# Lehninger
# PRINCIPLES OF BIOCHEMISTRY
## *fourth edition*

**DAVID L. NELSON**
Professor of Biochemistry
University of Wisconsin–Madison

**MICHAEL M. COX**
Professor of Biochemistry
University of Wisconsin–Madison

W. H. Freeman and Company
*New York*

Publisher: Sara Tenney
Acquisitions Editor: Katherine Ahr
Development Editor: Morgan Ryan
Marketing Manager: Sarah Martin
Marketing Director: John Britch
Project Editor: Jane O'Neill
Design Manager: Blake Logan
Text Designer: Rae Grant
Cover Designer: Yuichiro Nishizawa
Page Makeup: Paul Lacy
Illustration Coordinator: Shawn Churchman
Illustrations: Fine Line Illustrations
Molecular Graphics/Cover Illustration: Jean-Yves Sgro
Photo Editor: Vikii Wong
Production Coordinator: Paul Rohloff
Media & Supplements Editors: Jeffrey Ciprioni, Melanie Mays, Nick Tymoczko
Media Developers: Sumanas, Inc.
Composition: TechBooks
Manufacturing: RR Donnelley & Sons Company

On the cover: The $F_1$ ATPase, part of a complex responsible for ATP synthesis in eukaryotic mitochondria. See Chapter 19.

Library of Congress Control Number: 2004101716

ISBN: 0-7167-4339-6
EAN: 9 780716 743392

Printed in the United States of America

First printing
W. H. Freeman and Company
41 Madison Avenue
New York, NY 10010
Houndmills, Basingstoke RG21 6XS, England

www.whfreeman.com

# TO OUR TEACHERS:

PAUL R. BURTON

ALBERT FINHOLT

WILLIAM P. JENCKS

EUGENE P. KENNEDY

HOMER KNOSS

ARTHUR KORNBERG

I. ROBERT LEHMAN

EARL K. NELSON

DAVID E. SHEPPARD

HAROLD B. WHITE

# About the Authors

**David L. Nelson (left) and Michael M. Cox**

**David L. Nelson**, born in Fairmont, Minnesota, received his BS in Chemistry and Biology from St. Olaf College in 1964 and earned his PhD in Biochemistry at Stanford Medical School under Arthur Kornberg. He was a postdoctoral fellow at the Harvard Medical School with Eugene P. Kennedy, who was one of Albert Lehninger's first graduate students. Nelson went to the University of Wisconsin–Madison in 1971 and became a full professor of biochemistry in 1982.

Nelson's thesis research at Stanford was on the intermediary metabolism of sporulating and germinating bacteria. At Harvard he studied the energetics, genetics, and biochemistry of ion transport in *E. coli*. At Wisconsin his research has focused on the signal transductions that regulate ciliary motion and exocytosis in the protozoan *Paramecium*. The enzymes of signal transductions, including a variety of protein kinases, are primary targets of study. His research group uses enzyme purification, immunological techniques, electron microscopy, genetics, molecular biology, and electrophysiology to study these processes.

Dave Nelson has a distinguished record as a lecturer and research supervisor. For 32 years he has taught an intensive survey of biochemistry for advanced biochemistry undergraduates and graduate students in the life sciences (using Lehninger's *Biochemistry* and *Principles of Biochemistry* for much of that time). He has also taught a survey of biochemistry for nursing students, a graduate course on membrane structure and function, and a graduate seminar on membranes and sensory transductions. He has sponsored numerous PhD, MS, and undergraduate honors theses, and has received awards for his outstanding teaching, including the Dreyfus Teacher–Scholar Award and the Atwood Distinguished Professorship. In 1991–1992 he was a visiting professor of chemistry and biology at Spelman College. In 2002 he was appointed Director of the University of Wisconsin Center for Biology Education.

**Michael M. Cox** was born in Wilmington, Delaware. In his first biochemistry course, Lehninger's *Biochemistry* was a major influence in refocusing his fascination with biology and inspiring him to pursue a career in biochemistry. After graduating from the University of Delaware in 1974, Cox went to Brandeis University to do his doctoral work with William Jencks, and then to Stanford in 1979 for postdoctoral study with I. Robert Lehman. He moved to the University of Wisconsin–Madison in 1983, and became a full professor of biochemistry in 1992.

His doctoral research was on general acid and base catalysis as a model for enzyme-catalyzed reactions. At Stanford, Cox began work on the enzymes involved in genetic recombination. The work focused particularly on the RecA protein, designing purification and assay methods that are still in use, and illuminating the process of DNA branch migration. Exploration of the enzymes of genetic recombination has remained the central theme of his research.

Mike Cox has coordinated a large and active research team at Wisconsin, investigating the enzymology, topology, and energetics of genetic recombination. A primary focus has been the mechanism of RecA protein–mediated DNA strand exchange and the role of ATP in the RecA system. More recently, part of the research program has focused more generally on recombinational DNA repair processes in *E. coli* and *Deinococcus radiodurans*. For the past 20 years he has taught (with Dave Nelson) the survey of biochemistry to undergraduates and has lectured in graduate courses on DNA structure and topology, protein-DNA interactions, and the biochemistry of recombination. He has received awards for both his teaching and his research, including the Dreyfus Teacher–Scholar Award and the 1989 Eli Lilly Award in Biological Chemistry. His hobbies include gardening, wine collecting, and assisting in the design of laboratory buildings.

# PREFACE

*Lehninger Principles of Biochemistry*, Fourth Edition, is an introduction to our favorite subject—and our attempt to make it yours. In the four years since the publication of the third edition, both the pace and the scope of discovery in biochemistry have been breathtaking. We are learning not only what occurs within a cell but where and when it occurs, what determines location, and what controls timing. From the draft sequence of the human genome to elucidation of the high-resolution structure of the ribosome, major recent advances deepening our understanding of biochemistry fill the pages of this book.

While including these exciting new developments, we have also tried to remain focused on our mission: to communicate to students the fundamentals of biochemistry in a way that reflects the field as it is understood today. As always, our efforts to integrate new discoveries are balanced by our efforts to improve the accuracy and pedagogy of the core material of the field.

## Coverage of Major Recent Advances in Biochemistry

Every chapter has been thoroughly revised and updated. Some of the major recent advances incorporated into the Fourth Edition include:

- A revised mechanism for the enzyme lysozyme (Chapter 6)
- Functions of heparan sulfates in extracellular matrix (Chapter 7)
- The human genome (Chapter 9)
- The structure and roles of membrane rafts and caveolae (Chapter 11)

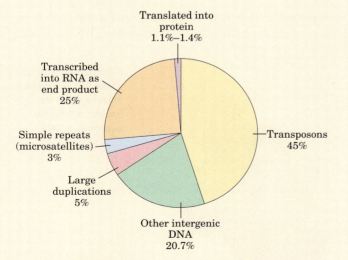

**Figure 9–19 Snapshot of the human genome. p. 324**

- The structure and function of ABC transporters (Chapter 11)
- Assembly of signaling complexes by the interaction of protein domains with phosphoserine and phosphotyrosine residues in partner proteins (Chapter 12)
- A two-component signaling system in bacteria and plants (Chapter 12)
- Receptorlike protein kinases in plant signaling (Chapter 12)
- Nonclassical actions of steroid hormones at plasma membrane receptors (Chapter 12)
- New developments in metabolic regulation and in the structure of enzymes and enzyme complexes, and new examples of human diseases that result from defective metabolism (throughout Part II)
- Inorganic polyphosphate metabolism (Chapter 13)
- Mechanisms by which metabolism of carbohydrates and fats is integrated in mammalian tissues (Chapters 14, 16, 17, and 23)
- Prevention of ATP hydrolysis by ATP synthase inhibitory protein during ischemia (Chapter 19)
- The role of mitochondria in apoptosis and oxidative stress (Chapter 19)
- The glycine decarboxylase complex and photorespiration (Chapter 20)
- A proposed lipid primer in cellulose synthesis (Chapter 20)
- A proposed new mechanism for starch synthesis from its reducing end (Chapter 20)
- The pathway for glyceroneogenesis, discussed in the context of its role in type II diabetes (Chapter 21)
- Hormonal regulation of body weight and the medical problems posed by the current epidemic of obesity in industrialized countries (Chapter 23)
- The role of cohesins and condensins in chromosome structure (Chapter 24)
- Coordination of multiple replication forks in replication factories (Chapter 25)
- New information on the mechanism and structure of bacterial RNA polymerase (Chapter 26)
- The high-resolution structure of the bacterial ribosome (Chapter 27)
- RNA interference (Chapter 28)

## New Methodologies

As scientists and as teachers, we are interested in new technology not just for its own sake but for what it can teach us—how it can provide us with clues to the biochemical mysteries around us. The following list includes just a few of the new experimental methodologies and their applications that are introduced or expanded in this edition:

- **High throughput methodologies** for determining protein structure, which have revealed many new structures that give us insight into the function of proteins and their interactions with each other and with ligands/substrates (Chapter 5)

- An introduction to modern **genomics** and **proteomics** (Chapter 9)

- Improvements in **DNA microarrays** that allow the detailed and comprehensive study of changes in gene expression caused by hormones and drugs and by mutations that lead to defects in function, including human diseases (Chapter 9)

- Visualization at the level of subcellular organelles of proteins fused genetically with green fluorescent protein (**GFP**) (Chapters 9, 12), providing information on cellular location and function

- **Atomic force microscopy,** used to visualize protein complexes and membrane surfaces (Chapter 11)

- Use of **fluorescence** for following the movement of a single lipid molecule in the plasma membrane (Chapter 11)

- Fluorescence resonance energy transfer (**FRET**), used to measure transient changes in second messenger or protein kinase activity in living cells (Chapter 12)

- Improved methods for **purifying and crystallizing membrane proteins,** which have provided a wealth of information about the details of such processes as oxidative phosphorylation and photophosphorylation (Chapter 19)

- RNA interference (**RNAi**), used to manipulate specific gene products in living cells, and the technology for creating knockout mice; these methods provide windows into the function of specific gene products in a whole organism. One strain of knockout mice has opened up a large field of research aimed at understanding the control of body weight—a subject of great interest, given the increasing incidence of obesity and its many medical consequences (Chapter 28).

## More Molecular Graphics

Our understanding of molecular structure has advanced significantly in the past few years, as has the interest on the part of students and instructors in working with this new structural information. As in the last edition, all the molecular structures you will find here are unique to this book, the result of close collaboration with our colleague Jean-Yves Sgro, biophysicist and master of molecular graphics. Jean-Yves has produced more than 70 new images for this edition. We have also added the Protein Data Bank identification code (PDB ID) to the legend of every figure where a structure appears. Students can use the PDB ID to access data about the molecule from the Protein Data Bank website and follow links to a variety of other resources. Structures new to this edition include:

- Aquaporin and voltage-gated $K^+$ channels—structures revealed by 2003 Nobel Prize–winning research (Chapter 11)

- The lactose permease of *E. coli* (Chapter 11)

- Extraordinary views of photosystem II and the cytochrome $b_6f$ complex (Chapter 19)

- Bacterial RNA polymerase (Chapter 26)

- Multiple images of the bacterial ribosome in all its detail (Chapter 27)

## Redesigned and Expanded Treatment of Enzyme Reaction Mechanisms

In response to instructor feedback, we have expanded our coverage of reaction mechanisms. We have also

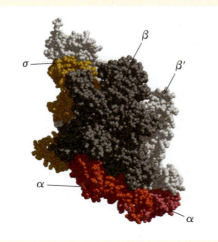
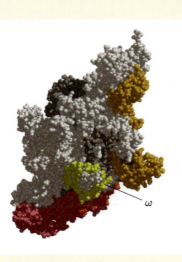

**Figure 26–4 Structure of the RNA polymerase holoenzyme of the bacterium *Thermus aquaticus.* p. 999**

worked to address the concern that students often enter biochemistry without a solid grasp of how to read reaction mechanisms. So, new in this edition:

- An expansive, two-page figure on the mechanism of chymotrypsin (the first reaction mechanism treated in the book), which serves as the framework for an illuminating discussion of how to read and extract information from mechanism diagrams (Fig. 6–21). Conventions are presented and explained.

- Building on the information communicated in the chymotrypsin figure, we've designed new **Mechanism Figures** that walk students through complex mechanisms step by step.

- Many new mechanisms; every major coenzyme is represented, with a discussion of its characteristic chemical role. Additional mechanisms highlight key classes of metabolic transformations and illuminate the chemical logic of reaction sequences in pathways.

## Detailed Mechanisms in the Fourth Edition

Alcohol dehydrogenase (Chapter 14) NEW
Aminoacyl-tRNA synthetase (Chapter 27)
Bacterial topoisomerase (Chapter 24) NEW
Chymotrypsin (Chapter 6) SIGNIFICANTLY REVISED
Citrate synthase (Chapter 16) NEW

Class I aldolase (Chapter 14) NEW
Coenzyme $B_{12}$ reactions (Chapter 17)
DNA photolyase (Chapter 25) SIGNIFICANTLY REVISED
DNA polymerases (Chapter 25) SIGNIFICANTLY REVISED
Enolase (Chapter 6)
Fatty acyl–CoA synthetase (Chapter 17)
Glutamine amidotransferases (Chapter 22) NEW
Glyceraldehyde 3-phosphate dehydrogenase (Chapter 14) SIGNIFICANTLY REVISED
Isocitrate dehydrogenase (Chapter 16) NEW
Lysozyme (Chapter 6) NEW
Phosphohexose isomerase (Chapter 14) NEW
Pyridoxal phosphate (PLP) reactions (Chapter 18) NEW
Pyruvate carboxylase (Chapter 16) NEW
Ribonucleotide reductase (Chapter 22) SIGNIFICANTLY REVISED
RNA polymerase (Chapter 26) SIGNIFICANTLY REVISED
Rubisco (Chapter 20) SIGNIFICANTLY REVISED
Thiamine pyrophosphate (TPP) reactions (Chapters 14, 16)
Thymidylate synthase (Chapter 22)
Tryptophan synthase (Chapter 22)
Urea production: origin of amino groups (multiple reactions) (Chapter 18) NEW

Many more mechanisms are provided in abbreviated form.

**Mechanism Figure 6–21** Hydrolytic cleavage of a peptide bond by chymotrypsin. pp. 216–217

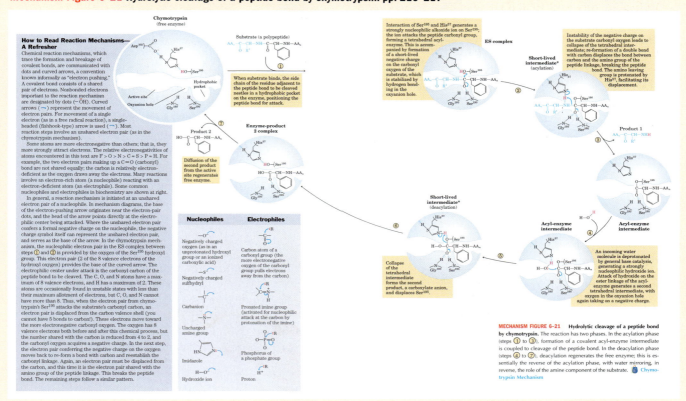

MECHANISM FIGURE 6–21 Hydrolytic cleavage of a peptide bond by chymotrypsin. The reaction has two phases. In the acylation phase (steps ① to ③), formation of a covalent acyl-enzyme intermediate is coupled to cleavage of the peptide bond. In the deacylation phase (steps ④ to ⑦), deacylation regenerates the free enzyme; this is essentially the reverse of the acylation phase, with water mirroring, in reverse, the role of the amine component of the substrate. Chymotrypsin Mechanism

# More Medically Relevant Examples

*This icon is used throughout the book to denote material of special medical interest.*

We have added many new and fascinating examples relating biochemistry to medicine to help students see the relevance of biochemistry to their lives. Some of the medical examples we have added in this edition:

- Scurvy (Chapter 4)
- Carbon monoxide poisoning (Chapter 5)
- Role of ABC transporters in multidrug resistance (Chapter 11)
- Adverse physiological consequences of defective ion channels (Chapter 11)
- Defect in the PTEN gene that causes tumors (Chapter 12)
- Treatment of erectile dysfunction with sildenafil (Viagra) (Chapter 12)
- Lactose intolerance (Chapter 14)
- Role of the pentose phosphate pathway in prevention of oxidative damage (Chapter 14)
- Exacerbation of Wernicke-Korsakoff syndrome by a defect in transketolase (Chapter 14)
- Glycogen storage diseases (Chapter 15)
- Disease caused by genetic defects in fatty acyl–CoA dehydrogenases (Chapter 17)
- Zellweger syndrome and X-linked adrenoleukodystrophy (Chapter 17)
- Refsum's disease (Chapter 17)
- Kidney stones (Chapter 18)
- Diabetes and glyceroneogenesis (Chapter 21)
- Heme as the source of bile pigments and its colorful metabolism in bruising (Chapter 22)
- Obesity as a medical problem, and the hormonal regulation of body weight (Chapter 23)
- Adiponectin and insulin resistance in mice (Chapter 23)
- RNAi and its potential uses in medicine (Chapter 28)

# Updated Organization

While the overall structure of *Lehninger Principles of Biochemistry* remains consistent with previous editions, we have made a few key changes in the organization of the fourth edition, in consultation with many teachers and students.

- Earlier placement of the thoroughly revised chapter on **DNA-based information technologies** (Chapter 9). This includes recent advances in genomics and proteomics and introduces the human and other genomes. Providing this material early in the text allows us to refer to important findings from molecular genetics throughout subsequent chapters.

- **Glycolysis and gluconeogenesis** presented in a single chapter (Chapter 14); the parallels between these pathways are thus more easily seen.

- A new chapter on the **regulation of metabolism** (Chapter 15). Using the pathways of glycolysis, gluconeogenesis, glycogenolysis, and glycogenesis to illustrate general principles, we discuss the various mechanisms by which cells and organisms regulate their metabolic affairs. We also introduce the principles of **metabolic control analysis,** a powerful way to determine which enzymes limit the flux through a pathway, and say goodbye to the concept of a single limiting step.

- Reorganization of the introductory material previously presented in three chapters and its condensation into one chapter, in which we describe the cellular, chemical, physical, genetic, and evolutionary **foundations of biochemistry** (Chapter 1). For students who would benefit from more background information on biology and chemistry, third edition Chapters 1 to 3 are available on the book's website, www.whfreeman.com/lehninger.

# Focus on Fundamental Principles

As in the third edition, we have carefully selected new material that illustrates or clarifies a broader theme. Our major objective has been to highlight principles that rationalize and systematize large bodies of information.

To help students navigate what can seem like an ocean of information, we wrote with these goals in mind:

- to introduce the language of biochemistry, with careful explanations of the meaning, origin, and significance of terms;

- to provide a balanced understanding of the physical, chemical, and biological context in which each biomolecule, reaction, or pathway operates; and

- to project a clear and repeated emphasis on major themes, especially those relating to evolution, thermodynamics, regulation, and the relationship between structure and function.

In keeping with these goals, we have numbered the major sections of each chapter and provided an overview and a summary of each section to help students master and review the information more efficiently.

# Supplements

## Instructor's Resource CD-ROM with Test Bank; 0-7167-5953-5

To help instructors create their own websites and orchestrate dynamic lectures, the two-disc IRCD contains:

- **Animated enzyme mechanisms,** which show key mechanisms step-by-step, in Flash format

- **Living Graphs,** which bring to life important equations from the text by allowing the user to see the graphic result of changing the parameters

- **Test Bank,** organized by chapter in both PDF files and editable Word files

- **All the figures** and tables from the text in JPEG and PowerPoint formats, **optimized for projection** with enhanced colors, higher resolution, and enlarged fonts for easy reading in the lecture hall

## Overhead Transparency Set; 0-7167-5956-X

The full-color transparency set contains 150 key illustrations from the text that have been **optimized** with enlarged labels, bolder lines, and more vivid colors that project more clearly for lecture hall presentation.

## Printed *Test Bank*, Fourth Edition, by Terry Platt and Eugene Barber (University of Rochester Medical Center) and David L. Nelson and Brook Chase Soltvedt (University of Wisconsin–Madison); 0-7167-5952-7

The new *Test Bank* contains many new multiple-choice and short-answer problems and solutions—approximately 50 problems and solutions per chapter. Each problem is keyed to the corresponding chapter of the text and rated by level of difficulty. The *Test Bank* is also available on the IRCD, organized by chapter, as PDF and editable Word files.

## Website at www.whfreeman.com/lehninger

Created with both instructors and students in mind, the website features:

- **All the figures** and tables from the book **optimized for projection,** available in PowerPoint and JPEG format; also available on the IRCD

- **Animated enzyme mechanisms,** showing key mechanisms step-by-step, in Flash format

- **Living Graphs,** dynamically illustrating equations and graphed material featured in the text

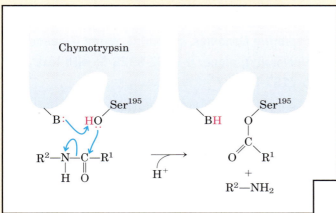

**Figure 6–10 Covalent and general acid-base catalysis. p. 202**

**Figure 6–10 Optimized for projection**

- **Biochemistry in 3D molecular structure tutorials,** self-paced interactive tutorials based on the Chemscape Chime molecular visualization browser plug-in; allow students to explore the basic and advanced topics covered in the textbook

- **Online quizzing** for each chapter, a new way for students to review material and prepare for exams

- **Flashcards** on key terms from the text

- Bonus material from *Lehninger Principles of Biochemistry*, Third Edition, fundamental Chapters 1, 2, and 3, which instructors find useful for students requiring a basis for the study of biochemistry

 Throughout the text, this icon alerts the reader to a website connection

### The Absolute, Ultimate Guide to Lehninger Principles of Biochemistry, Fourth Edition, Study Guide and Solutions Manual, by Marcy Osgood (University of New Mexico School of Medicine) and Karen Ocorr (University of California, San Diego); 0-7167-5955-1

*The Absolute, Ultimate Guide* combines an innovative study guide with a reliable solutions manual in one convenient volume. Thoroughly class-tested, it includes for each chapter:

- **Major Concepts:** a roadmap through the chapter

- **What to Review:** questions that recap key points from previous chapters

- **Discussion Questions:** provided for each section; designed for individual review, study groups, or classroom discussions

- **A Self-Test:** "Do you know the terms?"; crossword puzzles; multiple-choice, fact-driven questions; and questions that ask students to apply their new knowledge in new directions—plus answers!

- **Biochemistry on the Internet problems:** provided for many chapters, to encourage students to use the Web to solve problems

- **Complete and Detailed Solutions:** solutions for all end-of-chapter problems

A poster-size **Cellular Metabolic Map** is packaged with the Guide, on which students can draw the reactions and pathways of metabolism in their proper cellular compartments. To order the Cell Map separately, use ISBN 0-7167-9950-2.

### Lecture Notebook; 0-7167-5954-3

For students who find that they are too busy copying drawings to keep up with the lecture, the *Notebook* is an indispensable classroom companion, offering

- **Illustrations** in the order in which they appear in the textbook, with plenty of room to take notes

- **Essential reaction equations** and **mathematical equations** with identifying labels

- Complete **pathway diagrams** and **individual reaction diagrams** for all metabolic pathways in the book

- **References** that key the material in the text to the book's website.

- **Three-hole-punched, perforated pages** so that students can reorganize pages in any order necessary to follow instructor's lectures, and can insert instructor handouts

### Exploring Genomes, by Paul G. Young (Queens University); alone: 0-7167-9928-6; with text: 0-7167-8995-7

Used in conjunction with the online tutorials at www.whfreeman.com/young, *Exploring Genomes* guides students through live searches and analyses on the most commonly used National Center for Biotechnology Information (NCBI) databases.

# Acknowledgments

Our writing has been supported by many people whose advice and encouragement were critical throughout the years of revision, and we are indebted to all of them. At W. H. Freeman and Company, the organization of the project was overseen by Kate Ahr and Sara Tenney and developed by Morgan Ryan. Kate, Sara, and Morgan were part of our immediate production team. Their creative input and constructive challenges inspired many of the improvements in this edition. The final manuscript benefited from the superb copyediting of Linda Strange (who has been involved in the writing of every edition—this is number 6—of Lehninger's biochemistry texts!). Our project editor, Jane O'Neill, has kept us all on track if not always on schedule, showing inordinate patience. Our indexer, Ellen Brennan, has compiled a great index on a very tight schedule. We thank our production coordinator, Paul Rohloff, for helping us to produce the book on time. In Madison, we relied heavily on Brook Soltvedt for her invaluable editorial and organizational input, involving every phase of the project. Brook helped the two authors find a similar voice and avoid redundancies, while keeping the countless pieces of the project moving. Shelley Lusetti read and expertly critiqued the page proofs for every chapter. It was a great privilege to work with so talented and energetic a production team, and they have all had much to do with the quality of the book before you.

The new art for this edition includes work from Fine Line Illustrations, coordinated by Shawn Churchman. We have been impressed and gratified to witness the process by which our often indecipherable scribbles have been transformed into the elegant figures that grace this book. We thank Rae Grant and Blake Logan for the beautiful work they have done on the design for the book. Paul Lacy worked magic getting the chapters to lay out in the new two-column format. We have already noted the invaluable contribution of Jean-Yves Sgro, who prepared all the molecular graphics. The photographs used in this edition were carefully tracked down by Vikii Wong. Melanie Mays and Nick Tymoczko arranged for the many reviewers of the text and, with Jeff Ciprioni, coordinated the development of the supplemental materials—the website, *Test Bank, Lecture Notebook,* overhead transparencies, PowerPoint files, and *Absolute, Ultimate Guide.* Marcy Osgood of the University of New Mexico School of Medicine and Karen Ocorr of the University of California, San Diego, made helpful comments on the textbook manuscript as they revised their study guide for this edition. Five of the boxes in the book were written by Lou Pech of Carroll College.

One of the special advantages we have had is our access to the extraordinary community of researchers at the University of Wisconsin–Madison. More often than not, the expertise we needed to tie down a loose end or develop a new topic was right at hand. Our colleagues provided us with timely data, critiqued sections of the text, helped us develop figures, answered our questions, and helped in innumerable other ways. Space precludes a detailing of every contribution, but we are especially grateful to the following colleagues at the University of Wisconsin–Madison:

Rick Amasino, Laurens Anderson, Alan Attie, Wayne Becker, Sebastian Bednarek, Paul Bertics, Susan Eichhorn, Rick Eisenstein, Jerry Ensign, Ray Evert, Bob Fillingame, Brian Fox, Perry Frey, Rick Gourse, Colleen Hayes, Hazel Holden, Judith Kimble, Bob Landick, Paul Ludden, John Markley, Anant Menon, James Ntambi, Ann Palmenberg, George Phillips, Ron Raines, Ivan Rayment, Tom Record, George Reed, Bill Reznikoff, Ruth Saecker, Heinrich Schnoes, Tom Sharkey, Sue Smith, Gary Splitter, Bill Sugden, Mike Sussman, and Marv Wickens.

We were also able to draw on contacts from around the United States and around the world. For their patience in answering our questions, often including an informal reading and critique of passages, we thank:

Andre Adoutte, *Université d'Orsay*
Marlene Belfort, *Wadsworth Center, New York State Department of Health*
Terry Beveridge, *University of Guelph*
Pat Brown, *Stanford University*
Peter Burgers, *Washington University*
Joseph Clark, *Oxford University*
Ron Conaway, *University of Oklahoma*
Seth Darst, *Rockefeller University*
David Fell, *Oxford Brookes University*
Herbert Friedmann, *University of Chicago*
Barry Ganong, *Mansfield University*
Myron Goodman, *University of Southern California*
Jack Griffith, *University of North Carolina*
Carol Gross, *University of California, San Francisco*
Dick Hanson, *Case Western Reserve University*
Peter Hinkle, *Cornell University*
So Iwata, *Imperial College, London*
Ron Kabak, *University of California, Los Angeles*
Jack Kirsch, *University of California, Berkeley*
Denis Lynn, *University of Guelph*
Lynn Margulis, *University of Massachusetts–Amherst*
Ken Marians, *Memorial Sloan-Kettering Cancer Center*
Stanley Miller, *University of California, San Diego*
Paul Modrich, *Duke University*
Mike O'Donnell, *Rockefeller University*
Lea Reshef, *Hebrew University–Hadassah School of Medicine, Jerusalem*
Aziz Sancar, *University of North Carolina*
William Schopf, *University of California, Los Angeles*
JoAnne Stubbe, *Massachusetts Institute of Technology*
Sue Travis, *University of Iowa*
Brian Volkman, *Medical College of Wisconsin*
Claire Walczak, *Indiana University*
Richard Wolfenden, *University of North Carolina*

The book has benefited enormously from the combined wisdom and experience of countless reviewers. Every chapter was critiqued at one stage or another by biochemists with interests and expertise in both the chapter topic and biochemical education. These formal reviews helped shape this new edition in many ways. For their indispensable advice and criticism, we thank them.

## Fourth Edition Text Reviewers

Roger A. Acey, *California State University, Long Beach*

Loranne Agius, *University of Newcastle upon Tyne*

Richard Amasino, *University of Wisconsin–Madison*

Laurens Anderson, *University of Wisconsin–Madison*

Alan D. Attie, *University of Wisconsin–Madison*

Paul Azari, *Colorado State University*

John Martyn Bailey, *George Washington University*

Kenneth Balazovich, *University of Michigan*

Thurston E. Banks, *Tennessee Technological University*

Michael Barbush, *Baker University*

Paul J. Bertics, *University of Wisconsin–Madison*

Michael R. Borenstein, *Temple University School of Pharmacy*

Maureen A. Brandon, *Idaho State University*

Roger K. Bretthauer, *University of Notre Dame*

Alexander C. Brownie, *FRSE, University at Buffalo–School of Medicine and Biomedical Sciences*

John D. Brunstein, *University of British Columbia*

J. Thomas Buckley, *University of Victoria*

Brian P. Callahan, *University of North Carolina at Chapel Hill*

Marion L. Carroll, *Xavier University of Louisiana*

Guillaume Chanfreau, *University of California, Los Angeles*

Kim D. Collins, *University of Maryland Medical School*

Erik Conrad, *Northwestern University* (student)

John Crandall, *University of Wisconsin–Stout*

Tanya E. S. Dahms, *University of Regina*

Seth Darst, *Rockefeller University*

S. Todd Deal, *Georgia Southern University*

Rick Eisenstein, *University of Wisconsin–Madison*

Craig D. Fairchild, *Worcester Polytechnic Institute*

David Fell, *Oxford Brookes University*

Gregg B. Fields, *Florida Atlantic University*

Steven M. Firestine, *Duquesne University*

Paul Fitzpatrick, *Texas A&M University*

Brian G. Fox, *University of Wisconsin–Madison*

Gerald D. Frenkel, *Rutgers University*

George Gassner, *San Francisco State University*

Susan P. Gilbert, *University of Pittsburgh*

Donald L. Gill, *University of Maryland School of Medicine*

Burt Goldberg, *St. Francis College and New York University*

Rick Gourse, *University of Wisconsin–Madison*

E. M. Gregory, *Virginia Tech*

Anne A. Grippo, *Arkansas State University*

Richard I. Gumport, *University of Illinois at Urbana-Champaign*

J. Norman Hansen, *University of Maryland, College Park*

Richard W. Hanson, *Case Western Reserve University School of Medicine*

Sonja E. Hicks, *Wellesley College*

Anna M. Holmes, *University of Alabama in Huntsville*

Ralph A. Jacobson, *California Polytechnic State University*

Shamsa Jessa, *Douglas College*

Warren V. Johnson, *University of Wisconsin–Green Bay*

Harold E. Kasinsky, *University of British Columbia*

Anita S. Klein, *University of New Hampshire*

Rachel E. Klevit, *University of Washington*

Michael R. Koelle, *Yale University School of Medicine*

Gary R. Kunkel, *Texas A&M University*

Robert Landick, *University of Wisconsin–Madison*

Andrea Lee, *Northwestern University (student)*

Janet E. Lindsley, *University of Utah School of Medicine*

Robert C. MacDonald, *Northwestern University*

Colin T. Mant, *University of Kent at Canterbury*

John C. Matthews, *University of Mississippi School of Pharmacy*

Douglas D. McAbee, *California State University, Long Beach*

Ronald W. McCune, *Idaho State University*

Katherine D. McReynolds, *California State University, Sacramento*

Anant Menon, *University of Wisconsin–Madison*

Christopher R. Meyer, *California State University, Fullerton*

Kimberly Mowry, *Brown University*

Martin Nemeroff, *Rutgers University*

Anthony W. Norman, *University of California, Riverside*

Mike O'Donnell, *Rockefeller University*

Marcy P. Osgood, *University of New Mexico School of Medicine*

Annette Pasternak, *Loyola Marymount University*

Susan Pedigo, *University of Mississippi*

Darrell L. Peterson, *Virginia Commonwealth University*

Terry Platt, *University of Rochester Medical Center*

B. Franklin Pugh, *Pennsylvania State University*

S. Ramaswamy, *University of Iowa*

Kevin E. Redding, *University of Alabama*

George Reed, *University of Wisconsin–Madison*

David Reibstein, *Princeton University*

Ronald C. Reitz, *University of Nevada at Reno*

Lea Reshef, *Hebrew University–Hadassah Medical School, Jerusalem*

Margaret S. Rice, *California Polytechnic State University, San Luis Obispo*

Daniel M. Roberts, *University of Tennessee*

Anne G. Rosenwald, *Georgetown University*

Marvin L. Salin, *Mississippi State University*

Merlyn D. Schuh, *Davidson College*

David W. Seybert, *Duquesne University*

Thomas D. Sharkey, *University of Wisconsin–Madison*

Ram Singhal, *Wichita State University*

Christine M. Smith, *Reed College*

Ron G. Smith, *University College of the Cariboo*

Ronald Somerville, *Purdue University*
Jeffry Stock, *Princeton University*
Julie M. Stone, *University of Nebraska–Lincoln*
Koni Stone, *California State University, Stanislaus*
JoAnne Stubbe, *Massachusetts Institute of Technology*
Michael A. Sypes, *Pennsylvania State University*
John J. G. Tesmer, *University of Texas at Austin*
Jeremy Thorner, *University of California, Berkeley*
Cuong Tieu, *Northwestern University* (student)
Bik K. Tye, *Cornell University*
Jack Y. Vanderhoek, *George Washington University School of Medicine*
Manuel Varela, *Eastern New Mexico University*
John Voss, *University of California, Davis, School of Medicine*
Graham Walker, *Massachusetts Institute of Technology*
Peng George Wang, *Wayne State University*
Adele J. Wolfson, *Wellesley College*
Beulah M. Woodfin, *University of New Mexico*

### Media/Supplements Reviewers

Jim Blankenship, *Cornell University*
Larry Brown, *Appalachian State University*
Roger Brownsey, *University of British Columbia*
John Brunstein, *University of British Columbia*
Guillaume Chanfreau, *University of California, Los Angeles*
David W. Emerich, *University of Missouri, Columbia*
Partho Ghosh, *University of California, San Diego*
Burt Goldberg, *St. Francis College and New York University*
Pam Hay, *Davidson College*
Douglas Julin, *University of Maryland, College Park*
Roger Kaspar, *Brigham Young University*
Ramji Khandelwal, *University of Saskatchewan*
Rachel E. Klevit, *University of Washington*
Mark Lee, *Spelman College*
Ashley B. Mahoney, *Muhlenberg College*
John Makemson, *Florida International University*
Bjorn Olesen, *Southeast Missouri State University*
Dennis Pederson, *California State University, San Bernardino*

Michel Roberge, *University of British Columbia*
Christopher Rohlman, *Albion College*
Erik Sontheimer, *Northwestern University*
Koni Stone, *California State University, Stanislaus*
Miriam Ziegler, *University of Arizona*

We lack the space here to acknowledge all the other individuals whose special efforts went into this book. We offer instead our sincere thanks and the finished book that they helped guide to completion. We, of course, assume full responsibility for errors of fact or emphasis.

We are grateful to our students at the University of Wisconsin–Madison, who provided inspiration and numerous ideas for improvement; to the students and staff of our research groups, who helped us balance the competing demands of research, teaching, administration, and textbook writing; and to our colleagues in the Department of Biochemistry at Madison, who patiently answered our questions, corrected our misconceptions, and in many cases reviewed chapters. We have received many letters over the past four years from students and professors who used our book, often pointing out places it could be made better, for which we are grateful. (We hope future users will continue to tell us how we can improve the book.) We also wish to thank the entire staff of W. H. Freeman and Company, who allowed us the freedom we needed and exhibited extraordinary patience as we added late-breaking advances to "final" chapters, and whose dedication to producing a fine-quality book inspired our best efforts and made this a rewarding experience.

Finally, we express our deepest appreciation to our wives, Brook and Beth, and our children, who endured with extraordinary grace the long evenings and weekends we devoted to book writing and provided constant encouragement.

David L. Nelson          *Madison, Wisconsin*
Michael M. Cox          *March 2004*

# Contents in Brief

# CONTENTS

## 1  The Foundations of Biochemistry  1

### 1.1  Cellular Foundations  3
Cells Are the Structural and Functional Units of All Living Organisms  3
Cellular Dimensions Are Limited by Oxygen Diffusion  4
There Are Three Distinct Domains of Life  4
*Escherichia coli* Is the Most-Studied Prokaryotic Cell  5
Eukaryotic Cells Have a Variety of Membranous Organelles, Which Can Be Isolated for Study  6
The Cytoplasm Is Organized by the Cytoskeleton and Is Highly Dynamic  9
Cells Build Supramolecular Structures  10
In Vitro Studies May Overlook Important Interactions among Molecules  11

### 1.2  Chemical Foundations  12
Biomolecules Are Compounds of Carbon with a Variety of Functional Groups  13
Cells Contain a Universal Set of Small Molecules  14
Macromolecules Are the Major Constituents of Cells  15
**Box 1–1 Molecular Weight, Molecular Mass, and Their Correct Units  15**
Three-Dimensional Structure Is Described by Configuration and Conformation  16
**Box 1–2 Louis Pasteur and Optical Activity: *In Vino, Veritas*  19**
Interactions between Biomolecules Are Stereospecific  20

### 1.3  Physical Foundations  21
Living Organisms Exist in a Dynamic Steady State, Never at Equilibrium with Their Surroundings  21
Organisms Transform Energy and Matter from Their Surroundings  22
The Flow of Electrons Provides Energy for Organisms  22
Creating and Maintaining Order Requires Work and Energy  23
Energy Coupling Links Reactions in Biology  23
**Box 1–3 Entropy: The Advantages of Being Disorganized  24**
$K_{eq}$ and $\Delta G$ Are Measures of a Reaction's Tendency to Proceed Spontaneously  26
Enzymes Promote Sequences of Chemical Reactions  26
Metabolism Is Regulated to Achieve Balance and Economy  27

### 1.4  Genetic Foundations  28
Genetic Continuity Is Vested in Single DNA Molecules  29
The Structure of DNA Allows for Its Replication and Repair with Near-Perfect Fidelity  29
The Linear Sequence in DNA Encodes Proteins with Three-Dimensional Structures  29

### 1.5  Evolutionary Foundations  31
Changes in the Hereditary Instructions Allow Evolution  31
Biomolecules First Arose by Chemical Evolution  32
Chemical Evolution Can Be Simulated in the Laboratory  32
RNA or Related Precursors May Have Been the First Genes and Catalysts  32
Biological Evolution Began More Than Three and a Half Billion Years Ago  34
The First Cell Was Probably a Chemoheterotroph  34
Eukaryotic Cells Evolved from Prokaryotes in Several Stages  34
Molecular Anatomy Reveals Evolutionary Relationships  36

## 22 Biosynthesis of Amino Acids, Nucleotides, and Related Molecules   833

## 23 Hormonal Regulation and Integration of Mammalian Metabolism   881

# THE FOUNDATIONS OF BIOCHEMISTRY

With the cell, biology discovered its atom . . . To characterize life, it was henceforth essential to study the cell and analyze its structure: to single out the common denominators, necessary for the life of every cell; alternatively, to identify differences associated with the performance of special functions.

—*François Jacob,* La logique du vivant: une histoire de l'hérédité (The Logic of Life: A History of Heredity), *1970*

We must, however, acknowledge, as it seems to me, that man with all his noble qualities . . . still bears in his bodily frame the indelible stamp of his lowly origin.

—*Charles Darwin,* The Descent of Man, *1871*

Fifteen to twenty billion years ago, the universe arose as a cataclysmic eruption of hot, energy-rich subatomic particles. Within seconds, the simplest elements (hydrogen and helium) were formed. As the universe expanded and cooled, material condensed under the influence of gravity to form stars. Some stars became enormous and then exploded as supernovae, releasing the energy needed to fuse simpler atomic nuclei into the more complex elements. Thus were produced, over billions of years, the Earth itself and the chemical elements found on the Earth today. About four billion years ago, life arose—simple microorganisms with the ability to extract energy from organic compounds or from sunlight, which they used to make a vast array of more complex **biomolecules** from the simple elements and compounds on the Earth's surface.

Biochemistry asks how the remarkable properties of living organisms arise from the thousands of different lifeless biomolecules. When these molecules are isolated and examined individually, they conform to all the physical and chemical laws that describe the behavior of inanimate matter—as do all the processes occurring in living organisms. The study of biochemistry shows how the collections of inanimate molecules that constitute living organisms interact to maintain and perpetuate life animated solely by the physical and chemical laws that govern the nonliving universe.

Yet organisms possess extraordinary attributes, properties that distinguish them from other collections of matter. What are these distinguishing features of living organisms?

**A high degree of chemical complexity and microscopic organization.** Thousands of different molecules make up a cell's intricate internal structures (Fig. 1–1a). Each has its characteristic sequence of subunits, its unique three-dimensional structure, and its highly specific selection of binding partners in the cell.

**Systems for extracting, transforming, and using energy from the environment** (Fig. 1–1b), enabling organisms to build and maintain their intricate structures and to do mechanical, chemical, osmotic, and electrical work. Inanimate matter tends, rather, to decay toward a more disordered state, to come to equilibrium with its surroundings.

**(a)**

**(b)**

**(c)**

**FIGURE 1–1 Some characteristics of living matter. (a)** Microscopic complexity and organization are apparent in this colorized thin section of vertebrate muscle tissue, viewed with the electron microscope. **(b)** A prairie falcon acquires nutrients by consuming a smaller bird. **(c)** Biological reproduction occurs with near-perfect fidelity.

**A capacity for precise self-replication and self-assembly** (Fig. 1–1c). A single bacterial cell placed in a sterile nutrient medium can give rise to a billion identical "daughter" cells in 24 hours. Each cell contains thousands of different molecules, some extremely complex; yet each bacterium is a faithful copy of the original, its construction directed entirely from information contained within the genetic material of the original cell.

**Mechanisms for sensing and responding to alterations in their surroundings,** constantly adjusting to these changes by adapting their internal chemistry.

**Defined functions for each of their components and regulated interactions among them.**

This is true not only of macroscopic structures, such as leaves and stems or hearts and lungs, but also of microscopic intracellular structures and individual chemical compounds. The interplay among the chemical components of a living organism is dynamic; changes in one component cause coordinating or compensating changes in another, with the whole ensemble displaying a character beyond that of its individual parts. The collection of molecules carries out a program, the end result of which is reproduction of the program and self-perpetuation of that collection of molecules—in short, life.

**A history of evolutionary change.** Organisms change their inherited life strategies to survive in new circumstances. The result of eons of evolution is an enormous diversity of life forms, superficially very different (Fig. 1–2) but fundamentally related through their shared ancestry.

Despite these common properties, and the fundamental unity of life they reveal, very few generalizations about living organisms are absolutely correct for every organism under every condition; there is enormous diversity. The range of habitats in which organisms live, from hot springs to Arctic tundra, from animal intestines to college dormitories, is matched by a correspondingly wide range of specific biochemical adaptations, achieved

**FIGURE 1–2 Diverse living organisms share common chemical features.** Birds, beasts, plants, and soil microorganisms share with humans the same basic structural units (cells) and the same kinds of macromolecules (DNA, RNA, proteins) made up of the same kinds of monomeric subunits (nucleotides, amino acids). They utilize the same pathways for synthesis of cellular components, share the same genetic code, and derive from the same evolutionary ancestors. Shown here is a detail from "The Garden of Eden," by Jan van Kessel the Younger (1626–1679).

within a common chemical framework. For the sake of clarity, in this book we sometimes risk certain generalizations, which, though not perfect, remain useful; we also frequently point out the exceptions that illuminate scientific generalizations.

Biochemistry describes in molecular terms the structures, mechanisms, and chemical processes shared by all organisms and provides organizing principles that underlie life in all its diverse forms, principles we refer to collectively as *the molecular logic of life*. Although biochemistry provides important insights and practical applications in medicine, agriculture, nutrition, and industry, its ultimate concern is with the wonder of life itself.

In this introductory chapter, then, we describe (briefly!) the cellular, chemical, physical (thermodynamic), and genetic backgrounds to biochemistry and the overarching principle of evolution—the development over generations of the properties of living cells. As you read through the book, you may find it helpful to refer back to this chapter at intervals to refresh your memory of this background material.

## 1.1 Cellular Foundations

The unity and diversity of organisms become apparent even at the cellular level. The smallest organisms consist of single cells and are microscopic. Larger, multicellular organisms contain many different types of cells, which vary in size, shape, and specialized function. Despite these obvious differences, all cells of the simplest and most complex organisms share certain fundamental properties, which can be seen at the biochemical level.

### Cells Are the Structural and Functional Units of All Living Organisms

Cells of all kinds share certain structural features (Fig. 1–3). The **plasma membrane** defines the periphery of the cell, separating its contents from the surroundings. It is composed of lipid and protein molecules that form a thin, tough, pliable, hydrophobic barrier around the cell. The membrane is a barrier to the free passage of inorganic ions and most other charged or polar compounds. Transport proteins in the plasma membrane allow the passage of certain ions and molecules; receptor proteins transmit signals into the cell; and membrane enzymes participate in some reaction pathways. Because the individual lipids and proteins of the plasma membrane are not covalently linked, the entire structure is remarkably flexible, allowing changes in the shape and size of the cell. As a cell grows, newly made lipid and protein molecules are inserted into its plasma membrane; cell division produces two cells, each with its own membrane. This growth and cell division (fission) occurs without loss of membrane integrity.

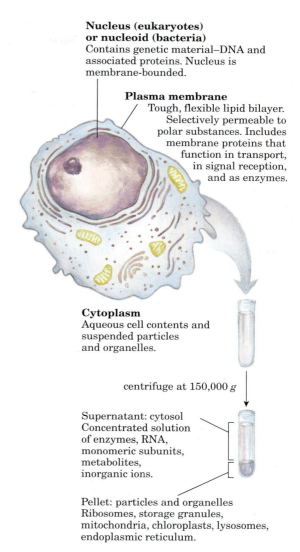

**Nucleus (eukaryotes) or nucleoid (bacteria)**
Contains genetic material–DNA and associated proteins. Nucleus is membrane-bounded.

**Plasma membrane**
Tough, flexible lipid bilayer. Selectively permeable to polar substances. Includes membrane proteins that function in transport, in signal reception, and as enzymes.

**Cytoplasm**
Aqueous cell contents and suspended particles and organelles.

centrifuge at 150,000 $g$

Supernatant: cytosol
Concentrated solution of enzymes, RNA, monomeric subunits, metabolites, inorganic ions.

Pellet: particles and organelles
Ribosomes, storage granules, mitochondria, chloroplasts, lysosomes, endoplasmic reticulum.

**FIGURE 1–3 The universal features of living cells.** All cells have a nucleus or nucleoid, a plasma membrane, and cytoplasm. The cytosol is defined as that portion of the cytoplasm that remains in the supernatant after centrifugation of a cell extract at 150,000 $g$ for 1 hour.

The internal volume bounded by the plasma membrane, the **cytoplasm** (Fig. 1–3), is composed of an aqueous solution, the **cytosol,** and a variety of suspended particles with specific functions. The cytosol is a highly concentrated solution containing enzymes and the RNA molecules that encode them; the components (amino acids and nucleotides) from which these macromolecules are assembled; hundreds of small organic molecules called **metabolites,** intermediates in biosynthetic and degradative pathways; **coenzymes,** compounds essential to many enzyme-catalyzed reactions; inorganic ions; and **ribosomes,** small particles (composed of protein and RNA molecules) that are the sites of protein synthesis.

All cells have, for at least some part of their life, either a **nucleus** or a **nucleoid,** in which the **genome**—

the complete set of genes, composed of DNA—is stored and replicated. The nucleoid, in bacteria, is not separated from the cytoplasm by a membrane; the nucleus, in higher organisms, consists of nuclear material enclosed within a double membrane, the nuclear envelope. Cells with nuclear envelopes are called **eukaryotes** (Greek *eu,* "true," and *karyon,* "nucleus"); those without nuclear envelopes—bacterial cells—are **prokaryotes** (Greek *pro,* "before").

## Cellular Dimensions Are Limited by Oxygen Diffusion

Most cells are microscopic, invisible to the unaided eye. Animal and plant cells are typically 5 to 100 $\mu$m in diameter, and many bacteria are only 1 to 2 $\mu$m long (see the inside back cover for information on units and their abbreviations). What limits the dimensions of a cell? The lower limit is probably set by the minimum number of each type of biomolecule required by the cell. The smallest cells, certain bacteria known as mycoplasmas, are 300 nm in diameter and have a volume of about $10^{-14}$ mL. A single bacterial ribosome is about 20 nm in its longest dimension, so a few ribosomes take up a substantial fraction of the volume in a mycoplasmal cell.

The upper limit of cell size is probably set by the rate of diffusion of solute molecules in aqueous systems. For example, a bacterial cell that depends upon oxygen-consuming reactions for energy production must obtain molecular oxygen by diffusion from the surrounding medium through its plasma membrane. The cell is so small, and the ratio of its surface area to its volume is so large, that every part of its cytoplasm is easily reached by $O_2$ diffusing into the cell. As cell size increases, however, surface-to-volume ratio decreases, until metabolism consumes $O_2$ faster than diffusion can supply it. Metabolism that requires $O_2$ thus becomes impossible as cell size increases beyond a certain point, placing a theoretical upper limit on the size of the cell.

## There Are Three Distinct Domains of Life

All living organisms fall into one of three large groups (kingdoms, or domains) that define three branches of evolution from a common progenitor (Fig. 1–4). Two large groups of prokaryotes can be distinguished on biochemical grounds: **archaebacteria** (Greek *archē,* "origin") and **eubacteria** (again, from Greek *eu,* "true"). Eubacteria inhabit soils, surface waters, and the tissues of other living or decaying organisms. Most of the well-studied bacteria, including *Escherichia coli,* are eubacteria. The archaebacteria, more recently discovered, are less well characterized biochemically; most inhabit extreme environments—salt lakes, hot springs, highly acidic bogs, and the ocean depths. The available evidence suggests that the archaebacteria and eubacteria diverged early in evolution and constitute two separate

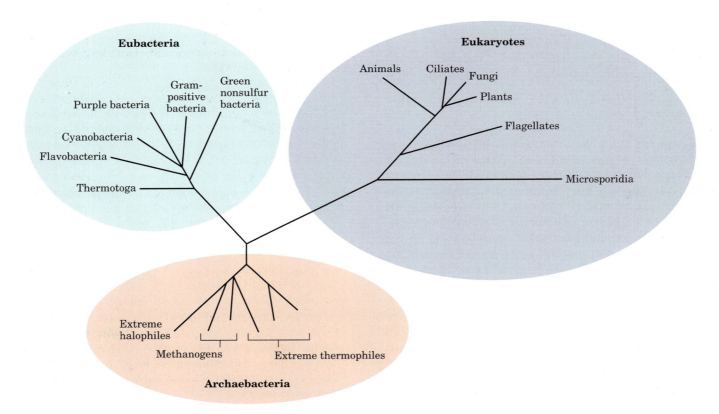

**FIGURE 1–4 Phylogeny of the three domains of life.** Phylogenetic relationships are often illustrated by a "family tree" of this type. The fewer the branch points between any two organisms, the closer is their evolutionary relationship.

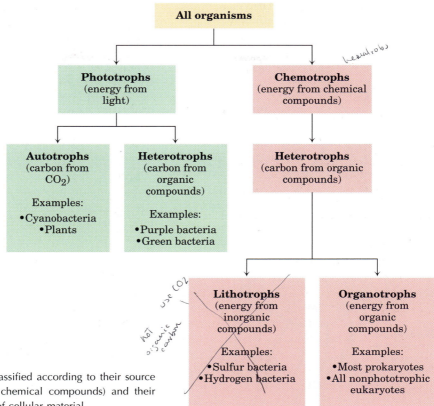

**FIGURE 1–5** Organisms can be classified according to their source of energy (sunlight or oxidizable chemical compounds) and their source of carbon for the synthesis of cellular material.

domains, sometimes called Archaea and Bacteria. All eukaryotic organisms, which make up the third domain, Eukarya, evolved from the same branch that gave rise to the Archaea; archaebacteria are therefore more closely related to eukaryotes than to eubacteria.

Within the domains of Archaea and Bacteria are subgroups distinguished by the habitats in which they live. In **aerobic** habitats with a plentiful supply of oxygen, some resident organisms derive energy from the transfer of electrons from fuel molecules to oxygen. Other environments are **anaerobic,** virtually devoid of oxygen, and microorganisms adapted to these environments obtain energy by transferring electrons to nitrate (forming $N_2$), sulfate (forming $H_2S$), or $CO_2$ (forming $CH_4$). Many organisms that have evolved in anaerobic environments are *obligate anaerobes:* they die when exposed to oxygen.

We can classify organisms according to how they obtain the energy and carbon they need for synthesizing cellular material (as summarized in Fig. 1–5). There are two broad categories based on energy sources: **phototrophs** (Greek *trophē,* "nourishment") trap and use sunlight, and **chemotrophs** derive their energy from oxidation of a fuel. All chemotrophs require a source of organic nutrients; they cannot fix $CO_2$ into organic compounds. The phototrophs can be further divided into those that can obtain all needed carbon from $CO_2$ (**autotrophs**) and those that require organic nutrients (**heterotrophs**). No chemotroph can get its carbon

atoms exclusively from $CO_2$ (that is, no chemotrophs are autotrophs), but the chemotrophs may be further classified according to a different criterion: whether the fuels they oxidize are inorganic (**lithotrophs**) or organic (**organotrophs**).

Most known organisms fall within one of these four broad categories—autotrophs or heterotrophs among the photosynthesizers, lithotrophs or organotrophs among the chemical oxidizers. The prokaryotes have several general modes of obtaining carbon and energy. *Escherichia coli,* for example, is a chemoorganoheterotroph; it requires organic compounds from its environment as fuel and as a source of carbon. Cyanobacteria are photolithoautotrophs; they use sunlight as an energy source and convert $CO_2$ into biomolecules. We humans, like *E. coli,* are chemoorganoheterotrophs.

### *Escherichia coli* Is the Most-Studied Prokaryotic Cell

Bacterial cells share certain common structural features, but also show group-specific specializations (Fig. 1–6). *E. coli* is a usually harmless inhabitant of the human intestinal tract. The *E. coli* cell is about 2 $\mu$m long and a little less than 1 $\mu$m in diameter. It has a protective outer membrane and an inner plasma membrane that encloses the cytoplasm and the nucleoid. Between the inner and outer membranes is a thin but strong layer of polymers called peptidoglycans, which gives the cell its shape and rigidity. The plasma membrane and the

**Ribosomes** Bacterial ribosomes are smaller than eukaryotic ribosomes, but serve the same function—protein synthesis from an RNA message.

**Nucleoid** Contains a single, simple, long circular DNA molecule.

**Pili** Provide points of adhesion to surface of other cells.

**Flagella** Propel cell through its surroundings.

**Cell envelope** Structure varies with type of bacteria.

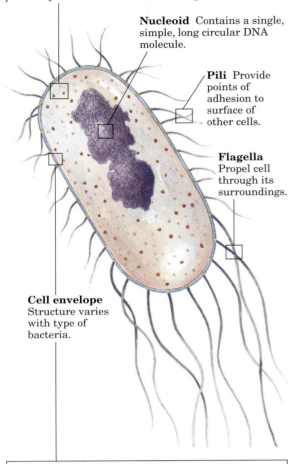

**Gram-negative bacteria**
Outer membrane;
peptidoglycan layer

Outer membrane
Peptidoglycan layer
Inner membrane

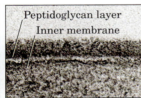

**Gram-positive bacteria**
No outer membrane;
thicker peptidoglycan layer

Peptidoglycan layer
Inner membrane

**Cyanobacteria**
Gram-negative; tougher
peptidoglycan layer;
extensive internal
membrane system with
photosynthetic pigments

**Archaebacteria**
No outer membrane;
peptidoglycan layer outside
plasma membrane

**FIGURE 1-6 Common structural features of bacterial cells.** Because of differences in the cell envelope structure, some eubacteria (gram-positive bacteria) retain Gram's stain, and others (gram-negative bacteria) do not. *E. coli* is gram-negative. Cyanobacteria are also eubacteria but are distinguished by their extensive internal membrane system, in which photosynthetic pigments are localized. Although the cell envelopes of archaebacteria and gram-positive eubacteria look similar under the electron microscope, the structures of the membrane lipids and the polysaccharides of the cell envelope are distinctly different in these organisms.

layers outside it constitute the **cell envelope.** In the Archaea, rigidity is conferred by a different type of polymer (pseudopeptidoglycan). The plasma membranes of eubacteria consist of a thin bilayer of lipid molecules penetrated by proteins. Archaebacterial membranes have a similar architecture, although their lipids differ strikingly from those of the eubacteria.

The cytoplasm of *E. coli* contains about 15,000 ribosomes, thousands of copies each of about 1,000 different enzymes, numerous metabolites and cofactors, and a variety of inorganic ions. The nucleoid contains a single, circular molecule of DNA, and the cytoplasm (like that of most bacteria) contains one or more smaller, circular segments of DNA called **plasmids.** In nature, some plasmids confer resistance to toxins and antibiotics in the environment. In the laboratory, these DNA segments are especially amenable to experimental manipulation and are extremely useful to molecular geneticists.

Most bacteria (including *E. coli*) lead existences as individual cells, but in some bacterial species cells tend to associate in clusters or filaments, and a few (the myxobacteria, for example) demonstrate simple social behavior.

## Eukaryotic Cells Have a Variety of Membranous Organelles, Which Can Be Isolated for Study

Typical eukaryotic cells (Fig. 1–7) are much larger than prokaryotic cells—commonly 5 to 100 $\mu$m in diameter, with cell volumes a thousand to a million times larger than those of bacteria. The distinguishing characteristics of eukaryotes are the nucleus and a variety of membrane-bounded organelles with specific functions: mitochondria, endoplasmic reticulum, Golgi complexes, and lysosomes. Plant cells also contain vacuoles and chloroplasts (Fig. 1–7). Also present in the cytoplasm of many cells are granules or droplets containing stored nutrients such as starch and fat.

In a major advance in biochemistry, Albert Claude, Christian de Duve, and George Palade developed methods for separating organelles from the cytosol and from each other—an essential step in isolating biomolecules and larger cell components and investigating their

**(a) Animal cell**

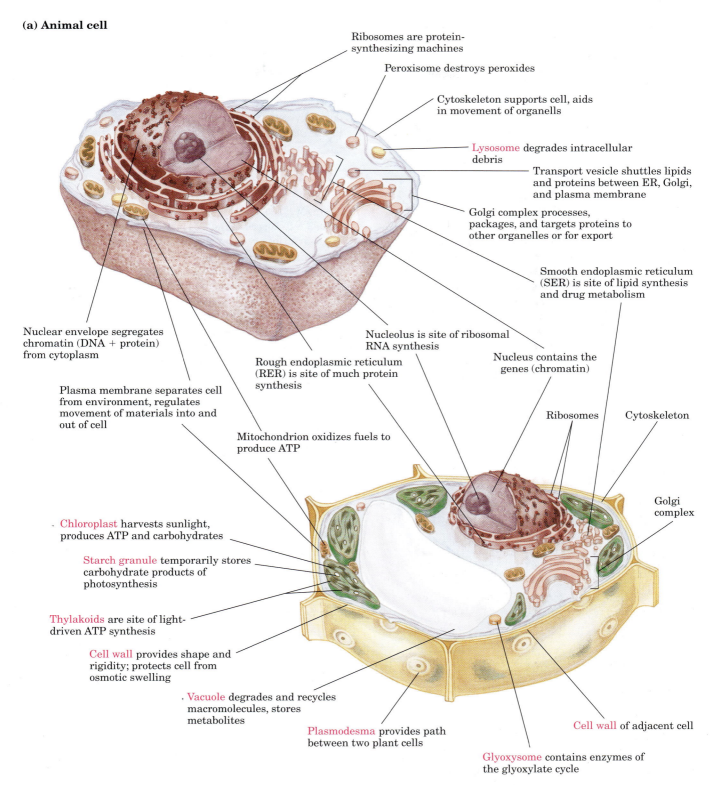

Ribosomes are protein-synthesizing machines

Peroxisome destroys peroxides

Cytoskeleton supports cell, aids in movement of organells

Lysosome degrades intracellular debris

Transport vesicle shuttles lipids and proteins between ER, Golgi, and plasma membrane

Golgi complex processes, packages, and targets proteins to other organelles or for export

Smooth endoplasmic reticulum (SER) is site of lipid synthesis and drug metabolism

Nuclear envelope segregates chromatin (DNA + protein) from cytoplasm

Nucleolus is site of ribosomal RNA synthesis

Nucleus contains the genes (chromatin)

Plasma membrane separates cell from environment, regulates movement of materials into and out of cell

Rough endoplasmic reticulum (RER) is site of much protein synthesis

Ribosomes

Cytoskeleton

Mitochondrion oxidizes fuels to produce ATP

Golgi complex

Chloroplast harvests sunlight, produces ATP and carbohydrates

Starch granule temporarily stores carbohydrate products of photosynthesis

Thylakoids are site of light-driven ATP synthesis

Cell wall provides shape and rigidity; protects cell from osmotic swelling

Vacuole degrades and recycles macromolecules, stores metabolites

Cell wall of adjacent cell

Plasmodesma provides path between two plant cells

Glyoxysome contains enzymes of the glyoxylate cycle

**(b) Plant cell**

**FIGURE 1–7 Eukaryotic cell structure.** Schematic illustrations of the two major types of eukaryotic cell: **(a)** a representative animal cell and **(b)** a representative plant cell. Plant cells are usually 10 to 100 $\mu$m in diameter—larger than animal cells, which typically range from 5 to 30 $\mu$m. Structures labeled in red are unique to either animal or plant cells.

structures and functions. In a typical cell fractionation (Fig. 1–8), cells or tissues in solution are disrupted by gentle homogenization. This treatment ruptures the plasma membrane but leaves most of the organelles intact. The homogenate is then centrifuged; organelles such as nuclei, mitochondria, and lysosomes differ in size and therefore sediment at different rates. They also differ in specific gravity, and they "float" at different levels in a density gradient.

Differential centrifugation results in a rough fractionation of the cytoplasmic contents, which may be further purified by isopycnic ("same density") centrifugation. In this procedure, organelles of different buoyant densities (the result of different ratios of lipid and protein in each type of organelle) are separated on a density gradient. By carefully removing material from each region of the gradient and observing it with a microscope, the biochemist can establish the sedimentation position of each organelle

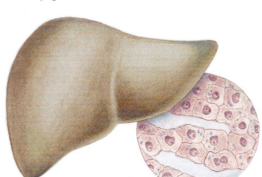

**FIGURE 1–8 Subcellular fractionation of tissue.** A tissue such as liver is first mechanically homogenized to break cells and disperse their contents in an aqueous buffer. The sucrose medium has an osmotic pressure similar to that in organelles, thus preventing diffusion of water into the organelles, which would swell and burst. **(a)** The large and small particles in the suspension can be separated by centrifugation at different speeds, or **(b)** particles of different density can be separated by isopycnic centrifugation. In isopycnic centrifugation, a centrifuge tube is filled with a solution, the density of which increases from top to bottom; a solute such as sucrose is dissolved at different concentrations to produce the density gradient. When a mixture of organelles is layered on top of the density gradient and the tube is centrifuged at high speed, individual organelles sediment until their buoyant density exactly matches that in the gradient. Each layer can be collected separately.

**(a) Differential centrifugation**

Tissue homogenization

Low-speed centrifugation (1,000 *g*, 10 min)

Tissue homogenate

Pellet contains whole cells, nuclei, cytoskeletons, plasma membranes

Supernatant subjected to medium-speed centrifugation (20,000 *g*, 20 min)

Pellet contains →mitochondria, lysosomes, peroxisomes

Supernatant subjected to high-speed centrifugation (80,000 *g*, 1 h)

Pellet contains → microsomes (fragments of ER), small vesicles

Supernatant subjected to very high-speed centrifugation (150,000 *g*, 3 h)

Supernatant contains soluble proteins

Pellet contains ribosomes, large macromolecules

**(b) Isopycnic (sucrose-density) centrifugation**

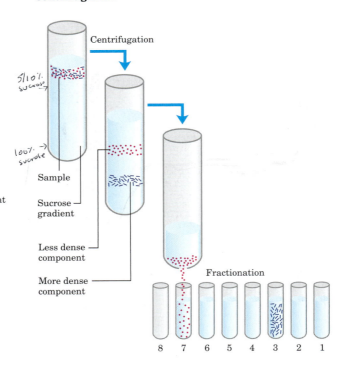

Centrifugation

5/10% sucrose

100% sucrose

Sample

Sucrose gradient

Less dense component

More dense component

Fractionation

8   7   6   5   4   3   2   1

and obtain purified organelles for further study. For example, these methods were used to establish that lysosomes contain degradative enzymes, mitochondria contain oxidative enzymes, and chloroplasts contain photosynthetic pigments. The isolation of an organelle enriched in a certain enzyme is often the first step in the purification of that enzyme.

## The Cytoplasm Is Organized by the Cytoskeleton and Is Highly Dynamic

Electron microscopy reveals several types of protein filaments crisscrossing the eukaryotic cell, forming an interlocking three-dimensional meshwork, the **cytoskeleton.** There are three general types of cytoplasmic filaments—actin filaments, microtubules, and intermediate filaments (Fig. 1–9)—differing in width (from about 6 to 22 nm), composition, and specific function. All types provide structure and organization to the cytoplasm and shape to the cell. Actin filaments and microtubules also help to produce the motion of organelles or of the whole cell.

Each type of cytoskeletal component is composed of simple protein subunits that polymerize to form filaments of uniform thickness. These filaments are not permanent structures; they undergo constant disassembly into their protein subunits and reassembly into filaments. Their locations in cells are not rigidly fixed but may change dramatically with mitosis, cytokinesis, amoeboid motion, or changes in cell shape. The assembly, disassembly, and location of all types of filaments are regulated by other proteins, which serve to link or bundle the filaments or to move cytoplasmic organelles along the filaments.

The picture that emerges from this brief survey of cell structure is that of a eukaryotic cell with a meshwork of structural fibers and a complex system of membrane-bounded compartments (Fig. 1–7). The filaments disassemble and then reassemble elsewhere. Membranous vesicles bud from one organelle and fuse with another. Organelles move through the cytoplasm along protein filaments, their motion powered by energy dependent motor proteins. The **endomembrane system** segregates specific metabolic processes and provides surfaces on which certain enzyme-catalyzed reactions occur. **Exocytosis** and **endocytosis,** mechanisms of transport (out of and into cells, respectively) that involve membrane fusion and fission, provide paths between the cytoplasm and surrounding medium, allowing for secretion of substances produced within the cell and uptake of extracellular materials.

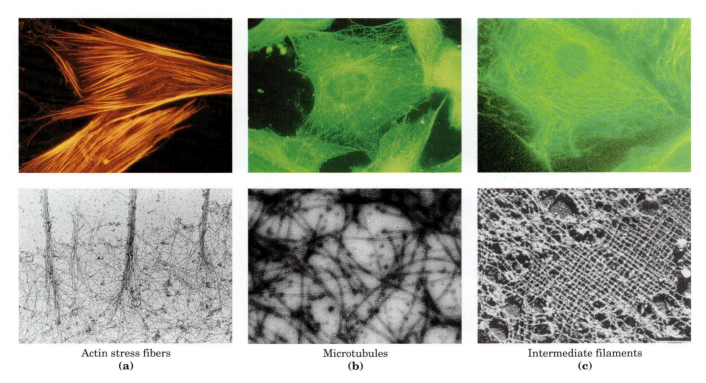

Actin stress fibers
(a)

Microtubules
(b)

Intermediate filaments
(c)

**FIGURE 1–9 The three types of cytoskeletal filaments.** The upper panels show epithelial cells photographed after treatment with antibodies that bind to and specifically stain (a) actin filaments bundled together to form "stress fibers," (b) microtubules radiating from the cell center, and (c) intermediate filaments extending throughout the cytoplasm. For these experiments, antibodies that specifically recognize actin, tubulin, or intermediate filament proteins are covalently attached to a fluorescent compound. When the cell is viewed with a fluorescence microscope, only the stained structures are visible. The lower panels show each type of filament as visualized by (a, b) transmission or (c) scanning electron microscopy.

Although complex, this organization of the cytoplasm is far from random. The motion and the positioning of organelles and cytoskeletal elements are under tight regulation, and at certain stages in a eukaryotic cell's life, dramatic, finely orchestrated reorganizations, such as the events of mitosis, occur. The interactions between the cytoskeleton and organelles are noncovalent, reversible, and subject to regulation in response to various intracellular and extracellular signals.

## Cells Build Supramolecular Structures

Macromolecules and their monomeric subunits differ greatly in size (Fig. 1–10). A molecule of alanine is less than 0.5 nm long. Hemoglobin, the oxygen-carrying protein of erythrocytes (red blood cells), consists of nearly 600 amino acid subunits in four long chains, folded into globular shapes and associated in a structure 5.5 nm in diameter. In turn, proteins are much smaller than ribosomes (about 20 nm in diameter), which are in turn much smaller than organelles such as mitochondria, typically 1,000 nm in diameter. It is a long jump from simple biomolecules to cellular structures that can be seen

**FIGURE 1–10 The organic compounds from which most cellular materials are constructed: the ABCs of biochemistry.** Shown here are **(a)** six of the 20 amino acids from which all proteins are built (the side chains are shaded pink); **(b)** the five nitrogenous bases, two five-carbon sugars, and phosphoric acid from which all nucleic acids are built; **(c)** five components of membrane lipids; and **(d)** D-glucose, the parent sugar from which most carbohydrates are derived. Note that phosphoric acid is a component of both nucleic acids and membrane lipids.

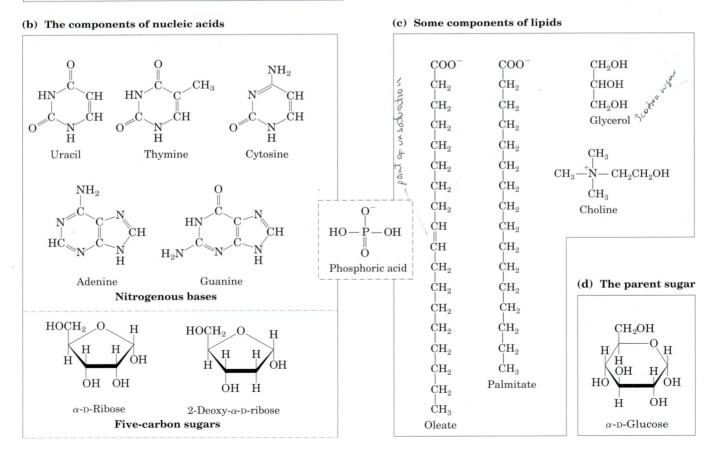

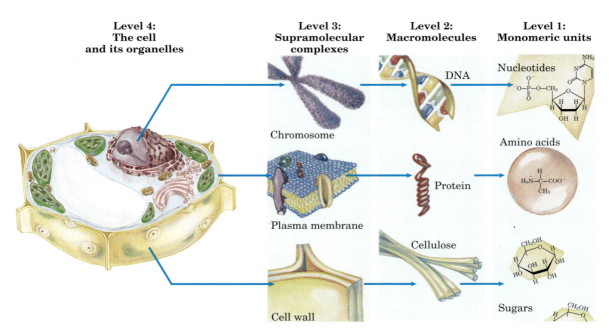

Level 4:
The cell
and its organelles

Level 3:
Supramolecular
complexes

Level 2:
Macromolecules

Level 1:
Monomeric units

Chromosome

Plasma membrane

Cell wall

DNA

Protein

Cellulose

Nucleotides

Amino acids

Sugars

**FIGURE 1–11 Structural hierarchy in the molecular organization of cells.** In this plant cell, the nucleus is an organelle containing several types of supramolecular complexes, including chromosomes. Chro-mosomes consist of macromolecules of DNA and many different proteins. Each type of macromolecule is made up of simple subunits—DNA of nucleotides (deoxyribonucleotides), for example.

with the light microscope. Figure 1–11 illustrates the structural hierarchy in cellular organization.

The monomeric subunits in proteins, nucleic acids, and polysaccharides are joined by covalent bonds. In supramolecular complexes, however, macromolecules are held together by noncovalent interactions—much weaker, individually, than covalent bonds. Among these noncovalent interactions are hydrogen bonds (between polar groups), ionic interactions (between charged groups), hydrophobic interactions (among nonpolar groups in aqueous solution), and van der Waals interactions—all of which have energies substantially smaller than those of covalent bonds (Table 1–1). The nature of these noncovalent interactions is described in Chapter 2. The large numbers of weak interactions between macromolecules in supramolecular complexes stabilize these assemblies, producing their unique structures.

### In Vitro Studies May Overlook Important Interactions among Molecules

One approach to understanding a biological process is to study purified molecules in vitro ("in glass"—in the test tube), without interference from other molecules present in the intact cell—that is, in vivo ("in the living"). Although this approach has been remarkably revealing, we must keep in mind that the inside of a cell is quite different from the inside of a test tube. The "interfering" components eliminated by purification may be critical to the biological function or regulation of the molecule purified. For example, in vitro studies of pure

enzymes are commonly done at very low enzyme concentrations in thoroughly stirred aqueous solutions. In the cell, an enzyme is dissolved or suspended in a gel-like cytosol with thousands of other proteins, some of which bind to that enzyme and influence its activity.

**TABLE 1–1   Strengths of Bonds Common in Biomolecules**

| Type of bond | Bond dissociation energy* (kJ/mol) | Type of bond | Bond dissociation energy (kJ/mol) |
|---|---|---|---|
| **Single bonds** | | **Double bonds** | |
| O—H | 470 | C=O | 712 |
| H—H | 435 | C=N | 615 |
| P—O | 419 | C=C | 611 |
| C—H | 414 | P=O | 502 |
| N—H | 389 | | |
| C—O | 352 | **Triple bonds** | |
| C—C | 348 | C≡C | 816 |
| S—H | 339 | N≡N | 930 |
| C—N | 293 | | |
| C—S | 260 | | |
| N—O | 222 | | |
| S—S | 214 | | |

*The greater the energy required for bond dissociation (breakage), the stronger the bond.

Some enzymes are parts of multienzyme complexes in which reactants are channeled from one enzyme to another without ever entering the bulk solvent. Diffusion is hindered in the gel-like cytosol, and the cytosolic composition varies in different regions of the cell. In short, a given molecule may function quite differently in the cell than in vitro. A central challenge of biochemistry is to understand the influences of cellular organization and macromolecular associations on the function of individual enzymes and other biomolecules—to understand function in vivo as well as in vitro.

### SUMMARY 1.1 Cellular Foundations

- All cells are bounded by a plasma membrane; have a cytosol containing metabolites, coenzymes, inorganic ions, and enzymes; and have a set of genes contained within a nucleoid (prokaryotes) or nucleus (eukaryotes).

- Phototrophs use sunlight to do work; chemotrophs oxidize fuels, passing electrons to good electron acceptors: inorganic compounds, organic compounds, or molecular oxygen.

- Bacterial cells contain cytosol, a nucleoid, and plasmids. Eukaryotic cells have a nucleus and are multicompartmented, segregating certain processes in specific organelles, which can be separated and studied in isolation.

- Cytoskeletal proteins assemble into long filaments that give cells shape and rigidity and serve as rails along which cellular organelles move throughout the cell.

- Supramolecular complexes are held together by noncovalent interactions and form a hierarchy of structures, some visible with the light microscope. When individual molecules are removed from these complexes to be studied in vitro, interactions important in the living cell may be lost.

## 1.2 Chemical Foundations

Biochemistry aims to explain biological form and function in chemical terms. As we noted earlier, one of the most fruitful approaches to understanding biological phenomena has been to purify an individual chemical component, such as a protein, from a living organism and to characterize its structural and chemical characteristics. By the late eighteenth century, chemists had concluded that the composition of living matter is strikingly different from that of the inanimate world. Antoine Lavoisier (1743–1794) noted the relative chemical simplicity of the "mineral world" and contrasted it with the complexity of the "plant and animal worlds"; the latter, he knew, were composed of compounds rich in the elements carbon, oxygen, nitrogen, and phosphorus.

During the first half of the twentieth century, parallel biochemical investigations of glucose breakdown in yeast and in animal muscle cells revealed remarkable chemical similarities in these two apparently very different cell types; the breakdown of glucose in yeast and muscle cells involved the same ten chemical intermediates. Subsequent studies of many other biochemical processes in many different organisms have confirmed the generality of this observation, neatly summarized by Jacques Monod: "What is true of *E. coli* is true of the elephant." The current understanding that all organisms share a common evolutionary origin is based in part on this observed universality of chemical intermediates and transformations.

Only about 30 of the more than 90 naturally occurring chemical elements are essential to organisms. Most of the elements in living matter have relatively low atomic numbers; only five have atomic numbers above that of selenium, 34 (Fig. 1–12). The four most abundant elements in living organisms, in terms of percentage of total number of atoms, are hydrogen, oxygen, nitrogen, and carbon, which together make up more than 99% of the mass of most cells. They are the lightest elements capable of forming one, two, three, and four bonds, respectively; in general, the lightest elements

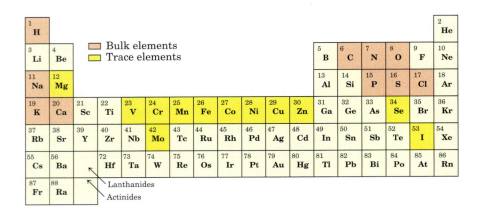

**FIGURE 1-12 Elements essential to animal life and health.** Bulk elements (shaded orange) are structural components of cells and tissues and are required in the diet in gram quantities daily. For trace elements (shaded bright yellow), the requirements are much smaller: for humans, a few milligrams per day of Fe, Cu, and Zn, even less of the others. The elemental requirements for plants and microorganisms are similar to those shown here; the ways in which they acquire these elements vary.

form the strongest bonds. The trace elements (Fig. 1–12) represent a miniscule fraction of the weight of the human body, but all are essential to life, usually because they are essential to the function of specific proteins, including enzymes. The oxygen-transporting capacity of the hemoglobin molecule, for example, is absolutely dependent on four iron ions that make up only 0.3% of its mass.

## Biomolecules Are Compounds of Carbon with a Variety of Functional Groups

The chemistry of living organisms is organized around carbon, which accounts for more than half the dry weight of cells. Carbon can form single bonds with hydrogen atoms, and both single and double bonds with oxygen and nitrogen atoms (Fig. 1–13). Of greatest significance in biology is the ability of carbon atoms to form very stable carbon–carbon single bonds. Each carbon atom can form single bonds with up to four other carbon atoms. Two carbon atoms also can share two (or three) electron pairs, thus forming double (or triple) bonds.

The four single bonds that can be formed by a carbon atom are arranged tetrahedrally, with an angle of

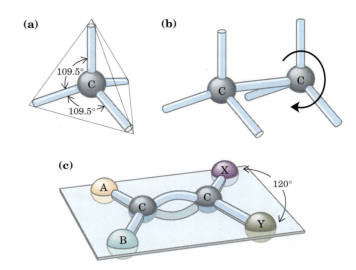

**FIGURE 1–14 Geometry of carbon bonding. (a)** Carbon atoms have a characteristic tetrahedral arrangement of their four single bonds. **(b)** Carbon–carbon single bonds have freedom of rotation, as shown for the compound ethane ($CH_3$—$CH_3$). **(c)** Double bonds are shorter and do not allow free rotation. The two doubly bonded carbons and the atoms designated A, B, X, and Y all lie in the same rigid plane.

about 109.5° between any two bonds (Fig. 1–14) and an average length of 0.154 nm. There is free rotation around each single bond, unless very large or highly charged groups are attached to both carbon atoms, in which case rotation may be restricted. A double bond is shorter (about 0.134 nm) and rigid and allows little rotation about its axis.

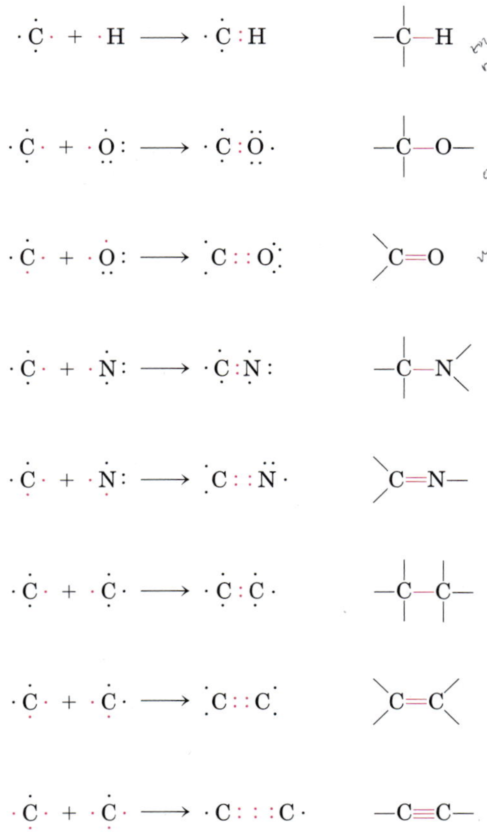

Covalently linked carbon atoms in biomolecules can form linear chains, branched chains, and cyclic structures. To these carbon skeletons are added groups of other atoms, called **functional groups,** which confer specific chemical properties on the molecule. It seems likely that the bonding versatility of carbon was a major factor in the selection of carbon compounds for the molecular machinery of cells during the origin and evolution of living organisms. No other chemical element can form molecules of such widely different sizes and shapes or with such a variety of functional groups.

Most biomolecules can be regarded as derivatives of hydrocarbons, with hydrogen atoms replaced by a variety of functional groups to yield different families of organic compounds. Typical of these are alcohols, which have one or more hydroxyl groups; amines, with amino groups; aldehydes and ketones, with carbonyl groups; and carboxylic acids, with carboxyl groups (Fig. 1–15). Many biomolecules are polyfunctional, containing two or more different kinds of functional groups (Fig. 1–16), each with its own chemical characteristics and reactions. The chemical "personality" of a compound is determined by the chemistry of its functional groups and their disposition in three-dimensional space.

**FIGURE 1–13 Versatility of carbon bonding.** Carbon can form covalent single, double, and triple bonds (in red), particularly with other carbon atoms. Triple bonds are rare in biomolecules.

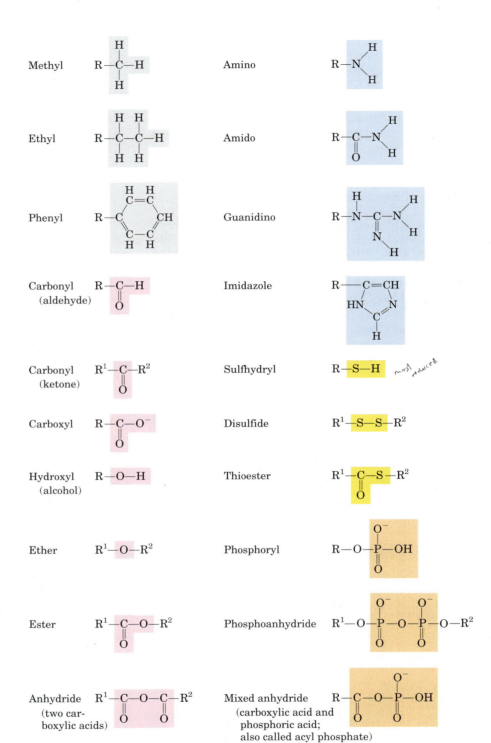

**FIGURE 1-15  Some common functional groups of biomolecules.** In this figure and throughout the book, we use R to represent "any substituent." It may be as simple as a hydrogen atom, but typically it is a carbon-containing moiety. When two or more substituents are shown in a molecule, we designate them $R^1$, $R^2$, and so forth.

## Cells Contain a Universal Set of Small Molecules

Dissolved in the aqueous phase (cytosol) of all cells is a collection of 100 to 200 different small organic molecules ($M_r$ ~100 to ~500), the central metabolites in the major pathways occurring in nearly every cell—the metabolites and pathways that have been conserved throughout the course of evolution. (See Box 1–1 for an explanation of the various ways of referring to molecu-

lar weight.) This collection of molecules includes the common amino acids, nucleotides, sugars and their phosphorylated derivatives, and a number of mono-, di-, and tricarboxylic acids. The molecules are polar or charged, water soluble, and present in micromolar to millimolar concentrations. They are trapped within the cell because the plasma membrane is impermeable to them—although specific membrane transporters can catalyze the movement of some molecules into and out

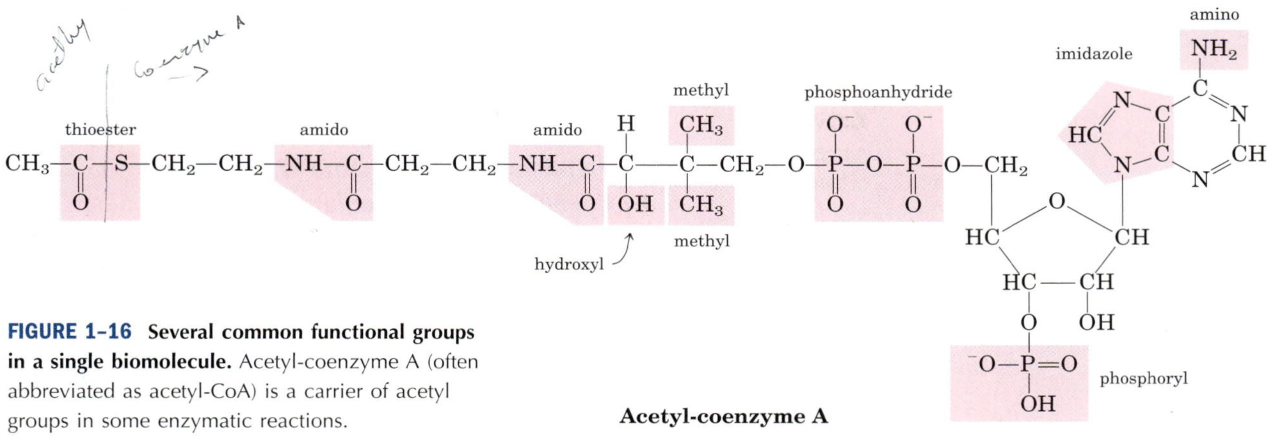

**FIGURE 1–16** **Several common functional groups in a single biomolecule.** Acetyl-coenzyme A (often abbreviated as acetyl-CoA) is a carrier of acetyl groups in some enzymatic reactions.

**Acetyl-coenzyme A**

of the cell or between compartments in eukaryotic cells. The universal occurrence of the same set of compounds in living cells is a manifestation of the universality of metabolic design, reflecting the evolutionary conservation of metabolic pathways that developed in the earliest cells.

There are other small biomolecules, specific to certain types of cells or organisms. For example, vascular plants contain, in addition to the universal set, small molecules called **secondary metabolites,** which play a role specific to plant life. These metabolites include compounds that give plants their characteristic scents, and compounds such as morphine, quinine, nicotine, and caffeine that are valued for their physiological effects on humans but used for other purposes by plants. The entire collection of small molecules in a given cell has been called that cell's **metabolome,** in parallel with the term "genome" (defined earlier and expanded on in Section 1.4). If we knew the composition of a cell's metabolome, we could predict which enzymes and metabolic pathways were active in that cell.

### Macromolecules Are the Major Constituents of Cells

Many biological molecules are macromolecules, polymers of high molecular weight assembled from relatively simple precursors. Proteins, nucleic acids, and polysaccharides are produced by the polymerization of relatively small compounds with molecular weights of 500 or less. The number of polymerized units can range from tens to millions. Synthesis of macromolecules is a major energy-consuming activity of cells. Macromolecules themselves may be further assembled into supramolecular complexes, forming functional units such as ribosomes. Table 1–2 shows the major classes of biomolecules in the bacterium *E. coli.*

---

### BOX 1–1 WORKING IN BIOCHEMISTRY

### Molecular Weight, Molecular Mass, and Their Correct Units

There are two common (and equivalent) ways to describe molecular mass; both are used in this text. The first is *molecular weight,* or *relative molecular mass,* denoted $M_r$. The molecular weight of a substance is defined as the ratio of the mass of a molecule of that substance to one-twelfth the mass of carbon-12 ($^{12}C$). Since $M_r$ is a ratio, it is dimensionless—it has no associated units. The second is *molecular mass,* denoted *m*. This is simply the mass of one molecule, or the molar mass divided by Avogadro's number. The molecular mass, *m*, is expressed in daltons (abbreviated Da). One dalton is equivalent to one-twelfth the mass of carbon-12; a kilodalton (kDa) is 1,000 daltons; a megadalton (MDa) is 1 million daltons.

Consider, for example, a molecule with a mass 1,000 times that of water. We can say of this molecule either $M_r = 18,000$ or $m = 18,000$ daltons. We can also describe it as an "18 kDa molecule." However, the expression $M_r = 18,000$ daltons is incorrect.

Another convenient unit for describing the mass of a single atom or molecule is the atomic mass unit (formerly amu, now commonly denoted u). One atomic mass unit (1 u) is defined as one-twelfth the mass of an atom of carbon-12. Since the experimentally measured mass of an atom of carbon-12 is $1.9926 \times 10^{-23}$ g, 1 u $= 1.6606 \times 10^{-24}$ g. The atomic mass unit is convenient for describing the mass of a peak observed by mass spectrometry (see Box 3–2).

**TABLE 1–2    Molecular Components of an E. coli Cell**

|  | Percentage of total weight of cell | Approximate number of different molecular species |
|---|---|---|
| Water | 70 | 1 |
| Proteins | 15 | 3,000 |
| Nucleic acids |  |  |
| DNA | 1 | 1 |
| RNA | 6 | >3,000 |
| Polysaccharides | 3 | 5 |
| Lipids | 2 | 20 |
| Monomeric subunits and intermediates | 2 | 500 |
| Inorganic ions | 1 | 20 |

**Proteins,** long polymers of amino acids, constitute the largest fraction (besides water) of cells. Some proteins have catalytic activity and function as enzymes; others serve as structural elements, signal receptors, or transporters that carry specific substances into or out of cells. Proteins are perhaps the most versatile of all biomolecules. The **nucleic acids,** DNA and RNA, are polymers of nucleotides. They store and transmit genetic information, and some RNA molecules have structural and catalytic roles in supramolecular complexes. The **polysaccharides,** polymers of simple sugars such as glucose, have two major functions: as energy-yielding fuel stores and as extracellular structural elements with specific binding sites for particular proteins. Shorter polymers of sugars (oligosaccharides) attached to proteins or lipids at the cell surface serve as specific cellular signals. The **lipids,** greasy or oily hydrocarbon derivatives, serve as structural components of membranes, energy-rich fuel stores, pigments, and intracellular signals. In proteins, nucleotides, polysaccharides, and lipids, the number of monomeric subunits is very large: molecular weights in the range of 5,000 to more than 1 million for proteins, up to several billion for nucleic acids, and in the millions for polysaccharides such as starch. Individual lipid molecules are much smaller ($M_r$ 750 to 1,500) and are not classified as macromolecules. However, large numbers of lipid molecules can associate noncovalently into very large structures. Cellular membranes are built of enormous noncovalent aggregates of lipid and protein molecules.

Proteins and nucleic acids are **informational macromolecules:** each protein and each nucleic acid has a characteristic information-rich subunit sequence. Some oligosaccharides, with six or more different sug-

ars connected in branched chains, also carry information; on the outer surface of cells they serve as highly specific points of recognition in many cellular processes (as described in Chapter 7).

## Three-Dimensional Structure Is Described by Configuration and Conformation

The covalent bonds and functional groups of a biomolecule are, of course, central to its function, but so also is the arrangement of the molecule's constituent atoms in three-dimensional space—its stereochemistry. A carbon-containing compound commonly exists as **stereoisomers,** molecules with the same chemical bonds but different stereochemistry—that is, different **configuration,** the fixed spatial arrangement of atoms. Interactions between biomolecules are invariably **stereospecific,** requiring specific stereochemistry in the interacting molecules.

Figure 1–17 shows three ways to illustrate the stereochemical structures of simple molecules. The perspective diagram specifies stereochemistry unambiguously, but bond angles and center-to-center bond lengths are better represented with ball-and-stick models. In space-

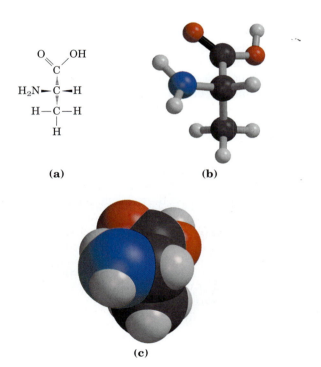

**FIGURE 1–17  Representations of molecules.** Three ways to represent the structure of the amino acid alanine. **(a)** Structural formula in perspective form: a solid wedge (—▶) represents a bond in which the atom at the wide end projects out of the plane of the paper, toward the reader; a dashed wedge (┅┅) represents a bond extending behind the plane of the paper. **(b)** Ball-and-stick model, showing relative bond lengths and the bond angles. **(c)** Space-filling model, in which each atom is shown with its correct relative van der Waals radius.

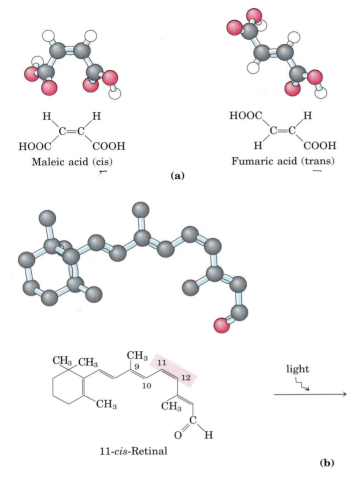

**Maleic acid (cis)**

**Fumaric acid (trans)**

**(a)**

**FIGURE 1-18** **Configurations of geometric isomers. (a)** Isomers such as maleic acid and fumaric acid cannot be interconverted without breaking covalent bonds, which requires the input of much energy. **(b)** In the vertebrate retina, the initial event in light detection is the absorption of visible light by 11-*cis*-retinal. The energy of the absorbed light (about 250 kJ/mol) converts 11-*cis*-retinal to all-*trans*-retinal, triggering electrical changes in the retinal cell that lead to a nerve impulse. (Note that the hydrogen atoms are omitted from the ball-and-stick models for the retinals.)

light

11-*cis*-Retinal

All-*trans*-Retinal

**(b)**

filling models, the radius of each atom is proportional to its van der Waals radius, and the contours of the model define the space occupied by the molecule (the volume of space from which atoms of other molecules are excluded).

Configuration is conferred by the presence of either (1) double bonds, around which there is no freedom of rotation, or (2) chiral centers, around which substituent groups are arranged in a specific sequence. The identifying characteristic of configurational isomers is that they cannot be interconverted without temporarily breaking one or more covalent bonds. Figure 1–18a shows the configurations of maleic acid and its isomer, fumaric acid. These compounds are **geometric, or cis-trans, isomers;** they differ in the arrangement of their substituent groups with respect to the nonrotating double bond (Latin *cis*, "on this side"—groups on the same side of the double bond; *trans*, "across"—groups on opposite sides). Maleic acid is the cis isomer and fumaric acid the trans isomer; each is a well-defined compound that can be separated from the other, and each has its own unique chemical properties. A binding site (on an enzyme, for example) that is complementary to one of these molecules would not be a suitable binding site for the other, which explains why the two compounds have distinct biological roles despite their similar chemistry.

In the second type of configurational isomer, four different substituents bonded to a tetrahedral carbon atom may be arranged two different ways in space—that is, have two configurations (Fig. 1–19)—yielding two stereoisomers with similar or identical chemical properties but differing in certain physical and biological properties. A carbon atom with four different substituents is said to be asymmetric, and asymmetric carbons are called **chiral centers** (Greek *chiros*, "hand"; some stereoisomers are related structurally as the right hand is to the left). A molecule with only one chiral carbon can have two stereoisomers; when two or more ($n$) chiral carbons are present, there can be $2^n$ stereoisomers. Some stereoisomers are mirror images of each other; they are called **enantiomers** (Fig. 1–19). Pairs of stereoisomers that are not mirror images of each other are called **diastereomers** (Fig. 1–20).

As Louis Pasteur first observed (Box 1–2), enantiomers have nearly identical chemical properties but differ in a characteristic physical property, their interaction with plane-polarized light. In separate solutions, two enantiomers rotate the plane of plane-polarized light in opposite directions, but an equimolar solution of the two enantiomers (a **racemic mixture**) shows no optical rotation. Compounds without chiral centers do not rotate the plane of plane-polarized light.

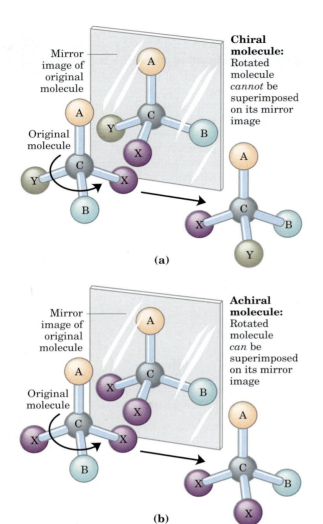

**(a)**

**(b)**

Mirror image of original molecule

Original molecule

**Chiral molecule:** Rotated molecule *cannot* be superimposed on its mirror image

Mirror image of original molecule

Original molecule

**Achiral molecule:** Rotated molecule *can* be superimposed on its mirror image

**FIGURE 1-19  Molecular asymmetry: chiral and achiral molecules.** **(a)** When a carbon atom has four different substituent groups (A, B, X, Y), they can be arranged in two ways that represent nonsuperimposable mirror images of each other (enantiomers). This asymmetric carbon atom is called a chiral atom or chiral center. **(b)** When a tetrahedral carbon has only three dissimilar groups (i.e., the same group occurs twice), only one configuration is possible and the molecule is symmetric, or achiral. In this case the molecule is superimposable on its mirror image: the molecule on the left can be rotated counterclockwise (when looking down the vertical bond from A to C) to create the molecule in the mirror.

Given the importance of stereochemistry in reactions between biomolecules (see below), biochemists must name and represent the structure of each biomolecule so that its stereochemistry is unambiguous. For compounds with more than one chiral center, the most useful system of nomenclature is the RS system. In this system, each group attached to a chiral carbon is assigned a *priority*. The priorities of some common substituents are

$$-OCH_2 > -OH > -NH_2 > -COOH > -CHO >$$
$$-CH_2OH > -CH_3 > -H$$

For naming in the RS system, the chiral atom is viewed with the group of lowest priority (4 in the diagram on the next page) pointing away from the viewer. If the priority of the other three groups (1 to 3) decreases in clockwise order, the configuration is (*R*) (Latin *rectus*, "right"); if in counterclockwise order, the configuration

Enantiomers (mirror images)

Enantiomers (mirror images)

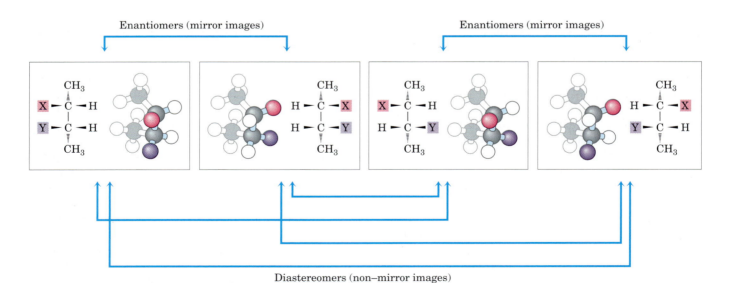

Diastereomers (non–mirror images)

**FIGURE 1-20  Two types of stereoisomers.** There are four different 2,3-disubstituted butanes ($n = 2$ asymmetric carbons, hence $2^n = 4$ stereoisomers). Each is shown in a box as a perspective formula and a ball-and-stick model, which has been rotated to allow the reader to view all the groups. Some pairs of stereoisomers are mirror images of each other, or enantiomers. Other pairs are not mirror images; these are diastereomers.

## BOX 1–2    WORKING IN BIOCHEMISTRY

### Louis Pasteur and Optical Activity:
### *In Vino, Veritas*

Louis Pasteur encountered the phenomenon of **optical activity** in 1843, during his investigation of the crystalline sediment that accumulated in wine casks (a form of tartaric acid called paratartaric acid—also called racemic acid, from Latin *racemus*, "bunch of grapes"). He used fine forceps to separate two types of crystals identical in shape but mirror images of each other. Both types proved to have all the chemical properties of tartaric acid, but in solution one type rotated polarized light to the left (levorotatory), the other to the right (dextrorotatory). Pasteur later described the experiment and its interpretation:

Louis Pasteur
1822–1895

> In isomeric bodies, the elements and the proportions in which they are combined are the same, only the arrangement of the atoms is different . . . We know, on the one hand, that the molecular arrangements of the two tartaric acids are asymmetric, and, on the other hand, that these arrangements are absolutely identical, excepting that they exhibit asymmetry in opposite directions. Are the atoms of the dextro acid grouped in the form of a right-handed spiral, or are they placed at the apex of an irregular tetrahedron, or are they disposed according to this or that asymmetric arrangement? We do not know.*

Now we do know. X-ray crystallographic studies in 1951 confirmed that the levorotatory and dextrorotatory forms of tartaric acid are mirror images of each other at the molecular level and established the absolute configuration of each (Fig. 1). The same approach has been used to demonstrate that although the amino acid alanine has two stereoisomeric forms (designated D and L), alanine in proteins exists exclusively in one form (the L isomer; see Chapter 3).

$$HOOC^1 \quad \quad {}^4COOH$$
$$C{-}C$$
$$H \quad OH \quad H \quad OH$$

**(2R,3R)-Tartaric acid**
(dextrorotatory)

$$HOOC^1 \quad \quad {}^4COOH$$
$$C{-}C$$
$$HO \quad H \quad H \quad OH$$

**(2S,3S)-Tartaric acid**
(levorotatory)

**FIGURE 1**  Pasteur separated crystals of two stereoisomers of tartaric acid and showed that solutions of the separated forms rotated polarized light to the same extent but in opposite directions. These dextrorotatory and levorotatory forms were later shown to be the (R,R) and (S,S) isomers represented here. The RS system of nomenclature is explained in the text.

*From Pasteur's lecture to the Société Chimique de Paris in 1883, quoted in DuBos, R. (1976) *Louis Pasteur: Free Lance of Science,* p. 95, Charles Scribner's Sons, New York.

---

is (S) (Latin *sinister,* "left"). In this way each chiral carbon is designated either (R) or (S), and the inclusion of these designations in the name of the compound provides an unambiguous description of the stereochemistry at each chiral center.

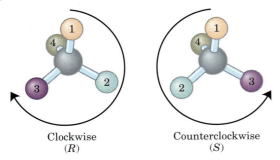

Clockwise
(R)

Counterclockwise
(S)

Another naming system for stereoisomers, the D and L system, is described in Chapter 3. A molecule with a single chiral center (glyceraldehydes, for example) can be named unambiguously by either system.

$$CHO$$
$$HO{-}C{-}H$$
$$CH_2OH$$

L-Glyceraldehyde

$$\equiv$$

$$CHO_{(2)}$$
$$H_{(4)} {-} OH_{(1)}$$
$$CH_2OH_{(3)}$$

(S)-Glyceraldehyde

Distinct from configuration is molecular **conformation,** the spatial arrangement of substituent groups that, without breaking any bonds, are free to assume different positions in space because of the freedom of rotation about single bonds. In the simple hydrocarbon ethane, for example, there is nearly complete freedom of rotation around the C—C bond. Many different, interconvertible conformations of ethane are possible, depending on the degree of rotation (Fig. 1–21). Two conformations are of special interest: the staggered, which is more stable than all others and thus predominates, and the eclipsed, which is least stable. We cannot isolate either of these conformational forms, because

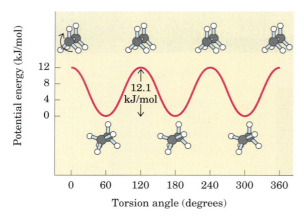

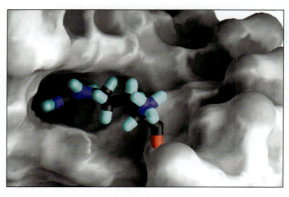

**FIGURE 1–21 Conformations.** Many conformations of ethane are possible because of freedom of rotation around the C—C bond. In the ball-and-stick model, when the front carbon atom (as viewed by the reader) with its three attached hydrogens is rotated relative to the rear carbon atom, the potential energy of the molecule rises to a maximum in the fully eclipsed conformation (torsion angle 0°, 120°, etc.), then falls to a minimum in the fully staggered conformation (torsion angle 60°, 180°, etc.). Because the energy differences are small enough to allow rapid interconversion of the two forms (millions of times per second), the eclipsed and staggered forms cannot be separately isolated.

**FIGURE 1–22 Complementary fit between a macromolecule and a small molecule.** A segment of RNA from the regulatory region TAR of the human immunodeficiency virus genome (gray) with a bound argininamide molecule (colored), representing one residue of a protein that binds to this region. The argininamide fits into a pocket on the RNA surface and is held in this orientation by several noncovalent interactions with the RNA. This representation of the RNA molecule is produced with the computer program GRASP, which can calculate the shape of the outer surface of a macromolecule, defined either by the van der Waals radii of all the atoms in the molecule or by the "solvent exclusion volume," into which a water molecule cannot penetrate.

they are freely interconvertible. However, when one or more of the hydrogen atoms on each carbon is replaced by a functional group that is either very large or electrically charged, freedom of rotation around the C—C bond is hindered. This limits the number of stable conformations of the ethane derivative.

### Interactions between Biomolecules Are Stereospecific

Biological interactions between molecules are stereospecific: the "fit" in such interactions must be stereochemically correct. The three-dimensional structure of biomolecules large and small—the combination of configuration and conformation—is of the utmost importance in their biological interactions: reactant with enzyme, hormone with its receptor on a cell surface, antigen with its specific antibody, for example (Fig. 1–22). The study of biomolecular stereochemistry with precise physical methods is an important part of modern research on cell structure and biochemical function.

In living organisms, chiral molecules are usually present in only one of their chiral forms. For example, the amino acids in proteins occur only as their L isomers; glucose occurs only as its D isomer. (The conventions for naming stereoisomers of the amino acids are described in Chapter 3; those for sugars, in Chapter 7; the RS system, described above, is the most useful for some biomolecules.) In contrast, when a compound with an asymmetric carbon atom is chemically synthesized in the laboratory, the reaction usually produces all possible chiral forms: a mixture of the D and L forms, for example. Living cells produce only one chiral form of biomolecules because the enzymes that synthesize them are also chiral.

Stereospecificity, the ability to distinguish between stereoisomers, is a property of enzymes and other proteins and a characteristic feature of the molecular logic of living cells. If the binding site on a protein is complementary to one isomer of a chiral compound, it will not be complementary to the other isomer, for the same reason that a left glove does not fit a right hand. Two striking examples of the ability of biological systems to distinguish stereoisomers are shown in Figure 1–23.

### SUMMARY 1.2 Chemical Foundations

■ Because of its bonding versatility, carbon can produce a broad array of carbon–carbon skeletons with a variety of functional groups; these groups give biomolecules their biological and chemical personalities.

■ A nearly universal set of several hundred small molecules is found in living cells; the interconversions of these molecules in the central metabolic pathways have been conserved in evolution.

■ Proteins and nucleic acids are linear polymers of simple monomeric subunits; their sequences contain the information that gives each molecule its three-dimensional structure and its biological functions.

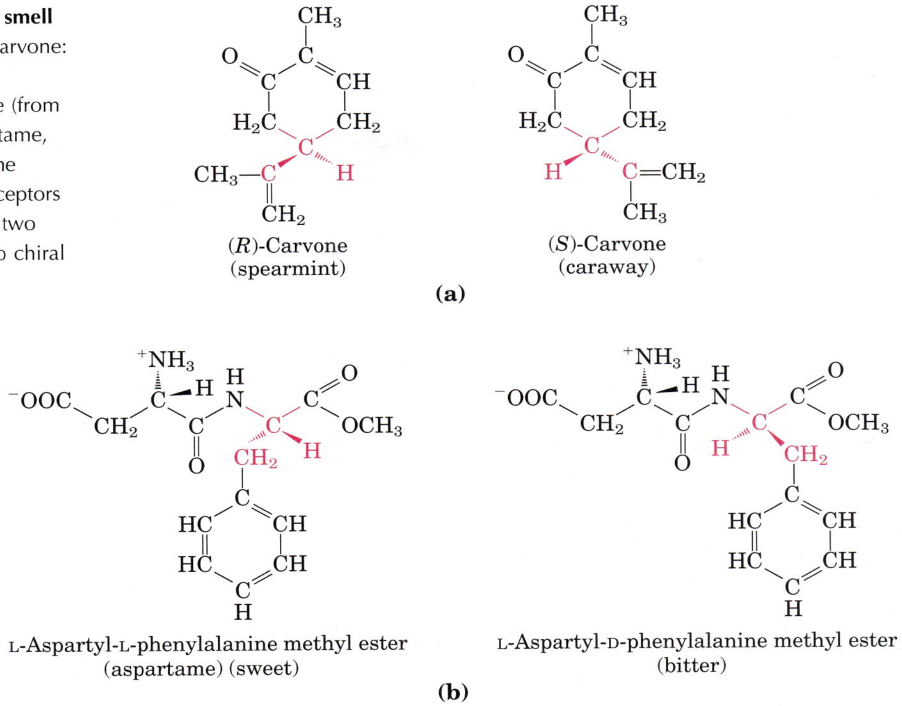

**FIGURE 1–23 Stereoisomers distinguishable by smell and taste in humans. (a)** Two stereoisomers of carvone: (R)-carvone (isolated from spearmint oil) has the characteristic fragrance of spearmint; (S)-carvone (from caraway seed oil) smells like caraway. **(b)** Aspartame, the artificial sweetener sold under the trade name NutraSweet, is easily distinguishable by taste receptors from its bitter-tasting stereoisomer, although the two differ only in the configuration at one of the two chiral carbon atoms.

(R)-Carvone (spearmint)    (S)-Carvone (caraway)

**(a)**

L-Aspartyl-L-phenylalanine methyl ester (aspartame) (sweet)    L-Aspartyl-D-phenylalanine methyl ester (bitter)

**(b)**

■ Molecular configuration can be changed only by breaking covalent bonds. For a carbon atom with four different substituents (a chiral carbon), the substituent groups can be arranged in two different ways, generating stereoisomers with distinct properties. Only one stereoisomer is biologically active. Molecular conformation is the position of atoms in space that can be changed by rotation about single bonds, without breaking covalent bonds.

■ Interactions between biological molecules are almost invariably stereospecific: they require a complementary match between the interacting molecules.

# 1.3 Physical Foundations

Living cells and organisms must perform work to stay alive and to reproduce themselves. The synthetic reactions that occur within cells, like the synthetic processes in any factory, require the input of energy. Energy is also consumed in the motion of a bacterium or an Olympic sprinter, in the flashing of a firefly or the electrical discharge of an eel. And the storage and expression of information require energy, without which structures rich in information inevitably become disordered and meaningless.

In the course of evolution, cells have developed highly efficient mechanisms for coupling the energy obtained from sunlight or fuels to the many energy-consuming processes they must carry out. One goal of biochemistry is to understand, in quantitative and chemical terms, the means by which energy is extracted, channeled, and consumed in living cells. We can consider cellular energy conversions—like all other energy conversions—in the context of the laws of thermodynamics.

## Living Organisms Exist in a Dynamic Steady State, Never at Equilibrium with Their Surroundings

The molecules and ions contained within a living organism differ in kind and in concentration from those in the organism's surroundings. A *Paramecium* in a pond, a shark in the ocean, an erythrocyte in the human bloodstream, an apple tree in an orchard—all are different in composition from their surroundings and, once they have reached maturity, all (to a first approximation) maintain a constant composition in the face of constantly changing surroundings.

Although the characteristic composition of an organism changes little through time, the population of molecules within the organism is far from static. Small molecules, macromolecules, and supramolecular complexes are continuously synthesized and then broken down in chemical reactions that involve a constant flux of mass and energy through the system. The hemoglobin molecules carrying oxygen from your lungs to your brain at this moment were synthesized within the past month; by next month they will have been degraded and entirely replaced by new hemoglobin molecules. The glucose you ingested with your most recent meal is now circulating in your bloodstream; before the day is over these particular glucose molecules will have been converted into something else—carbon dioxide or fat, perhaps—and will have been replaced with a fresh supply of glucose, so that your blood glucose concentration is more or less constant over the whole day. The amounts of hemoglobin and glucose in the blood remain nearly constant because the rate of synthesis or intake of each just balances the rate of its breakdown, consumption, or conversion into some other product. The constancy of concentration is the result of a *dynamic steady state,* a steady state that is far from equilibrium. Maintaining this steady state requires the constant investment of energy; when the cell can no longer generate energy, it dies and begins to decay toward equilibrium with its surroundings. We consider below exactly what is meant by "steady state" and "equilibrium."

## Organisms Transform Energy and Matter from Their Surroundings

For chemical reactions occurring in solution, we can define a **system** as all the reactants and products present, the solvent that contains them, and the immediate atmosphere—in short, everything within a defined region of space. The system and its surroundings together constitute the **universe.** If the system exchanges neither matter nor energy with its surroundings, it is said to be **isolated.** If the system exchanges energy but not matter with its surroundings, it is a **closed** system; if it exchanges both energy and matter with its surroundings, it is an **open** system.

A living organism is an open system; it exchanges both matter and energy with its surroundings. Living organisms derive energy from their surroundings in two ways: (1) they take up chemical fuels (such as glucose) from the environment and extract energy by oxidizing them (see Box 1–3, Case 2); or (2) they absorb energy from sunlight.

The first law of thermodynamics, developed from physics and chemistry but fully valid for biological systems as well, describes the principle of the conservation of energy: *in any physical or chemical change, the total amount of energy in the universe remains constant, although the form of the energy may change.* Cells are consummate transducers of energy, capable of interconverting chemical, electromagnetic, mechanical, and osmotic energy with great efficiency (Fig. 1–24).

## The Flow of Electrons Provides Energy for Organisms

Nearly all living organisms derive their energy, directly or indirectly, from the radiant energy of sunlight, which arises from thermonuclear fusion reactions carried out in the sun. Photosynthetic cells absorb light energy and use it to drive electrons from water to carbon dioxide, forming energy-rich products such as glucose ($C_6H_{12}O_6$), starch, and sucrose and releasing $O_2$ into the atmosphere:

$$6CO_2 + 6H_2O \xrightarrow{\text{light}} C_6H_{12}O_6 + 6O_2$$
(light-driven reduction of $CO_2$)

Nonphotosynthetic cells and organisms obtain the energy they need by oxidizing the energy-rich products of photosynthesis and then passing electrons to atmos-

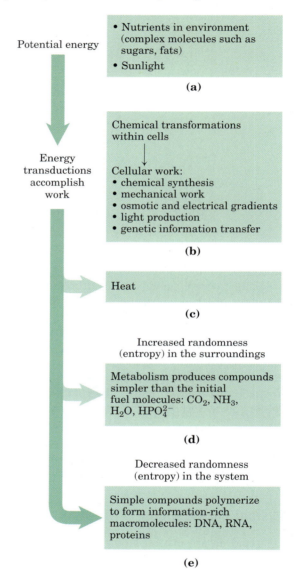

**FIGURE 1–24   Some energy interconversion in living organisms.** During metabolic energy transductions, the randomness of the system plus surroundings (expressed quantitatively as entropy) increases as the potential energy of complex nutrient molecules decreases. **(a)** Living organisms extract energy from their surroundings; **(b)** convert some of it into useful forms of energy to produce work; **(c)** return some energy to the surroundings as heat; and **(d)** release end-product molecules that are less well organized than the starting fuel, increasing the entropy of the universe. One effect of all these transformations is **(e)** increased order (decreased randomness) in the system in the form of complex macromolecules. We return to a quantitative treatment of entropy in Chapter 13.

pheric $O_2$ to form water, carbon dioxide, and other end products, which are recycled in the environment:

$$C_6H_{12}O_6 + O_2 \longrightarrow 6CO_2 + 6H_2O + \text{energy}$$
(energy-yielding oxidation of glucose)

Virtually all energy transductions in cells can be traced to this flow of electrons from one molecule to another, in a "downhill" flow from higher to lower electrochemical potential; as such, this is formally analogous to the flow of electrons in a battery-driven electric circuit. All these reactions involving electron flow are **oxidation-reduction reactions:** one reactant is oxidized (loses electrons) as another is reduced (gains electrons).

## Creating and Maintaining Order Requires Work and Energy

DNA, RNA, and proteins are informational macromolecules. In addition to using chemical energy to form the covalent bonds between the subunits in these polymers, the cell must invest energy to order the subunits in their correct sequence. It is extremely improbable that amino acids in a mixture would spontaneously condense into a single type of protein, with a unique sequence. This would represent increased order in a population of molecules; but according to the second law of thermodynamics, the tendency in nature is toward ever-greater disorder in the universe: *the total entropy of the universe is continually increasing.* To bring about the synthesis of macromolecules from their monomeric units, free energy must be supplied to the system (in this case, the cell).

The randomness or disorder of the components of a chemical system is expressed as **entropy, $S$** (Box 1–3).

J. Willard Gibbs, 1839–1903

Any change in randomness of the system is expressed as entropy change, $\Delta S$, which by convention has a positive value when randomness increases. J. Willard Gibbs, who developed the theory of energy changes during chemical reactions, showed that the **free-energy content, $G$,** of any closed system can be defined in terms of three quantities: **enthalpy, $H$,** reflecting the number and kinds of bonds; entropy, $S$; and the absolute temperature, $T$ (in degrees Kelvin). The definition of free energy is $G = H - TS$. When a chemical reaction occurs at constant temperature, the **free-energy change, $\Delta G$,** is determined by the enthalpy change, $\Delta H$, reflecting the kinds and numbers of chemical bonds and noncovalent interactions broken and formed, and the entropy change, $\Delta S$, describing the change in the system's randomness:

$$\Delta G = \Delta H - T\,\Delta S$$

**FIGURE 1–25 Adenosine triphosphate (ATP).** The removal of the terminal phosphoryl group (shaded pink) of ATP, by breakage of a phosphoanhydride bond, is highly exergonic, and this reaction is coupled to many endergonic reactions in the cell (as in the example in Fig. 1–26b).

A process tends to occur spontaneously only if $\Delta G$ is negative. Yet cell function depends largely on molecules, such as proteins and nucleic acids, for which the free energy of formation is positive: the molecules are less stable and more highly ordered than a mixture of their monomeric components. To carry out these thermodynamically unfavorable, energy-requiring **(endergonic)** reactions, cells couple them to other reactions that liberate free energy (**exergonic** reactions), so that the overall process is exergonic: the *sum* of the free-energy changes is negative. The usual source of free energy in coupled biological reactions is the energy released by hydrolysis of phosphoanhydride bonds such as those in adenosine triphosphate (ATP; Fig. 1–25; see also Fig. 1–15). Here, each ⓟ represents a phosphoryl group:

| | |
|---|---|
| Amino acids $\longrightarrow$ polymer | $\Delta G_1$ is positive (endergonic) |
| —ⓟ—ⓟ $\longrightarrow$ —ⓟ + ⓟ | $\Delta G_2$ is negative (exergonic) |

When these reactions are coupled, the sum of $\Delta G_1$ and $\Delta G_2$ is negative—the overall process is exergonic. By this coupling strategy, cells are able to synthesize and maintain the information-rich polymers essential to life.

## Energy Coupling Links Reactions in Biology

The central issue in *bioenergetics* (the study of energy transformations in living systems) is the means by which energy from fuel metabolism or light capture is coupled to a cell's energy-requiring reactions. In thinking about energy coupling, it is useful to consider a simple mechanical example, as shown in Figure 1–26a. An object at the top of an inclined plane has a certain amount of potential energy as a result of its elevation. It tends to slide down the plane, losing its potential energy of position as it approaches the ground. When an appropriate string-and-pulley device couples the falling object to another, smaller object, the spontaneous downward motion of the larger can lift the smaller, accomplishing a

## BOX 1–3   WORKING IN BIOCHEMISTRY

### Entropy: The Advantages of Being Disorganized

The term "entropy," which literally means "a change within," was first used in 1851 by Rudolf Clausius, one of the formulators of the second law of thermodynamics. A rigorous quantitative definition of entropy involves statistical and probability considerations. However, its nature can be illustrated qualitatively by three simple examples, each demonstrating one aspect of entropy. The key descriptors of entropy are *randomness* and *disorder,* manifested in different ways.

### Case 1: The Teakettle and the Randomization of Heat

We know that steam generated from boiling water can do useful work. But suppose we turn off the burner under a teakettle full of water at 100 °C (the "system") in the kitchen (the "surroundings") and allow the teakettle to cool. As it cools, no work is done, but heat passes from the teakettle to the surroundings, raising the temperature of the surroundings (the kitchen) by an infinitesimally small amount until complete equilibrium is attained. At this point all parts of the teakettle and the kitchen are at precisely the same temperature. The free energy that was once concentrated in the teakettle of hot water at 100 °C, *potentially* capable of doing work, has disappeared. Its equivalent in heat energy is still present in the teakettle + kitchen (i.e., the "universe") but has become completely randomized throughout. This energy is no longer available to do work because there is no temperature differential within the kitchen. Moreover, the increase in entropy of the kitchen (the surroundings) is irreversible. We know from everyday experience that heat never spontaneously passes back from the kitchen into the teakettle to raise the temperature of the water to 100 °C again.

### Case 2: The Oxidation of Glucose

Entropy is a state not only of energy but of matter. Aerobic (heterotrophic) organisms extract free en-

ergy from glucose obtained from their surroundings by oxidizing the glucose with $O_2$, also obtained from the surroundings. The end products of this oxidative metabolism, $CO_2$ and $H_2O$, are returned to the surroundings. In this process the surroundings undergo an increase in entropy, whereas the organism itself remains in a steady state and undergoes no change in its internal order. Although some entropy arises from the dissipation of heat, entropy also arises from another kind of disorder, illustrated by the equation for the oxidation of glucose:

$$C_6H_{12}O_6 + 6O_2 \longrightarrow 6CO_2 + 6H_2O$$

We can represent this schematically as

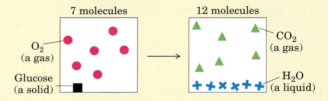

The atoms contained in 1 molecule of glucose plus 6 molecules of oxygen, a total of 7 molecules, are more randomly dispersed by the oxidation reaction and are now present in a total of 12 molecules ($6CO_2 + 6H_2O$).

Whenever a chemical reaction results in an increase in the number of molecules—or when a solid substance is converted into liquid or gaseous products, which allow more freedom of molecular movement than solids—molecular disorder, and thus entropy, increases.

### Case 3: Information and Entropy

The following short passage from *Julius Caesar,* Act IV, Scene 3, is spoken by Brutus, when he realizes that he must face Mark Antony's army. It is an information-rich nonrandom arrangement of 125 letters of the English alphabet:

certain amount of work. The amount of energy available to do work is the **free-energy change, $\Delta G$;** this is always somewhat less than the theoretical amount of energy released, because some energy is dissipated as the heat of friction. The greater the elevation of the larger object, the greater the energy released ($\Delta G$) as the object slides downward and the greater the amount of work that can be accomplished.

How does this apply in chemical reactions? In closed systems, chemical reactions proceed spontaneously until **equilibrium** is reached. When a system is at equilibrium, the rate of product formation exactly equals the

rate at which product is converted to reactant. Thus there is no net change in the concentration of reactants and products; a *steady state* is achieved. The energy change as the system moves from its initial state to equilibrium, with no changes in temperature or pressure, is given by the free-energy change, $\Delta G$. The magnitude of $\Delta G$ depends on the particular chemical reaction and on how far from equilibrium the system is initially. Each compound involved in a chemical reaction contains a certain amount of potential energy, related to the kind and number of its bonds. In reactions that occur spontaneously, the products have less free energy than the re-

There is a tide in the affairs of men,
Which, taken at the flood, leads on to fortune;
Omitted, all the voyage of their life
Is bound in shallows and in miseries.

In addition to what this passage says overtly, it has many hidden meanings. It not only reflects a complex sequence of events in the play, it also echoes the play's ideas on conflict, ambition, and the demands of leadership. Permeated with Shakespeare's understanding of human nature, it is very rich in information.

However, if the 125 letters making up this quotation were allowed to fall into a completely random, chaotic pattern, as shown in the following box, they would have no meaning whatsoever.

In this form the 125 letters contain little or no information, but they are very rich in entropy. Such considerations have led to the conclusion that information is a form of energy; information has been called "negative entropy." In fact, the branch of mathematics called information theory, which is basic to the programming logic of computers, is closely related to thermodynamic theory. Living organisms are highly ordered, nonrandom structures, immensely rich in information and thus entropy-poor.

## (a) Mechanical example

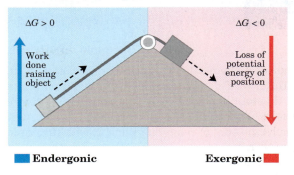

## (b) Chemical example

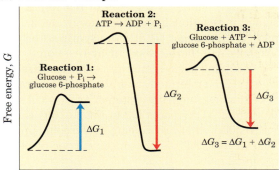

**FIGURE 1–26 Energy coupling in mechanical and chemical processes. (a)** The downward motion of an object releases potential energy that can do mechanical work. The potential energy made available by spontaneous downward motion, an exergonic process (pink), can be coupled to the endergonic upward movement of another object (blue). **(b)** In reaction 1, the formation of glucose 6-phosphate from glucose and inorganic phosphate ($P_i$) yields a product of higher energy than the two reactants. For this endergonic reaction, $\Delta G$ is positive. In reaction 2, the exergonic breakdown of adenosine triphosphate (ATP) can drive an endergonic reaction when the two reactions are coupled. The exergonic reaction has a large, negative free-energy change ($\Delta G_2$), and the endergonic reaction has a smaller, positive free-energy change ($\Delta G_1$). The third reaction accomplishes the sum of reactions 1 and 2, and the free-energy change, $\Delta G_3$, is the arithmetic sum of $\Delta G_1$ and $\Delta G_2$. Because $\Delta G_3$ is negative, the overall reaction is exergonic and proceeds spontaneously.

actants, thus the reaction releases free energy, which is then available to do work. Such reactions are exergonic; the decline in free energy from reactants to products is expressed as a negative value. Endergonic reactions require an input of energy, and their $\Delta G$ values are positive. As in mechanical processes, only part of the energy released in exergonic chemical reactions can be used to accomplish work. In living systems some energy is dissipated as heat or lost to increasing entropy.

In living organisms, as in the mechanical example in Figure 1–26a, an exergonic reaction can be coupled to an endergonic reaction to drive otherwise unfavorable

reactions. Figure 1–26b (a type of graph called a reaction coordinate diagram) illustrates this principle for the conversion of glucose to glucose 6-phosphate, the first step in the pathway for oxidation of glucose. The simplest way to produce glucose 6-phosphate would be:

Reaction 1:     Glucose + $P_i$ $\longrightarrow$ glucose 6-phosphate
(endergonic; $\Delta G_1$ is positive)

($P_i$ is an abbreviation for inorganic phosphate, $HPO_4^{2-}$. Don't be concerned about the structure of these compounds now; we describe them in detail later in the book.) This reaction does not occur spontaneously; $\Delta G$

is positive. A second, very exergonic reaction can occur in all cells:

Reaction 2:     $ATP \longrightarrow ADP + P_i$

(exergonic; $\Delta G_2$ is negative)

These two chemical reactions share a common intermediate, $P_i$, which is consumed in reaction 1 and produced in reaction 2. The two reactions can therefore be coupled in the form of a third reaction, which we can write as the sum of reactions 1 and 2, with the common intermediate, $P_i$, omitted from both sides of the equation:

Reaction 3:     $Glucose + ATP \longrightarrow$

glucose 6-phosphate + ADP

Because more energy is released in reaction 2 than is consumed in reaction 1, the free-energy change for reaction 3, $\Delta G_3$, is negative, and the synthesis of glucose 6-phosphate can therefore occur by reaction 3.

The coupling of exergonic and endergonic reactions through a shared intermediate is absolutely central to the energy exchanges in living systems. As we shall see, the breakdown of ATP (reaction 2 in Fig. 1–26b) is the exergonic reaction that drives many endergonic processes in cells. In fact, ATP is the major carrier of chemical energy in all cells.

## $K_{eq}$ and $\Delta G°$ Are Measures of a Reaction's Tendency to Proceed Spontaneously

The tendency of a chemical reaction to go to completion can be expressed as an equilibrium constant. For the reaction

$$aA + bB \longrightarrow cC + dD$$

the equilibrium constant, $K_{eq}$, is given by

$$K_{eq} = \frac{[C_{eq}]^c[D_{eq}]^d}{[A_{eq}]^a[B_{eq}]^b}$$

where $[A_{eq}]$ is the concentration of A, $[B_{eq}]$ the concentration of B, and so on, *when the system has reached equilibrium.* A large value of $K_{eq}$ means the reaction tends to proceed until the reactants have been almost completely converted into the products.

Gibbs showed that $\Delta G$ for any chemical reaction is a function of the **standard free-energy change, $\Delta G°$**— a constant that is characteristic of each specific reaction—and a term that expresses the initial concentrations of reactants and products:

$$\Delta G = \Delta G° + RT \ln \frac{[C_i]^c[D_i]^d}{[A_i]^a[B_i]^b} \qquad (1-1)$$

where $[A_i]$ is the initial concentration of A, and so forth; $R$ is the gas constant; and $T$ is the absolute temperature.

When a reaction has reached equilibrium, no driving force remains and it can do no work: $\Delta G = 0$. For this special case, $[A_i] = [A_{eq}]$, and so on, for all reactants and products, and

$$\frac{[C_i]^c[D_i]^d}{[A_i]^a[B_i]^b} = \frac{[C_{eq}]^c[D_{eq}]^d}{[A_{eq}]^a[B_{eq}]^b} = K_{eq}$$

Substituting 0 for $\Delta G$ and $K_{eq}$ for $[C_i]^c[D_i]^d/[A_i]^a[B_i]^b$ in Equation 1–1, we obtain the relationship

$$\Delta G° = -RT \ln K_{eq}$$

from which we see that $\Delta G°$ is simply a second way (besides $K_{eq}$) of expressing the driving force on a reaction. Because $K_{eq}$ is experimentally measurable, we have a way of determining $\Delta G°$, the thermodynamic constant characteristic of each reaction.

The units of $\Delta G°$ and $\Delta G$ are joules per mole (or calories per mole). When $K_{eq} \gg 1$, $\Delta G°$ is large and negative; when $K_{eq} \ll 1$, $\Delta G°$ is large and positive. From a table of experimentally determined values of either $K_{eq}$ or $\Delta G°$, we can see at a glance which reactions tend to go to completion and which do not.

One caution about the interpretation of $\Delta G°$: *thermodynamic* constants such as this show where the final equilibrium for a reaction lies but tell us nothing about how fast that equilibrium will be achieved. The rates of reactions are governed by the parameters of *kinetics*, a topic we consider in detail in Chapter 6.

## Enzymes Promote Sequences of Chemical Reactions

All biological macromolecules are much less thermodynamically stable than their monomeric subunits, yet they are *kinetically stable:* their *uncatalyzed* breakdown occurs so slowly (over years rather than seconds) that, on a time scale that matters for the organism, these molecules are stable. Virtually every chemical reaction in a cell occurs at a significant rate only because of the presence of **enzymes**—biocatalysts that, like all other catalysts, greatly enhance the rate of specific chemical reactions without being consumed in the process.

The path from reactant(s) to product(s) almost invariably involves an energy barrier, called the activation barrier (Fig. 1–27), that must be surmounted for any reaction to proceed. The breaking of existing bonds and formation of new ones generally requires, first, the distortion of the existing bonds, creating a **transition state** of higher free energy than either reactant or product. The highest point in the reaction coordinate diagram represents the transition state, and the difference in energy between the reactant in its ground state and in its transition state is the **activation energy, $\Delta G^{\ddagger}$**. An enzyme catalyzes a reaction by providing a more comfortable fit for the transition state: a surface that complements the transition state in stereochemistry, polarity, and charge. The binding of enzyme to the transition state is exergonic, and the energy released by this binding reduces the activation energy for the reaction and greatly increases the reaction rate.

A further contribution to catalysis occurs when two or more reactants bind to the enzyme's surface close to each other and with stereospecific orientations that favor the reaction. This increases by orders of magnitude the probability of productive collisions between reactants. As a result of these factors and several others, discussed in Chapter 6, enzyme-catalyzed reactions commonly proceed at rates greater than $10^{12}$ times faster than the uncatalyzed reactions.

Cellular catalysts are, with a few exceptions, proteins. (In some cases, RNA molecules have catalytic roles, as discussed in Chapters 26 and 27.) Again with a few exceptions, each enzyme catalyzes a specific reaction, and each reaction in a cell is catalyzed by a different enzyme. Thousands of different enzymes are therefore required by each cell. The multiplicity of enzymes, their specificity (the ability to discriminate between reactants), and their susceptibility to regulation give cells the capacity to lower activation barriers selectively. This selectivity is crucial for the effective regulation of cellular processes. By allowing specific reactions to proceed at significant rates at particular times, enzymes determine how matter and energy are channeled into cellular activities.

The thousands of enzyme-catalyzed chemical reactions in cells are functionally organized into many sequences of consecutive reactions, called **pathways,** in which the product of one reaction becomes the reactant in the next. Some pathways degrade organic nutrients into simple end products in order to extract chemical energy and convert it into a form useful to the cell; together these degradative, free-energy-yielding reactions are designated **catabolism.** Other pathways start with small precursor molecules and convert them to progressively larger and more complex molecules, including proteins and nucleic acids. Such synthetic pathways,

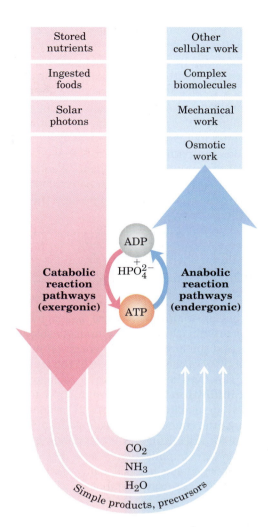

**FIGURE 1–28 The central role of ATP in metabolism.** ATP is the shared chemical intermediate linking energy-releasing to energy-requiring cell processes. Its role in the cell is analogous to that of money in an economy: it is "earned/produced" in exergonic reactions and "spent/consumed" in endergonic ones.

which invariably require the input of energy, are collectively designated **anabolism.** The overall network of enzyme-catalyzed pathways constitutes cellular **metabolism.** ATP is the major connecting link (the shared intermediate) between the catabolic and anabolic components of this network (shown schematically in Fig. 1–28). The pathways of enzyme-catalyzed reactions that act on the main constituents of cells—proteins, fats, sugars, and nucleic acids—are virtually identical in all living organisms.

## Metabolism Is Regulated to Achieve Balance and Economy

Not only do living cells simultaneously synthesize thousands of different kinds of carbohydrate, fat, protein, and nucleic acid molecules and their simpler subunits, but they do so in the precise proportions required by

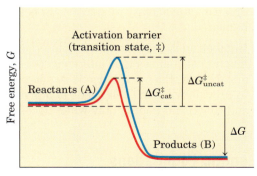

**FIGURE 1–27 Energy changes during a chemical reaction.** An activation barrier, representing the transition state, must be overcome in the conversion of reactants (A) into products (B), even though the products are more stable than the reactants, as indicated by a large, negative free-energy change ($\Delta G$). The energy required to overcome the activation barrier is the activation energy ($\Delta G^{\ddagger}$). Enzymes catalyze reactions by lowering the activation barrier. They bind the transition-state intermediates tightly, and the binding energy of this interaction effectively reduces the activation energy from $\Delta G^{\ddagger}_{uncat}$ to $\Delta G^{\ddagger}_{cat}$. (Note that activation energy is *not* related to free-energy change, $\Delta G$.)

the cell under any given circumstance. For example, during rapid cell growth the precursors of proteins and nucleic acids must be made in large quantities, whereas in nongrowing cells the requirement for these precursors is much reduced. Key enzymes in each metabolic pathway are regulated so that each type of precursor molecule is produced in a quantity appropriate to the current requirements of the cell.

Consider the pathway in *E. coli* that leads to the synthesis of the amino acid isoleucine, a constituent of proteins. The pathway has five steps catalyzed by five different enzymes (A through F represent the intermediates in the pathway):

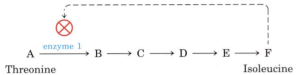

If a cell begins to produce more isoleucine than is needed for protein synthesis, the unused isoleucine accumulates and the increased concentration inhibits the catalytic activity of the first enzyme in the pathway, immediately slowing the production of isoleucine. Such **feedback inhibition** keeps the production and utilization of each metabolic intermediate in balance.

Although the concept of discrete pathways is an important tool for organizing our understanding of metabolism, it is an oversimplification. There are thousands of metabolic intermediates in a cell, many of which are part of more than one pathway. Metabolism would be better represented as a meshwork of interconnected and interdependent pathways. A change in the concentration of any one metabolite would have an impact on other pathways, starting a ripple effect that would influence the flow of materials through other sectors of the cellular economy. The task of understanding these complex interactions among intermediates and pathways in quantitative terms is daunting, but new approaches, discussed in Chapter 15, have begun to offer important insights into the overall regulation of metabolism.

Cells also regulate the synthesis of their own catalysts, the enzymes, in response to increased or diminished need for a metabolic product; this is the substance of Chapter 28. The expression of genes (the translation of information in DNA to active protein in the cell) and synthesis of enzymes are other layers of metabolic control in the cell. All layers must be taken into account when describing the overall control of cellular metabolism. Assembling the complete picture of how the cell regulates itself is a huge job that has only just begun.

**SUMMARY 1.3** Physical Foundations

■ Living cells are open systems, exchanging matter and energy with their surroundings, extracting and channeling energy to maintain themselves in a dynamic steady state distant from equilibrium. Energy is obtained from sunlight or fuels by converting the energy from electron flow into the chemical bonds of ATP.

■ The tendency for a chemical reaction to proceed toward equilibrium can be expressed as the free-energy change, $\Delta G$, which has two components: enthalpy change, $\Delta H$, and entropy change, $\Delta S$. These variables are related by the equation $\Delta G = \Delta H - T\,\Delta S$.

■ When $\Delta G$ of a reaction is negative, the reaction is exergonic and tends to go toward completion; when $\Delta G$ is positive, the reaction is endergonic and tends to go in the reverse direction. When two reactions can be summed to yield a third reaction, the $\Delta G$ for this overall reaction is the sum of the $\Delta G$s of the two separate reactions. This provides a way to couple reactions.

■ The conversion of ATP to $P_i$ and ADP is highly exergonic (large negative $\Delta G$), and many endergonic cellular reactions are driven by coupling them, through a common intermediate, to this reaction.

■ The standard free-energy change for a reaction, $\Delta G°$, is a physical constant that is related to the equilibrium constant by the equation $\Delta G° = -RT \ln K_{eq}$.

■ Most exergonic cellular reactions proceed at useful rates only because enzymes are present to catalyze them. Enzymes act in part by stabilizing the transition state, reducing the activation energy, $\Delta G^{\ddagger}$, and increasing the reaction rate by many orders of magnitude. The catalytic activity of enzymes in cells is regulated.

■ Metabolism is the sum of many interconnected reaction sequences that interconvert cellular metabolites. Each sequence is regulated so as to provide what the cell needs at a given time and to expend energy only when necessary.

## 1.4 Genetic Foundations

Perhaps the most remarkable property of living cells and organisms is their ability to reproduce themselves for countless generations with nearly perfect fidelity. This continuity of inherited traits implies constancy, over millions of years, in the structure of the molecules that contain the genetic information. Very few historical records of civilization, even those etched in copper or carved in stone (Fig. 1–29), have survived for a thousand years. But there is good evidence that the genetic instructions in living organisms have remained nearly unchanged over very much longer periods; many bacteria have nearly

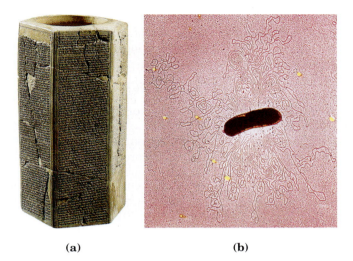

**(a)**                                              **(b)**

**FIGURE 1–29   Two ancient scripts. (a)** The Prism of Sennacherib, inscribed in about 700 B.C.E., describes in characters of the Assyrian language some historical events during the reign of King Sennacherib. The Prism contains about 20,000 characters, weighs about 50 kg, and has survived almost intact for about 2,700 years. **(b)** The single DNA molecule of the bacterium *E. coli,* seen leaking out of a disrupted cell, is hundreds of times longer than the cell itself and contains all the encoded information necessary to specify the cell's structure and functions. The bacterial DNA contains about 10 million characters (nucleotides), weighs less than $10^{-10}$ g, and has undergone only relatively minor changes during the past several million years. (The yellow spots and dark specks in this colorized electron micrograph are artifacts of the preparation.)

the same size, shape, and internal structure and contain the same kinds of precursor molecules and enzymes as bacteria that lived nearly four billion years ago.

Among the seminal discoveries in biology in the twentieth century were the chemical nature and the three-dimensional structure of the genetic material, **deoxyribonucleic acid, DNA.** The sequence of the monomeric subunits, the nucleotides (strictly, deoxyribonucleotides, as discussed below), in this linear polymer encodes the instructions for forming all other cellular components and provides a template for the production of identical DNA molecules to be distributed to progeny when a cell divides. The continued existence of a biological species requires its genetic information to be maintained in a stable form, expressed accurately in the form of gene products, and reproduced with a minimum of errors. Effective storage, expression, and reproduction of the genetic message defines individual species, distinguishes them from one another, and assures their continuity over successive generations.

### Genetic Continuity Is Vested in Single DNA Molecules

DNA is a long, thin organic polymer, the rare molecule that is constructed on the atomic scale in one dimension (width) and the human scale in another (length: a

molecule of DNA can be many centimeters long). A human sperm or egg, carrying the accumulated hereditary information of billions of years of evolution, transmits this inheritance in the form of DNA molecules, in which the linear sequence of covalently linked nucleotide subunits encodes the genetic message.

Usually when we describe the properties of a chemical species, we describe the average behavior of a very large number of identical molecules. While it is difficult to predict the behavior of any single molecule in a collection of, say, a picomole (about $6 \times 10^{11}$ molecules) of a compound, the *average* behavior of the molecules is predictable because so many molecules enter into the average. Cellular DNA is a remarkable exception. The DNA that is the entire genetic material of *E. coli* is a *single molecule* containing 4.64 million nucleotide pairs. That single molecule must be replicated perfectly in every detail if an *E. coli* cell is to give rise to identical progeny by cell division; there is no room for averaging in this process! The same is true for all cells. A human sperm brings to the egg that it fertilizes just one molecule of DNA in each of its 23 different chromosomes, to combine with just one DNA molecule in each corresponding chromosome in the egg. The result of this union is very highly predictable: an embryo with all of its 35,000 genes, constructed of 3 billion nucleotide pairs, intact. An amazing chemical feat!

### The Structure of DNA Allows for Its Replication and Repair with Near-Perfect Fidelity

The capacity of living cells to preserve their genetic material and to duplicate it for the next generation results from the structural complementarity between the two halves of the DNA molecule (Fig. 1–30). The basic unit of DNA is a linear polymer of four different monomeric subunits, **deoxyribonucleotides,** arranged in a precise linear sequence. It is this linear sequence that encodes the genetic information. Two of these polymeric strands are twisted about each other to form the DNA double helix, in which each deoxyribonucleotide in one strand pairs specifically with a complementary deoxyribonucleotide in the opposite strand. Before a cell divides, the two DNA strands separate and each serves as a template for the synthesis of a new complementary strand, generating two identical double-helical molecules, one for each daughter cell. If one strand is damaged, continuity of information is assured by the information present in the other strand, which acts as a template for repair of the damage.

### The Linear Sequence in DNA Encodes Proteins with Three-Dimensional Structures

The information in DNA is encoded in its linear (one-dimensional) sequence of deoxyribonucleotide subunits, but the expression of this information results in

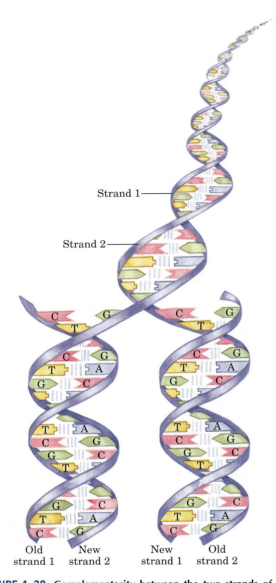

Strand 1

Strand 2

Old strand 1    New strand 2    New strand 1    Old strand 2

**FIGURE 1–30 Complementarity between the two strands of DNA.** DNA is a linear polymer of covalently joined deoxyribonucleotides, of four types: deoxyadenylate (A), deoxyguanylate (G), deoxycytidylate (C), and deoxythymidylate (T). Each nucleotide, with its unique three-dimensional structure, can associate very specifically but noncovalently with one other nucleotide in the complementary chain: A always associates with T, and G with C. Thus, in the double-stranded DNA molecule, the entire sequence of nucleotides in one strand is *complementary* to the sequence in the other. The two strands, held together by hydrogen bonds (represented here by vertical blue lines) between each pair of complementary nucleotides, twist about each other to form the DNA double helix. In DNA replication, the two strands separate and two new strands are synthesized, each with a sequence complementary to one of the original strands. The result is two double-helical molecules, each identical to the original DNA.

a three-dimensional cell. This change from one to three dimensions occurs in two phases. A linear sequence of deoxyribonucleotides in DNA codes (through an intermediary, RNA) for the production of a protein with a corresponding linear sequence of amino acids (Fig. 1–31). The protein folds into a particular three-dimensional

shape, determined by its amino acid sequence and stabilized primarily by noncovalent interactions. Although the final shape of the folded protein is dictated by its amino acid sequence, the folding process is aided by "molecular chaperones," which catalyze the process by discouraging incorrect folding. The precise three-dimensional structure, or **native conformation,** of the protein is crucial to its function.

Once in its native conformation, a protein may associate noncovalently with other proteins, or with nucleic acids or lipids, to form supramolecular complexes such as chromosomes, ribosomes, and membranes. The individual molecules of these complexes have specific, high-affinity binding sites for each other, and within the cell they spontaneously form functional complexes.

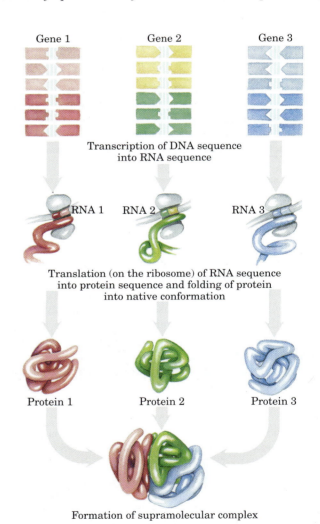

Gene 1    Gene 2    Gene 3

Transcription of DNA sequence into RNA sequence

RNA 1    RNA 2    RNA 3

Translation (on the ribosome) of RNA sequence into protein sequence and folding of protein into native conformation

Protein 1    Protein 2    Protein 3

Formation of supramolecular complex

**FIGURE 1–31 DNA to RNA to protein.** Linear sequences of deoxyribonucleotides in DNA, arranged into units known as genes, are transcribed into ribonucleic acid (RNA) molecules with complementary ribonucleotide sequences. The RNA sequences are then translated into linear protein chains, which fold into their native three-dimensional shapes, often aided by molecular chaperones. Individual proteins commonly associate with other proteins to form supramolecular complexes, stabilized by numerous weak interactions.

Although protein sequences carry all necessary information for the folding into their native conformation, this correct folding requires the right environment—pH, ionic strength, metal ion concentrations, and so forth. Self-assembly therefore requires both information (provided by the DNA sequence) and environment (the interior of a living cell), and in this sense the DNA sequence alone is not enough to dictate the formation of a cell. As Rudolph Virchow, the nineteenth-century Prussian pathologist and researcher, concluded, "Omnis cellula e cellula": every cell comes from another cell.

## SUMMARY 1.4 Genetic Foundations

- Genetic information is encoded in the linear sequence of four deoxyribonucleotides in DNA.

- The double-helical DNA molecule contains an internal template for its own replication and repair.

- The linear sequence of amino acids in a protein, which is encoded in the DNA of the gene for that protein, produces a protein's unique three-dimensional structure.

- Individual macromolecules with specific affinity for other macromolecules self-assemble into supramolecular complexes.

## 1.5 Evolutionary Foundations

Nothing in biology makes sense except in the light of evolution.

> —*Theodosius Dobzhansky*, The American Biology Teacher,
> *March 1973*

Great progress in biochemistry and molecular biology during the decades since Dobzhansky made this striking generalization has amply confirmed its validity. The remarkable similarity of metabolic pathways and gene sequences in organisms across the phyla argues strongly that all modern organisms share a common evolutionary progenitor and were derived from it by a series of small changes (mutations), each of which conferred a selective advantage to some organism in some ecological niche.

### Changes in the Hereditary Instructions Allow Evolution

Despite the near-perfect fidelity of genetic replication, infrequent, unrepaired mistakes in the DNA replication process lead to changes in the nucleotide sequence of DNA, producing a genetic **mutation** (Fig. 1–32) and changing the instructions for some cellular component. Incorrectly repaired damage to one of the DNA strands has the same effect. Mutations in the DNA handed down

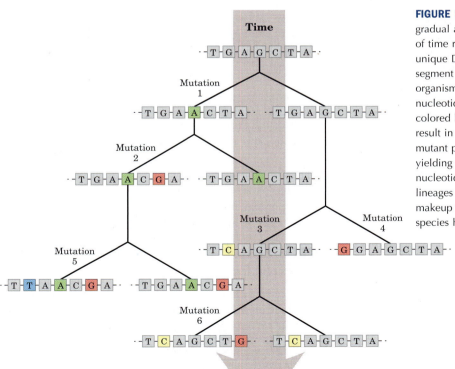

**FIGURE 1–32 Role of mutation in evolution.** The gradual accumulation of mutations over long periods of time results in new biological species, each with a unique DNA sequence. At the top is shown a short segment of a gene in a hypothetical progenitor organism. With the passage of time, changes in nucleotide sequence (mutations, indicated here by colored boxes), *occurring one nucleotide at a time*, result in progeny with different DNA sequences. These mutant progeny also undergo occasional mutations, yielding their own progeny that differ by two or more nucleotides from the progenitor sequence. When two lineages have diverged so much in their genetic makeup that they can no longer interbreed, a new species has been created.

to offspring—that is, mutations that are carried in the reproductive cells—may be harmful or even lethal to the organism; they may, for example, cause the synthesis of a defective enzyme that is not able to catalyze an essential metabolic reaction. Occasionally, however, a mutation better equips an organism or cell to survive in its environment. The mutant enzyme might have acquired a slightly different specificity, for example, so that it is now able to use some compound that the cell was previously unable to metabolize. If a population of cells were to find itself in an environment where that compound was the only or the most abundant available source of fuel, the mutant cell would have a selective advantage over the other, unmutated **(wild-type)** cells in the population. The mutant cell and its progeny would survive and prosper in the new environment, whereas wild-type cells would starve and be eliminated. This is what Darwin meant by "survival of the fittest under selective pressure."

Occasionally, a whole gene is duplicated. The second copy is superfluous, and mutations in this gene will not be deleterious; it becomes a means by which the cell may evolve: by producing a new gene with a new function while retaining the original gene and gene function. Seen in this light, the DNA molecules of modern organisms are historical documents, records of the long journey from the earliest cells to modern organisms. The historical accounts in DNA are not complete; in the course of evolution, many mutations must have been erased or written over. But DNA molecules are the best source of biological history that we have.

Several billion years of adaptive selection have refined cellular systems to take maximum advantage of the chemical and physical properties of the molecular raw materials for carrying out the basic energy-transforming and self-replicating activities of a living cell. Chance genetic variations in individuals in a population, combined with natural selection (survival and reproduction of the fittest individuals in a challenging or changing environment), have resulted in the evolution of an enormous variety of organisms, each adapted to life in its particular ecological niche.

## Biomolecules First Arose by Chemical Evolution

In our account thus far we have passed over the first chapter of the story of evolution: the appearance of the first living cell. Apart from their occurrence in living organisms, organic compounds, including the basic biomolecules such as amino acids and carbohydrates, are found in only trace amounts in the earth's crust, the sea, and the atmosphere. How did the first living organisms acquire their characteristic organic building blocks? In 1922, the biochemist Aleksandr I. Oparin proposed a theory for the origin of life early in the history of Earth, postulating that the atmosphere was very different from that of today. Rich in methane, ammonia, and water, and

essentially devoid of oxygen, it was a reducing atmosphere, in contrast to the oxidizing environment of our era. In Oparin's theory, electrical energy from lightning discharges or heat energy from volcanoes caused ammonia, methane, water vapor, and other components of the primitive atmosphere to react, forming simple organic compounds. These compounds then dissolved in the ancient seas, which over many millennia became enriched with a large variety of simple organic substances. In this warm solution (the "primordial soup"), some organic molecules had a greater tendency than others to associate into larger complexes. Over millions of years, these in turn assembled spontaneously to form membranes and catalysts (enzymes), which came together to become precursors of the earliest cells. Oparin's views remained speculative for many years and appeared untestable—until a surprising experiment was conducted using simple equipment on a desktop.

## Chemical Evolution Can Be Simulated in the Laboratory

The classic experiment on the abiotic (nonbiological) origin of organic biomolecules was carried out in 1953 by Stanley Miller in the laboratory of Harold Urey. Miller subjected gaseous mixtures of $NH_3$, $CH_4$, $H_2O$, and $H_2$ to electrical sparks produced across a pair of electrodes (to simulate lightning) for periods of a week or more, then analyzed the contents of the closed reaction vessel (Fig. 1–33). The gas phase of the resulting mixture contained CO and $CO_2$, as well as the starting materials. The water phase contained a variety of organic compounds, including some amino acids, hydroxy acids, aldehydes, and hydrogen cyanide (HCN). This experiment established the possibility of abiotic production of biomolecules in relatively short times under relatively mild conditions.

More refined laboratory experiments have provided good evidence that many of the chemical components of living cells, including polypeptides and RNA-like molecules, can form under these conditions. Polymers of RNA can act as catalysts in biologically significant reactions (as we discuss in Chapters 26 and 27), and RNA probably played a crucial role in prebiotic evolution, both as catalyst and as information repository.

## RNA or Related Precursors May Have Been the First Genes and Catalysts

In modern organisms, nucleic acids encode the genetic information that specifies the structure of enzymes, and enzymes catalyze the replication and repair of nucleic acids. The mutual dependence of these two classes of biomolecules brings up the perplexing question: which came first, DNA or protein?

The answer may be: neither. The discovery that RNA molecules can act as catalysts in their own forma-

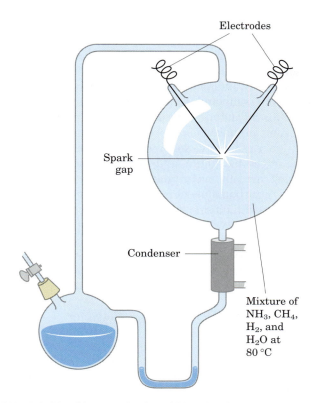

**FIGURE 1–33 Abiotic production of biomolecules.** Spark-discharge apparatus of the type used by Miller and Urey in experiments demonstrating abiotic formation of organic compounds under primitive atmospheric conditions. After subjection of the gaseous contents of the system to electrical sparks, products were collected by condensation. Biomolecules such as amino acids were among the products.

tion suggests that RNA or a similar molecule may have been the first gene *and* the first catalyst. According to this scenario (Fig. 1–34), one of the earliest stages of biological evolution was the chance formation, in the primordial soup, of an RNA molecule that could catalyze the formation of other RNA molecules of the same sequence—a self-replicating, self-perpetuating RNA. The concentration of a self-replicating RNA molecule would increase exponentially, as one molecule formed two, two formed four, and so on. The fidelity of self-replication was presumably less than perfect, so the process would generate variants of the RNA, some of which might be even better able to self-replicate. In the competition for nucleotides, the most efficient of the self-replicating sequences would win, and less efficient replicators would fade from the population.

The division of function between DNA (genetic information storage) and protein (catalysis) was, according to the "RNA world" hypothesis, a later development. New variants of self-replicating RNA molecules developed, with the additional ability to catalyze the condensation of amino acids into peptides. Occasionally, the peptide(s) thus formed would reinforce the self-replicating ability of the RNA, and the pair—RNA

molecule and helping peptide—could undergo further modifications in sequence, generating even more efficient self-replicating systems. The recent, remarkable discovery that, in the protein-synthesizing machinery of modern cells (ribosomes), RNA molecules, not proteins, catalyze the formation of peptide bonds is certainly consistent with the RNA world hypothesis.

Some time after the evolution of this primitive protein-synthesizing system, there was a further development: DNA molecules with sequences complementary to the self-replicating RNA molecules took over the function of conserving the "genetic" information, and RNA molecules evolved to play roles in protein synthesis. (We explain in Chapter 8 why DNA is a more stable molecule than RNA and thus a better repository of inheritable information.) Proteins proved to be versatile catalysts and, over time, took over that function. Lipidlike compounds in the primordial soup formed relatively impermeable layers around self-replicating collections of molecules. The concentration of proteins and nucleic acids within these lipid enclosures favored the molecular interactions required in self-replication.

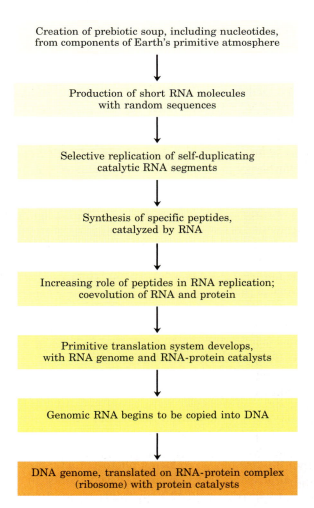

**FIGURE 1–34 A possible "RNA world" scenario.**

## Biological Evolution Began More Than Three and a Half Billion Years Ago

Earth was formed about 4.5 billion years ago, and the first evidence of life dates to more than 3.5 billion years ago. In 1996, scientists working in Greenland found not fossil remains but chemical evidence of life from as far back as 3.85 billion years ago, forms of carbon embedded in rock that appear to have a distinctly biological origin. Somewhere on Earth during its first billion years there arose the first simple organism, capable of replicating its own structure from a template (RNA?) that was the first genetic material. Because the terrestrial atmosphere at the dawn of life was nearly devoid of oxygen, and because there were few microorganisms to scavenge organic compounds formed by natural processes, these compounds were relatively stable. Given this stability and eons of time, the improbable became inevitable: the organic compounds were incorporated into evolving cells to produce increasingly effective self-reproducing catalysts. The process of biological evolution had begun.

## The First Cell Was Probably a Chemoheterotroph

The earliest cells that arose in the rich mixture of organic compounds, the primordial soup of prebiotic times, were almost certainly chemoheterotrophs (Fig. 1–5). The organic compounds they required were originally synthesized from components of the early atmosphere—$CO$, $CO_2$, $N_2$, $CH_4$, and such—by the nonbiological actions of volcanic heat and lightning. Early heterotrophs gradually acquired the ability to derive energy from compounds in their environment and to use that energy to synthesize more of their own precursor molecules, thereby becoming less dependent on outside sources. A very significant evolutionary event was the development of pigments capable of capturing the energy of light from the sun, which could be used to reduce, or "fix," $CO_2$ to form more complex, organic compounds. The original electron donor for these **photosynthetic** processes was probably $H_2S$, yielding elemental sulfur or sulfate ($SO_4^{2-}$) as the by-product, but later cells developed the enzymatic capacity to use $H_2O$ as the electron donor in photosynthetic reactions, eliminating $O_2$ as waste. Cyanobacteria are the modern descendants of these early photosynthetic oxygen-producers.

Because the atmosphere of Earth in the earliest stages of biological evolution was nearly devoid of oxygen, the earliest cells were anaerobic. Under these conditions, chemoheterotrophs could oxidize organic compounds to $CO_2$ by passing electrons not to $O_2$ but to acceptors such as $SO_4^{2-}$, yielding $H_2S$ as the product. With the rise of $O_2$-producing photosynthetic bacteria, the atmosphere became progressively richer in oxygen—a powerful oxidant and deadly poison to anaerobes. Responding to the evolutionary pressure of the "oxygen holocaust," some lineages of microorganisms gave rise to aerobes that obtained energy by passing electrons from fuel molecules to oxygen. Because the transfer of electrons from organic molecules to $O_2$ releases a great deal of energy, aerobic organisms had an energetic advantage over their anaerobic counterparts when both competed in an environment containing oxygen. This advantage translated into the predominance of aerobic organisms in $O_2$-rich environments.

Modern bacteria inhabit almost every ecological niche in the biosphere, and there are bacteria capable of using virtually every type of organic compound as a source of carbon and energy. Photosynthetic bacteria in both fresh and marine waters trap solar energy and use it to generate carbohydrates and all other cell constituents, which are in turn used as food by other forms of life. The process of evolution continues—and in rapidly reproducing bacterial cells, on a time scale that allows us to witness it in the laboratory.

## Eukaryotic Cells Evolved from Prokaryotes in Several Stages

Starting about 1.5 billion years ago, the fossil record begins to show evidence of larger and more complex organisms, probably the earliest eukaryotic cells (Fig. 1–35).

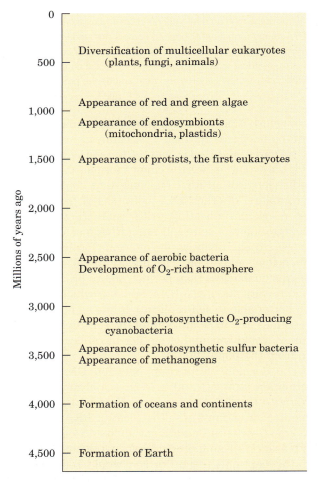

**FIGURE 1–35  Landmarks in the evolution of life on Earth.**

Details of the evolutionary path from prokaryotes to eukaryotes cannot be deduced from the fossil record alone, but morphological and biochemical comparisons of modern organisms have suggested a sequence of events consistent with the fossil evidence.

Three major changes must have occurred as prokaryotes gave rise to eukaryotes. First, as cells acquired more DNA, the mechanisms required to fold it compactly into discrete complexes with specific proteins and to divide it equally between daughter cells at cell division became more elaborate. For this, specialized proteins were required to stabilize folded DNA and to pull the resulting DNA-protein complexes (*chromosomes*) apart during cell division. Second, as cells became larger, a system of intracellular membranes developed, including a double membrane surrounding the DNA. This membrane segregated the nuclear process of RNA synthesis on a DNA template from the cytoplasmic process of protein synthesis on ribosomes. Finally, early eukaryotic cells, which were incapable of photosynthesis or aerobic metabolism, enveloped aerobic bacteria or photosynthetic bacteria to form **endosymbiotic** associations that became permanent (Fig. 1–36). Some aerobic bacteria evolved into the mitochondria of modern eukaryotes, and some photosynthetic cyanobacteria became the plastids, such as the chloroplasts of green algae, the likely ancestors of modern plant cells. Prokaryotic and eukaryotic cells are compared in Table 1–3.

At some later stage of evolution, unicellular organisms found it advantageous to cluster together, thereby acquiring greater motility, efficiency, or reproductive success than their free-living single-celled competitors. Further evolution of such clustered organisms led to permanent associations among individual cells and eventually to specialization within the colony—to cellular differentiation.

The advantages of cellular specialization led to the evolution of ever more complex and highly differentiated organisms, in which some cells carried out the sensory functions, others the digestive, photosynthetic, or reproductive functions, and so forth. Many modern multicellular organisms contain hundreds of different cell types, each specialized for some function that supports the entire organism. Fundamental mechanisms that evolved early have been further refined and embellished through evolution. The same basic structures and mechanisms that underlie the beating motion of cilia in *Paramecium* and of flagella in *Chlamydomonas* are employed by the highly differentiated vertebrate sperm cell.

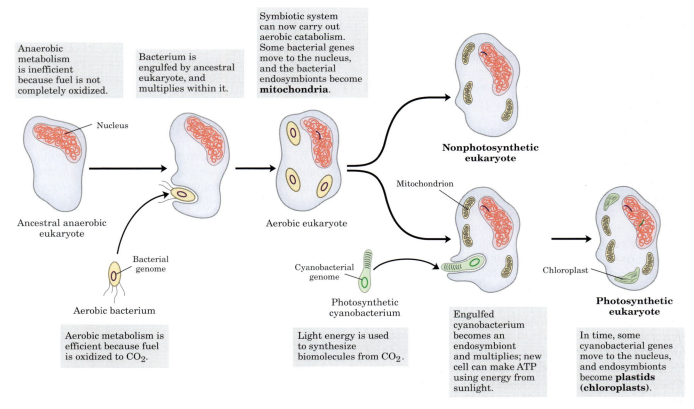

**FIGURE 1–36 Evolution of eukaryotes through endosymbiosis.** The earliest eukaryote, an anaerobe, acquired endosymbiotic purple bacteria (yellow), which carried with them their capacity for aerobic catabolism and became, over time, mitochondria. When photosynthetic cyanobacteria (green) subsequently became endosymbionts of some aerobic eukaryotes, these cells became the photosynthetic precursors of modern green algae and plants.

**TABLE 1-3**  **Comparison of Prokaryotic and Eukaryotic Cells**

| Characteristic | Prokaryotic cell | Eukaryotic cell |
|---|---|---|
| Size | Generally small (1–10 $\mu$m) | Generally large (5–100 $\mu$m) |
| Genome | DNA with nonhistone protein; genome in nucleoid, not surrounded by membrane | DNA complexed with histone and nonhistone proteins in chromosomes; chromosomes in nucleus with membranous envelope |
| Cell division | Fission or budding; no mitosis | Mitosis, including mitotic spindle; centrioles in many species |
| Membrane-bounded organelles | Absent | Mitochondria, chloroplasts (in plants, some algae), endoplasmic reticulum, Golgi complexes, lysosomes (in animals), etc. |
| Nutrition | Absorption; some photosynthesis | Absorption, ingestion; photosynthesis in some species |
| Energy metabolism | No mitochondria; oxidative enzymes bound to plasma membrane; great variation in metabolic pattern | Oxidative enzymes packaged in mitochondria; more unified pattern of oxidative metabolism |
| Cytoskeleton | None | Complex, with microtubules, intermediate filaments, actin filaments |
| Intracellular movement | None | Cytoplasmic streaming, endocytosis, phagocytosis, mitosis, vesicle transport |

**Source:** Modified from Hickman, C.P., Roberts, L.S., & Hickman, F.M. (1990) *Biology of Animals,* 5th edn, p. 30, Mosby-Yearbook, Inc., St. Louis, MO.

## Molecular Anatomy Reveals Evolutionary Relationships

The eighteenth-century naturalist Carolus Linnaeus recognized the anatomic similarities and differences among living organisms and used them to provide a framework for assessing the relatedness of species. Charles Darwin, in the nineteenth century, gave us a unifying hypothesis to explain the phylogeny of modern organisms—the origin of different species from a common ancestor. Biochemical research in the twentieth century revealed the molecular anatomy of cells of different species—the monomeric subunit sequences and the three-dimensional structures of individual nucleic acids and proteins. Biochemists now have an enormously rich and increasing treasury of evidence that can be used to analyze evolutionary relationships and to refine evolutionary theory.

The sequence of the **genome** (the complete genetic endowment of an organism) has been entirely determined for numerous eubacteria and for several archaebacteria; for the eukaryotic microorganisms *Saccharomyces cerevisiae* and *Plasmodium* sp.; for the plants *Arabidopsis thaliana* and rice; and for the multicellular animals *Caenorhabditis elegans* (a roundworm), *Drosophila melanogaster* (the fruit fly), mice, rats, and *Homo sapiens* (you*)* (Table 1–4). More sequences are being added to this list regularly. With such sequences in hand, detailed and quantitative comparisons among species can provide deep insight into the evolutionary process. Thus far, the molecular phylogeny derived from gene sequences is consistent with, but in many cases more precise than, the classical phylogeny based on macroscopic structures. Although organisms have continuously diverged at the level of gross anatomy, at the molecular level the basic unity of life is readily apparent; molecular structures and mechanisms are remarkably similar from the simplest to the most complex organisms. These similarities are most easily seen at the level of sequences, either the DNA sequences that encode proteins or the protein sequences themselves.

When two genes share readily detectable sequence similarities (nucleotide sequence in DNA or amino acid sequence in the proteins they encode), their sequences

Carolus Linnaeus,
1701–1778

Charles Darwin,
1809–1882

**TABLE 1–4  Some Organisms Whose Genomes Have Been Completely Sequenced**

| Organism | Genome size (millions of nucleotide pairs) | Biological interest |
|---|---|---|
| *Mycoplasma pneumoniae* | 0.8 | Causes pneumonia |
| *Treponema pallidum* | 1.1 | Causes syphilis |
| *Borrelia burgdorferi* | 1.3 | Causes Lyme disease |
| *Helicobacter pylori* | 1.7 | Causes gastric ulcers |
| *Methanococcus jannaschii* | 1.7 | Grows at 85 °C! |
| *Haemophilus influenzae* | 1.8 | Causes bacterial influenza |
| *Methanobacterium thermo- autotrophicum* | 1.8 | Member of the Archaea |
| *Archaeoglobus fulgidus* | 2.2 | High-temperature methanogen |
| *Synechocystis* sp. | 3.6 | Cyanobacterium |
| *Bacillus subtilis* | 4.2 | Common soil bacterium |
| *Escherichia coli* | 4.6 | Some strains cause toxic shock syndrome |
| *Saccharomyces cerevisiae* | 12.1 | Unicellular eukaryote |
| *Plasmodium falciparum* | 23 | Causes human malaria |
| *Caenorhabditis elegans* | 97.1 | Multicellular roundworm |
| *Anopheles gambiae* | 278 | Malaria vector |
| *Mus musculus domesticus* | $2.5 \times 10^3$ | Laboratory mouse |
| *Homo sapiens* | $2.9 \times 10^3$ ~ 30 million | Human |

are said to be homologous and the proteins they encode are **homologs.** If two homologous genes occur in the *same* species, they are said to be paralogous and their protein products are **paralogs.** Paralogous genes are presumed to have been derived by gene duplication followed by gradual changes in the sequences of both copies (Fig. 1–37). Typically, paralogous proteins are similar not only in sequence but also in three-dimensional structure, although they commonly have acquired different functions during their evolution.

Two homologous genes (or proteins) found in *different* species are said to be orthologous, and their protein products are **orthologs.** Orthologs are commonly found to have the same function in both organisms, and when a newly sequenced gene in one species is found to be strongly orthologous with a gene in another, this gene is presumed to encode a protein with the same function in both species. By this means, the function of gene products can be deduced from the genomic sequence, without any biochemical characterization of the gene product. An **annotated genome** includes, in addition to the DNA sequence itself, a description of the likely function of each gene product, deduced from comparisons with other genomic sequences and established protein functions. In principle, by identifying the pathways (sets of enzymes) encoded in a genome, we can deduce from the genomic sequence alone the organism's metabolic capabilities.

The sequence differences between homologous genes may be taken as a rough measure of the degree to which the two species have diverged during evolution—of how long ago their common evolutionary precursor gave rise to two lines with different evolutionary fates. The larger the number of sequence differences, the earlier the divergence in evolutionary history. One can construct a phylogeny (family tree) in which the evolutionary distance between any two species is represented by their proximity on the tree (Fig. 1–4 is an example).

As evolution advances, new structures, processes, or regulatory mechanisms are acquired, reflections of the changing genomes of the evolving organisms. The genome of a simple eukaryote such as yeast should have genes related to formation of the nuclear membrane, genes not present in prokaryotes. The genome of an insect should contain genes that encode proteins involved in specifying the characteristic insect segmented body plan, genes not present in yeast. The genomes of all vertebrate animals should share genes that specify the development of a spinal column, and those of mammals should have unique genes necessary for the development of the placenta, a characteristic of mammals—and so on. Comparisons of the whole genomes of species in each phylum may lead to the identification of genes critical to fundamental evolutionary changes in body plan and development.

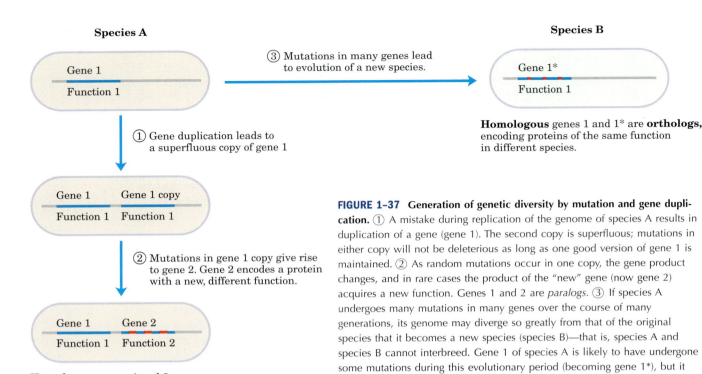

**Homologous** genes 1 and 1* are **orthologs,** encoding proteins of the same function in different species.

**Homologous** genes 1 and 2 are **paralogs,** related in sequence but encoding proteins of different function in the same species.

**FIGURE 1–37 Generation of genetic diversity by mutation and gene duplication.** ① A mistake during replication of the genome of species A results in duplication of a gene (gene 1). The second copy is superfluous; mutations in either copy will not be deleterious as long as one good version of gene 1 is maintained. ② As random mutations occur in one copy, the gene product changes, and in rare cases the product of the "new" gene (now gene 2) acquires a new function. Genes 1 and 2 are *paralogs*. ③ If species A undergoes many mutations in many genes over the course of many generations, its genome may diverge so greatly from that of the original species that it becomes a new species (species B)—that is, species A and species B cannot interbreed. Gene 1 of species A is likely to have undergone some mutations during this evolutionary period (becoming gene 1*), but it may retain enough of the original gene 1 sequence to be recognized as *homologous* with it, and its product may have the same function as (or similar function to) the product of gene 1. Gene 1* is an *ortholog* of gene 1.

## Functional Genomics Shows the Allocations of Genes to Specific Cellular Processes

When the sequence of a genome is fully determined and each gene is annotated (that is, assigned a function), molecular geneticists can group genes according to the processes (DNA synthesis, protein synthesis, generation of ATP, and so forth) in which they function and thus find what fraction of the genome is allocated to each of a cell's activities. The largest category of genes in *E. coli*, *A. thaliana*, and *H. sapiens* consists of genes of as yet unknown function, which make up more than 40% of the genes in each species. The transporters that move ions and small molecules across plasma membranes take up a significant proportion of the genes in all three species, more in the bacterium and plant than in the mammal (10% of the 4,269 genes of *E. coli*, ~8% of the 25,706 genes of *A. thaliana*, and ~4% of the ~35,000 genes of *H. sapiens*). Genes that encode the proteins and RNA required for protein synthesis make up 3% to 4% of the *E. coli* genome, but in the more complex cells of *A. thaliana*, more genes are needed for targeting proteins to their final location in the cell than are needed to synthesize those proteins (about 6% and 2%, respectively). In general, the more complex the organism, the greater the proportion of its genome that encodes genes involved in the *regulation* of cellular processes and the smaller the proportion dedicated to the basic processes themselves, such as ATP generation and protein synthesis.

## Genomic Comparisons Will Have Increasing Importance in Human Biology and Medicine

The genomes of chimpanzees and humans are 99.9% identical, yet the differences between the two species are vast. The relatively few differences in genetic endowment must explain the possession of language by humans, the extraordinary athleticism of chimpanzees, and myriad other differences. Genomic comparison will allow researchers to identify candidate genes linked to divergences in the developmental programs of humans and the other primates and to the emergence of complex functions such as language. The picture will become clearer only as more primate genomes become available for comparison with the human genome.

Similarly, the differences in genetic endowment among humans are vanishingly small compared with the differences between humans and chimpanzees, yet these differences account for the variety among us—including differences in health and in susceptibility to chronic diseases. We have much to learn about the variability in sequence among humans, and during the next decade the availability of genomic information will almost certainly transform medical diagnosis and treatment. We may expect that for some genetic diseases, palliatives will be replaced by cures; and that for disease susceptibilities associated with particular genetic markers, forewarning and perhaps increased preventive measures will prevail. Today's "medical history" may be replaced by a "medical forecast." ■

## SUMMARY 1.5 Evolutionary Foundations

- Occasional inheritable mutations yield an organism that is better suited for survival in an ecological niche and progeny that are preferentially selected. This process of mutation and selection is the basis for the Darwinian evolution that led from the first cell to all the organisms that now exist, and it explains the fundamental similarity of all living organisms.

- Life originated about 3.5 billion years ago, most likely with the formation of a membrane-enclosed compartment containing a self-replicating RNA molecule. The components for the first cell were produced by the action of lightning and high temperature on simple atmospheric molecules such as $CO_2$ and $NH_3$.

- The catalytic and genetic roles of the early RNA genome were separated over time, with DNA becoming the genomic material and proteins the major catalytic species.

- Eukaryotic cells acquired the capacity for photosynthesis and for oxidative phosphorylation from endosymbiotic bacteria. In multicellular organisms, differentiated cell types specialize in one or more of the functions essential to the organism's survival.

- Knowledge of the complete genomic nucleotide sequences of organisms from different branches of the phylogenetic tree provides insights into the evolution and function of extant organisms and offers great opportunities in human medicine.

## Key Terms

*All terms are defined in the glossary.*

| | | |
|---|---|---|
| metabolite   3 | stereoisomers   16 | exergonic reaction   23 |
| nucleus   3 | configuration   16 | equilibrium   24 |
| genome   3 | chiral center   17 | standard free-energy change, $\Delta G^\circ$   26 |
| eukaryote   4 | conformation   19 | activation energy, $\Delta G^{\ddagger}$   26 |
| prokaryote   4 | entropy, $S$   23 | catabolism   27 |
| archaebacteria   4 | enthalpy, $H$   23 | anabolism   27 |
| eubacteria   4 | free-energy change, $\Delta G$   23 | metabolism   27 |
| cytoskeleton   9 | endergonic reaction   23 | mutation   31 |

## Further Reading

### General

**Fruton, J.S.** (1999) *Proteins, Enzymes, Genes: The Interplay of Chemistry and Biochemistry,* Yale University Press, New Haven.
A distinguished historian of biochemistry traces the development of this science and discusses its impact on medicine, pharmacy, and agriculture.

**Harold, F.M.** (2001) *The Way of the Cell: Molecules, Organisms, and the Order of Life,* Oxford University Press, Oxford.

**Judson, H.F.** (1996) *The Eighth Day of Creation: The Makers of the Revolution in Biology,* expanded edn. Cold Spring Harbor Laboratory Press, Cold Spring Harbor, NY.
A highly readable and authoritative account of the rise of biochemistry and molecular biology in the twentieth century.

**Kornberg, A.** (1987) The two cultures: chemistry and biology. *Biochemistry* **26,** 6888–6891.
The importance of applying chemical tools to biological problems, described by an eminent practitioner.

**Monod, J.** (1971) *Chance and Necessity,* Alfred A. Knopf, Inc., New York. [Paperback edition, Vintage Books, 1972.] Originally published (1970) as *Le hasard et la nécessité,* Editions du Seuil, Paris.
An exploration of the philosophical implications of biological knowledge.

**Pace, N.R.** (2001) The universal nature of biochemistry. *Proc. Natl. Acad. Sci. USA* **98,** 805–808.
A short discussion of the minimal definition of life, on Earth and elsewhere.

**Schrödinger, E.** (1944) *What Is Life?* Cambridge University Press, New York. [Reprinted (1956) in *What Is Life? and Other Scientific Essays,* Doubleday Anchor Books, Garden City, NY.]
A thought-provoking look at life, written by a prominent physical chemist.

### Cellular Foundations

**Alberts, B., Johnson, A., Bray, D., Lewis, J., Raff, M., Roberts, K., & Walter, P.** (2002) *Molecular Biology of the Cell,* 4th edn, Garland Publishing, Inc., New York.
A superb textbook on cell structure and function, covering the topics considered in this chapter, and a useful reference for many of the following chapters.

**Becker, W.M., Kleinsmith, L.J., & Hardin, J.** (2000) *The World of the Cell,* 5th edn, The Benjamin/Cummings Publishing Company, Redwood City, CA.
An excellent introductory textbook of cell biology.

**Lodish, H., Berk, A., Matsudaira, P., Kaiser, C.A., Krieger, M., Scott, M.R., Zipursky, S.L., & Darnell, J.** (2003) *Molecular Cell Biology,* 5th edn, W. H. Freeman and Company, New York.

Like the book by Alberts and coauthors, a superb text useful for this and later chapters.

**Purves, W.K., Sadava, D., Orians, G.H., & Heller, H.C.** (2001) *Life: The Science of Biology,* 6th edn, W. H. Freeman and Company, New York.

## Chemical Foundations

**Barta, N.S. & Stille, J.R.** (1994) Grasping the concepts of stereochemistry. *J. Chem. Educ.* **71,** 20–23.

A clear description of the RS system for naming stereoisomers, with practical suggestions for determining and remembering the "handedness" of isomers.

**Brewster, J.H.** (1986) Stereochemistry and the origins of life. *J. Chem. Educ.* **63,** 667–670.

An interesting and lucid discussion of the ways in which evolution could have selected only one of two stereoisomers for the construction of proteins and other molecules.

**Kotz, J.C. & Treichel, P., Jr.** (1998) *Chemistry and Chemical Reactivity,* Saunders College Publishing, Fort Worth, TX.

An excellent, comprehensive introduction to chemistry.

**Vollhardt, K.P.C. & Shore, N.E.** (2002) *Organic Chemistry: Structure and Function,* W. H. Freeman and Company, New York.

Up-to-date discussions of stereochemistry, functional groups, reactivity, and the chemistry of the principal classes of biomolecules.

## Physical Foundations

**Atkins, P. W. & de Paula, J.** (2001) *Physical Chemistry,* 7th edn, W. H. Freeman and Company, New York.

**Atkins, P.W. & Jones, L.** (1999) *Chemical Principles: The Quest for Insight,* W. H. Freeman and Company, New York.

**Blum, H.F.** (1968) *Time's Arrow and Evolution,* 3rd edn, Princeton University Press, Princeton.

An excellent discussion of the way the second law of thermodynamics has influenced biological evolution.

## Genetic Foundations

**Adams, M.D., Celniker, S.E., Holt, R.A., Evans, C.A., Gocayne, J.D., Amanatides, P.G., Scherer, S.E., Li, P.W., Hoskins, R.A., Galle, R.F., et al.** (2000) The genome sequence of *Drosophila melanogaster. Science* **287,** 2185–2195.

Determination of the entire genome sequence of the fruit fly.

***Arabidopsis* Genome Initiative.** (2000) Analysis of the genome sequence of the flowering plant *Arabidopsis thaliana. Nature* **408,** 796–815.

***C. elegans* Sequencing Consortium.** (1998) Genome sequence of the nematode *C. elegans:* a platform for investigating biology. *Science* **282,** 2012–2018.

**Griffiths, A.J.F., Gelbart, W.M., Lewinton, R.C., & Miller, J.H.** (2002) *Modern Genetic Analysis: Integrating Genes and Genomes,* W. H. Freeman and Company, New York.

**International Human Genome Sequencing Consortium.** (2001) Initial sequencing and analysis of the human genome. *Nature* **409,** 860–921.

**Jacob, F.** (1973) *The Logic of Life: A History of Heredity,* Pantheon Books, Inc., New York. Originally published (1970) as *La logique du vivant: une histoire de l'hérédité,* Editions Gallimard, Paris.

A fascinating historical and philosophical account of the route by which we came to the present molecular understanding of life.

**Pierce, B.** (2002) *Genetics: A Conceptual Approach,* W. H. Freeman and Company, New York.

**Venter, J.C., Adams, M.D., Myers, E.W., Li, P.W., Mural, R.J., Sutton, G.G., Smith, H.O., Yandell, M., Evans, C.A., Holt, R.A., et al.** (2001) The sequence of the human genome. *Science* **291,** 1304–1351.

## Evolutionary Foundations

**Brow, J.R. & Doolittle, W.F.** (1997) Archaea and the prokaryote-to-eukaryote transition. *Microbiol. Mol. Biol. Rev.* **61,** 456–502.

A very thorough discussion of the arguments for placing the Archaea on the phylogenetic branch that led to multicellular organisms.

**Darwin, C.** (1964) *On the Origin of Species: A Facsimile of the First Edition (published in 1859),* Harvard University Press, Cambridge.

One of the most influential scientific works ever published.

**de Duve, C.** (1995) The beginnings of life on earth. *Am. Sci.* **83,** 428–437.

One scenario for the succession of chemical steps that led to the first living organism.

**de Duve, C.** (1996) The birth of complex cells. *Sci. Am.* **274** (April), 50–57.

**Dyer, B.D. & Obar, R.A.** (1994) *Tracing the History of Eukaryotic Cells: The Enigmatic Smile,* Columbia University Press, New York.

Evolution of Catalytic Function. (1987) *Cold Spring Harb. Symp. Quant. Biol.* **52.**

A collection of almost 100 articles on all aspects of prebiotic and early biological evolution; probably the single best source on molecular evolution.

**Fenchel, T. & Finlay, B.J.** (1994) The evolution of life without oxygen. *Am. Sci.* **82,** 22–29.

Discussion of the endosymbiotic hypothesis in the light of modern endosymbiotic anaerobic organisms.

**Gesteland, R.F. & Atkins, J.F. (eds)** (1993) *The RNA World,* Cold Spring Harbor Laboratory Press, Cold Spring Harbor, NY.

A collection of stimulating reviews on a wide range of topics related to the RNA world scenario.

**Hall, B.G.** (1982) Evolution on a Petri dish: the evolved $\beta$-galactosidase system as a model for studying acquisitive evolution in the laboratory. *Evolutionary Biol.* **15,** 85–150.

**Knoll, A.H.** (1991) End of the Proterozoic eon. *Sci. Am.* **265** (October), 64–73.

Discussion of the evidence that an increase in atmospheric oxygen led to the development of multicellular organisms, including large animals.

**Lazcano, A. & Miller, S.L.** (1996) The origin and early evolution of life: prebiotic chemistry, the pre-RNA world, and time. *Cell* **85,** 793–798.

Brief review of developments in studies of the origin of life: primitive atmospheres, submarine vents, autotrophic versus heterotrophic origin, the RNA and pre-RNA worlds, and the time required for life to arise.

**Margulis, L.** (1996) Archaeal-eubacterial mergers in the origin of Eukarya: phylogenetic classification of life. *Proc. Natl. Acad. Sci. USA* **93,** 1071–1076.

The arguments for dividing all living creatures into five king-doms: Monera, Protoctista, Fungi, Animalia, Plantae. (Compare the Woese et al. paper below.)

**Margulis, L., Gould, S.J., Schwartz, K.V., & Margulis, A.R.** (1998) *Five Kingdoms: An Illustrated Guide to the Phyla of Life on Earth,* 3rd edn, W. H. Freeman and Company, New York.

Description of all major groups of organisms, beautifully illustrated with electron micrographs and drawings.

**Mayr, E.** (1997) *This Is Biology: The Science of the Living World,* Belknap Press, Cambridge, MA.

A history of the development of science, with special emphasis on Darwinian evolution, by an eminent Darwin scholar.

**Miller, S.L.** (1987) Which organic compounds could have occurred on the prebiotic earth? *Cold Spring Harb. Symp. Quant. Biol.* **52,** 17–27.

Summary of laboratory experiments on chemical evolution, by the person who did the original Miller-Urey experiment.

**Morowitz, H.J.** (1992) *Beginnings of Cellular Life: Metabolism Recapitulates Biogenesis,* Yale University Press, New Haven.

**Schopf, J.W.** (1992) *Major Events in the History of Life,* Jones and Bartlett Publishers, Boston.

**Smith, J.M. & Szathmáry, E.** (1995) *The Major Transitions in Evolution,* W. H. Freeman and Company, New York.

**Woese, C.R., Kandler, O., & Wheelis, M.L.** (1990) Towards a natural system of organisms: proposal for the domains Archaea, Bacteria, and Eucarya. *Proc. Natl. Acad. Sci. USA* **87,** 4576–4579.

The arguments for dividing all living creatures into three kingdoms. (Compare the Margulis (1996) paper above.)

# Problems

Some problems related to the contents of the chapter follow. (In solving end-of-chapter problems, you may wish to refer to the tables on the inside of the back cover.) Each problem has a title for easy reference and discussion.

### 1. The Size of Cells and Their Components

(a) If you were to magnify a cell 10,000 fold (typical of the magnification achieved using an electron microscope), how big would it appear? Assume you are viewing a "typical" eukaryotic cell with a cellular diameter of 50 $\mu$m.

(b) If this cell were a muscle cell (myocyte), how many molecules of actin could it hold? (Assume the cell is spherical and no other cellular components are present; actin molecules are spherical, with a diameter of 3.6 nm. The volume of a sphere is 4/3 $\pi r^3$.)

(c) If this were a liver cell (hepatocyte) of the same dimensions, how many mitochondria could it hold? (Assume the cell is spherical; no other cellular components are present; and the mitochondria are spherical, with a diameter of 1.5 $\mu$m.)

(d) Glucose is the major energy-yielding nutrient for most cells. Assuming a cellular concentration of 1 mM, calculate how many molecules of glucose would be present in our hypothetical (and spherical) eukaryotic cell. (Avogadro's number, the number of molecules in 1 mol of a nonionized substance, is $6.02 \times 10^{23}$.)

(e) Hexokinase is an important enzyme in the metabolism of glucose. If the concentration of hexokinase in our eukaryotic cell is 20 $\mu$M, how many glucose molecules are present per hexokinase molecule?

### 2. Components of E. coli

*E. coli* cells are rod-shaped, about 2 $\mu$m long and 0.8 $\mu$m in diameter. The volume of a cylinder is $\pi r^2 h$, where $h$ is the height of the cylinder.

(a) If the average density of *E. coli* (mostly water) is $1.1 \times 10^3$ g/L, what is the mass of a single cell?

(b) *E. coli* has a protective cell envelope 10 nm thick. What percentage of the total volume of the bacterium does the cell envelope occupy?

(c) *E. coli* is capable of growing and multiplying rapidly because it contains some 15,000 spherical ribosomes (diameter 18 nm), which carry out protein synthesis. What percentage of the cell volume do the ribosomes occupy?

### 3. Genetic Information in E. coli DNA

The genetic information contained in DNA consists of a linear sequence of coding units, known as codons. Each codon is a specific sequence of three deoxyribonucleotides (three deoxyribonucleotide pairs in double-stranded DNA), and each codon codes for a single amino acid unit in a protein. The molecular weight of an *E. coli* DNA molecule is about $3.1 \times 10^9$ g/mol. The average molecular weight of a nucleotide pair is 660 g/mol, and each nucleotide pair contributes 0.34 nm to the length of DNA.

(a) Calculate the length of an *E. coli* DNA molecule. Compare the length of the DNA molecule with the cell dimensions (see Problem 2). How does the DNA molecule fit into the cell?

(b) Assume that the average protein in *E. coli* consists of a chain of 400 amino acids. What is the maximum number of proteins that can be coded by an *E. coli* DNA molecule?

### 4. The High Rate of Bacterial Metabolism

Bacterial cells have a much higher rate of metabolism than animal cells. Under ideal conditions some bacteria double in size and divide every 20 min, whereas most animal cells under rapid growth conditions require 24 hours. The high rate of bacterial metabolism requires a high ratio of surface area to cell volume.

(a) Why does surface-to-volume ratio affect the maximum rate of metabolism?

(b) Calculate the surface-to-volume ratio for the spherical bacterium *Neisseria gonorrhoeae* (diameter 0.5 $\mu$m), responsible for the disease gonorrhea. Compare it with the surface-to-volume ratio for a globular amoeba, a large eukaryotic cell (diameter 150 $\mu$m). The surface area of a sphere is $4\pi r^2$.

### 5. Fast Axonal Transport

Neurons have long thin processes called axons, structures specialized for conducting signals throughout the organism's nervous system. Some axonal processes can be as long as 2 m—for example, the axons that originate in your spinal cord and terminate in the muscles of your toes. Small membrane-enclosed vesicles carrying materials essential to axonal function move along microtubules of the cytoskeleton, from the cell body to the tips of the axons.

(a) If the average velocity of a vesicle is 1 $\mu$m/s, how long does it take a vesicle to move from a cell body in the spinal cord to the axonal tip in the toes?

(b) Movement of large molecules by diffusion occurs relatively slowly in cells. (For example, hemoglobin diffuses at a rate of approximately 5 $\mu$m/s.) However, the diffusion of sucrose in an aqueous solution occurs at a rate approaching that of fast cellular transport mechanisms (about 4 $\mu$m/s). What are some advantages to a cell or an organism of fast, directed transport mechanisms, compared with diffusion alone?

**6. Vitamin C: Is the Synthetic Vitamin as Good as the Natural One?** A claim put forth by some purveyors of health foods is that vitamins obtained from natural sources are more healthful than those obtained by chemical synthesis. For example, pure L-ascorbic acid (vitamin C) extracted from rose hips is better than pure L-ascorbic acid manufactured in a chemical plant. Are the vitamins from the two sources different? Can the body distinguish a vitamin's source?

**7. Identification of Functional Groups** Figures 1–15 and 1–16 show some common functional groups of biomolecules. Because the properties and biological activities of biomolecules are largely determined by their functional groups, it is important to be able to identify them. In each of the compounds below, circle and identify by name each functional group.

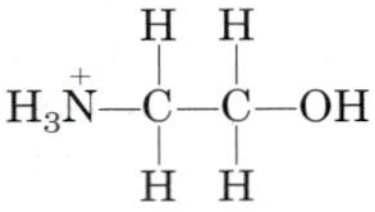

Ethanolamine (a)

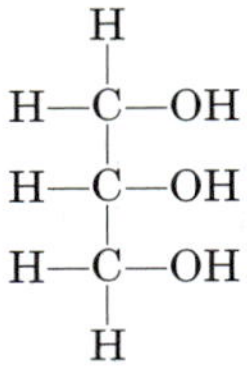

Glycerol (b)

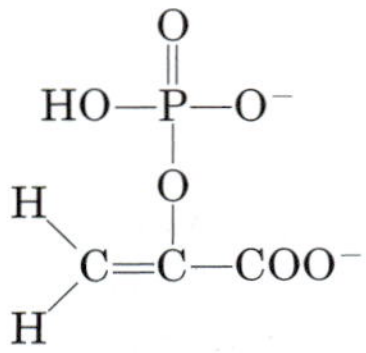

Phosphoenolpyruvate, an intermediate in glucose metabolism (c)

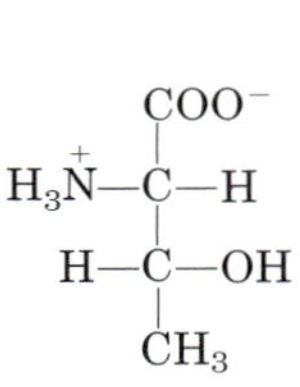

Threonine, an amino acid (d)

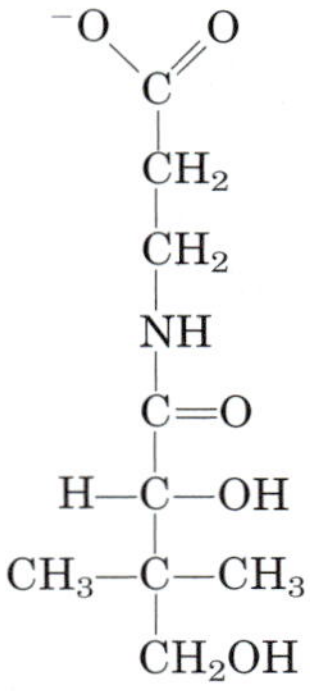

Pantothenate, a vitamin (e)

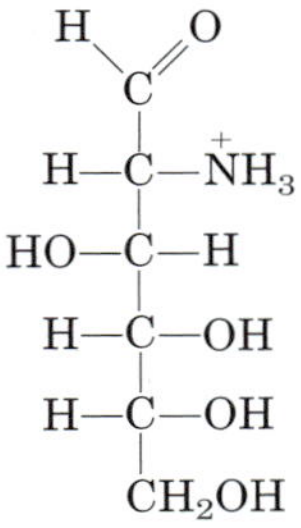

D-Glucosamine (f)

**8. Drug Activity and Stereochemistry** The quantitative differences in biological activity between the two enantiomers of a compound are sometimes quite large. For example, the D isomer of the drug isoproterenol, used to treat mild asthma, is 50 to 80 times more effective as a bronchodilator than the L isomer. Identify the chiral center in isoproterenol. Why do the two enantiomers have such radically different bioactivity?

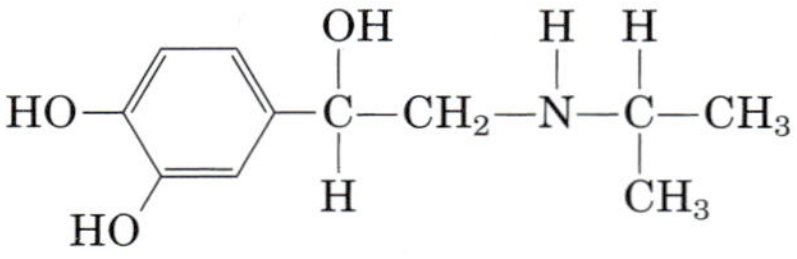

Isoproterenol

**9. Separating Biomolecules** In studying a particular biomolecule (a protein, nucleic acid, carbohydrate, or lipid) in the laboratory, the biochemist first needs to separate it from other biomolecules in the sample—that is, to *purify* it. Specific purification techniques are described later in the text. However, by looking at the monomeric subunits of a biomolecule, you should have some ideas about the characteristics of the molecule that would allow you to separate it from other molecules. For example, how would you separate (a) amino acids from fatty acids and (b) nucleotides from glucose?

**10. Silicon-Based Life?** Silicon is in the same group of the periodic table as carbon and, like carbon, can form up to four single bonds. Many science fiction stories have been based on the premise of silicon-based life. Is this realistic? What characteristics of silicon make it *less* well adapted than carbon as the central organizing element for life? To answer this question, consider what you have learned about carbon's bonding versatility, and refer to a beginning inorganic chemistry textbook for silicon's bonding properties.

**11. Drug Action and Shape of Molecules** Some years ago two drug companies marketed a drug under the trade names Dexedrine and Benzedrine. The structure of the drug is shown below.

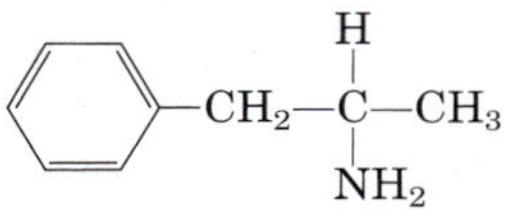

The physical properties (C, H, and N analysis, melting point, solubility, etc.) of Dexedrine and Benzedrine were identical. The recommended oral dosage of Dexedrine (which is still available) was 5 mg/day, but the recommended dosage of Benzedrine (no longer available) was twice that. Apparently it required considerably more Benzedrine than Dexedrine to yield the same physiological response. Explain this apparent contradiction.

**12. Components of Complex Biomolecules** Figure 1–10 shows the major components of complex biomolecules. For each of the three important biomolecules below (shown in their ionized forms at physiological pH), identify the constituents.

(a) Guanosine triphosphate (GTP), an energy-rich nucleotide that serves as a precursor to RNA:

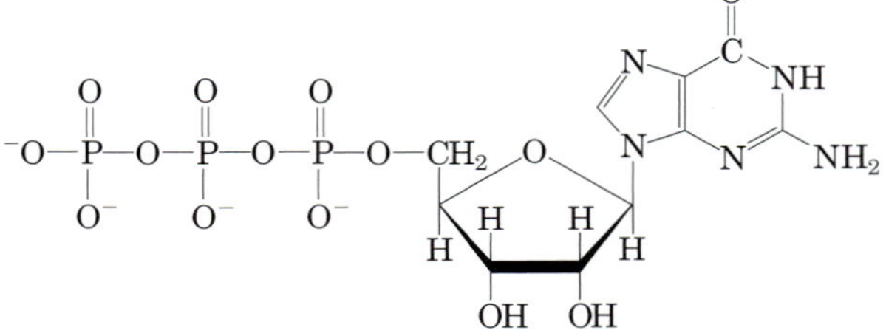

(b) Phosphatidylcholine, a component of many membranes:

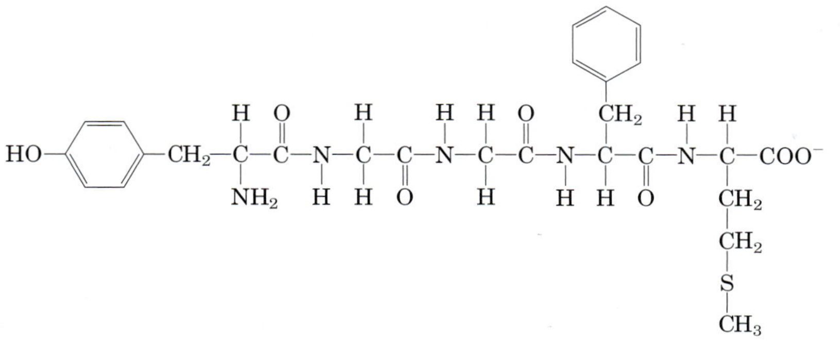

HO—⟨benzene ring⟩—CH₂—C—C—N—C—C—N—C—C—N—C—COO⁻

(c) Methionine enkephalin, the brain's own opiate:

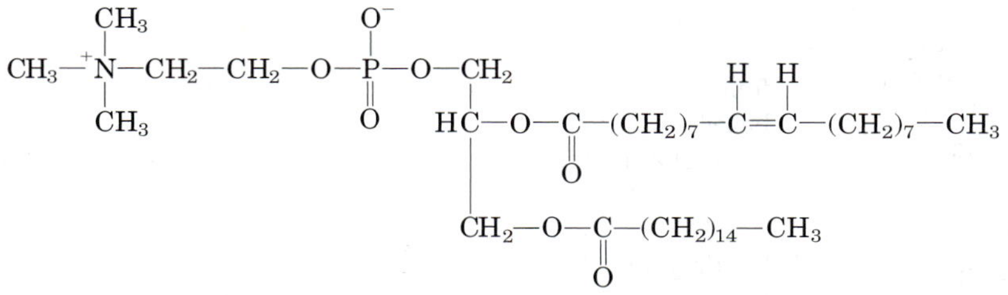

### 13. Determination of the Structure of a Biomolecule

An unknown substance, X, was isolated from rabbit muscle. Its structure was determined from the following observations and experiments. Qualitative analysis showed that X was composed entirely of C, H, and O. A weighed sample of X was completely oxidized, and the $H_2O$ and $CO_2$ produced were measured; this quantitative analysis revealed that X contained 40.00% C, 6.71% H, and 53.29% O by weight. The molecular mass of X, determined by mass spectrometry, was 90.00 u (atomic mass units; see Box 1–1). Infrared spectroscopy showed that X contained one double bond. X dissolved readily in water to give an acidic solution; the solution demonstrated optical activity when tested in a polarimeter.

(a) Determine the empirical and molecular formula of X.

(b) Draw the possible structures of X that fit the molecular formula and contain one double bond. Consider *only* linear or branched structures and disregard cyclic structures. Note that oxygen makes very poor bonds to itself.

(c) What is the structural significance of the observed optical activity? Which structures in (b) are consistent with the observation?

(d) What is the structural significance of the observation that a solution of X was acidic? Which structures in (b) are consistent with the observation?

(e) What is the structure of X? Is more than one structure consistent with all the data?

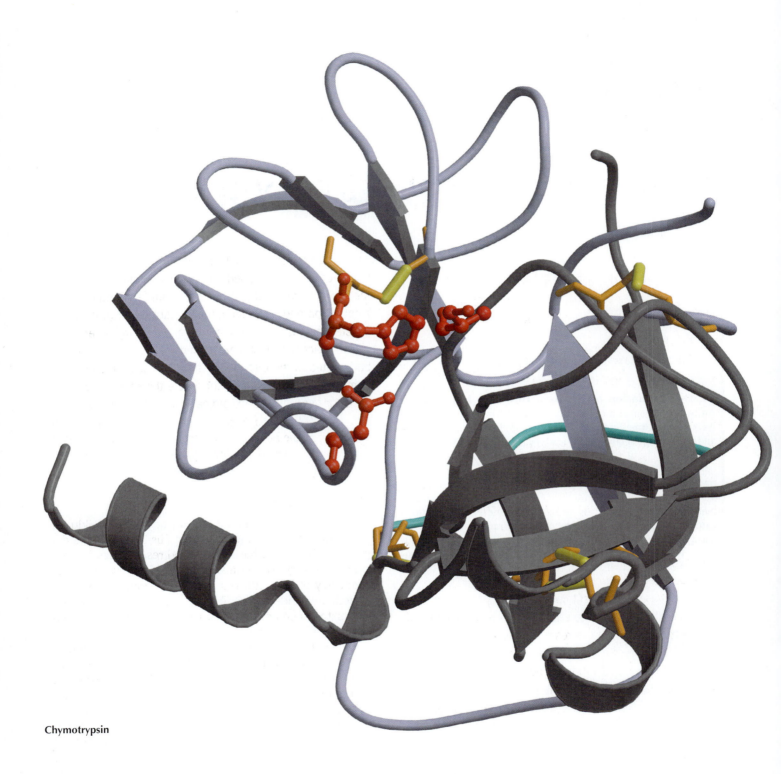

Chymotrypsin

# STRUCTURE AND CATALYSIS

was removed from the cells into the chemists' laboratories, to be studied there by the chemists' methods. It proved, too, that, apart from fermentation, combustion and respiration, the splitting up of protein substances, fats and carbohydrates, and many other similar reactions which characterise the living cell, could be imitated in the test tube without any cooperation at all from the cells, and that on the whole the same laws held for these reactions as for ordinary chemical processes.

—A. Tiselius, in presentation speech for the award of the Nobel Prize in Chemistry to James B. Sumner, John H. Northrop, and Wendell M. Stanley, 1946

In 1897 Eduard Buchner, the German research worker, discovered that sugar can be made to ferment, not only with ordinary yeast, but also with the help of the expressed juices of yeast which contain none of the cells of the Saccharomyces . . . Why was this apparently somewhat trivial experiment considered to be of such significance? The answer to this question is self-evident, if the development within the research work directed on the elucidation of the chemical nature of (life) is followed . . . there, more than in most fields, a tendency has showed itself to consider the unexplained as inexplicable . . . Thus ordinary yeast consists of living cells, and fermentation was considered by the majority of research workers—among them Pasteur—to be a manifestation of life, i.e. to be inextricably associated with the vital processes in these cells. Buchner's discovery showed that this was not the case. It may be said that thereby, at a blow, an important class of vital processes

The science of biochemistry can be dated to Eduard Buchner's pioneering discovery. His finding opened a world of chemistry that has inspired researchers for well over a century. Biochemistry is nothing less than the chemistry of life, and, yes, life can be investigated, analyzed, and understood. To begin, every student of biochemistry needs both a language and some fundamentals; these are provided in Part I.

The chapters of Part I are devoted to the structure and function of the major classes of cellular constituents: water (Chapter 2), amino acids and proteins (Chapters 3 through 6), sugars and polysaccharides (Chapter 7), nucleotides and nucleic acids (Chapter 8), fatty acids and lipids (Chapter 10), and, finally, membranes and membrane signaling proteins (Chapters 11 and 12). We supplement this discourse on molecules with information about the technologies used to study them. Some of the techniques sections are woven throughout the molecular descriptions, although one entire chapter (Chapter 9) is devoted to an integrated suite of modern advances in biotechnology that have greatly accelerated the pace of discovery.

The molecules found in a cell are a major part of the language of biochemistry; familiarity with them is a prerequisite for understanding more advanced topics covered in this book and for appreciating the rapidly growing and exciting literature of biochemistry. We begin with water because its properties affect the structure and function of all other cellular constituents. For each class of organic molecules, we first consider the covalent chemistry of the monomeric units (amino acids, monosaccharides, nucleotides, and fatty acids) and then describe the structure of the macromolecules and supramolecular complexes derived from them. An overriding theme is that the polymeric macromolecules in living systems, though large, are highly ordered chemical entities, with specific sequences of monomeric subunits giving rise to discrete structures and functions. This fundamental theme can be broken down into three interrelated principles: (1) the unique structure of each macromolecule determines its function; (2) noncovalent interactions play a critical role in the structure and thus the function of macromolecules; and (3) the monomeric subunits in polymeric macromolecules occur in specific sequences, representing a form of information upon which the ordered living state depends.

The relationship between structure and function is especially evident in proteins, which exhibit an extraordinary diversity of functions. One particular polymeric sequence of amino acids produces a strong, fibrous structure found in hair and wool; another produces a protein that transports oxygen in the blood; a third binds other proteins and catalyzes the cleavage of the bonds between their amino acids. Similarly, the special functions of polysaccharides, nucleic acids, and lipids can be understood as a direct manifestation of their chemical structure, with their characteristic monomeric subunits linked in precise functional polymers. Sugars linked together become energy stores, structural fibers, and points of specific molecular recognition; nucleotides strung together in DNA or RNA provide the blueprint for an entire organism; and aggregated lipids form membranes. Chapter 12 unifies the discussion of biomolecule function, describing how specific signaling systems regulate the activities of biomolecules—within a cell, within an organ, and among organs—to keep an organism in homeostasis.

As we move from monomeric units to larger and larger polymers, the chemical focus shifts from covalent bonds to noncovalent interactions. The properties of covalent bonds, both in the monomeric subunits and in the bonds that connect them in polymers, place constraints on the shapes assumed by large molecules. It is the numerous noncovalent interactions, however, that dictate the stable native conformations of large molecules while permitting the flexibility necessary for their biological function. As we shall see, noncovalent interactions are essential to the catalytic power of enzymes, the critical interaction of complementary base pairs in nucleic acids, the arrangement and properties of lipids in membranes, and the interaction of a hormone or growth factor with its membrane receptor.

The principle that sequences of monomeric subunits are rich in information emerges most fully in the discussion of nucleic acids (Chapter 8). However, proteins and some short polymers of sugars (oligosaccharides) are also information-rich molecules. The amino acid sequence is a form of information that directs the folding of the protein into its unique three-dimensional structure, and ultimately determines the function of the protein. Some oligosaccharides also have unique sequences and three-dimensional structures that are recognized by other macromolecules.

Each class of molecules has a similar structural hierarchy: subunits of fixed structure are connected by bonds of limited flexibility to form macromolecules with three-dimensional structures determined by noncovalent interactions. These macromolecules then interact to form the supramolecular structures and organelles that allow a cell to carry out its many metabolic functions. Together, the molecules described in Part I are the stuff of life. We begin with water.

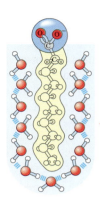

# WATER

I believe that as the methods of structural chemistry are further applied to physiological problems, it will be found that the significance of the hydrogen bond for physiology is greater than that of any other single structural feature.

—*Linus Pauling,* The Nature of the Chemical Bond, *1939*

What in water did Bloom, water lover, drawer of water, water carrier returning to the range, admire? Its universality, its democratic quality.

—*James Joyce,* Ulysses, *1922*

Water is the most abundant substance in living systems, making up 70% or more of the weight of most organisms. The first living organisms doubtless arose in an aqueous environment, and the course of evolution has been shaped by the properties of the aqueous medium in which life began.

This chapter begins with descriptions of the physical and chemical properties of water, to which all aspects of cell structure and function are adapted. The attractive forces between water molecules and the slight tendency of water to ionize are of crucial importance to the structure and function of biomolecules. We review the topic of ionization in terms of equilibrium constants, pH, and titration curves, and consider how aqueous solutions of weak acids or bases and their salts act as buffers against pH changes in biological systems. The water molecule and its ionization products, $H^+$ and $OH^-$, profoundly influence the structure, self-assembly, and properties of all cellular components, including proteins, nucleic acids, and lipids. The noncovalent interactions responsible for the strength and specificity of "recognition" among biomolecules are decisively influenced by the solvent properties of water, including its ability to form hydrogen bonds with itself and with solutes.

## 2.1 Weak Interactions in Aqueous Systems

Hydrogen bonds between water molecules provide the cohesive forces that make water a liquid at room temperature and that favor the extreme ordering of molecules that is typical of crystalline water (ice). Polar biomolecules dissolve readily in water because they can replace water-water interactions with more energetically favorable water-solute interactions. In contrast, nonpolar biomolecules interfere with water-water interactions but are unable to form water-solute interactions—consequently, nonpolar molecules are poorly soluble in water. In aqueous solutions, nonpolar molecules tend to cluster together.

Hydrogen bonds and ionic, hydrophobic (Greek, "water-fearing"), and van der Waals interactions are individually weak, but collectively they have a very significant influence on the three-dimensional structures of proteins, nucleic acids, polysaccharides, and membrane lipids.

### Hydrogen Bonding Gives Water Its Unusual Properties

Water has a higher melting point, boiling point, and heat of vaporization than most other common solvents (Table 2–1). These unusual properties are a consequence of

**TABLE 2–1    Melting Point, Boiling Point, and Heat of Vaporization of Some Common Solvents**

|  | Melting point (°C) | Boiling point (°C) | Heat of vaporization (J/g)* |
|---|---|---|---|
| Water | 0 | 100 | 2,260 |
| Methanol ($CH_3OH$) | −98 | 65 | 1,100 |
| Ethanol ($CH_3CH_2OH$) | −117 | 78 | 854 |
| Propanol ($CH_3CH_2CH_2OH$) | −127 | 97 | 687 |
| Butanol ($CH_3(CH_2)_2CH_2OH$) | −90 | 117 | 590 |
| Acetone ($CH_3COCH_3$) | −95 | 56 | 523 |
| Hexane ($CH_3(CH_2)_4CH_3$) | −98 | 69 | 423 |
| Benzene ($C_6H_6$) | 6 | 80 | 394 |
| Butane ($CH_3(CH_2)_2CH_3$) | −135 | −0.5 | 381 |
| Chloroform ($CHCl_3$) | −63 | 61 | 247 |

*The heat energy required to convert 1.0 g of a liquid at its boiling point, at atmospheric pressure, into its gaseous state at the same temperature. It is a direct measure of the energy required to overcome attractive forces between molecules in the liquid phase.

attractions between adjacent water molecules that give liquid water great internal cohesion. A look at the electron structure of the $H_2O$ molecule reveals the cause of these intermolecular attractions.

Each hydrogen atom of a water molecule shares an electron pair with the central oxygen atom. The geometry of the molecule is dictated by the shapes of the outer electron orbitals of the oxygen atom, which are similar to the $sp^3$ bonding orbitals of carbon (see Fig. 1–14). These orbitals describe a rough tetrahedron, with a hydrogen atom at each of two corners and unshared electron pairs at the other two corners (Fig. 2–1a). The H—O—H bond angle is 104.5°, slightly less than the 109.5° of a perfect tetrahedron because of crowding by the nonbonding orbitals of the oxygen atom.

The oxygen nucleus attracts electrons more strongly than does the hydrogen nucleus (a proton); that is, oxygen is more electronegative. The sharing of electrons between H and O is therefore unequal; the electrons are more often in the vicinity of the oxygen atom than of the hydrogen. The result of this unequal electron sharing is two electric dipoles in the water molecule, one along each of the H—O bonds; each hydrogen bears a partial positive charge ($\delta^+$) and the oxygen atom bears a partial negative charge equal to the sum of the two partial positives ($2\delta^-$) (Fig. 2–1b). As a result, there is an electrostatic attraction between the oxygen atom of one water molecule and the hydrogen of another (Fig. 2–1c), called a **hydrogen bond.** Throughout this book, we represent hydrogen bonds with three parallel blue lines, as in Figure 2–1c.

Hydrogen bonds are relatively weak. Those in liquid water have a **bond dissociation energy** (the energy required to break a bond) of about 23 kJ/mol, compared with 470 kJ/mol for the covalent O—H bond in

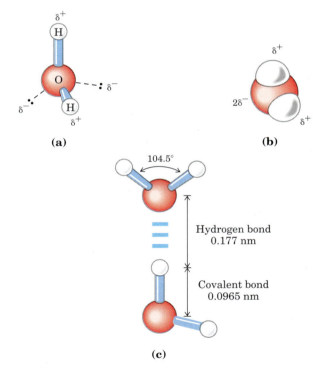

**FIGURE 2–1 Structure of the water molecule.** The dipolar nature of the $H_2O$ molecule is shown by **(a)** ball-and-stick and **(b)** space-filling models. The dashed lines in **(a)** represent the nonbonding orbitals. There is a nearly tetrahedral arrangement of the outer-shell electron pairs around the oxygen atom; the two hydrogen atoms have localized partial positive charges ($\delta^+$) and the oxygen atom has a partial negative charge ($2\delta^-$). **(c)** Two $H_2O$ molecules joined by a hydrogen bond (designated here, and throughout this book, by three blue lines) between the oxygen atom of the upper molecule and a hydrogen atom of the lower one. Hydrogen bonds are longer and weaker than covalent O—H bonds.

water or 348 kJ/mol for a covalent C—C bond. The hydrogen bond is about 10% covalent, due to overlaps in the bonding orbitals, and about 90% electrostatic. At room temperature, the thermal energy of an aqueous solution (the kinetic energy of motion of the individual atoms and molecules) is of the same order of magnitude as that required to break hydrogen bonds. When water is heated, the increase in temperature reflects the faster motion of individual water molecules. At any given time, most of the molecules in liquid water are engaged in hydrogen bonding, but the lifetime of each hydrogen bond is just 1 to 20 picoseconds (1 ps = $10^{-12}$ s); upon breakage of one hydrogen bond, another hydrogen bond forms, with the same partner or a new one, within 0.1 ps. The apt phrase "flickering clusters" has been applied to the short-lived groups of water molecules interlinked by hydrogen bonds in liquid water. The sum of all the hydrogen bonds between $H_2O$ molecules confers great internal cohesion on liquid water. Extended networks of hydrogen-bonded water molecules also form bridges between solutes (proteins and nucleic acids, for example) that allow the larger molecules to interact with each other over distances of several nanometers without physically touching.

The nearly tetrahedral arrangement of the orbitals about the oxygen atom (Fig. 2–1a) allows each water molecule to form hydrogen bonds with as many as four neighboring water molecules. In liquid water at room temperature and atmospheric pressure, however, water molecules are disorganized and in continuous motion, so that each molecule forms hydrogen bonds with an average of only 3.4 other molecules. In ice, on the other hand, each water molecule is fixed in space and forms hydrogen bonds with a full complement of four other water molecules to yield a regular lattice structure (Fig. 2–2). Breaking a sufficient proportion of hydrogen bonds to destabilize the crystal lattice of ice requires much thermal energy, which accounts for the relatively high melting point of water (Table 2–1). When ice melts or water evaporates, heat is taken up by the system:

$$H_2O(\text{solid}) \longrightarrow H_2O(\text{liquid}) \qquad \Delta H = +5.9 \text{ kJ/mol}$$

$$H_2O(\text{liquid}) \longrightarrow H_2O(\text{gas}) \qquad \Delta H = +44.0 \text{ kJ/mol}$$

During melting or evaporation, the entropy of the aqueous system increases as more highly ordered arrays of water molecules relax into the less orderly hydrogen-bonded arrays in liquid water or the wholly disordered gaseous state. At room temperature, both the melting of ice and the evaporation of water occur spontaneously; the tendency of the water molecules to associate through hydrogen bonds is outweighed by the energetic push toward randomness. Recall that the free-energy change ($\Delta G$) must have a negative value for a process to occur spontaneously: $\Delta G = \Delta H - T \Delta S$, where $\Delta G$ represents the driving force, $\Delta H$ the enthalpy change from making

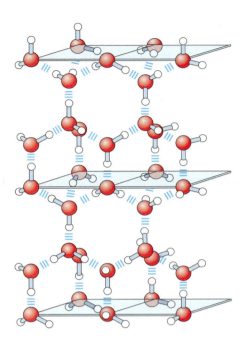

**FIGURE 2–2 Hydrogen bonding in ice.** In ice, each water molecule forms the maximum of four hydrogen bonds, creating a regular crystal lattice. By contrast, in liquid water at room temperature and atmospheric pressure, each water molecule hydrogen-bonds with an average of 3.4 other water molecules. This crystal lattice of ice makes it less dense than liquid water, and thus ice floats on liquid water.

and breaking bonds, and $\Delta S$ the change in randomness. Because $\Delta H$ is positive for melting and evaporation, it is clearly the increase in entropy ($\Delta S$) that makes $\Delta G$ negative and drives these transformations.

## Water Forms Hydrogen Bonds with Polar Solutes

Hydrogen bonds are not unique to water. They readily form between an electronegative atom (the hydrogen acceptor, usually oxygen or nitrogen with a lone pair of electrons) and a hydrogen atom covalently bonded to another electronegative atom (the hydrogen donor) in the same or another molecule (Fig. 2–3). Hydrogen atoms covalently bonded to carbon atoms do not participate in hydrogen bonding, because carbon is only

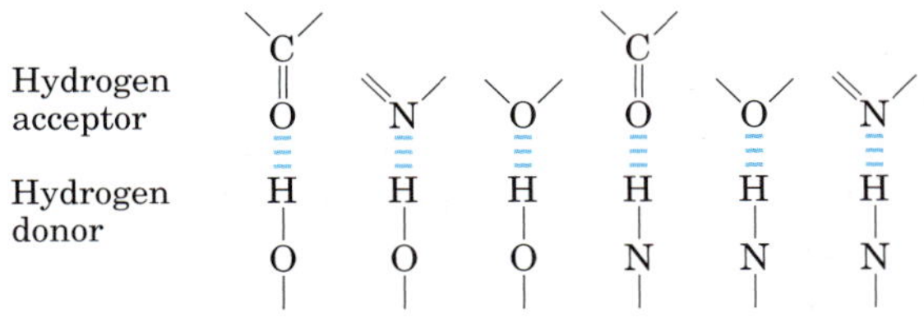

Hydrogen acceptor

Hydrogen donor

**FIGURE 2–3 Common hydrogen bonds in biological systems.** The hydrogen acceptor is usually oxygen or nitrogen; the hydrogen donor is another electronegative atom.

slightly more electronegative than hydrogen and thus the C—H bond is only very weakly polar. The distinction explains why butanol ($CH_3(CH_2)_2CH_2OH$) has a relatively high boiling point of 117 °C, whereas butane ($CH_3(CH_2)_2CH_3$) has a boiling point of only −0.5 °C. Butanol has a polar hydroxyl group and thus can form intermolecular hydrogen bonds. Uncharged but polar biomolecules such as sugars dissolve readily in water because of the stabilizing effect of hydrogen bonds between the hydroxyl groups or carbonyl oxygen of the sugar and the polar water molecules. Alcohols, aldehydes, ketones, and compounds containing N—H bonds all form hydrogen bonds with water molecules (Fig. 2–4) and tend to be soluble in water.

Hydrogen bonds are strongest when the bonded molecules are oriented to maximize electrostatic interaction, which occurs when the hydrogen atom and the two atoms that share it are in a straight line—that is, when the acceptor atom is in line with the covalent bond between the donor atom and H (Fig. 2–5). Hydrogen bonds are thus highly directional and capable of hold-

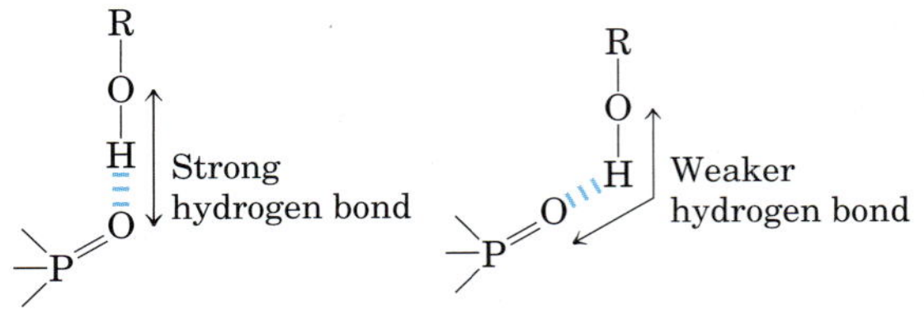

**FIGURE 2–5 Directionality of the hydrogen bond.** The attraction between the partial electric charges (see Fig. 2–1) is greatest when the three atoms involved (in this case O, H, and O) lie in a straight line. When the hydrogen-bonded moieties are structurally constrained (as when they are parts of a single protein molecule, for example), this ideal geometry may not be possible and the resulting hydrogen bond is weaker.

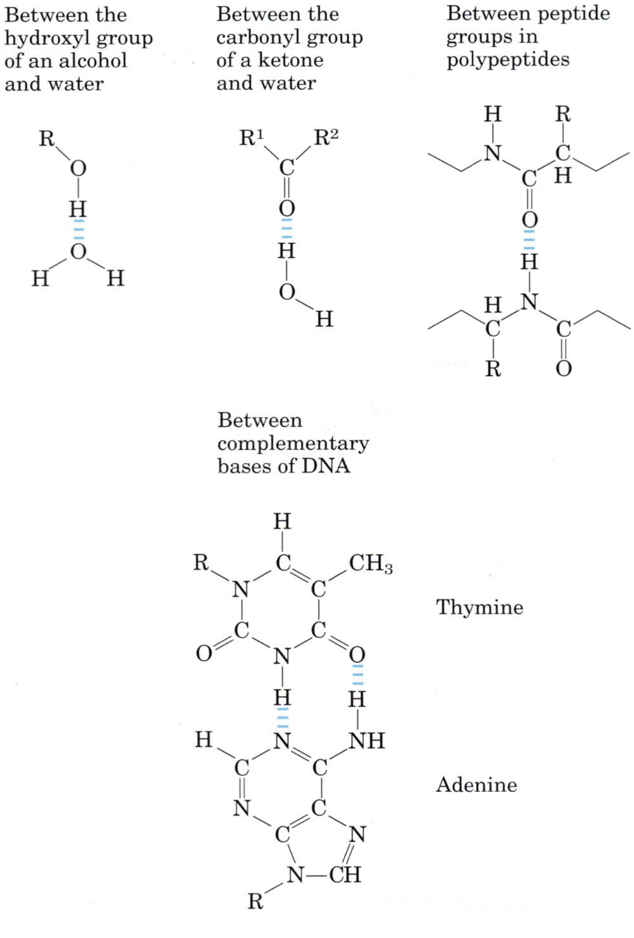

ing two hydrogen-bonded molecules or groups in a specific geometric arrangement. As we shall see later, this property of hydrogen bonds confers very precise three-dimensional structures on protein and nucleic acid molecules, which have many intramolecular hydrogen bonds.

## Water Interacts Electrostatically with Charged Solutes

Water is a polar solvent. It readily dissolves most biomolecules, which are generally charged or polar compounds (Table 2–2); compounds that dissolve easily in water are **hydrophilic** (Greek, "water-loving"). In contrast, nonpolar solvents such as chloroform and benzene are poor solvents for polar biomolecules but easily dissolve those that are **hydrophobic**—nonpolar molecules such as lipids and waxes.

Water dissolves salts such as NaCl by hydrating and stabilizing the $Na^+$ and $Cl^-$ ions, weakening the electrostatic interactions between them and thus counteracting their tendency to associate in a crystalline lattice (Fig. 2–6). The same factors apply to charged biomolecules, compounds with functional groups such as ionized carboxylic acids ($—COO^-$), protonated amines ($—NH_3^+$), and phosphate esters or anhydrides. Water readily dissolves such compounds by replacing solute-solute hydrogen bonds with solute-water hydrogen bonds, thus screening the electrostatic interactions between solute molecules.

Water is especially effective in screening the electrostatic interactions between dissolved ions because it has a high dielectric constant, a physical property reflecting the number of dipoles in a solvent. The strength, or force ($F$), of ionic interactions in a solution depends upon the magnitude of the charges ($Q$), the distance between the charged groups ($r$), and the dielectric constant ($\varepsilon$) of the solvent in which the interactions occur:

$$F = \frac{Q_1Q_2}{\varepsilon r^2}$$

**FIGURE 2–4 Some biologically important hydrogen bonds.**

**TABLE 2-2**  Some Examples of Polar, Nonpolar, and Amphipathic Biomolecules (Shown as Ionic Forms at pH 7)

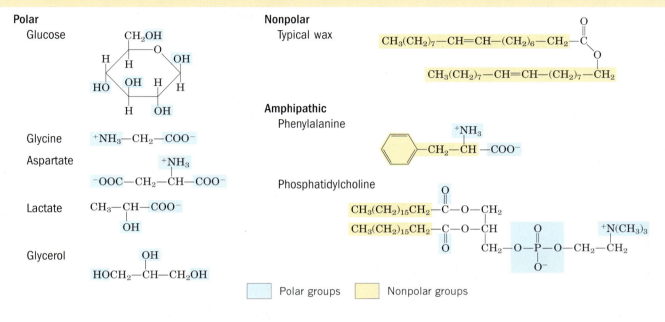

**Polar**

Glucose

Glycine $^+NH_3-CH_2-COO^-$

Aspartate $^-OOC-CH_2-\overset{\overset{\displaystyle +NH_3}{|}}{CH}-COO^-$

Lactate $CH_3-\overset{\overset{\displaystyle}{|}}{\underset{\underset{\displaystyle OH}{|}}{CH}}-COO^-$

Glycerol $HOCH_2-\overset{\overset{\displaystyle OH}{|}}{CH}-CH_2OH$

**Nonpolar**

Typical wax

**Amphipathic**

Phenylalanine

Phosphatidylcholine

☐ Polar groups    ☐ Nonpolar groups

For water at 25 °C, $\varepsilon$ (which is dimensionless) is 78.5, and for the very nonpolar solvent benzene, $\varepsilon$ is 4.6. Thus, ionic interactions are much stronger in less polar environments. The dependence on $r^2$ is such that ionic attractions or repulsions operate only over short distances—in the range of 10 to 40 nm (depending on the electrolyte concentration) when the solvent is water.

### Entropy Increases as Crystalline Substances Dissolve

As a salt such as NaCl dissolves, the $Na^+$ and $Cl^-$ ions leaving the crystal lattice acquire far greater freedom of motion (Fig. 2–6). The resulting increase in entropy (randomness) of the system is largely responsible for the ease of dissolving salts such as NaCl in water. In

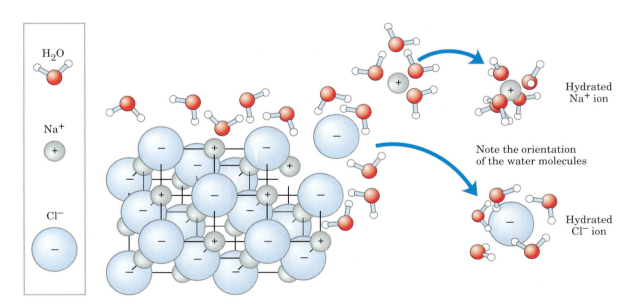

**FIGURE 2–6  Water as solvent.** Water dissolves many crystalline salts by hydrating their component ions. The NaCl crystal lattice is disrupted as water molecules cluster about the $Cl^-$ and $Na^+$ ions. The ionic charges are partially neutralized, and the electrostatic attractions necessary for lattice formation are weakened.

thermodynamic terms, formation of the solution occurs with a favorable free-energy change: $\Delta G = \Delta H - T\,\Delta S$, where $\Delta H$ has a small positive value and $T\,\Delta S$ a large positive value; thus $\Delta G$ is negative.

### Nonpolar Gases Are Poorly Soluble in Water

The molecules of the biologically important gases $CO_2$, $O_2$, and $N_2$ are nonpolar. In $O_2$ and $N_2$, electrons are shared equally by both atoms. In $CO_2$, each C=O bond is polar, but the two dipoles are oppositely directed and cancel each other (Table 2–3). The movement of molecules from the disordered gas phase into aqueous solution constrains their motion and the motion of water molecules and therefore represents a decrease in entropy. The nonpolar nature of these gases and the decrease in entropy when they enter solution combine to make them very poorly soluble in water (Table 2–3). Some organisms have water-soluble carrier proteins (hemoglobin and myoglobin, for example) that facilitate the transport of $O_2$. Carbon dioxide forms carbonic acid ($H_2CO_3$) in aqueous solution and is transported as the $HCO_3^-$ (bicarbonate) ion, either free—bicarbonate is very soluble in water (~100 g/L at 25 °C)—or bound to hemoglobin. Two other gases, $NH_3$ and $H_2S$, also have biological roles in some organisms; these gases are polar and dissolve readily in water.

### Nonpolar Compounds Force Energetically Unfavorable Changes in the Structure of Water

When water is mixed with benzene or hexane, two phases form; neither liquid is soluble in the other. Nonpolar compounds such as benzene and hexane are hydrophobic—they are unable to undergo energetically favorable interactions with water molecules, and they interfere with the hydrogen bonding among water molecules. All molecules or ions in aqueous solution interfere with the hydrogen bonding of some water molecules in their immediate vicinity, but polar or charged solutes (such as NaCl) compensate for lost water-water hydrogen bonds by forming new solute-water interactions. The net change in enthalpy ($\Delta H$) for dissolving these solutes is generally small. Hydrophobic solutes, however, offer no such compensation, and their addition to water may therefore result in a small gain of enthalpy; the breaking of hydrogen bonds between water molecules takes up energy from the system. Furthermore, dissolving hydrophobic compounds in water produces a measurable decrease in entropy. Water molecules in the immediate vicinity of a nonpolar solute are constrained in their possible orientations as they form a highly ordered cagelike shell around each solute molecule. These water molecules are not as highly oriented as those in **clathrates,** crystalline compounds of nonpolar solutes and water, but the effect is the same in both cases: the ordering of water molecules reduces entropy. The number of ordered water molecules, and therefore the magnitude of the entropy decrease, is proportional to the surface area of the hydrophobic solute enclosed within the cage of water molecules. The free-energy change for dissolving a nonpolar solute in water is thus unfavorable: $\Delta G = \Delta H - T\,\Delta S$, where $\Delta H$ has a positive value, $\Delta S$ has a negative value, and $\Delta G$ is positive.

**Amphipathic** compounds contain regions that are polar (or charged) and regions that are nonpolar (Table 2–2). When an amphipathic compound is mixed with

---

**TABLE 2–3   Solubilities of Some Gases in Water**

| Gas | Structure* | Polarity | Solubility in water (g/L)[†] |
|---|---|---|---|
| Nitrogen | N≡N | Nonpolar | 0.018 (40 °C) |
| Oxygen | O=O | Nonpolar | 0.035 (50 °C) |
| Carbon dioxide | 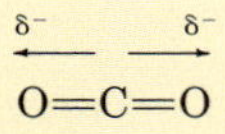 $\overset{\delta^-}{\underset{O=C=O}{\longleftarrow}}\ \overset{\delta^-}{\longrightarrow}$ | Nonpolar | 0.97 (45 °C) |
| Ammonia | 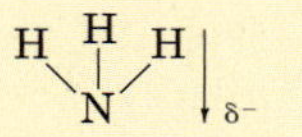 $\underset{N}{H\overset{H}{\diagup\!\!\diagdown}H}\ \downarrow\delta^-$ | Polar | 900 (10 °C) |
| Hydrogen sulfide | 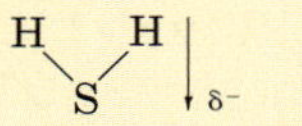 $\underset{S}{H\diagup\!\!\diagdown H}\ \downarrow\delta^-$ | Polar | 1,860 (40 °C) |

*The arrows represent electric dipoles; there is a partial negative charge ($\delta^-$) at the head of the arrow, a partial positive charge ($\delta^+$; not shown here) at the tail.

[†]Note that polar molecules dissolve far better even at low temperatures than do nonpolar molecules at relatively high temperatures.

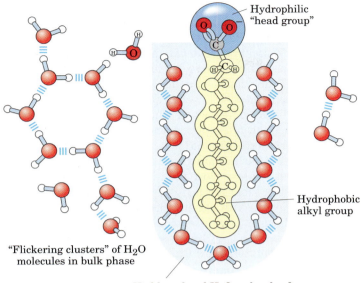

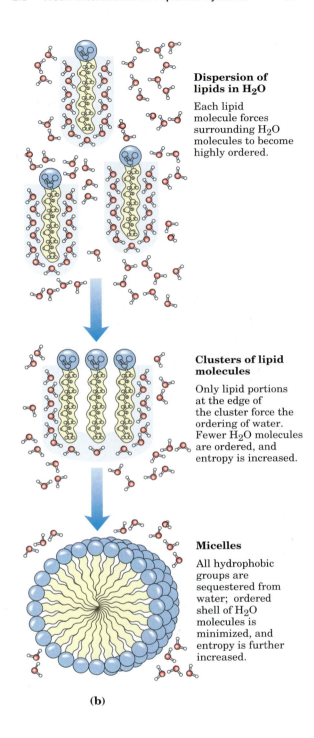

**Dispersion of lipids in $H_2O$**

Each lipid molecule forces surrounding $H_2O$ molecules to become highly ordered.

**Clusters of lipid molecules**

Only lipid portions at the edge of the cluster force the ordering of water. Fewer $H_2O$ molecules are ordered, and entropy is increased.

**Micelles**

All hydrophobic groups are sequestered from water; ordered shell of $H_2O$ molecules is minimized, and entropy is further increased.

**(b)**

Hydrophilic "head group"

Hydrophobic alkyl group

"Flickering clusters" of $H_2O$ molecules in bulk phase

Highly ordered $H_2O$ molecules form "cages" around the hydrophobic alkyl chains

**(a)**

**FIGURE 2–7 Amphipathic compounds in aqueous solution. (a)** Long-chain fatty acids have very hydrophobic alkyl chains, each of which is surrounded by a layer of highly ordered water molecules. **(b)** By clustering together in micelles, the fatty acid molecules expose the smallest possible hydrophobic surface area to the water, and fewer water molecules are required in the shell of ordered water. The energy gained by freeing immobilized water molecules stabilizes the micelle.

water, the polar, hydrophilic region interacts favorably with the solvent and tends to dissolve, but the nonpolar, hydrophobic region tends to avoid contact with the water (Fig. 2–7a). The nonpolar regions of the molecules cluster together to present the smallest hydrophobic area to the aqueous solvent, and the polar regions are arranged to maximize their interaction with the solvent (Fig. 2–7b). These stable structures of amphipathic compounds in water, called **micelles,** may contain hundreds or thousands of molecules. The forces that hold the nonpolar regions of the molecules together are called **hydrophobic interactions.** The strength of hydrophobic interactions is not due to any intrinsic attraction between nonpolar moieties. Rather, it results from the system's achieving greatest thermodynamic stability by minimizing the number of ordered water molecules required to surround hydrophobic portions of the solute molecules.

Many biomolecules are amphipathic; proteins, pigments, certain vitamins, and the sterols and phospholipids of membranes all have polar and nonpolar surface regions. Structures composed of these molecules are stabilized by hydrophobic interactions among the non-

polar regions. Hydrophobic interactions among lipids, and between lipids and proteins, are the most important determinants of structure in biological membranes. Hydrophobic interactions between nonpolar amino acids also stabilize the three-dimensional structures of proteins.

Hydrogen bonding between water and polar solutes also causes some ordering of water molecules, but the effect is less significant than with nonpolar solutes. Part

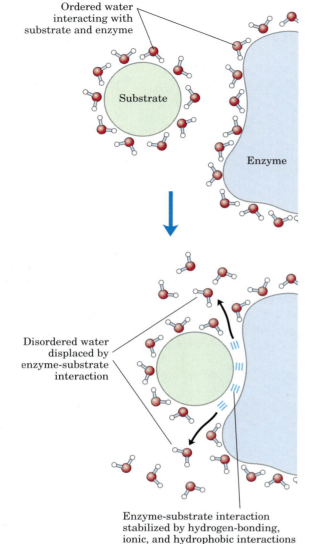

**FIGURE 2–8 Release of ordered water favors formation of an enzyme-substrate complex.** While separate, both enzyme and substrate force neighboring water molecules into an ordered shell. Binding of substrate to enzyme releases some of the ordered water, and the resulting increase in entropy provides a thermodynamic push toward formation of the enzyme-substrate complex.

other. Random variations in the positions of the electrons around one nucleus may create a transient electric dipole, which induces a transient, opposite electric dipole in the nearby atom. The two dipoles weakly attract each other, bringing the two nuclei closer. These weak attractions are called **van der Waals interactions.** As the two nuclei draw closer together, their electron clouds begin to repel each other. At the point where the van der Waals attraction exactly balances this repulsive force, the nuclei are said to be in van der Waals contact. Each atom has a characteristic **van der Waals radius,** a measure of how close that atom will allow another to approach (Table 2–4). In the "space-filling" molecular models shown throughout this book, the atoms are depicted in sizes proportional to their van der Waals radii.

## Weak Interactions Are Crucial to Macromolecular Structure and Function

The noncovalent interactions we have described (hydrogen bonds and ionic, hydrophobic, and van der Waals interactions) (Table 2–5) are much weaker than covalent bonds. An input of about 350 kJ of energy is required to break a mole of ($6 \times 10^{23}$) C—C single bonds, and about 410 kJ to break a mole of C—H bonds, but as little as 4 kJ is sufficient to disrupt a mole of typical van der Waals interactions. Hydrophobic interactions are also much weaker than covalent bonds, although they are substantially strengthened by a highly polar solvent (a concentrated salt solution, for example). Ionic interactions and hydrogen bonds are variable in strength, depending on the polarity of the solvent and

**TABLE 2–4   van der Waals Radii and Covalent (Single-Bond) Radii of Some Elements**

| Element | van der Waals radius (nm) | Covalent radius for single bond (nm) |
|---------|---------------------------|--------------------------------------|
| H | 0.11 | 0.030 |
| O | 0.15 | 0.066 |
| N | 0.15 | 0.070 |
| C | 0.17 | 0.077 |
| S | 0.18 | 0.104 |
| P | 0.19 | 0.110 |
| I | 0.21 | 0.133 |

**Sources:** For van der Waals radii, Chauvin, R. (1992) Explicit periodic trend of van der Waals radii. *J. Phys. Chem.* **96,** 9194–9197. For covalent radii, Pauling, L. (1960) *Nature of the Chemical Bond,* 3rd edn, Cornell University Press, Ithaca, NY.

**Note:** van der Waals radii describe the space-filling dimensions of atoms. When two atoms are joined covalently, the atomic radii at the point of bonding are less than the van der Waals radii, because the joined atoms are pulled together by the shared electron pair. The distance between nuclei in a van der Waals interaction or a covalent bond is about equal to the sum of the van der Waals or covalent radii, respectively, for the two atoms. Thus the length of a carbon-carbon single bond is about 0.077 nm + 0.077 nm = 0.154 nm.

of the driving force for binding of a polar substrate (reactant) to the complementary polar surface of an enzyme is the entropy increase as the enzyme displaces ordered water from the substrate (Fig. 2–8).

## van der Waals Interactions Are Weak Interatomic Attractions

When two uncharged atoms are brought very close together, their surrounding electron clouds influence each

**TABLE 2-5  Four Types of Noncovalent ("Weak") Interactions among Biomolecules in Aqueous Solvent**

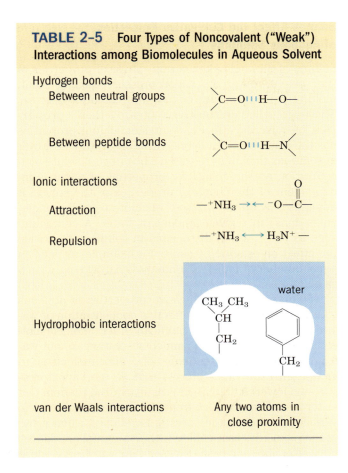

Hydrogen bonds
  Between neutral groups

  Between peptide bonds

Ionic interactions

  Attraction

  Repulsion

Hydrophobic interactions

                      water

van der Waals interactions      Any two atoms in
                                close proximity

through multiple weak interactions requires all these interactions to be disrupted at the same time. Because the interactions fluctuate randomly, such simultaneous disruptions are very unlikely. The molecular stability bestowed by 5 or 20 weak interactions is therefore much greater than would be expected intuitively from a simple summation of small binding energies.

Macromolecules such as proteins, DNA, and RNA contain so many sites of potential hydrogen bonding or ionic, van der Waals, or hydrophobic interactions that the cumulative effect of the many small binding forces can be enormous. For macromolecules, the most stable (that is, the native) structure is usually that in which weak-bonding possibilities are maximized. The folding of a single polypeptide or polynucleotide chain into its three-dimensional shape is determined by this principle. The binding of an antigen to a specific antibody depends on the cumulative effects of many weak interactions. As noted earlier, the energy released when an enzyme binds noncovalently to its substrate is the main source of the enzyme's catalytic power. The binding of a hormone or a neurotransmitter to its cellular receptor protein is the result of weak interactions. One consequence of the large size of enzymes and receptors is that their extensive surfaces provide many opportunities for weak interactions. At the molecular level, the complementarity between interacting biomolecules reflects the complementarity and weak interactions between polar, charged, and hydrophobic groups on the surfaces of the molecules.

When the structure of a protein such as hemoglobin (Fig. 2–9) is determined by x-ray crystallography (see

the alignment of the hydrogen-bonded atoms, but they are always significantly weaker than covalent bonds. In aqueous solvent at 25 °C, the available thermal energy can be of the same order of magnitude as the strength of these weak interactions, and the interaction between solute and solvent (water) molecules is nearly as favorable as solute-solute interactions. Consequently, hydrogen bonds and ionic, hydrophobic, and van der Waals interactions are continually formed and broken.

Although these four types of interactions are individually weak relative to covalent bonds, the cumulative effect of many such interactions can be very significant. For example, the noncovalent binding of an enzyme to its substrate may involve several hydrogen bonds and one or more ionic interactions, as well as hydrophobic and van der Waals interactions. The formation of each of these weak bonds contributes to a net decrease in the free energy of the system. We can calculate the stability of a noncovalent interaction, such as that of a small molecule hydrogen-bonded to its macromolecular partner, from the binding energy. Stability, as measured by the equilibrium constant (see below) of the binding reaction, varies *exponentially* with binding energy. The dissociation of two biomolecules (such as an enzyme and its bound substrate) associated noncovalently

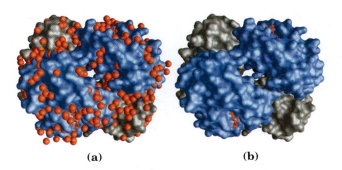

**FIGURE 2-9  Water binding in hemoglobin.** The crystal structure of hemoglobin, shown **(a)** with bound water molecules (red spheres) and **(b)** without the water molecules. These water molecules are so firmly bound to the protein that they affect the x-ray diffraction pattern as though they were fixed parts of the crystal. The two $\alpha$ subunits of hemoglobin are shown in gray, the two $\beta$ subunits in blue. Each subunit has a bound heme group (red stick structure), visible only in the $\beta$ subunits in this view. The structure and function of hemoglobin are discussed in detail in Chapter 5.

Box 4–4, p. 136), water molecules are often found to be bound so tightly as to be part of the crystal structure; the same is true for water in crystals of RNA or DNA. These bound water molecules, which can also be detected in aqueous solutions by nuclear magnetic resonance, have distinctly different properties from those of the "bulk" water of the solvent. They are, for example, not osmotically active (see below). For many proteins, tightly bound water molecules are essential to their function. In a reaction central to the process of photosynthesis, for example, light drives protons across a biological membrane as electrons flow through a series of electron-carrying proteins (see Fig. 19–54). One of these proteins, cytochrome *f,* has a chain of five bound water molecules (Fig. 2–10) that may provide a path for protons to move through the membrane by a process known as "proton hopping" (described below). Another such light-driven proton pump, bacteriorhodopsin, almost certainly uses a chain of precisely oriented bound water molecules in the transmembrane movement of protons (see Fig. 19–59).

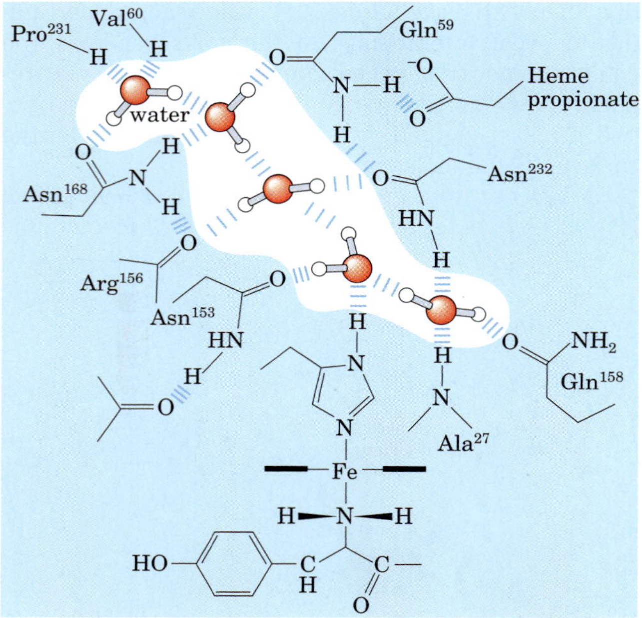

**FIGURE 2–10 Water chain in cytochrome f.** Water is bound in a proton channel of the membrane protein cytochrome *f,* which is part of the energy-trapping machinery of photosynthesis in chloroplasts (see Fig. 19–57). Five water molecules are hydrogen-bonded to each other and to functional groups of the protein: the peptide backbone atoms of valine, proline, arginine, and alanine residues, and the side chains of three asparagine and two glutamine residues. The protein has a bound heme (see Fig. 5–1), its iron ion facilitating electron flow during photosynthesis. Electron flow is coupled to the movement of protons across the membrane, which probably involves "proton hopping" (see Fig. 2–14) through this chain of bound water molecules.

## Solutes Affect the Colligative Properties of Aqueous Solutions

Solutes of all kinds alter certain physical properties of the solvent, water: its vapor pressure, boiling point, melting point (freezing point), and osmotic pressure. These are called **colligative** ("tied together") **properties,** because the effect of solutes on all four properties has the same basis: the concentration of water is lower in solutions than in pure water. The effect of solute concentration on the colligative properties of water is independent of the chemical properties of the solute; it depends only on the *number* of solute particles (molecules, ions) in a given amount of water. A compound such as NaCl, which dissociates in solution, has twice the effect on osmotic pressure, for example, as does an equal number of moles of a nondissociating solute such as glucose.

Solutes alter the colligative properties of aqueous solutions by lowering the effective concentration of water. For example, when a significant fraction of the molecules at the surface of an aqueous solution are not water but solute, the tendency of water molecules to escape into the vapor phase—that is, the vapor pressure—is lowered (Fig. 2–11). Similarly, the tendency of water molecules to move from the aqueous phase to the surface of a forming ice crystal is reduced when some of the molecules that collide with the crystal are solute, not water. In that case, the solution will freeze more slowly than pure water and at a lower temperature. For a 1.00 molal aqueous solution (1.00 mol of solute per 1,000 g of water) of an ideal, nonvolatile, and nondissociating solute at 101 kPa (1 atm) of pressure, the freezing point is 1.86 °C lower and the boiling point is 0.543 °C higher than for pure water. For a 0.100 molal solution of the same solute, the changes are one-tenth as large.

Water molecules tend to move from a region of higher water concentration to one of lower water concentration. When two different aqueous solutions are separated by a semipermeable membrane (one that allows the passage of water but not solute molecules), water molecules diffusing from the region of higher water concentration to that of lower water concentration produce osmotic pressure (Fig. 2–12). This pressure, $\Pi$, measured as the force necessary to resist water movement (Fig. 2–12c), is approximated by the van't Hoff equation:

$$\Pi = icRT$$

in which $R$ is the gas constant and $T$ is the absolute temperature. The term $ic$ is the **osmolarity** of the solution, the product of the solute's molar concentration $c$ and the van't Hoff factor $i$, which is a measure of the extent to which the solute dissociates into two or more ionic species. In dilute NaCl solutions, the solute completely

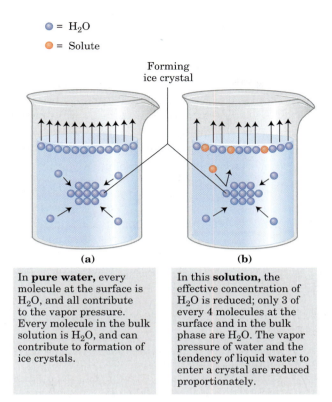

● = H₂O
● = Solute

Forming ice crystal

**(a)**

In **pure water,** every molecule at the surface is H₂O, and all contribute to the vapor pressure. Every molecule in the bulk solution is H₂O, and can contribute to formation of ice crystals.

**(b)**

In this **solution,** the effective concentration of H₂O is reduced; only 3 of every 4 molecules at the surface and in the bulk phase are H₂O. The vapor pressure of water and the tendency of liquid water to enter a crystal are reduced proportionately.

**FIGURE 2-11 Solutes alter the colligative properties of aqueous solutions. (a)** At 101 kPa (1 atm) pressure, pure water boils at 100 °C and freezes at 0 °C. **(b)** The presence of solute molecules reduces the probability of a water molecule leaving the solution and entering the gas phase, thereby reducing the vapor pressure of the solution and increasing the boiling point. Similarly, the probability of a water molecule colliding with and joining a forming ice crystal is reduced when some of the molecules colliding with the crystal are solute, not water, molecules. The effect is depression of the freezing point.

dissociates into $Na^+$ and $Cl^-$, doubling the number of solute particles, and thus $i = 2$. For nonionizing solutes, $i$ is always 1. For solutions of several ($n$) solutes, $\Pi$ is the sum of the contributions of each species:

$$\Pi = RT(i_1c_1 + i_2c_2 + \cdots + i_nc_n)$$

**Osmosis,** water movement across a semipermeable membrane driven by differences in osmotic pressure, is an important factor in the life of most cells. Plasma membranes are more permeable to water than to most other small molecules, ions, and macromolecules. This permeability is due partly to simple diffusion of water through the lipid bilayer and partly to protein channels (aquaporins; see Fig. 11–46) in the membrane that selectively permit the passage of water. Solutions of equal osmolarity are said to be **isotonic.** Surrounded by an isotonic solution, a cell neither gains nor loses water (Fig. 2–13). In a **hypertonic** solution, one with higher

osmolarity than the cytosol, the cell shrinks as water flows out. In a **hypotonic** solution, with lower osmolarity than the cytosol, the cell swells as water enters. In their natural environments, cells generally contain higher concentrations of biomolecules and ions than their surroundings, so osmotic pressure tends to drive water into cells. If not somehow counterbalanced, this inward movement of water would distend the plasma membrane and eventually cause bursting of the cell (osmotic lysis).

Several mechanisms have evolved to prevent this catastrophe. In bacteria and plants, the plasma membrane is surrounded by a nonexpandable cell wall of sufficient rigidity and strength to resist osmotic pressure and prevent osmotic lysis. Certain freshwater protists that live in a highly hypotonic medium have an organelle (contractile vacuole) that pumps water out of the cell. In multicellular animals, blood plasma and interstitial fluid (the extracellular fluid of tissues) are maintained at an osmolarity close to that of the cytosol. The high concentration of albumin and other proteins in blood plasma contributes to its osmolarity. Cells also actively pump out ions such as $Na^+$ into the interstitial fluid to stay in osmotic balance with their surroundings.

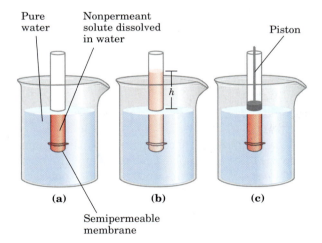

Pure water    Nonpermeant solute dissolved in water    Piston

$h$

Semipermeable membrane

**(a)**    **(b)**    **(c)**

**FIGURE 2-12 Osmosis and the measurement of osmotic pressure. (a)** The initial state. The tube contains an aqueous solution, the beaker contains pure water, and the semipermeable membrane allows the passage of water but not solute. Water flows from the beaker into the tube to equalize its concentration across the membrane. **(b)** The final state. Water has moved into the solution of the nonpermeant compound, diluting it and raising the column of water within the tube. At equilibrium, the force of gravity operating on the solution in the tube exactly balances the tendency of water to move into the tube, where its concentration is lower. **(c)** Osmotic pressure ($\Pi$) is measured as the force that must be applied to return the solution in the tube to the level of that in the beaker. This force is proportional to the height, $h$, of the column in **(b)**.

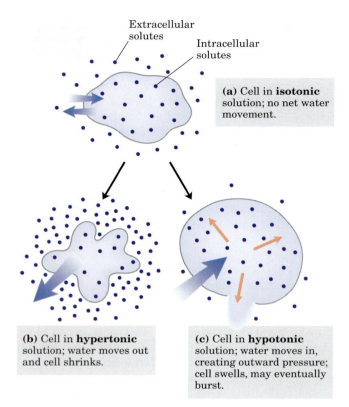

**(a)** Cell in **isotonic** solution; no net water movement.

**(b)** Cell in **hypertonic** solution; water moves out and cell shrinks.

**(c)** Cell in **hypotonic** solution; water moves in, creating outward pressure; cell swells, may eventually burst.

**FIGURE 2–13** **Effect of extracellular osmolarity on water movement across a plasma membrane.** When a cell in osmotic balance with its surrounding medium (that is, in an isotonic medium) **(a)** is transferred into a hypertonic solution **(b)** or hypotonic solution **(c),** water moves across the plasma membrane in the direction that tends to equalize osmolarity outside and inside the cell.

Because the effect of solutes on osmolarity depends on the *number* of dissolved particles, not their *mass,* macromolecules (proteins, nucleic acids, polysaccharides) have far less effect on the osmolarity of a solution than would an equal mass of their monomeric components. For example, a *gram* of a polysaccharide composed of 1,000 glucose units has the same effect on osmolarity as a *milligram* of glucose. One effect of storing fuel as polysaccharides (starch or glycogen) rather than as glucose or other simple sugars is prevention of an enormous increase in osmotic pressure within the storage cell.

Plants use osmotic pressure to achieve mechanical rigidity. The very high solute concentration in the plant cell vacuole draws water into the cell (Fig. 2–13). The resulting osmotic pressure against the cell wall (turgor pressure) stiffens the cell, the tissue, and the plant body. When the lettuce in your salad wilts, it is because loss of water has reduced turgor pressure. Sudden alterations in turgor pressure produce the movement of plant parts seen in touch-sensitive plants such as the Venus flytrap and mimosa (Box 2–1).

Osmosis also has consequences for laboratory protocols. Mitochondria, chloroplasts, and lysosomes, for example, are bounded by semipermeable membranes. In isolating these organelles from broken cells, biochemists must perform the fractionations in isotonic solutions (see Fig. 1–8). Buffers used in cellular fractionations commonly contain sufficient concentrations (about 0.2 M) of sucrose or some other inert solute to protect the organelles from osmotic lysis.

## SUMMARY 2.1 Weak Interactions in Aqueous Systems

- The very different electronegativities of H and O make water a highly polar molecule, capable of forming hydrogen bonds with itself and with solutes. Hydrogen bonds are fleeting, primarily electrostatic, and weaker than covalent bonds. Water is a good solvent for polar (hydrophilic) solutes, with which it forms hydrogen bonds, and for charged solutes, with which it interacts electrostatically.

- Nonpolar (hydrophobic) compounds dissolve poorly in water; they cannot hydrogen-bond with the solvent, and their presence forces an energetically unfavorable ordering of water molecules at their hydrophobic surfaces. To minimize the surface exposed to water, nonpolar compounds such as lipids form aggregates (micelles) in which the hydrophobic moieties are sequestered in the interior, associating through hydrophobic interactions, and only the more polar moieties interact with water.

- Numerous weak, noncovalent interactions decisively influence the folding of macromolecules such as proteins and nucleic acids. The most stable macromolecular conformations are those in which hydrogen bonding is maximized within the molecule and between the molecule and the solvent, and in which hydrophobic moieties cluster in the interior of the molecule away from the aqueous solvent.

- The physical properties of aqueous solutions are strongly influenced by the concentrations of solutes. When two aqueous compartments are separated by a semipermeable membrane (such as the plasma membrane separating a cell from its surroundings), water moves across that membrane to equalize the osmolarity in the two compartments. This tendency for water to move across a semipermeable membrane is the osmotic pressure.

BOX 2–1    THE WORLD OF BIOCHEMISTRY

## Touch Response in Plants: An Osmotic Event

The highly specialized leaves of the Venus flytrap (*Dionaea muscipula*) rapidly fold together in response to a light touch by an unsuspecting insect, entrapping the insect for later digestion. Attracted by nectar on the leaf surface, the insect touches three mechanically sensitive hairs, triggering the traplike closing of the leaf (Fig. 1). This leaf movement is produced by sudden (within 0.5 s) changes of turgor pressure in mesophyll cells (the inner cells of the leaf), probably achieved by the release of $K^+$ ions from the cells and the resulting efflux, by osmosis, of water. Digestive glands in the leaf's surface release enzymes that extract nutrients from the insect.

The sensitive plant (*Mimosa pudica*) also undergoes a remarkable change in leaf shape triggered by mechanical touch (Fig. 2). A light touch or vibration produces a sudden drooping of the leaves, the result of a dramatic reduction in turgor pressure in cells at the base of each leaflet and leaf. As in the Venus flytrap, the drop in turgor pressure results from $K^+$ release followed by the efflux of water.

(a)

(b)

**FIGURE 1**  Touch response in the Venus flytrap. A fly approaching an open leaf (a) is trapped for digestion by the plant (b).

(a)

(b)

**FIGURE 2**  The feathery leaflets of the sensitive plant (a) close and drop (b) to protect the plant from structural damage by wind.

## 2.2 Ionization of Water, Weak Acids, and Weak Bases

Although many of the solvent properties of water can be explained in terms of the uncharged $H_2O$ molecule, the small degree of ionization of water to hydrogen ions ($H^+$) and hydroxide ions ($OH^-$) must also be taken into account. Like all reversible reactions, the ionization of water can be described by an equilibrium constant. When weak acids are dissolved in water, they contribute $H^+$ by ionizing; weak bases consume $H^+$ by becoming protonated. These processes are also governed by equilibrium constants. The total hydrogen ion concentration from all sources is experimentally measurable and is expressed as the pH of the solution. To predict the state of ionization of solutes in water, we must take into account the relevant equilibrium constants for each ionization reaction. We therefore turn now to a brief discussion of the ionization of water and of weak acids and bases dissolved in water.

### Pure Water Is Slightly Ionized

Water molecules have a slight tendency to undergo reversible ionization to yield a hydrogen ion (a proton) and a hydroxide ion, giving the equilibrium

$$H_2O \rightleftharpoons H^+ + OH^- \qquad (2\text{–}1)$$

Although we commonly show the dissociation product of water as $H^+$, free protons do not exist in solution; hydrogen ions formed in water are immediately hydrated to **hydronium ions** ($H_3O^+$). Hydrogen bonding between water molecules makes the hydration of dissociating protons virtually instantaneous:

$$H-O\,{\scriptstyle|||}\,H-O \rightleftharpoons H-O^+-H + OH^-$$

The ionization of water can be measured by its electrical conductivity; pure water carries electrical current as $H^+$ migrates toward the cathode and $OH^-$ toward the anode. The movement of hydronium and hydroxide ions in the electric field is anomalously fast compared with that of other ions such as $Na^+$, $K^+$, and $Cl^-$. This high ionic mobility results from the kind of "proton hopping" shown in Figure 2–14. No individual proton moves very far through the bulk solution, but a series of proton hops between hydrogen-bonded water molecules causes the net movement of a proton over a long distance in a remarkably short time. As a result of the high ionic mobility of $H^+$ (and of $OH^-$, which also moves rapidly by proton hopping, but in the opposite direction), acid-base reactions in aqueous solutions are generally exceptionally fast. As noted above, proton hopping very likely also plays a role in biological proton-transfer reactions (Fig. 2–10; see also Fig. 19–59).

Because reversible ionization is crucial to the role of water in cellular function, we must have a means of

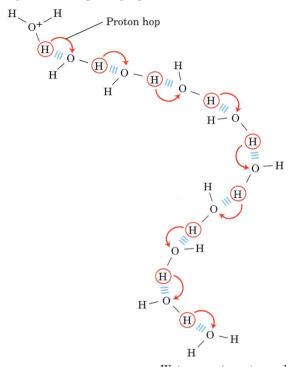

Hydronium ion gives up a proton

Water accepts proton and becomes a hydronium ion

**FIGURE 2–14 Proton hopping.** Short "hops" of protons between a series of hydrogen-bonded water molecules effect an extremely rapid net movement of a proton over a long distance. As a hydronium ion (upper left) gives up a proton, a water molecule some distance away (lower right) acquires one, becoming a hydronium ion. Proton hopping is much faster than true diffusion and explains the remarkably high ionic mobility of $H^+$ ions compared with other monovalent cations such as $Na^+$ or $K^+$.

expressing the extent of ionization of water in quantitative terms. A brief review of some properties of reversible chemical reactions shows how this can be done.

The position of equilibrium of any chemical reaction is given by its **equilibrium constant, $K_{eq}$** (sometimes expressed simply as $K$). For the generalized reaction

$$A + B \rightleftharpoons C + D \qquad (2\text{–}2)$$

an equilibrium constant can be defined in terms of the concentrations of reactants (A and B) and products (C and D) at equilibrium:

$$K_{eq} = \frac{[C][D]}{[A][B]}$$

Strictly speaking, the concentration terms should be the *activities*, or effective concentrations in nonideal solutions, of each species. Except in very accurate work, however, the equilibrium constant may be approximated by measuring the *concentrations* at equilibrium. For reasons beyond the scope of this discussion, equilibrium constants are dimensionless. Nonetheless, we have generally retained the concentration units (M) in the equilibrium expressions used in this book to remind you that molarity is the unit of concentration used in calculating $K_{eq}$.

The equilibrium constant is fixed and characteristic for any given chemical reaction at a specified temperature. It defines the composition of the final equilibrium mixture, regardless of the starting amounts of reactants and products. Conversely, we can calculate the equilibrium constant for a given reaction at a given temperature if the equilibrium concentrations of all its reactants and products are known. As we showed in Chapter 1 (p. 26), the standard free-energy change ($\Delta G°$) is directly related to $K_{eq}$.

### The Ionization of Water Is Expressed by an Equilibrium Constant

The degree of ionization of water at equilibrium (Eqn 2–1) is small; at 25 °C only about two of every $10^9$ molecules in pure water are ionized at any instant. The equilibrium constant for the reversible ionization of water (Eqn 2–1) is

$$K_{eq} = \frac{[H^+][OH^-]}{[H_2O]} \tag{2–3}$$

In pure water at 25 °C, the concentration of water is 55.5 M (grams of $H_2O$ in 1 L divided by its gram molecular weight: $(1{,}000 \text{ g/L})/(18.015 \text{ g/mol})$) and is essentially constant in relation to the very low concentrations of $H^+$ and $OH^-$, namely, $1 \times 10^{-7}$ M. Accordingly, we can substitute 55.5 M in the equilibrium constant expression (Eqn 2–3) to yield

$$K_{eq} = \frac{[H^+][OH^-]}{55.5 \text{ M}},$$

which, on rearranging, becomes

$$(55.5 \text{ M})(K_{eq}) = [H^+][OH^-] = K_w \tag{2–4}$$

where $K_w$ designates the product $(55.5 \text{ M})(K_{eq})$, the **ion product of water** at 25 °C.

The value for $K_{eq}$, determined by electrical-conductivity measurements of pure water, is $1.8 \times 10^{-16}$ M at 25 °C. Substituting this value for $K_{eq}$ in Equation 2–4 gives the value of the ion product of water:

$$K_w = [H^+][OH^-] = (55.5 \text{ M})(1.8 \times 10^{-16} \text{ M})$$
$$= 1.0 \times 10^{-14} \text{ M}^2$$

Thus the product $[H^+][OH^-]$ in aqueous solutions at 25 °C always equals $1 \times 10^{-14}$ M². When there are exactly equal concentrations of $H^+$ and $OH^-$, as in pure water, the solution is said to be at **neutral pH.** At this pH, the concentration of $H^+$ and $OH^-$ can be calculated from the ion product of water as follows:

$$K_w = [H^+][OH^-] = [H^+]^2$$

Solving for $[H^+]$ gives

$$[H^+] = \sqrt{K_w} = \sqrt{1 \times 10^{-14} \text{ M}^2}$$
$$[H^+] = [OH^-] = 10^{-7} \text{ M}$$

As the ion product of water is constant, whenever $[H^+]$ is greater than $1 \times 10^{-7}$ M, $[OH^-]$ must become less than $1 \times 10^{-7}$ M, and vice versa. When $[H^+]$ is very high, as in a solution of hydrochloric acid, $[OH^-]$ must be very low. From the ion product of water we can calculate $[H^+]$ if we know $[OH^-]$, and vice versa (Box 2–2).

### The pH Scale Designates the $H^+$ and $OH^-$ Concentrations

The ion product of water, $K_w$, is the basis for the **pH scale** (Table 2–6). It is a convenient means of designating the concentration of $H^+$ (and thus of $OH^-$) in any aqueous solution in the range between 1.0 M $H^+$ and 1.0 M $OH^-$. The term **pH** is defined by the expression

$$pH = \log \frac{1}{[H^+]} = -\log [H^+]$$

The symbol p denotes "negative logarithm of." For a precisely neutral solution at 25 °C, in which the concentration of hydrogen ions is $1.0 \times 10^{-7}$ M, the pH can be calculated as follows:

$$pH = \log \frac{1}{1.0 \times 10^{-7}} = \log (1.0 \times 10^7)$$
$$= \log 1.0 + \log 10^7 = 0 + 7 = 7$$

### TABLE 2–6 The pH Scale

| $[H^+]$ (M) | pH | $[OH^-]$ (M) | pOH* |
|---|---|---|---|
| $10^0$ (1) | 0 | $10^{-14}$ | 14 |
| $10^{-1}$ | 1 | $10^{-13}$ | 13 |
| $10^{-2}$ | 2 | $10^{-12}$ | 12 |
| $10^{-3}$ | 3 | $10^{-11}$ | 11 |
| $10^{-4}$ | 4 | $10^{-10}$ | 10 |
| $10^{-5}$ | 5 | $10^{-9}$ | 9 |
| $10^{-6}$ | 6 | $10^{-8}$ | 8 |
| $10^{-7}$ | 7 | $10^{-7}$ | 7 |
| $10^{-8}$ | 8 | $10^{-6}$ | 6 |
| $10^{-9}$ | 9 | $10^{-5}$ | 5 |
| $10^{-10}$ | 10 | $10^{-4}$ | 4 |
| $10^{-11}$ | 11 | $10^{-3}$ | 3 |
| $10^{-12}$ | 12 | $10^{-2}$ | 2 |
| $10^{-13}$ | 13 | $10^{-1}$ | 1 |
| $10^{-14}$ | 14 | $10^0$ (1) | 0 |

*The expression pOH is sometimes used to describe the basicity, or $OH^-$ concentration, of a solution; pOH is defined by the expression $pOH = -\log [OH^-]$, which is analogous to the expression for pH. Note that in all cases, $pH + pOH = 14$.

BOX 2–2     WORKING IN BIOCHEMISTRY

### The Ion Product of Water: Two Illustrative Problems

The ion product of water makes it possible to calculate the concentration of $H^+$, given the concentration of $OH^-$, and vice versa; the following problems demonstrate this.

**1.** What is the concentration of $H^+$ in a solution of 0.1 M NaOH?

$$K_w = [H^+][OH^-]$$

Solving for $[H^+]$ gives

$$[H^+] = \frac{K_w}{[OH^-]} = \frac{1 \times 10^{-14}\ M^2}{0.1\ M} = \frac{10^{-14}\ M^2}{10^{-1}\ M}$$

$$= 10^{-13}\ M \quad (answer)$$

**2.** What is the concentration of $OH^-$ in a solution with an $H^+$ concentration of $1.3 \times 10^{-4}$ M?

$$K_w = [H^+][OH^-]$$

Solving for $[OH^-]$ gives

$$[OH^-] = \frac{K_w}{[H^+]} = \frac{1.0 \times 10^{-14}\ M^2}{1.3 \times 10^{-4}\ M}$$

$$= 7.7 \times 10^{-11}\ M \quad (answer)$$

When doing these or any other calculations, be sure to round your answers to the correct number of significant figures.

---

The value of 7 for the pH of a precisely neutral solution is not an arbitrarily chosen figure; it is derived from the absolute value of the ion product of water at 25 °C, which by convenient coincidence is a round number. Solutions having a pH greater than 7 are alkaline or basic; the concentration of $OH^-$ is greater than that of $H^+$. Conversely, solutions having a pH less than 7 are acidic.

Note that the pH scale is logarithmic, not arithmetic. To say that two solutions differ in pH by 1 pH unit means that one solution has ten times the $H^+$ concentration of the other, but it does not tell us the absolute magnitude of the difference. Figure 2–15 gives the pH of some common aqueous fluids. A cola drink (pH 3.0) or red wine (pH 3.7) has an $H^+$ concentration approximately 10,000 times that of blood (pH 7.4).

The pH of an aqueous solution can be approximately measured using various indicator dyes, including litmus, phenolphthalein, and phenol red, which undergo color changes as a proton dissociates from the dye molecule. Accurate determinations of pH in the chemical or clinical laboratory are made with a glass electrode that is selectively sensitive to $H^+$ concentration but insensitive to $Na^+$, $K^+$, and other cations. In a pH meter the signal from such an electrode is amplified and compared with the signal generated by a solution of accurately known pH.

Measurement of pH is one of the most important and frequently used procedures in biochemistry. The pH affects the structure and activity of biological macromolecules; for example, the catalytic activity of enzymes is strongly dependent on pH (see Fig. 2–21). Measurements of the pH of blood and urine are commonly used in medical diagnoses. The pH of the blood plasma of people

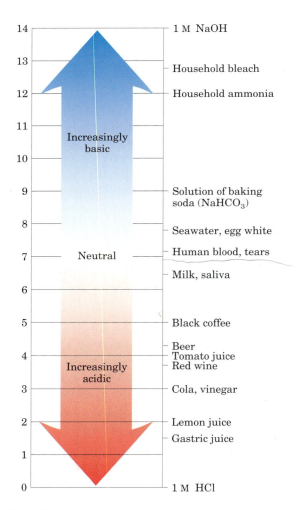

**FIGURE 2–15** The pH of some aqueous fluids.

with severe, uncontrolled diabetes, for example, is often below the normal value of 7.4; this condition is called acidosis. In certain other disease states the pH of the blood is higher than normal, the condition of alkalosis.

## Weak Acids and Bases Have Characteristic Dissociation Constants

Hydrochloric, sulfuric, and nitric acids, commonly called strong acids, are completely ionized in dilute aqueous solutions; the strong bases NaOH and KOH are also completely ionized. Of more interest to biochemists is the behavior of weak acids and bases—those not completely ionized when dissolved in water. These are common in biological systems and play important roles in metabolism and its regulation. The behavior of aqueous solutions of weak acids and bases is best understood if we first define some terms.

Acids may be defined as proton donors and bases as proton acceptors. A proton donor and its corresponding proton acceptor make up a **conjugate acid-base pair** (Fig. 2–16). Acetic acid ($CH_3COOH$), a proton donor, and the acetate anion ($CH_3COO^-$), the corre-

sponding proton acceptor, constitute a conjugate acid-base pair, related by the reversible reaction

$$CH_3COOH \rightleftharpoons H^+ + CH_3COO^-$$

Each acid has a characteristic tendency to lose its proton in an aqueous solution. The stronger the acid, the greater its tendency to lose its proton. The tendency of any acid (HA) to lose a proton and form its conjugate base ($A^-$) is defined by the equilibrium constant ($K_{eq}$) for the reversible reaction

$$HA \rightleftharpoons H^+ + A^-,$$

which is

$$K_{eq} = \frac{[H^+][A^-]}{[HA]} = K_a$$

Equilibrium constants for ionization reactions are usually called ionization or **dissociation constants,** often designated $K_a$. The dissociation constants of some acids are given in Figure 2–16. Stronger acids, such as phosphoric and carbonic acids, have larger dissociation constants; weaker acids, such as monohydrogen phosphate ($HPO_4^{2-}$), have smaller dissociation constants.

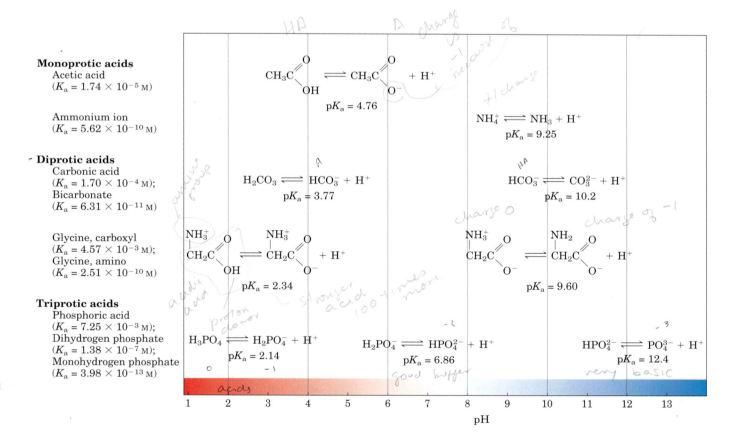

**FIGURE 2–16 Conjugate acid-base pairs consist of a proton donor and a proton acceptor.** Some compounds, such as acetic acid and ammonium ion, are monoprotic; they can give up only one proton. Others are diprotic ($H_2CO_3$ (carbonic acid) and glycine) or triprotic

($H_3PO_4$ (phosphoric acid)). The dissociation reactions for each pair are shown where they occur along a pH gradient. The equilibrium or dissociation constant ($K_a$) and its negative logarithm, the $pK_a$, are shown for each reaction.

Also included in Figure 2–16 are values of **$pK_a$,** which is analogous to pH and is defined by the equation

$$pK_a = \log \frac{1}{K_a} = -\log K_a$$

The stronger the tendency to dissociate a proton, the stronger is the acid and the lower its $pK_a$. As we shall now see, the $pK_a$ of any weak acid can be determined quite easily.

## Titration Curves Reveal the $pK_a$ of Weak Acids

Titration is used to determine the amount of an acid in a given solution. A measured volume of the acid is titrated with a solution of a strong base, usually sodium hydroxide (NaOH), of known concentration. The NaOH is added in small increments until the acid is consumed (neutralized), as determined with an indicator dye or a pH meter. The concentration of the acid in the original solution can be calculated from the volume and concentration of NaOH added.

A plot of pH against the amount of NaOH added (a **titration curve**) reveals the $pK_a$ of the weak acid. Consider the titration of a 0.1 M solution of acetic acid (for simplicity denoted as HAc) with 0.1 M NaOH at 25 °C (Fig. 2–17). Two reversible equilibria are involved in the process:

$$H_2O \rightleftharpoons H^+ + OH^- \tag{2–5}$$

$$HAc \rightleftharpoons H^+ + Ac^- \tag{2–6}$$

The equilibria must simultaneously conform to their characteristic equilibrium constants, which are, respectively,

$$K_w = [H^+][OH^-] = 1 \times 10^{-14}\ M^2 \tag{2–7}$$

$$K_a = \frac{[H^+][Ac^-]}{[HAc]} = 1.74 \times 10^5\ M \tag{2–8}$$

At the beginning of the titration, before any NaOH is added, the acetic acid is already slightly ionized, to an extent that can be calculated from its dissociation constant (Eqn 2–8).

As NaOH is gradually introduced, the added $OH^-$ combines with the free $H^+$ in the solution to form $H_2O$, to an extent that satisfies the equilibrium relationship in Equation 2–7. As free $H^+$ is removed, HAc dissociates further to satisfy its own equilibrium constant (Eqn 2–8). The net result as the titration proceeds is that more and more HAc ionizes, forming $Ac^-$, as the NaOH is added. At the midpoint of the titration, at which exactly 0.5 equivalent of NaOH has been added, one-half of the original acetic acid has undergone dissociation, so that the concentration of the proton donor, [HAc], now equals that of the proton acceptor, [Ac⁻]. At this midpoint a very important relationship holds: the pH of the equimolar solution of acetic acid and acetate is ex-

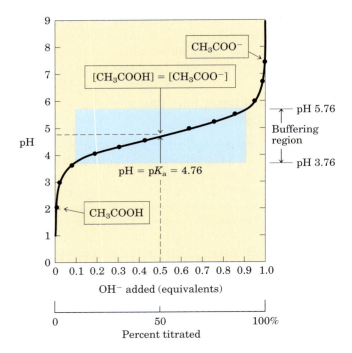

**FIGURE 2–17  The titration curve of acetic acid.** After addition of each increment of NaOH to the acetic acid solution, the pH of the mixture is measured. This value is plotted against the amount of NaOH expressed as a fraction of the total NaOH required to convert all the acetic acid to its deprotonated form, acetate. The points so obtained yield the titration curve. Shown in the boxes are the predominant ionic forms at the points designated. At the midpoint of the titration, the concentrations of the proton donor and proton acceptor are equal, and the pH is numerically equal to the $pK_a$. The shaded zone is the useful region of buffering power, generally between 10% and 90% titration of the weak acid.

actly equal to the $pK_a$ of acetic acid ($pK_a = 4.76$; Figs 2–16, 2–17). The basis for this relationship, which holds for all weak acids, will soon become clear.

As the titration is continued by adding further increments of NaOH, the remaining nondissociated acetic acid is gradually converted into acetate. The end point of the titration occurs at about pH 7.0: all the acetic acid has lost its protons to $OH^-$, to form $H_2O$ and acetate. Throughout the titration the two equilibria (Eqns 2–5, 2–6) coexist, each always conforming to its equilibrium constant.

Figure 2–18 compares the titration curves of three weak acids with very different dissociation constants: acetic acid ($pK_a = 4.76$); dihydrogen phosphate, $H_2PO_4^-$ ($pK_a = 6.86$); and ammonium ion, $NH_4^+$ ($pK_a = 9.25$). Although the titration curves of these acids have the same shape, they are displaced along the pH axis because the three acids have different strengths. Acetic acid, with the highest $K_a$ (lowest $pK_a$) of the three, is the strongest (loses its proton most readily); it is al-

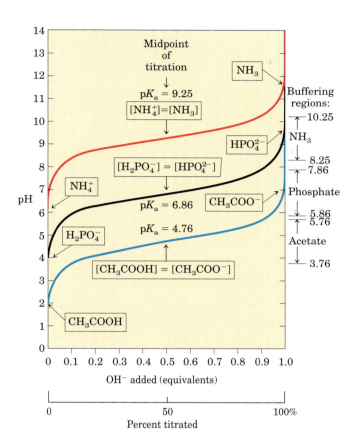

**FIGURE 2–18 Comparison of the titration curves of three weak acids.** Shown here are the titration curves for $CH_3COOH$, $H_2PO_4^-$, and $NH_4^+$. The predominant ionic forms at designated points in the titration are given in boxes. The regions of buffering capacity are indicated at the right. Conjugate acid-base pairs are effective buffers between approximately 10% and 90% neutralization of the proton-donor species.

ready half dissociated at pH 4.76. Dihydrogen phosphate loses a proton less readily, being half dissociated at pH 6.86. Ammonium ion is the weakest acid of the three and does not become half dissociated until pH 9.25.

The most important point about the titration curve of a weak acid is that it shows graphically that a weak acid and its anion—a conjugate acid-base pair—can act as a buffer.

## SUMMARY 2.2 Ionization of Water, Weak Acids, and Weak Bases

- Pure water ionizes slightly, forming equal numbers of hydrogen ions (hydronium ions, $H_3O^+$) and hydroxide ions. The extent of ionization is described by an equilibrium constant, $K_{eq} = \dfrac{[H^+][OH^-]}{[H_2O]}$, from which the ion product of

water, $K_w$, is derived. At 25 °C, $K_w = [H^+][OH^-] = (55.5 \text{ M})(K_{eq}) = 10^{-14} \text{ M}^2$.

- The pH of an aqueous solution reflects, on a logarithmic scale, the concentration of hydrogen ions: $pH = \log \dfrac{1}{[H^+]} = -\log [H^+]$.

- The greater the acidity of a solution, the lower its pH. Weak acids partially ionize to release a hydrogen ion, thus lowering the pH of the aqueous solution. Weak bases accept a hydrogen ion, increasing the pH. The extent of these processes is characteristic of each particular weak acid or base and is expressed as a dissociation constant, $K_a$: $K_{eq} = \dfrac{[H^+][A^-]}{[HA]} = K_a$.

- The $pK_a$ expresses, on a logarithmic scale, the relative strength of a weak acid or base: $pK_a = \log \dfrac{1}{K_a} = -\log K_a$.

- The stronger the acid, the lower its $pK_a$; the stronger the base, the higher its $pK_a$. The $pK_a$ can be determined experimentally; it is the pH at the midpoint of the titration curve for the acid or base.

## 2.3 Buffering against pH Changes in Biological Systems

Almost every biological process is pH dependent; a small change in pH produces a large change in the rate of the process. This is true not only for the many reactions in which the $H^+$ ion is a direct participant, but also for those in which there is no apparent role for $H^+$ ions. The enzymes that catalyze cellular reactions, and many of the molecules on which they act, contain ionizable groups with characteristic $pK_a$ values. The protonated amino and carboxyl groups of amino acids and the phosphate groups of nucleotides, for example, function as weak acids; their ionic state depends on the pH of the surrounding medium. As we noted above, ionic interactions are among the forces that stabilize a protein molecule and allow an enzyme to recognize and bind its substrate.

Cells and organisms maintain a specific and constant cytosolic pH, keeping biomolecules in their optimal ionic state, usually near pH 7. In multicellular organisms, the pH of extracellular fluids is also tightly regulated. Constancy of pH is achieved primarily by biological buffers: mixtures of weak acids and their conjugate bases.

We describe here the ionization equilibria that account for buffering, and we show the quantitative relationship between the pH of a buffered solution and the $pK_a$ of the buffer. Biological buffering is illustrated by the phosphate and carbonate buffering systems of humans.

## Buffers Are Mixtures of Weak Acids and Their Conjugate Bases

**Buffers** are aqueous systems that tend to resist changes in pH when small amounts of acid ($H^+$) or base ($OH^-$) are added. A buffer system consists of a weak acid (the proton donor) and its conjugate base (the proton acceptor). As an example, a mixture of equal concentrations of acetic acid and acetate ion, found at the midpoint of the titration curve in Figure 2–17, is a buffer system. The titration curve of acetic acid has a relatively flat zone extending about 1 pH unit on either side of its midpoint pH of 4.76. In this zone, an amount of $H^+$ or $OH^-$ added to the system has much less effect on pH than the same amount added outside the buffer range. This relatively flat zone is the buffering region of the acetic acid–acetate buffer pair. At the midpoint of the buffering region, where the concentration of the proton donor (acetic acid) exactly equals that of the proton acceptor (acetate), the buffering power of the system is maximal; that is, its pH changes least on addition of $H^+$ or $OH^-$. The pH at this point in the titration curve of acetic acid is equal to its $pK_a$. The pH of the acetate buffer system does change slightly when a small amount of $H^+$ or $OH^-$ is added, but this change is very small compared with the pH change that would result if the same amount of $H^+$ or $OH^-$ were added to pure water or to a solution of the salt of a strong acid and strong base, such as NaCl, which has no buffering power.

Buffering results from two reversible reaction equilibria occurring in a solution of nearly equal concentrations of a proton donor and its conjugate proton acceptor. Figure 2–19 explains how a buffer system works. Whenever $H^+$ or $OH^-$ is added to a buffer, the result is a small change in the ratio of the relative concentrations of the weak acid and its anion and thus a small change in pH. The decrease in concentration of one component of the system is balanced exactly by an increase in the other. The sum of the buffer components does not change, only their ratio.

Each conjugate acid-base pair has a characteristic pH zone in which it is an effective buffer (Fig. 2–18). The $H_2PO_4^-/HPO_4^{2-}$ pair has a $pK_a$ of 6.86 and thus can serve as an effective buffer system between approximately pH 5.9 and pH 7.9; the $NH_4^+/NH_3$ pair, with a $pK_a$ of 9.25, can act as a buffer between approximately pH 8.3 and pH 10.3.

## A Simple Expression Relates pH, p$K_a$, and Buffer Concentration

The titration curves of acetic acid, $H_2PO_4^-$, and $NH_4^+$ (Fig. 2–18) have nearly identical shapes, suggesting that these curves reflect a fundamental law or relationship. This is indeed the case. The shape of the titration curve of any weak acid is described by the Henderson-

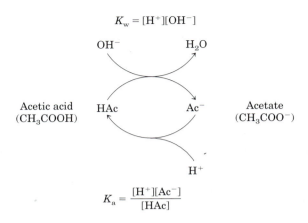

$$K_w = [H^+][OH^-]$$

$$K_a = \frac{[H^+][Ac^-]}{[HAc]}$$

**FIGURE 2–19 The acetic acid–acetate pair as a buffer system.** The system is capable of absorbing either $H^+$ or $OH^-$ through the reversibility of the dissociation of acetic acid. The proton donor, acetic acid (HAc), contains a reserve of bound $H^+$, which can be released to neutralize an addition of $OH^-$ to the system, forming $H_2O$. This happens because the product $[H^+][OH^-]$ transiently exceeds $K_w$ ($1 \times 10^{-14}$ M²). The equilibrium quickly adjusts so that this product equals $1 \times 10^{-14}$ M² (at 25 °C), thus transiently reducing the concentration of $H^+$. But now the quotient $[H^+][Ac^-] / [HAc]$ is less than $K_a$, so HAc dissociates further to restore equilibrium. Similarly, the conjugate base, $Ac^-$, can react with $H^+$ ions added to the system; again, the two ionization reactions simultaneously come to equilibrium. Thus a conjugate acid-base pair, such as acetic acid and acetate ion, tends to resist a change in pH when small amounts of acid or base are added. Buffering action is simply the consequence of two reversible reactions taking place simultaneously and reaching their points of equilibrium as governed by their equilibrium constants, $K_W$ and $K_a$.

Hasselbalch equation, which is important for understanding buffer action and acid-base balance in the blood and tissues of vertebrates. This equation is simply a useful way of restating the expression for the dissociation constant of an acid. For the dissociation of a weak acid HA into $H^+$ and $A^-$, the Henderson-Hasselbalch equation can be derived as follows:

$$K_a = \frac{[H^+][A^-]}{[HA]}$$

First solve for $[H^+]$:

$$[H^+] = K_a\frac{[HA]}{[A^-]}$$

Then take the negative logarithm of both sides:

$$-\log [H^+] = -\log K_a - \log \frac{[HA]}{[A^-]}$$

Substitute pH for $-\log [H^+]$ and $pK_a$ for $-\log K_a$:

$$pH = pK_a - \log \frac{[HA]}{[A^-]}$$

Now invert $-\log [HA]/[A^-]$, which involves changing its sign, to obtain the **Henderson-Hasselbalch equation:**

$$pH = pK_a + \log \frac{[A^-]}{[HA]} \qquad (2\text{–}9)$$

Stated more generally,

$$pH = pK_a + \log \frac{[\text{proton acceptor}]}{[\text{proton donor}]}$$

This equation fits the titration curve of all weak acids and enables us to deduce a number of important quantitative relationships. For example, it shows why the $pK_a$ of a weak acid is equal to the pH of the solution at the midpoint of its titration. At that point, [HA] equals $[A^-]$, and

$$pH = pK_a + \log 1 = pK_a + 0 = pK_a$$

As shown in Box 2–3, the Henderson-Hasselbalch equation also allows us to (1) calculate $pK_a$, given pH and the molar ratio of proton donor and acceptor; (2) calculate pH, given $pK_a$ and the molar ratio of proton donor and acceptor; and (3) calculate the molar ratio of proton donor and acceptor, given pH and $pK_a$.

## Weak Acids or Bases Buffer Cells and Tissues against pH Changes

The intracellular and extracellular fluids of multicellular organisms have a characteristic and nearly constant

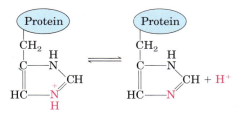

**FIGURE 2–20** The amino acid histidine, a component of proteins, is a weak acid. The $pK_a$ of the protonated nitrogen of the side chain is 6.0.

pH. The organism's first line of defense against changes in internal pH is provided by buffer systems. The cytoplasm of most cells contains high concentrations of proteins, which contain many amino acids with functional groups that are weak acids or weak bases. For example, the side chain of histidine (Fig. 2–20) has a $pK_a$ of 6.0; proteins containing histidine residues therefore buffer effectively near neutral pH. Nucleotides such as ATP, as well as many low molecular weight metabolites, contain ionizable groups that can contribute buffering power to the cytoplasm. Some highly specialized organelles and extracellular compartments have high concentrations of compounds that contribute buffering capacity: organic acids buffer the vacuoles of plant cells; ammonia buffers urine.

---

## BOX 2–3 WORKING IN BIOCHEMISTRY

### Solving Problems Using the Henderson-Hasselbalch Equation

1. Calculate the $pK_a$ of lactic acid, given that when the concentration of lactic acid is 0.010 M and the concentration of lactate is 0.087 M, the pH is 4.80.

$$pH = pK_a + \log \frac{[\text{lactate}]}{[\text{lactic acid}]}$$

$$pK_a = pH - \log \frac{[\text{lactate}]}{[\text{lactic acid}]}$$

$$= 4.80 - \log \frac{0.087}{0.010} = 4.80 - \log 8.7$$

$$= 4.80 - 0.94 = 3.9 \quad (answer)$$

2. Calculate the pH of a mixture of 0.10 M acetic acid and 0.20 M sodium acetate. The $pK_a$ of acetic acid is 4.76.

$$pH = pK_a + \log \frac{[\text{acetate}]}{[\text{acetic acid}]}$$

$$= 4.76 + \log \frac{0.20}{0.10} = 4.76 + 0.30$$

$$= 5.1 \quad (answer)$$

3. Calculate the ratio of the concentrations of acetate and acetic acid required in a buffer system of pH 5.30.

$$pH = pK_a + \log \frac{[\text{acetate}]}{[\text{acetic acid}]}$$

$$\log \frac{[\text{acetate}]}{[\text{acetic acid}]} = pH - pK_a$$

$$= 5.30 - 4.76 = 0.54$$

$$\frac{[\text{acetate}]}{[\text{acetic acid}]} = \text{antilog } 0.54 = 3.5 \quad (answer)$$

To see the effect of pH on the degree of ionization of a weak acid, see the Living Graph for Equation 2–9.

Two especially important biological buffers are the phosphate and bicarbonate systems. The phosphate buffer system, which acts in the cytoplasm of all cells, consists of $H_2PO_4^-$ as proton donor and $HPO_4^{2-}$ as proton acceptor:

$$H_2PO_4^- \rightleftharpoons H^+ + HPO_4^{2-}$$

The phosphate buffer system is maximally effective at a pH close to its $pK_a$ of 6.86 (Figs 2–16, 2–18) and thus tends to resist pH changes in the range between about 5.9 and 7.9. It is therefore an effective buffer in biological fluids; in mammals, for example, extracellular fluids and most cytoplasmic compartments have a pH in the range of 6.9 to 7.4.

Blood plasma is buffered in part by the bicarbonate system, consisting of carbonic acid ($H_2CO_3$) as proton donor and bicarbonate ($HCO_3^-$) as proton acceptor:

$$H_2CO_3 \rightleftharpoons H^+ + HCO_3^-$$

$$K_1 = \frac{[H^+][HCO_3^-]}{[H_2CO_3]}$$

This buffer system is more complex than other conjugate acid-base pairs because one of its components, carbonic acid ($H_2CO_3$), is formed from dissolved (d) carbon dioxide and water, in a reversible reaction:

$$CO_2(d) + H_2O \rightleftharpoons H_2CO_3$$

$$K_2 = \frac{[H_2CO_3]}{[CO_2(d)][H_2O]}$$

Carbon dioxide is a gas under normal conditions, and the concentration of dissolved $CO_2$ is the result of equilibration with $CO_2$ of the gas (g) phase:

$$CO_2(g) \rightleftharpoons CO_2(d)$$

$$K_3 = \frac{[CO_2(d)]}{[CO_2(g)]}$$

The pH of a bicarbonate buffer system depends on the concentration of $H_2CO_3$ and $HCO_3^-$, the proton donor and acceptor components. The concentration of $H_2CO_3$ in turn depends on the concentration of dissolved $CO_2$, which in turn depends on the concentration of $CO_2$ in the gas phase, called the **partial pressure** of $CO_2$. Thus the pH of a bicarbonate buffer exposed to a gas phase is ultimately determined by the concentration of $HCO_3^-$ in the aqueous phase and the partial pressure of $CO_2$ in the gas phase (Box 2–4).

Human blood plasma normally has a pH close to 7.4. Should the pH-regulating mechanisms fail or be overwhelmed, as may happen in severe uncontrolled diabetes when an overproduction of metabolic acids causes acidosis, the pH of the blood can fall to 6.8 or below, leading to irreparable cell damage and death. In other diseases the pH may rise to lethal levels.

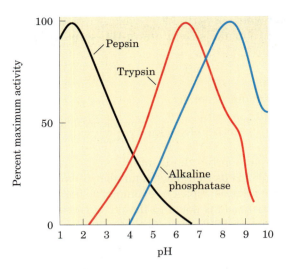

**FIGURE 2–21** **The pH optima of some enzymes.** Pepsin is a digestive enzyme secreted into gastric juice; trypsin, a digestive enzyme that acts in the small intestine; alkaline phosphatase of bone tissue, a hydrolytic enzyme thought to aid in bone mineralization.

Although many aspects of cell structure and function are influenced by pH, it is the catalytic activity of enzymes that is especially sensitive. Enzymes typically show maximal catalytic activity at a characteristic pH, called the **pH optimum** (Fig. 2–21). On either side of the optimum pH their catalytic activity often declines sharply. Thus, a small change in pH can make a large difference in the rate of some crucial enzyme-catalyzed reactions. Biological control of the pH of cells and body fluids is therefore of central importance in all aspects of metabolism and cellular activities.

## SUMMARY 2.3 Buffering against pH Changes in Biological Systems

- A mixture of a weak acid (or base) and its salt resists changes in pH caused by the addition of $H^+$ or $OH^-$. The mixture thus functions as a buffer.

- The pH of a solution of a weak acid (or base) and its salt is given by the Henderson-Hasselbalch equation: $pH = pK_a - \log \frac{[HA]}{[A^-]}$.

- In cells and tissues, phosphate and bicarbonate buffer systems maintain intracellular and extracellular fluids at their optimum (physiological) pH, which is usually close to pH 7. Enzymes generally work optimally at this pH.

BOX 2–4    BIOCHEMISTRY IN MEDICINE

## Blood, Lungs, and Buffer: The Bicarbonate Buffer System

In animals with lungs, the bicarbonate buffer system is an effective physiological buffer near pH 7.4, because the $H_2CO_3$ of blood plasma is in equilibrium with a large reserve capacity of $CO_2(g)$ in the air space of the lungs. This buffer system involves three reversible equilibria between gaseous $CO_2$ in the lungs and bicarbonate ($HCO_3^-$) in the blood plasma (Fig. 1).

When $H^+$ (from lactic acid produced in muscle tissue during vigorous exercise, for example) is added to blood as it passes through the tissues, reaction 1 proceeds toward a new equilibrium, in which the concentration of $H_2CO_3$ is increased. This increases the concentration of $CO_2(d)$ in the blood plasma (reaction 2) and thus increases the pressure of $CO_2(g)$ in the air space of the lungs (reaction 3); the extra $CO_2$ is exhaled. Conversely, when the pH of blood plasma is raised (by $NH_3$ production during protein catabolism, for example), the opposite events occur: the $H^+$ concentration of blood plasma is lowered, causing more $H_2CO_3$ to dissociate into $H^+$ and $HCO_3^-$. This in turn causes more $CO_2(g)$ from the lungs to dissolve in the blood plasma. The rate of breathing—that is, the rate of inhaling and exhaling $CO_2$—can quickly adjust these equilibria to keep the blood pH nearly constant.

**FIGURE 1** The $CO_2$ in the air space of the lungs is in equilibrium with the bicarbonate buffer in the blood plasma passing through the lung capillaries. Because the concentration of dissolved $CO_2$ can be adjusted rapidly through changes in the rate of breathing, the bicarbonate buffer system of the blood is in near-equilibrium with a large potential reservoir of $CO_2$.

## 2.4 Water as a Reactant

Water is not just the solvent in which the chemical reactions of living cells occur; it is very often a direct participant in those reactions. The formation of ATP from ADP and inorganic phosphate is an example of a **condensation reaction** in which the elements of water are eliminated (Fig. 2–22a). The reverse of this reaction—cleavage accompanied by the addition of the elements of water—is a **hydrolysis reaction.** Hydrolysis reactions are also responsible for the enzymatic depolymerization of proteins, carbohydrates, and nucleic acids. Hydrolysis reactions, catalyzed by enzymes called

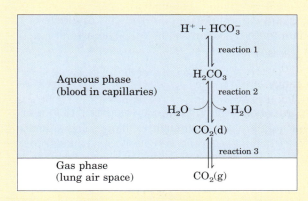

**FIGURE 2–22 Participation of water in biological reactions. (a)** ATP is a phosphoanhydride formed by a condensation reaction (loss of the elements of water) between ADP and phosphate. R represents adenosine monophosphate (AMP). This condensation reaction requires energy. The hydrolysis of (addition of the elements of water to) ATP to form ADP and phosphate releases an equivalent amount of energy. Also shown are some other condensation and hydrolysis reactions common in biological systems **(b)**, **(c)**, **(d)**.

**hydrolases,** are almost invariably exergonic. The formation of cellular polymers from their subunits by simple reversal of hydrolysis (that is, by condensation reactions) would be endergonic and therefore does not occur. As we shall see, cells circumvent this thermodynamic obstacle by coupling endergonic condensation reactions to exergonic processes, such as breakage of the anhydride bond in ATP.

You are (we hope!) consuming oxygen as you read. Water and carbon dioxide are the end products of the oxidation of fuels such as glucose. The overall reaction can be summarized as

$$C_6H_{12}O_6 + 6O_2 \longrightarrow 6CO_2 + 6H_2O$$
Glucose

The "metabolic water" formed by oxidation of foods and stored fats is actually enough to allow some animals in very dry habitats (gerbils, kangaroo rats, camels) to survive for extended periods without drinking water.

The $CO_2$ produced by glucose oxidation is converted in erythrocytes to the more soluble $HCO_3^-$, in a reaction catalyzed by the enzyme carbonic anhydrase:

$$CO_2 + H_2O \rightleftharpoons HCO_3^- + H^+$$

In this reaction, water not only is a substrate but also functions in proton transfer by forming a network of hydrogen-bonded water molecules through which proton hopping occurs (Fig. 2–14).

Green plants and algae use the energy of sunlight to split water in the process of photosynthesis:

$$2H_2O + 2A \xrightarrow{\text{light}} O_2 + 2AH_2$$

In this reaction, A is an electron-accepting species, which varies with the type of photosynthetic organism, and water serves as the electron donor in an oxidation-reduction sequence (see Fig. 19–49) that is fundamental to all life.

### SUMMARY 2.4 Water as a Reactant

■ Water is both the solvent in which metabolic reactions occur and a reactant in many biochemical processes, including hydrolysis, condensation, and oxidation-reduction reactions.

## 2.5 The Fitness of the Aqueous Environment for Living Organisms

Organisms have effectively adapted to their aqueous environment and have evolved means of exploiting the unusual properties of water. The high specific heat of water (the heat energy required to raise the temperature of 1 g of water by 1 °C) is useful to cells and organisms because it allows water to act as a "heat buffer," keeping the temperature of an organism relatively constant as the temperature of the surroundings fluctuates and as heat is generated as a byproduct of metabolism. Furthermore, some vertebrates exploit the high heat of vaporization of water (Table 2–1) by using (thus losing) excess body heat to evaporate sweat. The high degree of internal cohesion of liquid water, due to hydrogen bonding, is exploited by plants as a means of transporting dissolved nutrients from the roots to the leaves during the process of transpiration. Even the density of ice, lower than that of liquid water, has important biological consequences in the life cycles of aquatic organisms. Ponds freeze from the top down, and the layer of ice at the top insulates the water below from frigid air, preventing the pond (and the organisms in it) from freezing solid. Most fundamental to all living organisms is the fact that many physical and biological properties of cell macromolecules, particularly the proteins and nucleic acids, derive from their interactions with water molecules of the surrounding medium. The influence of water on the course of biological evolution has been profound and determinative. If life forms have evolved elsewhere in the universe, they are unlikely to resemble those of Earth unless their extraterrestrial origin is also a place in which plentiful liquid water is available.

Aqueous environments support countless species. Soft corals, sponges, bryozoans, and algae compete for space on this reef substrate off the Philippine Islands.

## *Key Terms*

**Terms in bold are defined in the glossary.**

**hydrogen bond** 48
**bond energy** 48
**hydrophilic** 50
**hydrophobic** 50
**amphipathic** 52
**micelle** 53
**hydrophobic interactions** 53
van der Waals interactions 54
osmolarity 56

**osmosis** 57
isotonic 57
hypertonic 57
hypotonic 57
**equilibrium constant ($K_{eq}$)** 60
**ion product of water ($K_w$)** 61
**pH** 61
**conjugate acid-base pair** 63
**dissociation constant ($K_a$)** 63

**p$K_a$** 64
**titration curve** 64
**buffer** 66
**Henderson-Hasselbalch equation** 67
**condensation** 69
**hydrolysis** 69

## *Further Reading*

### General

**Belton, P.S.** (2000) Nuclear magnetic resonance studies of the hydration of proteins and DNA. *Cell. Mol. Life Sci.* **57,** 993–998.

**Denny, M.W.** (1993) *Air and Water: The Biology and Physics of Life's Media,* Princeton University Press, Princeton, NJ.

A wonderful investigation of the biological relevance of the properties of water.

**Eisenberg, D. & Kauzmann, W.** (1969) *The Structure and Properties of Water,* Oxford University Press, New York.

An advanced, classic treatment of the physical chemistry of water and hydrophobic interactions.

**Franks, F. & Mathias, S.F. (eds)** (1982) *Biophysics of Water,* John Wiley & Sons, Inc., New York.

A large collection of papers on the structure of pure water and of the cytoplasm.

**Gerstein, M. & Levitt, M.** (1998) Simulating water and the molecules of life. *Sci. Am.* **279** (November), 100–105.

A well-illustrated description of the use of computer simulation to study the biologically important association of water with proteins and nucleic acids.

**Gronenborn, A. & Clore, M.** (1997) Water in and around proteins. *The Biochemist* **19** (3), 18–21.

A brief discussion of protein-bound water as detected by crystallography and NMR.

**Kandori, H.** (2000) Role of internal water molecules in bacteriorhodopsin. *Biochim. Biophys. Acta* **1460,** 177–191.

Intermediate-level review of the role of an internal chain of water molecules in proton movement through this protein.

**Kornblatt, J. & Kornblatt, J.** (1997) The role of water in recognition and catalysis by enzymes. *The Biochemist* **19** (3), 14–17.

A short, useful summary of the ways in which bound water influences the structure and activity of proteins.

**Kuntz, I.D. & Zipp, A.** (1977) Water in biological systems. *N. Engl. J. Med.* **297,** 262–266.

A brief review of the physical state of cytosolic water and its interactions with dissolved biomolecules.

**Ladbury, J.** (1996) Just add water! The effect of water on the specificity of protein-ligand binding sites and its potential application to drug design. *Chem. Biol.* **3,** 973–980.

**Luecke, H.** (2000) Atomic resolution structures of bacteriorhodopsin photocycle intermediates: the role of discrete water molecules in the function of this light-driven ion pump. *Biochim. Biophys. Acta* **1460,** 133–156.

Advanced review of a proton pump that employs an internal chain of water molecules.

**Nicolls, P.** (2000) Introduction: the biology of the water molecule. *Cell. Mol. Life Sci.* **57,** 987–992.

A short review of the properties of water, introducing several excellent advanced reviews published in the same issue (see especially Pocker and Rand et al., listed below).

**Pocker, Y.** (2000) Water in enzyme reactions: biophysical aspects of hydration-dehydration processes. *Cell. Mol. Life Sci.* **57,** 1008–1017.

Review of the role of water in enzyme catalysis, with carbonic anhydrase as the featured example.

**Rand, R.P., Parsegian, V.A., & Rau, D.C.** (2000) Intracellular osmotic action. *Cell. Mol. Life Sci.* **57,** 1018–1032.

Review of the roles of water in enzyme catalysis as revealed by studies in water-poor solutes.

**Record, M.T., Jr., Courtenay, E.S., Cayley, D.S., & Guttman, H.J.** (1998) Responses of *E. coli* to osmotic stress: large changes in amounts of cytoplasmic solutes and water. *Trends Biochem. Sci.* **23,** 143–148.

Intermediate-level review of the ways in which a bacterial cell counters changes in the osmolarity of its surroundings.

**Stillinger, F.H.** (1980) Water revisited. *Science* **209,** 451–457.

A short review of the physical structure of water, including the importance of hydrogen bonding and the nature of hydrophobic interactions.

**Symons, M.C.** (2000) Spectroscopy of aqueous solutions: protein and DNA interactions with water. *Cell. Mol. Life Sci.* **57,** 999–1007.

**Westhof, E.** (ed.) (1993) *Water and Biological Macromolecules,* CRC Press, Inc., Boca Raton, FL.

Fourteen chapters, each by a different author, cover (at an advanced level) the structure of water and its interactions with proteins, nucleic acids, polysaccharides, and lipids.

**Wiggins, P.M.** (1990) Role of water in some biological processes. *Microbiol. Rev.* **54,** 432–449.

A review of water in biology, including discussion of the physical structure of liquid water, its interaction with biomolecules, and the state of water in living cells.

## Weak Interactions in Aqueous Systems

**Fersht, A.R.** (1987) The hydrogen bond in molecular recognition. *Trends Biochem. Sci.* **12,** 301–304.

A clear, brief, quantitative discussion of the contribution of hydrogen bonding to molecular recognition and enzyme catalysis.

**Frieden, E.** (1975) Non-covalent interactions: key to biological flexibility and specificity. *J. Chem. Educ.* **52,** 754–761.

Review of the four kinds of weak interactions that stabilize macromolecules and confer biological specificity, with clear examples.

**Jeffrey, G.A.** (1997) *An Introduction to Hydrogen Bonding,* Oxford University Press, New York.

A detailed, advanced discussion of the structure and properties of hydrogen bonds, including those in water and biomolecules.

**Martin, T.W. & Derewenda, Z.S.** (1999) The name is bond—H bond. *Nat. Struct. Biol.* **6,** 403–406.

Brief review of the evidence that hydrogen bonds have some covalent character.

**Schwabe, J.W.R.** (1997) The role of water in protein-DNA interactions. *Curr. Opin. Struct. Biol.* **7,** 126–134.

An examination of the important role of water in both the specificity and the affinity of protein-DNA interactions.

**Tanford, C.** (1978) The hydrophobic effect and the organization of living matter. *Science* **200,** 1012–1018.

A review of the chemical and energetic bases for hydrophobic interactions between biomolecules in aqueous solutions.

## Weak Acids, Weak Bases, and Buffers: Problems for Practice

**Segel, I.H.** (1976) *Biochemical Calculations,* 2nd edn, John Wiley & Sons, Inc., New York.

# Problems

**1. Simulated Vinegar** One way to make vinegar (*not* the preferred way) is to prepare a solution of acetic acid, the sole acid component of vinegar, at the proper pH (see Fig. 2–15) and add appropriate flavoring agents. Acetic acid ($M_r$ 60) is a liquid at 25 °C, with a density of 1.049 g/mL. Calculate the volume that must be added to distilled water to make 1 L of simulated vinegar (see Fig. 2–16).

**2. Acidity of Gastric HCl** In a hospital laboratory, a 10.0 mL sample of gastric juice, obtained several hours after a meal, was titrated with 0.1 M NaOH to neutrality; 7.2 mL of NaOH was required. The patient's stomach contained no ingested food or drink, thus assume that no buffers were present. What was the pH of the gastric juice?

**3. Measurement of Acetylcholine Levels by pH Changes** The concentration of acetylcholine (a neurotransmitter) in a sample can be determined from the pH changes that accompany its hydrolysis. When the sample is incubated with the enzyme acetylcholinesterase, acetylcholine is quantitatively converted into choline and acetic acid, which dissociates to yield acetate and a hydrogen ion:

$$CH_3-\overset{\overset{\textstyle O}{\|}}{C}-O-CH_2-CH_2-\overset{\overset{\textstyle CH_3}{|}}{\underset{\underset{\textstyle CH_3}{|}}{{}^+N}}-CH_3 \xrightarrow{H_2O}$$

Acetylcholine

$$HO-CH_2-CH_2-\overset{\overset{\textstyle CH_3}{|}}{\underset{\underset{\textstyle CH_3}{|}}{{}^+N}}-CH_3 + CH_3-\overset{\overset{\textstyle }{}}{\underset{\underset{\textstyle O}{\|}}{C}}-O^- + H^+$$

Choline                    Acetate

In a typical analysis, 15 mL of an aqueous solution containing an unknown amount of acetylcholine had a pH of 7.65. When incubated with acetylcholinesterase, the pH of the solution decreased to 6.87. Assuming that there was no buffer in the assay mixture, determine the number of moles of acetylcholine in the 15 mL sample.

**4. Osmotic Balance in a Marine Frog** The crab-eating frog of Southeast Asia, *Rana cancrivora*, develops and matures in fresh water but searches for its food in coastal mangrove swamps (composed of 80% to full-strength seawater). When the frog moves from its freshwater home to seawater it experiences a large change in the osmolarity of its environment (from hypotonic to hypertonic).

(a) Eighty percent seawater contains 460 mM NaCl, 10 mM KCl, 10 mM CaCl₂, and 50 mM MgCl₂. What are the concentrations of the various ionic species in this seawater? Assuming that these salts account for nearly all the solutes in seawater, calculate the osmolarity of the seawater.

(b) The chart below lists the cytoplasmic concentrations of ions in *R. cancrivora*. Ignoring dissolved proteins, amino acids, nucleic acids, and other small metabolites, calculate the osmolarity of the frog's cells based solely on the ionic concentrations given below.

|  | $Na^+$ (mM) | $K^+$ (mM) | $Cl^-$ (mM) | $Ca^{2+}$ (mM) | $Mg^{2+}$ (mM) |
|---|---|---|---|---|---|
| *R. cancrivora* | 122 | 10 | 100 | 2 | 1 |

(c) Like all frogs, the crab-eating frog can exchange gases through its permeable skin, allowing it to stay underwater for long periods of time without breathing. How does the high permeability of frog skin affect the frog's cells when it moves from fresh water to seawater?

(d) The crab-eating frog uses two mechanisms to maintain its cells in osmotic balance with its environment. First, it allows the Na$^+$ and Cl$^-$ concentrations in its cells to increase slowly as the ions diffuse down their concentration gradients. Second, like many elasmobranchs (sharks), it retains the waste product urea in its cells. The addition of both NaCl and urea increases the osmolarity of the cytosol to a level nearly equal to that of the surrounding environment.

Urea (CH$_4$N$_2$O)

Assuming the volume of water in a typical frog is 100 mL, calculate how many grams of NaCl (formula weight (FW) 58.44) the frog must take up to make its tissues isotonic with seawater.

(e) How many grams of urea (FW 60) must it retain to accomplish the same thing?

**5. Properties of a Buffer**   The amino acid glycine is often used as the main ingredient of a buffer in biochemical experiments. The amino group of glycine, which has a pK$_a$ of 9.6, can exist either in the protonated form (—NH$_3^+$) or as the free base (—NH$_2$), because of the reversible equilibrium

$$R—NH_3^+ \rightleftharpoons R—NH_2 + H^+$$

(a) In what pH range can glycine be used as an effective buffer due to its amino group?

(b) In a 0.1 M solution of glycine at pH 9.0, what fraction of glycine has its amino group in the —NH$_3^+$ form?

(c) How much 5 M KOH must be added to 1.0 L of 0.1 M glycine at pH 9.0 to bring its pH to exactly 10.0?

(d) When 99% of the glycine is in its —NH$_3^+$ form, what is the numerical relation between the pH of the solution and the pK$_a$ of the amino group?

**6. The Effect of pH on Solubility**   The strongly polar, hydrogen-bonding properties of water make it an excellent solvent for ionic (charged) species. By contrast, nonionized, nonpolar organic molecules, such as benzene, are relatively insoluble in water. In principle, the aqueous solubility of any organic acid or base can be increased by converting the molecules to charged species. For example, the solubility of benzoic acid in water is low. The addition of sodium bicarbonate to a mixture of water and benzoic acid raises the pH and deprotonates the benzoic acid to form benzoate ion, which is quite soluble in water.

Benzoic acid
pK$_a \approx 5$

Benzoate ion

Are the following compounds more soluble in an aqueous solution of 0.1 M NaOH or 0.1 M HCl? (The dissociable protons are shown in red.)

Pyridine ion
pK$_a \approx 5$

**(a)**

β-Naphthol
pK$_a \approx 10$

**(b)**

N-Acetyltyrosine methyl ester
pK$_a \approx 10$

**(c)**

**7. Treatment of Poison Ivy Rash**   The components of poison ivy and poison oak that produce the characteristic itchy rash are catechols substituted with long-chain alkyl groups.

pK$_a \approx 8$

If you were exposed to poison ivy, which of the treatments below would you apply to the affected area? Justify your choice.

(a) Wash the area with cold water.

(b) Wash the area with dilute vinegar or lemon juice.

(c) Wash the area with soap and water.

(d) Wash the area with soap, water, and baking soda (sodium bicarbonate).

**8. pH and Drug Absorption**   Aspirin is a weak acid with a pK$_a$ of 3.5.

It is absorbed into the blood through the cells lining the stomach and the small intestine. Absorption requires passage through the plasma membrane, the rate of which is determined by the polarity of the molecule: charged and highly polar molecules pass slowly, whereas neutral hydrophobic ones pass rapidly. The pH of the stomach contents is about 1.5, and the pH of the contents of the small intestine is about 6. Is more aspirin absorbed into the bloodstream from the stomach or from the small intestine? Clearly justify your choice.

**9. Preparation of Standard Buffer for Calibration of a pH Meter**  The glass electrode used in commercial pH meters gives an electrical response proportional to the concentration of hydrogen ion. To convert these responses into pH, glass electrodes must be calibrated against standard solutions of known $H^+$ concentration. Determine the weight in grams of sodium dihydrogen phosphate ($NaH_2PO_4 \cdot H_2O$; FW 138.01) and disodium hydrogen phosphate ($Na_2HPO_4$; FW 141.98) needed to prepare 1 L of a standard buffer at pH 7.00 with a total phosphate concentration of 0.100 M (see Fig. 2–16).

**10. Calculating pH from Hydrogen Ion Concentration**  What is the pH of a solution that has an $H^+$ concentration of (a) $1.75 \times 10^{-5}$ mol/L; (b) $6.50 \times 10^{-10}$ mol/L; (c) $1.0 \times 10^{-4}$ mol/L; (d) $1.50 \times 10^{-5}$ mol/L?

**11. Calculating Hydrogen Ion Concentration from pH**  What is the $H^+$ concentration of a solution with pH of (a) 3.82; (b) 6.52; (c) 11.11?

**12. Calculating pH from Molar Ratios**  Calculate the pH of a dilute solution that contains a molar ratio of potassium acetate to acetic acid ($pK_a = 4.76$) of (a) 2:1; (b) 1:3; (c) 5:1; (d) 1:1; (e) 1:10.

**13. Working with Buffers**  A buffer contains 0.010 mol of lactic acid ($pK_a = 3.86$) and 0.050 mol of sodium lactate per liter. (a) Calculate the pH of the buffer. (b) Calculate the change in pH when 5 mL of 0.5 M HCl is added to 1 L of the buffer. (c) What pH change would you expect if you added the same quantity of HCl to 1 L of pure water?

**14. Calculating pH from Concentrations**  What is the pH of a solution containing 0.12 mol/L of $NH_4Cl$ and 0.03 mol/L of NaOH ($pK_a$ of $NH_4^+/NH_3$ is 9.25)?

**15. Calculating $pK_a$**  An unknown compound, X, is thought to have a carboxyl group with a $pK_a$ of 2.0 and another ionizable group with a $pK_a$ between 5 and 8. When 75 mL of 0.1 M NaOH was added to 100 mL of a 0.1 M solution of X at pH 2.0, the pH increased to 6.72. Calculate the $pK_a$ of the second ionizable group of X.

**16. Control of Blood pH by Respiration Rate**
(a) The partial pressure of $CO_2$ in the lungs can be varied rapidly by the rate and depth of breathing. For example, a common remedy to alleviate hiccups is to increase the concentration of $CO_2$ in the lungs. This can be achieved by holding one's breath, by very slow and shallow breathing (hypoventilation), or by breathing in and out of a paper bag. Under such conditions, the partial pressure of $CO_2$ in the air space of the lungs rises above normal. Qualitatively explain the effect of these procedures on the blood pH.
(b) A common practice of competitive short-distance runners is to breathe rapidly and deeply (hyperventilate) for about half a minute to remove $CO_2$ from their lungs just before running in, say, a 100 m dash. Blood pH may rise to 7.60. Explain why the blood pH increases.
(c) During a short-distance run the muscles produce a large amount of lactic acid ($CH_3CH(OH)COOH$, $K_a = 1.38 \times 10^{-4}$) from their glucose stores. In view of this fact, why might hyperventilation before a dash be useful?

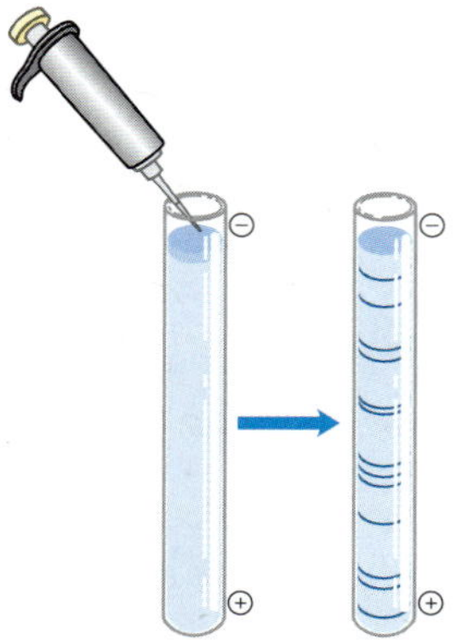

# AMINO ACIDS, PEPTIDES, AND PROTEINS

The word protein that I propose to you . . . I would wish to derive from *proteios,* because it appears to be the primitive or principal substance of animal nutrition that plants prepare for the herbivores, and which the latter then furnish to the carnivores.

—*J. J. Berzelius,* letter to G. J. Mulder, 1838

Proteins are the most abundant biological macromolecules, occurring in all cells and all parts of cells. Proteins also occur in great variety; thousands of different kinds, ranging in size from relatively small peptides to huge polymers with molecular weights in the millions, may be found in a single cell. Moreover, proteins exhibit enormous diversity of biological function and are the most important final products of the information pathways discussed in Part III of this book. Proteins are the molecular instruments through which genetic information is expressed.

Relatively simple monomeric subunits provide the key to the structure of the thousands of different proteins. All proteins, whether from the most ancient lines of bacteria or from the most complex forms of life, are constructed from the same ubiquitous set of 20 amino acids, covalently linked in characteristic linear sequences. Because each of these amino acids has a side chain with distinctive chemical properties, this group of 20 precursor molecules may be regarded as the alphabet in which the language of protein structure is written.

What is most remarkable is that cells can produce proteins with strikingly different properties and activities by joining the same 20 amino acids in many different combinations and sequences. From these building blocks different organisms can make such widely diverse products as enzymes, hormones, antibodies, transporters, muscle fibers, the lens protein of the eye, feathers, spider webs, rhinoceros horn, milk proteins, antibiotics, mushroom poisons, and myriad other substances having distinct biological activities (Fig. 3–1). Among these protein products, the enzymes are the most varied and specialized. Virtually all cellular reactions are catalyzed by enzymes.

Protein structure and function are the topics of this and the next three chapters. We begin with a description of the fundamental chemical properties of amino acids, peptides, and proteins.

## 3.1 Amino Acids

### Protein Architecture—Amino Acids

Proteins are polymers of amino acids, with each **amino acid residue** joined to its neighbor by a specific type of covalent bond. (The term "residue" reflects the loss of the elements of water when one amino acid is joined to another.) Proteins can be broken down (hydrolyzed) to their constituent amino acids by a variety of methods, and the earliest studies of proteins naturally focused on

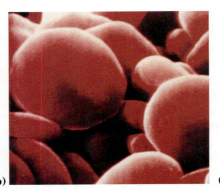

(a)  (b)  (c)

**FIGURE 3–1 Some functions of proteins. (a)** The light produced by fireflies is the result of a reaction involving the protein luciferin and ATP, catalyzed by the enzyme luciferase (see Box 13–2). **(b)** Erythrocytes contain large amounts of the oxygen-transporting protein hemoglobin. **(c)** The protein keratin, formed by all vertebrates, is the chief structural component of hair, scales, horn, wool, nails, and feath-ers. The black rhinoceros is nearing extinction in the wild because of the belief prevalent in some parts of the world that a powder derived from its horn has aphrodisiac properties. In reality, the chemical properties of powdered rhinoceros horn are no different from those of powdered bovine hooves or human fingernails.

the free amino acids derived from them. Twenty different amino acids are commonly found in proteins. The first to be discovered was asparagine, in 1806. The last of the 20 to be found, threonine, was not identified until 1938. All the amino acids have trivial or common names, in some cases derived from the source from which they were first isolated. Asparagine was first found in asparagus, and glutamate in wheat gluten; tyrosine was first isolated from cheese (its name is derived from the Greek *tyros,* "cheese"); and glycine (Greek *glykos,* "sweet") was so named because of its sweet taste.

## Amino Acids Share Common Structural Features

All 20 of the common amino acids are $\alpha$-amino acids. They have a carboxyl group and an amino group bonded to the same carbon atom (the $\alpha$ carbon) (Fig. 3–2). They differ from each other in their side chains, or **R groups,** which vary in structure, size, and electric charge, and which influence the solubility of the amino acids in water. In addition to these 20 amino acids there are many less common ones. Some are residues modified after a protein has been synthesized; others are amino acids present in living organisms but not as constituents of proteins. The common amino acids of proteins have been assigned three-letter abbreviations and one-letter

symbols (Table 3–1), which are used as shorthand to indicate the composition and sequence of amino acids polymerized in proteins.

Two conventions are used to identify the carbons in an amino acid—a practice that can be confusing. The additional carbons in an R group are commonly designated $\beta$, $\gamma$, $\delta$, $\varepsilon$, and so forth, proceeding out from the $\alpha$ carbon. For most other organic molecules, carbon atoms are simply numbered from one end, giving highest priority (C-1) to the carbon with the substituent containing the atom of highest atomic number. Within this latter convention, the carboxyl carbon of an amino acid would be C-1 and the $\alpha$ carbon would be C-2. In some cases, such as amino acids with heterocyclic R groups, the Greek lettering system is ambiguous and the numbering convention is therefore used.

$$\overset{\overset{\varepsilon}{6}}{CH_2} - \overset{\overset{\delta}{5}}{CH_2} - \overset{\overset{\gamma}{4}}{CH_2} - \overset{\overset{\beta}{3}}{CH_2} - \overset{\overset{\alpha}{2}}{CH} - \overset{1}{COO^-}$$
$$|\qquad\qquad\qquad\qquad\qquad\qquad |$$
$$^+NH_3 \qquad\qquad\qquad\qquad\quad ^+NH_3$$

Lysine

For all the common amino acids except glycine, the $\alpha$ carbon is bonded to four different groups: a carboxyl group, an amino group, an R group, and a hydrogen atom (Fig. 3–2; in glycine, the R group is another hydrogen atom). The $\alpha$-carbon atom is thus a **chiral center** (p. 17). Because of the tetrahedral arrangement of the bonding orbitals around the $\alpha$-carbon atom, the four different groups can occupy two unique spatial arrangements, and thus amino acids have two possible stereoisomers. Since they are nonsuperimposable mirror images of each other (Fig. 3–3), the two forms represent a class of stereoisomers called **enantiomers** (see Fig. 1–19). All molecules with a chiral center are also **optically active**—that is, they rotate plane-polarized light (see Box 1–2).

$$\begin{array}{c} COO^- \\ | \\ H_3\overset{+}{N} - C - H \\ | \\ R \end{array}$$

**FIGURE 3–2 General structure of an amino acid.** This structure is common to all but one of the $\alpha$-amino acids. (Proline, a cyclic amino acid, is the exception.) The R group or side chain (red) attached to the $\alpha$ carbon (blue) is different in each amino acid.

Special nomenclature has been developed to specify the **absolute configuration** of the four substituents of asymmetric carbon atoms. The absolute configurations of simple sugars and amino acids are specified by the **D, L system** (Fig. 3–4), based on the absolute configuration of the three-carbon sugar glyceraldehyde, a convention proposed by Emil Fischer in 1891. (Fischer knew what groups surrounded the asymmetric carbon of glyceraldehyde but had to guess at their absolute configuration; his guess was later confirmed by x-ray diffraction analysis.) For all chiral compounds, stereoisomers having a configuration related to that of L-glyceraldehyde are designated L, and stereoisomers related to D-glyceraldehyde are designated D. The functional groups of L-alanine are matched with those of L-glyceraldehyde by aligning those that can be interconverted by simple, one-step chemical reactions. Thus the carboxyl group of L-alanine occupies the same position about the chiral carbon as does the aldehyde group of L-glyceraldehyde, because an aldehyde is readily converted to a carboxyl group via a one-step oxidation. Historically, the similar *l* and *d* designations were used for levorotatory (rotating light to the left) and dextrorotatory (rotating light to the right). However, not all

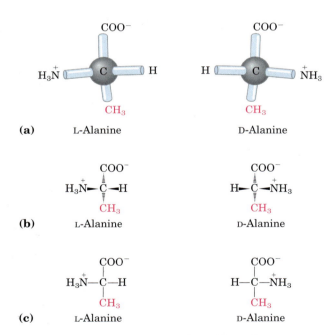

**FIGURE 3–3 Stereoisomerism in α-amino acids. (a)** The two stereoisomers of alanine, L- and D-alanine, are nonsuperimposable mirror images of each other (enantiomers). **(b, c)** Two different conventions for showing the configurations in space of stereoisomers. In perspective formulas **(b)** the solid wedge-shaped bonds project out of the plane of the paper, the dashed bonds behind it. In projection formulas **(c)** the horizontal bonds are assumed to project out of the plane of the paper, the vertical bonds behind. However, projection formulas are often used casually and are not always intended to portray a specific stereochemical configuration.

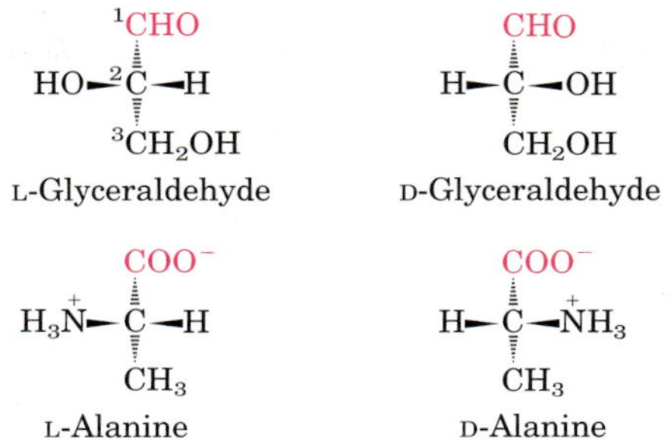

**FIGURE 3–4 Steric relationship of the stereoisomers of alanine to the absolute configuration of L- and D-glyceraldehyde.** In these perspective formulas, the carbons are lined up vertically, with the chiral atom in the center. The carbons in these molecules are numbered beginning with the terminal aldehyde or carboxyl carbon (red), 1 to 3 from top to bottom as shown. When presented in this way, the R group of the amino acid (in this case the methyl group of alanine) is always below the α carbon. L-Amino acids are those with the α-amino group on the left, and D-amino acids have the α-amino group on the right.

L-amino acids are levorotatory, and the convention shown in Figure 3–4 was needed to avoid potential ambiguities about absolute configuration. By Fischer's convention, L and D refer *only* to the absolute configuration of the four substituents around the chiral carbon, not to optical properties of the molecule.

Another system of specifying configuration around a chiral center is the **RS system,** which is used in the systematic nomenclature of organic chemistry and describes more precisely the configuration of molecules with more than one chiral center (see p. 18).

## The Amino Acid Residues in Proteins Are L Stereoisomers

Nearly all biological compounds with a chiral center occur naturally in only one stereoisomeric form, either D or L. The amino acid residues in protein molecules are exclusively L stereoisomers. D-Amino acid residues have been found only in a few, generally small peptides, including some peptides of bacterial cell walls and certain peptide antibiotics.

It is remarkable that virtually all amino acid residues in proteins are L stereoisomers. When chiral compounds are formed by ordinary chemical reactions, the result is a racemic mixture of D and L isomers, which are difficult for a chemist to distinguish and separate. But to a living system, D and L isomers are as different as the right hand and the left. The formation of stable, repeating substructures in proteins (Chapter 4) generally requires that their constituent amino acids be of one stereochemical series. Cells are able to specifically synthesize the L isomers of amino acids because the active sites of enzymes are asymmetric, causing the reactions they catalyze to be stereospecific.

**TABLE 3–1  Properties and Conventions Associated with the Common Amino Acids Found in Proteins**

| Amino acid | Abbreviation/ symbol | $M_r$ | $pK_1$ (—COOH) | $pK_2$ (—NH$_3^+$) | $pK_R$ (R group) | pI | Hydropathy index* | Occurrence in proteins (%)[†] |
|---|---|---|---|---|---|---|---|---|
| **Nonpolar, aliphatic R groups** | | | | | | | | |
| Glycine | Gly G | 75 | 2.34 | 9.60 | | 5.97 | −0.4 | 7.2 |
| Alanine | Ala A | 89 | 2.34 | 9.69 | | 6.01 | 1.8 | 7.8 |
| Proline | Pro P | 115 | 1.99 | 10.96 | | 6.48 | 1.6 | 5.2 |
| Valine | Val V | 117 | 2.32 | 9.62 | | 5.97 | 4.2 | 6.6 |
| Leucine | Leu L | 131 | 2.36 | 9.60 | | 5.98 | 3.8 | 9.1 |
| Isoleucine | Ile I | 131 | 2.36 | 9.68 | | 6.02 | 4.5 | 5.3 |
| Methionine | Met M | 149 | 2.28 | 9.21 | | 5.74 | 1.9 | 2.3 |
| **Aromatic R groups** | | | | | | | | |
| Phenylalanine | Phe F | 165 | 1.83 | 9.13 | | 5.48 | 2.8 | 3.9 |
| Tyrosine | Tyr Y | 181 | 2.20 | 9.11 | 10.07 | 5.66 | −1.3 | 3.2 |
| Tryptophan | Trp W | 204 | 2.38 | 9.39 | | 5.89 | −0.9 | 1.4 |
| **Polar, uncharged R groups** | | | | | | | | |
| Serine | Ser S | 105 | 2.21 | 9.15 | | 5.68 | −0.8 | 6.8 |
| Threonine | Thr T | 119 | 2.11 | 9.62 | | 5.87 | −0.7 | 5.9 |
| Cysteine | Cys C | 121 | 1.96 | 10.28 | 8.18 | 5.07 | 2.5 | 1.9 |
| Asparagine | Asn N | 132 | 2.02 | 8.80 | | 5.41 | −3.5 | 4.3 |
| Glutamine | Gln Q | 146 | 2.17 | 9.13 | | 5.65 | −3.5 | 4.2 |
| **Positively charged R groups** | | | | | | | | |
| Lysine | Lys K | 146 | 2.18 | 8.95 | 10.53 | 9.74 | −3.9 | 5.9 |
| Histidine | His H | 155 | 1.82 | 9.17 | 6.00 | 7.59 | −3.2 | 2.3 |
| Arginine | Arg R | 174 | 2.17 | 9.04 | 12.48 | 10.76 | −4.5 | 5.1 |
| **Negatively charged R groups** | | | | | | | | |
| Aspartate | Asp D | 133 | 1.88 | 9.60 | 3.65 | 2.77 | −3.5 | 5.3 |
| Glutamate | Glu E | 147 | 2.19 | 9.67 | 4.25 | 3.22 | −3.5 | 6.3 |

*A scale combining hydrophobicity and hydrophilicity of R groups; it can be used to measure the tendency of an amino acid to seek an aqueous environment (− values) or a hydrophobic environment (+ values). See Chapter 11. From Kyte, J. & Doolittle, R.F. (1982) A simple method for displaying the hydropathic character of a protein. *J. Mol. Biol.* **157,** 105–132.

[†]Average occurrence in more than 1,150 proteins. From Doolittle, R.F. (1989) Redundancies in protein sequences. In *Prediction of Protein Structure and the Principles of Protein Conformation* (Fasman, G.D., ed.), pp. 599–623, Plenum Press, New York.

## Amino Acids Can Be Classified by R Group

Knowledge of the chemical properties of the common amino acids is central to an understanding of biochemistry. The topic can be simplified by grouping the amino acids into five main classes based on the properties of their R groups (Table 3–1), in particular, their **polarity,** or tendency to interact with water at biological pH (near pH 7.0). The polarity of the R groups varies widely, from nonpolar and hydrophobic (water-insoluble) to highly polar and hydrophilic (water-soluble).

The structures of the 20 common amino acids are shown in Figure 3–5, and some of their properties are listed in Table 3–1. Within each class there are gradations of polarity, size, and shape of the R groups.

*Nonpolar, Aliphatic R Groups*    The R groups in this class of amino acids are nonpolar and hydrophobic. The side chains of **alanine, valine, leucine,** and **isoleucine** tend to cluster together within proteins, stabilizing protein structure by means of hydrophobic interactions. **Glycine** has the simplest structure. Although it is formally nonpolar, its very small side chain makes no real contribution to hydrophobic interactions. **Methionine,** one of the two sulfur-containing amino acids, has a nonpolar thioether group in its side chain. **Proline** has an

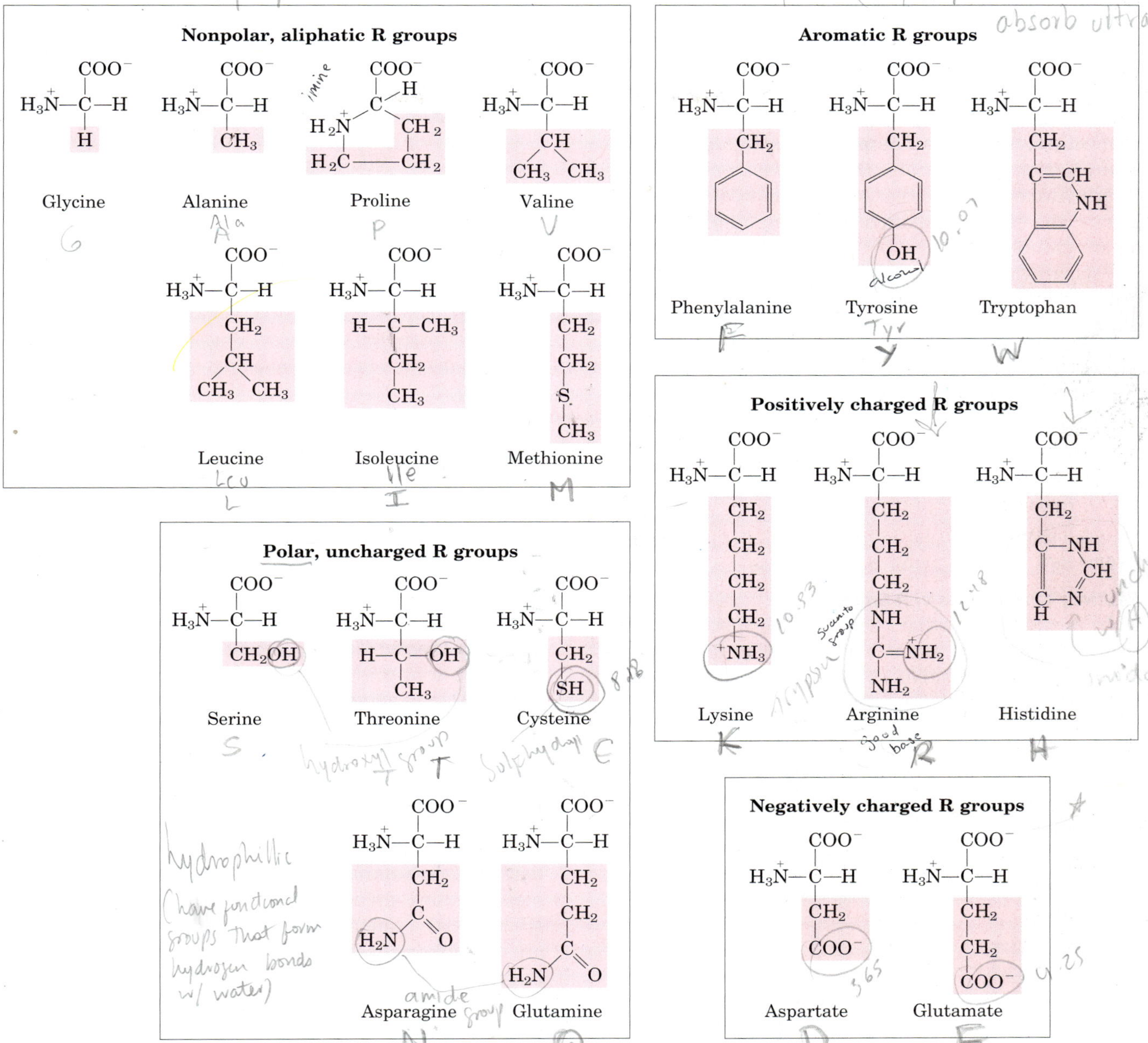

*hydrophobic*

## Nonpolar, aliphatic R groups

Glycine  
Alanine  
Proline  
Valine  
Leucine  
Isoleucine  
Methionine

*nonpolar (hydrophobic)* — *absorb ultraviolet light*

## Aromatic R groups

Phenylalanine  
Tyrosine  
Tryptophan

## Positively charged R groups

Lysine  
Arginine  
Histidine

## Polar, uncharged R groups

Serine  
Threonine  
Cysteine  
Asparagine  
Glutamine

*hydrophillic (have functional groups that form hydrogen bonds w/ water)*

## Negatively charged R groups

Aspartate  
Glutamate

**FIGURE 3–5 The 20 common amino acids of proteins.** The structural formulas show the state of ionization that would predominate at pH 7.0. The unshaded portions are those common to all the amino acids; the portions shaded in red are the R groups. Although the R group of histidine is shown uncharged, its $pK_a$ (see Table 3–1) is such that a small but significant fraction of these groups are positively charged at pH 7.0.

aliphatic side chain with a distinctive cyclic structure. The secondary amino (imino) group of proline residues is held in a rigid conformation that reduces the structural flexibility of polypeptide regions containing proline.

**Aromatic R Groups** **Phenylalanine, tyrosine,** and **tryptophan,** with their aromatic side chains, are relatively nonpolar (hydrophobic). All can participate in hydrophobic interactions. The hydroxyl group of tyrosine can form hydrogen bonds, and it is an important func-

tional group in some enzymes. Tyrosine and tryptophan are significantly more polar than phenylalanine, because of the tyrosine hydroxyl group and the nitrogen of the tryptophan indole ring.

Tryptophan and tyrosine, and to a much lesser extent phenylalanine, absorb ultraviolet light (Fig. 3–6; Box 3–1). This accounts for the characteristic strong absorbance of light by most proteins at a wavelength of 280 nm, a property exploited by researchers in the characterization of proteins.

***Polar, Uncharged R Groups***    The R groups of these amino acids are more soluble in water, or more hydrophilic, than those of the nonpolar amino acids, because they contain functional groups that form hydrogen bonds with water. This class of amino acids includes **serine, threonine, cysteine, asparagine,** and **glutamine.** The polarity of serine and threonine is contributed by their hydroxyl groups; that of cysteine by its sulfhydryl group; and that of asparagine and glutamine by their amide groups.

Asparagine and glutamine are the amides of two other amino acids also found in proteins, aspartate and glutamate, respectively, to which asparagine and glutamine are easily hydrolyzed by acid or base. Cysteine is readily oxidized to form a covalently linked dimeric amino acid called **cystine,** in which two cysteine molecules or residues are joined by a disulfide bond (Fig. 3–7). The disulfide-linked residues are strongly hydrophobic (nonpolar). Disulfide bonds play a special role in the structures of many proteins by forming covalent links between parts of a protein molecule or between two different polypeptide chains.

***Positively Charged (Basic) R Groups***    The most hydrophilic R groups are those that are either positively or negatively charged. The amino acids in which the R groups have significant positive charge at pH 7.0 are **lysine,** which has a second primary amino group at the $\varepsilon$ posi-

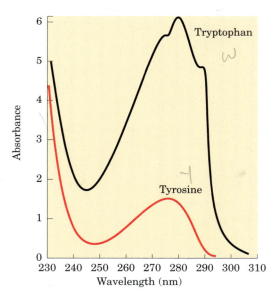

**FIGURE 3–6**  **Absorption of ultraviolet light by aromatic amino acids.** Comparison of the light absorption spectra of the aromatic amino acids tryptophan and tyrosine at pH 6.0. The amino acids are present in equimolar amounts ($10^{-3}$ M) under identical conditions. The measured absorbance of tryptophan is as much as four times that of tyrosine. Note that the maximum light absorption for both tryptophan and tyrosine occurs near a wavelength of 280 nm. Light absorption by the third aromatic amino acid, phenylalanine (not shown), generally contributes little to the spectroscopic properties of proteins.

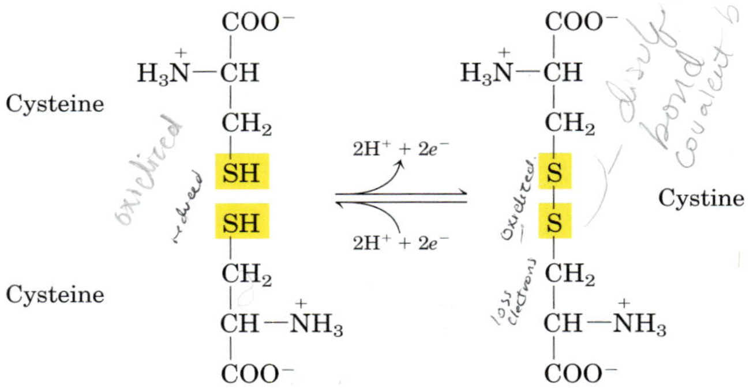

**FIGURE 3–7**  Reversible formation of a disulfide bond by the oxidation of two molecules of cysteine. Disulfide bonds between Cys residues stabilize the structures of many proteins.

tion on its aliphatic chain; **arginine,** which has a positively charged guanidino group; and **histidine,** which has an imidazole group. Histidine is the only common amino acid having an ionizable side chain with a $pK_a$ near neutrality. In many enzyme-catalyzed reactions, a His residue facilitates the reaction by serving as a proton donor/acceptor.

***Negatively Charged (Acidic) R Groups***    The two amino acids having R groups with a net negative charge at pH 7.0 are **aspartate** and **glutamate,** each of which has a second carboxyl group.

## Uncommon Amino Acids Also Have Important Functions

In addition to the 20 common amino acids, proteins may contain residues created by modification of common residues already incorporated into a polypeptide (Fig. 3–8a). Among these uncommon amino acids are **4-hydroxyproline,** a derivative of proline, and **5-hydroxylysine,** derived from lysine. The former is found in plant cell wall proteins, and both are found in collagen, a fibrous protein of connective tissues. **6-N-Methyllysine** is a constituent of myosin, a contractile protein of muscle. Another important uncommon amino acid is **γ-carboxyglutamate,** found in the blood-clotting protein prothrombin and in certain other proteins that bind $Ca^{2+}$ as part of their biological function. More complex is **desmosine,** a derivative of four Lys residues, which is found in the fibrous protein elastin.

**Selenocysteine** is a special case. This rare amino acid residue is introduced during protein synthesis rather than created through a postsynthetic modification. It contains selenium rather than the sulfur of cysteine. Actually derived from serine, selenocysteine is a constituent of just a few known proteins.

Some 300 additional amino acids have been found in cells. They have a variety of functions but are not constituents of proteins. **Ornithine** and **citrulline**

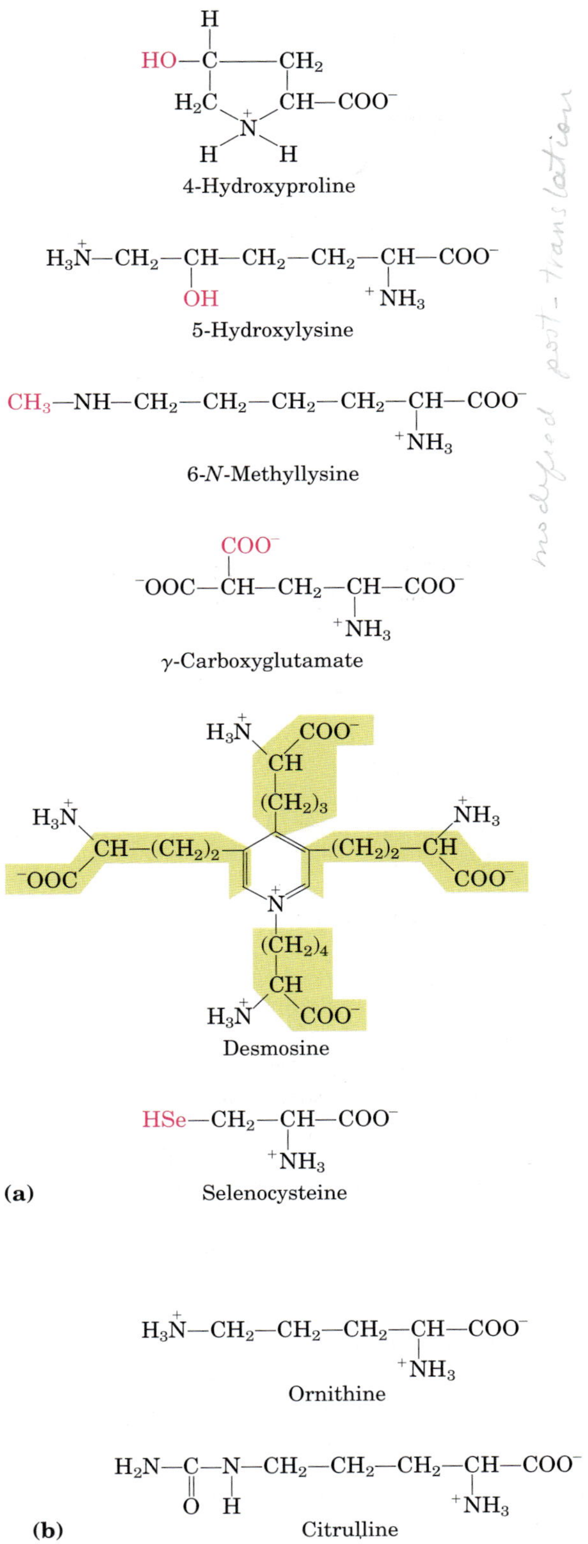
4-Hydroxyproline
5-Hydroxylysine
6-N-Methyllysine
γ-Carboxyglutamate
Desmosine
(a)
Selenocysteine
Ornithine
(b)
Citrulline

**FIGURE 3–8 Uncommon amino acids. (a)** Some uncommon amino acids found in proteins. All are derived from common amino acids. Extra functional groups added by modification reactions are shown in red. Desmosine is formed from four Lys residues (the four carbon backbones are shaded in yellow). Note the use of either numbers or Greek letters to identify the carbon atoms in these structures. **(b)** Ornithine and citrulline, which are not found in proteins, are intermediates in the biosynthesis of arginine and in the urea cycle.

(Fig. 3–8b) deserve special note because they are key intermediates (metabolites) in the biosynthesis of arginine (Chapter 22) and in the urea cycle (Chapter 18).

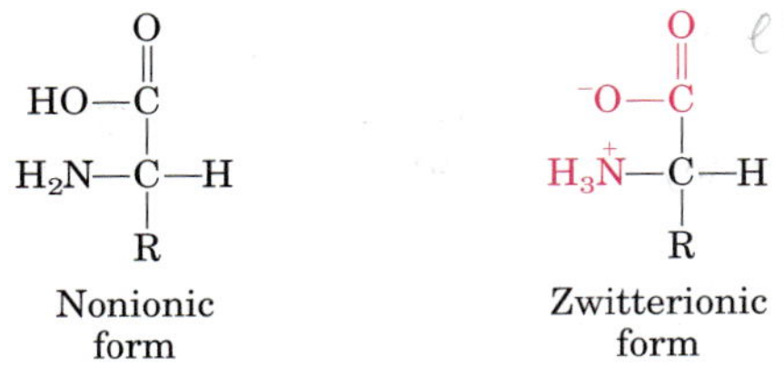
Nonionic form
Zwitterionic form

**FIGURE 3–9 Nonionic and zwitterionic forms of amino acids.** The nonionic form does not occur in significant amounts in aqueous solutions. The zwitterion predominates at neutral pH.

## Amino Acids Can Act as Acids and Bases

When an amino acid is dissolved in water, it exists in solution as the dipolar ion, or **zwitterion** (German for "hybrid ion"), shown in Figure 3–9. A zwitterion can act as either an acid (proton donor):

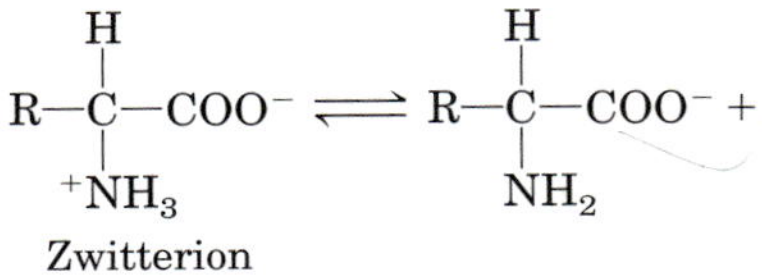
Zwitterion

or a base (proton acceptor):

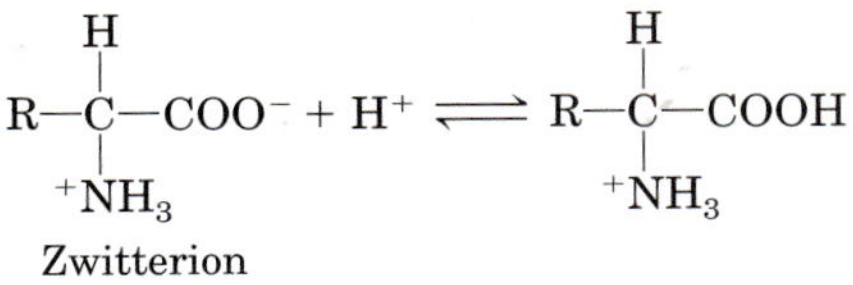
Zwitterion

Substances having this dual nature are **amphoteric** and are often called **ampholytes** (from "amphoteric electrolytes"). A simple monoamino monocarboxylic α-amino acid, such as alanine, is a diprotic acid when fully protonated—it has two groups, the —COOH group and the $—NH_3^+$ group, that can yield protons:

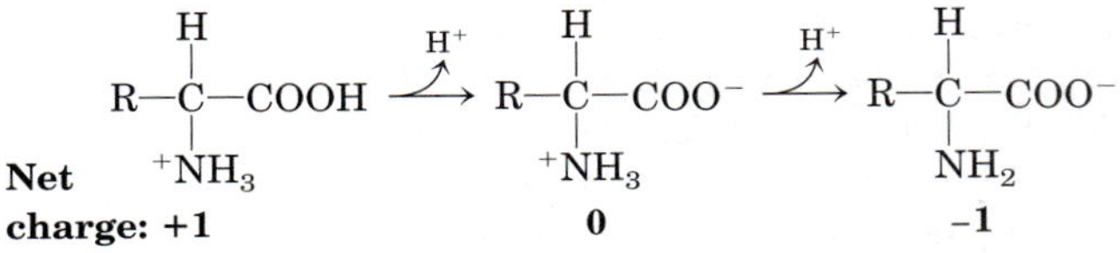
Net charge: +1
0
–1

## BOX 3–1    WORKING IN BIOCHEMISTRY

### Absorption of Light by Molecules: The Lambert-Beer Law

A wide range of biomolecules absorb light at characteristic wavelengths, just as tryptophan absorbs light at 280 nm (see Fig. 3–6). Measurement of light absorption by a spectrophotometer is used to detect and identify molecules and to measure their concentration in solution. The fraction of the incident light absorbed by a solution at a given wavelength is related to the thickness of the absorbing layer (path length) and the concentration of the absorbing species (Fig. 1). These two relationships are combined into the Lambert-Beer law,

$$\log \frac{I_0}{I} = \varepsilon c l$$

where $I_0$ is the intensity of the incident light, $I$ is the intensity of the transmitted light, $\varepsilon$ is the molar extinction coefficient (in units of liters per mole-centimeter), $c$ is the concentration of the absorbing species (in

moles per liter), and $l$ is the path length of the light-absorbing sample (in centimeters). The Lambert-Beer law assumes that the incident light is parallel and monochromatic (of a single wavelength) and that the solvent and solute molecules are randomly oriented. The expression $\log (I_0/I)$ is called the **absorbance,** designated $A$.

It is important to note that each successive millimeter of path length of absorbing solution in a 1.0 cm cell absorbs not a constant amount but a constant fraction of the light that is incident upon it. However, with an absorbing layer of fixed path length, *the absorbance,* A, *is directly proportional to the concentration of the absorbing solute.*

The molar extinction coefficient varies with the nature of the absorbing compound, the solvent, and the wavelength, and also with pH if the light-absorbing species is in equilibrium with an ionization state that has different absorbance properties.

**FIGURE 1**  The principal components of a spectrophotometer. A light source emits light along a broad spectrum, then the monochromator selects and transmits light of a particular wavelength. The monochromatic light passes through the sample in a cuvette of path length *l* and is absorbed by the sample in proportion to the concentration of the absorbing species. The transmitted light is measured by a detector.

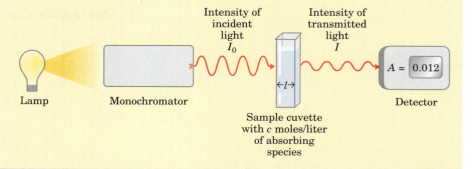

## Amino Acids Have Characteristic Titration Curves

Acid-base titration involves the gradual addition or removal of protons (Chapter 2). Figure 3–10 shows the titration curve of the diprotic form of glycine. The plot has two distinct stages, corresponding to deprotonation of two different groups on glycine. Each of the two stages resembles in shape the titration curve of a monoprotic acid, such as acetic acid (see Fig. 2–17), and can be analyzed in the same way. At very low pH, the predominant ionic species of glycine is the fully protonated form, $^+H_3N—CH_2—COOH$. At the midpoint in the first stage of the titration, in which the —COOH group of glycine loses its proton, equimolar concentrations of the proton-donor ($^+H_3N—CH_2—COOH$) and proton-acceptor ($^+H_3N—CH_2—COO^-$) species are present. At the midpoint of any titration, a point of inflection is reached where the pH is equal to the $pK_a$ of the protonated group being titrated (see Fig. 2–18). For glycine, the pH at the midpoint is 2.34, thus its —COOH group has a $pK_a$ (labeled $pK_1$ in Fig. 3–10) of 2.34.

(Recall from Chapter 2 that pH and $pK_a$ are simply convenient notations for proton concentration and the equilibrium constant for ionization, respectively. The $pK_a$ is a measure of the tendency of a group to give up a proton, with that tendency decreasing tenfold as the $pK_a$ increases by one unit.) As the titration proceeds, another important point is reached at pH 5.97. Here there is another point of inflection, at which removal of the first proton is essentially complete and removal of the second has just begun. At this pH glycine is present largely as the dipolar ion $^+H_3N—CH_2—COO^-$. We shall return to the significance of this inflection point in the titration curve (labeled pI in Fig. 3–10) shortly.

The second stage of the titration corresponds to the removal of a proton from the —$NH_3^+$ group of glycine. The pH at the midpoint of this stage is 9.60, equal to the $pK_a$ (labeled $pK_2$ in Fig. 3–10) for the —$NH_3^+$ group. The titration is essentially complete at a pH of about 12, at which point the predominant form of glycine is $H_2N—CH_2—COO^-$.

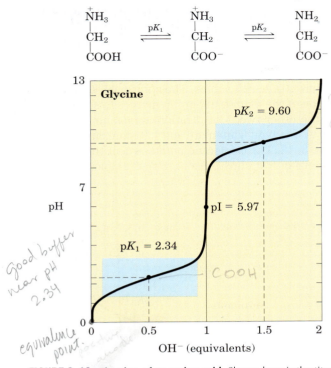

**FIGURE 3–10 Titration of an amino acid.** Shown here is the titration curve of 0.1 M glycine at 25 °C. The ionic species predominating at key points in the titration are shown above the graph. The shaded boxes, centered at about $pK_1 = 2.34$ and $pK_2 = 9.60$, indicate the regions of greatest buffering power.

From the titration curve of glycine we can derive several important pieces of information. First, it gives a quantitative measure of the $pK_a$ of each of the two ionizing groups: 2.34 for the —COOH group and 9.60 for the —$NH_3^+$ group. Note that the carboxyl group of glycine is over 100 times more acidic (more easily ionized) than the carboxyl group of acetic acid, which, as we saw in Chapter 2, has a $pK_a$ of 4.76—about average for a carboxyl group attached to an otherwise unsubstituted aliphatic hydrocarbon. The perturbed $pK_a$ of glycine is caused by repulsion between the departing proton and the nearby positively charged amino group on the $\alpha$-carbon atom, as described in Figure 3–11. The opposite charges on the resulting zwitterion are stabilizing, nudging the equilibrium farther to the right. Similarly, the $pK_a$ of the amino group in glycine is perturbed downward relative to the average $pK_a$ of an amino group. This effect is due partly to the electronegative oxygen atoms in the carboxyl groups, which tend to pull electrons toward them, increasing the tendency of the amino group to give up a proton. Hence, the $\alpha$-amino group has a $pK_a$ that is lower than that of an aliphatic amine such as methylamine (Fig. 3–11). In short, the $pK_a$ of any functional group is greatly affected by its chemical environment, a phenomenon sometimes exploited in the active sites of enzymes to promote exquisitely adapted reaction mechanisms that depend on the perturbed $pK_a$ values of proton donor/acceptor groups of specific residues.

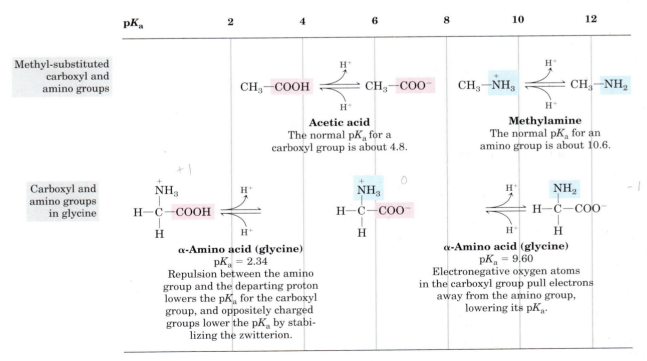

**FIGURE 3–11 Effect of the chemical environment on $pK_a$.** The $pK_a$ values for the ionizable groups in glycine are lower than those for simple, methyl-substituted amino and carboxyl groups. These downward perturbations of $pK_a$ are due to intramolecular interactions. Similar effects can be caused by chemical groups that happen to be positioned nearby—for example, in the active site of an enzyme.

The second piece of information provided by the titration curve of glycine is that this amino acid has *two* regions of buffering power. One of these is the relatively flat portion of the curve, extending for approximately 1 pH unit on either side of the first p$K_a$ of 2.34, indicating that glycine is a good buffer near this pH. The other buffering zone is centered around pH 9.60. (Note that glycine is not a good buffer at the pH of intracellular fluid or blood, about 7.4.) Within the buffering ranges of glycine, the Henderson-Hasselbalch equation (see Box 2–3) can be used to calculate the proportions of proton-donor and proton-acceptor species of glycine required to make a buffer at a given pH.

## Titration Curves Predict the Electric Charge of Amino Acids

Another important piece of information derived from the titration curve of an amino acid is the relationship between its net electric charge and the pH of the solution. At pH 5.97, the point of inflection between the two stages in its titration curve, glycine is present predominantly as its dipolar form, fully ionized but with no *net* electric charge (Fig. 3–10). The characteristic pH at which the net electric charge is zero is called the **isoelectric point** or **isoelectric pH,** designated **pI.** For glycine, which has no ionizable group in its side chain, the isoelectric point is simply the arithmetic mean of the two p$K_a$ values:

$$pI = \frac{1}{2} (pK_1 + pK_2) = \frac{1}{2} (2.34 + 9.60) = 5.97$$

As is evident in Figure 3–10, glycine has a net negative charge at any pH above its pI and will thus move toward the positive electrode (the anode) when placed in an electric field. At any pH below its pI, glycine has a net positive charge and will move toward the negative electrode (the cathode). The farther the pH of a glycine solution is from its isoelectric point, the greater the net electric charge of the population of glycine molecules. At pH 1.0, for example, glycine exists almost entirely as the form $^+H_3N—CH_2—COOH$, with a net positive charge of 1.0. At pH 2.34, where there is an equal mixture of $^+H_3N—CH_2—COOH$ and $^+H_3N—CH_2—COO^-$, the average or net positive charge is 0.5. The sign and the magnitude of the net charge of any amino acid at any pH can be predicted in the same way.

## Amino Acids Differ in Their Acid-Base Properties

The shared properties of many amino acids permit some simplifying generalizations about their acid-base behaviors. First, all amino acids with a single α-amino group, a single α-carboxyl group, and an R group that does not ionize have titration curves resembling that of glycine (Fig. 3–10). These amino acids have very similar, although not identical, p$K_a$ values: p$K_a$ of the —COOH

group in the range of 1.8 to 2.4, and p$K_a$ of the —$NH_3^+$ group in the range of 8.8 to 11.0 (Table 3–1).

Second, amino acids with an ionizable R group have more complex titration curves, with *three* stages corresponding to the three possible ionization steps; thus they have three p$K_a$ values. The additional stage for the titration of the ionizable R group merges to some extent with the other two. The titration curves for two amino acids of this type, glutamate and histidine, are shown in Figure 3–12. The isoelectric points reflect the nature of the ionizing R groups present. For example, glutamate

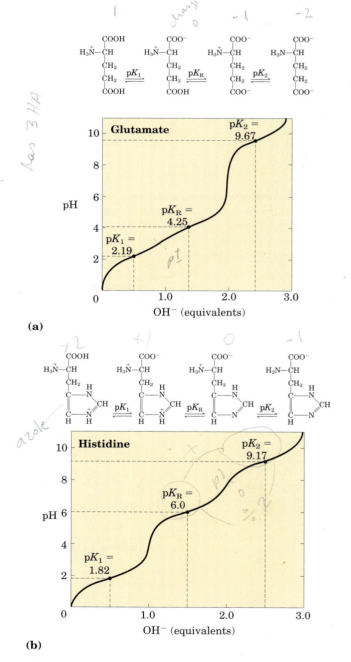

**FIGURE 3–12** Titration curves for **(a)** glutamate and **(b)** histidine. The p$K_a$ of the R group is designated here as p$K_R$.

has a pI of 3.22, considerably lower than that of glycine. This is due to the presence of two carboxyl groups, which, at the average of their $pK_a$ values (3.22), contribute a net charge of $-1$ that balances the $+1$ contributed by the amino group. Similarly, the pI of histidine, with two groups that are positively charged when protonated, is 7.59 (the average of the $pK_a$ values of the amino and imidazole groups), much higher than that of glycine.

Finally, as pointed out earlier, under the general condition of free and open exposure to the aqueous environment, only histidine has an R group ($pK_a = 6.0$) providing significant buffering power near the neutral pH usually found in the intracellular and extracellular fluids of most animals and bacteria (Table 3–1).

### SUMMARY 3.1 Amino Acids

- The 20 amino acids commonly found as residues in proteins contain an $\alpha$-carboxyl group, an $\alpha$-amino group, and a distinctive R group substituted on the $\alpha$-carbon atom. The $\alpha$-carbon atom of all amino acids except glycine is asymmetric, and thus amino acids can exist in at least two stereoisomeric forms. Only the L stereoisomers, with a configuration related to the absolute configuration of the reference molecule L-glyceraldehyde, are found in proteins.

- Other, less common amino acids also occur, either as constituents of proteins (through modification of common amino acid residues after protein synthesis) or as free metabolites.

- Amino acids are classified into five types on the basis of the polarity and charge (at pH 7) of their R groups.

- Amino acids vary in their acid-base properties and have characteristic titration curves. Monoamino monocarboxylic amino acids (with nonionizable R groups) are diprotic acids ($^+H_3NCH(R)COOH$) at low pH and exist in several different ionic forms as the pH is increased. Amino acids with ionizable R groups have additional ionic species, depending on the pH of the medium and the $pK_a$ of the R group.

## 3.2 Peptides and Proteins

We now turn to polymers of amino acids, the **peptides** and **proteins.** Biologically occurring polypeptides range in size from small to very large, consisting of two or three to thousands of linked amino acid residues. Our focus is on the fundamental chemical properties of these polymers.

### Peptides Are Chains of Amino Acids

Two amino acid molecules can be covalently joined through a substituted amide linkage, termed a **peptide bond,** to yield a dipeptide. Such a linkage is formed by removal of the elements of water (dehydration) from the $\alpha$-carboxyl group of one amino acid and the $\alpha$-amino group of another (Fig. 3–13). Peptide bond formation is an example of a condensation reaction, a common class of reactions in living cells. Under standard biochemical conditions, the equilibrium for the reaction shown in Figure 3–13 favors the amino acids over the dipeptide. To make the reaction thermodynamically more favorable, the carboxyl group must be chemically modified or activated so that the hydroxyl group can be more readily eliminated. A chemical approach to this problem is outlined later in this chapter. The biological approach to peptide bond formation is a major topic of Chapter 27.

Three amino acids can be joined by two peptide bonds to form a tripeptide; similarly, amino acids can be linked to form tetrapeptides, pentapeptides, and so forth. When a few amino acids are joined in this fashion, the structure is called an **oligopeptide.** When many amino acids are joined, the product is called a **polypeptide.** Proteins may have thousands of amino acid residues. Although the terms "protein" and "polypeptide" are sometimes used interchangeably, molecules referred to as polypeptides generally have molecular weights below 10,000, and those called proteins have higher molecular weights.

Figure 3–14 shows the structure of a pentapeptide. As already noted, an amino acid unit in a peptide is often called a residue (the part left over after losing a hydrogen atom from its amino group and the hydroxyl moiety from its carboxyl group). In a peptide, the amino acid residue at the end with a free $\alpha$-amino group is the **amino-terminal** (or N-terminal) residue; the residue

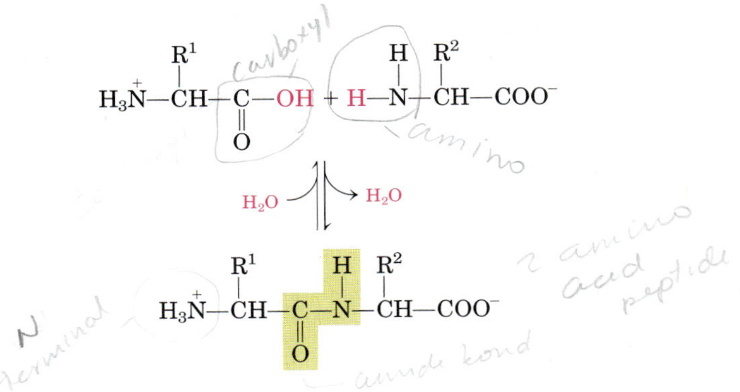

**FIGURE 3–13 Formation of a peptide bond by condensation.** The $\alpha$-amino group of one amino acid (with $R^2$ group) acts as a nucleophile to displace the hydroxyl group of another amino acid (with $R^1$ group), forming a peptide bond (shaded in yellow). Amino groups are good nucleophiles, but the hydroxyl group is a poor leaving group and is not readily displaced. At physiological pH, the reaction shown does not occur to any appreciable extent.

*end terminol* Ser - Gly - Tyr
Ala - Leu  SGYAL

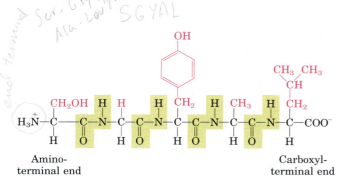

**FIGURE 3–14** The pentapeptide serylglycyltyrosylalanylleucine, or Ser–Gly–Tyr–Ala–Leu. Peptides are named beginning with the amino-terminal residue, which by convention is placed at the left. The peptide bonds are shaded in yellow; the R groups are in red.

at the other end, which has a free carboxyl group, is the **carboxyl-terminal** (*C*-terminal) residue.

Although hydrolysis of a peptide bond is an exergonic reaction, it occurs slowly because of its high activation energy. As a result, the peptide bonds in proteins are quite stable, with an average half-life ($t_{1/2}$) of about 7 years under most intracellular conditions.

## Peptides Can Be Distinguished by Their Ionization Behavior

Peptides contain only one free α-amino group and one free α-carboxyl group, at opposite ends of the chain (Fig. 3–15). These groups ionize as they do in free amino acids, although the ionization constants are different because an oppositely charged group is no longer linked to the α carbon. The α-amino and α-carboxyl groups of all nonterminal amino acids are covalently joined in the peptide bonds, which do not ionize and thus do not contribute to the total acid-base behavior of peptides. How-

ever, the R groups of some amino acids can ionize (Table 3–1), and in a peptide these contribute to the overall acid-base properties of the molecule (Fig. 3–15). Thus the acid-base behavior of a peptide can be predicted from its free α-amino and α-carboxyl groups as well as the nature and number of its ionizable R groups.

Like free amino acids, peptides have characteristic titration curves and a characteristic isoelectric pH (pI) at which they do not move in an electric field. These properties are exploited in some of the techniques used to separate peptides and proteins, as we shall see later in the chapter. It should be emphasized that the $pK_a$ value for an ionizable R group can change somewhat when an amino acid becomes a residue in a peptide. The loss of charge in the α-carboxyl and α-amino groups, the interactions with other peptide R groups, and other environmental factors can affect the $pK_a$. The $pK_a$ values for R groups listed in Table 3–1 can be a useful guide to the pH range in which a given group will ionize, but they cannot be strictly applied to peptides.

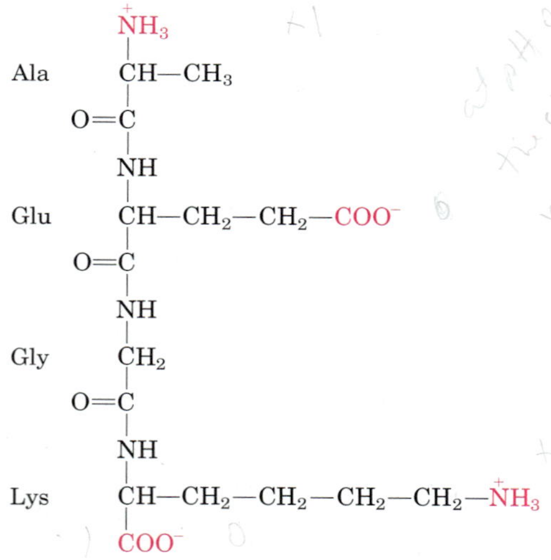

## Biologically Active Peptides and Polypeptides Occur in a Vast Range of Sizes

No generalizations can be made about the molecular weights of biologically active peptides and proteins in relation to their functions. Naturally occurring peptides range in length from two to many thousands of amino acid residues. Even the smallest peptides can have biologically important effects. Consider the commercially synthesized dipeptide L-aspartyl-L-phenylalanine methyl ester, the artificial sweetener better known as aspartame or NutraSweet.

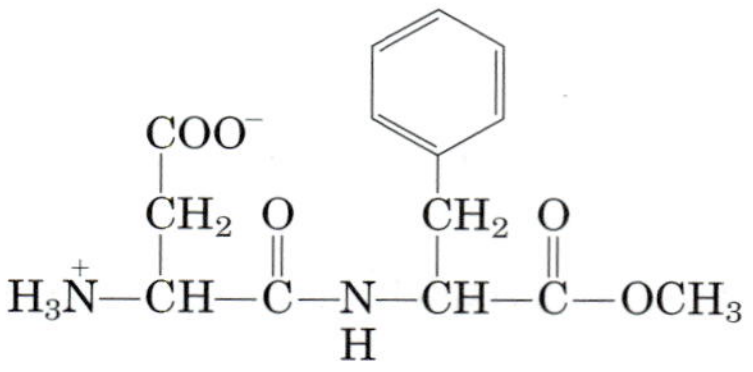

L-Aspartyl-L-phenylalanine methyl ester
(aspartame)

Many small peptides exert their effects at very low concentrations. For example, a number of vertebrate hormones (Chapter 23) are small peptides. These include oxytocin (nine amino acid residues), which is secreted by the posterior pituitary and stimulates uterine contractions; bradykinin (nine residues), which inhibits inflammation of tissues; and thyrotropin-releasing factor (three residues), which is formed in the hypothalamus and stimulates the release of another hormone, thyrotropin, from the anterior pituitary gland. Some extremely toxic mushroom poisons, such as amanitin, are also small peptides, as are many antibiotics.

Slightly larger are small polypeptides and oligopeptides such as the pancreatic hormone insulin, which contains two polypeptide chains, one having 30 amino acid

**FIGURE 3–15 Alanylglutamylglycyllysine.** This tetrapeptide has one free α-amino group, one free α-carboxyl group, and two ionizable R groups. The groups ionized at pH 7.0 are in red.

residues and the other 21. Glucagon, another pancreatic hormone, has 29 residues; it opposes the action of insulin. Corticotropin is a 39-residue hormone of the anterior pituitary gland that stimulates the adrenal cortex.

How long are the polypeptide chains in proteins? As Table 3–2 shows, lengths vary considerably. Human cytochrome c has 104 amino acid residues linked in a single chain; bovine chymotrypsinogen has 245 residues. At the extreme is titin, a constituent of vertebrate muscle, which has nearly 27,000 amino acid residues and a molecular weight of about 3,000,000. The vast majority of naturally occurring proteins are much smaller than this, containing fewer than 2,000 amino acid residues.

Some proteins consist of a single polypeptide chain, but others, called **multisubunit** proteins, have two or more polypeptides associated noncovalently (Table 3–2). The individual polypeptide chains in a multisubunit protein may be identical or different. If at least two are identical the protein is said to be **oligomeric,** and the identical units (consisting of one or more polypeptide chains) are referred to as **protomers.** Hemoglobin, for example, has four polypeptide subunits: two identical $\alpha$ chains and two identical $\beta$ chains, all four held together by noncovalent interactions. Each $\alpha$ subunit is paired in an identical way with a $\beta$ subunit within the structure of this multisubunit protein, so that hemoglobin can be considered either a tetramer of four polypeptide subunits or a dimer of $\alpha\beta$ protomers.

A few proteins contain two or more polypeptide chains linked covalently. For example, the two polypeptide chains of insulin are linked by disulfide bonds. In such cases, the individual polypeptides are not considered subunits but are commonly referred to simply as chains.

We can calculate the approximate number of amino acid residues in a simple protein containing no other chemical constituents by dividing its molecular weight by 110. Although the average molecular weight of the 20 common amino acids is about 138, the smaller amino acids predominate in most proteins. If we take into account the proportions in which the various amino acids occur in proteins (Table 3–1), the average molecular weight of protein amino acids is nearer to 128. Because a molecule of water ($M_r$ 18) is removed to create each peptide bond, the average molecular weight of an amino acid residue in a protein is about $128 - 18 = 110$.

## Polypeptides Have Characteristic Amino Acid Compositions

Hydrolysis of peptides or proteins with acid yields a mixture of free $\alpha$-amino acids. When completely hydrolyzed, each type of protein yields a characteristic proportion or mixture of the different amino acids. The 20 common amino acids almost never occur in equal amounts in a protein. Some amino acids may occur only once or not at all in a given type of protein; others may occur in large numbers. Table 3–3 shows the composition of the amino acid mixtures obtained on complete hydrolysis of bovine cytochrome c and chymotrypsinogen, the inactive precursor of the digestive enzyme chymotrypsin. These two proteins, with very different functions, also differ significantly in the relative numbers of each kind of amino acid they contain.

Complete hydrolysis alone is not sufficient for an exact analysis of amino acid composition, however, because some side reactions occur during the procedure. For example, the amide bonds in the side chains of asparagine and glutamine are cleaved by acid treatment, yielding aspartate and glutamate, respectively. The side chain of tryptophan is almost completely degraded by acid hydrolysis, and small amounts of serine, threonine,

**TABLE 3–2 Molecular Data on Some Proteins**

| | Molecular weight | Number of residues | Number of polypeptide chains |
|---|---|---|---|
| Cytochrome c (human) | 13,000 | 104 | 1 |
| Ribonuclease A (bovine pancreas) | 13,700 | 124 | 1 |
| Lysozyme (chicken egg white) | 13,930 | 129 | 1 |
| Myoglobin (equine heart) | 16,890 | 153 | 1 |
| Chymotrypsin (bovine pancreas) | 21,600 | 241 | 3 |
| Chymotrypsinogen (bovine) | 22,000 | 245 | 1 |
| Hemoglobin (human) | 64,500 | 574 | 4 |
| Serum albumin (human) | 68,500 | 609 | 1 |
| Hexokinase (yeast) | 102,000 | 972 | 2 |
| RNA polymerase (E. coli) | 450,000 | 4,158 | 5 |
| Apolipoprotein B (human) | 513,000 | 4,536 | 1 |
| Glutamine synthetase (E. coli) | 619,000 | 5,628 | 12 |
| Titin (human) | 2,993,000 | 26,926 | 1 |

## TABLE 3-3    Amino Acid Composition of Two Proteins

| Amino acid | Number of residues per molecule of protein* | |
|---|---|---|
| | Bovine cytochrome c | Bovine chymotrypsinogen |
| Ala | 6 | 22 |
| Arg | 2 | 4 |
| Asn | 5 | 15 |
| Asp | 3 | 8 |
| Cys | 2 | 10 |
| Gln | 3 | 10 |
| Glu | 9 | 5 |
| Gly | 14 | 23 |
| His | 3 | 2 |
| Ile | 6 | 10 |
| Leu | 6 | 19 |
| Lys | 18 | 14 |
| Met | 2 | 2 |
| Phe | 4 | 6 |
| Pro | 4 | 9 |
| Ser | 1 | 28 |
| Thr | 8 | 23 |
| Trp | 1 | 8 |
| Tyr | 4 | 4 |
| Val | 3 | 23 |
| Total | 104 | 245 |

*In some common analyses, such as acid hydrolysis, Asp and Asn are not readily distinguished from each other and are together designated Asx (or B). Similarly, when Glu and Gln cannot be distinguished, they are together designated Glx (or Z). In addition, Trp is destroyed. Additional procedures must be employed to obtain an accurate assessment of complete amino acid content.

## TABLE 3-4    Conjugated Proteins

| Class | Prosthetic group | Example |
|---|---|---|
| Lipoproteins | Lipids | $\beta_1$-Lipoprotein of blood |
| Glycoproteins | Carbohydrates | Immunoglobulin G |
| Phosphoproteins | Phosphate groups | Casein of milk |
| Hemoproteins | Heme (iron porphyrin) | Hemoglobin |
| Flavoproteins | Flavin nucleotides | Succinate dehydrogenase |
| Metalloproteins | Iron | Ferritin |
| | Zinc | Alcohol dehydrogenase |
| | Calcium | Calmodulin |
| | Molybdenum | Dinitrogenase |
| | Copper | Plastocyanin |

metal. A number of proteins contain more than one prosthetic group. Usually the prosthetic group plays an important role in the protein's biological function.

### There Are Several Levels of Protein Structure

For large macromolecules such as proteins, the tasks of describing and understanding structure are approached at several levels of complexity, arranged in a kind of conceptual hierarchy. Four levels of protein structure are commonly defined (Fig. 3–16). A description of all covalent bonds (mainly peptide bonds and disulfide bonds) linking amino acid residues in a polypeptide chain is its **primary structure.** The most important element of primary structure is the *sequence* of amino acid residues. **Secondary structure** refers to particularly stable arrangements of amino acid residues giving rise to recurring structural patterns. **Tertiary structure** describes all aspects of the three-dimensional folding of a polypeptide. When a protein has two or more polypeptide subunits, their arrangement in space is referred to as **quaternary structure.** Primary structure is the focus of Section 3.4; the higher levels of structure are discussed in Chapter 4.

### SUMMARY 3.2    Peptides and Proteins

■ Amino acids can be joined covalently through peptide bonds to form peptides and proteins. Cells generally contain thousands of different proteins, each with a different biological activity.

■ Proteins can be very long polypeptide chains of 100 to several thousand amino acid residues. However, some naturally occurring peptides have only a few amino acid residues. Some proteins are composed of several noncovalently

and tyrosine are also lost. When a precise amino acid composition is required, biochemists use additional procedures to resolve the ambiguities that remain from acid hydrolysis.

### Some Proteins Contain Chemical Groups Other Than Amino Acids

Many proteins, for example the enzymes ribonuclease A and chymotrypsinogen, contain only amino acid residues and no other chemical constituents; these are considered simple proteins. However, some proteins contain permanently associated chemical components in addition to amino acids; these are called **conjugated proteins.** The non–amino acid part of a conjugated protein is usually called its **prosthetic group.** Conjugated proteins are classified on the basis of the chemical nature of their prosthetic groups (Table 3–4); for example, **lipoproteins** contain lipids, **glycoproteins** contain sugar groups, and **metalloproteins** contain a specific

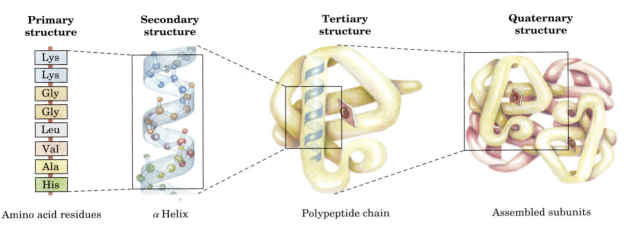

| Primary structure | Secondary structure | Tertiary structure | Quaternary structure |
|---|---|---|---|
| Lys Lys Gly Gly Leu Val Ala His | | | |
| Amino acid residues | α Helix | Polypeptide chain | Assembled subunits |

**FIGURE 3–16 Levels of structure in proteins.** The *primary structure* consists of a sequence of amino acids linked together by peptide bonds and includes any disulfide bonds. The resulting polypeptide can be coiled into units of *secondary structure,* such as an α helix. The he-lix is a part of the *tertiary structure* of the folded polypeptide, which is itself one of the subunits that make up the *quaternary structure* of the multisubunit protein, in this case hemoglobin.

associated polypeptide chains, called subunits. Simple proteins yield only amino acids on hydrolysis; conjugated proteins contain in addition some other component, such as a metal or organic prosthetic group.

■ The sequence of amino acids in a protein is characteristic of that protein and is called its primary structure. This is one of four generally recognized levels of protein structure.

## 3.3 Working with Proteins

Our understanding of protein structure and function has been derived from the study of many individual proteins. To study a protein in detail, the researcher must be able to separate it from other proteins and must have the techniques to determine its properties. The necessary methods come from protein chemistry, a discipline as old as biochemistry itself and one that retains a central position in biochemical research.

### Proteins Can Be Separated and Purified

A pure preparation is essential before a protein's properties and activities can be determined. Given that cells contain thousands of different kinds of proteins, how can one protein be purified? Methods for separating proteins take advantage of properties that vary from one protein to the next, including size, charge, and binding properties.

The source of a protein is generally tissue or microbial cells. The first step in any protein purification procedure is to break open these cells, releasing their proteins into a solution called a **crude extract.** If necessary, differential centrifugation can be used to pre-pare subcellular fractions or to isolate specific organelles (see Fig. 1–8).

Once the extract or organelle preparation is ready, various methods are available for purifying one or more of the proteins it contains. Commonly, the extract is subjected to treatments that separate the proteins into different fractions based on a property such as size or charge, a process referred to as **fractionation.** Early fractionation steps in a purification utilize differences in protein solubility, which is a complex function of pH, temperature, salt concentration, and other factors. The solubility of proteins is generally lowered at high salt concentrations, an effect called "salting out." The addition of a salt in the right amount can selectively precipitate some proteins, while others remain in solution. Ammonium sulfate $((NH_4)_2SO_4)$ is often used for this purpose because of its high solubility in water.

A solution containing the protein of interest often must be further altered before subsequent purification steps are possible. For example, **dialysis** is a procedure that separates proteins from solvents by taking advantage of the proteins' larger size. The partially purified extract is placed in a bag or tube made of a semipermeable membrane. When this is suspended in a much larger volume of buffered solution of appropriate ionic strength, the membrane allows the exchange of salt and buffer but not proteins. Thus dialysis retains large proteins within the membranous bag or tube while allowing the concentration of other solutes in the protein preparation to change until they come into equilibrium with the solution outside the membrane. Dialysis might be used, for example, to remove ammonium sulfate from the protein preparation.

The most powerful methods for fractionating proteins make use of **column chromatography,** which takes advantage of differences in protein charge, size, binding affinity, and other properties (Fig. 3–17). A porous solid material with appropriate chemical properties (the stationary phase) is held in a column, and a buffered solution (the mobile phase) percolates through it. The protein-containing solution, layered on the top of the column, percolates through the solid matrix as an ever-expanding band within the larger mobile phase (Fig. 3–17). Individual proteins migrate faster or more slowly through the column depending on their properties. For example, in **cation-exchange chromatography** (Fig. 3–18a), the solid matrix has negatively charged groups. In the mobile phase, proteins with a net positive charge migrate through the matrix more slowly than those with a net negative charge, because the migration of the former is retarded more by interaction with the stationary phase. The two types of protein can separate into two distinct bands. The expansion of the protein band in the mobile phase (the protein solution) is caused both by separation of proteins with different properties and by diffusional spreading. As the length of the column increases, the resolution of two types of protein with different net charges generally improves. However, the rate at which the protein solution can flow through the column usually decreases with column

length. And as the length of time spent on the column increases, the resolution can decline as a result of diffusional spreading within each protein band.

Figure 3–18 shows two other variations of column chromatography in addition to ion exchange. **Size-exclusion chromatography** separates proteins according to size. In this method, large proteins emerge from the column sooner than small ones—a somewhat counterintuitive result. The solid phase consists of beads with engineered pores or cavities of a particular size. Large proteins cannot enter the cavities, and so take a short (and rapid) path through the column, around the beads. Small proteins enter the cavities, and migrate through the column more slowly as a result (Fig. 3–18b). **Affinity chromatography** is based on the binding affinity of a protein. The beads in the column have a covalently attached chemical group. A protein with affinity for this particular chemical group will bind to the beads in the column, and its migration will be retarded as a result (Fig. 3–18c).

A modern refinement in chromatographic methods is **HPLC,** or **high-performance liquid chromatography.** HPLC makes use of high-pressure pumps that speed the movement of the protein molecules down the column, as well as higher-quality chromatographic materials that can withstand the crushing force of the pressurized flow. By reducing the transit time on the column, HPLC can limit diffusional spreading of protein bands and thus greatly improve resolution.

The approach to purification of a protein that has not previously been isolated is guided both by established precedents and by common sense. In most cases, several different methods must be used sequentially to purify a protein completely. The choice of method is

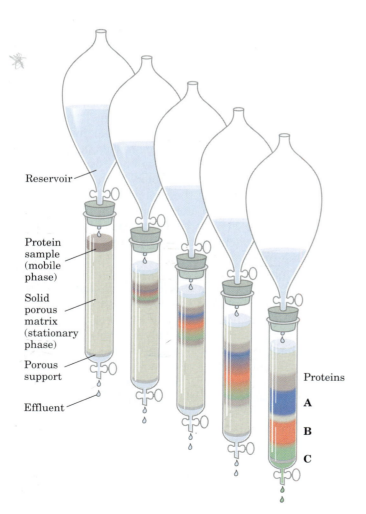

Reservoir

Protein sample (mobile phase)

Solid porous matrix (stationary phase)

Porous support

Effluent

Proteins

A

B

C

**FIGURE 3–17   Column chromatography.** The standard elements of a chromatographic column include a solid, porous material supported inside a column, generally made of plastic or glass. The solid material (matrix) makes up the stationary phase through which flows a solution, the mobile phase. The solution that passes out of the column at the bottom (the effluent) is constantly replaced by solution supplied from a reservoir at the top. The protein solution to be separated is layered on top of the column and allowed to percolate into the solid matrix. Additional solution is added on top. The protein solution forms a band within the mobile phase that is initially the depth of the protein solution applied to the column. As proteins migrate through the column, they are retarded to different degrees by their different interactions with the matrix material. The overall protein band thus widens as it moves through the column. Individual types of proteins (such as A, B, and C, shown in blue, red, and green) gradually separate from each other, forming bands within the broader protein band. Separation improves (resolution increases) as the length of the column increases. However, each individual protein band also broadens with time due to diffusional spreading, a process that decreases resolution. In this example, protein A is well separated from B and C, but diffusional spreading prevents complete separation of B and C under these conditions.

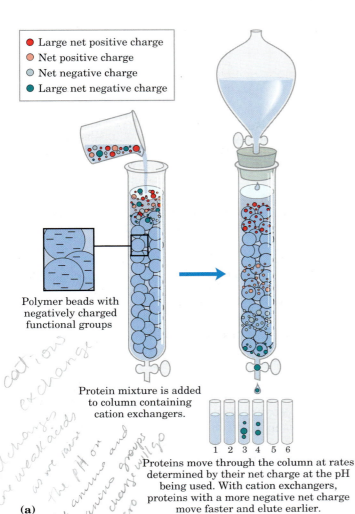

Large net positive charge
Net positive charge
Net negative charge
Large net negative charge

Polymer beads with negatively charged functional groups

Protein mixture is added to column containing cation exchangers.

1  2  3  4  5  6

Proteins move through the column at rates determined by their net charge at the pH being used. With cation exchangers, proteins with a more negative net charge move faster and elute earlier.

**(a)**

*[handwritten marginal notes:] cation exchanges all charges are weak acids as we raise the pH or the amino and guanino groups and their charge will go to zero*

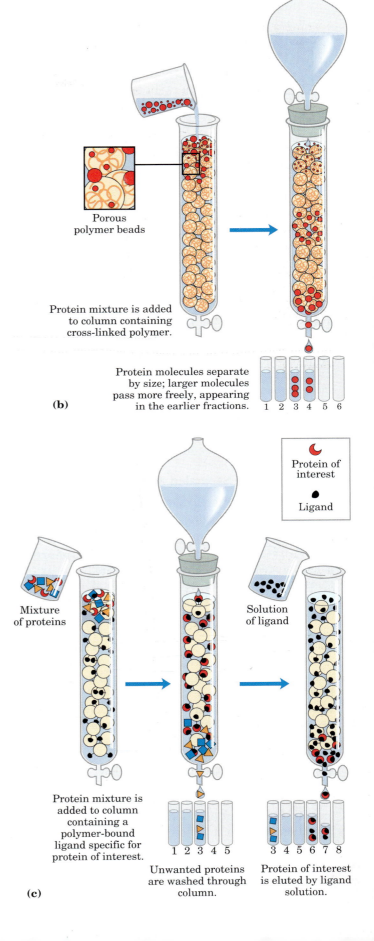

Porous polymer beads

Protein mixture is added to column containing cross-linked polymer.

Protein molecules separate by size; larger molecules pass more freely, appearing in the earlier fractions.

1  2  3  4  5  6

**(b)**

Protein of interest

Ligand

Mixture of proteins

Solution of ligand

Protein mixture is added to column containing a polymer-bound ligand specific for protein of interest.

Unwanted proteins are washed through column.

1  2  3  4  5

Protein of interest is eluted by ligand solution.

3  4  5  6  7  8

**(c)**

**FIGURE 3–18 Three chromatographic methods used in protein purification. (a) Ion-exchange chromatography** exploits differences in the sign and magnitude of the net electric charges of proteins at a given pH. The column matrix is a synthetic polymer containing bound charged groups; those with bound anionic groups are called **cation exchangers,** and those with bound cationic groups are called **anion exchangers.** Ion-exchange chromatography on a cation exchanger is shown here. The affinity of each protein for the charged groups on the column is affected by the pH (which determines the ionization state of the molecule) and the concentration of competing free salt ions in the surrounding solution. Separation can be optimized by gradually changing the pH and/or salt concentration of the mobile phase so as to create a pH or salt gradient. **(b) Size-exclusion chromatography,** also called gel filtration, separates proteins according to size. The column matrix is a cross-linked polymer with pores of selected size. Larger proteins migrate faster than smaller ones, because they are too large to enter the pores in the beads and hence take a more direct route through the column. The smaller proteins enter the pores and are slowed by their more labyrinthine path through the column. **(c) Affinity chromatography** separates proteins by their binding specificities. The proteins retained on the column are those that bind specifically to a ligand cross-linked to the beads. (In biochemistry, the term "ligand" is used to refer to a group or molecule that binds to a macromolecule such as a protein.) After proteins that do not bind to the ligand are washed through the column, the bound protein of particular interest is eluted (washed out of the column) by a solution containing free ligand.

## TABLE 3-5   A Purification Table for a Hypothetical Enzyme

| Procedure or step | Fraction volume (ml) | Total protein (mg) | Activity (units) | Specific activity (units/mg) |
|---|---|---|---|---|
| 1. Crude cellular extract | 1,400 | 10,000 | 100,000 | 10 |
| 2. Precipitation with ammonium sulfate | 280 | 3,000 | 96,000 | 32 |
| 3. Ion-exchange chromatography | 90 | 400 | 80,000 | 200 |
| 4. Size-exclusion chromatography | 80 | 100 | 60,000 | 600 |
| 5. Affinity chromatography | 6 | 3 | 45,000 | 15,000 |

**Note:** All data represent the status of the sample *after* the designated procedure has been carried out. Activity and specific activity are defined on page 94.

somewhat empirical, and many protocols may be tried before the most effective one is found. Trial and error can often be minimized by basing the procedure on purification techniques developed for similar proteins. Published purification protocols are available for many thousands of proteins. Common sense dictates that inexpensive procedures such as salting out be used first, when the total volume and the number of contaminants are greatest. Chromatographic methods are often impractical at early stages, because the amount of chromatographic medium needed increases with sample size. As each purification step is completed, the sample size generally becomes smaller (Table 3–5), making it feasible to use more sophisticated (and expensive) chromatographic procedures at later stages.

### Proteins Can Be Separated and Characterized by Electrophoresis

Another important technique for the separation of proteins is based on the migration of charged proteins in an electric field, a process called **electrophoresis.** These procedures are not generally used to purify proteins in large amounts, because simpler alternatives are usually available and electrophoretic methods often adversely affect the structure and thus the function of proteins. Electrophoresis is, however, especially useful as an analytical method. Its advantage is that proteins can be visualized as well as separated, permitting a researcher to estimate quickly the number of different proteins in a mixture or the degree of purity of a particular protein preparation. Also, electrophoresis allows determination of crucial properties of a protein such as its isoelectric point and approximate molecular weight.

Electrophoresis of proteins is generally carried out in gels made up of the cross-linked polymer polyacrylamide (Fig. 3–19). The polyacrylamide gel acts as a molecular sieve, slowing the migration of proteins approximately in proportion to their charge-to-mass ratio. Migration may also be affected by protein shape. In electrophoresis, the force moving the macromolecule is the electrical potential, $E$. The electrophoretic mobility of the molecule, $\mu$, is the ratio of the velocity of the par-

ticle molecule, $V$, to the electrical potential. Electrophoretic mobility is also equal to the net charge of the molecule, $Z$, divided by the frictional coefficient, $f$, which reflects in part a protein's shape. Thus:

$$\mu = \frac{V}{E} = \frac{Z}{f}$$

The migration of a protein in a gel during electrophoresis is therefore a function of its size and its shape.

An electrophoretic method commonly employed for estimation of purity and molecular weight makes use of the detergent **sodium dodecyl sulfate (SDS).**

$$Na^+ \, {}^-O - \overset{\overset{\displaystyle O}{\|}}{\underset{\underset{\displaystyle O}{\|}}{S}} - O - (CH_2)_{11}CH_3$$

Sodium dodecyl sulfate
(SDS)

SDS binds to most proteins in amounts roughly proportional to the molecular weight of the protein, about one molecule of SDS for every two amino acid residues. The bound SDS contributes a large net negative charge, rendering the intrinsic charge of the protein insignificant and conferring on each protein a similar charge-to-mass ratio. In addition, the native conformation of a protein is altered when SDS is bound, and most proteins assume a similar shape. Electrophoresis in the presence of SDS therefore separates proteins almost exclusively on the basis of mass (molecular weight), with smaller polypeptides migrating more rapidly. After electrophoresis, the proteins are visualized by adding a dye such as Coomassie blue, which binds to proteins but not to the gel itself (Fig. 3–19b). Thus, a researcher can monitor the progress of a protein purification procedure as the number of protein bands visible on the gel decreases after each new fractionation step. When compared with the positions to which proteins of known molecular weight migrate in the gel, the position of an unidentified protein can provide an excellent measure of its molecular weight (Fig. 3–20). If the protein has two or more different subunits, the subunits will generally be separated by the SDS treatment and a separate band will appear for each. 🖱 SDS Gel Electrophoresis

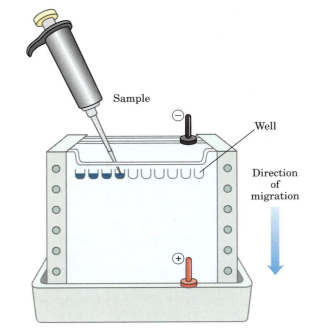

**(a)**

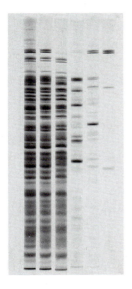

**(b)**

**FIGURE 3–19 Electrophoresis. (a)** Different samples are loaded in wells or depressions at the top of the polyacrylamide gel. The proteins move into the gel when an electric field is applied. The gel minimizes convection currents caused by small temperature gradients, as well as protein movements other than those induced by the electric field. **(b)** Proteins can be visualized after electrophoresis by treating the gel with a stain such as Coomassie blue, which binds to the proteins but not to the gel itself. Each band on the gel represents a different pro-

tein (or protein subunit); smaller proteins move through the gel more rapidly than larger proteins and therefore are found nearer the bottom of the gel. This gel illustrates the purification of the enzyme RNA polymerase from *E. coli*. The first lane shows the proteins present in the crude cellular extract. Successive lanes (left to right) show the proteins present after each purification step. The purified protein contains four subunits, as seen in the last lane on the right.

**Isoelectric focusing** is a procedure used to determine the isoelectric point (pI) of a protein (Fig. 3–21). A pH gradient is established by allowing a mixture of low molecular weight organic acids and bases (ampholytes; p. 81) to distribute themselves in an electric field generated across the gel. When a protein mix-

ture is applied, each protein migrates until it reaches the pH that matches its pI (Table 3–6). Proteins with different isoelectric points are thus distributed differently throughout the gel.

Combining isoelectric focusing and SDS electrophoresis sequentially in a process called **two-dimensional**

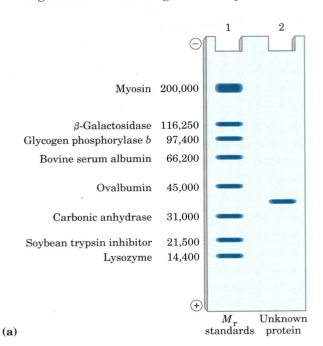

**(a)**

| | $M_r$ |
|---|---|
| Myosin | 200,000 |
| β-Galactosidase | 116,250 |
| Glycogen phosphorylase *b* | 97,400 |
| Bovine serum albumin | 66,200 |
| Ovalbumin | 45,000 |
| Carbonic anhydrase | 31,000 |
| Soybean trypsin inhibitor | 21,500 |
| Lysozyme | 14,400 |

$M_r$ standards   Unknown protein

**FIGURE 3–20 Estimating the molecular weight of a protein.** The electrophoretic mobility of a protein on an SDS polyacrylamide gel is related to its molecular weight, $M_r$. **(a)** Standard proteins of known molecular weight are subjected to electrophoresis (lane 1). These marker proteins can be used to estimate the molecular weight of an unknown protein (lane 2). **(b)** A plot of log $M_r$ of the marker proteins versus relative migration during electrophoresis is linear, which allows the molecular weight of the unknown protein to be read from the graph.

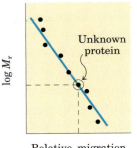

**(b)**   Relative migration

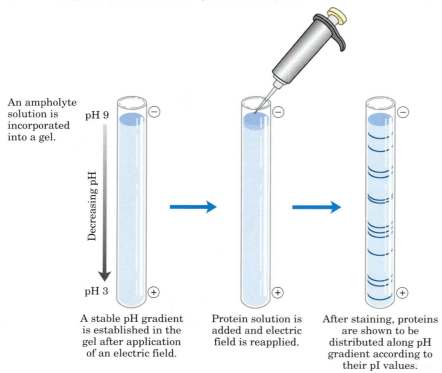

An ampholyte solution is incorporated into a gel.

Decreasing pH

pH 9

pH 3

A stable pH gradient is established in the gel after application of an electric field.

Protein solution is added and electric field is reapplied.

After staining, proteins are shown to be distributed along pH gradient according to their pI values.

**FIGURE 3–21 Isoelectric focusing.** This technique separates proteins according to their isoelectric points. A stable pH gradient is established in the gel by the addition of appropriate ampholytes. A protein mixture is placed in a well on the gel. With an applied electric field, proteins enter the gel and migrate until each reaches a pH equivalent to its pI. Remember that when pH = pI, the net charge of a protein is zero.

**electrophoresis** permits the resolution of complex mixtures of proteins (Fig. 3–22). This is a more sensitive analytical method than either electrophoretic method alone. Two-dimensional electrophoresis separates proteins of identical molecular weight that differ in pI, or proteins with similar pI values but different molecular weights.

## Unseparated Proteins Can Be Quantified

To purify a protein, it is essential to have a way of detecting and quantifying that protein in the presence of many other proteins at each stage of the procedure. Often, purification must proceed in the absence of any information about the size and physical properties of the protein or about the fraction of the total protein mass it represents in the extract. For proteins that are enzymes, the amount in a given solution or tissue extract can be measured, or assayed, in terms of the catalytic effect the enzyme produces—that is, the *increase* in the rate at which its substrate is converted to reaction products when the enzyme is present. For this purpose one must know (1) the overall equation of the reaction catalyzed, (2) an analytical procedure for determining the disappearance of the substrate or the appearance of a reaction product, (3) whether the enzyme requires cofactors such as metal ions or coenzymes, (4) the dependence of the enzyme activity on substrate concentration, (5) the optimum pH, and (6) a temperature zone in which the enzyme is stable and has high activity. Enzymes are usually assayed at their optimum pH and at some convenient temperature within the range

25 to 38 °C. Also, very high substrate concentrations are generally used so that the initial reaction rate, measured experimentally, is proportional to enzyme concentration (Chapter 6).

By international agreement, 1.0 unit of enzyme activity is defined as the amount of enzyme causing transformation of 1.0 $\mu$mol of substrate per minute at 25 °C under optimal conditions of measurement. The term **activity** refers to the total units of enzyme in a solution. The **specific activity** is the number of enzyme units per milligram of total protein (Fig. 3–23). The specific activity is a measure of enzyme purity: it increases during purification of an enzyme and becomes maximal and constant when the enzyme is pure (Table 3–5).

**TABLE 3–6    The Isoelectric Points of Some Proteins**

| Protein | pI |
|---|---|
| Pepsin | <1.0 |
| Egg albumin | 4.6 |
| Serum albumin | 4.9 |
| Urease | 5.0 |
| $\beta$-Lactoglobulin | 5.2 |
| Hemoglobin | 6.8 |
| Myoglobin | 7.0 |
| Chymotrypsinogen | 9.5 |
| Cytochrome $c$ | 10.7 |
| Lysozyme | 11.0 |

**First dimension**

Isoelectric focusing

Decreasing pI

Isoelectric focusing gel is placed on SDS polyacrylamide gel.

**Second dimension**

SDS polyacrylamide gel electrophoresis

Decreasing $M_r$

Decreasing pI

**(a)**

**(b)**

**FIGURE 3–22  Two-dimensional electrophoresis. (a)** Proteins are first separated by isoelectric focusing in a cylindrical gel. The gel is then laid horizontally on a second, slab-shaped gel, and the proteins are separated by SDS polyacrylamide gel electrophoresis. Horizontal separation reflects differences in pI; vertical separation reflects differences in molecular weight. **(b)** More than 1,000 different proteins from *E. coli* can be resolved using this technique.

when further purification steps fail to increase specific activity and when only a single protein species can be detected (for example, by electrophoresis).

For proteins that are not enzymes, other quantification methods are required. Transport proteins can be assayed by their binding to the molecule they transport, and hormones and toxins by the biological effect they produce; for example, growth hormones will stimulate the growth of certain cultured cells. Some structural proteins represent such a large fraction of a tissue mass that they can be readily extracted and purified without a functional assay. The approaches are as varied as the proteins themselves.

After each purification step, the activity of the preparation (in units of enzyme activity) is assayed, the total amount of protein is determined independently, and the ratio of the two gives the specific activity. Activity and total protein generally decrease with each step. Activity decreases because some loss always occurs due to inactivation or nonideal interactions with chromatographic materials or other molecules in the solution. Total protein decreases because the objective is to remove as much unwanted or nonspecific protein as possible. In a successful step, the loss of nonspecific protein is much greater than the loss of activity; therefore, specific activity increases even as total activity falls. The data are then assembled in a purification table similar to Table 3–5. A protein is generally considered pure

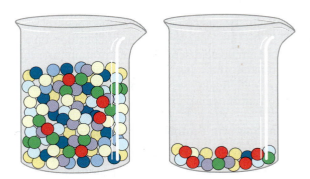

**FIGURE 3–23  Activity versus specific activity.** The difference between these two terms can be illustrated by considering two beakers of marbles. The beakers contain the same number of red marbles, but different numbers of marbles of other colors. If the marbles represent proteins, both beakers contain the same *activity* of the protein represented by the red marbles. The second beaker, however, has the higher *specific activity* because here the red marbles represent a much higher fraction of the total.

## SUMMARY 3.3 Working with Proteins

- Proteins are separated and purified by taking advantage of differences in their properties. Proteins can be selectively precipitated by the addition of certain salts. A wide range of chromatographic procedures makes use of differences in size, binding affinities, charge, and other properties. These include ion-exchange, size-exclusion, affinity, and high-performance liquid chromatography.

- Electrophoresis separates proteins on the basis of mass or charge. SDS gel electrophoresis and isoelectric focusing can be used separately or in combination for higher resolution.

- All purification procedures require a method for quantifying or assaying the protein of interest in the presence of other proteins. Purification can be monitored by assaying specific activity.

## 3.4 The Covalent Structure of Proteins

Purification of a protein is usually only a prelude to a detailed biochemical dissection of its structure and function. What is it that makes one protein an enzyme, another a hormone, another a structural protein, and still another an antibody? How do they differ chemically? The most obvious distinctions are structural, and these distinctions can be approached at every level of structure defined in Figure 3–16.

The differences in primary structure can be especially informative. Each protein has a distinctive number and sequence of amino acid residues. As we shall see in Chapter 4, the primary structure of a protein determines how it folds up into a unique three-dimensional structure, and this in turn determines the function of the protein. Primary structure is the focus of the remainder of this chapter. We first consider empirical clues that amino acid sequence and protein function are closely linked, then describe how amino acid sequence is determined; finally, we outline the many uses to which this information can be put.

### The Function of a Protein Depends on Its Amino Acid Sequence

The bacterium *Escherichia coli* produces more than 3,000 different proteins; a human produces 25,000 to 35,000. In both cases, each type of protein has a unique three-dimensional structure and this structure confers a unique function. Each type of protein also has a unique amino acid sequence. Intuition suggests that the amino acid sequence must play a fundamental role in determining the three-dimensional structure of the protein, and ultimately its function, but is this supposition cor-

rect? A quick survey of proteins and how they vary in amino acid sequence provides a number of empirical clues that help substantiate the important relationship between amino acid sequence and biological function.

First, as we have already noted, proteins with different functions always have different amino acid sequences. Second, thousands of human genetic diseases have been traced to the production of defective proteins. Perhaps one-third of these proteins are defective because of a single change in their amino acid sequence; hence, if the primary structure is altered, the function of the protein may also be changed. Finally, on comparing functionally similar proteins from different species, we find that these proteins often have similar amino acid sequences. An extreme case is ubiquitin, a 76-residue protein involved in regulating the degradation of other proteins. The amino acid sequence of ubiquitin is identical in species as disparate as fruit flies and humans.

Is the amino acid sequence absolutely fixed, or invariant, for a particular protein? No; some flexibility is possible. An estimated 20% to 30% of the proteins in humans are **polymorphic,** having amino acid sequence variants in the human population. Many of these variations in sequence have little or no effect on the function of the protein. Furthermore, proteins that carry out a broadly similar function in distantly related species can differ greatly in overall size and amino acid sequence.

Although the amino acid sequence in some regions of the primary structure might vary considerably without affecting biological function, most proteins contain crucial regions that are essential to their function and whose sequence is therefore conserved. The fraction of the overall sequence that is critical varies from protein to protein, complicating the task of relating sequence to three-dimensional structure, and structure to function. Before we can consider this problem further, however, we must examine how sequence information is obtained.

### The Amino Acid Sequences of Millions of Proteins Have Been Determined

Two major discoveries in 1953 were of crucial importance in the history of biochemistry. In that year James D. Watson and Francis Crick deduced the double-helical structure of DNA and proposed a structural basis for its precise replication (Chapter 8). Their proposal illuminated the molecular reality behind the idea of a gene. In that same year, Frederick Sanger worked out the sequence of amino acid residues in the polypeptide chains of the hormone insulin (Fig. 3–24), surprising many researchers who had long thought that elucidation of the amino acid sequence of a polypeptide would be a hopelessly difficult task. It quickly became evident that the nucleotide sequence in DNA and the amino acid sequence in proteins were somehow related. Barely a decade after these discoveries, the role of the nucleotide sequence of DNA in determining the amino acid sequence of protein molecules was revealed (Chapter 27). An enormous number of protein sequences can now be derived indirectly from the DNA sequences in the rapidly growing genome databases. However, many are still deduced by traditional methods of polypeptide sequencing.

The amino acid sequences of thousands of different proteins from many species have been determined using principles first developed by Sanger. These methods are still in use, although with many variations and improvements in detail. Chemical protein sequencing now

Frederick Sanger

complements a growing list of newer methods, providing multiple avenues to obtain amino acid sequence data. Such data are now critical to every area of biochemical investigation.

## Short Polypeptides Are Sequenced Using Automated Procedures

Various procedures are used to analyze protein primary structure. Several protocols are available to label and identify the amino-terminal amino acid residue (Fig. 3–25a). Sanger developed the reagent 1-fluoro-2,4-dinitrobenzene (FDNB) for this purpose; other reagents used to label the amino-terminal residue, dansyl chloride and dabsyl chloride, yield derivatives that are more easily detectable than the dinitrophenyl derivatives. After the amino-terminal residue is labeled with one of these reagents, the polypeptide is hydrolyzed to its constituent amino acids and the labeled amino acid is identified. Because the hydrolysis stage destroys the polypeptide, this procedure cannot be used to sequence a polypeptide beyond its amino-terminal residue. However, it can help determine the number of chemically distinct polypeptides in a protein, provided each has a different amino-terminal residue. For example, two residues—Phe and Gly—would be labeled if insulin (Fig. 3–24) were subjected to this procedure.

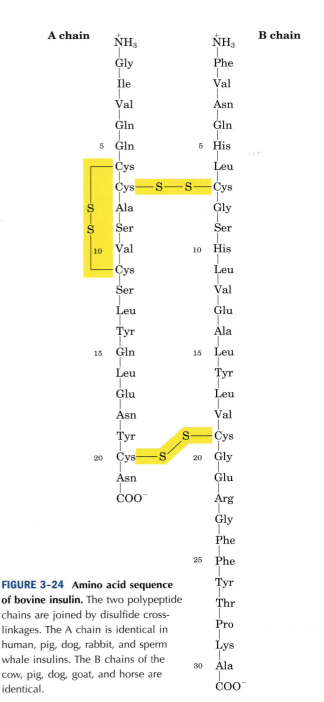

**FIGURE 3–24  Amino acid sequence of bovine insulin.** The two polypeptide chains are joined by disulfide cross-linkages. The A chain is identical in human, pig, dog, rabbit, and sperm whale insulins. The B chains of the cow, pig, dog, goat, and horse are identical.

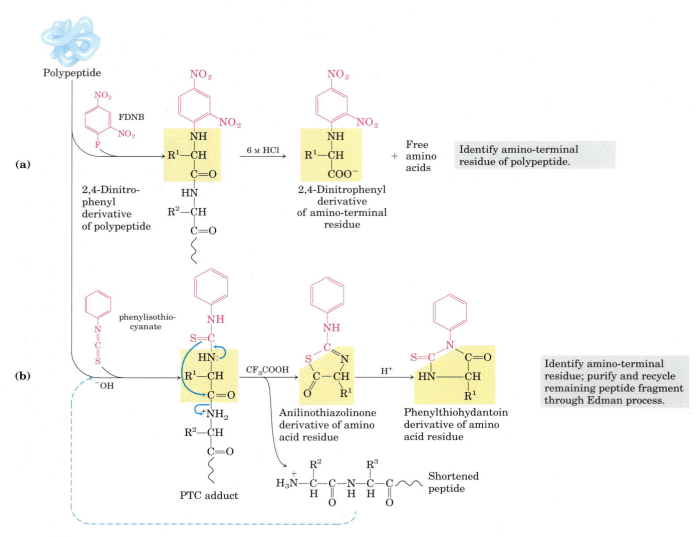

**FIGURE 3–25    Steps in sequencing a polypeptide. (a)** Identification of the amino-terminal residue can be the first step in sequencing a polypeptide. Sanger's method for identifying the amino-terminal residue is shown here. **(b)** The Edman degradation procedure reveals the entire sequence of a peptide. For shorter peptides, this method alone readily yields the entire sequence, and step **(a)** is often omitted. Step **(a)** is useful in the case of larger polypeptides, which are often fragmented into smaller peptides for sequencing (see Fig. 3–27).

To sequence an entire polypeptide, a chemical method devised by Pehr Edman is usually employed. The **Edman degradation** procedure labels and removes only the amino-terminal residue from a peptide, leaving all other peptide bonds intact (Fig. 3–25b). The peptide is reacted with phenylisothiocyanate under mildly alkaline conditions, which converts the amino-terminal amino acid to a phenylthiocarbamoyl (PTC) adduct. The peptide bond next to the PTC adduct is then cleaved in a step carried out in anhydrous trifluoroacetic acid, with removal of the amino-terminal amino acid as an anilinothiazolinone derivative. The derivatized amino acid is extracted with organic solvents, converted to the more stable phenylthiohydantoin derivative by treatment with aqueous acid, and then identified. The use of sequential reactions carried out under first basic and then acidic conditions provides control over the entire process. Each reaction with the amino-terminal amino acid can go essentially to completion without affecting any of the other peptide bonds in the peptide. After removal and identification of the amino-terminal residue, the *new* amino-terminal residue so exposed can be labeled, removed, and identified through the same series of reactions. This procedure is repeated until the entire sequence is determined. The Edman degradation is carried out on a machine, called a **sequenator,** that mixes reagents in the proper proportions, separates the products, identifies them, and records the results. These methods are extremely sensitive. Often, the complete amino acid sequence can be determined starting with only a few micrograms of protein.

The length of polypeptide that can be accurately sequenced by the Edman degradation depends on the efficiency of the individual chemical steps. Consider a peptide beginning with the sequence Gly–Pro–Lys– at its amino terminus. If glycine were removed with 97% efficiency, 3% of the polypeptide molecules in the solution would retain a Gly residue at their amino terminus. In the second Edman cycle, 97% of the liberated amino acids would be proline, and 3% glycine, while 3% of the polypeptide molecules would retain Gly (0.1%) or Pro (2.9%) residues at their amino terminus. At each cycle, peptides that did not react in earlier cycles would contribute amino acids to an ever-increasing background, eventually making it impossible to determine which amino acid is next in the original peptide sequence. Modern sequenators achieve efficiencies of better than 99% per cycle, permitting the sequencing of more than 50 contiguous amino acid residues in a polypeptide. The primary structure of insulin, worked out by Sanger and colleagues over a period of 10 years, could now be completely determined in a day or two.

## Large Proteins Must Be Sequenced in Smaller Segments

The overall accuracy of amino acid sequencing generally declines as the length of the polypeptide increases. The very large polypeptides found in proteins must be broken down into smaller pieces to be sequenced efficiently. There are several steps in this process. First, the protein is cleaved into a set of specific fragments by chemical or enzymatic methods. If any disulfide bonds are present, they must be broken. Each fragment is purified, then sequenced by the Edman procedure. Finally, the order in which the fragments appear in the original protein is determined and disulfide bonds (if any) are located.

**Breaking Disulfide Bonds**  Disulfide bonds interfere with the sequencing procedure. A cystine residue (Fig. 3–7) that has one of its peptide bonds cleaved by the Edman procedure may remain attached to another polypeptide strand via its disulfide bond. Disulfide bonds also interfere with the enzymatic or chemical cleavage of the polypeptide. Two approaches to irreversible breakage of disulfide bonds are outlined in Figure 3–26.

**Cleaving the Polypeptide Chain**  Several methods can be used for fragmenting the polypeptide chain. Enzymes called **proteases** catalyze the hydrolytic cleavage of peptide bonds. Some proteases cleave only the peptide bond adjacent to particular amino acid residues (Table 3–7) and thus fragment a polypeptide chain in a predictable and reproducible way. A number of chemical reagents also cleave the peptide bond adjacent to specific residues.

Among proteases, the digestive enzyme trypsin catalyzes the hydrolysis of only those peptide bonds in which the carbonyl group is contributed by either a Lys or an Arg residue, regardless of the length or amino acid sequence of the chain. The number of smaller peptides produced by trypsin cleavage can thus be predicted

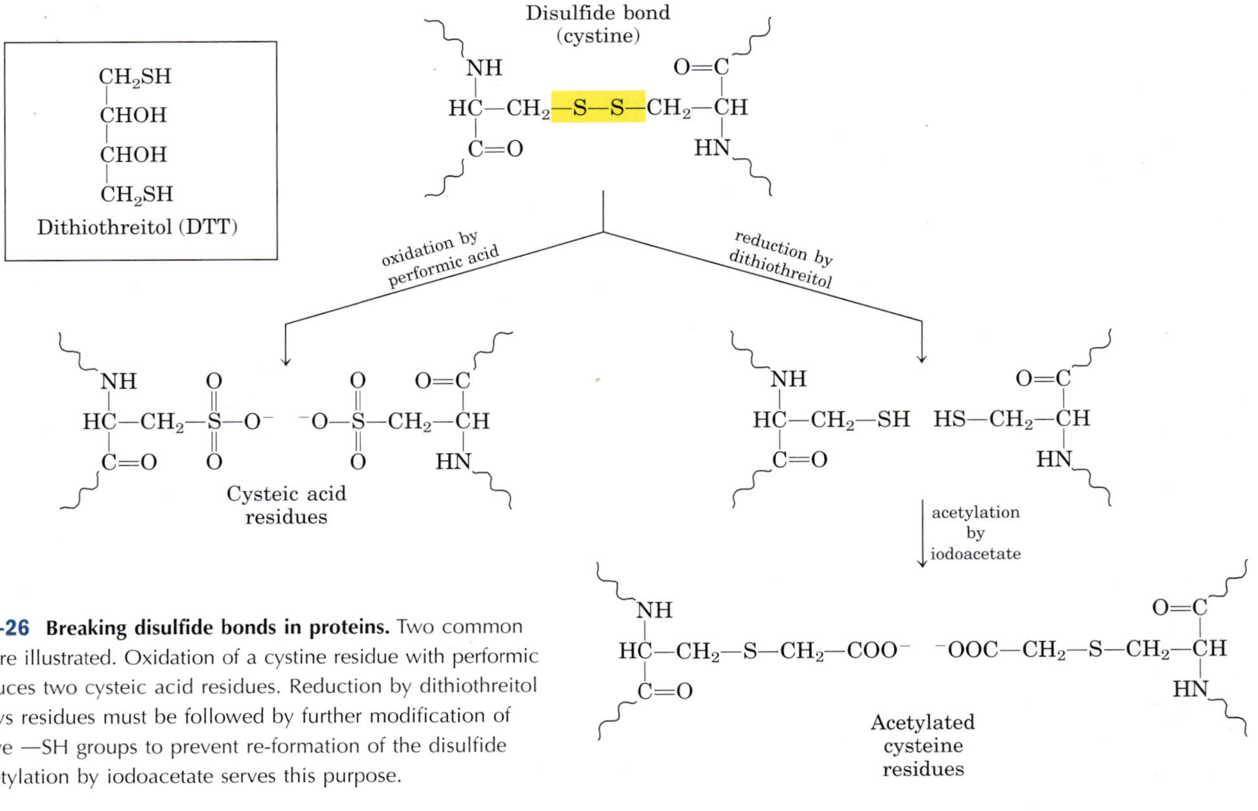

**FIGURE 3–26 Breaking disulfide bonds in proteins.** Two common methods are illustrated. Oxidation of a cystine residue with performic acid produces two cysteic acid residues. Reduction by dithiothreitol to form Cys residues must be followed by further modification of the reactive —SH groups to prevent re-formation of the disulfide bond. Acetylation by iodoacetate serves this purpose.

**TABLE 3–7   The Specificity of Some Common Methods for Fragmenting Polypeptide Chains**

| Reagent (biological source)* | Cleavage points† |
|---|---|
| Trypsin (bovine pancreas) | Lys, Arg (C) |
| Submaxillarus protease (mouse submaxillary gland) | Arg (C) |
| Chymotrypsin (bovine pancreas) | Phe, Trp, Tyr (C) |
| Staphylococcus aureus V8 protease (bacterium S. aureus) | Asp, Glu (C) |
| Asp-N-protease (bacterium Pseudomonas fragi) | Asp, Glu (N) |
| Pepsin (porcine stomach) | Phe, Trp, Tyr (N) |
| Endoproteinase Lys C (bacterium Lysobacter enzymogenes) | Lys (C) |
| Cyanogen bromide | Met (C) |

*All reagents except cyanogen bromide are proteases. All are available from commercial sources.

†Residues furnishing the primary recognition point for the protease or reagent; peptide bond cleavage occurs on either the carbonyl (C) or the amino (N) side of the indicated amino acid residues.

from the total number of Lys or Arg residues in the original polypeptide, as determined by hydrolysis of an intact sample (Fig. 3–27). A polypeptide with five Lys and/or Arg residues will usually yield six smaller peptides on cleavage with trypsin. Moreover, all except one of these will have a carboxyl-terminal Lys or Arg. The fragments produced by trypsin (or other enzyme or chemical) action are then separated by chromatographic or electrophoretic methods.

**Sequencing of Peptides**   Each peptide fragment resulting from the action of trypsin is sequenced separately by the Edman procedure.

**Ordering Peptide Fragments**   The order of the "trypsin fragments" in the original polypeptide chain must now be determined. Another sample of the intact polypeptide is cleaved into fragments using a different enzyme or reagent, one that cleaves peptide bonds at points other than those cleaved by trypsin. For example, cyanogen bromide cleaves only those peptide bonds in which the carbonyl group is contributed by Met. The fragments resulting from this second procedure are then separated and sequenced as before.

The amino acid sequences of each fragment obtained by the two cleavage procedures are examined, with the objective of finding peptides from the second procedure whose sequences establish continuity, be-

cause of overlaps, between the fragments obtained by the first cleavage procedure (Fig. 3–27). Overlapping peptides obtained from the second fragmentation yield the correct order of the peptide fragments produced in the first. If the amino-terminal amino acid has been identified before the original cleavage of the protein, this information can be used to establish which fragment is derived from the amino terminus. The two sets of fragments can be compared for possible errors in determining the amino acid sequence of each fragment. If the second cleavage procedure fails to establish continuity between all peptides from the first cleavage, a third or even a fourth cleavage method must be used to obtain a set of peptides that can provide the necessary overlap(s).

**Locating Disulfide Bonds**   If the primary structure includes disulfide bonds, their locations are determined in an additional step after sequencing is completed. A sample of the protein is again cleaved with a reagent such as trypsin, this time without first breaking the disulfide bonds. The resulting peptides are separated by electrophoresis and compared with the original set of peptides generated by trypsin. For each disulfide bond, two of the original peptides will be missing and a new, larger peptide will appear. The two missing peptides represent the regions of the intact polypeptide that are linked by the disulfide bond.

## Amino Acid Sequences Can Also Be Deduced by Other Methods

The approach outlined above is not the only way to determine amino acid sequences. New methods based on mass spectrometry permit the sequencing of short polypeptides (20 to 30 amino acid residues) in just a few minutes (Box 3–2). In addition, with the development of rapid DNA sequencing methods (Chapter 8), the elucidation of the genetic code (Chapter 27), and the development of techniques for isolating genes (Chapter 9), researchers can deduce the sequence of a polypeptide by determining the sequence of nucleotides in the gene that codes for it (Fig. 3–28). The techniques used to determine protein and DNA sequences are complementary. When the gene is available, sequencing the DNA can be faster and more accurate than sequencing the protein. Most proteins are now sequenced in this indirect way. If the gene has not been isolated, direct sequencing of peptides is necessary, and this can provide information (the location of disulfide bonds, for example) not available in a DNA sequence. In addition, a knowledge of the amino acid sequence of even a part of a polypeptide can greatly facilitate the isolation of the corresponding gene (Chapter 9).

The array of methods now available to analyze both proteins and nucleic acids is ushering in a new disci-

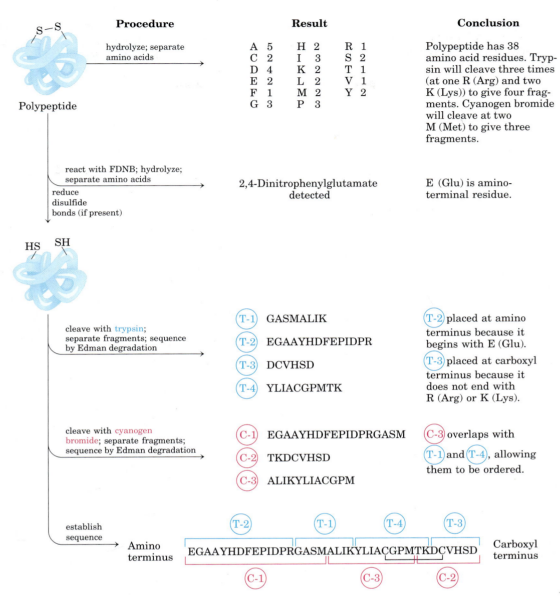

**FIGURE 3–27 Cleaving proteins and sequencing and ordering the peptide fragments.** First, the amino acid composition and amino-terminal residue of an intact sample are determined. Then any disulfide bonds are broken before fragmenting so that sequencing can proceed efficiently. In this example, there are only two Cys (C) residues and thus only one possibility for location of the disulfide bond. In polypeptides with three or more Cys residues, the position of disulfide bonds can be determined as described in the text. (The one-letter symbols for amino, acids are given in Table 3–1.)

pline of "whole cell biochemistry." The complete sequence of an organism's DNA, its genome, is now available for organisms ranging from viruses to bacteria to multicellular eukaryotes (see Table 1–4). Genes are being discovered by the millions, including many that encode proteins with no known function. To describe the entire protein complement encoded by an organism's DNA, researchers have coined the term **proteome.** As described in Chapter 9, the new disciplines of genomics and proteomics are complementing work carried out on cellular intermediary metabolism and nucleic acid

metabolism to provide a new and increasingly complete picture of biochemistry at the level of cells and even organisms.

| Amino acid sequence (protein) | Gln–Tyr–Pro–Thr–Ile–Trp |
|---|---|
| DNA sequence (gene) | CAGTATCCTACGATTTGG |

**FIGURE 3–28 Correspondence of DNA and amino acid sequences.** Each amino acid is encoded by a specific sequence of three nucleotides in DNA. The genetic code is described in detail in Chapter 27.

## BOX 3–2 WORKING IN BIOCHEMISTRY

### Investigating Proteins with Mass Spectrometry

The mass spectrometer has long been an indispensable tool in chemistry. Molecules to be analyzed, referred to as **analytes,** are first ionized in a vacuum. When the newly charged molecules are introduced into an electric and/or magnetic field, their paths through the field are a function of their mass-to-charge ratio, $m/z$. This measured property of the ionized species can be used to deduce the mass ($M$) of the analyte with very high precision.

Although mass spectrometry has been in use for many years, it could not be applied to macromolecules such as proteins and nucleic acids. The $m/z$ measurements are made on molecules in the gas phase, and the heating or other treatment needed to transfer a macromolecule to the gas phase usually caused its rapid decomposition. In 1988, two different techniques were developed to overcome this problem. In one, proteins are placed in a light-absorbing matrix. With a short pulse of laser light, the proteins are ionized and then desorbed from the matrix into the vacuum system. This process, known as **matrix-assisted laser desorption/ionization mass spectrometry,** or **MALDI MS,** has been successfully used to measure the mass of a wide range of macromolecules. In a second and equally successful method, macromolecules in solution are forced directly from the liquid to gas phase. A solution of analytes is passed through a charged needle that is kept at a high electrical potential, dispersing the solution into a fine mist of charged microdroplets. The solvent surrounding the macromolecules rapidly evaporates, and the resulting multiply charged macromolecular ions are thus introduced nondestructively into the gas phase. This technique is called **electrospray ionization mass spectrometry,** or **ESI MS.** Protons added during passage through the needle give additional charge to the macromolecule. The $m/z$ of the molecule can be analyzed in the vacuum chamber.

Mass spectrometry provides a wealth of information for proteomics research, enzymology, and protein chemistry in general. The techniques require only miniscule amounts of sample, so they can be readily applied to the small amounts of protein that can be extracted from a two-dimensional electrophoretic gel. The accurately measured molecular mass of a protein is one of the critical parameters in its identification. Once the mass of a protein is accurately known, mass spectrometry is a convenient and accurate method for detecting changes in mass due to the presence of bound cofactors, bound metal ions, covalent modifications, and so on.

The process for determining the molecular mass of a protein with ESI MS is illustrated in Figure 1. As it is injected into the gas phase, a protein acquires a variable number of protons, and thus positive charges, from the solvent. This creates a spectrum of species with different mass-to-charge ratios. Each successive peak corresponds to a species that differs from that

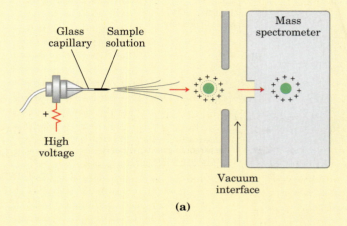

**(a)**

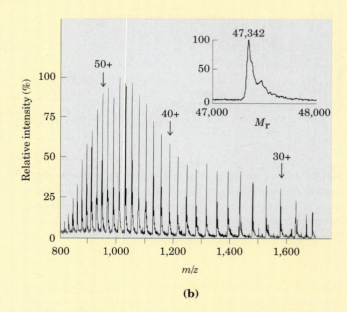

**(b)**

**FIGURE 1 Electrospray mass spectrometry of a protein. (a)** A protein solution is dispersed into highly charged droplets by passage through a needle under the influence of a high-voltage electric field. The droplets evaporate, and the ions (with added protons in this case) enter the mass spectrometer for $m/z$ measurement. The spectrum generated **(b)** is a family of peaks, with each successive peak (from right to left) corresponding to a charged species increased by 1 in both mass and charge. A computer-generated transformation of this spectrum is shown in the inset.

of its neighboring peak by a charge difference of 1 and a mass difference of 1 (1 proton). The mass of the protein can be determined from any two neighboring peaks. The measured $m/z$ of one peak is

$$(m/z)_2 = \frac{M + n_2 X}{n_2}$$

where $M$ is the mass of the protein, $n_2$ is the number of charges, and $X$ is the mass of the added groups (protons in this case). Similarly for the neighboring peak,

$$(m/z)_1 = \frac{M + (n_2 + 1)X}{n_2 + 1}$$

We now have two unknowns ($M$ and $n_2$) and two equations. We can solve first for $n_2$ and then for $M$:

$$n_2 = \frac{(m/z)_2 - X}{(m/z)_2 - (m/z)_1}$$

$$M = n_2\,[(m/z)_2 - X]$$

This calculation using the $m/z$ values for any two peaks in a spectrum such as that shown in Figure 1b usually provides the mass of the protein (in this case, aerolysin k; 47,342 Da) with an error of only ±0.01%. Generating several sets of peaks, repeating the calculation, and averaging the results generally provides an even more accurate value for $M$. Computer algorithms can transform the $m/z$ spectrum into a single peak that

**FIGURE 2 Obtaining protein sequence information with tandem MS. (a)** After proteolytic hydrolysis, a protein solution is injected into a mass spectrometer (MS-1). The different peptides are sorted so that only one type is selected for further analysis. The selected peptide is further fragmented in a chamber between the two mass spectrometers, and $m/z$ for each fragment is measured in the second mass spectrometer (MS-2). Many of the ions generated during this second fragmentation result from breakage of the peptide bond, as shown. These are called b-type or y-type ions, depending on whether the charge is retained on the amino- or carboxyl-terminal side, respectively. **(b)** A typical spectrum with peaks representing the peptide fragments generated from a sample of one small peptide (10 residues). The labeled peaks are y-type ions. The large peak next to $y_5''$ is a doubly charged ion and is not part of the y set. The successive peaks differ by the mass of a particular amino acid in the original peptide. In this case, the deduced sequence was Phe–Pro–Gly–Gln–(Ile/Leu)–Asn–Ala–Asp–(Ile/Leu)–Arg. Note the ambiguity about Ile and Leu residues, because they have the same molecular mass. In this example, the set of peaks derived from y-type ions predominates, and the spectrum is greatly simplified as a result. This is because an Arg residue occurs at the carboxyl terminus of the peptide, and most of the positive charges are retained on this residue.

also provides a very accurate mass measurement (Fig. 1b, inset).

Mass spectrometry can also be used to sequence short stretches of polypeptide, an application that has emerged as an invaluable tool for quickly identifying unknown proteins. Sequence information is extracted using a technique called **tandem MS, or MS/MS.** A solution containing the protein under investigation is first treated with a protease or chemical reagent to hydrolyze it to a mixture of shorter peptides. The mixture is then injected into a device that is essentially two mass spectrometers in tandem (Fig. 2a, top). In the first, the peptide mixture is sorted and the ionized fragments are manipulated so that only one of the several types of peptides produced by cleavage emerges at the other end. The sample of the selected

(continued on next page)

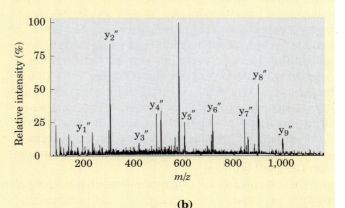

BOX 3–2    WORKING IN BIOCHEMISTRY (continued from previous page)

peptide, each molecule of which has a charge some-where along its length, then travels through a vacuum chamber between the two mass spectrometers. In this collision cell, the peptide is further fragmented by high-energy impact with a "collision gas," a small amount of a noble gas such as helium or argon that is bled into the vacuum chamber. This procedure is de-signed to fragment many of the peptide molecules in the sample, with each individual peptide broken in only one place, on average. Most breaks occur at pep-tide bonds. This fragmentation does not involve the addition of water (it is done in a near-vacuum), so the products may include molecular ion radicals such as carbonyl radicals (Fig. 2a, bottom). The charge on the original peptide is retained on one of the fragments generated from it.

The second mass spectrometer then measures the $m/z$ ratios of all the charged fragments (uncharged fragments are not detected). This generates one or more sets of peaks. A given set of peaks (Fig. 2b) con-sists of all the charged fragments that were generated by breaking the same type of bond (but at different points in the peptide) and are derived from the same side of the bond breakage, either the carboxyl- or amino-terminal side. Each successive peak in a given set has one less amino acid than the peak before. The difference in mass from peak to peak identifies the amino acid that was lost in each case, thus revealing the sequence of the peptide. The only ambiguities in-volve leucine and isoleucine, which have the same mass.

The charge on the peptide can be retained on ei-ther the carboxyl- or amino-terminal fragment, and

bonds other than the peptide bond can be broken in the fragmentation process, with the result that multi-ple sets of peaks are usually generated. The two most prominent sets generally consist of charged fragments derived from breakage of the peptide bonds. The set consisting of the carboxyl-terminal fragments can be unambiguously distinguished from that consisting of the amino-terminal fragments. Because the bond breaks generated between the spectrometers (in the collision cell) do not yield full carboxyl and amino groups at the sites of the breaks, the only intact $\alpha$-amino and $\alpha$-carboxyl groups on the peptide frag-ments are those at the very ends (Fig. 2a). The two sets of fragments can thereby be identified by the re-sulting slight differences in mass. The amino acid se-quence derived from one set can be confirmed by the other, improving the confidence in the sequence in-formation obtained.

Even a short sequence is often enough to permit unambiguous association of a protein with its gene, if the gene sequence is known. Sequencing by mass spectrometry cannot replace the Edman degradation procedure for the sequencing of long polypeptides, but it is ideal for proteomics research aimed at cata-loging the hundreds of cellular proteins that might be separated on a two-dimensional gel. In the coming decades, detailed genomic sequence data will be avail-able from hundreds, eventually thousands, of organ-isms. The ability to rapidly associate proteins with genes using mass spectrometry will greatly facilitate the exploitation of this extraordinary information resource.

## Small Peptides and Proteins Can Be Chemically Synthesized

Many peptides are potentially useful as pharmacologic agents, and their production is of considerable com-mercial importance. There are three ways to obtain a peptide: (1) purification from tissue, a task often made difficult by the vanishingly low concentrations of some peptides; (2) genetic engineering (Chapter 9); or (3) di-rect chemical synthesis. Powerful techniques now make direct chemical synthesis an attractive option in many cases. In addition to commercial applications, the syn-thesis of specific peptide portions of larger proteins is an increasingly important tool for the study of protein structure and function.

The complexity of proteins makes the traditional synthetic approaches of organic chemistry impractical for peptides with more than four or five amino acid

residues. One problem is the difficulty of purifying the product after each step.

The major breakthrough in this technology was provided by R. Bruce Merrifield in 1962. His innovation involved synthesizing a peptide while keeping it at-tached at one end to a solid support. The support is an insoluble polymer (resin) contained within a column, similar to that used for chromatographic procedures. The peptide is built up on this support one amino acid at a time using a standard set of reactions in a repeat-ing cycle (Fig. 3–29). At each successive step in the cycle, protective chemical groups block unwanted reactions.

The technology for chemical peptide synthesis is now automated. As in the sequencing reactions already considered, the most important limitation of the process is the efficiency of each chemical cycle, as can be seen by calculating the overall yields of peptides of various lengths when the yield for addition of each new amino acid is 96.0% versus 99.8% (Table 3–8). Incomplete reaction at one stage can lead to formation of an impurity (in the form of a shorter peptide) in the next. The chemistry has been optimized to permit the synthesis of proteins of 100 amino acid residues in a few days in reasonable yield. A very similar approach is used to synthesize nucleic acids (see Fig. 8–38). It is worth noting that this technology, impressive as it is, still pales when compared with biological processes. The same

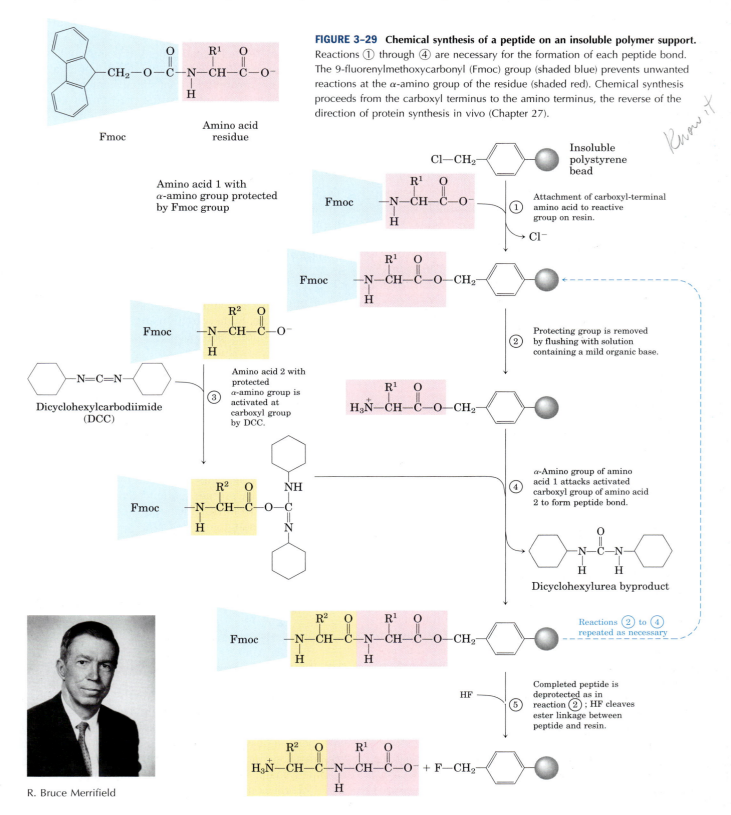

**FIGURE 3-29 Chemical synthesis of a peptide on an insoluble polymer support.** Reactions ① through ④ are necessary for the formation of each peptide bond. The 9-fluorenylmethoxycarbonyl (Fmoc) group (shaded blue) prevents unwanted reactions at the α-amino group of the residue (shaded red). Chemical synthesis proceeds from the carboxyl terminus to the amino terminus, the reverse of the direction of protein synthesis in vivo (Chapter 27).

R. Bruce Merrifield

**TABLE 3-8**    Effect of Stepwise Yield on Overall Yield in Peptide Synthesis

| Number of residues in the final polypeptide | Overall yield of final peptide (%) when the yield of each step is: | |
|---|---|---|
| | 96.0% | 99.8% |
| 11 | 66 | 98 |
| 21 | 44 | 96 |
| 31 | 29 | 94 |
| 51 | 13 | 90 |
| 100 | 1.7 | 82 |

100-amino-acid protein would be synthesized with exquisite fidelity in about 5 seconds in a bacterial cell.

A variety of new methods for the efficient ligation (joining together) of peptides has made possible the assembly of synthetic peptides into larger proteins. With these methods, novel forms of proteins can be created with precisely positioned chemical groups, including those that might not normally be found in a cellular protein. These novel forms provide new ways to test theories of enzyme catalysis, to create proteins with new chemical properties, and to design protein sequences that will fold into particular structures. This last application provides the ultimate test of our increasing ability to relate the primary structure of a peptide to the three-dimensional structure that it takes up in solution.

### Amino Acid Sequences Provide Important Biochemical Information

Knowledge of the sequence of amino acids in a protein can offer insights into its three-dimensional structure and its function, cellular location, and evolution. Most of these insights are derived by searching for similarities with other known sequences. Thousands of sequences are known and available in databases accessible through the Internet. A comparison of a newly obtained sequence with this large bank of stored sequences often reveals relationships both surprising and enlightening.

Exactly how the amino acid sequence determines three-dimensional structure is not understood in detail, nor can we always predict function from sequence. However, protein families that have some shared structural or functional features can be readily identified on the basis of amino acid sequence similarities. Individual proteins are assigned to families based on the degree of similarity in amino acid sequence. Members of a family are usually identical across 25% or more of their sequences, and proteins in these families generally share at least some structural and functional characteristics. Some families are defined, however, by identities involving only a few amino acid residues that are critical

to a certain function. A number of similar substructures (to be defined in Chapter 4 as "domains") occur in many functionally unrelated proteins. These domains often fold into structural configurations that have an unusual degree of stability or that are specialized for a certain environment. Evolutionary relationships can also be inferred from the structural and functional similarities within protein families.

Certain amino acid sequences serve as signals that determine the cellular location, chemical modification, and half-life of a protein. Special signal sequences, usually at the amino terminus, are used to target certain proteins for export from the cell; other proteins are targeted for distribution to the nucleus, the cell surface, the cytosol, and other cellular locations. Other sequences act as attachment sites for prosthetic groups, such as sugar groups in glycoproteins and lipids in lipoproteins. Some of these signals are well characterized and are easily recognized in the sequence of a newly characterized protein (Chapter 27).

### SUMMARY 3.4 The Covalent Structure of Proteins

- Differences in protein function result from differences in amino acid composition and sequence. Some variations in sequence are possible for a particular protein, with little or no effect on function.

- Amino acid sequences are deduced by fragmenting polypeptides into smaller peptides using reagents known to cleave specific peptide bonds; determining the amino acid sequence of each fragment by the automated Edman degradation procedure; then ordering the peptide fragments by finding sequence overlaps between fragments generated by different reagents. A protein sequence can also be deduced from the nucleotide sequence of its corresponding gene in DNA.

- Short proteins and peptides (up to about 100 residues) can be chemically synthesized. The peptide is built up, one amino acid residue at a time, while remaining tethered to a solid support.

## 3.5 Protein Sequences and Evolution

The simple string of letters denoting the amino acid sequence of a given protein belies the wealth of information this sequence holds. As more protein sequences have become available, the development of more powerful methods for extracting information from them has become a major biochemical enterprise. Each protein's function relies on its three-dimensional structure, which in turn is determined largely by its primary structure. Thus, the biochemical information conveyed by a protein sequence is in principle limited only by our own understanding of structural and functional principles. On a different level of inquiry, protein sequences are beginning to tell us how the proteins evolved and, ultimately, how life evolved on this planet.

## Protein Sequences Can Elucidate the History of Life on Earth

The field of molecular evolution is often traced to Emile Zuckerkandl and Linus Pauling, whose work in the mid-1960s advanced the use of nucleotide and protein sequences to explore evolution. The premise is deceptively straightforward. If two organisms are closely related, the sequences of their genes and proteins should be similar. The sequences increasingly diverge as the evolutionary distance between two organisms increases. The promise of this approach began to be realized in the 1970s, when Carl Woese used ribosomal RNA sequences to define archaebacteria as a group of living organisms distinct from other bacteria and eukaryotes (see Fig. 1–4). Protein sequences offer an opportunity to greatly refine the available information. With the advent of genome projects investigating organisms from bacteria to humans, the number of available sequences is growing at an enormous rate. This information can be used to trace biological history. The challenge is in learning to read the genetic hieroglyphics.

Evolution has not taken a simple linear path. Complexities abound in any attempt to mine the evolutionary information stored in protein sequences. For a given protein, the amino acid residues essential for the activity of the protein are conserved over evolutionary time. The residues that are less important to function may vary over time—that is, one amino acid may substitute for another—and these variable residues can provide the information used to trace evolution. Amino acid substitutions are not always random, however. At some positions in the primary structure, the need to maintain protein function may mean that only particular amino acid substitutions can be tolerated. Some proteins have more variable amino acid residues than others. For these and other reasons, proteins can evolve at different rates.

Another complicating factor in tracing evolutionary history is the rare transfer of a gene or group of genes from one organism to another, a process called **lateral gene transfer.** The transferred genes may be quite similar to the genes they were derived from in the original organism, whereas most other genes in the same two organisms may be quite distantly related. An example of lateral gene transfer is the recent rapid spread of antibiotic-resistance genes in bacterial populations. The proteins derived from these transferred genes would not be good candidates for the study of bacterial evolution, because they share only a very limited evolutionary history with their "host" organisms.

The study of molecular evolution generally focuses on families of closely related proteins. In most cases, the families chosen for analysis have essential functions in cellular metabolism that must have been present in the earliest viable cells, thus greatly reducing the chance that they were introduced relatively recently by lateral gene transfer. For example, a protein called EF-1$\alpha$ (elongation factor 1$\alpha$) is involved in the synthesis of proteins in all eukaryotes. A similar protein, EF-Tu, with the same function, is found in bacteria. Similarities in sequence and function indicate that EF-1$\alpha$ and EF-Tu are members of a family of proteins that share a common ancestor. The members of protein families are called **homologous proteins, or homologs.** The concept of a homolog can be further refined. If two proteins within a family (that is, two homologs) are present in the same species, they are referred to as **paralogs.** Homologs from different species are called **orthologs** (see Fig. 1–37). The process of tracing evolution involves first identifying suitable families of homologous proteins and then using them to reconstruct evolutionary paths.

Homologs are identified using increasingly powerful computer programs that can directly compare two or more chosen protein sequences, or can search vast databases to find the evolutionary relatives of one selected protein sequence. The electronic search process can be thought of as sliding one sequence past the other until a section with a good match is found. Within this sequence alignment, a positive score is assigned for each position where the amino acid residues in the two sequences are identical—the value of the score varying from one program to the next—to provide a measure of the quality of the alignment. The process has some complications. Sometimes the proteins being compared match well at, say, two sequence segments, and these segments are connected by less related sequences of different lengths. Thus the two matching segments cannot be aligned at the same time. To handle this, the computer program introduces "gaps" in one of the sequences to bring the matching segments into register (Fig. 3–30).

```
E. coli     TGNRTIAVYDLGGGTFDISIIEIDEVDGEKTFEVLATNGDTHLGGEDFDSRLIHYL
B. subtilis DEDQTILLYDLGGGTFDVSILELGDG    TFEVRSTAGDNRLGGDDFDQVIIDHL
                                      └──┘
                                      Gap
```

**FIGURE 3–30 Aligning protein sequences with the use of gaps.** Shown here is the sequence alignment of a short section of the EF-Tu protein from two well-studied bacterial species, *E. coli* and *Bacillus* *subtilis*. Introduction of a gap in the *B. subtilis* sequence allows a better alignment of amino acid residues on either side of the gap. Identical amino acid residues are shaded.

Of course, if a sufficient number of gaps are introduced, almost any two sequences could be brought into some sort of alignment. To avoid uninformative alignments, the programs include penalties for each gap introduced, thus lowering the overall alignment score. With electronic trial and error, the program selects the alignment with the optimal score that maximizes identical amino acid residues while minimizing the introduction of gaps.

Identical amino acids are often inadequate to identify related proteins or, more importantly, to determine how closely related the proteins are on an evolutionary time scale. A more useful analysis includes a consideration of the chemical properties of substituted amino acids. When amino acid substitutions are found within a protein family, many of the differences may be conservative—that is, an amino acid residue is replaced by a residue having similar chemical properties. For example, a Glu residue may substitute in one family member for the Asp residue found in another; both amino acids are negatively charged. Such a conservative substitution should logically garner a higher score in a sequence alignment than does a nonconservative substitution, such as the replacement of the Asp residue with a hydrophobic Phe residue.

To determine what scores to assign to the many different amino acid substitutions, Steven Henikoff and Jorja Henikoff examined the aligned sequences from a variety of different proteins. They did not analyze entire protein sequences, focusing instead on thousands of short conserved blocks where the fraction of identical amino acids was high and the alignments were thus reliable. Looking at the aligned sequence blocks, the Henikoffs analyzed the nonidentical amino acid residues within the blocks. Higher scores were given to nonidentical residues that occurred frequently than to those that appeared rarely. Even the identical residues were given scores based on how often they were replaced, such that amino acids with unique chemical properties (such as Cys and Trp) received higher scores than those more conservatively replaced (such as Asp and Glu). The result of this scoring system is a Blosum (*blocks substitution matrix*) table. The table in Figure 3–31 was generated from sequences that were identical in at least 62% of their amino acid residues, and it is thus referred to as Blosum62. Similar tables have been generated for blocks of homologous sequences that are 50% or 80% identical. When higher levels of identity are required, the most conservative amino acid substitutions can be

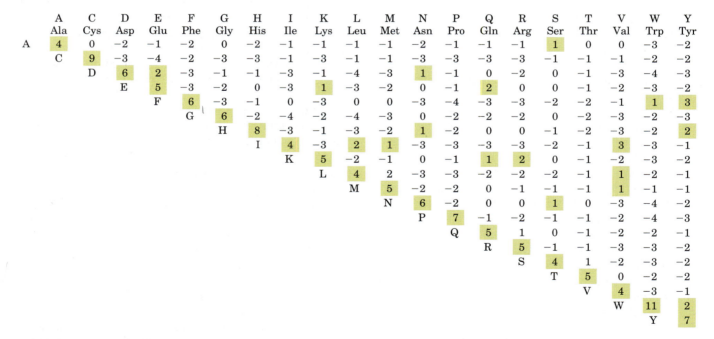

**FIGURE 3–31 The Blosum62 table.** This *blocks substitution matrix* was created by comparing thousands of short blocks of aligned sequences that were identical in at least 62% of their amino acid residues. The nonidentical residues were assigned scores based on how frequently they were replaced by each of the other amino acids. Each substitution contributes to the score given to a particular alignment. Positive numbers (shaded yellow) add to the score for a particular alignment; negative numbers subtract from the score. Identical residues in sequences being compared (the shaded diagonal from top left to bottom right in the matrix) receive scores based on how often they are replaced, such that amino acids with unique chemical properties (e.g., Cys and Trp) receive higher scores (9 and 11, respectively) than those more easily replaced in conservative substitutions (e.g., Asp (6) and Glu (5)). Many computer programs use Blosum62 to assign scores to new sequence alignments.

Signature sequence

| | | Conserved region | | |
|---|---|---|---|---|
| Archaebacteria | { Halobacterium halobium | IGHVDHGKST MVGR LLYETGSVPEHV IEQH | |
| | Sulfolobus solfataricus | IGHVDHGKST LVGR LLMDRGFIDEKT VKEA | |
| Eukaryotes | { Saccharomyces cerevisiae | IGHVDSGKST TTGH LIYKCGGIDKRT IEKF | |
| | Homo sapiens | IGHVDSGKST TTGH LIYKCGGIDKRT IEKF | |
| Gram-positive bacterium | Bacillus subtilis | IGHVDHGKST MVGR | ITTV |
| Gram-negative bacterium | Escherichia coli | IGHVDHGKTT LTAA | ITTV |

**FIGURE 3–32 A signature sequence in the EF-1α/EF-Tu protein family.** The signature sequence (boxed) is a 12-amino-acid insertion near the amino terminus of the sequence. Residues that align in all species are shaded yellow. Both archaebacteria and eukaryotes have the signature, although the sequences of the insertions are quite distinct for the two groups. The variation in the signature sequence reflects the significant evolutionary divergence that has occurred at this site since it first appeared in a common ancestor of both groups.

overrepresented, which limits the usefulness of the matrix in identifying homologs that are somewhat distantly related. Tests have shown that the Blosum62 table provides the most reliable alignments over a wide range of protein families, and it is the default table in many sequence alignment programs.

For most efforts to find homologies and explore evolutionary relationships, protein sequences (derived either directly from protein sequencing or from the sequencing of the DNA encoding the protein) are superior to nongenic nucleic acid sequences (those that do not encode a protein or functional RNA). For a nucleic acid, with its four different types of residues, random alignment of nonhomologous sequences will generally yield matches for at least 25% of the positions. Introduction of a few gaps can often increase the fraction of matched residues to 40% or more, and the probability of chance alignment of unrelated sequences becomes quite high. The 20 different amino acid residues in proteins greatly lower the probability of uninformative chance alignments of this type.

The programs used to generate a sequence alignment are complemented by methods that test the reliability of the alignments. A common computerized test is to shuffle the amino acid sequence of one of the proteins being compared to produce a random sequence, then instruct the program to align the shuffled sequence with the other, unshuffled one. Scores are assigned to the new alignment, and the shuffling and alignment process is repeated many times. The original alignment, before shuffling, should have a score significantly higher than any of those within the distribution of scores generated by the random alignments; this increases the confidence that the sequence alignment has identified a pair of homologs. Note that the *absence* of a significant alignment score does not necessarily mean that no evolutionary relationship exists between two proteins. As we shall see in Chapter 4, three-dimensional structural similarities sometimes reveal evolutionary relationships where sequence homology has been wiped away by time.

Using a protein family to explore evolution requires the identification of family members with similar molecular functions in the widest possible range of organisms. Information from the family can then be used to trace the evolution of those organisms. By analyzing the sequence divergence in selected protein families, investigators can segregate organisms into classes based on their evolutionary relationships. This information must be reconciled with more classical examinations of the physiology and biochemistry of the organisms.

Certain segments of a protein sequence may be found in the organisms of one taxonomic group but not in other groups; these segments can be used as **signature sequences** for the group in which they are found. An example of a signature sequence is an insertion of 12 amino acids near the amino terminus of the EF-1α/EF-Tu proteins in all archaebacteria and eukaryotes but not in other types of bacteria (Fig. 3–32). The signature is one of many biochemical clues that can help establish the evolutionary relatedness of eukaryotes and archaebacteria. For example, the major taxa of bacteria can be distinguished by signature sequences in several different proteins. The β and γ proteobacteria have signature sequences in the Hsp70 and DNA gyrase protein families (families of proteins involved in protein folding and DNA replication, respectively) that are not present in any other bacteria, including the other proteobacteria. The other types of proteobacteria (α, δ, ε), along with the β and γ proteobacteria, have a separate Hsp70 signature sequence and a signature in alanyl-tRNA synthetase (an enzyme of protein synthesis) that are not present in other bacteria. The appearance of unique signatures in the β and γ proteobacteria suggests the α, δ, and ε proteobacteria arose before their β and γ cousins.

By considering the entire sequence of a protein, researchers can now construct more elaborate evolutionary trees with many species in each taxonomic group. Figure 3–33 presents one such tree for bacteria, based on sequence divergence in the protein GroEL (a protein present in all bacteria that assists in the proper folding of proteins). The tree can be refined by basing it on the sequences of multiple proteins and by supplementing the sequence information with data on the unique biochemical and physiological properties of each species. There are many methods for generating trees, each with its own advantages and shortcomings, and many ways to represent the resulting evolutionary relationships. In Figure 3–33, the free end points of lines are called "external nodes"; each represents an extant species, and each is so labeled. The points where two lines come together, the "internal nodes," represent extinct ancestor species. In most representations (including Fig. 3–33), the lengths of the lines connecting the nodes are proportional to the number of amino acid substitutions separating one species from another. If we trace two extant species to a common internal node (representing the common ancestor of the two species), the length of the branch connecting each external node to the internal node represents the number of amino acid substitutions separating one extant species from this ancestor. The sum of the lengths of all the line segments that connect an extant species to another extant species through a common ancestor reflects the number of substitutions separating the two extant species. To determine how much time was needed for the various species to diverge, the tree must be calibrated by comparing it with information from the fossil record and other sources.

As more sequence information is made available in databases, we can generate evolutionary trees based on a variety of different proteins. Some proteins evolve faster than others, or change faster within one group of species than another. A large protein, with many variable amino acid residues, may exhibit a few differences between two closely related species. Another, smaller protein may be identical in the same two species. For many reasons, some details of an evolutionary tree based on the sequences of one protein may differ from those of a tree based on the sequences of another protein. Increasingly sophisticated analyses using the sequences of many different proteins can provide an exquisitely detailed and accurate picture of evolutionary relationships. The story is a work in progress, and the questions being asked and answered are fundamental to how humans view themselves and the world around them. The field of molecular evolution promises to be among the most vibrant of the scientific frontiers in the twenty-first century.

## SUMMARY 3.5 Protein Sequences and Evolution

■ Protein sequences are a rich source of information about protein structure and function, as well as the evolution of life on this planet. Sophisticated methods are being developed to trace evolution by analyzing the resultant slow changes in the amino acid sequences of homologous proteins.

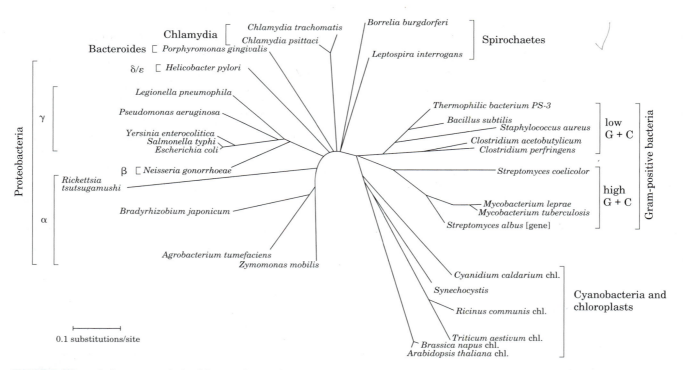

**FIGURE 3–33 Evolutionary tree derived from amino acid sequence comparisons.** A bacterial evolutionary tree, based on the sequence divergence observed in the GroEL family of proteins. Also included in this tree (lower right) are the chloroplasts (chl.) of some nonbacterial species.

# Key Terms

*Terms in bold are defined in the glossary.*

**amino acids**  75
**R group**  76
**chiral center**  76
**enantiomers**  76
**absolute
   configuration**  77
D, L system  77
**polarity**  78
**zwitterion**  81
absorbance, *A*  82
**isoelectric pH (isoelec-
   tric point, pI)**  84
**peptide**  85

**protein**  85
**peptide bond**  85
**oligopeptide**  85
**polypeptide**  85
**oligomeric protein**  87
**protomer**  87
**conjugated
   protein**  88
**prosthetic group**  88
**primary structure**  88
**secondary
   structure**  88
**tertiary structure**  88

**quaternary
   structure**  88
crude extract  89
**fractionation**  89
**dialysis**  89
column
   chromatography  89
**high-performance liquid
   chromatography
   (HPLC)**  90
**electrophoresis**  92
sodium dodecyl sulfate
   (SDS)  92

**isoelectric
   focusing**  93
Edman degradation  98
proteases  99
proteome  101
lateral gene transfer  107
**homologous
   proteins**  107
**homolog**  107
**paralog**  107
**ortholog**  107
signature sequence  109

# Further Reading

## Amino Acids

**Dougherty, D.A.** (2000) Unnatural amino acids as probes of protein structure and function. *Curr. Opin. Chem. Biol.* **4,** 645–652.

**Greenstein, J.P. & Winitz, M.** (1961) *Chemistry of the Amino Acids,* 3 Vols, John Wiley & Sons, New York.

**Kreil, G.** (1997) D-Amino acids in animal peptides. *Annu. Rev. Biochem.* **66,** 337–345.
   An update on the occurrence of these unusual stereoisomers of amino acids.

**Meister, A.** (1965) *Biochemistry of the Amino Acids,* 2nd edn, Vols 1 and 2, Academic Press, Inc., New York.
   Encyclopedic treatment of the properties, occurrence, and metabolism of amino acids.

## Peptides and Proteins

**Creighton, T.E.** (1992) *Proteins: Structures and Molecular Properties,* 2nd edn, W. H. Freeman and Company, New York.
   Very useful general source.

## Working with Proteins

**Dunn, M.J. & Corbett, J.M.** (1996) Two-dimensional polyacrylamide gel electrophoresis. *Methods Enzymol.* **271,** 177–203.
   A detailed description of the technology.

**Kornberg, A.** (1990) Why purify enzymes? *Methods Enzymol.* **182,** 1–5.
   The critical role of classical biochemical methods in a new age.

**Scopes, R.K.** (1994) *Protein Purification: Principles and Practice,* 3rd edn, Springer-Verlag, New York.
   A good source for more complete descriptions of the principles underlying chromatography and other methods.

## Covalent Structure of Proteins

**Andersson, L., Blomberg, L., Flegel, M., Lepsa, L., Nilsson, B., & Verlander, M.** (2000) Large-scale synthesis of peptides. *Biopolymers* **55,** 227–250.
   A discussion of approaches used to manufacture peptides as pharmaceuticals.

**Dell, A. & Morris, H.R.** (2001) Glycoprotein structure determination by mass spectrometry. *Science* **291,** 2351–2356.
   Glycoproteins can be complex; mass spectrometry is a method of choice for sorting things out.

**Dongre, A.R., Eng, J.K., & Yates, J.R. III** (1997) Emerging tandem-mass-spectrometry techniques for the rapid identification of proteins. *Trends Biotechnol.* **15,** 418–425.
   A detailed description of mass spectrometry methods.

**Gygi, S.P. & Aebersold, R.** (2000) Mass spectrometry and proteomics. *Curr. Opin. Chem. Biol.* **4,** 489–494.
   Uses of mass spectrometry to identify and study cellular proteins.

**Koonin, E.V., Tatusov, R.L., & Galperin, M.Y.** (1998) Beyond complete genomes: from sequence to structure and function. *Curr. Opin. Struct. Biol.* **8,** 355–363.
   A good discussion about the possible uses of the tremendous amount of protein sequence information becoming available.

**Mann, M. & Wilm, M.** (1995) Electrospray mass spectrometry for protein characterization. *Trends Biochem. Sci.* **20,** 219–224.
   An approachable summary of this technique for beginners.

**Mayo, K.H.** (2000) Recent advances in the design and construction of synthetic peptides: for the love of basics or just for the technology of it. *Trends Biotechnol.* **18,** 212–217.

**Miranda, L.P. & Alewood, P.F.** (2000) Challenges for protein chemical synthesis in the 21st century: bridging genomics and proteomics. *Biopolymers* **55,** 217–226.

This and the Mayo article describe how to make peptides and splice them together to address a wide range of problems in protein biochemistry.

**Sanger, F.** (1988) Sequences, sequences, sequences. *Annu. Rev. Biochem.* **57,** 1–28.

A nice historical account of the development of sequencing methods.

**Protein Sequences and Evolution**

**Gupta, R.S.** (1998) Protein phylogenies and signal sequences: a reappraisal of evolutionary relationships among Archaebacteria, Eubacteria, and Eukaryotes. *Microbiol. Mol. Biol. Rev.* **62,** 1435–1491.

An almost encyclopedic but very readable report of how protein sequences are used to explore evolution, introducing many in-

teresting ideas and supporting them with detailed sequence comparisons.

**Li, W.-H. & Graur, D.** (2000) *Fundamentals of Molecular Evolution*, 2nd edn, Sinauer Associates, Inc., Sunderland, MA.

A very readable text describing methods used to analyze protein and nucleic acid sequences. Chapter 5 provides one of the best available descriptions of how evolutionary trees are constructed from sequence data.

**Rokas, A., Williams, B.L., King, N., & Carroll, S.B.** (2003) Genome-scale approaches to resolving incongruence in molecular phylogenies. *Nature* **425,** 798–804.

How sequence comparisons of multiple proteins can yield accurate evolutionary information.

**Zuckerkandl, E. & Pauling, L.** (1965) Molecules as documents of evolutionary history. *J. Theor. Biol.* **8,** 357–366.

Considered by many the founding paper in the field of molecular evolution.

## Problems

**1. Absolute Configuration of Citrulline**   The citrulline isolated from watermelons has the structure shown below. Is it a D- or L-amino acid? Explain.

$$\begin{array}{c} CH_2(CH_2)_2NH\!-\!\overset{\displaystyle O}{\underset{\displaystyle \|}{C}}\!-\!NH_2 \\[1mm] H\!-\!\overset{\vdots}{\underset{\vdots}{C}}\!-\!\overset{+}{N}H_3 \\[1mm] COO^- \end{array}$$

**2. Relationship between the Titration Curve and the Acid-Base Properties of Glycine**   A 100 mL solution of 0.1 M glycine at pH 1.72 was titrated with 2 M NaOH solution. The pH was monitored and the results were plotted on a graph, as shown at right. The key points in the titration are designated I to V. For each of the statements (a) to (o), *identify* the appropriate key point in the titration and *justify* your choice.

(a) Glycine is present predominantly as the species $^+H_3N\!-\!CH_2\!-\!COOH$.

(b) The *average* net charge of glycine is $+\frac{1}{2}$.

(c) Half of the amino groups are ionized.

(d) The pH is equal to the $pK_a$ of the carboxyl group.

(e) The pH is equal to the $pK_a$ of the protonated amino group.

(f) Glycine has its maximum buffering capacity.

(g) The *average* net charge of glycine is zero.

(h) The carboxyl group has been completely titrated (first equivalence point).

(i) Glycine is completely titrated (second equivalence point).

(j) The predominant species is $^+H_3N\!-\!CH_2\!-\!COO^-$.

(k) The *average* net charge of glycine is $-1$.

(l) Glycine is present predominantly as a 50:50 mixture of $^+H_3N\!-\!CH_2\!-\!COOH$ and $^+H_3N\!-\!CH_2\!-\!COO^-$.

(m) This is the isoelectric point.

(n) This is the end of the titration.

(o) These are the *worst* pH regions for buffering power.

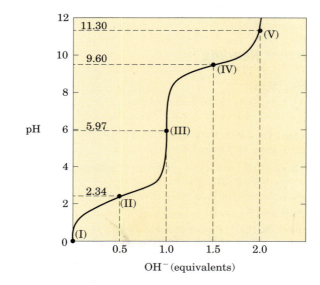

**3. How Much Alanine Is Present as the Completely Uncharged Species?**   At a pH equal to the isoelectric point of alanine, the *net* charge on alanine is zero. Two structures can be drawn that have a net charge of zero, but the predominant form of alanine at its pI is zwitterionic.

$$\begin{array}{cc} \underset{\text{Zwitterionic}}{H_3\overset{+}{N}\!-\!\overset{\overset{\textstyle CH_3}{|}}{\underset{\underset{\textstyle H}{|}}{C}}\!-\!\overset{\overset{\textstyle O}{\|}}{C}\!-\!O^-} & \underset{\text{Uncharged}}{H_2N\!-\!\overset{\overset{\textstyle CH_3}{|}}{\underset{\underset{\textstyle H}{|}}{C}}\!-\!\overset{\overset{\textstyle O}{\|}}{C}\!-\!OH} \end{array}$$

(a) Why is alanine predominantly zwitterionic rather than completely uncharged at its pI?

(b) What fraction of alanine is in the completely uncharged form at its pI? Justify your assumptions.

**4. Ionization State of Amino Acids** Each ionizable group of an amino acid can exist in one of two states, charged or neutral. The electric charge on the functional group is determined by the relationship between its $pK_a$ and the pH of the solution. This relationship is described by the Henderson-Hasselbalch equation.

(a) Histidine has three ionizable functional groups. Write the equilibrium equations for its three ionizations and assign the proper $pK_a$ for each ionization. Draw the structure of histidine in each ionization state. What is the net charge on the histidine molecule in each ionization state?

(b) Draw the structures of the predominant ionization state of histidine at pH 1, 4, 8, and 12. Note that the ionization state can be approximated by treating each ionizable group independently.

(c) What is the net charge of histidine at pH 1, 4, 8, and 12? For each pH, will histidine migrate toward the anode (+) or cathode (−) when placed in an electric field?

**5. Separation of Amino Acids by Ion-Exchange Chromatography** Mixtures of amino acids are analyzed by first separating the mixture into its components through ion-exchange chromatography. Amino acids placed on a cation-exchange resin containing sulfonate groups (see Fig. 3–18a) flow down the column at different rates because of two factors that influence their movement: (1) ionic attraction between the $-SO_3^-$ residues on the column and positively charged functional groups on the amino acids, and (2) hydrophobic interactions between amino acid side chains and the strongly hydrophobic backbone of the polystyrene resin. For each pair of amino acids listed, determine which will be eluted first from an ion-exchange column using a pH 7.0 buffer.

(a) Asp and Lys
(b) Arg and Met
(c) Glu and Val
(d) Gly and Leu
(e) Ser and Ala

**6. Naming the Stereoisomers of Isoleucine** The structure of the amino acid isoleucine is

$$
\begin{array}{c}
COO^- \\
| \\
H_3\overset{+}{N}-C-H \\
| \\
H-C-CH_3 \\
| \\
CH_2 \\
| \\
CH_3
\end{array}
$$

(a) How many chiral centers does it have? 2
(b) How many optical isomers?
(c) Draw perspective formulas for all the optical isomers of isoleucine.

**7. Comparing the $pK_a$ Values of Alanine and Polyalanine** The titration curve of alanine shows the ionization of two functional groups with $pK_a$ values of 2.34 and 9.69, corresponding to the ionization of the carboxyl and the protonated amino groups, respectively. The titration of di-, tri-, and larger oligopeptides of alanine also shows the ionization of only two functional groups, although the experimental $pK_a$ values are different. The trend in $pK_a$ values is summarized in the table.

| Amino acid or peptide | $pK_1$ | $pK_2$ |
|---|---|---|
| Ala | 2.34 | 9.69 |
| Ala–Ala | 3.12 | 8.30 |
| Ala–Ala–Ala | 3.39 | 8.03 |
| Ala–(Ala)$_n$–Ala, $n \geq 4$ | 3.42 | 7.94 |

(a) Draw the structure of Ala–Ala–Ala. Identify the functional groups associated with $pK_1$ and $pK_2$.

(b) Why does the value of $pK_1$ *increase* with each addition of an Ala residue to the Ala oligopeptide?

(c) Why does the value of $pK_2$ *decrease* with each addition of an Ala residue to the Ala oligopeptide?

**8. The Size of Proteins** What is the approximate molecular weight of a protein with 682 amino acid residues in a single polypeptide chain?

**9. The Number of Tryptophan Residues in Bovine Serum Albumin** A quantitative amino acid analysis reveals that bovine serum albumin (BSA) contains 0.58% tryptophan ($M_r$ 204) by weight.

(a) Calculate the *minimum* molecular weight of BSA (i.e., assuming there is only one tryptophan residue per protein molecule).

(b) Gel filtration of BSA gives a molecular weight estimate of 70,000. How many tryptophan residues are present in a molecule of serum albumin?

**10. Net Electric Charge of Peptides** A peptide has the sequence

Glu–His–Trp–Ser–Gly–Leu–Arg–Pro–Gly

(a) What is the net charge of the molecule at pH 3, 8, and 11? (Use $pK_a$ values for side chains and terminal amino and carboxyl groups as given in Table 3–1.)

(b) Estimate the pI for this peptide.

**11. Isoelectric Point of Pepsin** Pepsin is the name given to several digestive enzymes secreted (as larger precursor proteins) by glands that line the stomach. These glands also secrete hydrochloric acid, which dissolves the particulate matter in food, allowing pepsin to enzymatically cleave individual protein molecules. The resulting mixture of food, HCl, and digestive enzymes is known as chyme and has a pH near 1.5. What pI would you predict for the pepsin proteins? What functional groups must be present to confer this pI on pepsin? Which amino acids in the proteins would contribute such groups?

**12. The Isoelectric Point of Histones** Histones are proteins found in eukaryotic cell nuclei, tightly bound to DNA, which has many phosphate groups. The pI of histones is very high, about 10.8. What amino acid residues must be present in relatively large numbers in histones? In what way do these residues contribute to the strong binding of histones to DNA?

**13. Solubility of Polypeptides** One method for separating polypeptides makes use of their differential solubilities. The solubility of large polypeptides in water depends upon the relative polarity of their R groups, particularly on the number of ionized groups: the more ionized groups there are, the more soluble the polypeptide. Which of each pair of the polypeptides that follow is more soluble at the indicated pH?

(a) $(Gly)_{20}$ or $(Glu)_{20}$ at pH 7.0

(b) $(Lys–Ala)_3$ or $(Phe–Met)_3$ at pH 7.0

(c) $(Ala–Ser–Gly)_5$ or $(Asn–Ser–His)_5$ at pH 6.0

(d) $(Ala–Asp–Gly)_5$ or $(Asn–Ser–His)_5$ at pH 3.0

**14. Purification of an Enzyme**  A biochemist discovers and purifies a new enzyme, generating the purification table below.

| Procedure | Total protein (mg) | Activity (units) |
|---|---|---|
| 1. Crude extract | 20,000 | 4,000,000 |
| 2. Precipitation (salt) | 5,000 | 3,000,000 |
| 3. Precipitation (pH) | 4,000 | 1,000,000 |
| 4. Ion-exchange chromatography | 200 | 800,000 |
| 5. Affinity chromatography | 50 | 750,000 |
| 6. Size-exclusion chromatography | 45 | 675,000 |

(a) From the information given in the table, calculate the specific activity of the enzyme solution after each purification procedure.

(b) Which of the purification procedures used for this enzyme is most effective (i.e., gives the greatest relative increase in purity)?

(c) Which of the purification procedures is least effective?

(d) Is there any indication based on the results shown in the table that the enzyme after step 6 is now pure? What else could be done to estimate the purity of the enzyme preparation?

**15. Sequence Determination of the Brain Peptide Leucine Enkephalin**  A group of peptides that influence nerve transmission in certain parts of the brain has been isolated from normal brain tissue. These peptides are known as opioids, because they bind to specific receptors that also bind opiate drugs, such as morphine and naloxone. Opioids thus mimic some of the properties of opiates. Some researchers consider these peptides to be the brain's own pain killers. Using the information below, determine the amino acid sequence of the opioid leucine enkephalin. Explain how your structure is consistent with each piece of information.

(a) Complete hydrolysis by 6 M HCl at 110 °C followed by amino acid analysis indicated the presence of Gly, Leu, Phe, and Tyr, in a 2:1:1:1 molar ratio.

(b) Treatment of the peptide with 1-fluoro-2,4-dinitrobenzene followed by complete hydrolysis and chromatography indicated the presence of the 2,4-dinitrophenyl derivative of tyrosine. No free tyrosine could be found.

(c) Complete digestion of the peptide with pepsin followed by chromatography yielded a dipeptide containing Phe and Leu, plus a tripeptide containing Tyr and Gly in a 1:2 ratio.

**16. Structure of a Peptide Antibiotic from *Bacillus brevis***  Extracts from the bacterium *Bacillus brevis* contain a peptide with antibiotic properties. This peptide forms complexes with metal ions and apparently disrupts ion transport across the cell membranes of other bacterial species, killing them. The structure of the peptide has been determined from the following observations.

(a) Complete acid hydrolysis of the peptide followed by amino acid analysis yielded equimolar amounts of Leu, Orn,

Phe, Pro, and Val. Orn is ornithine, an amino acid not present in proteins but present in some peptides. It has the structure

$$\overset{+}{H_3N}-CH_2-CH_2-CH_2-\underset{\underset{NH_3}{\overset{|}{|}}}{\overset{\overset{\displaystyle H}{|}}{C}}-COO^-$$

(b) The molecular weight of the peptide was estimated as about 1,200.

(c) The peptide failed to undergo hydrolysis when treated with the enzyme carboxypeptidase. This enzyme catalyzes the hydrolysis of the carboxyl-terminal residue of a polypeptide unless the residue is Pro or, for some reason, does not contain a free carboxyl group.

(d) Treatment of the intact peptide with 1-fluoro-2,4-dinitrobenzene, followed by complete hydrolysis and chromatography, yielded only free amino acids and the following derivative:

$$O_2N-\underset{}{\overset{\overset{\displaystyle NO_2}{|}}{\bigcirc}}-NH-CH_2-CH_2-CH_2-\underset{\underset{NH_3}{\overset{|}{|}}}{\overset{\overset{\displaystyle H}{|}}{C}}-COO^-$$

(Hint: Note that the 2,4-dinitrophenyl derivative involves the amino group of a side chain rather than the $\alpha$-amino group.)

(e) Partial hydrolysis of the peptide followed by chromatographic separation and sequence analysis yielded the following di- and tripeptides (the amino-terminal amino acid is always at the left):

Leu–Phe    Phe–Pro    Orn–Leu    Val–Orn

Val–Orn–Leu    Phe–Pro–Val    Pro–Val–Orn

Given the above information, deduce the amino acid sequence of the peptide antibiotic. Show your reasoning. When you have arrived at a structure, demonstrate that it is consistent with *each* experimental observation.

**17. Efficiency in Peptide Sequencing**  A peptide with the primary structure Lys–Arg–Pro–Leu–Ile–Asp–Gly–Ala is sequenced by the Edman procedure. If each Edman cycle is 96% efficient, what percentage of the amino acids liberated in the fourth cycle will be leucine? Do the calculation a second time, but assume a 99% efficiency for each cycle.

**18. Biochemistry Protocols: Your First Protein Purification**  As the newest and least experienced student in a biochemistry research lab, your first few weeks are spent washing glassware and labeling test tubes. You then graduate to making buffers and stock solutions for use in various laboratory procedures. Finally, you are given responsibility for purifying a protein. It is a citric acid cycle enzyme, citrate synthase, located in the mitochondrial matrix. Following a protocol for the purification, you proceed through the steps below. As you work, a more experienced student questions you about the rationale for each procedure. Supply the answers. (Hint: See Chapter 2 for information about osmolarity; see p. 6 for information on separation of organelles from cells.)

(a) You pick up 20 kg of beef hearts from a nearby slaughterhouse. You transport the hearts on ice, and perform each step of the purification on ice or in a walk-in cold room. You homogenize the beef heart tissue in a high-speed blender in a medium containing 0.2 M sucrose, buffered to a pH of 7.2. *Why do you use beef heart tissue, and in such large quantity? What is the purpose of keeping the tissue cold and suspending it in 0.2 M sucrose, at pH 7.2? What happens to the tissue when it is homogenized?*

(b) You subject the resulting heart homogenate, which is dense and opaque, to a series of differential centrifugation steps. *What does this accomplish?*

(c) You proceed with the purification using the supernatant fraction that contains mostly intact mitochondria. Next you osmotically lyse the mitochondria. The lysate, which is less dense than the homogenate, but still opaque, consists primarily of mitochondrial membranes and internal mitochondrial contents. To this lysate you add ammonium sulfate, a highly soluble salt, to a specific concentration. You centrifuge the solution, decant the supernatant, and discard the pellet. To the supernatant, which is clearer than the lysate, you add *more* ammonium sulfate. Once again, you centrifuge the sample, but this time you save the pellet because it contains the protein of interest. *What is the rationale for the two-step addition of the salt?*

(d) You solubilize the ammonium sulfate pellet containing the mitochondrial proteins and dialyze it overnight against large volumes of buffered (pH 7.2) solution. *Why isn't ammonium sulfate included in the dialysis buffer? Why do you use the buffer solution instead of water?*

(e) You run the dialyzed solution over a size-exclusion chromatographic column. Following the protocol, you collect the *first* protein fraction that exits the column, and discard the rest of the fractions that elute from the column later. You detect the protein by measuring UV absorbance (at 280 nm) in the fractions. *What does the instruction to collect the first fraction tell you about the protein? Why is UV absorbance at 280 nm a good way to monitor for the presence of protein in the eluted fractions?*

(f) You place the fraction collected in (e) on a cation-exchange chromatographic column. After discarding the initial solution that exits the column (the flowthrough), you add a washing solution of higher pH to the column and collect the protein fraction that immediately elutes. *Explain what you are doing.*

(g) You run a small sample of your fraction, now very reduced in volume and quite clear (though tinged pink), on an isoelectric focusing gel. When stained, the gel shows three sharp bands. According to the protocol, the protein of interest is the one with the pI of 5.6, but you decide to do one more assay of the protein's purity. You cut out the pI 5.6 band and subject it to SDS polyacrylamide gel electrophoresis. The protein resolves as a single band. *Why were you unconvinced of the purity of the "single" protein band on your isoelectric focusing gel? What did the results of the SDS gel tell you? Why is it important to do the SDS gel electrophoresis after the isoelectric focusing?*

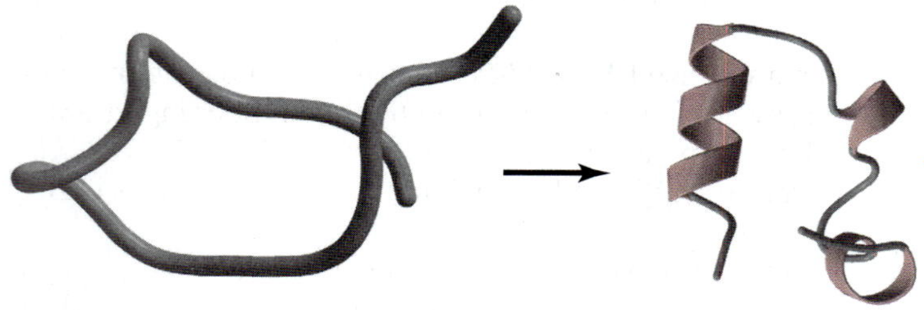

# THE THREE-DIMENSIONAL STRUCTURE OF PROTEINS

Perhaps the more remarkable features of [myoglobin] are its complexity and its lack of symmetry. The arrangement seems to be almost totally lacking in the kind of regularities which one instinctively anticipates, and it is more complicated than has been predicted by any theory of protein structure.

—*John Kendrew,* article in Nature, 1958

The covalent backbone of a typical protein contains hundreds of individual bonds. Because free rotation is possible around many of these bonds, the protein can assume an unlimited number of conformations. However, each protein has a specific chemical or structural function, strongly suggesting that each has a unique three-dimensional structure (Fig. 4–1). By the late 1920s, several proteins had been crystallized, including hemoglobin ($M_r$ 64,500) and the enzyme urease ($M_r$ 483,000). Given that the ordered array of molecules in a crystal can generally form only if the molecular units are identical, the simple fact that many proteins can be crystallized provides strong evidence that even very large proteins are discrete chemical entities with unique structures. This conclusion revolutionized thinking about proteins and their functions.

In this chapter, we explore the three-dimensional structure of proteins, emphasizing five themes. First, the three-dimensional structure of a protein is determined by its amino acid sequence. Second, the function of a protein depends on its structure. Third, an isolated protein usually exists in one or a small number of stable structural forms. Fourth, the most important forces stabilizing the specific structures maintained by a given protein are noncovalent interactions. Finally, amid the huge number of unique protein structures, we can recognize some common structural patterns that help us organize our understanding of protein architecture.

These themes should not be taken to imply that proteins have static, unchanging three-dimensional structures. Protein function often entails an interconversion between two or more structural forms. The dynamic aspects of protein structure will be explored in Chapters 5 and 6.

The relationship between the amino acid sequence of a protein and its three-dimensional structure is an intricate puzzle that is gradually yielding to techniques used in modern biochemistry. An understanding of structure, in turn, is essential to the discussion of function in succeeding chapters. We can find and understand the patterns within the biochemical labyrinth of protein structure by applying fundamental principles of chemistry and physics.

## 4.1 Overview of Protein Structure

The spatial arrangement of atoms in a protein is called its **conformation.** The possible conformations of a protein include any structural state that can be achieved without breaking covalent bonds. A change in conformation could occur, for example, by rotation about single bonds. Of the numerous conformations that are theoretically possible in a protein containing hundreds of single bonds, one or (more commonly) a few generally predominate under biological conditions. The need for multiple stable conformations reflects the changes that must occur in most proteins as they bind to other

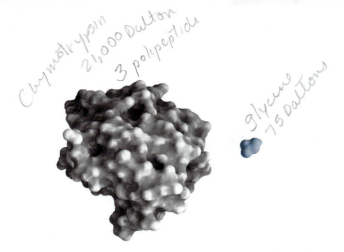

*Chymotrypsin*
*21,000 Dalton*
*3 polypeptide*

*glycine*
*75 Daltons*

**FIGURE 4–1 Structure of the enzyme chymotrypsin, a globular protein.** Proteins are large molecules and, as we shall see, each has a unique structure. A molecule of glycine (blue) is shown for size comparison. The known three-dimensional structures of proteins are archived in the Protein Data Bank, or PDB (www.rcsb.org/pdb). Each structure is assigned a unique four-character identifier, or PDB ID. Where appropriate, we will provide the PDB IDs for molecular graphic images in the figure captions. The image shown here was made using data from the PDB file 6GCH. The data from the PDB files provide only a series of coordinates detailing the location of atoms and their connectivity. Viewing the images requires easy-to-use graphics programs such as RasMol and Chime that convert the coordinates into an image and allow the viewer to manipulate the structure in three dimensions. You will find instructions for downloading Chime with the structure tutorials on the textbook website (www.whfreeman.com/lehninger). The PDB website has instructions for downloading other viewers. We encourage all students to take advantage of the resources of the PDB and the free molecular graphics programs.

molecules or catalyze reactions. The conformations existing under a given set of conditions are usually the ones that are thermodynamically the most stable, having the lowest Gibbs free energy ($G$). Proteins in any of their functional, folded conformations are called **native** proteins.

What principles determine the most stable conformations of a protein? An understanding of protein conformation can be built stepwise from the discussion of primary structure in Chapter 3 through a consideration of secondary, tertiary, and quaternary structures. To this traditional approach must be added a new emphasis on supersecondary structures, a growing set of known and classifiable protein folding patterns that provides an important organizational context to this complex endeavor. We begin by introducing some guiding principles.

## A Protein's Conformation Is Stabilized Largely by Weak Interactions

In the context of protein structure, the term **stability** can be defined as the tendency to maintain a native conformation. Native proteins are only marginally stable; the $\Delta G$ separating the folded and unfolded states in typical proteins under physiological conditions is in the range of only 20 to 65 kJ/mol. A given polypeptide chain can theoretically assume countless different conformations, and as a result the unfolded state of a protein is characterized by a high degree of conformational entropy. This entropy, and the hydrogen-bonding interactions of many groups in the polypeptide chain with solvent (water), tend to maintain the unfolded state. The chemical interactions that counteract these effects and stabilize the native conformation include disulfide bonds and the weak (noncovalent) interactions described in Chapter 2: hydrogen bonds, and hydrophobic and ionic interactions. An appreciation of the role of these weak interactions is especially important to our understanding of how polypeptide chains fold into specific secondary and tertiary structures, and how they combine with other polypeptides to form quaternary structures.

About 200 to 460 kJ/mol are required to break a single covalent bond, whereas weak interactions can be disrupted by a mere 4 to 30 kJ/mol. Individual covalent bonds that contribute to the native conformations of proteins, such as disulfide bonds linking separate parts of a single polypeptide chain, are clearly much stronger than individual weak interactions. Yet, because they are so numerous, it is weak interactions that predominate as a stabilizing force in protein structure. In general, the protein conformation with the lowest free energy (that is, the most stable conformation) is the one with the maximum number of weak interactions.

The stability of a protein is not simply the sum of the free energies of formation of the many weak interactions within it. Every hydrogen-bonding group in a folded polypeptide chain was hydrogen-bonded to water prior to folding, and for every hydrogen bond formed in a protein, a hydrogen bond (of similar strength) between the same group and water was broken. The net stability contributed by a given weak interaction, or the *difference* in free energies of the folded and unfolded states, may be close to zero. We must therefore look elsewhere to explain why the native conformation of a protein is favored.

We find that the contribution of weak interactions to protein stability can be understood in terms of the properties of water (Chapter 2). Pure water contains a network of hydrogen-bonded $H_2O$ molecules. No other molecule has the hydrogen-bonding potential of water, and other molecules present in an aqueous solution disrupt the hydrogen bonding of water. When water surrounds a hydrophobic molecule, the optimal arrangement of hydrogen bonds results in a highly structured shell, or **solvation layer,** of water in the immediate vicinity. The increased order of the water molecules in the solvation layer correlates with an unfavorable decrease in the entropy of the water. However, when non-polar groups are clustered together, there is a decrease in the extent of the solvation layer because each group no longer presents its entire surface to the solution. The result is a favorable increase in entropy. As described in

Chapter 2, this entropy term is the major thermodynamic driving force for the association of hydrophobic groups in aqueous solution. Hydrophobic amino acid side chains therefore tend to be clustered in a protein's interior, away from water.

Under physiological conditions, the formation of hydrogen bonds and ionic interactions in a protein is driven largely by this same entropic effect. Polar groups can generally form hydrogen bonds with water and hence are soluble in water. However, the number of hydrogen bonds per unit mass is generally greater for pure water than for any other liquid or solution, and there are limits to the solubility of even the most polar molecules as their presence causes a net decrease in hydrogen bonding per unit mass. Therefore, a solvation shell of structured water will also form to some extent around polar molecules. Even though the energy of formation of an intramolecular hydrogen bond or ionic interaction between two polar groups in a macromolecule is largely canceled out by the elimination of such interactions between the same groups and water, the release of structured water when the intramolecular interaction is formed provides an entropic driving force for folding. Most of the net change in free energy that occurs when weak interactions are formed within a protein is therefore derived from the increased entropy in the surrounding aqueous solution resulting from the burial of hydrophobic surfaces. This more than counterbalances the large loss of conformational entropy as a polypeptide is constrained into a single folded conformation.

Hydrophobic interactions are clearly important in stabilizing a protein conformation; the interior of a protein is generally a densely packed core of hydrophobic amino acid side chains. It is also important that any polar or charged groups in the protein interior have suitable partners for hydrogen bonding or ionic interactions. One hydrogen bond seems to contribute little to the stability of a native structure, but the presence of hydrogen-bonding or charged groups without partners in the hydrophobic core of a protein can be so *destabilizing* that conformations containing these groups are often thermodynamically untenable. The favorable free-energy change realized by combining such a group with a partner in the surrounding solution can be greater than the difference in free energy between the folded and unfolded states. In addition, hydrogen bonds between groups in proteins form cooperatively. Formation of one hydrogen bond facilitates the formation of additional hydrogen bonds. The overall contribution of hydrogen bonds and other noncovalent interactions to the stabilization of protein conformation is still being evaluated. The interaction of oppositely charged groups that form an ion pair (salt bridge) may also have a stabilizing effect on one or more native conformations of some proteins.

Most of the structural patterns outlined in this chapter reflect two simple rules: (1) hydrophobic residues are largely buried in the protein interior, away from water; and (2) the number of hydrogen bonds within the protein is maximized. Insoluble proteins and proteins within membranes (which we examine in Chapter 11) follow somewhat different rules because of their function or their environment, but weak interactions are still critical structural elements.

## The Peptide Bond Is Rigid and Planar

**Protein Architecture—Primary Structure** Covalent bonds also place important constraints on the conformation of a polypeptide. In the late 1930s, Linus Pauling and Robert Corey embarked on a series of studies that laid the foundation for our present understanding of protein structure. They began with a careful analysis of the peptide bond. The $\alpha$ carbons of adjacent amino acid residues are separated by three covalent bonds, arranged as $C_\alpha$—C—N—$C_\alpha$. X-ray diffraction studies of crystals of amino acids and of simple dipeptides and tripeptides demonstrated that the peptide C—N bond is somewhat shorter than the C—N bond in a simple amine and that the atoms associated with the peptide bond are coplanar. This indicated a resonance or partial sharing of two pairs of electrons between the carbonyl oxygen and the amide nitrogen (Fig. 4–2a). The oxygen has a partial negative charge and the nitrogen a partial positive charge, setting up a small electric dipole. The six atoms of the **peptide group** lie in a single plane, with the oxygen atom of the carbonyl group and the hydrogen atom of the amide nitrogen trans to each other. From these findings Pauling and Corey concluded that the peptide C—N bonds are unable to rotate freely because of their partial double-bond character. Rotation is permitted about the N—$C_\alpha$ and the $C_\alpha$—C bonds. The backbone of a polypeptide chain can thus be pictured as a series of rigid planes with consecutive planes sharing a common point of rotation at $C_\alpha$ (Fig. 4–2b). The rigid peptide bonds limit the range of conformations that can be assumed by a polypeptide chain.

By convention, the bond angles resulting from rotations at $C_\alpha$ are labeled $\phi$ (phi) for the N—$C_\alpha$ bond and $\psi$ (psi) for the $C_\alpha$—C bond. Again by convention, both $\phi$ and $\psi$ are defined as 180° when the polypeptide is in its fully extended conformation and all peptide groups are in the same plane (Fig. 4–2b). In principle, $\phi$ and $\psi$ can have any value between −180° and +180°, but many values are prohibited by steric interference between atoms in the polypeptide backbone and amino acid side chains. The conformation in which both $\phi$ and $\psi$ are 0° (Fig. 4–2c) is prohibited for this reason; this conformation is used merely as a reference point for describing the angles of rotation. Allowed values for $\phi$ and $\psi$ are graphically revealed when $\psi$ is plotted versus $\phi$ in a **Ramachandran plot** (Fig. 4–3), introduced by G. N. Ramachandran.

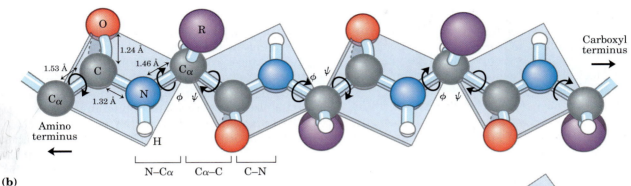

**(a)**

The carbonyl oxygen has a partial negative charge and the amide nitrogen a partial positive charge, setting up a small electric dipole. Virtually all peptide bonds in proteins occur in this trans configuration; an exception is noted in Figure 4–8b.

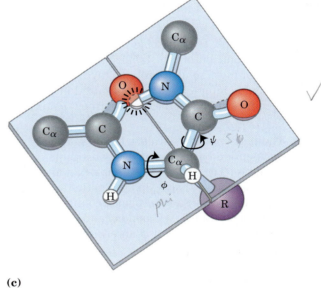

**(b)**

**FIGURE 4–2 The planar peptide group. (a)** Each peptide bond has some double-bond character due to resonance and cannot rotate. **(b)** Three bonds separate sequential α carbons in a polypeptide chain. The N—C$_\alpha$ and C$_\alpha$—C bonds can rotate, with bond angles designated φ and ψ, respectively. The peptide C—N bond is not free to rotate. Other single bonds in the backbone may also be rotationally hindered, depending on the size and charge of the R groups. In the conformation shown, φ and ψ are 180° (or −180°). As one looks out from the α carbon, the ψ and φ angles increase as the carbonyl or amide nitrogens (respectively) rotate clockwise. **(c)** By convention, both φ and ψ are defined as 0° when the two peptide bonds flanking that α carbon are in the same plane and positioned as shown. In a protein, this conformation is prohibited by steric overlap between an α-carbonyl oxygen and an α-amino hydrogen atom. To illustrate the bonds between atoms, the balls representing each atom are smaller than the van der Waals radii for this scale. 1 Å = 0.1 nm.

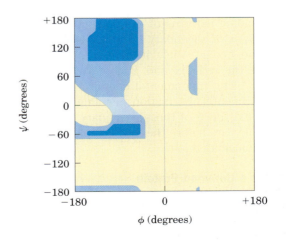

**(c)**

**FIGURE 4–3 Ramachandran plot for L-Ala residues.** The conformations of peptides are defined by the values of φ and ψ. Conformations deemed possible are those that involve little or no steric interference, based on calculations using known van der Waals radii and bond angles. The areas shaded dark blue reflect conformations that involve no steric overlap and thus are fully allowed; medium blue indicates conformations allowed at the extreme limits for unfavorable atomic contacts; the lightest blue area reflects conformations that are permissible if a little flexibility is allowed in the bond angles. The asymmetry of the plot results from the L stereochemistry of the amino acid residues. The plots for other L-amino acid residues with unbranched side chains are nearly identical. The allowed ranges for branched amino acid residues such as Val, Ile, and Thr are somewhat smaller than for Ala. The Gly residue, which is less sterically hindered, exhibits a much broader range of allowed conformations. The range for Pro residues is greatly restricted because φ is limited by the cyclic side chain to the range of −35° to −85°.

Linus Pauling, 1901–1994

Robert Corey, 1897–1971

### SUMMARY 4.1 Overview of Protein Structure

- Every protein has a three-dimensional structure that reflects its function.

- Protein structure is stabilized by multiple weak interactions. Hydrophobic interactions are the major contributors to stabilizing the globular form of most soluble proteins; hydrogen bonds and ionic interactions are optimized in the specific structures that are thermodynamically most stable.

- The nature of the covalent bonds in the polypeptide backbone places constraints on structure. The peptide bond has a partial double-bond character that keeps the entire six-atom peptide group in a rigid planar configuration. The N—$C_\alpha$ and $C_\alpha$—C bonds can rotate to assume bond angles of $\phi$ and $\psi$, respectively.

## 4.2 Protein Secondary Structure

The term **secondary structure** refers to the local conformation of some part of a polypeptide. The discussion of secondary structure most usefully focuses on common regular folding patterns of the polypeptide backbone. A few types of secondary structure are particularly stable and occur widely in proteins. The most prominent are the $\alpha$ helix and $\beta$ conformations described below. Using fundamental chemical principles and a few experimental observations, Pauling and Corey predicted the existence of these secondary structures in 1951, several years before the first complete protein structure was elucidated.

### The $\alpha$ Helix Is a Common Protein Secondary Structure

**Protein Architecture—$\alpha$ Helix** Pauling and Corey were aware of the importance of hydrogen bonds in orient-

ing polar chemical groups such as the C=O and N—H groups of the peptide bond. They also had the experimental results of William Astbury, who in the 1930s had conducted pioneering x-ray studies of proteins. Astbury demonstrated that the protein that makes up hair and porcupine quills (the fibrous protein $\alpha$-keratin) has a regular structure that repeats every 5.15 to 5.2 Å. (The angstrom, Å, named after the physicist Anders J. Ångström, is equal to 0.1 nm. Although not an SI unit, it is used universally by structural biologists to describe atomic distances.) With this information and their data on the peptide bond, and with the help of precisely constructed models, Pauling and Corey set out to determine the likely conformations of protein molecules.

The simplest arrangement the polypeptide chain could assume with its rigid peptide bonds (but other single bonds free to rotate) is a helical structure, which Pauling and Corey called the **$\alpha$ helix** (Fig. 4–4). In this structure the polypeptide backbone is tightly wound around an imaginary axis drawn longitudinally through the middle of the helix, and the R groups of the amino acid residues protrude outward from the helical backbone. The repeating unit is a single turn of the helix, which extends about 5.4 Å along the long axis, slightly greater than the periodicity Astbury observed on x-ray analysis of hair keratin. The amino acid residues in an $\alpha$ helix have conformations with $\psi = -45°$ to $-50°$ and $\phi = -60°$, and each helical turn includes 3.6 amino acid residues. The helical twist of the $\alpha$ helix found in all proteins is right-handed (Box 4–1). The $\alpha$ helix proved to be the predominant structure in $\alpha$-keratins. More generally, about one-fourth of all amino acid residues in polypeptides are found in $\alpha$ helices, the exact fraction varying greatly from one protein to the next.

Why does the $\alpha$ helix form more readily than many other possible conformations? The answer is, in part, that an $\alpha$ helix makes optimal use of internal hydrogen bonds. The structure is stabilized by a hydrogen bond between the hydrogen atom attached to the electronegative nitrogen atom of a peptide linkage and the electronegative carbonyl oxygen atom of the fourth amino acid on the amino-terminal side of that peptide bond (Fig. 4–4b). Within the $\alpha$ helix, every peptide bond (except those close to each end of the helix) participates in such hydrogen bonding. Each successive turn of the $\alpha$ helix is held to adjacent turns by three to four hydrogen bonds. All the hydrogen bonds combined give the entire helical structure considerable stability.

Further model-building experiments have shown that an $\alpha$ helix can form in polypeptides consisting of either L- or D-amino acids. However, all residues must be of one stereoisomeric series; a D-amino acid will disrupt a regular structure consisting of L-amino acids, and vice versa. Naturally occurring L-amino acids can form either right- or left-handed $\alpha$ helices, but extended left-handed helices have not been observed in proteins.

*Alpha helix*

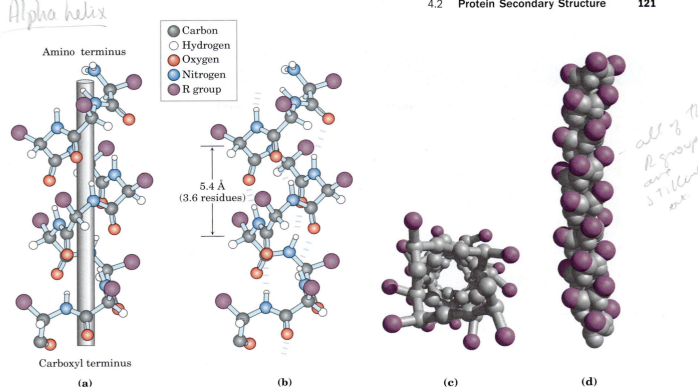

Carbon
Hydrogen
Oxygen
Nitrogen
R group

Amino terminus

5.4 Å
(3.6 residues)

Carboxyl terminus

*all of the R groups are sticking out.*

(a)      (b)      (c)      (d)

**FIGURE 4–4 Four models of the α helix, showing different aspects of its structure. (a)** Formation of a right-handed α helix. The planes of the rigid peptide bonds are parallel to the long axis of the helix, depicted here as a vertical rod. **(b)** Ball-and-stick model of a right-handed α helix, showing the intrachain hydrogen bonds. The repeat unit is a single turn of the helix, 3.6 residues. **(c)** The α helix as viewed from one end, looking down the longitudinal axis (derived from PDB ID 4TNC). Note the positions of the R groups, represented by purple spheres. This ball-and-stick model, used to emphasize the helical arrangement, gives the false impression that the helix is hollow, because the balls do not represent the van der Waals radii of the individual atoms. As the space-filling model **(d)** shows, the atoms in the center of the α helix are in very close contact.

## Amino Acid Sequence Affects α Helix Stability

Not all polypeptides can form a stable α helix. Interactions between amino acid side chains can stabilize or destabilize this structure. For example, if a polypeptide chain has a long block of Glu residues, this segment of the chain will not form an α helix at pH 7.0. The negatively charged carboxyl groups of adjacent Glu residues repel each other so strongly that they prevent formation of the α helix. For the same reason, if there are many adjacent Lys and/or Arg residues, which have positively charged R groups at pH 7.0, they will also repel each other and prevent formation of the α helix. The bulk and shape of Asn, Ser, Thr, and Cys residues can also destabilize an α helix if they are close together in the chain.

The twist of an α helix ensures that critical interactions occur between an amino acid side chain and the side chain three (and sometimes four) residues away on either side of it (Fig. 4–5). Positively charged amino acids are often found three residues away from negatively charged amino acids, permitting the formation of an ion pair. Two aromatic amino acid residues are often similarly spaced, resulting in a hydrophobic interaction.

*Interaction of R groups in an Alpha Helix*

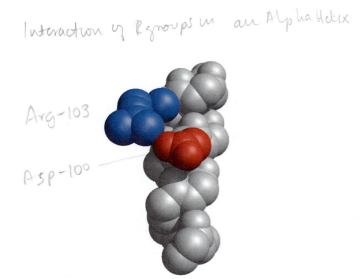

Arg-103

Asp-100

**FIGURE 4–5 Interactions between R groups of amino acids three residues apart in an α helix.** An ionic interaction between Asp$^{100}$ and Arg$^{103}$ in an α-helical region of the protein troponin C, a calcium-binding protein associated with muscle, is shown in this space-filling model (derived from PDB ID 4TNC). The polypeptide backbone (carbons, α-amino nitrogens, and α-carbonyl oxygens) is shown in gray for a helix segment 13 residues long. The only side chains represented here are the interacting Asp (red) and Arg (blue) side chains.

## BOX 4–1    WORKING IN BIOCHEMISTRY

### Knowing the Right Hand from the Left

There is a simple method for determining whether a helical structure is right-handed or left-handed. Make fists of your two hands with thumbs outstretched and pointing straight up. Looking at your right hand, think of a helix spiraling up your right thumb in the direction in which the other four fingers are curled as shown (counterclockwise). The resulting helix is right-handed. Your left hand will demonstrate a left-handed helix, which rotates in the clockwise direction as it spirals up your thumb.

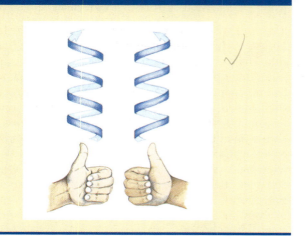

A constraint on the formation of the $\alpha$ helix is the presence of Pro or Gly residues. In proline, the nitrogen atom is part of a rigid ring (see Fig. 4–8b), and rotation about the N—C$_\alpha$ bond is not possible. Thus, a Pro residue introduces a destabilizing kink in an $\alpha$ helix. In addition, the nitrogen atom of a Pro residue in peptide linkage has no substituent hydrogen to participate in hydrogen bonds with other residues. For these reasons, proline is only rarely found within an $\alpha$ helix. Glycine occurs infrequently in $\alpha$ helices for a different reason: it has more conformational flexibility than the other amino acid residues. Polymers of glycine tend to take up coiled structures quite different from an $\alpha$ helix.

A final factor affecting the stability of an $\alpha$ helix in a polypeptide is the identity of the amino acid residues near the ends of the $\alpha$-helical segment. A small electric dipole exists in each peptide bond (Fig. 4–2a). These dipoles are connected through the hydrogen bonds of the helix, resulting in a net dipole extending along the helix that increases with helix length (Fig. 4–6). The four amino acid residues at each end of the helix do not participate fully in the helix hydrogen bonds. The partial positive and negative charges of the helix dipole actually reside on the peptide amino and carbonyl groups near the amino-terminal and carboxyl-terminal ends of the helix, respectively. For this reason, negatively charged amino acids are often found near the amino terminus of the helical segment, where they have a stabilizing interaction with the positive charge of the helix dipole; a positively charged amino acid at the amino-terminal end is destabilizing. The opposite is true at the carboxyl-terminal end of the helical segment.

Thus, five different kinds of constraints affect the stability of an $\alpha$ helix: (1) the electrostatic repulsion (or attraction) between successive amino acid residues with charged R groups, (2) the bulkiness of adjacent R groups, (3) the interactions between R groups spaced three (or four) residues apart, (4) the occurrence of Pro and Gly residues, and (5) the interaction between amino acid residues at the ends of the helical segment and the electric dipole inherent to the $\alpha$ helix. The tendency of a given segment of a polypeptide chain to fold up as an $\alpha$ helix therefore depends on the identity and sequence of amino acid residues within the segment.

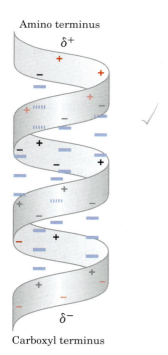

**FIGURE 4–6 Helix dipole.** The electric dipole of a peptide bond (see Fig. 4–2a) is transmitted along an $\alpha$-helical segment through the intrachain hydrogen bonds, resulting in an overall helix dipole. In this illustration, the amino and carbonyl constituents of each peptide bond are indicated by + and − symbols, respectively. Non-hydrogen-bonded amino and carbonyl constituents in the peptide bonds near each end of the $\alpha$-helical region are shown in red.

## The β Conformation Organizes Polypeptide Chains into Sheets

**Protein Architecture—β Sheet** Pauling and Corey predicted a second type of repetitive structure, the **β conformation.** This is a more extended conformation of polypeptide chains, and its structure has been confirmed by x-ray analysis. In the β conformation, the backbone of the polypeptide chain is extended into a zigzag rather than helical structure (Fig. 4–7). The zigzag polypeptide chains can be arranged side by side to form a structure resembling a series of pleats. In this arrangement, called a **β sheet,** hydrogen bonds are formed between adjacent segments of polypeptide chain. The individual segments that form a β sheet are usually nearby on the polypeptide chain, but can also be quite distant from each other in the linear sequence of the polypeptide; they may even be segments in different polypeptide chains. The R groups of adjacent amino acids protrude from the zigzag structure in opposite directions, creating the alternating pattern seen in the side views in Figure 4–7.

The adjacent polypeptide chains in a β sheet can be either parallel or antiparallel (having the same or opposite amino-to-carboxyl orientations, respectively). The structures are somewhat similar, although the repeat period is shorter for the parallel conformation (6.5 Å, versus 7 Å for antiparallel) and the hydrogen-bonding patterns are different.

Some protein structures limit the kinds of amino acids that can occur in the β sheet. When two or more β sheets are layered close together within a protein, the R groups of the amino acid residues on the touching surfaces must be relatively small. β-Keratins such as silk fibroin and the fibroin of spider webs have a very high content of Gly and Ala residues, the two amino acids with the smallest R groups. Indeed, in silk fibroin Gly and Ala alternate over large parts of the sequence.

## β Turns Are Common in Proteins

**Protein Architecture—β Turn** In globular proteins, which have a compact folded structure, nearly one-third of the amino acid residues are in turns or loops where the polypeptide chain reverses direction (Fig. 4–8). These are the connecting elements that link successive runs of α helix or β conformation. Particularly common are **β turns** that connect the ends of two adjacent segments of an antiparallel β sheet. The structure is a 180° turn involving four amino acid residues, with the carbonyl oxygen of the first residue forming a hydrogen bond with the amino-group hydrogen of the fourth. The peptide groups of the central two residues do not participate in any interresidue hydrogen bonding. Gly and Pro residues often occur in β turns, the former because it is small and flexible, the latter because peptide bonds

**(a) Antiparallel**

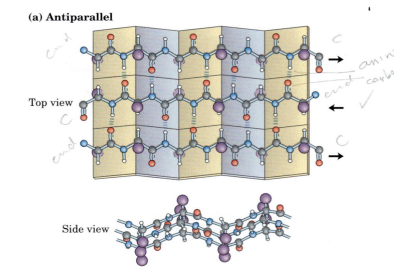

**(b) Parallel**

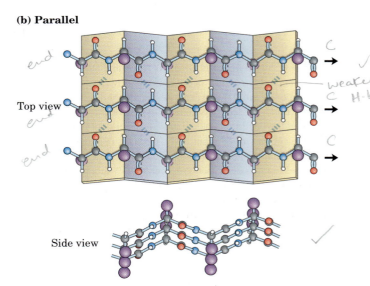

**FIGURE 4–7 The β conformation of polypeptide chains.** These top and side views reveal the R groups extending out from the β sheet and emphasize the pleated shape described by the planes of the peptide bonds. (An alternative name for this structure is β-pleated sheet.) Hydrogen-bond cross-links between adjacent chains are also shown. **(a)** Antiparallel β sheet, in which the amino-terminal to carboxyl-terminal orientation of adjacent chains (arrows) is inverse. **(b)** Parallel β sheet.

involving the imino nitrogen of proline readily assume the cis configuration (Fig. 4–8b), a form that is particularly amenable to a tight turn. Of the several types of β turns, the two shown in Figure 4–8a are the most common. Beta turns are often found near the surface of a protein, where the peptide groups of the central two amino acid residues in the turn can hydrogen-bond with water. Considerably less common is the γ turn, a three-residue turn with a hydrogen bond between the first and third residues.

**(a) β Turns**

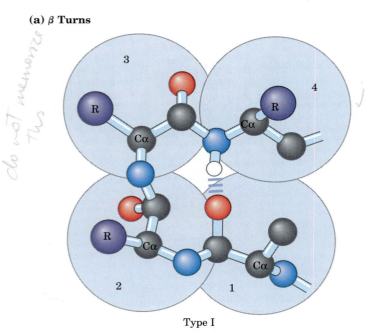

Type I

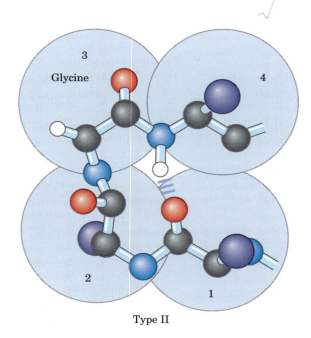

Type II

**FIGURE 4–8** Structures of β turns. **(a)** Type I and type II β turns are most common; type I turns occur more than twice as frequently as type II. Type II β turns always have Gly as the third residue. Note the hydrogen bond between the peptide groups of the first and fourth residues of the bends. (Individual amino acid residues are framed by large blue circles.) **(b)** The trans and cis isomers of a peptide bond involving the imino nitrogen of proline. Of the peptide bonds between amino acid residues other than Pro, over 99.95% are in the trans configuration. For peptide bonds involving the imino nitrogen of proline, however, about 6% are in the cis configuration; many of these occur at β turns.

**(b) Proline isomers**

trans          cis

## Common Secondary Structures Have Characteristic Bond Angles and Amino Acid Content

The α helix and the β conformation are the major repetitive secondary structures in a wide variety of proteins, although other repetitive structures do exist in some specialized proteins (an example is collagen; see Fig. 4–13 on page 128). Every type of secondary structure can be completely described by the bond angles $\phi$ and $\psi$ at each residue. As shown by a Ramachandran plot, the α helix and β conformation fall within a relatively restricted range of sterically allowed structures (Fig. 4–9a). Most values of $\phi$ and $\psi$ taken from known protein structures fall into the expected regions, with high concentrations near the α helix and β conformation values as predicted (Fig. 4–9b). The only amino acid residue often found in a conformation outside these regions is glycine. Because its side chain, a single hydrogen atom, is small, a Gly residue can take part in many conformations that are sterically forbidden for other amino acids.

Some amino acids are accommodated better than others in the different types of secondary structures. An overall summary is presented in Figure 4–10. Some biases, such as the common presence of Pro and Gly residues in β turns and their relative absence in α helices, are readily explained by the known constraints on the different secondary structures. Other evident biases may be explained by taking into account the sizes or charges of side chains, but not all the trends shown in Figure 4–10 are understood.

## SUMMARY 4.2 Protein Secondary Structure

- Secondary structure is the regular arrangement of amino acid residues in a segment of a polypeptide chain, in which each residue is spatially related to its neighbors in the same way.

- The most common secondary structures are the α helix, the β conformation, and β turns.

- The secondary structure of a polypeptide segment can be completely defined if the $\phi$ and $\psi$ angles are known for all amino acid residues in that segment.

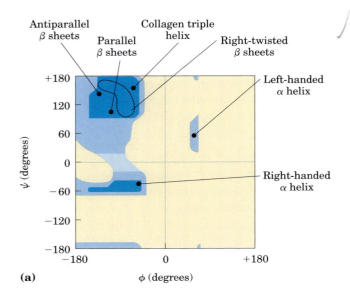

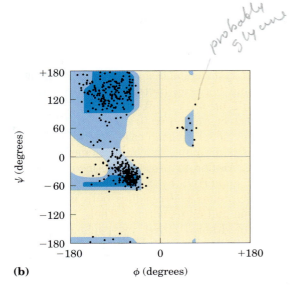

*probably glycine*

**FIGURE 4–9** **Ramachandran plots for a variety of structures. (a)** The values of $\phi$ and $\psi$ for various allowed secondary structures are overlaid on the plot from Figure 4–3. Although left-handed $\alpha$ helices extending over several amino acid residues are theoretically possible, they have not been observed in proteins. **(b)** The values of $\phi$ and $\psi$ for all the amino acid residues except Gly in the enzyme pyruvate kinase (isolated from rabbit) are overlaid on the plot of theoretically allowed conformations (Fig. 4–3). The small, flexible Gly residues were excluded because they frequently fall outside the expected ranges (blue).

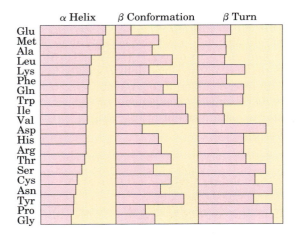

**FIGURE 4–10** Relative probabilities that a given amino acid will occur in the three common types of secondary structure.

## 4.3 Protein Tertiary and Quaternary Structures

**Protein Architecture—Introduction to Tertiary Structure** The overall three-dimensional arrangement of all atoms in a protein is referred to as the protein's **tertiary structure.** Whereas the term "secondary structure" refers to the spatial arrangement of amino acid residues that are adjacent in the primary structure, tertiary structure includes *longer-range* aspects of amino acid sequence. Amino acids that are far apart in the polypeptide sequence and that reside in different types of secondary structure may interact within the completely folded structure of a protein. The location of bends (including

$\beta$ turns) in the polypeptide chain and the direction and angle of these bends are determined by the number and location of specific bend-producing residues, such as Pro, Thr, Ser, and Gly. Interacting segments of polypeptide chains are held in their characteristic tertiary positions by different kinds of weak bonding interactions (and sometimes by covalent bonds such as disulfide cross-links) between the segments.

Some proteins contain two or more separate polypeptide chains, or subunits, which may be identical or different. The arrangement of these protein subunits in three-dimensional complexes constitutes **quaternary structure.**

In considering these higher levels of structure, it is useful to classify proteins into two major groups: **fibrous proteins,** having polypeptide chains arranged in long strands or sheets, and **globular proteins,** having polypeptide chains folded into a spherical or globular shape. The two groups are structurally distinct: fibrous proteins usually consist largely of a single type of secondary structure; globular proteins often contain several types of secondary structure. The two groups differ functionally in that the structures that provide support, shape, and external protection to vertebrates are made of fibrous proteins, whereas most enzymes and regulatory proteins are globular proteins. Certain fibrous proteins played a key role in the development of our modern understanding of protein structure and provide particularly clear examples of the relationship between structure and function. We begin our discussion with fibrous proteins, before turning to the more complex folding patterns observed in globular proteins.

## Fibrous Proteins Are Adapted for a Structural Function

🔵 **Protein Architecture—Tertiary Structure of Fibrous Proteins**
α-Keratin, collagen, and silk fibroin nicely illustrate the relationship between protein structure and biological function (Table 4–1). Fibrous proteins share properties that give strength and/or flexibility to the structures in which they occur. In each case, the fundamental structural unit is a simple repeating element of secondary structure. All fibrous proteins are insoluble in water, a property conferred by a high concentration of hydrophobic amino acid residues both in the interior of the protein and on its surface. These hydrophobic surfaces are largely buried by packing many similar polypeptide chains together to form elaborate supramolecular complexes. The underlying structural simplicity of fibrous proteins makes them particularly useful for illustrating some of the fundamental principles of protein structure discussed above.

**α-Keratin**    The α-keratins have evolved for strength. Found in mammals, these proteins constitute almost the entire dry weight of hair, wool, nails, claws, quills, horns, hooves, and much of the outer layer of skin. The α-keratins are part of a broader family of proteins called intermediate filament (IF) proteins. Other IF proteins are found in the cytoskeletons of animal cells. All IF proteins have a structural function and share structural features exemplified by the α-keratins.

The α-keratin helix is a right-handed α helix, the same helix found in many other proteins. Francis Crick

and Linus Pauling in the early 1950s independently suggested that the α helices of keratin were arranged as a coiled coil. Two strands of α-keratin, oriented in parallel (with their amino termini at the same end), are wrapped about each other to form a supertwisted coiled coil. The supertwisting amplifies the strength of the overall structure, just as strands are twisted to make a strong rope (Fig. 4–11). The twisting of the axis of an α helix to form a coiled coil explains the discrepancy between the 5.4 Å per turn predicted for an α helix by Pauling and Corey and the 5.15 to 5.2 Å repeating structure observed in the x-ray diffraction of hair (p. 120). The helical path of the supertwists is left-handed, opposite in sense to the α helix. The surfaces where the two α helices touch are made up of hydrophobic amino acid residues, their R groups meshed together in a regular interlocking pattern. This permits a close packing of the polypeptide chains within the left-handed supertwist. Not surprisingly, α-keratin is rich in the hydrophobic residues Ala, Val, Leu, Ile, Met, and Phe.

An individual polypeptide in the α-keratin coiled coil has a relatively simple tertiary structure, dominated by an α-helical secondary structure with its helical axis twisted in a left-handed superhelix. The intertwining of the two α-helical polypeptides is an example of quaternary structure. Coiled coils of this type are common structural elements in filamentous proteins and in the muscle protein myosin (see Fig. 5–29). The quaternary structure of α-keratin can be quite complex. Many coiled coils can be assembled into large supramolecular complexes, such as the arrangement of α-keratin to form the intermediate filament of hair (Fig. 4–11b).

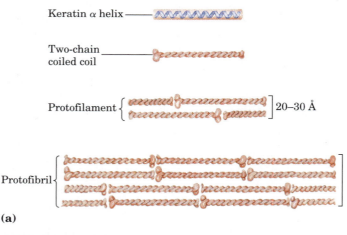

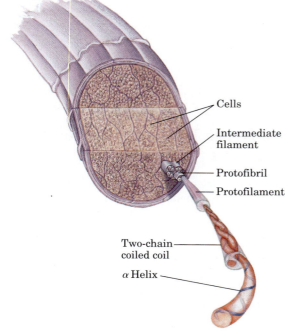

**(b) Cross section of a hair**

**FIGURE 4–11   Structure of hair. (a)** Hair α-keratin is an elongated α helix with somewhat thicker elements near the amino and carboxyl termini. Pairs of these helices are interwound in a left-handed sense to form two-chain coiled coils. These then combine in higher-order structures called protofilaments and protofibrils. About four protofibrils—32 strands of α-keratin altogether—combine to form an intermediate filament. The individual two-chain coiled coils in the various substructures also appear to be interwound, but the handedness of the interwinding and other structural details are unknown. **(b)** A hair is an array of many α-keratin filaments, made up of the substructures shown in **(a)**.

### TABLE 4–1   Secondary Structures and Properties of Fibrous Proteins

| Structure | Characteristics | Examples of occurrence |
|---|---|---|
| α Helix, cross-linked by disulfide bonds | Tough, insoluble protective structures of varying hardness and flexibility | α-Keratin of hair, feathers, and nails |
| β Conformation | Soft, flexible filaments | Silk fibroin |
| Collagen triple helix | High tensile strength, without stretch | Collagen of tendons, bone matrix |

The strength of fibrous proteins is enhanced by covalent cross-links between polypeptide chains within the multihelical "ropes" and between adjacent chains in a supramolecular assembly. In α-keratins, the cross-links stabilizing quaternary structure are disulfide bonds (Box 4–2). In the hardest and toughest α-keratins, such as those of rhinoceros horn, up to 18% of the residues are cysteines involved in disulfide bonds.

**Collagen**   Like the α-keratins, collagen has evolved to provide strength. It is found in connective tissue such as tendons, cartilage, the organic matrix of bone, and the cornea of the eye. The collagen helix is a unique secondary structure quite distinct from the α helix. It is left-handed and has three amino acid residues per turn (Fig. 4–12). Collagen is also a coiled coil, but one with distinct tertiary and quaternary structures: three separate polypeptides, called α chains (not to be confused with α helices), are supertwisted about each other (Fig. 4–12c). The superhelical twisting is right-handed in collagen, opposite in sense to the left-handed helix of the α chains.

There are many types of vertebrate collagen. Typically they contain about 35% Gly, 11% Ala, and 21% Pro and 4-Hyp (4-hydroxyproline, an uncommon amino acid; see Fig. 3–8a). The food product gelatin is derived

---

## BOX 4–2   THE WORLD OF BIOCHEMISTRY

### Permanent Waving Is Biochemical Engineering

When hair is exposed to moist heat, it can be stretched. At the molecular level, the α helices in the α-keratin of hair are stretched out until they arrive at the fully extended β conformation. On cooling they spontaneously revert to the α-helical conformation. The characteristic "stretchability" of α-keratins, and their numerous disulfide cross-linkages, are the basis of permanent waving. The hair to be waved or curled is first bent around a form of appropriate shape. A solution of a reducing agent, usually a compound containing a thiol or sulfhydryl group (—SH), is then applied with heat. The reducing agent cleaves the cross-linkages by reducing each disulfide bond to form two Cys residues. The moist heat breaks hydrogen bonds and causes the α-helical structure of the polypeptide chains to uncoil. After a time the reducing solution is removed, and an oxidizing agent is added to establish *new* disulfide bonds between pairs of Cys residues of adjacent polypeptide chains, but not the same pairs as before the treatment. After the hair is washed and cooled, the polypeptide chains revert to their α-helical conformation. The hair fibers now curl in the desired fashion because the new disulfide cross-linkages exert some torsion or twist on the bundles of α-helical coils in the hair fibers. A permanent wave is not truly permanent, because the hair grows; in the new hair replacing the old, the α-keratin has the natural, nonwavy pattern of disulfide bonds.

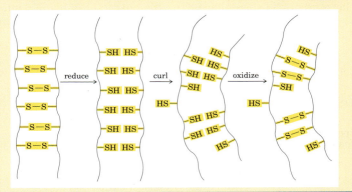

Collagen Triple helix

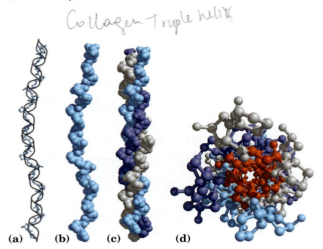

**(a)** **(b)** **(c)** **(d)**

**FIGURE 4–12** **Structure of collagen.** (Derived from PDB ID 1CGD.) **(a)** The α chain of collagen has a repeating secondary structure unique to this protein. The repeating tripeptide sequence Gly–X–Pro or Gly–X–4-Hyp adopts a left-handed helical structure with three residues per turn. The repeating sequence used to generate this model is Gly–Pro–4-Hyp. **(b)** Space-filling model of the same α chain. **(c)** Three of these helices (shown here in gray, blue, and purple) wrap around one another with a right-handed twist. **(d)** The three-stranded collagen superhelix shown from one end, in a ball-and-stick representation. Gly residues are shown in red. Glycine, because of its small size, is required at the tight junction where the three chains are in contact. The balls in this illustration do not represent the van der Waals radii of the individual atoms. The center of the three-stranded superhelix is not hollow, as it appears here, but is very tightly packed.

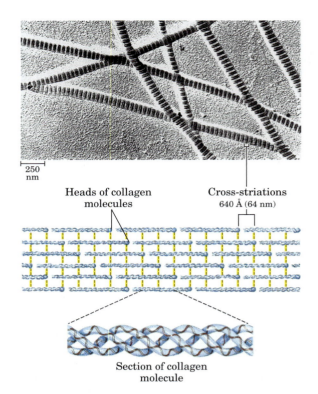

**FIGURE 4–13** **Structure of collagen fibrils.** Collagen ($M_r$ 300,000) is a rod-shaped molecule, about 3,000 Å long and only 15 Å thick. Its three helically intertwined α chains may have different sequences, but each has about 1,000 amino acid residues. Collagen fibrils are made up of collagen molecules aligned in a staggered fashion and cross-linked for strength. The specific alignment and degree of cross-linking vary with the tissue and produce characteristic cross-striations in an electron micrograph. In the example shown here, alignment of the head groups of every fourth molecule produces striations 640 Å apart.

from collagen; it has little nutritional value as a protein, because collagen is extremely low in many amino acids that are essential in the human diet. The unusual amino acid content of collagen is related to structural constraints unique to the collagen helix. The amino acid sequence in collagen is generally a repeating tripeptide unit, Gly–X–Y, where X is often Pro, and Y is often 4-Hyp. Only Gly residues can be accommodated at the very tight junctions between the individual α chains (Fig. 4–12d); The Pro and 4-Hyp residues permit the sharp twisting of the collagen helix. The amino acid sequence and the supertwisted quaternary structure of collagen allow a very close packing of its three polypeptides. 4-Hydroxyproline has a special role in the structure of collagen—and in human history (Box 4–3).

The tight wrapping of the α chains in the collagen triple helix provides tensile strength greater than that of a steel wire of equal cross section. Collagen fibrils (Fig. 4–13) are supramolecular assemblies consisting of triple-helical collagen molecules (sometimes referred to as tropocollagen molecules) associated in a variety of ways to provide different degrees of tensile strength. The α chains of collagen molecules and the collagen molecules of fibrils are cross-linked by unusual types of covalent bonds involving Lys, HyLys (5-hydroxylysine; see Fig. 3–8a), or His residues that are present at a few of the X and Y positions in collagens. These links create uncommon amino acid residues such as dehydrohydroxylysinonorleucine. The increasingly rigid and brittle character of aging connective tissue results from accumulated covalent cross-links in collagen fibrils.

H—N
‖
CH—CH₂—CH₂—CH₂—CH=N—CH₂—CH—CH₂—CH₂—CH
O=C                                          OH              N—H
                                                              C=O

Polypeptide          Lys residue              HyLys          Polypeptide
chain                minus ε-amino            residue        chain
                     group (norleucine)

Dehydrohydroxylysinonorleucine

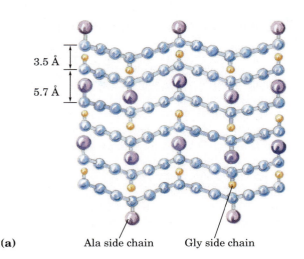

(a)                    Ala side chain    Gly side chain

(b)                                                  ⊢——⊣ 70 μm

**FIGURE 4–14 Structure of silk.** The fibers used to make silk cloth or a spider web are made up of the protein fibroin. **(a)** Fibroin consists of layers of antiparallel β sheets rich in Ala (purple) and Gly (yellow) residues. The small side chains interdigitate and allow close packing of each layered sheet, as shown in this side view. **(b)** Strands of fibroin (blue) emerge from the spinnerets of a spider in this colorized electron micrograph.

A typical mammal has more than 30 structural variants of collagen, particular to certain tissues and each somewhat different in sequence and function. Some human genetic defects in collagen structure illustrate the close relationship between amino acid sequence and three-dimensional structure in this protein. Osteogenesis imperfecta is characterized by abnormal bone formation in babies; Ehlers-Danlos syndrome is characterized by loose joints. Both conditions can be lethal, and both result from the substitution of an amino acid residue with a larger R group (such as Cys or Ser) for a single Gly residue in each α chain (a different Gly residue in each disorder). These single-residue substitutions have a catastrophic effect on collagen function because they disrupt the Gly–X–Y repeat that gives collagen its unique helical structure. Given its role in the collagen triple helix (Fig. 4–12d), Gly cannot be replaced by another amino acid residue without substantial deleterious effects on collagen structure. ∎

***Silk Fibroin*** Fibroin, the protein of silk, is produced by insects and spiders. Its polypeptide chains are predominantly in the β conformation. Fibroin is rich in Ala and Gly residues, permitting a close packing of β sheets and an interlocking arrangement of R groups (Fig. 4–14). The overall structure is stabilized by extensive hydrogen bonding between all peptide linkages in the polypeptides of each β sheet and by the optimization of van der Waals interactions between sheets. Silk does not stretch, because the β conformation is already highly extended (Fig. 4–7; see also Fig. 4–15). However, the structure is flexible because the sheets are held together by numerous weak interactions rather than by covalent bonds such as the disulfide bonds in α-keratins.

## Structural Diversity Reflects Functional Diversity in Globular Proteins

In a globular protein, different segments of a polypeptide chain (or multiple polypeptide chains) fold back on each other. As illustrated in Figure 4–15, this folding generates a compact form relative to polypeptides in a fully extended conformation. The folding also provides the structural diversity necessary for proteins to carry out a wide array of biological functions. Globular proteins include enzymes, transport proteins, motor proteins, regulatory proteins, immunoglobulins, and proteins with many other functions.

As a new millennium begins, the number of known three-dimensional protein structures is in the thousands and more than doubles every two years. This wealth of structural information is revolutionizing our understanding of protein structure, the relation of structure

β Conformation
2,000 × 5 Å

α Helix                          Native globular form
900 × 11 Å                       100 × 60 Å

**FIGURE 4–15 Globular protein structures are compact and varied.** Human serum albumin ($M_r$ 64,500) has 585 residues in a single chain. Given here are the approximate dimensions its single polypeptide chain would have if it occurred entirely in extended β conformation or as an α helix. Also shown is the size of the protein in its native globular form, as determined by X-ray crystallography; the polypeptide chain must be very compactly folded to fit into these dimensions.

**BOX 4–3** **BIOCHEMISTRY IN MEDICINE**

## Why Sailors, Explorers, and College Students Should Eat Their Fresh Fruits and Vegetables

. . . from this misfortune, together with the unhealthiness of the country, where there never falls a drop of rain, we were stricken with the "camp-sickness," which was such that the flesh of our limbs all shrivelled up, and the skin of our legs became all blotched with black, mouldy patches, like an old jack-boot, and proud flesh came upon the gums of those of us who had the sickness, and none escaped from this sickness save through the jaws of death. The signal was this: when the nose began to bleed, then death was at hand . . .

—*from* The Memoirs of the Lord of Joinville, *ca. 1300*

This excerpt describes the plight of Louis IX's army toward the end of the Seventh Crusade (1248–1254), immediately preceding the battle of Fariskur, where the scurvy-weakened Crusader army was destroyed by the Egyptians. What was the nature of the malady afflicting these thirteenth-century soldiers?

Scurvy is caused by lack of vitamin C, or ascorbic acid (ascorbate). Vitamin C is required for, among other things, the hydroxylation of proline and lysine in collagen; scurvy is a deficiency disease characterized by general degeneration of connective tissue. Manifestations of advanced scurvy include numerous small hemorrhages caused by fragile blood vessels, tooth loss, poor wound healing and the reopening of old wounds, bone pain and degeneration, and eventually heart failure. Despondency and oversensitivity to stimuli of many kinds are also observed. Milder cases of vitamin C deficiency are accompanied by fatigue, irritability, and an increased severity of respiratory tract infections. Most animals make large amounts of vitamin C, converting glucose to ascorbate in four enzymatic steps. But in the course of evolution, humans and some other animals—gorillas, guinea pigs, and fruit bats—have lost the last enzyme in this pathway and must obtain ascorbate in their diet. Vitamin C is available in a wide range of fruits and vegetables. Until 1800, however, it was often absent in the dried foods and other food supplies stored for winter or for extended travel.

Scurvy was recorded by the Egyptians in 1500 BCE, and it is described in the fifth century BCE writings of Hippocrates. Although scurvy played a critical role in medieval wars and made regular winter appearances in northern climates, it did not come to wide public notice until the European voyages of discovery from 1500 to 1800. The first circumnavigation of the globe, led by Ferdinand Magellan (1520), was accomplished only with the loss of more than 80% of his crew to scurvy. Vasco da Gama lost two-thirds of his crew to the same fate during his first exploration of trade routes to India (1499). During Jacques Cartier's second voyage to explore the St. Lawrence River (1535–1536), his band suffered numerous fatalities and was threatened with complete disaster until the native Americans taught the men to make a cedar tea that cured and prevented scurvy (it contained vitamin C) (Fig. 1). It is estimated that a million sailors died of scurvy in the years 1600 to 1800. Winter outbreaks of scurvy in Europe were gradually eliminated in the nineteenth century as the cultivation of the potato, introduced from South America, became widespread.

In 1747, James Lind, a Scottish surgeon in the Royal Navy (Fig. 2), carried out the first controlled clinical study in recorded history. During an extended voyage on the 50-gun warship *HMS Salisbury*, Lind selected 12 sailors suffering from scurvy and separated them into groups of two. All 12 received the same diet, except that each group was given a different remedy for scurvy from among those recommended at the time. The sailors given lemons and oranges recovered and returned to duty. The sailors given boiled apple juice improved slightly. The remainder continued to deteriorate. Lind's *Treatise on the Scurvy* was published in 1753, but inaction persisted in the Royal Navy for another 40 years.

**FIGURE 1** Iroquois showing Jacques Cartier how to make cedar tea as a remedy for scurvy.

**FIGURE 2** James Lind, 1716–1794; naval surgeon, 1739–1748.

In 1795 the British admiralty finally mandated a ration of concentrated lime or lemon juice for all British sailors (hence the name "limeys"). Scurvy continued to be a problem in some other parts of the world until 1932, when Hungarian scientist Albert Szent-Györgyi, and W. A. Waugh and C. G. King at the University of Pittsburgh, isolated and synthesized ascorbic acid.

L-Ascorbic acid (vitamin C) is a white, odorless, crystalline powder. It is freely soluble in water and relatively insoluble in organic solvents. In a dry state, away from light, it is stable for a considerable length of time. The appropriate daily intake of this vitamin is still in dispute. The recommended daily allowance in the United States is 60 mg (Australia and the United Kingdom recommend 30 to 40 mg; Russia recommends 100 mg). Higher doses of vitamin C are sometimes recommended, although the benefit of such a regimen is disputed. Notably, animals that synthesize their own vitamin C maintain levels found in humans only if they consume hundreds of times the recommended daily allowance. Along with citrus fruits and almost all other fresh fruits, other good sources of vitamin C include peppers, tomatoes, potatoes, and broccoli. The vitamin C of fruits and vegetables is destroyed by overcooking or prolonged storage.

So why is ascorbate so necessary to good health? Of particular interest to us here is its role in the formation of collagen. The proline derivative 4(R)-L-hydroxyproline (4-Hyp) plays an essential role in the folding of collagen and in maintaining its structure. As noted in the text, collagen is constructed of the repeating tripeptide unit Gly–X–Y, where X and Y are generally Pro or 4-Hyp. A constructed peptide with 10 Gly–Pro–Pro repeats will fold to form a collagen triple helix, but the structure melts at 41 °C. If the 10 repeats are changed to Gly–Pro–4-Hyp, the melting temperature jumps to 69 °C. The stability of collagen arises from the detailed structure of the collagen helix, determined independently by Helen Berman and Adriana Zagari and their colleagues. The proline ring is normally found as a mixture of two puckered conformations, called $C_\gamma$-endo and $C_\gamma$-exo (Fig. 3). The collagen helix structure requires the Pro residue in the Y positions to be in the $C_\gamma$-exo conformation, and it is this conformation that is enforced by the hydroxyl substitution at C-4 in 4-hydroxyproline. However, the collagen structure requires the Pro residue in the X positions to have the $C_\gamma$-endo conformation, and introduction of 4-Hyp here can destabilize the helix. The inability to hydroxylate the Pro at the Y positions when vitamin C is absent leads to collagen instability and the connective tissue problems seen in scurvy.

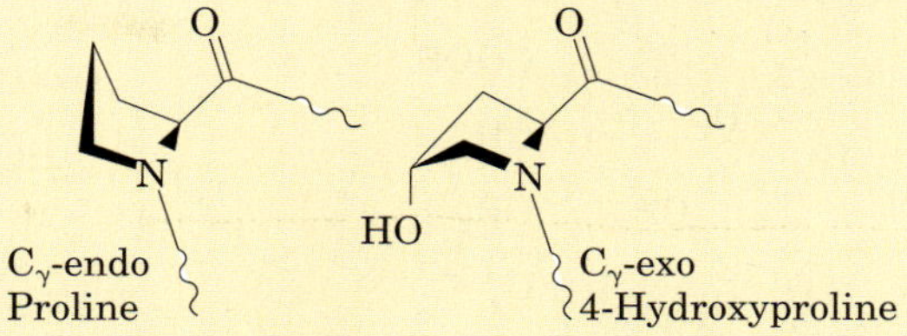

FIGURE 3  The $C_\gamma$-endo conformation of proline and the $C_\gamma$-exo conformation of 4-hydroxyproline.

The hydroxylation of specific Pro residues in procollagen, the precursor of collagen, requires the action of the enzyme prolyl 4-hydroxylase. This enzyme ($M_r$ 240,000) is an $\alpha_2\beta_2$ tetramer in all vertebrate sources. The proline-hydroxylating activity is found in the $\alpha$ subunits. (Researchers were surprised to find that the $\beta$ subunits are identical to the enzyme protein disulfide isomerase (PDI; p. 152); these subunits do not participate in the prolyl hydroxylation activity.) Each $\alpha$ subunit contains one atom of nonheme iron ($Fe^{2+}$), and the enzyme is one of a class of hydroxylases that require $\alpha$-ketoglutarate in their reactions.

In the normal prolyl 4-hydroxylase reaction (Fig. 4a), one molecule of $\alpha$-ketoglutarate and one of $O_2$ bind to the enzyme. The $\alpha$-ketoglutarate is oxidatively decarboxylated to form $CO_2$ and succinate. The remaining oxygen atom is then used to hydroxylate an appropriate Pro residue in procollagen. No ascorbate is needed in this reaction. However, prolyl 4-hydroxylase also catalyzes an oxidative decarboxylation of $\alpha$-ketoglutarate that is not coupled to proline hydroxylation—and this is the reaction that requires ascorbate (Fig. 4b). During this reaction, the heme $Fe^{2+}$ becomes oxidized, and the oxidized form of the enzyme is inactive—unable to hydroxylate proline. The ascorbate consumed in the reaction presumably functions to reduce the heme iron and restore enzyme activity.

But there is more to the vitamin C story than proline hydroxylation. Very similar hydroxylation reactions generate the less abundant 3-hydroxyproline and 5-hydroxylysine residues that also occur in collagen. The enzymes that catalyze these reactions are members of the same $\alpha$-ketoglutarate-dependent dioxygenase family, and for all these enzymes ascorbate plays the same role. These dioxygenases are just a few of the dozens of closely related enzymes that play a variety of metabolic roles in different classes of organisms. Ascorbate serves other roles too. It is an antioxidant, reacting enzymatically and nonenzymatically with reactive oxygen species, which in mammals play an important role in aging and cancer.

*(continued on next page)*

## BOX 4–3    BIOCHEMISTRY IN MEDICINE (continued from previous page)

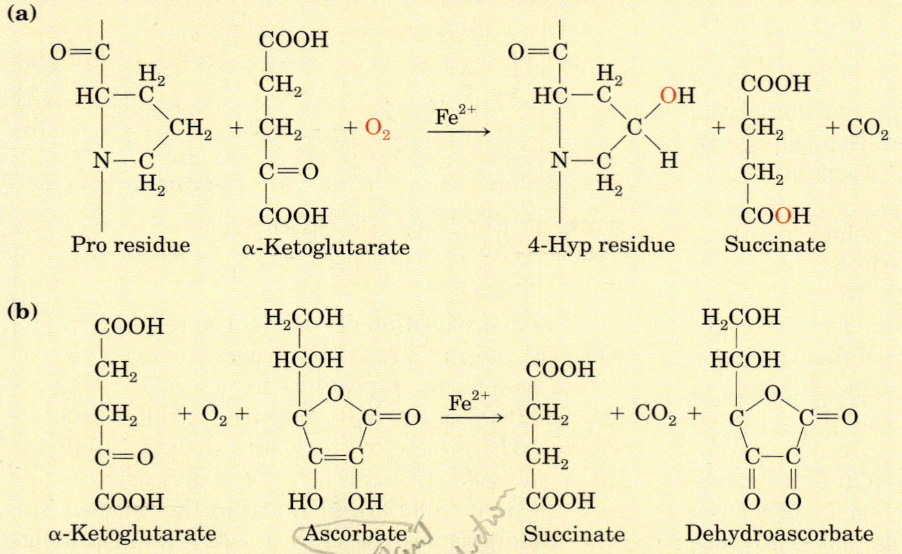

**(a)**

Pro residue   α-Ketoglutarate    4-Hyp residue   Succinate

**(b)**

α-Ketoglutarate   Ascorbate   Succinate   Dehydroascorbate

**FIGURE 4** The reactions catalyzed by prolyl 4-hydroxylase. **(a)** The normal reaction, coupled to proline hydroxylation, which does not require ascorbate. The fate of the two oxygen atoms from $O_2$ is shown in red. **(b)** The uncoupled reaction, in which α-ketoglutarate is oxidatively decarboxylated without hydroxylation of proline. Ascorbate is consumed stoichiometrically in this process as it is converted to dehydroascorbate.

In plants, ascorbate is required as a substrate for the enzyme ascorbate peroxidase, which converts $H_2O_2$ to water. The peroxide is generated from the $O_2$ produced in photosynthesis, an unavoidable consequence of generating $O_2$ in a compartment laden with powerful oxidation-reduction systems (Chapter 19). Ascorbate is a also a precursor of oxalate and tartrate in plants, and is involved in the hydroxylation of Pro residues in cell wall proteins called extensins. Ascorbate is found in all subcellular compartments of plants, at concentrations of 2 to 25 mM—which is why plants are such good sources of vitamin C.

Scurvy remains a problem today. The malady is still encountered not only in remote regions where nutritious food is scarce but, surprisingly, on U.S. college campuses. The only vegetables consumed by some students are those in tossed salads, and days go by without these young adults consuming fruit. A 1998 study of 230 students at Arizona State University revealed that 10% had serious vitamin C deficiencies, and 2 students had vitamin C levels so low that they probably had scurvy. Only half the students in the study consumed the recommended daily allowance of vitamin C.

Eat your fresh fruit and vegetables.

---

to function, and even the evolutionary paths by which proteins arrived at their present state, which can be glimpsed in the family resemblances that are revealed as protein databases are sifted and sorted. The sheer variety of structures can seem daunting. Yet as new protein structures become available it is becoming increasingly clear that they are manifestations of a finite set of recognizable, stable folding patterns.

Our discussion of globular protein structure begins with the principles gleaned from the earliest protein structures to be elucidated. This is followed by a detailed description of protein substructure and comparative categorization. Such discussions are possible only because of the vast amount of information available over the Internet from resources such as the Protein Data Bank (PDB; www.rcsb.org/pdb), an archive of experimentally determined three-dimensional structures of biological macromolecules.

## Myoglobin Provided Early Clues about the Complexity of Globular Protein Structure

**Protein Architecture—Tertiary Structure of Small Globular Proteins, II. Myoglobin** The first breakthrough in understanding the three-dimensional structure of a globular protein came from x-ray diffraction studies of myoglobin carried out by John Kendrew and his colleagues in the 1950s. Myoglobin is a relatively small ($M_r$ 16,700), oxygen-binding protein of muscle cells. It functions both to store oxygen and to facilitate oxygen diffusion in rapidly contracting muscle tissue. Myoglobin contains a single polypeptide chain of 153 amino acid residues of known sequence and a single iron protoporphyrin, or heme, group. The same heme group is found in hemoglobin, the oxygen-binding protein of erythrocytes, and is responsible for the deep red-brown color of both myoglobin and hemoglobin. Myoglobin is particularly abundant in the muscles of diving mammals such as the whale, seal, and porpoise, whose muscles are so rich in this protein that they are brown. Storage and distribution of oxygen by muscle myoglobin permit these animals to remain submerged for long periods of time. The activities of myoglobin and other globin molecules are investigated in greater detail in Chapter 5.

Figure 4–16 shows several structural representations of myoglobin, illustrating how the polypeptide chain is folded in three dimensions—its tertiary structure. The red group surrounded by protein is heme. The backbone of the myoglobin molecule is made up of eight relatively straight segments of α helix interrupted by bends, some of which are β turns. The longest α helix has 23 amino acid residues and the shortest only 7; all helices are right-handed. More than 70% of the residues in myoglobin are in these α-helical regions. X-ray analysis has revealed the precise position of each of the R groups, which occupy nearly all the space within the folded chain.

Many important conclusions were drawn from the structure of myoglobin. The positioning of amino acid side chains reflects a structure that derives much of its stability from hydrophobic interactions. Most of the hydrophobic R groups are in the interior of the myoglobin molecule, hidden from exposure to water. All but two of the polar R groups are located on the outer surface of the molecule, and all are hydrated. The myoglobin molecule is so compact that its interior has room for only four molecules of water. This dense hydrophobic core is typical of globular proteins. The fraction of space occupied by atoms in an organic liquid is 0.4 to 0.6; in a typical crystal the fraction is 0.70 to 0.78, near the theoretical maximum. In a globular protein the fraction is about 0.75, comparable to that in a crystal. In this packed environment, weak interactions strengthen and reinforce each other. For example, the nonpolar side chains in the core are so close together that short-range van der Waals interactions make a significant contribution to stabilizing hydrophobic interactions.

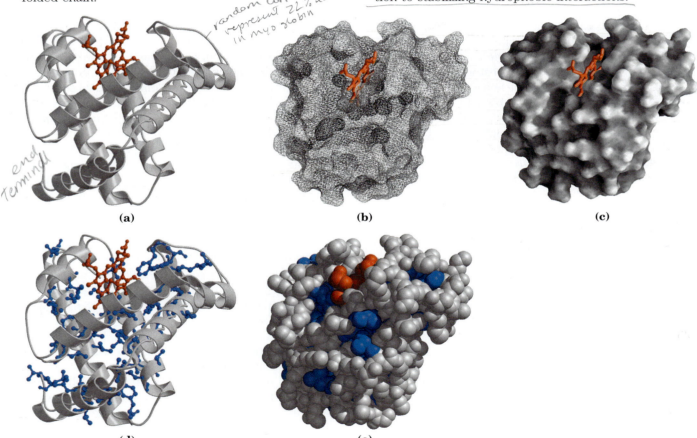

**FIGURE 4–16 Tertiary structure of sperm whale myoglobin.** (PDB ID 1MBO) The orientation of the protein is similar in all panels; the heme group is shown in red. In addition to illustrating the myoglobin structure, this figure provides examples of several different ways to display protein structure. **(a)** The polypeptide backbone, shown in a ribbon representation of a type introduced by Jane Richardson, which highlights regions of secondary structure. The α-helical regions are evident. **(b)** A "mesh" image emphasizes the protein surface. **(c)** A surface contour image is useful for visualizing pockets in the protein where other molecules might bind. **(d)** A ribbon representation, including side chains (blue) for the hydrophobic residues Leu, Ile, Val, and Phe. **(e)** A space-filling model with all amino acid side chains. Each atom is represented by a sphere encompassing its van der Waals radius. The hydrophobic residues are again shown in blue; most are not visible, because they are buried in the interior of the protein.

Deduction of the structure of myoglobin confirmed some expectations and introduced some new elements of secondary structure. As predicted by Pauling and Corey, all the peptide bonds are in the planar trans configuration. The $\alpha$ helices in myoglobin provided the first direct experimental evidence for the existence of this type of secondary structure. Three of the four Pro residues of myoglobin are found at bends (recall that proline, with its fixed $\phi$ bond angle and lack of a peptide-bond N—H group for participation in hydrogen bonds, is largely incompatible with $\alpha$-helical structure). The fourth Pro residue occurs within an $\alpha$ helix, where it creates a kink necessary for tight helix packing. Other bends contain Ser, Thr, and Asn residues, which are among the amino acids whose bulk and shape tend to make them incompatible with $\alpha$-helical structure if they are in close proximity in the amino acid sequence (p. 121).

The flat heme group rests in a crevice, or pocket, in the myoglobin molecule. The iron atom in the center of the heme group has two bonding (coordination) positions perpendicular to the plane of the heme (Fig. 4–17). One of these is bound to the R group of the His residue at position 93; the other is the site at which an $O_2$ molecule binds. Within this pocket, the accessibility of the heme group to solvent is highly restricted. This is important for function, because free heme groups in an oxygenated solution are rapidly oxidized from the ferrous $(Fe^{2+})$ form, which is active in the reversible binding of $O_2$, to the ferric $(Fe^{3+})$ form, which does not bind $O_2$.

Knowledge of the structure of myoglobin allowed researchers for the first time to understand in detail the

correlation between the structure and function of a protein. Many different myoglobin structures have been elucidated, allowing investigators to see how the structure changes when oxygen or other molecules bind to it. Hundreds of proteins have been subjected to similar analysis since then. Today, techniques such as NMR spectroscopy supplement x-ray diffraction data, providing more information on a protein's structure (Box 4–4). The ongoing sequencing of genomic DNA from many organisms (Chapter 9) has identified thousands of genes that encode proteins of known sequence but unknown function. Our first insight into what these proteins do often comes from our still-limited understanding of how primary structure determines tertiary structure, and how tertiary structure determines function.

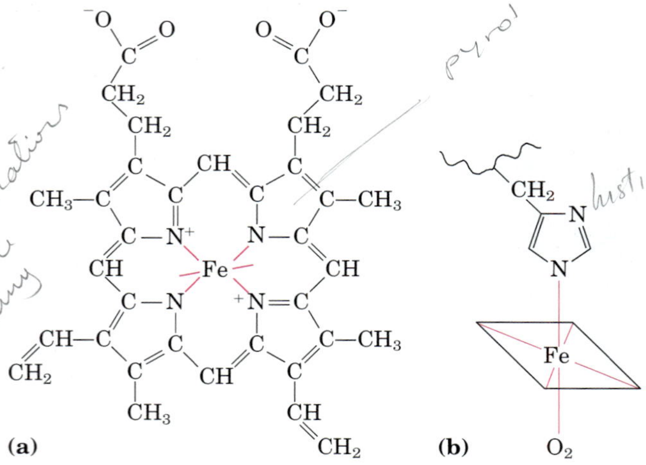

## Globular Proteins Have a Variety of Tertiary Structures

With elucidation of the tertiary structures of hundreds of other globular proteins by x-ray analysis, it became clear that myoglobin illustrates only one of many ways in which a polypeptide chain can be folded. In Figure 4–18 the structures of cytochrome $c$, lysozyme, and ribonuclease are compared. These proteins have different amino acid sequences and different tertiary structures, reflecting differences in function. All are relatively small and easy to work with, facilitating structural analysis. Cytochrome $c$ is a component of the respiratory chain of mitochondria (Chapter 19). Like myoglobin, cytochrome $c$ is a heme protein. It contains a single polypeptide chain of about 100 residues ($M_r$ 12,400) and a single heme group. In this case, the protoporphyrin of the heme group is covalently attached to the polypeptide. Only about 40% of the polypeptide is in $\alpha$-helical segments, compared with 70% of the myoglobin chain. The rest of the cytochrome $c$ chain contains $\beta$ turns and irregularly coiled and extended segments.

Lysozyme ($M_r$ 14,600) is an enzyme abundant in egg white and human tears that catalyzes the hydrolytic cleavage of polysaccharides in the protective cell walls of some families of bacteria. Lysozyme, because it can lyse, or degrade, bacterial cell walls, serves as a bactericidal agent. As in cytochrome $c$, about 40% of its 129 amino acid residues are in $\alpha$-helical segments, but the arrangement is different and some $\beta$-sheet structure is also present (Fig. 4–18). Four disulfide bonds contribute stability to this structure. The $\alpha$ helices line a long crevice in the side of the molecule, called the active site, which is the site of substrate binding and catalysis. The bacterial polysaccharide that is the substrate for lysozyme fits into this crevice. 🔵 Protein Architecture—Tertiary Structure of Small Globular Proteins, III. Lysozyme

Ribonuclease, another small globular protein ($M_r$ 13,700), is an enzyme secreted by the pancreas into the small intestine, where it catalyzes the hydrolysis of certain bonds in the ribonucleic acids present in ingested

**FIGURE 4–17 The heme group.** This group is present in myoglobin, hemoglobin, cytochromes, and many other heme proteins. **(a)** Heme consists of a complex organic ring structure, protoporphyrin, to which is bound an iron atom in its ferrous ($Fe^{2+}$) state. The iron atom has six coordination bonds, four in the plane of, and bonded to, the flat porphyrin molecule and two perpendicular to it. **(b)** In myoglobin and hemoglobin, one of the perpendicular coordination bonds is bound to a nitrogen atom of a His residue. The other is "open" and serves as the binding site for an $O_2$ molecule.

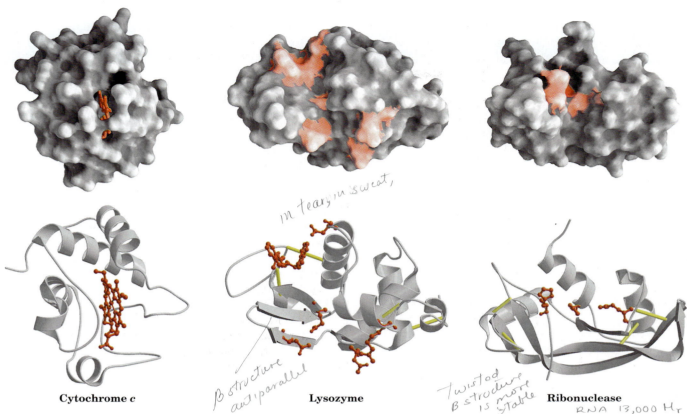

**Cytochrome c**    **Lysozyme**    **Ribonuclease**

*(handwritten annotations: "in tearyin sweat,", "β structure antiparallel", "Twisted β structure is more stable", "RNA 13,000 Mr")*

**FIGURE 4–18 Three-dimensional structures of some small proteins.** Shown here are cytochrome *c* (PDB ID 1CCR), lysozyme (PDB ID 3LYM), and ribonuclease (PDB ID 3RN3). Each protein is shown in surface contour and in a ribbon representation, in the same orientation. In the ribbon depictions, regions in the β conformation are represented by flat arrows and the α helices are represented by spiral ribbons. Key functional groups (the heme in cytochrome *c*; amino acid side chains in the active site of lysozyme and ribonuclease) are shown in red. Disulfide bonds are shown (in the ribbon representations) in yellow.

food. Its tertiary structure, determined by x-ray analysis, shows that little of its 124 amino acid polypeptide chain is in an α-helical conformation, but it contains many segments in the β conformation (Fig. 4–18). Like lysozyme, ribonuclease has four disulfide bonds between loops of the polypeptide chain.

In small proteins, hydrophobic residues are less likely to be sheltered in a hydrophobic interior—simple geometry dictates that the smaller the protein, the lower the ratio of volume to surface area. Small proteins also have fewer potential weak interactions available to stabilize them. This explains why many smaller proteins such as those in Figure 4–18 are stabilized by a number of covalent bonds. Lysozyme and ribonuclease, for example, have disulfide linkages, and the heme group in cytochrome *c* is covalently linked to the protein on two sides, providing significant stabilization of the entire protein structure.

Table 4–2 shows the proportions of α helix and β conformation (expressed as percentage of residues in each secondary structure) in several small, single-chain, globular proteins. Each of these proteins has a distinct structure, adapted for its particular biological function, but together they share several important properties. Each is folded compactly, and in each case the hydro-

phobic amino acid side chains are oriented toward the interior (away from water) and the hydrophilic side chains are on the surface. The structures are also stabilized by a multitude of hydrogen bonds and some ionic interactions.

**TABLE 4–2 Approximate Amounts of α Helix and β Conformation in Some Single-Chain Proteins**

| Protein (total residues) | Residues (%)* | |
|---|---|---|
| | α Helix | β Conformation |
| Chymotrypsin (247) | 14 | 45 |
| Ribonuclease (124) | 26 | 35 |
| Carboxypeptidase (307) | 38 | 17 |
| Cytochrome *c* (104) | 39 | 0 |
| Lysozyme (129) | 40 | 12 |
| Myoglobin (153) | 78 | 0 |

Source: Data from Cantor, C.R. & Schimmel, P.R. (1980) *Biophysical Chemistry*, Part I: *The Conformation of Biological Macromolecules*, p. 100, W. H. Freeman and Company, New York.

*Portions of the polypeptide chains that are not accounted for by α helix or β conformation consist of bends and irregularly coiled or extended stretches. Segments of α helix and β conformation sometimes deviate slightly from their normal dimensions and geometry.

BOX 4–4     WORKING IN BIOCHEMISTRY

## Methods for Determining the Three-Dimensional Structure of a Protein

### X-Ray Diffraction

The spacing of atoms in a crystal lattice can be determined by measuring the locations and intensities of spots produced on photographic film by a beam of x rays of given wavelength, after the beam has been diffracted by the electrons of the atoms. For example, x-ray analysis of sodium chloride crystals shows that $Na^+$ and $Cl^-$ ions are arranged in a simple cubic lattice. The spacing of the different kinds of atoms in complex organic molecules, even very large ones such as proteins, can also be analyzed by x-ray diffraction methods. However, the technique for analyzing crystals of complex molecules is far more laborious than for simple salt crystals. When the repeating pattern of the crystal is a molecule as large as, say, a protein, the numerous atoms in the molecule yield thousands of diffraction spots that must be analyzed by computer.

The process may be understood at an elementary level by considering how images are generated in a light microscope. Light from a point source is focused on an object. The light waves are scattered by the object, and these scattered waves are recombined by a series of lenses to generate an enlarged image of the object. The smallest object whose structure can be determined by such a system—that is, the resolving power of the microscope—is determined by the wavelength of the light, in this case visible light, with

wavelengths in the range of 400 to 700 nm. Objects smaller than half the wavelength of the incident light cannot be resolved. To resolve objects as small as proteins we must use x rays, with wavelengths in the range of 0.7 to 1.5 Å (0.07 to 0.15 nm). However, there are no lenses that can recombine x rays to form an image; instead the pattern of diffracted x rays is collected directly and an image is reconstructed by mathematical techniques.

The amount of information obtained from x-ray crystallography depends on the degree of structural order in the sample. Some important structural parameters were obtained from early studies of the diffraction patterns of the fibrous proteins arranged in fairly regular arrays in hair and wool. However, the orderly bundles formed by fibrous proteins are not crystals—the molecules are aligned side by side, but not all are oriented in the same direction. More detailed three-dimensional structural information about proteins requires a highly ordered protein crystal. Protein crystallization is something of an empirical science, and the structures of many important proteins are not yet known, simply because they have proved difficult to crystallize. Practitioners have compared making protein crystals to holding together a stack of bowling balls with cellophane tape.

Operationally, there are several steps in x-ray structural analysis (Fig. 1). Once a crystal is obtained, it is placed in an x-ray beam between the x-ray source and a detector, and a regular array of spots called re-

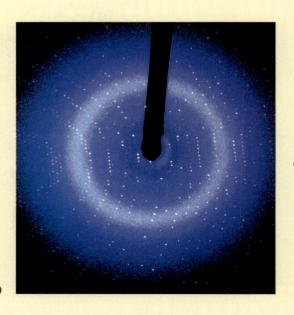

(a)

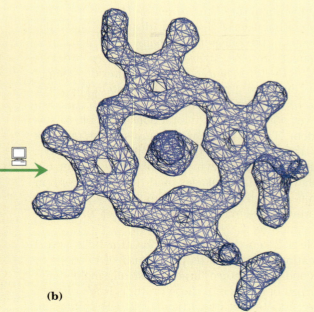

(b)

flections is generated. The spots are created by the diffracted x-ray beam, and each atom in a molecule makes a contribution to each spot. An electron-density map of the protein is reconstructed from the overall diffraction pattern of spots by using a mathematical technique called a Fourier transform. In effect, the computer acts as a "computational lens." A model for the structure is then built that is consistent with the electron-density map.

John Kendrew found that the x-ray diffraction pattern of crystalline myoglobin (isolated from muscles of the sperm whale) is very complex, with nearly 25,000 reflections. Computer analysis of these reflections took place in stages. The resolution improved at each stage, until in 1959 the positions of virtually all the non-hydrogen atoms in the protein had been determined. The amino acid sequence of the protein, obtained by chemical analysis, was consistent with the molecular model. The structures of thousands of proteins, many of them much more complex than myoglobin, have since been determined to a similar level of resolution.

The physical environment within a crystal, of course, is not identical to that in solution or in a living cell. A crystal imposes a space and time average on the structure deduced from its analysis, and x-ray diffraction studies provide little information about molecular motion within the protein. The conformation of proteins in a crystal could in principle also be affected by nonphysiological factors such as incidental protein-protein contacts within the crystal. However, when structures derived from the analysis of crystals are compared with structural information obtained by other means (such as NMR, as described below), the crystal-derived structure almost always represents a functional conformation of the protein. X-ray crystallography can be applied successfully to proteins too large to be structurally analyzed by NMR.

**Nuclear Magnetic Resonance**

An important complementary method for determining the three-dimensional structures of macromolecules is nuclear magnetic resonance (NMR). Modern NMR techniques are being used to determine the structures of ever-larger macromolecules, including carbohydrates, nucleic acids, and small to average-sized proteins. An advantage of NMR studies is that they are

*(continued on next page)*

**FIGURE 1** Steps in the determination of the structure of sperm whale myoglobin by x-ray crystallography. **(a)** X-ray diffraction patterns are generated from a crystal of the protein. **(b)** Data extracted from the diffraction patterns are used to calculate a three-dimensional electron-density map of the protein. The electron density of only part of the structure, the heme, is shown. **(c)** Regions of greatest electron density reveal the location of atomic nuclei, and this information is used to piece together the final structure. Here, the heme structure is modeled into its electron-density map. **(d)** The completed structure of sperm whale myoglobin, including the heme (PDB ID 2MBW).

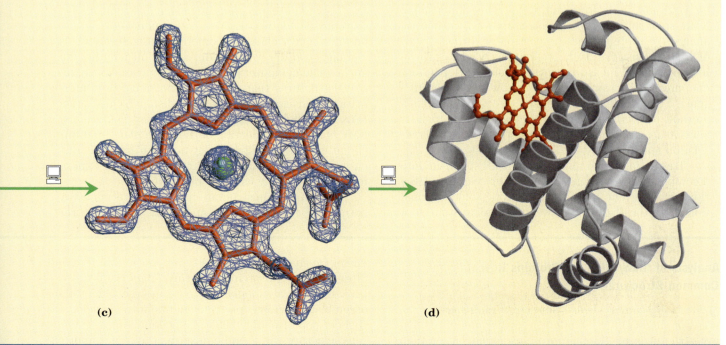

(c)

(d)

BOX 4–4    WORKING IN BIOCHEMISTRY (continued from previous page)

carried out on macromolecules in solution, whereas x-ray crystallography is limited to molecules that can be crystallized. NMR can also illuminate the dynamic side of protein structure, including conformational changes, protein folding, and interactions with other molecules.

NMR is a manifestation of nuclear spin angular momentum, a quantum mechanical property of atomic nuclei. Only certain atoms, including $^1H$, $^{13}C$, $^{15}N$, $^{19}F$, and $^{31}P$, possess the kind of nuclear spin that gives rise to an NMR signal. Nuclear spin generates a magnetic dipole. When a strong, static magnetic field is applied to a solution containing a single type of macromolecule, the magnetic dipoles are aligned in the field in one of two orientations, parallel (low energy) or antiparallel (high energy). A short (~10 $\mu s$) pulse of electromagnetic energy of suitable frequency (the resonant frequency, which is in the radio frequency range) is applied at right angles to the nuclei aligned in the magnetic field. Some energy is absorbed as nuclei switch to the high-energy state, and the absorption spectrum that results contains information about the identity of the nuclei and their immediate chemical environment. The data from many such experiments performed on a sample are averaged, increasing the signal-to-noise ratio, and an NMR spectrum such as that in Figure 2 is generated.

$^1H$ is particularly important in NMR experiments because of its high sensitivity and natural abundance. For macromolecules, $^1H$ NMR spectra can become quite complicated. Even a small protein has hundreds of $^1H$ atoms, typically resulting in a one-dimensional NMR spectrum too complex for analysis. Structural analysis of proteins became possible with the advent of two-dimensional NMR techniques (Fig. 3). These methods allow measurement of distance-dependent coupling of nuclear spins in nearby atoms through space (the nuclear Overhauser effect (NOE), in a method dubbed NOESY) or the coupling of nuclear spins in atoms connected by covalent bonds (total correlation spectroscopy, or TOCSY).

Translating a two-dimensional NMR spectrum into a complete three-dimensional structure can be a laborious process. The NOE signals provide some information about the distances between individual atoms, but

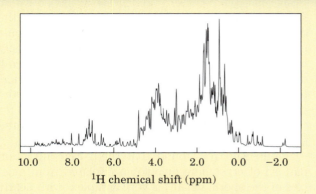

**FIGURE 2** A one-dimensional NMR spectrum of a globin from a marine blood worm. This protein and sperm whale myoglobin are very close structural analogs, belonging to the same protein structural family and sharing an oxygen-transport function.

for these distance constraints to be useful, the atoms giving rise to each signal must be identified. Complementary TOCSY experiments can help identify which NOE signals reflect atoms that are linked by covalent bonds. Certain patterns of NOE signals have been associated with secondary structures such as $\alpha$ helices. Modern genetic engineering (Chapter 9) can be used to prepare proteins that contain the rare isotopes $^{13}C$ or $^{15}N$. The new NMR signals produced by these atoms, and the coupling with $^1H$ signals resulting from these substitutions, help in the assignment of individual $^1H$ NOE signals. The process is also aided by a knowledge of the amino acid sequence of the polypeptide.

To generate a three-dimensional structure, researchers feed the distance constraints into a computer along with known geometric constraints such as chirality, van der Waals radii, and bond lengths and angles. The computer generates a family of closely related structures that represent the range of conformations consistent with the NOE distance constraints (Fig. 3c). The uncertainty in structures generated by NMR is in part a reflection of the molecular vibrations (breathing) within a protein structure in solution, discussed in more detail in Chapter 5. Normal experimental uncertainty can also play a role.

When a protein structure has been determined by both x-ray crystallography and NMR, the structures

## Analysis of Many Globular Proteins Reveals Common Structural Patterns

**Protein Architecture—Tertiary Structure of Large Globular Proteins** For the beginning student, the very complex tertiary structures of globular proteins much larger than those shown in Figure 4–18 are best approached by fo-

cusing on structural patterns that recur in different and often unrelated proteins. The three-dimensional structure of a typical globular protein can be considered an assemblage of polypeptide segments in the $\alpha$-helix and $\beta$-sheet conformations, linked by connecting segments. The structure can then be described to a first approximation by defining how these segments stack on one

generally agree well. In some cases, the precise locations of particular amino acid side chains on the protein exterior are different, often because of effects related to the packing of adjacent protein molecules in

a crystal. The two techniques together are at the heart of the rapid increase in the availability of structural information about the macromolecules of living cells.

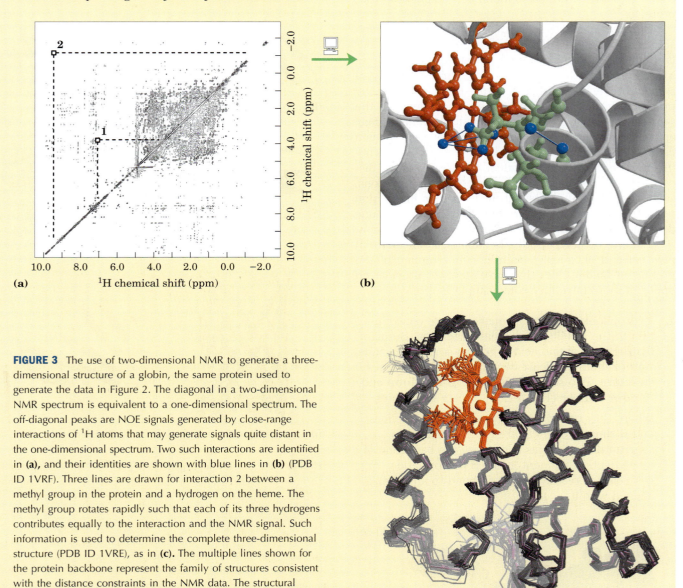

**(a)**

**(b)**

**(c)**

**FIGURE 3** The use of two-dimensional NMR to generate a three-dimensional structure of a globin, the same protein used to generate the data in Figure 2. The diagonal in a two-dimensional NMR spectrum is equivalent to a one-dimensional spectrum. The off-diagonal peaks are NOE signals generated by close-range interactions of $^1H$ atoms that may generate signals quite distant in the one-dimensional spectrum. Two such interactions are identified in **(a),** and their identities are shown with blue lines in **(b)** (PDB ID 1VRF). Three lines are drawn for interaction 2 between a methyl group in the protein and a hydrogen on the heme. The methyl group rotates rapidly such that each of its three hydrogens contributes equally to the interaction and the NMR signal. Such information is used to determine the complete three-dimensional structure (PDB ID 1VRE), as in **(c).** The multiple lines shown for the protein backbone represent the family of structures consistent with the distance constraints in the NMR data. The structural similarity with myoglobin (Fig. 1) is evident. The proteins are oriented in the same way in both figures.

another and how the segments that connect them are arranged. This formalism has led to the development of databases that allow informative comparisons of protein structures, complementing other databases that permit comparisons of protein sequences.

An understanding of a complete three-dimensional structure is built upon an analysis of its parts. We begin

by defining terms used to describe protein substructures, then turn to the folding rules elucidated from analysis of the structures of many proteins.

**Supersecondary structures,** also called **motifs** or simply **folds,** are particularly stable arrangements of several elements of secondary structure and the connections between them. There is no universal agreement among biochemists on the application of the three terms, and they are often used interchangeably. The terms are also applied to a wide range of structures. Recognized motifs range from simple to complex, sometimes appearing in repeating units or combinations. A single large motif may comprise the entire protein. We have already encountered one well-studied motif, the coiled coil of $\alpha$-keratin, also found in a number of other proteins.

Polypeptides with more than a few hundred amino acid residues often fold into two or more stable, globular units called **domains.** In many cases, a domain from a large protein will retain its correct three-dimensional structure even when it is separated (for example, by proteolytic cleavage) from the remainder of the polypeptide chain. A protein with multiple domains may appear to have a distinct globular lobe for each domain (Fig. 4–19), but, more commonly, extensive contacts between domains make individual domains hard to discern. Different domains often have distinct functions, such as the binding of small molecules or interaction with other proteins. Small proteins usually have only one domain (the domain *is* the protein).

Folding of polypeptides is subject to an array of physical and chemical constraints. A sampling of the prominent folding rules that have emerged provides an opportunity to introduce some simple motifs.

1.  Hydrophobic interactions make a large contribution to the stability of protein structures. Burial of hydrophobic amino acid R groups so as to exclude water requires at least two layers of secondary structure. Two simple motifs, the **$\beta$-$\alpha$-$\beta$ loop** and the **$\alpha$-$\alpha$ corner** (Fig. 4–20a), create two layers.

2.  Where they occur together in proteins, $\alpha$ helices and $\beta$ sheets generally are found in different structural layers. This is because the backbone of a polypeptide segment in the $\beta$ conformation (Fig. 4–7) cannot readily hydrogen-bond to an $\alpha$ helix aligned with it.

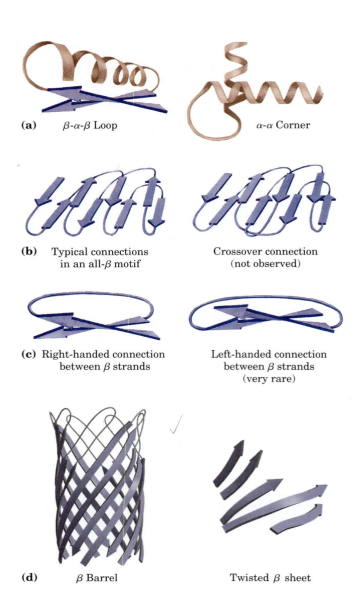

**(a)** $\beta$-$\alpha$-$\beta$ Loop      $\alpha$-$\alpha$ Corner

**(b)** Typical connections in an all-$\beta$ motif      Crossover connection (not observed)

**(c)** Right-handed connection between $\beta$ strands      Left-handed connection between $\beta$ strands (very rare)

**(d)** $\beta$ Barrel      Twisted $\beta$ sheet

**FIGURE 4–20 Stable folding patterns in proteins. (a)** Two simple and common motifs that provide two layers of secondary structure. Amino acid side chains at the interface between elements of secondary structure are shielded from water. Note that the $\beta$ strands in the $\beta$-$\alpha$-$\beta$ loop tend to twist in a right-handed fashion. **(b)** Connections between $\beta$ strands in layered $\beta$ sheets. The strands are shown from one end, with no twisting included in the schematic. Thick lines represent connections at the ends nearest the viewer; thin lines are connections at the far ends of the $\beta$ strands. The connections on a given end (e.g., near the viewer) do not cross each other. **(c)** Because of the twist in $\beta$ strands, connections between strands are generally right-handed. Left-handed connections must traverse sharper angles and are harder to form. **(d)** Two arrangements of $\beta$ strands stabilized by the tendency of the strands to twist. This $\beta$ barrel is a single domain of $\alpha$-hemolysin (a pore-forming toxin that kills a cell by creating a hole in its membrane) from the bacterium *Staphylococcus aureus* (derived from PDB ID 7AHL). The twisted $\beta$ sheet is from a domain of photolyase (a protein that repairs certain types of DNA damage) from *E. coli* (derived from PDB ID 1DNP).

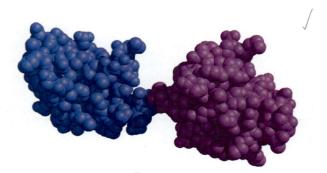

**FIGURE 4–19 Structural domains in the polypeptide troponin C.** (PDB ID 4TNC) This calcium-binding protein associated with muscle has separate calcium-binding domains, indicated in blue and purple.

**3.** Polypeptide segments adjacent to each other in the primary sequence are usually stacked adjacent to each other in the folded structure. Although distant segments of a polypeptide may come together in the tertiary structure, this is not the norm.

**4.** Connections between elements of secondary structure cannot cross or form knots (Fig. 4–20b).

**5.** The $\beta$ conformation is most stable when the individual segments are twisted slightly in a right-handed sense. This influences both the arrangement of $\beta$ sheets relative to one another and the path of the polypeptide connection between them. Two parallel $\beta$ strands, for example, must be connected by a crossover strand (Fig. 4–20c). In principle, this crossover could have a right- or left-handed conformation, but in proteins it is almost always right-handed. Right-handed connections tend to be shorter than left-handed connections and tend to bend through smaller angles, making them easier to form. The twisting of $\beta$ sheets also leads to a characteristic twisting of the structure formed when many segments are put together. Two examples of resulting structures are the $\beta$ barrel and twisted $\beta$ sheet (Fig. 4–20d), which form the core of many larger structures.

Following these rules, complex motifs can be built up from simple ones. For example, a series of $\beta$-$\alpha$-$\beta$ loops, arranged so that the $\beta$ strands form a barrel, creates a particularly stable and common motif called the **$\alpha/\beta$ barrel** (Fig. 4–21). In this structure, each parallel $\beta$ segment is attached to its neighbor by an $\alpha$-helical segment. All connections are right-handed. The $\alpha/\beta$ barrel is found in many enzymes, often with a binding site for a

cofactor or substrate in the form of a pocket near one end of the barrel. Note that domains exhibiting similar folding patterns are said to have the same motif even though their constituent $\alpha$ helices and $\beta$ sheets may differ in length.

## Protein Motifs Are the Basis for Protein Structural Classification

🔵 **Protein Architecture—Tertiary Structure of Large Globular Proteins, IV. Structural Classification of Proteins** As we have seen, the complexities of tertiary structure are decreased by considering substructures. Taking this idea further, researchers have organized the complete contents of databases according to hierarchical levels of structure. The Structural Classification of Proteins (SCOP) database offers a good example of this very important trend in biochemistry. At the highest level of classification, the SCOP database (http://scop.mrc-lmb.cam.ac.uk/scop) borrows a scheme already in common use, in which protein structures are divided into four classes: all $\alpha$, all $\beta$, $\alpha/\beta$ (in which the $\alpha$ and $\beta$ segments are interspersed or alternate), and $\alpha + \beta$ (in which the $\alpha$ and $\beta$ regions are somewhat segregated) (Fig. 4–22). Within each class are tens to hundreds of different folding arrangements, built up from increasingly identifiable substructures. Some of the substructure arrangements are very common, others have been found in just one protein. Figure 4–22 displays a variety of motifs arrayed among the four classes of protein structure. Those illustrated are just a minute sample of the hundreds of known motifs. The number of folding patterns is not infinite, however. As the rate at which new protein structures are elucidated has increased, the fraction of those structures containing a new motif has steadily declined. Fewer than 1,000 different folds or motifs may exist in all proteins. Figure 4–22 also shows how proteins can be organized based on the presence of the various motifs. The top two levels of organization, **class** and **fold,** are purely structural. Below the fold level, categorization is based on evolutionary relationships.

Many examples of recurring domain or motif structures are available, and these reveal that protein tertiary structure is more reliably conserved than primary sequence. The comparison of protein structures can thus provide much information about evolution. Proteins with significant primary sequence similarity, and/or with demonstrably similar structure and function, are said to be in the same **protein family.** A strong evolutionary relationship is usually evident within a protein family. For example, the globin family has many different proteins with both structural and sequence similarity to myoglobin (as seen in the proteins used as examples in Box 4–4 and again in the next chapter). Two or more families with little primary sequence similarity sometimes make use of the same major structural

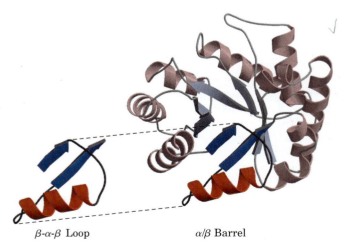

$\beta$-$\alpha$-$\beta$ Loop            $\alpha/\beta$ Barrel

**FIGURE 4–21 Constructing large motifs from smaller ones.** The $\alpha/\beta$ barrel is a common motif constructed from repetitions of the simpler $\beta$-$\alpha$-$\beta$ loop motif. This $\alpha/\beta$ barrel is a domain of the pyruvate kinase (a glycolytic enzyme) from rabbit (derived from PDB ID 1PKN).

**All α**

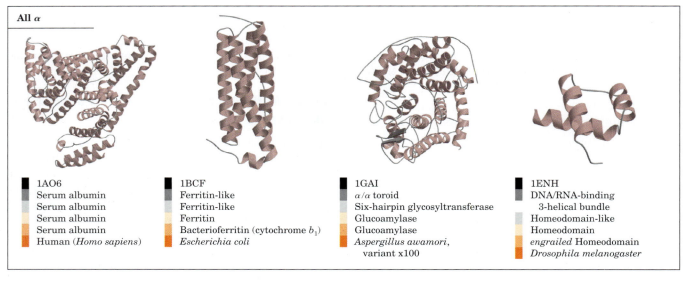

**1AO6**
Serum albumin
Serum albumin
Serum albumin
Serum albumin
Human (*Homo sapiens*)

**1BCF**
Ferritin-like
Ferritin-like
Ferritin
Bacterioferritin (cytochrome *b₁*)
*Escherichia coli*

**1GAI**
α/α toroid
Six-hairpin glycosyltransferase
Glucoamylase
Glucoamylase
*Aspergillus awamori*,
    variant x100

**1ENH**
DNA/RNA-binding
    3-helical bundle
Homeodomain-like
Homeodomain
*engrailed* Homeodomain
*Drosophila melanogaster*

**All β**

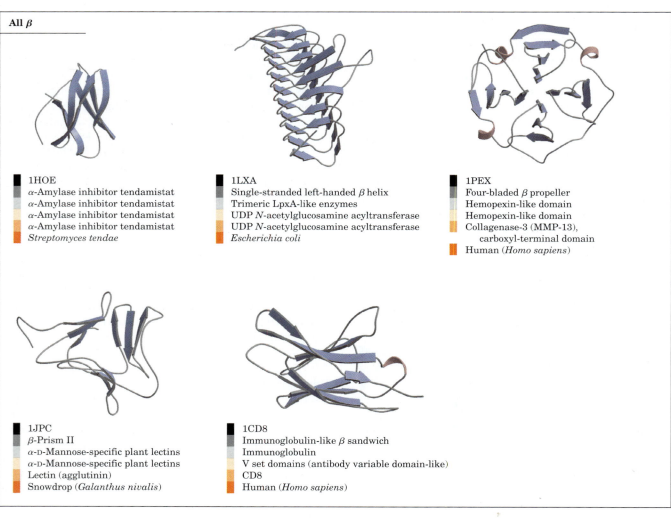

**1HOE**
α-Amylase inhibitor tendamistat
α-Amylase inhibitor tendamistat
α-Amylase inhibitor tendamistat
α-Amylase inhibitor tendamistat
*Streptomyces tendae*

**1LXA**
Single-stranded left-handed β helix
Trimeric LpxA-like enzymes
UDP *N*-acetylglucosamine acyltransferase
UDP *N*-acetylglucosamine acyltransferase
*Escherichia coli*

**1PEX**
Four-bladed β propeller
Hemopexin-like domain
Hemopexin-like domain
Collagenase-3 (MMP-13),
    carboxyl-terminal domain
Human (*Homo sapiens*)

**1JPC**
β-Prism II
α-D-Mannose-specific plant lectins
α-D-Mannose-specific plant lectins
Lectin (agglutinin)
Snowdrop (*Galanthus nivalis*)

**1CD8**
Immunoglobulin-like β sandwich
Immunoglobulin
V set domains (antibody variable domain-like)
CD8
Human (*Homo sapiens*)

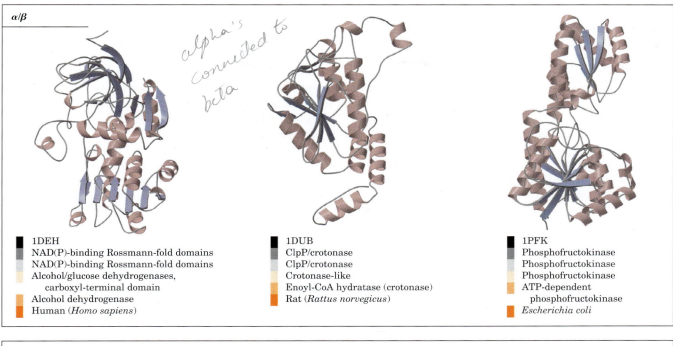

$\alpha/\beta$

*alpha's connected to beta*

1DEH
NAD(P)-binding Rossmann-fold domains
NAD(P)-binding Rossmann-fold domains
Alcohol/glucose dehydrogenases,
    carboxyl-terminal domain
Alcohol dehydrogenase
Human (*Homo sapiens*)

1DUB
ClpP/crotonase
ClpP/crotonase
Crotonase-like
Enoyl-CoA hydratase (crotonase)
Rat (*Rattus norvegicus*)

1PFK
Phosphofructokinase
Phosphofructokinase
Phosphofructokinase
ATP-dependent
    phosphofructokinase
*Escherichia coli*

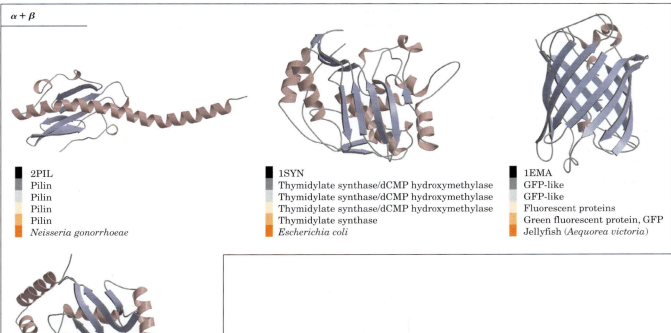

$\alpha + \beta$

2PIL
Pilin
Pilin
Pilin
Pilin
*Neisseria gonorrhoeae*

1SYN
Thymidylate synthase/dCMP hydroxymethylase
Thymidylate synthase/dCMP hydroxymethylase
Thymidylate synthase/dCMP hydroxymethylase
Thymidylate synthase
*Escherichia coli*

1EMA
GFP-like
GFP-like
Fluorescent proteins
Green fluorescent protein, GFP
Jellyfish (*Aequorea victoria*)

1U9A
UBC-like
UBC-like
Ubibuitin-conjugating enzyme, UBC
Ubiquitin-conjugating enzyme, UBC
Human (*Homo sapiens*) ubc9

PDB identifier
Fold
Superfamily
Family
Protein
Species

**FIGURE 4–22 Organization of proteins based on motifs.** Shown here are just a small number of the hundreds of known stable motifs. They are divided into four classes: all $\alpha$, all $\beta$, $\alpha/\beta$, and $\alpha + \beta$. Structural classification data from the SCOP (Structural Classification of Proteins) database (http://scop.mrc-lmb.cam.ac.uk/scop) are also provided. The PDB identifier is the unique number given to each structure archived in the Protein Data Bank (www.rcsb.org/pdb). The $\alpha/\beta$ barrel, shown in Figure 4–21, is another particularly common $\alpha/\beta$ motif.

motif and have functional similarities; these families are grouped as **superfamilies.** An evolutionary relationship between the families in a superfamily is considered probable, even though time and functional distinctions—hence different adaptive pressures—may have erased many of the telltale sequence relationships. A protein family may be widespread in all three domains of cellular life, the Bacteria, Archaea, and Eukarya, suggesting a very ancient origin. Other families may be present in only a small group of organisms, indicating that the structure arose more recently. Tracing the natural history of structural motifs, using structural classifications in databases such as SCOP, provides a powerful complement to sequence analyses in tracing many evolutionary relationships.

The SCOP database is curated manually, with the objective of placing proteins in the correct evolutionary framework based on conserved structural features. Two similar enterprises, the CATH (*c*lass, *a*rchitecture, *t*opology, and *h*omologous superfamily) and FSSP (*f*old classification based on *s*tructure-*s*tructure alignment of *p*roteins) databases, make use of more automated methods and can provide additional information.

Structural motifs become especially important in defining protein families and superfamilies. Improved classification and comparison systems for proteins lead inevitably to the elucidation of new functional relationships. Given the central role of proteins in living systems, these structural comparisons can help illuminate every aspect of biochemistry, from the evolution of individual proteins to the evolutionary history of complete metabolic pathways.

## Protein Quaternary Structures Range from Simple Dimers to Large Complexes

**Protein Architecture—Quaternary Structure** Many proteins have multiple polypeptide subunits. The association of polypeptide chains can serve a variety of functions. Many multisubunit proteins have regulatory roles; the binding of small molecules may affect the interaction between subunits, causing large changes in the protein's activity in response to small changes in the concentration of substrate or regulatory molecules (Chapter 6). In other cases, separate subunits can take on separate but related functions, such as catalysis and regulation. Some associations, such as the fibrous proteins considered earlier in this chapter and the coat proteins of viruses, serve primarily structural roles. Some very large protein assemblies are the site of complex, multistep reactions. One example is the ribosome, site of protein synthesis, which incorporates dozens of protein subunits along with a number of RNA molecules.

A multisubunit protein is also referred to as a **multimer.** Multimeric proteins can have from two to hundreds of subunits. A multimer with just a few subunits is often called an **oligomer.** If a multimer is composed of a number of nonidentical subunits, the overall structure of the protein can be asymmetric and quite complicated. However, most multimers have identical subunits or repeating groups of nonidentical subunits, usually in symmetric arrangements. As noted in Chapter 3, the repeating structural unit in such a multimeric protein, whether it is a single subunit or a group of subunits, is called a **protomer.**

The first oligomeric protein for which the three-dimensional structure was determined was hemoglobin ($M_r$ 64,500), which contains four polypeptide chains and four heme prosthetic groups, in which the iron atoms are in the ferrous ($Fe^{2+}$) state (Fig. 4–17). The protein portion, called globin, consists of two $\alpha$ chains (141 residues each) and two $\beta$ chains (146 residues each). Note that in this case $\alpha$ and $\beta$ do not refer to secondary structures. Because hemoglobin is four times as large as myoglobin, much more time and effort were required to solve its three-dimensional structure by x-ray analysis, finally achieved by Max Perutz, John Kendrew, and their colleagues in 1959. The subunits of hemoglobin are arranged in symmetric pairs (Fig. 4–23), each pair having one $\alpha$ and one $\beta$ subunit. Hemoglobin can therefore be described either as a tetramer or as a dimer of $\alpha\beta$ protomers.

Identical subunits of multimeric proteins are generally arranged in one or a limited set of symmetric patterns. A description of the structure of these proteins requires an understanding of conventions used to define symmetries. Oligomers can have either **rotational symmetry** or **helical symmetry;** that is, individual subunits can be superimposed on others (brought to coincidence) by rotation about one or more rotational axes, or by a helical rotation. In proteins with rotational symmetry, the subunits pack about the rotational axes to form closed structures. Proteins with helical symme-

Max Perutz, 1914–2002 (left)
John Kendrew, 1917–1997 (right)

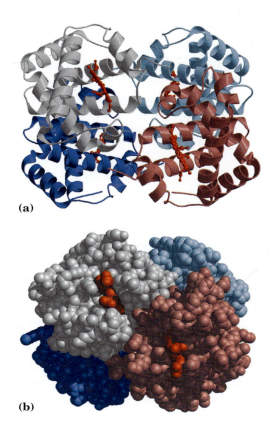

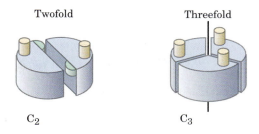

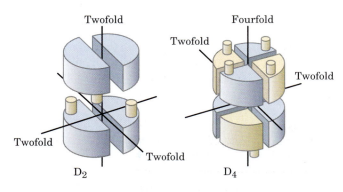

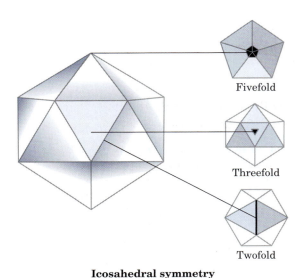

**FIGURE 4–23 Quaternary structure of deoxyhemoglobin.** (PDB ID 2HHB) X-ray diffraction analysis of deoxyhemoglobin (hemoglobin without oxygen molecules bound to the heme groups) shows how the four polypeptide subunits are packed together. **(a)** A ribbon representation. **(b)** A space-filling model. The $\alpha$ subunits are shown in gray and light blue; the $\beta$ subunits in pink and dark blue. Note that the heme groups (red) are relatively far apart.

try tend to form structures that are more open-ended, with subunits added in a spiraling array.

There are several forms of rotational symmetry. The simplest is **cyclic symmetry,** involving rotation about a single axis (Fig. 4–24a). If subunits can be superimposed by rotation about a single axis, the protein has a symmetry defined by convention as $C_n$ (C for cyclic, $n$ for the number of subunits related by the axis). The axis itself is described as an $n$-fold rotational axis. The $\alpha\beta$ protomers of hemoglobin (Fig. 4–23) are related by $C_2$ symmetry. A somewhat more complicated rotational symmetry is **dihedral symmetry,** in which a twofold rotational axis intersects an $n$-fold axis at right angles. The symmetry is defined as $D_n$ (Fig. 4–24b). A protein with dihedral symmetry has $2n$ protomers.

Proteins with cyclic or dihedral symmetry are particularly common. More complex rotational symmetries are possible, but only a few are regularly encountered. One example is **icosahedral symmetry.** An icosahedron is a regular 12-cornered polyhedron having 20 equilateral triangular faces (Fig. 4–24c). Each face can

**FIGURE 4–24 Rotational symmetry in proteins. (a)** In cyclic symmetry, subunits are related by rotation about a single $n$-fold axis, where $n$ is the number of subunits so related. The axes are shown as black lines; the numbers are values of $n$. Only two of many possible $C_n$ arrangements are shown. **(b)** In dihedral symmetry, all subunits can be related by rotation about one or both of two axes, one of which is twofold. $D_2$ symmetry is most common. **(c)** Icosahedral symmetry. Relating all 20 triangular faces of an icosahedron requires rotation about one or more of three separate rotational axes: twofold, threefold, and fivefold. An end-on view of each of these axes is shown at the right.

be brought to coincidence with another by rotation about one or more of three rotational axes. This is a common structure in virus coats, or capsids. The human poliovirus has an icosahedral capsid (Fig. 4–25a). Each triangular face is made up of three protomers, each protomer containing single copies of four different polypeptide chains, three of which are accessible at the outer surface. Sixty protomers form the 20 faces of the icosahedral shell enclosing the genetic material (RNA).

The other major type of symmetry found in oligomers, helical symmetry, also occurs in capsids. Tobacco mosaic virus is a right-handed helical filament made up of 2,130 identical subunits (Fig. 4–25b). This cylindrical structure encloses the viral RNA. Proteins with subunits arranged in helical filaments can also form long, fibrous structures such as the actin filaments of muscle (see Fig. 5–30).

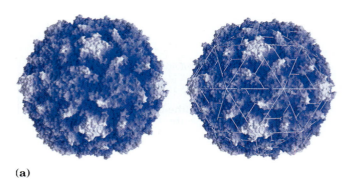

**(a)**

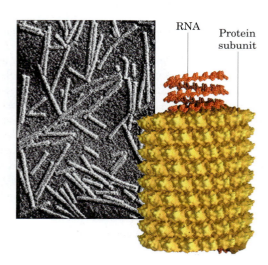

RNA    Protein subunit

**(b)**

**FIGURE 4–25 · Viral capsids. (a)** Poliovirus (derived from PDB ID 2PLV). The coat proteins of poliovirus assemble into an icosahedron 300 Å in diameter. Icosahedral symmetry is a type of rotational symmetry (see Fig. 4–24c). On the left is a surface contour image of the poliovirus capsid. In the image on the right, lines have been superimposed to show the axes of symmetry. **(b)** Tobacco mosaic virus (derived from PDB ID 1VTM). This rod-shaped virus (as shown in the electron micrograph) is 3,000 Å long and 180 Å in diameter; it has helical symmetry.

## There Are Limits to the Size of Proteins

The relatively large size of proteins reflects their functions. The function of an enzyme, for example, requires a stable structure containing a pocket large enough to bind its substrate and catalyze a reaction. Protein size has limits, however, imposed by two factors: the genetic coding capacity of nucleic acids and the accuracy of the protein biosynthetic process. The use of many copies of one or a few proteins to make a large enclosing structure (capsid) is important for viruses because this strategy conserves genetic material. Remember that there is a linear correspondence between the sequence of a gene in the nucleic acid and the amino acid sequence of the protein for which it codes (see Fig. 1–31). The nucleic acids of viruses are much too small to encode the information required for a protein shell made of a single polypeptide. By using many copies of much smaller polypeptides, a much shorter nucleic acid is needed for coding the capsid subunits, and this nucleic acid can be efficiently used over and over again. Cells also use large complexes of polypeptides in muscle, cilia, the cytoskeleton, and other structures. It is simply more efficient to make many copies of a small polypeptide than one copy of a very large protein. In fact, most proteins with a molecular weight greater than 100,000 have multiple subunits, identical or different.

The second factor limiting the size of proteins is the error frequency during protein biosynthesis. The error frequency is low (about 1 mistake per 10,000 amino acid residues added), but even this low rate results in a high probability of a damaged protein if the protein is very large. Simply put, the potential for incorporating a "wrong" amino acid in a protein is greater for a large protein than for a small one.

### SUMMARY 4.3 Protein Tertiary and Quaternary Structures

- Tertiary structure is the complete three-dimensional structure of a polypeptide chain. There are two general classes of proteins based on tertiary structure: fibrous and globular.

- Fibrous proteins, which serve mainly structural roles, have simple repeating elements of secondary structure.

- Globular proteins have more complicated tertiary structures, often containing several types of secondary structure in the same polypeptide chain. The first globular protein structure to be determined, using x-ray diffraction methods, was that of myoglobin.

- The complex structures of globular proteins can be analyzed by examining stable substructures called supersecondary structures, motifs, or folds. The thousands of known protein structures are generally assembled from a repertoire of only a few hundred motifs. Regions of a polypeptide chain that can fold stably and independently are called **domains**.

■ Quaternary structure results from interactions between the subunits of multisubunit (multimeric) proteins or large protein assemblies. Some multimeric proteins have a repeated unit consisting of a single subunit or a group of subunits referred to as a protomer. Protomers are usually related by rotational or helical symmetry.

## 4.4 Protein Denaturation and Folding

All proteins begin their existence on a ribosome as a linear sequence of amino acid residues (Chapter 27). This polypeptide must fold during and following synthesis to take up its native conformation. We have seen that a native protein conformation is only marginally stable. Modest changes in the protein's environment can bring about structural changes that can affect function. We now explore the transition that occurs between the folded and unfolded states.

### Loss of Protein Structure Results in Loss of Function

Protein structures have evolved to function in particular cellular environments. Conditions different from those in the cell can result in protein structural changes, large and small. A loss of three-dimensional structure sufficient to cause loss of function is called **denaturation.** The denatured state does not necessarily equate with complete unfolding of the protein and randomization of conformation. Under most conditions, denatured proteins exist in a set of partially folded states that are poorly understood.

Most proteins can be denatured by heat, which affects the weak interactions in a protein (primarily hydrogen bonds) in a complex manner. If the temperature is increased slowly, a protein's conformation generally remains intact until an abrupt loss of structure (and function) occurs over a narrow temperature range (Fig. 4–26). The abruptness of the change suggests that unfolding is a cooperative process: loss of structure in one part of the protein destabilizes other parts. The effects of heat on proteins are not readily predictable. The very heat-stable proteins of thermophilic bacteria have evolved to function at the temperature of hot springs (~100 °C). Yet the structures of these proteins often differ only slightly from those of homologous proteins derived from bacteria such as *Escherichia coli*. How these small differences promote structural stability at high temperatures is not yet understood.

Proteins can be denatured not only by heat but by extremes of pH, by certain miscible organic solvents

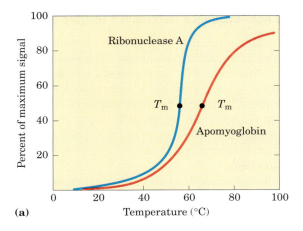

**(a)**

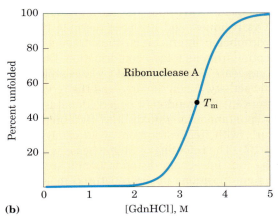

**(b)**

**FIGURE 4–26 Protein denaturation.** Results are shown for proteins denatured by two different environmental changes. In each case, the transition from the folded to unfolded state is fairly abrupt, suggesting cooperativity in the unfolding process. **(a)** Thermal denaturation of horse apomyoglobin (myoglobin without the heme prosthetic group) and ribonuclease A (with its disulfide bonds intact; see Fig. 4–27). The midpoint of the temperature range over which denaturation occurs is called the melting temperature, or $T_m$. The denaturation of apomyoglobin was monitored by circular dichroism, a technique that measures the amount of helical structure in a macromolecule. Denaturation of ribonuclease A was tracked by monitoring changes in the intrinsic fluorescence of the protein, which is affected by changes in the environment of Trp residues. **(b)** Denaturation of disulfide-intact ribonuclease A by guanidine hydrochloride (GdnHCl), monitored by circular dichroism.

such as alcohol or acetone, by certain solutes such as urea and guanidine hydrochloride, or by detergents. Each of these denaturing agents represents a relatively mild treatment in the sense that no covalent bonds in the polypeptide chain are broken. Organic solvents, urea, and detergents act primarily by disrupting the hydrophobic interactions that make up the stable core of globular proteins; extremes of pH alter the net charge on the protein, causing electrostatic repulsion and the disruption of some hydrogen bonding. The denatured states obtained with these various treatments need not be equivalent.

## Amino Acid Sequence Determines Tertiary Structure

The tertiary structure of a globular protein is determined by its amino acid sequence. The most important proof of this came from experiments showing that denaturation of some proteins is reversible. Certain globular proteins denatured by heat, extremes of pH, or denaturing reagents will regain their native structure and their biological activity if returned to conditions in which the native conformation is stable. This process is called **renaturation.**

A classic example is the denaturation and renaturation of ribonuclease. Purified ribonuclease can be completely denatured by exposure to a concentrated urea solution in the presence of a reducing agent. The reducing agent cleaves the four disulfide bonds to yield eight Cys residues, and the urea disrupts the stabilizing hydrophobic interactions, thus freeing the entire polypeptide from its folded conformation. Denaturation of ribonuclease is accompanied by a complete loss of catalytic activity. When the urea and the reducing agent are removed, the randomly coiled, denatured ribonuclease spontaneously refolds into its correct tertiary structure, with full restoration of its catalytic activity (Fig. 4–27). The refolding of ribonuclease is so accurate that the four intrachain disulfide bonds are re-formed in the same positions in the renatured molecule as in the native ribonuclease. As calculated mathematically, the eight Cys residues could recombine at random to form up to four disulfide bonds in 105 different ways. In fact, an essentially random distribution of disulfide bonds is obtained when the disulfides are allowed to re-form in the presence of denaturant, indicating that weak bonding interactions are required for correct positioning of disulfide bonds and assumption of the native conformation.

This classic experiment, carried out by Christian Anfinsen in the 1950s, provided the first evidence that the amino acid sequence of a polypeptide chain contains all the information required to fold the chain into its native, three-dimensional structure. Later, similar results were obtained using chemically synthesized, catalytically active ribonuclease. This eliminated the possibility that some minor contaminant in Anfinsen's purified ribonuclease preparation might have contributed to the renaturation of the enzyme, thus dispelling any remaining doubt that this enzyme folds spontaneously.

## Polypeptides Fold Rapidly by a Stepwise Process

In living cells, proteins are assembled from amino acids at a very high rate. For example, *E. coli* cells can make a complete, biologically active protein molecule containing 100 amino acid residues in about 5 seconds at 37 °C. How does such a polypeptide chain arrive at its native conformation? Let's assume conservatively that each of the amino acid residues could take up 10 dif-

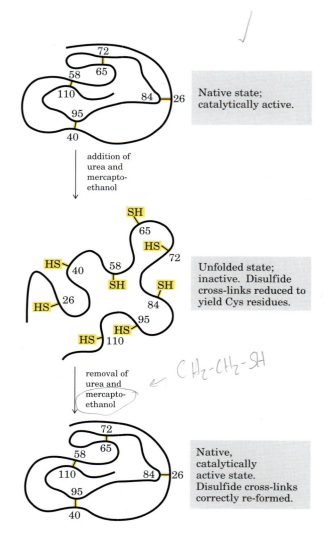

**FIGURE 4–27 Renaturation of unfolded, denatured ribonuclease.** Urea is used to denature ribonuclease, and mercaptoethanol ($HOCH_2CH_2SH$) to reduce and thus cleave the disulfide bonds to yield eight Cys residues. Renaturation involves reestablishment of the correct disulfide cross-links.

ferent conformations on average, giving $10^{100}$ different conformations for the polypeptide. Let's also assume that the protein folds itself spontaneously by a random process in which it tries out all possible conformations around every single bond in its backbone until it finds its native, biologically active form. If each conformation were sampled in the shortest possible time ($\sim 10^{-13}$ second, or the time required for a single molecular vibration), it would take about $10^{77}$ years to sample all possible conformations. Thus protein folding cannot be a completely random, trial-and-error process. There must be shortcuts. This problem was first pointed out by Cyrus Levinthal in 1968 and is sometimes called Levinthal's paradox.

The folding pathway of a large polypeptide chain is unquestionably complicated, and not all the principles that guide the process have been worked out. However, extensive study has led to the development of several

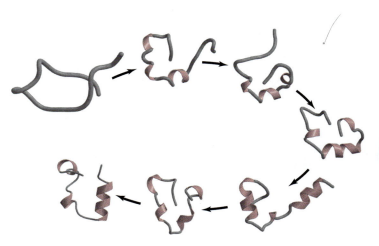

**FIGURE 4–28  A simulated folding pathway.** The folding pathway of a 36-residue segment of the protein villin (an actin-binding protein found principally in the microvilli lining the intestine) was simulated by computer. The process started with the randomly coiled peptide and 3,000 surrounding water molecules in a virtual "water box." The molecular motions of the peptide and the effects of the water molecules were taken into account in mapping the most likely paths to the final structure among the countless alternatives. The simulated folding took place in a theoretical time span of 1 ms; however, the calculation required half a billion integration steps on two Cray supercomputers, each running for two months.

plausible models. In one, the folding process is envisioned as hierarchical. Local secondary structures form first. Certain amino acid sequences fold readily into $\alpha$ helices or $\beta$ sheets, guided by constraints we have reviewed in our discussion of secondary structure. This is followed by longer-range interactions between, say, two $\alpha$ helices that come together to form stable supersecondary structures. The process continues until complete domains form and the entire polypeptide is folded (Fig. 4–28). In an alternative model, folding is initiated by a spontaneous collapse of the polypeptide into a compact state, mediated by hydrophobic interactions among nonpolar residues. The state resulting from this "hydrophobic collapse" may have a high content of secondary structure, but many amino acid side chains are not entirely fixed. The collapsed state is often referred to as a **molten globule.** Most proteins probably fold by a process that incorporates features of both models. Instead of following a single pathway, a population of peptide molecules may take a variety of routes to the same end point, with the number of different partly folded conformational species decreasing as folding nears completion.

Thermodynamically, the folding process can be viewed as a kind of free-energy funnel (Fig. 4–29). The unfolded states are characterized by a high degree of conformational entropy and relatively high free energy. As folding proceeds, the narrowing of the funnel repre-

sents a decrease in the number of conformational species present. Small depressions along the sides of the free-energy funnel represent semistable intermediates that can briefly slow the folding process. At the bottom of the funnel, an ensemble of folding intermediates has been reduced to a single native conformation (or one of a small set of native conformations).

Defects in protein folding may be the molecular basis for a wide range of human genetic disorders. For example, cystic fibrosis is caused by defects in a membrane-bound protein called *cystic fibrosis trans-membrane conductance regulator* (CFTR), which acts as a channel for chloride ions. The most common cystic

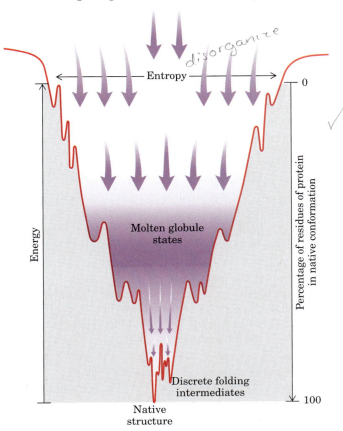

**FIGURE 4–29  The thermodynamics of protein folding depicted as a free-energy funnel.** At the top, the number of conformations, and hence the conformational entropy, is large. Only a small fraction of the intramolecular interactions that will exist in the native conformation are present. As folding progresses, the thermodynamic path down the funnel reduces the number of states present (decreases entropy), increases the amount of protein in the native conformation, and decreases the free energy. Depressions on the sides of the funnel represent semistable folding intermediates, which may, in some cases, slow the folding process.

fibrosis–causing mutation is the deletion of a Phe residue at position 508 in CFTR, which causes improper protein folding (see Box 11–3). Many of the disease-related mutations in collagen (p. 129) also cause defective folding. An improved understanding of protein folding may lead to new therapies for these and many other diseases (Box 4–5). ∎

Thermodynamic stability is not evenly distributed over the structure of a protein—the molecule has regions of high and low stability. For example, a protein

**BOX 4–5    BIOCHEMISTRY IN MEDICINE**

### Death by Misfolding: The Prion Diseases

A misfolded protein appears to be the causative agent of a number of rare degenerative brain diseases in mammals. Perhaps the best known of these is mad cow disease (bovine spongiform encephalopathy, BSE), an outbreak of which made international headlines in the spring of 1996. Related diseases include kuru and Creutzfeldt-Jakob disease in humans, scrapie in sheep, and chronic wasting disease in deer and elk. These diseases are also referred to as spongiform encephalopathies, because the diseased brain frequently becomes riddled with holes (Fig. 1). Typical symptoms include dementia and loss of coordination. The diseases are fatal.

In the 1960s, investigators found that preparations of the disease-causing agents appeared to lack nucleic acids. At this time, Tikvah Alper suggested that the agent was a protein. Initially, the idea seemed heretical. All disease-causing agents known up to that time—viruses, bacteria, fungi, and so on—contained nucleic acids, and their virulence was related to genetic reproduction and propagation. However, four decades of investigations, pursued most notably by Stanley Prusiner, have provided evidence that spongiform encephalopathies are different.

The infectious agent has been traced to a single protein ($M_r$ 28,000), which Prusiner dubbed **prion** (from *pr*oteinaceous *i*nfectious *only*) protein (PrP). Prion protein is a normal constituent of brain tissue in all mammals. Its role in the mammalian brain is not known in detail, but it appears to have a molecular signaling function. Strains of mice lacking the gene for PrP (and thus the protein itself) suffer no obvious ill effects. Illness occurs only when the normal cellular PrP, or PrP$^C$, occurs in an altered conformation called PrP$^{Sc}$ (Sc denotes scrapie). The interaction of PrP$^{Sc}$ with PrP$^C$ converts the latter to PrP$^{Sc}$, initiating a domino effect in which more and more of the brain protein converts to the disease-causing form. The mechanism by which the presence of PrP$^{Sc}$ leads to spongiform encephalopathy is not understood.

In inherited forms of prion diseases, a mutation in the gene encoding PrP produces a change in one amino acid residue that is believed to make the conversion of PrP$^C$ to PrP$^{Sc}$ more likely. A complete understanding of prion diseases awaits new information about how prion protein affects brain function. Structural information about PrP is beginning to provide insights into the molecular process that allows the prion proteins to interact so as to alter their conformation (Fig. 2).

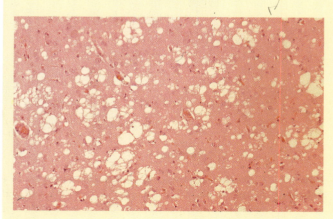

**FIGURE 1** A stained section of the cerebral cortex from a patient with Creutzfeldt-Jakob disease shows spongiform (vacuolar) degeneration, the most characteristic neurohistological feature. The yellowish vacuoles are intracellular and occur mostly in pre- and postsynaptic processes of neurons. The vacuoles in this section vary in diameter from 20 to 100 $\mu$m.

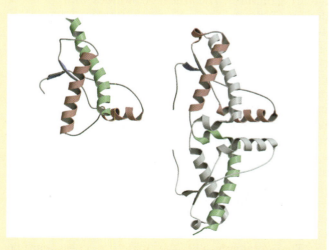

**FIGURE 2** The structure of the globular domain of human PrP in monomeric (left) and dimeric (right) forms. The second subunit is gray to highlight the dramatic conformational change in the green α helix when the dimer is formed.

may have two stable domains joined by a segment with lower structural stability, or one small part of a domain may have a lower stability than the remainder. The regions of low stability allow a protein to alter its conformation between two or more states. As we shall see in the next two chapters, variations in the stability of regions within a given protein are often essential to protein function.

## Some Proteins Undergo Assisted Folding

Not all proteins fold spontaneously as they are synthesized in the cell. Folding for many proteins is facilitated by the action of specialized proteins. **Molecular chaperones** are proteins that interact with partially folded or improperly folded polypeptides, facilitating correct folding pathways or providing microenvironments in which folding can occur. Two classes of molecular chaperones have been well studied. Both are found in organisms ranging from bacteria to humans. The first class, a family of proteins called **Hsp70**, generally have

a molecular weight near 70,000 and are more abundant in cells stressed by elevated temperatures (hence, *heat shock* proteins of $M_r$ 70,000, or Hsp70). Hsp70 proteins bind to regions of unfolded polypeptides that are rich in hydrophobic residues, preventing inappropriate aggregation. These chaperones thus "protect" proteins that have been denatured by heat and peptides that are being synthesized (and are not yet folded). Hsp70 proteins also block the folding of certain proteins that must remain unfolded until they have been translocated across membranes (as described in Chapter 27). Some chaperones also facilitate the quaternary assembly of oligomeric proteins. The Hsp70 proteins bind to and release polypeptides in a cycle that also involves several other proteins (including a class called Hsp40) and ATP hydrolysis. Figure 4–30 illustrates chaperone-assisted folding as elucidated for the chaperones DnaK and DnaJ in *E. coli*, homologs of the eukaryotic Hsp70 and Hsp40. DnaK and DnaJ were first identified as proteins required for in vitro replication of certain viral DNA molecules (hence the "Dna" designation).

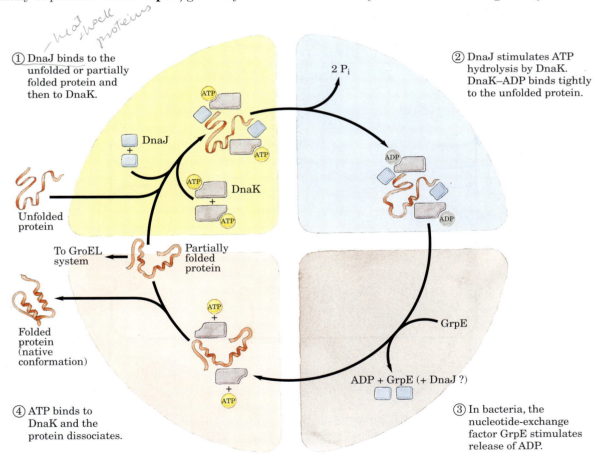

① DnaJ binds to the unfolded or partially folded protein and then to DnaK.

DnaJ

2 $P_i$

DnaK

Unfolded protein

To GroEL system

Partially folded protein

Folded protein (native conformation)

④ ATP binds to DnaK and the protein dissociates.

② DnaJ stimulates ATP hydrolysis by DnaK. DnaK–ADP binds tightly to the unfolded protein.

GrpE

ADP + GrpE (+ DnaJ ?)

③ In bacteria, the nucleotide-exchange factor GrpE stimulates release of ADP.

**FIGURE 4–30 Chaperones in protein folding.** The cyclic pathway by which chaperones bind and release polypeptides is illustrated for the *E. coli* chaperone proteins DnaK and DnaJ, homologs of the eukaryotic chaperones Hsp70 and Hsp40. The chaperones do not actively promote the folding of the substrate protein, but instead prevent aggregation of unfolded peptides. For a population of polypeptides, some fraction of the polypeptides released at the end of the cycle are in the native conformation. The remainder are rebound by DnaK or are diverted to the chaperonin system (GroEL; see Fig. 4–31). In bacteria, a protein called GrpE interacts transiently with DnaK late in the cycle (step ③), promoting dissociation of ADP and possibly DnaJ. No eukaryotic analog of GrpE is known.

The second class of chaperones is called **chaperonins.** These are elaborate protein complexes required for the folding of a number of cellular proteins that do not fold spontaneously. In *E. coli* an estimated 10% to 15% of cellular proteins require the resident chaperonin system, called GroEL/GroES, for folding under normal conditions (up to 30% require this assistance when the cells are heat stressed). These proteins first became known when they were found to be necessary for the growth of certain bacterial viruses (hence the designation "Gro"). Unfolded proteins are bound within pockets in the GroEL complex, and the pockets are capped transiently by the GroES "lid" (Fig. 4–31). GroEL un-

dergoes substantial conformational changes, coupled to ATP hydrolysis and the binding and release of GroES, which promote folding of the bound polypeptide. Although the structure of the GroEL/GroES chaperonin is known, many details of its mechanism of action remain unresolved.

Finally, the folding pathways of a number of proteins require two enzymes that catalyze isomerization reactions. **Protein disulfide isomerase (PDI)** is a widely distributed enzyme that catalyzes the interchange or shuffling of disulfide bonds until the bonds of the native conformation are formed. Among its functions, PDI catalyzes the elimination of folding interme-

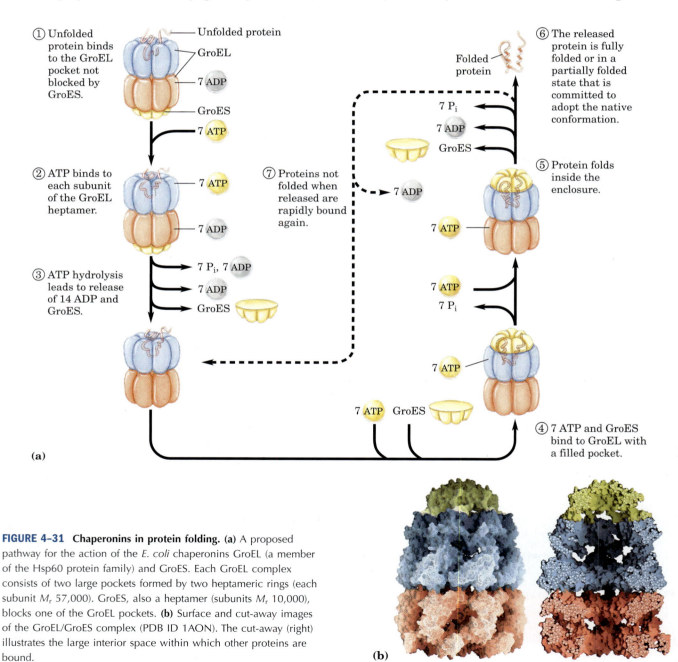

**FIGURE 4–31 Chaperonins in protein folding. (a)** A proposed pathway for the action of the *E. coli* chaperonins GroEL (a member of the Hsp60 protein family) and GroES. Each GroEL complex consists of two large pockets formed by two heptameric rings (each subunit $M_r$ 57,000). GroES, also a heptamer (subunits $M_r$ 10,000), blocks one of the GroEL pockets. **(b)** Surface and cut-away images of the GroEL/GroES complex (PDB ID 1AON). The cut-away (right) illustrates the large interior space within which other proteins are bound.

diates with inappropriate disulfide cross-links. **Peptide prolyl cis-trans isomerase (PPI)** catalyzes the interconversion of the cis and trans isomers of Pro peptide bonds (Fig. 4–8b), which can be a slow step in the folding of proteins that contain some Pro residue peptide bonds in the cis conformation.

Protein folding is likely to be a more complex process in the densely packed cellular environment than in the test tube. More classes of proteins that facilitate protein folding may be discovered as the biochemical dissection of the folding process continues.

### SUMMARY 4.4 Protein Denaturation and Folding

- The three-dimensional structure and the function of proteins can be destroyed by denaturation, demonstrating a relationship between structure and function. Some denatured proteins can renature spontaneously to form biologically active protein, showing that protein tertiary structure is determined by amino acid sequence.

- Protein folding in cells probably involves multiple pathways. Initially, regions of secondary structure may form, followed by folding into supersecondary structures. Large ensembles of folding intermediates are rapidly brought to a single native conformation.

- For many proteins, folding is facilitated by Hsp70 chaperones and by chaperonins. Disulfide bond formation and the cis-trans isomerization of Pro peptide bonds are catalyzed by specific enzymes.

## Key Terms

*Terms in bold are defined in the glossary.*

| | | | |
|---|---|---|---|
| **conformation** 116 | **β conformation** 123 | collagen 127 | **protomer** 144 |
| **native conformation** 117 | β sheet 123 | silk fibroin 129 | symmetry 144 |
| solvation layer 117 | **β turn** 123 | **supersecondary structures** 139 | **denaturation** 147 |
| peptide group 118 | **tertiary structure** 125 | **motif** 139 | molten globule 149 |
| Ramachandran plot 118 | **quaternary structure** 125 | **fold** 139 | prion 150 |
| **secondary structure** 120 | **fibrous proteins** 125 | **domain** 140 | molecular chaperone 151 |
| **α helix** 120 | **globular proteins** 125 | protein family 141 | Hsp70 151 |
| | α-keratin 126 | multimer 144 | chaperonin 152 |
| | | **oligomer** 144 | |

## Further Reading

### General

**Anfinsen, C.B.** (1973) Principles that govern the folding of protein chains. *Science* **181,** 223–230.

The author reviews his classic work on ribonuclease.

**Branden, C. & Tooze, J.** (1991) *Introduction to Protein Structure,* Garland Publishing, Inc., New York.

**Creighton, T.E.** (1993) *Proteins: Structures and Molecular Properties,* 2nd edn, W. H. Freeman and Company, New York.

A comprehensive and authoritative source.

Evolution of Catalytic Function. (1987) *Cold Spring Harb. Symp. Quant. Biol.* **52.**

A collection of excellent articles on many topics, including protein structure, folding, and function.

**Kendrew, J.C.** (1961) The three-dimensional structure of a protein molecule. *Sci. Am.* **205** (December), 96–111.

Describes how the structure of myoglobin was determined and what was learned from it.

**Richardson, J.S.** (1981) The anatomy and taxonomy of protein structure. *Adv. Prot. Chem.* **34,** 167–339.

An outstanding summary of protein structural patterns and principles; the author originated the very useful "ribbon" representations of protein structure.

### Secondary, Tertiary, and Quaternary Structures

**Berman, H.M.** (1999) The past and future of structure databases. *Curr. Opin. Biotechnol.* **10,** 76–80.

A broad summary of the different approaches being used to catalog protein structures.

**Brenner, S.E., Chothia, C., & Hubbard, T.J.P.** (1997) Population statistics of protein structures: lessons from structural classifications. *Curr. Opin. Struct. Biol.* **7,** 369–376.

**Fuchs, E. & Cleveland, D.W.** (1998) A structural scaffolding of intermediate filaments in health and disease. *Science* **279,** 514–519.

**McPherson, A.** (1989) Macromolecular crystals. *Sci. Am.* **260** (March), 62–69.

    A description of how macromolecules such as proteins are crystallized.

**Ponting, C.P. & Russell, R.R.** (2002) The natural history of protein domains. *Annu. Rev. Biophys. Biomol. Struct.* **31,** 45–71.

    An explanation of how structural databases can be used to explore evolution.

**Prockop, D.J. & Kivirikko, K.I.** (1995) Collagens, molecular biology, diseases, and potentials for therapy. *Annu. Rev. Biochem.* **64,** 403–434.

### Protein Denaturation and Folding

**Baldwin, R.L.** (1994) Matching speed and stability. *Nature* **369,** 183–184.

**Bukau, B., Deuerling, E., Pfund, C., & Craig, E.A.** (2000) Getting newly synthesized proteins into shape. *Cell* **101,** 119–122.

    A good summary of chaperone mechanisms.

**Collinge, J.** (2001) Prion diseases of humans and animals: their causes and molecular basis. *Annu. Rev. Neurosci.* **24,** 519–550.

**Creighton, T.E., Darby, N.J., & Kemmink, J.** (1996) The roles of partly folded intermediates in protein folding. *FASEB J.* **10,** 110–118.

**Daggett, V., & Fersht, A.R.** (2003) Is there a unifying mechanism for protein folding? *Trends Biochem. Sci.* **28,** 18–25.

**Dill, K.A. & Chan, H.S.** (1997) From Levinthal to pathways to funnels. *Nat. Struct. Biol.* **4,** 10–19.

**Luque, I., Leavitt, S.A., & Freire, E.** (2002) The linkage between protein folding and functional cooperativity: two sides of the same coin? *Annu. Rev. Biophys. Biomol. Struct.* **31,** 235–256.

    A review of how variations in structural stability within one protein contribute to function.

**Nicotera, P.** (2001) A route for prion neuroinvasion. *Neuron* **31,** 345–348.

**Prusiner, S.B.** (1995) The prion diseases. *Sci. Am.* **272** (January), 48–57.

    A good summary of the evidence leading to the prion hypothesis.

**Richardson, A., Landry, S.J., & Georgopolous, C.** (1998) The ins and outs of a molecular chaperone machine. *Trends Biochem. Sci.* **23,** 138–143.

**Thomas, P.J., Qu, B.-H., & Pederson, P.L.** (1995) Defective protein folding as a basis of human disease. *Trends Biochem. Sci.* **20,** 456–459.

**Westaway, D. & Carlson, G.A.** (2002) Mammalian prion proteins: enigma, variation and vaccination. *Trends Biochem. Sci.* **27,** 301–307.

    A good update.

## Problems

**1. Properties of the Peptide Bond**  In x-ray studies of crystalline peptides, Linus Pauling and Robert Corey found that the C—N bond in the peptide link is intermediate in length (1.32 Å) between a typical C—N single bond (1.49 Å) and a C=N double bond (1.27 Å). They also found that the peptide bond is planar (all four atoms attached to the C—N group are located in the same plane) and that the two α-carbon atoms attached to the C—N are always trans to each other (on opposite sides of the peptide bond):

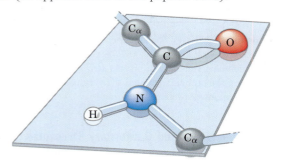

    (a)  What does the length of the C—N bond in the peptide linkage indicate about its strength and its bond order (i.e., whether it is single, double, or triple)?

    (b)  What do the observations of Pauling and Corey tell us about the ease of rotation about the C—N peptide bond?

**2. Structural and Functional Relationships in Fibrous Proteins**  William Astbury discovered that the x-ray pattern of wool shows a repeating structural unit spaced about 5.2 Å along the length of the wool fiber. When he steamed and

stretched the wool, the x-ray pattern showed a new repeating structural unit at a spacing of 7.0 Å. Steaming and stretching the wool and then letting it shrink gave an x-ray pattern consistent with the original spacing of about 5.2 Å. Although these observations provided important clues to the molecular structure of wool, Astbury was unable to interpret them at the time.

    (a)  Given our current understanding of the structure of wool, interpret Astbury's observations.

    (b)  When wool sweaters or socks are washed in hot water or heated in a dryer, they shrink. Silk, on the other hand, does not shrink under the same conditions. Explain.

**3. Rate of Synthesis of Hair α-Keratin**  Hair grows at a rate of 15 to 20 cm/yr. All this growth is concentrated at the base of the hair fiber, where α-keratin filaments are synthesized inside living epidermal cells and assembled into ropelike structures (see Fig. 4–11). The fundamental structural element of α-keratin is the α helix, which has 3.6 amino acid residues per turn and a rise of 5.4 Å per turn (see Fig. 4–4b). Assuming that the biosynthesis of α-helical keratin chains is the rate-limiting factor in the growth of hair, calculate the rate at which peptide bonds of α-keratin chains must be synthesized (peptide bonds per second) to account for the observed yearly growth of hair.

**4. Effect of pH on the Conformation of α-Helical Secondary Structures**  The unfolding of the α helix of a polypeptide to a randomly coiled conformation is accompanied by a large decrease in a property called its specific rotation, a measure of a solution's capacity to rotate plane-polarized light. Polyglutamate, a polypeptide made up of only L-Glu residues, has the $\alpha$-helical conformation at pH 3. When the pH is raised to 7, there is a large decrease in the specific rotation of the solution. Similarly, polylysine (L-Lys residues) is an $\alpha$ helix at pH 10, but when the pH is lowered to 7 the specific rotation also decreases, as shown by the following graph.

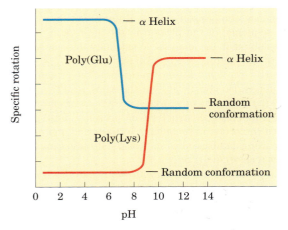

What is the explanation for the effect of the pH changes on the conformations of poly(Glu) and poly(Lys)? Why does the transition occur over such a narrow range of pH?

**5. Disulfide Bonds Determine the Properties of Many Proteins**   A number of natural proteins are very rich in disulfide bonds, and their mechanical properties (tensile strength, viscosity, hardness, etc.) are correlated with the degree of disulfide bonding. For example, glutenin, a wheat protein rich in disulfide bonds, is responsible for the cohesive and elastic character of dough made from wheat flour. Similarly, the hard, tough nature of tortoise shell is due to the extensive disulfide bonding in its $\alpha$-keratin.

(a)  What is the molecular basis for the correlation between disulfide-bond content and mechanical properties of the protein?

(b)  Most globular proteins are denatured and lose their activity when briefly heated to 65 °C. However, globular proteins that contain multiple disulfide bonds often must be heated longer at higher temperatures to denature them. One such protein is bovine pancreatic trypsin inhibitor (BPTI), which has 58 amino acid residues in a single chain and contains three disulfide bonds. On cooling a solution of denatured BPTI, the activity of the protein is restored. What is the molecular basis for this property?

**6. Amino Acid Sequence and Protein Structure**   Our growing understanding of how proteins fold allows researchers to make predictions about protein structure based on primary amino acid sequence data.

| 1 | 2 | 3 | 4 | 5 | 6 | 7 | 8 | 9 | 10 |
|---|---|---|---|---|---|---|---|---|---|
| Ile | Ala | His | Thr | Tyr | Gly | Pro | Phe | Glu | Ala – |

| 11 | 12 | 13 | 14 | 15 | 16 | 17 | 18 | 19 | 20 |
|---|---|---|---|---|---|---|---|---|---|
| Ala | Met | Cys | Lys | Trp | Glu | Ala | Gln | Pro | Asp – |

| 21 | 22 | 23 | 24 | 25 | 26 | 27 | 28 |
|---|---|---|---|---|---|---|---|
| Gly | Met | Glu | Cys | Ala | Phe | His | Arg |

(a)  In the amino acid sequence above, where would you predict that bends or $\beta$ turns would occur?

(b)  Where might intrachain disulfide cross-linkages be formed?

(c)  Assuming that this sequence is part of a larger globular protein, indicate the probable location (the external surface or interior of the protein) of the following amino acid residues: Asp, Ile, Thr, Ala, Gln, Lys. Explain your reasoning. (Hint: See the hydropathy index in Table 3–1.)

**7. Bacteriorhodopsin in Purple Membrane Proteins**   Under the proper environmental conditions, the salt-loving bacterium *Halobacterium halobium* synthesizes a membrane protein ($M_r$ 26,000) known as bacteriorhodopsin, which is purple because it contains retinal (see Fig. 10–21). Molecules of this protein aggregate into "purple patches" in the cell membrane. Bacteriorhodopsin acts as a light-activated proton pump that provides energy for cell functions. X-ray analysis of this protein reveals that it consists of seven parallel $\alpha$-helical segments, each of which traverses the bacterial cell membrane (thickness 45 Å). Calculate the minimum number of amino acid residues necessary for one segment of $\alpha$ helix to traverse the membrane completely. Estimate the fraction of the bacteriorhodopsin protein that is involved in membrane-spanning helices. (Use an average amino acid residue weight of 110.)

**8. Pathogenic Action of Bacteria That Cause Gas Gangrene**   The highly pathogenic anaerobic bacterium *Clostridium perfringens* is responsible for gas gangrene, a condition in which animal tissue structure is destroyed. This bacterium secretes an enzyme that efficiently catalyzes the hydrolysis of the peptide bond indicated in red:

$$-X-Gly-Pro-Y- \xrightarrow{H_2O}$$
$$-X-COO^- + H_3\overset{+}{N}-Gly-Pro-Y-$$

where X and Y are any of the 20 common amino acids. How does the secretion of this enzyme contribute to the invasiveness of this bacterium in human tissues? Why does this enzyme not affect the bacterium itself?

**9. Number of Polypeptide Chains in a Multisubunit Protein**   A sample (660 mg) of an oligomeric protein of $M_r$ 132,000 was treated with an excess of 1-fluoro-2,4-dinitrobenzene (Sanger's reagent) under slightly alkaline conditions until the chemical reaction was complete. The peptide bonds of the protein were then completely hydrolyzed by heating it with concentrated HCl. The hydrolysate was found to contain 5.5 mg of the following compound:

$$O_2N-\phantom{x}-NH-\underset{H}{\overset{NO_2 \quad CH_3 \;\; CH_3}{\underset{|}{\overset{|}{C}}\overset{CH}{\underset{|}{C}}}-COOH}$$

2,4-Dinitrophenyl derivatives of the $\alpha$-amino groups of other amino acids could not be found.

(a)  Explain how this information can be used to determine the number of polypeptide chains in an oligomeric protein.

(b)  Calculate the number of polypeptide chains in this protein.

(c)  What other protein analysis technique could you employ to determine whether the polypeptide chains in this protein are similar or different?

## Biochemistry on the Internet

**10. Protein Modeling on the Internet**   A group of patients suffering from Crohn's disease (an inflammatory bowel disease) underwent biopsies of their intestinal mucosa in an attempt to identify the causative agent. A protein was identified that was expressed at higher levels in patients with Crohn's disease than in patients with an unrelated inflammatory bowel disease or in unaffected controls. The protein was isolated and the following *partial* amino acid sequence was obtained (reads left to right):

| | | |
|---|---|---|
| EAELCPDRCI | HSFQNLGIQC | VKKRDLEQAI |
| SQRIQTNNNP | FQVPIEEQRG | DYDLNAVRLC |
| FQVTVRDPSG | RPLRLPPVLP | HPIFDNRAPN |
| TAELKICRVN | RNSGSCLGGD | EIFLLCDKVQ |
| KEDIEVYFTG | PGWEARGSFS | QADVHRQVAI |
| VFRTPPYADP | SLQAPVRVSM | QLRRPSDREL |
| SEPMEFQYLP | DTDDRHRIEE | KRKRTYETFK |
| SIMKKSPFSG | PTDPRPPPRR | IAVPSRSSAS |
| VPKPAPQPYP | | |

(a) You can identify this protein using a protein database on the Internet. Some good places to start include Protein Information Resource (PIR; pir.georgetown.edu/pirwww), Structural Classification of Proteins (SCOP; http://scop.berkeley.edu), and Prosite (http://us.expasy.org/prosite).

At your selected database site, follow links to locate the sequence comparison engine. Enter about 30 residues from the sequence of the protein in the appropriate search field and submit it for analysis. What does this analysis tell you about the identity of the protein?

(b) Try using different portions of the protein amino acid sequence. Do you always get the same result?

(c) A variety of websites provide information about the three-dimensional structure of proteins. Find information about the protein's secondary, tertiary, and quaternary structure using database sites such as the Protein Data Bank (PDB; www.rcsb.org/pdb) or SCOP.

(d) In the course of your Web searches try to find information about the cellular function of the protein.

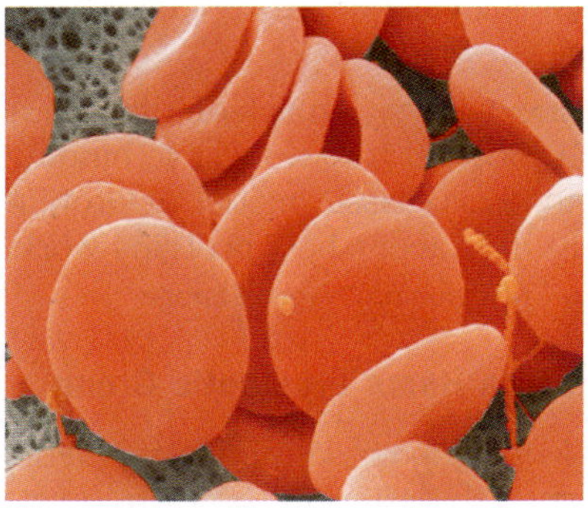

# PROTEIN FUNCTION

I have occasionally seen in almost dried blood, placed between glass plates in a desiccator, rectangular crystalline structures, which under the microscope had sharp edges and were bright red.

> —*Friedrich Ludwig Hünefeld,* Der Chemismus in der thierischen Organisation, *1840 (one of the first observations of hemoglobin)*

Since the proteins participate in one way or another in all chemical processes in the living organism, one may expect highly significant information for biological chemistry from the elucidation of their structure and their transformations.

> —*Emil Fischer, article in* Berichte der deutschen chemischen Gesellschaft zu Berlin, *1906*

Knowing the three-dimensional structure of a protein is an important part of understanding how the protein functions. However, the structure shown in two dimensions on a page is deceptively static. Proteins are dynamic molecules whose functions almost invariably depend on interactions with other molecules, and these interactions are affected in physiologically important ways by sometimes subtle, sometimes striking changes in protein conformation. In this chapter, we explore how proteins interact with other molecules and how their interactions are related to dynamic protein structure. The importance of molecular interactions to a protein's function can hardly be overemphasized. In Chapter 4, we saw that the function of fibrous proteins as structural elements of cells and tissues depends on stable, long-term quaternary interactions between identical polypeptide chains. As we shall see in this chapter, the functions of many other proteins involve interactions with a variety of different molecules. Most of these interactions are fleeting, though they may be the basis of complex physiological processes such as oxygen transport, immune function, and muscle contraction—the topics we examine in detail in this chapter. The proteins that carry out these processes illustrate the following key principles of protein function, some of which will be familiar from the previous chapter:

The functions of many proteins involve the reversible binding of other molecules. A molecule bound reversibly by a protein is called a **ligand.** A ligand may be any kind of molecule, including another protein. The transient nature of protein-ligand interactions is critical to life, allowing an organism to respond rapidly and reversibly to changing environmental and metabolic circumstances.

A ligand binds at a site on the protein called the **binding site,** which is complementary to the ligand in size, shape, charge, and hydrophobic or hydrophilic character. Furthermore, the interaction is specific: the protein can discriminate among the thousands of different molecules in its environment and selectively bind only one or a few. A given protein may have separate binding sites for several different ligands. These specific molecular interactions are crucial in maintaining the high degree of order in a living system. (This discussion excludes the binding of water, which may interact weakly and nonspecifically with many parts of a protein. In Chapter 6, we consider water as a specific ligand for many enzymes.)

Proteins are flexible. Changes in conformation may be subtle, reflecting molecular vibrations and small movements of amino acid residues throughout the protein. A protein flexing in this way is sometimes said to "breathe." Changes in conformation may also be quite dramatic, with major segments of the protein structure moving as much as several nanometers. Specific conformational changes are frequently essential to a protein's function.

The binding of a protein and ligand is often coupled to a conformational change in the protein that makes the binding site more complementary to the ligand, permitting tighter binding. The structural adaptation that occurs between protein and ligand is called **induced fit.**

In a multisubunit protein, a conformational change in one subunit often affects the conformation of other subunits.

Interactions between ligands and proteins may be regulated, usually through specific interactions with one or more additional ligands. These other ligands may cause conformational changes in the protein that affect the binding of the first ligand.

Enzymes represent a special case of protein function. Enzymes bind and chemically transform other molecules—they catalyze reactions. The molecules acted upon by enzymes are called reaction **substrates** rather than ligands, and the ligand-binding site is called the **catalytic site** or **active site.** In this chapter we emphasize the noncatalytic functions of proteins. In Chapter 6 we consider catalysis by enzymes, a central topic in biochemistry. You will see that the themes of this chapter—binding, specificity, and conformational change—are continued in the next chapter, with the added element of proteins participating in chemical transformations.

## 5.1 Reversible Binding of a Protein to a Ligand: Oxygen-Binding Proteins

Myoglobin and hemoglobin may be the most-studied and best-understood proteins. They were the first proteins for which three-dimensional structures were determined, and our current understanding of myoglobin and hemoglobin is garnered from the work of thousands of biochemists over several decades. Most important, these molecules illustrate almost every aspect of that most central of biochemical processes: the reversible binding of a ligand to a protein. This classic model of protein function tells us a great deal about how proteins work.

**Oxygen-Binding Proteins—Myoglobin: Oxygen Storage**

### Oxygen Can Be Bound to a Heme Prosthetic Group

Oxygen is poorly soluble in aqueous solutions (see Table 2–3) and cannot be carried to tissues in sufficient quantity if it is simply dissolved in blood serum. Diffusion of oxygen through tissues is also ineffective over distances greater than a few millimeters. The evolution of larger, multicellular animals depended on the evolution of proteins that could transport and store oxygen. However, none of the amino acid side chains in proteins is suited for the reversible binding of oxygen molecules. This role is filled by certain transition metals, among them iron and copper, that have a strong tendency to bind oxygen. Multicellular organisms exploit the properties of metals, most commonly iron, for oxygen transport. However, free iron promotes the formation of highly reactive oxygen species such as hydroxyl radicals that can damage DNA and other macromolecules. Iron used in cells is therefore bound in forms that sequester it and/or make it less reactive. In multicellular organisms—especially those in which iron, in its oxygen-carrying capacity, must be transported over large distances—iron is often incorporated into a protein-bound prosthetic group called **heme.** (Recall from Chapter 3 that a prosthetic group is a compound permanently associated with a protein that contributes to the protein's function.)

Heme (or haem) consists of a complex organic ring structure, **protoporphyrin,** to which is bound a single iron atom in its ferrous ($Fe^{2+}$) state (Fig. 5–1). The iron atom has six coordination bonds, four to nitrogen atoms that are part of the flat **porphyrin ring** system and two perpendicular to the porphyrin. The coordinated nitrogen atoms (which have an electron-donating character) help prevent conversion of the heme iron to the ferric ($Fe^{3+}$) state. Iron in the $Fe^{2+}$ state binds oxygen reversibly; in the $Fe^{3+}$ state it does not bind oxygen. Heme is found in a number of oxygen-transporting proteins, as well as in some proteins, such as the cytochromes, that participate in oxidation-reduction (electron-transfer) reactions (Chapter 19).

Free heme molecules (heme not bound to protein), leave $Fe^{2+}$ with two "open" coordination bonds. Simultaneous reaction of one $O_2$ molecule with two free heme molecules (or two free $Fe^{2+}$) can result in irreversible conversion of $Fe^{2+}$ to $Fe^{3+}$. In heme-containing proteins, this reaction is prevented by sequestering each heme deep within the protein structure where access to the two open coordination bonds is restricted. One of these two coordination bonds is occupied by a side-chain nitrogen of a His residue. The other is the bind-

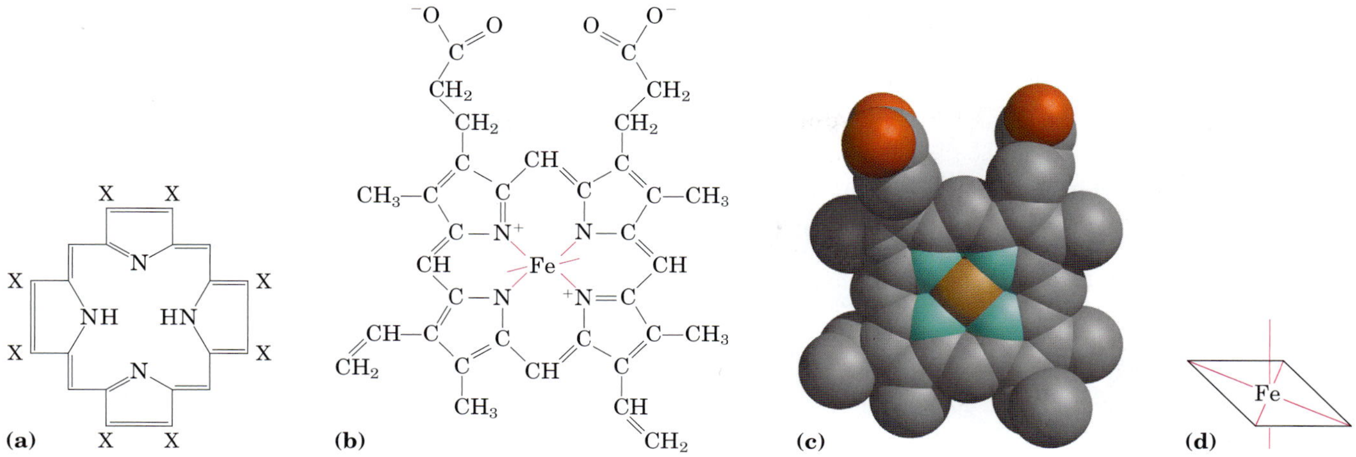

**FIGURE 5–1 Heme.** The heme group is present in myoglobin, hemoglobin, and many other proteins, designated heme proteins. Heme consists of a complex organic ring structure, protoporphyrin IX, to which is bound an iron atom in its ferrous ($Fe^{2+}$) state. **(a)** Porphyrins, of which protoporphyrin IX is only one example, consist of four pyr-role rings linked by methene bridges, with substitutions at one or more of the positions denoted X. **(b, c)** Two representations of heme. (Derived from PDB ID 1CCR.) The iron atom of heme has six coordination bonds: four in the plane of, and bonded to, the flat porphyrin ring system, and **(d)** two perpendicular to it.

ing site for molecular oxygen ($O_2$) (Fig. 5–2). When oxygen binds, the electronic properties of heme iron change; this accounts for the change in color from the dark purple of oxygen-depleted venous blood to the bright red of oxygen-rich arterial blood. Some small molecules, such as carbon monoxide (CO) and nitric oxide (NO), coordinate to heme iron with greater affinity than does $O_2$. When a molecule of CO is bound to heme, $O_2$ is excluded, which is why CO is highly toxic to aerobic organisms (a topic explored later, in Box 5–1). By surrounding and sequestering heme, oxygen-binding proteins regulate the access of CO and other small molecules to the heme iron.

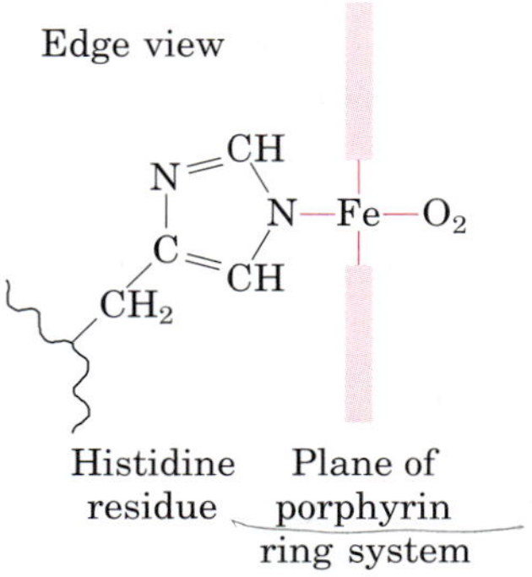

**FIGURE 5–2 The heme group viewed from the side.** This view shows the two coordination bonds to $Fe^{2+}$ perpendicular to the porphyrin ring system. One of these two bonds is occupied by a His residue, sometimes called the proximal His. The other bond is the binding site for oxygen. The remaining four coordination bonds are in the plane of, and bonded to, the flat porphyrin ring system.

## Myoglobin Has a Single Binding Site for Oxygen

Myoglobin ($M_r$ 16,700; abbreviated Mb) is a relatively simple oxygen-binding protein found in almost all mammals, primarily in muscle tissue. As a transport protein, it facilitates oxygen diffusion in muscle. Myoglobin is particularly abundant in the muscles of diving mammals such as seals and whales, where it also has an oxygen-storage function for prolonged excursions undersea. Proteins very similar to myoglobin are widely distributed, occurring even in some single-celled organisms.

Myoglobin is a single polypeptide of 153 amino acid residues with one molecule of heme. It is typical of the family of proteins called **globins,** all of which have similar primary and tertiary structures. The polypeptide is made up of eight α-helical segments connected by bends (Fig. 5–3). About 78% of the amino acid residues in the protein are found in these α helices.

Any detailed discussion of protein function inevitably involves protein structure. To facilitate our treatment of myoglobin, we first introduce some structural conventions peculiar to globins. As seen in Figure 5–3, the helical segments are named A through H. An individual amino acid residue is designated either by its position in the amino acid sequence or by its location within the sequence of a particular α-helical segment. For example, the His residue coordinated to the heme in myoglobin, His[93] (the 93rd amino acid residue from the amino-terminal end of the myoglobin polypeptide sequence), is also called His F8 (the 8th residue in α helix F). The bends in the structure are designated AB, CD, EF, FG, and so forth, reflecting the α-helical segments they connect.

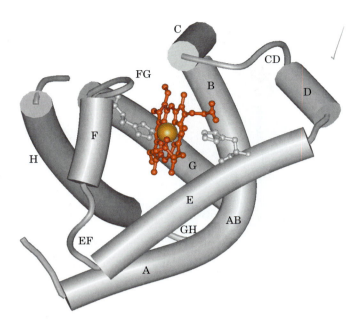

**FIGURE 5–3 The structure of myoglobin.** (PDB ID 1MBO) The eight α-helical segments (shown here as cylinders) are labeled A through H. Nonhelical residues in the bends that connect them are labeled AB, CD, EF, and so forth, indicating the segments they interconnect. A few bends, including BC and DE, are abrupt and do not contain any residues; these are not normally labeled. (The short segment visible between D and E is an artifact of the computer representation.) The heme is bound in a pocket made up largely of the E and F helices, although amino acid residues from other segments of the protein also participate.

## Protein-Ligand Interactions Can Be Described Quantitatively

The function of myoglobin depends on the protein's ability not only to bind oxygen but also to release it when and where it is needed. Function in biochemistry often revolves around a reversible protein-ligand interaction of this type. A quantitative description of this interaction is therefore a central part of many biochemical investigations.

In general, the reversible binding of a protein (P) to a ligand (L) can be described by a simple **equilibrium expression:**

$$P + L \rightleftharpoons PL \tag{5–1}$$

The reaction is characterized by an equilibrium constant, $K_a$, such that

$$K_a = \frac{[PL]}{[P][L]} \tag{5–2}$$

The term $\boldsymbol{K_a}$ is an **association constant** (not to be confused with the $K_a$ that denotes an acid dissociation constant; p. 63). The association constant provides a measure of the affinity of the ligand L for the protein. $K_a$ has units of $M^{-1}$; a higher value of $K_a$ corresponds to

a higher affinity of the ligand for the protein. A rearrangement of Equation 5–2 shows that the ratio of bound to free protein is directly proportional to the concentration of free ligand:

$$K_a[L] = \frac{[PL]}{[P]} \tag{5–3}$$

When the concentration of the ligand is much greater than the concentration of ligand-binding sites, the binding of the ligand by the protein does not appreciably change the concentration of free (unbound) ligand—that is, [L] remains constant. This condition is broadly applicable to most ligands that bind to proteins in cells and simplifies our description of the binding equilibrium.

We can now consider the binding equilibrium from the standpoint of the fraction, $\theta$ (theta), of ligand-binding sites on the protein that are occupied by ligand:

$$\theta = \frac{\text{binding sites occupied}}{\text{total binding sites}} = \frac{[PL]}{[PL] + [P]} \tag{5–4}$$

Substituting $K_a[L][P]$ for [PL] (see Eqn 5–3) and rearranging terms gives

$$\theta = \frac{K_a[L][P]}{K_a[L][P] + [P]} = \frac{K_a[L]}{K_a[L] + 1} = \frac{[L]}{[L] + \frac{1}{K_a}} \tag{5–5}$$

The value of $K_a$ can be determined from a plot of $\theta$ versus the concentration of free ligand, [L] (Fig. 5–4a). Any equation of the form $x = y/(y + z)$ describes a hyperbola, and $\theta$ is thus found to be a hyperbolic function of [L]. The fraction of ligand-binding sites occupied approaches saturation asymptotically as [L] increases. The [L] at which half of the available ligand-binding sites are occupied (at $\theta = 0.5$) corresponds to $1/K_a$.

It is more common (and intuitively simpler), however, to consider the **dissociation constant, $K_d$,** which is the reciprocal of $K_a$ ($K_d = 1/K_a$) and is given in units of molar concentration (M). $K_d$ is the equilibrium constant for the release of ligand. The relevant expressions change to

$$K_d = \frac{[P][L]}{[PL]} \tag{5–6}$$

$$[PL] = \frac{[P][L]}{K_d} \tag{5–7}$$

$$\theta = \frac{[L]}{[L] + K_d} \tag{5–8}$$

When [L] is equal to $K_d$, half of the ligand-binding sites are occupied. As [L] falls below $K_d$, progressively less of the protein has ligand bound to it. In order for 90% of the available ligand-binding sites to be occupied, [L] must be nine times greater than $K_d$.

In practice, $K_d$ is used much more often than $K_a$ to express the affinity of a protein for a ligand. Note that

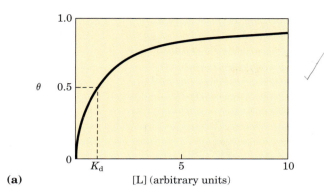

**(a)** [L] (arbitrary units)

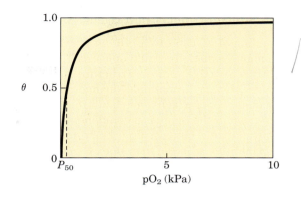

**(b)** pO$_2$ (kPa)

**FIGURE 5–4 Graphical representations of ligand binding.** The fraction of ligand-binding sites occupied, $\theta$, is plotted against the concentration of free ligand. Both curves are rectangular hyperbolas. **(a)** A hypothetical binding curve for a ligand L. The [L] at which half of the available ligand-binding sites are occupied is equivalent to $1/K_a$, or $K_d$. The curve has a horizontal asymptote at $\theta = 1$ and a vertical asymptote (not shown) at [L] $= -1/K_a$. **(b)** A curve describing the binding of oxygen to myoglobin. The partial pressure of O$_2$ in the air above the solution is expressed in kilopascals (kPa). Oxygen binds tightly to myoglobin, with a $P_{50}$ of only 0.26 kPa.

a lower value of $K_d$ corresponds to a higher affinity of ligand for the protein. The mathematics can be reduced to simple statements: $K_d$ is equivalent to the molar concentration of ligand at which half of the available ligand-binding sites are occupied. At this point, the protein is said to have reached half-saturation with respect to ligand binding. The more tightly a protein binds a ligand, the lower the concentration of ligand required for half the binding sites to be occupied, and thus the lower the value of $K_d$. Some representative dissociation constants are given in Table 5–1.

The binding of oxygen to myoglobin follows the patterns discussed above. However, because oxygen is a gas, we must make some minor adjustments to the equations so that laboratory experiments can be carried out more conveniently. We first substitute the concentration of dissolved oxygen for [L] in Equation 5–8 to give

$$\theta = \frac{[O_2]}{[O_2] + K_d} \qquad (5\text{–}9)$$

As for any ligand, $K_d$ is equal to the [O$_2$] at which half of the available ligand-binding sites are occupied, or $[O_2]_{0.5}$. Equation 5–9 thus becomes

$$\theta = \frac{[O_2]}{[O_2] + [O_2]_{0.5}} \qquad (5\text{–}10)$$

In experiments using oxygen as a ligand, it is the partial pressure of oxygen in the gas phase above the solution, pO$_2$, that is varied, because this is easier to measure than the concentration of oxygen dissolved in the solution. The concentration of a volatile substance in solution is always proportional to the local partial pressure of the gas. So, if we define the partial pressure of oxygen at $[O_2]_{0.5}$ as $P_{50}$, substitution in Equation 5–10 gives

$$\theta = \frac{pO_2}{pO_2 + P_{50}} \qquad (5\text{–}11)$$

A binding curve for myoglobin that relates $\theta$ to pO$_2$ is shown in Figure 5–4b.

### TABLE 5–1 Some Protein Dissociation Constants

| Protein | Ligand | $K_d$ (M)* |
|---|---|---|
| Avidin (egg white)[†] | Biotin | $1 \times 10^{-15}$ |
| Insulin receptor (human) | Insulin | $1 \times 10^{-10}$ |
| Anti-HIV immunoglobulin (human)[‡] | gp41 (HIV-1 surface protein) | $4 \times 10^{-10}$ |
| Nickel-binding protein (E. coli) | Ni$^{2+}$ | $1 \times 10^{-7}$ |
| Calmodulin (rat)[§] | Ca$^{2+}$ | $3 \times 10^{-6}$ |
| | | $2 \times 10^{-5}$ |

*A reported dissociation constant is valid only for the particular solution conditions under which it was measured. $K_d$ values for a protein-ligand interaction can be altered, sometimes by several orders of magnitude, by changes in the solution's salt concentration, pH, or other variables.

[†]Interaction of avidin with biotin, an enzyme cofactor, is among the strongest noncovalent biochemical interactions known.

[‡]This immunoglobulin was isolated as part of an effort to develop a vaccine against HIV. Immunoglobulins (described later in the chapter) are highly variable, and the $K_d$ reported here should not be considered characteristic of all immunoglobulins.

[§]Calmodulin has four binding sites for calcium. The values shown reflect the highest- and lowest-affinity binding sites observed in one set of measurements.

## Protein Structure Affects How Ligands Bind

The binding of a ligand to a protein is rarely as simple as the above equations would suggest. The interaction is greatly affected by protein structure and is often accompanied by conformational changes. For example, the specificity with which heme binds its various ligands is altered when the heme is a component of myoglobin. Carbon monoxide binds to free heme molecules more than 20,000 times better than does $O_2$ (that is, the $K_d$ or $P_{50}$ for CO binding to free heme is more than 20,000 times lower than that for $O_2$), but it binds only about 200 times better when the heme is bound in myoglobin. The difference may be partly explained by steric hindrance. When $O_2$ binds to free heme, the axis of the oxygen molecule is positioned at an angle to the Fe—O bond (Fig. 5–5a). In contrast, when CO binds to free heme, the Fe, C, and O atoms lie in a straight line (Fig. 5–5b). In both cases, the binding reflects the geometry of hybrid orbitals in each ligand. In myoglobin, His[64] (His E7), on the $O_2$-binding side of the heme, is too far away to coordinate with the heme iron, but it does interact with a ligand bound to heme. This residue, called the *distal His,* forms a hydrogen bond with $O_2$ (Fig. 5–5c) but may preclude the linear binding of CO, providing one explanation for the selectively diminished binding of CO to heme in myoglobin (and hemoglobin). A reduction in CO binding is physiologically important, because CO is a low-level byproduct of cellular metabolism. Other factors, not yet well-defined, also may modulate the interaction of heme with CO in these proteins.

The binding of $O_2$ to the heme in myoglobin also depends on molecular motions, or "breathing," in the protein structure. The heme molecule is deeply buried in the folded polypeptide, with no direct path for oxygen to move from the surrounding solution to the ligand-binding site. If the protein were rigid, $O_2$ could not enter or leave the heme pocket at a measurable rate. However, rapid molecular flexing of the amino acid side chains produces transient cavities in the protein structure, and $O_2$ evidently makes its way in and out by moving through these cavities. Computer simulations of rapid structural fluctuations in myoglobin suggest that there are many such pathways. One major route is provided by rotation of the side chain of the distal His (His[64]), which occurs on a nanosecond ($10^{-9}$ s) time scale. Even subtle conformational changes can be critical for protein activity.

## Oxygen Is Transported in Blood by Hemoglobin

**◉ Oxygen-Binding Proteins—Hemoglobin: Oxygen Transport**

Nearly all the oxygen carried by whole blood in animals is bound and transported by hemoglobin in erythrocytes (red blood cells). Normal human erythrocytes are small (6 to 9 $\mu$m in diameter), biconcave disks. They are formed from precursor stem cells called **hemocytoblasts.** In

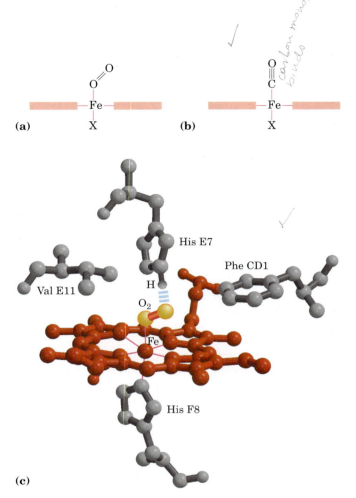

**FIGURE 5–5 Steric effects on the binding of ligands to the heme of myoglobin. (a)** Oxygen binds to heme with the $O_2$ axis at an angle, a binding conformation readily accommodated by myoglobin. **(b)** Carbon monoxide binds to free heme with the CO axis perpendicular to the plane of the porphyrin ring. When binding to the heme in myoglobin, CO is forced to adopt a slight angle because the perpendicular arrangement is sterically blocked by His E7, the distal His. This effect weakens the binding of CO to myoglobin. **(c)** Another view (derived from PDB ID 1MBO), showing the arrangement of key amino acid residues around the heme of myoglobin. The bound $O_2$ is hydrogen-bonded to the distal His, His E7 (His[64]), further facilitating the binding of $O_2$.

the maturation process, the stem cell produces daughter cells that form large amounts of hemoglobin and then lose their intracellular organelles—nucleus, mitochondria, and endoplasmic reticulum. Erythrocytes are thus incomplete, vestigial cells, unable to reproduce and, in humans, destined to survive for only about 120 days. Their main function is to carry hemoglobin, which is dissolved in the cytosol at a very high concentration (~34% by weight).

In arterial blood passing from the lungs through the heart to the peripheral tissues, hemoglobin is about 96% saturated with oxygen. In the venous blood returning to the heart, hemoglobin is only about 64% saturated. Thus, each 100 mL of blood passing through a tissue releases about one-third of the oxygen it carries, or 6.5 mL of $O_2$ gas at atmospheric pressure and body temperature.

Myoglobin, with its hyperbolic binding curve for oxygen (Fig. 5–4b), is relatively insensitive to small changes in the concentration of dissolved oxygen and so functions well as an oxygen-storage protein. Hemoglobin, with its multiple subunits and $O_2$-binding sites, is better suited to oxygen transport. As we shall see, interactions between the subunits of a multimeric protein can permit a highly sensitive response to small changes in ligand concentration. Interactions among the subunits in hemoglobin cause conformational changes that alter the affinity of the protein for oxygen. The modulation of oxygen binding allows the $O_2$-transport protein to respond to changes in oxygen demand by tissues.

## Hemoglobin Subunits Are Structurally Similar to Myoglobin

Hemoglobin ($M_r$ 64,500; abbreviated Hb) is roughly spherical, with a diameter of nearly 5.5 nm. It is a tetrameric protein containing four heme prosthetic groups, one associated with each polypeptide chain. Adult hemoglobin contains two types of globin, two $\alpha$ chains (141 residues each) and two $\beta$ chains (146 residues each). Although fewer than half of the amino acid residues in the polypeptide sequences of the $\alpha$ and $\beta$ subunits are identical, the three-dimensional structures of the two types of subunits are very similar. Furthermore, their structures are very similar to that of myoglobin (Fig. 5–6), even though the amino acid sequences of the three polypeptides are identical at only 27 positions (Fig. 5–7). All three polypeptides are members of the globin family of proteins. The helix-naming convention described for myoglobin is also applied to the hemoglobin polypeptides, except that the $\alpha$ subunit lacks the short D helix. The heme-binding pocket is made up largely of the E and F helices.

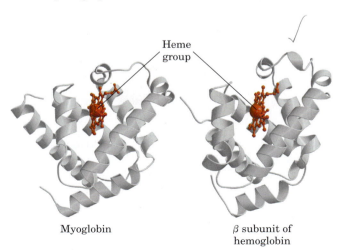

**FIGURE 5–6** A comparison of the structures of myoglobin (PDB ID 1MBO) and the $\beta$ subunit of hemoglobin (derived from PDB ID 1HGA).

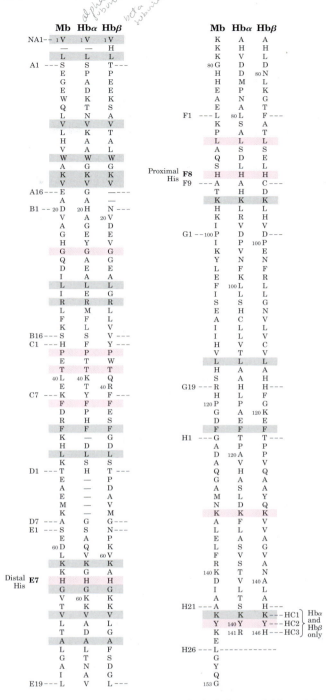

**FIGURE 5–7** **The amino acid sequences of whale myoglobin and the $\alpha$ and $\beta$ chains of human hemoglobin.** Dashed lines mark helix boundaries. To align the sequences optimally, short gaps must be introduced into both Hb sequences where a few amino acids are present in the compared sequences. With the exception of the missing D helix in Hb$\alpha$, this alignment permits the use of the helix lettering convention that emphasizes the common positioning of amino acid residues that are identical in all three structures (shaded). Residues shaded in pink are conserved in all known globins. Note that the common helix-letter-and-number designation for amino acids does not necessarily correspond to a common position in the linear sequence of amino acids in the polypeptides. For example, the distal His residue is His E7 in all three structures, but corresponds to His[64], His[58], and His[63] in the linear sequences of Mb, Hb$\alpha$, and Hb$\beta$, respectively. Nonhelical residues at the amino and carboxyl termini, beyond the first (A) and last (H) $\alpha$-helical segments, are labeled NA and HC, respectively.

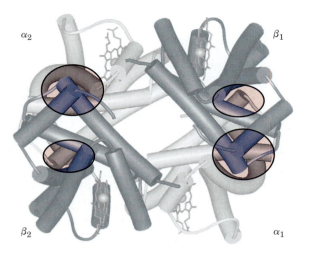

**FIGURE 5–8 Dominant interactions between hemoglobin subunits.** In this representation, $\alpha$ subunits are light and $\beta$ subunits are dark. The strongest subunit interactions (highlighted) occur between unlike subunits. When oxygen binds, the $\alpha_1\beta_1$ contact changes little, but there is a large change at the $\alpha_1\beta_2$ contact, with several ion pairs broken (PDB ID 1HGA).

The quaternary structure of hemoglobin features strong interactions between unlike subunits. The $\alpha_1\beta_1$ interface (and its $\alpha_2\beta_2$ counterpart) involves more than 30 residues, and its interaction is sufficiently strong that although mild treatment of hemoglobin with urea tends to cause the tetramer to disassemble into $\alpha\beta$ dimers, these dimers remain intact. The $\alpha_1\beta_2$ (and $\alpha_2\beta_1$) interface involves 19 residues (Fig. 5–8). Hydrophobic interactions predominate at the interfaces, but there are also many hydrogen bonds and a few ion pairs (sometimes referred to as salt bridges), whose importance is discussed below.

## Hemoglobin Undergoes a Structural Change on Binding Oxygen

X-ray analysis has revealed two major conformations of hemoglobin: the **R state** and the **T state.** Although oxygen binds to hemoglobin in either state, it has a significantly higher affinity for hemoglobin in the R state. Oxygen binding stabilizes the R state. When oxygen is absent experimentally, the T state is more stable and is thus the predominant conformation of **deoxyhemoglobin.** T and R originally denoted "tense" and "relaxed," respectively, because the T state is stabilized by a greater number of ion pairs, many of which lie at the $\alpha_1\beta_2$ (and $\alpha_2\beta_1$) interface (Fig. 5–9). The binding of $O_2$ to a hemoglobin subunit in the T state triggers a change in conformation to the R state. When the entire protein undergoes this transition, the structures of the individual subunits change little, but the $\alpha\beta$ subunit pairs slide past each other and rotate, narrowing the pocket between the $\beta$ subunits (Fig. 5–10). In this process, some

of the ion pairs that stabilize the T state are broken and some new ones are formed.

Max Perutz proposed that the $T \rightarrow R$ transition is triggered by changes in the positions of key amino acid side chains surrounding the heme. In the T state, the porphyrin is slightly puckered, causing the heme iron to protrude somewhat on the proximal His (His F8) side. The binding of $O_2$ causes the heme to assume a more planar conformation, shifting the position of the proximal His and the attached F helix (Fig. 5–11). These changes lead to adjustments in the ion pairs at the $\alpha_1\beta_2$ interface.

## Hemoglobin Binds Oxygen Cooperatively

Hemoglobin must bind oxygen efficiently in the lungs, where the $pO_2$ is about 13.3 kPa, and release oxygen in the tissues, where the $pO_2$ is about 4 kPa. Myoglobin, or any protein that binds oxygen with a hyperbolic binding curve, would be ill-suited to this function, for the reason illustrated in Figure 5–12. A protein that bound

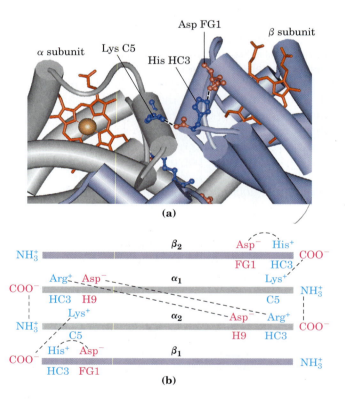

**FIGURE 5–9 Some ion pairs that stabilize the T state of deoxyhemoglobin. (a)** A close-up view of a portion of a deoxyhemoglobin molecule in the T state (PDB ID 1HGA). Interactions between the ion pairs His HC3 and Asp FG1 of the $\beta$ subunit (blue) and between Lys C5 of the $\alpha$ subunit (gray) and His HC3 (its $\alpha$-carboxyl group) of the $\beta$ subunit are shown with dashed lines. (Recall that HC3 is the carboxyl-terminal residue of the $\beta$ subunit.) **(b)** The interactions between these ion pairs, and between others not shown in **(a)**, are schematized in this representation of the extended polypeptide chains of hemoglobin.

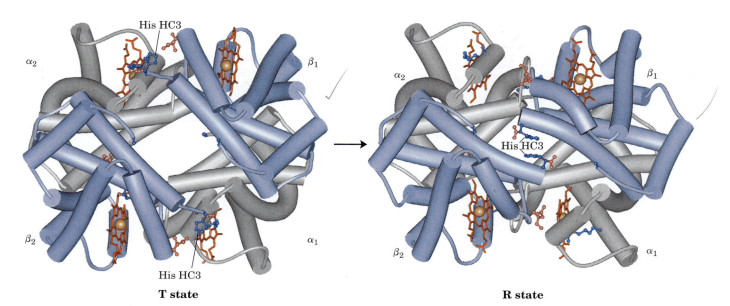

**T state** **R state**

**FIGURE 5–10** **The T → R transition.** (PDB ID 1HGA and 1BBB) In these depictions of deoxyhemoglobin, as in Figure 5–9, the β subunits are blue and the α subunits are gray. Positively charged side chains and chain termini involved in ion pairs are shown in blue, their negatively charged partners in red. The Lys C5 of each α subunit and Asp FG1 of each β subunit are visible but not labeled (compare Fig. 5–9a). Note that the molecule is oriented slightly differently than in Figure 5–9. The transition from the T state to the R state shifts the subunit pairs substantially, affecting certain ion pairs. Most noticeably, the His HC3 residues at the carboxyl termini of the β subunits, which are involved in ion pairs in the T state, rotate in the R state toward the center of the molecule, where they are no longer in ion pairs. Another dramatic result of the T → R transition is a narrowing of the pocket between the β subunits.

$O_2$ with high affinity would bind it efficiently in the lungs but would not release much of it in the tissues. If the protein bound oxygen with a sufficiently low affinity to release it in the tissues, it would not pick up much oxygen in the lungs.

Hemoglobin solves the problem by undergoing a transition from a low-affinity state (the T state) to a high-affinity state (the R state) as more $O_2$ molecules are bound. As a result, hemoglobin has a hybrid S-shaped, or sigmoid, binding curve for oxygen (Fig. 5–12). A single-subunit protein with a single ligand-binding site cannot produce a sigmoid binding curve—even if binding elicits a conformational change—because each molecule of ligand binds independently and cannot affect the binding of another molecule. In contrast, $O_2$ binding to individual subunits of hemoglobin can alter the affinity for $O_2$ in adjacent subunits. The first molecule of $O_2$ that interacts with deoxyhemoglobin binds weakly, because it binds to a subunit in the T state. Its binding, however, leads to conformational changes that are communicated to adjacent subunits, making it easier for additional molecules of $O_2$ to bind. In effect, the T → R transition occurs more readily in the second subunit once $O_2$ is bound to the first subunit. The last (fourth) $O_2$ molecule binds to a heme in a subunit that is already in the R state, and hence it binds with much higher affinity than the first molecule.

An **allosteric protein** is one in which the binding of a ligand to one site affects the binding properties of another site on the same protein. The term "allosteric" derives from the Greek *allos*, "other," and *stereos*, "solid" or "shape." Allosteric proteins are those having "other shapes," or conformations, induced by the binding of ligands referred to as modulators. The conformational changes induced by the modulator(s) interconvert more-active and less-active forms of the protein. The modulators for allosteric proteins may be either inhibitors or activators. When the normal ligand and

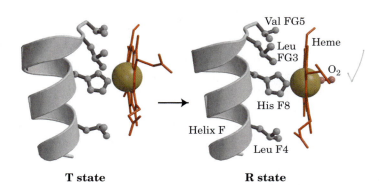

**T state** **R state**

**FIGURE 5–11** **Changes in conformation near heme on $O_2$ binding to deoxyhemoglobin.** (Derived from PDB ID 1HGA and 1BBB.) The shift in the position of the F helix when heme binds $O_2$ is thought to be one of the adjustments that triggers the T → R transition.

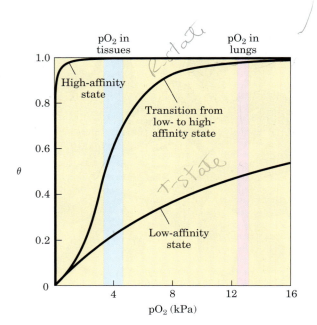

**FIGURE 5–12  A sigmoid (cooperative) binding curve.** A sigmoid binding curve can be viewed as a hybrid curve reflecting a transition from a low-affinity to a high-affinity state. Cooperative binding, as manifested by a sigmoid binding curve, renders hemoglobin more sensitive to the small differences in $O_2$ concentration between the tissues and the lungs, allowing hemoglobin to bind oxygen in the lungs (where $pO_2$ is high) and release it in the tissues (where $pO_2$ is low).

modulator are identical, the interaction is termed **homotropic.** When the modulator is a molecule other than the normal ligand the interaction is **heterotropic.** Some proteins have two or more modulators and therefore can have both homotropic and heterotropic interactions.

Cooperative binding of a ligand to a multimeric protein, such as we observe with the binding of $O_2$ to hemoglobin, is a form of allosteric binding often observed in multimeric proteins. The binding of one ligand affects

the affinities of any remaining unfilled binding sites, and $O_2$ can be considered as both a ligand and an activating homotropic modulator. There is only one binding site for $O_2$ on each subunit, so the allosteric effects giving rise to cooperativity are mediated by conformational changes transmitted from one subunit to another by subunit-subunit interactions. A sigmoid binding curve is diagnostic of cooperative binding. It permits a much more sensitive response to ligand concentration and is important to the function of many multisubunit proteins. The principle of allostery extends readily to regulatory enzymes, as we shall see in Chapter 6.

Cooperative conformational changes depend on variations in the structural stability of different parts of a protein, as described in Chapter 4. The binding sites of an allosteric protein typically consist of stable segments in proximity to relatively unstable segments, with the latter capable of frequent changes in conformation or disorganized motion (Fig. 5–13). When a ligand binds, the moving parts of the protein's binding site may be stabilized in a particular conformation, affecting the conformation of adjacent polypeptide subunits. If the

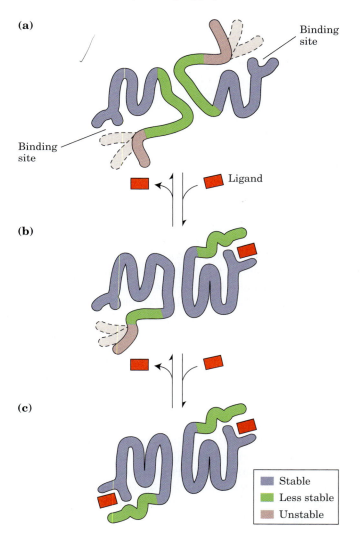

**FIGURE 5–13  Structural changes in a multisubunit protein undergoing cooperative binding to ligand.** Structural stability is not uniform throughout a protein molecule. Shown here is a hypothetical dimeric protein, with regions of high (blue), medium (green), and low (red) stability. The ligand-binding sites are composed of both high- and low-stability segments, so affinity for ligand is relatively low. **(a)** In the absence of ligand, the red segments are quite flexible and take up a variety of conformations, few of which facilitate ligand binding. The green segments are most stable in the low-affinity state. **(b)** The binding of ligand to one subunit stabilizes a high-affinity conformation of the nearby red segment (now shown in green), inducing a conformational change in the rest of the polypeptide. This is a form of induced fit. The conformational change is transmitted to the other subunit through protein-protein interactions, such that a higher-affinity conformation of the binding site is stabilized in the other subunit. **(c)** A second ligand molecule can now bind to the second subunit, with a higher affinity than the binding of the first, giving rise to the observed positive cooperativity.

entire binding site were highly stable, then few structural changes could occur in this site or be propagated to other parts of the protein when a ligand binds.

As is the case with myoglobin, ligands other than oxygen can bind to hemoglobin. An important example is carbon monoxide, which binds to hemoglobin about 250 times better than does oxygen. Human exposure to CO can have tragic consequences (Box 5–1).

## Cooperative Ligand Binding Can Be Described Quantitatively

Cooperative binding of oxygen by hemoglobin was first analyzed by Archibald Hill in 1910. From this work came a general approach to the study of cooperative ligand binding to multisubunit proteins.

For a protein with $n$ binding sites, the equilibrium of Equation 5–1 becomes

$$P + nL \rightleftharpoons PL_n \tag{5–12}$$

and the expression for the association constant becomes

$$K_a = \frac{[PL_n]}{[P][L]^n} \tag{5–13}$$

The expression for $\theta$ (see Eqn 5–8) is

$$\theta = \frac{[L]^n}{[L]^n + K_d} \tag{5–14}$$

Rearranging, then taking the log of both sides, yields

$$\frac{\theta}{1 - \theta} = \frac{[L]^n}{K_d} \tag{5–15}$$

$$\log\left(\frac{\theta}{1 - \theta}\right) = n \log [L] - \log K_d \tag{5–16}$$

where $K_d = [L]_{0.5}^n$.

Equation 5–16 is the **Hill equation,** and a plot of $\log [\theta/(1 - \theta)]$ versus $\log [L]$ is called a **Hill plot.** Based on the equation, the Hill plot should have a slope of $n$. However, the experimentally determined slope actually reflects not the number of binding sites but the degree of interaction between them. The slope of a Hill plot is therefore denoted by $n_H$, the **Hill coefficient,** which is a measure of the degree of cooperativity. If $n_H$ equals 1, ligand binding is not cooperative, a situation that can arise even in a multisubunit protein if the subunits do not communicate. An $n_H$ of greater than 1 indicates positive cooperativity in ligand binding. This is the situation observed in hemoglobin, in which the binding of one molecule of ligand facilitates the binding of others. The theoretical upper limit for $n_H$ is reached when $n_H = n$. In this case the binding would be completely cooperative: all binding sites on the protein would bind ligand simultaneously, and no protein molecules partially saturated with ligand would be present under any conditions. This limit is never reached in

practice, and the measured value of $n_H$ is always less than the actual number of ligand-binding sites in the protein.

An $n_H$ of less than 1 indicates negative cooperativity, in which the binding of one molecule of ligand *impedes* the binding of others. Well-documented cases of negative cooperativity are rare.

To adapt the Hill equation to the binding of oxygen to hemoglobin we must again substitute $pO_2$ for [L] and $P_{50}^n$ for $K_d$:

$$\log\left(\frac{\theta}{1 - \theta}\right) = n \log pO_2 - n \log P_{50}^n \tag{5–17}$$

Hill plots for myoglobin and hemoglobin are given in Figure 5–14.

## Two Models Suggest Mechanisms for Cooperative Binding

Biochemists now know a great deal about the T and R states of hemoglobin, but much remains to be learned about how the $T \rightarrow R$ transition occurs. Two models for the cooperative binding of ligands to proteins with multiple binding sites have greatly influenced thinking about this problem.

The first model was proposed by Jacques Monod, Jeffries Wyman, and Jean-Pierre Changeux in 1965, and is called the **MWC model** or the **concerted model** (Fig. 5–15a). The concerted model assumes that the subunits of a cooperatively binding protein are functionally identical, that each subunit can exist in (at

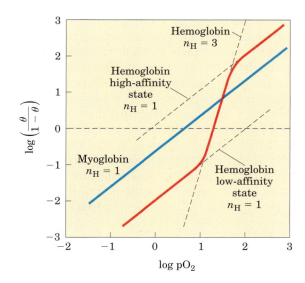

**FIGURE 5–14 Hill plots for the binding of oxygen to myoglobin and hemoglobin.** When $n_H = 1$, there is no evident cooperativity. The maximum degree of cooperativity observed for hemoglobin corresponds approximately to $n_H = 3$. Note that while this indicates a high level of cooperativity, $n_H$ is less than $n$, the number of $O_2$-binding sites in hemoglobin. This is normal for a protein that exhibits allosteric binding behavior.

BOX 5–1    BIOCHEMISTRY IN MEDICINE

## Carbon Monoxide: A Stealthy Killer

Lake Powell, Arizona, August 2000. A family was vacationing in a rented houseboat. They turned on the electrical generator to power an air conditioner and a television. About 15 minutes later, two brothers, aged 8 and 11, jumped off the swim deck at the stern. Situated immediately below the deck was the exhaust port for the generator. Within two minutes, both boys were overcome by the carbon monoxide in the exhaust, which had become concentrated in the space under the deck. Both drowned. These deaths, along with a series of deaths in the 1990s linked to houseboats of similar design, eventually led to the recall and redesign of the generator exhaust assembly.

Carbon monoxide (CO), a colorless, odorless gas, is responsible for more than half of yearly deaths due to poisoning worldwide. CO has an approximately 250-fold greater affinity for hemoglobin than does oxygen. Consequently, relatively low levels of CO can have substantial and tragic effects. When CO combines with hemoglobin, the complex is referred to as carboxyhemoglobin, or COHb.

Some CO is produced by natural processes, but locally high levels generally result only from human activities. Engine and furnace exhausts are important sources, as CO is a byproduct of the incomplete combustion of fossil fuels. In the United States alone, nearly 4,000 people succumb to CO poisoning each year, both accidentally and intentionally. Many of the accidental deaths involve undetected CO buildup in enclosed spaces, such as when a household furnace malfunctions or leaks, venting CO into a home. However, CO poisoning can also occur in open spaces, as unsuspecting people at work or play inhale the exhaust from generators, outboard motors, tractor engines, recreational vehicles, or lawn mowers.

Carbon monoxide levels in the atmosphere are rarely dangerous, ranging from less than 0.05 parts per million (ppm) in remote and uninhabited areas to 3 to 4 ppm in some cities of the northern hemisphere. In the United States, the government-mandated (Occupational Safety and Health Administration, OSHA) limit for CO at worksites is 50 ppm for people working an eight-hour shift. The tight binding of CO to hemoglobin means that COHb can accumulate over time as people are exposed to a constant low-level source of CO.

In an average, healthy individual, 1% or less of the total hemoglobin is complexed as COHb. Since CO is a product of tobacco smoke, many smokers have COHb levels in the range of 3% to 8% of total hemoglobin, and the levels can rise to 15% for chain-smokers. COHb levels equilibrate at 50% in people who breathe air containing 570 ppm of CO for several hours. Reliable methods have been developed that relate CO content in the atmosphere to COHb levels in the blood (Fig. 1). In tests of houseboats with a generator exhaust like the one responsible for the Lake Powell deaths, CO levels reached 6,000 to 30,000 ppm under the swim deck, and atmospheric $O_2$ levels under the deck declined from 21% to 12%. Even above the swim deck, CO levels of up to 7,200 ppm were detected, high enough to cause death within a few minutes.

How is a human affected by COHb? At levels of less than 10% of total hemoglobin, symptoms are rarely observed. At 15%, the individual experiences mild headaches. At 20% to 30%, the headache is severe and

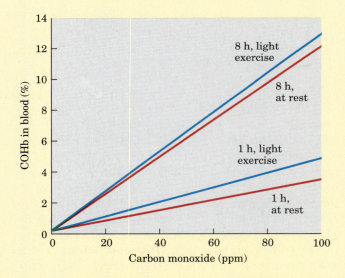

**FIGURE 1** Relationship between the levels of COHb in blood and the concentration of CO in the surrounding air. Four different conditions of exposure are shown, comparing the effects of short versus extended exposure, and exposure at rest versus exposure during light exercise.

least) two conformations, and that all subunits undergo the transition from one conformation to the other simultaneously. In this model, no protein has individual subunits in different conformations. The two conformations are in equilibrium. The ligand can bind to either conformation, but binds each with different affinity. Successive binding of ligand molecules to the low-affinity conformation (which is more stable in the absence of ligand) makes a transition to the high-affinity conformation more likely.

is generally accompanied by nausea, dizziness, confusion, disorientation, and some visual disturbances; these symptoms are generally reversed if the individual is treated with oxygen. At COHb levels of 30% to 50%, the neurological symptoms become more severe, and at levels near 50%, the individual loses consciousness and can sink into coma. Respiratory failure may follow. With prolonged exposure, some damage becomes permanent. Death normally occurs when COHb levels rise above 60%. Autopsy on the boys who died at Lake Powell revealed COHb levels of 59% and 52%.

Binding of CO to hemoglobin is affected by many factors, including exercise (Fig. 1) and changes in air pressure related to altitude. Because of their higher base levels of COHb, smokers exposed to a source of CO often develop symptoms faster than nonsmokers. Individuals with heart and lung conditions or blood diseases that reduce the availability of oxygen to tissues may also experience symptoms at lower levels of CO exposure. Fetuses are at particular risk for CO poisoning, because fetal hemoglobin has a somewhat higher affinity for CO than adult hemoglobin. Cases of CO exposure have been recorded in which the fetus died but the mother recovered.

It may seem surprising that the loss of half of one's hemoglobin to COHb can prove fatal—we know that people with any of several anemic conditions manage to function reasonably well with half the usual complement of active hemoglobin. However, the binding of CO to hemoglobin does more than remove protein from the pool available to bind oxygen. It also affects the affinity of the remaining hemoglobin subunits for oxygen. As CO binds to one or two subunits of a hemoglobin tetramer, the affinity for $O_2$ is increased substantially in the remaining subunits (Fig. 2). Thus, a hemoglobin tetramer with two bound CO molecules can efficiently bind $O_2$ in the lungs—but it releases very little of it in the tissues. Oxygen deprivation in the tissues rapidly becomes severe. To add to the problem, the effects of CO are not limited to interference with hemoglobin function. CO binds to other heme proteins and a variety of metalloproteins. The effects of these interactions are not yet well understood, but they may be responsible for some of the longer-term effects of acute but nonfatal CO poisoning.

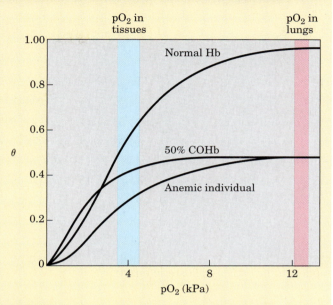

**FIGURE 2** Several oxygen-binding curves: for normal hemoglobin, hemoglobin from an anemic individual with only 50% of her hemoglobin functional, and hemoglobin from an individual with 50% of his hemoglobin subunits complexed with CO. The $pO_2$ in human lungs and tissues is indicated.

When CO poisoning is suspected, rapid evacuation of the person away from the CO source is essential, but this does not always result in rapid recovery. When an individual is moved from the CO-polluted site to a normal, outdoor atmosphere, $O_2$ begins to replace the CO in hemoglobin. The COHb levels drop rather slowly, however; the half-time is 2 to 6.5 hours, depending on individual and environmental factors. If 100% oxygen is administered with a mask, the rate of exchange can be increased about fourfold; the half-time for $O_2$-CO exchange can be reduced to tens of minutes if 100% oxygen at a pressure of 3 atm (303 kPa) is supplied. Thus, rapid treatment by a properly equipped medical team is critical.

Carbon monoxide detectors in all homes are highly recommended. This is a simple and inexpensive measure to avoid possible tragedy. After completing the research for this box, we immediately purchased several new CO detectors for our homes.

---

In the second model, the **sequential model** (Fig. 5–15b), proposed in 1966 by Daniel Koshland and colleagues, ligand binding can induce a change of conformation in an individual subunit. A conformational change in one subunit makes a similar change in an adjacent subunit, as well as the binding of a second ligand molecule, more likely. There are more potential intermediate states in this model than in the concerted model. The two models are not mutually exclusive; the concerted model may be viewed as the "all-or-none" limiting

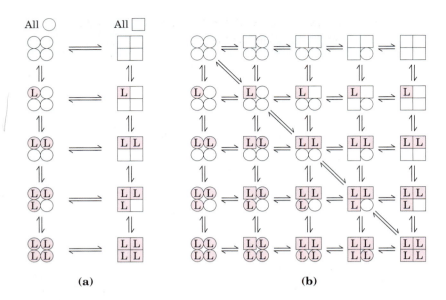

**(a)**                                             **(b)**

**FIGURE 5–15** **Two general models for the interconversion of inactive and active forms of cooperative ligand-binding proteins.** Although the models may be applied to any protein—including any enzyme (Chapter 6)—that exhibits cooperative binding, we show here four subunits because the model was originally proposed for hemoglobin. In the concerted, or all-or-none, model (MWC model) **(a)** all the subunits are postulated to be in the same conformation, either all ○ (low affinity or inactive) or all □ (high affinity or active). Depending on the equilibrium, $K_1$, between ○ and □ forms, the binding of one or more ligand molecules (L) will pull the equilibrium toward the □ form. Subunits with bound L are shaded. In the sequential model **(b)**, each individual subunit can be in either the ○ or □ form. A very large number of conformations is thus possible.

case of the sequential model. In Chapter 6 we use these models to investigate allosteric enzymes.

## Hemoglobin Also Transports H⁺ and CO₂

In addition to carrying nearly all the oxygen required by cells from the lungs to the tissues, hemoglobin carries two end products of cellular respiration—$H^+$ and $CO_2$—from the tissues to the lungs and the kidneys, where they are excreted. The $CO_2$, produced by oxidation of organic fuels in mitochondria, is hydrated to form bicarbonate:

$$CO_2 + H_2O \rightleftharpoons H^+ + HCO_3^-$$

This reaction is catalyzed by **carbonic anhydrase,** an enzyme particularly abundant in erythrocytes. Carbon dioxide is not very soluble in aqueous solution, and bubbles of $CO_2$ would form in the tissues and blood if it were not converted to bicarbonate. As you can see from the equation, the hydration of $CO_2$ results in an increase in the $H^+$ concentration (a decrease in pH) in the tissues. The binding of oxygen by hemoglobin is profoundly influenced by pH and $CO_2$ concentration, so the interconversion of $CO_2$ and bicarbonate is of great importance to the regulation of oxygen binding and release in the blood.

Hemoglobin transports about 40% of the total $H^+$ and 15% to 20% of the $CO_2$ formed in the tissues to the lungs and the kidneys. (The remainder of the $H^+$ is absorbed by the plasma's bicarbonate buffer; the remainder of the $CO_2$ is transported as dissolved $HCO_3^-$ and $CO_2$.) The binding of $H^+$ and $CO_2$ is inversely related to the binding of oxygen. At the relatively low pH and high $CO_2$ concentration of peripheral tissues, the affinity of hemoglobin for oxygen decreases as $H^+$ and $CO_2$ are bound, and $O_2$ is released to the tissues. Conversely, in the capillaries of the lung, as $CO_2$ is excreted

and the blood pH consequently rises, the affinity of hemoglobin for oxygen increases and the protein binds more $O_2$ for transport to the peripheral tissues. This effect of pH and $CO_2$ concentration on the binding and release of oxygen by hemoglobin is called the **Bohr effect,** after Christian Bohr, the Danish physiologist (and father of physicist Niels Bohr) who discovered it in 1904.

The binding equilibrium for hemoglobin and one molecule of oxygen can be designated by the reaction

$$Hb + O_2 \rightleftharpoons HbO_2$$

but this is not a complete statement. To account for the effect of $H^+$ concentration on this binding equilibrium, we rewrite the reaction as

$$HHb^+ + O_2 \rightleftharpoons HbO_2 + H^+$$

where $HHb^+$ denotes a protonated form of hemoglobin. This equation tells us that the $O_2$-saturation curve of hemoglobin is influenced by the $H^+$ concentration (Fig. 5–16). Both $O_2$ and $H^+$ are bound by hemoglobin, but with inverse affinity. When the oxygen concentration is high, as in the lungs, hemoglobin binds $O_2$ and releases protons. When the oxygen concentration is low, as in the peripheral tissues, $H^+$ is bound and $O_2$ is released.

Oxygen and $H^+$ are not bound at the same sites in hemoglobin. Oxygen binds to the iron atoms of the hemes, whereas $H^+$ binds to any of several amino acid residues in the protein. A major contribution to the Bohr effect is made by His146 (His HC3) of the $\beta$ subunits. When protonated, this residue forms one of the ion pairs—to Asp94 (Asp FG1)—that helps stabilize deoxyhemoglobin in the T state (Fig. 5–9). The ion pair stabilizes the protonated form of His HC3, giving this residue an abnormally high $pK_a$ in the T state. The $pK_a$ falls to its normal value of 6.0 in the R state because the ion pair cannot form, and this residue is largely unpro-

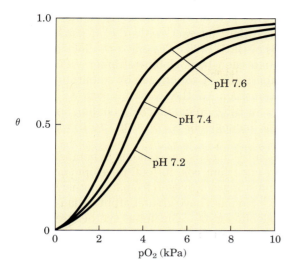

**FIGURE 5–16   Effect of pH on the binding of oxygen to hemoglobin.** The pH of blood is 7.6 in the lungs and 7.2 in the tissues. Experimental measurements on hemoglobin binding are often performed at pH 7.4.

tonated in oxyhemoglobin at pH 7.6, the blood pH in the lungs. As the concentration of $H^+$ rises, protonation of His HC3 promotes release of oxygen by favoring a transition to the T state. Protonation of the amino-terminal residues of the $\alpha$ subunits, certain other His residues, and perhaps other groups has a similar effect.

Thus we see that the four polypeptide chains of hemoglobin communicate with each other about not only $O_2$ binding to their heme groups but also $H^+$ binding to specific amino acid residues. And there is still more to the story. Hemoglobin also binds $CO_2$, again in a manner inversely related to the binding of oxygen. Carbon dioxide binds as a carbamate group to the $\alpha$-amino group at the amino-terminal end of each globin chain, forming carbaminohemoglobin:

$$CO_2 + H_2N-CH \rightarrow CO_2-NH-CH + H^+$$

Amino-terminal residue → Carbamino-terminal residue

This reaction produces $H^+$, contributing to the Bohr effect. The bound carbamates also form additional salt bridges (not shown in Fig. 5–9) that help to stabilize the T state and promote the release of oxygen.

When the concentration of carbon dioxide is high, as in peripheral tissues, some $CO_2$ binds to hemoglobin and the affinity for $O_2$ decreases, causing its release. Conversely, when hemoglobin reaches the lungs, the high oxygen concentration promotes binding of $O_2$ and release of $CO_2$. It is the capacity to communicate ligand-binding information from one polypeptide subunit to the others that makes the hemoglobin molecule so beautifully adapted to integrating the transport of $O_2$, $CO_2$, and $H^+$ by erythrocytes.

## Oxygen Binding to Hemoglobin Is Regulated by 2,3-Bisphosphoglycerate

The interaction of **2,3-bisphosphoglycerate (BPG)** with hemoglobin provides an example of heterotropic allosteric modulation.

2,3-Bisphosphoglycerate

BPG is present in relatively high concentrations in erythrocytes. When hemoglobin is isolated, it contains substantial amounts of bound BPG, which can be difficult to remove completely. In fact, the $O_2$-binding curves for hemoglobin that we have examined to this point were obtained in the presence of bound BPG. 2,3-Bisphosphoglycerate is known to greatly reduce the affinity of hemoglobin for oxygen—there is an inverse relationship between the binding of $O_2$ and the binding of BPG. We can therefore describe another binding process for hemoglobin:

$$HbBPG + O_2 \rightleftharpoons HbO_2 + BPG$$

BPG binds at a site distant from the oxygen-binding site and regulates the $O_2$-binding affinity of hemoglobin in relation to the $pO_2$ in the lungs. BPG plays an important role in the physiological adaptation to the lower $pO_2$ available at high altitudes. For a healthy human strolling by the ocean, the binding of $O_2$ to hemoglobin is regulated such that the amount of $O_2$ delivered to the tissues is equivalent to nearly 40% of the maximum that could be carried by the blood (Fig. 5–17). Imagine that this person is quickly transported to a mountainside at an altitude of 4,500 meters, where the $pO_2$ is considerably lower. The delivery of $O_2$ to the tissues is now reduced. However, after just a few hours at the higher altitude, the BPG concentration in the blood has begun to rise, leading to a decrease in the affinity of hemoglobin for oxygen. This adjustment in the BPG level has only a small effect on the binding of $O_2$ in the lungs but a considerable effect on the release of $O_2$ in the tissues. As a result, the delivery of oxygen to the tissues is restored to nearly 40% of that which can be transported by the blood. The situation is reversed when the person returns to sea level. The BPG concentration in erythrocytes also increases in people suffering from **hypoxia,** lowered oxygenation of peripheral tissues due to inadequate functioning of the lungs or circulatory system.

The site of BPG binding to hemoglobin is the cavity between the $\beta$ subunits in the T state (Fig. 5–18).

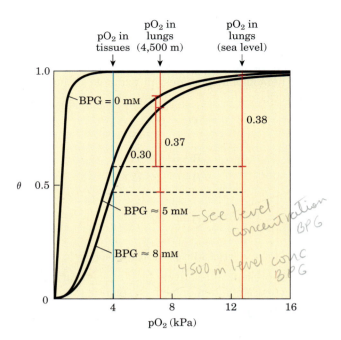

pO₂ in
tissues

pO₂ in
lungs
(4,500 m)

pO₂ in
lungs
(sea level)

*see level concentration BPG*

*4500 m level conc BPG*

**FIGURE 5–17 Effect of BPG on the binding of oxygen to hemoglobin.** The BPG concentration in normal human blood is about 5 mM at sea level and about 8 mM at high altitudes. Note that hemoglobin binds to oxygen quite tightly when BPG is entirely absent, and the binding curve appears to be hyperbolic. In reality, the measured Hill coefficient for O₂-binding cooperativity decreases only slightly (from 3 to about 2.5) when BPG is removed from hemoglobin, but the rising part of the sigmoid curve is confined to a very small region close to the origin. At sea level, hemoglobin is nearly saturated with O₂ in the lungs, but only 60% saturated in the tissues, so the amount of oxygen released in the tissues is close to 40% of the maximum that can be carried in the blood. At high altitudes, O₂ delivery declines by about one-fourth, to 30% of maximum. An increase in BPG concentration, however, decreases the affinity of hemoglobin for O₂, so nearly 40% of what can be carried is again delivered to the tissues.

This cavity is lined with positively charged amino acid residues that interact with the negatively charged groups of BPG. Unlike O₂, only one molecule of BPG is bound to each hemoglobin tetramer. BPG lowers hemoglobin's affinity for oxygen by stabilizing the T state. The transition to the R state narrows the binding pocket for BPG, precluding BPG binding. In the absence of BPG, hemoglobin is converted to the R state more easily.

Regulation of oxygen binding to hemoglobin by BPG has an important role in fetal development. Because a fetus must extract oxygen from its mother's blood, fetal hemoglobin must have greater affinity than the maternal hemoglobin for O₂. The fetus synthesizes $\gamma$ subunits rather than $\beta$ subunits, forming $\alpha_2\gamma_2$ hemoglobin. This tetramer has a much lower affinity for BPG than normal adult hemoglobin, and a correspondingly higher affinity for O₂. 🖱 **Oxygen-Binding Proteins—Hemoglobin Is Susceptible to Allosteric Regulation**

## Sickle-Cell Anemia Is a Molecular Disease of Hemoglobin

The great importance of the amino acid sequence in determining the secondary, tertiary, and quaternary structures of globular proteins, and thus their biological functions, is strikingly demonstrated by the hereditary human disease sickle-cell anemia. Almost 500 genetic variants of hemoglobin are known to occur in the human population; all but a few are quite rare. Most variations consist of differences in a single amino acid residue. The effects on hemoglobin structure and function are often minor but can sometimes be extraordinary. Each hemoglobin variation is the product of an altered gene. The variant genes are called alleles. Because humans generally have two copies of each gene, an in-

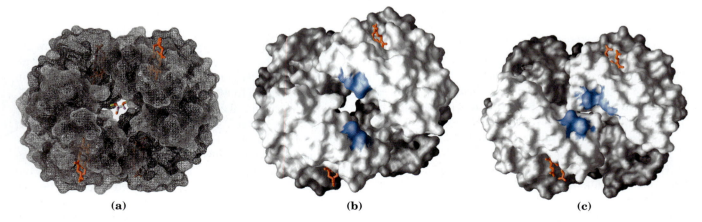

(a)                          (b)                          (c)

**FIGURE 5–18 Binding of BPG to deoxyhemoglobin. (a)** BPG binding stabilizes the T state of deoxyhemoglobin (PDB ID 1HGA), shown here as a mesh surface image. **(b)** The negative charges of BPG interact with several positively charged groups (shown in blue in this surface contour image) that surround the pocket between the $\beta$ subunits in the T state. **(c)** The binding pocket for BPG disappears on oxygenation, following transition to the R state (PDB ID 1BBB). (Compare **(b)** and **(c)** with Fig. 5–10.)

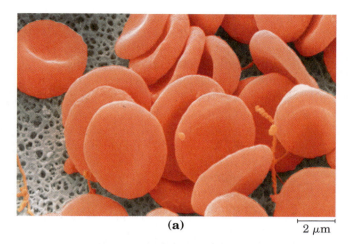

**(a)**                                    2 μm

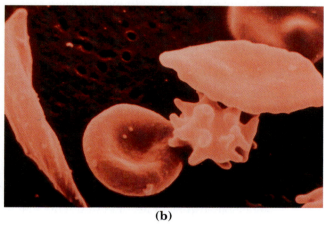

**(b)**

**FIGURE 5–19** A comparison of uniform, cup-shaped, normal erythrocytes **(a)** with the variably shaped erythrocytes seen in sickle-cell anemia **(b)**, which range from normal to spiny or sickle-shaped.

dividual may have two copies of one allele (thus being homozygous for that gene) or one copy of each of two different alleles (thus heterozygous).

Sickle-cell anemia is a genetic disease in which an individual has inherited the allele for sickle-cell hemoglobin from both parents. The erythrocytes of these individuals are fewer and also abnormal. In addition to an unusually large number of immature cells, the blood contains many long, thin, crescent-shaped erythrocytes that look like the blade of a sickle (Fig. 5–19). When hemoglobin from sickle cells (called hemoglobin S) is deoxygenated, it becomes insoluble and forms polymers that aggregate into tubular fibers (Fig. 5–20). Normal hemoglobin (hemoglobin A) remains soluble on deoxygenation. The insoluble fibers of deoxygenated hemoglobin S are responsible for the deformed sickle shape of the erythrocytes, and the proportion of sickled cells increases greatly as blood is deoxygenated.

The altered properties of hemoglobin S result from a single amino acid substitution, a Val instead of a Glu residue at position 6 in the two β chains. The R group

of valine has no electric charge, whereas glutamate has a negative charge at pH 7.4. Hemoglobin S therefore has two fewer negative charges than hemoglobin A, one for each of the two β chains. Replacement of the Glu residue by Val creates a "sticky" hydrophobic contact point at position 6 of the β chain, which is on the outer surface

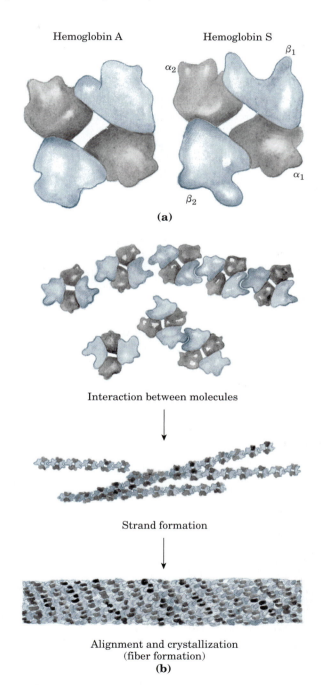

Hemoglobin A          Hemoglobin S

$\alpha_2$                  $\beta_1$

$\beta_2$                  $\alpha_1$

**(a)**

Interaction between molecules

Strand formation

Alignment and crystallization
(fiber formation)
**(b)**

**FIGURE 5–20 Normal and sickle-cell hemoglobin. (a)** Subtle differences between the conformations of hemoglobin A and hemoglobin S result from a single amino acid change in the β chains. **(b)** As a result of this change, deoxyhemoglobin S has a hydrophobic patch on its surface, which causes the molecules to aggregate into strands that align into insoluble fibers.

of the molecule. These sticky spots cause deoxyhemoglobin S molecules to associate abnormally with each other, forming the long, fibrous aggregates characteristic of this disorder. 🖱 **Oxygen-Binding Proteins—Defects in Hb Lead to Serious Genetic Disease**

Sickle-cell anemia, as we have noted, occurs in individuals homozygous for the sickle-cell allele of the gene encoding the $\beta$ subunit of hemoglobin. Individuals who receive the sickle-cell allele from only one parent and are thus heterozygous experience a milder condition called sickle-cell trait; only about 1% of their erythrocytes become sickled on deoxygenation. These individuals may live completely normal lives if they avoid vigorous exercise or other stresses on the circulatory system.

Sickle-cell anemia is a life-threatening and painful disease. People with sickle-cell anemia suffer from repeated crises brought on by physical exertion. They become weak, dizzy, and short of breath, and they also experience heart murmurs and an increased pulse rate. The hemoglobin content of their blood is only about half the normal value of 15 to 16 g/100 mL, because sickled cells are very fragile and rupture easily; this results in anemia ("lack of blood"). An even more serious consequence is that capillaries become blocked by the long, abnormally shaped cells, causing severe pain and interfering with normal organ function—a major factor in the early death of many people with the disease.

Without medical treatment, people with sickle-cell anemia usually die in childhood. Nevertheless, the sickle-cell allele is surprisingly common in certain parts of Africa. Investigation into the persistence of an allele that is so obviously deleterious in homozygous individuals led to the finding that in heterozygous individuals, the allele confers a small but significant resistance to lethal forms of malaria. Natural selection has resulted in an allele population that balances the deleterious effects of the homozygous condition against the resistance to malaria afforded by the heterozygous condition. ■

## SUMMARY 5.1  Reversible Binding of a Protein to a Ligand: Oxygen-Binding Proteins

- Protein function often entails interactions with other molecules. A molecule bound by a protein is called a ligand, and the site to which it binds is called the binding site. Proteins may undergo conformational changes when a ligand binds, a process called induced fit. In a multi-subunit protein, the binding of a ligand to one subunit may affect ligand binding to other subunits. Ligand binding can be regulated.

- Myoglobin contains a heme prosthetic group, which binds oxygen. Heme consists of a single atom of $Fe^{2+}$ coordinated within a porphyrin. Oxygen binds to myoglobin reversibly; this simple reversible binding can be described by an association constant $K_a$ or a dissociation constant $K_d$. For a monomeric protein such as myoglobin, the fraction of binding sites occupied by a ligand is a hyperbolic function of ligand concentration.

- Normal adult hemoglobin has four heme-containing subunits, two $\alpha$ and two $\beta$, similar in structure to each other and to myoglobin. Hemoglobin exists in two interchangeable structural states, T and R. The T state is most stable when oxygen is not bound. Oxygen binding promotes transition to the R state.

- Oxygen binding to hemoglobin is both allosteric and cooperative. As $O_2$ binds to one binding site, the hemoglobin undergoes conformational changes that affect the other binding sites—an example of allosteric behavior. Conformational changes between the T and R states, mediated by subunit-subunit interactions, result in cooperative binding; this is described by a sigmoid binding curve and can be analyzed by a Hill plot.

- Two major models have been proposed to explain the cooperative binding of ligands to multisubunit proteins: the concerted model and the sequential model.

- Hemoglobin also binds $H^+$ and $CO_2$, resulting in the formation of ion pairs that stabilize the T state and lessen the protein's affinity for $O_2$ (the Bohr effect). Oxygen binding to hemoglobin is also modulated by 2,3-bisphosphoglycerate, which binds to and stabilizes the T state.

- Sickle-cell anemia is a genetic disease caused by a single amino acid substitution ($Glu^6$ to $Val^6$) in each $\beta$ chain of hemoglobin. The change produces a hydrophobic patch on the surface of the hemoglobin that causes the molecules to aggregate into bundles of fibers. This homozygous condition results in serious medical complications.

## 5.2 Complementary Interactions between Proteins and Ligands: The Immune System and Immunoglobulins

Our discussion of oxygen-binding proteins showed how the conformations of these proteins affect and are affected by the binding of small ligands ($O_2$ or CO) to the heme group. However, most protein-ligand interactions do not involve a prosthetic group. Instead, the binding site for a ligand is more often like the hemoglobin binding site for BPG—a cleft in the protein lined with amino acid residues, arranged to render the binding interaction highly specific. Effective discrimination between ligands is the norm at binding sites, even when the ligands have only minor structural differences.

All vertebrates have an immune system capable of distinguishing molecular "self" from "nonself" and then destroying those entities identified as nonself. In this way, the immune system eliminates viruses, bacteria, and other pathogens and molecules that may pose a threat to the organism. On a physiological level, the response of the immune system to an invader is an intricate and coordinated set of interactions among many classes of proteins, molecules, and cell types. However, at the level of individual proteins, the immune response demonstrates how an acutely sensitive and specific biochemical system is built upon the reversible binding of ligands to proteins.

### The Immune Response Features a Specialized Array of Cells and Proteins

Immunity is brought about by a variety of **leukocytes** (white blood cells), including **macrophages** and **lymphocytes,** all developing from undifferentiated stem cells in the bone marrow. Leukocytes can leave the bloodstream and patrol the tissues, each cell producing one or more proteins capable of recognizing and binding to molecules that might signal an infection.

The immune response consists of two complementary systems, the humoral and cellular immune systems. The **humoral immune system** (Latin *humor,* "fluid") is directed at bacterial infections and extracellular viruses (those found in the body fluids), but can also respond to individual proteins introduced into the organism. The **cellular immune system** destroys host cells infected by viruses and also destroys some parasites and foreign tissues.

The proteins at the heart of the humoral immune response are soluble proteins called **antibodies** or **immunoglobulins,** often abbreviated **Ig.** Immunoglobulins bind bacteria, viruses, or large molecules identified as foreign and target them for destruction. Making up 20% of blood protein, the immunoglobulins are produced by **B lymphocytes,** or **B cells,** so named because they complete their development in the *b*one marrow.

The agents at the heart of the cellular immune response are a class of **T lymphocytes,** or **T cells** (so called because the latter stages of their development occur in the *t*hymus), known as **cytotoxic T cells (T$_C$ cells,** also called killer T cells). Recognition of infected cells or parasites involves proteins called **T-cell receptors** on the surface of T$_C$ cells. Receptors are proteins, usually found on the outer surface of cells and extending through the plasma membrane; they recognize and bind extracellular ligands, triggering changes inside the cell.

In addition to cytotoxic T cells, there are **helper T cells (T$_H$ cells),** whose function it is to produce soluble signaling proteins called cytokines, which include the interleukins. T$_H$ cells interact with macrophages. Table 5–2 summarizes the functions of the various leukocytes of the immune system.

Each recognition protein of the immune system, either an antibody produced by a B cell or a receptor on the surface of a T cell, specifically binds some particular chemical structure, distinguishing it from virtually all others. Humans are capable of producing more than $10^8$ different antibodies with distinct binding specificities. This extraordinary diversity makes it likely that any chemical structure on the surface of a virus or invading cell will be recognized and bound by one or more antibodies. Antibody diversity is derived from random reassembly of a set of immunoglobulin gene segments through genetic recombination mechanisms that are discussed in Chapter 25 (see Fig. 25–44).

Some properties of the interactions between antibodies or T-cell receptors and the molecules they bind are unique to the immune system, and a specialized lexicon is used to describe them. Any molecule or pathogen capable of eliciting an immune response is called an **antigen.** An antigen may be a virus, a bacterial cell wall, or an individual protein or other macromolecule. A complex antigen may be bound by a number of different antibodies. An individual antibody or T-cell receptor binds only a particular molecular structure within the antigen, called its **antigenic determinant** or **epitope.**

It would be unproductive for the immune system to respond to small molecules that are common intermediates and products of cellular metabolism. Molecules of $M_r$ <5,000 are generally not antigenic. However, small

**TABLE 5–2 Some Types of Leukocytes Associated with the Immune System**

| Cell type | Function |
|---|---|
| Macrophages | Ingest large particles and cells by phagocytosis |
| B lymphocytes (B cells) | Produce and secrete antibodies |
| T lymphocytes (T cells) | |
| Cytotoxic (killer) T cells (T$_C$) | Interact with infected host cells through receptors on T-cell surface |
| Helper T cells (T$_H$) | Interact with macrophages and secrete cytokines (interleukins) that stimulate T$_C$, T$_H$, and B cells to proliferate. |

molecules can be covalently attached to large proteins in the laboratory, and in this form they may elicit an immune response. These small molecules are called **haptens.** The antibodies produced in response to protein-linked haptens will then bind to the same small molecules when they are free. Such antibodies are sometimes used in the development of analytical tests described later in this chapter or as catalytic antibodies (see Box 6–3).

The interactions of antibody and antigen are much better understood than are the binding properties of T-cell receptors. However, before focusing on antibodies, we need to look at the humoral and cellular immune systems in more detail to put the fundamental biochemical interactions into their proper context.

### Self Is Distinguished from Nonself by the Display of Peptides on Cell Surfaces

The immune system must identify and destroy pathogens, but it must also recognize and *not* destroy the normal proteins and cells of the host organism—the "self." Detection of protein antigens in the host is mediated by **MHC (major histocompatibility complex) proteins.** MHC proteins bind peptide fragments of proteins digested in the cell and present them on the outside surface of the cell. These peptides normally come from the digestion of typical cellular proteins, but during a viral infection viral proteins are also digested and presented on the cell surface by MHC proteins. Peptide

fragments from foreign proteins that are displayed by MHC proteins are the antigens the immune system recognizes as nonself. T-cell receptors bind these fragments and launch the subsequent steps of the immune response. There are two classes of MHC proteins (Fig. 5–21), which differ in their distribution among cell types and in the source of digested proteins whose peptides they display.

**Class I MHC** proteins (Fig. 5–22) are found on the surface of virtually all vertebrate cells. There are countless variants in the human population, placing them among the most polymorphic of proteins. Because each individual produces up to six class I MHC protein variants, any two individuals are unlikely to have the same set. Class I MHC proteins bind and display peptides derived from the proteolytic degradation and turnover of proteins that occurs randomly within the cell. These complexes of peptides and class I MHC proteins are the recognition targets of the T-cell receptors of the $T_C$ cells in the cellular immune system. The general pattern of immune system recognition was first described by Rolf Zinkernagel and Peter Doherty in 1974.

Each $T_C$ cell has many copies of only one T-cell receptor that is specific for a particular class I MHC protein–peptide complex. To avoid creating a legion of $T_C$ cells that would set upon and destroy normal cells, the maturation of $T_C$ cells in the thymus includes a stringent selection process that eliminates more than 95% of the developing $T_C$ cells, including those that might recognize and bind class I MHC proteins displaying pep-

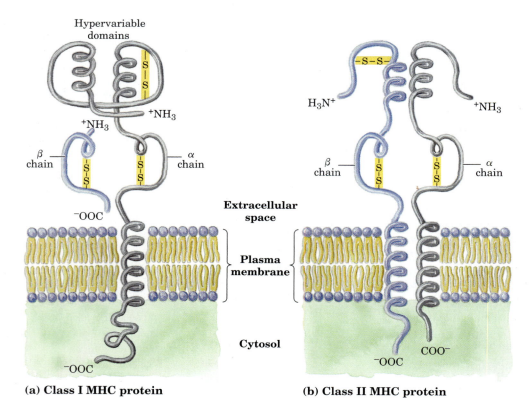

**(a) Class I MHC protein**

**(b) Class II MHC protein**

**FIGURE 5–21  MHC proteins.** These proteins consist of $\alpha$ and $\beta$ chains. **(a)** In class I MHC proteins, the small $\beta$ chain is invariant but the amino acid sequence of the $\alpha$ chain exhibits a high degree of variability, localized in specific domains of the protein that appear on the outside of the cell. Each human produces up to six different $\alpha$ chains for class I MHC proteins. **(b)** In class II MHC proteins, both the $\alpha$ and $\beta$ chains have regions of relatively high variability near their amino-terminal ends.

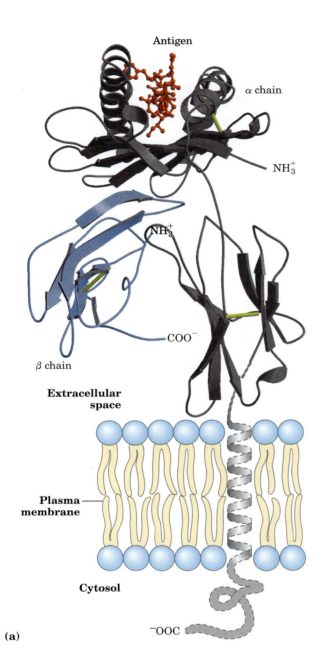

**(a)**

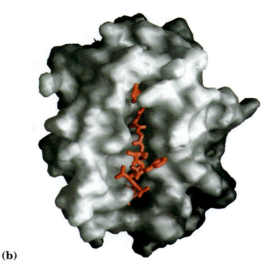

**(b)**

**FIGURE 5–22 Structure of a human class I MHC protein. (a)** This model is derived in part from the known structure of the extracellular portion of the protein (PDB ID 1DDH). The α chain of MHC is shown in gray; the small β chain is blue; the disulfide bonds are yellow. A bound ligand, a peptide derived from HIV, is shown in red. **(b)** Top view of the protein, showing a surface contour image of the site where peptides are bound and displayed. The HIV peptide (red) occupies the site. This part of the class I MHC protein interacts with T-cell receptors.

tion. Each human is capable of producing up to 12 variants, and thus it is unlikely that any two individuals have an identical set. The class II MHC proteins bind and display peptides derived not from cellular proteins but from external proteins ingested by the cells. The resulting class II MHC protein–peptide complexes are the binding targets of the T-cell receptors of the various helper T cells. $T_H$ cells, like $T_C$ cells, undergo a stringent selection process in the thymus, eliminating those that recognize the individual's own cellular proteins. 🖱 **MHC Molecules**

Despite the elimination of most $T_C$ and $T_H$ cells during the selection process in the thymus, a very large number survive, and these provide the immune response. Each survivor has a single type of T-cell receptor that can bind to one particular chemical structure. The T cells patrolling the bloodstream and the tissues carry millions of different binding specificities in the T-cell receptors. Within the highly varied T-cell population there is almost always a contingent of cells that can specifically bind any antigen that might appear. The vast majority of these cells never encounter a foreign antigen to which they can bind, and they typically die within a few days, replaced by new generations of T cells endlessly patrolling in search of the interaction that will launch the full immune response.

The $T_H$ cells participate only indirectly in the destruction of infected cells and pathogens, stimulating the selective proliferation of those $T_C$ and B cells that

tides from cellular proteins of the organism itself. The $T_C$ cells that survive and mature are those with T-cell receptors that do not bind to the organism's own proteins. The result is a population of cells that bind foreign peptides bound to class I MHC proteins of the host cell. These binding interactions lead to the destruction of parasites and virus-infected cells. Following organ transplantation, the donor's class I MHC proteins, recognized as foreign, are bound by the recipient's $T_C$ cells, leading to tissue rejection.

**Class II MHC** proteins occur on the surfaces of a few types of specialized cells, including macrophages and B lymphocytes that take up foreign antigens. Like class I MHC proteins, the class II proteins are highly polymorphic, with many variants in the human popula-

can bind to a particular antigen. This process, called **clonal selection,** increases the number of immune system cells that can respond to a particular pathogen. The importance of $T_H$ cells is dramatically illustrated by the epidemic produced by HIV (human immunodeficiency virus), the virus that causes AIDS (acquired immune deficiency syndrome). The primary targets of HIV infection are $T_H$ cells. Elimination of these cells progressively incapacitates the entire immune system.

### Antibodies Have Two Identical Antigen-Binding Sites

**Immunoglobulin G (IgG)** is the major class of antibody molecule and one of the most abundant proteins in the blood serum. IgG has four polypeptide chains: two large ones, called heavy chains, and two light chains, linked by noncovalent and disulfide bonds into a complex of $M_r$ 150,000. The heavy chains of an IgG molecule interact at one end, then branch to interact separately with the light chains, forming a Y-shaped molecule (Fig. 5–23). At the "hinges" separating the base of an IgG molecule from its branches, the immunoglobulin can be cleaved with proteases. Cleavage with the protease papain liberates the basal fragment, called **Fc** because it usually *crystallizes* readily, and the two branches, called

**Fab,** the *antigen-binding* fragments. Each branch has a single antigen-binding site.

The fundamental structure of immunoglobulins was first established by Gerald Edelman and Rodney Porter. Each chain is made up of identifiable domains; some are constant in sequence and structure from one IgG to the next, others are variable. The constant domains have a characteristic structure known as the **immunoglobulin fold,** a well-conserved structural motif in the all β class of proteins (Chapter 4). There are three of these constant domains in each heavy chain and one in each light chain. The heavy and light chains also have one variable domain each, in which most of the variability in amino acid residue sequence is found. The variable domains associate to create the antigen-binding site (Fig. 5–24).

In many vertebrates, IgG is but one of five classes of immunoglobulins. Each class has a characteristic type of heavy chain, denoted $\alpha$, $\delta$, $\varepsilon$, $\gamma$, and $\mu$ for IgA, IgD, IgE, IgG, and IgM, respectively. Two types of light chain, $\kappa$ and $\lambda$, occur in all classes of immunoglobulins. The overall structures of **IgD** and **IgE** are similar to that of IgG. **IgM** occurs either in a monomeric, membrane-bound form or in a secreted form that is a cross-linked pentamer of this basic structure (Fig. 5–25). **IgA,** found

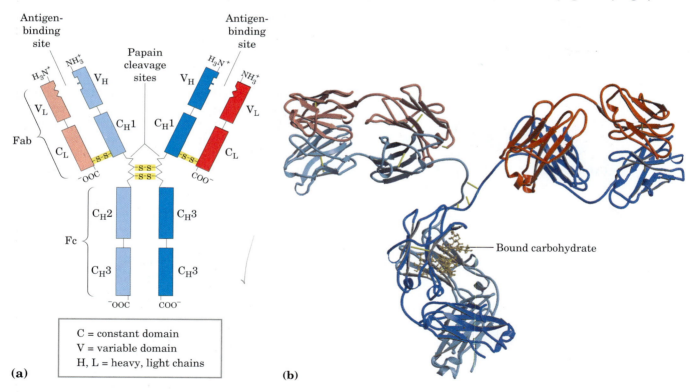

**(a)**

C = constant domain
V = variable domain
H, L = heavy, light chains

**(b)**

— Bound carbohydrate

**FIGURE 5–23 The structure of immunoglobulin G. (a)** Pairs of heavy and light chains combine to form a Y-shaped molecule. Two antigen-binding sites are formed by the combination of variable domains from one light ($V_L$) and one heavy ($V_H$) chain. Cleavage with papain separates the Fab and Fc portions of the protein in the hinge region. The Fc portion of the molecule also contains bound carbohydrate.

**(b)** A ribbon model of the first complete IgG molecule to be crystallized and structurally analyzed (PDB ID 1IGT). Although the molecule contains two identical heavy chains (two shades of blue) and two identical light chains (two shades of red), it crystallized in the asymmetric conformation shown. Conformational flexibility may be important to the function of immunoglobulins.

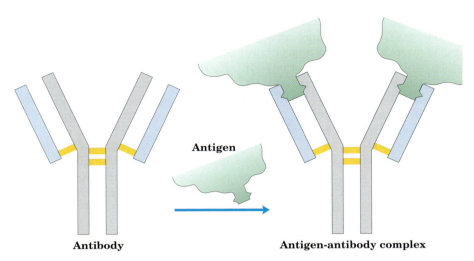

**FIGURE 5–24 Binding of IgG to an antigen.** To generate an optimal fit for the antigen, the binding sites of IgG often undergo slight conformational changes. Such induced fit is common to many protein-ligand interactions.

principally in secretions such as saliva, tears, and milk, can be a monomer, dimer, or trimer. IgM is the first antibody to be made by B lymphocytes and is the major antibody in the early stages of a primary immune response. Some B cells soon begin to produce IgD (with the same antigen-binding site as the IgM produced by the same cell), but the unique function of IgD is less clear.

The IgG described above is the major antibody in secondary immune responses, which are initiated by memory B cells. As part of the organism's ongoing immunity to antigens already encountered and dealt with, IgG is the most abundant immunoglobulin in the blood. When IgG binds to an invading bacterium or virus, it

activates certain leukocytes such as macrophages to engulf and destroy the invader, and also activates some other parts of the immune response. Yet another class of receptors on the cell surface of macrophages recognizes and binds the Fc region of IgG. When these Fc receptors bind an antibody-pathogen complex, the macrophage engulfs the complex by phagocytosis (Fig. 5–26).

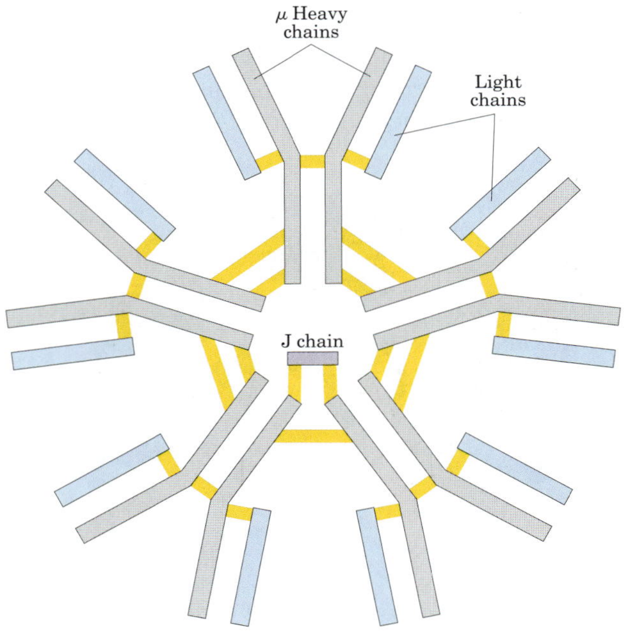

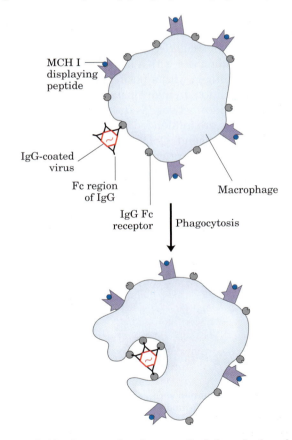

**FIGURE 5–26 Phagocytosis of an antibody-bound virus by a macrophage.** The Fc regions of the antibodies bind to Fc receptors on the surface of the macrophage, triggering the macrophage to engulf and destroy the virus.

**FIGURE 5–25 IgM pentamer of immunoglobulin units.** The pentamer is cross-linked with disulfide bonds (yellow). The J chain is a polypeptide of $M_r$ 20,000 found in both IgA and IgM.

IgE plays an important role in the allergic response, interacting with basophils (phagocytic leukocytes) in the blood and histamine-secreting cells called mast cells that are widely distributed in tissues. This immunoglobulin binds, through its Fc region, to special Fc receptors on the basophils or mast cells. In this form, IgE serves as a kind of receptor for antigen. If antigen is bound, the cells are induced to secrete histamine and other biologically active amines that cause dilation and increased permeability of blood vessels. These effects on the blood vessels are thought to facilitate the movement of immune system cells and proteins to sites of inflammation. They also produce the symptoms normally associated with allergies. Pollen or other allergens are recognized as foreign, triggering an immune response normally reserved for pathogens.

### Antibodies Bind Tightly and Specifically to Antigen

The binding specificity of an antibody is determined by the amino acid residues in the variable domains of its heavy and light chains. Many residues in these domains are variable, but not equally so. Some, particularly those lining the antigen-binding site, are hypervariable—especially likely to differ. Specificity is conferred by chemical complementarity between the antigen and its specific binding site, in terms of shape and the location of charged, nonpolar, and hydrogen-bonding groups. For example, a binding site with a negatively charged group may bind an antigen with a positive charge in the complementary position. In many instances, complementarity is achieved interactively as the structures of antigen and binding site are influenced by each other during the approach of the ligand. Conformational changes in the antibody and/or the antigen then occur that allow the complementary groups to interact fully. This is an example of induced fit (Fig. 5–27).

A typical antibody-antigen interaction is quite strong, characterized by $K_d$ values as low as $10^{-10}$ M (recall that a lower $K_d$ corresponds to a stronger binding interaction). The $K_d$ reflects the energy derived from the various ionic, hydrogen-bonding, hydrophobic, and van der Waals interactions that stabilize the binding. The binding energy required to produce a $K_d$ of $10^{-10}$ M is about 65 kJ/mol.

The complex of a peptide derived from HIV (a model antigen) and an Fab molecule, shown in Figure 5–27, illustrates some of these properties. The changes in structure observed on antigen binding are particularly striking in this example.

### The Antibody-Antigen Interaction Is the Basis for a Variety of Important Analytical Procedures

The extraordinary binding affinity and specificity of antibodies make them valuable analytical reagents. Two types of antibody preparations are in use: polyclonal and monoclonal. **Polyclonal antibodies** are those produced by many different B lymphocytes responding to one antigen, such as a protein injected into an animal. Cells in the population of B lymphocytes produce antibodies that bind specific, different epitopes within the antigen. Thus, polyclonal preparations contain a mixture of antibodies that recognize different parts of the protein. **Monoclonal antibodies,** in contrast, are synthesized by a population of identical B cells (a **clone**) grown in cell culture. These antibodies are homogeneous, all recognizing the same epitope. The techniques for producing monoclonal antibodies were developed by Georges Köhler and Cesar Milstein.

The specificity of antibodies has practical uses. A selected antibody can be covalently attached to a resin and used in a chromatography column of the type shown in Figure 3–18c. When a mixture of proteins is added to

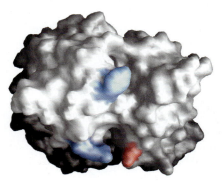 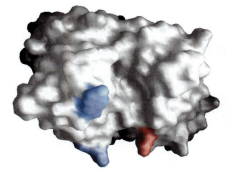 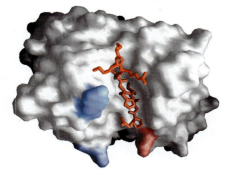

|  |  |  |
|---|---|---|
| **(a)** Conformation with no antigen bound | **(b)** Antigen bound (hidden) | **(c)** Antigen bound (shown) |

**FIGURE 5–27 Induced fit in the binding of an antigen to IgG.** The molecule, shown in surface contour, is the Fab fragment of an IgG. The antigen bound by this IgG is a small peptide derived from HIV. Two residues from the heavy chain (blue) and one from the light chain (pink) are colored to provide visual points of reference. **(a)** View of the Fab fragment, looking down on the antigen-binding site (PDB ID 1GGC). **(b)** The same view, but here the Fab fragment is in the "bound" conformation (PDB ID 1GGI); the antigen has been omitted from the image to provide an unobstructed view of the altered binding site. Note how the binding cavity has enlarged and several groups have shifted position. **(c)** The same view as in **(b)**, but with the antigen in the binding site, pictured as a red stick structure.

Georges Köhler, 1946–1995    Cesar Milstein, 1927–2002

the column, the antibody specifically binds its target protein and retains it on the column while other proteins are washed through. The target protein can then be eluted from the resin by a salt solution or some other agent. This is a powerful tool for protein purification.

In another versatile analytical technique, an antibody is attached to a radioactive label or some other reagent that makes it easy to detect. When the antibody binds the target protein, the label reveals the presence of the protein in a solution or its location in a gel or even a living cell. Several variations of this procedure are illustrated in Figure 5–28.

An **ELISA** (enzyme-linked immunosorbent assay) allows for rapid screening and quantification of the presence of an antigen in a sample (Fig. 5–28b). Proteins in a sample are adsorbed to an inert surface, usually a 96-well polystyrene plate. The surface is washed with a solution of an inexpensive nonspecific protein (often casein from nonfat dry milk powder) to block proteins introduced in subsequent steps from also adsorbing to these surfaces. The surface is then treated with a solution containing the primary antibody—an antibody against the protein of interest. Unbound antibody is washed away and the surface is treated with a solution containing antibodies against the primary antibody. These secondary antibodies have been linked to an enzyme that catalyzes a reaction that forms a colored product. After unbound secondary antibody is washed away, the substrate of the antibody-linked enzyme is added. Product formation (monitored as color intensity) is proportional to the concentration of the protein of interest in the sample.

In an **immunoblot assay** (Fig. 5–28c), proteins that have been separated by gel electrophoresis are transferred electrophoretically to a nitrocellulose membrane. The membrane is blocked (as described above for ELISA), then treated successively with primary

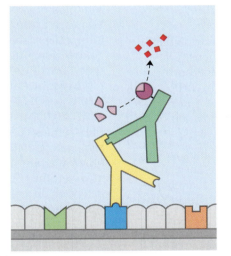

① Coat surface with sample (antigens).

② Block unoccupied sites with nonspecific protein.

③ Incubate with primary antibody against specific antigen.

④ Incubate with antibody-enzyme complex that binds primary antibody.

⑤ Add substrate.

⑥ Formation of colored product indicates presence of specific antigen.

**(a)**

**FIGURE 5–28 Antibody techniques.** The specific reaction of an antibody with its antigen is the basis of several techniques that identify and quantify a specific protein in a complex sample. **(a)** A schematic representation of the general method. **(b)** An ELISA to test for the presence of herpes simplex virus (HSV) antibodies in blood samples. Wells were coated with an HSV antigen, to which antibodies against HSV will bind. The second antibody is anti-human IgG linked to horseradish peroxidase. Blood samples with greater amounts of HSV antibody turn brighter yellow. **(c)** An immunoblot. Lanes 1 to 3 are from an SDS gel; samples from successive stages in the purification of a protein kinase have been separated and stained with Coomassie blue. Lanes 4 to 6 show the same samples, but these were electrophoretically transferred to a nitrocellulose membrane after separation on an SDS gel. The membrane was then "probed" with antibody against the protein kinase. The numbers between the SDS gel and the immunoblot indicate $M_r$ in thousands.

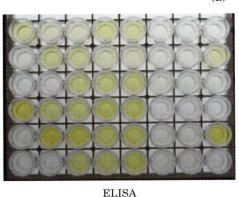

ELISA

**(b)**

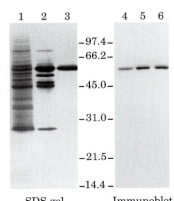

SDS gel    Immunoblot

**(c)**

antibody, secondary antibody linked to enzyme, and substrate. A colored precipitate forms only along the band containing the protein of interest. The immunoblot allows the detection of a minor component in a sample and provides an approximation of its molecular weight.

🖱 **Immunoblotting**

We will encounter other aspects of antibodies in later chapters. They are extremely important in medicine and can tell us much about the structure of proteins and the action of genes.

### SUMMARY 5.2 Complementary Interactions between Proteins and Ligands: The Immune System and Immunoglobulins

- The immune response is mediated by interactions among an array of specialized leukocytes and their associated proteins. T lymphocytes produce T-cell receptors. B lymphocytes produce immunoglobulins. All cells produce MHC proteins, which display host (self) or antigenic (nonself) peptides on the cell surface. In a process called clonal selection, helper T cells induce the proliferation of B cells and cytotoxic T cells that produce immunoglobulins or of T-cell receptors that bind to a specific antigen.

- Humans have five classes of immunoglobulins, each with different biological functions. The most abundant class is IgG, a Y-shaped protein with two heavy and two light chains. The domains near the upper ends of the Y are hypervariable within the broad population of IgGs and form two antigen-binding sites.

- A given immunoglobulin generally binds to only a part, called the epitope, of a large antigen. Binding often involves a conformational change in the IgG, an induced fit to the antigen.

## 5.3 Protein Interactions Modulated by Chemical Energy: Actin, Myosin, and Molecular Motors

Organisms move. Cells move. Organelles and macromolecules within cells move. Most of these movements arise from the activity of a fascinating class of protein-based molecular motors. Fueled by chemical energy, usually derived from ATP, large aggregates of motor proteins undergo cyclic conformational changes that accumulate into a unified, directional force—the tiny force that pulls apart chromosomes in a dividing cell, and the immense force that levers a pouncing, quarter-ton jungle cat into the air.

The interactions among motor proteins, as you might predict, feature complementary arrangements of ionic, hydrogen-bonding, hydrophobic, and van der Waals interactions at protein binding sites. In motor proteins, however, these interactions achieve exceptionally high levels of spatial and temporal organization.

Motor proteins underlie the contraction of muscles, the migration of organelles along microtubules, the rotation of bacterial flagella, and the movement of some proteins along DNA. Proteins called kinesins and dyneins move along microtubules in cells, pulling along organelles or reorganizing chromosomes during cell division. An interaction of dynein with microtubules brings about the motion of eukaryotic flagella and cilia. Flagellar motion in bacteria involves a complex rotational motor at the base of the flagellum (see Fig. 19–34). Helicases, polymerases, and other proteins move along DNA as they carry out their functions in DNA metabolism (Chapter 25). Here, we focus on the well-studied example of the contractile proteins of vertebrate skeletal muscle as a paradigm for how proteins translate chemical energy into motion.

### The Major Proteins of Muscle Are Myosin and Actin

The contractile force of muscle is generated by the interaction of two proteins, myosin and actin. These proteins are arranged in filaments that undergo transient interactions and slide past each other to bring about contraction. Together, actin and myosin make up more than 80% of the protein mass of muscle.

**Myosin** ($M_r$ 540,000) has six subunits: two heavy chains (each of $M_r$ 220,000) and four light chains (each of $M_r$ 20,000). The heavy chains account for much of the overall structure. At their carboxyl termini, they are arranged as extended $\alpha$ helices, wrapped around each other in a fibrous, left-handed coiled coil similar to that of $\alpha$-keratin (Fig. 5–29a). At its amino terminus, each heavy chain has a large globular domain containing a site where ATP is hydrolyzed. The light chains are associated with the globular domains. When myosin is treated briefly with the protease trypsin, much of the fibrous tail is cleaved off, dividing the protein into components called light and heavy meromyosin (Fig. 5–29b). The globular domain, called myosin subfragment 1, or S1, or simply the myosin head group, is liberated from heavy meromyosin by cleavage with papain. The S1 fragment produced by this procedure is the motor domain that makes muscle contraction possible. S1 fragments can be crystallized and their structure has been determined. The overall structure of the S1 fragment as determined by Ivan Rayment and Hazel Holden is shown in Figure 5–29c.

In muscle cells, molecules of myosin aggregate to form structures called **thick filaments** (Fig. 5–30a). These rodlike structures serve as the core of the contractile unit. Within a thick filament, several hundred myosin molecules are arranged with their fibrous "tails" associated to form a long bipolar structure. The globular domains project from either end of this structure, in regular stacked arrays.

The second major muscle protein, **actin,** is abundant in almost all eukaryotic cells. In muscle, molecules

**(a)**

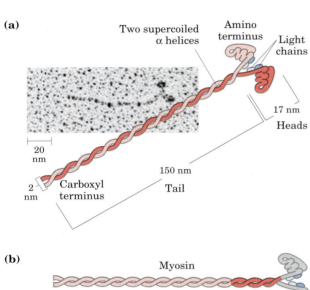

**(b)**

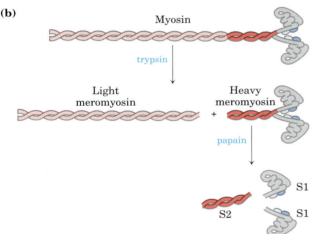

**(c)**

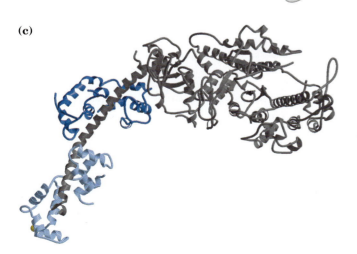

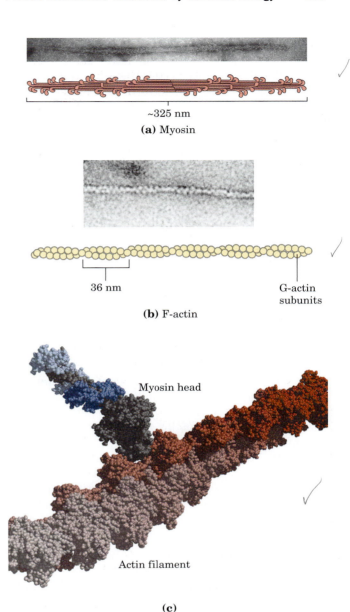

**FIGURE 5–30 The major components of muscle. (a)** Myosin aggregates to form a bipolar structure called a thick filament. **(b)** F-actin is a filamentous assemblage of G-actin monomers that polymerize two by two, giving the appearance of two filaments spiraling about one another in a right-handed fashion. **(c)** Space-filling model of an actin filament (shades of red) with one myosin head (gray and two shades of blue) bound to an actin monomer within the filament. (From coordinates supplied by Ivan Rayment.)

◀ **FIGURE 5–29 (at left) Myosin. (a)** Myosin has two heavy chains (in two shades of pink), the carboxyl termini forming an extended coiled coil (tail) and the amino termini having globular domains (heads). Two light chains (blue) are associated with each myosin head. **(b)** Cleavage with trypsin and papain separates the myosin heads (S1 fragments) from the tails. **(c)** Ribbon representation of the myosin S1 fragment. The heavy chain is in gray, the two light chains in two shades of blue. (From coordinates supplied by Ivan Rayment.)

of monomeric actin, called G-actin (*g*lobular actin; $M_r$ 42,000), associate to form a long polymer called F-actin (*f*ilamentous actin). The **thin filament** (Fig. 5–30b) consists of F-actin, along with the proteins troponin and tropomyosin. The filamentous parts of thin filaments assemble as successive monomeric actin molecules add to one end. On addition, each monomer binds ATP, then hydrolyzes it to ADP, so every actin molecule in the filament is complexed to ADP. This ATP hydrolysis by actin functions only in the assembly of the filaments; it does not contribute directly to the energy expended in muscle contraction. Each actin monomer in the thin filament can bind tightly and specifically to one myosin head group (Fig. 5–30c).

## Additional Proteins Organize the Thin and Thick Filaments into Ordered Structures

Skeletal muscle consists of parallel bundles of **muscle fibers,** each fiber a single, very large, multinucleated cell, 20 to 100 $\mu$m in diameter, formed from many cells fused together and often spanning the length of the muscle. Each fiber, in turn, contains about 1,000 **myofibrils,** 2 $\mu$m in diameter, each consisting of a vast number of regularly arrayed thick and thin filaments complexed to other proteins (Fig. 5–31). A system of flat membranous vesicles called the **sarcoplasmic reticulum** surrounds each myofibril. Examined under the electron microscope, muscle fibers reveal alternating regions of high and low electron density, called the **A bands** and **I bands** (Fig. 5–31b, c). The A and I bands arise from the arrangement of thick and thin filaments, which are aligned and partially overlapping. The I band is the region of the bundle that in cross section would contain only thin filaments. The darker A band stretches the length of the thick filament and includes the region where parallel thick and thin filaments overlap. Bisecting the I band is a thin structure called the **Z disk,** perpendicular to the thin filaments and serving as an anchor to which the thin filaments are attached. The A band too is bisected by a thin line, the **M line** or M disk, a region of high electron density in the middle of the thick filaments. The entire contractile unit, consisting of bundles of thick filaments interleaved at either end with bundles of thin filaments, is called the **sarcomere.** The arrangement of interleaved bundles allows the thick and thin filaments to slide past each other (by a mechanism discussed below), causing a progressive shortening of each sarcomere (Fig. 5–32).

The thin actin filaments are attached at one end to the Z disk in a regular pattern. The assembly includes the minor muscle proteins **α-actinin, desmin,** and **vimentin.** Thin filaments also contain a large protein called **nebulin** (~7,000 amino acid residues), thought to be structured as an $\alpha$ helix that is long enough to span the length of the filament. The M line similarly organizes the thick filaments. It contains the proteins **paramyosin, C-protein,** and **M-protein.** Another class of proteins called **titins,** the largest single polypeptide chains discovered thus far (the titin of human cardiac muscle has 26,926 amino acid residues), link the thick filaments to the Z disk, providing additional organization to the overall structure. Among their structural functions, the proteins nebulin and titin are believed to act as "molecular

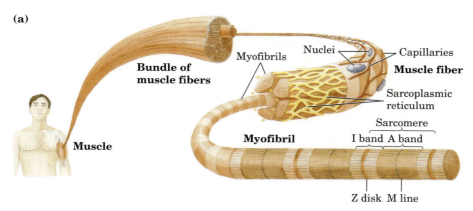

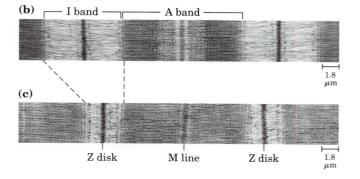

**FIGURE 5–31 Structure of skeletal muscle. (a)** Muscle fibers consist of single, elongated, multinucleated cells that arise from the fusion of many precursor cells. Within the fibers are many myofibrils (only six are shown here for simplicity) surrounded by the membranous sarcoplasmic reticulum. The organization of thick and thin filaments in the myofibril gives it a striated appearance. When muscle contracts, the I bands narrow and the Z disks come closer together, as seen in electron micrographs of **(b)** relaxed and **(c)** contracted muscle.

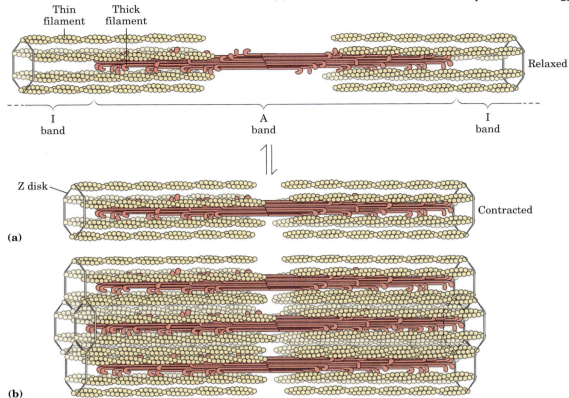

Thin filament
Thick filament

Relaxed

I band
A band
I band

Z disk

Contracted

**(a)**

**(b)**

**FIGURE 5–32 Muscle contraction.** Thick filaments are bipolar structures created by the association of many myosin molecules. **(a)** Muscle contraction occurs by the sliding of the thick and thin filaments past each other so that the Z disks in neighboring I bands approach each other. **(b)** The thick and thin filaments are interleaved such that each thick filament is surrounded by six thin filaments.

rulers," regulating the length of the thin and thick filaments, respectively. Titin extends from the Z disk to the M line, regulating the length of the sarcomere itself and preventing overextension of the muscle. The characteristic sarcomere length varies from one muscle tissue to the next in a vertebrate organism, a finding attributed in large part to the different titin variants in the tissues.

## Myosin Thick Filaments Slide along Actin Thin Filaments

The interaction between actin and myosin, like that between all proteins and ligands, involves weak bonds. When ATP is not bound to myosin, a face on the myosin head group binds tightly to actin (Fig. 5–33). When ATP binds to myosin and is hydrolyzed to ADP and phosphate, a coordinated and cyclic series of conformational changes occurs in which myosin releases the F-actin subunit and binds another subunit farther along the thin filament.

The cycle has four major steps (Fig. 5–33). In step ①, ATP binds to myosin and a cleft in the myosin molecule opens, disrupting the actin-myosin interaction so that the bound actin is released. ATP is then hydrolyzed in step ②, causing a conformational change in the protein to a "high-energy" state that moves the myosin head and changes its orientation in relation to the actin thin filament. Myosin then binds weakly to an F-actin subunit

closer to the Z disk than the one just released. As the phosphate product of ATP hydrolysis is released from myosin in step ③, another conformational change occurs in which the myosin cleft closes, strengthening the myosin-actin binding. This is followed quickly by step ④, a "power stroke" during which the conformation of the myosin head returns to the original resting state, its orientation relative to the bound actin changing so as to pull the tail of the myosin toward the Z disk. ADP is then released to complete the cycle. Each cycle generates about 3 to 4 pN (piconewtons) of force and moves the thick filament 5 to 10 nm relative to the thin filament.

Because there are many myosin heads in a thick filament, at any given moment some (probably 1% to 3%) are bound to the thin filaments. This prevents the thick filaments from slipping backward when an individual myosin head releases the actin subunit to which it was bound. The thick filament thus actively slides forward past the adjacent thin filaments. This process, coordinated among the many sarcomeres in a muscle fiber, brings about muscle contraction.

The interaction between actin and myosin must be regulated so that contraction occurs only in response to appropriate signals from the nervous system. The regulation is mediated by a complex of two proteins, **tropomyosin** and **troponin.** Tropomyosin binds to the thin filament, blocking the attachment sites for the myosin head groups. Troponin is a $Ca^{2+}$-binding protein.

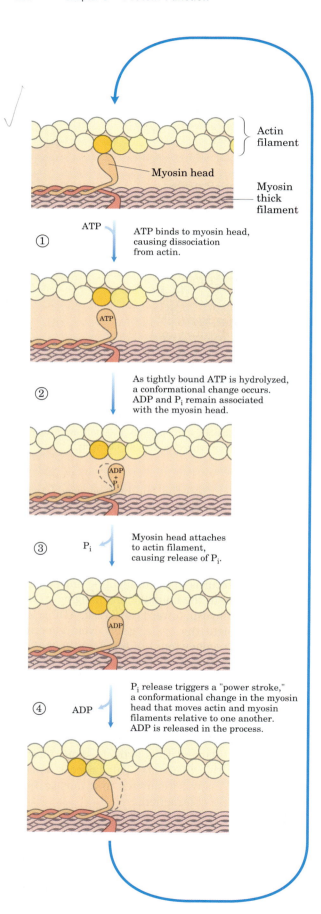

① ATP ATP binds to myosin head, causing dissociation from actin.

② As tightly bound ATP is hydrolyzed, a conformational change occurs. ADP and $P_i$ remain associated with the myosin head.

③ $P_i$ Myosin head attaches to actin filament, causing release of $P_i$.

④ ADP $P_i$ release triggers a "power stroke," a conformational change in the myosin head that moves actin and myosin filaments relative to one another. ADP is released in the process.

Actin filament

Myosin head

Myosin thick filament

A nerve impulse causes release of $Ca^{2+}$ from the sarcoplasmic reticulum. The released $Ca^{2+}$ binds to troponin (another protein-ligand interaction) and causes a conformational change in the tropomyosin-troponin complexes, exposing the myosin-binding sites on the thin filaments. Contraction follows.

Working skeletal muscle requires two types of molecular functions that are common in proteins—binding and catalysis. The actin-myosin interaction, a protein-ligand interaction like that of immunoglobulins with antigens, is reversible and leaves the participants unchanged. When ATP binds myosin, however, it is hydrolyzed to ADP and $P_i$. Myosin is not only an actin-binding protein, it is also an ATPase—an enzyme. The function of enzymes in catalyzing chemical transformations is the topic of the next chapter.

### SUMMARY 5.3 Protein Interactions Modulated by Chemical Energy: Actin, Myosin, and Molecular Motors

- Protein-ligand interactions achieve a special degree of spatial and temporal organization in motor proteins. Muscle contraction results from choreographed interactions between myosin and actin, coupled to the hydrolysis of ATP by myosin.

- Myosin consists of two heavy and four light chains, forming a fibrous coiled coil (tail) domain and a globular (head) domain. Myosin molecules are organized into thick filaments, which slide past thin filaments composed largely of actin. ATP hydrolysis in myosin is coupled to a series of conformational changes in the myosin head, leading to dissociation of myosin from one F-actin subunit and its eventual reassociation with another, farther along the thin filament. The myosin thus slides along the actin filaments.

- Muscle contraction is stimulated by the release of $Ca^{2+}$ from the sarcoplasmic reticulum. The $Ca^{2+}$ binds to the protein troponin, leading to a conformational change in a troponin-tropomyosin complex that triggers the cycle of actin-myosin interactions.

**FIGURE 5–33 Molecular mechanism of muscle contraction.** Conformational changes in the myosin head that are coupled to stages in the ATP hydrolytic cycle cause myosin to successively dissociate from one actin subunit, then associate with another farther along the actin filament. In this way the myosin heads slide along the thin filaments, drawing the thick filament array into the thin filament array (see Fig. 5–32).

## Key Terms

*Terms in bold are defined in the glossary.*

**ligand**   157
**binding site**   157
**induced fit**   158
**heme**   158
**porphyrin**   158
globins   159
equilibrium expression   160
association constant, $K_a$   160
**dissociation constant, $K_d$**   160

**allosteric protein**   165
Hill equation   167
Bohr effect   170
**lymphocytes**   175
**antibody**   175
**immunoglobulin**   175
**B lymphocytes** or **B cells**   175
**T lymphocytes** or **T cells**   175
**antigen**   175

**epitope**   175
**hapten**   176
immunoglobulin fold   178
**polyclonal antibodies**   180
**monoclonal antibodies**   180
ELISA   181
**myosin**   182
**actin**   183
**sarcomere**   184

## Further Reading

### Oxygen-Binding Proteins

**Ackers, G.K. & Hazzard, J.H.** (1993) Transduction of binding energy into hemoglobin cooperativity. *Trends Biochem. Sci.* **18**, 385–390.

**Changeux, J.-P.** (1993) Allosteric proteins: from regulatory enzymes to receptors—personal recollections. *Bioessays* **15**, 625–634.

An interesting perspective from a leader in the field.

**Dickerson, R.E. & Geis, I.** (1982) *Hemoglobin: Structure, Function, Evolution, and Pathology,* The Benjamin/Cummings Publishing Company, Redwood City, CA.

**di Prisco, G., Condò, S.G., Tamburrini, M., & Giardina, B.** (1991) Oxygen transport in extreme environments. *Trends Biochem. Sci.* **16**, 471–474.

A revealing comparison of the oxygen-binding properties of hemoglobins from polar species.

**Koshland, D.E., Jr., Nemethy, G., & Filmer, D.** (1966) Comparison of experimental binding data and theoretical models in proteins containing subunits. *Biochemistry* **6**, 365–385.

The paper that introduced the sequential model.

**Monod, J., Wyman, J., & Changeux, J.-P.** (1965) On the nature of allosteric transitions: a plausible model. *J. Mol. Biol.* **12**, 88–118.

The concerted model was first proposed in this landmark paper.

**Olson, J.S. & Phillips, G.N., Jr.** (1996) Kinetic pathways and barriers for ligand binding to myoglobin. *J. Biol. Chem.* **271**, 17,593–17,596.

**Perutz, M.F.** (1989) Myoglobin and haemoglobin: role of distal residues in reactions with haem ligands. *Trends Biochem. Sci.* **14**, 42–44.

**Perutz, M.F., Wilkinson, A.J., Paoli, M., & Dodson, G.G.** (1998) The stereochemical mechanism of the cooperative effects in hemoglobin revisited. *Annu. Rev. Biophys. Biomol. Struct.* **27**, 1–34.

### Immune System Proteins

**Blom, B., Res, P.C., & Spits, H.** (1998) T cell precursors in man and mice. *Crit. Rev. Immunol.* **18**, 371–388.

**Davies, D.R. & Chacko, S.** (1993) Antibody structure. *Acc. Chem. Res.* **26**, 421–427.

**Davies, D.R., Padlan, E.A., & Sheriff, S.** (1990) Antibody-antigen complexes. *Annu. Rev. Biochem.* **59**, 439–473.

**Davis, M.M.** (1990) T cell receptor gene diversity and selection. *Annu. Rev. Biochem.* **59**, 475–496.

**Dutton, R.W., Bradley, L.M., & Swain, S.L.** (1998) T cell memory. *Annu. Rev. Immunol.* **16**, 201–223.

Life, Death and the Immune System. (1993) *Sci. Am.* **269** (September).

A special issue on the immune system.

**Goldsby, R.A., Kindt, T.J., Osborne, B.A., & Kuby, J.** (2003) *Immunology,* 5th ed. W. H. Freeman and Company, New York.

**Marrack, P. & Kappler, J.** (1987) The T cell receptor. *Science* **238**, 1073–1079.

**Parham, P. & Ohta, T.** (1996) Population biology of antigen presentation by MHC class I molecules. *Science* **272**, 67–74.

**Ploegh, H.L.** (1998) Viral strategies of immune evasion. *Science* **280**, 248–253.

**Thomsen, A.R., Nansen, A., & Christensen, J.P.** (1998) Virus-induced T cell activation and the inflammatory response. *Curr. Top. Microbiol. Immunol.* **231**, 99–123.

**Van Parjis, L. & Abbas, A.K.** (1998) Homeostasis and self-tolerance in the immune system: turning lymphocytes off. *Science* **280**, 243–248.

**York, I.A. & Rock, K.L.** (1996) Antigen processing and presentation by the class-I major histocompatibility complex. *Annu. Rev. Immunol.* **14**, 369–396.

### Molecular Motors

**Finer, J.T., Simmons, R.M., & Spudich, J.A.** (1994) Single myosin molecule mechanics: piconewton forces and nanometre steps. *Nature* **368**, 113–119.

Modern techniques reveal the forces affecting individual motor proteins.

**Geeves, M.A. & Holmes, K.C.** (1999) Structural mechanism of muscle contraction. *Annu. Rev. Biochem.* **68**, 687–728.

**Goldman, Y.E.** (1998) Wag the tail: structural dynamics of actomyosin. *Cell* **93**, 1–4.

**Huxley, H.E.** (1998) Getting to grips with contraction: the interplay of structure and biochemistry. *Trends Biochem. Sci.* **23**, 84–87.

An interesting historical perspective on deciphering the mechanism of muscle contraction.

**Labeit, S. & Kolmerer, B.** (1995) Titins: giant proteins in charge of muscle ultrastructure and elasticity. *Science* **270**, 293–296.

A structural and functional description of some of the largest proteins.

**Molloy, J.E. & Veigel, C.** (2003) Myosin motors walk the walk. *Science* **300**, 2045–2046.

**Rayment, I.** (1996) The structural basis of the myosin ATPase activity. *J. Biol. Chem.* **271**, 15,850–15,853.

Examines the muscle-contraction mechanism from a structural perspective.

**Rayment, I. & Holden, H.M.** (1994) The three-dimensional structure of a molecular motor. *Trends Biochem. Sci.* **19**, 129–134.

**Spudich, J.A.** (1994) How molecular motors work. *Nature* **372**, 515–518.

**Vale, R.D.** (2003) The molecular motor toolbox for intracellular transport. *Cell* **112**, 467–480.

## Problems

**1. Relationship between Affinity and Dissociation Constant**    Protein A has a binding site for ligand X with a $K_d$ of $10^{-6}$ M. Protein B has a binding site for ligand X with a $K_d$ of $10^{-9}$ M. Which protein has a higher affinity for ligand X? Explain your reasoning. Convert the $K_d$ to $K_a$ for both proteins.

**2. Negative Cooperativity**    Which of the following situations would produce a Hill plot with $n_H < 1.0$? Explain your reasoning in each case.

(a) The protein has multiple subunits, each with a single ligand-binding site. Binding of ligand to one site decreases the binding affinity of other sites for the ligand.

(b) The protein is a single polypeptide with two ligand-binding sites, each having a different affinity for the ligand.

(c) The protein is a single polypeptide with a single ligand-binding site. As purified, the protein preparation is heterogeneous, containing some protein molecules that are partially denatured and thus have a lower binding affinity for the ligand.

**3. Affinity for Oxygen in Myoglobin and Hemoglobin**    What is the effect of the following changes on the $O_2$ affinity of myoglobin and hemoglobin? (a) A drop in the pH of blood plasma from 7.4 to 7.2. (b) A decrease in the partial pressure of $CO_2$ in the lungs from 6 kPa (holding one's breath) to 2 kPa (normal). (c) An increase in the BPG level from 5 mM (normal altitudes) to 8 mM (high altitudes).

**4. Cooperativity in Hemoglobin**    Under appropriate conditions, hemoglobin dissociates into its four subunits. The isolated $\alpha$ subunit binds oxygen, but the $O_2$-saturation curve is hyperbolic rather than sigmoid. In addition, the binding of oxygen to the isolated $\alpha$ subunit is not affected by the presence of $H^+$, $CO_2$, or BPG. What do these observations indicate about the source of the cooperativity in hemoglobin?

**5. Comparison of Fetal and Maternal Hemoglobins**    Studies of oxygen transport in pregnant mammals have shown that the $O_2$-saturation curves of fetal and maternal blood are markedly different when measured under the same conditions. Fetal erythrocytes contain a structural variant of hemoglobin, HbF, consisting of two $\alpha$ and two $\gamma$ subunits ($\alpha_2\gamma_2$), whereas maternal erythrocytes contain HbA ($\alpha_2\beta_2$).

(a) Which hemoglobin has a higher affinity for oxygen under physiological conditions, HbA or HbF? Explain.

(b) What is the physiological significance of the different $O_2$ affinities?

(c) When all the BPG is carefully removed from samples of HbA and HbF, the measured $O_2$-saturation curves (and consequently the $O_2$ affinities) are displaced to the left. However,

HbA now has a greater affinity for oxygen than does HbF. When BPG is reintroduced, the $O_2$-saturation curves return to normal, as shown in the graph. What is the effect of BPG on the $O_2$ affinity of hemoglobin? How can the above information be used to explain the different $O_2$ affinities of fetal and maternal hemoglobin?

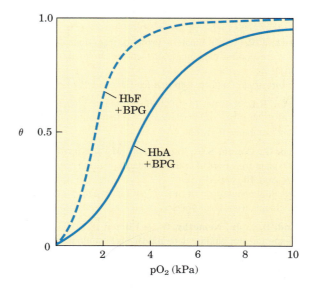

**6. Hemoglobin Variants**    There are almost 500 naturally occurring variants of hemoglobin. Most are the result of a single amino acid substitution in a globin polypeptide chain. Some variants produce clinical illness, though not all variants have deleterious effects. A brief sample is presented below.

HbS (sickle-cell Hb): substitutes a Val for a Glu on the surface

Hb Cowtown: eliminates an ion pair involved in T-state stabilization

Hb Memphis: substitutes one uncharged polar residue for another of similar size on the surface

Hb Bibba: substitutes a Pro for a Leu involved in an $\alpha$ helix

Hb Milwaukee: substitutes a Glu for a Val

Hb Providence: substitutes an Asn for a Lys that normally projects into the central cavity of the tetramer

Hb Philly: substitutes a Phe for a Tyr, disrupting hydrogen bonding at the $\alpha_1\beta_1$ interface

Explain your choices for each of the following:

(a) The Hb variant *least* likely to cause pathological symptoms.

(b) The variant(s) most likely to show pI values different from that of HbA when run on an isoelectric focusing gel.

(c) The variant(s) most likely to show a decrease in BPG binding and an increase in the overall affinity of the hemoglobin for oxygen.

**7. Reversible (but Tight) Binding to an Antibody**  An antibody binds to an antigen with a $K_d$ of $5 \times 10^{-8}$ M. At what concentration of antigen will $\theta$ be (a) 0.2, (b) 0.5, (c) 0.6, (d) 0.8?

**8. Using Antibodies to Probe Structure-Function Relationships in Proteins**  A monoclonal antibody binds to G-actin but not to F-actin. What does this tell you about the epitope recognized by the antibody?

**9. The Immune System and Vaccines**  A host organism needs time, often days, to mount an immune response against a new antigen, but memory cells permit a rapid response to pathogens previously encountered. A vaccine to protect against a particular viral infection often consists of weakened or killed virus or isolated proteins from a viral protein coat. When injected into a human patient, the vaccine generally does not cause an infection and illness, but it effectively "teaches" the immune system what the viral particles look like, stimulating the production of memory cells. On subsequent infection, these cells can bind to the virus and trigger a rapid immune response. Some pathogens, including HIV, have developed mechanisms to evade the immune system, making it difficult or impossible to develop effective vaccines against them. What strategy could a pathogen use to evade the immune system? Assume that antibodies and/or T-cell receptors are available to bind to any structure that might appear on the surface of a pathogen and that, once bound, the pathogen is destroyed.

**10. How We Become a "Stiff"**  When a higher vertebrate dies, its muscles stiffen as they are deprived of ATP, a state called rigor mortis. Explain the molecular basis of the rigor state.

**11. Sarcomeres from Another Point of View**  The symmetry of thick and thin filaments in a sarcomere is such that six thin filaments ordinarily surround each thick filament in a hexagonal array. Draw a cross section (transverse cut) of a myofibril at the following points: (a) at the M line; (b) through the I band; (c) through the dense region of the A band; (d) through the less dense region of the A band, adjacent to the M line (see Fig. 5–31b, c).

**Biochemistry on the Internet**

**12. Lysozyme and Antibodies**  To fully appreciate how proteins function in a cell, it is helpful to have a three-dimensional view of how proteins interact with other cellular components. Fortunately, this is possible using on-line protein databases and the three-dimensional molecular viewing utilities RasMol and Protein Explorer. If you have not yet installed both programs on your computer, go to www.umass.edu/microbio/rasmol and follow the instructions for your operating system and browser. Then, go to the Protein Data Bank (www.rcsb.org/pdb).

In this exercise you will examine the interactions between the enzyme lysozyme (Chapter 4) and the Fab portion of the anti-lysozyme antibody. Use the PDB identifier 1FDL to explore the structure of the IgG1 Fab fragment–lysozyme complex (antibody-antigen complex). View the structure using Protein Explorer, and also use the information on the structure summary page at the PDB (click on "explore" to get there) to answer the following questions.

(a) Which chains in the three-dimensional model correspond to the antibody fragment and which correspond to the antigen, lysozyme?

(b) What secondary structure predominates in this Fab fragment?

(c) How many amino acid residues are in the heavy and light chains of the Fab fragment? In lysozyme? Estimate the percentage of the lysozyme that interacts with the antigen-binding site of the antibody fragment.

(d) Identify the specific amino acid residues in lysozyme and in the variable regions of the Fab heavy and light chains that appear to be situated at the antigen-antibody interface. Are the residues contiguous in the primary sequence of the polypeptide chains?

**13. Exploring Reversible Interactions of Proteins and Ligands with Living Graphs**  Use the living graphs for Equations 5–8, 5–11, 5–14, and 5–16 to work through the following exercises.

(a) Reversible binding of a ligand to a simple protein, without cooperativity. For Equation 5–8, set up a plot of $\theta$ versus [L] (vertical and horizontal axes, respectively). Examine the plots generated when $K_d$ is set at 5 $\mu$M, 10 $\mu$M, 20 $\mu$M, and 100 $\mu$M. Higher affinity of the protein for the ligand means more binding at lower ligand concentrations. Suppose that four different proteins exhibit these four different $K_d$ values for ligand L. Which protein would have the highest affinity for L?

Examine the plot generated when $K_d = 10$ $\mu$M. How much does $\theta$ increase when [L] increases from 0.2 $\mu$M to 0.4 $\mu$M? How much does $\theta$ increase when [L] increases from 40 $\mu$M to 80 $\mu$M?

You can do the same exercise for Equation 5–11. Convert [L] to $pO_2$ and $K_d$ to $P_{50}$. Examine the curves generated when $P_{50}$ is set at 0.5 kPa, 1 kPa, 2 kPa, and 10 kPa. For the curve generated when $P_{50} = 1$ kPa, how much does $\theta$ change when the $pO_2$ increases from 0.02 kPa to 0.04 kPa? From 4 kPa to 8 kPa?

(b) Cooperative binding of a ligand to a multisubunit protein. Using Equation 5–14, generate a binding curve for a protein and ligand with $K_d = 10$ $\mu$M and $n = 3$. Note the altered definition of $K_d$ in Equation 5–16. On the same plot, add a curve for a protein with $K_d = 20$ $\mu$M and $n = 3$. Now see how both curves change when you change to $n = 4$. Generate Hill plots (Eqn 5–16) for each of these cases. For $K_d = 10$ $\mu$M and $n = 3$, what is $\theta$ when [L] = 20 $\mu$M?

(c) Explore these equations further by varying all the parameters used above.

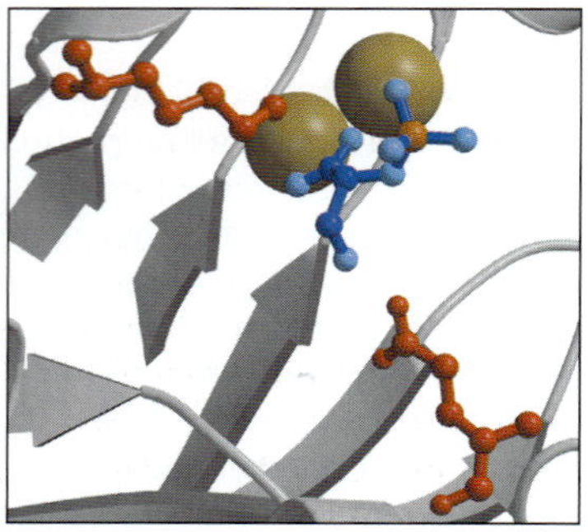

# chapter 6

# ENZYMES

One way in which this condition might be fulfilled would be if the molecules when combined with the enzyme, lay slightly further apart than their equilibrium distance when [covalently joined], but nearer than their equilibrium distance when free. . . . Using Fischer's lock and key simile, the key does not fit the lock quite perfectly but exercises a certain strain on it.

—*J. B. S. Haldane,* Enzymes, *1930*

Catalysis can be described formally in terms of a stabilization of the transition state through tight binding to the catalyst.

—*William P. Jencks, article in* Advances in Enzymology, *1975*

There are two fundamental conditions for life. First, the living entity must be able to self-replicate (a topic considered in Part III); second, the organism must be able to catalyze chemical reactions efficiently and selectively. The central importance of catalysis may surprise some beginning students of biochemistry, but it is easy to demonstrate. As described in Chapter 1, living systems make use of energy from the environment. Many of us, for example, consume substantial amounts of sucrose—common table sugar—as a kind of fuel, whether in the form of sweetened foods and drinks or as sugar itself. The conversion of sucrose to $CO_2$ and $H_2O$ in the presence of oxygen is a highly exergonic process, releasing free energy that we can use to think, move, taste, and see. However, a bag of sugar can remain on the shelf for years without any obvious conversion to $CO_2$ and $H_2O$. Although this chemical process is thermodynamically favorable, it is very slow! Yet when sucrose is consumed by a human (or almost any other organism), it releases its chemical energy in seconds. The difference is catalysis. Without catalysis, chemical reactions such as sucrose oxidation could not occur on a useful time scale, and thus could not sustain life.

In this chapter, then, we turn our attention to the reaction catalysts of biological systems: the enzymes, the most remarkable and highly specialized proteins. Enzymes have extraordinary catalytic power, often far greater than that of synthetic or inorganic catalysts. They have a high degree of specificity for their substrates, they accelerate chemical reactions tremendously, and they function in aqueous solutions under very mild conditions of temperature and pH. Few nonbiological catalysts have all these properties.

Enzymes are central to every biochemical process. Acting in organized sequences, they catalyze the hundreds of stepwise reactions that degrade nutrient molecules, conserve and transform chemical energy, and make biological macromolecules from simple precursors. Through the action of regulatory enzymes, metabolic pathways are highly coordinated to yield a harmonious interplay among the many activities necessary to sustain life.

The study of enzymes has immense practical importance. In some diseases, especially inheritable genetic disorders, there may be a deficiency or even a total absence of one or more enzymes. For other disease conditions, excessive activity of an enzyme may be the cause. Measurements of the activities of enzymes in blood plasma, erythrocytes, or tissue samples are important in diagnosing certain illnesses. Many drugs exert their biological effects through interactions with enzymes. And enzymes are important practical tools, not only in medicine but in the chemical industry, food processing, and agriculture.

We begin with descriptions of the properties of enzymes and the principles underlying their catalytic power, then introduce enzyme kinetics, a discipline that provides much of the framework for any discussion of enzymes. Specific examples of enzyme mechanisms are then provided, illustrating principles introduced earlier in the chapter. We end with a discussion of how enzyme activity is regulated.

## 6.1 An Introduction to Enzymes

Much of the history of biochemistry is the history of enzyme research. Biological catalysis was first recognized and described in the late 1700s, in studies on the digestion of meat by secretions of the stomach, and research continued in the 1800s with examinations of the conversion of starch to sugar by saliva and various plant extracts. In the 1850s, Louis Pasteur concluded that fermentation of sugar into alcohol by yeast is catalyzed by "ferments." He postulated that these ferments were inseparable from the structure of living yeast cells; this view, called vitalism, prevailed for decades. Then in 1897 Eduard Buchner discovered that yeast extracts could ferment sugar to alcohol, proving that fermentation was promoted by molecules that continued to function when removed from cells. Frederick W. Kühne called these molecules **enzymes.** As vitalistic notions of life were disproved, the isolation of new enzymes and the investigation of their properties advanced the science of biochemistry.

The isolation and crystallization of urease by James Sumner in 1926 provided a breakthrough in early enzyme studies. Sumner found that urease crystals consisted entirely of protein, and he postulated that all enzymes are proteins. In the absence of other examples, this idea remained controversial for some time. Only in the 1930s was Sumner's conclusion widely accepted, after John Northrop and Moses Kunitz crystallized pepsin, trypsin, and other digestive enzymes and found them also to be proteins. During this period, J. B. S. Haldane wrote a treatise entitled *Enzymes.* Although the molecular nature of enzymes was not yet fully appreciated, Haldane made the remarkable suggestion that weak bonding interactions between an enzyme and its substrate might be used to catalyze a reaction. This insight lies at the heart of our current understanding of enzymatic catalysis.

Since the latter part of the twentieth century, research on enzymes has been intensive. It has led to the purification of thousands of enzymes, elucidation of the

structure and chemical mechanism of many of them, and a general understanding of how enzymes work.

### Most Enzymes Are Proteins

With the exception of a small group of catalytic RNA molecules (Chapter 26), all enzymes are proteins. Their catalytic activity depends on the integrity of their native protein conformation. If an enzyme is denatured or dissociated into its subunits, catalytic activity is usually lost. If an enzyme is broken down into its component amino acids, its catalytic activity is always destroyed. Thus the primary, secondary, tertiary, and quaternary structures of protein enzymes are essential to their catalytic activity.

Enzymes, like other proteins, have molecular weights ranging from about 12,000 to more than 1 million. Some enzymes require no chemical groups for activity other than their amino acid residues. Others require an additional chemical component called a **cofactor**—either one or more inorganic ions, such as $Fe^{2+}$, $Mg^{2+}$, $Mn^{2+}$, or $Zn^{2+}$ (Table 6–1), or a complex organic or metalloorganic molecule called a **coenzyme** (Table 6–2). Some enzymes require *both* a coenzyme

**TABLE 6–1  Some Inorganic Elements That Serve as Cofactors for Enzymes**

| | |
|---|---|
| $Cu^{2+}$ | Cytochrome oxidase |
| $Fe^{2+}$ or $Fe^{3+}$ | Cytochrome oxidase, catalase, peroxidase |
| $K^+$ | Pyruvate kinase |
| $Mg^{2+}$ | Hexokinase, glucose 6-phosphatase, pyruvate kinase |
| $Mn^{2+}$ | Arginase, ribonucleotide reductase |
| Mo | Dinitrogenase |
| $Ni^{2+}$ | Urease |
| Se | Glutathione peroxidase |
| $Zn^{2+}$ | Carbonic anhydrase, alcohol dehydrogenase, carboxypeptidases A and B |

Eduard Buchner, 1860–1917

James Sumner, 1887–1955

J. B. S. Haldane, 1892–1964

**TABLE 6-2    Some Coenzymes That Serve as Transient Carriers of Specific Atoms or Functional Groups**

| Coenzyme | Examples of chemical groups transferred | Dietary precursor in mammals |
|---|---|---|
| Biocytin | $CO_2$ | Biotin |
| Coenzyme A | Acyl groups | Pantothenic acid and other compounds |
| 5'-Deoxyadenosylcobalamin (coenzyme $B_{12}$) | H atoms and alkyl groups | Vitamin $B_{12}$ |
| Flavin adenine dinucleotide | Electrons | Riboflavin (vitamin $B_2$) |
| Lipoate | Electrons and acyl groups | Not required in diet |
| Nicotinamide adenine dinucleotide | Hydride ion ($:H^-$) | Nicotinic acid (niacin) |
| Pyridoxal phosphate | Amino groups | Pyridoxine (vitamin $B_6$) |
| Tetrahydrofolate | One-carbon groups | Folate |
| Thiamine pyrophosphate | Aldehydes | Thiamine (vitamin $B_1$) |

Note: The structures and modes of action of these coenzymes are described in Part II.

and one or more metal ions for activity. A coenzyme or metal ion that is very tightly or even covalently bound to the enzyme protein is called a **prosthetic group.** A complete, catalytically active enzyme together with its bound coenzyme and/or metal ions is called a **holoenzyme.** The protein part of such an enzyme is called the **apoenzyme** or **apoprotein.** Coenzymes act as transient carriers of specific functional groups. Most are derived from vitamins, organic nutrients required in small amounts in the diet. We consider coenzymes in more detail as we encounter them in the metabolic pathways discussed in Part II. Finally, some enzyme proteins are modified covalently by phosphorylation, glycosylation, and other processes. Many of these alterations are involved in the regulation of enzyme activity.

## Enzymes Are Classified by the Reactions They Catalyze

Many enzymes have been named by adding the suffix "-ase" to the name of their substrate or to a word or phrase describing their activity. Thus urease catalyzes hydrolysis of urea, and DNA polymerase catalyzes the polymerization of nucleotides to form DNA. Other enzymes were named by their discovers for a broad func-

tion, before the specific reaction catalyzed was known. For example, an enzyme known to act in the digestion of foods was named pepsin, from the Greek *pepsis,* "digestion," and lysozyme was named for its ability to lyse bacterial cell walls. Still others were named for their source: trypsin, named in part from the Greek *tryein,* "to wear down," was obtained by rubbing pancreatic tissue with glycerin. Sometimes the same enzyme has two or more names, or two different enzymes have the same name. Because of such ambiguities, and the ever-increasing number of newly discovered enzymes, biochemists, by international agreement, have adopted a system for naming and classifying enzymes. This system divides enzymes into six classes, each with subclasses, based on the type of reaction catalyzed (Table 6–3). Each enzyme is assigned a four-part classification number and a systematic name, which identifies the reaction it catalyzes. As an example, the formal systematic name of the enzyme catalyzing the reaction

$$ATP + \text{D-glucose} \longrightarrow ADP + \text{D-glucose 6-phosphate}$$

is ATP:glucose phosphotransferase, which indicates that it catalyzes the transfer of a phosphoryl group from ATP to glucose. Its Enzyme Commission number (E.C. number) is 2.7.1.1. The first number (2) denotes the

**TABLE 6-3    International Classification of Enzymes**

| No. | Class | Type of reaction catalyzed |
|---|---|---|
| 1 | Oxidoreductases | Transfer of electrons (hydride ions or H atoms) |
| 2 | Transferases | Group transfer reactions |
| 3 | Hydrolases | Hydrolysis reactions (transfer of functional groups to water) |
| 4 | Lyases | Addition of groups to double bonds, or formation of double bonds by removal of groups |
| 5 | Isomerases | Transfer of groups within molecules to yield isomeric forms |
| 6 | Ligases | Formation of C—C, C—S, C—O, and C—N bonds by condensation reactions coupled to ATP cleavage |

Note: Most enzymes catalyze the transfer of electrons, atoms, or functional groups. They are therefore classified, given code numbers, and assigned names according to the type of transfer reaction, the group donor, and the group acceptor.

class name (transferase); the second number (7), the subclass (phosphotransferase); the third number (1), a phosphotransferase with a hydroxyl group as acceptor; and the fourth number (1), D-glucose as the phosphoryl group acceptor. For many enzymes, a trivial name is more commonly used—in this case hexokinase. A complete list and description of the thousands of known enzymes is maintained by the Nomenclature Committee of the International Union of Biochemistry and Molecular Biology (www.chem.qmul.ac.uk/iubmb/enzyme). This chapter is devoted primarily to principles and properties common to all enzymes.

### SUMMARY 6.1 An Introduction to Enzymes

- Life depends on the existence of powerful and specific catalysts: the enzymes. Almost every biochemical reaction is catalyzed by an enzyme.

- With the exception of a few catalytic RNAs, all known enzymes are proteins. Many require nonprotein coenzymes or cofactors for their catalytic function.

- Enzymes are classified according to the type of reaction they catalyze. All enzymes have formal E.C. numbers and names, and most have trivial names.

## 6.2 How Enzymes Work

The enzymatic catalysis of reactions is essential to living systems. Under biologically relevant conditions, uncatalyzed reactions tend to be slow—most biological molecules are quite stable in the neutral-pH, mild-temperature, aqueous environment inside cells. Furthermore, many common reactions in biochemistry entail chemical events that are unfavorable or unlikely in the cellular environment, such as the transient formation of unstable charged intermediates or the collision of two or more molecules in the precise orientation required for reaction. Reactions required to digest food, send nerve signals, or contract a muscle simply do not occur at a useful rate without catalysis.

An enzyme circumvents these problems by providing a specific environment within which a given reaction can occur more rapidly. The distinguishing feature of an enzyme-catalyzed reaction is that it takes place within the confines of a pocket on the enzyme called the **active site** (Fig. 6–1). The molecule that is bound in the active site and acted upon by the enzyme is called the **substrate.** The surface of the active site is lined with amino acid residues with substituent groups that bind the substrate and catalyze its chemical transformation. Often, the active site encloses a substrate, sequestering it completely from solution. The enzyme-

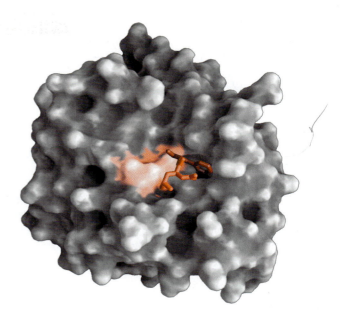

**FIGURE 6–1 Binding of a substrate to an enzyme at the active site.** The enzyme chymotrypsin, with bound substrate in red (PDB ID 7GCH). Some key active-site amino acid residues appear as a red splotch on the enzyme surface.

substrate complex, whose existence was first proposed by Charles-Adolphe Wurtz in 1880, is central to the action of enzymes. It is also the starting point for mathematical treatments that define the kinetic behavior of enzyme-catalyzed reactions and for theoretical descriptions of enzyme mechanisms.

### Enzymes Affect Reaction Rates, Not Equilibria

A simple enzymatic reaction might be written

$$E + S \rightleftharpoons ES \rightleftharpoons EP \rightleftharpoons E + P \qquad (6\text{–}1)$$

where E, S, and P represent the enzyme, substrate, and product; ES and EP are transient complexes of the enzyme with the substrate and with the product.

To understand catalysis, we must first appreciate the important distinction between reaction equilibria and reaction rates. The function of a catalyst is to increase the *rate* of a reaction. Catalysts do not affect reaction *equilibria.* Any reaction, such as S $\rightleftharpoons$ P, can be described by a reaction coordinate diagram (Fig. 6–2), a picture of the energy changes during the reaction. As discussed in Chapter 1, energy in biological systems is described in terms of free energy, $G$. In the coordinate diagram, the free energy of the system is plotted against the progress of the reaction (the reaction coordinate). The starting point for either the forward or the reverse reaction is called the **ground state,** the contribution to the free energy of the system by an average molecule (S or P) under a given set of conditions. To describe the free-energy changes for reactions, chemists define a standard set of conditions (temperature 298 K; partial pressure of each gas 1 atm, or 101.3 kPa; concentration

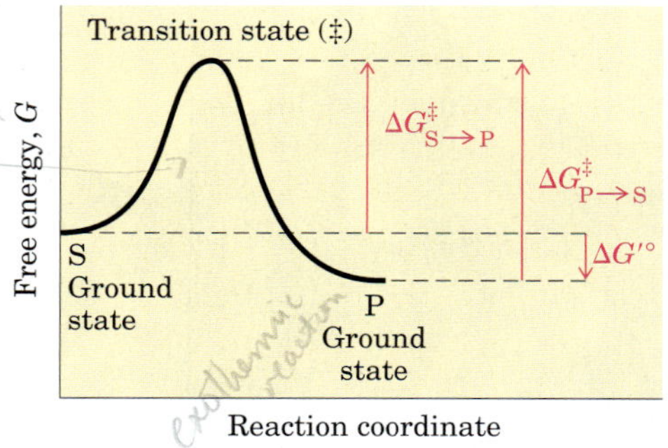

*(handwritten: binding reaction)*
*(handwritten: S + E ⇌ → ES → E + P)*

*(handwritten, left margin vertical: activation energy)*

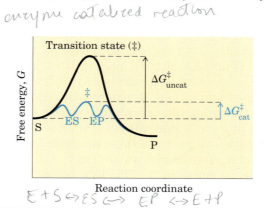

*(handwritten, vertical: exothermic reaction)*

**FIGURE 6–2 Reaction coordinate diagram for a chemical reaction.** The free energy of the system is plotted against the progress of the reaction S → P. A diagram of this kind is a description of the energy changes during the reaction, and the horizontal axis (reaction coordinate) reflects the progressive chemical changes (e.g., bond breakage or formation) as S is converted to P. The activation energies, $\Delta G^{\ddagger}$, for the S → P and P → S reactions are indicated. $\Delta G'^{\circ}$ is the overall standard free-energy change in the direction S → P.

of each solute 1 M) and express the free-energy change for this reacting system as $\Delta G^{\circ}$, the **standard free-energy change.** Because biochemical systems commonly involve $H^+$ concentrations far below 1 M, biochemists define a **biochemical standard free-energy change,** $\Delta G'^{\circ}$, the standard free-energy change *at pH 7.0;* we employ this definition throughout the book. A more complete definition of $\Delta G'^{\circ}$ is given in Chapter 13.

The equilibrium between S and P reflects the difference in the free energies of their ground states. In the example shown in Figure 6–2, the free energy of the ground state of P is lower than that of S, so $\Delta G'^{\circ}$ for the reaction is negative and the equilibrium favors P. The position and direction of equilibrium are *not* affected by any catalyst.

A favorable equilibrium does not mean that the S → P conversion will occur at a detectable rate. The *rate* of a reaction is dependent on an entirely different parameter. There is an energy barrier between S and P: the energy required for alignment of reacting groups, formation of transient unstable charges, bond rearrangements, and other transformations required for the reaction to proceed in either direction. This is illustrated by the energy "hill" in Figures 6–2 and 6–3. To undergo reaction, the molecules must overcome this barrier and therefore must be raised to a higher energy level. At the top of the energy hill is a point at which decay to the S or P state is equally probable (it is downhill either way). This is called the **transition state.** The transition state is not a chemical species with any significant stability and should not be confused with a reaction intermediate (such as ES or EP). It is simply a fleeting molecular moment in which events such as bond breakage, bond formation, and charge development have proceeded to the precise point at which decay to

either substrate or product is equally likely. The difference between the energy levels of the ground state and the transition state is the **activation energy, $\Delta G^{\ddagger}$.** The rate of a reaction reflects this activation energy: a higher activation energy corresponds to a slower reaction. Reaction rates can be increased by raising the temperature, thereby increasing the number of molecules with sufficient energy to overcome the energy barrier. Alternatively, the activation energy can be lowered by adding a catalyst (Fig. 6–3). *Catalysts enhance reaction rates by lowering activation energies.*

Enzymes are no exception to the rule that catalysts do not affect reaction equilibria. The bidirectional arrows in Equation 6–1 make this point: any enzyme that catalyzes the reaction S → P also catalyzes the reaction P → S. The role of enzymes is to *accelerate* the interconversion of S and P. The enzyme is not used up in the process, and the equilibrium point is unaffected. However, the reaction reaches equilibrium much faster when the appropriate enzyme is present, because the rate of the reaction is increased.

This general principle can be illustrated by considering the conversion of sucrose and oxygen to carbon dioxide and water:

$$C_{12}H_{22}O_{11} + 12O_2 \longrightarrow 12CO_2 + 11H_2O$$

This conversion, which takes place through a series of separate reactions, has a very large and negative $\Delta G'^{\circ}$, and at equilibrium the amount of sucrose present is negligible. Yet sucrose is a stable compound, because the activation energy barrier that must be overcome before sucrose reacts with oxygen is quite high. Sucrose can be stored in a container with oxygen almost indefinitely without reacting. In cells, however, sucrose is readily broken down to $CO_2$ and $H_2O$ in a series of reactions catalyzed by enzymes. These enzymes not only accel-

*(handwritten: enzyme catalyzed reaction)*

*(handwritten: E + S ⇌ ES ⇌ EP ⇌ E + P)*

**FIGURE 6–3 Reaction coordinate diagram comparing enzyme-catalyzed and uncatalyzed reactions.** In the reaction S → P, the ES and EP intermediates occupy minima in the energy progress curve of the enzyme-catalyzed reaction. The terms $\Delta G^{\ddagger}_{uncat}$ and $\Delta G^{\ddagger}_{cat}$ correspond to the activation energy for the uncatalyzed reaction and the overall activation energy for the catalyzed reaction, respectively. The activation energy is lower when the enzyme catalyzes the reaction.

erate the reactions, they organize and control them so that much of the energy released is recovered in other chemical forms and made available to the cell for other tasks. The reaction pathway by which sucrose (and other sugars) is broken down is the primary energy-yielding pathway for cells, and the enzymes of this pathway allow the reaction sequence to proceed on a biologically useful time scale.

Any reaction may have several steps, involving the formation and decay of transient chemical species called **reaction intermediates.**[*] A reaction intermediate is any species on the reaction pathway that has a finite chemical lifetime (longer than a molecular vibration, $\sim 10^{-13}$ seconds). When the S $\rightleftharpoons$ P reaction is catalyzed by an enzyme, the ES and EP complexes can be considered intermediates, even though S and P are stable chemical species (Eqn 6–1); the ES and EP complexes occupy valleys in the reaction coordinate diagram (Fig. 6–3). Additional, less stable chemical intermediates often exist in the course of an enzyme-catalyzed reaction. The interconversion of two sequential reaction intermediates thus constitutes a reaction step. When several steps occur in a reaction, the overall rate is determined by the step (or steps) with the highest activation energy; this is called the **rate-limiting step.** In a simple case, the rate-limiting step is the highest-energy point in the diagram for interconversion of S and P. In practice, the rate-limiting step can vary with reaction conditions, and for many enzymes several steps may have similar activation energies, which means they are all partially rate-limiting.

Activation energies are energy barriers to chemical reactions. These barriers are crucial to life itself. The rate at which a molecule undergoes a particular reaction decreases as the activation barrier for that reaction increases. Without such energy barriers, complex macromolecules would revert spontaneously to much simpler molecular forms, and the complex and highly ordered structures and metabolic processes of cells could not exist. Over the course of evolution, enzymes have developed lower activation energies *selectively* for reactions that are needed for cell survival.

## Reaction Rates and Equilibria Have Precise Thermodynamic Definitions

Reaction *equilibria* are inextricably linked to the standard free-energy change for the reaction, $\Delta G'^{\circ}$, and re-

action *rates* are linked to the activation energy, $\Delta G^{\ddagger}$. A basic introduction to these thermodynamic relationships is the next step in understanding how enzymes work.

An equilibrium such as S $\rightleftharpoons$ P is described by an **equilibrium constant, $K_{eq}$,** or simply $K$ (p. 26). Under the standard conditions used to compare biochemical processes, an equilibrium constant is denoted $K'_{eq}$ (or $K'$):

$$K'_{eq} = \frac{[P]}{[S]} \qquad (6\text{–}2)$$

From thermodynamics, the relationship between $K'_{eq}$ and $\Delta G'^{\circ}$ can be described by the expression

$$\Delta G'^{\circ} = -RT \ln K'_{eq} \qquad (6\text{–}3)$$

where $R$ is the gas constant, 8.315 J/mol $\cdot$ K, and $T$ is the absolute temperature, 298 K (25 °C). Equation 6–3 is developed and discussed in more detail in Chapter 13. The important point here is that the equilibrium constant is directly related to the overall standard free-energy change for the reaction (Table 6–4). A large negative value for $\Delta G'^{\circ}$ reflects a favorable reaction equilibrium—but as already noted, this does not mean the reaction will proceed at a rapid rate.

The rate of any reaction is determined by the concentration of the reactant (or reactants) and by a **rate constant,** usually denoted by $k$. For the unimolecular reaction S $\rightarrow$ P, the rate (or velocity) of the reaction, $V$—representing the amount of S that reacts per unit time—is expressed by a **rate equation:**

$$V = k[S] \qquad (6\text{–}4)$$

In this reaction, the rate depends only on the concentration of S. This is called a first-order reaction. The factor $k$ is a proportionality constant that reflects the probability of reaction under a given set of conditions (pH, temperature, and so forth). Here, $k$ is a first-order rate constant and has units of reciprocal time, such as s$^{-1}$. If a first-order reaction has a rate constant $k$ of 0.03 s$^{-1}$,

---

[*]In this chapter, *step* and *intermediate* refer to chemical species in the reaction pathway of a single enzyme-catalyzed reaction. In the context of metabolic pathways involving many enzymes (discussed in Part II), these terms are used somewhat differently. An entire enzymatic reaction is often referred to as a "step" in a pathway, and the product of one enzymatic reaction (which is the substrate for the next enzyme in the pathway) is referred to as an "intermediate."

**TABLE 6–4  Relationship between $K'_{eq}$ and $\Delta G'^{\circ}$**

| $K'_{eq}$ | $\Delta G'^{\circ}$ (kJ/mol) |
|---|---|
| $10^{-6}$ | 34.2 |
| $10^{-5}$ | 28.5 |
| $10^{-4}$ | 22.8 |
| $10^{-3}$ | 17.1 |
| $10^{-2}$ | 11.4 |
| $10^{-1}$ | 5.7 |
| 1 | 0.0 |
| $10^{1}$ | −5.7 |
| $10^{2}$ | −11.4 |
| $10^{3}$ | −17.1 |

*Note:* The relationship is calculated from $\Delta G'^{\circ} = -RT \ln K'_{eq}$ (Eqn 6-3).

this may be interpreted (qualitatively) to mean that 3% of the available S will be converted to P in 1 s. A reaction with a rate constant of 2,000 s$^{-1}$ will be over in a small fraction of a second. If a reaction rate depends on the concentration of two different compounds, or if the reaction is between two molecules of the same compound, the reaction is second order and $k$ is a second-order rate constant, with units of M$^{-1}$s$^{-1}$. The rate equation then becomes

$$V = k[S_1][S_2] \qquad (6\text{--}5)$$

From transition-state theory we can derive an expression that relates the magnitude of a rate constant to the activation energy:

$$k = \frac{\mathbf{k}T}{h}e^{-\Delta G^{\ddagger}/RT} \qquad (6\text{--}6)$$

where **k** is the Boltzmann constant and $h$ is Planck's constant. The important point here is that the relationship between the rate constant $k$ and the activation energy $\Delta G^{\ddagger}$ is inverse and exponential. In simplified terms, this is the basis for the statement that a lower activation energy means a faster reaction rate.

Now we turn from *what* enzymes do to *how* they do it.

## A Few Principles Explain the Catalytic Power and Specificity of Enzymes

Enzymes are extraordinary catalysts. The rate enhancements they bring about are in the range of 5 to 17 orders of magnitude (Table 6–5). Enzymes are also very specific, readily discriminating between substrates with quite similar structures. How can these enormous and highly selective rate enhancements be explained? What is the source of the energy for the dramatic lowering of the activation energies for specific reactions?

The answer to these questions has two distinct but interwoven parts. The first lies in the rearrangements of covalent bonds during an enzyme-catalyzed reaction. Chemical reactions of many types take place between substrates and enzymes' functional groups (specific amino acid side chains, metal ions, and coenzymes). Catalytic functional groups on an enzyme may form a transient covalent bond with a substrate and activate it for reaction, or a group may be transiently transferred from the substrate to the enzyme. In many cases, these reactions occur only in the enzyme active site. Covalent interactions between enzymes and substrates lower the activation energy (and thereby accelerate the reaction) by providing an alternative, lower-energy reaction path. The specific types of rearrangements that occur are described in Section 6.4.

The second part of the explanation lies in the *noncovalent* interactions between enzyme and substrate. Much of the energy required to lower activation energies is derived from weak, noncovalent interactions between substrate and enzyme. What really sets enzymes apart from most other catalysts is the formation of a specific ES complex. The interaction between substrate and enzyme in this complex is mediated by the same forces that stabilize protein structure, including hydrogen bonds and hydrophobic and ionic interactions (Chapter 4). Formation of each weak interaction in the ES complex is accompanied by release of a small amount of free energy that provides a degree of stability to the interaction. The energy derived from enzyme-substrate interaction is called **binding energy, $\Delta G_B$**. Its significance extends beyond a simple stabilization of the enzyme-substrate interaction. *Binding energy is a major source of free energy used by enzymes to lower the activation energies of reactions.*

Two fundamental and interrelated principles provide a general explanation for how enzymes use noncovalent binding energy:

1. Much of the catalytic power of enzymes is ultimately derived from the free energy released in forming many weak bonds and interactions between an enzyme and its substrate. This binding energy contributes to specificity as well as to catalysis.

2. Weak interactions are optimized in the reaction transition state; enzyme active sites are complementary not to the substrates per se but to the transition states through which substrates pass as they are converted to products during an enzymatic reaction.

These themes are critical to an understanding of enzymes, and they now become our primary focus.

## Weak Interactions between Enzyme and Substrate Are Optimized in the Transition State

How does an enzyme use binding energy to lower the activation energy for a reaction? Formation of the ES complex is not the explanation in itself, although some

---

**TABLE 6–5    Some Rate Enhancements Produced by Enzymes**

| | |
|---|---|
| Cyclophilin | $10^5$ |
| Carbonic anhydrase | $10^7$ |
| Triose phosphate isomerase | $10^9$ |
| Carboxypeptidase A | $10^{11}$ |
| Phosphoglucomutase | $10^{12}$ |
| Succinyl-CoA transferase | $10^{13}$ |
| Urease | $10^{14}$ |
| Orotidine monophosphate decarboxylase | $10^{17}$ |

of the earliest considerations of enzyme mechanisms began with this idea. Studies on enzyme specificity carried out by Emil Fischer led him to propose, in 1894, that enzymes were structurally complementary to their substrates, so that they fit together like a lock and key (Fig. 6–4). This elegant idea, that a specific (exclusive) interaction between two biological molecules is mediated by molecular surfaces with complementary shapes, has greatly influenced the development of biochemistry, and such interactions lie at the heart of many biochemical processes. However, the "lock and key" hypothesis can be misleading when applied to enzymatic catalysis. An enzyme completely complementary to its substrate would be a very poor enzyme, as we can demonstrate.

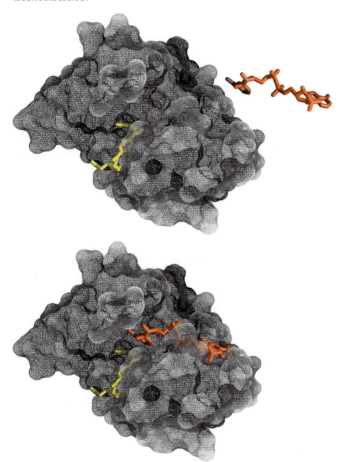

**FIGURE 6–4 Complementary shapes of a substrate and its binding site on an enzyme.** The enzyme dihydrofolate reductase with its substrate NADP$^+$ (red), unbound (top) and bound (bottom). Another bound substrate, tetrahydrofolate (yellow), is also visible (PDB ID 1RA2). The NADP$^+$ binds to a pocket that is complementary to it in shape and ionic properties. In reality, the complementarity between protein and ligand (in this case substrate) is rarely perfect, as we saw in Chapter 5. The interaction of a protein with a ligand often involves changes in the conformation of one or both molecules, a process called induced fit. This *lack* of perfect complementarity between enzyme and substrate (not evident in this figure) is important to enzymatic catalysis.

Consider an imaginary reaction, the breaking of a magnetized metal stick. The uncatalyzed reaction is shown in Figure 6–5a. Let's examine two imaginary enzymes—two "stickases"—that could catalyze this reaction, both of which employ magnetic forces as a paradigm for the binding energy used by real enzymes. We first design an enzyme perfectly complementary to the substrate (Fig. 6–5b). The active site of this stickase is a pocket lined with magnets. To react (break), the stick must reach the transition state of the reaction, but the stick fits so tightly in the active site that it cannot bend, because bending would eliminate some of the magnetic interactions between stick and enzyme. Such an enzyme *impedes* the reaction, stabilizing the substrate instead. In a reaction coordinate diagram (Fig. 6–5b), this kind of ES complex would correspond to an energy trough from which the substrate would have difficulty escaping. Such an enzyme would be useless.

The modern notion of enzymatic catalysis, first proposed by Michael Polanyi (1921) and Haldane (1930), was elaborated by Linus Pauling in 1946: in order to catalyze reactions, an enzyme must be complementary to the *reaction transition state.* This means that optimal interactions between substrate and enzyme occur only in the transition state. Figure 6–5c demonstrates how such an enzyme can work. The metal stick binds to the stickase, but only a subset of the possible magnetic interactions are used in forming the ES complex. The bound substrate must still undergo the increase in free energy needed to reach the transition state. Now, however, the increase in free energy required to draw the stick into a bent and partially broken conformation is offset, or "paid for," by the magnetic interactions (binding energy) that form between the enzyme and substrate in the transition state. Many of these interactions involve parts of the stick that are distant from the point of breakage; thus interactions between the stickase and nonreacting parts of the stick provide some of the energy needed to catalyze stick breakage. This "energy payment" translates into a lower net activation energy and a faster reaction rate.

Real enzymes work on an analogous principle. Some weak interactions are formed in the ES complex, but the full complement of such interactions between substrate and enzyme is formed only when the substrate reaches the transition state. The free energy (binding energy) released by the formation of these interactions partially offsets the energy required to reach the top of the energy hill. The summation of the unfavorable (positive) activation energy $\Delta G^{\ddagger}$ and the favorable (negative) binding energy $\Delta G_B$ results in a lower *net* activation energy (Fig. 6–6). Even on the enzyme, the transition state is not a stable species but a brief point in time that the substrate spends atop an energy hill. The enzyme-catalyzed reaction is much faster than the uncatalyzed process, however, because the hill is much smaller. The

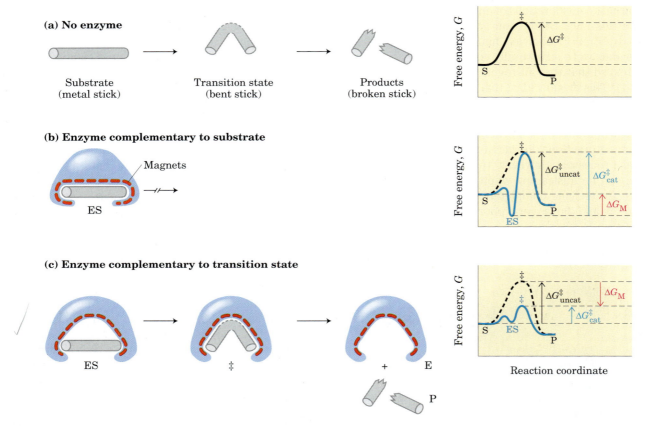

**FIGURE 6–5 An imaginary enzyme (stickase) designed to catalyze breakage of a metal stick.** (a) Before the stick is broken, it must first be bent (the transition state). In both stickase examples, magnetic interactions take the place of weak bonding interactions between enzyme and substrate. (b) A stickase with a magnet-lined pocket complementary in structure to the stick (the substrate) stabilizes the substrate. Bending is impeded by the magnetic attraction between stick and stickase. (c) An enzyme with a pocket complementary to the reaction transition state helps to destabilize the stick, contributing to catalysis of the reaction. The binding energy of the magnetic interactions compensates for the increase in free energy required to bend the stick. Reaction coordinate diagrams (right) show the energy consequences of complementarity to substrate versus complementarity to transition state (EP complexes are omitted). $\Delta G_M$, the difference between the transition-state energies of the uncatalyzed and catalyzed reactions, is contributed by the magnetic interactions between the stick and stickase. When the enzyme is complementary to the substrate (b), the ES complex is more stable and has less free energy in the ground state than substrate alone. The result is an *increase* in the activation energy.

important principle is that *weak binding interactions between the enzyme and the substrate provide a substantial driving force for enzymatic catalysis*. The groups on the substrate that are involved in these weak interactions can be at some distance from the bonds that are broken or changed. The weak interactions formed only in the transition state are those that make the primary contribution to catalysis.

The requirement for multiple weak interactions to drive catalysis is one reason why enzymes (and some coenzymes) are so large. An enzyme must provide functional groups for ionic, hydrogen-bond, and other interactions, and also must precisely position these groups so that binding energy is optimized in the transition state. Adequate binding is accomplished most readily by positioning a substrate in a cavity (the active site) where it is effectively removed from water. The size of proteins reflects the need for superstructure to keep interacting groups properly positioned and to keep the cavity from collapsing.

## Binding Energy Contributes to Reaction Specificity and Catalysis

Can we demonstrate quantitatively that binding energy accounts for the huge rate accelerations brought about by enzymes? Yes. As a point of reference, Equation 6–6 allows us to calculate that $\Delta G^{\ddagger}$ must be lowered by about 5.7 kJ/mol to accelerate a first-order reaction by a factor of ten, under conditions commonly found in cells. The energy available from formation of a single weak interaction is generally estimated to be 4 to 30 kJ/mol. The overall energy available from a number of such interactions is therefore sufficient to lower activation en-

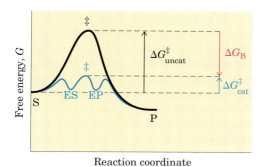

**FIGURE 6–6  Role of binding energy in catalysis.** To lower the activation energy for a reaction, the system must acquire an amount of energy equivalent to the amount by which $\Delta G^{\ddagger}$ is lowered. Much of this energy comes from binding energy ($\Delta G_B$) contributed by formation of weak noncovalent interactions between substrate and enzyme in the transition state. The role of $\Delta G_B$ is analogous to that of $\Delta G_M$ in Figure 6–5.

ergies by the 60 to 100 kJ/mol required to explain the large rate enhancements observed for many enzymes.

The same binding energy that provides energy for catalysis also gives an enzyme its **specificity,** the ability to discriminate between a substrate and a competing molecule. Conceptually, specificity is easy to distinguish from catalysis, but this distinction is much more difficult to make experimentally, because catalysis and specificity arise from the same phenomenon. If an enzyme active site has functional groups arranged optimally to form a variety of weak interactions with a particular substrate in the transition state, the enzyme will not be able to interact to the same degree with any other molecule. For example, if the substrate has a hydroxyl group that forms a hydrogen bond with a specific Glu residue on the enzyme, any molecule lacking a hydroxyl group at that particular position will be a poorer substrate for the enzyme. In addition, any molecule with an extra functional group for which the enzyme has no pocket or binding site is likely to be excluded from the enzyme. In general, *specificity* is derived from the formation of many weak interactions between the enzyme and its specific substrate molecule.

The importance of binding energy to catalysis can be readily demonstrated. For example, the glycolytic enzyme triose phosphate isomerase catalyzes the interconversion of glyceraldehyde 3-phosphate and dihydroxyacetone phosphate:

This reaction rearranges the carbonyl and hydroxyl groups on carbons 1 and 2. However, more than 80% of the enzymatic rate acceleration has been traced to enzyme-substrate interactions involving the phosphate group on carbon 3 of the substrate. This was determined by a careful comparison of the enzyme-catalyzed reactions with glyceraldehyde 3-phosphate and with glyceraldehyde (no phosphate group at position 3) as substrate.

The general principles outlined above can be illustrated by a variety of recognized catalytic mechanisms. These mechanisms are not mutually exclusive, and a given enzyme might incorporate several types in its overall mechanism of action. For most enzymes, it is difficult to quantify the contribution of any one catalytic mechanism to the rate and/or specificity of a particular enzyme-catalyzed reaction.

As we have noted, binding energy makes an important, and sometimes the dominant, contribution to catalysis. Consider what needs to occur for a reaction to take place. Prominent physical and thermodynamic factors contributing to $\Delta G^{\ddagger}$, the barrier to reaction, might include (1) a reduction in entropy, in the form of decreased freedom of motion of two molecules in solution; (2) the solvation shell of hydrogen-bonded water that surrounds and helps to stabilize most biomolecules in aqueous solution; (3) the distortion of substrates that must occur in many reactions; and (4) the need for proper alignment of catalytic functional groups on the enzyme. Binding energy can be used to overcome all these barriers.

First, a large restriction in the relative motions of two substrates that are to react, or **entropy reduction,** is one obvious benefit of binding them to an enzyme. Binding energy holds the substrates in the proper orientation to react—a substantial contribution to catalysis, because productive collisions between molecules in solution can be exceedingly rare. Substrates can be precisely aligned on the enzyme, with many weak interactions between each substrate and strategically located groups on the enzyme clamping the substrate molecules into the proper positions. Studies have shown that constraining the motion of two reactants can produce rate enhancements of many orders of magnitude (Fig. 6–7).

Second, formation of weak bonds between substrate and enzyme also results in **desolvation** of the substrate. Enzyme-substrate interactions replace most or all of the hydrogen bonds between the substrate and water. Third, binding energy involving weak interactions formed only in the reaction transition state helps to compensate thermodynamically for any distortion, primarily electron redistribution, that the substrate must undergo to react.

Finally, the enzyme itself usually undergoes a change in conformation when the substrate binds, induced by multiple weak interactions with the substrate.

$$\overset{1}{H}C{=}O \qquad\qquad H_2C{-}OH$$
$$\overset{2}{H}C{-}OH \qquad\quad C{=}O$$
$$\overset{3}{C}H_2OPO_3^{2-} \qquad\quad CH_2OPO_3^{2-}$$

triose phosphate isomerase

Glyceraldehyde 3-phosphate          Dihydroxyacetone phosphate

**Reaction**    **Rate enhancement**

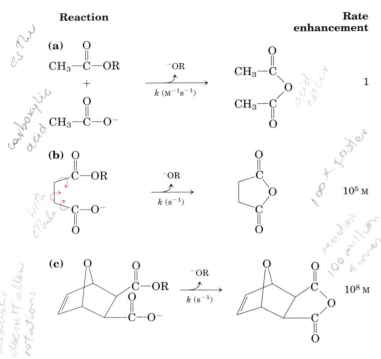

*Es The*

*carboxylic acid*

*with Thes*

*aromatic doesn't allow rotations*

(a)

$$CH_3-C-OR$$

$$+$$

$$CH_3-C-O^-$$

→ $k$ ($M^{-1}s^{-1}$)

$CH_3-C$ ... $CH_3-C$    *acid ester*    1

(b)

→ $k$ ($s^{-1}$)    $10^5$ M    *100 x faster*

(c)

→ $k$ ($s^{-1}$)    $10^8$ M    *100 million reaction times faster*

**FIGURE 6–7 Rate enhancement by entropy reduction.** Shown here are reactions of an ester with a carboxylate group to form an anhydride. The R group is the same in each case. **(a)** For this bimolecular reaction, the rate constant $k$ is second order, with units of $M^{-1}s^{-1}$. **(b)** When the two reacting groups are in a single molecule, the reaction is much faster. For this unimolecular reaction, $k$ has units of $s^{-1}$. Dividing the rate constant for **(b)** by the rate constant for **(a)** gives a rate enhancement of about $10^5$ M. (The enhancement has units of molarity because we are comparing a unimolecular and a bimolecular reaction.) Put another way, if the reactant in **(b)** were present at a concentration of 1 M, the reacting groups would *behave* as though they were present at a concentration of $10^5$ M. Note that the reactant in **(b)** has freedom of rotation about three bonds (shown with curved arrows), but this still represents a substantial reduction of entropy over **(a)**. If the bonds that rotate in **(b)** are constrained as in **(c)**, the entropy is reduced further and the reaction exhibits a rate enhancement of $10^8$ M relative to **(a)**.

This is referred to as **induced fit,** a mechanism postulated by Daniel Koshland in 1958. Induced fit serves to bring specific functional groups on the enzyme into the proper position to catalyze the reaction. The conformational change also permits formation of additional weak bonding interactions in the transition state. In either case, the new enzyme conformation has enhanced catalytic properties. As we have seen, induced fit is a common feature of the reversible binding of ligands to proteins (Chapter 5). Induced fit is also important in the interaction of almost every enzyme with its substrate.

## Specific Catalytic Groups Contribute to Catalysis

In most enzymes, the binding energy used to form the ES complex is just one of several contributors to the overall catalytic mechanism. Once a substrate is bound to an enzyme, properly positioned catalytic functional groups aid in the cleavage and formation of bonds by a variety of mechanisms, including general acid-base catalysis, covalent catalysis, and metal ion catalysis. These are distinct from mechanisms based on binding energy, because they generally involve transient *covalent* interaction with a substrate or group transfer to or from a substrate.

**General Acid-Base Catalysis**    Many biochemical reactions involve the formation of unstable charged intermediates that tend to break down rapidly to their constituent reactant species, thus impeding the reaction (Fig. 6–8). Charged intermediates can often be stabilized by the transfer of protons to or from the substrate or intermediate to form a species that breaks down more readily to products. For nonenzymatic reactions, the proton transfers can involve either the constituents of water alone or other weak proton donors or acceptors. Catalysis of this type that uses only the $H^+$ ($H_3O^+$) or $OH^-$ ions present in water is referred to as **specific acid-base catalysis.** If protons are transferred between the intermediate and water faster than the intermediate breaks down to reactants, the intermediate is effectively stabilized every time it forms. No additional catalysis mediated by other proton acceptors or donors will occur. In many cases, however, water is not enough. The term **general acid-base catalysis** refers to proton transfers mediated by other classes of molecules. For nonenzymatic reactions in aqueous solutions, this occurs only when the unstable reaction intermediate breaks down to reactants faster than protons can be transferred to or from water. Many weak organic acids can supplement water as proton donors in this situation, or weak organic bases can serve as proton acceptors.

In the active site of an enzyme, a number of amino acid side chains can similarly act as proton donors and acceptors (Fig. 6–9). These groups can be precisely positioned in an enzyme active site to allow proton transfers, providing rate enhancements of the order of $10^2$ to $10^5$. This type of catalysis occurs on the vast majority of enzymes. In fact, proton transfers are the most common biochemical reactions.

**Covalent Catalysis**    In covalent catalysis, a transient covalent bond is formed between the enzyme and the substrate. Consider the hydrolysis of a bond between groups A and B:

$$A-B \xrightarrow{H_2O} A + B$$

In the presence of a covalent catalyst (an enzyme with a nucleophilic group X:) the reaction becomes

$$A-B + X: \longrightarrow A-X + B \xrightarrow{H_2O} A + X: + B$$

This alters the pathway of the reaction, and it results in catalysis only when the new pathway has a lower activation energy than the uncatalyzed pathway. Both of the new steps must be faster than the uncatalyzed

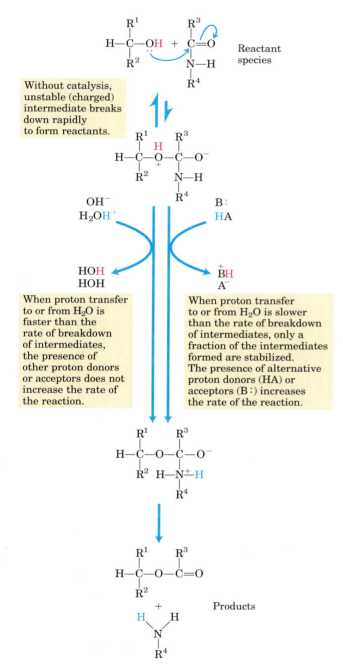

**FIGURE 6–8 How a catalyst circumvents unfavorable charge development during cleavage of an amide.** The hydrolysis of an amide bond, shown here, is the same reaction as that catalyzed by chymotrypsin and other proteases. Charge development is unfavorable and can be circumvented by donation of a proton by $H_3O^+$ (specific acid catalysis) or HA (general acid catalysis), where HA represents any acid. Similarly, charge can be neutralized by proton abstraction by $OH^-$ (specific base catalysis) or B: (general base catalysis), where B: represents any base.

reaction. A number of amino acid side chains, including all those in Figure 6–9, and the functional groups of some enzyme cofactors can serve as nucleophiles in the formation of covalent bonds with substrates. These covalent complexes always undergo further reaction to regenerate the free enzyme. The covalent bond formed between the enzyme and the substrate can activate a substrate for further reaction in a manner that is usually specific to the particular group or coenzyme.

**Metal Ion Catalysis** Metals, whether tightly bound to the enzyme or taken up from solution along with the substrate, can participate in catalysis in several ways. Ionic interactions between an enzyme-bound metal and a substrate can help orient the substrate for reaction or stabilize charged reaction transition states. This use of weak bonding interactions between metal and substrate is similar to some of the uses of enzyme-substrate binding energy described earlier. Metals can also mediate oxidation-reduction reactions by reversible changes in the metal ion's oxidation state. Nearly a third of all known enzymes require one or more metal ions for catalytic activity.

Most enzymes employ a combination of several catalytic strategies to bring about a rate enhancement. A good example of the use of both covalent catalysis and general acid-base catalysis is the reaction catalyzed by chymotrypsin. The first step is cleavage of a peptide bond, which is accompanied by formation of a covalent linkage between a Ser residue on the enzyme and part

| Amino acid residues | General acid form (proton donor) | General base form (proton acceptor) |
|---|---|---|
| **Glu, Asp** | $R-COOH$ | $R-COO^-$ |
| **Lys, Arg** | $R-\overset{H}{\underset{H}{\overset{+}{N}}}H$ | $R-\ddot{N}H_2$ |
| **Cys** | $R-SH$ | $R-S^-$ |
| **His** | 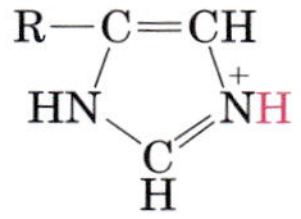 $R-C=CH$ $HN \quad \overset{+}{N}H$ $\quad C$ $\quad H$ | 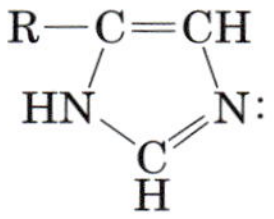 $R-C=CH$ $HN \quad N:$ $\quad C$ $\quad H$ |
| **Ser** | $R-OH$ | $R-O^-$ |
| **Tyr** | 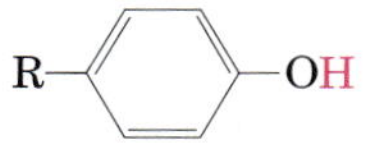 $R-\text{(ring)}-OH$ |  $R-\text{(ring)}-O^-$ |

**FIGURE 6–9 Amino acids in general acid-base catalysis.** Many organic reactions are promoted by proton donors (general acids) or proton acceptors (general bases). The active sites of some enzymes contain amino acid functional groups, such as those shown here, that can participate in the catalytic process as proton donors or proton acceptors.

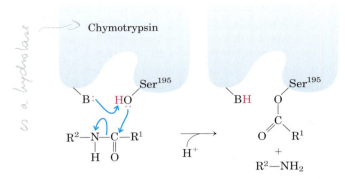

*is a hydrolase*

**FIGURE 6–10 Covalent and general acid-base catalysis.** The first step in the reaction catalyzed by chymotrypsin is the acylation step. The hydroxyl group of Ser[195] is the nucleophile in a reaction aided by general base catalysis (the base is the side chain of His[57]). This provides a new pathway for the hydrolytic cleavage of a peptide bond. Catalysis occurs only if each step in the new pathway is faster than the uncatalyzed reaction. The chymotrypsin reaction is described in more detail in Figure 6–21.

of the substrate; the reaction is enhanced by general base catalysis by other groups on the enzyme (Fig. 6–10). The chymotrypsin reaction is described in more detail in Section 6.4.

### SUMMARY 6.2 How Enzymes Work

- Enzymes are highly effective catalysts, commonly enhancing reaction rates by a factor of $10^5$ to $10^{17}$.

- Enzyme-catalyzed reactions are characterized by the formation of a complex between substrate and enzyme (an ES complex). Substrate binding occurs in a pocket on the enzyme called the active site.

- The function of enzymes and other catalysts is to lower the activation energy, $\Delta G^{\ddagger}$, for a reaction and thereby enhance the reaction rate. The equilibrium of a reaction is unaffected by the enzyme.

- A significant part of the energy used for enzymatic rate enhancements is derived from weak interactions (hydrogen bonds and hydrophobic and ionic interactions) between substrate and enzyme. The enzyme active site is structured so that some of these weak interactions occur preferentially in the reaction transition state, thus stabilizing the transition state. The need for multiple interactions is one reason for the large size of enzymes. The binding energy, $\Delta G_B$, can be used to lower substrate entropy or to cause a conformational change in the enzyme (induced fit). Binding energy also accounts for the exquisite specificity of enzymes for their substrates.

- Additional catalytic mechanisms employed by enzymes include general acid-base catalysis, covalent catalysis, and metal ion catalysis. Catalysis often involves transient covalent interactions between the substrate and the enzyme, or group transfers to and from the enzyme, so as to provide a new, lower-energy reaction path.

## 6.3 Enzyme Kinetics as an Approach to Understanding Mechanism

Biochemists commonly use several approaches to study the mechanism of action of purified enzymes. A knowledge of the three-dimensional structure of the protein provides important information, and the value of structural information is greatly enhanced by classical protein chemistry and modern methods of site-directed mutagenesis (changing the amino acid sequence of a protein by genetic engineering; see Fig. 9–12). These technologies permit enzymologists to examine the role of individual amino acids in enzyme structure and action. However, the central approach to studying the mechanism of an enzyme-catalyzed reaction is to determine the *rate* of the reaction and how it changes in response to changes in experimental parameters, a discipline known as **enzyme kinetics.** This is the oldest approach to understanding enzyme mechanisms and remains the most important. We provide here a basic introduction to the kinetics of enzyme-catalyzed reactions. More advanced treatments are available in the sources cited at the end of the chapter.

### Substrate Concentration Affects the Rate of Enzyme-Catalyzed Reactions

A key factor affecting the rate of a reaction catalyzed by an enzyme is the concentration of substrate, [S]. However, studying the effects of substrate concentration is complicated by the fact that [S] changes during the course of an in vitro reaction as substrate is converted to product. One simplifying approach in kinetics experiments is to measure the **initial rate** (or **initial velocity**), designated $V_0$, when [S] is much greater than the concentration of enzyme, [E]. In a typical reaction, the enzyme may be present in nanomolar quantities, whereas [S] may be five or six orders of magnitude higher. If only the beginning of the reaction is monitored (often the first 60 seconds or less), changes in [S] can be limited to a few percent, and [S] can be regarded as constant. $V_0$ can then be explored as a function of [S], which is adjusted by the investigator. The effect on $V_0$ of varying [S] when the enzyme concentration is held constant is shown in Figure 6–11. At relatively low concentrations of substrate, $V_0$ increases almost linearly

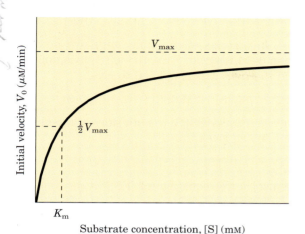

*micromolar per minute* ↓

**FIGURE 6-11 Effect of substrate concentration on the initial velocity of an enzyme-catalyzed reaction.** $V_{max}$ is extrapolated from the plot, because $V_0$ approaches but never quite reaches $V_{max}$. The substrate concentration at which $V_0$ is half maximal is $K_m$, the Michaelis constant. The concentration of enzyme in an experiment such as this is generally so low that $[S] \gg [E]$ even when $[S]$ is described as low or relatively low. The units shown are typical for enzyme-catalyzed reactions and are given only to help illustrate the meaning of $V_0$ and $[S]$. (Note that the curve describes *part* of a rectangular hyperbola, with one asymptote at $V_{max}$. If the curve were continued below $[S] = 0$, it would approach a vertical asymptote at $[S] = -K_m$.)

with an increase in $[S]$. At higher substrate concentrations, $V_0$ increases by smaller and smaller amounts in response to increases in $[S]$. Finally, a point is reached beyond which increases in $V_0$ are vanishingly small as $[S]$ increases. This plateau-like $V_0$ region is close to the **maximum velocity, $V_{max}$.**

The ES complex is the key to understanding this kinetic behavior, just as it was a starting point for our discussion of catalysis. The kinetic pattern in Figure 6–11 led Victor Henri, following the lead of Wurtz, to propose in 1903 that the combination of an enzyme with its substrate molecule to form an ES complex is a necessary step in enzymatic catalysis. This idea was expanded into a general theory of enzyme action, particularly by Leonor Michaelis and Maud Menten in 1913. They postulated that the enzyme first combines reversibly with

Leonor Michaelis,
1875–1949

Maud Menten,
1879–1960

its substrate to form an enzyme-substrate complex in a relatively fast reversible step:

$$E + S \underset{k_{-1}}{\overset{k_1}{\rightleftharpoons}} ES \qquad (6-7)$$

The ES complex then breaks down in a slower second step to yield the free enzyme and the reaction product P:

$$ES \underset{k_{-2}}{\overset{k_2}{\rightleftharpoons}} E + P \qquad (6-8)$$

Because the slower second reaction (Eqn 6–8) must limit the rate of the overall reaction, the overall rate must be proportional to the concentration of the species that reacts in the second step, that is, ES.

At any given instant in an enzyme-catalyzed reaction, the enzyme exists in two forms, the free or uncombined form E and the combined form ES. At low $[S]$, most of the enzyme is in the uncombined form E. Here, the rate is proportional to $[S]$ because the equilibrium of Equation 6–7 is pushed toward formation of more ES as $[S]$ increases. The maximum initial rate of the catalyzed reaction ($V_{max}$) is observed when virtually all the enzyme is present as the ES complex and $[E]$ is vanishingly small. Under these conditions, the enzyme is "saturated" with its substrate, so that further increases in $[S]$ have no effect on rate. This condition exists when $[S]$ is sufficiently high that essentially all the free enzyme has been converted to the ES form. After the ES complex breaks down to yield the product P, the enzyme is free to catalyze reaction of another molecule of substrate. The saturation effect is a distinguishing characteristic of enzymatic catalysts and is responsible for the plateau observed in Figure 6–11. The pattern seen in Figure 6–11 is sometimes referred to as saturation kinetics.

When the enzyme is first mixed with a large excess of substrate, there is an initial period, the **pre–steady state,** during which the concentration of ES builds up. This period is usually too short to be easily observed, lasting just microseconds. The reaction quickly achieves a **steady state** in which $[ES]$ (and the concentrations of any other intermediates) remains approximately constant over time. The concept of a steady state was introduced by G. E. Briggs and Haldane in 1925. The measured $V_0$ generally reflects the steady state, even though $V_0$ is limited to the early part of the reaction, and analysis of these initial rates is referred to as **steady-state kinetics.**

## The Relationship between Substrate Concentration and Reaction Rate Can Be Expressed Quantitatively

The curve expressing the relationship between $[S]$ and $V_0$ (Fig. 6–11) has the same general shape for most enzymes (it approaches a rectangular hyperbola), which can be expressed algebraically by the Michaelis-Menten equation. Michaelis and Menten derived this equation starting from their basic hypothesis that the rate-limiting step in enzymatic reactions is the breakdown of the ES complex to product and free enzyme. The equation is

$$V_0 = \frac{V_{max}[S]}{K_m + [S]} \qquad (6-9)$$

The important terms are [S], $V_0$, $V_{max}$, and a constant called the Michaelis constant, $K_m$. All these terms are readily measured experimentally.

Here we develop the basic logic and the algebraic steps in a modern derivation of the Michaelis-Menten equation, which includes the steady-state assumption introduced by Briggs and Haldane. The derivation starts with the two basic steps of the formation and breakdown of ES (Eqns 6–7 and 6–8). Early in the reaction, the concentration of the product, [P], is negligible, and we make the simplifying assumption that the reverse reaction, $P \rightarrow S$ (described by $k_{-2}$), can be ignored. This assumption is not critical but it simplifies our task. The overall reaction then reduces to

$$E + S \underset{k_{-1}}{\overset{k_1}{\rightleftharpoons}} ES \xrightarrow{k_2} E + P \qquad (6-10)$$

$V_0$ is determined by the breakdown of ES to form product, which is determined by [ES]:

$$V_0 = k_2[ES] \qquad (6-11)$$

Because [ES] in Equation 6–11 is not easily measured experimentally, we must begin by finding an alternative expression for this term. First, we introduce the term [$E_t$], representing the total enzyme concentration (the sum of free and substrate-bound enzyme). Free or unbound enzyme can then be represented by [$E_t$] − [ES]. Also, because [S] is ordinarily far greater than [$E_t$], the amount of substrate bound by the enzyme at any given time is negligible compared with the total [S]. With these conditions in mind, the following steps lead us to an expression for $V_0$ in terms of easily measurable parameters.

*Step 1* The rates of formation and breakdown of ES are determined by the steps governed by the rate constants $k_1$ (formation) and $k_{-1} + k_2$ (breakdown), according to the expressions

**Rate of ES formation** = $k_1([E_t] - [ES])[S]$ $\qquad$ (6–12)

**Rate of ES breakdown** = $k_{-1}[ES] + k_2[ES]$ $\qquad$ (6–13)

*Step 2* We now make an important assumption: that the initial rate of reaction reflects a steady state in which [ES] is constant—that is, the rate of formation of ES is equal to the rate of its breakdown. This is called the **steady-state assumption.** The expressions in Equations 6–12 and 6–13 can be equated for the steady state, giving

$$k_1([E_t] - [ES])[S] = k_{-1}[ES] + k_2[ES] \qquad (6-14)$$

*Step 3* In a series of algebraic steps, we now solve Equation 6–14 for [ES]. First, the left side is multiplied out and the right side simplified to give

$$k_1[E_t][S] - k_1[ES][S] = (k_{-1} + k_2)[ES] \qquad (6-15)$$

Adding the term $k_1[ES][S]$ to both sides of the equation and simplifying gives

$$k_1[E_t][S] = (k_1[S] + k_{-1} + k_2)[ES] \qquad (6-16)$$

We then solve this equation for [ES]:

$$[ES] = \frac{k_1[E_t][S]}{k_1[S] + k_{-1} + k_2} \qquad (6-17)$$

This can now be simplified further, combining the rate constants into one expression:

$$[ES] = \frac{[E_t][S]}{[S] + (k_2 + k_{-1})/k_1} \qquad (6-18)$$

The term $(k_2 + k_{-1})/k_1$ **is defined as the Michaelis constant, $K_m$.** Substituting this into Equation 6–18 simplifies the expression to

$$[ES] = \frac{[E_t][S]}{K_m + [S]} \qquad (6-19)$$

*Step 4* We can now express $V_0$ in terms of [ES]. Substituting the right side of Equation 6–19 for [ES] in Equation 6–11 gives

$$V_0 = \frac{k_2[E_t][S]}{K_m + [S]} \qquad (6-20)$$

This equation can be further simplified. Because the maximum velocity occurs when the enzyme is saturated (that is, with [ES] = [$E_t$]) $V_{max}$ can be defined as $k_2[E_t]$. Substituting this in Equation 6–20 gives Equation 6–9:

$$V_0 = \frac{V_{max}[S]}{K_m + [S]}$$

This is the **Michaelis-Menten equation,** the **rate equation** for a one-substrate enzyme-catalyzed reaction. It is a statement of the quantitative relationship between the initial velocity $V_0$, the maximum velocity $V_{max}$, and the initial substrate concentration [S], all related through the Michaelis constant $K_m$. Note that $K_m$ has units of concentration. Does the equation fit experimental observations? Yes; we can confirm this by considering the limiting situations where [S] is very high or very low, as shown in Figure 6–12.

An important numerical relationship emerges from the Michaelis-Menten equation in the special case when $V_0$ is exactly one-half $V_{max}$ (Fig. 6–12). Then

$$\frac{V_{max}}{2} = \frac{V_{max}[S]}{K_m + [S]} \qquad (6-21)$$

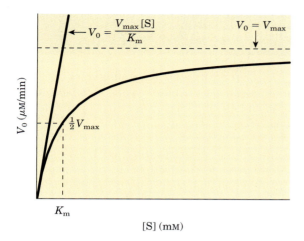

**FIGURE 6–12 Dependence of initial velocity on substrate concentration.** This graph shows the kinetic parameters that define the limits of the curve at high and low [S]. At low [S], $K_m >>$ [S] and the [S] term in the denominator of the Michaelis-Menten equation (Eqn 6–9) becomes insignificant. The equation simplifies to $V_0 = V_{max}[S]/K_m$ and $V_0$ exhibits a linear dependence on [S], as observed here. At high [S], where [S] $>> K_m$, the $K_m$ term in the denominator of the Michaelis-Menten equation becomes insignificant and the equation simplifies to $V_0 = V_{max}$; this is consistent with the plateau observed at high [S]. The Michaelis-Menten equation is therefore consistent with the observed dependence of $V_0$ on [S], and the shape of the curve is defined by the terms $V_{max}/K_m$ at low [S] and $V_{max}$ at high [S].

On dividing by $V_{max}$, we obtain

$$\frac{1}{2} = \frac{[S]}{K_m + [S]} \tag{6–22}$$

Solving for $K_m$, we get $K_m + [S] = 2[S]$, or

$$K_m = [S], \text{ when } V_0 = \frac{1}{2}V_{max} \tag{6–23}$$

This is a very useful, practical definition of $K_m$: $K_m$ is equivalent to the substrate concentration at which $V_0$ is one-half $V_{max}$.

The Michaelis-Menten equation (Eqn 6–9) can be algebraically transformed into versions that are useful in the practical determination of $K_m$ and $V_{max}$ (Box 6–1) and, as we describe later, in the analysis of inhibitor action (see Box 6–2 on page 210).

## Kinetic Parameters Are Used to Compare Enzyme Activities

It is important to distinguish between the Michaelis-Menten equation and the specific kinetic mechanism on which it was originally based. The equation describes the kinetic behavior of a great many enzymes, and all enzymes that exhibit a hyperbolic dependence of $V_0$ on [S] are said to follow **Michaelis-Menten kinetics.** The practical rule that

$K_m = [S]$ when $V_0 = \frac{1}{2}V_{max}$ (Eqn 6–23) holds for all enzymes that follow Michaelis-Menten kinetics. (The most important exceptions to Michaelis-Menten kinetics are the regulatory enzymes, discussed in Section 6.5.) However, the Michaelis-Menten equation does not depend on the relatively simple two-step reaction mechanism proposed by Michaelis and Menten (Eqn 6–10). Many enzymes that follow Michaelis-Menten kinetics have quite different reaction mechanisms, and enzymes that catalyze reactions with six or eight identifiable steps often exhibit the same steady-state kinetic behavior. Even though Equation 6–23 holds true for many enzymes, both the magnitude and the real meaning of $V_{max}$ and $K_m$ can differ from one enzyme to the next. This is an important limitation of the steady-state approach to enzyme kinetics. The parameters $V_{max}$ and $K_m$ can be obtained experimentally for any given enzyme, but by themselves they provide little information about the number, rates, or chemical nature of discrete steps in the reaction. Steady-state kinetics nevertheless is the standard language by which biochemists compare and characterize the catalytic efficiencies of enzymes.

***Interpreting*** $V_{max}$ **and** $K_m$ Figure 6–12 shows a simple graphical method for obtaining an approximate value for $K_m$. A more convenient procedure, using a **double-reciprocal plot,** is presented in Box 6–1. The $K_m$ can vary greatly from enzyme to enzyme, and even for different substrates of the same enzyme (Table 6–6). The term is sometimes used (often inappropriately) as an indicator of the affinity of an enzyme for its substrate. The actual meaning of $K_m$ depends on specific aspects of the reaction mechanism such as the number and relative rates of the individual steps. For reactions with two steps,

$$K_m = \frac{k_2 + k_{-1}}{k_1} \tag{6–24}$$

When $k_2$ is rate-limiting, $k_2 << k_{-1}$ and $K_m$ reduces to $k_{-1}/k_1$, which is defined as the **dissociation constant, $K_d$,** of the ES complex. Where these conditions hold, $K_m$ does represent a measure of the affinity of the enzyme

## TABLE 6–6 $K_m$ for Some Enzymes and Substrates

| Enzyme | Substrate | $K_m$ (mM) |
|---|---|---|
| Hexokinase (brain) | ATP | 0.4 |
| | D-Glucose | 0.05 |
| | D-Fructose | 1.5 |
| Carbonic anhydrase | $HCO_3^-$ | 26 |
| Chymotrypsin | Glycyltyrosinylglycine | 108 |
| | N-Benzoyltyrosinamide | 2.5 |
| β-Galactosidase | D-Lactose | 4.0 |
| Threonine dehydratase | L-Threonine | 5.0 |

BOX 6–1    WORKING IN BIOCHEMISTRY

## Transformations of the Michaelis-Menten Equation: The Double-Reciprocal Plot

The Michaelis-Menten equation

$$V_0 = \frac{V_{max}[S]}{K_m + [S]}$$

can be algebraically transformed into equations that are more useful in plotting experimental data. One common transformation is derived simply by taking the reciprocal of both sides of the Michaelis-Menten equation:

$$\frac{1}{V_0} = \frac{K_m + [S]}{V_{max}[S]}$$

Separating the components of the numerator on the right side of the equation gives

$$\frac{1}{V_0} = \frac{K_m}{V_{max}[S]} + \frac{[S]}{V_{max}[S]}$$

which simplifies to

$$\frac{1}{V_0} = \frac{K_m}{V_{max}[S]} + \frac{1}{V_{max}}$$

This form of the Michaelis-Menten equation is called the **Lineweaver-Burk equation.** For enzymes obeying the Michaelis-Menten relationship, a plot of $1/V_0$ versus $1/[S]$ (the "double reciprocal" of the $V_0$ versus [S] plot we have been using to this point) yields a straight line (Fig. 1). This line has a slope of $K_m/V_{max}$, an intercept of $1/V_{max}$ on the $1/V_0$ axis, and an intercept

of $-1/K_m$ on the $1/[S]$ axis. The double-reciprocal presentation, also called a Lineweaver-Burk plot, has the great advantage of allowing a more accurate determination of $V_{max}$, which can only be *approximated* from a simple plot of $V_0$ versus [S] (see Fig. 6–12).

Other transformations of the Michaelis-Menten equation have been derived, each with some particular advantage in analyzing enzyme kinetic data. (See Problem 11 at the end of this chapter.)

The double-reciprocal plot of enzyme reaction rates is very useful in distinguishing between certain types of enzymatic reaction mechanisms (see Fig. 6–14) and in analyzing enzyme inhibition (see Box 6–2).

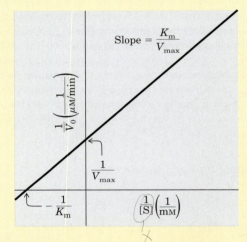

**FIGURE 1**  A double-reciprocal or Lineweaver-Burk plot.

---

for its substrate in the ES complex. However, this scenario does not apply for most enzymes. Sometimes $k_2 \gg k_{-1}$, and then $K_m = k_2/k_1$. In other cases, $k_2$ and $k_{-1}$ are comparable and $K_m$ remains a more complex function of all three rate constants (Eqn 6–24). The Michaelis-Menten equation and the characteristic saturation behavior of the enzyme still apply, but $K_m$ cannot be considered a simple measure of substrate affinity. Even more common are cases in which the reaction goes through several steps after formation of ES; $K_m$ can then become a very complex function of many rate constants.

The quantity $V_{max}$ also varies greatly from one enzyme to the next. If an enzyme reacts by the two-step Michaelis-Menten mechanism, $V_{max} = k_2[E_t]$, where $k_2$ is rate-limiting. However, the number of reaction steps and the identity of the rate-limiting step(s) can vary from enzyme to enzyme. For example, consider the quite common situation where product release, $EP \rightarrow E + P$, is rate-limiting. Early in the reaction (when [P] is low), the overall reaction can be described by the scheme

$$E + S \underset{k_{-1}}{\overset{k_1}{\rightleftharpoons}} ES \underset{k_{-2}}{\overset{k_2}{\rightleftharpoons}} EP \overset{k_3}{\rightleftharpoons} E + P \qquad (6\text{–}25)$$

In this case, most of the enzyme is in the EP form at saturation, and $V_{max} = k_3[E_t]$. It is useful to define a more general rate constant, $k_{cat}$, to describe the limiting rate of any enzyme-catalyzed reaction at saturation. If the reaction has several steps and one is clearly rate-limiting, $k_{cat}$ is equivalent to the rate constant for that limiting step. For the simple reaction of Equation 6–10, $k_{cat} = k_2$. For the reaction of Equation 6–25, $k_{cat} = k_3$. When several steps are partially rate-limiting, $k_{cat}$ can become a complex function of several of the rate constants that define each individual reaction step. In the Michaelis-Menten equation, $k_{cat} = V_{max}/[E_t]$, and Equation 6–9 becomes

$$V_0 = \frac{k_{cat}[E_t][S]}{K_m + [S]} \qquad (6\text{–}26)$$

The constant $k_{cat}$ is a first-order rate constant and hence has units of reciprocal time. It is also called the

*To gen an intrinsic catalytic constant from Vmax*

*catalytic constant*

**turnover number.** It is equivalent to the number of substrate molecules converted to product in a given unit of time on a single enzyme molecule when the enzyme is saturated with substrate. The turnover numbers of several enzymes are given in Table 6–7.

### Comparing Catalytic Mechanisms and Efficiencies

The kinetic parameters $k_{cat}$ and $K_m$ are generally useful for the study and comparison of different enzymes, whether their reaction mechanisms are simple or complex. Each enzyme has values of $k_{cat}$ and $K_m$ that reflect the cellular environment, the concentration of substrate normally encountered in vivo by the enzyme, and the chemistry of the reaction being catalyzed.

The parameters $k_{cat}$ and $K_m$ also allow us to evaluate the kinetic efficiency of enzymes, but either parameter alone is insufficient for this task. Two enzymes catalyzing different reactions may have the same $k_{cat}$ (turnover number), yet the rates of the uncatalyzed reactions may be different and thus the rate enhancements brought about by the enzymes may differ greatly. Experimentally, the $K_m$ for an enzyme tends to be similar to the cellular concentration of its substrate. An enzyme that acts on a substrate present at a very low concentration in the cell usually has a lower $K_m$ than an enzyme that acts on a substrate that is more abundant.

The best way to compare the catalytic efficiencies of different enzymes or the turnover of different substrates by the same enzyme is to compare the ratio $k_{cat}/K_m$ for the two reactions. This parameter, sometimes called the **specificity constant,** is the rate constant for the conversion of E + S to E + P. When [S] << $K_m$, Equation 6–26 reduces to the form

$$V_0 = \frac{k_{cat}}{K_m}[E_t][S] \qquad (6\text{–}27)$$

$V_0$ in this case depends on the concentration of two reactants, $[E_t]$ and $[S]$; therefore this is a second-order rate equation and the constant $k_{cat}/K_m$ is a second-order rate

**TABLE 6–7 Turnover Numbers, $k_{cat}$, of Some Enzymes**

| Enzyme | Substrate | $k_{cat}$ (s$^{-1}$) |
|---|---|---|
| Catalase | $H_2O_2$ | 40,000,000 |
| Carbonic anhydrase | $HCO_3^-$ | 400,000 |
| Acetylcholinesterase | Acetylcholine | 14,000 |
| β-Lactamase | Benzylpenicillin | 2,000 |
| Fumarase | Fumarate | 800 |
| RecA protein (an ATPase) | ATP | 0.4 |

constant with units of $M^{-1}s^{-1}$. There is an upper limit to $k_{cat}/K_m$, imposed by the rate at which E and S can diffuse together in an aqueous solution. This diffusion-controlled limit is $10^8$ to $10^9$ $M^{-1}s^{-1}$, and many enzymes have a $k_{cat}/K_m$ near this range (Table 6–8). Such enzymes are said to have achieved catalytic perfection. Note that different values of $k_{cat}$ and $K_m$ can produce the maximum ratio.

## Many Enzymes Catalyze Reactions with Two or More Substrates

We have seen how [S] affects the rate of a simple enzymatic reaction (S → P) with only one substrate molecule. In most enzymatic reactions, however, two (and sometimes more) different substrate molecules bind to the enzyme and participate in the reaction. For example, in the reaction catalyzed by hexokinase, ATP and glucose are the substrate molecules, and ADP and glucose 6-phosphate are the products:

$$\text{ATP} + \text{glucose} \longrightarrow \text{ADP} + \text{glucose 6-phosphate}$$

The rates of such bisubstrate reactions can also be analyzed by the Michaelis-Menten approach. Hexokinase has a characteristic $K_m$ for each of its substrates (Table 6–6).

Enzymatic reactions with two substrates usually involve transfer of an atom or a functional group from one substrate to the other. These reactions proceed by one

**TABLE 6–8 Enzymes for Which $k_{cat}/K_m$ Is Close to the Diffusion-Controlled Limit ($10^8$ to $10^9$ $M^{-1}s^{-1}$)**

*per mole per second. units of diffusion*

| Enzyme | Substrate | $k_{cat}$ (s$^{-1}$) | $K_m$ (M) | $k_{cat}/K_m$ (M$^{-1}$s$^{-1}$) |
|---|---|---|---|---|
| Acetylcholinesterase | Acetylcholine | $1.4 \times 10^4$ | $9 \times 10^{-5}$ | $1.6 \times 10^8$ |
| Carbonic anhydrase | $CO_2$ | $1 \times 10^6$ | $1.2 \times 10^{-2}$ | $8.3 \times 10^7$ |
| | $HCO_3^-$ | $4 \times 10^5$ | $2.6 \times 10^{-2}$ | $1.5 \times 10^7$ |
| Catalase | $H_2O_2$ | $4 \times 10^7$ | $1.1 \times 10^0$ | $4 \times 10^7$ |
| Crotonase | Crotonyl-CoA | $5.7 \times 10^3$ | $2 \times 10^{-5}$ | $2.8 \times 10^8$ |
| Fumarase | Fumarate | $8 \times 10^2$ | $5 \times 10^{-6}$ | $1.6 \times 10^8$ |
| | Malate | $9 \times 10^2$ | $2.5 \times 10^{-5}$ | $3.6 \times 10^7$ |
| β-Lactamase | Benzylpenicillin | $2.0 \times 10^3$ | $2 \times 10^{-5}$ | $1 \times 10^8$ |

*almost 10⁹*

**Source:** Fersht, A. (1999) *Structure and Mechanism in Protein Science*, p. 166, W. H. Freeman and Company, New York.

*almost as fast as the diffusion of H₂O*

**(a) Enzyme reaction involving a ternary complex**

Random order

$$E \underset{\searrow ES_2 \nearrow}{\overset{\nearrow ES_1 \searrow}{}} ES_1S_2 \longrightarrow E + P_1 + P_2$$

Ordered

$$E + S_1 \rightleftharpoons ES_1 \overset{S_2}{\rightleftharpoons} ES_1S_2 \longrightarrow E + P_1 + P_2$$

**(b) Enzyme reaction in which no ternary complex is formed**

$$E + S_1 \rightleftharpoons ES_1 \rightleftharpoons E'P_1 \overset{P_1}{\rightleftharpoons} E' \overset{S_2}{\rightleftharpoons} E'S_2 \longrightarrow E + P_2$$

**FIGURE 6–13** **Common mechanisms for enzyme-catalyzed bisubstrate reactions.** **(a)** The enzyme and both substrates come together to form a ternary complex. In ordered binding, substrate 1 must bind before substrate 2 can bind productively. In random binding, the substrates can bind in either order. **(b)** An enzyme-substrate complex forms, a product leaves the complex, the altered enzyme forms a second complex with another substrate molecule, and the second product leaves, regenerating the enzyme. Substrate 1 may transfer a functional group to the enzyme (to form the covalently modified E′), which is subsequently transferred to substrate 2. This is called a Ping-Pong or double-displacement mechanism.

of several different pathways. In some cases, both substrates are bound to the enzyme concurrently at some point in the course of the reaction, forming a noncovalent ternary complex (Fig. 6–13a); the substrates bind in a random sequence or in a specific order. In other cases, the first substrate is converted to product and dissociates before the second substrate binds, so no ternary complex is formed. An example of this is the Ping-Pong, or double-displacement, mechanism (Fig. 6–13b). Steady-state kinetics can often help distinguish among these possibilities (Fig. 6–14).

### Pre–Steady State Kinetics Can Provide Evidence for Specific Reaction Steps

We have introduced kinetics as the primary method for studying the steps in an enzymatic reaction, and we have also outlined the limitations of the most common kinetic parameters in providing such information. The two most important experimental parameters obtained from steady-state kinetics are $k_{cat}$ and $k_{cat}/K_m$. Variation in $k_{cat}$ and $k_{cat}/K_m$ with changes in pH or temperature can provide additional information about steps in a reaction pathway. In the case of bisubstrate reactions, steady-state kinetics can help determine whether a ternary complex is formed during the reaction (Fig. 6–14). A more complete picture generally requires more sophisticated kinetic methods that go beyond the scope of an introductory text. Here, we briefly introduce one of the most important kinetic approaches for studying reaction mechanisms, pre–steady state kinetics.

A complete description of an enzyme-catalyzed reaction requires direct measurement of the rates of individual reaction steps—for example, measurement of the association of enzyme and substrate to form the ES complex. It is during the pre–steady state that the rates of many reaction steps can be measured independently. Experimenters adjust reaction conditions so that they can observe events during reaction of a single substrate molecule. Because the pre–steady state phase is gener-

ally very short, the experiments often require specialized techniques for very rapid mixing and sampling. One objective is to gain a complete and quantitative picture of the energy changes during the reaction. As we have already noted, reaction rates and equilibria are related to the free-energy changes during a reaction. Measur-

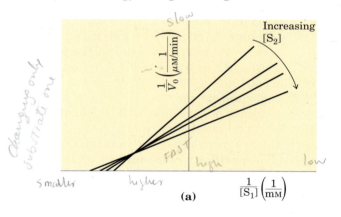

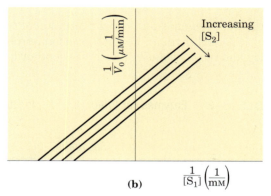

**FIGURE 6–14** **Steady-state kinetic analysis of bisubstrate reactions.** In these double-reciprocal plots (see Box 6–1), the concentration of substrate 1 is varied while the concentration of substrate 2 is held constant. This is repeated for several values of $[S_2]$, generating several separate lines. **(a)** Intersecting lines indicate that a ternary complex is formed in the reaction; **(b)** parallel lines indicate a Ping-Pong (double-displacement) pathway.

ing the rate of individual reaction steps reveals how energy is used by a specific enzyme, which is an important component of the overall reaction mechanism. In a number of cases investigators have been able to record the rates of every individual step in a multistep enzymatic reaction. Some examples of the application of pre–steady state kinetics are included in the descriptions of specific enzymes in Section 6.4.

### Enzymes Are Subject to Reversible or Irreversible Inhibition

Enzyme inhibitors are molecular agents that interfere with catalysis, slowing or halting enzymatic reactions. Enzymes catalyze virtually all cellular processes, so it should not be surprising that enzyme inhibitors are among the most important pharmaceutical agents known. For example, aspirin (acetylsalicylate) inhibits the enzyme that catalyzes the first step in the synthesis of prostaglandins, compounds involved in many processes, including some that produce pain. The study of enzyme inhibitors also has provided valuable information about enzyme mechanisms and has helped define some metabolic pathways. There are two broad classes of enzyme inhibitors: reversible and irreversible.

**Reversible Inhibition**  One common type of **reversible inhibition** is called competitive (Fig. 6–15a). A **competitive inhibitor** competes with the substrate for the active site of an enzyme. While the inhibitor (I) occupies the active site it prevents binding of the substrate to the enzyme. Many competitive inhibitors are compounds that resemble the substrate and combine with the enzyme to form an EI complex, but without leading to catalysis. Even fleeting combinations of this type will reduce the efficiency of the enzyme. By taking into account the molecular geometry of inhibitors that resemble the substrate, we can reach conclusions about which parts of the normal substrate bind to the enzyme. Competitive inhibition can be analyzed quantitatively by steady-state kinetics. In the presence of a competitive inhibitor, the Michaelis-Menten equation (Eqn 6–9) becomes

$$V_0 = \frac{V_{max}[S]}{\alpha K_m + [S]} \qquad (6\text{–}28)$$

where

$$\alpha = 1 + \frac{[I]}{K_I} \qquad \text{and} \qquad K_I = \frac{[E][I]}{[EI]}$$

Equation 6–28 describes the important features of competitive inhibition. The experimentally determined variable $\alpha K_m$, the $K_m$ observed in the presence of the inhibitor, is often called the "apparent" $K_m$.

Because the inhibitor binds reversibly to the enzyme, the competition can be biased to favor the substrate sim-

**(a) Competitive inhibition**

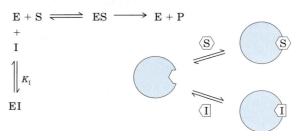

**(b) Uncompetitive inhibition**

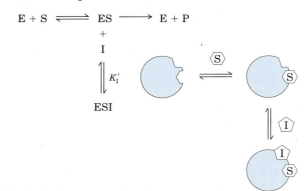

**(c) Mixed inhibition**

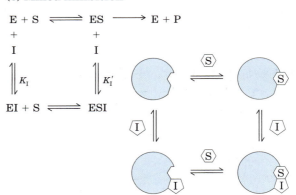

**FIGURE 6–15  Three types of reversible inhibition. (a)** Competitive inhibitors bind to the enzyme's active site. **(b)** Uncompetitive inhibitors bind at a separate site, but bind only to the ES complex. $K_I$ is the equilibrium constant for inhibitor binding to E; $K_I'$ is the equilibrium constant for inhibitor binding to ES. **(c)** Mixed inhibitors bind at a separate site, but may bind to either E or ES.

ply by adding more substrate. When [S] far exceeds [I], the probability that an inhibitor molecule will bind to the enzyme is minimized and the reaction exhibits a normal $V_{max}$. However, the [S] at which $V_0 = \frac{1}{2}V_{max}$, the apparent $K_m$, increases in the presence of inhibitor by the factor $\alpha$. This effect on apparent $K_m$, combined with the absence of an effect on $V_{max}$, is diagnostic of competitive inhibition and is readily revealed in a double-reciprocal plot (Box 6–2). The equilibrium constant for inhibitor binding, $K_I$, can be obtained from the same plot.

## Kinetic Tests for Determining Inhibition Mechanisms

The double-reciprocal plot (see Box 6–1) offers an easy way of determining whether an enzyme inhibitor is competitive, uncompetitive, or mixed. Two sets of rate experiments are carried out, with the enzyme concentration held constant in each set. In the first set, [S] is also held constant, permitting measurement of the effect of increasing inhibitor concentration [I] on the initial rate $V_0$ (not shown). In the second set, [I] is held constant but [S] is varied. The results are plotted as $1/V_0$ versus $1/[S]$.

Figure 1 shows a set of double-reciprocal plots, one obtained in the absence of inhibitor and two at different concentrations of a competitive inhibitor. Increasing [I] results in a family of lines with a common intercept on the $1/V_0$ axis but with different slopes. Because the intercept on the $1/V_0$ axis equals $1/V_{max}$, we know that $V_{max}$ is unchanged by the presence of a competitive inhibitor. That is, regardless of the concentration of a competitive inhibitor, a sufficiently high substrate concentration will always displace the inhibitor from the enzyme's active site. Above the graph is the rearrangement of Equation 6–28 on which the plot is based. The value of $\alpha$ can be calculated

from the change in slope at any given [I]. Knowing [I] and $\alpha$, we can calculate $K_I$ from the expression

$$\alpha = 1 + \frac{[I]}{K_I}$$

*when increase* *it increases* *inhibition* *(it gets slower)*

For uncompetitive and mixed inhibition, similar plots of rate data give the families of lines shown in Figures 2 and 3. Changes in axis intercepts signal changes in $V_{max}$ and $K_m$.

$$\frac{1}{V_0} = \left(\frac{K_m}{V_{max}}\right)\frac{1}{[S]} + \frac{\alpha'}{V_{max}}$$

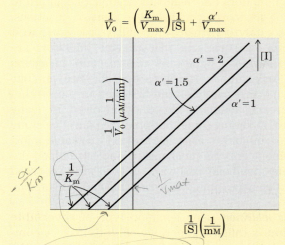

**FIGURE 2** Uncompetitive inhibition.

$$\frac{1}{V_0} = \left(\frac{\alpha K_m}{V_{max}}\right)\frac{1}{[S]} + \frac{1}{V_{max}}$$

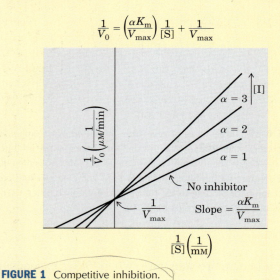

**FIGURE 1** Competitive inhibition.

$$\frac{1}{V_0} = \left(\frac{\alpha K_m}{V_{max}}\right)\frac{1}{[S]} + \frac{\alpha'}{V_{max}}$$

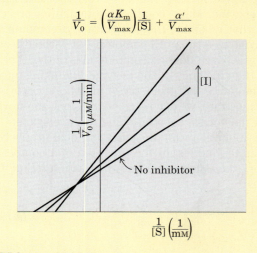

**FIGURE 3** Mixed inhibition.

A medical therapy based on competition at the active site is used to treat patients who have ingested methanol, a solvent found in gas-line antifreeze. The liver enzyme alcohol dehydrogenase converts methanol to formaldehyde, which is damaging to many tissues. Blindness is a common result of methanol ingestion, because

the eyes are particularly sensitive to formaldehyde. Ethanol competes effectively with methanol as an alternative substrate for alcohol dehydrogenase. The effect of ethanol is much like that of a competitive inhibitor, with the distinction that ethanol is also a substrate for alcohol dehydrogenase and its concentration will decrease over

*[handwritten margin notes: $K_m = 3.33 \, mM$ / $K_i = -.5 \, mM$ / inhibitor binds / enzyme better / than substrate]*

time as the enzyme converts it to acetaldehyde. The therapy for methanol poisoning is slow intravenous infusion of ethanol, at a rate that maintains a controlled concentration in the bloodstream for several hours. This slows the formation of formaldehyde, lessening the danger while the kidneys filter out the methanol to be excreted harmlessly in the urine. ■

Two other types of reversible inhibition, uncompetitive and mixed, though often defined in terms of one-substrate enzymes, are in practice observed only with enzymes having two or more substrates. An **uncompetitive inhibitor** (Fig. 6–15b) binds at a site distinct from the substrate active site and, unlike a competitive inhibitor, binds only to the ES complex. In the presence of an uncompetitive inhibitor, the Michaelis-Menten equation is altered to

$$V_0 = \frac{V_{max}[S]}{K_m + \alpha'[S]} \qquad (6\text{--}29)$$

where

$$\alpha' = 1 + \frac{[I]}{K'_I} \quad \text{and} \quad K'_I = \frac{[ES][I]}{[ESI]}$$

As described by Equation 6–29, at high concentrations of substrate, $V_0$ approaches $V_{max}/\alpha'$. Thus, an uncompetitive inhibitor lowers the measured $V_{max}$. Apparent $K_m$ also decreases, because the [S] required to reach one-half $V_{max}$ decreases by the factor $\alpha'$.

A **mixed inhibitor** (Fig. 6–15c) also binds at a site distinct from the substrate active site, but it binds to either E or ES. The rate equation describing mixed inhibition is

$$V_0 = \frac{V_{max}[S]}{\alpha K_m + \alpha'[S]} \qquad (6\text{--}30)$$

where $\alpha$ and $\alpha'$ are defined as above. A mixed inhibitor usually affects both $K_m$ and $V_{max}$. The special case of $\alpha = \alpha'$, rarely encountered in experiments, classically has been defined as **noncompetitive inhibition.** Examine Equation 6–30 to see why a noncompetitive inhibitor would affect the $V_{max}$ but not the $K_m$.

Equation 6–30 serves as a general expression for the effects of reversible inhibitors, simplifying to the expressions for competitive and uncompetitive inhibition when $\alpha' = 1.0$ or $\alpha = 1.0$, respectively. From this expression we can summarize the effects of inhibitors on individual kinetic parameters. For all reversible inhibitors, apparent $V_{max} = V_{max}/\alpha'$, because the right side of Equation 6–30 always simplifies to $V_{max}/\alpha'$ at sufficiently high substrate concentrations. For competitive inhibitors, $\alpha' = 1.0$ and can thus be ignored. Taking this expression for apparent $V_{max}$, we can also derive a general expression for apparent $K_m$ to show how this parameter changes in the presence of reversible inhibitors. Apparent $K_m$, as always, equals the [S] at which $V_0$ is one-half apparent $V_{max}$ or, more generally, when $V_0 = V_{max}/2\alpha'$. This condition is

**TABLE 6–9 Effects of Reversible Inhibitors on Apparent $V_{max}$ and Apparent $K_m$**

| Inhibitor type | Apparent $V_{max}$ | Apparent $K_m$ |
|---|---|---|
| None | $V_{max}$ | $K_m$ |
| Competitive | $V_{max}$ | $\alpha K_m$ |
| Uncompetitive | $V_{max}/\alpha'$ | $K_m/\alpha'$ |
| Mixed | $V_{max}/\alpha'$ | $\alpha K_m/\alpha'$ |

met when $[S] = \alpha K_m/\alpha'$. Thus, apparent $K_m = \alpha K_m/\alpha'$. This expression is simpler when either $\alpha$ or $\alpha'$ is 1.0 (for uncompetitive or competitive inhibitors), as summarized in Table 6–9.

In practice, uncompetitive and mixed inhibition are observed only for enzymes with two or more substrates—say, $S_1$ and $S_2$—and are very important in the experimental analysis of such enzymes. If an inhibitor binds to the site normally occupied by $S_1$, it may act as a competitive inhibitor in experiments in which $[S_1]$ is varied. If an inhibitor binds to the site normally occupied by $S_2$, it may act as a mixed or uncompetitive inhibitor of $S_1$. The actual inhibition patterns observed depend on whether the $S_1$- and $S_2$-binding events are ordered or random, and thus the order in which substrates bind and products leave the active site can be determined. Use of one of the reaction products as an inhibitor is often particularly informative. If only one of two reaction products is present, no reverse reaction can take place. However, a product generally binds to some part of the active site, thus serving as an inhibitor. Enzymologists can use elaborate kinetic studies involving different combinations and amounts of products and inhibitors to develop a detailed picture of the mechanism of a bisubstrate reaction.

**Irreversible Inhibition** The **irreversible inhibitors** are those that bind covalently with or destroy a functional group on an enzyme that is essential for the enzyme's activity, or those that form a particularly stable noncovalent association. Formation of a covalent link between an irreversible inhibitor and an enzyme is common. Irreversible inhibitors are another useful tool for studying reaction mechanisms. Amino acids with key catalytic functions in the active site can sometimes be identified by determining which residue is covalently linked to an inhibitor after the enzyme is inactivated. An example is shown in Figure 6–16.

A special class of irreversible inhibitors is the **suicide inactivators.** These compounds are relatively unreactive until they bind to the active site of a specific enzyme. A suicide inactivator undergoes the first few chemical steps of the normal enzymatic reaction, but instead of being transformed into the normal product, the

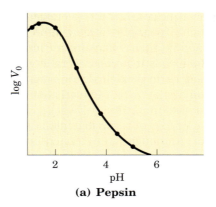

**(a) Pepsin**

**FIGURE 6–16** **Irreversible inhibition.** Reaction of chymotrypsin with diisopropylfluorophosphate (DIFP) irreversibly inhibits the enzyme. This has led to the conclusion that Ser[195] is the key active-site Ser residue in chymotrypsin.

inactivator is converted to a very reactive compound that combines irreversibly with the enzyme. These compounds are also called **mechanism-based inactivators,** because they hijack the normal enzyme reaction mechanism to inactivate the enzyme. Suicide inactivators play a significant role in *rational drug design,* a modern approach to obtaining new pharmaceutical agents in which chemists synthesize novel substrates based on knowledge of substrates and reaction mechanisms. A well-designed suicide inactivator is specific for a single enzyme and is unreactive until within that enzyme's active site, so drugs based on this approach can offer the important advantage of few side effects (see Box 22–2).

### Enzyme Activity Depends on pH

Enzymes have an optimum pH (or pH range) at which their activity is maximal (Fig. 6–17); at higher or lower pH, activity decreases. This is not surprising. Amino acid side chains in the active site may act as weak acids and bases with critical functions that depend on their maintaining a certain state of ionization, and elsewhere in the protein ionized side chains may play an essential role in the interactions that maintain protein structure. Removing a proton from a His residue, for example, might eliminate an ionic interaction essential for stabilizing the active conformation of the enzyme. A less common cause of pH sensitivity is titration of a group on the substrate.

The pH range over which an enzyme undergoes changes in activity can provide a clue to the type of amino acid residue involved (see Table 3–1). A change in activity near pH 7.0, for example, often reflects titration of a His residue. The effects of pH must be interpreted with some caution, however. In the closely packed environment of a protein, the $pK_a$ of amino acid

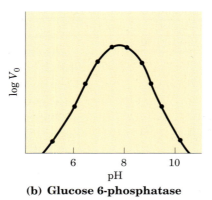

**(b) Glucose 6-phosphatase**

**FIGURE 6–17** **The pH-activity profiles of two enzymes.** These curves are constructed from measurements of initial velocities when the reaction is carried out in buffers of different pH. Because pH is a logarithmic scale reflecting tenfold changes in $[H^+]$, the changes in $V_0$ are also plotted on a logarithmic scale. The pH optimum for the activity of an enzyme is generally close to the pH of the environment in which the enzyme is normally found. **(a)** Pepsin, which hydrolyzes certain peptide bonds of proteins during digestion in the stomach, has a pH optimum of about 1.6. The pH of gastric juice is between 1 and 2. **(b)** Glucose 6-phosphatase of hepatocytes (liver cells), with a pH optimum of about 7.8, is responsible for releasing glucose into the blood. The normal pH of the cytosol of hepatocytes is about 7.2.

side chains can be significantly altered. For example, a nearby positive charge can lower the $pK_a$ of a Lys residue, and a nearby negative charge can increase it. Such effects sometimes result in a $pK_a$ that is shifted by several pH units from its value in the free amino acid. In the enzyme acetoacetate decarboxylase, for example, one Lys residue has a $pK_a$ of 6.6 (compared with 10.5 in free lysine) due to electrostatic effects of nearby positive charges.

### SUMMARY 6.3 Enzyme Kinetics As an Approach to Understanding Mechanism

■ Most enzymes have certain kinetic properties in common. When substrate is added to an enzyme, the reaction rapidly achieves a steady state in which the rate at which the ES complex forms balances the rate at which it reacts. As [S] increases, the steady-state activity of a fixed concentration of enzyme increases in a hyperbolic fashion to approach a characteristic maximum rate, $V_{max}$, at which essentially all the enzyme has formed a complex with substrate.

■ The substrate concentration that results in a reaction rate equal to one-half $V_{max}$ is the Michaelis constant $K_m$, which is characteristic for each enzyme acting on a given substrate. The Michaelis-Menten equation

$$V_0 = \frac{V_{max}[S]}{K_m + [S]}$$

relates initial velocity to [S] and $V_{max}$ through the constant $K_m$. Michaelis-Menten kinetics is also called steady-state kinetics.

■ $K_m$ and $V_{max}$ have different meanings for different enzymes. The limiting rate of an enzyme-catalyzed reaction at saturation is described by the constant $k_{cat}$, the turnover number. The ratio $k_{cat}/K_m$ provides a good measure of catalytic efficiency. The Michaelis-Menten equation is also applicable to bisubstrate reactions, which occur by ternary-complex or Ping-Pong (double-displacement) pathways.

■ Reversible inhibition of an enzyme is competitive, uncompetitive, or mixed. Competitive inhibitors compete with substrate by binding reversibly to the active site, but they are not transformed by the enzyme. Uncompetitive inhibitors bind only to the ES complex, at a site distinct from the active site. Mixed inhibitors bind to either E or ES, again at a site distinct from the active site. In irreversible inhibition an inhibitor binds permanently to an active site by forming a covalent bond or a very stable noncovalent interaction.

■ Every enzyme has an optimum pH (or pH range) at which it has maximal activity.

## 6.4 Examples of Enzymatic Reactions

Thus far we have focused on the general principles of catalysis and on introducing some of the kinetic parameters used to describe enzyme action. We now turn to several examples of specific enzyme reaction mechanisms.

An understanding of the complete mechanism of action of a purified enzyme requires identification of all substrates, cofactors, products, and regulators. Moreover, it requires a knowledge of (1) the temporal sequence in which enzyme-bound reaction intermediates

form, (2) the structure of each intermediate and each transition state, (3) the rates of interconversion between intermediates, (4) the structural relationship of the enzyme to each intermediate, and (5) the energy contributed by all reacting and interacting groups to intermediate complexes and transition states. As yet, there is probably no enzyme for which we have an understanding that meets all these requirements. Many decades of research, however, have produced mechanistic information about hundreds of enzymes, and in some cases this information is highly detailed.

We present here the mechanisms for four enzymes: chymotrypsin, hexokinase, enolase, and lysozyme. These examples are not intended to cover all possible classes of enzyme chemistry. They are chosen in part because they are among the best understood enzymes, and in part because they clearly illustrate some general principles outlined in this chapter. The discussion concentrates on selected principles, along with some key experiments that have helped to bring these principles into focus. We use the chymotrypsin example to review some of the conventions used to depict enzyme mechanisms. Much mechanistic detail and experimental evidence is necessarily omitted; no one book could completely document the rich experimental history of these enzymes. Also absent from these discussions is the special contribution of coenzymes to the catalytic activity of many enzymes. The function of coenzymes is chemically varied, and we describe each as it is encountered in Part II.

### The Chymotrypsin Mechanism Involves Acylation and Deacylation of a Ser Residue

Bovine pancreatic chymotrypsin ($M_r$ 25,191) is a protease, an enzyme that catalyzes the hydrolytic cleavage of peptide bonds. This protease is specific for peptide bonds adjacent to aromatic amino acid residues (Trp, Phe, Tyr). The three-dimensional structure of chymotrypsin is shown in Figure 6–18, with functional groups in the active site emphasized. The reaction catalyzed by this enzyme illustrates the principle of transition-state stabilization and also provides a classic example of general acid-base catalysis and covalent catalysis.

Chymotrypsin enhances the rate of peptide bond hydrolysis by a factor of at least $10^9$. It does not catalyze a direct attack of water on the peptide bond; instead, a transient covalent acyl-enzyme intermediate is formed. The reaction thus has two distinct phases. In the acylation phase, the peptide bond is cleaved and an ester linkage is formed between the peptide carbonyl carbon and the enzyme. In the deacylation phase, the ester linkage is hydrolyzed and the nonacylated enzyme regenerated.

The first evidence for a covalent acyl-enzyme intermediate came from a classic application of pre–steady state kinetics. In addition to its action on polypeptides,

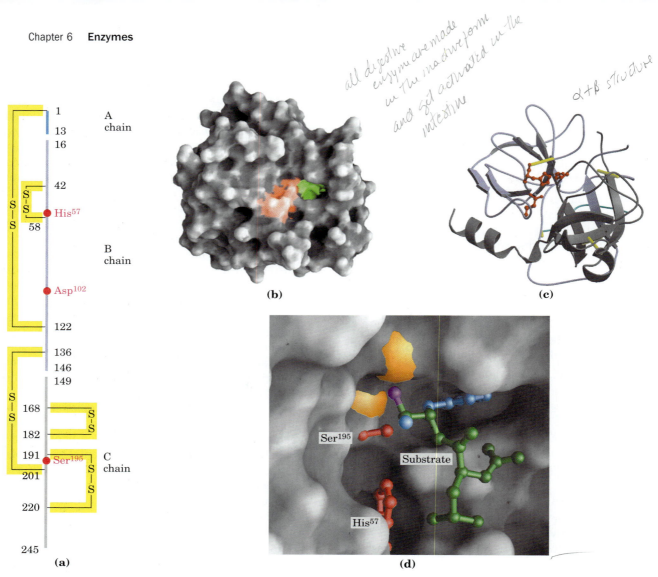

*(handwritten notes in margin:* all digestive enzymes are made in the inactive form and get activated in the intestine ... α+β structure*)*

**FIGURE 6–18  Structure of chymotrypsin.** (PDB ID 7GCH) **(a)** A representation of primary structure, showing disulfide bonds and the amino acid residues crucial to catalysis. The protein consists of three polypeptide chains linked by disulfide bonds. (The numbering of residues in chymotrypsin, with "missing" residues 14, 15, 147, and 148, is explained in Fig. 6–33.) The active-site amino acid residues are grouped together in the three-dimensional structure. **(b)** A depiction of the enzyme emphasizing its surface. The pocket in which the aromatic amino acid side chain of the substrate is bound is shown in green. Key active-site residues, including Ser[195], His[57], and Asp[102], are red. The roles of these residues in catalysis are illustrated in Figure 6–21. **(c)** The polypeptide backbone as a ribbon structure. Disulfide bonds are yellow; the three chains are colored as in part **(a)**. **(d)** A close-up of the active site with a substrate (mostly green) bound. Two of the active-site residues, Ser[195] and His[57] (both red), are partly visible. The hydroxyl of Ser[195] attacks the carbonyl group of the substrate (the oxygen is purple); the developing negative charge on the oxygen is stabilized by the oxyanion hole (amide nitrogens, including one from Ser[195] in orange), as explained in Figure 6–21. In the substrate, the aromatic amino acid side chain and the amide nitrogen of the peptide bond to be cleaved (protruding toward the viewer and projecting the path of the rest of the substrate polypeptide chain) are in blue.

chymotrypsin also catalyzes the hydrolysis of small esters and amides. These reactions are much slower than hydrolysis of peptides because less binding energy is available with smaller substrates, and they are therefore easier to study. Investigations by B. S. Hartley and B. A. Kilby in 1954 found that chymotrypsin hydrolysis of the ester *p*-nitrophenylacetate, as measured by release of *p*-nitrophenol, proceeded with a rapid burst before leveling off to a slower rate (Fig. 6–19). By extrapolating back to zero time, they concluded that the burst phase corresponded to just under one molecule of *p*-nitrophenol released for every enzyme molecule present. Hartley and Kilby suggested that this reflected a rapid acylation of all the enzyme molecules (with release of *p*-nitrophenol), with the rate for subsequent turnover of the enzyme limited by a slow deacylation step. Similar results have since been obtained with many other enzymes. The observation of a burst phase provides yet another example of the use of kinetics to break down a reaction into its constituent steps.

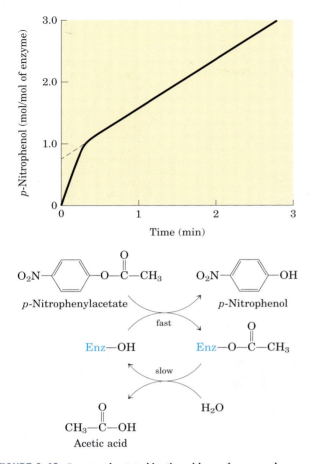

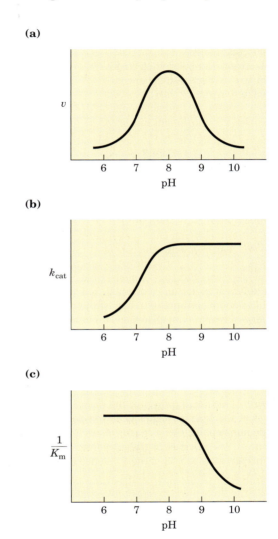

FIGURE 6–19 Pre–steady state kinetic evidence for an acyl-enzyme intermediate. The hydrolysis of $p$-nitrophenylacetate by chymotrypsin is measured by release of $p$-nitrophenol (a colored product). Initially, the reaction releases a rapid burst of $p$-nitrophenol nearly stoichiometric with the amount of enzyme present. This reflects the fast acylation phase of the reaction. The subsequent rate is slower, because enzyme turnover is limited by the rate of the slower deacylation phase.

$1/K_m$ term reflect the ionization of the $\alpha$-amino group of Ile[16] (at the amino-terminal end of one of chymotrypsin's three polypeptide chains). This group forms a salt bridge to Asp[194], stabilizing the active conformation of the enzyme. When this group loses its proton at high pH, the salt bridge is eliminated and a conformational change closes the hydrophobic pocket where the

Additional features of the chymotrypsin mechanism have been elucidated by analyzing the dependence of the reaction on pH. The rate of chymotrypsin-catalyzed cleavage generally exhibits a bell-shaped pH-rate profile (Fig. 6–20). The rates plotted in Figure 6–20a are obtained at low (subsaturating) substrate concentrations and therefore represent $k_{cat}/K_m$. The plot can be dissected further by first obtaining the maximum rates at each pH, and then plotting $k_{cat}$ alone versus pH (Fig. 6–20b); after obtaining the $K_m$ at each pH, researchers can then plot $1/K_m$ (Fig. 6–20c). Kinetic and structural analyses have revealed that the change in $k_{cat}$ reflects the ionization state of His[57]. The decline in $k_{cat}$ at low pH results from protonation of His[57] (so that it cannot extract a proton from Ser[195] in step ① of the reaction; see Fig. 6–21). This rate reduction illustrates the importance of general acid and general base catalysis in the mechanism for chymotrypsin. The changes in the

FIGURE 6–20 The pH dependence of chymotrypsin-catalyzed reactions. (a) The rates of chymotrypsin-mediated cleavage produce a bell-shaped pH-rate profile with an optimum at pH 8.0. The rate ($v$) being plotted is that at low substrate concentrations and thus reflects the term $k_{cat}/K_m$. The plot can be broken down to its components by using kinetic methods to determine the terms $k_{cat}$ and $K_m$ separately at each pH. When this is done (b and c), it becomes clear that the transition just above pH 7 is due to changes in $k_{cat}$, whereas the transition above pH 8.5 is due to changes in $1/K_m$. Kinetic and structural studies have shown that the transitions illustrated in (b) and (c) reflect the ionization states of the His[57] side chain (when substrate is not bound) and the $\alpha$-amino group of Ile[16] (at the amino terminus of the B chain), respectively. For optimal activity, His[57] must be unprotonated and Ile[16] must be protonated.

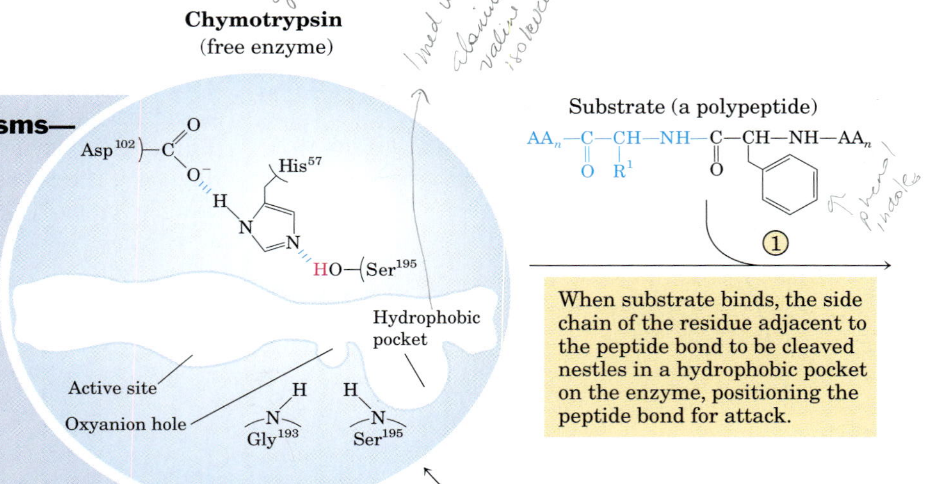

*(handwritten annotations at top: "cuts C-terminal of Phe-Trp-Tyr", "lined with alanine valine isoleucine) does not attract water")*

## How to Read Reaction Mechanisms— A Refresher

Chemical reaction mechanisms, which trace the formation and breakage of covalent bonds, are communicated with dots and curved arrows, a convention known informally as "electron pushing." A covalent bond consists of a shared pair of electrons. Nonbonded electrons important to the reaction mechanism are designated by dots (—Ö̈H). Curved arrows (⤻) represent the movement of electron pairs. For movement of a single electron (as in a free radical reaction), a single-headed (fishhook-type) arrow is used (⇀). Most reaction steps involve an unshared electron pair (as in the chymotrypsin mechanism).

Some atoms are more electronegative than others; that is, they more strongly attract electrons. The relative electronegativities of atoms encountered in this text are F > O > N > C ≈ S > P ≈ H. For example, the two electron pairs making up a C=O (carbonyl) bond are not shared equally; the carbon is relatively electron-deficient as the oxygen draws away the electrons. Many reactions involve an electron-rich atom (a nucleophile) reacting with an electron-deficient atom (an electrophile). Some common nucleophiles and electrophiles in biochemistry are shown at right. In general, a reaction mechanism is initiated at an unshared electron pair of a nucleophile. In mechanism diagrams, the base of the electron-pushing arrow originates near the electron-pair dots, and the head of the arrow points directly at the electro-philic center being attacked. Where the unshared electron pair confers a formal negative charge on the nucleophile, the negative charge symbol itself can represent the unshared electron pair, and serves as the base of the arrow. In the chymotrypsin mech-anism, the nucleophilic electron pair in the ES complex between steps ① and ② is provided by the oxygen of the Ser[195] hydroxyl group. This electron pair (2 of the 8 valence electrons of the hydroxyl oxygen) provides the base of the curved arrow. The electrophilic center under attack is the carbonyl carbon of the peptide bond to be cleaved. The C, O, and N atoms have a max-imum of 8 valence electrons, and H has a maximum of 2. These atoms are occasionally found in unstable states with less than their maximum allotment of electrons, but C, O, and N cannot have more than 8. Thus, when the electron pair from chymo-trypsin's Ser[195] attacks the substrate's carbonyl carbon, an electron pair is displaced from the carbon valence shell (you cannot have 5 bonds to carbon!). These electrons move toward the more electronegative carbonyl oxygen. The oxygen has 8 valence electrons both before and after this chemical process, but the number shared with the carbon is reduced from 4 to 2, and the carbonyl oxygen acquires a negative charge. In the next step, the electron pair conferring the negative charge on the oxygen moves back to re-form a bond with carbon and reestablish the carbonyl linkage. Again, an electron pair must be displaced from the carbon, and this time it is the electron pair shared with the amino group of the peptide linkage. This breaks the peptide bond. The remaining steps follow a similar pattern.

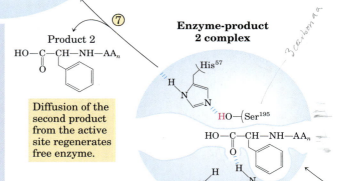

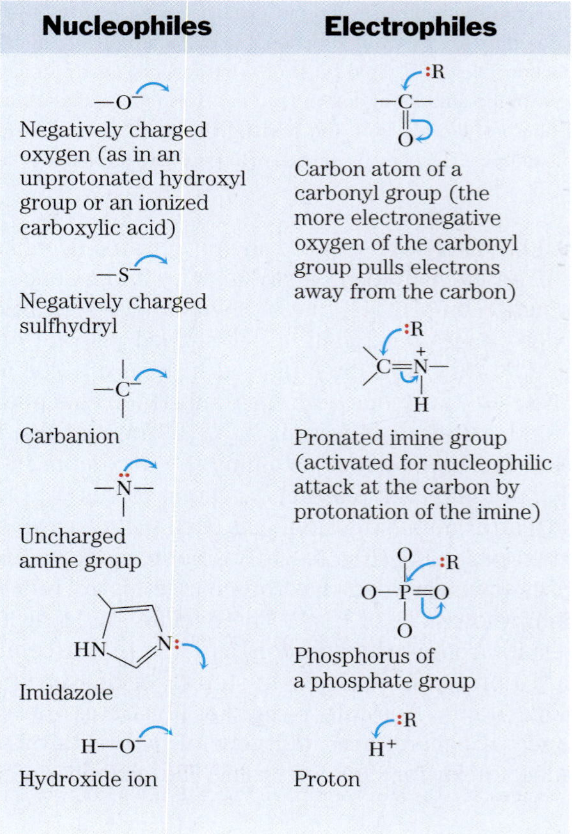

**Chymotrypsin**
(free enzyme)

Asp[102]

His[57]

HO—(Ser[195]

Active site

Oxyanion hole

Gly[193]   Ser[195]

Hydrophobic pocket

**Substrate (a polypeptide)**

$AA_n$—C—CH—NH—C—CH—NH—$AA_n$
       O   $R^1$

①

> When substrate binds, the side chain of the residue adjacent to the peptide bond to be cleaved nestles in a hydrophobic pocket on the enzyme, positioning the peptide bond for attack.

⑦

**Product 2**

HO—C—CH—NH—$AA_n$
     O

> Diffusion of the second product from the active site regenerates free enzyme.

**Enzyme–product 2 complex**

His[57]

HO—(Ser[195]

HO—C—CH—NH—$AA_n$
     O

Gly[193]   Ser[195]

| **Nucleophiles** | **Electrophiles** |
|---|---|
| —O⁻  Negatively charged oxygen (as in an unprotonated hydroxyl group or an ionized carboxylic acid) | :R / —C— / O  Carbon atom of a carbonyl group (the more electronegative oxygen of the carbonyl group pulls electrons away from the carbon) |
| —S⁻  Negatively charged sulfhydryl | :R / C=N— / H  Pronated imine group (activated for nucleophilic attack at the carbon by protonation of the imine) |
| —C⁻  Carbanion | :R / O—P=O / O  Phosphorus of a phosphate group |
| —N̈—  Uncharged amine group | |
| HN⟍N:  Imidazole | |
| H—O⁻  Hydroxide ion | :R / H⁺  Proton |

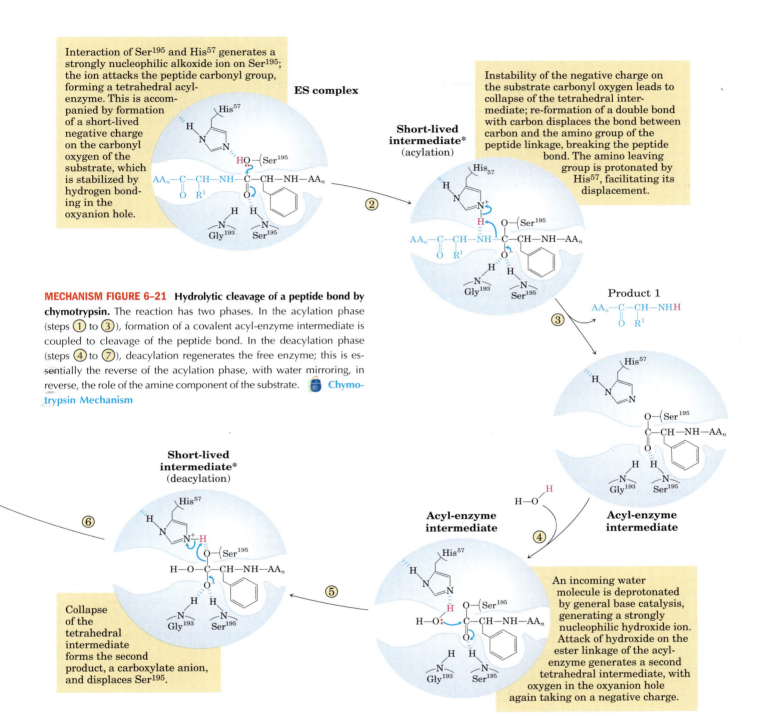

Interaction of Ser¹⁹⁵ and His⁵⁷ generates a strongly nucleophilic alkoxide ion on Ser¹⁹⁵; the ion attacks the peptide carbonyl group, forming a tetrahedral acyl-enzyme. This is accompanied by formation of a short-lived negative charge on the carbonyl oxygen of the substrate, which is stabilized by hydrogen bonding in the oxyanion hole.

**ES complex**

**Short-lived intermediate\*** (acylation)

Instability of the negative charge on the substrate carbonyl oxygen leads to collapse of the tetrahedral intermediate; re-formation of a double bond with carbon displaces the bond between carbon and the amino group of the peptide linkage, breaking the peptide bond. The amino leaving group is protonated by His⁵⁷, facilitating its displacement.

**MECHANISM FIGURE 6–21** **Hydrolytic cleavage of a peptide bond by chymotrypsin.** The reaction has two phases. In the acylation phase (steps ① to ③), formation of a covalent acyl-enzyme intermediate is coupled to cleavage of the peptide bond. In the deacylation phase (steps ④ to ⑦), deacylation regenerates the free enzyme; this is essentially the reverse of the acylation phase, with water mirroring, in reverse, the role of the amine component of the substrate. 🖱 **Chymotrypsin Mechanism**

**Product 1**

**Short-lived intermediate\*** (deacylation)

**Acyl-enzyme intermediate**

**Acyl-enzyme intermediate**

Collapse of the tetrahedral intermediate forms the second product, a carboxylate anion, and displaces Ser¹⁹⁵.

An incoming water molecule is deprotonated by general base catalysis, generating a strongly nucleophilic hydroxide ion. Attack of hydroxide on the ester linkage of the acyl-enzyme generates a second tetrahedral intermediate, with oxygen in the oxyanion hole again taking on a negative charge.

---

\*The tetrahedral intermediate in the chymotrypsin reaction pathway, and the second tetrahedral intermediate that forms later, are sometimes referred to as transition states, which can lead to confusion. An *intermediate* is any chemical species with a finite lifetime, "finite" being defined as longer than the time required for a molecular vibration (~$10^{-13}$ seconds). A *transition state* is simply the maximum-energy species formed on the reaction coordinate and does not have a finite lifetime. The tetrahedral intermediates formed in the chymotrypsin reaction closely resemble, both energetically and structurally, the transition states leading to their formation and breakdown. However, the intermediate represents a committed stage of completed bond formation, whereas the transition state is part of the process of reaction. In the case of chymotrypsin, given the close relationship between the intermediate and the actual transition state the distinction between them is routinely glossed over. Furthermore, the interaction of the negatively charged oxygen with the amide nitrogens in the oxyanion hole, often referred to as transition-state stabilization, also serves to stabilize the intermediate in this case. Not all intermediates are so short-lived that they resemble transition states. The chymotrypsin acyl-enzyme intermediate is much more stable and more readily detected and studied, and it is never confused with a transition state.

aromatic amino acid side chain of the substrate inserts (Fig. 6–18). Substrates can no longer bind properly, which is measured kinetically as an increase in $K_m$.

The nucleophile in the acylation phase is the oxygen of Ser$^{195}$. (Proteases with a Ser residue that plays this role in reaction mechanisms are called serine proteases.) The $pK_a$ of a Ser hydroxyl group is generally too high for the unprotonated form to be present in significant concentrations at physiological pH. However, in chymotrypsin, Ser$^{195}$ is linked to His$^{57}$ and Asp$^{102}$ in a hydrogen-bonding network referred to as the **catalytic triad.** When a peptide substrate binds to chymotrypsin, a subtle change in conformation compresses the hydrogen bond between His$^{57}$ and Asp$^{102}$, resulting in a stronger interaction, called a low-barrier hydrogen bond. This enhanced interaction increases the $pK_a$ of His$^{57}$ from ~7 (for free histidine) to >12, allowing the His residue to act as an enhanced general base that can remove the proton from the Ser$^{195}$ hydroxyl group. Deprotonation prevents development of a very unstable positive charge on the Ser$^{195}$ hydroxyl and makes the Ser side chain a stronger nucleophile. At later reaction stages, His$^{57}$ also acts as a proton donor, protonating the amino group in the displaced portion of the substrate (the leaving group).

As the Ser$^{195}$ oxygen attacks the carbonyl group of the substrate, a very short-lived tetrahedral intermediate is formed in which the carbonyl oxygen acquires a negative charge (Fig 6-21). This charge, forming within a pocket on the enzyme called the oxyanion hole, is stabilized by hydrogen bonds contributed by the amide groups of two peptide bonds in the chymotrypsin backbone. One of these hydrogen bonds (contributed by Gly$^{193}$) is present only in this intermediate and in the transition states for its formation and breakdown; it reduces the energy required to reach these states. This is an example of the use of binding energy in catalysis.

The role of transition state complementarity in enzyme catalysis is further explored in Box 6-3.

## Hexokinase Undergoes Induced Fit on Substrate Binding

Yeast hexokinase ($M_r$ 107,862) is a bisubstrate enzyme that catalyzes the reversible reaction

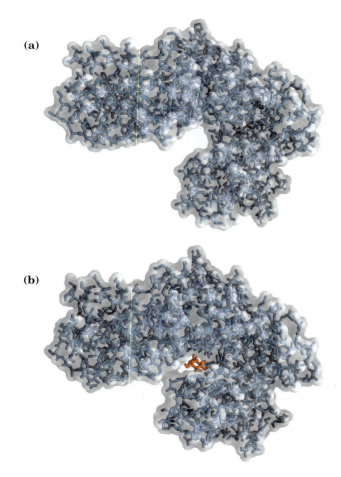

**(a)**

**(b)**

**FIGURE 6–22 Induced fit in hexokinase. (a)** Hexokinase has a U-shaped structure (PDB ID 2YHX). **(b)** The ends pinch toward each other in a conformational change induced by binding of D-glucose (red) (derived from PDB ID 1HKG and PDB ID 1GLK).

The hydroxyl at C-6 of glucose (to which the γ-phosphoryl of ATP is transferred in the hexokinase reaction) is similar in chemical reactivity to water, and water freely enters the enzyme active site. Yet hexokinase favors the reaction with glucose by a factor of $10^6$. The enzyme can discriminate between glucose and water because of a conformational change in the enzyme when the correct substrates binds (Fig. 6–22). Hexokinase thus provides a good example of induced fit. When glucose is not present, the enzyme is in an inactive conformation with the active-site amino acid side chains out of position for reaction. When glucose (but not water) and Mg · ATP bind, the binding energy derived from this interaction induces a conformational change in hexokinase to the catalytically active form.

This model has been reinforced by kinetic studies. The five-carbon sugar xylose, stereochemically similar to glucose but one carbon shorter, binds to hexokinase but in a position where it cannot be phosphorylated. Nevertheless, addition of xylose to the reaction mixture increases the rate of ATP hydrolysis. Evidently, the binding of xylose is sufficient to induce a change in

β-D-Glucose + Mg · ATP → Glucose 6-phosphate + Mg · ADP (hexokinase)

ATP and ADP always bind to enzymes as a complex with the metal ion $Mg^{2+}$.

hexokinase to its active conformation, and the enzyme is thereby "tricked" into phosphorylating water. The hexokinase reaction also illustrates that enzyme specificity is not always a simple matter of binding one compound but not another. In the case of hexokinase, specificity is observed not in the formation of the ES complex but in the relative rates of subsequent catalytic steps. Water is not excluded from the active site, but reaction rates increase greatly in the presence of the functional phosphoryl group acceptor (glucose).

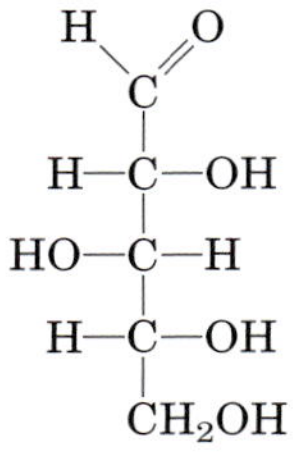

Xylose

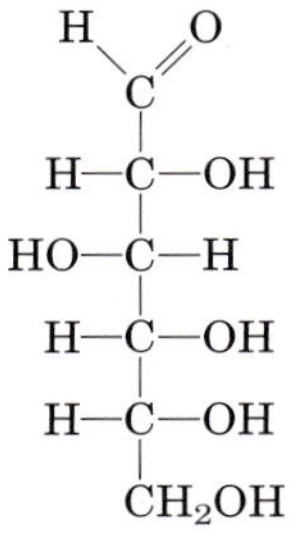

Glucose

Induced fit is only one aspect of the catalytic mechanism of hexokinase—like chymotrypsin, hexokinase uses several catalytic strategies. For example, the active-site amino acid residues (those brought into position by the conformational change that follows substrate binding) participate in general acid-base catalysis and transition-state stabilization.

## The Enolase Reaction Mechanism Requires Metal Ions

Another glycolytic enzyme, enolase, catalyzes the reversible dehydration of 2-phosphoglycerate to phosphoenolpyruvate:

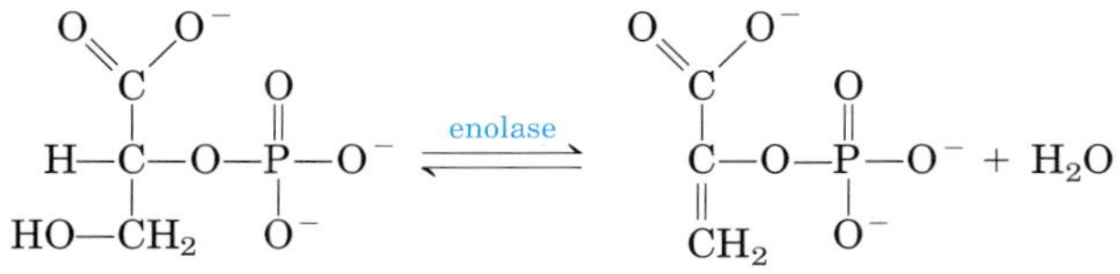

2-Phosphoglycerate            Phosphoenolpyruvate

Yeast enolase ($M_r$ 93,316) is a dimer with 436 amino acid residues per subunit. The enolase reaction illustrates one type of metal ion catalysis and provides an additional example of general acid-base catalysis and transition-state stabilization. The reaction occurs in two steps (Fig. 6–23a). First, Lys$^{345}$ acts as a general base catalyst,

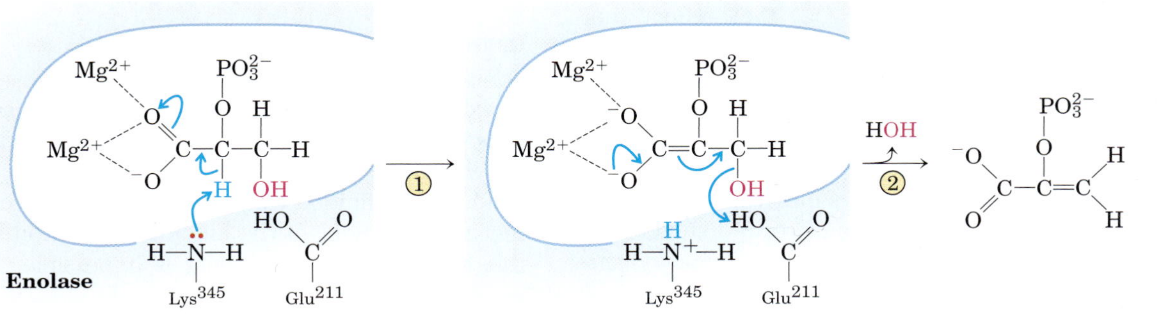

**(a)**    2-Phosphoglycerate bound to enzyme    Enolic intermediate    Phosphoenolpyruvate

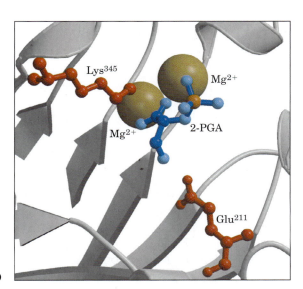

**(b)**

**MECHANISM FIGURE 6–23** **Two-step reaction catalyzed by enolase. (a)** The mechanism by which enolase converts 2-phosphoglycerate (2-PGA) to phosphoenolpyruvate. The carboxyl group of 2-PGA is coordinated by two magnesium ions at the active site. A proton is abstracted in step ① by general base catalysis (Lys$^{345}$), and the resulting enolic intermediate is stabilized by the two $Mg^{2+}$ ions. Elimination of the —OH in step ② is facilitated by general acid catalysis (Glu$^{211}$). **(b)** The substrate, 2-PGA, in relation to the $Mg^{2+}$ ions, Lys$^{345}$, and Glu$^{211}$ in the enolase active site. Hydrogen atoms are not shown. All the oxygen atoms of 2-PGA are light blue; phosphorus is orange (PDB ID 1ONE).

**BOX 6-3** **WORKING IN BIOCHEMISTRY**

## Evidence for Enzyme–Transition State Complementarity

The transition state of a reaction is difficult to study because it is so short-lived. To understand enzymatic catalysis, however, we must dissect the interaction between the enzyme and this ephemeral moment in the course of a reaction. Complementarity between an enzyme and the transition state is virtually a requirement for catalysis, because the energy hill upon which the transition state sits is what the enzyme must lower if catalysis is to occur. How can we obtain evidence for enzyme–transition state complementarity? Fortunately, we have a variety of approaches, old and new, to address this problem, each providing compelling evidence in support of this general principle of enzyme action.

### Structure-Activity Correlations

If enzymes are complementary to reaction transition states, then some functional groups in both the substrate and the enzyme must interact preferentially in the transition state rather than in the ES complex. Changing these groups should have little effect on formation of the ES complex and hence should not affect kinetic parameters (the dissociation constant, $K_d$; or sometimes $K_m$, if $K_d = K_m$) that reflect the $E + S \rightleftharpoons ES$ equilibrium. Changing these same groups should have a large effect on the overall rate ($k_{cat}$ or $k_{cat}/K_m$) of the reaction, however, because the bound substrate lacks potential binding interactions needed to lower the activation energy.

An excellent example of this effect is seen in the kinetics associated with a series of related substrates for the enzyme chymotrypsin (Fig. 1). Chymotrypsin normally catalyzes the hydrolysis of peptide bonds next to aromatic amino acids. The substrates shown in Figure 1 are convenient smaller models for the natural substrates (long polypeptides and proteins). The additional chemical groups added in each substrate (A to B to C) are shaded. As the table shows, the interaction between the enzyme and these added functional groups has a minimal effect on $K_m$ (taken here as a reflection of $K_d$) but a large, positive effect on $k_{cat}$ and $k_{cat}/K_m$. This is what we would expect if the interaction contributed largely to stabilization of the transition state. The results also demonstrate that the rate of a reaction can be affected greatly by enzyme-substrate interactions that are physically remote from the covalent bonds that are altered in the enzyme-catalyzed reaction. Chymotrypsin is described in more detail in the text.

A complementary experimental approach is to modify the enzyme, eliminating certain enzyme-substrate interactions by replacing specific amino acid residues through site-directed mutagenesis (see Fig. 9–12). Results from such experiments again demonstrate the importance of binding energy in stabilizing the transition state.

### Transition-State Analogs

Even though transition states cannot be observed directly, chemists can often predict the approximate structure of a transition state based on accumulated knowledge about reaction mechanisms. The transition state is by definition transient and so unstable that direct measurement of the binding interaction between this species and the enzyme is impossible. In some

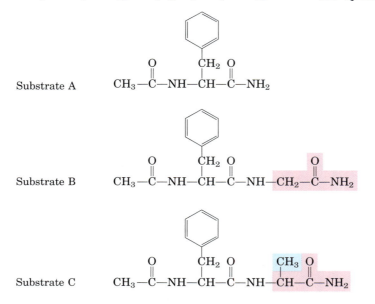

| | $k_{cat}$ (s$^{-1}$) | $K_m$ (mM) | $k_{cat}/K_m$ (M$^{-1}$s$^{-1}$) |
|---|---|---|---|
| Substrate A | 0.06 | 31 | 2 |
| Substrate B | 0.14 | 15 | 10 |
| Substrate C | 2.8 | 25 | 114 |

**FIGURE 1** Effects of small structural changes in the substrate on kinetic parameters for chymotrypsin-catalyzed amide hydrolysis.

cases, however, stable molecules can be designed that resemble transition states. These are called **transition-state analogs.** In principle, they should bind to an enzyme more tightly than does the substrate in the ES complex, because they should fit the active site better (that is, form a greater number of weak interactions) than the substrate itself. The idea of transition-state analogs was suggested by Pauling in the 1940s, and it has been explored using a number of enzymes. These experiments have the limitation that a transition-state analog cannot perfectly mimic a transition state. Some analogs, however, bind an enzyme $10^2$ to $10^6$ times more tightly than does the normal substrate, providing good evidence that enzyme active sites are indeed complementary to transition states. The same principle is now used in the pharmaceutical industry to design new drugs. The powerful anti-HIV drugs called protease inhibitors were designed in part as tight-binding transition-state analogs directed at the active site of HIV protease.

### Catalytic Antibodies

If a transition-state analog can be designed for the reaction S→P, then an antibody that binds tightly to this analog might be expected to catalyze S→P. Antibodies (immunoglobulins; see Fig. 5–23) are key components of the immune response. When a transition-state analog is used as a protein-bound epitope to stimulate the production of antibodies, the antibodies that bind it are potential catalysts of the corresponding reaction. This use of "catalytic antibodies," first suggested by William P. Jencks in 1969, has become practical with the development of laboratory techniques to produce quantities of identical antibodies that bind one specific antigen (monoclonal antibodies; see Chapter 5).

Pioneering work in the laboratories of Richard Lerner and Peter Schultz has resulted in the isolation of a number of monoclonal antibodies that catalyze the hydrolysis of esters or carbonates (Fig. 2). In these reactions, the attack by water (OH⁻) on the carbonyl carbon produces a tetrahedral transition state in which a partial negative charge has developed on the carbonyl oxygen. Phosphonate ester compounds mimic the structure and charge distribution of this transition state in ester hydrolysis, making them good transition-state analogs; phosphate ester compounds are used for carbonate hydrolysis reactions. Antibodies that bind the phosphonate or phosphate compound tightly have been found to accelerate the corresponding ester or carbonate hydrolysis reaction by factors of $10^3$ to $10^4$. Structural analyses of a few of these catalytic antibodies have shown that some catalytic amino acid side chains are arranged such that they could interact with the substrate in the transition state.

Catalytic antibodies generally do not approach the catalytic efficiency of enzymes, but medical and industrial uses for them are nevertheless emerging. For example, catalytic antibodies designed to degrade cocaine are being investigated as a potential aid in the treatment of cocaine addiction.

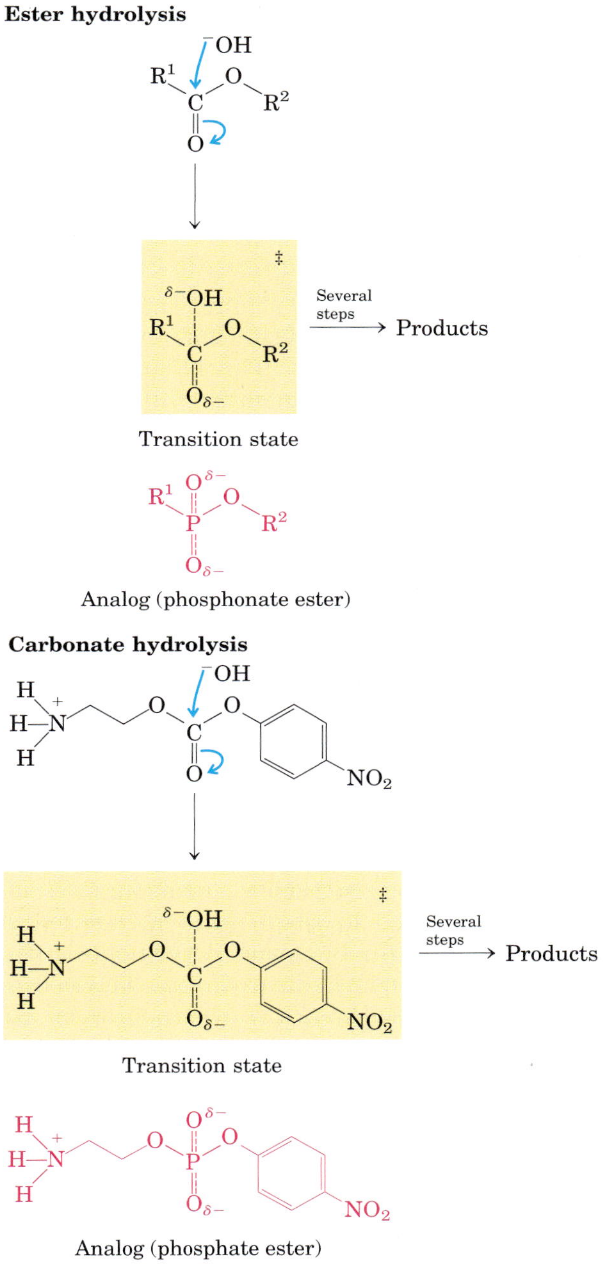

**Ester hydrolysis**

**Carbonate hydrolysis**

**FIGURE 2** The expected transition states for ester or carbonate hydrolysis reactions. Phosphonate ester and phosphate ester compounds, respectively, make good transition-state analogs for these reactions.

abstracting a proton from C-2 of 2-phosphoglycerate; then Glu$^{211}$ acts as a general acid catalyst, donating a proton to the —OH leaving group. The proton at C-2 of 2-phosphoglycerate is not very acidic and thus is not readily removed. However, in the enzyme active site, 2-phosphoglycerate undergoes strong ionic interactions with two bound $Mg^{2+}$ ions (Fig. 6–23b), making the C-2 proton more acidic (lowering the $pK_a$) and easier to abstract. Hydrogen bonding to other active-site amino acid residues also contributes to the overall mechanism. The various interactions effectively stabilize both the enolate intermediate and the transition state preceding its formation.

## Lysozyme Uses Two Successive Nucleophilic Displacement Reactions

Lysozyme is a natural antibacterial agent found in tears and egg whites. The hen egg white lysozyme ($M_r$ 14,296) is a monomer with 129 amino acid residues. This was the first enzyme to have its three-dimensional structure determined, by David Phillips and colleagues in 1965. The structure revealed four stabilizing disulfide bonds and a cleft containing the active site (Fig. 6–24a; see also Fig. 4–18). More than five decades of lysozyme investigations have provided a detailed picture of the structure and activity of the enzyme, and an interesting story of how biochemical science progresses.

The substrate of lysozyme is peptidoglycan, a carbohydrate found in many bacterial cell walls (see Fig. 7–22). Lysozyme cleaves the ($\beta1{\rightarrow}4$) glycosidic C—O bond between the two types of sugar residue in the molecule, $N$-acetylmuramic acid (Mur2Ac) and $N$-acetylglucosamine (GlcNAc), often referred to as NAM and NAG, respectively, in the research literature on enzymology (Fig. 6–24b). Six residues of the alternating Mur2Ac and GlcNAc in peptidoglycan bind in the active site, in binding sites labeled A through F. Model building has shown that the lactyl side chain of Mur2Ac cannot be accommodated in sites C and E, restricting Mur2Ac binding to sites B, D, and F. Only one of the bound glycosidic bonds is cleaved, that between a Mur2Ac residue in site D and a GlcNAc residue in site E. The key catalytic amino acid residues in the active site are Glu$^{35}$ and Asp$^{52}$ (Fig. 6–25a). The reaction is a nucleophilic substitution, with —OH from water replacing the GlcNAc at C-1 of Mur2Ac.

With the active site residues identified and a detailed structure of the enzyme available, the path to understanding the reaction mechanism seemed open in the 1960s. However, definitive evidence for a particular mechanism eluded investigators for nearly four decades. There are two chemically reasonable mechanisms that could generate the observed product of lysozyme-mediated cleavage of the glycosidic bond. Phillips and

colleagues proposed a dissociative ($S_N1$-type) mechanism (Fig. 6–25a, left), in which the GlcNAc initially dissociates in step ① to leave behind a glycosyl cation (a carbocation) intermediate. In this mechanism, the departing GlcNAc is protonated by general acid catalysis by Glu$^{35}$, located in a hydrophobic pocket that gives its carboxyl group an unusually high $pK_a$. The carbocation is stabilized by resonance involving the adjacent ring oxygen, as well as by electrostatic interaction with the negative charge on the nearby Asp$^{52}$. In step ②, water attacks at C-1 of Mur2Ac to yield the product. The alternative mechanism (Fig. 6–25a, right) involves two consecutive direct-displacement ($S_N2$-type) steps. In step ①, Asp$^{52}$ attacks C-1 of Mur2Ac to displace the GlcNAc. As in the first mechanism, Glu$^{35}$ acts as a general acid to protonate the departing GlcNAc. In step ②, water attacks at C-1 of Mur2Ac to displace the Asp$^{52}$ and generate product.

The Phillips mechanism ($S_N1$), based on structural considerations and bolstered by a variety of binding studies with artificial substrates, was widely accepted for more than three decades. However, some controversy persisted and tests continued. The scientific method sometimes advances an issue slowly, and a truly insightful experiment can be difficult to design. Some early arguments against the Phillips mechanism were suggestive but not completely persuasive. For example, the half-life of the proposed glycosyl cation was estimated to be $10^{-12}$ seconds, just longer than a molecular vibration and not long enough for the needed diffusion of other molecules. More important, lysozyme is a member of a family of enzymes called "retaining glycosidases," all of which catalyze reactions in which the product has the same anomeric configuration as the substrate (anomeric configurations of carbohydrates are examined in Chapter 7), and all of which are known to have reactive covalent intermediates like that envisioned in the alternative ($S_N2$) pathway. Hence, the Phillips mechanism ran counter to experimental findings for closely related enzymes.

A compelling experiment tipped the scales decidedly in favor of the $S_N2$ pathway, as reported by Stephen Withers and colleagues in 2001. Making use of a mutant enzyme (with residue 35 changed from Glu to Gln) and artificial substrates, which combined to slow the rate of key steps in the reaction, these workers were able to stabilize the elusive covalent intermediate. This in turn allowed them to observe the intermediate directly, using both mass spectrometry and x-ray crystallography (Fig. 6–25b).

Is the lysozyme mechanism now proven? No. A key feature of the scientific method, as Albert Einstein once summarized it, is "No amount of experimentation can ever prove me right; a single experiment can prove me wrong." In the case of the lysozyme mechanism,

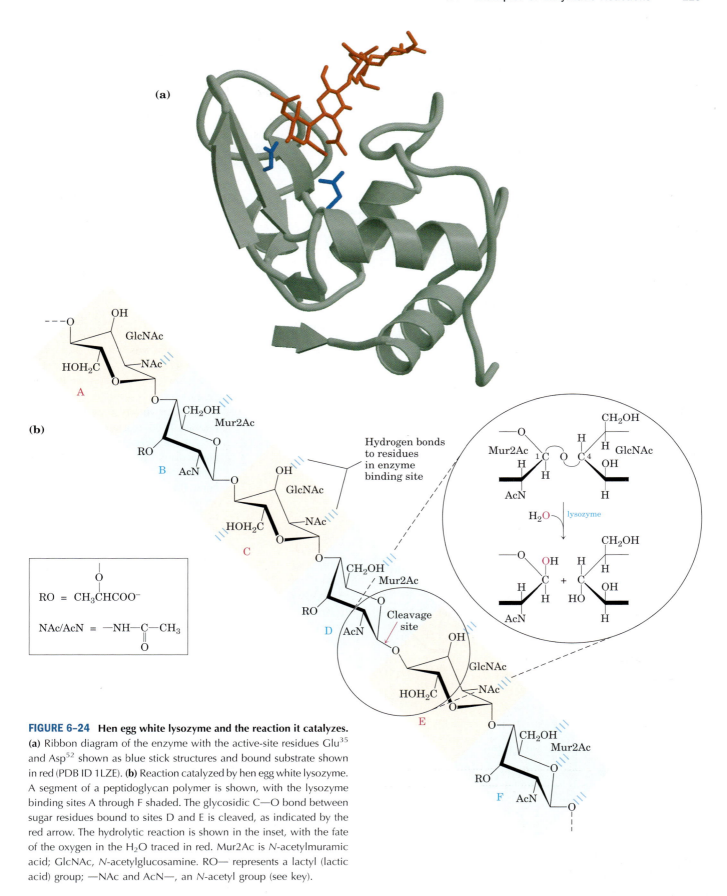

**FIGURE 6-24 Hen egg white lysozyme and the reaction it catalyzes.**
(a) Ribbon diagram of the enzyme with the active-site residues Glu[35] and Asp[52] shown as blue stick structures and bound substrate shown in red (PDB ID 1LZE). (b) Reaction catalyzed by hen egg white lysozyme. A segment of a peptidoglycan polymer is shown, with the lysozyme binding sites A through F shaded. The glycosidic C—O bond between sugar residues bound to sites D and E is cleaved, as indicated by the red arrow. The hydrolytic reaction is shown in the inset, with the fate of the oxygen in the $H_2O$ traced in red. Mur2Ac is *N*-acetylmuramic acid; GlcNAc, *N*-acetylglucosamine. RO— represents a lactyl (lactic acid) group; —NAc and AcN—, an *N*-acetyl group (see key).

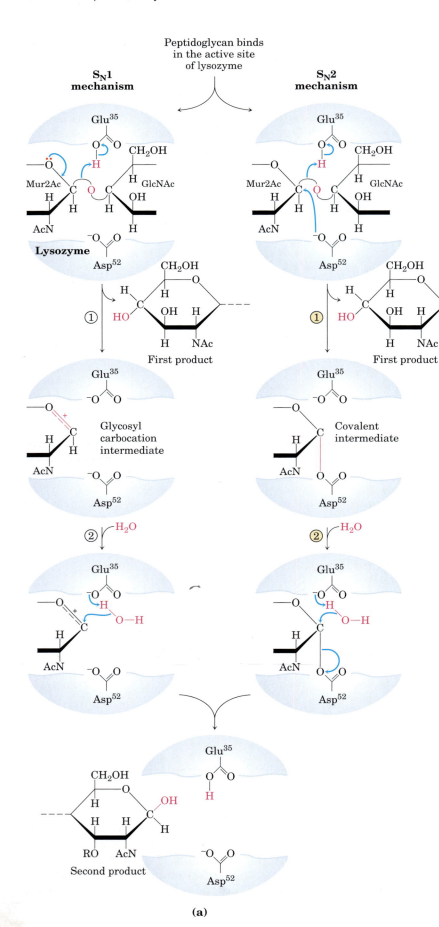

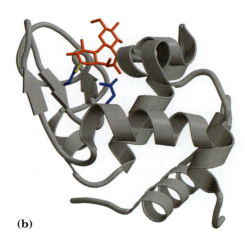

**MECHANISM FIGURE 6–25  Lysozyme reaction.** In this reaction (described on p. 222), the water introduced into the product at C-1 of Mur2Ac is in the same configuration as the original glycosidic bond. The reaction is thus a molecular substitution with retention of configuration. **(a)** Two proposed pathways potentially explain the overall reaction and its properties. The $S_N1$ pathway (left) is the original Phillips mechanism. The $S_N2$ pathway (right) is the mechanism most consistent with current data. **(b)** A ribbon diagram of the covalent enzyme-substrate intermediate with the active-site residues (blue) and bound substrate (red) shown as stick structures (PDB ID 1H6M).

**(a)**                                    **(b)**

one might argue (and some have) that the artificial substrates, with fluorine substitutions at C-1 and C-2, that were used to stabilize the covalent intermediate might have altered the reaction pathway. The highly electronegative fluorine could destabilize an already electron-deficient oxocarbenium ion in the glycosyl cation intermediate that might occur in an $S_N1$ pathway. However, the $S_N2$ pathway is now the mechanism most in concert with available data.

### SUMMARY 6.4 Examples of Enzymatic Reactions

- Chymotrypsin is a serine protease with a well-understood mechanism, featuring general acid-base catalysis, covalent catalysis, and transition-state stabilization.

- Hexokinase provides an excellent example of induced fit as a means of using substrate binding energy.

- The enolase reaction proceeds via metal ion catalysis.

- Lysozyme makes use of covalent catalysis and general acid catalysis as it promotes two successive nucleophilic displacement reactions.

## 6.5 Regulatory Enzymes

In cellular metabolism, groups of enzymes work together in sequential pathways to carry out a given metabolic process, such as the multireaction breakdown of glucose to lactate or the multireaction synthesis of an amino acid from simpler precursors. In such enzyme systems, the reaction product of one enzyme becomes the substrate of the next.

Most of the enzymes in each metabolic pathway follow the kinetic patterns we have already described. Each pathway, however, includes one or more enzymes that have a greater effect on the rate of the overall sequence. These **regulatory enzymes** exhibit increased or decreased catalytic activity in response to certain signals. Adjustments in the rate of reactions catalyzed by regulatory enzymes, and therefore in the rate of entire metabolic sequences, allow the cell to meet changing needs for energy and for biomolecules required in growth and repair.

In most multienzyme systems, the first enzyme of the sequence is a regulatory enzyme. This is an excellent place to regulate a pathway, because catalysis of even the first few reactions of a sequence that leads to an unneeded product diverts energy and metabolites from more important processes. Other enzymes in the sequence are usually present at levels that provide an excess of catalytic activity; they can generally promote their reactions as fast as their substrates are made available from preceding reactions.

The activities of regulatory enzymes are modulated in a variety of ways. **Allosteric enzymes** function through reversible, noncovalent binding of regulatory compounds called **allosteric modulators** or **allosteric effectors,** which are generally small metabolites or cofactors. Other enzymes are regulated by reversible **covalent modification.** Both classes of regulatory enzymes tend to be multisubunit proteins, and in some cases the regulatory site(s) and the active site are on separate subunits. Metabolic systems have at least two other mechanisms of enzyme regulation. Some enzymes are stimulated or inhibited when they are bound by separate regulatory proteins. Others are activated when peptide segments are removed by proteolytic cleavage; unlike effector-mediated regulation, regulation by proteolytic cleavage is irreversible. Important examples of both mechanisms are found in physiological processes such as digestion, blood clotting, hormone action, and vision.

Cell growth and survival depend on efficient use of resources, and this efficiency is made possible by regulatory enzymes. No single rule governs the occurrence of different types of regulation in different systems. To a degree, allosteric (noncovalent) regulation may permit fine-tuning of metabolic pathways that are required continuously but at different levels of activity as cellular conditions change. Regulation by covalent modification may be all or none—usually the case with proteolytic cleavage—or it may allow for subtle changes in activity. Several types of regulation may occur in a single regulatory enzyme. The remainder of this chapter is devoted to a discussion of these methods of enzyme regulation.

### Allosteric Enzymes Undergo Conformational Changes in Response to Modulator Binding

As we saw in Chapter 5, allosteric proteins are those having "other shapes" or conformations induced by the binding of modulators. The same concept applies to certain regulatory enzymes, as conformational changes induced by one or more modulators interconvert more-active and less-active forms of the enzyme. The modulators for allosteric enzymes may be inhibitory or stimulatory. Often the modulator is the substrate itself; regulatory enzymes for which substrate and modulator are identical are called homotropic. The effect is similar to that of $O_2$ binding to hemoglobin (Chapter 5): binding of the ligand—or substrate, in the case of enzymes—causes conformational changes that affect the subsequent activity of other sites on the protein. When the modulator is a molecule other than the substrate, the enzyme is said to be heterotropic. Note that allosteric modulators should not be confused with uncompetitive

*allosteric activation*

*#1*

*catalytic side*
*active side*
*regulatory*

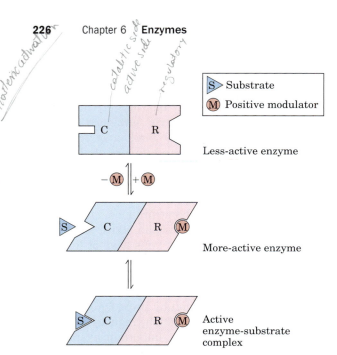

| S | Substrate |
| M | Positive modulator |

C  R  Less-active enzyme

$-$ Ⓜ ⇅ $+$ Ⓜ

Ⓢ C  R Ⓜ  More-active enzyme

Ⓢ C  R Ⓜ  Active enzyme-substrate complex

**FIGURE 6–26 Subunit interactions in an allosteric enzyme, and interactions with inhibitors and activators.** In many allosteric enzymes the substrate binding site and the modulator binding site(s) are on different subunits, the catalytic (C) and regulatory (R) subunits, respectively. Binding of the positive (stimulatory) modulator (M) to its specific site on the regulatory subunit is communicated to the catalytic subunit through a conformational change. This change renders the catalytic subunit active and capable of binding the substrate (S) with higher affinity. On dissociation of the modulator from the regulatory subunit, the enzyme reverts to its inactive or less active form.

and mixed inhibitors. Although the latter bind at a second site on the enzyme, they do not necessarily mediate conformational changes between active and inactive forms, and the kinetic effects are distinct.

The properties of allosteric enzymes are significantly different from those of simple nonregulatory enzymes. Some of the differences are structural. In addition to active sites, allosteric enzymes generally have one or more regulatory, or allosteric, sites for binding the modulator (Fig. 6–26). Just as an enzyme's active site is specific for its substrate, each regulatory site is specific for its modulator. Enzymes with several modulators generally have different specific binding sites for each. In homotropic enzymes, the active site and regulatory site are the same.

Allosteric enzymes are generally larger and more complex than nonallosteric enzymes. Most have two or more subunits. Aspartate transcarbamoylase, which catalyzes an early reaction in the biosynthesis of pyrimidine nucleotides (see Fig. 22–36), has 12 polypeptide chains organized into catalytic and regulatory subunits. Figure 6–27 shows the quaternary structure of this enzyme, deduced from x-ray analysis.

## In Many Pathways a Regulated Step Is Catalyzed by an Allosteric Enzyme

In some multienzyme systems, the regulatory enzyme is specifically inhibited by the end product of the pathway whenever the concentration of the end product exceeds the cell's requirements. When the regulatory enzyme reaction is slowed, all subsequent enzymes operate at reduced rates as their substrates are depleted. The rate

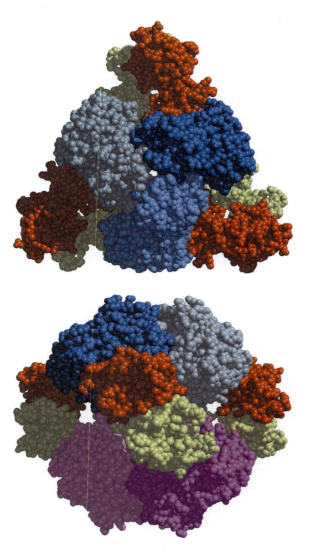

**FIGURE 6–27 Two views of the regulatory enzyme aspartate transcarbamoylase.** (Derived from PDB ID 2AT2.) This allosteric regulatory enzyme has two stacked catalytic clusters, each with three catalytic polypeptide chains (in shades of blue and purple), and three regulatory clusters, each with two regulatory polypeptide chains (in red and yellow). The regulatory clusters form the points of a triangle surrounding the catalytic subunits. Binding sites for allosteric modulators are on the regulatory subunits. Modulator binding produces large changes in enzyme conformation and activity. The role of this enzyme in nucleotide synthesis, and details of its regulation, are discussed in Chapter 22.

of production of the pathway's end product is thereby brought into balance with the cell's needs. This type of regulation is called **feedback inhibition.** Buildup of the end product ultimately slows the entire pathway.

One of the first known examples of allosteric feedback inhibition was the bacterial enzyme system that catalyzes the conversion of L-threonine to L-isoleucine in five steps (Fig. 6–28). In this system, the first enzyme, threonine dehydratase, is inhibited by isoleucine, the product of the last reaction of the series. This is an example of heterotropic allosteric inhibition. Isoleucine is quite specific as an inhibitor. No other intermediate in this sequence inhibits threonine dehydratase, nor is any other enzyme in the sequence inhibited by isoleucine. Isoleucine binds not to the active site but to another specific site on the enzyme molecule, the regulatory site. This binding is noncovalent and readily reversible; if the isoleucine concentration decreases, the rate of threonine dehydration increases. Thus threonine dehydratase activity responds rapidly and reversibly to fluctuations in the cellular concentration of isoleucine.

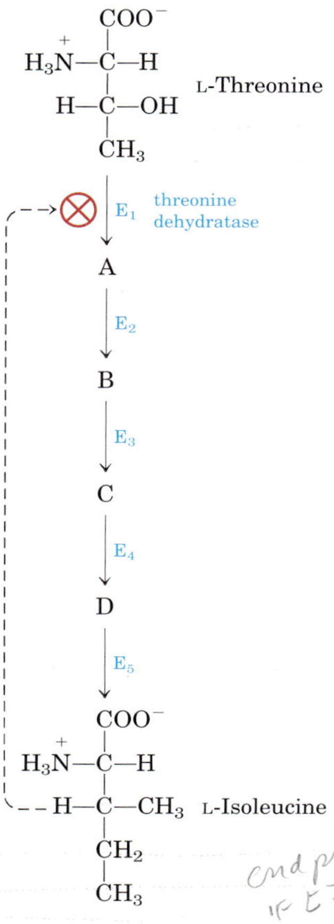

## The Kinetic Properties of Allosteric Enzymes Diverge from Michaelis-Menten Behavior

Allosteric enzymes show relationships between $V_0$ and [S] that differ from Michaelis-Menten kinetics. They do exhibit saturation with the substrate when [S] is sufficiently high, but for some allosteric enzymes, plots of $V_0$ versus [S] (Fig. 6–29) produce a sigmoid saturation curve, rather than the hyperbolic curve typical of nonregulatory enzymes. On the sigmoid saturation curve we can find a value of [S] at which $V_0$ is half-maximal, but we cannot refer to it with the designation $K_m$, because the enzyme does not follow the hyperbolic Michaelis-Menten relationship. Instead, the symbol $[S]_{0.5}$ or $K_{0.5}$ is often used to represent the substrate concentration giving half-maximal velocity of the reaction catalyzed by an allosteric enzyme (Fig. 6–29).

Sigmoid kinetic behavior generally reflects cooperative interactions between protein subunits. In other words, changes in the structure of one subunit are translated into structural changes in adjacent subunits, an effect mediated by noncovalent interactions at the interface between subunits. The principles are particularly well illustrated by a nonenzyme: $O_2$ binding to hemoglobin. Sigmoid kinetic behavior is explained by the concerted and sequential models for subunit interactions (see Fig. 5–15).

Homotropic allosteric enzymes generally are multisubunit proteins and, as noted earlier, the same binding site on each subunit functions as both the active site and the regulatory site. Most commonly, the substrate acts as a positive modulator (an activator), because the subunits act cooperatively: the binding of one molecule

*[handwritten annotation: end product. If [ ] gets too ↑ then isoleucine binds to the 1st enzyme in the pathway]*

**FIGURE 6–28 Feedback inhibition.** The conversion of L-threonine to L-isoleucine is catalyzed by a sequence of five enzymes ($E_1$ to $E_5$). Threonine dehydratase ($E_1$) is specifically inhibited allosterically by L-isoleucine, the end product of the sequence, but not by any of the four intermediates (A to D). Feedback inhibition is indicated by the dashed feedback line and the ⊗ symbol at the threonine dehydratase reaction arrow, a device used throughout this book.

of substrate to one binding site alters the enzyme's conformation and enhances the binding of subsequent substrate molecules. This accounts for the sigmoid rather than hyperbolic change in $V_0$ with increasing [S]. One characteristic of sigmoid kinetics is that small changes in the concentration of a modulator can be associated with large changes in activity. As is evident in Figure 6–29a, a relatively small increase in [S] in the steep part of the curve causes a comparatively large increase in $V_0$.

For heterotropic allosteric enzymes, those whose modulators are metabolites other than the normal substrate, it is difficult to generalize about the shape of the substrate-saturation curve. An activator may cause the curve to become more nearly hyperbolic, with a decrease in $K_{0.5}$ but no change in $V_{max}$, resulting in an increased reaction velocity at a fixed substrate concentration ($V_0$ is higher for any value of [S]; Fig. 6–29b, upper curve).

Other heterotropic allosteric enzymes respond to an activator by an increase in $V_{max}$ with little change in $K_{0.5}$ (Fig. 6–29c). A negative modulator (an inhibitor) may produce a *more* sigmoid substrate-saturation curve,

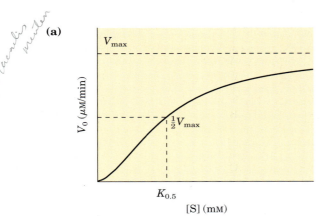

**(a)**

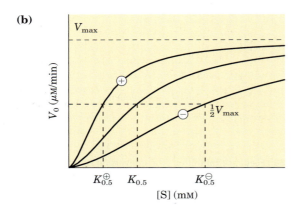

**(b)**

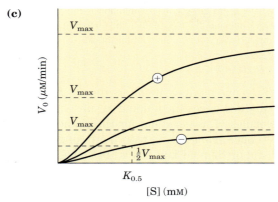

**(c)**

**FIGURE 6–29 Substrate-activity curves for representative allosteric enzymes.** Three examples of complex responses of allosteric enzymes to their modulators. **(a)** The sigmoid curve of a homotropic enzyme, in which the substrate also serves as a positive (stimulatory) modulator, or activator. Note the resemblance to the oxygen-saturation curve of hemoglobin (see Fig. 5–12). **(b)** The effects of a positive modulator (+) and a negative modulator (−) on an allosteric enzyme in which $K_{0.5}$ is altered without a change in $V_{max}$. The central curve shows the substrate-activity relationship without a modulator. **(c)** A less common type of modulation, in which $V_{max}$ is altered and $K_{0.5}$ is nearly constant.

with an increase in $K_{0.5}$ (Fig. 6–29b, lower curve). Heterotropic allosteric enzymes therefore show different kinds of responses in their substrate-activity curves, because some have inhibitory modulators, some have activating modulators, and some have both.

## Some Regulatory Enzymes Undergo Reversible Covalent Modification

In another important class of regulatory enzymes, activity is modulated by covalent modification of the enzyme molecule. Modifying groups include phosphoryl, adenylyl, uridylyl, methyl, and adenosine diphosphate ribosyl groups (Fig. 6–30). These groups are generally linked to and removed from the regulatory enzyme by separate enzymes.

An example of an enzyme regulated by methylation is the methyl-accepting chemotaxis protein of bacteria. This protein is part of a system that permits a bacterium to swim toward an attractant (such as a sugar) in solution and away from repellent chemicals. The methylating agent is *S*-adenosylmethionine (adoMet) (see Fig. 18–18b). ADP-ribosylation is an especially interesting reaction, observed in only a few proteins; the ADP-ribose is derived from nicotinamide adenine dinucleotide (NAD) (see Fig. 8–41). This type of modification occurs for the bacterial enzyme dinitrogenase reductase, resulting in regulation of the important process of biological nitrogen fixation. Diphtheria toxin and cholera toxin are enzymes that catalyze the ADP-ribosylation (and inactivation) of key cellular enzymes or proteins. Diphtheria toxin acts on and inhibits elongation factor 2, a protein involved in protein biosynthesis. Cholera toxin acts on a G protein that is part of a signaling pathway (see Fig. 12–39), leading to several physiological responses including a massive loss of body fluids and, sometimes, death.

Phosphorylation is the most common type of regulatory modification; one-third to one-half of all proteins in a eukaryotic cell are phosphorylated. Some proteins have only one phosphorylated residue, others have several, and a few have dozens of sites for phosphorylation. This mode of covalent modification is central to a large number of regulatory pathways, and we therefore discuss it in considerable detail.

## Phosphoryl Groups Affect the Structure and Catalytic Activity of Proteins

The attachment of phosphoryl groups to specific amino acid residues of a protein is catalyzed by **protein kinases;** removal of phosphoryl groups is catalyzed by **protein phosphatases.** The addition of a phosphoryl group to a Ser, Thr, or Tyr residue introduces a bulky, charged group into a region that was only moderately polar. The oxygen atoms of a phosphoryl group can hydrogen-bond with one or several groups in a protein, commonly the

*[handwritten note at top:]* not in the active site this will cause a change in the shape of the protein

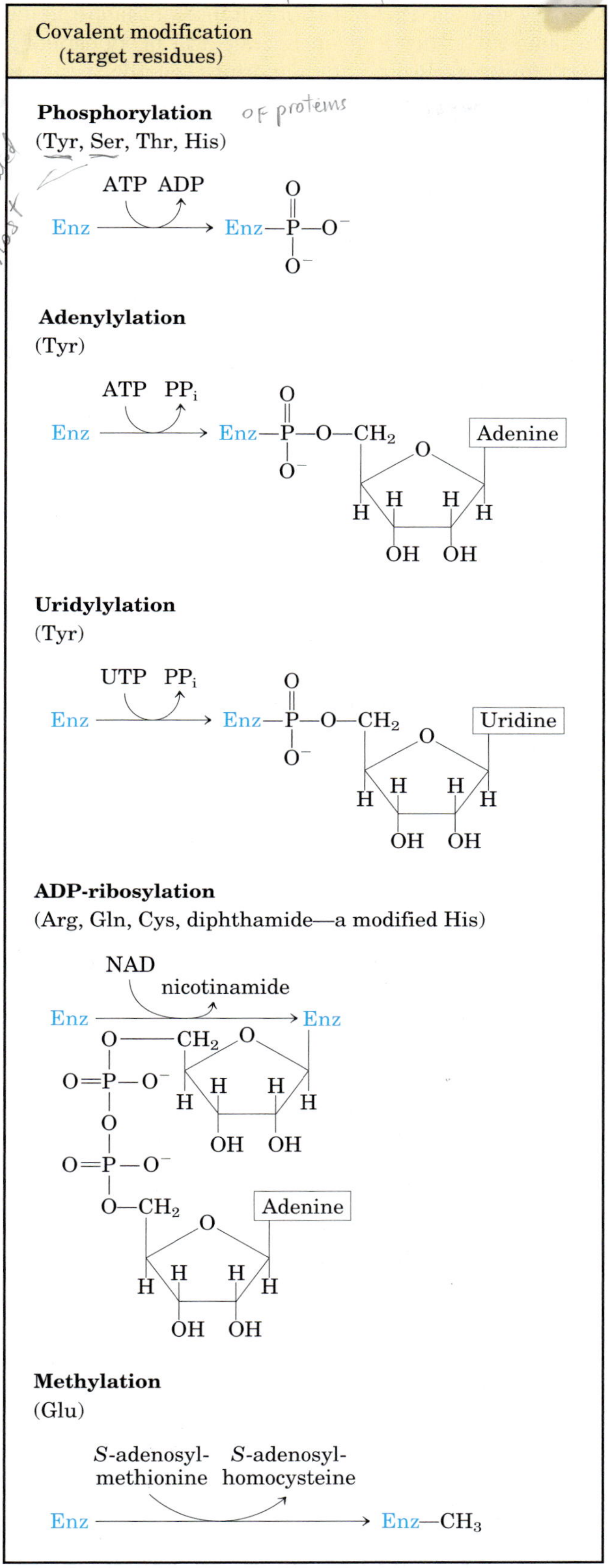

*[handwritten note at left margin:]* phosphorylated the most

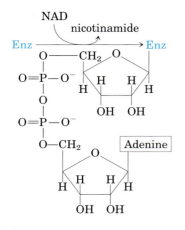

**Covalent modification**
(target residues)

**Phosphorylation** *of proteins*
(Tyr, Ser, Thr, His)

**Adenylylation**
(Tyr)

**Uridylylation**
(Tyr)

**ADP-ribosylation**
(Arg, Gln, Cys, diphthamide—a modified His)

**Methylation**
(Glu)

**FIGURE 6–30** Some enzyme modification reactions.

amide groups of the peptide backbone at the start of an $\alpha$ helix or the charged guanidinium group of an Arg residue. The two negative charges on a phosphorylated side chain can also repel neighboring negatively charged (Asp or Glu) residues. When the modified side chain is located in a region of the protein critical to its three-dimensional structure, phosphorylation can have dramatic effects on protein conformation and thus on substrate binding and catalysis.

An important example of regulation by phosphorylation is seen in glycogen phosphorylase ($M_r$ 94,500) of muscle and liver (Chapter 15), which catalyzes the reaction

$$\underset{\text{Glycogen}}{(\text{Glucose})_n} + P_i \longrightarrow \underset{\substack{\text{Shortened} \\ \text{glycogen} \\ \text{chain}}}{(\text{glucose})_{n-1}} + \text{glucose 1-phosphate}$$

The glucose 1-phosphate so formed can be used for ATP synthesis in muscle or converted to free glucose in the liver. Glycogen phosphorylase occurs in two forms: the more active phosphorylase *a* and the less active phosphorylase *b* (Fig. 6–31). Phosphorylase *a* has two subunits, each with a specific Ser residue that is phosphorylated at its hydroxyl group. These serine phosphate residues are required for maximal activity of the enzyme.

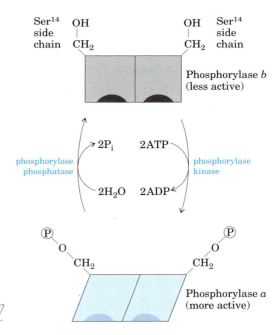

**FIGURE 6–31** **Regulation of glycogen phosphorylase activity by covalent modification.** In the more active form of the enzyme, phosphorylase *a*, specific Ser residues, one on each subunit, are phosphorylated. Phosphorylase *a* is converted to the less active phosphorylase *b* by enzymatic loss of these phosphoryl groups, promoted by phosphorylase phosphatase. Phosphorylase *b* can be reconverted (reactivated) to phosphorylase *a* by the action of phosphorylase kinase.

The phosphoryl groups can be hydrolytically removed by a separate enzyme called phosphorylase phosphatase:

$$\text{Phosphorylase } a + 2H_2O \longrightarrow \text{phosphorylase } b + 2P_i$$
$$\text{(more active)} \qquad\qquad\qquad \text{(less active)}$$

In this reaction, phosphorylase $a$ is converted to phosphorylase $b$ by the cleavage of two serine phosphate covalent bonds, one on each subunit of glycogen phosphorylase.

Phosphorylase $b$ can in turn be reactivated—covalently transformed back into active phosphorylase $a$—by another enzyme, phosphorylase kinase, which catalyzes the transfer of phosphoryl groups from ATP to the hydroxyl groups of the two specific Ser residues in phosphorylase $b$:

$$2ATP + \text{phosphorylase } b \longrightarrow 2ADP + \text{phosphorylase } a$$
$$\text{(less active)} \qquad\qquad\qquad \text{(more active)}$$

The breakdown of glycogen in skeletal muscles and the liver is regulated by variations in the ratio of the two forms of glycogen phosphorylase. The $a$ and $b$ forms differ in their secondary, tertiary, and quaternary structures; the active site undergoes changes in structure and, consequently, changes in catalytic activity as the two forms are interconverted.

The regulation of glycogen phosphorylase by phosphorylation illustrates the effects on both structure and catalytic activity of adding a phosphoryl group. In the unphosphorylated state, each subunit of this protein is folded so as to bring the 20 residues at its amino terminus, including a number of basic residues, into a region containing several acidic amino acids; this produces an electrostatic interaction that stabilizes the conformation. Phosphorylation of Ser[14] interferes with this interaction, forcing the amino-terminal domain out of the acidic environment and into a conformation that allows interaction between the ⓟ–Ser and several Arg side chains. In this conformation, the enzyme is much more active.

Phosphorylation of an enzyme can affect catalysis in another way: by altering substrate-binding affinity. For example, when isocitrate dehydrogenase (an enzyme of the citric acid cycle; Chapter 16) is phosphorylated, electrostatic repulsion by the phosphoryl group inhibits the binding of citrate (a tricarboxylic acid) at the active site.

## Multiple Phosphorylations Allow Exquisite Regulatory Control

The Ser, Thr, or Tyr residues that are phosphorylated in regulated proteins occur within common structural motifs, called consensus sequences, that are recognized by specific protein kinases (Table 6–10). Some kinases are basophilic, preferring to phosphorylate a residue having basic neighbors; others have different substrate preferences, such as for a residue near a Pro residue.

Primary sequence is not the only important factor in determining whether a given residue will be phosphorylated, however. Protein folding brings together residues that are distant in the primary sequence; the resulting three-dimensional structure can determine whether a protein kinase has access to a given residue and can recognize it as a substrate. Another factor influencing the substrate specificity of certain protein kinases is the proximity of other phosphorylated residues.

Regulation by phosphorylation is often complicated. Some proteins have consensus sequences recognized by several different protein kinases, each of which can phosphorylate the protein and alter its enzymatic activity. In some cases, phosphorylation is hierarchical: a certain residue can be phosphorylated only if a neighboring residue has already been phosphorylated. For example, glycogen synthase, the enzyme that catalyzes the condensation of glucose monomers to form glycogen (Chapter 15), is inactivated by phosphorylation of specific Ser residues and is also modulated by at least four other protein kinases that phosphorylate four other sites in the protein (Fig. 6–32). The protein is not a substrate for glycogen synthase kinase 3, for example, until one site has been phosphorylated by casein kinase II. Some phosphorylations inhibit glycogen synthase more than

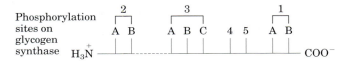

| Kinase | Phosphorylation sites | Degree of synthase inactivation |
|---|---|---|
| Protein kinase A | 1A, 1B, 2, 4 | + |
| Protein kinase G | 1A, 1B, 2 | + |
| Protein kinase C | 1A | + |
| $Ca^{2+}$/calmodulin kinase | 1B, 2 | + |
| Phosphorylase $b$ kinase | 2 | + |
| Casein kinase I | At least nine | + + + + |
| Casein kinase II | 5 | 0 |
| Glycogen synthase kinase 3 | 3A, 3B, 3C | + + + |
| Glycogen synthase kinase 4 | 2 | + |

**FIGURE 6–32 Multiple regulatory phosphorylations.** The enzyme glycogen synthase has at least nine separate sites in five designated regions susceptible to phosphorylation by one of the cellular protein kinases. Thus, regulation of this enzyme is a matter not of binary (on/off) switching but of finely tuned modulation of activity over a wide range in response to a variety of signals.

**TABLE 6–10    Consensus Sequences for Protein Kinases**

| Protein kinase | Consensus sequence and phosphorylated residue[*] |
|---|---|
| Protein kinase A | –X–R–(R/K)–X–(S/T)–B– |
| Protein kinase G | –X–R–(R/K)–X–(S/T)–X– |
| Protein kinase C | –(R/K)–(R/K)–X–(S/T)–B–(R/K)–(R/K)– |
| Protein kinase B | –X–R–X–(S/T)–X–K– |
| $Ca^{2+}$/calmodulin kinase I | –B–X–R–X–X–(S/T)–X–X–X–B– |
| $Ca^{2+}$/calmodulin kinase II | –B–X–(R/K)–X–X–(S/T)–X–X– |
| Myosin light chain kinase (smooth muscle) | –K–K–R–X–X–S–X–B–B– |
| Phosphorylase *b* kinase | *–K–R–K–Q–I–S–V–R–* |
| Extracellular signal–regulated kinase (ERK) | –P–X–(S/T)–P–P– |
| Cyclin-dependent protein kinase (cdc2) | –X–(S/T)–P–X–(K/R)– |
| Casein kinase I | –(Sp/Tp)–X–X–(X)–(S/T)–B |
| Casein kinase II | –X–(S/T)–X–X–(E/D/Sp/Yp)–X– |
| β-Adrenergic receptor kinase | –(D/E)$_n$–(S/T)–X–X–X– |
| Rhodopsin kinase | –X–X–(S/T)–(E)$_n$– |
| Insulin receptor kinase | –X–E–E–E–*Y*–M–M–M–M–*K–K–S–R–G–D–Y–M–T–M–Q–I–G–K–K–K–* <br> *L–P–A–T–G–D–Y–M–N–M–S–P–V–G–D–* |
| Epidermal growth factor (EGF) receptor kinase | *–E–E–E–E–Y–F–E–L–V–* |

**Sources:** Pinna, L.A. & Ruzzene, M.H. (1996) How do protein kinases recognize their substrates? *Biochim. Biophys. Acta* **1314,** 191–225; Kemp, B.E. & Pearson, R.B. (1990) Protein kinase recognition sequence motifs. *Trends Biochem. Sci.* **15,** 342–346; Kennelly, P.J. & Krebs, E.G. (1991) Consensus sequences as substrate specificity determinants for protein kinases and protein phosphatases. *J. Biol. Chem.* **266,** 15,555–15,558.

[*]Shown here are deduced consensus sequences (in roman type) and actual sequences from known substrates (italic). The Ser (S), Thr (T), or Tyr (Y) residue that undergoes phosphorylation is in red; all amino acid residues are shown as their one-letter abbreviations (see Table 3–1). X represents any amino acid; B, any hydrophobic amino acid; Sp, Tp, and Yp, already phosphorylated Ser, Thr, and Tyr residues.

others, and some combinations of phosphorylations are cumulative. These multiple regulatory phosphorylations provide the potential for extremely subtle modulation of enzyme activity.

To serve as an effective regulatory mechanism, phosphorylation must be reversible. In general, phosphoryl groups are added and removed by different enzymes, and the processes can therefore be separately regulated. Cells contain a family of phosphoprotein phosphatases that hydrolyze specific Ⓟ–Ser, Ⓟ–Thr, and Ⓟ–Tyr esters, releasing $P_i$. The phosphoprotein phosphatases we know of thus far act only on a subset of phosphoproteins, but they show less substrate specificity than protein kinases.

## Some Enzymes and Other Proteins Are Regulated by Proteolytic Cleavage of an Enzyme Precursor

For some enzymes, an inactive precursor called a **zymogen** is cleaved to form the active enzyme. Many proteolytic enzymes (proteases) of the stomach and pancreas are regulated in this way. Chymotrypsin and trypsin are initially synthesized as chymotrypsinogen and trypsinogen (Fig. 6–33). Specific cleavage causes conformational changes that expose the enzyme active site. Because this type of activation is irreversible, other

mechanisms are needed to inactivate these enzymes. Proteases are inactivated by inhibitor proteins that bind very tightly to the enzyme active site. For example, pancreatic trypsin inhibitor ($M_r$ 6,000) binds to and inhibits trypsin; α1-antiproteinase ($M_r$ 53,000) primarily inhibits neutrophil elastase (neutrophils are a type of leukocyte, or white blood cell; elastase is a protease acting on elastin, a component of some connective tissues). An insufficiency of $α_1$-antiproteinase, which can be caused by exposure to cigarette smoke, has been associated with lung damage, including emphysema.

Proteases are not the only proteins activated by proteolysis. In other cases, however, the precursors are called not zymogens but, more generally, **proproteins** or **proenzymes,** as appropriate. For example, the connective tissue protein collagen is initially synthesized as the soluble precursor procollagen. The blood clotting system provides many examples of the proteolytic activation of proteins. Fibrin, the protein of blood clots, is produced by proteolysis of fibrinogen, its inactive proprotein. The protease responsible for this activation is thrombin (similar in many respects to chymotrypsin), which itself is produced by proteolysis of a proprotein (in this case a zymogen), prothrombin. Blood clotting is mediated by a complicated cascade of proteolytic activations.

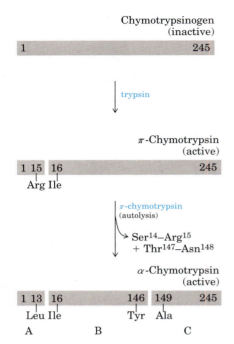

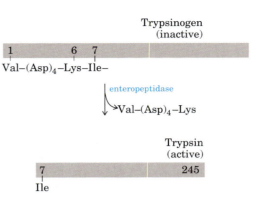

**FIGURE 6–33 Activation of zymogens by proteolytic cleavage.** Shown here is the formation of chymotrypsin and trypsin from their zymogens. The bars represent the primary sequences of the polypeptide chains. Amino acid residues at the termini of the polypeptide fragments generated by cleavage are indicated below the bars. The numbering of amino acid residues represents their positions in the primary sequence of the zymogens, chymotrypsinogen or trypsinogen (the amino-terminal residue is number 1). Thus, in the active forms, some numbered residues are missing. Recall that the three polypeptide chains (A, B, and C) of chymotrypsin are linked by disulfide bonds (see Fig. 6–18).

### Some Regulatory Enzymes Use Several Regulatory Mechanisms

Glycogen phosphorylase catalyzes the first reaction in a pathway that feeds stored glucose into energy-yielding carbohydrate metabolism (Chapters 14 and 15). This is an important metabolic step, and its regulation is correspondingly complex. Although its primary regulation is through covalent modification, as outlined in Figure 6–31, glycogen phosphorylase is also modulated allosterically by AMP, which is an activator of phosphorylase *b,* and by several other molecules that are inhibitors.

Other complex regulatory enzymes are found at key metabolic crossroads. Bacterial glutamine synthetase, which catalyzes a reaction that introduces reduced nitrogen into cellular metabolism (Chapter 22), is among the most complex regulatory enzymes known. It is regulated allosterically (with at least eight different modulators); by reversible covalent modification; and by the association of other regulatory proteins, a mechanism examined in detail when we consider the regulation of specific metabolic pathways.

What is the advantage of such complexity in the regulation of enzymatic activity? We began this chapter by stressing the central importance of catalysis to the very existence of life. The *control* of catalysis is also critical to life. If all possible reactions in a cell were catalyzed simultaneously, macromolecules and metabolites would quickly be broken down to much simpler chem-

ical forms. Instead, cells catalyze only the reactions they need at a given moment. When chemical resources are plentiful, cells synthesize and store glucose and other metabolites. When chemical resources are scarce, cells use these stores to fuel cellular metabolism. Chemical energy is used economically, parceled out to various metabolic pathways as cellular needs dictate. The availability of powerful catalysts, each specific for a given reaction, makes the regulation of these reactions possible. This in turn gives rise to the complex, highly regulated symphony we call life.

### SUMMARY 6.5 Regulatory Enzymes

- The activities of metabolic pathways in cells are regulated by control of the activities of certain enzymes.

- In feedback inhibition, the end product of a pathway inhibits the first enzyme of that pathway.

- The activity of allosteric enzymes is adjusted by reversible binding of a specific modulator to a regulatory site. Modulators may be the substrate itself or some other metabolite, and the effect of the modulator may be inhibitory or stimulatory. The kinetic behavior of allosteric enzymes reflects cooperative interactions among enzyme subunits.

- Other regulatory enzymes are modulated by covalent modification of a specific functional group necessary for activity. The phosphorylation of specific amino acid residues is a particularly common way to regulate enzyme activity.

- Many proteolytic enzymes are synthesized as inactive precursors called zymogens, which are activated by cleavage of small peptide fragments.

- Enzymes at important metabolic intersections may be regulated by complex combinations of effectors, allowing coordination of the activities of interconnected pathways.

## Key Terms

*Terms in bold are defined in the glossary.*

| | | |
|---|---|---|
| **enzyme**  191 | **rate constant**  195 | $k_{cat}$  206 |
| **cofactor**  191 | **binding energy ($\Delta G_B$)**  196 | **turnover number**  207 |
| **coenzyme**  191 | **specificity**  199 | reversible inhibition  209 |
| **prosthetic group**  192 | **induced fit**  200 | **competitive inhibition**  209 |
| **holoenzyme**  192 | **specific acid-base catalysis**  200 | **uncompetitive inhibition**  211 |
| **apoenzyme**  192 | **general acid-base catalysis**  200 | **mixed inhibition**  211 |
| **apoprotein**  192 | covalent catalysis  200 | noncompetitive inhibition  211 |
| **active site**  193 | enzyme kinetics  202 | irreversible inhibitors  211 |
| **substrate**  193 | initial rate (initial velocity), $V_0$  202 | **suicide inactivator**  211 |
| **ground state**  193 | $V_{max}$ 203 | transition state analogs  221 |
| **standard free-energy change** | pre–steady state  203 | **regulatory enzyme**  225 |
| **($\Delta G°$)**  194 | **steady state**  203 | **allosteric enzyme**  225 |
| **transition state**  194 | steady-state kinetics  203 | allosteric modulator  225 |
| **activation energy ($\Delta G^{\ddagger}$)**  194 | **Michaelis constant ($K_m$)**  204 | **feedback inhibition**  227 |
| **reaction intermediate**  195 | **Michaelis-Menten equation**  204 | **protein kinases**  228 |
| **rate-limiting step**  195 | **dissociation constant ($K_d$)**  205 | **zymogen**  231 |
| **equilibrium constant ($K_{eq}$)**  195 | **Lineweaver-Burk equation**  206 | |

## Further Reading

### General

Evolution of Catalytic Function. (1987) *Cold Spring Harb. Symp. Quant. Biol.* **52.**

A collection of excellent papers on fundamentals; continues to be very useful.

Fersht, A. (1999) *Structure and Mechanism in Protein Science: A Guide to Enzyme Catalysis and Protein Folding,* W. H. Freeman and Company, New York.

A clearly written, concise introduction. More advanced.

Friedmann, H. (ed.) (1981) *Benchmark Papers in Biochemistry,* Vol. 1: *Enzymes,* Hutchinson Ross Publishing Company, Stroudsburg, PA.

A collection of classic papers on enzyme chemistry, with historical commentaries by the editor. Extremely interesting.

Jencks, W.P. (1987) *Catalysis in Chemistry and Enzymology,* Dover Publications, Inc., New York.

An outstanding book on the subject. More advanced.

Kornberg, A. (1989) *For the Love of Enzymes: The Odyssey of a Biochemist,* Harvard University Press, Cambridge.

### Principles of Catalysis

Amyes, T.L., O'Donoghue, A.C., & Richard, J.P. (2001) Contribution of phosphate intrinsic binding energy to the enzymatic rate acceleration for triosephosphate isomerase. *J. Am. Chem. Soc.* **123,** 11,325–11,326.

Hansen, D.E. & Raines, R.T. (1990) Binding energy and enzymatic catalysis. *J. Chem. Educ.* **67,** 483–489.

A good place for the beginning student to acquire a better understanding of principles.

Harris, T.K. & Turner, G.J. (2002) Structural basis of perturbed $pK_a$ values of catalytic groups in enzyme active sites. *IUBMB Life* **53,** 85–98.

Kraut, J. (1988) How do enzymes work? *Science* **242,** 533–540.

Landry, D.W., Zhao, K., Yang, G.X.-Q., Glickman, M., & Georgiadis, T.M. (1993) Antibody degradation of cocaine. *Science* **259,** 1899–1901.

An interesting application of catalytic antibodies.

Lerner, R.A., Benkovic, S.J., & Schulz, P.G. (1991) At the crossroads of chemistry and immunology: catalytic antibodies. *Science* **252,** 659–667.

**Miller, B.G. & Wolfenden, R.** (2002) Catalytic proficiency: the unusual case of OMP decarboxylase. *Annu. Rev. Biochem.* **71,** 847–885.

Orotidine monophosphate decarboxylase seems to be a reigning champion of catalytic rate enhancement by an enzyme.

**Schramm, V.L.** (1998) Enzymatic transition states and transition state analog design. *Annu. Rev. Biochem.* **67,** 693–720.

Many good illustrations of the principles introduced in this chapter.

### Kinetics

**Cleland, W.W.** (1977) Determining the chemical mechanisms of enzyme-catalyzed reactions by kinetic studies. *Adv. Enzymol.* **45,** 273–387.

**Cleland, W.W.** (2002) Enzyme kinetics: steady state. In *Nature Encyclopedia of Life Sciences,* Vol. 6, pp. 421–425, Nature Publishing Group, London. Article originally published in 1998. *Encyclopedia* available online (2001), by subscription, at www.els.net.

A clear and concise presentation of the basics.

**Raines, R.T. & Hansen, D.E.** (1988) An intuitive approach to steady-state kinetics. *J. Chem. Educ.* **65,** 757–759.

**Segel, I.H.** (1975) *Enzyme Kinetics: Behavior and Analysis of Rapid Equilibrium and Steady State Enzyme Systems,* John Wiley & Sons, Inc., New York.

A more advanced treatment.

### Enzyme Examples

**Babbitt, P.C. & Gerlt, J.A.** (1997) Understanding enzyme superfamilies: chemistry as the fundamental determinant in the evolution of new catalytic activities. *J. Biol. Chem.* **27,** 30,591–30,594.

An interesting description of the evolution of enzymes with different catalytic specificities, and the use of a limited repertoire of protein structural motifs.

**Babbitt, P.C., Hasson, M.S., Wedekind, J.E., Palmer, D.R.J., Barrett, W.C., Reed, G.H., Rayment, I., Ringe, D., Kenyon, G.L., & Gerlt, J.A.** (1996) The enolase superfamily: a general strategy for enzyme-catalyzed abstraction of the α-protons of carboxylic acids. *Biochemistry* **35,** 16,489–16,501.

**Kirby, A.J.** (2001) The lysozyme mechanism sorted—after 50 years. *Nat. Struct. Biol.* **8,** 737–739.

A nice discussion of the catalytic power of enzymes and the principles underlying it.

**Warshel, A., Naray-Szabo, G., Sussman, F., & Hwang, J.-K.** (1989) How do serine proteases really work? *Biochemistry* **28,** 3629–3637.

### Regulatory Enzymes

**Barford, D., Das, A.K., & Egloff, M.-P.** (1998). The structure and mechanism of protein phosphatases: insights into catalysis and regulation. *Annu. Rev. Biophys. Biomol. Struct.* **27,** 133–164.

**Dische, Z.** (1976) The discovery of feedback inhibition. *Trends Biochem. Sci.* **1,** 269–270.

**Hunter, T. & Plowman, G.D.** (1997) The protein kinases of budding yeast: six score and more. *Trends Biochem. Sci.* **22,** 18–22.

Details of the variety of these important enzymes in a model eukaryote.

**Johnson, L.N. & Barford, D.** (1993) The effects of phosphorylation on the structure and function of proteins. *Annu. Rev. Biophys. Biomol. Struct.* **22,** 199–232.

**Koshland, D.E., Jr. & Neet, K.E.** (1968) The catalytic and regulatory properties of enzymes. *Annu. Rev. Biochem.* **37,** 359–410.

**Monod, J., Changeux, J.-P., & Jacob, F.** (1963) Allosteric proteins and cellular control systems. *J. Mol. Biol.* **6,** 306–329.

A classic paper introducing the concept of allosteric regulation.

# Problems

**1. Keeping the Sweet Taste of Corn** The sweet taste of freshly picked corn (maize) is due to the high level of sugar in the kernels. Store-bought corn (several days after picking) is not as sweet, because about 50% of the free sugar is converted to starch within one day of picking. To preserve the sweetness of fresh corn, the husked ears can be immersed in boiling water for a few minutes ("blanched") then cooled in cold water. Corn processed in this way and stored in a freezer maintains its sweetness. What is the biochemical basis for this procedure?

**2. Intracellular Concentration of Enzymes** To approximate the actual concentration of enzymes in a bacterial cell, assume that the cell contains equal concentrations of 1,000 different enzymes in solution in the cytosol and that each protein has a molecular weight of 100,000. Assume also that the bacterial cell is a cylinder (diameter 1.0 $\mu$m, height 2.0 $\mu$m), that the cytosol (specific gravity 1.20) is 20% soluble protein by weight, and that the soluble protein consists entirely of enzymes. Calculate the *average* molar concentration of each enzyme in this hypothetical cell.

**3. Rate Enhancement by Urease** The enzyme urease enhances the rate of urea hydrolysis at pH 8.0 and 20 °C by a factor of $10^{14}$. If a given quantity of urease can completely hydrolyze a given quantity of urea in 5.0 min at 20 °C and pH 8.0, how long would it take for this amount of urea to be hydrolyzed under the same conditions in the absence of urease? Assume that both reactions take place in sterile systems so that bacteria cannot attack the urea.

**4. Protection of an Enzyme against Denaturation by Heat** When enzyme solutions are heated, there is a progressive loss of catalytic activity over time due to denaturation of the enzyme. A solution of the enzyme hexokinase incubated at 45 °C lost 50% of its activity in 12 min, but when incubated at 45 °C in the presence of a very large concentration of one of its substrates, it lost only 3% of its activity in 12 min. Suggest why thermal denaturation of hexokinase was retarded in the presence of one of its substrates.

**5. Requirements of Active Sites in Enzymes** Carboxypeptidase, which sequentially removes carboxyl-terminal amino acid residues from its peptide substrates, is a single polypeptide of 307 amino acids. The two essential catalytic groups in the active site are furnished by $Arg^{145}$ and $Glu^{270}$.

(a) If the carboxypeptidase chain were a perfect $\alpha$ helix, how far apart (in Å) would $Arg^{145}$ and $Glu^{270}$ be? (Hint: See Fig. 4–4b.)

(b) Explain how the two amino acid residues can catalyze a reaction occurring in the space of a few angstroms.

**6. Quantitative Assay for Lactate Dehydrogenase** The muscle enzyme lactate dehydrogenase catalyzes the reaction

$$CH_3-\overset{\overset{\displaystyle O}{\|}}{C}-COO^- + NADH + H^+ \longrightarrow$$

Pyruvate

*enzyme is not working*

$$CH_3-\overset{\overset{\displaystyle OH}{|}}{\underset{\underset{\displaystyle H}{|}}{C}}-COO^- + NAD^+$$

Lactate

*reaction is working*

NADH and $NAD^+$ are the reduced and oxidized forms, respectively, of the coenzyme NAD. Solutions of NADH, but *not* $NAD^+$, absorb light at 340 nm. This property is used to determine the concentration of NADH in solution by measuring spectrophotometrically the amount of light absorbed at 340 nm by the solution. Explain how these properties of NADH can be used to design a quantitative assay for lactate dehydrogenase.

**7. Relation between Reaction Velocity and Substrate Concentration: Michaelis-Menten Equation** (a) At what substrate concentration would an enzyme with a $k_{cat}$ of 30.0 $s^{-1}$ and a $K_m$ of 0.0050 M operate at one-quarter of its maximum rate? (b) Determine the fraction of $V_{max}$ that would be obtained at the following substrate concentrations: $[S] = \frac{1}{2}K_m$, $2K_m$, and $10K_m$.

**8. Estimation of $V_{max}$ and $K_m$ by Inspection** Although graphical methods are available for accurate determination of the $V_{max}$ and $K_m$ of an enzyme-catalyzed reaction (see Box 6–1), sometimes these quantities can be quickly estimated by inspecting values of $V_0$ at increasing $[S]$. Estimate the $V_{max}$ and $K_m$ of the enzyme-catalyzed reaction for which the following data were obtained.

| [S] (M) | $V_0$ ($\mu M$/min) |
|---|---|
| $2.5 \times 10^{-6}$ | 28 |
| $4.0 \times 10^{-6}$ | 40 |
| $1 \times 10^{-5}$ | 70 |
| $2 \times 10^{-5}$ | 95 |
| $4 \times 10^{-5}$ | 112 |
| $1 \times 10^{-4}$ | 128 |
| $2 \times 10^{-3}$ | 139 |
| $1 \times 10^{-2}$ | 140 |

**9. Properties of an Enzyme of Prostaglandin Synthesis** Prostaglandins are a class of eicosanoids, fatty acid derivatives with a variety of extremely potent actions on vertebrate tissues. They are responsible for producing fever and inflammation and its associated pain. Prostaglandins are derived from the 20-carbon fatty acid arachidonic acid in a reaction catalyzed by the enzyme prostaglandin endoperoxide synthase. This enzyme, a cyclooxygenase, uses oxygen to convert arachidonic acid to $PGG_2$, the immediate precursor of many different prostaglandins (prostaglandin synthesis is described in Chapter 21).

(a) The kinetic data given below are for the reaction catalyzed by prostaglandin endoperoxide synthase. Focusing here on the first two columns, determine the $V_{max}$ and $K_m$ of the enzyme.

| [Arachidonic acid] (mM) | Rate of formation of $PGG_2$ (mM/min) | Rate of formation of $PGG_2$ with 10 mg/mL ibuprofen (mM/min) |
|---|---|---|
| 0.5 | 23.5 | 16.67 |
| 1.0 | 32.2 | 25.25 |
| 1.5 | 36.9 | 30.49 |
| 2.5 | 41.8 | 37.04 |
| 3.5 | 44.0 | 38.91 |

(b) Ibuprofen is an inhibitor of prostaglandin endoperoxide synthase. By inhibiting the synthesis of prostaglandins, ibuprofen reduces inflammation and pain. Using the data in the first and third columns of the table, determine the type of inhibition that ibuprofen exerts on prostaglandin endoperoxide synthase.

**10. Graphical Analysis of $V_{max}$ and $K_m$** The following experimental data were collected during a study of the catalytic activity of an intestinal peptidase with the substrate glycylglycine:

$$\text{Glycylglycine} + H_2O \longrightarrow 2 \text{ glycine}$$

| [S] (mM) | Product formed ($\mu$mol/min) |
|---|---|
| 1.5 | 0.21 |
| 2.0 | 0.24 |
| 3.0 | 0.28 |
| 4.0 | 0.33 |
| 8.0 | 0.40 |
| 16.0 | 0.45 |

Use graphical analysis (see Box 6–1 and its associated Living Graph) to determine the $K_m$ and $V_{max}$ for this enzyme preparation and substrate.

**11. The Eadie-Hofstee Equation** One transformation of the Michaelis-Menten equation is the Lineweaver-Burk, or double-reciprocal, equation. Multiplying both sides of the Lineweaver-Burk equation by $V_{max}$ and rearranging gives the Eadie-Hofstee equation:

$$V_0 = (-K_m)\frac{V_0}{[S]} + V_{max}$$

A plot of $V_0$ vs. $V_0/[S]$ for an enzyme-catalyzed reaction is shown below. The blue curve was obtained in the absence of inhibitor. Which of the other curves (A, B, or C) shows the enzyme activity when a competitive inhibitor is added to the reaction mixture? Hint: See Equation 6–30.

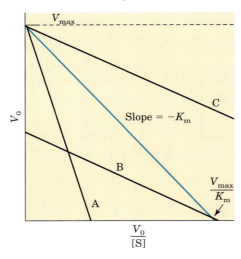

**12. The Turnover Number of Carbonic Anhydrase** Carbonic anhydrase of erythrocytes ($M_r$ 30,000) has one of the highest turnover numbers we know of. It catalyzes the reversible hydration of $CO_2$:

$$H_2O + CO_2 \rightleftharpoons H_2CO_3$$

This is an important process in the transport of $CO_2$ from the tissues to the lungs. If 10.0 $\mu$g of pure carbonic anhydrase catalyzes the hydration of 0.30 g of $CO_2$ in 1 min at 37 °C at $V_{max}$, what is the turnover number ($k_{cat}$) of carbonic anhydrase (in units of $min^{-1}$)?

**13. Deriving a Rate Equation for Competitive Inhibition** The rate equation for an enzyme subject to competitive inhibition is

$$V_0 = \frac{V_{max}[S]}{\alpha K_m + [S]}$$

Beginning with a new definition of total enzyme as

$$[E]_t = [E] + [EI] + [ES]$$

and the definitions of $\alpha$ and $K_I$ provided in the text, derive the rate equation above. Use the derivation of the Michaelis-Menten equation as a guide.

**14. Irreversible Inhibition of an Enzyme** Many enzymes are inhibited irreversibly by heavy metal ions such as $Hg^{2+}$, $Cu^{2+}$, or $Ag^+$, which can react with essential sulfhydryl groups to form mercaptides:

$$Enz\text{–}SH + Ag^+ \longrightarrow Enz\text{–}S\text{–}Ag + H^+$$

The affinity of $Ag^+$ for sulfhydryl groups is so great that $Ag^+$ can be used to titrate —SH groups quantitatively. To 10.0 mL of a solution containing 1.0 mg/mL of a pure enzyme, an investigator added just enough $AgNO_3$ to completely inactivate the enzyme. A total of 0.342 $\mu$mol of $AgNO_3$ was required.

Calculate the minimum molecular weight of the enzyme. Why does the value obtained in this way give only the *minimum* molecular weight?

**15. Clinical Application of Differential Enzyme Inhibition** Human blood serum contains a class of enzymes known as acid phosphatases, which hydrolyze biological phosphate esters under slightly acidic conditions (pH 5.0):

$$R\text{–}O\text{–}\overset{\overset{\displaystyle O^-}{|}}{\underset{\underset{\displaystyle O}{\|}}{P}}\text{–}O^- + H_2O \longrightarrow R\text{–}OH + HO\text{–}\overset{\overset{\displaystyle O^-}{|}}{\underset{\underset{\displaystyle O}{\|}}{P}}\text{–}O^-$$

Acid phosphatases are produced by erythrocytes, the liver, kidney, spleen, and prostate gland. The enzyme of the prostate gland is clinically important, because its increased activity in the blood can be an indication of prostate cancer. The phosphatase from the prostate gland is strongly inhibited by tartrate ion, but acid phosphatases from other tissues are not. How can this information be used to develop a specific procedure for measuring the activity of the acid phosphatase of the prostate gland in human blood serum?

**16. Inhibition of Carbonic Anhydrase by Acetazolamide** Carbonic anhydrase is strongly inhibited by the drug acetazolamide, which is used as a diuretic (i.e., to increase the production of urine) and to lower excessively high pressure in the eye (due to accumulation of intraocular fluid) in glaucoma. Carbonic anhydrase plays an important role in these and other secretory processes, because it participates in regulating the pH and bicarbonate content of several body fluids. The experimental curve of initial reaction velocity (as percentage of $V_{max}$) versus [S] for the carbonic anhydrase reaction is illustrated below (upper curve). When the experiment is repeated in the presence of acetazolamide, the lower curve is obtained. From an inspection of the curves and your knowledge of the kinetic properties of competitive and mixed enzyme inhibitors, determine the nature of the inhibition by acetazolamide. Explain your reasoning.

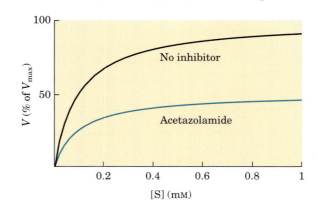

**17. The Effects of Reversible Inhibitors** Derive the expression for the effect of a reversible inhibitor on observed $K_m$ (apparent $K_m = \alpha K_m/\alpha'$). Start with Equation 6–30 and the statement that apparent $K_m$ is equivalent to the [S] at which $V_0 = V_{max}/2\alpha'$.

**18. pH Optimum of Lysozyme**   The active site of lysozyme contains two amino acid residues essential for catalysis: $Glu^{35}$ and $Asp^{52}$. The $pK_a$ values of the carboxyl side chains of these residues are 5.9 and 4.5, respectively. What is the ionization state (protonated or deprotonated) of each residue at pH 5.2, the pH optimum of lysozyme? How can the ionization states of these residues explain the pH-activity profile of lysozyme shown below?

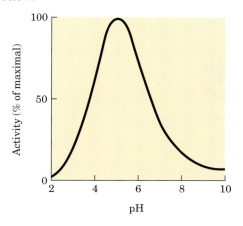

**19. Working with Kinetics**   Go to the Living Graphs for Chapter 6.

(a)  Using the Living Graph for Equation 6–9, create a $V$ versus [S] plot. Use $V_{max} = 100\ \mu M\ s^{-1}$, and $K_m = 10\ \mu M$. How much does $V_0$ increase when [S] is doubled, from 0.2 to 0.4 $\mu M$? What is $V_0$ when [S] = 10 $\mu M$? How much does the $V_0$ increase when [S] increases from 100 to 200 $\mu M$? Observe how the graph changes when the values for $V_{max}$ or $K_m$ are halved or doubled.

(b)  Using the Living Graph for Equation 6–30 and the kinetic parameters in (a), create a plot in which both $\alpha$ and $\alpha'$ are 1.0. Now observe how the plot changes when $\alpha = 2.0$; when $\alpha' = 3.0$; and when $\alpha = 2.0$ and $\alpha' = 3.0$.

(c)  Using the Living Graphs for Equation 6–30 and the Lineweaver-Burk equation in Box 6–1, create Lineweaver-Burk (double-reciprocal) plots for all the cases in (a) and (b). When $\alpha = 2.0$, does the $x$ intercept move to the right or to the left? If $\alpha = 2.0$ and $\alpha' = 3.0$, does the $x$ intercept move to the right or to the left?

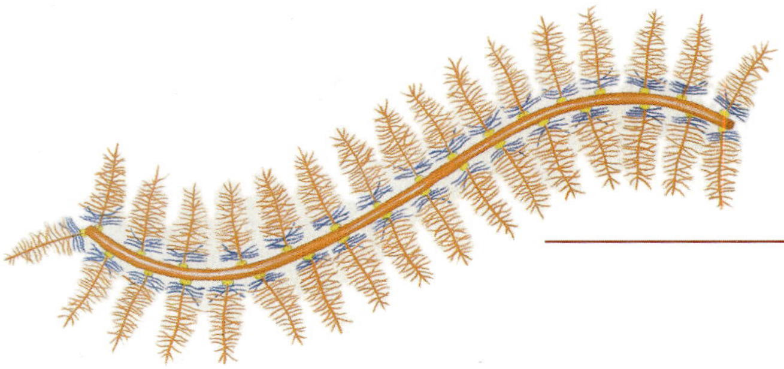

# CARBOHYDRATES AND GLYCOBIOLOGY

Ah! sweet mystery of life . . .

—Rida Johnson Young (lyrics) and Victor Herbert (music),
"Ah! Sweet Mystery of Life," 1910

I would feel more optimistic about a bright future for man if he spent less time proving that he can outwit Nature and more time tasting her sweetness and respecting her seniority.

—E. B. White, "Coon Tree," 1977

Carbohydrates are the most abundant biomolecules on Earth. Each year, photosynthesis converts more than 100 billion metric tons of $CO_2$ and $H_2O$ into cellulose and other plant products. Certain carbohydrates (sugar and starch) are a dietary staple in most parts of the world, and the oxidation of carbohydrates is the central energy-yielding pathway in most nonphotosynthetic cells. Insoluble carbohydrate polymers serve as structural and protective elements in the cell walls of bacteria and plants and in the connective tissues of animals. Other carbohydrate polymers lubricate skeletal joints and participate in recognition and adhesion between cells. More complex carbohydrate polymers covalently attached to proteins or lipids act as signals that determine the intracellular location or metabolic fate of these hybrid molecules, called **glycoconjugates.** This chapter introduces the major classes of carbohydrates and glycoconjugates and provides a few examples of their many structural and functional roles.

Carbohydrates are polyhydroxy aldehydes or ketones, or substances that yield such compounds on hydrolysis. Many, but not all, carbohydrates have the empirical formula $(CH_2O)_n$; some also contain nitrogen, phosphorus, or sulfur.

There are three major size classes of carbohydrates: monosaccharides, oligosaccharides, and polysaccharides (the word "saccharide" is derived from the Greek *sakcharon,* meaning "sugar"). **Monosaccharides,** or simple sugars, consist of a single polyhydroxy aldehyde or ketone unit. The most abundant monosaccharide in nature is the six-carbon sugar D-glucose, sometimes referred to as dextrose. Monosaccharides of more than four carbons tend to have cyclic structures.

**Oligosaccharides** consist of short chains of monosaccharide units, or residues, joined by characteristic linkages called glycosidic bonds. The most abundant are the **disaccharides,** with two monosaccharide units. Typical is sucrose (cane sugar), which consists of the six-carbon sugars D-glucose and D-fructose. All common monosaccharides and disaccharides have names ending with the suffix "-ose." In cells, most oligosaccharides consisting of three or more units do not occur as free entities but are joined to nonsugar molecules (lipids or proteins) in glycoconjugates.

The **polysaccharides** are sugar polymers containing more than 20 or so monosaccharide units, and some have hundreds or thousands of units. Some polysaccharides, such as cellulose, are linear chains; others, such as glycogen, are branched. Both glycogen and cellulose consist of recurring units of D-glucose, but they differ in the type of glycosidic linkage and consequently have strikingly different properties and biological roles.

## 7.1 Monosaccharides and Disaccharides

The simplest of the carbohydrates, the monosaccharides, are either aldehydes or ketones with two or more hydroxyl groups; the six-carbon monosaccharides glucose and fructose have five hydroxyl groups. Many of the carbon atoms to which hydroxyl groups are attached are chiral centers, which give rise to the many sugar stereoisomers found in nature. We begin by describing the families of monosaccharides with backbones of three to seven carbons—their structure and stereoisomeric forms, and the means of representing their three-dimensional structures on paper. We then discuss several chemical reactions of the carbonyl groups of monosaccharides. One such reaction, the addition of a hydroxyl group from within the same molecule, generates the cyclic forms of five- and six-carbon sugars (the forms that predominate in aqueous solution) and creates a new chiral center, adding further stereochemical complexity to this class of compounds. The nomenclature for unambiguously specifying the configuration about each carbon atom in a cyclic form and the means of representing these structures on paper are therefore described in some detail; this information will be useful as we discuss the metabolism of monosaccharides in Part II. We also introduce here some important monosaccharide derivatives encountered in later chapters.

### The Two Families of Monosaccharides Are Aldoses and Ketoses

Monosaccharides are colorless, crystalline solids that are freely soluble in water but insoluble in nonpolar solvents. Most have a sweet taste. The backbones of common monosaccharide molecules are unbranched carbon chains in which all the carbon atoms are linked by single bonds. In the open-chain form, one of the carbon atoms is double-bonded to an oxygen atom to form a carbonyl group; each of the other carbon atoms has a hydroxyl group. If the carbonyl group is at an end of the carbon chain (that is, in an aldehyde group) the monosaccharide is an **aldose;** if the carbonyl group is at any other position (in a ketone group) the monosaccharide is a **ketose.** The simplest monosaccharides are the two three-carbon trioses: glyceraldehyde, an aldotriose, and dihydroxyacetone, a ketotriose (Fig. 7–1a).

Monosaccharides with four, five, six, and seven carbon atoms in their backbones are called, respectively, tetroses, pentoses, hexoses, and heptoses. There are aldoses and ketoses of each of these chain lengths:

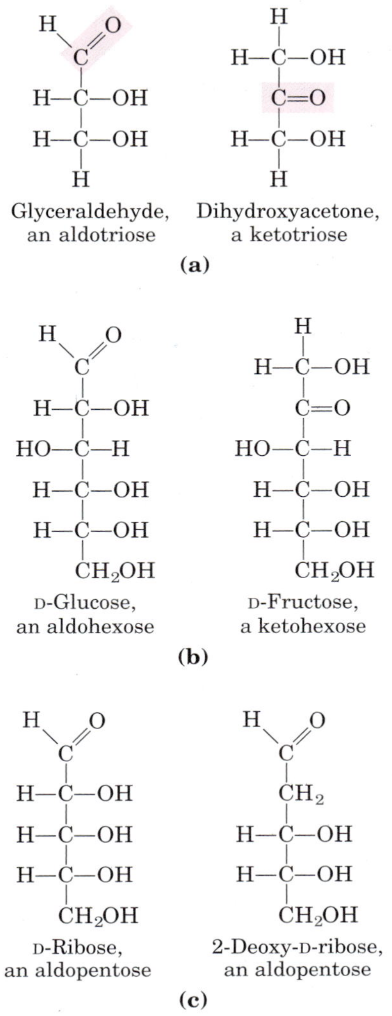

**FIGURE 7–1  Representative monosaccharides. (a)** Two trioses, an aldose and a ketose. The carbonyl group in each is shaded. **(b)** Two common hexoses. **(c)** The pentose components of nucleic acids. D-Ribose is a component of ribonucleic acid (RNA), and 2-deoxy-D-ribose is a component of deoxyribonucleic acid (DNA).

aldotetroses and ketotetroses, aldopentoses and ketopentoses, and so on. The hexoses, which include the aldohexose D-glucose and the ketohexose D-fructose (Fig. 7–1b), are the most common monosaccharides in nature. The aldopentoses D-ribose and 2-deoxy-D-ribose (Fig. 7–1c) are components of nucleotides and nucleic acids (Chapter 8).

### Monosaccharides Have Asymmetric Centers

All the monosaccharides except dihydroxyacetone contain one or more asymmetric (chiral) carbon atoms and thus occur in optically active isomeric forms (pp. 17–19). The simplest aldose, glyceraldehyde, contains one chiral center (the middle carbon atom) and therefore has two different optical isomers, or enantiomers (Fig. 7–2).

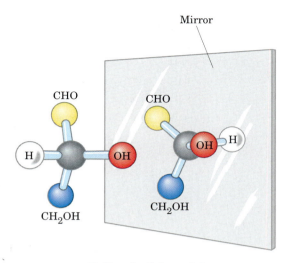

**Ball-and-stick models**

| | |
|---|---|
| CHO | CHO |
| H—C—OH | HO—C—H |
| CH$_2$OH | CH$_2$OH |
| D-Glyceraldehyde | L-Glyceraldehyde |

**Fischer projection formulas**

| | |
|---|---|
| CHO | CHO |
| H—C—OH | HO—C—H |
| CH$_2$OH | CH$_2$OH |
| D-Glyceraldehyde | L-Glyceraldehyde |

**Perspective formulas**

**FIGURE 7–2    Three ways to represent the two stereoisomers of glyc-eraldehyde.** The stereoisomers are mirror images of each other. Ball-and-stick models show the actual configuration of molecules. By convention, in Fischer projection formulas, horizontal bonds project out of the plane of the paper, toward the reader; vertical bonds project behind the plane of the paper, away from the reader. Recall (see Fig. 1–17) that in perspective formulas, solid wedge-shaped bonds point toward the reader, dashed wedges point away.

By convention, one of these two forms is designated the D isomer, the other the L isomer. As for other biomolecules with chiral centers, the absolute configurations of sugars are known from x-ray crystallography. To represent three-dimensional sugar structures on paper, we often use **Fischer projection formulas** (Fig. 7–2).

In general, a molecule with $n$ chiral centers can have $2^n$ stereoisomers. Glyceraldehyde has $2^1 = 2$; the aldohexoses, with four chiral centers, have $2^4 = 16$ stereoisomers. The stereoisomers of monosaccharides of each carbon-chain length can be divided into two groups that differ in the configuration about the chiral center *most distant* from the carbonyl carbon. Those in which the configuration at this reference carbon is the same as that of D-glyceraldehyde are designated D isomers, and those with the same configuration as L-glyceraldehyde are L isomers. When the hydroxyl group on the reference carbon is on the right in the projection formula, the sugar is the D isomer; when on the left, it is the L isomer. Of the 16 possible aldohexoses, eight are D forms and eight are L. Most of the hexoses of living organisms are D isomers.

Figure 7–3 shows the structures of the D stereoisomers of all the aldoses and ketoses having three to six carbon atoms. The carbons of a sugar are numbered beginning at the end of the chain nearest the carbonyl group. Each of the eight D-aldohexoses, which differ in the stereochemistry at C-2, C-3, or C-4, has its own name: D-glucose, D-galactose, D-mannose, and so forth (Fig. 7–3a). The four- and five-carbon ketoses are designated by inserting "ul" into the name of a corresponding aldose; for example, D-ribulose is the ketopentose corresponding to the aldopentose D-ribose. The keto-hexoses are named otherwise: for example, fructose (from the Latin *fructus*, "fruit"; fruits are rich in this sugar) and sorbose (from *Sorbus*, the genus of mountain ash, which has berries rich in the related sugar alcohol sorbitol). Two sugars that differ only in the configuration around one carbon atom are called **epimers;** D-glucose and D-mannose, which differ only in the stereochemistry at C-2, are epimers, as are D-glucose and D-galactose (which differ at C-4) (Fig. 7–4).

Some sugars occur naturally in their L form; examples are L-arabinose and the L isomers of some sugar derivatives that are common components of glycoconjugates (Section 7.3).

| | |
|---|---|
| H    O | |
| \\  // | |
| C | |
| H—C—OH | |
| HO—C—H | |
| HO—C—H | |
| CH$_2$OH | |

L-Arabinose

## The Common Monosaccharides Have Cyclic Structures

For simplicity, we have thus far represented the structures of aldoses and ketoses as straight-chain molecules (Figs 7–3, 7–4). In fact, in aqueous solution, aldotetroses and all monosaccharides with five or more carbon atoms in the backbone occur predominantly as cyclic (ring) structures in which the carbonyl group has formed a covalent bond with the oxygen of a hydroxyl

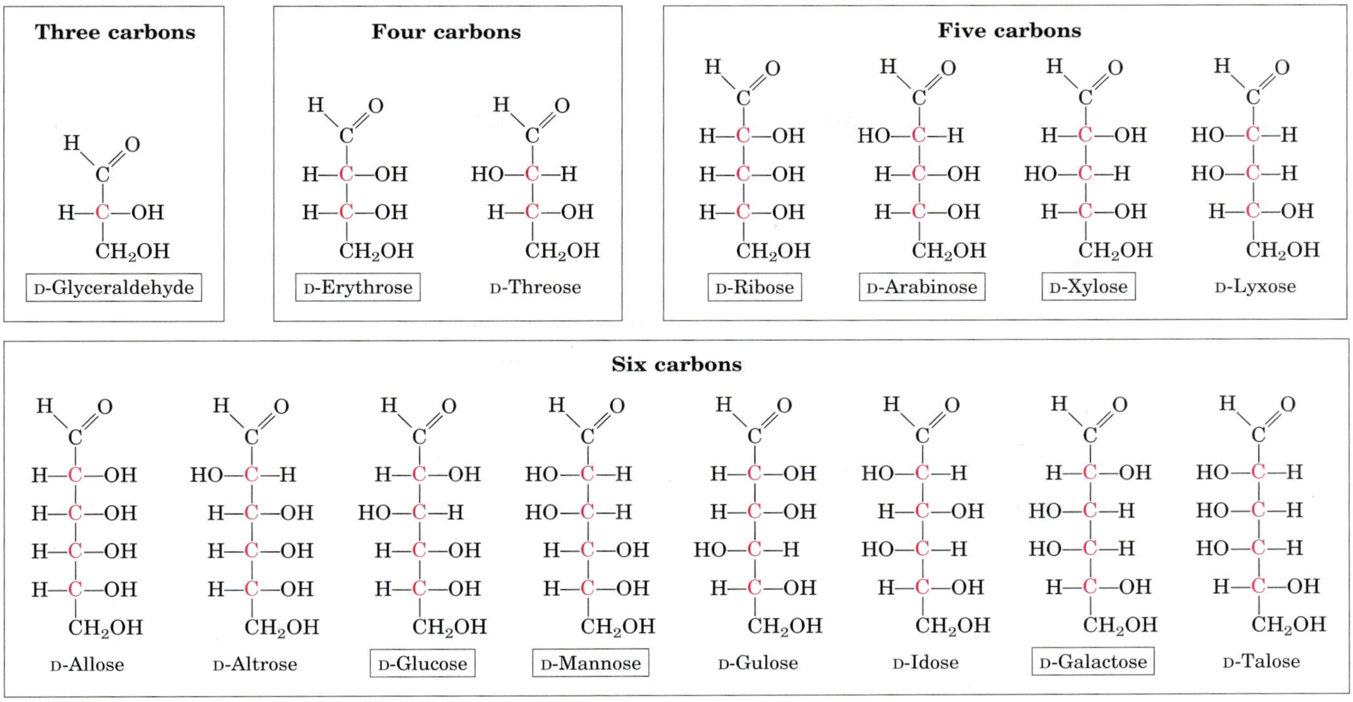

**Three carbons**

D-Glyceraldehyde

**Four carbons**

D-Erythrose

D-Threose

**Five carbons**

D-Ribose

D-Arabinose

D-Xylose

D-Lyxose

**Six carbons**

D-Allose

D-Altrose

D-Glucose

D-Mannose

D-Gulose

D-Idose

D-Galactose

D-Talose

**D-Aldoses**

**(a)**

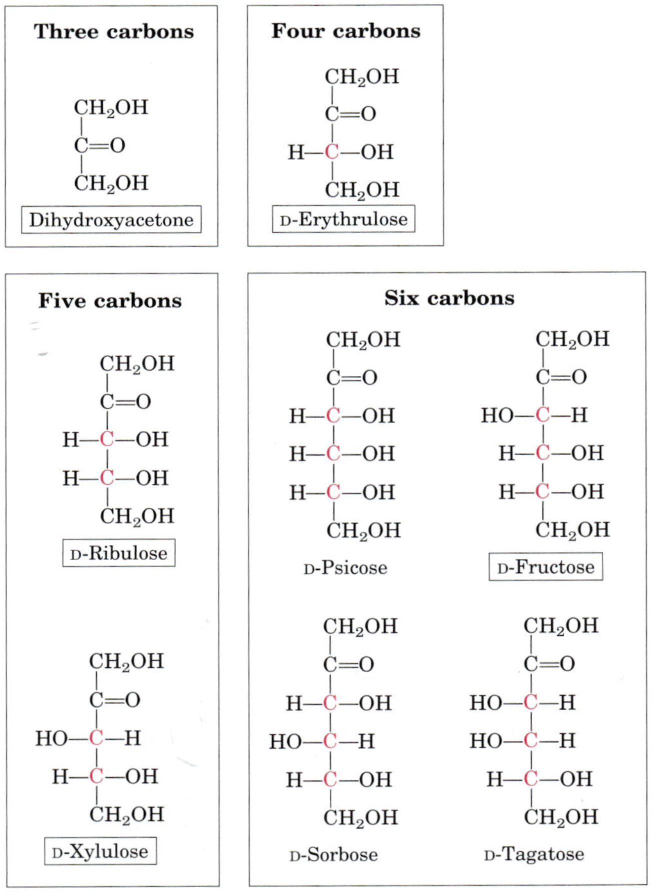

**Three carbons**

Dihydroxyacetone

**Four carbons**

D-Erythrulose

**Five carbons**

D-Ribulose

D-Xylulose

**Six carbons**

D-Psicose

D-Fructose

D-Sorbose

D-Tagatose

**D-Ketoses**

**(b)**

**FIGURE 7–3 Aldoses and ketoses.** The series of **(a)** D-aldoses and **(b)** D-ketoses having from three to six carbon atoms, shown as projection formulas. The carbon atoms in red are chiral centers. In all these D isomers, the chiral carbon *most distant from the carbonyl carbon* has the same configuration as the chiral carbon in D-glyceraldehyde. The sugars named in boxes are the most common in nature; you will encounter these again in this and later chapters.

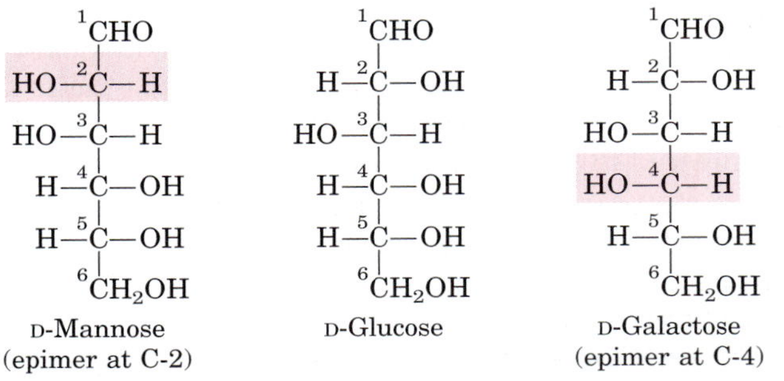

D-Mannose
(epimer at C-2)

D-Glucose

D-Galactose
(epimer at C-4)

**FIGURE 7–4 Epimers.** D-Glucose and two of its epimers are shown as projection formulas. Each epimer differs from D-glucose in the configuration at one chiral center (shaded pink).

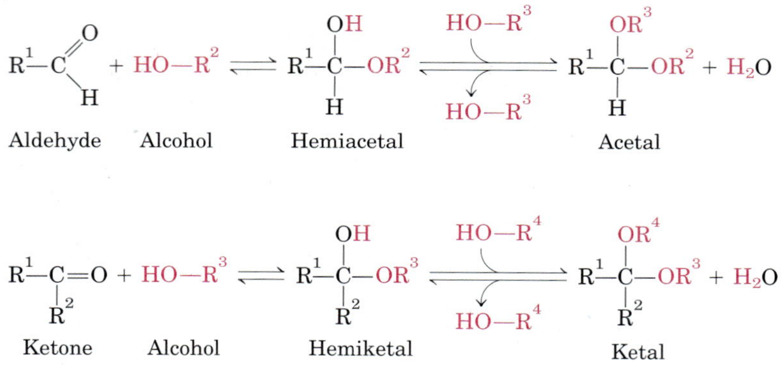

Aldehyde    Alcohol    Hemiacetal    Acetal

Ketone    Alcohol    Hemiketal    Ketal

**FIGURE 7–5 Formation of hemiacetals and hemiketals.** An aldehyde or ketone can react with an alcohol in a 1:1 ratio to yield a hemiacetal or hemiketal, respectively, creating a new chiral center at the carbonyl carbon. Substitution of a second alcohol molecule produces an acetal or ketal. When the second alcohol is part of another sugar molecule, the bond produced is a glycosidic bond (p. 245).

group along the chain. The formation of these ring structures is the result of a general reaction between alcohols and aldehydes or ketones to form derivatives called **hemiacetals** or **hemiketals** (Fig. 7–5), which contain an additional asymmetric carbon atom and thus can exist in two stereoisomeric forms. For example, D-glucose exists in solution as an intramolecular hemiacetal in which the free hydroxyl group at C-5 has reacted with the aldehydic C-1, rendering the latter carbon asymmetric and producing two stereoisomers, designated $\alpha$ and $\beta$ (Fig. 7–6). These six-membered ring compounds are called **pyranoses** because they resemble the six-membered ring compound pyran (Fig. 7–7). The systematic names for the two ring forms of D-glucose are $\alpha$-D-glucopyranose and $\beta$-D-glucopyranose.

Aldohexoses also exist in cyclic forms having five-membered rings, which, because they resemble the five-membered ring compound furan, are called **furanoses.** However, the six-membered aldopyranose ring is much more stable than the aldofuranose ring and predominates in aldohexose solutions. Only aldoses having five or more carbon atoms can form pyranose rings.

Isomeric forms of monosaccharides that differ only in their configuration about the hemiacetal or hemiketal carbon atom are called **anomers.** The hemiacetal (or carbonyl) carbon atom is called the **anomeric carbon.** The $\alpha$ and $\beta$ anomers of D-glucose interconvert in aqueous solution by a process called **mutarotation** (Fig. 7–6). Thus, a solution of $\alpha$-D-glucose and a solution of $\beta$-D-glucose eventually form identical equilibrium mixtures having identical optical properties. This mixture consists of about one-third $\alpha$-D-glucose, two-thirds $\beta$-D-glucose, and very small amounts of the linear and five-membered ring (glucofuranose) forms.

Ketohexoses also occur in $\alpha$ and $\beta$ anomeric forms. In these compounds the hydroxyl group at C-5 (or C-6) reacts with the keto group at C-2, forming a furanose (or pyranose) ring containing a hemiketal linkage (Fig. 7–5). D-Fructose readily forms the furanose ring (Fig. 7–7); the more common anomer of this sugar in combined forms or in derivatives is $\beta$-D-fructofuranose.

**Haworth perspective formulas** like those in Figure 7–7 are commonly used to show the stereochem-

istry of ring forms of monosaccharides. However, the six-membered pyranose ring is not planar, as Haworth perspectives suggest, but tends to assume either of two "chair" conformations (Fig. 7–8). Recall from Chapter 1 (p. 19) that two *conformations* of a molecule are interconvertible without the breakage of covalent bonds,

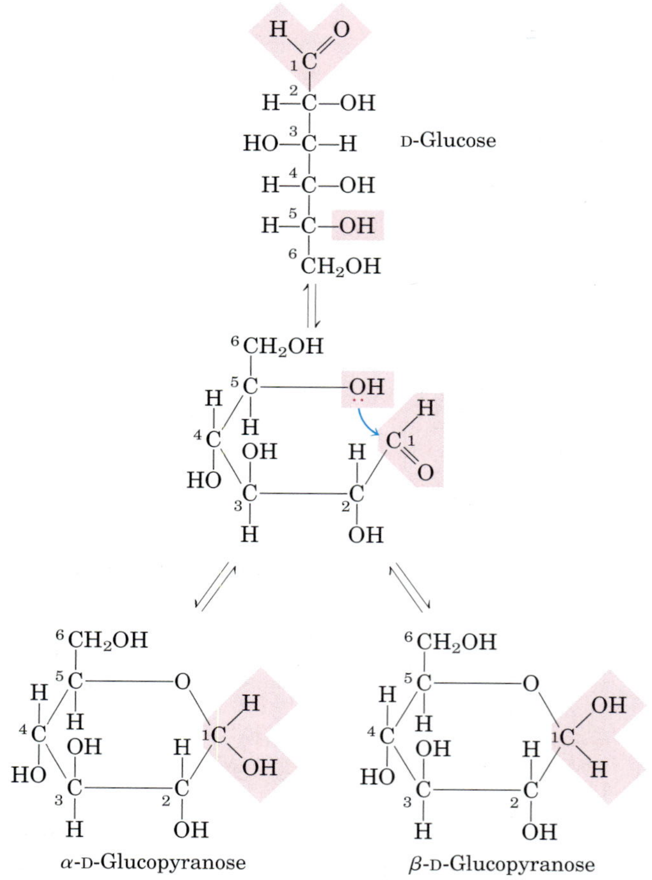

D-Glucose

$\alpha$-D-Glucopyranose    $\beta$-D-Glucopyranose

**FIGURE 7–6 Formation of the two cyclic forms of D-glucose.** Reaction between the aldehyde group at C-1 and the hydroxyl group at C-5 forms a hemiacetal linkage, producing either of two stereoisomers, the $\alpha$ and $\beta$ anomers, which differ only in the stereochemistry around the hemiacetal carbon. The interconversion of $\alpha$ and $\beta$ anomers is called mutarotation.

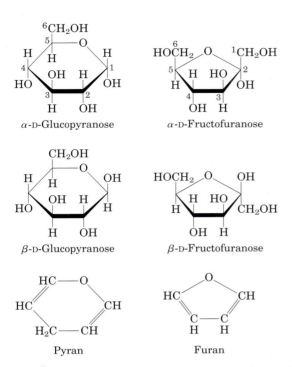

**FIGURE 7-7 Pyranoses and furanoses.** The pyranose forms of D-glucose and the furanose forms of D-fructose are shown here as Haworth perspective formulas. The edges of the ring nearest the reader are represented by bold lines. Hydroxyl groups below the plane of the ring in these Haworth perspectives would appear at the right side of a Fischer projection (compare with Fig. 7–6). Pyran and furan are shown for comparison.

whereas two *configurations* can be interconverted only by breaking a covalent bond—for example, in the case of α and β configurations, the bond involving the ring oxygen atom. The specific three-dimensional conformations of the monosaccharide units are important in determining the biological properties and functions of some polysaccharides, as we shall see.

## Organisms Contain a Variety of Hexose Derivatives

In addition to simple hexoses such as glucose, galactose, and mannose, there are a number of sugar derivatives in which a hydroxyl group in the parent compound is replaced with another substituent, or a carbon atom is oxidized to a carboxyl group (Fig. 7–9). In glucosamine, galactosamine, and mannosamine, the hydroxyl at C-2 of the parent compound is replaced with an amino group. The amino group is nearly always condensed with acetic acid, as in *N*-acetylglucosamine. This glucosamine derivative is part of many structural polymers, including those of the bacterial cell wall. Bacterial cell walls also contain a derivative of glucosamine, *N*-acetylmuramic acid, in which lactic acid (a three-carbon carboxylic acid) is ether-linked to the oxygen at C-3 of *N*-acetylglucosamine. The substitution of a hydrogen for

the hydroxyl group at C-6 of L-galactose or L-mannose produces L-fucose or L-rhamnose, respectively; these deoxy sugars are found in plant polysaccharides and in the complex oligosaccharide components of glycoproteins and glycolipids.

Oxidation of the carbonyl (aldehyde) carbon of glucose to the carboxyl level produces gluconic acid; other aldoses yield other **aldonic acids.** Oxidation of the carbon at the other end of the carbon chain—C-6 of glucose, galactose, or mannose—forms the corresponding **uronic acid:** glucuronic, galacturonic, or mannuronic acid. Both aldonic and uronic acids form stable intramolecular esters called lactones (Fig. 7–9, lower left). In addition to these acidic hexose derivatives, one nine-carbon acidic sugar deserves mention: *N*-acetylneuraminic acid (a sialic acid, but often referred to simply as "sialic acid"), a derivative of *N*-acetylmannosamine, is a component of many glycoproteins and glycolipids in animals. The carboxylic acid groups of the acidic sugar derivatives are ionized at pH 7, and the compounds are therefore correctly named as the carboxylates—glucuronate, galacturonate, and so forth.

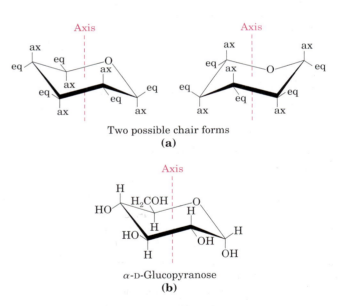

**FIGURE 7-8 Conformational formulas of pyranoses. (a)** Two chair forms of the pyranose ring. Substituents on the ring carbons may be either axial (ax), projecting parallel to the vertical axis through the ring, or equatorial (eq), projecting roughly perpendicular to this axis. Two *conformers* such are these are not readily interconvertible without breaking the ring. However, when the molecule is "stretched" (by atomic force microscopy; see Box 11–1), an input of about 46 kJ of energy per mole of sugar can force the interconversion of chair forms. Generally, substituents in the equatorial positions are less sterically hindered by neighboring substituents, and conformers with bulky substituents in equatorial positions are favored. Another conformation, the "boat" (not shown), is seen only in derivatives with very bulky substituents. **(b)** A chair conformation of α-D-glucopyranose.

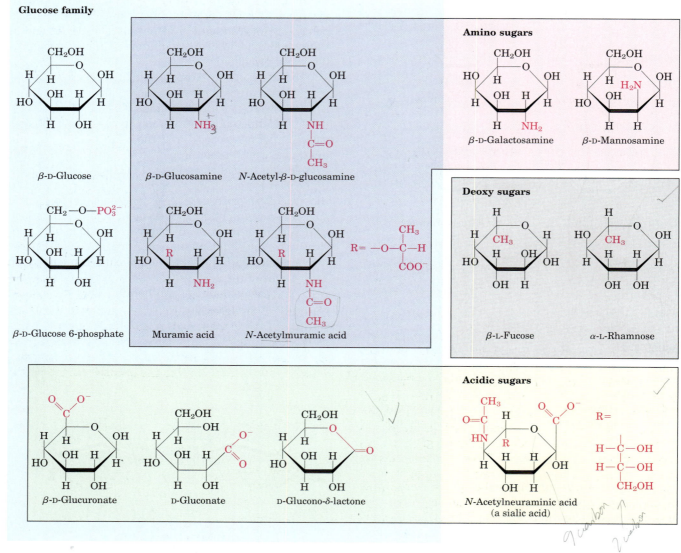

**FIGURE 7–9  Some hexose derivatives important in biology.** In amino sugars, an —NH₂ group replaces one of the —OH groups in the parent hexose. Substitution of —H for —OH produces a deoxy sugar; note that the deoxy sugars shown here occur in nature as the L isomers. The acidic sugars contain a carboxylate group, which confers a negative charge at neutral pH. D-Glucono-δ-lactone results from formation of an ester linkage between the C-1 carboxylate group and the C-5 (also known as the δ carbon) hydroxyl group of D-gluconate.

In the synthesis and metabolism of carbohydrates, the intermediates are very often not the sugars themselves but their phosphorylated derivatives. Condensation of phosphoric acid with one of the hydroxyl groups of a sugar forms a phosphate ester, as in glucose 6-phosphate (Fig. 7–9). Sugar phosphates are relatively stable at neutral pH and bear a negative charge. One effect of sugar phosphorylation within cells is to trap the sugar inside the cell; most cells do not have plasma membrane transporters for phosphorylated sugars. Phosphorylation also activates sugars for subsequent chemical transformation. Several important phosphorylated derivatives of sugars are components of nucleotides (discussed in the next chapter).

## Monosaccharides Are Reducing Agents

Monosaccharides can be oxidized by relatively mild oxidizing agents such as ferric ($Fe^{3+}$) or cupric ($Cu^{2+}$) ion (Fig. 7–10a). The carbonyl carbon is oxidized to a carboxyl group. Glucose and other sugars capable of reducing ferric or cupric ion are called **reducing sugars.** This property is the basis of Fehling's reaction, a qualitative test for the presence of reducing sugar. By measuring the amount of oxidizing agent reduced by a solution of a sugar, it is also possible to estimate the concentration of that sugar. For many years this test was used to detect and measure elevated glucose levels in blood and urine in the diagnosis of dia-

**(a)**

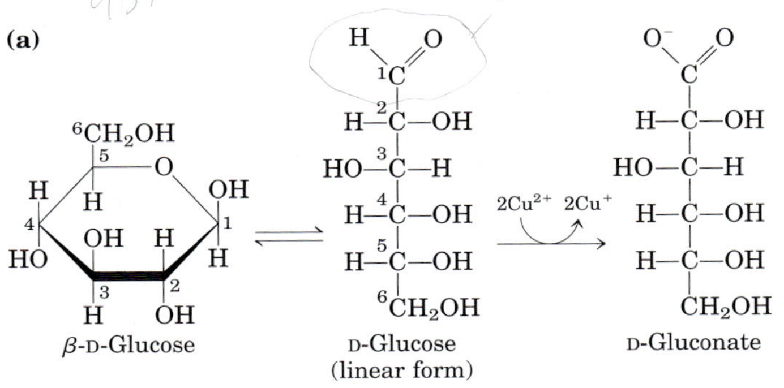

*[handwritten: 95%.]*  *[handwritten: 5%]*  *[handwritten: aldehyde]*

β-D-Glucose ⇌ D-Glucose (linear form) →($2Cu^{2+}$ →$2Cu^+$)→ D-Gluconate

**(b)** D-Glucose + $O_2$ →(glucose oxidase)→ D-Glucono-δ-lactone + $H_2O_2$

**FIGURE 7–10** **Sugars as reducing agents.** **(a)** Oxidation of the anomeric carbon of glucose and other sugars is the basis for Fehling's reaction. The cuprous ion ($Cu^+$) produced under alkaline conditions forms a red cuprous oxide precipitate. In the hemi-acetal (ring) form, C-1 of glucose cannot be oxidized by $Cu^{2+}$. However, the open-chain form is in equilibrium with the ring form, and eventually the oxidation reaction goes to completion. The reaction with $Cu^{2+}$ is not as simple as the equation here implies; in addition to D-gluconate, a number of shorter-chain acids are produced by the fragmentation of glucose. **(b)** Blood glucose concentration is commonly determined by measuring the amount of $H_2O_2$ produced in the reaction catalyzed by glucose oxidase. In the reaction mixture, a second enzyme, peroxidase, catalyzes reaction of the $H_2O_2$ with a colorless compound to produce a colored compound, the amount of which is then measured spectrophotometrically.

betes mellitus. Now, more sensitive methods for measuring blood glucose employ an enzyme, glucose oxidase (Fig. 7–10b). ∎

## Disaccharides Contain a Glycosidic Bond

Disaccharides (such as maltose, lactose, and sucrose) consist of two monosaccharides joined covalently by an **O-glycosidic bond**, which is formed when a hydroxyl group of one sugar reacts with the anomeric carbon of the other (Fig. 7–11). This reaction represents the formation of an acetal from a hemiacetal (such as glucopyranose) and an alcohol (a hydroxyl group of the second sugar molecule) (Fig. 7–5). Glycosidic bonds are readily hydrolyzed by acid but resist cleavage by base. Thus disaccharides can be hydrolyzed to yield their free monosaccharide components by boiling with dilute acid. N-glycosyl bonds join the anomeric carbon of a sugar to a nitrogen atom in glycoproteins (see Fig. 7–31) and nucleotides (see Fig. 8–1).

The oxidation of a sugar's anomeric carbon by cupric or ferric ion (the reaction that defines a reducing sugar) occurs only with the linear form, which exists in equilibrium with the cyclic form(s). When the anomeric carbon is involved in a glycosidic bond, that sugar residue cannot take the linear form and therefore becomes a nonreducing sugar. In describing disaccharides or polysaccharides, the end of a chain with a free anomeric carbon (one not involved in a glycosidic bond) is commonly called the **reducing end.**

The disaccharide maltose (Fig. 7–11) contains two D-glucose residues joined by a glycosidic linkage between C-1 (the anomeric carbon) of one glucose residue and C-4 of the other. Because the disaccharide retains a free anomeric carbon (C-1 of the glucose residue on the right in Fig. 7–11), maltose is a reducing sugar. The configuration of the anomeric carbon atom in the glycosidic linkage is α. The glucose residue with the free anomeric carbon is capable of existing in α- and β-pyranose forms.

To name reducing disaccharides such as maltose unambiguously, and especially to name more complex oligosaccharides, several rules are followed. By convention, the name describes the compound with its nonreducing end to the left, and we can "build up" the name in the following order. (1) Give the configuration (α or β) at the anomeric carbon joining the first monosaccharide unit (on the left) to the second. (2) Name the

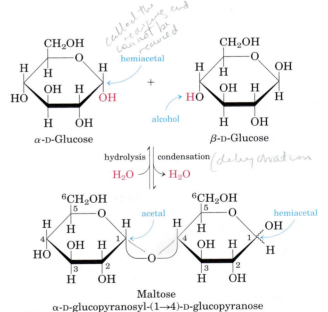

*[handwritten annotations: "Called the reducing end, can not be reduced"; "hemiacetal"; "alcohol"; "condensation (dehydration)"; "acetal"; "hemiacetal"]*

α-D-Glucose + β-D-Glucose

hydrolysis $H_2O$ ⇅ condensation $H_2O$

Maltose
α-D-glucopyranosyl-(1→4)-D-glucopyranose

**FIGURE 7–11** **Formation of maltose.** A disaccharide is formed from two monosaccharides (here, two molecules of D-glucose) when an —OH (alcohol) of one glucose molecule (right) condenses with the intramolecular hemiacetal of the other glucose molecule (left), with elimination of $H_2O$ and formation of an O-glycosidic bond. The reversal of this reaction is hydrolysis—attack by $H_2O$ on the glycosidic bond. The maltose molecule retains a reducing hemiacetal at the C-1 not involved in the glycosidic bond. Because mutarotation interconverts the α and β forms of the hemiacetal, the bonds at this position are sometimes depicted with wavy lines, as shown here, to indicate that the structure may be either α or β.

nonreducing residue; to distinguish five- and six-membered ring structures, insert "furano" or "pyrano" into the name. (3) Indicate in parentheses the two carbon atoms joined by the glycosidic bond, with an arrow connecting the two numbers; for example, (1→4) shows that C-1 of the first-named sugar residue is joined to C-4 of the second. (4) Name the second residue. If there is a third residue, describe the second glycosidic bond by the same conventions. (To shorten the description of complex polysaccharides, three-letter abbreviations for the monosaccharides are often used, as given in Table 7–1.) Following this convention for naming oligosaccharides, maltose is α-D-glucopyranosyl-(1→4)-D-glucopyranose. Because most sugars encountered in this book are the D enantiomers and the pyranose form of hexoses predominates, we generally use a shortened version of the formal name of such compounds, giving the configuration of the anomeric carbon and naming the carbons joined by the glycosidic bond. In this abbreviated nomenclature, maltose is Glc(α1→4)Glc.

The disaccharide lactose (Fig. 7–12), which yields D-galactose and D-glucose on hydrolysis, occurs naturally only in milk. The anomeric carbon of the glucose residue is available for oxidation, and thus lactose is a reducing disaccharide. Its abbreviated name is Gal(β1→4)Glc. Sucrose (table sugar) is a disaccharide of glucose and fructose. It is formed by plants but not by animals. In contrast to maltose and lactose, sucrose contains no free anomeric carbon atom; the anomeric carbons of both monosaccharide units are involved in the glycosidic bond (Fig. 7–12). Sucrose is therefore a nonreducing sugar. Nonreducing disaccharides are named as glycosides; in this case, the positions joined are the anomeric carbons. In the abbreviated nomenclature, a double-headed arrow connects the symbols specifying the anomeric carbons and their configurations. For example, the abbreviated name of sucrose is either Glc(α1↔2β)Fru or Fru(β2↔1α)Glc. Sucrose is a major intermediate product of photosynthesis; in

many plants it is the principal form in which sugar is transported from the leaves to other parts of the plant body. Trehalose, Glc(α1↔1α)Glc (Fig. 7–12)—a disaccharide of D-glucose that, like sucrose, is a nonreducing sugar—is a major constituent of the circulating fluid (hemolymph) of insects, serving as an energy-storage compound.

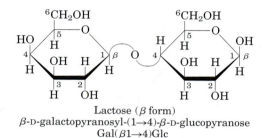

Lactose (β form)
β-D-galactopyranosyl-(1→4)-β-D-glucopyranose
Gal(β1→4)Glc

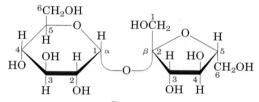

Sucrose
α-D-glucopyranosyl β-D-fructofuranoside
Glc(α1↔2β)Fru

Trehalose
α-D-glucopyranosyl α-D-glucopyranoside
Glc(α1↔1α)Glc

**FIGURE 7–12 Some common disaccharides.** Like maltose in Figure 7–11, these are shown as Haworth perspectives. The common name, full systematic name, and abbreviation are given for each disaccharide.

| TABLE 7–1 | Abbreviations for Common Monosaccharides and Some of Their Derivatives | | |
|---|---|---|---|
| Abequose | Abe | Glucuronic acid | GlcA |
| Arabinose | Ara | Galactosamine | GalN |
| Fructose | Fru | Glucosamine | GlcN |
| Fucose | Fuc | N-Acetylgalactosamine | GalNAc |
| Galactose | Gal | N-Acetylglucosamine | GlcNAc |
| Glucose | Glc | Iduronic acid | IdoA |
| Mannose | Man | Muramic acid | Mur |
| Rhamnose | Rha | N-Acetylmuramic acid | Mur2Ac |
| Ribose | Rib | N-Acetylneuraminic acid | Neu5Ac |
| Xylose | Xyl | (a sialic acid) | |

## SUMMARY 7.1 Monosaccharides and Disaccharides

■ Sugars (also called saccharides) are compounds containing an aldehyde or ketone group and two or more hydroxyl groups.

■ Monosaccharides generally contain several chiral carbons and therefore exist in a variety of stereochemical forms, which may be represented on paper as Fischer projections. Epimers are sugars that differ in configuration at only one carbon atom.

■ Monosaccharides commonly form internal hemiacetals or hemiketals, in which the aldehyde or ketone group joins with a hydroxyl group of the same molecule, creating a cyclic structure; this can be represented as a Haworth perspective formula. The carbon atom originally found in the aldehyde or ketone group (the anomeric carbon) can assume either of two configurations, $\alpha$ and $\beta$, which are interconvertible by mutarotation. In the linear form, which is in equilibrium with the cyclized forms, the anomeric carbon is easily oxidized.

■ A hydroxyl group of one monosaccharide can add to the anomeric carbon of a second monosaccharide to form an acetal. In this disaccharide, the glycosidic bond protects the anomeric carbon from oxidation.

■ Oligosaccharides are short polymers of several monosaccharides joined by glycosidic bonds. At one end of the chain, the reducing end, is a monosaccharide unit whose anomeric carbon is not involved in a glycosidic bond.

■ The common nomenclature for di- or oligosaccharides specifies the order of monosaccharide units, the configuration at each anomeric carbon, and the carbon atoms involved in the glycosidic linkage(s).

## 7.2 Polysaccharides

Most carbohydrates found in nature occur as polysaccharides, polymers of medium to high molecular weight. Polysaccharides, also called **glycans,** differ from each other in the identity of their recurring monosaccharide units, in the length of their chains, in the types of bonds linking the units, and in the degree of branching. **Homopolysaccharides** contain only a single type of monomer; **heteropolysaccharides** contain two or more different kinds (Fig. 7–13). Some homopolysaccharides serve as storage forms of monosaccharides that are used as fuels; starch and glycogen are homopolysaccharides of this type. Other homopolysaccharides (cellulose and chitin,

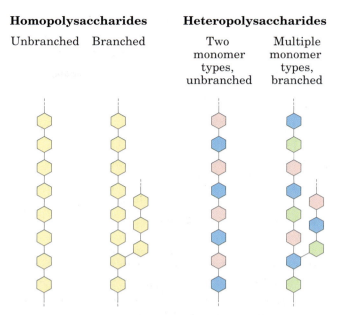

| Homopolysaccharides | | Heteropolysaccharides | |
|---|---|---|---|
| Unbranched | Branched | Two monomer types, unbranched | Multiple monomer types, branched |

**FIGURE 7–13 Homo- and heteropolysaccharides.** Polysaccharides may be composed of one, two, or several different monosaccharides, in straight or branched chains of varying length.

for example) serve as structural elements in plant cell walls and animal exoskeletons. Heteropolysaccharides provide extracellular support for organisms of all kingdoms. For example, the rigid layer of the bacterial cell envelope (the peptidoglycan) is composed in part of a heteropolysaccharide built from two alternating monosaccharide units. In animal tissues, the extracellular space is occupied by several types of heteropolysaccharides, which form a matrix that holds individual cells together and provides protection, shape, and support to cells, tissues, and organs.

Unlike proteins, polysaccharides generally do not have definite molecular weights. This difference is a consequence of the mechanisms of assembly of the two types of polymers. As we shall see in Chapter 27, proteins are synthesized on a template (messenger RNA) of defined sequence and length, by enzymes that follow the template exactly. For polysaccharide synthesis there is no template; rather, the program for polysaccharide synthesis is intrinsic to the enzymes that catalyze the polymerization of the monomeric units, and there is no specific stopping point in the synthetic process.

### Some Homopolysaccharides Are Stored Forms of Fuel

The most important storage polysaccharides are starch in plant cells and glycogen in animal cells. Both polysaccharides occur intracellularly as large clusters or granules (Fig. 7–14). Starch and glycogen molecules are heavily hydrated, because they have many exposed hydroxyl groups available to hydrogen-bond with water. Most plant cells have the ability to form starch, but it is

Starch granules

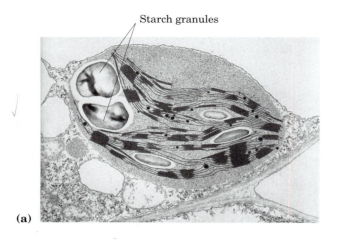

**(a)**

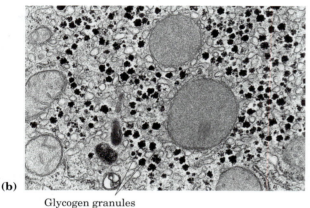

**(b)**

Glycogen granules

**FIGURE 7–14 Electron micrographs of starch and glycogen granules.**
**(a)** Large starch granules in a single chloroplast. Starch is made in the chloroplast from D-glucose formed photosynthetically. **(b)** Glycogen granules in a hepatocyte. These granules form in the cytosol and are much smaller (~0.1 $\mu$m) than starch granules (~1.0 $\mu$m).

especially abundant in tubers, such as potatoes, and in seeds.

**Starch** contains two types of glucose polymer, amylose and amylopectin (Fig. 7–15). The former consists of long, unbranched chains of D-glucose residues connected by ($\alpha1{\rightarrow}4$) linkages. Such chains vary in molecular weight from a few thousand to more than a million. Amylopectin also has a high molecular weight (up to 100 million) but unlike amylose is highly branched. The glycosidic linkages joining successive glucose residues in amylopectin chains are ($\alpha1{\rightarrow}4$); the branch points (occurring every 24 to 30 residues) are ($\alpha1{\rightarrow}6$) linkages.

**Glycogen** is the main storage polysaccharide of animal cells. Like amylopectin, glycogen is a polymer of ($\alpha1{\rightarrow}4$)-linked subunits of glucose, with ($\alpha1{\rightarrow}6$)-linked branches, but glycogen is more extensively branched (on average, every 8 to 12 residues) and more compact than starch. Glycogen is especially abundant in the liver,

where it may constitute as much as 7% of the wet weight; it is also present in skeletal muscle. In hepatocytes glycogen is found in large granules (Fig. 7–14b), which are themselves clusters of smaller granules composed of single, highly branched glycogen molecules with an average molecular weight of several million. Such glycogen granules also contain, in tightly bound form, the enzymes responsible for the synthesis and degradation of glycogen.

Because each branch in glycogen ends with a nonreducing sugar unit, a glycogen molecule has as many nonreducing ends as it has branches, but only one reducing end. When glycogen is used as an energy source, glucose units are removed one at a time from the nonreducing ends. Degradative enzymes that act only at nonreducing ends can work simultaneously on the many branches, speeding the conversion of the polymer to monosaccharides.

Why not store glucose in its monomeric form? It has been calculated that hepatocytes store glycogen equivalent to a glucose concentration of 0.4 M. The actual concentration of glycogen, which is insoluble and contributes little to the osmolarity of the cytosol, is about 0.01 $\mu$M. If the cytosol contained 0.4 M glucose, the osmolarity would be threateningly elevated, leading to osmotic entry of water that might rupture the cell (see Fig. 2–13). Furthermore, with an intracellular glucose concentration of 0.4 M and an external concentration of about 5 mM (the concentration in the blood of a mammal), the free-energy change for glucose uptake into cells against this very high concentration gradient would be prohibitively large.

**Dextrans** are bacterial and yeast polysaccharides made up of ($\alpha1{\rightarrow}6$)-linked poly-D-glucose; all have ($\alpha1{\rightarrow}3$) branches, and some also have ($\alpha1{\rightarrow}2$) or ($\alpha1{\rightarrow}4$) branches. Dental plaque, formed by bacteria growing on the surface of teeth, is rich in dextrans. Synthetic dextrans are used in several commercial products (for example, Sephadex) that serve in the fractionation of proteins by size-exclusion chromatography (see Fig. 3–18b). The dextrans in these products are chemically cross-linked to form insoluble materials of various porosities, admitting macromolecules of various sizes.

## Some Homopolysaccharides Serve Structural Roles

Cellulose, a fibrous, tough, water-insoluble substance, is found in the cell walls of plants, particularly in stalks, stems, trunks, and all the woody portions of the plant body. Cellulose constitutes much of the mass of wood, and cotton is almost pure cellulose. Like amylose and the main chains of amylopectin and glycogen, the cellulose molecule is a linear, unbranched homopolysaccharide, consisting of 10,000 to 15,000 D-glucose units. But there is a very important difference: in cellulose the glucose residues have the $\beta$ configuration (Fig. 7–16),

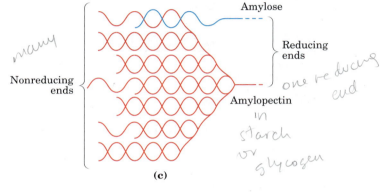

**(a) Amylose** — *starch*

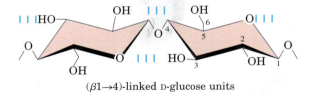

*many* Nonreducing ends

Amylose
Reducing ends
*one reducing end*
Amylopectin *in starch or glycogen*

**(b)**

**(c)**

**FIGURE 7–15 Amylose and amylopectin, the polysaccharides of starch. (a)** A short segment of amylose, a linear polymer of D-glucose residues in ($\alpha1\rightarrow4$) linkage. A single chain can contain several thousand glucose residues. Amylopectin has stretches of similarly linked residues between branch points. **(b)** An ($\alpha1\rightarrow6$) branch point of amylopectin. **(c)** A cluster of amylose and amylopectin like that believed to occur in starch granules. Strands of amylopectin (red) form double-helical structures with each other or with amylose strands (blue). Glucose residues at the nonreducing ends of the outer branches are removed enzymatically during the mobilization of starch for energy production. Glycogen has a similar structure but is more highly branched and more compact.

whereas in amylose, amylopectin, and glycogen the glucose is in the $\alpha$ configuration. The glucose residues in cellulose are linked by ($\beta1\rightarrow4$) glycosidic bonds, in contrast to the ($\alpha1\rightarrow4$) bonds of amylose, starch, and glycogen. This difference gives cellulose and amylose very different structures and physical properties.

Glycogen and starch ingested in the diet are hydrolyzed by $\alpha$-amylases, enzymes in saliva and intestinal secretions that break ($\alpha1\rightarrow4$) glycosidic bonds between glucose units. Most animals cannot use cellulose as a fuel source, because they lack an enzyme to hydrolyze the ($\beta1\rightarrow4$) linkages. Termites readily digest cellulose

**FIGURE 7–16 The structure of cellulose. (a)** Two units of a cellulose chain; the D-glucose residues are in ($\beta1\rightarrow4$) linkage. The rigid chair structures can rotate relative to one another. **(b)** Scale drawing of segments of two parallel cellulose chains, showing the conformation of the D-glucose residues and the hydrogen-bond cross-links. In the hexose unit at the lower left, all hydrogen atoms are shown; in the other three hexose units, the hydrogens attached to carbon have been omitted for clarity as they do not participate in hydrogen bonding.

($\beta1\rightarrow4$)-linked D-glucose units

**(a)**

**(b)**

**FIGURE 7-17 Cellulose breakdown by wood fungi.** A wood fungus growing on an oak log. All wood fungi have the enzyme cellulase, which breaks the ($\beta1{\rightarrow}4$) glycosidic bonds in cellulose, such that wood is a source of metabolizable sugar (glucose) for the fungus. The only vertebrates able to use cellulose as food are cattle and other ruminants (sheep, goats, camels, giraffes). The extra stomach compartment (rumen) of a ruminant teems with bacteria and protists that secrete cellulase.

(and therefore wood), but only because their intestinal tract harbors a symbiotic microorganism, *Trichonympha*, that secretes cellulase, which hydrolyzes the ($\beta1{\rightarrow}4$) linkages. Wood-rot fungi and bacteria also produce cellulase (Fig. 7–17).

**Chitin** is a linear homopolysaccharide composed of *N*-acetylglucosamine residues in $\beta$ linkage (Fig. 7–18). The only chemical difference from cellulose is the replacement of the hydroxyl group at C-2 with an acetylated amino group. Chitin forms extended fibers similar to those of cellulose, and like cellulose cannot be digested by vertebrates. Chitin is the principal component of the hard exoskeletons of nearly a million species of arthropods—insects, lobsters, and crabs, for example—and is probably the second most abundant polysaccharide, next to cellulose, in nature.

## Steric Factors and Hydrogen Bonding Influence Homopolysaccharide Folding

The folding of polysaccharides in three dimensions follows the same principles as those governing polypeptide structure: subunits with a more-or-less rigid structure dictated by covalent bonds form three-dimensional macromolecular structures that are stabilized by weak interactions within or between molecules: hydrogen-bond, hydrophobic, and van der Waals interactions, and, for polymers with charged subunits, electrostatic interactions. Because polysaccharides have so many hydroxyl groups, hydrogen bonding has an especially important influence on their structure. Glycogen, starch, and cellulose are composed of pyranoside subunits (having six-membered rings), as are the oligosaccharides of glycoproteins and glycolipids to be discussed later. Such molecules can be represented as a series of rigid pyranose rings connected by an oxygen atom bridging two carbon atoms (the glycosidic bond). There is, in princi-

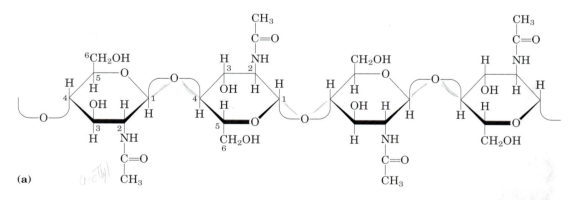

**(a)**

**FIGURE 7-18 Chitin. (a)** A short segment of chitin, a homopolymer of *N*-acetyl-D-glucosamine units in ($\beta1{\rightarrow}4$) linkage. **(b)** A spotted June beetle (*Pelidnota punctata*), showing its surface armor (exoskeleton) of chitin.

**(b)**

ple, free rotation about both C—O bonds linking the residues (Fig. 7–16a), but as in polypeptides (see Figs 4–2, 4–9), rotation about each bond is limited by steric hindrance by substituents. The three-dimensional structures of these molecules can be described in terms of the dihedral angles, $\phi$ and $\psi$, made with the glycosidic bond (Fig. 7–19), analogous to angles $\phi$ and $\psi$ made by the peptide bond (see Fig. 4–2). Because of the bulkiness of the pyranose ring and its substituents, their size and shape place constraints on the angles $\phi$ and $\psi$; certain conformations are much more stable than others, as can be shown on a map of energy as a function of $\phi$ and $\psi$ (Fig. 7–20).

The most stable three-dimensional structure for starch and glycogen is a tightly coiled helix (Fig. 7–21), stabilized by interchain hydrogen bonds. In amylose (with no branches) this structure is regular enough to allow crystallization and thus determination of the structure by x-ray diffraction. Each residue along the amylose chain forms a 60° angle with the preceding residue, so the helical structure has six residues per turn. For amylose, the core of the helix is of precisely the right dimensions to accommodate iodine in the form $I^{3-}$ or $I^{5-}$ (iodide ions), and this interaction with iodine is a common qualitative test for amylose.

For cellulose, the most stable conformation is that in which each chair is turned 180° relative to its neighbors, yielding a straight, extended chain. All —OH groups are available for hydrogen bonding with neighboring chains. With several chains lying side by side, a stabilizing network of interchain and intrachain hydrogen bonds produces straight, stable supramolecular

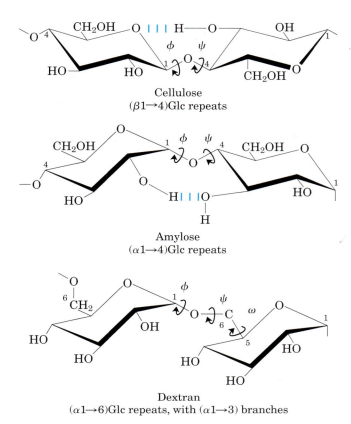

Cellulose
($\beta 1{\rightarrow}4$)Glc repeats

Amylose
($\alpha 1{\rightarrow}4$)Glc repeats

Dextran
($\alpha 1{\rightarrow}6$)Glc repeats, with ($\alpha 1{\rightarrow}3$) branches

**FIGURE 7–19 Conformation at the glycosidic bonds of cellulose, amylose, and dextran.** The polymers are depicted as rigid pyranose rings joined by glycosidic bonds, with free rotation about these bonds. Note that in dextran there is also free rotation about the bond between C-5 and C-6 (torsion angle $\omega$ (omega)).

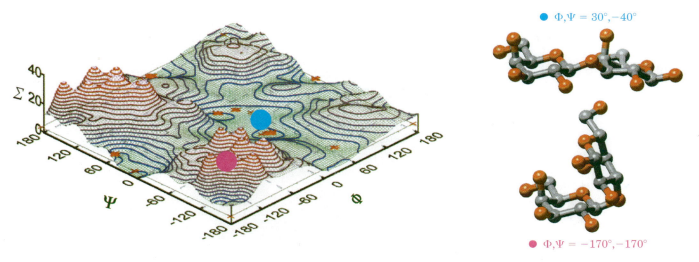

**FIGURE 7–20 A map of favored conformations for oligosaccharides and polysaccharides.** The torsion angles $\psi$ and $\phi$ (see Fig. 7–19), which define the spatial relationship between adjacent rings, can in principle have any value from 0° to 360°. In fact, some of the torsion angles would give conformations that are sterically hindered, whereas others give conformations that maximize hydrogen bonding. When the relative energy is plotted for each value of $\phi$ and $\psi$, with isoenergy ("same energy") contours drawn at intervals of 1 kcal/mol above the minimum energy state, the result is a map of preferred conformations. This is analogous to the Ramachandran plot for peptides (see Figs 4–3, 4–9). Two energetic extremes shown for the disaccharide gal($\beta 1{\rightarrow}3$)gal fall on the energy diagram as shown with dots. The known conformations of the three polysaccharides shown in Figure 7–19 have been determined by x-ray crystallography, and all fall within the lowest-energy regions of the map.

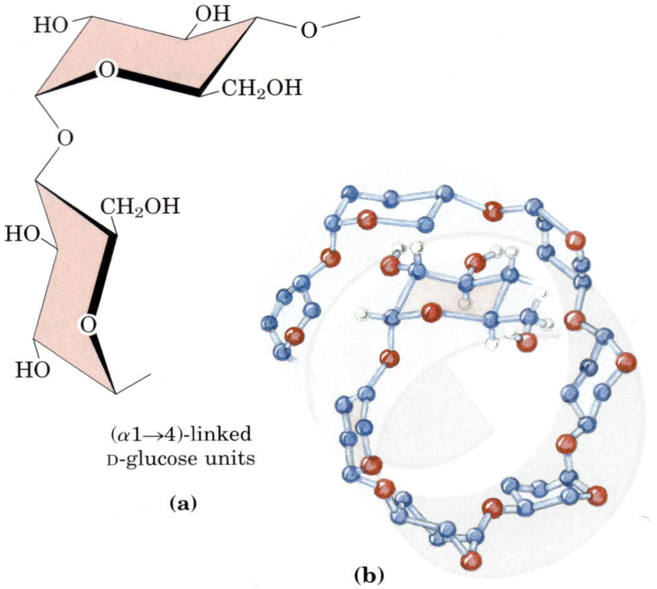

(α1→4)-linked
D-glucose units

**(a)**

**(b)**

**FIGURE 7-21 The structure of starch (amylose). (a)** In the most stable conformation, with adjacent rigid chairs, the polysaccharide chain is curved, rather than linear as in cellulose (see Fig. 7–16). **(b)** Scale drawing of a segment of amylose. The conformation of (α1→4) linkages in amylose, amylopectin, and glycogen causes these polymers to assume tightly coiled helical structures. These compact structures produce the dense granules of stored starch or glycogen seen in many cells (see Fig. 7–14).

fibers of great tensile strength (Fig. 7–16b). This property of cellulose has made it a useful substance to civilizations for millennia. Many manufactured products, including papyrus, paper, cardboard, rayon, insulating tiles, and a variety of other useful materials, are derived from cellulose. The water content of these materials is low because extensive interchain hydrogen bonding between cellulose molecules satisfies their capacity for hydrogen-bond formation.

### Bacterial and Algal Cell Walls Contain Structural Heteropolysaccharides

The rigid component of bacterial cell walls is a heteropolymer of alternating (β1→4)-linked N-acetylglucosamine and N-acetylmuramic acid residues (Fig. 7–22). The linear polymers lie side by side in the cell wall, cross-linked by short peptides, the exact structure of which depends on the bacterial species. The peptide cross-links weld the polysaccharide chains into a strong sheath that envelops the entire cell and prevents cellular swelling and lysis due to the osmotic entry of water. The enzyme lysozyme kills bacteria by hydrolyzing the (β1→4) glycosidic bond between N-acetylglucosamine and N-acetylmuramic acid (see Fig. 6–24). Lysozyme is notably present in tears, presumably as a defense against bacterial infections of the eye. It is also produced by certain bacterial viruses to ensure their release from the host

bacterial cell, an essential step of the viral infection cycle. Penicillin and related antibiotics kill bacteria by preventing synthesis of the cross-links, leaving the cell wall too weak to resist osmotic lysis (see Box 20–1).

Certain marine red algae, including some of the seaweeds, have cell walls that contain **agar,** a mixture of sulfated heteropolysaccharides made up of D-galactose and an L-galactose derivative ether-linked between C-3 and C-6 (Fig. 7–23). The two major components of agar are the unbranched polymer **agarose** ($M_r$ ~120,000) and a branched component, agaropectin. The remarkable gel-forming property of agarose makes it useful in the biochemistry laboratory. When a suspension of agarose in water is heated and cooled, the agarose forms a double helix: two molecules in parallel orientation twist together with a helix repeat of three residues; water molecules are trapped in the central cavity. These struc-

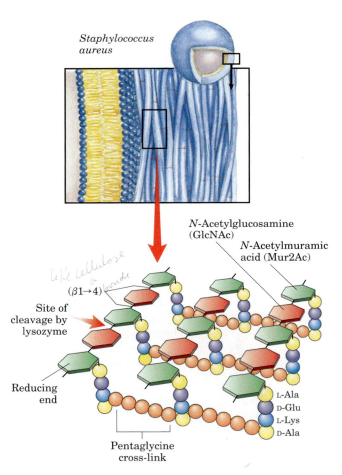

**FIGURE 7-22 Peptidoglycan.** Shown here is the peptidoglycan of the cell wall of *Staphylococcus aureus*, a gram-positive bacterium. Peptides (strings of colored spheres) covalently link N-acetylmuramic acid residues in neighboring polysaccharide chains. Note the mixture of L and D amino acids in the peptides. Gram-positive bacteria have a pentaglycine chain in the cross-link. Gram-negative bacteria, such as *E. coli,* lack the pentaglycine; instead, the terminal D-Ala residue of one tetrapeptide is attached directly to a neighboring tetrapeptide through either L-Lys or a lysine-like amino acid, diaminopimelic acid.

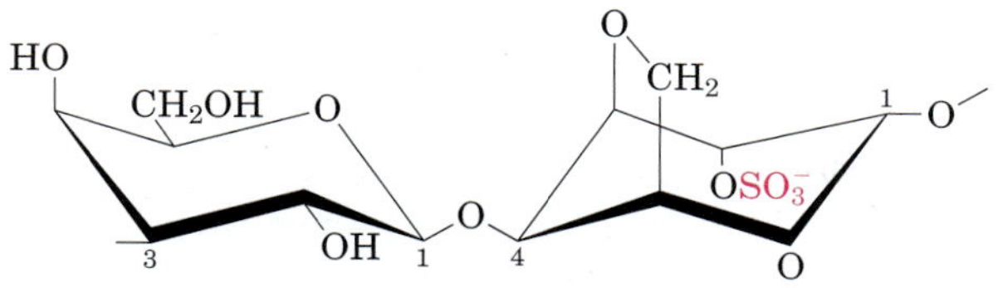

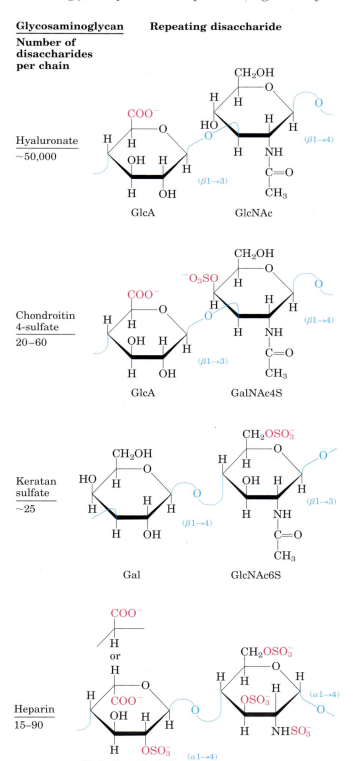

**Agarose**
3)D-Gal(β1→4)3,6-anhydro-L-Gal2S(α1 repeats

**FIGURE 7–23 The structure of agarose.** The repeating unit consists of D-galactose (β1→4)-linked to 3,6-anhydro-L-galactose (in which an ether ring connects C-3 and C-6). These units are joined by (α1→3) glycosidic links to form a polymer 600 to 700 residues long. A small fraction of the 3,6-anhydrogalactose residues have a sulfate ester at C-2 (as shown here).

tures in turn associate with each other to form a gel—a three-dimensional matrix that traps large amounts of water. Agarose gels are used as inert supports for the electrophoretic separation of nucleic acids, an essential part of the DNA sequencing process (p. 296). Agar is also used to form a surface for the growth of bacterial colonies. Another commercial use of agar is for the capsules in which some vitamins and drugs are packaged; the dried agar material dissolves readily in the stomach and is metabolically inert.

## Glycosaminoglycans Are Heteropolysaccharides of the Extracellular Matrix

The extracellular space in the tissues of multicellular animals is filled with a gel-like material, the **extracellular matrix,** also called ground substance, which holds the cells together and provides a porous pathway for the diffusion of nutrients and oxygen to individual cells. The extracellular matrix is composed of an interlocking meshwork of heteropolysaccharides and fibrous proteins such as collagen, elastin, fibronectin, and laminin. These heteropolysaccharides, the **glycosaminoglycans,** are a family of linear polymers composed of repeating disaccharide units (Fig. 7–24). One of the two monosaccharides is always either *N*-acetylglucosamine or *N*-acetylgalactosamine; the other is in most cases a uronic acid, usually D-glucuronic or L-iduronic acid. In some glycosaminoglycans, one or more of the hydroxyls of the amino sugar are esterified with sulfate. The combination

**FIGURE 7–24 Repeating units of some common glycosaminoglycans of extracellular matrix.** The molecules are copolymers of alternating uronic acid and amino sugar residues, with sulfate esters in any of several positions. The ionized carboxylate and sulfate groups (red) give these polymers their characteristic high negative charge. Heparin contains primarily iduronic acid (IdoA) and a smaller proportion of glucuronic acid (GlcA), and is generally highly sulfated and heterogeneous in length. Heparan sulfate (not shown) is similar to heparin but has a higher proportion of GlcA and fewer sulfate groups, arranged in a less regular pattern.

of sulfate groups and the carboxylate groups of the uronic acid residues gives glycosaminoglycans a very high density of negative charge. To minimize the repulsive forces among neighboring charged groups, these molecules assume an extended conformation in solution. The specific patterns of sulfated and nonsulfated sugar residues in glycosaminoglycans provide for specific recognition by a variety of protein ligands that bind electrostatically to these molecules. Glycosaminoglycans are attached to extracellular proteins to form proteoglycans (Section 7.3).

The glycosaminoglycan **hyaluronic acid** (hyaluronate at physiological pH) contains alternating residues of D-glucuronic acid and N-acetylglucosamine (Fig. 7–24). With up to 50,000 repeats of the basic disaccharide unit, hyaluronates have molecular weights greater than 1 million; they form clear, highly viscous solutions that serve as lubricants in the synovial fluid of joints and give the vitreous humor of the vertebrate eye its jelly-like consistency (the Greek *hyalos* means "glass"; hyaluronates can have a glassy or translucent appearance). Hyaluronate is also an essential component of the extracellular matrix of cartilage and tendons, to which it contributes tensile strength and elasticity as a result of its strong interactions with other components of the matrix. Hyaluronidase, an enzyme secreted by some pathogenic bacteria, can hydrolyze the glycosidic linkages of hyaluronate, rendering tissues more susceptible to bacterial invasion. In many organisms, a similar enzyme in sperm hydrolyzes an outer glycosaminoglycan coat around the ovum, allowing sperm penetration.

Other glycosaminoglycans differ from hyaluronate in two respects: they are generally much shorter polymers and they are covalently linked to specific proteins (proteoglycans). Chondroitin sulfate (Greek *chondros*, "cartilage") contributes to the tensile strength of cartilage, tendons, ligaments, and the walls of the aorta. Dermatan sulfate (Greek *derma*, "skin") contributes to the pliability of skin and is also present in blood vessels and heart valves. In this polymer, many of the glucuronate (GlcA) residues present in chondroitin sulfate are replaced by their epimer, iduronate (IdoA).

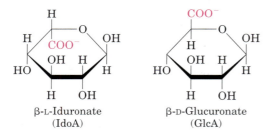

β-L-Iduronate
(IdoA)

β-D-Glucuronate
(GlcA)

Keratan sulfates (Greek *keras*, "horn") have no uronic acid and their sulfate content is variable. They are present in cornea, cartilage, bone, and a variety of horny structures formed of dead cells: horn, hair, hoofs, nails, and claws. Heparin (Greek *hēpar*, "liver") is a natural anticoagulant made in mast cells (a type of leukocyte) and released into the blood, where it inhibits blood coagulation by binding to the protein antithrombin. Heparin binding causes antithrombin to bind to and inhibit thrombin, a protease essential to blood clotting. The interaction is strongly electrostatic; heparin has the highest negative charge density of any known biological macromolecule (Fig. 7–25). Purified heparin is routinely

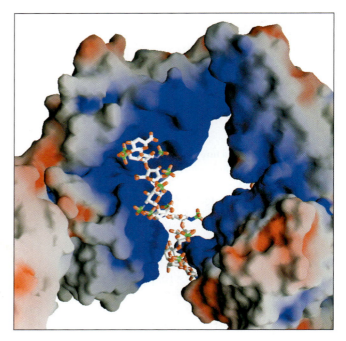

**FIGURE 7–25  Interaction between a glycosaminoglycan and its binding protein.** Fibroblast growth factor (FGF1), its cell surface receptor (FGFR), and a short segment of a glycosaminoglycan (heparin) were co-crystallized to yield the structure shown here (PDB ID 1E0O). The proteins are represented as surface contour images, with color to represent surface electrostatic potential: red, predominantly negative charge; blue, predominantly positive charge. Heparin is shown in a ball-and-stick representation, with the negative charges ($-SO_3^-$ and $-COO^-$) attracted to the positive (blue) surface of the FGF protein. Heparin was used in this experiment, but, in vivo, the glycosaminoglycan that binds FGF is heparan sulfate on the cell surface.

added to blood samples obtained for clinical analysis, and to blood donated for transfusion, to prevent clotting.

Table 7–2 summarizes the composition, properties, roles, and occurrence of the polysaccharides described in Section 7.2.

## SUMMARY 7.2 Polysaccharides

- Polysaccharides (glycans) serve as stored fuel and as structural components of cell walls and extracellular matrix.

- The homopolysaccharides starch and glycogen are stored fuels in plant, animal, and bacterial cells. They consist of D-glucose with ($\alpha1{\rightarrow}4$) linkages, and all three contain some branches.

- The homopolysaccharides cellulose, chitin, and dextran serve structural roles. Cellulose, composed of ($\beta1{\rightarrow}4$)-linked D-glucose residues, lends strength and rigidity to plant cell walls. Chitin, a polymer of ($\beta1{\rightarrow}4$)-linked N-acetylglucosamine, strengthens the

## TABLE 7–2  Structures and Roles of Some Polysaccharides

| Polymer | Type* | Repeating unit† | Size (number of monosaccharide units) | Roles/significance |
|---|---|---|---|---|
| Starch | | | | Energy storage: in plants |
|   Amylose | Homo- | $(\alpha1\rightarrow4)$Glc, linear | 50–5,000 | |
|   Amylopectin | Homo- | $(\alpha1\rightarrow4)$Glc, with $(\alpha1\rightarrow6)$Glc branches every 24–30 residues | Up to $10^6$ | |
| Glycogen | Homo- | $(\alpha1\rightarrow4)$Glc, with $(\alpha1\rightarrow6)$Glc branches every 8–12 residues | Up to 50,000 | Energy storage: in bacteria and animal cells |
| Cellulose | Homo- | $(\beta1\rightarrow4)$Glc | Up to 15,000 | Structural: in plants, gives rigidity and strength to cell walls |
| Chitin | Homo- | $(\beta1\rightarrow4)$GlcNAc | Very large | Structural: in insects, spiders, crustaceans, gives rigidity and strength to exoskeletons |
| Dextran | Homo- | $(\alpha1\rightarrow6)$Glc, with $(\alpha1\rightarrow3)$ branches | Wide range | Structural: in bacteria, extracellular adhesive |
| Peptidoglycan | Hetero-; peptides attached | 4)Mur2Ac$(\beta1\rightarrow4)$ GlcNAc$(\beta1$ | Very large | Structural: in bacteria, gives rigidity and strength to cell envelope |
| Agarose | Hetero- | 3)D-Gal$(\beta1\rightarrow4)$3,6-anhydro-L-Gal$(\alpha1$ | 1,000 | Structural: in algae, cell wall material |
| Hyaluronate (a glycosaminoglycan) | Hetero-; acidic | 4)GlcA$(\beta1\rightarrow3)$ GlcNAc$(\beta1$ | Up to 100,000 | Structural: in vertebrates, extracellular matrix of skin and connective tissue; viscosity and lubrication in joints |

*Each polymer is classified as a homopolysaccharide (homo-) or heteropolysaccharide (hetero-).

†The abbreviated names for the peptidoglycan, agarose, and hyaluronate repeating units indicate that the polymer contains repeats of this disaccharide unit. For example, in peptidoglycan, the GlcNAc of one disaccharide unit is $(\beta1\rightarrow4)$-linked to the first residue of the next disaccharide unit.

exoskeletons of arthropods. Dextran forms an adhesive coat around certain bacteria.

■ Homopolysaccharides fold in three dimensions. The chair form of the pyranose ring is essentially rigid, so the conformation of the polymers is determined by rotation about the bonds to the oxygen on the anomeric carbon. Starch and glycogen form helical structures with intrachain hydrogen bonding; cellulose and chitin form long, straight strands that interact with neighboring strands.

■ Bacterial and algal cell walls are strengthened by heteropolysaccharides—peptidoglycan in bacteria, agar in red algae. The repeating disaccharide in peptidoglycan is GlcNAc$(\beta1\rightarrow4)$Mur2Ac; in agarose, it is D-Gal$(\beta1\rightarrow4)$3,6-anhydro-L-Gal.

■ Glycosaminoglycans are extracellular heteropolysaccharides in which one of the two monosaccharide units is a uronic acid and the other an N-acetylated amino sugar. Sulfate esters on some of the hydroxyl groups give these polymers a high density of negative charge, forcing them to assume extended conformations. These polymers (hyaluronate, chondroitin sulfate, dermatan sulfate, keratan sulfate, and heparin) provide viscosity, adhesiveness, and tensile strength to the extracellular matrix.

## 7.3 Glycoconjugates: Proteoglycans, Glycoproteins, and Glycolipids

In addition to their important roles as stored fuels (starch, glycogen, dextran) and as structural materials (cellulose, chitin, peptidoglycans), polysaccharides and oligosaccharides are information carriers: they serve as destination labels for some proteins and as mediators of specific cell-cell interactions and interactions between cells and the extracellular matrix. Specific carbohydrate-containing molecules act in cell-cell recognition and adhesion, cell migration during development, blood clotting, the immune response, and wound healing, to name but a few of their many roles. In most of these cases, the informational carbohydrate is covalently joined to a protein or a lipid to form a **glycoconjugate,** which is the biologically active molecule.

**Proteoglycans** are macromolecules of the cell surface or extracellular matrix in which one or more glycosaminoglycan chains are joined covalently to a membrane protein or a secreted protein. The glycosaminoglycan moiety commonly forms the greater fraction (by mass) of the proteoglycan molecule, dominates the structure, and is often the main site of biological activity. In many cases the biological activity is the provision of multiple binding sites, rich in opportunities for hydrogen bonding and electrostatic interactions with other proteins of the cell surface or the extracellular matrix. Proteoglycans are major components of connective tissue such as cartilage, in which their many noncovalent interactions with other proteoglycans, proteins, and glycosaminoglycans provide strength and resilience.

**Glycoproteins** have one or several oligosaccharides of varying complexity joined covalently to a protein. They are found on the outer face of the plasma membrane, in the extracellular matrix, and in the blood. Inside cells they are found in specific organelles such as Golgi complexes, secretory granules, and lysosomes. The oligosaccharide portions of glycoproteins are less monotonous than the glycosaminoglycan chains of proteoglycans; they are rich in information, forming highly specific sites for recognition and high-affinity binding by other proteins.

**Glycolipids** are membrane lipids in which the hydrophilic head groups are oligosaccharides, which, as in glycoproteins, act as specific sites for recognition by carbohydrate-binding proteins.

## Proteoglycans Are Glycosaminoglycan-Containing Macromolecules of the Cell Surface and Extracellular Matrix

Mammalian cells can produce at least 30 types of molecules that are members of the proteoglycan superfamily. These molecules act as tissue organizers, influence the development of specialized tissues, mediate the activities of various growth factors, and regulate the extracellular assembly of collagen fibrils. The basic proteoglycan unit consists of a "core protein" with covalently attached glycosaminoglycan(s). For example, the sheet-like extracellular matrix (basal lamina) that separates organized groups of cells contains a family of core proteins ($M_r$ 20,000 to 40,000), each with several covalently attached heparan sulfate chains. (Heparan sulfate is structurally similar to heparin but has a lower density of sulfate esters.) The point of attachment is commonly a

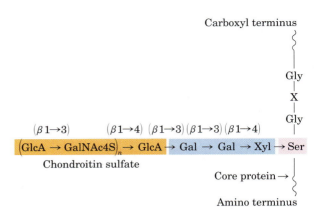

**FIGURE 7–26 Proteoglycan structure, showing the trisaccharide bridge.** A typical trisaccharide linker (blue) connects a glycosaminoglycan—in this case chondroitin sulfate (orange)—to a Ser residue (red) in the core protein. The xylose residue at the reducing end of the linker is joined by its anomeric carbon to the hydroxyl of the Ser residue.

Ser residue, to which the glycosaminoglycan is joined through a trisaccharide bridge (Fig. 7–26). The Ser residue is generally in the sequence –Ser–Gly–X–Gly– (where X is any amino acid residue), although not every protein with this sequence has an attached glycosaminoglycan. Many proteoglycans are secreted into the extracellular matrix, but some are integral membrane proteins (see Fig. 11–6). For example, syndecan core protein ($M_r$ 56,000) has a single transmembrane domain and an extracellular domain bearing three chains of heparan sulfate and two of chondroitin sulfate, each attached to a Ser residue (Fig. 7–27a). There are at least four members of the syndecan family in mammals. Another family of core proteins is the glypicans, with six members. These proteins are attached to the membrane by a lipid anchor, a derivative of the membrane lipid phosphatidylinositol (Chapter 11).

The heparan sulfate moieties in proteoglycans bind a variety of extracellular ligands and thereby modulate the ligands' interaction with specific receptors of the cell surface. Detailed examination of the glycan moiety of proteoglycans has revealed a sequence heterogeneity that is not random; some domains (typically 3 to 8 disaccharide units long) differ from neighboring domains in sequence and in ability to bind to specific proteins. Heparan sulfate, for example, is initially synthesized as a long polymer (50 to 200 disaccharide units) of alternating N-acetylglucosamine (GlcNAc) and glucuronic acid (GlcA) residues. This simple chain is acted on by a series of enzymes that introduce alterations in specific regions. First, an N-deacetylase: N-sulfotransferase replaces some acetyl groups of GlcNAc residues with sulfates, creating clusters of N-sulfated glucosamine (GlcN) residues. These clusters then attract enzymes that carry out further modifications: an epimerase con-

**(a) Syndecan**

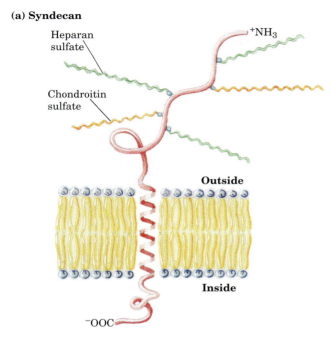

**(b) Heparan sulfate**

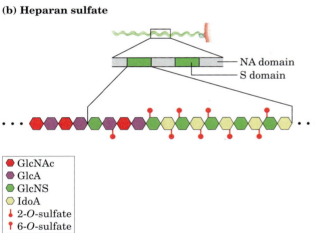

- ⬡ GlcNAc
- ⬡ GlcA
- ⬡ GlcNS
- ⬡ IdoA
- 🔴 2-*O*-sulfate
- 🔺 6-*O*-sulfate

**FIGURE 7–27 Proteoglycan structure of an integral membrane protein. (a)** Schematic diagram of syndecan, a core protein of the plasma membrane. The amino-terminal domain on the extracellular surface of the membrane is covalently attached (by trisaccharide linkers such as those in Fig. 7–26) to three heparan sulfate chains and two chondroitin sulfate chains. Some core proteins (syndecans, as here) are anchored by a single transmembrane helix; others (glypicans), by a covalently attached membrane glycolipid. In a third class of core proteins, the protein is released into the extracellular space, where it forms part of the basement membrane. **(b)** Along a heparan sulfate chain, regions rich in sulfated sugars, the S domains (green), alternate with regions with chiefly unmodified residues of GlcNAc and GlcA, the NA domains (gray). One of the S domains is shown in more detail, revealing a high density of modified residues: GlcA, with a sulfate ester at C-6; and IdoA, with a sulfate ester at C-2. The exact pattern of sulfation in the S domain differs among proteoglycans. Given all the possible modifications of the GlcNAc–IdoA dimer, at least 32 different disaccharide units are possible.

verts GlcA to IdoA; sulfotransferases then create sulfate esters at the C-2 hydroxyl of IdoA and the C-6 hydroxyl of *N*-sulfated GlcN, but only in regions that already have *N*-sulfated GlcN residues. The result is a polymer in which highly sulfated domains (S domains) alternate with domains having unmodified GlcNAc and GlcA residues (*N*-acetylated, or NA, domains) (Fig. 7–27b). The exact pattern of sulfation in the S domain differs in different proteoglycans; given the number of possible modifications of the GlcNAc–IdoA dimer, at least 32 different disaccharide units are possible. Furthermore, the same core protein can display different heparan sulfate structures when synthesized in different cell types.

The S domains bind specifically to extracellular proteins and signaling molecules to alter their activities. The change in activity may result from a conformational change in the protein that is induced by the binding (Fig. 7–28a), or it may be due to the ability of adjacent domains of heparan sulfate to bind to two different proteins, bringing them into close proximity and enhancing protein-protein interactions (Fig. 7–28b). A third general mechanism of action is the binding of extracellular signal molecules (growth factors, for example) to heparan sulfate, which increases their local concentrations and enhances their interaction with growth factor receptors in the cell surface; in this case, the heparan sulfate acts as a coreceptor (Fig. 7–28c). For example, fibroblast growth factor (FGF), an extracellular protein signal that stimulates cell division, first binds to heparan sulfate moieties of syndecan molecules in the target cell's plasma membrane. Syndecan presents FGF to the FGF plasma membrane receptor, and only then can FGF interact productively with its receptor to trigger cell division. Finally, the S domains interact—electrostatically and otherwise—with a variety of soluble molecules outside the cell, maintaining high local concentrations at the cell surface (Fig. 7–28d). The importance of correctly synthesizing sulfated domains in heparan sulfate is demonstrated in "knockout" mice that lack the enzyme that places sulfates at the C-2 hydroxyl of IdoA. Such animals are born without kidneys and with very severe abnormalities in development of the skeleton and eyes.

Some proteoglycans can form **proteoglycan aggregates,** enormous supramolecular assemblies of many core proteins all bound to a single molecule of hyaluronate. Aggrecan core protein ($M_r$ ~250,000) has multiple chains of chondroitin sulfate and keratan sulfate, joined to Ser residues in the core protein through trisaccharide linkers, to give an aggrecan monomer of $M_r$ ~$2 \times 10^6$. When a hundred or more of these "decorated" core proteins bind a single, extended molecule of hyaluronate (Fig. 7–29), the resulting proteoglycan aggregate ($M_r > 2 \times 10^8$) and its associated water of hydration occupy a volume about equal to that of a bacterial cell! Aggrecan interacts strongly with collagen in the

**(a) Conformational activation**

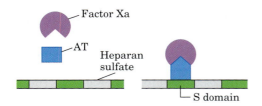

A conformational change induced in the protein antithrombin (AT) on binding a specific pentasaccharide S domain allows its interaction with Factor Xa, a blood clotting factor, preventing clotting.

**(b) Enhanced protein-protein interaction**

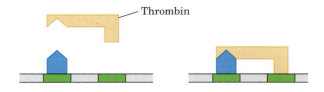

Binding of AT and thrombin to two adjacent S domains brings the two proteins into close proximity, favoring their interaction, which inhibits blood clotting.

**(c) Coreceptor for extracellular ligands**

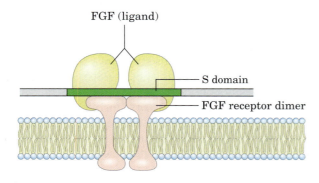

S domains interact with both the fibroblast growth factor (FGF) and its receptor, bringing the oligomeric complex together and increasing the effectiveness of a low concentration of FGF.

**(d) Cell surface localization/concentration**

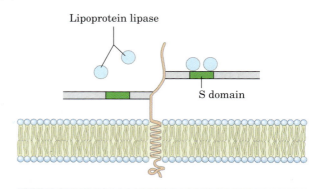

The high density of negative charges in heparan sulfate brings positively charged molecules of lipoprotein lipase into the vicinity and holds them by electrostatic interactions as well as by sequence-specific interactions with S domains. Such interactions are also central in the first step in the entry of certain viruses (such as herpes simplex viruses HSV-1 and HSV-2) into cells.

**FIGURE 7–28** Four types of protein interactions with S domains of heparan sulfate.

extracellular matrix of cartilage, contributing to the development and tensile strength of this connective tissue.

Interwoven with these enormous extracellular proteoglycans are fibrous matrix proteins such as collagen, elastin, and fibronectin, forming a cross-linked meshwork that gives the whole extracellular matrix strength and resilience. Some of these proteins are multiadhesive, a single protein having binding sites for several different matrix molecules. Fibronectin, for example, has separate domains that bind fibrin, heparan sulfate, collagen, and a family of plasma membrane proteins called integrins that mediate signaling between the cell interior and the extracellular matrix (see Fig. 11–22). Integrins, in turn, have binding sites for a number of other extracellular macromolecules. The overall picture of cell-matrix interactions that emerges (Fig. 7–30) shows an array of interactions between cellular and extracellular molecules. These interactions serve not merely to

anchor cells to the extracellular matrix but also to provide paths that direct the migration of cells in developing tissue and, through integrins, to convey information in both directions across the plasma membrane.

## Glycoproteins Have Covalently Attached Oligosaccharides

Glycoproteins are carbohydrate-protein conjugates in which the carbohydrate moieties are smaller and more structurally diverse than the glycosaminoglycans of proteoglycans. The carbohydrate is attached at its anomeric carbon through a glycosidic link to the —OH of a Ser or Thr residue (*O*-linked), or through an *N*-glycosyl link to the amide nitrogen of an Asn residue (*N*-linked) (Fig. 7–31). Some glycoproteins have a single oligosaccharide chain, but many have more than one; the carbohydrate may constitute from 1% to 70% or more of the glyco-

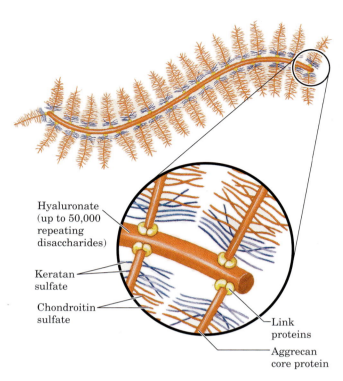

**FIGURE 7–29 Proteoglycan aggregate of the extracellular matrix.** One very long molecule of hyaluronate is associated noncovalently with about 100 molecules of the core protein aggrecan. Each aggrecan molecule contains many covalently bound chondroitin sulfate and keratan sulfate chains. Link proteins situated at the junction between each core protein and the hyaluronate backbone mediate the core protein–hyaluronate interaction.

protein by mass. The structures of a large number of *O*- and *N*-linked oligosaccharides from a variety of glycoproteins are known; Figure 7–31 shows a few typical examples.

As we shall see in Chapter 11, the external surface of the plasma membrane has many membrane glycoproteins with arrays of covalently attached oligosaccharides of varying complexity. One of the best-characterized membrane glycoproteins is glycophorin A of the erythrocyte membrane (see Fig. 11–7). It contains 60% carbohydrate by mass, in the form of 16 oligosaccharide chains (totaling 60 to 70 monosaccharide residues) covalently attached to amino acid residues near the amino terminus of the polypeptide chain. Fifteen of the oligosaccharide chains are *O*-linked to Ser or Thr residues, and one is *N*-linked to an Asn residue.

Many of the proteins secreted by eukaryotic cells are glycoproteins, including most of the proteins of blood. For example, immunoglobulins (antibodies) and certain hormones, such as follicle-stimulating hormone, luteinizing hormone, and thyroid-stimulating hormone, are glycoproteins. Many milk proteins, including lactalbumin, and some of the proteins secreted by the pancreas (such as ribonuclease) are glycosylated, as are most of the proteins contained in lysosomes.

A number of cases are known in which the same protein produced in two types of tissues has different glycosylation patterns. For example, the human protein interferon IFN-β1 has one set of oligosaccharide chains when produced in ovarian cells and a different set when produced in breast epithelial cells. The biological significance of these **tissue glycoforms** is not understood, but in some way the oligosaccharide chains represent a tissue-specific marker.

The biological advantages of adding oligosaccharides to proteins are not fully understood. The very hydrophilic clusters of carbohydrate alter the polarity and solubility of the proteins with which they are conjugated. Oligosaccharide chains that are attached to newly synthesized proteins in the endoplasmic reticulum and elaborated in the Golgi complex may also influence the sequence of polypeptide-folding events that determine the tertiary structure of the protein (see Fig. 27–34). Steric interactions between peptide and oligosaccharide may preclude one folding route and favor another. When numerous negatively charged oligosaccharide chains are clustered in a single region of a protein, the charge repulsion among them favors the formation of an extended, rodlike structure in that region. The bulkiness and negative charge of oligosaccharide chains also

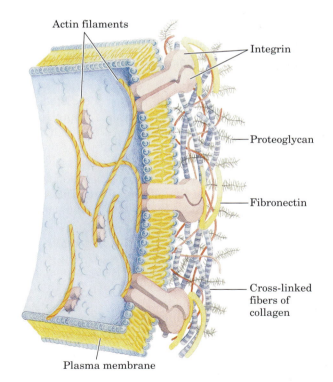

**FIGURE 7–30 Interactions between cells and the extracellular matrix.** The association between cells and the proteoglycan of the extracellular matrix is mediated by a membrane protein (integrin) and by an extracellular protein (fibronectin in this example) with binding sites for both integrin and the proteoglycan. Note the close association of collagen fibers with the fibronectin and proteoglycan.

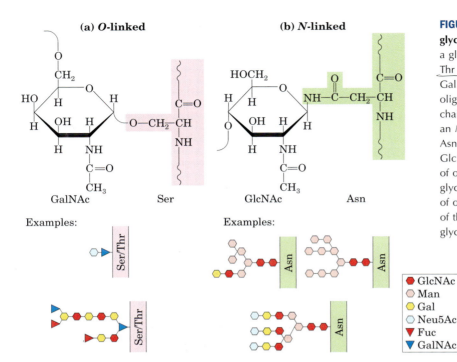

**(a) O-linked**

GalNAc        Ser

Examples:

**(b) N-linked**

GlcNAc        Asn

Examples:

**FIGURE 7–31 Oligosaccharide linkages in glycoproteins. (a)** *O*-linked oligosaccharides have a glycosidic bond to the hydroxyl group of Ser or Thr residues (shaded pink), illustrated here with GalNAc as the sugar at the reducing end of the oligosaccharide. One simple chain and one complex chain are shown. **(b)** *N*-linked oligosaccharides have an *N*-glycosyl bond to the amide nitrogen of an Asn residue (shaded green), illustrated here with GlcNAc as the terminal sugar. Three common types of oligosaccharide chains that are *N*-linked in glycoproteins are shown. A complete description of oligosaccharide structure requires specification of the position and stereochemistry (α or β) of each glycosidic linkage.

- GlcNAc
- Man
- Gal
- Neu5Ac
- Fuc
- GalNAc

protect some proteins from attack by proteolytic enzymes. Beyond these global physical effects on protein structure, there are also more specific biological effects of oligosaccharide chains in glycoproteins (Section 7.4).

### Glycolipids and Lipopolysaccharides Are Membrane Components

Glycoproteins are not the only cellular components that bear complex oligosaccharide chains; some lipids, too, have covalently bound oligosaccharides. **Gangliosides** are membrane lipids of eukaryotic cells in which the polar head group, the part of the lipid that forms the outer surface of the membrane, is a complex oligosaccharide containing sialic acid (Fig. 7–9) and other monosaccharide residues. Some of the oligosaccharide moieties of gangliosides, such as those that determine human blood groups (see Fig. 10–14), are identical with those found in certain glycoproteins, which therefore also contribute to blood group type determination. Like the oligosaccharide moieties of glycoproteins, those of membrane lipids are generally, perhaps always, found on the outer face of the plasma membrane.

**Lipopolysaccharides** are the dominant surface feature of the outer membrane of gram-negative bacteria such as *Escherichia coli* and *Salmonella typhimurium*. These molecules are prime targets of the antibodies produced by the vertebrate immune system in response to bacterial infection and are therefore important determinants of the serotype of bacterial strains (serotypes are strains that are distinguished on the basis of antigenic properties). The lipopolysaccharides of *S. typhimurium* contain six fatty acids bound to two glucosamine residues, one of which is the point of attachment for a complex oligosaccharide (Fig. 7–32). *E. coli* has similar but unique lipopolysaccharides. The lipopolysaccharides of some bacteria are toxic to humans and other animals; for example, they are responsible for the dangerously lowered blood pressure that occurs in toxic shock syndrome resulting from gram-negative bacterial infections. ■

### SUMMARY 7.3 Glycoconjugates: Proteoglycans, Glycoproteins, and Glycolipids

- Proteoglycans are glycoconjugates in which a core protein is attached covalently to one or more large glycans, such as heparan sulfate, chondroitin sulfate, or keratan sulfate. The glycan is the greater portion (by mass) of the molecule. Bound to the outside of the plasma membrane by a transmembrane peptide or a covalently attached lipid, proteoglycans provide points of adhesion, recognition, and information transfer between cells, or between the cell and the extracellular matrix.

- Glycoproteins contain covalently linked oligosaccharides that are smaller but more structurally complex, and therefore more information-rich, than glycosaminoglycans. Many cell surface or extracellular proteins are glycoproteins, as are most secreted proteins. The covalently attached oligosaccharides influence the folding and stability of the proteins, provide critical information about the

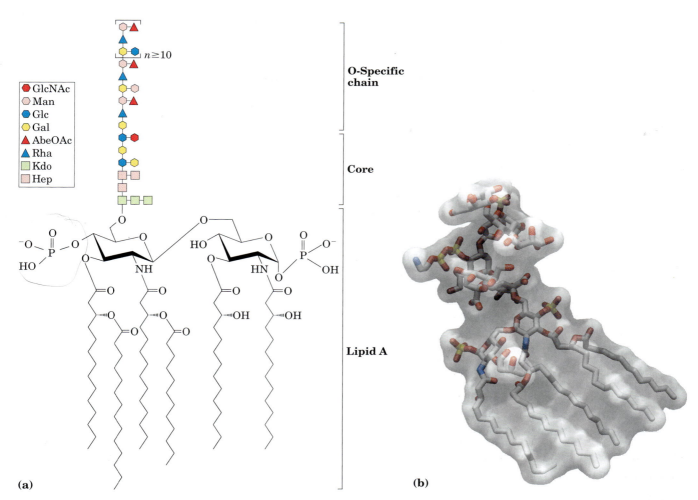

**FIGURE 7–32 Bacterial lipopolysaccharides. (a)** Schematic diagram of the lipopolysaccharide of the outer membrane of *Salmonella typhimurium*. Kdo is 3-deoxy-D-manno-octulosonic acid, previously called *keto*deoxyoctonic acid; Hep is L-glycero-D-mannoheptose; AbeOAc is abequose (a 3,6-dideoxyhexose) acetylated on one of its hydroxyls. There are six fatty acids in the lipid A portion of the molecule. Different bacterial species have subtly different lipopolysaccharide structures, but they have in common a lipid region (lipid A), a core oligosaccharide, and an "O-specific" chain, which is the prin-

cipal determinant of the serotype (immunological reactivity) of the bacterium. The outer membranes of the gram-negative bacteria *S. typhimurium* and *E. coli* contain so many lipopolysaccharide molecules that the cell surface is virtually covered with O-specific chains. **(b)** The stick structure of the lipopolysaccharide of *E. coli* is visible through a transparent surface contour model of the molecule. The position of the sixth fatty acyl chain was not defined in the crystallographic study, so it is not shown.

targeting of newly synthesized proteins, and allow for specific recognition by other proteins.

■ Glycolipids and lipopolysaccharides are components of the plasma membrane with covalently attached oligosaccharide chains exposed on the cell's outer surface.

## 7.4 Carbohydrates as Informational Molecules: The Sugar Code

Glycobiology, the study of the structure and function of glycoconjugates, is one of the most active and exciting areas of biochemistry and cell biology. As is becoming

increasingly clear, cells use specific oligosaccharides to encode important information about intracellular targeting of proteins, cell-cell interaction, tissue development, and extracellular signals. Our discussion uses just a few examples to illustrate the diversity of structure and the range of biological activity of the glycoconjugates. In Chapter 20 we discuss the biosynthesis of polysaccharides, including the peptidoglycans; and in Chapter 27, the assembly of oligosaccharide chains on glycoproteins.

Improved methods for the analysis of oligosaccharide and polysaccharide structure have revealed remarkable complexity and diversity in the oligosaccharides of glycoproteins and glycolipids. Consider the oligosaccharide chains in Figure 7–31, typical of those found in many glycoproteins. The most complex of those shown contains 14 monosaccharide residues of four different kinds, variously linked as $(1\rightarrow2)$, $(1\rightarrow3)$, $(1\rightarrow4)$, $(1\rightarrow6)$, $(2\rightarrow3)$, and $(2\rightarrow6)$, some with the $\alpha$ and some with the $\beta$ configuration. Branched structures, not found in nucleic acids or proteins, are common in oligosaccharides. With the reasonable assumption that 20 different monosaccharide subunits are available for construction of oligosaccharides, we can calculate that $1.44 \times 10^{15}$ different hexameric oligosaccharides are possible; this compares with $6.4 \times 10^7$ $(20^6)$ different hexapeptides possible with the 20 common amino acids, and 4,096 $(4^6)$ different hexanucleotides with the four nucleotide subunits. If we also allow for variations in oligosaccharides resulting from sulfation of one or more residues, the number of possible oligosaccharides increases by two orders of magnitude. Oligosaccharides are enormously rich in structural information, not merely rivaling but far surpassing nucleic acids in the density of information contained in a molecule of modest size. Each of the oligosaccharides represented in Figure 7–31 presents a unique, three-dimensional face—a word in the sugar code—readable by the proteins that interact with it.

## Lectins Are Proteins That Read the Sugar Code and Mediate Many Biological Processes

**Lectins,** found in all organisms, are proteins that bind carbohydrates with high affinity and specificity (Table 7–3). Lectins serve in a wide variety of cell-cell recognition, signaling, and adhesion processes and in intra-cellular targeting of newly synthesized proteins. In the laboratory, purified lectins are useful reagents for detecting and separating glycoproteins with different oligosaccharide moieties. Here we discuss just a few examples of the roles of lectins in cells.

Some peptide hormones that circulate in the blood have oligosaccharide moieties that strongly influence their circulatory half-life. Luteinizing hormone and thyrotropin (polypeptide hormones produced in the adrenal cortex) have $N$-linked oligosaccharides that end with the disaccharide GalNAc4S$(\beta1\rightarrow4)$GlcNAc, which is recognized by a lectin (receptor) of hepatocytes. (GalNAc4S is $N$-acetylgalactosamine sulfated on the —OH group of C-4.) Receptor-hormone interaction mediates the uptake and destruction of luteinizing hormone and thyrotropin, reducing their concentration in the blood. Thus the blood levels of these hormones undergo a periodic rise (due to secretion by the adrenal cortex) and fall (due to destruction by hepatocytes).

The importance of the oligosaccharide moiety of these hormones is apparent from studies of individuals with a defective enzyme in the pathway that produces this oligosaccharide. Females with this congenital defect often fail to undergo the sexual changes of puberty (although males with the same defect develop normally). ■

The residues of Neu5Ac (a sialic acid) situated at the ends of the oligosaccharide chains of many plasma glycoproteins (Fig. 7–31) protect those proteins from uptake and degradation in the liver. For example, ceruloplasmin, a copper-containing serum glycoprotein, has several oligosaccharide chains ending in Neu5Ac. Re-

**TABLE 7–3  Some Lectins and the Oligosaccharide Ligands They Bind**

| Lectin source and lectin | Abbreviation | Ligand(s) |
|---|---|---|
| **Plant** | | |
| Concanavalin A | ConA | Man$\alpha$1—OCH$_3$ |
| *Griffonia simplicifolia* lectin 4 | GS4 | Lewis b (Le$^b$) tetrasaccharide |
| Wheat germ agglutinin | WGA | Neu5Ac$(\alpha2\rightarrow3)$Gal$(\beta1\rightarrow4)$Glc |
| | | GlcNAc$(\beta1\rightarrow4)$GlcNAc |
| Ricin | | Gal$(\beta1\rightarrow4)$Glc |
| **Animal** | | |
| Galectin-1 | | Gal$(\beta1\rightarrow4)$Glc |
| Mannose-binding protein A | MBP-A | High-mannose octasaccharide |
| **Viral** | | |
| Influenza virus hemagglutinin | HA | Neu5Ac$(\alpha2\rightarrow6)$Gal$(\beta1\rightarrow4)$Glc |
| Polyoma virus protein 1 | VP1 | Neu5Ac$(\alpha2\rightarrow3)$Gal$(\beta1\rightarrow4)$Glc |
| **Bacterial** | | |
| Enterotoxin | LT | Gal |
| Cholera toxin | CT | GM1 pentasaccharide |

Source: Weiss, W.I. & Drickamer, K. (1996) Structural basis of lectin-carbohydrate recognition. *Annu. Rev. Biochem.* **65,** 441–473.

moval of the sialic acid residues by the enzyme sialidase (also called neuraminidase) is one way in which the body marks "old" proteins for destruction and replacement. The plasma membrane of hepatocytes has lectin molecules (asialoglycoprotein receptors; "asialo-" indicating "without sialic acid") that specifically bind oligosaccharide chains with galactose residues no longer "protected" by a terminal Neu5Ac residue. Receptor-ceruloplasmin interaction triggers endocytosis and destruction of the ceruloplasmin.

*N*-Acetylneuraminic acid (Neu5Ac)
(a sialic acid)

A similar mechanism is apparently responsible for removing old erythrocytes from the mammalian bloodstream. Newly synthesized erythrocytes have several membrane glycoproteins with oligosaccharide chains that end in Neu5Ac. When the sialic acid residues are removed by withdrawing a sample of blood, treating it with sialidase in vitro, and reintroducing it into the circulation, the treated erythrocytes disappear from the bloodstream within a few hours; those with intact oligosaccharides (erythrocytes withdrawn and reintroduced without sialidase treatment) continue to circulate for days.

Several animal viruses, including the influenza virus, attach to their host cells through interactions with oligosaccharides displayed on the host cell surface. The lectin of the influenza virus, the HA protein, is essential for viral entry and infection (see Fig. 11–24). After initial binding of the virus to a sialic acid–containing oligosaccharide on the host surface, a viral sialidase removes the terminal sialic acid residue, triggering the entry of the virus into the cell. Inhibitors of this enzyme are used clinically in the treatment of influenza. Lectins on the surface of the herpes simplex viruses HS-1 and HS-2 (the causative agents of oral and genital herpes, respectively) bind specifically to heparan sulfate on the cell surface as a first step in their infection cycle; infection requires precisely the right pattern of sulfation on this polymer.

**Selectins** are a family of plasma membrane lectins that mediate cell-cell recognition and adhesion in a wide range of cellular processes. One such process is the movement of immune cells (T lymphocytes) through the capillary wall, from blood to tissues, at sites of infection or inflammation (Fig. 7–33). At an infection site, P-selectin on the surface of capillary endothelial cells interacts with a specific oligosaccharide of the glycoproteins of circu-

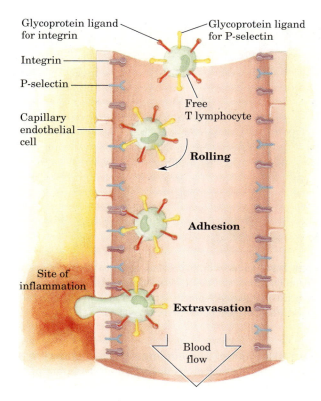

**FIGURE 7–33 Role of lectin-ligand interactions in lymphocyte movement to the site of an infection or injury.** A T lymphocyte circulating through a capillary is slowed by transient interactions between P-selectin molecules in the plasma membrane of the capillary endothelial cells and glycoprotein ligands for P-selectin on the T-cell surface. As it interacts with successive P-selectin molecules, the T cell rolls along the capillary surface. Near a site of inflammation, stronger interactions between integrin in the capillary surface and its ligand in the T-cell surface lead to tight adhesion. The T cell stops rolling and, under the influence of signals sent out from the site of inflammation, begins extravasation—escape through the capillary wall—as it moves toward the site of inflammation.

lating T cells. This interaction slows the T cells as they adhere to and roll along the endothelial lining of the capillaries. A second interaction, between integrin molecules (see p. 385) in the T-cell plasma membrane and an adhesion protein on the endothelial cell surface, now stops the T cell and allows it to move through the capillary wall into the infected tissues to initiate the immune attack. Two other selectins participate in this "lymphocyte homing": E-selectin on the endothelial cell and L-selectin on the T cell bind their cognate oligosaccharides on the T cell and endothelial cell, respectively.

Some microbial pathogens have lectins that mediate bacterial adhesion to host cells or toxin entry into cells. The bacterium believed responsible for most gastric ulcers, *Helicobacter pylori*, adheres to the inner surface of the stomach by interactions between bacterial membrane lectins and specific oligosaccharides of membrane glycoproteins of the gastric epithelial cells

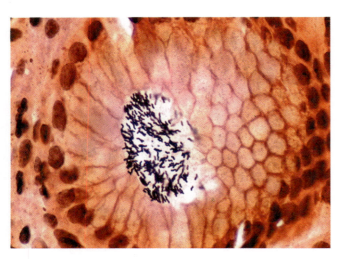

**FIGURE 7–34 An ulcer in the making.** *Helicobacter pylori* cells adhering to the gastric surface. This bacterium causes ulcers by interactions between a bacterial surface lectin and the Le[b] oligosaccharide (a blood group antigen) of the gastric epithelium.

(Fig. 7–34). Among the binding sites recognized by *H. pylori* is the oligosaccharide Le[b] when it is part of the type O blood group determinant. This observation helps to explain the severalfold greater incidence of gastric ulcers in people of blood type O than in those of type A or B. Chemically synthesized analogs of the Le[b] oligosaccharide may prove useful in treating this type of ulcer. Administered orally, they could prevent bacterial adhesion (and thus infection) by competing with the gastric glycoproteins for binding to the bacterial lectin.

The cholera toxin molecule (produced by *Vibrio cholerae*) triggers diarrhea after entering intestinal cells responsible for water absorption from the intestine. The toxin attaches to its target cell through the oligosaccharide of ganglioside GM1, a membrane phospholipid (for the structure of GM1 see Box 10–2, Fig. 1), on the surface of intestinal epithelial cells. Similarly, the pertussis toxin produced by *Bordetella pertussis,* the bacterium that causes whooping cough, enters target cells only after interacting with an oligosaccharide (or perhaps several oligosaccharides) with a terminal sialic acid residue. Understanding the details of the oligosaccharide-binding sites of these toxins (lectins) may allow the development of genetically engineered toxin analogs for use in vaccines. Toxin analogs engineered to lack the carbohydrate-binding site would be harmless because they could not bind to and enter cells, but they might elicit an immune response that would protect the recipient if later exposed to the natural toxin. It is also possible to imagine drugs that would act by mimicking the oligosaccharides of the cell surface, binding to the lectins of bacteria or toxins and preventing their productive binding to cell surfaces. ■

Lectins also act intracellularly. An oligosaccharide containing mannose 6-phosphate marks newly synthe-

sized proteins in the Golgi complex for transfer to the lysosome (see Fig. 27–36). A common structural feature on the surface of these glycoproteins, the signal patch, causes them to be recognized by an enzyme that phosphorylates a mannose residue at the terminus of an oligosaccharide chain. This mannose phosphate residue is recognized by the cation-dependent mannose 6-phosphate receptor, a membrane-associated lectin with its mannose phosphate binding site on the lumenal side of the Golgi complex. When a section of the Golgi complex containing this receptor buds off to form a transport vesicle, proteins containing mannose phosphate residues are dragged into the forming bud by interaction of their mannose phosphates with the receptor; the vesicle then moves to and fuses with a lysosome, depositing its cargo therein. Many, perhaps all, of the degradative enzymes (hydrolases) of the lysosome are targeted and delivered by this mechanism.

## Lectin-Carbohydrate Interactions Are Very Strong and Highly Specific

In all the functions of lectins described above, and in many more known to involve lectin-oligosaccharide interactions, it is essential that the oligosaccharide have a unique structure, so that recognition by the lectin is highly specific. The high density of information in oligosaccharides provides a sugar code with an essentially unlimited number of unique "words" small enough to be read by a single protein. In their carbohydrate-binding sites, lectins have a subtle molecular complementarity that allows interaction only with their correct carbohydrate cognates. The result is extraordinarily high specificity in these interactions.

X-ray crystallographic studies of the structures of several lectin-carbohydrate complexes have provided rich details of the lectin-sugar interaction. Sialoadhesin (also called siglec-1) is a membrane-bound lectin on the surface of mouse macrophages that recognizes certain sialic acid–containing oligosaccharides. This protein has a $\beta$ sandwich domain (see this motif in the CD8 protein in Fig. 4–22) that contains the sialic acid binding site (Fig. 7–35a). Each of the ring substituents unique to Neu5Ac is involved in the interaction between sugar and lectin; the acetyl group at C-5 undergoes both hydrogen-bond and van der Waals interactions with the protein; the carboxyl group makes a salt bridge with Arg[97]; and the hydroxyls of the glycerol moiety hydrogen-bond with the protein (Fig. 7–35b).

The structure of the mannose 6-phosphate receptor/lectin has also been resolved crystallographically, revealing details of its interaction with mannose 6-phosphate that explain the specificity of the binding and the necessity for a divalent cation in the lectin-sugar interaction (Fig. 7–35c). Arg[111] of the receptor is hydrogen-bonded to the C-2 hydroxyl of mannose and coordinated with $Mn^{2+}$. His[105] is hydrogen-bonded to

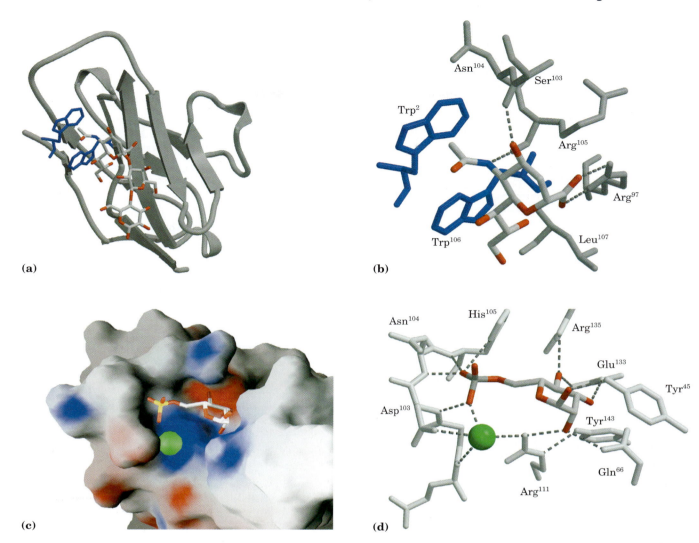

**FIGURE 7–35 Details of lectin-carbohydrate interaction. (a)** X-ray crystallographic studies of a sialic acid–specific lectin (derived from PDB ID 1QFO) show how a protein can recognize and bind to a sialic acid (Neu5Ac) residue. Sialoadhesin (also called siglec-1), a membrane-bound lectin of the surface of mouse macrophages, has a β sandwich domain (gray) that contains the Neu5Ac binding site (dark blue). Neu5Ac is shown as a stick structure. **(b)** Each ring substituent unique to Neu5Ac is involved in the interaction between sugar and lectin: the acetyl group at C-5 has both hydrogen-bond and van der Waals interactions with the protein; the carboxyl group makes a salt bridge with Arg$^{97}$; and the hydroxyls of the glycerol moiety hydrogen-bond with the protein. **(c)** Structure of the bovine mannose 6-phosphate receptor complexed with mannose 6-phosphate (PDB ID 1M6P). The protein is represented here as a surface contour image, with color to indicate the surface electrostatic potential: red, predominantly negative charge; blue, predominantly positive charge. Mannose 6-phosphate is shown as a stick structure; a manganese ion is shown in green. **(d)** In this complex, mannose 6-phosphate is hydrogen-bonded to Arg$^{111}$ and coordinated with the manganese ion (green). The His$^{105}$ hydrogen-bonded to a phosphate oxygen of mannose 6-phosphate may be the residue that, when protonated at low pH, causes the receptor to release mannose 6-phosphate into the lysosome.

one of the oxygen atoms of the phosphate (Fig. 7–35d). When the protein-tagged with mannose 6-phosphate reaches the lysosome (which has a lower internal pH than the Golgi complex), the receptor apparently loses its affinity for mannose 6-phosphate. Protonation of His$^{105}$ may be responsible for this change in binding.

In addition to these very specific interactions, there are more general interactions that contribute to the binding of many carbohydrates to their lectins. For example, many sugars have a more polar and a less polar side (Fig. 7–36); the more polar side hydrogen-bonds with the lectin, while the less polar undergoes hydrophobic interactions with nonpolar amino acid residues. The sum of all these interactions produces high-affinity binding ($K_d$ often $10^{-8}$ M or less) and high specificity of lectins for their carbohydrates. This represents a kind of information transfer that is clearly central in many processes within and between cells. Figure 7–37 summarizes some of the biological interactions mediated by the sugar code.

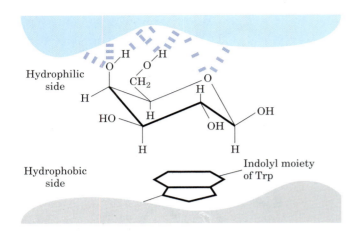

**FIGURE 7–36 Hydrophobic interactions of sugar residues.** Sugar units such as galactose have a more polar side (the top of the chair, with the ring oxygen and several hydroxyls), available to hydrogen-bond with the lectin, and a less polar side that can have hydrophobic interactions with nonpolar side chains in the protein, such as the indole ring of tryptophan.

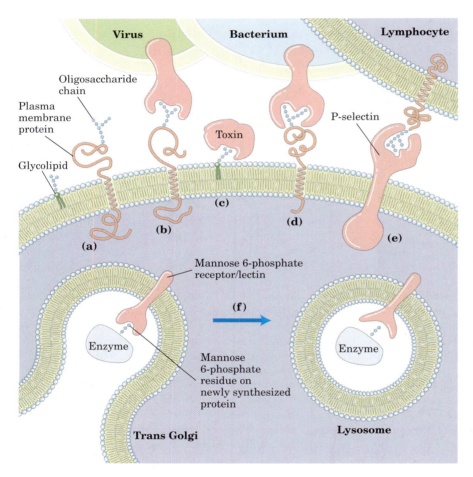

**FIGURE 7–37 Roles of oligosaccharides in recognition and adhesion at the cell surface.** (a) Oligosaccharides with unique structures (represented as strings of hexagons), components of a variety of glycoproteins or glycolipids on the outer surface of plasma membranes, interact with high specificity and affinity with lectins in the extracellular milieu. (b) Viruses that infect animal cells, such as the influenza virus, bind to cell surface glycoproteins as the first step in infection. (c) Bacterial toxins, such as the cholera and pertussis toxins, bind to a surface glycolipid before entering a cell. (d) Some bacteria, such as *H. pylori*, adhere to and then colonize or infect animal cells. (e) Selectins (lectins) in the plasma membrane of certain cells mediate cell-cell interactions, such as those of T lymphocytes with the endothelial cells of the capillary wall at an infection site. (f) The mannose 6-phosphate receptor/lectin of the trans Golgi complex binds to the oligosaccharide of lysosomal enzymes, targeting them for transfer into the lysosome.

## SUMMARY 7.4 Carbohydrates as Informational Molecules: The Sugar Code

■ Monosaccharides can be assembled into an almost limitless variety of oligosaccharides, which differ in the stereochemistry and position of glycosidic bonds, the type and orientation of substituent groups, and the number and type of branches. Oligosaccharides are far more information-dense than nucleic acids or proteins.

■ Lectins, proteins with highly specific carbohydrate-binding domains, are commonly found on the outer surface of cells, where they initiate interaction with other cells. In vertebrates, oligosaccharide tags "read" by lectins govern the rate of degradation of certain peptide hormones, circulating proteins, and blood cells.

■ The adhesion of bacterial and viral pathogens to their animal-cell targets occurs through binding of lectins in the pathogens to oligosaccharides in the target cell surface. Lectins are also present inside cells, where they mediate intracellular protein targeting.

■ X-ray crystallography of lectin-sugar complexes shows the detailed complementarity between the two molecules, which accounts for the strength and specificity of their interactions with carbohydrates.

■ Selectins are plasma membrane lectins that bind carbohydrate chains in the extracellular matrix or on the surfaces of other cells, thereby mediating the flow of information between cell and matrix or between cells.

## 7.5 Working with Carbohydrates

The growing appreciation of the importance of oligosaccharide structure in biological recognition has been the driving force behind the development of methods for analyzing the structure and stereochemistry of complex oligosaccharides. Oligosaccharide analysis is complicated by the fact that, unlike nucleic acids and proteins, oligosaccharides can be branched and are joined by a variety of linkages. Oligosaccharides are generally removed from their protein or lipid conjugates before analysis, then subjected to stepwise degradation with specific reagents that reveal bond position or stereochemistry. Mass spectrometry and NMR spectroscopy have also become invaluable in deciphering oligosaccharide structure.

The oligosaccharide moieties of glycoproteins or glycolipids can be released by purified enzymes—glycosidases that specifically cleave *O*- or *N*-linked oligosaccharides or lipases that remove lipid head groups. Mixtures of carbohydrates are resolved into their individual components (Fig. 7–38) by some of the same techniques useful in protein and amino acid separation: fractional precipitation by solvents, and ion-exchange and size-exclusion chromatography (see Fig. 3–18). Highly purified lectins, attached covalently to an insoluble support, are commonly used in affinity chromatography of carbohydrates (see Fig. 3–18c). Hydrolysis of oligosaccharides and polysaccharides in strong acid yields a mixture of monosaccharides, which, after conversion to suitable volatile derivatives, may be separated, identified, and quantified by gas-liquid chromatography (p. 365) to yield the overall composition of the polymer.

For simple, linear polymers such as amylose, the positions of the glycosidic bonds are determined by treating the intact polysaccharide with methyl iodide in a strongly basic medium to convert all free hydroxyls to acid-stable methyl ethers, then hydrolyzing the methylated polysaccharide in acid. The only free hydroxyls present in the monosaccharide derivatives so produced are those that were involved in glycosidic bonds. To de-termine the sequence of monosaccharide residues, including branches if they are present, exoglycosidases of known specificity are used to remove residues one at a time from the nonreducing end(s). The specificity of these exoglycosidases often allows deduction of the position and stereochemistry of the linkages. Polysaccharides and large oligosaccharides can be treated chemically or with endoglycosidases to split specific internal glycosidic bonds, producing several smaller, more easily analyzable oligosaccharides.

Oligosaccharide analysis relies increasingly on mass spectrometry and high-resolution NMR spectroscopy (see Box 4–4). Matrix-assisted laser desorption/ionization mass spectrometry (MALDI MS) and tandem mass spectrometry (MS/MS) (described in Box 3–2), are readily applicable to polar compounds like oligosaccharides. MALDI MS is a very sensitive method for determining the mass of the molecular ion (the entire oligosaccharide chain). Tandem MS reveals the mass of the molecular ion and many of its fragments, which are usually the result of breakage of the glycosidic bonds. A comparison of the masses of each fragment therefore gives information about the sequence of monosaccharide units. NMR analysis alone, especially for oligosaccharides of moderate size, can yield much information about sequence, linkage position, and anomeric carbon configuration. Automated procedures and commercial instruments are used for the routine determination of oligosaccharide structure, but the sequencing of branched oligosaccharides joined by more than one type of bond remains a far more formidable task than determining the linear sequences of proteins and nucleic acids, with monomers joined by a single bond type.

### SUMMARY 7.5 Working with Carbohydrates

■ Establishing the complete structure of oligosaccharides and polysaccharides requires determination of branching positions, the sequence in each branch, the configuration of each monosaccharide unit, and the positions of the glycosidic links—a more complex problem than protein and nucleic acid analysis.

■ The structures of oligosaccharides and polysaccharides are usually determined by a combination of methods: specific enzymatic hydrolysis to determine stereochemistry and produce smaller fragments for further analysis; methylation analysis to locate glycosidic bonds; and stepwise degradation to determine sequence and configuration of anomeric carbons.

■ Mass spectrometry and high-resolution NMR spectroscopy, applicable to small samples of carbohydrate, yield essential information about sequence, configuration at anomeric and other carbons, and positions of glycosidic bonds.

**FIGURE 7–38  Methods of carbohydrate analysis.**
A carbohydrate purified in the first stage of the
analysis often requires all four analytical routes for
its complete characterization.

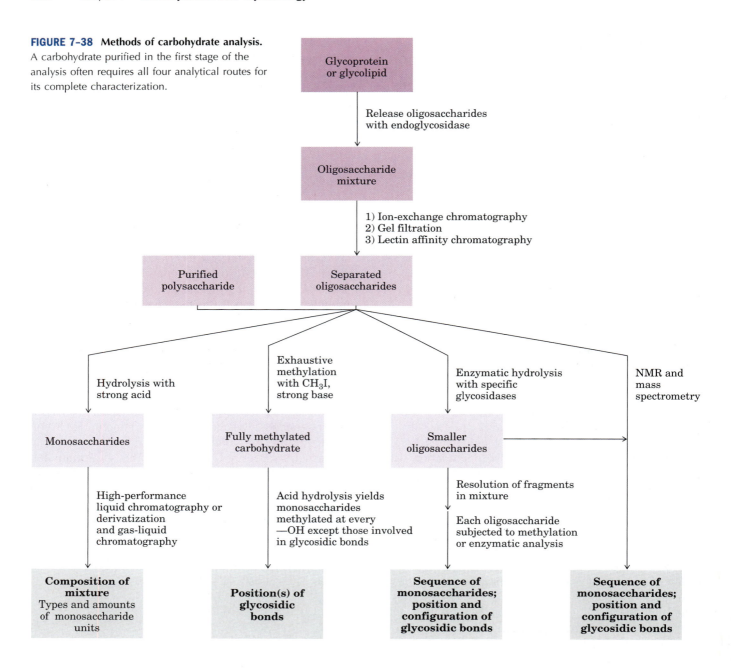

## Key Terms

*Terms in bold are defined in the glossary.*

| | | | |
|---|---|---|---|
| **glycoconjugate**  238 | hemiacetal  242 | **glycosidic bonds**  245 | **glycoprotein**  256 |
| **monosaccharide**  238 | hemiketal  242 | **reducing end**  245 | **glycolipid**  256 |
| **oligosaccharide**  238 | **pyranose**  242 | **glycan**  247 | **lectin**  262 |
| **disaccharide**  238 | **furanose**  242 | starch  248 | **selectins**  263 |
| **polysaccharide**  238 | **anomers**  242 | glycogen  248 | |
| **aldose**  239 | anomeric carbon  242 | extracellular matrix  253 | |
| **ketose**  239 | **mutarotation**  242 | **glycosaminoglycan** | |
| **Fischer projection** | **Haworth perspective** | 253 | |
| formulas  240 | formulas  242 | **hyaluronic acid**  254 | |
| **epimers**  240 | **reducing sugar**  244 | **proteoglycan**  256 | |

# *Further Reading*

## General Background on Carbohydrate Chemistry

**Aspinall, G.O. (ed.)** (1982, 1983, 1985) *The Polysaccharides,* Vols 1-3, Academic Press, Inc., New York.

**Collins, P.M. & Ferrier, R.J.** (1995) *Monosaccharides: Their Chemistry and Their Roles in Natural Products,* John Wiley & Sons, Chichester, England.

A comprehensive text at the graduate level.

**Fukuda, M. & Hindsgaul, O.** (1994) *Molecular Glycobiology,* IRL Press at Oxford University Press, Inc., New York.

Thorough, advanced treatment of the chemistry and biology of cell surface carbohydrates. Good chapters on lectins, carbohydrate recognition in cell-cell interactions, and chemical synthesis of oligosaccharides.

**Lehmann, J. (Haines, A.H., trans.)** (1998) *Carbohydrates: Structure and Biology,* G. Thieme Verlag, New York.

The fundamentals of carbohydrate chemistry and biology, presented at a level suitable for advanced undergraduates and graduate students.

**Morrison, R.T. & Boyd, R.N.** (1992) *Organic Chemistry,* 6th edn, Benjamin Cummings, San Francisco.

Chapters 34 and 35 cover the structure, stereochemistry, nomenclature, and chemical reactions of carbohydrates.

**Pigman, W. & Horton, D. (eds)** (1970, 1972, 1980) *The Carbohydrates: Chemistry and Biochemistry,* Vols IA, IB, IIA, and IIB, Academic Press, Inc., New York.

Comprehensive treatise on carbohydrate chemistry.

**Varki, A., Cummings, R., Esko, J., Freeze, H., Hart, G., & Marth, J.** (1999) *Essentials of Glycobiology,* Cold Spring Harbor Laboratory Press, Cold Spring Harbor, NY.

Structure, biosynthesis, metabolism, and function of glycosaminoglycans, proteoglycans, glycoproteins, and glycolipids, all presented at an intermediate level and very well illustrated.

## Glycosaminoglycans and Proteoglycans

**Esko, J.D. & Lindahl, U.** (2001) Molecular diversity of heparan sulfate. *J. Clin. Invest.* **108,** 169–173.

**Esko, J.D. & Selleck, S.B.** (2002) Order out of chaos: assembly of ligand binding sites in heparan sulfate. *Annu. Rev. Biochem.* **71,** 435–471.

**Iozzo, R.V.** (1998) Matrix proteoglycans: from molecular design to cellular function. *Annu. Rev. Biochem.* **67,** 609–652.

A review focusing on recent genetic and molecular biological studies of the matrix proteoglycans. The structure-function relationships of some paradigmatic proteoglycans are discussed in depth, and novel aspects of their biology are examined.

**Jackson, R.L., Busch, S.J., & Cardin, A.D.** (1991) Glycosaminoglycans: molecular properties, protein interactions, and role in physiological processes. *Physiol. Rev.* **71,** 481–539.

An advanced review of the chemistry and biology of glycosaminoglycans.

**Roseman, S.** (2001) Reflections on glycobiology. *J. Biol. Chem.* **276,** 41,527–41,542.

A masterful review of the history of carbohydrate and glycosaminoglycan studies, by one of the major contributors to this field.

**Turnbull, J., Powell, A., & Guimond, S.** (2001) Heparan sulfate: decoding a dynamic multifunctional cell regulator. *Trends Cell Biol.* **11,** 75–82.

Review of the chemistry and biology of high-sulfate domains in heparan sulfate.

## Glycoproteins

**Gahmberg, C.G. & Tolvanen, M.** (1996) Why mammalian cell surface proteins are glycoproteins. *Trends Biochem. Sci.* **21,** 308–311.

**Opdenakker, G., Rudd, P., Ponting, C., & Dwek, R.** (1993) Concepts and principles of glycobiology. *FASEB J.* **7,** 1330–1337.

This review considers the genesis of glycoforms, functional roles for glycosylation, and structure-function relationships for several glycoproteins.

**Varki, A.** (1993) Biological roles of oligosaccharides: all of the theories are correct. *Glycobiology* **3,** 97–130.

## Glycobiology and the Sugar Code

**Angata, T. & Brinkman-Van der Linden, E.** (2002) I-type lectins. *Biochim. Biophys. Acta* **1572,** 294–316.

**Aplin, A.E., Howe, A., Alahari, S.K., & Juliano, R.L.** (1998) Signal transduction and signal modulation by cell adhesion receptors: the role of integrins, cadherins, immunoglobulin-cell adhesion molecules, and selectins. *Pharmacol. Rev.* **50,** 197–263.

**Bernfield, M., Götte, M., Park, P.W., Reizes, O., Fitzgerald, M.L., Lincecum, J., & Zako, M.** (1999) Functions of cell surface heparan sulfate proteoglycans. *Annu. Rev. Biochem.* **68,** 729–777.

Extensive review of the biological roles of heparan sulfate.

**Bertozzi, C.R. & Kiessling, L.L.** (2001) Chemical glycobiology. *Science* **291,** 2357–2363.

A review of applications of chemical synthesis of carbohydrates to an understanding of the biological roles of oligosaccharides.

**Borén, T., Normark, S., & Falk, P.** (1994) *Helicobacter pylori:* molecular basis for host recognition and bacterial adherence. *Trends Microbiol.* **2,** 221–228.

A look at the role of the oligosaccharides that determine blood type in the adhesion of *H. pylori* to the stomach lining, producing ulcers.

**Cooper, D.N.** (2002) Galectinomics: finding themes in complexity. *Biochim. Biophys. Acta* **1572,** 209–231.

A review of the genomic evidence for the conservation of the galectins, a family of lectins.

**Cornejo, C.J., Winn, R.K., & Harlan, J.M.** (1997) Anti-adhesion therapy. *Adv. Pharmacol.* **39,** 99–142.

Analogs of recognition oligosaccharides are used to block adhesion of a pathogen to its host-cell target.

**Dahms, N.M. & Hancock, M.K.** (2002) P-type lectins. *Biochim. Biophys. Acta* **1572,** 317–340.

**Gabius, H.-J.** (2000) Biological information transfer beyond the genetic code: the sugar code. *Naturwissenschaften* **87,** 108–121.

Description of the basis for the high information density in oligosaccharides, with examples of the importance of the sugar code.

**Gabius, H.-J., Andre, S., Kaltner, H., & Siebert, H.C.** (2002) The sugar code: functional lectinomics. *Biochim. Biophys. Acta* **1572**, 165–177.

This review examines the reasons for the relatively late appreciation of the informational roles of oligosaccharides and polysaccharides.

**Ghosh, P., Dahms, N.M., & Kornfeld, S.** (2003) Mannose 6-phosphate receptors: new twists in the tale. *Nat. Rev. Mol. Cell Biol.* **4**, 202–212.

**Helenius, A. & Aebi, M.** (2001) Intracellular functions of *N*-linked glycans. *Science* **291**, 2364–2369.

Review of the synthesis of *N*-linked oligosaccharides and their targeting functions.

**Hooper, L.A., Manzella, S.M., & Baenziger, J.U.** (1996) From legumes to leukocytes: biological roles for sulfated carbohydrates. *FASEB J.* **10**, 1137–1146.

Evidence for roles of sulfated oligosaccharides in peptide hormone half-life, symbiont interactions in nitrogen-fixing legumes, and lymphocyte homing.

**Horwitz, A.F.** (1997) Integrins and health. *Sci. Am.* **276** (May), 68–75.

Article on the role of integrins in cell-cell adhesion, and possible roles in arthritis, heart disease, stroke, osteoporosis, and the spread of cancer.

**Iozzo, R.V.** (2001) Heparan sulfate proteoglycans: intricate molecules with intriguing functions. *J. Clin. Invest.* **108**, 165–167.

Introduction to a series of papers on heparan sulfates published in this issue; all are rewarding reading.

**Kilpatrick, D.C.** (2002) Animal lectins: a historical introduction and overview. *Biochim. Biophys. Acta* **1572**, 187–197.

Introduction to a series of excellent reviews on lectins and their biological roles, all published in this issue.

**Loris, R.** (2002) Principles of structures of animal and plant lectins. *Biochim. Biophys. Acta* **1572**, 198–208.

**McEver, R.P., Moore, K.L., & Cummings, R.D.** (1995) Leukocyte trafficking mediated by selectin-carbohydrate interactions. *J. Biol. Chem.* **270**, 11,025–11,028.

This short review focuses on the interaction of selectins with their carbohydrate ligands.

**Reuter, G. & Gabius, H.-J.** (1999) Eukaryotic glycosylation: whim of nature or multipurpose tool? *Cell. Mol. Life Sci.* **55**, 368–422.

Excellent review of the chemical diversity of oligosaccharides and polysaccharides and of biological processes dependent upon protein-carbohydrate recognition.

**Selleck, S.** (2000) Proteoglycans and pattern formation: sugar biochemistry meets developmental genetics. *Trends Genet.* **16**, 206–212.

A short, intermediate-level review of the genetic evidence for proteoglycans as determinants of development.

**Weigel, P.H. & Yik, J.H.** (2002) Glycans as endocytosis signals: the cases of the asialoglycoprotein and hyaluronan/chondroitin sulfate receptors. *Biochim. Biophys. Acta* **1572**, 341–363.

**Weiss, W.I. & Drickamer, K.** (1996) Structural basis of lectin-carbohydrate recognition. *Annu. Rev. Biochem.* **65**, 441–473.

Good treatment of the chemical basis of carbohydrate-protein interactions.

**Working with Carbohydrates**

**Chaplin, M.F. & Kennedy, J.F. (eds)** (1994) *Carbohydrate Analysis: A Practical Approach,* 2nd edn, IRL Press, Oxford.

Very useful manual for analysis of all types of sugar-containing molecules—monosaccharides, polysaccharides and glycosaminoglycans, glycoproteins, proteoglycans, and glycolipids.

**Dell, A. & Morris, H.R.** (2001) Glycoprotein structure determination by mass spectroscopy. *Science* **291**, 2351–2356.

Short review of the uses of MALDI MS and tandem MS in oligosaccharide structure determination.

**Dwek, R.A., Edge, C.J., Harvey, D.J., & Wormald, M.R.** (1993) Analysis of glycoprotein-associated oligosaccharides. *Annu. Rev. Biochem.* **62**, 65–100.

Excellent survey of the uses of NMR, mass spectrometry, and enzymatic reagents to determine oligosaccharide structure.

**Fukuda, M. & Kobata, A.** (1993) *Glycobiology: A Practical Approach,* IRL Press, Oxford.

A how-to manual for the isolation and characterization of the oligosaccharide moieties of glycoproteins, using the whole range of modern techniques. Available as part of the IRL Press *Practical Approach Series* on CD-ROM, from Oxford University Press (www.oup-usa.org/acadsci/pasbooks.html).

**Jay, A.** (1996) The methylation reaction in carbohydrate analysis. *J. Carbohydr. Chem.* **15**, 897–923.

Thorough description of methylation analysis of carbohydrates.

**Lennarz, W.J. & Hart, G.W. (eds)** (1994) *Guide to Techniques in Glycobiology,* Methods in Enzymology, Vol. 230, Academic Press, Inc., New York.

Practical guide to working with oligosaccharides.

**McCleary, B.V. & Matheson, N.K.** (1986) Enzymic analysis of polysaccharide structure. *Adv. Carbohydr. Chem. Biochem.* **44**, 147–276.

On the use of purified enzymes in analysis of structure and stereochemistry.

**Rudd, P.M., Guile, G.R., Kuester, B., Harvey, D.J., Opdenakker, G., & Dwek, R.A.** (1997) Oligosaccharide sequencing technology. *Nature* **388**, 205–207.

## Problems

**1. Determination of an Empirical Formula** An unknown substance containing only C, H, and O was isolated from goose liver. A 0.423 g sample produced 0.620 g of $CO_2$ and 0.254 g of $H_2O$ after complete combustion in excess oxygen. Is the empirical formula of this substance consistent with its being a carbohydrate? Explain.

**2. Sugar Alcohols** In the monosaccharide derivatives known as sugar alcohols, the carbonyl oxygen is reduced to a hydroxyl group. For example, D-glyceraldehyde can be reduced to glycerol. However, this sugar alcohol is no longer designated D or L. Why?

**3. Melting Points of Monosaccharide Osazone Derivatives** Many carbohydrates react with phenylhydrazine ($C_6H_5NHNH_2$) to form bright yellow crystalline derivatives known as osazones:

Glucose → Osazone derivative of glucose

The melting temperatures of these derivatives are easily determined and are characteristic for each osazone. This information was used to help identify monosaccharides before the development of HPLC or gas-liquid chromatography. Listed below are the melting points (MPs) of some aldose-osazone derivatives:

| Monosaccharide | MP of anhydrous monosaccharide (°C) | MP of osazone derivative (°C) |
|---|---|---|
| Glucose | 146 | 205 |
| Mannose | 132 | 205 |
| Galactose | 165–168 | 201 |
| Talose | 128–130 | 201 |

As the table shows, certain pairs of derivatives have the same melting points, although the underivatized monosaccharides do not. Why do glucose and mannose, and galactose and talose, form osazone derivatives with the same melting points?

**4. Interconversion of D-Glucose Forms** A solution of one stereoisomer of a given monosaccharide rotates plane-polarized light to the left (counterclockwise) and is called the levorotatory isomer, designated (−); the other stereoisomer rotates plane-polarized light to the same extent but to the right (clockwise) and is called the dextrorotatory isomer, designated (+). An equimolar mixture of the (+) and (−) forms does not rotate plane-polarized light.

The optical activity of a stereoisomer is expressed quantitatively by its *optical rotation,* the number of degrees by which plane-polarized light is rotated on passage through a given path length of a solution of the compound at a given concentration. The *specific rotation* $[\alpha]_D^{25°C}$ of an optically active compound is defined thus:

$$[\alpha]_D^{25°C} = \frac{\text{observed optical rotation (°)}}{\text{optical path length (dm)} \times \text{concentration (g/mL)}}$$

The temperature and the wavelength of the light employed (usually the D line of sodium, 589 nm) must be specified in the definition.

A freshly prepared solution of $\alpha$-D-glucose shows a specific rotation of +112°. Over time, the rotation of the solution gradually decreases and reaches an equilibrium value corresponding to $[\alpha]_D^{25°C} = +52.5°$. In contrast, a freshly pre-

pared solution of $\beta$-D-glucose has a specific rotation of +19°. The rotation of this solution increases over time to the same equilibrium value as that shown by the $\alpha$ anomer.

(a) Draw the Haworth perspective formulas of the $\alpha$ and $\beta$ forms of D-glucose. What feature distinguishes the two forms?

(b) Why does the specific rotation of a freshly prepared solution of the $\alpha$ form gradually decrease with time? Why do solutions of the $\alpha$ and $\beta$ forms reach the same specific rotation at equilibrium?

(c) Calculate the percentage of each of the two forms of D-glucose present at equilibrium.

**5. A Taste of Honey** The fructose in honey is mainly in the $\beta$-D-pyranose form. This is one of the sweetest carbohydrates known, about twice as sweet as glucose. The $\beta$-D-furanose form of fructose is much less sweet. The sweetness of honey gradually decreases at a high temperature. Also, high-fructose corn syrup (a commercial product in which much of the glucose in corn syrup is converted to fructose) is used for sweetening *cold* but not *hot* drinks. What chemical property of fructose could account for both these observations?

**6. Glucose Oxidase in Determination of Blood Glucose** The enzyme glucose oxidase isolated from the mold *Penicillium notatum* catalyzes the oxidation of $\beta$-D-glucose to D-glucono-$\delta$-lactone. This enzyme is highly specific for the $\beta$ anomer of glucose and does not affect the $\alpha$ anomer. In spite of this specificity, the reaction catalyzed by glucose oxidase is commonly used in a clinical assay for total blood glucose—that is, for solutions consisting of a mixture of $\beta$- and $\alpha$-D-glucose. How is this possible? Aside from allowing the detection of smaller quantities of glucose, what advantage does glucose oxidase offer over Fehling's reagent for the determination of blood glucose?

**7. Invertase "Inverts" Sucrose** The hydrolysis of sucrose (specific rotation +66.5°) yields an equimolar mixture of D-glucose (specific rotation +52.5°) and D-fructose (specific rotation −92°). (See Problem 4 for details of specific rotation.)

(a) Suggest a convenient way to determine the rate of hydrolysis of sucrose by an enzyme preparation extracted from the lining of the small intestine.

(b) Explain why an equimolar mixture of D-glucose and D-fructose formed by hydrolysis of sucrose is called invert sugar in the food industry.

(c) The enzyme invertase (now commonly called sucrase) is allowed to act on a 10% (0.1 g/mL) solution of sucrose until hydrolysis is complete. What will be the observed optical rotation of the solution in a 10 cm cell? (Ignore a possible small contribution from the enzyme.)

**8. Manufacture of Liquid-Filled Chocolates** The manufacture of chocolates containing a liquid center is an interesting application of enzyme engineering. The flavored liquid center consists largely of an aqueous solution of sugars rich in fructose to provide sweetness. The technical dilemma is the following: the chocolate coating must be prepared by pouring hot melted chocolate over a solid (or almost solid) core, yet the final product must have a liquid, fructose-rich center. Suggest a way to solve this problem. (Hint: Sucrose is much less soluble than a mixture of glucose and fructose.)

**9. Anomers of Sucrose?** Although lactose exists in two anomeric forms, no anomeric forms of sucrose have been reported. Why?

**10. Physical Properties of Cellulose and Glycogen** The almost pure cellulose obtained from the seed threads of *Gossypium* (cotton) is tough, fibrous, and completely insoluble in water. In contrast, glycogen obtained from muscle or liver disperses readily in hot water to make a turbid solution. Although they have markedly different physical properties, both substances are composed of (1→4)-linked D-glucose polymers of comparable molecular weight. What structural features of these two polysaccharides underlie their different physical properties? Explain the biological advantages of their respective properties.

**11. Growth Rate of Bamboo** The stems of bamboo, a tropical grass, can grow at the phenomenal rate of 0.3 m/day under optimal conditions. Given that the stems are composed almost entirely of cellulose fibers oriented in the direction of growth, calculate the number of sugar residues per second that must be added enzymatically to growing cellulose chains to account for the growth rate. Each D-glucose unit contributes ~0.5 nm to the length of a cellulose molecule.

**12. Glycogen as Energy Storage: How Long Can a Game Bird Fly?** Since ancient times it has been observed that certain game birds, such as grouse, quail, and pheasants, are easily fatigued. The Greek historian Xenophon wrote, "The bustards . . . can be caught if one is quick in starting them up, for they will fly only a short distance, like partridges, and soon tire; and their flesh is delicious." The flight muscles of game birds rely almost entirely on the use of glucose 1-phosphate for energy, in the form of ATP (Chapter 14). In game birds, glucose 1-phosphate is formed by the breakdown of stored muscle glycogen, catalyzed by the enzyme glycogen phosphorylase. The rate of ATP production is limited by the rate at which glycogen can be broken down. During a "panic flight," the game bird's rate of glycogen breakdown is quite high, approximately 120 $\mu$mol/min of glucose 1-phosphate produced per gram of fresh tissue. Given that the flight muscles usually contain about 0.35% glycogen by weight, calculate how long a game bird can fly. (Assume the average molecular weight of a glucose residue in glycogen is 160 g/mol.)

**13. Volume of Chondroitin Sulfate in Solution** One critical function of chondroitin sulfate is to act as a lubricant in skeletal joints by creating a gel-like medium that is resilient to friction and shock. This function appears to be related to a distinctive property of chondroitin sulfate: the volume occupied by the molecule is much greater in solution than in the dehydrated solid. Why is the volume occupied by the molecule so much larger in solution?

**14. Heparin Interactions** Heparin, a highly negatively charged glycosaminoglycan, is used clinically as an anticoagulant. It acts by binding several plasma proteins, including antithrombin III, an inhibitor of blood clotting. The 1:1 binding of heparin to antithrombin III appears to cause a conformational change in the protein that greatly increases its ability to inhibit clotting. What amino acid residues of antithrombin III are likely to interact with heparin?

**15. Information Content of Oligosaccharides** The carbohydrate portion of some glycoproteins may serve as a cellular recognition site. In order to perform this function, the oligosaccharide moiety of glycoproteins must have the potential to exist in a large variety of forms. Which can produce a greater variety of structures: oligopeptides composed of five different amino acid residues or oligosaccharides composed of five different monosaccharide residues? Explain.

**16. Determination of the Extent of Branching in Amylopectin** The extent of branching (number of ($\alpha$1→6) glycosidic bonds) in amylopectin can be determined by the following procedure. A sample of amylopectin is exhaustively methylated—treated with a methylating agent (methyl iodide) that replaces all the hydrogens of the sugar hydroxyls with methyl groups, converting —OH to —OCH$_3$. All the glycosidic bonds in the treated sample are then hydrolyzed in aqueous acid. The amount of 2,3-di-*O*-methylglucose in the hydrolyzed sample is determined.

(a) Explain the basis of this procedure for determining the number of ($\alpha$1→6) branch points in amylopectin. What happens to the unbranched glucose residues in amylopectin during the methylation and hydrolysis procedure?

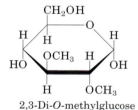

2,3-Di-*O*-methylglucose

(b) A 258 mg sample of amylopectin treated as described above yielded 12.4 mg of 2,3-di-*O*-methylglucose. Determine what percentage of the glucose residues in amylopectin contain an ($\alpha$1→6) branch. (Assume that the average molecular weight of a glucose residue in amylopectin is 162 g/mol.)

**17. Structural Analysis of a Polysaccharide** A polysaccharide of unknown structure was isolated, subjected to exhaustive methylation, and hydrolyzed. Analysis of the products revealed three methylated sugars in the ratio 20:1:1. The sugars were 2,3,4-tri-*O*-methyl-D-glucose; 2,4-di-*O*-methyl-D-glucose; and 2,3,4,6-tetra-*O*-methyl-D-glucose. What is the structure of the polysaccharide?

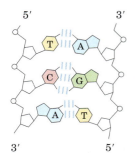

5′      3′

3′      5′

# NUCLEOTIDES AND NUCLEIC ACIDS

A structure this pretty just had to exist.

—*James Watson,* The Double Helix, *1968*

Nucleotides have a variety of roles in cellular metabolism. They are the energy currency in metabolic transactions, the essential chemical links in the response of cells to hormones and other extracellular stimuli, and the structural components of an array of enzyme cofactors and metabolic intermediates. And, last but certainly not least, they are the constituents of nucleic acids: deoxyribonucleic acid (DNA) and ribonucleic acid (RNA), the molecular repositories of genetic information. The structure of every protein, and ultimately of every biomolecule and cellular component, is a product of information programmed into the nucleotide sequence of a cell's nucleic acids. The ability to store and transmit genetic information from one generation to the next is a fundamental condition for life.

This chapter provides an overview of the chemical nature of the nucleotides and nucleic acids found in most cells; a more detailed examination of the function of nucleic acids is the focus of Part III of this text.

## 8.1 Some Basics

**Nucleotides, Building Blocks of Nucleic Acids** The amino acid sequence of every protein in a cell, and the nucleotide sequence of every RNA, is specified by a nucleotide sequence in the cell's DNA. A segment of a DNA molecule that contains the information required for the synthesis of a functional biological product, whether protein or RNA, is referred to as a **gene.** A cell typically has many thousands of genes, and DNA molecules, not surprisingly, tend to be very large. The storage and transmission of biological information are the only known functions of DNA.

RNAs have a broader range of functions, and several classes are found in cells. **Ribosomal RNAs (rRNAs)** are components of ribosomes, the complexes that carry out the synthesis of proteins. **Messenger RNAs (mRNAs)** are intermediaries, carrying genetic information from one or a few genes to a ribosome, where the corresponding proteins can be synthesized. **Transfer RNAs (tRNAs)** are adapter molecules that faithfully translate the information in mRNA into a specific sequence of amino acids. In addition to these major classes there is a wide variety of RNAs with special functions, described in depth in Part III.

### Nucleotides and Nucleic Acids Have Characteristic Bases and Pentoses

**Nucleotides** have three characteristic components: (1) a nitrogenous (nitrogen-containing) base, (2) a pentose, and (3) a phosphate (Fig. 8–1). The molecule without the phosphate group is called a **nucleoside.** The nitrogenous bases are derivatives of two parent compounds, **pyrimidine** and **purine.** The bases and pentoses of the common nucleotides are heterocyclic compounds. The carbon and nitrogen atoms in the parent structures are conventionally numbered to facilitate the naming and identification of the many derivative compounds. The convention for the pentose ring follows rules outlined in Chapter 7, but in the pentoses of nucleotides

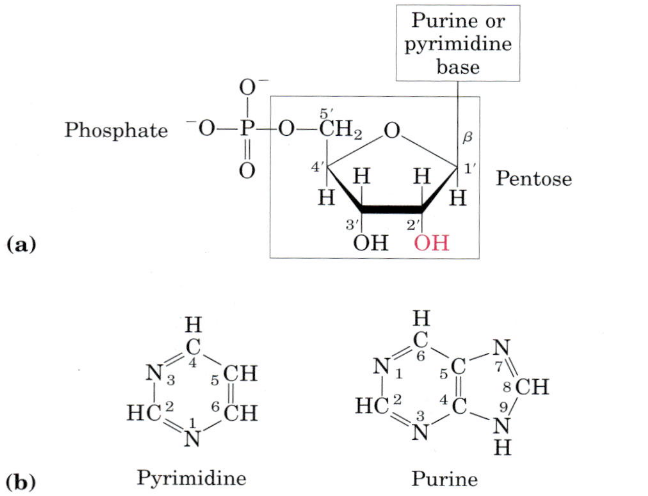

**(a)**

Phosphate

Pentose

**(b)**  Pyrimidine          Purine

**FIGURE 8–1 Structure of nucleotides. (a)** General structure showing the numbering convention for the pentose ring. This is a ribonucleotide. In deoxyribonucleotides the —OH group on the 2′ carbon (in red) is replaced with —H. **(b)** The parent compounds of the pyrimidine and purine bases of nucleotides and nucleic acids, showing the numbering conventions.

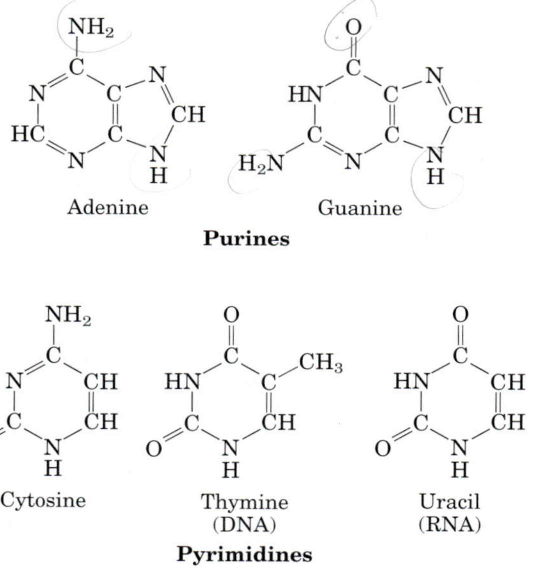

Adenine                  Guanine
**Purines**

Cytosine          Thymine          Uracil
                   (DNA)            (RNA)
**Pyrimidines**

**FIGURE 8–2 Major purine and pyrimidine bases of nucleic acids.** Some of the common names of these bases reflect the circumstances of their discovery. Guanine, for example, was first isolated from guano (bird manure), and thymine was first isolated from thymus tissue.

and nucleosides the carbon numbers are given a prime (′) designation to distinguish them from the numbered atoms of the nitrogenous bases.

The base of a nucleotide is joined covalently (at N-1 of pyrimidines and N-9 of purines) in an *N*-β-glycosyl bond to the 1′ carbon of the pentose, and the phosphate is esterified to the 5′ carbon. The *N*-β-glycosyl bond is formed by removal of the elements of water (a hydroxyl group from the pentose and hydrogen from the base), as in *O*-glycosidic bond formation (see Fig. 7–31).

Both DNA and RNA contain two major purine bases, **adenine** (A) and **guanine** (G), and two major pyrimidines. In both DNA and RNA one of the pyrimidines is **cytosine** (C), but the second major pyrimidine is not the same in both: it is **thymine** (T) in DNA and **uracil** (U) in RNA. Only rarely does thymine occur in RNA or uracil in DNA. The structures of the five major bases are shown in Figure 8–2, and the nomenclature of their corresponding nucleotides and nucleosides is summarized in Table 8–1.

Nucleic acids have two kinds of pentoses. The recurring deoxyribonucleotide units of DNA contain 2′-deoxy-D-ribose, and the ribonucleotide units of RNA contain D-ribose. In nucleotides, both types of pentoses are in their β-furanose (closed five-membered ring) form. As Figure 8–3 shows, the pentose ring is not planar but occurs in one of a variety of conformations generally described as "puckered."

Figure 8–4 gives the structures and names of the four major **deoxyribonucleotides** (deoxyribonucleoside 5′-monophosphates), the structural units of DNAs, and the four major **ribonucleotides** (ribonucleoside 5′-monophosphates), the structural units of RNAs. Specific

long sequences of A, T, G, and C nucleotides in DNA are the repository of genetic information.

Although nucleotides bearing the major purines and pyrimidines are most common, both DNA and RNA also

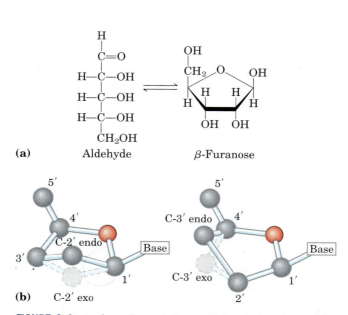

**(a)**  Aldehyde          β-Furanose

**(b)**  C-2′ exo

**FIGURE 8–3 Conformations of ribose. (a)** In solution, the straight-chain (aldehyde) and ring (β-furanose) forms of free ribose are in equilibrium. RNA contains only the ring form, β-D-ribofuranose. Deoxyribose undergoes a similar interconversion in solution, but in DNA exists solely as β-2′-deoxy-D-ribofuranose. **(b)** Ribofuranose rings in nucleotides can exist in four different puckered conformations. In all cases, four of the five atoms are in a single plane. The fifth atom (C-2′ or C-3′) is on either the same (endo) or the opposite (exo) side of the plane relative to the C-5′ atom.

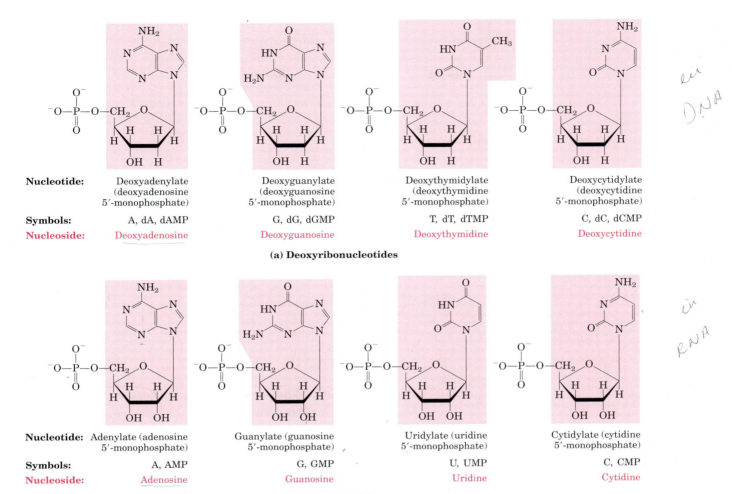

**FIGURE 8–4 Deoxyribonucleotides and ribonucleotides of nucleic acids.** All nucleotides are shown in their free form at pH 7.0. The nucleotide units of DNA **(a)** are usually symbolized as A, G, T, and C, sometimes as dA, dG, dT, and dC; those of RNA **(b)** as A, G, U, and C. In their free form the deoxyribonucleotides are commonly abbreviated dAMP, dGMP, dTMP, and dCMP; the ribonucleotides, AMP, GMP, UMP, and CMP. For each nucleotide, the more common name is followed by the complete name in parentheses. All abbreviations assume that the phosphate group is at the 5' position. The nucleoside portion of each molecule is shaded in light red. In this and the following illustrations, the ring carbons are not shown.

**TABLE 8–1  Nucleotide and Nucleic Acid Nomenclature**

| Base | Nucleoside | Nucleotide | Nucleic acid |
|------|-----------|-----------|--------------|
| **Purines** | | | |
| Adenine | Adenosine | Adenylate | RNA |
|  | Deoxyadenosine | Deoxyadenylate | DNA |
| Guanine | Guanosine | Guanylate | RNA |
|  | Deoxyguanosine | Deoxyguanylate | DNA |
| **Pyrimidines** | | | |
| Cytosine | Cytidine | Cytidylate | RNA |
|  | Deoxycytidine | Deoxycytidylate | DNA |
| Thymine | Thymidine or deoxythymidine | Thymidylate or deoxythymidylate | DNA |
| Uracil | Uridine | Uridylate | RNA |

**Note:** "Nucleoside" and "nucleotide" are generic terms that include both ribo- and deoxyribo- forms. Also, ribonucleosides and ribonucleotides are here designated simply as nucleosides and nucleotides (e.g., ribo-adenosine as adenosine), and deoxyribo-nucleosides and deoxyribonucleotides as deoxynucleosides and deoxynucleotides (e.g., deoxyriboadenosine as deoxyadenosine). Both forms of naming are acceptable, but the shortened names are more commonly used. Thymine is an exception; "ribothymidine" is used to describe its unusual occurrence in RNA.

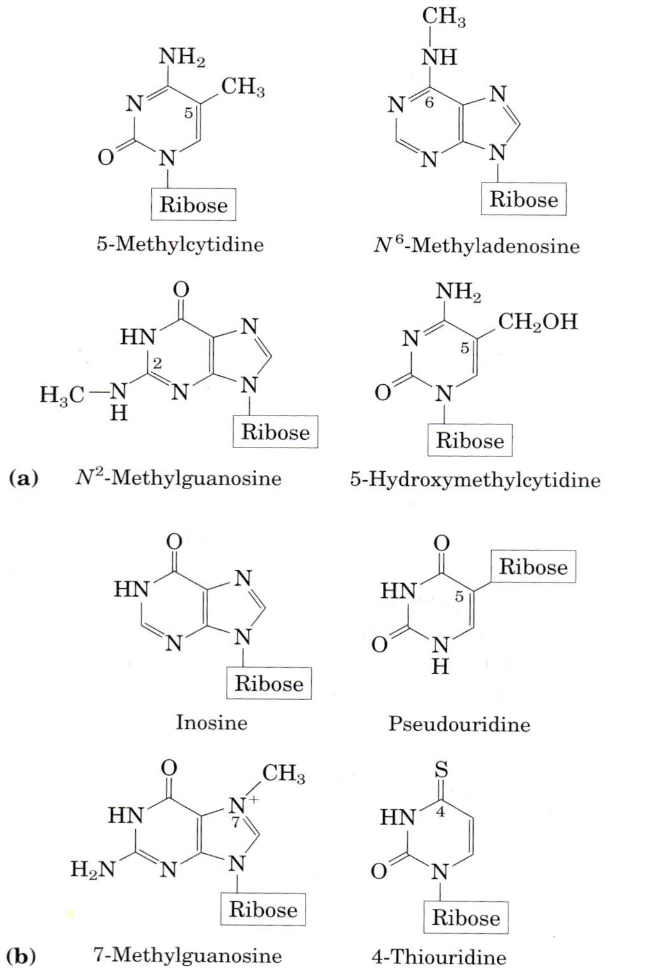

**(a)**

5-Methylcytidine

$N^6$-Methyladenosine

$N^2$-Methylguanosine

5-Hydroxymethylcytidine

Inosine

Pseudouridine

**(b)**

7-Methylguanosine

4-Thiouridine

**FIGURE 8–5** **Some minor purine and pyrimidine bases, shown as the nucleosides.** **(a)** Minor bases of DNA. 5-Methylcytidine occurs in the DNA of animals and higher plants, $N^6$-methyladenosine in bacterial DNA, and 5-hydroxymethylcytidine in the DNA of bacteria infected with certain bacteriophages. **(b)** Some minor bases of tRNAs. Inosine contains the base hypoxanthine. Note that pseudouridine, like uridine, contains uracil; they are distinct in the point of attachment to the ribose—in uridine, uracil is attached through N-1, the usual attachment point for pyrimidines; in pseudouridine, through C-5.

here) is simply to indicate the ring position of the substituent by its number—for example, 5-methylcytosine, 7-methylguanine, and 5-hydroxymethylcytosine (shown as the nucleosides in Fig. 8–5). The element to which the substituent is attached (N, C, O) is not identified. The convention changes when the substituted atom is exocyclic (not within the ring structure), in which case the type of atom is identified and the ring position to which it is attached is denoted with a superscript. The amino nitrogen attached to C-6 of adenine is $N^6$; similarly, the carbonyl oxygen and amino nitrogen at C-6 and C-2 of guanine are $O^6$ and $N^2$, respectively. Examples of this nomenclature are $N^6$-methyladenosine and $N^2$-methylguanosine (Fig. 8–5).

Cells also contain nucleotides with phosphate groups in positions other than on the 5' carbon (Fig. 8–6). **Ribonucleoside 2′,3′-cyclic monophosphates** are isolatable intermediates, and **ribonucleoside 3′-monophosphates** are end products of the hydrolysis of RNA by certain ribonucleases. Other variations are adenosine 3′,5′-cyclic monophosphate (cAMP) and guanosine 3′,5′-cyclic monophosphate (cGMP), considered at the end of this chapter.

## Phosphodiester Bonds Link Successive Nucleotides in Nucleic Acids

The successive nucleotides of both DNA and RNA are covalently linked through phosphate-group "bridges," in which the 5'-phosphate group of one nucleotide unit is

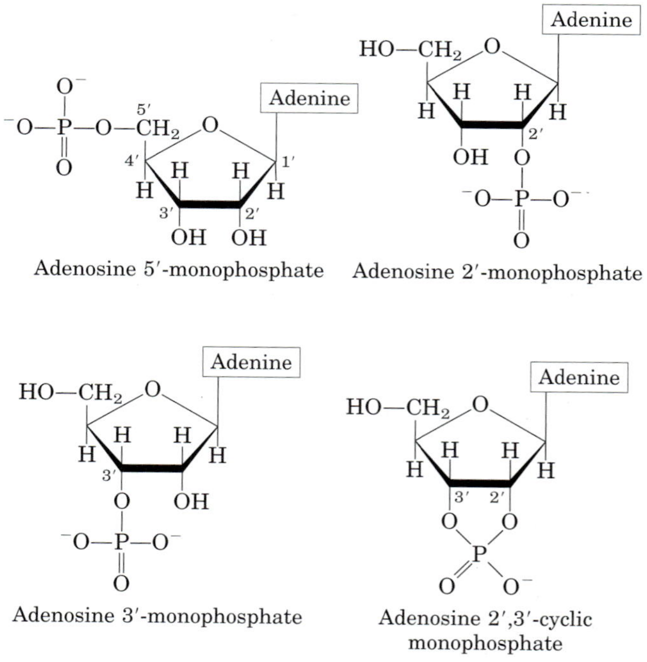

contain some minor bases (Fig. 8–5). In DNA the most common of these are methylated forms of the major bases; in some viral DNAs, certain bases may be hydroxymethylated or glucosylated. Altered or unusual bases in DNA molecules often have roles in regulating or protecting the genetic information. Minor bases of many types are also found in RNAs, especially in tRNAs (see Fig. 26–24).

The nomenclature for the minor bases can be confusing. Like the major bases, many have common names—hypoxanthine, for example, shown as its nucleoside inosine in Figure 8–5. When an atom in the purine or pyrimidine ring is substituted, the usual convention (used

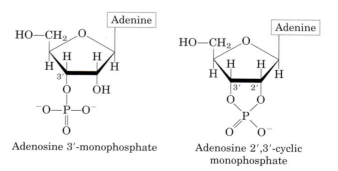

Adenosine 5′-monophosphate

Adenosine 2′-monophosphate

Adenosine 3′-monophosphate

Adenosine 2′,3′-cyclic monophosphate

**FIGURE 8–6** **Some adenosine monophosphates.** Adenosine 2′-monophosphate, 3′-monophosphate, and 2′,3′-cyclic monophosphate are formed by enzymatic and alkaline hydrolysis of RNA.

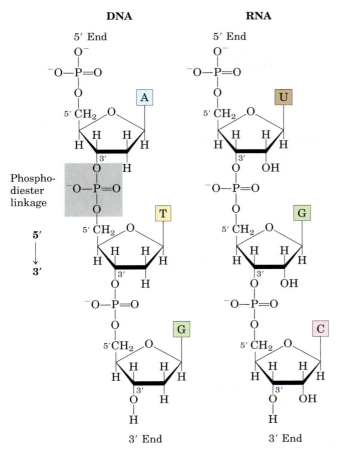

**DNA**

**RNA**

**FIGURE 8-7 Phosphodiester linkages in the covalent backbone of DNA and RNA.** The phosphodiester bonds (one of which is shaded in the DNA) link successive nucleotide units. The backbone of alternating pentose and phosphate groups in both types of nucleic acid is highly polar. The 5′ end of the macromolecule lacks a nucleotide at the 5′ position, and the 3′ end lacks a nucleotide at the 3′ position.

joined to the 3′-hydroxyl group of the next nucleotide, creating a **phosphodiester linkage** (Fig. 8–7). Thus the covalent backbones of nucleic acids consist of alternating phosphate and pentose residues, and the nitrogenous bases may be regarded as side groups joined to the backbone at regular intervals. The backbones of both DNA and RNA are hydrophilic. The hydroxyl groups of the sugar residues form hydrogen bonds with water. The phosphate groups, with a $pK_a$ near 0, are completely ionized and negatively charged at pH 7, and the negative charges are generally neutralized by ionic interactions with positive charges on proteins, metal ions, and polyamines.

All the phosphodiester linkages have the same orientation along the chain (Fig. 8–7), giving each linear nucleic acid strand a specific polarity and distinct 5′ and 3′ ends. By definition, the **5′ end** lacks a nucleotide at the 5′ position and the **3′ end** lacks a nucleotide at the 3′ position. Other groups (most often one or more phosphates) may be present on one or both ends.

The covalent backbone of DNA and RNA is subject to slow, nonenzymatic hydrolysis of the phosphodiester bonds. In the test tube, RNA is hydrolyzed rapidly under alkaline conditions, but DNA is not; the 2′-hydroxyl groups in RNA (absent in DNA) are directly involved in the process. Cyclic 2′,3′-monophosphate nucleotides are the first products of the action of alkali on RNA and are rapidly hydrolyzed further to yield a mixture of 2′- and 3′-nucleoside monophosphates (Fig. 8–8).

The nucleotide sequences of nucleic acids can be represented schematically, as illustrated on the following page by a segment of DNA with five nucleotide units. The phosphate groups are symbolized by ⓟ, and each deoxyribose is symbolized by a vertical line, from C-1′ at the top to C-5′ at the bottom (but keep in mind that

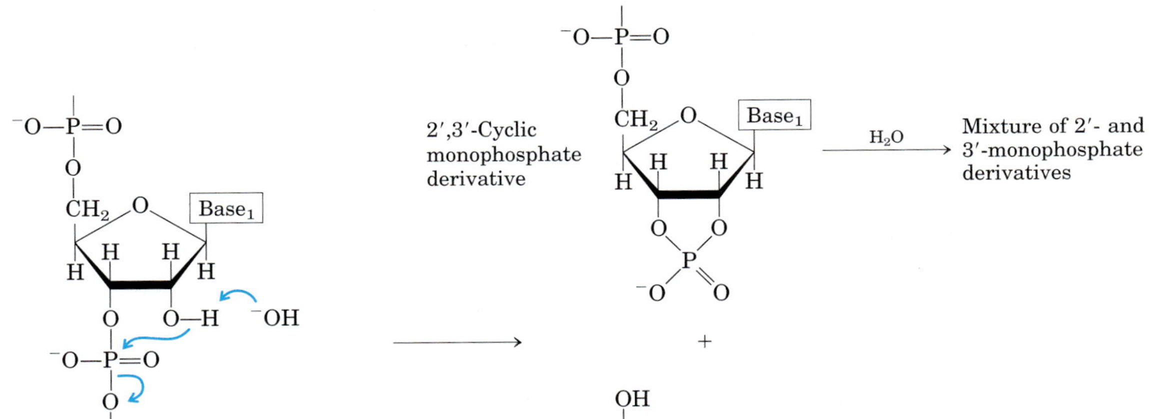

**2′,3′-Cyclic monophosphate derivative**

Mixture of 2′- and 3′-monophosphate derivatives

**Shortened RNA**

**FIGURE 8-8 Hydrolysis of RNA under alkaline conditions.** The 2′ hydroxyl acts as a nucleophile in an intramolecular displacement. The 2′,3′-cyclic monophosphate derivative is further hydrolyzed to a mixture of 2′- and 3′-monophosphates. DNA, which lacks 2′ hydroxyls, is stable under similar conditions.

the sugar is always in its closed-ring β-furanose form in nucleic acids). The connecting lines between nucleotides (which pass through Ⓟ) are drawn diagonally from the middle (C-3′) of the deoxyribose of one nucleotide to the bottom (C-5′) of the next.

By convention, the structure of a single strand of nucleic acid is always written with the 5′ end at the left and the 3′ end at the right—that is, in the 5′ → 3′ direction. Some simpler representations of this pentade-oxyribonucleotide are pA-C-G-T-A$_{OH}$, pApCpGpTpA, and pACGTA.

A short nucleic acid is referred to as an **oligonucleotide.** The definition of "short" is somewhat arbitrary, but polymers containing 50 or fewer nucleotides are generally called oligonucleotides. A longer nucleic acid is called a **polynucleotide.**

## The Properties of Nucleotide Bases Affect the Three-Dimensional Structure of Nucleic Acids

Free pyrimidines and purines are weakly basic compounds and are thus called bases. They have a variety of chemical properties that affect the structure, and ultimately the function, of nucleic acids. The purines and pyrimidines common in DNA and RNA are highly conjugated molecules (Fig. 8–2), a property with important consequences for the structure, electron distribution, and light absorption of nucleic acids. Resonance among atoms in the ring gives most of the bonds partial double-bond character. One result is that pyrimidines are planar molecules; purines are very nearly

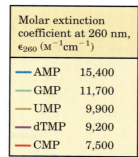

**Uracil**

**FIGURE 8-9  Tautomeric forms of uracil.** The lactam form predominates at pH 7.0; the other forms become more prominent as pH decreases. The other free pyrimidines and the free purines also have tautomeric forms, but they are more rarely encountered.

planar, with a slight pucker. Free pyrimidine and purine bases may exist in two or more tautomeric forms depending on the pH. Uracil, for example, occurs in lactam, lactim, and double lactim forms (Fig. 8–9). The structures shown in Figure 8–2 are the tautomers that predominate at pH 7.0. As a result of resonance, all nucleotide bases absorb UV light, and nucleic acids are characterized by a strong absorption at wavelengths near 260 nm (Fig. 8–10).

The purine and pyrimidine bases are hydrophobic and relatively insoluble in water at the near-neutral pH of the cell. At acidic or alkaline pH the bases become charged and their solubility in water increases. Hydrophobic stacking interactions in which two or more bases are positioned with the planes of their rings parallel (like a stack of coins) are one of two important modes of interaction between bases in nucleic acids. The stacking also involves a combination of van der Waals and dipole-dipole interactions between the bases. Base stacking helps to minimize contact of the bases with water, and base-stacking interactions are very important in stabilizing the three-dimensional structure of nucleic acids, as described later.

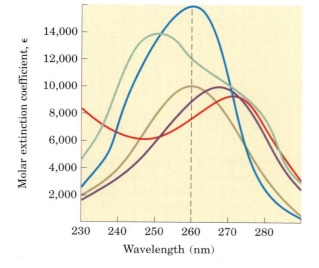

| Molar extinction coefficient at 260 nm, $\epsilon_{260}$ ($M^{-1}cm^{-1}$) | |
|---|---|
| AMP | 15,400 |
| GMP | 11,700 |
| UMP | 9,900 |
| dTMP | 9,200 |
| CMP | 7,500 |

**FIGURE 8-10  Absorption spectra of the common nucleotides.** The spectra are shown as the variation in molar extinction coefficient with wavelength. The molar extinction coefficients at 260 nm and pH 7.0 ($\varepsilon_{260}$) are listed in the table. The spectra of corresponding ribonucleotides and deoxyribonucleotides, as well as the nucleosides, are essentially identical. For mixtures of nucleotides, a wavelength of 260 nm (dashed vertical line) is used for absorption measurements.

**FIGURE 8–11** **Hydrogen-bonding patterns in the base pairs defined by Watson and Crick.** Here as elsewhere, hydrogen bonds are represented by three blue lines.

The most important functional groups of pyrimidines and purines are ring nitrogens, carbonyl groups, and exocyclic amino groups. Hydrogen bonds involving the amino and carbonyl groups are the second important mode of interaction between bases in nucleic acid molecules. Hydrogen bonds between bases permit a complementary association of two (and occasionally three or four) strands of nucleic acid. The most important hydrogen-bonding patterns are those defined by James D. Watson and Francis Crick in 1953, in which A bonds specifically to T (or U) and G bonds to C (Fig. 8–11). These two types of **base pairs** predominate in double-stranded DNA and RNA, and the tautomers shown in Figure 8–2 are responsible for these patterns. It is this specific pairing of bases that permits the duplication of genetic information, as we shall discuss later in this chapter.

### SUMMARY 8.1 Some Basics

■ A nucleotide consists of a nitrogenous base (purine or pyrimidine), a pentose sugar, and one or more phosphate groups. Nucleic acids are polymers of nucleotides, joined together by phosphodiester linkages between the 5'-hydroxyl group of one pentose and the 3'-hydroxyl group of the next.

■ There are two types of nucleic acid: RNA and DNA. The nucleotides in RNA contain ribose, and the common pyrimidine bases are uracil and cytosine. In DNA, the nucleotides contain 2'-deoxyribose, and the common pyrimidine bases are thymine and cytosine. The primary purines are adenine and guanine in both RNA and DNA.

## 8.2 Nucleic Acid Structure

The discovery of the structure of DNA by Watson and Crick in 1953 was a momentous event in science, an event that gave rise to entirely new disciplines and influenced the course of many established ones. Our present understanding of the storage and utilization of a cell's genetic information is based on work made possible by this discovery, and an outline of how genetic information is processed by the cell is now a prerequisite for the discussion of any area of biochemistry. Here, we concern ourselves with DNA structure itself, the events

James Watson

Francis Crick

that led to its discovery, and more recent refinements in our understanding. RNA structure is also introduced.

As in the case of protein structure (Chapter 4), it is sometimes useful to describe nucleic acid structure in terms of hierarchical levels of complexity (primary, secondary, tertiary). The primary structure of a nucleic acid is its covalent structure and nucleotide sequence. Any regular, stable structure taken up by some or all of the nucleotides in a nucleic acid can be referred to as secondary structure. All structures considered in the remainder of this chapter fall under the heading of secondary structure. The complex folding of large chromosomes within eukaryotic chromatin and bacterial nucleoids is generally considered tertiary structure; this is discussed in Chapter 24.

## DNA Stores Genetic Information

The biochemical investigation of DNA began with Friedrich Miescher, who carried out the first systematic chemical studies of cell nuclei. In 1868 Miescher isolated a phosphorus-containing substance, which he called "nuclein," from the nuclei of pus cells (leukocytes) obtained from discarded surgical bandages. He found nuclein to consist of an acidic portion, which we know today as DNA, and a basic portion, protein. Miescher later found a similar acidic substance in the heads of sperm cells from salmon. Although he partially purified nuclein and studied its properties, the covalent (primary) structure of DNA (as shown in Fig. 8–7) was not known with certainty until the late 1940s.

Miescher and many others suspected that nuclein (nucleic acid) was associated in some way with cell inheritance, but the first direct evidence that DNA is the bearer of genetic information came in 1944 through a discovery made by Oswald T. Avery, Colin MacLeod, and Maclyn McCarty. These investigators found that DNA extracted from a virulent (disease-causing) strain of the bacterium *Streptococcus pneumoniae*, also known as pneumococcus, genetically transformed a nonvirulent strain of this organism into a virulent form (Fig. 8–12).

**FIGURE 8–12 The Avery-MacLeod-McCarty experiment. (a)** When injected into mice, the encapsulated strain of pneumococcus is lethal, **(b)** whereas the nonencapsulated strain, **(c)** like the heat-killed encapsulated strain, is harmless. **(d)** Earlier research by the bacteriologist Frederick Griffith had shown that adding heat-killed virulent bacteria (harmless to mice) to a live nonvirulent strain permanently transformed the latter into lethal, virulent, encapsulated bacteria. **(e)** Avery and his colleagues extracted the DNA from heat-killed virulent pneumococci, removing the protein as completely as possible, and added this DNA to nonvirulent bacteria. The DNA gained entrance into the nonvirulent bacteria, which were permanently transformed into a virulent strain.

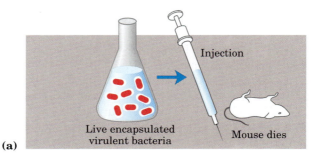

(a)

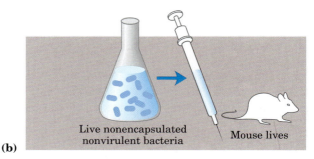

(b)

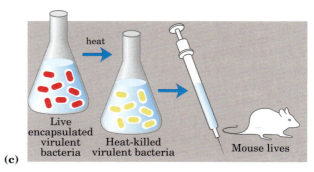

(c)

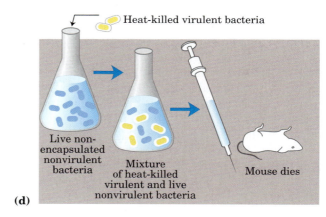

(d)

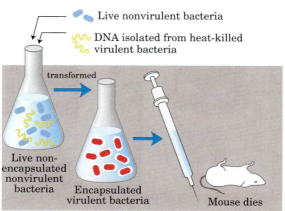

(e)

Avery and his colleagues concluded that the DNA extracted from the virulent strain carried the inheritable genetic message for virulence. Not everyone accepted these conclusions, because protein impurities present in the DNA could have been the carrier of the genetic information. This possibility was soon eliminated by the finding that treatment of the DNA with proteolytic enzymes did not destroy the transforming activity, but treatment with deoxyribonucleases (DNA-hydrolyzing enzymes) did.

A second important experiment provided independent evidence that DNA carries genetic information. In 1952 Alfred D. Hershey and Martha Chase used radioactive phosphorus ($^{32}$P) and radioactive sulfur ($^{35}$S) tracers to show that when the bacterial virus (bacteriophage) T2 infects its host cell, *Escherichia coli,* it is the phosphorus-containing DNA of the viral particle, not the sulfur-containing protein of the viral coat, that enters the host cell and furnishes the genetic information for viral replication (Fig. 8–13). These important early experiments and many other lines of evidence have shown that DNA is the exclusive chromosomal component bearing the genetic information of living cells.

## DNA Molecules Have Distinctive Base Compositions

A most important clue to the structure of DNA came from the work of Erwin Chargaff and his colleagues in the late 1940s. They found that the four nucleotide bases of DNA occur in different ratios in the DNAs of different organisms and that the amounts of certain bases are closely related. These data, collected from DNAs of a great many different species, led Chargaff to the following conclusions:

1. The base composition of DNA generally varies from one species to another.

2. DNA specimens isolated from different tissues of the same species have the same base composition.

3. The base composition of DNA in a given species does not change with an organism's age, nutritional state, or changing environment.

4. In *all* cellular DNAs, regardless of the species, the number of adenosine residues is equal to the number of thymidine residues (that is, A = T), and the number of guanosine residues is equal to the number of cytidine residues (G = C). From these relationships it follows that the sum of the purine residues equals the sum of the pyrimidine residues; that is, A + G = T + C.

These quantitative relationships, sometimes called "Chargaff's rules," were confirmed by many subsequent researchers. They were a key to establishing the three-dimensional structure of DNA and yielded clues to how genetic information is encoded in DNA and passed from one generation to the next.

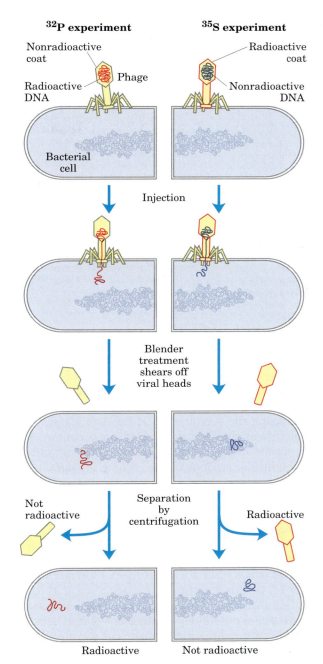

**FIGURE 8–13 The Hershey-Chase experiment.** Two batches of isotopically labeled bacteriophage T2 particles were prepared. One was labeled with $^{32}$P in the phosphate groups of the DNA, the other with $^{35}$S in the sulfur-containing amino acids of the protein coats (capsids). (Note that DNA contains no sulfur and viral protein contains no phosphorus.) The two batches of labeled phage were then allowed to infect separate suspensions of unlabeled bacteria. Each suspension of phage-infected cells was agitated in a blender to shear the viral capsids from the bacteria. The bacteria and empty viral coats (called "ghosts") were then separated by centrifugation. The cells infected with the $^{32}$P-labeled phage were found to contain $^{32}$P, indicating that the labeled viral DNA had entered the cells; the viral ghosts contained no radioactivity. The cells infected with $^{35}$S-labeled phage were found to have no radioactivity after blender treatment, but the viral ghosts contained $^{35}$S. Progeny virus particles (not shown) were produced in both batches of bacteria some time after the viral coats were removed, indicating that the genetic message for their replication had been introduced by viral DNA, not by viral protein.

## DNA Is a Double Helix

To shed more light on the structure of DNA, Rosalind Franklin and Maurice Wilkins used the powerful method of x-ray diffraction (see Box 4–4) to analyze DNA fibers. They showed in the early 1950s that DNA produces a characteristic x-ray diffraction pattern (Fig. 8–14). From this pattern it was deduced that DNA molecules are helical with two periodicities along their long axis, a primary one of 3.4 Å and a secondary one of 34 Å. The problem then was to formulate a three-dimensional model of the DNA molecule that could account not only for the x-ray diffraction data but also for the specific A = T and G = C base equivalences discovered by Chargaff and for the other chemical properties of DNA.

In 1953 Watson and Crick postulated a three-dimensional model of DNA structure that accounted for all the available data. It consists of two helical DNA chains wound around the same axis to form a right-handed double helix (see Box 4–1 for an explanation of the right- or left-handed sense of a helical structure). The hydrophilic backbones of alternating deoxyribose and phosphate groups are on the outside of the double helix, facing the surrounding water. The furanose ring of each deoxyribose is in the C-2′ endo conformation. The purine and pyrimidine bases of both strands are stacked inside the double helix, with their hydrophobic and nearly planar ring structures very close together and perpendicular to the long axis. The offset pairing of the two strands creates a **major groove** and **minor groove** on the surface of the duplex (Fig. 8–15). Each nucleotide base of one strand is paired in the same plane with a base of the other strand. Watson and Crick found that the hydrogen-bonded base pairs illustrated in Figure 8–11, G with C and A with T, are those that fit best within the structure, providing a rationale for Chargaff's rule that in any DNA, G = C and A = T. It is important to note that three hydrogen bonds can form between G and C, symbolized G≡C, but only two can form between A and T, symbolized A=T. This is one reason for the

Rosalind Franklin, 1920–1958      Maurice Wilkins

finding that separation of paired DNA strands is more difficult the higher the ratio of G≡C to A=T base pairs. Other pairings of bases tend (to varying degrees) to destabilize the double-helical structure.

When Watson and Crick constructed their model, they had to decide at the outset whether the strands of DNA should be **parallel** or **antiparallel**—whether their 5′,3′-phosphodiester bonds should run in the same or opposite directions. An antiparallel orientation produced the most convincing model, and later work with DNA polymerases (Chapter 25) provided experimental evidence that the strands are indeed antiparallel, a finding ultimately confirmed by x-ray analysis.

To account for the periodicities observed in the x-ray diffraction patterns of DNA fibers, Watson and Crick manipulated molecular models to arrive at a structure

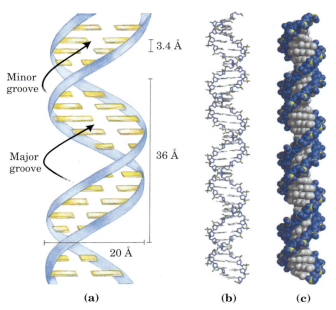

**FIGURE 8–15 Watson-Crick model for the structure of DNA.** The original model proposed by Watson and Crick had 10 base pairs, or 34 Å (3.4 nm), per turn of the helix; subsequent measurements revealed 10.5 base pairs, or 36 Å (3.6 nm), per turn. **(a)** Schematic representation, showing dimensions of the helix. **(b)** Stick representation showing the backbone and stacking of the bases. **(c)** Space-filling model.

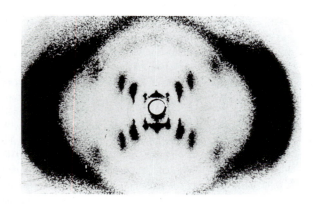

**FIGURE 8–14 X-ray diffraction pattern of DNA.** The spots forming a cross in the center denote a helical structure. The heavy bands at the left and right arise from the recurring bases.

in which the vertically stacked bases inside the double helix would be 3.4 Å apart; the secondary repeat distance of about 34 Å was accounted for by the presence of 10 base pairs in each complete turn of the double helix. In aqueous solution the structure differs slightly from that in fibers, having 10.5 base pairs per helical turn (Fig. 8–15).

As Figure 8–16 shows, the two antiparallel polynucleotide chains of double-helical DNA are not identical in either base sequence or composition. Instead they are **complementary** to each other. Wherever adenine occurs in one chain, thymine is found in the other; similarly, wherever guanine occurs in one chain, cytosine is found in the other.

The DNA double helix, or duplex, is held together by two forces, as described earlier: hydrogen bonding between complementary base pairs (Fig. 8–11) and base-stacking interactions. The complementarity between the DNA strands is attributable to the hydrogen bonding between base pairs. The base-stacking interactions, which are largely nonspecific with respect to the identity of the stacked bases, make the major contribution to the stability of the double helix.

The important features of the double-helical model of DNA structure are supported by much chemical and

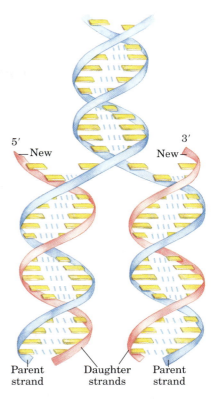

**FIGURE 8–17 Replication of DNA as suggested by Watson and Crick.** The preexisting or "parent" strands become separated, and each is the template for biosynthesis of a complementary "daughter" strand (in red).

biological evidence. Moreover, the model immediately suggested a mechanism for the transmission of genetic information. The essential feature of the model is the complementarity of the two DNA strands. As Watson and Crick were able to see, well before confirmatory data became available, this structure could logically be replicated by (1) separating the two strands and (2) synthesizing a complementary strand for each. Because nucleotides in each new strand are joined in a sequence specified by the base-pairing rules stated above, each preexisting strand functions as a template to guide the synthesis of one complementary strand (Fig. 8–17). These expectations were experimentally confirmed, inaugurating a revolution in our understanding of biological inheritance.

## DNA Can Occur in Different Three-Dimensional Forms

DNA is a remarkably flexible molecule. Considerable rotation is possible around a number of bonds in the sugar–phosphate (phosphodeoxyribose) backbone, and thermal fluctuation can produce bending, stretching, and unpairing (melting) of the strands. Many significant deviations from the Watson-Crick DNA structure are found in cellular DNA, some or all of which may play important roles in DNA metabolism. These structural variations generally do not affect the key properties of DNA defined by Watson and Crick: strand complementarity,

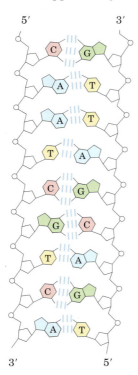

**FIGURE 8–16 Complementarity of strands in the DNA double helix.** The complementary antiparallel strands of DNA follow the pairing rules proposed by Watson and Crick. The base-paired antiparallel strands differ in base composition: the left strand has the composition $A_3 T_2 G_1 C_3$; the right, $A_2 T_3 G_3 C_1$. They also differ in sequence when each chain is read in the $5' \rightarrow 3'$ direction. Note the base equivalences: A = T and G = C in the duplex.

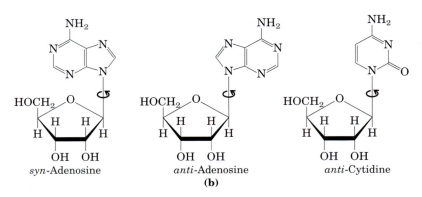

*syn*-Adenosine          *anti*-Adenosine          *anti*-Cytidine

**(b)**

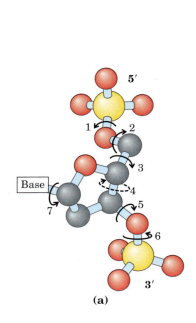

**(a)**

**FIGURE 8–18 Structural variation in DNA. (a)** The conformation of a nucleotide in DNA is affected by rotation about seven different bonds. Six of the bonds rotate freely. The limited rotation about bond 4 gives rise to ring pucker, in which one of the atoms in the five-membered furanose ring is out of the plane described by the other four. This conformation is endo or exo, depending on whether the atom is displaced to the same side of the plane as C-5′ or to the opposite side (see Fig. 8–3b). **(b)** For purine bases in nucleotides, only two conformations with respect to the attached ribose units are sterically permitted, anti or syn. Pyrimidines generally occur in the anti conformation.

antiparallel strands, and the requirement for A=T and G≡C base pairs.

Structural variation in DNA reflects three things: the different possible conformations of the deoxyribose, rotation about the contiguous bonds that make up the phosphodeoxyribose backbone (Fig. 8–18a), and free rotation about the C-1′–*N*-glycosyl bond (Fig. 8–18b). Because of steric constraints, purines in purine nucleotides are restricted to two stable conformations with respect to deoxyribose, called syn and anti (Fig. 8–18b). Pyrimidines are generally restricted to the anti conformation because of steric interference between the sugar and the carbonyl oxygen at C-2 of the pyrimidine.

The Watson-Crick structure is also referred to as **B-form DNA,** or B-DNA. The B form is the most stable structure for a random-sequence DNA molecule under physiological conditions and is therefore the standard point of reference in any study of the properties of DNA. Two structural variants that have been well characterized in crystal structures are the **A** and **Z forms.** These three DNA conformations are shown in Figure 8–19, with a summary of their properties. The A form is favored in many solutions that are relatively devoid of water. The DNA is still arranged in a right-handed double helix, but the helix is wider and the number of base pairs per helical turn is 11, rather than 10.5 as in B-DNA. The

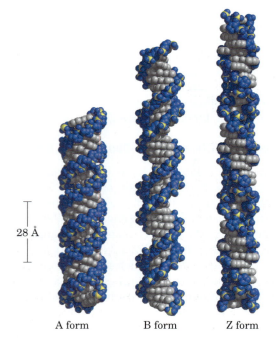

28 Å

A form          B form          Z form

| | A form | B form | Z form |
|---|---|---|---|
| Helical sense | Right handed | Right handed | Left handed |
| Diameter | ~26 Å | ~20 Å | ~18 Å |
| Base pairs per helical turn | 11 | 10.5 | 12 |
| Helix rise per base pair | 2.6 Å | 3.4 Å | 3.7 Å |
| Base tilt normal to the helix axis | 20° | 6° | 7° |
| Sugar pucker conformation | C-3′ endo | C-2′ endo | C-2′ endo for pyrimidines; C-3′ endo for purines |
| Glycosyl bond conformation | Anti | Anti | Anti for pyrimidines; syn for purines |

**FIGURE 8–19 Comparison of A, B, and Z forms of DNA.** Each structure shown here has 36 base pairs. The bases are shown in gray, the phosphate atoms in yellow, and the riboses and phosphate oxygens in blue. Blue is the color used to represent DNA strands in later chapters. The table summarizes some properties of the three forms of DNA.

plane of the base pairs in A-DNA is tilted about 20° with respect to the helix axis. These structural changes deepen the major groove while making the minor groove shallower. The reagents used to promote crystallization of DNA tend to dehydrate it, and thus most short DNA molecules tend to crystallize in the A form.

Z-form DNA is a more radical departure from the B structure; the most obvious distinction is the left-handed helical rotation. There are 12 base pairs per helical turn, and the structure appears more slender and elongated. The DNA backbone takes on a zigzag appearance. Certain nucleotide sequences fold into left-handed Z helices much more readily than others. Prominent examples are sequences in which pyrimidines alternate with purines, especially alternating C and G or 5-methyl-C and G residues. To form the left-handed helix in Z-DNA, the purine residues flip to the syn conformation, alternating with pyrimidines in the anti conformation. The major groove is barely apparent in Z-DNA, and the minor groove is narrow and deep.

Whether A-DNA occurs in cells is uncertain, but there is evidence for some short stretches (tracts) of Z-DNA in both prokaryotes and eukaryotes. These Z-DNA tracts may play a role (as yet undefined) in regulating the expression of some genes or in genetic recombination.

### Certain DNA Sequences Adopt Unusual Structures

A number of other sequence-dependent structural variations have been detected within larger chromosomes that may affect the function and metabolism of the DNA segments in their immediate vicinity. For example, bends occur in the DNA helix wherever four or more adenosine residues appear sequentially in one strand. Six adenosines in a row produce a bend of about 18°.

The bending observed with this and other sequences may be important in the binding of some proteins to DNA.

A rather common type of DNA sequence is a **palindrome.** A palindrome is a word, phrase, or sentence that is spelled identically read either forward or backward; two examples are ROTATOR and NURSES RUN. The term is applied to regions of DNA with **inverted repeats** of base sequence having twofold symmetry over two strands of DNA (Fig. 8–20). Such sequences are self-complementary within each strand and therefore have the potential to form **hairpin** or **cruciform** (cross-shaped) structures (Fig. 8–21). When the inverted repeat occurs within each individual strand of the DNA, the sequence is called a **mirror repeat.** Mirror repeats do not have complementary sequences within the same strand and cannot form hairpin or cruciform structures. Sequences of these types are found

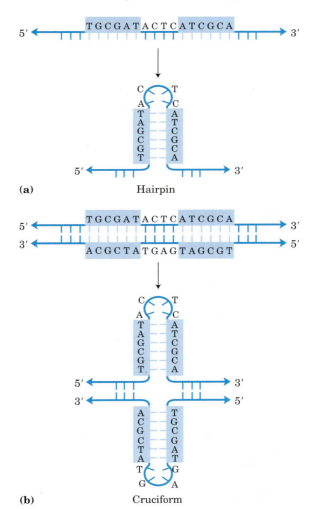

**(a)** Hairpin

**(b)** Cruciform

**FIGURE 8–21 Hairpins and cruciforms.** Palindromic DNA (or RNA) sequences can form alternative structures with intrastrand base pairing. **(a)** When only a single DNA (or RNA) strand is involved, the structure is called a hairpin. **(b)** When both strands of a duplex DNA are involved, it is called a cruciform. Blue shading highlights asymmetric sequences that can pair with the complementary sequence either in the same strand or in the complementary strand.

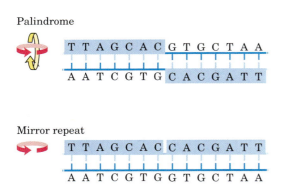

**FIGURE 8–20 Palindromes and mirror repeats.** Palindromes are sequences of double-stranded nucleic acids with twofold symmetry. In order to superimpose one repeat (shaded sequence) on the other, it must be rotated 180° about the horizontal axis then 180° about the vertical axis, as shown by the colored arrows. A mirror repeat, on the other hand, has a symmetric sequence within each strand. Superimposing one repeat on the other requires only a single 180° rotation about the vertical axis.

in virtually every large DNA molecule and can encompass a few base pairs or thousands. The extent to which palindromes occur as cruciforms in cells is not known, although some cruciform structures have been demonstrated in vivo in *E. coli*. Self-complementary sequences cause isolated single strands of DNA (or RNA) in solution to fold into complex structures containing multiple hairpins.

Several unusual DNA structures involve three or even four DNA strands. These structural variations merit investigation because there is a tendency for many of them to appear at sites where important events in DNA metabolism (replication, recombination, transcription) are initiated or regulated. Nucleotides participating in a

Watson-Crick base pair (Fig. 8–11) can form a number of additional hydrogen bonds, particularly with functional groups arrayed in the major groove. For example, a cytidine residue (if protonated) can pair with the guanosine residue of a G≡C nucleotide pair, and a thymidine can pair with the adenosine of an A=T pair (Fig. 8–22). The N-7, $O^6$, and $N^6$ of purines, the atoms that participate in the hydrogen bonding of triplex DNA, are often referred to as **Hoogsteen positions,** and the non-Watson-Crick pairing is called **Hoogsteen pairing,** after Karst Hoogsteen, who in 1963 first recognized the potential for these unusual pairings. Hoogsteen pairing allows the formation of **triplex DNAs.** The triplexes shown in Figure 8–22 (a, b) are most stable at low pH

T=A•T

C≡G•C⁺

**(a)**

**(b)**

Guanosine tetraplex

**(c)**

**(d)**

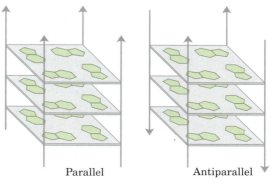

Parallel          Antiparallel

**(e)**

**FIGURE 8–22 DNA structures containing three or four DNA strands.**
**(a)** Base-pairing patterns in one well-characterized form of triplex DNA. The Hoogsteen pair in each case is shown in red. **(b)** Triple-helical DNA containing two pyrimidine strands (poly(T)) and one purine strand (poly(A)) (derived from PDB ID 1BCE). The dark blue and light blue strands are antiparallel and paired by normal Watson-Crick base-pairing patterns. The third (all-pyrimidine) strand (purple) is parallel to the purine strand and paired through non-Watson-Crick hydrogen bonds. The triplex is viewed end-on, with five triplets shown. Only the triplet closest to the viewer is colored. **(c)** Base-pairing pattern in the guanosine tetraplex structure. **(d)** Two successive tetraplets from a G tetraplex structure (derived from PDB ID 1QDG), viewed end-on with the one closest to the viewer in color. **(e)** Possible variants in the orientation of strands in a G tetraplex.

because the C≡G • C⁺ triplet requires a protonated cytosine. In the triplex, the p$K_a$ of this cytosine is >7.5, altered from its normal value of 4.2. The triplexes also form most readily within long sequences containing only pyrimidines or only purines in a given strand. Some triplex DNAs contain two pyrimidine strands and one purine strand; others contain two purine strands and one pyrimidine strand.

Four DNA strands can also pair to form a tetraplex (quadruplex), but this occurs readily only for DNA sequences with a very high proportion of guanosine residues (Fig. 8–22c, d). The guanosine tetraplex, or **G tetraplex**, is quite stable over a wide range of conditions. The orientation of strands in the tetraplex can vary as shown in Figure 8–22e.

A particularly exotic DNA structure, known as **H-DNA**, is found in polypyrimidine or polypurine tracts that also incorporate a mirror repeat. A simple example is a long stretch of alternating T and C residues (Fig. 8–23). The H-DNA structure features the triple-stranded form illustrated in Figure 8–22 (a, b). Two of the three strands in the H-DNA triple helix contain pyrimidines and the third contains purines.

In the DNA of living cells, sites recognized by many sequence-specific DNA-binding proteins (Chapter 28) are arranged as palindromes, and polypyrimidine or polypurine sequences that can form triple helices or even H-DNA are found within regions involved in the regulation of expression of some eukaryotic genes. In principle, synthetic DNA strands designed to pair with these sequences to form triplex DNA could disrupt gene expression. This approach to controlling cellular metabolism is of growing commercial interest for its potential application in medicine and agriculture.

### Messenger RNAs Code for Polypeptide Chains

We now turn our attention briefly from DNA structure to the expression of the genetic information that it contains. RNA, the second major form of nucleic acid in cells, has many functions. In gene expression, RNA acts as an intermediary by using the information encoded in DNA to specify the amino acid sequence of a functional protein.

Given that the DNA of eukaryotes is largely confined to the nucleus whereas protein synthesis occurs on ribosomes in the cytoplasm, some molecule other than DNA must carry the genetic message from the nucleus to the cytoplasm. As early as the 1950s, RNA was considered the logical candidate: RNA is found in both the nucleus and the cytoplasm, and an increase in protein synthesis is accompanied by an increase in the amount of cytoplasmic RNA and an increase in its rate of turnover. These and other observations led several researchers to suggest that RNA carries genetic information from DNA to the protein biosynthetic machin-

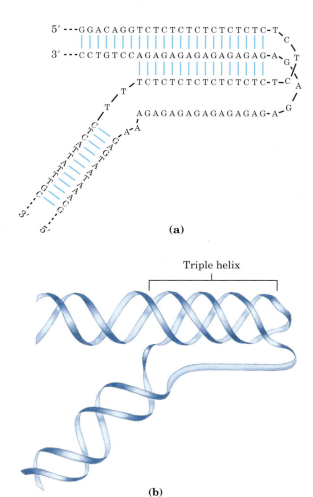

**(a)**

**(b)**

**FIGURE 8–23  H-DNA. (a)** A sequence of alternating T and C residues can be considered a mirror repeat centered about a central T or C. **(b)** These sequences form an unusual structure in which the strands in one half of the mirror repeat are separated and the pyrimidine-containing strand (alternating T and C residues) folds back on the other half of the repeat to form a triple helix. The purine strand (alternating A and G residues) is left unpaired. This structure produces a sharp bend in the DNA.

ery of the ribosome. In 1961 François Jacob and Jacques Monod presented a unified (and essentially correct) picture of many aspects of this process. They proposed the name "messenger RNA" (mRNA) for that portion of the total cellular RNA carrying the genetic information from DNA to the ribosomes, where the messengers provide the templates that specify amino acid sequences in polypeptide chains. Although mRNAs from different genes can vary greatly in length, the mRNAs from a particular gene generally have a defined size. The process of forming mRNA on a DNA template is known as transcription.

In prokaryotes, a single mRNA molecule may code for one or several polypeptide chains. If it carries the code for only one polypeptide, the mRNA is **monocistronic;**

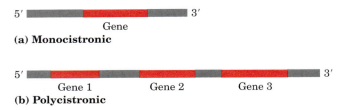

**(a) Monocistronic**

5′ Gene 3′

**(b) Polycistronic**

5′ Gene 1 Gene 2 Gene 3 3′

**FIGURE 8–24 Prokaryotic mRNA.** Schematic diagrams show **(a)** monocistronic and **(b)** polycistronic mRNAs of prokaryotes. Red segments represent RNA coding for a gene product; gray segments represent noncoding RNA. In the polycistronic transcript, noncoding RNA separates the three genes.

if it codes for two or more different polypeptides, the mRNA is **polycistronic.** In eukaryotes, most mRNAs are monocistronic. (For the purposes of this discussion, "cistron" refers to a gene. The term itself has historical roots in the science of genetics, and its formal genetic definition is beyond the scope of this text.) The minimum length of an mRNA is set by the length of the polypeptide chain for which it codes. For example, a polypeptide chain of 100 amino acid residues requires an RNA coding sequence of at least 300 nucleotides, because each amino acid is coded by a nucleotide triplet (this and other details of protein synthesis are discussed in Chapter 27). However, mRNAs transcribed from DNA are always somewhat longer than the length needed simply to code for a polypeptide sequence (or sequences). The additional, noncoding RNA includes sequences that regulate protein synthesis. Figure 8–24 summarizes the general structure of prokaryotic mRNAs.

## Many RNAs Have More Complex Three-Dimensional Structures

Messenger RNA is only one of several classes of cellular RNA. Transfer RNAs serve as adapter molecules in protein synthesis; covalently linked to an amino acid at one end, they pair with the mRNA in such a way that amino acids are joined to a growing polypeptide in the correct sequence. Ribosomal RNAs are components of ribosomes. There is also a wide variety of special-function RNAs, including some (called ribozymes) that have enzymatic activity. All the RNAs are considered in detail in Chapter 26. The diverse and often complex functions of these RNAs reflect a diversity of structure much richer than that observed in DNA molecules.

The product of transcription of DNA is always single-stranded RNA. The single strand tends to assume a right-handed helical conformation dominated by base-stacking interactions (Fig. 8–25), which are stronger between two purines than between a purine and pyrimidine or between two pyrimidines. The purine-purine interaction is so strong that a pyrimidine separating two purines is often displaced from the stacking pattern so

that the purines can interact. Any self-complementary sequences in the molecule produce more complex structures. RNA can base-pair with complementary regions of either RNA or DNA. Base pairing matches the pattern for DNA: G pairs with C and A pairs with U (or with the occasional T residue in some RNAs). One difference is that base pairing between G and U residues—unusual in DNA—is fairly common in RNA (see Fig. 8–27). The paired strands in RNA or RNA-DNA duplexes are antiparallel, as in DNA.

RNA has no simple, regular secondary structure that serves as a reference point, as does the double helix for DNA. The three-dimensional structures of many RNAs, like those of proteins, are complex and unique. Weak interactions, especially base-stacking interactions, play a major role in stabilizing RNA structures, just as they do in DNA. Where complementary sequences are present, the predominant double-stranded structure is an A-form right-handed double helix. Z-form helices have been made in the laboratory (under very high-salt or high-temperature conditions). The B form of RNA has not been observed. Breaks in the regular A-form helix caused by mismatched or unmatched bases in one or both strands are common and result in bulges or internal loops (Fig. 8–26). Hairpin loops form between nearby self-complementary sequences. The potential for base-paired helical structures in many RNAs is extensive (Fig. 8–27), and the resulting hairpins are the most common type of secondary structure in RNA. Specific

**FIGURE 8–25 Typical right-handed stacking pattern of single-stranded RNA.** The bases are shown in gray, the phosphate atoms in yellow, and the riboses and phosphate oxygens in green. Green is used to represent RNA strands in succeeding chapters, just as blue is used for DNA.

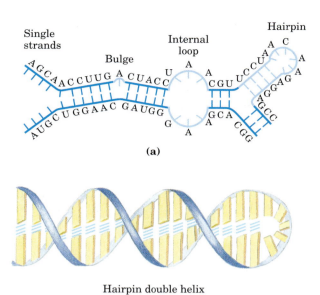

**FIGURE 8–26 Secondary structure of RNAs. (a)** Bulge, internal loop, and hairpin loop. **(b)** The paired regions generally have an A-form right-handed helix, as shown for a hairpin.

short base sequences (such as UUCG) are often found at the ends of RNA hairpins and are known to form particularly tight and stable loops. Such sequences may act as starting points for the folding of an RNA molecule into its precise three-dimensional structure. Important additional structural contributions are made by hydrogen bonds that are not part of standard Watson-Crick base pairs. For example, the 2′-hydroxyl group of ribose can hydrogen-bond with other groups. Some of these properties are evident in the structure of the phenylalanine transfer RNA of yeast—the tRNA responsible for inserting Phe residues into polypeptides—and in two RNA enzymes, or ribozymes, whose functions, like those of protein enzymes, depend on their three-dimensional structures (Fig. 8–28).

The analysis of RNA structure and the relationship between structure and function is an emerging field of inquiry that has many of the same complexities as the analysis of protein structure. The importance of understanding RNA structure grows as we become increasingly aware of the large number of functional roles for RNA molecules.

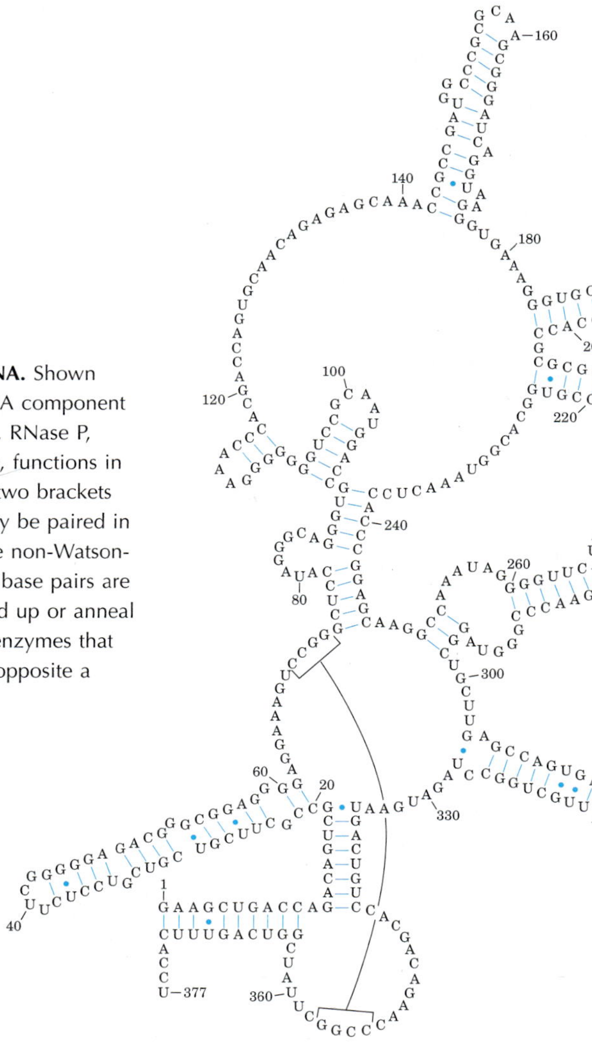

**FIGURE 8–27 Base-paired helical structures in an RNA.** Shown here is the possible secondary structure of the M1 RNA component of the enzyme RNase P of *E. coli*, with many hairpins. RNase P, which also contains a protein component (not shown), functions in the processing of transfer RNAs (see Fig. 26–23). The two brackets indicate additional complementary sequences that may be paired in the three-dimensional structure. The blue dots indicate non-Watson-Crick G≡U base pairs (boxed inset). Note that G≡U base pairs are allowed only when presynthesized strands of RNA fold up or anneal with each other. There are no RNA polymerases (the enzymes that synthesize RNAs on a DNA template) that insert a U opposite a template G, or vice versa, during RNA synthesis.

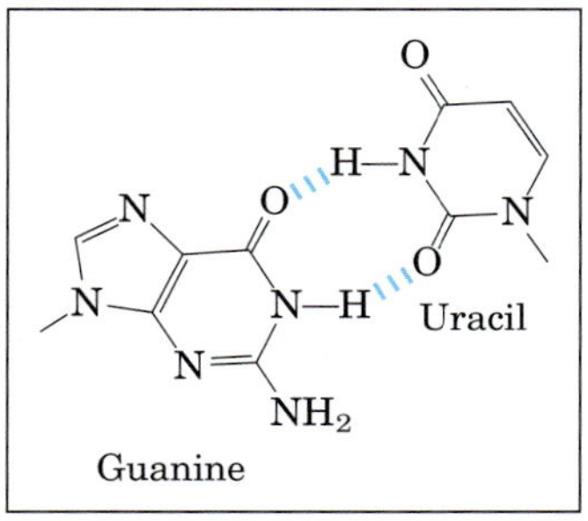

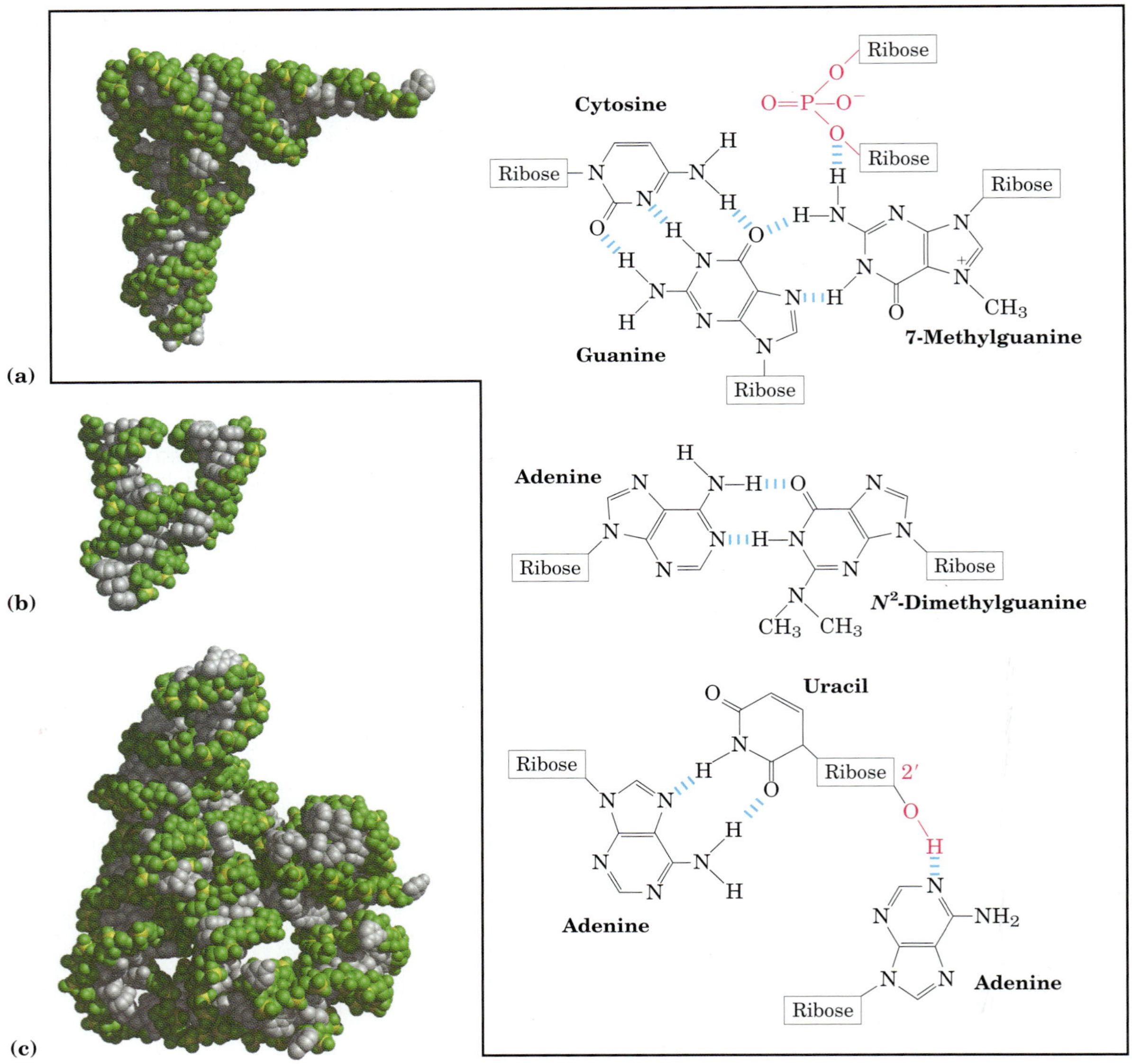

**FIGURE 8–28 Three-dimensional structure in RNA. (a)** Three-dimensional structure of phenylalanine tRNA of yeast (PDB ID 1TRA). Some unusual base-pairing patterns found in this tRNA are shown. Note also the involvement of the oxygen of a ribose phosphodiester bond in one hydrogen-bonding arrangement, and a ribose 2′-hydroxyl group in another (both in red). **(b)** A hammerhead ribozyme (so named because the secondary structure at the active site looks like the head of a hammer), derived from certain plant viruses (derived from PDB ID 1MME). Ribozymes, or RNA enzymes, catalyze a variety of reactions, primarily in RNA metabolism and protein synthesis. The complex three-dimensional structures of these RNAs reflect the complexity inherent in catalysis, as described for protein enzymes in Chapter 6. **(c)** A segment of mRNA known as an intron, from the ciliated protozoan *Tetrahymena thermophila* (derived from PDB ID 1GRZ). This intron (a ribozyme) catalyzes its own excision from between exons in an mRNA strand (discussed in Chapter 26).

## SUMMARY 8.2 Nucleic Acid Structure

■ Many lines of evidence show that DNA bears genetic information. In particular, the Avery-MacLeod-McCarty experiment showed that DNA isolated from one bacterial strain can enter and transform the cells of another strain, endowing it with some of the inheritable characteristics of the donor. The Hershey-Chase experiment showed that the DNA of a bacterial virus, but not its protein coat, carries the genetic message for replication of the virus in a host cell.

■ Putting together much published data, Watson and Crick postulated that native DNA consists of two antiparallel chains in a right-handed double-helical arrangement. Complementary base pairs, A=T and G≡C, are formed by hydrogen bonding within the helix. The base pairs are stacked perpendicular to the long axis of the double helix, 3.4 Å apart, with 10.5 base pairs per turn.

- DNA can exist in several structural forms. Two variations of the Watson-Crick form, or B-DNA, are A- and Z-DNA. Some sequence-dependent structural variations cause bends in the DNA molecule. DNA strands with appropriate sequences can form hairpin/cruciform structures or triplex or tetraplex DNA.

- Messenger RNA transfers genetic information from DNA to ribosomes for protein synthesis. Transfer RNA and ribosomal RNA are also involved in protein synthesis. RNA can be structurally complex; single RNA strands can be folded into hairpins, double-stranded regions, or complex loops.

## 8.3 Nucleic Acid Chemistry

To understand how nucleic acids function, we must understand their chemical properties as well as their structures. The role of DNA as a repository of genetic information depends in part on its inherent stability. The chemical transformations that do occur are generally very slow in the absence of an enzyme catalyst. The long-term storage of information without alteration is so important to a cell, however, that even very slow reactions that alter DNA structure can be physiologically significant. Processes such as carcinogenesis and aging may be intimately linked to slowly accumulating, irreversible alterations of DNA. Other, nondestructive alterations also occur and are essential to function, such as the strand separation that must precede DNA replication or transcription. In addition to providing insights into physiological processes, our understanding of nucleic acid chemistry has given us a powerful array of technologies that have applications in molecular biology, medicine, and forensic science. We now examine the chemical properties of DNA and some of these technologies.

### Double-Helical DNA and RNA Can Be Denatured

Solutions of carefully isolated, native DNA are highly viscous at pH 7.0 and room temperature (25 °C). When such a solution is subjected to extremes of pH or to temperatures above 80 °C, its viscosity decreases sharply, indicating that the DNA has undergone a physical change. Just as heat and extremes of pH denature globular proteins, they also cause denaturation, or melting, of double-helical DNA. Disruption of the hydrogen bonds between paired bases and of base stacking causes unwinding of the double helix to form two single strands, completely separate from each other along the entire

length or part of the length (partial denaturation) of the molecule. No covalent bonds in the DNA are broken (Fig. 8–29).

Renaturation of a DNA molecule is a rapid one-step process, as long as a double-helical segment of a dozen or more residues still unites the two strands. When the temperature or pH is returned to the range in which most organisms live, the unwound segments of the two strands spontaneously rewind, or **anneal,** to yield the intact duplex (Fig. 8–29). However, if the two strands are completely separated, renaturation occurs in two steps. In the first, relatively slow step, the two strands "find" each other by random collisions and form a short segment of complementary double helix. The second step is much faster: the remaining unpaired bases successively come into register as base pairs, and the two strands "zipper" themselves together to form the double helix.

The close interaction between stacked bases in a nucleic acid has the effect of decreasing its absorption of UV light relative to that of a solution with the same concentration of free nucleotides, and the absorption is decreased further when two complementary nucleic acids strands are paired. This is called the hypochromic effect. Denaturation of a double-stranded nucleic acid produces the opposite result: an increase in absorption

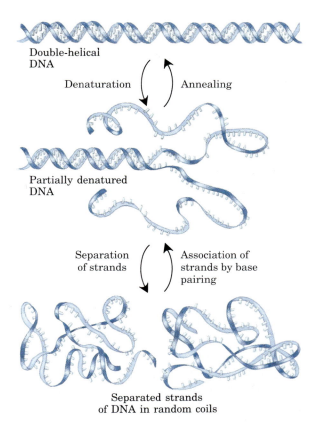

**Double-helical DNA**

Denaturation ⟳ Annealing

**Partially denatured DNA**

Separation of strands ⟳ Association of strands by base pairing

**Separated strands of DNA in random coils**

**FIGURE 8–29  Reversible denaturation and annealing (renaturation) of DNA.**

called the hyperchromic effect. The transition from double-stranded DNA to the single-stranded, denatured form can thus be detected by monitoring the absorption of UV light.

Viral or bacterial DNA molecules in solution denature when they are heated slowly (Fig. 8–30). Each species of DNA has a characteristic denaturation temperature, or melting point ($t_m$): the higher its content of G≡C base pairs, the higher the melting point of the DNA. This is because G≡C base pairs, with three hydrogen bonds, require more heat energy to dissociate than A=T base pairs. Careful determination of the melting point of a DNA specimen, under fixed conditions of pH and ionic strength, can yield an estimate of its base composition. If denaturation conditions are carefully controlled, regions that are rich in A=T base pairs will specifically denature while most of the DNA remains

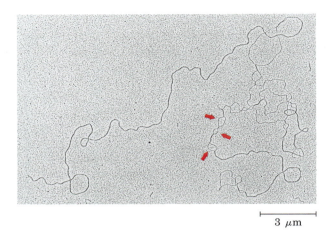

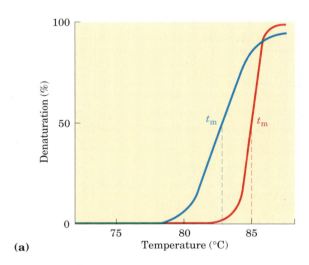

**(a)**

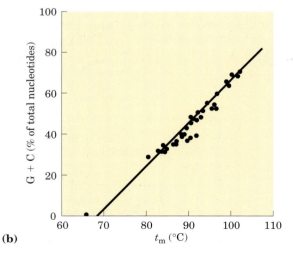

**(b)**

**FIGURE 8–30 Heat denaturation of DNA. (a)** The denaturation, or melting, curves of two DNA specimens. The temperature at the midpoint of the transition ($t_m$) is the melting point; it depends on pH and ionic strength and on the size and base composition of the DNA. **(b)** Relationship between $t_m$ and the G≡C content of a DNA.

$\overline{\quad\quad\quad} 3 \ \mu m$

**FIGURE 8–31 Partially denatured DNA.** This DNA was partially denatured, then fixed to prevent renaturation during sample preparation. The shadowing method used to visualize the DNA in this electron micrograph increases its diameter approximately fivefold and obliterates most details of the helix. However, length measurements can be obtained, and single-stranded regions are readily distinguishable from double-stranded regions. The arrows point to some single-stranded bubbles where denaturation has occurred. The regions that denature are highly reproducible and are rich in A=T base pairs.

double-stranded. Such denatured regions (called bubbles) can be visualized with electron microscopy (Fig. 8–31). Strand separation of DNA *must* occur in vivo during processes such as DNA replication and transcription. As we shall see, the DNA sites where these processes are initiated are often rich in A=T base pairs.

Duplexes of two RNA strands or of one RNA strand and one DNA strand (RNA-DNA hybrids) can also be denatured. Notably, RNA duplexes are more stable than DNA duplexes. At neutral pH, denaturation of a double-helical RNA often requires temperatures 20 °C or more higher than those required for denaturation of a DNA molecule with a comparable sequence. The stability of an RNA-DNA hybrid is generally intermediate between that of RNA and that of DNA. The physical basis for these differences in thermal stability is not known.

## Nucleic Acids from Different Species Can Form Hybrids

The ability of two complementary DNA strands to pair with one another can be used to detect similar DNA sequences in two different species or within the genome of a single species. If duplex DNAs isolated from human cells and from mouse cells are completely denatured by heating, then mixed and kept at 65 °C for many hours, much of the DNA will anneal. Most of the mouse DNA strands anneal with complementary mouse DNA strands to form mouse duplex DNA; similarly, most human DNA strands anneal with complementary human DNA strands. However, some strands of the mouse DNA will associate with human DNA strands to yield **hybrid**

**duplexes,** in which segments of a mouse DNA strand form base-paired regions with segments of a human DNA strand (Fig. 8–32). This reflects a common evolutionary heritage; different organisms generally have some proteins and RNAs with similar functions and, often, similar structures. In many cases, the DNAs encoding these proteins and RNAs have similar sequences. The closer the evolutionary relationship between two species, the more extensively their DNAs will hybridize. For example, human DNA hybridizes much more extensively with mouse DNA than with DNA from yeast.

The hybridization of DNA strands from different sources forms the basis for a powerful set of techniques essential to the practice of modern molecular genetics. A specific DNA sequence or gene can be detected in the presence of many other sequences, if one already has an appropriate complementary DNA strand (usually labeled in some way) to hybridize with it (Chapter 9). The complementary DNA can be from a different species or from the same species, or it can be synthesized chemically in the laboratory using techniques described later

in this chapter. Hybridization techniques can be varied to detect a specific RNA rather than DNA. The isolation and identification of specific genes and RNAs rely on these hybridization techniques. Applications of this technology make possible the identification of an individual on the basis of a single hair left at the scene of a crime or the prediction of the onset of a disease decades before symptoms appear (see Box 9–1).

## Nucleotides and Nucleic Acids Undergo Nonenzymatic Transformations

Purines and pyrimidines, along with the nucleotides of which they are a part, undergo a number of spontaneous alterations in their covalent structure. The rate of these reactions is generally *very slow,* but they are physiologically significant because of the cell's very low tolerance for alterations in its genetic information. Alterations in DNA structure that produce permanent changes in the genetic information encoded therein are called **mutations,** and much evidence suggests an intimate link between the accumulation of mutations in an individual organism and the processes of aging and carcinogenesis.

Several nucleotide bases undergo spontaneous loss of their exocyclic amino groups (deamination) (Fig. 8–33a). For example, under typical cellular conditions, deamination of cytosine (in DNA) to uracil occurs in about one of every $10^7$ cytidine residues in 24 hours. This corresponds to about 100 spontaneous events per day, on average, in a mammalian cell. Deamination of adenine and guanine occurs at about 1/100th this rate.

The slow cytosine deamination reaction seems innocuous enough, but is almost certainly the reason why DNA contains thymine rather than uracil. The product of cytosine deamination (uracil) is readily recognized as foreign in DNA and is removed by a repair system (Chapter 25). If DNA normally contained uracil, recognition of uracils resulting from cytosine deamination would be more difficult, and unrepaired uracils would lead to permanent sequence changes as they were paired with adenines during replication. Cytosine deamination would gradually lead to a decrease in G≡C base pairs and an increase in A=U base pairs in the DNA of all cells. Over the millennia, cytosine deamination could eliminate G≡C base pairs and the genetic code that depends on them. Establishing thymine as one of the four bases in DNA may well have been one of the crucial turning points in evolution, making the long-term storage of genetic information possible.

Another important reaction in deoxyribonucleotides is the hydrolysis of the *N*-β-glycosyl bond between the base and the pentose (Fig. 8–33b). This occurs at a higher rate for purines than for pyrimidines. As many as one in $10^5$ purines (10,000 per mammalian cell) are lost from DNA every 24 hours under typical

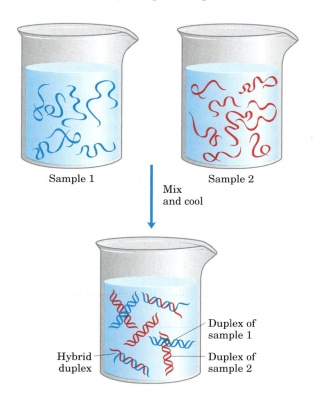

Sample 1        Sample 2

Mix and cool

Hybrid duplex

Duplex of sample 1

Duplex of sample 2

**FIGURE 8–32 DNA hybridization.** Two DNA samples to be compared are completely denatured by heating. When the two solutions are mixed and slowly cooled, DNA strands of each sample associate with their normal complementary partner and anneal to form duplexes. If the two DNAs have significant sequence similarity, they also tend to form partial duplexes or hybrids with each other: the greater the sequence similarity between the two DNAs, the greater the number of hybrids formed. Hybrid formation can be measured in several ways. One of the DNAs is usually labeled with a radioactive isotope to simplify the measurements.

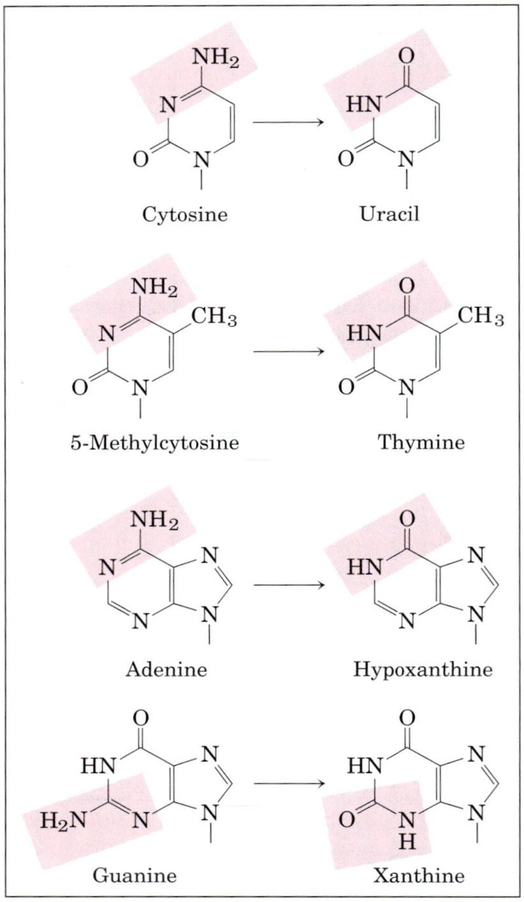

**(a) Deamination**

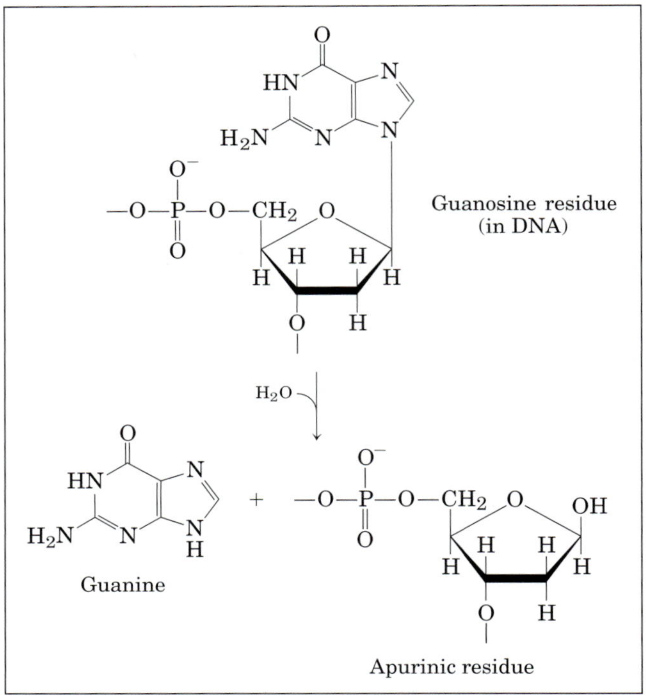

**(b) Depurination**

**FIGURE 8–33 Some well-characterized nonenzymatic reactions of nucleotides. (a)** Deamination reactions. Only the base is shown. **(b)** Depurination, in which a purine is lost by hydrolysis of the *N*-β-glycosyl bond. The deoxyribose remaining after depurination is readily converted from the β-furanose to the aldehyde form (see Fig. 8–3). Further nonenzymatic reactions are illustrated in Figures 8–34 and 8–35.

cellular conditions. Depurination of ribonucleotides and RNA is much slower and generally is not considered physiologically significant. In the test tube, loss of purines can be accelerated by dilute acid. Incubation of DNA at pH 3 causes selective removal of the purine bases, resulting in a derivative called apurinic acid.

Other reactions are promoted by radiation. UV light induces the condensation of two ethylene groups to form a cyclobutane ring. In the cell, the same reaction between adjacent pyrimidine bases in nucleic acids forms cyclobutane pyrimidine dimers. This happens most frequently between adjacent thymidine residues on the same DNA strand (Fig. 8–34). A second type of pyrimidine dimer, called a 6-4 photoproduct, is also formed during UV irradiation. Ionizing radiation (x rays and gamma rays) can cause ring opening and fragmentation of bases as well as breaks in the covalent backbone of nucleic acids.

Virtually all forms of life are exposed to energy-rich radiation capable of causing chemical changes in DNA. Near-UV radiation (with wavelengths of 200 to 400 nm), which makes up a significant portion of the solar spectrum, is known to cause pyrimidine dimer formation and other chemical changes in the DNA of bacteria and of

human skin cells. We are subject to a constant field of ionizing radiation in the form of cosmic rays, which can penetrate deep into the earth, as well as radiation emitted from radioactive elements, such as radium, plutonium, uranium, radon, $^{14}C$, and $^{3}H$. X rays used in medical and dental examinations and in radiation therapy of cancer and other diseases are another form of ionizing radiation. It is estimated that UV and ionizing radiations are responsible for about 10% of all DNA damage caused by environmental agents.

DNA also may be damaged by reactive chemicals introduced into the environment as products of industrial activity. Such products may not be injurious per se but may be metabolized by cells into forms that are. Two prominent classes of such agents (Fig. 8–35) are (1) deaminating agents, particularly nitrous acid ($HNO_2$) or compounds that can be metabolized to nitrous acid or nitrites, and (2) alkylating agents.

Nitrous acid, formed from organic precursors such as nitrosamines and from nitrite and nitrate salts, is a potent accelerator of the deamination of bases. Bisulfite has similar effects. Both agents are used as preservatives in processed foods to prevent the growth of toxic bacteria. They do not appear to increase cancer risks

**FIGURE 8–34 Formation of pyrimidine dimers induced by UV light.** (a) One type of reaction (on the left) results in the formation of a cyclobutyl ring involving C-5 and C-6 of adjacent pyrimidine residues. An alternative reaction (on the right) results in a 6-4 photoproduct, with a linkage between C-6 of one pyrimidine and C-4 of its neighbor. (b) Formation of a cyclobutane pyrimidine dimer introduces a bend or kink into the DNA.

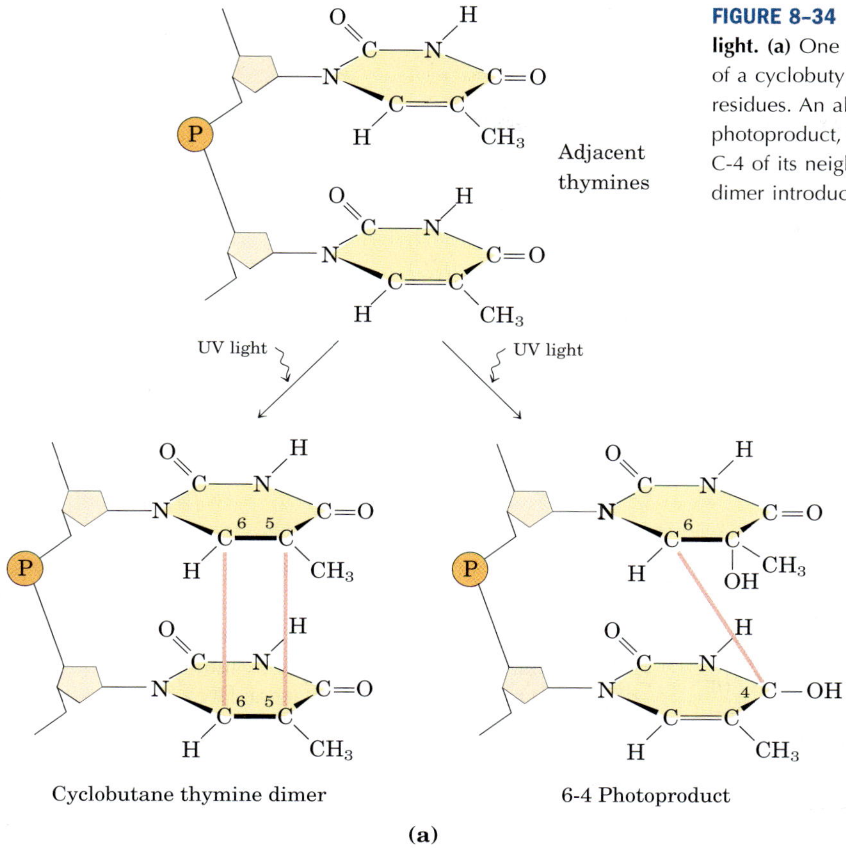

Cyclobutane thymine dimer

6-4 Photoproduct

**(a)**

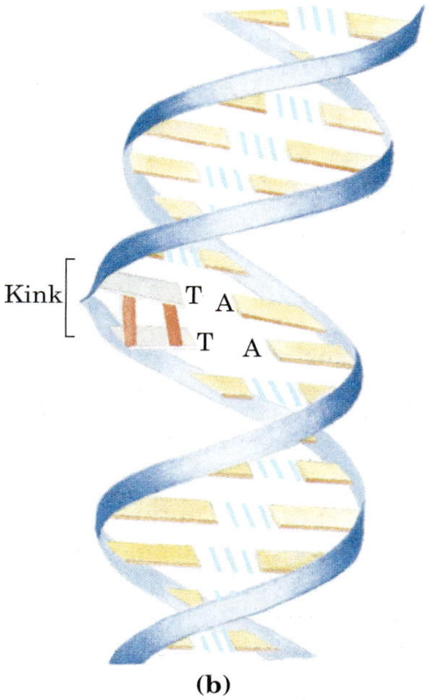

**(b)**

significantly when used in this way, perhaps because they are used in small amounts and make only a minor contribution to the overall levels of DNA damage. (The potential health risk from food spoilage if these preservatives were not used is much greater.)

Alkylating agents can alter certain bases of DNA. For example, the highly reactive chemical dimethylsulfate (Fig. 8–35b) can methylate a guanine to yield $O^6$-methylguanine, which cannot base-pair with cytosine.

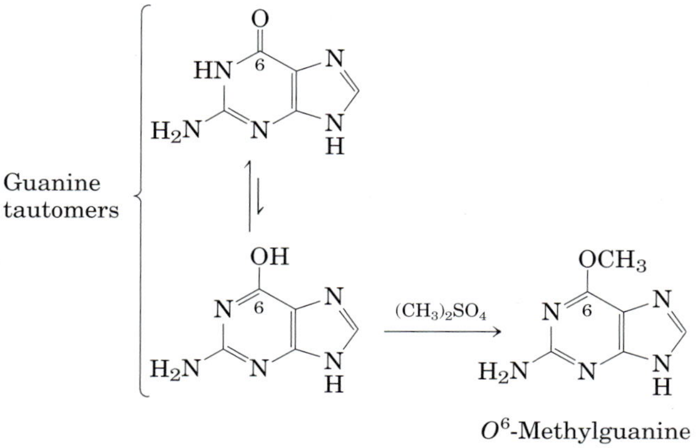

Guanine tautomers

$O^6$-Methylguanine

Many similar reactions are brought about by alkylating agents normally present in cells, such as $S$-adenosylmethionine.

Possibly the most important source of mutagenic alterations in DNA is oxidative damage. Excited-oxygen species such as hydrogen peroxide, hydroxyl radicals, and superoxide radicals arise during irradiation or as a byproduct of aerobic metabolism. Of these species, the hydroxyl radicals are responsible for most oxidative DNA damage. Cells have an elaborate defense system to destroy reactive oxygen species, including enzymes such as catalase and superoxide dismutase that convert reactive oxygen species to harmless products. A fraction of these oxidants inevitably escape cellular defenses, however, and damage to DNA occurs through any of a large, complex group of reactions ranging from oxidation of deoxyribose and base moieties to strand breaks. Accurate estimates for the extent of this damage are not yet available, but every day the DNA of each human cell is subjected to thousands of damaging oxidative reactions.

This is merely a sampling of the best-understood reactions that damage DNA. Many carcinogenic compounds in food, water, or air exert their cancer-causing effects by modifying bases in DNA. Nevertheless, the integrity of DNA as a polymer is better maintained than that of either RNA or protein, because DNA is the only macromolecule that has the benefit of biochemical repair systems. These repair processes (described in Chapter 25) greatly lessen the impact of damage to DNA.

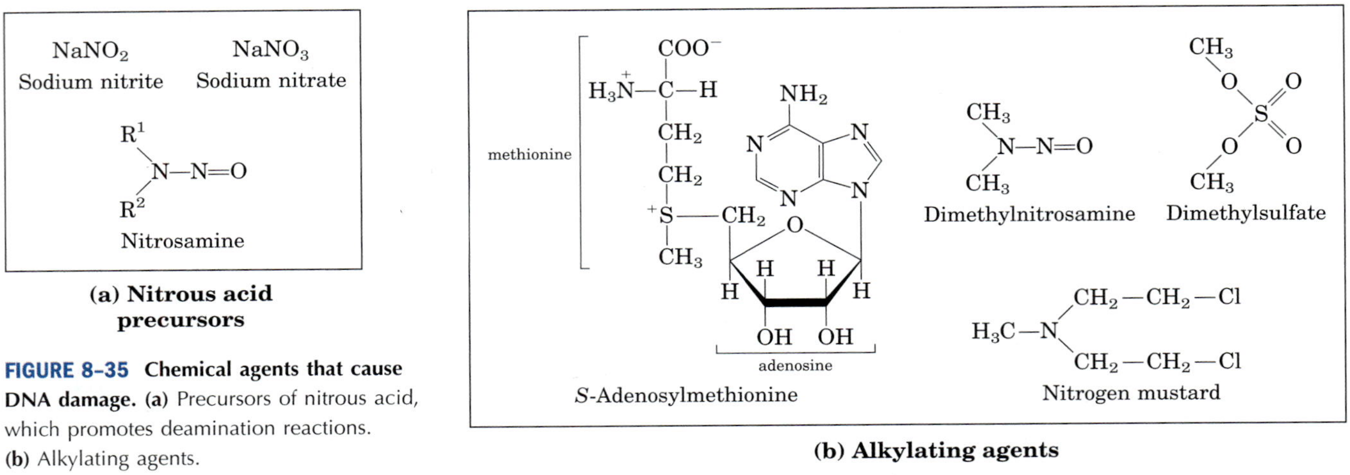

**(a) Nitrous acid precursors**

**FIGURE 8–35 Chemical agents that cause DNA damage. (a)** Precursors of nitrous acid, which promotes deamination reactions. **(b)** Alkylating agents.

**(b) Alkylating agents**

## Some Bases of DNA Are Methylated

Certain nucleotide bases in DNA molecules are enzymatically methylated. Adenine and cytosine are methylated more often than guanine and thymine. Methylation is generally confined to certain sequences or regions of a DNA molecule. In some cases the function of methylation is well understood; in others the function remains unclear. All known DNA methylases use $S$-adenosylmethionine as a methyl group donor. $E.$ $coli$ has two prominent methylation systems. One serves as part of a defense mechanism that helps the cell to distinguish its DNA from foreign DNA by marking its own DNA with methyl groups and destroying (foreign) DNA without the methyl groups (this is known as a restriction-modification system; see Chapter 9). The other system methylates adenosine residues within the sequence (5′)GATC(3′) to $N^6$-methyladenosine (Fig. 8–5a). This is mediated by the Dam (*DNA adenine methylation*) methylase, a component of a system that repairs mismatched base pairs formed occasionally during DNA replication (see Fig. 25–20).

In eukaryotic cells, about 5% of cytidine residues in DNA are methylated to 5-methylcytidine (Fig. 8–5a). Methylation is most common at CpG sequences, producing methyl-CpG symmetrically on both strands of the DNA. The extent of methylation of CpG sequences varies by molecular region in large eukaryotic DNA molecules. Methylation suppresses the migration of segments of DNA called transposons, described in Chapter 25. These methylations of cytosine also have structural significance. The presence of 5-methylcytosine in an alternating CpG sequence markedly increases the tendency for that segment of DNA to assume the Z form.

## The Sequences of Long DNA Strands Can Be Determined

In its capacity as a repository of information, a DNA molecule's most important property is its nucleotide se-

quence. Until the late 1970s, determining the sequence of a nucleic acid containing even five or ten nucleotides was difficult and very laborious. The development of two new techniques in 1977, one by Alan Maxam and Walter Gilbert and the other by Frederick Sanger, has made possible the sequencing of ever larger DNA molecules with an ease unimagined just a few decades ago. The techniques depend on an improved understanding of nucleotide chemistry and DNA metabolism, and on electrophoretic methods for separating DNA strands differing in size by only one nucleotide. Electrophoresis of DNA is similar to that of proteins (see Fig. 3–19). Polyacrylamide is often used as the gel matrix in work with short DNA molecules (up to a few hundred nucleotides); agarose is generally used for longer pieces of DNA.

In both Sanger and Maxam-Gilbert sequencing, the general principle is to reduce the DNA to four sets of labeled fragments. The reaction producing each set is base-specific, so the lengths of the fragments correspond to positions in the DNA sequence where a certain base occurs. For example, for an oligonucleotide with the sequence pAATCGACT, labeled at the 5′ end (the left end), a reaction that breaks the DNA after each C residue will generate two labeled fragments: a four-nucleotide and a seven-nucleotide fragment; a reaction that breaks the DNA after each G will produce only one labeled, five-nucleotide fragment. Because the fragments are radioactively labeled at their 5′ ends, only the fragment to the 5′ side of the break is visualized. The fragment sizes correspond to the relative positions of C and G residues in the sequence. When the sets of fragments corresponding to each of the four bases are electrophoretically separated side by side, they produce a ladder of bands from which the sequence can be read directly (Fig. 8–36). We illustrate only the Sanger method, because it has proven to be technically easier and is in more widespread use. It requires the enzymatic synthesis of a DNA strand complementary to the strand under analysis, using a radioactively labeled "primer" and dideoxynucleotides.

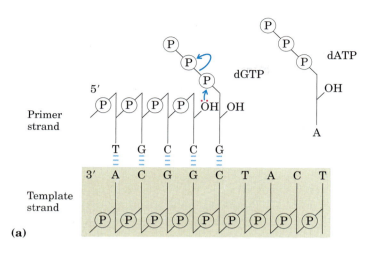

**(b)** ddNTP analog

**FIGURE 8–36 DNA sequencing by the Sanger method.** This method makes use of the mechanism of DNA synthesis by DNA polymerases (Chapter 25). **(a)** DNA polymerases require both a primer (a short oligonucleotide strand), to which nucleotides are added, and a template strand to guide selection of each new nucleotide. In cells, the 3'-hydroxyl group of the primer reacts with an incoming deoxynucleoside triphosphate (dNTP) to form a new phosphodiester bond. **(b)** The Sanger sequencing procedure uses dideoxynucleoside triphosphate (ddNTP) analogs to interrupt DNA synthesis. (The Sanger method is also known as the dideoxy method.) When a ddNTP is inserted in place of a dNTP, strand elongation is halted after the analog is added, because it lacks the 3'-hydroxyl group needed for the next step.

**(c)** The DNA to be sequenced is used as the template strand, and a short primer, radioactively or fluorescently labeled, is annealed to it. By addition of small amounts of a single ddNTP, for example ddCTP, to an otherwise normal reaction system, the synthesized strands will be prematurely terminated at some locations where dC normally occurs. Given the excess of dCTP over ddCTP, the chance that the analog will be incorporated whenever a dC is to be added is small. However, ddCTP is present in sufficient amounts to ensure that each new strand has a high probability of acquiring at least one ddC at some point during synthesis. The result is a solution containing a mixture of labeled fragments, each ending with a C residue. Each C residue in the sequence generates a set of fragments of a particular length, such that the different-sized fragments, separated by electrophoresis, reveal the location of C residues. This procedure is repeated separately for each of the four ddNTPs, and the sequence can be read directly from an autoradiogram of the gel. Because shorter DNA fragments migrate faster, the fragments near the bottom of the gel represent the nucleotide positions closest to the primer (the 5' end), and the sequence is read (in the 5' → 3' direction) from bottom to top. Note that the sequence obtained is that of the strand *complementary* to the strand being analyzed.

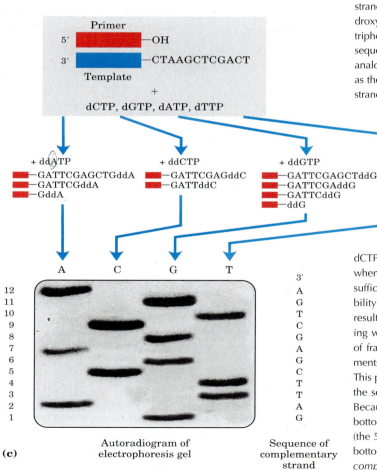

**(c)** Autoradiogram of electrophoresis gel | Sequence of complementary strand

DNA sequencing is readily automated by a variation of Sanger's sequencing method in which the dideoxynucleotides used for each reaction are labeled with a differently colored fluorescent tag (Fig. 8–37). This technology allows DNA sequences containing thousands of nucleotides to be determined in a few hours. Entire genomes of many organisms have now been sequenced (see Table 1–4), and many very large DNA-sequencing projects are in progress. Perhaps the most ambitious of these is the Human Genome Project, in which researchers have sequenced all 3.2 billion base pairs of the DNA in a human cell (Chapter 9). 🖱 **Dideoxy Sequencing of DNA**

## The Chemical Synthesis of DNA Has Been Automated

Another technology that has paved the way for many biochemical advances is the chemical synthesis of oligonucleotides with any chosen sequence. The chemical methods for synthesizing nucleic acids were developed primarily by H. Gobind Khorana and his colleagues in the 1970s. Refinement and automation of these methods have made possible the rapid and accurate synthesis of DNA strands. The synthesis is carried out with the growing strand attached to a solid support (Fig. 8–38), using principles similar to those used by Merrifield in peptide synthesis (see Fig. 3–29). The efficiency of each addition step is very high, allowing the routine laboratory synthesis of polymers containing 70 or 80 nucleotides and, in some laboratories, much longer strands. The availability of relatively inexpensive DNA polymers with predesigned sequences is having a powerful impact on all areas of biochemistry (Chapter 9).

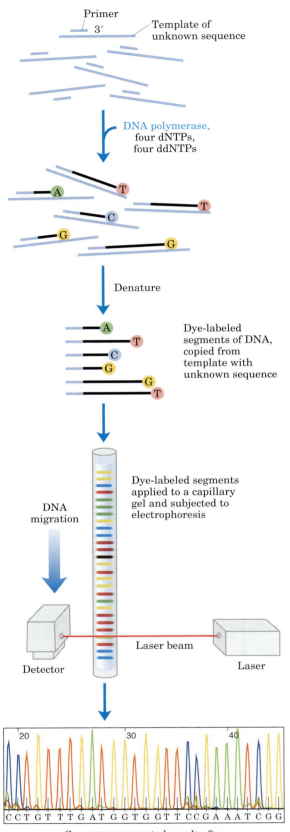

**FIGURE 8–37 Strategy for automating DNA sequencing reactions.** Each dideoxynucleotide used in the Sanger method can be linked to a fluorescent molecule that gives all the fragments terminating in that nucleotide a particular color. All four labeled ddNTPs are added to a single tube. The resulting colored DNA fragments are then separated by size in a single electrophoretic gel contained in a capillary tube (a refinement of gel electrophoresis that allows for faster separations). All fragments of a given length migrate through the capillary gel in a single peak, and the color associated with each peak is detected using a laser beam. The DNA sequence is read by determining the sequence of colors in the peaks as they pass the detector. This information is fed directly to a computer, which determines the sequence.

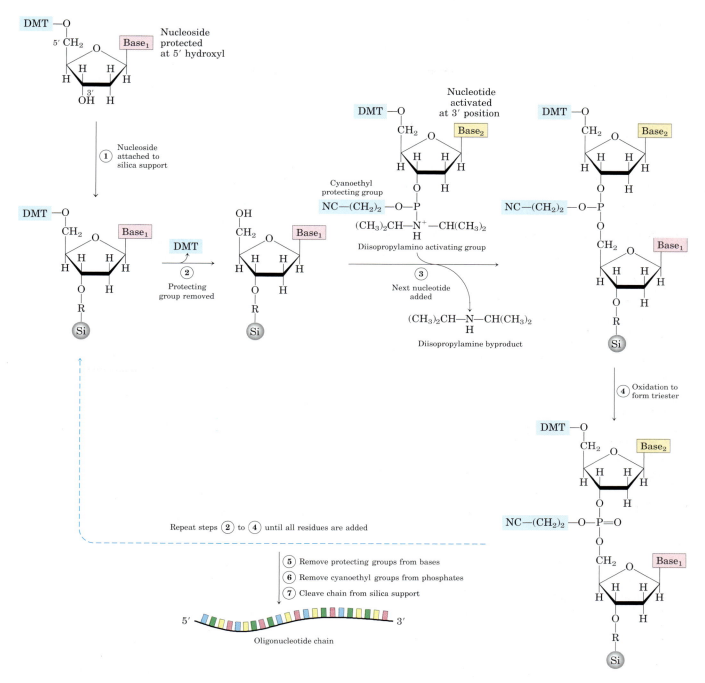

**FIGURE 8–38 Chemical synthesis of DNA.** Automated DNA synthesis is conceptually similar to the synthesis of polypeptides on a solid support. The oligonucleotide is built up on the solid support (silica), one nucleotide at a time, in a repeated series of chemical reactions with suitably protected nucleotide precursors. ① The first nucleoside (which will be the 3′ end) is attached to the silica support at the 3′ hydroxyl (through a linking group, R) and is protected at the 5′ hydroxyl with an acid-labile dimethoxytrityl group (DMT). The reactive groups on all bases are also chemically protected. ② The protecting DMT group is removed by washing the column with acid (the DMT group is colored, so this reaction can be followed spectrophotometrically). ③ The next nucleotide is activated with a diisopropylamino group and reacted with the bound nucleotide to form a 5′,3′ linkage, which in step ④ is oxidized with iodine to produce a phosphotriester linkage. (One of the phosphate oxygens carries a cyanoethyl protecting group.) Reactions ② through ④ are repeated until all nucleotides are added. At each step, excess nucleotide is removed before addition of the next nucleotide. In steps ⑤ and ⑥ the remaining protecting groups on the bases and the phosphates are removed, and in ⑦ the oligonucleotide is separated from the solid support and purified. The chemical synthesis of RNA is somewhat more complicated because of the need to protect the 2′ hydroxyl of ribose without adversely affecting the reactivity of the 3′ hydroxyl.

## SUMMARY 8.3 Nucleic Acid Chemistry

■ Native DNA undergoes reversible unwinding and separation of strands (melting) on heating or at extremes of pH. DNAs rich in G≡C pairs have higher melting points than DNAs rich in A=T pairs.

■ Denatured single-stranded DNAs from two species can form a hybrid duplex, the degree of hybridization depending on the extent of sequence similarity. Hybridization is the basis for important techniques used to study and isolate specific genes and RNAs.

■ DNA is a relatively stable polymer. Spontaneous reactions such as deamination of certain bases, hydrolysis of base-sugar N-glycosyl bonds, radiation-induced formation of pyrimidine dimers, and oxidative damage occur at very low rates, yet are important because of cells' very low tolerance for changes in genetic material.

■ DNA sequences can be determined and DNA polymers synthesized with simple, automated protocols involving chemical and enzymatic methods.

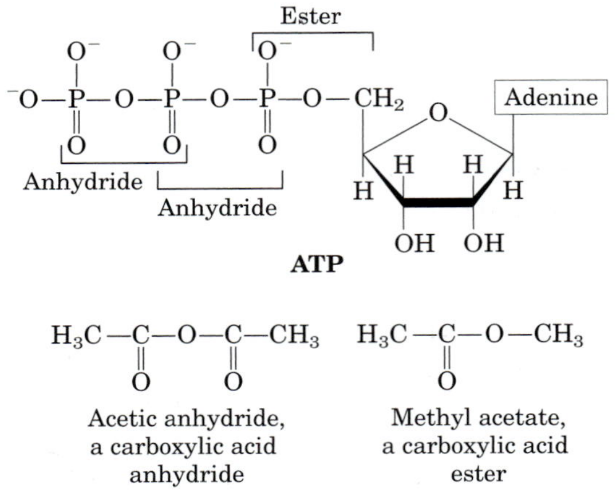

## 8.4 Other Functions of Nucleotides

In addition to their roles as the subunits of nucleic acids, nucleotides have a variety of other functions in every cell: as energy carriers, components of enzyme cofactors, and chemical messengers.

### Nucleotides Carry Chemical Energy in Cells

The phosphate group covalently linked at the 5′ hydroxyl of a ribonucleotide may have one or two additional phosphates attached. The resulting molecules are referred to as nucleoside mono-, di-, and triphosphates (Fig. 8–39). Starting from the ribose, the three phos-

**FIGURE 8–40 The phosphate ester and phosphoanhydride bonds of ATP.** Hydrolysis of an anhydride bond yields more energy than hydrolysis of the ester. A carboxylic acid anhydride and carboxylic acid ester are shown for comparison.

phates are generally labeled $\alpha$, $\beta$, and $\gamma$. Hydrolysis of nucleoside triphosphates provides the chemical energy to drive a wide variety of cellular reactions. Adenosine 5′-triphosphate, ATP, is by far the most widely used for this purpose, but UTP, GTP, and CTP are also used in some reactions. Nucleoside triphosphates also serve as the activated precursors of DNA and RNA synthesis, as described in Chapters 25 and 26.

The energy released by hydrolysis of ATP and the other nucleoside triphosphates is accounted for by the structure of the triphosphate group. The bond between the ribose and the $\alpha$ phosphate is an ester linkage. The $\alpha,\beta$ and $\beta,\gamma$ linkages are phosphoanhydrides (Fig. 8–40). Hydrolysis of the ester linkage yields about 14 kJ/mol under standard conditions, whereas hydrolysis of each anhydride bond yields about 30 kJ/mol. ATP hydrolysis often plays an important thermodynamic role in biosynthesis. When coupled to a reaction with a positive free-energy change, ATP hydrolysis shifts the equilibrium of the overall process to favor product forma-

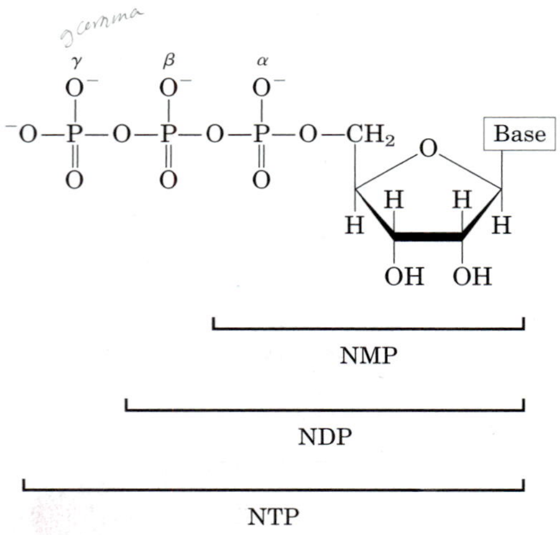

| Abbreviations of ribonucleoside 5′-phosphates | | | |
|---|---|---|---|
| Base | Mono- | Di- | Tri- |
| Adenine | AMP | ADP | ATP |
| Guanine | GMP | GDP | GTP |
| Cytosine | CMP | CDP | CTP |
| Uracil | UMP | UDP | UTP |

| Abbreviations of deoxyribonucleoside 5′-phosphates | | | |
|---|---|---|---|
| Base | Mono- | Di- | Tri- |
| Adenine | dAMP | dADP | dATP |
| Guanine | dGMP | dGDP | dGTP |
| Cytosine | dCMP | dCDP | dCTP |
| Thymine | dTMP | dTDP | dTTP |

**FIGURE 8–39 Nucleoside phosphates.** General structure of the nucleoside 5′-mono-, di-, and triphosphates (NMPs, NDPs, and NTPs) and their standard abbreviations. In the deoxyribonucleoside phosphates (dNMPs, dNDPs, and dNTPs), the pentose is 2′-deoxy-D-ribose.

tion (recall the relationship between equilibrium constant and free-energy change described by Eqn 6–3 on p. 195).

## Adenine Nucleotides Are Components of Many Enzyme Cofactors

A variety of enzyme cofactors serving a wide range of chemical functions include adenosine as part of their structure (Fig. 8–41). They are unrelated structurally except for the presence of adenosine. In none of these cofactors does the adenosine portion participate directly in the primary function, but removal of adenosine generally results in a drastic reduction of cofactor activities. For example, removal of the adenine nucleotide (3′-phosphoadenosine diphosphate) from acetoacetyl-

CoA, the coenzyme A derivative of acetoacetate, reduces its reactivity as a substrate for $\beta$-ketoacyl-CoA transferase (an enzyme of lipid metabolism) by a factor of $10^6$. Although this requirement for adenosine has not been investigated in detail, it must involve the binding energy between enzyme and substrate (or cofactor) that is used both in catalysis and in stabilizing the initial enzyme-substrate complex (Chapter 6). In the case of $\beta$-ketoacyl-CoA transferase, the nucleotide moiety of coenzyme A appears to be a binding "handle" that helps to pull the substrate (acetoacetyl-CoA) into the active site. Similar roles may be found for the nucleoside portion of other nucleotide cofactors.

Why is adenosine, rather than some other large molecule, used in these structures? The answer here may involve a form of evolutionary economy. Adenosine is

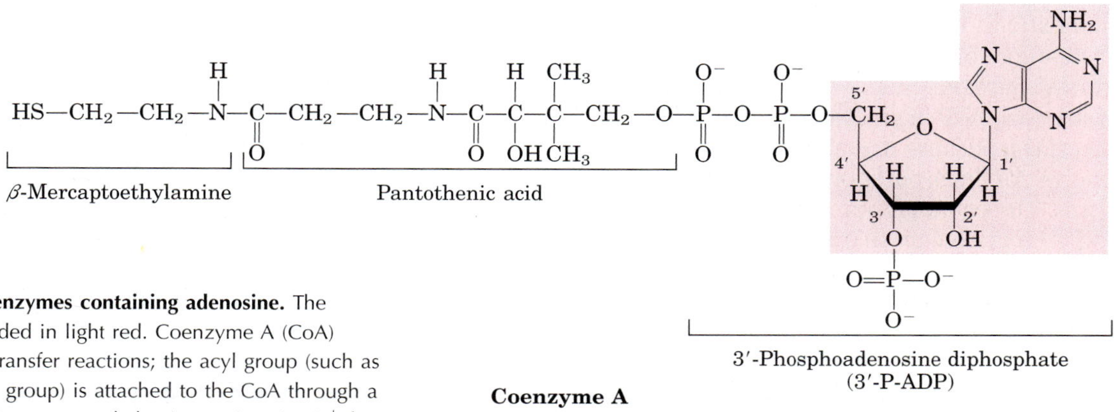

**FIGURE 8–41 Some coenzymes containing adenosine.** The adenosine portion is shaded in light red. Coenzyme A (CoA) functions in acyl group transfer reactions; the acyl group (such as the acetyl or acetoacetyl group) is attached to the CoA through a thioester linkage to the $\beta$-mercaptoethylamine moiety. $NAD^+$ functions in hydride transfers, and FAD, the active form of vitamin $B_2$ (riboflavin), in electron transfers. Another coenzyme incorporating adenosine is 5′-deoxyadenosylcobalamin, the active form of vitamin $B_{12}$ (see Box 17-2), which participates in intramolecular group transfers between adjacent carbons.

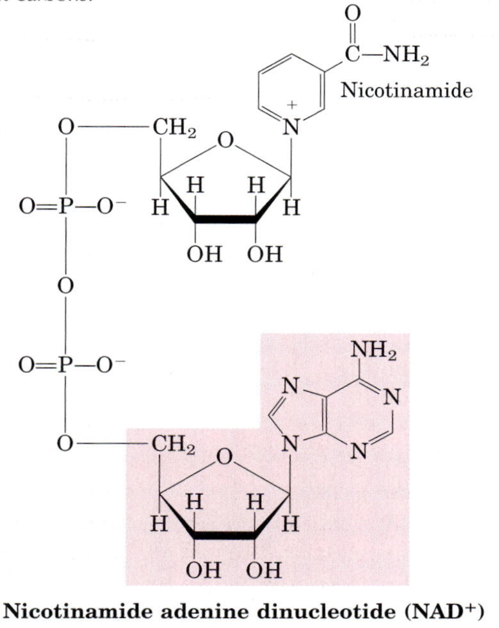

$\beta$-Mercaptoethylamine    Pantothenic acid

3′-Phosphoadenosine diphosphate (3′-P-ADP)

**Coenzyme A**

Nicotinamide

**Nicotinamide adenine dinucleotide (NAD$^+$)**

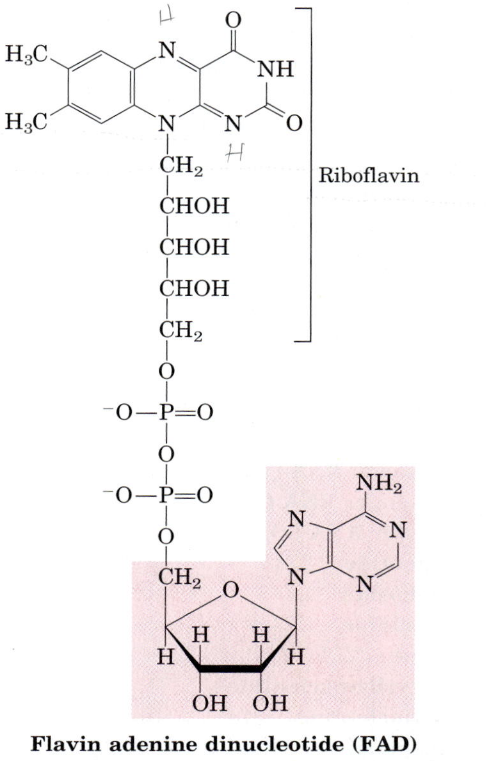

Riboflavin

**Flavin adenine dinucleotide (FAD)**

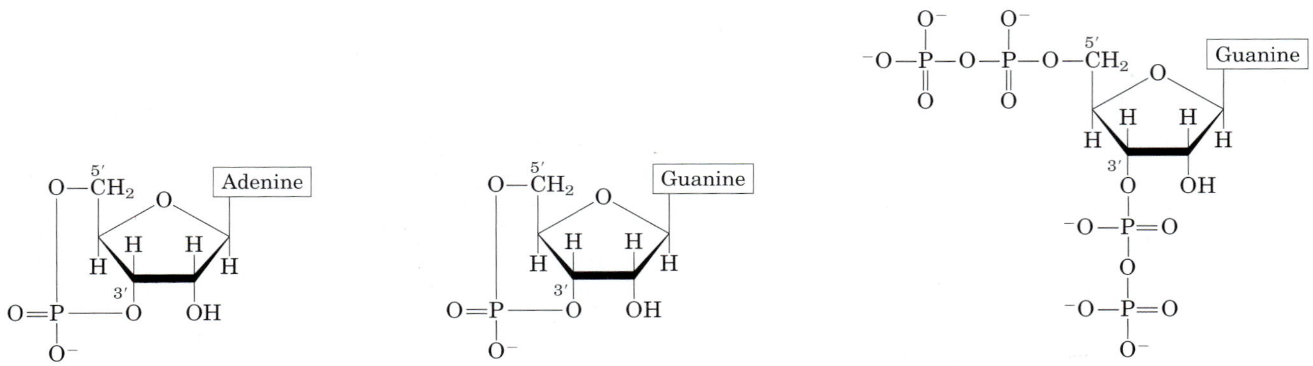


Adenosine 3′,5′-cyclic monophosphate (cyclic AMP; cAMP)

Guanosine 3′,5′-cyclic monophosphate (cyclic GMP; cGMP)

Guanosine 5′-diphosphate,3′-diphosphate (guanosine tetraphosphate) (ppGpp)

**FIGURE 8–42** **Three regulatory nucleotides.**

certainly not unique in the amount of potential binding energy it can contribute. The importance of adenosine probably lies not so much in some special chemical characteristic as in the evolutionary advantage of using one compound for multiple roles. Once ATP became the universal source of chemical energy, systems developed to synthesize ATP in greater abundance than the other nucleotides; because it is abundant, it becomes the logical choice for incorporation into a wide variety of structures. The economy extends to protein structure. A single protein domain that binds adenosine can be used in a wide variety of enzymes. Such a domain, called a **nucleotide-binding fold,** is found in many enzymes that bind ATP and nucleotide cofactors.

## Some Nucleotides Are Regulatory Molecules

Cells respond to their environment by taking cues from hormones or other external chemical signals. The interaction of these extracellular chemical signals ("first messengers") with receptors on the cell surface often leads to the production of **second messengers** inside the cell, which in turn leads to adaptive changes in the cell interior (Chapter 12). Often, the second messenger is a nucleotide (Fig. 8–42). One of the most common is **adenosine 3′,5′-cyclic monophosphate (cyclic AMP,** or **cAMP),** formed from ATP in a reaction catalyzed by adenylyl cyclase, an enzyme associated with the inner face of the plasma membrane. Cyclic AMP serves regulatory functions in virtually every cell outside the plant kingdom. Guanosine 3′,5′-cyclic monophosphate (cGMP) occurs in many cells and also has regulatory functions.

Another regulatory nucleotide, ppGpp (Fig. 8–42), is produced in bacteria in response to a slowdown in protein synthesis during amino acid starvation. This nucleotide inhibits the synthesis of the rRNA and tRNA molecules (see Fig. 28–24) needed for protein synthesis, preventing the unnecessary production of nucleic acids.

## *SUMMARY 8.4* Other Functions of Nucleotides

- ATP is the central carrier of chemical energy in cells. The presence of an adenosine moiety in a variety of enzyme cofactors may be related to binding-energy requirements.
- Cyclic AMP, formed from ATP in a reaction catalyzed by adenylyl cyclase, is a common second messenger produced in response to hormones and other chemical signals.

# *Key Terms*

***Terms in bold are defined in the glossary.***

**gene** 273
**ribosomal RNA (rRNA)** 273
**messenger RNA (mRNA)** 273
**transfer RNA (tRNA)** 273
**nucleotide** 273
**nucleoside** 273
**pyrimidine** 273
**purine** 273
**deoxyribonucleotides** 274
**ribonucleotide** 274
**phosphodiester linkage** 277
**5′ end** 277
**3′ end** 277
**oligonucleotide** 278
**polynucleotide** 278
**base pair** 279
major groove 282
minor groove 282
B-form DNA 284
A-form DNA 284
Z-form DNA 284
**palindrome** 285
**hairpin** 285
**cruciform** 285
triplex DNA 286
G tetraplex 287
H-DNA 287
**monocistronic mRNA** 287
**polycistronic mRNA** 288
**mutation** 293
**second messenger** 302
**adenosine 3′,5′-cyclic monophosphate (cyclic AMP, cAMP)** 302

# Further Reading

## General

**Chang, K.Y. & Varani, G.** (1997) Nucleic acids structure and recognition. *Nat. Struct. Biol.* **4** (Suppl.), 854–858.

> Describes the application of NMR to determination of nucleic acid structure.

**Friedberg, E.C., Walker, G.C., & Siede, W.** (1995) *DNA Repair and Mutagenesis,* W. H. Freeman and Company, New York.

> A good source for more information on the chemistry of nucleotides and nucleic acids.

**Hecht, S.M. (ed.)** (1996) *Bioorganic Chemistry: Nucleic Acids,* Oxford University Press, Oxford.

> A very useful set of articles.

**Kornberg, A. & Baker, T.A.** (1991) *DNA Replication,* 2nd edn, W. H. Freeman and Company, New York.

> The best place to start to learn more about DNA structure.

**Sinden, R.R.** (1994) *DNA Structure and Function,* Academic Press, Inc., San Diego.

> Good discussion of many topics covered in this chapter.

## Historical

**Judson, H.F.** (1996) *The Eighth Day of Creation: Makers of the Revolution in Biology,* expanded edn, Cold Spring Harbor Laboratory Press, Cold Spring Harbor, NY.

**Olby, R.C.** (1994) *The Path to the Double Helix: The Discovery of DNA,* Dover Publications, Inc., New York.

**Sayre, A.** (1978) *Rosalind Franklin and DNA,* W. W. Norton & Co., Inc., New York.

**Watson, J.D.** (1968) *The Double Helix: A Personal Account of the Discovery of the Structure of DNA,* Atheneum, New York. [Paperback edition, Touchstone Books, 2001.]

## Variations in DNA Structure

**Frank-Kamenetskii, M.D. & Mirkin, S.M.** (1995) Triplex DNA structures. *Annu. Rev. Biochem.* **64,** 65–95.

**Herbert, A. & Rich, A.** (1996) The biology of left-handed Z-DNA. *J. Biol. Chem.* **271,** 11,595–11,598.

**Htun, H. & Dahlberg, J.E.** (1989) Topology and formation of triple-stranded H-DNA. *Science* **243,** 1571–1576.

**Keniry, M.A.** (2000) Quadruplex structures in nucleic acids. *Biopolymers* **56,** 123–146.

> Good summary of the structural properties of quadruplexes.

**Moore, P.B.** (1999) Structural motifs in RNA. *Annu. Rev. Biochem.* **68,** 287–300.

**Shafer, R.H.** (1998) Stability and structure of model DNA triplexes and quadruplexes and their interactions with small ligands. *Prog. Nucleic Acid Res. Mol. Biol.* **59,** 55–94.

**Wells, R.D.** (1988) Unusual DNA structures. *J. Biol. Chem.* **263,** 1095–1098.

> Minireview; a concise summary.

## Nucleic Acid Chemistry

**Collins, A.R.** (1999) Oxidative DNA damage, antioxidants, and cancer. *Bioessays* **21,** 238–246.

**Marnett, L.J. & Plastaras, J.P.** (2001) Endogenous DNA damage and mutation. *Trends Genet.* **17,** 214–221.

## ATP As Energy Carrier

**Jencks, W.P.** (1987) Economics of enzyme catalysis. *Cold Spring Harb. Symp. Quant. Biol.* **52,** 65–73.

> A relatively short article, full of insights.

# Problems

**1. Nucleotide Structure** Which positions in a purine ring of a purine nucleotide in DNA have the potential to form hydrogen bonds but are not involved in Watson-Crick base pairing?

**2. Base Sequence of Complementary DNA Strands** One strand of a double-helical DNA has the sequence (5′)GCGCAATATTTCTCAAAATATTGCGC(3′). Write the base sequence of the complementary strand. What special type of sequence is contained in this DNA segment? Does the double-stranded DNA have the potential to form any alternative structures?

**3. DNA of the Human Body** Calculate the weight in grams of a double-helical DNA molecule stretching from the earth to the moon (~320,000 km). The DNA double helix weighs about $1 \times 10^{-18}$ g per 1,000 nucleotide pairs; each base pair extends 3.4 Å. For an interesting comparison, your body contains about 0.5 g of DNA!

**4. DNA Bending** Assume that a poly(A) tract five base pairs long produces a 20° bend in a DNA strand. Calculate the total (net) bend produced in a DNA if the center base pairs (the third of five) of two successive $(dA)_5$ tracts are located (a) 10 base pairs apart; (b) 15 base pairs apart. Assume 10 base pairs per turn in the DNA double helix.

**5. Distinction between DNA Structure and RNA Structure** Hairpins may form at palindromic sequences in single strands of either RNA or DNA. How is the helical structure of a long and fully base-paired (except at the end) hairpin in RNA different from that of a similar hairpin in DNA?

**6. Nucleotide Chemistry** The cells of many eukaryotic organisms have highly specialized systems that specifically repair G–T mismatches in DNA. The mismatch is repaired to form a G≡C (not A=T) base pair. This G–T mismatch repair mechanism occurs in addition to a more general system that repairs virtually all mismatches. Can you suggest why cells might require a specialized system to repair G–T mismatches?

**7. Nucleic Acid Structure** Explain why the absorption of UV light by double-stranded DNA increases (hyperchromic effect) when the DNA is denatured.

**8. Determination of Protein Concentration in a Solution Containing Proteins and Nucleic Acids**  The concentration of protein or nucleic acid in a solution containing both can be estimated by using their different light absorption properties: proteins absorb most strongly at 280 nm and nucleic acids at 260 nm. Their respective concentrations in a mixture can be estimated by measuring the absorbance (A) of the solution at 280 nm and 260 nm and using the table below, which gives $R_{280/260}$, the ratio of absorbances at 280 and 260 nm; the percentage of total mass that is nucleic acid; and a factor, $F$, that corrects the $A_{280}$ reading and gives a more accurate protein estimate. The protein concentration (in mg/ml) = $F \times A_{280}$ (assuming the cuvette is 1 cm wide). Calculate the protein concentration in a solution of $A_{280} = 0.69$ and $A_{260} = 0.94$.

| $R_{280/260}$ | Proportion of nucleic acid (%) | $F$ |
|---|---|---|
| 1.75 | 0.00 | 1.116 |
| 1.63 | 0.25 | 1.081 |
| 1.52 | 0.50 | 1.054 |
| 1.40 | 0.75 | 1.023 |
| 1.36 | 1.00 | 0.994 |
| 1.30 | 1.25 | 0.970 |
| 1.25 | 1.50 | 0.944 |
| 1.16 | 2.00 | 0.899 |
| 1.09 | 2.50 | 0.852 |
| 1.03 | 3.00 | 0.814 |
| 0.979 | 3.50 | 0.776 |
| 0.939 | 4.00 | 0.743 |
| 0.874 | 5.00 | 0.682 |
| 0.846 | 5.50 | 0.656 |
| 0.822 | 6.00 | 0.632 |
| 0.804 | 6.50 | 0.607 |
| 0.784 | 7.00 | 0.585 |
| 0.767 | 7.50 | 0.565 |
| 0.753 | 8.00 | 0.545 |
| 0.730 | 9.00 | 0.508 |
| 0.705 | 10.00 | 0.478 |
| 0.671 | 12.00 | 0.422 |
| 0.644 | 14.00 | 0.377 |
| 0.615 | 17.00 | 0.322 |
| 0.595 | 20.00 | 0.278 |

**9. Base Pairing in DNA**  In samples of DNA isolated from two unidentified species of bacteria, X and Y, adenine makes up 32% and 17%, respectively, of the total bases. What relative proportions of adenine, guanine, thymine, and cytosine would you expect to find in the two DNA samples? What assumptions have you made? One of these species was isolated from a hot spring (64 °C). Suggest which species is the thermophilic bacterium. What is the basis for your answer?

**10. Solubility of the Components of DNA**  Draw the following structures and rate their relative solubilities in water (most soluble to least soluble): deoxyribose, guanine, phosphate. How are these solubilities consistent with the three-dimensional structure of double-stranded DNA?

**11. DNA Sequencing**  The following DNA fragment was sequenced by the Sanger method. The red asterisk indicates a fluorescent label.

*5′ ———— 3′-OH
3′ ———— ATTACGCAAGGACATTAGAC---5′

A sample of the DNA was reacted with DNA polymerase and each of the nucleotide mixtures (in an appropriate buffer) listed below. Dideoxynucleotides (ddNTPs) were added in relatively small amounts.

*1.* dATP, dTTP, dCTP, dGTP, ddTTP
*2.* dATP, dTTP, dCTP, dGTP, ddGTP
*3.* dATP, dCTP, dGTP, ddTTP
*4.* dATP, dTTP, dCTP, dGTP

The resulting DNA was separated by electrophoresis on an agarose gel, and the fluorescent bands on the gel were located. The band pattern resulting from nucleotide mixture 1 is shown below. Assuming that all mixtures were run on the same gel, what did the remaining lanes of the gel look like?

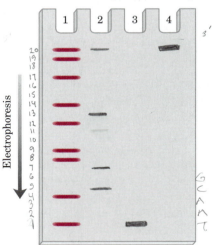

**12. Snake Venom Phosphodiesterase**  An exonuclease is an enzyme that sequentially cleaves nucleotides from the end of a polynucleotide strand. Snake venom phosphodiesterase, which hydrolyzes nucleotides from the 3′ end of any oligonucleotide with a free 3′-hydroxyl group, cleaves between the 3′ hydroxyl of the ribose or deoxyribose and the phosphoryl group of the next nucleotide. It acts on single-stranded DNA or RNA and has no base specificity. This enzyme was used in sequence determination experiments before the development of modern nucleic acid sequencing techniques. What are the products of partial digestion by snake venom phosphodiesterase of an oligonucleotide with the following sequence?

(5′)GCGCCAUUGC(3′)–OH

**13. Preserving DNA in Bacterial Endospores**  Bacterial endospores form when the environment is no longer conducive to active cell metabolism. The soil bacterium *Bacillus subtilis,* for example, begins the process of sporulation when one or more nutrients are depleted. The end product is a small, metabolically dormant structure that can survive almost indefinitely with no detectable metabolism. Spores have mechanisms to prevent accumulation of potentially lethal mutations in their DNA over periods of dormancy that can exceed 1,000 years. *B. subtilis* spores are much more resistant than the organism's growing cells to heat, UV radiation, and oxidizing agents, all of which promote mutations.

(a) One factor that prevents potential DNA damage in spores is their greatly decreased water content. How would this affect some types of mutations?

(b) Endospores have a category of proteins called small acid-soluble proteins (SASPs) that bind to their DNA, preventing formation of cyclobutane-type dimers. What causes cyclobutane dimers, and why do bacterial endospores need mechanisms to prevent their formation?

### Biochemistry on the Internet

**14. The Structure of DNA** Elucidation of the three-dimensional structure of DNA helped researchers understand how this molecule conveys information that can be faithfully replicated from one generation to the next. To see the secondary structure of double-stranded DNA, go to the Protein Data Bank website (www.rcsb.org/pdb). Use the PDB identifiers listed below to retrieve the data pages for the two forms of DNA.

Open the structures using RasMol or Chime, and use the different viewing options to complete the following exercises.

(a) Obtain the file for 141D, a highly conserved, repeated DNA sequence from the end of the HIV-1 (the virus that causes AIDS) genome. Display the molecule as a stick or ball-and-stick structure. Identify the sugar–phosphate backbone for each strand of the DNA duplex. Locate and identify individual bases. Which is the 5′ end of this molecule? Locate the major and minor grooves. Is this a right- or left-handed helix?

(b) Obtain the file for 145D, a DNA with the Z conformation. Display the molecule as a stick or ball-and-stick structure. Identify the sugar–phosphate backbone for each strand of the DNA duplex. Is this a right- or left-handed helix?

(c) To fully appreciate the secondary structure of DNA, select "Stereo" in the Options menu in the viewer. You will see two images of the DNA molecule. Sit with your nose approximately 10 inches from the monitor and focus on the tip of your nose. In the background you should see three images of the DNA helix. Shift your focus from the tip of your nose to the middle image, which should appear three-dimensional. (Note that only one of the two authors can make this work.) For additional tips, see the Study Guide or the textbook website (www.whfreeman.com/lehninger).

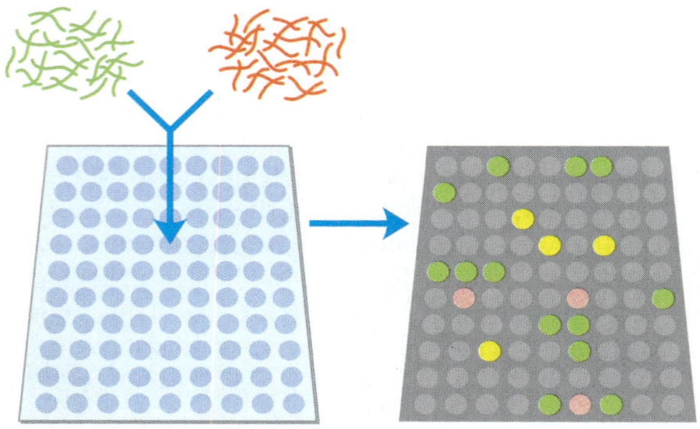

# DNA-BASED INFORMATION TECHNOLOGIES

Of all the natural systems, living matter is the one which, in the face of great transformations, preserves inscribed in its organization the largest amount of its own past history.

—Emile Zuckerkandl and Linus Pauling, *article in* Journal of Theoretical Biology, *1965*

We now turn to a technology that is fundamental to the advance of modern biological sciences, defining present and future biochemical frontiers and illustrating many important principles of biochemistry. Elucidation of the laws governing enzymatic catalysis, macromolecular structure, cellular metabolism, and information pathways allows research to be directed at increasingly complex biochemical processes. Cell division, immunity, embryogenesis, vision, taste, oncogenesis, cognition—all are orchestrated in an elaborate symphony of molecular and macromolecular interactions that we are now beginning to understand with increasing clarity. The real implications of the biochemical journey begun in the nineteenth century are found in the ever-increasing power to analyze and alter living systems.

To understand a complex biological process, a biochemist isolates and studies the individual components in vitro, then pieces together the parts to get a coherent picture of the overall process. A major source of molecular insights is the cell's own information archive, its DNA. The sheer size of chromosomes, however, presents an enormous challenge: how does one find and study a particular gene among the tens of thousands of genes nested in the billions of base pairs of a mammalian genome? Solutions began to emerge in the 1970s.

Decades of advances by thousands of scientists working in genetics, biochemistry, cell biology, and physical chemistry came together in the laboratories of Paul Berg, Herbert Boyer, and Stanley Cohen to yield techniques for locating, isolating, preparing, and studying small segments of DNA derived from much larger chromosomes. Techniques for DNA cloning paved the way to the modern fields of **genomics** and **proteomics,** the study of genes and proteins on the scale of whole cells and organisms. These new methods are transforming basic research, agriculture, medicine, ecology, forensics, and many other fields, while occasionally presenting society with difficult choices and ethical dilemmas.

We begin this chapter with an outline of the fundamental biochemical principles of the now-classic discipline of DNA cloning. Next, after laying the groundwork for a discussion of genomics, we illustrate the range of applications and the potential of these technologies, with a broad emphasis on modern advances in genomics and proteomics.

## 9.1 DNA Cloning: The Basics

A *clone* is an identical copy. This term originally applied to cells of a single type, isolated and allowed to reproduce to create a population of identical cells. **DNA cloning** involves separating a specific gene or DNA segment from a larger chromosome, attaching it to a small molecule of carrier DNA, and then replicating this modified DNA thousands or millions of times through both an increase in cell number and the creation of multiple copies of the cloned DNA in each cell. The result is selective amplification of a particular gene or DNA segment. Cloning of DNA from any organism entails five general procedures:

1. *Cutting DNA at precise locations.* Sequence-specific endonucleases (restriction endonucleases) provide the necessary molecular scissors.

2. *Selecting a small molecule of DNA capable of self-replication.* These DNAs are called **cloning vectors** (a vector is a delivery agent). They are typically plasmids or viral DNAs.

3. *Joining two DNA fragments covalently.* The enzyme DNA ligase links the cloning vector and DNA to be cloned. Composite DNA molecules comprising covalently linked segments from two or more sources are called **recombinant DNAs.**

4. *Moving recombinant DNA from the test tube to a host cell* that will provide the enzymatic machinery for DNA replication.

5. *Selecting or identifying host cells that contain recombinant DNA.*

The methods used to accomplish these and related tasks are collectively referred to as **recombinant DNA technology** or, more informally, **genetic engineering.**

Much of our initial discussion will focus on DNA cloning in the bacterium *Escherichia coli,* the first organism used for recombinant DNA work and still the most common host cell. *E. coli* has many advantages: its DNA metabolism (like many other of its biochemical processes) is well understood; many naturally occurring cloning vectors associated with *E. coli,* such as plasmids and bacteriophages (bacterial viruses; also called phages), are well characterized; and techniques are available for moving DNA expeditiously from one bac-

Paul Berg

Herbert Boyer

Stanley N. Cohen

terial cell to another. We also address DNA cloning in other organisms, a topic discussed more fully later in the chapter.

## Restriction Endonucleases and DNA Ligase Yield Recombinant DNA

Particularly important to recombinant DNA technology is a set of enzymes (Table 9–1) made available through decades of research on nucleic acid metabolism. Two classes of enzymes lie at the heart of the general approach to generating and propagating a recombinant DNA molecule (Fig. 9–1). First, **restriction endonucleases** (also called restriction enzymes) recognize and cleave DNA at specific DNA sequences (recognition sequences or restriction sites) to generate a set of smaller fragments. Second, the DNA fragment to be cloned can be joined to a suitable cloning vector by using **DNA ligases** to link the DNA molecules together. The recombinant vector is then introduced into a host cell, which amplifies the fragment in the course of many generations of cell division.

Restriction endonucleases are found in a wide range of bacterial species. Werner Arber discovered in the early 1960s that their biological function is to recognize and cleave foreign DNA (the DNA of an infecting virus, for example); such DNA is said to be *restricted.* In the host cell's DNA, the sequence that would be recognized

| TABLE 9–1 | Some Enzymes Used in Recombinant DNA Technology |
|---|---|
| *Enzyme(s)* | *Function* |
| Type II restriction endonucleases | Cleave DNAs at specific base sequences |
| DNA ligase | Joins two DNA molecules or fragments |
| DNA polymerase I (*E. coli*) | Fills gaps in duplexes by stepwise addition of nucleotides to 3′ ends |
| Reverse transcriptase | Makes a DNA copy of an RNA molecule |
| Polynucleotide kinase | Adds a phosphate to the 5′-OH end of a polynucleotide to label it or permit ligation |
| Terminal transferase | Adds homopolymer tails to the 3′-OH ends of a linear duplex |
| Exonuclease III | Removes nucleotide residues from the 3′ ends of a DNA strand |
| Bacteriophage λ exonuclease | Removes nucleotides from the 5′ ends of a duplex to expose single-stranded 3′ ends |
| Alkaline phosphatase | Removes terminal phosphates from either the 5′ or 3′ end (or both) |

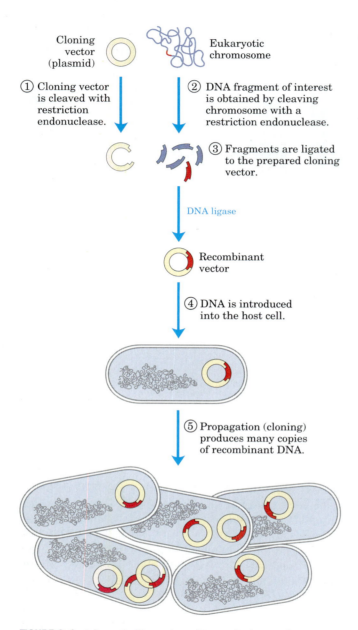

Cloning vector (plasmid)

Eukaryotic chromosome

① Cloning vector is cleaved with restriction endonuclease.

② DNA fragment of interest is obtained by cleaving chromosome with a restriction endonuclease.

③ Fragments are ligated to the prepared cloning vector.

DNA ligase

Recombinant vector

④ DNA is introduced into the host cell.

⑤ Propagation (cloning) produces many copies of recombinant DNA.

**FIGURE 9–1 Schematic illustration of DNA cloning.** A cloning vector and eukaryotic chromosomes are separately cleaved with the same restriction endonuclease. The fragments to be cloned are then ligated to the cloning vector. The resulting recombinant DNA (only one recombinant vector is shown here) is introduced into a host cell where it can be propagated (cloned). Note that this drawing is not to scale: the size of the *E. coli* chromosome relative to that of a typical cloning vector (such as a plasmid) is much greater than depicted here.

by its own restriction endonuclease is protected from digestion by methylation of the DNA, catalyzed by a specific DNA methylase. The restriction endonuclease and the corresponding methylase are sometimes referred to as a **restriction-modification system.**

There are three types of restriction endonucleases, designated I, II, and III. Types I and III are generally large, multisubunit complexes containing both the endonucle-

ase and methylase activities. Type I restriction endonucleases cleave DNA at random sites that can be more than 1,000 base pairs (bp) from the recognition sequence. Type III restriction endonucleases cleave the DNA about 25 bp from the recognition sequence. Both types move along the DNA in a reaction that requires the energy of ATP. **Type II restriction endonucleases,** first isolated by Hamilton Smith in 1970, are simpler, require no ATP, and cleave the DNA within the recognition sequence itself. The extraordinary utility of this group of restriction endonucleases was demonstrated by Daniel Nathans, who first used them to develop novel methods for mapping and analyzing genes and genomes.

Thousands of restriction endonucleases have been discovered in different bacterial species, and more than 100 different DNA sequences are recognized by one or more of these enzymes. The recognition sequences are usually 4 to 6 bp long and palindromic (see Fig. 8–20). Table 9–2 lists sequences recognized by a few type II restriction endonucleases. In some cases, the interaction between a restriction endonuclease and its target sequence has been elucidated in exquisite molecular detail; for example, Figure 9–2 shows the complex of the type II restriction endonuclease *Eco*RV and its target sequence.

Some restriction endonucleases make staggered cuts on the two DNA strands, leaving two to four nucleotides of one strand unpaired at each resulting end. These unpaired strands are referred to as **sticky ends** (Fig. 9–3a), because they can base-pair with each other or with complementary sticky ends of other DNA fragments. Other restriction endonucleases cleave both strands of DNA at the opposing phosphodiester bonds, leaving no unpaired bases on the ends, often called **blunt ends** (Fig. 9–3b).

The average size of the DNA fragments produced by cleaving genomic DNA with a restriction endonuclease depends on the frequency with which a particular restriction site occurs in the DNA molecule; this in turn depends largely on the size of the recognition sequence. In a DNA molecule with a random sequence in which all four nucleotides were equally abundant, a 6 bp sequence recognized by a restriction endonuclease such as *Bam*HI would occur on average once every $4^6$ (4,096) bp, assuming the DNA had a 50% G≡C content. Enzymes that recognize a 4 bp sequence would produce smaller DNA fragments from a random-sequence DNA molecule; a recognition sequence of this size would be expected to occur about once every $4^4$ (256) bp. In natural DNA molecules, particular recognition sequences tend to occur less frequently than this because nucleotide sequences in DNA are not random and the four nucleotides are not equally abundant. In laboratory experiments, the average size of the fragments produced by restriction endonuclease cleavage of a large DNA can be increased by simply terminating the reaction before completion; the result is called a partial digest. Fragment size can also

**TABLE 9–2** Recognition Sequences for Some Type II Restriction Endonucleases

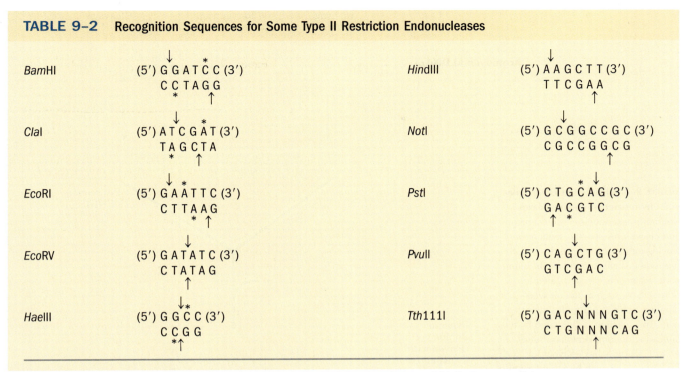

Arrows indicate the phosphodiester bonds cleaved by each restriction endonuclease. Asterisks indicate bases that are methylated by the corresponding methylase (where known). N denotes any base. Note that the name of each enzyme consists of a three-letter abbreviation (in italics) of the bacterial species from which it is derived, sometimes followed by a strain designation and Roman numerals to distinguish different restriction endonucleases isolated from the same bacterial species. Thus *Bam*HI is the first (I) restriction endonuclease characterized from B*acillus amyloliquefaciens,* strain H.

be increased by using a special class of endonucleases called homing endonucleases (see Fig. 26–34). These recognize and cleave much longer DNA sequences (14 to 20 bp).

Once a DNA molecule has been cleaved into fragments, a particular fragment of known size can be enriched by agarose or acrylamide gel electrophoresis or by HPLC (pp. 92, 90). For a typical mammalian genome, however, cleavage by a restriction endonuclease usually yields too many different DNA fragments to permit isolation of a particular fragment by electrophoresis or HPLC. A common intermediate step in the cloning of a specific gene or DNA segment is the construction of a DNA library (as described in Section 9.2).

After the target DNA fragment is isolated, DNA ligase can be used to join it to a similarly digested cloning vector—that is, a vector digested by the *same* restriction endonuclease; a fragment generated by *Eco*RI, for example, generally will not link to a fragment generated by *Bam*HI. As described in more detail in Chapter 25 (see Fig. 25–16), DNA ligase catalyzes the formation of new phosphodiester bonds in a reaction that uses ATP

**FIGURE 9–2 Interaction of *Eco*RV restriction endonuclease with its target sequence. (a)** The dimeric *Eco*RV endonuclease (its two subunits in light blue and gray) is bound to the products of DNA cleavage at the sequence recognized by the enzyme. The DNA backbone is shown in two shades of blue to distinguish the segments separated by cleavage (PDB ID 1RVC). **(b)** In this view, showing just the DNA, the DNA segment has been turned 180°. The enzyme creates blunt ends; the cleavage points appear staggered on the two DNA strands because the DNA is kinked. Bound magnesium ions (orange) play a role in catalysis of the cleavage reaction.

**Restriction Endonucleases**

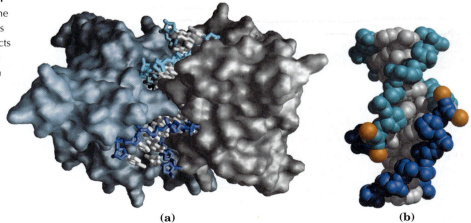

(a)

(b)

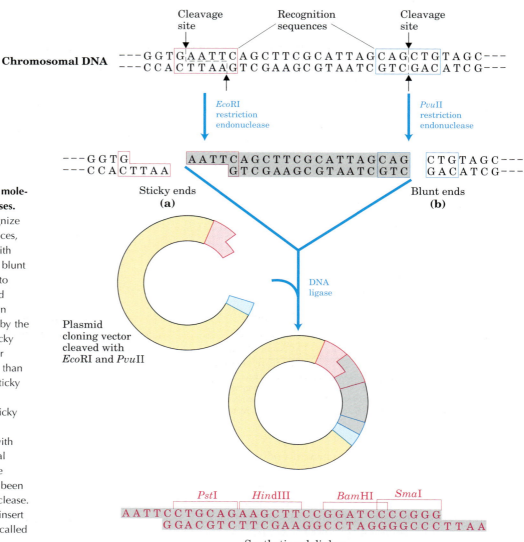

**FIGURE 9–3 Cleavage of DNA molecules by restriction endonucleases.** Restriction endonucleases recognize and cleave only specific sequences, leaving either **(a)** sticky ends (with protruding single strands) or **(b)** blunt ends. Fragments can be ligated to other DNAs, such as the cleaved cloning vector (a plasmid) shown here. This reaction is facilitated by the annealing of complementary sticky ends. Ligation is less efficient for DNA fragments with blunt ends than for those with complementary sticky ends, and DNA fragments with different (noncomplementary) sticky ends generally are not ligated. **(c)** A synthetic DNA fragment with recognition sequences for several restriction endonucleases can be inserted into a plasmid that has been cleaved by a restriction endonuclease. The insert is called a linker; an insert with multiple restriction sites is called a polylinker.

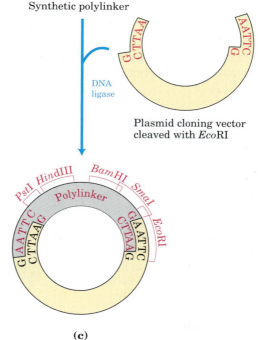

or a similar cofactor. The base-pairing of complementary sticky ends greatly facilitates the ligation reaction (Fig. 9–3a). Blunt ends can also be ligated, albeit less efficiently. Researchers can create new DNA sequences by inserting synthetic DNA fragments (called **linkers**) between the ends that are being ligated. Inserted DNA fragments with multiple recognition sequences for restriction endonucleases (often useful later as points for inserting additional DNA by cleavage and ligation) are called **polylinkers** (Fig. 9–3c).

The effectiveness of sticky ends in selectively joining two DNA fragments was apparent in the earliest recombinant DNA experiments. Before restriction endonucleases were widely available, some workers found they could generate sticky ends by the combined action of the bacteriophage λ exonuclease and terminal transferase (Table 9–1). The fragments to be joined were given complementary homopolymeric tails. Peter Lobban and Dale Kaiser used this method in 1971 in the first experiments to join naturally occurring DNA fragments.

Similar methods were used soon after in the laboratory of Paul Berg to join DNA segments from simian virus 40 (SV40) to DNA derived from bacteriophage λ, thereby creating the first recombinant DNA molecule with DNA segments from different species.

## Cloning Vectors Allow Amplification of Inserted DNA Segments

The principles that govern the delivery of recombinant DNA in clonable form to a host cell, and its subsequent amplification in the host, are well illustrated by considering three popular cloning vectors commonly used in experiments with *E. coli*—plasmids, bacteriophages, and bacterial artificial chromosomes—and a vector used to clone large DNA segments in yeast.

**Plasmids**   Plasmids are circular DNA molecules that replicate separately from the host chromosome. Naturally occurring bacterial plasmids range in size from 5,000 to 400,000 bp. They can be introduced into bacterial cells by a process called **transformation.** The cells (generally *E. coli*) and plasmid DNA are incubated together at 0 °C in a calcium chloride solution, then subjected to a shock by rapidly shifting the temperature to 37 to 43 °C. For reasons not well understood, some of the cells treated in this way take up the plasmid DNA. Some species of bacteria are naturally competent for DNA uptake and do not require the calcium chloride treatment. In an alternative method, cells incubated with the plasmid DNA are subjected to a high-voltage pulse. This approach, called **electroporation,** transiently renders the bacterial membrane permeable to large molecules.

Regardless of the approach, few cells actually take up the plasmid DNA, so a method is needed to select those that do. The usual strategy is to use a plasmid that includes a gene that the host cell requires for growth under specific conditions, such as a gene that confers resistance to an antibiotic. Only cells transformed by the recombinant plasmid can grow in the presence of that antibiotic, making any cell that contains the plasmid "selectable" under those growth conditions. Such a gene is called a selectable marker.

Investigators have developed many different plasmid vectors suitable for cloning by modifying naturally occurring plasmids. The *E. coli* plasmid pBR322 offers a good example of the features useful in a cloning vector (Fig. 9–4):

1.  pBR322 has an origin of replication, ori, a sequence where replication is initiated by cellular enzymes (Chapter 25). This sequence is required to propagate the plasmid and maintain it at a level of 10 to 20 copies per cell.

2.  The plasmid contains two genes that confer resistance to different antibiotics ($tet^R$, $amp^R$),

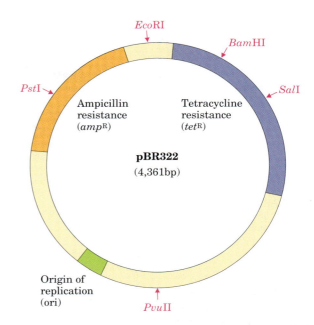

**FIGURE 9-4  The constructed *E. coli* plasmid pBR322.** Note the location of some important restriction sites—for *Pst*I, *Eco*RI, *Bam*HI, *Sal*I, and *Pvu*II; ampicillin- and tetracycline-resistance genes; and the replication origin (ori). Constructed in 1977, this was one of the early plasmids designed expressly for cloning in *E. coli*.

allowing the identification of cells that contain the intact plasmid or a recombinant version of the plasmid (Fig. 9–5).

3.  Several unique recognition sequences in pBR322 (*Pst*I, *Eco*RI, *Bam*HI, *Sal*I, *Pvu*II) are targets for different restriction endonucleases, providing sites where the plasmid can later be cut to insert foreign DNA.

4.  The small size of the plasmid (4,361 bp) facilitates its entry into cells and the biochemical manipulation of the DNA.

Transformation of typical bacterial cells with purified DNA (never a very efficient process) becomes less successful as plasmid size increases, and it is difficult to clone DNA segments longer than about 15,000 bp when plasmids are used as the vector.

**Bacteriophages**   Bacteriophage λ has a very efficient mechanism for delivering its 48,502 bp of DNA into a bacterium, and it can be used as a vector to clone somewhat larger DNA segments (Fig. 9–6). Two key features contribute to its utility:

1.  About one-third of the λ genome is nonessential and can be replaced with foreign DNA.

2.  DNA is packaged into infectious phage particles only if it is between 40,000 and 53,000 bp long, a constraint that can be used to ensure packaging of recombinant DNA only.

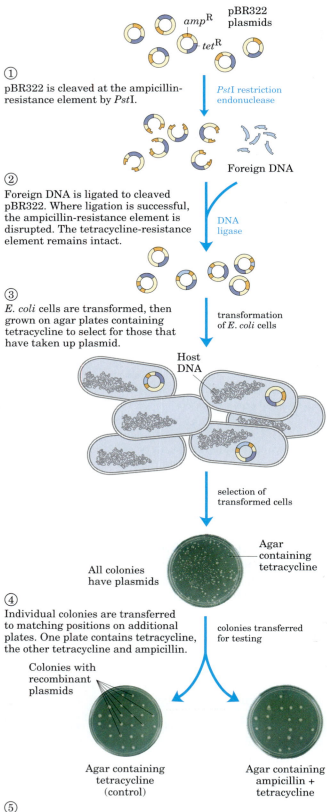

① pBR322 is cleaved at the ampicillin-resistance element by *Pst*I.

*Pst*I restriction endonuclease

Foreign DNA

② Foreign DNA is ligated to cleaved pBR322. Where ligation is successful, the ampicillin-resistance element is disrupted. The tetracycline-resistance element remains intact.

DNA ligase

③ *E. coli* cells are transformed, then grown on agar plates containing tetracycline to select for those that have taken up plasmid.

transformation of *E. coli* cells

Host DNA

selection of transformed cells

All colonies have plasmids

Agar containing tetracycline

④ Individual colonies are transferred to matching positions on additional plates. One plate contains tetracycline, the other tetracycline and ampicillin.

colonies transferred for testing

Colonies with recombinant plasmids

Agar containing tetracycline (control)

Agar containing ampicillin + tetracycline

⑤ Cells that grow on tetracycline but not on tetracycline + ampicillin contain recombinant plasmids with disrupted ampicillin resistance, hence the foreign DNA. Cells with pBR322 without foreign DNA retain ampicillin resistance and grow on both plates.

**FIGURE 9–5** Use of pBR322 to clone and identify foreign DNA in *E. coli*. 🔵 **Plasmid Cloning**

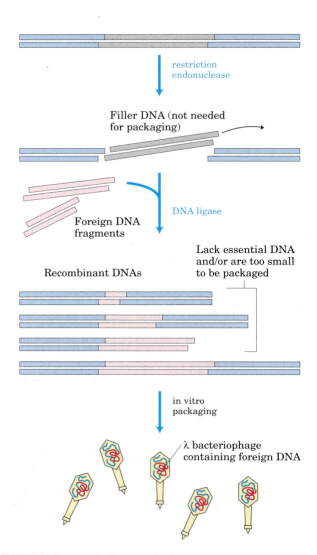

**FIGURE 9–6 Bacteriophage λ cloning vectors.** Recombinant DNA methods are used to modify the bacteriophage λ genome, removing the genes not needed for phage production and replacing them with "filler" DNA to make the phage DNA large enough for packaging into phage particles. As shown here, the filler is replaced with foreign DNA in cloning experiments. Recombinants are packaged into viable phage particles in vitro only if they include an appropriately sized foreign DNA fragment as well as both of the essential λ DNA end fragments.

Researchers have developed bacteriophage λ vectors that can be readily cleaved into three pieces, two of which contain essential genes but which together are only about 30,000 bp long. The third piece, "filler" DNA, is discarded when the vector is to be used for cloning, and additional DNA is inserted between the two essential segments to generate ligated DNA molecules long enough to produce viable phage particles. In effect, the packaging mechanism *selects for* recombinant viral DNAs.

Bacteriophage λ vectors permit the cloning of DNA fragments of up to 23,000 bp. Once the bacteriophage λ fragments are ligated to foreign DNA fragments of suitable size, the resulting recombinant DNAs can be packaged into phage particles by adding them to crude bacterial cell extracts that contain all the proteins needed to assemble a complete phage. This is called **in vitro packaging** (Fig. 9–6). All viable phage particles will contain a foreign DNA fragment. The subsequent transmission of the recombinant DNA into *E. coli* cells is highly efficient.

***Bacterial Artificial Chromosomes (BACs)*** Bacterial artificial chromosomes are simply plasmids designed for the cloning of very long segments (typically 100,000 to 300,000 bp) of DNA (Fig. 9–7). They generally include selectable markers such as resistance to the antibiotic chloramphenicol ($Cm^R$), as well as a very stable origin of replication (ori) that maintains the plasmid at one or two copies per cell. DNA fragments of several hundred thousand base pairs are cloned into the BAC vector. The large circular DNAs are then introduced into host bacteria by electroporation. These procedures use host bacteria with mutations that compromise the structure of their cell wall, permitting the uptake of the large DNA molecules.

***Yeast Artificial Chromosomes (YACs)*** *E. coli* cells are by no means the only hosts for genetic engineering. Yeasts are particularly convenient eukaryotic organisms for this work. As with *E. coli*, yeast genetics is a well-developed discipline. The genome of the most commonly used yeast, *Saccharomyces cerevisiae*, contains only $14 \times 10^6$ bp (a simple genome by eukaryotic standards, less than four times the size of the *E. coli* chromosome), and its entire sequence is known. Yeast is also very easy to maintain and grow on a large scale in the laboratory. Plasmid vectors have been constructed for yeast, employing the same principles that govern the use of *E. coli* vectors described above. Convenient methods are now available for moving DNA into and out of yeast cells, facilitating the study of many aspects of eukaryotic cell biochemistry. Some recombinant plasmids incorporate multiple replication origins and other elements that allow them to be used in more than one species (for example, yeast or *E. coli*). Plasmids that can be propagated in cells of two or more different species are called **shuttle vectors.**

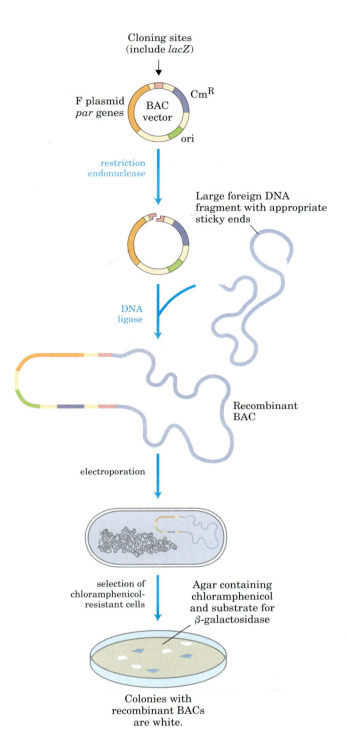

**FIGURE 9–7 (above right) Bacterial artificial chromosomes (BACs) as cloning vectors.** The vector is a relatively simple plasmid, with a replication origin (ori) that directs replication. The *par* genes, derived from a type of plasmid called an F plasmid, assist in the even distribution of plasmids to daughter cells at cell division. This increases the likelihood of each daughter cell carrying one copy of the plasmid, even when few copies are present. The low number of copies is useful in cloning large segments of DNA because it limits the opportunities for unwanted recombination reactions that can unpredictably alter large cloned DNAs over time. The BAC includes selectable markers. A *lacZ* gene (required for the production of the enzyme β-galactosidase) is situated in the cloning region such that it is inactivated by cloned DNA inserts. Introduction of recombinant BACs into cells by electroporation is promoted by the use of cells with an altered (more porous) cell wall. Recombinant DNAs are screened for resistance to the antibiotic chloramphenicol ($Cm^R$). Plates also contain a substrate for β-galactosidase that yields a colored product. Colonies with active β-galactosidase and hence no DNA insert in the BAC vector turn blue; colonies without β-galactosidase activity—and thus with the desired DNA inserts—are white.

Research work with large genomes and the associated need for high-capacity cloning vectors led to the development of **yeast artificial chromosomes** (**YACS;** Fig. 9–8). YAC vectors contain all the elements needed to maintain a eukaryotic chromosome in the yeast nucleus: a yeast origin of replication, two selectable markers, and specialized sequences (derived from the telomeres and centromere, regions of the chromosome discussed in Chapter 24) needed for stability and

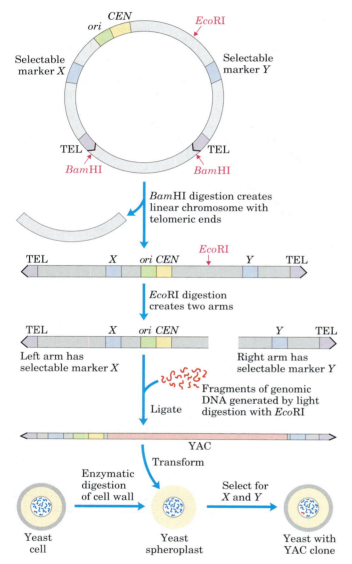

**FIGURE 9–8 Construction of a yeast artificial chromosome (YAC).** A YAC vector includes an origin of replication (ori), a centromere (*CEN*), two telomeres (TEL), and selectable markers (*X* and *Y*). Digestion with *Bam*H1 and *Eco*RI generates two separate DNA arms, each with a telomeric end and one selectable marker. A large segment of DNA (e.g., up to $2 \times 10^6$ bp from the human genome) is ligated to the two arms to create a yeast artificial chromosome. The YAC transforms yeast cells (prepared by removal of the cell wall to form spheroplasts), and the cells are selected for *X* and *Y*; the surviving cells propagate the DNA insert.

proper segregation of the chromosomes at cell division. Before being used in cloning, the vector is propagated as a circular bacterial plasmid. Cleavage with a restriction endonuclease (*Bam*H1 in Fig. 9–8) removes a length of DNA between two telomere sequences (TEL), leaving the telomeres at the ends of the linearized DNA. Cleavage at another internal site (*Eco*RI in Fig. 9–8) divides the vector into two DNA segments, referred to as vector arms, each with a different selectable marker.

The genomic DNA is prepared by partial digestion with restriction endonucleases (*Eco*RI in Fig. 9–8) to obtain a suitable fragment size. Genomic fragments are then separated by **pulsed field gel electrophoresis,** a variation of gel electrophoresis (see Fig. 3–19) that allows the separation of very large DNA segments. The DNA fragments of appropriate size (up to about $2 \times 10^6$ bp) are mixed with the prepared vector arms and ligated. The ligation mixture is then used to transform treated yeast cells with very large DNA molecules. Culture on a medium that requires the presence of both selectable marker genes ensures the growth of only those yeast cells that contain an artificial chromosome with a large insert sandwiched between the two vector arms (Fig. 9–8). The stability of YAC clones increases with size (up to a point). Those with inserts of more than 150,000 bp are nearly as stable as normal cellular chromosomes, whereas those with inserts of less than 100,000 bp are gradually lost during mitosis (so generally there are no yeast cell clones carrying only the two vector ends ligated together or with only short inserts). YACs that lack a telomere at either end are rapidly degraded.

## Specific DNA Sequences Are Detectable by Hybridization

DNA hybridization, a process outlined in Chapter 8 (see Fig. 8–32), is the most common sequence-based process for detecting a particular gene or segment of nucleic acid. There are many variations of the basic method, most making use of a labeled (such as radioactive) DNA or RNA fragment, known as a **probe,** complementary to the DNA being sought. In one classic approach to detect a particular DNA sequence within a DNA library (a collection of DNA clones), nitrocellulose paper is pressed onto an agar plate containing many individual bacterial colonies from the library, each colony with a different recombinant DNA. Some cells from each colony adhere to the paper, forming a replica of the plate. The paper is treated with alkali to disrupt the cells and denature the DNA within, which remains bound to the region of the paper around the colony from which it came. Added radioactive DNA probe anneals only to its complementary DNA. After any unannealed probe DNA is washed away, the hybridized DNA can be detected by autoradiography (Fig. 9–9).

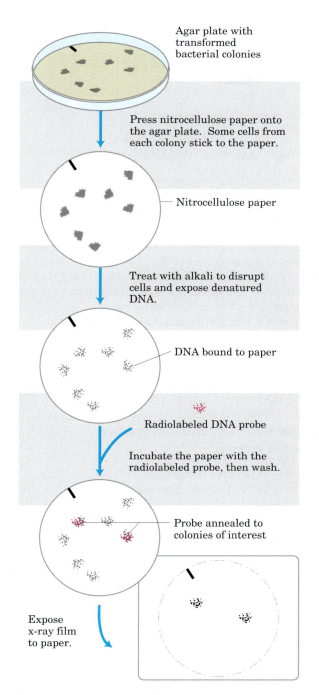

Agar plate with transformed bacterial colonies

Press nitrocellulose paper onto the agar plate. Some cells from each colony stick to the paper.

Nitrocellulose paper

Treat with alkali to disrupt cells and expose denatured DNA.

DNA bound to paper

Radiolabeled DNA probe

Incubate the paper with the radiolabeled probe, then wash.

Probe annealed to colonies of interest

Expose x-ray film to paper.

**FIGURE 9–9 Use of hybridization to identify a clone with a particular DNA segment.** The radioactive DNA probe hybridizes to complementary DNA and is revealed by autoradiography. Once the labeled colonies have been identified, the corresponding colonies on the original agar plate can be used as a source of cloned DNA for further study.

A common limiting step in detecting and cloning a gene is the generation of a complementary strand of nucleic acid to use as a probe. The origin of a probe depends on what is known about the gene under investigation. Sometimes a homologous gene cloned from another species makes a suitable probe. Or, if the protein product of a gene has been purified, probes can be designed and synthesized by working backward from the amino acid sequence, deducing the DNA sequence that would code for it (Fig. 9–10). Now, researchers typically obtain the necessary DNA sequence information from sequence databases that detail the structure of millions of genes from a wide range of organisms.

## Expression of Cloned Genes Produces Large Quantities of Protein

Frequently it is the product of the cloned gene, rather than the gene itself, that is of primary interest—particularly when the protein has commercial, therapeutic, or research value. With an increased understanding of the fundamentals of DNA, RNA, and protein metabolism and their regulation in *E. coli,* investigators can now manipulate cells to express cloned genes in order to study their protein products.

Most eukaryotic genes lack the DNA sequence elements—such as promoters, sequences that instruct RNA polymerase where to bind—required for their expression in *E. coli* cells, so bacterial regulatory sequences for transcription and translation must be inserted at appropriate positions relative to the eukaryotic gene in the vector DNA. (Promoters, regulatory sequences, and other aspects of the regulation of gene expression are discussed in Chapter 28.) In some cases cloned genes are so efficiently expressed that their protein product represents 10% or more of the cellular protein; they are said to be overexpressed. At these concentrations some foreign proteins can kill an *E. coli* cell, so gene expression must be limited to the few hours before the planned harvest of the cells.

Cloning vectors with the transcription and translation signals needed for the regulated expression of a cloned gene are often called **expression vectors.** The rate of expression of the cloned gene is controlled by replacing the gene's own promoter and regulatory sequences with more efficient and convenient versions supplied by the vector. Generally, a well-characterized promoter and its regulatory elements are positioned near several unique restriction sites for cloning, so that genes inserted at the restriction sites will be expressed from the regulated promoter element (Fig. 9–11). Some of these vectors incorporate other features, such as a bacterial ribosome binding site to enhance translation of the mRNA derived from the gene, or a transcription termination sequence.

Genes can similarly be cloned and expressed in eukaryotic cells, with various species of yeast as the usual hosts. A eukaryotic host can sometimes promote posttranslational modifications (changes in protein structure made after synthesis on the ribosomes) that might be required for the function of a cloned eukaryotic protein.

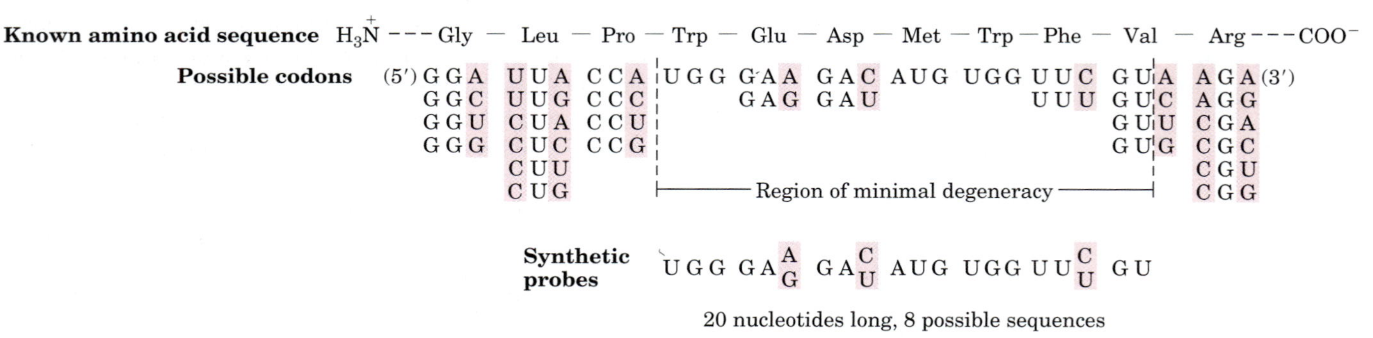

**Known amino acid sequence**   $H_3\overset{+}{N}$ --- Gly — Leu — Pro — Trp — Glu — Asp — Met — Trp — Phe — Val — Arg --- COO⁻

**Possible codons**

| (5′) G G A | U U A | C C A | U G G | G A A | G A C | A U G | U G G | U U C | G U A | A G A (3′) |
|---|---|---|---|---|---|---|---|---|---|---|
| G G C | U U G | C C C | | G A G | G A U | | | U U U | G U C | A G G |
| G G U | C U A | C C U | | | | | | | G U U | C G A |
| G G G | C U C | C C G | | | | | | | G U G | C G C |
| | C U U | | | | | | | | | C G U |
| | C U G | | | | | | | | | C G G |

├─────── Region of minimal degeneracy ───────┤

**Synthetic probes**   U G G  G A $\substack{A\\G}$  G A $\substack{C\\U}$  A U G  U G G  U U $\substack{C\\U}$  G U

20 nucleotides long, 8 possible sequences

**FIGURE 9–10  Probe to detect the gene for a protein of known amino acid sequence.** Because more than one DNA sequence can code for any given amino acid sequence, the genetic code is said to be "degenerate." (As described in Chapter 27, an amino acid is coded for by a set of three nucleotides called a *codon*. Most amino acids have two or more codons; see Fig. 27–7.) Thus the correct DNA sequence for a known amino acid sequence cannot be known in advance. The probe is designed to be complementary to a region of the gene with minimal degeneracy, that is, a region with the fewest possible codons for the amino acids—two codons at most in the example shown here. Oligonucleotides are synthesized with selectively randomized sequences, so that they contain either of the two possible nucleotides at each position of potential degeneracy (shaded in pink). The oligonucleotide shown here represents a mixture of eight different sequences: one of the eight will complement the gene perfectly, and all eight will match at least 17 of the 20 positions.

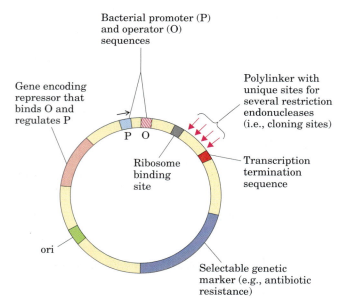

**FIGURE 9–11  DNA sequences in a typical *E. coli* expression vector.** The gene to be expressed is inserted into one of the restriction sites in the polylinker, near the promoter (P), with the end encoding the amino terminus proximal to the promoter. The promoter allows efficient transcription of the inserted gene, and the transcription termination sequence sometimes improves the amount and stability of the mRNA produced. The operator (O) permits regulation by means of a repressor that binds to it (Chapter 28). The ribosome binding site provides sequence signals needed for efficient translation of the mRNA derived from the gene. The selectable marker allows the selection of cells containing the recombinant DNA.

## Alterations in Cloned Genes Produce Modified Proteins

Cloning techniques can be used not only to overproduce proteins but to produce protein products subtly altered from their native forms. Specific amino acids may be replaced individually by **site-directed mutagenesis.** This powerful approach to studying protein structure and function changes the amino acid sequence of a protein by altering the DNA sequence of the cloned gene. If appropriate restriction sites flank the sequence to be altered, researchers can simply remove a DNA segment and replace it with a synthetic one that is identical to the original except for the desired change (Fig. 9–12a). When suitably located restriction sites are not present, an approach called **oligonucleotide-directed mutagenesis** (Fig. 9–12b) can create a specific DNA sequence change. A short synthetic DNA strand with a specific base change is annealed to a single-stranded copy of the cloned gene within a suitable vector. The mismatch of a single base pair in 15 to 20 bp does not prevent annealing if it is done at an appropriate temperature. The annealed strand serves as a primer for the synthesis of a strand complementary to the plasmid vector. This slightly mismatched duplex recombinant plasmid is then used to transform bacteria, where the mismatch is repaired by cellular DNA repair enzymes (Chapter 25). About half of the repair events will remove and replace the altered base and restore the gene to its original sequence; the other half will remove and

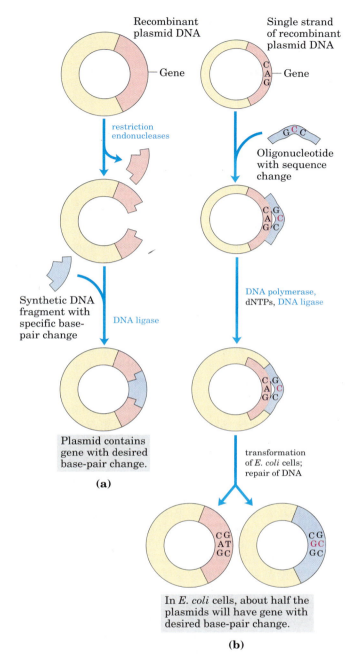

**FIGURE 9–12 Two approaches to site-directed mutagenesis. (a)** A synthetic DNA segment replaces a DNA fragment that has been removed by cleavage with a restriction endonuclease. **(b)** A synthetic oligonucleotide with a desired sequence change at one position is hybridized to a single-stranded copy of the gene to be altered. This acts as primer for synthesis of a duplex DNA (with one mismatch), which is then used to transform cells. Cellular DNA repair systems will convert about 50% of the mismatches to reflect the desired sequence change.

Changes can also be introduced that involve more than one base pair. Large parts of a gene can be deleted by cutting out a segment with restriction endonucleases and ligating the remaining portions to form a smaller gene. Parts of two different genes can be ligated to create new combinations. The product of such a fused gene is called a **fusion protein.**

Researchers now have ingenious methods to bring about virtually any genetic alteration in vitro. Reintroduction of the altered DNA into the cell permits investigation of the consequences of the alteration. Site-directed mutagenesis has greatly facilitated research on proteins by allowing investigators to make specific changes in the primary structure of a protein and to examine the effects of these changes on the folding, three-dimensional structure, and activity of the protein.

### SUMMARY 9.1 DNA Cloning: The Basics

- DNA cloning and genetic engineering involve the cleavage of DNA and assembly of DNA segments in new combinations—recombinant DNA.

- Cloning entails cutting DNA into fragments with enzymes; selecting and possibly modifying a fragment of interest; inserting the DNA fragment into a suitable cloning vector; transferring the vector with the DNA insert into a host cell for replication; and identifying and selecting cells that contain the DNA fragment.

- Key enzymes in gene cloning include restriction endonucleases (especially the type II enzymes) and DNA ligase.

- Cloning vectors include plasmids, bacteriophages, and, for the longest DNA inserts, bacterial artificial chromosomes (BACs) and yeast artificial chromosomes (YACs).

- Cells containing particular DNA sequences can be identified by DNA hybridization methods.

- Genetic engineering techniques manipulate cells to express and/or alter cloned genes.

## 9.2 From Genes to Genomes

The modern science of **genomics** now permits the study of DNA on a cellular scale, from individual genes to the entire genetic complement of an organism—its genome. Genomic databases are growing rapidly, as one sequencing milestone is superseded by the next. Biology in the twenty-first century will move forward with the aid of informational resources undreamed of only a few years ago. We now turn to a consideration of some of the technologies fueling these advances.

replace the *normal* base, retaining the desired mutation. Transformants are screened (often by sequencing their plasmid DNA) until a bacterial colony containing a plasmid with the altered sequence is found.

## DNA Libraries Provide Specialized Catalogs of Genetic Information

A DNA library is a collection of DNA clones, gathered together as a source of DNA for sequencing, gene discovery, or gene function studies. The library can take a variety of forms, depending on the source of the DNA. Among the largest types of DNA library is a **genomic library,** produced when the complete genome of a particular organism is cleaved into thousands of fragments, and *all* the fragments are cloned by insertion into a cloning vector.

The first step in preparing a genomic library is partial digestion of the DNA by restriction endonucleases, such that any given sequence will appear in fragments of a range of sizes—a range that is compatible with the cloning vector and ensures that virtually all sequences are represented among the clones in the library. Fragments that are too large or too small for cloning are removed by centrifugation or electrophoresis. The cloning vector, such as a BAC or YAC plasmid, is cleaved with the same restriction endonuclease and ligated to the genomic DNA fragments. The ligated DNA mixture is then used to transform bacterial or yeast cells to produce a library of cell types, each type harboring a different recombinant DNA molecule. Ideally, all the DNA in the genome under study will be represented in the library. Each transformed bacterium or yeast cell grows into a colony, or "clone," of identical cells, each cell bearing the same recombinant plasmid.

Using hybridization methods, researchers can order individual clones in a library by identifying clones with overlapping sequences. A set of overlapping clones represents a catalog for a long contiguous segment of a genome, often referred to as a **contig** (Fig. 9–13). Previously studied sequences or entire genes can be located within the library using hybridization methods to determine which library clones harbor the known sequence. If the sequence has already been mapped on a chromosome, investigators can determine the location (in the genome) of the cloned DNA and any contig of which it is a part. A well-characterized library may contain thousands of long contigs, all assigned to and ordered on particular chromosomes to form a detailed physical map. The known sequences within the library (each called a **sequence-tagged site,** or **STS**) can provide landmarks for genomic sequencing projects.

As more and more genome sequences become available, the utility of genomic libraries is diminishing and investigators are constructing more specialized libraries designed to study gene function. An example is a library that includes only those genes that are *expressed*—that is, are transcribed into RNA—in a given organism or even in certain cells or tissues. Such a library lacks the noncoding DNA that makes up a large portion of many eukaryotic genomes. The researcher first extracts mRNA from an organism or from specific cells of an or-

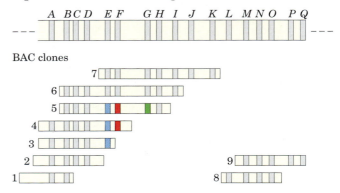

Segment of chromosome from organism X

**FIGURE 9–13 Ordering of the clones in a DNA library.** Shown here is a segment of a chromosome from a hypothetical organism X, with markers *A* through *Q* representing sequence-tagged sites (STSs—DNA segments of known sequence, including known genes). Below the chromosome is an array of ordered BAC clones, numbered 1 to 9. Ordering the clones on the genetic map is a many-stage process. The presence or absence of an STS on an individual clone can be determined by hybridization—for example, by probing each clone with PCR-amplified DNA from the STS. Once the STSs on each BAC clone are identified, the clones (and the STSs themselves, if their location is not yet known) can be ordered on the map. For example, compare clones 3, 4, and 5. Marker *E* (blue) is found on all three clones; *F* (red) on clones 4 and 5, but not on 3; and *G* (green) only on clone 5. This indicates that the order of the sites is *E, F, G*. The clones partially overlap and their order must be 3, 4, 5. The resulting ordered series of clones is called a contig.

ganism and then prepares **complementary DNAs (cDNAs)** from the RNA in a multistep reaction catalyzed by the enzyme reverse transcriptase (Fig. 9–14). The resulting double-stranded DNA fragments are then inserted into a suitable vector and cloned, creating a population of clones called a **cDNA library.** The search for a particular gene is made easier by focusing on a cDNA library generated from the mRNAs of a cell known to express that gene. For example, if we wished to clone globin genes, we could first generate a cDNA library from erythrocyte precursor cells, in which about half the mRNAs code for globins. To aid in the mapping of large genomes, cDNAs in a library can be partially sequenced at random to produce a useful type of STS called an **expressed sequence tag (EST).** ESTs, ranging in size from a few dozen to several hundred base pairs, can be positioned within the larger genome map, providing markers for expressed genes. Hundreds of thousands of ESTs were included in the detailed physical maps used as a guide to sequencing the human genome.

A cDNA library can be made even more specialized by cloning a cDNA or cDNA fragment into a vector that fuses the cDNA sequence with the sequence for a marker, or reporter gene; the fused genes form a "reporter construct." Two useful markers are the genes for green fluorescent protein and epitope tags. A target

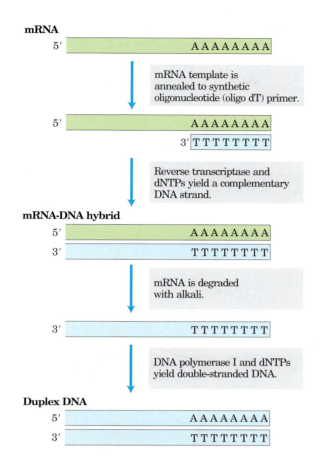

**mRNA**

mRNA template is annealed to synthetic oligonucleotide (oligo dT) primer.

Reverse transcriptase and dNTPs yield a complementary DNA strand.

**mRNA-DNA hybrid**

mRNA is degraded with alkali.

DNA polymerase I and dNTPs yield double-stranded DNA.

**Duplex DNA**

**FIGURE 9–14 Construction of a cDNA library from mRNA.** A cell's mRNA includes transcripts from thousands of genes, and the cDNAs generated are correspondingly heterogeneous. The duplex DNA produced by this method is inserted into an appropriate cloning vector. Reverse transcriptase can synthesize DNA on an RNA or a DNA template (see Fig. 26–29).

gene fused with a gene for **green fluorescent protein (GFP)** generates a fusion protein that is highly fluorescent—it literally lights up (Fig. 9–15a). Just a few molecules of this protein can be observed microscopically, allowing the study of its location and movements in a cell. An **epitope tag** is a short protein sequence that is bound tightly by a well-characterized monoclonal antibody (Chapter 5). The tagged protein can be specifically precipitated from a crude protein extract by interaction with the antibody (Fig. 9–15b). If any other proteins bind to the tagged protein, those will precipitate as well, providing information about protein-protein interactions in a cell. The diversity and utility of specialized DNA libraries are growing every year.

## The Polymerase Chain Reaction Amplifies Specific DNA Sequences

The Human Genome Project, along with the many associated efforts to sequence the genomes of organisms of every type, is providing unprecedented access to gene sequence information. This in turn is simplifying the

process of cloning individual genes for more detailed biochemical analysis. If we know the sequence of at least the flanking parts of a DNA segment to be cloned, we can hugely amplify the number of copies of that DNA segment, using the **polymerase chain reaction (PCR),** a process conceived by Kary Mullis in 1983. The amplified DNA can be cloned directly or used in a variety of analytical procedures.

The PCR procedure has an elegant simplicity. Two synthetic oligonucleotides are prepared, complementary to sequences on opposite strands of the target DNA at positions just beyond the ends of the segment to be amplified. The oligonucleotides serve as replication primers that can be extended by DNA polymerase. The 3' ends of the hybridized probes are oriented toward each other and positioned to prime DNA synthesis across the desired DNA segment (Fig. 9–16). (DNA polymerases

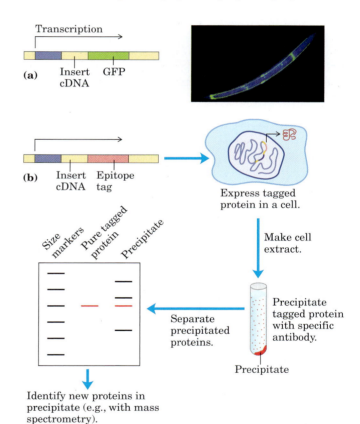

**FIGURE 9–15 Specialized DNA libraries. (a)** Cloning of cDNA next to a gene for green fluorescent protein (GFP) creates a reporter construct. RNA transcription proceeds through the gene of interest (insert DNA) and the reporter gene, and the mRNA transcript is then expressed as a fusion protein. The GFP part of the protein is visible in the fluorescence microscope. The photograph shows a nematode worm containing a GFP fusion protein expressed only in the four "touch" neurons that run the length of its body. 🔵 **Reporter Constructs (b)** If the cDNA is cloned next to a gene for an epitope tag, the resulting fusion protein can be precipitated by antibodies to the epitope. Any other proteins that interact with the tagged protein also precipitate, helping to elucidate protein-protein interactions.

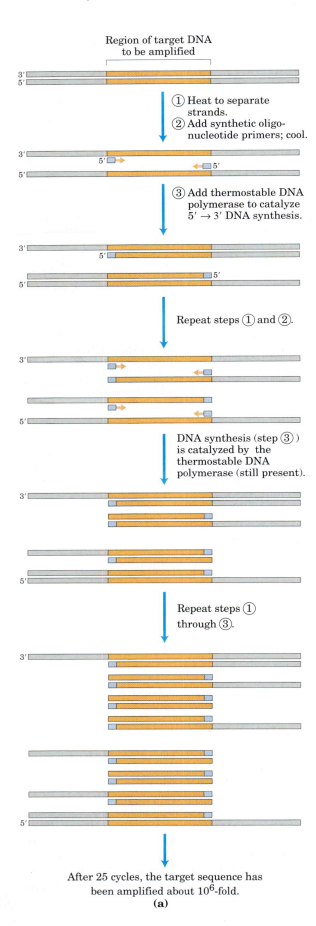

**(a)**

Region of target DNA to be amplified

① Heat to separate strands.
② Add synthetic oligo-nucleotide primers; cool.

③ Add thermostable DNA polymerase to catalyze 5′ → 3′ DNA synthesis.

Repeat steps ① and ②.

DNA synthesis (step ③) is catalyzed by the thermostable DNA polymerase (still present).

Repeat steps ① through ③.

After 25 cycles, the target sequence has been amplified about $10^6$-fold.

**FIGURE 9–16 Amplification of a DNA segment by the polymerase chain reaction. (a)** The PCR procedure has three steps. DNA strands are ① separated by heating, then ② annealed to an excess of short synthetic DNA primers (blue) that flank the region to be amplified; ③ new DNA is synthesized by polymerization. The three steps are repeated for 25 or 30 cycles. The thermostable DNA polymerase *Taq*I (from *Thermus aquaticus,* a bacterial species that grows in hot springs) is not denatured by the heating steps. **(b)** DNA amplified by PCR can be cloned. The primers can include noncomplementary ends that have a site for cleavage by a restriction endonuclease. Although these parts of the primers do not anneal to the target DNA, the PCR process incorporates them into the DNA that is amplified. Cleavage of the amplified fragments at these sites creates sticky ends, used in ligation of the amplified DNA to a cloning vector. 🖱 **Polymerase Chain Reaction**

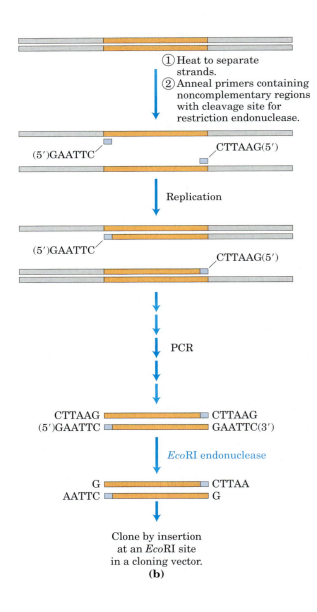

① Heat to separate strands.
② Anneal primers containing noncomplementary regions with cleavage site for restriction endonuclease.

(5′)GAATTC          CTTAAG(5′)

Replication

(5′)GAATTC          CTTAAG(5′)

PCR

CTTAAG        CTTAAG
(5′)GAATTC        GAATTC(3′)

*Eco*RI endonuclease

G        CTTAA
AATTC        G

Clone by insertion at an *Eco*RI site in a cloning vector.

**(b)**

synthesize DNA strands from deoxyribonucleotides, using a DNA template, as described in Chapter 25.) Isolated DNA containing the segment to be amplified is heated briefly to denature it, and then cooled in the presence of a large excess of the synthetic oligonucleotide primers. The four deoxynucleoside triphosphates are then added, and the primed DNA segment is replicated selectively. The cycle of heating, cooling, and replication is repeated 25 or 30 times over a few hours in an automated process, amplifying the DNA segment flanked by the primers until it can be readily analyzed or cloned. PCR uses a heat-stable DNA polymerase, such as the *Taq* polymerase (derived from a bacterium that lives at 90 °C), which remains active after every heating step and does not have to be replenished. Careful design of the primers used for PCR, such as including restriction endonuclease cleavage sites, can facilitate the subsequent cloning of the amplified DNA (Fig. 9–16b).

This technology is highly sensitive: PCR can detect and amplify as little as one DNA molecule in almost any type of sample. Although DNA degrades over time (p. 293), PCR has allowed successful cloning of DNA from samples more than 40,000 years old. Investigators have used the technique to clone DNA fragments from the mummified remains of humans and extinct animals such as the woolly mammoth, creating the new fields of molecular archaeology and molecular paleontology. DNA from burial sites has been amplified by PCR and used to trace ancient human migrations. Epidemiologists can use PCR-enhanced DNA samples from human remains to trace the evolution of human pathogenic viruses. Thus, in addition to its usefulness for cloning DNA, PCR is a potent tool in forensic medicine (Box 9–1). It is also being used for detection of viral infections before they cause symptoms and for prenatal diagnosis of a wide array of genetic diseases.

The PCR method is also important in advancing the goal of whole genome sequencing. For example, the mapping of expressed sequence tags to particular chromosomes often involves amplification of the EST by PCR, followed by hybridization of the amplified DNA to clones in an ordered library. Investigators found many other applications of PCR in the Human Genome Project, to which we now turn.

### Genome Sequences Provide the Ultimate Genetic Libraries

The genome is the ultimate source of information about an organism, and there is no genome we are more interested in than our own. Less than 10 years after the development of practical DNA sequencing methods, serious discussions began about the prospects for sequencing the entire 3 billion base pairs of the human genome. The international Human Genome Project got underway with substantial funding in the late 1980s. The effort eventually included significant contributions from

20 sequencing centers distributed among six nations: the United States, Great Britain, Japan, France, China, and Germany. General coordination was provided by the Office of Genome Research at the National Institutes of Health, led first by James Watson and after 1992 by Francis Collins. At the outset, the task of sequencing a $3 \times 10^9$ bp genome seemed to be a titanic job, but it gradually yielded to advances in technology. The completed sequence of the human genome was published in April 2003, several years ahead of schedule.

This advance was the product of a carefully planned international effort spanning 14 years. Research teams first generated a detailed physical map of the human genome, with clones derived from each chromosome organized into a series of long contigs (Fig. 9–17). Each contig contained landmarks in the form of STSs at a distance of every 100,000 bp or less. The genome thus mapped could be divided up between the international sequencing centers, each center sequencing the mapped BAC or YAC clones corresponding to its particular segments of the genome. Because many of the

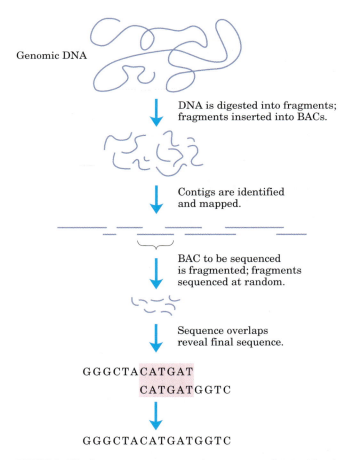

Genomic DNA

DNA is digested into fragments; fragments inserted into BACs.

Contigs are identified and mapped.

BAC to be sequenced is fragmented; fragments sequenced at random.

Sequence overlaps reveal final sequence.

GGGCTACATGAT
CATGATGGTC

GGGCTACATGATGGTC

**FIGURE 9–17 The Human Genome Project strategy.** Clones isolated from a genomic library were ordered into a detailed physical map, then individual clones were sequenced by shotgun sequencing protocols. The strategy used by the commercial sequencing effort eliminated the step of creating the physical map and sequenced the entire genome by shotgun cloning.

## BOX 9–1    WORKING IN BIOCHEMISTRY

### A Potent Weapon in Forensic Medicine

Traditionally, one of the most accurate methods for placing an individual at the scene of a crime has been a fingerprint. With the advent of recombinant DNA technology, a more powerful tool is now available: **DNA fingerprinting** (also called DNA typing or DNA profiling).

DNA fingerprinting is based on **sequence polymorphisms,** slight sequence differences (usually single base-pair changes) between individuals, 1 bp in every 1,000 bp, on average. Each difference from the prototype human genome sequence (the first one obtained) occurs in some fraction of the human population; every individual has some differences. Some of the sequence changes affect recognition sites for restriction enzymes, resulting in variation in the size of DNA fragments produced by digestion with a particular restriction enzyme. These variations are **restriction fragment length polymorphisms (RFLPs).**

The detection of RFLPs relies on a specialized hybridization procedure called **Southern blotting** (Fig. 1). DNA fragments from digestion of genomic DNA by restriction endonucleases are separated by size electrophoretically, denatured by soaking the agarose gel in alkali, and then blotted onto a nylon membrane to reproduce the distribution of fragments in the gel. The membrane is immersed in a solution containing a radioactively labeled DNA probe. A probe for a sequence that is repeated several times in the human genome generally identifies a few of the thousands of DNA fragments generated when the human genome is digested with a restriction endonuclease. Autoradiography reveals the fragments to which the probe hybridizes, as in Figure 9–9.

The genomic DNA sequences used in these tests are generally regions containing repetitive DNA (short sequences repeated thousands of times in tandem), which are common in the genomes of higher eukaryotes (see Fig. 24–8). The number of repeated units in these DNA regions varies among individuals (except between identical twins). With a suitable probe, the pattern of bands produced by DNA fingerprinting is distinctive for each individual. Combining the use of several probes makes the test so selective that it can positively identify a single individual in the entire human population. However, the Southern blot procedure requires relatively fresh DNA samples and larger amounts of DNA than are generally present at a crime scene. RFLP analysis sensitivity is augmented by using PCR (see Fig. 9–16a) to amplify vanishingly small amounts of DNA. This allows investigators to obtain DNA fingerprints from a single hair follicle, a drop of blood, a small semen sample from a rape victim, or samples that might be months or even many years old.

These methods are proving decisive in court cases worldwide. In the example in Figure 1, the DNA from a semen sample obtained from a rape and murder victim was compared with DNA samples from the victim and two suspects. Each sample was cleaved into fragments and separated by gel electrophoresis. Radioactive DNA probes were used to identify a small subset of fragments that contained sequences complementary to the probe. The sizes of the identified fragments varied from one individual to the next, as seen here in the different patterns for the three individuals (victim and two suspects) tested. One suspect's DNA exhibits a banding pattern identical to that of a semen sample taken from the victim. This test used a single probe, but three or four different probes would be used (in separate experiments) to achieve an unambiguous positive identification.

---

clones were more than 100,000 bp long, and modern sequencing techniques can resolve only 600 to 750 bp of sequence at a time, each clone had to be sequenced in pieces. The sequencing strategy used a shotgun approach, in which researchers used powerful new automated sequencers to sequence random segments of a given clone, then assembled the sequence of the entire clone by computerized identification of overlaps. The number of clone pieces sequenced was determined statistically so that the entire length of the clone was sequenced four to six times on average. The sequenced DNA was then made available in a database covering the entire genome. Construction of the physical map was a time-consuming task, and its progress was followed in annual reports in major journals throughout the 1990s— by the end of which the map was largely in place. Completion of the entire sequencing project was initially projected for the year 2005, but circumstances and technology intervened to accelerate the process.

A competing commercial effort to sequence the human genome was initiated by the newly established Celera Corporation in 1997. Led by J. Craig Venter, the Celera group made use of a different strategy called "whole genome shotgun sequencing," which eliminates the step of assembling a physical map of the genome. Instead, teams sequenced DNA segments from through-

Such results have been used to both convict and acquit suspects and, in other cases, to establish paternity with an extraordinary degree of certainty. The impact of these procedures on court cases will continue to grow as societies agree on the standards and as formal methods become widely established in forensic laboratories. Even decades-old murder mysteries can be solved: in 1996, DNA fingerprinting helped to confirm the identification of the bones of the last Russian czar and his family, who were assassinated in 1918.

**FIGURE 1** The Southern blot procedure, as applied to DNA fingerprinting. This procedure was named after Jeremy Southern, who developed the technique.

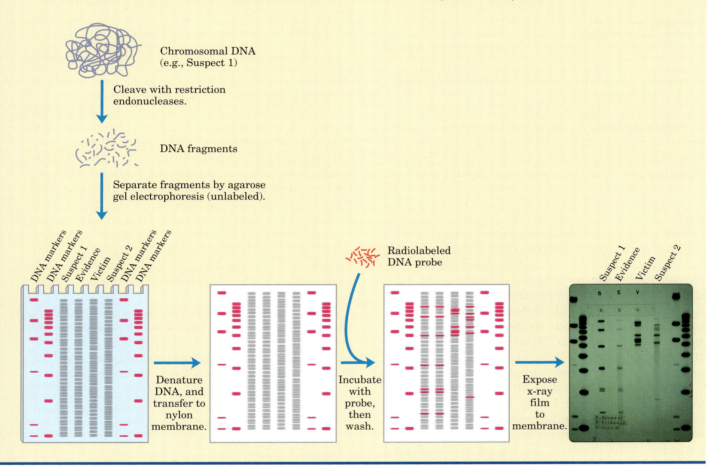

Francis S. Collins

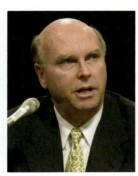
J. Craig Venter

out the genome at random. The sequenced segments were ordered by the computerized identification of sequence overlaps (with some reference to the public project's detailed physical map). At the outset of the Human Genome Project, shotgun sequencing on this scale had been deemed impractical. However, advances in computer software and sequencing automation had made the approach feasible by 1997. The ensuing race between the private and public sequencing efforts substantially advanced the timeline for completion of the project. Publication of the draft human genome sequence in 2001 was followed by two years of follow-up work to eliminate nearly a thousand discontinuities and

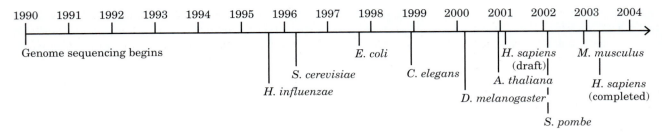

**FIGURE 9–18 Genomic sequencing timeline.** Discussions in the mid-1980s led to initiation of the project in 1989. Preparatory work, including extensive mapping to provide genome landmarks, occupied much of the 1990s. Separate projects were launched to sequence the genomes of other organisms important to research. The first sequencing efforts to be completed included many bacterial species (such as *Haemophilus influenzae*), yeast (*S. cerevisiae*), a nematode worm (*C. elegans*), the fruit fly (*D. melanogaster*), and a plant (*A. thaliana*). Completed sequences for mammalian genomes, including the human genome, began to emerge in 2000. Each genome project has a website that serves as a central repository for the latest data.

to provide high-quality sequence data that are contiguous throughout the genome.

The Human Genome Project marks the culmination of twentieth-century biology and promises a vastly changed scientific landscape for the new century. The human genome is only part of the story, as the genomes of many other species are also being (or have been) sequenced, including the yeasts *Saccharomyces cerevisiae* (completed in 1996) and *Schizosaccharomyces pombe* (2002), the nematode *Caenorhabditis elegans* (1998), the fruit fly *Drosophila melanogaster* (2000), the plant *Arabidopsis thaliana* (2000), the mouse *Mus musculus* (2002), zebrafish, and dozens of bacterial and archaebacterial species (Fig. 9–18). Most of the early efforts have been focused on species commonly used in laboratories. However, genome sequencing is destined to branch out to many other species as experience grows and technology improves. Broad efforts to map genes, attempts to identify new proteins and disease genes, and many other initiatives are currently under way.

The result is a database with the potential not only to fuel rapid advances in biology but to change the way that humans think about themselves. Early insights provided by the human genome sequence range from the intriguing to the profound. We are not as complicated as we thought. Decades-old estimates that humans possessed about 100,000 genes within the approximately $3.2 \times 10^9$ bp in the human genome have been supplanted by the discovery that we have only 30,000 to 35,000 genes. This is perhaps three times more genes than a fruit fly (with 13,000) and twice as many as a nematode worm (18,000). Although humans evolved relatively recently, the human genome is very old. Of 1,278 protein families identified in one early screen, only 94 were unique to vertebrates. However, while we share many protein domain types with plants, worms, and flies, we use these domains in more complex arrangements. Alternative modes of gene expression (Chapter 26) allow the production of more than one protein from a single gene—a process that humans and other vertebrates engage in more than do bacteria, worms, or any

other forms of life. This allows for greater complexity in the proteins generated from our gene complement.

We now know that only 1.1% to 1.4% of our DNA actually encodes proteins (Fig. 9–19). More than 50% of our genome consists of short, repeated sequences, the vast majority of which—about 45% of our genome in all—come from transposons, short movable DNA sequences that are molecular parasites (Chapter 25). Many of the transposons have been there a long time, now altered so that they can no longer move to new genomic locations. Others are still actively moving at low frequencies, helping to make the genome an ever-dynamic and evolving entity. At least a few transposons have been co-opted by their host and appear to serve useful cellular functions.

What does all this information tell us about how much one human differs from another? Within the human population are millions of single-base differences, called **single nucleotide polymorphisms,** or **SNPs** (pronounced "snips"). Each human differs from the next

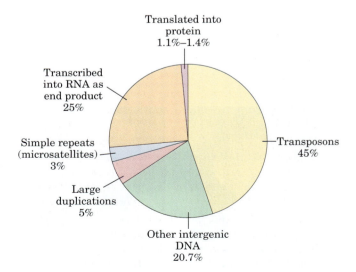

**FIGURE 9–19 Snapshot of the human genome.** The chart shows the proportions of our genome made up of various types of sequences.

by about 1 bp in every 1,000 bp. From these small genetic differences arises the human variety we are all aware of—differences in hair color, eyesight, allergies to medication, foot size, and even (to some unknown degree) behavior. Some of the SNPs are linked to particular human populations and can provide important information about human migrations that occurred thousands of years ago and about our more distant evolutionary past.

As spectacular as this advance is, the sequencing of the human genome is easy compared with what comes next—the effort to understand all the information in each genome. The genome sequences being added monthly to international databases are roadmaps, parts of which are written in a language we do not yet understand. However, they have great utility in catalyzing the discovery of new proteins and processes affecting every aspect of biochemistry, as will become apparent in chapters to come.

### SUMMARY 9.2   From Genes to Genomes

- The science of genomics broadly encompasses the study of genomes and their gene content.

- Genomic DNA segments can be organized in libraries—such as genomic libraries and cDNA libraries—with a wide range of designs and purposes.

- The polymerase chain reaction (PCR) can be used to amplify selected DNA segments from a DNA library or an entire genome.

- In an international cooperative research effort, the genomes of many organisms, including that of humans, have been sequenced in their entirety and are now available in public databases.

## 9.3 From Genomes to Proteomes

A gene is not simply a DNA sequence; it is information that is converted to a useful product—a protein or functional RNA molecule—when and if needed by the cell. The first and most obvious step in exploring a large sequenced genome is to catalog the products of the genes within that genome. Genes that encode RNA as their final product are somewhat harder to identify than are protein-encoding genes, and even the latter can be very difficult to spot in a vertebrate genome. The explosion of DNA sequence information has also revealed a sobering truth. Despite many years of biochemical advances, there are still thousands of proteins in every eukaryotic cell (and quite a few in bacteria) that we know nothing about. These proteins may have functions in processes not yet discovered, or may contribute in unexpected

ways to processes we think we understand. In addition, the genomic sequences tell us nothing about the three-dimensional structure of proteins or how proteins are modified after they are synthesized. The proteins, with their myriad critical functions in every cell, are now becoming the focus of new strategies for whole cell biochemistry.

The complement of proteins expressed by a genome is called its **proteome**, a term that first appeared in the research literature in 1995. This concept rapidly evolved into a separate field of investigation, called **proteomics.** The problem addressed by proteomics research is straightforward, although the solution is not. Each genome presents us with thousands of genes encoding proteins, and ideally we want to know the structure and function of all those proteins. Given that many proteins offer surprises even after years of study, the investigation of an entire proteome is a daunting enterprise. Simply discovering the function of new proteins requires intensive work. Biochemists can now apply shortcuts in the form of a broad array of new and updated technologies.

Protein function can be described on three levels. **Phenotypic function** describes the effects of a protein on the entire organism. For example, the loss of the protein may lead to slower growth of the organism, an altered development pattern, or even death. **Cellular function** is a description of the network of interactions engaged in by a protein at the cellular level. Interactions with other proteins in the cell can help define the kinds of metabolic processes in which the protein participates. Finally, **molecular function** refers to the precise biochemical activity of a protein, including details such as the reactions an enzyme catalyzes or the ligands a receptor binds.

For several genomes, such as those of the yeast *Saccharomyces cerevisiae* and the plant *Arabidopsis*, a massive effort is underway to inactivate each gene by genetic engineering and to investigate the effect on the organism. If the growth patterns or other properties of the organism change (or if it does not grow at all), this provides information on the phenotypic function of the protein product of the gene.

There are three other main paths to investigating protein function: (1) sequence and structural comparisons with genes and proteins of known function, (2) determination of when and where a gene is expressed, and (3) investigation of the interactions of the protein with other proteins. We discuss each of these approaches in turn.

### Sequence or Structural Relationships Provide Information on Protein Function

One of the important reasons to sequence many genomes is to provide a database that can be used to assign gene functions by genome comparisons, an enterprise referred to as **comparative genomics.** Sometimes a newly discovered gene is related by sequence homologies to a gene previously studied in another or the same species, and its function can be entirely or partly defined by that relationship. Such genes—of different species but possessing a clear sequence and functional relationship to each other—are called **orthologs.** Genes similarly related to each other within a single species are called **paralogs** (see Fig. 1–37). If the function of a gene has been characterized for one species, this information can be used to assign gene function to the ortholog found in the second species. The identity is easiest to make when comparing genomes from relatively closely related species, such as mouse and human, although many clearly orthologous genes have been identified in species as distant as bacteria and humans. Sometimes even the order of genes on a chromosome is conserved over large segments of the genomes of closely related species (Fig. 9–20). Conserved gene order, called **synteny,** provides additional evidence for an orthologous relationship between genes at identical locations within the related segments.

Alternatively, certain sequences associated with particular structural motifs (Chapter 4) may be identified within a protein. The presence of a structural motif may suggest that it, say, catalyzes ATP hydrolysis, binds to DNA, or forms a complex with zinc ions, helping to define molecular function. These relationships are determined with the aid of increasingly sophisticated computer programs, limited only by the current information on gene and protein structure and our capacity to associate sequences with particular structural motifs.

To further the assignment of function based on structural relationships, a large-scale structural proteomics project has been initiated. The goal is to crystallize and determine the structure of as many proteins and protein domains as possible, in many cases with little or no existing information about protein function. The project has been assisted by the automation of some of the tedious steps of protein crystallization (see Box 4–4). As these structures are revealed, they will be made available in the structural databases described in Chapter 4. The effort should help define the extent of variation in structural motifs. When a newly discovered protein is found to have structural folds that are clearly related to motifs with known functions in the databases, this information can suggest a molecular function for the protein.

## Cellular Expression Patterns Can Reveal the Cellular Function of a Gene

In every newly sequenced genome, researchers find genes that encode proteins with no evident structural relationships to known genes or proteins. In these cases, other approaches must be used to generate information about gene function. Determining which tissues a gene is expressed in, or what circumstances trigger the appearance of the gene product, can provide valuable clues. Many different approaches have been developed to study these patterns.

**Two-Dimensional Gel Electrophoresis** As shown in Figure 3–22, two-dimensional gel electrophoresis allows the separation and display of up to 1,000 different proteins on a single gel. Mass spectrometry (see Box 3–2) can then be used to partially sequence individual protein spots and assign each to a gene. The appearance and nonappearance (or disappearance) of particular protein spots in samples from different tissues, from similar tissues at different stages of development, or from tissues treated in ways that simulate a variety of biological conditions can help define cellular function.

**DNA Microarrays** Major refinements of the technology underlying DNA libraries, PCR, and hybridization have come together in the development of **DNA microarrays** (sometimes called **DNA chips**), which allow the rapid and simultaneous screening of many thousands of genes. DNA segments from known genes, a few dozen to hundreds of nucleotides long, are amplified by PCR and placed on a solid surface, using robotic devices that accurately deposit nanoliter quantities of DNA solution. Many thousands of such spots are deposited in a predesigned array on a surface area of just a few square centimeters. An alternative strategy is to synthesize DNA directly on the solid surface, using photolithography (Fig. 9–21). Once the chip is constructed, it can be probed with mRNAs or cDNAs from a particular cell type

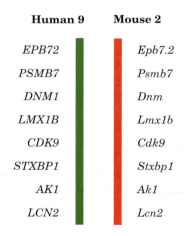

| Human 9 | Mouse 2 |
|---------|---------|
| EPB72 | Epb7.2 |
| PSMB7 | Psmb7 |
| DNM1 | Dnm |
| LMX1B | Lmx1b |
| CDK9 | Cdk9 |
| STXBP1 | Stxbp1 |
| AK1 | Ak1 |
| LCN2 | Lcn2 |

**FIGURE 9–20 Synteny in the mouse and human genomes.** Large segments of the mouse and human genomes have closely related genes aligned in the same order on chromosomes, a relationship called synteny. This diagram shows segments of human chromosome 9 and mouse chromosome 2. The genes in these segments exhibit a very high degree of homology as well as the same gene order. The different lettering schemes for the gene names reflect different naming conventions in the two organisms.

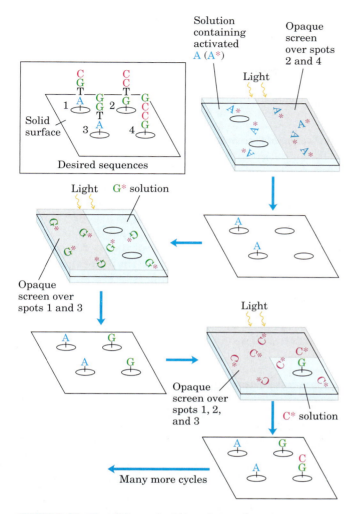

**FIGURE 9–21 Photolithography.** This technique for preparing a DNA microarray makes use of nucleotide precursors that are activated by light, joining one nucleotide to the next in a photoreaction (as opposed to the chemical process illustrated in Fig. 8–38). A computer is programmed with the oligonucleotide sequences to be synthesized at each point on a solid surface. The surface is washed successively with solutions containing one type of activated nucleotide (A*, G*, etc.). As in the chemical synthesis of DNA, the activated nucleotides are blocked so that only one can be added to a chain in each cycle. A screen covering the surface is opened over the areas programmed to receive a particular nucleotide, and a flash of light joins the nucleotide to the polymers in the uncovered areas. This continues until the required sequences are built up on each spot on the surface. Many polymers with the same sequence are generated on each spot, not just the single polymer shown. Also, the surfaces have thousands of spots with different sequences (see Fig. 9–22); this array shows just four spots, to illustrate the strategy.

or cell culture to identify the genes being expressed in those cells.

A microarray can answer such questions as which genes are expressed at a given stage in the development of an organism. The total complement of mRNA is isolated from cells at two different stages of development

and converted to cDNA, using reverse transcriptase and fluorescently labeled deoxynucleotides. The fluorescent cDNAs are then mixed and used as probes, each hybridizing to complementary sequences on the microarray. In Figure 9–22, for example, the labeled nucleotides used to make the cDNA for each sample fluoresce in two different colors. The cDNA from the two samples is mixed and used to probe the microarray. Spots that fluoresce green represent mRNAs more abundant at the single-cell stage; those that fluoresce red represent sequences more abundant later in development. The mRNAs that are equally abundant at both stages of development fluoresce yellow. By using a mixture of two samples to measure relative rather than absolute abundance of sequences, the method corrects for variations in the amounts of DNA originally deposited in each spot on the grid and other possible inconsistencies among spots in the microarray. The spots that fluoresce provide a snapshot of all the genes being expressed in the cells at the moment they were harvested—gene expression examined on a genome-wide scale. For a gene of unknown function, the time and circumstances of its expression can provide important clues about its role in the cell.

An example of this technique is illustrated in Figure 9–23, showing the dramatic results this technique can produce. Segments from each of the more than 6,000 genes in the completely sequenced yeast genome were separately amplified by PCR, and each segment was deposited in a defined pattern to create the illustrated microarray. In a sense, this array provides a snapshot of the entire yeast genome.

**Protein Chips** Proteins, too, can be immobilized on a solid surface and used to help define the presence or absence of other proteins in a sample. For example, researchers prepare an array of antibodies to particular proteins by immobilizing them as individual spots on a solid surface. A sample of proteins is added, and if the protein that binds any of the antibodies is present in the sample, it can be detected by a solid-state form of the ELISA assay (see Fig. 5–28). Many other types and applications of protein chips are being developed.

### Detection of Protein-Protein Interactions Helps to Define Cellular and Molecular Function

A key to defining the function of any protein is to determine what it binds to. In the case of protein-protein interactions, the association of a protein of unknown function with one whose function is known can provide a useful and compelling "guilt by association." The techniques used in this effort are quite varied.

**Comparisons of Genome Composition** Although not evidence of direct association, the mere presence of combinations of genes in particular genomes can hint at

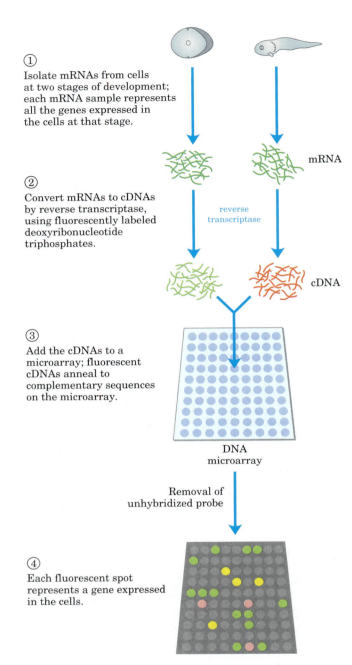

① Isolate mRNAs from cells at two stages of development; each mRNA sample represents all the genes expressed in the cells at that stage.

mRNA

② Convert mRNAs to cDNAs by reverse transcriptase, using fluorescently labeled deoxyribonucleotide triphosphates.

reverse transcriptase

cDNA

③ Add the cDNAs to a microarray; fluorescent cDNAs anneal to complementary sequences on the microarray.

DNA microarray

Removal of unhybridized probe

④ Each fluorescent spot represents a gene expressed in the cells.

**FIGURE 9–22  DNA microarray.** A microarray can be prepared from any known DNA sequence, from any source, generated by chemical synthesis or by PCR. The DNA is positioned on a solid surface (usually specially treated glass slides) with the aid of a robotic device capable of depositing very small (nanoliter) drops in precise patterns. UV light cross-links the DNA to the glass slides. Once the DNA is attached to the surface, the microarray can be probed with other fluorescently labeled nucleic acids. Here, mRNA samples are collected from cells at two different stages in the development of a frog. The cDNA probes are made with nucleotides that fluoresce in different colors for each sample; a mixture of the cDNAs is used to probe the microarray. Green spots represent mRNAs more abundant at the single-cell stage; red spots, sequences more abundant later in development. The yellow spots indicate approximately equal abundance at both stages.  🖱 **Synthesizing an Oligonucleotide Array**

protein function. We can simply search the genomic databases for particular genes, then determine what other genes are present in the same genomes (Fig. 9–24). When two genes always appear together in a genome, it suggests that the proteins they encode may be functionally related. Such correlations are most useful if the function of at least one of the proteins is known.

**Purification of Protein Complexes**    With the construction of cDNA libraries in which each gene is contiguous with (fused to) an epitope tag, workers can immunoprecipitate the protein product of a gene by using the antibody that binds to the epitope (Fig. 9–15b). If the tagged protein is expressed in cells, other proteins that bind to it may also be precipitated with it. Identification of the associated proteins reveals some of the protein-protein interactions of the tagged protein. There are many variations of this process. For example, a crude extract of cells that express a similarly tagged protein is added to a column containing immobilized antibody. The tagged protein binds to the antibody, and proteins that interact with the tagged protein are sometimes also retained

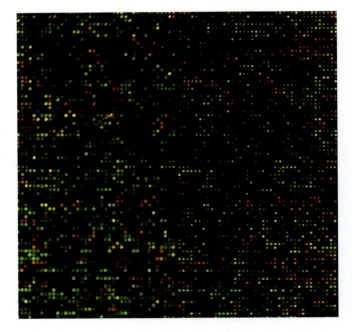

**FIGURE 9–23  Enlarged image of a DNA microarray.** Each glowing spot in this microarray contains DNA from one of the 6,200 genes of the yeast (*S. cerevisiae*) genome, with every gene represented in the array. The microarray has been probed with fluorescently labeled nucleic acid derived from the mRNAs obtained (1) when the cells were growing normally in culture and (2) five hours after the cells began to form spores. The green spots represent genes expressed at higher levels during normal growth; the red spots, genes expressed at higher levels during sporulation. The yellow spots represent genes that do not change their levels of expression during sporulation. This image is enlarged; the microarray actually measures only 1.8 × 1.8 cm.  🖱 **Screening Oligonucleotide Array for Patterns of Gene Expression**

|         | Species |     |     |     |
| :-----: | :-----: | :-: | :-: | :-: |
| Protein |    1    |  2  |  3  |  4  |
|   P1    |    +    |  −  |  +  |  +  |
|   P2    |    −    |  −  |  +  |  −  |
|   P3    |    +    |  +  |  −  |  +  |
|   P4    |    +    |  −  |  +  |  −  |
|   P5    |    +    |  −  |  −  |  −  |
|   P6    |    +    |  +  |  −  |  +  |
|   P7    |    +    |  +  |  +  |  −  |

**FIGURE 9–24  Use of comparative genomics to identify functionally related genes.** One use of comparative genomics is to prepare phylogenetic profiles in order to identify genes that always appear together in a genome. This example shows a comparison of genes from four organisms, but in practice, computer searches can look at dozens of species. The designations P1, P2, and so forth refer to proteins encoded by each species. This technique does not require homologous proteins. In this example, because proteins P3 and P6 always appear together in a genome they may be functionally related. Further testing would be needed to confirm this inference.

on the column. The connection between the protein and the tag is cleaved with a specific protease, and the protein complexes are eluted from the column and analyzed. Researchers can use these methods to define complex networks of interactions within a cell.

A variety of useful protein tags are available. A common one is a histidine tag, often just a string of six His residues. A poly-His sequence binds quite tightly to metals such as nickel. If a protein is cloned so that its sequence is contiguous with a His tag, it will have the extra His residues at its carboxyl terminus. The protein can then be purified by chromatography on columns with immobilized nickel. These procedures are convenient but require caution, because the additional amino acid residues in an epitope or His tag can affect protein activity.

**Yeast Two-Hybrid Analysis**  A sophisticated genetic approach to defining protein-protein interactions is based on the properties of the Gal4 protein (Gal4p), which activates transcription of certain genes in yeast (see Fig. 28–28). Gal4p has two domains, one that binds to a specific DNA sequence and another that activates the RNA polymerase that synthesizes mRNA from an adjacent reporter gene. The domains are stable when separated, but activation of the RNA polymerase requires interaction with the activation domain, which in turn requires positioning by the DNA-binding domain. Hence, the domains must be brought together to function correctly (Fig. 9–25a).

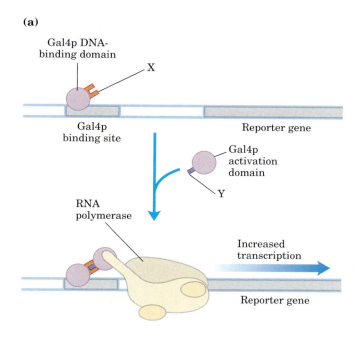

**(a)**

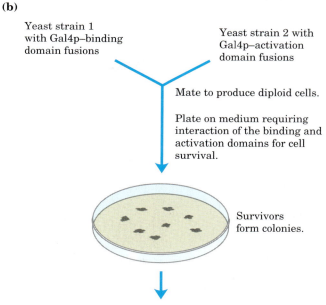

**(b)**

**FIGURE 9–25  The yeast two-hybrid system. (a)** In this system for detecting protein-protein interactions, the aim is to bring together the DNA-binding domain and the activation domain of the yeast Gal4 protein through the interaction of two proteins, X and Y, to which each domain is fused. This interaction is accompanied by the expression of a reporter gene. **(b)** The two fusions are created in separate yeast strains, which are then mated. The mated mixture is plated on a medium on which the yeast cannot survive unless the reporter gene is expressed. Thus, all surviving colonies have interacting protein fusion pairs. Sequencing of the fusion proteins in the survivors reveals which proteins are interacting.  **Yeast Two-Hybrid Systems**

In this method, the protein-coding regions of genes to be analyzed are fused to the coding sequences of either the DNA-binding domain or the activation domain of Gal4p, and the resulting genes express a series of fusion proteins. If a protein fused to the DNA-binding domain interacts with a protein fused to the activation domain, transcription is activated. The reporter gene transcribed by this activation is generally one that yields a protein required for growth, or is an enzyme that catalyzes a reaction with a colored product. Thus, when grown on the proper medium, cells that contain such a pair of interacting proteins are easily distinguished from those that do not. Typically, many genes are fused to the Gal4p DNA-binding domain gene in one yeast strain, and many other genes are fused to the Gal4p activation domain gene in another yeast strain, then the yeast strains are mated and individual diploid cells grown into colonies (Fig. 9–25b). This allows for large-scale screening for proteins that interact in the cell.

All these techniques provide important clues to protein function. However, they do not replace classical biochemistry. They simply provide researchers with an expedited entrée into important new biological problems. In the end, a detailed functional understanding of any new protein requires traditional biochemical analyses—such as were used for the many well-studied proteins described in this text. When paired with the simultaneously evolving tools of biochemistry and molecular biology, genomics and proteomics are speeding the discovery not only of new proteins but of new biological processes and mechanisms.

### SUMMARY 9.3 From Genomes to Proteomes

- A proteome is the complement of proteins produced by a cell's genome. The new field of proteomics encompasses an effort to catalog and determine the functions of all the proteins in a cell.

- One of the most effective ways to determine the function of a new gene is by comparative genomics, the search of databases for genes with similar sequences. Paralogs and orthologs are proteins (and their genes) with clear functional and sequence relationships in the same or in different species. In some cases, the presence of a gene in combination with certain other genes, observed as a pattern in several genomes, can point toward a possible function.

- Cellular proteomes can be displayed by two-dimensional gel electrophoresis and explored with the aid of mass spectrometry.

- The cellular function of a protein can sometimes be inferred by determining when

and where its gene is expressed. Researchers use DNA microarrays (chips) and protein chips to explore gene expression at the cellular level.

- Several new techniques, including comparative genomics, immunoprecipitation, and yeast two-hybrid analysis, can identify protein-protein interactions. These interactions provide important clues to protein function.

## 9.4 Genome Alterations and New Products of Biotechnology

We don't need to look far to find practical applications for the new biotechnologies or to find new opportunities for breakthroughs in basic research. Herein lie both the promise and the challenge of genomics. As our knowledge of the genome increases, we will improve our understanding of every aspect of biological function. We will enhance our capacity to engineer organisms and produce new pharmaceutical agents and, as a consequence, will improve human nutrition and health. This promise can be realized only if practical safeguards are in place to ensure responsible application of these techniques.

### A Bacterial Plant Parasite Aids Cloning in Plants

We not only can understand genomes, we can change them. This is perhaps the ultimate manifestation of the new technologies. The introduction of recombinant DNA into plants has enormous implications for agriculture, making possible the alteration of the nutritional profile or yield of crops or their resistance to environmental stresses, such as insect pests, diseases, cold, salinity, and drought. Fertile plants of some species may be generated from a single transformed cell, so that an introduced gene passes to progeny through the seeds.

As yet, researchers have not found any naturally occurring plant cell plasmids to facilitate cloning in plants, so the biggest technical challenge is getting DNA into plant cells. An important and adaptable ally in this effort is the soil bacterium *Agrobacterium tumefaciens*. This bacterium can invade plants at the site of a wound, transform nearby cells, and induce them to form a tumor called a crown gall. *Agrobacterium* contains the large (200,000 bp) **Ti plasmid** (Fig. 9–26a). When the bacterium is in contact with a damaged plant cell, a 23,000 bp segment of the Ti plasmid called T DNA is transferred from the plasmid and integrated at a random position in one of the plant cell chromosomes (Fig. 9–26b). The transfer of T DNA from *Agrobacterium* to the plant cell chromosome depends on two 25 bp repeats that flank the T DNA and on the products of the virulence (*vir*) genes on the Ti plasmid (Fig. 9–26a).

The T DNA encodes enzymes that convert plant metabolites to two classes of compounds that benefit

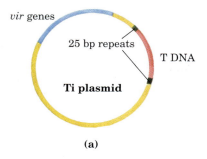

**(a)**

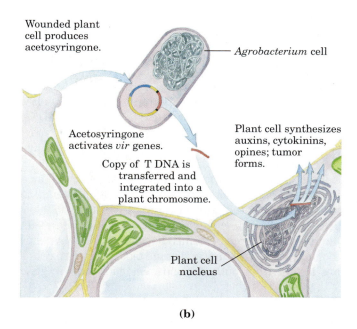

**(b)**

**FIGURE 9-26 Transfer of DNA to plant cells by a bacterial parasite.** **(a)** The Ti (tumor-inducing) plasmid of *Agrobacterium tumefaciens*. **(b)** Wounded plant cells produce and release the phenolic compound acetosyringone. When *Agrobacterium* detects this compound, the virulence (*vir*) genes on the Ti plasmid are expressed. The *vir* genes encode enzymes needed to introduce the T DNA segment of the Ti plasmid into the genome of nearby plant cells. A single-stranded copy of the T DNA is synthesized and transferred to the plant cell, where it is converted to duplex DNA and integrated into a plant cell chromosome. The T DNA encodes enzymes that synthesize both plant growth hormones and opines (see Fig. 9–27); the latter compounds are metabolized (as a nutrient source) only by *Agrobacterium*. Expression of the T DNA genes by transformed plant cells thus leads to both aberrant plant cell growth (tumor formation) and the diversion of plant cell nutrients to the invading bacteria.

the bacterium (Fig. 9–27). The first group consists of plant growth hormones (auxins and cytokinins) that stimulate growth of the transformed plant cells to form the crown gall tumor. The second constitutes a series of unusual amino acids called opines, which serve as a food source for the bacterium. The opines are produced in high concentrations in the tumor cells and secreted to the surroundings, where they can be metabolized only by *Agrobacterium*, using enzymes encoded elsewhere on the Ti plasmid. The bacterium thereby diverts plant resources by converting them to a form that benefits only itself.

**FIGURE 9-27 Metabolites produced in *Agrobacterium*-infected plant cells.** Auxins and cytokinins are plant growth hormones. The most common auxin, indoleacetate, is derived from tryptophan. Cytokinins are adenine derivatives. Opines generally are derived from amino acid precursors; at least 14 different opines are produced by enzymes encoded by the Ti plasmids of different *Agrobacterium* species.

This rare example of DNA transfer from a prokaryote to a eukaryotic cell is a natural genetic engineering process—one that researchers can harness to transfer recombinant DNA (instead of T DNA) to the plant genome. A common cloning strategy employs an *Agrobacterium* with two different recombinant plasmids. The first is a Ti plasmid from which the T DNA segment has been removed in the laboratory (Fig. 9–28a). The second is an *Agrobacterium–E. coli* shuttle vector in which the 25 bp repeats of the T DNA flank a foreign gene that the researcher wants to introduce into the plant cell, along with a selectable marker such as resistance to the antibiotic kanamycin (Fig. 9–28b). The engineered *Agrobacterium* is used to infect a leaf, but crown galls are not formed because the T DNA genes for the auxin, cytokinin, and opine biosynthetic enzymes are absent from both plasmids. Instead, the *vir* gene

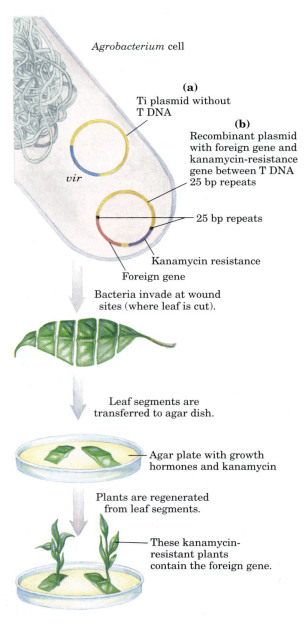

**(a)**
Ti plasmid without
T DNA

*Agrobacterium* cell

*vir*

**(b)**
Recombinant plasmid
with foreign gene and
kanamycin-resistance
gene between T DNA
25 bp repeats

25 bp repeats

Kanamycin resistance

Foreign gene

Bacteria invade at wound
sites (where leaf is cut).

Leaf segments are
transferred to agar dish.

Agar plate with growth
hormones and kanamycin

Plants are regenerated
from leaf segments.

These kanamycin-
resistant plants
contain the foreign gene.

products from the altered Ti plasmid direct the transformation of the plant cells by the foreign gene—the gene flanked by the T DNA 25 bp repeats in the second plasmid. The transformed plant cells can be selected by growth on agar plates that contain kanamycin, and addition of growth hormones induces the formation of new plants that contain the foreign gene in every cell.

The successful transfer of recombinant DNA into plants was vividly illustrated by an experiment in which the luciferase gene from fireflies was introduced into the cells of a tobacco plant (Fig. 9–29)—a favorite plant for transformation experiments because its cells are particularly easy to transform with *Agrobacterium*. The potential of this technology is not limited to the production of glow-in-the-dark plants, of course. The same approach has been used to produce crop plants that are resistant to herbicides, plant viruses, and insect pests (Fig. 9–30). Potential benefits include increased yields and less need for environmentally harmful agricultural chemicals.

Biotechnology can introduce new traits into a plant much faster than traditional methods of plant breeding. A prominent example is the development of soybeans that are resistant to the general herbicide glyphosate (the active ingredient in the product RoundUp). Glyphosate breaks down rapidly in the environment (glyphosate-sensitive plants can be planted in a treated area after as little as 48 hours), and its use does not generally lead to contamination of groundwater or carryover from one year to the next. A field of glyphosate-resistant soybeans can be treated once with glyphosate during a summer growing season to eliminate essentially all weeds in the field, while leaving the soybeans unaffected (Fig. 9–31). Potential pitfalls of the technology, such as the evolution of glyphosate-resistant weeds or the escape of difficult-to-control recombinant plants, remain a concern of researchers and the public.

**FIGURE 9–28 A two-plasmid strategy to create a recombinant plant.** **(a)** One plasmid is a modified Ti plasmid that contains the *vir* genes but lacks T DNA. **(b)** The other plasmid contains a segment of DNA that bears both a foreign gene (the gene of interest, e.g., the gene for the insecticidal protein described in Fig. 9–30) and an antibiotic-resistance element (here, kanamycin resistance), flanked by the two 25 bp repeats of T DNA that are required for transfer of the plasmid genes to the plant chromosome. This plasmid also contains the replication origin needed for propagation in *Agrobacterium*.

When bacteria invade at the site of a wound (the edge of the cut leaf), the *vir* genes on the first plasmid mediate transfer into the plant genome of the segment of the second plasmid that is flanked by the 25 bp repeats. Leaf segments are placed on an agar dish that contains both kanamycin and appropriate levels of plant growth hormones, and new plants are generated from segments with the transformed cells. Nontransformed cells are killed by the kanamycin. The foreign gene and the antibiotic-resistance element are normally transferred together, so plant cells that grow in this medium generally contain the foreign gene.

FIGURE 9–29 **A tobacco plant expressing the gene for firefly luciferase.** Light was produced after the plant was watered with a solution containing luciferin, the substrate for the light-producing luciferase enzyme (see Box 13–2). Don't expect glow-in-the-dark ornamental plants at your local plant nursery anytime soon. The light is actually quite weak; this photograph required a 24-hour exposure. The real point—that this technology allows the introduction of new traits into plants—is nevertheless elegantly made.

FIGURE 9–30 **Tomato plants engineered to be resistant to insect larvae.** Two tomato plants were exposed to equal numbers of moth larvae. The plant on the left has not been genetically altered. The plant on the right expresses a gene for a protein toxin derived from the bacterium *Bacillus thuringiensis*. This protein, introduced by a protocol similar to that depicted in Figure 9–28, is toxic to the larvae of some moth species while being harmless to humans and other organisms. Insect resistance has also been genetically engineered in cotton and other plants.

(a)

(b)

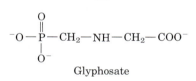

Glyphosate

FIGURE 9–31 **Glyphosate-resistant soybean plants.** The photographs show two areas of a soybean field in Wisconsin. **(a)** Without glyphosate treatment, this part of the field is overrun with weeds. **(b)** Glyphosate-resistant soybean plants thrive in the glyphosate-treated section of the field. Glyphosate breaks down rapidly in the environment. Agricultural use of engineered plants such as these proceeds only after considerable deliberation, balancing the extraordinary promise of the technology with the need to select new traits with care. Both science and society as a whole have a stake in ensuring that the use of the resultant plants has no adverse impact on the environment or on human health.

## Manipulation of Animal Cell Genomes Provides Information on Chromosome Structure and Gene Expression

The transformation of animal cells by foreign genetic material offers an important mechanism for expanding our knowledge of the structure and function of animal genomes, as well as for the generation of animals with new traits. This potential has stimulated intensive research into more-sophisticated means of cloning animals.

Most work of this kind requires a source of cells into which DNA can be introduced. Although intact tissues are often difficult to maintain and manipulate in vitro, many types of animal cells can be isolated and grown in the laboratory if their growth requirements are carefully met. Cells derived from a particular animal tissue and grown under appropriate **tissue culture** conditions can maintain their differentiated properties (for example, a hepatocyte (liver cell) remains a hepatocyte) for weeks or even months.

No suitable plasmidlike vector is available for introducing DNA into an animal cell, so transformation usually requires the integration of the DNA into a host-cell chromosome. The efficient delivery of DNA to a cell nucleus and integration of this DNA into a chromosome without disrupting any critical genes remain the major technical problems in the genetic engineering of animal cells.

Available methods for carrying DNA into an animal cell vary in efficiency and convenience. Some success has been achieved with spontaneous uptake of DNA or electroporation, techniques roughly comparable to the common methods used to transform bacteria. They are inefficient in animal cells, however, transforming only 1 in 100 to 10,000 cells. **Microinjection**—the injection of DNA directly into a nucleus, using a very fine needle—has a high success rate for skilled practitioners, but the total number of cells that can be treated is small, because each must be injected individually.

The most efficient and widely used methods for transforming animal cells rely on liposomes or viral vectors. Liposomes are small vesicles consisting of a lipid bilayer that encloses an aqueous compartment (see Fig. 11–4). Liposomes that enclose a recombinant DNA molecule can be fused with the membranes of target cells to deliver DNA into the cell. The DNA sometimes reaches the nucleus, where it can integrate into a chromosome (mostly at random locations). **Viral vectors** are even more efficient at delivering DNA. Animal viruses have effective mechanisms for introducing their nucleic acids into cells, and several types also have mechanisms to integrate their DNA into a host-cell chromosome. Some of these, such as retroviruses (see Fig. 26–30) and adenoviruses, have been modified to serve as viral vectors to introduce foreign DNA into mammalian cells.

The work on retroviral vectors illustrates some of the strategies being used (Fig. 9–32). When an engineered retrovirus enters a cell, its RNA genome is transcribed to DNA by reverse transcriptase and then integrated into the host genome by the enzyme viral integrase. Special regions of DNA are required for this

procedure: long terminal repeat (LTR) sequences to integrate retroviral DNA into the host chromosome and the ψ (psi) sequence to package the viral RNA in viral particles (see Fig. 26–30).

The *gag*, *pol*, and *env* genes of the retroviral genome, required for retroviral replication and assembly of viral particles, can be replaced with foreign DNA. To assemble viruses that contain the recombinant genetic information, researchers must introduce the DNA into cultured cells that are simultaneously infected with a "helper virus" that has the genes to produce viral particles but lacks the ψ sequence required for packaging. Thus the recombinant DNA can be transcribed and its

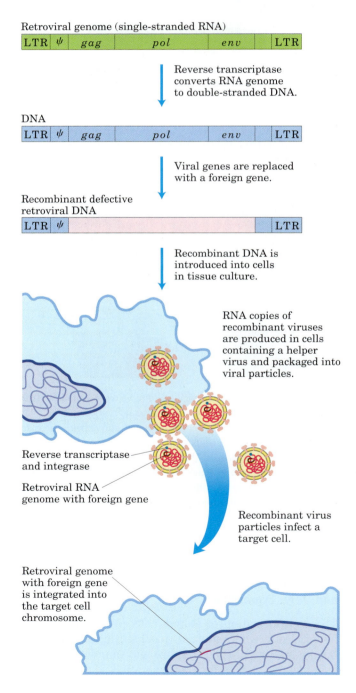

**FIGURE 9–32 Use of retroviral vectors in mammalian cell cloning.** A typical retroviral genome (somewhat simplified here), engineered to carry a foreign gene (pink), is added to a host-cell tissue culture. The helper virus (not shown) lacks the packaging sequence, ψ, so its RNA transcripts cannot be packaged into viral particles, but it provides the *gag*, *pol*, and *env* gene products needed to package the engineered retrovirus into functional viral particles. This enables the foreign gene in the recombinant retroviral genome to be introduced efficiently into the target cells.

RNA packaged into viral particles. These particles can act as vectors to introduce the recombinant RNA into target cells. Viral reverse transcriptase and integrase enzymes (produced by the helper virus) are also packaged in the viral particle and introduced into the target cells. Once the engineered viral genome is inside a cell, these enzymes create a DNA copy of the recombinant viral RNA genome and integrate it into a host chromosome. The integrated recombinant DNA then becomes a permanent part of the target cell's chromosome and is replicated with the chromosome at every cell division. The cell itself is not endangered by the integrated viral DNA, because the recombinant virus lacks the genes needed to produce RNA copies of its genome and package them into new viral particles. The use of recombinant retroviruses is often the best method for introducing DNA into large numbers of mammalian cells.

Each type of virus has different attributes, so several classes of animal viruses are being engineered as vectors to transform mammalian cells. Adenoviruses, for example, lack a mechanism for integrating DNA into a chromosome. Recombinant DNA introduced via an adenoviral vector is therefore expressed for only a short time and then destroyed. This can be useful if the objective is transient expression of a gene.

Transformation of animal cells by any of the above techniques has its problems. Introduced DNA is generally integrated into chromosomes at random locations. Even when the foreign DNA contains a sequence similar to a sequence in a host chromosome, allowing targeting to that position, nonhomologous integrants still outnumber the targeted ones by several orders of magnitude. If these integration events disrupt essential genes, they can sometimes alter cellular functions (most cells are diploid or polyploid, however, so an integration usually leaves at least one unaffected copy of any given gene). A particularly poor outcome would involve an integration event that inadvertently activated a gene that stimulated cell division, potentially creating a cancer cell. Although such an event was once thought to be rare, recent trials suggest it is a significant hazard (Box 9–2). Finally, the site of an integration can determine the level of expression of the integrated gene, because integrants are not transcribed equally well everywhere in the genome.

Despite these challenges, the transformation of animal cells has been used extensively to study chromosome structure and the function, regulation, and expression of genes. The successful introduction of recombinant DNA into an animal can be illustrated by an experiment that permanently altered an easily observable inheritable physical trait. Microinjection of DNA into the nuclei of fertilized mouse eggs can produce efficient transformation (chromosomal integration). When the injected eggs are introduced into a female mouse and allowed to develop, the new gene is often expressed in some of the newborn mice. Those in which the germ line has been

**FIGURE 9–33 Cloning in mice.** The gene for human growth hormone was introduced into the genome of the mouse on the right. Expression of the gene resulted in the unusually large size of this mouse.

altered can be identified by testing *their* offspring. By careful breeding of these mice, researchers can establish a **transgenic** mouse line in which all the mice are homozygous for the new gene or genes. This technology was used to introduce into mice the gene for human growth hormone, under the control of an inducible promoter. When the mice were fed a diet that included the inducer, some of the mice that developed from injected embryos grew to an unusually large size (Fig. 9–33). Transgenic mice have now been produced with a wide range of genetic variations, including many relevant to human diseases and their control, pointing the way to human gene therapy (Box 9–2). A very similar approach is used to generate mice in which a particular gene has been inactivated ("knockout mice"), a way of establishing the function of the inactivated gene. 🖱 **Creating a Transgenic Mouse**

## New Technologies Promise to Expedite the Discovery of New Pharmaceuticals

It is difficult to summarize all the ways in which genomics and proteomics might affect the development of pharmaceutical agents, but a few examples illustrate the potential. Hypertension, congestive heart failure, hypercholesterolemia, and obesity are treated by pharmaceutical drugs that alter human physiology. Therapies are arrived at by identifying an enzyme or receptor involved in the process and discovering an inhibitor that interferes with its action. Proteomics will play an increasing role in identifying such potential drug targets. For example, the most potent vasoconstrictor known is the peptide hormone urotensin II. First discovered in fish spinal fluid, urotensin II is a small cyclic peptide, with 11 amino acid residues in humans and 12 or 13 in some other organisms. The vasoconstriction it induces can cause or exacerbate hypertension, congestive heart failure, and coronary artery disease. Some of the methods described in Section 9.3 for elucidating

BOX 9–2   BIOCHEMISTRY IN MEDICINE

## The Human Genome and Human Gene Therapy

As biotechnology gained momentum in the 1980s, a rational approach to the treatment of genetic diseases became increasingly attractive. In principle, DNA can be introduced into human cells to correct inherited genetic deficiencies. Genetic correction may even be targeted to a specific tissue by inoculating an individual with a genetically engineered, tissue-specific virus carrying a payload of DNA to be incorporated into deficient cells. The goal is entrancing, but the research path is strewn with impediments.

Altering chromosomal DNA entails substantial risk—a risk that cannot be quantified in the early stages of discovery. Consequently, early efforts at human gene therapy were directed at only a small subset of genetic diseases. Panels of scientists and ethicists developed a list of several conditions that should be satisfied to justify the risk involved, including the following. (1) The genetic defect must be a well-characterized, single-gene disorder. (2) Both the mutant and the normal gene must be cloned and sequenced. (3) In the absence of a technique for eliminating the existing mutant gene, the functional gene must function well in the presence of the mutant gene. (4) Finally, and most important, the risks inherent in a new technology must be outweighed by the seriousness of the disease. Protocols for human clinical trials were submitted by scientists in several nations and reviewed for scientific rigor and ethical compliance by carefully selected advisory panels in each country; then human trials commenced.

Early targets of gene therapy included cancer and genetic diseases affecting the immune system. Immunity is mediated by leukocytes (white blood cells) of several different types, all arising from undifferentiated stem cells in the bone marrow. These cells divide quickly and have special metabolic requirements. Differentiation can become blocked in several ways, resulting in a condition called severe combined immune deficiency (SCID). One form of SCID results from genetically inherited defects in the gene encoding adenosine deaminase (ADA), an enzyme involved in nucleotide biosynthesis (discussed in Chapter 22). Another form of SCID arises from a defect in a cell-surface receptor protein that binds chemical signals called cytokines, which trigger differentiation. In both cases, the progenitor stem cells cannot differentiate into the mature immune system cells, such as T and B lymphocytes (Chapter 5). Children with these rare human diseases are highly susceptible to bacterial and viral infections, and often suffer from a range of related physiological and neurological problems. In the absence of an effective therapy, the children must be confined in a sterile environment. About 20% of these children have a human leukocyte antigen (HLA)–identical sibling who can serve as a bone marrow transplant donor, a procedure that can cure the disease. The remaining children need a different approach.

The first human gene therapy trial was carried out at the National Institutes of Health in Bethesda, Maryland, in 1990. The patient was a four-year-old girl crippled by ADA deficiency. Bone marrow cells from the child were transformed with an engineered retrovirus containing a functional ADA gene; when the alteration of cells is done in this way—in the laboratory rather than in the living patient—the procedure is said to be done ex vivo. The treated cells were reintroduced into the patient's marrow. Four years later, the child was leading a normal life, going to school, and even testifying about her experiences before Congress. However, her recovery cannot be uniquely attributed to gene therapy. Before the gene therapy clinical trials began, researchers had developed a new treatment for ADA deficiency, in which synthetic ADA was administered in a complex with polyethylene glycol (PEG). For many ADA-SCID patients, injection of the ADA-PEG complex allowed some immune system development, with weight gain and reduced infection, although not full immune reconstitution. The new gene therapy was risky, and withdrawing the inoculation treatment from patients in the gene therapy trial was judged unethical. So trial participants received both treatments at once, making it unclear which treatment was primarily responsible for the positive clinical outcome. Nevertheless, the clinical trial provided important information: it was feasible to transfer genes ex vivo to large numbers of leukocytes, and cells bearing the transferred gene were still detectable years after treatment, suggesting that long-term correction was possible. In addition, the risk associated with use of the retroviral vectors appeared to be low.

Through the 1990s, hundreds of human gene therapy clinical trials were carried out, targeting a variety of genetic diseases, but the results in most cases were

protein-protein interactions have been used to demonstrate that urotensin II is bound by a G-protein-coupled receptor called GPR14. As we shall see in Chapter 12, G proteins play an important role in many signaling pathways. However, GPR14 was an "orphan" receptor, in that human genome sequencing had identified it as a G-protein-coupled receptor, but with no known function. The association of urotensin II with GPR14 now makes the latter protein a key target for drug therapies aimed at interfering with the action of urotensin II.

discouraging. One major impediment proved to be the inefficiency of introducing new genes into cells. Transformation failed in many cells, and the number of transformed cells often proved insufficient to reverse the disorder. In the ADA trials, achieving a sufficient population of transformed cells was particularly difficult, because of the ongoing ADA-PEG therapy. Normally, stem cells with the correct ADA gene would have a growth advantage over the untreated cells, expanding their population and gradually predominating in the bone marrow. However, the injections of ADA-PEG in the same patients allowed the untransformed (ADA-deficient) cells to live and develop, and the transformed cells did not have the needed growth advantage to expand their population at the expense of the others.

A gene therapy trial initiated in 1999 was successful in correcting a form of SCID caused by defective cytokine receptors (in particular a subunit called $\gamma c$), as reported in 2000 by physician researchers in France, Italy, and Britain. These researchers introduced the corrected gene for the $\gamma c$ cytokine–receptor subunit into $CD34^+$ cells. (The stem cells that give rise to immune system cells have a protein called CD34 on their surface; these cells can be separated from other bone marrow cells by antibodies to CD34.) The transformed cells were placed back into the patients' bone marrow. In this trial, introduction of the corrected gene clearly conferred a growth advantage over the untreated cells. A functioning immune system was detected in four of the first five patients within 6 to 12 weeks, and levels of mature immune system T lymphocytes reached the levels found in age-matched control subjects (who did not have SCID) within 6 to 8 months. Immune system function was restored, and nearly 4 years later (mid-2003) most of the children are leading normal lives. Similar results have been obtained with four additional patients. This provided dramatic confirmation that human gene therapy could cure a serious genetic disease.

In early 2003 came a setback. One of the original four patients who had received cells with the correct cytokine receptor gene developed a severe form of leukemia. During the gene therapy treatment, one of the introduced retroviruses had by chance inserted itself into a chromosome of one $CD34^+$ cell, resulting in abnormally high expression of a gene called LMO-2. The affected cell differentiated into an immune system T cell, and the elevated expression of LMO-2 led to uncontrolled growth of the cell, giving rise to the leukemia. As of mid-2003 the patient had responded well to chemotherapy, but there may be more chapters to write. The incident shows that early worries about the risk associated with retroviral vectors were well founded. After a review of the gene therapy trial protocols, including consultations with ethicists and parents of children affected by these diseases, further gene therapy trials are still planned for children who are not candidates for bone marrow transplants. The reason is simple enough. The potential benefit to the children with these debilitating conditions has been judged to outweigh the demonstrated risk.

Human gene therapy is not limited to genetic diseases. Cancer cells are being targeted by delivering genes for proteins that might destroy the cell or restore the normal control of cell division. Immune system cells associated with tumors, called tumor-infiltrating lymphocytes, can be genetically modified to produce tumor necrosis factor (TNF; see Fig. 12–50). When these lymphocytes are taken from a cancer patient, modified, and reintroduced, the engineered cells target the tumor, and the TNF they produce causes tumor shrinkage. AIDS may also be treatable with gene therapy; DNA that encodes an RNA molecule complementary to a vital HIV mRNA could be introduced into immune system cells (the targets of HIV). The RNA transcribed from the introduced DNA would pair with the HIV mRNA, preventing its translation and interfering with the virus's life cycle. Alternatively, a gene could be introduced that encodes an inactive form of one subunit of a multisubunit HIV enzyme; with one nonfunctional subunit, the entire enzyme might be inactivated.

Our growing understanding of the human genome and the genetic basis for some diseases brings the promise of early diagnosis and constructive intervention. As the early results demonstrate, however, the road to effective therapies will be a long one, with many detours. We need to learn more about cellular metabolism, more about how genes interact, and more about how to manage the dangers. The prospect of vanquishing life-destroying genetic defects and other debilitating diseases provides the motivation to press on.

Glu–Thr–Pro–Asp–Cys–S–S–Cys–Val
/ \
Phe Tyr
\ Trp–Lys /

Urotensin II

Another objective of medical research is to identify new agents that can treat the diseases caused by human pathogens. This now means identifying enzymatic targets in microbial pathogens that can be inactivated with a new drug. The ideal microbial target enzyme should be (1) essential to the pathogen cell's survival, (2) well-conserved among a wide range of pathogens, and (3) absent or significantly different in humans. The task of identifying metabolic processes that are critical to microorganisms but absent in humans is made much easier by comparative genomics, augmented by the functional information available from genomics and proteomics. ■

## Recombinant DNA Technology Yields New Products and Challenges

The products of recombinant DNA technology range from proteins to engineered organisms. The technology can produce large amounts of commercially useful proteins, can design microorganisms for special tasks, and can engineer plants or animals with traits that are useful in agriculture or medicine. Some products of this technology have been approved for consumer or professional use, and many more are in development. Genetic engineering has been transformed over a few years from a promising new technology to a multibillion-dollar industry, with much of the growth occurring in the pharmaceutical industry. Some major classes of new products are listed in Table 9–3.

Erythropoietin is typical of the newer products. This protein hormone ($M_r$ 51,000) stimulates erythrocyte production. People with diseases that com-

promise kidney function often have a deficiency of this protein, resulting in anemia. Erythropoietin produced by recombinant DNA technology can be used to treat these individuals, reducing the need for repeated blood transfusions. ■

Other applications of this technology continue to emerge. Enzymes produced by recombinant DNA technology are already used in the production of detergents, sugars, and cheese. Engineered proteins are being used as food additives to supplement nutrition, flavor, and fragrance. Microorganisms are being engineered with altered or entirely novel metabolic pathways to extract oil and minerals from ground deposits, to digest oil spills, and to detoxify hazardous waste dumps and sewage. Engineered plants with improved resistance to drought, frost, pests, and disease are increasing crop yields and reducing the need for agricultural chemicals. Complete animals can be cloned by moving an entire nucleus and all of its genetic material to a prepared egg from which the nucleus has been removed.

The extraordinary promise of modern biotechnology does not come without controversy. The cloning of mammals challenges societal mores and may be accompanied by serious deficiencies in the health and longevity of the cloned animal. If useful pharmaceutical agents can be produced, so can toxins suitable for biological warfare. The potential for hazards posed by the release of engineered plants and other organisms into

| TABLE 9–3 | Some Recombinant DNA Products in Medicine |
|---|---|
| Product category | Examples/uses |
| Anticoagulants | Tissue plasminogen activator (TPA); activates plasmin, an enzyme involved in dissolving clots; effective in treating heart attack patients. |
| Blood factors | Factor VIII; promotes clotting; it is deficient in hemophiliacs; treatment with factor VIII produced by recombinant DNA technology eliminates infection risks associated with blood transfusions. |
| Colony-stimulating factors | Immune system growth factors that stimulate leukocyte production; treatment of immune deficiencies and infections. |
| Erythropoietin | Stimulates erythrocyte production; treatment of anemia in patients with kidney disease. |
| Growth factors | Stimulate differentiation and growth of various cell types; promote wound healing. |
| Human growth hormone | Treatment of dwarfism. |
| Human insulin | Treatment of diabetes. |
| Interferons | Interfere with viral reproduction; used to treat some cancers. |
| Interleukins | Activate and stimulate different classes of leukocytes; possible uses in treatment of wounds, HIV infection, cancer, and immune deficiencies. |
| Monoclonal antibodies | Extraordinary binding specificity is used in: diagnostic tests; targeted transport of drugs, toxins, or radioactive compounds to tumors as a cancer therapy; many other applications. |
| Superoxide dismutase | Prevents tissue damage from reactive oxygen species when tissues briefly deprived of $O_2$ during surgery suddenly have blood flow restored. |
| Vaccines | Proteins derived from viral coats are as effective in "priming" an immune system as is the killed virus more traditionally used for vaccines, and are safer; first developed was the vaccine for hepatitis B. |

the biosphere continues to be monitored carefully. The full range of the long-term consequences of this technology for our species and for the global environment is impossible to foresee, but will certainly demand our increasing understanding of both cellular metabolism and ecology.

## SUMMARY 9.4 Genome Alterations and New Products of Biotechnology

■ Advances in whole genome sequencing and genetic engineering methods are enhancing our ability to modify genomes in all species.

■ Cloning in plants, which makes use of the Ti plasmid vector from *Agrobacterium,* allows the introduction of new plant traits.

■ In animal cloning, researchers introduce foreign DNA primarily with the use of viral vectors or microinjection. These techniques can produce transgenic animals and provide new methods for human gene therapy.

■ The use of genomics and proteomics in basic and pharmaceutical research is greatly advancing the discovery of new drugs. Biotechnology is also generating an ever-expanding range of other products and technologies.

## Key Terms

*Terms in bold are defined in the glossary.*

| | | |
|---|---|---|
| **cloning** 306 | **genomics** 317 | **restriction fragment length** |
| **vector** 307 | **genomic library** 318 | **polymorphisms (RFLPs)** 322 |
| **recombinant DNA** 307 | **contig** 318 | **Southern blot** 322 |
| **restriction endonucleases** 307 | **sequence-tagged site (STS)** 318 | single nucleotide polymorphisms |
| **DNA ligase** 307 | **complementary DNA** | (SNPs) 324 |
| **plasmid** 311 | **(cDNA)** 318 | **proteome** 325 |
| bacterial artificial chromosome | **cDNA library** 318 | **proteomics** 325 |
| (BAC) 313 | expressed sequence tag (EST) 318 | **orthologs** 326 |
| yeast artificial chromosome | epitope tag 319 | **synteny** 326 |
| (YAC) 314 | **polymerase chain reaction** | **DNA microarray** 326 |
| **site-directed mutagenesis** 316 | **(PCR)** 319 | Ti plasmid 330 |
| **fusion protein** 317 | DNA fingerprinting 322 | **transgenic** 335 |

## Further Reading

### General

**Jackson, D.A., Symons, R.H., & Berg, P.** (1972) Biochemical method for inserting new genetic information into DNA of simian virus 40: circular SV40 DNA molecules containing lambda phage genes and the galactose operon of *Escherichia coli. Proc. Natl. Acad. Sci. USA* **69,** 2904–2909.

The first recombinant DNA experiment linking DNA from two species.

**Lobban, P.E. & Kaiser, A.D.** (1973) Enzymatic end-to-end joining of DNA molecules. *J. Mol. Biol.* **78,** 453–471.

Report of the first recombinant DNA experiment.

**Sambrook, J., Fritsch, E.F., & Maniatis, T.** (1989) *Molecular Cloning: A Laboratory Manual,* 2nd edn, Cold Spring Harbor Laboratory Press, Cold Spring Harbor, NY.

Although supplanted by more recent manuals, this three-volume set includes much useful background information on the biological, chemical, and physical principles underlying both classic and still-current techniques.

### Gene Cloning

**Arnheim, N. & Erlich, H.** (1992) Polymerase chain reaction strategy. *Annu. Rev. Biochem.* **61,** 131–156.

**Hofreiter, M., Serre, D., Poinar, H.N., Kuch, M., & Paabo, S.** (2001) Ancient DNA. *Nat. Rev. Genet.* **2,** 353–359.

Successes and pitfalls in the retrieval of DNA from very old samples.

**Ivanov, P.L., Wadhams, M.J., Roby, R.K., Holland, M.M., Weedn, V.W., & Parsons, T.J.** (1996) Mitochondrial DNA sequence heteroplasmy in the Grand Duke of Russia Georgij Romanov establishes the authenticity of the remains of Tsar Nicholas II. *Nat. Genet.* **12,** 417–420.

**Lindahl, T.** (1997) Facts and artifacts of ancient DNA. *Cell* **90,** 1–3.

Good description of how nucleic acid chemistry affects the retrieval of DNA in archaeology.

### Genomics

**Adams, M.D., Kelley, J.M., Gocayne, J.D., Dubnick, M., Polymeropoulos, M.H., Xiao, H., Merril, C.R., Wu, A., Olde, B., Moreno, R.F., et al.** (1991) Complementary DNA sequencing: expressed sequence tags and Human Genome Project. *Science* **252,** 1651–1656.

The paper that introduced expressed sequence tags (ESTs).

**Bamshad, M. & Wooding, S.P.** (2003) Signatures of natural selection in the human genome. *Nat. Rev. Genet.* **4,** 99A–111A.

Use of the human genome to trace human evolution.

**Brenner, S.** (2004) Genes to genomics. *Annu. Rev. Genet.* **38,** in press.

**Carroll, S.B.** (2003) Genetics and the making of *Homo sapiens. Nature* **422**, 849–857.

**Clark, M.S.** (1999) Comparative genomics: the key to understanding the Human Genome Project. *Bioessays* **21**, 121–130.

Useful background on some reasons for the importance of sequencing the genomes of many organisms.

**Collins, F.S., Green, E.D., Guttmacher, A.E., & Guyer, M.S.** (2003) A vision for the future of genomics research. *Nature* **422**, 835–847.

A wide-ranging overview of the enormous potential of genomics research.

**Lander, E.S., Linton, L.M., Birren, B., Nusbaum, C., Zody, M.C., Baldwin, J., Devon, K., Dewar, K., Doyle, M., FitzHugh, W., et al.** (2001) Initial sequencing and analysis of the human genome. *Nature* **409**, 860–921.

Discussion of the draft genome sequence put together by the international Human Genome Project. Many other useful articles are to be found in this issue.

**Venter, J.C., Adams, M.D., Myers, E.W., Li, P.W., Mural, R.J., Sutton, G.G., Smith, H.O., Yandell, M., Evans, C.A., Holt, R.A., et al.** (2001) The sequence of the human genome. *Science* **291**, 1304–1351.

Description of the draft of the human genome sequence produced by Celera Corporation. Many other articles in the same issue provide insight and additional information.

## Proteomics

**Brown, P.O. & Botstein, D.** (1999) Exploring the new world of the genome with DNA microarrays. *Nat. Genet.* **21**, 33–37.

**Eisenberg, D., Marcotte, E.M., Xenarios, I., & Yeates, T.O.** (2000) Protein function in the post-genomic era. *Nature* **405**, 823–826.

**Pandey, A. & Mann, M.** (2000) Proteomics to study genes and genomes. *Nature* **405**, 837–846.

An especially good description of the various strategies and methods used to identify proteins and their functions.

**Zhu, H., Bilgin, M., & Snyder, M.** (2003) Proteomics. *Annu. Rev. Biochem.* **72**, 783–812.

## Applying Biotechnology

**Foster, E.A., Jobling, M.A., Taylor, P.G., Donnelly, P., de Knijff, P., Mieremet, R., Zerjal, T., & Tyler-Smith, C.** (1999) The Thomas Jefferson paternity case. *Nature* **397**, 32.

Last article of a series in an interesting case study of the uses of biotechnology to address historical questions.

**Hansen, G. & Wright M.S.** (1999) Recent advances in the transformation of plants. *Trends Plant Sci.* **4**, 226–231.

**Koopman, P., Gubbay, J., Vivian, N., Goodfellow, P., & Lovell-Badge, R.** (1991) Male development of chromosomally female mice transgenic for *Sry. Nature* **351**, 117–121.

Recombinant DNA technology shows that a single gene directs development of chromosomally female mice into males.

**Lapham, E.V., Kozma, C., & Weiss, J.** (1996) Genetic discrimination: perspectives of consumers. *Science* **274**, 621–624.

The upside and downside of knowing what is in your genome.

**Mahowald, M.B., Verp, M.S., & Anderson, R.R.** (1998) Genetic counseling: clinical and ethical challenges. *Annu. Rev. Genet.* **32**, 547–559.

**Ohlstein, E.H., Ruffolo, R.R., Jr., & Elliott, J.D.** (2000) Drug discovery in the next millennium. *Annu. Rev. Pharmacol. Toxicol.* **40**, 177–191.

**Palmiter, R.D., Brinster, R.L., Hammer, R.E., Trumbauer, M.E., Rosenfeld, M.G., Birnberg, N.C., & Evans, R.M.** (1982) Dramatic growth of mice that develop from eggs microinjected with metallothionein-growth hormone fusion genes. *Nature* **300**, 611–615.

A description of how to make giant mice.

**Pfeifer, A. & Verma, I. M.** (2001) Gene therapy: promises and problems. *Annu. Rev. Genomics Hum. Genet.* **2**, 177–211.

**Thompson, J. & Donkersloot, J.A.** (1992) *N*-(Carboxyalkyl) amino acids: occurrence, synthesis, and functions. *Annu. Rev. Biochem.* **61**, 517–557.

A summary of the structure and biological functions of opines.

**Wadhwa, P.D., Zielske, S.P., Roth, J.C., Ballas, C.B., Bowman, J.E., & Gerson, S.L.** (2002) Cancer gene therapy: scientific basis. *Annu. Rev. Med.* **53**, 437–452.

## Problems

**1. Cloning**  When joining two or more DNA fragments, a researcher can adjust the sequence at the junction in a variety of subtle ways, as seen in the following exercises.

(a) Draw the structure of each end of a linear DNA fragment produced by an *Eco*RI restriction digest (include those sequences remaining from the *Eco*RI recognition sequence).

(b) Draw the structure resulting from the reaction of this end sequence with DNA polymerase I and the four deoxynucleoside triphosphates (see Fig. 8–36).

(c) Draw the sequence produced at the junction that arises if two ends with the structure derived in (b) are ligated (see Fig. 25–16).

(d) Draw the structure produced if the structure derived in (a) is treated with a nuclease that degrades only single-stranded DNA.

(e) Draw the sequence of the junction produced if an end with structure (b) is ligated to an end with structure (d).

(f) Draw the structure of the end of a linear DNA fragment that was produced by a *Pvu*II restriction digest (include those sequences remaining from the *Pvu*II recognition sequence).

(g) Draw the sequence of the junction produced if an end with structure (b) is ligated to an end with structure (f).

(h) Suppose you can synthesize a short duplex DNA fragment with any sequence you desire. With this synthetic fragment and the procedures described in (a) through (g), design a protocol that would remove an *Eco*RI restriction site from a DNA molecule and incorporate a new *Bam*HI restriction site at approximately the same location. (See Fig. 9–3.)

(i) Design four different short synthetic double-stranded DNA fragments that would permit ligation of structure (a) with a DNA fragment produced by a *Pst*I restriction digest. In one of these fragments, design the sequence so that the final junction contains the recognition sequences for both *Eco*RI and *Pst*I. In the second and third fragments, design the sequence so that the junction contains only the *Eco*RI and only the *Pst*I recognition sequence, respectively. Design the sequence of the fourth fragment so that neither the *Eco*RI nor the *Pst*I sequence appears in the junction.

**2. Selecting for Recombinant Plasmids**   When cloning a foreign DNA fragment into a plasmid, it is often useful to insert the fragment at a site that interrupts a selectable marker (such as the tetracycline-resistance gene of pBR322). The loss of function of the interrupted gene can be used to identify clones containing recombinant plasmids with foreign DNA. With a bacteriophage λ vector it is not necessary to do this, yet one can easily distinguish vectors that incorporate large foreign DNA fragments from those that do not. How are these recombinant vectors identified?

**3. DNA Cloning**   The plasmid cloning vector pBR322 (see Fig. 9–4) is cleaved with the restriction endonuclease *Pst*I. An isolated DNA fragment from a eukaryotic genome (also produced by *Pst*I cleavage) is added to the prepared vector and ligated. The mixture of ligated DNAs is then used to transform bacteria, and plasmid-containing bacteria are selected by growth in the presence of tetracycline.

(a) In addition to the desired recombinant plasmid, what other types of plasmids might be found among the transformed bacteria that are tetracycline resistant? How can the types be distinguished?

(b) The cloned DNA fragment is 1,000 bp long and has an *Eco*RI site 250 bp from one end. Three different recombinant plasmids are cleaved with *Eco*RI and analyzed by gel electrophoresis, giving the patterns shown. What does each pattern say about the cloned DNA? Note that in pBR322, the *Pst*I and *Eco*RI restriction sites are about 750 bp apart. The entire plasmid with no cloned insert is 4,361 bp. Size markers in lane 4 have the number of nucleotides noted.

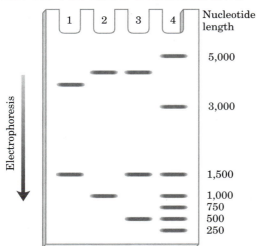

**4. Identifying the Gene for a Protein with a Known Amino Acid Sequence**   Using Figure 27–7 to translate the genetic code, design a DNA probe that would allow you to identify the gene for a protein with the following amino-terminal amino acid sequence. The probe should be 18 to 20 nucleotides long, a size that provides adequate specificity if there is sufficient homology between the probe and the gene.

H₃N⁺–Ala–Pro–Met–Thr–Trp–Tyr–Cys–Met–
Asp–Trp–Ile–Ala–Gly–Gly–Pro–Trp–Phe–Arg–
Lys–Asn–Thr–Lys–

**5. Designing a Diagnostic Test for a Genetic Disease**   Huntington's disease (HD) is an inherited neurodegenerative disorder, characterized by the gradual, irreversible impairment of psychological, motor, and cognitive functions. Symptoms typically appear in middle age, but onset can occur at almost any age. The course of the disease can last 15 to 20 years. The molecular basis of the disease is becoming better understood. The genetic mutation underlying HD has been traced to a gene encoding a protein ($M_r$ 350,000) of unknown function. In individuals who will not develop HD, a region of the gene that encodes the amino terminus of the protein has a sequence of CAG codons (for glutamine) that is repeated 6 to 39 times in succession. In individuals with adult-onset HD, this codon is typically repeated 40 to 55 times. In individuals with childhood-onset HD, this codon is repeated more than 70 times. The length of this simple trinucleotide repeat indicates whether an individual will develop HD, and at approximately what age the first symptoms will occur.

A small portion of the amino-terminal coding sequence of the 3,143-codon HD gene is given below. The nucleotide sequence of the DNA is shown in black, the amino acid sequence corresponding to the gene is shown in blue, and the CAG repeat is shaded. Using Figure 27–7 to translate the genetic code, outline a PCR-based test for HD that could be carried out using a blood sample. Assume the PCR primer must be 25 nucleotides long. By convention, unless otherwise specified a DNA sequence encoding a protein is displayed with the coding strand (the sequence identical to the mRNA transcribed from the gene) on top such that it is read 5′ to 3′, left to right.

```
307  ATGGCGACCCTGGAAAAGCTGATGAAGGCCTTCGAGTCCCTCAAGTCCTTC
1     M  A  T  L  E  K  L  M  K  A  F  E  S  L  K  S  F

358  CAGCAGTTCCAGCAGCAGCAGCAGCAGCAGCAGCAGCAGCAGCAGCAGCAG
18    Q  Q  F  Q  Q  Q  Q  Q  Q  Q  Q  Q  Q  Q  Q  Q  Q

409  CAGCAGCAGCAGCAGCAGCAGCAACAGCCGCCACCGCCGCCGCCGCCGCCG
35    Q  Q  Q  Q  Q  Q  Q  Q  Q  P  P  P  P  P  P  P  P

460  CCGCCTCCTCAGCTTCCTCAGCCGCCGCCG
52    P  P  P  Q  L  P  Q  P  P  P
```

Source: The Huntington's Disease Collaborative Research Group. (1993) A novel gene containing a trinucleotide repeat that is expanded and unstable on Huntington's disease chromosomes. *Cell* **72**, 971–983.

**6. Using PCR to Detect Circular DNA Molecules**   In a species of ciliated protist, a segment of genomic DNA is sometimes deleted. The deletion is a genetically programmed reaction associated with cellular mating. A researcher proposes that the DNA is deleted in a type of recombination called site-specific recombination, with the DNA on either end of the segment joined together and the deleted DNA ending up as a circular DNA reaction product.

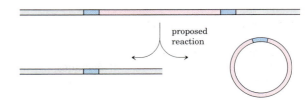

Suggest how the researcher might use the polymerase chain reaction (PCR) to detect the presence of the circular form of the deleted DNA in an extract of the protist.

**7. RFLP Analysis for Paternity Testing** DNA fingerprinting and RFLP analysis are often used to test for paternity. A child inherits chromosomes from both mother and father, so DNA from a child displays restriction fragments derived from each parent. In the gel shown here, which child, if any, can be excluded as being the biological offspring of the putative father? Explain your reasoning. Lane M is the sample from the mother, F from the putative father, and C1, C2, and C3 from the children.

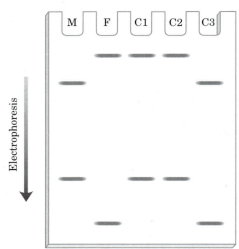

**8. Mapping a Chromosome Segment** A group of overlapping clones, designated A through F, is isolated from one region of a chromosome. Each of the clones is separately cleaved by a restriction enzyme and the pieces resolved by agarose gel electrophoresis, with the results shown in the figure below. There are nine different restriction fragments in this chromosomal region, with a subset appearing in each clone. Using this information, deduce the order of the restriction fragments in the chromosome.

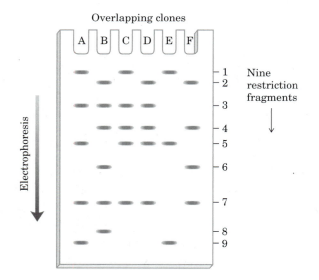

**9. Cloning in Plants** The strategy outlined in Figure 9–28 employs *Agrobacterium* cells that contain two separate plasmids. Suggest why the sequences on the two plasmids are not combined on one plasmid.

**10. DNA Fingerprinting and RFLP Analysis** DNA is extracted from the blood cells of two different humans, individuals 1 and 2. In separate experiments, the DNA from each individual is cleaved by restriction endonucleases A, B, and C, and the fragments separated by electrophoresis. A hypothetical map of a 10,000 bp segment of a human chromosome is shown (1 kbp = 1,000 bp). Individual 2 has point mutations that eliminate restriction recognition sites B* and C*. You probe the gel with a radioactive oligonucleotide complementary to the indicated sequence and expose a piece of x-ray film to the gel. Indicate where you would expect to see bands on the film. The lanes of the gel are marked in the accompanying diagram.

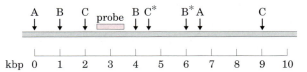

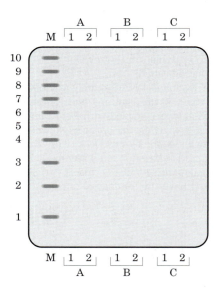

**11. Use of Photolithography to Make a DNA Microarray** Figure 9–21 shows the first steps in the process of making a DNA microarray, or DNA chip, using photolithography. Describe the remaining steps needed to obtain the desired sequences (a different four-nucleotide sequence on each of the four spots) shown in the first panel of the figure. After each step, give the resulting nucleotide sequence attached at each spot.

**12. Cloning in Mammals** The retroviral vectors described in Figure 9–32 make possible the efficient integration of foreign DNA into a mammalian genome. Explain how these vectors, which lack genes for replication and viral packaging (*gag, pol, env*), are assembled into infectious viral particles. Suggest why it is important that these vectors lack the replication and packaging genes.

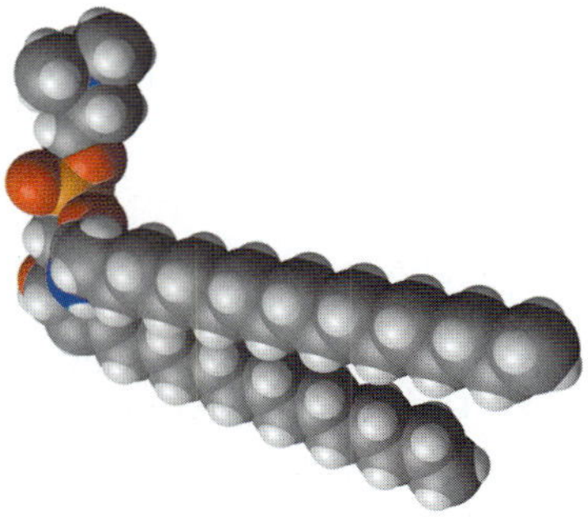

# chapter 10

# LIPIDS

The fatty substance, separated from the salifiable bases, was dissolved in boiling alcohol. On cooling, it was obtained crystallized and very pure, and in this state it was examined. As it has not been hitherto described . . . I purpose to call it margarine, from the Greek word signifying pearl, because one of its characters is to have the appearance of mother of pearl, which it communicates to several of the combinations of which it forms with the salifiable bases.

— *Michel-Eugène Chevreul,*
*article in* Philosophical Magazine, *1814*

**B**iological lipids are a chemically diverse group of compounds, the common and defining feature of which is their insolubility in water. The biological functions of the lipids are as diverse as their chemistry. Fats and oils are the principal stored forms of energy in many organisms. Phospholipids and sterols are major structural elements of biological membranes. Other lipids, although present in relatively small quantities, play crucial roles as enzyme cofactors, electron carriers, light-absorbing pigments, hydrophobic anchors for proteins, "chaperones" to help membrane proteins fold, emulsifying agents in the digestive tract, hormones, and intracellular messengers. This chapter introduces representative lipids of each type, with emphasis on their chemical structure and physical properties. We discuss the energy-yielding oxidation of lipids in Chapter 17 and their synthesis in Chapter 21.

## 10.1 Storage Lipids

The fats and oils used almost universally as stored forms of energy in living organisms are derivatives of **fatty acids.** The fatty acids are hydrocarbon derivatives, at about the same low oxidation state (that is, as highly reduced) as the hydrocarbons in fossil fuels. The cellular oxidation of fatty acids (to $CO_2$ and $H_2O$), like the controlled, rapid burning of fossil fuels in internal combustion engines, is highly exergonic.

We introduce here the structures and nomenclature of the fatty acids most commonly found in living organisms. Two types of fatty acid–containing compounds, triacylglycerols and waxes, are described to illustrate the diversity of structure and physical properties in this family of compounds.

### Fatty Acids Are Hydrocarbon Derivatives

Fatty acids are carboxylic acids with hydrocarbon chains ranging from 4 to 36 carbons long ($C_4$ to $C_{36}$). In some fatty acids, this chain is unbranched and fully saturated (contains no double bonds); in others the chain contains one or more double bonds (Table 10–1). A few contain three-carbon rings, hydroxyl groups, or methyl-group branches. A simplified nomenclature for these compounds specifies the chain length and number of double bonds, separated by a colon; for example, the 16-carbon saturated palmitic acid is abbreviated 16:0, and the 18-carbon oleic acid, with one double bond, is 18:1. The positions of any double bonds are specified by superscript numbers following $\Delta$ (delta); a 20-carbon fatty acid with one double bond between C-9 and C-10 (C-1 being the carboxyl carbon) and another between C-12 and C-13 is designated $20:2(\Delta^{9,12})$. The most commonly occurring fatty acids have even numbers of carbon atoms in an unbranched chain of 12 to 24 carbons (Table 10–1). As we shall see in Chapter 21, the even number of carbons results from the mode of synthesis of these compounds, which involves condensation of two-carbon (acetate) units.

There is also a common pattern in the location of double bonds; in most monounsaturated fatty acids the double bond is between C-9 and C-10 ($\Delta^9$), and the other double bonds of polyunsaturated fatty acids are generally $\Delta^{12}$ and $\Delta^{15}$. (Arachidonic acid is an exception to this generalization.) The double bonds of polyunsaturated fatty acids are almost never conjugated (alternating single and double bonds, as in —CH=CH—CH=CH—), but are separated by a methylene group: —CH=CH—CH$_2$—CH=CH—. In nearly all naturally occurring unsaturated fatty acids, the double bonds are in the cis configuration. Trans fatty acids are produced by fermentation in the rumen of dairy animals and are obtained from dairy products and meat.

They are also produced during hydrogenation of fish or vegetable oils. Because diets high in trans fatty acids correlate with increased blood levels of LDL (bad cholesterol) and decreased HDL (good cholesterol), it is generally recommended that one avoid large amounts of these fatty acids. Unfortunately, French fries, doughnuts, and cookies tend to be high in trans fatty acids.

The physical properties of the fatty acids, and of compounds that contain them, are largely determined by the length and degree of unsaturation of the hydrocarbon chain. The nonpolar hydrocarbon chain accounts for the poor solubility of fatty acids in water. Lauric acid (12:0, $M_r$ 200), for example, has a solubility in water of 0.063 mg/g—much less than that of glucose ($M_r$ 180), which is 1,100 mg/g. The longer the fatty acyl chain and the fewer the double bonds, the lower is the solubility

**TABLE 10–1 Some Naturally Occurring Fatty Acids: Structure, Properties, and Nomenclature**

| Carbon skeleton | Structure* | Systematic name[†] | Common name (derivation) | Melting point (°C) | Solubility at 30°C (mg/g solvent) Water | Benzene |
|---|---|---|---|---|---|---|
| 12:0 | CH$_3$(CH$_2$)$_{10}$COOH | n-Dodecanoic acid | Lauric acid (Latin *laurus*, "laurel plant") | 44.2 | 0.063 | 2,600 |
| 14:0 | CH$_3$(CH$_2$)$_{12}$COOH | n-Tetradecanoic acid | Myristic acid (Latin *Myristica*, nutmeg genus) | 53.9 | 0.024 | 874 |
| 16:0 | CH$_3$(CH$_2$)$_{14}$COOH | n-Hexadecanoic acid | Palmitic acid (Latin *palma*, "palm tree") | 63.1 | 0.0083 | 348 |
| 18:0 | CH$_3$(CH$_2$)$_{16}$COOH | n-Octadecanoic acid | Stearic acid (Greek *stear*, "hard fat") | 69.6 | 0.0034 | 124 |
| 20:0 | CH$_3$(CH$_2$)$_{18}$COOH | n-Eicosanoic acid | Arachidic acid (Latin *Arachis*, legume genus) | 76.5 | | |
| 24:0 | CH$_3$(CH$_2$)$_{22}$COOH | n-Tetracosanoic acid | Lignoceric acid (Latin *lignum*, "wood" + *cera*, "wax") | 86.0 | | |
| 16:1($\Delta^9$) | CH$_3$(CH$_2$)$_5$CH=CH(CH$_2$)$_7$COOH | cis-9-Hexadecenoic acid | Palmitoleic acid | 1–0.5 | | |
| 18:1($\Delta^9$) | CH$_3$(CH$_2$)$_7$CH=CH(CH$_2$)$_7$COOH | cis-9-Octadecenoic acid | Oleic acid (Latin *oleum*, "oil") | 13.4 | | |
| 18:2($\Delta^{9,12}$) | CH$_3$(CH$_2$)$_4$CH=CHCH$_2$CH=CH(CH$_2$)$_7$COOH | cis-,cis-9,12-Octadecadienoic acid | Linoleic acid (Greek *linon*, "flax") | 1–5 | | |
| 18:3($\Delta^{9,12,15}$) | CH$_3$CH$_2$CH=CHCH$_2$CH=CHCH$_2$CH=CH(CH$_2$)$_7$COOH | cis-,cis-,cis-9,12,15-Octadecatrienoic acid | α-Linolenic acid | −11 | | |
| 20:4($\Delta^{5,8,11,14}$) | CH$_3$(CH$_2$)$_4$CH=CHCH$_2$CH=CHCH$_2$CH=CHCH$_2$CH=CH(CH$_2$)$_3$COOH | cis-,cis-,cis-,cis-5,8,11,14-Icosatetraenoic acid | Arachidonic acid | −49.5 | | |

*All acids are shown in their nonionized form. At pH 7, all free fatty acids have an ionized carboxylate. Note that numbering of carbon atoms begins at the carboxyl carbon.

[†]The prefix *n-* indicates the "normal" unbranched structure. For instance, "dodecanoic" simply indicates 12 carbon atoms, which could be arranged in a variety of branched forms; "*n*-dodecanoic" specifies the linear, unbranched form. For unsaturated fatty acids, the configuration of each double bond is indicated; in biological fatty acids the configuration is almost always cis.

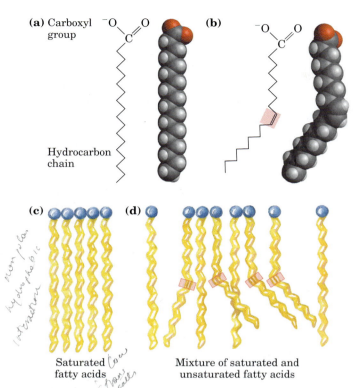

**(a)** Carboxyl group

Hydrocarbon chain

**(b)**

**(c)** **(d)**

Saturated fatty acids

Mixture of saturated and unsaturated fatty acids

**FIGURE 10–1 The packing of fatty acids into stable aggregates.** The extent of packing depends on the degree of saturation. **(a)** Two representations of the fully saturated acid stearic acid (stearate at pH 7) in its usual extended conformation. Each line segment of the zigzag represents a single bond between adjacent carbons. **(b)** The cis double bond (shaded) in oleic acid (oleate) does not permit rotation and introduces a rigid bend in the hydrocarbon tail. All other bonds in the chain are free to rotate. **(c)** Fully saturated fatty acids in the extended form pack into nearly crystalline arrays, stabilized by many hydrophobic interactions. **(d)** The presence of one or more cis double bonds interferes with this tight packing and results in less stable aggregates.

in water. The carboxylic acid group is polar (and ionized at neutral pH) and accounts for the slight solubility of short-chain fatty acids in water.

Melting points are also strongly influenced by the length and degree of unsaturation of the hydrocarbon chain. At room temperature (25 °C), the saturated fatty acids from 12:0 to 24:0 have a waxy consistency, whereas unsaturated fatty acids of these lengths are oily liquids. This difference in melting points is due to different degrees of packing of the fatty acid molecules (Fig. 10–1). In the fully saturated compounds, free rotation around each carbon–carbon bond gives the hydrocarbon chain great flexibility; the most stable conformation is the fully extended form, in which the steric hindrance of neighboring atoms is minimized. These molecules can pack together tightly in nearly crystalline arrays, with atoms all along their lengths in van der Waals contact with the atoms of neighboring molecules. In unsaturated fatty acids, a cis double bond forces a kink in the hydrocarbon chain. Fatty acids with one or several such kinks

cannot pack together as tightly as fully saturated fatty acids, and their interactions with each other are therefore weaker. Because it takes less thermal energy to disorder these poorly ordered arrays of unsaturated fatty acids, they have markedly lower melting points than saturated fatty acids of the same chain length (Table 10–1).

In vertebrates, free fatty acids (unesterified fatty acids, with a free carboxylate group) circulate in the blood bound noncovalently to a protein carrier, serum albumin. However, fatty acids are present in blood plasma mostly as carboxylic acid derivatives such as esters or amides. Lacking the charged carboxylate group, these fatty acid derivatives are generally even less soluble in water than are the free fatty acids.

### Triacylglycerols Are Fatty Acid Esters of Glycerol

The simplest lipids constructed from fatty acids are the **triacylglycerols,** also referred to as triglycerides, fats, or neutral fats. Triacylglycerols are composed of three fatty acids each in ester linkage with a single glycerol (Fig. 10–2). Those containing the same kind of fatty acid

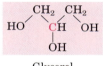

Glycerol

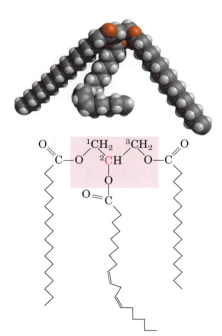

1-Stearoyl, 2-linoleoyl, 3-palmitoyl glycerol, a mixed triacylglycerol

**FIGURE 10–2 Glycerol and a triacylglycerol.** The mixed triacylglycerol shown here has three different fatty acids attached to the glycerol backbone. When glycerol has two different fatty acids at C-1 and C-3, the C-2 is a chiral center (p. 76).

in all three positions are called simple triacylglycerols and are named after the fatty acid they contain. Simple triacylglycerols of 16:0, 18:0, and 18:1, for example, are tristearin, tripalmitin, and triolein, respectively. Most naturally occurring triacylglycerols are mixed; they contain two or more different fatty acids. To name these compounds unambiguously, the name and position of each fatty acid must be specified.

Because the polar hydroxyls of glycerol and the polar carboxylates of the fatty acids are bound in ester linkages, triacylglycerols are nonpolar, hydrophobic molecules, essentially insoluble in water. Lipids have lower specific gravities than water, which explains why mixtures of oil and water (oil-and-vinegar salad dressing, for example) have two phases: oil, with the lower specific gravity, floats on the aqueous phase.

## Triacylglycerols Provide Stored Energy and Insulation

In most eukaryotic cells, triacylglycerols form a separate phase of microscopic, oily droplets in the aqueous cytosol, serving as depots of metabolic fuel. In vertebrates, specialized cells called adipocytes, or fat cells, store large amounts of triacylglycerols as fat droplets that nearly fill the cell (Fig. 10–3a). Triacylglycerols are also stored as oils in the seeds of many types of plants, providing energy and biosynthetic precursors during seed germination (Fig. 10–3b). Adipocytes and germinating seeds contain **lipases,** enzymes that catalyze the hydrolysis of stored triacylglycerols, releasing fatty acids for export to sites where they are required as fuel.

There are two significant advantages to using triacylglycerols as stored fuels, rather than polysaccharides such as glycogen and starch. First, because the carbon atoms of fatty acids are more reduced than those of sugars, oxidation of triacylglycerols yields more than twice as much energy, gram for gram, as the oxidation of carbohydrates. Second, because triacylglycerols are hydrophobic and therefore unhydrated, the organism that carries fat as fuel does not have to carry the extra weight of water of hydration that is associated with stored polysaccharides (2 g per gram of polysaccharide). Humans have fat tissue (composed primarily of adipocytes) under the skin, in the abdominal cavity, and in the mammary glands. Moderately obese people with 15 to 20 kg of triacylglycerols deposited in their adipocytes could meet their energy needs for months by drawing on their fat stores. In contrast, the human body can store less than a day's energy supply in the form of glycogen. Carbohydrates such as glucose and glycogen do offer certain advantages as quick sources of metabolic energy, one of which is their ready solubility in water.

In some animals, triacylglycerols stored under the skin serve not only as energy stores but as insulation against low temperatures. Seals, walruses, penguins, and other warm-blooded polar animals are amply padded with triacylglycerols. In hibernating animals (bears, for

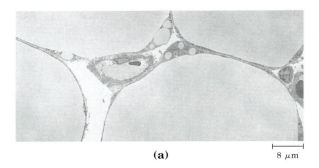

**(a)**　　　　　　　　　　8 μm

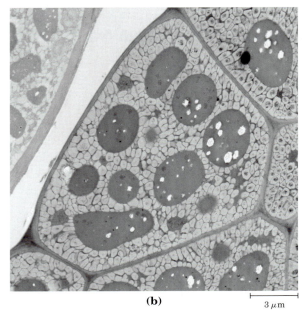

**(b)**　　　　　　　3 μm

**FIGURE 10–3　Fat stores in cells. (a)** Cross section of four guinea pig adipocytes, showing huge fat droplets that virtually fill the cells. Also visible are several capillaries in cross section. **(b)** Cross section of a cotyledon cell from a seed of the plant *Arabidopsis*. The large dark structures are protein bodies, which are surrounded by stored oils in the light-colored oil bodies.

example), the huge fat reserves accumulated before hibernation serve the dual purposes of insulation and energy storage (see Box 17–1). The low density of triacylglycerols is the basis for another remarkable function of these compounds. In sperm whales, a store of triacylglycerols and waxes allows the animals to match the buoyancy of their bodies to that of their surroundings during deep dives in cold water (Box 10–1).

## Many Foods Contain Triacylglycerols

Most natural fats, such as those in vegetable oils, dairy products, and animal fat, are complex mixtures of simple and mixed triacylglycerols. These contain a variety of fatty acids differing in chain length and degree of saturation (Fig. 10–4). Vegetable oils such as corn (maize) and olive oil are composed largely of triacylglycerols with unsaturated fatty acids and thus are liquids at room temperature. They are converted industrially into solid

## BOX 10–1 THE WORLD OF BIOCHEMISTRY

### Sperm Whales: Fatheads of the Deep

Studies of sperm whales have uncovered another way in which triacylglycerols are biologically useful. The sperm whale's head is very large, accounting for over one-third of its total body weight. About 90% of the weight of the head is made up of the spermaceti organ, a blubbery mass that contains up to 3,600 kg (about 4 tons) of spermaceti oil, a mixture of triacylglycerols and waxes containing an abundance of unsaturated fatty acids. This mixture is liquid at the normal resting body temperature of the whale, about 37 °C, but it begins to crystallize at about 31 °C and becomes solid when the temperature drops several more degrees.

The probable biological function of spermaceti oil has been deduced from research on the anatomy and feeding behavior of the sperm whale. These mammals feed almost exclusively on squid in very deep water. In their feeding dives they descend 1,000 m or more; the deepest recorded dive is 3,000 m (almost 2 miles). At these depths, there are no competitors for the very plentiful squid; the sperm whale rests quietly, waiting for schools of squid to pass.

For a marine animal to remain at a given depth without a constant swimming effort, it must have the same density as the surrounding water. The sperm whale undergoes changes in buoyancy to match the density of its surroundings—from the tropical ocean surface to great depths where the water is much

colder and thus denser. The key is the freezing point of spermaceti oil. When the temperature of the oil is lowered several degrees during a deep dive, it congeals or crystallizes and becomes denser. Thus the buoyancy of the whale changes to match the density of seawater. Various physiological mechanisms promote rapid cooling of the oil during a dive. During the return to the surface, the congealed spermaceti oil warms and melts, decreasing its density to match that of the surface water. Thus we see in the sperm whale a remarkable anatomical and biochemical adaptation. The triacylglycerols and waxes synthesized by the sperm whale contain fatty acids of the necessary chain length and degree of unsaturation to give the spermaceti oil the proper melting point for the animal's diving habits.

Unfortunately for the sperm whale population, spermaceti oil was at one time considered the finest lamp oil and continues to be commercially valuable as a lubricant. Several centuries of intensive hunting of these mammals have driven sperm whales onto the endangered species list.

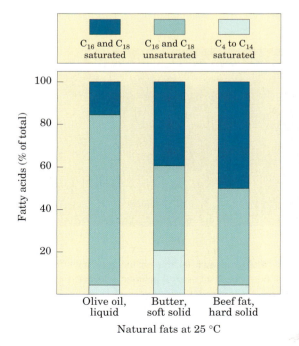

fats by catalytic hydrogenation, which reduces some of their double bonds to single bonds and converts others to trans double bonds. Triacylglycerols containing only saturated fatty acids, such as tristearin, the major component of beef fat, are white, greasy solids at room temperature.

When lipid-rich foods are exposed too long to the oxygen in air, they may spoil and become rancid. The unpleasant taste and smell associated with rancidity result from the oxidative cleavage of the double bonds in

**FIGURE 10–4 Fatty acid composition of three food fats.** Olive oil, butter, and beef fat consist of mixtures of triacylglycerols, differing in their fatty acid composition. The melting points of these fats—and hence their physical state at room temperature (25 °C)—are a direct function of their fatty acid composition. Olive oil has a high proportion of long-chain ($C_{16}$ and $C_{18}$) unsaturated fatty acids, which accounts for its liquid state at 25 °C. The higher proportion of long-chain ($C_{16}$ and $C_{18}$) saturated fatty acids in butter increases its melting point, so butter is a soft solid at room temperature. Beef fat, with an even higher proportion of long-chain saturated fatty acids, is a hard solid.

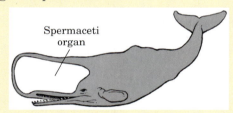

unsaturated fatty acids, which produces aldehydes and carboxylic acids of shorter chain length and therefore higher volatility.

### Waxes Serve as Energy Stores and Water Repellents

Biological waxes are esters of long-chain ($C_{14}$ to $C_{36}$) saturated and unsaturated fatty acids with long-chain ($C_{16}$ to $C_{30}$) alcohols (Fig. 10–5). Their melting points (60 to 100 °C) are generally higher than those of triacylglycerols. In plankton, the free-floating microorganisms at the bottom of the food chain for marine animals, waxes are the chief storage form of metabolic fuel.

Waxes also serve a diversity of other functions related to their water-repellent properties and their firm consistency. Certain skin glands of vertebrates secrete waxes to protect hair and skin and keep it pliable, lubricated, and waterproof. Birds, particularly waterfowl, secrete waxes from their preen glands to keep their feathers water-repellent. The shiny leaves of holly, rhododendrons, poison ivy, and many tropical plants are coated with a thick layer of waxes, which prevents excessive evaporation of water and protects against parasites.

Biological waxes find a variety of applications in the pharmaceutical, cosmetic, and other industries. Lanolin (from lamb's wool), beeswax (Fig. 10–5), carnauba wax (from a Brazilian palm tree), and wax extracted from spermaceti oil (from whales; see Box 10–1) are widely used in the manufacture of lotions, ointments, and polishes.

$$CH_3(CH_2)_{14}\!-\!\overset{\overset{\displaystyle O}{\|}}{C}\!-\!O\!-\!CH_2\!-\!(CH_2)_{28}\!-\!CH_3$$

$\underbrace{\hphantom{CH_3(CH_2)_{14}C}}_{\text{Palmitic acid}}$   $\underbrace{\hphantom{OCH_2(CH_2)_{28}CH_3}}_{\text{1-Triacontanol}}$

**(a)**

**(b)**

**FIGURE 10–5 Biological wax. (a)** Triacontanoylpalmitate, the major component of beeswax, is an ester of palmitic acid with the alcohol triacontanol. **(b)** A honeycomb, constructed of beeswax, is firm at 25 °C and completely impervious to water. The term "wax" originates in the Old English *weax,* meaning "the material of the honeycomb."

### SUMMARY 10.1  Storage Lipids

- Lipids are water-insoluble cellular components of diverse structure that can be extracted by nonpolar solvents.

- Almost all fatty acids, the hydrocarbon components of many lipids, have an even number of carbon atoms (usually 12 to 24); they are either saturated or unsaturated, with double bonds almost always in the cis configuration.

- Triacylglycerols contain three fatty acid molecules esterified to the three hydroxyl groups of glycerol. Simple triacylglycerols contain only one type of fatty acid; mixed triacylglycerols, two or three types. Triacylglycerols are primarily storage fats; they are present in many foods.

## 10.2 Structural Lipids in Membranes

The central architectural feature of biological membranes is a double layer of lipids, which acts as a barrier to the passage of polar molecules and ions. Membrane lipids are amphipathic: one end of the molecule is hydrophobic, the other hydrophilic. Their hydrophobic interactions with each other and their hydrophilic interactions with water direct their packing into sheets called membrane bilayers. In this section we describe five general types of membrane lipids: glycerophospholipids, in which the hydrophobic regions are composed of two fatty acids joined to glycerol; galactolipids and sulfolipids, which also contain two fatty acids esterified to glycerol, but lack the characteristic phosphate of phospholipids; archaebacterial tetraether lipids, in which two very long alkyl chains are ether-linked to glycerol at both ends; sphingolipids, in which a single fatty acid is joined to a fatty amine, sphingosine; and sterols, compounds characterized by a rigid system of four fused hydrocarbon rings.

The hydrophilic moieties in these amphipathic compounds may be as simple as a single —OH group at one end of the sterol ring system, or they may be much more complex. In glycerophospholipids and some sphingolipids, a polar head group is joined to the hydrophobic moiety by a phosphodiester linkage; these are the **phospholipids.** Other sphingolipids lack phosphate but have a simple sugar or complex oligosaccharide at their polar ends; these are the **glycolipids** (Fig. 10–6). Within these groups of membrane lipids, enormous diversity results from various combinations of fatty acid "tails" and polar "heads." The arrangement of these lipids in membranes, and their structural and functional roles therein, are considered in the next chapter.

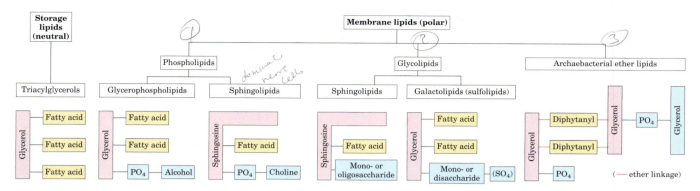

**FIGURE 10–6 Some common types of storage and membrane lipids.** All the lipid types shown here have either glycerol or sphingosine as the backbone (pink screen), to which are attached one or more long-chain alkyl groups (yellow) and a polar head group (blue). In triacylglycerols, glycerophospholipids, galactolipids, and sulfolipids, the alkyl groups are fatty acids in ester linkage. Sphingolipids contain a single fatty acid, in amide linkage to the sphingosine backbone. The membrane lipids of archaebacteria are variable; that shown here has two very long, branched alkyl chains, each end in ether linkage with a glycerol moiety. In phospholipids the polar head group is joined through a phosphodiester, whereas glycolipids have a direct glycosidic linkage between the head-group sugar and the backbone glycerol.

## Glycerophospholipids Are Derivatives of Phosphatidic Acid

**Glycerophospholipids,** also called phosphoglycerides, are membrane lipids in which two fatty acids are attached in ester linkage to the first and second carbons of glycerol, and a highly polar or charged group is attached through a phosphodiester linkage to the third carbon. Glycerol is prochiral; it has no asymmetric carbons, but attachment of phosphate at one end converts it into a chiral compound, which can be correctly named either L-glycerol 3-phosphate, D-glycerol 1-phosphate, or *sn*-glycerol 3-phosphate (Fig. 10–7). Glycerophospholipids are named as derivatives of the parent compound, phosphatidic acid (Fig. 10–8), according to the polar alcohol in the head group. Phosphatidylcholine and phosphatidylethanolamine have choline and ethanolamine in their polar head groups, for example. In all these compounds, the head group is joined to glycerol through a phosphodiester bond, in which the phosphate group bears a negative charge at neutral pH. The polar alcohol may be negatively charged (as in phosphatidylinositol 4,5-bisphosphate), neutral (phosphatidylserine), or positively charged (phosphatidylcholine, phosphatidylethanolamine). As we shall see in Chapter 11, these charges contribute greatly to the surface properties of membranes.

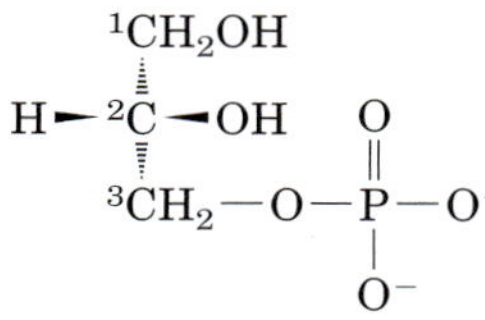

The fatty acids in glycerophospholipids can be any of a wide variety, so a given phospholipid (phosphatidylcholine, for example) may consist of a number of molecular species, each with its unique complement of fatty acids. The distribution of molecular species is specific for different organisms, different tissues of the same organism, and different glycerophospholipids in the same cell or tissue. In general, glycerophospholipids contain a $C_{16}$ or $C_{18}$ saturated fatty acid at C-1 and a $C_{18}$ to $C_{20}$ unsaturated fatty acid at C-2. With few exceptions, the biological significance of the variation in fatty acids and head groups is not yet understood.

$$^1CH_2OH$$
$$H-^2C-OH \quad O$$
$$^3CH_2-O-\overset{\displaystyle \|}{P}-O^-$$
$$\underset{\displaystyle |}{} O^-$$

L-Glycerol 3-phosphate
(*sn*-glycerol 3-phosphate)

**FIGURE 10–7 L-Glycerol 3-phosphate, the backbone of phospholipids.** Glycerol itself is not chiral, as it has a plane of symmetry through C-2. However, glycerol can be converted to a chiral compound by adding a substituent such as phosphate to either of the —CH₂OH groups; that is, glycerol is prochiral. One unambiguous nomenclature for glycerol phosphate is the DL system (described on p. 77), in which the isomers are named according to their stereochemical relationships to glyceraldehyde isomers. By this system, the stereoisomer of glycerol phosphate found in most lipids is correctly named either L-glycerol 3-phosphate or D-glycerol 1-phosphate. Another way to specify stereoisomers is the *stereospecific numbering* (*sn*) system, in which C-1 is, by definition, that group of the prochiral compound that occupies the pro-S position. The common form of glycerol phosphate in phospholipids is, by this system, *sn*-glycerol 3-phosphate.

## Some Phospholipids Have Ether-Linked Fatty Acids

Some animal tissues and some unicellular organisms are rich in **ether lipids,** in which one of the two acyl chains is attached to glycerol in ether, rather than ester, linkage. The ether-linked chain may be saturated, as in the alkyl ether lipids, or may contain a double bond between C-1 and C-2, as in **plasmalogens** (Fig. 10–9). Vertebrate heart tissue is uniquely enriched in ether lipids; about half of the heart phospholipids are plasmalogens. The membranes of halophilic bacteria, ciliated protists, and certain invertebrates also contain high proportions of

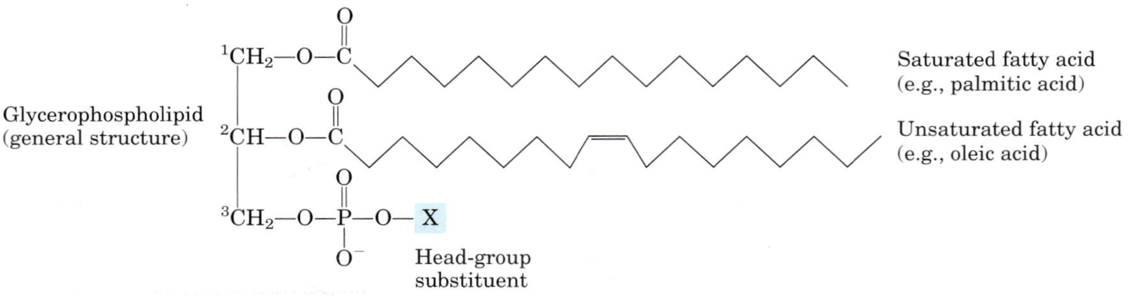

Glycerophospholipid (general structure)

Saturated fatty acid (e.g., palmitic acid)

Unsaturated fatty acid (e.g., oleic acid)

Head-group substituent

| Name of glycerophospholipid | Name of X | Formula of X | Net charge (at pH 7) |
|---|---|---|---|
| Phosphatidic acid | — | $-H$ | $-1$ |
| Phosphatidylethanolamine | Ethanolamine | $-CH_2-CH_2-\overset{+}{N}H_3$ | $0$ |
| Phosphatidylcholine | Choline | $-CH_2-CH_2-\overset{+}{N}(CH_3)_3$ | $0$ |
| Phosphatidylserine | Serine | $-CH_2-\underset{COO^-}{\overset{+}{\underset{|}{CH}}}-\overset{+}{N}H_3$ | $-1$ |
| Phosphatidylglycerol | Glycerol | $-CH_2-\underset{OH}{\overset{|}{CH}}-CH_2-OH$ | $-1$ |
| Phosphatidylinositol 4,5-bisphosphate | *myo*-Inositol 4,5-bisphosphate | 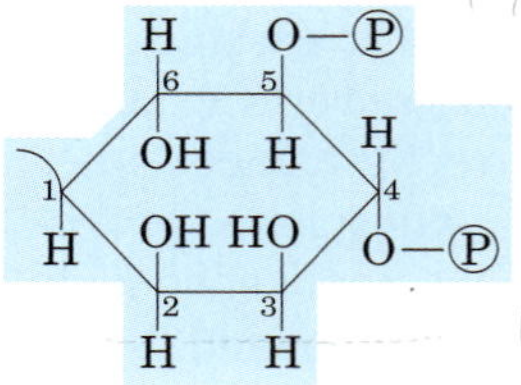 | $-4$ |
| Cardiolipin | Phosphatidyl-glycerol | 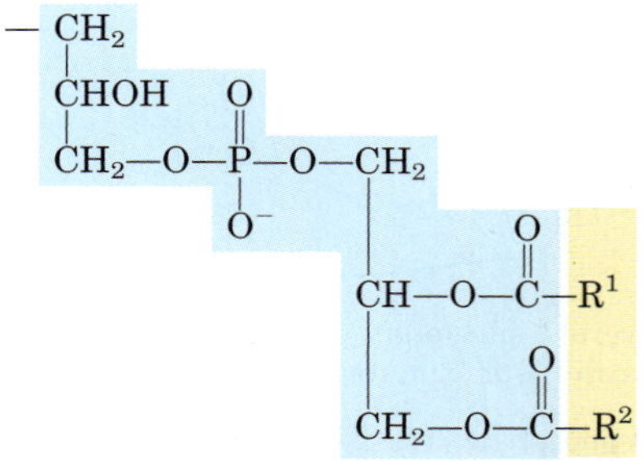 | $-2$ |

FIGURE 10–8  **Glycerophospholipids.** The common glycerophospholipids are diacylglycerols linked to head-group alcohols through a phosphodiester bond. Phosphatidic acid, a phosphomonoester, is the parent compound. Each derivative is named for the head-group alcohol (X), with the prefix "phosphatidyl-." In cardiolipin, two phosphatidic acids share a single glycerol.

ether lipids. The functional significance of ether lipids in these membranes is unknown; perhaps their resistance to the phospholipases that cleave ester-linked fatty acids from membrane lipids is important in some roles.

At least one ether lipid, **platelet-activating factor,** is a potent molecular signal. It is released from leukocytes called basophils and stimulates platelet aggregation and the release of serotonin (a vasoconstrictor) from platelets. It also exerts a variety of effects on liver, smooth muscle, heart, uterine, and lung tissues and plays an important role in inflammation and the allergic response. ■

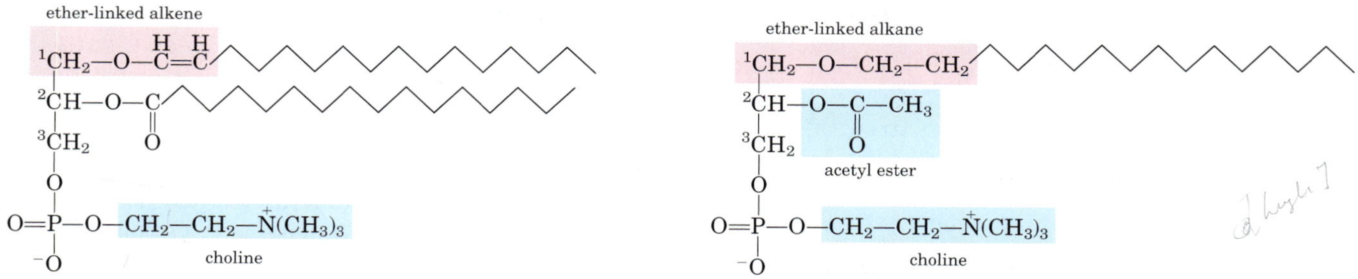

**FIGURE 10-9   Ether lipids.** Plasmalogens have an ether-linked alkenyl chain where most glycerophospholipids have an ester-linked fatty acid (compare Fig. 10–8). Platelet-activating factor has a long ether-linked alkyl chain at C-1 of glycerol, but C-2 is ester-linked to acetic acid, which makes the compound much more water-soluble than most glycerophospholipids and plasmalogens. The head-group alcohol is choline in plasmalogens and in platelet-activating factor.

## Chloroplasts Contain Galactolipids and Sulfolipids

The second group of membrane lipids are those that predominate in plant cells: the **galactolipids,** in which one or two galactose residues are connected by a glycosidic linkage to C-3 of a 1,2-diacylglycerol (Fig. 10–10; see also Fig. 10–6). Galactolipids are localized in the thylakoid membranes (internal membranes) of chloroplasts; they make up 70% to 80% of the total membrane lipids of a vascular plant. They are probably the most abundant membrane lipids in the biosphere. Phosphate is often the limiting plant nutrient in soil, and perhaps the evolutionary pressure to conserve phosphate for more critical roles favored plants that made phosphate-free lipids. Plant membranes also contain sulfolipids, in which a sulfonated glucose residue is joined to a diacylglycerol in glycosidic linkage. In sulfolipids, the sulfonate on the head group bears a fixed negative charge like that of the phosphate group in phospholipids (Fig. 10–10).

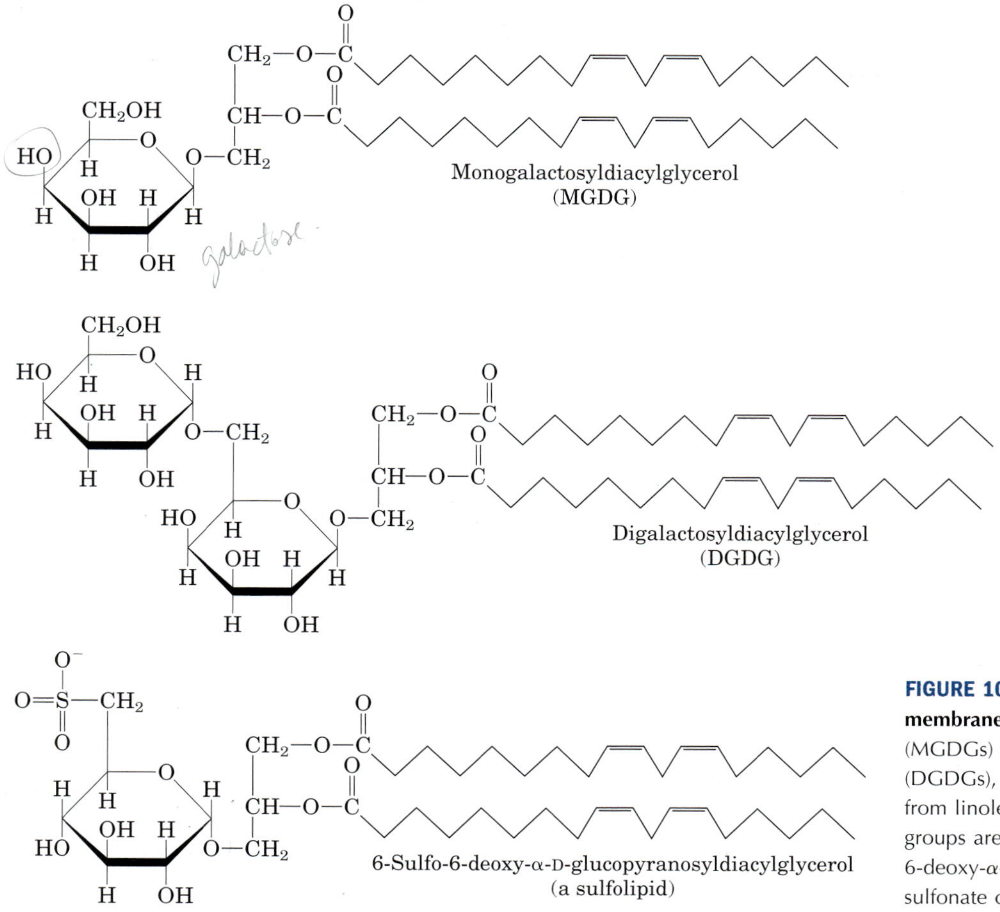

**FIGURE 10-10   Three glycolipids of chloroplast membranes.** In monogalactosyldiacylglycerols (MGDGs) and digalactosyldiacylglycerols (DGDGs), almost all the acyl groups are derived from linoleic acid (18:2($\Delta^{9,12}$)) and the head groups are uncharged. In the sulfolipid 6-sulfo-6-deoxy-$\alpha$-D-glucopyranosyldiacylglycerol, the sulfonate carries a fixed negative charge.

## Archaebacteria Contain Unique Membrane Lipids

The archaebacteria, most of which live in ecological niches with extreme conditions—high temperatures (boiling water), low pH, high ionic strength, for example—have membrane lipids containing long-chain (32 carbons) branched hydrocarbons linked at each end to glycerol (Fig. 10–11). These linkages are through ether bonds, which are much more stable to hydrolysis at low pH and high temperature than are the ester bonds found in the lipids of eubacteria and eukaryotes. In their fully extended form, these archaebacterial lipids are twice the length of phospholipids and sphingolipids and span the width of the surface membrane. At each end of the extended molecule is a polar head consisting of glycerol linked to either phosphate or sugar residues. The general name for these compounds, glycerol dialkyl glycerol tetraethers (GDGTs), reflects their unique structure. The glycerol moiety of the archaebacterial lipids is not the same stereoisomer as that in the lipids of eubacteria and eukaryotes; the central carbon is in the R configuration in archaebacteria, in the S configuration in the other kingdoms (Fig. 10–7).

## Sphingolipids Are Derivatives of Sphingosine

**Sphingolipids,** the fourth large class of membrane lipids, also have a polar head group and two nonpolar tails, but unlike glycerophospholipids and galactolipids they contain no glycerol. Sphingolipids are composed of one molecule of the long-chain amino alcohol sphingosine (also called 4-sphingenine) or one of its derivatives, one molecule of a long-chain fatty acid, and a polar head group that is joined by a glycosidic linkage in some cases and by a phosphodiester in others (Fig. 10–12).

Carbons C-1, C-2, and C-3 of the sphingosine molecule are structurally analogous to the three carbons of glycerol in glycerophospholipids. When a fatty acid is attached in amide linkage to the —NH$_2$ on C-2, the resulting compound is a **ceramide,** which is structurally similar to a diacylglycerol. Ceramide is the structural parent of all sphingolipids.

There are three subclasses of sphingolipids, all derivatives of ceramide but differing in their head groups: sphingomyelins, neutral (uncharged) glycolipids, and gangliosides. **Sphingomyelins** contain phosphocholine or phosphoethanolamine as their polar head group and are therefore classified along with glycerophospholipids as phospholipids (Fig. 10–6). Indeed, sphingomyelins resemble phosphatidylcholines in their general properties and three-dimensional structure, and in having no net charge on their head groups (Fig. 10–13). Sphingomyelins are present in the plasma membranes of animal cells and are especially prominent in myelin, a membranous sheath that surrounds and insulates the axons of some neurons—thus the name "sphingomyelins."

**Glycosphingolipids,** which occur largely in the outer face of plasma membranes, have head groups with one or more sugars connected directly to the —OH at C-1 of the ceramide moiety; they do not contain phosphate. **Cerebrosides** have a single sugar linked to ceramide; those with galactose are characteristically found in the plasma membranes of cells in neural tissue, and those with glucose in the plasma membranes of cells in nonneural tissues. **Globosides** are neutral (uncharged) glycosphingolipids with two or more sugars, usually D-glucose, D-galactose, or N-acetyl-D-galactosamine. Cerebrosides and globosides are sometimes called **neutral glycolipids,** as they have no charge at pH 7.

**Gangliosides,** the most complex sphingolipids, have oligosaccharides as their polar head groups and one or more residues of N-acetylneuraminic acid (Neu5Ac), a sialic acid (often simply called "sialic acid"), at the

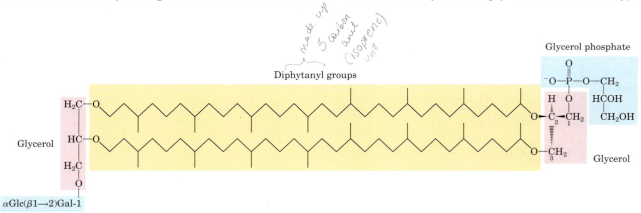

**FIGURE 10–11 A typical membrane lipid of archaebacteria.** In this diphytanyl tetraether lipid, the diphytanyl moieties (yellow) are long hydrocarbons composed of eight five-carbon isoprene groups condensed end-to-end (on the condensation of isoprene units, see Fig. 21–36; also, compare the diphytanyl groups with the 20-carbon phytol side chain of chlorophylls in Fig. 19–40a). In this extended form, the diphytanyl groups are about twice the length of a 16-carbon fatty acid typically found in the membrane lipids of eubacteria and eukaryotes. The glycerol moieties in the archaebacterial lipids are in the R configuration, in contrast to those of eubacteria and eukaryotes, which have the S configuration. Archaebacterial lipids differ in the substituents on the glycerols. In the molecule shown here, one glycerol is linked to the disaccharide α-glucopyranosyl-(1→2)-β-galactofuranose; the other glycerol is linked to a glycerol phosphate head group.

Sphingosine

$$HO-^3CH-CH=CH-(CH_2)_{12}-CH_3$$

Fatty acid

Sphingolipid
(general
structure)

$$^2CH-N-\overset{\overset{\displaystyle O}{\|}}{C}$$

H

$$^1CH_2-O-X$$

| Name of sphingolipid | Name of X | Formula of X |
|---|---|---|
| Ceramide | — | —H |
| Sphingomyelin | Phosphocholine | $-\overset{\overset{\displaystyle O}{\|}}{\underset{\underset{\displaystyle O^-}{\|}}{P}}-O-CH_2-CH_2-\overset{+}{N}(CH_3)_3$ |
| Neutral glycolipids Glucosylcerebroside | Glucose | (structure of glucose) |
| Lactosylceramide (a globoside) | Di-, tri-, or tetrasaccharide | Glc—Gal |
| Ganglioside GM2 | Complex oligosaccharide | Glc—Gal—GalNAc with Neu5Ac |

Johann Thudichum,
1829–1901

**FIGURE 10–12 Sphingolipids.** The first three carbons at the polar end of sphingosine are analogous to the three carbons of glycerol in glycerophospholipids. The amino group at C-2 bears a fatty acid in amide linkage. The fatty acid is usually saturated or monounsaturated, with 16, 18, 22, or 24 carbon atoms. Ceramide is the parent compound for this group. Other sphingolipids differ in the polar head group (X) attached at C-1. Gangliosides have very complex oligosaccharide head groups. Standard abbreviations for sugars are used in this figure: Glc, D-glucose; Gal, D-galactose; GalNAc, N-acetyl-D-galactosamine; Neu5Ac, N-acetylneuraminic acid (sialic acid).

termini. Sialic acid gives gangliosides the negative charge at pH 7 that distinguishes them from globosides. Gangliosides with one sialic acid residue are in the GM (M for mono-) series, those with two are in the GD (D for di-) series, and so on (GT, three sialic acid residues; GQ, four).

*N*-Acetylneuraminic acid (a sialic acid)
(Neu5Ac)

## Sphingolipids at Cell Surfaces Are Sites of Biological Recognition

When sphingolipids were discovered a century ago by the physician-chemist Johann Thudichum, their biological role seemed as enigmatic as the Sphinx, for which he therefore named them. In humans, at least 60 different sphingolipids have been identified in cellular membranes. Many of these are especially prominent in the plasma membranes of neurons, and some are clearly recognition sites on the cell surface, but a specific function for only a few sphingolipids has been discovered thus far. The carbohydrate moieties of certain sphingolipids define the human blood groups and therefore determine the type of blood that individuals can safely receive in blood transfusions (Fig. 10–14).

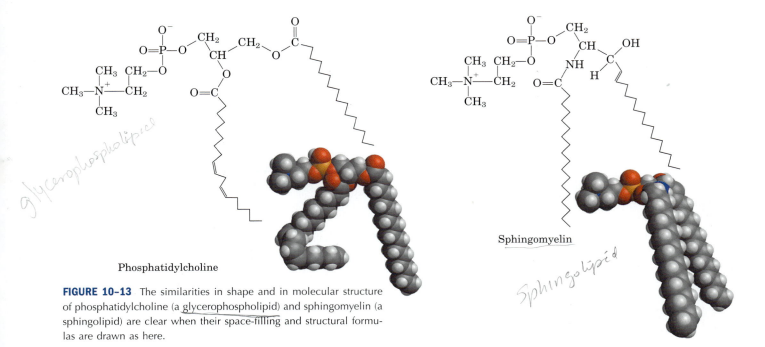

*glycerophospholipid*

Phosphatidylcholine

Sphingomyelin

*Sphingolipid*

**FIGURE 10–13** The similarities in shape and in molecular structure of phosphatidylcholine (a glycerophospholipid) and sphingomyelin (a sphingolipid) are clear when their space-filling and structural formulas are drawn as here.

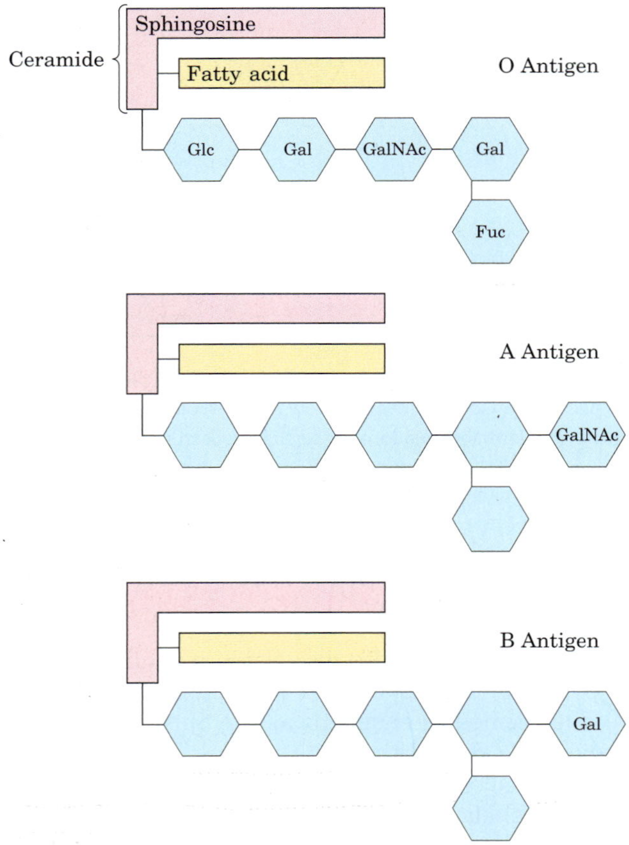

Ceramide {

Sphingosine

Fatty acid

O Antigen

Glc — Gal — GalNAc — Gal

Fuc

A Antigen

GalNAc

B Antigen

Gal

**FIGURE 10–14** **Glycosphingolipids as determinants of blood groups.** The human blood groups (O, A, B) are determined in part by the oligosaccharide head groups (blue) of these glycosphingolipids. The same three oligosaccharides are also found attached to certain blood proteins of individuals of blood types O, A, and B, respectively. (Fuc represents the sugar fucose.)

Gangliosides are concentrated in the outer surface of cells, where they present points of recognition for extracellular molecules or surfaces of neighboring cells. The kinds and amounts of gangliosides in the plasma membrane change dramatically during embryonic development. Tumor formation induces the synthesis of a new complement of gangliosides, and very low concentrations of a specific ganglioside have been found to induce differentiation of cultured neuronal tumor cells. Investigation of the biological roles of diverse gangliosides remains fertile ground for future research.

### Phospholipids and Sphingolipids Are Degraded in Lysosomes

Most cells continually degrade and replace their membrane lipids. For each hydrolyzable bond in a glycerophospholipid, there is a specific hydrolytic enzyme in the lysosome (Fig. 10–15). Phospholipases of the A type remove one of the two fatty acids, producing a lysophospholipid. (These esterases do not attack the ether link of plasmalogens.) Lysophospholipases remove the remaining fatty acid.

Gangliosides are degraded by a set of lysosomal enzymes that catalyze the stepwise removal of sugar units, finally yielding a ceramide. A genetic defect in any of these hydrolytic enzymes leads to the accumulation of gangliosides in the cell, with severe medical consequences (Box 10–2).

### Sterols Have Four Fused Carbon Rings

**Sterols** are structural lipids present in the membranes of most eukaryotic cells. The characteristic structure of this fifth group of membrane lipids is the steroid nu-

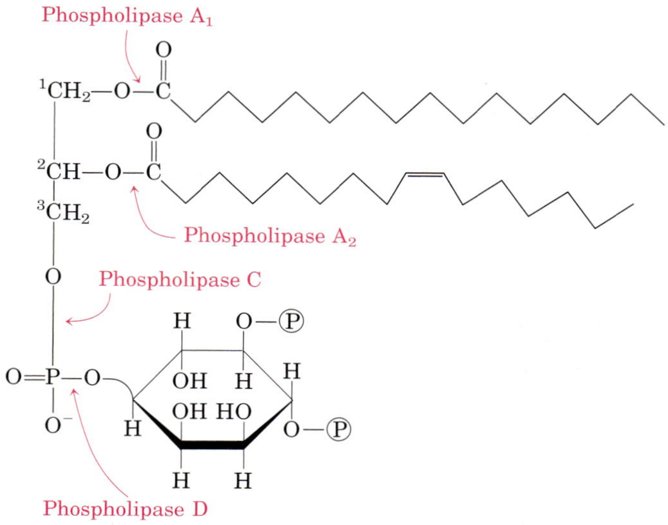

Phospholipase A₁

Phospholipase A₂

Phospholipase C

Phospholipase D

**FIGURE 10–15 The specificities of phospholipases.** Phospholipases A₁ and A₂ hydrolyze the ester bonds of intact glycerophospholipids at C-1 and C-2 of glycerol, respectively. Phospholipases C and D each split one of the phosphodiester bonds in the head group. Some phospholipases act on only one type of glycerophospholipid, such as phosphatidylinositol 4,5-bisphosphate (shown here) or phosphatidylcholine; others are less specific. When one of the fatty acids has been removed by a type A phospholipase, the second fatty acid is cleaved from the molecule by a lysophospholipase (not shown).

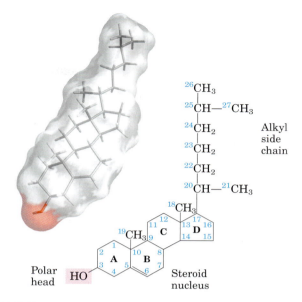

Polar head

Alkyl side chain

Steroid nucleus

**FIGURE 10–16 Cholesterol.** The stick structure of cholesterol is visible through a transparent surface contour model of the molecule (from coordinates supplied by Dave Woodcock). In the chemical structure, the rings are labeled A through D to simplify reference to derivatives of the steroid nucleus, and the carbon atoms are numbered in blue. The C-3 hydroxyl group (pink in both representations) is the polar head group. For storage and transport of the sterol, this hydroxyl group condenses with a fatty acid to form a sterol ester.

cleus, consisting of four fused rings, three with six carbons and one with five (Fig. 10–16). The steroid nucleus is almost planar and is relatively rigid; the fused rings do not allow rotation about C—C bonds. **Cholesterol,** the major sterol in animal tissues, is amphipathic, with a polar head group (the hydroxyl group at C-3) and a nonpolar hydrocarbon body (the steroid nucleus and the hydrocarbon side chain at C-17), about as long as a 16-carbon fatty acid in its extended form. Similar sterols are found in other eukaryotes: stigmasterol in plants and ergosterol in fungi, for example. Bacteria cannot synthesize sterols; a few bacterial species, however, can incorporate exogenous sterols into their membranes. The sterols of all eukaryotes are synthesized from simple five-carbon isoprene subunits, as are the fat-soluble vitamins, quinones, and dolichols described in Section 10.3.

In addition to their roles as membrane constituents, the sterols serve as precursors for a variety of products with specific biological activities. Steroid hormones, for example, are potent biological signals that regulate gene expression. Bile acids are polar derivatives of cholesterol that act as detergents in the intestine, emulsifying dietary fats to make them more readily accessible to digestive lipases. We return to cholesterol and other sterols in later chapters, to consider the structural role of cholesterol in biological membranes (Chapter 11), signaling by steroid hormones (Chapter 12), the remarkable biosynthetic pathway to cholesterol, and the transport of cholesterol by lipoprotein carriers (Chapter 21).

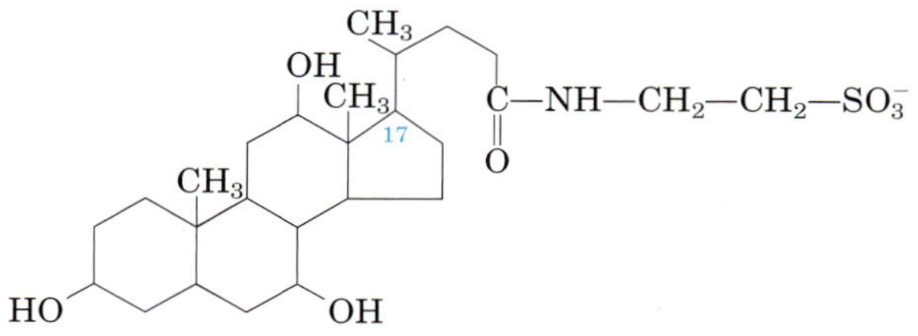

Taurocholic acid
(a bile acid)

## SUMMARY 10.2 Structural Lipids in Membranes

■ The polar lipids, with polar heads and nonpolar tails, are major components of membranes. The most abundant are the glycerophospholipids, which contain fatty acids esterified to two of the hydroxyl groups of glycerol, and a second alcohol, the head group, esterified to the third hydroxyl of glycerol via a phosphodiester bond. Other polar lipids are the sterols.

■ Glycerophospholipids differ in the structure of their head group; common glycerophospholipids are phosphatidylethanolamine and phosphatidylcholine. The polar heads of the glycerophospholipids carry electric charges at pH near 7.

■ Chloroplast membranes are remarkably rich in galactolipids, composed of a diacylglycerol with

**BOX 10–2    BIOCHEMISTRY IN MEDICINE**

## Inherited Human Diseases Resulting from Abnormal Accumulations of Membrane Lipids

The polar lipids of membranes undergo constant metabolic turnover, the rate of their synthesis normally counterbalanced by the rate of breakdown. The breakdown of lipids is promoted by hydrolytic enzymes in lysosomes, each enzyme capable of hydrolyzing a specific bond. When sphingolipid degradation is impaired by a defect in one of these enzymes (Fig. 1), partial breakdown products accumulate in the tissues, causing serious disease.

For example, Niemann-Pick disease is caused by a rare genetic defect in the enzyme sphingomyelinase, which cleaves phosphocholine from sphingomyelin. Sphingomyelin accumulates in the brain, spleen, and liver. The disease becomes evident in infants, and causes mental retardation and early death. More common is Tay-Sachs disease, in which ganglioside GM2 accumulates in the brain and spleen (Fig. 2) owing to lack of the enzyme hexosaminidase A. The symptoms of Tay-Sachs disease are progressive retardation in development, paralysis, blindness, and death by the age of 3 or 4 years.

Genetic counseling can predict and avert many inheritable diseases. Tests on prospective parents can detect abnormal enzymes, then DNA testing can determine the exact nature of the defect and the risk it poses for offspring. Once a pregnancy occurs, fetal cells obtained by sampling a part of the placenta (chorionic villus sampling) or the fluid surrounding the fetus (amniocentesis) can be tested in the same way.

**FIGURE 1** Pathways for the breakdown of GM1, globoside, and sphingomyelin to ceramide. A defect in the enzyme hydrolyzing a particular step is indicated by ⊗, and the disease that results from accumulation of the partial breakdown product is noted.

**FIGURE 2** Electron micrograph of a portion of a brain cell from an infant with Tay-Sachs disease, showing abnormal ganglioside deposits in the lysosomes.

Ceramide — GM1
β-galactosidase ⊗ → Generalized gangliosidosis

Ceramide — GM2
hexosaminidase A ⊗ → Tay-Sachs disease

Ceramide — GM3
ganglioside neuraminidase     α-galactosidase A

Ceramide
β-galactosidase

Ceramide
glucocerebrosidase ⊗ → Gaucher's disease

Ceramide

Ceramide — Globoside
hexosaminidase A and B ⊗ → Sandhoff's disease

Ceramide
α-galactosidase A ⊗ → Fabry's disease

Ceramide — Phosphocholine
Sphingomyelin

sphingomyelinase ⊗ → Niemann-Pick disease
Phosphocholine

1 μm

⬡ Glc     ⬡ GalNAc
⬡ Gal     ⬡ Neu5Ac

one or two linked galactose residues, and sulfolipids, diacylglycerols with a linked sulfonated sugar residue and thus a negatively charged head group.

- Archaebacteria have unique membrane lipids, with long-chain alkyl groups ether-linked to glycerol at both ends and with sugar residues and/or phosphate joined to the glycerol to provide a polar or charged head group. These lipids are stable under the harsh conditions in which archaebacteria live.

- The sphingolipids contain sphingosine, a long-chain aliphatic amino alcohol, but no glycerol. Sphingomyelin has, in addition to phosphoric acid and choline, two long hydrocarbon chains, one contributed by a fatty acid and the other by sphingosine. Three other classes of sphingolipids are cerebrosides, globosides, and gangliosides, which contain sugar components.

- Sterols have four fused rings and a hydroxyl group. Cholesterol, the major sterol in animals, is both a structural component of membranes and precursor to a wide variety of steroids.

## 10.3 Lipids as Signals, Cofactors, and Pigments

The two functional classes of lipids considered thus far (storage lipids and structural lipids) are major cellular components; membrane lipids make up 5% to 10% of the dry mass of most cells, and storage lipids more than 80% of the mass of an adipocyte. With some important exceptions, these lipids play a *passive* role in the cell; lipid fuels are stored until oxidized by enzymes, and membrane lipids form impermeable barriers around cells and cellular compartments. Another group of lipids, present in much smaller amounts, have *active* roles in the metabolic traffic as metabolites and messengers. Some serve as potent signals—as hormones, carried in the blood from one tissue to another, or as intracellular messengers generated in response to an extracellular signal (hormone or growth factor). Others function as enzyme cofactors in electron-transfer reactions in chloroplasts and mitochondria, or in the transfer of sugar moieties in a variety of glycosylation (addition of sugar) reactions. A third group consists of lipids with a system of conjugated double bonds: pigment molecules that absorb visible light. Some of these act as light-capturing pigments in vision and photosynthesis; others produce natural colorations, such as the orange of pumpkins and carrots and the yellow of canary feathers. Specialized lipids such as these are derived from lipids of the plasma membrane or from the fat-soluble vitamins A, D, E, and K. We describe in this section a few of these biologically active lipids. In later chapters, their synthesis and biological roles are considered in more detail.

### Phosphatidylinositols and Sphingosine Derivatives Act as Intracellular Signals

Phosphatidylinositol and its phosphorylated derivatives act at several levels to regulate cell structure and metabolism (Fig. 10–17). Phosphatidylinositol 4,5-bisphosphate (Fig. 10–8) in the cytoplasmic (inner) face of plasma membranes serves as a specific binding site for certain cytoskeletal proteins and for some soluble proteins involved in membrane fusion during exocytosis. It also serves as a reservoir of messenger molecules that are released inside the cell in response to extracellular signals interacting with specific receptors on the outer surface of the plasma membrane. The signals act through a series of steps (Fig. 10–17) that begins with enzymatic removal of a phospholipid head group and ends with activation of an enzyme (protein kinase C). For example, when the hormone vasopressin binds to plasma membrane receptors on the epithelial cells of the renal collecting duct, a specific phospholipase C is activated.

Phospholipase C hydrolyzes the bond between glycerol and phosphate in phosphatidylinositol 4,5-bisphosphate, releasing two products: inositol 1,4,5-trisphosphate ($IP_3$), which is water-soluble, and diacylglycerol, which remains associated with the plasma membrane. $IP_3$ triggers release of $Ca^{2+}$ from the endoplasmic reticulum, and the combination of diacylglycerol and elevated cytosolic $Ca^{2+}$ activates the enzyme protein kinase C.

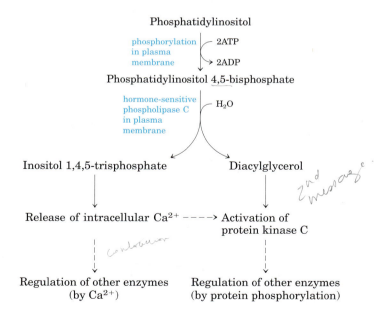

**FIGURE 10–17 Phosphatidylinositols in cellular regulation.** Phosphatidylinositol 4,5-bisphosphate in the plasma membrane is hydrolyzed by a specific phospholipase C in response to hormonal signals. Both products of hydrolysis act as intracellular messengers.

This enzyme catalyzes the transfer of a phosphoryl group from ATP to a specific residue in one or more target proteins, thereby altering their activity and consequently the cell's metabolism. This signaling mechanism is described more fully in Chapter 12 (see Fig. 12–19).

Inositol phospholipids also serve as points of nucleation for certain supramolecular complexes involved in signaling or in exocytosis. Proteins that contain certain structural motifs, called PH and PX domains (for *p*leckstrin *h*omology and *Phox* homology, respectively), bind phosphatidylinositols in the membrane with high specificity and affinity, initiating the formation of multienzyme complexes at the membrane's cytosolic surface. A number of proteins bind specifically to phosphatidylinositol 3,4,5-trisphosphate, and the formation of this phospholipid in response to extracellular signals brings the proteins together at the surface of the plasma membrane (see Fig. 12–8).

Membrane sphingolipids also can serve as sources of intracellular messengers. Both ceramide and sphingomyelin (Fig. 10–12) are potent regulators of protein kinases, and ceramide or its derivatives are known to be involved in the regulation of cell division, differentiation, migration, and programmed cell death (also called apoptosis; see Chapter 12).

## Eicosanoids Carry Messages to Nearby Cells

Eicosanoids are paracrine hormones, substances that act only on cells near the point of hormone synthesis instead of being transported in the blood to act on cells in other tissues or organs. These fatty acid derivatives have a variety of dramatic effects on vertebrate tissues. They are known to be involved in reproductive function; in the inflammation, fever, and pain associated with injury or disease; in the formation of blood clots and the regulation of blood pressure; in gastric acid secretion; and in a variety of other processes important in human health or disease.

All eicosanoids are derived from arachidonic acid ($20{:}4(\Delta^{5,8,11,14})$) (Fig. 10–18), the 20-carbon polyunsaturated fatty acid from which they take their gen-

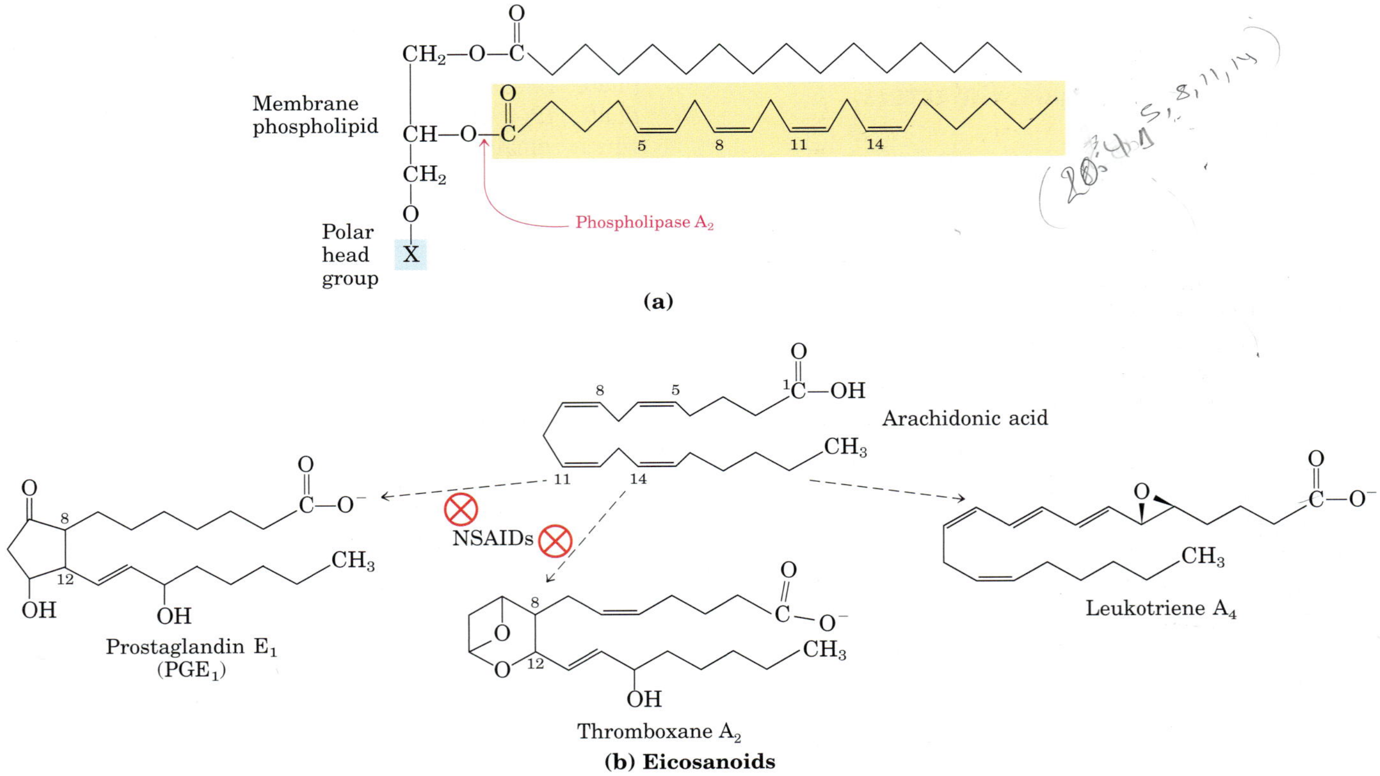

**(a)**

**(b) Eicosanoids**

**FIGURE 10–18 Arachidonic acid and some eicosanoid derivatives. (a)** In response to hormonal signals, phospholipase $A_2$ cleaves arachidonic acid–containing membrane phospholipids to release arachidonic acid (arachidonate at pH 7), the precursor to various eicosanoids. **(b)** These compounds include prostaglandins such as PGE$_1$, in which C-8 and C-12 of arachidonate are joined to form the characteristic five-membered ring. In thromboxane A$_2$, the C-8 and C-12 are joined and an oxygen atom is added to form the six-membered ring. Leukotriene A$_4$ has a series of three conjugated double bonds. Nonsteroidal antiinflammatory drugs (NSAIDs) such as aspirin and ibuprofen block the formation of prostaglandins and thromboxanes from arachidonate by inhibiting the enzyme cyclooxygenase (prostaglandin H$_2$ synthase).

John Vane, Sune Bergström, and Bengt Samuelsson

eral name (Greek *eikosi*, "twenty"). There are three classes of eicosanoids: prostaglandins, thromboxanes, and leukotrienes.

**Prostaglandins** (PG) contain a five-carbon ring originating from the chain of arachidonic acid. Their name derives from the prostate gland, the tissue from which they were first isolated by Bengt Samuelsson and Sune Bergström. Two groups of prostaglandins were originally defined: PGE, for *ether*-soluble, and PGF, for phosphate ( *fosfat* in Swedish) buffer–soluble. Each group contains numerous subtypes, named $PGE_1$, $PGE_2$, and so forth. Prostaglandins act in many tissues by regulating the synthesis of the intracellular messenger 3′,5′-cyclic AMP (cAMP). Because cAMP mediates the action of diverse hormones, the prostaglandins affect a wide range of cellular and tissue functions. Some prostaglandins stimulate contraction of the smooth muscle of the uterus during menstruation and labor. Others affect blood flow to specific organs, the wake-sleep cycle, and the responsiveness of certain tissues to hormones such as epinephrine and glucagon. Prostaglandins in a third group elevate body temperature ( producing fever) and cause inflammation and pain.

The **thromboxanes** have a six-membered ring containing an ether. They are produced by platelets (also called thrombocytes) and act in the formation of blood clots and the reduction of blood flow to the site of a clot. The nonsteroidal antiinflammatory drugs (NSAIDs)— aspirin, ibuprofen, and meclofenamate, for example— were shown by John Vane to inhibit the enzyme prostaglandin $H_2$ synthase (also called cyclooxygenase or COX), which catalyzes an early step in the pathway from arachidonate to prostaglandins and thromboxanes (Fig. 10–18; see also Box 21–2).

**Leukotrienes,** first found in leukocytes, contain three conjugated double bonds. They are powerful biological signals. For example, leukotriene $D_4$, derived from leukotriene $A_4$, induces contraction of the muscle lining the airways to the lung. Overproduction of leukotrienes causes asthmatic attacks, and leukotriene synthesis is one target of antiasthmatic drugs such as prednisone. The strong contraction of the smooth muscles of the lung that occurs during anaphylactic shock is part of the potentially fatal allergic reaction in individuals hypersensitive to bee stings, penicillin, or other agents. ■

## Steroid Hormones Carry Messages between Tissues

Steroids are oxidized derivatives of sterols; they have the sterol nucleus but lack the alkyl chain attached to ring D of cholesterol, and they are more polar than cholesterol. Steroid hormones move through the bloodstream (on protein carriers) from their site of production to target tissues, where they enter cells, bind to highly specific receptor proteins in the nucleus, and trigger changes in gene expression and metabolism. Because hormones have very high affinity for their receptors, very low concentrations of hormones (nanomolar or less) are sufficient to produce responses in target tissues. The major groups of steroid hormones are the male and female sex hormones and the hormones produced by the adrenal cortex, cortisol and aldosterone (Fig. 10–19). Prednisone and prednisolone are steroid drugs with potent antiinflammatory activities, mediated in part by the inhibition of arachidonate release by phospholipase $A_2$ (Fig. 10–18) and consequent inhibition of the

Testosterone

Estradiol

Cortisol

Aldosterone

Prednisolone

Prednisone

**FIGURE 10–19  Steroids derived from cholesterol.** Testosterone, the male sex hormone, is produced in the testes. Estradiol, one of the female sex hormones, is produced in the ovaries and placenta. Cortisol and aldosterone are hormones synthesized in the cortex of the adrenal gland; they regulate glucose metabolism and salt excretion, respectively. Prednisolone and prednisone are synthetic steroids used as antiinflammatory agents.

synthesis of leukotrienes, prostaglandins, and thromboxanes. They have a variety of medical applications, including the treatment of asthma and rheumatoid arthritis. ■

## Plants Use Phosphatidylinositols, Steroids, and Eicosanoidlike Compounds in Signaling

Vascular plants contain phosphatidylinositol 4,5-bisphosphate, as well as the phospholipase that releases IP$_3$, and they use IP$_3$ to regulate the intracellular concentration of Ca$^{2+}$. Brassinolide and the related group of brassinosteroids are potent growth regulators in plants, increasing the rate of stem elongation and influencing the orientation of cellulose microfibrils in the cell wall during growth. Jasmonate, derived from the fatty acid 18:3($\Delta^{9,12,15}$) in membrane lipids, is chemically similar to the eicosanoids of animal tissues and also serves as a powerful signal, triggering the plant's defenses in response to insect-inflicted damage. The methyl ester of jasmonate gives the characteristic fragrance of jasmine oil, which is widely used in the perfume industry.

Brassinolide
(a brassinosteroid)

Jasmonate

## Vitamins A and D Are Hormone Precursors

During the first third of the twentieth century, a major focus of research in physiological chemistry was the identification of **vitamins,** compounds that are essential to the health of humans and other vertebrates but cannot be synthesized by these animals and must therefore be obtained in the diet. Early nutritional

**(a)**

7-Dehydrocholesterol

UV light
⟿
2 steps (in skin)

Cholecalciferol (vitamin D$_3$)

1 step in the liver
1 step in the kidney

1,25-Dihydroxycholecalciferol
(1,25-dihydroxyvitamin D$_3$)

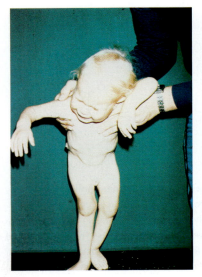

Before vitamin D treatment

After 14 months of vitamin D treatment

**(b)**

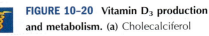

**FIGURE 10–20 Vitamin D$_3$ production and metabolism. (a)** Cholecalciferol (vitamin D$_3$) is produced in the skin by UV irradiation of 7-dehydrocholesterol, which breaks the bond shaded pink. In the liver, a hydroxyl group is added at C-25 (pink); in the kidney, a second hydroxylation at C-1 (pink) produces the active hormone, 1,25-dihydroxycholecalciferol. This hormone regulates the metabolism of Ca$^{2+}$ in kidney, intestine, and bone. **(b)** Dietary vitamin D prevents rickets, a disease once common in cold climates where heavy clothing blocks the UV component of sunlight necessary for the production of vitamin D$_3$ in skin. On the left is a 2½-year-old boy with severe rickets; on the right, the same boy at age 5, after 14 months of vitamin D therapy.

studies identified two general classes of such compounds: those soluble in nonpolar organic solvents (fat-soluble vitamins) and those that could be extracted from foods with aqueous solvents (water-soluble vitamins). Eventually the fat-soluble group was resolved into the four vitamin groups A, D, E, and K, all of which are isoprenoid compounds synthesized by the condensation of multiple isoprene units. Two of these (D and A) serve as hormone precursors.

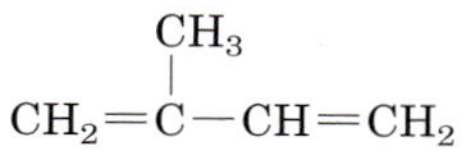

$$CH_2{=}\underset{\underset{CH_3}{|}}{C}{-}CH{=}CH_2$$

Isoprene

**Vitamin $D_3$,** also called **cholecalciferol,** is normally formed in the skin from 7-dehydrocholesterol in a photochemical reaction driven by the UV component of sunlight (Fig. 10–20). Vitamin $D_3$ is not itself biologically active, but it is converted by enzymes in the liver and kidney to 1,25-dihydroxycholecalciferol, a hormone that regulates calcium uptake in the intestine and calcium levels in kidney and bone. Deficiency of vitamin D leads to defective bone formation and the disease rickets, for which administration of vitamin D produces a dramatic cure. Vitamin $D_2$ (ergocalciferol) is a commercial product formed by UV irradiation of the ergosterol of yeast. Vitamin $D_2$ is structurally similar to $D_3$, with slight modification to the side chain attached to the sterol D ring. Both have the same biological effects, and $D_2$ is commonly added to milk and butter as a dietary supplement. Like steroid hormones, the product of vitamin D metabolism, 1,25-dihydroxycholecalciferol, regulates gene expression—for example, turning on the synthesis of an intestinal $Ca^{2+}$-binding protein.

**Vitamin A (retinol)** in its various forms functions as a hormone and as the visual pigment of the vertebrate eye (Fig. 10–21). Acting through receptor proteins in the cell nucleus, the vitamin A derivative retinoic acid regulates gene expression in the development of epithelial tissue, including skin. Retinoic acid is the active ingredient in the drug tretinoin (Retin-A), used in the treatment of severe acne and wrinkled skin. The vitamin A derivative retinal is the pigment that initiates the

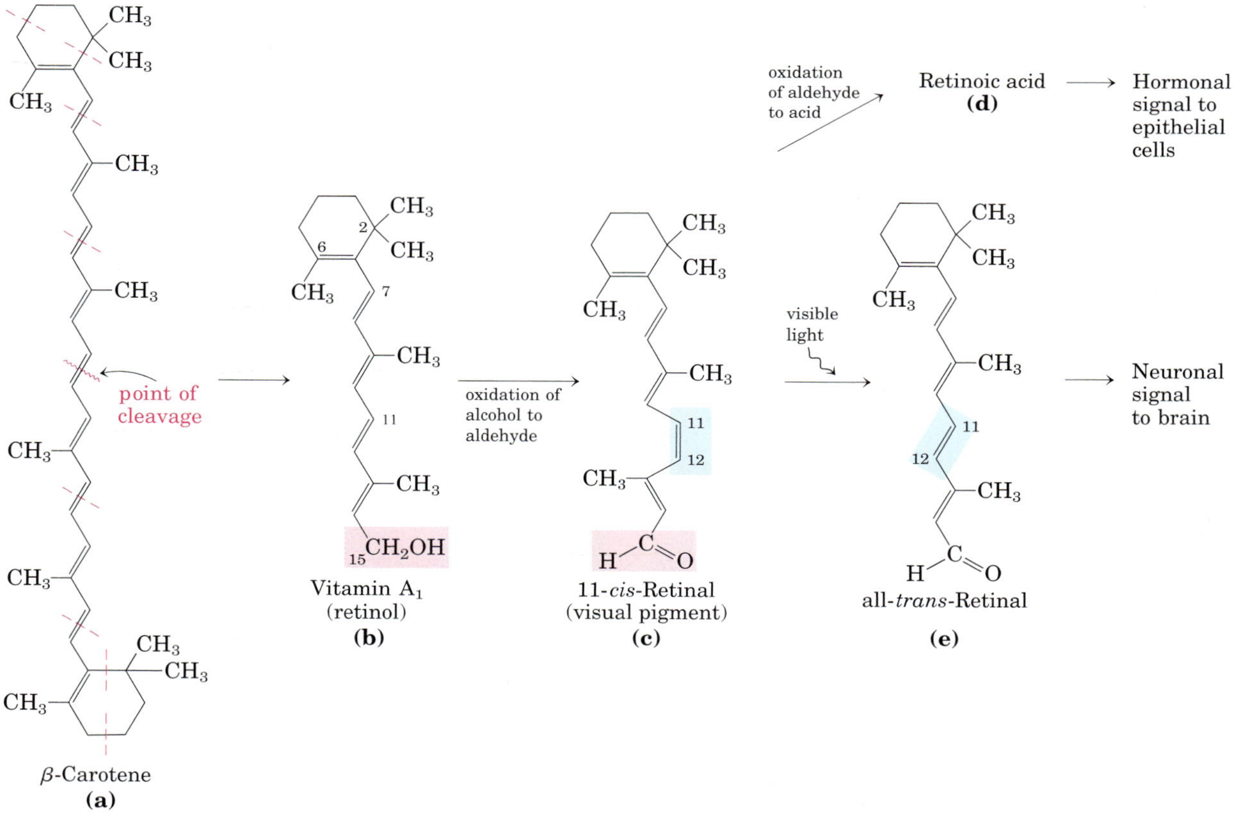

**FIGURE 10–21 Vitamin $A_1$ and its precursor and derivatives.**
**(a)** β-Carotene is the precursor of vitamin $A_1$. Isoprene structural units are set off by dashed red lines. Cleavage of β-carotene yields two molecules of vitamin $A_1$ (retinol) **(b).** Oxidation at C-15 converts retinol to the aldehyde, retinal **(c),** and further oxidation produces retinoic acid **(d),** a hormone that regulates gene expression. Retinal combines with the protein opsin to form rhodopsin (not shown), a visual pigment widespread in nature. In the dark, retinal of rhodopsin is in the 11-cis form **(c).** When a rhodopsin molecule is excited by visible light, the 11-cis-retinal undergoes a series of photochemical reactions that convert it to all-trans-retinal **(e),** forcing a change in the shape of the entire rhodopsin molecule. This transformation in the rod cell of the vertebrate retina sends an electrical signal to the brain that is the basis of visual transduction, a topic we address in more detail in Chapter 12.

response of rod and cone cells of the retina to light, producing a neuronal signal to the brain. This role of retinal is described in detail in Chapter 12.

Vitamin A was first isolated from fish liver oils; liver, eggs, whole milk, and butter are good dietary sources. In vertebrates, β-carotene, the pigment that gives carrots, sweet potatoes, and other yellow vegetables their characteristic color, can be enzymatically converted to vitamin A. Deficiency of vitamin A leads to a variety of symptoms in humans, including dryness of the skin, eyes, and mucous membranes; retarded development and growth; and night blindness, an early symptom commonly used in diagnosing vitamin A deficiency. ■

## Vitamins E and K and the Lipid Quinones Are Oxidation-Reduction Cofactors

**Vitamin E** is the collective name for a group of closely related lipids called **tocopherols,** all of which contain a substituted aromatic ring and a long isoprenoid side chain (Fig. 10–22a). Because they are hydrophobic, tocopherols associate with cell membranes, lipid deposits, and lipoproteins in the blood. Tocopherols are biological antioxidants. The aromatic ring reacts with and destroys the most reactive forms of oxygen radicals and other free radicals, protecting unsaturated fatty acids from oxidation and preventing oxidative

**(a)**
Vitamin E: an antioxidant

**(b)**
Vitamin $K_1$: a blood-clotting cofactor (phylloquinone)

**(c)**
Warfarin: a blood anticoagulant

**(d)**
Ubiquinone: a mitochondrial electron carrier (coenzyme Q)
($n$ = 4 to 8)

**(e)**
Plastoquinone: a chloroplast electron carrier ($n$ = 4 to 8)

**(f)**
Dolichol: a sugar carrier
($n$ = 9 to 22)

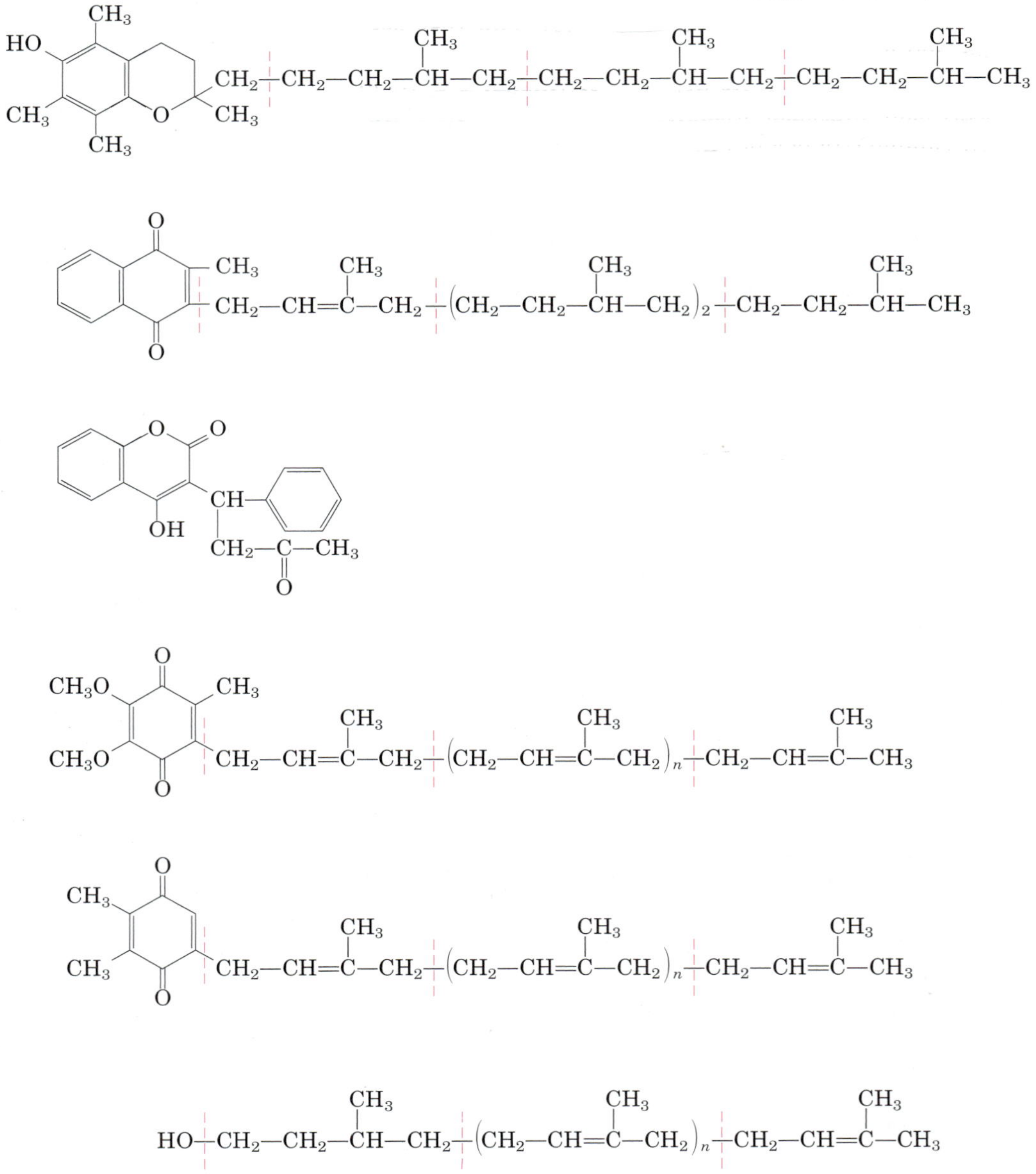

**FIGURE 10–22 Some other biologically active isoprenoid compounds or derivatives.** Isoprene structural units are set off by dashed red lines. In most mammalian tissues, ubiquinone (also called coenzyme Q) has 10 isoprene units. Dolichols of animals have 17 to 21 isoprene units (85 to 105 carbon atoms), bacterial dolichols have 11, and those of plants and fungi have 14 to 24.

Edward A. Doisy,
1893–1986

Henrik Dam,
1895–1976

damage to membrane lipids, which can cause cell fragility. Tocopherols are found in eggs and vegetable oils and are especially abundant in wheat germ. Laboratory animals fed diets depleted of vitamin E develop scaly skin, muscular weakness and wasting, and sterility. Vitamin E deficiency in humans is very rare; the principal symptom is fragile erythrocytes.

The aromatic ring of **vitamin K** (Fig. 10–22b) undergoes a cycle of oxidation and reduction during the formation of active prothrombin, a blood plasma protein essential in blood clot formation. Prothrombin is a proteolytic enzyme that splits peptide bonds in the blood protein fibrinogen to convert it to fibrin, the insoluble fibrous protein that holds blood clots together. Henrik Dam and Edward A. Doisy independently discovered that vitamin K deficiency slows blood clotting, which can be fatal. Vitamin K deficiency is very uncommon in humans, aside from a small percentage of infants who suffer from hemorrhagic disease of the newborn, a potentially fatal disorder. In the United States, newborns are routinely given a 1 mg injection of vitamin K. Vitamin K$_1$ (phylloquinone) is found in green plant leaves; a related form, vitamin K$_2$ (menaquinone), is formed by bacteria residing in the vertebrate intestine.

Warfarin (Fig. 10–22c) is a synthetic compound that inhibits the formation of active prothrombin. It is particularly poisonous to rats, causing death by internal bleeding. Ironically, this potent rodenticide is also an invaluable anticoagulant drug for treating humans at risk for excessive blood clotting, such as surgical patients and those with coronary thrombosis. ■

Ubiquinone (also called coenzyme Q) and plastoquinone (Fig. 10–22d, e) are isoprenoids that function as lipophilic electron carriers in the oxidation-reduction reactions that drive ATP synthesis in mitochondria and chloroplasts, respectively. Both ubiquinone and plastoquinone can accept either one or two electrons and either one or two protons (see Fig. 19–54).

### Dolichols Activate Sugar Precursors for Biosynthesis

During assembly of the complex carbohydrates of bacterial cell walls, and during the addition of polysaccha-
ride units to certain proteins (glycoproteins) and lipids (glycolipids) in eukaryotes, the sugar units to be added are chemically activated by attachment to isoprenoid alcohols called **dolichols** (Fig. 10–22f). These compounds have strong hydrophobic interactions with membrane lipids, anchoring the attached sugars to the membrane, where they participate in sugar-transfer reactions.

### SUMMARY 10.3  Lipids as Signals, Cofactors, and Pigments

- Some types of lipids, although present in relatively small quantities, play critical roles as cofactors or signals.

- Phosphatidylinositol bisphosphate is hydrolyzed to yield two intracellular messengers, diacylglycerol and inositol 1,4,5-trisphosphate. Phosphatidylinositol 3,4,5-trisphosphate is a nucleation point for supramolecular protein complexes involved in biological signaling.

- Prostaglandins, thromboxanes, and leukotrienes (the eicosanoids), derived from arachidonate, are extremely potent hormones.

- Steroid hormones, derived from sterols, serve as powerful biological signals, such as the sex hormones.

- Vitamins D, A, E, and K are fat-soluble compounds made up of isoprene units. All play essential roles in the metabolism or physiology of animals. Vitamin D is precursor to a hormone that regulates calcium metabolism. Vitamin A furnishes the visual pigment of the vertebrate eye and is a regulator of gene expression during epithelial cell growth. Vitamin E functions in the protection of membrane lipids from oxidative damage, and vitamin K is essential in the blood-clotting process.

- Ubiquinones and plastoquinones, also isoprenoid derivatives, function as electron carriers in mitochondria and chloroplasts, respectively.

- Dolichols activate and anchor sugars on cellular membranes for use in the synthesis of certain complex carbohydrates, glycolipids, and glycoproteins.

## 10.4 Working with Lipids

In exploring the biological role of lipids in cells and tissues, it is essential to know which lipids are present and in what proportions. Because lipids are insoluble in water, their extraction and subsequent fractionation require the use of organic solvents and some techniques not commonly used in the purification of water-soluble molecules such as proteins and carbohydrates. In general, complex mixtures of lipids are separated by differences in the polarity or solubility of the components in nonpolar solvents. Lipids that contain ester- or amide-linked fatty acids can be hydrolyzed by treatment with acid or alkali or with highly specific hydrolytic enzymes (phospholipases, glycosidases) to yield their component parts for analysis. Some methods commonly used in lipid analysis are shown in Figure 10–23 and discussed below.

## Lipid Extraction Requires Organic Solvents

Neutral lipids (triacylglycerols, waxes, pigments, and so forth) are readily extracted from tissues with ethyl ether, chloroform, or benzene, solvents that do not permit lipid clustering driven by hydrophobic interactions. Membrane lipids are more effectively extracted by more polar organic solvents, such as ethanol or methanol, which reduce the hydrophobic interactions among lipid molecules while also weakening the hydrogen bonds and electrostatic interactions that bind membrane lipids to membrane proteins. A commonly used extractant is a mixture of chloroform, methanol, and water, initially in volume proportions (1:2:0.8) that are miscible, producing a single phase. After tissue is homogenized in this solvent to extract all lipids, more water is added to the resulting extract and the mixture separates into two phases, methanol/water (top phase) and chloroform (bottom phase). The lipids remain in the chloroform layer, and more polar molecules such as proteins and sugars partition into the methanol/water layer.

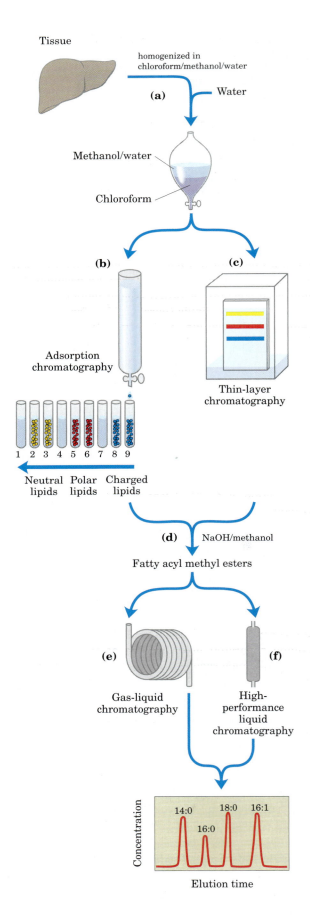

**FIGURE 10–23 Common procedures in the extraction, separation, and identification of cellular lipids. (a)** Tissue is homogenized in a chloroform/methanol/water mixture, which on addition of water and removal of unextractable sediment by centrifugation yields two phases. Different types of extracted lipids in the chloroform phase may be separated by **(b)** adsorption chromatography on a column of silica gel, through which solvents of increasing polarity are passed, or **(c)** thin-layer chromatography (TLC), in which lipids are carried up a silica gel-coated plate by a rising solvent front, less polar lipids traveling farther than more polar or charged lipids. TLC with appropriate solvents can also be used to separate closely related lipid species; for example, the charged lipids phosphatidylserine, phosphatidylglycerol, and phosphatidylinositol are easily separated by TLC.

For the determination of fatty acid composition, a lipid fraction containing ester-linked fatty acids is transesterified in a warm aqueous solution of NaOH and methanol **(d)**, producing a mixture of fatty acyl methyl esters. These methyl esters are then separated on the basis of chain length and degree of saturation by **(e)** gas-liquid chromatography (GLC) or **(f)** high-performance liquid chromatography (HPLC). Precise determination of molecular mass by mass spectrometry allows unambiguous identification of individual lipids.

## Adsorption Chromatography Separates Lipids of Different Polarity

Complex mixtures of tissue lipids can be fractionated by chromatographic procedures based on the different polarities of each class of lipid. In adsorption chromatography (Fig. 10–23b), an insoluble, polar material such as silica gel (a form of silicic acid, $Si(OH)_4$) is packed into a glass column, and the lipid mixture (in chloroform solution) is applied to the top of the column. (In high-performance liquid chromatography, the column is of smaller diameter and solvents are forced through the column under high pressure.) The polar lipids bind tightly to the polar silicic acid, but the neutral lipids pass directly through the column and emerge in the first chloroform wash. The polar lipids are then eluted, in order of increasing polarity, by washing the column with solvents of progressively higher polarity. Uncharged but polar lipids (cerebrosides, for example) are eluted with acetone, and very polar or charged lipids (such as glycerophospholipids) are eluted with methanol.

Thin-layer chromatography on silicic acid employs the same principle (Fig. 10–23c). A thin layer of silica gel is spread onto a glass plate, to which it adheres. A small sample of lipids dissolved in chloroform is applied near one edge of the plate, which is dipped in a shallow container of an organic solvent or solvent mixture—all of which is enclosed within a chamber saturated with the solvent vapor. As the solvent rises on the plate by capillary action, it carries lipids with it. The less polar lipids move farthest, as they have less tendency to bind to the silicic acid. The separated lipids can be detected by spraying the plate with a dye (rhodamine) that fluoresces when associated with lipids or by exposing the plate to iodine fumes. Iodine reacts reversibly with the double bonds in fatty acids, such that lipids containing unsaturated fatty acids develop a yellow or brown color. A number of other spray reagents are also useful in detecting specific lipids. For subsequent analysis, regions containing separated lipids can be scraped from the plate and the lipids recovered by extraction with an organic solvent.

## Gas-Liquid Chromatography Resolves Mixtures of Volatile Lipid Derivatives

Gas-liquid chromatography separates volatile components of a mixture according to their relative tendencies to dissolve in the inert material packed in the chromatography column and to volatilize and move through the column, carried by a current of an inert gas such as helium. Some lipids are naturally volatile, but most must first be derivatized to increase their volatility (that is, lower their boiling point). For an analysis of the fatty acids in a sample of phospholipids, the lipids are first heated in a methanol/HCl or methanol/NaOH mixture, which converts fatty acids esterified to glycerol into their methyl esters (in a process of transesterification; Fig. 10–23d). These fatty acyl methyl esters are then loaded onto the gas-liquid chromatography column, and the column is heated to volatilize the compounds. Those fatty acyl esters most soluble in the column material partition into (dissolve in) that material; the less soluble lipids are carried by the stream of inert gas and emerge first from the column. The order of elution depends on the nature of the solid adsorbant in the column and on the boiling point of the components of the lipid mixture. Using these techniques, mixtures of fatty acids of various chain lengths and various degrees of unsaturation can be completely resolved (Fig. 10–23e).

## Specific Hydrolysis Aids in Determination of Lipid Structure

Certain classes of lipids are susceptible to degradation under specific conditions. For example, all ester-linked fatty acids in triacylglycerols, phospholipids, and sterol esters are released by mild acid or alkaline treatment, and somewhat harsher hydrolysis conditions release amide-bound fatty acids from sphingolipids. Enzymes that specifically hydrolyze certain lipids are also useful in the determination of lipid structure. Phospholipases A, C, and D (Fig. 10–15) each split particular bonds in phospholipids and yield products with characteristic solubilities and chromatographic behaviors. Phospholipase C, for example, releases a water-soluble phosphoryl alcohol (such as phosphocholine from phosphatidylcholine) and a chloroform-soluble diacylglycerol, each of which can be characterized separately to determine the structure of the intact phospholipid. The combination of specific hydrolysis with characterization of the products by thin-layer, gas-liquid, or high-performance liquid chromatography often allows determination of a lipid structure.

## Mass Spectrometry Reveals Complete Lipid Structure

To establish unambiguously the length of a hydrocarbon chain or the position of double bonds, mass spectral analysis of lipids or their volatile derivatives is invaluable. The chemical properties of similar lipids (for example, two fatty acids of similar length unsaturated at different positions, or two isoprenoids with different numbers of isoprene units) are very much alike, and their positions of elution from the various chromatographic procedures often do not distinguish between them. When the effluent from a chromatography column is sampled by mass spectrometry, however, the components of a lipid mixture can be simultaneously separated and identified by their unique pattern of fragmentation (Fig. 10–24).

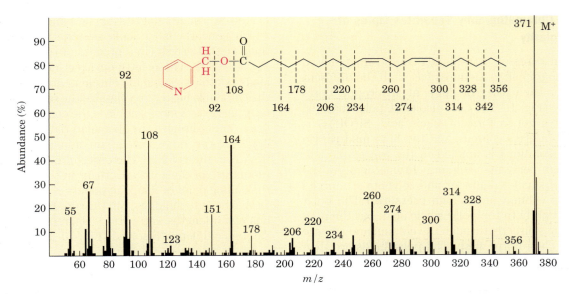

**FIGURE 10–24** **Determination of the structure of a fatty acid by mass spectrometry.** The fatty acid is first converted to a derivative that minimizes migration of the double bonds when the molecule is fragmented by electron bombardment. The derivative shown here is a picolinyl ester of linoleic acid—18:2($\Delta^{9,12}$) ($M_r$ 371)—in which the alcohol is picolinol (red). When bombarded with a stream of electrons, this molecule is volatilized and converted to a parent ion (M$^+$; $M_r$ 371), in which the N atom bears the positive charge, and a series of smaller fragments produced by breakage of C—C bonds in the fatty acid. The mass spectrometer separates these charged fragments according to their mass/charge ratio (m/z). (To review the principles of mass spectrometry, see Box 3–2.)

The prominent ions at m/z = 92, 108, 151, and 164 contain the pyridine ring of the picolinol and various fragments of the carboxyl group, showing that the compound is indeed a picolinyl ester. The molecular ion (m/z = 371) confirms the presence of a C-18 fatty acid with two double bonds. The uniform series of ions 14 atomic mass units (amu) apart represents loss of each successive methyl and methylene group from the right end of the molecule (C-18 of the fatty acid), until the ion at m/z = 300 is reached. This is followed by a gap of 26 amu for the carbons of the terminal double bond, at m/z = 274; a further gap of 14 amu for the C-11 methylene group, at m/z = 260, and so forth. By this means the entire structure is determined, although these data alone do not reveal the configuration (cis or trans) of the double bonds.

## SUMMARY 10.4  Working with Lipids

- In the determination of lipid composition, the lipids are first extracted from tissues with organic solvents and separated by thin-layer, gas-liquid, or high-performance liquid chromatography.

- Phospholipases specific for one of the bonds in a phospholipid can be used to generate simpler compounds for subsequent analysis.

- Individual lipids are identified by their chromatographic behavior, their susceptibility to hydrolysis by specific enzymes, or mass spectrometry.

## Key Terms

*Terms in bold are defined in the glossary.*

| | | | |
|---|---|---|---|
| **fatty acid**  343 | **plasmalogen**  349 | neutral glycolipids  352 | vitamin D$_3$  361 |
| **triacylglycerol**  345 | galactolipid  351 | **gangliosides**  352 | cholecalciferol  361 |
| **lipases**  346 | **sphingolipid**  352 | **sterols**  354 | vitamin A (retinol)  361 |
| **phospholipid**  348 | ceramide  352 | cholesterol  355 | vitamin E  362 |
| **glycolipid**  348 | sphingomyelin  352 | **prostaglandins**  359 | **tocopherols**  362 |
| **glycerophospholipid** | glycosphingolipid  352 | **thromboxanes**  359 | vitamin K  363 |
| 349 | **cerebroside**  352 | **leukotrienes**  359 | dolichol  363 |
| ether lipid  349 | globoside  352 | **vitamin**  360 | |

# Further Reading

## General
**Gurr, M.I. & Harwood, J.L.** (1991) *Lipid Biochemistry: An Introduction,* 4th edn, Chapman & Hall, London.

A good general resource on lipid structure and metabolism, at the intermediate level.

**Vance, D.E. & Vance, J.E. (eds)** (2002) *Biochemistry of Lipids, Lipoproteins, and Membranes,* New Comprehensive Biochemistry, Vol. 36, Elsevier Science Publishing Co., Inc., New York.

An excellent collection of reviews on various aspects of lipid structure, biosynthesis, and function.

## Structural Lipids in Membranes
**Bogdanov, M. & Dowhan, W.** (1999) Lipid-assisted protein folding. *J. Biol. Chem.* **274,** 36,827–36,830.

A minireview of the role of membrane lipids in the folding of membrane proteins.

**De Rosa, M. & Gambacorta, A.** (1988) The lipids of archaebacteria. *Prog. Lipid Res.* **27,** 153–175.

**Dowhan, W.** (1997) Molecular basis for membrane phospholipid diversity: why are there so many lipids? *Annu. Rev. Biochem.* **66,** 199–232.

**Gravel, R.A., Kaback, M.M., Proia, R., Sandhoff, K., Suzuki, K., & Suzuki, K.** (2001) The GM$_2$ gangliosidoses. In *The Metabolic and Molecular Bases of Inherited Disease,* 8th edn (Scriver, C.R., Sly, W.S., Childs, B., Beaudet, A.L., Valle, D., Kinzler, K.W., & Vogelstein, B., eds), pp. 3827–3876, McGraw-Hill, Inc., New York.

This article is one of many in a four-volume set that contains definitive descriptions of the clinical, biochemical, and genetic aspects of hundreds of human metabolic diseases—an authoritative source and fascinating reading.

**Hoekstra, D. (ed.)** (1994) *Cell Lipids,* Current Topics in Membranes, Vol. 4, Academic Press, Inc., San Diego.

## Lipids as Signals, Cofactors, and Pigments
**Bell, R.M., Exton, J.H., & Prescott, S.M. (eds)** (1996) *Lipid Second Messengers,* Handbook of Lipid Research, Vol. 8, Plenum Press, New York.

**Binkley, N.C. & Suttie, J.W.** (1995) Vitamin K nutrition and osteoporosis. *J. Nutr.* **125,** 1812–1821.

**Brigelius-Flohé, R. & Traber, M.G.** (1999) Vitamin E: function and metabolism. *FASEB J.* **13,** 1145–1155.

**Chojnacki, T. & Dallner, G.** (1988) The biological role of dolichol. *Biochem. J.* **251,** 1–9.

**Clouse, S.D.** (2002) Brassinosteroid signal transduction: clarifying the pathway from ligand perception to gene expression. *Mol. Cell* **10,** 973–982.

**Lemmon, M.A. & Ferguson, K.M.** (2000) Signal-dependent membrane targeting by pleckstrin homology (PH) domains. *Biochem. J.* **350,** 1–18.

**Prescott, S.M., Zimmerman, G.A., Stafforini, D.M., & McIntyre, T.M.** (2000) Platelet-activating factor and related lipid mediators. *Annu. Rev. Biochem.* **69,** 419–445.

**Schneiter, R.** (1999) Brave little yeast, please guide us to Thebes: sphingolipid function in *S. cerevisiae. BioEssays* **21,** 1004–1010.

**Suttie, J.W.** (1993) Synthesis of vitamin K-dependent proteins. *FASEB J.* **7,** 445–452.

**Vermeer, C.** (1990) γ-Carboxyglutamate-containing proteins and the vitamin K-dependent carboxylase. *Biochem. J.* **266,** 625–636.

Describes the biochemical basis for the requirement of vitamin K in blood clotting and the importance of carboxylation in the synthesis of the blood-clotting protein thrombin.

**Viitala, J. & Järnefelt, J.** (1985) The red cell surface revisited. *Trends Biochem. Sci.* **10,** 392–395.

Includes discussion of the human A, B, and O blood type determinants.

**Weber, H.** (2002) Fatty acid-derived signals in plants. *Trends Plant Sci.* **7,** 217–224.

**Zittermann, A.** (2001) Effects of vitamin K on calcium and bone metabolism. *Curr. Opin. Clin. Nutr. Metab. Care* **4,** 483–487.

## Working with Lipids
**Christie, W.W.** (1998) Gas chromatography-mass spectrometry methods for structural analysis of fatty acids. *Lipids* **33,** 343–353.

A detailed description of the methods used to obtain data such as those presented in Figure 10–24.

**Christie, W.W.** (2003) *Lipid Analysis,* 3rd edn, The Oily Press, Bridgwater, England.

**Hamilton, R.J. & Hamilton, S. (eds)** (1992) *Lipid Analysis: A Practical Approach,* IRL Press at Oxford University Press, New York.

This text, though out of print, is available as part of the IRL Press *Practical Approach Series* on CD-ROM, from Oxford University Press (www.oup-usa/acadsci/pasbooks.html).

**Matsubara, T. & Hagashi, A.** (1991) FAB/mass spectrometry of lipids. *Prog. Lipid Res.* **30,** 301–322.

An advanced discussion of the identification of lipids by fast atom bombardment (FAB) mass spectrometry, a powerful technique for structure determination.

# Problems

**1. Operational Definition of Lipids** How is the definition of "lipid" different from the types of definitions used for other biomolecules that we have considered, such as amino acids, nucleic acids, and proteins?

**2. Melting Points of Lipids** The melting points of a series of 18-carbon fatty acids are: stearic acid, 69.6 °C; oleic acid, 13.4 °C; linoleic acid, −5 °C; and linolenic acid, −11 °C.

(a) What structural aspect of these 18-carbon fatty acids

can be correlated with the melting point? Provide a molecular explanation for the trend in melting points.

(b) Draw all the possible triacylglycerols that can be constructed from glycerol, palmitic acid, and oleic acid. Rank them in order of increasing melting point.

(c) Branched-chain fatty acids are found in some bacterial membrane lipids. Would their presence increase or decrease the fluidity of the membranes (that is, give them a lower or higher melting point)? Why?

**3. Preparation of Béarnaise Sauce** During the preparation of béarnaise sauce, egg yolks are incorporated into melted butter to stabilize the sauce and avoid separation. The stabilizing agent in the egg yolks is lecithin (phosphatidylcholine). Suggest why this works.

**4. Hydrophobic and Hydrophilic Components of Membrane Lipids** A common structural feature of membrane lipids is their amphipathic nature. For example, in phosphatidylcholine, the two fatty acid chains are hydrophobic and the phosphocholine head group is hydrophilic. For each of the following membrane lipids, name the components that serve as the hydrophobic and hydrophilic units: (a) phosphatidylethanolamine; (b) sphingomyelin; (c) galactosylcerebroside; (d) ganglioside; (e) cholesterol.

**5. Alkali Lability of Triacylglycerols** A common procedure for cleaning the grease trap in a sink is to add a product that contains sodium hydroxide. Explain why this works.

**6. The Action of Phospholipases** The venom of the Eastern diamondback rattler and the Indian cobra contains phospholipase $A_2$, which catalyzes the hydrolysis of fatty acids at the C-2 position of glycerophospholipids. The phospholipid breakdown product of this reaction is lysolecithin (lecithin is phosphatidylcholine). At high concentrations, this and other lysophospholipids act as detergents, dissolving the membranes of erythrocytes and lysing the cells. Extensive hemolysis may be life-threatening.

(a) Detergents are amphipathic. What are the hydrophilic and hydrophobic portions of lysolecithin?

(b) The pain and inflammation caused by a snake bite can be treated with certain steroids. What is the basis of this treatment?

(c) Though high levels of phospholipase $A_2$ can be deadly, this enzyme is necessary for a variety of normal metabolic processes. What are these processes?

**7. Intracellular Messengers from Phosphatidylinositols** When the hormone vasopressin stimulates cleavage of phosphatidylinositol 4,5-bisphosphate by hormone-sensitive phospholipase C, two products are formed. What are they? Compare their properties and their solubilities in water, and predict whether either would diffuse readily through the cytosol.

**8. Storage of Fat-Soluble Vitamins** In contrast to water-soluble vitamins, which must be a part of our daily diet, fat-soluble vitamins can be stored in the body in amounts sufficient for many months. Suggest an explanation for this difference, based on solubilities.

**9. Hydrolysis of Lipids** Name the products of mild hydrolysis with dilute NaOH of (a) 1-stearoyl-2,3-dipalmitoylglycerol; (b) 1-palmitoyl-2-oleoylphosphatidylcholine.

**10. Effect of Polarity on Solubility** Rank the following in order of increasing solubility in water: a triacylglycerol, a diacylglycerol, and a monoacylglycerol, all containing only palmitic acid.

**11. Chromatographic Separation of Lipids** A mixture of lipids is applied to a silica gel column, and the column is then washed with increasingly polar solvents. The mixture consists of phosphatidylserine, phosphatidylethanolamine, phosphatidylcholine, cholesteryl palmitate (a sterol ester), sphingomyelin, palmitate, *n*-tetradecanol, triacylglycerol, and cholesterol. In what order do you expect the lipids to elute from the column? Explain your reasoning.

**12. Identification of Unknown Lipids** Johann Thudichum, who practiced medicine in London about 100 years ago, also dabbled in lipid chemistry in his spare time. He isolated a variety of lipids from neural tissue, and characterized and named many of them. His carefully sealed and labeled vials of isolated lipids were rediscovered many years later.

(a) How would you confirm, using techniques not available to Thudichum, that the vials labeled "sphingomyelin" and "cerebroside" actually contain these compounds?

(b) How would you distinguish sphingomyelin from phosphatidylcholine by chemical, physical, or enzymatic tests?

**13. Ninhydrin to Detect Lipids on TLC Plates** Ninhydrin reacts specifically with primary amines to form a purplish-blue product. A thin-layer chromatogram of rat liver phospholipids is sprayed with ninhydrin, and the color is allowed to develop. Which phospholipids can be detected in this way?

# BIOLOGICAL MEMBRANES AND TRANSPORT

Good fences make good neighbors.
—*Robert Frost, "Mending Wall," in* North of Boston, *1914*

The first cell probably came into being when a membrane formed, enclosing a small volume of aqueous solution and separating it from the rest of the universe. Membranes define the external boundaries of cells and regulate the molecular traffic across that boundary (Fig. 11–1); in eukaryotic cells, they divide the internal space into discrete compartments to segregate processes and components. They organize complex reaction sequences and are central to both biological energy conservation and cell-to-cell communication. The biological activities of membranes flow from their remarkable physical properties. Membranes are flexible, self-sealing, and selectively permeable to polar solutes. Their flexibility permits the shape changes that accompany cell growth and movement (such as amoeboid movement). With their ability to break and reseal, two membranes can fuse, as in exocytosis, or a single membrane-enclosed compartment can undergo fission to yield two sealed compartments, as in endocytosis or cell division, without creating gross leaks through cellular surfaces. Because membranes are selectively permeable, they retain certain compounds and ions within cells and within specific cellular compartments, while excluding others.

Membranes are not merely passive barriers. They include an array of proteins specialized for promoting or catalyzing various cellular processes. At the cell surface, transporters move specific organic solutes and inorganic ions across the membrane; receptors sense extracellular signals and trigger molecular changes in the cell; adhesion molecules hold neighboring cells together. Within the cell, membranes organize cellular processes such as the synthesis of lipids and certain proteins, and the energy transductions in mitochondria and chloroplasts. Because membranes consist of just two layers of molecules, they are very thin—essentially two-dimensional. Intermolecular collisions are far more probable in this two-dimensional space than in three-dimensional space, so the efficiency of enzyme-catalyzed processes organized within membranes is vastly increased.

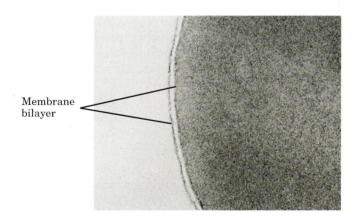

**FIGURE 11–1 Biological membranes.** Viewed in cross section, all cell membranes share a characteristic trilaminar appearance. When an erythrocyte is stained with osmium tetroxide and viewed with an electron microscope, the plasma membrane appears as a three-layer structure, 5 to 8 nm (50 to 80 Å) thick. The trilaminar image consists of two electron-dense layers (the osmium, bound to the inner and outer surfaces of the membrane) separated by a less dense central region.

Membrane bilayer

In this chapter we first describe the composition of cellular membranes and their chemical architecture—the molecular structures that underlie their biological functions. Next, we consider the remarkable dynamic features of membranes, in which lipids and proteins move relative to each other. Cell adhesion, endocytosis, and the membrane fusion accompanying neurotransmitter secretion illustrate the dynamic role of membrane proteins. We then turn to the protein-mediated passage of solutes across membranes via transporters and ion channels. In later chapters we discuss the role of membranes in signal transduction (Chapters 12 and 23), energy transduction (Chapter 19), lipid synthesis (Chapter 21), and protein synthesis (Chapter 27).

## 11.1 The Composition and Architecture of Membranes

One approach to understanding membrane function is to study membrane composition—to determine, for example, which components are common to all membranes and which are unique to membranes with specific functions. So before describing membrane structure and function we consider the molecular components of membranes: proteins and polar lipids, which account for almost all the mass of biological membranes, and carbohydrates, present as part of glycoproteins and glycolipids.

### Each Type of Membrane Has Characteristic Lipids and Proteins

The relative proportions of protein and lipid vary with the type of membrane (Table 11–1), reflecting the diversity of biological roles. For example, certain neurons have a myelin sheath, an extended plasma membrane that wraps around the cell many times and acts as a passive electrical insulator. The myelin sheath consists primarily of lipids, whereas the plasma membranes of bacteria and the membranes of mitochondria and chloroplasts, the sites of many enzyme-catalyzed processes, contain more protein than lipid (in mass per total mass).

For studies of membrane composition, the first task is to isolate a selected membrane. When eukaryotic cells are subjected to mechanical shear, their plasma membranes are torn and fragmented, releasing cytoplasmic components and membrane-bounded organelles such as mitochondria, chloroplasts, lysosomes, and nuclei. Plasma membrane fragments and intact organelles can be isolated by centrifugal techniques described in Chapter 1 (see Fig. 1–8).

Chemical analyses of membranes isolated from various sources reveal certain common properties. Each kingdom, each species, each tissue or cell type, and the organelles of each cell type have a characteristic set of membrane lipids. Plasma membranes, for example, are enriched in cholesterol and contain no detectable cardiolipin (Fig. 11–2); in the inner mitochondrial membrane of the hepatocyte, this distribution is reversed: very low cholesterol and high cardiolipin. Cardiolipin is essential to the function of certain proteins of the inner mitochondrial membrane. Cells clearly have mechanisms to control the kinds and amounts of membrane lipids they synthesize and to target specific lipids to particular organelles. In many cases, we can surmise the adaptive advantages of distinct combinations of membrane lipids; in other cases, the functional significance of these combinations is as yet unknown.

The protein composition of membranes from different sources varies even more widely than their lipid composition, reflecting functional specialization. In a rod cell of the vertebrate retina, one portion of the cell is highly specialized for the reception of light; more than 90% of the plasma membrane protein in this region is the light-absorbing glycoprotein rhodopsin. The less-specialized plasma membrane of the erythrocyte has about 20 prominent types of proteins as well as scores of minor ones; many of these are transporters, each moving a specific solute across the membrane. The plasma membrane of *Escherichia coli* contains hun-

---

**TABLE 11–1  Major Components of Plasma Membranes in Various Organisms**

| | Components (% by weight) | | | | |
| | Protein | Phospholipid | Sterol | Sterol type | Other lipids |
|---|---|---|---|---|---|
| Human myelin sheath | 30 | 30 | 19 | Cholesterol | Galactolipids, plasmalogens |
| Mouse liver | 45 | 27 | 25 | Cholesterol | — |
| Maize leaf | 47 | 26 | 7 | Sitosterol | Galactolipids |
| Yeast | 52 | 7 | 4 | Ergosterol | Triacylglycerols, steryl esters |
| *Paramecium* (ciliated protist) | 56 | 40 | 4 | Stigmasterol | — |
| *E. coli* | 75 | 25 | 0 | — | — |

**Note:** Values do not add up to 100% in every case, because there are components other than protein, phospholipids, and sterol; plants, for example, have high levels of glycolipids.

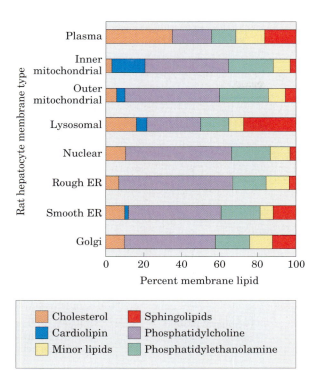

**FIGURE 11–2 Lipid composition of the plasma membrane and organelle membranes of a rat hepatocyte.** The functional specialization of each membrane type is reflected in its unique lipid composition. Cholesterol is prominent in plasma membranes but barely detectable in mitochondrial membranes. Cardiolipin is a major component of the inner mitochondrial membrane but not of the plasma membrane. Phosphatidylserine, phosphatidylinositol, and phosphatidylglycerol are relatively minor components (yellow) of most membranes but serve critical functions; phosphatidylinositol and its derivatives, for example, are important in signal transductions triggered by hormones. Sphingolipids, phosphatidylcholine, and phosphatidylethanolamine are present in most membranes, but in varying proportions. Glycolipids, which are major components of the chloroplast membranes of plants, are virtually absent from animal cells.

dreds of different proteins, including transporters and many enzymes involved in energy-conserving metabolism, lipid synthesis, protein export, and cell division. The outer membrane of *E. coli,* which encloses the plasma membrane, has a different function (protection) and a different set of proteins.

Some membrane proteins are covalently linked to complex arrays of carbohydrate. For example, in glycophorin, a glycoprotein of the erythrocyte plasma membrane, 60% of the mass consists of complex oligosaccharide units covalently attached to specific amino acid residues. Ser, Thr, and Asn residues are the most common points of attachment (see Fig. 7–31). At the other end of the scale is rhodopsin of the rod cell plasma membrane, which contains just one hexasaccharide. The sugar moieties of surface glycoproteins influence the folding of the proteins, as well as their sta-

bilities and intracellular destinations, and they play a significant role in the specific binding of ligands to glycoprotein surface receptors (see Fig. 7–37).

Some membrane proteins are covalently attached to one or more lipids, which serve as hydrophobic anchors that hold the proteins to the membrane, as we shall see.

## All Biological Membranes Share Some Fundamental Properties

Membranes are impermeable to most polar or charged solutes, but permeable to nonpolar compounds; they are 5 to 8 nm (50 to 80 Å) thick and appear trilaminar when viewed in cross section with the electron microscope (Fig. 11–1). The combined evidence from electron microscopy and studies of chemical composition, as well as physical studies of permeability and the motion of individual protein and lipid molecules within membranes, led to the development of the **fluid mosaic model** for the structure of biological membranes (Fig. 11–3). Phospholipids form a bilayer in which the nonpolar regions of the lipid molecules in each layer face the core of the bilayer and their polar head groups face outward, interacting with the aqueous phase on either side. Proteins are embedded in this bilayer sheet, held by hydrophobic interactions between the membrane lipids and hydrophobic domains in the proteins. Some proteins protrude from only one side of the membrane; others have domains exposed on both sides. The orientation of proteins in the bilayer is asymmetric, giving the membrane "sidedness": the protein domains exposed on one side of the bilayer are different from those exposed on the other side, reflecting functional asymmetry. The individual lipid and protein units in a membrane form a fluid mosaic with a pattern that, unlike a mosaic of ceramic tile and mortar, is free to change constantly. The membrane mosaic is fluid because most of the interactions among its components are noncovalent, leaving individual lipid and protein molecules free to move laterally in the plane of the membrane.

We now look at some of these features of the fluid mosaic model in more detail and consider the experimental evidence that supports the basic model but has necessitated its refinement in several ways.

## A Lipid Bilayer Is the Basic Structural Element of Membranes

Glycerophospholipids, sphingolipids, and sterols are virtually insoluble in water. When mixed with water, they spontaneously form microscopic lipid aggregates in a phase separate from their aqueous surroundings, clustering together, with their hydrophobic moieties in contact with each other and their hydrophilic groups interacting with the surrounding water. Recall that lipid clustering reduces the amount of hydrophobic surface

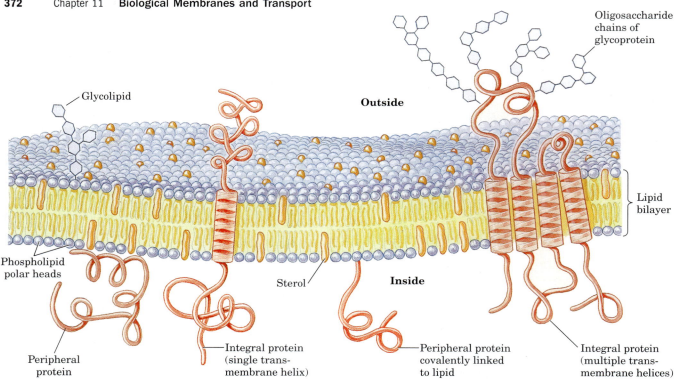

**FIGURE 11–3 Fluid mosaic model for membrane structure.** The fatty acyl chains in the interior of the membrane form a fluid, hydrophobic region. Integral proteins float in this sea of lipid, held by hydrophobic interactions with their nonpolar amino acid side chains. Both proteins and lipids are free to move laterally in the plane of the bilayer, but movement of either from one face of the bilayer to the other is restricted. The carbohydrate moieties attached to some proteins and lipids of the plasma membrane are exposed on the extracellular surface of the membrane.

exposed to water and thus minimizes the number of molecules in the shell of ordered water at the lipid-water interface (see Fig. 2–7), resulting in an increase in entropy. Hydrophobic interactions among lipid molecules provide the thermodynamic driving force for the formation and maintenance of these clusters.

Depending on the precise conditions and the nature of the lipids, three types of lipid aggregates can form when amphipathic lipids are mixed with water (Fig. 11–4). **Micelles** are spherical structures that contain anywhere from a few dozen to a few thousand amphipathic molecules. These molecules are arranged with

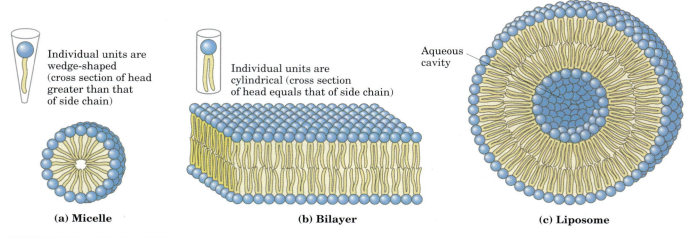

**FIGURE 11–4 Amphipathic lipid aggregates that form in water. (a)** In micelles, the hydrophobic chains of the fatty acids are sequestered at the core of the sphere. There is virtually no water in the hydrophobic interior. **(b)** In an open bilayer, all acyl side chains except those at the edges of the sheet are protected from interaction with water. **(c)** When a two-dimensional bilayer folds on itself, it forms a closed bilayer, a three-dimensional hollow vesicle (liposome) enclosing an aqueous cavity.

their hydrophobic regions aggregated in the interior, where water is excluded, and their hydrophilic head groups at the surface, in contact with water. Micelle formation is favored when the cross-sectional area of the head group is greater than that of the acyl side chain(s), as in free fatty acids, lysophospholipids (phospholipids lacking one fatty acid), and detergents such as sodium dodecyl sulfate (SDS; p. 92).

A second type of lipid aggregate in water is the **bilayer,** in which two lipid monolayers (leaflets) form a two-dimensional sheet. Bilayer formation occurs most readily when the cross-sectional areas of the head group and acyl side chain(s) are similar, as in glycerophospholipids and sphingolipids. The hydrophobic portions in each monolayer, excluded from water, interact with each other. The hydrophilic head groups interact with water at each surface of the bilayer. Because the hydrophobic regions at its edges (Fig. 11–4b) are transiently in contact with water, the bilayer sheet is relatively unstable and spontaneously forms a third type of aggregate: it folds back on itself to form a hollow sphere, a vesicle or **liposome** (Fig. 11–4c). By forming vesicles, bilayers lose their hydrophobic edge regions, achieving maximal stability in their aqueous environment. These bilayer vesicles enclose water, creating a separate aqueous compartment. It is likely that the precursors to the first living cells resembled liposomes, their aqueous contents segregated from the rest of the world by a hydrophobic shell.

Biological membranes are constructed of lipid bilayers 3 nm (30 Å) thick, with proteins protruding on each side. The hydrocarbon core of the membrane, made up of the $-CH_2-$ and $-CH_3$ of the fatty acyl groups, is about as nonpolar as decane, and liposomes formed in the laboratory from pure lipids are essentially impermeable to polar solutes, as are biological membranes (although the latter, as we shall see, are permeable to solutes for which they have specific transporters).

Plasma membrane lipids are asymmetrically distributed between the two monolayers of the bilayer, although the asymmetry, unlike that of membrane proteins, is not absolute. In the plasma membrane of the erythrocyte, for example, choline-containing lipids (phosphatidylcholine and sphingomyelin) are typically found in the outer (extracellular or exoplasmic) leaflet (Fig. 11–5), whereas phosphatidylserine, phosphatidylethanolamine, and the phosphatidylinositols are much more common in the inner (cytoplasmic) leaflet. Changes in the distribution of lipids between plasma membrane leaflets have biological consequences. For example, only when the phosphatidylserine in the plasma membrane moves into the outer leaflet is a platelet able to play its role in formation of a blood clot. For many other cells types, phosphatidylserine exposure on the outer surface marks a cell for destruction by programmed cell death.

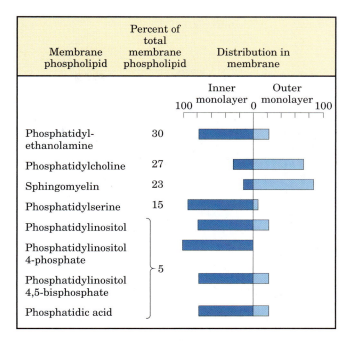

**FIGURE 11–5 Asymmetric distribution of phospholipids between the inner and outer monolayers of the erythrocyte plasma membrane.** The distribution of a specific phospholipid is determined by treating the intact cell with phospholipase C, which cannot reach lipids in the inner monolayer (leaflet) but removes the head groups of lipids in the outer monolayer. The proportion of each head group released provides an estimate of the fraction of each lipid in the outer monolayer.

## Peripheral Membrane Proteins Are Easily Solubilized

Membrane proteins may be divided operationally into two groups (Fig. 11–6). **Integral proteins** are very firmly associated with the membrane, removable only by agents that interfere with hydrophobic interactions, such as detergents, organic solvents, or denaturants. **Peripheral proteins** associate with the membrane through electrostatic interactions and hydrogen bonding with the hydrophilic domains of integral proteins and with the polar head groups of membrane lipids. They can be released by relatively mild treatments that interfere with electrostatic interactions or break hydrogen bonds; a commonly used agent is carbonate at high pH. Peripheral proteins may serve as regulators of membrane-bound enzymes or may limit the mobility of integral proteins by tethering them to intracellular structures.

## Many Membrane Proteins Span the Lipid Bilayer

Membrane protein topology (localization relative to the lipid bilayer) can be determined with reagents that react with protein side chains but cannot cross membranes—polar chemical reagents that react with primary amines of Lys residues, for example, or enzymes

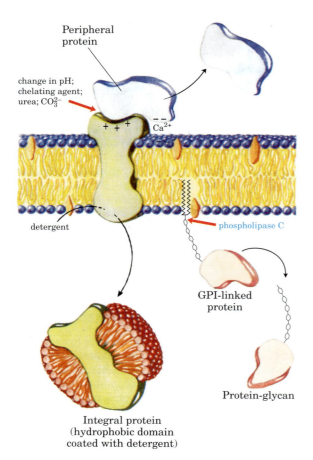

**FIGURE 11-6 Peripheral and integral proteins.** Membrane proteins can be operationally distinguished by the conditions required to release them from the membrane. Most peripheral proteins are released by changes in pH or ionic strength, removal of $Ca^{2+}$ by a chelating agent, or addition of urea or carbonate. Integral proteins are extractable with detergents, which disrupt the hydrophobic interactions with the lipid bilayer and form micelle-like clusters around individual protein molecules. Integral proteins covalently attached to a membrane lipid, such as a glycosyl phosphatidylinositol (GPI; see Fig. 11–14), can be released by treatment with phospholipase C.

like trypsin that cleave proteins but cannot cross the membrane. The human erythrocyte is convenient for such studies because it has no membrane-bounded organelles; the plasma membrane is the only membrane present. If a membrane protein in an intact erythrocyte reacts with a membrane-impermeant reagent, that protein must have at least one domain exposed on the outer (extracellular) face of the membrane. Trypsin is found to cleave extracellular domains but does not affect domains buried within the bilayer or exposed on the inner surface only, unless the plasma membrane is broken to make these domains accessible to the enzyme.

Experiments with such topology-specific reagents show that the erythrocyte glycoprotein **glycophorin** spans the plasma membrane. Its amino-terminal domain (bearing the carbohydrate chains) is on the outer sur-

face and is cleaved by trypsin. The carboxyl terminus protrudes on the inside of the cell, where it cannot react with impermeant reagents. Both the amino-terminal and carboxyl-terminal domains contain many polar or charged amino acid residues and are therefore quite hydrophilic. However, a segment in the center of the protein (residues 75 to 93) contains mainly hydrophobic amino acid residues, suggesting that glycophorin has a transmembrane segment arranged as shown in Figure 11–7.

One further fact may be deduced from the results of experiments with glycophorin: its disposition in the membrane is asymmetric. Similar studies of other membrane proteins show that each has a specific orientation

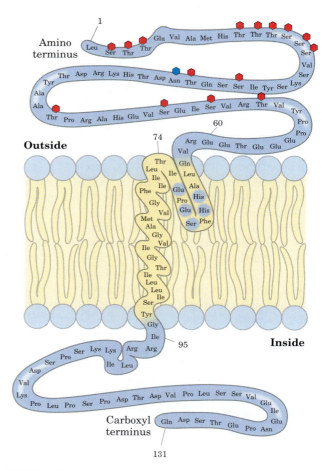

**FIGURE 11-7 Transbilayer disposition of glycophorin in an erythrocyte.** One hydrophilic domain, containing all the sugar residues, is on the outer surface, and another hydrophilic domain protrudes from the inner face of the membrane. Each red hexagon represents a tetrasaccharide (containing two Neu5Ac (sialic acid), Gal, and GalNAc) O-linked to a Ser or Thr residue; the blue hexagon represents an oligosaccharide chain N-linked to an Asn residue. The relative size of the oligosaccharide units is larger than shown here. A segment of 19 hydrophobic residues (residues 75 to 93) forms an α helix that traverses the membrane bilayer (see Fig. 11–11a). The segment from residues 64 to 74 has some hydrophobic residues and probably penetrates into the outer face of the lipid bilayer, as shown.

in the bilayer; one domain of a transmembrane protein always faces out, the other always faces in. Furthermore, glycoproteins of the plasma membrane are invariably situated with their sugar residues on the outer surface of the cell. As we shall see, the asymmetric arrangement of membrane proteins results in functional asymmetry. All the molecules of a given ion pump, for example, have the same orientation in the membrane and therefore pump in the same direction.

## Integral Proteins Are Held in the Membrane by Hydrophobic Interactions with Lipids

The firm attachment of integral proteins to membranes is the result of hydrophobic interactions between membrane lipids and hydrophobic domains of the protein. Some proteins have a single hydrophobic sequence in the middle (as in glycophorin) or at the amino or carboxyl terminus. Others have multiple hydrophobic sequences, each of which, when in the $\alpha$-helical conformation, is long enough to span the lipid bilayer (Fig. 11–8). The same techniques used to determine the three-dimensional structures of soluble proteins can, in principle, be applied to membrane proteins. In practice, however, membrane proteins have until recently proved difficult to crystallize. New techniques are overcoming this obstacle, and crystallographic structures of membrane proteins are regularly becoming available, yielding deep insights into membrane events at the molecular level.

One of the best-studied membrane-spanning proteins, bacteriorhodopsin, has seven very hydrophobic internal sequences and crosses the lipid bilayer seven times. Bacteriorhodopsin is a light-driven proton pump densely packed in regular arrays in the purple membrane of the bacterium *Halobacterium salinarum.* X-ray crystallography reveals a structure with seven $\alpha$-helical segments, each traversing the lipid bilayer, connected by nonhelical loops at the inner and outer face of the membrane (Fig. 11–9). In the amino acid sequence of bacteriorhodopsin, seven segments of about 20 hydrophobic residues can be identified, each segment just long enough to form an $\alpha$ helix that spans the bilayer. Hydrophobic interactions between the nonpolar amino acids and the fatty acyl groups of the membrane lipids firmly anchor the protein in the membrane. The seven helices are clustered together and oriented not quite perpendicular to the bilayer plane, providing a transmembrane pathway for proton movement. As we shall see in Chapter 12, this pattern of seven hydrophobic membrane-spanning helices is a common motif in membrane proteins involved in signal reception.

The photosynthetic reaction center of a purple bacterium was the first membrane protein structure solved by crystallography. Although a more complex membrane protein than bacteriorhodopsin, it is constructed on the same principles. The reaction center has four protein

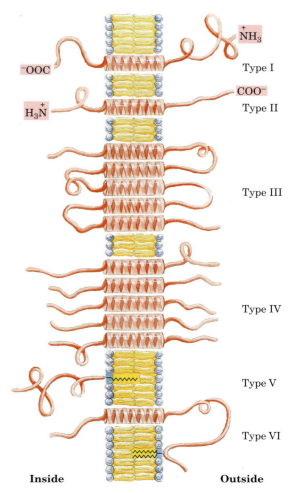

**FIGURE 11–8 Integral membrane proteins.** For known proteins of the plasma membrane, the spatial relationships of protein domains to the lipid bilayer fall into six categories. Types I and II have only one transmembrane helix; the amino-terminal domain is outside the cell in type I proteins and inside in type II. Type III proteins have multiple transmembrane helices in a single polypeptide. In type IV proteins, transmembrane domains of several different polypeptides assemble to form a channel through the membrane. Type V proteins are held to the bilayer primarily by covalently linked lipids (see Fig. 11–14), and type VI proteins have both transmembrane helices and lipid (GPI) anchors.

In this figure, and in figures throughout the book, we represent transmembrane protein segments in their most likely conformations: as $\alpha$ helices of six to seven turns. Sometimes these helices are shown simply as cylinders. As relatively few membrane protein structures have been deduced by x-ray crystallography, our representation of the extramembrane domains is arbitrary and not necessarily to scale.

subunits, three of which contain $\alpha$-helical segments that span the membrane (Fig. 11–10). These segments are rich in nonpolar amino acids, their hydrophobic side chains oriented toward the outside of the molecule where they interact with membrane lipids. The architecture of the reaction center protein is therefore the inverse of that seen in most water-soluble proteins, in

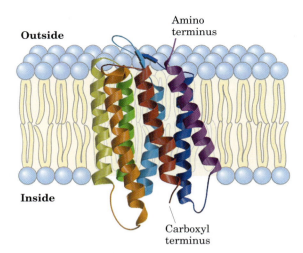

**FIGURE 11–9  Bacteriorhodopsin, a membrane-spanning protein.**
(PDB ID 2AT9) The single polypeptide chain folds into seven hydrophobic α helices, each of which traverses the lipid bilayer roughly
perpendicular to the plane of the membrane. The seven transmembrane helices are clustered, and the space around and between them
is filled with the acyl chains of membrane lipids. The light-absorbing
pigment retinal (see Fig. 10–21) is buried deep in the membrane in
contact with several of the helical segments (not shown). The helices
are colored to correspond with the hydropathy plot in Figure 11–11b.

which hydrophobic residues are buried within the
protein core and hydrophilic residues are on the
surface (recall the structures of myoglobin and hemoglobin, for example). In Chapter 19 we will encounter
several complex membrane proteins having multiple
transmembrane helical segments in which hydrophobic
chains are positioned to interact with the lipid bilayer.

## The Topology of an Integral Membrane Protein Can Be Predicted from Its Sequence

Determination of the three-dimensional structure of a
membrane protein, or its topology, is generally much
more difficult than determining its amino acid sequence,
which can be accomplished by sequencing the protein
or its gene. Thousands of sequences are known for membrane proteins, but relatively few three-dimensional
structures have been established by crystallography or
NMR spectroscopy. The presence of unbroken sequences
of more than 20 hydrophobic residues in a membrane
protein is commonly taken as evidence that these sequences traverse the lipid bilayer, acting as hydrophobic anchors or forming transmembrane channels. Virtually all integral proteins have at least one such sequence.
Application of this logic to entire genomic sequences
leads to the conclusion that in many species, 10% to
20% of all proteins are integral membrane proteins.

What can we predict about the secondary structure
of the membrane-spanning portions of integral proteins?
An α-helical sequence of 20 to 25 residues is just long

enough to span the thickness (30 Å) of the lipid bilayer
(recall that the length of an α helix is 1.5 Å (0.15 nm)
per amino acid residue). A polypeptide chain surrounded by lipids, having no water molecules with which
to hydrogen-bond, will tend to form α helices or β
sheets, in which intrachain hydrogen bonding is maximized. If the side chains of all amino acids in a helix are
nonpolar, hydrophobic interactions with the surrounding lipids further stabilize the helix.

Several simple methods of analyzing amino acid sequences yield reasonably accurate predictions of secondary structure for transmembrane proteins. The relative polarity of each amino acid has been determined
experimentally by measuring the free-energy change accompanying the movement of that amino acid side chain
from a hydrophobic solvent into water. This free energy
of transfer ranges from very exergonic for charged or
polar residues to very endergonic for amino acids with
aromatic or aliphatic hydrocarbon side chains. The
overall hydrophobicity of a sequence of amino acids is
estimated by summing the free energies of transfer for

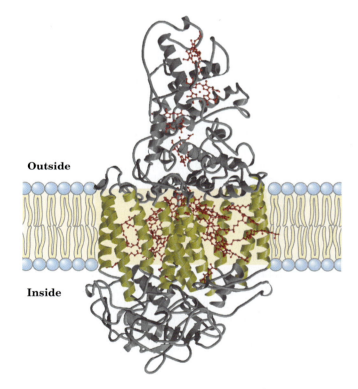

**FIGURE 11–10  Three-dimensional structure of the photosynthetic
reaction center of *Rhodopseudomonas viridis,* a purple bacterium.**
This was the first integral membrane protein to have its atomic structure determined by x-ray diffraction methods (PDB ID 1PRC). Eleven
α-helical segments from three of the four subunits span the lipid bilayer, forming a cylinder 45 Å (4.5 nm) long; hydrophobic residues on
the exterior of the cylinder interact with lipids of the bilayer. In this
ribbon representation, residues that are part of the transmembrane helices are shown in yellow. The prosthetic groups (light-absorbing pigments and electron carriers; see Fig. 19–45) are red.

the residues in the sequence, which yields a **hydropathy index** for that region (see Table 3–1). To scan a polypeptide sequence for potential membrane-spanning segments, an investigator calculates the hydropathy index for successive segments (called windows) of a given size, from 7 to 20 residues. For a window of seven residues, for example, the indices for residues 1 to 7, 2 to 8, 3 to 9, and so on, are plotted as in Figure 11–11 (plotted for the middle residue in each window— residue 4 for residues 1 to 7, for example). A region with more than 20 residues of high hydropathy index is presumed to be a transmembrane segment. When the sequences of membrane proteins of known three-dimensional structure are scanned in this way, we find a reasonably good correspondence between predicted and known membrane-spanning segments. Hydropathy analysis predicts a single hydrophobic helix for glycophorin (Fig. 11–11a) and seven transmembrane segments for bacteriorhodopsin (Fig. 11–11b)—in agreement with experimental studies.

On the basis of their amino acid sequences and hydropathy plots, many of the transport proteins described in this chapter are believed to have multiple membrane-spanning helical regions—that is, they are type III or type IV integral proteins (Fig. 11–8). When predictions are consistent with chemical studies of protein localization (such as those described above for glycophorin and bacteriorhodopsin), the assumption that hydrophobic regions correspond to membrane-spanning domains is much better justified.

A further remarkable feature of many transmembrane proteins of known structure is the presence of Tyr and Trp residues at the interface between lipid and water (Fig. 11–12). The side chains of these residues apparently serve as membrane interface anchors, able to interact simultaneously with the central lipid phase and the aqueous phases on either side of the membrane.

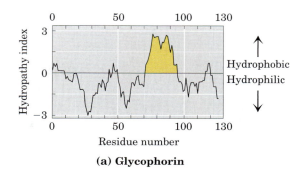

**(a) Glycophorin**

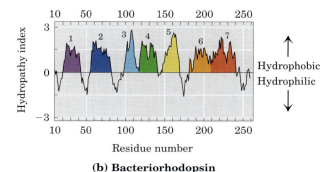

**(b) Bacteriorhodopsin**

**FIGURE 11–11 Hydropathy plots.** Hydropathy index (see Table 3–1) is plotted against residue number for two integral membrane proteins. The hydropathy index for each amino acid residue in a sequence of defined length (called the window) is used to calculate the average hydropathy for the residues in that window. The horizontal axis shows the residue number in the middle of the window. **(a)** Glycophorin from human erythrocytes has a single hydrophobic sequence between residues 75 and 93 (yellow); compare this with Figure 11–7. **(b)** Bacteriorhodopsin, known from independent physical studies to have seven transmembrane helices (see Fig. 11–9), has seven hydrophobic regions. Note, however, that the hydropathy plot is ambiguous in the region of segments 6 and 7. Physical studies have confirmed that this region has two transmembrane segments.

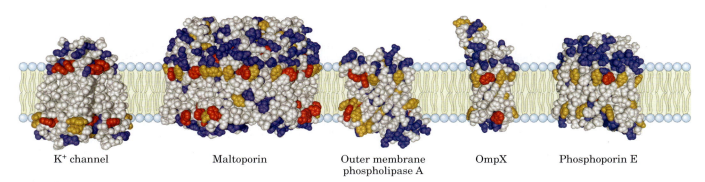

K⁺ channel     Maltoporin     Outer membrane phospholipase A     OmpX     Phosphoporin E

**FIGURE 11–12 Tyr and Trp residues of membrane proteins clustering at the water-lipid interface.** The detailed structures of these five integral membrane proteins are known from crystallographic studies. The K⁺ channel (PDB ID 1BL8) is from the bacterium *Streptomyces lividans* (see Fig. 11–48); maltoporin (PDB ID 1AF6), outer membrane phospholipase A (PDB ID 1QD5), OmpX (PDB ID 1QJ9), and phosphoporin E (PDB ID 1PHO) are proteins of the outer membrane of *E. coli*. Residues of Tyr (orange) and Trp (red) are found predominantly where the nonpolar region of acyl chains meets the polar head group region. Charged residues (Lys, Arg, Glu, Asp) are shown in blue; they are found almost exclusively in the aqueous phases.

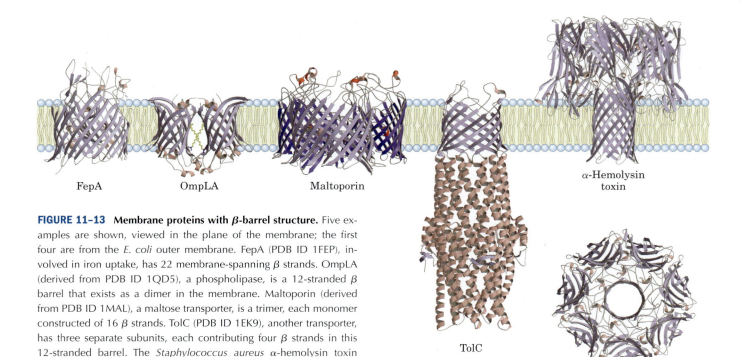

FepA          OmpLA          Maltoporin

α-Hemolysin
toxin

TolC

Top view

**FIGURE 11–13 Membrane proteins with β-barrel structure.** Five examples are shown, viewed in the plane of the membrane; the first four are from the *E. coli* outer membrane. FepA (PDB ID 1FEP), involved in iron uptake, has 22 membrane-spanning β strands. OmpLA (derived from PDB ID 1QD5), a phospholipase, is a 12-stranded β barrel that exists as a dimer in the membrane. Maltoporin (derived from PDB ID 1MAL), a maltose transporter, is a trimer, each monomer constructed of 16 β strands. TolC (PDB ID 1EK9), another transporter, has three separate subunits, each contributing four β strands in this 12-stranded barrel. The *Staphylococcus aureus* α-hemolysin toxin (PDB ID 7AHL; top view below) is composed of seven identical subunits, each contributing one hairpin-shaped pair of β strands to the 14-stranded barrel.

The hydrophobic domains of some integral membrane proteins penetrate only one leaflet of the bilayer. Cyclooxygenase, the target of aspirin action, is an example; its hydrophobic helices do not span the whole membrane but interact strongly with the acyl groups on one side of the bilayer (see Box 21–2, Fig. 1a).

Not all integral membrane proteins are composed of transmembrane α helices. Another structural motif common in membrane proteins is the **β barrel** (see Fig. 4–20d), in which 20 or more transmembrane segments form β sheets that line a cylinder (Fig. 11–13). The same factors that favor α-helix formation in the hydrophobic interior of a lipid bilayer also stabilize β barrels. When no water molecules are available to hydrogen-bond with the carbonyl oxygen and nitrogen of the peptide bond, maximal intrachain hydrogen bonding gives the most stable conformation. Planar β sheets do not maximize these interactions and are generally not found in the membrane interior; β barrels do allow all possible hydrogen bonds and are apparently common among membrane proteins. **Porins,** proteins that allow certain polar solutes to cross the outer membrane of gram-negative bacteria such as *E. coli*, have many-stranded β barrels lining the polar transmembrane passage.

A polypeptide is more extended in the β conformation than in an α helix; just seven to nine residues of β conformation are needed to span a membrane. Recall that in the β conformation, alternating side chains

project above and below the sheet (see Fig. 4–7). In β strands of membrane proteins, every second residue in the membrane-spanning segment is hydrophobic and interacts with the lipid bilayer; aromatic side chains are commonly found at the lipid-protein interface. The other residues may or may not be hydrophilic. The hydropathy plot is not useful in predicting transmembrane segments for proteins with β barrel motifs, but as the database of known β barrel motifs increases, sequence-based predictions of transmembrane β conformations have become feasible. For example, a number of outer membrane proteins of gram-negative bacteria (Fig. 11–13) have been correctly predicted, by sequence analysis, to contain β barrels.

## Covalently Attached Lipids Anchor Some Membrane Proteins

Some membrane proteins contain one or more covalently linked lipids of several types: long-chain fatty acids, isoprenoids, sterols, or *g*lycosylated derivatives of *p*hosphatidyl*i*nositol, GPI (Fig. 11–14). The attached lipid provides a hydrophobic anchor that inserts into the lipid bilayer and holds the protein at the membrane surface. The strength of the hydrophobic interaction between a bilayer and a single hydrocarbon chain linked to a protein is barely enough to anchor the protein securely, but many proteins have more than one attached

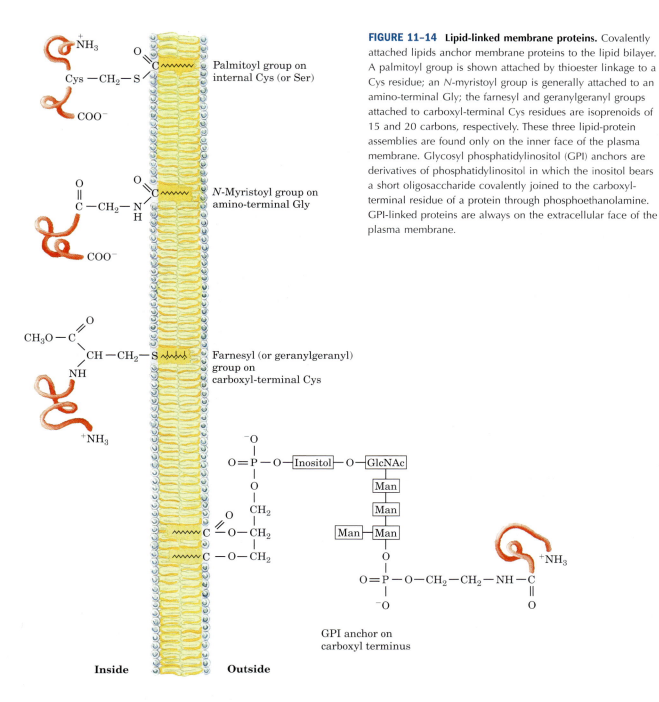

**Palmitoyl group on internal Cys (or Ser)**

**N-Myristoyl group on amino-terminal Gly**

**Farnesyl (or geranylgeranyl) group on carboxyl-terminal Cys**

**GPI anchor on carboxyl terminus**

**Inside**     **Outside**

**FIGURE 11–14 Lipid-linked membrane proteins.** Covalently attached lipids anchor membrane proteins to the lipid bilayer. A palmitoyl group is shown attached by thioester linkage to a Cys residue; an *N*-myristoyl group is generally attached to an amino-terminal Gly; the farnesyl and geranylgeranyl groups attached to carboxyl-terminal Cys residues are isoprenoids of 15 and 20 carbons, respectively. These three lipid-protein assemblies are found only on the inner face of the plasma membrane. Glycosyl phosphatidylinositol (GPI) anchors are derivatives of phosphatidylinositol in which the inositol bears a short oligosaccharide covalently joined to the carboxyl-terminal residue of a protein through phosphoethanolamine. GPI-linked proteins are always on the extracellular face of the plasma membrane.

lipid moiety. Other interactions, such as ionic attractions between positively charged Lys residues in the protein and negatively charged lipid head groups, probably contribute to the stability of the attachment. The association of these lipid-linked proteins with the membrane is certainly weaker than that for integral membrane proteins and is, in at least some cases, reversible. But treatment with alkaline carbonate does not release GPI-linked proteins, which are therefore, by the working definition, integral proteins.

Beyond merely anchoring a protein to the membrane, the attached lipid may have a specific role. In the plasma membrane, proteins with GPI anchors are exclusively on the outer face and are confined within clusters, as we shall see below, whereas other types of lipid-linked proteins (with farnesyl or geranylgeranyl groups attached; Fig. 11–14) are exclusively on the inner face. Attachment of a specific lipid to a newly synthesized membrane protein therefore has a targeting function, directing the protein to its correct membrane location.

## SUMMARY 11.1 The Composition and Architecture of Membranes

■ Biological membranes define cellular boundaries, divide cells into discrete compartments, organize complex reaction sequences, and act in signal reception and energy transformations.

■ Membranes are composed of lipids and proteins in varying combinations particular to each species, cell type, and organelle. The fluid mosaic model describes features common to all biological membranes. The lipid bilayer is the basic structural unit. Fatty acyl chains of phospholipids and the steroid nucleus of sterols are oriented toward the interior of the bilayer; their hydrophobic interactions stabilize the bilayer but give it flexibility.

■ Peripheral proteins are loosely associated with the membrane through electrostatic interactions and hydrogen bonds or by covalently attached lipid anchors. Integral proteins associate firmly with membranes by hydrophobic interactions between the lipid bilayer and their nonpolar amino acid side chains, which are oriented toward the outside of the protein molecule.

■ Some membrane proteins span the lipid bilayer several times, with hydrophobic sequences of about 20 amino acid residues forming transmembrane α helices. Detection of such hydrophobic sequences in proteins can be used to predict their secondary structure and transmembrane disposition. Multistranded β barrels are also common in integral membrane proteins. Tyr and Trp residues of transmembrane proteins are commonly found at the lipid-water interface.

■ The lipids and proteins of membranes are inserted into the bilayer with specific sidedness; thus membranes are structurally and functionally asymmetric. Many membrane proteins contain covalently attached oligosaccharides. Plasma membrane glycoproteins are always oriented with the carbohydrate-bearing domain on the extracellular surface.

## 11.2 Membrane Dynamics

One remarkable feature of all biological membranes is their flexibility—their ability to change shape without losing their integrity and becoming leaky. The basis for this property is the noncovalent interactions among lipids in the bilayer and the motions allowed to individual lipids because they are not covalently anchored to one another. We turn now to the dynamics of membranes: the motions that occur and the transient structures allowed by these motions.

### Acyl Groups in the Bilayer Interior Are Ordered to Varying Degrees

Although the lipid bilayer structure is quite stable, its individual phospholipid and sterol molecules have some freedom of motion (Fig. 11–15). The structure and flexibility of the lipid bilayer depend on temperature and on the kinds of lipids present. At relatively low temperatures, the lipids in a bilayer form a semisolid **gel phase,** in which all types of motion of individual lipid molecules are strongly constrained; the bilayer is paracrystalline (Fig. 11–15a). At relatively high temperatures, individual hydrocarbon chains of fatty acids are in constant motion produced by rotation about the carbon–carbon bonds of the long acyl side chains. In this **liquid-disordered state,** or fluid state (Fig. 11–15b), the interior of the bilayer is more fluid than solid and the bilayer is like a sea of constantly moving lipid. At intermediate temperatures, the lipids exist in a **liquid-ordered state;** there is less thermal motion in the acyl chains of the lipid bilayer, but lateral movement in the plane of the bilayer still takes place. These differences in bilayer state are easily observed in liposomes composed of a single lipid,

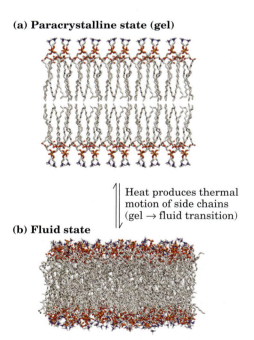

**(a) Paracrystalline state (gel)**

Heat produces thermal motion of side chains (gel → fluid transition)

**(b) Fluid state**

**FIGURE 11–15 Two states of bilayer lipids. (a)** In the paracrystalline state, or gel phase, polar head groups are uniformly arrayed at the surface, and the acyl chains are nearly motionless and packed with regular geometry; **(b)** in the liquid disordered state, or fluid state, acyl chains undergo much thermal motion and have no regular organization. Intermediate between these extremes is the liquid-ordered state, in which individual phospholipid molecules can diffuse laterally but the acyl groups remain extended and more or less ordered.

but biological membranes contain many lipids with a variety of fatty acyl chains and thus do not show sharp phase changes with temperature.

At temperatures in the physiological range (about 20 to 40 °C), long-chain saturated fatty acids (such as 16:0 and 18:0) pack well into a liquid-ordered array, but the kinks in unsaturated fatty acids (see Fig. 10–1) interfere with this packing, favoring the liquid-disordered state. Shorter-chain fatty acyl groups have the same effect. The sterol content of a membrane (which varies greatly with organism and organelle; Table 11–1) is another important determinant of lipid state. The rigid planar structure of the steroid nucleus, inserted between fatty acyl side chains, reduces the freedom of neighboring fatty acyl chains to move by rotation about their carbon–carbon bonds, forcing acyl chains into their fully extended conformation. The presence of sterols therefore reduces the fluidity in the core of the bilayer, thus favoring the liquid-ordered phase, and increases the thickness of the lipid leaflet (as described below).

Cells regulate their lipid composition to achieve a constant membrane fluidity under various growth conditions. For example, bacteria synthesize more unsaturated fatty acids and fewer saturated ones when cultured at low temperatures than when cultured at higher temperatures (Table 11–2). As a result of this adjustment in lipid composition, membranes of bacteria cultured at high or low temperatures have about the same degree of fluidity.

## Transbilayer Movement of Lipids Requires Catalysis

At physiological temperature, transbilayer—or "flip-flop"—diffusion of a lipid molecule from one leaflet of the bilayer to the other (Fig. 11–16a) occurs very slowly if at all in most membranes. Transbilayer movement requires that a polar or charged head group leave its

**(a) Uncatalyzed transverse ("flip-flop") diffusion**

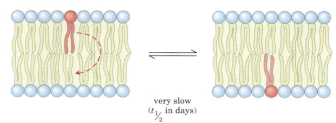

very slow
($t_{1/2}$ in days)

**(b) Transverse diffusion catalyzed by flippase**

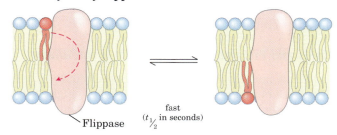

Flippase

fast
($t_{1/2}$ in seconds)

**(c) Uncatalyzed lateral diffusion**

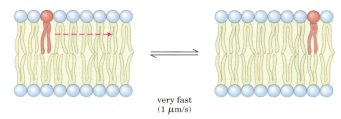

very fast
(1 $\mu$m/s)

**FIGURE 11–16 Motion of single phospholipids in a bilayer. (a)** Movement from one leaflet to the other is very slow, unless **(b)** catalyzed by a flippase; in contrast, lateral diffusion within the leaflet **(c)** is very rapid and requires no protein catalysis.

**TABLE 11–2 Fatty Acid Composition of *E. coli* Cells Cultured at Different Temperatures**

| | Percentage of total fatty acids[*] | | | |
|---|---|---|---|---|
| | 10 °C | 20 °C | 30 °C | 40 °C |
| Myristic acid (14:0) | 4 | 4 | 4 | 8 |
| Palmitic acid (16:0) | 18 | 25 | 29 | 48 |
| Palmitoleic acid (16:1) | 26 | 24 | 23 | 9 |
| Oleic acid (18:1) | 38 | 34 | 30 | 12 |
| Hydroxymyristic acid | 13 | 10 | 10 | 8 |
| Ratio of unsaturated to saturated[†] | 2.9 | 2.0 | 1.6 | 0.38 |

**Source:** Data from Marr, A.G. & Ingraham, J.L. (1962) Effect of temperature on the composition of fatty acids in *Escherichia coli*. *J. Bacteriol.* **84,** 1260.

[*]The exact fatty acid composition depends not only on growth temperature but on growth stage and growth medium composition.

[†]Ratios calculated as the total percentage of 16:1 plus 18:1 divided by the total percentage of 14:0 plus 16:0. Hydroxymyristic acid was omitted from this calculation.

aqueous environment and move into the hydrophobic interior of the bilayer, a process with a large, positive free-energy change. There are, however, situations in which such movement is essential. For example, during synthesis of the bacterial plasma membrane, phospholipids are produced on the inside surface of the membrane and must undergo flip-flop diffusion to enter the outer leaflet of the bilayer. Similar transbilayer diffusion must also take place in eukaryotic cells as membrane lipids synthesized in one organelle move from the inner to the outer leaflet and into other organelles. A family of proteins, the **flippases** (Fig. 11–16b), facilitates flip-flop diffusion, providing a transmembrane path that is energetically more favorable and much faster than the uncatalyzed movement.

### Lipids and Proteins Diffuse Laterally in the Bilayer

Individual lipid molecules can move laterally in the plane of the membrane by changing places with neighboring lipid molecules (Fig. 11–16c). A molecule in one monolayer, or leaflet, of the lipid bilayer—the outer leaflet of the erythrocyte plasma membrane, for example—can diffuse laterally so fast that it circumnavigates the erythrocyte in seconds. This rapid lateral diffusion within the plane of the bilayer tends to randomize the positions of individual molecules in a few seconds.

Lateral diffusion can be shown experimentally by attaching fluorescent probes to the head groups of lipids and using fluorescence microscopy to follow the probes over time (Fig. 11–17). In one technique, a small region (5 $\mu$m$^2$) of a cell surface with fluorescence-tagged lipids is bleached by intense laser radiation so that the irradiated patch no longer fluoresces when viewed in the much dimmer light of the fluorescence microscope. However, within milliseconds, the region recovers its fluorescence as unbleached lipid molecules diffuse into the bleached patch and bleached lipid molecules diffuse away from it. The rate of *f*luorescence *r*ecovery *a*fter *p*hotobleaching, or **FRAP,** is a measure of the rate of lateral diffusion of the lipids. Using the FRAP technique,

researchers have shown that some membrane lipids diffuse laterally by up to 1 $\mu$m/s.

Another technique, single particle tracking, allows one to follow the movement of a *single* lipid molecule in the plasma membrane on a much shorter time scale. Results from these studies confirm the rapid lateral diffusion within small, discrete regions of the cell sur-

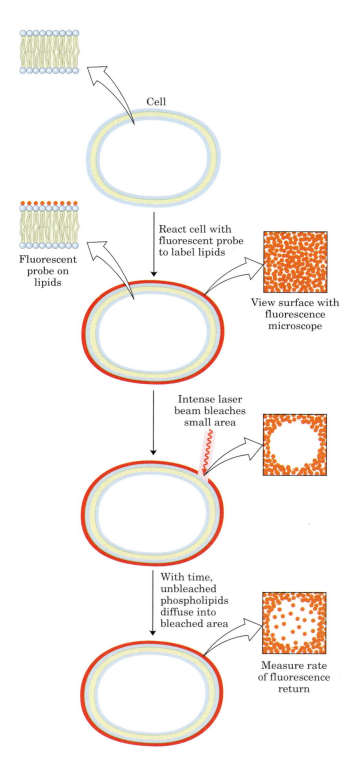

**FIGURE 11–17 Measurement of lateral diffusion rates of lipids by fluorescence recovery after photobleaching (FRAP).** The lipids in the outer leaflet of the plasma membrane are labeled by reaction with a membrane-impermeant fluorescent probe (red), so the surface is uniformly labeled when viewed with a fluorescence microscope. A small area is bleached by irradiation with an intense laser beam, leaving that area nonfluorescent. With the passage of time, labeled lipid molecules diffuse into the bleached region, and it again becomes fluorescent. From the time course of fluorescence return to this area, the diffusion coefficient for the labeled lipid is determined. The rates are typically high; a lipid moving at this speed could circumnavigate *E. coli* in one second. (The FRAP method can also be used to measure the lateral diffusion of membrane proteins.)

Labels in figure:
Cell
Fluorescent probe on lipids
React cell with fluorescent probe to label lipids
View surface with fluorescence microscope
Intense laser beam bleaches small area
With time, unbleached phospholipids diffuse into bleached area
Measure rate of fluorescence return

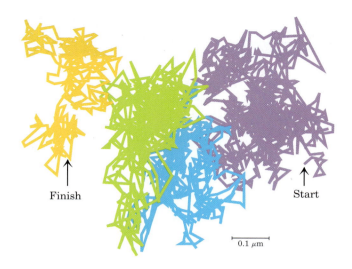

Finish

Start

0.1 μm

**FIGURE 11–18 Hop diffusion of individual lipid molecules.** The motion of a single fluorescent lipid molecule in a cell surface is recorded on video by fluorescence microscopy, with a time resolution of 25 μs (equivalent to 40,000 frames/s). The track shown here represents a molecule followed for 56 ms (a total of 2,250 frames); the trace begins in the purple area and continues through blue, green, and orange. The pattern of movement indicates rapid diffusion within a confined region (about 250 nm in diameter, shown by a single color), with occasional hops into an adjoining region. This finding suggests that the lipids are corralled by molecular fences that they occasionally jump.

face and show that movement from one such region to a nearby region is inhibited; lipids behave as though corralled by fences that they can occasionally jump (Fig. 11–18).

Many membrane proteins seem to be afloat in a sea of lipids. Like membrane lipids, these proteins are free to diffuse laterally in the plane of the bilayer and are in constant motion, as shown by the FRAP technique with fluorescence-tagged surface proteins. Some membrane

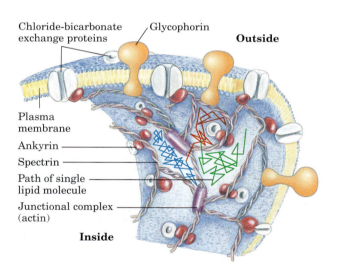

Chloride-bicarbonate exchange proteins — Glycophorin — **Outside**

Plasma membrane

Ankyrin

Spectrin

Path of single lipid molecule

Junctional complex (actin)

**Inside**

**FIGURE 11–19 Restricted motion of the erythrocyte chloride-bicarbonate exchanger and glycophorin.** The proteins span the membrane and are tethered to spectrin, a cytoskeletal protein, by another protein, ankyrin, limiting their lateral mobilities. Ankyrin is anchored in the membrane by a covalently bound palmitoyl side chain (see Fig. 11–14). Spectrin, a long, filamentous protein, is cross-linked at junctional complexes containing actin. A network of cross-linked spectrin molecules attached to the cytoplasmic face of the plasma membrane stabilizes the membrane against deformation. This network of anchored membrane proteins may be the "corral" suggested by the experiment shown in Figure 11–18; the lipid tracks shown here are confined to subregions defined by the tethered membrane proteins.

proteins associate to form large aggregates ("patches") on the surface of a cell or organelle in which individual protein molecules do not move relative to one another; for example, acetylcholine receptors (see Fig. 11–51) form dense patches on neuron plasma membranes at synapses. Other membrane proteins are anchored to internal structures that prevent their free diffusion. In the erythrocyte membrane, both glycophorin and the chloride-bicarbonate exchanger (p. 395) are tethered to spectrin, a filamentous cytoskeletal protein (Fig. 11–19). One possible explanation for the pattern of lateral diffusion of lipid molecules shown in Figure 11–18 is that membrane proteins immobilized by their association with spectrin are the "fences" that define the regions of relatively unrestricted lipid motion.

## Sphingolipids and Cholesterol Cluster Together in Membrane Rafts

We have seen that diffusion of membrane lipids from one bilayer leaflet into the other is very slow unless catalyzed, and that the different lipid species of the plasma membrane are asymmetrically distributed in the two leaflets of the bilayer (Fig. 11–5). Even within a single leaflet, the lipid distribution is not random. Glycosphingolipids (cerebrosides and gangliosides), which typically contain long-chain saturated fatty acids, form transient clusters in the outer leaflet that largely exclude glycerophospholipids, which typically contain one unsaturated fatty acyl group and a shorter saturated fatty acyl group. The long, saturated acyl groups of sphingolipids can form more compact, more stable associations with the long ring system of cholesterol than can the shorter, often unsaturated, chains of phospholipids. The cholesterol-sphingolipid **microdomains** in the outer monolayer of the plasma membrane, visible with atomic-force microscopy (Box 11–1), are slightly thicker and more ordered (less fluid) than neighboring microdomains rich in phospholipids (Fig. 11–20) and are more difficult

BOX 11–1    WORKING IN BIOCHEMISTRY

## Atomic Force Microscopy to Visualize Membrane Proteins

In atomic force microscopy (AFM), the sharp tip of a microscopic probe attached to a flexible cantilever is drawn across an uneven surface such as a membrane (Fig. 1). Electrostatic and van der Waals interactions between the tip and the sample produce a force that moves the probe up and down (in the $z$ dimension) as it encounters hills and valleys in the sample. A laser beam reflected from the cantilever detects motions of as little as 1 Å. In one type of atomic force microscope, the force on the probe is held constant (relative to a standard force, on the order of piconewtons) by a feedback circuit that causes the platform holding the sample to rise or fall to keep the force constant. A series of scans in the $x$ and $y$ dimensions (the plane of the membrane) yields a three-dimensional contour map of the surface with resolution near the atomic scale—0.1 nm in the vertical dimension, 0.5 to 1.0 nm in the lateral dimensions. The membrane rafts shown in Figure 11–20b were visualized by this technique.

In favorable cases, AFM can be used to study single membrane protein molecules. Single molecules of bacteriorhodopsin in the purple membranes of the bacterium *Halobacterium salinarum* (see Fig. 11–9) are seen as highly regular structures (Fig. 2a). When a number of images of individual units are superimposed with the help of a computer, the real parts of the image reinforce each other and the noise in individual images is averaged out, yielding a high-resolution image of the protein (inset in Fig. 2a). AFM of purified *E. coli* aquaporin, reconstituted into

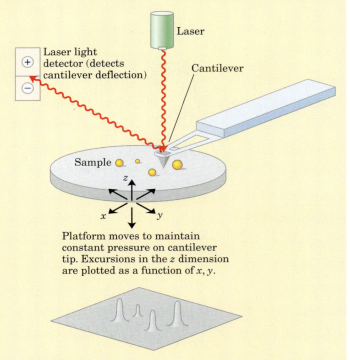

**FIGURE 1**

lipid bilayers and viewed as if from the outside of a cell, shows the fine details of the protein's periplasmic domains (Fig. 2b). And AFM reveals that $F_o$, the proton-driven rotor of the chloroplast ATP synthase (p. 742), is composed of many subunits (14 in Fig. 2c) arranged in a circle.

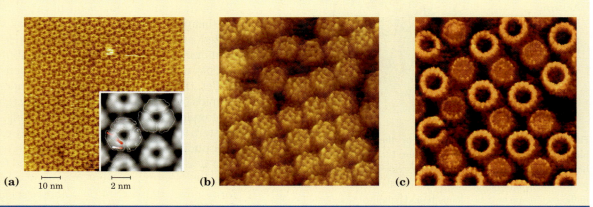

**FIGURE 2**    **(a)**   10 nm    2 nm      **(b)**                **(c)**

to dissolve with nonionic detergents; they behave like liquid-ordered sphingolipid **rafts** adrift on an ocean of liquid-disordered phospholipids.

These lipid rafts are remarkably enriched in two classes of integral membrane proteins: those anchored to the membrane by two covalently attached long-chain saturated fatty acids (two palmitoyl groups or a palmitoyl and a myristoyl group) and GPI-anchored proteins (Fig. 11–14). Presumably these lipid anchors, like the acyl chains of sphingolipids, form more stable associations with the cholesterol and long acyl groups in rafts than with the surrounding phospholipids. (It is notable

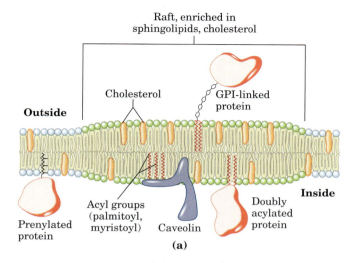

Raft, enriched in
sphingolipids, cholesterol

Cholesterol

GPI-linked
protein

**Outside**

**Inside**

Acyl groups
(palmitoyl,
myristoyl)

Caveolin

Doubly
acylated
protein

Prenylated
protein

**(a)**

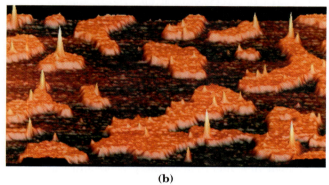

**(b)**

**FIGURE 11–20 Microdomains (rafts) in the plasma membrane.**
**(a)** Stable associations of sphingolipids and cholesterol in the outer
leaflet produce a microdomain, slightly thicker than other membrane
regions, that is enriched with specific types of membrane proteins.
GPI-linked proteins are commonly found in the outer leaflet of such
rafts, and proteins with one or several covalently attached long-chain
acyl groups are common in the inner leaflet. Caveolin is especially
common in inwardly curved rafts called caveolae (see Fig. 11–21).
Proteins with attached prenyl groups (such as Ras; see Fig. 12–6) tend
to be excluded from rafts. **(b)** The greater thickness of raft regions can
be visualized by atomic force microscopy (see Box 11–1). In this view
of a membrane region, we can see the rafts protruding from a lipid
bilayer ocean; in the rafts, sharp peaks represent GPI-linked proteins.
Note that these peaks are found almost exclusively in rafts.

brane that resists detergent solubilization, which can be
as high as 50% in some cases: the rafts cover half of the
ocean (Fig. 11–20b). Indirect measurements in cultured
fibroblasts suggest a diameter of roughly 50 nm for an
individual raft, which corresponds to a patch containing
a few thousand sphingolipids and perhaps 10 to 50
membrane proteins. Because most cells express more
than 50 different kinds of plasma membrane proteins, it
is likely that a single raft contains only a subset of membrane proteins and that this segregation of membrane
proteins is functionally significant. For a process that
involves interaction of two membrane proteins, their
presence in a single raft would hugely increase the
likelihood of their collision. Certain membrane receptors and signaling proteins, for example, appear to be
segregated together in membrane rafts. Experiments
show that signaling through these proteins can be disrupted by manipulations that deplete the plasma membrane of cholesterol and destroy lipid rafts.

### Caveolins Define a Special Class of Membrane Rafts

**Caveolin** is an integral membrane protein with two
globular domains connected by a hairpin-shaped
hydrophobic domain, which binds the protein to the cytoplasmic leaflet of the plasma membrane. Three palmitoyl groups attached to the carboxyl-terminal globular
domain further anchor it to the membrane. Caveolin
(actually, a family of related caveolins) binds cholesterol
in the membrane, and the presence of caveolin forces
the associated lipid bilayer to curve inward, forming
**caveolae** ("little caves") in the surface of the cell (Fig.
11–21). Caveolae are unusual rafts: they involve *both*
leaflets of the bilayer—the cytoplasmic leaflet, from
which the caveolin globular domains project, and the exoplasmic leaflet, a typical sphingolipid/cholesterol raft
with associated GPI-anchored proteins. Caveolae are implicated in a variety of cellular functions, including membrane trafficking within cells and the transduction of
external signals into cellular responses. The receptors
for insulin and other growth factors, as well as certain
GTP-binding proteins and protein kinases associated
with transmembrane signaling, appear to be localized in
rafts and perhaps in caveolae. We discuss some possible roles of rafts in signaling in Chapter 12.

### Certain Integral Proteins Mediate Cell-Cell Interactions and Adhesion

Several families of integral proteins in the plasma membrane provide specific points of attachment between
cells, or between a cell and extracellular matrix proteins.
**Integrins** are heterodimeric proteins (two unlike subunits, $\alpha$ and $\beta$) anchored to the plasma membrane by a
single hydrophobic transmembrane helix in each subunit (Fig. 11–22; see also Fig. 7–30). The large extracellular domains of the $\alpha$ and $\beta$ subunits combine to
form a specific binding site for extracellular proteins

that other lipid-linked proteins, those with covalently
attached isoprenyl groups such as farnesyl, are *not*
preferentially associated with the outer leaflet of
sphingolipid/cholesterol rafts (Fig. 11–20a).) The "raft"
and "sea" domains of the plasma membrane are not
rigidly separated; membrane proteins can move into and
out of lipid rafts on a time scale of seconds. But in the
shorter time scale (microseconds) more relevant to
many membrane-mediated biochemical processes, many
of these proteins reside primarily in a raft.

We can estimate the fraction of the cell surface
occupied by rafts from the fraction of the plasma mem-

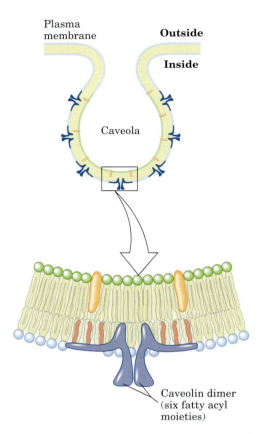

**FIGURE 11–21 Caveolin forces inward curvature in membranes.** The protein caveolin has a central hydrophobic domain and three long-chain acyl groups on each monomeric unit, which hold the molecule to the inside of the plasma membrane. When a number of caveolin dimers are concentrated in a small region (a raft), they force a curvature in the lipid bilayer, forming a caveola.

such as collagen and fibronectin. As there are 18 different $\alpha$ subunits and at least 8 different $\beta$ subunits, a wide variety of specificities may be generated from various combinations of $\alpha$ and $\beta$. One common determinant of integrin binding in several extracellular partners of integrins is the sequence Arg–Gly–Asp (RGD).

Integrins are not merely adhesives; they serve as receptors and signal transducers, conveying information across the plasma membrane in both directions. Integrins regulate many processes, including platelet aggregation at the site of a wound, tissue repair, the activity of immune cells, and the invasion of tissue by a tumor. Mutation in an integrin gene encoding the $\beta$ subunit known as CD18 is the cause of leukocyte adhesion deficiency in humans, a rare genetic disease in which leukocytes fail to pass out of blood vessels to reach sites of infection (see Fig. 7–33). Infants with a severe defect in CD18 commonly die of infections before the age of two. ■

At least three other families of plasma membrane proteins are also involved in surface adhesion (Fig. 11–22). **Cadherins** undergo homophilic ("with same kind") interactions with identical cadherins in an adjacent cell. **Immunoglobulin-like proteins** can undergo either homophilic interactions with their identical counterparts on another cell or heterophilic interactions with an integrin on a neighboring cell. **Selectins** have extracellular domains that, in the presence of $Ca^{2+}$, bind specific polysaccharides on the surface of an adjacent cell. Selectins are present primarily in the various types of blood cells and in the endothelial cells that line blood vessels (see Fig. 7–33). They are an essential part of the blood-clotting process.

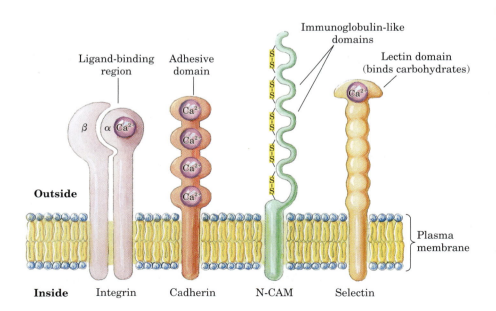

**FIGURE 11–22 Four examples of integral protein types that function in cell-cell interactions.** Integrins consist of $\alpha$ and $\beta$ transmembrane polypeptides; their extracellular domains combine to form binding sites for divalent metal ions and proteins of the extracellular matrix (such as collagen and fibronectin) or for specific surface proteins of other cells. Cadherin has four extracellular $Ca^{2+}$-binding domains, the most distal of which contains the site that binds to cadherin on another cell surface. N-CAM (*neuronal cell adhesion molecule*) is one of a family of immunoglobulin-like proteins that mediate $Ca^{2+}$-independent interactions with surface proteins of nearby cells. Selectins bind tightly to carbohydrate moieties in neighboring cells; this binding is $Ca^{2+}$-dependent.

Integral proteins play a role in many other cellular processes. They serve as transporters and ion channels (discussed in Section 11.3) and as receptors for hormones, neurotransmitters, and growth factors (Chapter 12). They are central to oxidative phosphorylation and photosynthesis (Chapter 19) and to cell-cell and antigen-cell recognition in the immune system (Chapter 5). Integral proteins are also important players in the membrane fusion that accompanies exocytosis, endocytosis, and the entry of many types of viruses into host cells.

### Membrane Fusion Is Central to Many Biological Processes

A remarkable feature of the biological membrane is its ability to undergo fusion with another membrane without losing its continuity. Although membranes are stable, they are by no means static. Within the eukaryotic endomembrane system (which includes the nuclear membrane, endoplasmic reticulum, Golgi, and various small vesicles), the membranous compartments constantly reorganize. Vesicles bud from the endoplasmic reticulum to carry newly synthesized lipids and proteins to other organelles and to the plasma membrane. Exocytosis, endocytosis, cell division, fusion of egg and sperm cells, and entry of a membrane-enveloped virus into its host cell all involve membrane reorganization in which the fundamental operation is fusion of two membrane segments without loss of continuity (Fig. 11–23).

Specific fusion of two membranes requires that (1) they recognize each other; (2) their surfaces become closely apposed, which requires the removal of water molecules normally associated with the polar head groups of lipids; (3) their bilayer structures become locally disrupted, resulting in fusion of the outer leaflet of each membrane (hemifusion); and (4) their bilayers fuse to form a single continuous bilayer. Receptor-mediated endocytosis, or regulated secretion, also requires that (5) the fusion process is triggered at the appropriate time or in response to a specific signal. Integral proteins called **fusion proteins** mediate these events, bringing about specific recognition and a transient local distortion of the bilayer structure that favors membrane fusion. (Note that these fusion proteins are unrelated to the products of two fused genes, also called fusion proteins, discussed in Chapter 9.)

Two cases of membrane fusion are especially well studied: the entry into a host cell of an enveloped virus such as influenza virus, and the release of neurotransmitters by exocytosis. Both processes involve complexes of fusion proteins that undergo dramatic conformational changes.

The influenza virus is surrounded by a membrane containing, among other proteins, many molecules of the hemagglutination (HA) protein (named for its abil-

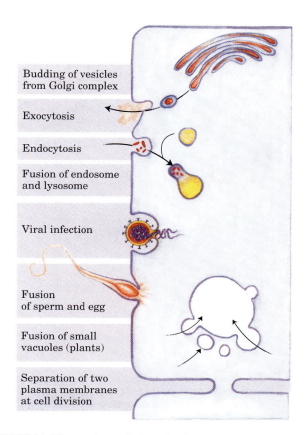

Budding of vesicles from Golgi complex

Exocytosis

Endocytosis

Fusion of endosome and lysosome

Viral infection

Fusion of sperm and egg

Fusion of small vacuoles (plants)

Separation of two plasma membranes at cell division

**FIGURE 11–23 Membrane fusion.** The fusion of two membranes is central to a variety of cellular processes involving both organelles and the plasma membrane.

ity to cause erythrocytes to clump together). The virus enters a host cell by inducing endocytosis, which encloses the virus in an endosome, a small membrane vesicle with a pH of about 5 (Fig. 11–24). At this pH, a conformational change in the HA protein occurs, exposing a sequence within the HA protein called the **fusion peptide** and enabling the protein to penetrate the endosomal membrane. The endosomal membrane and the viral membrane are now connected through the HA protein. Next, the HA protein bends at its middle to form a hairpin shape, bringing its two ends together. This pulls the two membranes into close apposition and causes fusion of the viral membrane and the endosomal membrane. The HA protein functions as a trimer (Fig. 11–24). In its low-pH form, three HA domains at the closed end of the hairpin twist about each other to form a stable, coiled structure. The fusion process involves an intermediate stage (hemifusion) in which the outer leaflet of the viral membrane is fused with the inner leaflet of the endosomal membrane, while the other two leaflets maintain their continuity. At the point of hemifusion, the lipid bilayer must be temporarily disorganized, presumably caused by the HA fusion peptide

**Cytosol**

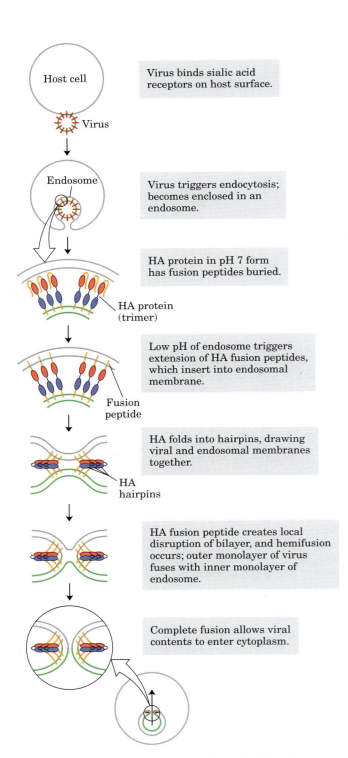

Virus binds sialic acid receptors on host surface.

Virus triggers endocytosis; becomes enclosed in an endosome.

HA protein in pH 7 form has fusion peptides buried.

Low pH of endosome triggers extension of HA fusion peptides, which insert into endosomal membrane.

HA folds into hairpins, drawing viral and endosomal membranes together.

HA fusion peptide creates local disruption of bilayer, and hemifusion occurs; outer monolayer of virus fuses with inner monolayer of endosome.

Complete fusion allows viral contents to enter cytoplasm.

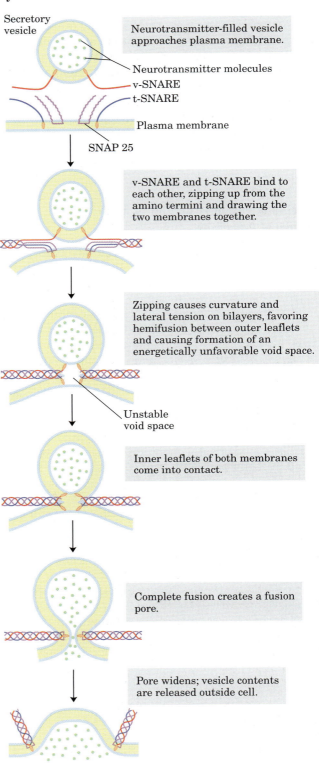

Neurotransmitter-filled vesicle approaches plasma membrane.

v-SNARE and t-SNARE bind to each other, zipping up from the amino termini and drawing the two membranes together.

Zipping causes curvature and lateral tension on bilayers, favoring hemifusion between outer leaflets and causing formation of an energetically unfavorable void space.

Inner leaflets of both membranes come into contact.

Complete fusion creates a fusion pore.

Pore widens; vesicle contents are released outside cell.

**FIGURE 11–24 Fusion induced by the hemagglutinin (HA) protein during viral infection.** HA protein is exposed on the membrane surface of the influenza virus. When the virus moves from the neutral pH of the interstitial fluid to the low-pH compartment (endosome) in the host cell, HA undergoes dramatic shape changes that mediate fusion of the viral and endosomal membranes, releasing the viral contents into the cytoplasm.

**FIGURE 11–25 Fusion during neurotransmitter release at a synapse.** The membrane of the secretory vesicle contains the v-SNARE synaptobrevin (red). The target (plasma) membrane contains the t-SNAREs syntaxin (blue) and SNAP25 (violet). When a local increase in [Ca$^{2+}$] signals release of neurotransmitter, the v-SNARE, SNAP25, and t-SNARE interact, forming a coiled bundle of four $\alpha$ helices, pulling the two membranes together and disrupting the bilayer locally, which leads to membrane fusion and neurotransmitter release.

domains. Complete fusion results in release of the viral contents into the host cell cytoplasm.

Neurotransmitters are released at synapses when intracellular vesicles loaded with neurotransmitter fuse with the plasma membrane. This process involves a family of proteins called SNARES (Fig. 11–25). SNAREs in the cytoplasmic face of the intracellular vesicles are called **v-SNAREs;** those in the target membranes with which the vesicles fuse (the plasma membrane during exocytosis) are **t-SNAREs.** Two other proteins, SNAP25 and NSF, are also involved. During fusion, v- and t-SNAREs bind to each other and undergo a structural change that produces a bundle of long thin rods made up of helices from both v- and t-SNARES and two helices from SNAP25 (Fig. 11–25). The two SNAREs initially interact at their ends, then zip up into the bundle of helices. This structural change pulls the two membranes into contact and initiates the fusion of their lipid bilayers.

The complex of SNAREs and SNAP25 is the target of the powerful *Clostridium botulinum* toxin, a protease that cleaves specific bonds in these proteins, preventing neurotransmission and causing the death of the organism. Because of its very high specificity for these proteins, purified botulinum toxin has served as a powerful tool for dissecting the mechanism of neurotransmitter release in vivo and in vitro.

### SUMMARY 11.2 Membrane Dynamics

- Lipids in a biological membrane can exist in liquid-ordered or liquid-disordered states; in the latter state, thermal motion of acyl chains makes the interior of the bilayer fluid. Fluidity is affected by temperature, fatty acid composition, and sterol content.

- Flip-flop diffusion of lipids between the inner and outer leaflets of a membrane is very slow except when specifically catalyzed by flippases.

- Lipids and proteins can diffuse laterally within the plane of the membrane, but this mobility is limited by interactions of membrane proteins with internal cytoskeletal structures and interactions of lipids with lipid rafts. One class of lipid rafts consists of sphingolipids and cholesterol with a subset of membrane proteins that are GPI-linked or attached to several long-chain fatty acyl moieties.

- Caveolin is an integral membrane protein that associates with the inner leaflet of the plasma membrane, forcing it to curve inward to form caveolae, probably involved in membrane transport and signaling.

- Integrins are transmembrane proteins of the plasma membrane that act both to attach cells to each other and to carry messages between the extracellular matrix and the cytoplasm.

- Specific proteins mediate the fusion of two membranes, which accompanies processes such as viral invasion and endocytosis and exocytosis.

## 11.3 Solute Transport across Membranes

Every living cell must acquire from its surroundings the raw materials for biosynthesis and for energy production, and must release to its environment the byproducts of metabolism. A few nonpolar compounds can dissolve in the lipid bilayer and cross the membrane unassisted, but for polar or charged compounds or ions, a membrane protein is essential for transmembrane movement. In some cases a membrane protein simply facilitates the diffusion of a solute down its concentration gradient, but transport often occurs against a gradient of concentration, electrical charge, or both, in which case solutes must be "pumped" in a process that requires energy (Fig. 11–26). The energy may come directly from ATP hydrolysis or may be supplied in the form of movement of another solute down its electrochemical gradient with enough energy to carry another solute up its gradient. Ions may also move across membranes via ion channels formed by proteins, or they may be carried across by ionophores, small molecules that mask the charge of the ions and allow them to diffuse through the lipid bilayer. With very few exceptions, the traffic of small molecules across the plasma membrane is mediated by proteins such as transmembrane channels, carriers, or pumps. Within the eukaryotic cell, different compartments have different concentrations of metabolic intermediates and products and of ions, and these, too, must move across intracellular membranes in tightly regulated, protein-mediated processes.

### Passive Transport Is Facilitated by Membrane Proteins

When two aqueous compartments containing unequal concentrations of a soluble compound or ion are separated by a permeable divider (membrane), the solute moves by **simple diffusion** from the region of higher concentration, through the membrane, to the region of lower concentration, until the two compartments have equal solute concentrations (Fig. 11–27a). When ions of opposite charge are separated by a permeable membrane, there is a transmembrane electrical gradient, a **membrane potential, $V_m$** (expressed in volts or millivolts). This membrane potential produces a force opposing ion movements that increase $V_m$ and driving ion movements that reduce $V_m$ (Fig. 11–27b). Thus the direction in which a charged solute tends to move spontaneously across a membrane depends on both the

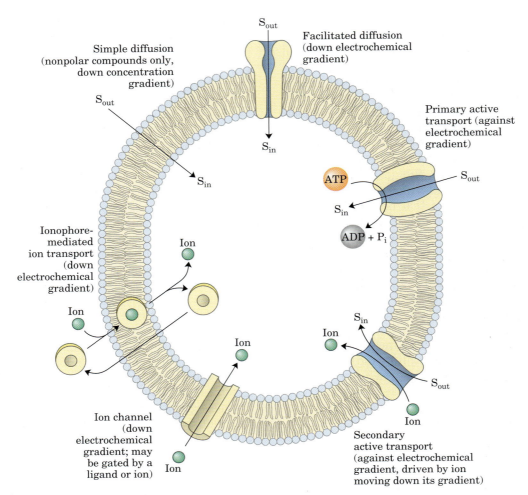

**FIGURE 11–26 Summary of transport types.**

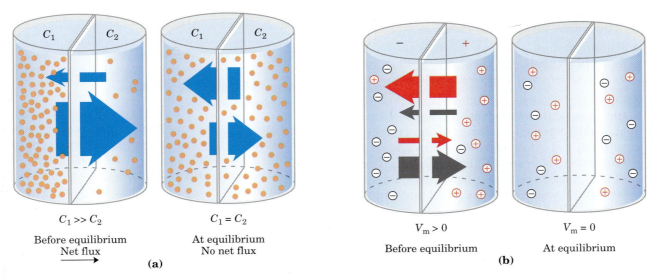

**FIGURE 11–27 Movement of solutes across a permeable membrane.**
**(a)** Net movement of electrically neutral solutes is toward the side of lower solute concentration until equilibrium is achieved. The solute concentrations on the left and right sides of the membrane are designated $C_1$ and $C_2$. The rate of transmembrane movement (indicated by the large arrows) is proportional to the concentration gradient, $C_1/C_2$. **(b)** Net movement of electrically charged solutes is dictated by a combination of the electrical potential ($V_m$) and the chemical concentration difference across the membrane; net ion movement continues until this electrochemical potential reaches zero.

chemical gradient (the difference in solute concentration) and the electrical gradient ($V_m$) across the membrane. Together, these two factors are referred to as the **electrochemical gradient** or **electrochemical potential.** This behavior of solutes is in accord with the second law of thermodynamics: molecules tend to spontaneously assume the distribution of greatest randomness and lowest energy.

To pass through a lipid bilayer, a polar or charged solute must first give up its interactions with the water molecules in its hydration shell, then diffuse about 3 nm (30 Å) through a solvent (lipid) in which it is poorly soluble (Fig. 11–28). The energy used to strip away the hydration shell and to move the polar compound from water into and through lipid is regained as the compound leaves the membrane on the other side and is rehydrated. However, the intermediate stage of transmembrane passage is a high-energy state comparable to the transition state in an enzyme-catalyzed chemical reaction. In both cases, an activation barrier must be overcome to reach the intermediate stage (Fig. 11–28; compare with Fig. 6–3). The energy of activation ($\Delta G^{\ddagger}$) for

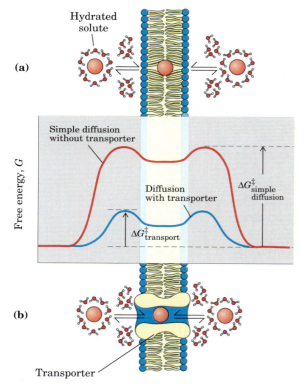

**(a)**

**(b)**

**FIGURE 11–28 Energy changes accompanying passage of a hydrophilic solute through the lipid bilayer of a biological membrane. (a)** In simple diffusion, removal of the hydration shell is highly endergonic, and the energy of activation ($\Delta G^{\ddagger}$) for diffusion through the bilayer is very high. **(b)** A transporter protein reduces the $\Delta G^{\ddagger}$ for transmembrane diffusion of the solute. It does this by forming noncovalent interactions with the dehydrated solute to replace the hydrogen bonding with water and by providing a hydrophilic transmembrane passageway.

translocation of a polar solute across the bilayer is so large that pure lipid bilayers are virtually impermeable to polar and charged species over periods of time relevant to cell growth and division.

Membrane proteins lower the activation energy for transport of polar compounds and ions by providing an alternative path through the bilayer for specific solutes. Proteins that bring about this **facilitated diffusion,** or **passive transport,** are not enzymes in the usual sense; their "substrates" are moved from one compartment to another, but are not chemically altered. Membrane proteins that speed the movement of a solute across a membrane by facilitating diffusion are called **transporters** or **permeases.**

Like enzymes, transporters bind their substrates with stereochemical specificity through multiple weak, noncovalent interactions. The negative free-energy change associated with these weak interactions, $\Delta G_{\text{binding}}$, counterbalances the positive free-energy change that accompanies loss of the water of hydration from the substrate, $\Delta G_{\text{dehydration}}$, thereby lowering $\Delta G^{\ddagger}$ for transmembrane passage (Fig. 11–28). Transporters span the lipid bilayer several times, forming a transmembrane channel lined with hydrophilic amino acid side chains. The channel provides an alternative path for a specific substrate to move across the lipid bilayer without its having to dissolve in the bilayer, further lowering $\Delta G^{\ddagger}$ for transmembrane diffusion. The result is an increase of several orders of magnitude in the rate of transmembrane passage of the substrate.

## Transporters Can Be Grouped into Superfamilies Based on Their Structures

We know from genomic studies that transporters constitute a significant fraction of all proteins encoded in the genomes of both simple and complex organisms. There are probably a thousand or more different transporters in the human genome. A few hundred transporters from various species have been studied with biochemical, genetic, and electrophysiological tools, but investigators have determined the three-dimensional structures for only a handful of these. Examination of the many transporter genes reveals obvious sequence similarities among subsets of transporters. And as experience has shown, similar amino acid sequences in proteins generally reflect similar three-dimensional structures and, often, similar mechanisms of action. It is reasonable to hope that by determining the structure and mechanism of action of at least one member of each transporter family, we can learn much about the other members of the family—about their structures, substrate specificities, transport rates, and mechanisms of energy coupling. A phylogenetic tree in which proteins are grouped together based on sequence homologies has the potential to tell us much about the transport properties of individual proteins on that tree. When this

**TABLE 11–3    The Transporter Classification (TC) System**

1.A. α Helix type channels
  1.A.1.    Voltage-gated ion channel VIC superfamily
            Voltage-gated $K^+$ channel
  1.A.3.    Ryanodine/$IP_3$ receptor $Ca^{2+}$ channel
  1.A.8.    Major intrinsic protein family
            Aquaporins
  1.A.9.    Ligand-gated ion channel (LIC) of neurotransmitter receptors
            Acetylcholine receptor/channel
1.B. β Barrel porins
  1.B.1.    General bacterial porin (GBP) family
1.C. Pore-forming toxins
  1.C.7.    Diphtheria toxin family
  1.C.18.   Mellitin family (bee venoms)

2.A. Porters: uniporters, symporters, and antiporters
  2.A.1.    Major facilitator superfamily (MFS)
            Lactose transporter/permease of *E. coli*
  2.A.1.1.  Sugar porter family
            GLUT1 glucose transporter of erythrocyte
  2.A.1.9.  $P_i$-$H^+$ symporter
  2.A.12.   ATP-ADP antiporter (AAA) family
  2.A.13.   $C_4$-dicarboxylate uptake (Dcu) family
  2.A.21.   Solute-$Na^+$ symporter (SSS) family
            $Na^+$-glucose symporter in epithelial cells
  2.A.73.   $HCO_3^-$ transporters
            $HCO_3^-$-$Cl^-$ antiporter
2.B. Nonribosomally synthesized porters
  2.B.1.    Valinomycin carrier family
            Valinomycin

3.A. Diphosphate bond hydrolysis–driven transporters (use $PP_i$, not ATP)
  3.A.1.    ATP-binding cassette (ABC) superfamily
            CFTR $Cl^-$ channel; multidrug transporter MDR1
  3.A.2.    $H^+$- or $Na^+$-translocating F-type, V-type, A-type ATPase superfamily
            $F_oF_1$ ATPase proton pump; $V_oV_1$ ATPase; $A_oA_1$ ATPase
  3.A.3.    P-type ATPase superfamily
            $Na^+K^+$ ATPase antiporter; SERCA $Ca^{2+}$ pump

**Note:** The three broad groups correspond to groups 1, 2, and 3 in Figure 11-29. The individual transporters listed here (screened in yellow) are discussed in this chapter.

phylogeny is combined with knowledge of structure, specificity, or mechanism, we have a very useful and relatively simple representation of the huge group of transporters (Table 11–3).

Transporters can usefully be classified into superfamilies, whose members have considerable similarity of sequence and might therefore be expected to share structural and functional properties. There are two very broad categories of transporters: carriers and channels (Fig. 11–29). **Carriers** bind their substrates with high stereospecificity, catalyze transport at rates well below the limits of free diffusion, and are saturable in the same sense as are enzymes: there is some substrate concentration above which further increases will not produce a greater rate of activity. **Channels** generally allow transmembrane movement at rates several orders of magnitude greater than those typical of carriers, rates approaching the limit of unhindered diffusion. Channels typically show less stereospecificity than carriers and are usually not saturable. Most channels are oligomeric complexes of several, often identical, subunits, whereas many carriers function as monomeric proteins. The classification as carrier or channel is the broadest distinction among transporters. Within each of these categories

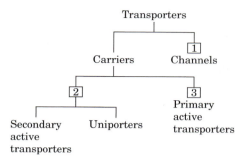

**FIGURE 11–29 Classification of transporters.** The numbers here correspond to the main subdivisions in Table 11–3.

are superfamilies of various types, defined not only by their primary sequences but by their secondary structures. Some channels are constructed primarily of helical transmembrane segments, others have β-barrel structures (Table 11–3). Among the carriers, some simply facilitate diffusion down a concentration gradient; they are the uniporter superfamily. Others (active transporters) can drive substrates across the membrane against a concentration gradient, some using energy provided directly by a chemical reaction (primary active transporters) and some coupling uphill transport of one substrate with the downhill transport of another (secondary active transporters). We now consider some well-studied representatives of the main transporter su-

perfamilies. You will encounter some of these transporters again in later chapters in the context of the metabolic pathways in which they participate.

## The Glucose Transporter of Erythrocytes Mediates Passive Transport

Energy-yielding metabolism in erythrocytes depends on a constant supply of glucose from the blood plasma, where the glucose concentration is maintained at about 5 mM. Glucose enters the erythrocyte by facilitated diffusion via a specific glucose transporter, at a rate about 50,000 times greater than the uncatalyzed diffusion rate. The glucose transporter of erythrocytes (called GLUT1 to distinguish it from related glucose transporters in other tissues) is a type III integral protein ($M_r$ ~45,000) with 12 hydrophobic segments, each of which is believed to form a membrane-spanning helix. The detailed structure of GLUT1 is not yet known, but one plausible model suggests that the side-by-side assembly of several helices produces a transmembrane channel lined with hydrophilic residues that can hydrogen-bond with glucose as it moves through the channel (Fig. 11–30).

The process of glucose transport can be described by analogy with an enzymatic reaction in which the "substrate" is glucose outside the cell ($S_{out}$), the "product" is glucose inside ($S_{in}$), and the "enzyme" is the transporter, T. When the rate of glucose uptake is measured

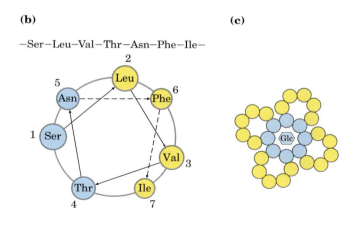

**FIGURE 11–30 Proposed structure of GLUT1. (a)** Transmembrane helices are represented as oblique (angled) rows of three or four amino acid residues, each row depicting one turn of the α helix. Nine of the 12 helices contain three or more polar or charged amino acid residues, often separated by several hydrophobic residues. **(b)** A helical wheel diagram shows the distribution of polar and nonpolar residues on the surface of a helical segment. The helix is diagrammed as though observed along its axis from the amino terminus. Adjacent residues in the linear sequence are connected with arrows, and each residue is placed around the wheel in the position it occupies in the helix; recall that 3.6 residues are required to make one complete turn of the

α helix. In this example, the polar residues (blue) are on one side of the helix and the hydrophobic residues (yellow) on the other. This is, by definition, an amphipathic helix. **(c)** Side-by-side association of five or six amphipathic helices, each with its polar face oriented toward the central cavity, can produce a transmembrane channel lined with polar and charged residues. This channel provides many opportunities for hydrogen bonding with glucose as it moves through the transporter. The three-dimensional structure of GLUT1 has not yet been determined by x-ray crystallography, but researchers expect that the hydrophilic transmembrane channels of this and many other transporters and ion channels will resemble this model.

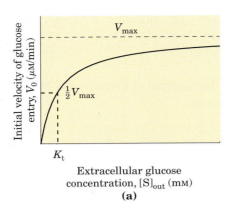

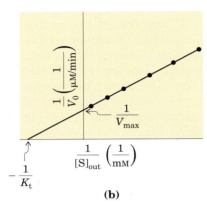

**FIGURE 11–31 Kinetics of glucose transport into erythrocytes. (a)** The initial rate of glucose entry into an erythrocyte, $V_0$, depends upon the initial concentration of glucose on the outside, $[S]_{out}$. **(b)** Double-reciprocal plot of the data in **(a).** The kinetics of facilitated diffusion is analogous to the kinetics of an enzyme-catalyzed reaction. Compare these plots with Figure 6–11, and Figure 1 in Box 6–1. Note that $K_t$ is analogous to $K_m$, the Michaelis constant.

as a function of external glucose concentration (Fig. 11–31), the resulting plot is hyperbolic; at high external glucose concentrations the rate of uptake approaches $V_{max}$. Formally, such a transport process can be described by the equations

$$S_{out} + T_1 \underset{k_{-1}}{\overset{k_1}{\rightleftharpoons}} S_{out} \cdot T_1$$

$$k_{-4} \Big\| k_4 \qquad k_{-2} \Big\| k_2$$

$$S_{in} + T_2 \underset{k_{-3}}{\overset{k_3}{\rightleftharpoons}} S_{in} \cdot T_2$$

in which $k_1$, $k_{-1}$, and so forth, are the forward and reverse rate constants for each step; $T_2$ is the transporter conformation that faces out, and $T_2$ the one that faces in. The steps are summarized in Figure 11–32.

The rate equations for this process can be derived exactly as for enzyme-catalyzed reactions (Chapter 6), yielding an expression analogous to the Michaelis-Menten equation:

$$V_0 = \frac{V_{max}[S]_{out}}{K_t + [S]_{out}}$$

in which $V_0$ is the initial velocity of accumulation of glucose inside the cell when its concentration in the surrounding medium is $[S]_{out}$, and $K_t$ ($K_{transport}$) is a constant analogous to the Michaelis constant, a combination of rate constants that is characteristic of each transport system. This equation describes the *initial* velocity, the rate observed when $[S]_{in} = 0$. As is the case for enzyme-catalyzed reactions, the slope-intercept form of the equation describes a linear plot of $1/V_0$ against $1/[S]_{out}$, from which we can obtain values of $K_t$ and $V_{max}$ (Fig. 11–31b). When $[S] = K_t$, the rate of uptake is $\frac{1}{2} V_{max}$; the transport process is half-saturated. The concentration of blood glucose, 4.5 to 5 mM, is about

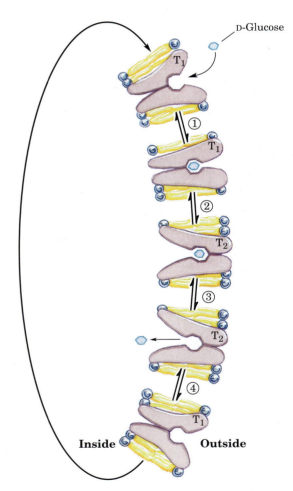

**FIGURE 11–32 Model of glucose transport into erythrocytes by GLUT1.** The transporter exists in two conformations: $T_1$, with the glucose-binding site exposed on the outer surface of the plasma membrane, and $T_2$, with the binding site exposed on the inner surface. Glucose transport occurs in four steps. ① Glucose in blood plasma binds to a stereospecific site on $T_1$; this lowers the activation energy for ② a conformational change from $S_{out} \cdot T_1$ to $S_{in} \cdot T_2$, effecting the transmembrane passage of the glucose. ③ Glucose is now released from $T_2$ into the cytoplasm, and ④ the transporter returns to the $T_1$ conformation, ready to transport another glucose molecule.

three times $K_t$, which ensures that GLUT1 is nearly saturated with substrate and operates near $V_{max}$.

Because no chemical bonds are made or broken in the conversion of $S_{out}$ to $S_{in}$, neither "substrate" nor "product" is intrinsically more stable, and the process of entry is therefore fully reversible. As $[S]_{in}$ approaches $[S]_{out}$, the rates of entry and exit become equal. Such a system is therefore incapable of accumulating the substrate (glucose) within a cell at concentrations above that in the surrounding medium; it simply achieves equilibration of glucose on the two sides of the membrane much faster than would occur in the absence of a specific transporter. GLUT1 is specific for D-glucose, having a measured $K_t$ of 1.5 mM. For the close analogs D-mannose and D-galactose, which differ only in the position of one hydroxyl group, the values of $K_t$ are 20 and 30 mM, respectively; and for L-glucose, $K_t$ exceeds 3,000 mM. Thus GLUT1 shows the three hallmarks of passive transport: high rates of diffusion down a concentration gradient, saturability, and specificity.

Twelve glucose transporters are encoded in the human genome, each with unique kinetic properties, patterns of tissue distribution, and function (Table 11–4). In liver, GLUT2 transports glucose out of hepatocytes when liver glycogen is broken down to replenish blood glucose. GLUT2 has a $K_t$ of about 66 mM and can therefore respond to increased levels of intracellular glucose (produced by glycogen breakdown) by increasing outward transport. Skeletal muscle and adipose tissue have yet another glucose transporter, GLUT4 ($K_t = 5$ mM), which is distinguished by its stimulation by insulin: its activity increases when release of insulin signals a high blood glucose concentration, thus increasing the rate of glucose uptake into muscle and adipose tissue (Box 11–2 describes some malfunctions of this transporter).

## The Chloride-Bicarbonate Exchanger Catalyzes Electroneutral Cotransport of Anions across the Plasma Membrane

The erythrocyte contains another facilitated diffusion system, an anion exchanger that is essential in $CO_2$ transport to the lungs from tissues such as skeletal muscle and liver. Waste $CO_2$ released from respiring tissues into the blood plasma enters the erythrocyte, where it is converted to bicarbonate ($HCO_3^-$) by the enzyme carbonic anhydrase. (Recall that $HCO_3^-$ is the primary buffer of blood pH; see Box 2–4). The $HCO_3^-$ reenters the blood plasma for transport to the lungs (Fig. 11–33). Because $HCO_3^-$ is much more soluble in blood plasma than is $CO_2$, this roundabout route increases the capacity of the blood to carry carbon dioxide from the tissues to the lungs. In the lungs, $HCO_3^-$ reenters the erythrocyte and is converted to $CO_2$, which is eventually released into the lung space and exhaled. To be effective, this shuttle requires very rapid movement of $HCO_3^-$ across the erythrocyte membrane.

The **chloride-bicarbonate exchanger,** also called the **anion exchange (AE) protein,** increases the permeability of the erythrocyte membrane to $HCO_3^-$ more than a millionfold. Like the glucose transporter, it is an integral protein that probably spans the membrane at least 12 times. This protein mediates the simultaneous movement of two anions: for each $HCO_3^-$ ion that moves in one direction, one $Cl^-$ ion moves in the opposite direction (Fig. 11–33), with no net transfer of charge; the exchange is **electroneutral.** The coupling of $Cl^-$ and $HCO_3^-$ movements is obligatory; in the absence of chloride, bicarbonate transport stops. In this respect, the anion exchanger is typical of all systems, called **cotransport systems,** that simultaneously carry

---

| TABLE 11–4 | Glucose Transporters in the Human Genome | | |
|---|---|---|---|
| Transporter | Tissue(s) where expressed | Gene | Role[*] |
| GLUT1 | Ubiquitous | SLC2A1 | Basal glucose uptake |
| GLUT2 | Liver, pancreatic islets, intestine | SLC2A2 | In liver, removal of excess glucose from blood; in pancreas, regulation of insulin release |
| GLUT3 | Brain (neuronal) | SLC2A3 | Basal glucose uptake |
| GLUT4 | Muscle, fat, heart | SLC2A4 | Activity increased by insulin |
| GLUT5 | Intestine, testis, kidney, sperm | SLC2A5 | Primarily fructose transport |
| GLUT6 | Spleen, leukocytes, brain | SLC2A6 | Possibly no transporter function |
| GLUT7 | Liver microsomes | SLC2A7 | – |
| GLUT8 | Testis, blastocyst, brain | SLC2A8 | – |
| GLUT9 | Liver, kidney | SLC2A9 | – |
| GLUT10 | Liver, pancreas | SLC2A10 | – |
| GLUT11 | Heart, skeletal muscle | SLC2A11 | – |
| GLUT12 | Skeletal muscle, adipose, small intestine | SLC2A12 | – |

*Dash indicates role uncertain.

BOX 11–2 **BIOCHEMISTRY IN MEDICINE**

## Defective Glucose and Water Transport in Two Forms of Diabetes

When ingestion of a carbohydrate-rich meal causes blood glucose to exceed the usual concentration between meals (about 5 mM), excess glucose is taken up by the myocytes of cardiac and skeletal muscle (which store it as glycogen) and by adipocytes (which convert it to triacylglycerols). Glucose uptake into myocytes and adipocytes is mediated by the glucose transporter GLUT4. Between meals, some GLUT4 is present in the plasma membrane, but most is sequestered in the membranes of small intracellular vesicles (Fig. 1). Insulin released from the pancreas in response to high blood glucose triggers the movement of these intracellular vesicles to the plasma membrane, where they fuse, thus exposing GLUT4 molecules on the outer surface of the cell (see Fig. 12–8). With more GLUT4 molecules in action, the rate of glucose uptake increases 15-fold or more. When blood glucose levels return to normal, insulin release slows and most GLUT4 molecules are removed from the plasma membrane and stored in vesicles.

In type I (juvenile onset) diabetes mellitus, the inability to release insulin (and thus to mobilize glucose transporters) results in low rates of glucose uptake into muscle and adipose tissue. One consequence is a prolonged period of high blood glucose after a carbohydrate-rich meal. This condition is the basis for the glucose tolerance test used to diagnose diabetes (Chapter 23).

The water permeability of epithelial cells lining the renal collecting duct in the kidney is due to the presence of an aquaporin (AQP-2) in their apical plasma membranes (facing the lumen of the duct). Antidiuretic hormone (ADH) regulates the retention of water by mobilizing AQP-2 molecules stored in vesicle membranes within the epithelial cells, much as insulin mobilizes GLUT4 in muscle and adipose tissue. When the vesicles fuse with the epithelial cell plasma membrane, water permeability greatly increases and more water is reabsorbed from the collecting duct and returned to the blood. When the ADH level drops, AQP-2 is resequestered within vesicles, reducing water retention. In the relatively rare human disease diabetes insipidus, a genetic defect in AQP-2 leads to impaired water reabsorption by the kidney. The result is excretion of copious volumes of very dilute urine.

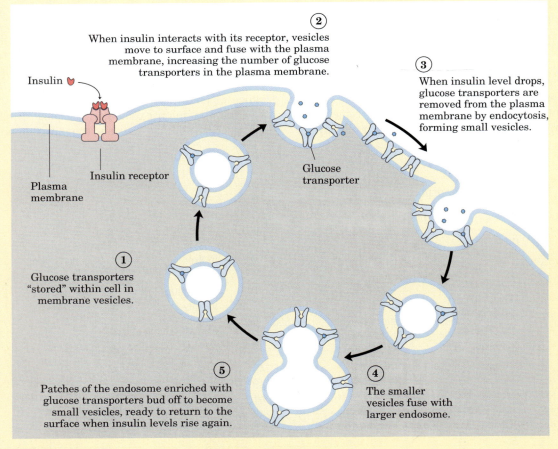

**FIGURE 1** Regulation by insulin of glucose transport by GLUT4 into a myocyte.

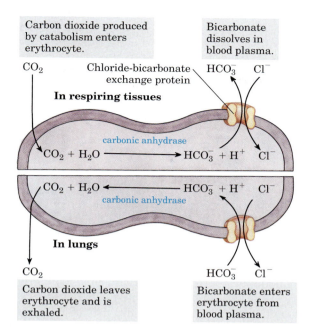

Carbon dioxide produced by catabolism enters erythrocyte.

Bicarbonate dissolves in blood plasma.

**In respiring tissues**

carbonic anhydrase

carbonic anhydrase

**In lungs**

Carbon dioxide leaves erythrocyte and is exhaled.

Bicarbonate enters erythrocyte from blood plasma.

**FIGURE 11–33  Chloride-bicarbonate exchanger of the erythrocyte membrane.** This cotransport system allows the entry and exit of $HCO_3^-$ without changes in the transmembrane electrical potential. Its role is to increase the $CO_2$-carrying capacity of the blood.

two solutes across a membrane. When, as in this case, the two substrates move in opposite directions, the process is **antiport.** In **symport,** two substrates are moved simultaneously in the same direction. As we noted earlier, transporters that carry only one substrate, such as the erythrocyte glucose transporter, are **uniport** systems (Fig. 11–34).

The human genome has genes for three closely related chloride-bicarbonate exchangers, all with the same predicted transmembrane topology. Erythrocytes

contain the AE1 transporter, AE2 is prominent in liver, and AE3 is present in plasma membranes of the brain, heart, and retina. Similar anion exchangers are also found in plants and microorganisms.

### Active Transport Results in Solute Movement against a Concentration or Electrochemical Gradient

In passive transport, the transported species always moves down its electrochemical gradient and is not accumulated above the equilibrium concentration. Active transport, by contrast, results in the accumulation of a solute above the equilibrium point. Active transport is thermodynamically unfavorable (endergonic) and takes place only when coupled (directly or indirectly) to an exergonic process such as the absorption of sunlight, an oxidation reaction, the breakdown of ATP, or the concomitant flow of some other chemical species down its electrochemical gradient. In **primary active transport,** solute accumulation is coupled directly to an exergonic chemical reaction, such as conversion of ATP to $ADP + P_i$ (Fig. 11–35). **Secondary active transport** occurs when endergonic (uphill) transport of one solute is coupled to the exergonic (downhill) flow of a different solute that was originally pumped uphill by primary active transport.

The amount of energy needed for the transport of a solute against a gradient can be calculated from the initial concentration gradient. The general equation for the free-energy change in the chemical process that converts S to P is

$$\Delta G = \Delta G'^\circ + RT \ln [P]/[S] \qquad (11\text{–}1)$$

where $R$ is the gas constant, 8.315 J/mol · K, and $T$ is the absolute temperature. When the "reaction" is simply

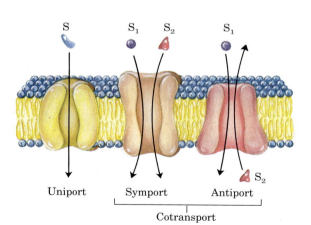

Uniport      Symport      Antiport

Cotransport

**FIGURE 11–34  Three general classes of transport systems.** Transporters differ in the number of solutes (substrates) transported and the direction in which each is transported. Examples of all three types of transporters are discussed in the text. Note that this classification tells us nothing about whether these are energy-requiring (active transport) or energy-independent (passive transport) processes.

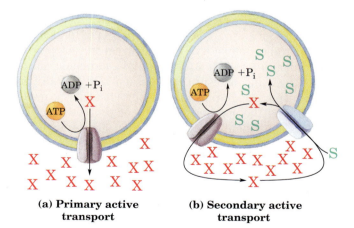

**(a) Primary active transport**

**(b) Secondary active transport**

**FIGURE 11–35  Two types of active transport. (a)** In primary active transport, the energy released by ATP hydrolysis drives solute movement against an electrochemical gradient. **(b)** In secondary active transport, a gradient of ion X (often $Na^+$) has been established by primary active transport. Movement of X down its electrochemical gradient now provides the energy to drive cotransport of a second solute (S) against its electrochemical gradient.

transport of a solute from a region where its concentration is $C_1$ to a region where its concentration is $C_2$, no bonds are made or broken and the standard free-energy change, $\Delta G'^\circ$, is zero. The free-energy change for transport, $\Delta G_t$, is then

$$\Delta G_t = RT \ln \frac{C_2}{C_1} \qquad (11\text{–}2)$$

If there is a tenfold difference in concentration between two compartments, the cost of moving 1 mol of an uncharged solute at 25 °C across a membrane separating the compartments is therefore

$$\Delta G_t = (8.315 \text{ J/mol} \cdot \text{K})(298 \text{ K})(\ln 10/1) = 5{,}700 \text{ J/mol}$$
$$= 5.7 \text{ kJ/mol}$$

Equation 11–2 holds for all uncharged solutes.

When the solute is an ion, its movement without an accompanying counterion results in the endergonic separation of positive and negative charges, producing an electrical potential; such a transport process is said to be **electrogenic.** The energetic cost of moving an ion depends on the electrochemical potential (p. 391), the sum of the chemical and electrical gradients:

$$\Delta G_t = RT \ln \left( \frac{C_2}{C_1} \right) + Z \mathcal{F} \Delta \psi \qquad (11\text{–}3)$$

where $Z$ is the charge on the ion, $\mathcal{F}$ is the Faraday constant (96,480 J/V · mol), and $\Delta \psi$ is the transmembrane electrical potential (in volts). Eukaryotic cells typically have electrical potentials across their plasma membranes of about 0.05 to 0.1 V (with the inside negative relative to the outside), so the second term of Equation 11–3 can make a significant contribution to the total free-energy change for transporting an ion. Most cells maintain more than tenfold differences in ion concentrations across their plasma or intracellular membranes, and for many cells and tissues active transport is therefore a major energy-consuming process.

The mechanism of active transport is of fundamental importance in biology. As we shall see in Chapter 19, the formation of ATP in mitochondria and chloroplasts occurs by a mechanism that is essentially ATP-driven ion transport operating in reverse. The energy made available by the spontaneous flow of protons across a membrane is calculable from Equation 11–3; remember that $\Delta G$ for flow *down* an electrochemical gradient has a negative value, and $\Delta G$ for transport of ions *against* an electrochemical gradient has a positive value.

### P-Type ATPases Undergo Phosphorylation during Their Catalytic Cycles

The family of active transporters called **P-type ATPases** are ATP-driven cation transporters that are reversibly phosphorylated by ATP as part of the transport cycle; phosphorylation forces a conformational change that is central to moving the cation across the membrane. All P-type transport ATPases have similarities in amino acid sequence, especially near the Asp residue that undergoes phosphorylation, and all are sensitive to inhibition by the phosphate analog **vanadate.**

$$
\begin{array}{cc}
\text{O}^- & \text{O}^- \\
| & | \\
\text{O=P—O}^- & \text{O=V—O}^- \\
| & | \\
\text{OH} & \text{OH} \\
\text{Phosphate} & \text{Vanadate}
\end{array}
$$

Each P-type ATPase transporter is an integral protein with ten predicted membrane-spanning regions in a single polypeptide; some also have a second subunit. The P-type transporters are very widely distributed. In animal tissues, the $Na^+K^+$ ATPase (an antiporter for $Na^+$ and $K^+$) and the $Ca^{2+}$ ATPase (a uniporter for $Ca^{2+}$) are ubiquitous P-type ATPases that maintain differences in the ionic composition of the cytosol and the extracellular medium. Parietal cells in the lining of the mammalian stomach have a P-type ATPase that pumps $H^+$ and $K^+$ across the plasma membrane, thereby acidifying the stomach contents. In vascular plants, a P-type ATPase pumps protons out of the cell, establishing an electrochemical difference of as much as 2 pH units and 250 mV across the plasma membrane. A similar P-type ATPase in the bread mold *Neurospora* pumps protons out of cells to establish an inside-negative membrane potential, which is used to drive the uptake of substrates and ions from the surrounding medium by secondary active transport. Bacteria use P-type ATPases to pump out toxic heavy metal ions such as $Cd^{2+}$ and $Cu^{2+}$.

In virtually every animal cell type, the concentration of $Na^+$ is lower in the cell than in the surrounding medium, and the concentration of $K^+$ is higher (Fig. 11–36). This imbalance is maintained by a primary active transport system in the plasma membrane. The enzyme **$Na^+K^+$ ATPase,** discovered by Jens Skou in 1957, couples breakdown of ATP to the simultaneous movement of both $Na^+$ and $K^+$ against their electrochemical gradients. For each molecule of ATP converted to ADP and $P_i$, the transporter moves two $K^+$ ions inward and three $Na^+$ ions outward across the plasma membrane. The $Na^+K^+$ ATPase is an integral protein with two subunits ($M_r$ ~50,000 and ~110,000), both of which span the membrane.

Jens Skou

The detailed mechanism by which ATP hydrolysis is coupled to transport awaits determination of the protein's three-dimensional structure, but a current model (Fig. 11–37) proposes that the ATPase cycles between two forms, a phosphorylated form (designated P-Enz$_\text{II}$) with high affinity for $K^+$ and low affinity for $Na^+$, and a dephosphorylated form (Enz$_\text{I}$) with high affinity for $Na^+$

and low affinity for $K^+$. The conversion of ATP to ADP and $P_i$ takes place in two steps catalyzed by the enzyme, involving formation then hydrolysis of the phospho-enzyme:

$$(1) \quad ATP + Enz_I \longrightarrow ADP + P\text{-}Enz_{II}$$

$$(2) \quad P\text{-}Enz_{II} + H_2O \longrightarrow Enz_I + P_i$$

$$Sum: \quad ATP + H_2O \longrightarrow ADP + P_i$$

Because three $Na^+$ ions move outward for every two $K^+$ ions that move inward, the process is electrogenic—it creates a net separation of charge across the membrane. The result is a transmembrane potential of $-50$ to $-70$ mV (inside negative relative to outside), which is characteristic of most animal cells and essential to the conduction of action potentials in neurons. The central role of the $Na^+K^+$ ATPase is reflected in the energy invested in this single reaction: about 25% of the total energy consumption of a human at rest!

The steroid derivative **ouabain** (pronounced wah'-bane; from *waa bayyo*, Somali for "arrow poison") is a potent and specific inhibitor of the $Na^+K^+$ ATPase. Oubain binds preferentially to the form of the enzyme that is open to the extracellular side, locking in two $Na^+$ ions and preventing the changes of conformation necessary to ion transport. Another very potent toxin, palytoxin (produced by a coral on the Hawaiian shoreline), also targets the $Na^+K^+$ ATPase, but it binds

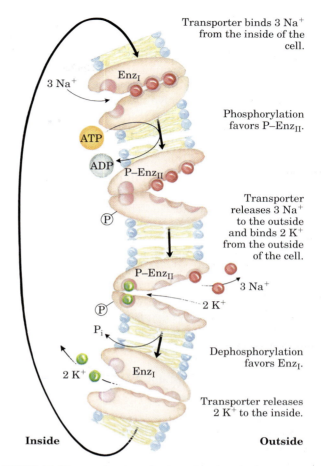

**FIGURE 11–37** Postulated mechanism of $Na^+$ and $K^+$ transport by the $Na^+K^+$ ATPase.

to the protein so as to lock it into a position in which the ion-binding sites are permanently accessible from both sides, converting the transporter into a nonspecific ion channel. This allows exit of $K^+$ from cells and deflates the (essential) ion gradient across the plasma membrane, which accounts for the high toxicity of this compound.

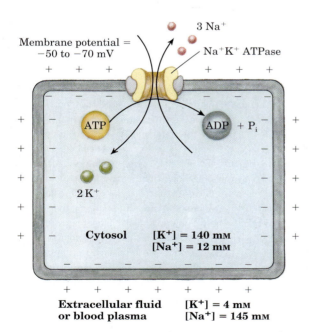

**FIGURE 11–36** $Na^+K^+$ ATPase. In animal cells, this active transport system is primarily responsible for setting and maintaining the intracellular concentrations of $Na^+$ and $K^+$ and for generating the transmembrane electrical potential. It does this by moving three $Na^+$ out of the cell for every two $K^+$ it moves in. The electrical potential is central to electrical signaling in neurons, and the gradient of $Na^+$ is used to drive the uphill cotransport of solutes in many cell types.

Ouabain

Ouabain and another steroid derivative, **digitoxigenin,** are the active ingredients of digitalis, an extract of the leaves of the foxglove plant. (Ouabain is found in lower concentrations in a number of other plants, presumably serving to discourage herbivores.) Digitalis has been used to treat congestive heart failure since its introduction for that purpose (treatment of "dropsy") by the British physician William Withering in 1785. It strengthens heart muscle contractions without increasing the heart rate and thus increases the efficiency of the heart. Digitalis inhibits the efflux of $Na^+$, raising the intracellular $[Na^+]$ enough to activate a $Na^+$-$Ca^{2+}$ antiporter in cardiac muscle. The increased influx of $Ca^{2+}$ through this antiporter produces elevated cytosolic $[Ca^{2+}]$, which strengthens the contractions of the heart. The potency of ouabain in animals led to the suggestion (50 years ago) that this plant product might act by mimicking a normal regulator of the $Na^+K^+$ ATPase produced in animals, and it now appears that this may be so. Ouabain itself has been isolated from bovine adrenal glands and has been detected in the blood plasma and hypothalamus of mammals. ■

## P-Type $Ca^{2+}$ Pumps Maintain a Low Concentration of Calcium in the Cytosol

The cytosolic concentration of free $Ca^{2+}$ is generally at or below 100 nM, far lower than that in the surrounding medium, whether pond water or blood plasma. The ubiquitous occurrence of inorganic phosphates ($P_i$ and $PP_i$) at millimolar concentrations in the cytosol necessitates a low cytosolic $Ca^{2+}$ concentration, because inorganic phosphate combines with calcium to form relatively insoluble calcium phosphates. Calcium ions are pumped out of the cytosol by a P-type ATPase, the **plasma membrane $Ca^{2+}$ pump.** Another P-type $Ca^{2+}$ pump in the endoplasmic reticulum moves $Ca^{2+}$ into the ER lumen, a compartment separate from the cytosol. In myocytes, $Ca^{2+}$ is normally sequestered in a specialized form of endoplasmic reticulum called the sarcoplasmic reticulum. **The sarcoplasmic and endoplasmic reticulum calcium (SERCA) pumps** are closely related in structure and mechanism, and both are inhibited by the tumor-promoting agent thapsigargin, which does not affect the plasma membrane $Ca^{2+}$ pump.

The plasma membrane $Ca^{2+}$ pump and SERCA pumps are integral proteins that cycle between phosphorylated and dephosphorylated conformations in a mechanism similar to that for $Na^+K^+$ ATPase (Fig. 11–37). Phosphorylation favors a conformation with a high-affinity $Ca^{2+}$-binding site exposed on the cytoplasmic side, and dephosphorylation favors one with a low-affinity $Ca^{2+}$-binding site on the lumenal side. By this mechanism, the energy released by hydrolysis of ATP during one phosphorylation-dephosphorylation cycle drives $Ca^{2+}$ across the membrane against a large electrochemical gradient.

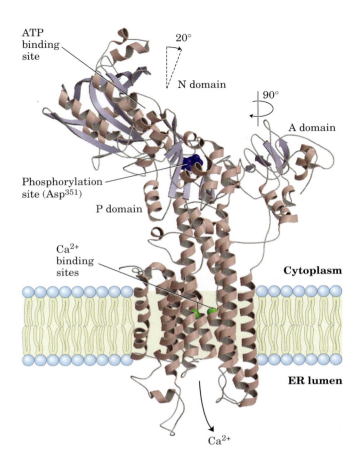

**FIGURE 11–38 Structure of the $Ca^{2+}$ pump of sarcoplasmic reticulum.** (PDB ID 1EUL) Ten transmembrane helices surround the path for $Ca^{2+}$ movement through the membrane. Two of the helices are interrupted near the middle of the bilayer, and their nonhelical regions form the binding sites for two $Ca^{2+}$ ions (green). The carboxylate groups of an Asp residue in one helix and a Glu residue in another are central to the $Ca^{2+}$-binding sites. Three globular domains extend from the cytoplasmic side: the N (nucleotide-binding) domain has the binding site for ATP; the P (phosphorylation) domain contains the $Asp^{351}$ residue (blue) that undergoes reversible phosphorylation, and the A (actuator) domain somehow mediates the structural changes that alter the $Ca^{2+}$ affinity of the $Ca^{2+}$-binding site and its exposure to cytoplasm or lumen. Note the long distance between the phosphorylation site and the $Ca^{2+}$-binding site. There is strong evidence that during one transport cycle, the N domain tips about 20° to the right, bringing the ATP site close to $Asp^{351}$, and that during each catalytic cycle the A domain twists by about 90° around the normal (perpendicular) to the membrane. These conformational changes must expose the $Ca^{2+}$-binding site first on one side of the membrane, then on the other, changing the $Ca^{2+}$ affinity of the site from high on the cytoplasmic side to lower on the lumenal side. A complete understanding of the coupling between phosphorylation and $Ca^{2+}$ transport awaits determination of all the conformations involved in the cycle.

The $Ca^{2+}$ pump of the sarcoplasmic reticulum, which comprises 80% of the protein in that membrane, consists of a single polypeptide ($M_r$ ~100,000) that spans the membrane ten times and has three cytoplasmic domains formed by loops that connect the transmembrane helices (Fig. 11–38). The two $Ca^{2+}$-binding sites are located near the middle of the membrane bilayer, 40 to 50 Å from the phosphorylated Asp residue characteristic of all P-type ATPases, so the effects of Asp phosphorylation are not direct. They must be mediated by conformational changes that alter the affinity for $Ca^{2+}$ and open a path for $Ca^{2+}$ release on the lumenal side of the membrane.

The amino acid sequences of the SERCA pumps and the $Na^+K^+$ ATPase share 30% identity and 65% sequence similarity, and their topology relative to the membrane is also the same. Thus it seems likely that the $Na^+K^+$ ATPase structure is similar to that of the SERCA pumps and that all P-type ATPase transporters share the same basic structure.

## F-Type ATPases Are Reversible, ATP-Driven Proton Pumps

The **F-type ATPase** active transporters play a central role in energy-conserving reactions in mitochondria, bacteria, and chloroplasts; we discuss that role in detail in our description of oxidative phosphorylation and photophosphorylation in Chapter 19. The F-type ATPases catalyze the uphill transmembrane passage of protons driven by ATP hydrolysis ("F-type" originated in the identification of these ATPases as energy-coupling factors). The $F_o$ integral membrane protein complex (Fig. 11–39; subscript $o$ denoting its inhibition by the drug oligomycin) provides a transmembrane pore for protons, and the peripheral protein $F_1$ (subscript $1$ indicating that it was the first of several factors isolated from mitochondria) is a molecular machine that uses the energy

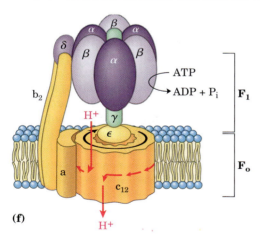

**(f)**

**FIGURE 11–39 Structure of the $F_oF_1$ ATPase/ATP synthase.** F-type ATPases have a peripheral domain, $F_1$, consisting of three α subunits, three β subunits, one δ subunit (purple), and a central shaft (the γ subunit, green). The integral portion of F-type ATPases, $F_o$ (yellow), has multiple copies of c, one a, and two b subunits. $F_o$ provides a transmembrane channel through which about four protons are pumped (red arrows) for each ATP hydrolyzed on the β subunits of $F_1$. The remarkable mechanism by which these two events are coupled is described in detail in Chapter 19. It involves rotation of $F_o$ relative to $F_1$ (black arrow). The structures of $V_oV_1$ and $A_oA_1$ are essentially similar to that of $F_oF_1$, and the mechanisms are probably similar, too.

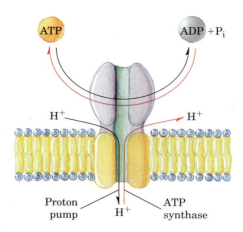

**FIGURE 11–40 Reversibility of F-type ATPases.** An ATP-driven proton transporter also can catalyze ATP synthesis (red arrows) as protons flow *down* their electrochemical gradient. This is the central reaction in the processes of oxidative phosphorylation and photophosphorylation, both described in detail in Chapter 19.

of ATP to drive protons uphill (into a region of higher $H^+$ concentration). The $F_oF_1$ organization of proton-pumping transporters must have developed very early in evolution. Eubacteria such as *E. coli* use an $F_oF_1$ ATPase complex in their plasma membrane to pump protons outward, and archaebacteria have a closely homologous proton pump, the $A_oA_1$ ATPase.

The reaction catalyzed by F-type ATPases is reversible, so a proton gradient can supply the energy to drive the reverse reaction, ATP synthesis (Fig. 11–40). When functioning in this direction, the F-type ATPases are more appropriately named **ATP synthases.** ATP synthases are central to ATP production in mitochondria during oxidative phosphorylation and in chloroplasts during photophosphorylation, as well as in eubacteria and archaebacteria. The proton gradient needed to drive ATP synthesis is produced by other types of proton pumps powered by substrate oxidation or sunlight. As noted above, we return to a detailed description of these processes in Chapter 19.

**V-type ATPases,** a class of proton-transporting ATPases structurally (and possibly mechanistically) related to the F-type ATPases, are responsible for acidifying intracellular compartments in many organisms (thus *V* for *v*acuolar). Proton pumps of this type maintain the vacuoles of fungi and higher plants at a pH between 3 and 6, well below that of the surrounding cytosol (pH 7.5). V-type ATPases are also responsible for the acidification of lysosomes, endosomes, the Golgi complex, and secretory vesicles in animal cells. All V-type ATPases have a similar complex structure, with an integral (transmembrane) domain ($V_o$) that serves as a proton channel and a peripheral domain ($V_1$) that contains the ATP-binding site and the ATPase activity. The mechanism by which V-type ATPases couple ATP hydrolysis to the uphill transport of protons is not understood in detail.

## ABC Transporters Use ATP to Drive the Active Transport of a Wide Variety of Substrates

**ABC transporters** (Fig. 11–41) constitute a large family of ATP-dependent transporters that pump amino acids, peptides, proteins, metal ions, various lipids, bile salts, and many hydrophobic compounds, including drugs, out of cells against a concentration gradient. One ABC transporter in humans, the **multidrug transporter (MDR1),** is responsible for the striking resistance of certain tumors to some generally effective antitumor drugs. MDR1 has a broad substrate specificity for hydrophobic compounds, including, for example, the chemotherapeutic drugs adriamycin, doxorubicin, and vinblastine. By pumping these drugs out of the cell, the transporter prevents their accumulation within a tumor and thus blocks their therapeutic effects. MDR1 is an integral membrane protein ($M_r$ 170,000) with 12 transmembrane segments and two ATP-binding domains ("cassettes"), which give the family its name: *A*TP-*b*inding *c*assette transporters.

All ABC transporters have two nucleotide-binding domains (NBDs) and two transmembrane domains (Fig. 11–41). In some cases, all these domains are in a single long polypeptide; other ABC transporters have two subunits, each contributing an NBD and a domain with six (or in some cases ten) transmembrane helices. Although many of the ABC transporters are in the plasma membrane, some types are also found in the endoplasmic reticulum and in the membranes of mitochondria and lysosomes. Most ABC transporters act as pumps, but at least some members of the superfamily act as ion channels that are opened and closed by ATP hydrolysis. The CFTR transporter (Box 11–3) is a Cl⁻ channel operated by ATP hydrolysis.

The NBDs of all ABC proteins are similar in sequence and presumably in three-dimensional structure; they are the conserved molecular motor that can be coupled to a wide variety of pumps and channels. When coupled with a pump, the ATP-driven motor moves solutes against a concentration gradient; when coupled with an ion channel, the motor opens and closes the channel using ATP as energy source. The stoichiometry of ABC pumps is about one ATP hydrolyzed per molecule of substrate transported, but neither the mechanism of coupling nor the site of substrate binding are known.

Some ABC transporters have very high specificity for a single substrate; others are more promiscuous. The human genome contains at least 48 genes that encode ABC transporters, many of which are involved in maintaining the lipid bilayer and in transporting sterols, sterol derivatives, and fatty acids throughout the body. The flippases that move membrane lipids from one leaflet of the bilayer to the other are ABC transporters, and the cellular machinery for exporting excess cholesterol includes an ABC transporter. Mutations in the genes that encode some of these proteins contribute to several genetic diseases, including cystic fibrosis (Box 11–3), Tangier disease (p. 827), retinal degeneration, anemia, and liver failure.

ABC transporters are also present in simpler animals and in plants and microorganisms. Yeast has 31 genes that encode ABC transporters, *Drosophila* has 56, and *E. coli* has 80, representing 2% of its entire genome. The presence of ABC transporters that confer antibiotic resistance in pathogenic microbes (*Pseudomonas aeruginosa, Staphylococcus aureus, Candida albicans, Neisseria gonorrhoeae,* and *Plasmodium falciparum*) is a serious public health concern and makes these transporters attractive targets for drug design. ■

## Ion Gradients Provide the Energy for Secondary Active Transport

The ion gradients formed by primary transport of Na⁺ or H⁺ can in turn provide the driving force for cotransport of other solutes. Many cell types contain transport

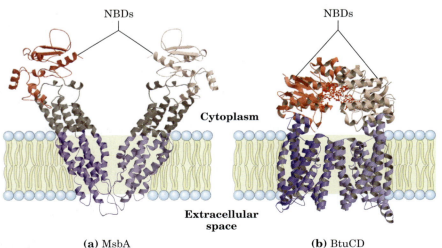

**(a)** MsbA          **(b)** BtuCD

**FIGURE 11–41 Structures of two ABC transporters of *E. coli*. (a)** The lipid A flippase MsbA (PDB ID 1JSQ) and **(b)** the vitamin B₁₂ importer BtuCD (PDB ID 1L7V). Both structures are homodimers. The two nucleotide-binding domains (NBDs, in red) extend into the cytoplasm. In **(b)**, residues involved in ATP binding and hydrolysis are shown as ball-and-stick structures. Each monomer of MsbA has six transmembrane helical segments (blue), and each monomer of BtuCD has ten.

## BOX 11-3   BIOCHEMISTRY IN MEDICINE

### A Defective Ion Channel in Cystic Fibrosis

Cystic fibrosis (CF) is a serious and relatively common hereditary disease of humans. About 5% of white Americans are carriers, having one defective and one normal copy of the gene. Only individuals with two defective copies show the severe symptoms of the disease: obstruction of the gastrointestinal and respiratory tracts, commonly leading to bacterial infection of the airways and death due to respiratory insufficiency before the age of 30. In CF, the thin layer of mucus that normally coats the internal surfaces of the lungs is abnormally thick, obstructing air flow and providing a haven for pathogenic bacteria, particularly *Staphylococcus aureus* and *Pseudomonas aeruginosa.*

The defective gene in CF patients was discovered in 1989. It encodes a membrane protein called *cystic fibrosis transmembrane conductance regula-*tor, or CFTR. Hydropathy analysis predicted that CFTR has 12 transmembrane helices and is structurally related to the multidrug (MDR1) transporters of drug-resistant tumors (Fig. 1). The normal CFTR protein proved to be an ion channel specific for $Cl^-$ ions. The $Cl^-$ channel activity increases greatly when phosphoryl groups are transferred from ATP to several side chains of the protein, catalyzed by cAMP-dependent protein kinase (Chapter 12). The mutation responsible for CF in 70% of cases results in deletion of a Phe residue at position 508, with the effect that the mutant protein is not correctly folded and inserted in the plasma membrane. Other mutations yield a protein that is inserted properly but cannot be activated by phosphorylation. In each case, the fundamental problem is a nonfunctional $Cl^-$ channel in the epithelial cells that line the airways (Fig. 2), the digestive tract, and exocrine glands (pancreas, sweat glands, bile ducts, and vas deferens).

Normally, epithelial cells that line the inner surface of the lungs secrete a substance that traps and kills bacteria, and the cilia on the epithelial cells constantly sweep away the resulting debris. When CFTR is defective or missing, this process is less efficient, and frequent infections by bacteria such as *S. aureus* and *P. aeruginosa* progressively damage the lungs and reduce respiratory efficiency.

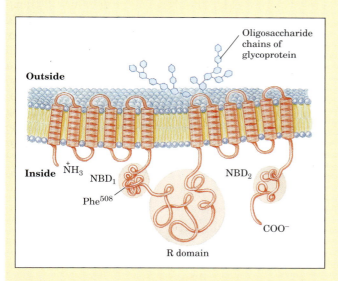

**FIGURE 1**   Topology of the cystic fibrosis transmembrane conductance regulator, CFTR. It has 12 transmembrane helices, and three functionally significant domains extend from the cytoplasmic surface: $NBD_1$ and $NBD_2$ are nucleotide-binding domains to which ATP binds, and a regulatory domain (R domain) is the site of phosphorylation by cAMP-dependent protein kinase. Oligosaccharide chains are attached to several residues on the outer surface of the segment between helices 7 and 8. The most commonly occurring mutation leading to CF is the deletion of Phe$^{508}$, in the $NBD_1$ domain. The structure of CFTR is very similar to that of the multidrug transporter of tumors, described in the text.

**FIGURE 2**   Mucus lining the surface of the lungs traps bacteria. In healthy lungs, these bacteria are killed and swept away by the action of cilia. In CF, the bactericidal activity is impaired, resulting in recurring infections and progressive damage to the lungs.

systems that couple the spontaneous, downhill flow of these ions to the simultaneous uphill pumping of another ion, sugar, or amino acid (Table 11–5). The **lactose transporter (lactose permease)** of *E. coli* is the well-studied prototype for proton-driven cotransporters. This protein consists of a single polypeptide chain (417 residues) that functions as a monomer to transport one proton and one lactose molecule into the

**TABLE 11-5    Cotransport Systems Driven by Gradients of Na⁺ or H⁺**

| Organism/tissue/cell type | Transported solute (moving against its gradient) | Cotransported solute (moving down its gradient) | Type of transport |
|---|---|---|---|
| E. coli | Lactose | $H^+$ | Symport |
| | Proline | $H^+$ | Symport |
| | Dicarboxylic acids | $H^+$ | Symport |
| Intestine, kidney (vertebrates) | Glucose | $Na^+$ | Symport |
| | Amino acids | $Na^+$ | Symport |
| Vertebrate cells (many types) | $Ca^{2+}$ | $Na^+$ | Antiport |
| Higher plants | $K^+$ | $H^+$ | Antiport |
| Fungi (Neurospora) | $K^+$ | $H^+$ | Antiport |

cell, with the net accumulation of lactose (Fig. 11–42). *E. coli* normally produces a gradient of protons and charge across its plasma membrane by oxidizing fuels and using the energy of oxidation to pump protons outward. (This mechanism is discussed in detail in Chapter 19.) The lipid bilayer is impermeable to protons, but the lactose transporter provides a route for proton reentry, and lactose is simultaneously carried into the cell by symport. The endergonic accumulation of lactose is thereby coupled to the exergonic flow of protons into the cell, with a negative overall free-energy change.

The lactose transporter is one member of the **major facilitator superfamily (MFS)** of transporters, which comprises 28 families. Almost all proteins in this superfamily have 12 transmembrane domains (the few exceptions have 14). The proteins share relatively little sequence homology, but the similarity of their secondary structures and topology suggests a common tertiary structure. The crystallographic solution of the *E. coli* lactose transporter by Ron Kaback and So Iwata in 2003 may provide a glimpse of this general structure (Fig. 11–43a). The protein has 12 transmembrane helices, and connecting loops that protrude into the cytoplasm or the periplasmic space. All six amino-terminal and six carboxyl-terminal helices form very similar domains, to produce a structure with a rough twofold symmetry. In the crystallized form of the protein, a large aqueous cavity is exposed on the cytoplasmic side of the membrane. The substrate-binding site is in this cavity, more or less in the middle of the membrane. The side of the transporter facing outward (the periplasmic face) is closed tightly, with no channel big enough for lactose to enter. The proposed mecha-

**(a)**

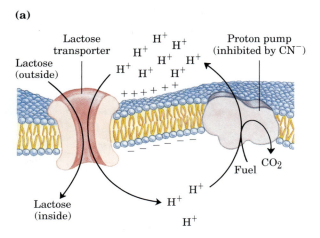
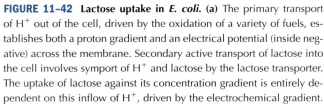

**(b)**

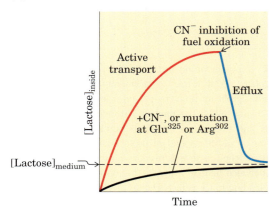

**FIGURE 11–42 Lactose uptake in *E. coli*. (a)** The primary transport of H⁺ out of the cell, driven by the oxidation of a variety of fuels, establishes both a proton gradient and an electrical potential (inside negative) across the membrane. Secondary active transport of lactose into the cell involves symport of H⁺ and lactose by the lactose transporter. The uptake of lactose against its concentration gradient is entirely dependent on this inflow of H⁺, driven by the electrochemical gradient.

**(b)** When the energy-yielding oxidation reactions of metabolism are blocked by cyanide (CN⁻), the lactose transporter allows equilibration of lactose inside and outside the cell via passive transport. Mutations that affect Glu³²⁵ or Arg³⁰² have the same effect as cyanide. The dashed line represents the concentration of lactose in the surrounding medium.

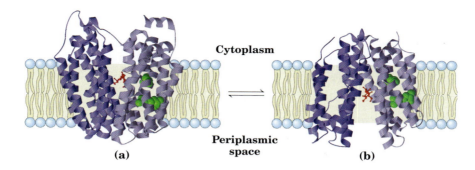

**Cytoplasm**

**(a)**          **(b)**

**Periplasmic space**

**FIGURE 11–43 Structure of the lactose transporter (lactose permease) of *E. coli*. (a)** Ribbon representation viewed parallel to the plane of the membrane shows the 12 transmembrane helices arranged in two nearly symmetrical domains shown in different shades of blue. In the form of the protein for which the crystal structure was determined, the substrate sugar (red) is bound near the middle of the membrane where it is exposed to the cytoplasm (derived from PDB ID 1PV7). **(b)** The structural changes postulated to take place during one transport cycle. The two halves of the transporter undergo a large, reversible conformational change in which the two domains tilt relative to each other, exposing the substrate-binding site first to the periplasm (structure on the right), where lactose is picked up, then to the cytoplasm (left), where the lactose is released. The interconversion of the two forms is driven by changes in the pairing of charged (protonatable) side chains such as those of Glu$^{325}$ and Arg$^{302}$ (green), which is affected by the transmembrane proton gradient.

nism for transmembrane passage of the substrate (Fig. 11–43b) involves a rocking motion between the two domains, driven by substrate binding and proton movement, alternately exposing the substrate-binding domain to the cytoplasm and to the periplasm. This so-called rocking banana model is similar to that shown in Figure 11–32 for GLUT1.

How is proton movement into the cell coupled with lactose uptake? Extensive genetic studies of the lactose transporter have established that of the 417 residues in the protein, only 6 are absolutely essential for cotransport of H$^+$ and lactose—some for lactose binding, others for proton transport. Mutation in either of two residues (Glu$^{325}$ and Arg$^{302}$; Fig. 11–43) results in a protein still able to catalyze facilitated diffusion of lactose

but incapable of coupling H$^+$ flow to uphill lactose transport. A similar effect is seen in wild-type (unmutated) cells when their ability to generate a proton gradient is blocked with CN$^-$: the transporter carries out facilitated diffusion normally, but it cannot pump lactose against a concentration gradient (Fig. 11–42b). The balance between the two conformations of the lactose transporter is affected by changes in charge pairing between side chains.

In intestinal epithelial cells, glucose and certain amino acids are accumulated by symport with Na$^+$, down the Na$^+$ gradient established by the Na$^+$K$^+$ ATPase of the plasma membrane (Fig. 11–44). The apical surface of the intestinal epithelial cell is covered with microvilli, long thin projections of the plasma membrane

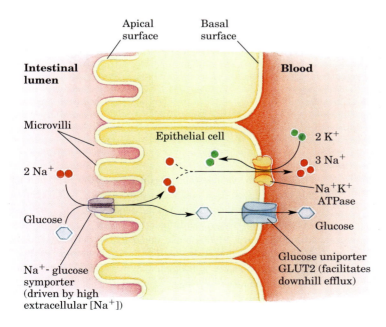

**FIGURE 11–44 Glucose transport in intestinal epithelial cells.** Glucose is cotransported with Na$^+$ across the apical plasma membrane into the epithelial cell. It moves through the cell to the basal surface, where it passes into the blood via GLUT2, a passive glucose transporter. The Na$^+$K$^+$ ATPase continues to pump Na$^+$ outward to maintain the Na$^+$ gradient that drives glucose uptake.

that greatly increase the surface area exposed to the intestinal contents. **Na$^+$-glucose symporters** in the apical plasma membrane take up glucose from the intestine in a process driven by the downhill flow of Na$^+$:

$$2Na^+_{out} + glucose_{out} \longrightarrow 2Na^+_{in} + glucose_{in}$$

The energy required for this process comes from two sources: the greater concentration of Na$^+$ outside than inside (the chemical potential) and the transmembrane potential (the electrical potential), which is inside-negative and therefore draws Na$^+$ inward. The electro-chemical potential of Na$^+$ is

$$\Delta G = RT \ln \frac{[Na+]_{in}}{[Na^+]_{out}} + n\,\mathcal{J}\,\Delta E$$

where $n = 2$, the number of Na$^+$ ions cotransported with each glucose molecule. Given the typical membrane potential of $-50$ mV, an intracellular [Na$^+$] of 12 mM, and an extracellular [Na$^+$] of 145 mM, the energy, $\Delta G$, made available as two Na$^+$ ions reenter the cell is 22.5 kJ, enough to pump glucose against a large concentration gradient:

$$\Delta G_t = -22.5 \text{ kJ} = RT \ln \frac{[glucose]_{in}}{[glucose]_{out}}$$

and thus

$$\frac{[Glucose]_{in}}{[Glucose]_{out}} \approx 9,000$$

That is, the cotransporter can pump glucose inward until its concentration within the epithelial cell is about 9,000 times that in the intestine. As glucose is pumped from the intestine into the epithelial cell at the apical surface, it is simultaneously moved from the cell into the blood by passive transport through a glucose transporter (GLUT2) in the basal surface (Fig. 11–44). The crucial role of Na$^+$ in symport and antiport systems such as these requires the continued outward pumping of Na$^+$ to maintain the transmembrane Na$^+$ gradient.

Because of the essential role of ion gradients in active transport and energy conservation, compounds that collapse ion gradients across cellular membranes are effective poisons, and those that are specific for infectious microorganisms can serve as antibiotics. One such substance is valinomycin, a small cyclic peptide that neutralizes the K$^+$ charge by surrounding it with six carbonyl oxygens (Fig. 11–45). The hydrophobic peptide then acts as a shuttle, carrying K$^+$ across membranes down its concentration gradient and deflating that gradient. Compounds that shuttle ions across membranes in this way are called **ionophores** ("ion bearers"). Both valinomycin and monensin (a Na$^+$-carrying ionophore) are antibiotics; they kill microbial cells by disrupting secondary transport processes and energy-conserving reactions.

### Aquaporins Form Hydrophilic Transmembrane Channels for the Passage of Water

A family of integral proteins discovered by Peter Agre, the **aquaporins (AQPs),** provide channels for rapid movement of water molecules across all plasma membranes (Table 11–6 lists a few examples). Ten aquaporins are known in humans, each with its specialized role. Erythrocytes, which swell or shrink rapidly in response to abrupt changes in extracellular osmolarity as blood travels through the renal medulla, have a high density of aquaporin in their plasma membranes ($2 \times 10^5$ copies of AQP-1 per cell). In the nephron (the functional unit of the kidney), the plasma membranes of proximal renal tubule cells have five different aquaporin types.

Peter Agre

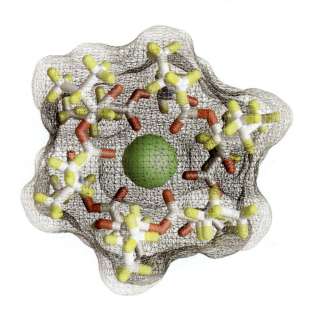

**FIGURE 11–45 Valinomycin, a peptide ionophore that binds K$^+$.** In this image, the surface contours are shown as a transparent mesh, through which a stick structure of the peptide and a K$^+$ atom (green) are visible. The oxygen atoms (red) that bind K$^+$ are part of a central hydrophilic cavity. Hydrophobic amino acid side chains (yellow) coat the outside of the molecule. Because the exterior of the K$^+$-valinomycin complex is hydrophobic, the complex readily diffuses through membranes, carrying K$^+$ down its concentration gradient. The resulting dissipation of the transmembrane ion gradient kills microbial cells, making valinomycin a potent antibiotic.

These cells reabsorb water during urine formation, a process for which water movement across membranes is essential (Box 11–3). The plant *Arabidopsis thaliana* has 38 genes that encode various types of aquaporins, reflecting the critical roles of water movement in plant physiology. Changes in turgor pressure, for example, require rapid movement of water across a membrane.

Water molecules flow through an AQP-1 channel at the rate of about $10^9$ $s^{-1}$. For comparison, the highest known turnover number for an enzyme is that for catalase, $4 \times 10^7$ $s^{-1}$, and many enzymes have turnover numbers between $1$ $s^{-1}$ and $10^4$ $s^{-1}$ (see Table 6–7). The low activation energy for passage of water through aquaporin channels ($\Delta G^{\ddagger} < 15$ kJ/mol) suggests that water moves through the channels in a continuous stream, in the direction dictated by the osmotic gradient. (For a discussion of osmosis, see p. 57.) It is essential that aquaporins not allow passage of protons (hydronium ions, $H_3O^+$), which would collapse membrane electrochemical potentials. And they do not. What is the basis for this extraordinary selectivity?

We find an answer in the structure of AQP-1, as determined by x-ray diffraction analysis (Fig. 11–46). AQP-1 has four monomers (each $M_r$ 28,000) associated in a tetramer, each monomer forming a transmembrane pore with a diameter (2 to 3 Å) sufficient to allow passage of water molecules in single file. Each monomer consists of six transmembrane helical segments and two shorter helices, each of which contains the sequence Asn–Pro–Ala (NPA). The NPA-containing short helices extend toward the middle of the bilayer from opposite

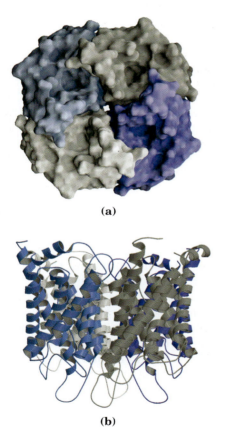

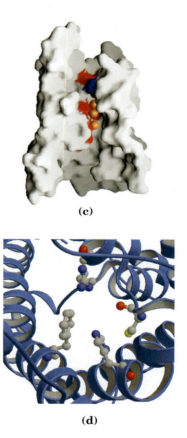

**FIGURE 11–46 Structure of an aquaporin, AQP-1.** The protein is a tetramer of identical monomeric units, each of which forms a transmembrane pore (derived from PBD ID 1J4N). **(a)** Surface model viewed perpendicular to the plane of the membrane. The protein contains four pores, one in each subunit. (The opening at the junction of the subunits is not a pore.) **(b)** An AQP-1 tetramer, viewed in the plane of the membrane. The helices of each subunit cluster around a central transmembrane pore. In each monomer, two short helical loops, one between helices 2 and 3 and the other between 5 and 6, contain the Asn–Pro–Ala (NPA) sequences found in all aquaporins, and form part of the water channel. **(c)** Surface representation of a single subunit, viewed in the plane of the membrane. The near side of the AQP-1 monomer has been cut away to reveal the channel running from top to bottom. The series of water molecules (orange spheres) shows the likely path of water molecules through the aquaporin channel, as predicted by molecular dynamics simulations in which investigators use the properties of water and aquaporin to calculate the lowest energy states. Hydrophilic atoms that provide selective interactions with water in the channel are colored red. A Phe residue (Phe[58]) at the constriction is shown in blue. **(d)** A view down the channel, showing the constriction region of the specificity pore, which lets only a molecule as small as water pass. The side chains of Phe[58], His[182], Cys[191], and Arg[197] create this constriction.

**TABLE 11–6** **Aquaporins**

| Aquaporin | Roles and/or location |
|---|---|
| AQP-1 | Fluid reabsorption in proximal renal tubule; secretion of aqueous humor in eye and cerebrospinal fluid in central nervous system; water homeostasis in lung |
| AQP-2 | Water permeability in renal collecting duct (mutations produce nephrogenic diabetes insipidus) |
| AQP-3 | Water retention in renal collecting duct |
| AQP-4 | Cerebrospinal fluid reabsorption in central nervous system; regulation of brain edema |
| AQP-5 | Fluid secretion in salivary glands, lachrymal glands, and alveolar epithelium of lung |
| AQP-6 | Kidney |
| AQP-7 | Renal proximal tubule, intestine |
| AQP-8 | Liver, pancreas, colon, placenta |
| AQP-9 | Liver, leukocytes |
| TIP | Regulation of turgor pressure in plant tonoplast |
| PIP | Plant plasma membrane |
| AQY | Yeast plasma membrane |

sides, with their NPA regions overlapping in the middle of the membrane to form part of the specificity filter—the structure that allows only water to pass.

The residues that line the channel of each AQP-1 monomer are generally nonpolar, but carbonyl oxygens in the peptide backbone, projecting into the narrow part of the channel at intervals, can form hydrogen bonds with individual water molecules as they pass through; the two Asn residues ($Asn^{76}$ and $Asn^{192}$) in the NPA loops also hydrogen-bond with the water. The structure does not admit closely spaced water molecules that might form a chain to allow proton hopping (see Fig. 2–14), which would effectively move protons across the membrane. Critical Arg and His residues and electric dipoles formed by the short helices of the NPA loops provide positive charges in positions that repel any protons that might leak through the pore.

### Ion-Selective Channels Allow Rapid Movement of Ions across Membranes

**Ion-selective channels**—first recognized in neurons and now known to be present in the plasma membranes of all cells, as well as in the intracellular membranes of eukaryotes—provide another mechanism for moving inorganic ions across membranes. Ion channels, together with ion pumps such as the $Na^+K^+$ ATPase, determine a plasma membrane's permeability to specific ions and regulate the cytosolic concentration of ions and the membrane potential. In neurons, very rapid changes in the activity of ion channels cause the changes in membrane potential (the action potentials) that carry signals from one end of a neuron to the other. In myocytes,

rapid opening of $Ca^{2+}$ channels in the sarcoplasmic reticulum releases the $Ca^{2+}$ that triggers muscle contraction. We discuss the signaling functions of ion channels in Chapter 12. Here we describe the structural basis for ion-channel function, using as examples a bacterial $K^+$ channel, the neuronal $Na^+$ channel, and the acetylcholine receptor ion channel.

Ion channels are distinguished from ion transporters in at least three ways. First, the rate of flux through channels can be several orders of magnitude greater than the turnover number for a transporter—$10^7$ to $10^8$ ions/s for an ion channel, near the theoretical maximum for unrestricted diffusion. Second, ion channels are not saturable: rates do not approach a maximum at high substrate concentration. Third, they are "gated"—opened or closed in response to some cellular event. In **ligand-gated channels** (which are generally oligomeric), binding of an extracellular or intracellular small molecule forces an allosteric transition in the protein, which opens or closes the channel. In **voltage-gated ion channels,** a change in transmembrane electrical potential ($V_m$) causes a charged protein domain to move relative to the membrane, opening or closing the ion channel. Both types of gating can be very fast. A channel typically opens in a fraction of a millisecond and may remain open for only milliseconds, making these molecular devices effective for very fast signal transmission in the nervous system.

### Ion-Channel Function Is Measured Electrically

Because a single ion channel typically remains open for only a few milliseconds, monitoring this process is be-

Erwin Neher

Bert Sakmann

fication of the initial signal; for example, only two acetylcholine molecules are needed to open an acetylcholine receptor channel (as described below).

## The Structure of a K⁺ Channel Reveals the Basis for Its Specificity

The structure of a potassium channel from the bacterium *Streptomyces lividans*, determined crystallographically by Roderick MacKinnon in 1998, provides much insight into the way ion channels work. This bacterial ion channel is related in sequence to all other known $K^+$ channels and serves as the prototype for such channels, including the voltage-gated $K^+$ channel of neurons. Among the members of this protein family, the similarities in sequence are greatest in the "pore region," which contains the ion selectivity filter that allows $K^+$ (radius 1.33 Å) to pass 10,000 times more readily than $Na^+$ (radius 0.95 Å)—at a rate (about $10^8$ ions/s) approaching the theoretical limit for unrestricted diffusion.

Roderick MacKinnon

The $K^+$ channel consists of four identical subunits that span the membrane and form a cone within a cone surrounding the ion channel, with the wide end of the double cone facing the extracellular space (Fig. 11–48).

yond the limit of most biochemical measurements. Ion fluxes must therefore be measured electrically, either as changes in $V_m$ (in the millivolt range) or as electric currents $I$ (in the microampere or picoampere range), using microelectrodes and appropriate amplifiers. In patch-clamping, a technique developed by Erwin Neher and Bert Sakmann in 1976, very small currents are measured through a tiny region of the membrane surface containing only one or a few ion-channel molecules (Fig. 11–47). The researcher can measure the size and duration of the current that flows during one opening of an ion channel and can determine how often a channel opens and how that frequency is affected by transmembrane potential, regulatory ligands, toxins, and other agents. Patch-clamp studies have revealed that as many as $10^4$ ions can move through a single ion channel in 1 ms. Such an ion flux represents a huge ampli-

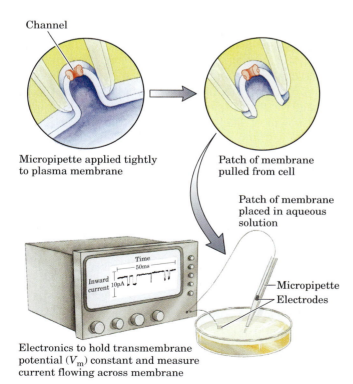

Channel

Micropipette applied tightly to plasma membrane

Patch of membrane pulled from cell

Patch of membrane placed in aqueous solution

Time

50ms

Inward current  10pA

Micropipette
Electrodes

Electronics to hold transmembrane potential ($V_m$) constant and measure current flowing across membrane

**FIGURE 11–47  Electrical measurements of ion-channel function.** The "activity" of an ion channel is estimated by measuring the flow of ions through it, using the patch-clamp technique. A finely drawn-out pipette (micropipette) is pressed against the cell surface, and negative pressure in the pipette forms a pressure seal between pipette and membrane. As the pipette is pulled away from the cell, it pulls off a tiny patch of membrane (which may contain one or a few ion channels). After placing the pipette and attached patch in an aqueous solution, the researcher can measure channel activity as the electric current that flows between the contents of the pipette and the aqueous solution. In practice, a circuit is set up that "clamps" the transmembrane potential at a given value and measures the current that must flow to maintain this voltage. With highly sensitive current detectors, researchers can measure the current flowing through a single ion channel, typically a few picoamperes. The trace showing the current as a function of time (in milliseconds) reveals how fast the channel opens and closes, how frequently it opens, and how long it stays open. Clamping the $V_m$ at different values permits determination of the effect of membrane potential on these parameters of channel function.

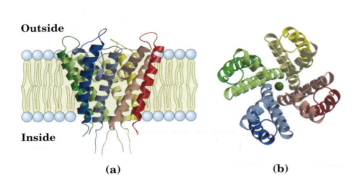

Outside

Inside

**(a)**          **(b)**

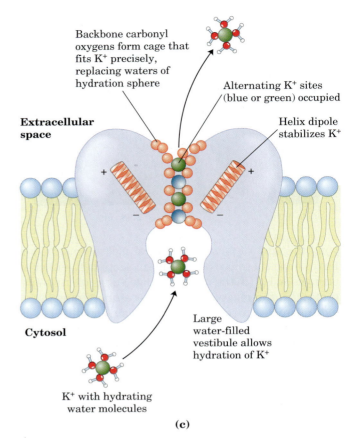

Backbone carbonyl
oxygens form cage that
fits K+ precisely,
replacing waters of
hydration sphere

Alternating K+ sites
(blue or green) occupied

Helix dipole
stabilizes K+

**Extracellular
space**

Large
water-filled
vestibule allows
hydration of K+

**Cytosol**

K+ with hydrating
water molecules

**(c)**

**FIGURE 11–48** **Structure and function of the K+ channel of Streptomyces lividans.** (PDB ID 1BL8) **(a)** Viewed in the plane of the membrane, the channel consists of eight transmembrane helices (two from
each of the four identical subunits), forming a cone with its wide end
toward the extracellular space. The inner helices of the cone (lighter
colored) line the transmembrane channel, and the outer helices interact with the lipid bilayer. Short segments of each subunit converge
in the open end of the cone to make a selectivity filter. **(b)** This view
perpendicular to the plane of the membrane shows the four subunits
arranged around a central channel just wide enough for a single K+
ion to pass. **(c)** Diagram of a K+ channel in cross section, showing
the structural features critical to function. (See also Fig. 11–49.)

Each subunit has two transmembrane $\alpha$ helices as well
as a third, shorter helix that contributes to the pore region. The outer cone is formed by one of the transmembrane helices of each subunit. The inner cone,
formed by the other four transmembrane helices, surrounds the ion channel and cradles the ion selectivity
filter.

Both the ion specificity and the high flux through
the channel are understandable from what we know of
the channel's structure. At the inner and outer plasma
membrane surfaces, the entryways to the channel have
several negatively charged amino acid residues, which
presumably increase the local concentration of cations
such as K+ and Na+. The ion path through the membrane begins (on the inner surface) as a wide, water-
filled channel in which the ion can retain its hydration
sphere. Further stabilization is provided by the short $\alpha$
helices in the pore region of each subunit, with the partial negative charges of their electric dipoles pointed at
K+ in the channel. About two-thirds of the way through
the membrane, this channel narrows in the region of the
selectivity filter, forcing the ion to give up its hydrating
water molecules. Carbonyl oxygen atoms in the backbone of the selectivity filter replace the water molecules
in the hydration sphere, forming a series of perfect coordination shells through which the K+ moves. This favorable interaction with the filter is not possible for Na+,
which is too small to make contact with all the poten-

tial oxygen ligands. The preferential stabilization of K+
is the basis for the ion selectivity of the filter, and mutations that change residues in this part of the protein
eliminate the channel's ion selectivity.

There are four potential K+-binding sites along the
selectivity filter, each composed of an oxygen "cage"
that provides ligands for the K+ ions (Fig. 11–49). In
the crystal structure, two K+ ions are visible within the
selectivity filter, about 7.5 Å apart, and two water molecules occupy the unfilled positions. K+ ions pass
through the filter in single file; their mutual electrostatic
repulsion most likely just balances the interaction of
each ion with the selectivity filter and keeps them moving. Movement of the two K+ ions is concerted: first they
occupy positions 1 and 3, then they hop to positions 2
and 4 (Fig. 11–48c). The energetic difference between
these two configurations (1, 3 and 2, 4) is very small;
energetically, the selectivity pore is not a series of hills
and valleys but a flat surface, which is ideal for rapid
ion movement through the channel. The structure of the
channel appears to have been optimized during evolution to give maximal flow rates and high specificity.

### The Neuronal Na+ Channel Is a Voltage-Gated
### Ion Channel

Sodium ion channels in the plasma membranes of neurons and of myocytes of heart and skeletal muscle sense

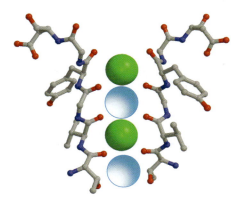

**FIGURE 11–49 K⁺ binding sites in the selectivity pore of the K⁺ channel.** (PDB ID 1J95) Carbonyl oxygens (red) of the peptide backbone in the selectivity filter protrude into the channel, interacting with and stabilizing a $K^+$ ion passing through. These ligands are perfectly positioned to interact with each of four $K^+$ ions, but not with the smaller $Na^+$ ions. This preferential interaction with $K^+$ is the basis for the ion selectivity. The mutual repulsion between $K^+$ ions results in occupation of only two of the four $K^+$ sites at a time (both green or both blue) and counteracts the tendency for a lone $K^+$ to stay bound in one site. The combined effect of $K^+$ binding to carbonyl oxygens and repulsion between $K^+$ ions ensures that an ion keeps moving, changing positions within 10 to 100 ns, and that there are no large energy barriers to ion flow along the path through the membrane.

electrical gradients across the membrane and respond by opening or closing. These voltage-gated ion channels are typically very selective for $Na^+$ over other monovalent or divalent cations (by factors of 100 or more) and have a very high flux rate ($>10^7$ ions/s). Normally (in the resting state) in the closed conformation, $Na^+$ channels are opened—activated—by a reduction in the transmembrane electrical potential, then they undergo very rapid inactivation. Within milliseconds of the opening, the channel closes and remains inactive for many milliseconds. Activation followed by inactivation of $Na^+$ channels is the basis for signaling by neurons (see Fig. 12–5).

The essential component of a $Na^+$ channel is a single, large polypeptide (1,840 amino acid residues) organized into four domains clustered around a central channel (Fig. 11–50a, b), providing a path for $Na^+$ through the membrane. The path is made $Na^+$-specific by a "pore region" composed of the segments between transmembrane helices 5 and 6 of each domain, which fold into the channel. Helix 4 of each domain has a high density of positively charged residues; this segment is believed to move within the membrane in response to changes in the transmembrane voltage, from the "resting" potential of about $-60$ mV (inside negative) to about $+30$ mV. The movement of helix 4 triggers opening of the channel, and this is the basis for voltage gating (Fig. 11–50c).

Inactivation of the channel is thought to occur by a ball-and-chain mechanism. A protein domain on the cytosolic surface of the $Na^+$ channel, the inactivation gate (the ball), is tethered to the channel by a short segment of the polypeptide (the chain) (Fig. 11–50b). This domain is free to move about when the channel is closed, but when it opens, a site on the inner face of the channel becomes available for the tethered ball to bind, blocking the channel. The length of the tether appears to determine how long an ion channel stays open; the longer the tether, the longer the open period. Inactivation of other ion channels may proceed by a similar mechanism.

## The Acetylcholine Receptor Is a Ligand-Gated Ion Channel

Another very well-studied ion channel is the **nicotinic acetylcholine receptor,** essential in the passage of an electrical signal from a motor neuron to a muscle fiber at the neuromuscular junction (signaling the muscle to contract). (Nicotinic receptors were originally distinguished from muscarinic receptors by the sensitivity of the former to nicotine, the latter to the mushroom alkaloid muscarine. They are structurally and functionally different.) Acetylcholine released by the motor neuron diffuses a few micrometers to the plasma membrane of a myocyte, where it binds to the acetylcholine receptor. This forces a conformational change in the receptor, causing its ion channel to open. The resulting inward movement of positive charges depolarizes the plasma membrane, triggering contraction. The acetylcholine receptor allows $Na^+$, $Ca^{2+}$, and $K^+$ to pass through with equal ease, but other cations and all anions are unable to pass. Movement of $Na^+$ through an acetylcholine receptor ion channel is unsaturable (its rate is linear with respect to extracellular $[Na^+]$) and very fast—about $2 \times 10^7$ ions/s under physiological conditions.

$$CH_3-C \overset{\displaystyle O}{\underset{\displaystyle O-CH_2-CH_2-\overset{+}{\underset{CH_3}{\overset{CH_3}{N}}}-CH_3}{}}$$

Acetylcholine

This receptor channel is typical of many other ion channels that produce or respond to electrical signals: it has a "gate" that opens in response to stimulation by a signal molecule (in this case acetylcholine) and an intrinsic timing mechanism that closes the gate after a split second. Thus the acetylcholine signal is transient—an essential feature of electrical signal conduction. We understand the structural changes underlying gating in the acetylcholine receptor, but not the exact mechanism of "desensitization"—of closing the gate even in the continued presence of acetylcholine.

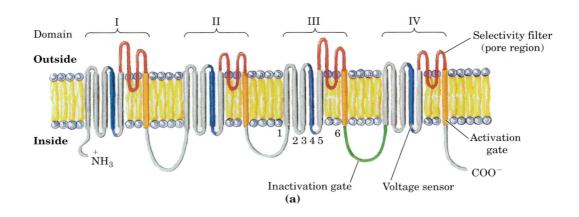

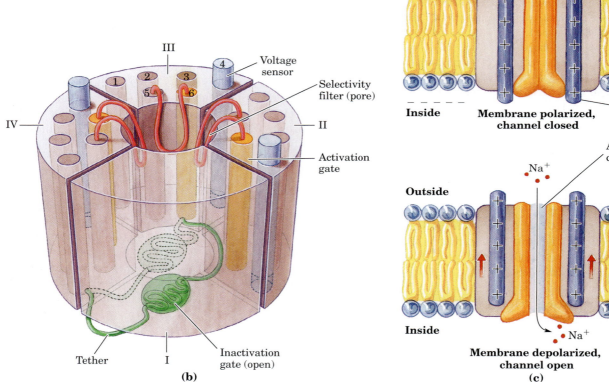

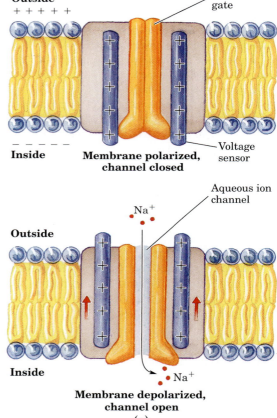

**FIGURE 11–50 Voltage-gated Na$^+$ channels of neurons.** Sodium channels of different tissues and organisms have a variety of subunits, but only the principal subunit ($\alpha$) is essential. **(a)** The $\alpha$ subunit is a large protein with four homologous domains (I to IV), each containing six transmembrane helices (1 to 6). Helix 4 in each domain (blue) is the voltage sensor; helix 6 (orange) is thought to be the activation gate. The segments between helices 5 and 6, the pore region (red), form the selectivity filter, and the segment connecting domains III and IV (green) is the inactivation gate. **(b)** The four domains are wrapped about a central transmembrane channel lined with polar amino acid residues. The segments linking helices 5 and 6 (red) in each domain come together near the extracellular surface to form the selectivity fil-

ter, which is conserved in all Na$^+$ channels. The filter gives the channel its ability to discriminate between Na$^+$ and other ions of similar size. The inactivation gate (green) closes (dotted lines) soon after the activation gate opens. **(c)** The voltage-sensing mechanism involves movement of helix 4 (blue) perpendicular to the plane of the membrane in response to a change in transmembrane potential. As shown at the top, the strong positive charge on helix 4 allows it to be pulled inward in response to the inside-negative membrane potential ($V_m$). Depolarization lessens this pull, and helix 4 relaxes by moving outward (bottom). This movement is communicated to the activation gate (orange), inducing conformational changes that open the channel in response to depolarization.

The nicotinic acetylcholine receptor has five subunits: single copies of subunits $\beta$, $\gamma$, and $\delta$, and two identical $\alpha$ subunits each with an acetylcholine-binding site. All five subunits are related in sequence and tertiary structure, each having four transmembrane helical segments (M1 to M4) (Fig. 11–51a). The five subunits surround a central pore, which is lined with their M2 helices. The pore is about 20 Å wide in the parts of the channel that protrude on the cytoplasmic and extracellular surfaces, but narrows as it passes through the

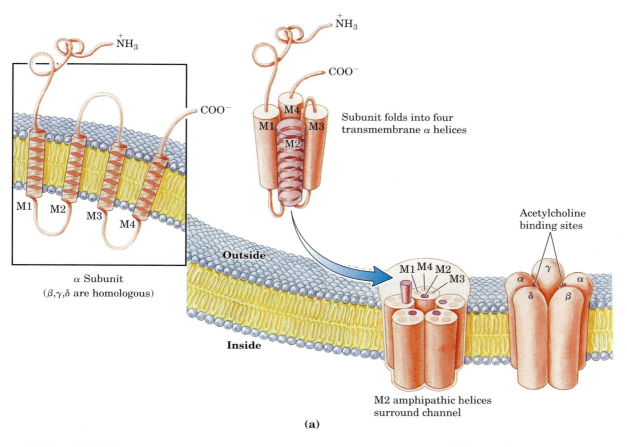

**(a)**

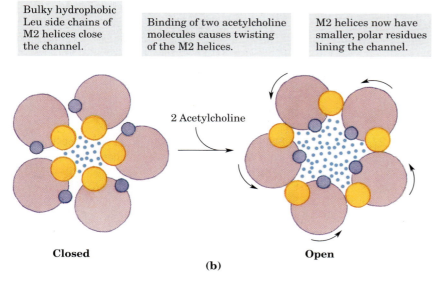

| Bulky hydrophobic Leu side chains of M2 helices close the channel. | Binding of two acetylcholine molecules causes twisting of the M2 helices. | M2 helices now have smaller, polar residues lining the channel. |

2 Acetylcholine

**Closed**                          **Open**

**(b)**

**FIGURE 11–51    Structure of the acetylcholine receptor ion channel.** **(a)** Each of the five subunits ($\alpha_2\beta\gamma\delta$) has four transmembrane helices, M1 to M4. The M2 helices are amphipathic; the others have mainly hydrophobic residues. The five subunits are arranged around a central transmembrane channel, which is lined with the polar sides of the M2 helices. At the top and bottom of the channel are rings of negatively charged amino acid residues. **(b)** This top view of a cross section through the center of the M2 helices shows five Leu side chains (one from each M2 helix) protruding into the channel, constricting it to a diameter too small to allow passage of ions such as $Ca^{2+}$, $Na^+$, and $K^+$. When both acetylcholine receptor sites (one on each $\alpha$ subunit) are occupied, a conformational change occurs. As the M2 helices twist slightly, the five Leu residues (yellow) rotate away from the channel and are replaced by smaller, polar residues (blue). This gating mechanism opens the channel, allowing the passage of $Ca^{2+}$, $Na^+$, or $K^+$.

lipid bilayer. Near the center of the bilayer is a ring of bulky hydrophobic side chains of Leu residues in the M2 helices, positioned so close together that they prevent ions from passing through the channel. Allosteric conformational changes induced by acetylcholine binding to the two $\alpha$ subunits include a slight twisting of the M2

**TABLE 11–7**   **Transport Systems Described Elsewhere in This Text**

| Transport system and location | Figure number | Role |
|---|---|---|
| Adenine nucleotide antiporter of mitochondrial inner membrane | 19–26 | Imports substrate ADP for oxidative phosphorylation, and exports product ATP |
| Acyl-carnitine/carnitine transporter of mitochondrial inner membrane | 17–6 | Imports fatty acids into matrix for $\beta$ oxidation |
| $P_i$-$H^+$ symporter of mitochondrial inner membrane | 19–26 | Supplies $P_i$ for oxidative phosphorylation |
| Malate–$\alpha$-ketoglutarate transporter of mitochondrial inner membrane | 19–27 | Shuttles reducing equivalents (as malate) from matrix to cytosol |
| Glutamate-aspartate transporter of mitochondrial inner membrane | 19–27 | Completes shuttling begun by malate–$\alpha$-ketoglutarate shuttle |
| Citrate transporter of mitochondrial inner membrane | 21–10 | Provides cytosolic citrate as source of acetyl-CoA for lipid synthesis |
| Pyruvate transporter of mitochondrial inner membrane | 21–10 | Is part of mechanism for shuttling citrate from matrix to cytosol |
| Fatty acid transporter of myocyte plasma membrane | 17–3 | Imports fatty acids for fuel |
| Complex I, III, and IV proton transporters of mitochondrial inner membrane | 19–15 | Acts as energy-conserving mechanism in oxidative phosphorylation, converting electron flow into proton gradient |
| Thermogenin (uncoupler protein), a proton pore of mitochondrial inner membrane | 19–30, 23–22 | Allows dissipation of proton gradient in mitochondria as means of thermogenesis and/or disposal of excess fuel |
| Cytochrome $bf$ complex, a proton transporter of chloroplast thylakoid | 19–50, 19–54 | Acts as proton pump, driven by electron flow through the Z scheme; source of proton gradient for photosynthetic ATP synthesis |
| Bacteriorhodopsin, a light-driven proton pump | 19–59 | Is light-driven source of proton gradient for ATP synthesis in halophilic bacterium |
| $F_oF_1$ ATPase/ATP synthase of mitochondrial inner membrane, chloroplast thylakoid, and bacterial plasma membrane | 19–58 | Interconverts energy of proton gradient and ATP during oxidative phosphorylation and photophosphorylation |
| $P_i$-triose phosphate antiporter of chloroplast inner membrane | 20–15, 20–16 | Exports photosynthetic product from stroma; imports $P_i$ for ATP synthesis |
| Bacterial protein transporter | 27–39 | Exports secreted proteins through plasma membrane |
| Protein translocase of ER | 27–33 | Transports into ER proteins destined for plasma membrane, secretion, or organelles |
| Nuclear pore protein translocase | 27–37 | Shuttles proteins between nucleus and cytoplasm |
| LDL receptor in animal cell plasma membrane | 21–42 | Imports, by receptor-mediated endocytosis, lipid carrying particles |
| Glucose transporter of animal cell plasma membrane; regulated by insulin | 12–8 | Increases capacity of muscle and adipose tissue to take up excess glucose from blood |
| $IP_3$-gated $Ca^{2+}$ channel of endoplasmic reticulum | 12–19 | Allows signaling via changes of cytosolic $Ca^{2+}$ concentration |
| cGMP-gated $Ca^{2+}$ channel of retinal rod and cone cells | 12–32 | Allows signaling via rhodopsin linked to cAMP phosphodiesterase in vertebrate eye |
| Voltage-gated $Na^+$ channel of neuron | 12–5 | Creates action potentials in neuronal signal transmission |

helices (Fig. 11–51b), which draws these hydrophobic side chains away from the center of the channel, opening it to the passage of ions.

Based on similarities between the amino acid sequences of other ligand-gated ion channels and the acetylcholine receptor, the receptor channels that respond to the extracellular signals γ-aminobutyric acid (GABA), glycine, and serotonin have been classified in the acetylcholine receptor superfamily, and probably share three-dimensional structure and gating mechanisms. The GABA$_A$ and glycine receptors are anion channels specific for $Cl^-$ or $HCO_3^-$, whereas the serotonin receptor, like the acetylcholine receptor, is cation-specific. The subunits of each of these channels, like those of the acetylcholine receptor, have four transmembrane helical segments and form oligomeric channels.

A second class of ligand-gated ion channels respond to *intracellular* ligands: 3′,5′-cyclic guanosine mononucleotide (cGMP) in the vertebrate eye, cGMP and cAMP in olfactory neurons, and ATP and inositol 1,4,5-trisphosphate (IP3) in many cell types. These channels are composed of multiple subunits, each with six transmembrane helical domains. We discuss the signaling functions of these ion channels in Chapter 12.

Table 11–7 shows a number of transporters not discussed in this chapter but encountered later in the book in the context of the paths in which they act.

## Defective Ion Channels Can Have Adverse Physiological Consequences

The importance of ion channels to physiological processes is clear from the effects of mutations in specific ion-channel proteins (Table 11–8). Genetic defects in the voltage-gated $Na^+$ channel of the myocyte plasma membrane result in diseases in which muscles are periodically either paralyzed (as in hyperkalemic pe-

riodic paralysis) or stiff (as in paramyotonia congenita). As noted earlier, cystic fibrosis is the result of a mutation that changes one amino acid in the protein CFTR, a $Cl^-$ ion channel; the defective process here is not neurotransmission but secretion by various exocrine gland cells whose activities are tied to $Cl^-$ ion fluxes.

Many naturally occurring toxins act on ion channels, and the potency of these toxins further illustrates the importance of normal ion-channel function. Tetrodotoxin (produced by the puffer fish, *Sphaeroides rubripes*) and saxitoxin (produced by the marine dinoflagellate *Gonyaulax,* which causes "red tides") act by binding to the voltage-gated $Na^+$ channels of neurons and preventing normal action potentials. Puffer fish is an ingredient of the Japanese delicacy fugu, which may be prepared only by chefs specially trained to separate

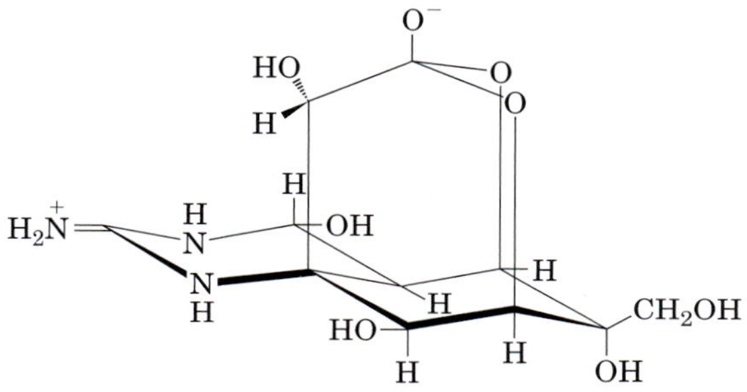

Tetrodotoxin

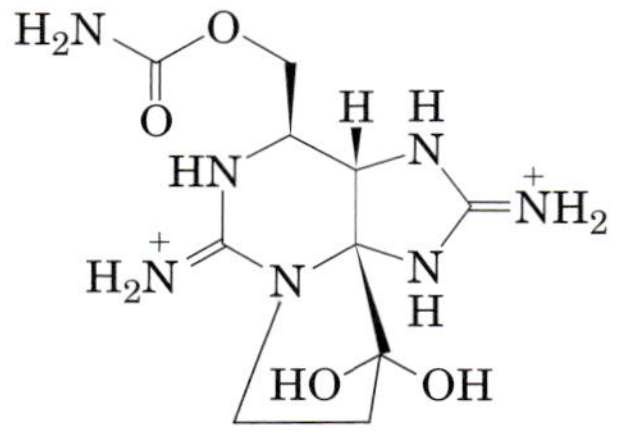

Saxitoxin

## TABLE 11–8  Some Diseases Resulting from Ion Channel Defects

| Ion channel | Affected gene | Disease |
|---|---|---|
| $Na^+$ (voltage-gated, skeletal muscle) | SCN4A | Hyperkalemic periodic paralysis (or paramyotonia congenita) |
| $Na^+$ (voltage-gated, neuronal ) | SCN1A | Generalized epilepsy with febrile seizures |
| $Na^+$ (voltage-gated, cardiac muscle) | SCN5A | Long QT syndrome 3 |
| $Ca^{2+}$ (neuronal) | CACNA1A | Familial hemiplegic migraine |
| $Ca^{2+}$ (voltage-gated, retina) | CACNA1F | Congenital stationary night blindness |
| $Ca^{2+}$ (polycystin-1) | PKD1 | Polycystic kidney disease |
| $K^+$ (neuronal) | KCNQ4 | Dominant deafness |
| $K^+$ (voltage-gated, neuronal) | KCNQ2 | Benign familial neonatal convulsions |
| Nonspecific cation (cGMP-gated, retinal) | CNCG1 | Retinitis pigmentosa |
| Acetylcholine receptor (skeletal muscle) | CHRNA1 | Congenital myasthenic syndrome |
| $Cl^-$ | CFTR | Cystic fibrosis |

D-Tubocurarine chloride

succulent morsel from deadly poison. Eating shellfish that have fed on *Gonyaulax* can also be fatal; shellfish are not sensitive to saxitoxin, but they concentrate it in their muscles, which become highly poisonous to organisms higher up the food chain. The venom of the black mamba snake contains dendrotoxin, which interferes with voltage-gated $K^+$ channels. Tubocurarine, the active component of curare (used as an arrow poison in the Amazon), and two other toxins from snake venoms, cobrotoxin and bungarotoxin, block the acetylcholine receptor or prevent the opening of its ion channel. By blocking signals from nerves to muscles, all these toxins cause paralysis and possibly death. On the positive side, the extremely high affinity of bungarotoxin for the acetylcholine receptor ($K_d = 10^{-15}$ M) has proved useful experimentally: the radiolabeled toxin was used to quantify the receptor during its purification. ■

## SUMMARY 11.3 Solute Transport across Membranes

- Movement of polar compounds and ions across biological membranes requires protein transporters. Some transporters simply facilitate passive diffusion across the membrane from the side with higher concentration to the side with lower. Others bring about active movement of solutes against an electrochemical gradient; such transport must be coupled to a source of metabolic energy.

- Carriers, like enzymes, show saturation and stereospecificity for their substrates. Transport via these systems may be passive or active. Primary active transport is driven by ATP or electron-transfer reactions; secondary active transport, by coupled flow of two solutes, one of which (often $H^+$ or $Na^+$) flows down its electrochemical gradient as the other is pulled up its gradient.

- The GLUT transporters, such as GLUT1 of erythrocytes, carry glucose into cells by facilitated diffusion. These transporters are uniporters, carrying only one substrate. Symporters permit simultaneous passage of two

substances in the same direction; examples are the lactose transporter of *E. coli*, driven by the energy of a proton gradient (lactose-$H^+$ symport), and the glucose transporter of intestinal epithelial cells, driven by a $Na^+$ gradient (glucose-$Na^+$ symport). Antiporters mediate simultaneous passage of two substances in opposite directions; examples are the chloride-bicarbonate exchanger of erythrocytes and the ubiquitous $Na^+K^+$ ATPase.

- In animal cells, $Na^+K^+$ ATPase maintains the differences in cytosolic and extracellular concentrations of $Na^+$ and $K^+$, and the resulting $Na^+$ gradient is used as the energy source for a variety of secondary active transport processes.

- The $Na^+K^+$ ATPase of the plasma membrane and the $Ca^{2+}$ transporters of the sarcoplasmic and endoplasmic reticulums (the SERCA pumps) are examples of P-type ATPases; they undergo reversible phosphorylation during their catalytic cycle and are inhibited by the phosphate analog vanadate. F-type ATPase proton pumps (ATP synthases) are central to energy-conserving mechanisms in mitochondria and chloroplasts. V-type ATPases produce gradients of protons across some intracellular membranes, including plant vacuolar membranes.

- ABC transporters carry a variety of substrates, including many drugs, out of cells, using ATP as energy source.

- Ionophores are lipid-soluble molecules that bind specific ions and carry them passively across membranes, dissipating the energy of electrochemical ion gradients.

- Water moves across membranes through aquaporins.

- Ion channels provide hydrophilic pores through which select ions can diffuse, moving down their electrical or chemical concentration gradients; they are characteristically unsaturable and have very high flux rates. Many ion channels are highly specific for one ion, and most are gated by either voltage or a ligand. In bacterial $K^+$ channels, a selectivity filter provides ligands with the right geometry to replace the water of hydration of a $K^+$ ion as it crosses the membrane. Some $K^+$ channels are voltage gated. The acetylcholine receptor/channel is gated by acetylcholine, which triggers subtle conformational changes that open and close the transmembrane path.

# Key Terms

*Terms in bold are defined in the glossary.*

**fluid mosaic model** 371

**micelle** 372

**bilayer** 373

**integral proteins** 373

**peripheral proteins** 373

**hydropathy index** 377

$\beta$ barrel 378

gel phase 380

liquid-disordered state 380

liquid-ordered state 380

**flippases** 382

**FRAP** 382

microdomains 383

rafts 384

caveolin 385

caveolae 385

**fusion proteins** 387

**SNAREs** 389

**simple diffusion** 389

**membrane potential ($V_m$)** 389

**electrochemical gradient** 391

**electrochemical potential** 391

**facilitated diffusion** 391

passive transport 391

**transporters** 391

carriers 392

channels 392

electroneutral 395

**cotransport systems** 395

**antiport** 397

**symport** 397

**uniport** 397

**active transport** 397

**electrogenic** 398

P-type ATPases 398

SERCA pump 400

F-type ATPases 401

**ATP synthase** 401

V-type ATPases 401

**ABC transporters** 402

multidrug transporter 402

**ionophores** 406

aquaporins (AQPs) 406

**ion channel** 408

# Further Reading

## Composition and Architecture of Membranes

**Boon, J.M. & Smith, B.D.** (2002) Chemical control of phospholipid distribution across bilayer membranes. *Med. Res. Rev.* **22,** 251–281.

Intermediate-level review of phospholipid asymmetry and factors that influence it.

**Dowhan, W.** (1997) Molecular basis for membrane phospholipids diversity: why are there so many lipids? *Annu. Rev. Biochem.* **66,** 199–232.

**Ediden, M.** (2002) Lipids on the frontier: a century of cell-membrane bilayers. *Nat. Rev. Mol. Cell Biol.* **4,** 414–418.

Short review of how the notion of a lipid bilayer membrane was developed and confirmed.

**Haltia, T. & Freire, E.** (1995) Forces and factors that contribute to the structural stability of membrane proteins. *Biochim. Biophys. Acta* **1241,** 295–322.

Good discussion of the secondary and tertiary structures of membrane proteins and the factors that stabilize them.

**von Heijne, G.** (1994) Membrane proteins: from sequence to structure. *Annu. Rev. Biophys. Biomol. Struct.* **23,** 167–192.

A review of the steps required to predict the structure of an integral protein from its sequence.

**White, S.H., Ladokhin, A.S., Jayasinghe, S., & Hristova, K.** (2001) How membranes shape protein structure. *J. Biol. Chem.* **276,** 32,395–32,398.

Brief, intermediate-level review of the forces that shape transmembrane helices.

**Wimley, W.C.** (2003) The versatile $\beta$ barrel membrane protein. *Curr. Opin. Struct. Biol.* **13,** 1–8.

Intermediate-level review.

## Membrane Dynamics

**Brown, D.A. & London, E.** (1998) Functions of lipid rafts in biological membranes. *Annu. Rev. Cell Dev. Biol.* **14,** 111–136.

**Chen, Y.A. & Scheller, R.H.** (2001) SNARE-mediated membrane fusion. *Nat. Rev. Mol. Cell Biol.* **2,** 98–106.

Intermediate-level review.

**Edidin, M.** (2003) The state of lipid rafts: from model membranes to cells. *Annu. Rev. Biophys. Biomol. Struct.* **32,** 257–283.

Advanced review.

**Frye, L.D. & Ediden, M.** (1970) The rapid intermixing of cell-surface antigens after formation of mouse-human heterokaryons. *J. Cell Sci.* **7,** 319–335.

The classic demonstration of membrane protein mobility.

**Mayer, A.** (2002) Membrane fusion in eukaryotic cells. *Annu. Rev. Cell Dev. Biol.* **18,** 289–314.

Advanced review of membrane fusion, with emphasis on the conserved general features.

**Parton, R.G.** (2003) Caveolae—from ultrastructure to molecular mechanisms. *Nat. Rev. Mol. Cell Biol.* **4,** 162–167.

A concise historical review of caveolae, caveolin, and rafts.

**Sprong, H., van der Sluijs, P., & van Meer, G.** (2001) How proteins move lipids and lipids move proteins. *Nat. Rev. Mol. Cell Biol.* **2,** 504–513.

Intermediate-level review.

**van Deurs, B., Roepstorff, K., Hommelgaard, A.M., & Sandvig, K.** (2003) Caveolae: anchored, multifunctional platforms in the lipid ocean. *Trends Cell Biol.* **13,** 92–100.

**Vereb, G., Szöllósi, J., Matcó, J., Nagy, P., Farkas, T., Vigh, L., Mátyus, L., Waldmann, T.A., & Damjanovich, S.** (2003) Dynamic, yet structured: the cell membrane three decades after the Singer-Nicolson model. *Proc. Natl. Acad. Sci. USA* **100,** 8053–8058.

Intermediate-level review of membrane structure and dynamics.

## Transporters

**Abramson, J., Smirnova, I., Kasho, V., Verner, G., Kaback, H.R., & Iwata, S.** (2003) Structure and mechanism of the lactose permease of *Escherichia coli. Science* **301,** 610–615.

Fujiyoshi, Y., Mitsuoka, K., de Groot, B.L., Philippsen, A., Grubmüller, H., Agre, P., & Engel, A. (2002) Structure and function of water channels. *Curr. Opin. Struct. Biol.* **12**, 509–515.

Jorgensen, P.L., Håkansson, K.O., & Karlish, S.J.D. (2003) Structure and mechanism of Na,K-ATPase: functional sites and their interactions. *Annu. Rev. Physiol.* **65**, 817–849.

Kjellbom, P., Larsson, C., Johansson, I., Karlsson, M., & Johanson, U. (1999) Aquaporins and water homeostasis in plants. *Trends Plant Sci.* **4**, 308–314.

    Intermediate-level review.

Mueckler, M. (1994) Facilitative glucose transporters. *Eur. J. Biochem.* **219**, 713–725.

Saier, M.H., Jr. (2000) A functional-phylogenetic classification system for transmembrane solute transporters. *Microbiol. Mol. Biol. Rev.* **64**, 354–411.

Schmitt, L. & Tampé, R. (2002) Structure and mechanism of ABC transporters. *Curr. Opin. Struct. Biol.* **12**, 754–760.

Sheppard, D.N. & Welsh, M.J. (1999) Structure and function of the CFTR chloride channel. *Physiol. Rev.* **79**, S23–S46.

    This issue of the journal has 11 reviews on the CFTR chloride channel, covering its structure, activity, regulation, biosynthesis, and pathophysiology.

Stokes, D.L. & Green, N.M. (2003) Structure and function of the calcium pump. *Annu. Rev. Biophys. Biomol. Struct.* **32**, 445–468.

    Advanced review.

Sui, H., Han, B.-G., Lee, J.K., Walian, P., & Jap, B.K. (2001) Structural basis of water-specific transport through the AQP1 water channel. *Nature* **414**, 872–878.

    High-resolution solution of the aquaporin structure by x-ray crystallography.

### Ion Channels

Changeux, J.P. (1993) Chemical signaling in the brain. *Sci. Am.* **269** (November), 58–62.

Discussion of structure and function of the acetylcholine receptor channel.

Doyle, D.A., Cabral, K.M., Pfuetzner, R.A., Kuo, A., Gulbis, J.M., Cohen, S.L., Chait, B.T., & MacKinnon, R. (1998) The structure of the potassium channel: molecular basis of $K^+$ conduction and selectivity. *Science* **280**, 69–77.

    The first crystal structure of an ion channel is described.

Edelstein, S.J. & Changeux, J.P. (1998) Allosteric transitions of the acetylcholine receptor. *Adv. Prot. Chem.* **51**, 121–184.

    Advanced discussion of the conformational changes induced by acetylcholine.

Hille, B. (2001) *Ion Channels of Excitable Membranes,* 3rd edn, Sinauer Associates, Sunderland, MA.

    Intermediate-level text emphasizing the function of ion channels.

Jiang, Y., Lee, A., Chen, J., Ruta, V., Cadene, M., Chait, B.T., & MacKinnon, R. (2003) X-ray structure of a voltage-dependent $K^+$ channel. *Nature* **423**, 33–41.

Lee, A.G. & East, J.M. (2001) What the structure of a calcium pump tells us about its mechanism. *Biochem J.* **356**, 665–683.

Miyazawa, A., Fujiyoshi, Y., & Unwin, N. (2003) Structure and gating mechanism of the acetylcholine receptor pore. *Nature* **423**, 949–955.

    Intermediate-level review.

Neher, E. & Sakmann, B. (1992) The patch clamp technique. *Sci. Am.* (March) **266**, 44–51.

    Clear description of the electrophysiological methods used to measure the activity of single ion channels, by the Nobel Prize–winning developers of this technique.

Yellen, G. (2002) The voltage-gated potassium channels and their relatives. *Nature* **419**, 35–42.

Zhou, Y., Morias-Cabral, J.H., Kaufman, A., & MacKinnon, R. (2001) Chemistry of ion coordination and hydration revealed by a $K^+$ channel–Fab complex at 2.0 Å resolution. *Nature* **414**, 43–48.

## *Problems*

**1. Determining the Cross-Sectional Area of a Lipid Molecule** When phospholipids are layered gently onto the surface of water, they orient at the air-water interface with their head groups in the water and their hydrophobic tails in the air. An experimental apparatus **(a)** has been devised that reduces the surface area available to a layer of lipids. By measuring the force necessary to push the lipids together, it is possible to determine when the molecules are packed tightly in a continuous monolayer; as that area is approached, the force needed to further reduce the surface area increases sharply **(b)**. How would you use this apparatus to determine the average area occupied by a single lipid molecule in the monolayer?

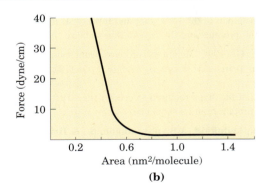

**(b)**

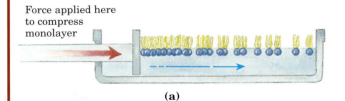

Force applied here to compress monolayer

**(a)**

**2. Evidence for a Lipid Bilayer** In 1925, E. Gorter and F. Grendel used an apparatus like that described in Problem 1 to determine the surface area of a lipid monolayer formed by lipids extracted from erythrocytes of several animal species. They used a microscope to measure the dimensions of individual cells, from which they calculated the average surface area of one erythrocyte. They obtained the data shown in the table. Were these investigators justified in concluding that "chromocytes [erythrocytes] are covered by a layer of fatty substances that is two molecules thick" (i.e., a lipid bilayer)?

| Animal | Volume of packed cells (mL) | Number of cells (per $mm^3$) | Total surface area of lipid monolayer from cells ($m^2$) | Total surface area of one cell ($\mu m^2$) |
|---|---|---|---|---|
| Dog | 40 | 8,000,000 | 62 | 98 |
| Sheep | 10 | 9,900,000 | 6.0 | 29.8 |
| Human | 1 | 4,740,000 | 0.92 | 99.4 |

*Source:* Data from Gorter, E. & Grendel, F. (1925) On bimolecular layers of lipoids on the chromocytes of the blood. *J. Exp. Med.* **41**, 439–443.

**3. Number of Detergent Molecules per Micelle**  When a small amount of sodium dodecyl sulfate (SDS; $Na^+CH_3(CH_2)_{11}OSO_3^-$) is dissolved in water, the detergent ions enter the solution as monomeric species. As more detergent is added, a concentration is reached (the critical micelle concentration) at which the monomers associate to form micelles. The critical micelle concentration of SDS is 8.2 mM. The micelles have an average particle weight (the sum of the molecular weights of the constituent monomers) of 18,000. Calculate the number of detergent molecules in the average micelle.

**4. Properties of Lipids and Lipid Bilayers**  Lipid bilayers formed between two aqueous phases have this important property: they form two-dimensional sheets, the edges of which close upon each other and undergo self-sealing to form liposomes.

(a) What properties of lipids are responsible for this property of bilayers? Explain.

(b) What are the consequences of this property for the structure of biological membranes?

**5. Length of a Fatty Acid Molecule**  The carbon–carbon bond distance for single-bonded carbons such as those in a saturated fatty acyl chain is about 1.5 Å. Estimate the length of a single molecule of palmitate in its fully extended form. If two molecules of palmitate were placed end to end, how would their total length compare with the thickness of the lipid bilayer in a biological membrane?

**6. Temperature Dependence of Lateral Diffusion**  The experiment described in Figure 11–17 was performed at 37 °C. If the experiment were carried out at 10 °C, what effect would you expect on the rate of diffusion? Why?

**7. Synthesis of Gastric Juice: Energetics**  Gastric juice (pH 1.5) is produced by pumping HCl from blood plasma (pH 7.4) into the stomach. Calculate the amount of free energy required to concentrate the $H^+$ in 1 L of gastric juice at 37 °C. Under cellular conditions, how many moles of ATP must be hydrolyzed to provide this amount of free energy? The free-energy change for ATP hydrolysis under cellular conditions is about −58 kJ/mol (as explained in Chapter 13). Ignore the effects of the transmembrane electrical potential.

**8. Energetics of the $Na^+K^+$ ATPase**  For a typical vertebrate cell with a transmembrane potential of −0.070 V (inside negative), what is the free-energy change for transporting 1 mol of $Na^+$ out of the cell and into the blood at 37 °C? Assume the concentration of $Na^+$ inside the cell is 12 mM, and that in blood plasma is 145 mM.

**9. Action of Ouabain on Kidney Tissue**  Ouabain specifically inhibits the $Na^+K^+$ ATPase activity of animal tissues but is not known to inhibit any other enzyme. When ouabain is added to thin slices of living kidney tissue, it inhibits oxygen consumption by 66%. Why? What does this observation tell us about the use of respiratory energy by kidney tissue?

**10. Energetics of Symport**  Suppose that you determined experimentally that a cellular transport system for glucose, driven by symport of $Na^+$, could accumulate glucose to concentrations 25 times greater than in the external medium, while the external $[Na^+]$ was only 10 times greater than the intracellular $[Na^+]$. Would this violate the laws of thermodynamics? If not, how could you explain this observation?

**11. Location of a Membrane Protein**  The following observations are made on an unknown membrane protein, X. It can be extracted from disrupted erythrocyte membranes into a concentrated salt solution, and it can be cleaved into fragments by proteolytic enzymes. Treatment of erythrocytes with proteolytic enzymes followed by disruption and extraction of membrane components yields intact X. However, treatment of erythrocyte "ghosts" (which consist of just plasma membranes, produced by disrupting the cells and washing out the hemoglobin) with proteolytic enzymes followed by disruption and extraction yields extensively fragmented X. What do these observations indicate about the location of X in the plasma membrane? Do the properties of X resemble those of an integral or peripheral membrane protein?

**12. Membrane Self-sealing**  Cellular membranes are self-sealing—if they are punctured or disrupted mechanically, they quickly and automatically reseal. What properties of membranes are responsible for this important feature?

**13. Lipid Melting Temperatures**  Membrane lipids in tissue samples obtained from different parts of the leg of a reindeer have different fatty acid compositions. Membrane lipids from tissue near the hooves contain a larger proportion of unsaturated fatty acids than those from tissue in the upper leg. What is the significance of this observation?

**14. Flip-Flop Diffusion**  The inner leaflet (monolayer) of the human erythrocyte membrane consists predominantly of phosphatidylethanolamine and phosphatidylserine. The outer leaflet consists predominantly of phosphatidylcholine and sphingomyelin. Although the phospholipid components of the membrane can diffuse in the fluid bilayer, this sidedness is preserved at all times. How?

**15. Membrane Permeability**  At pH 7, tryptophan crosses a lipid bilayer at about one-thousandth the rate of the closely related substance indole:

Suggest an explanation for this observation.

**16. Water Flow through an Aquaporin** Each human erythrocyte has about $2 \times 10^5$ AQP-1 monomers. If water molecules flow through the plasma membrane at a rate of $5 \times 10^8$ per AQP-1 tetramer per second, and the volume of an erythrocyte is $5 \times 10^{-11}$ mL, how rapidly could an erythrocyte halve its volume as it encounters the high osmolarity (1 M) in the interstitial fluid of the renal medulla? Assume that the erythrocyte consists entirely of water.

**17. Labeling the Lactose Transporter** A bacterial lactose transporter, which is highly specific for its substrate lactose, contains a Cys residue that is essential to its transport activity. Covalent reaction of N-ethylmaleimide (NEM) with this Cys residue irreversibly inactivates the transporter. A high concentration of lactose in the medium prevents inactivation by NEM, presumably by sterically protecting the Cys residue, which is in or near the lactose-binding site. You know nothing else about the transporter protein. Suggest an experiment that might allow you to determine the $M_r$ of the Cys-containing transporter polypeptide.

**18. Predicting Membrane Protein Topology from Sequence** You have cloned the gene for a human erythrocyte protein, which you suspect is a membrane protein. From the nucleotide sequence of the gene, you know the amino acid sequence. From this sequence alone, how would you evaluate the possibility that the protein is an integral protein? Suppose the protein proves to be an integral protein, either type I or type II. Suggest biochemical or chemical experiments that might allow you to determine which type it is.

**19. Intestinal Uptake of Leucine** You are studying the uptake of L-leucine by epithelial cells of the mouse intestine. Measurements of the rate of uptake of L-leucine and several of its analogs, with and without $Na^+$ in the assay buffer, yield the results given in the table. What can you conclude about the properties and mechanism of the leucine transporter? Would you expect L-leucine uptake to be inhibited by ouabain?

| Substrate | Uptake in presence of $Na^+$ | | Uptake in absence of $Na^+$ | |
|---|---|---|---|---|
| | $V_{max}$ | $K_t$ (mM) | $V_{max}$ | $K_t$ (mM) |
| L-Leucine | 420 | 0.24 | 23 | 0.24 |
| D-Leucine | 310 | 4.7 | 5 | 4.7 |
| L-Valine | 225 | 0.31 | 19 | 0.31 |

**20. Effect of an Ionophore on Active Transport** Consider the leucine transporter described in Problem 19. Would $V_{max}$ and/or $K_t$ change if you added a $Na^+$ ionophore to the assay solution containing $Na^+$? Explain.

**21. Surface Density of a Membrane Protein** E. coli can be induced to make about 10,000 copies of the lactose transporter ($M_r$ 31,000) per cell. Assume that E. coli is a cylinder 1 $\mu$m in diameter and 2 $\mu$m long. What fraction of the plasma membrane surface is occupied by the lactose transporter molecules? Explain how you arrived at this conclusion.

### Biochemistry on the Internet

**22. Membrane Protein Topology** The receptor for the hormone epinephrine in animal cells is an integral membrane protein ($M_r$ 64,000) that is believed to have seven membrane-spanning regions.

(a) Show that a protein of this size is capable of spanning the membrane seven times.

(b) Given the amino acid sequence of this protein, how would you predict which regions of the protein form the membrane-spanning helices?

(c) Go to the Protein Data Bank (www.rcsb.org/pdb). Use the PDB identifier 1DEP to retrieve the data page for a portion of the β-adrenergic receptor (one type of epinephrine receptor) from a turkey. Using Chime to explore the structure, predict where this portion of the receptor is located: within the membrane or at the membrane surface. Explain.

(d) Retrieve the data for a portion of another receptor, the acetylcholine receptor of neurons and myocytes, using the PDB identifier 1A11. As in (c), predict where this portion of the receptor is located and explain your answer.

If you have not used the PDB or Chemscape Chime, you can find instructions at www.whfreeman.com/lehninger.

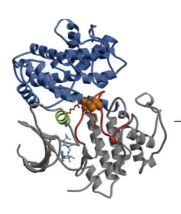

# BIOSIGNALING

When I first entered the study of hormone action, some 25 years ago, there was a widespread feeling among biologists that hormone action could not be studied meaningfully in the absence of organized cell structure. However, as I reflected on the history of biochemistry, it seemed to me there was a real possibility that hormones might act at the molecular level.

—Earl W. Sutherland, Nobel Address, 1971

The ability of cells to receive and act on signals from beyond the plasma membrane is fundamental to life. Bacterial cells receive constant input from membrane proteins that act as information receptors, sampling the surrounding medium for pH, osmotic strength, the availability of food, oxygen, and light, and the presence of noxious chemicals, predators, or competitors for food.

chapter **12**

These signals elicit appropriate responses, such as motion toward food or away from toxic substances or the formation of dormant spores in a nutrient-depleted medium. In multicellular organisms, cells with different functions exchange a wide variety of signals. Plant cells respond to growth hormones and to variations in sunlight. Animal cells exchange information about the concentrations of ions and glucose in extracellular fluids, the interdependent metabolic activities taking place in different tissues, and, in an embryo, the correct placement of cells during development. In all these cases, the signal represents *information* that is detected by specific receptors and converted to a cellular response, which always involves a *chemical* process. This conversion of information into a chemical change, **signal transduction,** is a universal property of living cells.

The number of different biological signals is large (Table 12–1), as is the variety of biological responses to these signals, but organisms use just a few evolutionarily conserved mechanisms to detect extracellular signals and *transduce* them into intracellular changes. In this chapter we examine some examples of the major classes of signaling mechanisms, looking at how they are integrated in specific biological functions such as the transmission of nerve signals; responses to hormones and growth factors; the senses of sight, smell, and taste; and

**TABLE 12–1**

**Some Signals to Which Cells Respond**

| | |
|---|---|
| Antigens | Light |
| Cell surface glycoproteins/ oligosaccharides | Mechanical touch |
| Developmental signals | Neurotransmitters |
| Extracellular matrix components | Nutrients |
| Growth factors | Odorants |
| Hormones | Pheromones |
| | Tastants |

control of the cell cycle. Often, the end result of a signaling pathway is the phosphorylation of a few specific target-cell proteins, which changes their activities and thus the activities of the cell. Throughout our discussion we emphasize the conservation of fundamental mechanisms for the transduction of biological signals and the adaptation of these basic mechanisms to a wide range of signaling pathways.

## 12.1 Molecular Mechanisms of Signal Transduction

Signal transductions are remarkably specific and exquisitely sensitive. **Specificity** is achieved by precise molecular complementarity between the signal and receptor molecules (Fig. 12–1a), mediated by the same kinds of weak (noncovalent) forces that mediate enzyme-substrate and antigen-antibody interactions. Multicellular organisms have an additional level of specificity, because the receptors for a given signal, or the intracellular targets of a given signal pathway, are present only in certain cell types. Thyrotropin-releasing hormone, for example, triggers responses in the cells of the anterior pituitary but not in hepatocytes, which lack receptors for this hormone. Epinephrine alters glycogen metabolism in hepatocytes but not in erythrocytes; in this case, both cell types have receptors for the hormone, but whereas hepatocytes contain glycogen and the glycogen-metabolizing enzyme that is stimulated by epinephrine, erythrocytes contain neither.

Three factors account for the extraordinary sensitivity of signal transducers: the high affinity of receptors for signal molecules, cooperativity (often but not

always) in the ligand-receptor interaction, and amplification of the signal by enzyme cascades. The **affinity** between signal (ligand) and receptor can be expressed as the dissociation constant $K_d$, usually $10^{-10}$ M or less—meaning that the receptor detects picomolar concentrations of a signal molecule. Receptor-ligand interactions are quantified by Scatchard analysis, which yields a quantitative measure of affinity ($K_d$) and the number of ligand-binding sites in a receptor sample (Box 12–1).

**Cooperativity** in receptor-ligand interactions results in large changes in receptor activation with small changes in ligand concentration (recall the effect of cooperativity on oxygen binding to hemoglobin; see Fig. 5–12). **Amplification** by **enzyme cascades** results when an enzyme associated with a signal receptor is activated and, in turn, catalyzes the activation of many molecules of a second enzyme, each of which activates many molecules of a third enzyme, and so on (Fig. 12–1b). Such cascades can produce amplifications of several orders of magnitude within milliseconds.

The sensitivity of receptor systems is subject to modification. When a signal is present continuously, **desensitization** of the receptor system results (Fig. 12–1c); when the stimulus falls below a certain threshold, the system again becomes sensitive. Think of what happens to your visual transduction system when you walk from bright sunlight into a darkened room or from darkness into the light.

A final noteworthy feature of signal-transducing systems is **integration** (Fig. 12–1d), the ability of the system to receive multiple signals and produce a unified response appropriate to the needs of the cell or organism. Different signaling pathways converse with

**(a) Specificity**
Signal molecule fits binding site on its complementary receptor; other signals do not fit.

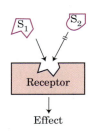

**(b) Amplification**
When enzymes activate enzymes, the number of affected molecules increases geometrically in an enzyme cascade.

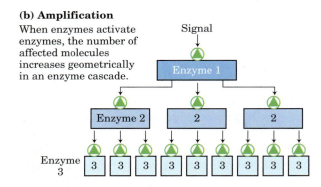

**(c) Desensitization/Adaptation**
Receptor activation triggers a feedback circuit that shuts off the receptor or removes it from the cell surface.

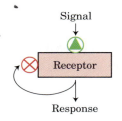

**(d) Integration**
When two signals have opposite effects on a metabolic characteristic such as the concentration of a second messenger X, or the membrane potential $V_m$, the regulatory outcome results from the integrated input from both receptors.

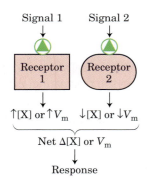

**FIGURE 12–1** Four features of signal-transducing systems.

# BOX 12-1    WORKING IN BIOCHEMISTRY

## Scatchard Analysis Quantifies the Receptor-Ligand Interaction

The cellular actions of a hormone begin when the hormone (ligand, L) binds specifically and tightly to its protein receptor (R) on or in the target cell. Binding is mediated by noncovalent interactions (hydrogen-bonding, hydrophobic, and electrostatic) between the complementary surfaces of ligand and receptor. Receptor-ligand interaction brings about a conformational change that alters the biological activity of the receptor, which may be an enzyme, an enzyme regulator, an ion channel, or a regulator of gene expression.

Receptor-ligand binding is described by the equation

$$\underset{\text{Receptor}}{R} + \underset{\text{Ligand}}{L} \rightleftharpoons \underset{\text{Receptor-ligand complex}}{RL}$$

This binding, like that of an enzyme to its substrate, depends on the concentrations of the interacting components and can be described by an equilibrium constant:

$$\underset{\text{Receptor}}{R} + \underset{\text{Ligand}}{L} \underset{k_{-1}}{\overset{k_{+1}}{\rightleftharpoons}} \underset{\substack{\text{Receptor-ligand} \\ \text{complex}}}{RL}$$

$$K_a = \frac{[RL]}{[R][L]} = \frac{k_{+1}}{k_{-1}} = 1/K_d$$

where $K_a$ is the association constant and $K_d$ is the dissociation constant.

Like enzyme-substrate binding, receptor-ligand binding is saturable. As more ligand is added to a fixed amount of receptor, an increasing fraction of receptor molecules is occupied by ligand (Fig. 1a). A rough measure of receptor-ligand affinity is given by the concentration of ligand needed to give half-saturation of the receptor. Using **Scatchard analysis** of receptor-ligand binding, we can estimate both the dissociation constant $K_d$ and the number of receptor-binding sites in a given preparation. When binding has reached equilibrium, the total number of possible binding sites, $B_{max}$, equals the number of unoccupied sites, represented by [R], plus the number of occupied or ligand-bound sites, [RL]; that is, $B_{max} = [R] + [RL]$. The number of unbound sites can be expressed in terms of total sites minus occupied sites: $[R] = B_{max} - [RL]$. The equilibrium expression can now be written

$$K_a = \frac{[RL]}{[L](B_{max} - [RL])}$$

Rearranging to obtain the ratio of receptor-bound ligand to free (unbound) ligand, we get

$$\frac{[Bound]}{[Free]} = \frac{[RL]}{[L]} = K_a(B_{max} - [RL])$$

$$= \frac{1}{K_d}(B_{max} - [RL])$$

From this slope-intercept form of the equation, we can see that a plot of [bound ligand]/[free ligand] versus [bound ligand] should give a straight line with a slope of $-K_a$ ($-1/K_d$) and an intercept on the abscissa of $B_{max}$, the total number of binding sites (Fig. 1b). Hormone-ligand interactions typically have $K_d$ values of $10^{-9}$ to $10^{-11}$ M, corresponding to very tight binding.

Scatchard analysis is reliable for the simplest cases, but as with Lineweaver-Burk plots for enzymes, when the receptor is an allosteric protein, the plots deviate from linearity.

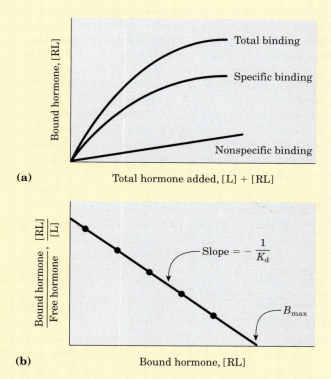

(a)

(b)

**FIGURE 1** Scatchard analysis of a receptor-ligand interaction. A radiolabeled ligand (L)—a hormone, for example—is added at several concentrations to a fixed amount of receptor (R), and the fraction of the hormone bound to receptor is determined by separating the receptor-hormone complex (RL) from free hormone. **(a)** A plot of [RL] versus [L] + [RL] (total hormone added) is hyperbolic, rising toward a maximum for [RL] as the receptor sites become saturated. To control for nonsaturable, nonspecific binding sites (eicosanoid hormones bind nonspecifically to the lipid bilayer, for example), a separate series of binding experiments is also necessary. A large excess of unlabeled hormone is added along with the dilute solution of labeled hormone. The unlabeled molecules compete with the labeled molecules for specific binding to the saturable site on the receptor, but not for the nonspecific binding. The true value for specific binding is obtained by subtracting nonspecific binding from total binding. **(b)** A linear plot of [RL]/[L] versus [RL] gives $K_d$ and $B_{max}$ for the receptor-hormone complex. Compare these plots with those of $V_0$ versus [S] and $1/V_0$ versus $1/[S]$ for an enzyme-substrate complex (see Fig. 6-12, Box 6-1).

each other at several levels, generating a wealth of interactions that maintain homeostasis in the cell and the organism.

We consider here the molecular details of several representative signal-transduction systems. The trigger for each system is different, but the general features of signal transduction are common to all: a signal interacts with a receptor; the activated receptor interacts with cellular machinery, producing a second signal or a change in the activity of a cellular protein; the metabolic activity (broadly defined to include metabolism of RNA, DNA, and protein) of the target cell undergoes a change; and finally, the transduction event ends and the cell returns to its prestimulus state. To illustrate these general features of signaling systems, we provide examples of each of six basic signaling mechanisms (Fig. 12–2).

1.  Gated ion channels of the plasma membrane that open and close (hence the term "gating") in response to the binding of chemical ligands or changes in transmembrane potential. These are the simplest signal transducers. The acetylcholine receptor ion channel is an example of this mechanism (Section 12.2).

2.  Receptor enzymes, plasma membrane receptors that are also enzymes. When one of these receptors is activated by its extracellular ligand, it catalyzes the production of an intracellular second messenger. An example is the insulin receptor (Section 12.3).

3.  Receptor proteins (serpentine receptors) that *indirectly* activate (through GTP-binding proteins, or G proteins) enzymes that generate intracellular second messengers. This is illustrated by the β-adrenergic receptor system that detects epinephrine (adrenaline) (Section 12.4).

4.  Nuclear receptors (steroid receptors) that, when bound to their specific ligand (such as the hormone estrogen), alter the rate at which specific genes are transcribed and translated into cellular proteins. Because steroid hormones function through mechanisms intimately related to the regulation of gene expression, we consider them here only briefly (Section 12.8) and defer a detailed discussion of their action until Chapter 28.

5.  Receptors that lack enzymatic activity but attract and activate cytoplasmic enzymes that act on downstream proteins, either by directly converting them to gene-regulating proteins or by activating a cascade of enzymes that finally activates a gene regulator. The JAK-STAT system exemplifies the first mechanism (Section 12.3); and the TLR4 (Toll) signaling system in humans, the second (Section 12.6).

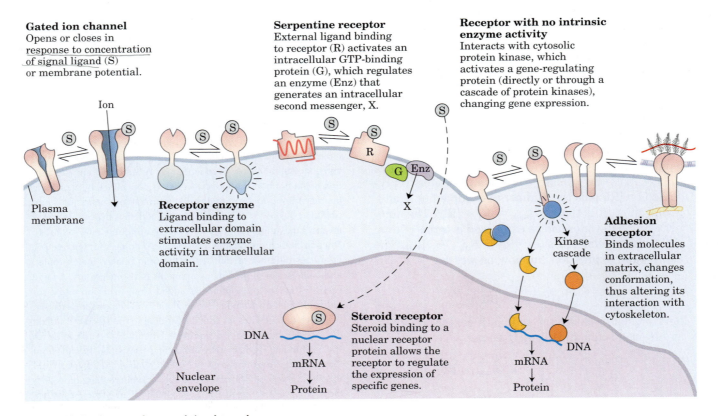

**Gated ion channel**
Opens or closes in response to concentration of signal ligand (S) or membrane potential.

**Serpentine receptor**
External ligand binding to receptor (R) activates an intracellular GTP-binding protein (G), which regulates an enzyme (Enz) that generates an intracellular second messenger, X.

**Receptor with no intrinsic enzyme activity**
Interacts with cytosolic protein kinase, which activates a gene-regulating protein (directly or through a cascade of protein kinases), changing gene expression.

Ion

Plasma membrane

**Receptor enzyme**
Ligand binding to extracellular domain stimulates enzyme activity in intracellular domain.

Kinase cascade

**Adhesion receptor**
Binds molecules in extracellular matrix, changes conformation, thus altering its interaction with cytoskeleton.

Nuclear envelope

DNA → mRNA → Protein

**Steroid receptor**
Steroid binding to a nuclear receptor protein allows the receptor to regulate the expression of specific genes.

DNA → mRNA → Protein

**FIGURE 12–2** Six general types of signal transducers.

**6.** Receptors (adhesion receptors) that interact with macromolecular components of the extracellular matrix (such as collagen) and convey to the cytoskeletal system instructions on cell migration or adherence to the matrix. Integrins (discussed in Chapter 10) illustrate this general type of transduction mechanism.

As we shall see, transductions of all six types commonly require the activation of protein kinases, enzymes that transfer a phosphoryl group from ATP to a protein side chain.

## SUMMARY 12.1 Molecular Mechanisms of Signal Transduction

- All cells have specific and highly sensitive signal-transducing mechanisms, which have been conserved during evolution.

- A wide variety of stimuli, including hormones, neurotransmitters, and growth factors, act through specific protein receptors in the plasma membrane.

- The receptors bind the signal molecule, amplify the signal, integrate it with input from other receptors, and transmit it into the cell. If the signal persists, receptor desensitization reduces or ends the response.

- Eukaryotic cells have six general types of signaling mechanisms: gated ion channels; receptor enzymes; membrane proteins that act through G proteins; nuclear proteins that bind steroids and act as transcription factors; membrane proteins that attract and activate soluble protein kinases; and adhesion receptors that carry information between the extracellular matrix and the cytoskeleton.

## 12.2 Gated Ion Channels

### Ion Channels Underlie Electrical Signaling in Excitable Cells

The excitability of sensory cells, neurons, and myocytes depends on ion channels, signal transducers that provide a regulated path for the movement of inorganic ions such as $Na^+$, $K^+$, $Ca^{2+}$, and $Cl^-$ across the plasma membrane in response to various stimuli. Recall from Chapter 11 that these ion channels are "gated"; they may be open or closed, depending on whether the associated receptor has been activated by the binding of its specific ligand (a neurotransmitter, for example) or by a change in the transmembrane electrical potential, $V_m$. The $Na^+K^+$ ATPase creates a charge imbalance across

the plasma membrane by carrying 3 $Na^+$ out of the cell for every 2 $K^+$ carried in (Fig. 12–3a), making the inside negative relative to the outside. The membrane is said to be polarized. By convention, $V_m$ is negative when the inside of the cell is negative relative to the outside. For a typical animal cell, $V_m = -60$ to $-70$ mV.

Because ion channels generally allow passage of either anions or cations but not both, ion flux through a channel causes a redistribution of charge on the two sides of the membrane, changing $V_m$. Influx of a positively charged ion such as $Na^+$, or efflux of a negatively charged ion such as $Cl^-$, depolarizes the membrane and brings $V_m$ closer to zero. Conversely, efflux of $K^+$ hyperpolarizes the membrane and $V_m$ becomes more negative. These ion fluxes through channels are passive, in contrast to active transport by the $Na^+K^+$ ATPase.

The direction of spontaneous ion flow across a polarized membrane is dictated by the electrochemical

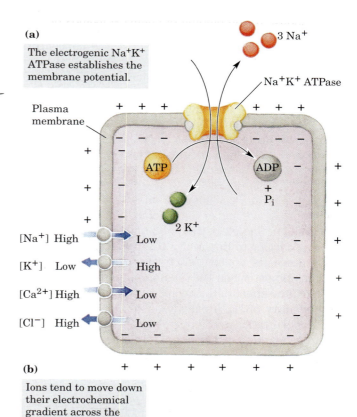

**(a)**

The electrogenic $Na^+K^+$ ATPase establishes the membrane potential.

3 $Na^+$

$Na^+K^+$ ATPase

Plasma membrane

ATP

ADP

$P_i$

2 $K^+$

[$Na^+$] High → Low

[$K^+$] Low → High

[$Ca^{2+}$] High → Low

[$Cl^-$] High ← Low

**(b)**

Ions tend to move down their electrochemical gradient across the polarized membrane.

**FIGURE 12–3 Transmembrane electrical potential. (a)** The electrogenic $Na^+K^+$ ATPase produces a transmembrane electrical potential of $-60$ mV (inside negative). **(b)** Blue arrows show the direction in which ions tend to move spontaneously across the plasma membrane in an animal cell, driven by the combination of chemical and electrical gradients. The chemical gradient drives $Na^+$ and $Ca^{2+}$ inward (producing depolarization) and $K^+$ outward (producing hyperpolarization). The electrical gradient drives $Cl^-$ outward, against its concentration gradient (producing depolarization).

**TABLE 12–2    Ion Concentrations in Cells and Extracellular Fluids (mM)**

| Cell type | K⁺ | | Na⁺ | | Ca²⁺ | | Cl⁻ | |
|---|---|---|---|---|---|---|---|---|
| | In | Out | In | Out | In | Out | In | Out |
| Squid axon | 400 | 20 | 50 | 440 | $\leq 0.4$ | 10 | 40–150 | 560 |
| Frog muscle | 124 | 2.3 | 10.4 | 109 | $< 0.1$ | 2.1 | 1.5 | 78 |

potential of that ion across the membrane. The force ($\Delta G$) that causes a cation (say, $Na^+$) to pass spontaneously inward through an ion channel is a function of the ratio of its concentrations on the two sides of the membrane ($C_{in}/C_{out}$) and of the difference in electrical potential ($\Delta \psi$ or $V_m$):

$$\Delta G = RT \ln \left( \frac{C_{in}}{C_{out}} \right) + Z \mathcal{F} V_m \qquad (12\text{–}1)$$

where $R$ is the gas constant, $T$ the absolute temperature, $Z$ the charge on the ion, and $\mathcal{F}$ the Faraday constant. In a typical neuron or myocyte, the concentrations of $Na^+$, $K^+$, $Ca^{2+}$, and $Cl^-$ in the cytosol are very different from those in the extracellular fluid (Table 12–2). Given these concentration differences, the resting $V_m$ of −60 mV, and the relationship shown in Equation 12–1, opening of a $Na^+$ or $Ca^{2+}$ channel will result in a spontaneous inward flow of $Na^+$ or $Ca^{2+}$ (and depolarization), whereas opening of a $K^+$ channel will result in a spontaneous outward flux of $K^+$ (and hyperpolarization) (Fig. 12–3b).

A given ionic species continues to flow through a channel only as long as the combination of concentration gradient and electrical potential provides a driving force, according to Equation 12–1. For example, as $Na^+$ flows down its concentration gradient it depolarizes the membrane. When the membrane potential reaches +70 mV, the effect of this membrane potential (to resist further entry of $Na^+$) exactly equals the effect of the $Na^+$ concentration gradient (to cause more $Na^+$ to flow inward). At this equilibrium potential ($E$), the driving force ($\Delta G$) tending to move an ion is zero. The equilibrium potential is different for each ionic species because the concentration gradients differ for each ion.

The number of ions that must flow to change the membrane potential significantly is negligible relative to the concentrations of $Na^+$, $K^+$, and $Cl^-$ in cells and extracellular fluid, so the ion fluxes that occur during signaling in excitable cells have essentially no effect on the concentrations of those ions. However, because the intracellular concentration of $Ca^{2+}$ is generally very low (~$10^{-7}$ M), inward flow of $Ca^{2+}$ can significantly alter the cytosolic $[Ca^{2+}]$.

The membrane potential of a cell at a given time is the result of the types and numbers of ion channels open at that instant. In most cells at rest, more $K^+$ channels than $Na^+$, $Cl^-$, or $Ca^{2+}$ channels are open and thus the resting potential is closer to the $E$ for $K^+$ (−98 mV) than that for any other ion. When channels for $Na^+$, $Ca^{2+}$, or $Cl^-$ open, the membrane potential moves toward the $E$ for that ion. The precisely timed opening and closing of ion channels and the resulting transient changes in membrane potential underlie the electrical signaling by which the nervous system stimulates the skeletal muscles to contract, the heart to beat, or secretory cells to release their contents. Moreover, many hormones exert their effects by altering the membrane potentials of their target cells. These mechanisms are not limited to complex animals; ion channels play important roles in the responses of bacteria, protists, and plants to environmental signals.

To illustrate the action of ion channels in cell-to-cell signaling, we describe the mechanisms by which a neuron passes a signal along its length and across a synapse to the next neuron (or to a myocyte) in a cellular circuit, using acetylcholine as the neurotransmitter.

## The Nicotinic Acetylcholine Receptor Is a Ligand-Gated Ion Channel

One of the best-understood examples of a **ligand-gated receptor channel** is the **nicotinic acetylcholine receptor** (see Fig. 11–51). The receptor channel opens in response to the neurotransmitter acetylcholine (and to nicotine, hence the name). This receptor is found in the postsynaptic membrane of neurons at certain synapses and in muscle fibers (myocytes) at neuromuscular junctions.

$$CH_3-\overset{\displaystyle \overset{CH_3}{\overset{+|}{}}}{N}-CH_2CH_2O-\overset{\displaystyle \overset{O}{\|}}{C}-CH_3$$
$$\underset{\displaystyle CH_3}{|}$$

Acetylcholine (Ach)

Acetylcholine released by an excited neuron diffuses a few micrometers across the synaptic cleft or neuromuscular junction to the postsynaptic neuron or myocyte, where it interacts with the acetylcholine receptor and triggers electrical excitation (depolarization) of the receiving cell. The acetylcholine receptor is an allosteric protein with two high-affinity binding sites for acetylcholine, about 3.0 nm from the ion gate, on the two α

*ligand-gated receptor channel.*

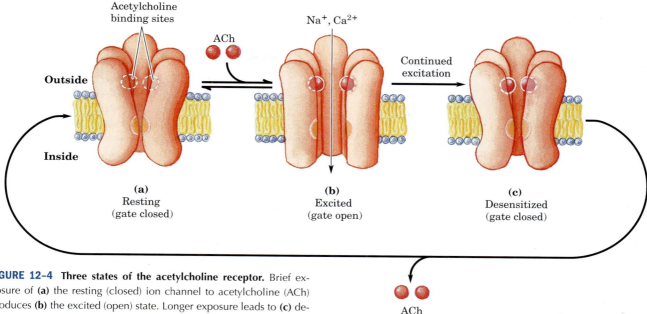

**FIGURE 12–4 Three states of the acetylcholine receptor.** Brief exposure of **(a)** the resting (closed) ion channel to acetylcholine (ACh) produces **(b)** the excited (open) state. Longer exposure leads to **(c)** desensitization and channel closure.

subunits. The binding of acetylcholine causes a change from the closed to the open conformation. The process is positively cooperative: binding of acetylcholine to the first site increases the acetylcholine-binding affinity of the second site. When the presynaptic cell releases a brief pulse of acetylcholine, both sites on the postsynaptic cell receptor are occupied briefly and the channel opens (Fig. 12–4). Either $Na^+$ or $Ca^{2+}$ can now pass, and the inward flux of these ions depolarizes the plasma membrane, initiating subsequent events that vary with the type of tissue. In a postsynaptic neuron, depolarization initiates an action potential (see below); at a neuromuscular junction, depolarization of the muscle fiber triggers muscle contraction.

Normally, the acetylcholine concentration in the synaptic cleft is quickly lowered by the enzyme acetylcholinesterase, present in the cleft. When acetylcholine levels remain high for more than a few milliseconds, the receptor is desensitized (Fig. 12–1c). The receptor channel is converted to a third conformation (Fig. 12–4c) in which the channel is closed and the acetylcholine is very tightly bound. The slow release (in tens of milliseconds) of acetylcholine from its binding sites eventually allows the receptor to return to its resting state—closed and resensitized to acetylcholine levels.

## Voltage-Gated Ion Channels Produce Neuronal Action Potentials

Signaling in the nervous system is accomplished by networks of neurons, specialized cells that carry an electrical impulse (action potential) from one end of the cell (the cell body) through an elongated cytoplasmic ex-

tension (the axon). The electrical signal triggers release of neurotransmitter molecules at the synapse, carrying the signal to the next cell in the circuit. Three types of **voltage-gated ion channels** are essential to this signaling mechanism. Along the entire length of the axon are **voltage-gated $Na^+$ channels** (Fig. 12–5; see also Fig. 11–50), which are closed when the membrane is at rest ($V_m = -60$ mV) but open briefly when the membrane is depolarized locally in response to acetylcholine (or some other neurotransmitter). The depolarization induced by the opening of $Na^+$ channels causes **voltage-gated $K^+$ channels** to open, and the resulting efflux of $K^+$ repolarizes the membrane locally. A brief pulse of depolarization traverses the axon as local depolarization triggers the brief opening of neighboring $Na^+$ channels, then $K^+$ channels. After each opening of a $Na^+$ channel, a short refractory period follows during which that channel cannot open again, and thus a unidirectional wave of depolarization sweeps from the nerve cell body toward the end of the axon. The voltage sensitivity of ion channels is due to the presence at critical positions in the channel protein of charged amino acid side chains that interact with the electric field across the membrane. Changes in transmembrane potential produce subtle conformational changes in the channel protein (see Fig. 11–50).

At the distal tip of the axon are **voltage-gated $Ca^{2+}$ channels.** When the wave of depolarization reaches these channels, they open, and $Ca^{2+}$ enters from the extracellular space. The rise in cytoplasmic $[Ca^{2+}]$ then triggers release of acetylcholine by exocytosis into the synaptic cleft (step ③ in Fig. 12–5). Acetylcholine diffuses to the postsynaptic cell (another

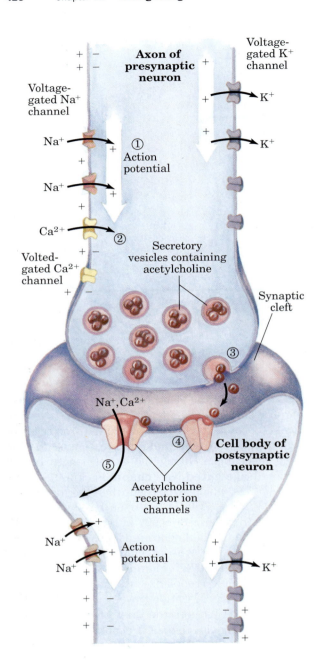

**FIGURE 12–5 Role of voltage-gated and ligand-gated ion channels in neural transmission.** Initially, the plasma membrane of the presynaptic neuron is polarized (inside negative) through the action of the electrogenic $Na^+K^+$ ATPase, which pumps 3 $Na^+$ out for every 2 $K^+$ pumped into the neuron (see Fig. 12–3). ① A stimulus to this neuron causes an action potential to move along the axon (white arrow), away from the cell body. The opening of one voltage-gated $Na^+$ channel allows $Na^+$ entry, and the resulting local depolarization causes the adjacent $Na^+$ channel to open, and so on. The directionality of movement of the action potential is ensured by the brief refractory period that follows the opening of each voltage-gated $Na^+$ channel. ② When the wave of depolarization reaches the axon tip, voltage-gated $Ca^{2+}$ channels open, allowing $Ca^{2+}$ entry into the presynaptic neuron. ③ The resulting increase in internal $[Ca^{2+}]$ triggers exocytic release of the neurotransmitter acetylcholine into the synaptic cleft. ④ Acetylcholine binds to a receptor on the postsynaptic neuron, causing its ligand-gated ion channel to open. ⑤ Extracellular $Na^+$ and $Ca^{2+}$ enter through this channel, depolarizing the postsynaptic cell. The electrical signal has thus passed to the cell body of the postsynaptic neuron and will move along its axon to a third neuron by this same sequence of events.

## Neurons Have Receptor Channels That Respond to Different Neurotransmitters

Animal cells, especially those of the nervous system, contain a variety of ion channels gated by ligands, voltage, or both. The neurotransmitters 5-hydroxytryptamine (serotonin), glutamate, and glycine can all act through receptor channels that are structurally related to the acetylcholine receptor. Serotonin and glutamate trigger the opening of cation ($K^+$, $Na^+$, $Ca^{2+}$) channels, whereas glycine opens $Cl^-$-specific channels. Cation and anion channels are distinguished by subtle differences in the amino acid residues that line the hydrophilic channel. Cation channels have negatively charged Glu and Asp side chains at crucial positions. When a few of these acidic residues are experimentally replaced with basic residues, the cation channel is converted to an anion channel.

Depending on which ion passes through a channel, the ligand (neurotransmitter) for that channel either depolarizes or hyperpolarizes the target cell. A single neuron normally receives input from several (or many) other neurons, each releasing its own characteristic neurotransmitter with its characteristic depolarizing or hyperpolarizing effect. The target cell's $V_m$ therefore reflects the *integrated* input (Fig. 12–1d) from multi-

neuron or a myocyte), where it binds to acetylcholine receptors and triggers depolarization. Thus the message is passed to the next cell in the circuit.

We see, then, that gated ion channels convey signals in either of two ways: by changing the cytosolic concentration of an ion (such as $Ca^{2+}$), which then serves as an intracellular **second messenger** (the hormone or neurotransmitter is the first messenger), or by changing $V_m$ and affecting other membrane proteins that are sensitive to $V_m$. The passage of an electrical signal through one neuron and on to the next illustrates both types of mechanism.

Serotonin
(5-hydroxytryptamine)

Glutamate

ple neurons. The cell responds with an action potential only if the integrated input adds up to a net depolarization of sufficient size.

The receptor channels for acetylcholine, glycine, glutamate, and γ-aminobutyric acid (GABA) are gated by *extracellular* ligands. *Intracellular* second messengers—such as cAMP, cGMP (3′,5′-cyclic GMP, a close analog of cAMP), IP$_3$ (inositol 1,4,5-trisphosphate), Ca$^{2+}$, and ATP—regulate ion channels of another class, which, as we shall see in Section 12.7, participate in the sensory transductions of vision, olfaction, and gustation.

### SUMMARY 12.2 Gated Ion Channels

- Ion channels gated by ligands or membrane potential are central to signaling in neurons and other cells.

- The acetylcholine receptor of neurons and myocytes is a ligand-gated ion channel.

- The voltage-gated Na$^+$ and K$^+$ channels of neuronal membranes carry the action potential along the axon as a wave of depolarization (Na$^+$ influx) followed by repolarization (K$^+$ efflux).

- The arrival of an action potential triggers neurotransmitter release from the presynaptic cell. The neurotransmitter (acetylcholine, for example) diffuses to the postsynaptic cell, binds to specific receptors in the plasma membrane, and triggers a change in $V_m$.

## 12.3 Receptor Enzymes

A fundamentally different mechanism of signal transduction is carried out by the receptor enzymes. These proteins have a ligand-binding domain on the extracellular surface of the plasma membrane and an enzyme active site on the cytosolic side, with the two domains connected by a single transmembrane segment. Commonly, the receptor enzyme is a protein kinase that phosphorylates Tyr residues in specific target proteins; the insulin receptor is the prototype for this group. In plants, the protein kinase of receptors is specific for Ser or Thr residues. Other receptor enzymes synthesize the intracellular second messenger cGMP in response to extracellular signals. The receptor for atrial natriuretic factor is typical of this type.

### The Insulin Receptor Is a Tyrosine-Specific Protein Kinase

Insulin regulates both metabolism and gene expression: the insulin signal passes from the plasma membrane receptor to insulin-sensitive metabolic enzymes and to the nucleus, where it stimulates the transcription of specific genes. The active insulin receptor consists of two identical α chains protruding from the outer face of the plasma membrane and two transmembrane β subunits with their carboxyl termini protruding into the cytosol (Fig. 12–6, step ①). The α chains contain the insulin-binding domain, and the intracellular domains of the β chains contain the protein kinase activity that transfers a phosphoryl group from ATP to the hydroxyl group of Tyr residues in specific target proteins. Signaling through the insulin receptor begins (step ①) when binding of insulin to the α chains activates the Tyr kinase activity of the β chains, and each αβ monomer phosphorylates three critical Tyr residues near the carboxyl terminus of the β chain of its partner in the dimer. This **autophosphorylation** opens up the active site so that the enzyme can phosphorylate Tyr residues of other target proteins (Fig. 12–7).

One of these target proteins (Fig. 12–6, step ②) is insulin receptor substrate-1 (IRS-1). Once phosphorylated on its Tyr residues, IRS-1 becomes the point of nucleation for a complex of proteins (step ③) that carry the message from the insulin receptor to end targets in the cytosol and nucleus, through a long series of intermediate proteins. First, a Ⓟ–Tyr residue in IRS-1 is bound by the **SH2 domain** of the protein Grb2. (SH2 is an abbreviation of *Src homology 2*; the sequences of SH2 domains are similar to a domain in another protein Tyr kinase, Src (pronounced sark).) A number of signaling proteins contain SH2 domains, all of which bind Ⓟ–Tyr residues in a protein partner. Grb2 also contains a second protein-binding domain, SH3, that binds to regions rich in Pro residues. Grb2 binds to a proline-rich region of Sos, recruiting Sos to the growing receptor complex. When bound to Grb2, Sos catalyzes the replacement of bound GDP by GTP on Ras, one of a family of guanosine nucleotide–binding proteins (**G proteins**) that mediate a wide variety of signal transductions (Section 12.4). When GTP is bound, Ras can activate a protein kinase, Raf-1 (step ④), the first of three protein kinases—Raf-1, MEK, and ERK—that form a cascade in which each kinase activates the next by phosphorylation (step ⑤). The protein kinase ERK is activated by phosphorylation of both a Thr and a Tyr residue. When activated, it mediates some of the biological effects of insulin by entering the nucleus and phosphorylating proteins such as Elk1, which modulates the transcription of about 100 insulin-regulated genes (step ⑥).

The proteins Raf-1, MEK, and ERK are members of three larger families, for which several nomenclatures are employed. ERK is a member of the **MAPK** family (*m*itogen-*a*ctivated *p*rotein *k*inases; mitogens are signals that act from outside the cell to induce mitosis and cell division). Soon after discovery of the first MAPK, that enzyme was found to be activated by another protein kinase, which came to be called MAP kinase kinase (MEK

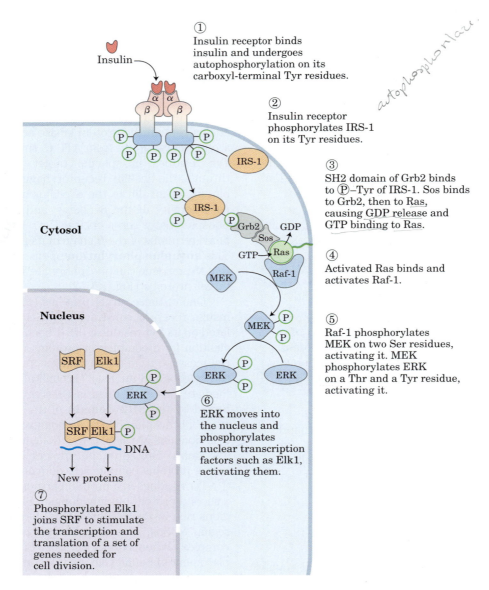

*autophosphorylase*

① Insulin receptor binds insulin and undergoes autophosphorylation on its carboxyl-terminal Tyr residues.

② Insulin receptor phosphorylates IRS-1 on its Tyr residues.

③ SH2 domain of Grb2 binds to ℗–Tyr of IRS-1. Sos binds to Grb2, then to Ras, causing GDP release and GTP binding to Ras.

④ Activated Ras binds and activates Raf-1.

⑤ Raf-1 phosphorylates MEK on two Ser residues, activating it. MEK phosphorylates ERK on a Thr and a Tyr residue, activating it.

⑥ ERK moves into the nucleus and phosphorylates nuclear transcription factors such as Elk1, activating them.

⑦ Phosphorylated Elk1 joins SRF to stimulate the transcription and translation of a set of genes needed for cell division.

**FIGURE 12–6  Regulation of gene expression by insulin.** The insulin receptor consists of two α chains on the outer face of the plasma membrane and two β chains that traverse the membrane and protrude from the cytoplasmic face. Binding of insulin to the α chains triggers a conformational change that allows the autophosphorylation of Tyr residues in the carboxyl-terminal domain of the β subunits. Autophosphorylation further activates the Tyr kinase domain, which then catalyzes phosphorylation of other target proteins. The signaling pathway by which insulin regulates the expression of specific genes consists of a cascade of protein kinases, each of which activates the next. The insulin receptor is a Tyr-specific kinase; the other kinases (all shown in blue) phosphorylate Ser or Thr residues. MEK is a dual-specificity kinase, which phosphorylates both a Thr and a Tyr residue in ERK (extracellular regulated kinase); MEK is mitogen-activated, ERK-activating kinase; SRF is serum response factor. Abbreviations for other components are explained in the text.

belongs to this family); and when a third kinase that activated MAP kinase kinase was discovered, it was given the slightly ludicrous family name MAP kinase kinase kinase (Raf-1 is a member of this family; Fig. 12–6). Slightly less cumbersome are the acronyms for these three families, MAPK, MAPKK, and MAPKKK. Kinases in the MAPK and MAPKKK families are specific for Ser or Thr residues, but MAPKKs (here, MEK) phosphorylate both a Ser and a Tyr residue in their substrate, a MAPK (here, ERK).

Biochemists now recognize the insulin pathway as but one instance of a more general theme in which hormone signals, via pathways similar to that shown in Figure 12–6, result in phosphorylation of target enzymes by protein kinases. The target of phosphorylation is often another protein kinase, which then phosphorylates a third protein kinase, and so on. The result is a cascade of reactions that amplifies the initial signal by many orders of magnitude (see Fig. 12–1b). Cascades such as that shown in Figure 12–6 are called **MAPK cascades.**

*enzyme part that's in the cytoplasm*

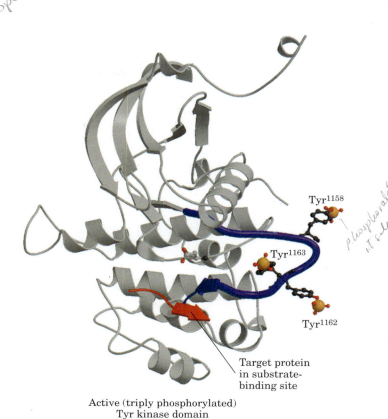

Tyr¹¹⁵⁸          Asp¹¹³²

Tyr¹¹⁶²

Tyr¹¹⁶³

Activation loop
blocks substrate-
binding site

Inactive (unphosphorylated)
Tyr kinase domain

**(a)**

Tyr¹¹⁵⁸

Tyr¹¹⁶³

*phosphorylates itself*

Tyr¹¹⁶²

Target protein
in substrate-
binding site

Active (triply phosphorylated)
Tyr kinase domain

**(b)**

**FIGURE 12–7 Activation of the insulin-receptor Tyr kinase by autophosphorylation. (a)** In the inactive form of the Tyr kinase domain (PDB ID 1IRK), the activation loop (blue) sits in the active site, and none of the critical Tyr residues (black and red ball-and-stick structures) are phosphorylated. This conformation is stabilized by hydrogen bonding between Tyr¹¹⁶² and Asp¹¹³². **(b)** When insulin binds to the α chains of insulin receptors, the Tyr kinase of each β subunit of the dimer phosphorylates three Tyr residues (Tyr¹¹⁵⁸, Tyr¹¹⁶², and Tyr¹¹⁶³) on the other β subunit (shown here; PDB ID 1IR3). (Phosphoryl groups are depicted here as an orange space-filling phosphorus atom and red ball-and-stick oxygen atoms.) The effect of introducing three highly charged Ⓟ–Tyr residues is to force a 30 Å change in the position of the activation loop, away from the substrate-binding site, which becomes available to bind to and phosphorylate a target protein, shown here as a red arrow.

Grb2 is not the only protein that associates with phosphorylated IRS-1. The enzyme phosphoinositide 3-kinase (PI-3K) binds IRS-1 through the former's SH2 domain (Fig. 12–8). Thus activated, PI-3K converts the membrane lipid phosphatidylinositol 4,5-bisphosphate (see Fig. 10–15), also called PIP₂, to phosphatidylinositol 3,4,5-trisphosphate (PIP₃). When bound to PIP₃, protein kinase B (PKB) is phosphorylated and activated by yet another protein kinase, PDK1. The activated PKB then phosphorylates Ser or Thr residues on its target proteins, one of which is glycogen synthase kinase 3 (GSK3). In its active, nonphosphorylated form, GSK3 phosphorylates glycogen synthase, inactivating it and thereby contributing to the slowing of glycogen synthesis. (This mechanism is believed to be only part of the explanation for the effects of insulin on glycogen metabolism.) When phosphorylated by PKB, GSK3 is inactivated. By thus preventing inactivation of glycogen synthase in liver and muscle, the cascade of protein phosphorylations initiated by insulin stimulates glycogen synthesis (Fig. 12–8). In muscle, PKB triggers the movement of glucose transporters (GLUT4) from internal vesicles to the plasma membrane, stimulating glucose uptake from the blood (Fig. 12–8; see also Box 11–2). PKB also functions in several other signaling pathways, including that triggered by Δ⁹-tetrahydrocannabinol (THC), the active ingredient of marijuana

Δ⁹-Tetrahydrocannabinol (THC)

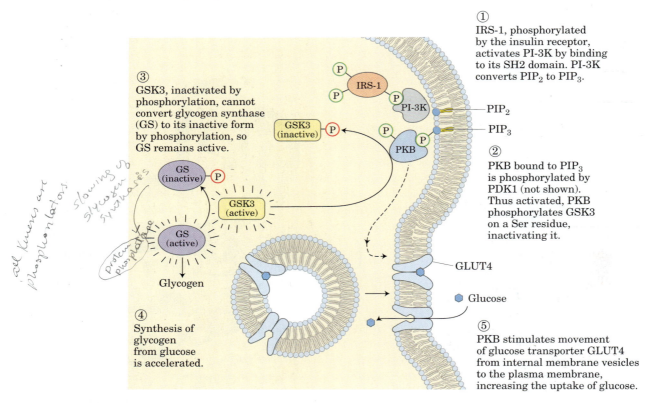

③ GSK3, inactivated by phosphorylation, cannot convert glycogen synthase (GS) to its inactive form by phosphorylation, so GS remains active.

① IRS-1, phosphorylated by the insulin receptor, activates PI-3K by binding to its SH2 domain. PI-3K converts $PIP_2$ to $PIP_3$.

② PKB bound to $PIP_3$ is phosphorylated by PDK1 (not shown). Thus activated, PKB phosphorylates GSK3 on a Ser residue, inactivating it.

④ Synthesis of glycogen from glucose is accelerated.

⑤ PKB stimulates movement of glucose transporter GLUT4 from internal membrane vesicles to the plasma membrane, increasing the uptake of glucose.

*all kinases are phosphorylators*
*slowing of glycogen synthases*
*protein phosphatase*

**FIGURE 12–8** **Activation of glycogen synthase by insulin.** Transmission of the signal is mediated by PI-3 kinase (PI-3K) and protein kinase B (PKB).

and hashish. THC activates the $CB_1$ receptor in the plasma membrane of neurons in the brain, triggering a signaling cascade that involves MAPKs. One consequence of $CB_1$ activation is the stimulation of appetite, one of the well-established effects of marijuana use. The normal ligands for the $CB_1$ receptor are endocannabinoids such as anandamide, which serve to protect the brain from the toxicity of excessive neuronal activity— as in an epileptic seizure, for example. Hashish has for centuries been used in the treatment of epilepsy.

**Anandamide (arachidonylethanolamide, an endogenous cannabinoid)**

As in all signaling pathways, there is a mechanism for terminating signaling through the PI-3K–PKB pathway. A $PIP_3$-specific phosphatase (PTEN in humans) removes the phosphate at the 3 position of $PIP_3$ to produce $PIP_2$, which no longer serves as a binding site for PKB, and the signaling chain is broken. In various types of advanced cancer, tumor cells often have a defect in the PTEN gene and thus have abnormally high levels of $PIP_3$ and of PKB activity. The result seems to be a continuing signal for cell division and thus tumor growth. ■

What spurred the evolution of such complicated regulatory machinery? This system allows one activated receptor to activate several IRS-1 molecules, amplifying the insulin signal, and it provides for the integration of signals from several receptors, each of which can phosphorylate IRS-1. Furthermore, because IRS-1 can activate any of several proteins that contain SH2 domains, a single receptor acting through IRS-1 can trigger two or more signaling pathways; insulin affects gene expression through the Grb2-Sos-Ras-MAPK pathway and glycogen metabolism through the PI-3K–PKB pathway.

The insulin receptor is the prototype for a number of receptor enzymes with a similar structure and **receptor Tyr kinase** activity. The receptors for epidermal growth factor and platelet-derived growth factor, for example, have structural and sequence similarities to the insulin receptor, and both have a protein Tyr kinase activity that phosphorylates IRS-1. Many of these receptors dimerize after binding ligand; the insulin receptor is already a dimer before insulin binds. The binding of adaptor proteins such as Grb2 to ℗–Tyr residues is a common mechanism for promoting protein-protein interactions, a subject to which we return in Section 12.5.

In addition to the many receptors that act as protein Tyr kinases, a number of receptorlike plasma membrane proteins have protein Tyr phosphatase activity. Based on the structures of these proteins, we can surmise that their ligands are components of the extracellular matrix or the surfaces of other cells. Although their signaling roles are not yet as well understood as those of the receptor Tyr kinases, they clearly have the potential to reverse the actions of signals that stimulate these kinases.

A variation on the basic theme of receptor Tyr kinases is seen in receptors that have no intrinsic protein kinase activity but, when occupied by their ligand, bind a *soluble* Tyr kinase. One example is the system that regulates the formation of erythrocytes in mammals. The **cytokine** (developmental signal) for this system is erythropoietin (EPO), a 165 amino acid protein produced in the kidneys. When EPO binds to its plasma membrane receptor (Fig. 12–9), the receptor dimerizes and can now bind the soluble protein kinase JAK (*Janus kinase*). This binding activates JAK, which phosphorylates several Tyr residues in the cytoplasmic domain of the EPO receptor. A family of transcription factors, collectively called STATs (signal *transducers* and *activators* of *transcription*), are also targets of the JAK kinase activity. An SH2 domain in STAT5 binds ⓟ–Tyr residues in the EPO receptor, positioning it for this phosphorylation by JAK. When STAT5 is phosphorylated in re-

sponse to EPO, it forms dimers, exposing a signal for its transport into the nucleus. There, STAT5 causes the expression (transcription) of specific genes essential for erythrocyte maturation. This JAK-STAT system operates in a number of other signaling pathways, including that for the hormone leptin, described in detail in Chapter 23 (see Fig. 23–34). Activated JAK can also trigger, through Grb2, the MAPK cascade (Fig. 12–6), which leads to altered expression of specific genes.

Src is another soluble protein Tyr kinase that associates with certain receptors when they bind their ligands. Src was the first protein found to have the characteristic ⓟ–Tyr-binding domain that was subsequently named the Src homology (SH2) domain. Yet another example of a receptor's association with a soluble protein kinase is the Toll-like receptor (TLR4) system through which mammals detect the bacterial lipopolysaccharide (LPS), a potent toxin. We return to the Toll-like receptor system in Section 12.6, in the context of the evolution of signaling proteins.

### Receptor Guanylyl Cyclases Generate the Second Messenger cGMP

Guanylyl cyclases (Fig. 12–10) are another type of receptor enzyme. When activated, a guanylyl cyclase produces **guanosine 3′,5′-cyclic monophosphate (cyclic GMP, cGMP)** from GTP:

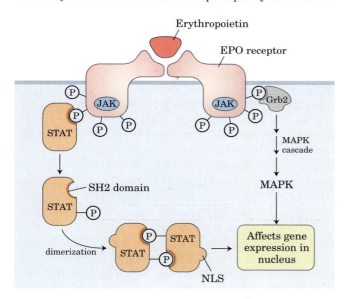

**FIGURE 12–9 The JAK-STAT transduction mechanism for the erythropoietin receptor.** Binding of erythropoietin (EPO) causes dimerization of the EPO receptor, which allows the soluble Tyr kinase JAK to bind to the internal domain of the receptor and phosphorylate it on several Tyr residues. The STAT protein STAT5 contains an SH2 domain and binds to the ⓟ–Tyr residues on the receptor, bringing the receptor into proximity with JAK. Phosphorylation of STAT5 by JAK allows two STAT molecules to dimerize, each binding the other's ⓟ–Tyr residue. Dimerization of STAT5 exposes a nuclear localization sequence (NLS) that targets STAT5 for transport into the nucleus. In the nucleus, STAT causes the expression of genes controlled by EPO. A second signaling pathway is also triggered by autophosphorylation of JAK that is associated with EPO binding to its receptor. The adaptor protein Grb2 binds ⓟ–Tyr in JAK and triggers the MAPK cascade, as in the insulin system (see Fig. 12–6).

GTP

PPᵢ

Guanosine 3′,5′-cyclic monophosphate
(cGMP)

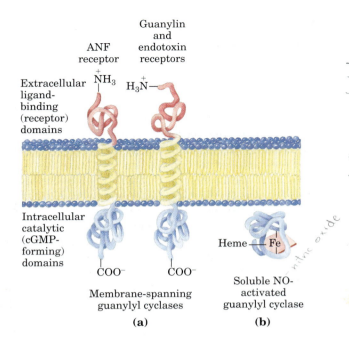

**FIGURE 12–10 Two types (isozymes) of guanylyl cyclase that participate in signal transduction. (a)** One isozyme exists in two similar membrane-spanning forms that are activated by their extracellular ligands: atrial natriuretic factor, ANF (receptors in cells of the renal collecting ducts and the smooth muscle of blood vessels), and guanylin (receptors in intestinal epithelial cells). The guanylin receptor is also the target of a type of bacterial endotoxin that triggers severe diarrhea. **(b)** The other isozyme is a soluble enzyme that is activated by intracellular nitric oxide (NO); this form is found in many tissues, including smooth muscle of the heart and blood vessels.

Cyclic GMP is a second messenger that carries different messages in different tissues. In the kidney and intestine it triggers changes in ion transport and water retention; in cardiac muscle (a type of smooth muscle) it signals relaxation; in the brain it may be involved both in development and in adult brain function. Guanylyl cyclase in the kidney is activated by the hormone **atrial natriuretic factor (ANF),** which is released by cells in the atrium of the heart when the heart is stretched by increased blood volume. Carried in the blood to the kidney, ANF activates guanylyl cyclase in cells of the collecting ducts (Fig. 12–10a). The resulting rise in [cGMP] triggers increased renal excretion of $Na^+$ and, consequently, of water, driven by the change in osmotic pressure. Water loss reduces the blood volume, countering the stimulus that initially led to ANF secretion. Vascular smooth muscle also has an ANF receptor—guanylyl cyclase; on binding to this receptor, ANF causes relaxation (vasodilation) of the blood vessel, which increases blood flow while decreasing blood pressure.

A similar receptor guanylyl cyclase in the plasma membrane of intestinal epithelial cells is activated by an intestinal peptide, **guanylin,** which regulates $Cl^-$ secretion in the intestine. This receptor is also the target of a

heat-stable peptide endotoxin produced by *Escherichia coli* and other gram-negative bacteria. The elevation in [cGMP] caused by the endotoxin increases $Cl^-$ secretion and consequently decreases reabsorption of water by the intestinal epithelium, producing diarrhea.

A distinctly different type of guanylyl cyclase is a cytosolic protein with a tightly associated heme group (Fig. 12–10b), an enzyme activated by nitric oxide (NO). Nitric oxide is produced from arginine by $Ca^{2+}$-dependent **NO synthase,** present in many mammalian tissues, and diffuses from its cell of origin into nearby cells. NO is sufficiently nonpolar to cross plasma membranes without a carrier. In the target cell, it binds to the heme group of guanylyl cyclase and activates cGMP production. In the heart, cGMP reduces the forcefulness of contractions by stimulating the ion pump(s) that expel $Ca^{2+}$ from the cytosol.

$$\underset{\text{Arginine}}{\begin{array}{c} NH_2 \\ | \\ C{=}\overset{+}{N}H_2 \\ | \\ NH \\ | \\ (CH_2)_3 \\ | \\ CH{-}COO^- \\ | \\ {}^+NH_3 \end{array}} \xrightarrow[\text{NO synthase}]{\substack{\text{NADPH} \quad \text{NADP}^+ \\ O_2 \qquad Ca^{2+}}} \underset{\text{Citrulline}}{\begin{array}{c} NH_2 \\ | \\ C{=}O \\ | \\ NH \\ | \\ (CH_2)_3 \\ | \\ CH{-}COO^- \\ | \\ {}^+NH_3 \end{array}} + NO$$

This NO-induced relaxation of cardiac muscle is the same response brought about by nitroglycerin tablets and other nitrovasodilators taken to relieve angina, the pain caused by contraction of a heart deprived of $O_2$ because of blocked coronary arteries. Nitric oxide is unstable and its action is brief; within seconds of its formation, it undergoes oxidation to nitrite or nitrate. Nitrovasodilators produce long-lasting relaxation of cardiac muscle because they break down over several hours, yielding a steady stream of NO. The value of nitroglycerin as a treatment for angina was discovered serendipitously in factories producing nitroglycerin as an explosive in the 1860s. Workers with angina reported that their condition was much improved during the work week but returned on weekends. The physicians treating these workers heard this story so often that they made the connection, and a drug was born.

The effects of increased cGMP synthesis diminish after the stimulus ceases, because a specific phosphodiesterase (cGMP PDE) converts cGMP to the inactive 5′-GMP. Humans have several isoforms of cGMP PDE, with different tissue distributions. The isoform in the blood vessels of the penis is inhibited by the drug sildenafil (Viagra), which therefore causes cGMP levels to remain elevated once raised by an appropriate stimulus, accounting for the usefulness of this drug in the treatment of erectile dysfunction. ■

Most of the actions of cGMP in animals are believed to be mediated by **cGMP-dependent protein kinase,** also called **protein kinase G** or **PKG,** which, when ac-

Sildenafil (Viagra)

tivated by cGMP, phosphorylates Ser and Thr residues in target proteins. The catalytic and regulatory domains of this enzyme are in a single polypeptide ($M_r$ ~80,000). Part of the regulatory domain fits snugly in the substrate-binding site. Binding of cGMP forces this part of the regulatory domain out of the binding site, activating the catalytic domain.

Cyclic GMP has a second mode of action in the vertebrate eye: it causes ion-specific channels to open in the retinal rod and cone cells. We return to this role of cGMP in the discussion of vision in Section 12.7.

### SUMMARY 12.3  Receptor Enzymes

- The insulin receptor is the prototype of receptor enzymes with Tyr kinase activity. When insulin binds to its receptor, each $\alpha\beta$ monomer of the receptor phosphorylates the $\beta$ chain of its partner, activating the receptor's Tyr kinase activity. The kinase catalyzes the phosphorylation of Tyr residues on other proteins such as IRS-1.

- Ⓟ–Tyr residues in IRS-1 serve as binding sites for proteins with SH2 domains. Some of these proteins, such as Grb2, have two or more protein-binding domains and can serve as adaptors that bring two proteins into proximity.

- Further protein-protein interactions result in GTP binding to and activation of the Ras protein, which in turn activates a protein kinase cascade that ends with the phosphorylation of target proteins in the cytosol and nucleus. The result is specific metabolic changes and altered gene expression.

- Several signals, including atrial natriuretic factor and the intestinal peptide guanylin, act through receptor enzymes with guanylyl cyclase activity. The cGMP produced acts as a second messenger, activating cGMP-dependent protein kinase (PKG). This enzyme alters metabolism by phosphorylating specific enzyme targets.

- Nitric oxide (NO) is a short-lived messenger that acts by stimulating a soluble guanylyl cyclase, raising [cGMP] and stimulating PKG.

## 12.4 G Protein–Coupled Receptors and Second Messengers

A third mechanism of signal transduction, distinct from gated ion channels and receptor enzymes, is defined by three essential components: a plasma membrane receptor with seven transmembrane helical segments, an enzyme in the plasma membrane that generates an intracellular second messenger, and a **guanosine nucleotide–binding protein (G protein).** The G protein, stimulated by the activated receptor, exchanges bound GDP for GTP; the GTP-protein dissociates from the occupied receptor and binds to a nearby enzyme, altering its activity. The human genome encodes more than 1,000 members of this family of receptors, specialized for transducing messages as diverse as light, smells, tastes, and hormones. The $\beta$-adrenergic receptor, which mediates the effects of epinephrine on many tissues, is the prototype for this type of transducing system.

### The $\beta$-Adrenergic Receptor System Acts through the Second Messenger cAMP

Epinephrine action begins when the hormone binds to a protein receptor in the plasma membrane of a hormone-sensitive cell. **Adrenergic receptors** ("adrenergic" reflects the alternative name for epinephrine, adrenaline) are of four general types, $\alpha_1$, $\alpha_2$, $\beta_1$, and $\beta_2$, defined by subtle differences in their affinities and responses to a group of agonists and antagonists. **Agonists** are structural analogs that bind to a receptor and mimic the effects of its natural ligand; **antagonists** are analogs that bind without triggering the normal effect and thereby block the effects of agonists. In some cases, the affinity of the synthetic agonist or antagonist for the receptor is greater than that of the natural agonist (Fig. 12–11). The four types of adrenergic receptors are found in different target tissues and mediate different responses to epinephrine. Here we focus on the $\beta$-adrenergic receptors of muscle, liver, and adipose tissue. These receptors mediate changes in fuel metabolism, as described in Chapter 23, including the increased breakdown of glycogen and fat. Adrenergic receptors of the $\beta_1$ and $\beta_2$ subtypes act through the same mechanism, so in our discussion, "$\beta$-adrenergic" applies to both types.

The $\beta$-adrenergic receptor is an integral protein with seven hydrophobic regions of 20 to 28 amino acid residues that "snake" back and forth across the plasma membrane seven times. This protein is a member of a very large family of receptors, all with seven transmembrane helices, that are commonly called **serpentine receptors, G protein–coupled receptors (GPCR),** or **7 transmembrane segment (7tm) receptors.** The binding of epinephrine to a site on the

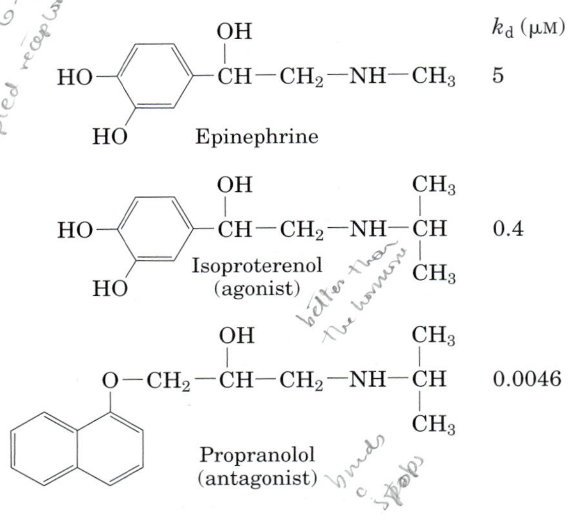

*(handwritten margin note: binds a G-protein Coupled receptor)*

OH

HO—⬡—CH—CH$_2$—NH—CH$_3$     $k_d$ (μM)

HO     Epinephrine     5

OH          CH$_3$

HO—⬡—CH—CH$_2$—NH—CH     0.4

HO     Isoproterenol     CH$_3$
       (agonist)

*(handwritten: better than the hormone)*

OH          CH$_3$

O—CH$_2$—CH—CH$_2$—NH—CH     0.0046

          CH$_3$

     Propranolol
     (antagonist)

*(handwritten: binds & stops)*

**FIGURE 12–11 Epinephrine and its synthetic analogs.** Epinephrine, also called adrenaline, is released from the adrenal gland and regulates energy-yielding metabolism in muscle, liver, and adipose tissue. It also serves as a neurotransmitter in adrenergic neurons. Its affinity for its receptor is expressed as a dissociation constant for the receptor-ligand complex. Isoproterenol and propranolol are synthetic analogs, one an agonist with an affinity for the receptor that is higher than that of epinephrine, and the other an antagonist with extremely high affinity.

receptor deep within the membrane (Fig. 12–12, step ① ) promotes a conformational change in the receptor's intracellular domain that affects its interaction with the second protein in the signal-transduction pathway, a heterotrimeric GTP-binding **stimulatory G protein,** or **G$_s$,** on the cytosolic side of the plasma membrane. Alfred G. Gilman and Martin Rodbell discovered that when GTP is bound to G$_s$, G$_s$ stimulates the production of cAMP by adenylyl cyclase (see below) in the plasma membrane. The function of G$_s$ as a molecular switch resembles that of another class of G proteins typified by Ras, discussed in Section 12.3 in the context of the insulin receptor. Structurally, G$_s$ and Ras are quite distinct; G proteins of the Ras type are monomers ($M_r$ ~20,000), whereas the G proteins that interact with serpentine

Alfred G. Gilman

Martin Rodbell, 1925–1998

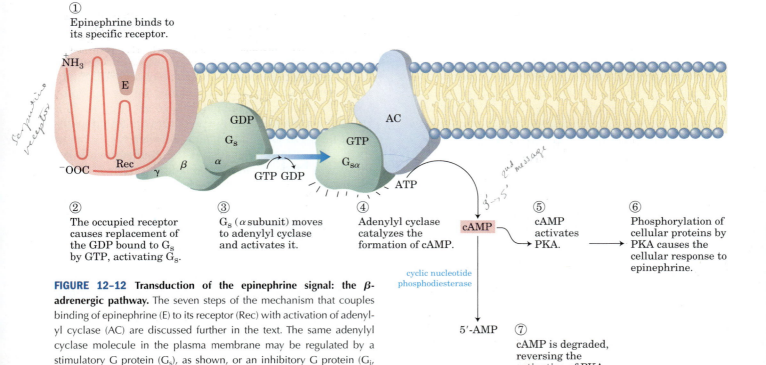

① Epinephrine binds to its specific receptor.

② The occupied receptor causes replacement of the GDP bound to G$_s$ by GTP, activating G$_s$.

③ G$_s$ (α subunit) moves to adenylyl cyclase and activates it.

④ Adenylyl cyclase catalyzes the formation of cAMP.

⑤ cAMP activates PKA.

⑥ Phosphorylation of cellular proteins by PKA causes the cellular response to epinephrine.

⑦ cAMP is degraded, reversing the activation of PKA.

**FIGURE 12–12 Transduction of the epinephrine signal: the β-adrenergic pathway.** The seven steps of the mechanism that couples binding of epinephrine (E) to its receptor (Rec) with activation of adenylyl cyclase (AC) are discussed further in the text. The same adenylyl cyclase molecule in the plasma membrane may be regulated by a stimulatory G protein (G$_s$), as shown, or an inhibitory G protein (G$_i$, not shown). G$_s$ and G$_i$ are under the influence of different hormones. Hormones that induce GTP binding to G$_i$ cause *inhibition* of adenylyl cyclase, resulting in lower cellular [cAMP].

receptors are trimers of three different subunits, $\alpha$ ($M_r$ 43,000), $\beta$ ($M_r$ 37,000), and $\gamma$ ($M_r$ 7,500 to 10,000).

When the nucleotide-binding site of $G_s$ (on the $\alpha$ subunit) is occupied by GTP, $G_s$ is active and can activate adenylyl cyclase (AC in Fig. 12–12); with GDP bound to the site, $G_s$ is inactive. Binding of epinephrine enables the receptor to catalyze displacement of bound GDP by GTP, converting $G_s$ to its active form (step ②). As this occurs, the $\beta$ and $\gamma$ subunits of $G_s$ dissociate from the $\alpha$ subunit, and $G_{s\alpha}$, with its bound GTP, moves in the plane of the membrane from the receptor to a nearby molecule of adenylyl cyclase (step ③). The $G_{s\alpha}$ is held to the membrane by a covalently attached palmitoyl group (see Fig. 11–14).

**Adenylyl cyclase** (Fig. 12–13) is an integral protein of the plasma membrane, with its active site on the cytosolic face. It catalyzes the synthesis of cAMP from ATP:

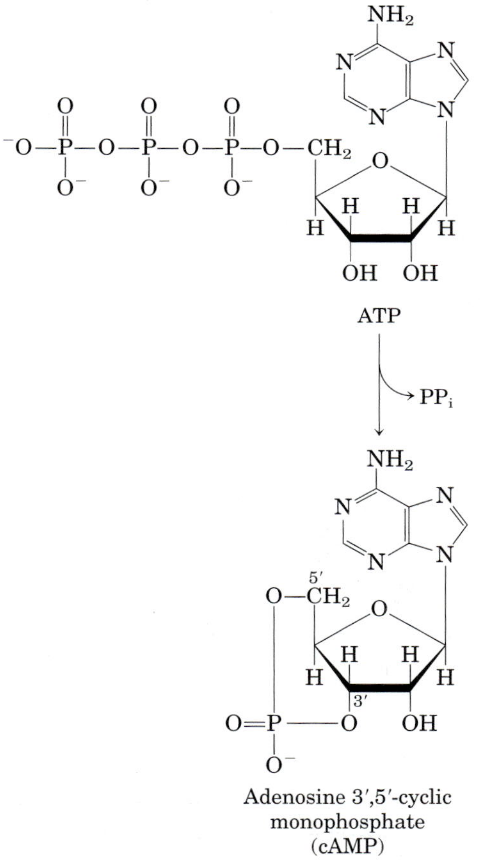

ATP

PP$_i$

Adenosine 3',5'-cyclic monophosphate (cAMP)

The association of active $G_{s\alpha}$ with adenylyl cyclase stimulates the cyclase to catalyze cAMP synthesis (Fig. 12–12, step ④), raising the cytosolic [cAMP]. This stimulation by $G_{s\alpha}$ is self-limiting; $G_{s\alpha}$ is a GTPase that turns itself off by converting its bound GTP to GDP (Fig. 12–14). The now inactive $G_{s\alpha}$ dissociates from adenylyl cyclase, rendering the cyclase inactive. After $G_{s\alpha}$ reassociates with the $\beta$ and $\gamma$ subunits ($G_{s\beta\gamma}$), $G_s$ is again available to interact with a hormone-bound receptor.

**Trimeric G Proteins: Molecular On/Off Switches**

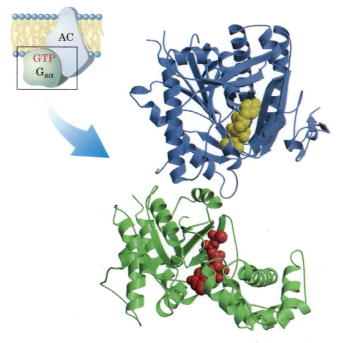

**FIGURE 12–13 Interaction of $G_{s\alpha}$ with adenylyl cyclase.** (PDB ID 1AZS) The soluble catalytic core of the adenylyl cyclase (AC, blue), severed from its membrane anchor, was cocrystallized with $G_{s\alpha}$ (green) to give this crystal structure. The plant terpene forskolin (yellow) is a drug that strongly stimulates the enzyme, and GTP (red) bound to $G_{s\alpha}$ triggers interaction of $G_{s\alpha}$ with adenylyl cyclase.

① $G_s$ with GDP bound is turned off; it cannot activate adenylyl cyclase.

② Contact of $G_s$ with hormone-receptor complex causes displacement of bound GDP by GTP.

③ $G_s$ with GTP bound dissociates into $\alpha$ and $\beta\gamma$ subunits. $G_{s\alpha}$-GTP is turned on; it can activate adenylyl cyclase.

④ GTP bound to $G_{s\alpha}$ is hydrolyzed by the protein's intrinsic GTPase; $G_{s\alpha}$ thereby turns itself off. The inactive $\alpha$ subunit reassociates with the $\beta\gamma$ subunit.

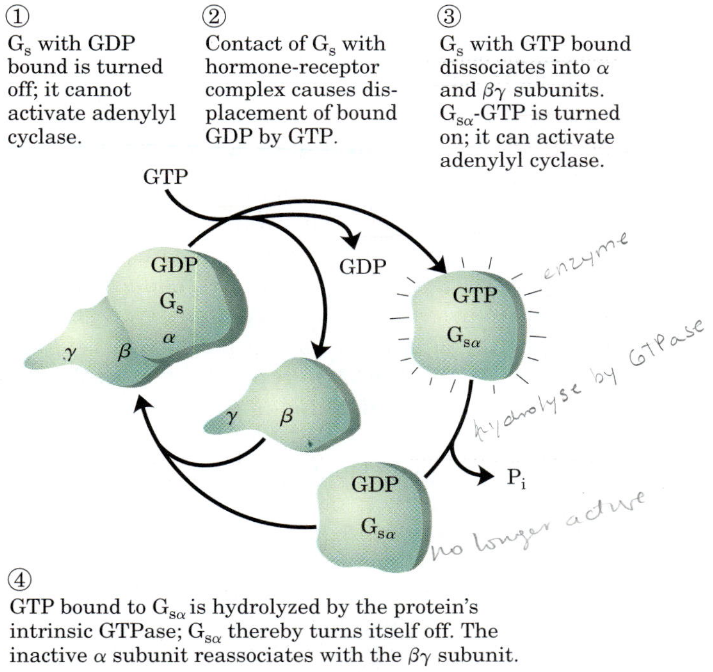

**FIGURE 12–14 Self-inactivation of $G_s$.** The steps are further described in the text. The protein's intrinsic GTPase activity, in many cases stimulated by RGS proteins (regulators of *G* protein signaling), determines how quickly bound GTP is hydrolyzed to GDP and thus how long the G protein remains active.

*enzyme regulation*

*c-AMP allosterically Activates protein Kinase A.*

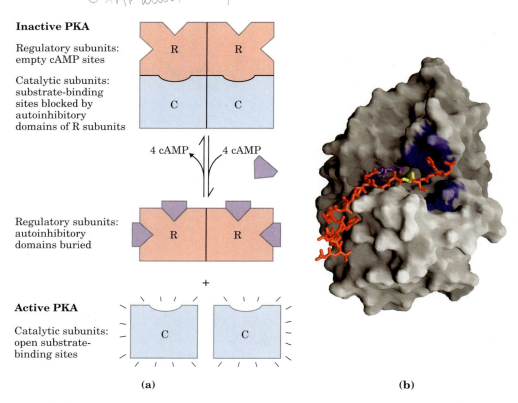

**Inactive PKA**

Regulatory subunits: empty cAMP sites

Catalytic subunits: substrate-binding sites blocked by autoinhibitory domains of R subunits

4 cAMP ⇌ 4 cAMP

Regulatory subunits: autoinhibitory domains buried

**+**

**Active PKA**

Catalytic subunits: open substrate-binding sites

**(a)**                                     **(b)**

**FIGURE 12–15  Activation of cAMP-dependent protein kinase, PKA. (a)** A schematic representation of the inactive $R_2C_2$ tetramer, in which the autoinhibitory domain of a regulatory (R) subunit occupies the substrate-binding site, inhibiting the activity of the catalytic (C) subunit. Cyclic AMP activates PKA by causing dissociation of the C subunits from the inhibitory R subunits. Activated PKA can phosphorylate a variety of protein substrates (Table 12–3) that contain the PKA consensus sequence (X–Arg–(Arg/Lys)–X–(Ser/Thr)–B, where X is any residue and B is any hydrophobic residue), including phosphorylase *b* kinase. **(b)** The substrate-binding region of a catalytic subunit revealed by x-ray crystallography (derived from PDB ID 1JBP). Enzyme side chains known to be critical in substrate binding and specificity are in blue. The peptide substrate (red) lies in a groove in the enzyme surface, with its Ser residue (yellow) positioned in the catalytic site. In the inactive $R_2C_2$ tetramer, the autoinhibitory domain of R lies in this groove, blocking access to the substrate.

One downstream effect of epinephrine is to activate glycogen phosphorylase *b*. This conversion is promoted by the enzyme phosphorylase *b* kinase, which catalyzes the phosphorylation of two specific Ser residues in phosphorylase *b*, converting it to phosphorylase *a* (see Fig. 6–31). Cyclic AMP does not affect phosphorylase *b* kinase directly. Rather, **cAMP-dependent protein kinase,** also called **protein kinase A** or **PKA,** which is allosterically activated by cAMP (Fig. 12–12, step ⑤), catalyzes the phosphorylation of inactive phosphorylase *b* kinase to yield the active form.

The inactive form of PKA contains two catalytic subunits (C) and two regulatory subunits (R) (Fig. 12–15a), which are similar in sequence to the catalytic and regulatory domains of PKG (cGMP-dependent protein kinase). The tetrameric $R_2C_2$ complex is catalytically inactive, because an autoinhibitory domain of each R subunit occupies the substrate-binding site of each C subunit. When cAMP binds to two sites on each R subunit, the R subunits undergo a conformational change and the $R_2C_2$ complex dissociates to yield two free,

catalytically active C subunits. This same basic mechanism—displacement of an autoinhibitory domain—mediates the allosteric activation of many types of protein kinases by their second messengers (as in Figs 12–7 and 12–23, for example).

As indicated in Figure 12–12 (step ⑥), PKA regulates a number of enzymes (Table 12–3). Although the proteins regulated by cAMP-dependent phosphorylation have diverse functions, they share a region of sequence similarity around the Ser or Thr residue that undergoes phosphorylation, a sequence that marks them for regulation by PKA. The catalytic site of PKA (Fig. 12–15b) interacts with several residues near the Thr or Ser residue in the target protein, and these interactions define the substrate specificity. Comparison of the sequences of a number of protein substrates for PKA has yielded the consensus sequence—the specific neighboring residues needed to mark a Ser or Thr residue for phosphorylation (see Table 12–3).

Signal transduction by adenylyl cyclase entails several steps that amplify the original hormone signal (Fig.

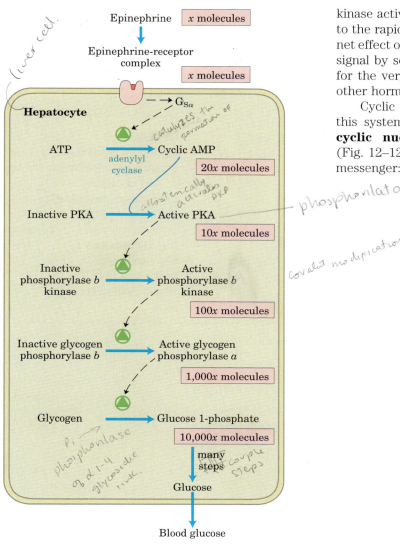

*(liver cell.)*

Epinephrine $x$ molecules

Epinephrine-receptor complex $x$ molecules

**Hepatocyte**

$G_{s\alpha}$

*catalyzes the formation of*

ATP → Cyclic AMP

adenylyl cyclase

20$x$ molecules

*allosterically activates PKA*

Inactive PKA → Active PKA

*phosphorilator*

10$x$ molecules

*covalent modification*

Inactive phosphorylase $b$ kinase → Active phosphorylase $b$ kinase

100$x$ molecules

Inactive glycogen phosphorylase $b$ → Active glycogen phosphorylase $a$

1,000$x$ molecules

Glycogen → Glucose 1-phosphate

*Pi, phosphorilase of α 1-4 glycosidic innk.*

10,000$x$ molecules

many steps *couple steps*

Glucose

Blood glucose

10,000$x$ molecules

**FIGURE 12–16 Epinephrine cascade.** Epinephrine triggers a series of reactions in hepatocytes in which catalysts activate catalysts, resulting in great amplification of the signal. Binding of a small number of molecules of epinephrine to specific β-adrenergic receptors on the cell surface activates adenylyl cyclase. To illustrate amplification, we show 20 molecules of cAMP produced by each molecule of adenylyl cyclase, the 20 cAMP molecules activating 10 molecules of PKA, each PKA molecule activating 10 molecules of the next enzyme (a total of 100), and so forth. These amplifications are probably gross underestimates.

12–16). First, the binding of one hormone molecule to one receptor catalytically activates several $G_s$ molecules. Next, by activating a molecule of adenylyl cyclase, each active $G_{s\alpha}$ molecule stimulates the catalytic synthesis of many molecules of cAMP. The second messenger cAMP now activates PKA, each molecule of which catalyzes the phosphorylation of many molecules of the target protein—phosphorylase $b$ kinase in Figure 12–16. This

kinase activates glycogen phosphorylase $b$, which leads to the rapid mobilization of glucose from glycogen. The net effect of the cascade is amplification of the hormonal signal by several orders of magnitude, which accounts for the very low concentration of epinephrine (or any other hormone) required for hormone activity.

Cyclic AMP, the intracellular second messenger in this system, is short-lived. It is quickly degraded by **cyclic nucleotide phosphodiesterase** to 5′-AMP (Fig. 12–12, step ⑦ ), which is not active as a second messenger:

Cyclic AMP

— $H_2O$  *cyclic nucleotide phosphodiesterase enzyme.*

Adenosine 5′-monophosphate (AMP)

The intracellular signal therefore persists only as long as the hormone receptor remains occupied by epinephrine. Methyl xanthines such as caffeine and theophylline (a component of tea) inhibit the phosphodiesterase, increasing the half-life of cAMP and thereby potentiating agents that act by stimulating adenylyl cyclase.

## The β-Adrenergic Receptor Is Desensitized by Phosphorylation

As noted earlier, signal-transducing systems undergo desensitization when the signal persists. Desensitization of the β-adrenergic receptor is mediated by a protein kinase that phosphorylates the receptor on the intracellular domain that normally interacts with $G_s$ (Fig. 12–17). When the receptor is occupied by epinephrine,

**TABLE 12–3 Some Enzymes and Other Proteins Regulated by cAMP-Dependent Phosphorylation (by PKA)**

| Enzyme/protein | Sequence phosphorylated* | Pathway/process regulated |
|---|---|---|
| Glycogen synthase | RASCTSSS | Glycogen synthesis |
| Phosphorylase *b* kinase | | |
| α subunit | VEFRRLSI | Glycogen breakdown |
| β subunit | RTKRSGSV | |
| Pyruvate kinase (rat liver) | GVLRRASVAZL | Glycolysis |
| Pyruvate dehydrogenase complex (type L) | GYLRRASV | Pyruvate to acetyl-CoA |
| Hormone-sensitive lipase | PMRRSV | Triacylglycerol mobilization and fatty acid oxidation |
| Phosphofructokinase-2/fructose 2,6-bisphosphatase | LQRRRGSSIPQ | Glycolysis/gluconeogenesis |
| Tyrosine hydroxylase | FIGRRQSL | Synthesis of L-DOPA, dopamine, norepinephrine, and epinephrine |
| Histone H1 | AKRKASGPPVS | DNA condensation |
| Histone H2B | KKAKASRKESYSVYVYK | DNA condensation |
| Cardiac phospholamban (cardiac pump regulator) | AIRRAST | Intracellular $[Ca^{2+}]$ |
| Protein phosphatase-1 inhibitor-1 | IRRRPTP | Protein dephosphorylation |
| PKA consensus sequence† | XR(R/K)X(S/T)B | Many |

*The phosphorylated S or T residue is shown in red. All residues are given as their one-letter abbreviations (see Table 3–1).
†X is any amino acid; B is any hydrophobic amino acid.

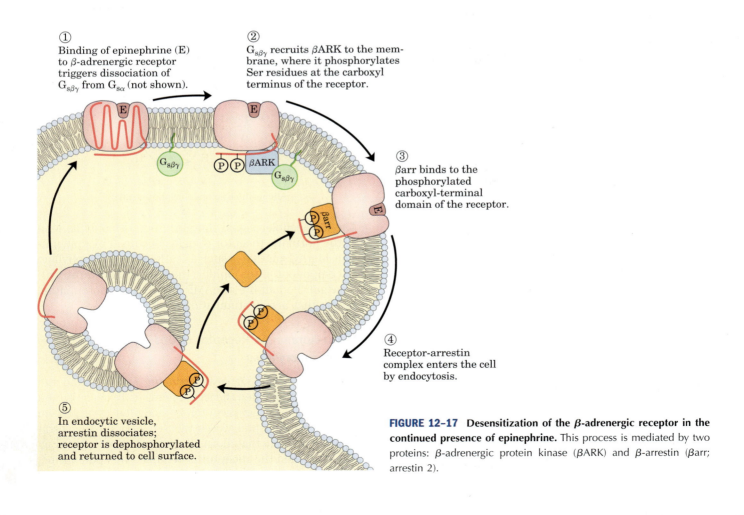

① Binding of epinephrine (E) to β-adrenergic receptor triggers dissociation of $G_{s\beta\gamma}$ from $G_{s\alpha}$ (not shown).

② $G_{s\beta\gamma}$ recruits βARK to the membrane, where it phosphorylates Ser residues at the carboxyl terminus of the receptor.

③ βarr binds to the phosphorylated carboxyl-terminal domain of the receptor.

④ Receptor-arrestin complex enters the cell by endocytosis.

⑤ In endocytic vesicle, arrestin dissociates; receptor is dephosphorylated and returned to cell surface.

**FIGURE 12–17 Desensitization of the β-adrenergic receptor in the continued presence of epinephrine.** This process is mediated by two proteins: β-adrenergic protein kinase (βARK) and β-arrestin (βarr; arrestin 2).

**β-adrenergic receptor kinase (βARK)** phosphorylates Ser residues near the carboxyl terminus of the receptor. Normally located in the cytosol, βARK is drawn to the plasma membrane by its association with th $G_{s\beta\gamma}$ subunits and is thus positioned to phosphorylate the receptor. The phosphorylation creates a binding site for the protein **β-arrestin (βarr)**, also called arrestin 2, and binding of β-arrestin effectively prevents interaction between the receptor and the G protein. The binding of β-arrestin also facilitates receptor sequestration, the removal of receptors from the plasma membrane by endocytosis into small intracellular vesicles. Receptors in the endocytic vesicles are dephosphorylated, then returned to the plasma membrane, completing the circuit and resensitizing the system to epinephrine. β-Adrenergic receptor kinase is a member of a family of **G protein–coupled receptor kinases (GRKs),** all of which phosphorylate serpentine receptors on their carboxyl-terminal cytosolic domains and play roles similar to that of βARK in desensitization and resensitization of their receptors. At least five different GRKs and four different arrestins are encoded in the human genome; each GRK is capable of desensitizing a subset of the serpentine receptors, and each arrestin can interact with many different types of phosphorylated receptors.

While preventing the signal from a serpentine receptor from reaching its associated G protein, arrestins can also initiate a second signaling cascade, by acting as **scaffold proteins** that bring together several protein kinases that function in a cascade. For example, the β-arrestin associated with the serpentine receptor for angiotensin, a potent regulator of blood pressure, binds the three protein kinases Raf-1, MEK1, and ERK (Fig. 12–18), serving as a scaffold that facilitates any signaling process, such as insulin signaling (Fig. 12–6), that requires these three protein kinases to interact. This is one of many known examples of cross-talk between systems triggered by different ligands (angiotensin and insulin, in this case).

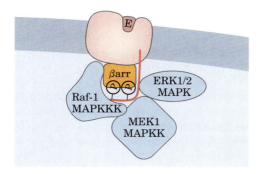

**FIGURE 12–18** β-Arrestin uncouples the serpentine receptor from its G protein and brings together the three enzymes of the MAPK cascade. The effect is that one stimulus triggers two distinct response pathways: the path activated by the G protein and the MAPK cascade.

## Cyclic AMP Acts as a Second Messenger for a Number of Regulatory Molecules

Epinephrine is only one of many hormones, growth factors, and other regulatory molecules that act by changing the intracellular [cAMP] and thus the activity of PKA (Table 12–4). For example, glucagon binds to its receptors in the plasma membrane of adipocytes, activating (via a $G_s$ protein) adenylyl cyclase. PKA, stimulated by the resulting rise in [cAMP], phosphorylates and activates two proteins critical to the conversion of stored fat to fatty acids (perilipin and hormone-sensitive triacylglycerol lipase; see Fig. 17–3), leading to the mobilization of fatty acids. Similarly, the peptide hormone ACTH (adrenocorticotropic hormone, also called corticotropin), produced by the anterior pituitary, binds to specific receptors in the adrenal cortex, activating adenylyl cyclase and raising the intracellular [cAMP]. PKA then phosphorylates and activates several of the enzymes required for the synthesis of cortisol and other steroid hormones. The catalytic subunit of PKA can also move into the nucleus, where it phosphorylates a protein that alters the expression of specific genes.

Some hormones act by *inhibiting* adenylyl cyclase, *lowering* cAMP levels, and *suppressing* protein phosphorylation. For example, the binding of somatostatin to its receptor leads to activation of an **inhibitory G protein,** or **$G_i$,** structurally homologous to $G_s$, that inhibits adenylyl cyclase and lowers [cAMP]. Somatostatin therefore counterbalances the effects of glucagon. In adipose tissue, prostaglandin $E_1$ (PGE$_1$; see Fig. 10–18b) inhibits adenylyl cyclase, thus lowering [cAMP] and slowing the

**TABLE 12–4  Some Signals That Use cAMP as Second Messenger**

Corticotropin (ACTH)
Corticotropin-releasing hormone (CRH)
Dopamine [D$_1$, D$_2$]*
Epinephrine (β-adrenergic)
Follicle-stimulating hormone (FSH)
Glucagon
Histamine [H$_2$]*
Luteinizing hormone (LH)
Melanocyte-stimulating hormone (MSH)
Odorants (many)
Parathyroid hormone
Prostaglandins E$_1$, E$_2$ (PGE$_1$, PGE$_2$)
Serotonin [5-HT-1a, 5-HT-2]*
Somatostatin
Tastants (sweet, bitter)
Thyroid-stimulating hormone (TSH)

*Receptor subtypes in square brackets. Subtypes may have different transduction mechanisms. For example, serotonin is detected in some tissues by receptor subtypes 5-HT-1a and 5-HT-1b, which act through adenylyl cyclase and cAMP, and in other tissues by receptor subtype 5-HT-1c, acting through the phospholipase C-IP$_3$ mechanism (see Table 12–5).

mobilization of lipid reserves triggered by epinephrine and glucagon. In certain other tissues PGE$_1$ stimulates cAMP synthesis, because its receptors are coupled to adenylyl cyclase through a stimulatory G protein, G$_s$. In tissues with $\alpha_2$-adrenergic receptors, epinephrine lowers [cAMP], because the $\alpha_2$ receptors are coupled to adenylyl cyclase through an inhibitory G protein, G$_i$. In short, an extracellular signal such as epinephrine or PGE$_1$ can have quite different effects on different tissues or cell types, depending on three factors: the type of receptor in each tissue, the type of G protein (G$_s$ or G$_i$) with which the receptor is coupled, and the set of PKA target enzymes in the cells.

A fourth factor that explains how so many signals can be mediated by a single second messenger (cAMP) is the confinement of the signaling process to a specific region of the cell by scaffold proteins. AKAPs (*A k*inase *a*nchoring *p*roteins) are bivalent; one part binds to the R subunit of PKA, and another to a specific structure within the cell, confining the PKA to the vicinity of that structure. For example, specific AKAPs bind PKA to microtubules, actin filaments, Ca$^{2+}$ channels, mitochondria, and the nucleus. Different types of cells have different AKAPs, so cAMP might stimulate phosphorylation of mitochondrial proteins in one cell and phosphorylation of actin filaments in another. In studies of the intracellular localization of biochemical changes, biochemistry meets cell biology, and techniques that cross this boundary become invaluable (Box 12–2).

## Two Second Messengers Are Derived from Phosphatidylinositols

A second class of serpentine receptors are coupled through a G protein to a plasma membrane **phospholipase C (PLC)** that is specific for the plasma membrane lipid phosphatidylinositol 4,5-bisphosphate (see Fig. 10–15). This hormone-sensitive enzyme catalyzes the formation of two potent second messengers: **diacylglycerol** and **inositol 1,4,5-trisphosphate**, or **IP$_3$** (not to be confused with PIP$_3$, p. 431).

Inositol 1,4,5-trisphosphate (IP$_3$)

When a hormone of this class (Table 12–5) binds its specific receptor in the plasma membrane (Fig. 12–19, step ①), the receptor-hormone complex catalyzes GTP-GDP exchange on an associated G protein, **G$_q$**

(step ②), activating it exactly as the $\beta$-adrenergic receptor activates G$_s$ (Fig. 12–12). The activated G$_q$ in turn activates a specific membrane-bound PLC (step ③), which catalyzes the production of the two second messengers diacylglycerol and IP$_3$ by hydrolysis of phosphatidylinositol 4,5-bisphosphate in the plasma membrane (step ④).

Inositol trisphosphate, a water-soluble compound, diffuses from the plasma membrane to the endoplasmic reticulum, where it binds to specific IP$_3$ receptors and causes Ca$^{2+}$ channels within the ER to open. Sequestered Ca$^{2+}$ is thus released into the cytosol (step ⑤), and the cytosolic [Ca$^{2+}$] rises sharply to about $10^{-6}$ M. One effect of elevated [Ca$^{2+}$] is the activation of **protein kinase C (PKC)**. Diacylglycerol cooperates with Ca$^{2+}$ in activating PKC, thus also acting as a second messenger (step ⑥). PKC phosphorylates Ser or Thr residues of specific target proteins, changing their catalytic activities (step ⑦). There are a number of isozymes of PKC, each with a characteristic tissue distribution, target protein specificity, and role.

The action of a group of compounds known as **tumor promoters** is attributable to their effects on PKC. The best understood of these are the phorbol esters, synthetic compounds that are potent activators of PKC. They apparently mimic cellular diacylglycerol as second messengers, but unlike naturally occurring diacylglycerols they are not rapidly metabolized. By continuously activating PKC, these synthetic tumor promoters interfere with the normal regulation of cell growth and division (discussed in Section 12.10). ■

Myristoylphorbol acetate (a phorbol ester)

## Calcium Is a Second Messenger in Many Signal Transductions

In many cells that respond to extracellular signals, Ca$^{2+}$ serves as a second messenger that triggers intracellular responses, such as exocytosis in neurons and endocrine cells, contraction in muscle, and cytoskeletal rearrangement during amoeboid movement. Normally, cytosolic [Ca$^{2+}$] is kept very low ($<10^{-7}$ M) by the action of Ca$^{2+}$ pumps in the ER, mitochondria, and plasma membrane. Hormonal, neural, or other stimuli cause either an influx of Ca$^{2+}$ into the cell through specific

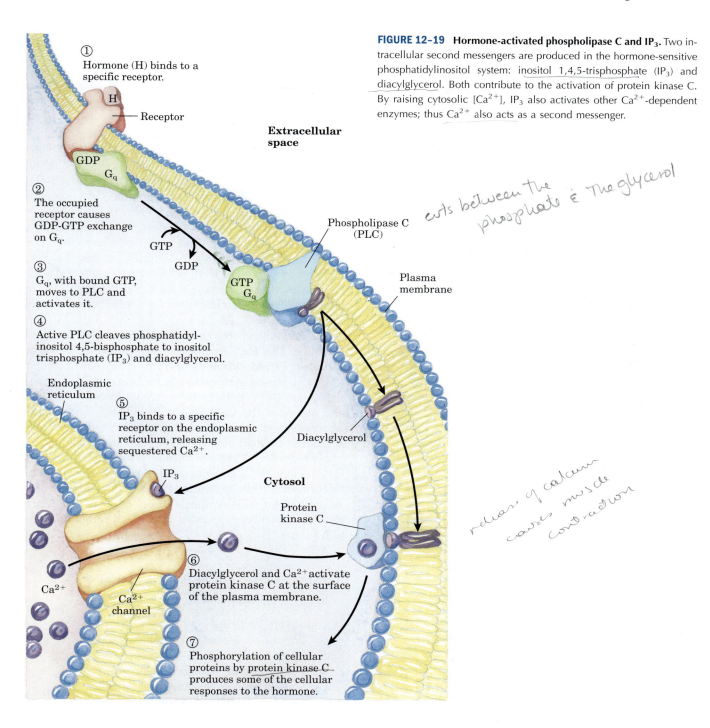

**FIGURE 12–19 Hormone-activated phospholipase C and IP₃.** Two intracellular second messengers are produced in the hormone-sensitive phosphatidylinositol system: inositol 1,4,5-trisphosphate (IP₃) and diacylglycerol. Both contribute to the activation of protein kinase C. By raising cytosolic [Ca²⁺], IP₃ also activates other Ca²⁺-dependent enzymes; thus Ca²⁺ also acts as a second messenger.

① Hormone (H) binds to a specific receptor.

② The occupied receptor causes GDP-GTP exchange on Gq.

③ Gq, with bound GTP, moves to PLC and activates it.

④ Active PLC cleaves phosphatidyl-inositol 4,5-bisphosphate to inositol trisphosphate (IP₃) and diacylglycerol.

⑤ IP₃ binds to a specific receptor on the endoplasmic reticulum, releasing sequestered Ca²⁺.

⑥ Diacylglycerol and Ca²⁺ activate protein kinase C at the surface of the plasma membrane.

⑦ Phosphorylation of cellular proteins by protein kinase C produces some of the cellular responses to the hormone.

*Handwritten annotations:* cuts between the phosphate & the glycerol; release of calcium causes muscle contraction

**TABLE 12–5 Some Signals That Act through Phospholipase C and IP₃**

| | | |
|---|---|---|
| Acetylcholine [muscarinic M₁] | Gastrin-releasing peptide | Platelet-derived growth factor (PDGF) |
| α₁-Adrenergic agonists | Glutamate | Serotonin [5-HT-1c]* |
| Angiogenin | Gonadotropin-releasing hormone (GRH) | Thyrotropin-releasing hormone (TRH) |
| Angiotensin II | Histamine [H₁]* | Vasopressin |
| ATP [P₂ₓ and P₂ᵧ]* | Light (*Drosophila*) | |
| Auxin | Oxytocin | |

*Receptor subtypes are in square brackets; see footnote to Table 12–4.

Ca²⁺ channels in the plasma membrane or the release of sequestered Ca²⁺ from the ER or mitochondria, in either case raising the cytosolic [Ca²⁺] and triggering a cellular response.

Very commonly, [Ca²⁺] does not simply rise and then decrease, but rather oscillates with a period of a few seconds (Fig. 12–20), even when the extracellular concentration of hormone remains constant. The mechanism underlying [Ca²⁺] oscillations presumably entails feedback regulation by Ca²⁺ of either the phospholipase

that generates IP₃ or the ion channel that regulates Ca²⁺ release from the ER, or both. Whatever the mechanism, the effect is that one kind of signal (hormone concentration, for example) is converted into another (frequency and amplitude of intracellular [Ca²⁺] "spikes").

Changes in intracellular [Ca²⁺] are detected by Ca²⁺-binding proteins that regulate a variety of Ca²⁺-dependent enzymes. **Calmodulin (CaM)** ($M_r$ 17,000) is an acidic protein with four high-affinity Ca²⁺-binding sites. When intracellular [Ca²⁺] rises to about $10^{-6}$ M (1 μM), the binding of Ca²⁺ to calmodulin drives a conformational change in the protein (Fig. 12–21). Calmodulin associates with a variety of proteins and, in its Ca²⁺-bound state, modulates their activities. Calmodulin is a member of a family of Ca²⁺-binding proteins that also includes troponin (p. 185), which triggers skeletal muscle contraction in response to increased [Ca²⁺]. This family shares a characteristic Ca²⁺-binding structure, the EF hand (Fig. 12–21c).

Calmodulin is also an integral subunit of a family of enzymes, the **Ca²⁺/calmodulin-dependent protein kinases (CaM kinases I–IV).** When intracellular [Ca²⁺] increases in response to some stimulus, calmodulin binds Ca²⁺, undergoes a change in conformation, and activates the CaM kinase. The kinase then phosphorylates a number of target enzymes, regulating their activities. Calmodulin is also a regulatory subunit of phosphorylase *b* kinase of muscle, which is activated by Ca²⁺. Thus Ca²⁺ triggers ATP-requiring muscle contractions while also activating glycogen breakdown, providing fuel for ATP synthesis. Many other enzymes are also known to be modulated by Ca²⁺ through calmodulin (Table 12–6).

**(a)**

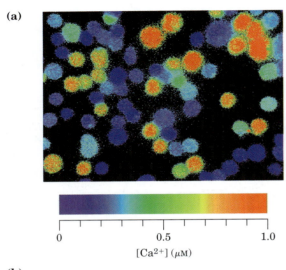

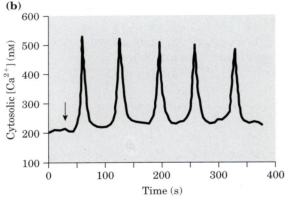

**(b)**

**FIGURE 12–20 Triggering of oscillations in intracellular [Ca²⁺] by extracellular signals. (a)** A dye (fura) that undergoes fluorescence changes when it binds Ca²⁺ is allowed to diffuse into cells, and its instantaneous light output is measured by fluorescence microscopy. Fluorescence intensity is represented by color; the color scale relates intensity of color to [Ca²⁺], allowing determination of the absolute [Ca²⁺]. In this case, thymocytes (cells of the thymus) have been stimulated with extracellular ATP, which raises their internal [Ca²⁺]. The cells are heterogeneous in their responses; some have high intracellular [Ca²⁺] (red), others much lower (blue). **(b)** When such a probe is used to measure [Ca²⁺] in a single hepatocyte, we observe that the agonist norepinephrine (added at the arrow) causes oscillations of [Ca²⁺] from 200 to 500 nM. Similar oscillations are induced in other cell types by other extracellular signals.

**TABLE 12–6    Some Proteins Regulated by Ca²⁺ and Calmodulin**

Adenylyl cyclase (brain)
Ca²⁺/calmodulin-dependent protein kinases (CaM kinases I to IV)
Ca²⁺-dependent Na⁺ channel (*Paramecium*)
Ca²⁺-release channel of sarcoplasmic reticulum
Calcineurin (phosphoprotein phosphatase 2B)
cAMP phosphodiesterase
cAMP-gated olfactory channel
cGMP-gated Na⁺, Ca²⁺ channels (rod and cone cells)
Glutamate decarboxylase
Myosin light chain kinases
NAD⁺ kinase
Nitric oxide synthase
Phosphoinositide 3-kinase
Plasma membrane Ca²⁺ ATPase (Ca²⁺ pump)
RNA helicase (p68)

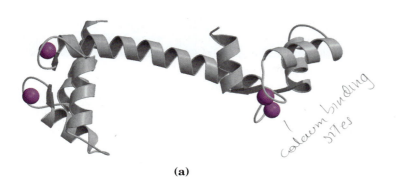

**(a)**

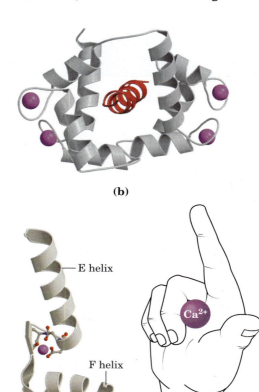

**(b)**

**FIGURE 12–21 Calmodulin.** This is the protein mediator of many $Ca^{2+}$-stimulated enzymatic reactions. Calmodulin has four high-affinity $Ca^{2+}$-binding sites ($K_d \approx 0.1$ to $1\ \mu M$). **(a)** A ribbon model of the crystal structure of calmodulin (PDB ID 1CLL). The four $Ca^{2+}$-binding sites are occupied by $Ca^{2+}$ (purple). The amino-terminal domain is on the left; the carboxyl-terminal domain on the right. **(b)** Calmodulin associated with a helical domain (red) of one of the many enzymes it regulates, calmodulin-dependent protein kinase II (PDB ID 1CDL). Notice that the long central $\alpha$ helix visible in **(a)** has bent back on itself in binding to the helical substrate domain. The central helix is clearly more flexible in solution than in the crystal. **(c)** Each of the four $Ca^{2+}$-binding sites occurs in a helix-loop-helix motif called the EF hand, also found in many other $Ca^{2+}$-binding proteins.

— E helix

— F helix

EF hand

**(c)**

## SUMMARY 12.4 G Protein–Coupled Receptors and Second Messengers

- A large family of plasma membrane receptors with seven transmembrane segments act through heterotrimeric G proteins. On ligand binding, these receptors catalyze the exchange of GTP for GDP bound to an associated G protein, forcing dissociation of the $\alpha$ subunit of the G protein. This subunit stimulates or inhibits the activity of a nearby membrane-bound enzyme, changing the level of its second messenger product.

- The $\beta$-adrenergic receptor binds epinephrine, then through a stimulatory G protein, $G_s$, activates adenylyl cyclase in the plasma membrane. The cAMP produced by adenylyl cyclase is an intracellular second messenger that stimulates cAMP-dependent protein kinase, which mediates the effects of epinephrine by phosphorylating key proteins, changing their enzymatic activities or structural features.

- The cascade of events in which a single molecule of hormone activates a catalyst that in turn activates another catalyst, and so on, results in large signal amplification; this is characteristic of most hormone-activated systems.

- Some receptors stimulate adenylyl cyclase through $G_s$; others inhibit it through $G_i$. Thus cellular [cAMP] reflects the integrated input of two (or more) signals.

- Cyclic AMP is eventually eliminated by cAMP phosphodiesterase, and $G_s$ turns itself off by hydrolysis of its bound GTP to GDP. When the epinephrine signal persists, $\beta$-adrenergic receptor–specific protein kinase and arrestin 2 temporarily desensitize the receptor and cause it to move into intracellular vesicles. In some cases, arrestin also acts as a scaffold protein, bringing together protein components of a signaling pathway such as the MAPK cascade.

- Some serpentine receptors are coupled to a plasma membrane phospholipase C that cleaves $PIP_2$ to diacylglycerol and $IP_3$. By opening $Ca^{2+}$ channels in the endoplasmic reticulum, $IP_3$ raises cytosolic [$Ca^{2+}$]. Diacylglycerol and $Ca^{2+}$ act together to activate protein kinase C, which phosphorylates and changes the activity of specific cellular proteins. Cellular [$Ca^{2+}$] also regulates a number of other enzymes, often through calmodulin.

## BOX 12–2    WORKING IN BIOCHEMISTRY

### FRET: Biochemistry Visualized in a Living Cell

Fluorescent probes are commonly used to detect rapid biochemical changes in single living cells. They can be designed to give an essentially instantaneous report (within nanoseconds) on the changes in intracellular concentration of a second messenger or in the activity of a protein kinase. Furthermore, fluorescence microscopy has sufficient resolution to reveal where in the cell such changes are occurring. In one widely used procedure, the fluorescent probes are derived from a naturally occurring fluorescent protein, the green fluorescent protein (GFP) of the jellyfish *Aequorea victoria* (Fig. 1).

When excited by absorption of a photon of light, GFP emits a photon (that is, it fluoresces) in the green region of the spectrum. GFP is an 11-stranded β barrel, and the light-absorbing/emitting center of the protein (its chromophore) comprises the tripeptide Ser$^{65}$–Tyr$^{66}$–Gly$^{67}$, located within the barrel (Fig. 2). Variants of this protein, with different fluorescence spectra, can be produced by genetic engineering of the GFP gene. For example, in the yellow fluorescent protein (YFP), Ala$^{206}$ in GFP is replaced by a Lys residue, changing the wavelength of light absorption

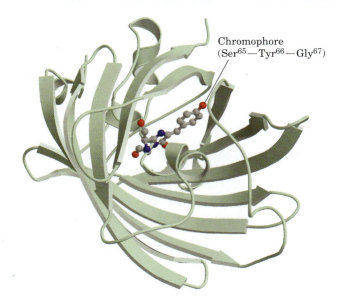

**FIGURE 2** Green fluorescent protein (GFP), with the fluorescent chromophore shown in ball-and-stick form (derived from PDB ID 1GFL).

and fluorescence. Other variants of GFP fluoresce blue (BFP) or cyan (CFP) light, and a related protein (mRFP1) fluoresces red light (Fig. 3). GFP and its variants are compact structures that retain their ability to fold into their native β-barrel conformation even when fused with another protein. Investigators are using these fluorescent hybrid proteins as spectroscopic rulers to measure distances between interacting components within a cell.

**FIGURE 1** *Aequorea victoria,* a jellyfish abundant in Puget Sound, Washington State.

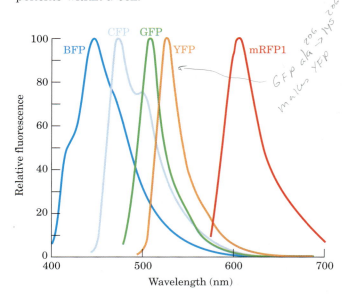

**FIGURE 3** Emission spectra of GFP variants.

An excited fluorescent molecule such as GFP or YFP can dispose of the energy from the absorbed photon in either of two ways: (1) by fluorescence, emitting a photon of slightly longer wavelength (lower energy) than the exciting light, or (2) by nonradiative **fluorescence resonance energy transfer (FRET),** in which the energy of the excited molecule (the donor) passes directly to a nearby molecule (the acceptor) *without emission of a photon,* exciting the acceptor (Fig. 4). The acceptor can now decay to its ground state by fluorescence; the emitted photon has a longer wavelength (lower energy) than both the original exciting light and the fluorescence emission of the donor. This second mode of decay (FRET) is possible only when donor and acceptor are close to each other (within 1 to 50 Å); the efficiency of FRET is inversely proportional to the *sixth power* of the distance between donor and acceptor. Thus very small changes in the distance between donor and acceptor register as very large changes in FRET, measured as the fluorescence of the acceptor molecule when the donor is excited. With sufficiently sensitive light detectors, this fluorescence signal can be located to specific regions of a single, living cell.

FRET has been used to measure [cAMP] in living cells. The gene for GFP is fused with that for the regulatory subunit (R) of cAMP-dependent protein kinase, and the gene for BFP is fused with that for the

catalytic subunit (C) (Fig. 5). When these two hybrid proteins are expressed in a cell, BFP (donor; excitation at 380 nm, emission at 460 nm) and GFP (acceptor; excitation at 475 nm, emission at 545 nm) in the inactive PKA ($R_2C_2$ tetramer) are close enough to undergo FRET. Wherever in the cell [cAMP] increases, the $R_2C_2$ complex dissociates into $R_2$ and 2C and the FRET signal is lost, because donor and acceptor are now too far apart for efficient FRET. Viewed in the fluorescence microscope, the region of higher [cAMP] has a minimal GFP signal and higher BFP signal. Measuring the ratio of emission at 460 nm and 545 nm gives a sensitive measure of the change in [cAMP]. By determining this ratio for all regions of the cell, the investigator can generate a false color image of the

*(continued on next page)*

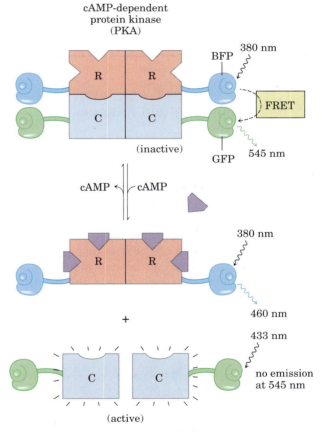

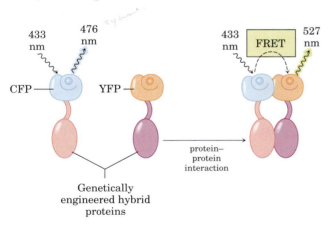

**FIGURE 4**  When the donor protein (CFP) is excited with monochromatic light of wavelength 433 nm, it emits fluorescent light at 476 nm (left). When the (red) protein fused with CFP interacts with the (purple) protein fused with YFP, that interaction brings CFP and YFP close enough to allow fluorescence resonance energy transfer (FRET) between them. Now, when CFP absorbs light of 433 nm, instead of fluorescing at 476 nm, it transfers energy directly to YFP, which then fluoresces at its characteristic emission wavelength, 527 nm. The ratio of light emission at 527 and 476 nm is therefore a measure of the interaction of the red and purple protein.

**FIGURE 5**  **Measuring [cAMP] with FRET.** Gene fusion creates hybrid proteins that exhibit FRET when the PKA regulatory and catalytic subunits are associated (low [cAMP]). When [cAMP] rises, the subunits dissociate, and FRET ceases. The ratio of emission at 460 nm (dissociated) and 545 nm (complexed) thus offers a sensitive measure of [cAMP].

BOX 12–2    WORKING IN BIOCHEMISTRY (continued from previous page)

cell in which the ratio, or relative [cAMP], is represented by the intensity of the color. Images recorded at timed intervals reveal changes in [cAMP] over time.

A variation of this technology has been used to measure the activity of PKA in a living cell (Fig. 6). Researchers create a phosphorylation target for PKA by producing a hybrid protein containing four elements: YFP (acceptor); a short peptide with a Ser residue surrounded by the consensus sequence for PKA; a ℗–Ser-binding domain (called 14-3-3); and CFP (donor). When the Ser residue is not phosphorylated, 14-3-3 has no affinity for the Ser residue and the hybrid protein exists in an extended form, with the donor and acceptor too far apart to generate a FRET signal. Wherever PKA is active in the cell, it phosphorylates the Ser residue of the hybrid protein, and 14-3-3 binds to the ℗–Ser. In doing so, it draws YFP and CFP together and a FRET signal is detected with the fluorescence microscope, revealing the presence of active PKA.

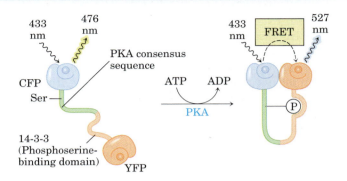

FIGURE 6 **Measuring the activity of PKA with FRET.** An engineered protein links YFP and CFP via a peptide that contains a Ser residue surrounded by the consensus sequence for phosphorylation by PKA, and the 14-3-3 phosphoserine binding domain. Active PKA phosphorylates the Ser residue, which docks with the 14-3-3 binding domain, bringing the fluorescence proteins close enough to allow FRET to occur, revealing the presence of active PKA.

## 12.5 Multivalent Scaffold Proteins and Membrane Rafts

About 10% of the 30,000 to 35,000 genes in the human genome encode signaling proteins—receptors, G proteins, enzymes that generate second messengers, protein kinases (>500), proteins involved in desensitization, and ion channels. Not every signaling protein is expressed in a given cell type, but most cells doubtless contain many such proteins. How does one protein find another in a signaling pathway, and how are their interactions regulated? As is becoming clear, the reversible phosphorylation of Tyr, Ser, and Thr residues in signaling proteins creates *docking sites* for other proteins, and many signaling proteins are *multivalent* in that they can interact with several different proteins simultaneously to form multiprotein signaling complexes. In this section we present a few examples to illustrate the general principles of protein interactions in signaling.

### Protein Modules Bind Phosphorylated Tyr, Ser, or Thr Residues in Partner Proteins

We have seen that the protein Grb2 in the insulin signaling pathway (Fig. 12–6) binds through its SH2 domain to other proteins that contain exposed ℗–Tyr residues. The human genome encodes at least 87 SH2-containing proteins, many already known to participate in signaling. The ℗–Tyr residue is bound in a deep

pocket in an SH2 domain, with each of its phosphate oxygens participating in hydrogen-bonding or electrostatic interactions; the positive charges on two Arg residues figure prominently in the binding. Subtle differences in the structure of SH2 domains in different proteins account for the specificities of their interactions with various ℗–Tyr-containing proteins. The three to five residues on the carboxyl-terminal side of the ℗–Tyr residue are critical in determining the specificity of interactions with SH2 domains (Fig. 12–22).

**PTB domains** (phosphotyrosine-binding domains) also bind ℗–Tyr in partner proteins, but their critical sequences and three-dimensional structures distinguish them from SH2 domains. The human genome encodes 24 proteins that contain PTB domains, including IRS-1, which we have already met in its role as a scaffold protein in insulin-signal transduction (Fig. 12–6).

Many of the signaling protein kinases, including PKA, PKC, PKG, and members of the MAPK cascade, phosphorylate Ser or Thr residues in their target proteins, which in some cases acquire the ability to interact with partner proteins through the phosphorylated residue, triggering a downstream process. An alphabet soup of domains that bind ℗–Ser or ℗–Thr residues has been identified, and more are sure to be found. Each domain favors a certain sequence around the phosphorylated residue, so the domains represent families of highly specific recognition sites, able to bind to a specific subset of phosphorylated proteins.

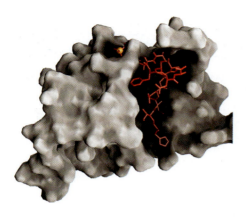

**FIGURE 12–22 Structure of an SH2 domain and its interaction with a ⓟ–Tyr residue in a partner protein.** (PDB ID 1SHC) The SH2 domain is shown as a gray surface contour representation. The phosphorus of the phosphate group in the interacting ⓟ–Tyr is visible as an orange sphere; most of the residue is obscured in this view. The next few residues toward the carboxyl end of the partner protein are shown in red. The SH2 domain interacts with ⓟ–Tyr (which, as the phosphorylated residue, is assigned the index position 0) and also with the next three residues toward the carboxyl terminus (designated +1, +2, +3). The residues important in the ⓟ–Tyr residue are conserved in all SH2 domains. Some SH2 domains (Src, Fyn, Hck, Nck) favor negatively charged residues in the +1 and +2 positions; others (PLC-γ1, SHP-2) have a long hydrophobic groove that selects for aliphatic residues in positions +1 to +5. These differences define subclasses of SH2 domains that have different partner specificities.

In some cases, the domain-binding partner is internal. Phosphorylation of some protein kinases inhibits their activity by favoring the interaction of an SH2 domain with a ⓟ–Tyr in another domain of the same enzyme. For example, the soluble protein Tyr kinase Src, when phosphorylated on a critical Tyr residue, is rendered inactive as an SH2 domain needed to bind to the substrate protein instead binds to an internal ⓟ–Tyr

(Fig. 12–23). Glycogen synthase kinase 3 (GSK3) is inactive when phosphorylated on a Ser residue in its autoinhibitory domain (Fig. 12–23b). Dephosphorylation of that domain frees the enzyme to bind and phosphorylate its target proteins. Similarly, the polar head group of the phospholipid $PIP_3$, protruding from the inner leaflet of the plasma membrane, provides points of attachment for proteins that contain SH3 and other domains.

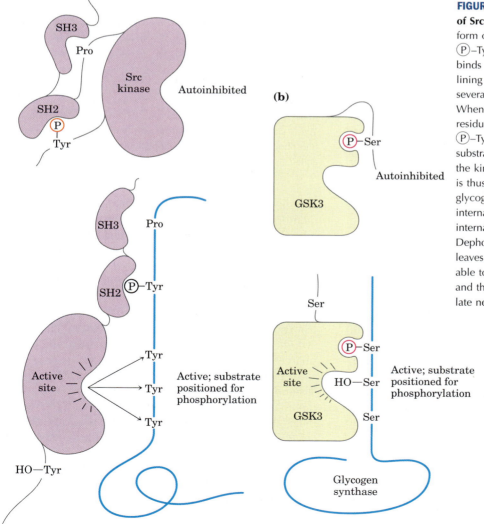

(a)

(b)

**FIGURE 12–23 Mechanism of autoinhibition of Src Tyr kinase and GSK3. (a)** In the active form of Src kinase, an SH2 domain binds a ⓟ–Tyr in the substrate, and an SH3 domain binds a proline-rich region of the substrate, lining up the active site of the kinase with several target Tyr residues in the substrate. When Src is phosphorylated on a specific Tyr residue, the SH2 domain binds to the internal ⓟ–Tyr instead of to the ⓟ–Tyr of the substrate, preventing productive binding of the kinase to its protein substrate; the enzyme is thus autoinhibited. **(b)** In the autoinhibited glycogen synthase kinase 3 (GSK3), an internal ⓟ–Ser residue is bound to an internal ⓟ–Ser-binding domain (top). Dephosphorylation of this internal Ser residue leaves the ⓟ–Ser-binding site of GSK3 available to bind ⓟ–Ser in a protein substrate, and thus to position the kinase to phosphorylate neighboring Ser residues (bottom).

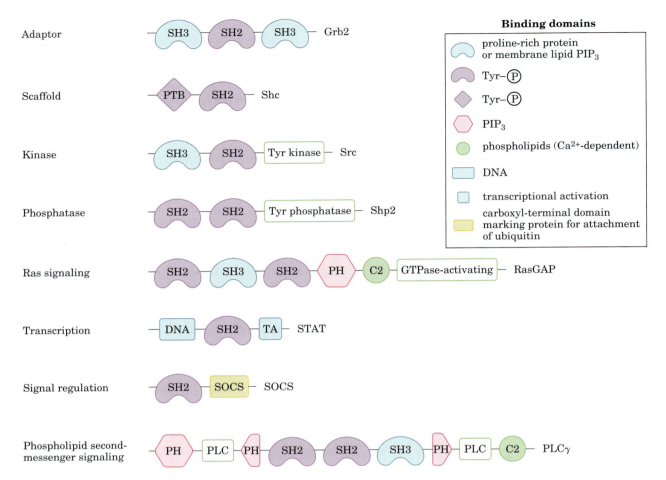

**FIGURE 12–24 Some binding modules of signaling proteins.** Each protein is represented by a line (with the amino terminus to the left); symbols indicate the location of conserved binding domains (with specificities as listed in the key; PH denotes plextrin homology; other abbreviations explained in the text); green boxes indicate catalytic ac- tivities. The name of each protein is given at its carboxyl-terminal end. These signaling proteins interact with phosphorylated proteins or phospholipids in many permutations and combinations to form inte- grated signaling complexes.

Most of the proteins involved in signaling at the plasma membrane have one or more protein- or phospholipid-binding domains; many have three or more, and thus are multivalent in their interactions with other signaling proteins. Figure 12–24 shows a few of the many multivalent proteins known to participate in signaling.

A remarkable picture of signaling pathways has emerged from studies of many signaling proteins and the multiple binding domains they contain (Fig. 12–25). An initial signal results in phosphorylation of the receptor or a target protein, triggering the assembly of large multiprotein complexes, held together on scaffolds made from adaptor proteins with multivalent binding capacities. Some of these complexes have several protein kinases that activate each other in turn, producing a cascade of phosphorylation and a great amplification of the initial signal. Animal cells also have phosphotyrosine phosphatases (PTPases), which remove the phosphate from $(P)$–Tyr residues, reversing the effect of phosphorylation. Some of these phosphatases are receptorlike membrane proteins, presumably controlled by extracellular factors not yet identified; other PTPases are soluble and contain SH2 domains. In addition, animal cells have protein phosphoserine and phosphothreonine phosphatases, which reverse the effects of Ser- and Thr-specific protein kinases. We can see, then, that signaling occurs in protein circuits, effectively hard-wired from signal receptor to response effector and able to be switched off instantly by the hydrolysis of a single phosphate ester bond.

The multivalency of signaling proteins allows for the assembly of many different combinations of signaling modules, each combination presumably suited to particular signals, cell types, and metabolic circumstances. The large variety of protein kinases and of phosphoprotein-binding domains, each with its own specificity (the consensus sequence required in its substrate), provides for many permutations and combinations and many different signaling circuits of extraordinary complexity. And given the variety of specific phosphatases that reverse the

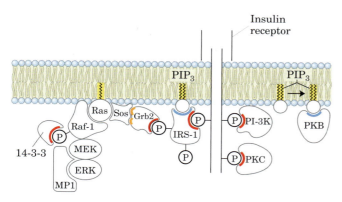

**FIGURE 12–25 Insulin-induced formation of supramolecular signaling complexes.** The binding of insulin to its receptor sets off a series of events that lead eventually to the formation of membrane-associated complexes involving the 12 signaling proteins shown here, as well as others. Phosphorylation of Tyr residues in the insulin receptor initiates complex formation, and dephosphorylation of any of the phosphoproteins breaks the circuit. Four general types of interaction hold the complex together: the binding of a protein to a second phosphoprotein through SH2 or PTB domains in the first (red); the binding of SH3 domains in the first with proline-rich domains in the second (orange); the binding of PH domains in one protein to the phospholipid PIP₃ in the plasma membrane (blue); or the association of a protein (RAS) with the plasma membrane through a lipid covalently bound to the protein (yellow). Two proteins shown here are not described in the text: 14-3-3, which binds a (P)–Ser in Raf and mediates its interaction with MEK; and MP1, a scaffold protein that cements the links between Raf, MEK, and ERK.

the action of protein kinases, some under specific types of external control, a cell can quickly "disconnect" the entire protein circuitry of a signaling pathway. Together, these mechanisms confer a huge capacity for cellular regulation in response to signals of many types.

## Membrane Rafts and Caveolae May Segregate Signaling Proteins

Membrane rafts are regions of the membrane bilayer enriched in sphingolipids, sterols, and certain proteins, including many attached to the bilayer by GPI anchors (Chapter 11). Some receptor Tyr kinases, such as the receptors for epidermal growth factor (EGF) and platelet-derived growth factor (PDGF), appear to be localized in rafts; other signaling proteins, such as the small G protein Ras (which is prenylated) and the heterotrimeric G protein Gₛ (also prenylated, on the α and γ subunits), are not. Growing evidence suggests that this sequestration of signaling proteins is functionally significant. When cholesterol is removed from rafts by treatment with cyclodextrin (which binds cholesterol and removes it from membranes), the rafts are disrupted and a number of signaling pathways become defective.

How might localization in rafts influence signaling through a receptor? There are several possibilities. If a receptor Tyr kinase in a raft is phosphorylated, and the

phosphotyrosine phosphatase that reverses this phosphorylation is in another raft, then dephosphorylation of the Tyr kinase will be slowed or prevented. If two signaling proteins must interact during transduction of a signal, the probability of encounters between these proteins is greatly enhanced if both are in the same raft. Interactions between scaffold proteins might be strong enough to pull into a raft a signaling protein not normally located there, or strong enough to pull receptors out of a raft. For example, the EGF receptor in isolated fibroblasts is normally concentrated in the specialized rafts called caveolae (see Fig. 11–21), but treatment with EGF causes the receptor to leave the raft. This migration depends on the receptor's protein kinase activity; mutant receptors lacking this activity remain in the rafts during treatment with EGF. Caveolin, an integral membrane protein localized in caveolae, is phosphorylated on Tyr residues in response to insulin, and phosphorylation may allow the now-activated EGF receptor to draw its binding partners into the raft. Finally, another example of the clustering of signaling proteins in rafts is the β-adrenergic receptor. This receptor is segregated in membrane rafts that also contain the G proteins, adenylyl cyclase, PKA, and a specific protein phosphatase, PP2, providing a highly integrated signaling unit. Spatial segregation of signaling proteins in rafts adds yet another dimension to the already complex processes initiated by extracellular signals.

## SUMMARY 12.5 Multivalent Scaffold Proteins and Membrane Rafts

- Many signaling proteins have domains that bind phosphorylated Tyr, Ser, or Thr residues in other proteins; the binding specificity for each domain is determined by sequences that adjoin the phosphorylated residue.

- SH2 and PTB domains bind to proteins containing (P)–Tyr residues; other domains bind (P)–Ser and (P)–Thr residues in various contexts.

- Plextrin homology domains bind the membrane phospholipid PIP₃.

- Many signaling proteins are multivalent, with several different binding modules. By combining the substrate specificities of various protein kinases with the specificities of domains that bind phosphorylated Ser, Thr, or Tyr residues, and with phosphatases that can rapidly inactivate a pathway, cells create a large number of multiprotein signaling complexes.

- Membrane rafts and caveolae sequester groups of signaling proteins in small regions of the plasma membrane, enhancing their interactions and making signaling more efficient.

## 12.6 Signaling in Microorganisms and Plants

Much of what we have said here about signaling relates to mammalian tissues or cultured cells from such tissues. Bacteria, eukaryotic microorganisms, and vascular plants must also respond to a variety of external signals, such as $O_2$, nutrients, light, noxious chemicals, and so on. We turn here to a brief consideration of the kinds of signaling machinery used by microorganisms and plants.

### Bacterial Signaling Entails Phosphorylation in a Two-Component System

*E. coli* responds to a number of nutrients in its environment, including sugars and amino acids, by swimming toward them, propelled by one or a few flagella. A family of membrane proteins have binding domains on the outside of the plasma membrane to which specific **attractants** (sugars or amino acids) bind (Fig. 12–26). Ligand binding causes another domain on the inside of the plasma membrane to phosphorylate itself on a His residue. This first component of the **two-component system,** the **receptor His kinase,** then catalyzes the transfer of the phosphoryl group from the His residue to an Asp residue on a second, soluble protein, the **response regulator;** this phosphoprotein moves to the base of the flagellum, carrying the signal from the membrane receptor. The flagellum is driven by a rotary motor that can propel the cell through its medium or cause it to stall, depending on the direction of the motor's rotation. Information from the receptor allows the cell to determine whether it is moving toward or away from the source of the attractant. If its motion is toward the attractant, the response regulator signals the cell to continue in a straight line; if away from it, the cell tumbles momentarily, acquiring a new direction. Repetition of this behavior results in a random path, biased toward movement in the direction of increasing attractant concentration.

*E. coli* detects not only sugars and amino acids but also $O_2$, extremes of temperature, and other environmental factors, using this basic two-component system. Two-component systems have been detected in many other bacteria, including gram-positive and gram-negative eubacteria and archaebacteria, as well as in protists and fungi. Clearly this signaling mechanism developed early in the course of cellular evolution and has been conserved.

Various signaling systems used by animal cells also have analogs in the prokaryotes. As the full genomic sequences of more, and more diverse, bacteria become known, researchers have discovered genes that encode proteins similar to protein Ser or Thr kinases, Ras-like proteins regulated by GTP binding, and proteins with SH3 domains. Receptor Tyr kinases have not been detected in bacteria, but (P)–Tyr residues do occur in some bacterial proteins, so there must be an enzyme that phosphorylates Tyr residues.

### Signaling Systems of Plants Have Some of the Same Components Used by Microbes and Mammals

Like animals, vascular plants must have a means of communication between tissues to coordinate and direct growth and development; to adapt to conditions of $O_2$, nutrients, light, and temperature; and to warn of the presence of noxious chemicals and damaging pathogens (Fig. 12–27). At least a billion years of evolution have passed since the plant and animal branches of the eukaryotes diverged, which is reflected in the differences in signaling mechanisms: some plant mechanisms are conserved—that is, are similar to those in animals (protein kinases, scaffold proteins, cyclic nucleotides, electrogenic ion pumps, and gated ion channels); some are similar to bacterial two-component systems; and some are unique to plants (light-sensing mechanisms, for ex-

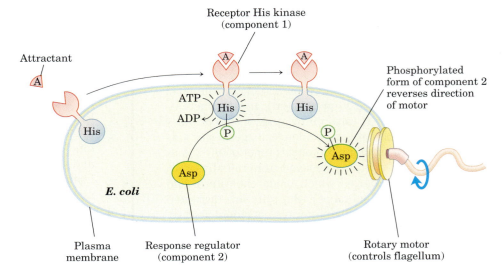

**FIGURE 12–26 The two-component signaling mechanism in bacterial chemotaxis.** When an attractant ligand (A) binds to the receptor domain of the membrane-bound receptor, a protein His kinase in the cytosolic domain (component 1) is activated and autophosphorylates on a His residue. This phosphoryl group is then transferred to an Asp residue on component 2 (in some cases a separate protein; in others, another domain of the receptor protein). After phosphorylation on Asp, component 2 moves to the base of the flagellum, where it determines the direction of rotation of the flagellar motor.

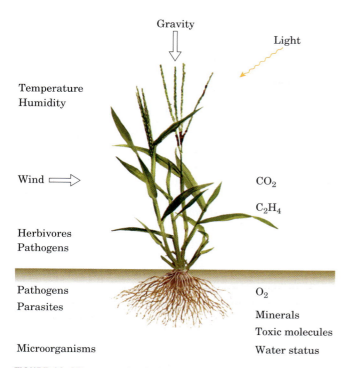

Gravity

Light

Temperature
Humidity

Wind

$CO_2$

$C_2H_4$

Herbivores
Pathogens

Pathogens
Parasites

$O_2$

Minerals

Toxic molecules

Microorganisms

Water status

**FIGURE 12–27** Some stimuli that produce responses in plants.

kinases that phosphorylate Ser or Thr residues; a variety of protein phosphatases; scaffold proteins that bring other proteins together in signaling complexes; enzymes for the synthesis and degradation of cyclic nucleotides; and 100 or more ion channels, including about 20 gated by cyclic nucleotides. Inositol phospholipids are present, as are kinases that interconvert them by phosphorylation of inositol head groups.

However, some types of signaling proteins common in animal tissues are not present in plants, or are represented by only a few genes. Cyclic nucleotide–dependent protein kinases (PKA and PKG) appear to be absent, for example. Heterotrimeric G proteins and protein Tyr kinase genes are much less prominent in the plant genome, and serpentine (G protein–coupled) receptors, the largest gene family in the human genome (>1,000 genes), are very sparsely represented in the plant genome. DNA-binding nuclear steroid receptors are certainly not prominent, and may be absent from plants. Although plants lack the most widely conserved light-sensing mechanism present in animals (rhodopsin, with retinal as pigment), they have a rich collection of other light-detecting mechanisms not found in animal tissues—phytochromes and cryptochromes, for example (Chapter 19).

The kinds of compounds that elicit signals in plants are similar to certain signaling molecules in mammals (Fig. 12–28). Instead of prostaglandins, plants have jasmonate; instead of steroid hormones, brassinosteroids.

ample) (Table 12–7). The genome of the widely studied plant *Arabidopsis thaliana,* for example, encodes about 1,000 protein Ser/Thr kinases, including about 60 MAPKs and nearly 400 membrane-associated receptor

**TABLE 12–7  Signaling Components Present in Mammals, Plants, or Bacteria**

| Signaling protein | Mammals | Plants | Bacteria |
|---|---|---|---|
| Ion channels | + | + | + |
| Electrogenic ion pumps | + | + | + |
| Two-component His kinases | + | + | + |
| Adenylyl cyclase | + | + | + |
| Guanylyl cyclase | + | + | ? |
| Receptor protein kinases (Ser/Thr) | + | + | ? |
| $Ca^{2+}$ as second messenger | + | + | ? |
| $Ca^{2+}$ channels | + | + | ? |
| Calmodulin, CaM-binding protein | + | + | − |
| MAPK cascade | + | + | − |
| Cyclic nucleotide–gated channels | + | + | − |
| $IP_3$-gated $Ca^{2+}$ channels | + | + | − |
| Phosphatidylinositol kinases | + | + | − |
| Serpentine receptors | + | +/− | + |
| Trimeric G proteins | + | +/− | − |
| PI-specific phospholipase C | + | ? | − |
| Tyrosine kinase receptors | + | ? | − |
| SH2 domains | + | ? | ? |
| Nuclear steroid receptors | + | − | − |
| Protein kinase A | + | − | − |
| Protein kinase G | + | − | − |

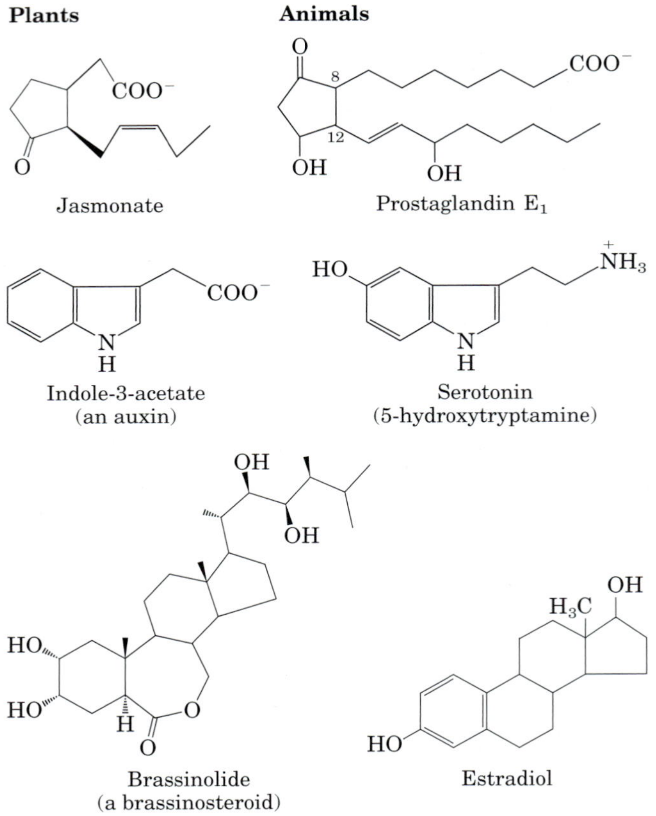

**Plants**                          **Animals**

Jasmonate

Prostaglandin E$_1$

Indole-3-acetate
(an auxin)

Serotonin
(5-hydroxytryptamine)

Brassinolide
(a brassinosteroid)

Estradiol

**FIGURE 12–28** **Structural similarities between plant and animal signals.** The plant signals jasmonate, indole-3-acetate, and brassinolide resemble the mammalian signals prostaglandin E$_1$, serotonin, and estradiol.

About 100 different small peptides serve as plant signals, and both plants and animals use compounds derived from aromatic amino acids as signals.

## Plants Detect Ethylene through a Two-Component System and a MAPK Cascade

The receptors for the plant hormone ethylene (CH$_2$=CH$_2$) are related in primary sequence to the receptor His kinases of the bacterial two-component systems and probably evolved from them; the cyanobacterial origin of chloroplasts (see Fig. 1–36) may have brought the bacterial signaling genes into the plant cell nucleus. In *Arabidopsis*, the two-component signaling system is contained within a single protein. The first downstream component affected by ethylene signaling is a protein Ser/Thr kinase (CTR-1; Fig. 12–29) with sequence homology to Raf, the protein kinase that begins the MAPK cascade in the mammalian response to insulin (see the comparison in Fig. 12–30). In plants, in the absence of ethylene, the CTR-1 kinase is active and *inhibits* the MAPK cascade, preventing transcription of ethylene-responsive genes. Exposure to ethylene *inac-*

*tivates* the CTR-1 kinase, thereby activating the MAPK cascade that leads to activation of the transcription factor EIN3. Active EIN3 stimulates the synthesis of a second transcription factor (ERF1), which in turn activates transcription of a number of ethylene-responsive genes; the gene products affect processes ranging from seedling development to fruit ripening.

Although apparently derived from the bacterial two-component signaling system, the ethylene system in *Arabidopsis* is different in that the His kinase activity that defines component 1 in bacteria is not essential to the transduction in *Arabidopsis*. The genome of the cyanobacterium *Anabaena* encodes proteins with both an ethylene-binding domain and an active His kinase domain. It seems likely that in the course of evolution, the ethylene receptor of vascular plants was derived from that of a cyanobacterial endosymbiont, and that the bacterial His kinase became a Ser/Thr kinase in the plant.

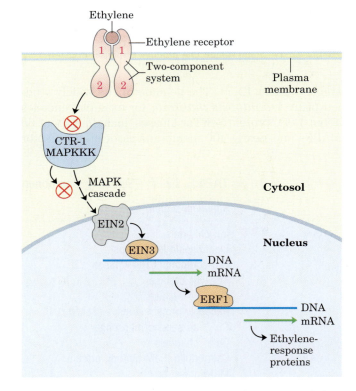

**FIGURE 12–29** **Transduction mechanism for detection of ethylene by plants.** The ethylene receptor in the plasma membrane (pink) is a two-component system contained within a single protein, which has both a receptor domain (component 1) and a response regulator domain (component 2). The receptor controls (in ways we do not yet understand) the activity of CTR1, a protein kinase similar to MAPKKKs and therefore presumed to be part of a MAPK cascade. CTR1 is a negative regulator of the ethylene response; when CTR1 is *inactive*, the ethylene signal passes through the gene product EIN2 (thought to be a nuclear envelope protein), which somehow causes increased synthesis of ERF1, a transcription factor; ERF1 in turn stimulates expression of proteins specific to the ethylene response.

### Receptorlike Protein Kinases Transduce Signals from Peptides and Brassinosteroids

One common motif in plant signaling involves **receptorlike kinases (RLKs)** with a single helical segment in the plasma membrane that connects a receptor domain on the outside of the membrane with a protein Ser/Thr kinase on the cytoplasmic side. This type of receptor participates in the defense mechanism triggered by infection with a bacterial pathogen (Fig. 12–30a). The signal to turn on the genes needed for defense against infection is a peptide (flg22) released by breakdown of flagellin, the major protein of the bacterial flagellum. Binding of flg22 to the FLS2 receptor of *Arabidopsis* induces receptor dimerization and autophosphorylation on Ser and Thr residues, and the downstream effect is activation of a MAPK cascade like that described above for insulin action (Fig. 12–6). The final kinase in this cascade activates a specific transcription factor, triggering synthesis of the proteins that defend against the bacterial infection. The steps between receptor phosphorylation and the MAPK cascade are not yet known. A phosphoprotein phosphatase (KAPP) associates with the active receptor protein and inactivates it by dephosphorylation to end the response.

The MAPK cascade in the plant's defense against bacterial pathogens is remarkably similar to the innate immune response triggered by bacterial lipopolysaccharide and mediated by the Toll-like receptors in mammals (Fig. 12–30b). Other membrane receptors use similar mechanisms to activate a MAPK cascade, ultimately activating transcription factors and turning on the genes essential to the defense response.

Most of the several hundred RLKs in plants are presumed to act in similar ways: ligand binding induces dimerization and autophosphorylation, and the

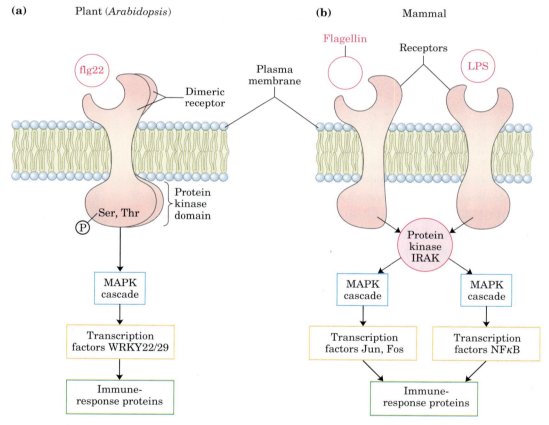

**FIGURE 12–30  Similarities between the signaling pathways that trigger immune responses in plants and animals. (a)** In the plant *Arabidopsis thaliana*, the peptide flg22, derived from the flagella of a bacterial pathogen, binds to its receptor in the plasma membrane, causing the receptors to form dimers and triggering autophosphorylation of the cytosolic protein kinase domain on a Ser or Thr residue (*not* a Tyr). Autophosphorylation activates the receptor protein kinase, which then phosphorylates downstream proteins. The activated receptor also activates (by means unknown) a MAPKKK. The resulting kinase cascade leads to phosphorylation of a nuclear protein that normally inhibits the transcription factors WRKY22 and 29, triggering proteolytic degradation of the inhibitor and freeing the transcription factors to stimulate gene expression related to the immune response. **(b)** In mammals, the toxic bacterial lipopolysaccharide (LPS; see Fig. 7–32) is detected by plasma membrane receptors that associate with and activate a soluble protein kinase (IRAK). The major flagellar protein of pathogenic bacteria acts through a similar receptor to activate IRAK. Then IRAK initiates two distinct MAPK cascades that end in the nucleus, causing the synthesis of proteins needed in the immune response. Jun, Fos, and NFκB are transcription factors.

activated receptor kinase triggers downstream responses by phosphorylating key proteins at Ser or Thr residues. The ligands for these kinases have been identified in only a few cases: brassinosteroids, the peptide trigger for the self-incompatibility response that prevents self-pollination, and CLV1 peptide, a factor involved in regulating the fate of stem cells (undifferentiated cells) in plant development.

### SUMMARY 12.6 Signaling in Microorganisms and Plants

- Bacteria and unicellular eukaryotes have a variety of sensory systems that allow them to sample and respond to their environment. In the two-component system, a receptor His kinase senses the signal and autophosphorylates a His residue, then phosphorylates the response regulator on an Asp residue.

- Plants respond to many environmental stimuli, and employ hormones and growth factors to coordinate the development and metabolic activities of their tissues. Plant genomes encode hundreds of signaling proteins, including some very similar to those used in signal transductions in mammalian cells.

- Two-component signaling mechanisms common in bacteria have been acquired in altered forms by plants. Cyanobacteria use typical two-component systems in the detection of chemical signals and light; plants use related proteins—which autophosphorylate on Ser/Thr, not His, residues—to detect ethylene.

- Plant receptorlike kinases (RLKs), with an extracellular ligand-binding domain, a single transmembrane segment, and a cytosolic protein kinase domain, participate in detecting a wide variety of stimuli, including peptides that originate from pathogens, brassinosteroid hormones, self-incompatible pollen, and developmental signals. RLKs autophosphorylate Ser/Thr residues, then activate downstream proteins that in some cases are MAPK cascades. The end result of many such signals is increased transcription of specific genes.

## 12.7 Sensory Transduction in Vision, Olfaction, and Gustation

The detection of light, smells, and tastes (vision, olfaction, and gustation, respectively) in animals is accomplished by specialized sensory neurons that use signal-transduction mechanisms fundamentally similar to those that detect hormones, neurotransmitters, and growth factors. An initial sensory signal is amplified greatly by mechanisms that include gated ion channels and intracellular second messengers; the system adapts to continued stimulation by changing its sensitivity to the stimulus (desensitization); and sensory input from several receptors is integrated before the final signal goes to the brain.

### Light Hyperpolarizes Rod and Cone Cells of the Vertebrate Eye

In the vertebrate eye, light entering through the pupil is focused on a highly organized collection of light-sensitive neurons (Fig. 12–31). The light-sensing cells are of two types: **rods** (about $10^9$ per retina), which sense low levels of light but cannot discriminate colors, and **cones** (about $3 \times 10^6$ per retina), which are less sensitive to light but can discriminate colors. Both cell types are long, narrow, specialized sensory neurons with two distinct cellular compartments: the **outer segment** contains dozens of membranous disks loaded with the membrane protein rhodopsin, and the **inner segment** contains the nucleus and many mitochondria, which produce the ATP essential to phototransduction.

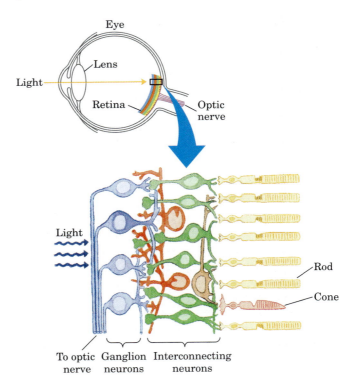

**FIGURE 12–31 Light reception in the vertebrate eye.** The lens of the eye focuses light on the retina, which is composed of layers of neurons. The primary photosensory neurons are rod cells (yellow), which are responsible for high-resolution and night vision, and cone cells of three subtypes (pink), which initiate color vision. The rods and cones form synapses with several ranks of interconnecting neurons that convey and integrate the electrical signals. The signals eventually pass from ganglion neurons through the optic nerve to the brain.

Like other neurons, rods and cones have a transmembrane electrical potential ($V_m$), produced by the electrogenic pumping of the $Na^+K^+$ ATPase in the plasma membrane of the inner segment (Fig. 12–32). Also contributing to the membrane potential is an ion channel in the outer segment that permits passage of either $Na^+$ or $Ca^{2+}$ and is gated (opened) by cGMP. In the dark, rod cells contain enough cGMP to keep this channel open. The membrane potential is therefore determined by the net difference between the $Na^+$ and $K^+$ pumped by the inner segment (which polarizes the membrane) and the influx of $Na^+$ through the ion channels of the outer segment (which tends to depolarize the membrane).

The essence of signaling in the retinal rod or cone cell is a light-induced decrease in the concentration of cGMP, which causes the cGMP-gated ion channel to close. The plasma membrane then becomes hyperpolarized by the $Na^+K^+$ ATPase. Rod and cone cells synapse with interconnecting neurons (Fig. 12–31) that carry information about the electrical activity to the ganglion neurons near the inner surface of the retina. The ganglion neurons integrate the output from many rod or cone cells and send the resulting signal through the optic nerve to the visual cortex of the brain.

## Light Triggers Conformational Changes in the Receptor Rhodopsin

Visual transduction begins when light falls on rhodopsin, many thousands of molecules of which are present in each disk of the outer segments of rod and cone cells. **Rhodopsin** ($M_r$ 40,000) is an integral protein with seven membrane-spanning $\alpha$ helices (Fig. 12–33), the characteristic serpentine architecture. The amino-terminal domain projects into the disk, and the carboxyl-terminal domain faces the cytosol of the outer segment. The light-absorbing pigment (chromophore) **11-*cis*-retinal** is covalently attached to **opsin,** the protein component of rhodopsin, through a Schiff base to a Lys residue. The retinal lies near the middle of the bilayer (Fig. 12–33), oriented with its long axis approximately in the plane of the membrane. When a photon is absorbed by the retinal component of rhodopsin, the energy causes a photochemical change; 11-*cis*-retinal is converted to **all-*trans*-retinal** (see Figs 1–18b, 10–21). This change in the structure of the chromophore causes conformational changes in the rhodopsin molecule—the first stage in visual transduction.

## Excited Rhodopsin Acts through the G Protein Transducin to Reduce the cGMP Concentration

In its excited conformation, rhodopsin interacts with a second protein, **transducin,** which hovers nearby on the cytoplasmic face of the disk membrane (Fig. 12–33). Transducin (T) belongs to the same family of heterotrimeric GTP-binding proteins as $G_s$ and $G_i$. Although

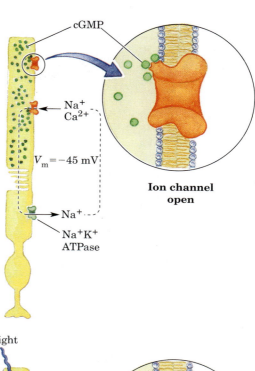

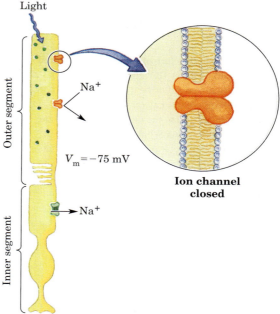

**FIGURE 12–32 Light-induced hyperpolarization of rod cells.** The rod cell consists of an outer segment that is filled with stacks of membranous disks (not shown) containing the photoreceptor rhodopsin and an inner segment that contains the nucleus and other organelles. Cones have a similar structure. ATP in the inner segment powers the $Na^+K^+$ ATPase, which creates a transmembrane electrical potential by pumping 3 $Na^+$ out for every 2 $K^+$ pumped in. The membrane potential is reduced by the flow of $Na^+$ and $Ca^{2+}$ into the cell through cGMP-gated cation channels in the plasma membrane of the outer segment. When rhodopsin absorbs light, it triggers degradation of cGMP (green dots) in the outer segment, causing closure of the cation channel. Without cation influx through this channel, the cell becomes hyperpolarized. This electrical signal is passed to the brain through the ranks of neurons shown in Figure 12–31.

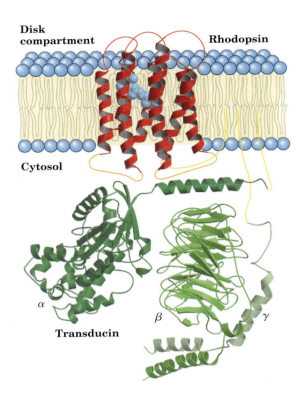

**Disk compartment**

**Rhodopsin**

**Cytosol**

α

**Transducin**

β

γ

**FIGURE 12–33** **Likely structure of rhodopsin complexed with the G protein transducin.** (PDB ID 1BAC) Rhodopsin (red) has seven transmembrane helices embedded in the disk membranes of rod outer segments and is oriented with its carboxyl terminus on the cytosolic side and its amino terminus inside the disk. The chromophore 11-*cis* retinal (blue), attached through a Schiff base linkage to Lys[256] of the seventh helix, lies near the center of the bilayer. (This location is similar to that of the epinephrine-binding site in the β-adrenergic receptor.) Several Ser and Thr residues near the carboxyl terminus are substrates for phosphorylations that are part of the desensitization mechanism for rhodopsin. Cytosolic loops that interact with the G protein transducin are shown in orange; their exact positions are not yet known. The three subunits of transducin (green) are shown in their likely arrangement. Rhodopsin is palmitoylated at its carboxyl terminus, and both the α and γ subunits of transducin have attached lipids (yellow) that assist in anchoring them to the membrane.

specialized for visual transduction, transducin shares many functional features with $G_s$ and $G_i$. It can bind either GDP or GTP. In the dark, GDP is bound, all three subunits of the protein ($T_\alpha$, $T_\beta$, and $T_\gamma$) remain together, and no signal is sent. When rhodopsin is excited by light, it interacts with transducin, catalyzing the replacement of bound GDP by GTP from the cytosol (Fig. 12–34, steps ① and ②). Transducin then dissociates into $T_\alpha$ and $T_{\beta\gamma}$, and the $T_\alpha$-GTP carries the signal from the excited receptor to the next element in the transduction pathway, cGMP phosphodiesterase (PDE); this enzyme converts cGMP to 5′-GMP (steps ③ and ④). Note that this is not the same cyclic nucleotide phosphodiesterase that hydrolyzes cAMP to terminate the β-adrenergic response. The cGMP-specific PDE is unique to the visual cells of the retina.

PDE is an integral protein with its active site on the cytoplasmic side of the disk membrane. In the dark, a tightly bound inhibitory subunit very effectively suppresses PDE activity. When $T_\alpha$-GTP encounters PDE, the inhibitory subunit is released, and the enzyme's activity immediately increases by several orders of magnitude. Each molecule of active PDE degrades many molecules of cGMP to the biologically inactive 5′-GMP, lowering [cGMP] in the outer segment within a fraction of a second. At the new, lower [cGMP], the cGMP-gated ion channels close, blocking reentry of $Na^+$ and $Ca^{2+}$ into the outer segment and hyperpolarizing the membrane of the rod or cone cell (step ⑤). Through this process, the initial stimulus—a photon—changes the $V_m$ of the cell.

### Amplification of the Visual Signal Occurs in the Rod and Cone Cells

Several steps in the visual-transduction process result in great amplification of the signal. Each excited rhodopsin molecule activates at least 500 molecules of transducin, each of which can activate a molecule of PDE. This phosphodiesterase has a remarkably high turnover number, each activated molecule hydrolyzing 4,200 molecules of cGMP per second. The binding of cGMP to cGMP-gated ion channels is cooperative (at least three cGMP molecules must be bound to open one channel), and a relatively small change in [cGMP] therefore registers as a large change in ion conductance. The result of these amplifications is exquisite sensitivity to light. Absorption of a single photon closes 1,000 or more ion channels and changes the cell's membrane potential by about 1 mV.

### The Visual Signal Is Quickly Terminated

As your eyes move across this line, the images of the first words disappear rapidly—before you see the next series of words. In that short interval, a great deal of biochemistry has taken place. Very shortly after illumination of the rod or cone cells stops, the photosensory system shuts off. The α subunit of transducin (with bound GTP) has intrinsic GTPase activity. Within milliseconds after the decrease in light intensity, GTP is hydrolyzed and $T_\alpha$ reassociates with $T_{\beta\gamma}$. The inhibitory subunit of PDE, which had been bound to $T_\alpha$-GTP, is released and reassociates with PDE, strongly inhibiting that enzyme. To return [cGMP] to its "dark" level, the enzyme guanylyl cyclase converts GTP to cGMP (step ⑦ in Fig. 12–34) in a reaction that is inhibited by high [$Ca^{2+}$] (>100 nM). Calcium levels drop during illumination, because the steady-state [$Ca^{2+}$] in the outer segment is the result of outward pumping of $Ca^{2+}$ through the $Na^+$-$Ca^{2+}$ exchanger of the plasma membrane and inward movement of $Ca^{2+}$ through open cGMP-gated channels. In the dark, this produces a [$Ca^{2+}$] of about 500 nM—enough to inhibit cGMP synthesis. After brief illumination, $Ca^{2+}$ entry slows and [$Ca^{2+}$] declines (step

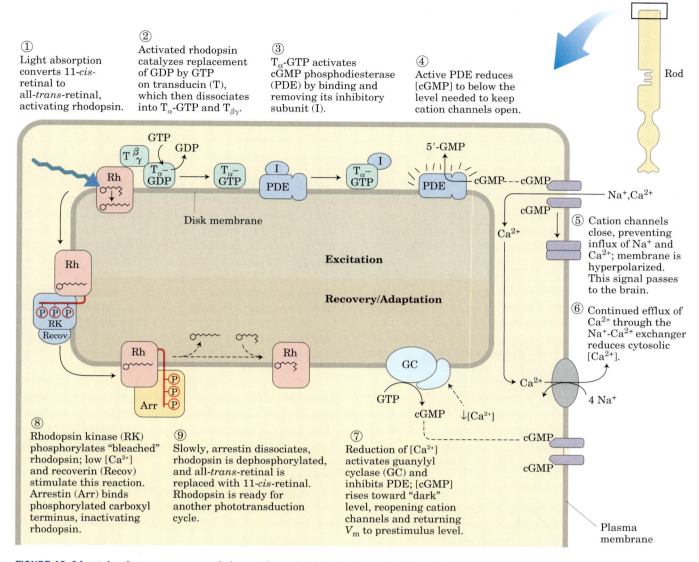

① Light absorption converts 11-*cis*-retinal to all-*trans*-retinal, activating rhodopsin.

② Activated rhodopsin catalyzes replacement of GDP by GTP on transducin (T), which then dissociates into $T_\alpha$-GTP and $T_{\beta\gamma}$.

③ $T_\alpha$-GTP activates cGMP phosphodiesterase (PDE) by binding and removing its inhibitory subunit (I).

④ Active PDE reduces [cGMP] to below the level needed to keep cation channels open.

Rod

⑤ Cation channels close, preventing influx of $Na^+$ and $Ca^{2+}$; membrane is hyperpolarized. This signal passes to the brain.

⑥ Continued efflux of $Ca^{2+}$ through the $Na^+$-$Ca^{2+}$ exchanger reduces cytosolic [$Ca^{2+}$].

⑧ Rhodopsin kinase (RK) phosphorylates "bleached" rhodopsin; low [$Ca^{2+}$] and recoverin (Recov) stimulate this reaction. Arrestin (Arr) binds phosphorylated carboxyl terminus, inactivating rhodopsin.

⑨ Slowly, arrestin dissociates, rhodopsin is dephosphorylated, and all-*trans*-retinal is replaced with 11-*cis*-retinal. Rhodopsin is ready for another phototransduction cycle.

⑦ Reduction of [$Ca^{2+}$] activates guanylyl cyclase (GC) and inhibits PDE; [cGMP] rises toward "dark" level, reopening cation channels and returning $V_m$ to prestimulus level.

**FIGURE 12–34 Molecular consequences of photon absorption by rhodopsin in the rod outer segment.** The top half of the figure (steps ① to ⑤) describes excitation; the bottom (steps ⑥ to ⑨), recovery and adaptation after illumination.

⑥). The inhibition of guanylyl cyclase by $Ca^{2+}$ is relieved, and the cyclase converts GTP to cGMP to return the system to its prestimulus state (step ⑦).

## Rhodopsin Is Desensitized by Phosphorylation

Rhodopsin itself also undergoes changes in response to prolonged illumination. The conformational change induced by light absorption exposes several Thr and Ser residues in the carboxyl-terminal domain. These residues are quickly phosphorylated by **rhodopsin kinase** (step ⑧ in Fig. 12–34), which is functionally and structurally homologous to the β-adrenergic kinase (βARK) that desensitizes the β-adrenergic receptor (Fig. 12–17). The $Ca^{2+}$-binding protein **recoverin** inhibits rhodopsin kinase at high [$Ca^{2+}$], but the inhibi-

tion is relieved when [$Ca^{2+}$] drops after illumination, as described above. The phosphorylated carboxyl-terminal domain of rhodopsin is bound by the protein **arrestin 1,** preventing further interaction between activated rhodopsin and transducin. Arrestin 1 is a close homolog of arrestin 2 (βarr; Fig. 12–17). On a relatively long time scale (seconds to minutes), the all-*trans*-retinal of an excited rhodopsin molecule is removed and replaced by 11-*cis*-retinal, to produce rhodopsin that is ready for another round of excitation (step ⑨ in Fig. 12–34).

Humans cannot synthesize retinal from simpler precursors and must obtain it in the diet in the form of vitamin A (see Fig. 10–21). Given the role of retinal in the process of vision, it is not surprising that dietary deficiency of vitamin A causes night blindness (poor vision at night or in dim light). ∎

## Cone Cells Specialize in Color Vision

Color vision in cone cells involves a path of sensory transduction essentially identical to that described above, but triggered by slightly different light receptors. Three types of cone cells are specialized to detect light from different regions of the spectrum, using three related photoreceptor proteins (opsins). Each cone cell expresses only one kind of opsin, but each type is closely related to rhodopsin in size, amino acid sequence, and presumably three-dimensional structure. The differences among the opsins, however, are great enough to place the chromophore, 11-*cis*-retinal, in three slightly different environments, with the result that the three photoreceptors have different absorption spectra (Fig. 12–35). We discriminate colors and hues by integrating the output from the three types of cone cells, each containing one of the three photoreceptors.

Color blindness, such as the inability to distinguish red from green, is a fairly common, genetically inherited trait in humans. The various types of color blindness result from different opsin mutations. One form is due to loss of the red photoreceptor; affected individuals are **red⁻ dichromats** (they see only two primary colors). Others lack the green pigment and are **green⁻ dichromats.** In some cases, the red and green photoreceptors are present but have a changed amino acid sequence that causes a change in their absorption spectra, resulting in abnormal color vision. Depending on which pigment is altered, such individuals are **red-anomalous trichromats** or **green-anomalous trichromats.** Examination of the genes for the visual receptors has allowed the diagnosis of color blindness in a famous "patient" more than a century after his death (Box 12–3)! ■

### Vertebrate Olfaction and Gustation Use Mechanisms Similar to the Visual System

The sensory cells used to detect odors and tastes have much in common with the rod and cone cells that detect light. Olfactory neurons have a number of long thin cilia extending from one end of the cell into a mucous layer that overlays the cell. These cilia present a large surface area for interaction with olfactory signals. The receptors for olfactory stimuli are ciliary membrane proteins with the familiar serpentine structure of seven transmembrane α helices. The olfactory signal can be any one of the many volatile compounds for which there are specific receptor proteins. Our ability to discriminate odors stems from hundreds of different olfactory receptors in the tongue and nasal passages and from the brain's ability to integrate input from different types of olfactory receptors to recognize a "hybrid" pattern, extending our range of discrimination far beyond the number of receptors.

The olfactory stimulus arrives at the sensory cells by diffusion through the air. In the mucous layer overlaying the olfactory neurons, the odorant molecule binds directly to an olfactory receptor or to a specific binding protein that carries the odorant to a receptor (Fig. 12–36). Interaction between odorant and receptor triggers a change in receptor conformation that results in the replacement of bound GDP by GTP on a G protein, $G_{olf}$, analogous to transducin and to $G_s$ of the β-adrenergic system. The activated $G_{olf}$ then activates adenylyl cyclase of the ciliary membrane, which synthesizes cAMP from ATP, raising the local [cAMP]. The cAMP-gated $Na^+$ and $Ca^{2+}$ channels of the ciliary membrane open, and the influx of $Na^+$ and $Ca^{2+}$ produces a small depolarization called the **receptor potential.** If a sufficient number of odorant molecules encounter receptors, the receptor potential is strong enough to cause the neuron to fire an action potential. This is relayed to the brain in several stages and registers as a specific smell. All these events occur within 100 to 200 ms.

Some olfactory neurons may use a second transduction mechanism. They have receptors coupled through G proteins to PLC rather than to adenylyl cyclase. Signal reception in these cells triggers production of $IP_3$ (Fig. 12–19), which opens $IP_3$-gated $Ca^{2+}$ channels in the ciliary membrane. Influx of $Ca^{2+}$ then depolarizes the ciliary membrane and generates a receptor potential or regulates $Ca^{2+}$-dependent enzymes in the olfactory pathway.

In either type of olfactory neuron, when the stimulus is no longer present, the transducing machinery shuts

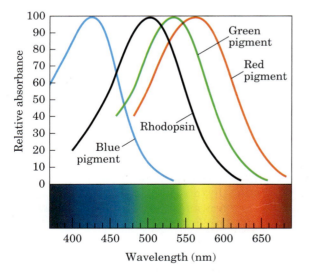

**FIGURE 12–35 Absorption spectra of purified rhodopsin and the red, green, and blue receptors of cone cells.** The spectra, obtained from individual cone cells isolated from cadavers, peak at about 420, 530, and 560 nm, and the maximum absorption for rhodopsin is at about 500 nm. For reference, the visible spectrum for humans is about 380 to 750 nm.

## BOX 12–3 BIOCHEMISTRY IN MEDICINE

### Color Blindness: John Dalton's Experiment from the Grave

The chemist John Dalton (of atomic theory fame) was color-blind. He thought it probable that the vitreous humor of his eyes (the fluid that fills the eyeball behind the lens) was tinted blue, unlike the colorless fluid of normal eyes. He proposed that after his death, his eyes should be dissected and the color of the vitreous humor determined. His wish was honored. The day after Dalton's death in July 1844, Joseph Ransome dissected his eyes and found the vitreous humor to be perfectly colorless. Ransome, like many scientists, was reluctant to throw samples away. He placed Dalton's eyes in a jar of preservative (Fig. 1), where they stayed for a century and a half.

Then, in the mid-1990s, molecular biologists in England took small samples of Dalton's retinas and extracted DNA. Using the known gene sequences for the opsins of the red and green photopigments, they am-

plified the relevant sequences (using techniques described in Chapter 9) and determined that Dalton had the opsin gene for the red photopigment but lacked the opsin gene for the green photopigment. Dalton was a green⁻ dichromat. So, 150 years after his death, the experiment Dalton started—by hypothesizing about the cause of his color blindness—was finally finished.

**FIGURE 1** Dalton's eyes.

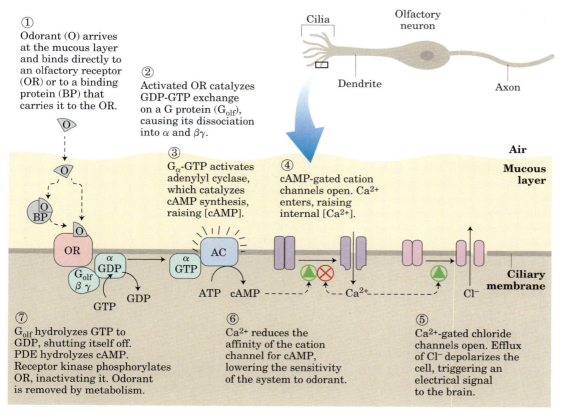

① Odorant (O) arrives at the mucous layer and binds directly to an olfactory receptor (OR) or to a binding protein (BP) that carries it to the OR.

② Activated OR catalyzes GDP-GTP exchange on a G protein ($G_{olf}$), causing its dissociation into $\alpha$ and $\beta\gamma$.

③ $G_\alpha$-GTP activates adenylyl cyclase, which catalyzes cAMP synthesis, raising [cAMP].

④ cAMP-gated cation channels open. $Ca^{2+}$ enters, raising internal [$Ca^{2+}$].

⑦ $G_{olf}$ hydrolyzes GTP to GDP, shutting itself off. PDE hydrolyzes cAMP. Receptor kinase phosphorylates OR, inactivating it. Odorant is removed by metabolism.

⑥ $Ca^{2+}$ reduces the affinity of the cation channel for cAMP, lowering the sensitivity of the system to odorant.

⑤ $Ca^{2+}$-gated chloride channels open. Efflux of $Cl^-$ depolarizes the cell, triggering an electrical signal to the brain.

**FIGURE 12–36 Molecular events of olfaction.** These interactions occur in the cilia of olfactory receptor cells.

itself off in several ways. A cAMP phosphodiesterase returns [cAMP] to the prestimulus level. $G_{olf}$ hydrolyzes its bound GTP to GDP, thereby inactivating itself. Phosphorylation of the receptor by a specific kinase prevents its interaction with $G_{olf}$, by a mechanism analogous to that used to desensitize the $\beta$-adrenergic receptor and rhodopsin. And lastly, some odorants are enzymatically destroyed by oxidases.

The sense of taste in vertebrates reflects the activity of gustatory neurons clustered in taste buds on the surface of the tongue. In these sensory neurons, serpentine receptors are coupled to the heterotrimeric G protein **gustducin** (very similar to the transducin of rod and cone cells). Sweet-tasting molecules are those that bind receptors in "sweet" taste buds. When the molecule (tastant) binds, gustducin is activated by replacement of bound GDP with GTP and then stimulates cAMP production by adenylyl cyclase. The resulting elevation of [cAMP] activates PKA, which phosphorylates $K^+$ channels in the plasma membrane, causing them to close. Reduced efflux of $K^+$ depolarizes the cell (Fig. 12–37). Other taste buds specialize in detecting bitter, sour, or salty tastants, using various combinations of second messengers and ion channels in the transduction mechanisms.

## G Protein–Coupled Serpentine Receptor Systems Share Several Features

We have now looked at four systems (hormone signaling, vision, olfaction, and gustation) in which membrane receptors are coupled to second messenger–generating enzymes through G proteins. It is clear that signaling mechanisms arose early in evolution; serpentine receptors, heterotrimeric G proteins, and adenylyl cyclase are found in virtually all eukaryotic organisms. Even the common brewer's yeast *Saccharomyces* uses serpentine receptors and G proteins to detect the opposite mating type. Overall patterns have been conserved, and the introduction of variety has given modern organisms the ability to respond to a wide range of stimuli (Table 12–8). Of the 35,000 or so genes in the human genome, as many as 1,000 encode serpentine receptors, including hundreds for olfactory stimuli and a number of "orphan receptors" for which the natural ligand is not yet known.

All well-studied transducing systems that act through heterotrimeric G proteins share certain common features (Fig. 12–38). The receptors have seven transmembrane segments, a domain (generally the loop between transmembrane helices 6 and 7) that interacts with a G protein, and a carboxyl-terminal cytoplasmic domain that undergoes reversible phosphorylation on several Ser or Thr residues. The ligand-binding site (or, in the case of light reception, the light receptor) is buried deep in the membrane and includes residues from several of the transmembrane segments. Ligand binding (or light) induces a conformational change in the receptor, exposing a domain that can interact with a G protein. Heterotrimeric G proteins activate or inhibit effector enzymes (adenylyl cyclase, PDE, or PLC), which change the concentration of a second messenger (cAMP, cGMP, $IP_3$, or $Ca^{2+}$). In the hormone-detecting

---

**TABLE 12–8   Some Signals Transduced by G Protein–Coupled Serpentine Receptors**

| | |
|---|---|
| Acetylcholine (muscarinic) | Leukotrienes |
| Adenosine | Light |
| Angiotensin | Luteinizing hormone (LH) |
| ATP (extracellular) | Melatonin |
| Bradykinin | Odorants |
| Calcitonin | Opioids |
| Cannabinoids | Oxytocin |
| Catecholamines | Platelet-activating factor |
| Cholecystokinin | Prostaglandins |
| Corticotropin-releasing factor (CRF) | Secretin |
| Cyclic AMP (*Dictyostelium discoideum*) | Serotonin |
| Dopamine | Somatostatin |
| Follicle-stimulating hormone (FSH) | Tastants |
| $\gamma$-Aminobutyric acid (GABA) | Thyrotropin |
| Glucagon | Thyrotropin-releasing hormone (TRH) |
| Glutamate | Vasoactive intestinal peptide |
| Growth hormone–releasing hormone (GHRH) | Vasopressin |
| Histamine | Yeast mating factors |

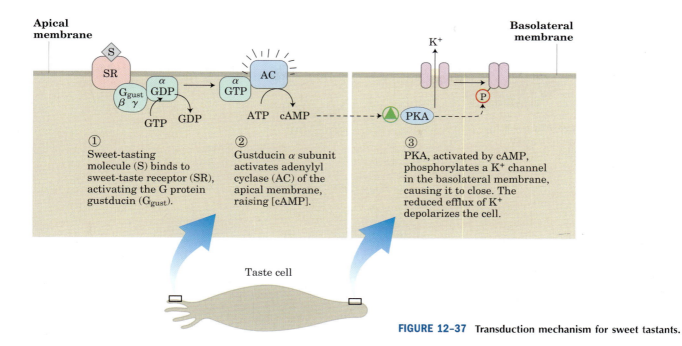

**Apical membrane**

**Basolateral membrane**

① Sweet-tasting molecule (S) binds to sweet-taste receptor (SR), activating the G protein gustducin (G$_{gust}$).

② Gustducin $\alpha$ subunit activates adenylyl cyclase (AC) of the apical membrane, raising [cAMP].

③ PKA, activated by cAMP, phosphorylates a K$^+$ channel in the basolateral membrane, causing it to close. The reduced efflux of K$^+$ depolarizes the cell.

Taste cell

**FIGURE 12–37** Transduction mechanism for sweet tastants.

systems, the final output is an activated protein kinase that regulates some cellular process by phosphorylating a protein critical to that process. In sensory neurons, the output is a change in membrane potential and a consequent electrical signal that passes to another neuron in the pathway connecting the sensory cell to the brain.

All these systems self-inactivate. Bound GTP is converted to GDP by the intrinsic GTPase activity of G proteins, often augmented by GTPase-activating proteins (GAPs) or RGS proteins (*regulators of G-protein signaling*). In some cases, the effector enzymes that are the targets of G protein modulation also serve as GAPs.

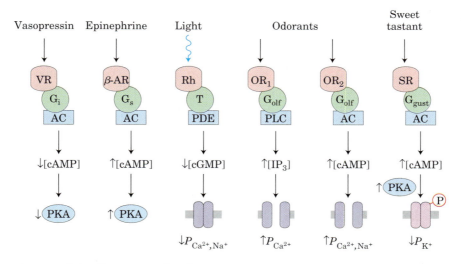

**FIGURE 12–38 Common features of signaling systems that detect hormones, light, smells, and tastes.** Serpentine receptors provide signal specificity, and their interaction with G proteins provides signal amplification. Heterotrimeric G proteins activate effector enzymes: adenylyl cyclase (AC), phospholipase C (PLC), and phosphodiesterases (PDE) that degrade cAMP or cGMP. Changes in concentration of the second messengers (cAMP, cGMP, IP$_3$) result in alterations of enzymatic activities by phosphorylation or alterations in the permeability

(P) of surface membranes to Ca$^{2+}$, Na$^+$, and K$^+$. The resulting depolarization or hyperpolarization of the sensory cell (the signal) is passed through relay neurons to sensory centers in the brain. In the best-studied cases, desensitization includes phosphorylation of the receptor and binding of a protein (arrestin) that interrupts receptor–G protein interactions. VR is the vasopressin receptor; other receptor and G protein abbreviations are as used in earlier illustrations.

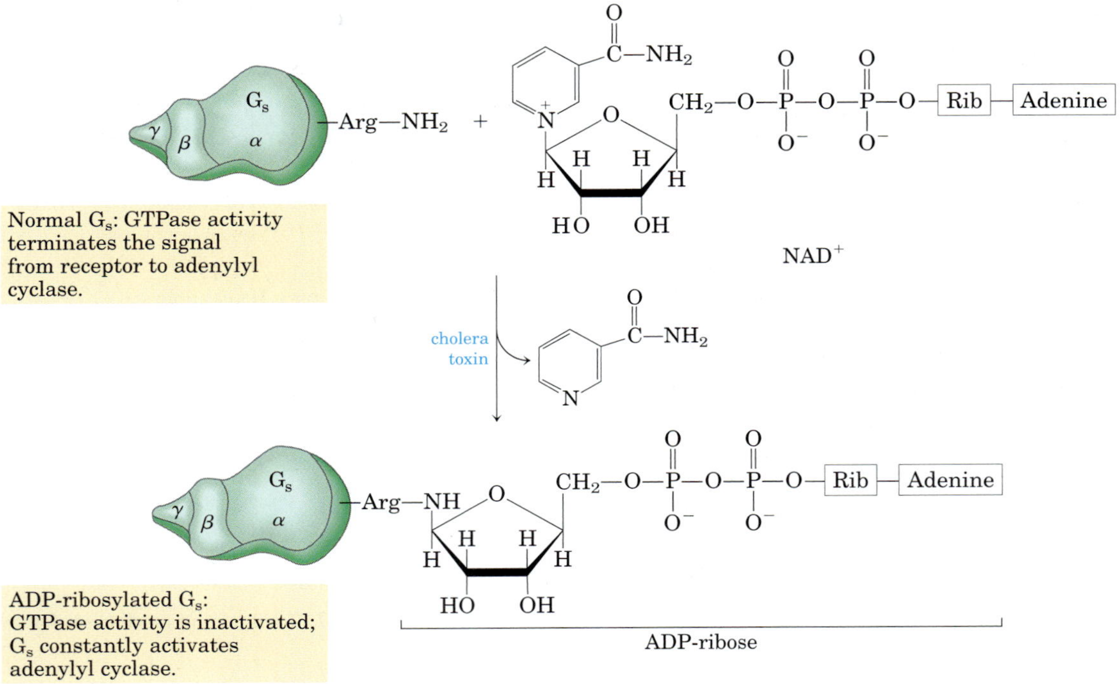

**FIGURE 12–39** **Toxins produced by bacteria that cause cholera and whooping cough (pertussis).** These toxins are enzymes that catalyze transfer of the ADP-ribose moiety of $NAD^+$ to an Arg residue (cholera toxin) or a Cys residue (pertussis toxin) of G proteins: $G_s$ in the case of cholera (as shown here) and $G_i$ in whooping cough. The G proteins thus modified fail to respond to normal hormonal stimuli. The pathology of both diseases results from defective regulation of adenylyl cyclase and overproduction of cAMP.

## Disruption of G-Protein Signaling Causes Disease

Biochemical studies of signal transductions have led to an improved understanding of the pathological effects of toxins produced by the bacteria that cause cholera and pertussis (whooping cough). Both toxins are enzymes that interfere with normal signal transductions in the host animal. **Cholera toxin,** secreted by *Vibrio cholerae* found in contaminated drinking water, catalyzes the transfer of ADP-ribose from $NAD^+$ to the $\alpha$ subunit of $G_s$ (Fig. 12–39), blocking its GTPase activity and thereby rendering $G_s$ permanently activated. This results in continuous activation of the adenylyl cyclase of intestinal epithelial cells and chronically high [cAMP], which triggers constant secretion of $Cl^-$, $HCO_3^-$, and water into the intestinal lumen. The resulting dehydration and electrolyte loss are the major pathologies in cholera. The **pertussis toxin,** produced by *Bordetella pertussis,* catalyzes ADP-ribosylation of $G_i$, preventing displacement of GDP by GTP and blocking inhibition of adenylyl cyclase by $G_i$. ■

## SUMMARY 12.7 Sensory Transduction in Vision, Olfaction, and Gustation

- Vision, olfaction, and gustation in vertebrates employ serpentine receptors, which act through heterotrimeric G proteins to change the $V_m$ of the sensory neuron.

- In rod and cone cells of the retina, light activates rhodopsin, which stimulates replacement of GDP by GTP on the G protein transducin. The freed $\alpha$ subunit of transducin activates cGMP phosphodiesterase, which lowers [cGMP] and thus closes cGMP-dependent ion channels in the outer segment of the neuron. The resulting hyperpolarization of the rod or cone cell carries the signal to the next neuron in the pathway, and eventually to the brain.

- In olfactory neurons, olfactory stimuli, acting through serpentine receptors and G proteins, trigger either an increase in [cAMP] (by activating adenylyl cyclase) or an increase in $[Ca^{2+}]$ (by activating PLC). These second messengers affect ion channels and thus the $V_m$.

- Gustatory neurons have serpentine receptors that respond to tastants by altering [cAMP], which in turn changes $V_m$ by gating ion channels.

- There is a high degree of conservation of signaling proteins and transduction mechanisms across species.

# 12.8 Regulation of Transcription by Steroid Hormones

The large group of steroid, retinoic acid (retinoid), and thyroid hormones exert at least part of their effects by a mechanism fundamentally different from that of other hormones: they act in the nucleus to alter gene expression. We therefore discuss their mode of action in detail in Chapter 28, along with other mechanisms for regulating gene expression. Here we give a brief overview.

Steroid hormones (estrogen, progesterone, and cortisol, for example), too hydrophobic to dissolve readily in the blood, are carried on specific carrier proteins from their point of release to their target tissues. In target cells, these hormones pass through the plasma membranes by simple diffusion and bind to specific receptor proteins in the nucleus (Fig. 12–40). Hormone binding triggers changes in the conformation of the receptor proteins so that they become capable of interacting with specific regulatory sequences in DNA called **hormone response elements (HREs),** thus altering gene expression (see Fig. 28–31). The bound receptor-hormone complex can either enhance or suppress the expression of specific genes adjacent to HREs. Hours or days are required for these regulators to have their full effect—the time required for the changes in RNA synthesis and subsequent protein synthesis to become evident in altered metabolism.

The specificity of the steroid-receptor interaction is exploited in the use of the drug **tamoxifen** to treat breast cancer. In some types of breast cancer, division of the cancerous cells depends on the continued presence of the hormone estrogen. Tamoxifen competes with estrogen for binding to the estrogen receptor, but the tamoxifen-receptor complex has little or no effect on gene expression; tamoxifen is an antagonist of estrogen. Consequently, tamoxifen administered after surgery or during chemotherapy for hormone-dependent breast cancer slows or stops the growth of remaining cancerous cells.

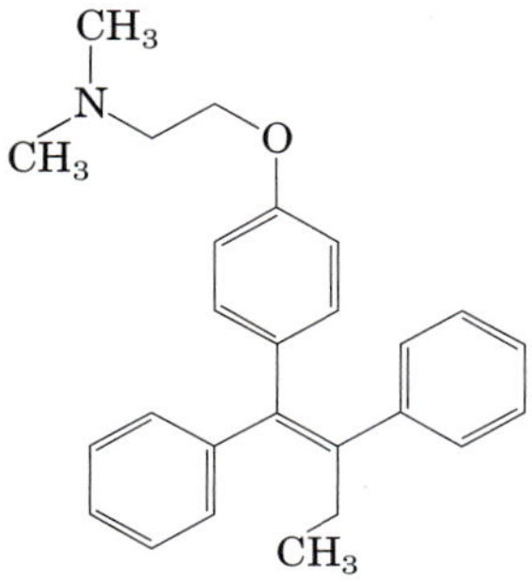

Tamoxifen

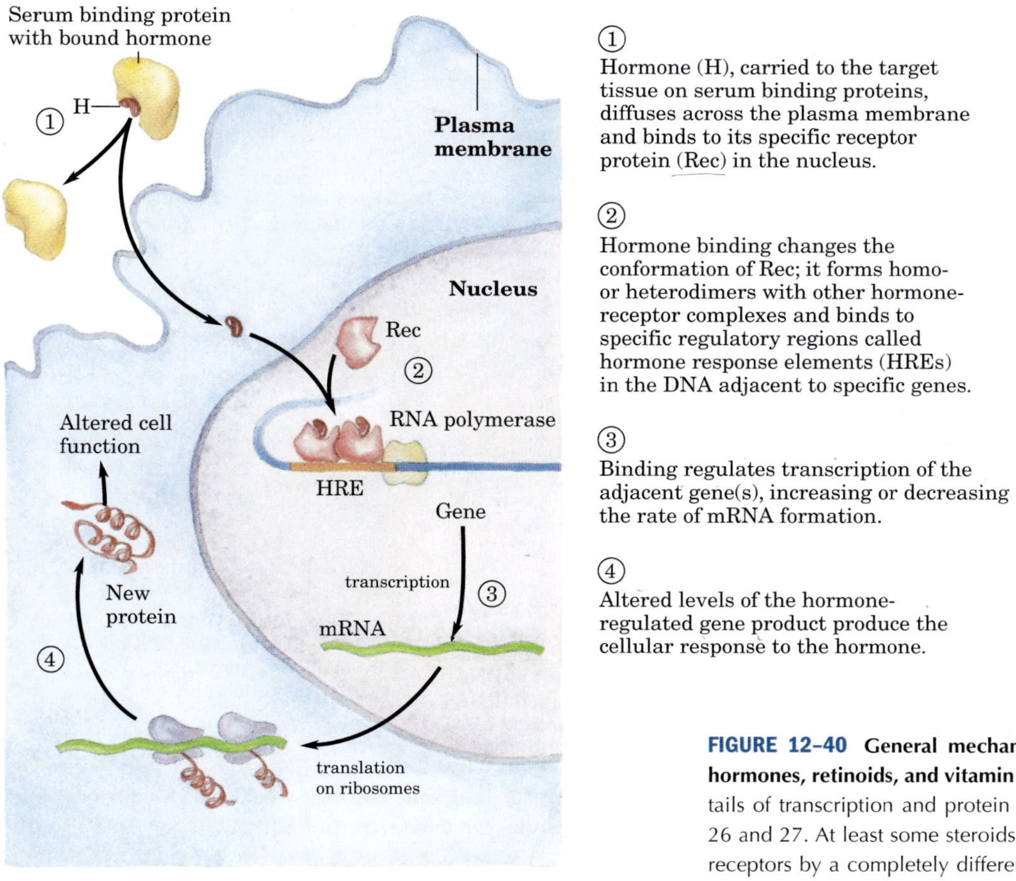

① Hormone (H), carried to the target tissue on serum binding proteins, diffuses across the plasma membrane and binds to its specific receptor protein (Rec) in the nucleus.

② Hormone binding changes the conformation of Rec; it forms homo- or heterodimers with other hormone-receptor complexes and binds to specific regulatory regions called hormone response elements (HREs) in the DNA adjacent to specific genes.

③ Binding regulates transcription of the adjacent gene(s), increasing or decreasing the rate of mRNA formation.

④ Altered levels of the hormone-regulated gene product produce the cellular response to the hormone.

**FIGURE 12–40 General mechanism by which steroid and thyroid hormones, retinoids, and vitamin D regulate gene expression.** The details of transcription and protein synthesis are discussed in Chapters 26 and 27. At least some steroids also act through plasma membrane receptors by a completely different mechanism.

RU486
(mifepristone)

Another steroid analog, the drug **RU486,** is used to terminate early (preimplantation) pregnancies. An antagonist of the hormone progesterone, RU486 binds to the progesterone receptor and blocks hormone actions essential to implantation of the fertilized ovum in the uterus. ■

The classic mechanism for steroid hormone action through nuclear receptors does not explain certain effects of steroids that are too fast to be the result of altered protein synthesis. For example, the estrogen-mediated dilation of blood vessels is known to be independent of gene transcription or protein synthesis, as is the steroid-induced decrease in cellular [cAMP]. Another transduction mechanism is probably responsible for some of these effects. A plasma membrane protein predicted to have seven transmembrane helical segments binds progesterone with very high affinity and mediates the inhibition of adenylyl cyclase by that hormone, accounting for the decrease in [cAMP]. A second nonclassical mechanism involves the rapid activation of the MAPK cascade by progesterone, acting through the soluble progesterone receptor. This is the same receptor that, in the nucleus, causes the much slower changes in gene expression that constitute the classic mechanism of progesterone action. How the MAPK cascade is activated is not yet clear.

### SUMMARY 12.8 Regulation of Transcription by Steroid Hormones

- Steroid hormones enter cells and bind to specific receptor proteins.

- The hormone-receptor complex binds specific regions of DNA, the hormone response elements, and regulates the expression of nearby genes by interacting with transcription factors.

- Two other, faster-acting mechanisms produce some of the effects of steroids. Progesterone triggers a rapid drop in [cAMP], mediated by a plasma membrane receptor, and binding of progesterone to the classic soluble steroid receptor activates a MAPK cascade.

## 12.9 Regulation of the Cell Cycle by Protein Kinases

One of the most dramatic roles for protein phosphorylation is the regulation of the eukaryotic cell cycle. During embryonic growth and later development, cell division occurs in virtually every tissue. In the adult organism most tissues become quiescent. A cell's "decision" to divide or not is of crucial importance to the organism. When the regulatory mechanisms that limit cell division are defective and cells undergo unregulated division, the result is catastrophic—cancer. Proper cell division requires a precisely ordered sequence of biochemical events that assures every daughter cell a full complement of the molecules required for life. Investigations into the control of cell division in diverse eukaryotic cells have revealed universal regulatory mechanisms. Protein kinases and protein phosphorylation are central to the timing mechanism that determines entry into cell division and ensures orderly passage through these events.

### The Cell Cycle Has Four Stages

Cell division in eukaryotes occurs in four well-defined stages (Fig. 12–41). In the S (synthesis) phase, the DNA is replicated to produce copies for both daughter

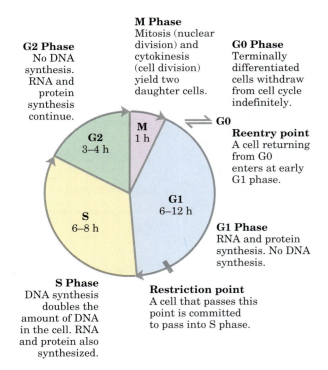

**FIGURE 12–41 Eukaryotic cell cycle.** The durations (in hours) of the four stages vary, but those shown are typical.

cells. In the G2 phase (G indicates the gap between divisions), new proteins are synthesized and the cell approximately doubles in size. In the M phase (mitosis), the maternal nuclear envelope breaks down, matching chromosomes are pulled to opposite poles of the cell, each set of daughter chromosomes is surrounded by a newly formed nuclear envelope, and cytokinesis pinches the cell in half, producing two daughter cells. In embryonic or rapidly proliferating tissue, each daughter cell divides again, but only after a waiting period (G1). In cultured animal cells the entire process takes about 24 hours.

After passing through mitosis and into G1, a cell either continues through another division or ceases to divide, entering a quiescent phase (G0) that may last hours, days, or the lifetime of the cell. When a cell in G0 begins to divide again, it reenters the division cycle through the G1 phase. Differentiated cells such as hepatocytes or adipocytes have acquired their specialized function and form; they remain in the G0 phase.

## Levels of Cyclin-Dependent Protein Kinases Oscillate

The timing of the cell cycle is controlled by a family of protein kinases with activities that change in response to cellular signals. By phosphorylating specific proteins at precisely timed intervals, these protein kinases orchestrate the metabolic activities of the cell to produce orderly cell division. The kinases are heterodimers with a regulatory subunit, **cyclin,** and a catalytic subunit, **cyclin-dependent protein kinase (CDK).** In the absence of cyclin, the catalytic subunit is virtually inactive. When cyclin binds, the catalytic site opens up, a residue essential to catalysis becomes accessible (Fig. 12–42), and the activity of the catalytic subunit increases 10,000-fold. Animal cells have at least ten different cyclins (designated A, B, and so forth) and at least eight cyclin-dependent kinases (CDK1 through CDK8), which act in various combinations at specific points in the cell cycle. Plants also use a family of CDKs to regulate their cell division.

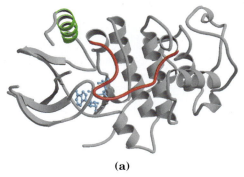

(a)

**FIGURE 12–42 Activation of cyclin-dependent protein kinases (CDKs) by cyclin and phosphorylation.** CDKs, a family of related enzymes, are active only when associated with cyclins, another protein family. The crystal structure of CDK2 with and without cyclin reveals the basis for this activation. **(a)** Without cyclin (PDB ID 1HCK), CDK2 folds so that one segment, the T loop (red), obstructs the binding site for protein substrates and thus inhibits protein kinase activity. The binding site for ATP (blue) is also near the T loop. **(b)** When cyclin binds (PDB ID 1FIN), it forces conformational changes that move the T loop away from the active site and reorient an amino-terminal helix (green), bringing a residue critical to catalysis (Glu[51]) into the active site. **(c)** Phosphorylation of a Thr residue (dark orange space-filling structure) in the T loop produces a negatively charged residue that is stabilized by interaction with three Arg residues (red ball-and-stick structures), holding CDK in its active conformation (PDB ID 1JST).

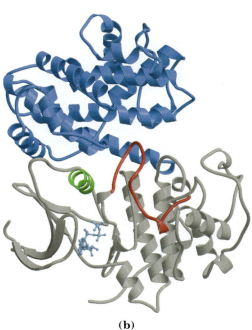

(b)

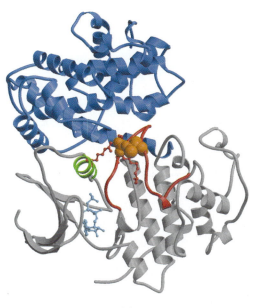

(c)

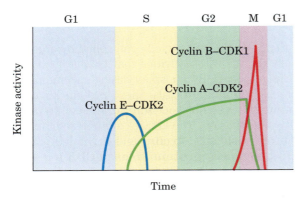

**FIGURE 12–43 Variations in the activities of specific CDKs during the cell cycle in animals.** Cyclin E–CDK2 activity peaks near the G1 phase–S phase boundary, when the active enzyme triggers synthesis of enzymes required for DNA synthesis (see Fig. 12–46). Cyclin A–CDK2 activity rises during the S and G2 phases, then drops sharply in the M phase, as cyclin B–CDK1 peaks.

In a population of animal cells undergoing synchronous division, some CDK activities show striking oscillations (Fig. 12–43). These oscillations are the result of four mechanisms for regulating CDK activity: phosphorylation or dephosphorylation of the CDK, controlled degradation of the cyclin subunit, periodic synthesis of CDKs and cyclins, and the action of specific CDK-inhibiting proteins.

**Regulation of CDKs by Phosphorylation** The activity of a CDK is strikingly affected by phosphorylation and dephosphorylation of two critical residues in the protein (Fig. 12–44a). Phosphorylation of $Tyr^{15}$ near the amino terminus renders CDK2 inactive; the Ⓟ–Tyr residue is in the ATP-binding site of the kinase, and the negatively charged phosphate group blocks the entry of ATP. A specific phosphatase dephosphorylates this Ⓟ–Tyr residue, permitting the binding of ATP. Phosphorylation

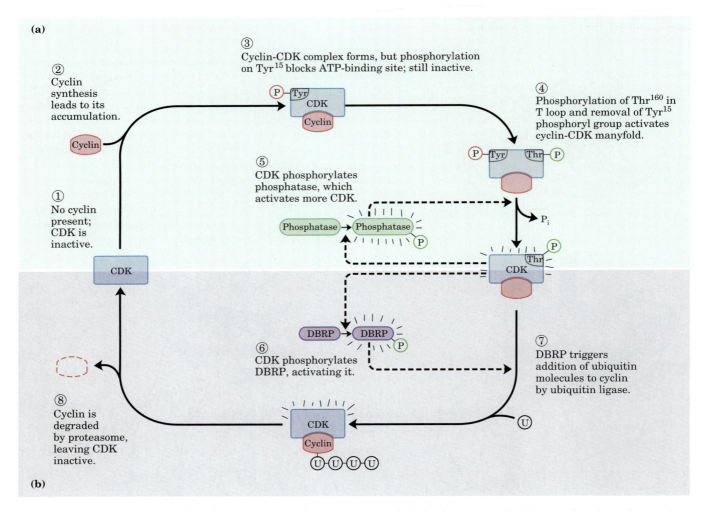

**FIGURE 12–44 Regulation of CDK by phosphorylation and proteolysis. (a)** The cyclin-dependent protein kinase activated at the time of mitosis (the M phase CDK) has a "T loop" that can fold into the substrate-binding site. When $Thr^{160}$ in the T loop is phosphorylated, the loop moves out of the substrate-binding site, activating the CDK manyfold. **(b)** The active cyclin-CDK complex triggers its own inactivation by phosphorylation of DBRP (destruction box recognizing protein). DBRP and ubiquitin ligase then attach several molecules of ubiquitin (U) to cyclin, targeting it for destruction by proteasomes, proteolytic enzyme complexes.

of Thr$^{160}$ in the "T loop" of CDK, catalyzed by the CDK-activating kinase, forces the T loop out of the substrate-binding cleft, permitting substrate binding and catalytic activity.

One circumstance that triggers this control mechanism is the presence of single-strand breaks in DNA, which leads to arrest of the cell cycle in G2. A specific protein kinase (called Rad3 in yeast), which is activated by single-strand breaks, triggers a cascade leading to the inactivation of the phosphatase that dephosphorylates Tyr$^{15}$ of CDK. The CDK remains inactive and the cell is arrested in G2. The cell will not divide until the DNA is repaired and the effects of the cascade are reversed.

***Controlled Degradation of Cyclin***  Highly specific and precisely timed proteolytic breakdown of mitotic cyclins regulates CDK activity throughout the cell cycle. Progress through mitosis requires first the activation then the destruction of cyclins A and B, which activate the catalytic subunit of the M-phase CDK. These cyclins contain near their amino terminus the sequence Arg–Thr–Ala–Leu–Gly–Asp–Ile–Gly–Asn, the "destruction box," which targets them for degradation. (This usage of "box" derives from the common practice, in diagramming the sequence of a nucleic acid or protein, of enclosing within a box a short sequence of nucleotide or amino acid residues with some specific function. It does not imply any three-dimensional structure.) The protein DBRP (destruction box recognizing protein) recognizes this sequence and initiates the process of cyclin degradation by bringing together the cyclin and another protein, **ubiquitin.** Cyclin and activated ubiquitin are covalently joined by the enzyme ubiquitin ligase (Fig. 12–44b). Several more ubiquitin molecules are then appended, providing the signal for a proteolytic enzyme complex, or **proteasome,** to degrade cyclin.

What controls the timing of cyclin breakdown? A feedback loop occurs in the overall process shown in Figure 12–44. Increased CDK activity activates cyclin proteolysis. Newly synthesized cyclin associates with and activates CDK, which phosphorylates and activates DBRP. Active DBRP then causes proteolysis of cyclin. Lowered [cyclin] causes a decline in CDK activity, and the activity of DBRP also drops through slow, constant dephosphorylation and inactivation by a DBRP phosphatase. The cyclin level is ultimately restored by synthesis of new cyclin molecules.

The role of ubiquitin and proteasomes is not limited to the regulation of cyclin; as we shall see in Chapter 27, both also take part in the turnover of cellular proteins, a process fundamental to cellular housekeeping.

***Regulated Synthesis of CDKs and Cyclins***  The third mechanism for changing CDK activity is regulation of the rate of synthesis of cyclin or CDK or both. For example, cyclin D, cyclin E, CDK2, and CDK4 are synthesized only when a specific transcription factor, E2F, is present in

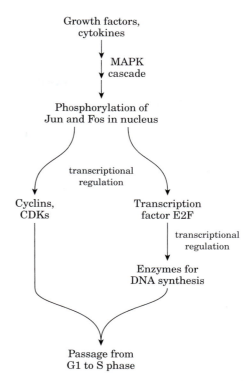

**FIGURE 12–45  Regulation of cell division by growth factors.** The path from growth factors to cell division leads through the enzyme cascade that activates MAPK; phosphorylation of the nuclear transcription factors Jun and Fos; and the activity of the transcription factor E2F, which promotes synthesis of several enzymes essential for DNA synthesis.

the nucleus to activate transcription of their genes. Synthesis of E2F is in turn regulated by extracellular signals such as **growth factors** and **cytokines** (inducers of cell division), compounds found to be essential for the division of mammalian cells in culture. These growth factors induce the synthesis of specific nuclear transcription factors essential to the production of the enzymes of DNA synthesis. Growth factors trigger phosphorylation of the nuclear proteins Jun and Fos, transcription factors that promote the synthesis of a variety of gene products, including cyclins, CDKs, and E2F. In turn, E2F controls production of several enzymes essential for the synthesis of deoxynucleotides and DNA, enabling cells to enter the S phase (Fig. 12–45).

***Inhibition of CDKs***  Finally, specific protein inhibitors bind to and inactivate specific CDKs. One such protein is p21, which we discuss below.

These four control mechanisms modulate the activity of specific CDKs that, in turn, control whether a cell will divide, differentiate, become permanently quiescent, or begin a new cycle of division after a period of quiescence. The details of cell cycle regulation, such as the number of different cyclins and kinases and the combinations in which they act, differ from species to species, but the basic mechanism has been conserved in the evolution of all eukaryotic cells.

## CDKs Regulate Cell Division by Phosphorylating Critical Proteins

We have examined how cells maintain close control of CDK activity, but how does the activity of CDK control the cell cycle? The list of target proteins that CDKs are known to act upon continues to grow, and much remains to be learned. But we can see a general pattern behind CDK regulation by inspecting the effect of CDKs on the structures of laminin and myosin and on the activity of retinoblastoma protein.

The structure of the nuclear envelope is maintained in part by highly organized meshworks of intermediate filaments composed of the protein laminin. Breakdown of the nuclear envelope before segregation of the sister chromatids in mitosis is partly due to the phosphorylation of laminin by a CDK, which causes laminin filaments to depolymerize.

A second kinase target is the ATP-driven actin-myosin contractile machinery that pinches a dividing cell into two equal parts during cytokinesis. After the division, CDK phosphorylates a small regulatory subunit of myosin, causing dissociation of myosin from actin filaments and inactivating the contractile machinery. Subsequent dephosphorylation allows reassembly of the contractile apparatus for the next round of cytokinesis.

A third and very important CDK substrate is the **retinoblastoma protein, pRb;** when DNA damage is detected, this protein participates in a mechanism that arrests cell division in G1 (Fig. 12–46). Named for the retinal tumor cell line in which it was discovered, pRb functions in most, perhaps all, cell types to regulate cell division in response to a variety of stimuli. Unphosphorylated pRb binds the transcription factor E2F; while bound to pRb, E2F cannot promote transcription of a group of genes necessary for DNA synthesis (the genes for DNA polymerase $\alpha$, ribonucleotide reductase, and other proteins; Chapter 25). In this state, the cell cycle cannot proceed from the G1 to the S phase, the step that commits a cell to mitosis and cell division. The pRb-E2F blocking mechanism is relieved when pRb is phosphorylated by cyclin E–CDK2, which occurs in response to a signal for cell division to proceed.

When the protein kinases ATM and ATR detect damage to DNA, such as a single-strand break, they activate p53 to serve as a transcription factor that stimulates the synthesis of the protein p21 (Fig. 12–46). This protein inhibits the protein kinase activity of cyclin E–CDK2. In the presence of p21, pRb remains unphosphorylated and bound to E2F, blocking the activity of this transcription factor, and the cell cycle is arrested in G1. This gives the cell time to repair its DNA before entering the S

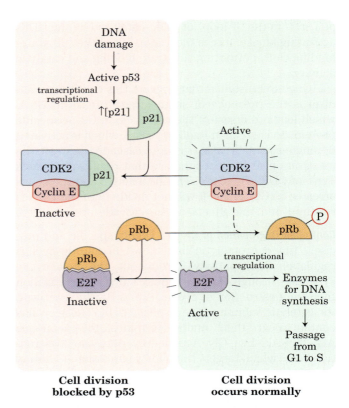

**FIGURE 12–46 Regulation of passage from G1 to S by phosphorylation of pRb.** When the retinoblastoma protein, pRb, is phosphorylated, it cannot bind and inactivate EF2, a transcription factor that promotes synthesis of enzymes essential to DNA synthesis. If the regulatory protein p53 is activated by ATM and ATR, protein kinases that detect damaged DNA, it stimulates the synthesis of p21, which can bind to and inhibit cyclin E–CDK2 and thus prevent phosphorylation of pRb. Unphosphorylated pRb binds and inactivates E2F, blocking passage from G1 to S until the DNA has been repaired.

phase, thereby avoiding the potentially disastrous transfer of a defective genome to one or both daughter cells.

## SUMMARY 12.9 Regulation of the Cell Cycle by Protein Kinases

- Progression through the cell cycle is regulated by the cyclin-dependent protein kinases (CDKs), which act at specific points in the cycle, phosphorylating key proteins and modulating their activities. The catalytic subunit of CDKs is inactive unless associated with the regulatory cyclin subunit.

- The activity of a cyclin-CDK complex changes during the cell cycle through differential synthesis of CDKs, specific degradation of cyclin, phosphorylation and dephosphorylation of critical residues in CDKs, and binding of inhibitory proteins to specific cyclin-CDKs.

# 12.10 Oncogenes, Tumor Suppressor Genes, and Programmed Cell Death

Tumors and cancer are the result of uncontrolled cell division. Normally, cell division is regulated by a family of extracellular growth factors, proteins that cause resting cells to divide and, in some cases, differentiate. Defects in the synthesis, regulation, or recognition of growth factors can lead to cancer.

## Oncogenes Are Mutant Forms of the Genes for Proteins That Regulate the Cell Cycle

**Oncogenes** were originally discovered in tumor-causing viruses, then later found to be closely similar to or derived from genes in the animal host cells, proto-oncogenes, which encode growth-regulating proteins. During viral infections, the DNA sequence of a proto-oncogene is sometimes copied by the virus and incorporated into its genome (Fig. 12–47). At some point during the viral infection cycle, the gene can become defective by truncation or mutation. When this viral oncogene is expressed in its host cell during a subsequent infection, the abnormal protein product interferes with normal regulation of cell growth, sometimes resulting in a tumor.

Proto-oncogenes can become oncogenes without a viral intermediary. Chromosomal rearrangements, chemical agents, and radiation are among the factors that can cause oncogenic mutations. The mutations that produce oncogenes are genetically dominant; if either of a pair of chromosomes contains a defective gene, that gene product sends the signal "divide" and a tumor will result. The oncogenic defect can be in any of the proteins involved in communicating the "divide" signal. We know of oncogenes that encode secreted proteins, growth factors, transmembrane proteins (receptors), cytoplasmic proteins (G proteins and protein kinases), and the nuclear transcription factors that control the expression of genes essential for cell division (Jun, Fos).

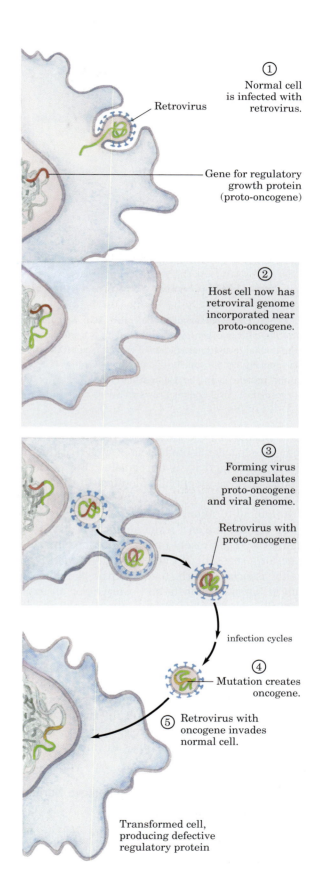

**FIGURE 12–47  Conversion of a regulatory gene to a viral oncogene.** ① A normal cell is infected by a retrovirus (Chapter 26), which ② inserts its own genome into the chromosome of the host cell, near the gene for a regulatory protein (the proto-oncogene). ③ Viral particles released from the infected cell sometimes "capture" a host gene, in this case a proto-oncogene. ④ During several cycles of infection, a mutation occurs in the viral proto-oncogene, converting it to an oncogene. ⑤ When the virus subsequently infects a cell, it introduces the oncogene into the cell's DNA. Transcription of the oncogene leads to the production of a defective regulatory protein that continuously gives the signal for cell division, overriding normal regulatory mechanisms. Host cells infected with oncogene-carrying viruses undergo unregulated cell division—they form tumors. Proto-oncogenes can also undergo mutation to oncogenes without the intervention of a retrovirus, as described in the text.

① Normal cell is infected with retrovirus.

Retrovirus

Gene for regulatory growth protein (proto-oncogene)

② Host cell now has retroviral genome incorporated near proto-oncogene.

③ Forming virus encapsulates proto-oncogene and viral genome.

Retrovirus with proto-oncogene

infection cycles

④ Mutation creates oncogene.

⑤ Retrovirus with oncogene invades normal cell.

Transformed cell, producing defective regulatory protein

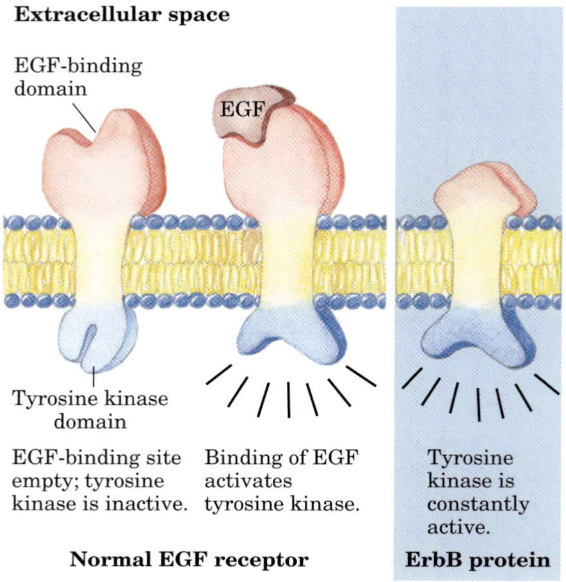

**Extracellular space**

EGF-binding
domain

EGF

Tyrosine kinase
domain

| EGF-binding site empty; tyrosine kinase is inactive. | Binding of EGF activates tyrosine kinase. | Tyrosine kinase is constantly active. |

**Normal EGF receptor** | **ErbB protein**

**FIGURE 12–48 Oncogene-encoded defective EGF receptor.** The product of the *erbB* oncogene (the ErbB protein) is a truncated version of the normal receptor for epidermal growth factor (EGF). Its intracellular domain has the structure normally induced by EGF binding, but the protein lacks the extracellular binding site for EGF. Unregulated by EGF, ErbB continuously signals cell division.

Some oncogenes encode surface receptors with defective or missing signal-binding sites such that their intrinsic Tyr kinase activity is unregulated. For example, the protein ErbB is essentially identical to the normal receptor for epidermal growth factor, except that ErbB lacks the amino-terminal domain that normally binds EGF (Fig. 12–48) and as a result sends the "divide" signal whether EGF is present or not. Mutations in *erbB2*, the gene for a receptor Tyr kinase related to ErbB, are commonly associated with cancers of the glandular epithelium in breast, stomach, and ovary. (For an explanation of the use of abbreviations in naming genes and their products, see Chapter 25.)

Mutant forms of the G protein Ras are common in tumor cells. The *ras* oncogene encodes a protein with normal GTP binding but no GTPase activity. The mutant Ras protein is therefore always in its activated (GTP-bound) form, regardless of the signals arriving through normal receptors. The result can be unregulated growth. Mutations in *ras* are associated with 30% to 50% of lung and colon carcinomas and more than 90% of pancreatic carcinomas.

## Defects in Tumor Suppressor Genes Remove Normal Restraints on Cell Division

**Tumor suppressor genes** encode proteins that normally restrain cell division. Mutation in one or more of these genes can lead to tumor formation.

Unregulated growth due to defective tumor suppressor genes, unlike that due to oncogenes, is genetically recessive; tumors form only if *both* chromosomes of a pair contain a defective gene. In a person who inherits one correct copy and one defective copy, every cell has one defective copy of the gene. If any one of those $10^{12}$ somatic cells undergoes mutation in the one good copy, a tumor may grow from that doubly mutant cell. Mutations in both copies of the genes for pRb, p53, or p21 yield cells in which the normal restraint on cell division is lost and a tumor forms.

Retinoblastoma is a cancer of the retina that occurs in children who have two defective *Rb* alleles. Very young children who develop retinoblastoma commonly have multiple tumors in both eyes. Each tumor is derived from a single retinal cell that has undergone a mutation in its one good copy of the *Rb* gene. (A fetus with two mutant alleles in every cell is nonviable.) Retinoblastoma patients also have a high incidence of cancers of the lung, prostate, and breast.

A far less likely event is that a person born with two good copies of a gene will have two independent mutations in the *same* gene in the *same* cell, but this does occur. Some individuals develop retinoblastomas later in childhood, usually with only one tumor in only one eye. These individuals were presumably born with two good copies of *Rb* in every cell, but both *Rb* genes in a single retinal cell have undergone mutation, leading to a tumor.

Mutations in the gene for p53 also cause tumors; in more than 90% of human cutaneous squamous cell carcinomas (skin cancers) and about 50% of all other human cancers, *p53* is defective. Those very rare individuals who *inherit* one defective copy of *p53* commonly have the Li-Fraumeni cancer syndrome, in which multiple cancers (of the breast, brain, bone, blood, lung, and skin) occur at high frequency and at an early age. The explanation for multiple tumors in this case is the same as that for *Rb* mutations: an individual born with one defective copy of *p53* in every somatic cell is likely to suffer a second *p53* mutation in more than one cell in his or her lifetime.

Mutations in oncogenes and tumor suppressor genes do not have an all-or-none effect. In some cancers, perhaps in all, the progression from a normal cell to a malignant tumor requires an accumulation of mutations (sometimes over several decades), none of which, alone, is responsible for the end effect. For example, the development of colorectal cancer has several recognizable stages, each associated with a mutation (Fig. 12–49). If a normal epithelial cell in the colon undergoes mutation of both copies of the tumor suppressor gene *APC* (adenomatous polyposis coli), it begins to divide faster than normal and produces a clone of itself, a benign polyp (early adenoma). For reasons not yet known, the *APC* mutation results in chromosomal instability; whole regions of a chromosome are lost or re-

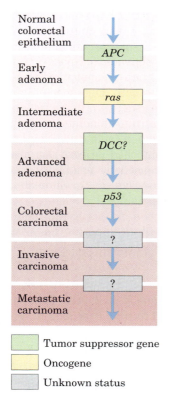

Normal colorectal epithelium

APC

Early adenoma

ras

Intermediate adenoma

DCC?

Advanced adenoma

p53

Colorectal carcinoma

?

Invasive carcinoma

?

Metastatic carcinoma

Tumor suppressor gene

Oncogene

Unknown status

**FIGURE 12–49   From normal epithelial cell to colorectal cancer.** In the colon, mutations in both copies of the tumor suppressor gene *APC* lead to benign clusters of epithelial cells that multiply too rapidly (early adenoma). If a cell already defective in *APC* suffers a second mutation in the proto-oncogene *ras,* the doubly mutant cell gives rise to an intermediate adenoma, forming a benign polyp of the colon. When one of these cells undergoes further mutations in the tumor suppressor genes *DCC* (probably) and *p53,* increasingly aggressive tumors form. Finally, mutations in genes not yet characterized lead to a malignant tumor and finally to a metastatic tumor that can spread to other tissues. Most malignant tumors probably result from a series of mutations such as this.

arranged during cell division. This instability can lead to another mutation, commonly in *ras,* that converts the clone into an intermediate adenoma. A third mutation (probably in the tumor suppressor gene *DCC*) leads to a late adenoma. Only when both copies of *p53* become defective does this cell mass become a carcinoma, a malignant, life-threatening cancer. The full sequence therefore requires at least seven genetic "hits": two on each of three tumor suppressor genes (*APC, DCC,* and *p53*) and one on the protooncogene *ras.* There are probably several other routes to colorectal cancer as well, but the principle that full malignancy results only from multiple mutations is likely to hold. When a polyp is detected in the early adenoma stage and the cells containing the first mutations are removed surgically, late adenomas and carcinomas will not develop; hence the importance of early detection. ■

## Apoptosis Is Programmed Cell Suicide

Many cells can precisely control the time of their own death by the process of **programmed cell death,** or **apoptosis** (app′-a-toe′-sis; from the Greek for "dropping off," as in leaves dropping in the fall). In the development of an embryo, for example, some cells must die. Carving fingers from stubby limb buds requires the precisely timed death of cells between developing finger bones. During development of the nematode *Caenorhabditis elegans* from a fertilized egg, exactly 131 cells (of a total of 1,090 somatic cells in the embryo) must undergo programmed death in order to construct the adult body.

Apoptosis also has roles in processes other than development. When an antibody-producing cell begins to make antibodies against an antigen normally present in the body, that cell undergoes programmed death in the thymus gland—an essential mechanism for eliminating anti-self antibodies. The monthly sloughing of cells of the uterine wall (menstruation) is another case of apoptosis mediating normal cell death. Sometimes cell suicide is not programmed but occurs in response to biological circumstances that threaten the rest of the organism. For example, a virus-infected cell that dies before completion of the infection cycle prevents spread of the virus to nearby cells. Severe stresses such as heat, hyperosmolarity, UV light, and gamma irradiation also trigger cell suicide; presumably the organism is better off with aberrant cells dead.

The regulatory mechanisms that trigger apoptosis involve some of the same proteins that regulate the cell cycle. The signal for suicide often comes from outside, through a surface receptor. Tumor necrosis factor (TNF), produced by cells of the immune system, interacts with cells through specific TNF receptors. These receptors have TNF-binding sites on the outer face of the plasma membrane and a "death domain" of about 80 amino acid residues that passes the self-destruct signal through the membrane to cytosolic proteins such as TRADD (*T*NF *r*eceptor-*a*ssociated *d*eath *d*omain) (Fig. 12–50). Another receptor, Fas, has a similar death domain that allows it to interact with the cytosolic protein FADD (*F*as-*a*ssociated *d*eath *d*omain), which activates a cytosolic protease called caspase 8. This enzyme belongs to a family of proteases that participate in apoptosis; all are synthesized as inactive proenzymes, all have a critical *Cys* residue at the active site, and all hydrolyze their target proteins on the carboxyl-terminal side of specific *Asp* residues (hence the name caspase).

When caspase 8, an "initiator" caspase, is activated by an apoptotic signal carried through FADD, it further self-activates by cleaving its own proenzyme form. Mitochondria are one target of active caspase 8. The protease causes the release of certain proteins contained between the inner and outer mitochondrial membranes:

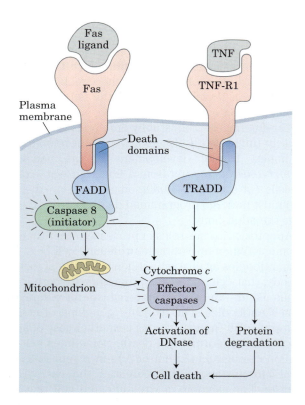

**FIGURE 12–50 Initial events of apoptosis.** Receptors in the plasma membrane (Fas, TNF-R1) receive signals from outside the cell (the Fas ligand or tumor necrosis factor (TNF), respectively). Activated receptors foster interaction between the "death domain" (an 80 amino acid sequence) in Fas or TNF-R1 and a similar death domain in the cytosolic proteins FADD or TRADD. FADD activates a cytosolic protease, caspase 8, that proteolytically activates other cellular proteases. TRADD also activates proteases. The resulting proteolysis is a primary factor in cell death.

cytochrome *c* (Chapter 19) and several "effector" caspases. Cytochrome *c* binds to the proenzyme form of the effector enzyme caspase 9 and stimulates its proteolytic activation. The activated caspase 9 in turn catalyzes wholesale destruction of cellular proteins—a major cause of apoptotic cell death. One specific target of caspase action is a caspase-activated deoxyribonuclease.

In apoptosis, the monomeric products of protein and DNA degradation (amino acids and nucleotides) are released in a controlled process that allows them to be taken up and reused by neighboring cells. Apoptosis thus allows the organism to eliminate a cell without wasting its components.

### SUMMARY 12.10 Oncogenes, Tumor Suppressor Genes, and Programmed Cell Death

- Oncogenes encode defective signaling proteins. By continually giving the signal for cell division, they lead to tumor formation. Oncogenes are genetically dominant and may encode defective growth factors, receptors, G proteins, protein kinases, or nuclear regulators of transcription.

- Tumor suppressor genes encode regulatory proteins that normally inhibit cell division; mutations in these genes are genetically recessive but can lead to tumor formation.

- Cancer is generally the result of an accumulation of mutations in oncogenes and tumor suppressor genes.

- Apoptosis can be triggered by extracellular signals such as TNF through plasma membrane receptors.

## Key Terms

*Terms in bold are defined in the glossary.*

| | | |
|---|---|---|
| **signal transduction**  421 | **stimulatory G protein (G$_s$)**  436 | **hormone response element (HRE)**  465 |
| **enzyme cascade**  422 | β-adrenergic receptor kinase (βARK)  441 | tamoxifen  465 |
| **desensitization**  422 | β-arrestin (βarr; arrestin 2)  441 | RU486  466 |
| ligand-gated receptor channel  426 | G protein–coupled receptor kinases (GRKs)  441 | **cyclin**  467 |
| voltage-gated ion channel  427 | scaffold proteins  441 | cyclin-dependent protein kinase (CDK)  467 |
| **second messenger**  428 | **inhibitory G protein (G$_i$)**  441 | **ubiquitin**  469 |
| **autophosphorylation**  429 | calmodulin (CaM)  444 | **proteasome**  469 |
| **SH2 domain**  429 | Ca$^{2+}$/calmodulin-dependent protein kinases (CaM kinases I–IV)  444 | **growth factors**  469 |
| **G proteins**  429 | **two-component signaling systems**  452 | **cytokine**  469 |
| MAPK cascade  430 | receptor His kinase  452 | retinoblastoma protein (pRb)  470 |
| receptor Tyr kinase  432 | response regulator  452 | **oncogene**  471 |
| **serpentine receptors**  435 | receptorlike kinase (RLK)  455 | **tumor suppressor genes**  472 |
| **G protein–coupled receptors (GPCR)**  435 | | programmed cell death  473 |
| **7 transmembrane segment (7tm) receptors**  435 | | **apoptosis**  473 |

# Further Reading

## General

**Cohen, P.** (2001) The role of protein phosphorylation in human health and disease: the Sir Hans Krebs Medal Lecture. *Eur. J. Biochem.* **268**, 5001–5010.

An intermediate-level review of the role of protein kinases, their alteration in disease, and the drugs that affect their activity.

**Cohen, P.** (2000) The regulation of protein function by multisite phosphorylation—a 25 year update. *Trends Biochem. Sci.* **25**, 596–601.

Historical account of the developments in protein phosphorylation.

**Heilmeyer, L. & Friedrich, P. (eds)** (2001) *Protein Modules in Cellular Signalling,* IOS Press, Washington, DC.

## Receptor Ion Channels

See also Chapter 11, Further Reading, Ion Channels.

**Aidley, D.J. & Stanfield, P.R.** (1996) *Ion Channels: Molecules in Action,* Cambridge University Press, Cambridge.

Clear, concise introduction to the physics, chemistry, and molecular biology used in research on ion channels; emphasis is on molecular approaches.

**Lehmann-Horn, F. & Jurkat-Rott, K.** (1999) Voltage-gated ion channels and hereditary disease. *Physiol. Rev.* **79**, 1317–1372.

Advanced review of the structure and function of ion channels, with special emphasis on cases in which ion-channel defects produce human diseases.

## Receptor Enzymes

**Foster, D.C., Wedel, B.J., Robinson, S.W., & Garbers, D.L.** (1999) Mechanisms of regulation and functions of guanylyl cyclases. *Rev. Physiol. Biochem. Pharmacol.* **135**, 1–39.

Advanced review of the structure and function of signal-transducing guanylyl cyclases.

**Saltiel, A.R. & Pessin, J.E.** (2002) Insulin signaling pathways in time and space. *Trends Biochem Sci.* **12**, 65–71.

Short, intermediate-level review.

**Schaeffer, H.J. & Weber, M.J.** (1999) Mitogen-activated protein kinases: specific messages from ubiquitous messengers. *Mol. Cell. Biol.* **19**, 2435–2444.

Intermediate-level review of MAPKs and the basis for specific signaling through these general signaling proteins.

**Schlessinger, J.** (2000) Cell signaling by receptor tyrosine kinases. *Cell* **103**, 211–225.

**Shepherd, P.R., Withers, D.J., & Siddle, K.** (1998) Phospho-inositide 3-kinase: the key switch mechanism in insulin signalling. *Biochem. J.* **333**, 471–490.

Intermediate-level review of the importance of PKB and PI-3K in metabolic regulation by insulin.

**Shields, J.M., Pruitt, K., McFall, A., Shaub, A., & Der, C.J.** (2000) Understanding Ras: it ain't over 'til it's over. *Trends Cell Biol.* **10**, 147–154.

Intermediate-level review of the monomeric G protein Ras.

**Widmann, C., Gibson, S., Jarpe, M.B., & Johnson, G.L.** (1999) Mitogen-activated protein kinase: conservation of a three-kinase module from yeast to human. *Physiol. Rev.* **79**, 143–180.

Advanced review of the roles of MAPKs in diverse organisms, from yeast, slime mold, and nematode to vertebrates and plants.

**Zajchowski, L.D. & Robbins, S.M.** (2002) Lipid rafts and little caves: compartmentalized signalling in membrane microdomains. *Eur. J. Biochem.* **269**, 737–752.

## Serpentine Receptors

**Hahn, K. & Toutchkine, A.** (2002) Live-cell fluorescent biosensors for activated signaling proteins. *Curr. Opin. Cell Biol.* **14**, 167–172.

Brief, intermediate-level review.

**Hamm, H.E.** (1998) The many faces of G protein signaling. *J. Biol. Chem.* **273**, 669–672.

Introduction to a series of short reviews on G proteins.

**Lodowski, D.T., Pitcher, J.A., Capel, W.D., Lefkowitz, R.J., & Tesmer, J.J.G.** (2003) Keeping G proteins at bay: a complex between G protein–coupled receptor kinase 2 and $G_{\beta\gamma}$. *Science* **300**, 1256–1262.

**Martin, T.F.J.** (1998) Phosphoinositide lipids as signaling molecules. *Annu. Rev. Cell Dev. Biol.* **14**, 231–264.

Discussion of the roles of phosphatidylinositol derivatives in signal transduction, cytoskeletal regulation, and membrane trafficking.

**Perry, S.J. & Lefkowitz, R.J.** (2002) Arresting developments in heptahelical receptor signaling and regulation. *Trends Cell Biol.* **12**, 130–138.

Role of arrestins in serpentine receptor adaptation.

**Pinna, L.A. & Ruzzene, M.** (1996) How do protein kinases recognize their substrates? *Biochim. Biophys. Acta* **1314**, 191–225.

Advanced review of the factors, including consensus sequences, that give protein kinases their specificity.

**Roach, P.J.** (1991) Multisite and hierarchal protein phosphorylation. *J. Biol. Chem.* **266**, 14,139–14,142.

Report on the importance of multiple phosphorylation sites in the fine regulation of protein function.

**Skiba, N.P. & Hamm, H.E.** (1998) How $G_{s\alpha}$ activates adenylyl cyclase. *Nat. Struct. Biol.* **5**, 88–92.

Intermediate-level review of the mechanism of G-protein action, based on structural studies.

**Zaccolo, M., De Giorgi, F., Cho, C.Y., Feng, L., Knapp, T., Negulescu, P.A., Taylor, S.S., Tsien, R.Y., & Pozzan, T.** (2000) A genetically encoded, fluorescent indicator for cyclic AMP in living cells. *Nat. Cell Biol.* **2**, 25–29.

Describes the technique shown in Figure 5 of Box 12–2.

**Zhang, J., Campbell, R.E., Ting, A.Y., & Tsien, R.Y.** (2002) Creating new fluorescent probes for cell biology. *Nat. Rev. Molec. Cell Biol.* **3**, 906–918.

The basis for techniques such as those described in Box 12–2.

## Scaffold Proteins and Membrane Rafts

See also Chapter 11, Further Reading, Membrane Dynamics.

**Filipek, S., Stenkamp, R.E., Teller, D.C., & Palczewski, K.** (2003) G protein-coupled receptor rhodopsin: A prospectus. *Annu. Rev. Physiol.* **65**, 851–879.

Advanced review.

**Lim, W.A.** (2002) The modular logic of signaling proteins: building allosteric switches from simple binding domains. *Curr. Opin. Struct. Biol.* **12,** 61–68.

**Michel, J.J.C. & Scott, J.D.** (2002) AKAP mediated signal transduction. *Annu. Rev. Pharmacol. Toxicol.* **42,** 235–257.
>Advanced review of the proteins that target PKA to subcellular compartments.

**Neel, B.G., Gu, H., & Pao, L.** (2003) The "Shp"ing news: SH2 domain–containing tyrosine phosphatases in cell signaling. *Trends Biochem. Sci.* **28,** 284–293.

**Pawson, T. & Nash, P.** (2003) Assembly of cell regulatory systems through protein interaction domains. *Science* **300,** 445–452.
>Intermediate-level review of multivalent proteins and the complexes they form.

**Saltiel, A.R. & Pessin, J.E.** (2002) Insulin signaling pathways in time and space. *Trends Cell Biol.* **12,** 65–71.

**Siegal, G.** (1999) The surprisingly flexible PTB domain. *Nat. Struct. Biol.* **6,** 7–10.

**Tsui-Pierchala, B.A., Encinas, M., Milbrandt, J., & Johnson, E.M., Jr.** (2002) Lipid rafts in neuronal signaling and function. *Trends Neurosci.* **25,** 412–417.

**Yaffe, M.D. & Elia, A.E.H.** (2001) Phosphoserine/threonine-binding domains. *Curr. Opin. Cell Biol.* **13,** 131–138.

### Calcium Ions in Signaling

**Berridge, M.J.** (1993) Inositol triphosphate and calcium signalling. *Nature* **361,** 315–325.
>Classic description of the IP$_3$ signaling system.

**Berridge, M.J., Lipp, P., & Bootman, M.D.** (2000) The versatility and universality of calcium signaling. *Nat. Rev. Molec. Cell Biol.* **1,** 11–21.
>Intermediate review.

**Chin, D. & Means, A.R.** (2000) Calmodulin: a prototypical calcium receptor. *Trends Cell Biol.* **10,** 322–328.

**Nowycky, M.C. & Thomas, A.P.** (2002) Intracellular calcium signaling. *J. Cell Sci.* **115,** 3715–3716.
>A comprehensive poster of all elements of Ca$^{2+}$ signaling.

**Takahashi, A., Camacho, P., Lechleiter, J.D., & Herman, B.** (1999) Measurement of intracellular calcium. *Physiol. Rev.* **79,** 1089–1125.
>Advanced review of methods for estimating intracellular Ca$^{2+}$ levels in real time.

**Thomas, A.P., Bird, G.S.J., Hajnoczky, G., Robb-Gaspers, L.D., & Putney, J.W., Jr.** (1996) Spatial and temporal aspects of cellular calcium signaling. *FASEB J.* **10,** 1505–1517.

### Signaling in Plants and Bacteria

**Bakal, C.J. & Davies, J.E.** (2000) No longer an exclusive club: eukaryotic signaling domains in bacteria. *Trends Cell Biol.* **10,** 32–38.
>Intermediate review.

**Becraft, P.W.** (2002) Receptor kinase signaling in plant development. *Annu. Rev. Cell Dev. Biol.* **18,** 163–192.
>Advanced review.

**Chang, C. & Stadler, R.** (2001) Ethylene hormone receptor action in *Arabidopsis*. *Bioessays* **23,** 619–627.

**Cock, J.M., Vanoosthuyse, V., & Gaude, T.** (2002) Receptor kinase signalling in plants and animals: distinct molecular systems with mechanistic similarities. *Curr. Opin. Cell Biol.* **14,** 230–236.

**Hwang, I., Chen, H.-C., & Sheen, J.** (2002) Two-component circuitry in *Arabidopsis* cytokinin signal transduction. *Nature* **413,** 383–389.

**Jones, A.M.** (2002) G-protein–coupled signaling in *Arabidopsis*. *Curr. Opin. Plant Biol.* **5,** 402–407.

**Matsubayashi, Y., Yang, H., & Sakagami, Y.** (2001) Peptide signals and their receptors in higher plants. *Trends Plant Sci.* **6,** 573–577.

**McCarty, D.R. & Chory, J.** (2000) Conservation and innovation in plant signaling pathways. *Cell* **103,** 201–209.

**Meijer, H.J.G. & Munnik, T.** (2003) Phospholipid-based signaling in plants. *Annu. Rev. Plant Biol.* **54,** 265–306.
>Advanced review.

**Ouaked, F., Rozhon, W., Lecourieus, D., & Hirt, H.** (2003) A MAPK pathway mediates ethylene signaling in plants. *EMBO J.* **22,** 1282–1288.

**Talke, I.N., Blaudez, D., Maathuis, F.J.M., & Sanders, D.** (2003) CNGCs: prime targets of plant cyclic nucleotide signalling? *Trends Plant Sci.* **8,** 286–293.

**Tichtinsky, G., Vanoosthuyse, V., Cock, J.M., & Gaude, T.** (2003) Making inroads into plant receptor kinase signalling pathways. *Trends Plant Sci.* **8,** 231–237.

### Vision, Olfaction, and Gustation

**Baylor, D.** (1996) How photons start vision. *Proc. Natl. Acad. Sci. USA* **93,** 560–565.
>One of six short reviews on vision in this journal issue.

**Margolskee, R.F.** (2002) Molecular mechanisms of bitter and sweet taste transduction. *J. Biol. Chem.* **277,** 1–4.

**Menon, S.T., Han, M., & Sakmar, T.P.** (2001) Rhodopsin: structural basis of molecular physiology. *Physiol. Rev.* **81,** 1659–1688.
>Advanced review.

**Mombaerts, P.** (2001) The human repertoire of odorant receptor genes and pseudogenes. *Annu. Rev. Genomics Hum. Genet.* **2,** 493–510.
>Advanced review.

**Nathans, J.** (1989) The genes for color vision. *Sci. Am.* **260** (February), 42–49.

**Ronnett, G.V. & Moon, C.** (2002) G proteins and olfactory signal transduction. *Annu. Rev. Physiol.* **64,** 189–222.
>Advanced review.

**Scott, K. & Zuker, C.** (1997) Lights out: deactivation of the phototransduction cascade. *Trends Biochem. Sci.* **22,** 350–354.

### Steroid Hormone Receptors and Action

**Carson-Jurica, M.A., Schrader, W.T., & O'Malley, B.W.** (1990) Steroid receptor family—structure and functions. *Endocr. Rev.* **11,** 201–220.
>Advanced discussion of the structure of nuclear hormone receptors and the mechanisms of their action.

**Hall, J.M., Couse, J.F., & Korach, K.S.** (2001) The multifaceted mechanisms of estradiol and estrogen receptor signaling. *J. Biol. Chem.* **276,** 36,869–36,872.
>Brief, intermediate-level review.

**Jordan, V.C.** (1998) Designer estrogens. *Sci. Am.* **279** (October), 60–67.

Introductory review of the mechanism of estrogen action and the effects of estrogenlike compounds in medicine.

**Lösel, R.M., Falkenstein, E., Feuring, M., Schultz, A., Tillmann, H.-C., Rossol-Haseroth, K., & Wehling, M.** (2003) Nongenomic steroid action: controversies, questions, and answers. *Physiol. Rev.* **83**, 965–1016.

Detailed review of the evidence for steroid hormone action through plasma membrane receptors.

## Cell Cycle and Cancer

**Cavenee, W.K. & White, R.L.** (1995) The genetic basis of cancer. *Sci. Am.* **272** (March), 72–79.

**Chau, B.N. & Wang, J.Y.J.** (2003) Coordinated regulation of life and death by RB. *Nat. Rev. Cancer* **3**, 130–138.

**Fearon, E.R.** (1997) Human cancer syndromes—clues to the origin and nature of cancer. *Science* **278**, 1043–1050.

Intermediate-level review of the role of inherited mutations in the development of cancer.

**Herwig, S. & Strauss, M.** (1997) The retinoblastoma protein: a master regulator of cell cycle, differentiation and apoptosis. *Eur. J. Biochem.* **246**, 581–601.

**Hunt, M. & Hunt, T.** (1993) *The Cell Cycle: An Introduction,* W. H. Freeman and Company/Oxford University Press, New York/Oxford.

**Kinzler, K.W. & Vogelstein, B.** (1996) Lessons from hereditary colorectal cancer. *Cell* **87**, 159–170.

Evidence for multistep processes in the development of cancer.

**Levine, A.J.** (1997) p53, the cellular gatekeeper for growth and division. *Cell* **88**, 323–331.

Intermediate coverage of the function of protein p53 in the normal cell cycle and in cancer.

**Morgan, D.O.** (1997) Cyclin-dependent kinases: engines, clocks, and microprocessors. *Annu. Rev. Cell Dev. Biol.* **13**, 261–291.

Advanced review.

**Obaya, A.J. & Sedivy, J.M.** (2002) Regulation of cyclin-Cdk activity in mammalian cells. *Cell. Mol. Life Sci.* **59**, 126–142.

**Rajagopalan, H., Nowak, M.A., Vogelstein, B., & Lengauer, C.** (2003) The significance of unstable chromosomes in colorectal cancer. *Nat. Rev. Cancer* **3**, 695–701.

**Sherr, C.J. & McCormick, F.** (2002) The RB and p53 pathways in cancer. *Cancer Cell* **2**, 102–112.

**Weinberg, R.A.** (1996) How cancer arises. *Sci. Am.* **275** (September), 62–70.

**Yamasaki, L.** (2003) Role of the RB tumor suppressor in cancer. *Cancer Treatment Res.* **115**, 209–239.

Advanced review.

## Apoptosis

**Anderson, P.** (1997) Kinase cascades regulating entry into apoptosis. *Microbiol. Mol. Biol. Rev.* **61**, 33–46.

**Ashkenazi, A. & Dixit, V.M.** (1998) Death receptors: signaling and modulation. *Science* **281**, 1305–1308.

This and the papers by Green and Reed and by Thornberry and Lazebnik (below) are in an issue of *Science* devoted to apoptosis.

**Duke, R.C., Ojcius, D.M., & Young, J.D.-E.** (1996) Cell suicide in health and disease. *Sci. Am.* **275** (December), 80–87.

**Green, D.R. & Reed, J.C.** (1998) Mitochondria and apoptosis. *Science* **281**, 1309–1312.

**Jacobson, M.D., Weil, M., & Raff, M.C.** (1997) Programmed cell death in animal development. *Cell* **88**, 347–354.

**Lawen, A.** (2003) Apoptosis—an introduction. *Bioessays* **25**, 888–896.

**Thornberry, N.A. & Lazebnik, Y.** (1998) Caspases: enemies within. *Science* **281**, 1312.

# Problems

**1. Therapeutic Effects of Albuterol** The respiratory symptoms of asthma result from constriction of the bronchi and bronchioles of the lungs due to contraction of the smooth muscle of their walls. This constriction can be reversed by raising the [cAMP] in the smooth muscle. Explain the therapeutic effects of albuterol, a β-adrenergic agonist taken (by inhalation) for asthma. Would you expect this drug to have any side effects? How might one design a better drug that did not have these effects?

**2. Amplification of Hormonal Signals** Describe all the sources of amplification in the insulin receptor system.

**3. Termination of Hormonal Signals** Signals carried by hormones must eventually be terminated. Describe several different mechanisms for signal termination.

**4. Specificity of a Signal for a Single Cell Type** Discuss the validity of the following proposition. A signaling molecule (hormone, growth factor, or neurotransmitter) elicits identical responses in different types of target cells if they contain identical receptors.

**5. Resting Membrane Potential** A variety of unusual invertebrates, including giant clams, mussels, and polychaete worms, live on the fringes of hydrothermal vents on the ocean bottom, where the temperature is 60 °C.

(a) The adductor muscle of a deep-sea giant clam has a resting membrane potential of −95 mV. Given the intracellular and extracellular ionic compositions shown below, would you have predicted this membrane potential? Why or why not?

| Ion | Concentration (mM) | |
| --- | --- | --- |
| | Intracellular | Extracellular |
| $Na^+$ | 50 | 440 |
| $K^+$ | 400 | 20 |
| $Cl^-$ | 21 | 560 |
| $Ca^{2+}$ | 0.4 | 10 |

(b) Assume that the adductor muscle membrane is permeable to only one of the ions listed above. Which ion could determine the $V_m$?

**6. Membrane Potentials in Frog Eggs** Fertilization of a frog oocyte by a sperm cell triggers ionic changes similar to those observed in neurons (during movement of the action potential) and initiates the events that result in cell division and development of the embryo. Oocytes can be stimulated to divide without fertilization by suspending them in 80 mM KCl (normal pond water contains 9 mM KCl).

(a) Calculate how much the change in extracellular [KCl] changes the resting membrane potential of the oocyte. (Hint: Assume the oocyte contains 120 mM $K^+$ and is permeable *only* to $K^+$.) Assume a temperature of 20 °C.

(b) When the experiment is repeated in $Ca^{2+}$-free water, elevated [KCl] has no effect. What does this suggest about the mechanism of the KCl effect?

**7. Excitation Triggered by Hyperpolarization** In most neurons, membrane *depolarization* leads to the opening of voltage-dependent ion channels, generation of an action potential, and ultimately an influx of $Ca^{2+}$, which causes release of neurotransmitter at the axon terminus. Devise a cellular strategy by which *hyperpolarization* in rod cells could produce excitation of the visual pathway and passage of visual signals to the brain. (Hint: The neuronal signaling pathway in higher organisms consists of a *series* of neurons that relay information to the brain (see Fig. 12–31). The signal released by one neuron can be either excitatory or inhibitory to the following, postsynaptic neuron.)

**8. Hormone Experiments in Cell-Free Systems** In the 1950s, Earl W. Sutherland, Jr., and his colleagues carried out pioneering experiments to elucidate the mechanism of action of epinephrine and glucagon. Given what you have learned in this chapter about hormone action, interpret each of the experiments described below. Identify substance X and indicate the significance of the results.

(a) Addition of epinephrine to a homogenate of normal liver resulted in an increase in the activity of glycogen phosphorylase. However, if the homogenate was first centrifuged at a high speed and epinephrine or glucagon was added to the clear supernatant fraction that contains phosphorylase, no increase in the phosphorylase activity occurred.

(b) When the particulate fraction from the centrifugation in (a) was treated with epinephrine, substance X was produced. The substance was isolated and purified. Unlike epinephrine, substance X activated glycogen phosphorylase when added to the clear supernatant fraction of the centrifuged homogenate.

(c) Substance X was heat-stable; that is, heat treatment did not affect its capacity to activate phosphorylase. (Hint: Would this be the case if substance X were a protein?) Substance X was nearly identical to a compound obtained when pure ATP was treated with barium hydroxide. (Fig. 8–6 will be helpful.)

**9. Effect of Cholera Toxin on Adenylyl Cyclase** The gram-negative bacterium *Vibrio cholerae* produces a protein, cholera toxin ($M_r$ 90,000), that is responsible for the characteristic symptoms of cholera: extensive loss of body water and $Na^+$ through continuous, debilitating diarrhea. If body fluids and $Na^+$ are not replaced, severe dehydration results; untreated, the disease is often fatal. When the cholera toxin gains access to the human intestinal tract it binds tightly to specific sites in the plasma membrane of the epithelial cells lining the small intestine, causing adenylyl cyclase to undergo prolonged activation (hours or days).

(a) What is the effect of cholera toxin on [cAMP] in the intestinal cells?

(b) Based on the information above, suggest how cAMP normally functions in intestinal epithelial cells.

(c) Suggest a possible treatment for cholera.

**10. Effect of Dibutyryl cAMP versus cAMP on Intact Cells** The physiological effects of epinephrine should in principle be mimicked by addition of cAMP to the target cells. In practice, addition of cAMP to intact target cells elicits only a minimal physiological response. Why? When the structurally related derivative dibutyryl cAMP (shown below) is added to intact cells, the expected physiological response is readily apparent. Explain the basis for the difference in cellular response to these two substances. Dibutyryl cAMP is widely used in studies of cAMP function.

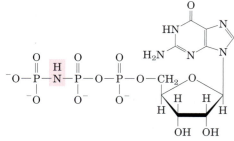

Dibutyryl cAMP
($N^6,O^{2'}$-Dibutyryl adenosine 3′,5′-cyclic monophosphate)

**11. Nonhydrolyzable GTP Analogs** Many enzymes can hydrolyze GTP between the $\beta$ and $\gamma$ phosphates. The GTP analog $\beta,\gamma$-imidoguanosine 5′-triphosphate Gpp(NH)p, shown below, cannot be hydrolyzed between the $\beta$ and $\gamma$ phosphates. Predict the effect of microinjection of Gpp(NH)p into a myocyte on the cell's response to $\beta$-adrenergic stimulation.

Gpp(NH)p
($\beta,\gamma$-imidoguanosine 5′-triphosphate)

**12. G Protein Differences** Compare the G proteins $G_s$, which acts in transducing the signal from $\beta$-adrenergic receptors, and Ras. What properties do they share? How do they differ? What is the functional difference between $G_s$ and $G_I$?

**13. EGTA Injection**  EGTA (ethylene glycol-bis($\beta$-amino-ethyl ether)-$N,N,N',N'$-tetraacetic acid) is a chelating agent with high affinity and specificity for $Ca^{2+}$. By microinjecting a cell with an appropriate $Ca^{2+}$-EDTA solution, an experimenter can prevent cytosolic $[Ca^{2+}]$ from rising above $10^{-7}$ M. How would EGTA microinjection affect a cell's response to vasopressin (see Table 12–5)? To glucagon?

**14. Visual Desensitization**  Oguchi's disease is an inherited form of night blindness. Affected individuals are slow to recover vision after a flash of bright light against a dark background, such as the headlights of a car on the freeway. Suggest what the molecular defect(s) might be in Oguchi's disease. Explain in molecular terms how this defect accounts for the night blindness.

**15. Mutations in PKA**  Explain how mutations in the R or C subunit of cAMP-dependent protein kinase (PKA) might lead to (a) a constantly active PKA or (b) a constantly inactive PKA.

**16. Mechanisms for Regulating Protein Kinases**  Identify eight general types of protein kinases found in eukaryotic cells, and explain what factor is *directly* responsible for activating each type.

**17. Mutations in Tumor Suppressor Genes and Oncogenes**  Explain why mutations in tumor suppressor genes are recessive (both copies of the gene must be defective for the regulation of cell division to be defective) whereas mutations in oncogenes are dominant.

**18. Retinoblastoma in Children**  Explain why some children with retinoblastoma develop multiple tumors of the retina in both eyes, whereas others have a single tumor in only one eye.

**19. Mutations in *ras***  How does a mutation in the *ras* gene that leads to formation of a Ras protein with no GTPase activity affect a cell's response to insulin?

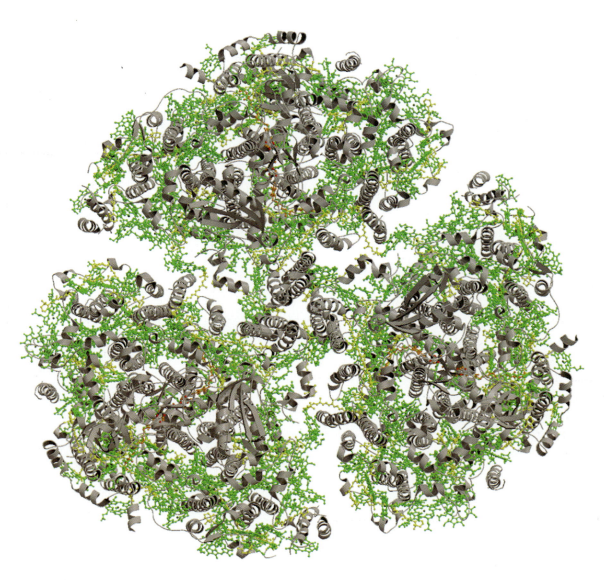

Photosystem I

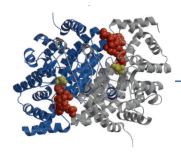

# BIOENERGETICS AND METABOLISM

Metabolism is a highly coordinated cellular activity in which many multienzyme systems (metabolic pathways) cooperate to (1) obtain chemical energy by capturing solar energy or degrading energy-rich nutrients from the environment; (2) convert nutrient molecules into the cell's own characteristic molecules, including precursors of macromolecules; (3) polymerize monomeric precursors into macromolecules: proteins, nucleic acids, and polysaccharides; and (4) synthesize and degrade biomolecules required for specialized cellular functions, such as membrane lipids, intracellular messengers, and pigments.

Although metabolism embraces hundreds of different enzyme-catalyzed reactions, our major concern in Part II is the central metabolic pathways, which are few in number and remarkably similar in all forms of life. Living organisms can be divided into two large groups according to the chemical form in which they obtain carbon from the environment. **Autotrophs** (such as photosynthetic bacteria and vascular plants) can use carbon dioxide from the atmosphere as their sole source of carbon, from which they construct all their carbon-containing biomolecules (see Fig. 1–5). Some autotrophic organisms, such as cyanobacteria, can also use atmospheric nitrogen to generate all their nitrogenous components. **Heterotrophs** cannot use atmospheric carbon dioxide and must obtain carbon from their environment in the form of relatively complex organic molecules such as glucose. Multicellular animals and most microorganisms are heterotrophic. Autotrophic cells and organisms are relatively self-sufficient, whereas heterotrophic cells and organisms, with their requirements for carbon in more complex forms, must subsist on the products of other organisms.

Many autotrophic organisms are photosynthetic and obtain their energy from sunlight, whereas heterotrophic organisms obtain their energy from the degradation of organic nutrients produced by autotrophs. In our biosphere, autotrophs and heterotrophs live together in a vast, interdependent cycle in which autotrophic organisms use atmospheric carbon dioxide to build their organic biomolecules, some of them generating oxygen from water in the process. Heterotrophs in turn use the organic products of autotrophs as nutrients and return carbon dioxide to the atmosphere. Some of the oxidation reactions that produce carbon dioxide also consume oxygen, converting it to water. Thus carbon, oxygen, and water are constantly cycled between the heterotrophic and autotrophic worlds, with solar energy as the driving force for this global process (Fig. 1).

All living organisms also require a source of nitrogen, which is necessary for the synthesis of amino acids, nucleotides, and other compounds. Plants can generally use either ammonia or nitrate as their sole source of nitrogen, but vertebrates must obtain nitrogen in the form of amino acids or other organic compounds. Only a few organisms—the cyanobacteria and many species of soil bacteria that live symbiotically on the roots of some plants—are capable of converting ("fixing") atmospheric nitrogen ($N_2$) into ammonia. Other bacteria (the nitrifying bacteria) oxidize ammonia to nitrites and nitrates; yet others convert nitrate to $N_2$. Thus, in addition to the global carbon and oxygen cycle, a nitrogen cycle operates in the biosphere, turning over huge amounts of nitrogen (Fig. 2). The cycling of carbon, oxygen, and nitrogen, which ultimately involves all species, depends on a proper balance between the activities of the producers (autotrophs) and consumers (heterotrophs) in our biosphere.

These cycles of matter are driven by an enormous flow of energy into and through the biosphere, beginning with the capture of solar energy by photosynthetic organisms and use of this energy to generate energy-rich carbohydrates and other organic nutrients; these nutrients are then used as energy sources by heterotrophic organisms. In metabolic processes, and in all energy transformations, there is a loss of useful energy (free energy) and an inevitable increase in the amount of unusable energy (heat and entropy). In contrast to the cycling of matter, therefore, energy flows one way

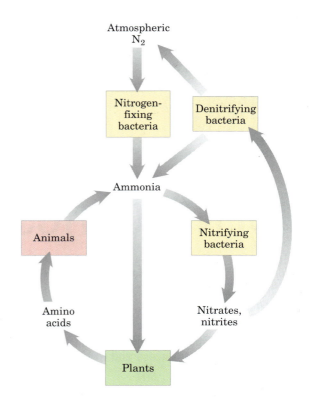

**FIGURE 2  Cycling of nitrogen in the biosphere.** Gaseous nitrogen ($N_2$) makes up 80% of the earth's atmosphere.

through the biosphere; organisms cannot regenerate useful energy from energy dissipated as heat and entropy. Carbon, oxygen, and nitrogen recycle continuously, but energy is constantly transformed into unusable forms such as heat.

**Metabolism,** the sum of all the chemical transformations taking place in a cell or organism, occurs through a series of enzyme-catalyzed reactions that constitute **metabolic pathways.** Each of the consecutive steps in a metabolic pathway brings about a specific, small chemical change, usually the removal, transfer, or addition of a particular atom or functional group. The precursor is converted into a product through a series of metabolic intermediates called **metabolites.** The term **intermediary metabolism** is often applied to the combined activities of all the metabolic pathways that interconvert precursors, metabolites, and products of low molecular weight (generally, $M_r <1,000$).

**Catabolism** is the degradative phase of metabolism in which organic nutrient molecules (carbohydrates, fats, and proteins) are converted into smaller, simpler end products (such as lactic acid, $CO_2$, $NH_3$). Catabolic pathways release energy, some of which is conserved in the formation of ATP and reduced electron carriers (NADH, NADPH, and $FADH_2$); the rest is lost as heat. In **anabolism,** also called biosynthesis, small, simple precursors are built up into larger and more complex

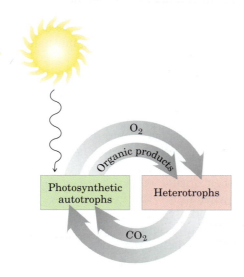

**FIGURE 1  Cycling of carbon dioxide and oxygen between the autotrophic (photosynthetic) and heterotrophic domains in the biosphere.** The flow of mass through this cycle is enormous; about $4 \times 10^{11}$ metric tons of carbon are turned over in the biosphere annually.

molecules, including lipids, polysaccharides, proteins, and nucleic acids. Anabolic reactions require an input of energy, generally in the form of the phosphoryl group transfer potential of ATP and the reducing power of NADH, NADPH, and $FADH_2$ (Fig. 3).

Some metabolic pathways are linear, and some are branched, yielding multiple useful end products from a single precursor or converting several starting materials into a single product. In general, catabolic pathways are convergent and anabolic pathways divergent (Fig. 4). Some pathways are cyclic: one starting component of the pathway is regenerated in a series of reactions that converts another starting component into a product. We shall see examples of each type of pathway in the following chapters.

Most cells have the enzymes to carry out both the degradation and the synthesis of the important categories of biomolecules—fatty acids, for example. The

simultaneous synthesis and degradation of fatty acids would be wasteful, however, and this is prevented by reciprocally regulating the anabolic and catabolic reaction sequences: when one sequence is active, the other is suppressed. Such regulation could not occur if anabolic and catabolic pathways were catalyzed by exactly the same set of enzymes, operating in one direction for anabolism, the opposite direction for catabolism: inhibition of an enzyme involved in catabolism would also inhibit the reaction sequence in the anabolic direction. Catabolic and anabolic pathways that connect the same two end points (glucose → → pyruvate and pyruvate → → glucose, for example) may employ many of the same enzymes, but invariably at least one of the steps is catalyzed by different enzymes in the catabolic and anabolic directions, and these enzymes are the sites of separate regulation. Moreover, for both anabolic and catabolic pathways to be essentially irreversible, the reactions unique to each direction must include at least one that is thermodynamically very favorable—in other words, a reaction for which the reverse reaction is very unfavorable. As a further contribution to the separate regulation of catabolic and anabolic reaction sequences, paired catabolic and anabolic pathways commonly take place in different cellular compartments: for example, fatty acid catabolism in mitochondria, fatty acid synthesis in the cytosol. The concentrations of intermediates, enzymes, and regulators can be maintained at different levels in these different compartments. Because metabolic pathways are subject to kinetic control by substrate concentration, separate pools of anabolic and catabolic intermediates also contribute to the control of metabolic rates. Devices that separate anabolic and catabolic processes will be of particular interest in our discussions of metabolism.

Metabolic pathways are regulated at several levels, from within the cell and from outside. The most immediate regulation is by the availability of substrate; when the intracellular concentration of an enzyme's substrate is near or below $K_m$ (as is commonly the case), the rate of the reaction depends strongly upon substrate concentration (see Fig. 6–11). A second type of rapid control from within is allosteric regulation (p. 225) by a metabolic intermediate or coenzyme—an amino acid or ATP, for example—that signals the cell's internal metabolic state. When the cell contains an amount of, say, aspartate sufficient for its immediate needs, or when the cellular level of ATP indicates that further fuel consumption is unnecessary at the moment, these signals allosterically inhibit the activity of one or more enzymes in the relevant pathway. In multicellular organisms the metabolic activities of different tissues are regulated and integrated by growth factors and hormones that act from outside the cell. In some cases this regulation occurs virtually instantaneously (sometimes in less than a millisecond) through changes in the levels of intracellular

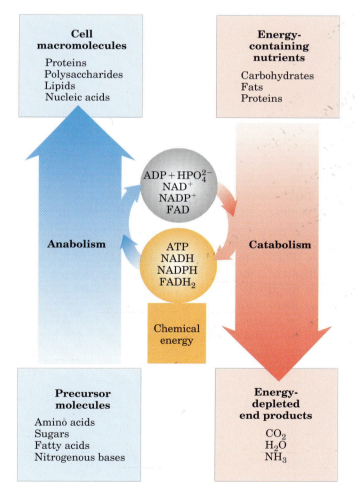

**FIGURE 3   Energy relationships between catabolic and anabolic pathways.** Catabolic pathways deliver chemical energy in the form of ATP, NADH, NADPH, and $FADH_2$. These energy carriers are used in anabolic pathways to convert small precursor molecules into cell macromolecules.

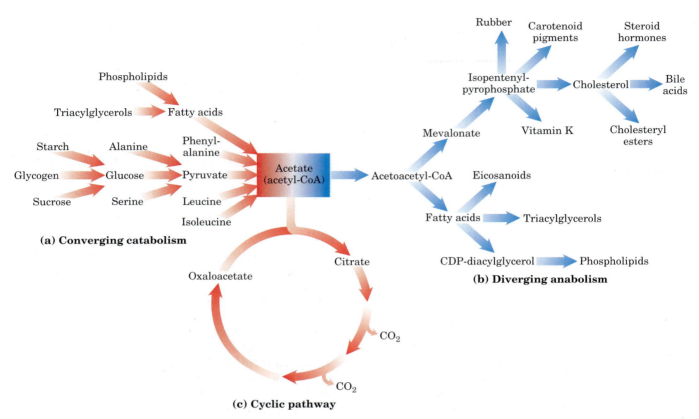

**(a) Converging catabolism**

**(b) Diverging anabolism**

**(c) Cyclic pathway**

**FIGURE 4  Three types of nonlinear metabolic pathways. (a)** Converging, catabolic; **(b)** diverging, anabolic; and **(c)** cyclic, in which one of the starting materials (oxaloacetate in this case) is regenerated and reenters the pathway. Acetate, a key metabolic intermediate, is the breakdown product of a variety of fuels **(a),** serves as the precursor for an array of products **(b),** and is consumed in the catabolic pathway known as the citric acid cycle **(c).**

messengers that modify the activity of existing enzyme molecules by allosteric mechanisms or by covalent modification such as phosphorylation. In other cases, the extracellular signal changes the cellular concentration of an enzyme by altering the rate of its synthesis or degradation, so the effect is seen only after minutes or hours.

The number of metabolic transformations taking place in a typical cell can seem overwhelming to a beginning student. Most cells have the capacity to carry out thousands of specific, enzyme-catalyzed reactions: for example, transformation of a simple nutrient such as glucose into amino acids, nucleotides, or lipids; extraction of energy from fuels by oxidation; or polymerization of monomeric subunits into macromolecules. Fortunately for the student of biochemistry, there are patterns within this multitude of reactions; you do not need to learn all these reactions to comprehend the molecular logic of biochemistry. Most of the reactions in living cells fall into one of five general categories: (1) oxidation-reductions; (2) reactions that make or break carbon–carbon bonds; (3) internal rearrangements, isomerizations, and eliminations; (4) group transfers; and (5) free radical reactions. Reactions within each general category usually proceed by a limited set of mechanisms and often employ characteristic cofactors.

Before reviewing the five main reaction classes of biochemistry, let's consider two basic chemical principles. First, a covalent bond consists of a shared pair of electrons, and the bond can be broken in two general ways (Fig. 5). In homolytic cleavage, each atom leaves the bond as a radical, carrying one of the two electrons (now unpaired) that held the bonded atoms together. In the more common, heterolytic cleavage, one atom retains both bonding electrons. The species generated when C—C and C—H bonds are cleaved are illustrated in Figure 5. Carbanions, carbocations, and hydride ions are highly unstable; this instability shapes the chemistry of these ions, as described further below.

The second chemical principle of interest here is that many biochemical reactions involve interactions between nucleophiles (functional groups rich in electrons and capable of donating them) and electrophiles (electron-deficient functional groups that seek electrons). Nucleophiles combine with, and give up electrons to, electrophiles. Common nucleophiles and electrophiles are listed in Figure 6–21. Note that a carbon atom can act as either a nucleophile or an electrophile, depending on which bonds and functional groups surround it.

We now consider the five main reaction classes you will encounter in upcoming chapters.

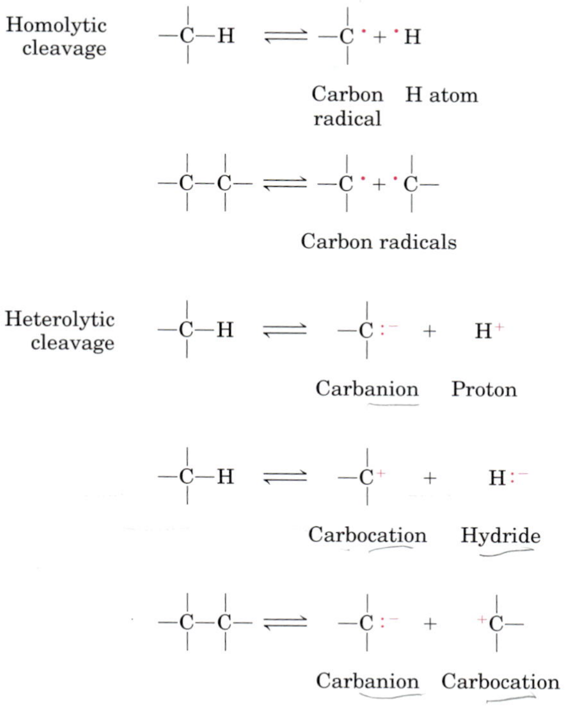

Homolytic cleavage

$$-\overset{|}{\underset{|}{C}}-H \;\rightleftharpoons\; -\overset{|}{\underset{|}{C}}\cdot + \cdot H$$

Carbon    H atom
radical

$$-\overset{|}{\underset{|}{C}}-\overset{|}{\underset{|}{C}}- \;\rightleftharpoons\; -\overset{|}{\underset{|}{C}}\cdot + \cdot\overset{|}{\underset{|}{C}}-$$

Carbon radicals

Heterolytic cleavage

$$-\overset{|}{\underset{|}{C}}-H \;\rightleftharpoons\; -\overset{|}{\underset{|}{C}}{:}^- + H^+$$

Carbanion    Proton

$$-\overset{|}{\underset{|}{C}}-H \;\rightleftharpoons\; -\overset{|}{\underset{|}{C}}{}^+ + H{:}^-$$

Carbocation    Hydride

$$-\overset{|}{\underset{|}{C}}-\overset{|}{\underset{|}{C}}- \;\rightleftharpoons\; -\overset{|}{\underset{|}{C}}{:}^- + {}^+\overset{|}{\underset{|}{C}}-$$

Carbanion    Carbocation

**FIGURE 5**  **Two mechanisms for cleavage of a C—C or C—H bond.** In homolytic cleavages, each atom keeps one of the bonding electrons, resulting in the formation of carbon radicals (carbons having unpaired electrons) or uncharged hydrogen atoms. In heterolytic cleavages, one of the atoms retains both bonding electrons. This can result in the formation of carbanions, carbocations, protons, or hydride ions.

**1. Oxidation-reduction reactions**    Carbon atoms encountered in biochemistry can exist in five oxidation states, depending on the elements with which carbon shares electrons (Fig. 6). In many biological oxidations, a compound loses two electrons and two hydrogen ions (that is, two hydrogen atoms); these reactions are commonly called dehydrogenations and the enzymes that catalyze them are called dehydrogenases (Fig. 7). In some, but not all, biological oxidations, a carbon atom becomes covalently bonded to an oxygen atom. The enzymes that

catalyze these oxidations are generally called oxidases or, if the oxygen atom is derived directly from molecular oxygen ($O_2$), oxygenases.

Every oxidation must be accompanied by a reduction, in which an electron acceptor acquires the electrons removed by oxidation. Oxidation reactions generally release energy (think of camp fires: the compounds in wood are oxidized by oxygen molecules in the air). Most living cells obtain the energy needed for cellular work by oxidizing metabolic fuels such as carbohydrates or fat; photosynthetic organisms can also trap and use the energy of sunlight. The catabolic (energy-yielding) pathways described in Chapters 14 through 19 are oxidative reaction sequences that result in the transfer of electrons from fuel molecules, through a series of electron carriers, to oxygen. The high affinity of $O_2$ for electrons makes the overall electron-transfer process highly exergonic, providing the energy that drives ATP synthesis—the central goal of catabolism.

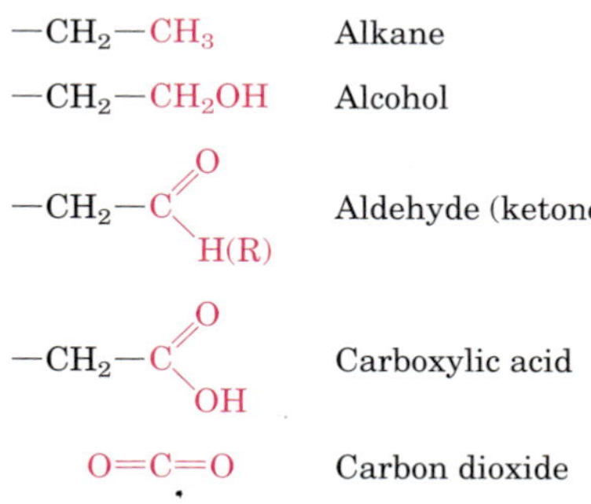

**2. Reactions that make or break carbon–carbon bonds**    Heterolytic cleavage of a C—C bond yields a carbanion and a carbocation (Fig. 5). Conversely, the formation of a C—C bond involves the combination of a nucleophilic carbanion and an electrophilic carbocation. Groups with electronegative atoms play key roles in these reactions. Carbonyl groups are particularly important in the chemical transformations of metabolic pathways. As noted above, the carbon of a carbonyl group has a partial positive charge due to the electron-withdrawing nature of the adjacent bonded oxygen, and thus is an electrophilic carbon. The presence of a carbonyl group can also facilitate the formation of a carbanion on an adjoining carbon, because the carbonyl group can delocalize electrons through resonance (Fig. 8a, b). The importance of a carbonyl group is evident in three major classes of reactions in which C—C bonds are formed or broken (Fig 8c): aldol condensations (such as the aldolase reaction; see Fig. 14–5), Claisen condensations (as in the citrate synthase reaction; see Fig. 16–9), and

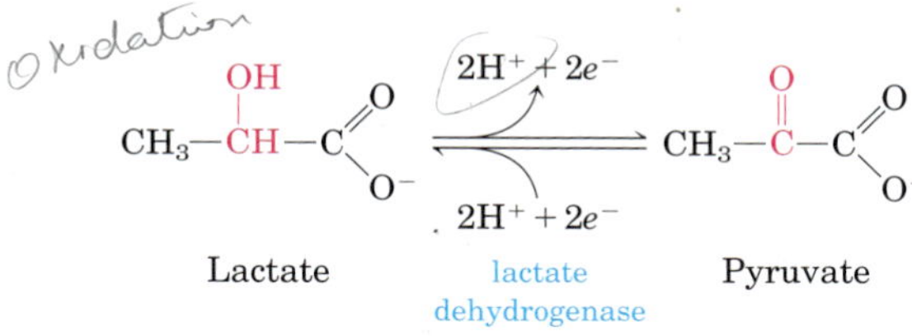

$-CH_2-CH_3$    Alkane

$-CH_2-CH_2OH$    Alcohol

$-CH_2-\overset{O}{\underset{H(R)}{C}}$    Aldehyde (ketone)

$-CH_2-\overset{O}{\underset{OH}{C}}$    Carboxylic acid

$O{=}C{=}O$    Carbon dioxide

**FIGURE 6**  **The oxidation states of carbon in biomolecules.** Each compound is formed by oxidation of the red carbon in the compound listed above it. Carbon dioxide is the most highly oxidized form of carbon found in living systems.

$$CH_3-\overset{OH}{\underset{\underset{O^-}{}}{\overset{|}{CH}}}-\overset{O}{C} \;\underset{2H^+ + 2e^-}{\overset{2H^+ + 2e^-}{\rightleftharpoons}}\; CH_3-\overset{O}{C}-\overset{O}{C}\underset{O^-}{}$$

Lactate          lactate          Pyruvate
dehydrogenase

**FIGURE 7**  **An oxidation-reduction reaction.** Shown here is the oxidation of lactate to pyruvate. In this dehydrogenation, two electrons and two hydrogen ions (the equivalent of two hydrogen atoms) are removed from C-2 of lactate, an alcohol, to form pyruvate, a ketone. In cells the reaction is catalyzed by lactate dehydrogenase and the electrons are transferred to a cofactor called nicotinamide adenine dinucleotide. This reaction is fully reversible; pyruvate can be reduced by electrons from the cofactor. In Chapter 13 we discuss the factors that determine the direction of a reaction.

decarboxylations (as in the acetoacetate decarboxylase reaction; see Fig. 17–18). Entire metabolic pathways are organized around the introduction of a carbonyl group in a particular location so that a nearby carbon–carbon bond can be formed or cleaved. In some reactions, this role is played by an imine group or a specialized cofactor such as pyridoxal phosphate, rather than by a carbonyl group.

**3. Internal rearrangements, isomerizations, and eliminations** Another common type of cellular reaction is an intramolecular rearrangement, in which redistribution of

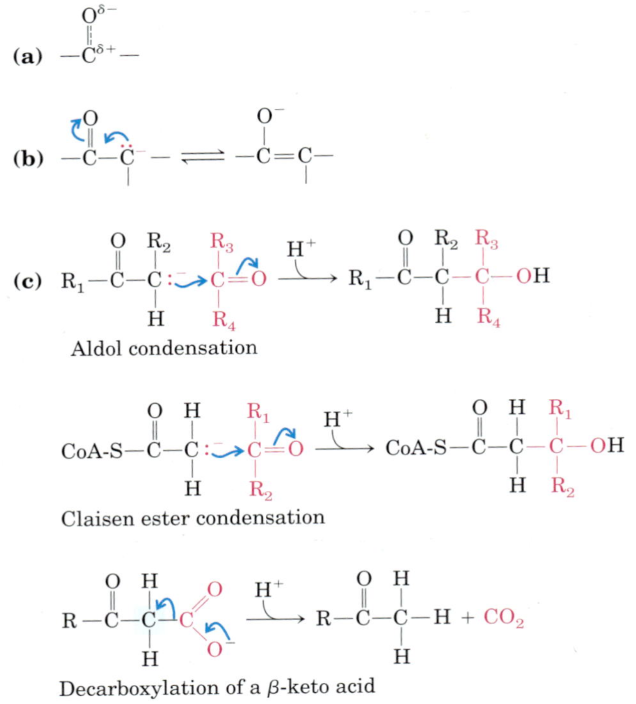

**FIGURE 8 Carbon–carbon bond formation reactions. (a)** The carbon atom of a carbonyl group is an electrophile by virtue of the electron-withdrawing capacity of the electronegative oxygen atom, which results in a resonance hybrid structure in which the carbon has a partial positive charge. **(b)** Within a molecule, delocalization of electrons into a carbonyl group facilitates the transient formation of a carbanion on an adjacent carbon. **(c)** Some of the major reactions involved in the formation and breakage of C—C bonds in biological systems. For both the aldol condensation and the Claisen condensation, a carbanion serves as nucleophile and the carbon of a carbonyl group serves as electrophile. The carbanion is stabilized in each case by another carbonyl at the carbon adjoining the carbanion carbon. In the decarboxylation reaction, a carbanion is formed on the carbon shaded blue as the $CO_2$ leaves. The reaction would not occur at an appreciable rate but for the stabilizing effect of the carbonyl adjacent to the carbanion carbon. Wherever a carbanion is shown, a stabilizing resonance with the adjacent carbonyl, as shown in **(a)**, is assumed. The formation of the carbanion is highly disfavored unless the stabilizing carbonyl group, or a group of similar function such as an imine, is present.

electrons results in isomerization, transposition of double bonds, or cis-trans rearrangements of double bonds. An example of isomerization is the formation of fructose 6-phosphate from glucose 6-phosphate during sugar metabolism (Fig 9a; this reaction is discussed in detail in Chapter 14). Carbon-1 is reduced (from aldehyde to alcohol) and C-2 is oxidized (from alcohol to ketone). Figure 9b shows the details of the electron movements that result in isomerization.

A simple transposition of a C=C bond occurs during metabolism of the common fatty acid oleic acid (see Fig. 17–9), and you will encounter some spectacular examples of double-bond repositioning in the synthesis of cholesterol (see Fig. 21–33).

Elimination of water introduces a C=C bond between two carbons that previously were saturated (as in the enolase reaction; see Fig. 6–23). Similar reactions can result in the elimination of alcohols and amines.

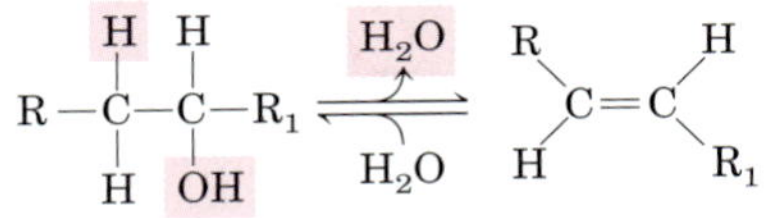

**4. Group transfer reactions** The transfer of acyl, glycosyl, and phosphoryl groups from one nucleophile to another is common in living cells. Acyl group transfer generally involves the addition of a nucleophile to the carbonyl carbon of an acyl group to form a tetrahedral intermediate.

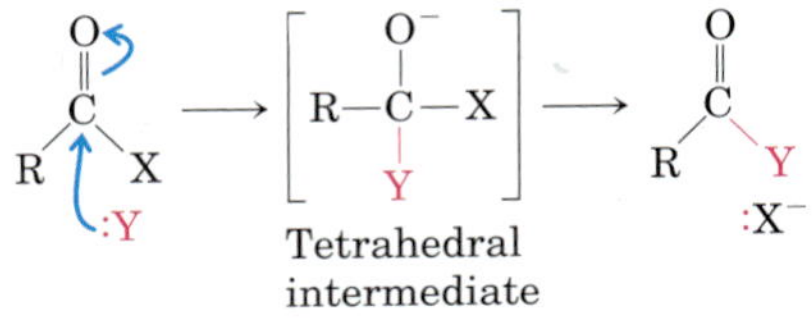

The chymotrypsin reaction is one example of acyl group transfer (see Fig. 6–21). Glycosyl group transfers involve nucleophilic substitution at C-1 of a sugar ring, which is the central atom of an acetal. In principle, the substitution could proceed by an $S_N1$ or $S_N2$ path, as described for the enzyme lysozyme (see Fig. 6–25).

Phosphoryl group transfers play a special role in metabolic pathways. A general theme in metabolism is the attachment of a good leaving group to a metabolic intermediate to "activate" the intermediate for subsequent reaction. Among the better leaving groups in nucleophilic substitution reactions are inorganic orthophosphate (the ionized form of $H_3PO_4$ at neutral pH, a mixture of $H_2PO_4^-$ and $HPO_4^{2-}$, commonly abbreviated $P_i$) and inorganic pyrophosphate ($P_2O_7^{4-}$, abbreviated $PP_i$); esters and anhydrides of phosphoric acid are effectively activated for reaction. Nucleophilic substitution is made more favorable by the attachment of a phosphoryl group to an otherwise poor leaving group such as —OH. Nucleophilic substitutions in which the

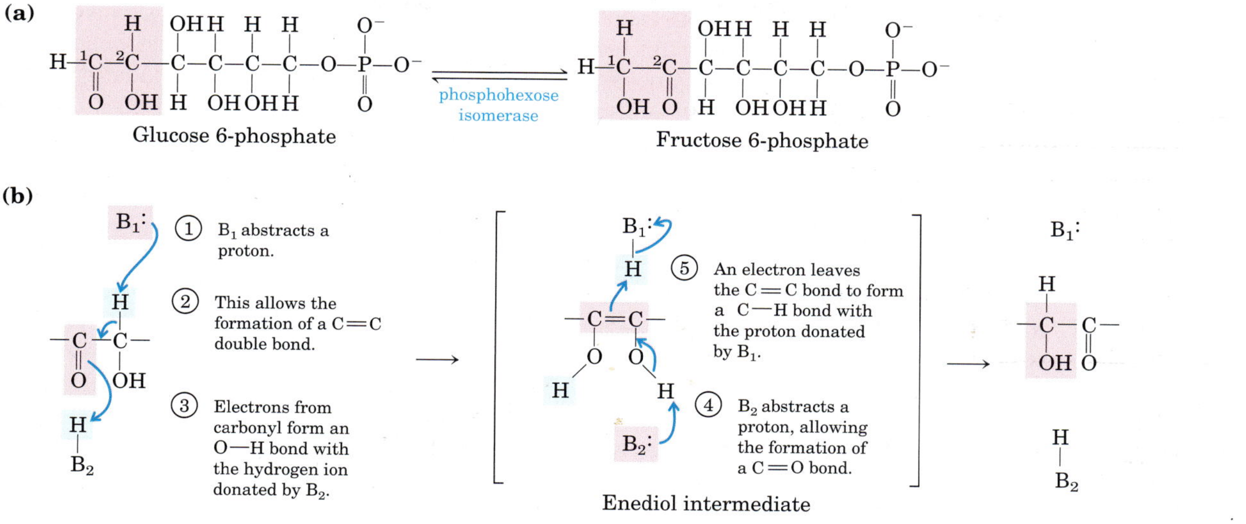

**(a)**

Glucose 6-phosphate     phosphohexose isomerase     Fructose 6-phosphate

**(b)**

① $B_1$ abstracts a proton.

② This allows the formation of a C=C double bond.

③ Electrons from carbonyl form an O—H bond with the hydrogen ion donated by $B_2$.

⑤ An electron leaves the C=C bond to form a C—H bond with the proton donated by $B_1$.

④ $B_2$ abstracts a proton, allowing the formation of a C=O bond.

Enediol intermediate

**FIGURE 9  Isomerization and elimination reactions. (a)** The conversion of glucose 6-phosphate to fructose 6-phosphate, a reaction of sugar metabolism catalyzed by phosphohexose isomerase. **(b)** This reaction proceeds through an enediol intermediate. The curved blue arrows represent the movement of bonding electrons from nucleophile (pink) to electrophile (blue). $B_1$ and $B_2$ are basic groups on the enzyme; they are capable of donating and accepting hydrogen ions (protons) as the reaction progresses.

phosphoryl group ($-PO_3^{2-}$) serves as a leaving group occur in hundreds of metabolic reactions.

Phosphorus can form five covalent bonds. The conventional representation of $P_i$ (Fig. 10a), with three P—O bonds and one P=O bond, is not an accurate picture. In $P_i$, four equivalent phosphorus–oxygen bonds share some double-bond character, and the anion has a tetrahedral structure (Fig. 10b). As oxygen is more electronegative than phosphorus, the sharing of electrons is unequal: the central phosphorus bears a partial positive charge and can therefore act as an electrophile. In a very large number of metabolic reactions, a phosphoryl group ($-PO_3^{2-}$) is transferred from ATP to an alcohol (forming a phosphate ester) (Fig. 10c) or to a carboxylic acid (forming a mixed anhydride). When a nucleophile attacks the electrophilic phosphorus atom in ATP, a relatively stable pentacovalent structure is formed as a reaction intermediate (Fig. 10d). With departure of the leaving group (ADP), the transfer of a phosphoryl group is complete. The large family of enzymes that catalyze

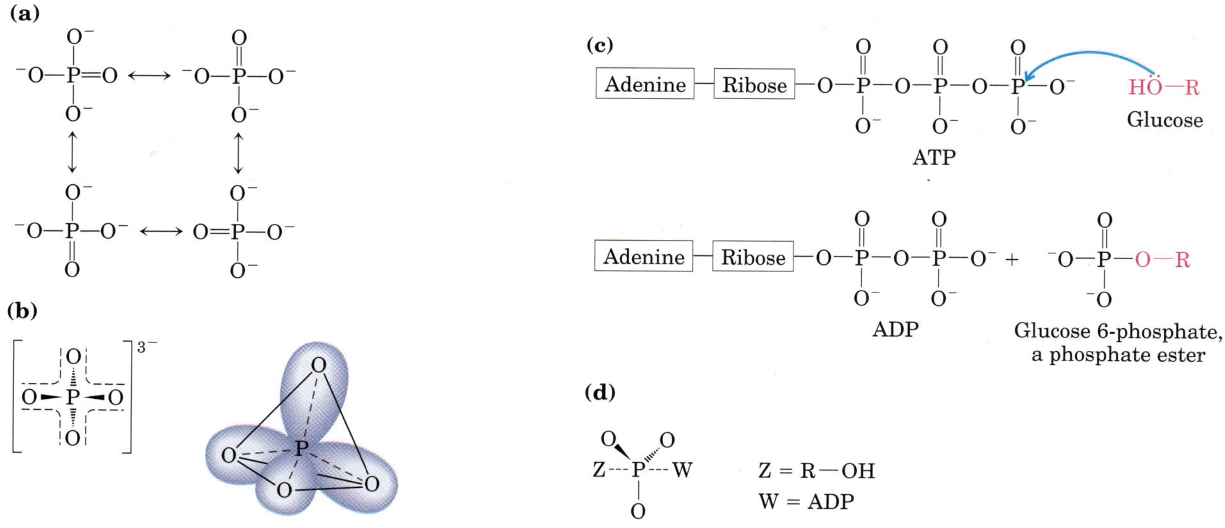

**(a)**

**(b)**

**(c)**

ATP

ADP     Glucose 6-phosphate, a phosphate ester

**(d)**

Z = R—OH
W = ADP

**FIGURE 10  Alternative ways of showing the structure of inorganic orthophosphate. (a)** In one (inadequate) representation, three oxygens are single-bonded to phosphorus, and the fourth is double-bonded, allowing the four different resonance structures shown. **(b)** The four resonance structures can be represented more accurately by showing all four phosphorus–oxygen bonds with some double-bond character; the hybrid orbitals so represented are arranged in a tetrahedron with P at its center. **(c)** When a nucleophile Z (in this case, the —OH on C-6 of glucose) attacks ATP, it displaces ADP (W). In this $S_N2$ reaction, a pentacovalent intermediate **(d)** forms transiently.

phosphoryl group transfers with ATP as donor are called kinases (Greek *kinein*, "to move"). Hexokinase, for example, "moves" a phosphoryl group from ATP to glucose.

Phosphoryl groups are not the only activators of this type. Thioalcohols (thiols), in which the oxygen atom of an alcohol is replaced with a sulfur atom, are also good leaving groups. Thiols activate carboxylic acids by forming thioesters (thiol esters) with them. We will discuss a number of cases, including the reactions catalyzed by the fatty acyl transferases in lipid synthesis (see Fig. 21–2), in which nucleophilic substitution at the carbonyl carbon of a thioester results in transfer of the acyl group to another moiety.

**5. Free radical reactions**   Once thought to be rare, the homolytic cleavage of covalent bonds to generate free radicals has now been found in a range of biochemical processes. Some examples are the reactions of methylmalonyl-CoA mutase (see Box 17–2), ribonucleotide reductase (see Fig. 22–41), and DNA photolyase (see Fig. 25–25).

We begin Part II with a discussion of the basic energetic principles that govern all metabolism (Chapter 13). We then consider the major catabolic pathways by which cells obtain energy from the oxidation of various fuels (Chapters 14 through 19). Chapter 19 is the pivotal point of our discussion of metabolism; it concerns chemiosmotic energy coupling, a universal mechanism in which a transmembrane electrochemical potential, produced either by substrate oxidation or by light absorption, drives the synthesis of ATP.

Chapters 20 through 22 describe the major anabolic pathways by which cells use the energy in ATP to produce carbohydrates, lipids, amino acids, and nucleotides from simpler precursors. In Chapter 23 we step back from our detailed look at the metabolic pathways—as they occur in all organisms, from *Escherichia coli* to humans—and consider how they are regulated and integrated in mammals by hormonal mechanisms.

As we undertake our study of intermediary metabolism, a final word. Keep in mind that the myriad reactions described in these pages take place in, and play crucial roles in, living organisms. As you encounter each reaction and each pathway ask, What does this chemical transformation do for the organism? How does this pathway interconnect with the other pathways operating simultaneously in the same cell to produce the energy and products required for cell maintenance and growth? How do the multilayered regulatory mechanisms cooperate to balance metabolic and energy inputs and outputs, achieving the dynamic steady state of life? Studied with this perspective, metabolism provides fascinating and revealing insights into life, with countless applications in medicine, agriculture, and biotechnology.

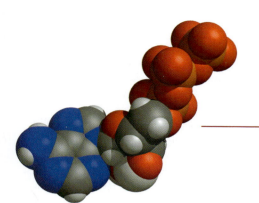

chapter **13**

# PRINCIPLES OF BIOENERGETICS

The total energy of the universe is constant; the total entropy is continually increasing.

—*Rudolf Clausius,* The Mechanical Theory of Heat with Its Applications to the Steam-Engine and to the Physical Properties of Bodies, *1865 (trans. 1867)*

The isomorphism of entropy and information establishes a link between the two forms of power: the power to do and the power to direct what is done.

—*François Jacob,* La logique du vivant: une histoire de l'hérédité (The Logic of Life: A History of Heredity), *1970*

Living cells and organisms must perform work to stay alive, to grow, and to reproduce. The ability to harness energy and to channel it into biological work is a fundamental property of all living organisms; it must have been acquired very early in cellular evolution. Modern organisms carry out a remarkable variety of energy transductions, conversions of one form of energy to another. They use the chemical energy in fuels to bring about the synthesis of complex, highly ordered macromolecules from simple precursors. They also convert the chemical energy of fuels into concentration gradients and electrical gradients, into motion and heat, and, in a few organisms such as fireflies and some deep-sea fish, into light. Photosynthetic organisms transduce light energy into all these other forms of energy.

The chemical mechanisms that underlie biological energy transductions have fascinated and challenged biologists for centuries. Antoine Lavoisier, before he lost his head in the French Revolution, recognized that animals somehow transform chemical fuels (foods) into

heat and that this process of respiration is essential to life. He observed that

> . . . in general, respiration is nothing but a slow combustion of carbon and hydrogen, which is entirely similar to that which occurs in a lighted lamp or candle, and that, from this point of view, animals that respire are true combustible bodies that burn and consume themselves . . . One may say that this analogy between combustion and respiration has not escaped the notice of the poets, or rather the philosophers of antiquity, and which they had expounded and interpreted. This fire stolen from heaven, this torch of Prometheus, does not only represent an ingenious and poetic idea, it is a faithful picture of the operations of nature, at least for animals that breathe; one may therefore say, with the ancients, that the torch of life lights itself at the moment the infant breathes for the first time, and it does not extinguish itself except at death.[*]

Antoine Lavoisier, 1743–1794

In this century, biochemical studies have revealed much of the chemistry underlying that "torch of life." Biological energy transductions obey the same physical laws that govern all other natural processes. It is therefore essential for a student of biochemistry to understand these laws and how they apply to the flow of energy in the biosphere. In this chapter we first review the laws of thermodynamics and the quantitative relationships among free energy, enthalpy, and entropy. We then describe the special role of ATP in biological

[*]From a memoir by Armand Seguin and Antoine Lavoisier, dated 1789, quoted in Lavoisier, A. (1862) *Oeuvres de Lavoisier,* Imprimerie Impériale, Paris.

energy exchanges. Finally, we consider the importance of oxidation-reduction reactions in living cells, the energetics of electron-transfer reactions, and the electron carriers commonly employed as cofactors of the enzymes that catalyze these reactions.

## 13.1 Bioenergetics and Thermodynamics

Bioenergetics is the quantitative study of the energy transductions that occur in living cells and of the nature and function of the chemical processes underlying these transductions. Although many of the principles of thermodynamics have been introduced in earlier chapters and may be familiar to you, a review of the quantitative aspects of these principles is useful here.

### Biological Energy Transformations Obey the Laws of Thermodynamics

Many quantitative observations made by physicists and chemists on the interconversion of different forms of energy led, in the nineteenth century, to the formulation of two fundamental laws of thermodynamics. The first law is the principle of the conservation of energy: *for any physical or chemical change, the total amount of energy in the universe remains constant; energy may change form or it may be transported from one region to another, but it cannot be created or destroyed.* The second law of thermodynamics, which can be stated in several forms, says that the universe always tends toward increasing disorder: *in all natural processes, the entropy of the universe increases.*

Living organisms consist of collections of molecules much more highly organized than the surrounding materials from which they are constructed, and organisms maintain and produce order, seemingly oblivious to the second law of thermodynamics. But living organisms do

not violate the second law; they operate strictly within it. To discuss the application of the second law to biological systems, we must first define those systems and their surroundings.

The reacting system is the collection of matter that is undergoing a particular chemical or physical process; it may be an organism, a cell, or two reacting compounds. The reacting system and its surroundings together constitute the universe. In the laboratory, some chemical or physical processes can be carried out in isolated or closed systems, in which no material or energy is exchanged with the surroundings. Living cells and organisms, however, are open systems, exchanging both material and energy with their surroundings; living systems are never at equilibrium with their surroundings, and the constant transactions between system and surroundings explain how organisms can create order within themselves while operating within the second law of thermodynamics.

In Chapter 1 (p. 23) we defined three thermodynamic quantities that describe the energy changes occurring in a chemical reaction:

**Gibbs free energy, G,** expresses the amount of energy capable of doing work during a reaction at constant temperature and pressure. When a reaction proceeds with the release of free energy (that is, when the system changes so as to possess less free energy), the free-energy change, $\Delta G$, has a negative value and the reaction is said to be exergonic. In endergonic reactions, the system gains free energy and $\Delta G$ is positive.

**Enthalpy, H,** is the heat content of the reacting system. It reflects the number and kinds of chemical bonds in the reactants and products. When a chemical reaction releases heat, it is said to be exothermic; the heat content of the products is less than that of the reactants and $\Delta H$ has, by convention, a negative value. Reacting systems that take up heat from their surroundings are endothermic and have positive values of $\Delta H$.

**Entropy, S,** is a quantitative expression for the randomness or disorder in a system (see Box 1–3). When the products of a reaction are less complex and more disordered than the reactants, the reaction is said to proceed with a gain in entropy.

The units of $\Delta G$ and $\Delta H$ are joules/mole or calories/mole (recall that 1 cal = 4.184 J); units of entropy are joules/mole · Kelvin (J/mol · K) (Table 13–1).

Under the conditions existing in biological systems (including constant temperature and pressure), changes in free energy, enthalpy, and entropy are related to each other quantitatively by the equation

$$\Delta G = \Delta H - T \Delta S \qquad (13\text{–}1)$$

"Now, in the *second* law of thermodynamics . . . "

in which $\Delta G$ is the change in Gibbs free energy of the reacting system, $\Delta H$ is the change in enthalpy of the system, $T$ is the absolute temperature, and $\Delta S$ is the change in entropy of the system. By convention, $\Delta S$ has a positive sign when entropy increases and $\Delta H$, as noted above, has a negative sign when heat is released by the system to its surroundings. Either of these conditions, which are typical of favorable processes, tend to make $\Delta G$ negative. In fact, $\Delta G$ of a spontaneously reacting system is always negative.

The second law of thermodynamics states that the entropy *of the universe* increases during all chemical and physical processes, but it does not require that the entropy increase take place *in the reacting system* itself. The order produced within cells as they grow and divide is more than compensated for by the disorder they create in their surroundings in the course of growth and division (see Box 1–3, case 2). In short, living organisms preserve their internal order by taking from the surroundings free energy in the form of nutrients or sunlight, and returning to their surroundings an equal amount of energy as heat and entropy.

## Cells Require Sources of Free Energy

Cells are isothermal systems—they function at essentially constant temperature (they also function at constant pressure). Heat flow is not a source of energy for cells, because heat can do work only as it passes to a zone or object at a lower temperature. The energy that cells can and must use is free energy, described by the Gibbs free-energy function $G$, which allows prediction of the direction of chemical reactions, their exact equilibrium position, and the amount of work they can in theory perform at constant temperature and pressure. Heterotrophic cells acquire free energy from nutrient molecules, and photosynthetic cells acquire it from absorbed solar radiation. Both kinds of cells transform this free energy into ATP and other energy-rich compounds capable of providing energy for biological work at constant temperature.

## The Standard Free-Energy Change Is Directly Related to the Equilibrium Constant

The composition of a reacting system (a mixture of chemical reactants and products) tends to continue changing until equilibrium is reached. At the equilibrium concentration of reactants and products, the rates of the forward and reverse reactions are exactly equal and no further net change occurs in the system. The concentrations of reactants and products *at equilibrium* define the equilibrium constant, $K_{eq}$ (p. 26). In the general reaction $aA + bB \rightleftharpoons cC + dD$, where $a$, $b$, $c$, and $d$ are the number of molecules of A, B, C, and D participating, the equilibrium constant is given by

$$K_{eq} = \frac{[C]^c[D]^d}{[A]^a[B]^b} \qquad (13-2)$$

where [A], [B], [C], and [D] are the molar concentrations of the reaction components at the point of equilibrium.

When a reacting system is not at equilibrium, the tendency to move toward equilibrium represents a driving force, the magnitude of which can be expressed as the free-energy change for the reaction, $\Delta G$. Under standard conditions (298 K = 25 °C), when reactants and products are initially present at 1 M concentrations or, for gases, at partial pressures of 101.3 kilopascals (kPa), or 1 atm, the force driving the system toward equilibrium is defined as the standard free-energy change, $\Delta G°$. By this definition, the standard state for reactions that involve hydrogen ions is [H$^+$] = 1 M, or pH 0. Most biochemical reactions, however, occur in well-buffered aqueous solutions near pH 7; both the pH and the concentration of water (55.5 M) are essentially constant. For convenience of calculations, biochemists therefore define a different standard state, in which the concentration of H$^+$ is $10^{-7}$ M (pH 7) and that of water is 55.5 M; for reactions that involve Mg$^{2+}$ (including most in which ATP is a reactant), its concentration in solution is commonly taken to be constant at 1 mM. Physical constants based on this biochemical standard state are called **standard transformed constants** and are written with a prime (such as $\Delta G'°$ and $K'_{eq}$) to distinguish them from the untransformed constants used by chemists and physicists. (Notice that most other textbooks use the symbol $\Delta G°'$ rather than $\Delta G'°$. Our use of $\Delta G'°$, recommended by an international committee of chemists and biochemists, is intended to emphasize that the transformed free energy $G'$ is the criterion for equilibrium.) By convention, when H$_2$O, H$^+$, and/or Mg$^{2+}$ are reactants or products, their concentrations are not included in equations such as Equation 13–2 but are instead incorporated into the constants $K'_{eq}$ and $\Delta G'°$.

Just as $K'_{eq}$ is a physical constant characteristic for each reaction, so too is $\Delta G'^\circ$ a constant. As we noted in Chapter 6, there is a simple relationship between $K'_{eq}$ and $\Delta G'^\circ$:

$$\Delta G'^\circ = -RT \ln K'_{eq}$$

*The standard free-energy change of a chemical reaction is simply an alternative mathematical way of expressing its equilibrium constant.* Table 13–2 shows the relationship between $\Delta G'^\circ$ and $K'_{eq}$. If the equilibrium constant for a given chemical reaction is 1.0, the standard free-energy change of that reaction is 0.0 (the natural logarithm of 1.0 is zero). If $K'_{eq}$ of a reaction is greater than 1.0, its $\Delta G'^\circ$ is negative. If $K'_{eq}$ is less than 1.0, $\Delta G'^\circ$ is positive. Because the relationship between $\Delta G'^\circ$ and $K'_{eq}$ is exponential, relatively small changes in $\Delta G'^\circ$ correspond to large changes in $K'_{eq}$.

It may be helpful to think of the standard free-energy change in another way. $\Delta G'^\circ$ is the difference between the free-energy content of the products and the free-energy content of the reactants, under standard conditions. When $\Delta G'^\circ$ is negative, the products contain less free energy than the reactants and the reaction will proceed spontaneously under standard conditions; all chemical reactions tend to go in the direction that results in a decrease in the free energy of the system. A positive value of $\Delta G'^\circ$ means that the products of the reaction contain more free energy than the reactants, and this reaction will tend to go in the reverse direction if we start with 1.0 M concentrations of all components (standard conditions). Table 13–3 summarizes these points.

**TABLE 13–2** **Relationship between the Equilibrium Constants and Standard Free-Energy Changes of Chemical Reactions**

| $K'_{eq}$ | $\Delta G'^\circ$ (kJ/mol) | $\Delta G'^\circ$ (kcal/mol)* |
|---|---|---|
| $10^3$ | −17.1 | −4.1 |
| $10^2$ | −11.4 | −2.7 |
| $10^1$ | −5.7 | −1.4 |
| 1 | 0.0 | 0.0 |
| $10^{-1}$ | 5.7 | 1.4 |
| $10^{-2}$ | 11.4 | 2.7 |
| $10^{-3}$ | 17.1 | 4.1 |
| $10^{-4}$ | 22.8 | 5.5 |
| $10^{-5}$ | 28.5 | 6.8 |
| $10^{-6}$ | 34.2 | 8.2 |

*Although joules and kilojoules are the standard units of energy and are used throughout this text, biochemists sometimes express $\Delta G'^\circ$ values in kilocalories per mole. We have therefore included values in both kilojoules and kilocalories in this table and in Tables 13–4 and 13–6. To convert kilojoules to kilocalories, divide the number of kilojoules by 4.184.

**TABLE 13–3** **Relationships among $K'_{eq}$, $\Delta G'^\circ$, and the Direction of Chemical Reactions under Standard Conditions**

| When $K'_{eq}$ is . . . | $\Delta G'^\circ$ is . . . | Starting with all components at 1 M, the reaction . . . |
|---|---|---|
| >1.0 | negative | proceeds forward |
| 1.0 | zero | is at equilibrium |
| <1.0 | positive | proceeds in reverse |

As an example, let's make a simple calculation of the standard free-energy change of the reaction catalyzed by the enzyme phosphoglucomutase:

Glucose 1-phosphate $\rightleftharpoons$ glucose 6-phosphate

Chemical analysis shows that whether we start with, say, 20 mM glucose 1-phosphate (but no glucose 6-phosphate) or with 20 mM glucose 6-phosphate (but no glucose 1-phosphate), the final equilibrium mixture at 25 °C and pH 7.0 will be the same: 1 mM glucose 1-phosphate and 19 mM glucose 6-phosphate. (Remember that enzymes do not affect the point of equilibrium of a reaction; they merely hasten its attainment.) From these data we can calculate the equilibrium constant:

$$K'_{eq} = \frac{[\text{glucose 6-phosphate}]}{[\text{glucose 1-phosphate}]} = \frac{19 \text{ mM}}{1 \text{ mM}} = 19$$

From this value of $K'_{eq}$ we can calculate the standard free-energy change:

$$\begin{aligned} \Delta G'^\circ &= -RT \ln K'_{eq} \\ &= -(8.315 \text{ J/mol} \cdot \text{K})(298 \text{ K})(\ln 19) \\ &= -7.3 \text{ kJ/mol} \end{aligned}$$

Because the standard free-energy change is negative, when the reaction starts with 1.0 M glucose 1-phosphate and 1.0 M glucose 6-phosphate, the conversion of glucose 1-phosphate to glucose 6-phosphate proceeds with a loss (release) of free energy. For the reverse reaction (the conversion of glucose 6-phosphate to glucose 1-phosphate), $\Delta G'^\circ$ has the same magnitude but the opposite sign.

Table 13–4 gives the standard free-energy changes for some representative chemical reactions. Note that hydrolysis of simple esters, amides, peptides, and glycosides, as well as rearrangements and eliminations, proceed with relatively small standard free-energy changes, whereas hydrolysis of acid anhydrides is accompanied by relatively large decreases in standard free energy. The complete oxidation of organic compounds such as glucose or palmitate to $CO_2$ and $H_2O$, which in cells requires many steps, results in very large decreases in standard free energy. However, standard free-energy

## TABLE 13–4 Standard Free-Energy Changes of Some Chemical Reactions at pH 7.0 and 25 °C (298 K)

| Reaction type | $\Delta G'^{\circ}$ (kJ/mol) | (kcal/mol) |
|---|---|---|
| **Hydrolysis reactions** | | |
| Acid anhydrides | | |
| Acetic anhydride + $H_2O$ ⟶ 2 acetate | −91.1 | −21.8 |
| ATP + $H_2O$ ⟶ ADP + $P_i$ | −30.5 | −7.3 |
| ATP + $H_2O$ ⟶ AMP + $PP_i$ | −45.6 | −10.9 |
| $PP_i$ + $H_2O$ ⟶ $2P_i$ | −19.2 | −4.6 |
| UDP-glucose + $H_2O$ ⟶ UMP + glucose 1-phosphate | −43.0 | −10.3 |
| Esters | | |
| Ethyl acetate + $H_2O$ ⟶ ethanol + acetate | −19.6 | −4.7 |
| Glucose 6-phosphate + $H_2O$ ⟶ glucose + $P_i$ | −13.8 | −3.3 |
| Amides and peptides | | |
| Glutamine + $H_2O$ ⟶ glutamate + $NH_4^+$ | −14.2 | −3.4 |
| Glycylglycine + $H_2O$ ⟶ 2 glycine | −9.2 | −2.2 |
| Glycosides | | |
| Maltose + $H_2O$ ⟶ 2 glucose | −15.5 | −3.7 |
| Lactose + $H_2O$ ⟶ glucose + galactose | −15.9 | −3.8 |
| **Rearrangements** | | |
| Glucose 1-phosphate ⟶ glucose 6-phosphate | −7.3 | −1.7 |
| Fructose 6-phosphate ⟶ glucose 6-phosphate | −1.7 | −0.4 |
| **Elimination of water** | | |
| Malate ⟶ fumarate + $H_2O$ | 3.1 | 0.8 |
| **Oxidations with molecular oxygen** | | |
| Glucose + $6O_2$ ⟶ $6CO_2$ + $6H_2O$ | −2,840 | −686 |
| Palmitate + $23O_2$ ⟶ $16CO_2$ + $16H_2O$ | −9,770 | −2,338 |

changes such as those in Table 13–4 indicate how much free energy is available from a reaction under *standard conditions*. To describe the energy released under the conditions existing in cells, an expression for the *actual* free-energy change is essential.

## Actual Free-Energy Changes Depend on Reactant and Product Concentrations

We must be careful to distinguish between two different quantities: the free-energy change, $\Delta G$, and the standard free-energy change, $\Delta G'^{\circ}$. Each chemical reaction has a characteristic standard free-energy change, which may be positive, negative, or zero, depending on the equilibrium constant of the reaction. The standard free-energy change tells us in which direction and how far a given reaction must go to reach equilibrium *when the initial concentration of each component is 1.0 M,* the pH is 7.0, the temperature is 25 °C, and the pressure is 101.3 kPa. Thus $\Delta G'^{\circ}$ is a constant: it has a characteristic, unchanging value for a given reaction. But the *actual* free-energy change, $\Delta G$, is a function of reactant and product concentrations and of the temperature prevailing during the reaction, which will not necessarily match the standard conditions as defined above. Moreover, the $\Delta G$ of any reaction proceeding spontaneously toward its equilibrium is always negative, becomes less negative as the reaction proceeds, and is zero at the point of equilibrium, indicating that no more work can be done by the reaction.

$\Delta G$ and $\Delta G'^\circ$ for any reaction A + B $\rightleftharpoons$ C + D are related by the equation

$$\Delta G = \Delta G'^\circ + RT \ln \frac{[C][D]}{[A][B]} \qquad \text{🖱} \quad (13\text{–}3)$$

in which the terms in red are those *actually prevailing* in the system under observation. The concentration terms in this equation express the effects commonly called mass action, and the term [C][D]/[A][B] is called the **mass-action ratio, *Q*.** As an example, let us suppose that the reaction A + B $\rightleftharpoons$ C + D is taking place at the standard conditions of temperature (25 °C) and pressure (101.3 kPa) but that the concentrations of A, B, C, and D are *not* equal and none of the components is present at the standard concentration of 1.0 M. To determine the actual free-energy change, $\Delta G$, under these nonstandard conditions of concentration as the reaction proceeds from left to right, we simply enter the *actual* concentrations of A, B, C, and D in Equation 13–3; the values of $R$, $T$, and $\Delta G'^\circ$ are the standard values. $\Delta G$ is negative and approaches zero as the reaction proceeds because the actual concentrations of A and B decrease and the concentrations of C and D increase. Notice that when a reaction is at equilibrium—when there is no force driving the reaction in either direction and $\Delta G$ is zero—Equation 13–3 reduces to

$$0 = \Delta G = \Delta G'^\circ + RT \ln \frac{[C]_{eq}[D]_{eq}}{[A]_{eq}[B]_{eq}}$$

or

$$\Delta G'^\circ = -RT \ln K'_{eq}$$

which is the equation relating the standard free-energy change and equilibrium constant given earlier.

The criterion for spontaneity of a reaction is the value of $\Delta G$, not $\Delta G'^\circ$. A reaction with a positive $\Delta G'^\circ$ can go in the forward direction *if $\Delta G$ is negative.* This is possible if the term $RT \ln$ ([products]/[reactants]) in Equation 13–3 is negative and has a larger absolute value than $\Delta G'^\circ$. For example, the immediate removal of the products of a reaction can keep the ratio [products]/[reactants] well below 1, such that the term $RT \ln$ ([products]/[reactants]) has a large, negative value.

$\Delta G'^\circ$ and $\Delta G$ are expressions of the *maximum* amount of free energy that a given reaction can *theoretically* deliver—an amount of energy that could be realized only if a perfectly efficient device were available to trap or harness it. Given that no such device is possible (some free energy is always lost to entropy during any process), the amount of work done by the reaction at constant temperature and pressure is always less than the theoretical amount.

Another important point is that some thermodynamically favorable reactions (that is, reactions for which $\Delta G'^\circ$ is large and negative) do not occur at measurable rates. For example, combustion of firewood to $CO_2$ and $H_2O$ is very favorable thermodynamically, but firewood remains stable for years because the activation energy (see Figs 6–2 and 6–3) for the combustion reaction is higher than the energy available at room temperature. If the necessary activation energy is provided (with a lighted match, for example), combustion will begin, converting the wood to the more stable products $CO_2$ and $H_2O$ and releasing energy as heat and light. The heat released by this exothermic reaction provides the activation energy for combustion of neighboring regions of the firewood; the process is self-perpetuating.

In living cells, reactions that would be extremely slow *if uncatalyzed* are caused to proceed, not by supplying additional heat but by lowering the activation energy with an enzyme. An enzyme provides an alternative reaction pathway with a lower activation energy than the uncatalyzed reaction, so that at room temperature a large fraction of the substrate molecules have enough thermal energy to overcome the activation barrier, and the reaction rate increases dramatically. *The free-energy change for a reaction is independent of the pathway by which the reaction occurs;* it depends only on the nature and concentration of the initial reactants and the final products. Enzymes cannot, therefore, change equilibrium constants; but they can and do increase the *rate* at which a reaction proceeds in the direction dictated by thermodynamics.

## Standard Free-Energy Changes Are Additive

In the case of two sequential chemical reactions, A $\rightleftharpoons$ B and B $\rightleftharpoons$ C, each reaction has its own equilibrium constant and each has its characteristic standard free-energy change, $\Delta G_1'^\circ$ and $\Delta G_2'^\circ$. As the two reactions are sequential, B cancels out to give the overall reaction A $\rightleftharpoons$ C, which has its own equilibrium constant and thus its own standard free-energy change, $\Delta G_{total}'^\circ$. *The $\Delta G'^\circ$ values of sequential chemical reactions are additive.* For the overall reaction A $\rightleftharpoons$ C, $\Delta G_{total}'^\circ$ is the sum of the individual standard free-energy changes, $\Delta G_1'^\circ$ and $\Delta G_2'^\circ$, of the two reactions: $\Delta G_{total}'^\circ = \Delta G_1'^\circ + \Delta G_2'^\circ$.

| (1) | A $\longrightarrow$ B | $\Delta G_1'^\circ$ |
|---|---|---|
| (2) | B $\longrightarrow$ C | $\Delta G_2'^\circ$ |
| *Sum:* | A $\longrightarrow$ C | $\Delta G_1'^\circ + \Delta G_2'^\circ$ |

This principle of bioenergetics explains how a thermodynamically unfavorable (endergonic) reaction can be driven in the forward direction by coupling it to a highly exergonic reaction through a common intermediate. For example, the synthesis of glucose 6-phosphate is the first step in the utilization of glucose by many organisms:

$$\text{Glucose} + P_i \longrightarrow \text{glucose 6-phosphate} + H_2O$$
$$\Delta G'^\circ = 13.8 \text{ kJ/mol}$$

The positive value of $\Delta G'^{\circ}$ predicts that under standard conditions the reaction will tend not to proceed spontaneously in the direction written. Another cellular reaction, the hydrolysis of ATP to ADP and $P_i$, is very exergonic:

$$\text{ATP} + H_2O \longrightarrow \text{ADP} + P_i \qquad \Delta G'^{\circ} = -30.5 \text{ kJ/mol}$$

These two reactions share the common intermediates $P_i$ and $H_2O$ and may be expressed as sequential reactions:

$$
\begin{array}{ll}
(1) & \text{Glucose} + P_i \longrightarrow \text{glucose 6-phosphate} + H_2O \\
(2) & \text{ATP} + H_2O \longrightarrow \text{ADP} + P_i \\
\hline
Sum: & \text{ATP} + \text{glucose} \longrightarrow \text{ADP} + \text{glucose 6-phosphate}
\end{array}
$$

The overall standard free-energy change is obtained by adding the $\Delta G'^{\circ}$ values for individual reactions:

$$\Delta G'^{\circ} = 13.8 \text{ kJ/mol} + (-30.5 \text{ kJ/mol}) = -16.7 \text{ kJ/mol}$$

The overall reaction is exergonic. In this case, energy stored in ATP is used to drive the synthesis of glucose 6-phosphate, even though its formation from glucose and inorganic phosphate ($P_i$) is endergonic. The *pathway* of glucose 6-phosphate formation by phosphoryl transfer from ATP is different from reactions (1) and (2) above, but the net result is the same as the sum of the two reactions. In thermodynamic calculations, all that matters is the state of the system at the beginning of the process and its state at the end; the route between the initial and final states is immaterial.

We have said that $\Delta G'^{\circ}$ is a way of expressing the equilibrium constant for a reaction. For reaction (1) above,

$$K'_{eq_1} = \frac{[\text{glucose 6-phosphate}]}{[\text{glucose}][P_i]} = 3.9 \times 10^{-3} \text{ M}^{-1}$$

Notice that $H_2O$ is not included in this expression, as its concentration (55.5 M) is assumed to remain unchanged by the reaction. The equilibrium constant for the hydrolysis of ATP is

$$K'_{eq_2} = \frac{[\text{ADP}][P_i]}{[\text{ATP}]} = 2.0 \times 10^5 \text{ M}$$

The equilibrium constant for the two coupled reactions is

$$
\begin{aligned}
K'_{eq_3} &= \frac{[\text{glucose 6-phosphate}][\text{ADP}][P_i]}{[\text{glucose}][P_i][\text{ATP}]} \\
&= (K'_{eq_1})(K'_{eq_2}) = (3.9 \times 10^{-3} \text{ M}^{-1})(2.0 \times 10^5 \text{ M}) \\
&= 7.8 \times 10^2
\end{aligned}
$$

This calculation illustrates an important point about equilibrium constants: although the $\Delta G'^{\circ}$ values for two reactions that sum to a third are *additive*, the $K'_{eq}$ for a reaction that is the sum of two reactions is the *product* of their individual $K'_{eq}$ values. Equilibrium constants are *multiplicative*. By coupling ATP hydrolysis to glu-

cose 6-phosphate synthesis, the $K'_{eq}$ for formation of glucose 6-phosphate has been raised by a factor of about $2 \times 10^5$.

This common-intermediate strategy is employed by all living cells in the synthesis of metabolic intermediates and cellular components. Obviously, the strategy works only if compounds such as ATP are continuously available. In the following chapters we consider several of the most important cellular pathways for producing ATP.

## SUMMARY 13.1 Bioenergetics and Thermodynamics

- Living cells constantly perform work. They require energy for maintaining their highly organized structures, synthesizing cellular components, generating electric currents, and many other processes.

- Bioenergetics is the quantitative study of energy relationships and energy conversions in biological systems. Biological energy transformations obey the laws of thermodynamics.

- All chemical reactions are influenced by two forces: the tendency to achieve the most stable bonding state (for which enthalpy, $H$, is a useful expression) and the tendency to achieve the highest degree of randomness, expressed as entropy, $S$. The net driving force in a reaction is $\Delta G$, the free-energy change, which represents the net effect of these two factors: $\Delta G = \Delta H - T \Delta S$.

- The standard transformed free-energy change, $\Delta G'^{\circ}$, is a physical constant that is characteristic for a given reaction and can be calculated from the equilibrium constant for the reaction: $\Delta G'^{\circ} = -RT \ln K'_{eq}$.

- The actual free-energy change, $\Delta G$, is a variable that depends on $\Delta G'^{\circ}$ and on the concentrations of reactants and products: $\Delta G = \Delta G'^{\circ} + RT \ln ([\text{products}]/[\text{reactants}])$.

- When $\Delta G$ is large and negative, the reaction tends to go in the forward direction; when $\Delta G$ is large and positive, the reaction tends to go in the reverse direction; and when $\Delta G = 0$, the system is at equilibrium.

- The free-energy change for a reaction is independent of the pathway by which the reaction occurs. Free-energy changes are additive; the net chemical reaction that results from successive reactions sharing a common intermediate has an overall free-energy change that is the sum of the $\Delta G$ values for the individual reactions.

## 13.2 Phosphoryl Group Transfers and ATP

Having developed some fundamental principles of energy changes in chemical systems, we can now examine the energy cycle in cells and the special role of ATP as the energy currency that links catabolism and anabolism (see Fig. 1–28). Heterotrophic cells obtain free energy in a chemical form by the catabolism of nutrient molecules, and they use that energy to make ATP from ADP and $P_i$. ATP then donates some of its chemical energy to endergonic processes such as the synthesis of metabolic intermediates and macromolecules from smaller precursors, the transport of substances across membranes against concentration gradients, and mechanical motion. This donation of energy from ATP generally involves the covalent participation of ATP in the reaction that is to be driven, with the eventual result that ATP is converted to ADP and $P_i$ or, in some reactions, to AMP and $2 P_i$. We discuss here the chemical basis for the large free-energy changes that accompany hydrolysis of ATP and other high-energy phosphate compounds, and we show that most cases of energy donation by ATP involve group transfer, not simple hydrolysis of ATP. To illustrate the range of energy transductions in which ATP provides the energy, we consider the synthesis of information-rich macromolecules, the transport of solutes across membranes, and motion produced by muscle contraction.

### The Free-Energy Change for ATP Hydrolysis Is Large and Negative

Figure 13–1 summarizes the chemical basis for the relatively large, negative, standard free energy of hydrolysis of ATP. The hydrolytic cleavage of the terminal phosphoric acid anhydride (phosphoanhydride) bond in ATP separates one of the three negatively charged phosphates and thus relieves some of the electrostatic repulsion in ATP; the $P_i$ ($HPO_4^{2-}$) released is stabilized by the formation of several resonance forms not possible in ATP; and $ADP^{2-}$, the other direct product of hydrolysis, immediately ionizes, releasing $H^+$ into a medium of very low $[H^+]$ ($\sim 10^{-7}$ M). Because the concentrations of the direct products of ATP hydrolysis are, in the cell, far below the concentrations at equilibrium (Table 13–5), mass action favors the hydrolysis reaction in the cell.

Although the hydrolysis of ATP is highly exergonic ($\Delta G'^\circ = -30.5$ kJ/mol), the molecule is kinetically stable at pH 7 because the activation energy for ATP hydrolysis is relatively high. Rapid cleavage of the phosphoanhydride bonds occurs only when catalyzed by an enzyme.

The free-energy change for ATP hydrolysis is $-30.5$ kJ/mol under standard conditions, but the *actual* free energy of hydrolysis ($\Delta G$) of ATP in living cells is very different: the cellular concentrations of ATP, ADP,

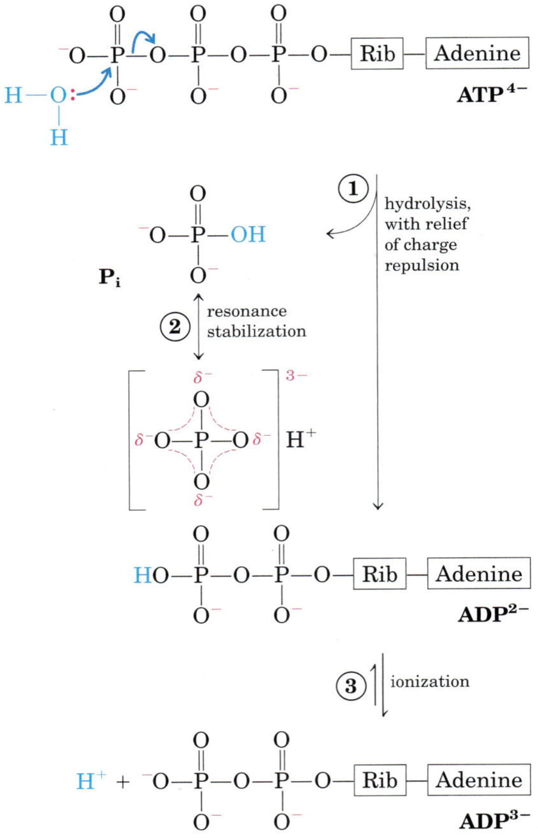

$$ATP^{4-} + H_2O \longrightarrow ADP^{3-} + P_i^{2-} + H^+$$
$$\Delta G'^\circ = -30.5 \text{ kJ/mol}$$

**FIGURE 13–1 Chemical basis for the large free-energy change associated with ATP hydrolysis.** ① The charge separation that results from hydrolysis relieves electrostatic repulsion among the four negative charges on ATP. ② The product inorganic phosphate ($P_i$) is stabilized by formation of a resonance hybrid, in which each of the four phosphorus–oxygen bonds has the same degree of double-bond character and the hydrogen ion is not permanently associated with any one of the oxygens. (Some degree of resonance stabilization also occurs in phosphates involved in ester or anhydride linkages, but fewer resonance forms are possible than for $P_i$.) ③ The product $ADP^{2-}$ immediately ionizes, releasing a proton into a medium of very low $[H^+]$ (pH 7). A fourth factor (not shown) that favors ATP hydrolysis is the greater degree of solvation (hydration) of the products $P_i$ and ADP relative to ATP, which further stabilizes the products relative to the reactants.

and $P_i$ are not identical and are much lower than the 1.0 M of standard conditions (Table 13–5). Furthermore, $Mg^{2+}$ in the cytosol binds to ATP and ADP (Fig. 13–2), and for most enzymatic reactions that involve ATP as phosphoryl group donor, the true substrate is $MgATP^{2-}$. The relevant $\Delta G'^\circ$ is therefore that for $MgATP^{2-}$ hydrolysis. Box 13–1 shows how $\Delta G$ for ATP hydrolysis in the intact erythrocyte can be calculated from the data in Table 13–5. In intact cells, $\Delta G$ for ATP hydrolysis, usually designated $\Delta G_p$, is much more negative than

**TABLE 13–5  Adenine Nucleotide, Inorganic Phosphate, and Phosphocreatine Concentrations in Some Cells**

| | Concentration (mM)* | | | | |
|---|---|---|---|---|---|
| | ATP | ADP† | AMP | $P_i$ | PCr |
| Rat hepatocyte | 3.38 | 1.32 | 0.29 | 4.8 | 0 |
| Rat myocyte | 8.05 | 0.93 | 0.04 | 8.05 | 28 |
| Rat neuron | 2.59 | 0.73 | 0.06 | 2.72 | 4.7 |
| Human erythrocyte | 2.25 | 0.25 | 0.02 | 1.65 | 0 |
| *E. coli* cell | 7.90 | 1.04 | 0.82 | 7.9 | 0 |

*For erythrocytes the concentrations are those of the cytosol (human erythrocytes lack a nucleus and mitochondria). In the other types of cells the data are for the entire cell contents, although the cytosol and the mitochondria have very different concentrations of ADP. PCr is phosphocreatine, discussed on p. 505.

†This value reflects total concentration; the true value for free ADP may be much lower (see Box 13–1).

$\Delta G'^\circ$, ranging from $-50$ to $-65$ kJ/mol. $\Delta G_p$ is often called the **phosphorylation potential.** In the following discussions we use the standard free-energy change for ATP hydrolysis, because this allows comparison, on the same basis, with the energetics of other cellular reactions. Remember, however, that in living cells $\Delta G$ is the relevant quantity—for ATP hydrolysis and all other reactions—and may be quite different from $\Delta G'^\circ$.

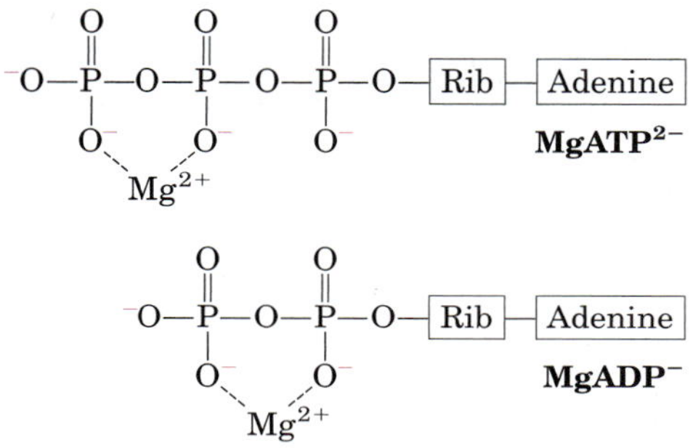

## Other Phosphorylated Compounds and Thioesters Also Have Large Free Energies of Hydrolysis

Phosphoenolpyruvate (Fig. 13–3) contains a phosphate ester bond that undergoes hydrolysis to yield the enol form of pyruvate, and this direct product can immediately tautomerize to the more stable keto form of pyruvate. Because the reactant (phosphoenolpyruvate) has only one form (enol) and the product (pyruvate) has two possible forms, the product is stabilized relative to the reactant. This is the greatest contributing factor to the high standard free energy of hydrolysis of phosphoenolpyruvate: $\Delta G'^\circ = -61.9$ kJ/mol.

Another three-carbon compound, 1,3-bisphosphoglycerate (Fig. 13–4), contains an anhydride bond between the carboxyl group at C-1 and phosphoric acid. Hydrolysis of this acyl phosphate is accompanied by a large, negative, standard free-energy change ($\Delta G'^\circ =$

**FIGURE 13–2  $Mg^{2+}$ and ATP.** Formation of $Mg^{2+}$ complexes partially shields the negative charges and influences the conformation of the phosphate groups in nucleotides such as ATP and ADP.

$-49.3$ kJ/mol), which can, again, be explained in terms of the structure of reactant and products. When $H_2O$ is added across the anhydride bond of 1,3-bisphosphoglycerate, one of the direct products, 3-phosphoglyceric acid, can immediately lose a proton to give the carboxylate ion, 3-phosphoglycerate, which has two equally probable resonance forms (Fig. 13–4). Removal of the direct product (3-phosphoglyceric acid) and formation of the resonance-stabilized ion favor the forward reaction.

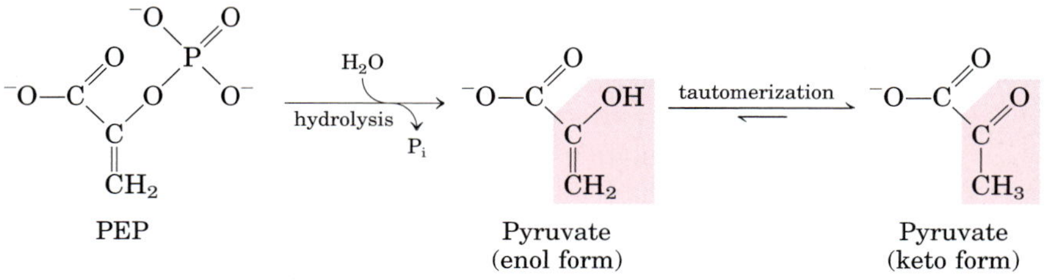

$$PEP^{3-} + H_2O \longrightarrow pyruvate^- + P_i^{2-}$$
$$\Delta G'^\circ = -61.9 \text{ kJ/mol}$$

**FIGURE 13–3  Hydrolysis of phosphoenolpyruvate (PEP).** Catalyzed by pyruvate kinase, this reaction is followed by spontaneous tautomerization of the product, pyruvate. Tautomerization is not possible in PEP, and thus the products of hydrolysis are stabilized relative to the reactants. Resonance stabilization of $P_i$ also occurs, as shown in Figure 13–1.

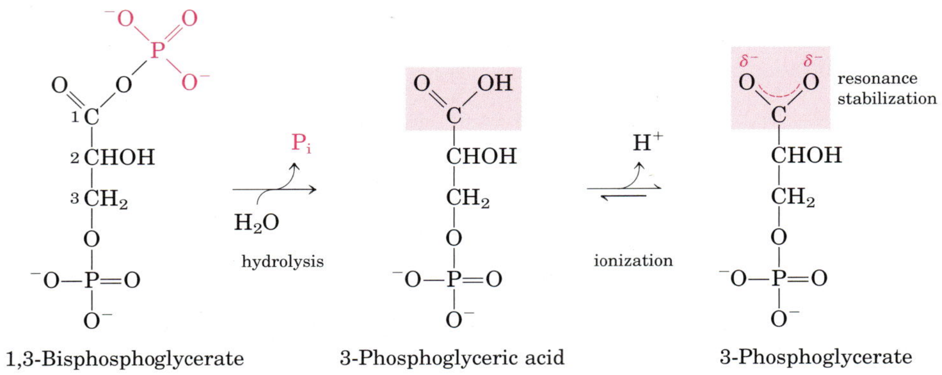

**FIGURE 13-4 Hydrolysis of 1,3-bisphosphoglycerate.** The direct product of hydrolysis is 3-phospho-glyceric acid, with an undissociated carboxylic acid group, but dissociation occurs immediately. This ionization and the resonance structures it makes possible stabilize the product relative to the reactants. Resonance stabilization of $P_i$ further contributes to the negative free-energy change.

$$\text{1,3-Bisphosphoglycerate}^{4-} + H_2O \longrightarrow \text{3-phosphoglycerate}^{3-} + P_i^{2-} + H^+$$
$$\Delta G'^\circ = -49.3 \text{ kJ/mol}$$

## BOX 13-1    WORKING IN BIOCHEMISTRY

### The Free Energy of Hydrolysis of ATP within Cells: The Real Cost of Doing Metabolic Business

The standard free energy of hydrolysis of ATP is −30.5 kJ/mol. In the cell, however, the concentrations of ATP, ADP, and $P_i$ are not only unequal but much lower than the standard 1 M concentrations (see Table 13–5). Moreover, the cellular pH may differ somewhat from the standard pH of 7.0. Thus the *actual* free energy of hydrolysis of ATP under intracellular conditions ($\Delta G_p$) differs from the standard free-energy change, $\Delta G'^\circ$. We can easily calculate $\Delta G_p$.

In human erythrocytes, for example, the concentrations of ATP, ADP, and $P_i$ are 2.25, 0.25, and 1.65 mM, respectively. Let us assume for simplicity that the pH is 7.0 and the temperature is 25 °C, the standard pH and temperature. The actual free energy of hydrolysis of ATP in the erythrocyte under these conditions is given by the relationship

$$\Delta G_p = \Delta G'^\circ + RT \ln \frac{[\text{ADP}][P_i]}{[\text{ATP}]}$$

Substituting the appropriate values we obtain

$$\Delta G_p = -30.5 \text{ kJ/mol} +$$

$$\left[ (8.315 \text{ J/mol} \cdot \text{K})(298 \text{ K}) \ln \frac{(0.25 \times 10^{-3})(1.65 \times 10^{-3})}{2.25 \times 10^{-3}} \right]$$

$$= -30.5 \text{ kJ/mol} + (2.48 \text{ kJ/mol}) \ln 1.8 \times 10^{-4}$$
$$= -30.5 \text{ kJ/mol} - 21 \text{ kJ/mol}$$
$$= -52 \text{ kJ/mol}$$

Thus $\Delta G_p$, the actual free-energy change for ATP hydrolysis in the intact erythrocyte (−52 kJ/mol), is much larger than the standard free-energy change (−30.5 kJ/mol). By the same token, the free energy required to *synthesize* ATP from ADP and $P_i$ under the conditions prevailing in the erythrocyte would be 52 kJ/mol.

Because the concentrations of ATP, ADP, and $P_i$ differ from one cell type to another (see Table 13–5), $\Delta G_p$ for ATP hydrolysis likewise differs among cells. Moreover, in any given cell, $\Delta G_p$ can vary from time to time, depending on the metabolic conditions in the cell and how they influence the concentrations of ATP, ADP, $P_i$, and $H^+$ (pH). We can calculate the actual free-energy change for any given metabolic reaction as it occurs in the cell, providing we know the concentrations of all the reactants and products of the reaction and know about the other factors (such as pH, temperature, and concentration of $Mg^{2+}$) that may affect the $\Delta G'^\circ$ and thus the calculated free-energy change, $\Delta G_p$.

To further complicate the issue, the total concentrations of ATP, ADP, $P_i$, and $H^+$ may be substantially higher than the *free* concentrations, which are the thermodynamically relevant values. The difference is due to tight binding of ATP, ADP, and $P_i$ to cellular proteins. For example, the concentration of free ADP in resting muscle has been variously estimated at between 1 and 37 μM. Using the value 25 μM in the calculation outlined above, we get a $\Delta G_p$ of −58 kJ/mol.

Calculation of the exact value of $\Delta G_p$ is perhaps less instructive than the generalization we can make about actual free-energy changes: in vivo, the energy released by ATP hydrolysis is greater than the standard free-energy change, $\Delta G'^\circ$.

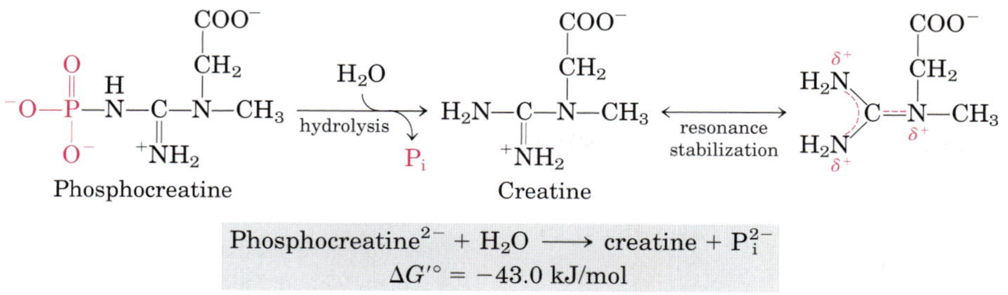

FIGURE 13-5 Hydrolysis of phosphocreatine. Breakage of the P—N bond in phosphocreatine produces creatine, which is stabilized by formation of a resonance hybrid. The other product, $P_i$, is also resonance stabilized.

$$\text{Phosphocreatine}^{2-} + H_2O \longrightarrow \text{creatine} + P_i^{2-}$$
$$\Delta G'^\circ = -43.0 \text{ kJ/mol}$$

In phosphocreatine (Fig. 13–5), the P—N bond can be hydrolyzed to generate free creatine and $P_i$. The release of $P_i$ and the resonance stabilization of creatine favor the forward reaction. The standard free-energy change of phosphocreatine hydrolysis is again large, −43.0 kJ/mol.

In all these phosphate-releasing reactions, the several resonance forms available to $P_i$ (Fig. 13–1) stabilize this product relative to the reactant, contributing to an already negative free-energy change. Table 13–6 lists the standard free energies of hydrolysis for a number of phosphorylated compounds.

**Thioesters,** in which a sulfur atom replaces the usual oxygen in the ester bond, also have large, negative, standard free energies of hydrolysis. Acetyl-coenzyme A, or acetyl-CoA (Fig. 13–6), is one of many thioesters important in metabolism. The acyl group in

these compounds is activated for transacylation, condensation, or oxidation-reduction reactions. Thioesters undergo much less resonance stabilization than do oxygen esters; consequently, the difference in free energy between the reactant and its hydrolysis products, which *are* resonance-stabilized, is greater for thioesters than for comparable oxygen esters (Fig. 13–7). In both cases, hydrolysis of the ester generates a carboxylic acid, which can ionize and assume several resonance forms. Together, these factors result in the large, negative $\Delta G'^\circ$ (−31.4 kJ/mol) for acetyl-CoA hydrolysis.

To summarize, for hydrolysis reactions with large, negative, standard free-energy changes, the products are more stable than the reactants for one or more of the following reasons: (1) the bond strain in reactants due to electrostatic repulsion is relieved by charge separation, as for ATP; (2) the products are stabilized by

**TABLE 13–6  Standard Free Energies of Hydrolysis of Some Phosphorylated Compounds and Acetyl-CoA (a Thioester)**

| | $\Delta G'^\circ$ | |
|---|---|---|
| | (kJ/mol) | (kcal/mol) |
| Phosphoenolpyruvate | −61.9 | −14.8 |
| 1,3-bisphosphoglycerate | | |
| ($\rightarrow$ 3-phosphoglycerate + $P_i$) | −49.3 | −11.8 |
| Phosphocreatine | −43.0 | −10.3 |
| ADP ($\rightarrow$ AMP + $P_i$) | −32.8 | −7.8 |
| ATP ($\rightarrow$ ADP + $P_i$) | −30.5 | −7.3 |
| ATP ($\rightarrow$ AMP + $PP_i$) | −45.6 | −10.9 |
| AMP ($\rightarrow$ adenosine + $P_i$) | −14.2 | −3.4 |
| $PP_i$ ($\rightarrow$ $2P_i$) | −19.2 | −4.0 |
| Glucose 1-phosphate | −20.9 | −5.0 |
| Fructose 6-phosphate | −15.9 | −3.8 |
| Glucose 6-phosphate | −13.8 | −3.3 |
| Glycerol 1-phosphate | −9.2 | −2.2 |
| Acetyl-CoA | −31.4 | −7.5 |

Source: Data mostly from Jencks, W.P. (1976) in *Handbook of Biochemistry and Molecular Biology*, 3rd edn (Fasman, G.D., ed.), *Physical and Chemical Data*, Vol. I, pp. 296–304, CRC Press, Boca Raton, FL. The value for the free energy of hydrolysis of $PP_i$ is from Frey, P.A. & Arabshahi, A. (1995) Standard free-energy change for the hydrolysis of the α–β-phosphoanhydride bridge in ATP. *Biochemistry* **34,** 11,307–11,310.

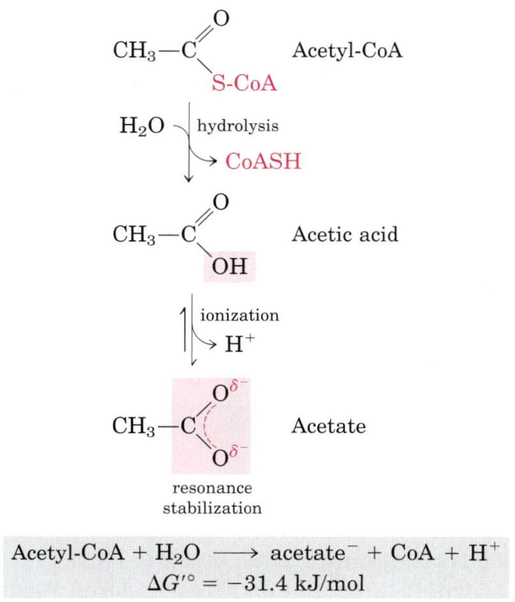

FIGURE 13-6 Hydrolysis of acetyl-coenzyme A. Acetyl-CoA is a thioester with a large, negative, standard free energy of hydrolysis. Thioesters contain a sulfur atom in the position occupied by an oxygen atom in oxygen esters. The complete structure of coenzyme A (CoA, or CoASH) is shown in Figure 8–41.

$$\text{Acetyl-CoA} + H_2O \longrightarrow \text{acetate}^- + \text{CoA} + H^+$$
$$\Delta G'^\circ = -31.4 \text{ kJ/mol}$$

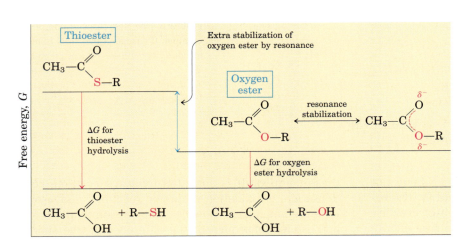

*(handwritten note in top margin: * Overall chemistry of the reaction)*

**FIGURE 13–7 Free energy of hydrolysis for thioesters and oxygen esters.** The *products* of both types of hydrolysis reaction have about the same free-energy content (*G*), but the thioester has a higher free-energy content than the oxygen ester. Orbital overlap between the O and C atoms allows resonance stabilization in oxygen esters; orbital overlap between S and C atoms is poorer and provides little resonance stabilization.

ionization, as for ATP, acyl phosphates, and thioesters; (3) the products are stabilized by isomerization (tautomerization), as for phosphoenolpyruvate; and/or (4) the products are stabilized by resonance, as for creatine released from phosphocreatine, carboxylate ion released from acyl phosphates and thioesters, and phosphate ($P_i$) released from anhydride or ester linkages.

## ATP Provides Energy by Group Transfers, Not by Simple Hydrolysis

Throughout this book you will encounter reactions or processes for which ATP supplies energy, and the contribution of ATP to these reactions is commonly indicated as in Figure 13–8a, with a single arrow showing the conversion of ATP to ADP and $P_i$ (or, in some cases, of ATP to AMP and pyrophosphate, $PP_i$). When written this way, these reactions of ATP appear to be simple hydrolysis reactions in which water displaces $P_i$ (or $PP_i$), and one is tempted to say that an ATP-dependent reaction is "driven by the hydrolysis of ATP." This is *not* the case. ATP hydrolysis per se usually accomplishes nothing but the liberation of heat, which cannot drive a chemical process in an isothermal system. A single reaction arrow such as that in Figure 13–8a almost invariably represents a two-step process (Fig. 13–8b) in which part of the ATP molecule, a phosphoryl or pyrophosphoryl group or the adenylate moiety (AMP), is first transferred to a substrate molecule or to an amino acid residue in an enzyme, becoming covalently attached to the substrate or the enzyme and raising its free-energy content. Then, in a second step, the phosphate-containing moiety transferred in the first step is displaced, generating $P_i$, $PP_i$, or AMP. Thus ATP participates *covalently* in the enzyme-catalyzed reaction to which it contributes free energy.

Some processes *do* involve direct hydrolysis of ATP (or GTP), however. For example, noncovalent binding of ATP (or of GTP), followed by its hydrolysis to ADP (or GDP) and $P_i$, can provide the energy to cycle some proteins between two conformations, producing me-

chanical motion. This occurs in muscle contraction and in the movement of enzymes along DNA or of ribosomes along messenger RNA. The energy-dependent reactions catalyzed by helicases, RecA protein, and some topoisomerases (Chapter 25) also involve direct hydrolysis of phosphoanhydride bonds. GTP-binding proteins that act in signaling pathways directly hydrolyze GTP to drive conformational changes that terminate signals

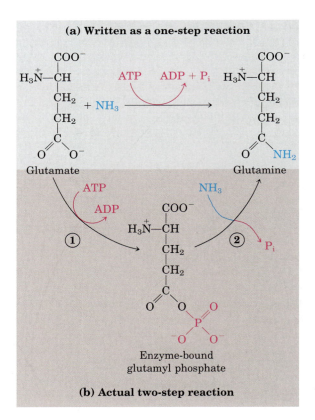

**FIGURE 13–8 ATP hydrolysis in two steps. (a)** The contribution of ATP to a reaction is often shown as a single step, but is almost always a two-step process. **(b)** Shown here is the reaction catalyzed by ATP-dependent glutamine synthetase. ① A phosphoryl group is transferred from ATP to glutamate, then ② the phosphoryl group is displaced by $NH_3$ and released as $P_i$.

triggered by hormones or by other extracellular factors (Chapter 12).

The phosphate compounds found in living organisms can be divided somewhat arbitrarily into two groups, based on their standard free energies of hydrolysis (Fig. 13–9). "High-energy" compounds have a $\Delta G'^{\circ}$ of hydrolysis more negative than $-25$ kJ/mol; "low-energy" compounds have a less negative $\Delta G'^{\circ}$. Based on this criterion, ATP, with a $\Delta G'^{\circ}$ of hydrolysis of $-30.5$ kJ/mol ($-7.3$ kcal/mol), is a high-energy compound; glucose 6-phosphate, with a $\Delta G'^{\circ}$ of hydrolysis of $-13.8$ kJ/mol ($-3.3$ kcal/mol), is a low-energy compound.

The term "high-energy phosphate bond," long used by biochemists to describe the P—O bond broken in hydrolysis reactions, is incorrect and misleading as it wrongly suggests that the bond itself contains the energy. In fact, the breaking of all chemical bonds requires an *input* of energy. The free energy released by hydrolysis of phosphate compounds does not come from the specific bond that is broken; it results from the products of the reaction having a lower free-energy content than the reactants. For simplicity, we will sometimes use the term "high-energy phosphate compound" when referring to ATP or other phosphate compounds with a large, negative, standard free energy of hydrolysis.

As is evident from the additivity of free-energy changes of sequential reactions, any phosphorylated compound can be synthesized by coupling the synthesis to the breakdown of another phosphorylated compound with a more negative free energy of hydrolysis. For example, because cleavage of $P_i$ from phosphoenolpyruvate (PEP) releases more energy than is needed to drive the condensation of $P_i$ with ADP, the direct donation of a phosphoryl group from PEP to ADP is thermodynamically feasible:

|  |  | $\Delta G'^{\circ}$ (kJ/mol) |
|---|---|---|
| (1) | $PEP + H_2O \longrightarrow pyruvate + P_i$ | $-61.9$ |
| (2) | $ADP + P_i \longrightarrow ATP + H_2O$ | $+30.5$ |
| *Sum:* | $PEP + ADP \longrightarrow pyruvate + ATP$ | $-31.4$ |

Notice that while the overall reaction above is represented as the algebraic sum of the first two reactions, the overall reaction is actually a third, distinct reaction that does not involve $P_i$; PEP donates a *phosphoryl* group *directly* to ADP. We can describe phosphorylated compounds as having a high or low phosphoryl group transfer potential, on the basis of their standard free energies of hydrolysis (as listed in Table 13–6). The phosphoryl group transfer potential of phosphoenolpyruvate is very high, that of ATP is high, and that of glucose 6-phosphate is low (Fig. 13–9).

Much of catabolism is directed toward the synthesis of high-energy phosphate compounds, but their formation is not an end in itself; they are the means of activating a very wide variety of compounds for further chemical transformation. The transfer of a phosphoryl group to a compound effectively puts free energy into that compound, so that it has more free energy to give up during subsequent metabolic transformations. We described above how the synthesis of glucose 6-phosphate is accomplished by phosphoryl group transfer from ATP. In the next chapter we see how this phosphorylation of glucose activates, or "primes," the glucose for catabolic reactions that occur in nearly every living cell. Because of its intermediate position on the scale of group transfer potential, ATP can carry energy from high-energy

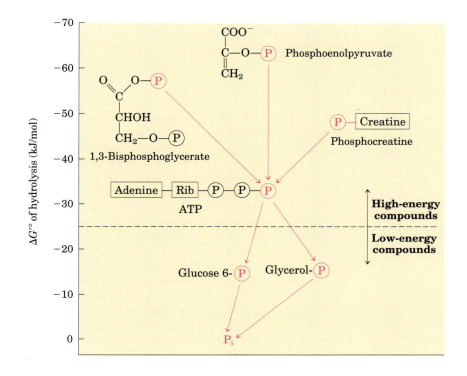

**FIGURE 13–9 Ranking of biological phosphate compounds by standard free energies of hydrolysis.** This shows the flow of phosphoryl groups, represented by Ⓟ, from high-energy phosphoryl donors via ATP to acceptor molecules (such as glucose and glycerol) to form their low-energy phosphate derivatives. This flow of phosphoryl groups, catalyzed by enzymes called kinases, proceeds with an overall loss of free energy under intracellular conditions. Hydrolysis of low-energy phosphate compounds releases $P_i$, which has an even lower phosphoryl group transfer potential (as defined in the text).

phosphate compounds produced by catabolism to compounds such as glucose, converting them into more reactive species. ATP thus serves as the universal energy currency in all living cells.

One more chemical feature of ATP is crucial to its role in metabolism: although in aqueous solution ATP is thermodynamically unstable and is therefore a good phosphoryl group donor, it is *kinetically* stable. Because of the huge activation energies (200 to 400 kJ/mol) required for uncatalyzed cleavage of its phosphoanhydride bonds, ATP does not spontaneously donate phosphoryl groups to water or to the hundreds of other potential acceptors in the cell. Only when specific enzymes are present to lower the energy of activation does phosphoryl group transfer from ATP proceed. The cell is therefore able to regulate the disposition of the energy carried by ATP by regulating the various enzymes that act on it.

### ATP Donates Phosphoryl, Pyrophosphoryl, and Adenylyl Groups

The reactions of ATP are generally $S_N2$ nucleophilic displacements (p. 486), in which the nucleophile may be, for example, the oxygen of an alcohol or carboxylate, or a nitrogen of creatine or of the side chain of arginine or histidine. Each of the three phosphates of ATP is susceptible to nucleophilic attack (Fig. 13–10), and each position of attack yields a different type of product.

Nucleophilic attack by an alcohol on the $\gamma$ phosphate (Fig. 13–10a) displaces ADP and produces a new phosphate ester. Studies with $^{18}O$-labeled reactants have shown that the bridge oxygen in the new compound is derived from the alcohol, not from ATP; the group transferred from ATP is a phosphoryl ($-PO_3^{2-}$), not a phosphate ($-OPO_3^{2-}$). Phosphoryl group transfer from ATP to glutamate (Fig. 13–8) or to glucose

(p. 218) involves attack at the $\gamma$ position of the ATP molecule.

Attack at the $\beta$ phosphate of ATP displaces AMP and transfers a pyrophosphoryl (not pyrophosphate) group to the attacking nucleophile (Fig. 13–10b). For example, the formation of 5′-phosphoribosyl-1-pyrophosphate (p. 842), a key intermediate in nucleotide synthesis, results from attack of an —OH of the ribose on the $\beta$ phosphate.

Nucleophilic attack at the $\alpha$ position of ATP displaces $PP_i$ and transfers adenylate (5′-AMP) as an adenylyl group (Fig. 13–10c); the reaction is an **adenylylation** (a-den′-i-li-lā′-shun, probably the most ungainly word in the biochemical language). Notice that hydrolysis of the $\alpha$–$\beta$ phosphoanhydride bond releases considerably more energy (~46 kJ/mol) than hydrolysis of the $\beta$–$\gamma$ bond (~31 kJ/mol) (Table 13–6). Furthermore, the $PP_i$ formed as a byproduct of the adenylylation is hydrolyzed to two $P_i$ by the ubiquitous enzyme **inorganic pyrophosphatase,** releasing 19 kJ/mol and thereby providing a further energy "push" for the adenylylation reaction. In effect, both phosphoanhydride bonds of ATP are split in the overall reaction. Adenylylation reactions are therefore thermodynamically very favorable. When the energy of ATP is used to drive a particularly unfavorable metabolic reaction, adenylylation is often the mechanism of energy coupling. Fatty acid activation is a good example of this energy-coupling strategy.

The first step in the activation of a fatty acid—either for energy-yielding oxidation or for use in the synthesis of more complex lipids—is the formation of its thiol ester (see Fig. 17–5). The direct condensation of a fatty acid with coenzyme A is endergonic, but the formation of fatty acyl–CoA is made exergonic by stepwise removal of *two* phosphoryl groups from ATP. First, adenylate (AMP) is transferred from ATP to the carboxyl group of the fatty acid, forming a mixed anhydride

**Three positions on ATP for attack by the nucleophile $R^{18}\overset{..}{\underset{..}{O}}$**

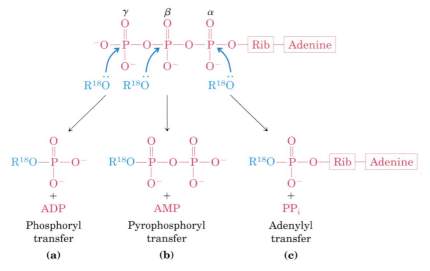

**FIGURE 13–10 Nucleophilic displacement reactions of ATP.** Any of the three P atoms ($\alpha$, $\beta$, or $\gamma$) may serve as the electrophilic target for nucleophilic attack—in this case, by the labeled nucleophile R—$^{18}O$:. The nucleophile may be an alcohol (ROH), a carboxyl group (RCOO⁻), or a phosphoanhydride (a nucleoside mono- or diphosphate, for example). **(a)** When the oxygen of the nucleophile attacks the $\gamma$ position, the bridge oxygen of the product is labeled, indicating that the group transferred from ATP is a phosphoryl ($-PO_3^{2-}$), not a phosphate ($-OPO_3^{2-}$). **(b)** Attack on the $\beta$ position displaces AMP and leads to the transfer of a pyrophosphoryl (not pyrophosphate) group to the nucleophile. **(c)** Attack on the $\alpha$ position displaces $PP_i$ and transfers the adenylyl group to the nucleophile.

(fatty acyl adenylate) and liberating PP$_i$. The thiol group of coenzyme A then displaces the adenylate group and forms a thioester with the fatty acid. The sum of these two reactions is energetically equivalent to the exergonic hydrolysis of ATP to AMP and PP$_i$ ($\Delta G'^\circ = -45.6$ kJ/mol) and the endergonic formation of fatty acyl–CoA ($\Delta G'^\circ = 31.4$ kJ/mol). The formation of fatty acyl–CoA is made energetically favorable by hydrolysis of the PP$_i$ by inorganic pyrophosphatase. Thus, in the activation of a fatty acid, both phosphoanhydride bonds of ATP are broken. The resulting $\Delta G'^\circ$ is the sum of the $\Delta G'^\circ$ values

for the breakage of these bonds, or $-45.6$ kJ/mol + $(-19.2)$ kJ/mol:

$$\text{ATP} + 2\text{H}_2\text{O} \longrightarrow \text{AMP} + 2\text{Pi} \qquad \Delta G'^\circ = -64.8 \text{ kJ/mol}$$

The activation of amino acids before their polymerization into proteins (see Fig. 27–14) is accomplished by an analogous set of reactions in which a transfer RNA molecule takes the place of coenzyme A. An interesting use of the cleavage of ATP to AMP and PP$_i$ occurs in the firefly, which uses ATP as an energy source to produce light flashes (Box 13–2).

---

**BOX 13–2    THE WORLD OF BIOCHEMISTRY**

## Firefly Flashes: Glowing Reports of ATP

Bioluminescence requires considerable amounts of energy. In the firefly, ATP is used in a set of reactions that converts chemical energy into light energy. In the 1950s, from many thousands of fireflies collected by children in and around Baltimore, William McElroy and his colleagues at The Johns Hopkins University isolated the principal biochemical components: luciferin, a complex carboxylic acid, and luciferase, an enzyme. The generation of a light flash requires activation of luciferin by an enzymatic reaction involving

pyrophosphate cleavage of ATP to form luciferyl adenylate. In the presence of molecular oxygen and luciferase, the luciferin undergoes a multistep oxidative decarboxylation to oxyluciferin. This process is accompanied by emission of light. The color of the light flash differs with the firefly species and appears to be determined by differences in the structure of the luciferase. Luciferin is regenerated from oxyluciferin in a subsequent series of reactions.

In the laboratory, pure firefly luciferin and luciferase are used to measure minute quantities of ATP by the intensity of the light flash produced. As little as a few picomoles ($10^{-12}$ mol) of ATP can be measured in this way. An enlightening extension of the studies in luciferase was the cloning of the luciferase gene into tobacco plants. When watered with a solution containing luciferin, the plants glowed in the dark (see Fig. 9–29).

The firefly, a beetle of the Lampyridae family.

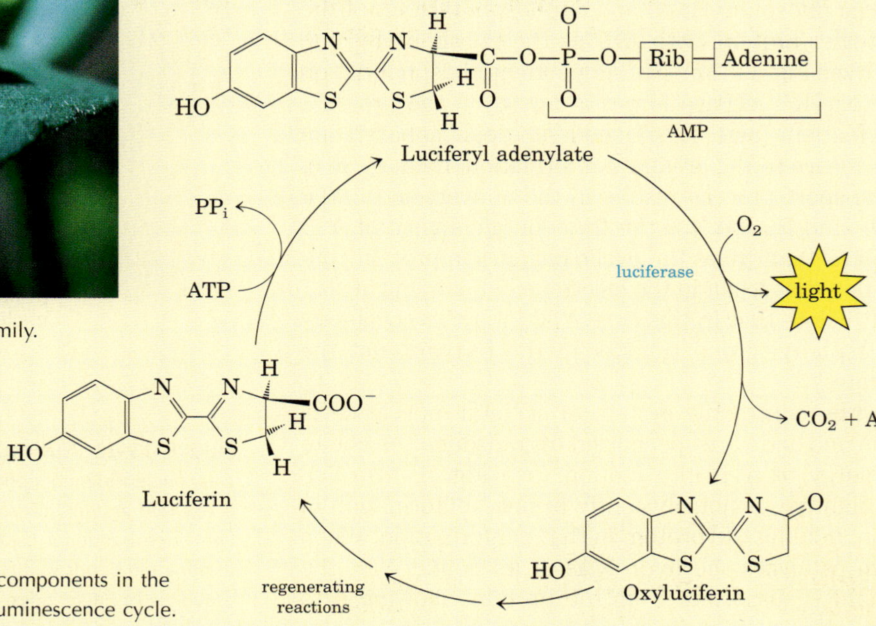

Important components in the firefly bioluminescence cycle.

## Assembly of Informational Macromolecules Requires Energy

When simple precursors are assembled into high molecular weight polymers with defined sequences (DNA, RNA, proteins), as described in detail in Part III, energy is required both for the condensation of monomeric units and for the creation of *ordered* sequences. The precursors for DNA and RNA synthesis are nucleoside triphosphates, and polymerization is accompanied by cleavage of the phosphoanhydride linkage between the $\alpha$ and $\beta$ phosphates, with the release of $PP_i$ (Fig. 13–11). The moieties transferred to the growing polymer in these reactions are adenylate (AMP), guanylate (GMP), cytidylate (CMP), or uridylate (UMP) for RNA synthesis, and their deoxy analogs (with TMP in place of UMP) for DNA synthesis. As noted above, the activation of amino acids for protein synthesis involves the donation of adenylate groups from ATP, and we shall see in Chapter 27 that several steps of protein synthesis on the ribosome are also accompanied by GTP hydrolysis. In all these cases, the exergonic breakdown of a nucleoside triphosphate is coupled to the endergonic process of synthesizing a polymer of a specific sequence.

## ATP Energizes Active Transport and Muscle Contraction

ATP can supply the energy for transporting an ion or a molecule across a membrane into another aqueous compartment where its concentration is higher (see Fig. 11–36). Transport processes are major consumers of energy; in human kidney and brain, for example, as much as two-thirds of the energy consumed at rest is used to pump $Na^+$ and $K^+$ across plasma membranes via the $Na^+K^+$ ATPase. The transport of $Na^+$ and $K^+$ is driven by cyclic phosphorylation and dephosphorylation of the transporter protein, with ATP as the phosphoryl group donor (see Fig. 11–37). $Na^+$-dependent phosphorylation of the $Na^+K^+$ ATPase forces a change in the protein's conformation, and $K^+$-dependent dephosphorylation favors return to the original conformation. Each cycle in the transport process results in the conversion of ATP to ADP and $P_i$, and it is the free-energy change of ATP hydrolysis that drives the cyclic changes in protein conformation that result in the electrogenic pumping of $Na^+$ and $K^+$. Note that in this case ATP interacts covalently by phosphoryl group transfer to the enzyme, not the substrate.

In the contractile system of skeletal muscle cells, myosin and actin are specialized to transduce the chemical energy of ATP into motion (see Fig. 5–33). ATP binds tightly but noncovalently to one conformation of myosin, holding the protein in that conformation. When myosin catalyzes the hydrolysis of its bound ATP, the ADP and $P_i$ dissociate from the protein, allowing it to

relax into a second conformation until another molecule of ATP binds. The binding and subsequent hydrolysis of ATP (by myosin ATPase) provide the energy that forces cyclic changes in the conformation of the myosin head. The change in conformation of many individual myosin

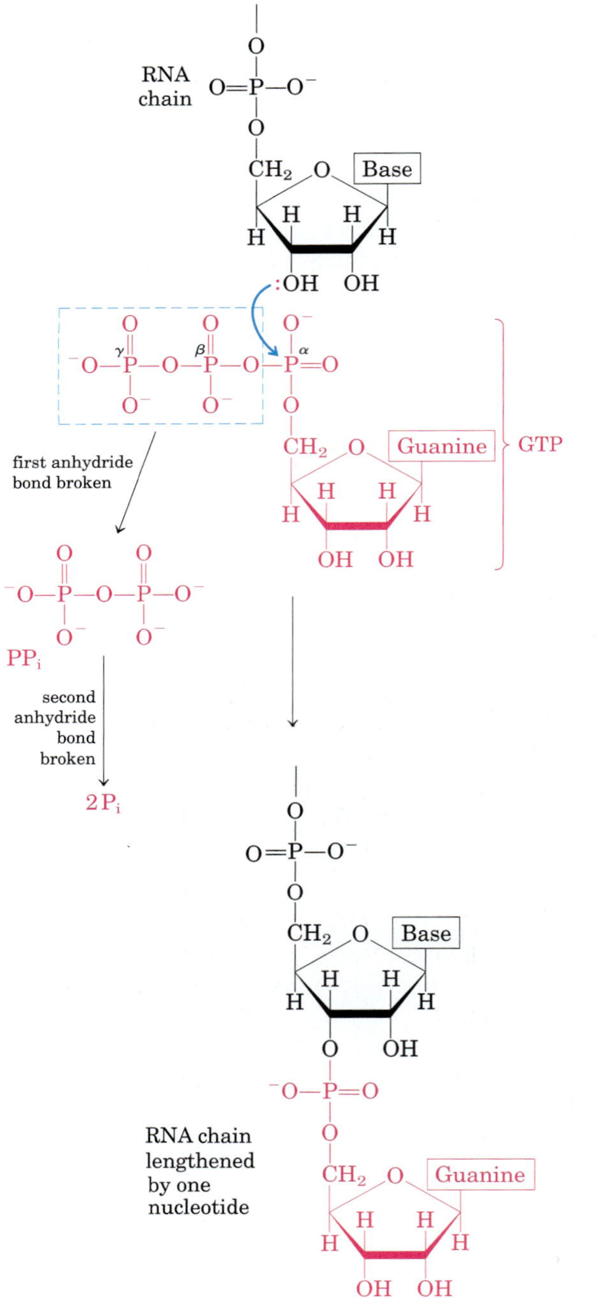

**FIGURE 13–11 Nucleoside triphosphates in RNA synthesis.** With each nucleoside monophosphate added to the growing chain, one $PP_i$ is released and hydrolyzed to two $P_i$. The hydrolysis of two phosphoanhydride bonds for each nucleotide added provides the energy for forming the bonds in the RNA polymer and for assembling a specific sequence of nucleotides.

molecules results in the sliding of myosin fibrils along actin filaments (see Fig. 5–32), which translates into macroscopic contraction of the muscle fiber.

As we noted earlier, this production of mechanical motion at the expense of ATP is one of the few cases in which ATP hydrolysis per se, rather than group transfer from ATP, is the source of the chemical energy in a coupled process.

## Transphosphorylations between Nucleotides Occur in All Cell Types

Although we have focused on ATP as the cell's energy currency and donor of phosphoryl groups, all other nucleoside triphosphates (GTP, UTP, and CTP) and all the deoxynucleoside triphosphates (dATP, dGTP, dTTP, and dCTP) are energetically equivalent to ATP. The free-energy changes associated with hydrolysis of their phosphoanhydride linkages are very nearly identical with those shown in Table 13–6 for ATP. In preparation for their various biological roles, these other nucleotides are generated and maintained as the nucleoside triphosphate (NTP) forms by phosphoryl group transfer to the corresponding nucleoside diphosphates (NDPs) and monophosphates (NMPs).

ATP is the primary high-energy phosphate compound produced by catabolism, in the processes of glycolysis, oxidative phosphorylation, and, in photosynthetic cells, photophosphorylation. Several enzymes then carry phosphoryl groups from ATP to the other nucleotides. **Nucleoside diphosphate kinase,** found in all cells, catalyzes the reaction

$$\text{ATP} + \text{NDP (or dNDP)} \xrightleftharpoons{\text{Mg}^{2+}} \text{ADP} + \text{NTP (or dNTP)}$$
$$\Delta G'^{\circ} \approx 0$$

Although this reaction is fully reversible, the relatively high [ATP]/[ADP] ratio in cells normally drives the reaction to the right, with the net formation of NTPs and dNTPs. The enzyme actually catalyzes a two-step phosphoryl transfer, which is a classic case of a double-displacement (Ping-Pong) mechanism (Fig. 13–12; see also Fig. 6–13b). First, phosphoryl group transfer from ATP to an active-site His residue produces a phospho-

enzyme intermediate; then the phosphoryl group is transferred from the P–His residue to an NDP acceptor. Because the enzyme is nonspecific for the base in the NDP and works equally well on dNDPs and NDPs, it can synthesize all NTPs and dNTPs, given the corresponding NDPs and a supply of ATP.

Phosphoryl group transfers from ATP result in an accumulation of ADP; for example, when muscle is contracting vigorously, ADP accumulates and interferes with ATP-dependent contraction. During periods of intense demand for ATP, the cell lowers the ADP concentration, and at the same time acquires ATP, by the action of **adenylate kinase:**

$$2\text{ADP} \xrightleftharpoons{\text{Mg}^{2+}} \text{ATP} + \text{AMP} \qquad \Delta G'^{\circ} \approx 0$$

This reaction is fully reversible, so after the intense demand for ATP ends, the enzyme can recycle AMP by converting it to ADP, which can then be phosphorylated to ATP in mitochondria. A similar enzyme, guanylate kinase, converts GMP to GDP at the expense of ATP. By pathways such as these, energy conserved in the catabolic production of ATP is used to supply the cell with all required NTPs and dNTPs.

Phosphocreatine (Fig. 13–5), also called creatine phosphate, serves as a ready source of phosphoryl groups for the quick synthesis of ATP from ADP. The phosphocreatine (PCr) concentration in skeletal muscle is approximately 30 mM, nearly ten times the concentration of ATP, and in other tissues such as smooth muscle, brain, and kidney [PCr] is 5 to 10 mM. The enzyme **creatine kinase** catalyzes the reversible reaction

$$\text{ADP} + \text{PCr} \xrightleftharpoons{\text{Mg}^{2+}} \text{ATP} + \text{Cr} \qquad \Delta G'^{\circ} = -12.5 \text{ kJ/mol}$$

When a sudden demand for energy depletes ATP, the PCr reservoir is used to replenish ATP at a rate considerably faster than ATP can be synthesized by catabolic pathways. When the demand for energy slackens, ATP produced by catabolism is used to replenish the PCr reservoir by reversal of the creatine kinase reaction. Organisms in the lower phyla employ other PCr-like molecules (collectively called **phosphagens**) as phosphoryl reservoirs.

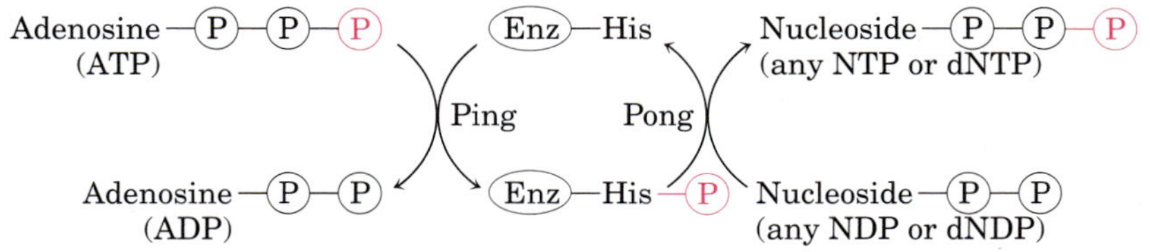

**FIGURE 13–12 Ping-Pong mechanism of nucleoside diphosphate kinase.** The enzyme binds its first substrate (ATP in our example), and a phosphoryl group is transferred to the side chain of a His residue. ADP departs, and another nucleoside (or deoxynucleoside) diphosphate replaces it, and this is converted to the corresponding triphosphate by transfer of the phosphoryl group from the phosphohistidine residue.

## Inorganic Polyphosphate Is a Potential Phosphoryl Group Donor

Inorganic polyphosphate (polyP) is a linear polymer composed of many tens or hundreds of $P_i$ residues linked through phosphoanhydride bonds. This polymer, present in all organisms, may accumulate to high levels in some cells. In yeast, for example, the amount of polyP that accumulates in the vacuoles would represent, if distributed uniformly throughout the cell, a concentration of 200 mM! (Compare this with the concentrations of other phosphoryl donors listed in Table 13–5.)

$$-O-\overset{O}{\underset{O^-}{\overset{||}{P}}}-O-\overset{O}{\underset{O^-}{\overset{||}{P}}}-O-\overset{O}{\underset{O^-}{\overset{||}{P}}}-O-\overset{O}{\underset{O^-}{\overset{||}{P}}}-O-\overset{O}{\underset{O^-}{\overset{||}{P}}}-O-$$

Inorganic polyphosphate (polyP)

One potential role for polyP is to serve as a phosphagen, a reservoir of phosphoryl groups that can be used to generate ATP, as creatine phosphate is used in muscle. PolyP has about the same phosphoryl group transfer potential as $PP_i$. The shortest polyphosphate, $PP_i$ ($n = 2$), can serve as the energy source for active transport of $H^+$ in plant vacuoles. For at least one form of the enzyme phosphofructokinase in plants, $PP_i$ is the phosphoryl group donor, a role played by ATP in animals and microbes (p. 527). The finding of high concentrations of polyP in volcanic condensates and steam vents suggests that it could have served as an energy source in prebiotic and early cellular evolution.

In prokaryotes, the enzyme **polyphosphate kinase-1** (PPK-1) catalyzes the reversible reaction

$$ATP + polyP_n \underset{}{\overset{Mg^{2+}}{\rightleftharpoons}} ADP + polyP_{n+1}$$
$$\Delta G'^{\circ} = -20 \text{ kJ/mol}$$

by a mechanism involving an enzyme-bound phosphohistidine intermediate (recall the mechanism of nucleoside diphosphate kinase, described above). A second enzyme, **polyphosphate kinase-2** (PPK-2), catalyzes the reversible synthesis of GTP (or ATP) from polyphosphate and GDP (or ADP):

$$GDP + polyP_{n+1} \underset{}{\overset{Mn^{2+}}{\rightleftharpoons}} GTP + polyP_n$$

PPK-2 is believed to act primarily in the direction of GTP and ATP synthesis, and PPK-1 in the direction of polyphosphate synthesis. PPK-1 and PPK-2 are present in a wide variety of prokaryotes, including many pathogenic bacteria.

In prokaryotes, elevated levels of polyP have been shown to promote expression of a number of genes involved in adaptation of the organism to conditions of starvation or other threats to survival. In *Escherichia coli*, for example, polyP accumulates when cells are starved for amino acids or $P_i$, and this accumulation con-

fers a survival advantage. Deletion of the genes for polyphosphate kinases diminishes the ability of certain pathogenic bacteria to invade animal tissues. The enzymes may therefore prove to be vulnerable targets in the development of new antimicrobial drugs.

No gene in yeast encodes a PPK-like protein, but four genes—unrelated to bacterial PPK genes—are necessary for the synthesis of polyphosphate. The mechanism for polyphosphate synthesis in eukaryotes seems to be quite different from that in prokaryotes.

## Biochemical and Chemical Equations Are Not Identical

Biochemists write metabolic equations in a simplified way, and this is particularly evident for reactions involving ATP. Phosphorylated compounds can exist in several ionization states and, as we have noted, the different species can bind $Mg^{2+}$. For example, at pH 7 and 2 mM $Mg^{2+}$, ATP exists in the forms $ATP^{4-}$, $HATP^{3-}$, $H_2ATP^{2-}$, $MgHATP^-$, and $Mg_2ATP$. In thinking about the biological role of ATP, however, we are not always interested in all this detail, and so we consider ATP as an entity made up of a sum of species, and we write its hydrolysis as the biochemical equation

$$ATP + H_2O \longrightarrow ADP + P_i$$

where ATP, ADP, and $P_i$ are sums of species. The corresponding apparent equilibrium constant, $K'_{eq} = [ADP][P_i]/[ATP]$, depends on the pH and the concentration of free $Mg^{2+}$. Note that $H^+$ and $Mg^{2+}$ do not appear in the biochemical equation because they are held constant. Thus a biochemical equation does not balance H, Mg, or charge, although it does balance all other elements involved in the reaction (C, N, O, and P in the equation above).

We can write a chemical equation that *does* balance for all elements and for charge. For example, when ATP is hydrolyzed at a pH above 8.5 in the absence of $Mg^{2+}$, the chemical reaction is represented by

$$ATP^{4-} + H_2O \longrightarrow ADP^{3-} + HPO_4^{2-} + H^+$$

The corresponding equilibrium constant, $K'_{eq} = [ADP^{3-}][HPO_4^{2-}][H^+]/[ATP^{4-}]$, depends only on temperature, pressure, and ionic strength.

Both ways of writing a metabolic reaction have value in biochemistry. Chemical equations are needed when we want to account for all atoms and charges in a reaction, as when we are considering the mechanism of a chemical reaction. Biochemical equations are used to determine in which direction a reaction will proceed spontaneously, given a specified pH and $[Mg^{2+}]$, or to calculate the equilibrium constant of such a reaction.

Throughout this book we use biochemical equations, unless the focus is on chemical mechanism, and we use values of $\Delta G'^{\circ}$ and $K'_{eq}$ as determined at pH 7 and 1 mM $Mg^{2+}$.

## SUMMARY 13.2 Phosphoryl Group Transfers and ATP

- ATP is the chemical link between catabolism and anabolism. It is the energy currency of the living cell. The exergonic conversion of ATP to ADP and $P_i$, or to AMP and $PP_i$, is coupled to many endergonic reactions and processes.

- Direct hydrolysis of ATP is the source of energy in the conformational changes that produce muscle contraction but, in general, it is not ATP hydrolysis but the transfer of a phosphoryl, pyrophosphoryl, or adenylyl group from ATP to a substrate or enzyme molecule that couples the energy of ATP breakdown to endergonic transformations of substrates.

- Through these group transfer reactions, ATP provides the energy for anabolic reactions, including the synthesis of informational molecules, and for the transport of molecules and ions across membranes against concentration gradients and electrical potential gradients.

- Cells contain other metabolites with large, negative, free energies of hydrolysis, including phosphoenolpyruvate, 1,3-bisphosphoglycerate, and phosphocreatine. These high-energy compounds, like ATP, have a high phosphoryl group transfer potential; they are good donors of the phosphoryl group. Thioesters also have high free energies of hydrolysis.

- Inorganic polyphosphate, present in all cells, may serve as a reservoir of phosphoryl groups with high group transfer potential.

## 13.3 Biological Oxidation-Reduction Reactions

The transfer of phosphoryl groups is a central feature of metabolism. Equally important is another kind of transfer, electron transfer in oxidation-reduction reactions. These reactions involve the loss of electrons by one chemical species, which is thereby oxidized, and the gain of electrons by another, which is reduced. The flow of electrons in oxidation-reduction reactions is responsible, directly or indirectly, for all work done by living organisms. In nonphotosynthetic organisms, the sources of electrons are reduced compounds (foods); in photosynthetic organisms, the initial electron donor is a chemical species excited by the absorption of light. The path of electron flow in metabolism is complex. Electrons move from various metabolic intermediates to specialized electron carriers in enzyme-catalyzed reactions.

The carriers in turn donate electrons to acceptors with higher electron affinities, with the release of energy. Cells contain a variety of molecular energy transducers, which convert the energy of electron flow into useful work.

We begin our discussion with a description of the general types of metabolic reactions in which electrons are transferred. After considering the theoretical and experimental basis for measuring the energy changes in oxidation reactions in terms of electromotive force, we discuss the relationship between this force, expressed in volts, and the free-energy change, expressed in joules. We conclude by describing the structures and oxidation-reduction chemistry of the most common of the specialized electron carriers, which you will encounter repeatedly in later chapters.

### The Flow of Electrons Can Do Biological Work

Every time we use a motor, an electric light or heater, or a spark to ignite gasoline in a car engine, we use the flow of electrons to accomplish work. In the circuit that powers a motor, the source of electrons can be a battery containing two chemical species that differ in affinity for electrons. Electrical wires provide a pathway for electron flow from the chemical species at one pole of the battery, through the motor, to the chemical species at the other pole of the battery. Because the two chemical species differ in their affinity for electrons, electrons flow spontaneously through the circuit, driven by a force proportional to the difference in electron affinity, the **electromotive force (emf).** The electromotive force (typically a few volts) can accomplish work if an appropriate energy transducer—in this case a motor—is placed in the circuit. The motor can be coupled to a variety of mechanical devices to accomplish useful work.

Living cells have an analogous biological "circuit," with a relatively reduced compound such as glucose as the source of electrons. As glucose is enzymatically oxidized, the released electrons flow spontaneously through a series of electron-carrier intermediates to another chemical species, such as $O_2$. This electron flow is exergonic, because $O_2$ has a higher affinity for electrons than do the electron-carrier intermediates. The resulting electromotive force provides energy to a variety of molecular energy transducers (enzymes and other proteins) that do biological work. In the mitochondrion, for example, membrane-bound enzymes couple electron flow to the production of a transmembrane pH difference, accomplishing osmotic and electrical work. The proton gradient thus formed has potential energy, sometimes called the proton-motive force by analogy with electromotive force. Another enzyme, ATP synthase in the inner mitochondrial membrane, uses the proton-motive force to do chemical work: synthesis of ATP from ADP and $P_i$ as protons flow spontaneously across the membrane. Similarly, membrane-localized enzymes in

*E. coli* convert electromotive force to proton-motive force, which is then used to power flagellar motion.

The principles of electrochemistry that govern energy changes in the macroscopic circuit with a motor and battery apply with equal validity to the molecular processes accompanying electron flow in living cells. We turn now to a discussion of those principles.

## Oxidation-Reductions Can Be Described as Half-Reactions

Although oxidation and reduction must occur together, it is convenient when describing electron transfers to consider the two halves of an oxidation-reduction reaction separately. For example, the oxidation of ferrous ion by cupric ion,

$$Fe^{2+} + Cu^{2+} \rightleftharpoons Fe^{3+} + Cu^+$$

can be described in terms of two half-reactions:

$$(1) \quad Fe^{2+} \rightleftharpoons Fe^{3+} + e^-$$
$$(2) \quad Cu^{2+} + e^- \rightleftharpoons Cu^+$$

The electron-donating molecule in an oxidation-reduction reaction is called the reducing agent or reductant; the electron-accepting molecule is the oxidizing agent or oxidant. A given agent, such as an iron cation existing in the ferrous ($Fe^{2+}$) or ferric ($Fe^{3+}$) state, functions as a conjugate reductant-oxidant pair (redox pair), just as an acid and corresponding base function as a conjugate acid-base pair. Recall from Chapter 2 that in acid-base reactions we can write a general equation: proton donor $\rightleftharpoons$ H$^+$ + proton acceptor. In redox reactions we can write a similar general equation: electron donor $\rightleftharpoons$ $e^-$ + electron acceptor. In the reversible half-reaction (1) above, $Fe^{2+}$ is the electron donor and $Fe^{3+}$ is the electron acceptor; together, $Fe^{2+}$ and $Fe^{3+}$ constitute a **conjugate redox pair.**

The electron transfers in the oxidation-reduction reactions of organic compounds are not fundamentally different from those of inorganic species. In Chapter 7 we considered the oxidation of a reducing sugar (an aldehyde or ketone) by cupric ion (see Fig. 7–10a):

$$R-\overset{\displaystyle O}{\underset{\displaystyle H}{C}} + 4OH^- + 2Cu^{2+} \rightleftharpoons R-\overset{\displaystyle O}{\underset{\displaystyle OH}{C}} + Cu_2O + 2H_2O$$

This overall reaction can be expressed as two half-reactions:

$$(1) \quad R-\overset{\displaystyle O}{\underset{\displaystyle H}{C}} + 2OH^- \rightleftharpoons R-\overset{\displaystyle O}{\underset{\displaystyle OH}{C}} + 2e^- + H_2O$$

$$(2) \quad 2Cu^{2+} + 2e^- + 2OH^- \rightleftharpoons Cu_2O + H_2O$$

Because two electrons are removed from the aldehyde carbon, the second half-reaction (the one-electron reduction of cupric to cuprous ion) must be doubled to balance the overall equation.

## Biological Oxidations Often Involve Dehydrogenation

The carbon in living cells exists in a range of oxidation states (Fig. 13–13). When a carbon atom shares an electron pair with another atom (typically H, C, S, N, or O), the sharing is unequal in favor of the more electronegative atom. The order of increasing electronegativity is H < C < S < N < O. In oversimplified but useful terms, the more electronegative atom "owns" the bonding electrons it shares with another atom. For example, in methane ($CH_4$), carbon is more electronegative than the four hydrogens bonded to it, and the C atom therefore "owns" all eight bonding electrons (Fig. 13–13). In ethane, the electrons in the C—C bond are shared equally, so each C atom owns only seven of its eight bonding electrons. In ethanol, C-1 is less electronegative than the oxygen to which it is bonded, and the O atom therefore "owns" both electrons of the C—O bond, leaving C-1 with only five bonding electrons. With each formal loss of electrons, the carbon atom has undergone oxidation—even when no oxygen is involved, as in the conversion of an alkane (—CH$_2$—CH$_2$—) to an alkene (—CH=CH—). In this case, oxidation (loss of electrons) is coincident with the loss of hydrogen. In biological systems, oxidation is often synonymous with **dehydrogenation,** and many enzymes that catalyze oxidation reactions are **dehydrogenases.** Notice that the more reduced compounds in Figure 13–13 (top) are richer in hydrogen than in oxygen, whereas the more oxidized compounds (bottom) have more oxygen and less hydrogen.

Not all biological oxidation-reduction reactions involve carbon. For example, in the conversion of molecular nitrogen to ammonia, $6H^+ + 6e^- + N_2 \rightarrow 2NH_3$, the nitrogen atoms are reduced.

Electrons are transferred from one molecule (electron donor) to another (electron acceptor) in one of four different ways:

1. Directly as *electrons*. For example, the $Fe^{2+}/Fe^{3+}$ redox pair can transfer an electron to the $Cu^+/Cu^{2+}$ redox pair:

$$Fe^{2+} + Cu^{2+} \rightleftharpoons Fe^{3+} + Cu^+$$

2. As *hydrogen atoms*. Recall that a hydrogen atom consists of a proton (H$^+$) and a single electron ($e^-$). In this case we can write the general equation

$$AH_2 \rightleftharpoons A + 2e^- + 2H^+$$

where $AH_2$ is the hydrogen/electron donor. (Do not mistake the above reaction for an acid dissociation; the H$^+$ arises from the removal of a hydrogen atom, H$^+$ + $e^-$.) $AH_2$ and A together constitute a conjugate redox pair ($A/AH_2$), which can reduce another compound B (or redox pair, $B/BH_2$) by transfer of hydrogen atoms:

$$AH_2 + B \rightleftharpoons A + BH_2$$

**3.** As a *hydride ion* ($:H^-$), which has two electrons. This occurs in the case of NAD-linked dehydrogenases, described below.

**4.** Through direct *combination with oxygen*. In this case, oxygen combines with an organic reductant and is covalently incorporated in the product, as in the oxidation of a hydrocarbon to an alcohol:

$$R{-}CH_3 + \tfrac{1}{2}O_2 \longrightarrow R{-}CH_2{-}OH$$

The hydrocarbon is the electron donor and the oxygen atom is the electron acceptor.

All four types of electron transfer occur in cells. The neutral term **reducing equivalent** is commonly used to designate a single electron equivalent participating in an oxidation-reduction reaction, no matter whether this equivalent is an electron per se, a hydrogen atom, or a hydride ion, or whether the electron transfer takes place in a reaction with oxygen to yield an oxygenated product. Because biological fuel molecules are usually enzymatically dehydrogenated to lose *two* reducing equivalents at a time, and because each oxygen atom can accept two reducing equivalents, biochemists by convention regard the unit of biological oxidations as two reducing equivalents passing from substrate to oxygen.

## Reduction Potentials Measure Affinity for Electrons

When two conjugate redox pairs are together in solution, electron transfer from the electron donor of one pair to the electron acceptor of the other may proceed spontaneously. The tendency for such a reaction depends on the relative affinity of the electron acceptor of each redox pair for electrons. The **standard reduction potential, $E°$,** a measure (in volts) of this affinity, can be determined in an experiment such as that described in Figure 13–14. Electrochemists have chosen as a standard of reference the half-reaction

$$H^+ + e^- \longrightarrow \tfrac{1}{2}H_2$$

The electrode at which this half-reaction occurs (called a half-cell) is arbitrarily assigned a standard reduction

potential of 0.00 V. When this hydrogen electrode is connected through an external circuit to another half-cell in which an oxidized species and its corresponding reduced species are present at standard concentrations (each solute at 1 M, each gas at 101.3 kPa), electrons tend to flow through the external circuit from the half-cell of

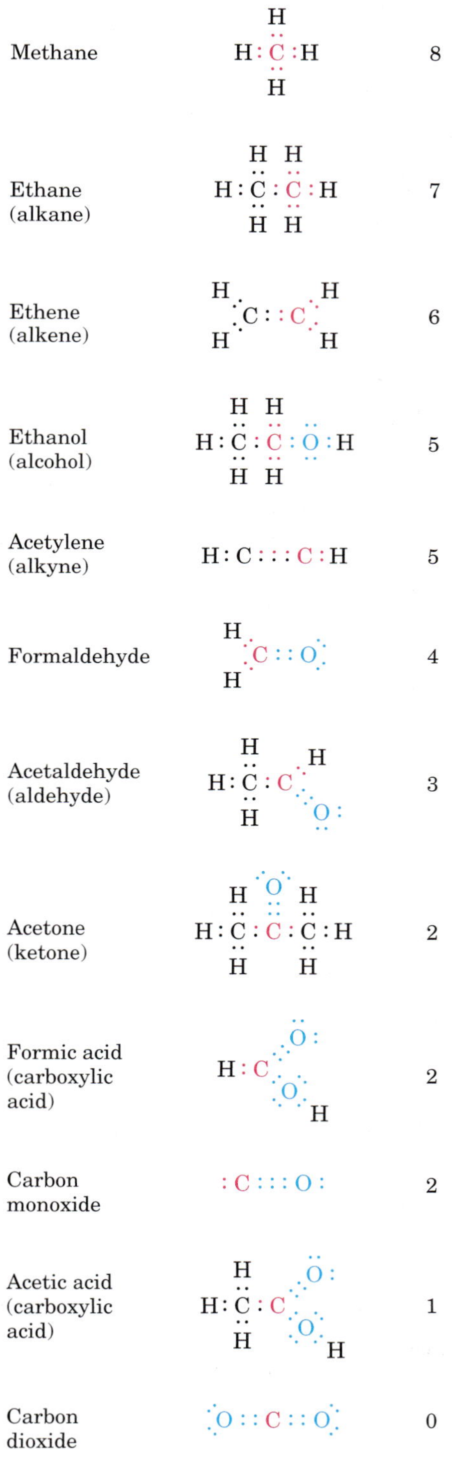

**FIGURE 13–13 Oxidation states of carbon in the biosphere.** The oxidation states are illustrated with some representative compounds. Focus on the red carbon atom and its bonding electrons. When this carbon is bonded to the less electronegative H atom, both bonding electrons (red) are assigned to the carbon. When carbon is bonded to another carbon, bonding electrons are shared equally, so one of the two electrons is assigned to the red carbon. When the red carbon is bonded to the more electronegative O atom, the bonding electrons are assigned to the oxygen. The number to the right of each compound is the number of electrons "owned" by the red carbon, a rough expression of the oxidation state of that carbon. When the red carbon undergoes oxidation (loses electrons), the number gets smaller. Thus the oxidation state increases from top to bottom of the list.

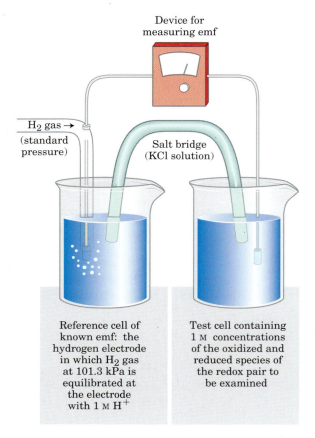

**FIGURE 13–14 Measurement of the standard reduction potential ($E'°$) of a redox pair.** Electrons flow from the test electrode to the reference electrode, or vice versa. The ultimate reference half-cell is the hydrogen electrode, as shown here, at pH 0. The electromotive force (emf) of this electrode is designated 0.00 V. At pH 7 in the test cell, $E'°$ for the hydrogen electrode is −0.414 V. The direction of electron flow depends on the relative electron "pressure" or potential of the two cells. A salt bridge containing a saturated KCl solution provides a path for counter-ion movement between the test cell and the reference cell. From the observed emf and the known emf of the reference cell, the experimenter can find the emf of the test cell containing the redox pair. The cell that gains electrons has, by convention, the more positive reduction potential.

lower standard reduction potential to the half-cell of higher standard reduction potential. By convention, the half-cell with the stronger tendency to acquire electrons is assigned a positive value of $E°$.

The reduction potential of a half-cell depends not only on the chemical species present but also on their activities, approximated by their concentrations. About a century ago, Walther Nernst derived an equation that relates standard reduction potential ($E°$) to the reduction potential ($E$) at any concentration of oxidized and reduced species in the cell:

$$E = E° + \frac{RT}{n\mathcal{F}} \ln \frac{[\text{electron acceptor}]}{[\text{electron donor}]} \quad (13\text{–}4)$$

where $R$ and $T$ have their usual meanings, $n$ is the number of electrons transferred per molecule, and $\mathcal{F}$ is the Faraday constant (Table 13–1). At 298 K (25 °C), this expression reduces to

$$E = E° + \frac{0.026\ \text{V}}{n} \ln \frac{[\text{electron acceptor}]}{[\text{electron donor}]} \quad (13\text{–}5)$$

Many half-reactions of interest to biochemists involve protons. As in the definition of $\Delta G'°$, biochemists define the standard state for oxidation-reduction reactions as pH 7 and express reduction potential as $E'°$, the standard reduction potential at pH 7. The standard reduction potentials given in Table 13–7 and used throughout this book are values for $E'°$ and are therefore valid only for systems at neutral pH. Each value represents the potential difference when the conjugate redox pair, at 1 M concentrations and pH 7, is connected with the standard (pH 0) hydrogen electrode. Notice in Table 13–7 that when the conjugate pair $2H^+/H_2$ at pH 7 is connected with the standard hydrogen electrode (pH 0), electrons tend to flow from the pH 7 cell to the standard (pH 0) cell; the measured $E'°$ for the $2H^+/H_2$ pair is −0.414 V.

## Standard Reduction Potentials Can Be Used to Calculate the Free-Energy Change

The usefulness of reduction potentials stems from the fact that when $E$ values have been determined for any two half-cells, relative to the standard hydrogen electrode, their reduction potentials relative to each other are also known. We can then predict the direction in which electrons will tend to flow when the two half-cells are connected through an external circuit or when components of both half-cells are present in the same solution. Electrons tend to flow to the half-cell with the more positive $E$, and the strength of that tendency is proportional to the difference in reduction potentials, $\Delta E$.

The energy made available by this spontaneous electron flow (the free-energy change for the oxidation-reduction reaction) is proportional to $\Delta E$:

$$\Delta G = -n\mathcal{F}\,\Delta E \quad \text{or} \quad \Delta G'° = -n\mathcal{F}\,\Delta E'° \quad (13\text{–}6)$$

Here $n$ represents the number of electrons transferred in the reaction. With this equation we can calculate the free-energy change for any oxidation-reduction reaction from the values of $E'°$ in a table of reduction potentials (Table 13–7) and the concentrations of the species participating in the reaction.

Consider the reaction in which acetaldehyde is reduced by the biological electron carrier NADH:

$$\text{Acetaldehyde} + \text{NADH} + \text{H}^+ \longrightarrow \text{ethanol} + \text{NAD}^+$$

The relevant half-reactions and their $E'°$ values are:

(1)  Acetaldehyde + $2H^+$ + $2e^-$ $\longrightarrow$ ethanol

$$E'° = -0.197\ \text{V}$$

(2) $NAD^+ + 2H^+ + 2e^- \longrightarrow NADH + H^+$

$$E'^\circ = -0.320 \text{ V}$$

By convention, $\Delta E'^\circ$ is expressed as $E'^\circ$ of the electron acceptor minus $E'^\circ$ of the electron donor. Because acetaldehyde is accepting electrons from NADH in our example, $\Delta E'^\circ = -0.197 \text{ V} - (-0.320 \text{ V}) = 0.123 \text{ V}$, and $n$ is 2. Therefore,

$$\Delta G'^\circ = -n\mathcal{F}\,\Delta E'^\circ = -2(96.5 \text{ kJ/V} \cdot \text{mol})(0.123 \text{ V})$$
$$= -23.7 \text{ kJ/mol}$$

This is the free-energy change for the oxidation-reduction reaction at pH 7, when acetaldehyde, ethanol, $NAD^+$, and NADH are all present at 1.00 M concentrations. If, instead, acetaldehyde and NADH were present at 1.00 M but ethanol and $NAD^+$ were present at 0.100 M, the value for $\Delta G$ would be calculated as follows. First,

the values of $E$ for both reductants are determined (Eqn 13–4):

$$E_{\text{acetaldehyde}} = E^\circ + \frac{RT}{n\mathcal{F}} \ln \frac{[\text{acetaldehyde}]}{[\text{ethanol}]}$$

$$= -0.197 \text{ V} + \frac{0.026 \text{ V}}{2} \ln \frac{1.00}{0.100} = -0.167 \text{ V}$$

$$E_{\text{NADH}} = E^\circ + \frac{RT}{n\mathcal{F}} \ln \frac{[\text{NAD}^+]}{[\text{NADH}]}$$

$$= -0.320 \text{ V} + \frac{0.026 \text{ V}}{2} \ln \frac{1.00}{0.100} = -0.350 \text{ V}$$

Then $\Delta E$ is used to calculate $\Delta G$ (Eqn 13–5):

$$\Delta E = -0.167 \text{ V} - (-0.350) \text{ V} = 0.183 \text{ V}$$
$$\Delta G = -n\mathcal{F}\,\Delta E$$
$$= -2(96.5 \text{ kJ/V} \cdot \text{mol})(0.183 \text{ V})$$
$$= -35.3 \text{ kJ/mol}$$

**TABLE 13–7** Standard Reduction Potentials of Some Biologically Important Half-Reactions, at pH 7.0 and 25 °C (298 K)

| Half-reaction | $E'^\circ$ (V) |
|---|---|
| $\frac{1}{2}O_2 + 2H^+ + 2e^- \longrightarrow H_2O$ | 0.816 |
| $Fe^{3+} + e^- \longrightarrow Fe^{2+}$ | 0.771 |
| $NO_3^- + 2H^+ + 2e^- \longrightarrow NO_2^- + H_2O$ | 0.421 |
| Cytochrome $f$ $(Fe^{3+}) + e^- \longrightarrow$ cytochrome $f$ $(Fe^{2+})$ | 0.365 |
| $Fe(CN)_6^{3-}$ (ferricyanide) $+ e^- \longrightarrow Fe(CN)_6^{4-}$ | 0.36 |
| Cytochrome $a_3$ $(Fe^{3+}) + e^- \longrightarrow$ cytochrome $a_3$ $(Fe^{2+})$ | 0.35 |
| $O_2 + 2H^+ + 2e^- \longrightarrow H_2O_2$ | 0.295 |
| Cytochrome $a$ $(Fe^{3+}) + e^- \longrightarrow$ cytochrome $a$ $(Fe^{2+})$ | 0.29 |
| Cytochrome $c$ $(Fe^{3+}) + e^- \longrightarrow$ cytochrome $c$ $(Fe^{2+})$ | 0.254 |
| Cytochrome $c_1$ $(Fe^{3+}) + e^- \longrightarrow$ cytochrome $c_1$ $(Fe^{2+})$ | 0.22 |
| Cytochrome $b$ $(Fe^{3+}) + e^- \longrightarrow$ cytochrome $b$ $(Fe^{2+})$ | 0.077 |
| Ubiquinone $+ 2H^+ + 2e^- \longrightarrow$ ubiquinol $+ H_2$ | 0.045 |
| Fumarate$^{2-}$ $+ 2H^+ + 2e^- \longrightarrow$ succinate$^{2-}$ | 0.031 |
| $2H^+ + 2e^- \longrightarrow H_2$ (at standard conditions, pH 0) | 0.000 |
| Crotonyl-CoA $+ 2H^+ + 2e^- \longrightarrow$ butyryl-CoA | −0.015 |
| Oxaloacetate$^{2-}$ $+ 2H^+ + 2e^- \longrightarrow$ malate$^{2-}$ | −0.166 |
| Pyruvate$^-$ $+ 2H^+ + 2e^- \longrightarrow$ lactate$^-$ | −0.185 |
| Acetaldehyde $+ 2H^+ + 2e^- \longrightarrow$ ethanol | −0.197 |
| FAD $+ 2H^+ + 2e^- \longrightarrow FADH_2$ | −0.219* |
| Glutathione $+ 2H^+ + 2e^- \longrightarrow$ 2 reduced glutathione | −0.23 |
| $S + 2H^+ + 2e^- \longrightarrow H_2S$ | −0.243 |
| Lipoic acid $+ 2H^+ + 2e^- \longrightarrow$ dihydrolipoic acid | −0.29 |
| $NAD^+ + H^+ + 2e^- \longrightarrow$ NADH | −0.320 |
| $NADP^+ + H^+ + 2e^- \longrightarrow$ NADPH | −0.324 |
| Acetoacetate $+ 2H^+ + 2e^- \longrightarrow \beta$-hydroxybutyrate | −0.346 |
| $\alpha$-Ketoglutarate $+ CO_2 + 2H^+ + 2e^- \longrightarrow$ isocitrate | −0.38 |
| $2H^+ + 2e^- \longrightarrow H_2$ (at pH 7) | −0.414 |
| Ferredoxin $(Fe^{3+}) + e^- \longrightarrow$ ferredoxin $(Fe^{2+})$ | −0.432 |

Source: Data mostly from Loach, P.A. (1976) In *Handbook of Biochemistry and Molecular Biology,* 3rd edn (Fasman, G.D., ed.), *Physical and Chemical Data,* Vol. I, pp. 122–130, CRC Press, Boca Raton, FL.

* This is the value for free FAD; FAD bound to a specific flavoprotein (for example succinate dehydrogenase) has a different $E'^\circ$ that depends on its protein environments.

It is thus possible to calculate the free-energy change for any biological redox reaction at any concentrations of the redox pairs.

## Cellular Oxidation of Glucose to Carbon Dioxide Requires Specialized Electron Carriers

The principles of oxidation-reduction energetics described above apply to the many metabolic reactions that involve electron transfers. For example, in many organisms, the oxidation of glucose supplies energy for the production of ATP. The complete oxidation of glucose:

$$C_6H_{12}O_6 + 6O_2 \longrightarrow 6CO_2 + 6H_2O$$

has a $\Delta G'^\circ$ of $-2,840$ kJ/mol. This is a much larger release of free energy than is required for ATP synthesis (50 to 60 kJ/mol; see Box 13–1). Cells convert glucose to $CO_2$ not in a single, high-energy-releasing reaction, but rather in a series of controlled reactions, some of which are oxidations. The free energy released in these oxidation steps is of the same order of magnitude as that required for ATP synthesis from ADP, with some energy to spare. Electrons removed in these oxidation steps are transferred to coenzymes specialized for carrying electrons, such as $NAD^+$ and FAD (described below).

## A Few Types of Coenzymes and Proteins Serve as Universal Electron Carriers

The multitude of enzymes that catalyze cellular oxidations channel electrons from their hundreds of different substrates into just a few types of universal electron carriers. The reduction of these carriers in catabolic processes results in the conservation of free energy released by substrate oxidation. $NAD^+$, $NADP^+$, FMN, and FAD are water-soluble coenzymes that undergo reversible oxidation and reduction in many of the electron-transfer reactions of metabolism. The nucleotides $NAD^+$ and $NADP^+$ move readily from one enzyme to another; the flavin nucleotides FMN and FAD are usually very tightly bound to the enzymes, called flavoproteins, for which they serve as prosthetic groups. Lipid-soluble quinones such as ubiquinone and plastoquinone act as electron carriers and proton donors in the nonaqueous environment of membranes. Iron-sulfur proteins and cytochromes, which have tightly bound prosthetic groups that undergo reversible oxidation and reduction, also serve as electron carriers in many oxidation-reduction reactions. Some of these proteins are water-soluble, but others are peripheral or integral membrane proteins (see Fig. 11–6).

We conclude this chapter by describing some chemical features of nucleotide coenzymes and some of the enzymes (dehydrogenases and flavoproteins) that use them. The oxidation-reduction chemistry of quinones, iron-sulfur proteins, and cytochromes is discussed in Chapter 19.

## NADH and NADPH Act with Dehydrogenases as Soluble Electron Carriers

Nicotinamide adenine dinucleotide ($NAD^+$ in its oxidized form) and its close analog nicotinamide adenine dinucleotide phosphate ($NADP^+$) are composed of two nucleotides joined through their phosphate groups by a phosphoanhydride bond (Fig. 13–15a). Because the nicotinamide ring resembles pyridine, these compounds are sometimes called **pyridine nucleotides.** The vitamin niacin is the source of the nicotinamide moiety in nicotinamide nucleotides.

Both coenzymes undergo reversible reduction of the nicotinamide ring (Fig. 13–15). As a substrate molecule undergoes oxidation (dehydrogenation), giving up two hydrogen atoms, the oxidized form of the nucleotide ($NAD^+$ or $NADP^+$) accepts a hydride ion ($:H^-$, the equivalent of a proton and two electrons) and is transformed into the reduced form (NADH or NADPH). The second proton removed from the substrate is released to the aqueous solvent. The half-reaction for each type of nucleotide is therefore

$$NAD^+ + 2e^- + 2H^+ \longrightarrow NADH + H^+$$
$$NADP^+ + 2e^- + 2H^+ \longrightarrow NADPH + H^+$$

Reduction of $NAD^+$ or $NADP^+$ converts the benzenoid ring of the nicotinamide moiety (with a fixed positive charge on the ring nitrogen) to the quinonoid form (with no charge on the nitrogen). Note that the reduced nucleotides absorb light at 340 nm; the oxidized forms do not (Fig. 13–15b). The plus sign in the abbreviations $NAD^+$ and $NADP^+$ does *not* indicate the net charge on these molecules (they are both negatively charged); rather, it indicates that the nicotinamide ring is in its oxidized form, with a positive charge on the nitrogen atom. In the abbreviations NADH and NADPH, the "H" denotes the added hydride ion. To refer to these nucleotides without specifying their oxidation state, we use NAD and NADP.

The total concentration of $NAD^+$ + NADH in most tissues is about $10^{-5}$ M; that of $NADP^+$ + NADPH is about $10^{-6}$ M. In many cells and tissues, the ratio of $NAD^+$ (oxidized) to NADH (reduced) is high, favoring hydride transfer from a substrate *to* $NAD^+$ to form NADH. By contrast, NADPH (reduced) is generally present in greater amounts than its oxidized form, $NADP^+$, favoring hydride transfer *from* NADPH to a substrate. This reflects the specialized metabolic roles of the two coenzymes: $NAD^+$ generally functions in oxidations—usually as part of a catabolic reaction; and NADPH is the usual coenzyme in reductions—nearly always as part of an anabolic reaction. A few enzymes can use either coenzyme, but most show a strong preference for one over the other. The processes in which these two cofactors function are also segregated in specific organelles of eukaryotic cells: oxidations of fuels such as pyruvate, fatty acids, and $\alpha$-keto acids derived from

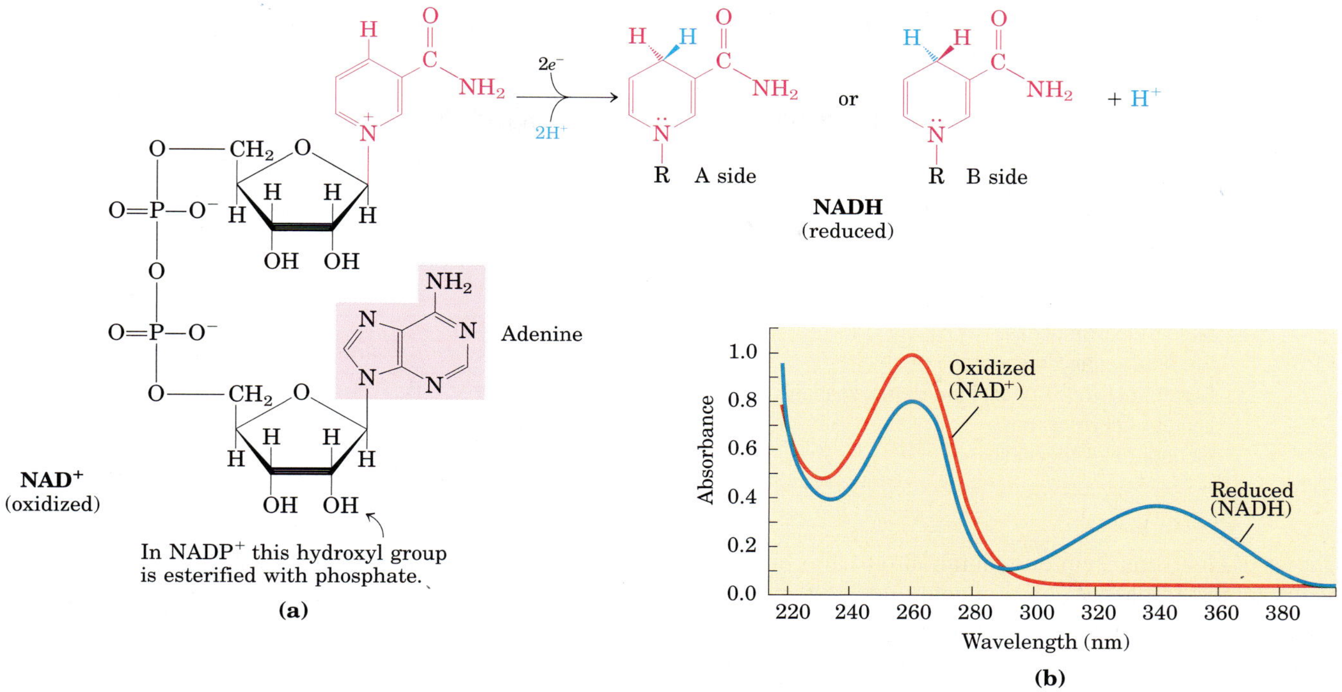

**(a)**

**(b)**

**FIGURE 13–15 NAD and NADP. (a)** Nicotinamide adenine dinucleotide, $NAD^+$, and its phosphorylated analog $NADP^+$ undergo reduction to NADH and NADPH, accepting a hydride ion (two electrons and one proton) from an oxidizable substrate. The hydride ion is added to either the front (the A side) or the back (the B side) of the planar nicotinamide ring (see Table 13–8). **(b)** The UV absorption spectra of $NAD^+$ and NADH. Reduction of the nicotinamide ring produces a new, broad absorption band with a maximum at 340 nm. The production of NADH during an enzyme-catalyzed reaction can be conveniently followed by observing the appearance of the absorbance at 340 nm (the molar extinction coefficient $\varepsilon_{340} = 6{,}200$ $M^{-1}cm^{-1}$).

amino acids occur in the mitochondrial matrix, whereas reductive biosynthesis processes such as fatty acid synthesis take place in the cytosol. This functional and spatial specialization allows a cell to maintain two distinct pools of electron carriers, with two distinct functions.

More than 200 enzymes are known to catalyze reactions in which $NAD^+$ (or $NADP^+$) accepts a hydride ion from a reduced substrate, or NADPH (or NADH) donates a hydride ion to an oxidized substrate. The general reactions are

$$AH_2 + NAD^+ \longrightarrow A + NADH + H^+$$
$$A + NADPH + H^+ \longrightarrow AH_2 + NADP^+$$

where $AH_2$ is the reduced substrate and A the oxidized substrate. The general name for an enzyme of this type is **oxidoreductase;** they are also commonly called **dehydrogenases.** For example, alcohol dehydrogenase catalyzes the first step in the catabolism of ethanol, in which ethanol is oxidized to acetaldehyde:

$$\underset{\text{Ethanol}}{CH_3CH_2OH} + NAD^+ \longrightarrow \underset{\text{Acetaldehyde}}{CH_3CHO} + NADH + H^+$$

Notice that one of the carbon atoms in ethanol has lost a hydrogen; the compound has been oxidized from an alcohol to an aldehyde (refer again to Fig. 13–13 for the oxidation states of carbon).

When $NAD^+$ or $NADP^+$ is reduced, the hydride ion could in principle be transferred to either side of the nicotinamide ring: the front (A side) or the back (B side), as represented in Figure 13–15a. Studies with isotopically labeled substrates have shown that a given enzyme catalyzes either an A-type or a B-type transfer, but not both. For example, yeast alcohol dehydrogenase and lactate dehydrogenase of vertebrate heart transfer a hydride ion to (or remove a hydride ion from) the A side of the nicotinamide ring; they are classed as type A dehydrogenases to distinguish them from another group of enzymes that transfer a hydride ion to (or remove a hydride ion from) the B side of the nicotinamide ring (Table 13–8). The specificity for one side or another can be very striking; lactate dehydrogenase, for example, prefers the A side over the B side by a factor of $5 \times 10^7$!

Most dehydrogenases that use NAD or NADP bind the cofactor in a conserved protein domain called the Rossmann fold (named for Michael Rossmann, who deduced the structure of lactate dehydrogenase and first described this structural motif). The Rossmann fold typically consists of a six-stranded parallel $\beta$ sheet and four associated $\alpha$ helices (Fig. 13–16).

The association between a dehydrogenase and NAD or NADP is relatively loose; the coenzyme readily diffuses from one enzyme to another, acting as a water-soluble

**TABLE 13–8 Stereospecificity of Dehydrogenases That Employ NAD$^+$ or NADP$^+$ as Coenzymes**

| Enzyme | Coenzyme | Stereochemical specificity for nicotinamide ring (A or B) | Text page(s) |
|---|---|---|---|
| Isocitrate dehydrogenase | NAD$^+$ | A | 610 |
| α-Ketoglutarate dehydrogenase | NAD$^+$ | B | 610 |
| Glucose 6-phosphate dehydrogenase | NADP$^+$ | B | 540 |
| Malate dehydrogenase | NAD$^+$ | A | 612 |
| Glutamate dehydrogenase | NAD$^+$ or NADP$^+$ | B | 665 |
| Glyceraldehyde 3-phosphate dehydrogenase | NAD$^+$ | B | 530 |
| Lactate dehydrogenase | NAD$^+$ | A | 538 |
| Alcohol dehydrogenase | NAD$^+$ | A | 540 |

carrier of electrons from one metabolite to another. For example, in the production of alcohol during fermentation of glucose by yeast cells, a hydride ion is removed from glyceraldehyde 3-phosphate by one enzyme (glyceraldehyde 3-phosphate dehydrogenase, a type B enzyme) and transferred to NAD$^+$. The NADH produced then leaves the enzyme surface and diffuses to another enzyme (alcohol dehydrogenase, a type A enzyme), which transfers a hydride ion to acetaldehyde, producing ethanol:

(1) Glyceraldehyde 3-phosphate + NAD$^+$ $\longrightarrow$ 3-phosphoglycerate + NADH + H$^+$

(2) Acetaldehyde + NADH + H$^+$ $\longrightarrow$ ethanol + NAD$^+$

*Sum:* Glyceraldehyde 3-phosphate + acetaldehyde $\longrightarrow$ 3-phosphoglycerate + ethanol

Notice that in the overall reaction there is no net production or consumption of NAD$^+$ or NADH; the coenzymes function catalytically and are recycled repeatedly without a net change in the concentration of NAD$^+$ + NADH.

## Dietary Deficiency of Niacin, the Vitamin Form of NAD and NADP, Causes Pellagra

The pyridine-like rings of NAD and NADP are derived from the vitamin **niacin** (nicotinic acid; Fig. 13–17), which is synthesized from tryptophan. Humans generally cannot synthesize niacin in sufficient quantities, and this is especially so for those with diets low in tryptophan (maize, for example, has a low tryptophan content). Niacin deficiency, which affects all the NAD(P)-dependent dehydrogenases, causes the serious human disease pellagra (Italian for "rough skin") and a related disease in dogs, blacktongue. These diseases are characterized by the "three Ds": dermatitis, diarrhea, and dementia, followed in many cases by death. A century ago, pellagra was a common human disease; in the southern United States, where maize was a dietary staple, about 100,000 people were afflicted and about 10,000 died between 1912 and 1916. In 1920 Joseph Goldberger showed pellagra to be caused by a dietary insufficiency, and in 1937 Frank Strong, D. Wayne Woolley, and Conrad Elvehjem identified niacin as the curative agent for blacktongue. Supplementation of the human diet with this inexpensive compound led to the eradication of pellagra in the populations of the developed world—with one significant exception. Pellagra is still found among alcoholics, whose intestinal absorption of niacin is much

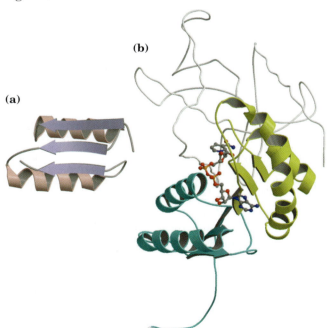

**(b)**

**(a)**

**FIGURE 13–16 The nucleotide binding domain of the enzyme lactate dehydrogenase. (a)** The Rossmann fold is a structural motif found in the NAD-binding site of many dehydrogenases. It consists of a six-stranded parallel β sheet and four α helices; inspection reveals the arrangement to be a pair of structurally similar β-α-β-α-β motifs. **(b)** The dinucleotide NAD binds in an extended conformation through hydrogen bonds and salt bridges (derived from PDB ID 3LDH).

## FIGURE 13-17 Structures of niacin (nicotinic acid) and its derivative nicotinamide.

**FIGURE 13-17 Structures of niacin (nicotinic acid) and its derivative nicotinamide.** The biosynthetic precursor of these compounds is tryptophan. In the laboratory, nicotinic acid was first produced by oxidation of the natural product nicotine—thus the name. Both nicotinic acid and nicotinamide cure pellagra, but nicotine (from cigarettes or elsewhere) has no curative activity.

reduced, and whose caloric needs are often met with distilled spirits that are virtually devoid of vitamins, including niacin. In a few places, including the Deccan Plateau in India, pellagra still occurs, especially among the poor. ■

## Flavin Nucleotides Are Tightly Bound in Flavoproteins

**Flavoproteins** (Table 13–9) are enzymes that catalyze oxidation-reduction reactions using either flavin mononucleotide (FMN) or flavin adenine dinucleotide (FAD) as coenzyme (Fig. 13–18). These coenzymes, the **flavin nucleotides,** are derived from the vitamin riboflavin. The fused ring structure of flavin nucleotides (the isoalloxazine ring) undergoes reversible reduction, accepting either one or two electrons in the form of one or two hydrogen atoms (each atom an electron plus a proton) from a reduced substrate. The fully reduced forms are abbreviated $FADH_2$ and $FMNH_2$. When a fully oxidized flavin nucleotide accepts only one electron (one hydrogen atom), the semiquinone form of the isoal-

**TABLE 13-9 Some Enzymes (Flavoproteins) That Employ Flavin Nucleotide Coenzymes**

| Enzyme | Flavin nucleotide | Text page(s) |
|---|---|---|
| Acyl–CoA dehydrogenase | FAD | 638 |
| Dihydrolipoyl dehydrogenase | FAD | 605 |
| Succinate dehydrogenase | FAD | 612 |
| Glycerol 3-phosphate dehydrogenase | FAD | 714–715 |
| Thioredoxin reductase | FAD | 869 |
| NADH dehydrogenase (Complex I) | FMN | 696–697 |
| Glycolate oxidase | FMN | 767 |

loxazine ring is produced, abbreviated FADH$^\bullet$ and FMNH$^\bullet$. Because flavoproteins can participate in either one- or two-electron transfers, this class of proteins is involved in a greater diversity of reactions than the NAD(P)-linked dehydrogenases.

Like the nicotinamide coenzymes (Fig. 13–15), the flavin nucleotides undergo a shift in a major absorption band on reduction. Flavoproteins that are fully reduced (two electrons accepted) generally have an absorption maximum near 360 nm. When partially reduced (one electron), they acquire another absorption maximum at about 450 nm; when fully oxidized, the flavin has maxima at 370 and 440 nm. The intermediate radical form, reduced by one electron, has absorption maxima at 380, 480, 580, and 625 nm. These changes can be used to assay reactions involving a flavoprotein.

The flavin nucleotide in most flavoproteins is bound rather tightly to the protein, and in some enzymes, such as succinate dehydrogenase, it is bound covalently. Such tightly bound coenzymes are properly called prosthetic groups. They do not transfer electrons by diffusing from one enzyme to another; rather, they provide a means by which the flavoprotein can temporarily hold electrons while it catalyzes electron transfer from a reduced substrate to an electron acceptor. One important feature of the flavoproteins is the variability in the standard reduction potential ($E'^\circ$) of the bound flavin nucleotide. Tight association between the enzyme and prosthetic group confers on the flavin ring a reduction potential typical of that particular flavoprotein, sometimes quite different from the reduction potential of the free flavin nucleotide. FAD bound to succinate dehydrogenase, for example, has an $E'^\circ$ close to 0.0 V, compared with $-0.219$ V for free FAD; $E'^\circ$ for other flavoproteins ranges from $-0.40$ V to $+0.06$ V.

Frank Strong,
1908–1993

D. Wayne Woolley,
1914–1966

Conrad Elvehjem,
1901–1962

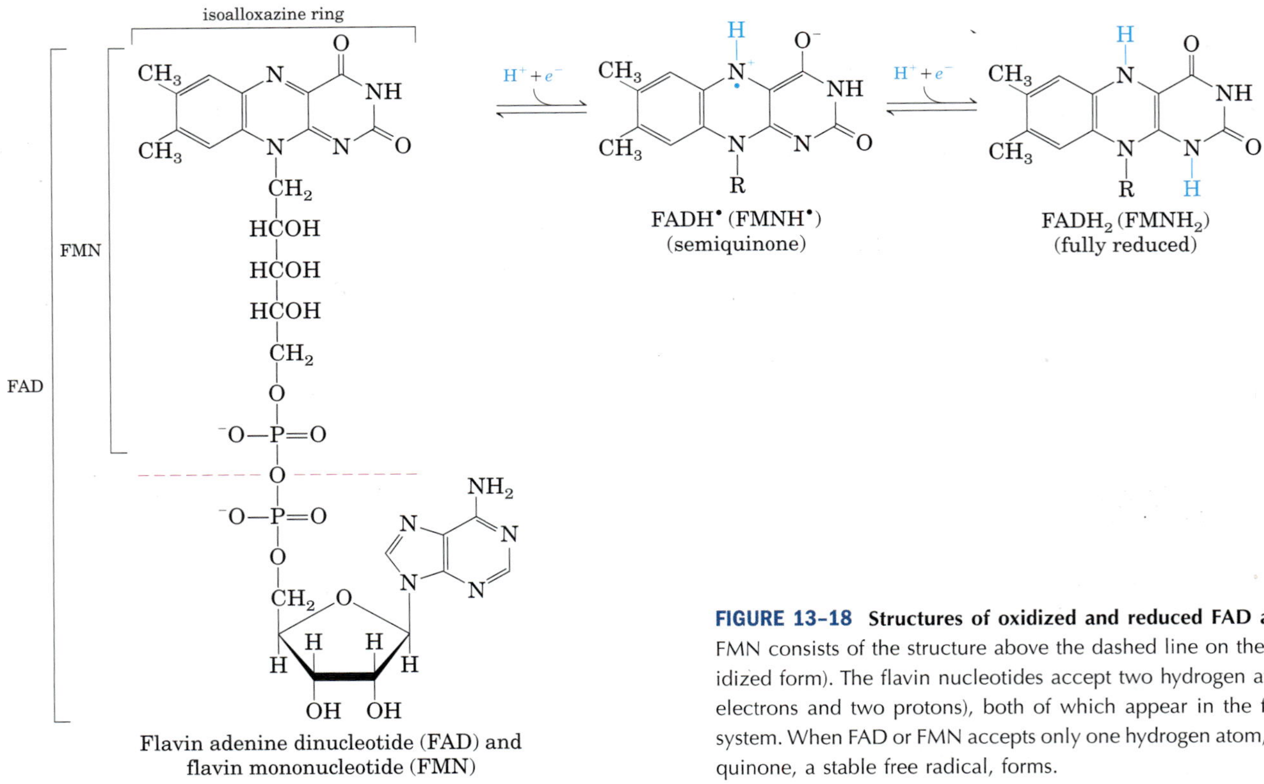

Flavin adenine dinucleotide (FAD) and
flavin mononucleotide (FMN)

**FIGURE 13–18    Structures of oxidized and reduced FAD and FMN.**
FMN consists of the structure above the dashed line on the FAD (oxidized form). The flavin nucleotides accept two hydrogen atoms (two electrons and two protons), both of which appear in the flavin ring system. When FAD or FMN accepts only one hydrogen atom, the semiquinone, a stable free radical, forms.

Flavoproteins are often very complex; some have, in addition to a flavin nucleotide, tightly bound inorganic ions (iron or molybdenum, for example) capable of participating in electron transfers.

Certain flavoproteins act in a quite different role as light receptors. **Cryptochromes** are a family of flavoproteins, widely distributed in the eukaryotic phyla, that mediate the effects of blue light on plant development and the effects of light on mammalian circadian rhythms (oscillations in physiology and biochemistry, with a 24-hour period). The cryptochromes are homologs of another family of flavoproteins, the photolyases. Found in both prokaryotes and eukaryotes, **photolyases** use the energy of absorbed light to repair chemical defects in DNA.

We examine the function of flavoproteins as electron carriers in Chapter 19, when we consider their roles in oxidative phosphorylation (in mitochondria) and photophosphorylation (in chloroplasts), and we describe the photolyase reactions in Chapter 25.

## SUMMARY 13.3 Biological Oxidation-Reduction Reactions

- In many organisms, a central energy-conserving process is the stepwise oxidation of glucose to $CO_2$, in which some of the energy of oxidation is conserved in ATP as electrons are passed to $O_2$.

- Biological oxidation-reduction reactions can be described in terms of two half-reactions, each with a characteristic standard reduction potential, $E'^\circ$.

- When two electrochemical half-cells, each containing the components of a half-reaction, are connected, electrons tend to flow to the half-cell with the higher reduction potential. The strength of this tendency is proportional to the difference between the two reduction potentials ($\Delta E$) and is a function of the concentrations of oxidized and reduced species.

- The standard free-energy change for an oxidation-reduction reaction is directly proportional to the difference in standard reduction potentials of the two half-cells: $\Delta G'^\circ = -n\mathcal{F}\,\Delta E'^\circ$.

- Many biological oxidation reactions are dehydrogenations in which one or two hydrogen atoms ($H^+ + e^-$) are transferred from a substrate to a hydrogen acceptor. Oxidation-reduction reactions in living cells involve specialized electron carriers.

- NAD and NADP are the freely diffusible coenzymes of many dehydrogenases. Both $NAD^+$ and $NADP^+$ accept two electrons and one proton. NAD and NADP are bound to dehydrogenases in a widely conserved structural motif called the Rossmann fold.

✳ ■ FAD and FMN, the flavin nucleotides, serve as tightly bound prosthetic groups of

flavoproteins. They can accept either one or two electrons and one or two protons. Flavoproteins also serve as light receptors in cryptochromes and photolyases.

## Key Terms

*Terms in bold are defined in the glossary.*

## Further Reading

### Bioenergetics and Thermodynamics

**Atkins, P.W.** (1984) *The Second Law,* Scientific American Books, Inc., New York.

A well-illustrated and elementary discussion of the second law and its implications.

**Becker, W.M.** (1977) *Energy and the Living Cell: An Introduction to Bioenergetics,* J. B. Lippincott Company, Philadelphia.

A clear introductory account of cellular metabolism, in terms of energetics.

**Bergethon, P.R.** (1998) *The Physical Basis of Biochemistry,* Springer Verlag, New York.

Chapters 11 through 13 of this book, and the books by Tinoco et al. and van Holde et al. (below), are excellent general references for physical biochemistry, with good discussions of the applications of thermodynamics to biochemistry.

**Edsall, J.T. & Gutfreund, H.** (1983) *Biothermodynamics: The Study of Biochemical Processes at Equilibrium,* John Wiley & Sons, Inc., New York.

**Harold, F.M.** (1986) *The Vital Force: A Study of Bioenergetics,* W. H. Freeman and Company, New York.

A beautifully clear discussion of thermodynamics in biological processes.

**Harris, D.A.** (1995) *Bioenergetics at a Glance,* Blackwell Science, Oxford.

A short, clearly written account of cellular energetics, including introductory chapters on thermodynamics.

**Loewenstein, W.R.** (1999) *The Touchstone of Life: Molecular Information, Cell Communication, and the Foundations of Life,* Oxford University Press, New York.

Beautifully written discussion of the relationship between entropy and information.

**Morowitz, H.J.** (1978) *Foundations of Bioenergetics,* Academic Press, Inc., New York. [Out of print.]

Clear, rigorous description of thermodynamics in biology.

**Nicholls, D.G. & Ferguson, S.J.** (2002) *Bioenergetics 3,* Academic Press, Inc., New York.

Clear, well-illustrated intermediate-level discussion of the theory of bioenergetics and the mechanisms of energy transductions.

**Tinoco, I., Jr., Sauer, K., & Wang, J.C.** (1996) *Physical Chemistry: Principles and Applications in Biological Sciences,* 3rd edn, Prentice-Hall, Inc., Upper Saddle River, NJ.

Chapters 2 through 5 cover thermodynamics.

**van Holde, K.E., Johnson, W.C., & Ho, P.S.** (1998) *Principles of Physical Biochemistry,* Prentice-Hall, Inc., Upper Saddle River, NJ.

Chapters 2 and 3 are especially relevant.

### Phosphoryl Group Transfers and ATP

**Alberty, R.A.** (1994) Biochemical thermodynamics. *Biochim. Biophys. Acta* **1207,** 1–11.

Explains the distinction between biochemical and chemical equations, and the calculation and meaning of transformed thermodynamic properties for ATP and other phosphorylated compounds.

**Bridger, W.A. & Henderson, J.F.** (1983) *Cell ATP,* John Wiley & Sons, Inc., New York.

The chemistry of ATP, its role in metabolic regulation, and its catabolic and anabolic roles.

**Frey, P.A. & Arabshahi, A.** (1995) Standard free-energy change for the hydrolysis of the α–β-phosphoanhydride bridge in ATP. *Biochemistry* **34,** 11,307–11,310.

**Hanson, R.W.** (1989) The role of ATP in metabolism. *Biochem. Educ.* **17,** 86–92.

Excellent summary of the chemistry and biology of ATP.

**Kornberg, A.** (1999) Inorganic polyphosphate: a molecule of many functions. *Annu. Rev. Biochem.* **68,** 89–125.

**Lipmann, F.** (1941) Metabolic generation and utilization of phosphate bond energy. *Adv. Enzymol.* **11,** 99–162.

The classic description of the role of high-energy phosphate compounds in biology.

**Pullman, B. & Pullman, A.** (1960) Electronic structure of energy-rich phosphates. *Radiat. Res.,* Suppl. 2, 160–181.

An advanced discussion of the chemistry of ATP and other "energy-rich" compounds.

**Veech, R.L., Lawson, J.W.R., Cornell, N.W., & Krebs, H.A.** (1979) Cytosolic phosphorylation potential. *J. Biol. Chem.* **254,** 6538–6547.

Experimental determination of ATP, ADP, and $P_i$ concentrations in brain, muscle, and liver, and a discussion of the problems in determining the real free-energy change for ATP synthesis in cells.

**Westheimer, F.H.** (1987) Why nature chose phosphates. *Science* **235,** 1173–1178.

A chemist's description of the unique suitability of phosphate esters and anhydrides for metabolic transformations.

**Biological Oxidation-Reduction Reactions**

**Cashmore, A.R., Jarillo, J.A., Wu, Y.J., & Liu D.** (1999) Cryptochromes: blue light receptors for plants and animals. *Science* **284,** 760–765.

**Dolphin, D., Avramovic, O., & Poulson, R. (eds)** (1987) *Pyridine Nucleotide Coenzymes: Chemical, Biochemical, and Medical Aspects,* John Wiley & Sons, Inc., New York.

An excellent two-volume collection of authoritative reviews. Among the most useful are the chapters by Kaplan, Westheimer, Veech, and Ohno and Ushio.

**Fraaije, M.W. & Mattevi, A.** (2000) Flavoenzymes: diverse catalysts with recurrent features. *Trends Biochem. Sci.* **25,** 126–132.

**Massey, V.** (1994) Activation of molecular oxygen by flavins and flavoproteins. *J. Biol. Chem.* **269,** 22,459–22,462.

A short review of the chemistry of flavin–oxygen interactions in flavoproteins.

**Rees, D.C.** (2002) Great metalloclusters in enzymology. *Annu. Rev. Biochem.* **71,** 221–246.

Advanced review of the types of metal ion clusters found in enzymes and their modes of action.

**Williams, R.E. & Bruce, N.C.** (2002) New uses for an old enzyme—the old yellow enzyme family of flavoenzymes. *Microbiology* **148,** 1607–1614.

## Problems

**1. Entropy Changes during Egg Development** Consider a system consisting of an egg in an incubator. The white and yolk of the egg contain proteins, carbohydrates, and lipids. If fertilized, the egg is transformed from a single cell to a complex organism. Discuss this irreversible process in terms of the entropy changes in the system, surroundings, and universe. Be sure that you first clearly define the system and surroundings.

**2. Calculation of $\Delta G'^\circ$ from an Equilibrium Constant** Calculate the standard free-energy changes of the following metabolically important enzyme-catalyzed reactions at 25 °C and pH 7.0, using the equilibrium constants given.

(a) Glutamate + oxaloacetate $\underset{\text{aminotransferase}}{\overset{\text{aspartate}}{\rightleftharpoons}}$

aspartate + α-ketoglutarate    $K'_{eq} = 6.8$

(b) Dihydroxyacetone phosphate $\underset{\text{isomerase}}{\overset{\text{triose phosphate}}{\rightleftharpoons}}$

glyceraldehyde 3-phosphate    $K'_{eq} = 0.0475$

(c) Fructose 6-phosphate + ATP $\overset{\text{phosphofructokinase}}{\rightleftharpoons}$

fructose 1,6-bisphosphate + ADP    $K'_{eq} = 254$

**3. Calculation of the Equilibrium Constant from $\Delta G'^\circ$** Calculate the equilibrium constants $K'_{eq}$ for each of the following reactions at pH 7.0 and 25 °C, using the $\Delta G'^\circ$ values in Table 13–4.

(a) Glucose 6-phosphate + $H_2O$ $\underset{\text{6-phosphatase}}{\overset{\text{glucose}}{\rightleftharpoons}}$

glucose + $P_i$

(b) Lactose + $H_2O$ $\overset{\beta\text{-galactosidase}}{\rightleftharpoons}$ glucose + galactose

(c) Malate $\overset{\text{fumarase}}{\rightleftharpoons}$ fumarate + $H_2O$

**4. Experimental Determination of $K'_{eq}$ and $\Delta G'^\circ$** If a 0.1 M solution of glucose 1-phosphate is incubated with a catalytic amount of phosphoglucomutase, the glucose 1-phosphate is transformed to glucose 6-phosphate. At equilibrium, the concentrations of the reaction components are

Glucose 1-phosphate $\rightleftharpoons$ glucose 6-phosphate
$4.5 \times 10^{-3}$ M           $9.6 \times 10^{-2}$ M

Calculate $K'_{eq}$ and $\Delta G'^\circ$ for this reaction at 25 °C.

**5. Experimental Determination of $\Delta G'^\circ$ for ATP Hydrolysis** A direct measurement of the standard free-energy change associated with the hydrolysis of ATP is technically demanding because the minute amount of ATP remaining at equilibrium is difficult to measure accurately. The value of $\Delta G'^\circ$ can be calculated indirectly, however, from the equilibrium constants of two other enzymatic reactions having less favorable equilibrium constants:

Glucose 6-phosphate + $H_2O \longrightarrow$ glucose + $P_i$ $\qquad K'_{eq} = 270$

ATP + glucose $\longrightarrow$ ADP + glucose 6-phosphate

$$K'_{eq} = 890$$

Using this information, calculate the standard free energy of hydrolysis of ATP at 25 °C.

**6. Difference between $\Delta G'^\circ$ and $\Delta G$** Consider the following interconversion, which occurs in glycolysis (Chapter 14):

Fructose 6-phosphate $\rightleftharpoons$ glucose 6-phosphate

$$K'_{eq} = 1.97$$

(a) What is $\Delta G'^\circ$ for the reaction (at 25 °C)?

(b) If the concentration of fructose 6-phosphate is adjusted to 1.5 M and that of glucose 6-phosphate is adjusted to 0.50 M, what is $\Delta G$?

(c) Why are $\Delta G'^\circ$ and $\Delta G$ different?

**7. Dependence of $\Delta G$ on pH** The free energy released by the hydrolysis of ATP under standard conditions at pH 7.0 is $-30.5$ kJ/mol. If ATP is hydrolyzed under standard conditions but at pH 5.0, is more or less free energy released? Explain. Use the Living Graph to explore this relationship. 🖱

**8. The $\Delta G'^\circ$ for Coupled Reactions** Glucose 1-phosphate is converted into fructose 6-phosphate in two successive reactions:

Glucose 1-phosphate $\longrightarrow$ glucose 6-phosphate
Glucose 6-phosphate $\longrightarrow$ fructose 6-phosphate

Using the $\Delta G'^\circ$ values in Table 13–4, calculate the equilibrium constant, $K'_{eq}$, for the sum of the two reactions at 25 °C:

Glucose 1-phosphate $\longrightarrow$ fructose 6-phosphate

**9. Strategy for Overcoming an Unfavorable Reaction: ATP-Dependent Chemical Coupling** The phosphorylation of glucose to glucose 6-phosphate is the initial step in the catabolism of glucose. The direct phosphorylation of glucose by $P_i$ is described by the equation

Glucose + $P_i \longrightarrow$ glucose 6-phosphate + $H_2O$

$$\Delta G'^\circ = 13.8 \text{ kJ/mol}$$

(a) Calculate the equilibrium constant for the above reaction. In the rat hepatocyte the physiological concentrations of glucose and $P_i$ are maintained at approximately 4.8 mM. What is the equilibrium concentration of glucose 6-phosphate obtained by the direct phosphorylation of glucose by $P_i$? Does this reaction represent a reasonable metabolic step for the catabolism of glucose? Explain.

(b) In principle, at least, one way to increase the concentration of glucose 6-phosphate is to drive the equilibrium reaction to the right by increasing the intracellular concentrations of glucose and $P_i$. Assuming a fixed concentration of $P_i$ at 4.8 mM, how high would the intracellular concentration of glucose have to be to give an equilibrium concentration of glucose 6-phosphate of 250 μM (the normal physiological concentration)? Would this route be physiologically reasonable, given that the maximum solubility of glucose is less than 1 M?

(c) The phosphorylation of glucose in the cell is coupled to the hydrolysis of ATP; that is, part of the free energy of ATP hydrolysis is used to phosphorylate glucose:

(1) $\qquad$ Glucose + $P_i \longrightarrow$ glucose 6-phosphate + $H_2O$

$$\Delta G'^\circ = 13.8 \text{ kJ/mol}$$

(2) $\qquad$ ATP + $H_2O \longrightarrow$ ADP + $P_i$

$$\Delta G'^\circ = -30.5 \text{ kJ/mol}$$

*Sum:* $\quad$ Glucose + ATP $\longrightarrow$ glucose 6-phosphate + ADP

Calculate $K'_{eq}$ for the overall reaction. For the ATP-dependent phosphorylation of glucose, what concentration of glucose is needed to achieve a 250 μM intracellular concentration of glucose 6-phosphate when the concentrations of ATP and ADP are 3.38 mM and 1.32 mM, respectively? Does this coupling process provide a feasible route, at least in principle, for the phosphorylation of glucose in the cell? Explain.

(d) Although coupling ATP hydrolysis to glucose phosphorylation makes thermodynamic sense, we have not yet specified how this coupling is to take place. Given that coupling requires a common intermediate, one conceivable route is to use ATP hydrolysis to raise the intracellular concentration of $P_i$ and thus drive the unfavorable phosphorylation of glucose by $P_i$. Is this a reasonable route? (Think about the solubility products of metabolic intermediates.)

(e) The ATP-coupled phosphorylation of glucose is catalyzed in hepatocytes by the enzyme glucokinase. This enzyme binds ATP and glucose to form a glucose-ATP-enzyme complex, and the phosphoryl group is transferred directly from ATP to glucose. Explain the advantages of this route.

**10. Calculations of $\Delta G'^\circ$ for ATP-Coupled Reactions** From data in Table 13–6 calculate the $\Delta G'^\circ$ value for the reactions

(a) Phosphocreatine + ADP $\longrightarrow$ creatine + ATP

(b) ATP + fructose $\longrightarrow$ ADP + fructose 6-phosphate

**11. Coupling ATP Cleavage to an Unfavorable Reaction** To explore the consequences of coupling ATP hydrolysis under physiological conditions to a thermodynamically unfavorable biochemical reaction, consider the hypothetical transformation $X \longrightarrow Y$, for which $\Delta G'^\circ = 20$ kJ/mol.

(a) What is the ratio [Y]/[X] at equilibrium?

(b) Suppose X and Y participate in a sequence of reactions during which ATP is hydrolyzed to ADP and $P_i$. The overall reaction is

$$X + ATP + H_2O \longrightarrow Y + ADP + P_i$$

Calculate [Y]/[X] for this reaction at equilibrium. Assume that the equilibrium concentrations of ATP, ADP, and $P_i$ are 1 M.

(c) We know that [ATP], [ADP], and [$P_i$] are *not* 1 M under physiological conditions. Calculate [Y]/[X] for the ATP-coupled reaction when the values of [ATP], [ADP], and [$P_i$] are those found in rat myocytes (Table 13–5).

**12. Calculations of $\Delta G$ at Physiological Concentrations** Calculate the physiological $\Delta G$ (not $\Delta G'^\circ$) for the reaction

Phosphocreatine + ADP $\longrightarrow$ creatine + ATP

at 25 °C, as it occurs in the cytosol of neurons, with phosphocreatine at 4.7 mM, creatine at 1.0 mM, ADP at 0.73 mM, and ATP at 2.6 mM.

**13. Free Energy Required for ATP Synthesis under Physiological Conditions**    In the cytosol of rat hepatocytes, the mass-action ratio, $Q$, is

$$\frac{[ATP]}{[ADP][P_i]} = 5.33 \times 10^2 \ \text{M}^{-1}$$

Calculate the free energy required to synthesize ATP in a rat hepatocyte.

**14. Daily ATP Utilization by Human Adults**

(a) A total of 30.5 kJ/mol of free energy is needed to synthesize ATP from ADP and $P_i$ when the reactants and products are at 1 M concentrations (standard state). Because the actual physiological concentrations of ATP, ADP, and $P_i$ are not 1 M, the free energy required to synthesize ATP under physiological conditions is different from $\Delta G'^\circ$. Calculate the free energy required to synthesize ATP in the human hepatocyte when the physiological concentrations of ATP, ADP, and $P_i$ are 3.5, 1.50, and 5.0 mM, respectively.

(b) A 68 kg (150 lb) adult requires a caloric intake of 2,000 kcal (8,360 kJ) of food per day (24 h). The food is metabolized and the free energy is used to synthesize ATP, which then provides energy for the body's daily chemical and mechanical work. Assuming that the efficiency of converting food energy into ATP is 50%, calculate the weight of ATP used by a human adult in 24 h. What percentage of the body weight does this represent?

(c) Although adults synthesize large amounts of ATP daily, their body weight, structure, and composition do not change significantly during this period. Explain this apparent contradiction.

**15. Rates of Turnover of $\gamma$ and $\beta$ Phosphates of ATP**    If a small amount of ATP labeled with radioactive phosphorus in the terminal position, [$\gamma$-$^{32}$P]ATP, is added to a yeast extract, about half of the $^{32}$P activity is found in $P_i$ within a few minutes, but the concentration of ATP remains unchanged. Explain. If the same experiment is carried out using ATP labeled with $^{32}$P in the central position, [$\beta$-$^{32}$P]ATP, the $^{32}$P does not appear in $P_i$ within such a short time. Why?

**16. Cleavage of ATP to AMP and $PP_i$ during Metabolism**    The synthesis of the activated form of acetate (acetyl-CoA) is carried out in an ATP-dependent process:

Acetate + CoA + ATP $\longrightarrow$ acetyl-CoA + AMP + $PP_i$

(a) The $\Delta G'^\circ$ for the hydrolysis of acetyl-CoA to acetate and CoA is $-32.2$ kJ/mol and that for hydrolysis of ATP to AMP and $PP_i$ is $-30.5$ kJ/mol. Calculate $\Delta G'^\circ$ for the ATP-dependent synthesis of acetyl-CoA.

(b) Almost all cells contain the enzyme inorganic pyrophosphatase, which catalyzes the hydrolysis of $PP_i$ to $P_i$. What effect does the presence of this enzyme have on the synthesis of acetyl-CoA? Explain.

**17. Energy for H$^+$ Pumping**    The parietal cells of the stomach lining contain membrane "pumps" that transport hydrogen ions from the cytosol of these cells (pH 7.0) into the stomach, contributing to the acidity of gastric juice (pH 1.0). Calculate the free energy required to transport 1 mol of hydrogen ions through these pumps. (Hint: See Chapter 11.) Assume a temperature of 25 °C.

**18. Standard Reduction Potentials**    The standard reduction potential, $E'^\circ$, of any redox pair is defined for the half-cell reaction:

Oxidizing agent + $n$ electrons $\longrightarrow$ reducing agent

The $E'^\circ$ values for the NAD$^+$/NADH and pyruvate/lactate conjugate redox pairs are $-0.32$ V and $-0.19$ V, respectively.

(a) Which conjugate pair has the greater tendency to lose electrons? Explain.

(b) Which is the stronger oxidizing agent? Explain.

(c) Beginning with 1 M concentrations of each reactant and product at pH 7, in which direction will the following reaction proceed?

Pyruvate + NADH + H$^+$ $\rightleftharpoons$ lactate + NAD$^+$

(d) What is the standard free-energy change ($\Delta G'^\circ$) at 25 °C for the conversion of pyruvate to lactate?

(e) What is the equilibrium constant ($K'_{eq}$) for this reaction?

**19. Energy Span of the Respiratory Chain**    Electron transfer in the mitochondrial respiratory chain may be represented by the net reaction equation

NADH + H$^+$ + $\tfrac{1}{2}$ O$_2$ $\rightleftharpoons$ H$_2$O + NAD$^+$

(a) Calculate the value of $\Delta E'^\circ$ for the net reaction of mitochondrial electron transfer. Use $E'^\circ$ values from Table 13–7.

(b) Calculate $\Delta G'^\circ$ for this reaction.

(c) How many ATP molecules can *theoretically* be generated by this reaction if the free energy of ATP synthesis under cellular conditions is 52 kJ/mol?

**20. Dependence of Electromotive Force on Concentrations**    Calculate the electromotive force (in volts) registered by an electrode immersed in a solution containing the following mixtures of NAD$^+$ and NADH at pH 7.0 and 25 °C, with reference to a half-cell of $E'^\circ$ 0.00 V.

(a) 1.0 mM NAD$^+$ and 10 mM NADH

(b) 1.0 mM NAD$^+$ and 1.0 mM NADH

(c) 10 mM NAD$^+$ and 1.0 mM NADH

**21. Electron Affinity of Compounds**    List the following substances in order of increasing tendency to accept electrons: (a) $\alpha$-ketoglutarate + CO$_2$ (yielding isocitrate); (b) oxaloacetate; (c) O$_2$; (d) NADP$^+$.

**22. Direction of Oxidation-Reduction Reactions**    Which of the following reactions would you expect to proceed in the direction shown, under standard conditions, assuming that the appropriate enzymes are present to catalyze them?

(a) Malate + NAD$^+$ $\longrightarrow$ oxaloacetate + NADH + H$^+$

(b) Acetoacetate + NADH + H$^+$ $\longrightarrow$

$\beta$-hydroxybutyrate + NAD$^+$

(c) Pyruvate + NADH + H$^+$ $\longrightarrow$ lactate + NAD$^+$

(d) Pyruvate + $\beta$-hydroxybutyrate $\longrightarrow$

lactate + acetoacetate

(e) Malate + pyruvate $\longrightarrow$ oxaloacetate + lactate

(f) Acetaldehyde + succinate $\longrightarrow$ ethanol + fumarate

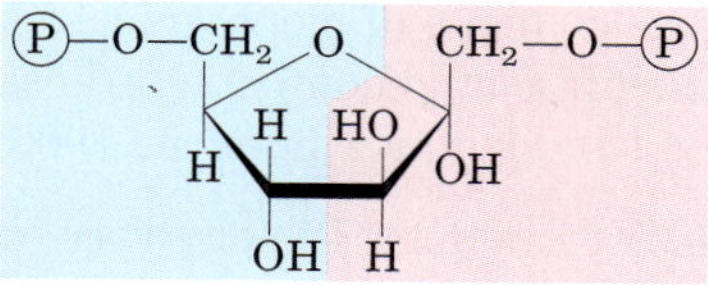

## chapter 14

# GLYCOLYSIS, GLUCONEOGENESIS, AND THE PENTOSE PHOSPHATE PATHWAY

The problem of alcoholic fermentation, of the origin and nature of that mysterious and apparently spontaneous change, which converted the insipid juice of the grape into stimulating wine, seems to have exerted a fascination over the minds of natural philosophers from the very earliest times.

—*Arthur Harden, Alcoholic Fermentation, 1923*

Glucose occupies a central position in the metabolism of plants, animals, and many microorganisms. It is relatively rich in potential energy, and thus a good fuel; the complete oxidation of glucose to carbon dioxide and water proceeds with a standard free-energy change of −2,840 kJ/mol. By storing glucose as a high molecular weight polymer such as starch or glycogen, a cell can stockpile large quantities of hexose units while maintaining a relatively low cytosolic osmolarity. When energy demands increase, glucose can be released from these intracellular storage polymers and used to produce ATP either aerobically or anaerobically.

Glucose is not only an excellent fuel, it is also a remarkably versatile precursor, capable of supplying a huge array of metabolic intermediates for biosynthetic reactions. A bacterium such as *Escherichia coli* can obtain from glucose the carbon skeletons for every amino acid, nucleotide, coenzyme, fatty acid, or other metabolic intermediate it needs for growth. A comprehensive study of the metabolic fates of glucose would encompass hundreds or thousands of transformations. In animals and vascular plants, glucose has three major fates: it may be stored (as a polysaccharide or as sucrose); oxidized to a three-carbon compound (pyruvate) via glycolysis to provide ATP and metabolic intermediates; or oxidized via the pentose phosphate (phosphogluconate) pathway to yield ribose 5-phosphate for nucleic acid synthesis and NADPH for reductive biosynthetic processes (Fig. 14–1).

Organisms that do not have access to glucose from other sources must make it. Photosynthetic organisms make glucose by first reducing atmospheric $CO_2$ to trioses, then converting the trioses to glucose. Nonphotosynthetic cells make glucose from simpler three- and four-carbon precursors by the process of gluconeogenesis, effectively reversing glycolysis in a pathway that uses many of the glycolytic enzymes.

In this chapter we describe the individual reactions of glycolysis, gluconeogenesis, and the pentose phosphate pathway and the functional significance of each pathway. We also describe the various fates of the pyruvate produced by glycolysis; they include the fermentations that are used by many organisms in anaerobic niches to produce ATP and that are exploited industrially as sources of ethanol, lactic acid, and other

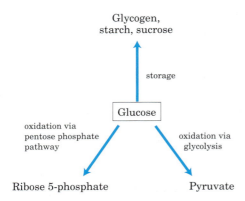

**FIGURE 14-1 Major pathways of glucose utilization.** Although not the only possible fates for glucose, these three pathways are the most significant in terms of the amount of glucose that flows through them in most cells.

commercially useful products. And we look at the pathways that feed various sugars from mono-, di-, and polysaccharides into the glycolytic pathway. The discussion of glucose metabolism continues in Chapter 15, where we describe the opposing anabolic and catabolic pathways that connect glucose and glycogen, and use the processes of carbohydrate synthesis and degradation as examples of the many mechanisms by which organisms regulate metabolic pathways.

## 14.1 Glycolysis

In **glycolysis** (from the Greek *glykys,* meaning "sweet," and *lysis,* meaning "splitting"), a molecule of glucose is degraded in a series of enzyme-catalyzed reactions to yield two molecules of the three-carbon compound pyruvate. During the sequential reactions of glycolysis, some of the free energy released from glucose is conserved in the form of ATP and NADH. Glycolysis was the first metabolic pathway to be elucidated and is probably the best understood. From Eduard Buchner's discovery in 1897 of fermentation in broken extracts of yeast cells until the elucidation of the whole pathway in yeast (by Otto Warburg and Hans von Euler-Chelpin)

and in muscle (by Gustav Embden and Otto Meyerhof) in the 1930s, the reactions of glycolysis in extracts of yeast and muscle were a major focus of biochemical research. The philosophical shift that accompanied these discoveries was announced by Jacques Loeb in 1906:

> Through the discovery of Buchner, Biology was relieved of another fragment of mysticism. The splitting up of sugar into $CO_2$ and alcohol is no more the effect of a "vital principle" than the splitting up of cane sugar by invertase. The history of this problem is instructive, as it warns us against considering problems as beyond our reach because they have not yet found their solution.

The development of methods of enzyme purification, the discovery and recognition of the importance of coenzymes such as NAD, and the discovery of the pivotal metabolic role of ATP and other phosphorylated compounds all came out of studies of glycolysis. The glycolytic enzymes of many species have long since been purified and thoroughly studied.

Glycolysis is an almost universal central pathway of glucose catabolism, the pathway with the largest flux of carbon in most cells. The glycolytic breakdown of glucose is the sole source of metabolic energy in some mammalian tissues and cell types (erythrocytes, renal medulla, brain, and sperm, for example). Some plant tissues that are modified to store starch (such as potato tubers) and some aquatic plants (watercress, for example) derive most of their energy from glycolysis; many anaerobic microorganisms are entirely dependent on glycolysis.

**Fermentation** is a general term for the *anaerobic* degradation of glucose or other organic nutrients to obtain energy, conserved as ATP. Because living organisms first arose in an atmosphere without oxygen, anaerobic breakdown of glucose is probably the most ancient biological mechanism for obtaining energy from organic fuel molecules. In the course of evolution, the chemistry of this reaction sequence has been completely conserved; the glycolytic enzymes of vertebrates are closely similar, in amino acid sequence and three-dimensional structure, to their homologs in yeast and spinach. Glycolysis differs among species only in the details of its regulation and in the subsequent metabolic fate of the pyruvate formed. The thermodynamic principles and the types of regulatory mechanisms that govern glycolysis are common to all pathways of cell metabolism. A study of glycolysis can therefore serve as a model for many aspects of the pathways discussed throughout this book.

Hans von Euler-Chelpin, 1873–1964

Gustav Embden, 1874–1933

Otto Meyerhof, 1884–1951

Before examining each step of the pathway in some detail, we take a look at glycolysis as a whole.

## An Overview: Glycolysis Has Two Phases

The breakdown of the six-carbon glucose into two molecules of the three-carbon pyruvate occurs in ten steps, the first five of which constitute the *preparatory phase* (Fig. 14–2a). In these reactions, glucose is first phosphorylated at the hydroxyl group on C-6 (step ①). The D-glucose 6-phosphate thus formed is converted to D-fructose 6-phosphate (step ②), which is again phosphorylated, this time at C-1, to yield D-fructose 1,6-bisphosphate (step ③). For both phosphorylations, ATP is the phosphoryl group donor. As all sugar derivatives in glycolysis are the D isomers, we will usually omit the D designation except when emphasizing stereochemistry.

Fructose 1,6-bisphosphate is split to yield two three-carbon molecules, dihydroxyacetone phosphate and glyceraldehyde 3-phosphate (step ④); this is the "lysis" step that gives the pathway its name. The dihydroxyacetone phosphate is isomerized to a second molecule of glyceraldehyde 3-phosphate (step ⑤), ending the first phase of glycolysis. From a chemical perspective, the isomerization in step ② is critical for setting up the phosphorylation and C—C bond cleavage reactions in steps ③ and ④, as detailed later. Note that two molecules of ATP are invested before the cleavage of glucose into two three-carbon pieces; later there will be a good return on this investment. To summarize: in the preparatory phase of glycolysis the energy of ATP is invested, raising the free-energy content of the intermediates, and the carbon chains of all the metabolized hexoses are converted into a common product, glyceraldehyde 3-phosphate.

The energy gain comes in the *payoff phase* of glycolysis (Fig. 14–2b). Each molecule of glyceraldehyde 3-phosphate is oxidized and phosphorylated by inorganic phosphate (*not* by ATP) to form 1,3-bisphosphoglycerate (step ⑥). Energy is then released as the two molecules of 1,3-bisphosphoglycerate are converted to two molecules of pyruvate (steps ⑦ through ⑩). Much of this energy is conserved by the coupled phosphorylation of four molecules of ADP to ATP. The net yield is two molecules of ATP per molecule of glucose used, because two molecules of ATP were invested in the preparatory phase. Energy is also conserved in the payoff phase in the formation of two molecules of NADH per molecule of glucose.

In the sequential reactions of glycolysis, three types of chemical transformations are particularly noteworthy: (1) degradation of the carbon skeleton of glucose to yield pyruvate, (2) phosphorylation of ADP to ATP by high-energy phosphate compounds formed during glycolysis, and (3) transfer of a hydride ion to $NAD^+$, forming NADH.

***Fates of Pyruvate*** With the exception of some interesting variations in the bacterial realm, the pyruvate formed by glycolysis is further metabolized via one of three catabolic routes. In aerobic organisms or tissues, under aerobic conditions, glycolysis is only the first stage in the complete degradation of glucose (Fig. 14–3). Pyruvate is oxidized, with loss of its carboxyl group as $CO_2$, to yield the acetyl group of acetyl-coenzyme A; the acetyl group is then oxidized completely to $CO_2$ by the citric acid cycle (Chapter 16). The electrons from these oxidations are passed to $O_2$ through a chain of carriers in the mitochondrion, to form $H_2O$. The energy from the electron-transfer reactions drives the synthesis of ATP in the mitochondrion (Chapter 19).

The second route for pyruvate is its reduction to lactate via **lactic acid fermentation.** When vigorously contracting skeletal muscle must function under low-oxygen conditions **(hypoxia),** NADH cannot be reoxidized to $NAD^+$, but $NAD^+$ is required as an electron acceptor for the further oxidation of pyruvate. Under these conditions pyruvate is reduced to lactate, accepting electrons from NADH and thereby regenerating the $NAD^+$ necessary for glycolysis to continue. Certain tissues and cell types (retina and erythrocytes, for example) convert glucose to lactate even under aerobic conditions, and lactate is also the product of glycolysis under anaerobic conditions in some microorganisms (Fig. 14–3).

The third major route of pyruvate catabolism leads to ethanol. In some plant tissues and in certain invertebrates, protists, and microorganisms such as brewer's yeast, pyruvate is converted under hypoxic or anaerobic conditions into ethanol and $CO_2$, a process called **ethanol (alcohol) fermentation** (Fig. 14–3).

The oxidation of pyruvate is an important catabolic process, but pyruvate has anabolic fates as well. It can, for example, provide the carbon skeleton for the synthesis of the amino acid alanine. We return to these anabolic reactions of pyruvate in later chapters.

***ATP Formation Coupled to Glycolysis*** During glycolysis some of the energy of the glucose molecule is conserved in ATP, while much remains in the product, pyruvate. The overall equation for glycolysis is

$$\text{Glucose} + 2NAD^+ + 2ADP + 2P_i \longrightarrow$$
$$2 \text{ pyruvate} + 2NADH + 2H^+ + 2ATP + 2H_2O \quad (14\text{–}1)$$

For each molecule of glucose degraded to pyruvate, two molecules of ATP are generated from ADP and $P_i$. We can now resolve the equation of glycolysis into two processes—the conversion of glucose to pyruvate, which is exergonic:

$$\text{Glucose} + 2NAD^+ \longrightarrow 2 \text{ pyruvate} + 2NADH + 2H^+ \quad (14\text{–}2)$$
$$\Delta G_1'^\circ = -146 \text{ kJ/mol}$$

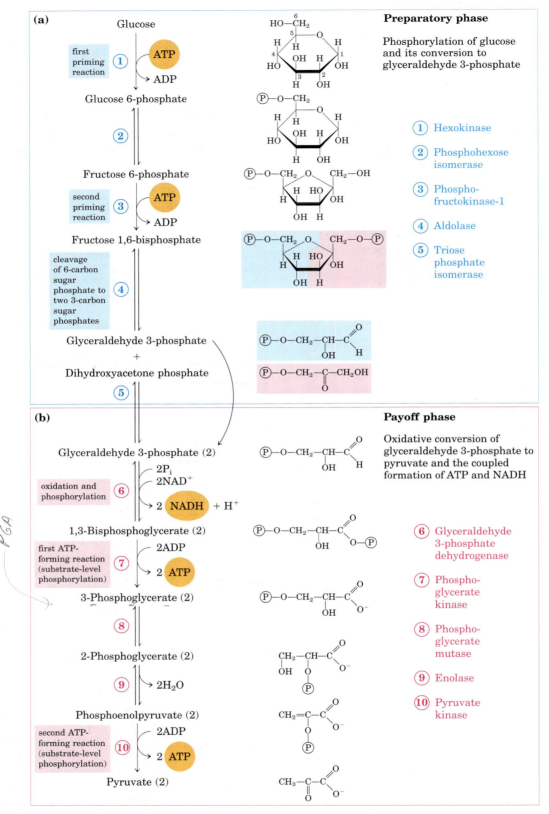

**FIGURE 14–2 The two phases of glycolysis.** For each molecule of glucose that passes through the preparatory phase **(a),** two molecules of glyceraldehyde 3-phosphate are formed; both pass through the payoff phase **(b).** Pyruvate is the end product of the second phase of glycolysis. For each glucose molecule, two ATP are consumed in the preparatory phase and four ATP are produced in the payoff phase, giving a net yield of two ATP per molecule of glucose converted to pyruvate. The numbered reaction steps are catalyzed by the enzymes listed on the right, and also correspond to the numbered headings in the text discussion. Keep in mind that each phosphoryl group, represented here as $P$, has two negative charges ($-PO_3^{2-}$).

and the formation of ATP from ADP and $P_i$, which is endergonic:

$$2ADP + 2P_i \longrightarrow 2ATP + 2H_2O \qquad (14\text{--}3)$$
$$\Delta G_2'^\circ = 2(30.5 \text{ kJ/mol}) = 61.0 \text{ kJ/mol}$$

The sum of Equations 14–2 and 14–3 gives the overall standard free-energy change of glycolysis, $\Delta G_s'^\circ$:

$$\Delta G_s'^\circ = \Delta G_1'^\circ + \Delta G_2'^\circ = -146 \text{ kJ/mol} + 61.0 \text{ kJ/mol}$$
$$= -85 \text{ kJ/mol}$$

Under standard conditions and in the cell, glycolysis is an essentially irreversible process, driven to completion by a large net decrease in free energy. At the actual intracellular concentrations of ATP, ADP, and $P_i$ (see Box 13–1) and of glucose and pyruvate, the energy released in glycolysis (with pyruvate as the end product) is recovered as ATP with an efficiency of more than 60%.

**Energy Remaining in Pyruvate**  Glycolysis releases only a small fraction of the total available energy of the glucose molecule; the two molecules of pyruvate formed by glycolysis still contain most of the chemical potential energy of glucose, energy that can be extracted by oxidative reactions in the citric acid cycle (Chapter 16) and oxidative phosphorylation (Chapter 19).

**Importance of Phosphorylated Intermediates**  Each of the nine glycolytic intermediates between glucose and pyruvate is phosphorylated (Fig. 14–2). The phosphoryl groups appear to have three functions.

1.  Because the plasma membrane generally lacks transporters for phosphorylated sugars, the phosphorylated glycolytic intermediates cannot leave the cell. After the initial phosphorylation, no further energy is necessary to retain phosphorylated intermediates in the cell, despite the large difference in their intracellular and extracellular concentrations.

2.  Phosphoryl groups are essential components in the enzymatic conservation of metabolic energy. Energy released in the breakage of phosphoanhydride bonds (such as those in ATP) is partially conserved in the formation of phosphate esters such as glucose 6-phosphate. High-energy phosphate compounds formed in glycolysis (1,3-bisphosphoglycerate and phosphoenolpyruvate) donate phosphoryl groups to ADP to form ATP.

3.  Binding energy resulting from the binding of phosphate groups to the active sites of enzymes lowers the activation energy and increases the specificity of the enzymatic reactions (Chapter 6). The phosphate groups of ADP, ATP, and the glycolytic intermediates form complexes with $Mg^{2+}$, and the substrate binding sites of many glycolytic enzymes are specific for these $Mg^{2+}$ complexes. Most glycolytic enzymes require $Mg^{2+}$ for activity.

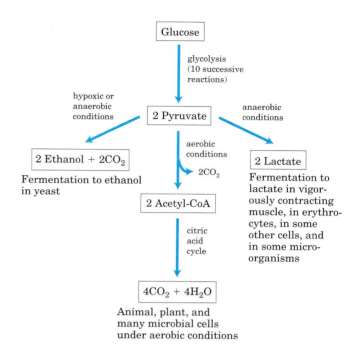

**FIGURE 14–3  Three possible catabolic fates of the pyruvate formed in glycolysis.** Pyruvate also serves as a precursor in many anabolic reactions, not shown here.

## The Preparatory Phase of Glycolysis Requires ATP

In the preparatory phase of glycolysis, two molecules of ATP are invested and the hexose chain is cleaved into two triose phosphates. The realization that *phosphorylated* hexoses were intermediates in glycolysis came slowly and serendipitously. In 1906, Arthur Harden and William Young tested their hypothesis that inhibitors of proteolytic enzymes would stabilize the glucose-fermenting enzymes in yeast extract. They added blood serum (known to contain inhibitors of proteolytic enzymes) to yeast extracts and observed the predicted stimulation of glucose metabolism. However, in a control experiment intended to show that boiling the serum destroyed the stimulatory activity, they discovered that boiled serum was just as effective at stimulating glycolysis. Careful examination and testing of the contents of

Arthur Harden, 1865–1940

William Young, 1878–1942

the boiled serum revealed that inorganic phosphate was responsible for the stimulation. Harden and Young soon discovered that glucose added to their yeast extract was converted to a hexose bisphosphate (the "Harden-Young ester," eventually identified as fructose 1,6-bisphosphate). This was the beginning of a long series of investigations on the role of organic esters of phosphate in biochemistry, which has led to our current understanding of the central role of phosphoryl group transfer in biology.

① **Phosphorylation of Glucose** In the first step of glycolysis, glucose is activated for subsequent reactions by its phosphorylation at C-6 to yield **glucose 6-phosphate,** with ATP as the phosphoryl donor:

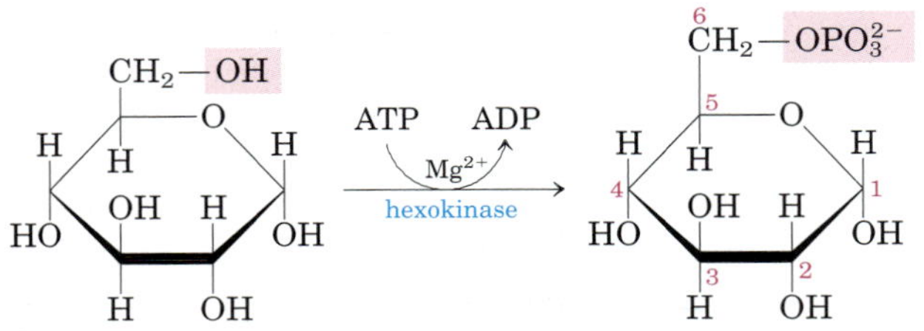

Glucose → Glucose 6-phosphate

$$\Delta G'^\circ = -16.7 \text{ kJ/mol}$$

This reaction, which is irreversible under intracellular conditions, is catalyzed by **hexokinase.** Recall that kinases are enzymes that catalyze the transfer of the terminal phosphoryl group from ATP to an acceptor nucleophile (see Fig. 13–10). Kinases are a subclass of transferases (see Table 6–3). The acceptor in the case of hexokinase is a hexose, normally D-glucose, although hexokinase also catalyzes the phosphorylation of other common hexoses, such as D-fructose and D-mannose.

Hexokinase, like many other kinases, requires $Mg^{2+}$ for its activity, because the true substrate of the enzyme is not $ATP^{4-}$ but the $MgATP^{2-}$ complex (see Fig. 13–2). $Mg^{2+}$ shields the negative charges of the phosphoryl groups in ATP, making the terminal phosphorus atom an easier target for nucleophilic attack by an —OH of glucose. Hexokinase undergoes a profound change in shape, an induced fit, when it binds glucose; two domains of the protein move about 8 Å closer to each other when ATP binds (see Fig. 6–22). This movement brings bound ATP closer to a molecule of glucose also bound to the enzyme and blocks the access of water (from the solvent), which might otherwise enter the active site and attack (hydrolyze) the phosphoanhydride bonds of ATP. Like the other nine enzymes of glycolysis, hexokinase is a soluble, cytosolic protein.

Hexokinase is present in all cells of all organisms. Hepatocytes also contain a form of hexokinase called hexokinase IV or glucokinase, which differs from other forms of hexokinase in kinetic and regulatory properties (see Box 15–2). Two enzymes that catalyze the

same reaction but are encoded in different genes are called **isozymes.**

② **Conversion of Glucose 6-Phosphate to Fructose 6-Phosphate** The enzyme **phosphohexose isomerase (phosphoglucose isomerase)** catalyzes the reversible isomerization of glucose 6-phosphate, an aldose, to **fructose 6-phosphate,** a ketose:

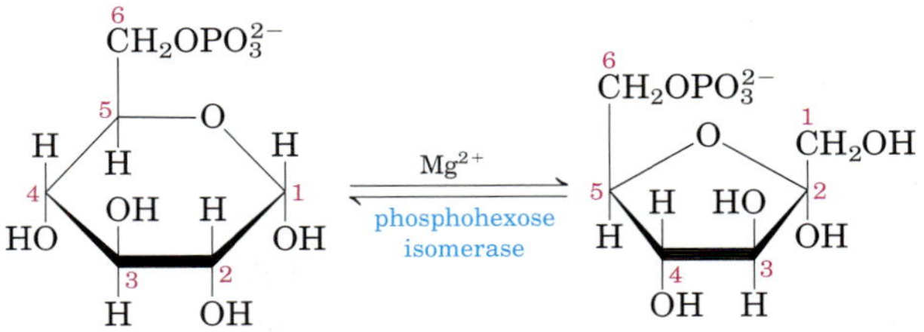

Glucose 6-phosphate → Fructose 6-phosphate

$$\Delta G'^\circ = 1.7 \text{ kJ/mol}$$

The mechanism for this reaction is shown in Figure 14–4. The reaction proceeds readily in either direction, as might be expected from the relatively small change in standard free energy. This isomerization has a critical role in the overall chemistry of the glycolytic pathway, as the rearrangement of the carbonyl and hydroxyl groups at C-1 and C-2 is a necessary prelude to the next two steps. The phosphorylation that occurs in the next reaction (step ③) requires that the group at C-1 first be converted from a carbonyl to an alcohol, and in the subsequent reaction (step ④) cleavage of the bond between C-3 and C-4 requires a carbonyl group at C-2 (p. 485).

③ **Phosphorylation of Fructose 6-Phosphate to Fructose 1,6-Bisphosphate** In the second of the two priming reactions of glycolysis, **phosphofructokinase-1 (PFK-1)** catalyzes the transfer of a phosphoryl group from ATP to fructose 6-phosphate to yield **fructose 1,6-bisphosphate:**

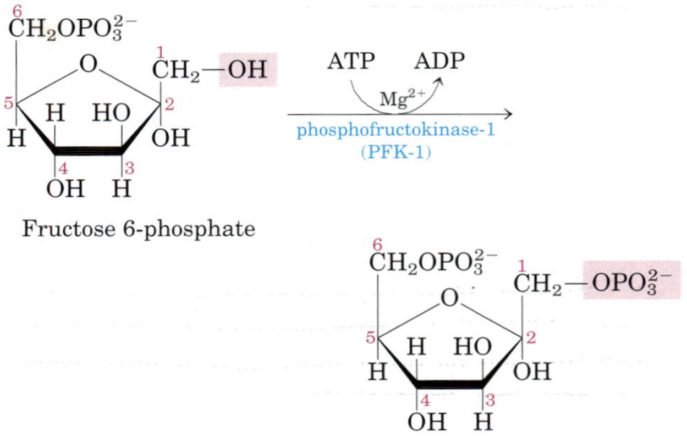

Fructose 6-phosphate → Fructose 1,6-bisphosphate

$$\Delta G'^\circ = -14.2 \text{ kJ/mol}$$

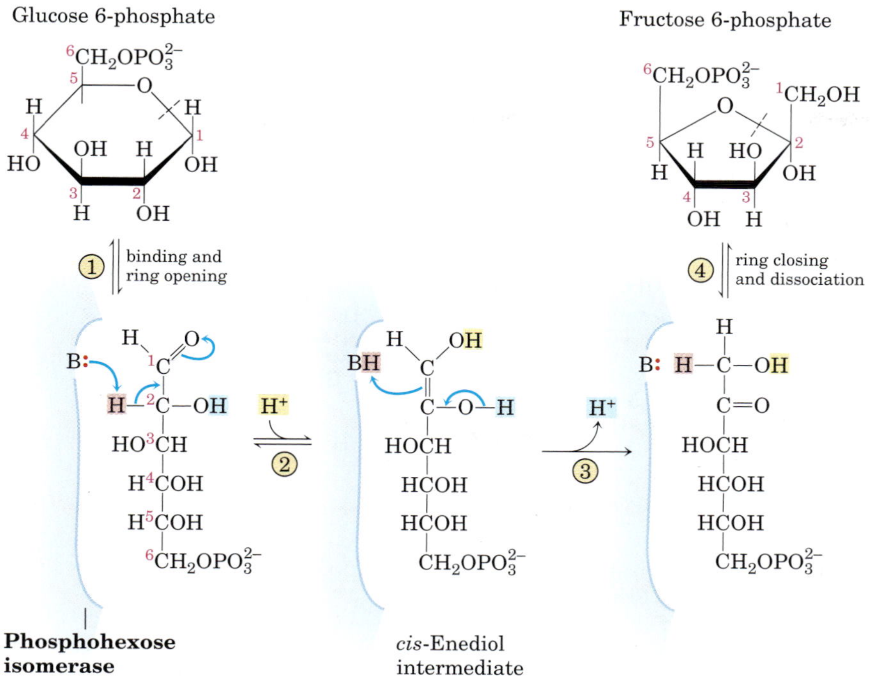

Glucose 6-phosphate

$^6CH_2OPO_3^{2-}$

Fructose 6-phosphate

$^6CH_2OPO_3^{2-}$ $^1CH_2OH$

**MECHANISM FIGURE 14-4** The **phosphohexose isomerase reaction.** The ring opening and closing reactions (steps ① and ④) are catalyzed by an active-site His residue, by mechanisms omitted here for simplicity. The movement of the proton between C-2 and C-1 (steps ② and ③) is base-catalyzed by an active-site Glu residue (shown as B:). The proton (pink) initially at C-2 is made more easily abstractable by electron withdrawal by the adjacent carbonyl and the nearby hydroxyl group. After its transfer from C-2 to the active-site Glu residue, the proton is freely exchanged with the surrounding solution; that is, the proton abstracted from C-2 in step ② is not necessarily the same one that is added to C-1 in step ③. (The additional exchange of protons (yellow and blue) between the hydroxyl groups and solvent is shown for completeness. The hydroxyl groups are weak acids and can exchange protons with the surrounding water whether the isomerization reaction is underway or not.)

🔹 **Phosphohexose Isomerase Mechanism**

① binding and ring opening

④ ring closing and dissociation

**Phosphohexose isomerase**

*cis*-Enediol intermediate

This enzyme is called PFK-1 to distinguish it from a second enzyme (PFK-2) that catalyzes the formation of fructose 2,6-bisphosphate from fructose 6-phosphate in a separate pathway. The PFK-1 reaction is essentially irreversible under cellular conditions, and it is the first "committed" step in the glycolytic pathway; glucose 6-phosphate and fructose 6-phosphate have other possible fates, but fructose 1,6-bisphosphate is targeted for glycolysis.

Some bacteria and protists and perhaps all plants have a phosphofructokinase that uses pyrophosphate (PP$_i$), not ATP, as the phosphoryl group donor in the synthesis of fructose 1,6-bisphosphate:

$$\text{Fructose 6-phosphate} + PP_i \xrightarrow{Mg^{2+}}$$
$$\text{fructose 1,6-bisphosphate} + P_i$$
$$\Delta G'^{\circ} = -14 \text{ kJ/mol}$$

Phosphofructokinase-1 is a regulatory enzyme (Chapter 6), one of the most complex known. It is the major point of regulation in glycolysis. The activity of PFK-1 is increased whenever the cell's ATP supply is depleted or when the ATP breakdown products, ADP and AMP (particularly the latter), are in excess. The enzyme is inhibited whenever the cell has ample ATP and is well supplied by other fuels such as fatty acids. In some organisms, fructose 2,6-bisphosphate (not to be confused with the PFK-1 reaction product, fructose 1,6-bisphosphate) is a potent allosteric activator of PFK-1. The regulation of this step in glycolysis is discussed in greater detail in Chapter 15.

④ **Cleavage of Fructose 1,6-Bisphosphate** The enzyme **fructose 1,6-bisphosphate aldolase,** often called simply **aldolase,** catalyzes a reversible aldol condensation (p. 485). Fructose 1,6-bisphosphate is cleaved to yield two different triose phosphates, **glyceraldehyde 3-phosphate,** an aldose, and **dihydroxyacetone phosphate,** a ketose:

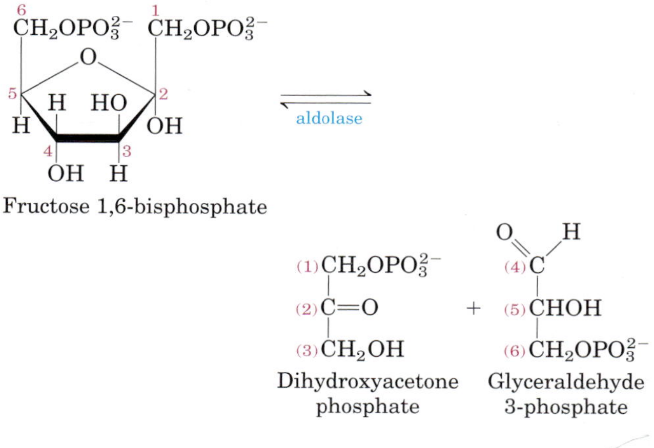

$^6CH_2OPO_3^{2-}$ $^1CH_2OPO_3^{2-}$

Fructose 1,6-bisphosphate

aldolase

(1)$CH_2OPO_3^{2-}$
(2)$C=O$
(3)$CH_2OH$
Dihydroxyacetone phosphate

$+$

(4)$C$ (with $O$, $H$)
(5)$CHOH$
(6)$CH_2OPO_3^{2-}$
Glyceraldehyde 3-phosphate

$$\Delta G'^{\circ} = 23.8 \text{ kJ/mol}$$

There are two classes of aldolases. Class I aldolases, found in animals and plants, use the mechanism shown in Figure 14-5. Class II enzymes, in fungi and bacteria, do not form the Schiff base intermediate. Instead, a zinc ion at the active site is coordinated with the carbonyl oxygen at C-2; the $Zn^{2+}$ polarizes the carbonyl group

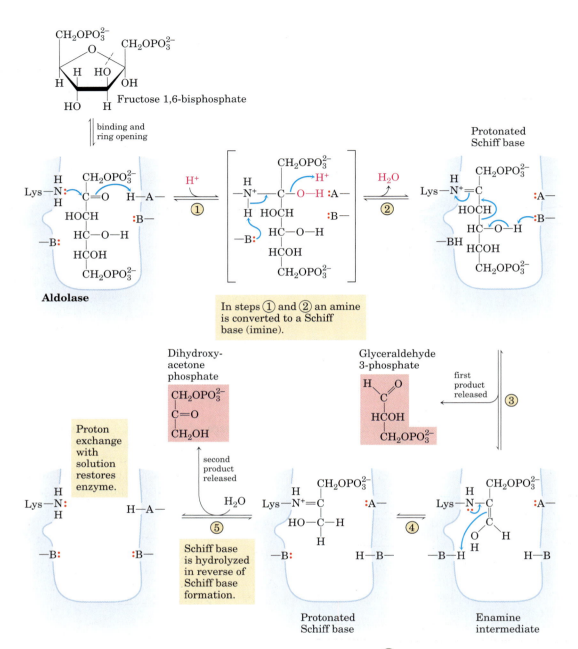

**MECHANISM FIGURE 14–5 The class I aldolase reaction.** The reaction shown here is the reverse of an aldol condensation. Note that cleavage between C-3 and C-4 depends on the presence of the carbonyl group at C-2. ① and ② The carbonyl reacts with an active-site Lys residue to form an imine, which stabilizes the carbanion generated by the bond cleavage—an imine delocalizes electrons even better than does a carbonyl. ③ Bond cleavage releases glyceraldeyde 3-phosphate as the first product. ④ The resulting enamine covalently linked to the enzyme is isomerized to a protonated Schiff base, and ⑤ hydrolysis of the Schiff base generates dihydroxyacetone phosphate as the second product. A and B represent amino acid residues that serve as general acid (A) or base (B).

and stabilizes the enolate intermediate created in the C—C bond cleavage step.

Although the aldolase reaction has a strongly positive standard free-energy change in the direction of fructose 1,6-bisphosphate cleavage, at the lower concentrations of reactants present in cells, the actual free-energy change is small and the aldolase reaction is readily reversible. We shall see later that aldolase acts in the reverse direction during the process of gluconeogenesis (see Fig. 14–16).

**⑤ Interconversion of the Triose Phosphates** Only one of the two triose phosphates formed by aldolase, glyceraldehyde 3-phosphate, can be directly degraded in the subsequent steps of glycolysis. The other product, dihydroxyacetone phosphate, is rapidly and reversibly

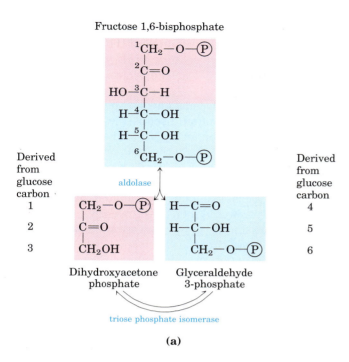

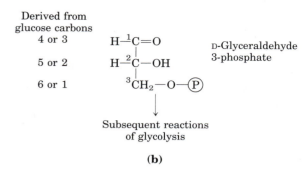

**(b)**

**FIGURE 14–6 Fate of the glucose carbons in the formation of glyceraldehyde 3-phosphate. (a)** The origin of the carbons in the two three-carbon products of the aldolase and triose phosphate isomerase reactions. The end product of the two reactions is glyceraldehyde 3-phosphate (two molecules). **(b)** Each carbon of glyceraldehyde 3-phosphate is derived from either of two specific carbons of glucose. Note that the numbering of the carbon atoms of glyceraldehyde 3-phosphate differs from that of the glucose from which it is derived. In glyceraldehyde 3-phosphate, the most complex functional group (the carbonyl) is specified as C-1. This numbering change is important for interpreting experiments with glucose in which a single carbon is labeled with a radioisotope. (See Problems 3 and 5 at the end of this chapter.)

converted to glyceraldehyde 3-phosphate by the fifth enzyme of the glycolitic sequence, **triose phosphate isomerase:**

$$\Delta G'^\circ = 7.5 \text{ kJ/mol}$$

The reaction mechanism is similar to the reaction promoted by phosphohexose isomerase in step ② of glycolysis (Fig. 14–4). After the triose phosphate isomerase reaction, C-1, C-2, and C-3 of the starting glucose are chemically indistinguishable from C-6, C-5, and C-4, respectively (Fig. 14–6), setting up the efficient metabolism of the entire six-carbon glucose molecule.

This reaction completes the preparatory phase of glycolysis. The hexose molecule has been phosphorylated at C-1 and C-6 and then cleaved to form two molecules of glyceraldehyde 3-phosphate.

## The Payoff Phase of Glycolysis Yields ATP and NADH

The payoff phase of glycolysis (Fig. 14–2b) includes the energy-conserving phosphorylation steps in which some of the free energy of the glucose molecule is conserved in the form of ATP. Remember that one molecule of glucose yields two molecules of glyceraldehyde 3-phosphate; both halves of the glucose molecule follow the same pathway in the second phase of glycolysis. The conversion of two molecules of glyceraldehyde 3-phosphate to two molecules of pyruvate is accompanied by the formation of four molecules of ATP from ADP. However, the net yield of ATP per molecule of glucose degraded is only two, because two ATP were invested in the preparatory phase of glycolysis to phosphorylate the two ends of the hexose molecule.

⑥ **Oxidation of Glyceraldehyde 3-Phosphate to 1,3-Bisphosphoglycerate** The first step in the payoff phase is the oxidation of glyceraldehyde 3-phosphate to **1,3-bisphosphoglycerate,** catalyzed by **glyceraldehyde 3-phosphate dehydrogenase:**

$$\Delta G'^\circ = 6.3 \text{ kJ/mol}$$

This is the first of the two energy-conserving reactions of glycolysis that eventually lead to the formation of ATP. The aldehyde group of glyceraldehyde 3-phosphate is oxidized, not to a free carboxyl group but to a carboxylic acid anhydride with phosphoric acid. This type of anhydride, called an **acyl phosphate,** has a very high standard free energy of hydrolysis ($\Delta G'° = -49.3$ kJ/mol; see Fig. 13–4, Table 13–6). Much of the free energy of oxidation of the aldehyde group of glyceraldehyde 3-phosphate is conserved by formation of the acyl phosphate group at C-1 of 1,3-bisphosphoglycerate.

The acceptor of hydrogen in the glyceraldehyde 3-phosphate dehydrogenase reaction is $NAD^+$ (see Fig. 13–15), bound to a Rossmann fold as shown in Figure 13–16. The reduction of $NAD^+$ proceeds by the enzymatic transfer of a hydride ion ($:H^-$) from the aldehyde group of glyceraldehyde 3-phosphate to the nicoti-

namide ring of $NAD^+$, yielding the reduced coenzyme NADH. The other hydrogen atom of the substrate molecule is released to the solution as $H^+$.

Glyceraldehyde 3-phosphate is covalently bound to the dehydrogenase during the reaction (Fig. 14–7). The aldehyde group of glyceraldehyde 3-phosphate reacts with the —SH group of an essential Cys residue in the active site, in a reaction analogous to the formation of a hemiacetal (see Fig. 7–5), in this case producing a *thio*hemiacetal. Reaction of the essential Cys residue with a heavy metal such as $Hg^{2+}$ irreversibly inhibits the enzyme.

Because cells maintain only limited amounts of $NAD^+$, glycolysis would soon come to a halt if the NADH formed in this step of glycolysis were not continuously reoxidized. The reactions in which $NAD^+$ is regenerated anaerobically are described in detail in Section 14.3, in our discussion of the alternative fates of pyruvate.

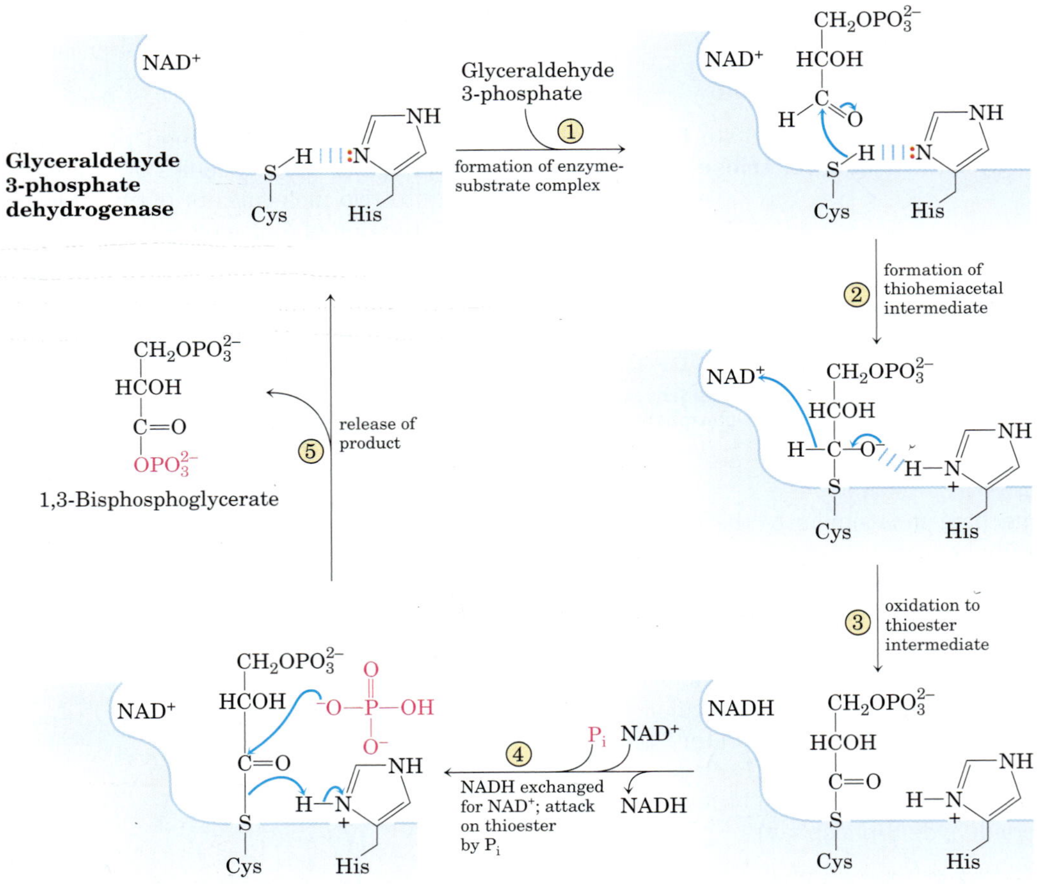

**MECHANISM FIGURE 14–7 The glyceraldehyde 3-phosphate dehydrogenase reaction.** After ① formation of the enzyme-substrate complex, ② a covalent thiohemiacetal linkage forms between the substrate and the —SH group of a Cys residue—facilitated by acid-base catalysis with a neighboring base catalyst, probably a His residue. ③ This enzyme-substrate intermediate is oxidized by $NAD^+$ bound to the active site, forming a covalent acyl-enzyme intermediate, a thioester. ④ The newly formed NADH leaves the active site and is replaced by another $NAD^+$ molecule. The bond between the acyl group and the thiol group of the enzyme has a very high standard free energy of hydrolysis. ⑤ This bond undergoes phosphorolysis (attack by $P_i$), releasing the acyl phosphate product, 1,3-bisphosphoglycerate. Formation of this product conserves much of the free energy liberated during oxidation of the aldehyde group of glyceraldehyde 3-phosphate.

**(7) Phosphoryl Transfer from 1,3-Bisphosphoglycerate to ADP** The enzyme **phosphoglycerate kinase** transfers the high-energy phosphoryl group from the carboxyl group of 1,3-bisphosphoglycerate to ADP, forming ATP and **3-phosphoglycerate:**

1,3-Bisphosphoglycerate ADP

$Mg^{2+}$  phosphoglycerate kinase

3-Phosphoglycerate ATP

$$\Delta G'^{\circ} = -18.5 \text{ kJ/mol}$$

Notice that phosphoglycerate kinase is named for the reverse reaction. Like all enzymes, it catalyzes the reaction in both directions. This enzyme acts in the direction suggested by its name during gluconeogenesis (see Fig. 14–16) and during photosynthetic $CO_2$ assimilation (see Fig. 20–4).

Steps (6) and (7) of glycolysis together constitute an energy-coupling process in which 1,3-bisphosphoglycerate is the common intermediate; it is formed in the first reaction (which would be endergonic in isolation), and its acyl phosphate group is transferred to ADP in the second reaction (which is strongly exergonic). The sum of these two reactions is

Glyceraldehyde 3-phosphate + ADP + $P_i$ + $NAD^+$ ⇌
    3-phosphoglycerate + ATP + NADH + $H^+$
    $$\Delta G'^{\circ} = -12.5 \text{ kJ/mol}$$

Thus the overall reaction is exergonic.

Recall from Chapter 13 that the actual free-energy change, $\Delta G$, is determined by the standard free-energy change, $\Delta G'^{\circ}$, and the mass-action ratio, $Q$, which is the ratio [products]/[reactants] (see Eqn 13–3). For step (6)

$$\Delta G = \Delta G'^{\circ} + RT \ln Q$$

$$= \Delta G'^{\circ} + RT \ln \frac{[\text{1,3-bisphosphoglycerate}][\text{NADH}]}{[\text{glyceraldehyde 3-phosphate}][P_i][\text{NAD}^+]}$$

Notice that $[H^+]$ is not included in $Q$. In biochemical calculations, $[H^+]$ is assumed to be a constant ($10^{-7}$ M), and this constant is included in the definition of $\Delta G'^{\circ}$ (p. 491).

When the mass-action ratio is less than 1.0, its natural logarithm has a negative sign. Step (7), by consuming the product of step (6) (1,3-bisphosphoglycerate), keeps [1,3-bisphosphoglycerate] relatively low in the steady state and thereby keeps $Q$ for the overall energy-coupling process small. When $Q$ is small, the contribution of $\ln Q$ can make $\Delta G$ strongly negative. This is simply another way of showing how the two reactions, steps (6) and (7), are coupled through a common intermediate.

The outcome of these coupled reactions, both reversible under cellular conditions, is that the energy released on oxidation of an aldehyde to a carboxylate group is conserved by the coupled formation of ATP from ADP and $P_i$. The formation of ATP by phosphoryl group transfer from a substrate such as 1,3-bisphosphoglycerate is referred to as a **substrate-level phosphorylation,** to distinguish this mechanism from **respiration-linked phosphorylation.** Substrate-level phosphorylations involve soluble enzymes and chemical intermediates (1,3-bisphosphoglycerate in this case). Respiration-linked phosphorylations, on the other hand, involve membrane-bound enzymes and transmembrane gradients of protons (Chapter 19).

**(8) Conversion of 3-Phosphoglycerate to 2-Phosphoglycerate** The enzyme **phosphoglycerate mutase** catalyzes a reversible shift of the phosphoryl group between C-2 and C-3 of glycerate; $Mg^{2+}$ is essential for this reaction:

3-Phosphoglycerate          2-Phosphoglycerate

$$\Delta G'^{\circ} = 4.4 \text{ kJ/mol}$$

The reaction occurs in two steps (Fig. 14–8). A phosphoryl group initially attached to a His residue of the mutase is transferred to the hydroxyl group at C-2 of 3-phosphoglycerate, forming 2,3-bisphosphoglycerate (2,3-BPG). The phosphoryl group at C-3 of 2,3-BPG is then transferred to the same His residue, producing 2-phosphoglycerate and regenerating the phosphorylated enzyme. Phosphoglycerate mutase is initially phosphorylated by phosphoryl transfer from 2,3-BPG, which is required in small quantities to initiate the catalytic cycle and is continuously regenerated by that cycle. Although in most cells 2,3-BPG is present in only trace amounts, it is a major component (~5 mM) of erythrocytes, where it regulates the affinity of hemoglobin for

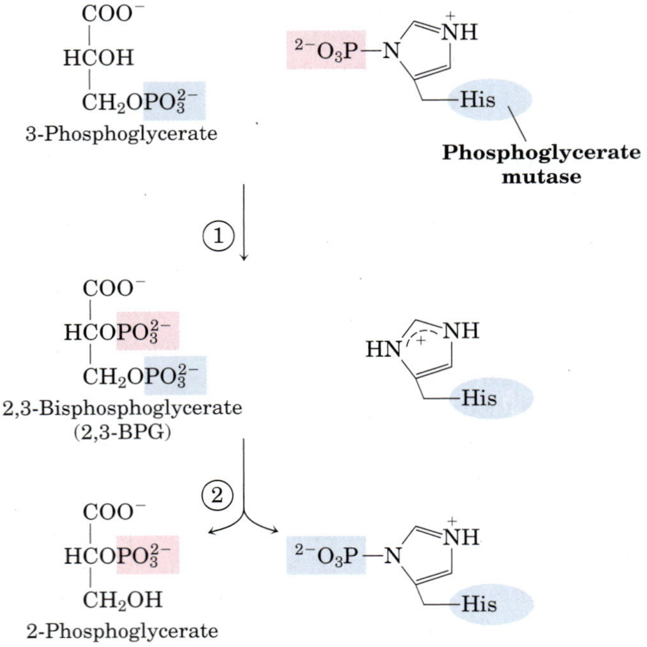

**3-Phosphoglycerate**

**Phosphoglycerate mutase**

**2,3-Bisphosphoglycerate (2,3-BPG)**

**2-Phosphoglycerate**

**FIGURE 14–8 The phosphoglycerate mutase reaction.** The enzyme is initially phosphorylated on a His residue. ① The phosphoenzyme transfers its phosphoryl group to 3-phosphoglycerate, forming 2,3-BPG. ② The phosphoryl group at C-3 of 2,3-BPG is transferred to the same His residue on the enzyme, producing 2-phosphoglycerate and regenerating the phosphoenzyme.

oxygen (see Fig. 5–17; note that in the context of hemoglobin regulation, 2,3-bisphosphoglycerate is usually abbreviated as simply BPG).

⑨ **Dehydration of 2-Phosphoglycerate to Phosphoenolpyruvate** In the second glycolytic reaction that generates a compound with high phosphoryl group transfer potential, **enolase** promotes reversible removal of a molecule of water from 2-phosphoglycerate to yield **phosphoenolpyruvate (PEP):**

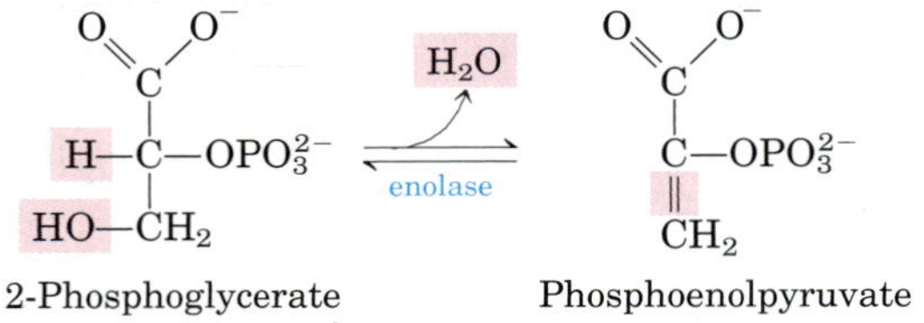

**2-Phosphoglycerate** → **Phosphoenolpyruvate**

$$\Delta G'^\circ = 7.5 \text{ kJ/mol}$$

The mechanism of the enolase reaction is presented in Figure 6–23. Despite the relatively small standard free-energy change of this reaction, there is a very large difference in the standard free energy of hydrolysis of

the phosphoryl groups of the reactant and product: $-17.6$ kJ/mol for 2-phosphoglycerate (a low-energy phosphate ester) and $-61.9$ kJ/mol for phosphoenolpyruvate (a compound with a very high standard free energy of hydrolysis) (see Fig. 13–3, Table 13–6). Although 2-phosphoglycerate and phosphoenolpyruvate contain nearly the same *total* amount of energy, the loss of the water molecule from 2-phosphoglycerate causes a redistribution of energy within the molecule, greatly increasing the standard free energy of hydrolysis of the phosphoryl group.

⑩ **Transfer of the Phosphoryl Group from Phosphoenolpyruvate to ADP** The last step in glycolysis is the transfer of the phosphoryl group from phosphoenolpyruvate to ADP, catalyzed by **pyruvate kinase,** which requires $K^+$ and either $Mg^{2+}$ or $Mn^{2+}$:

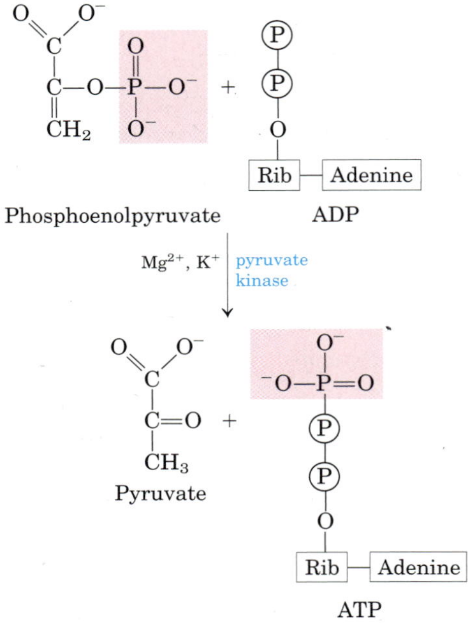

**Phosphoenolpyruvate**   **ADP**

$Mg^{2+}$, $K^+$ | pyruvate kinase

**Pyruvate**

**ATP**

$$\Delta G'^\circ = -31.4 \text{ kJ/mol}$$

In this substrate-level phosphorylation, the product **pyruvate** first appears in its enol form, then tautomerizes rapidly and nonenzymatically to its keto form, which predominates at pH 7:

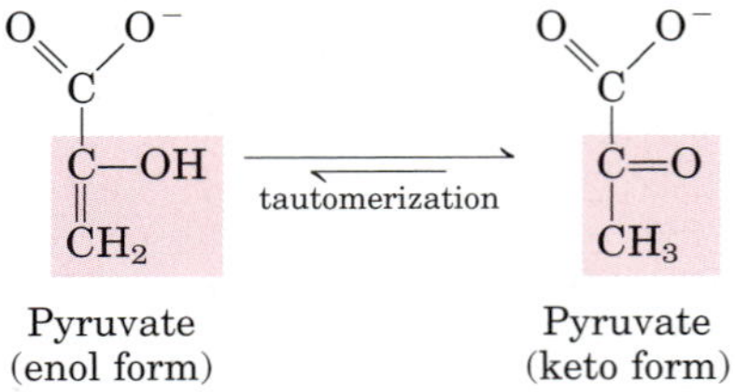

**Pyruvate (enol form)**   tautomerization   **Pyruvate (keto form)**

The overall reaction has a large, negative standard free-energy change, due in large part to the spontaneous conversion of the enol form of pyruvate to the keto form (see Fig. 13–3). The $\Delta G'^\circ$ of phosphoenolpyruvate hydrolysis is $-61.9$ kJ/mol; about half of this energy is conserved in the formation of the phosphoanhydride bond of ATP ($\Delta G'^{\circ} = -30.5$ kJ/mol), and the rest ($-31.4$ kJ/mol) constitutes a large driving force pushing the reaction toward ATP synthesis. The pyruvate kinase reaction is essentially irreversible under intracellular conditions and is an important site of regulation, as described in Chapter 15.

## The Overall Balance Sheet Shows a Net Gain of ATP

We can now construct a balance sheet for glycolysis to account for (1) the fate of the carbon skeleton of glucose, (2) the input of $P_i$ and ADP and the output of ATP, and (3) the pathway of electrons in the oxidation-reduction reactions. The left-hand side of the following equation shows all the inputs of ATP, $NAD^+$, ADP, and $P_i$ (consult Fig. 14–2), and the right-hand side shows all the outputs (keep in mind that each molecule of glucose yields two molecules of pyruvate):

$$\text{Glucose} + 2ATP + 2NAD^+ + 4ADP + 2P_i \longrightarrow$$
$$2 \text{ pyruvate} + 2ADP + 2NADH + 2H^+ + 4ATP + 2H_2O$$

Canceling out common terms on both sides of the equation gives the overall equation for glycolysis under aerobic conditions:

$$\text{Glucose} + 2NAD^+ + 2ADP + 2P_i \longrightarrow$$
$$2 \text{ pyruvate} + 2NADH + 2H^+ + 2ATP + 2H_2O$$

The two molecules of NADH formed by glycolysis in the cytosol are, under aerobic conditions, reoxidized to $NAD^+$ by transfer of their electrons to the electron-transfer chain, which in eukaryotic cells is located in the mitochondria. The electron-transfer chain passes these electrons to their ultimate destination, $O_2$:

$$2NADH + 2H^+ + O_2 \longrightarrow 2NAD^+ + 2H_2O$$

Electron transfer from NADH to $O_2$ in mitochondria provides the energy for synthesis of ATP by respiration-linked phosphorylation (Chapter 19).

In the overall glycolytic process, one molecule of glucose is converted to two molecules of pyruvate (the pathway of carbon). Two molecules of ADP and two of $P_i$ are converted to two molecules of ATP (the pathway of phosphoryl groups). Four electrons, as two hydride ions, are transferred from two molecules of glyceraldehyde 3-phosphate to two of $NAD^+$ (the pathway of electrons).

## Glycolysis Is under Tight Regulation

During his studies on the fermentation of glucose by yeast, Louis Pasteur discovered that both the rate and the total amount of glucose consumption were many times greater under anaerobic than aerobic conditions. Later studies of muscle showed the same large difference in the rates of anaerobic and aerobic glycolysis. The biochemical basis of this "Pasteur effect" is now clear. The ATP yield from glycolysis under anaerobic conditions (2 ATP per molecule of glucose) is much smaller than that from the complete oxidation of glucose to $CO_2$ under aerobic conditions (30 or 32 ATP per glucose; see Table 19–5). About 15 times as much glucose must therefore be consumed anaerobically as aerobically to yield the same amount of ATP.

The flux of glucose through the glycolytic pathway is regulated to maintain nearly constant ATP levels (as well as adequate supplies of glycolytic intermediates that serve biosynthetic roles). The required adjustment in the rate of glycolysis is achieved by a complex interplay among ATP consumption, NADH regeneration, and allosteric regulation of several glycolytic enzymes—including hexokinase, PFK-1, and pyruvate kinase—and by second-to-second fluctuations in the concentration of key metabolites that reflect the cellular balance between ATP production and consumption. On a slightly longer time scale, glycolysis is regulated by the hormones glucagon, epinephrine, and insulin, and by changes in the expression of the genes for several glycolytic enzymes. We return to a more detailed discussion of the regulation of glycolysis in Chapter 15.

## Cancerous Tissue Has Deranged Glucose Catabolism

Glucose uptake and glycolysis proceed about ten times faster in most solid tumors than in noncancerous tissues. Tumor cells commonly experience hypoxia (limited oxygen supply), because they initially lack an extensive capillary network to supply the tumor with oxygen. As a result, cancer cells more than 100 to 200 $\mu$m from the nearest capillaries depend on anaerobic glycolysis for much of their ATP production. They take up more glucose than normal cells, converting it to pyruvate and then to lactate as they recycle NADH. The high glycolytic rate may also result in part from smaller numbers of mitochondria in tumor cells; less ATP made by respiration-linked phosphorylation in mitochondria means more ATP is needed from glycolysis. In addition, some tumor cells overproduce several glycolytic enzymes, including an isozyme of hexokinase that associates with the cytosolic face of the mitochondrial inner membrane and is insensitive to feedback inhibition by glucose 6-phosphate. This enzyme may monopolize the ATP produced in mitochondria, using it to convert glucose to glucose 6-phosphate and committing the cell to continued glycolysis. The hypoxia-inducible transcription factor (HIF-1) is a protein that acts at the level of mRNA synthesis to stimulate the synthesis of at least eight of the glycolytic enzymes. This gives the tumor cell the capacity to survive anaerobic conditions until the supply of blood vessels has caught up with tumor growth.

The German biochemist Otto Warburg was the first to show, as early as 1928, that tumors have a higher rate of glucose metabolism than other tissues. With his associates, Warburg purified and crystallized seven of the enzymes of glycolysis. In these studies he developed and used an experimental tool that revolutionized biochemical studies of oxidative metabolism: the Warburg manometer, which measured directly the consumption of oxygen by monitoring changes in gas volume, and therefore allowed quantitative measurement of any enzyme with oxidase activity.

Warburg, considered by many the preeminent biochemist of the first half of the twentieth century, made

seminal contributions to many other areas of biochemistry, including respiration, photosynthesis, and the enzymology of intermediary metabolism. Trained in carbohydrate chemistry in the laboratory of the great Emil Fischer (who won the Nobel Prize in Chemistry in 1902), Warburg himself won the Nobel Prize in Physiology or Medicine in 1931. A number of Warburg's students and colleagues also were awarded Nobel Prizes:

Otto Warburg, 1883–1970

Otto Meyerhof in 1922, Hans Krebs and Fritz Lipmann in 1953, and Hugo Theorell in 1955. Meyerhof's laboratory provided training for Lipmann, and for several other Nobel Prize winners: Severo Ochoa (1959), Andre Lwoff (1965), and George Wald (1967). ■

## SUMMARY 14.1 Glycolysis

- Glycolysis is a near-universal pathway by which a glucose molecule is oxidized to two molecules of pyruvate, with energy conserved as ATP and NADH.

- All ten glycolytic enzymes are in the cytosol, and all ten intermediates are phosphorylated compounds of three or six carbons.

- In the preparatory phase of glycolysis, ATP is invested to convert glucose to fructose 1,6-bisphosphate. The bond between C-3 and C-4 is then broken to yield two molecules of triose phosphate.

- In the payoff phase, each of the two molecules of glyceraldehyde 3-phosphate derived from glucose undergoes oxidation at C-1; the energy of this oxidation reaction is conserved in the formation of one NADH and two ATP per triose phosphate oxidized. The net equation for the overall process is

$$\text{Glucose} + 2NAD^+ + 2ADP + 2P_i \longrightarrow$$
$$2 \text{ pyruvate} + 2NADH + 2H^+ + 2ATP + 2H_2O$$

- Glycolysis is tightly regulated in coordination with other energy-yielding pathways to assure a steady supply of ATP. Hexokinase, PFK-1, and pyruvate kinase are all subject to allosteric regulation that controls the flow of carbon through the pathway and maintains constant levels of metabolic intermediates.

## 14.2 Feeder Pathways for Glycolysis

Many carbohydrates besides glucose meet their catabolic fate in glycolysis, after being transformed into one of the glycolytic intermediates. The most significant are the storage polysaccharides glycogen and starch; the disaccharides maltose, lactose, trehalose, and sucrose; and the monosaccharides fructose, mannose, and galactose (Fig. 14–9).

### Glycogen and Starch Are Degraded by Phosphorolysis

Glycogen in animal tissues and in microorganisms (and starch in plants) can be mobilized for use within the same cell by a phosphorolytic reaction catalyzed by **glycogen phosphorylase** (**starch phosphorylase** in plants). These enzymes catalyze an attack by $P_i$ on the ($\alpha1\rightarrow4$) glycosidic linkage that joins the last two glucose residues at a nonreducing end, generating glucose 1-phosphate and a polymer one glucose unit shorter (Fig. 14–10). *Phosphorolysis* preserves some of the energy of the glycosidic bond in the phosphate ester glucose 1-phosphate. Glycogen phosphorylase (or starch phosphorylase) acts repetitively until it approaches an ($\alpha1\rightarrow6$) branch point (see Fig. 7–15), where its action stops. A **debranching enzyme** removes the branches. The mechanisms and control of glycogen degradation are described in detail in Chapter 15.

Glucose 1-phosphate produced by glycogen phosphorylase is converted to glucose 6-phosphate by **phosphoglucomutase,** which catalyzes the reversible reaction

$$\text{Glucose 1-phosphate} \rightleftharpoons \text{glucose 6-phosphate}$$

The glucose 6-phosphate thus formed can enter glycolysis or another pathway such as the pentose phosphate pathway, described in Section 14.5. Phosphoglucomutase employs essentially the same mechanism as phosphoglycerate mutase (p. 531). The general name **mutase** is given to enzymes that catalyze the transfer of a functional group from one position to another in the same molecule. Mutases are a subclass of **isomerases,** enzymes that interconvert stereoisomers or structural or positional isomers (see Table 6–3).

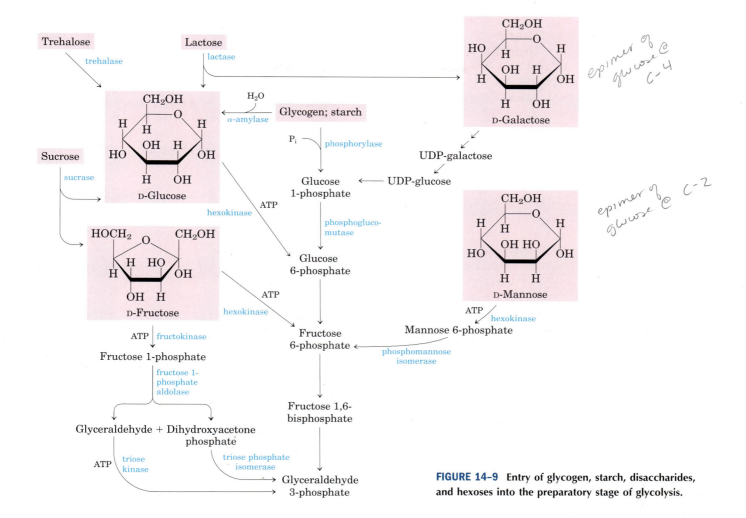

FIGURE 14–9  Entry of glycogen, starch, disaccharides, and hexoses into the preparatory stage of glycolysis.

## Dietary Polysaccharides and Disaccharides Undergo Hydrolysis to Monosaccharides

For most humans, starch is the major source of carbohydrates in the diet. Digestion begins in the mouth, where salivary **α-amylase** (Fig. 14–9) hydrolyzes the internal glycosidic linkages of starch, producing short polysaccharide fragments or oligosaccharides. (Note that in this *hydrolysis* reaction, water, not $P_i$, is the attacking species.) In the stomach, salivary α-amylase is inactivated by the low pH, but a second form of α-amylase, secreted by the pancreas into the small intestine, continues the breakdown process. Pancreatic α-amylase yields mainly maltose and maltotriose (the di- and trisaccharides of $α(1→4)$ glucose) and oligosaccharides called limit dextrins, fragments of amylopectin containing $α(1→6)$ branch points. Maltose and dextrins are degraded by enzymes of the intestinal brush border (the fingerlike microvilli of intestinal epithelial cells, which greatly increase the area of the intestinal surface). Dietary glycogen has essentially the same structure as starch, and its digestion proceeds by the same pathway.

Disaccharides must be hydrolyzed to monosaccharides before entering cells. Intestinal disaccharides and dextrins are hydrolyzed by enzymes attached to the outer surface of the intestinal epithelial cells:

$$Dextrin + n\,H_2O \xrightarrow{\text{dextrinase}} n\ \text{D-glucose}$$

$$Maltose + H_2O \xrightarrow{\text{maltase}} 2\ \text{D-glucose}$$

$$Lactose + H_2O \xrightarrow{\text{lactase}} \text{D-galactose} + \text{D-glucose}$$

$$Sucrose + H_2O \xrightarrow{\text{sucrase}} \text{D-fructose} + \text{D-glucose}$$

$$Trehalose + H_2O \xrightarrow{\text{trehalase}} 2\ \text{D-glucose}$$

The monosaccharides so formed are actively transported into the epithelial cells (see Fig. 11–44), then passed into the blood to be carried to various tissues, where they are phosphorylated and funneled into the glycolytic sequence.

 **Lactose intolerance,** common among adults of most human populations except those originating

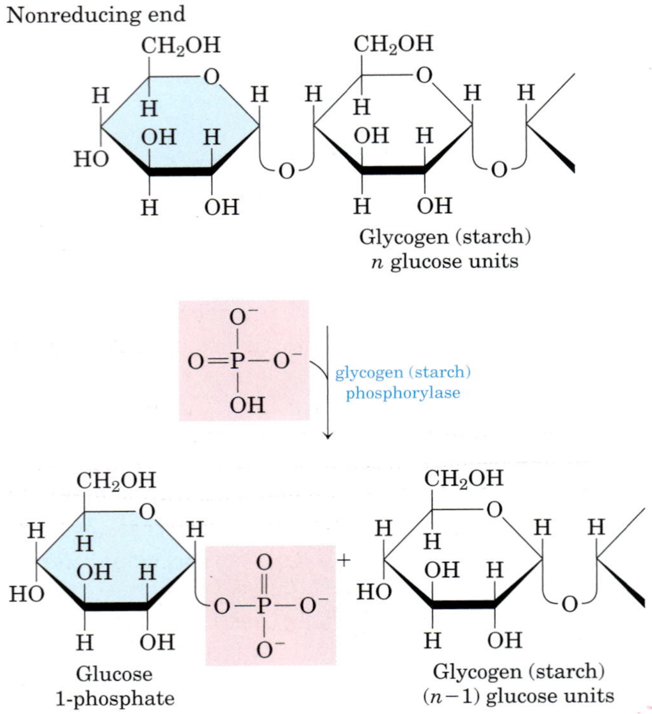

Nonreducing end

Glycogen (starch)
*n* glucose units

glycogen (starch)
phosphorylase

Glucose
1-phosphate

Glycogen (starch)
(*n* − 1) glucose units

**FIGURE 14–10 Glycogen breakdown by glycogen phosphorylase.** The enzyme catalyzes attack by inorganic phosphate (pink) on the terminal glucosyl residue (blue) at the nonreducing end of a glycogen molecule, releasing glucose 1-phosphate and generating a glycogen molecule shortened by one glucose residue. The reaction is a *phosphorolysis* (not hydrolysis).

in Northern Europe and some parts of Africa, is due to the disappearance after childhood of most or all of the lactase activity of the intestinal cells. Lactose cannot be completely digested and absorbed in the small intestine and passes into the large intestine, where bacteria convert it to toxic products that cause abdominal cramps and diarrhea. The problem is further complicated because undigested lactose and its metabolites increase the osmolarity of the intestinal contents, favoring the retention of water in the intestine. In most parts of the world where lactose intolerance is prevalent, milk is not used as a food by adults, although milk products predigested with lactase are commercially available in some countries. In certain human disorders, several or all of the intestinal disaccharidases are missing. In these cases, the digestive disturbances triggered by dietary disaccharides can sometimes be minimized by a controlled diet. ■

## Other Monosaccharides Enter the Glycolytic Pathway at Several Points

In most organisms, hexoses other than glucose can undergo glycolysis after conversion to a phosphorylated

derivative. D-Fructose, present in free form in many fruits and formed by hydrolysis of sucrose in the small intestine of vertebrates, is phosphorylated by hexokinase:

$$\text{Fructose} + \text{ATP} \xrightarrow{\text{Mg}^{2+}} \text{fructose 6-phosphate} + \text{ADP}$$

This is a major pathway of fructose entry into glycolysis in the muscles and kidney. In the liver, however, fructose enters by a different pathway. The liver enzyme **fructokinase** catalyzes the phosphorylation of fructose at C-1 rather than C-6:

$$\text{Fructose} + \text{ATP} \xrightarrow{\text{Mg}^{2+}} \text{fructose 1-phosphate} + \text{ADP}$$

The fructose 1-phosphate is then cleaved to glyceraldehyde and dihydroxyacetone phosphate by **fructose 1-phosphate aldolase:**

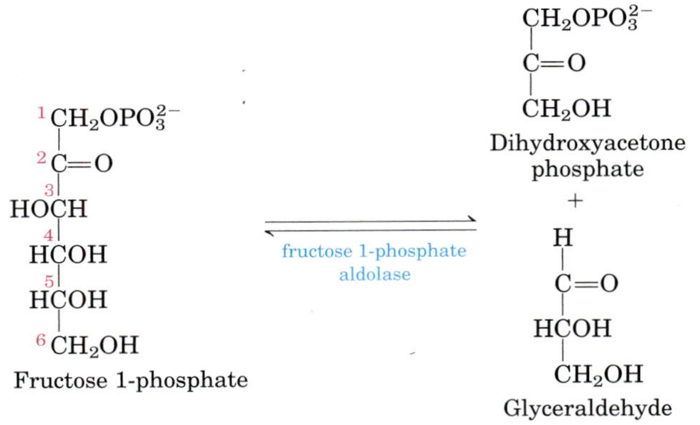

$$^{1}\text{CH}_2\text{OPO}_3^{2-}$$
$$^{2}\text{C}=\text{O}$$
$$^{3}\text{HOCH}$$
$$^{4}\text{HCOH}$$
$$^{5}\text{HCOH}$$
$$^{6}\text{CH}_2\text{OH}$$
Fructose 1-phosphate

fructose 1-phosphate aldolase

$$\text{CH}_2\text{OPO}_3^{2-}$$
$$\text{C}=\text{O}$$
$$\text{CH}_2\text{OH}$$
Dihydroxyacetone phosphate

+

$$\text{H}$$
$$\text{C}=\text{O}$$
$$\text{HCOH}$$
$$\text{CH}_2\text{OH}$$
Glyceraldehyde

Dihydroxyacetone phosphate is converted to glyceraldehyde 3-phosphate by the glycolytic enzyme triose phosphate isomerase. Glyceraldehyde is phosphorylated by ATP and **triose kinase** to glyceraldehyde 3-phosphate:

$$\text{Glyceraldehyde} + \text{ATP} \xrightarrow{\text{Mg}^{2+}}$$
$$\text{glyceraldehyde 3-phosphate} + \text{ADP}$$

Thus both products of fructose 1-phosphate hydrolysis enter the glycolytic pathway as glyceraldehyde 3-phosphate.

D-Galactose, a product of hydrolysis of the disaccharide lactose (milk sugar), passes in the blood from the intestine to the liver, where it is first phosphorylated at C-1, at the expense of ATP, by the enzyme **galactokinase:**

$$\text{Galactose} + \text{ATP} \xrightarrow{\text{Mg}^{2+}} \text{galactose 1-phosphate} + \text{ADP}$$

The galactose 1-phosphate is then converted to its epimer at C-4, glucose 1-phosphate, by a set of reactions in which **uridine diphosphate** (UDP) functions as a coenzyme-like carrier of hexose groups (Fig. 14–11). The epimerization involves first the oxidation of the C-4 —OH group to a ketone, then reduction of the ketone to an —OH, with inversion of the configuration at C-4. NAD is the cofactor for both the oxidation and the reduction.

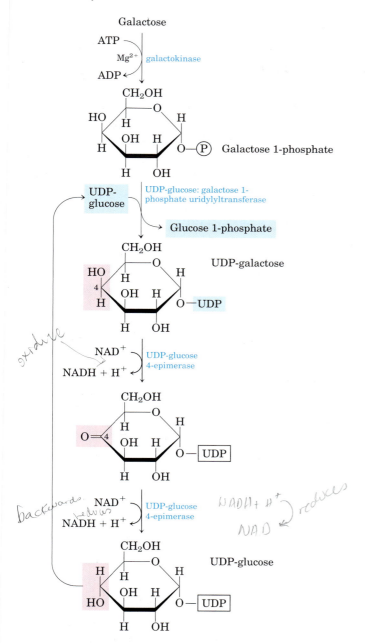

**FIGURE 14–11** **Conversion of galactose to glucose 1-phosphate.** The conversion proceeds through a sugar-nucleotide derivative, UDP-galactose, which is formed when galactose 1-phosphate displaces glucose 1-phosphate from UDP-glucose. UDP-galactose is then converted by UDP-glucose 4-epimerase to UDP-glucose, in a reaction that involves oxidation of C-4 (pink) by $NAD^+$, then reduction of C-4 by NADH; the result is inversion of the configuration at C-4. The UDP-glucose is recycled through another round of the same reaction. The net effect of this cycle is the conversion of galactose 1-phosphate to glucose 1-phosphate; there is no net production or consumption of UDP-galactose or UDP-glucose.

Defects in any of the three enzymes in this pathway cause **galactosemia** in humans. In galactokinase-deficiency galactosemia, high galactose concentrations are found in blood and urine. Infants develop cataracts, caused by deposition of the galactose metabolite galactitol in the lens.

The symptoms in this disorder are relatively mild, and strict limitation of galactose in the diet greatly diminishes their severity.

Transferase-deficiency galactosemia is more serious; it is characterized by poor growth in children, speech abnormality, mental deficiency, and liver damage that may be fatal, even when galactose is withheld from the diet. Epimerase-deficiency galactosemia leads to similar symptoms, but is less severe when dietary galactose is carefully controlled. ■

D-Mannose, released in the digestion of various polysaccharides and glycoproteins of foods, can be phosphorylated at C-6 by hexokinase:

$$\text{Mannose} + \text{ATP} \xrightarrow{\text{Mg}^{2+}} \text{mannose 6-phosphate} + \text{ADP}$$

Mannose 6-phosphate is isomerized by **phosphomannose isomerase** to yield fructose 6-phosphate, an intermediate of glycolysis.

### SUMMARY 14.2  Feeder Pathways for Glycolysis

- Glycogen and starch, polymeric storage forms of glucose, enter glycolysis in a two-step process. Phosphorolytic cleavage of a glucose residue from an end of the polymer, forming glucose 1-phosphate, is catalyzed by glycogen phosphorylase or starch phosphorylase. Phosphoglucomutase then converts the glucose 1-phosphate to glucose 6-phosphate, which can enter glycolysis.

- Ingested polysaccharides and disaccharides are converted to monosaccharides by intestinal hydrolytic enzymes, and the monosaccharides then enter intestinal cells and are transported to the liver or other tissues.

- A variety of D-hexoses, including fructose, galactose, and mannose, can be funneled into glycolysis. Each is phosphorylated and converted to either glucose 6-phosphate or fructose 6-phosphate.

- Conversion of galactose 1-phosphate to glucose 1-phosphate involves two nucleotide derivatives: UDP-galactose and UDP-glucose. Genetic defects in any of the three enzymes that catalyze conversion of galactose to glucose 1-phosphate result in galactosemias of varying severity.

## 14.3 Fates of Pyruvate under Anaerobic Conditions: Fermentation

Pyruvate occupies an important junction in carbohydrate catabolism (Fig. 14–3). Under aerobic conditions pyruvate is oxidized to acetate, which enters the citric acid cycle and is oxidized to $CO_2$ and $H_2O$, and NADH formed by the dehydrogenation of glyceraldehyde 3-phosphate is ultimately reoxidized to $NAD^+$ by passage of its electrons to $O_2$ in mitochondrial respiration. However, under hypoxic conditions, as in very active skeletal muscle, in submerged plant tissues, or in lactic acid bacteria, NADH generated by glycolysis cannot be reoxidized by $O_2$. Failure to regenerate $NAD^+$ would leave the cell with no electron acceptor for the oxidation of glyceraldehyde 3-phosphate, and the energy-yielding reactions of glycolysis would stop. $NAD^+$ must therefore be regenerated in some other way.

The earliest cells lived in an atmosphere almost devoid of oxygen and had to develop strategies for deriving energy from fuel molecules under anaerobic conditions. Most modern organisms have retained the ability to constantly regenerate $NAD^+$ during anaerobic glycolysis by transferring electrons from NADH to form a reduced end product such as lactate or ethanol.

### Pyruvate Is the Terminal Electron Acceptor in Lactic Acid Fermentation

When animal tissues cannot be supplied with sufficient oxygen to support aerobic oxidation of the pyruvate and NADH produced in glycolysis, $NAD^+$ is regenerated from NADH by the reduction of pyruvate to **lactate.** As mentioned earlier, some tissues and cell types (such as erythrocytes, which have no mitochondria and thus cannot oxidize pyruvate to $CO_2$) produce lactate from glucose even under aerobic conditions. The reduction of pyruvate is catalyzed by **lactate dehydrogenase,** which forms the L isomer of lactate at pH 7:

$$\Delta G'^{\circ} = -25.1 \text{ kJ/mol}$$

Pyruvate → L-Lactate, via lactate dehydrogenase (NADH + H⁺ → NAD⁺)

The overall equilibrium of this reaction strongly favors lactate formation, as shown by the large negative standard free-energy change.

In glycolysis, dehydrogenation of the two molecules of glyceraldehyde 3-phosphate derived from each molecule of glucose converts two molecules of $NAD^+$ to two of NADH. Because the reduction of two molecules of pyruvate to two of lactate regenerates two molecules of $NAD^+$, there is no net change in $NAD^+$ or NADH:

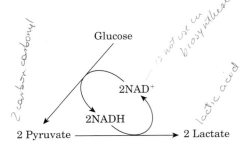

The lactate formed by active skeletal muscles (or by erythrocytes) can be recycled; it is carried in the blood to the liver, where it is converted to glucose during the recovery from strenuous muscular activity. When lactate is produced in large quantities during vigorous muscle contraction (during a sprint, for example), the acidification that results from ionization of lactic acid in muscle and blood limits the period of vigorous activity. The best-conditioned athletes can sprint at top speed for no more than a minute (Box 14–1).

Although conversion of glucose to lactate includes two oxidation-reduction steps, there is no net change in the oxidation state of carbon; in glucose ($C_6H_{12}O_6$) and lactic acid ($C_3H_6O_3$), the H:C ratio is the same. Nevertheless, some of the energy of the glucose molecule has been extracted by its conversion to lactate—enough to give a net yield of two molecules of ATP for every glucose molecule consumed. **Fermentation** is the general term for such processes, which extract energy (as ATP) but do not consume oxygen or change the concentrations of $NAD^+$ or NADH. Fermentations are carried out by a wide range of organisms, many of which occupy anaerobic niches, and they yield a variety of end products, some of which find commercial uses.

### Ethanol Is the Reduced Product in Ethanol Fermentation

Yeast and other microorganisms ferment glucose to ethanol and $CO_2$, rather than to lactate. Glucose is converted to pyruvate by glycolysis, and the pyruvate is converted to ethanol and $CO_2$ in a two-step process:

Pyruvate → Acetaldehyde (via pyruvate decarboxylase, TPP, $Mg^{2+}$, releasing $CO_2$) → Ethanol (via alcohol dehydrogenase, NADH + H⁺ → NAD⁺)

In the first step, pyruvate is decarboxylated in an irreversible reaction catalyzed by **pyruvate decarboxylase.** This reaction is a simple decarboxylation and does not involve the net oxidation of pyruvate. Pyruvate decarboxylase requires $Mg^{2+}$ and has a tightly bound coenzyme, thiamine pyrophosphate, discussed below. In the second step, acetaldehyde is reduced to ethanol through the action of **alcohol dehydrogenase,** with

BOX 14–1    THE WORLD OF BIOCHEMISTRY

## Athletes, Alligators, and Coelacanths: Glycolysis at Limiting Concentrations of Oxygen

Most vertebrates are essentially aerobic organisms; they convert glucose to pyruvate by glycolysis, then use molecular oxygen to oxidize the pyruvate completely to $CO_2$ and $H_2O$. Anaerobic catabolism of glucose to lactate occurs during short bursts of extreme muscular activity, for example in a 100 m sprint, during which oxygen cannot be carried to the muscles fast enough to oxidize pyruvate. Instead, the muscles use their stored glucose (glycogen) as fuel to generate ATP by fermentation, with lactate as the end product. In a sprint, lactate in the blood builds up to high concentrations. It is slowly converted back to glucose by gluconeogenesis in the liver in the subsequent rest or recovery period, during which oxygen is consumed at a gradually diminishing rate until the breathing rate returns to normal. The excess oxygen consumed in the recovery period represents a repayment of the oxygen debt. This is the amount of oxygen required to supply ATP for gluconeogenesis during recovery respiration, in order to regenerate the glycogen "borrowed" from liver and muscle to carry out intense muscular activity in the sprint. The cycle of reactions that includes glucose conversion to lactate in muscle and lactate conversion to glucose in liver is called the Cori cycle, for Carl and Gerty Cori, whose studies in the 1930s and 1940s clarified the pathway and its role (see Box 15–1).

The circulatory systems of most small vertebrates can carry oxygen to their muscles fast enough to avoid having to use muscle glycogen anaerobically. For example, migrating birds often fly great distances at high speeds without rest and without incurring an oxygen debt. Many running animals of moderate size also maintain an essentially aerobic metabolism in their skeletal muscle. However, the circulatory systems of larger animals, including humans, cannot completely sustain aerobic metabolism in skeletal muscles over long periods of intense muscular activity. These animals generally are slow-moving under normal circumstances and engage in intense muscular activity only in the gravest emergencies, because such bursts of activity require long recovery periods to repay the oxygen debt.

Alligators and crocodiles, for example, are normally sluggish animals. Yet when provoked they are capable of lightning-fast charges and dangerous lashings of their powerful tails. Such intense bursts of activity are short and must be followed by long periods of recovery. The fast emergency movements require

lactic acid fermentation to generate ATP in skeletal muscles. The stores of muscle glycogen are rapidly expended in intense muscular activity, and lactate reaches very high concentrations in muscles and extracellular fluid. Whereas a trained athlete can recover from a 100 m sprint in 30 min or less, an alligator may require many hours of rest and extra oxygen consumption to clear the excess lactate from its blood and regenerate muscle glycogen after a burst of activity.

Other large animals, such as the elephant and rhinoceros, have similar metabolic characteristics, as do diving mammals such as whales and seals. Dinosaurs and other huge, now-extinct animals probably had to depend on lactic acid fermentation to supply energy for muscular activity, followed by very long recovery periods during which they were vulnerable to attack by smaller predators better able to use oxygen and thus better adapted to continuous, sustained muscular activity.

Deep-sea explorations have revealed many species of marine life at great ocean depths, where the oxygen concentration is near zero. For example, the primitive coelacanth, a large fish recovered from depths of 4,000 m or more off the coast of South Africa, has an essentially anaerobic metabolism in virtually all its tissues. It converts carbohydrates to lactate and other products, most of which must be excreted. Some marine vertebrates ferment glucose to ethanol and $CO_2$ in order to generate ATP.

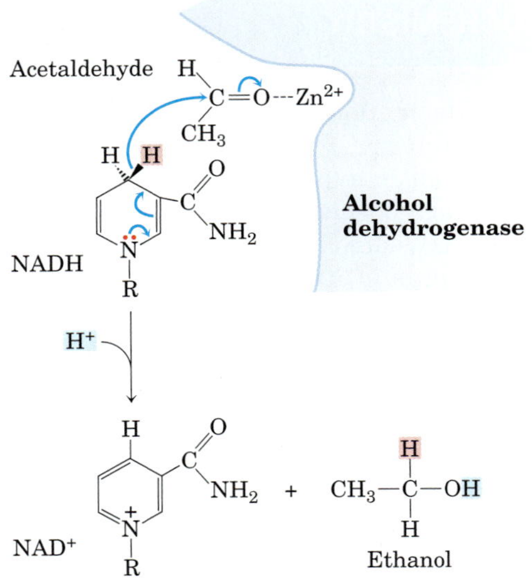

**MECHANISM FIGURE 14–12 The alcohol dehydrogenase reaction.**
A $Zn^{2+}$ at the active site polarizes the carbonyl oxygen of acetaldehyde, allowing transfer of a hydride ion (red) from the reduced cofactor NADH. The reduced intermediate acquires a proton from the medium (blue) to form ethanol. 🐭 Alcohol Dehydrogenase Mechanism

the reducing power furnished by NADH derived from the dehydrogenation of glyceraldehyde 3-phosphate. This reaction is a well-studied case of hydride transfer from NADH (Fig. 14–12). Ethanol and $CO_2$ are thus the end products of ethanol fermentation, and the overall equation is

$$Glucose + 2ADP + 2P_i \longrightarrow$$
$$2\ ethanol + 2CO_2 + 2ATP + 2H_2O$$

As in lactic acid fermentation, there is no net change in the ratio of hydrogen to carbon atoms when glucose (H:C ratio = 12/6 = 2) is fermented to two ethanol and

two $CO_2$ (combined H:C ratio = 12/6 = 2). In all fermentations, the H:C ratio of the reactants and products remains the same.

Pyruvate decarboxylase is present in brewer's and baker's yeast and in all other organisms that ferment glucose to ethanol, including some plants. The $CO_2$ produced by pyruvate decarboxylation in brewer's yeast is responsible for the characteristic carbonation of champagne. The ancient art of brewing beer involves a number of enzymatic processes in addition to the reactions of ethanol fermentation (Box 14–2). In baking, $CO_2$ released by pyruvate decarboxylase when yeast is mixed with a fermentable sugar causes dough to rise. The enzyme is absent in vertebrate tissues and in other organisms that carry out lactic acid fermentation.

Alcohol dehydrogenase is present in many organisms that metabolize ethanol, including humans. In human liver it catalyzes the oxidation of ethanol, either ingested or produced by intestinal microorganisms, with the concomitant reduction of $NAD^+$ to NADH.

### Thiamine Pyrophosphate Carries "Active Acetaldehyde" Groups

The pyruvate decarboxylase reaction provides our first encounter with **thiamine pyrophosphate (TPP)** (Fig. 14–13), a coenzyme derived from vitamin $B_1$. Lack of vitamin $B_1$ in the human diet leads to the condition known as beriberi, characterized by an accumulation of body fluids (swelling), pain, paralysis, and ultimately death.

Thiamine pyrophosphate plays an important role in the cleavage of bonds adjacent to a carbonyl group, such as the decarboxylation of $\alpha$-keto acids, and in chemical rearrangements in which an activated acetaldehyde group is transferred from one carbon atom to another (Table 14–1). The functional part of TPP, the thiazolium ring, has a relatively acidic proton at C-2. Loss of this

| TABLE 14–1 | Some TPP-Dependent Reactions | | |
|---|---|---|---|
| Enzyme | Pathway(s) | Bond cleaved | Bond formed |
| Pyruvate decarboxylase | Ethanol fermentation | $R^1-C(=O)-C(=O)-O^-$ | $R^1-C(=O)-H$ |
| Pyruvate dehydrogenase $\alpha$-Ketoglutarate dehydrogenase | Synthesis of acetyl-CoA Citric acid cycle | $R^2-C(=O)-C(=O)-O^-$ | $R^2-C(=O)-S\text{-}CoA$ |
| Transketolase | Carbon-assimilation reactions Pentose phosphate pathway | $R^3-C(=O)-C(OH)(H)-R^4$ | $R^3-C(=O)-C(OH)(H)-R^5$ |

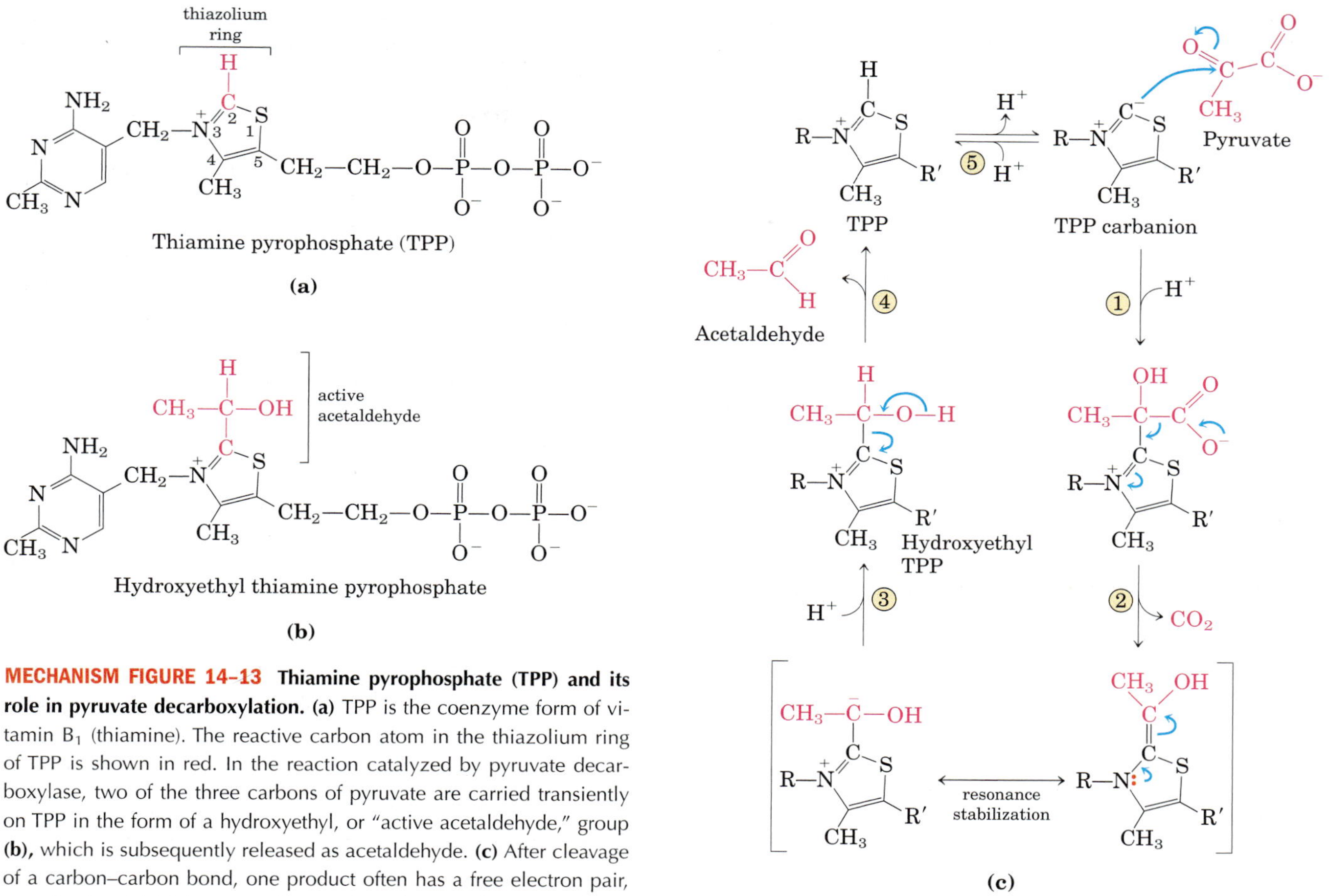

Thiamine pyrophosphate (TPP)

**(a)**

Hydroxyethyl thiamine pyrophosphate

**(b)**

**MECHANISM FIGURE 14–13   Thiamine pyrophosphate (TPP) and its role in pyruvate decarboxylation. (a)** TPP is the coenzyme form of vitamin B$_1$ (thiamine). The reactive carbon atom in the thiazolium ring of TPP is shown in red. In the reaction catalyzed by pyruvate decarboxylase, two of the three carbons of pyruvate are carried transiently on TPP in the form of a hydroxyethyl, or "active acetaldehyde," group **(b)**, which is subsequently released as acetaldehyde. **(c)** After cleavage of a carbon–carbon bond, one product often has a free electron pair, or carbanion, which because of its strong tendency to form a new bond is generally unstable. The thiazolium ring of TPP stabilizes carbanion intermediates by providing an electrophilic (electron-deficient) structure into which the carbanion electrons can be delocalized by resonance. Structures with this property, often called "electron sinks," play a role in many biochemical reactions. This principle is illustrated here for the reaction catalyzed by pyruvate decarboxylase. ① The TPP carbanion acts as a nucleophile, attacking the carbonyl group of pyruvate. ② Decarboxylation produces a carbanion that is stabilized by the thiazolium ring. ③ Protonation to form hydroxyethyl TPP is followed by ④ release of acetaldehyde. ⑤ A proton dissociates to regenerate the carbanion. 🔵 **Thiamine Pyrophosphate Mechanism**

**(c)**

proton produces a carbanion that is the active species in TPP-dependent reactions (Fig. 14–13). The carbanion readily adds to carbonyl groups, and the thiazolium ring is thereby positioned to act as an "electron sink" that greatly facilitates reactions such as the decarboxylation catalyzed by pyruvate decarboxylase.

## Fermentations Yield a Variety of Common Foods and Industrial Chemicals

Our progenitors learned millennia ago to use fermentation in the production and preservation of foods. Certain microorganisms present in raw food products fer-
ment the carbohydrates and yield metabolic products that give the foods their characteristic forms, textures, and tastes. Yogurt, already known in Biblical times, is produced when the bacterium *Lactobacillus bulgaricus* ferments the carbohydrate in milk, producing lactic acid; the resulting drop in pH causes the milk proteins to precipitate, producing the thick texture and sour taste of unsweetened yogurt. Another bacterium, *Propionibacterium freudenreichii*, ferments milk to produce propionic acid and CO$_2$; the propionic acid precipitates milk proteins, and bubbles of CO$_2$ cause the holes characteristic of Swiss cheese. Many other food products are the result of fermentations: pickles, sauerkraut, sausage, soy sauce, and a variety of national favorites, such as kimchi (Korea), tempoyak (Indonesia), kefir (Russia), dahi (India), and pozol (Mexico). The drop in pH associated with fermentation also helps to preserve foods, because most of the microorganisms that cause food spoilage cannot grow at low pH. In agriculture, plant byproducts such as corn stalks are preserved for use as animal feed by packing them into a large container (a silo) with limited access to air; microbial fermentation produces acids that lower the pH. The silage that results from this fermentation

## BOX 14–2    THE WORLD OF BIOCHEMISTRY

### Brewing Beer

Brewers prepare beer by ethanol fermentation of the carbohydrates in cereal grains (seeds) such as barley, carried out by yeast glycolytic enzymes. The carbohydrates, largely polysaccharides, must first be degraded to disaccharides and monosaccharides. In a process called malting, the barley seeds are allowed to germinate until they form the hydrolytic enzymes required to break down their polysaccharides, at which point germination is stopped by controlled heating. The product is malt, which contains enzymes that catalyze the hydrolysis of the $\beta$ linkages of cellulose and other cell wall polysaccharides of the barley husks, and enzymes such as $\alpha$-amylase and maltase.

The brewer next prepares the wort, the nutrient medium required for fermentation by yeast cells. The malt is mixed with water and then mashed or crushed. This allows the enzymes formed in the malting process to act on the cereal polysaccharides to form maltose, glucose, and other simple sugars, which are soluble in the aqueous medium. The remaining cell matter is then separated, and the liquid wort is boiled with hops to give flavor. The wort is cooled and then aerated.

Now the yeast cells are added. In the aerobic wort the yeast grows and reproduces very rapidly, using energy obtained from available sugars. No ethanol forms during this stage, because the yeast, amply supplied with oxygen, oxidizes the pyruvate formed by glycolysis to $CO_2$ and $H_2O$ via the citric acid cycle. When all the dissolved oxygen in the vat of wort has been consumed, the yeast cells switch to anaerobic metabolism, and from this point they ferment the sugars into ethanol and $CO_2$. The fermentation process is controlled in part by the concentration of the ethanol formed, by the pH, and by the amount of remaining sugar. After fermentation has been stopped, the cells are removed and the "raw" beer is ready for final processing.

In the final steps of brewing, the amount of foam or head on the beer, which results from dissolved proteins, is adjusted. Normally this is controlled by proteolytic enzymes that arise in the malting process. If these enzymes act on the proteins too long, the beer will have very little head and will be flat; if they do not act long enough, the beer will not be clear when it is cold. Sometimes proteolytic enzymes from other sources are added to control the head.

---

process can be kept as animal feed for long periods without spoilage.

In 1910 Chaim Weizmann (later to become the first president of Israel) discovered that the bacterium *Clostridium acetobutyricum* ferments starch to butanol and acetone. This discovery opened the field of industrial fermentations, in which some readily available material rich in carbohydrate (corn starch or molasses, for example) is supplied to a pure culture of a specific microorganism, which ferments it into a product of greater value. The methanol used to make "gasohol" is produced by microbial fermentation, as are formic, acetic, propionic, butyric, and succinic acids, and glycerol, ethanol, isopropanol, butanol, and butanediol. These fermentations are generally carried out in huge closed vats in which temperature and access to air are adjusted to favor the multiplication of the desired microorganism and to exclude contaminating organisms (Fig. 14–14). The beauty of industrial fermentations is that complicated, multistep chemical transformations are carried out in high yields and with few side products by chemical factories that reproduce themselves—microbial cells. For some industrial fermentations, technology has been developed to immobilize the cells in an inert support, to pass the starting material continuously through the bed of immobilized cells, and to collect the desired product in the effluent—an engineer's dream!

**FIGURE 14–14 Industrial-scale fermentation.** Microorganisms are cultured in a sterilizable vessel containing thousands of liters of growth medium—an inexpensive source of both carbon and energy—under carefully controlled conditions, including low oxygen concentration and constant temperature. After centrifugal separation of the cells from the growth medium, the valuable products of the fermentation are recovered from the cells or from the supernatant fluid.

## SUMMARY 14.3 Fates of Pyruvate under Anaerobic Conditions: Fermentation

- The NADH formed in glycolysis must be recycled to regenerate NAD$^+$, which is required as an electron acceptor in the first step of the payoff phase. Under aerobic conditions, electrons pass from NADH to O$_2$ in mitochondrial respiration.

- Under anaerobic or hypoxic conditions, many organisms regenerate NAD$^+$ by transferring electrons from NADH to pyruvate, forming lactate. Other organisms, such as yeast, regenerate NAD$^+$ by reducing pyruvate to ethanol and CO$_2$. In these anaerobic processes (fermentations), there is no *net* oxidation or reduction of the carbons of glucose.

- A variety of microorganisms can ferment sugar in fresh foods, resulting in changes in pH, taste, and texture, and preserving food from spoilage. Fermentations are used in industry to produce a wide variety of commercially valuable organic compounds from inexpensive starting materials.

## 14.4 Gluconeogenesis

The central role of glucose in metabolism arose early in evolution, and this sugar remains the nearly universal fuel and building block in modern organisms, from microbes to humans. In mammals, some tissues depend almost completely on glucose for their metabolic energy. For the human brain and nervous system, as well as the erythrocytes, testes, renal medulla, and embryonic tissues, glucose from the blood is the sole or major fuel source. The brain alone requires about 120 g of glucose each day—more than half of all the glucose stored as glycogen in muscle and liver. However, the supply of glucose from these stores is not always sufficient; between meals and during longer fasts, or after vigorous exercise, glycogen is depleted. For these times, organisms need a method for synthesizing glucose from noncarbohydrate precursors. This is accomplished by a pathway called **gluconeogenesis** ("formation of new sugar"), which converts pyruvate and related three- and four-carbon compounds to glucose.

Gluconeogenesis occurs in all animals, plants, fungi, and microorganisms. The reactions are essentially the same in all tissues and all species. The important precursors of glucose in animals are three-carbon compounds such as lactate, pyruvate, and glycerol, as well as certain amino acids (Fig. 14–15). In mammals, gluconeogenesis takes place mainly in the liver, and to a lesser extent in renal cortex. The glucose produced passes into the blood to supply other tissues. After vigorous exercise, lactate produced by anaerobic glycolysis in skeletal muscle returns to the liver and is converted to glucose, which moves back to muscle and is converted to glycogen—a circuit called the Cori cycle (Box 14–1; see also Fig. 23–18). In plant seedlings, stored fats and proteins are converted, via paths that include gluconeogenesis, to the disaccharide sucrose for transport throughout the developing plant. Glucose and its derivatives are precursors for the synthesis of plant cell walls, nucleotides and coenzymes, and a variety of other essential metabolites. In many microorganisms, gluconeogenesis starts from simple organic compounds of two or three carbons, such as acetate, lactate, and propionate, in their growth medium.

Although the reactions of gluconeogenesis are the same in all organisms, the metabolic context and the regulation of the pathway differ from one species to another and from tissue to tissue. In this section we focus on gluconeogenesis as it occurs in the mammalian liver. In Chapter 20 we show how photosynthetic organisms use this pathway to convert the primary products of photosynthesis into glucose, to be stored as sucrose or starch.

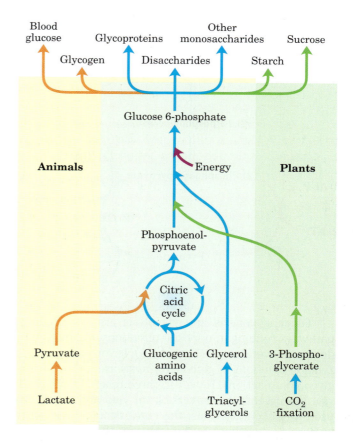

**FIGURE 14–15 Carbohydrate synthesis from simple precursors.** The pathway from phosphoenolpyruvate to glucose 6-phosphate is common to the biosynthetic conversion of many different precursors of carbohydrates in animals and plants. Plants and photosynthetic bacteria are uniquely able to convert CO$_2$ to carbohydrates.

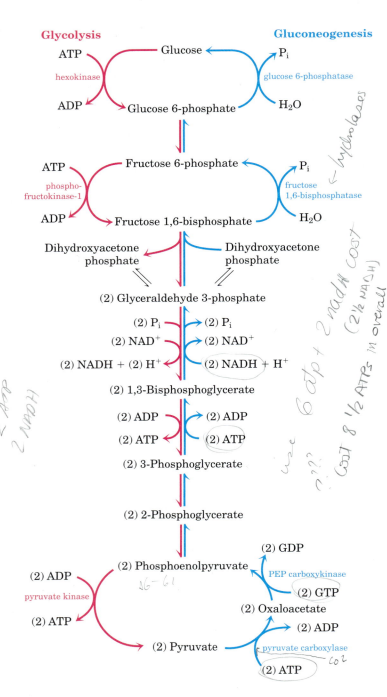

**Glycolysis**

**Gluconeogenesis**

*(handwritten notes in margin:)* ~ hydrolaoses, (2½ NADH) 6 atp + 2 nadH (ost (2 nadH cost) 8 ½ ATPs in overall, use atp? (cost 8 ½ ace), NG-61, CO2, Produce 2 ATP 2 NADH

**FIGURE 14–16 Opposing pathways of glycolysis and gluconeogenesis in rat liver.** The reactions of glycolysis are shown on the left side in blue; the opposing pathway of gluconeogenesis is shown on the right in red. The major sites of regulation of gluconeogenesis shown here are discussed later in this chapter, and in detail in Chapter 15. Figure 14–19 illustrates an alternative route for oxaloacetate produced in mitochondria.

terized by a large negative free-energy change, $\Delta G$, whereas other glycolytic reactions have a $\Delta G$ near 0 (Table 14–2). In gluconeogenesis, the three irreversible steps are bypassed by a separate set of enzymes, catalyzing reactions that are sufficiently exergonic to be effectively irreversible in the direction of glucose synthesis. Thus, both glycolysis and gluconeogenesis are irreversible processes in cells. In animals, both pathways occur largely in the cytosol, necessitating their reciprocal and coordinated regulation. Separate regulation of the two pathways is brought about through controls exerted on the enzymatic steps unique to each.

We begin by considering the three bypass reactions of gluconeogenesis. (Keep in mind that "bypass" refers throughout to the bypass of irreversible glycolytic reactions.)

## Conversion of Pyruvate to Phosphoenolpyruvate Requires Two Exergonic Reactions

The first of the bypass reactions in gluconeogenesis is the conversion of pyruvate to phosphoenolpyruvate (PEP). This reaction cannot occur by reversal of the pyruvate kinase reaction of glycolysis (p. 532), which has a large, negative standard free-energy change and is irreversible under the conditions prevailing in intact cells (Table 14–2, step ⑩). Instead, the phosphorylation of pyruvate is achieved by a roundabout sequence of reactions that in eukaryotes requires enzymes in both the cytosol and mitochondria. As we shall see, the pathway shown in Figure 14–16 and described in detail here is one of two routes from pyruvate to PEP; it is the predominant path when pyruvate or alanine is the glucogenic precursor. A second pathway, described later, predominates when lactate is the glucogenic precursor.

Pyruvate is first transported from the cytosol into mitochondria or is generated from alanine within mitochondria by transamination, in which the $\alpha$-amino group is removed from alanine (leaving pyruvate) and added to an $\alpha$-keto carboxylic acid (transamination reactions are discussed in detail in Chapter 18). Then **pyruvate carboxylase,** a mitochondrial enzyme that requires the coenzyme **biotin,** converts the pyruvate to oxaloacetate (Fig. 14–17):

$$\text{Pyruvate} + \text{HCO}_3^- + \text{ATP} \longrightarrow$$

$$\text{oxaloacetate} + \text{ADP} + \text{P}_i \quad (14\text{–}4)$$

Gluconeogenesis and glycolysis are not identical pathways running in opposite directions, although they do share several steps (Fig. 14–16); seven of the ten enzymatic reactions of gluconeogenesis are the reverse of glycolytic reactions. However, three reactions of glycolysis are essentially irreversible in vivo and cannot be used in gluconeogenesis: the conversion of glucose to glucose 6-phosphate by hexokinase, the phosphorylation of fructose 6-phosphate to fructose 1,6-bisphosphate by phosphofructokinase-1, and the conversion of phosphoenolpyruvate to pyruvate by pyruvate kinase (Fig. 14–16). In cells, these three reactions are charac-

**TABLE 14–2 Free-Energy Changes of Glycolytic Reactions in Erythrocytes**

| Glycolytic reaction step | $\Delta G'^{\circ}$ (kJ/mol) | $\Delta G$ (kJ/mol) |
|---|---|---|
| ① Glucose + ATP $\longrightarrow$ glucose 6-phosphate + ADP | $-16.7$ | $-33.4$ |
| ② Glucose 6-phosphate $\rightleftharpoons$ fructose 6-phosphate | 1.7 | 0 to 25 |
| ③ Fructose 6-phosphate + ATP $\longrightarrow$ fructose 1,6-bisphosphate + ADP | $-14.2$ | $-22.2$ |
| ④ Fructose 1,6-bisphosphate $\rightleftharpoons$ dihydroxyacetone phosphate + glyceraldehyde 3-phosphate | 23.8 | 0 to $-6$ |
| ⑤ Dihydroxyacetone phosphate $\rightleftharpoons$ glyceraldehyde 3-phosphate | 7.5 | 0 to 4 |
| ⑥ Glyceraldehyde 3-phosphate + $P_i$ + $NAD^+$ $\rightleftharpoons$ 1,3-bisphosphoglycerate + NADH + $H^+$ | 6.3 | $-2$ to 2 |
| ⑦ 1,3-Bisphosphoglycerate + ADP $\rightleftharpoons$ 3-phosphoglycerate + ATP | $-18.8$ | 0 to 2 |
| ⑧ 3-Phosphoglycerate $\rightleftharpoons$ 2-phosphoglycerate | 4.4 | 0 to 0.8 |
| ⑨ 2-Phosphoglycerate $\rightleftharpoons$ phosphoenolpyruvate + $H_2O$ | 7.5 | 0 to 3.3 |
| ⑩ Phosphoenolpyruvate + ADP $\longrightarrow$ pyruvate + ATP | $-31.4$ | $-16.7$ |

Note: $\Delta G'^{\circ}$ is the standard free-energy change, as defined in Chapter 13 (p. 491). $\Delta G$ is the free-energy change calculated from the actual concentrations of glycolytic intermediates present under physiological conditions in erythrocytes, at pH 7. The glycolytic reactions bypassed in gluconeogenesis are shown in red. Biochemical equations are not necessarily balanced for H or charge (p. 506).

The reaction involves biotin as a carrier of activated $HCO_3^-$ (Fig. 14–18). The reaction mechanism is shown in Figure 16–16. Pyruvate carboxylase is the first regulatory enzyme in the gluconeogenic pathway, requiring acetyl-CoA as a positive effector. (Acetyl-CoA is produced by fatty acid oxidation (Chapter 17), and its accumulation signals the availability of fatty acids as fuel.) As we shall see in Chapter 16 (see Fig. 16–15), the pyruvate carboxylase reaction can replenish intermediates in another central metabolic pathway, the citric acid cycle.

Because the mitochondrial membrane has no transporter for oxaloacetate, before export to the cytosol the oxaloacetate formed from pyruvate must be reduced to malate by mitochondrial **malate dehydrogenase,** at the expense of NADH:

Oxaloacetate + NADH + $H^+$ $\rightleftharpoons$ L-malate + $NAD^+$ (14–5)

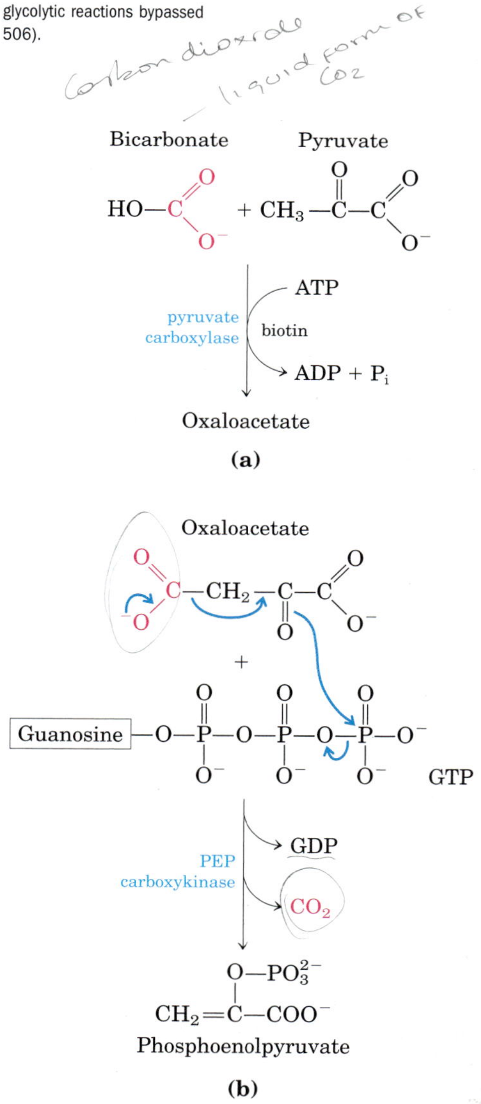

**FIGURE 14–17 Synthesis of phosphoenolpyruvate from pyruvate.**
**(a)** In mitochondria, pyruvate is converted to oxaloacetate in a biotin-requiring reaction catalyzed by pyruvate carboxylase. **(b)** In the cytosol, oxaloacetate is converted to phosphoenolpyruvate by PEP carboxykinase. The $CO_2$ incorporated in the pyruvate carboxylase reaction is lost here as $CO_2$. The decarboxylation leads to a rearrangement of electrons that facilitates attack of the carbonyl oxygen of the pyruvate moiety on the $\gamma$ phosphate of GTP.

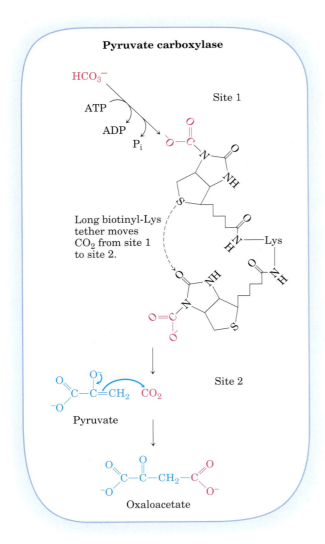

**FIGURE 14–18 Role of biotin in the pyruvate carboxylase reaction.** The cofactor biotin is covalently attached to the enzyme through an amide linkage to the $\varepsilon$-amino group of a Lys residue, forming a biotinyl-enzyme. The reaction occurs in two phases, which occur at two different sites in the enzyme. At catalytic site 1, bicarbonate ion is converted to $CO_2$ at the expense of ATP. Then $CO_2$ reacts with biotin, forming carboxybiotinyl-enzyme. The long arm composed of biotin and the side chain of the Lys to which it is attached then carry the $CO_2$ of carboxybiotinyl-enzyme to catalytic site 2 on the enzyme surface, where $CO_2$ is released and reacts with the pyruvate, forming oxaloacetate and regenerating the biotinyl-enzyme. The general role of flexible arms in carrying reaction intermediates between enzyme active sites is described in Figure 16–17, and the mechanistic details of the pyruvate carboxylase reaction are shown in Figure 16–16. Similar mechanisms occur in other biotin-dependent carboxylation reactions, such as those catalyzed by propionyl-CoA carboxylase (see Fig. 17–11) and acetyl-CoA carboxylase (see Fig. 21–1).

The standard free-energy change for this reaction is quite high, but under physiological conditions (including a very low concentration of oxaloacetate) $\Delta G \approx 0$ and the reaction is readily reversible. Mitochondrial malate dehydrogenase functions in both gluconeogenesis and the citric acid cycle, but the overall flow of metabolites in the two processes is in opposite directions.

Malate leaves the mitochondrion through a specific transporter in the inner mitochondrial membrane (see Fig. 19–27), and in the cytosol it is reoxidized to oxaloacetate, with the production of cytosolic NADH:

$$\text{Malate} + NAD^+ \longrightarrow \text{oxaloacetate} + NADH + H^+ \quad (14\text{–}6)$$

The oxaloacetate is then converted to PEP by **phosphoenolpyruvate carboxykinase** (Fig. 14–17). This $Mg^{2+}$-dependent reaction requires GTP as the phosphoryl group donor :

$$\text{Oxaloacetate} + GTP \rightleftharpoons PEP + CO_2 + GDP \quad (14\text{–}7)$$

The reaction is reversible under intracellular conditions; the formation of one high-energy phosphate compound (PEP) is balanced by the hydrolysis of another (GTP).

The overall equation for this set of bypass reactions, the sum of Equations 14–4 through 14–7, is

$$\text{Pyruvate} + ATP + GTP + HCO_3^- \longrightarrow$$
$$PEP + ADP + GDP + P_i + CO_2$$
$$\Delta G'^\circ = 0.9 \text{ kJ/mol} \quad (14\text{–}8)$$

Two high-energy phosphate equivalents (one from ATP and one from GTP), each yielding about 50 kJ/mol under cellular conditions, must be expended to phosphorylate one molecule of pyruvate to PEP. In contrast, when PEP is converted to pyruvate during glycolysis, only one ATP is generated from ADP. Although the standard free-energy change ($\Delta G'^\circ$) of the two-step path from pyruvate to PEP is 0.9 kJ/mol, the actual free-energy change ($\Delta G$), calculated from measured cellular concentrations of intermediates, is very strongly negative ($-25$ kJ/mol); this results from the ready consumption of PEP in other reactions such that its concentration remains relatively low. The reaction is thus effectively irreversible in the cell.

Note that the $CO_2$ added to pyruvate in the pyruvate carboxylase step is the same molecule that is lost in the PEP carboxykinase reaction (Fig. 14–17). This carboxylation-decarboxylation sequence represents a way of "activating" pyruvate, in that the decarboxylation of oxaloacetate facilitates PEP formation. In Chapter 21 we shall see how a similar carboxylation-decarboxylation sequence is used to activate acetyl-CoA for fatty acid biosynthesis (see Fig. 21–1).

There is a logic to the route of these reactions through the mitochondrion. The [NADH]/[NAD$^+$] ratio in the cytosol is $8 \times 10^{-4}$, about $10^5$ times lower than in mitochondria. Because cytosolic NADH is consumed in gluconeogenesis (in the conversion of 1,3-bisphosphoglycerate to glyceraldehyde 3-phosphate; Fig. 14–16), glucose biosynthesis cannot proceed unless NADH is available. The transport of malate from the mitochondrion to the cytosol and its reconversion there to oxaloacetate effectively moves reducing equivalents to the cytosol, where they are scarce. This path from pyruvate to PEP therefore provides an important balance between NADH produced and consumed in the cytosol during gluconeogenesis.

A second pyruvate → PEP bypass predominates when lactate is the glucogenic precursor (Fig. 14–19). This pathway makes use of lactate produced by glycolysis in erythrocytes or anaerobic muscle, for example, and it is particularly important in large vertebrates after vigorous exercise (Box 14–1). The conversion of lactate to pyruvate in the cytosol of hepatocytes yields NADH, and the export of reducing equivalents (as malate) from mitochondria is therefore unnecessary. After the pyruvate produced by the lactate dehydrogenase reaction is transported into the mitochondrion, it is converted to oxaloacetate by pyruvate carboxylase, as described above. This oxaloacetate, however, is converted directly to PEP by a mitochondrial isozyme of PEP carboxykinase, and the PEP is transported out of the mitochondrion to continue on the gluconeogenic path. The mitochondrial and cytosolic isozymes of PEP carboxykinase are encoded by separate genes in the nuclear chromosomes, providing another example of two distinct enzymes catalyzing the same reaction but having different cellular locations or metabolic roles (recall the isozymes of hexokinase).

## Conversion of Fructose 1,6-Bisphosphate to Fructose 6-Phosphate Is the Second Bypass

The second glycolytic reaction that cannot participate in gluconeogenesis is the phosphorylation of fructose 6-phosphate by PFK-1 (Table 14–2, step ③). Because this reaction is highly exergonic and therefore irreversible in intact cells, the generation of fructose 6-phosphate from fructose 1,6-bisphosphate (Fig. 14–16) is catalyzed by a different enzyme, $Mg^{2+}$-dependent **fructose 1,6-bisphosphatase (FBPase-1),** which promotes the essentially irreversible _hydrolysis_ of the C-1 phosphate (_not_ phosphoryl group transfer to ADP):

Fructose 1,6-bisphosphate + $H_2O$ ⟶

fructose 6-phosphate + $P_i$
$\Delta G'^\circ = -16.3$ kJ/mol

## Conversion of Glucose 6-Phosphate to Glucose Is the Third Bypass

The third bypass is the final reaction of gluconeogenesis, the dephosphorylation of glucose 6-phosphate to yield glucose (Fig. 14–16). Reversal of the hexokinase reaction (p. 526) would require phosphoryl group trans-

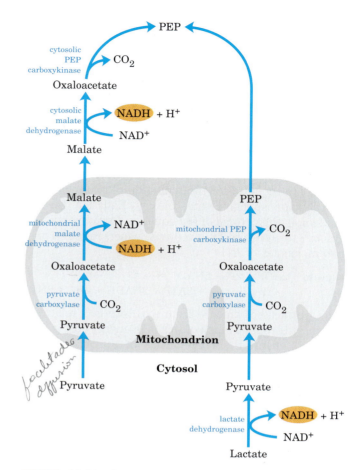

**FIGURE 14–19 Alternative paths from pyruvate to phosphoenolpyruvate.** The path that predominates depends on the glucogenic precursor (lactate or pyruvate). The path on the right predominates when lactate is the precursor, because cytosolic NADH is generated in the lactate dehydrogenase reaction and does not have to be shuttled out of the mitochondrion (see text). The relative importance of the two pathways depends on the availability of lactate and the cytosolic requirements for NADH by gluconeogenesis.

fer from glucose 6-phosphate to ADP, forming ATP, an energetically unfavorable reaction (Table 14–2, step ①). The reaction catalyzed by **glucose 6-phosphatase** does not require synthesis of ATP; it is a simple hydrolysis of a phosphate ester:

Glucose 6-phosphate + $H_2O$ ⟶ glucose + $P_i$
$\Delta G'^\circ = -13.8$ kJ/mol

This $Mg^{2+}$-activated enzyme is found on the lumenal side of the endoplasmic reticulum of hepatocytes and renal cells (see Fig. 15–6). Muscle and brain tissue do not contain this enzyme and so cannot carry out gluconeogenesis. Glucose produced by gluconeogenesis in the liver or kidney or ingested in the diet is delivered to brain and muscle through the bloodstream.

## Gluconeogenesis Is Energetically Expensive, but Essential

The sum of the biosynthetic reactions leading from pyruvate to free blood glucose (Table 14–3) is

$$\text{2 Pyruvate} + 4ATP + 2GTP + 2NADH + 2H^+ + 4H_2O \longrightarrow$$
$$\text{glucose} + 4ADP + 2GDP + 6P_i + 2NAD^+ \quad (14\text{--}9)$$

For each molecule of glucose formed from pyruvate, six high-energy phosphate groups are required, four from ATP and two from GTP. In addition, two molecules of NADH are required for the reduction of two molecules of 1,3-bisphosphoglycerate. Clearly, Equation 14–9 is not simply the reverse of the equation for conversion of glucose to pyruvate by glycolysis, which requires only two molecules of ATP:

$$\text{Glucose} + 2ADP + 2P_i + 2NAD^+ \longrightarrow$$
$$\text{2 pyruvate} + 2ATP + 2NADH + 2H^+ + 2H_2O$$

The synthesis of glucose from pyruvate is a relatively expensive process. Much of this high energy cost is necessary to ensure the irreversibility of gluconeogenesis. Under intracellular conditions, the overall free-energy change of glycolysis is at least $-63$ kJ/mol. Under the same conditions the overall $\Delta G$ of gluconeogenesis is $-16$ kJ/mol. Thus both glycolysis and gluconeogenesis are essentially irreversible processes in cells.

## Citric Acid Cycle Intermediates and Many Amino Acids Are Glucogenic

The biosynthetic pathway to glucose described above allows the net synthesis of glucose not only from pyruvate but also from the four-, five-, and six-carbon intermediates of the citric acid cycle (Chapter 16). Citrate, isocitrate, $\alpha$-ketoglutarate, succinyl-CoA, succinate, fumarate, and malate—all are citric acid cycle intermediates that can undergo oxidation to oxaloacetate (see Fig. 16–7). Some or all of the carbon atoms of most amino acids derived from proteins are ultimately catabolized to pyruvate or to intermediates of the citric acid cycle. Such amino acids can therefore undergo net conversion to glucose and are said to be **glucogenic** (Table 14–4). Alanine and glutamine, the principal molecules that transport amino groups from extrahepatic tissues to the liver (see Fig. 18–9), are particularly important glucogenic amino acids in mammals. After removal of their amino groups in liver mitochondria, the carbon skeletons remaining (pyruvate and $\alpha$-ketoglutarate, respectively) are readily funneled into gluconeogenesis.

In contrast, no net conversion of fatty acids to glucose occurs in mammals. As we shall see in Chapter 17, the catabolism of most fatty acids yields only acetyl-CoA. Mammals cannot use acetyl-CoA as a precursor of glucose, because the pyruvate dehydrogenase reaction is irreversible and cells have no other pathway to convert acetyl-CoA to pyruvate. Plants, yeast, and many bacteria do have a pathway (the glyoxylate cycle; see Fig. 16–20) for converting acetyl-CoA to oxaloacetate, so these organisms can use fatty acids as the starting material for gluconeogenesis. This is especially important during the germination of seedlings, before photosynthesis can serve as a source of glucose.

## Glycolysis and Gluconeogenesis Are Regulated Reciprocally

If glycolysis (the conversion of glucose to pyruvate) and gluconeogenesis (the conversion of pyruvate to glucose) were allowed to proceed simultaneously at high rates,

---

**TABLE 14–3 Sequential Reactions in Gluconeogenesis Starting from Pyruvate**

| Reaction | |
|---|---|
| Pyruvate + $HCO_3^-$ + ATP $\longrightarrow$ oxaloacetate + ADP + $P_i$ | $\times 2$ |
| Oxaloacetate + GTP $\rightleftharpoons$ phosphoenolpyruvate + $CO_2$ + GDP | $\times 2$ |
| Phosphoenolpyruvate + $H_2O$ $\rightleftharpoons$ 2-phosphoglycerate | $\times 2$ |
| 2-Phosphoglycerate $\rightleftharpoons$ 3-phosphoglycerate | $\times 2$ |
| 3-Phosphoglycerate + ATP $\rightleftharpoons$ 1,3-bisphosphoglycerate + ADP | $\times 2$ |
| 1,3-Bisphosphoglycerate + NADH + $H^+$ $\rightleftharpoons$ glyceraldehyde 3-phosphate + $NAD^+$ + $P_i$ | $\times 2$ |
| Glyceraldehyde 3-phosphate $\rightleftharpoons$ dihydroxyacetone phosphate | |
| Glyceraldehyde 3-phosphate + dihydroxyacetone phosphate $\rightleftharpoons$ fructose 1,6-bisphosphate | |
| Fructose 1,6-bisphosphate $\longrightarrow$ fructose 6-phosphate + $P_i$ | |
| Fructose 6-phosphate $\rightleftharpoons$ glucose 6-phosphate | |
| Glucose 6-phosphate + $H_2O$ $\longrightarrow$ glucose + $P_i$ | |
| *Sum:* 2 Pyruvate + 4ATP + 2GTP + 2NADH + $2H^+$ + $4H_2O$ $\longrightarrow$ glucose + 4ADP + 2GDP + 6$P_i$ + 2$NAD^+$ | |

Note: The bypass reactions are in red; all other reactions are reversible steps of glycolysis. The figures at the right indicate that the reaction is to be counted twice, because two three-carbon precursors are required to make a molecule of glucose. The reactions required to replace the cytosolic NADH consumed in the glyceraldehyde 3-phosphate dehydrogenase reaction (the conversion of lactate to pyruvate in the cytosol or the transport of reducing equivalents from mitochondria to the cytosol in the form of malate) are not considered in this summary. Biochemical equations are not necessarily balanced for H and charge (p. 506).

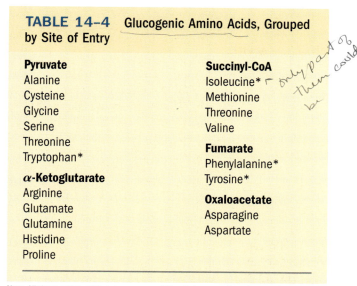

**TABLE 14–4 Glucogenic Amino Acids, Grouped by Site of Entry**

| Pyruvate | Succinyl-CoA |
|---|---|
| Alanine | Isoleucine* |
| Cysteine | Methionine |
| Glycine | Threonine |
| Serine | Valine |
| Threonine | |
| Tryptophan* | **Fumarate** |
| | Phenylalanine* |
| **α-Ketoglutarate** | Tyrosine* |
| Arginine | |
| Glutamate | **Oxaloacetate** |
| Glutamine | Asparagine |
| Histidine | Aspartate |
| Proline | |

*[handwritten note: only part of them could be]*

Note: All these amino acids are precursors of blood glucose or liver glycogen, because they can be converted to pyruvate or citric acid cycle intermediates. Of the 20 common amino acids, only leucine and lysine are unable to furnish carbon for net glucose synthesis.

*These amino acids are also ketogenic (see Fig. 18–21).

the result would be the consumption of ATP and the production of heat. For example, PFK-1 and FBPase-1 catalyze opposing reactions:

$$ATP + \text{fructose 6-phosphate} \xrightarrow{\text{PFK-1}} ADP + \text{fructose 1,6-bisphosphate}$$

$$\text{Fructose 1,6-bisphosphate} + H_2O \xrightarrow{\text{FBPase-1}} \text{fructose 6-phosphate} + P_i$$

The sum of these two reactions is

$$ATP + H_2O \longrightarrow ADP + P_i + \text{heat}$$

These two enzymatic reactions, and a number of others in the two pathways, are regulated allosterically and by covalent modification (phosphorylation). In Chapter 15 we take up the mechanisms of this regulation in detail. For now, suffice it to say that the pathways are regulated so that when the flux of glucose through glycolysis goes up, the flux of pyruvate toward glucose goes down, and vice versa.

### SUMMARY 14.4 Gluconeogenesis

- Gluconeogenesis is a ubiquitous multistep process in which pyruvate or a related three-carbon compound (lactate, alanine) is converted to glucose. Seven of the steps in gluconeogenesis are catalyzed by the same enzymes used in glycolysis; these are the reversible reactions.

- Three irreversible steps in the glycolytic pathway are bypassed by reactions catalyzed by gluconeogenic enzymes: (1) conversion of pyruvate to PEP via oxaloacetate, catalyzed by pyruvate carboxylase and PEP carboxykinase; (2) dephosphorylation of fructose 1,6-bisphosphate by FBPase-1; and (3) dephosphorylation of glucose 6-phosphate by glucose 6-phosphatase.

- Formation of one molecule of glucose from pyruvate requires 4 ATP, 2 GTP, and 2 NADH; it is expensive.

- In mammals, gluconeogenesis in the liver and kidney provides glucose for use by the brain, muscles, and erythrocytes.

- Pyruvate carboxylase is stimulated by acetyl-CoA, increasing the rate of gluconeogenesis when the cell already has adequate supplies of other substrates (fatty acids) for energy production.

- Animals cannot convert acetyl-CoA derived from fatty acids into glucose; plants and microorganisms can.

- Glycolysis and gluconeogenesis are reciprocally regulated to prevent wasteful operation of both pathways at the same time.

## 14.5 Pentose Phosphate Pathway of Glucose Oxidation

In most animal tissues, the major catabolic fate of glucose 6-phosphate is glycolytic breakdown to pyruvate, much of which is then oxidized via the citric acid cycle, ultimately leading to the formation of ATP. Glucose 6-phosphate does have other catabolic fates, however, which lead to specialized products needed by the cell. Of particular importance in some tissues is the oxidation of glucose 6-phosphate to pentose phosphates by the **pentose phosphate pathway** (also called the **phosphogluconate pathway** or the **hexose monophosphate pathway;** Fig. 14–20). In this oxidative pathway, NADP$^+$ is the electron acceptor, yielding NADPH. Rapidly dividing cells, such as those of bone marrow, skin, and intestinal mucosa, use the pentoses to make RNA, DNA, and such coenzymes as ATP, NADH, FADH$_2$, and coenzyme A.

In other tissues, the essential product of the pentose phosphate pathway is not the pentoses but the electron donor NADPH, needed for reductive biosynthesis or to counter the damaging effects of oxygen radicals. Tissues that carry out extensive fatty acid synthesis (liver, adipose, lactating mammary gland) or very active synthesis of cholesterol and steroid hormones (liver, adrenal gland, gonads) require the NADPH provided by the pathway. Erythrocytes and the cells of the lens and cornea are directly exposed to oxygen and thus to the damaging free radicals generated by oxygen.

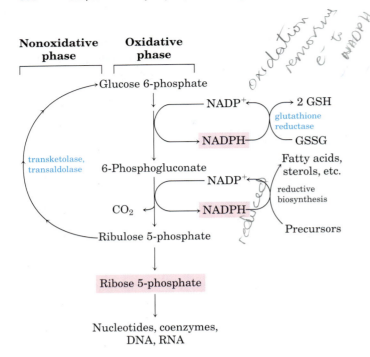

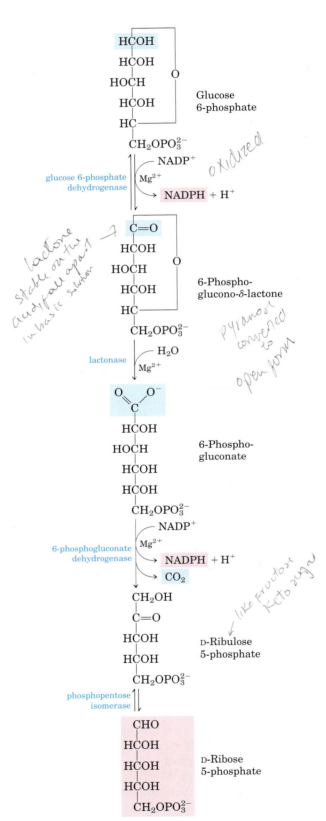

**FIGURE 14–20 General scheme of the pentose phosphate pathway.** NADPH formed in the oxidative phase is used to reduce glutathione, GSSG (see Box 14–3) and to support reductive biosynthesis. The other product of the oxidative phase is ribose 5-phosphate, which serves as precursor for nucleotides, coenzymes, and nucleic acids. In cells that are not using ribose 5-phosphate for biosynthesis, the nonoxidative phase recycles six molecules of the pentose into five molecules of the hexose glucose 6-phosphate, allowing continued production of NADPH and converting glucose 6-phosphate (in six cycles) to $CO_2$.

By maintaining a reducing atmosphere (a high ratio of NADPH to $NADP^+$ and a high ratio of reduced to oxidized glutathione), they can prevent or undo oxidative damage to proteins, lipids, and other sensitive molecules. In erythrocytes, the NADPH produced by the pentose phosphate pathway is so important in preventing oxidative damage that a genetic defect in glucose 6-phosphate dehydrogenase, the first enzyme of the pathway, can have serious medical consequences (Box 14–3). ■

## The Oxidative Phase Produces Pentose Phosphates and NADPH

The first reaction of the pentose phosphate pathway (Fig. 14–21) is the oxidation of glucose 6-phosphate by **glucose 6-phosphate dehydrogenase (G6PD)** to form 6-phosphoglucono-δ-lactone, an intramolecular ester. $NADP^+$ is the electron acceptor, and the overall equilibrium lies far in the direction of NADPH formation. The lactone is hydrolyzed to the free acid 6-phosphogluconate by a specific **lactonase,** then 6-phosphogluconate undergoes oxidation and decarboxylation by **6-phosphogluconate dehydrogenase** to form the ketopentose ribulose 5-phosphate. This reaction generates

**FIGURE 14–21 Oxidative reactions of the pentose phosphate pathway.** The end products are ribose 5-phosphate, $CO_2$, and NADPH.

## Why Pythagoras Wouldn't Eat Falafel: Glucose 6-Phosphate Dehydrogenase Deficiency

Fava beans, an ingredient of falafel, have been an important food source in the Mediterranean and Middle East since antiquity. The Greek philosopher and mathematician Pythagoras prohibited his followers from dining on fava beans, perhaps because they make many people sick with a condition called favism, which can be fatal. In favism, erythrocytes begin to lyse 24 to 48 hours after ingestion of the beans, releasing free hemoglobin into the blood. Jaundice and sometimes kidney failure can result. Similar symptoms can occur with ingestion of the antimalarial drug primaquine or of sulfa antibiotics or following exposure to certain herbicides. These symptoms have a genetic basis: glucose 6-phosphate dehydrogenase (G6PD) deficiency, which affects about 400 million people. Most G6PD-deficient individuals are asymptomatic; only the combination of G6PD deficiency and certain environmental factors produces the clinical manifestations.

G6PD catalyzes the first step in the pentose phosphate pathway (see Fig. 14–21), which produces NADPH. This reductant, essential in many biosynthetic pathways, also protects cells from oxidative damage by hydrogen peroxide ($H_2O_2$) and superoxide free radicals, highly reactive oxidants generated as metabolic byproducts and through the actions of drugs such as primaquine and natural products such as divicine—the toxic ingredient of fava beans. During normal detoxification, $H_2O_2$ is converted to $H_2O$ by reduced glutathione and glutathione peroxidase, and the oxidized glutathione is converted back to the reduced form by glutathione reductase and NADPH (Fig. 1). $H_2O_2$ is also broken down to $H_2O$ and $O_2$ by catalase, which also requires NADPH. In G6PD-deficient individuals, the NADPH production is diminished and detoxification of $H_2O_2$ is inhibited. Cellular damage results: lipid peroxidation leading to breakdown of erythrocyte membranes and oxidation of proteins and DNA.

The geographic distribution of G6PD deficiency is instructive. Frequencies as high as 25% occur in tropical Africa, parts of the Middle East, and Southeast Asia, areas where malaria is most prevalent. In addition to such epidemiological observations, in vitro studies show that growth of one malaria parasite, *Plasmodium falciparum,* is inhibited in G6PD-deficient erythrocytes. The parasite is very sensitive to oxidative damage and is killed by a level of oxidative stress that is tolerable to a G6PD-deficient human host. Because the advantage of resistance to malaria balances the disadvantage of lowered resistance to oxidative damage, natural selection sustains the G6PD-deficient genotype in human populations where malaria is prevalent. Only under overwhelming oxidative stress, caused by drugs, herbicides, or divicine, does G6PD deficiency cause serious medical problems.

An antimalarial drug such as primaquine is believed to act by causing oxidative stress to the parasite. It is ironic that antimalarial drugs can cause illness through the same biochemical mechanism that provides resistance to malaria. Divicine also acts as an antimalarial drug, and ingestion of fava beans may protect against malaria. By refusing to eat falafel, many Pythagoreans with normal G6PD activity may have unwittingly increased their risk of malaria!

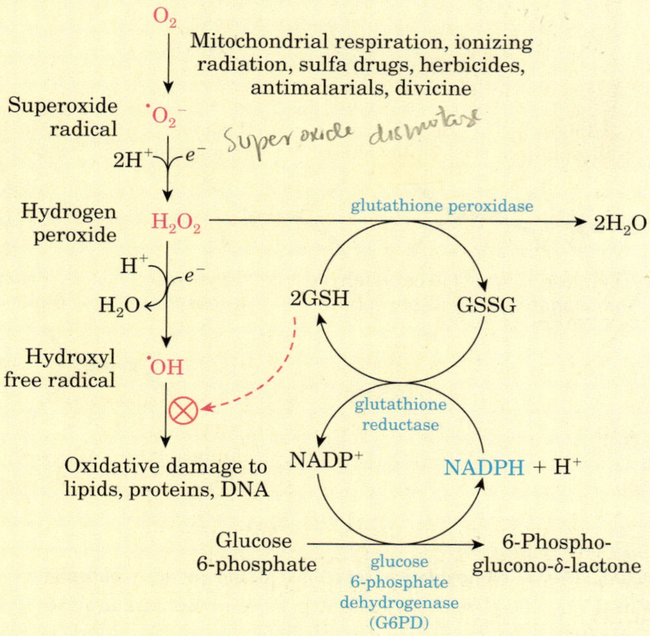

**FIGURE 1**  Role of NADPH and glutathione in protecting cells against highly reactive oxygen derivatives. Reduced glutathione (GSH) protects the cell by destroying hydrogen peroxide and hydroxyl free radicals. Regeneration of GSH from its oxidized form (GSSG) requires the NADPH produced in the glucose 6-phosphate dehydrogenase reaction.

a second molecule of NADPH. **Phosphopentose isomerase** converts ribulose 5-phosphate to its aldose isomer, ribose 5-phosphate. In some tissues, the pentose phosphate pathway ends at this point, and its overall equation is

$$\text{Glucose 6-phosphate} + 2NADP^+ + H_2O \longrightarrow$$
$$\text{ribose 5-phosphate} + CO_2 + 2NADPH + 2H^+$$

The net result is the production of NADPH, a reductant for biosynthetic reactions, and ribose 5-phosphate, a precursor for nucleotide synthesis.

## The Nonoxidative Phase Recycles Pentose Phosphates to Glucose 6-Phosphate

In tissues that require primarily NADPH, the pentose phosphates produced in the oxidative phase of the pathway are recycled into glucose 6-phosphate. In this nonoxidative phase, ribulose 5-phosphate is first epimerized to xylulose 5-phosphate:

$$
\begin{array}{ccc}
\text{CH}_2\text{OH} & & \text{CH}_2\text{OH} \\
| & & | \\
\text{C}=\text{O} & & \text{C}=\text{O} \\
| & & | \\
\text{H}-\text{C}-\text{OH} & \underset{\substack{\text{ribose} \\ \text{5-phosphate} \\ \text{epimerase}}}{\rightleftharpoons} & \text{HO}-\text{C}-\text{H} \\
| & & | \\
\text{H}-\text{C}-\text{OH} & & \text{H}-\text{C}-\text{OH} \\
| & & | \\
\text{CH}_2\text{OPO}_3^{2-} & & \text{CH}_2\text{OPO}_3^{2-} \\
\text{Ribulose} & & \text{Xylulose 5-phosphate} \\
\text{5-phosphate} & &
\end{array}
$$

Then, in a series of rearrangements of the carbon skeletons (Fig. 14–22), six five-carbon sugar phosphates are converted to five six-carbon sugar phosphates, completing the cycle and allowing continued oxidation of glucose 6-phosphate with production of NADPH. Continued recycling leads ultimately to the conversion of glucose 6-phosphate to six $CO_2$. Two enzymes unique to the pentose phosphate pathway act in these interconversions of sugars: transketolase and transaldolase. **Transketolase** catalyzes the transfer of a two-carbon fragment from a ketose donor to an aldose acceptor (Fig. 14–23a). In its first appearance in the pentose phosphate pathway, transketolase transfers C-1 and C-2 of xylulose 5-phosphate to ribose 5-phosphate, forming the seven-carbon product sedoheptulose 7-phosphate (Fig. 14–23b). The remaining three-carbon fragment from xylulose is glyceraldehyde 3-phosphate.

Next, **transaldolase** catalyzes a reaction similar to the aldolase reaction of glycolysis: a three-carbon fragment is removed from sedoheptulose 7-phosphate and condensed with glyceraldehyde 3-phosphate, forming fructose 6-phosphate and the tetrose erythrose 4-phosphate (Fig. 14–24). Now transketolase acts again, forming fructose 6-phosphate and glyceraldehyde 3-phosphate from erythrose 4-phosphate and xylulose 5-phosphate (Fig. 14–25). Two molecules of glyceraldehyde 3-phosphate formed by two iterations of these reactions can be converted to a molecule of fructose 1,6-bisphosphate as in gluconeogenesis (Fig. 14–16), and finally FBPase-1 and phosphohexose isomerase convert fructose 1,6-bisphosphate to glucose 6-phosphate. The cycle is complete: six pentose phosphates have been converted to five hexose phosphates (Fig. 14–22b).

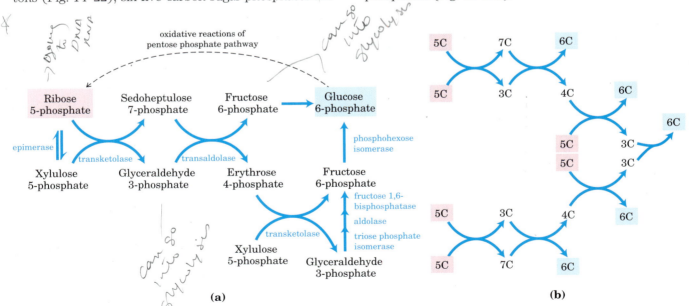

**FIGURE 14–22 Nonoxidative reactions of the pentose phosphate pathway. (a)** These reactions convert pentose phosphates to hexose phosphates, allowing the oxidative reactions (see Fig. 14–21) to continue. The enzymes transketolase and transaldolase are specific to this pathway; the other enzymes also serve in the glycolytic or gluconeogenic pathways. **(b)** A schematic diagram showing the pathway from six pentoses (5C) to five hexoses (6C). Note that this involves two sets of the interconversions shown in **(a)**. Every reaction shown here is reversible; unidirectional arrows are used only to make clear the direction of the reactions during continuous oxidation of glucose 6-phosphate. In the light-independent reactions of photosynthesis, the direction of these reactions is reversed (see Fig. 20–10).

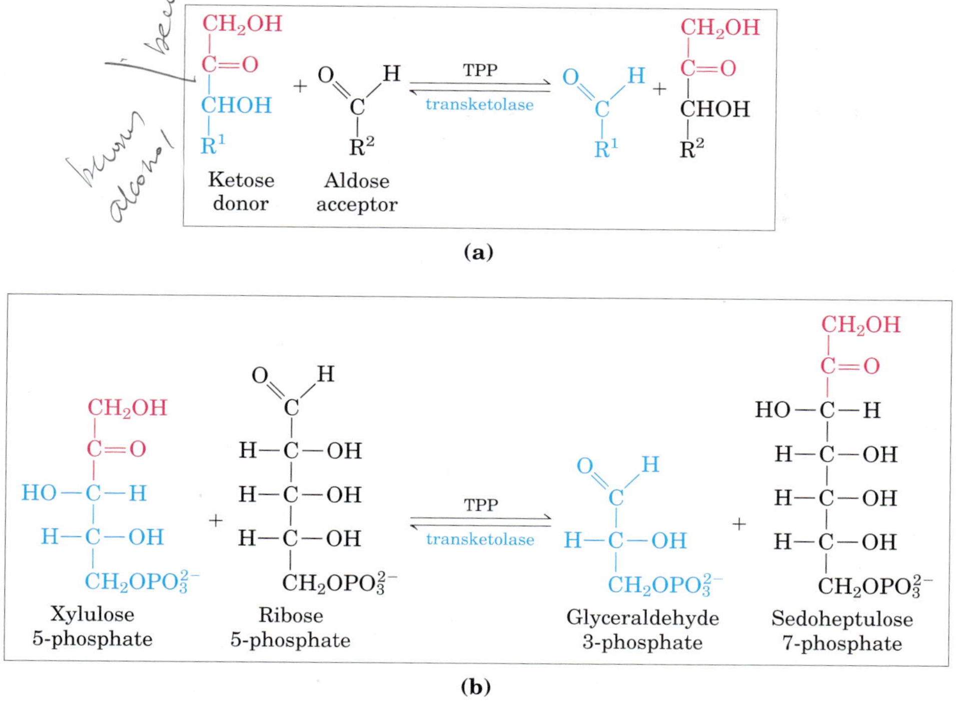

**(a)**

**(b)**

FIGURE 14–23 **The first reaction catalyzed by transketolase. (a)** The general reaction catalyzed by transketolase is the transfer of a two-carbon group, carried temporarily on enzyme-bound TPP, from a ketose donor to an aldose acceptor. **(b)** Conversion of two pentose phosphates to a triose phosphate and a seven-carbon sugar phosphate, sedoheptulose 7-phosphate.

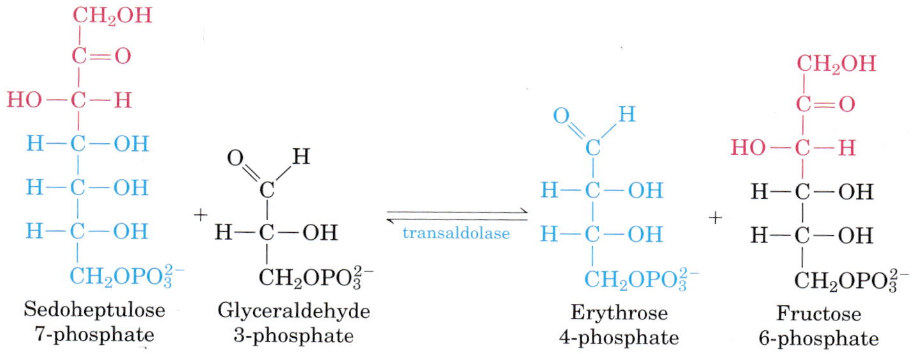

FIGURE 14–24 **The reaction catalyzed by transaldolase.**

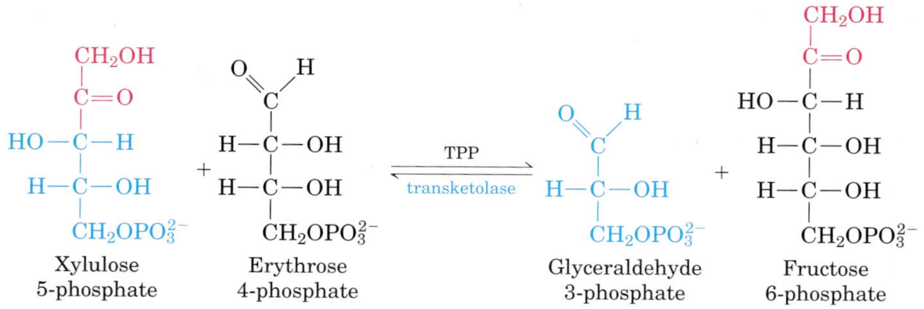

FIGURE 14–25 **The second reaction catalyzed by transketolase.**

Transketolase requires the cofactor thiamine pyrophosphate (TPP), which stabilizes a two-carbon carbanion in this reaction (Fig. 14–26a), just as it does in the pyruvate decarboxylase reaction (Fig. 14–13). Transaldolase uses a Lys side chain to form a Schiff base with the carbonyl group of its substrate, a ketose, thereby stabilizing a carbanion (Fig. 14–26b) that is central to the reaction mechanism.

The process described in Figure 14–21 is known as the **oxidative pentose phosphate pathway.** The first two steps are oxidations with large, negative standard free-energy changes and are essentially irreversible in

**(a) Transketolase**

TPP

**(b) Transaldolase**

Protonated
Schiff base

**FIGURE 14–26 Carbanion intermediates stabilized by covalent interactions with transketolase and transaldolase. (a)** The ring of TPP stabilizes the two-carbon carbanion carried by transketolase; see Fig. 14–13 for the chemistry of TPP action. **(b)** In the transaldolase reaction, the protonated Schiff base formed between the $\varepsilon$-amino group of a Lys side chain and the substrate stabilizes a three-carbon carbanion.

the cell. The reactions of the nonoxidative part of the pentose phosphate pathway (Fig. 14–22) are readily reversible and thus also provide a means of converting hexose phosphates to pentose phosphates. As we shall see in Chapter 20, a process that converts hexose phosphates to pentose phosphates is crucial to the photosynthetic assimilation of $CO_2$ by plants. That pathway, the **reductive pentose phosphate pathway,** is essentially the reversal of the reactions shown in Figure 14–22 and employs many of the same enzymes.

All the enzymes in the pentose phosphate pathway are located in the cytosol, like those of glycolysis and most of those of gluconeogenesis. In fact, these three pathways are connected through several shared intermediates and enzymes. The glyceraldehyde 3-phosphate formed by the action of transketolase is readily converted to dihydroxyacetone phosphate by the glycolytic enzyme triose phosphate isomerase, and these two trioses can be joined by the aldolase as in gluconeogenesis, forming fructose 1,6-bisphosphate. Alternatively, the triose phosphates can be oxidized to pyruvate by the glycolytic reactions. The fate of the trioses is determined by the cell's relative needs for pentose phosphates, NADPH, and ATP.

## Wernicke-Korsakoff Syndrome Is Exacerbated by a Defect in Transketolase

 In humans with Wernicke-Korsakoff syndrome, a mutation in the gene for transketolase results in

an enzyme having an affinity for its coenzyme TPP that is one-tenth that of the normal enzyme. Although moderate deficiencies in the vitamin thiamine have little effect on individuals with an unmutated transketolase gene, in those with the altered gene, thiamine deficiency drops the level of TPP below that needed to saturate the enzyme. The lowering of transketolase activity slows the whole pentose phosphate pathway, and the result is the Wernicke-Korsakoff syndrome: severe memory loss, mental confusion, and partial paralysis. The syndrome is more common among alcoholics than in the general population; chronic alcohol consumption interferes with the intestinal absorption of some vitamins, including thiamine. ■

## Glucose 6-Phosphate Is Partitioned between Glycolysis and the Pentose Phosphate Pathway

Whether glucose 6-phosphate enters glycolysis or the pentose phosphate pathway depends on the current needs of the cell and on the concentration of $NADP^+$ in the cytosol. Without this electron acceptor, the first reaction of the pentose phosphate pathway (catalyzed by G6PD) cannot proceed. When a cell is rapidly converting NADPH to $NADP^+$ in biosynthetic reductions, the level of $NADP^+$ rises, allosterically stimulating G6PD and thereby increasing the flux of glucose 6-phosphate through the pentose phosphate pathway (Fig. 14–27). When the demand for NADPH slows, the level of $NADP^+$ drops, the pentose phosphate pathway slows, and glucose 6-phosphate is instead used to fuel glycolysis.

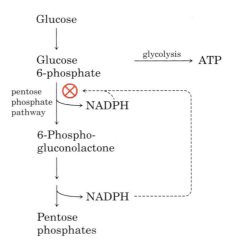

**FIGURE 14–27 Role of NADPH in regulating the partitioning of glucose 6-phosphate between glycolysis and the pentose phosphate pathway.** When NADPH is forming faster than it is being used for biosynthesis and glutathione reduction (see Fig. 14–20), [NADPH] rises and inhibits the first enzyme in the pentose phosphate pathway. As a result, more glucose 6-phosphate is available for glycolysis.

## *SUMMARY 14.5* Pentose Phosphate Pathway of Glucose Oxidation

- The *oxidative* pentose phosphate pathway (phosphogluconate pathway, or hexose monophosphate pathway) brings about oxidation and decarboxylation at C-1 of glucose 6-phosphate, reducing $NADP^+$ to NADPH and producing pentose phosphates.

- NADPH provides reducing power for biosynthetic reactions, and ribose 5-phosphate is a precursor for nucleotide and nucleic acid synthesis. Rapidly growing tissues and tissues carrying out active biosynthesis of fatty acids, cholesterol, or steroid hormones send more glucose 6-phosphate through the pentose phosphate pathway than do tissues with less demand for pentose phosphates and reducing power.

- The first phase of the pentose phosphate pathway consists of two oxidations that convert glucose 6-phosphate to ribulose 5-phosphate and reduce $NADP^+$ to NADPH. The second phase comprises nonoxidative steps that convert pentose phosphates to glucose 6-phosphate, which begins the cycle again.

- In the second phase, transaldolase (with TPP as cofactor) and transketolase catalyze the interconversion of three-, four-, five-, six-, and seven-carbon sugars, with the reversible conversion of six pentose phosphates to five hexose phosphates. In the carbon-assimilating reactions of photosynthesis, the same enzymes catalyze the reverse process, called the *reductive* pentose phosphate pathway: conversion of five hexose phosphates to six pentose phosphates.

- A genetic defect in transketolase that lowers its affinity for TPP exacerbates the Wernicke-Korsakoff syndrome.

- Entry of glucose 6-phosphate either into glycolysis or into the pentose phosphate pathway is largely determined by the relative concentrations of $NADP^+$ and NADPH.

## Key Terms

*Terms in bold are defined in the glossary.*

| | | | |
|---|---|---|---|
| **glycolysis**   522 | **acyl phosphate**   530 | **isomerases**   534 | **pentose phosphate** |
| **fermentation**   522 | **substrate-level phos-** | lactose intolerance | **pathway**   549 |
| lactic acid fermen- | **phorylation**   531 | galactosemia   537 | **phosphogluconate** |
| tation   523 | **respiration-linked phos-** | **thiamine pyrophos-** | **pathway**   549 |
| **hypoxia**   523 | **phorylation**   531 | **phate (TPP)**   540 | **hexose monophosphate** |
| **ethanol (alcohol)** | phosphoenolpyruvate | **gluconeogenesis**   543 | **pathway**   549 |
| **fermentation**   523 | (PEP)   532 | biotin   544 | |
| **isozymes**   526 | **mutases**   534 | | |

## Further Reading

### General

**Fruton, J.S.** (1999) *Proteins, Genes, and Enzymes: The Interplay of Chemistry and Biology,* Yale University Press, New Haven.
  This text includes a detailed historical account of research on glycolysis.

### Glycolysis

**Boiteux, A. & Hess, B.** (1981) Design of glycolysis. *Philos. Trans. R. Soc. Lond. Ser. B Biol. Sci.* **293,** 5–22.
  Intermediate-level review of the pathway and the classic view of its control.

**Dandekar, T., Schuster, S., Snel, B., Huynen, M., & Bork, P.** (1999) Pathway alignment: application to the comparative analysis of glycolytic enzymes. *Biochem. J.* **343,** 115–124.
  Intermediate-level review of the bioinformatic view of the evolution of glycolysis.

**Dang, C.V. & Semenza, G.L.** (1999) Oncogenic alterations of metabolism. *Trends Biochem. Sci.* **24,** 68–72.
  Brief review of the molecular basis for increased glycolysis in tumors.

**Erlandsen, H., Abola, E.E., & Stevens, R.C.** (2000) Combining structural genomics and enzymology: completing the picture in metabolic pathways and enzyme active sites. *Curr. Opin. Struct. Biol.* **10,** 719–730.
  Intermediate-level review of the structures of the glycolytic enzymes.

**Hardie, D.G.** (2000) Metabolic control: a new solution to an old problem. *Curr. Biol.* **10,** R757–R759.

**Harris, A.L.** (2002) Hypoxia—a key regulatory factor in tumour growth. *Nat. Rev. Cancer* **2,** 38–47.

Heinrich, R., Melendez-Hevia, E., Montero, F., Nuno, J.C., Stephani, A., & Waddell, T.D. (1999) The structural design of glycolysis: an evolutionary approach. *Biochem. Soc. Trans.* **27**, 294–298.

Knowles, J. & Albery, W.J. (1977) Perfection in enzyme catalysis: the energetics of triose phosphate isomerase. *Acc. Chem. Res.* **10**, 105–111.

Phillips, D., Blake, C.C.F., & Watson, H.C. (eds) (1981) The Enzymes of Glycolysis: Structure, Activity and Evolution. *Philos. Trans. R. Soc. Lond. Ser. B Biol. Sci.* **293**, 1–214.

A collection of excellent reviews on the enzymes of glycolysis, written at a level challenging but comprehensible to a beginning student of biochemistry.

Plaxton, W.C. (1996) The organization and regulation of plant glycolysis. *Annu. Rev. Plant Physiol. Plant Mol. Biol.* **47**, 185–214.

Very helpful review of the subcellular localization of glycolytic enzymes and the regulation of glycolysis in plants.

Rose, I. (1981) Chemistry of proton abstraction by glycolytic enzymes (aldolase, isomerases, and pyruvate kinase). *Philos. Trans. R. Soc. Lond. Ser. B Biol. Sci.* **293**, 131–144.

Intermediate-level review of the mechanisms of these enzymes.

Shirmer, T. & Evans, P.R. (1990) Structural basis for the allosteric behavior of phosphofructokinase. *Nature* **343**, 140–145.

Smith, T.A. (2000) Mammalian hexokinases and their abnormal expression in cancer. *Br. J. Biomed. Sci.* **57**, 170–178.

A review of the four hexokinase isozymes of mammals: their properties and tissue distributions and their expression during the development of tumors.

### Feeder Pathways for Glycolysis

Elsas, L.J. & Lai, K. (1998) The molecular biology of galactosemia. *Genet. Med.* **1**, 40–48.

Novelli, G. & Reichardt, J.K. (2000) Molecular basis of disorders of human galactose metabolism: past, present, and future. *Mol. Genet. Metab.* **71**, 62–65.

Petry, K.G. & Reichardt, J.K. (1998) The fundamental importance of human galactose metabolism: lessons from genetics and biochemistry. *Trends Genet.* **14**, 98–102.

Van Beers, E.H., Buller, H.A., Grand, R.J., Einerhand, A.W.C., & Dekker, J. (1995) Intestinal brush border glycohydrolases: structure, function, and development. *Crit. Rev. Biochem. Mol. Biol.* **30**, 197–262.

### Fermentations

Behal, R.H., Buxton, D.B., Robertson, J.G., & Olson, M.S. (1993) Regulation of the pyruvate dehydrogenase multienzyme complex. *Annu. Rev. Nutr.* **13**, 497–520.

Patel, M.S., Naik, S., Wexler, I.D., & Kerr, D.S. (1995) Gene regulation and genetic defects in the pyruvate dehydrogenase complex. *J. Nutr.* **125**, 1753S–1757S.

Patel, M.S. & Roche, T.E. (1990) Molecular biology and biochemistry of pyruvate dehydrogenase complexes. *FASEB J.* **4**, 3224–3233.

Robinson, B.H., MacKay, N., Chun, K., & Ling, M. (1996) Disorders of pyruvate carboxylase and the pyruvate dehydrogenase complex. *J. Inherit. Metab. Dis.* **19**, 452–462.

### Gluconeogenesis

Gerich, J.E., Meyer, C., Woerle, H.J., & Stumvoll, M. (2001) Renal gluconeogenesis: its importance in human glucose homeostasis. *Diabetes Care* **24**, 382–391.

Intermediate-level review of the contribution of kidney tissue to gluconeogenesis.

Gleeson, T. (1996) Post-exercise lactate metabolism: a comparative review of sites, pathways, and regulation. *Annu. Rev. Physiol.* **58**, 565–581.

Hers, H.G. & Hue, L. (1983) Gluconeogenesis and related aspects of glycolysis. *Annu. Rev. Biochem.* **52**, 617–653.

Matte, A., Tari, L.W., Goldie, H., & Delbaere, L.T.J. (1997) Structure and mechanism of phosphoenolpyruvate carboxykinase. *J. Biol. Chem.* **272**, 8105–8108.

### Oxidative Pentose Phosphate Pathway

Chayen, J., Howat, D.W., & Bitensky, L. (1986) Cellular biochemistry of glucose 6-phosphate and 6-phosphogluconate dehydrogenase activities. *Cell Biochem. Funct.* **4**, 249–253.

Horecker, B.L. (1976) Unraveling the pentose phosphate pathway. In *Reflections on Biochemistry* (Kornberg, A., Cornudella, L., Horecker, B.L., & Oro, J., eds), pp. 65–72, Pergamon Press, Inc., Oxford.

Kletzien, R.F., Harris, P.K., & Foellmi, L.A. (1994) Glucose 6-phosphate dehydrogenase: a "housekeeping" enzyme subject to tissue-specific regulation by hormones, nutrients, and oxidant stress. *FASEB J.* **8**, 174–181.

An intermediate-level review.

Luzzato, L., Mehta, A., & Vulliamy, T. (2001) Glucose 6-phosphate dehydrogenase deficiency. In *The Metabolic and Molecular Bases of Inherited Disease*, 8th edn (Scriver, C.R., Sly, W.S., Childs, B., Beaudet, A.L., Valle, D., Kinzler, K.W., & Vogelstein, B., eds), pp. 4517–4553, McGraw-Hill Inc., New York.

The four-volume treatise in which this article appears is filled with fascinating information about the clinical and biochemical features of hundreds of inherited diseases of metabolism.

Martini, G. & Ursini, M.V. (1996) A new lease on life for an old enzyme. *BioEssays* **18**, 631–637.

An intermediate-level review of glucose 6-phosphate dehydrogenase, the effects of mutations in this enzyme in humans, and the effects of knock-out mutations in mice.

Notaro, R., Afolayan, A., & Luzzatto, L. (2000) Human mutations in glucose 6-phosphate dehydrogenase reflect evolutionary history. *FASEB J.* **14**, 485–494.

Wood, T. (1985) *The Pentose Phosphate Pathway*, Academic Press, Inc., Orlando, FL.

Wood, T. (1986) Physiological functions of the pentose phosphate pathway. *Cell Biochem. Funct.* **4**, 241–247.

## Problems

**1. Equation for the Preparatory Phase of Glycolysis**
Write balanced biochemical equations for all the reactions in the catabolism of glucose to two molecules of glyceraldehyde 3-phosphate (the preparatory phase of glycolysis), including the standard free-energy change for each reaction. Then write the overall or net equation for the preparatory phase of glycolysis, with the net standard free-energy change.

**2. The Payoff Phase of Glycolysis in Skeletal Muscle**
In working skeletal muscle under anaerobic conditions, glyceraldehyde 3-phosphate is converted to pyruvate (the payoff phase of glycolysis), and the pyruvate is reduced to lactate. Write balanced biochemical equations for all the reactions in this process, with the standard free-energy change for each reaction. Then write the overall or net equation for the payoff phase of glycolysis (with lactate as the end product), including the net standard free-energy change.

**3. Pathway of Atoms in Fermentation**   A "pulse-chase" experiment using $^{14}C$-labeled carbon sources is carried out on a yeast extract maintained under strictly anaerobic conditions to produce ethanol. The experiment consists of incubating a small amount of $^{14}C$-labeled substrate (the pulse) with the yeast extract just long enough for each intermediate in the fermentation pathway to become labeled. The label is then "chased" through the pathway by the addition of excess unlabeled glucose. The chase effectively prevents any further entry of labeled glucose into the pathway.

(a) If $[1\text{-}^{14}C]$glucose (glucose labeled at C-1 with $^{14}C$) is used as a substrate, what is the location of $^{14}C$ in the product ethanol? Explain.

(b) Where would $^{14}C$ have to be located in the starting glucose to ensure that all the $^{14}C$ activity is liberated as $^{14}CO_2$ during fermentation to ethanol? Explain.

**4. Fermentation to Produce Soy Sauce**   Soy sauce is prepared by fermenting a salted mixture of soybeans and wheat with several microorganisms, including yeast, over a period of 8 to 12 months. The resulting sauce (after solids are removed) is rich in lactate and ethanol. How are these two compounds produced? To prevent the soy sauce from having a strong vinegar taste (vinegar is dilute acetic acid), oxygen must be kept out of the fermentation tank. Why?

**5. Equivalence of Triose Phosphates**   $^{14}C$-Labeled glyceraldehyde 3-phosphate was added to a yeast extract. After a short time, fructose 1,6-bisphosphate labeled with $^{14}C$ at C-3 and C-4 was isolated. What was the location of the $^{14}C$ label in the starting glyceraldehyde 3-phosphate? Where did the second $^{14}C$ label in fructose 1,6-bisphosphate come from? Explain.

**6. Glycolysis Shortcut**   Suppose you discovered a mutant yeast whose glycolytic pathway was shorter because of the presence of a new enzyme catalyzing the reaction:

$$\text{Glyceraldehyde 3-phosphate} + H_2 \xrightarrow{\quad NAD^+ \quad NADH + H^+ \quad} \text{3-phosphoglycerate}$$

Would shortening the glycolytic pathway in this way benefit the cell? Explain.

**7. Role of Lactate Dehydrogenase**   During strenuous activity, the demand for ATP in muscle tissue is vastly increased. In rabbit leg muscle or turkey flight muscle, the ATP is produced almost exclusively by lactic acid fermentation. ATP is formed in the payoff phase of glycolysis by two reactions, promoted by phosphoglycerate kinase and pyruvate kinase. Suppose skeletal muscle were devoid of lactate dehydrogenase. Could it carry out strenuous physical activity; that is, could it generate ATP at a high rate by glycolysis? Explain.

**8. Efficiency of ATP Production in Muscle**   The transformation of glucose to lactate in myocytes releases only about 7% of the free energy released when glucose is completely oxidized to $CO_2$ and $H_2O$. Does this mean that anaerobic glycolysis in muscle is a wasteful use of glucose? Explain.

**9. Free-Energy Change for Triose Phosphate Oxidation**   The oxidation of glyceraldehyde 3-phosphate to 1,3-bisphosphoglycerate, catalyzed by glyceraldehyde 3-phosphate dehydrogenase, proceeds with an unfavorable equilibrium constant ($K'_{eq} = 0.08$; $\Delta G'^\circ = 6.3$ kJ/mol), yet the flow through this point in the glycolytic pathway proceeds smoothly. How does the cell overcome the unfavorable equilibrium?

**10. Arsenate Poisoning**   Arsenate is structurally and chemically similar to inorganic phosphate ($P_i$), and many enzymes that require phosphate will also use arsenate. Organic compounds of arsenate are less stable than analogous phosphate compounds, however. For example, acyl *arsenates* decompose rapidly by hydrolysis:

$$R-\overset{\displaystyle O}{\overset{\displaystyle \|}{C}}-O-\overset{\displaystyle O}{\underset{\displaystyle O^-}{\overset{\displaystyle \|}{As}}}-O^- + H_2O \longrightarrow$$

$$R-\overset{\displaystyle O}{\overset{\displaystyle \|}{C}}-O^- + HO-\overset{\displaystyle O}{\underset{\displaystyle O^-}{\overset{\displaystyle \|}{As}}}-O^- + H^+$$

On the other hand, acyl *phosphates*, such as 1,3-bisphosphoglycerate, are more stable and undergo further enzyme-catalyzed transformation in cells.

(a) Predict the effect on the net reaction catalyzed by glyceraldehyde 3-phosphate dehydrogenase if phosphate were replaced by arsenate.

(b) What would be the consequence to an organism if arsenate were substituted for phosphate? Arsenate is very toxic to most organisms. Explain why.

**11. Requirement for Phosphate in Ethanol Fermentation**   In 1906 Harden and Young, in a series of classic studies on the fermentation of glucose to ethanol and $CO_2$ by extracts of brewer's yeast, made the following observations. (1) Inorganic phosphate was essential to fermentation; when the supply of phosphate was exhausted, fermentation ceased before all the glucose was used. (2) During fermentation under these conditions, ethanol, $CO_2$, and a hexose bisphosphate accumulated. (3) When arsenate was substituted for phosphate, no hexose bisphosphate accumulated, but the fermentation proceeded until all the glucose was converted to ethanol and $CO_2$.

(a) Why did fermentation cease when the supply of phosphate was exhausted?

(b) Why did ethanol and $CO_2$ accumulate? Was the conversion of pyruvate to ethanol and $CO_2$ essential? Why? Identify the hexose bisphosphate that accumulated. Why did it accumulate?

(c) Why did the substitution of arsenate for phosphate prevent the accumulation of the hexose bisphosphate yet allow fermentation to ethanol and $CO_2$ to go to completion? (See Problem 10.)

**12. Role of the Vitamin Niacin**    Adults engaged in strenuous physical activity require an intake of about 160 g of carbohydrate daily but only about 20 mg of niacin for optimal nutrition. Given the role of niacin in glycolysis, how do you explain the observation?

**13. Metabolism of Glycerol**    Glycerol obtained from the breakdown of fat is metabolized by conversion to dihydroxyacetone phosphate, a glycolytic intermediate, in two enzyme-catalyzed reactions. Propose a reaction sequence for glycerol metabolism. On which known enzyme-catalyzed reactions is your proposal based? Write the net equation for the conversion of glycerol to pyruvate according to your scheme.

$$HOCH_2-\overset{\overset{\displaystyle OH}{|}}{\underset{\underset{\displaystyle H}{|}}{C}}-CH_2OH$$

Glycerol

**14. Severity of Clinical Symptoms Due to Enzyme Deficiency**    The clinical symptoms of two forms of galactosemia—deficiency of galactokinase or of UDP-glucose:galactose 1-phosphate uridylyltransferase—show radically different severity. Although both types produce gastric discomfort after milk ingestion, deficiency of the transferase also leads to liver, kidney, spleen, and brain dysfunction and eventual death. What products accumulate in the blood and tissues with each type of enzyme deficiency? Estimate the relative toxicities of these products from the above information.

**15. Muscle Wasting in Starvation**    One consequence of starvation is a reduction in muscle mass. What happens to the muscle proteins?

**16. Pathway of Atoms in Gluconeogenesis**    A liver extract capable of carrying out all the normal metabolic reactions of the liver is briefly incubated in separate experiments with the following $^{14}C$-labeled precursors:

(a) [$^{14}C$]Bicarbonate, $HO-^{14}C\overset{\displaystyle O^-}{\underset{\displaystyle O}{\diagdown}}$

(b) [1-$^{14}C$]Pyruvate, $CH_3-\overset{\overset{\displaystyle}{}}{\underset{\underset{\displaystyle O}{||}}{C}}-^{14}COO^-$

Trace the pathway of each precursor through gluconeogenesis. Indicate the location of $^{14}C$ in all intermediates and in the product, glucose.

**17. Pathway of $CO_2$ in Gluconeogenesis**    In the first bypass step of gluconeogenesis, the conversion of pyruvate to phosphoenolpyruvate, pyruvate is carboxylated by pyruvate carboxylase to oxaloacetate, which is subsequently decarboxylated by PEP carboxykinase to yield phosphoenolpyruvate. The observation that the addition of $CO_2$ is directly followed by the loss of $CO_2$ suggests that $^{14}C$ of $^{14}CO_2$ would not be incorporated into PEP, glucose, or any intermediates in gluconeogenesis. However, when a rat liver preparation synthesizes glucose in the presence of $^{14}CO_2$, $^{14}C$ slowly appears in PEP and eventually at C-3 and C-4 of glucose. How does the $^{14}C$ label get into PEP and glucose? (Hint: During gluconeogenesis in the presence of $^{14}CO_2$, several of the four-carbon citric acid cycle intermediates also become labeled.)

**18. Energy Cost of a Cycle of Glycolysis and Gluconeogenesis**    What is the cost (in ATP equivalents) of transforming glucose to pyruvate via glycolysis and back again to glucose via gluconeogenesis?

**19. Glucogenic Substrates**    A common procedure for determining the effectiveness of compounds as precursors of glucose in mammals is to starve the animal until the liver glycogen stores are depleted and then administer the compound in question. A substrate that leads to a *net* increase in liver glycogen is termed glucogenic, because it must first be converted to glucose 6-phosphate. Show by means of known enzymatic reactions which of the following substances are glucogenic:

(a) Succinate, $^-OOC-CH_2-CH_2-COO^-$

(b) Glycerol, $\overset{\overset{\displaystyle OH\quad OH\,OH}{|\qquad |\ \ |}}{\underset{\underset{\displaystyle H}{|}}{CH_2-C-CH_2}}$

(c) Acetyl-CoA, $CH_3-\overset{\overset{\displaystyle O}{||}}{C}-S\text{-}CoA$

(d) Pyruvate, $CH_3-\overset{\overset{\displaystyle O}{||}}{C}-COO^-$

(e) Butyrate, $CH_3-CH_2-CH_2-COO^-$

**20. Ethanol Affects Blood Glucose Levels**    The consumption of alcohol (ethanol), especially after periods of strenuous activity or after not eating for several hours, results in a deficiency of glucose in the blood, a condition known as hypoglycemia. The first step in the metabolism of ethanol by the liver is oxidation to acetaldehyde, catalyzed by liver alcohol dehydrogenase:

$$CH_3CH_2OH + NAD^+ \longrightarrow CH_3CHO + NADH + H^+$$

Explain how this reaction inhibits the transformation of lactate to pyruvate. Why does this lead to hypoglycemia?

## 21. Blood Lactate Levels during Vigorous Exercise

The concentrations of lactate in blood plasma before, during, and after a 400 m sprint are shown in the graph.

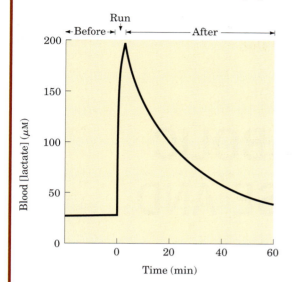

(a) What causes the rapid rise in lactate concentration?

(b) What causes the decline in lactate concentration after completion of the sprint? Why does the decline occur more slowly than the increase?

(c) Why is the concentration of lactate not zero during the resting state?

## 22. Relationship between Fructose 1,6-Bisphosphatase and Blood Lactate Levels

A congenital defect in the liver enzyme fructose 1,6-bisphosphatase results in abnormally high levels of lactate in the blood plasma. Explain.

## 23. Effect of Phloridzin on Carbohydrate Metabolism

Phloridzin, a toxic glycoside from the bark of the pear tree, blocks the normal reabsorption of glucose from the kidney tubule, thus causing blood glucose to be almost completely excreted in the urine. In an experiment, rats fed phloridzin and sodium succinate excreted about 0.5 mol of glucose (made by gluconeogenesis) for every 1 mol of sodium succinate ingested. How is the succinate transformed to glucose? Explain the stoichiometry.

Phloridzin

## 24. Excess $O_2$ Uptake during Gluconeogenesis

Lactate absorbed by the liver is converted to glucose, with the input of 6 mol of ATP for every mole of glucose produced. The extent of this process in a rat liver preparation can be monitored by administering [$^{14}$C]lactate and measuring the amount of [$^{14}$C]glucose produced. Because the stoichiometry of $O_2$ consumption and ATP production is known (about 5 ATP per $O_2$), we can predict the extra $O_2$ consumption above the normal rate when a given amount of lactate is administered. However, when the extra $O_2$ used in the synthesis of glucose from lactate is actually measured, it is always higher than predicted by known stoichiometric relationships. Suggest a possible explanation for this observation.

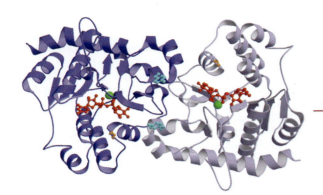

## chapter 15

# PRINCIPLES OF METABOLIC REGULATION: GLUCOSE AND GLYCOGEN

Formation of liver glycogen from lactic acid is thus seen to establish an important connection between the metabolism of the muscle and that of the liver. Muscle glycogen becomes available as blood sugar through the intervention of the liver, and blood sugar in turn is converted into muscle glycogen. There exists therefore a complete cycle of the glucose molecule in the body . . . Epinephrine was found to accelerate this cycle in the direction of muscle glycogen to liver glycogen . . . Insulin, on the other hand, was found to accelerate the cycle in the direction of blood glucose to muscle glycogen.

—C. F. Cori and G. T. Cori, article in Journal of Biological Chemistry, 1929

Metabolic regulation, a central theme in biochemistry, is one of the most remarkable features of a living cell. Of the thousands of enzyme-catalyzed reactions that can take place in a cell, there is probably not one that escapes some form of regulation. Although it is convenient (and perhaps essential) in writing a textbook to divide metabolic processes into "pathways" that play discrete roles in the cell's economy, no such separation exists inside the cell. Rather, each of the pathways we discuss in this book is inextricably intertwined with all the other cellular pathways in a multidimensional network of reactions (Fig. 15–1). For example, in Chapter 14 we discussed three possible fates for glucose 6-phosphate in a hepatocyte: passage into glycolysis for the production of ATP, passage into the pentose phosphate pathway for the production of NADPH and pentose phosphates, or hydrolysis to glucose and phosphate to replenish blood glucose. In fact, glucose 6-phosphate has a number of other possible fates; it may, for example, be used to synthesize other sugars, such as glucosamine, galactose, galactosamine, fucose, and neuraminic acid, for use in protein glycosylation, or it may be partially degraded to provide acetyl-CoA for fatty acid and sterol synthesis. In the extreme case, the bacterium *Escherichia coli* can use glucose to produce the carbon skeleton of every one of its molecules. When a cell "decides" to use glucose 6-phosphate for one purpose, that decision affects all the other pathways for which glucose 6-phosphate is a precursor or intermediate; any change in the allocation of glucose 6-phosphate to one pathway affects, directly or indirectly, the metabolite flow through all the others.

Such changes in allocation are common in the life of a cell. Louis Pasteur was the first to describe the large (greater than tenfold) increase in glucose consumption by a yeast culture when it was shifted from aerobic to anaerobic conditions. This phenomenon, called the

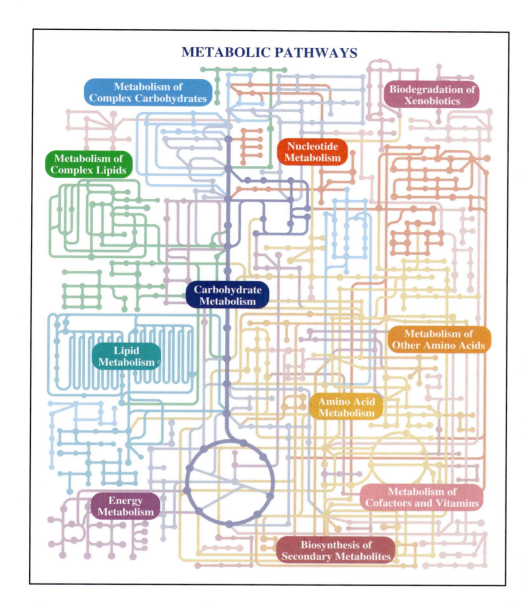

**FIGURE 15–1 Metabolism as a three-dimensional meshwork.** A typical eukaryotic cell has the capacity to make about 30,000 different proteins, which catalyze thousands of different reactions involving many hundreds of metabolites, most shared by more than one "pathway." This overview image of metabolic pathways is from the online KEGG (Kyoto Encyclopedia of Genes and Genomes) PATHWAY database (www.genome.ad.jp/kegg/pathway/map/map01100.html). Each area can be further expanded for increasingly detailed information, to the level of specific enzymes and intermediates.

Pasteur effect, occurs without a significant change in the concentration of ATP or any of the hundreds of metabolic intermediates and products derived from glucose. A similar change takes place in cells of skeletal muscle when a sprinter leaves the starting blocks. The ability of a cell to carry out all these interlocking metabolic processes simultaneously—obtaining every product in the amount needed and at the right time, in the face of major perturbations from outside, and without generating leftovers—is an *astounding* accomplishment.

In this chapter we look at mechanisms of metabolic regulation, using the pathways in which glucose is an intermediate to illustrate some general principles. First we consider the pathways by which glycogen is synthesized and broken down, a very well-studied case of metabolic regulation. Then we look at the general roles of regulation in achieving metabolic homeostasis. Focusing on the pathways that connect pyruvate with glycogen in both directions, we next consider the specific regulatory properties of the participating enzymes and the ways in which the cell accomplishes coordinated regulation of catabolic and anabolic pathways. Finally, we discuss metabolic control analysis, a system for treating complex metabolic interactions quantitatively, and consider some surprising results of its application.

In selecting carbohydrate metabolism to illustrate the principles of metabolic regulation, we have artificially separated the metabolism of fats and carbohydrates. In fact, these two activities are very tightly integrated, as we shall see in Chapter 23.

# 15.1 The Metabolism of Glycogen in Animals

In a wide range of organisms, excess glucose is converted to polymeric forms for storage—glycogen in vertebrates and many microorganisms, starch in plants. In vertebrates, glycogen is found primarily in the liver and skeletal muscle; it may represent up to 10% of the weight of liver and 1% to 2% of the weight of muscle. If this much glucose were dissolved in the cytosol of a hepatocyte, its concentration would be about 0.4 M, enough to dominate the osmotic properties of the cell. When stored as a long polymer (glycogen), however, the same mass of glucose has a concentration of only 0.01 $\mu$M. Glycogen is stored in large cytosolic granules. The elementary particle of glycogen, the $\beta$-particle, about 21 nm in diameter, consists of up to 55,000 glucose residues with about 2,000 nonreducing ends. Twenty to 40 of these particles cluster together to form $\alpha$-rosettes, easily seen with the microscope in tissue samples from well-fed animals (Fig. 15–2) but essentially absent after a 24-hour fast.

The glycogen in muscle is there to provide a quick source of energy for either aerobic or anaerobic metabolism. Muscle glycogen can be exhausted in less than an hour during vigorous activity. Liver glycogen serves as a reservoir of glucose for other tissues when dietary glucose is not available (between meals or during a fast); this is especially important for the neurons of the brain, which cannot use fatty acids as fuel. Liver glycogen can be depleted in 12 to 24 hours. In humans, the total amount of energy stored as glycogen is far less than the amount stored as fat (triacylglycerol) (see Table 23–5), but fats cannot be converted to glucose in mammals and cannot be catabolized anaerobically.

Glycogen granules are complex aggregates of glycogen and the enzymes that synthesize it and degrade it, as well as the machinery for regulating these enzymes. The general mechanisms for storing and mobilizing glycogen are the same in muscle and liver, but the enzymes differ in subtle yet important ways that reflect the different roles of glycogen in the two tissues. Glycogen is also obtained in the diet and broken down in the gut, and this involves a separate set of hydrolytic enzymes that convert glycogen (and starch) to free glucose.

The transformations of glucose discussed in this chapter and in Chapter 14 are central to the metabolism of most organisms, microbial, animal, or plant. We begin with a discussion of the *catabolic* pathways from glycogen to glucose 6-phosphate (**glycogenolysis**) and from glucose 6-phosphate to pyruvate (**glycolysis**), then turn to the *anabolic* pathways from pyruvate to glucose (**gluconeogenesis**) and from glucose to glycogen (**glycogenesis**).

## Glycogen Breakdown Is Catalyzed by Glycogen Phosphorylase

In skeletal muscle and liver, the glucose units of the outer branches of glycogen enter the glycolytic pathway through the action of three enzymes: glycogen phosphorylase, glycogen debranching enzyme, and phosphoglucomutase. Glycogen phosphorylase catalyzes the reaction in which an ($\alpha 1 \rightarrow 4$) glycosidic linkage between two glucose residues at a nonreducing end of glycogen undergoes attack by inorganic phosphate ($P_i$), removing the terminal glucose residue as $\alpha$-D-glucose 1-phosphate (Fig. 15–3). This *phosphorolysis* reaction is different from the *hydrolysis* of glycosidic bonds by amylase during intestinal degradation of dietary glycogen and starch. In phosphorolysis, some of the energy of the glycosidic bond is preserved in the formation of the phosphate ester, glucose 1-phosphate.

Pyridoxal phosphate is an essential cofactor in the glycogen phosphorylase reaction; its phosphate group acts as a general acid catalyst, promoting attack by $P_i$ on the glycosidic bond. (This is an unusual role for this cofactor; its more typical role is as a cofactor in amino acid metabolism; see Fig. 18–6.)

Glycogen phosphorylase acts repetitively on the nonreducing ends of glycogen branches until it reaches a point four glucose residues away from an ($\alpha 1 \rightarrow 6$) branch point (see Fig. 7–15), where its action stops. Further degradation by glycogen phosphorylase can occur only after the **debranching enzyme,** formally known as **oligo ($\alpha 1 \rightarrow 6$) to ($\alpha 1 \rightarrow 4$) glucantransferase,** catalyzes two successive reactions that transfer

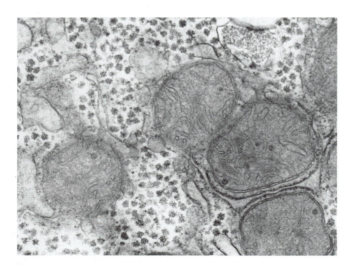

**FIGURE 15–2 Glycogen granules in a hepatocyte.** Glycogen is a storage form of carbohydrate in cells, especially hepatocytes, as illustrated here. Glycogen appears as electron-dense particles, often in aggregates or rosettes. In hepatocytes the glycogen is closely associated with tubules of the smooth endoplasmic reticulum. Many mitochondria are also present.

**FIGURE 15–3** Removal of a terminal glucose residue from the nonreducing end of a glycogen chain by glycogen phosphorylase. This process is repetitive; the enzyme removes successive glucose residues until it reaches the fourth glucose unit from a branch point (see Fig. 15–4).

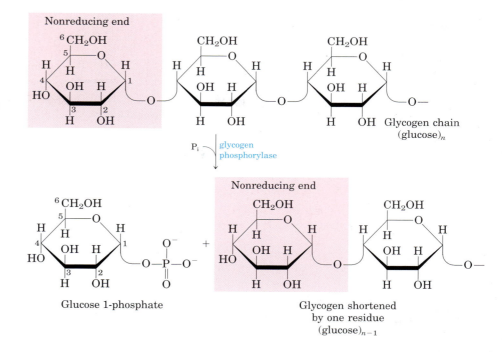

Glucose 1-phosphate

Glycogen shortened by one residue $(glucose)_{n-1}$

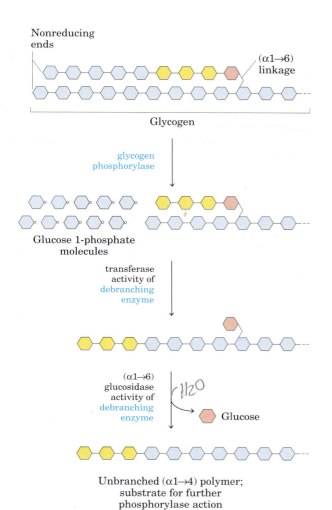

branches (Fig. 15–4). Once these branches are transferred and the glucosyl residue at C-6 is hydrolyzed, glycogen phosphorylase activity can continue.

## Glucose 1-Phosphate Can Enter Glycolysis or, in Liver, Replenish Blood Glucose

Glucose 1-phosphate, the end product of the glycogen phosphorylase reaction, is converted to glucose 6-phosphate by **phosphoglucomutase,** which catalyzes the reversible reaction

$$\text{Glucose 1-phosphate} \rightleftharpoons \text{glucose 6-phosphate}$$

Initially phosphorylated at a Ser residue, the enzyme donates a phosphoryl group to C-6 of the substrate, then accepts a phosphoryl group from C-1 (Fig. 15–5).

The glucose 6-phosphate formed from glycogen in skeletal muscle can enter glycolysis and serve as an energy source to support muscle contraction. In liver,

**FIGURE 15–4 Glycogen breakdown near an $(\alpha 1 \rightarrow 6)$ branch point.** Following sequential removal of terminal glucose residues by glycogen phosphorylase (see Fig. 15–3), glucose residues near a branch are removed in a two-step process that requires a bifunctional "debranching enzyme." First, the transferase activity of the enzyme shifts a block of three glucose residues from the branch to a nearby nonreducing end, to which they are reattached in $(\alpha 1 \rightarrow 4)$ linkage. The single glucose residue remaining at the branch point, in $(\alpha 1 \rightarrow 6)$ linkage, is then released as free glucose by the enzyme's $(\alpha 1 \rightarrow 6)$ glucosidase activity. The glucose residues are shown in shorthand form, which omits the —H, —OH, and —$CH_2OH$ groups from the pyranose rings.

**FIGURE 15–5 Reaction catalyzed by phosphogluco-mutase.** The reaction begins with the enzyme phosphorylated on a Ser residue. In step ①, the enzyme donates its phosphoryl group (green) to glucose 1-phosphate, producing glucose 1,6-bisphosphate. In step ②, the phosphoryl group at C-1 of glucose 1,6-bisphosphate (red) is transferred back to the enzyme, re-forming the phosphoenzyme and producing glucose 6-phosphate.

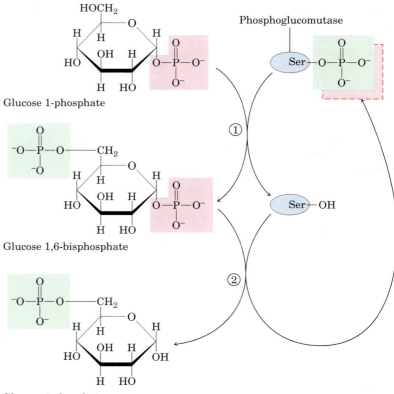

glycogen breakdown serves a different purpose: to release glucose into the blood when the blood glucose level drops, as it does between meals. This requires an enzyme, glucose 6-phosphatase, that is present in liver and kidney but not in other tissues. The enzyme is an integral membrane protein of the endoplasmic reticulum, predicted to contain nine transmembrane helices,

with its active site on the lumenal side of the ER. Glucose 6-phosphate formed in the cytosol is transported into the ER lumen by a specific transporter (T1) (Fig. 15–6) and hydrolyzed at the lumenal surface by the glucose 6-phosphatase. The resulting P$_i$ and glucose are thought to be carried back into the cytosol by two different transporters (T2 and T3), and the glucose leaves

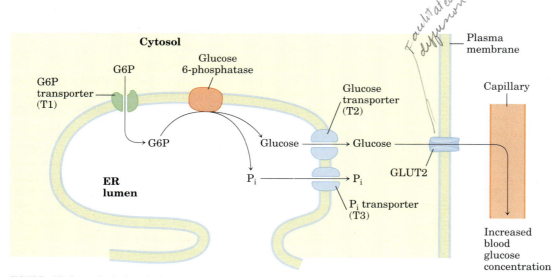

**FIGURE 15–6 Hydrolysis of glucose 6-phosphate by glucose 6-phosphatase of the ER.** The catalytic site of glucose 6-phosphatase faces the lumen of the ER. A glucose 6-phosphate (G6P) transporter (T1) carries the substrate from the cytosol to the lumen, and the products glucose and P$_i$ pass to the cytosol on specific transporters (T2 and T3). Glucose leaves the cell via the GLUT2 transporter in the plasma membrane.

the hepatocyte via yet another transporter in the plasma membrane (GLUT2). Notice that by having the active site of glucose 6-phosphatase inside the ER lumen, the cell separates this reaction from the process of glycolysis, which takes place in the cytosol and would be aborted by the action of glucose 6-phosphatase. Genetic defects in either glucose 6-phosphatase or T1 lead to serious derangement of glycogen metabolism, resulting in type Ia glycogen storage disease (Box 15–1).

Because muscle and adipose tissue lack glucose 6-phosphatase, they cannot convert the glucose 6-phosphate formed by glycogen breakdown to glucose, and these tissues therefore do not contribute glucose to the blood.

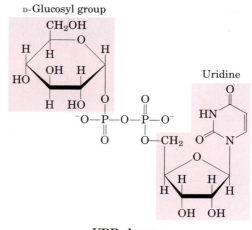

**UDP-glucose**
(a sugar nucleotide)

### The Sugar Nucleotide UDP-Glucose Donates Glucose for Glycogen Synthesis

Many of the reactions in which hexoses are transformed or polymerized involve **sugar nucleotides,** compounds in which the anomeric carbon of a sugar is activated by attachment to a nucleotide through a phosphate ester linkage. Sugar nucleotides are the substrates for polymerization of monosaccharides into disaccharides, glycogen, starch, cellulose, and more complex extracellular polysaccharides. They are also key intermediates in the production of the aminohexoses and deoxyhexoses found in some of these polysaccharides, and in the synthesis of vitamin C (L-ascorbic acid). The role of sugar nucleotides in the biosynthesis of glycogen and many other carbohydrate derivatives was first discovered by the Argentine biochemist Luis Leloir.

The suitability of sugar nucleotides for biosynthetic reactions stems from several properties:

1. Their formation is metabolically irreversible, contributing to the irreversibility of the synthetic pathways in which they are intermediates. The condensation of a nucleoside triphosphate with a hexose 1-phosphate to form a sugar nucleotide has a small positive free-energy change, but the reaction releases $PP_i$, which is rapidly hydrolyzed by inorganic pyrophosphatase in a reaction that is strongly exergonic ($\Delta G'^\circ = -19.2$ kJ/mol; Fig. 15–7). This keeps the cellular concentration of $PP_i$ low, ensuring that the actual free-energy change in

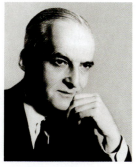

Luis Leloir, 1906–1987

**FIGURE 15–7   Formation of a sugar nucleotide.** A condensation reaction occurs between a nucleoside triphosphate (NTP) and a sugar phosphate. The negatively charged oxygen on the sugar phosphate serves as a nucleophile, attacking the $\alpha$ phosphate of the nucleoside triphosphate and displacing pyrophosphate. The reaction is pulled in the forward direction by the hydrolysis of $PP_i$ by inorganic pyrophosphatase.

Net reaction: Sugar phosphate $+$ NTP $\longrightarrow$ NDP-sugar $+$ $2P_i$

# BOX 15–1   WORKING IN BIOCHEMISTRY

## Carl and Gerty Cori: Pioneers in Glycogen Metabolism and Disease

Much of what is written in present-day biochemistry textbooks about the metabolism of glycogen was discovered between about 1925 and 1950 by the remarkable husband and wife team of Carl F. Cori and Gerty T. Cori. Both trained in medicine in Europe at the end of World War I (she completed premedical studies and medical school in one year!). They left Europe together in 1922 to establish research laboratories in the United States, first for nine years in Buffalo, New York, at what is now the Roswell Park Memorial Institute, then from 1931 until the end of their lives at Washington University in St. Louis.

The Coris in Gerty Cori's laboratory, around 1947.

In their early physiological studies of the origin and fate of glycogen in animal muscle, the Coris demonstrated the conversion of glycogen to lactate in tissues, movement of lactate in the blood to the liver, and, in the liver, reconversion of lactate to glycogen—a pathway that came to be known as the Cori cycle (see Fig. 23–18). Pursuing these observations at the biochemical level, they showed that glycogen was mobilized in a phosphorolysis reaction catalyzed by the enzyme they discovered, glycogen phosphorylase. They identified the product of this reaction (the "Cori ester") as glucose 1-phosphate and showed that it could be reincorporated into glycogen in the reverse reaction. Although this did not prove to be the reaction by which glycogen is synthesized in cells, it was the first in vitro demonstration of the synthesis of a macromolecule from simple monomeric subunits, and it inspired others to search for polymerizing enzymes. Arthur Kornberg, discoverer of the first DNA polymerase, has said of his experience in the Coris' lab, "Glycogen phosphorylase, not base pairing, was what led me to DNA polymerase."

Gerty Cori became interested in human genetic diseases in which too much glycogen is stored in the liver. She was able to identify the biochemical defect in several of these diseases and to show that these diseases could be diagnosed by assays of the enzymes of glycogen metabolism in small samples of tissue obtained by biopsy. Table 1 summarizes what we now know about 13 genetic diseases of this sort. ■

Carl and Gerty Cori shared the Nobel Prize in Physiology or Medicine in 1947 with Bernardo Houssay of Argentina, who was cited for his studies of hormonal regulation of carbohydrate metabolism. The Cori laboratories in St. Louis became an international center of biochemical research in the 1940s and 1950s, and at least six scientists who trained with the Coris became Nobel laureates: Arthur Kornberg (for DNA synthesis, 1959), Severo Ochoa (for RNA synthesis, 1959), Luis Leloir (for the role of sugar nucleotides in

the cell is favorable. In effect, rapid removal of the product, driven by the large, negative free-energy change of $PP_i$ hydrolysis, pulls the synthetic reaction forward, a common strategy in biological polymerization reactions.

2. Although the chemical transformations of sugar nucleotides do not involve the atoms of the nucleotide itself, the nucleotide moiety has many

groups that can undergo noncovalent interactions with enzymes; the additional free energy of binding can contribute significantly to catalytic activity (Chapter 6; see also p. 301).

3. Like phosphate, the nucleotidyl group (UMP or AMP, for example) is an excellent leaving group, facilitating nucleophilic attack by activating the sugar carbon to which it is attached.

polysaccharide synthesis, 1970), Earl Sutherland (for the discovery of cAMP in the regulation of carbohydrate metabolism, 1971), Christian de Duve (for subcellular fractionation, 1974), and Edwin Krebs (for the discovery of phosphorylase kinase, 1991).

**TABLE 1    Glycogen Storage Diseases of Humans**

| Type (name) | Enzyme affected | Primary organ affected | Symptoms |
|---|---|---|---|
| Type 0 | Glycogen synthase | Liver | Low blood glucose, high ketone bodies, early death |
| Type Ia (von Gierke's) | Glucose 6-phosphatase | Liver | Enlarged liver, kidney failure |
| Type Ib | Microsomal glucose 6-phosphate translocase | Liver | As in Ia; also high susceptibility to bacterial infections |
| Type Ic | Microsomal $P_i$ transporter | Liver | As in Ia |
| Type II (Pompe's) | Lysosomal glucosidase | Skeletal and cardiac muscle | Infantile form: death by age 2; juvenile form: muscle defects (myopathy); adult form: as in muscular dystrophy |
| Type IIIa (Cori's or Forbes's) | Debranching enzyme | Liver, skeletal and cardiac muscle | Enlarged liver in infants; myopathy |
| Type IIIb | Liver debranching enzyme (muscle enzyme normal) | Liver | Enlarged liver in infants |
| Type IV (Andersen's) | Branching enzyme | Liver, skeletal muscle | Enlarged liver and spleen, myoglobin in urine |
| Type V (McArdle's) | Muscle phosphorylase | Skeletal muscle | Exercise-induced cramps and pain; myoglobin in urine |
| Type VI (Hers's) | Liver phosphorylase | Liver | Enlarged liver |
| Type VII (Tarui's) | Muscle PFK-1 | Muscle, erythrocytes | As in V; also hemolytic anemia |
| Type VIb, VIII, or IX | Phosphorylase kinase | Liver, leukocytes, muscle | Enlarged liver |
| Type XI (Fanconi-Bickel) | Glucose transporter (GLUT2) | Liver | Failure to thrive, enlarged liver, rickets, kidney dysfunction |

4. By "tagging" some hexoses with nucleotidyl groups, cells can set them aside in a pool for one purpose (glycogen synthesis, for example), separate from hexose phosphates destined for another purpose (such as glycolysis).

Glycogen synthesis takes place in virtually all animal tissues but is especially prominent in the liver and skeletal muscles. The starting point for synthesis of glycogen is **glucose 6-phosphate.** As we saw in Chapter 14, this can be derived from free glucose in a reaction catalyzed by the isozymes **hexokinase I** and **hexokinase II** in muscle and **hexokinase IV** (glucokinase) in liver:

$$\text{D-Glucose} + \text{ATP} \longrightarrow \text{D-glucose 6-phosphate} + \text{ADP}$$

However, some ingested glucose takes a more roundabout path to glycogen. It is first taken up by erythrocytes and converted to lactate glycolytically; the lactate is then taken up by the liver and converted to glucose 6-phosphate by gluconeogenesis.

To initiate glycogen synthesis, the glucose 6-phosphate is converted to **glucose 1-phosphate** in the phosphoglucomutase reaction:

$$\text{Glucose 6-phosphate} \rightleftharpoons \text{glucose 1-phosphate}$$

The product of this reaction is converted to UDP-glucose by the action of **UDP-glucose pyrophosphorylase,** in a key step of glycogen biosynthesis:

$$\text{Glucose 1-phosphate} + \text{UTP} \longrightarrow \text{UDP-glucose} + \text{PP}_i$$

Notice that this enzyme is named for the reverse reaction; in the cell, the reaction proceeds in the direction of UDP-glucose formation, because pyrophosphate is rapidly hydrolyzed by inorganic pyrophosphatase (Fig. 15–7).

UDP-glucose is the immediate donor of glucose residues in the reaction catalyzed by **glycogen synthase,** which promotes the transfer of the glucose residue from UDP-glucose to a nonreducing end of a branched glyco-

gen molecule (Fig. 15–8). The overall equilibrium of the path from glucose 6-phosphate to lengthened glycogen greatly favors synthesis of glycogen.

Glycogen synthase cannot make the $(\alpha 1{\rightarrow}6)$ bonds found at the branch points of glycogen; these are formed by the glycogen-branching enzyme, also called **amylo $(1{\rightarrow}4)$ to $(1{\rightarrow}6)$ transglycosylase** or glycosyl-$(4{\rightarrow}6)$-transferase. The glycogen-branching enzyme catalyzes transfer of a terminal fragment of 6 or 7 glucose residues from the nonreducing end of a glycogen branch having at least 11 residues to the C-6 hydroxyl group of a glucose residue at a more interior position of the same or another glycogen chain, thus creating a new branch (Fig. 15–9). Further glucose residues may be added to the new branch by glycogen synthase. The biological effect of branching is to make the glycogen molecule more soluble and to increase the number of nonreducing ends. This increases the number of sites accessible to glycogen phosphorylase and glycogen synthase, both of which act only at nonreducing ends.

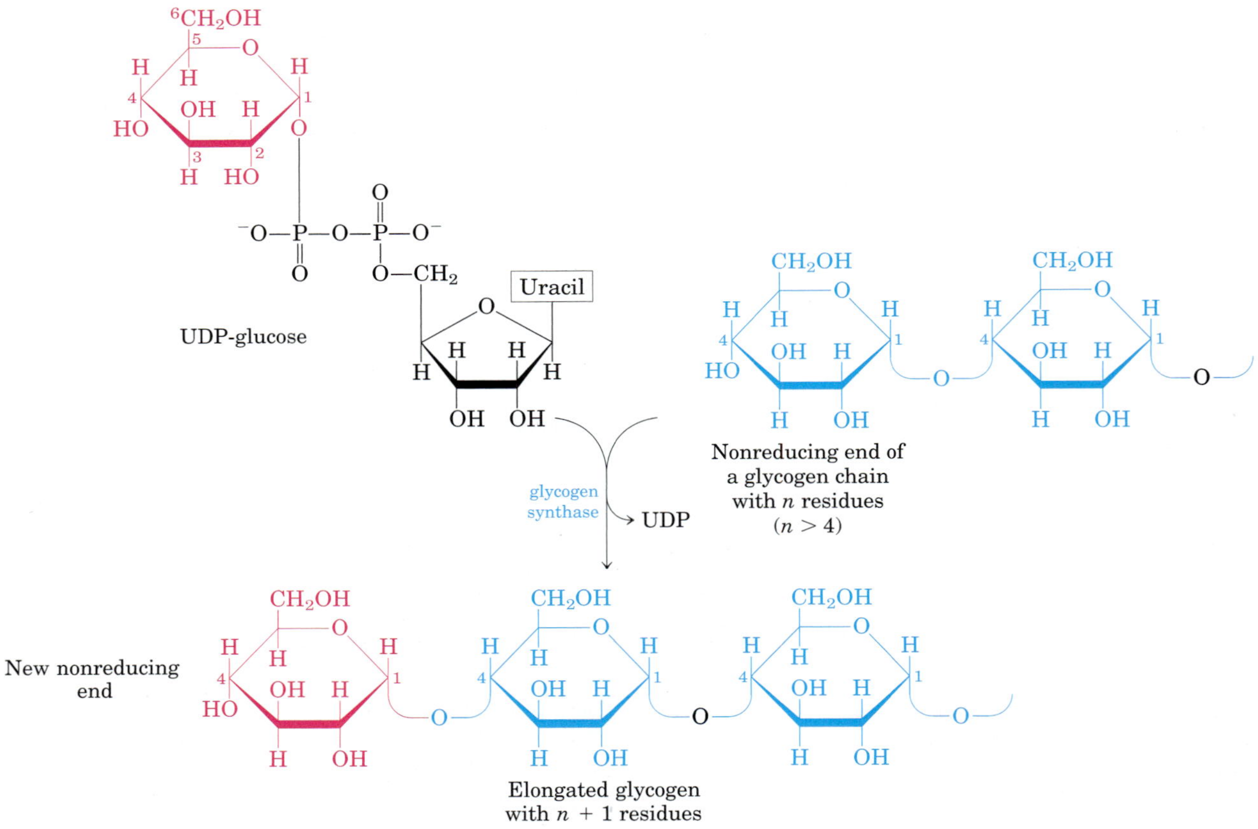

**FIGURE 15–8 Glycogen synthesis.** A glycogen chain is elongated by glycogen synthase. The enzyme transfers the glucose residue of UDP-glucose to the nonreducing end of a glycogen branch (see Fig. 7–15) to make a new $(\alpha 1{\rightarrow}4)$ linkage.

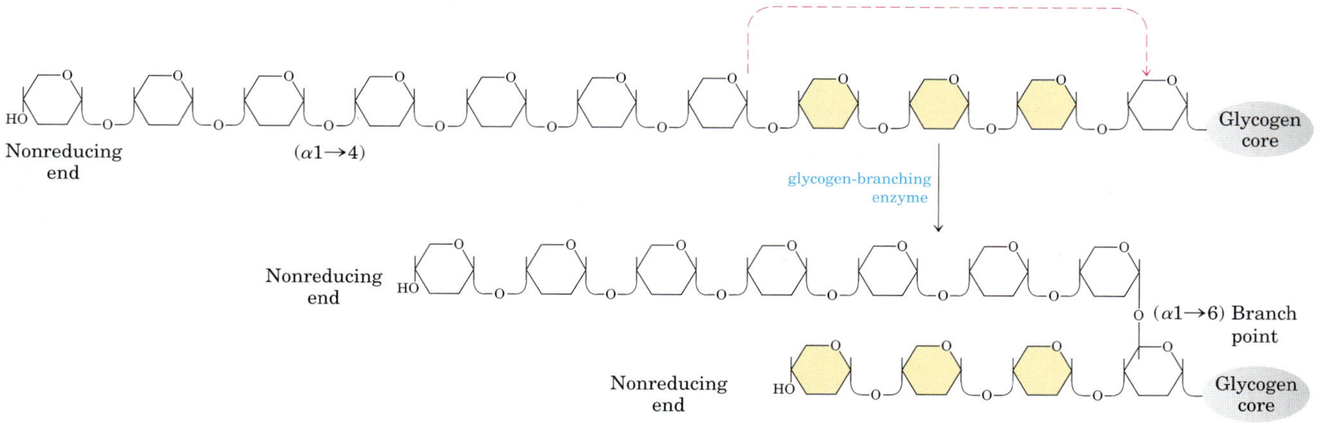

**FIGURE 15–9 Branch synthesis in glycogen.** The glycogen-branching enzyme (also called amylo (1→4) to (1→6) transglycosylase or glycosyl-(4→6)-transferase) forms a new branch point during glycogen synthesis.

## Glycogenin Primes the Initial Sugar Residues in Glycogen

Glycogen synthase cannot initiate a new glycogen chain de novo. It requires a primer, usually a preformed (α1→4) polyglucose chain or branch having at least eight glucose residues. How is a *new* glycogen molecule initiated? The intriguing protein **glycogenin** (Fig. 15–10) is both the primer on which new chains are assembled and the enzyme that catalyzes their assembly. The first step in the synthesis of a new glycogen mole-

cule is the transfer of a glucose residue from UDP-glucose to the hydroxyl group of $Tyr^{194}$ of glycogenin, catalyzed by the protein's intrinsic glucosyltransferase activity (Fig. 15–11a). The nascent chain is extended by the sequential addition of seven more glucose residues, each derived from UDP-glucose; the reactions are catalyzed by the chain-extending activity of glycogenin. At this point, glycogen synthase takes over, further extending the glycogen chain. Glycogenin remains buried within the particle, covalently attached to the single reducing end of the glycogen molecule (Fig. 15–11b).

**FIGURE 15–10 Glycogenin structure.** (PDB 1D 1772) Muscle glycogenin ($M_r$ 37,000) forms dimers in solution. Humans have a second isoform in liver, glycogenin-2. The substrate, UDP-glucose (shown as a red ball-and-stick structure), is bound to a Rossman fold near the amino terminus and is some distance from the $Tyr^{194}$ residues (turquoise)—15 Å from that in the same monomer, 12 Å from that in the dimeric partner. Each UDP-glucose is bound through its phosphates to a $Mn^{2+}$ ion (green) that is essential to catalysis. $Mn^{2+}$ is believed to function as an electron-pair acceptor (Lewis acid) to stabilize the leaving group, UDP. The glycosidic bond in the product has the same configuration about the C-1 of glucose as the substrate UDP-glucose, suggesting that the transfer of glucose from UDP to $Tyr^{194}$ occurs in two steps. The first step is probably a nucleophilic attack by $Asp^{162}$ (orange), forming a temporary intermediate with inverted configuration. A second nucleophilic attack by $Tyr^{194}$ then restores the starting configuration.

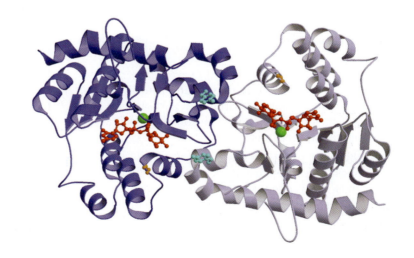

**(a)**                                 **(b)**

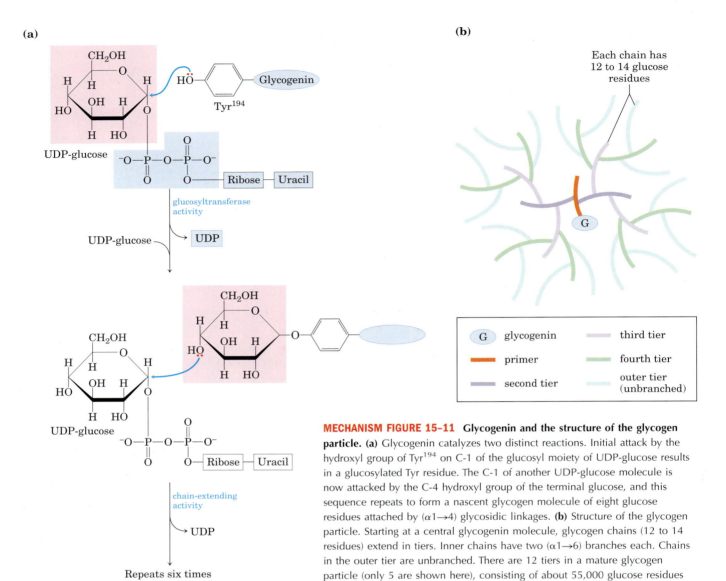

**MECHANISM FIGURE 15–11 Glycogenin and the structure of the glycogen particle. (a)** Glycogenin catalyzes two distinct reactions. Initial attack by the hydroxyl group of Tyr[194] on C-1 of the glucosyl moiety of UDP-glucose results in a glucosylated Tyr residue. The C-1 of another UDP-glucose molecule is now attacked by the C-4 hydroxyl group of the terminal glucose, and this sequence repeats to form a nascent glycogen molecule of eight glucose residues attached by $(\alpha 1{\rightarrow}4)$ glycosidic linkages. **(b)** Structure of the glycogen particle. Starting at a central glycogenin molecule, glycogen chains (12 to 14 residues) extend in tiers. Inner chains have two $(\alpha 1{\rightarrow}6)$ branches each. Chains in the outer tier are unbranched. There are 12 tiers in a mature glycogen particle (only 5 are shown here), consisting of about 55,000 glucose residues in a molecule of about 21 nm diameter and $M_r$ $10^7$.

## SUMMARY 15.1 The Metabolism of Glycogen in Animals

- Glycogen is stored in muscle and liver as large particles. Contained within the particles are the enzymes that metabolize glycogen, as well as regulatory enzymes.

- Glycogen phosphorylase catalyzes phosphorolytic cleavage at the nonreducing ends of glycogen chains, producing glucose 1-phosphate. The debranching enzyme transfers branches onto main chains and releases the residue at the $(\alpha 1{\rightarrow}6)$ branch as free glucose.

- Phosphoglucomutase interconverts glucose 1-phosphate and glucose 6-phosphate. Glucose

6-phosphate can enter glycolysis or, in liver, can be converted to free glucose by glucose 6-phosphatase in the endoplasmic reticulum, then released to replenish blood glucose.

- The sugar nucleotide UDP-glucose donates glucose residues to the nonreducing end of glycogen in the reaction catalyzed by glycogen synthase. A separate branching enzyme produces the $(\alpha 1{\rightarrow}6)$ linkages at branch points.

- New glycogen particles begin with the auto-catalytic formation of a glycosidic bond between the glucose of UDP-glucose and a Tyr residue in the protein glycogenin, followed by addition of several glucose residues to form a primer that can be acted upon by glycogen synthase.

## 15.2 Regulation of Metabolic Pathways

The pathways of glycogen metabolism provide, in the catabolic direction, the energy essential to oppose the forces of entropy and, in the anabolic direction, biosynthetic precursors and a storage form of metabolic energy. These reactions are so important to survival that very complex regulatory mechanisms have evolved to ensure that metabolites move through each pathway in the correct direction and at the correct rate to exactly match the cell's or the organism's current circumstances, and that appropriate adjustments are made in the rate of metabolite flow through the whole pathway if external circumstances change.

Circumstances do change, sometimes dramatically. The demand for ATP production in muscle may increase 100-fold in a few seconds in response to exercise. The availability of oxygen may decrease due to hypoxia (diminished delivery of oxygen to tissues) or ischemia (diminished flow of blood to tissues). The relative proportions of carbohydrate, fat, and protein in the diet vary from meal to meal, and the supply of fuels obtained in the diet is intermittent, requiring metabolic adjustments between meals and during starvation. Wound healing requires huge amounts of energy and biosynthetic precursors.

### Living Cells Maintain a Dynamic Steady State

Fuels such as glucose enter a cell, and waste products such as $CO_2$ leave, but the mass and the gross composition of a typical cell do not change appreciably over time; cells and organisms exist in a dynamic steady state, but not at equilibrium with their surroundings. At the molecular level, this means that for each metabolic reaction in a pathway, the substrate is provided by the preceding reaction at the same rate at which it is converted to product. Thus, although the rate of metabolite flow, or **flux,** through this step of the pathway may be high, the concentration of substrate, S, remains constant. For the reaction

$$A \xrightarrow{v_1} S \xrightarrow{v_2} P$$

when $v_1 = v_2$, [S] is constant.

When the steady state is disturbed by some change in external circumstances or energy supply, the temporarily altered fluxes through individual metabolic pathways trigger regulatory mechanisms intrinsic to each pathway. The net effect of all these adjustments is to return the organism to a new steady state—to achieve **homeostasis.**

### Regulatory Mechanisms Evolved under Strong Selective Pressures

In the course of evolution, organisms have acquired a remarkable collection of regulatory mechanisms for maintaining homeostasis at the molecular, cellular, and organismal level. The importance of metabolic regulation to an organism is reflected in the relative proportion of genes that encode regulatory machinery—in humans, about 4,000 genes (~12% of all genes) encode regulatory proteins, including a variety of receptors, regulators of gene expression, and about 500 different protein kinases! These regulatory mechanisms act over different time scales (from seconds to days) and have different sensitivities to external changes. In many cases, the mechanisms overlap: one enzyme is subject to regulation by several different mechanisms.

After the protection of its DNA from damage, perhaps nothing is more important to a cell than maintaining a constant supply and concentration of ATP. Many ATP-using enzymes have $K_m$ values between 0.1 and 1 mM, and the ATP concentration in a typical cell is about 5 mM. If [ATP] were to drop significantly, the rates of hundreds of reactions that involve ATP would decrease, and the cell would probably not survive. Furthermore, because ATP is converted to ADP or AMP when "spent" to accomplish cellular work, the [ATP]/[ADP] ratio profoundly affects all reactions that employ these cofactors. The same is true for other important cofactors, such as NADH/NAD$^+$ and NADPH/NADP$^+$. For example, consider the reaction catalyzed by hexokinase:

$$\text{ATP} + \text{glucose} \longrightarrow \text{ADP} + \text{glucose 6-phosphate}$$

$$K'_{eq} = \frac{[\text{ADP}]_{eq}[\text{glucose 6-phosphate}]_{eq}}{[\text{ATP}]_{eq}[\text{glucose}]_{eq}} = 2 \times 10^3$$

Note that this expression is true *only* when reactants and products are at their *equilibrium* concentrations, where $\Delta G' = 0$. At any other set of concentrations, $\Delta G'$ is not zero. Recall (from Chapter 13) that the ratio of products to substrates (the mass action ratio, $Q$) determines the magnitude and sign of $\Delta G'$ and therefore the amount of free energy released during the reaction:

$$\Delta G' = \Delta G'^\circ + RT \ln \frac{[\text{ADP}][\text{glucose 6-phosphate}]}{[\text{ATP}][\text{glucose}]}$$

Because an alteration of this mass action ratio profoundly influences every reaction that involves ATP, organisms have been under strong evolutionary pressure to develop regulatory mechanisms that respond to the [ATP]/[ADP] ratio. Similar arguments show the importance of maintaining appropriate [NADH]/[NAD$^+$] and [NADPH]/[NADP$^+$] ratios.

AMP concentration is a much more sensitive indicator of a cell's energetic state than is ATP. Normally cells have a far higher concentration of ATP (5 to 10 mM) than of AMP (<0.1 mM). When some process (say, muscle contraction) consumes ATP, AMP is produced in two steps. First, hydrolysis of ATP produces ADP, then the reaction catalyzed by **adenylate kinase** produces AMP:

$$2\,\text{ADP} \longrightarrow \text{AMP} + \text{ATP}$$

**TABLE 15–1    Relative Changes in [ATP] and [AMP] When ATP Is Consumed**

| Adenine nucleotide | Concentration before ATP depletion (mM) | Concentration after ATP depletion (mM) | Relative change |
|---|---|---|---|
| ATP | 5.0 | 4.5 | 10% |
| ADP | 1.0 | 1.0 | 0 |
| AMP | 0.1 | 0.6 | **600%** |

If [ATP] drops by 10%, producing ADP and AMP in the same amounts, the relative change in [AMP] is much greater (Table 15–1). It is not surprising, therefore, that many regulatory processes hinge on changes in [AMP]. One important mediator of regulation by AMP is **AMP-dependent protein kinase (AMPK),** which responds to an increase in [AMP] by phosphorylating key proteins, thereby regulating their activities. The rise in [AMP] may be caused by a reduced nutrient supply or increased exercise. The action of AMPK (not to be confused with the *cyclic* AMP–dependent protein kinase; see Section 15.4) increases glucose transport and activates glycolysis and fatty acid oxidation, while suppressing energy-requiring processes such as the synthesis of fatty acids, cholesterol, and protein. We discuss this enzyme further, and the detailed mechanisms by which it effects these changes, in Chapter 23.

In addition to the cofactors ATP, NADH, and NADPH, hundreds of metabolic intermediates also must be present at appropriate concentrations in the cell. For example, the glycolytic intermediates dihydroxyacetone phosphate and 3-phosphoglycerate are precursors of triacylglycerols and serine, respectively. When these products are needed, the rate of glycolysis must be adjusted to provide them without reducing the glycolytic production of ATP.

Priorities at the *organismal* level have also driven the evolution of regulatory mechanisms. In mammals, the brain has virtually no stored source of energy, depending instead on a constant supply of glucose from the blood. If glucose in the blood drops from its normal concentration of 4 to 5 mM to half that level, mental confusion results, and a fivefold reduction in blood glucose can lead to coma and death. To buffer against changes in blood glucose concentration, release of the hormones insulin and glucagon, elicited by high or low blood glucose, respectively, triggers metabolic changes that tend to return the blood glucose concentration to normal.

Other selective pressures must also have operated throughout evolution, selecting for regulatory mechanisms that

1. maximize the efficiency of fuel utilization by preventing the simultaneous operation of

pathways in opposite directions (such as glycolysis and gluconeogenesis);

2. partition metabolites appropriately between alternative pathways (such as glycolysis and the pentose phosphate pathway);

3. draw on the fuel best suited for the immediate needs of the organism (glucose, fatty acids, glycogen, or amino acids); and

4. shut down biosynthetic pathways when their products accumulate.

The importance of effective metabolic regulation is clear from the consequences of failed regulation: in many cases, serious disease (as described in Box 15–1, for example).

## Regulatory Enzymes Respond to Changes in Metabolite Concentration

Flux through a biochemical pathway depends on the activities of the enzymes that catalyze each reaction in that pathway. For some steps, the reaction is close to equilibrium within the cell (Fig. 15–12). The net flow of metabolites through these steps is the small difference between the rates of the forward and reverse reactions, rates that are very similar when the reaction is near equilibrium. Small changes in substrate or product concentration can produce large changes in the net rate, and can even change the direction of the net flow. We can identify these near-equilibrium reactions in a cell by comparing the **mass action ratio, $Q$,** with the equilibrium constant for the reaction, $K'_{eq}$. Recall that for the reaction $A + B \rightarrow C + D$, $Q = [C][D]/[A][B]$. When $Q$ and $K'_{eq}$ are within a few orders of magnitude, the reaction is near equilibrium. This is the case for six of the ten reactions in the glycolytic pathway (Table 15–2).

Other reactions are far from equilibrium in the cell. For example, $K'_{eq}$ for the phosphofructokinase-1 (PFK-1) reaction in glycolysis is about 1,000, but $Q$ ([fructose 1,6-bisphosphate][ADP]/[fructose 6-phosphate][ATP]) in a typical cell in the steady state is about 0.1 (Table 15–2). It is *because* the reaction is so far from equilibrium that the process is exergonic under cellular con-

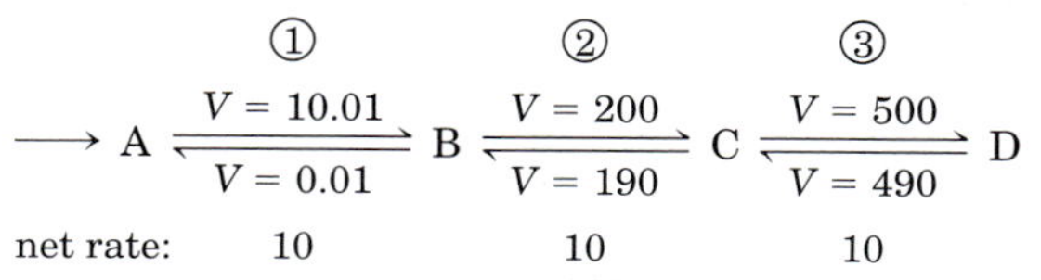

**FIGURE 15–12 Near-equilibrium and nonequilibrium steps in a metabolic pathway.** Steps 2 and 3 of this pathway are near equilibrium in the cell; their forward rates are only slightly greater than their reverse rates, so the net forward rates (10) are relatively low and the free-energy change $\Delta G'$ for each step is close to zero. An increase in the intracellular concentration of metabolite C or D can reverse the direction of these steps. Step 1 is maintained in the cell far from equilibrium; its forward rate greatly exceeds its reverse rate. The net rate of step 1 (10) is much larger than the reverse rate (0.01) and is identical to the net rates of steps 2 and 3 when the pathway is operating in the steady state. Step 1 has a large, negative $\Delta G'$.

ditions and tends to go in the forward direction. The reaction is held far from equilibrium because, under prevailing cellular conditions of substrate, product, and effector concentrations, the rate of conversion of fructose 6-phosphate to fructose 1,6-bisphosphate is limited by the activity of PFK-1, which is itself limited by the number of PFK-1 molecules present and by the actions of effectors. Thus the net forward rate of the enzyme-catalyzed reaction is equal to the net flow of glycolytic intermediates through other steps in the pathway, and the reverse flow through PFK-1 remains near zero.

The cell *cannot* allow reactions with large equilibrium constants to reach equilibrium. If [fructose 6-phosphate], [ATP], and [ADP] in the cell were held at their usual level (low millimolar) and the PFK-1 reaction were allowed to reach equilibrium by an increase in [fructose 1,6-bisphosphate], the concentration of fructose 1,6-bisphosphate would rise into the molar range, wreaking osmotic havoc on the cell. Consider another case: if the reaction ATP → ADP + $P_i$ were allowed to approach equilibrium in the cell, the actual free-energy change ($\Delta G'$) for that reaction (see Box 13–1) would approach zero, and ATP would lose the high phosphoryl group transfer potential that makes it valuable to the cell as an energy source. It is therefore essential that enzymes catalyzing ATP breakdown and other highly exergonic reactions in a cell be sensitive to regulation, so that when metabolic changes are forced by external circumstances, the flow through these enzymes will be adjusted to ensure that [ATP] remains far above its equilibrium level. When such metabolic changes occur, enzymatic activities in all interconnected pathways adjust to keep these critical steps away from equilibrium. Thus, not surprisingly, many enzymes that catalyze highly exergonic reactions (such as PFK-1) are subject to a variety of subtle regulatory mechanisms. The multiplicity of these adjustments is so great that we cannot predict by examining the properties of any one enzyme in a pathway whether that enzyme has a strong influence on net

**TABLE 15–2 Equilibrium Constants, Mass Action Coefficients, and Free-Energy Changes for Enzymes of Carbohydrate Metabolism**

| Enzyme | $K'_{eq}$ | Mass action ratio, Q | | Reaction near equilibrium in vivo?* | $\Delta G'^o$ (kJ/mol) | $\Delta G'$ (kJ/mol) in heart |
| | | Liver | Heart | | | |
|---|---|---|---|---|---|---|
| Hexokinase | $2 \times 10^3$ | $2 \times 10^{-2}$ | $8 \times 10^{-2}$ | No | $-17$ | $-27$ |
| PFK-1 | $1.0 \times 10^3$ | $9 \times 10^{-2}$ | $3 \times 10^{-2}$ | No | $-14.2$ | $-23$ |
| Aldolase | $1.0 \times 10^{-4}$ | $1.2 \times 10^{-6}$ | $9 \times 10^{-6}$ | Yes | $+23.8$ | $-6.0$ |
| Triose phosphate isomerase | $4 \times 10^{-2}$ | $-$ | $2.4 \times 10^{-1}$ | Yes | $+7.5$ | $+3.8$ |
| Glyceraldehyde 3-phosphate dehydrogenase + phosphoglycerate kinase | $2 \times 10^3$ | $6 \times 10^2$ | 9.0 | Yes | $-13$ | $+3.5$ |
| Phosphoglycerate mutase | $1 \times 10^{-1}$ | $1 \times 10^{-1}$ | $1.2 \times 10^{-1}$ | Yes | $+4.4$ | $+0.6$ |
| Enolase | 3 | 2.9 | 1.4 | Yes | $-3.2$ | $-0.5$ |
| Pyruvate kinase | $2 \times 10^4$ | $7 \times 10^{-1}$ | 40 | No | $-31.0$ | $-17$ |
| Phosphoglucose isomerase | $4 \times 10^{-1}$ | $3.1 \times 10^{-1}$ | $2.4 \times 10^{-1}$ | Yes | $+2.2$ | $-1.4$ |
| Pyruvate carboxylase + PEP carboxykinase | 7 | $1 \times 10^{-3}$ | $-$ | No | $-5.0$ | $-22.8$ |
| Glucose 6-phosphatase | $8.5 \times 10^2$ | $1.2 \times 10^2$ | $-$ | Yes | $-17.3$ | $-5.0$ |

*For simplicity, any reaction for which the absolute value of the calculated $\Delta G'$ is less than 6 is considered near equilibrium.

**Source:** $K'_{eq}$ and Q from Newsholme, E.A. & Start, C. (1973) *Regulation in Metabolism*, Wiley Press, New York, pp. 97, 263. $\Delta G'$ and $\Delta G'^o$ were calculated from these data.

**FIGURE 15–13 Factors that determine the activity of an enzyme.** Blue arrows represent processes that determine the number of enzyme molecules in the cell; red arrows show factors that determine the enzymatic activity of an existing enzyme molecule. Each arrow represents a point at which regulation can occur.

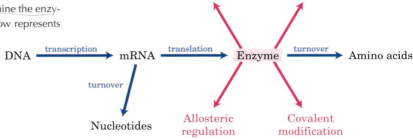

### Enzyme Activity Can Be Altered in Several Ways

The activity of an enzyme can be modulated by changes in the number of enzyme molecules in the cell or by changes in the catalytic activity of each enzyme molecule already present, for example through allosteric regulation or covalent alteration (Fig. 15–13). The number of enzyme molecules in the cell is a function of the rates of synthesis and degradation, both of which, for many enzymes, are tightly controlled. The rate of synthesis of a protein can be adjusted by the production or alteration (in response to some outside signal) of a transcription factor, a protein that binds to a regulatory region of DNA adjacent to the gene in question and increases the likelihood of its transcription into mRNA (Chapter 28). The stability of mRNAs—their resistance to degradation by a ribonuclease—varies, so the amount of a given mRNA in the cell is a function of its rates of synthesis and degradation (Chapter 26). Finally, the rate at which an mRNA is translated on the ribosome depends on several factors, described in detail in Chapter 27.

The rate of protein degradation also differs from enzyme to enzyme and depends on the conditions in the cell; protein half-lives vary from a few minutes to many days. Some proteins are tagged for degradation in proteasomes (discussed in Chapter 28) by the covalent attachment of ubiquitin (recall the case of cyclin; see Fig. 12–44). Some proteins are synthesized as inactive forms, or proenzymes, that become active only when a proteolytic event removes an inhibitory sequence in the proenzyme.

As a result of these several mechanisms of regulating enzyme level, cells can change their complement of enzymes in response to changes in metabolic circumstances. In vertebrates, liver is the most adaptable tissue; a change in diet from high carbohydrate to high lipid, for example, affects the transcription into mRNA of hundreds of genes and thus the levels of hundreds of proteins. These global changes in gene expression can be quantified by the use of DNA microarrays (see Fig. 9–22) or two-dimensional gel electrophoresis (see Fig.

3–22) to display the protein complement of a tissue. Both techniques provide great insights into metabolic regulation.

Changes in the number of molecules of an enzyme are generally relatively slow, occurring over seconds to hours. Covalent modifications of existing proteins are faster, taking seconds to minutes. Various types of covalent modifications are known, such as adenylylation, methylation, or attachment of lipids (p. 228). By far the most common type is phosphorylation and dephosphorylation (Fig. 15–14); up to half of a eukaryotic cell's proteins are targets of phosphorylation under some circumstances. Phosphorylation may alter the electrostatic features of the active site, cause movement of an inhibitory region of the protein out of the active site, alter the protein's interaction with other proteins, or force conformational changes that translate into changes in $V_{max}$ or $K_m$. For covalent modification to be useful in regulation, the cell must be able to restore the altered enzyme to its original condition. A family of phosphoprotein phosphatases, at least some of which are themselves under regulation, catalyze dephosphorylation of proteins that have been phosphorylated by protein kinases.

Alteration of the number of enzyme molecules and covalent modifications are generally triggered by some signal from outside the cell—a hormone or growth factor, for example—and result in a change of metabolite

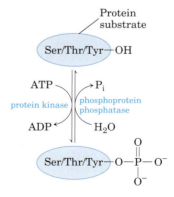

**FIGURE 15–14 Protein phosphorylation and dephosphorylation.** Protein kinases transfer a phosphoryl group from ATP to a Ser, Thr, or Tyr residue in a protein. Protein phosphatases remove the phosphoryl group as $P_i$.

flux through one or more pathways. In contrast, the very rapid (milliseconds) allosteric changes in enzyme activity are generally triggered locally, by changes in the level of a metabolite within the cell. The allosteric effector may be a substrate of the affected pathway (glucose for glycolysis, for example), a product of a pathway (ATP from glycolysis), or a key metabolite or cofactor (such as NADH) that indicates the cell's metabolic state. A single enzyme is commonly regulated in several ways—for example, by modulation of its synthesis, by covalent alteration, and by allosteric effectors.

Yet another way to alter the effective activity of an enzyme is to change the accessibility of its substrate. The hexokinase of muscle cannot act on glucose until the sugar enters the myocyte from the blood, and the rate at which it enters depends on the activity of glucose transporters in the plasma membrane. Within cells, membrane-bounded compartments segregate certain enzymes and enzyme systems, and the transport of substrate into these compartments may be the limiting factor in enzyme action.

In the discussion that follows, it is useful to think of changes in enzymatic activity as serving two distinct though complementary roles. We use the term **metabolic regulation** to refer to processes that serve to maintain homeostasis at the molecular level—to hold some cellular parameter (concentration of a metabolite, for example) at a steady level over time, even as the flow of metabolites through the pathway changes. The term **metabolic control** refers to a process that leads to a change in the output of a metabolic pathway over time, in response to some outside signal or change in circumstances. The distinction, although useful, is not always easy to make.

## SUMMARY 15.2 Regulation of Metabolic Pathways

- In a metabolically active cell in a steady state, intermediates are formed and consumed at equal rates. When a perturbation alters the rate of formation or consumption of a metabolite, compensating changes in enzyme activities return the system to the steady state.

- Regulatory mechanisms maintain nearly constant levels of key metabolites such as ATP and NADH in cells and glucose in the blood, while matching the use or storage of glycogen to the organism's changing needs.

- In multistep processes such as glycolysis, certain reactions are essentially at equilibrium in the steady state; the rates of these substrate-limited reactions rise and fall with substrate concentration. Other reactions are far from equilibrium; their rates are too slow to produce instant equilibration of substrate and product. These enzyme-limited reactions are

often highly exergonic and therefore metabolically irreversible, and the enzymes that catalyze them are commonly the points at which flux through the pathway is regulated.

- The activity of an enzyme can be regulated by changing the rate of its synthesis or degradation, by allosteric or covalent alteration of existing enzyme molecules, or by separating the enzyme from its substrate in subcellular compartments.

- Fast metabolic adjustments (on the time scale of seconds or less) at the intracellular level are generally allosteric. The effects of hormones and growth factors are generally slower (seconds to hours) and are commonly achieved by covalent modification or changes in enzyme synthesis.

## 15.3 Coordinated Regulation of Glycolysis and Gluconeogenesis

In mammals, gluconeogenesis occurs primarily in the liver, where its role is to provide glucose for export to other tissues when glycogen stores are exhausted. Gluconeogenesis employs most of the enzymes that act in glycolysis, but it is not simply the reversal of glycolysis. Seven of the glycolytic reactions are freely reversible, and the enzymes that catalyze these reactions also function in gluconeogenesis (Fig. 15–15). Three reactions of glycolysis are so exergonic as to be essentially irreversible: those catalyzed by hexokinase, PFK-1, and pyruvate kinase. Notice in Table 15–2 that all three reactions have a large, negative $\Delta G'$. Gluconeogenesis uses detours around each of these irreversible steps; for example, the conversion of fructose 1,6-bisphosphate to fructose 6-phosphate is catalyzed by fructose 1,6-bisphosphatase (FBPase-1; Fig. 15–15). Note that each of these bypass reactions also has a large, negative $\Delta G'$.

At each of the three points where glycolytic reactions are bypassed by alternative, gluconeogenic reactions, simultaneous operation of both pathways would consume ATP without accomplishing any chemical or biological work. For example, PFK-1 and FBPase-1 catalyze opposing reactions:

$$\text{ATP} + \text{fructose 6-phosphate} \xrightarrow{\text{PFK-1}}$$
$$\text{ADP} + \text{fructose 1,6-bisphosphate}$$

$$\text{Fructose 1,6-bisphosphate} + \text{H}_2\text{O} \xrightarrow{\text{FBPase-1}}$$
$$\text{fructose 6-phosphate} + \text{P}_i$$

The sum of these two reactions is

$$\text{ATP} + \text{H}_2\text{O} \longrightarrow \text{ADP} + \text{P}_i + \text{heat}$$

that is, hydrolysis of ATP without any useful metabolic work being done. Clearly, if these two reactions were

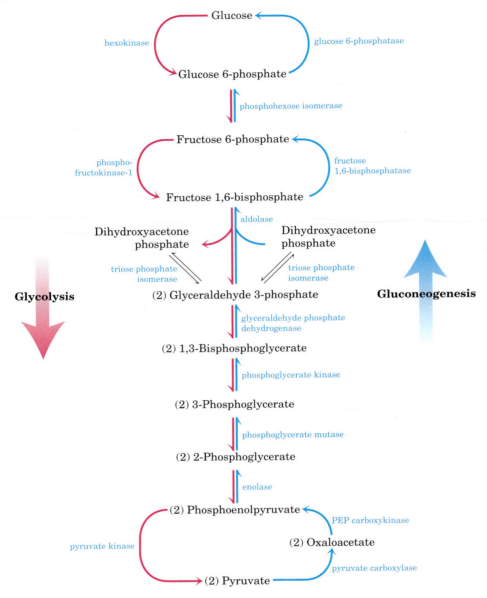

**FIGURE 15–15 Glycolysis and gluconeogenesis.** Opposing pathways of glycolysis (pink) and gluconeogenesis (blue) in rat liver. Three steps are catalyzed by different enzymes in gluconeogenesis (the "bypass reactions") and glycolysis; seven steps are catalyzed by the same enzymes in the two pathways. Cofactors have been omitted for simplicity.

allowed to proceed simultaneously at a high rate in the same cell, a large amount of chemical energy would be dissipated as heat. This uneconomical process has been called a **futile cycle.** However, as we shall see later, such cycles may provide advantages for controlling pathways, and the term **substrate cycle** is a better description. Similar substrate cycles also occur with the other two sets of bypass reactions of gluconeogenesis (Fig. 15–15).

We begin our examination of the coordinated regulation of glycolysis and gluconeogenesis by considering the regulatory patterns seen at the three main control points of glycolysis. We then look at the regulation of the enzymes of gluconeogenesis, leading to a consideration of how the regulation of both pathways is tightly, reciprocally coordinated.

## Hexokinase Isozymes of Muscle and Liver Are Affected Differently by Their Product, Glucose 6-Phosphate

Hexokinase, which catalyzes the entry of free glucose into the glycolytic pathway, is a regulatory enzyme. There are four isozymes (designated I to IV), encoded by four different genes. Isozymes are different proteins that catalyze the same reaction (Box 15–2). The predominant hexokinase isozyme of myocytes (hexokinase II) has a high affinity for glucose—it is half-saturated at about 0.1 mM. Because glucose entering myocytes from the blood (where the glucose concentration is 4 to 5 mM) produces an intracellular glucose concentration high enough to saturate hexokinase II, the enzyme normally acts at or near its maximal rate. Muscle hexokinases I and II are allosterically inhibited by their product, glucose 6-phosphate, so whenever the cellular concentra-

tion of glucose 6-phosphate rises above its normal level, these isozymes are temporarily and reversibly inhibited, bringing the rate of glucose 6-phosphate formation into balance with the rate of its utilization and reestablishing the steady state.

The different hexokinase isozymes of liver and muscle reflect the different roles of these organs in carbohydrate metabolism: muscle consumes glucose, using it for energy production, whereas liver maintains blood glucose homeostasis by removing or producing glucose, depending on the prevailing glucose concentration. The

---

## BOX 15–2    WORKING IN BIOCHEMISTRY

### Isozymes: Different Proteins That Catalyze the Same Reaction

The four forms of hexokinase found in mammalian tissues are but one example of a common biological situation: the same reaction catalyzed by two or more different molecular forms of an enzyme. These multiple forms, called isozymes or isoenzymes, may occur in the same species, in the same tissue, or even in the same cell. The different forms of the enzyme generally differ in kinetic or regulatory properties, in the cofactor they use (NADH or NADPH for dehydrogenase isozymes, for example), or in their subcellular distribution (soluble or membrane-bound). Isozymes may have similar, but not identical, amino acid sequences, and in many cases they clearly share a common evolutionary origin.

One of the first enzymes found to have isozymes was lactate dehydrogenase (LDH) (p. 538), which, in vertebrate tissues, exists as at least five different isozymes separable by electrophoresis. All LDH isozymes contain four polypeptide chains (each of $M_r$ 33,500), each type containing a different ratio of two kinds of polypeptides. The M (for muscle) chain and the H (for heart) chain are encoded by two different genes.

In skeletal muscle the predominant isozyme contains four M chains, and in heart the predominant isozyme contains four H chains. Other tissues have some combination of the five possible types of LDH isozymes:

| Type | Composition | Location |
|------|-------------|----------|
| LDH$_1$ | HHHH | Heart and erythrocyte |
| LDH$_2$ | HHHM | Heart and erythrocyte |
| LDH$_3$ | HHMM | Brain and kidney |
| LDH$_4$ | HMMM | Skeletal muscle and liver |
| LDH$_5$ | MMMM | Skeletal muscle and liver |

These differences in the isozyme content of tissues can be used to assess the timing and extent of heart damage due to myocardial infarction (heart attack). Damage to heart tissue results in the release of heart LDH into the blood. Shortly after a heart attack, the blood level of total LDH increases, and there is more LDH$_2$ than LDH$_1$. After 12 hours the amounts of LDH$_1$ and LDH$_2$ are very similar, and after 24 hours there is more LDH$_1$ than LDH$_2$. This switch in the LDH$_1$/LDH$_2$ ratio, combined with increased concentrations in the blood of another heart enzyme, creatine kinase, is very strong evidence of a recent myocardial infarction. ■

The different LDH isozymes have significantly different values of $V_{max}$ and $K_M$, particularly for pyruvate. The properties of LDH$_4$ favor rapid reduction of very low concentrations of pyruvate to lactate in skeletal muscle, whereas those of isozyme LDH$_1$ favor rapid oxidation of lactate to pyruvate in the heart.

In general, the distribution of different isozymes of a given enzyme reflects at least four factors:

1. *Different metabolic patterns in different organs.* For glycogen phosphorylase, the isozymes in skeletal muscle and liver have different regulatory properties, reflecting the different roles of glycogen breakdown in these two tissues.

2. *Different locations and metabolic roles for isozymes in the same cell.* The isocitrate dehydrogenase isozymes of the cytosol and the mitochondrion are an example (Chapter 16).

3. *Different stages of development in embryonic or fetal tissues and in adult tissues.* For example, the fetal liver has a characteristic isozyme distribution of LDH, which changes as the organ develops into its adult form. Some enzymes of glucose catabolism in malignant (cancer) cells occur as their fetal, not adult, isozymes.

4. *Different responses of isozymes to allosteric modulators.* This difference is useful in fine-tuning metabolic rates. Hexokinase IV (glucokinase) of liver and the hexokinase isozymes of other tissues differ in their sensitivity to inhibition by glucose 6-phosphate.

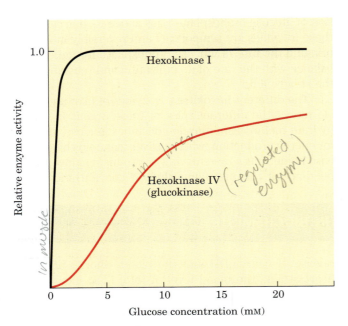

FIGURE 15–16 **Comparison of the kinetic properties of hexokinase IV (glucokinase) and hexokinase I.** Note the sigmoidicity for hexokinase IV and the much lower $K_m$ for hexokinase I. When blood glucose rises above 5 mM, hexokinase IV activity increases, but hexokinase I is already operating near $V_{max}$ at 5 mM glucose and cannot respond to an increase in glucose concentration. Hexokinases I, II, and III have similar kinetic properties.

predominant hexokinase isozyme of liver is hexokinase IV (glucokinase), which differs in three important respects from hexokinases I–III of muscle. First, the glucose concentration at which hexokinase IV is half-saturated (about 10 mM) is higher than the usual concentration of glucose in the blood. Because an efficient glucose transporter in hepatocytes (**GLUT2;** see Fig. 11–31) rapidly equilibrates the glucose concentrations in cytosol and blood, the high $K_m$ of hexokinase IV allows its direct regulation by the level of blood glucose (Fig. 15–16). When the blood glucose concentration is high, as it is after a meal rich in carbohydrates, excess glucose

is transported into hepatocytes, where hexokinase IV converts it to glucose 6-phosphate. Because hexokinase IV is not saturated at 10 mM glucose, its activity continues to increase as the glucose concentration rises to 10 mM or more.

Second, hexokinase IV is subject to inhibition by the reversible binding of a regulatory protein specific to liver (Fig. 15–17). The binding is much tighter in the presence of the allosteric effector fructose 6-phosphate. Glucose competes with fructose 6-phosphate for binding and causes dissociation of the regulatory protein from hexokinase IV, relieving the inhibition. Immediately after a carbohydrate-rich meal, when blood glucose is high, glucose enters the hepatocyte via GLUT2 and activates hexokinase IV by this mechanism. During a fast, when blood glucose drops below 5 mM, fructose 6-phosphate triggers the inhibition of hexokinase IV by the regulatory protein, so the liver does not compete with other organs for the scarce glucose. The mechanism of inhibition by the regulatory protein is interesting: the protein anchors hexokinase IV inside the nucleus, where it is segregated from the other enzymes of glycolysis in the cytosol (Fig. 15–17). When the glucose concentration in the cell rises, it equilibrates with glucose in the nucleus by transport through the nuclear pores. Glucose causes dissociation of the regulatory protein, and hexokinase IV enters the cytosol and begins to phosphorylate glucose.

Third, hexokinase IV is not inhibited by glucose 6-phosphate, and it can therefore continue to operate when the accumulation of glucose 6-phosphate completely inhibits hexokinases I–III.

## Phosphofructokinase-1 Is under Complex Allosteric Regulation

As we have noted, glucose 6-phosphate can flow either into glycolysis or through any of several other pathways, including glycogen synthesis and the pentose phosphate pathway. The metabolically irreversible reaction catalyzed by PFK-1 is the step that commits glucose to gly-

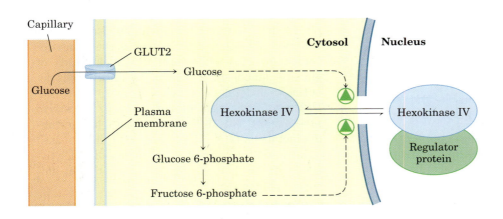

FIGURE 15–17 **Regulation of hexokinase IV (glucokinase) by sequestration in the nucleus.** The protein inhibitor of hexokinase IV is a nuclear binding protein that draws hexokinase IV into the nucleus when the fructose 6-phosphate concentration in liver is high and releases it to the cytosol when the glucose concentration is high.

colysis. In addition to its substrate-binding sites, this complex enzyme has several regulatory sites at which allosteric activators or inhibitors bind.

ATP is not only a substrate for PFK-1 but also an end product of the glycolytic pathway. When high cellular [ATP] signals that ATP is being produced faster

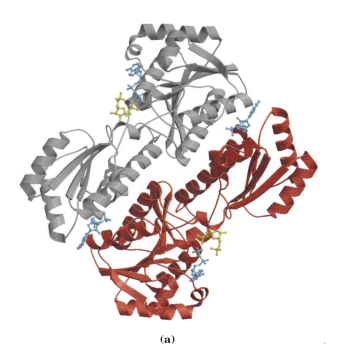

**(a)**

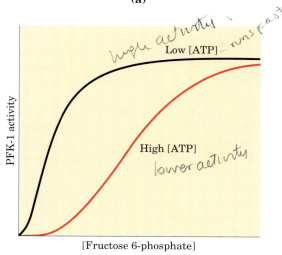

*(handwritten) high activity — runs faster*

Low [ATP]

High [ATP]

*(handwritten) lower activity*

PFK-1 activity

[Fructose 6-phosphate]

**(b)**

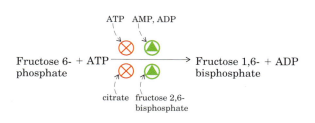

ATP  AMP, ADP

Fructose 6-   + ATP  →  Fructose 1,6-  + ADP
phosphate                    bisphosphate

citrate  fructose 2,6-
bisphosphate

**(c)**

than it is being consumed, ATP inhibits PFK-1 by binding to an allosteric site and lowering the affinity of the enzyme for fructose 6-phosphate (Fig. 15–18). ADP and AMP, which increase in concentration as consumption of ATP outpaces production, act allosterically to relieve this inhibition by ATP. These effects combine to produce higher enzyme activity when ADP or AMP accumulates and lower activity when ATP accumulates.

Citrate (the ionized form of citric acid), a key intermediate in the aerobic oxidation of pyruvate, fatty acids, and amino acids, is also an allosteric regulator of PFK-1; high citrate concentration increases the inhibitory effect of ATP, further reducing the flow of glucose through glycolysis. In this case, as in several others encountered later, citrate serves as an intracellular signal that the cell is meeting its current needs for energy-yielding metabolism by the oxidation of fats and proteins.

The most significant allosteric regulator of PFK-1 is fructose 2,6-bisphosphate, which strongly activates the enzyme. We return to this role of fructose 2,6-bisphosphate later.

### Pyruvate Kinase Is Allosterically Inhibited by ATP

At least three isozymes of pyruvate kinase are found in vertebrates, differing in their tissue distribution and their response to modulators. High concentrations of ATP, acetyl-CoA, and long-chain fatty acids (signs of abundant energy supply) allosterically inhibit all isozymes of pyruvate kinase (Fig. 15–19). The liver isozyme (L form), but not the muscle isozyme (M form), is subject to further regulation by phosphorylation. When low blood glucose causes glucagon release, cAMP-dependent protein kinase phosphorylates the L isozyme of pyruvate kinase, inactivating it. This slows the use of glucose as a fuel in liver, sparing it for export to the brain and other organs. In muscle, the effect of increased [cAMP] is quite different. In response to epinephrine, cAMP activates glycogen breakdown and glycolysis and provides the fuel needed for the fight-or-flight response.

**FIGURE 15–18 Phosphofructokinase-1 (PFK-1) and its regulation. (a)** Ribbon diagram of *E. coli* phosphofructokinase-1, showing two of its four identical subunits (PDB ID 1PFK). Each subunit has its own catalytic site, where ADP (blue) and fructose 1,6-bisphosphate (yellow) are almost in contact, and its own binding sites for the allosteric regulator ADP (blue), located at the interface between subunits. **(b)** Allosteric regulation of muscle PFK-1 by ATP, shown by a substrate-activity curve. At low concentrations of ATP, the $K_{0.5}$ for fructose 6-phosphate is relatively low, enabling the enzyme to function at a high rate at relatively low concentrations of fructose 6-phosphate. (Recall from Chapter 6 that $K_{0.5}$ or $K_m$ is equivalent to the substrate concentration at which half-maximal enzyme activity occurs.) When the concentration of ATP is high, $K_{0.5}$ for fructose 6-phosphate is greatly increased, as indicated by the sigmoid relationship between substrate concentration and enzyme activity. **(c)** Summary of the regulators affecting PFK-1 activity.

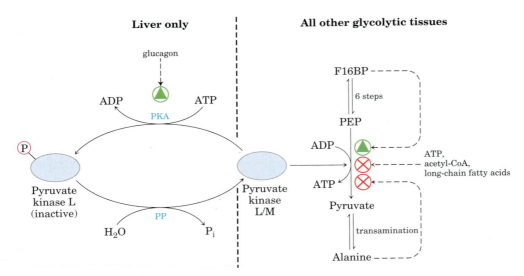

**FIGURE 15–19 Regulation of pyruvate kinase.** The enzyme is allosterically inhibited by ATP, acetyl-CoA, and long-chain fatty acids (all signs of an abundant energy supply), and the accumulation of fructose 1,6-bisphosphate triggers its activation. Accumulation of alanine, which can be synthesized from pyruvate in one step, allosterically inhibits pyruvate kinase, slowing the production of pyruvate by glycolysis. The liver isozyme (L form) is also regulated hormonally; glucagon activates cAMP-dependent protein kinase (PKA; see Fig. 15–25), which phosphorylates the pyruvate kinase L isozyme, inactivating it. When the glucagon level drops, a protein phosphatase (PP) dephosphorylates pyruvate kinase, activating it. This mechanism prevents the liver from consuming glucose by glycolysis when the blood glucose concentration is low; instead, liver exports glucose. The muscle isozyme (M form) is not affected by this phosphorylation mechanism.

## Gluconeogenesis Is Regulated at Several Steps

In the pathway leading from pyruvate to glucose, the first control point determines the fate of pyruvate in the mitochondrion. Pyruvate can be converted either to acetyl-CoA (by the pyruvate dehydrogenase complex; Chapter 16) to fuel the citric acid cycle, or to oxaloacetate (by pyruvate carboxylase) to start the process of gluconeogenesis (Fig. 15–20). When fatty acids are readily available as fuels, their breakdown in liver mitochondria yields acetyl-CoA, a signal that further oxidation of glucose for fuel is not necessary. Acetyl-CoA is a positive allosteric modulator of pyruvate carboxylase and a negative modulator of pyruvate dehydrogenase, through stimulation of a protein kinase that inactivates the dehydrogenase. When the cell's energetic needs are being met, oxidative phosphorylation slows, NADH rises relative to $NAD^+$ and inhibits the citric acid cycle, and acetyl-CoA accumulates. The increased concentration of acetyl-CoA inhibits the pyruvate dehydrogenase complex, slowing the formation of acetyl-CoA from pyruvate, and stimulates gluconeogenesis by activating pyruvate carboxylase, allowing excess pyruvate to be converted to glucose.

The second control point in gluconeogenesis is the reaction catalyzed by FBPase-1 (Fig. 15–21), which is strongly inhibited by AMP. The corresponding glycolytic enzyme, PFK-1, is stimulated by AMP and ADP but inhibited by citrate and ATP. Thus these opposing steps

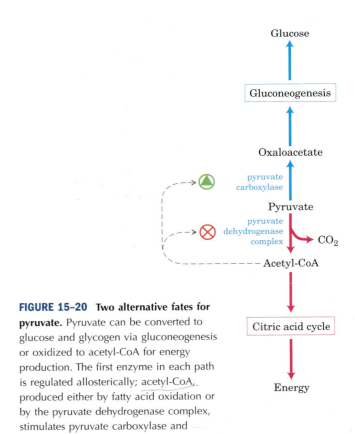

**FIGURE 15–20 Two alternative fates for pyruvate.** Pyruvate can be converted to glucose and glycogen via gluconeogenesis or oxidized to acetyl-CoA for energy production. The first enzyme in each path is regulated allosterically; acetyl-CoA, produced either by fatty acid oxidation or by the pyruvate dehydrogenase complex, stimulates pyruvate carboxylase and inhibits pyruvate dehydrogenase.

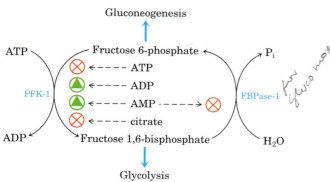

**FIGURE 15–21 Regulation of fructose 1,6-bisphosphatase-1 (FBPase-1) and phosphofructokinase-1 (PFK-1).** The important role of fructose 2,6-bisphosphate in the regulation of this substrate cycle is detailed in subsequent figures.

in the two pathways are regulated in a coordinated and reciprocal manner. In general, when sufficient concentrations of acetyl-CoA or citrate (the product of acetyl-CoA condensation with oxaloacetate) are present, or when a high proportion of the cell's adenylate is in the form of ATP, gluconeogenesis is favored. AMP promotes glycogen degradation and glycolysis by activating glycogen phosphorylase (via activation of phosphorylase kinase) and stimulating the activity of PFK-1.

All the regulatory actions discussed here are triggered by changes inside the cell and are mediated by very rapid, instantly reversible, allosteric mechanisms. Another set of regulatory processes is triggered from outside the cell by the hormones insulin and glucagon, which signal too much or too little glucose in the blood, respectively, or by epinephrine, which signals the impending need for fuel for a fight-or-flight response. These hormonal signals bring about covalent modification (phosphorylation or dephosphorylation) of target proteins inside the cell; this takes place on a somewhat longer time scale than the internally driven allosteric mechanisms—seconds or minutes, rather than milliseconds.

### Fructose 2,6-Bisphosphate Is a Potent Regulator of Glycolysis and Gluconeogenesis

The special role of liver in maintaining a constant blood glucose level requires additional regulatory mechanisms to coordinate glucose production and consumption. When the blood glucose level decreases, the hormone **glucagon** signals the liver to produce and release more glucose and to stop consuming it for its own needs. One source of glucose is glycogen stored in the liver; another source is gluconeogenesis.

The hormonal regulation of glycolysis and gluconeogenesis is mediated by **fructose 2,6-bisphosphate,** an allosteric effector for the enzymes PFK-1 and FBPase-1 (Fig. 15–22):

Fructose 2,6-bisphosphate

When fructose 2,6-bisphosphate binds to its allosteric site on PFK-1, it increases that enzyme's affinity for its substrate, fructose 6-phosphate, and reduces its affinity for the allosteric inhibitors ATP and citrate. At the physiological concentrations of its substrates ATP and fructose 6-phosphate and of its other positive and negative effectors (ATP, AMP, citrate), PFK-1 is virtually inactive in the absence of fructose 2,6-bisphosphate. Fructose 2,6-bisphosphate *activates* PFK-1 and stimulates glycolysis in liver and, at the same time, *inhibits* FBPase-1, thereby slowing gluconeogenesis.

Although structurally related to fructose 1,6-bisphosphate, fructose 2,6-bisphosphate is not an intermediate in gluconeogenesis or glycolysis; it is a *regulator* whose cellular level reflects the level of glucagon in the blood, which rises when blood glucose falls. The cellular concentration of fructose 2,6-bisphosphate is set by the relative rates of its formation and breakdown (Fig. 15–23a). It is formed by phosphorylation of fructose 6-phosphate, catalyzed by **phosphofructokinase-2 (PFK-2),** and is broken down by **fructose 2,6-bisphosphatase (FBPase-2).** (Note that these enzymes are distinct from PFK-1 and FBPase-1, which catalyze the formation and breakdown, respectively, of fructose 1,6-bisphosphate.) PFK-2 and FBPase-2 are two distinct enzymatic activities of a single, bifunctional protein. The balance of these two activities in the liver, which determines the cellular level of fructose 2,6-bisphosphate, is regulated by glucagon and insulin (Fig. 15–23b). As we saw in Chapter 12 (p. 441), glucagon stimulates the adenylyl cyclase of liver to synthesize 3′,5′-cyclic AMP (cAMP) from ATP. Then cAMP activates cAMP-dependent protein kinase, which transfers a phosphoryl group from ATP to the bifunctional protein PFK-2/FBPase-2. Phosphorylation of this protein enhances its FBPase-2 activity and inhibits its PFK-2 activity. Glucagon thereby lowers the cellular level of fructose 2,6-bisphosphate, inhibiting glycolysis and stimulating gluconeogenesis. The resulting production of more glucose enables the liver to replenish blood glucose in response to glucagon. Insulin has the opposite effect, stimulating the activity of a phosphoprotein phosphatase that catalyzes removal of the phosphoryl group from the bifunctional protein PFK-2/FBPase-2, activating its PFK-2 activity, increasing the level of fructose 2,6-bisphosphate, stimulating glycolysis, and inhibiting gluconeogenesis.

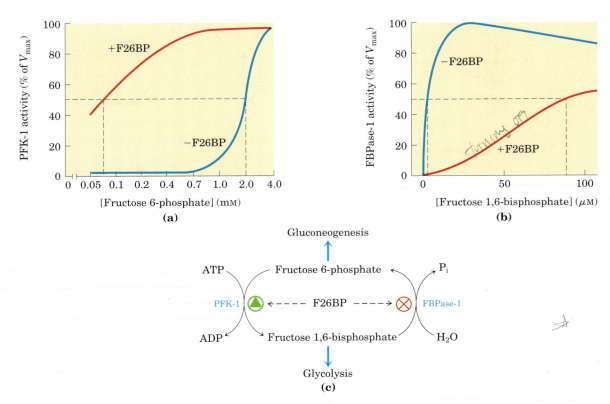

**FIGURE 15–22 Role of fructose 2,6-bisphosphate in regulation of glycolysis and gluconeogenesis.** Fructose 2,6-bisphosphate (F26BP) has opposite effects on the enzymatic activities of phosphofructokinase-1 (PFK-1, a glycolytic enzyme) and fructose 1,6-bisphosphatase (FBPase-1, a gluconeogenic enzyme). **(a)** PFK-1 activity in the absence of F26BP (blue curve) is half-maximal when the concentration of fructose 6-phosphate is 2 mM (that is, $K_{0.5} = 2$ mM). When 0.13 μM F26BP is present (red curve), the $K_{0.5}$ for fructose 6-phosphate is only 0.08 mM. Thus F26BP activates PFK-1 by increasing its apparent affinity (Fig. 15–18) for fructose 6-phosphate. **(b)** FBPase-1 activity is inhibited by as little as 1 μM F26BP and is strongly inhibited by 25 μM. In the absence of this inhibitor (blue curve) the $K_{0.5}$ for fructose 1,6-bisphosphate is 5 μM, but in the presence of 25 μM F26BP (red curve) the $K_{0.5}$ is >70 μM. Fructose 2,6-bisphosphate also makes FBPase-1 more sensitive to inhibition by another allosteric regulator, AMP. **(c)** Summary of regulation by F26BP.

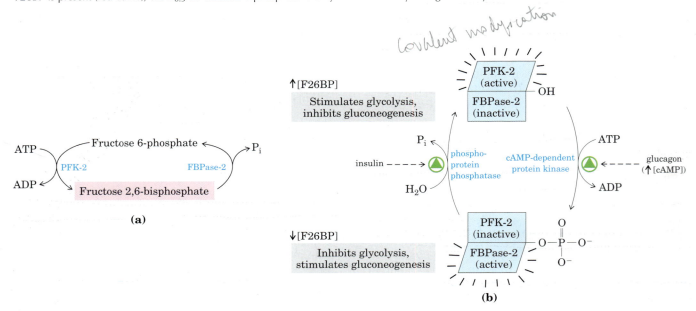

**FIGURE 15–23 Regulation of fructose 2,6-bisphosphate level.** **(a)** The cellular concentration of the regulator fructose 2,6-bisphosphate (F26BP) is determined by the rates of its synthesis by phosphofructokinase-2 (PFK-2) and breakdown by fructose 2,6-bisphosphatase (FBPase-2). **(b)** Both enzymes are part of the same polypeptide chain, and both are regulated, in a reciprocal fashion, by insulin and glucagon. Here and elsewhere, arrows are used to indicate increasing (↑) and decreasing (↓) levels of metabolites.

## Are Substrate Cycles Futile?

We noted above that substrate cycles (sometimes called futile cycles) occur at several points in the pathways that interconnect glycogen and pyruvate. For reactions such as those catalyzed by PFK-1 and FBPase-1 (Fig. 15–22c) to take place at the same time, each must be exergonic under the conditions prevailing in the cell. The PFK-1 reaction is exergonic because it involves a phosphoryl group transfer from ATP, and the FBPase-1 reaction is exergonic because it entails hydrolysis of a phosphate ester. Because the cycle involves two different enzymes, not simply one working in both directions, each activity can be regulated separately: fructose 2,6-bisphosphate activates PFK-1, favoring glycolysis, and inhibits FBPase-1, inhibiting gluconeogenesis. The two-enzyme cycle thus provides a means of controlling the *direction* of net metabolite flow. The apparent energetic disadvantage of the "futile" cycle is evidently outweighed by the advantage of allowing this type of control of pathway direction.

## Xylulose 5-Phosphate Is a Key Regulator of Carbohydrate and Fat Metabolism

Another recently discovered regulatory mechanism also acts by controlling the level of fructose 2,6-bisphosphate. In the mammalian liver, xylulose 5-phosphate (see Fig. 14–23), a product of the hexose monophosphate pathway, mediates the increase in glycolysis that follows ingestion of a high-carbohydrate meal. The xylulose 5-phosphate concentration rises as glucose entering the liver is converted to glucose 6-phosphate and enters both the glycolytic and hexose monophosphate pathways. Xylulose 5-phosphate activates a phosphoprotein phosphatase, PP2A, that dephosphorylates the bifunctional PFK-2/FBPase-2 enzyme. Dephosphorylation activates PFK-2 and inhibits FBPase-2, and the resulting rise in [fructose 2,6-bisphosphate] stimulates glycolysis and inhibits gluconeogenesis. The increased glycolysis boosts the production of acetyl-CoA, while the increased flow of hexose through the hexose monophosphate pathway generates NADPH. Acetyl-CoA and NADPH are the starting materials for fatty acid synthesis, which has long been known to increase dramatically in response to intake of a high-carbohydrate meal. Xylulose 5-phosphate also increases the synthesis of all the enzymes required for fatty acid synthesis; we shall return to this effect in our discussion of the integration of carbohydrate and lipid metabolism (Chapter 23).

## SUMMARY 15.3 Coordinated Regulation of Glycolysis and Gluconeogenesis

- Three glycolytic enzymes are subject to allosteric regulation: hexokinase IV, phosphofructokinase-1 (PFK-1), and pyruvate kinase.

- Hexokinase IV (glucokinase) is sequestered in the nucleus of the hepatocyte, but is released when the cytosolic glucose concentration rises.

- PFK-1 is allosterically inhibited by ATP and citrate. In most mammalian tissues, including liver, PFK-1 is allosterically activated by fructose 2,6-bisphosphate.

- Pyruvate kinase is allosterically inhibited by ATP, and the liver isozyme is inhibited by cAMP-dependent phosphorylation.

- Gluconeogenesis is regulated at the level of pyruvate carboxylase (which is activated by acetyl-CoA) and FBPase-1 (which is inhibited by fructose 2,6-bisphosphate and AMP).

- To limit futile cycling between glycolysis and gluconeogenesis, the two pathways are under reciprocal allosteric control, mainly achieved by the opposite effects of fructose 2,6-bisphosphate on PFK-1 and FBPase-1.

- Glucagon or epinephrine decreases [fructose 2,6-bisphosphate]. The hormones do this by raising [cAMP] and bringing about phosphorylation of the bifunctional enzyme that makes and breaks down fructose 2,6-bisphosphate. Phosphorylation inactivates PFK-2 and activates FBPase-2, leading to breakdown of fructose 2,6-bisphosphate. Insulin increases [fructose 2,6-bisphosphate] by activating a phosphoprotein phosphatase that dephosphorylates (activates) PFK-2.

# 15.4 Coordinated Regulation of Glycogen Synthesis and Breakdown

As we have seen, the mobilization of stored glycogen is brought about by glycogen phosphorylase, which degrades glycogen to glucose 1-phosphate (Fig. 15–3). Glycogen phosphorylase provides an especially instructive case of enzyme regulation. It was one of the first known examples of an allosterically regulated enzyme and the first enzyme shown to be controlled by reversible phosphorylation. It was also one of the first allosteric enzymes for which the detailed three-dimensional structures of the active and inactive forms were revealed by x-ray crystallographic studies. Glycogen phosphorylase also illustrates how isozymes play their tissue-specific roles.

## Glycogen Phosphorylase Is Regulated Allosterically and Hormonally

In the late 1930s, Carl and Gerty Cori (Box 15–1) discovered that the glycogen phosphorylase of skeletal muscle exists in two interconvertible forms: **glycogen phosphorylase *a*,** which is catalytically active, and

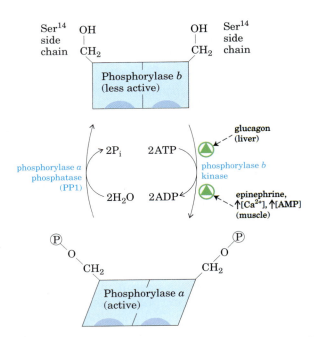

**FIGURE 15–24 Regulation of muscle glycogen phosphorylase by covalent modification.** In the more active form of the enzyme, phosphorylase *a*, Ser[14] residues, one on each subunit, are phosphorylated. Phosphorylase *a* is converted to the less active form, phosphorylase *b*, by enzymatic loss of these phosphoryl groups, catalyzed by phosphorylase *a* phosphatase (PP1). Phosphorylase *b* can be reconverted (reactivated) to phosphorylase *a* by the action of phosphorylase *b* kinase.

Earl W. Sutherland, Jr., 1915–1974

**glycogen phosphorylase *b*,** which is less active (Fig. 15–24). Subsequent studies by Earl Sutherland showed that phosphorylase *b* predominates in resting muscle, but during vigorous muscular activity the hormone epinephrine triggers phosphorylation of a specific Ser residue in phosphorylase *b*, converting it to its more active form, phosphorylase *a*. (Note that glycogen phosphorylase is often referred to simply as phosphorylase—so honored because it was the first phosphorylase to be discovered; the shortened name has persisted in common usage and in the literature.)

The enzyme (phosphorylase *b* kinase) responsible for activating phosphorylase by transferring a phosphoryl group to its Ser residue is itself activated by epinephrine or glucagon through a series of steps shown in Figure 15–25. Sutherland discovered the second messenger cAMP, which increases in concentration in response to stimulation by epinephrine (in muscle) or glucagon (in liver). Elevated [cAMP] ini-

tiates an **enzyme cascade,** in which a catalyst activates a catalyst, which activates a catalyst. Such cascades allow for large amplification of the initial signal (see pink boxes in Fig. 15–25). The rise in [cAMP] activates cAMP-dependent protein kinase, also called protein kinase A (PKA). PKA then phosphorylates and activates **phosphorylase *b* kinase,** which catalyzes the phosphorylation of Ser residues in each of the two identical subunits of glycogen phosphorylase, activating it and thus stimulating glycogen breakdown. In muscle, this provides fuel for glycolysis to sustain muscle contraction for the fight-or-flight response signaled by epinephrine. In liver, glycogen breakdown counters the low blood glucose signaled by glucagon, releasing glucose. These different roles are reflected in subtle differences in the regulatory mechanisms in muscle and liver. The glycogen phosphorylases of liver and muscle are isozymes, encoded by different genes and differing in their regulatory properties.

In muscle, superimposed on the regulation of phosphorylase by covalent modification are two allosteric control mechanisms (Fig. 15–25). $Ca^{2+}$, the signal for muscle contraction, binds to and activates phosphorylase *b* kinase, promoting conversion of phosphorylase *b* to the active *a* form. $Ca^{2+}$ binds to phosphorylase *b* kinase through its δ subunit, which is calmodulin (see Fig. 12–21). AMP, which accumulates in vigorously contracting muscle as a result of ATP breakdown, binds to and activates phosphorylase, speeding the release of glucose 1-phosphate from glycogen. When ATP levels are adequate, ATP blocks the allosteric site to which AMP binds, inactivating phosphorylase.

When the muscle returns to rest, a second enzyme, **phosphorylase *a* phosphatase,** also called **phosphoprotein phosphatase 1 (PP1),** removes the phosphoryl groups from phosphorylase *a*, converting it to the less active form, phosphorylase *b*.

Like the enzyme of muscle, the glycogen phosphorylase of liver is regulated hormonally (by phosphorylation/dephosphorylation) and allosterically. The dephosphorylated form is essentially inactive. When the blood glucose level is too low, glucagon (acting by the same cascade mechanism shown in Fig. 15–25) activates phosphorylase *b* kinase, which in turn converts phosphorylase *b* to its active *a* form, initiating the release of glucose into the blood. When blood glucose levels return to normal, glucose enters hepatocytes and binds to an inhibitory allosteric site on phosphorylase *a*. This binding also produces a conformational change that exposes the phosphorylated Ser residues to PP1, which catalyzes their dephosphorylation and inactivates the phosphorylase (Fig. 15–26). The allosteric site for glucose allows liver glycogen phosphorylase to act as its own glucose sensor and to respond appropriately to changes in blood glucose.

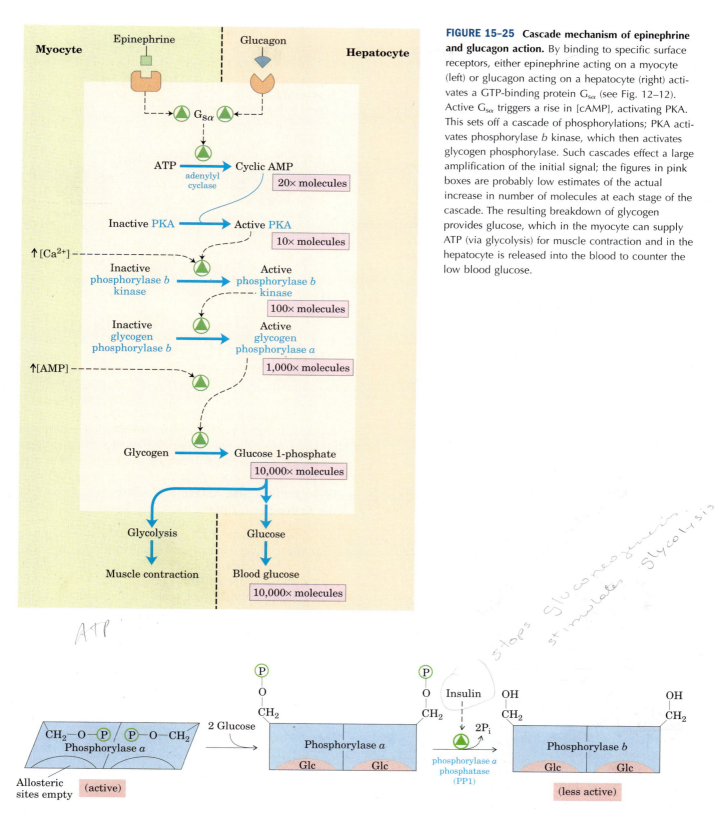

**FIGURE 15-25 Cascade mechanism of epinephrine and glucagon action.** By binding to specific surface receptors, either epinephrine acting on a myocyte (left) or glucagon acting on a hepatocyte (right) activates a GTP-binding protein $G_{s\alpha}$ (see Fig. 12–12). Active $G_{s\alpha}$ triggers a rise in [cAMP], activating PKA. This sets off a cascade of phosphorylations; PKA activates phosphorylase b kinase, which then activates glycogen phosphorylase. Such cascades effect a large amplification of the initial signal; the figures in pink boxes are probably low estimates of the actual increase in number of molecules at each stage of the cascade. The resulting breakdown of glycogen provides glucose, which in the myocyte can supply ATP (via glycolysis) for muscle contraction and in the hepatocyte is released into the blood to counter the low blood glucose.

**FIGURE 15-26 Glycogen phosphorylase of liver as a glucose sensor.** Glucose binding to an allosteric site of the phosphorylase a isozyme of liver induces a conformational change that exposes its phosphorylated Ser residues to the action of phosphorylase a phosphatase 1(PP1). This phosphatase converts phosphorylase a to phosphorylase b, sharply reducing the activity of phosphorylase and slowing glycogen breakdown in response to high blood glucose. Insulin also acts indirectly to stimulate PP1 and slow glycogen breakdown.

**FIGURE 15–27 Effects of GSK3 on glycogen synthase activity.**
Glycogen synthase *a*, the active form, has three Ser residues near its carboxyl terminus, which are phosphorylated by glycogen synthase kinase 3 (GSK3). This converts glycogen synthase to the inactive (*b*) form (GS*b*). GSK3 action requires prior phosphorylation (priming) by casein kinase (CKII). Insulin triggers activation of glycogen synthase *b* by blocking the activity of GSK3 (see the pathway for this action in Fig. 12–8) and activating a phosphoprotein phosphatase (PP1 in muscle, another phosphatase in liver). In muscle, epinephrine activates PKA, which phosphorylates the glycogen-targeting protein G$_M$ (see Fig. 15–30) on a site that causes dissociation of PP1 from glycogen. Glucose 6-phosphate favors dephosphorylation of glycogen synthase by binding to it and promoting a conformation that is a good substrate for PP1. Glucose also promotes dephosphorylation; the binding of glucose to glycogen phosphorylase *a* forces a conformational change that favors dephosphorylation to glycogen phosphorylase *b*, thus relieving its inhibition of PP1 (see Fig. 15–29).

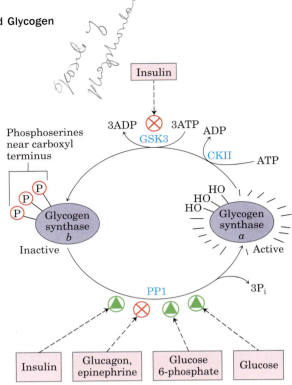

## Glycogen Synthase Is Also Regulated by Phosphorylation and Dephosphorylation

Like glycogen phosphorylase, glycogen synthase can exist in phosphorylated and dephosphorylated forms (Fig. 15–27). Its active form, **glycogen synthase *a*,** is unphosphorylated. Phosphorylation of the hydroxyl side chains of several Ser residues of both subunits converts glycogen synthase *a* to **glycogen synthase *b*,** which is inactive unless its allosteric activator, glucose 6-phosphate, is present. Glycogen synthase is remarkable for its ability to be phosphorylated on various residues by at least 11 different protein kinases. The most important regulatory kinase is **glycogen synthase kinase 3 (GSK3),** which adds phosphoryl groups to three Ser residues near the carboxyl terminus of glycogen synthase, strongly inactivating it. The action of GSK3 is hierarchical; it cannot phosphorylate glycogen synthase until another protein kinase, **casein kinase II (CKII),** has first phosphorylated the glycogen synthase on a nearby residue, an event called **priming** (Fig. 15–28a).

In liver, conversion of glycogen synthase *b* to the active form is promoted by PP1, which is bound to the glycogen particle. PP1 removes the phosphoryl groups from the three Ser residues phosphorylated by GSK3. Glucose 6-phosphate binds to an allosteric site on glycogen synthase *b*, making the enzyme a better substrate for dephosphorylation by PP1 and causing its activation. By analogy with glycogen phosphorylase, which acts as a glucose sensor, glycogen synthase can be regarded as a glucose 6-phosphate sensor. In muscle, a different phosphatase may have the role played by PP1 in liver, activating glycogen synthase by dephosphorylating it.

## Glycogen Synthase Kinase 3 Mediates the Actions of Insulin

As we saw in Chapter 12, one way in which insulin triggers intracellular changes is by activating a protein kinase (protein kinase B, or PKB) that in turn phosphorylates and inactivates GSK3 (Fig. 15–29; see also Fig. 12–8). Phosphorylation of a Ser residue near the amino terminus of GSK3 converts that region of the protein to a pseudosubstrate, which folds into the site at which the priming phosphorylated Ser residue normally binds (Fig. 15–28b). This prevents GSK3 from binding the priming site of a real substrate, thereby inactivating the enzyme and tipping the balance in favor of dephosphorylation of glycogen synthase by PP1. Glycogen phosphorylase can also affect the phosphorylation of glycogen synthase: active glycogen phosphorylase directly inhibits PP1, preventing it from activating glycogen synthase (Fig. 15–27).

Although first discovered in its role in glycogen metabolism (hence the name glycogen synthase kinase), GSK3 clearly has a much broader role than the regulation of glycogen synthase. It mediates signaling by insulin and other growth factors and nutrients, and it acts in the specification of cell fates during embryonic development. Among its targets are cytoskeletal proteins

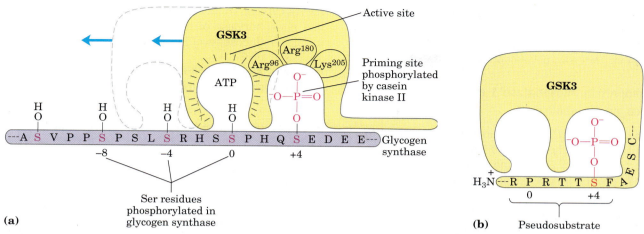

**(a)**

**(b)** Pseudosubstrate

**FIGURE 15–28 Priming of GSK3 phosphorylation of glycogen synthase. (a)** Glycogen synthase kinase 3 first associates with its substrate (glycogen synthase) by interaction between three positively charged residues (Arg[96], Arg[180], Lys[205]) and a phosphoserine residue at position +4 in the substrate. (For orientation, the Ser or Thr residue to be phosphorylated in the substrate is assigned the index 0. Residues on the amino-terminal side of this residue are numbered −1, −2, and so forth; residues on the carboxyl-terminal side are numbered +1, +2, and so forth.) This association aligns the active site of the enzyme with a Ser residue at position 0, which it phosphorylates. This creates a new priming site, and the enzyme moves down the protein to phosphorylate the Ser residue at position −4, and then the Ser at −8. **(b)** GSK3 has a Ser residue near its amino terminus that can be phosphorylated by PKA or PKB (see Fig. 15–29). This produces a "pseudosubstrate" region in GSK3 that folds into the priming site and makes the active site inaccessible to another protein substrate, inhibiting GSK3 until the priming phosphoryl group of its pseudosubstrate region is removed by PP1. Other proteins that are substrates for GSK3 also have a priming site at position +4, which must be phosphorylated by another protein kinase before GSK3 can act on them.

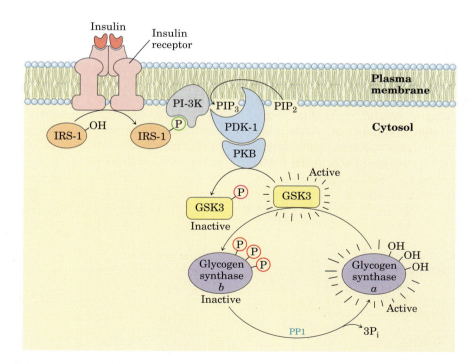

**FIGURE 15–29 The path from insulin to GSK3 and glycogen synthase.** Insulin binding to its receptor activates a tyrosine protein kinase in the receptor, which phosphorylates insulin receptor substrate-1 (IRS-1). The phosphotyrosine in this protein is then bound by phosphatidylinositol 3-kinase (PI-3K), which converts phosphatidylinositol 4,5-bisphosphate (PIP₂) in the membrane to phosphatidylinositol 3,4,5-trisphosphate (PIP₃). A protein kinase (PDK-1) that is activated when bound to PIP₃ activates a second protein kinase (PKB), which phosphorylates glycogen synthase kinase 3 (GSK3) in its pseudosubstrate region, inactivating it by the mechanisms shown in Figure 15–28b. The inactivation of GSK3 allows phosphoprotein phosphatase 1 (PP1) to dephosphorylate glycogen synthase, converting it to its active form. In this way, insulin stimulates glycogen synthesis. (See Fig. 12–8 for more details on insulin action.)

and proteins essential for mRNA and protein synthesis. These targets, like glycogen synthase, must first undergo a priming phosphorylation by another protein kinase before they can be phosphorylated by GSK3.

## Phosphoprotein Phosphatase 1 Is Central to Glycogen Metabolism

A single enzyme, PP1, can remove phosphoryl groups from all three of the enzymes phosphorylated in response to glucagon (liver) and epinephrine (liver and muscle): phosphorylase kinase, glycogen phosphorylase, and glycogen synthase. Insulin stimulates glycogen synthesis by activating PP1 and by inactivating GSK3.

PP1 does not exist free in the cytosol, but is tightly bound to its target proteins by one of a family of **glycogen-targeting proteins** that bind glycogen and each of the three enzymes, glycogen phosphorylase, phosphorylase kinase, and glycogen synthase (Fig. 15–30). PP1 is itself subject to covalent and allosteric regulation; it is inactivated when phosphorylated by PKA and is allosterically activated by glucose 6-phosphate.

## Transport into Cells Can Limit Glucose Utilization

The passive uptake of glucose by muscle and adipose tissue is catalyzed by the GLUT4 transporter described in Box 11–2. In the absence of insulin, most GLUT4 molecules are sequestered in membrane vesicles within the cell, but when blood glucose rises, release of insulin triggers GLUT4 movement to the plasma membrane. Glucose transport into hepatocytes involves a different, high-capacity transporter, GLUT2, which is always present in the plasma membrane. It catalyzes facilitated diffusion of glucose in both directions, at a rate high enough to ensure virtually instantaneous equilibration of glucose concentration in the blood and in the hepatocyte cytosol. In its role as a glucose sensor, the glycogen phosphorylase of hepatocytes is essentially measuring the glucose level in blood.

## Allosteric and Hormonal Signals Coordinate Carbohydrate Metabolism

Having looked at the mechanisms that regulate individual enzymes, we can now consider the overall shifts in carbohydrate metabolism that occur in the well-fed state, during fasting, and in the fight-or-flight response—signaled by insulin, glucagon, and epinephrine, respectively. We need to contrast two cases in which regulation serves different ends: (1) the role of hepatocytes in supplying glucose to the blood, and (2) the selfish use of carbohydrate fuels by nonhepatic tissues, typified by skeletal muscle (the myocyte), to support their own activities.

After ingestion of a carbohydrate-rich meal, the elevation of blood glucose triggers insulin release (Fig. 15–31, top). In a hepatocyte, insulin has two immediate effects: it inactivates GSK3, acting through the cascade shown in Figure 15–29, and activates a protein phosphatase, perhaps PP1. These two actions fully activate glycogen synthase. PP1 also inactivates glycogen phosphorylase $a$ and phosphorylase kinase by dephosphorylating both, effectively stopping glycogen breakdown. Glucose enters the hepatocyte through the high-capacity transporter GLUT2, always present in the plasma membrane, and the elevated intracellular glucose leads to dissociation of hexokinase IV (glucokinase) from its nuclear regulator protein. Hexokinase IV enters the cytosol and phosphorylates glucose, stimulating glycolysis and supplying the precursor for glycogen synthesis. Under these conditions, hepatocytes use the excess glucose in the blood to synthesize glycogen, up to the limit of about 10% of the total weight of the liver.

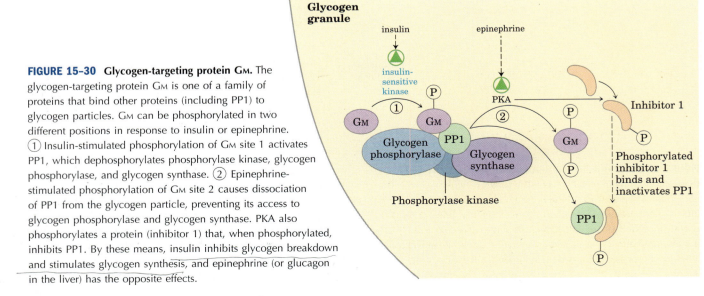

**FIGURE 15–30 Glycogen-targeting protein G$_M$.** The glycogen-targeting protein G$_M$ is one of a family of proteins that bind other proteins (including PP1) to glycogen particles. G$_M$ can be phosphorylated in two different positions in response to insulin or epinephrine. ① Insulin-stimulated phosphorylation of G$_M$ site 1 activates PP1, which dephosphorylates phosphorylase kinase, glycogen phosphorylase, and glycogen synthase. ② Epinephrine-stimulated phosphorylation of G$_M$ site 2 causes dissociation of PP1 from the glycogen particle, preventing its access to glycogen phosphorylase and glycogen synthase. PKA also phosphorylates a protein (inhibitor 1) that, when phosphorylated, inhibits PP1. By these means, insulin inhibits glycogen breakdown and stimulates glycogen synthesis, and epinephrine (or glucagon in the liver) has the opposite effects.

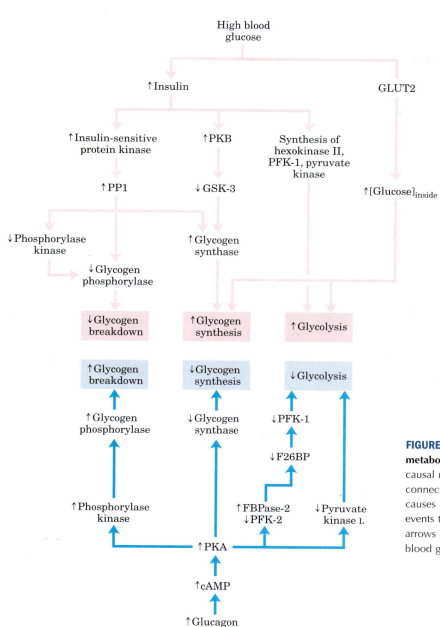

**FIGURE 15–31 Regulation of carbohydrate metabolism in the hepatocyte.** Arrows indicate causal relationships between the changes they connect. $\downarrow$A → $\uparrow$B means that a decrease in A causes an increase in B. Pink arrows connect events that result from high blood glucose; blue arrows connect events that result from low blood glucose.

Between meals, or during an extended fast, the drop in blood glucose triggers the release of glucagon, which, acting through the cascade shown in Figure 15–25, activates PKA. PKA mediates all the effects of glucagon (Fig. 15–31, bottom). It phosphorylates phosphorylase kinase, activating it and leading to the activation of glycogen phosphorylase. It phosphorylates glycogen synthase, inactivating it and blocking glycogen synthesis. It phosphorylates PFK-2/FBPase-2, leading to a drop in the concentration of the regulator fructose 2,6-bisphosphate, which has the effect of inactivating the glycolytic enzyme PFK-1 and activating the gluconeogenic enzyme FBPase-1. And it phosphorylates and inactivates the glycolytic enzyme pyruvate kinase. Under these conditions, the liver produces glucose 6-phosphate by glycogen breakdown and by gluconeogenesis, and it stops using glucose to fuel glycolysis or make glycogen, maximizing the amount of glucose it can release to the blood. This release of glucose is possible only in liver, because other tissues lack glucose 6-phosphatase (Fig. 15–6).

The physiology of skeletal muscle differs from that of liver in three ways important to our discussion of metabolic regulation (Fig. 15–32): (1) muscle uses its stored glycogen only for its own needs; (2) as it goes from rest to vigorous contraction, muscle undergoes very large changes in its demand for ATP, which is supported by glycolysis; (3) muscle lacks the enzymatic machinery for gluconeogenesis. The regulation of carbohydrate

**TABLE 15–3    Some of the Genes Regulated by Insulin**

| Change in gene expression | Pathway |
| --- | --- |
| **Increased expression** | |
| Hexokinase II | Glycolysis |
| Hexokinase IV | Glycolysis |
| Phosphofructokinase-1 (PFK-1) | Glycolysis |
| Pyruvate kinase | Glycolysis |
| PFK-2/FBPase-2 | Regulation of glycolysis/gluconeogenesis |
| Glucose 6-phosphate dehydrogenase | Pentose phosphate pathway (NADPH) |
| 6-Phosphogluconate dehydrogenase | Pentose phosphate pathway (NADPH) |
| Pyruvate dehydrogenase | Fatty acid synthesis |
| Acetyl-CoA carboxylase | Fatty acid synthesis |
| Malic enzyme | Fatty acid synthesis (NADPH) |
| ATP-citrate lyase | Fatty acid synthesis (provides acetyl-CoA) |
| Fatty acid synthase complex | Fatty acid synthesis |
| Stearoyl-CoA dehydrogenase | Fatty acid desaturation |
| Acyl-CoA–glycerol transferases | Triacylglycerol synthesis |
| **Decreased expression** | |
| PEP carboxykinase | Gluconeogenesis |
| Glucose 6-phosphatase (catalytic subunit) | Glucose release to blood |

metabolism in muscle reflects these differences from liver. First, myocytes lack receptors for glucagon. Second, the muscle isozyme of pyruvate kinase is not phosphorylated by PKA, so glycolysis is not turned off when [cAMP] is high. In fact, cAMP *increases* the rate of glycolysis in muscle, probably by activating glycogen phosphorylase. When epinephrine is released into the blood in a fight-or-flight situation, PKA is activated by the rise in [cAMP], and phosphorylates and activates glycogen phosphorylase kinase. The resulting phosphorylation and activation of glycogen phosphorylase results in faster glycogen breakdown. Epinephrine is not released under low-stress conditions, but with each neuronal stimulation of muscle contraction, cytosolic [$Ca^{2+}$] rises briefly and activates phosphorylase kinase through its calmodulin subunit.

Elevated insulin triggers increased glycogen synthesis in myocytes by activating PP1 and inactivating GSK3. Unlike hepatocytes, myocytes have a reserve of GLUT4 sequestered in intracellular vesicles. Insulin triggers their movement to the plasma membrane, where they allow increased glucose uptake. In response to insulin, therefore, myocytes help to lower blood glucose by increasing their rates of glucose uptake, glycogen synthesis, and glycolysis.

## Insulin Changes the Expression of Many Genes Involved in Carbohydrate and Fat Metabolism

In addition to its effects on the activity of existing enzymes, insulin also regulates the expression of as many as 150 genes, including some related to fuel metabolism

(Fig. 15–31; Table 15–3). Insulin stimulates the transcription of the genes that encode hexokinases II and IV, PFK-1, pyruvate kinase, and the bifunctional enzyme PFK-2/FBPase-2 (all involved in glycolysis and its regulation), several enzymes involved in fatty acid synthesis, and two enzymes that generate the reductant for fatty acid synthesis (NADPH) via the pentose phosphate pathway (glucose 6-phosphate dehydrogenase and 6-phosphogluconate dehydrogenase). Insulin also slows the expression of the genes for two enzymes of gluconeogenesis (PEP carboxykinase and glucose 6-phosphatase). These effects take place on a longer time scale (minutes to hours) than those mediated by covalent alteration of enzymes, but the impact on metabolism can be very significant. When the diet provides an excess of glucose, the resulting rise in insulin increases the synthesis of glucose-metabolizing proteins, and glucose becomes the fuel of choice (via glycolysis) for liver, adipose tissue, and muscle. In liver and adipose tissue, glucose is converted to glycogen and triacylglycerols for temporary storage.

## Carbohydrate and Lipid Metabolism Are Integrated by Hormonal and Allosteric Mechanisms

As complex as the regulation of carbohydrate metabolism is, it is far from the whole story of fuel metabolism. The metabolism of fats and fatty acids is very closely tied to that of carbohydrates. Hormonal signals such as insulin and changes in diet or exercise are equally important in regulating fat metabolism and integrating it with that of carbohydrates. We shall return to this overall metabolic integration in mammals in Chapter 23,

**FIGURE 15–32 Difference in the regulation of carbohydrate metabolism in liver and muscle.** In liver, either glucagon (indicating low blood glucose) or epinephrine (signaling the need to fight or flee) has the effect of maximizing the output of glucose into the bloodstream. In muscle, epinephrine increases glycogen breakdown and glycolysis, which together provide fuel to produce the ATP needed for muscle contraction.

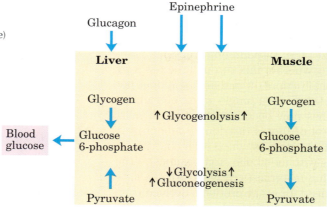

after first considering the metabolic pathways for fats and amino acids (Chapters 17 and 18). The message we wish to convey here is that metabolic pathways are over-laid with complex regulatory controls that are exquisitely sensitive to changes in metabolic circumstances. These mechanisms act to adjust the flow of metabolites through various metabolic pathways as needed by the cell and organism, and do so without causing major changes in the concentrations of intermediates shared with other pathways.

## SUMMARY 15.4 Coordinated Regulation of Glycogen Synthesis and Breakdown

■ Glycogen phosphorylase is activated in response to glucagon or epinephrine, which raise [cAMP] and activate PKA. PKA phosphorylates and activates phosphorylase kinase, which converts glycogen phosphorylase *b* to its active *a* form. Phosphoprotein phosphatase 1 (PP1) reverses the phosphorylation of glycogen phosphorylase *a*, inactivating it. Glucose binds to the liver isozyme of glycogen phosphorylase *a*, favoring its dephosphorylation and inactivation.

■ Glycogen synthase *a* is inactivated by phosphorylation catalyzed by GSK3. Insulin blocks GSK3. PP1, which is activated by insulin, reverses the inhibition by dephosphorylating glycogen synthase *b*.

■ Insulin increases glucose uptake into myocytes and adipocytes by triggering movement of the glucose transporter GLUT4 to the plasma membrane.

■ Insulin stimulates the synthesis of hexokinases II and IV, PFK-1, pyruvate kinase, and several enzymes involved in lipid synthesis. Insulin stimulates glycogen synthesis in muscle and liver.

■ In liver, glucagon stimulates glycogen breakdown and gluconeogenesis while blocking glycolysis, thereby sparing glucose for export to the brain and other tissues.

■ In muscle, epinephrine stimulates glycogen breakdown and glycolysis, providing ATP to support contraction.

## 15.5 Analysis of Metabolic Control

For every complex problem there is a simple solution. And it is always wrong.

—H. L. Mencken, A Mencken Chrestomathy, 1949

Beginning with Eduard Buchner's discovery (c. 1900) that an extract of broken yeast cells could convert glucose to ethanol and $CO_2$, a major thrust of biochemical research was to deduce the steps by which this transformation occurred and to purify and characterize the enzymes that catalyzed each step. By the middle of the twentieth century, all ten enzymes of the glycolytic pathway had been purified and characterized. In the next 50 years much was learned about the regulation of these enzymes by intracellular and extracellular signals, through the kinds of allosteric and covalent mechanisms we have described in this chapter. The conventional wisdom was that in a linear pathway such as glycolysis, catalysis by one enzyme must be the slowest and must therefore determine the rate of metabolite flow, or flux, through the whole pathway. For glycolysis, PFK-1 was considered the rate-limiting enzyme, because it was known to be closely regulated by fructose 2,6-bisphosphate and other allosteric effectors.

Eduard Buchner, 1860–1917

With the advent of genetic engineering technology, it became possible to test this "single rate-determining step" hypothesis by increasing the concentration of the enzyme that catalyzes the "rate-limiting step" in a pathway and determining whether flux through the pathway increases proportionally. More often than not, it does not do so: the simple solution (a single rate-determining step) is wrong. It has now become clear that in most pathways the control of flux is distributed among several enzymes, and the extent to which each contributes to the control varies with metabolic circumstances—the supply of the starting material (say, glucose), the supply of oxygen, the need for other products derived from intermediates of the pathway (say, glucose 6-phosphate for the pentose phosphate pathway in cells synthesizing large amounts of nucleotides), the effects of metabolites with regulatory roles, and the hormonal status of the organism (the levels of insulin and glucagon), among other factors.

Why are we interested in what limits the flux through a pathway? To understand the action of hormones or drugs, or the pathology that results from a failure of metabolic regulation, we must know where control is exercised. If researchers wish to develop a drug that stimulates or inhibits a pathway, the logical target is the enzyme that has the greatest impact on the flux through that pathway. And the bioengineering of a microorganism to overproduce a product of commercial value (p. 315) requires a knowledge of what limits the flux of metabolites toward that product.

### The Contribution of Each Enzyme to Flux through a Pathway Is Experimentally Measurable

There are several ways to determine experimentally how a change in the activity of one enzyme in a pathway affects metabolite flux through that pathway. Consider the experimental results shown in Figure 15–33. When a sample of rat liver was homogenized to release all soluble enzymes, the extract carried out the glycolytic conversion of glucose to fructose 1,6-bisphosphate at a measurable rate. (This experiment, for simplicity, focused on just the first part of the glycolytic pathway.) When increasing amounts of purified hexokinase IV were added to the extract, the rate of glycolysis progressively increased. The addition of purified PFK-1 to the extract also increased the rate of glycolysis, but not as dramatically as did hexokinase. Purified phosphohexose isomerase was without effect. These results suggest that hexokinase and PFK-1 both contribute to setting the flux through the pathway (hexokinase more than PFK-1), and that phosphohexose isomerase does not.

Similar experiments can be done on intact cells or organisms, using specific inhibitors or activators to change the activity of one enzyme while observing the effect on flux through the pathway. The amount of an

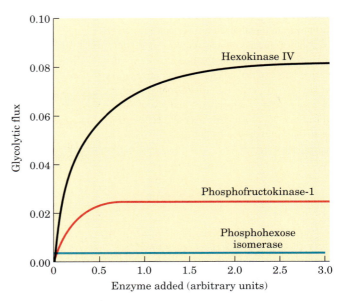

**FIGURE 15–33 Dependence of glycolytic flux in a rat liver homogenate on added enzymes.** Purified enzymes in the amounts shown on the *x* axis were added to an extract of liver carrying out glycolysis in vitro. The increase in flux through the pathway is shown on the *y* axis.

enzyme can also be altered genetically; bioengineering can produce a cell that makes extra copies of the enzyme under investigation or has a version of the enzyme that is less active than the normal enzyme. Increasing the concentration of an enzyme genetically sometimes has significant effects on flux; sometimes it has no effect.

Three critical parameters, which together describe the responsiveness of a pathway to changes in metabolic circumstances, lie at the center of **metabolic control analysis.** We turn now to a qualitative description of these parameters and their meaning in the context of a living cell. In Box 15–3 we will provide a more rigorous quantitative discussion.

### The Control Coefficient Quantifies the Effect of a Change in Enzyme Activity on Metabolite Flux through a Pathway

Quantitative data obtained as described in Figure 15–33 can be used to calculate a **flux control coefficient, *C*,** for each enzyme in a pathway. This coefficient expresses the relative contribution of each enzyme to setting the rate at which metabolites flow through the pathway—that is, the **flux, *J*.** *C* can have any value from 0.0 (for an enzyme with no impact on the flux) to 1.0 (for an enzyme that wholly determines the flux). An enzyme can also have a *negative* flux control coefficient. In a branched pathway, an enzyme in one branch, by drawing intermediates away from the other branch, can have a negative impact on the flux through that other branch (Fig. 15–34). *C* is not a constant, and it is not

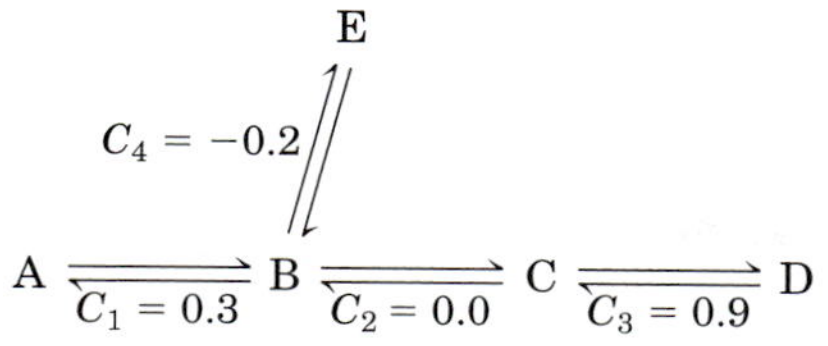

$$E$$

$$C_4 = -0.2$$

$$A \xrightleftharpoons[C_1 = 0.3]{} B \xrightleftharpoons[C_2 = 0.0]{} C \xrightleftharpoons[C_3 = 0.9]{} D$$

**FIGURE 15–34 Flux control coefficient, C, in a branched metabolic pathway.** In this simple pathway, the intermediate B has two alternative fates. To the extent that reaction B → E draws B away from the pathway A → D, it controls that pathway, which will result in a *negative* flux control coefficient for the enzyme that catalyzes step B → E. Note that the sum of all four coefficients equals 1.0, as it must.

intrinsic to a single enzyme; it is a function of the whole system of enzymes, and its value depends on the concentrations of substrates and effectors.

When real data from the experiment on glycolysis in a rat liver extract (Fig. 15–33) were subjected to this kind of analysis, investigators found flux control coefficients (for enzymes at the concentrations found in the extract) of 0.79 for hexokinase, 0.21 for PFK-1, and 0.0 for phosphohexose isomerase. It is not just fortuitous that these values add up to 1.0; we can show that for any complete pathway, the sum of the flux control coefficients must equal unity.

### The Elasticity Coefficient Is Related to an Enzyme's Responsiveness to Changes in Metabolite or Regulator Concentrations

A second parameter, the **elasticity coefficient, $\varepsilon$,** expresses quantitatively the responsiveness of a single enzyme to changes in the concentration of a metabolite or regulator; it is a function of the enzyme's intrinsic kinetic properties. For example, an enzyme with typical Michaelis-Menten kinetics shows a hyperbolic response to increasing substrate concentration (Fig. 15–35). At low concentrations of substrate (say, $0.1\ K_m$) each increment in substrate concentration results in a comparable increase in enzymatic activity, yielding an $\varepsilon$ near 1.0. At relatively high substrate concentrations (say, 10 $K_m$), increasing the substrate concentration has little effect on the reaction rate, because the enzyme is already saturated with substrate. The elasticity in this case approaches zero. For allosteric enzymes that show positive cooperativity, $\varepsilon$ may exceed 1.0, but it cannot exceed the Hill coefficient. Recall that the Hill coefficient is a measure of the degree of cooperativity, typically between 1.0 and 4.0 (p. 167).

### The Response Coefficient Expresses the Effect of an Outside Controller on Flux through a Pathway

We can also derive a quantitative expression for the relative impact of an outside factor (such as a hormone or growth factor), which is neither a metabolite nor an en-

zyme in the pathway, on the flux through the pathway. The experiment would measure the flux through the pathway (glycolysis, in this case) at various levels of the parameter $P$ (the insulin concentration, for example) to obtain the **response coefficient, $R$,** which expresses the change in pathway flux when $P$ ([insulin]) changes.

The three coefficients $C$, $\varepsilon$, and $R$ are related in a simple way: the responsiveness ($R$) of a pathway to an outside factor that affects a certain enzyme is a function of (1) how sensitive the pathway is to changes in the activity of that enzyme (the control coefficient, $C$) and (2) how sensitive that specific enzyme is to changes in the outside factor (the elasticity, $\varepsilon$):

$$R = C \cdot \varepsilon$$

Each enzyme in the pathway can be examined in this way, and the effects of any of several outside factors on flux through the pathway can be separately determined. Thus, in principle, we can predict how the flux of substrate through a series of enzymatic steps will change when there is a change in one or more controlling factors external to the pathway. Box 15–3 shows how these qualitative concepts are treated quantitatively.

### Metabolic Control Analysis Has Been Applied to Carbohydrate Metabolism, with Surprising Results

Metabolic control analysis provides a framework within which we can think quantitatively about regulation, interpret the significance of the regulatory properties of each enzyme in a pathway, identify the steps that most affect the flux through the pathway, and distinguish between *regulatory* mechanisms that act to maintain metabolite concentrations and *control* mechanisms that actually alter the flux through the pathway. Analysis of the glycolytic pathway in yeast, for example, has

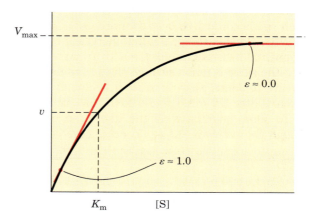

**FIGURE 15–35 Elasticity coefficient, $\varepsilon$, of an enzyme with typical Michaelis-Menten kinetics.** At substrate concentrations far below the $K_m$, each increase in [S] produces a correspondingly large increase in the reaction velocity, $v$. For this region of the curve, the enzyme has an elasticity, $\varepsilon$, of about 1.0. At [S] >> $K_m$, increasing [S] has little effect on $v$; $\varepsilon$ here is close to 0.0.

## BOX 15–3    WORKING IN BIOCHEMISTRY

### Metabolic Control Analysis: Quantitative Aspects

The factors that influence the flow of intermediates (flux) through a pathway may be determined quantitatively by experiment and expressed in terms useful for predicting the change in flux when some factor involved in the pathway changes. Consider the simple reaction sequence in Figure 1, in which a substrate X (say, glucose) is converted in several steps to a product Z (perhaps pyruvate, formed glycolytically). A later enzyme in the pathway is a dehydrogenase (ydh) that acts on substrate Y. Because the action of a dehydrogenase is easily measured (see Fig. 13–15), we can use the flux ($J$) through this step ($J_{ydh}$) to measure the flux through the whole path. We manipulate experimentally the level of an early enzyme in the pathway (xase, which acts on the substrate X) and measure the flux through the path ($J_{ydh}$) for several levels of the enzyme xase.

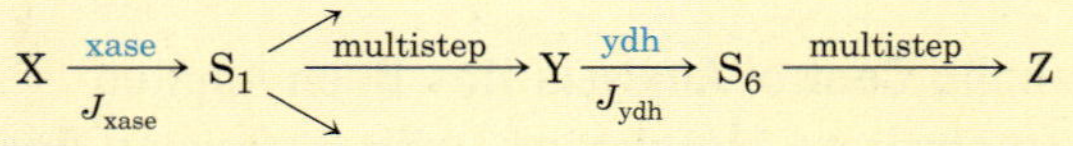

$$X \xrightarrow[J_{xase}]{xase} S_1 \overset{\text{multistep}}{\underset{}{\diagup\diagdown}} Y \xrightarrow[J_{ydh}]{ydh} S_6 \xrightarrow{\text{multistep}} Z$$

**FIGURE 1**   Flux through a hypothetical multienzyme pathway.

The relationship between the flux through the pathway from X to Z in the intact cell and the concentration of each enzyme in the path should be hyperbolic, with virtually no flux at infinitely low enzyme and near-maximum flux at very high enzyme activity. In a plot of $J_{ydh}$ against the concentration of xase, $E_{xase}$, the change of flux with a small change of enzyme is $\partial J_{ydh}/\partial E_{xase}$, which is simply the slope of the tangent to the curve at any concentration of enzyme, $E_{xase}$, and which tends toward zero at saturating $E_{xase}$. At low $E_{xase}$, the slope is steep; the flux increases with each incremental increase in enzyme activity. At very high $E_{xase}$, the slope is much smaller; the system is less responsive to added xase, because it is already present in excess over the other enzymes in the pathway.

To show quantitatively the dependence of flux through the pathway, $\partial J_{ydh}$, on $\partial E_{xase}$, we could use the ratio $\partial J_{ydh}/\partial E_{xase}$. However, its usefulness is limited because its value depends on the units used to express flux and enzyme activity. By expressing the *fractional* changes in flux and enzyme activity, $\partial J_{ydh}/J_{ydh}$, and $\partial E_{xase}/E_{xase}$, we obtain a unitless expression for the flux control coefficient, $C_{xase}^{Jydh}$:

$$C_{xase}^{Jydh} \approx \frac{\partial J_{ydh}}{J_{ydh}} \bigg/ \frac{\partial E_{xase}}{E_{xase}} \tag{1}$$

This can be rearranged to

$$C_{xase}^{Jydh} \approx \frac{\partial J_{ydh}}{\partial E_{xase}} \cdot \frac{E_{xase}}{J_{ydh}}$$

which is mathematically identical to

$$C_{xase}^{Jydh} = \frac{\partial \ln J_{ydh}}{\partial \ln E_{xase}}$$

This equation suggests a simple graphical means for determining the flux control coefficient: $C_{xase}^{Jydh}$ is the slope of the tangent to the plot of $\ln J_{ydh}$ versus $\ln E_{xase}$, which can be obtained by replotting the experimental data in Figure 2a to obtain Figure 2b. Notice that $C_{xase}^{Jydh}$ is not a constant; it depends on the starting $E_{xase}$ from which the change in enzyme level takes place. For the cases shown in Figure 2, $C_{xase}^{Jydh}$ is about 1.0 at the lowest $E_{xase}$, but only about 0.2 at high $E_{xase}$. A value near 1.0 for $C_{xase}^{Jydh}$ means that the enzyme's concentration wholly determines the flux through the pathway; a value near 0.0 means that the enzyme's concentration does not limit the flux through the path. Unless the flux control coefficient is greater than about 0.5, changes in the activity of the enzyme will not have a strong effect on the flux.

The **elasticity, $\varepsilon$,** of an enzyme is a measure of how that enzyme's catalytic activity changes when the concentration of a metabolite—substrate, product, or effector—changes. It is obtained from an experimental plot of the rate of the reaction catalyzed by the enzyme versus the concentration of the metabolite, at metabolite concentrations that prevail in the cell. By arguments analogous to those used to derive $C$, we can show $\varepsilon$ to be the slope of the tangent to a plot of

---

revealed an unexpectedly low flux control coefficient for PFK-1, which, because of its known elaborate allosteric regulation, has been viewed as the main point of flux control—the "rate-determining step"—in glycolysis. Experimentally raising the level of PFK-1 fivefold led to a change in flux through glycolysis of less than 10%, suggesting that the real role of PFK-1 regulation is not to control flux through glycolysis but to mediate metabolite homeostasis—to prevent large changes in metabolite concentrations when the flux through glycolysis increases in response to elevated blood glucose or insulin. Recall that the study of glycolysis in a liver

ln $V$ versus ln [substrate, or product, or effector]:

$$\varepsilon_S^{\text{xase}} = \frac{\partial V_{\text{xase}}}{\partial S} \cdot \frac{S}{V_{\text{xase}}}$$

$$= \frac{\partial \ln |V_{\text{xase}}|}{\partial \ln S}$$

For an enzyme with typical Michaelis-Menten kinetics, the value of $\varepsilon$ ranges from about 1 at substrate concentrations far below $K_m$ to near 0 as $V_{\text{max}}$ is approached. Allosteric enzymes can have elasticities greater than 1.0, but not larger than their Hill coefficients (p. 167).

Finally, the effect of controllers outside the pathway itself (that is, not metabolites) can be measured and expressed as the **response coefficient, R.** The change in flux through the pathway is measured for changes in the concentration of the controlling parameter $P$, and $R$ is defined in a form analogous to that of Equation 1, yielding the expression

$$R_P^{J_{\text{ydh}}} = \frac{\partial J_{\text{ydh}}}{\partial P} \cdot \frac{P}{J_{\text{ydh}}}$$

Using the same logic and graphical methods as described above for determining $C$, we can obtain $R$ as the slope of the tangent to the plot of ln $J$ versus ln $P$.

The three coefficients we have described are related in this simple way:

$$R_P^{J_{\text{ydh}}} = C_{\text{xase}}^{J_{\text{ydh}}} \cdot \varepsilon_P^{\text{xase}}$$

Thus the responsiveness of each enzyme in a pathway to a change in an outside controlling factor is a simple function of two things: the control coefficient, a variable that expresses the extent to which that enzyme influences the flux under a given set of conditions, and the elasticity, an intrinsic property of the enzyme that reflects its sensitivity to substrate and effector concentrations.

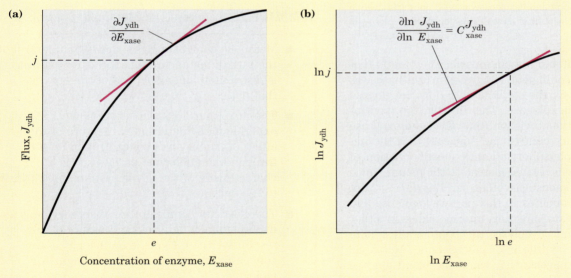

**FIGURE 2** **The flux control coefficient. (a)** Typical variation of the pathway flux, $J_{\text{ydh}}$, measured at the step catalyzed by the enzyme ydh, as a function of the amount of the enzyme xase, $E_{\text{xase}}$, which catalyzes an earlier step in the pathway. The flux control coefficient at $(e,j)$ is the slope of the product of the tangent to the curve, $\partial J_{\text{ydh}}/\partial E_{\text{xase}}$, and the ratio (scaling factor), $e/j$. **(b)** On a double-logarithmic plot of the same curve, the flux control coefficient is the slope of the tangent to the curve.

extract (Fig. 15–33) also yielded a flux control coefficient that contradicted the conventional wisdom; it showed that hexokinase, not PFK-1, is most influential in setting the flux through glycolysis. We must note here that a liver extract is far from equivalent to a hepatocyte; the ideal way to study flux control is by manipulating one enzyme at a time in the living cell. This is already feasible in some cases.

Investigators have used nuclear magnetic resonance (NMR) as a noninvasive means to determine the concentration of glycogen and metabolites in the five-step pathway from glucose in the blood to glycogen in myocytes

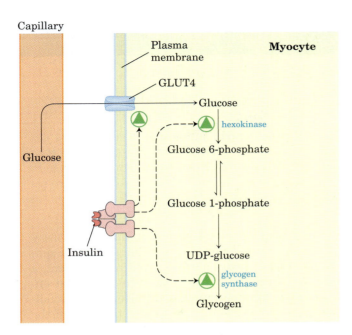

**FIGURE 15–36** **Control of glycogen synthesis from blood glucose in myocytes.** Insulin affects three of the five steps in this pathway, but it is the effects on transport and hexokinase activity, not the change in glycogen synthase activity, that increase the flux toward glycogen.

(Fig. 15–36) in rat and human muscle. They found that the flux control coefficient for glycogen synthase was smaller than that for the steps catalyzed by the glucose transporter and hexokinase. This finding, too, contradicts the conventional wisdom that glycogen synthase is the locus of flux control and suggests that the importance of the phosphorylation/dephosphorylation of glycogen synthase is related instead to the maintenance of metabolite homeostasis—that is, *regulation,* not *control.* Two metabolites in this pathway, glucose and glucose 6-phosphate, are key intermediates in other pathways, including glycolysis, the pentose phosphate pathway, and the synthesis of glucosamine. Metabolic control analysis suggests that when the blood glucose level rises, insulin acts in muscle to (1) increase glucose transport into cells by bringing GLUT4 to the plasma membrane, (2) induce the synthesis of hexokinase, and (3) activate glycogen synthase by covalent alteration (Fig. 15–29). The first two effects of insulin increase glucose flux through the pathway (control), and the third serves to adapt the activity of glycogen synthase so that metabolite levels (glucose 6-phosphate, for example) will not change dramatically with the increased flux (regulation).

## Metabolic Control Analysis Suggests a General Method for Increasing Flux through a Pathway

How could an investigator engineer a cell to increase the flux through one pathway without altering the concen-

trations of other metabolites or the fluxes through other pathways? More than two decades ago Henrik Kacser predicted, on the basis of metabolic control analysis, that this could be accomplished by increasing the concentrations of every enzyme in a pathway. The prediction has been confirmed in several experimental tests, and it also fits with the way cells normally control fluxes through a pathway. For example, when rats are fed a high-protein diet, they dispose of excess amino groups by converting them to urea in the urea cycle (Chapter 18). After such a dietary shift, the urea output increases fourfold, and the amount of all eight enzymes in the urea cycle increases two- to threefold. Similarly, when increased fatty acid oxidation is triggered by activation of the enzyme peroxisome proliferator-activated receptor γ (PPARγ; see Fig. 21–22), synthesis of the *whole set* of oxidative enzymes is increased. With the growing use of DNA microarrays to study the expression of whole sets of genes in response to various perturbations, we should soon learn whether this is the general mechanism by which cells make long-term adjustments in the fluxes through specific pathways.

## SUMMARY 15.5 Analysis of Metabolic Control

- Metabolic control analysis shows that control of the rate of metabolite flux through a pathway is distributed among several of the enzymes in that path.

- The flux control coefficient, $C$, is an experimentally determined measure of the effect of an enzyme's concentration on flux through a multienzyme pathway. It is characteristic of the whole system, not intrinsic to the enzyme.

- The elasticity coefficient, $\varepsilon$, of an enzyme is an experimentally determined measure of how responsive the enzyme is to changes in the concentration of a metabolite or regulator molecule.

- The response coefficient, $R$, is the expression for the experimentally determined change in flux through a pathway in response to a regulatory hormone or second messenger. It is a function of $C$ and $\varepsilon$: $R = C \cdot \varepsilon$.

- Some regulated enzymes control the flux through a pathway, while others rebalance the level of metabolites in response to the change in flux. This latter, rebalancing activity is *regulation;* the former activity is *control.*

- Metabolic control analysis predicts that flux toward a desired product is most effectively increased by raising the concentration of all enzymes in the pathway.

## Key Terms

*Terms in bold are defined in the glossary.*

glycogenolysis  562
**glycolysis**  562
**gluconeogenesis**  562
**glycogenesis**  562
debranching enzyme  562
sugar nucleotides  565
**glycogenin**  569
**homeostasis**  571
adenylate kinase  571
**mass action ratio, Q**  572
**metabolic regulation**  575

**metabolic control**  575
**futile cycle**  576
substrate cycle  576
**GLUT**  578
glucagon  581
fructose 2,6-bisphosphate  581
glycogen phosphorylase $a$  583
glycogen phosphorylase $b$  584
**enzyme cascade**  584
phosphoprotein phosphatase 1
  (PP1)  584

glycogen synthase $a$  586
glycogen synthase $b$  586
glycogen synthase kinase 3
  (GSK3)  586
**priming**  586
glycogen-targeting proteins  588
flux control coefficient, $C$  592
flux, $J$  592
elasticity coefficient, $\varepsilon$  593
response coefficient, $R$  593

## Further Reading

### General and Historical

**Gibson, D.M. & Harris, R.A.** (2002) *Metabolic Regulation in Mammals,* Taylor and Francis, New York.

An excellent introduction to the regulation of metabolism in each of the major organs.

**Kornberg, A.** (2001) Remembering our teachers. *J. Biol. Chem.* **276,** 3–11.

An appreciative description of the Coris' laboratories and coworkers.

**Ochs, R.S., Hanson, R.W., & Hall, J. (eds)** (1985) *Metabolic Regulation,* Elsevier Science Publishing Co. Inc., New York.

A collection of short essays first published in *Trends in Biochemical Sciences,* better known as *TIBS.*

**Simoni, R.D., Hill, R.L., & Vaughan, M.** (2002) Carbohydrate metabolism: glycogen phosphorylase and the work of Carl F. and Gerty T. Cori. *J. Biol. Chem.* **277** (www.jbc.org/cgi/content/full/277/29/e18).

A brief historical note with references to five classic papers by the Coris (online journal only).

### Metabolism of Glycogen in Animals

**Gibbons, B.J., Roach, P.J., & Hurley, T.D.** (2002) Crystal structure of the autocatalytic initiator of glycogen biosynthesis, glycogenin. *J. Mol. Biol.* **319,** 463–477.

**Melendez-Hevia, E., Waddell, T.G., & Shelton, E.D.** (1993) Optimization of molecular design in the evolution of metabolism: the glycogen molecule. *Biochem. J.* **295,** 477–483.

Comparison of theoretical and experimental aspects of glycogen structure.

### Regulation of Metabolic Pathways

**Barford, D.** (1999) Structural studies of reversible protein phosphorylation and protein phosphatases. *Biochem. Soc. Trans.* **27,** 751–766.

An intermediate-level review.

### Coordinated Regulation of Glycolysis and Gluconeogenesis

**de la Iglesia, N., Mukhtar, M., Seoane, J., Guinovart, J.J., & Agius, L.** (2000) The role of the regulatory protein of glucokinase

in the glucose sensory mechanism of the hepatocyte. *J. Biol. Chem.* **275,** 10,597–10,603.

Report of the experimental determination of the flux control coefficients for glucokinase and the glucokinase regulatory protein in hepatocytes.

**Hue, L. & Rider, M.H.** (1987) Role of fructose 2,6-bisphosphate in the control of glycolysis in mammalian tissues. *Biochem. J.* **245,** 313–324.

**Nordlie, R.C., Foster, J.D., & Lange, A.J.** (1999) Regulation of glucose production by the liver. *Annu. Rev. Nutr.* **19,** 379–406.

Advanced review.

**Okar, D.A., Manzano, A., Navarro-Sabate, A., Riera, L., Bartrons, R., & Lange, A.J.** (2001) PFK-2/FBPase-2: maker and breaker of the essential biofactor fructose-2,6-bisphosphate. *Trends Biochem. Sci.* **26,** 30–35.

A brief review of the bifunctional kinase/phosphatase.

**Pilkis, S.J. & Granner, D.K.** (1992) Molecular physiology of the regulation of hepatic gluconeogenesis and glycolysis. *Annu. Rev. Physiol.* **54,** 885–909.

**Schirmer, T. & Evans, P.R.** (1990) Structural basis of the allosteric behavior of phosphofructokinase. *Nature* **343,** 140–145.

**van Shaftingen, E. & Gerin, I.** (2002) The glucose-6-phosphatase system. *Biochem. J.* **362,** 513–532.

**Veech, R.L.** (2003) A humble hexose monophosphate pathway metabolite regulates short- and long-term control of lipogenesis. *Proc. Natl. Acad. Sci. USA* **100,** 5578–5580.

Short review of the work from K. Uyeda's laboratory on the role of xylulose 5-phosphate in carbohydrate and fat metabolism; Uyeda's papers are cited here.

**Yamada, K. & Noguchi, T.** (1999) Nutrient and hormonal regulation of pyruvate kinase gene expression. *Biochem. J.* **337,** 1–11.

Detailed review of recent work on the genes and proteins of this system and their regulation.

### Coordinated Regulation of Glycogen Synthesis and Breakdown

**Barford, D., Hu, S.-H., & Johnson, L.N.** (1991) Structural mechanism for glycogen phosphorylase control by phosphorylation and AMP. *J. Mol. Biol.* **218,** 233–260.

Clear discussion of the regulatory changes in the structure of glycogen phosphorylase, based on the structures (from x-ray diffraction studies) of the active and less active forms of the enzyme.

**Frame, S. & Cohen, P.** (2001) GSK3 takes centre stage more than 20 years after its discovery. *Biochem. J.* **359,** 1–16.
Review of the roles of GSK3 in carbohydrate metabolism and in other regulatory phenomena.

**Harwood, A.J.** (2001) Regulation of GSK-3: a cellular multiprocessor. *Cell* **105,** 821–824.
Short review of several regulatory roles of GSK3.

**Hudson, J.W., Golding, G.B., & Crerar, M.M.** (1993) Evolution of allosteric control in glycogen phosphorylase. *J. Mol. Biol.* **234,** 700–721.

**Newgard, C.B., Brady, M.J., O'Doherty, R.M., & Saltiel, A.R.** (2000) Organizing glucose disposal: emerging roles of the glycogen targeting subunits of protein phosphatase-1. *Diabetes* **49,** 1967–1977.
Intermediate-level review.

**Radziuk, J. & Pye, S.** (2001) Hepatic glucose uptake, gluconeogenesis and the regulation of glycogen synthesis. *Diabetes/Metab. Res. Rev.* **17,** 250–272.
Advanced review.

**Stalmans, W., Keppens, S., & Bollen, M.** (1998) Specific features of glycogen metabolism in the liver. *Biochem. J.* **336,** 19–31.
A review that goes into greater depth than this chapter.

### Metabolic Control Analysis

**Aiston, S., Hampson, L., Gomez-Foix, A.M., Guinovart, J.J., & Agius, L.** (2001) Hepatic glycogen synthesis is highly sensitive to phosphorylase activity: evidence from metabolic control analysis. *J. Biol. Chem.* **276,** 23,858–23,866.

**Fell, D.A.** (1992) Metabolic control analysis: a survey of its theoretical and experimental development. *Biochem. J.* **286,** 313–330.
Clear statement of the principles of metabolic control analysis.

**Fell, D.A.** (1997) *Understanding the Control of Metabolism,* Portland Press, Ltd., London.
An excellent, clear exposition of metabolic regulation, from the point of view of metabolic control analysis. If you read only one treatment on metabolic control analysis, this should be it.

**Jeffrey, F.M.H., Rajagopal, A., Maloy, C.R., & Sherry, A.D.** (1991) [13]C-NMR: a simple yet comprehensive method for analysis of intermediary metabolism. *Trends Biochem. Sci.* **16,** 5–10.
Brief, intermediate-level review.

**Kacser, H. & Burns, J.A.** (1973) The control of flux. *Symp. Soc. Exp. Biol.* **32,** 65–104.
A classic paper in the field.

**Kacser, H., Burns, J.A., & Fell, D.A.** (1995) The control of flux: 21 years on. *Biochem. Soc. Trans.* **23,** 341–366.

**Schilling, C.H., Schuster, S., Palsson, B.O., & Heinrich, R.** (1999) Metabolic pathway analysis: basic concepts and scientific applications in the post-genomic era. *Biotechnol. Prog.* **15,** 296–303.
Short, advanced discussion of theoretical treatments that attempt to find ways of manipulating metabolism to optimize the formation of metabolic products.

**Schuster, S., Fell, D.A., & Dandekar, T.** (2000) A general definition of metabolic pathways useful for systematic organization and analysis of complex metabolic networks. *Nat. Biotechnol.* **18,** 326–332.
An interesting and provocative analysis of the interplay between the pentose phosphate pathway and glycolysis, from a theoretical standpoint.

**Shulman, R.G., Block, G., & Rothman, D.L.** (1995) In vivo regulation of muscle glycogen synthase and the control of glycogen synthesis. *Proc. Natl. Acad. Sci. USA* **92,** 8535–8542.
Review of the use of NMR to measure metabolite concentrations during glycogen synthesis, interpreted by metabolic control analysis.

**Westerhoff, H.V., Hofmeyr, J.-H.S., & Kholodenko, B.N.** (1994) Getting to the inside of cells using metabolic control analysis. *Biophys. Chem.* **50,** 273–283.

## Problems

**1. Measurement of Intracellular Metabolite Concentrations** Measuring the concentrations of metabolic intermediates in a living cell presents great experimental difficulties—usually a cell must be destroyed before metabolite concentrations can be measured. Yet enzymes catalyze metabolic interconversions very rapidly, so a common problem associated with these types of measurements is that the findings reflect not the physiological concentrations of metabolites but the equilibrium concentrations. A reliable experimental technique requires all enzyme-catalyzed reactions to be instantaneously stopped in the intact tissue so that the metabolic intermediates do not undergo change. This objective is accomplished by rapidly compressing the tissue between large aluminum plates cooled with liquid nitrogen ($-190\ ^\circ$C), a process called **freeze-clamping.** After freezing, which stops enzyme action instantly, the tissue is powdered and the enzymes are inactivated by precipitation with perchloric acid. The precipitate is removed by centrifugation, and the clear supernatant extract is analyzed for metabolites. To calculate intracellular concentrations, the intracellular volume is determined from the total water content of the tissue and a measurement of the extracellular volume.

The intracellular concentrations of the substrates and products of the phosphofructokinase-1 reaction in isolated rat heart tissue are given in the table below.

| Metabolite | Concentration ($\mu$M)[*] |
|---|---|
| Fructose 6-phosphate | 87.0 |
| Fructose 1,6-bisphosphate | 22.0 |
| ATP | 11,400 |
| ADP | 1,320 |

**Source:** From Williamson, J.R. (1965) Glycolytic control mechanisms I: inhibition of glycolysis by acetate and pyruvate in the isolated, perfused rat heart. *J. Biol. Chem.* **240,** 2308-2321.

[*]Calculated as $\mu$mol/mL of intracellular water.

(a) Calculate $Q$, [fructose 1,6-bisphosphate] [ADP] / [fructose 6-phosphate][ATP], for the PFK-1 reaction under physiological conditions.

(b) Given a $\Delta G'^{\circ}$ for the PFK-1 reaction of $-14.2$ kJ/mol, calculate the equilibrium constant for this reaction.

(c) Compare the values of $Q$ and $K'_{eq}$. Is the physiological reaction near or far from equilibrium? Explain. What does this experiment suggest about the role of PFK-1 as a regulatory enzyme?

**2. Effect of O$_2$ Supply on Glycolytic Rates** The regulated steps of glycolysis in intact cells can be identified by studying the catabolism of glucose in whole tissues or organs. For example, the glucose consumption by heart muscle can be measured by artificially circulating blood through an isolated intact heart and measuring the concentration of glucose before and after the blood passes through the heart. If the circulating blood is deoxygenated, heart muscle consumes glucose at a steady rate. When oxygen is added to the blood, the rate of glucose consumption drops dramatically, then is maintained at the new, lower rate. Why?

**3. Regulation of PFK-1** The effect of ATP on the allosteric enzyme PFK-1 is shown below. For a given concentration of fructose 6-phosphate, the PFK-1 activity increases with increasing concentrations of ATP, but a point is reached beyond which increasing the concentration of ATP inhibits the enzyme.

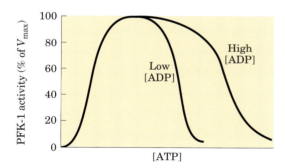

(a) Explain how ATP can be both a substrate and an inhibitor of PFK-1. How is the enzyme regulated by ATP?

(b) In what ways is glycolysis regulated by ATP levels?

(c) The inhibition of PFK-1 by ATP is diminished when the ADP concentration is high, as shown in the illustration. How can this observation be explained?

**4. Are All Metabolic Reactions at Equilibrium?**

(a) Phosphoenolpyruvate (PEP) is one of the two phosphoryl group donors in the synthesis of ATP during glycolysis. In human erythrocytes, the steady-state concentration of ATP is 2.24 mM, that of ADP is 0.25 mM, and that of pyruvate is 0.051 mM. Calculate the concentration of PEP at 25 °C, assuming that the pyruvate kinase reaction (see Fig. 13–3) is at equilibrium in the cell.

(b) The physiological concentration of PEP in human erythrocytes is 0.023 mM. Compare this with the value obtained in (a). Explain the significance of this difference.

**5. Cellular Glucose Concentration** The concentration of glucose in human blood plasma is maintained at about 5 mM. The concentration of free glucose inside a myocyte is much lower. Why is the concentration so low in the cell? What

happens to glucose after entry into the cell? Glucose is administered intravenously as a food source in certain clinical situations. Given that the transformation of glucose to glucose 6-phosphate consumes ATP, why not administer intravenous glucose 6-phosphate instead?

**6. Enzyme Activity and Physiological Function** The $V_{max}$ of the enzyme glycogen phosphorylase from skeletal muscle is much greater than the $V_{max}$ of the same enzyme from liver tissue.

(a) What is the physiological function of glycogen phosphorylase in skeletal muscle? In liver tissue?

(b) Why does the $V_{max}$ of the muscle enzyme need to be greater than that of the liver enzyme?

**7. Glycogen Phosphorylase Equilibrium** Glycogen phosphorylase catalyzes the removal of glucose from glycogen. The $\Delta G'^{\circ}$ for this reaction is 3.1 kJ/mol. (a) Calculate the ratio of [P$_i$] to [glucose 1-phosphate] when the reaction is at equilibrium. (Hint: The removal of glucose units from glycogen does not change the glycogen concentration.) (b) The measured ratio [P$_i$]/[glucose 1-phosphate] in myocytes under physiological conditions is more than 100:1. What does this indicate about the direction of metabolite flow through the glycogen phosphorylase reaction in muscle? (c) Why are the equilibrium and physiological ratios different? What is the possible significance of this difference?

**8. Regulation of Glycogen Phosphorylase** In muscle tissue, the rate of conversion of glycogen to glucose 6-phosphate is determined by the ratio of phosphorylase $a$ (active) to phosphorylase $b$ (less active). Determine what happens to the rate of glycogen breakdown if a muscle preparation containing glycogen phosphorylase is treated with (a) phosphorylase kinase and ATP; (b) PP1; (c) epinephrine.

**9. Glycogen Breakdown in Rabbit Muscle** The intracellular use of glucose and glycogen is tightly regulated at four points. In order to compare the regulation of glycolysis when oxygen is plentiful and when it is depleted, consider the utilization of glucose and glycogen by rabbit leg muscle in two physiological settings: a resting rabbit, with low ATP demands, and a rabbit that sights its mortal enemy, the coyote, and dashes into its burrow. For each setting, determine the relative levels (high, intermediate, or low) of AMP, ATP, citrate, and acetyl-CoA and how these levels affect the flow of metabolites through glycolysis by regulating specific enzymes. In periods of stress, rabbit leg muscle produces much of its ATP by anaerobic glycolysis (lactate fermentation) and very little by oxidation of acetyl-CoA derived from fat breakdown.

**10. Glycogen Breakdown in Migrating Birds** Unlike the rabbit with its short dash, migratory birds require energy for extended periods of time. For example, ducks generally fly several thousand miles during their annual migration. The flight muscles of migratory birds have a high oxidative capacity and obtain the necessary ATP through the oxidation of acetyl-CoA (obtained from fats) via the citric acid cycle. Compare the regulation of muscle glycolysis during short-term intense activity, as in the fleeing rabbit, and during extended activity, as in the migrating duck. Why must the regulation in these two settings be different?

**11. Enzyme Defects in Carbohydrate Metabolism** Summaries of four clinical case studies follow. For each case determine which enzyme is defective and designate the appropriate treatment, from the lists provided at the end of the problem. Justify your choices. Answer the questions contained in each case study. (You may need to refer to information in Chapter 14.)

*Case A* The patient develops vomiting and diarrhea shortly after milk ingestion. A lactose tolerance test is administered. (The patient ingests a standard amount of lactose, and the glucose and galactose concentrations of blood plasma are measured at intervals. In normal individuals the levels increase to a maximum in about 1 hour, then decline.) The patient's blood glucose and galactose concentrations do not increase during the test. Why do blood glucose and galactose increase and then decrease during the test in normal individuals? Why do they fail to rise in the patient?

*Case B* The patient develops vomiting and diarrhea after ingestion of milk. His blood is found to have a low concentration of glucose but a much higher than normal concentration of reducing sugars. The urine test for galactose is positive. Why is the concentration of reducing sugar in the blood high? Why does galactose appear in the urine?

*Case C* The patient complains of painful muscle cramps when performing strenuous physical exercise but has no other symptoms. A muscle biopsy indicates a muscle glycogen concentration much higher than normal. Why does glycogen accumulate?

*Case D* The patient is lethargic, her liver is enlarged, and a biopsy of the liver shows large amounts of excess glycogen. She also has a lower than normal blood glucose level. What is the reason for the low blood glucose in this patient?

*Defective Enzyme*
(a) Muscle PFK-1
(b) Phosphomannose isomerase
(c) Galactose 1-phosphate uridylyltransferase
(d) Liver glycogen phosphorylase
(e) Triose kinase
(f) Lactase in intestinal mucosa
(g) Maltase in intestinal mucosa
(h) Muscle-debranching enzyme

*Treatment*
1. Jogging 5 km each day
2. Fat-free diet
3. Low-lactose diet
4. Avoiding strenuous exercise
5. Large doses of niacin (the precursor of $NAD^+$)
6. Frequent regular feedings

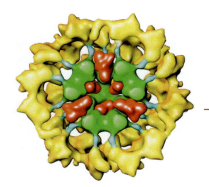

# THE CITRIC ACID CYCLE

If citrate is added the rate of respiration is often increased . . . the extra oxygen uptake is by far greater than can be accounted for by the complete oxidation of citrate . . . Since citric acid reacts catalytically in the tissue it is probable that it is removed by a primary reaction but regenerated by a subsequent reaction.

—*H. A. Krebs and W. A. Johnson, article in* Enzymologia, *1937*

As we saw in Chapter 14, some cells obtain energy (ATP) by fermentation, breaking down glucose in the absence of oxygen. For most eukaryotic cells and many bacteria, which live under aerobic conditions and oxidize their organic fuels to carbon dioxide and water, glycolysis is but the first stage in the complete oxidation of glucose. Rather than being reduced to lactate, ethanol, or some other fermentation product, the pyruvate produced by glycolysis is further oxidized to $H_2O$ and $CO_2$. This aerobic phase of catabolism is called **respiration.** In the broader physiological or macroscopic sense, respiration refers to a multicellular organism's uptake of $O_2$ and release of $CO_2$. Biochemists and cell biologists, however, use the term in a narrower sense to refer to the molecular processes by which *cells* consume $O_2$ and produce $CO_2$—processes more precisely termed **cellular respiration.**

Cellular respiration occurs in three major stages (Fig. 16–1). In the first, organic fuel molecules—glucose, fatty acids, and some amino acids—are oxidized to yield two-carbon fragments in the form of the acetyl group of acetyl-coenzyme A (acetyl-CoA). In the sec-

ond stage, the acetyl groups are fed into the citric acid cycle, which enzymatically oxidizes them to $CO_2$; the energy released is conserved in the reduced electron carriers NADH and $FADH_2$. In the third stage of respiration, these reduced coenzymes are themselves oxidized, giving up protons ($H^+$) and electrons. The electrons are transferred to $O_2$—the final electron acceptor—via a chain of electron-carrying molecules known as the respiratory chain. In the course of electron transfer, the large amount of energy released is conserved in the form of ATP, by a process called oxidative phosphorylation (Chapter 19). Respiration is more complex than glycolysis and is believed to have evolved much later, after the appearance of cyanobacteria. The metabolic activities of cyanobacteria account for the rise of oxygen levels in the earth's atmosphere, a dramatic turning point in evolutionary history.

We consider first the conversion of pyruvate to acetyl groups, then the entry of those groups into the **citric acid cycle,** also called the **tricarboxylic acid (TCA) cycle** or the **Krebs cycle** (after its discoverer, Hans Krebs). We next examine the cycle reactions and the enzymes that catalyze them. Because intermediates of the citric acid cycle are also siphoned off as biosynthetic precursors, we go on to consider some ways in which these intermediates are replenished. The citric acid cycle is a hub in metabolism, with degradative pathways leading in and anabolic pathways leading out, and it is closely regulated in coordination with other pathways. The chapter ends with a description of the glyoxylate pathway, a metabolic sequence in some organisms that employs several of the same enzymes and reactions used in the citric acid cycle, bringing about the net synthesis of glucose from stored triacylglycerols.

Hans Krebs, 1900–1981

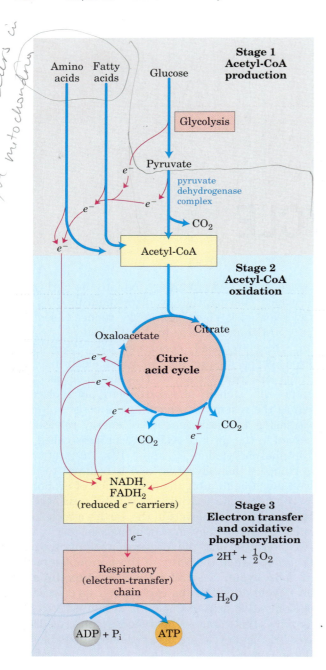

*Catabolism also occurs in the mitochondria* [handwritten annotation]

**FIGURE 16–1  Catabolism of proteins, fats, and carbohydrates in the three stages of cellular respiration.** Stage 1: oxidation of fatty acids, glucose, and some amino acids yields acetyl-CoA. Stage 2: oxidation of acetyl groups in the citric acid cycle includes four steps in which electrons are abstracted. Stage 3: electrons carried by NADH and $FADH_2$ are funneled into a chain of mitochondrial (or, in bacteria, plasma membrane–bound) electron carriers—the respiratory chain—ultimately reducing $O_2$ to $H_2O$. This electron flow drives the production of ATP.

and other sugars by glycolysis, is oxidized to acetyl-CoA and $CO_2$ by the **pyruvate dehydrogenase (PDH) complex,** a cluster of enzymes—multiple copies of each of three enzymes—located in the mitochondria of eukaryotic cells and in the cytosol of prokaryotes.

A careful examination of this enzyme complex is rewarding in several respects. The PDH complex is a classic, much-studied example of a multienzyme complex in which a series of chemical intermediates remain bound to the enzyme molecules as a substrate is transformed into the final product. Five cofactors, four derived from vitamins, participate in the reaction mechanism. The regulation of this enzyme complex also illustrates how a combination of covalent modification and allosteric regulation results in precisely regulated flux through a metabolic step. Finally, the PDH complex is the prototype for two other important enzyme complexes: α-ketoglutarate dehydrogenase, of the citric acid cycle, and the branched-chain α-keto acid dehydrogenase, of the oxidative pathways of several amino acids (see Fig. 18–28). The remarkable similarity in the protein structure, cofactor requirements, and reaction mechanisms of these three complexes doubtless reflects a common evolutionary origin.

### Pyruvate Is Oxidized to Acetyl-CoA and $CO_2$

The overall reaction catalyzed by the pyruvate dehydrogenase complex is an **oxidative decarboxylation,** an irreversible oxidation process in which the carboxyl group is removed from pyruvate as a molecule of $CO_2$

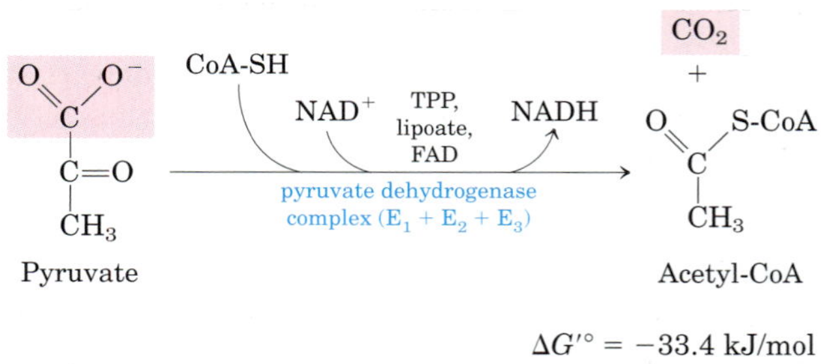

## 16.1 Production of Acetyl-CoA (Activated Acetate)

In aerobic organisms, glucose and other sugars, fatty acids, and most amino acids are ultimately oxidized to $CO_2$ and $H_2O$ via the citric acid cycle and the respiratory chain. Before entering the citric acid cycle, the carbon skeletons of sugars and fatty acids are degraded to the acetyl group of acetyl-CoA, the form in which the cycle accepts most of its fuel input. Many amino acid carbons also enter the cycle this way, although several amino acids are degraded to other cycle intermediates. Here we focus on how pyruvate, derived from glucose

$$\Delta G'^\circ = -33.4 \text{ kJ/mol}$$

**FIGURE 16–2  Overall reaction catalyzed by the pyruvate dehydrogenase complex.** The five coenzymes participating in this reaction, and the three enzymes that make up the enzyme complex, are discussed in the text.

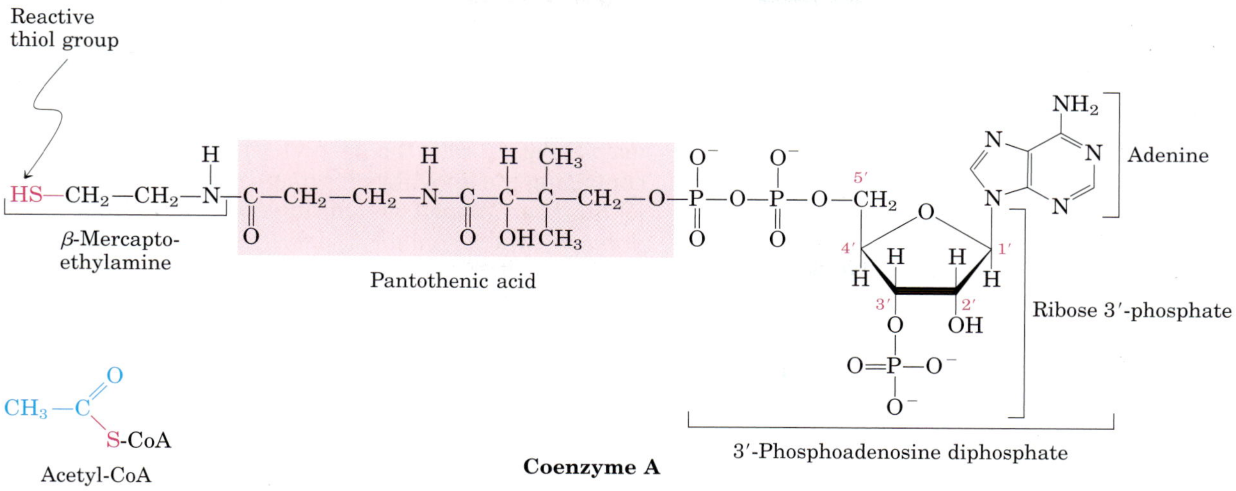

**FIGURE 16-3 Coenzyme A (CoA).** A hydroxyl group of pantothenic acid is joined to a modified ADP moiety by a phosphate ester bond, and its carboxyl group is attached to β-mercaptoethylamine in amide linkage. The hydroxyl group at the 3′ position of the ADP moiety has a phosphoryl group not present in free ADP. The —SH group of the mercaptoethylamine moiety forms a thioester with acetate in acetyl-coenzyme A (acetyl-CoA) (lower left).

and the two remaining carbons become the acetyl group of acetyl-CoA (Fig. 16–2). The NADH formed in this reaction gives up a hydride ion ($:H^-$) to the respiratory chain (Fig. 16–1), which carries the two electrons to oxygen or, in anaerobic microorganisms, to an alternative electron acceptor such as nitrate or sulfate. The transfer of electrons from NADH to oxygen ultimately generates 2.5 molecules of ATP per pair of electrons. The irreversibility of the PDH complex reaction has been demonstrated by isotopic labeling experiments: the complex cannot reattach radioactively labeled $CO_2$ to acetyl-CoA to yield carboxyl-labeled pyruvate.

## The Pyruvate Dehydrogenase Complex Requires Five Coenzymes

The combined dehydrogenation and decarboxylation of pyruvate to the acetyl group of acetyl-CoA (Fig. 16–2) requires the sequential action of three different enzymes and five different coenzymes or prosthetic groups—thiamine pyrophosphate (TPP), flavin adenine dinucleotide (FAD), coenzyme A (CoA, sometimes denoted CoA-SH, to emphasize the role of the —SH group), nicotinamide adenine dinucleotide (NAD), and lipoate. Four different vitamins required in human nutrition are vital components of this system: thiamine (in TPP), riboflavin (in FAD), niacin (in NAD), and pantothenate (in CoA). We have already described the roles of FAD and NAD as electron carriers (Chapter 13), and we have encountered TPP as the coenzyme of pyruvate decarboxylase (see Fig. 14–13).

Coenzyme A (Fig. 16–3) has a reactive thiol (—SH) group that is critical to the role of CoA as an acyl car-

rier in a number of metabolic reactions. Acyl groups are covalently linked to the thiol group, forming **thioesters.** Because of their relatively high standard free energies of hydrolysis (see Figs 13–6, 13–7), thioesters have a high acyl group transfer potential and can donate their acyl groups to a variety of acceptor molecules. The acyl group attached to coenzyme A may thus be thought of as "activated" for group transfer.

The fifth cofactor of the PDH complex, **lipoate** (Fig. 16–4), has two thiol groups that can undergo reversible oxidation to a disulfide bond (—S—S—), similar to that between two Cys residues in a protein. Because of its capacity to undergo oxidation-reduction reactions, lipoate can serve both as an electron hydrogen carrier and as an acyl carrier, as we shall see.

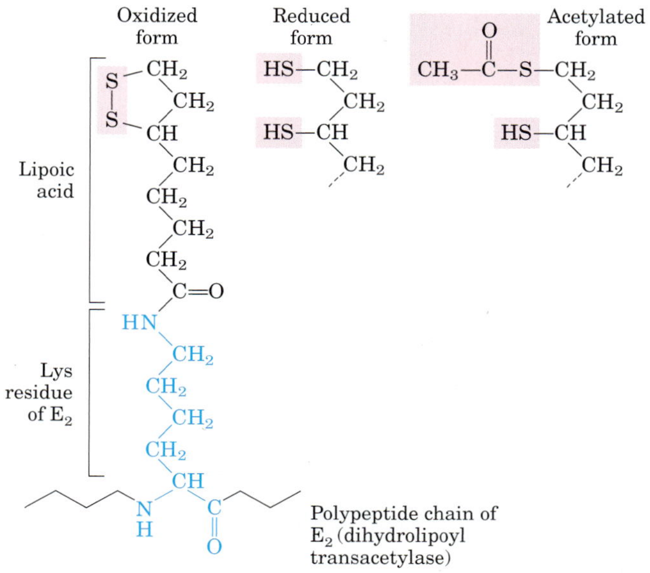

**FIGURE 16–4 Lipoic acid (lipoate) in amide linkage with a Lys residue.** The lipoyllysyl moiety is the prosthetic group of dihydrolipoyl transacetylase ($E_2$ of the PDH complex). The lipoyl group occurs in oxidized (disulfide) and reduced (dithiol) forms and acts as a carrier of both hydrogen and an acetyl (or other acyl) group.

## The Pyruvate Dehydrogenase Complex Consists of Three Distinct Enzymes

The PDH complex contains three enzymes—**pyruvate dehydrogenase** ($E_1$), **dihydrolipoyl transacetylase** ($E_2$), and **dihydrolipoyl dehydrogenase** ($E_3$)—each present in multiple copies. The number of copies of each enzyme and therefore the size of the complex varies among species. The PDH complex isolated from mammals is about 50 nm in diameter—more than five times the size of an entire ribosome and big enough to be visualized with the electron microscope (Fig. 16–5a). In the bovine enzyme, 60 identical copies of $E_2$ form a pentagonal dodecahedron (the core) with a diameter of about 25 nm (Fig. 16–5b). (The core of the *Escherichia coli* enzyme contains 24 copies of $E_2$.) $E_2$ is the point of

connection for the prosthetic group lipoate, attached through an amide bond to the ε-amino group of a Lys residue (Fig. 16–4). $E_2$ has three functionally distinct domains (Fig. 16–5c): the amino-terminal *lipoyl domain*, containing the lipoyl-Lys residue(s); the central $E_1$- and $E_3$-*binding domain*; and the inner-core *acyltransferase domain*, which contains the acyltransferase active site. The yeast PDH complex has a single lipoyl domain with a lipoate attached, but the mammalian complex has two, and *E. coli* has three (Fig. 16–5c). The domains of $E_2$ are separated by linkers, sequences of 20 to 30 amino acid residues, rich in Ala and Pro and interspersed with charged residues; these linkers tend to assume their extended forms, holding the three domains apart.

The active site of $E_1$ has bound TPP, and that of $E_3$ has bound FAD. Also part of the complex are two reg-

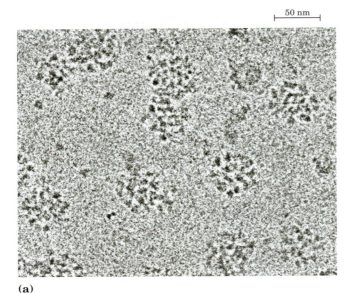

(a)

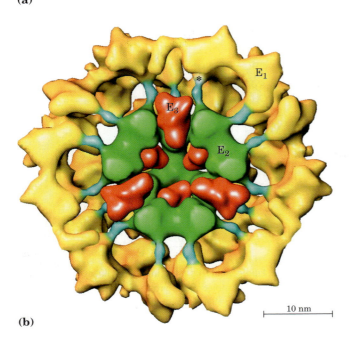

(b)

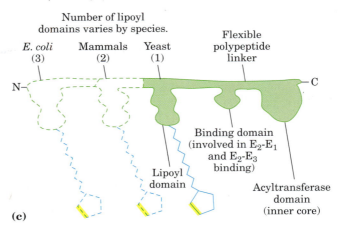

(c)

**FIGURE 16-5 Structure of the pyruvate dehydrogenase complex (a)** Cryoelectron micrograph of PDH complexes isolated from bovine kidney. In cryoelectron microscopy, biological samples are viewed at extremely low temperatures; this avoids potential artifacts introduced by the usual process of dehydrating, fixing, and staining. **(b)** Three-dimensional image of PDH complex, showing the subunit structure: $E_1$, pyruvate dehydrogenase; $E_2$, dihydrolipoyl transacetylase; and $E_3$, dihydrolipoyl dehydrogenase. This image is reconstructed by analysis of a large number of images such as those in **(a)**, combined with crystallographic studies of individual subunits. The core (green) consists of 60 molecules of $E_2$, arranged in 20 trimers to form a pentagonal dodecahedron. The lipoyl domain of $E_2$ (blue) reaches outward to touch the active sites of $E_1$ molecules (yellow) arranged on the $E_2$ core. A number of $E_3$ subunits (red) are also bound to the core, where the swinging arm on $E_2$ can reach their active sites. An asterisk marks the site where a lipoyl group is attached to the lipoyl domain of $E_2$. To make the structure clearer, about half of the complex has been cut away from the front. This model was prepared by Z. H. Zhou et al. (2001); in another model, proposed by J. L. S. Milne et al. (2002), the $E_3$ subunits are located more toward the periphery (see Further Reading). **(c)** $E_2$ consists of three types of domains linked by short polypeptide linkers: a catalytic acyltransferase domain; a binding domain, involved in the binding of $E_2$ to $E_1$ and $E_3$; and one or more (depending on the species) lipoyl domains.

ulatory proteins, a protein kinase and a phosphoprotein phosphatase, discussed below. This basic $E_1$-$E_2$-$E_3$ structure has been conserved during evolution and used in a number of similar metabolic reactions, including the oxidation of $\alpha$-ketoglutarate in the citric acid cycle (described below) and the oxidation of $\alpha$-keto acids derived from the breakdown of the branched-chain amino acids valine, isoleucine, and leucine (see Fig. 18–28). Within a given species, $E_3$ of PDH is identical to $E_3$ of the other two enzyme complexes. The attachment of lipoate to the end of a Lys side chain in $E_2$ produces a long, flexible arm that can move from the active site of $E_1$ to the active sites of $E_2$ and $E_3$, a distance of perhaps 5 nm or more.

## In Substrate Channeling, Intermediates Never Leave the Enzyme Surface

Figure 16–6 shows schematically how the pyruvate dehydrogenase complex carries out the five consecutive reactions in the decarboxylation and dehydrogenation of pyruvate. Step ① is essentially identical to the reaction catalyzed by pyruvate decarboxylase (see Fig. 14–13c); C-1 of pyruvate is released as $CO_2$, and C-2, which in pyruvate has the oxidation state of an aldehyde, is attached to TPP as a hydroxyethyl group. This first step is the slowest and therefore limits the rate of the overall reaction. It is also the point at which the PDH complex exercises its substrate specificity. In step ② the hydroxyethyl group is oxidized to the level of a car-

boxylic acid (acetate). The two electrons removed in this reaction reduce the —S—S— of a lipoyl group on $E_2$ to two thiol (—SH) groups. The acetyl moiety produced in this oxidation-reduction reaction is first esterified to one of the lipoyl —SH groups, then transesterified to CoA to form acetyl-CoA (step ③). Thus the energy of oxidation drives the formation of a high-energy thioester of acetate. The remaining reactions catalyzed by the PDH complex (by $E_3$, in steps ④ and ⑤) are electron transfers necessary to regenerate the oxidized (disulfide) form of the lipoyl group of $E_2$ to prepare the enzyme complex for another round of oxidation. The electrons removed from the hydroxyethyl group derived from pyruvate pass through FAD to $NAD^+$.

Central to the mechanism of the PDH complex are the swinging lipoyllysyl arms of $E_2$, which accept from $E_1$ the two electrons and the acetyl group derived from pyruvate, passing them to $E_3$. All these enzymes and coenzymes are clustered, allowing the intermediates to react quickly without diffusing away from the surface of the enzyme complex. The five-reaction sequence shown in Figure 16–6 is thus an example of **substrate channeling.** The intermediates of the multistep sequence never leave the complex, and the local concentration of the substrate of $E_2$ is kept very high. Channeling also prevents theft of the activated acetyl group by other enzymes that use this group as substrate. As we shall see, a similar tethering mechanism for the channeling of substrate between active

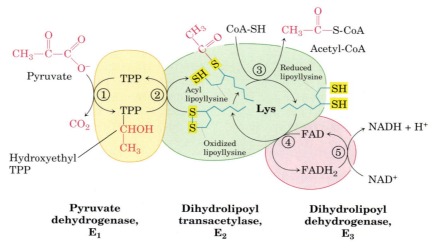

**FIGURE 16–6 Oxidative decarboxylation of pyruvate to acetyl-CoA by the PDH complex.** The fate of pyruvate is traced in red. In step ① pyruvate reacts with the bound thiamine pyrophosphate (TPP) of pyruvate dehydrogenase ($E_1$), undergoing decarboxylation to the hydroxyethyl derivative (see Fig. 14–13). Pyruvate dehydrogenase also carries out step ②, the transfer of two electrons and the acetyl group from TPP to the oxidized form of the lipoyllysyl group of the core enzyme, dihydrolipoyl transacetylase ($E_2$), to form the acetyl thioester of the reduced lipoyl group. Step ③ is a transesterification in which the

—SH group of CoA replaces the —SH group of $E_2$ to yield acetyl-CoA and the fully reduced (dithiol) form of the lipoyl group. In step ④ dihydrolipoyl dehydrogenase ($E_3$) promotes transfer of two hydrogen atoms from the reduced lipoyl groups of $E_2$ to the FAD prosthetic group of $E_3$, restoring the oxidized form of the lipoyllysyl group of $E_2$. In step ⑤ the reduced $FADH_2$ of $E_3$ transfers a hydride ion to $NAD^+$, forming NADH. The enzyme complex is now ready for another catalytic cycle. (Subunit colors correspond to those in Fig. 16–5b.)

sites is used in some other enzymes, with lipoate, biotin, or a CoA-like moiety serving as cofactors.

As one might predict, mutations in the genes for the subunits of the PDH complex, or a dietary thiamine deficiency, can have severe consequences. Thiamine-deficient animals are unable to oxidize pyruvate normally. This is of particular importance to the brain, which usually obtains all its energy from the aerobic oxidation of glucose in a pathway that necessarily includes the oxidation of pyruvate. Beriberi, a disease that results from thiamine deficiency, is characterized by loss of neural function. This disease occurs primarily in populations that rely on a diet consisting mainly of white (polished) rice, which lacks the hulls in which most of the thiamine of rice is found. People who habitually consume large amounts of alcohol can also develop thiamine deficiency, because much of their dietary intake consists of the vitamin-free "empty calories" of distilled spirits. An elevated level of pyruvate in the blood is often an indicator of defects in pyruvate oxidation due to one of these causes. ■

## SUMMARY 16.1 Production of Acetyl-CoA (Activated Acetate)

- Pyruvate, the product of glycolysis, is converted to acetyl-CoA, the starting material for the citric acid cycle, by the pyruvate dehydrogenase complex.

- The PDH complex is composed of multiple copies of three enzymes: pyruvate dehydrogenase, $E_1$ (with its bound cofactor TPP); dihydrolipoyl transacetylase, $E_2$ (with its covalently bound lipoyl group); and dihydrolipoyl dehydrogenase, $E_3$ (with its cofactors FAD and NAD).

- $E_1$ catalyzes first the decarboxylation of pyruvate, producing hydroxyethyl-TPP, and then the oxidation of the hydroxyethyl group to an acetyl group. The electrons from this oxidation reduce the disulfide of lipoate bound to $E_2$, and the acetyl group is transferred into thioester linkage with one —SH group of reduced lipoate.

- $E_2$ catalyzes the transfer of the acetyl group to coenzyme A, forming acetyl-CoA.

- $E_3$ catalyzes the regeneration of the disulfide (oxidized) form of lipoate; electrons pass first to FAD, then to $NAD^+$.

- The long lipoyllysine arm swings from the active site of $E_1$ to $E_2$ to $E_3$, tethering the intermediates to the enzyme complex to allow substrate channeling.

- The organization of the PDH complex is very similar to that of the enzyme complexes that catalyze the oxidation of $\alpha$-ketoglutarate and the branched-chain $\alpha$-keto acids.

## 16.2 Reactions of the Citric Acid Cycle

We are now ready to trace the process by which acetyl-CoA undergoes oxidation. This chemical transformation is carried out by the citric acid cycle, the first *cyclic* pathway we have encountered (Fig. 16–7). To begin a turn of the cycle, acetyl-CoA donates its acetyl group to the four-carbon compound oxaloacetate to form the six-carbon citrate. Citrate is then transformed into isocitrate, also a six-carbon molecule, which is dehydrogenated with loss of $CO_2$ to yield the five-carbon compound $\alpha$-ketoglutarate (also called oxoglutarate). $\alpha$-Ketoglutarate undergoes loss of a second molecule of $CO_2$ and ultimately yields the four-carbon compound succinate. Succinate is then enzymatically converted in three steps into the four-carbon oxaloacetate—which is then ready to react with another molecule of acetyl-CoA. In each turn of the cycle, one acetyl group (two carbons) enters as acetyl-CoA and two molecules of $CO_2$ leave; one molecule of oxaloacetate is used to form citrate and one molecule of oxaloacetate is regenerated. No net removal of oxaloacetate occurs; one molecule of oxaloacetate can theoretically bring about oxidation of an infinite number of acetyl groups, and, in fact, oxaloacetate is present in cells in very low concentrations. Four of the eight steps in this process are oxidations, in which the energy of oxidation is very efficiently conserved in the form of the reduced coenzymes NADH and $FADH_2$.

As noted earlier, although the citric acid cycle is central to energy-yielding metabolism its role is not limited to energy conservation. Four- and five-carbon intermediates of the cycle serve as precursors for a wide variety of products. To replace intermediates removed for this purpose, cells employ anaplerotic (replenishing) reactions, which are described below.

Eugene Kennedy and Albert Lehninger showed in 1948 that, in eukaryotes, the entire set of reactions of the citric acid cycle takes place in mitochondria. Isolated mitochondria were found to contain not only all the enzymes and coenzymes required for the citric acid cycle, but also all the enzymes and proteins necessary for the last stage of respiration—electron transfer and ATP synthesis by oxidative phosphorylation. As we shall see in later chapters, mitochondria also contain the enzymes for the oxidation of fatty acids and some amino acids to acetyl-CoA, and the oxidative degradation of other amino acids to $\alpha$-ketoglutarate, succinyl-CoA, or oxaloacetate. Thus, in nonphotosynthetic eukaryotes, the mitochondrion is the site of most energy-yielding

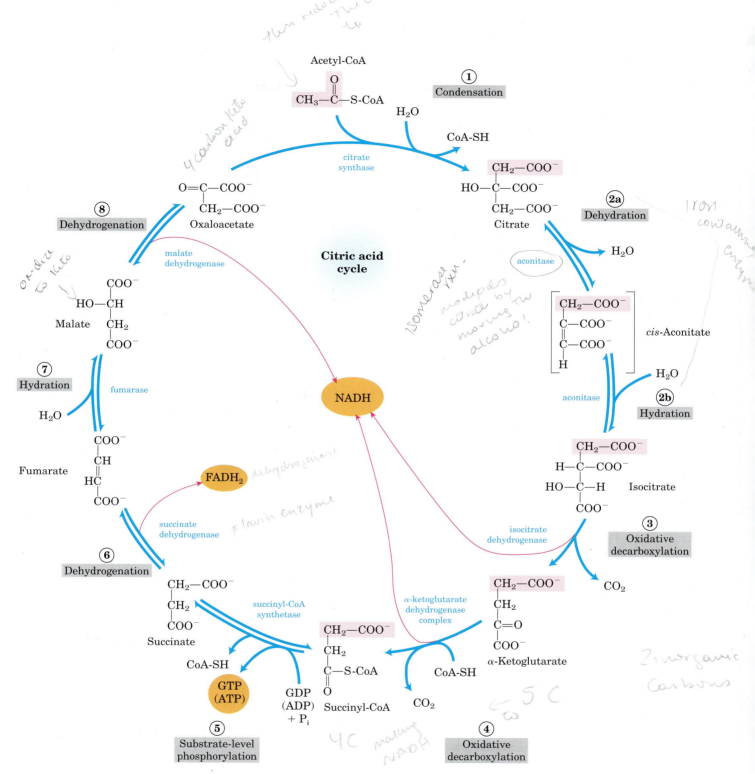

**FIGURE 16–7 Reactions of the citric acid cycle.** The carbon atoms shaded in pink are those derived from the acetate of acetyl-CoA in the first turn of the cycle; these are *not* the carbons released as CO₂ in the first turn. Note that in succinate and fumarate, the two-carbon group derived from acetate can no longer be specifically denoted; because succinate and fumarate are symmetric molecules, C-1 and C-2 are indistinguishable from C-4 and C-3. The number beside each reaction step corresponds to a numbered heading on pages 608–612. The red arrows show where energy is conserved by electron transfer to FAD or NAD⁺, forming FADH₂ or NADH + H⁺. Steps ①, ③, and ④ are essentially irreversible in the cell; all other steps are reversible. The product of step ⑤ may be either ATP or GTP, depending on which succinyl-CoA synthetase isozyme is the catalyst.

oxidative reactions and of the coupled synthesis of ATP. In photosynthetic eukaryotes, mitochondria are the major site of ATP production in the dark, but in daylight chloroplasts produce most of the organism's ATP. In most prokaryotes, the enzymes of the citric acid cycle are in the cytosol, and the plasma membrane plays a role analogous to that of the inner mitochondrial membrane in ATP synthesis (Chapter 19).

## The Citric Acid Cycle Has Eight Steps

In examining the eight successive reaction steps of the citric acid cycle, we place special emphasis on the chemical transformations taking place as citrate formed from acetyl-CoA and oxaloacetate is oxidized to yield $CO_2$ and the energy of this oxidation is conserved in the reduced coenzymes NADH and $FADH_2$.

(1) *Formation of Citrate* The first reaction of the cycle is the condensation of acetyl-CoA with **oxaloacetate** to form **citrate,** catalyzed by **citrate synthase:**

$$CH_3-\overset{O}{\underset{S-CoA}{C}} + \overset{O=C-COO^-}{\underset{CH_2-COO^-}{}} \xrightarrow[\text{synthase}]{\text{citrate}} \quad \overset{H_2O \quad CoA\text{-}SH}{}$$

Acetyl-CoA          Oxaloacetate

$$\begin{array}{c} CH_2-\overset{O}{\underset{O^-}{C}} \\ HO-C-COO^- \\ CH_2-COO^- \end{array}$$

Citrate          $\Delta G'^\circ = -32.2$ kJ/mol

In this reaction the methyl carbon of the acetyl group is joined to the carbonyl group (C-2) of oxaloacetate. Citroyl-CoA is a transient intermediate formed on the active site of the enzyme (see Fig. 16–9). It rapidly undergoes hydrolysis to free CoA and citrate, which are released from the active site. The hydrolysis of this high-energy thioester intermediate makes the forward reaction highly exergonic. The large, negative standard free-energy change of the citrate synthase reaction is essential to the operation of the cycle because, as noted earlier, the concentration of oxaloacetate is normally very low. The CoA liberated in this reaction is recycled to participate in the oxidative decarboxylation of another molecule of pyruvate by the PDH complex.

Citrate synthase from mitochondria has been crystallized and visualized by x-ray diffraction in the presence and absence of its substrates and inhibitors (Fig. 16–8). Each subunit of the homodimeric enzyme is a single polypeptide with two domains, one large and rigid, the other smaller and more flexible, with the active site between them. Oxaloacetate, the first substrate to bind to the enzyme, induces a large conformational

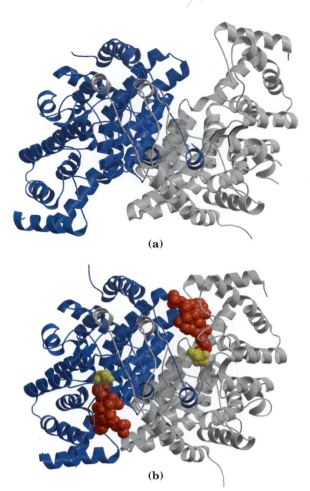

**(a)**

**(b)**

**FIGURE 16–8 Structure of citrate synthase.** The flexible domain of each subunit undergoes a large conformational change on binding oxaloacetate creating a binding site for acetyl-CoA. **(a)** Open form of the enzyme alone (PDB ID 5CSC); **(b)** closed form with bound oxaloacetate (yellow) and a stable analog of acetyl-CoA (carboxymethyl-CoA; red) (derived from PDB ID 5CTS).

change in the flexible domain, creating a binding site for the second substrate, acetyl-CoA. When citroyl-CoA has formed in the enzyme active site, another conformational change brings about thioester hydrolysis, releasing CoA-SH. This induced fit of the enzyme first to its substrate and then to its reaction intermediate decreases the likelihood of premature and unproductive cleavage of the thioester bond of acetyl-CoA. Kinetic studies of the enzyme are consistent with this ordered bisubstrate mechanism (see Fig. 6–13). The reaction catalyzed by citrate synthase is essentially a Claisen condensation (p. 485), involving a thioester (acetyl-CoA) and a ketone (oxaloacetate) (Fig. 16–9).

(2) *Formation of Isocitrate via cis-Aconitate* The enzyme **aconitase** (more formally, **aconitate hydratase**) catalyzes the reversible transformation of citrate to **isocitrate,** through the intermediary formation of the tricarboxylic acid *cis*-**aconitate,** which normally does

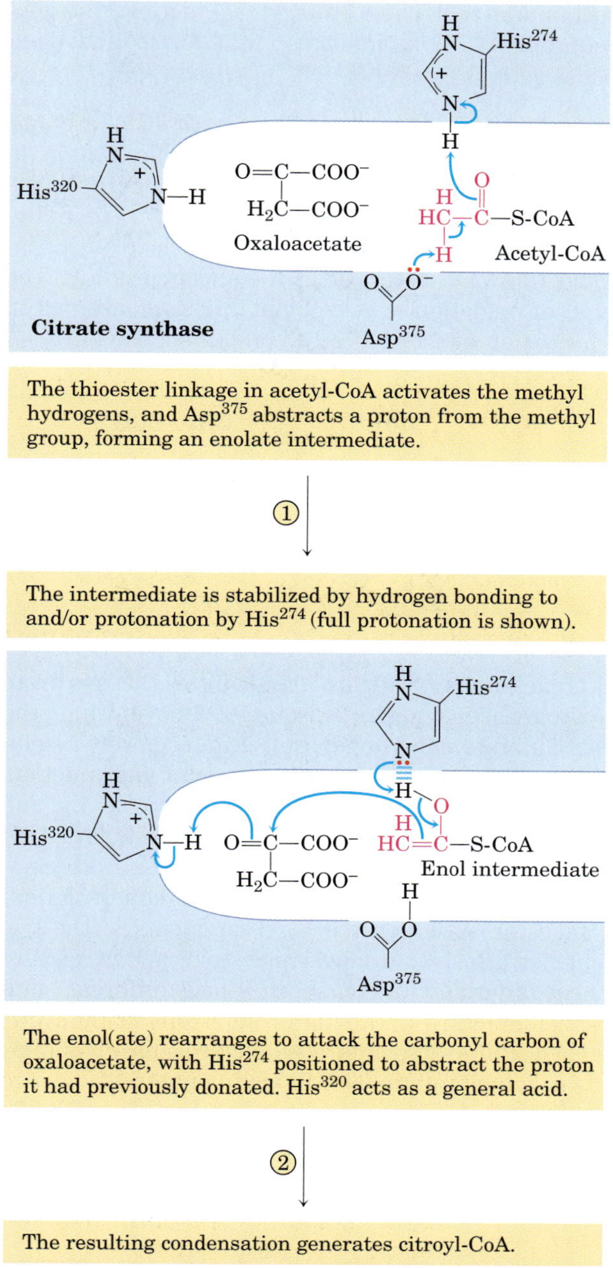
H
N
His274
H
N
H
His320
N—H
O=C—COO⁻
H₂C—COO⁻
Oxaloacetate
O
H
HC—C—S-CoA
H
Acetyl-CoA
O
O⁻
Asp375
Citrate synthase
The thioester linkage in acetyl-CoA activates the methyl hydrogens, and Asp375 abstracts a proton from the methyl group, forming an enolate intermediate.
1
The intermediate is stabilized by hydrogen bonding to and/or protonation by His274 (full protonation is shown).
H
N
His274
N
H
H
His320
N—H
O=C—COO⁻
H₂C—COO⁻
O
H
HC=C—S-CoA
Enol intermediate
H
O
O
Asp375
The enol(ate) rearranges to attack the carbonyl carbon of oxaloacetate, with His274 positioned to abstract the proton it had previously donated. His320 acts as a general acid.
2

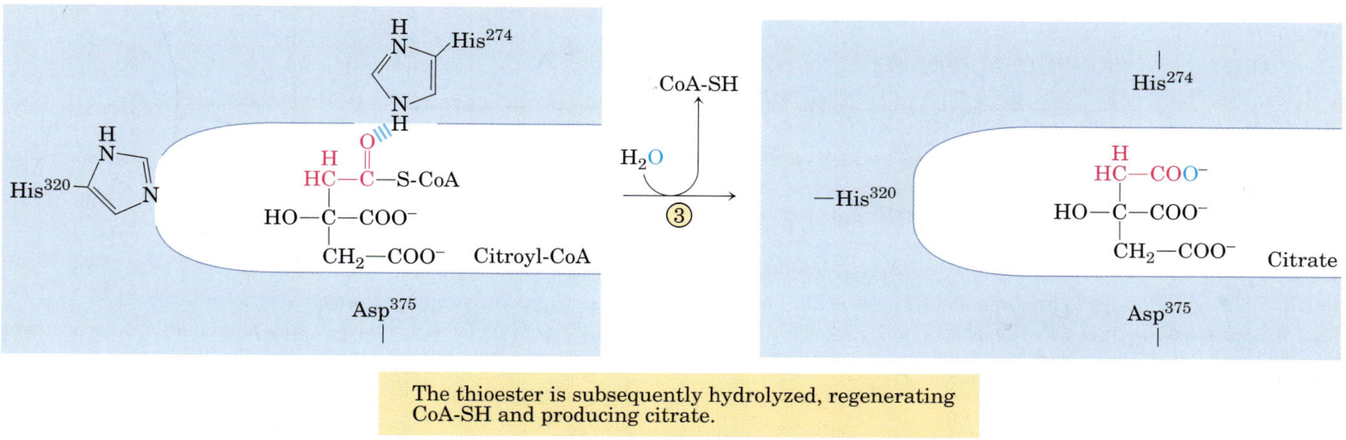
The resulting condensation generates citroyl-CoA.
H
N
His274
N
H
H
N
His320
N
O
H
HC—C—S-CoA
HO—C—COO⁻
CH₂—COO⁻
Citroyl-CoA
Asp375
CoA-SH
H₂O
3
His274
—His320
H
HC—COO⁻
HO—C—COO⁻
CH₂—COO⁻
Citrate
Asp375
The thioester is subsequently hydrolyzed, regenerating CoA-SH and producing citrate.

not dissociate from the active site. Aconitase can promote the reversible addition of $H_2O$ to the double bond of enzyme-bound *cis*-aconitate in two different ways, one leading to citrate and the other to isocitrate:

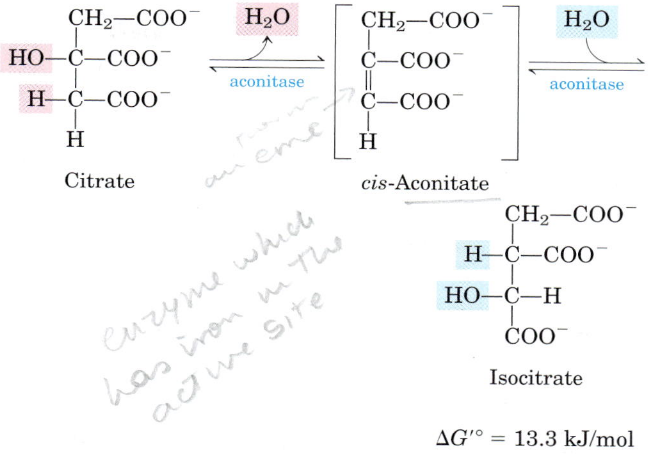
CH₂—COO⁻
HO—C—COO⁻
H—C—COO⁻
H
Citrate
H₂O
aconitase
CH₂—COO⁻
C—COO⁻
C—COO⁻
H
cis-Aconitate
H₂O
aconitase
CH₂—COO⁻
H—C—COO⁻
HO—C—H
COO⁻
Isocitrate
ΔG′° = 13.3 kJ/mol

Although the equilibrium mixture at pH 7.4 and 25 °C contains less than 10% isocitrate, in the cell the reaction is pulled to the right because isocitrate is rapidly consumed in the next step of the cycle, lowering its steady-state concentration. Aconitase contains an **iron-sulfur center** (Fig. 16–10), which acts both in the binding of the substrate at the active site and in the catalytic addition or removal of $H_2O$.

**MECHANISM FIGURE 16–9 Citrate synthase.** In the mammalian citrate synthase reaction, oxaloacetate binds first, in a strictly ordered reaction sequence. This binding triggers a conformation change that opens up the binding site for acetyl-CoA. Oxaloacetetate is specifically oriented in the active site of citrate synthase by interaction of its two carboxylates with two positively charged Arg residues (not shown here). The details of the mechanism are described in the figure. Citrate Synthase Mechanism

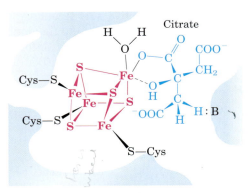

**FIGURE 16–10 Iron-sulfur center in aconitase.** The iron-sulfur center is in red, the citrate molecule in blue. Three Cys residues of the enzyme bind three iron atoms; the fourth iron is bound to one of the carboxyl groups of citrate and also interacts noncovalently with a hydroxyl group of citrate (dashed bond). A basic residue (:B) on the enzyme helps to position the citrate in the active site. The iron-sulfur center acts in both substrate binding and catalysis. The general properties of iron-sulfur proteins are discussed in Chapter 19 (see Fig. 19–5).

③ **Oxidation of Isocitrate to α-Ketoglutarate and CO₂** In the next step, **isocitrate dehydrogenase** catalyzes oxidative decarboxylation of isocitrate to form **α-ketoglutarate** (Fig. 16–11). $Mn^{2+}$ in the active site interacts with the carbonyl group of the intermediate oxalosuccinate, which is formed transiently but does not leave the binding site until decarboxylation converts it to α-ketoglutarate. $Mn^{2+}$ also stabilizes the enol formed transiently by decarboxylation.

There are two different forms of isocitrate dehydrogenase in all cells, one requiring $NAD^+$ as electron acceptor and the other requiring $NADP^+$. The overall reactions are otherwise identical. In eukaryotic cells, the NAD-dependent enzyme occurs in the mitochondrial matrix and serves in the citric acid cycle. The main function of the NADP-dependent enzyme, found in both the mitochondrial matrix and the cytosol, may be the generation of NADPH, which is essential for reductive anabolic reactions.

④ **Oxidation of α-Ketoglutarate to Succinyl-CoA and CO₂** The next step is another oxidative decarboxylation, in which α-ketoglutarate is converted to **succinyl-CoA** and $CO_2$ by the action of the **α-ketoglutarate dehydrogenase complex;** $NAD^+$ serves as electron acceptor and CoA as the carrier of the succinyl group. The energy of oxidation of α-ketoglutarate is conserved in the formation of the thioester bond of succinyl-CoA:

$$\text{CH}_2\text{—COO}^- \quad \xrightarrow[\substack{\alpha\text{-ketoglutarate} \\ \text{dehydrogenase} \\ \text{complex}}]{\substack{\text{CoA-SH} \\ NAD^+ \quad NADH}} \quad \text{CH}_2\text{—COO}^- + CO_2$$

α-Ketoglutarate → Succinyl-CoA

$$\Delta G'^\circ = -33.5 \text{ kJ/mol}$$

This reaction is virtually identical to the pyruvate dehydrogenase reaction discussed above, and the α-ketoglutarate dehydrogenase complex closely resembles the PDH complex in both structure and function. It includes three enzymes, homologous to $E_1$, $E_2$, and $E_3$ of the PDH complex, as well as enzyme-bound TPP, bound lipoate, FAD, NAD, and coenzyme A. Both complexes are certainly derived from a common evolutionary ancestor. Although the $E_1$ components of the two complexes are structurally similar, their amino acid sequences differ and, of course, they have different binding specificities: $E_1$ of the PDH complex binds pyruvate, and $E_1$ of the α-ketoglutarate dehydrogenase complex binds α-ketoglutarate. The $E_2$ components of the two complexes are also very similar, both having covalently bound lipoyl moieties. The subunits of $E_3$ are identical in the two enzyme complexes.

**MECHANISM FIGURE 16–11 Isocitrate dehydrogenase.** In this reaction, the substrate, isocitrate, loses one carbon by oxidative decarboxylation. In step ①, isocitrate binds to the enzyme and is oxidized by hydride transfer to $NAD^+$ or $NADP^+$, depending on the isocitrate dehydrogenase isozyme. (See Fig. 14–12 for more information on hydride transfer reactions involving $NAD^+$ and $NADP^+$.) The resulting carbonyl group sets up the molecule for decarboxylation in step ②. Interaction of the carbonyl oxygen with a bound $Mn^{2+}$ ion increases the electron-withdrawing capacity of the carbonyl group and facilitates the decarboxylation step. The reaction is completed in step ③ by rearrangement of the enol intermediate to generate α-ketoglutarate.

**⑤ Conversion of Succinyl-CoA to Succinate** Succinyl-CoA, like acetyl-CoA, has a thioester bond with a strongly negative standard free energy of hydrolysis ($\Delta G'^{\circ} \approx$ −36 kJ/mol). In the next step of the citric acid cycle, energy released in the breakage of this bond is used to drive the synthesis of a phosphoanhydride bond in either GTP or ATP, with a net $\Delta G'^{\circ}$ of only −2.9 kJ/mol. **Succinate** is formed in the process:

$$\Delta G'^{\circ} = -2.9 \text{ kJ/mol}$$

The enzyme that catalyzes this reversible reaction is called **succinyl-CoA synthetase** or **succinic thiokinase**; both names indicate the participation of a nucleoside triphosphate in the reaction (Box 16–1).

This energy-conserving reaction involves an intermediate step in which the enzyme molecule itself becomes phosphorylated at a His residue in the active site (Fig. 16–12a). This phosphoryl group, which has a high group transfer potential, is transferred to ADP (or GDP) to form ATP (or GTP). Animal cells have two isozymes of succinyl-CoA synthetase, one specific for ADP and the other for GDP. The enzyme has two subunits, $\alpha$ ($M_r$ 32,000), which has the ℗-His residue (His$^{246}$) and the binding site for CoA, and $\beta$ ($M_r$ 42,000), which confers specificity for either ADP or GDP. The active site is at the interface between subunits. The crystal structure of succinyl-CoA synthetase reveals two "power helices" (one from each subunit), oriented so that their electric dipoles situate partial positive charges close to the negatively charged ℗-His (Fig. 16–12b), stabilizing the phosphoenzyme intermediate. (Recall the similar role of helix dipoles in stabilizing K$^+$ ions in the K$^+$ channel (see Fig. 11–48).)

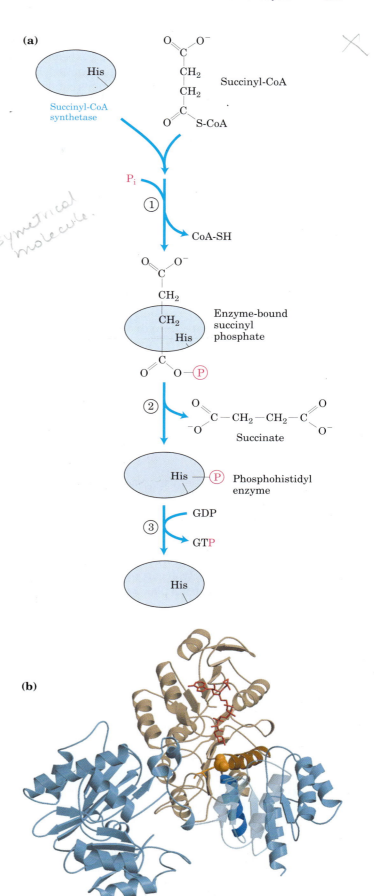

(a)

(b)

**FIGURE 16–12 The succinyl-CoA synthetase reaction. (a)** In step ① a phosphoryl group replaces the CoA of succinyl-CoA bound to the enzyme, forming a high-energy acyl phosphate. In step ② the succinyl phosphate donates its phosphoryl group to a His residue on the enzyme, forming a high-energy phosphohistidyl enzyme. In step ③ the phosphoryl group is transferred from the His residue to the terminal phosphate of GDP (or ADP), forming GTP (or ATP). **(b)** Succinyl-CoA synthetase of *E. coli* (derived from PDB ID 1SCU). The bacterial and mammalian enzymes have similar amino acid sequences and presumably have very similar three-dimensional structures. The active site includes part of both the $\alpha$ (blue) and $\beta$ (brown) subunits. The power helices (bright blue, dark brown) situate the partial positive charges of the helix dipole near the phosphate group (orange) on His$^{246}$ of the $\alpha$ chain, stabilizing the phosphohistidyl enzyme. Coenzyme A is shown here as a red stick structure. (To improve the visibility of the power helices, some nearby secondary structures have been made transparent.)

The formation of ATP (or GTP) at the expense of the energy released by the oxidative decarboxylation of $\alpha$-ketoglutarate is a substrate-level phosphorylation, like the synthesis of ATP in the glycolytic reactions catalyzed by glyceraldehyde 3-phosphate dehydrogenase and pyruvate kinase (see Fig. 14–2). The GTP formed by succinyl-CoA synthetase can donate its terminal phosphoryl group to ADP to form ATP, in a reversible reaction catalyzed by **nucleoside diphosphate kinase** (p. 505):

$$GTP + ADP \longrightarrow GDP + ATP \qquad \Delta G'^{\circ} = 0 \text{ kJ/mol}$$

Thus the net result of the activity of either isozyme of succinyl-CoA synthetase is the conservation of energy as ATP. There is no change in free energy for the nucleoside diphosphate kinase reaction; ATP and GTP are energetically equivalent.

⑥ **Oxidation of Succinate to Fumarate**  The succinate formed from succinyl-CoA is oxidized to **fumarate** by the flavoprotein **succinate dehydrogenase:**

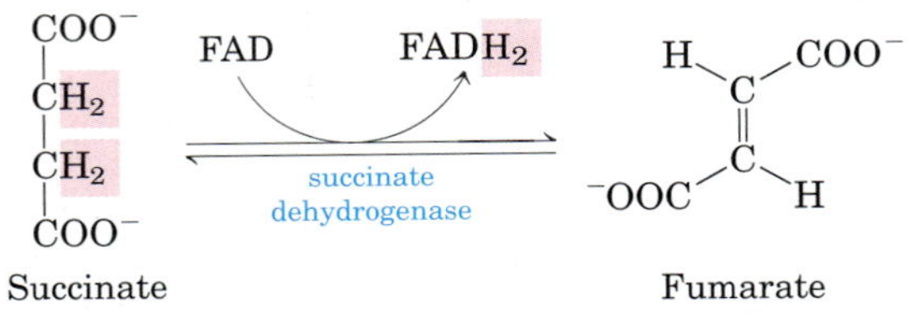

*[handwritten: equilibrium / max action driven]*

*[handwritten: Old yellow / L comes from the / Flavin FAD]*

$$\Delta G'^{\circ} = 0 \text{ kJ/mol}$$

*[handwritten: Trans]*

In eukaryotes, succinate dehydrogenase is tightly bound to the inner mitochondrial membrane; in prokaryotes, to the plasma membrane. The enzyme contains three different iron-sulfur clusters and one molecule of covalently bound FAD (see Fig. 19–8). Electrons pass from succinate through the FAD and iron-sulfur centers before entering the chain of electron carriers in the mitochondrial inner membrane (or the plasma membrane in bacteria). Electron flow from succinate through these carriers to the final electron acceptor, $O_2$, is coupled to the synthesis of about 1.5 ATP molecules per pair of electrons (respiration-linked phosphorylation). Malonate, an analog of succinate not normally present in cells, is a strong competitive inhibitor of succinate dehydrogenase, and its addition to mitochondria blocks the activity of the citric acid cycle.

*[handwritten: 1st inhibitor]*

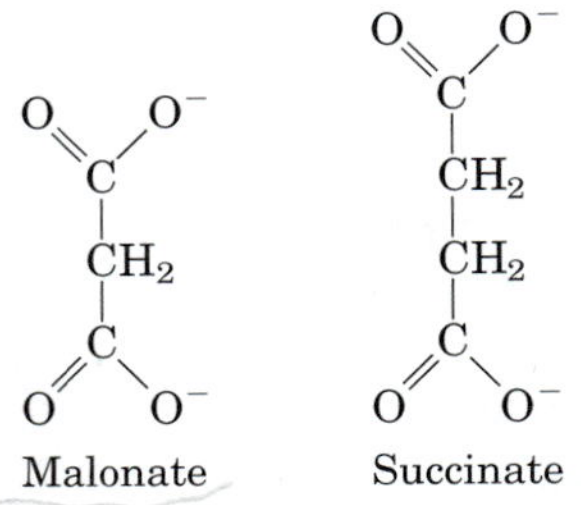

⑦ **Hydration of Fumarate to Malate**  The reversible hydration of fumarate to **L-malate** is catalyzed by **fumarase**

(formally, **fumarate hydratase**). The transition state in this reaction is a carbanion:

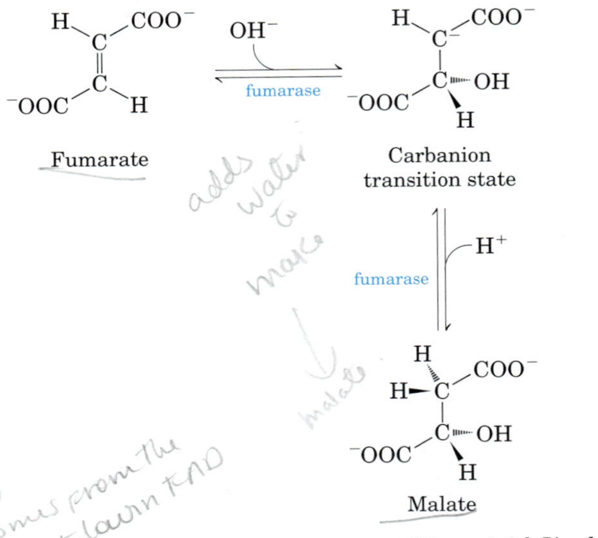

*[handwritten: adds water to make]*
*[handwritten: malate]*

$$\Delta G'^{\circ} = -3.8 \text{ kJ/mol}$$

This enzyme is highly stereospecific; it catalyzes hydration of the trans double bond of fumarate but not the cis double bond of maleate (the cis isomer of fumarate). In the reverse direction (from L-malate to fumarate), fumarase is equally stereospecific: D-malate is not a substrate.

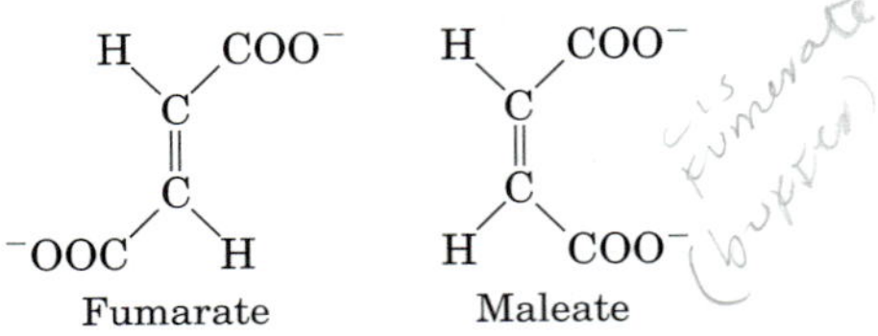

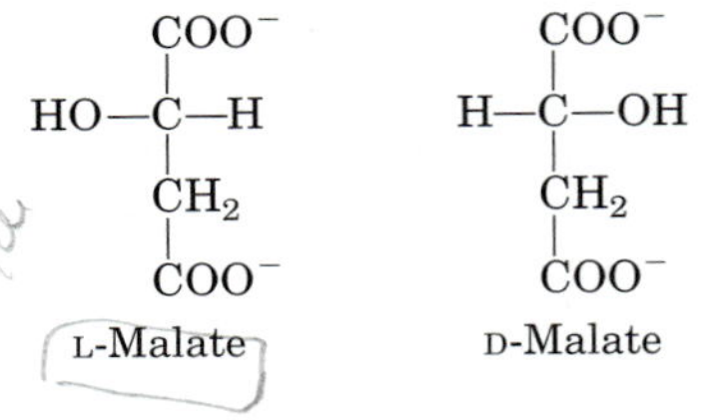

*[handwritten: cis fumerate (buptica)]*

*[handwritten: made in citric acid]*

⑧ **Oxidation of Malate to Oxaloacetate**  In the last reaction of the citric acid cycle, NAD-linked **L-malate dehydrogenase** catalyzes the oxidation of L-malate to oxaloacetate:

*[handwritten: Oxidize alcohol to carbonyl]*

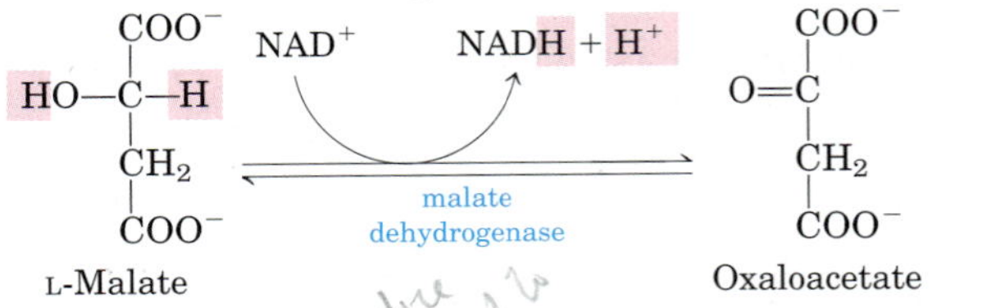

$$\Delta G'^{\circ} = 29.7 \text{ kJ/mol}$$

The equilibrium of this reaction lies far to the left under standard thermodynamic conditions, but in intact cells

## BOX 16–1    WORKING IN BIOCHEMISTRY

### Synthases and Synthetases; Ligases and Lyases; Kinases, Phosphatases, and Phosphorylases: Yes, the Names Are Confusing!

Citrate synthase is one of many enzymes that catalyze condensation reactions, yielding a product more chemically complex than its precursors. **Synthases** catalyze condensation reactions in which no nucleoside triphosphate (ATP, GTP, and so forth) is required as an energy source. **Synthetases** catalyze condensations that *do* use ATP or another nucleoside triphosphate as a source of energy for the synthetic reaction. Succinyl-CoA synthetase is such an enzyme. **Ligases** (from the Latin *ligare,* "to tie together") are enzymes that catalyze condensation reactions in which two atoms are joined using ATP or another energy source. (Thus synthetases are ligases.) DNA ligase, for example, closes breaks in DNA molecules, using energy supplied by either ATP or $NAD^+$; it is widely used in joining DNA pieces for genetic engineering. Ligases are not to be confused with **lyases,** enzymes that catalyze cleavages (or, in the reverse direction, additions) in which electronic rearrangements occur. The PDH complex, which oxidatively cleaves $CO_2$ from pyruvate, is a member of the large class of lyases.

The name **kinase** is applied to enzymes that transfer a phosphoryl group from a nucleoside triphosphate such as ATP to an acceptor molecule—a sugar (as in hexokinase and glucokinase), a protein (as in glycogen phosphorylase kinase), another nucleotide (as in nucleoside diphosphate kinase), or a metabolic intermediate such as oxaloacetate (as in PEP carboxykinase). The reaction catalyzed by a kinase is a *phosphorylation.* On the other hand, *phosphorolysis* is a displacement reaction in which phosphate is the attacking species and becomes covalently attached at the point of bond breakage. Such reactions are catalyzed by **phosphorylases.** Glycogen phosphorylase, for example, catalyzes the phosphorolysis of glycogen, producing glucose 1-phosphate. *Dephos-*

*phorylation,* the removal of a phosphoryl group from a phosphate ester, is catalyzed by **phosphatases,** with water as the attacking species. Fructose bisphosphatase-1 converts fructose 1,6-bisphosphate to fructose 6-phosphate in gluconeogenesis, and phosphorylase *a* phosphatase removes phosphoryl groups from phosphoserine in phosphorylated glycogen phosphorylase. Whew!

Unfortunately, these descriptions of enzyme types overlap, and many enzymes are commonly called by two or more names. Succinyl-CoA synthetase, for example, is also called succinate thiokinase; the enzyme is both a synthetase in the citric acid cycle and a kinase when acting in the direction of succinyl-CoA synthesis. This raises another source of confusion in the naming of enzymes. An enzyme may have been discovered by the use of an assay in which, say, A is converted to B. The enzyme is then named for that reaction. Later work may show, however, that in the cell, the enzyme functions primarily in converting B to A. Commonly, the first name continues to be used, although the metabolic role of the enzyme would be better described by naming it for the reverse reaction. The glycolytic enzyme pyruvate kinase illustrates this situation (p. 532). To a beginner in biochemistry, this duplication in nomenclature can be bewildering. International committees have made heroic efforts to systematize the nomenclature of enzymes (see Table 6–3 for a brief summary of the system), but some systematic names have proved too long and cumbersome and are not frequently used in biochemical conversation.

We have tried throughout this book to use the enzyme name most commonly used by working biochemists and to point out cases in which an enzyme has more than one widely used name. For current information on enzyme nomenclature, refer to the recommendations of the Nomenclature Committee of the International Union of Biochemistry and Molecular Biology (www.chem.qmw.ac.uk/iubmb/nomenclature/).

---

oxaloacetate is continually removed by the highly exergonic citrate synthase reaction (step ① of Fig. 16–7). This keeps the concentration of oxaloacetate in the cell extremely low ($<10^{-6}$ M), pulling the malate dehydrogenase reaction toward the formation of oxaloacetate.

Although the individual reactions of the citric acid cycle were initially worked out in vitro, using minced muscle tissue, the pathway and its regulation have also been studied extensively in vivo. By using radioactively la-

beled precursors such as [$^{14}$C]pyruvate and [$^{14}$C]acetate, researchers have traced the fate of individual carbon atoms through the citric acid cycle. Some of the earliest experiments with isotopes produced an unexpected result, however, which aroused considerable controversy about the pathway and mechanism of the citric acid cycle. In fact, these experiments at first seemed to show that citrate was not the first tricarboxylic acid to be formed. Box 16–2 gives some details of this episode in the history of citric acid cycle research. Metabolic flux through the cycle can now be monitored in living tissue by using $^{13}C$-labeled precursors and whole-tissue NMR spectroscopy. Because the NMR signal is unique to the compound containing the $^{13}C$, biochemists can trace the movement of precursor carbons into each cycle intermediate and into compounds derived from the intermediates. This technique has great promise for studies of regulation of the citric acid cycle and its interconnections with other metabolic pathways such as glycolysis.

## The Energy of Oxidations in the Cycle Is Efficiently Conserved

We have now covered one complete turn of the citric acid cycle (Fig. 16–13). A two-carbon acetyl group entered the cycle by combining with oxaloacetate. Two carbon atoms emerged from the cycle as $CO_2$ from the oxidation of isocitrate and $\alpha$-ketoglutarate. The energy released by these oxidations was conserved in the reduction of three $NAD^+$ and one FAD and the produc-

tion of one ATP or GTP. At the end of the cycle a molecule of oxaloacetate was regenerated. Note that the two carbon atoms appearing as $CO_2$ are not the same two carbons that entered in the form of the acetyl group; additional turns around the cycle are required to release these carbons as $CO_2$ (Fig. 16–7).

Although the citric acid cycle directly generates only one ATP per turn (in the conversion of succinyl-CoA to succinate), the four oxidation steps in the cycle provide a large flow of electrons into the respiratory chain via NADH and $FADH_2$ and thus lead to formation of a large number of ATP molecules during oxidative phosphorylation.

We saw in Chapter 14 that the energy yield from the production of two molecules of pyruvate from one molecule of glucose in glycolysis is 2 ATP and 2 NADH. In oxidative phosphorylation (Chapter 19), passage of two electrons from NADH to $O_2$ drives the formation of about 2.5 ATP, and passage of two electrons from $FADH_2$ to $O_2$ yields about 1.5 ATP. This stoichiometry allows us to calculate the overall yield of ATP from the complete

---

**BOX 16–2    WORKING IN BIOCHEMISTRY**

### Citrate: A Symmetrical Molecule That Reacts Asymmetrically

When compounds enriched in the heavy-carbon isotope $^{13}C$ and the radioactive carbon isotopes $^{11}C$ and $^{14}C$ became available about 60 years ago, they were soon put to use in tracing the pathway of carbon atoms through the citric acid cycle. One such experiment initiated the controversy over the role of citrate. Acetate labeled in the carboxyl group (designated [1-$^{14}C$] acetate) was incubated aerobically with an animal tissue preparation. Acetate is enzymatically converted to acetyl-CoA in animal tissues, and the pathway of the

labeled carboxyl carbon of the acetyl group in the cycle reactions could thus be traced. $\alpha$-Ketoglutarate was isolated from the tissue after incubation, then degraded by known chemical reactions to establish the position(s) of the isotopic carbon.

Condensation of unlabeled oxaloacetate with carboxyl-labeled acetate would be expected to produce citrate labeled in one of the two primary carboxyl groups. Citrate is a symmetric molecule, its two terminal carboxyl groups being chemically indistinguishable. Therefore, half the labeled citrate molecules were expected to yield $\alpha$-ketoglutarate labeled in

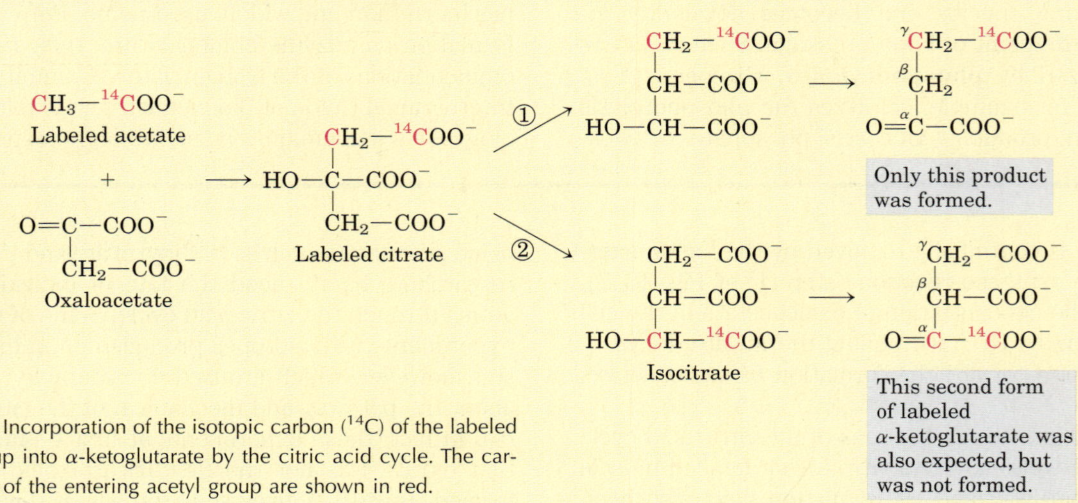

**FIGURE 1** Incorporation of the isotopic carbon ($^{14}C$) of the labeled acetyl group into $\alpha$-ketoglutarate by the citric acid cycle. The carbon atoms of the entering acetyl group are shown in red.

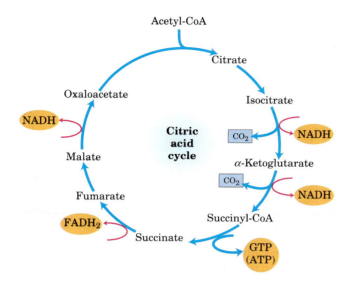

**FIGURE 16–13 Products of one turn of the citric acid cycle.** At each turn of the cycle, three NADH, one FADH$_2$, one GTP (or ATP), and two CO$_2$ are released in oxidative decarboxylation reactions. Here and in several following figures, all cycle reactions are shown as proceeding in one direction only, but keep in mind that most of the reactions are reversible (see Fig. 16–7).

oxidation of glucose. When both pyruvate molecules are oxidized to 6 CO$_2$ via the pyruvate dehydrogenase complex and the citric acid cycle, and the electrons are transferred to O$_2$ via oxidative phosphorylation, as many as 32 ATP are obtained per glucose (Table 16–1). In round numbers, this represents the conservation of 32 × 30.5 kJ/mol = 976 kJ/mol, or 34% of the theoretical maximum of about 2,840 kJ/mol available from the complete oxidation of glucose. These calculations employ the standard free-energy changes; when corrected for the actual free energy required to form ATP within cells (see Box 13–1), the calculated efficiency of the process is closer to 65%.

### Why Is the Oxidation of Acetate So Complicated?

The eight-step cyclic process for oxidation of simple two-carbon acetyl groups to CO$_2$ may seem unnecessarily cumbersome and not in keeping with the biological principle of maximum economy. The role of the citric acid cycle is not confined to the oxidation of acetate, however.

the α-carboxyl group and the other half to yield α-keto-glutarate labeled in the γ-carboxyl group; that is, the α-ketoglutarate isolated was expected to be a mixture of the two types of labeled molecules (Fig. 1, pathways ① and ②). Contrary to this expectation, the labeled α-ketoglutarate isolated from the tissue suspension contained $^{14}$C only in the γ-carboxyl group (Fig. 1, pathway ①). The investigators concluded that citrate (or any other symmetric molecule) could not be an intermediate in the pathway from acetate to α-ketoglutarate. Rather, an asymmetric tricarboxylic acid, presumably *cis*-aconitate or isocitrate, must be the first product formed from condensation of acetate and oxaloacetate.

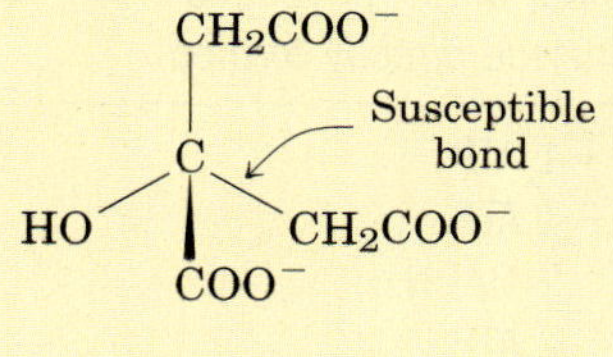

In 1948, however, Alexander Ogston pointed out that although citrate has no chiral center (see Fig. 1–19), it has the *potential* to react asymmetrically if an enzyme with which it interacts has an active site that is asymmetric. He suggested that the active site of aconitase may have three points to which the citrate must be bound and that the citrate must undergo a specific three-point attachment to these binding points. As seen in Figure 2, the binding of citrate to three such points could happen in only one way, and this would account for the formation of only one type of labeled α-ketoglutarate. Organic molecules such as citrate that have no chiral center but are potentially capable of reacting asymmetrically with an asymmetric active site are now called **prochiral molecules.**

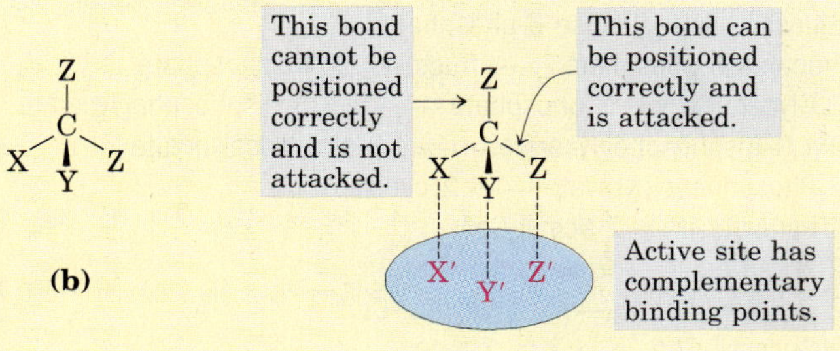

**FIGURE 2 The prochiral nature of citrate. (a)** Structure of citrate; **(b)** schematic representation of citrate: X = —OH; Y = —COO$^-$; Z = —CH$_2$COO$^-$. **(c)** Correct complementary fit of citrate to the binding site of aconitase. There is only one way in which the three specified groups of citrate can fit on the three points of the binding site. Thus only one of the two —CH$_2$COO$^-$ groups is bound by aconitase.

(a) — structure showing CH$_2$COO$^-$, C, HO, CH$_2$COO$^-$ (Susceptible bond), COO$^-$

(b) — schematic Z, C, X, Y, Z

(c) — This bond cannot be positioned correctly and is not attacked. → This bond can be positioned correctly and is attacked. Active site has complementary binding points.

This pathway is the hub of intermediary metabolism. Four- and five-carbon end products of many catabolic processes feed into the cycle to serve as fuels. Oxaloacetate and α-ketoglutarate, for example, are produced from aspartate and glutamate, respectively, when proteins are degraded. Under some metabolic circumstances, intermediates are drawn out of the cycle to be used as precursors in a variety of biosynthetic pathways.

The citric acid cycle, like all other metabolic pathways, is the product of evolution, and much of this evolution occurred before the advent of aerobic organisms. It does not necessarily represent the *shortest* pathway from acetate to CO$_2$, but it is the pathway that has, over time, conferred the greatest selective advantage. Early anaerobes most probably used some of the reactions of the citric acid cycle in linear biosynthetic processes. In fact, some modern anaerobic microorganisms use an incomplete citric acid cycle as a source of, not energy, but biosynthetic precursors (Fig. 16–14). These organisms use the first three reactions of the cycle to make α-ketoglutarate but, lacking α-ketoglutarate dehydrogenase, they cannot carry out the complete set of citric acid cycle reactions. They do have the four enzymes that catalyze the reversible conversion of oxaloacetate to succinyl-CoA and can produce malate, fumarate, succinate, and succinyl-CoA from oxaloacetate in a reversal of the "normal" (oxidative) direction of flow through the cycle. This pathway is a fermentation, with the NADH produced by isocitrate oxidation recycled to NAD$^+$ by reduction of oxaloacetate to succinate.

With the evolution of cyanobacteria that produced O$_2$ from water, the earth's atmosphere became aerobic and organisms were under selective pressure to develop aerobic metabolism, which, as we have seen, is much more efficient than anaerobic fermentation.

## Citric Acid Cycle Components Are Important Biosynthetic Intermediates

In aerobic organisms, the citric acid cycle is an **amphibolic pathway,** one that serves in both catabolic and anabolic processes. Besides its role in the oxidative catabolism of carbohydrates, fatty acids, and amino acids, the cycle provides precursors for many biosynthetic pathways (Fig. 16–15), through reactions that served the same purpose in anaerobic ancestors. α-Ketoglutarate and oxaloacetate can, for example, serve as precursors of the amino acids aspartate and glutamate by simple transamination (Chapter 22). Through aspartate and glutamate, the carbons of oxaloacetate and α-ketoglutarate are then used to build other amino acids, as well as purine and pyrimidine nucleotides. Oxaloacetate is converted to glucose in gluconeogenesis (see Fig. 15–15). Succinyl-CoA is a central intermediate in the synthesis of the porphyrin ring of heme groups, which serve as oxygen carriers (in hemoglobin and myoglobin) and electron carriers (in cytochromes) (see Fig. 22–23). And the citrate produced in some organisms is used commercially for a variety of purposes (Box 16–3).

## Anaplerotic Reactions Replenish Citric Acid Cycle Intermediates

As intermediates of the citric acid cycle are removed to serve as biosynthetic precursors, they are replenished by **anaplerotic reactions** (Fig. 16–15; Table 16–2). Under normal circumstances, the reactions by which cycle intermediates are siphoned off into other pathways and those by which they are replenished are in dynamic balance, so that the concentrations of the citric acid cycle intermediates remain almost constant.

**TABLE 16–1 Stoichiometry of Coenzyme Reduction and ATP Formation in the Aerobic Oxidation of Glucose via Glycolysis, the Pyruvate Dehydrogenase Complex Reaction, the Citric Acid Cycle, and Oxidative Phosphorylation**

| Reaction | Number of ATP or reduced coenzyme directly formed | Number of ATP ultimately formed* |
|---|---|---|
| Glucose ⟶ glucose 6-phosphate | −1 ATP | −1 |
| Fructose 6-phosphate ⟶ fructose 1,6-bisphosphate | −1 ATP | −1 |
| 2 Glyceraldehyde 3-phosphate ⟶ 2 1,3-bisphosphoglycerate | 2 NADH | 3 or 5[†] |
| 2 1,3-Bisphosphoglycerate ⟶ 2 3-phosphoglycerate | 2 ATP | 2 |
| 2 Phosphoenolpyruvate ⟶ 2 pyruvate | 2 ATP | 2 |
| 2 Pyruvate ⟶ 2 acetyl-CoA | 2 NADH | 5 |
| 2 Isocitrate ⟶ 2 α-ketoglutarate | 2 NADH | 5 |
| 2 α-Ketoglutarate ⟶ 2 succinyl-CoA | 2 NADH | 5 |
| 2 Succinyl-CoA ⟶ 2 succinate | 2 ATP (or 2 GTP) | 2 |
| 2 Succinate ⟶ 2 fumarate | 2 FADH$_2$ | 3 |
| 2 Malate ⟶ 2 oxaloacetate | 2 NADH | 5 |
| Total | | 30–32 |

*This is calculated as 2.5 ATP per NADH and 1.5 ATP per FADH$_2$. A negative value indicates consumption.

[†] This number is either 3 or 5, depending on the mechanism used to shuttle NADH equivalents from the cytosol to the mitochondrial matrix; see Figures 19–27 and 19–28.

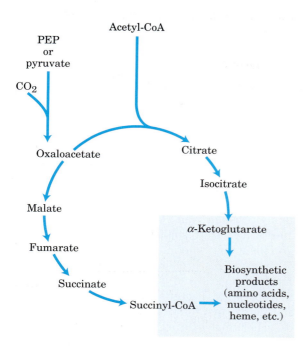

**FIGURE 16-14 Biosynthetic precursors produced by an incomplete citric acid cycle in anaerobic bacteria.** These anaerobes lack α-ketoglutarate dehydrogenase and therefore cannot carry out the complete citric acid cycle. α-Ketoglutarate and succinyl-CoA serve as precursors in a variety of biosynthetic pathways. (See Fig. 16–13 for the "normal" direction of these reactions in the citric acid cycle.)

Table 16–2 shows the most common anaplerotic reactions, all of which, in various tissues and organisms, convert either pyruvate or phosphoenolpyruvate to oxaloacetate or malate. The most important anaplerotic reaction in mammalian liver and kidney is the reversible carboxylation of pyruvate by $CO_2$ to form oxaloacetate, catalyzed by **pyruvate carboxylase.** When the citric acid cycle is deficient in oxaloacetate or any other intermediates, pyruvate is carboxylated to produce more oxaloacetate. The enzymatic addition of a carboxyl group to pyruvate requires energy, which is supplied by ATP—the free energy required to attach a carboxyl group to pyruvate is about equal to the free energy available from ATP.

Pyruvate carboxylase is a regulatory enzyme and is virtually inactive in the absence of acetyl-CoA, its positive allosteric modulator. Whenever acetyl-CoA, the fuel for the citric acid cycle, is present in excess, it stimulates the pyruvate carboxylase reaction to produce more oxaloacetate, enabling the cycle to use more acetyl-CoA in the citrate synthase reaction.

The other anaplerotic reactions shown in Table 16–2 are also regulated to keep the level of intermediates high enough to support the activity of the citric acid cycle. Phosphoenolpyruvate (PEP) carboxylase, for example, is activated by the glycolytic intermediate fructose 1,6-bisphosphate, which accumulates when the citric acid cycle operates too slowly to process the pyruvate generated by glycolysis.

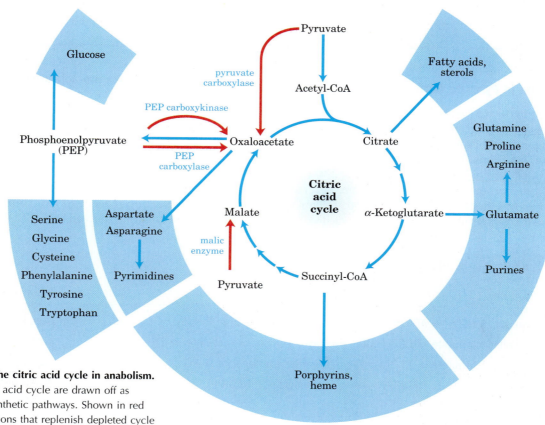

**FIGURE 16-15 Role of the citric acid cycle in anabolism.** Intermediates of the citric acid cycle are drawn off as precursors in many biosynthetic pathways. Shown in red are four anaplerotic reactions that replenish depleted cycle intermediates (see Table 16–2).

## BOX 16–3 THE WORLD OF BIOCHEMISTRY

### Citrate Synthase, Soda Pop, and the World Food Supply

Citrate has a number of important industrial applications. A quick examination of the ingredients in most soft drinks reveals the common use of citric acid to provide a tart or fruity flavor. Citric acid is also used as a plasticizer and foam inhibitor in the manufacture of certain resins, as a mordant to brighten colors, and as an antioxidant to preserve the flavors of foods. Citric acid is produced industrially by growing the fungus *Aspergillus niger* in the presence of an inexpensive sugar source, usually beet molasses. Culture conditions are designed to inhibit the reactions of the citric acid cycle such that citrate accumulates.

On a grander scale, citric acid may one day play a spectacular role in the alleviation of world hunger. With its three negatively charged carboxyl groups, citrate is a good chelator of metal ions, and some plants exploit this property by releasing citrate into the soil, where it binds metal ions and prevents their absorption by the plant. Of particular importance is the aluminum ion ($Al^{3+}$), which is toxic to many plants and causes decreased crop yields on 30% to 40% of the world's arable land. Aluminum is the most abundant metal in the earth's crust, yet it occurs mostly in chemical compounds, such as $Al(OH)_3$, that are biologically inert. However, when soil pH is less than 5, $Al^{3+}$ be-

comes soluble and thus can be absorbed by plant roots. Acidic soil and $Al^{3+}$ toxicity are most prevalent in the tropics, where maize yields can be depressed by as much as 80%. In Mexico, $Al^{3+}$ toxicity limits papaya production to 20,000 hectares, instead of the 3 million hectares that could theoretically be cultivated. One solution would be to raise soil pH with lime, but this is economically and environmentally unsound. An alternative would be to breed $Al^{3+}$-resistant plants. Naturally resistant plants do exist, and these provide the means for a third solution: transferring resistance to crop plants by genetic engineering.

A group of researchers in Mexico has genetically engineered tobacco and papaya plants to express elevated levels of bacterial citrate synthase. These plants secrete five to six times their normal amount of $Al^{3+}$-chelating citric acid and can grow in soils with $Al^{3+}$ levels ten times those at which control plants can grow. This degree of resistance would allow Mexico to grow papaya on the 3 million hectares of land currently rendered unsuitable by $Al^{3+}$.

Given projected levels of population growth, world food production must more than triple in the next 50 years to adequately feed 9.6 billion people. A long-term solution may turn on increasing crop productivity on the arable land affected by aluminum toxicity, and citric acid may play an important role in achieving this goal.

### Biotin in Pyruvate Carboxylase Carries CO₂ Groups

The pyruvate carboxylase reaction requires the vitamin **biotin** (Fig. 16–16), which is the prosthetic group of the enzyme. Biotin plays a key role in many carboxylation reactions. It is a specialized carrier of one-carbon groups in their most oxidized form: $CO_2$. (The transfer of one-carbon groups in more reduced forms is mediated by other cofactors, notably tetrahydrofolate and *S*-adenosylmethionine, as described in Chapter 18.)

Carboxyl groups are activated in a reaction that splits ATP and joins $CO_2$ to enzyme-bound biotin. This "activated" $CO_2$ is then passed to an acceptor (pyruvate in this case) in a carboxylation reaction.

Pyruvate carboxylase has four identical subunits, each containing a molecule of biotin covalently attached through an amide linkage to the $\varepsilon$-amino group of a specific Lys residue in the enzyme active site. Carboxylation of pyruvate proceeds in two steps (Fig. 16–16): first, a carboxyl group derived from $HCO_3^-$ is attached to biotin,

**TABLE 16–2 Anaplerotic Reactions**

| Reaction | Tissue(s)/organism(s) |
|---|---|
| Pyruvate + $HCO_3^-$ + ATP $\underset{\text{pyruvate carboxylase}}{\rightleftharpoons}$ oxaloacetate + ADP + $P_i$ | Liver, kidney |
| Phosphoenolpyruvate + $CO_2$ + GDP $\underset{\text{PEP carboxykinase}}{\rightleftharpoons}$ oxaloacetate + GTP | Heart, skeletal muscle |
| Phosphoenolpyruvate + $HCO_3^-$ $\underset{\text{PEP carboxylase}}{\rightleftharpoons}$ oxaloacetate + $P_i$ | Higher plants, yeast, bacteria |
| Pyruvate + $HCO_3^-$ + NAD(P)H $\underset{\text{malic enzyme}}{\rightleftharpoons}$ malate + NAD(P)$^+$ | Widely distributed in eukaryotes and prokaryotes |

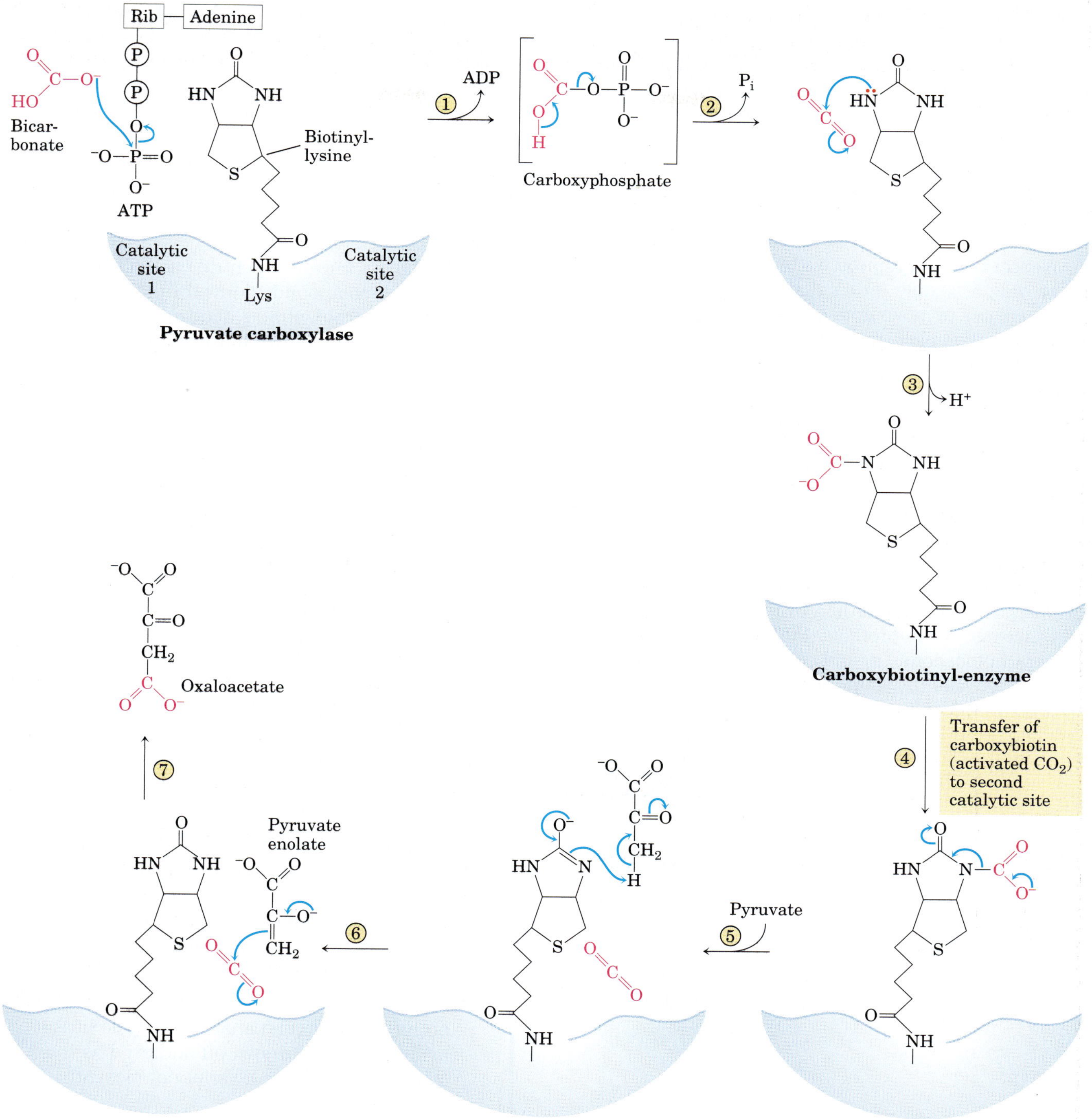

**MECHANISM FIGURE 16–16 The role of biotin in the reaction catalyzed by pyruvate carboxylase.** Biotin is attached to the enzyme through an amide bond with the $\epsilon$-amino group of a Lys residue, forming biotinyl-enzyme. Biotin-mediated carboxylation reactions occur in two phases, generally catalyzed by separate active sites on the enzyme as exemplified by the pyruvate carboxylase reaction. In the first phase (steps ① to ③), bicarbonate is converted to the more activated $CO_2$, and then used to carboxylate biotin. The bicarbonate is first activated by reaction with ATP to form carboxyphosphate (step ①), which breaks down to carbon dioxide (step ②). In effect, the bicarbonate is dehydrated by its reaction with ATP, and the $CO_2$ can react with biotin to form carboxybiotin (step ③). The biotin acts as a carrier to transport the $CO_2$ from one active site to another on the same enzyme (step ④). In the second phase of the reaction (steps ⑤ to ⑦), catalyzed in a second active site, the $CO_2$ reacts with pyruvate to form oxaloacetate. The $CO_2$ is released in the second active site (step ⑤). Pyruvate is converted to its enolate form in step ⑥, transferring a proton to biotin. The enolate then attacks the $CO_2$ to generate oxaloacetate in the final step of the reaction (step ⑦).

then the carboxyl group is transferred to pyruvate to form oxaloacetate. These two steps occur at separate active sites; the long flexible arm of biotin transfers activated carboxyl groups from the first active site to the second, functioning much like the long lipoyllysine arm of $E_2$ in the PDH complex (Fig. 16–6) and the long arm of the CoA-like moiety in the acyl carrier protein involved in fatty acid synthesis (see Fig. 21–5); these are compared in Figure 16–17. Lipoate, biotin, and pantothenate all enter cells on the same transporter; all become covalently attached to proteins by similar reactions; and all provide a flexible tether that allows bound reaction intermediates to move from one active site to another in an enzyme complex, without dissociating from it—all, that is, participate in substrate channeling.

Biotin is a vitamin required in the human diet; it is abundant in many foods and is synthesized by intestinal bacteria. Biotin deficiency is rare, but can sometimes be caused by a diet rich in raw eggs. Egg whites contain a large amount of the protein **avidin** ($M_r$ 70,000), which binds very tightly to biotin and prevents its absorption in the intestine. The avidin of egg whites may be a defense mechanism for the potential chick embryo, inhibiting the growth of bacteria. When eggs are cooked, avidin is denatured (and thereby inactivated) along with all other egg white proteins. Purified avidin is a useful reagent in biochemistry and cell biology. A protein that contains covalently bound biotin (derived experimentally or produced in vivo) can be recovered by affinity chromatography (see Fig. 3–18c) based on biotin's strong affinity for avidin. The protein is then eluted from the column with an excess of free biotin. The very high affinity of biotin for avidin is also used in the laboratory in the form of a molecular glue that can hold two structures together (see Fig. 19–25).

## SUMMARY 16.2 Reactions of the Citric Acid Cycle

■ The citric acid cycle (Krebs cycle, TCA cycle) is a nearly universal central catabolic pathway in which compounds derived from the breakdown of carbohydrates, fats, and proteins are oxidized to $CO_2$, with most of the energy of oxidation temporarily held in the electron carriers $FADH_2$ and NADH. During aerobic metabolism, these electrons are transferred to $O_2$ and the energy of electron flow is trapped as ATP.

■ Acetyl-CoA enters the citric acid cycle (in the mitochondria of eukaryotes, the cytosol of prokaryotes) as citrate synthase catalyzes its condensation with oxaloacetate to form citrate.

■ In seven sequential reactions, including two decarboxylations, the citric acid cycle converts citrate to oxaloacetate and releases two $CO_2$. The pathway is cyclic in that the intermediates of the cycle are not used up; for each oxaloacetate consumed in the path, one is produced.

■ For each acetyl-CoA oxidized by the citric acid cycle, the energy gain consists of three molecules of NADH, one $FADH_2$, and one nucleoside triphosphate (either ATP or GTP).

■ Besides acetyl-CoA, any compound that gives rise to a four- or five-carbon intermediate of the citric acid cycle—for example, the breakdown products of many amino acids—can be oxidized by the cycle.

■ The citric acid cycle is amphibolic, serving in both catabolism and anabolism; cycle intermediates can be drawn off and used as the starting material for a variety of biosynthetic products.

■ When intermediates are shunted from the citric acid cycle to other pathways, they are replenished by several anaplerotic reactions, which produce four-carbon intermediates by carboxylation of three-carbon compounds; these reactions are catalyzed by pyruvate carboxylase, PEP carboxykinase, PEP carboxylase, and malic enzyme. Enzymes that catalyze carboxylations commonly employ biotin to activate $CO_2$ and to carry it to acceptors such as pyruvate or phosphoenolpyruvate.

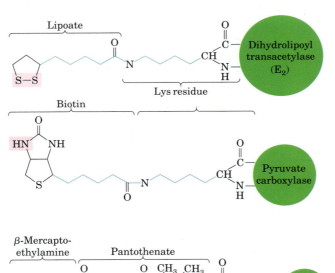

**FIGURE 16–17 Biological tethers.** The cofactors lipoate, biotin, and the combination of β-mercaptoethylamine and pantothenate form long, flexible arms in the enzymes to which they are covalently bound, acting as tethers that move intermediates from one active site to the next. The group shaded pink is in each case the point of attachment of the activated intermediate to the tether.

# 16.3 Regulation of the Citric Acid Cycle

As we have seen in Chapter 15, the regulation of key enzymes in metabolic pathways, by allosteric effectors and by covalent modification, ensures the production of intermediates at the rates required to keep the cell in a stable steady state while avoiding wasteful overproduction. The flow of carbon atoms from pyruvate into and through the citric acid cycle is under tight regulation at two levels: the conversion of pyruvate to acetyl-CoA, the starting material for the cycle (the pyruvate dehydrogenase complex reaction), and the entry of acetyl-CoA into the cycle (the citrate synthase reaction). Acetyl-CoA is also produced by pathways other than the PDH complex reaction—most cells produce acetyl-CoA from the oxidation of fatty acids and certain amino acids—and the availability of intermediates from these other pathways is important in the regulation of pyruvate oxidation and of the citric acid cycle. The cycle is also regulated at the isocitrate dehydrogenase and $\alpha$-ketoglutarate dehydrogenase reactions.

## Production of Acetyl-CoA by the Pyruvate Dehydrogenase Complex Is Regulated by Allosteric and Covalent Mechanisms

The PDH complex of mammals is strongly inhibited by ATP and by acetyl-CoA and NADH, the products of the reaction catalyzed by the complex (Fig. 16–18). The allosteric inhibition of pyruvate oxidation is greatly enhanced when long-chain fatty acids are available. AMP, CoA, and NAD$^+$, all of which accumulate when too little acetate flows into the citric acid cycle, allosterically activate the PDH complex. Thus, this enzyme activity is turned off when ample fuel is available in the form of fatty acids and acetyl-CoA and when the cell's [ATP]/[ADP] and [NADH]/[NAD$^+$] ratios are high, and it is turned on again when energy demands are high and the cell requires greater flux of acetyl-CoA into the citric acid cycle.

In mammals, these allosteric regulatory mechanisms are complemented by a second level of regulation: covalent protein modification. The PDH complex is inhibited by reversible phosphorylation of a specific Ser residue on one of the two subunits of E$_1$. As noted earlier, in addition to the enzymes E$_1$, E$_2$, and E$_3$, the mammalian PDH complex contains two regulatory proteins whose sole purpose is to regulate the activity of the complex. A specific protein kinase phosphorylates and thereby inactivates E$_1$, and a specific phosphoprotein phosphatase removes the phosphoryl group by hydrolysis and thereby activates E$_1$. The kinase is allosterically activated by ATP: when [ATP] is high (reflecting a sufficient supply of energy), the PDH complex is inactivated by phosphorylation of E$_1$. When [ATP] declines, kinase activity decreases and phosphatase action removes the phosphoryl groups from E$_1$, activating the complex.

The PDH complex of plants, located in the mitochondrial matrix and in plastids, is inhibited by its products, NADH and acetyl-CoA. The plant mitochondrial

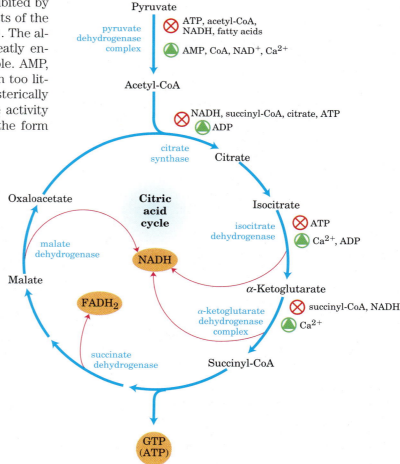

**FIGURE 16–18 Regulation of metabolite flow from the PDH complex through the citric acid cycle.** The PDH complex is allosterically inhibited when [ATP]/[ADP], [NADH]/[NAD$^+$], and [acetyl-CoA]/[CoA] ratios are high, indicating an energy-sufficient metabolic state. When these ratios decrease, allosteric activation of pyruvate oxidation results. The rate of flow through the citric acid cycle can be limited by the availability of the citrate synthase substrates, oxaloacetate and acetyl-CoA, or of NAD$^+$, which is depleted by its conversion to NADH, slowing the three NAD-dependent oxidation steps. Feedback inhibition by succinyl-CoA, citrate, and ATP also slows the cycle by inhibiting early steps. In muscle tissue, Ca$^{2+}$ signals contraction and, as shown here, stimulates energy-yielding metabolism to replace the ATP consumed by contraction.

enzyme is also regulated by reversible phosphorylation; pyruvate inhibits the kinase, thus activating the PDH complex, and $NH_4^+$ stimulates the kinase, causing inactivation of the complex. The PDH complex of *E. coli* is under allosteric regulation similar to that of the mammalian enzyme, but it does not seem to be regulated by phosphorylation.

## The Citric Acid Cycle Is Regulated at Its Three Exergonic Steps

The flow of metabolites through the citric acid cycle is under stringent regulation. Three factors govern the rate of flux through the cycle: substrate availability, inhibition by accumulating products, and allosteric feedback inhibition of the enzymes that catalyze early steps in the cycle.

Each of the three strongly exergonic steps in the cycle—those catalyzed by citrate synthase, isocitrate dehydrogenase, and $\alpha$-ketoglutarate dehydrogenase (Fig. 16–18)—can become the rate-limiting step under some circumstances. The availability of the substrates for citrate synthase (acetyl-CoA and oxaloacetate) varies with the metabolic state of the cell and sometimes limits the rate of citrate formation. NADH, a product of isocitrate and $\alpha$-ketoglutarate oxidation, accumulates under some conditions, and at high [NADH]/[NAD$^+$] both dehydrogenase reactions are severely inhibited by mass action. Similarly, in the cell, the malate dehydrogenase reaction is essentially at equilibrium (that is, it is substrate-limited, and when [NADH]/[NAD$^+$] is high the concentration of oxaloacetate is low, slowing the first step in the cycle. Product accumulation inhibits all three limiting steps of the cycle: succinyl-CoA inhibits $\alpha$-ketoglutarate dehydrogenase (and also citrate synthase); citrate blocks citrate synthase; and the end product, ATP, inhibits both citrate synthase and isocitrate dehydrogenase. The inhibition of citrate synthase by ATP is relieved by ADP, an allosteric activator of this enzyme. In vertebrate muscle, $Ca^{2+}$, the signal for contraction and for a concomitant increase in demand for ATP, activates both isocitrate dehydrogenase and $\alpha$-ketoglutarate dehydrogenase, as well as the PDH complex. In short, the concentrations of substrates and intermediates in the citric acid cycle set the flux through this pathway at a rate that provides optimal concentrations of ATP and NADH.

Under normal conditions, the rates of glycolysis and of the citric acid cycle are integrated so that only as much glucose is metabolized to pyruvate as is needed to supply the citric acid cycle with its fuel, the acetyl groups of acetyl-CoA. Pyruvate, lactate, and acetyl-CoA are normally maintained at steady-state concentrations. The rate of glycolysis is matched to the rate of the citric acid cycle not only through its inhibition by high levels of ATP and NADH, which are common to both the glycolytic and respiratory stages of glucose oxidation, but also by the concentration of citrate. Citrate, the product of the first step of the citric acid cycle, is an important allosteric inhibitor of phosphofructokinase-1 in the glycolytic pathway (see Fig. 15–18).

## Substrate Channeling through Multienzyme Complexes May Occur in the Citric Acid Cycle

Although the enzymes of the citric acid cycle are usually described as soluble components of the mitochondrial matrix (except for succinate dehydrogenase, which is membrane-bound), growing evidence suggests that within the mitochondrion these enzymes exist as multienzyme complexes. The classic approach of enzymology—purification of individual proteins from extracts of broken cells—was applied with great success to the citric acid cycle enzymes. However, the first casualty of cell breakage is higher-level organization within the cell—the noncovalent, reversible interaction of one protein with another, or of an enzyme with some structural component such as a membrane, microtubule, or microfilament. When cells are broken open, their contents, including enzymes, are diluted 100- or 1,000-fold (Fig. 16–19).

Several types of evidence suggest that, in cells, multienzyme complexes ensure efficient passage of the product of one enzyme reaction to the next enzyme in the pathway. Such complexes are called **metabolons.** Certain enzymes of the citric acid cycle have been isolated together as supramolecular aggregates, or have been found associated with the inner mitochondrial membrane, or have been shown to diffuse in the mitochondrial matrix more slowly than expected for the individual protein in solution. There is strong evidence for substrate channeling through multienzyme complexes in

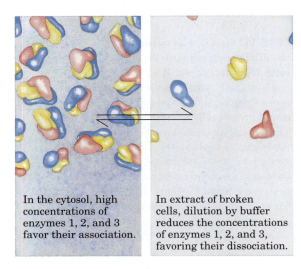

In the cytosol, high concentrations of enzymes 1, 2, and 3 favor their association.

In extract of broken cells, dilution by buffer reduces the concentrations of enzymes 1, 2, and 3, favoring their dissociation.

**FIGURE 16–19** Dilution of a solution containing a noncovalent protein complex—such as one consisting of three enzymes—favors dissociation of the complex into its constituents.

other metabolic pathways, and many enzymes thought of as "soluble" probably function in the cell as highly organized complexes that channel intermediates. We will encounter other examples of channeling when we discuss the biosynthesis of amino acids and nucleotides in Chapter 22.

### SUMMARY 16.3 Regulation of the Citric Acid Cycle

- The overall rate of the citric acid cycle is controlled by the rate of conversion of pyruvate to acetyl-CoA and by the flux through citrate synthase, isocitrate dehydrogenase, and $\alpha$-ketoglutarate dehydrogenase. These fluxes are largely determined by the concentrations of substrates and products: the end products ATP and NADH are inhibitory, and the substrates $NAD^+$ and ADP are stimulatory.

- The production of acetyl-CoA for the citric acid cycle by the PDH complex is inhibited allosterically by metabolites that signal a sufficiency of metabolic energy (ATP, acetyl-CoA, NADH, and fatty acids) and stimulated by metabolites that indicate a reduced energy supply (AMP, $NAD^+$, CoA).

## 16.4 The Glyoxylate Cycle

Vertebrates cannot convert fatty acids, or the acetate derived from them, to carbohydrates. Conversion of phosphoenolpyruvate to pyruvate (p. 532) and of pyruvate to acetyl-CoA (Fig. 16–2) are so exergonic as to be essentially irreversible. If a cell cannot convert acetate into phosphoenolpyruvate, acetate cannot serve as the starting material for the gluconeogenic pathway, which leads from phosphoenolpyruvate to glucose (see Fig. 15–15). Without this capacity, then, a cell or organism is unable to convert fuels or metabolites that are degraded to acetate (fatty acids and certain amino acids) into carbohydrates.

As noted in the discussion of anaplerotic reactions (Table 16–2), phosphoenolpyruvate can be synthesized from oxaloacetate in the reversible reaction catalyzed by PEP carboxykinase:

$$\text{Oxaloacetate} + \text{GTP} \rightleftharpoons$$
$$\text{phosphoenolpyruvate} + CO_2 + \text{GDP}$$

Because the carbon atoms of acetate molecules that enter the citric acid cycle appear eight steps later in oxaloacetate, it might seem that this pathway could generate oxaloacetate from acetate and thus generate phosphoenolpyruvate for gluconeogenesis. However, as an examination of the stoichiometry of the citric acid cycle shows, there is no *net* conversion of acetate to ox-

aloacetate; in vertebrates, for every two carbons that enter the cycle as acetyl-CoA, two leave as $CO_2$. In many organisms other than vertebrates, the glyoxylate cycle serves as a mechanism for converting acetate to carbohydrate.

### The Glyoxylate Cycle Produces Four-Carbon Compounds from Acetate

In plants, certain invertebrates, and some microorganisms (including *E. coli* and yeast) acetate can serve both as an energy-rich fuel and as a source of phosphoenolpyruvate for carbohydrate synthesis. In these organisms, enzymes of the **glyoxylate cycle** catalyze the net conversion of acetate to succinate or other four-carbon intermediates of the citric acid cycle:

$$2\,\text{Acetyl-CoA} + NAD^+ + 2H_2O \longrightarrow$$
$$\text{succinate} + 2\text{CoA} + \text{NADH} + H^+$$

In the glyoxylate cycle, acetyl-CoA condenses with oxaloacetate to form citrate, and citrate is converted to isocitrate, exactly as in the citric acid cycle. The next step, however, is not the breakdown of isocitrate by isocitrate dehydrogenase but the cleavage of isocitrate by **isocitrate lyase,** forming succinate and **glyoxylate.** The glyoxylate then condenses with a second molecule of acetyl-CoA to yield malate, in a reaction catalyzed by **malate synthase.** The malate is subsequently oxidized to oxaloacetate, which can condense with another molecule of acetyl-CoA to start another turn of the cycle (Fig. 16–20). Each turn of the glyoxylate cycle consumes two molecules of acetyl-CoA and produces one molecule of succinate, which is then available for biosynthetic purposes. The succinate may be converted through fumarate and malate into oxaloacetate, which can then be converted to phosphoenolpyruvate by PEP carboxykinase, and thus to glucose by gluconeogenesis. Vertebrates do not have the enzymes specific to the glyoxylate cycle (isocitrate lyase and malate synthase) and therefore cannot bring about the net synthesis of glucose from lipids.

In plants, the enzymes of the glyoxylate cycle are sequestered in membrane-bounded organelles called glyoxysomes, which are specialized peroxisomes (Fig. 16–21). Those enzymes common to the citric acid and glyoxylate cycles have two isozymes, one specific to mitochondria, the other to glyoxysomes. Glyoxysomes are not present in all plant tissues at all times. They develop in lipid-rich seeds during germination, before the developing plant acquires the ability to make glucose by photosynthesis. In addition to glyoxylate cycle enzymes, glyoxysomes contain all the enzymes needed for the degradation of the fatty acids stored in seed oils (see Fig. 17–13). Acetyl-CoA formed from lipid breakdown is converted to succinate via the glyoxylate cycle, and the succinate is exported to mitochondria, where citric

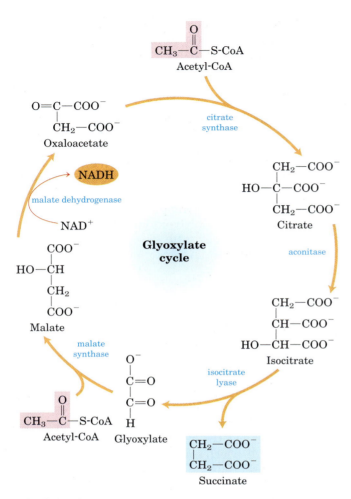

**FIGURE 16–20  Glyoxylate cycle.** The citrate synthase, aconitase, and malate dehydrogenase of the glyoxylate cycle are isozymes of the citric acid cycle enzymes; isocitrate lyase and malate synthase are unique to the glyoxylate cycle. Notice that two acetyl groups (pink) enter the cycle and four carbons leave as succinate (blue). The glyoxylate cycle was elucidated by Hans Kornberg and Neil Madsen in the laboratory of Hans Krebs.

acid cycle enzymes transform it to malate. A cytosolic isozyme of malate dehydrogenase oxidizes malate to oxaloacetate, a precursor for gluconeogenesis. Germinating seeds can therefore convert the carbon of stored lipids into glucose.

## The Citric Acid and Glyoxylate Cycles Are Coordinately Regulated

In germinating seeds, the enzymatic transformations of dicarboxylic and tricarboxylic acids occur in three intracellular compartments: mitochondria, glyoxysomes, and the cytosol. There is a continuous interchange of metabolites among these compartments (Fig. 16–22).

The carbon skeleton of oxaloacetate from the citric acid cycle (in the mitochondrion) is carried to the glyoxysome in the form of aspartate. Aspartate is converted

to oxaloacetate, which condenses with acetyl-CoA derived from fatty acid breakdown. The citrate thus formed is converted to isocitrate by aconitase, then split into glyoxylate and succinate by isocitrate lyase. The succinate returns to the mitochondrion, where it reenters the citric acid cycle and is transformed into malate, which enters the cytosol and is oxidized (by cytosolic malate dehydrogenase) to oxaloacetate. Oxaloacetate is converted via gluconeogenesis into hexoses and sucrose, which can be transported to the growing roots and shoot. Four distinct pathways participate in these conversions: fatty acid breakdown to acetyl-CoA (in glyoxysomes), the glyoxylate cycle (in glyoxysomes), the citric acid cycle (in mitochondria), and gluconeogenesis (in the cytosol).

The sharing of common intermediates requires that these pathways be coordinately regulated. Isocitrate is a crucial intermediate, at the branch point between the glyoxylate and citric acid cycles (Fig. 16–23). Isocitrate dehydrogenase is regulated by covalent modification: a specific protein kinase phosphorylates and thereby inactivates the dehydrogenase. This inactivation shunts isocitrate to the glyoxylate cycle, where it begins the synthetic route toward glucose. A phosphoprotein phosphatase removes the phosphoryl group from isocitrate dehydrogenase, reactivating the enzyme and sending more isocitrate through the energy-yielding citric acid cycle. The regulatory protein kinase and phosphoprotein phosphatase are separate enzymatic activities of a single polypeptide.

Some bacteria, including *E. coli,* have the full complement of enzymes for the glyoxylate and citric acid cycles in the cytosol and can therefore grow on acetate as their sole source of carbon and energy. The phosphoprotein phosphatase that activates isocitrate dehydrogenase is stimulated by intermediates of the citric acid cycle and glycolysis and by indicators of reduced cellular energy supply (Fig. 16–23). The same metabolites *inhibit* the protein kinase activity of the bifunctional polypeptide. Thus, the accumulation of intermediates of

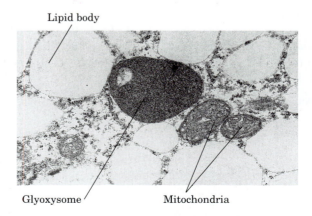

**FIGURE 16–21** Electron micrograph of a germinating cucumber seed, showing a glyoxysome, mitochondria, and surrounding lipid bodies.

The same intermediates of glycolysis and the citric acid cycle that activate isocitrate dehydrogenase are allosteric inhibitors of isocitrate lyase. When energy-yielding metabolism is sufficiently fast to keep the concentrations of glycolytic and citric acid cycle intermediates low, isocitrate dehydrogenase is inactivated, the inhibition of isocitrate lyase is relieved, and isocitrate flows into the glyoxylate pathway, to be used in the biosynthesis of carbohydrates, amino acids, and other cellular components.

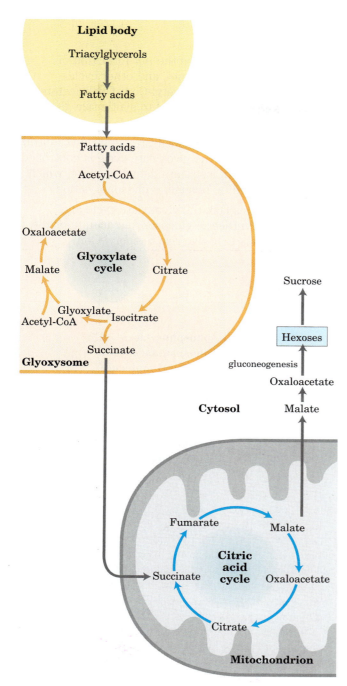

**FIGURE 16–22 Relationship between the glyoxylate and citric acid cycles.** The reactions of the glyoxylate cycle (in glyoxysomes) proceed simultaneously with, and mesh with, those of the citric acid cycle (in mitochondria), as intermediates pass between these compartments. The conversion of succinate to oxaloacetate is catalyzed by citric acid cycle enzymes. The oxidation of fatty acids to acetyl-CoA is described in Chapter 17; the synthesis of hexoses from oxaloacetate is described in Chapter 20.

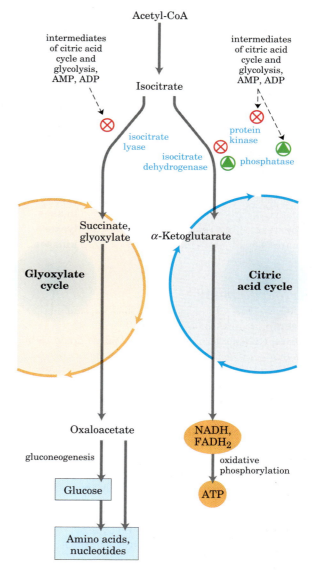

**FIGURE 16–23 Coordinated regulation of glyoxylate and citric acid cycles.** Regulation of isocitrate dehydrogenase activity determines the partitioning of isocitrate between the glyoxylate and citric acid cycles. When the enzyme is inactivated by phosphorylation (by a specific protein kinase), isocitrate is directed into biosynthetic reactions via the glyoxylate cycle. When the enzyme is activated by dephosphorylation (by a specific phosphatase), isocitrate enters the citric acid cycle and ATP is produced.

the central energy-yielding pathways—indicating energy depletion—results in the activation of isocitrate dehydrogenase. When the concentration of these regulators falls, signaling a sufficient flux through the energy-yielding citric acid cycle, isocitrate dehydrogenase is inactivated by the protein kinase.

$CH_3-COO^-$ / acetate

## SUMMARY 16.4 The Glyoxylate Cycle

- The glyoxylate cycle is active in the germinating seeds of some plants and in certain microorganisms that can live on acetate as the sole carbon source. In plants, the pathway takes place in glyoxysomes in seedlings. It involves several citric acid cycle enzymes and two additional enzymes: isocitrate lyase and malate synthase.

- In the glyoxylate cycle, the bypassing of the two decarboxylation steps of the citric acid cycle makes possible the *net* formation of succinate, oxaloacetate, and other cycle intermediates from acetyl-CoA. Oxaloacetate thus formed can be used to synthesize glucose via gluconeogenesis.

- The partitioning of isocitrate between the citric acid cycle and the glyoxylate cycle is controlled at the level of isocitrate dehydrogenase, which is regulated by reversible phosphorylation.

- Vertebrates lack the glyoxylate cycle and cannot synthesize glucose from acetate or the fatty acids that give rise to acetyl-CoA.

## Key Terms

*Terms in bold are defined in the glossary.*

**respiration** 601
cellular respiration 601
**citric acid cycle** 601
**tricarboxylic acid (TCA) cycle** 601
**Krebs cycle** 601
pyruvate dehydrogenase (PDH) complex 602
oxidative decarboxylation 602
**thioester** 603
**lipoate** 603

**substrate channeling** 605
**iron-sulfur center** 609
α-ketoglutarate dehydrogenase complex 610
**nucleoside diphosphate kinase** 612
**synthases** 613
**synthetases** 613
**ligases** 613
**lyases** 613
**kinases** 613

**phosphorylases** 613
**phosphatases** 613
**prochiral molecule** 615
**amphibolic pathway** 616
**anaplerotic reaction** 616
**biotin** 618
avidin 620
**metabolon** 622
**glyoxylate cycle** 623

## Further Reading

### General

**Holmes, F.L.** (1990, 1993) *Hans Krebs*, Vol 1: *Formation of a Scientific Life, 1900–1933;* Vol. 2: *Architect of Intermediary Metabolism, 1933–1937,* Oxford University Press, Oxford.

A scientific and personal biography of Krebs by an eminent historian of science, with a thorough description of the work that revealed the urea and citric acid cycles.

**Kay, J. & Weitzman, P.D.J. (eds)** (1987) *Krebs' Citric Acid Cycle: Half a Century and Still Turning,* Biochemical Society Symposium **54,** The Biochemical Society, London.

A multiauthor book on the citric acid cycle, including molecular genetics, regulatory mechanisms, variations on the cycle in microorganisms from unusual ecological niches, and evolution of the pathway. Especially relevant are the chapters by H. Gest (Evolutionary Roots of the Citric Acid Cycle in Prokaryotes), W. H. Holms (Control of Flux through the Citric Acid Cycle and the Glyoxylate Bypass in *Escherichia coli*), and R. N. Perham et al. (α-Keto Acid Dehydrogenase Complexes).

### Pyruvate Dehydrogenase Complex

**Harris, R.A., Bowker-Kinley, M.M., Huang, B., & Wu, P.** (2002) Regulation of the activity of the pyruvate dehydrogenase complex. *Adv. Enzyme Regul.* **42,** 249–259.

**Milne, J.L.S., Shi, D., Rosenthal, P.B., Sunshine, J.S., Domingo, G.J., Wu, X., Brooks, B.R., Perham, R.N., Henderson, R., & Subramaniam, S.** (2002) Molecular architecture and mechanism of an icosahedral pyruvate dehydrogenase complex: a multifunctional catalytic machine. *EMBO J.* **21,** 5587–5598.

Beautiful illustration of the power of image reconstruction methodology with cryoelectron microscopy, here used to develop a plausible model for the structure of the PDH complex. Compare this model with that in the paper by Zhou et al.

**Perham, R.N.** (2000) Swinging arms and swinging domains in multifunctional enzymes: catalytic machines for multistep reactions. *Annu. Rev. Biochem.* **69,** 961–1004.

Review of the roles of swinging arms containing lipoate, biotin, and pantothenate in substrate channeling through multienzyme complexes.

**Zhou, Z.H., McCarthy, D.B., O'Conner, C.M., Reed, L.J., & Stoops, J.K.** (2001) The remarkable structural and functional organization of the eukaryotic pyruvate dehydrogenase complexes. *Proc. Natl. Acad. Sci. USA* **98,** 14,802–14,807.

Another striking paper in which image reconstruction with cryoelectron microscopy yields a model of the PDH complex. Compare this model with that in the paper by Milne et al.

### Citric Acid Cycle Enzymes

**Fraser, M.D., James, M.N., Bridger, W.A., & Wolodko, W.T.** (1999) A detailed structural description of *Escherichia coli* succinyl-CoA synthetase. *J. Mol. Biol.* **285,** 1633–1653. (See also the erratum in *J. Mol. Biol.* **288,** 501 (1999).)

**Goward, C.R. & Nicholls, D.J.** (1994) Malate dehydrogenase: a model for structure, evolution, and catalysis. *Protein Sci.* **3**, 1883–1888.

A good, short review.

**Hagerhall, C.** (1997) Succinate:quinone oxidoreductases: variations on a conserved theme. *Biochim. Biophys. Acta* **1320**, 107–141.

A review of the structure and function of succinate dehydrogenases.

**Jitrapakdee, S. & Wallace, J.C.** (1999) Structure, function, and regulation of pyruvate carboxylase. *Biochem. J.* **340**, 1–16.

**Knowles, J.** (1989) The mechanism of biotin-dependent enzymes. *Annu. Rev. Biochem.* **58**, 195–221.

**Matte, A., Tari, L.W., Goldie, H., & Delbaere, L.T.J.** (1997) Structure and mechanism of phosphoenolpyruvate carboxykinase. *J. Biol. Chem.* **272**, 8105–8108.

**Ovadi, J. & Srere, P.** (2000) Macromolecular compartmentation and channeling. *Int. Rev. Cytol.* **192**, 255–280.

Advanced review of the evidence for channeling and metabolons.

**Remington, S.J.** (1992) Structure and mechanism of citrate synthase. *Curr. Top. Cell. Regul.* **33**, 209–229.

A thorough review of this enzyme.

**Singer, T.P. & Johnson, M.K.** (1985) The prosthetic groups of succinate dehydrogenase: 30 years from discovery to identification. *FEBS Lett.* **190**, 189–198.

A description of the structure and role of the iron-sulfur centers in this enzyme.

**Weigand, G. & Remington, S.J.** (1986) Citrate synthase: structure, control, and mechanism. *Annu. Rev. Biophys. Biophys. Chem.* **15**, 97–117.

**Wolodko, W.T., Fraser, M.E., James, M.N.G., & Bridger, W.A.** (1994) The crystal structure of succinyl-CoA synthetase from *Escherichia coli* at 2.5-Å resolution. *J. Biol. Chem.* **269**, 10,883–10,890.

### Regulation of the Citric Acid Cycle

**Hansford, R.G.** (1980) Control of mitochondrial substrate oxidation. *Curr. Top. Bioenerget.* **10**, 217–278.

A detailed review of the regulation of the citric acid cycle.

**Kaplan, N.O.** (1985) The role of pyridine nucleotides in regulating cellular metabolism. *Curr. Top. Cell. Regul.* **26**, 371–381.

An excellent general discussion of the importance of the [NADH]/[NAD$^+$] ratio in cellular regulation.

**Reed, L.J., Damuni, Z., & Merryfield, M.L.** (1985) Regulation of mammalian pyruvate and branched-chain $\alpha$-keto acid dehydrogenase complexes by phosphorylation-dephosphorylation. *Curr. Top. Cell. Regul.* **27**, 41–49.

### Glyoxylate Cycle

**Eastmond, P.J. & Graham, I.A.** (2001) Re-examining the role of the glyoxylate cycle in oilseeds. *Trends Plant Sci.* **6**, 72–77.

Intermediate-level review of studies of the glyoxylate cycle in *Arabidopsis*.

**Holms, W.H.** (1986) The central metabolic pathways of *Escherichia coli*: relationship between flux and control at a branch point, efficiency of conversion to biomass, and excretion of acetate. *Curr. Top. Cell. Regul.* **28**, 69–106.

## Problems

**1. Balance Sheet for the Citric Acid Cycle** The citric acid cycle has eight enzymes: citrate synthase, aconitase, isocitrate dehydrogenase, $\alpha$-ketoglutarate dehydrogenase, succinyl-CoA synthetase, succinate dehydrogenase, fumarase, and malate dehydrogenase.

(a) Write a balanced equation for the reaction catalyzed by each enzyme.

(b) Name the cofactor(s) required by each enzyme reaction.

(c) For each enzyme determine which of the following describes the type of reaction(s) catalyzed: condensation (carbon–carbon bond formation); dehydration (loss of water); hydration (addition of water); decarboxylation (loss of $CO_2$); oxidation-reduction; substrate-level phosphorylation; isomerization.

(d) Write a balanced net equation for the catabolism of acetyl-CoA to $CO_2$.

**2. Recognizing Oxidation and Reduction Reactions** One biochemical strategy of many living organisms is the stepwise oxidation of organic compounds to $CO_2$ and $H_2O$ and the conservation of a major part of the energy thus produced in the form of ATP. It is important to be able to recognize oxidation-reduction processes in metabolism. Reduction of an organic molecule results from the hydrogenation of a double bond (Eqn 1, below) or of a single bond with accompanying cleavage (Eqn 2). Conversely, oxidation results from dehydrogenation. In biochemical redox reactions, the coenzymes NAD and FAD dehydrogenate/hydrogenate organic molecules in the presence of the proper enzymes.

$$(1)$$

$$(2)$$

For each of the metabolic transformations in (a) through (h), determine whether oxidation or reduction has occurred. Balance each transformation by inserting H—H and, where necessary, $H_2O$.

(a) $CH_3$—OH $\longrightarrow$ H—$\overset{\overset{\displaystyle O}{\|}}{C}$—H

    Methanol      Formaldehyde

(b) H—$\overset{\overset{\displaystyle O}{\|}}{C}$—H $\longrightarrow$ H—$\overset{\overset{\displaystyle O}{\|}}{C}$$\underset{O^-}{\diagup}$ + $H^+$

    Formaldehyde      Formate

(c) O=C=O $\longrightarrow$ H—$\overset{\overset{\displaystyle O}{\|}}{C}$$\underset{O^-}{\diagup}$ + $H^+$

    Carbon dioxide    Formate

(d) $\overset{OH}{\overset{|}{CH_2}}$—$\overset{OH}{\overset{|}{\underset{|}{\underset{H}{C}}}}$—$C$$\underset{O^-}{\overset{\overset{\displaystyle O}{\|}}{\diagup}}$ + $H^+$ $\longrightarrow$ $\overset{OH}{\overset{|}{CH_2}}$—$\overset{OH}{\overset{|}{\underset{|}{\underset{H}{C}}}}$—$C$$\underset{H}{\overset{\overset{\displaystyle O}{\|}}{\diagup}}$

    Glycerate       Glyceraldehyde

(e) $\overset{OH}{\overset{|}{CH_2}}$—$\overset{OH}{\overset{|}{\underset{|}{\underset{H}{C}}}}$—$\overset{OH}{\overset{|}{CH_2}}$ $\longrightarrow$ $\overset{OH}{\overset{|}{CH_2}}$—$\overset{\overset{\displaystyle O}{\|}}{C}$—$\overset{OH}{\overset{|}{CH_2}}$

    Glycerol      Dihydroxyacetone

(f) [benzene ring]—$\overset{H}{\overset{|}{\underset{|}{\underset{H}{C}}}}$—H $\longrightarrow$ [benzene ring]—$C$$\underset{O^-}{\overset{\overset{\displaystyle O}{\|}}{\diagup}}$ + $H^+$

    Toluene      Benzoate

(g) $\overset{\overset{\displaystyle O}{\|}}{\underset{^-O}{C}}$—$CH_2$—$CH_2$—$C$$\underset{O}{\overset{O^-}{\diagup}}$ $\longrightarrow$ $\overset{H}{\underset{^-O—C}{\overset{|}{C}}}$=$\overset{C}{\underset{H}{\overset{|}{C}}}$

    Succinate      Fumarate

(h) $CH_3$—$\overset{\overset{\displaystyle O}{\|}}{\underset{\underset{O}{\underset{\|}{C—O^-}}}{C}}$ $\longrightarrow$ $CH_3$—$C$$\underset{O^-}{\overset{\overset{\displaystyle O}{\|}}{\diagup}}$ + $CO_2$

    Pyruvate      Acetate

**3. Relationship between Energy Release and the Oxidation State of Carbon**    A eukaryotic cell can use glucose ($C_6H_{12}O_6$) and hexanoic acid ($C_6H_{14}O_2$) as fuels for cellular respiration. On the basis of their structural formulas, which substance releases more energy per gram on complete combustion to $CO_2$ and $H_2O$?

**4. Nicotinamide Coenzymes as Reversible Redox Carriers**    The nicotinamide coenzymes (see Fig. 13-15) can undergo reversible oxidation-reduction reactions with specific substrates in the presence of the appropriate dehydrogenase. In these reactions, NADH + $H^+$ serves as the hydrogen source, as described in Problem 2. Whenever the coenzyme is oxidized, a substrate must be simultaneously reduced:

$$\underset{\text{Oxidized}}{\text{Substrate}} + \underset{\text{Reduced}}{\text{NADH} + H^+} \rightleftharpoons \underset{\text{Reduced}}{\text{product}} + \underset{\text{Oxidized}}{NAD^+}$$

For each of the reactions in (a) through (f), determine whether the substrate has been oxidized or reduced or is unchanged in oxidation state (see Problem 2). If a redox change has occurred, balance the reaction with the necessary amount of $NAD^+$, NADH, $H^+$, and $H_2O$. The objective is to recognize when a redox coenzyme is necessary in a metabolic reaction.

(a) $CH_3CH_2OH \longrightarrow CH_3$—$C$$\underset{H}{\overset{\overset{\displaystyle O}{\|}}{\diagup}}$

    Ethanol      Acetaldehyde

(b) $\overset{OPO_3^{2-}}{\overset{|}{CH_2}}$—$\overset{OH}{\overset{|}{\underset{|}{\underset{H}{C}}}}$—$C$$\underset{OPO_3^{2-}}{\overset{\overset{\displaystyle O}{\|}}{\diagup}}$ $\longrightarrow$ $\overset{OPO_3^{2-}}{\overset{|}{CH_2}}$—$\overset{OH}{\overset{|}{\underset{|}{\underset{H}{C}}}}$—$C$$\underset{H}{\overset{\overset{\displaystyle O}{\|}}{\diagup}}$ + $HPO_4^{2-}$

    1,3-Bisphosphoglycerate    Glyceraldehyde 3-phosphate

(c) $CH_3$—$\overset{\overset{\displaystyle O}{\|}}{C}$—$C$$\underset{O}{\overset{O^-}{\diagup}}$ $\longrightarrow$ $CH_3$—$C$$\underset{H}{\overset{O^-}{\diagup}}$ + $CO_2$

    Pyruvate      Acetaldehyde

(d) $CH_3$—$\overset{\overset{\displaystyle O}{\|}}{C}$—$C$$\underset{O}{\overset{O^-}{\diagup}}$ $\longrightarrow$ $CH_3$—$C$$\underset{O^-}{\overset{\overset{\displaystyle O}{\|}}{\diagup}}$ + $CO_2$

    Pyruvate·      Acetate

(e) $^-OOC$—$CH_2$—$\overset{\overset{\displaystyle O}{\|}}{C}$—$COO^-$ $\longrightarrow$ $^-OOC$—$CH_2$—$\overset{OH}{\overset{|}{\underset{|}{\underset{H}{C}}}}$—$COO^-$

    Oxaloacetate      Malate

(f) $CH_3$—$\overset{\overset{\displaystyle O}{\|}}{C}$—$CH_2$—$C$$\underset{O^-}{\overset{\overset{\displaystyle O}{\|}}{\diagup}}$ + $H^+$ $\longrightarrow$ $CH_3$—$\overset{\overset{\displaystyle O}{\|}}{C}$—$CH_3$ + $CO_2$

    Acetoacetate      Acetone

**5. Stimulation of Oxygen Consumption by Oxaloacetate and Malate**    In the early 1930s, Albert Szent György reported the interesting observation that the addition of small amounts of oxaloacetate or malate to suspensions of minced pigeon-breast muscle stimulated the oxygen consumption of the preparation. Surprisingly, the amount of oxygen consumed was about seven times more than the amount necessary for complete oxidation (to $CO_2$ and $H_2O$) of the added oxaloacetate or malate. Why did the addition of oxaloacetate or malate stimulate oxygen consumption? Why was the amount of oxygen consumed so much greater than the amount necessary to completely oxidize the added oxaloacetate or malate?

**6. Formation of Oxaloacetate in a Mitochondrion**   In the last reaction of the citric acid cycle, malate is dehydrogenated to regenerate the oxaloacetate necessary for the entry of acetyl-CoA into the cycle:

$$\text{L-Malate} + \text{NAD}^+ \longrightarrow \text{oxaloacetate} + \text{NADH} + \text{H}^+$$
$$\Delta G'^\circ = 30.0 \text{ kJ/mol}$$

(a) Calculate the equilibrium constant for this reaction at 25 °C.

(b) Because $\Delta G'^\circ$ assumes a standard pH of 7, the equilibrium constant calculated in (a) corresponds to

$$K'_{eq} = \frac{[\text{oxaloacetate}][\text{NADH}]}{[\text{L-malate}][\text{NAD}^+]}$$

The measured concentration of L-malate in rat liver mitochondria is about 0.20 mM when $[\text{NAD}^+]/[\text{NADH}]$ is 10. Calculate the concentration of oxaloacetate at pH 7 in these mitochondria.

(c) To appreciate the magnitude of the mitochondrial oxaloacetate concentration, calculate the number of oxaloacetate molecules in a single rat liver mitochondrion. Assume the mitochondrion is a sphere of diameter 2.0 $\mu$m.

**7. Energy Yield from the Citric Acid Cycle**   The reaction catalyzed by succinyl-CoA synthetase produces the high-energy compound GTP. How is the free energy contained in GTP incorporated into the cellular ATP pool?

**8. Respiration Studies in Isolated Mitochondria**   Cellular respiration can be studied in isolated mitochondria by measuring oxygen consumption under different conditions. If 0.01 M sodium malonate is added to actively respiring mitochondria that are using pyruvate as fuel source, respiration soon stops and a metabolic intermediate accumulates.

(a) What is the structure of this intermediate?

(b) Explain why it accumulates.

(c) Explain why oxygen consumption stops.

(d) Aside from removal of the malonate, how can this inhibition of respiration be overcome? Explain.

**9. Labeling Studies in Isolated Mitochondria**   The metabolic pathways of organic compounds have often been delineated by using a radioactively labeled substrate and following the fate of the label.

(a) How can you determine whether glucose added to a suspension of isolated mitochondria is metabolized to $CO_2$ and $H_2O$?

(b) Suppose you add a brief pulse of [3-$^{14}$C]pyruvate (labeled in the methyl position) to the mitochondria. After one turn of the citric acid cycle, what is the location of the $^{14}$C in the oxaloacetate? Explain by tracing the $^{14}$C label through the pathway. How many turns of the cycle are required to release all the [3-$^{14}$C]pyruvate as $CO_2$?

**10. [1-$^{14}$C]Glucose Catabolism**   An actively respiring bacterial culture is briefly incubated with [1-$^{14}$C] glucose, and the glycolytic and citric acid cycle intermediates are isolated. Where is the $^{14}$C in each of the intermediates listed below? Consider only the initial incorporation of $^{14}$C, in the first pass of labeled glucose through the pathways.

(a) Fructose 1,6-bisphosphate

(b) Glyceraldehyde 3-phosphate

(c) Phosphoenolpyruvate

(d) Acetyl-CoA

(e) Citrate

(f) $\alpha$-Ketoglutarate

(g) Oxaloacetate

**11. Role of the Vitamin Thiamine**   People with beriberi, a disease caused by thiamine deficiency, have elevated levels of blood pyruvate and $\alpha$-ketoglutarate, especially after consuming a meal rich in glucose. How are these effects related to a deficiency of thiamine?

**12. Synthesis of Oxaloacetate by the Citric Acid Cycle**   Oxaloacetate is formed in the last step of the citric acid cycle by the $\text{NAD}^+$-dependent oxidation of L-malate. Can a net synthesis of oxaloacetate from acetyl-CoA occur using only the enzymes and cofactors of the citric acid cycle, without depleting the intermediates of the cycle? Explain. How is oxaloacetate that is lost from the cycle (to biosynthetic reactions) replenished?

**13. Mode of Action of the Rodenticide Fluoroacetate**   Fluoroacetate, prepared commercially for rodent control, is also produced by a South African plant. After entering a cell, fluoroacetate is converted to fluoroacetyl-CoA in a reaction catalyzed by the enzyme acetate thiokinase:

$$\text{F—CH}_2\text{COO}^- + \text{CoA-SH} + \text{ATP} \longrightarrow$$
$$\text{F—CH}_2\overset{\text{O}}{\underset{\|}{\text{C}}}\text{—S-CoA} + \text{AMP} + \text{PP}_i$$

The toxic effect of fluoroacetate was studied in an experiment using intact isolated rat heart. After the heart was perfused with 0.22 mM fluoroacetate, the measured rate of glucose uptake and glycolysis decreased, and glucose 6-phosphate and fructose 6-phosphate accumulated. Examination of the citric acid cycle intermediates revealed that their concentrations were below normal, except for citrate, with a concentration 10 times higher than normal.

(a) Where did the block in the citric acid cycle occur? What caused citrate to accumulate and the other cycle intermediates to be depleted?

(b) Fluoroacetyl-CoA is enzymatically transformed in the citric acid cycle. What is the structure of the end product of fluoroacetate metabolism? Why does it block the citric acid cycle? How might the inhibition be overcome?

(c) In the heart perfusion experiments, why did glucose uptake and glycolysis decrease? Why did hexose monophosphates accumulate?

(d) Why is fluoroacetate poisoning fatal?

**14. Synthesis of L-Malate in Wine Making**   The tartness of some wines is due to high concentrations of L-malate. Write a sequence of reactions showing how yeast cells synthesize L-malate from glucose under anaerobic conditions in the presence of dissolved $CO_2$ ($HCO_3^-$). Note that the overall reaction for this fermentation cannot involve the consumption of nicotinamide coenzymes or citric acid cycle intermediates.

**15. Net Synthesis of $\alpha$-Ketoglutarate**   $\alpha$-Ketoglutarate plays a central role in the biosynthesis of several amino acids. Write a sequence of enzymatic reactions that could result in the net synthesis of $\alpha$-ketoglutarate from pyruvate. Your proposed sequence must not involve the net consumption of other citric acid cycle intermediates. Write an equation for the overall reaction and identify the source of each reactant.

**16. Regulation of the Pyruvate Dehydrogenase Complex** In animal tissues, the rate of conversion of pyruvate to acetyl-CoA is regulated by the ratio of active, phosphorylated to inactive, unphosphorylated PDH complex. Determine what happens to the rate of this reaction when a preparation of rabbit muscle mitochondria containing the PDH complex is treated with (a) pyruvate dehydrogenase kinase, ATP, and NADH; (b) pyruvate dehydrogenase phosphatase and $Ca^{2+}$; (c) malonate.

**17. Commercial Synthesis of Citric Acid** Citric acid is used as a flavoring agent in soft drinks, fruit juices, and many other foods. Worldwide, the market for citric acid is valued at hundreds of millions of dollars per year. Commercial production uses the mold *Aspergillus niger*, which metabolizes sucrose under carefully controlled conditions.

(a) The yield of citric acid is strongly dependent on the concentration of $FeCl_3$ in the culture medium, as indicated in the graph. Why does the yield decrease when the concentration of $Fe^{3+}$ is above or below the optimal value of 0.5 mg/L?

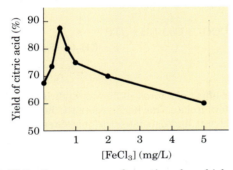

(b) Write the sequence of reactions by which *A. niger* synthesizes citric acid from sucrose. Write an equation for the overall reaction.

(c) Does the commercial process require the culture medium to be aerated—that is, is this a fermentation or an aerobic process? Explain.

**18. Regulation of Citrate Synthase** In the presence of saturating amounts of oxaloacetate, the activity of citrate synthase from pig heart tissue shows a sigmoid dependence on the concentration of acetyl-CoA, as shown in the graph. When succinyl-CoA is added, the curve shifts to the right and the sigmoid dependence is more pronounced.

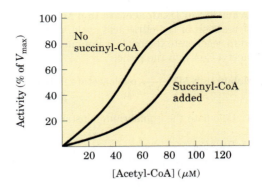

On the basis of these observations, suggest how succinyl-CoA regulates the activity of citrate synthase. (Hint: See Fig. 6–29.) Why is succinyl-CoA an appropriate signal for regulation of the citric acid cycle? How does the regulation of citrate synthase control the rate of cellular respiration in pig heart tissue?

**19. Regulation of Pyruvate Carboxylase** The carboxylation of pyruvate by pyruvate carboxylase occurs at a very low rate unless acetyl-CoA, a positive allosteric modulator, is present. If you have just eaten a meal rich in fatty acids (triacylglycerols) but low in carbohydrates (glucose), how does this regulatory property shut down the oxidation of glucose to $CO_2$ and $H_2O$ but increase the oxidation of acetyl-CoA derived from fatty acids?

**20. Relationship between Respiration and the Citric Acid Cycle** Although oxygen does not participate directly in the citric acid cycle, the cycle operates only when $O_2$ is present. Why?

**21. Thermodynamics of Citrate Synthase Reaction in Cells** Citrate is formed by the condensation of acetyl-CoA with oxaloacetate, catalyzed by citrate synthase:

$$\text{Oxaloacetate} + \text{acetyl-CoA} + H_2O \longrightarrow \text{citrate} + \text{CoA} + H^+$$

In rat heart mitochondria at pH 7.0 and 25 °C, the concentrations of reactants and products are: oxaloacetate, 1 $\mu M$; acetyl-CoA, 1 $\mu M$; citrate, 220 $\mu M$; and CoA, 65 $\mu M$. The standard free-energy change for the citrate synthase reaction is $-32.2$ kJ/mol. What is the direction of metabolite flow through the citrate synthase reaction in rat heart cells? Explain.

**22. Reactions of the Pyruvate Dehydrogenase Complex** Two of the steps in the oxidative decarboxylation of pyruvate (steps ④ and ⑤ in Fig. 16–6) do not involve any of the three carbons of pyruvate yet are essential to the operation of the PDH complex. Explain.

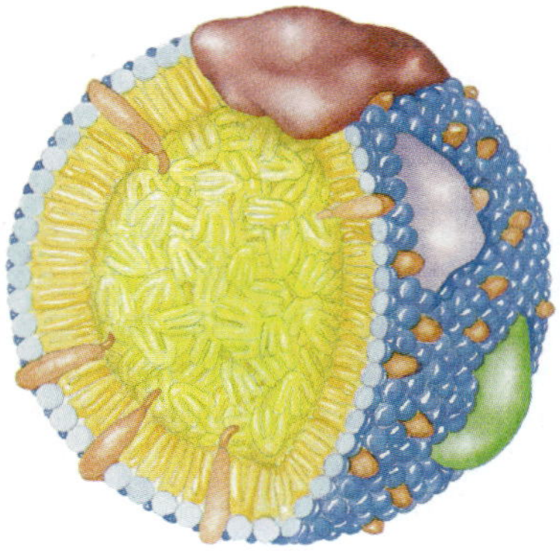

# FATTY ACID CATABOLISM

Jack Sprat could eat no fat,
His wife could eat no lean,
And so between them both you see,
They licked the platter clean.

—*John Clarke,* Paroemiologia Anglo-Latina
(Proverbs English and Latin), *1639*

The oxidation of long-chain fatty acids to acetyl-CoA is a central energy-yielding pathway in many organisms and tissues. In mammalian heart and liver, for example, it provides as much as 80% of the energetic needs under all physiological circumstances. The electrons removed from fatty acids during oxidation pass through the respiratory chain, driving ATP synthesis; the acetyl-CoA produced from the fatty acids may be completely oxidized to $CO_2$ in the citric acid cycle, resulting in further energy conservation. In some species and in some tissues, the acetyl-CoA has alternative fates. In liver, acetyl-CoA may be converted to ketone bodies—water-soluble fuels exported to the brain and other tissues when glucose is not available. In higher plants, acetyl-CoA serves primarily as a biosynthetic precursor, only secondarily as fuel. Although the biological role of fatty acid oxidation differs from organism to organism, the mechanism is essentially the same. The repetitive four-step process, called **β oxidation,** by which fatty acids are converted into acetyl-CoA is the main topic of this chapter.

In Chapter 10 we described the properties of triacylglycerols (also called triglycerides or neutral fats) that make them especially suitable as storage fuels. The long alkyl chains of their constituent fatty acids are essentially hydrocarbons, highly reduced structures with an energy of complete oxidation (~38 kJ/g) more than twice that for the same weight of carbohydrate or protein. This advantage is compounded by the extreme insolubility of lipids in water; cellular triacylglycerols aggregate in lipid droplets, which do not raise the osmolarity of the cytosol, and they are unsolvated. (In storage polysaccharides, by contrast, water of solvation can account for two-thirds of the overall weight of the stored molecules.) And because of their relative chemical inertness, triacylglycerols can be stored in large quantity in cells without the risk of undesired chemical reactions with other cellular constituents.

The properties that make triacylglycerols good storage compounds, however, present problems in their role as fuels. Because they are insoluble in water, ingested triacylglycerols must be emulsified before they can be digested by water-soluble enzymes in the intestine, and triacylglycerols absorbed in the intestine or mobilized from storage tissues must be carried in the blood bound to proteins that counteract their insolubility. To overcome the relative stability of the C—C bonds in a fatty acid, the carboxyl group at C-1 is activated by attachment to coenzyme A, which allows stepwise oxidation of the fatty acyl group at the C-3, or β, position—hence the name β oxidation.

We begin this chapter with a brief discussion of the sources of fatty acids and the routes by which they travel to the site of their oxidation, with special emphasis on the process in vertebrates. We then describe the chemical steps of fatty acid oxidation in mitochondria. The complete oxidation of fatty acids to $CO_2$ and $H_2O$ takes place in three stages: the oxidation of long-chain fatty acids to two-carbon fragments, in the form of acetyl-CoA (β oxidation); the oxidation of acetyl-CoA to $CO_2$ in the citric acid cycle (Chapter 16); and the transfer of electrons from reduced electron carriers to the mitochondrial respiratory chain (Chapter 19). In this chapter we focus on the first of these stages. We begin our discussion of $\beta$ oxidation with the simple case in which a fully saturated fatty acid with an even number of carbon atoms is degraded to acetyl-CoA. We then look briefly at the extra transformations necessary for the degradation of unsaturated fatty acids and fatty acids with an odd number of carbons. Finally, we discuss variations on the $\beta$-oxidation theme in specialized organelles—peroxisomes and glyoxysomes—and two less common pathways of fatty acid catabolism, $\omega$ and $\alpha$ oxidation. The chapter concludes with a description of an alternative fate for the acetyl-CoA formed by $\beta$ oxidation in vertebrates: the production of ketone bodies in the liver.

## 17.1 Digestion, Mobilization, and Transport of Fats

Cells can obtain fatty acid fuels from three sources: fats consumed in the diet, fats stored in cells as lipid droplets, and fats synthesized in one organ for export to another. Some species use all three sources under various circumstances, others use one or two. Vertebrates, for example, obtain fats in the diet, mobilize fats stored in specialized tissue (adipose tissue, consisting of cells called adipocytes), and, in the liver, convert excess dietary carbohydrates to fats for export to other tissues. On average, 40% or more of the daily energy requirement of humans in highly industrialized countries is supplied by dietary triacylglycerols (although most nutritional guidelines recommend no more than 30% of daily caloric intake from fats). Triacylglycerols provide more than half the energy requirements of some organs, particularly the liver, heart, and resting skeletal muscle. Stored triacylglycerols are virtually the sole source of energy in hibernating animals and migrating birds. Protists obtain fats by consuming organisms lower in the food chain, and some also store fats as cytosolic lipid droplets. Vascular plants mobilize fats stored in seeds during germination, but do not otherwise depend on fats for energy.

### Dietary Fats Are Absorbed in the Small Intestine

In vertebrates, before ingested triacylglycerols can be absorbed through the intestinal wall they must be converted from insoluble macroscopic fat particles to finely dispersed microscopic micelles. This solubilization is carried out by bile salts, such as taurocholic acid (p. 355), which are synthesized from cholesterol in the liver, stored in the gallbladder, and released into the small intestine after ingestion of a fatty meal. Bile salts are amphipathic compounds that act as biological detergents, converting dietary fats into mixed micelles of bile salts and triacylglycerols (Fig. 17–1, step ①).

Micelle formation enormously increases the fraction of lipid molecules accessible to the action of water-soluble lipases in the intestine, and lipase action converts triacylglycerols to monoacylglycerols (monoglycerides) and diacylglycerols (diglycerides), free fatty acids, and glycerol (step ②). These products of lipase action diffuse into the epithelial cells lining the intestinal surface (the intestinal mucosa) (step ③), where they are reconverted to triacylglycerols and packaged with dietary cholesterol and specific proteins into lipoprotein aggregates called **chylomicrons** (Fig. 17–2; see also Fig. 17–1, step ④).

**Apolipoproteins** are lipid-binding proteins in the blood, responsible for the transport of triacylglycerols, phospholipids, cholesterol, and cholesteryl esters between organs. Apolipoproteins ("apo" means "detached" or "separate," designating the protein in its lipid-free form) combine with lipids to form several classes of lipoprotein particles, spherical aggregates with hydrophobic lipids at the core and hydrophilic protein side chains and lipid head groups at the surface. Various combinations of lipid and protein produce particles of different densities, ranging from chylomicrons and very-low-density lipoproteins (VLDL) to very-high-density lipoproteins (VHDL), which can be separated by ultracentrifugation. The structures of these lipoprotein particles and their roles in lipid transport are detailed in Chapter 21.

The protein moieties of lipoproteins are recognized by receptors on cell surfaces. In lipid uptake from the intestine, chylomicrons, which contain apolipoprotein C-II (apoC-II), move from the intestinal mucosa into the lymphatic system, and then enter the blood, which carries them to muscle and adipose tissue (Fig. 17–1, step ⑤). In the capillaries of these tissues, the extracellular enzyme **lipoprotein lipase,** activated by apoC-II, hydrolyzes triacylglycerols to fatty acids and glycerol (step ⑥), which are taken up by cells in the target tissues (step ⑦). In muscle, the fatty acids are oxidized for energy; in adipose tissue, they are reesterified for storage as triacylglycerols (step ⑧).

The remnants of chylomicrons, depleted of most of their triacylglycerols but still containing cholesterol and apolipoproteins, travel in the blood to the liver, where they are taken up by endocytosis, mediated by receptors for their apolipoproteins. Triacylglycerols that enter the liver by this route may be oxidized to provide energy or to provide precursors for the synthesis of ketone bodies, as described in Section 17.3. When the diet contains more fatty acids than are needed immediately for fuel or as precursors, the liver converts them to triacylglycerols, which are packaged with specific apolipoproteins into VLDLs. The VLDLs are transported in the blood to adipose tissues, where the triacylglycerols are removed and stored in lipid droplets within adipocytes.

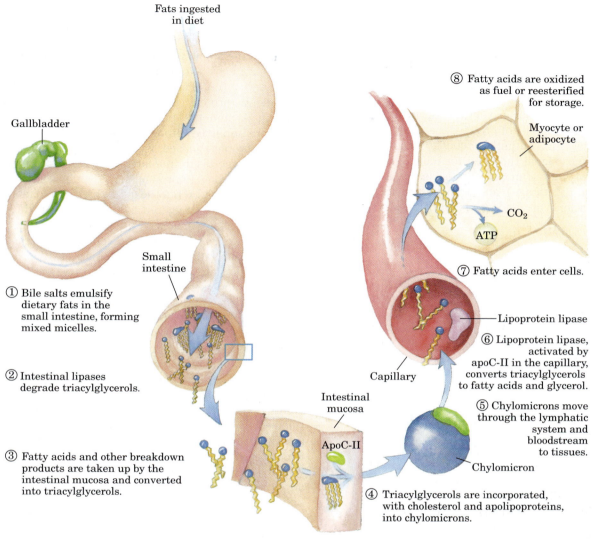

① Bile salts emulsify dietary fats in the small intestine, forming mixed micelles.

② Intestinal lipases degrade triacylglycerols.

③ Fatty acids and other breakdown products are taken up by the intestinal mucosa and converted into triacylglycerols.

④ Triacylglycerols are incorporated, with cholesterol and apolipoproteins, into chylomicrons.

⑤ Chylomicrons move through the lymphatic system and bloodstream to tissues.

⑥ Lipoprotein lipase, activated by apoC-II in the capillary, converts triacylglycerols to fatty acids and glycerol.

⑦ Fatty acids enter cells.

⑧ Fatty acids are oxidized as fuel or reesterified for storage.

**FIGURE 17–1 Processing of dietary lipids in vertebrates.** Digestion and absorption of dietary lipids occur in the small intestine, and the fatty acids released from triacylglycerols are packaged and delivered to muscle and adipose tissues. The eight steps are discussed in the text.

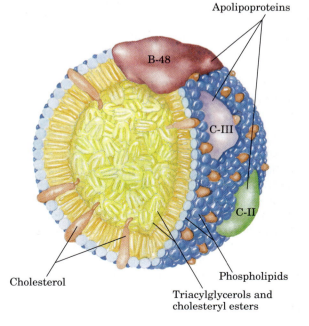

**FIGURE 17–2 Molecular structure of a chylomicron.** The surface is a layer of phospholipids, with head groups facing the aqueous phase. Triacylglycerols sequestered in the interior (yellow) make up more than 80% of the mass. Several apolipoproteins that protrude from the surface (B-48, C-III, C-II) act as signals in the uptake and metabolism of chylomicron contents. The diameter of chylomicrons ranges from about 100 to 500 nm.

## Hormones Trigger Mobilization of Stored Triacylglycerols

Neutral lipids are stored in adipocytes (and in steroid-synthesizing cells of the adrenal cortex, ovary, and testes) in the form of lipid droplets, with a core of sterol esters and triacylglycerols surrounded by a monolayer of phospholipids. The surface of these droplets is coated with **perilipins,** a family of proteins that restrict access to lipid droplets, preventing untimely lipid mobilization. When hormones signal the need for metabolic energy, triacylglycerols stored in adipose tissue are mobilized (brought out of storage) and transported to tissues (skeletal muscle, heart, and renal cortex) in which fatty

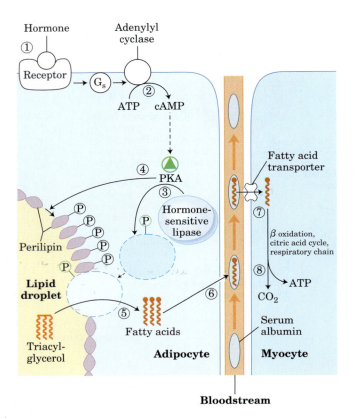

**FIGURE 17–3 Mobilization of triacylglycerols stored in adipose tissue.** When low levels of glucose in the blood trigger the release of glucagon, ① the hormone binds its receptor in the adipocyte membrane and thus ② stimulates adenylyl cyclase, via a G protein, to produce cAMP. This activates PKA, which phosphorylates ③ the hormone-sensitive lipase and ④ perilipin molecules on the surface of the lipid droplet. Phosphorylation of perilipin permits hormone-sensitive lipase access to the surface of the lipid droplet, where ⑤ it hydrolyzes triacylglycerols to free fatty acids. ⑥ Fatty acids leave the adipocyte, bind serum albumin in the blood, and are carried in the blood; they are released from the albumin and ⑦ enter a myocyte via a specific fatty acid transporter. ⑧ In the myocyte, fatty acids are oxidized to $CO_2$, and the energy of oxidation is conserved in ATP, which fuels muscle contraction and other energy-requiring metabolism in the myocyte.

acids can be oxidized for energy production. The hormones epinephrine and glucagon, secreted in response to low blood glucose levels, activate the enzyme adenylyl cyclase in the adipocyte plasma membrane (Fig. 17–3), which produces the intracellular second messenger cyclic AMP (cAMP; see Fig. 12–13). Cyclic AMP–dependent protein kinase (PKA) phosphorylates perilipin A, and the phosphorylated perilipin causes **hormone-sensitive lipase** in the cytosol to move to the lipid droplet surface, where it can begin hydrolyzing triacylglycerols to free fatty acids and glycerol. PKA also phosphorylates hormone-sensitive lipase, doubling or tripling its activity, but the more than 50-fold increase in fat mobilization triggered by epinephrine is due primarily to perilipin phosphorylation. Cells with defective perilipin genes have almost no response to increases in cAMP concentration; their hormone-sensitive lipase does not associate with lipid droplets.

As hormone-sensitive lipase hydrolyzes triacylglycerol in adipocytes, the fatty acids thus released **(free fatty acids, FFA)** pass from the adipocyte into the blood, where they bind to the blood protein **serum albumin.** This protein ($M_r$ 66,000), which makes up about half of the total serum protein, noncovalently binds as many as 10 fatty acids per protein monomer. Bound to this soluble protein, the otherwise insoluble fatty acids are carried to tissues such as skeletal muscle, heart, and renal cortex. In these target tissues, fatty acids dissociate from albumin and are moved by plasma membrane transporters into cells to serve as fuel.

About 95% of the biologically available energy of triacylglycerols resides in their three long-chain fatty acids; only 5% is contributed by the glycerol moiety. The glycerol released by lipase action is phosphorylated by **glycerol kinase** (Fig. 17–4), and the resulting glycerol 3-phosphate is oxidized to dihydroxyacetone phosphate. The glycolytic enzyme triose phosphate isomerase converts this compound to glyceraldehyde 3-phosphate, which is oxidized via glycolysis.

## Fatty Acids Are Activated and Transported into Mitochondria

The enzymes of fatty acid oxidation in animal cells are located in the mitochondrial matrix, as demonstrated in 1948 by Eugene P. Kennedy and Albert Lehninger. The fatty acids with chain lengths of 12 or fewer carbons enter mitochondria without the help of membrane transporters. Those with 14 or more carbons, which constitute the majority of the FFA obtained in the diet or released from adipose tissue, cannot pass directly through the mitochondrial membranes—they must first undergo the three enzymatic reactions of the **carnitine shuttle.** The first reaction is catalyzed by a family of isozymes (different isozymes specific for fatty acids having short, intermediate, or long carbon chains) present

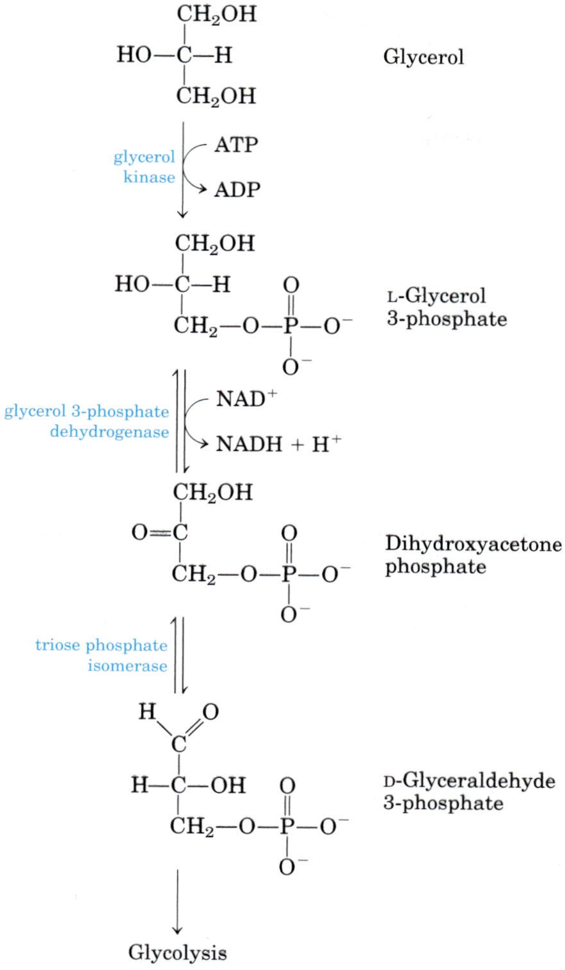

**FIGURE 17-4 Entry of glycerol into the glycolytic pathway.**

in the outer mitochondrial membrane, the **acyl-CoA synthetases,** which promote the general reaction

$$\text{Fatty acid} + \text{CoA} + \text{ATP} \rightleftharpoons \text{fatty acyl–CoA} + \text{AMP} + \text{PP}_i$$

Thus, acyl-CoA synthetases catalyze the formation of a thioester linkage between the fatty acid carboxyl group and the thiol group of coenzyme A to yield a **fatty acyl–CoA,** coupled to the cleavage of ATP to AMP and PP$_i$. (Recall the description of this reaction in Chapter 13, to illustrate how the free energy released by cleavage of phosphoanhydride bonds in ATP could be coupled to the formation of a high-energy compound; p. 502.) The reaction occurs in two steps and involves a fatty acyl–adenylate intermediate (Fig. 17–5).

Fatty acyl–CoAs, like acetyl-CoA, are high-energy compounds; their hydrolysis to FFA and CoA has a large, negative standard free-energy change ($\Delta G'^\circ \approx -31$ kJ/mol). The formation of a fatty acyl–CoA is made more favorable by the hydrolysis of *two* high-energy bonds in ATP; the pyrophosphate formed in the activation reaction is immediately hydrolyzed by inorganic pyrophosphatase (left side of Fig. 17–5), which pulls the preceding activation reaction in the direction of fatty acyl–CoA formation. The overall reaction is

$$\text{Fatty acid} + \text{CoA} + \text{ATP} \longrightarrow$$
$$\text{fatty acyl–CoA} + \text{AMP} + 2\text{P}_i \qquad (17\text{–}1)$$
$$\Delta G'^\circ = -34 \text{ kJ/mol}$$

Fatty acyl–CoA esters formed at the cytosolic side of the outer mitochondrial membrane can be transported into the mitochondrion and oxidized to produce ATP, or they can be used in the cytosol to synthesize

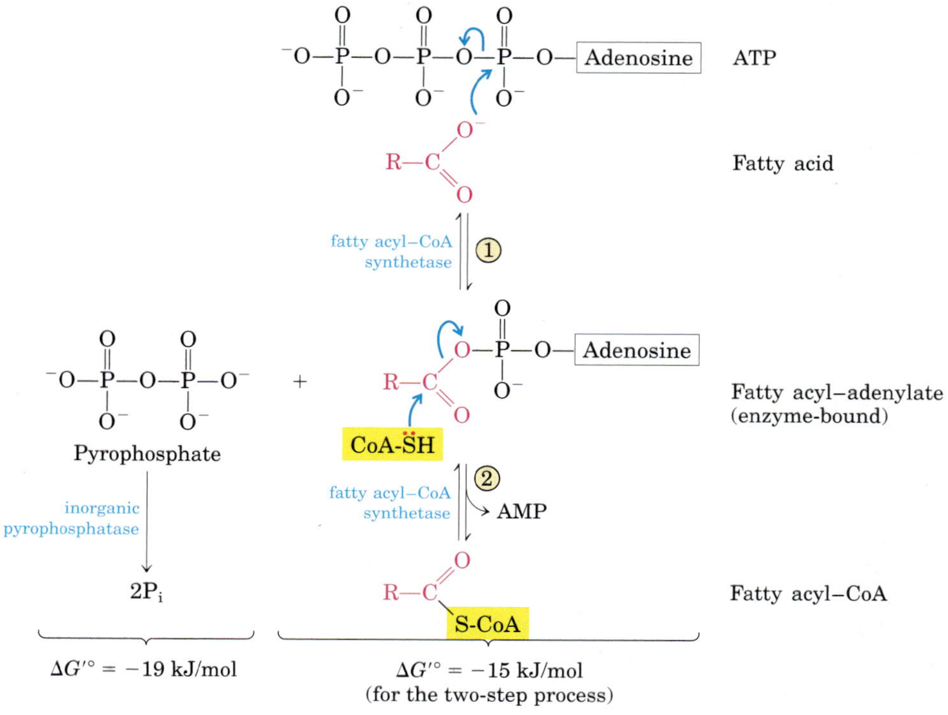

**MECHANISM FIGURE 17-5 Conversion of a fatty acid to a fatty acyl–CoA.** The conversion is catalyzed by fatty acyl–CoA synthetase and inorganic pyrophosphatase. Fatty acid activation by formation of the fatty acyl–CoA derivative occurs in two steps. In step ①, the carboxylate ion displaces the outer two (β and γ) phosphates of ATP to form a fatty acyl–adenylate, the mixed anhydride of a carboxylic acid and a phosphoric acid. The other product is PP$_i$, an excellent leaving group that is immediately hydrolyzed to two P$_i$, pulling the reaction in the forward direction. In step ②, the thiol group of coenzyme A carries out nucleophilic attack on the enzyme-bound mixed anhydride, displacing AMP and forming the thioester fatty acyl–CoA. The overall reaction is highly exergonic. 🖱 **Fatty Acyl–CoA Synthetase Mechanism**

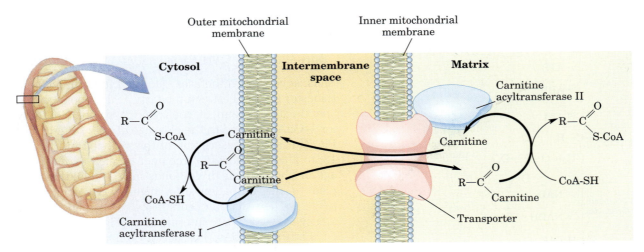

**FIGURE 17–6 Fatty acid entry into mitochondria via the acyl-carnitine/carnitine transporter.** After fatty acyl–carnitine is formed at the outer membrane or in the intermembrane space, it moves into the matrix by facilitated diffusion through the transporter in the inner membrane. In the matrix, the acyl group is transferred to mitochondrial coenzyme A, freeing carnitine to return to the intermembrane space through the same transporter. Acyltransferase I is inhibited by malonyl-CoA, the first intermediate in fatty acid synthesis (see Fig. 21–1). This inhibition prevents the simultaneous synthesis and degradation of fatty acids.

membrane lipids. Fatty acids destined for mitochondrial oxidation are transiently attached to the hydroxyl group of **carnitine** to form fatty acyl–carnitine—the second reaction of the shuttle. This transesterification is catalyzed by **carnitine acyltransferase I** ($M_r$ 88,000), in the outer membrane. Either the acyl-CoA passes through the outer membrane and is converted to the carnitine ester in the intermembrane space (Fig. 17–6), or the carnitine ester is formed on the cytosolic face of the outer membrane, then moved across the outer membrane to the intermembrane space—the current evidence does not reveal which. In either case, passage into the intermembrane space (the space between the outer and inner membranes) occurs through large pores (formed by the protein porin) in the outer membrane. The fatty acyl–carnitine ester then enters the matrix by facilitated diffusion through the **acyl-carnitine/carnitine transporter** of the inner mitochondrial membrane (Fig. 17–6).

$$
\begin{array}{c}
\underset{\substack{|\\ \mathrm{CH_3}}}{\overset{\mathrm{CH_3}}{|}} \\
\mathrm{CH_3{-}N^+{-}CH_2{-}CH{-}CH_2{-}COO^-} \\
\underset{\mathrm{CH_3}}{\phantom{CH_3{-}N^+{-}CH_2{-}}}\underset{\mathrm{OH}}{\phantom{}}
\end{array}
$$
Carnitine

In the third and final step of the carnitine shuttle, the fatty acyl group is enzymatically transferred from carnitine to intramitochondrial coenzyme A by **carnitine acyltransferase II.** This isozyme, located on the inner face of the inner mitochondrial membrane, regenerates fatty acyl–CoA and releases it, along with free carnitine, into the matrix (Fig. 17–6). Carnitine reenters the intermembrane space via the acyl-carnitine/carnitine transporter.

This three-step process for transferring fatty acids into the mitochondrion—esterification to CoA, transesterification to carnitine followed by transport, and transesterification back to CoA—links two separate pools of coenzyme A and of fatty acyl–CoA, one in the cytosol, the other in mitochondria. These pools have different functions. Coenzyme A in the mitochondrial matrix is largely used in oxidative degradation of pyruvate, fatty acids, and some amino acids, whereas cytosolic coenzyme A is used in the biosynthesis of fatty acids (see Fig. 21–10). Fatty acyl–CoA in the cytosolic pool can be used for membrane lipid synthesis or can be moved into the mitochondrial matrix for oxidation and ATP production. Conversion to the carnitine ester commits the fatty acyl moiety to the oxidative fate.

The carnitine-mediated entry process is the rate-limiting step for oxidation of fatty acids in mitochondria and, as discussed later, is a regulation point. Once inside the mitochondrion, the fatty acyl–CoA is acted upon by a set of enzymes in the matrix.

## SUMMARY 17.1 Digestion, Mobilization, and Transport of Fats

- The fatty acids of triacylglycerols furnish a large fraction of the oxidative energy in animals. Dietary triacylglycerols are emulsified in the small intestine by bile salts, hydrolyzed by intestinal lipases, absorbed by intestinal epithelial cells, reconverted into triacylglycerols, then formed into chylomicrons by combination with specific apolipoproteins.

■ Chylomicrons deliver triacylglycerols to tissues, where lipoprotein lipase releases free fatty acids for entry into cells. Triacylglycerols stored in adipose tissue are mobilized by a hormone-sensitive triacylglycerol lipase. The released fatty acids bind to serum albumin and are carried in the blood to the heart, skeletal muscle, and other tissues that use fatty acids for fuel.

■ Once inside cells, fatty acids are activated at the outer mitochondrial membrane by conversion to fatty acyl–CoA thioesters. Fatty acyl–CoA to be oxidized enters mitochondria in three steps, via the carnitine shuttle.

## 17.2 Oxidation of Fatty Acids

As noted earlier, mitochondrial oxidation of fatty acids takes place in three stages (Fig. 17–7). In the first stage—$\beta$ oxidation—fatty acids undergo oxidative removal of successive two-carbon units in the form of acetyl-CoA, starting from the carboxyl end of the fatty acyl chain. For example, the 16-carbon palmitic acid (palmitate at pH 7) undergoes seven passes through the oxidative sequence, in each pass losing two carbons as acetyl-CoA. At the end of seven cycles the last two carbons of palmitate (originally C-15 and C-16) remain as acetyl-CoA. The overall result is the conversion of the 16-carbon chain of palmitate to eight two-carbon acetyl groups of acetyl-CoA molecules. Formation of each acetyl-CoA requires removal of four hydrogen atoms (two pairs of electrons and four $H^+$) from the fatty acyl moiety by dehydrogenases.

In the second stage of fatty acid oxidation, the acetyl groups of acetyl-CoA are oxidized to $CO_2$ in the citric acid cycle, which also takes place in the mitochondrial matrix. Acetyl-CoA derived from fatty acids thus enters a final common pathway of oxidation with the acetyl-CoA derived from glucose via glycolysis and pyruvate oxidation (see Fig. 16–1). The first two stages of fatty acid oxidation produce the reduced electron carriers NADH and $FADH_2$, which in the third stage donate electrons to the mitochondrial respiratory chain, through which the electrons pass to oxygen with the concomitant phosphorylation of ADP to ATP (Fig. 17–7). The energy released by fatty acid oxidation is thus conserved as ATP.

We now take a closer look at the first stage of fatty acid oxidation, beginning with the simple case of a saturated fatty acyl chain with an even number of carbons, then turning to the slightly more complicated cases of unsaturated and odd-number chains. We also consider the regulation of fatty acid oxidation, the $\beta$-oxidative processes as they occur in organelles other than mitochondria, and, finally, two less-general modes of fatty acid catabolism, $\alpha$ oxidation and $\omega$ oxidation.

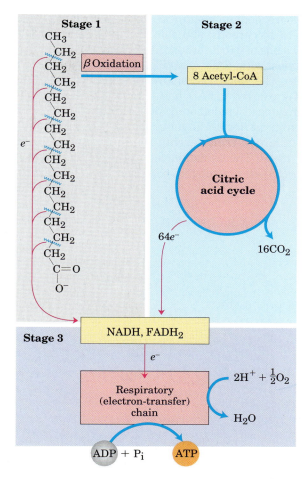

**FIGURE 17–7 Stages of fatty acid oxidation.** Stage 1: A long-chain fatty acid is oxidized to yield acetyl residues in the form of acetyl-CoA. This process is called $\beta$ oxidation. Stage 2: The acetyl groups are oxidized to $CO_2$ via the citric acid cycle. Stage 3: Electrons derived from the oxidations of stages 1 and 2 pass to $O_2$ via the mitochondrial respiratory chain, providing the energy for ATP synthesis by oxidative phosphorylation.

## The $\beta$ Oxidation of Saturated Fatty Acids Has Four Basic Steps

Four enzyme-catalyzed reactions make up the first stage of fatty acid oxidation (Fig. 17–8a). First, dehydrogenation of fatty acyl–CoA produces a double bond between the $\alpha$ and $\beta$ carbon atoms (C-2 and C-3), yielding a ***trans*-$\Delta^2$-enoyl-CoA** (the symbol $\Delta^2$ designates the position of the double bond; you may want to review fatty acid nomenclature, p. 343.) Note that the new double bond has the trans configuration, whereas the double bonds in naturally occurring unsaturated fatty acids are normally in the cis configuration. We consider the significance of this difference later.

This first step is catalyzed by three isozymes of **acyl-CoA dehydrogenase,** each specific for a range of fatty-acyl chain lengths: very-long-chain acyl-CoA dehydrogenase (VLCAD), acting on fatty acids of 12 to 18 carbons; medium-chain (MCAD), acting on fatty acids of 4 to 14 carbons; and short-chain (SCAD), acting on fatty acids of 4 to 8 carbons. All three isozymes are flavoproteins with FAD (see Fig. 13–18) as a prosthetic group. The electrons removed from the fatty acyl–CoA are transferred to FAD, and the reduced form of the dehydrogenase immediately donates its electrons to an electron carrier of the mitochondrial respiratory chain, the **electron-transferring flavoprotein (ETF)** (see Fig. 19–8). The oxidation catalyzed by an acyl-CoA dehydrogenase is analogous to succinate dehydrogenation

in the citric acid cycle (p. 612); in both reactions the enzyme is bound to the inner membrane, a double bond is introduced into a carboxylic acid between the $\alpha$ and $\beta$ carbons, FAD is the electron acceptor, and electrons from the reaction ultimately enter the respiratory chain and pass to $O_2$, with the concomitant synthesis of about 1.5 ATP molecules per electron pair.

In the second step of the $\beta$-oxidation cycle (Fig. 17–8a), water is added to the double bond of the *trans*-$\Delta^2$-enoyl-CoA to form the L stereoisomer of **$\beta$-hydroxyacyl-CoA (3-hydroxyacyl-CoA).** This reaction, catalyzed by **enoyl-CoA hydratase,** is formally analogous to the fumarase reaction in the citric acid cycle, in which $H_2O$ adds across an $\alpha$–$\beta$ double bond (p. 612).

In the third step, L-$\beta$-hydroxyacyl-CoA is dehydrogenated to form **$\beta$-ketoacyl-CoA,** by the action of **$\beta$-hydroxyacyl-CoA dehydrogenase;** $NAD^+$ is the electron acceptor. This enzyme is absolutely specific for the L stereoisomer of hydroxyacyl-CoA. The NADH formed in the reaction donates its electrons to **NADH dehydrogenase,** an electron carrier of the respiratory chain, and ATP is formed from ADP as the electrons pass to $O_2$. The reaction catalyzed by $\beta$-hydroxyacyl-CoA dehydrogenase is closely analogous to the malate dehydrogenase reaction of the citric acid cycle (p. 612).

The fourth and last step of the $\beta$-oxidation cycle is catalyzed by **acyl-CoA acetyltransferase,** more commonly called **thiolase,** which promotes reaction of $\beta$-ketoacyl-CoA with a molecule of free coenzyme A to split off the carboxyl-terminal two-carbon fragment of the original fatty acid as acetyl-CoA. The other product is the coenzyme A thioester of the fatty acid, now shortened by two carbon atoms (Fig. 17–8a). This reaction is called thiolysis, by analogy with the process of hydrolysis, because the $\beta$-ketoacyl-CoA is cleaved by reaction with the thiol group of coenzyme A.

The last three steps of this four-step sequence are catalyzed by either of two sets of enzymes, with the enzymes employed depending on the length of the fatty acyl chain. For fatty acyl chains of 12 or more carbons, the reactions are catalyzed by a multienzyme complex associated with the inner mitochondrial membrane, the **trifunctional protein (TFP).** TFP is a heterooctamer of $\alpha_4\beta_4$ subunits. Each $\alpha$ subunit contains two activities, the enoyl-CoA hydratase and the $\beta$-hydroxyacyl-CoA dehydrogenase; the $\beta$ subunits contain the thiolase activity. This tight association of three enzymes may allow efficient substrate channeling from one active site to the

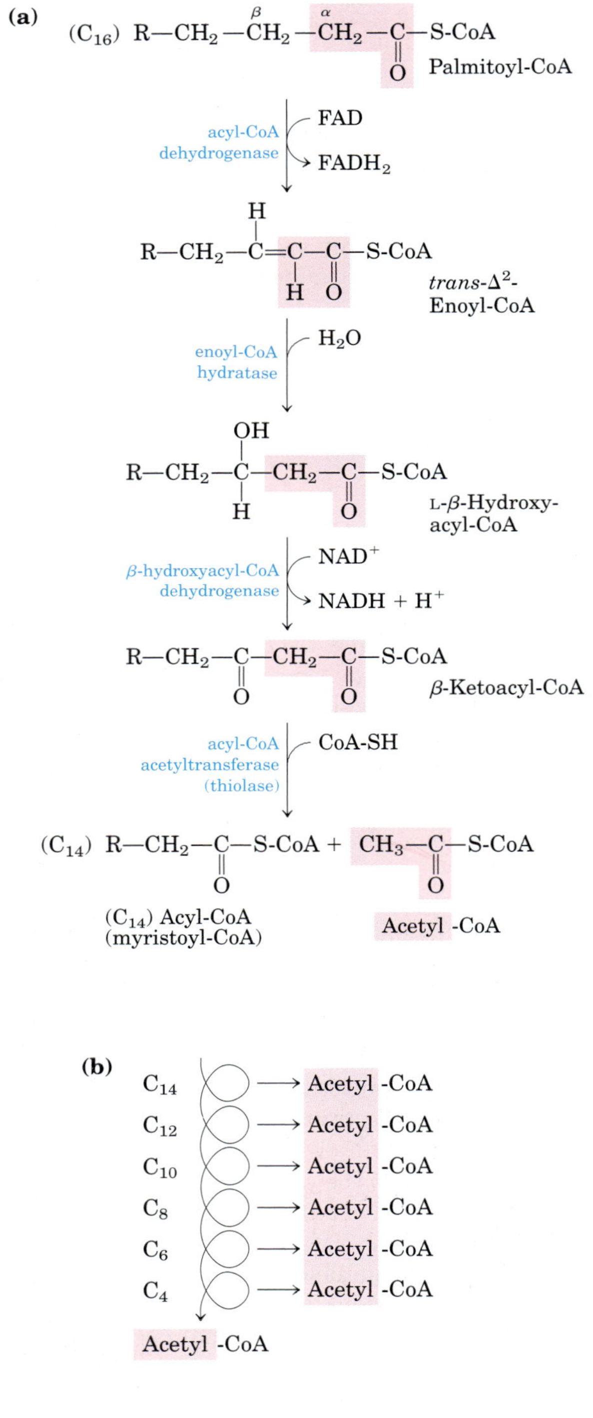

**FIGURE 17–8 The $\beta$-oxidation pathway. (a)** In each pass through this four-step sequence, one acetyl residue (shaded in pink) is removed in the form of acetyl-CoA from the carboxyl end of the fatty acyl chain—in this example palmitate ($C_{16}$), which enters as palmitoyl-CoA. **(b)** Six more passes through the pathway yield seven more molecules of acetyl-CoA, the seventh arising from the last two carbon atoms of the 16-carbon chain. Eight molecules of acetyl-CoA are formed in all.

next, without diffusion of the intermediates away from the enzyme surface. When TFP has shortened the fatty acyl chain to 12 or fewer carbons, further oxidations are catalyzed by a set of four soluble enzymes in the matrix.

As noted earlier, the single bond between methylene (—$CH_2$—) groups in fatty acids is relatively stable. The $\beta$-oxidation sequence is an elegant mechanism for destabilizing and breaking these bonds. The first three reactions of $\beta$ oxidation create a much less stable C—C bond, in which the $\alpha$ carbon (C-2) is bonded to *two* carbonyl carbons (the $\beta$-ketoacyl-CoA intermediate). The ketone function on the $\beta$ carbon (C-3) makes it a good target for nucleophilic attack by the —SH of coenzyme A, catalyzed by thiolase. The acidity of the $\alpha$ hydrogen and the resonance stabilization of the carbanion generated by the departure of this hydrogen make the terminal —$CH_2$—CO—S-CoA a good leaving group, facilitating breakage of the $\alpha$–$\beta$ bond.

### The Four $\beta$-Oxidation Steps Are Repeated to Yield Acetyl-CoA and ATP

In one pass through the $\beta$-oxidation sequence, one molecule of acetyl-CoA, two pairs of electrons, and four protons ($H^+$) are removed from the long-chain fatty acyl–CoA, shortening it by two carbon atoms. The equation for one pass, beginning with the coenzyme A ester of our example, palmitate, is

Palmitoyl-CoA + CoA + FAD + NAD$^+$ + $H_2O$ $\longrightarrow$
$\qquad$ myristoyl-CoA + acetyl-CoA +FADH$_2$ + NADH + H$^+$
$\qquad\qquad\qquad\qquad\qquad\qquad\qquad\qquad\qquad$ (17–2)

Following removal of one acetyl-CoA unit from palmitoyl-CoA, the coenzyme A thioester of the shortened fatty acid (now the 14-carbon myristate) remains. The myristoyl-CoA can now go through another set of four $\beta$-oxidation reactions, exactly analogous to the first, to yield a second molecule of acetyl-CoA and lauroyl-CoA, the coenzyme A thioester of the 12-carbon laurate. Altogether, seven passes through the $\beta$-oxidation sequence are required to oxidize one molecule of palmitoyl-CoA to eight molecules of acetyl-CoA (Fig. 17–8b). The overall equation is

Palmitoyl-CoA + 7CoA + 7FAD + 7NAD$^+$ + 7$H_2O$ $\longrightarrow$
$\qquad$ 8 acetyl-CoA + 7FADH$_2$ + 7NADH + 7H$^+$ (17–3)

Each molecule of FADH$_2$ formed during oxidation of the fatty acid donates a pair of electrons to ETF of the respiratory chain, and about 1.5 molecules of ATP are generated during the ensuing transfer of each electron pair to $O_2$. Similarly, each molecule of NADH formed delivers a pair of electrons to the mitochondrial NADH dehydrogenase, and the subsequent transfer of each pair of electrons to $O_2$ results in formation of about 2.5 molecules of ATP. Thus four molecules of ATP are formed for each two-carbon unit removed in one pass through the sequence. Note that water is also produced in this

process. Transfer of electrons from NADH or FADH$_2$ to $O_2$ yields one $H_2O$ per electron pair. Reduction of $O_2$ by NADH also consumes one $H^+$ per NADH molecule: NADH + H$^+$ + $\frac{1}{2}O_2$ $\longrightarrow$ NAD$^+$ + $H_2O$. In hibernating animals, fatty acid oxidation provides metabolic energy, heat, and water—all essential for survival of an animal that neither eats nor drinks for long periods (Box 17–1). Camels obtain water to supplement the meager supply available in their natural environment by oxidation of fats stored in their hump.

The overall equation for the oxidation of palmitoyl-CoA to eight molecules of acetyl-CoA, including the electron transfers and oxidative phosphorylations, is

Palmitoyl-CoA + 7CoA + 7$O_2$ + 28P$_i$ + 28ADP $\longrightarrow$
$\qquad$ 8 acetyl-CoA + 28ATP + 7$H_2O$ (17–4)

### Acetyl-CoA Can Be Further Oxidized in the Citric Acid Cycle

The acetyl-CoA produced from the oxidation of fatty acids can be oxidized to $CO_2$ and $H_2O$ by the citric acid cycle. The following equation represents the balance sheet for the second stage in the oxidation of palmitoyl-CoA, together with the coupled phosphorylations of the third stage:

8 Acetyl-CoA + 16$O_2$ + 80P$_i$ + 80ADP $\longrightarrow$
$\qquad$ 8CoA + 80ATP + 16$CO_2$ + 16$H_2O$ (17–5)

Combining Equations 17–4 and 17–5, we obtain the overall equation for the complete oxidation of palmitoyl-CoA to carbon dioxide and water:

Palmitoyl-CoA + 23$O_2$ + 108P$_i$ + 108ADP $\longrightarrow$
$\qquad$ CoA + 108ATP + 16$CO_2$ + 23$H_2O$ (17–6)

Table 17–1 summarizes the yields of NADH, FADH$_2$, and ATP in the successive steps of palmitoyl-CoA oxidation. Note that because the activation of palmitate to palmitoyl-CoA breaks both phosphoanhydride bonds in ATP (Fig. 17–5), the energetic cost of activating a fatty acid is equivalent to two ATP, and the net gain per molecule of palmitate is 106 ATP. The standard free-energy change for the oxidation of palmitate to $CO_2$ and $H_2O$ is about 9,800 kJ/mol. Under standard conditions, the energy recovered as the phosphate bond energy of ATP is 106 × 30.5 kJ/mol = 3,230 kJ/mol, about 33% of the theoretical maximum. However, when the free-energy changes are calculated from actual concentrations of reactants and products under intracellular conditions (see Box 13–1), the free-energy recovery is more than 60%; the energy conservation is remarkably efficient.

### Oxidation of Unsaturated Fatty Acids Requires Two Additional Reactions

The fatty acid oxidation sequence just described is typical when the incoming fatty acid is saturated (that is, has only single bonds in its carbon chain). However,

**BOX 17–1**     THE WORLD OF BIOCHEMISTRY

### Fat Bears Carry Out $\beta$ Oxidation in Their Sleep

Many animals depend on fat stores for energy during hibernation, during migratory periods, and in other situations involving radical metabolic adjustments. One of the most pronounced adjustments of fat metabolism occurs in hibernating grizzly bears. These animals remain in a continuous state of dormancy for periods as long as seven months. Unlike most hibernating species, the bear maintains a body temperature of between 32 and 35 °C, close to the normal (nonhiber-

A grizzly bear prepares its hibernation nest, near the McNeil River in Canada.

nating) level. Although expending about 25,000 kJ/day (6,000 kcal/day), the bear does not eat, drink, urinate, or defecate for months at a time.

Experimental studies have shown that hibernating grizzly bears use body fat as their sole fuel. Fat oxidation yields sufficient energy for maintenance of body temperature, active synthesis of amino acids and proteins, and other energy-requiring activities, such as membrane transport. Fat oxidation also releases large amounts of water, as described in the text, which replenishes water lost in breathing. The glycerol released by degradation of triacylglycerols is converted into blood glucose by gluconeogenesis. Urea formed during breakdown of amino acids is reabsorbed in the kidneys and recycled, the amino groups reused to make new amino acids for maintaining body proteins.

Bears store an enormous amount of body fat in preparation for their long sleep. An adult grizzly consumes about 38,000 kJ/day during the late spring and summer, but as winter approaches it feeds 20 hours a day, consuming up to 84,000 kJ daily. This change in feeding is a response to a seasonal change in hormone secretion. Large amounts of triacylglycerols are formed from the huge intake of carbohydrates during the fattening-up period. Other hibernating species, including the tiny dormouse, also accumulate large amounts of body fat.

most of the fatty acids in the triacylglycerols and phospholipids of animals and plants are unsaturated, having one or more double bonds. These bonds are in the cis configuration and cannot be acted upon by enoyl-CoA hydratase, the enzyme catalyzing the addition of $H_2O$ to

the trans double bond of the $\Delta^2$-enoyl-CoA generated during $\beta$ oxidation. Two auxiliary enzymes are needed for $\beta$ oxidation of the common unsaturated fatty acids: an isomerase and a reductase. We illustrate these auxiliary reactions with two examples.

**TABLE 17–1**    Yield of ATP during Oxidation of One Molecule of Palmitoyl-CoA to $CO_2$ and $H_2O$

| Enzyme catalyzing the oxidation step | Number of NADH or FADH₂ formed | Number of ATP ultimately formed* |
|---|---|---|
| Acyl-CoA dehydrogenase | 7 FADH₂ | 10.5 |
| $\beta$-Hydroxyacyl-CoA dehydrogenase | 7 NADH | 17.5 |
| Isocitrate dehydrogenase | 8 NADH | 20 |
| $\alpha$-Ketoglutarate dehydrogenase | 8 NADH | 20 |
| Succinyl-CoA synthetase | | 8† |
| Succinate dehydrogenase | 8 FADH₂ | 12 |
| Malate dehydrogenase | 8 NADH | 20 |
|   Total | | 108 |

*These calculations assume that mitochondrial oxidative phosphorylation produces 1.5 ATP per FADH₂ oxidized and 2.5 ATP per NADH oxidized.

†GTP produced directly in this step yields ATP in the reaction catalyzed by nucleoside diphosphate kinase (p. 612).

Oleate is an abundant 18-carbon monounsaturated fatty acid with a cis double bond between C-9 and C-10 (denoted $\Delta^9$). In the first step of oxidation, oleate is converted to oleoyl-CoA and, like the saturated fatty acids, enters the mitochondrial matrix via the carnitine shuttle (Fig. 17–6). Oleoyl-CoA then undergoes three passes through the fatty acid oxidation cycle to yield three molecules of acetyl-CoA and the coenzyme A ester of a $\Delta^3$, 12-carbon unsaturated fatty acid, $cis$-$\Delta^3$-dodecenoyl-CoA (Fig. 17–9). This product cannot serve as a substrate for enoyl-CoA hydratase, which acts only on trans double bonds. The auxiliary enzyme **$\Delta^3$,$\Delta^2$-enoyl-CoA isomerase** isomerizes the $cis$-$\Delta^3$-enoyl-CoA to the $trans$-$\Delta^2$-enoyl-CoA, which is converted by enoyl-CoA hydratase into the corresponding L-$\beta$-hydroxyacyl-CoA ($trans$-$\Delta^2$-dodecenoyl-CoA). This intermediate is now acted upon by the remaining enzymes of $\beta$ oxidation to yield acetyl-CoA and the coenzyme A ester of a 10-carbon saturated fatty acid, decanoyl-CoA. The latter undergoes four more passes through the pathway to yield five more molecules of acetyl-CoA. Altogether, nine acetyl-CoAs are produced from one molecule of the 18-carbon oleate.

The other auxiliary enzyme (a reductase) is required for oxidation of polyunsaturated fatty acids—for

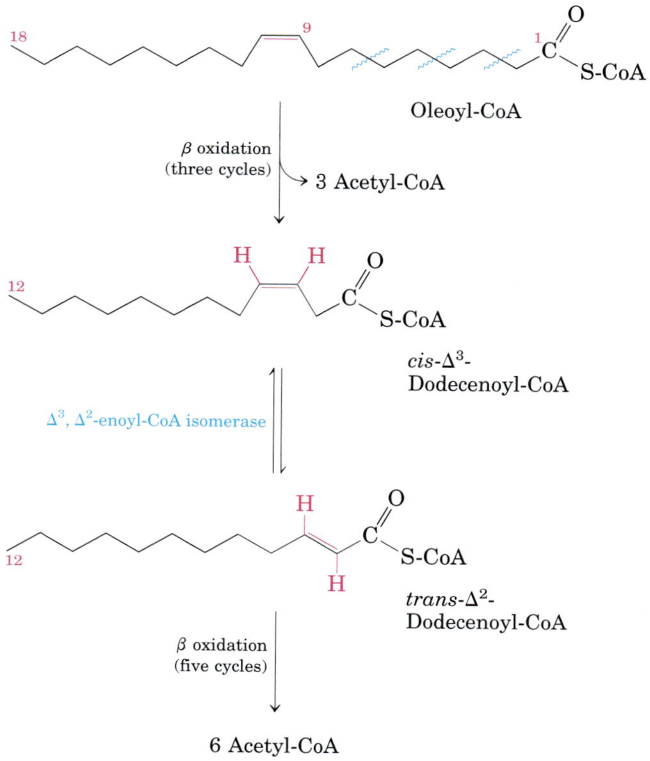

**FIGURE 17–9 Oxidation of a monounsaturated fatty acid.** Oleic acid, as oleoyl-CoA ($\Delta^9$), is the example used here. Oxidation requires an additional enzyme, enoyl-CoA isomerase, to reposition the double bond, converting the cis isomer to a trans isomer, a normal intermediate in $\beta$ oxidation.

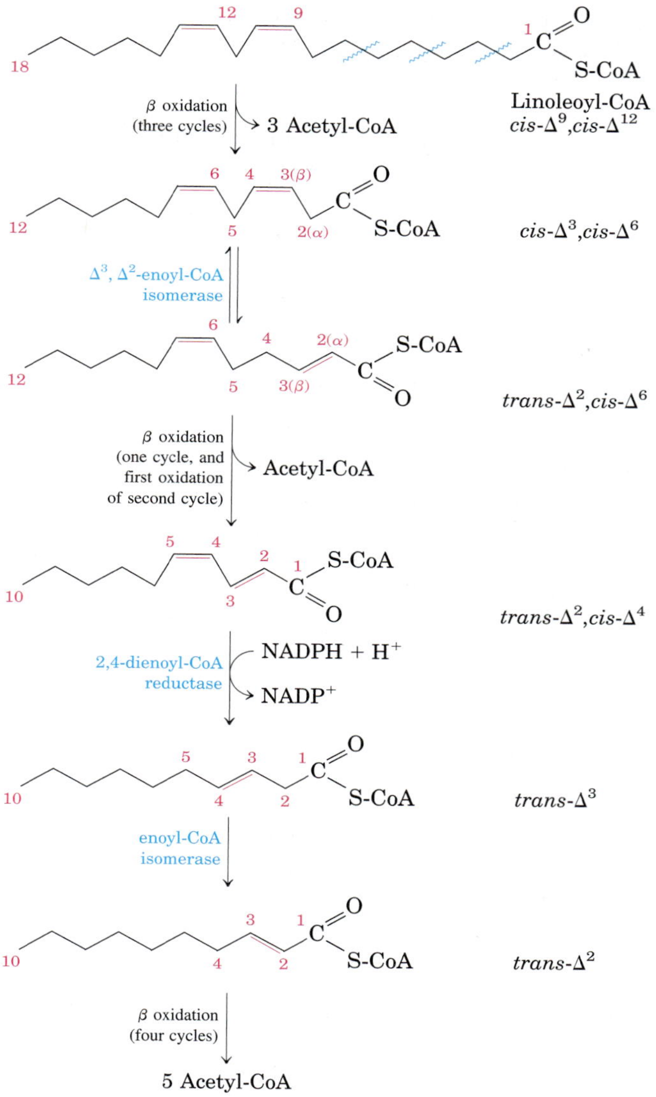

**FIGURE 17–10 Oxidation of a polyunsaturated fatty acid.** The example here is linoleic acid, as linoleoyl-CoA ($\Delta^{9,12}$). Oxidation requires a second auxiliary enzyme in addition to enoyl-CoA isomerase: NADPH-dependent 2,4-dienoyl-CoA reductase. The combined action of these two enzymes converts a $trans$-$\Delta^2$,$cis$-$\Delta^4$-dienoyl-CoA intermediate to the $trans$-$\Delta^2$-enoyl-CoA substrate necessary for $\beta$ oxidation.

example, the 18-carbon linoleate, which has a $cis$-$\Delta^9$,$cis$-$\Delta^{12}$ configuration (Fig. 17–10). Linoleoyl-CoA undergoes three passes through the $\beta$-oxidation sequence to yield three molecules of acetyl-CoA and the coenzyme A ester of a 12-carbon unsaturated fatty acid with a $cis$-$\Delta^3$,$cis$-$\Delta6$ configuration. This intermediate cannot be used by the enzymes of the $\beta$-oxidation pathway; its double bonds are in the wrong position and have the wrong configuration (cis, not trans). However, the combined action of enoyl-CoA isomerase and **2,4-dienoyl-CoA reductase,** as shown in Figure 17–10, allows reentry of this intermediate into the $\beta$-oxidation pathway and its degradation to six acetyl-CoAs. The overall result is conversion of linoleate to nine molecules of acetyl-CoA.

## Complete Oxidation of Odd-Number Fatty Acids Requires Three Extra Reactions

Although most naturally occurring lipids contain fatty acids with an even number of carbon atoms, fatty acids with an odd number of carbons are common in the lipids of many plants and some marine organisms. Cattle and other ruminant animals form large amounts of the three-carbon **propionate** ($CH_3$—$CH_2$—$COO^-$) during fermentation of carbohydrates in the rumen. The propionate is absorbed into the blood and oxidized by the liver and other tissues. And small quantities of propionate are added as a mold inhibitor to some breads and cereals, thus entering the human diet.

Long-chain odd-number fatty acids are oxidized in the same pathway as the even-number acids, beginning at the carboxyl end of the chain. However, the substrate for the last pass through the $\beta$-oxidation sequence is a fatty acyl–CoA with a five-carbon fatty acid. When this is oxidized and cleaved, the products are acetyl-CoA and **propionyl-CoA.** The acetyl-CoA can be oxidized in the citric acid cycle, of course, but propionyl-CoA enters a different pathway involving three enzymes.

Propionyl-CoA is first carboxylated to form the D stereoisomer of **methylmalonyl-CoA** (Fig. 17–11) by **propionyl-CoA carboxylase,** which contains the cofactor biotin. In this enzymatic reaction, as in the pyruvate carboxylase reaction (see Fig. 16–16), $CO_2$ (or its hydrated ion, $HCO_3^-$) is activated by attachment to biotin before its transfer to the substrate, in this case the propionate moiety. Formation of the carboxybiotin intermediate requires energy, which is provided by the cleavage of ATP to ADP and $P_i$. The D-methylmalonyl-CoA thus formed is enzymatically epimerized to its L stereoisomer by **methylmalonyl-CoA epimerase** (Fig. 17–11). The L-methylmalonyl-CoA then undergoes an intramolecular rearrangement to form succinyl-CoA, which can enter the citric acid cycle. This rearrangement is catalyzed by **methylmalonyl-CoA mutase,** which requires as its coenzyme **5′-deoxyadenosyl-cobalamin,** or **coenzyme $B_{12}$,** which is derived from vitamin $B_{12}$ (cobalamin). Box 17–2 describes the role of coenzyme $B_{12}$ in this remarkable exchange reaction.

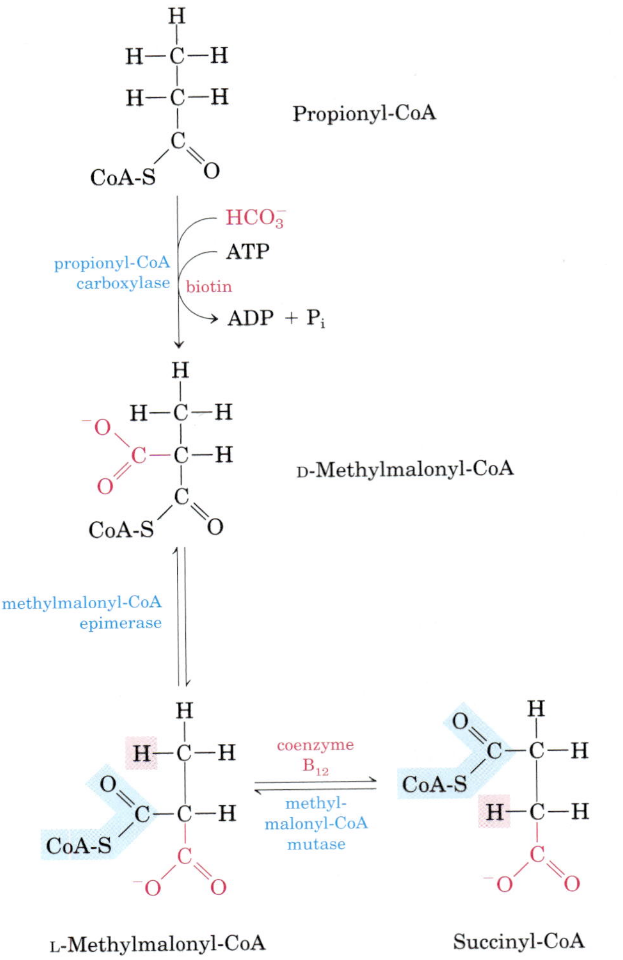

## Fatty Acid Oxidation Is Tightly Regulated

Oxidation of fatty acids consumes a precious fuel, and it is regulated so as to occur only when the need for energy requires it. In the liver, fatty acyl–CoA formed in the cytosol has two major pathways open to it: (1) $\beta$ oxidation by enzymes in mitochondria or (2) conversion into triacylglycerols and phospholipids by enzymes in the cytosol. The pathway taken depends on the rate of

**FIGURE 17–11 Oxidation of propionyl-CoA produced by $\beta$ oxidation of odd-number fatty acids.** The sequence involves the carboxylation of propionyl-CoA to D-methylmalonyl-CoA and conversion of the latter to succinyl-CoA. This conversion requires epimerization of D- to L-methylmalonyl-CoA, followed by a remarkable reaction in which substituents on adjacent carbon atoms exchange positions (see Box 17–2).

transfer of long-chain fatty acyl–CoA into mitochondria. The three-step process (carnitine shuttle) by which fatty acyl groups are carried from cytosolic fatty acyl–CoA into the mitochondrial matrix (Fig. 17–6) is rate-limiting for fatty acid oxidation and is an important point of regulation. Once fatty acyl groups have entered the mitochondrion, they are committed to oxidation to acetyl-CoA.

**Malonyl-CoA,** the first intermediate in the cytosolic biosynthesis of long-chain fatty acids from acetyl-CoA (see Fig. 21–1), increases in concentration whenever the animal is well supplied with carbohydrate; excess glucose that cannot be oxidized or stored as glycogen is converted in the cytosol into fatty acids for storage as triacylglycerol. The inhibition of carnitine acyltransferase I by malonyl-CoA ensures that the oxidation of fatty acids is inhibited whenever the liver is amply supplied with glucose as fuel and is actively making triacylglycerols from excess glucose.

$$^-OOC-CH_2-\overset{\overset{\displaystyle O}{\|}}{C}-S\text{-CoA}$$

Malonyl-CoA

Two of the enzymes of $\beta$ oxidation are also regulated by metabolites that signal energy sufficiency. When the [NADH]/[NAD$^+$] ratio is high, $\beta$-hydroxyacyl-CoA dehydrogenase is inhibited; in addition, high concentrations of acetyl-CoA inhibit thiolase (Fig. 17–12).

## Genetic Defects in Fatty Acyl–CoA Dehydrogenases Cause Serious Disease

Stored triacylglycerols are typically the chief source of energy for muscle contraction, and an inability to oxidize fatty acids from triacylglycerols has serious consequences for health. The most common genetic defect in fatty acid catabolism in U.S. and northern European populations is due to a mutation in the gene encoding the **medium-chain acyl-CoA dehydrogenase (MCAD)**. Among northern Europeans, the frequency of carriers (individuals with this recessive mutation on one of the two homologous chromosomes) is about 1 in 40, and about 1 individual in 10,000 has the disease—that is, has two copies of the mutant MCAD allele and is unable to oxidize fatty acids of 6 to 12 carbons. The disease is characterized by recurring episodes of a syndrome that includes fat accumulation in the liver, high blood levels of octanoic acid, low blood glucose (hypoglycemia), sleepiness, vomiting, and coma. The pattern of organic acids in the urine helps in the diagnosis of this disease: the urine commonly contains high levels of 6-carbon to 10-carbon dicarboxylic acids (produced by $\omega$ oxidation) and low levels of urinary ketone bodies (we discuss $\omega$ oxidation below and ketone bodies in Section 17.3). Although individuals may have no symptoms between episodes, the episodes are very serious; mortality from this disease is 25% to 60% in early childhood. If the genetic defect is detected shortly after birth, the infant can be started on a low-fat, high-carbohydrate diet. With early detection and careful management of the diet—including avoiding long intervals between meals, to prevent the body from turning to its fat reserves for energy—the prognosis for these individuals is good.

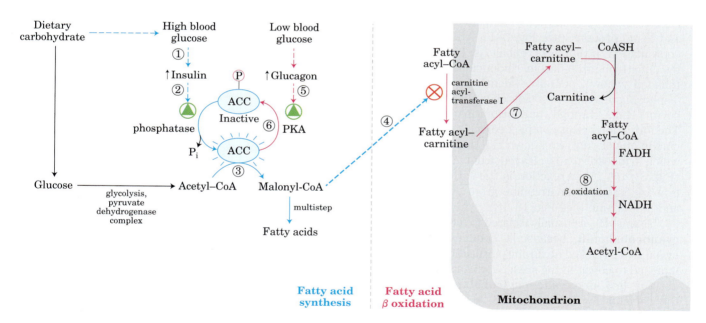

**FIGURE 17–12 Coordinated regulation of fatty acid synthesis and breakdown.** When the diet provides a ready source of carbohydrate as fuel, $\beta$ oxidation of fatty acids is unnecessary and is therefore downregulated. Two enzymes are key to the coordination of fatty acid metabolism: acetyl-CoA carboxylase (ACC), the first enzyme in the synthesis of fatty acids (see Fig. 21–1), and carnitine acyl transferase I, which limits the transport of fatty acids into the mitochondrial matrix for $\beta$ oxidation (see Fig. 17–6). Ingestion of a high-carbohydrate meal raises the blood glucose level and thus ① triggers the release of insulin. ② Insulin-dependent protein phosphatase dephosphorylates ACC, activating it. ③ ACC catalyzes the formation of malonyl-CoA (the first intermediate of fatty acid synthesis), and ④ malonyl-CoA inhibits carnitine acyltransferase I, thereby preventing fatty acid entry into the mitochondrial matrix.

When blood glucose levels drop between meals, ⑤ glucagon release activates cAMP-dependent protein kinase (PKA), which ⑥ phosphorylates and inactivates ACC. The concentration of malonyl-CoA falls, the inhibition of fatty acid entry into mitochondria is relieved, and ⑦ fatty acids enter the mitochondrial matrix and ⑧ become the major fuel. Because glucagon also triggers the mobilization of fatty acids in adipose tissue, a supply of fatty acids begins arriving in the blood.

BOX 17–2    WORKING IN BIOCHEMISTRY

## Coenzyme B₁₂: A Radical Solution to a Perplexing Problem

In the methylmalonyl-CoA mutase reaction (see Fig. 17–11), the group —CO—S-CoA at C-2 of the original propionate exchanges position with a hydrogen atom at C-3 of the original propionate (Fig. 1a). Coenzyme B₁₂ is the cofactor for this reaction, as it is for almost all enzymes that catalyze reactions of this general type (Fig. 1b). These coenzyme B₁₂–dependent processes are among the very few enzymatic reactions in biology in which there is an exchange of an alkyl or substituted alkyl group (X) with a hydrogen atom on an adjacent carbon, *with no mixing of the transferred hydrogen atom with the hydrogen of the solvent,* $H_2O$. How can the hydrogen atom move between two carbons without mixing with the enormous excess of hydrogen atoms in the solvent?

Coenzyme B₁₂ is the cofactor form of vitamin B₁₂, which is unique among all the vitamins in that it contains not only a complex organic molecule but an essential trace element, cobalt. The complex **corrin ring system** of vitamin B₁₂ (colored blue in Fig. 2), to which cobalt (as $Co^{3+}$) is coordinated, is chemically related to the porphyrin ring system of heme and heme proteins (see Fig. 5–1). A fifth coordination position of cobalt is filled by dimethylbenzimidazole ribonucleotide (shaded yellow), bound covalently by its 3′-phosphate group to a side chain of the corrin ring, through aminoisopropanol. The formation of this complex cofactor occurs in one of only two known reactions in which triphosphate is cleaved from ATP (Fig. 3); the other reaction is the formation of S-adenosylmethionine from ATP and methionine (see Fig. 18–18).

Vitamin B₁₂ as usually isolated is called **cyanocobalamin,** because it contains a cyano group (picked up during purification) attached to cobalt in the sixth coordination position. In **5′-deoxyadenosylcobalamin,** the cofactor for methylmalonyl-CoA mutase, the cyano group is replaced by the **5′-deoxyadenosyl** group (red in Fig. 2), covalently bound through C-5′ to the cobalt. The three-dimensional structure of the cofactor was determined by Dorothy Crowfoot Hodgkin in 1956, using x-ray crystallography.

The key to understanding how coenzyme B₁₂ catalyzes hydrogen exchange lies in the properties of the covalent bond between cobalt and C-5′ of the deoxyadeno-

**FIGURE 1**

syl group (Fig. 2). This is a relatively weak bond; its bond dissociation energy is about 110 kJ/mol, compared with 348 kJ/mol for a typical C—C bond or 414 kJ/mol for a C—H bond. Merely illuminating the compound with visible light is enough to break this Co—C bond. (This extreme photolability probably accounts for the absence of vitamin B₁₂ in plants.) Dissociation produces a 5′-deoxyadenosyl radical and the $Co^{2+}$

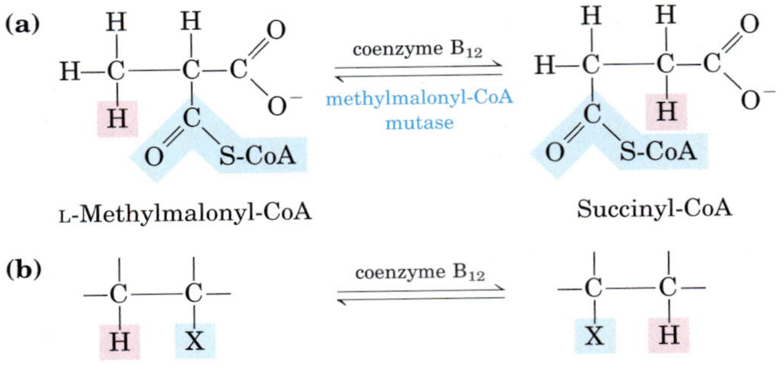

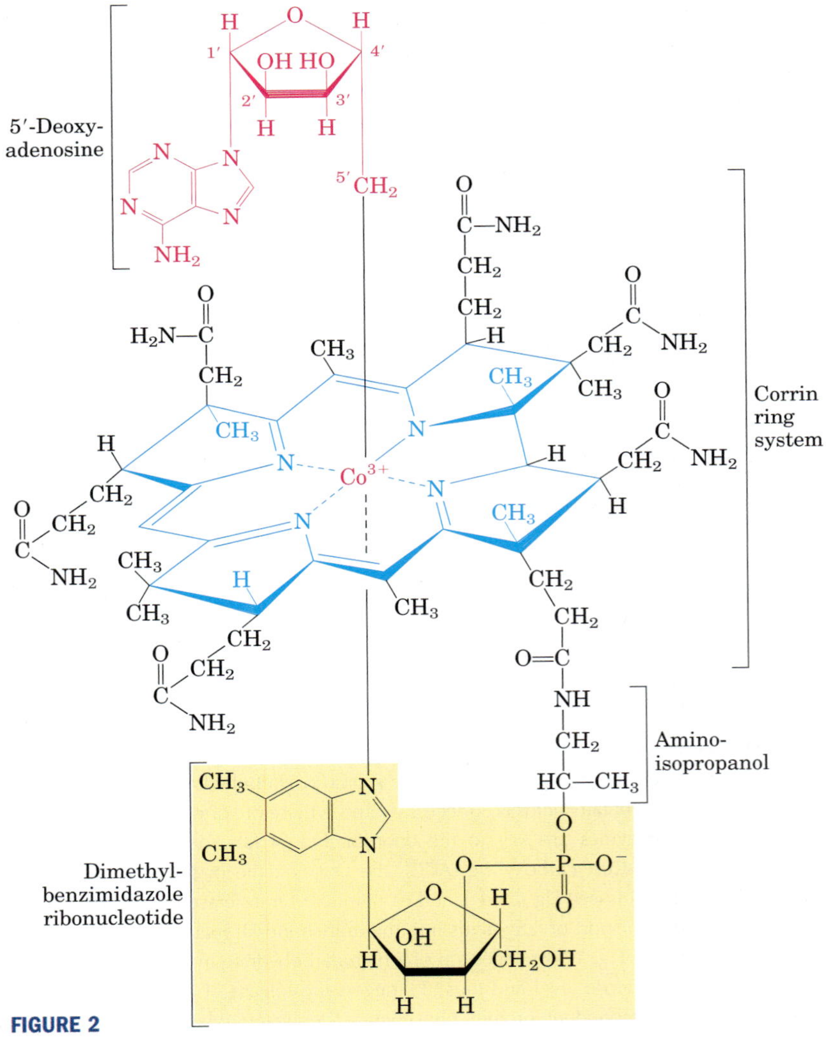

**FIGURE 2**

form of the vitamin. The chemical function of 5′-deoxyadenosylcobalamin is to generate free radicals in this way, thus initiating a series of transformations such as that illustrated in Figure 4— a postulated mechanism for the reaction catalyzed by methylmalonyl-CoA mutase and a number of other coenzyme $B_{12}$–dependent transformations.

① The enzyme first breaks the Co—C bond in the cofactor, leaving the coenzyme in its $Co^{2+}$ form and producing the 5′-deoxyadenosyl free radical. ② This radical now abstracts a hydrogen atom from the substrate, converting the substrate to a radical and producing 5′-deoxyadenosine. ③ Rearrangement of the substrate radical yields another radical, in which the migrating group X (—CO—S-CoA for methylmalonyl-CoA mutase) has moved to the adjacent carbon to form a radical that has the carbon skeleton of the eventual product (a four-carbon straight chain). The hydrogen atom initially abstracted from the substrate is now part of the —CH₃ group of 5′-deoxyadenosine. ④ One of the hydrogens from this same —CH₃ group (it can be the same one originally abstracted) is now returned to the productlike radical, generating the product and regenerating the deoxyadenosyl free radical. ⑤ The bond re-forms between cobalt and the —CH₂ group of the deoxyadenosyl radical, destroying the free radical and regenerating the cofactor in its $Co^{3+}$ form, ready to undergo another catalytic cycle. In this postulated mechanism, the migrating hydrogen atom never exists as a free species and is thus never free to exchange with the hydrogen of surrounding water molecules.

Vitamin $B_{12}$ deficiency results in serious disease. This vitamin is not made by plants or animals and can be synthesized only by a few species of microorganisms. It is required by healthy people in only minute amounts, about 3 $\mu$g/day. The severe disease **pernicious anemia** results from failure to absorb vitamin $B_{12}$ efficiently from the intestine, where it is synthesized by intestinal bacteria or obtained from digestion of meat. Individuals with this disease do not produce sufficient amounts of **intrinsic factor,** a glycoprotein essential to vitamin $B_{12}$ absorption. The pathology in pernicious anemia includes reduced production of erythrocytes, reduced levels of hemoglobin, and severe, progressive impairment of the central nervous system. Administration of large doses of vitamin $B_{12}$ alleviates these symptoms in at least some cases. ■

**FIGURE 3**

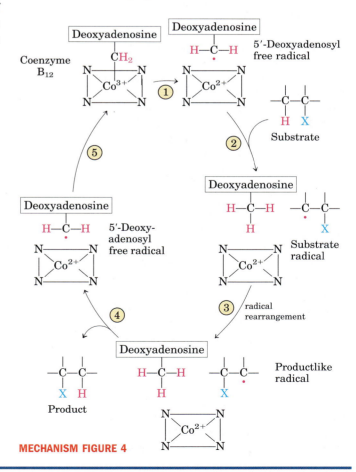

**MECHANISM FIGURE 4**

More than 20 other human genetic defects in fatty acid transport or oxidation have been documented, most much less common than the defect in MCAD. One of the most severe disorders results from loss of the long-chain β-hydroxyacyl-CoA dehydrogenase activity of the trifunctional protein, TFP. Other disorders include defects in the α or β subunits that affect all three activities of TFP and cause serious heart disease and abnormal skeletal muscle. ∎

## Peroxisomes Also Carry Out β Oxidation

The mitochondrial matrix is the major site of fatty acid oxidation in animal cells, but in certain cells other compartments also contain enzymes capable of oxidizing fatty acids to acetyl-CoA, by a pathway similar to, but not identical with, that in mitochondria. In plant cells, the major site of β oxidation is not mitochondria but peroxisomes.

In **peroxisomes,** membrane-enclosed organelles of animal and plant cells, the intermediates for β oxidation of fatty acids are coenzyme A derivatives, and the process consists of four steps, as in mitochondrial β oxidation (Fig. 17–13): (1) dehydrogenation, (2) addition of water to the resulting double bond, (3) oxidation of the β-hydroxyacyl-CoA to a ketone, and (4) thiolytic cleavage by coenzyme A. (The identical reactions also occur in glyoxysomes, as discussed below.)

One difference between the peroxisomal and mitochondrial pathways is in the chemistry of the first step. In peroxisomes, the flavoprotein acyl-CoA oxidase that introduces the double bond passes electrons directly to $O_2$, producing $H_2O_2$ (Fig. 17–13). This strong and potentially damaging oxidant is immediately cleaved to $H_2O$ and $O_2$ by **catalase.** Recall that in mitochondria, the electrons removed in the first oxidation step pass through the respiratory chain to $O_2$ to produce $H_2O$, and this process is accompanied by ATP synthesis. In peroxisomes, the energy released in the first oxidative step of fatty acid breakdown is not conserved as ATP, but is dissipated as heat.

A second important difference between mitochondrial and peroxisomal β oxidation in mammals is in the specificity for fatty acyl–CoAs; the peroxisomal system is much more active on very-long-chain fatty acids such as hexacosanoic acid (26:0) and on branched-chain fatty acids such as phytanic acid and pristanic acid (see Fig. 17–17). These less-common fatty acids are obtained in the diet from dairy products, the fat of ruminant animals, meat, and fish. Their catabolism in the peroxisome involves several auxiliary enzymes unique to this organelle. The inability to oxidize these compounds is responsible for several serious human diseases. Individuals with **Zellweger syndrome** are unable to make peroxisomes and therefore lack all the metabolism unique to that organelle. In **X-linked adrenoleukodystrophy (XALD),** peroxisomes fail to

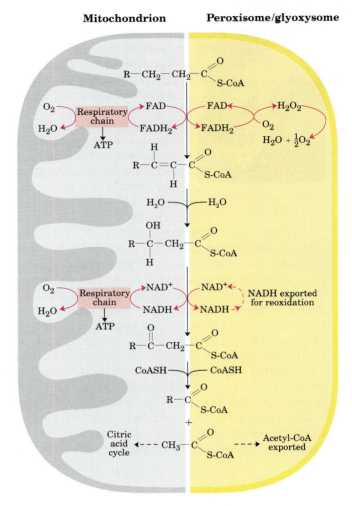

**FIGURE 17–13 Comparison of β oxidation in mitochondria and in peroxisomes and glyoxysomes.** The peroxisomal/glyoxysomal system differs from the mitochondrial system in two respects: (1) in the first oxidative step electrons pass directly to $O_2$, generating $H_2O_2$, and (2) the NADH formed in the second oxidative step cannot be reoxidized in the peroxisome or glyoxysome, so reducing equivalents are exported to the cytosol, eventually entering mitochondria. The acetyl-CoA produced by peroxisomes and glyoxysomes is also exported; the acetate from glyoxysomes (organelles found only in germinating seeds) serves as a biosynthetic precursor (see Fig. 17–14). Acetyl-CoA produced in mitochondria is further oxidized in the citric acid cycle.

oxidize very-long-chain fatty acids, apparently for lack of a functional transporter for these fatty acids in the peroxisomal membrane. Both defects lead to accumulation in the blood of very-long-chain fatty acids, especially 26:0. XALD affects young boys before the age of 10 years, causing loss of vision, behavioral disturbances, and death within a few years. ∎

In mammals, high concentrations of fats in the diet result in increased synthesis of the enzymes of peroxisomal β oxidation in the liver. Liver peroxisomes do not contain the enzymes of the citric acid cycle and cannot catalyze the oxidation of acetyl-CoA to $CO_2$. Instead, long-chain or branched fatty acids are catabolized to shorter-chain products, such as hexanoyl-CoA, which are exported to mitochondria and completely oxidized.

## Plant Peroxisomes and Glyoxysomes Use Acetyl-CoA from β Oxidation as a Biosynthetic Precursor

In plants, fatty acid oxidation does not occur primarily in mitochondria (as noted earlier) but in the peroxisomes of leaf tissue and in the glyoxysomes of germinating seeds. Plant peroxisomes and glyoxysomes are similar in structure and function; glyoxysomes, which occur only in germinating seeds, may be considered specialized peroxisomes. The biological role of β oxidation in these organelles is to use stored lipids primarily to provide biosynthetic precursors, not energy.

During seed germination, stored triacylglycerols are converted into glucose, sucrose, and a wide variety of essential metabolites (Fig. 17–14). Fatty acids released from the triacylglycerols are first activated to their coenzyme A derivatives and oxidized in glyoxysomes by the same four-step process that takes place in peroxisomes (Fig. 17–13). The acetyl-CoA produced is converted via the glyoxylate cycle (see Fig. 16–20) to four-carbon precursors for gluconeogenesis (see Fig. 14–18). Glyoxysomes, like peroxisomes, contain high concentrations of catalase, which converts the $H_2O_2$ produced by β oxidation to $H_2O$ and $O_2$.

## The β-Oxidation Enzymes of Different Organelles Have Diverged during Evolution

Although the β-oxidation reactions in mitochondria are essentially the same as those in peroxisomes and glyoxysomes, the enzymes (isozymes) differ significantly between the two types of organelles. The differences apparently reflect an evolutionary divergence that occurred very early, with the separation of gram-positive and gram-negative bacteria (see Fig. 1–6).

In mitochondria, the four β-oxidation enzymes that act on short-chain fatty acyl–CoAs are separate, soluble proteins (as noted earlier), similar in structure to the analogous enzymes of gram-positive bacteria (Fig. 17–15a). The gram-negative bacteria have four activities in three soluble subunits (Fig. 17–15b), and the eukaryotic enzyme system that acts on long-chain fatty acids—the trifunctional protein, TFP—has three enzyme activities in two subunits that are membrane-associated (Fig. 17–15c). The β-oxidation enzymes of plant peroxisomes and glyoxysomes, however, form a complex of proteins, one of which contains four enzymatic activities in a single polypeptide chain (Fig. 17–15d). The first enzyme, acyl-CoA oxidase, is a single polypeptide chain; the **multifunctional protein (MFP)** contains the second and third enzyme activities (enoyl-CoA hydratase and hydroxyacyl-CoA dehydrogenase) as well as two auxiliary activities needed for the oxidation of unsaturated

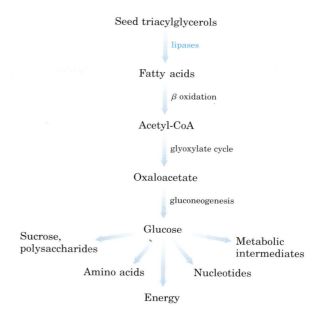

fatty acids (D-3-hydroxyacyl-CoA epimerase and $\Delta^3,\Delta^2$-enoyl-CoA isomerase); the fourth enzyme, thiolase, is a separate, soluble polypeptide.

**FIGURE 17–14 Triacylglycerols as glucose source in seeds.** β Oxidation is one stage in a pathway that converts stored triacylglycerols to glucose in germinating seeds. For more detail, see Figure 16–22.

It is interesting that the enzymes that catalyze essentially the reversal of β oxidation in the synthesis of fatty acids are also organized differently in prokaryotes and eukaryotes; in bacteria, the seven enzymes needed for fatty acid synthesis are separate polypeptides, but in mammals, all seven activities are part of a single, huge polypeptide chain (see Fig. 21–7). One advantage to the cell in having several enzymes of the same pathway encoded in a single polypeptide chain is that this solves the problem of regulating the synthesis of enzymes that must interact functionally; regulation of the expression of *one* gene ensures production of the same number of active sites for all enzymes in the path. When each enzyme activity is on a separate polypeptide, some mechanism is required to coordinate the synthesis of all the gene products. The *disadvantage* of having several activities on the same polypeptide is that the longer the polypeptide chain, the greater is the probability of a mistake in its synthesis: a single incorrect amino acid in the chain may make all the enzyme activities in that chain useless. Comparison of the gene structures for these proteins in many species may shed light on the reasons for the selection of one or the other strategy in evolution.

## The ω Oxidation of Fatty Acids Occurs in the Endoplasmic Reticulum

Although mitochondrial β oxidation, in which enzymes act at the carboxyl end of a fatty acid, is by far the most

**(a)** Gram-positive bacteria and mitochondrial short-chain-specific system

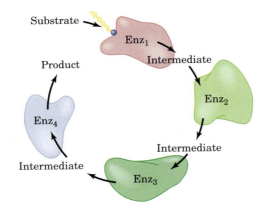

**(b)** Gram-negative bacteria

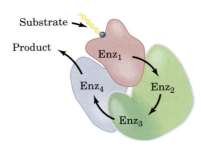

**(c)** Mitochondrial very-long-chain-specific system

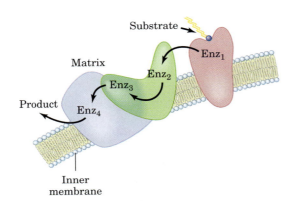

**(d)** Peroxisomal and glyoxysomal system of plants

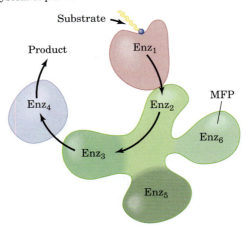

**FIGURE 17–15** **The enzymes of $\beta$ oxidation.** Shown here are the different subunit structures of the enzymes of $\beta$ oxidation in gram-positive and gram-negative bacteria, mitochondria, and plant peroxisomes and glyoxysomes. $Enz_1$ is acyl-CoA dehydrogenase; $Enz_2$, enoyl-CoA hydratase; $Enz_3$, L-$\beta$-hydroxyacyl-CoA dehydrogenase; $Enz_4$, thiolase; $Enz_5$, D-3-hydroxyacyl-CoA epimerase, and $Enz_6$, $\Delta^3,\Delta^2$-enoyl-CoA isomerase. **(a)** The four enzymes of $\beta$ oxidation in gram-positive bacteria are separate, soluble entities, as are those of the short-chain-specific system of mitochondria. **(b)** In gram-negative bacteria, the four enzyme activities reside in three polypeptides; enzymes 2 and 3 are parts of a single polypeptide chain. **(c)** The very-long-chain-specific system of mitochondria is also composed of three polypeptides, one of which includes the activities of enzymes 2 and 3; in this case, the system is bound to the inner mitochondrial membrane. **(d)** In the peroxisomal and glyoxysomal $\beta$-oxidation systems of plants, enzymes 1 and 4 are separate polypeptides, but enzymes 2 and 3, as well as two auxiliary enzymes, are part of a single polypeptide chain, the multifunctional protein, MFP.

important catabolic fate for fatty acids in animal cells, there is another pathway in some species, including vertebrates, that involves oxidation of the $\omega$ (omega) carbon—the carbon most distant from the carboxyl group. The enzymes unique to **$\omega$ oxidation** are located (in vertebrates) in the endoplasmic reticulum of liver and kidney, and the preferred substrates are fatty acids of 10 or 12 carbon atoms. In mammals $\omega$ oxidation is normally a minor pathway for fatty acid degradation, but when $\beta$ oxidation is defective (because of mutation or a carnitine deficiency, for example) it becomes more important.

The first step introduces a hydroxyl group onto the $\omega$ carbon (Fig. 17–16). The oxygen for this group comes from molecular oxygen ($O_2$) in a complex reaction that involves cytochrome P450 and the electron donor NADPH. Reactions of this type are catalyzed by **mixed-function oxidases,** described in Box 21–1. Two more enzymes now act on the $\omega$ carbon: **alcohol dehydrogenase** oxidizes the hydroxyl group to an aldehyde, and **aldehyde dehydrogenase** oxidizes the aldehyde group to a carboxylic acid, producing a fatty acid with a carboxyl group at each end. At this point, either end can be attached to coenzyme A, and the molecule can en-

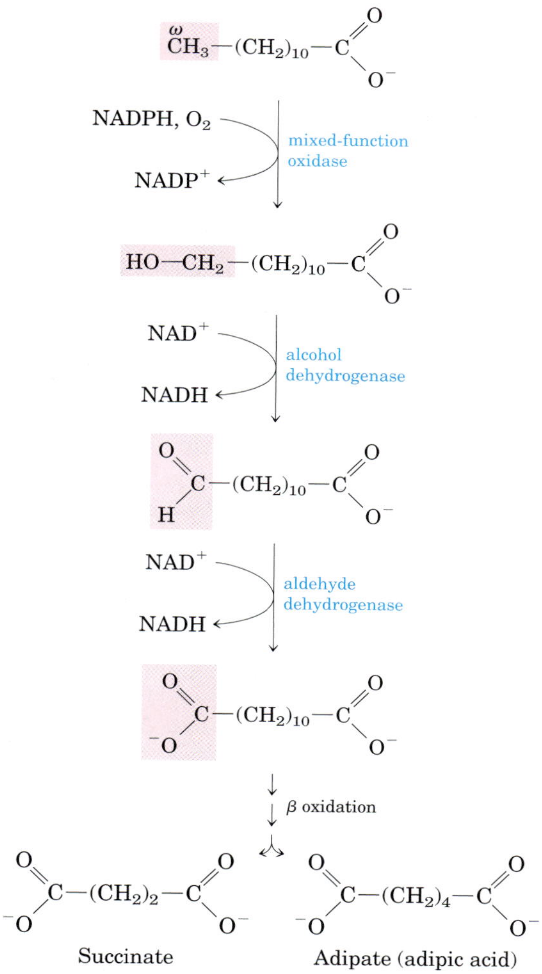

**FIGURE 17–16 The ω oxidation of fatty acids in the endoplasmic reticulum.** This alternative to β oxidation begins with oxidation of the carbon most distant from the α carbon—the ω (omega) carbon. The substrate is usually a medium-chain fatty acid; shown here is lauric acid (laurate). This pathway is generally not the major route for oxidative catabolism of fatty acids.

ter the mitochondrion and undergo β oxidation by the normal route. In each pass through the β-oxidation pathway, the "double-ended" fatty acid yields dicarboxylic acids such as succinic acid, which can enter the citric acid cycle, and adipic acid (Fig. 17–16).

## Phytanic Acid Undergoes α Oxidation in Peroxisomes

The presence of a methyl group on the β carbon of a fatty acid makes β oxidation impossible, and these branched fatty acids are catabolized in peroxisomes of animal cells by **α oxidation.** In the oxidation of phytanic acid, for example (Fig. 17–17), phytanoyl-CoA is hydroxylated on its α carbon, in a reaction that involves molecular oxygen; decarboxylated to form an aldehyde one carbon shorter; and then oxidized to the

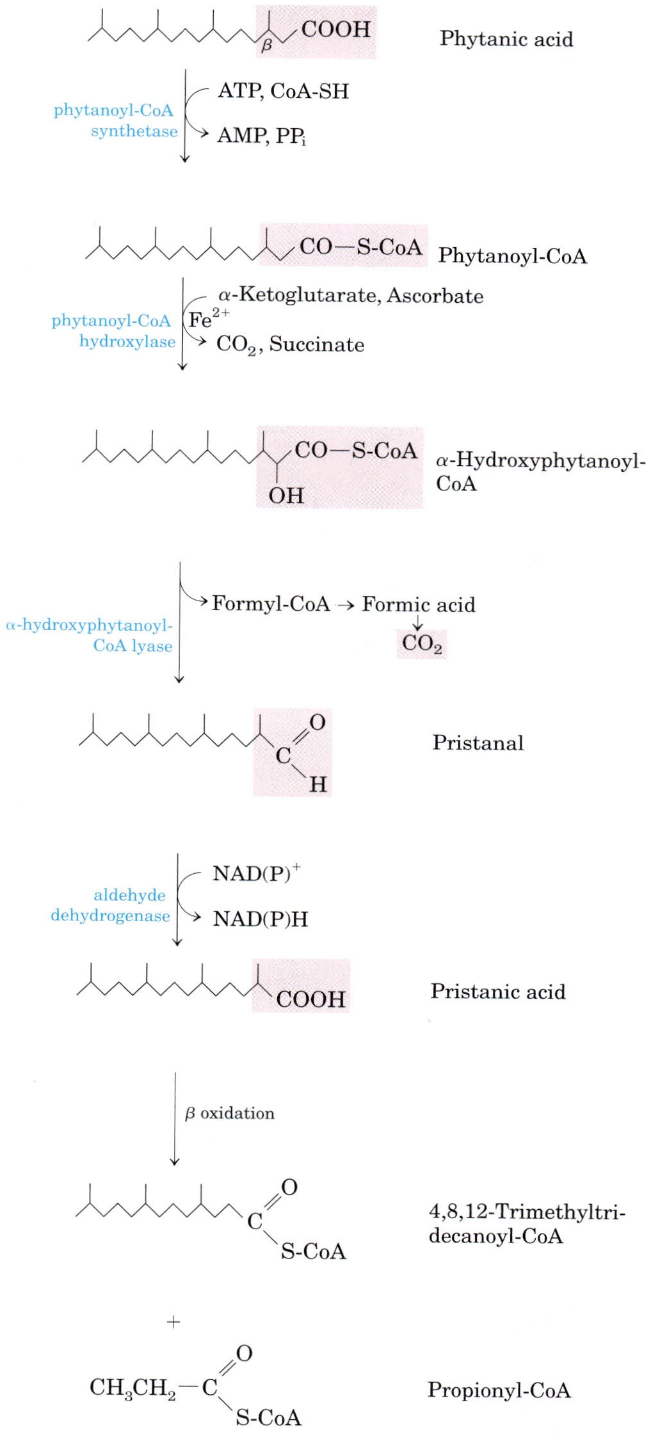

**FIGURE 17–17 The α oxidation of a branched-chain fatty acid (phytanic acid) in peroxisomes.** Phytanic acid has a methyl-substituted β carbon and therefore cannot undergo β oxidation. The combined action of the enzymes shown here removes the carboxyl carbon of phytanic acid, to produce pristanic acid, in which the β carbon is unsubstituted, allowing β oxidation. Notice that β oxidation of pristanic acid releases propionyl-CoA, not acetyl-CoA. This is further catabolized as in Figure 17–11. (The details of the reaction that produces pristanal remain controversial.)

corresponding carboxylic acid, which now has no substituent on the $\beta$ carbon and can be oxidized further by $\beta$ oxidation. **Refsum's disease,** resulting from a genetic defect in phytanoyl-CoA hydroxylase, leads to very high blood levels of phytanic acid and severe neurological problems including blindness and deafness. ■

### SUMMARY 17.2 Oxidation of Fatty Acids

■ In the first stage of $\beta$ oxidation, four reactions remove each acetyl-CoA unit from the carboxyl end of a saturated fatty acyl–CoA: (1) dehydrogenation of the $\alpha$ and $\beta$ carbons (C-2 and C-3) by FAD-linked acyl-CoA dehydrogenases, (2) hydration of the resulting $trans$-$\Delta^2$ double bond by enoyl-CoA hydratase, (3) dehydrogenation of the resulting L-$\beta$-hydroxyacyl-CoA by NAD-linked $\beta$-hydroxyacyl-CoA dehydrogenase, and (4) CoA-requiring cleavage of the resulting $\beta$-ketoacyl-CoA by thiolase, to form acetyl-CoA and a fatty acyl–CoA shortened by two carbons. The shortened fatty acyl–CoA then reenters the sequence.

■ In the second stage of fatty acid oxidation, the acetyl-CoA is oxidized to $CO_2$ in the citric acid cycle. A large fraction of the theoretical yield of free energy from fatty acid oxidation is recovered as ATP by oxidative phosphorylation, the final stage of the oxidative pathway.

■ Malonyl-CoA, an early intermediate of fatty acid synthesis, inhibits carnitine acyltransferase I, preventing fatty acid entry into mitochondria. This blocks fatty acid breakdown while synthesis is occurring.

■ Genetic defects in the medium-chain acyl-CoA dehydrogenase result in serious human disease, as do mutations in other components of the $\beta$-oxidation system.

■ Oxidation of unsaturated fatty acids requires two additional enzymes: enoyl-CoA isomerase and 2,4-dienoyl-CoA reductase. Odd-number fatty acids are oxidized by the $\beta$-oxidation pathway to yield acetyl-CoA and a molecule of propionyl-CoA. This is carboxylated to methylmalonyl-CoA, which is isomerized to succinyl-CoA in a reaction catalyzed by methylmalonyl-CoA mutase, an enzyme requiring coenzyme $B_{12}$.

■ Peroxisomes of plants and animals, and glyoxysomes of plants, carry out $\beta$ oxidation in four steps similar to those of the mitochondrial pathway in animals. The first oxidation step, however, transfers electrons directly to $O_2$, generating $H_2O_2$. Peroxisomes of animal tissues

specialize in the oxidation of very-long-chain fatty acids and branched fatty acids. In glyoxysomes, in germinating seeds, $\beta$ oxidation is one step in the conversion of stored lipids into a variety of intermediates and products.

■ The reactions of $\omega$ oxidation, occurring in the endoplasmic reticulum, produce dicarboxylic fatty acyl intermediates, which can undergo $\beta$ oxidation at either end to yield short dicarboxylic acids such as succinate.

## 17.3 Ketone Bodies

In humans and most other mammals, acetyl-CoA formed in the liver during oxidation of fatty acids can either enter the citric acid cycle (stage 2 of Fig. 17–7) or undergo conversion to the "ketone bodies," **acetone, acetoacetate,** and **D-$\beta$-hydroxybutyrate,** for export to other tissues. (The term "bodies" is a historical artifact; the term is occasionally applied to insoluble particles, but these compounds are quite soluble in blood and urine.) Acetone, produced in smaller quantities than the other ketone bodies, is exhaled. Acetoacetate and D-$\beta$-hydroxybutyrate are transported by the blood to tissues other than the liver (extrahepatic tissues), where they are converted to acetyl-CoA and oxidized in the citric acid cycle, providing much of the energy required by tissues such as skeletal and heart muscle and the renal cortex. The brain, which preferentially uses glucose as fuel, can adapt to the use of acetoacetate or D-$\beta$-hydroxybutyrate under starvation conditions, when glucose is unavailable. The production and export of ketone bodies from the liver to extrahepatic tissues allow continued oxidation of fatty acids in the liver when acetyl-CoA is not being oxidized in the citric acid cycle.

Acetone

Acetoacetate

D-$\beta$-Hydroxybutyrate

### Ketone Bodies, Formed in the Liver, Are Exported to Other Organs as Fuel

The first step in the formation of acetoacetate, occurring in the liver (Fig. 17–18), is the enzymatic condensation of two molecules of acetyl-CoA, catalyzed by thiolase;

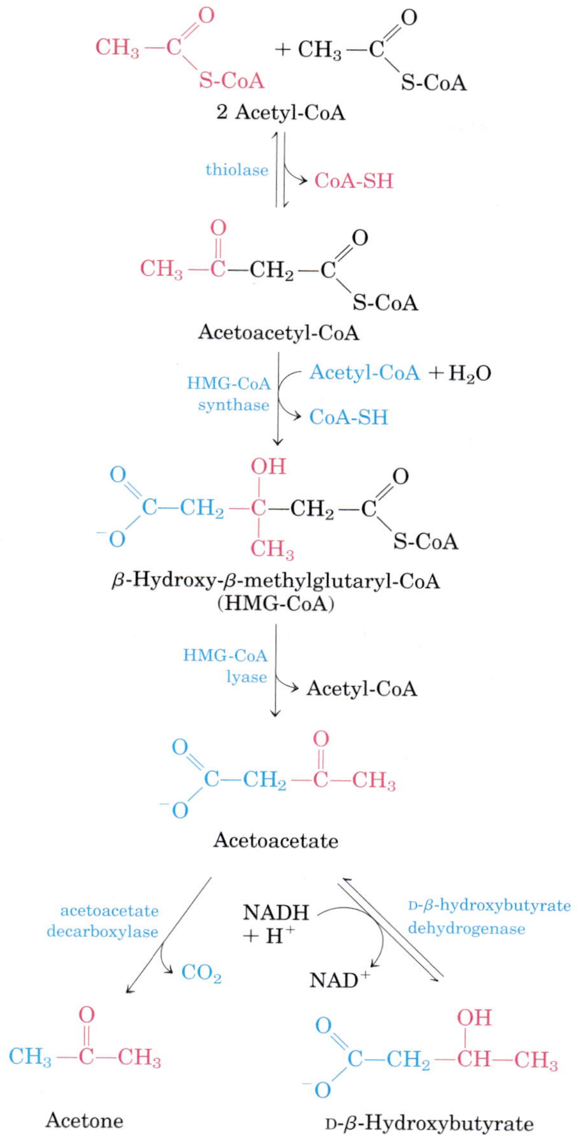

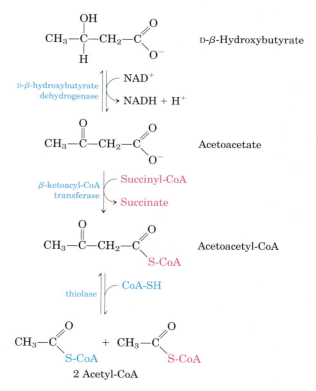

**FIGURE 17–18 Formation of ketone bodies from acetyl-CoA.** Healthy, well-nourished individuals produce ketone bodies at a relatively low rate. When acetyl-CoA accumulates (as in starvation or untreated diabetes, for example), thiolase catalyzes the condensation of two acetyl-CoA molecules to acetoacetyl-CoA, the parent compound of the three ketone bodies. The reactions of ketone body formation occur in the matrix of liver mitochondria. The six-carbon compound β-hydroxy-β-methylglutaryl-CoA (HMG-CoA) is also an intermediate of sterol biosynthesis, but the enzyme that forms HMG-CoA in that pathway is cytosolic. HMG-CoA lyase is present only in the mitochondrial matrix.

this is simply the reversal of the last step of β oxidation. The acetoacetyl-CoA then condenses with acetyl-CoA to form **β-hydroxy-β-methylglutaryl-CoA (HMG-CoA),** which is cleaved to free acetoacetate and acetyl-CoA. The acetoacetate is reversibly reduced by D-β-hydroxybutyrate dehydrogenase, a mitochondrial enzyme, to D-β-hydroxybutyrate. This enzyme is specific for the

D stereoisomer; it does not act on L-β-hydroxyacyl-CoAs and is not to be confused with L-β-hydroxyacyl-CoA dehydrogenase of the β-oxidation pathway.

In healthy people, acetone is formed in very small amounts from acetoacetate, which is easily decarboxylated, either spontaneously or by the action of **acetoacetate decarboxylase** (Fig. 17–18). Because individuals with untreated diabetes produce large quantities of acetoacetate, their blood contains significant amounts of acetone, which is toxic. Acetone is volatile and imparts a characteristic odor to the breath, which is sometimes useful in diagnosing diabetes. ■

In extrahepatic tissues, D-β-hydroxybutyrate is oxidized to acetoacetate by D-β-hydroxybutyrate dehydrogenase (Fig. 17–19). The acetoacetate is activated to its coenzyme A ester by transfer of CoA from succinyl-CoA, an intermediate of the citric acid cycle (see Fig. 16–7), in a reaction catalyzed by **β-ketoacyl-CoA transferase.** The acetoacetyl-CoA is then cleaved by thiolase to yield two acetyl-CoAs, which enter the citric acid cycle. Thus the ketone bodies are used as fuels.

The production and export of ketone bodies by the liver allow continued oxidation of fatty acids with only minimal oxidation of acetyl-CoA. When intermediates of the citric acid cycle are being siphoned off for glucose

**FIGURE 17–19 D-β-Hydroxybutyrate as a fuel.** D-β-Hydroxybutyrate, synthesized in the liver, passes into the blood and thus to other tissues, where it is converted in three steps to acetyl-CoA. It is first oxidized to acetoacetate, which is activated with coenzyme A donated from succinyl-CoA, then split by thiolase. The acetyl-CoA thus formed is used for energy production.

synthesis by gluconeogenesis, for example, oxidation of cycle intermediates slows—and so does acetyl-CoA oxidation. Moreover, the liver contains only a limited amount of coenzyme A, and when most of it is tied up in acetyl-CoA, β oxidation slows for want of the free coenzyme. The production and export of ketone bodies free coenzyme A, allowing continued fatty acid oxidation.

## Ketone Bodies Are Overproduced in Diabetes and during Starvation

Starvation and untreated diabetes mellitus lead to overproduction of ketone bodies, with several associated medical problems. During starvation, gluconeogenesis depletes citric acid cycle intermediates, diverting acetyl-CoA to ketone body production (Fig. 17–20). In untreated diabetes, when the insulin level is insufficient, extrahepatic tissues cannot take up glucose efficiently from the blood, either for fuel or for conversion to fat. Under these conditions, levels of malonyl-CoA (the starting material for fatty acid synthesis) fall, inhibition of carnitine acyltransferase I is relieved, and fatty acids enter mitochondria to be degraded to acetyl-CoA—which cannot pass through the citric acid cycle because cycle intermediates have been drawn off for use as substrates in gluconeogenesis. The resulting accumulation of acetyl-CoA accelerates the formation of ketone bodies beyond the capacity of extrahepatic tissues to oxidize them. The increased blood levels of acetoacetate and D-β-hydroxybutyrate lower the blood pH, causing the condition known as **acidosis.** Extreme acidosis can lead to coma and in some cases death. Ketone bodies in the blood and urine of untreated diabetics can reach extraordinary levels—a blood concentration of 90 mg/100 mL (compared with a normal level of <3 mg/100 mL) and urinary excretion of 5,000 mg/24 hr (compared with a normal rate of ≤125 mg/24 hr). This condition is called **ketosis.**

Individuals on very low-calorie diets, using the fats stored in adipose tissue as their major energy source, also have increased levels of ketone bodies in their blood and urine. These levels must be monitored to avoid the dangers of acidosis and ketosis (ketoacidosis). ■

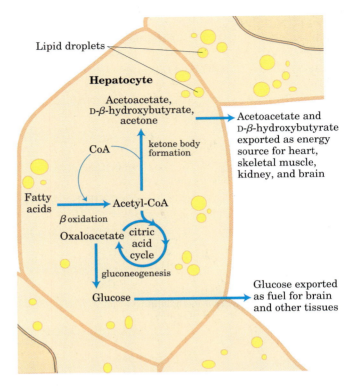

**FIGURE 17–20 Ketone body formation and export from the liver.** Conditions that promote gluconeogenesis (untreated diabetes, severely reduced food intake) slow the citric acid cycle (by drawing off oxaloacetate) and enhance the conversion of acetyl-CoA to acetoacetate. The released coenzyme A allows continued β oxidation of fatty acids.

## SUMMARY 17.3 Ketone Bodies

- The ketone bodies—acetone, acetoacetate, and D-β-hydroxybutyrate—are formed in the liver. The latter two compounds serve as fuel molecules in extrahepatic tissues, through oxidation to acetyl-CoA and entry into the citric acid cycle.

- Overproduction of ketone bodies in uncontrolled diabetes or severely reduced calorie intake can lead to acidosis or ketosis.

## Key Terms

*Terms in bold are defined in the glossary.*

| | | | |
|---|---|---|---|
| **β oxidation** 631 | **carnitine shuttle** 634 | methylmalonyl-CoA mutase 642 | multifunctional protein (MFP) 647 |
| **chylomicron** 632 | carnitine acyltransferase I 636 | **coenzyme B₁₂** 642 | **ω oxidation** 648 |
| **apolipoprotein** 632 | acyl-carnitine/carnitine transporter 636 | malonyl-CoA 642 | **mixed-function oxidases** 648 |
| **lipoprotein** 632 | carnitine acyltransferase II 636 | medium-chain acyl-CoA dehydrogenase (MCAD) 643 | **α oxidation** 649 |
| perilipins 634 | trifunctional protein (TFP) 638 | pernicious anemia 645 | **acidosis** 652 |
| hormone-sensitive lipase 634 | | intrinsic factor 645 | **ketosis** 652 |
| free fatty acids 634 | | | |
| serum albumin 634 | | | |

# Further Reading

## General

**Boyer, P.D.** (1983) *The Enzymes,* 3rd edn, Vol. 16: *Lipid Enzymology,* Academic Press, Inc., San Diego, CA.

**Ferry, G.** (1998) *Dorothy Hodgkin: A Life,* Cold Spring Harbor Laboratory Press, Cold Spring Harbor, NY.

Fascinating biography of an amazing woman.

**Gurr, M.I., Harwood, J.L., & Frayn, K.N.** (2002) *Lipid Biochemistry: An Introduction,* 5th edn, Blackwell Science, Oxford, UK.

**Langin, D., Holm, C., & Lafontan, M.** (1996) Adipocyte hormone-sensitive lipase: a major regulator of lipid metabolism. *Proc. Nutr. Soc.* **55,** 93–109.

**Ramsay, T.G.** (1996) Fat cells. *Endocrinol. Metab. Clin. N. Am.* **25,** 847–870.

A review of all aspects of fat storage and mobilization in adipocytes.

**Scheffler, I.E.** (1999) *Mitochondria,* Wiley-Liss, New York.

An excellent book on mitochondrial structure and function.

**Wang, C.S., Hartsuck, J., & McConathy, W.J.** (1992) Structure and functional properties of lipoprotein lipase. *Biochim. Biophys. Acta* **1123,** 1–17.

Advanced-level discussion of the enzyme that releases fatty acids from lipoproteins in the capillaries of muscle and adipose tissue.

## Mitochondrial $\beta$ Oxidation

**Bannerjee, R.** (1997) The yin-yang of cobalamin biochemistry. *Chem. Biol.* **4,** 175–186.

A review of the biochemistry of coenzyme $B_{12}$ reactions, including the methylmalonyl-CoA mutase reaction.

**Eaton, S., Bartlett, K., & Pourfarzam, M.** (1996) Mammalian mitochondrial $\beta$-oxidation. *Biochem. J.* **320,** 345–357.

A review of the enzymology of $\beta$ oxidation, inherited defects in this pathway, and regulation of the process in mitochondria.

**Eaton, S., Bursby, T., Middleton, B., Pourfarzam, M., Mills, K., Johnson, A.W., & Bartlett, K.** (2000) The mitochondrial trifunctional protein: centre of a $\beta$-oxidation metabolon? *Biochem. Soc. Trans.* **28,** 177–182.

Short, intermediate-level review.

**Harwood, J.L.** (1988) Fatty acid metabolism. *Annu. Rev. Plant Physiol. Plant Mol. Biol.* **39,** 101–138.

**Jeukendrup, A.E., Saris, W.H., & Wagenmakers, A.J.** (1998) Fat metabolism during exercise: a review. Part III: effects of nutritional interventions. *Int. J. Sports Med.* **19,** 371–379.

This paper is one of a series that reviews the factors that influence fat mobilization and utilization during exercise.

**Kerner, J. & Hoppel, C.** (1998) Genetic disorders of carnitine metabolism and their nutritional management. *Annu. Rev. Nutr.* **18,** 179–206.

**Kerner, J. & Hoppel, C.** (2000) Fatty acid import into mitochondria. *Biochim. Biophys. Acta* **1486,** 1–17.

**Kunau, W.H., Dommes, V., & Schulz, H.** (1995) $\beta$-Oxidation of fatty acids in mitochondria, peroxisomes, and bacteria: a century of continued progress. *Prog. Lipid Res.* **34,** 267–342.

A good historical account and a useful comparison of $\beta$ oxidation in different systems.

**Rinaldo, P., Matern, D., & Bennett, M.J.** (2002) Fatty acid oxidation disorders. *Annu. Rev. Physiol.* **64,** 477–502.

Advanced review of metabolic defects in fat oxidation, including MCAD mutations.

**Sherratt, H.S.** (1994) Introduction: the regulation of fatty acid oxidation in cells. *Biochem. Soc. Trans.* **22,** 421–422.

Introduction to reviews (in this journal issue) of various aspects of fatty acid oxidation and its regulation.

**Thorpe, C. & Kim, J.J.** (1995) Structure and mechanism of action of the acyl-CoA dehydrogenases. *FASEB J.* **9,** 718–725.

Short, clear description of the three-dimensional structure and catalytic mechanism of these enzymes.

## Peroxisomal $\beta$ Oxidation

**Graham, I.A. & Eastmond, P.J.** (2002) Pathways of straight and branched chain fatty acid catabolism in higher plants. *Prog. Lipid Res.* **41,** 156–181.

**Hashimoto, T.** (1996) Peroxisomal $\beta$-oxidation: enzymology and molecular biology. *Ann. N. Y. Acad. Sci.* **804,** 86–98.

**Mannaerts, G.P. & van Veldhoven, P.P.** (1996) Functions and organization of peroxisomal $\beta$-oxidation. *Ann. N. Y. Acad. Sci.* **804,** 99–115.

**Wanders, R.J.A., van Grunsven, E.G., & Jansen, G.A.** (2000) Lipid metabolism in peroxisomes: enzymology, functions and dysfunctions of the fatty acid $\alpha$- and $\beta$-oxidation systems in humans. *Biochem. Soc. Trans.* **28,** 141–148.

## Ketone Bodies

**Foster, D.W. & McGarry, J.D.** (1983) The metabolic derangements and treatment of diabetic ketoacidosis. *N. Engl. J. Med.* **309,** 159–169.

**McGarry, J.D. & Foster, D.W.** (1980) Regulation of hepatic fatty acid oxidation and ketone body production. *Annu. Rev. Biochem.* **49,** 395–420.

**Robinson, A.M. & Williamson, D.H.** (1980) Physiological roles of ketone bodies as substrates and signals in mammalian tissues. *Physiol. Rev.* **60,** 143–187.

# Problems

**1. Energy in Triacylglycerols** On a per-carbon basis, where does the largest amount of biologically available energy in triacylglycerols reside: in the fatty acid portions or the glycerol portion? Indicate how knowledge of the chemical structure of triacylglycerols provides the answer.

**2. Fuel Reserves in Adipose Tissue** Triacylglycerols, with their hydrocarbon-like fatty acids, have the highest energy content of the major nutrients.

(a) If 15% of the body mass of a 70.0 kg adult consists of triacylglycerols, what is the total available fuel reserve, in both kilojoules and kilocalories, in the form of triacylglycerols? Recall that 1.00 kcal = 4.18 kJ.

(b) If the basal energy requirement is approximately 8,400 kJ/day (2,000 kcal/day), how long could this person survive if the oxidation of fatty acids stored as triacylglycerols were the only source of energy?

(c) What would be the weight loss in pounds per day under such starvation conditions (1 lb = 0.454 kg)?

**3. Common Reaction Steps in the Fatty Acid Oxidation Cycle and Citric Acid Cycle** Cells often use the same enzyme reaction pattern for analogous metabolic conversions. For example, the steps in the oxidation of pyruvate to acetyl-CoA and of $\alpha$-ketoglutarate to succinyl-CoA, although catalyzed by different enzymes, are very similar. The first stage of fatty acid oxidation follows a reaction sequence closely resembling a sequence in the citric acid cycle. Use equations to show the analogous reaction sequences in the two pathways.

**4. Chemistry of the Acyl-CoA Synthetase Reaction** Fatty acids are converted to their coenzyme A esters in a reversible reaction catalyzed by acyl-CoA synthetase:

$$R\text{---}COO^- + ATP + CoA \rightleftharpoons$$
$$R\text{---}\overset{\overset{\displaystyle O}{\|}}{C}\text{---}CoA + AMP + PP_i$$

(a) The enzyme-bound intermediate in this reaction has been identified as the mixed anhydride of the fatty acid and adenosine monophosphate (AMP), acyl-AMP:

Write two equations corresponding to the two steps of the reaction catalyzed by acyl-CoA synthetase.

(b) The acyl-CoA synthetase reaction is readily reversible, with an equilibrium constant near 1. How can this reaction be made to favor formation of fatty acyl–CoA?

**5. Oxidation of Tritiated Palmitate** Palmitate uniformly labeled with tritium ($^3H$) to a specific activity of $2.48 \times 10^8$ counts per minute (cpm) per micromole of palmitate is added to a mitochondrial preparation that oxidizes it to acetyl-CoA. The acetyl-CoA is isolated and hydrolyzed to acetate. The specific activity of the isolated acetate is $1.00 \times 10^7$ cpm/$\mu$mol. Is this result consistent with the $\beta$-oxidation pathway? Explain. What is the final fate of the removed tritium?

**6. Compartmentation in $\beta$ Oxidation** Free palmitate is activated to its coenzyme A derivative (palmitoyl-CoA) in the cytosol before it can be oxidized in the mitochondrion. If palmitate and [$^{14}C$]coenzyme A are added to a liver homogenate, palmitoyl-CoA isolated from the cytosolic fraction is radioactive, but that isolated from the mitochondrial fraction is not. Explain.

**7. Comparative Biochemistry: Energy-Generating Pathways in Birds** One indication of the relative impor-

tance of various ATP-producing pathways is the $V_{max}$ of certain enzymes of these pathways. The values of $V_{max}$ of several enzymes from the pectoral muscles (chest muscles used for flying) of pigeon and pheasant are listed below.

| Enzyme | $V_{max}$ ($\mu$mol substrate/min/g tissue) | |
| --- | --- | --- |
| | Pigeon | Pheasant |
| Hexokinase | 3.0 | 2.3 |
| Glycogen phosphorylase | 18.0 | 120.0 |
| Phosphofructokinase-1 | 24.0 | 143.0 |
| Citrate synthase | 100.0 | 15.0 |
| Triacylglycerol lipase | 0.07 | 0.01 |

(a) Discuss the relative importance of glycogen metabolism and fat metabolism in generating ATP in the pectoral muscles of these birds.

(b) Compare oxygen consumption in the two birds.

(c) Judging from the data in the table, which bird is the long-distance flyer? Justify your answer.

(d) Why were these particular enzymes selected for comparison? Would the activities of triose phosphate isomerase and malate dehydrogenase be equally good bases for comparison? Explain.

**8. Effect of Carnitine Deficiency** An individual developed a condition characterized by progressive muscular weakness and aching muscle cramps. The symptoms were aggravated by fasting, exercise, and a high-fat diet. The homogenate of a skeletal muscle specimen from the patient oxidized added oleate more slowly than did control homogenates, consisting of muscle specimens from healthy individuals. When carnitine was added to the patient's muscle homogenate, the rate of oleate oxidation equaled that in the control homogenates. The patient was diagnosed as having a carnitine deficiency.

(a) Why did added carnitine increase the rate of oleate oxidation in the patient's muscle homogenate?

(b) Why were the patient's symptoms aggravated by fasting, exercise, and a high-fat diet?

(c) Suggest two possible reasons for the deficiency of muscle carnitine in this individual.

**9. Fatty Acids as a Source of Water** Contrary to legend, camels do not store water in their humps, which actually consist of large fat deposits. How can these fat deposits serve as a source of water? Calculate the amount of water (in liters) that a camel can produce from 1.0 kg of fat. Assume for simplicity that the fat consists entirely of tripalmitoylglycerol.

**10. Petroleum as a Microbial Food Source** Some microorganisms of the genera *Nocardia* and *Pseudomonas* can grow in an environment where hydrocarbons are the only food source. These bacteria oxidize straight-chain aliphatic hydrocarbons, such as octane, to their corresponding carboxylic acids:

$$CH_3(CH_2)_6CH_3 + NAD^+ + O_2 \rightleftharpoons$$
$$CH_3(CH_2)_6COOH + NADH + H^+$$

How could these bacteria be used to clean up oil spills? What would be some of the limiting factors to the efficiency of this process?

## 11. Metabolism of a Straight-Chain Phenylated Fatty Acid

A crystalline metabolite was isolated from the urine of a rabbit that had been fed a straight-chain fatty acid containing a terminal phenyl group:

$$\text{\ceCH_2-(CH_2)_n-COO^-}$$

A 302 mg sample of the metabolite in aqueous solution was completely neutralized by 22.2 mL of 0.100 M NaOH.

(a) What is the probable molecular weight and structure of the metabolite?

(b) Did the straight-chain fatty acid contain an even or an odd number of methylene ($-CH_2-$) groups (i.e., is $n$ even or odd)? Explain.

## 12. Fatty Acid Oxidation in Uncontrolled Diabetes

When the acetyl-CoA produced during $\beta$ oxidation in the liver exceeds the capacity of the citric acid cycle, the excess acetyl-CoA forms ketone bodies—acetone, acetoacetate, and D-$\beta$-hydroxybutyrate. This occurs in severe, uncontrolled diabetes: because the tissues cannot use glucose, they oxidize large amounts of fatty acids instead. Although acetyl-CoA is not toxic, the mitochondrion must divert the acetyl-CoA to ketone bodies. What problem would arise if acetyl-CoA were not converted to ketone bodies? How does the diversion to ketone bodies solve the problem?

## 13. Consequences of a High-Fat Diet with No Carbohydrates

Suppose you had to subsist on a diet of whale blubber and seal blubber, with little or no carbohydrate.

(a) What would be the effect of carbohydrate deprivation on the utilization of fats for energy?

(b) If your diet were totally devoid of carbohydrate, would it be better to consume odd- or even-numbered fatty acids? Explain.

## 14. Metabolic Consequences of Ingesting $\omega$-Fluorooleate

The shrub *Dichapetalum toxicarium,* native to Sierra Leone, produces $\omega$-fluorooleate, which is highly toxic to warm-blooded animals.

$$\overset{\displaystyle H \quad H}{\underset{\omega\text{-Fluorooleate}}{F-CH_2-(CH_2)_7-C=C-(CH_2)_7-COO^-}}$$

This substance has been used as an arrow poison, and powdered fruit from the plant is sometimes used as a rat poison (hence the plant's common name, ratsbane). Why is this substance so toxic? (Hint: Review Chapter 16, Problem 13.)

## 15. Role of FAD as Electron Acceptor

Acyl-CoA dehydrogenase uses enzyme-bound FAD as a prosthetic group to dehydrogenate the $\alpha$ and $\beta$ carbons of fatty acyl–CoA. What is the advantage of using FAD as an electron acceptor rather than NAD$^+$? Explain in terms of the standard reduction potentials for the Enz-FAD/FADH$_2$ ($E'^\circ = -0.219$ V) and NAD$^+$/NADH ($E'^\circ = -0.320$ V) half-reactions.

## 16. $\beta$ Oxidation of Arachidic Acid

How many turns of the fatty acid oxidation cycle are required for complete oxidation of arachidic acid (see Table 10–1) to acetyl-CoA?

## 17. Fate of Labeled Propionate

If [3-$^{14}$C]propionate ($^{14}$C in the methyl group) is added to a liver homogenate, $^{14}$C-labeled oxaloacetate is rapidly produced. Draw a flow chart for the pathway by which propionate is transformed to oxaloacetate, and indicate the location of the $^{14}$C in oxaloacetate.

## 18. Sources of H$_2$O Produced in $\beta$ Oxidation

The complete oxidation of palmitoyl-CoA to carbon dioxide and water is represented by the overall equation

$$\text{Palmitoyl-CoA} + 23O_2 + 108P_i + 108ADP \longrightarrow$$
$$\text{CoA} + 16CO_2 + 108ATP + 23H_2O$$

Water is also produced in the reaction

$$ADP + P_i \longrightarrow ATP + H_2O$$

but is not included as a product in the overall equation. Why?

## 19. Biological Importance of Cobalt

In cattle, deer, sheep, and other ruminant animals, large amounts of propionate are produced in the rumen through the bacterial fermentation of ingested plant matter. Propionate is the principal source of glucose for these animals, via the route propionate $\rightarrow$ oxaloacetate $\rightarrow$ glucose. In some areas of the world, notably Australia, ruminant animals sometimes show symptoms of anemia with concomitant loss of appetite and retarded growth, resulting from an inability to transform propionate to oxaloacetate. This condition is due to a cobalt deficiency caused by very low cobalt levels in the soil and thus in plant matter. Explain.

## 20. Fat Loss during Hibernation

Bears expend about $25 \times 10^6$ J/day during periods of hibernation, which may last as long as seven months. The energy required to sustain life is obtained from fatty acid oxidation. How much weight loss (in kilograms) has occurred after seven months? How might ketosis be minimized during hibernation? (Assume the oxidation of fat yields 38 kJ/g.)

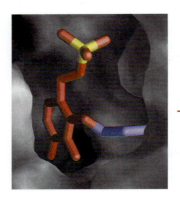

## chapter 18

# AMINO ACID OXIDATION AND THE PRODUCTION OF UREA

I chose the study of the synthesis of urea in the liver because it appeared to be a relatively simple problem.

—*Hans Krebs, article in* Perspectives in Biology and Medicine, *1970*

We now turn our attention to the amino acids, the final class of biomolecules that, through their oxidative degradation, make a significant contribution to the generation of metabolic energy. The fraction of metabolic energy obtained from amino acids, whether they are derived from dietary protein or from tissue protein, varies greatly with the type of organism and with metabolic conditions. Carnivores can obtain (immediately following a meal) up to 90% of their energy requirements from amino acid oxidation, whereas herbivores may fill only a small fraction of their energy needs by this route. Most microorganisms can scavenge amino acids from their environment and use them as fuel when required by metabolic conditions. Plants, however, rarely if ever oxidize amino acids to provide energy; the carbohydrate produced from $CO_2$ and $H_2O$ in photosynthesis is generally their sole energy source. Amino acid concentrations in plant tissues are carefully regulated to just meet the requirements for biosynthesis of proteins, nucleic acids, and other molecules needed to support growth. Amino acid catabolism does occur in plants, but its purpose is to produce metabolites for other biosynthetic pathways.

In animals, amino acids undergo oxidative degradation in three different metabolic circumstances:

1. During the normal synthesis and degradation of cellular proteins (protein turnover; Chapter 27), some amino acids that are released from protein breakdown and are not needed for new protein synthesis undergo oxidative degradation.

2. When a diet is rich in protein and the ingested amino acids exceed the body's needs for protein synthesis, the surplus is catabolized; amino acids cannot be stored.

3. During starvation or in uncontrolled diabetes mellitus, when carbohydrates are either unavailable or not properly utilized, cellular proteins are used as fuel.

Under all these metabolic conditions, amino acids lose their amino groups to form $\alpha$-keto acids, the "carbon skeletons" of amino acids. The $\alpha$-keto acids undergo oxidation to $CO_2$ and $H_2O$ or, often more importantly, provide three- and four-carbon units that can be converted by gluconeogenesis into glucose, the fuel for brain, skeletal muscle, and other tissues.

The pathways of amino acid catabolism are quite similar in most organisms. The focus of this chapter is on the pathways in vertebrates, because these have received the most research attention. As in carbohydrate and fatty acid catabolism, the processes of amino acid degradation converge on the central catabolic pathways, with the carbon skeletons of most amino acids finding their way to the citric acid cycle. In some cases the reaction pathways of amino acid breakdown closely parallel steps in the catabolism of fatty acids (Chapter 17).

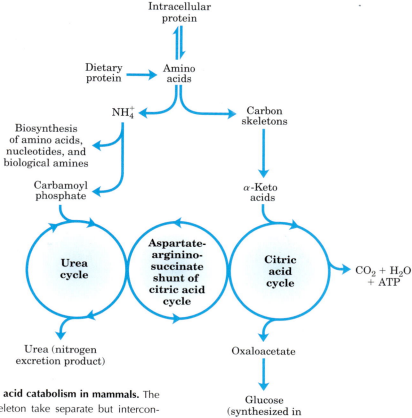

**FIGURE 18–1 Overview of amino acid catabolism in mammals.** The amino groups and the carbon skeleton take separate but interconnected pathways.

One important feature distinguishes amino acid degradation from other catabolic processes described to this point: every amino acid contains an amino group, and the pathways for amino acid degradation therefore include a key step in which the $\alpha$-amino group is separated from the carbon skeleton and shunted into the pathways of amino group metabolism (Fig. 18–1). We deal first with amino group metabolism and nitrogen excretion, then with the fate of the carbon skeletons derived from the amino acids; along the way we see how the pathways are interconnected.

## 18.1 Metabolic Fates of Amino Groups

Nitrogen, $N_2$, is abundant in the atmosphere but is too inert for use in most biochemical processes. Because only a few microorganisms can convert $N_2$ to biologically useful forms such as $NH_3$ (Chapter 22), amino groups are carefully husbanded in biological systems.

Figure 18–2a provides an overview of the catabolic pathways of ammonia and amino groups in vertebrates. Amino acids derived from dietary protein are the source of most amino groups. Most amino acids are metabolized in the liver. Some of the ammonia generated in this process is recycled and used in a variety of biosynthetic pathways; the excess is either excreted directly or converted to urea or uric acid for excretion, depending on the organism (Fig. 18–2b). Excess ammonia generated in other (extrahepatic) tissues travels to the liver (in the form of amino groups, as described below) for conversion to the excretory form.

Glutamate and glutamine play especially critical roles in nitrogen metabolism, acting as a kind of general collection point for amino groups. In the cytosol of hepatocytes, amino groups from most amino acids are transferred to $\alpha$-ketoglutarate to form glutamate, which enters mitochondria and gives up its amino group to form $NH_4^+$. Excess ammonia generated in most other tissues is converted to the amide nitrogen of glutamine, which passes to the liver, then into liver mitochondria. Glutamine or glutamate or both are present in higher concentrations than other amino acids in most tissues.

In skeletal muscle, excess amino groups are generally transferred to pyruvate to form alanine, another important molecule in the transport of amino groups to the liver.

We begin with a discussion of the breakdown of dietary proteins, then give a general description of the metabolic fates of amino groups.

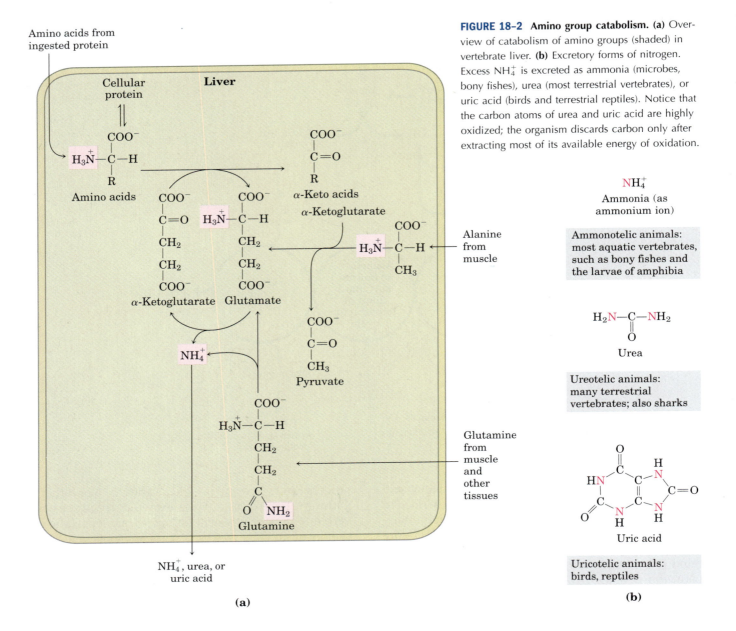

**FIGURE 18–2 Amino group catabolism. (a)** Overview of catabolism of amino groups (shaded) in vertebrate liver. **(b)** Excretory forms of nitrogen. Excess $NH_4^+$ is excreted as ammonia (microbes, bony fishes), urea (most terrestrial vertebrates), or uric acid (birds and terrestrial reptiles). Notice that the carbon atoms of urea and uric acid are highly oxidized; the organism discards carbon only after extracting most of its available energy of oxidation.

Ammonia (as ammonium ion)

Ammonotelic animals: most aquatic vertebrates, such as bony fishes and the larvae of amphibia

Urea

Ureotelic animals: many terrestrial vertebrates; also sharks

Uric acid

Uricotelic animals: birds, reptiles

## Dietary Protein Is Enzymatically Degraded to Amino Acids

In humans, the degradation of ingested proteins to their constituent amino acids occurs in the gastrointestinal tract. Entry of dietary protein into the stomach stimulates the gastric mucosa to secrete the hormone **gastrin,** which in turn stimulates the secretion of hydrochloric acid by the parietal cells and pepsinogen by the chief cells of the gastric glands (Fig. 18–3a). The acidic gastric juice (pH 1.0 to 2.5) is both an antiseptic, killing most bacteria and other foreign cells, and a denaturing agent, unfolding globular proteins and rendering their internal peptide bonds more accessible to enzymatic hydrolysis. **Pepsinogen** ($M_r$ 40,554), an inactive precursor, or zymogen (p. 231), is converted to active pepsin

($M_r$ 34,614) by the enzymatic action of pepsin itself. In the stomach, pepsin hydrolyzes ingested proteins at peptide bonds on the amino-terminal side of the aromatic amino acid residues Phe, Trp, and Tyr (see Table 3–7), cleaving long polypeptide chains into a mixture of smaller peptides.

As the acidic stomach contents pass into the small intestine, the low pH triggers secretion of the hormone **secretin** into the blood. Secretin stimulates the pancreas to secrete bicarbonate into the small intestine to neutralize the gastric HCl, abruptly increasing the pH to about 7. (All pancreatic secretions pass into the small intestine through the pancreatic duct.) The digestion of proteins now continues in the small intestine. Arrival of amino acids in the upper part of the intestine (duodenum) causes release into the blood of the hormone

**FIGURE 18–3 Part of the human digestive (gastrointestinal) tract. (a)** The parietal cells and chief cells of the gastric glands secrete their products in response to the hormone gastrin. Pepsin begins the process of protein degradation in the stomach. **(b)** The cytoplasm of exocrine cells is completely filled with rough endoplasmic reticulum, the site of synthesis of the zymogens of many digestive enzymes. The zymogens are concentrated in membrane-enclosed transport particles called zymogen granules. When an exocrine cell is stimulated, its plasma membrane fuses with the zymogen granule membrane and zymogens are released into the lumen of the collecting duct by exocytosis. The collecting ducts ultimately lead to the pancreatic duct and thence to the small intestine. **(c)** Amino acids are absorbed through the epithelial cell layer (intestinal mucosa) of the villi and enter the capillaries. Recall that the products of lipid hydrolysis in the small intestine enter the lymphatic system after their absorption by the intestinal mucosa (see Fig. 17–1).

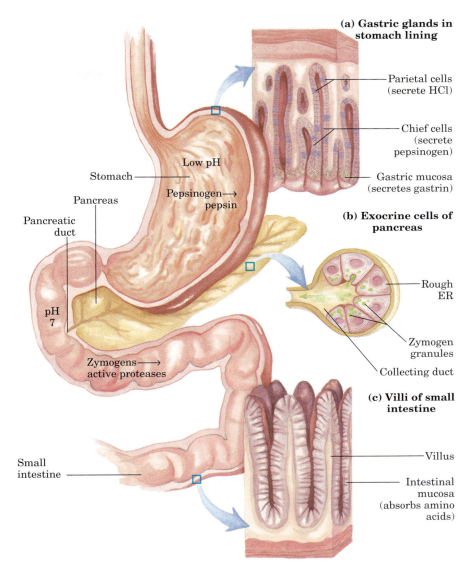

**(a) Gastric glands in stomach lining**

- Parietal cells (secrete HCl)
- Chief cells (secrete pepsinogen)
- Gastric mucosa (secretes gastrin)

Stomach
Low pH
Pepsinogen→pepsin

Pancreas
Pancreatic duct

pH 7

Zymogens→active proteases

**(b) Exocrine cells of pancreas**

- Rough ER
- Zymogen granules
- Collecting duct

**(c) Villi of small intestine**

Small intestine

- Villus
- Intestinal mucosa (absorbs amino acids)

**cholecystokinin,** which stimulates secretion of several pancreatic enzymes with activity optima at pH 7 to 8. **Trypsinogen, chymotrypsinogen,** and **procarboxypeptidases A** and **B,** the zymogens of **trypsin, chymotrypsin,** and **carboxypeptidases A** and **B,** are synthesized and secreted by the exocrine cells of the pancreas (Fig. 18–3b). Trypsinogen is converted to its active form, trypsin, by **enteropeptidase,** a proteolytic enzyme secreted by intestinal cells. Free trypsin then catalyzes the conversion of additional trypsinogen to trypsin (see Fig. 6–33). Trypsin also activates chymotrypsinogen, the procarboxypeptidases, and proelastase.

Why this elaborate mechanism for getting active digestive enzymes into the gastrointestinal tract? Synthesis of the enzymes as inactive precursors protects the exocrine cells from destructive proteolytic attack. The pancreas further protects itself against self-digestion by making a specific inhibitor, a protein called **pancreatic trypsin inhibitor** (p. 231), that effectively prevents premature production of active proteolytic enzymes within the pancreatic cells.

Trypsin and chymotrypsin further hydrolyze the peptides that were produced by pepsin in the stomach. This stage of protein digestion is accomplished very efficiently, because pepsin, trypsin, and chymotrypsin have different amino acid specificities (see Table 3–7). Degradation of the short peptides in the small intestine is then completed by other intestinal peptidases. These include carboxypeptidases A and B (both of which are zinc-containing enzymes), which remove successive carboxyl-terminal residues from peptides, and an **aminopeptidase** that hydrolyzes successive amino-terminal residues from short peptides. The resulting mixture of free amino acids is transported into the epithelial cells lining the small intestine (Fig. 18–3c), through which the amino acids enter the blood capillaries in the villi and travel to the liver. In humans, most globular proteins from animal sources are almost completely hydrolyzed to amino acids in the gastrointestinal tract, but some fibrous proteins, such as keratin, are only partly digested. In addition, the protein content of some plant foods is protected against breakdown by indigestible cellulose husks.

**Acute pancreatitis** is a disease caused by obstruction of the normal pathway by which pancreatic secretions enter the intestine. The zymogens of the proteolytic enzymes are converted to their catalytically active forms prematurely, *inside* the pancreatic cells, and attack the pancreatic tissue itself. This causes excruciating pain and damage to the organ that can prove fatal. ■

## Pyridoxal Phosphate Participates in the Transfer of $\alpha$-Amino Groups to $\alpha$-Ketoglutarate

The first step in the catabolism of most L-amino acids, once they have reached the liver, is removal of the $\alpha$-amino groups, promoted by enzymes called **aminotransferases** or **transaminases**. In these **transamination** reactions, the $\alpha$-amino group is transferred to the $\alpha$-carbon atom of $\alpha$-ketoglutarate, leaving behind the corresponding $\alpha$-keto acid analog of the amino acid (Fig. 18–4). There is no net deamination (loss of amino groups) in these reactions, because the $\alpha$-ketoglutarate becomes aminated as the $\alpha$-amino acid is deaminated. The effect of transamination reactions is to collect the amino groups from many different amino acids in the form of L-glutamate. The glutamate then functions as an amino group donor for biosynthetic pathways or for

excretion pathways that lead to the elimination of nitrogenous waste products.

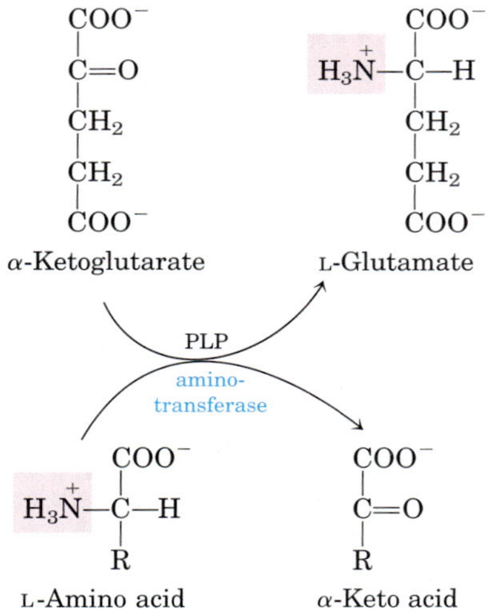

Cells contain different types of aminotransferases. Many are specific for $\alpha$-ketoglutarate as the amino group acceptor but differ in their specificity for the L-amino acid. The enzymes are named for the amino group donor (alanine aminotransferase, aspartate aminotransferase, for example). The reactions catalyzed by aminotransferases are freely reversible, having an equilibrium constant of about 1.0 ($\Delta G'^{\circ} \approx 0$ kJ/mol).

All aminotransferases have the same prosthetic group and the same reaction mechanism. The prosthetic group is **pyridoxal phosphate (PLP),** the coenzyme form of pyridoxine, or vitamin $B_6$. We encountered pyridoxal phosphate in Chapter 15, as a coenzyme in the glycogen phosphorylase reaction, but its role in that reaction is not representative of its usual coenzyme function. Its primary role in cells is in the metabolism of molecules with amino groups.

Pyridoxal phosphate functions as an intermediate carrier of amino groups at the active site of aminotransferases. It undergoes reversible transformations between its aldehyde form, pyridoxal phosphate, which can accept an amino group, and its aminated form, pyridoxamine phosphate, which can donate its amino group to an $\alpha$-keto acid (Fig. 18–5a). Pyridoxal phosphate is generally covalently bound to the enzyme's active site through an aldimine (Schiff base) linkage to the $\varepsilon$-amino group of a Lys residue (Fig. 18–5b, d).

Pyridoxal phosphate participates in a variety of reactions at the $\alpha$, $\beta$, and $\gamma$ carbons (C-2 to C-4) of amino acids. Reactions at the $\alpha$ carbon (Fig. 18–6) include racemizations (interconverting L- and D-amino acids) and decarboxylations, as well as transaminations. Pyridoxal phosphate plays the same chemical role in each of these reactions. A bond to the $\alpha$ carbon of the substrate is broken, removing either a proton or a carboxyl group. The electron pair left behind on the $\alpha$ carbon would form a highly unstable carbanion, but pyridoxal phosphate provides resonance stabilization of this intermediate (Fig. 18–6 inset). The highly conjugated structure of PLP (an electron sink) permits delocalization of the negative charge.

Aminotransferases (Fig. 18–5) are classic examples of enzymes catalyzing bimolecular Ping-Pong reactions (see Fig. 6–13b), in which the first substrate reacts and the product must leave the active site before the second substrate can bind. Thus the incoming amino acid binds to the active site, donates its amino group to pyridoxal phosphate, and departs in the form of an $\alpha$-keto acid. The incoming $\alpha$-keto acid then binds, accepts the amino group from pyridoxamine phosphate, and departs in the form of an amino acid. As described in Box 18–1 on page 664, measurement of the alanine aminotransferase and aspartate aminotransferase levels in blood serum is important in some medical diagnoses.

**FIGURE 18–4 Enzyme-catalyzed transaminations.** In many aminotransferase reactions, $\alpha$-ketoglutarate is the amino group acceptor. All aminotransferases have pyridoxal phosphate (PLP) as cofactor. Although the reaction is shown here in the direction of transfer of the amino group to $\alpha$-ketoglutarate, it is readily reversible.

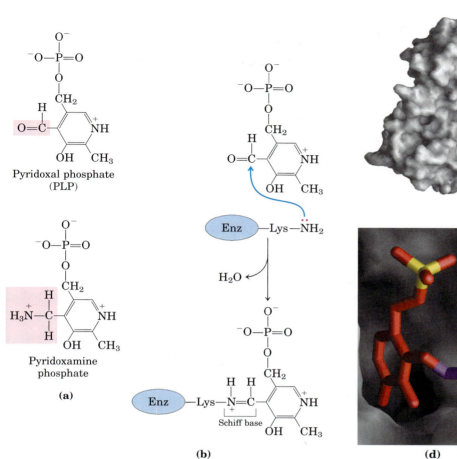

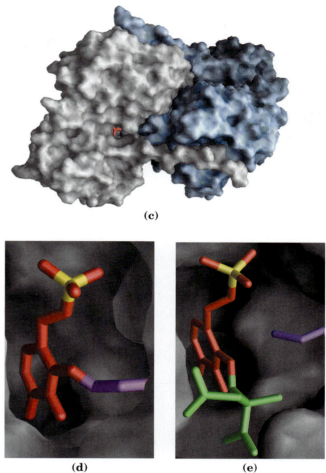

FIGURE 18–5 **Pyridoxal phosphate, the prosthetic group of aminotransferases.** (a) Pyridoxal phosphate (PLP) and its aminated form, pyridoxamine phosphate, are the tightly bound coenzymes of aminotransferases. The functional groups are shaded. (b) Pyridoxal phosphate is bound to the enzyme through noncovalent interactions and a Schiff-base linkage to a Lys residue at the active site. The steps in the formation of a Schiff base from a primary amine and a carbonyl group are detailed in Figure 14–5. (c) PLP (red) bound to one of the two active sites of the dimeric enzyme aspartate aminotransferase, a typical aminotransferase; (d) close-up view of the active site, with PLP (red, with yellow phosphorus) in aldimine linkage with the side chain of Lys$^{258}$ (purple); (e) another close-up view of the active site, with PLP linked to the substrate analog 2-methylaspartate (green) via a Schiff base (PDB ID 1AJS).

## Glutamate Releases Its Amino Group as Ammonia in the Liver

As we have seen, the amino groups from many of the $\alpha$-amino acids are collected in the liver in the form of the amino group of L-glutamate molecules. These amino groups must next be removed from glutamate to prepare them for excretion. In hepatocytes, glutamate is transported from the cytosol into mitochondria, where it undergoes **oxidative deamination** catalyzed by **L-glutamate dehydrogenase** ($M_r$ 330,000). In mammals, this enzyme is present in the mitochondrial matrix. It is the only enzyme that can use either NAD$^+$ or NADP$^+$ as the acceptor of reducing equivalents (Fig. 18–7).

The combined action of an aminotransferase and glutamate dehydrogenase is referred to as **transdeamination.** A few amino acids bypass the transdeamination pathway and undergo direct oxidative deamination. The fate of the NH$_4^+$ produced by any of these deamination processes is discussed in detail in Section 18.2. The $\alpha$-ketoglutarate formed from glutamate deamination can be used in the citric acid cycle and for glucose synthesis.

Glutamate dehydrogenase operates at an important intersection of carbon and nitrogen metabolism. An allosteric enzyme with six identical subunits, its activity is influenced by a complicated array of allosteric modulators. The best-studied of these are the positive modulator ADP and the negative modulator GTP. The metabolic rationale for this regulatory pattern has not been elucidated in detail. Mutations that alter the allosteric binding site for GTP or otherwise cause permanent activation of glutamate dehydrogenase lead to a human genetic disorder called hyperinsulinism-hyperammonemia

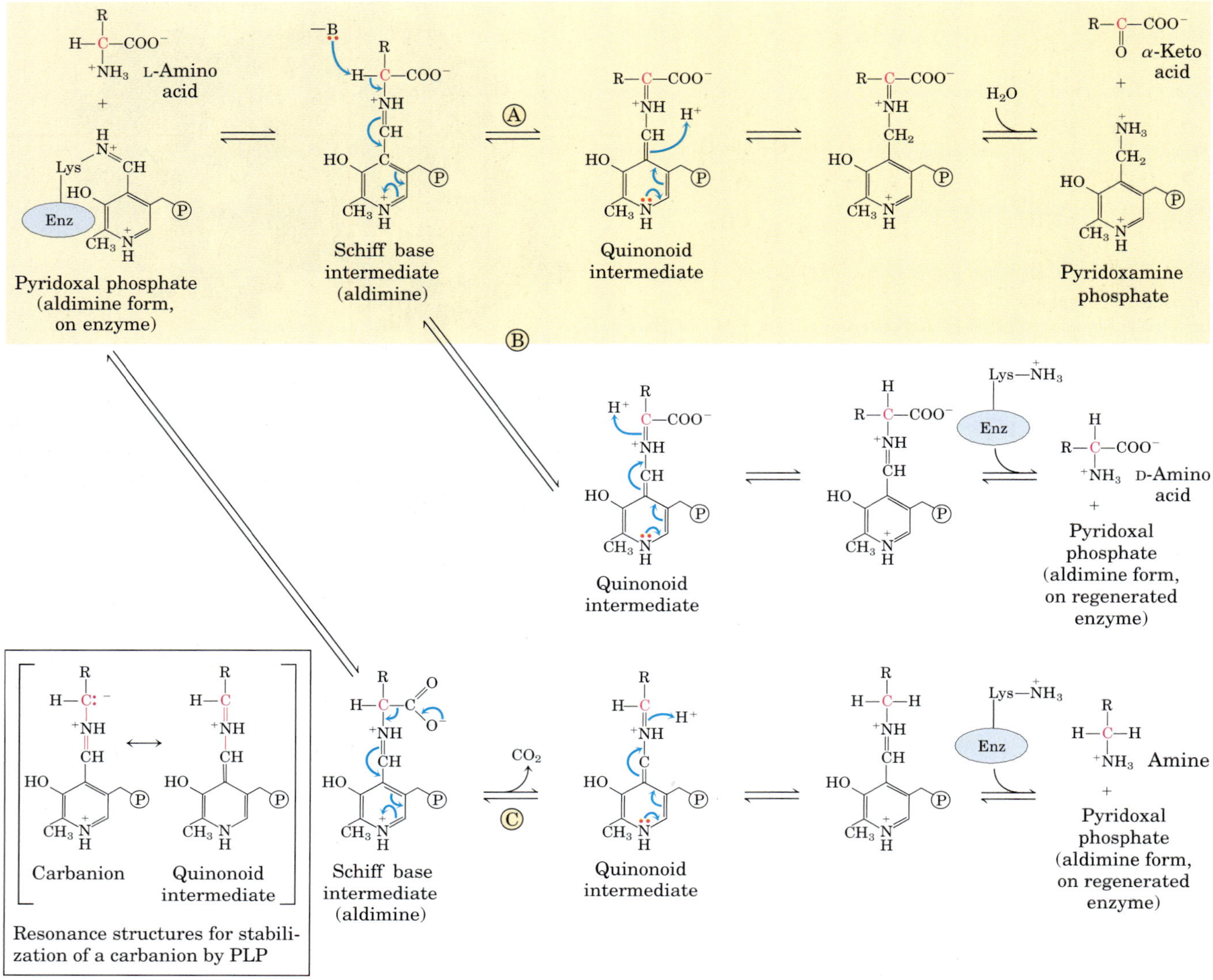

**MECHANISM FIGURE 18–6** **Some amino acid transformations at the α carbon that are facilitated by pyridoxal phosphate.** Pyridoxal phosphate is generally bonded to the enzyme through a Schiff base (see Fig. 18–5b, d). Reactions begin (top left) with formation of a new Schiff base (aldimine) between the α-amino group of the amino acid and PLP, which substitutes for the enzyme-PLP linkage. Three alternative fates for this Schiff base are shown: Ⓐ transamination, Ⓑ racemization, and Ⓒ decarboxylation. The Schiff base formed between PLP and the amino acid is in conjugation with the pyridine ring, an electron sink that permits delocalization of an electron pair to avoid formation of an unstable carbanion on the α carbon (inset). A quinonoid intermediate is involved in all three types of reactions. The transamination route (Ⓐ) is especially important in the pathways described in this chapter. The pathway highlighted here (shown left to right) represents only part of the overall reaction catalyzed by aminotransferases. To complete the process, a second α-keto acid replaces the one that is released, and this is converted to an amino acid in a reversal of the reaction steps (right to left). Pyridoxal phosphate is also involved in certain reactions at the β and γ carbons of some amino acids (not shown). 🔵 **Pyridoxal Phosphate Reaction Mechanisms**

syndrome, characterized by elevated levels of ammonia in the bloodstream and hypoglycemia.

## Glutamine Transports Ammonia in the Bloodstream

Ammonia is quite toxic to animal tissues (we examine some possible reasons for this toxicity later), and the levels present in blood are regulated. In many tissues, including the brain, some processes such as nucleotide

degradation generate free ammonia. In most animals much of the free ammonia is converted to a nontoxic compound before export from the extrahepatic tissues into the blood and transport to the liver or kidneys. For this transport function, glutamate, critical to *intracellular* amino group metabolism, is supplanted by L-glutamine. The free ammonia produced in tissues is combined with glutamate to yield glutamine by the action of **glutamine synthetase.** This reaction requires

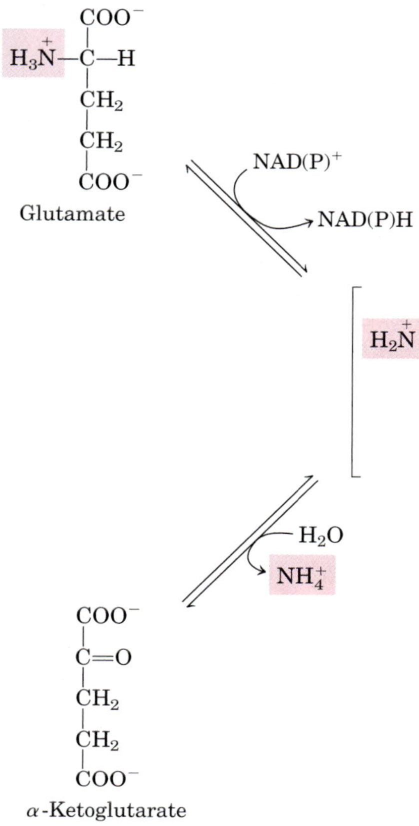

**FIGURE 18–7 Reaction catalyzed by glutamate dehydrogenase.** The glutamate dehydrogenase of mammalian liver has the unusual capacity to use either NAD$^+$ or NADP$^+$ as cofactor. The glutamate dehydrogenases of plants and microorganisms are generally specific for one or the other. The mammalian enzyme is allosterically regulated by GTP and ADP.

ATP and occurs in two steps (Fig. 18–8). First, glutamate and ATP react to form ADP and a γ-glutamyl phosphate intermediate, which then reacts with ammonia to produce glutamine and inorganic phosphate. Glutamine is a nontoxic transport form of ammonia; it is normally present in blood in much higher concentrations than other amino acids. Glutamine also serves as a source of amino groups in a variety of biosynthetic reactions. Glutamine synthetase is found in all organisms, always playing a central metabolic role. In microorganisms, the enzyme serves as an essential portal for the entry of fixed nitrogen into biological systems. (The roles of glutamine and glutamine synthetase in metabolism are further discussed in Chapter 22.)

In most terrestrial animals, glutamine in excess of that required for biosynthesis is transported in the blood to the intestine, liver, and kidneys for processing. In these tissues, the amide nitrogen is released as ammonium ion in the mitochondria, where the enzyme **glutaminase** converts glutamine to glutamate and NH$_4^+$ (Fig. 18–8). The NH$_4^+$ from intestine and kidney is transported in the

blood to the liver. In the liver, the ammonia from all sources is disposed of by urea synthesis. Some of the glutamate produced in the glutaminase reaction may be further processed in the liver by glutamate dehydrogenase, releasing more ammonia and producing carbon skeletons for metabolic fuel. However, most glutamate enters the transamination reactions required for amino acid biosynthesis and other processes (Chapter 22).

In metabolic acidosis (p. 652) there is an increase in glutamine processing by the kidneys. Not all the excess NH$_4^+$ thus produced is released into the bloodstream or converted to urea; some is excreted directly into the urine. In the kidney, the NH$_4^+$ forms salts with metabolic acids, facilitating their removal in the urine. Bicarbonate produced by the decarboxylation of α-ketoglutarate in the citric acid cycle can also serve as a buffer in blood plasma. Taken together, these effects of glutamine metabolism in the kidney tend to counteract acidosis. ■

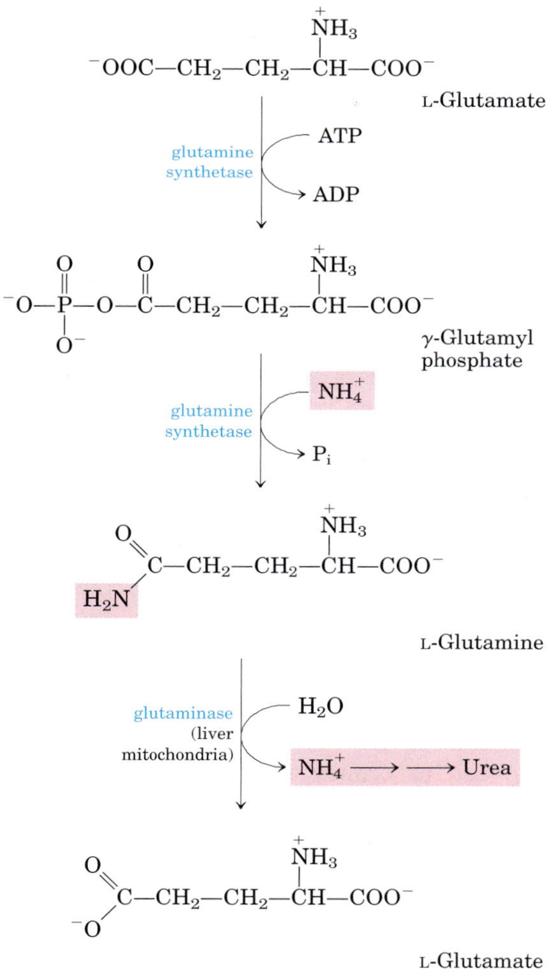

**FIGURE 18–8 Ammonia transport in the form of glutamine.** Excess ammonia in tissues is added to glutamate to form glutamine, a process catalyzed by glutamine synthetase. After transport in the bloodstream, the glutamine enters the liver and NH$_4^+$ is liberated in mitochondria by the enzyme glutaminase.

## Alanine Transports Ammonia from Skeletal Muscles to the Liver

Alanine also plays a special role in transporting amino groups to the liver in a nontoxic form, via a pathway called the **glucose-alanine cycle** (Fig. 18–9). In muscle and certain other tissues that degrade amino acids for fuel, amino groups are collected in the form of glutamate by transamination (Fig. 18–2a). Glutamate can be converted to glutamine for transport to the liver, as described above, or it can transfer its $\alpha$-amino group to pyruvate, a readily available product of muscle glycolysis, by the action of **alanine aminotransferase** (Fig. 18–9). The alanine so formed passes into the blood and travels to the liver. In the cytosol of hepatocytes, alanine aminotransferase transfers the amino group from alanine to $\alpha$-ketoglutarate, forming pyruvate and glutamate. Glutamate can then enter mitochondria, where the glutamate dehydrogenase reaction releases $NH_4^+$ (Fig. 18–7), or can undergo transamination with oxaloacetate to form aspartate, another nitrogen donor in urea synthesis, as we shall see.

The use of alanine to transport ammonia from skeletal muscles to the liver is another example of the intrinsic economy of living organisms. Vigorously contracting skeletal muscles operate anaerobically, producing pyruvate and lactate from glycolysis as well as

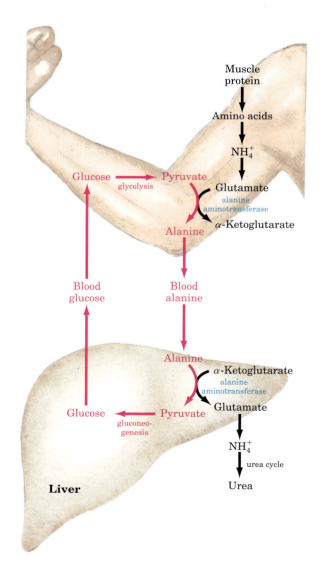

**FIGURE 18–9   Glucose-alanine cycle.** Alanine serves as a carrier of ammonia and of the carbon skeleton of pyruvate from skeletal muscle to liver. The ammonia is excreted and the pyruvate is used to produce glucose, which is returned to the muscle.

ammonia from protein breakdown. These products must find their way to the liver, where pyruvate and lactate are incorporated into glucose, which is returned to the muscles, and ammonia is converted to urea for excretion. The glucose-alanine cycle, in concert with the Cori cycle (see Box 14–1 and Fig. 23–18), accomplishes this transaction. The energetic burden of gluconeogenesis is thus imposed on the liver rather than the muscle, and all available ATP in muscle is devoted to muscle contraction.

### Ammonia Is Toxic to Animals

The catabolic production of ammonia poses a serious biochemical problem, because ammonia is very toxic. The molecular basis for this toxicity is not entirely understood. The terminal stages of ammonia intoxication in humans are characterized by onset of a comatose state accompanied by cerebral edema (an increase in the brain's water content) and increased cranial pressure, so research and speculation on ammonia toxicity have focused on this tissue. Speculation centers on a potential depletion of ATP in brain cells.

Ridding the cytosol of excess ammonia requires reductive amination of $\alpha$-ketoglutarate to glutamate by glutamate dehydrogenase (the reverse of the reaction described earlier; Fig. 18–7) and conversion of glutamate to glutamine by glutamine synthetase. Both enzymes are present at high levels in the brain, although the glutamine synthetase reaction is almost certainly the more important pathway for removal of ammonia. High levels of $NH_4^+$ lead to increased levels of glutamine, which acts as an osmotically active solute (osmolyte) in brain astrocytes, star-shaped cells of the nervous system that provide nutrients, support, and insulation for neurons. This triggers an uptake of water into the astrocytes to maintain osmotic balance, leading to swelling and the symptoms noted above.

Depletion of glutamate in the glutamine synthetase reaction may have additional effects on the brain. Glutamate and its derivative $\gamma$-aminobutyrate (GABA; see Fig. 22–29) are important neurotransmitters; the sensitivity of the brain to ammonia may reflect a depletion of neurotransmitters as well as changes in cellular osmotic balance. ∎

As we close this discussion of amino group metabolism, note that we have described several processes that deposit excess ammonia in the mitochondria of hepatocytes (Fig. 18–2). We now look at the fate of that ammonia.

### SUMMARY 18.1 Metabolic Fates of Amino Groups

- Humans derive a small fraction of their oxidative energy from the catabolism of amino acids. Amino acids are derived from the normal breakdown (recycling) of cellular proteins, degradation of ingested proteins, and breakdown of body proteins in lieu of other fuel sources during starvation or in uncontrolled diabetes mellitus.

- Proteases degrade ingested proteins in the stomach and small intestine. Most proteases are initially synthesized as inactive zymogens.

- An early step in the catabolism of amino acids is the separation of the amino group from the carbon skeleton. In most cases, the amino group is transferred to $\alpha$-ketoglutarate to form glutamate. This transamination reaction requires the coenzyme pyridoxal phosphate.

- Glutamate is transported to liver mitochondria, where glutamate dehydrogenase liberates the amino group as ammonium ion ($NH_4^+$). Ammonia formed in other tissues is transported to the liver as the amide nitrogen of glutamine or, in transport from skeletal muscle, as the amino group of alanine.

- The pyruvate produced by deamination of alanine in the liver is converted to glucose, which is transported back to muscle as part of the glucose-alanine cycle.

## 18.2 Nitrogen Excretion and the Urea Cycle

If not reused for the synthesis of new amino acids or other nitrogenous products, amino groups are channeled into a single excretory end product (Fig. 18–10). Most aquatic species, such as the bony fishes, are **ammonotelic,** excreting amino nitrogen as ammonia. The toxic ammonia is simply diluted in the surrounding water. Terrestrial animals require pathways for nitrogen excretion that minimize toxicity and water loss. Most terrestrial animals are **ureotelic,** excreting amino nitrogen in the form of urea; birds and reptiles are **uricotelic,** excreting amino nitrogen as uric acid. (The pathway of uric acid synthesis is described in Fig. 22–45.) Plants recycle virtually all amino groups, and nitrogen excretion occurs only under very unusual circumstances.

In ureotelic organisms, the ammonia deposited in the mitochondria of hepatocytes is converted to urea in the **urea cycle.** This pathway was discovered in 1932 by Hans Krebs (who later also discovered the citric acid cycle) and a medical student associate, Kurt Henseleit. Urea production occurs almost exclusively in the liver and is the fate of most of the ammonia channeled there. The urea passes into the bloodstream and thus to the kidneys and is excreted into the urine. The production of urea now becomes the focus of our discussion.

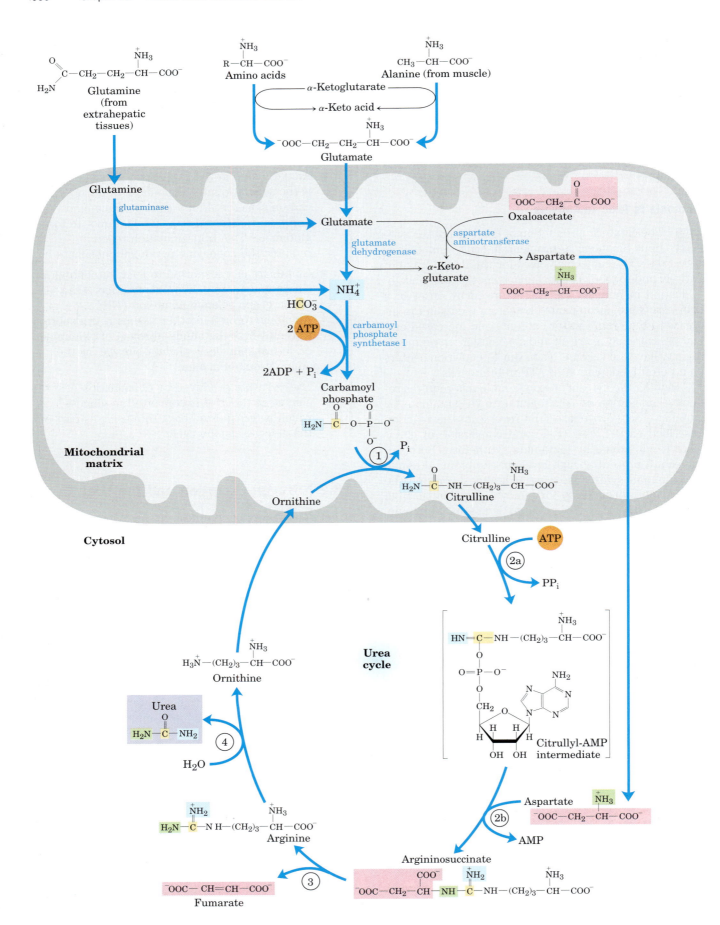

FIGURE 18–10 (facing page) **Urea cycle and reactions that feed amino groups into the cycle.** The enzymes catalyzing these reactions (named in the text) are distributed between the mitochondrial matrix and the cytosol. One amino group enters the urea cycle as carbamoyl phosphate, formed in the matrix; the other enters as aspartate, formed in the matrix by transamination of oxaloacetate and glutamate, catalyzed by aspartate aminotransferase. The urea cycle consists of four steps. ① Formation of citrulline from ornithine and carbamoyl phosphate (entry of the first amino group); the citrulline passes into the cytosol. ② Formation of argininosuccinate through a citrullyl-AMP intermediate (entry of the second amino group). ③ Formation of arginine from argininosuccinate; this reaction releases fumarate, which enters the citric acid cycle. ④ Formation of urea; this reaction also regenerates ornithine. The pathways by which $NH_4^+$ arrives in the mitochondrial matrix of hepatocytes were discussed in Section 18.1.

## Urea Is Produced from Ammonia in Five Enzymatic Steps

The urea cycle begins inside liver mitochondria, but three of the subsequent steps take place in the cytosol; the cycle thus spans two cellular compartments (Fig. 18–10). The first amino group to enter the urea cycle is derived from ammonia in the mitochondrial matrix—$NH_4^+$ arising by the pathways described above.

The liver also receives some ammonia via the portal vein from the intestine, from the bacterial oxidation of amino acids. Whatever its source, the $NH_4^+$ generated in liver mitochondria is immediately used, together with $CO_2$ (as $HCO_3^-$) produced by mitochondrial respiration, to form carbamoyl phosphate in the matrix (Fig. 18–11a; see also Fig. 18–10). This ATP-dependent reaction is catalyzed by **carbamoyl phosphate synthetase I,** a regulatory enzyme (see below). The mitochondrial form of the enzyme is distinct from the cytosolic (II) form, which has a separate function in pyrimidine biosynthesis (Chapter 22).

The carbamoyl phosphate, which functions as an activated carbamoyl group donor, now enters the urea cycle. The cycle has four enzymatic steps. First, carbamoyl phosphate donates its carbamoyl group to ornithine to form citrulline, with the release of $P_i$ (Fig. 18–10, step ①). Ornithine plays a role resembling that of oxaloacetate in the citric acid cycle, accepting material at each turn of the cycle. The reaction is catalyzed by **ornithine transcarbamoylase,** and the citrulline passes from the mitochondrion to the cytosol.

The second amino group now enters from aspartate (generated in mitochondria by transamination and transported into the cytosol) by a condensation reaction between the amino group of aspartate and the ureido

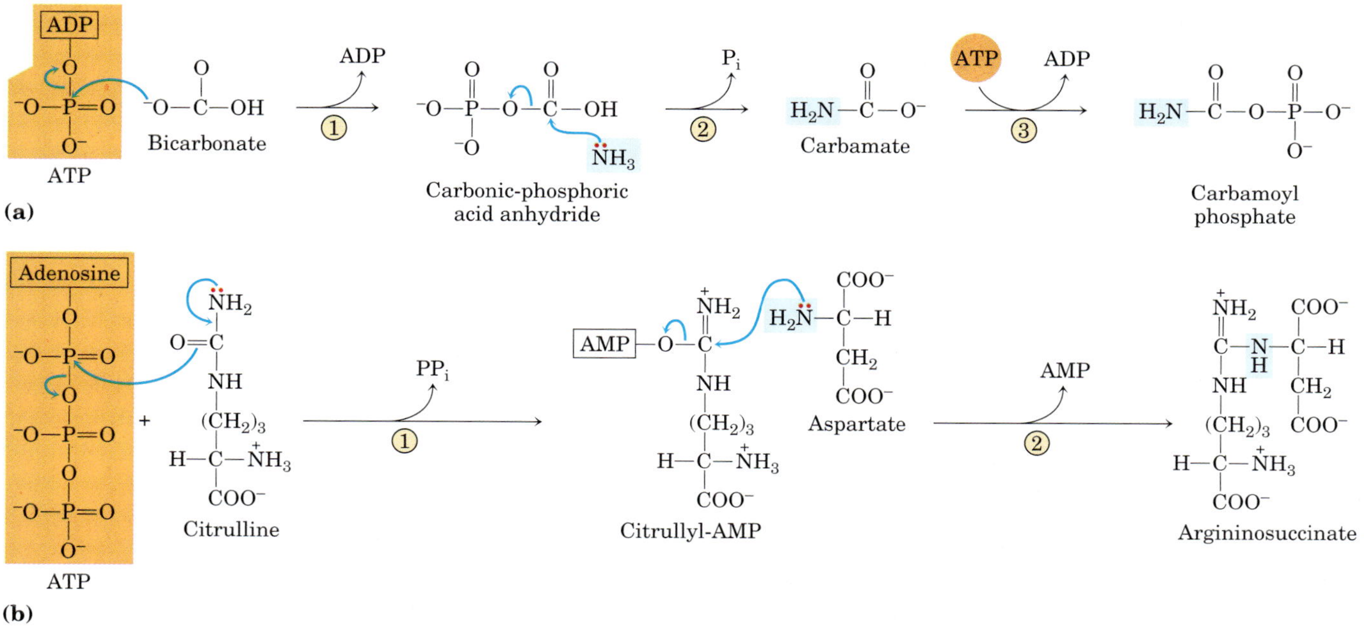

**(a)**

**(b)**

MECHANISM FIGURE 18–11 **Nitrogen-acquiring reactions in the synthesis of urea.** The urea nitrogens are acquired in two reactions, each requiring ATP. **(a)** In the reaction catalyzed by carbamoyl phosphate synthetase I, the first nitrogen enters from ammonia. The terminal phosphate groups of two molecules of ATP are used to form one molecule of carbamoyl phosphate. In other words, this reaction has two activa-

tion steps (① and ③). 🖱 Carbamoyl Phosphate Synthetase I Mechanism **(b)** In the reaction catalyzed by argininosuccinate synthetase, the second nitrogen enters from aspartate. The ureido oxygen of citrulline is activated by the addition of AMP in step ①; this sets up the addition of aspartate in step ②, with AMP (including the ureido oxygen) as the leaving group. 🖱 Argininosuccinate Synthetase Mechanism

(carbonyl) group of citrulline, forming argininosucci-nate (step ② in Fig. 18–10). This cytosolic reaction, catalyzed by **argininosuccinate synthetase,** requires ATP and proceeds through a citrullyl-AMP intermediate (Fig. 18–11b). The argininosuccinate is then cleaved by **argininosuccinase** (step ③ in Fig. 18–10) to form free arginine and fumarate, the latter entering mitochondria to join the pool of citric acid cycle intermediates. This is the only reversible step in the urea cycle. In the last reaction of the urea cycle (step ④), the cytosolic enzyme **arginase** cleaves arginine to yield **urea** and ornithine. Ornithine is transported into the mitochondrion to initiate another round of the urea cycle.

As we noted in Chapter 16, the enzymes of many metabolic pathways are clustered (p. 605), with the product of one enzyme reaction being channeled directly to the next enzyme in the pathway. In the urea cycle, the mitochondrial and cytosolic enzymes appear to be clustered in this way. The citrulline transported out of the mitochondrion is not diluted into the general pool of metabolites in the cytosol but is passed directly to the active site of argininosuccinate synthetase. This channeling between enzymes continues for argininosuccinate, arginine, and ornithine. Only urea is released into the general cytosolic pool of metabolites.

## The Citric Acid and Urea Cycles Can Be Linked

Because the fumarate produced in the argininosuccinase reaction is also an intermediate of the citric acid cycle, the cycles are, in principle, interconnected—in a process dubbed the "Krebs bicycle" (Fig. 18–12). However, each cycle can operate independently and communication between them depends on the transport of key intermediates between the mitochondrion and cytosol. Several enzymes of the citric acid cycle, including fumarase (fumarate hydratase) and malate dehydrogenase (p. 612), are also present as isozymes in the cytosol. The fumarate generated in cytosolic arginine synthesis can therefore be converted to malate in the cytosol, and these intermediates can be further metabolized in the cytosol or transported into mitochondria for use in the citric acid cycle. Aspartate formed in mitochondria by transamination between oxaloacetate and glutamate can be transported to the cytosol, where it serves as nitrogen donor in the urea cycle reaction catalyzed by argininosuccinate synthetase. These reactions, making up the **aspartate-argininosuccinate shunt,** provide metabolic links between the separate pathways by which the amino groups and carbon skeletons of amino acids are processed.

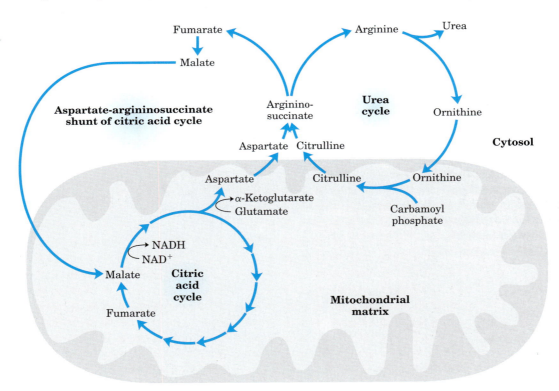

**FIGURE 18–12 Links between the urea cycle and citric acid cycle.** The interconnected cycles have been called the "Krebs bicycle." The pathways linking the citric acid and urea cycles are called the aspartate-argininosuccinate shunt; these effectively link the fates of the amino groups and the carbon skeletons of amino acids. The interconnections are even more elaborate than the arrows suggest. For example, some citric acid cycle enzymes, such as fumarase and malate dehydrogenase, have both cytosolic and mitochondrial isozymes. Fumarate produced in the cytosol—whether by the urea cycle, purine biosynthesis, or other processes—can be converted to cytosolic malate, which is used in the cytosol or transported into mitochondria (via the malate-aspartate shuttle; see Fig. 19–27) to enter the citric acid cycle.

## The Activity of the Urea Cycle Is Regulated at Two Levels

The flux of nitrogen through the urea cycle in an individual animal varies with diet. When the dietary intake is primarily protein, the carbon skeletons of amino acids are used for fuel, producing much urea from the excess amino groups. During prolonged starvation, when breakdown of muscle protein begins to supply much of the organism's metabolic energy, urea production also increases substantially.

These changes in demand for urea cycle activity are met over the long term by regulation of the rates of synthesis of the four urea cycle enzymes and carbamoyl phosphate synthetase I in the liver. All five enzymes are synthesized at higher rates in starving animals and in animals on very-high-protein diets than in well-fed animals eating primarily carbohydrates and fats. Animals on protein-free diets produce lower levels of urea cycle enzymes.

On a shorter time scale, allosteric regulation of at least one key enzyme adjusts the flux through the urea cycle. The first enzyme in the pathway, carbamoyl phosphate synthetase I, is allosterically activated by **N-acetylglutamate,** which is synthesized from acetyl-CoA and glutamate by **N-acetylglutamate synthase** (Fig. 18–13). In plants and microorganisms this enzyme catalyzes the first step in the de novo synthesis of arginine from glutamate (see Fig. 22–10), but in mammals N-acetylglutamate synthase activity in the liver has a purely regulatory function (mammals lack the other enzymes needed to convert glutamate to arginine). The steady-state levels of N-acetylglutamate are determined by the concentrations of glutamate and acetyl-CoA (the substrates for N-acetylglutamate synthase) and arginine (an activator of N-acetylglutamate synthase, and thus an activator of the urea cycle).

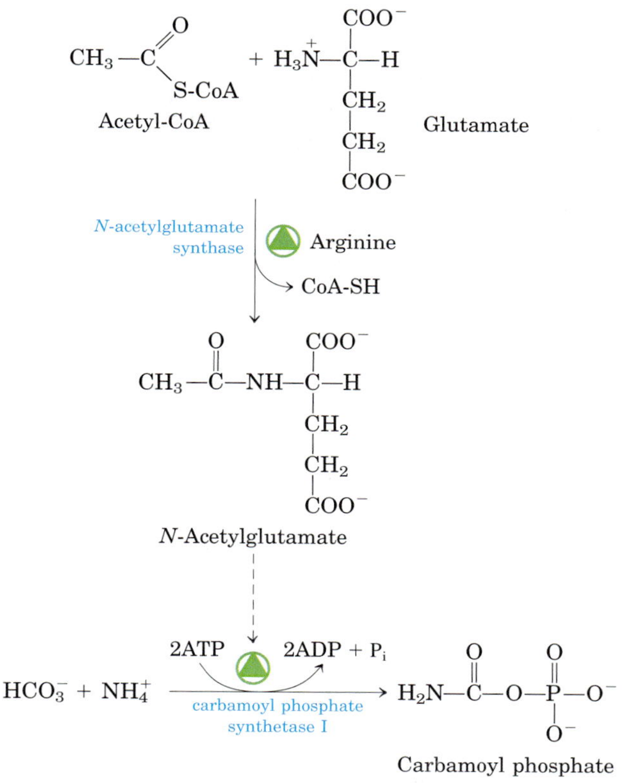

## Pathway Interconnections Reduce the Energetic Cost of Urea Synthesis

If we consider the urea cycle in isolation, we see that the synthesis of one molecule of urea requires four high-energy phosphate groups (Fig. 18–10). Two ATP molecules are required to make carbamoyl phosphate, and one ATP to make argininosuccinate—the latter ATP undergoing a pyrophosphate cleavage to AMP and $PP_i$, which is hydrolyzed to two $P_i$. The overall equation of the urea cycle is

$$2NH_4^+ + HCO_3^- + 3ATP^{4-} + H_2O \longrightarrow$$
$$\text{urea} + 2ADP^{3-} + 4P_i^{2-} + AMP^{2-} + 2H^+$$

However, the urea cycle also causes a net conversion of oxaloacetate to fumarate (via aspartate), and the regeneration of oxaloacetate (Fig. 18–12) produces NADH in the malate dehydrogenase reaction. Each NADH molecule can generate up to 2.5 ATP during mitochondrial

**FIGURE 18–13** Synthesis of N-acetylglutamate and its activation of carbamoyl phosphate synthetase I.

respiration (Chapter 19), greatly reducing the overall energetic cost of urea synthesis.

## Genetic Defects in the Urea Cycle Can Be Life-Threatening

People with genetic defects in any enzyme involved in urea formation cannot tolerate protein-rich diets. Amino acids ingested in excess of the minimum daily requirements for protein synthesis are deaminated in the liver, producing free ammonia that cannot be converted to urea and exported into the bloodstream, and, as we have seen, ammonia is highly toxic. The absence of a urea cycle enzyme can result in hyperammonemia or in the build-up of one or more urea cycle intermediates, depending on the enzyme that is missing. Given that most urea cycle steps are irreversible, the absent enzyme activity can often be identified by determining which cycle intermediate is present in especially elevated concentration in the blood and/or urine. Although the breakdown of amino acids can have serious health consequences in individuals with urea cycle deficiencies, a protein-free diet is not a treatment option. Humans are incapable of synthesizing half of the 20 common amino acids, and these **essential amino acids** (Table 18–1) must be provided in the diet.

**TABLE 18-1 Nonessential and Essential Amino Acids for Humans and the Albino Rat**

| Nonessential | Conditionally essential* | Essential |
|---|---|---|
| Alanine | Arginine | Histidine |
| Asparagine | Cysteine | Isoleucine |
| Aspartate | Glutamine | Leucine |
| Glutamate | Glycine | Lysine |
| Serine | Proline | Methionine |
| | Tyrosine | Phenylalanine |
| | | Threonine |
| | | Tryptophan |
| | | Valine |

*Required to some degree in young, growing animals, and/or sometimes during illness.

A variety of treatments are available for individuals with urea cycle defects. Careful administration of the aromatic acids benzoate or phenylbutyrate in the diet can help lower the level of ammonia in the blood. Benzoate is converted to benzoyl-CoA, which combines with glycine to form hippurate (Fig. 18–14, left). The glycine used up in this reaction must be regenerated, and ammonia is thus taken up in the glycine synthase reaction. Phenylbutyrate is converted to phenylacetate by $\beta$ oxidation. The phenylacetate is then converted to phenylacetyl-CoA, which combines with glutamine to form phenylacetylglutamine (Fig. 18–14, right). The resulting removal of glutamine triggers its further synthesis by glutamine synthetase (see Eqn 22–1) in a reaction that takes up ammonia. Both hippurate and phenylacetylglutamine are nontoxic compounds that are excreted in the urine. The pathways shown in Figure 18–14 make only minor contributions to normal metabolism, but they become prominent when aromatic acids are ingested.

Other therapies are more specific to a particular enzyme deficiency. Deficiency of N-acetylglutamate synthase results in the absence of the normal activator of carbamoyl phosphate synthetase I (Fig. 18–13). This condition can be treated by administering carbamoyl glutamate, an analog of N-acetylglutamate that is effective in activating carbamoyl phosphate synthetase I.

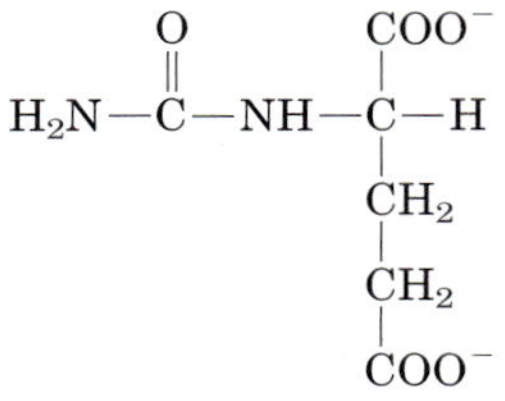

Carbamoyl glutamate

Supplementing the diet with arginine is useful in treating deficiencies of ornithine transcarbamoylase, argininosuccinate synthetase, and argininosuccinase. Many

of these treatments must be accompanied by strict dietary control and supplements of essential amino acids. In the rare cases of arginase deficiency, arginine, the substrate of the defective enzyme, must be excluded from the diet. ∎

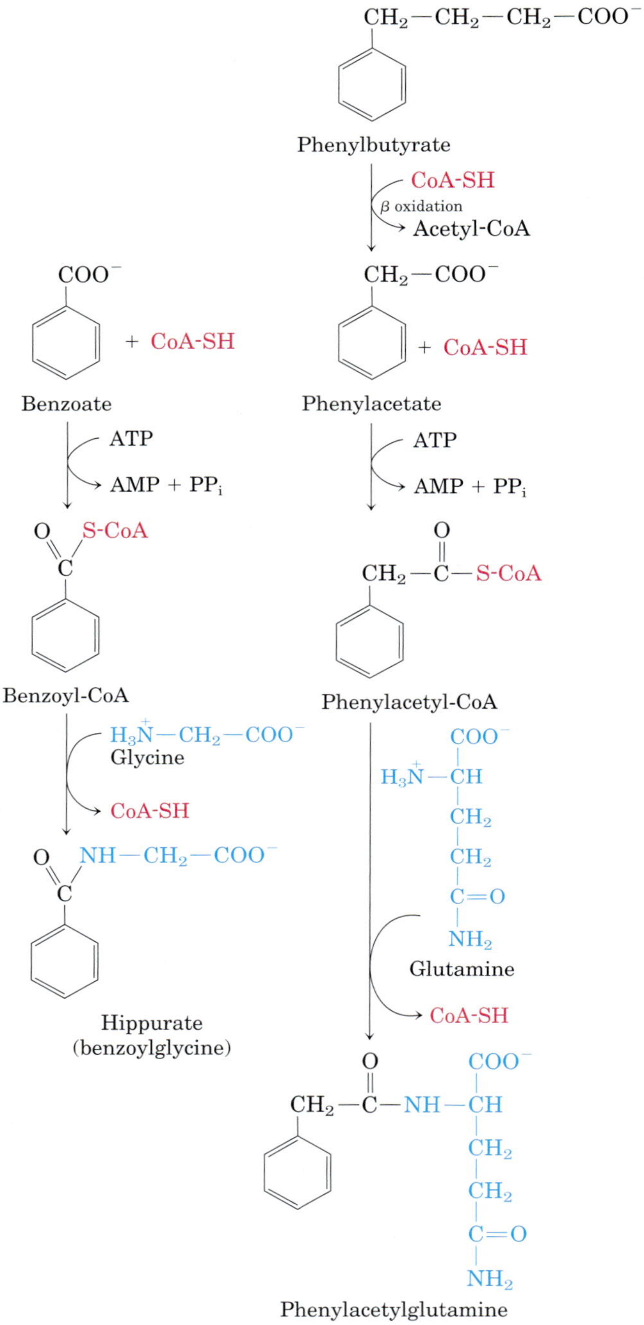

**FIGURE 18–14 Treatment for deficiencies in urea cycle enzymes.** The aromatic acids benzoate and phenylbutyrate, administered in the diet, are metabolized and combine with glycine and glutamine, respectively. The products are excreted in the urine. Subsequent synthesis of glycine and glutamine to replenish the pool of these intermediates removes ammonia from the bloodstream.

## SUMMARY 18.2 Nitrogen Excretion and the Urea Cycle

- Ammonia is highly toxic to animal tissues. In the urea cycle, ornithine combines with ammonia, in the form of carbamoyl phosphate, to form citrulline. A second amino group is transferred to citrulline from aspartate to form arginine—the immediate precursor of urea. Arginase catalyzes hydrolysis of arginine to urea and ornithine; thus ornithine is regenerated in each turn of the cycle.

- The urea cycle results in a net conversion of oxaloacetate to fumarate, both of which are intermediates in the citric acid cycle. The two cycles are thus interconnected.

- The activity of the urea cycle is regulated at the level of enzyme synthesis and by allosteric regulation of the enzyme that catalyzes the formation of carbamoyl phosphate.

## 18.3 Pathways of Amino Acid Degradation

The pathways of amino acid catabolism, taken together, normally account for only 10% to 15% of the human body's energy production; these pathways are not nearly as active as glycolysis and fatty acid oxidation. Flux through these catabolic routes also varies greatly, depending on the balance between requirements for bio-synthetic processes and the availability of a particular amino acid. The 20 catabolic pathways converge to form only six major products, all of which enter the citric acid cycle (Fig. 18–15). From here the carbon skeletons are diverted to gluconeogenesis or ketogenesis or are completely oxidized to $CO_2$ and $H_2O$.

All or part of the carbon skeletons of seven amino acids are ultimately broken down to acetyl-CoA. Five amino acids are converted to $\alpha$-ketoglutarate, four to succinyl-CoA, two to fumarate, and two to oxaloacetate. Parts or all of six amino acids are converted to pyruvate, which can be converted to either acetyl-CoA or oxaloacetate. We later summarize the individual pathways for the 20 amino acids in flow diagrams, each leading to a specific point of entry into the citric acid cycle. In these diagrams the carbon atoms that enter the citric acid cycle are shown in color. Note that some amino acids appear more than once, reflecting different fates for different parts of their carbon skeletons. Rather than examining every step of every pathway in amino acid catabolism, we single out for special discussion some enzymatic reactions that are particularly noteworthy for their mechanisms or their medical significance.

### Some Amino Acids Are Converted to Glucose, Others to Ketone Bodies

The seven amino acids that are degraded entirely or in part to acetoacetyl-CoA and/or acetyl-CoA—phenylalanine, tyrosine, isoleucine, leucine, tryptophan, threonine, and lysine—can yield ketone bodies in the liver,

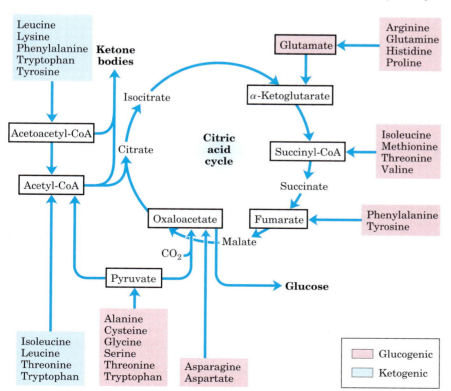

**FIGURE 18–15 Summary of amino acid catabolism.** Amino acids are grouped according to their major degradative end product. Some amino acids are listed more than once because different parts of their carbon skeletons are degraded to different end products. The figure shows the most important catabolic pathways in vertebrates, but there are minor variations among vertebrate species. Threonine, for instance, is degraded via at least two different pathways (see Figs 18–19, 18–27), and the importance of a given pathway can vary with the organism and its metabolic conditions. The glucogenic and ketogenic amino acids are also delineated in the figure, by color shading. Notice that five of the amino acids are both glucogenic and ketogenic. The amino acids degraded to pyruvate are also potentially ketogenic. Only two amino acids, leucine and lysine, are exclusively ketogenic.

where acetoacetyl-CoA is converted to acetoacetate and then to acetone and $\beta$-hydroxybutyrate (see Fig. 17–18). These are the **ketogenic** amino acids (Fig. 18–15). Their ability to form ketone bodies is particularly evident in uncontrolled diabetes mellitus, in which the liver produces large amounts of ketone bodies from both fatty acids and the ketogenic amino acids.

The amino acids that are degraded to pyruvate, $\alpha$-ketoglutarate, succinyl-CoA, fumarate, and/or oxaloacetate can be converted to glucose and glycogen by pathways described in Chapters 14 and 15. They are the **glucogenic** amino acids. The division between ketogenic and glucogenic amino acids is not sharp; five amino acids—tryptophan, phenylalanine, tyrosine, threonine, and isoleucine—are both ketogenic and glucogenic. Catabolism of amino acids is particularly critical to the survival of animals with high-protein diets or during starvation. Leucine is an exclusively ketogenic amino acid that is very common in proteins. Its degradation makes a substantial contribution to ketosis under starvation conditions.

## Several Enzyme Cofactors Play Important Roles in Amino Acid Catabolism

A variety of interesting chemical rearrangements occur in the catabolic pathways of amino acids. It is useful to begin our study of these pathways by noting the classes of reactions that recur and introducing their enzyme cofactors. We have already considered one important class: transamination reactions requiring pyridoxal phosphate. Another common type of reaction in amino acid catabolism is one-carbon transfers, which usually involve one of three cofactors: biotin, tetrahydrofolate, or $S$-adenosylmethionine (Fig. 18–16). These cofactors transfer one-carbon groups in different oxidation states: biotin transfers carbon in its most oxidized state, $CO_2$

(see Fig. 14–18); tetrahydrofolate transfers one-carbon groups in intermediate oxidation states and sometimes as methyl groups; and $S$-adenosylmethionine transfers methyl groups, the most reduced state of carbon. The latter two cofactors are especially important in amino acid and nucleotide metabolism.

**Tetrahydrofolate ($H_4$ folate),** synthesized in bacteria, consists of substituted pterin (6-methylpterin),

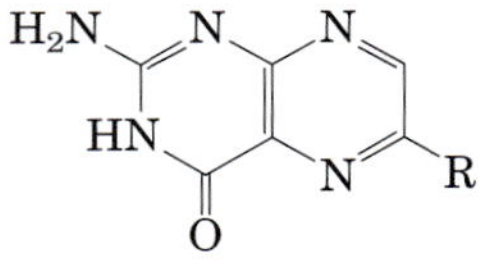

Pterin

$p$-aminobenzoate, and glutamate moieties (Fig. 18–16). The oxidized form, folate, is a vitamin for mammals; it is converted in two steps to tetrahydrofolate by the enzyme dihydrofolate reductase. The one-carbon group undergoing transfer, in any of three oxidation states, is bonded to N-5 or N-10 or both. The most reduced form of the cofactor carries a methyl group, a more oxidized form carries a methylene group, and the most oxidized forms carry a methenyl, formyl, or formimino group (Fig. 18–17). Most forms of tetrahydrofolate are interconvertible and serve as donors of one-carbon units in a variety of metabolic reactions. The primary source of one-carbon units for tetrahydrofolate is the carbon removed in the conversion of serine to glycine, producing $N^5,N^{10}$-methylenetetrahydrofolate.

Although tetrahydrofolate can carry a methyl group at N-5, the transfer potential of this methyl group is insufficient for most biosynthetic reactions. **$S$-Adenosylmethionine (adoMet)** is the preferred cofactor for biological methyl group transfers. It is synthesized from ATP and methionine by the action of **methionine**

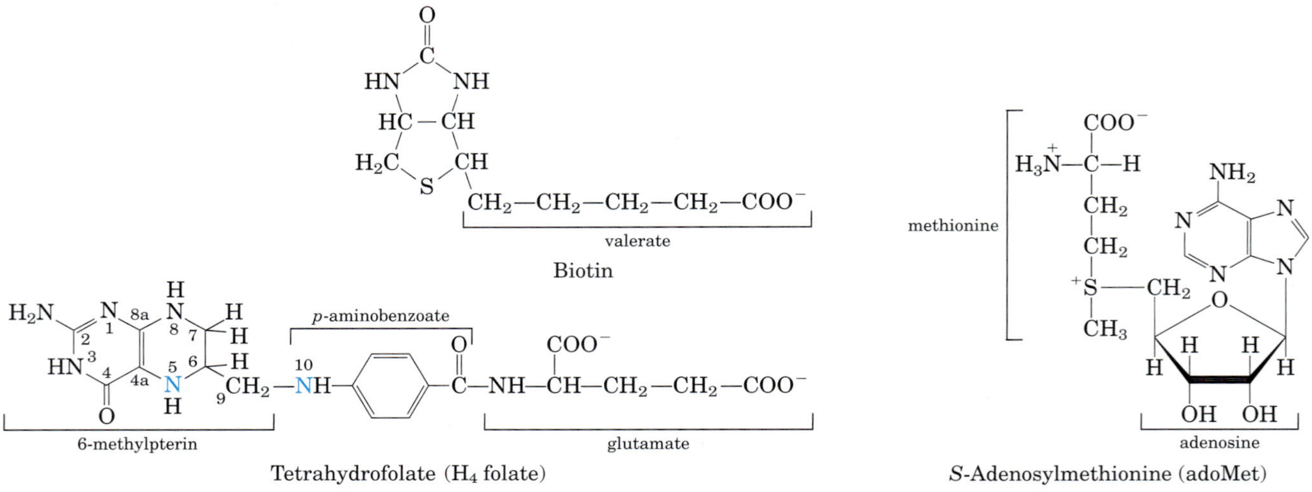

**FIGURE 18–16 Some enzyme cofactors important in one-carbon transfer reactions.** The nitrogen atoms to which one-carbon groups are attached in tetrahydrofolate are shown in blue.

**FIGURE 18–17 Conversions of one-carbon units on tetrahydrofolate.**
The different molecular species are grouped according to oxidation state, with the most reduced at the top and most oxidized at the bottom. All species within a single shaded box are at the same oxidation state. The conversion of $N^5,N^{10}$-methylenetetrahydrofolate to $N^5$-methyltetrahydrofolate is effectively irreversible. The enzymatic transfer of formyl groups, as in purine synthesis (see Fig. 22–33) and in the formation of formylmethionine in prokaryotes (Chapter 27), generally uses $N^{10}$-formyltetrahydrofolate rather than $N^5$-formyltetrahydrofolate. The latter species is significantly more stable and therefore a weaker donor of formyl groups. $N^5$-Formyltetrahydrofolate is a minor byproduct of the cyclohydrolase reaction, and can also form spontaneously. Conversion of $N^5$-formyltetrahydrofolate to $N^5,N^{10}$-methenyltetrahydrofolate requires ATP, because of an otherwise unfavorable equilibrium. Note that $N^5$-formiminotetrahydrofolate is derived from histidine in a pathway shown in Figure 18–26.

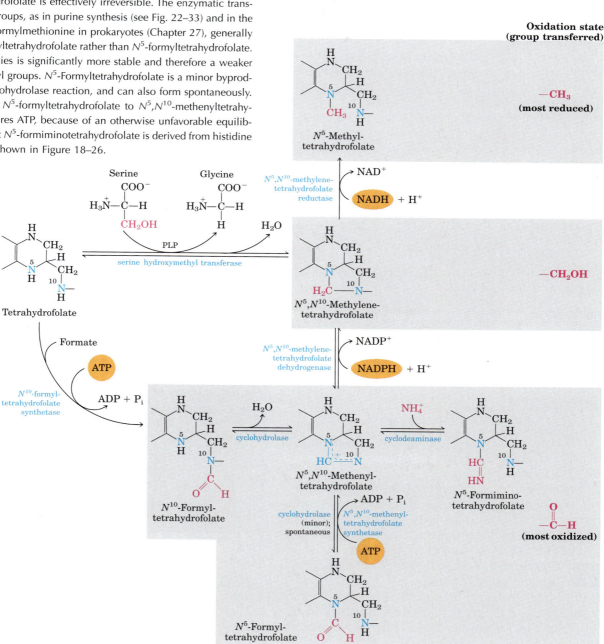

adenosyl transferase (Fig. 18–18, step ①). This reaction is unusual in that the nucleophilic sulfur atom of methionine attacks the 5′ carbon of the ribose moiety of ATP rather than one of the phosphorus atoms. Triphosphate is released and is cleaved to $P_i$ and $PP_i$ on the enzyme, and the $PP_i$ is cleaved by inorganic pyrophosphatase; thus three bonds, including two bonds of high-energy phosphate groups, are broken in this reaction. The only other known reaction in which triphosphate is displaced from ATP occurs in the synthesis of coenzyme $B_{12}$ (see Box 17–2, Fig. 3).

S-Adenosylmethionine is a potent alkylating agent by virtue of its destabilizing sulfonium ion. The methyl group is subject to attack by nucleophiles and is about

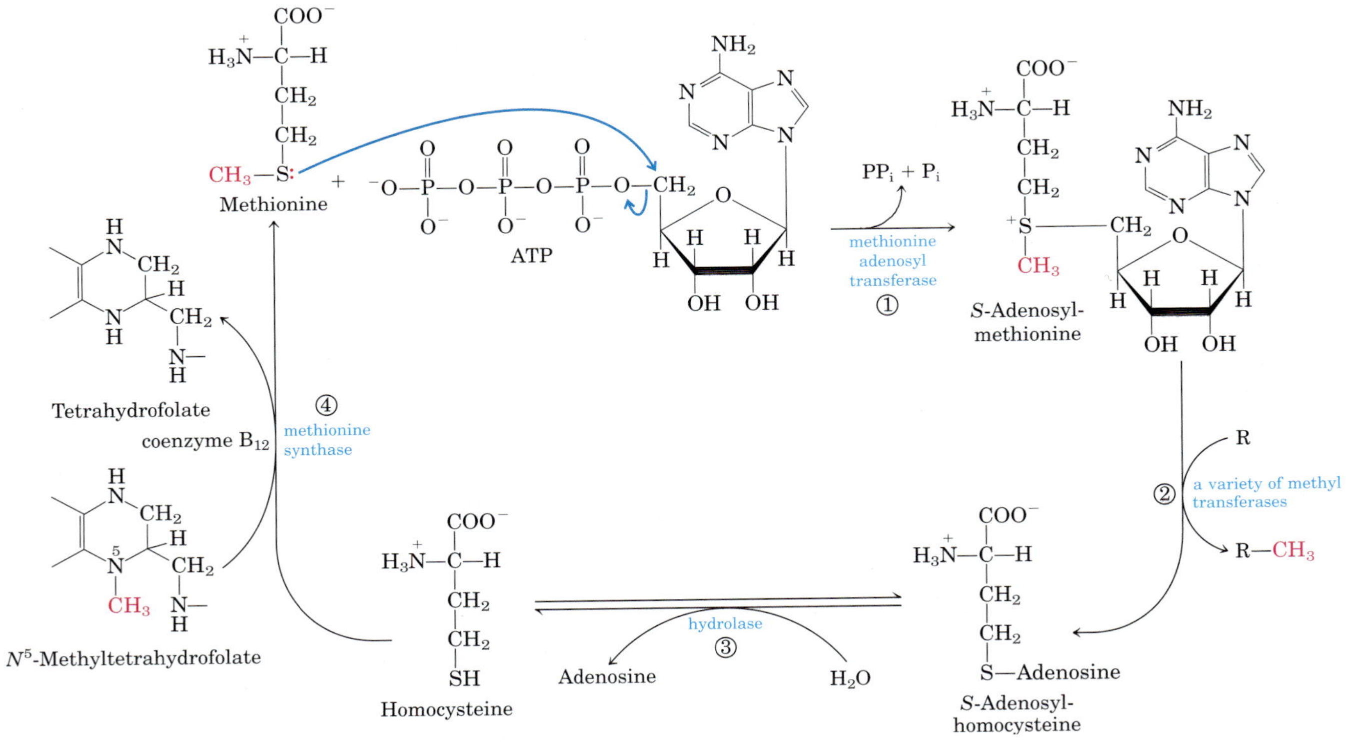

**FIGURE 18–18 Synthesis of methionine and S-adenosylmethionine in an activated-methyl cycle.** The steps are described in the text. In the methionine synthase reaction (step ④), the methyl group is transferred to cobalamin to form methylcobalamin, which in turn is the methyl donor in the formation of methionine. S-Adenosylmethionine, which has a positively charged sulfur (and is thus a sulfonium ion), is a powerful methylating agent in a number of biosynthetic reactions. The methyl group acceptor (step ②) is designated R.

1,000 times more reactive than the methyl group of $N^5$-methyltetrahydrofolate.

Transfer of the methyl group from S-adenosylmethionine to an acceptor yields **S-adenosylhomocysteine** (Fig. 18–18, step ②), which is subsequently broken down to homocysteine and adenosine (step ③). Methionine is regenerated by transfer of a methyl group to homocysteine in a reaction catalyzed by methionine synthase (step ④), and methionine is reconverted to S-adenosylmethionine to complete an activated-methyl cycle.

One form of methionine synthase common in bacteria uses $N^5$-methyltetrahydrofolate as a methyl donor. Another form of the enzyme present in some bacteria and mammals uses $N^5$-methyltetrahydrofolate, but the methyl group is first transferred to cobalamin, derived from coenzyme $B_{12}$, to form methylcobalamin as the methyl donor in methionine formation. This reaction and the rearrangement of L-methylmalonyl-CoA to succinyl-CoA (see Box 17–2, Fig. 1a) are the only known coenzyme $B_{12}$–dependent reactions in mammals. In cases of vitamin $B_{12}$ deficiency, some symptoms can be alleviated by administering not only vitamin $B_{12}$ but folate. As noted above, the methyl group of methylcobalamin is derived from $N^5$-methyltetrahydrofolate. Because the reaction converting the $N^5,N^{10}$-methylene form to the $N^5$-methyl form of tetrahydrofo-

late is irreversible (Fig. 18–17), if coenzyme $B_{12}$ is not available for the synthesis of methylcobalamin, then no acceptor is available for the methyl group of $N^5$-methyltetrahydrofolate and metabolic folates become trapped in the $N^5$-methyl form. This sequestering of folates in one form may be the cause of some symptoms of the vitamin $B_{12}$ deficiency disease pernicious anemia. However, we do not know whether this is the only effect of insufficient vitamin $B_{12}$. ∎

**Tetrahydrobiopterin,** another cofactor of amino acid catabolism, is similar to the pterin moiety of tetrahydrofolate, but it is not involved in one-carbon transfers; instead it participates in oxidation reactions. We consider its mode of action when we discuss phenylalanine degradation (see Fig. 18–24).

## Six Amino Acids Are Degraded to Pyruvate

The carbon skeletons of six amino acids are converted in whole or in part to pyruvate. The pyruvate can then be converted to either acetyl-CoA (a ketone body precursor) or oxaloacetate (a precursor for gluconeogenesis). Thus amino acids catabolized to pyruvate are both ketogenic and glucogenic. The six are alanine, tryptophan, cysteine, serine, glycine, and threonine (Fig. 18–19). **Alanine** yields pyruvate directly on transamination with

$\alpha$-ketoglutarate, and the side chain of **tryptophan** is cleaved to yield alanine and thus pyruvate. **Cysteine** is converted to pyruvate in two steps; one removes the sulfur atom, the other is a transamination. **Serine** is converted to pyruvate by serine dehydratase. Both the $\beta$-hydroxyl and the $\alpha$-amino groups of serine are removed in this single pyridoxal phosphate–dependent reaction (Fig. 18–20a).

**Glycine** is degraded via three pathways, only one of which leads to pyruvate. Glycine is converted to serine by enzymatic addition of a hydroxymethyl group (Figs 18–19 and 18–20b). This reaction, catalyzed by **serine hydroxymethyl transferase,** requires the coenzymes tetrahydrofolate and pyridoxal phosphate. The serine is converted to pyruvate as described above. In the second pathway, which predominates in animals, glycine undergoes oxidative cleavage to $CO_2$, $NH_4^+$, and a methylene group ($-CH_2-$) (Fig. 18–19). This readily reversible reaction, catalyzed by **glycine cleavage enzyme** (also called glycine synthase), also requires tetrahydrofolate, which accepts the methylene group. In this oxidative cleavage pathway the two carbon atoms of glycine do not enter the citric acid cycle. One carbon is lost as $CO_2$ and the other becomes the methylene group of $N^5,N^{10}$-methylenetetrahydrofolate (Fig. 18–17), a one-carbon group donor in certain biosynthetic pathways.

This second pathway for glycine degradation appears to be critical in mammals. Humans with serious defects in glycine cleavage enzyme activity suffer from a condition known as nonketotic hyperglycinemia. The condition is characterized by elevated serum levels of glycine, leading to severe mental deficiencies and death in very early childhood. At high levels, glycine is an inhibitory neurotransmitter, perhaps explaining the neurological effects of the disease. Many genetic defects of amino acid metabolism have been identified in humans (Table 18–2). We will encounter several more in this chapter. ■

In the third and final pathway of glycine degradation, the achiral glycine molecule is a substrate for the enzyme D-amino acid oxidase. The glycine is converted to glyoxylate, an alternative substrate for hepatic lactate

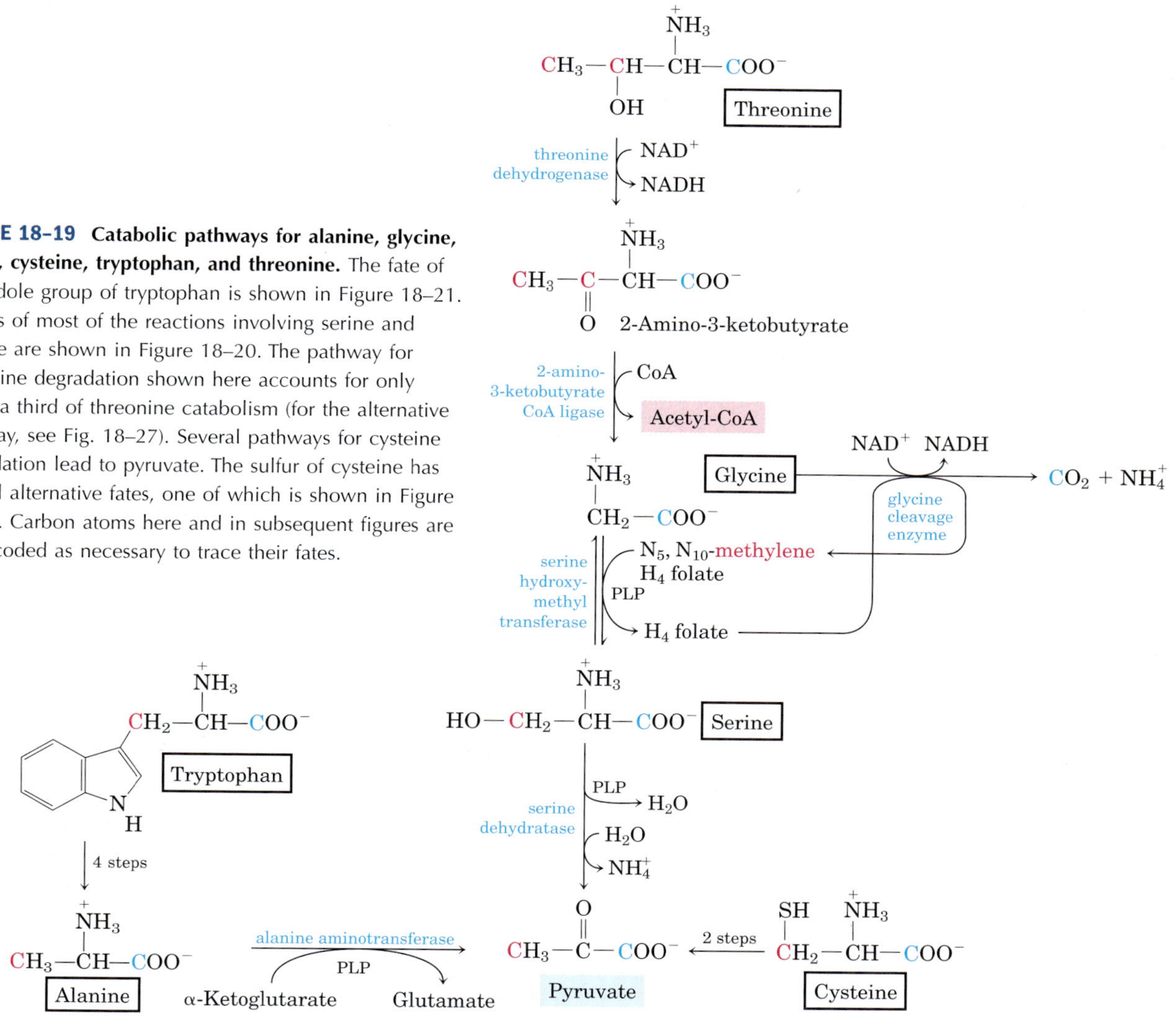

**FIGURE 18–19 Catabolic pathways for alanine, glycine, serine, cysteine, tryptophan, and threonine.** The fate of the indole group of tryptophan is shown in Figure 18–21. Details of most of the reactions involving serine and glycine are shown in Figure 18–20. The pathway for threonine degradation shown here accounts for only about a third of threonine catabolism (for the alternative pathway, see Fig. 18–27). Several pathways for cysteine degradation lead to pyruvate. The sulfur of cysteine has several alternative fates, one of which is shown in Figure 22–15. Carbon atoms here and in subsequent figures are color-coded as necessary to trace their fates.

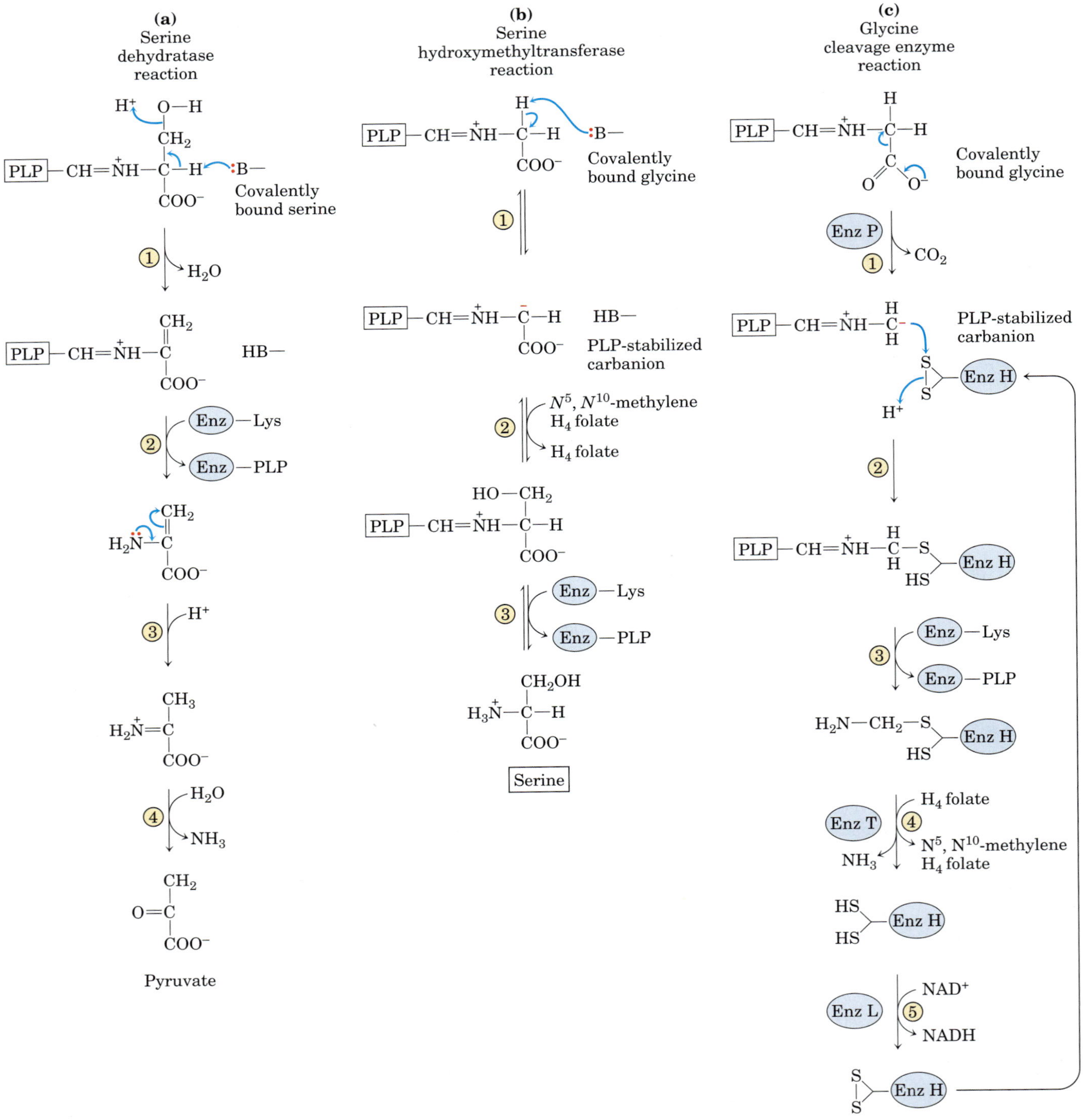

**(a)**
Serine
dehydratase
reaction

**(b)**
Serine
hydroxymethyltransferase
reaction

**(c)**
Glycine
cleavage enzyme
reaction

**MECHANISM FIGURE 18–20 Interplay of the pyridoxal phosphate and tetrahydrofolate cofactors in serine and glycine metabolism.** The first step in each of these reactions (not shown) involves the formation of a covalent imine linkage between enzyme-bound PLP and the substrate amino acid—serine in **(a)**, glycine in **(b)** and **(c)**. **(a)** The serine dehydratase reaction entails a PLP-catalyzed elimination of water across the bond between the $\alpha$ and $\beta$ carbons (step ①), leading eventually to the production of pyruvate (steps ② through ④). **(b)** In the serine hydroxymethyltransferase reaction, a PLP-stabilized carbanion on the $\alpha$ carbon of glycine (product of step ①) is a key intermediate in the transfer of the methylene group (as —$CH_2$—OH) from $N^5,N^{10}$-methylenetetrahydrofolate to form serine. This reaction is reversible. **(c)** The glycine cleavage enzyme is a multienzyme complex, with components P, H, T, and L. The overall reaction, which is reversible, converts glycine to $CO_2$ and $NH_4^+$, with the second glycine carbon taken up by tetrahydrofolate to form $N^5,N^{10}$-methylenetetrahydrofolate. Pyridoxal phosphate activates the $\alpha$ carbon of amino acids at critical stages in all these reactions, and tetrahydrofolate carries one-carbon units in two of them (see Figs 18–6, 18–17).

**TABLE 18–2 Some Human Genetic Disorders Affecting Amino Acid Catabolism**

| Medical condition | Approximate incidence (per 100,000 births) | Defective process | Defective enzyme | Symptoms and effects |
|---|---|---|---|---|
| Albinism | <3 | Melanin synthesis from tyrosine | Tyrosine 3-monooxygenase (tyrosinase) | Lack of pigmentation: white hair, pink skin |
| Alkaptonuria | <0.4 | Tyrosine degradation | Homogentisate 1,2-dioxygenase | Dark pigment in urine; late-developing arthritis |
| Argininemia | <0.5 | Urea synthesis | Arginase | Mental retardation |
| Argininosuccinic acidemia | <1.5 | Urea synthesis | Argininosuccinase | Vomiting; convulsions |
| Carbamoyl phosphate synthetase I deficiency | <0.5 | Urea synthesis | Carbamoyl phosphate synthetase I | Lethargy; convulsions; early death |
| Homocystinuria | <0.5 | Methionine degradation | Cystathionine $\beta$-synthase | Faulty bone development; mental retardation |
| Maple syrup urine disease (branched-chain ketoaciduria) | <0.4 | Isoleucine, leucine, and valine degradation | Branched-chain $\alpha$-keto acid dehydrogenase complex | Vomiting; convulsions; mental retardation; early death |
| Methylmalonic acidemia | <0.5 | Conversion of propionyl-CoA to succinyl-CoA | Methylmalonyl-CoA mutase | Vomiting; convulsions; mental retardation; early death |
| Phenylketonuria | <8 | Conversion of phenyl-alanine to tyrosine | Phenylalanine hydroxylase | Neonatal vomiting; mental retardation |

dehydrogenase (p. 538). Glyoxylate is oxidized in an $NAD^+$-dependent reaction to oxalate:

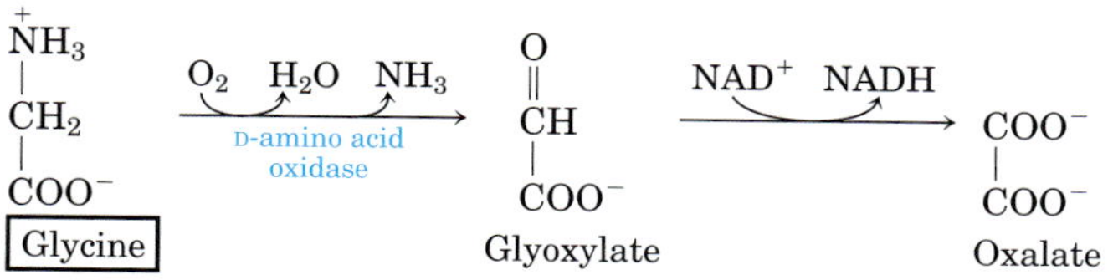

The primary function of D-amino acid oxidase, present at high levels in the kidney, is thought to be the detoxification of ingested D-amino acids derived from bacterial cell walls and from cooked foodstuffs (heat causes some spontaneous racemization of the L-amino acids in proteins). Oxalate, whether obtained in foods or produced enzymatically in the kidneys, has medical significance. Crystals of calcium oxalate account for up to 75% of all kidney stones. ■

There are two significant pathways for **threonine** degradation. One pathway leads to pyruvate via glycine (Fig. 18–19). The conversion to glycine occurs in two steps, with threonine first converted to 2-amino-3-ketobutyrate by the action of threonine dehydrogenase. This is a relatively minor pathway in humans, accounting for 10% to 30% of threonine catabolism, but is more important in some other mammals. The major pathway in humans leads to succinyl-CoA and is described later.

In the laboratory, serine hydroxymethyltransferase will catalyze the conversion of threonine to glycine and acetaldehyde in one step, but this is not a significant pathway for threonine degradation in mammals.

### Seven Amino Acids Are Degraded to Acetyl-CoA

Portions of the carbon skeletons of seven amino acids—**tryptophan, lysine, phenylalanine, tyrosine, leucine, isoleucine,** and **threonine**—yield acetyl-CoA and/or acetoacetyl-CoA, the latter being converted to acetyl-CoA (Fig. 18–21). Some of the final steps in the degradative pathways for leucine, lysine, and tryptophan resemble steps in the oxidation of fatty acids. Threonine (not shown in Fig. 18–21) yields some acetyl-CoA via the minor pathway illustrated in Figure 18–19.

The degradative pathways of two of these seven amino acids deserve special mention. Tryptophan breakdown is the most complex of all the pathways of amino

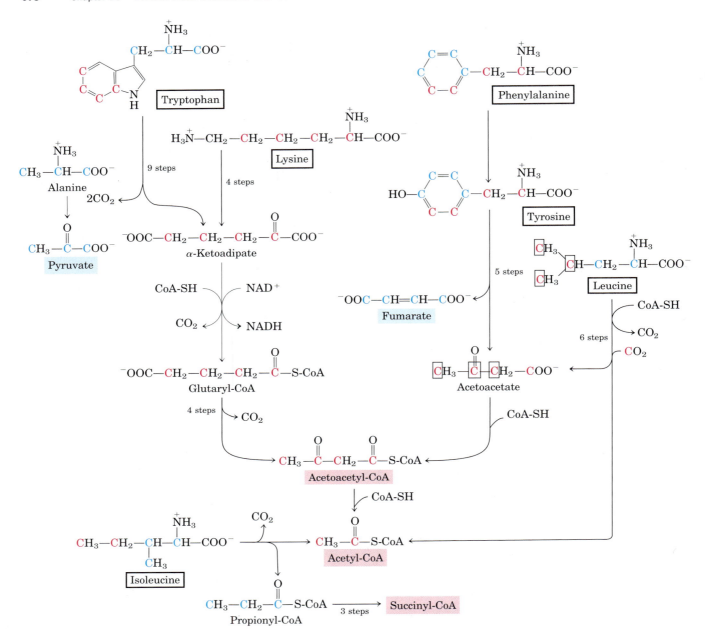

**FIGURE 18–21 Catabolic pathways for tryptophan, lysine, phenyl-alanine, tyrosine, leucine, and isoleucine.** These amino acids donate some of their carbons (red) to acetyl-CoA. Tryptophan, phenylalanine, tyrosine, and isoleucine also contribute carbons (blue) to pyruvate or citric acid cycle intermediates. The phenylalanine pathway is de-scribed in more detail in Figure 18–23. The fate of nitrogen atoms is not traced in this scheme; in most cases they are transferred to $\alpha$-ketoglutarate to form glutamate.

acid catabolism in animal tissues; portions of tryptophan (four of its carbons) yield acetyl-CoA via acetoacetyl-CoA. Some of the intermediates in tryptophan catabolism are precursors for the synthesis of other biomolecules (Fig. 18–22), including nicotinate, a precursor of NAD and NADP in animals; serotonin, a neurotransmitter in vertebrates; and indoleacetate, a growth factor in plants. Some of these biosynthetic pathways are described in more detail in Chapter 22 (see Figs 22–28, 22–29).

The breakdown of phenylalanine is noteworthy be-cause genetic defects in the enzymes of this pathway lead to several inheritable human diseases (Fig. 18–23), as discussed below. Phenylalanine and its oxidation

product tyrosine (both with nine carbons) are degraded into two fragments, both of which can enter the citric acid cycle: four of the nine carbon atoms yield free ace-toacetate, which is converted to acetoacetyl-CoA and thus acetyl-CoA, and a second four-carbon fragment is recovered as fumarate. Eight of the nine carbons of these two amino acids thus enter the citric acid cycle; the remaining carbon is lost as $CO_2$. Phenylalanine, af-ter its hydroxylation to tyrosine, is also the precursor of dopamine, a neurotransmitter, and of norepinephrine and epinephrine, hormones secreted by the adrenal medulla (see Fig. 22–29). Melanin, the black pigment of skin and hair, is also derived from tyrosine.

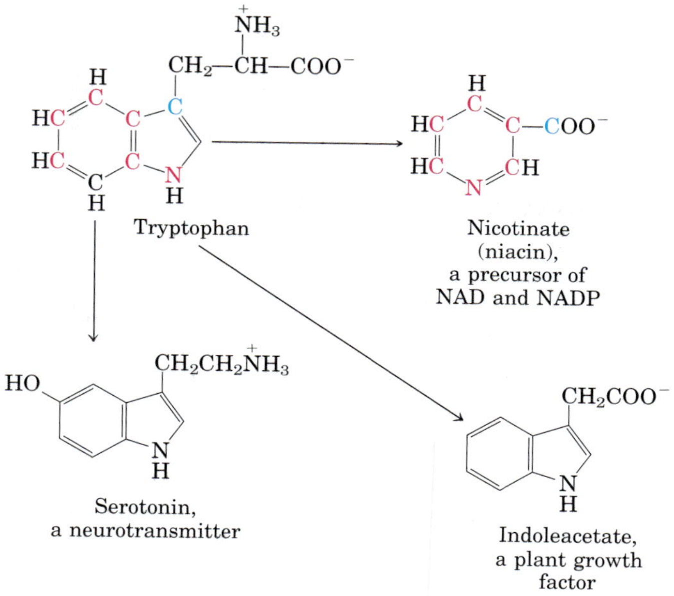

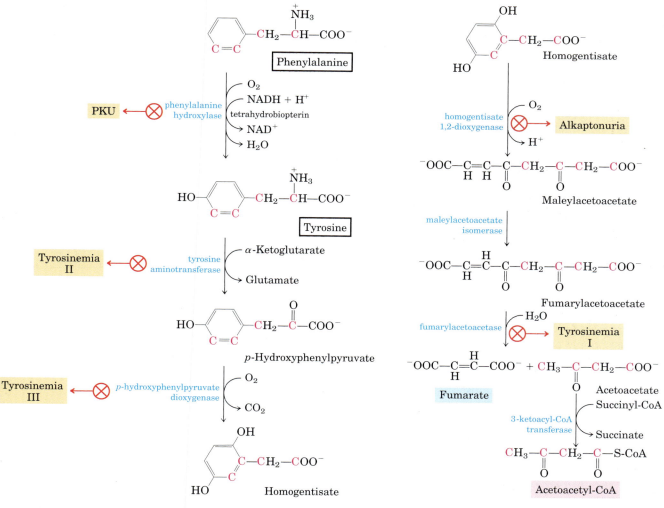

**FIGURE 18–22 Tryptophan as precursor.** The aromatic rings of tryptophan give rise to nicotinate, indoleacetate, and serotonin. Colored atoms trace the source of the ring atoms in nicotinate.

## Phenylalanine Catabolism Is Genetically Defective in Some People

Given that many amino acids are either neurotransmitters or precursors or antagonists of neutrotransmitters, genetic defects of amino acid metabolism can cause defective neural development and mental retardation. In most such diseases specific intermediates accumulate. For example, a genetic defect in **phenylalanine hydroxylase,** the first enzyme in the catabolic pathway for phenylalanine (Fig. 18–23), is responsible for the disease **phenylketonuria (PKU),** the most common cause of elevated levels of phenylalanine (hyperphenylalaninemia).

Phenylalanine hydroxylase (also called phenylalanine-4-monooxygenase) is one of a general class of enzymes called **mixed-function oxidases** (see Box 21–1), all of which catalyze simultaneous hydroxylation of a substrate by an oxygen atom of $O_2$ and reduction of the other oxygen atom to $H_2O$. Phenylalanine hydroxylase

**FIGURE 18–23 Catabolic pathways for phenylalanine and tyrosine.** In humans these amino acids are normally converted to acetoacetyl-CoA and fumarate. Genetic defects in many of these enzymes cause inheritable human diseases (shaded yellow).

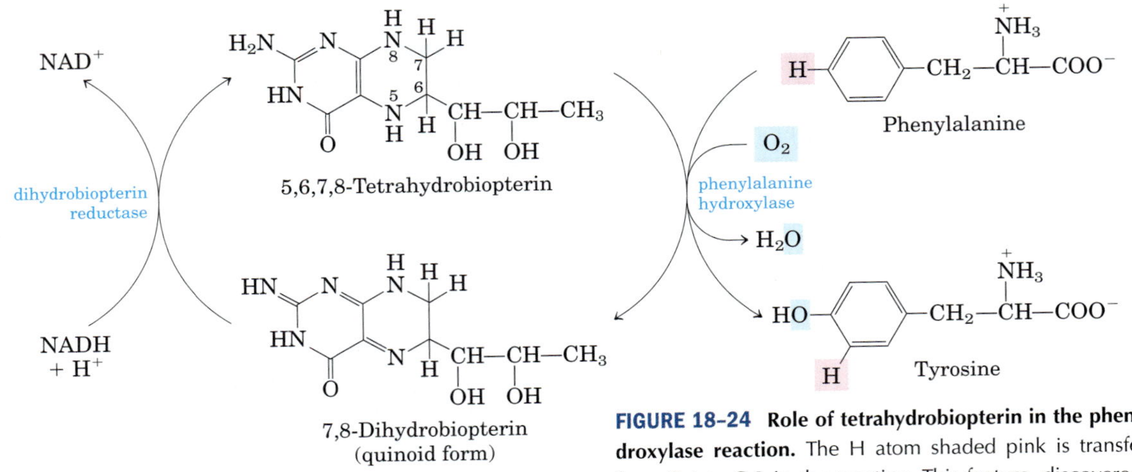

**FIGURE 18–24** **Role of tetrahydrobiopterin in the phenylalanine hydroxylase reaction.** The H atom shaded pink is transferred directly from C-4 to C-3 in the reaction. This feature, discovered at the NIH, is called the NIH Shift.

requires the cofactor tetrahydrobiopterin, which carries electrons from NADH to $O_2$ and becomes oxidized to dihydrobiopterin in the process (Fig. 18–24). It is subsequently reduced by the enzyme **dihydrobiopterin reductase** in a reaction that requires NADH.

In individuals with PKU, a secondary, normally little-used pathway of phenylalanine metabolism comes into play. In this pathway phenylalanine undergoes transamination with pyruvate to yield **phenylpyruvate** (Fig. 18–25). Phenylalanine and phenylpyruvate accumulate in the blood and tissues and are excreted in the urine—hence the name "phenylketonuria." Much of the phenylpyruvate, rather than being excreted as such, is either decarboxylated to phenylacetate or reduced to phenyllactate. Phenylacetate imparts a characteristic odor to the urine, which nurses have traditionally used to detect PKU in infants. The accumulation of phenylalanine or its metabolites in early life impairs normal development of the brain, causing severe mental retardation. This may be caused by excess phenylalanine competing with other amino acids for transport across the blood-brain barrier, resulting in a deficit of required metabolites.

Phenylketonuria was among the first inheritable metabolic defects discovered in humans. When this condition is recognized early in infancy, mental retardation can largely be prevented by rigid dietary control. The diet must supply only enough phenylalanine and tyrosine to meet the needs for protein synthesis. Consumption of protein-rich foods must be curtailed. Natural proteins, such as casein of milk, must first be hydrolyzed and much of the phenylalanine removed to provide an appropriate diet, at least through childhood. Because the artificial sweetener aspartame is a dipeptide of aspartate and the methyl ester of phenylalanine (see Fig. 1–23b), foods sweetened with aspartame bear warnings addressed to individuals on phenylalanine-controlled diets.

Phenylketonuria can also be caused by a defect in the enzyme that catalyzes the regeneration of tetrahydrobiopterin (Fig. 18–24). The treatment in this case is

more complex than restricting the intake of phenylalanine and tyrosine. Tetrahydrobiopterin is also required for the formation of L-3,4-dihydroxyphenylalanine (L-dopa) and 5-hydroxytryptophan—precursors of the neurotransmitters norepinephrine and serotonin, respectively—and in phenylketonuria of this type, these precursors must be supplied in the diet. Supplementing the diet with tetrahydrobiopterin itself is ineffective because it is unstable and does not cross the blood-brain barrier.

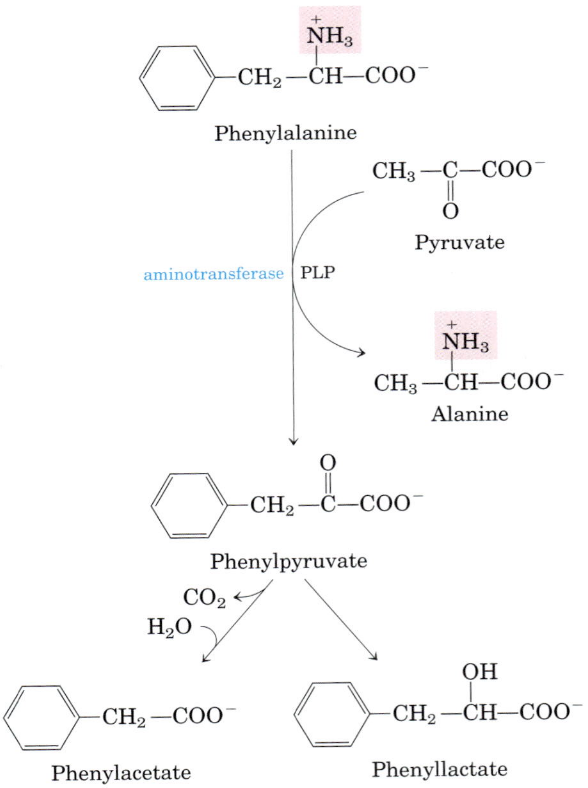

**FIGURE 18–25** **Alternative pathways for catabolism of phenylalanine in phenylketonuria.** In PKU, phenylpyruvate accumulates in the tissues, blood, and urine. The urine may also contain phenylacetate and phenyllactate.

Screening newborns for genetic diseases can be highly cost-effective, especially in the case of PKU. The tests (no longer relying on urine odor) are relatively inexpensive, and the detection and early treatment of PKU in infants (eight to ten cases per 100,000 newborns) saves millions of dollars in later health care costs each year. More importantly, the emotional trauma avoided by early detection with these simple tests is inestimable.

Another inheritable disease of phenylalanine catabolism is **alkaptonuria,** in which the defective enzyme is **homogentisate dioxygenase** (Fig. 18–23). Less serious than PKU, this condition produces few ill effects, although large amounts of homogentisate are excreted and its oxidation turns the urine black. Individuals with alkaptonuria are also prone to develop a form of arthri-

tis. Alkaptonuria is of considerable historical interest. Archibald Garrod discovered in the early 1900s that this condition is inherited, and he traced the cause to the absence of a single enzyme. Garrod was the first to make a connection between an inheritable trait and an enzyme, a great advance on the path that ultimately led to our current understanding of genes and the information pathways described in Part III. ■

## Five Amino Acids Are Converted to $\alpha$-Ketoglutarate

The carbon skeletons of five amino acids (proline, glutamate, glutamine, arginine, and histidine) enter the citric acid cycle as $\alpha$-ketoglutarate (Fig. 18–26). **Proline, glutamate,** and **glutamine** have five-carbon skeletons. The cyclic structure of proline is opened by oxidation

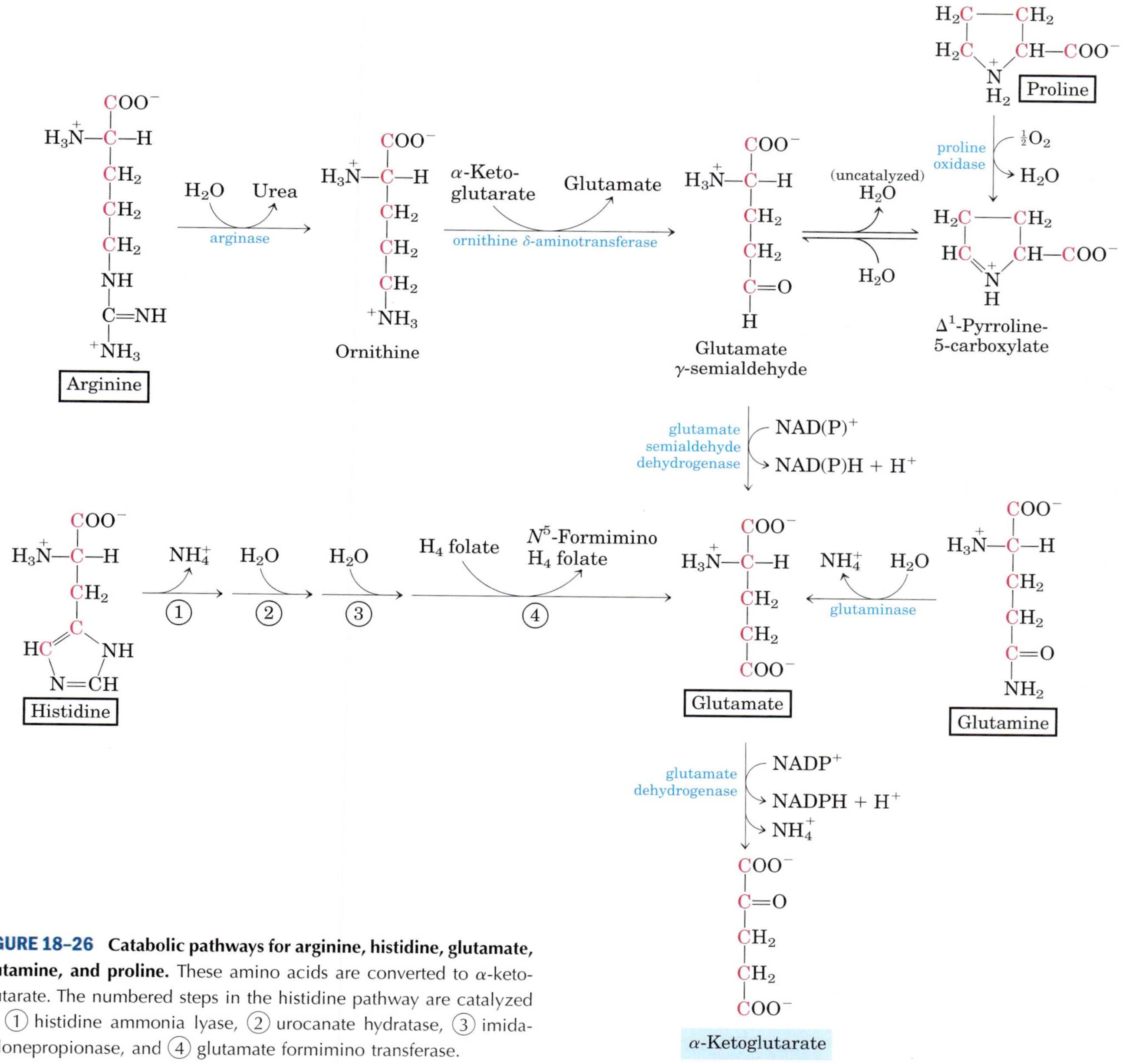

**FIGURE 18–26 Catabolic pathways for arginine, histidine, glutamate, glutamine, and proline.** These amino acids are converted to $\alpha$-ketoglutarate. The numbered steps in the histidine pathway are catalyzed by ① histidine ammonia lyase, ② urocanate hydratase, ③ imidazolonepropionase, and ④ glutamate formimino transferase.

of the carbon most distant from the carboxyl group to create a Schiff base, then hydrolysis of the Schiff base to a linear semialdehyde, glutamate γ-semialdehyde. This intermediate is further oxidized at the same carbon to produce glutamate. The action of glutaminase, or any of several enzyme reactions in which glutamine donates its amide nitrogen to an acceptor, converts glutamine to glutamate. Transamination or deamination of glutamate produces α-ketoglutarate.

**Arginine** and **histidine** contain five adjacent carbons and a sixth carbon attached through a nitrogen atom. The catabolic conversion of these amino acids to glutamate is therefore slightly more complex than the path from proline or glutamine (Fig. 18–26). Arginine is converted to the five-carbon skeleton of ornithine in the urea cycle (Fig. 18–10), and the ornithine is transaminated to glutamate γ-semialdehyde. Conversion of histidine to the five-carbon glutamate occurs in a multistep pathway; the extra carbon is removed in a step that uses tetrahydrofolate as cofactor.

## Four Amino Acids Are Converted to Succinyl-CoA

The carbon skeletons of methionine, isoleucine, threonine, and valine are degraded by pathways that yield succinyl-CoA (Fig. 18–27), an intermediate of the citric acid cycle. **Methionine** donates its methyl group to one of several possible acceptors through S-adenosylmethionine,

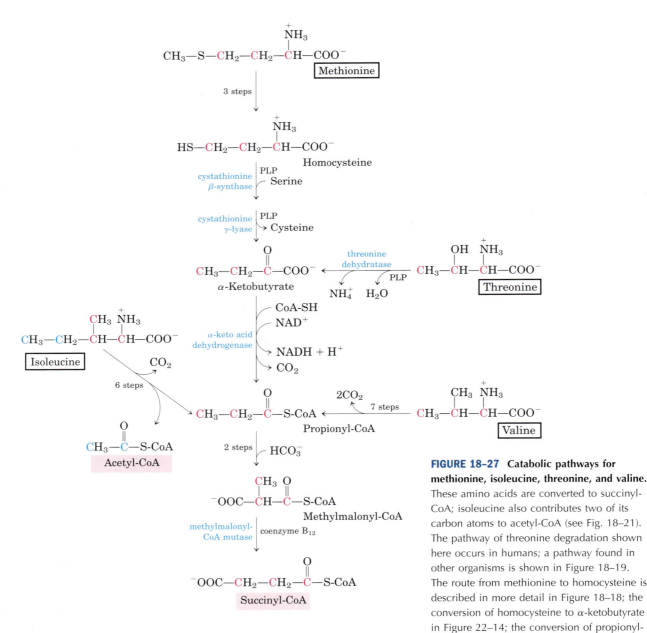

**FIGURE 18-27 Catabolic pathways for methionine, isoleucine, threonine, and valine.** These amino acids are converted to succinyl-CoA; isoleucine also contributes two of its carbon atoms to acetyl-CoA (see Fig. 18–21). The pathway of threonine degradation shown here occurs in humans; a pathway found in other organisms is shown in Figure 18–19. The route from methionine to homocysteine is described in more detail in Figure 18–18; the conversion of homocysteine to α-ketobutyrate in Figure 22–14; the conversion of propionyl-CoA to succinyl-CoA in Figure 17–11.

and three of its four remaining carbon atoms are converted to the propionate of propionyl-CoA, a precursor of succinyl-CoA. **Isoleucine** undergoes transamination, followed by oxidative decarboxylation of the resulting $\alpha$-keto acid. The remaining five-carbon skeleton is further oxidized to acetyl-CoA and propionyl-CoA. **Valine** undergoes transamination and decarboxylation, then a series of oxidation reactions that convert the remaining four carbons to propionyl-CoA. Some parts of the valine and isoleucine degradative pathways closely parallel steps in fatty acid degradation (see Fig. 17–8a). In human tissues, **threonine** is also converted in two steps to propionyl-CoA. This is the primary pathway for threonine degradation in humans (see Fig. 18–19 for the alternative pathway). The mechanism of the first step is analogous to that catalyzed by serine dehydratase, and the serine and threonine dehydratases may actually be the same enzyme.

The propionyl-CoA derived from these three amino acids is converted to succinyl-CoA by a pathway described in Chapter 17: carboxylation to methylmalonyl-CoA, epimerization of the methylmalonyl-CoA, and conversion to succinyl-CoA by the coenzyme $B_{12}$–dependent methylmalonyl-CoA mutase (see Fig. 17–11). In the rare genetic disease known as methylmalonic acidemia, methylmalonyl-CoA mutase is lacking—with serious metabolic consequences (Table 18–2; Box 18–2).

## Branched-Chain Amino Acids Are Not Degraded in the Liver

Although much of the catabolism of amino acids takes place in the liver, the three amino acids with branched side chains (leucine, isoleucine, and valine) are oxidized as fuels primarily in muscle, adipose, kidney, and brain tissue. These extrahepatic tissues contain an aminotransferase, absent in liver, that acts on all three branched-chain amino acids to produce the corresponding $\alpha$-keto acids (Fig. 18–28). The **branched-chain $\alpha$-keto acid dehydrogenase complex** then catalyzes oxidative decarboxylation of all three $\alpha$-keto acids, in each case releasing the carboxyl group as $CO_2$ and producing the acyl-CoA derivative. This reaction is formally analogous to two other oxidative decarboxylations encountered in Chapter 16: oxidation of pyruvate to acetyl-CoA by the pyruvate dehydrogenase complex (see Fig. 16–6) and oxidation of $\alpha$-ketoglutarate to succinyl-CoA by the $\alpha$-ketoglutarate dehydrogenase complex (p. 610). In fact, all three enzyme complexes are similar in structure and share essentially the same reaction mechanism. Five cofactors (thiamine pyrophosphate, FAD, NAD, lipoate, and coenzyme A) participate, and the three proteins in each complex catalyze homologous reactions. This is clearly a case in which enzymatic machinery that evolved to catalyze

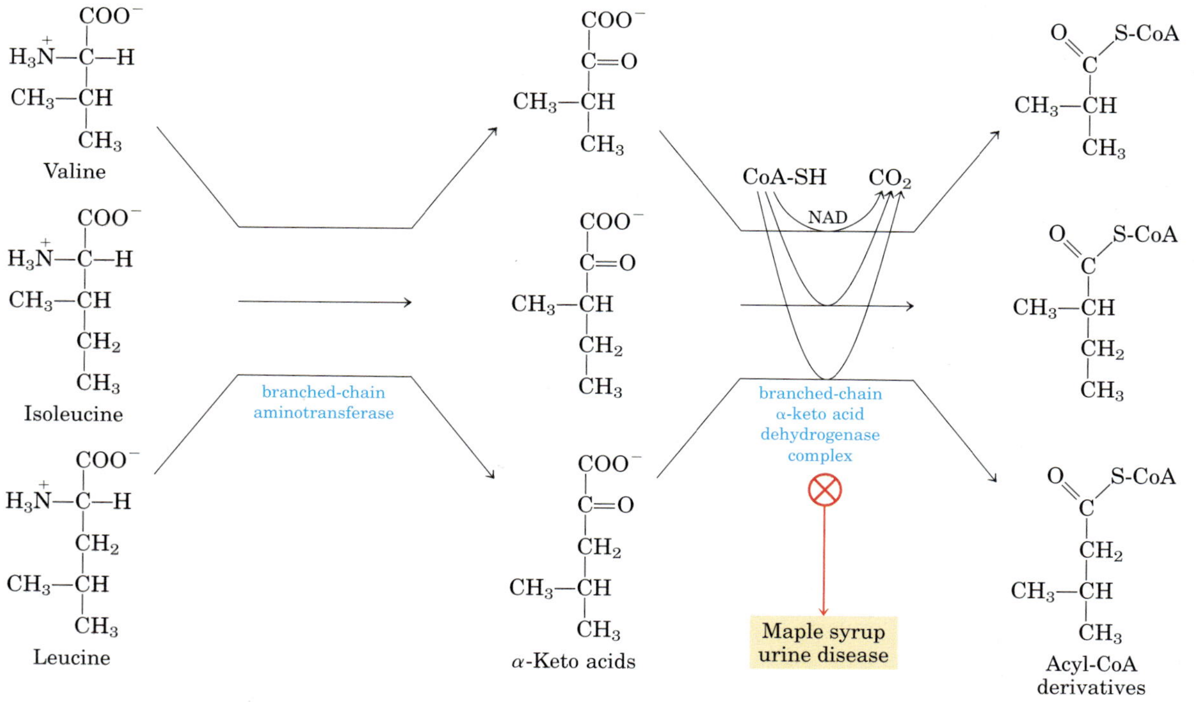

**FIGURE 18–28 Catabolic pathways for the three branched-chain amino acids: valine, isoleucine, and leucine.** The three pathways, which occur in extrahepatic tissues, share the first two enzymes, as shown here. The branched-chain $\alpha$-keto acid dehydrogenase complex is analogous to the pyruvate and $\alpha$-ketoglutarate dehydrogenase complexes and requires the same five cofactors (some not shown here). This enzyme is defective in people with maple syrup urine disease.

## BOX 18–2    BIOCHEMISTRY IN MEDICINE

### Scientific Sleuths Solve a Murder Mystery

Truth can sometimes be stranger than fiction—or at least as strange as a made-for-TV movie. Take, for example, the case of Patricia Stallings. Convicted of the murder of her infant son, she was sentenced to life in prison—but was later found innocent, thanks to the medical sleuthing of three persistent researchers.

The story began in the summer of 1989 when Stallings brought her three-month-old son, Ryan, to the emergency room of Cardinal Glennon Children's Hospital in St. Louis. The child had labored breathing, uncontrollable vomiting, and gastric distress. According to the attending physician, a toxicologist, the child's symptoms indicated that he had been poisoned with ethylene glycol, an ingredient of antifreeze, a conclusion apparently confirmed by analysis at a commercial lab.

After he recovered, the child was placed in a foster home, and Stallings and her husband, David, were allowed to see him in supervised visits. But when the infant became ill, and subsequently died, after a visit in which Stallings had been briefly left alone with him, she was charged with first-degree murder and held without bail. At the time, the evidence seemed compelling as both the commercial lab and the hospital lab found large amounts of ethylene glycol in the boy's blood and traces of it in a bottle of milk Stallings had fed her son during the visit.

But without knowing it, Stallings had performed a brilliant experiment. While in custody, she learned she was pregnant; she subsequently gave birth to another son, David Stallings Jr., in February 1990. He was placed immediately in a foster home, but within two weeks he started having symptoms similar to Ryan's. David was eventually diagnosed with a rare metabolic disorder called methylmalonic acidemia (MMA). A recessive genetic disorder of amino acid metabolism, MMA affects about 1 in 48,000 newborns and presents symptoms almost identical with those caused by ethylene glycol poisoning.

Stallings couldn't possibly have poisoned her second son, but the Missouri state prosecutor's office was not impressed by the new developments and pressed forward with her trial anyway. The court wouldn't allow the MMA diagnosis of the second child to be introduced as evidence, and in January 1991 Patricia Stallings was convicted of assault with a deadly weapon and sentenced to life in prison.

Fortunately for Stallings, however, William Sly, chairman of the Department of Biochemistry and Molecular Biology at St. Louis University, and James Shoemaker, head of a metabolic screening lab at the university, got interested in her case when they heard about it from a television broadcast. Shoemaker performed his own analysis of Ryan's blood and didn't detect ethylene glycol. He and Sly then contacted Piero Rinaldo, a metabolic disease expert at Yale University School of Medicine whose lab is equipped to diagnose MMA from blood samples.

When Rinaldo analyzed Ryan's blood serum, he found high concentrations of methylmalonic acid, a breakdown product of the branched-chain amino acids isoleucine and valine, which accumulates in MMA patients because the enzyme that should convert it to the next product in the metabolic pathway is defective. And particularly telling, he says, the child's blood and urine contained massive amounts of ketones, another metabolic consequence of the disease. Like Shoemaker, he did not find any ethylene glycol in a sample of the baby's bodily fluids. The bottle couldn't be tested, since it had mysteriously disappeared. Rinaldo's analyses convinced him that Ryan had died from MMA, but how to account for the results from two labs, indicating that the boy had ethylene glycol in his blood? Could they both be wrong?

When Rinaldo obtained the lab reports, what he saw was, he says, "scary." One lab said that Ryan Stallings' blood contained ethylene glycol, even though the blood sample analysis did not match the lab's own profile for a known sample containing ethylene glycol. "This was not just a matter of questionable interpretation. The quality of their analysis was unacceptable," Rinaldo says. And the second laboratory? According to Rinaldo, that lab detected an abnormal component in Ryan's blood and just "assumed it was ethylene glycol." Samples from the bottle had produced nothing unusual, says Rinaldo, yet the lab claimed evidence of ethylene glycol in that, too.

Rinaldo presented his findings to the case's prosecutor, George McElroy, who called a press conference the very next day. "I no longer believe the laboratory data," he told reporters. Having concluded that Ryan Stallings had died of MMA after all, McElroy dismissed all charges against Patricia Stallings on September 20, 1991.

By Michelle Hoffman (1991). *Science* **253,** 931. Copyright 1991 by the American Association for the Advancement of Science.

one reaction was "borrowed" by gene duplication and further evolved to catalyze similar reactions in other pathways.

Experiments with rats have shown that the branched-chain α-keto acid dehydrogenase complex is regulated by covalent modification in response to the content of branched-chain amino acids in the diet. With little or no excess dietary intake of branched-chain amino acids, the enzyme complex is phosphorylated and thereby inactivated by a protein kinase. Addition of excess branched-chain amino acids to the diet results in dephosphorylation and consequent activation of the enzyme. Recall that the pyruvate dehydrogenase complex is subject to similar regulation by phosphorylation and dephosphorylation (p. 621).

There is a relatively rare genetic disease in which the three branched-chain α-keto acids (as well as their precursor amino acids, especially leucine) accumulate in the blood and "spill over" into the urine. This condition, called **maple syrup urine disease** because of the characteristic odor imparted to the urine by the α-keto acids, results from a defective branched-chain α-keto acid dehydrogenase complex. Untreated, the disease results in abnormal development of the brain, mental retardation, and death in early infancy. Treatment entails rigid control of the diet, limiting the intake of valine, isoleucine, and leucine to the minimum required to permit normal growth. ∎

## Asparagine and Aspartate Are Degraded to Oxaloacetate

The carbon skeletons of **asparagine** and **aspartate** ultimately enter the citric acid cycle as oxaloacetate. The enzyme **asparaginase** catalyzes the hydrolysis of asparagine to aspartate, which undergoes transamination with α-ketoglutarate to yield glutamate and oxaloacetate (Fig. 18–29).

We have now seen how the 20 common amino acids, after losing their nitrogen atoms, are degraded by dehydrogenation, decarboxylation, and other reactions to yield portions of their carbon backbones in the form of six central metabolites that can enter the citric acid cycle. Those portions degraded to acetyl-CoA are completely oxidized to carbon dioxide and water, with generation of ATP by oxidative phosphorylation.

As was the case for carbohydrates and lipids, the degradation of amino acids results ultimately in the generation of reducing equivalents (NADH and $FADH_2$) through the action of the citric acid cycle. Our survey of catabolic processes concludes in the next chapter with a discussion of respiration, in which these reducing equivalents fuel the ultimate oxidative and energy-generating process in aerobic organisms.

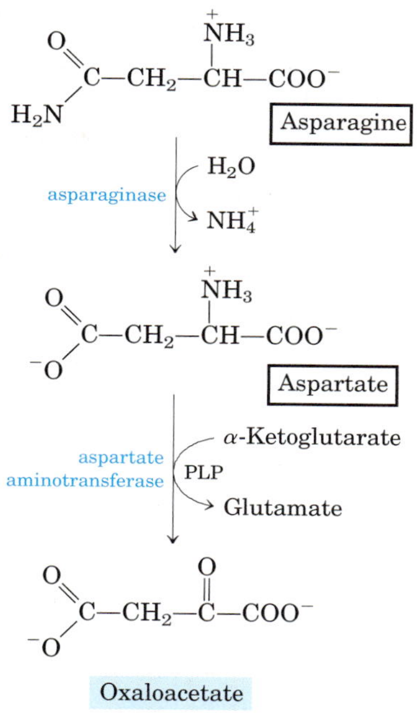

**FIGURE 18–29 Catabolic pathway for asparagine and aspartate.** Both amino acids are converted to oxaloacetate.

## SUMMARY 18.3 Pathways of Amino Acid Degradation

- After removal of their amino groups, the carbon skeletons of amino acids undergo oxidation to compounds that can enter the citric acid cycle for oxidation to $CO_2$ and $H_2O$. The reactions of these pathways require a number of cofactors, including tetrahydrofolate and S-adenosylmethionine in one-carbon transfer reactions and tetrahydrobiopterin in the oxidation of phenylalanine by phenylalanine hydroxylase.

- Depending on their degradative end product, some amino acids can be converted to ketone bodies, some to glucose, and some to both. Thus amino acid degradation is integrated into intermediary metabolism and can be critical to survival under conditions in which amino acids are a significant source of metabolic energy.

- The carbon skeletons of amino acids enter the citric acid cycle through five intermediates: acetyl-CoA, α-ketoglutarate, succinyl-CoA, fumarate, and oxaloacetate. Some are also degraded to pyruvate, which can be converted to either acetyl-CoA or oxaloacetate.

- The amino acids producing pyruvate are alanine, cysteine, glycine, serine, threonine, and tryptophan. Leucine, lysine, phenylalanine, and tryptophan yield acetyl-CoA via acetoacetyl-CoA. Isoleucine, leucine, threonine, and tryptophan also form acetyl-CoA directly.

- Arginine, glutamate, glutamine, histidine, and proline produce $\alpha$-ketoglutarate; isoleucine, methionine, threonine, and valine produce succinyl-CoA; four carbon atoms of

phenylalanine and tyrosine give rise to fumarate; and asparagine and aspartate produce oxaloacetate.

- The branched-chain amino acids (isoleucine, leucine, and valine), unlike the other amino acids, are degraded only in extrahepatic tissues.

- A number of serious human diseases can be traced to genetic defects in the enzymes of amino acid catabolism.

## Key Terms

*Terms in bold are defined in the glossary.*

| | | |
|---|---|---|
| **aminotransferases** 660 | glucose-alanine cycle 664 | **tetrahydrofolate** 672 |
| **transaminases** 660 | **ammonotelic** 665 | *S*-**adenosylmethionine (adoMet)** 672 |
| **transamination** 660 | **ureotelic** 665 | |
| **pyridoxal phosphate (PLP)** 660 | **uricotelic** 665 | **tetrahydrobiopterin** 674 |
| oxidative deamination 661 | **urea cycle** 665 | phenylketonuria (PKU) 679 |
| L-glutamate dehydrogenase 661 | urea 668 | **mixed-function oxidases** 679 |
| glutamine synthetase 662 | **essential amino acids** 669 | alkaptonuria 681 |
| glutaminase 663 | **ketogenic** 672 | maple syrup urine disease 685 |
| creatine kinase 664 | **glucogenic** 672 | |

## Further Reading

### General

**Arias, I.M., Boyer, J.L., Chisari, F.V., Fausto, N., Schachter, D., & Shafritz, D.A.** (2001) *The Liver: Biology and Pathobiology,* 4th edn, Lippincott Williams & Wilkins, Philadelphia.

**Bender, D.A.** (1985) *Amino Acid Metabolism,* 2nd edn, Wiley-Interscience, Inc., New York.

**Brosnan, J.T.** (2001) Amino acids, then and now—a reflection on Sir Hans Krebs' contribution to nitrogen metabolism. *IUBMB Life* **52,** 265–270.

An interesting tour through the life of this important biochemist.

**Campbell, J.W.** (1991) Excretory nitrogen metabolism. In *Environmental and Metabolic Animal Physiology,* 4th edn (Prosser, C.L., ed.), pp. 277–324, John Wiley & Sons, Inc., New York.

**Coomes, M.W.** (1997) Amino acid metabolism. In *Textbook of Biochemistry with Clinical Correlations,* 5th edn (Devlin, T.M., ed.), pp. 779–823, Wiley-Liss, New York.

**Hayashi, H.** (1995) Pyridoxal enzymes: mechanistic diversity and uniformity. *J. Biochem.* **118,** 463–473.

**Mazelis, M.** (1980) Amino acid catabolism. In *The Biochemistry of Plants: A Comprehensive Treatise* (Stumpf, P.K. & Conn, E.E., eds), Vol. 5: *Amino Acids and Derivatives* (Miflin, B.J., ed.), pp. 541–567, Academic Press, Inc., New York.

A discussion of the various fates of amino acids in plants.

**Walsh, C.** (1979) *Enzymatic Reaction Mechanisms,* W. H. Freeman and Company, San Francisco.

A good source for in-depth discussion of the classes of enzymatic reaction mechanisms described in the chapter.

### Amino Acid Metabolism

**Christen, P. & Metzler, D.E.** (1985) *Transaminases,* Wiley-Interscience, Inc., New York.

**Curthoys, N.P. & Watford, M.** (1995) Regulation of glutaminase activity and glutamine metabolism. *Annu. Rev. Nutr.* **15,** 133–159.

**Fitzpatrick, P.F.** (1999) Tetrahydropterin-dependent amino acid hydroxylases. *Annu. Rev. Biochem.* **68,** 355–382.

**Kirsch, J.F. & Eliot, A.C.** (2004) Pyridoxal phosphate enzymes: mechanistic, structural and evolutionary considerations. *Annu. Rev. Biochem.* **73** [in press].

**Pencharz, P.B. & Ball, R.O.** (2003) Different approaches to define individual amino acid requirements. *Annu. Rev. Nutr.* **23,** 101–116.

Determination of which amino acids are essential in the human diet is not a trivial problem, as this review relates.

### The Urea Cycle

**Brusilow, S.W. & Horwich, A.L.** (2001) Urea cycle enzymes. In *The Metabolic Bases of Inherited Disease,* 8th edn (Scriver, C.R., Beaudet, A.C., Sly, W.S., Valle, D., Childs, B., Kinzler, K., & Vogelstein, B., eds), pp. 1909–1963, McGraw-Hill Companies Inc., New York.

An authoritative source on this pathway.

Holmes, F.L. (1980) Hans Krebs and the discovery of the ornithine cycle. *Fed. Proc.* **39,** 216–225.

> A medical historian reconstructs the events leading to the discovery of the urea cycle.

Kirsch, J.F., Eichele, G., Ford, G.C., Vincent, M.G., Jansonius, J.N., Gehring, H., & Christen, P. (1984) Mechanism of action of aspartate aminotransferase proposed on the basis of its spatial structure. *J. Mol. Biol.* **174,** 497–525.

Morris, S.M. (2002) Regulation of enzymes of the urea cycle and arginine metabolism. *Annu. Rev. Nutr.* **22,** 87–105.

> This review details what is known about some levels of regulation not covered in the chapter, such as hormonal and nutritional regulation.

**Disorders of Amino Acid Degradation**

Ledley, F.D., Levy, H.L., & Woo, S.L.C. (1986) Molecular analysis of the inheritance of phenylketonuria and mild hyperphenylalaninemia in families with both disorders. *N. Engl. J. Med.* **314,** 1276–1280.

Nyhan, W.L. (1984) *Abnormalities in Amino Acid Metabolism in Clinical Medicine,* Appleton-Century-Crofts, Norwalk, CT.

Scriver, C.R., Beaudet, A.L., Sly, W.S., Valle, D., Childs, B., Kinzler, A.W., & Vogelstein, B. (eds) (2001) *The Metabolic and Molecular Bases of Inherited Disease,* 8th edn, Part 5: Amino Acids, McGraw-Hill, Inc., New York.

Scriver, C.R., Kaufman, S., & Woo, S.L.C. (1988) Mendelian hyperphenylalaninemia. *Annu. Rev. Genet.* **22,** 301–321.

## Problems

**1. Products of Amino Acid Transamination** Name and draw the structure of the $\alpha$-keto acid resulting when each of the following amino acids undergoes transamination with $\alpha$-ketoglutarate: (a) aspartate, (b) glutamate, (c) alanine, (d) phenylalanine.

**2. Measurement of Alanine Aminotransferase Activity** The activity (reaction rate) of alanine aminotransferase is usually measured by including an excess of pure lactate dehydrogenase and NADH in the reaction system. The rate of alanine disappearance is equal to the rate of NADH disappearance measured spectrophotometrically. Explain how this assay works.

**3. Distribution of Amino Nitrogen** If your diet is rich in alanine but deficient in aspartate, will you show signs of aspartate deficiency? Explain.

**4. A Genetic Defect in Amino Acid Metabolism: A Case History** A two-year-old child was taken to the hospital. His mother said that he vomited frequently, especially after feedings. The child's weight and physical development were below normal. His hair, although dark, contained patches of white. A urine sample treated with ferric chloride ($FeCl_3$) gave a green color characteristic of the presence of phenylpyruvate. Quantitative analysis of urine samples gave the results shown in the table.

| | Concentration (mM) | |
|---|---|---|
| Substance | Patient's urine | Normal urine |
| Phenylalanine | 7.0 | 0.01 |
| Phenylpyruvate | 4.8 | 0 |
| Phenyllactate | 10.3 | 0 |

(a) Suggest which enzyme might be deficient in this child. Propose a treatment.

(b) Why does phenylalanine appear in the urine in large amounts?

(c) What is the source of phenylpyruvate and phenyllactate? Why does this pathway (normally not functional) come into play when the concentration of phenylalanine rises?

(d) Why does the boy's hair contain patches of white?

**5. Role of Cobalamin in Amino Acid Catabolism** Pernicious anemia is caused by impaired absorption of vitamin $B_{12}$. What is the effect of this impairment on the catabolism of amino acids? Are all amino acids equally affected? (Hint: See Box 17–2.)

**6. Lactate versus Alanine as Metabolic Fuel: The Cost of Nitrogen Removal** The three carbons in lactate and alanine have identical oxidation states, and animals can use either carbon source as a metabolic fuel. Compare the net ATP yield (moles of ATP per mole of substrate) for the complete oxidation (to $CO_2$ and $H_2O$) of lactate versus alanine when the cost of nitrogen excretion as urea is included.

Lactate          Alanine

**7. Pathway of Carbon and Nitrogen in Glutamate Metabolism** When $[2\text{-}^{14}C, ^{15}N]$ glutamate undergoes oxidative degradation in the liver of a rat, in which atoms of the following metabolites will each isotope be found: (a) urea, (b) succinate, (c) arginine, (d) citrulline, (e) ornithine, (f) aspartate?

Labeled glutamate

**8. Chemical Strategy of Isoleucine Catabolism**
Isoleucine is degraded in six steps to propionyl-CoA and acetyl-CoA:

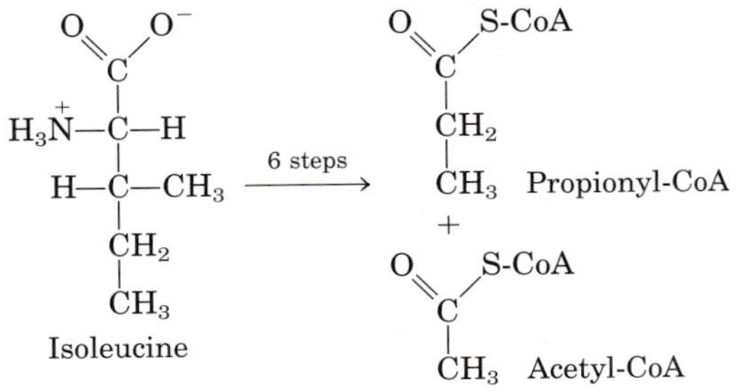

Isoleucine — 6 steps → Propionyl-CoA + Acetyl-CoA

(a) The chemical process of isoleucine degradation includes strategies analogous to those used in the citric acid cycle and the $\beta$ oxidation of fatty acids. The intermediates of isoleucine degradation (I to V) shown below are not in the proper order. Use your knowledge and understanding of the citric acid cycle and $\beta$-oxidation pathway to arrange the intermediates in the proper metabolic sequence for isoleucine degradation.

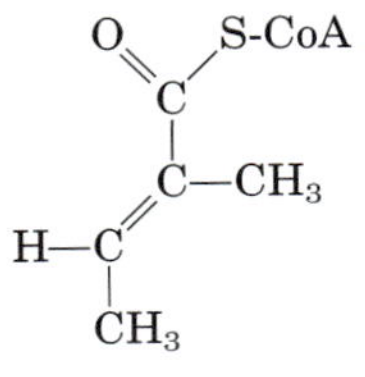

I 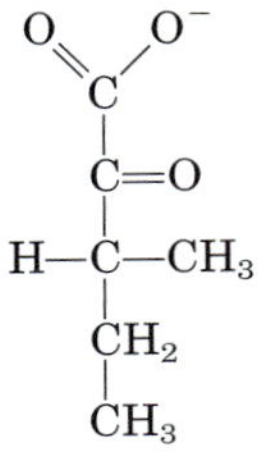   II

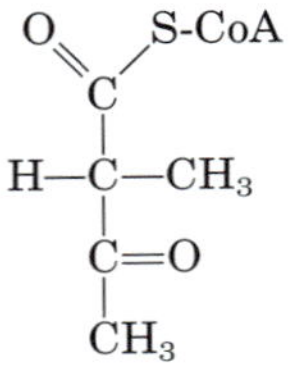

III 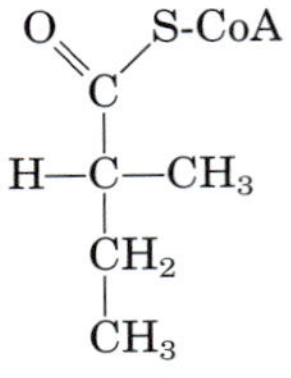   IV

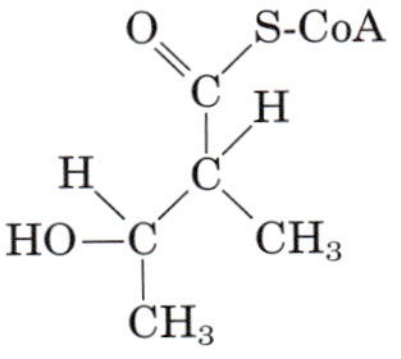

V

(b) For each step you propose, describe the chemical process, provide an analogous example from the citric acid cycle or $\beta$-oxidation pathway (where possible), and indicate any necessary cofactors.

**9. Role of Pyridoxal Phosphate in Glycine Metabolism**
The enzyme serine hydroxymethyltransferase requires pyridoxal phosphate as cofactor. Propose a mechanism for the reaction catalyzed by this enzyme, in the direction of serine degradation (glycine production). (Hint: See Figs 18–19 and 18–20b.)

**10. Parallel Pathways for Amino Acid and Fatty Acid Degradation** The carbon skeleton of leucine is degraded by a series of reactions closely analogous to those of the citric acid cycle and $\beta$ oxidation. For each reaction, (a) through (f), indicate its type, provide an analogous example from the citric acid cycle or $\beta$-oxidation pathway (where possible), and note any necessary cofactors.

Leucine

(a) ↓

$\alpha$-Ketoisocaproate

(b) — CoA-SH → $CO_2$

Isovaleryl-CoA

(c) ↓

$\beta$-Methylcrotonyl-CoA

(d) — $HCO_3^-$

$\beta$-Methylglutaconyl-CoA

(e) — $H_2O$

$\beta$-Hydroxy-$\beta$-methylglutaryl-CoA

(f) ↓

Acetoacetate + Acetyl-CoA

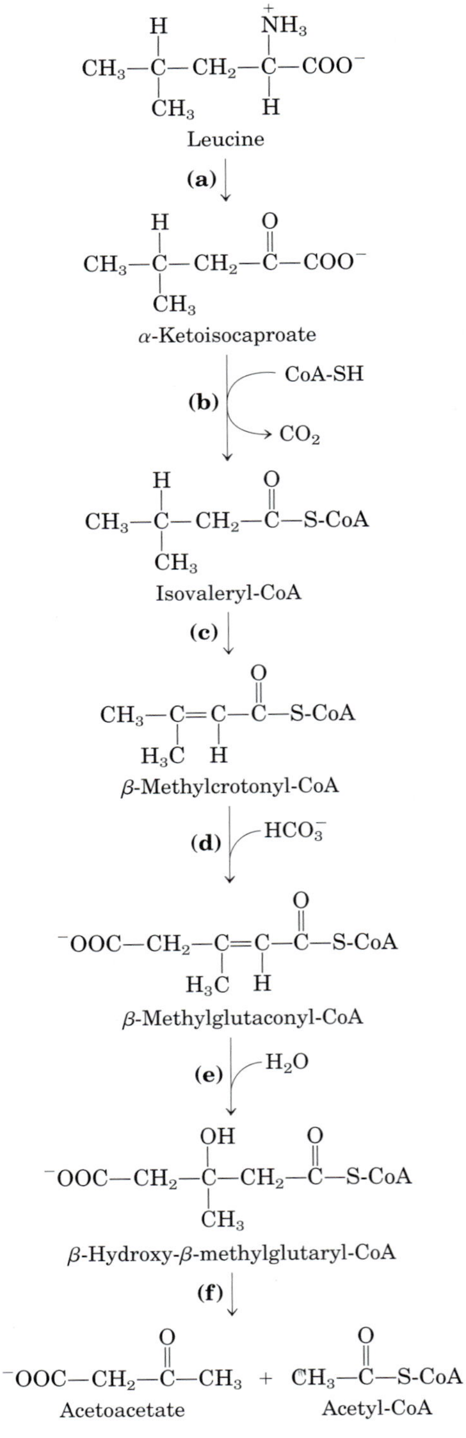

**11. Ammonia Toxicity Resulting from an Arginine-Deficient Diet**  In a study conducted some years ago, cats were fasted overnight then given a single meal complete in all amino acids except arginine. Within 2 hours, blood ammonia levels increased from a normal level of 18 $\mu$g/L to 140 $\mu$g/L, and the cats showed the clinical symptoms of ammonia toxicity. A control group fed a complete amino acid diet or an amino acid diet in which arginine was replaced by ornithine showed no unusual clinical symptoms.

(a) What was the role of fasting in the experiment?

(b) What caused the ammonia levels to rise in the experimental group? Why did the absence of arginine lead to ammonia toxicity? Is arginine an essential amino acid in cats? Why or why not?

(c) Why can ornithine be substituted for arginine?

**12. Oxidation of Glutamate**  Write a series of balanced equations, and an overall equation for the net reaction, describing the oxidation of 2 mol of glutamate to 2 mol of $\alpha$-ketoglutarate and 1 mol of urea.

**13. Transamination and the Urea Cycle**  Aspartate aminotransferase has the highest activity of all the mammalian liver aminotransferases. Why?

**14. The Case against the Liquid Protein Diet**  A weight-reducing diet heavily promoted some years ago required the daily intake of "liquid protein" (soup of hydrolyzed gelatin), water, and an assortment of vitamins. All other food and drink were to be avoided. People on this diet typically lost 10 to 14 lb in the first week.

(a) Opponents argued that the weight loss was almost entirely due to water loss and would be regained very soon after a normal diet was resumed. What is the biochemical basis for this argument?

(b) A number of people on this diet died. What are some of the dangers inherent in the diet, and how can they lead to death?

**15. Alanine and Glutamine in the Blood**  Normal human blood plasma contains all the amino acids required for the synthesis of body proteins, but not in equal concentrations. Alanine and glutamine are present in much higher concentrations than any other amino acids. Suggest why.

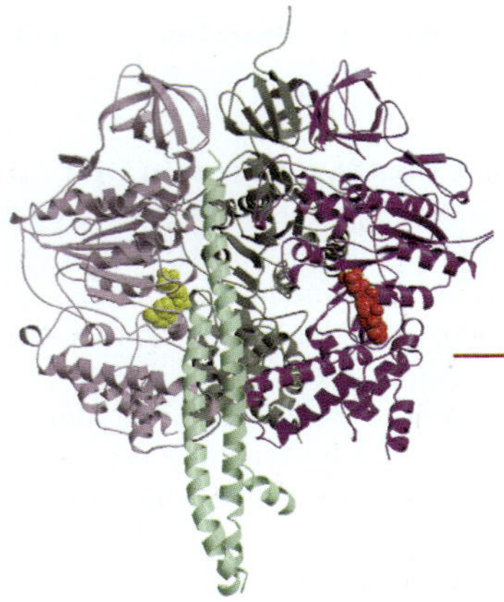

## chapter 19

# OXIDATIVE PHOSPHORYLATION AND PHOTOPHOSPHORYLATION

If an idea presents itself to us, we must not reject it simply because it does not agree with the logical deductions of a reigning theory.

—*Claude Bernard,* An Introduction to the Study of Experimental Medicine, *1813*

The aspect of the present position of consensus that I find most remarkable and admirable, is the altruism and generosity with which former opponents of the chemiosmotic hypothesis have not only come to accept it, but have actively promoted it to the status of a theory.

—*Peter Mitchell, Nobel Address, 1978*

Oxidative phosphorylation is the culmination of energy-yielding metabolism in aerobic organisms. All oxidative steps in the degradation of carbohydrates, fats, and amino acids converge at this final stage of cellular respiration, in which the energy of oxidation drives the synthesis of ATP. Photophosphorylation is the means by which photosynthetic organisms capture the energy of sunlight—the ultimate source of energy in the biosphere—and harness it to make ATP. Together, oxidative phosphorylation and photophosphorylation account for most of the ATP synthesized by most organisms most of the time.

In eukaryotes, oxidative phosphorylation occurs in mitochondria, photophosphorylation in chloroplasts. Oxidative phosphorylation involves the *reduction* of $O_2$ to $H_2O$ with electrons donated by NADH and $FADH_2$; it occurs equally well in light or darkness. Photophosphorylation involves the *oxidation* of $H_2O$ to $O_2$, with $NADP^+$ as ultimate electron acceptor; it is absolutely dependent on the energy of light. Despite their differences, these two highly efficient energy-converting processes have fundamentally similar mechanisms.

Our current understanding of ATP synthesis in mitochondria and chloroplasts is based on the hypothesis, introduced by Peter Mitchell in 1961, that transmembrane differences in proton concentration are the reservoir for the energy extracted from biological oxidation reactions. This **chemiosmotic theory** has been accepted as one of the great unifying principles of twentieth century biology. It provides insight into the processes of oxidative phosphorylation and photophosphorylation, and into such apparently disparate energy transductions as active transport across membranes and the motion of bacterial flagella.

Oxidative phosphorylation and photophosphorylation are mechanistically similar in three respects. (1) Both processes involve the flow of electrons through a chain of membrane-bound carriers. (2) The free energy made available by this "downhill" (exergonic) electron flow is coupled to the "uphill" transport of protons across a proton-impermeable membrane, conserving the free energy of fuel oxidation as a transmembrane electrochemical potential (p. 391). (3) The transmembrane flow of protons down their concentration gradient through specific protein channels provides the free energy for synthesis of ATP, catalyzed by a membrane protein complex (ATP synthase) that couples proton flow to phosphorylation of ADP.

We begin this chapter with oxidative phosphorylation. We first describe the components of the electron-transfer chain, their organization into large functional complexes in the inner mitochondrial membrane, the path of electron flow through them, and the proton movements that accompany this flow. We then consider the remarkable enzyme complex that, by "rotational catalysis," captures the energy of proton flow in ATP, and the regulatory mechanisms that coordinate oxidative phosphorylation with the many catabolic pathways by which fuels are oxidized. With this understanding of mitochondrial oxidative phosphorylation, we turn to photophosphorylation, looking first at the absorption of light by photosynthetic pigments, then at the light-driven flow of electrons from $H_2O$ to $NADP^+$ and the molecular basis for coupling electron and proton flow. We also consider the similarities of structure and mechanism between the ATP synthases of chloroplasts and mitochondria, and the evolutionary basis for this conservation of mechanism.

# OXIDATIVE PHOSPHORYLATION

## 19.1 Electron-Transfer Reactions in Mitochondria

The discovery in 1948 by Eugene Kennedy and Albert Lehninger that mitochondria are the site of oxidative phosphorylation in eukaryotes marked the beginning of the modern phase of studies in biological energy transductions. Mitochondria, like gram-negative bacteria, have two membranes (Fig. 19–1). The outer mitochondrial membrane is readily permeable to small molecules ($M_r$ <5,000) and ions, which move freely through transmembrane channels formed by a family of integral membrane proteins called porins. The inner membrane is impermeable to most small

Albert L. Lehninger,
1917–1986

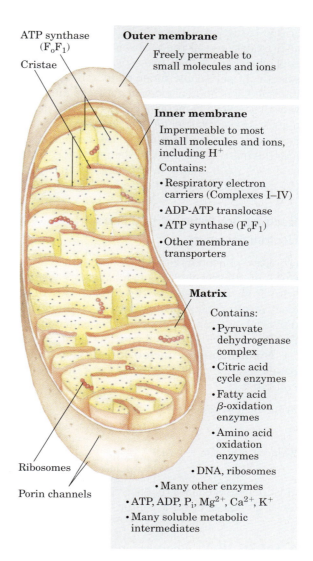

ATP synthase ($F_oF_1$)

Cristae

**Outer membrane**
Freely permeable to small molecules and ions

**Inner membrane**
Impermeable to most small molecules and ions, including $H^+$
Contains:
• Respiratory electron carriers (Complexes I–IV)
• ADP-ATP translocase
• ATP synthase ($F_oF_1$)
• Other membrane transporters

**Matrix**
Contains:
• Pyruvate dehydrogenase complex
• Citric acid cycle enzymes
• Fatty acid β-oxidation enzymes
• Amino acid oxidation enzymes
• DNA, ribosomes
• Many other enzymes
• ATP, ADP, $P_i$, $Mg^{2+}$, $Ca^{2+}$, $K^+$
• Many soluble metabolic intermediates

Ribosomes

Porin channels

**FIGURE 19–1 Biochemical anatomy of a mitochondrion.** The convolutions (cristae) of the inner membrane provide a very large surface area. The inner membrane of a single liver mitochondrion may have more than 10,000 sets of electron-transfer systems (respiratory chains) and ATP synthase molecules, distributed over the membrane surface. Heart mitochondria, which have more profuse cristae and thus a much larger area of inner membrane, contain more than three times as many sets of electron-transfer systems as liver mitochondria. The mitochondrial pool of coenzymes and intermediates is functionally separate from the cytosolic pool. The mitochondria of invertebrates, plants, and microbial eukaryotes are similar to those shown here, but with much variation in size, shape, and degree of convolution of the inner membrane.

molecules and ions, including protons ($H^+$); the only species that cross this membrane do so through specific transporters. The inner membrane bears the components of the respiratory chain and the ATP synthase.

The mitochondrial matrix, enclosed by the inner membrane, contains the pyruvate dehydrogenase complex and the enzymes of the citric acid cycle, the fatty acid $\beta$-oxidation pathway, and the pathways of amino acid oxidation—all the pathways of fuel oxidation except glycolysis, which takes place in the cytosol. The selectively permeable inner membrane segregates the intermediates and enzymes of cytosolic metabolic pathways from those of metabolic processes occurring in the matrix. However, specific transporters carry pyruvate, fatty acids, and amino acids or their $\alpha$-keto derivatives into the matrix for access to the machinery of the citric acid cycle. ADP and $P_i$ are specifically transported into the matrix as newly synthesized ATP is transported out.

## Electrons Are Funneled to Universal Electron Acceptors

Oxidative phosphorylation begins with the entry of electrons into the respiratory chain. Most of these electrons arise from the action of dehydrogenases that collect electrons from catabolic pathways and funnel them into universal electron acceptors—nicotinamide nucleotides ($NAD^+$ or $NADP^+$) or flavin nucleotides (FMN or FAD).

**Nicotinamide nucleotide–linked dehydrogenases** catalyze reversible reactions of the following general types:

$$\text{Reduced substrate} + NAD^+ \rightleftharpoons$$
$$\text{oxidized substrate} + NADH + H^+$$

$$\text{Reduced substrate} + NADP^+ \rightleftharpoons$$
$$\text{oxidized substrate} + NADPH + H^+$$

Most dehydrogenases that act in catabolism are specific for $NAD^+$ as electron acceptor (Table 19–1). Some are

in the cytosol, others are in mitochondria, and still others have mitochondrial and cytosolic isozymes.

NAD-linked dehydrogenases remove two hydrogen atoms from their substrates. One of these is transferred as a hydride ion ($:H^-$) to $NAD^+$; the other is released as $H^+$ in the medium (see Fig. 13–15). NADH and NADPH are water-soluble electron carriers that associate *reversibly* with dehydrogenases. NADH carries electrons from catabolic reactions to their point of entry into the respiratory chain, the NADH dehydrogenase complex described below. NADPH generally supplies electrons to anabolic reactions. Cells maintain separate pools of NADPH and NADH, with different redox potentials. This is accomplished by holding the ratios of [reduced form]/[oxidized form] relatively high for NADPH and relatively low for NADH. Neither NADH nor NADPH can cross the inner mitochondrial membrane, but the electrons they carry can be shuttled across indirectly, as we shall see.

**Flavoproteins** contain a very tightly, sometimes covalently, bound flavin nucleotide, either FMN or FAD (see Fig. 13–18). The oxidized flavin nucleotide can accept either one electron (yielding the semiquinone form) or two (yielding $FADH_2$ or $FMNH_2$). Electron transfer occurs because the flavoprotein has a higher reduction potential than the compound oxidized. The standard reduction potential of a flavin nucleotide, unlike that of NAD or NADP, depends on the protein with which it is associated. Local interactions with functional groups in the protein distort the electron orbitals in the flavin ring, changing the relative stabilities of oxidized and reduced forms. The relevant standard reduction

---

**TABLE 19–1 Some Important Reactions Catalyzed by NAD(P)H-Linked Dehydrogenases**

| Reaction[*] | Location[†] |
|---|---|
| **NAD-linked** | |
| $\alpha$-Ketoglutarate + CoA + $NAD^+$ $\rightleftharpoons$ succinyl-CoA + $CO_2$ + NADH + $H^+$ | M |
| L-Malate + $NAD^+$ $\rightleftharpoons$ oxaloacetate + NADH + $H^+$ | M and C |
| Pyruvate + CoA + $NAD^+$ $\rightleftharpoons$ acetyl-CoA + $CO_2$ + NADH + $H^+$ | M |
| Glyceraldehyde 3-phosphate + $P_i$ + $NAD^+$ $\rightleftharpoons$ 1,3-bisphosphoglycerate + NADH + $H^+$ | C |
| Lactate + $NAD^+$ $\rightleftharpoons$ pyruvate + NADH + $H^+$ | C |
| $\beta$-Hydroxyacyl-CoA + $NAD^+$ $\rightleftharpoons$ $\beta$-ketoacyl-CoA + NADH + $H^+$ | M |
| **NADP-linked** | |
| Glucose 6-phosphate + $NADP^+$ $\rightleftharpoons$ 6-phosphogluconate + NADPH + $H^+$ | C |
| **NAD- or NADP-linked** | |
| L-Glutamate + $H_2O$ + $NAD(P)^+$ $\rightleftharpoons$ $\alpha$-ketoglutarate + $NH_4^+$ + NAD(P)H | M |
| Isocitrate + $NAD(P)^+$ $\rightleftharpoons$ $\alpha$-ketoglutarate + $CO_2$ + NAD(P)H + $H^+$ | M and C |

[*]These reactions and their enzymes are discussed in Chapters 14 through 18.
[†]M designates mitochondria; C, cytosol.

potential is therefore that of the particular flavoprotein, not that of isolated FAD or FMN. The flavin nucleotide should be considered part of the flavoprotein's active site rather than a reactant or product in the electron-transfer reaction. Because flavoproteins can participate in either one- or two-electron transfers, they can serve as intermediates between reactions in which two electrons are donated (as in dehydrogenations) and those in which only one electron is accepted (as in the reduction of a quinone to a hydroquinone, described below).

## Electrons Pass through a Series of Membrane-Bound Carriers

The mitochondrial respiratory chain consists of a series of sequentially acting electron carriers, most of which are integral proteins with prosthetic groups capable of accepting and donating either one or two electrons. Three types of electron transfers occur in oxidative phosphorylation: (1) direct transfer of electrons, as in the reduction of $Fe^{3+}$ to $Fe^{2+}$; (2) transfer as a hydrogen atom ($H^+ + e^-$); and (3) transfer as a hydride ion ($:H^-$), which bears two electrons. The term **reducing equivalent** is used to designate a single electron equivalent transferred in an oxidation-reduction reaction.

In addition to NAD and flavoproteins, three other types of electron-carrying molecules function in the respiratory chain: a hydrophobic quinone (ubiquinone) and two different types of iron-containing proteins (cytochromes and iron-sulfur proteins). **Ubiquinone** (also called **coenzyme Q,** or simply **Q**) is a lipid-soluble benzoquinone with a long isoprenoid side chain (Fig. 19–2). The closely related compounds plastoquinone (of plant chloroplasts) and menaquinone (of bacteria) play roles analogous to that of ubiquinone, carrying electrons in membrane-associated electron-transfer chains. Ubiquinone can accept one electron to become the semiquinone radical ($^\cdot$QH) or two electrons to form ubiquinol ($QH_2$) (Fig. 19–2) and, like flavoprotein carriers, it can act at the junction between a two-electron donor and a one-electron acceptor. Because ubiquinone is both small and hydrophobic, it is freely diffusible within the lipid bilayer of the inner mitochondrial membrane and can shuttle reducing equivalents between other, less mobile electron carriers in the membrane. And because it carries both electrons and protons, it plays a central role in coupling electron flow to proton movement.

The **cytochromes** are proteins with characteristic strong absorption of visible light, due to their iron-containing heme prosthetic groups (Fig. 19–3). Mitochondria contain three classes of cytochromes, designated $a$, $b$, and $c$, which are distinguished by differences in their light-absorption spectra. Each type of cytochrome in its reduced ($Fe^{2+}$) state has three absorption bands in the visible range (Fig. 19–4). The longest-wavelength band is near 600 nm in type $a$ cytochromes,

**FIGURE 19–2 Ubiquinone (Q, or coenzyme Q).** Complete reduction of ubiquinone requires two electrons and two protons, and occurs in two steps through the semiquinone radical intermediate.

near 560 nm in type $b$, and near 550 nm in type $c$. To distinguish among closely related cytochromes of one type, the exact absorption maximum is sometimes used in the names, as in cytochrome $b_{562}$.

The heme cofactors of $a$ and $b$ cytochromes are tightly, but not covalently, bound to their associated proteins; the hemes of $c$-type cytochromes are covalently attached through Cys residues (Fig. 19–3). As with the flavoproteins, the standard reduction potential of the heme iron atom of a cytochrome depends on its interaction with protein side chains and is therefore different for each cytochrome. The cytochromes of type $a$ and $b$ and some of type $c$ are integral proteins of the inner mitochondrial membrane. One striking exception is the cytochrome $c$ of mitochondria, a soluble protein that associates through electrostatic interactions with the outer surface of the inner membrane. We encountered cytochrome $c$ in earlier discussions of protein structure (see Fig. 4–18).

In **iron-sulfur proteins,** first discovered by Helmut Beinert, the iron is present not in heme but in association with inorganic sulfur atoms or with the sulfur atoms of Cys residues in the protein, or both. These iron-sulfur (Fe-S)

Helmut Beinert

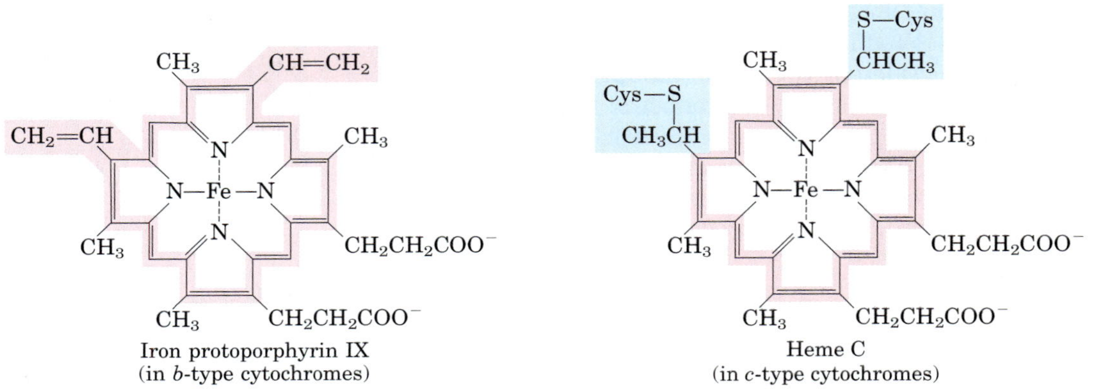

CH$_3$   CH=CH$_2$

CH$_2$=CH   CH$_3$

N—Fe—N

CH$_3$   CH$_2$CH$_2$COO$^-$

CH$_3$   CH$_2$CH$_2$COO$^-$

Iron protoporphyrin IX
(in *b*-type cytochromes)

S—Cys

CH$_3$   CHCH$_3$

Cys—S
CH$_3$CH   CH$_3$

N—Fe—N

CH$_3$   CH$_2$CH$_2$COO$^-$

CH$_3$   CH$_2$CH$_2$COO$^-$

Heme C
(in *c*-type cytochromes)

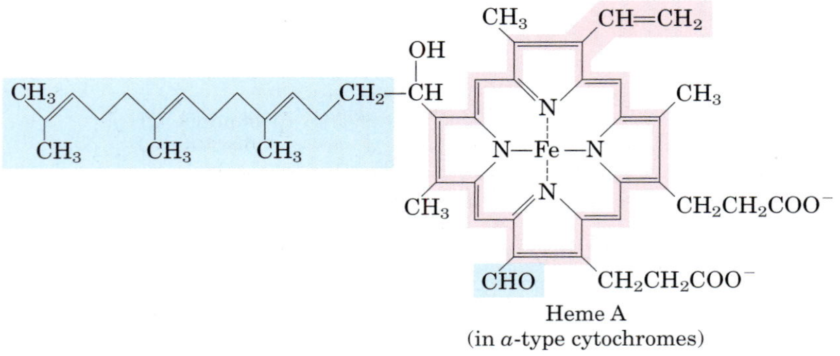

CH$_3$   CH=CH$_2$

OH

CH$_3$   CH$_2$—CH   CH$_3$

CH$_3$   CH$_3$   CH$_3$

N—Fe—N

CH$_3$   CH$_2$CH$_2$COO$^-$

CHO   CH$_2$CH$_2$COO$^-$

Heme A
(in *a*-type cytochromes)

**FIGURE 19–3 Prosthetic groups of cytochromes.** Each group consists of four five-membered, nitrogen-containing rings in a cyclic structure called a porphyrin. The four nitrogen atoms are coordinated with a central Fe ion, either Fe$^{2+}$ or Fe$^{3+}$. Iron protoporphyrin IX is found in *b*-type cytochromes and in hemoglobin and myoglobin (see Fig. 4–17). Heme *c* is covalently bound to the protein of cytochrome *c* through thioether bonds to two Cys residues. Heme *a*, found in the *a*-type cytochromes, has a long isoprenoid tail attached to one of the five-membered rings. The conjugated double-bond system (shaded pink) of the porphyrin ring accounts for the absorption of visible light by these hemes.

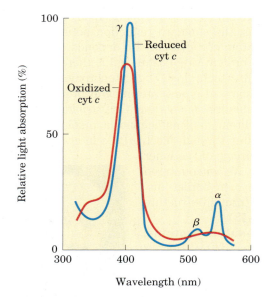

**FIGURE 19–4** Absorption spectra of cytochrome *c* (cyt *c*) in its oxidized (red) and reduced (blue) forms. Also labeled are the characteristic $\alpha$, $\beta$, and $\gamma$ bands of the reduced form.

centers range from simple structures with a single Fe atom coordinated to four Cys —SH groups to more complex Fe-S centers with two or four Fe atoms (Fig. 19–5). **Rieske iron-sulfur proteins** (named after their discoverer, John S. Rieske) are a variation on this theme, in which one Fe atom is coordinated to two His residues rather than two Cys residues. All iron-sulfur proteins participate in one-electron transfers in which one iron atom of the iron-sulfur cluster is oxidized or reduced. At least eight Fe-S proteins function in mitochondrial electron transfer. The reduction potential of Fe-S proteins varies from −0.65 V to +0.45 V, depending on the microenvironment of the iron within the protein.

In the overall reaction catalyzed by the mitochondrial respiratory chain, electrons move from NADH, succinate, or some other primary electron donor through flavoproteins, ubiquinone, iron-sulfur proteins, and cytochromes, and finally to O$_2$. A look at the methods used to determine the sequence in which the carriers act is instructive, as the same general approaches have been used to study other electron-transfer chains, such as those of chloroplasts.

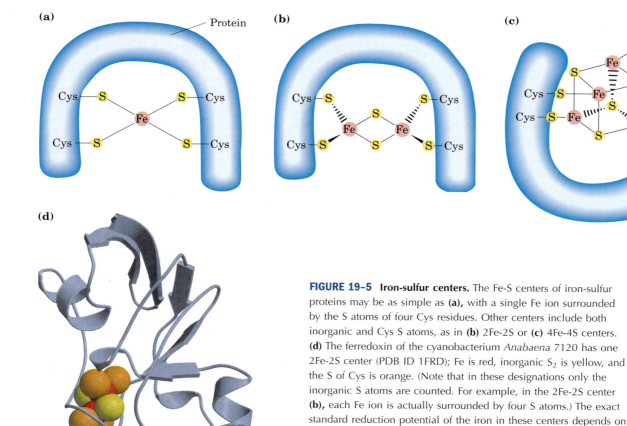

**FIGURE 19–5 Iron-sulfur centers.** The Fe-S centers of iron-sulfur proteins may be as simple as **(a)**, with a single Fe ion surrounded by the S atoms of four Cys residues. Other centers include both inorganic and Cys S atoms, as in **(b)** 2Fe-2S or **(c)** 4Fe-4S centers. **(d)** The ferredoxin of the cyanobacterium *Anabaena* 7120 has one 2Fe-2S center (PDB ID 1FRD); Fe is red, inorganic $S_2$ is yellow, and the S of Cys is orange. (Note that in these designations only the inorganic S atoms are counted. For example, in the 2Fe-2S center **(b)**, each Fe ion is actually surrounded by four S atoms.) The exact standard reduction potential of the iron in these centers depends on the type of center and its interaction with the associated protein.

First, the standard reduction potentials of the individual electron carriers have been determined experimentally (Table 19–2). We would expect the carriers to function in order of increasing reduction potential, because electrons tend to flow spontaneously from carriers of lower $E'^\circ$ to carriers of higher $E'^\circ$. The order of carriers deduced by this method is NADH →

Q → cytochrome $b$ → cytochrome $c_1$ → cytochrome $c$ → cytochrome $a$ → cytochrome $a_3$ → $O_2$. Note, however, that the order of standard reduction potentials is not necessarily the same as the order of *actual* reduction potentials under cellular conditions, which depend on the concentration of reduced and oxidized forms (p. 510). A second method for determining the sequence

**TABLE 19–2 Standard Reduction Potentials of Respiratory Chain and Related Electron Carriers**

| Redox reaction (half-reaction) | $E'^\circ$ (V) |
|---|---|
| $2H^+ + 2e^- \longrightarrow H_2$ | −0.414 |
| $NAD^+ + H^+ + 2e^- \longrightarrow NADH$ | −0.320 |
| $NADP^+ + H^+ + 2e^- \longrightarrow NADPH$ | −0.324 |
| NADH dehydrogenase (FMN) $+ 2H^+ + 2e^- \longrightarrow$ NADH dehydrogenase $(FMNH_2)$ | −0.30 |
| Ubiquinone $+ 2H^+ + 2e^- \longrightarrow$ ubiquinol | 0.045 |
| Cytochrome $b$ $(Fe^{3+}) + e^- \longrightarrow$ cytochrome $b$ $(Fe^{2+})$ | 0.077 |
| Cytochrome $c_1$ $(Fe^{3+}) + e^- \longrightarrow$ cytochrome $c_1$ $(Fe^{2+})$ | 0.22 |
| Cytochrome $c$ $(Fe^{3+}) + e^- \longrightarrow$ cytochrome $c$ $(Fe^{2+})$ | 0.254 |
| Cytochrome $a$ $(Fe^{3+}) + e^- \longrightarrow$ cytochrome $a$ $(Fe^{2+})$ | 0.29 |
| Cytochrome $a_3$ $(Fe^{3+}) + e^- \longrightarrow$ cytochrome $a_3$ $(Fe^{2+})$ | 0.35 |
| $\frac{1}{2}O_2 + 2H^+ + 2e^- \longrightarrow H_2O$ | 0.8166 |

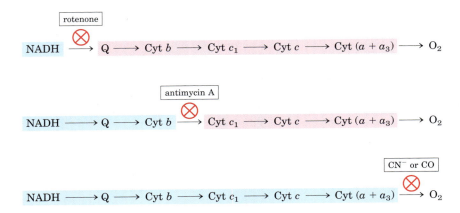

**FIGURE 19–6 Method for determining the sequence of electron carriers.** This method measures the effects of inhibitors of electron transfer on the oxidation state of each carrier. In the presence of an electron donor and $O_2$, each inhibitor causes a characteristic pattern of oxidized/reduced carriers: those before the block become reduced (blue), and those after the block become oxidized (pink).

of electron carriers involves reducing the entire chain of carriers experimentally by providing an electron source but no electron acceptor (no $O_2$). When $O_2$ is suddenly introduced into the system, the rate at which each electron carrier becomes oxidized (measured spectroscopically) reveals the order in which the carriers function. The carrier nearest $O_2$ (at the end of the chain) gives up its electrons first, the second carrier from the end is oxidized next, and so on. Such experiments have confirmed the sequence deduced from standard reduction potentials.

In a final confirmation, agents that inhibit the flow of electrons through the chain have been used in combination with measurements of the degree of oxidation of each carrier. In the presence of $O_2$ and an electron donor, carriers that function before the inhibited step become fully reduced, and those that function after this step are completely oxidized (Fig. 19–6). By using several inhibitors that block different steps in the chain, investigators have determined the entire sequence; it is the same as deduced in the first two approaches.

## Electron Carriers Function in Multienzyme Complexes

The electron carriers of the respiratory chain are organized into membrane-embedded supramolecular

complexes that can be physically separated. Gentle treatment of the inner mitochondrial membrane with detergents allows the resolution of four unique electron-carrier complexes, each capable of catalyzing electron transfer through a portion of the chain (Table 19–3; Fig. 19–7). Complexes I and II catalyze electron transfer to ubiquinone from two different electron donors: NADH (Complex I) and succinate (Complex II). Complex III carries electrons from reduced ubiquinone to cytochrome $c$, and Complex IV completes the sequence by transferring electrons from cytochrome $c$ to $O_2$.

We now look in more detail at the structure and function of each complex of the mitochondrial respiratory chain.

**Complex I: NADH to Ubiquinone** Figure 19–8 illustrates the relationship between Complexes I and II and ubiquinone. **Complex I,** also called **NADH:ubiquinone oxidoreductase** or **NADH dehydrogenase,** is a large enzyme composed of 42 different polypeptide chains, including an FMN-containing flavoprotein and at least six iron-sulfur centers. High-resolution electron microscopy shows Complex I to be L-shaped, with one arm of the L in the membrane and the other extending into the matrix. As shown in Figure 19–9, Complex I catalyzes two simultaneous and obligately coupled processes: (1) the

**TABLE 19–3 The Protein Components of the Mitochondrial Electron-Transfer Chain**

| Enzyme complex/protein | Mass (kDa) | Number of subunits[*] | Prosthetic group(s) |
|---|---|---|---|
| I NADH dehydrogenase | 850 | 43 (14) | FMN, Fe-S |
| II Succinate dehydrogenase | 140 | 4 | FAD, Fe-S |
| III Ubiquinone cytochrome $c$ oxidoreductase | 250 | 11 | Hemes, Fe-S |
| Cytochrome $c$[†] | 13 | 1 | Heme |
| IV Cytochrome oxidase | 160 | 13 (3–4) | Hemes; $Cu_A$, $Cu_B$ |

[*]Numbers of subunits in the bacterial equivalents in parentheses.
[†]Cytochrome $c$ is not part of an enzyme complex; it moves between Complexes III and IV as a freely soluble protein.

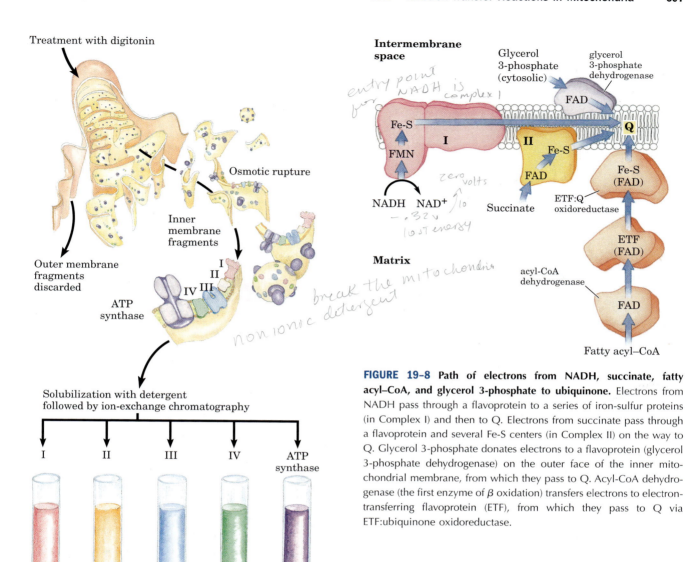

**FIGURE 19–7 Separation of functional complexes of the respiratory chain.** The outer mitochondrial membrane is first removed by treatment with the detergent digitonin. Fragments of inner membrane are then obtained by osmotic rupture of the mitochondria, and the fragments are gently dissolved in a second detergent. The resulting mixture of inner membrane proteins is resolved by ion-exchange chromatography into different complexes (I through IV) of the respiratory chain, each with its unique protein composition (see Table 19–3), and the enzyme ATP synthase (sometimes called Complex V). The isolated Complexes I through IV catalyze transfers between donors (NADH and succinate), intermediate carriers (Q and cytochrome $c$), and $O_2$, as shown. In vitro, isolated ATP synthase has only ATP-hydrolyzing (ATPase), not ATP-synthesizing, activity.

**FIGURE 19–8 Path of electrons from NADH, succinate, fatty acyl–CoA, and glycerol 3-phosphate to ubiquinone.** Electrons from NADH pass through a flavoprotein to a series of iron-sulfur proteins (in Complex I) and then to Q. Electrons from succinate pass through a flavoprotein and several Fe-S centers (in Complex II) on the way to Q. Glycerol 3-phosphate donates electrons to a flavoprotein (glycerol 3-phosphate dehydrogenase) on the outer face of the inner mitochondrial membrane, from which they pass to Q. Acyl-CoA dehydrogenase (the first enzyme of β oxidation) transfers electrons to electron-transferring flavoprotein (ETF), from which they pass to Q via ETF:ubiquinone oxidoreductase.

exergonic transfer to ubiquinone of a hydride ion from NADH and a proton from the matrix, expressed by

$$NADH + H^+ + Q \longrightarrow NAD^+ + QH_2 \qquad (19–1)$$

and (2) the endergonic transfer of four protons from the matrix to the intermembrane space. Complex I is therefore a proton pump driven by the energy of electron transfer, and the reaction it catalyzes is **vectorial**: it moves protons in a specific direction from one location (the matrix, which becomes negatively charged with the departure of protons) to another (the intermembrane space, which becomes positively charged). To emphasize the vectorial nature of the process, the overall reaction is often written with subscripts that indicate the location of the protons: P for the positive side of the inner membrane (the intermembrane space), N for the negative side (the matrix):

$$NADH + 5H_N^+ + Q \longrightarrow NAD^+ + QH_2 + 4H_P^+ \qquad (19–2)$$

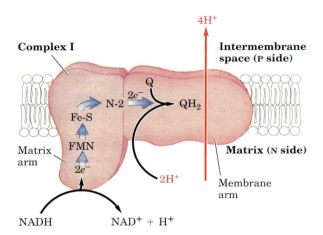

**FIGURE 19–9 NADH:ubiquinone oxidoreductase (Complex I).** Complex I catalyzes the transfer of a hydride ion from NADH to FMN, from which two electrons pass through a series of Fe-S centers to the iron-sulfur protein N-2 in the matrix arm of the complex. Electron transfer from N-2 to ubiquinone on the membrane arm forms QH₂, which diffuses into the lipid bilayer. This electron transfer also drives the expulsion from the matrix of four protons per pair of electrons. The detailed mechanism that couples electron and proton transfer in Complex I is not yet known, but probably involves a Q cycle similar to that in Complex III in which QH₂ participates twice per electron pair (see Fig. 19–12). Proton flux produces an electrochemical potential across the inner mitochondrial membrane (N side negative, P side positive), which conserves some of the energy released by the electron-transfer reactions. This electrochemical potential drives ATP synthesis.

Amytal (a barbiturate drug), rotenone (a plant product commonly used as an insecticide), and piericidin A (an antibiotic) inhibit electron flow from the Fe-S centers of Complex I to ubiquinone (Table 19–4) and therefore block the overall process of oxidative phosphorylation.

Ubiquinol (QH₂, the fully reduced form; Fig. 19–2) diffuses in the inner mitochondrial membrane from Complex I to Complex III, where it is oxidized to Q in a process that also involves the outward movement of H⁺.

***Complex II: Succinate to Ubiquinone*** We encountered **Complex II** in Chapter 16 as **succinate dehydrogenase,** the only membrane-bound enzyme in the citric acid cycle (p. 612). Although smaller and simpler than Complex I, it contains five prosthetic groups of two types and four different protein subunits (Fig. 19–10). Subunits C and D are integral membrane proteins, each with three transmembrane helices. They contain a heme group, heme *b,* and a binding site for ubiquinone, the final electron acceptor in the reaction catalyzed by Complex II. Subunits A and B extend into the matrix (or the cytosol of a bacterium); they contain three 2Fe-2S centers, bound FAD, and a binding site for the substrate, succinate. The path of electron transfer from the succinate-binding site to FAD, then through the Fe-S centers to the Q-binding site, is more than 40 Å long, but none of the individual electron-transfer distances exceeds about 11 Å—a reasonable distance for rapid electron transfer (Fig. 19–10).

**TABLE 19–4  Agents That Interfere with Oxidative Phosphorylation or Photophosphorylation**

| Type of interference | Compound* | Target/mode of action |
|---|---|---|
| Inhibition of electron transfer | Cyanide<br>Carbon monoxide | Inhibit cytochrome oxidase |
| | Antimycin A | Blocks electron transfer from cytochrome *b* to cytochrome $c_1$ |
| | Myxothiazol<br>Rotenone<br>Amytal<br>Piericidin A | Prevent electron transfer from Fe-S center to ubiquinone |
| | DCMU | Competes with $Q_B$ for binding site in PSII |
| Inhibition of ATP synthase | Aurovertin | Inhibits $F_1$ |
| | Oligomycin<br>Venturicidin | Inhibit $F_o$ and $CF_o$ |
| | DCCD | Blocks proton flow through $F_o$ and $CF_o$ |
| Uncoupling of phosphorylation from electron transfer | FCCP<br>DNP | Hydrophobic proton carriers |
| | Valinomycin | $K^+$ ionophore |
| | Thermogenin | In brown fat, forms proton-conducting pores in inner mitochondrial membrane |
| Inhibition of ATP-ADP exchange | Atractyloside | Inhibits adenine nucleotide translocase |

*DCMU is 3-(3,4-dichlorophenyl)-1,1-dimethylurea; DCCD, dicyclohexylcarbodiimide; FCCP, cyanide-*p*-trifluoromethoxyphenylhydrazone; DNP, 2,4-dinitrophenol.

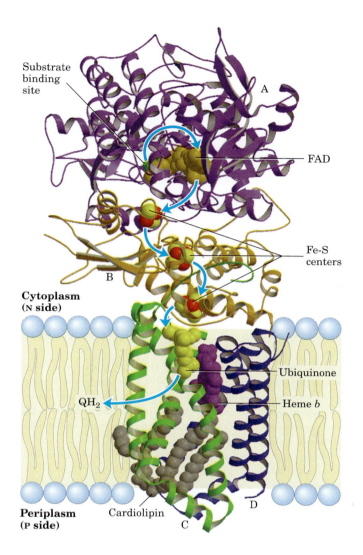

Substrate
binding
site

A

FAD

Fe-S
centers

B

**Cytoplasm**
(N side)

QH₂

Ubiquinone

Heme *b*

**Periplasm**
(P side)

Cardiolipin

C

D

**FIGURE 19–10 Structure of Complex II (succinate dehydrogenase) of *E. coli*** (PDB ID 1NEK). The enzyme has two transmembrane subunits, C (green) and D (blue); the cytoplasmic extensions contain subunits B (orange) and A (purple). Just behind the FAD in subunit A (gold) is the binding site for succinate (occupied in this crystal structure by the inhibitor oxaloacetate, green). Subunit B has three sets of Fe-S centers (yellow and red); ubiquinone (yellow) is bound to subunit C; and heme *b* (purple) is sandwiched between subunits C and D. A cardiolipin molecule is so tightly bound to subunit C that it shows up in the crystal structure (gray spacefilling). Electrons move (blue arrows) from succinate to FAD, then through the three Fe-S centers to ubiquinone. The heme *b* is not on the main path of electron transfer but protects against the formation of reactive oxygen species (ROS) by electrons that go astray.

flavoprotein **acyl-CoA dehydrogenase** (see Fig. 17–8), involves transfer of electrons from the substrate to the FAD of the dehydrogenase, then to electron-transferring flavoprotein (ETF), which in turn passes its electrons to **ETF:ubiquinone oxidoreductase** (Fig. 19–8). This enzyme transfers electrons into the respiratory chain by reducing ubiquinone. Glycerol 3-phosphate, formed either from glycerol released by triacylglycerol breakdown or by the reduction of dihydroxyacetone phosphate from glycolysis, is oxidized by **glycerol 3-phosphate dehydrogenase** (see Fig. 17–4). This enzyme is a flavoprotein located on the outer face of the inner mitochondrial membrane, and like succinate dehydrogenase and acyl-CoA dehydrogenase it channels electrons into the respiratory chain by reducing ubiquinone (Fig. 19–8). The important role of glycerol 3-phosphate dehydrogenase in shuttling reducing equivalents from cytosolic NADH into the mitochondrial matrix is described in Section 19.2 (see Fig. 19–28). The effect of each of these electron-transferring enzymes is to contribute to the pool of reduced ubiquinone. QH₂ from all these reactions is reoxidized by Complex III.

***Complex III: Ubiquinone to Cytochrome c*** The next respiratory complex, **Complex III**, also called **cytochrome *bc₁* complex** or **ubiquinone:cytochrome *c* oxidoreductase,** couples the transfer of electrons from ubiquinol (QH₂) to cytochrome *c* with the vectorial transport of protons from the matrix to the intermembrane space. The determination of the complete structure of this huge complex (Fig. 19–11) and of Complex IV (below) by x-ray crystallography, achieved between 1995 and 1998, were landmarks in the study of mitochondrial electron transfer, providing the structural framework to integrate the many biochemical observations on the functions of the respiratory complexes.

Based on the structure of Complex III and detailed biochemical studies of the redox reactions, a reasonable model has been proposed for the passage of electrons

The heme *b* of Complex II is apparently not in the direct path of electron transfer; it may serve instead to reduce the frequency with which electrons "leak" out of the system, moving from succinate to molecular oxygen to produce the **reactive oxygen species (ROS)** hydrogen peroxide (H₂O₂) and the **superoxide radical** (˙O₂⁻) described in Section 19.5. Humans with point mutations in Complex II subunits near heme *b* or the quinone-binding site suffer from hereditary paraganglioma. This inherited condition is characterized by benign tumors of the head and neck, commonly in the carotid body, an organ that senses O₂ levels in the blood. These mutations result in greater production of ROS and perhaps greater tissue damage during succinate oxidation. ■

Other substrates for mitochondrial dehydrogenases pass electrons into the respiratory chain at the level of ubiquinone, but not through Complex II. The first step in the β oxidation of fatty acyl–CoA, catalyzed by the

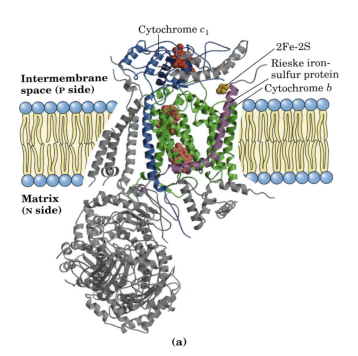

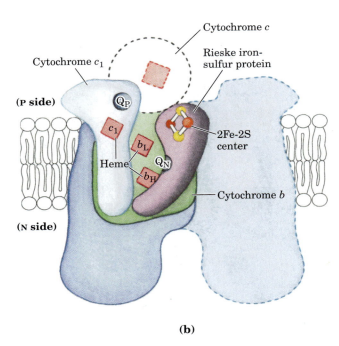

**(a)**                    **(b)**

**FIGURE 19–11 Cytochrome $bc_1$ complex (Complex III).** The complex is a dimer of identical monomers, each with 11 different subunits. **(a)** Structure of a monomer. The functional core is three subunits: cytochrome $b$ (green) with its two hemes ($b_H$ and $b_L$, light red); the Rieske iron-sulfur protein (purple) with its 2Fe-2S centers (yellow); and cytochrome $c_1$ (blue) with its heme (red) (PDB ID 1BGY). **(b)** The dimeric functional unit. Cytochrome $c_1$ and the Rieske iron-sulfur protein project from the P surface and can interact with cytochrome $c$ (not part of the functional complex) in the intermembrane space. The complex has two distinct binding sites for ubiquinone, $Q_N$ and $Q_P$, which correspond to the sites of inhibition by two drugs that block oxidative phosphorylation. Antimycin A, which blocks electron flow from heme $b_H$ to Q, binds at $Q_N$, close to heme $b_H$ on the N (matrix) side of the membrane. Myxothiazol, which prevents electron flow from

$QH_2$ to the Rieske iron-sulfur protein, binds at $Q_P$, near the 2Fe-2S center and heme $b_L$ on the P side. The dimeric structure is essential to the function of Complex III. The interface between monomers forms two pockets, each containing a $Q_P$ site from one monomer and a $Q_N$ site from the other. The ubiquinone intermediates move within these sheltered pockets.

Complex III crystallizes in two distinct conformations (not shown). In one, the Rieske Fe-S center is close to its electron acceptor, the heme of cytochrome $c_1$, but relatively distant from cytochrome $b$ and the $QH_2$-binding site at which the Rieske Fe-S center receives electrons. In the other, the Fe-S center has moved away from cytochrome $c_1$ and toward cytochrome $b$. The Rieske protein is thought to oscillate between these two conformations as it is reduced, then oxidized.

and protons through the complex. The net equation for the redox reactions of this **Q cycle** (Fig. 19–12) is

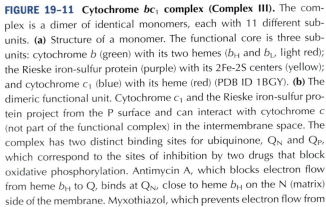

$$QH_2 + 2\text{ cyt }c_1(\text{oxidized}) + 2H_N^+ \longrightarrow$$
$$Q + 2\text{ cyt }c_1(\text{reduced}) + 4H_P^+ \quad (19\text{–}3)$$

The Q cycle accommodates the switch between the two-electron carrier ubiquinone and the one-electron carriers—cytochromes $b_{562}$, $b_{566}$, $c_1$, and $c$—and explains the measured stoichiometry of four protons translocated per pair of electrons passing through the Complex III to cytochrome $c$. Although the path of electrons through this segment of the respiratory chain is complicated, the net effect of the transfer is simple: $QH_2$ is oxidized to Q and two molecules of cytochrome $c$ are reduced.

Cytochrome $c$ (see Fig. 4–18) is a soluble protein of the intermembrane space. After its single heme accepts an electron from Complex III, cytochrome $c$ moves to Complex IV to donate the electron to a binuclear copper center.

**Complex IV: Cytochrome c to O₂**    In the final step of the respiratory chain, **Complex IV,** also called **cytochrome oxidase,** carries electrons from cytochrome $c$ to molecular oxygen, reducing it to $H_2O$. Complex IV is a large enzyme (13 subunits; $M_r$ 204,000) of the inner mitochondrial membrane. Bacteria contain a form that is much simpler, with only three or four subunits, but still capable of catalyzing both electron transfer and proton pumping. Comparison of the mitochondrial and bacterial complexes suggests that three subunits are critical to the function (Fig. 19–13).

Mitochondrial subunit II contains two Cu ions complexed with the —SH groups of two Cys residues in a binuclear center ($Cu_A$; Fig. 19–13b) that resembles the 2Fe-2S centers of iron-sulfur proteins. Subunit I contains two heme groups, designated $a$ and $a_3$, and another copper ion ($Cu_B$). Heme $a_3$ and $Cu_B$ form a second binuclear center that accepts electrons from heme $a$ and transfers them to $O_2$ bound to heme $a_3$.

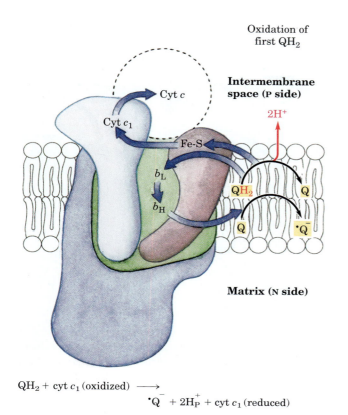

Oxidation of first QH$_2$

**Intermembrane space (P side)**

Cyt $c$

Cyt $c_1$

Fe-S

2H$^+$

$b_L$

QH$_2$    Q

$b_H$

Q    ·Q$^-$

**Matrix (N side)**

QH$_2$ + cyt $c_1$ (oxidized) $\longrightarrow$

·Q$^-$ + 2H$_P^+$ + cyt $c_1$ (reduced)

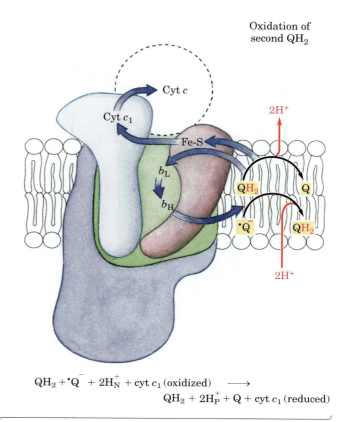

Oxidation of second QH$_2$

Cyt $c$

Cyt $c_1$

Fe-S

2H$^+$

$b_L$

QH$_2$    Q

$b_H$

·Q    QH$_2$

2H$^+$

QH$_2$ + ·Q$^-$ + 2H$_N^+$ + cyt $c_1$ (oxidized) $\longrightarrow$

QH$_2$ + 2H$_P^+$ + Q + cyt $c_1$ (reduced)

Net equation:

QH$_2$ + 2 cyt $c_1$ (oxidized) + 2H$_N^+$ $\longrightarrow$ Q + 2 cyt $c_1$ (reduced) + 4H$_P^+$

**FIGURE 19–12 The Q cycle.** The path of electrons through Complex III is shown by blue arrows. On the P side of the membrane, two molecules of QH$_2$ are oxidized to Q near the P side, releasing two protons per Q (four protons in all) into the intermembrane space. Each

QH$_2$ donates one electron (via the Rieske Fe-S center) to cytochrome $c_1$, and one electron (via cytochrome $b$) to a molecule of Q near the N side, reducing it in two steps to QH$_2$. This reduction also uses two protons per Q, which are taken up from the matrix.

Electron transfer through Complex IV is from cytochrome $c$ to the Cu$_A$ center, to heme $a$, to the heme $a_3$–Cu$_B$ center, and finally to O$_2$ (Fig. 19–14). For every four electrons passing through this complex, the enzyme consumes four "substrate" H$^+$ from the matrix (N side) in converting O$_2$ to 2H$_2$O. It also uses the energy of this redox reaction to pump one proton outward into the intermembrane space (P side) for each electron that passes through, adding to the electrochemical potential produced by redox-driven proton transport through Complexes I and III. The overall reaction catalyzed by Complex IV is

4 Cyt $c$ (reduced) + 8H$_N^+$ + O$_2$ $\longrightarrow$

4 cyt $c$ (oxidized) + 4H$_P^+$ + 2H$_2$O    (19–4)

This four-electron reduction of O$_2$ involves redox centers that carry only one electron at a time, and it must occur without the release of incompletely reduced intermediates such as hydrogen peroxide or hydroxyl free radicals—very reactive species that would damage cellular components. The intermediates remain tightly

bound to the complex until completely converted to water.

## The Energy of Electron Transfer Is Efficiently Conserved in a Proton Gradient

The transfer of two electrons from NADH through the respiratory chain to molecular oxygen can be written as

$$\text{NADH} + \text{H}^+ + \tfrac{1}{2}\text{O}_2 \longrightarrow \text{NAD}^+ + \text{H}_2\text{O} \qquad (19\text{–}5)$$

This net reaction is highly exergonic. For the redox pair NAD$^+$/NADH, $E'^{\circ}$ is $-0.320$ V, and for the pair O$_2$/H$_2$O, $E'^{\circ}$ is 0.816 V. The $\Delta E'^{\circ}$ for this reaction is therefore 1.14 V, and the standard free-energy change (see Eqn 13–6, p. 510) is

$$\Delta G'^{\circ} = -n\mathcal{F}\,\Delta E'^{\circ} \qquad (19\text{–}6)$$
$$= -2(96.5 \text{ kJ/V} \cdot \text{mol})(1.14 \text{ V})$$
$$= -220 \text{ kJ/mol (of NADH)}$$

This *standard* free-energy change is based on the assumption of equal concentrations (1 M) of NADH and

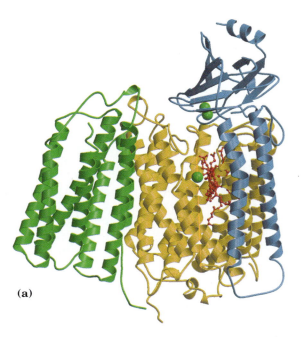

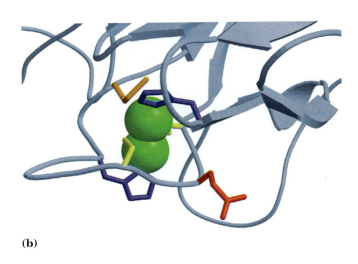

**(a)**

**(b)**

**FIGURE 19–13 Critical subunits of cytochrome oxidase (Complex IV).** The bovine complex is shown here (PDB ID 1OCC). **(a)** The core of Complex IV, with three subunits. Subunit I (yellow) has two heme groups, $a$ and $a_3$ (red), and a copper ion, $Cu_B$ (green sphere). Heme $a_3$ and $Cu_B$ form a binuclear Fe-Cu center. Subunit II (blue) contains two Cu ions (green spheres) complexed with the —SH groups of two Cys residues in a binuclear center, $Cu_A$, that resembles the 2Fe-2S centers of iron-sulfur proteins. This binuclear center and the cytochrome $c$–binding site are located in a domain of subunit II that protrudes from the P side of the inner membrane (into the intermembrane space). Subunit III (green) seems to be essential for Complex IV function, but its role is not well understood. **(b)** The binuclear center of $Cu_A$. The Cu ions (green spheres) share electrons equally. When the center is reduced they have the formal charges $Cu^{1+}Cu^{1+}$; when oxidized, $Cu^{1.5+}Cu^{1.5+}$. Ligands around the Cu ions include two His (dark blue), two Cys (yellow), an Asp (red), and Met (orange) residues.

$NAD^+$. In actively respiring mitochondria, the actions of many dehydrogenases keep the actual $[NADH]/[NAD^+]$ ratio well above unity, and the real free-energy change for the reaction shown in Equation 19–5 is therefore substantially greater (more negative) than $-220$ kJ/mol. A similar calculation for the oxidation of succinate shows that electron transfer from succinate ($E'^{\circ}$ for fumarate/succinate $= 0.031$ V) to $O_2$ has a smaller, but still negative, standard free-energy change of about $-150$ kJ/mol.

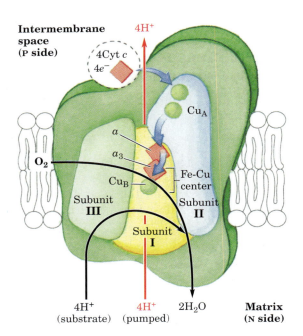

**FIGURE 19–14 Path of electrons through Complex IV.** The three proteins critical to electron flow are subunits I, II, and III. The larger green structure includes the other ten proteins in the complex. Electron transfer through Complex IV begins when two molecules of reduced cytochrome $c$ (top) each donate an electron to the binuclear center $Cu_A$. From here electrons pass through heme $a$ to the Fe-Cu center (cytochrome $a_3$ and $Cu_B$). Oxygen now binds to heme $a_3$ and is reduced to its peroxy derivative ($O_2^{2-}$) by two electrons from the Fe-Cu center. Delivery of two more electrons from cytochrome $c$ (making four electrons in all) converts the $O_2^{2-}$ to two molecules of water, with consumption of four "substrate" protons from the matrix. At the same time, four more protons are pumped from the matrix by an as yet unknown mechanism.

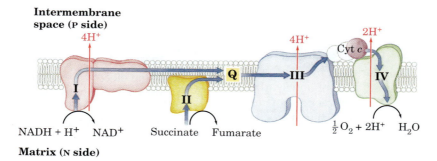

**Intermembrane space (P side)**

NADH + H$^+$    NAD$^+$     Succinate  Fumarate        $\frac{1}{2}O_2 + 2H^+$  H$_2$O

**Matrix (N side)**

**FIGURE 19–15 Summary of the flow of electrons and protons through the four complexes of the respiratory chain.** Electrons reach Q through Complexes I and II. QH$_2$ serves as a mobile carrier of electrons and protons. It passes electrons to Complex III, which passes them to another mobile connecting link, cytochrome $c$. Complex IV then transfers electrons from reduced cytochrome $c$ to O$_2$. Electron flow through Complexes I, III, and IV is accompanied by proton flow from the matrix to the intermembrane space. Recall that electrons from β oxidation of fatty acids can also enter the respiratory chain through Q (see Fig. 19–8).

Much of this energy is used to pump protons out of the matrix. For each pair of electrons transferred to O$_2$, four protons are pumped out by Complex I, four by Complex III, and two by Complex IV (Fig. 19–15). The *vectorial* equation for the process is therefore

$$\text{NADH} + 11\text{H}^+_N + \tfrac{1}{2}\text{O}_2 \longrightarrow \text{NAD}^+ + 10\text{H}^+_P + \text{H}_2\text{O} \quad (19\text{–}7)$$

The electrochemical energy inherent in this difference in proton concentration and separation of charge represents a temporary conservation of much of the energy of electron transfer. The energy stored in such a gradient, termed the **proton-motive force,** has two components: (1) the *chemical potential energy* due to the difference in concentration of a chemical species (H$^+$) in the two regions separated by the membrane, and (2) the *electrical potential energy* that results from the separation of charge when a proton moves across the membrane without a counterion (Fig. 19–16).

As we showed in Chapter 11, the free-energy change for the creation of an electrochemical gradient by an ion pump is

$$\Delta G = RT \ln\left(\frac{C_2}{C_1}\right) + Z\mathcal{F}\,\Delta\psi \qquad (19\text{–}8)$$

where $C_2$ and $C_1$ are the concentrations of an ion in two regions, and $C_2 > C_1$; $Z$ is the absolute value of its electrical charge (1 for a proton); and $\Delta\psi$ is the transmembrane difference in electrical potential, measured in volts.

For protons at 25 °C,

$$\ln\left(\frac{C_2}{C_1}\right) = 2.3(\log[\text{H}^+]_P - \log[\text{H}^+]_N)$$

$$= 2.3(\text{pH}_N - \text{pH}_P) = 2.3\,\Delta\text{pH}$$

and Equation 19–8 reduces to

$$\Delta G = 2.3RT\,\Delta\text{pH} + \mathcal{F}\,\Delta\psi \qquad (19\text{–}9)$$

$$= (5.70\ \text{kJ/mol})\Delta\text{pH} + (96.5\ \text{kJ/V}\cdot\text{mol})\Delta\psi$$

In actively respiring mitochondria, the measured $\Delta\psi$ is 0.15 to 0.20 V and the pH of the matrix is about 0.75

units more alkaline than that of the intermembrane space, so the calculated free-energy change for pumping protons outward is about 20 kJ/mol (of H$^+$), most of which is contributed by the electrical portion of the electrochemical potential. Because the transfer of two electrons from NADH to O$_2$ is accompanied by the outward pumping of 10 H$^+$ (Eqn 19–7), roughly 200 kJ of the 220 kJ released by oxidation of a mole of NADH is conserved in the proton gradient.

When protons flow spontaneously *down* their electrochemical gradient, energy is made available to do work. In mitochondria, chloroplasts, and aerobic bacteria, the electrochemical energy in the proton gradient drives the synthesis of ATP from ADP and P$_i$. We return to the energetics and stoichiometry of ATP synthesis driven by the electrochemical potential of the proton gradient in Section 19.2.

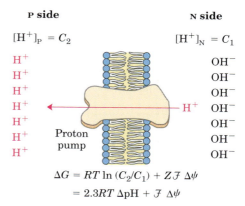

**P side**                          **N side**

$[\text{H}^+]_P = C_2$               $[\text{H}^+]_N = C_1$

$$\Delta G = RT\ln(C_2/C_1) + Z\mathcal{F}\,\Delta\psi$$
$$= 2.3RT\,\Delta\text{pH} + \mathcal{F}\,\Delta\psi$$

**FIGURE 19–16 Proton-motive force.** The inner mitochondrial membrane separates two compartments of different [H$^+$], resulting in differences in chemical concentration ($\Delta$pH) and charge distribution ($\Delta\psi$) across the membrane. The net effect is the proton-motive force ($\Delta G$), which can be calculated as shown here. This is explained more fully in the text.

## Plant Mitochondria Have Alternative Mechanisms for Oxidizing NADH

Plant mitochondria supply the cell with ATP during periods of low illumination or darkness by mechanisms entirely analogous to those used by nonphotosynthetic organisms. In the light, the principal source of mitochondrial NADH is a reaction in which glycine, produced by a process known as photorespiration, is converted to serine (see Fig. 20–21):

$$2 \text{ Glycine} + \text{NAD}^+ \longrightarrow \text{serine} + \text{CO}_2 + \text{NH}_3 + \text{NADH} + \text{H}^+$$

For reasons discussed in Chapter 20, plants must carry out this reaction even when they do not need NADH for ATP production. To regenerate $\text{NAD}^+$ from unneeded NADH, plant mitochondria transfer electrons from NADH directly to ubiquinone and from ubiquinone directly to $\text{O}_2$, bypassing Complexes III and IV and their proton pumps. In this process the energy in NADH is dissipated as heat, which can sometimes be of value to the plant (Box 19–1). Unlike cytochrome oxidase (Complex IV), the alternative $\text{QH}_2$ oxidase is not inhibited by cyanide. Cyanide-resistant NADH oxidation is therefore the hallmark of this unique plant electron-transfer pathway.

## SUMMARY 19.1 Electron-Transfer Reactions in Mitochondria

- Chemiosmotic theory provides the intellectual framework for understanding many biological energy transductions, including oxidative phosphorylation and photophosphorylation. The mechanism of energy coupling is similar in both cases: the energy of electron flow is conserved by the concomitant pumping of protons across the membrane, producing an electrochemical gradient, the proton-motive force.

- In mitochondria, hydride ions removed from substrates by NAD-linked dehydrogenases donate electrons to the respiratory (electron-transfer) chain, which transfers the electrons to molecular $\text{O}_2$, reducing it to $\text{H}_2\text{O}$.

- Shuttle systems convey reducing equivalents from cytosolic NADH to mitochondrial NADH. Reducing equivalents from all NAD-linked dehydrogenations are transferred to mitochondrial NADH dehydrogenase (Complex I).

- Reducing equivalents are then passed through a series of Fe-S centers to ubiquinone, which transfers the electrons to cytochrome $b$, the first carrier in Complex III. In this complex, electrons take two separate paths through two $b$-type cytochromes and cytochrome $c_1$ to an Fe-S center. The Fe-S center passes electrons, one at a time, through cytochrome $c$ and into

Complex IV, cytochrome oxidase. This copper-containing enzyme, which also contains cytochromes $a$ and $a_3$, accumulates electrons, then passes them to $\text{O}_2$, reducing it to $\text{H}_2\text{O}$.

- Some electrons enter this chain of carriers through alternative paths. Succinate is oxidized by succinate dehydrogenase (Complex II), which contains a flavoprotein that passes electrons through several Fe-S centers to ubiquinone. Electrons derived from the oxidation of fatty acids pass to ubiquinone via the electron-transferring flavoprotein.

- Plants also have an alternative, cyanide-resistant NADH oxidation pathway.

## 19.2 ATP Synthesis

How is a concentration gradient of protons transformed into ATP? We have seen that electron transfer releases, and the proton-motive force conserves, more than enough free energy (about 200 kJ) per "mole" of electron pairs to drive the formation of a mole of ATP, which requires about 50 kJ (see Box 13–1). Mitochondrial oxidative phosphorylation therefore poses no thermodynamic problem. But what is the chemical mechanism that couples proton flux with phosphorylation?

Peter Mitchell, 1920–1992

The **chemiosmotic model,** proposed by Peter Mitchell, is the paradigm for this mechanism. According to the model (Fig. 19–17), the electrochemical energy inherent in the difference in proton concentration and separation of charge across the inner mitochondrial membrane—the proton-motive force—drives the synthesis of ATP as protons flow passively back into the matrix through a proton pore associated with **ATP synthase.** To emphasize this crucial role of the proton-motive force, the equation for ATP synthesis is sometimes written

$$\text{ADP} + \text{P}_i + n\text{H}_\text{P}^+ \longrightarrow \text{ATP} + \text{H}_2\text{O} + n\text{H}_\text{N}^+ \quad (19\text{--}10)$$

Mitchell used "chemiosmotic" to describe enzymatic reactions that involve, simultaneously, a chemical reaction and a transport process. The operational definition of "coupling" is shown in Figure 19–18. When isolated mitochondria are suspended in a buffer containing ADP, $\text{P}_i$, and an oxidizable substrate such as succinate, three easily measured processes occur: (1) the substrate is oxidized (succinate yields fumarate), (2) $\text{O}_2$ is consumed, and (3) ATP is synthesized. Oxygen consumption and ATP synthesis depend on the presence of an oxidizable substrate (succinate in this case) as well as ADP and $\text{P}_i$.

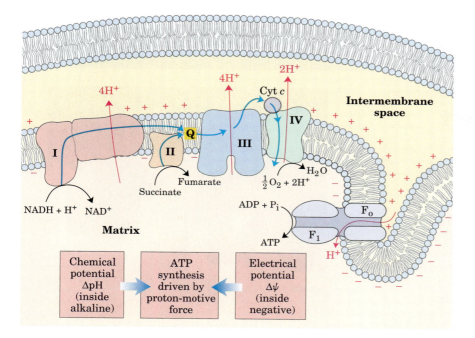

**FIGURE 19–17 Chemiosmotic model.** In this simple representation of the chemiosmotic theory applied to mitochondria, electrons from NADH and other oxidizable substrates pass through a chain of carriers arranged asymmetrically in the inner membrane. Electron flow is accompanied by proton transfer across the membrane, producing both a chemical gradient ($\Delta$pH) and an electrical gradient ($\Delta\psi$). The inner mitochondrial membrane is impermeable to protons; protons can reenter the matrix only through proton-specific channels ($F_o$). The proton-motive force that drives protons back into the matrix provides the energy for ATP synthesis, catalyzed by the $F_1$ complex associated with $F_o$.

Because the energy of substrate oxidation drives ATP synthesis in mitochondria, we would expect inhibitors of the passage of electrons to $O_2$ (such as cyanide, carbon monoxide, and antimycin A) to block ATP synthesis (Fig. 19–18a). More surprising is the finding that the converse is also true: inhibition of ATP synthesis blocks electron transfer in intact mitochondria. This obligatory coupling can be demonstrated in isolated mitochondria by providing $O_2$ and oxidizable substrates, but not ADP (Fig. 19–18b). Under these conditions, no ATP synthesis can occur and electron transfer to $O_2$ does not proceed. Coupling of oxidation and phosphorylation can also be demonstrated using oligomycin or venturicidin, toxic antibiotics that bind to the ATP synthase in mitochondria. These compounds are potent in-

hibitors of both ATP synthesis *and* the transfer of electrons through the chain of carriers to $O_2$ (Fig. 19–18b). Because oligomycin is known to interact not directly with the electron carriers but with ATP synthase, it follows that electron transfer and ATP synthesis are obligately coupled; neither reaction occurs without the other.

Chemiosmotic theory readily explains the dependence of electron transfer on ATP synthesis in mitochondria. When the flow of protons into the matrix through the proton channel of ATP synthase is blocked (with oligomycin, for example), no path exists for the return of protons to the matrix, and the continued extrusion of protons driven by the activity of the respiratory chain generates a large proton gradient. The proton-motive force builds up until the cost (free energy) of pumping

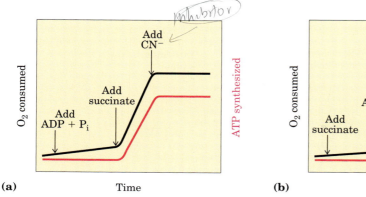

(a)

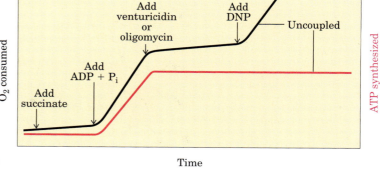

(b)

**FIGURE 19–18 Coupling of electron transfer and ATP synthesis in mitochondria.** In experiments to demonstrate coupling, mitochondria are suspended in a buffered medium and an $O_2$ electrode monitors $O_2$ consumption. At intervals, samples are removed and assayed for the presence of ATP. **(a)** Addition of ADP and $P_i$ alone results in little or no increase in either respiration ($O_2$ consumption; black) or ATP synthesis (red). When succinate is added, respiration begins immediately and

ATP is synthesized. Addition of cyanide ($CN^-$), which blocks electron transfer between cytochrome oxidase and $O_2$, inhibits both respiration and ATP synthesis. **(b)** Mitochondria provided with succinate respire and synthesize ATP only when ADP and $P_i$ are added. Subsequent addition of venturicidin or oligomycin, inhibitors of ATP synthase, blocks both ATP synthesis and respiration. Dinitrophenol (DNP) is an uncoupler, allowing respiration to continue without ATP synthesis.

**BOX 19–1   THE WORLD OF BIOCHEMISTRY**

## Hot, Stinking Plants and Alternative Respiratory Pathways

Many flowering plants attract insect pollinators by releasing odorant molecules that mimic an insect's natural food sources or potential egg-laying sites. Plants pollinated by flies or beetles that normally feed on or lay their eggs in dung or carrion sometimes use foul-smelling compounds to attract these insects.

One family of stinking plants is the Araceae, which includes philodendrons, arum lilies, and skunk cabbages. These plants have tiny flowers densely packed on an erect structure, the spadix, surrounded by a modified leaf, the spathe. The spadix releases odors of rotting flesh or dung. Before pollination the spadix also heats up, in some species to as much as 20 to 40 °C above the ambient temperature. Heat production (thermogenesis) helps evaporate odorant molecules for better dispersal, and because rotting flesh and dung are usually warm from the hyperactive metabolism of scavenging microbes, the heat itself might also attract insects. In the case of the eastern skunk cabbage (Fig. 1), which flowers in late winter or early spring when snow still covers the ground, thermogenesis allows the spadix to grow up through the snow.

How does a skunk cabbage heat its spadix? The mitochondria of plants, fungi, and unicellular eukaryotes have electron-transfer systems that are essentially the same as those in animals, but they also have an alternative respiratory pathway. A cyanide-resistant $QH_2$ oxidase transfers electrons from the ubiquinone pool directly to oxygen, bypassing the two proton-translocating steps of Complexes III and IV (Fig. 2). Energy that might have been conserved as

**FIGURE 1**   Eastern skunk cabbage.

ATP is instead released as heat. Plant mitochondria also have an alternative NADH dehydrogenase, insensitive to the Complex I inhibitor rotenone (see Table 19–4), that transfers electrons from NADH in the matrix directly to ubiquinone, bypassing Complex I and its associated proton pumping. And plant mitochondria have yet another NADH dehydrogenase, on the external face of the inner membrane, that transfers electrons from NADPH or NADH in the intermembrane space to ubiquinone, again bypassing Complex I. Thus when electrons enter the alternative respiratory pathway through the rotenone-insensitive NADH dehydrogenase, the external NADH dehydrogenase, or succinate dehydrogenase (Complex II), and pass to $O_2$ via the cyanide-resistant alternative oxidase, energy is not conserved as ATP but is released as heat. A skunk cabbage can use the heat to melt snow, produce a foul stench, or attract beetles or flies.

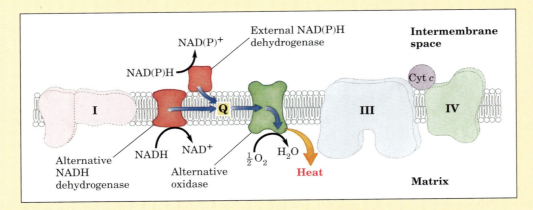

**FIGURE 2**   Electron carriers of the inner membrane of plant mitochondria. Electrons can flow through Complexes I, III, and IV, as in animal mitochondria, or through plant-specific alternative carriers by the paths shown with blue arrows.

protons out of the matrix against this gradient equals or exceeds the energy released by the transfer of electrons from NADH to $O_2$. At this point electron flow must stop; the free energy for the overall process of electron flow coupled to proton pumping becomes zero, and the system is at equilibrium.

Certain conditions and reagents, however, can uncouple oxidation from phosphorylation. When intact mitochondria are disrupted by treatment with detergent or by physical shear, the resulting membrane fragments can still catalyze electron transfer from succinate or NADH to $O_2$, but no ATP synthesis is coupled to this respiration. Certain chemical compounds cause uncoupling without disrupting mitochondrial structure. Chemical uncouplers include 2,4-dinitrophenol (DNP) and carbonylcyanide-$p$-trifluoromethoxyphenylhydrazone (FCCP) (Table 19–4; Fig. 19–19), weak acids with hydrophobic properties that permit them to diffuse readily across mitochondrial membranes. After entering the matrix in the protonated form, they can release a proton, thus dissipating the proton gradient. Ionophores such as valinomycin (see Fig. 11–45) allow inorganic ions to pass easily through membranes. Ionophores uncouple electron transfer from oxidative phosphorylation by dissipating the electrical contribution to the electrochemical gradient across the mitochondrial membrane.

A prediction of the chemiosmotic theory is that, because the role of electron transfer in mitochondrial ATP synthesis is simply to pump protons to create the electrochemical potential of the proton-motive force, an artificially created proton gradient should be able to replace electron transfer in driving ATP synthesis. This has been experimentally confirmed (Fig. 19–20). Mitochondria manipulated so as to impose a difference of proton concentration and a separation of charge across the inner membrane synthesize ATP *in the absence of an oxidizable substrate;* the proton-motive force alone suffices to drive ATP synthesis.

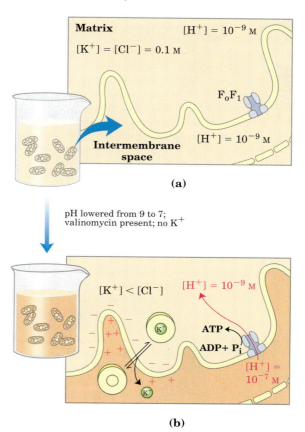

**FIGURE 19–20 Evidence for the role of a proton gradient in ATP synthesis.** An artificially imposed electrochemical gradient can drive ATP synthesis in the absence of an oxidizable substrate as electron donor. In this two-step experiment, **(a)** isolated mitochondria are first incubated in a pH 9 buffer containing 0.1 M KCl. Slow leakage of buffer and KCl into the mitochondria eventually brings the matrix into equilibrium with the surrounding medium. No oxidizable substrates are present. **(b)** Mitochondria are now separated from the pH 9 buffer and resuspended in pH 7 buffer containing valinomycin but no KCl. The change in buffer creates a difference of two pH units across the inner mitochondrial membrane. The outward flow of $K^+$, carried (by valinomycin) down its concentration gradient without a counterion, creates a charge imbalance across the membrane (matrix negative). The sum of the chemical potential provided by the pH difference and the electrical potential provided by the separation of charges is a proton-motive force large enough to support ATP synthesis in the absence of an oxidizable substrate.

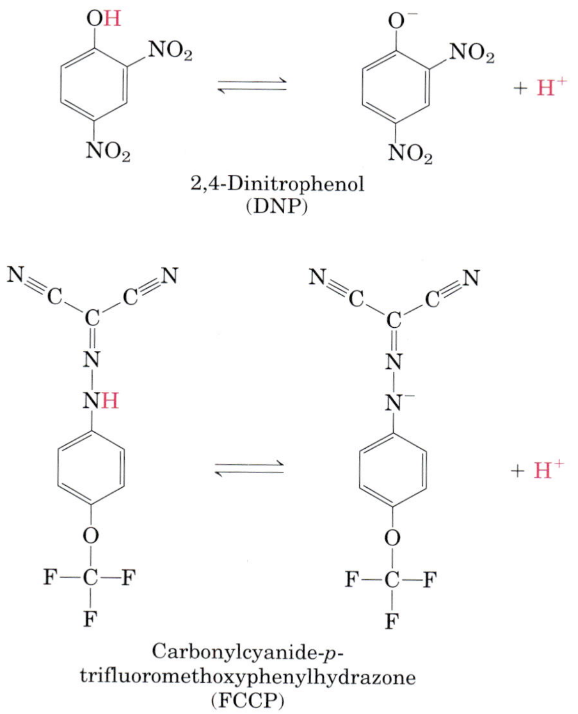

**FIGURE 19–19 Two chemical uncouplers of oxidative phosphorylation.** Both DNP and FCCP have a dissociable proton and are very hydrophobic. They carry protons across the inner mitochondrial membrane, dissipating the proton gradient. Both also uncouple photophosphorylation (see Fig. 19–57).

## ATP Synthase Has Two Functional Domains, $F_o$ and $F_1$

Mitochondrial **ATP synthase** is an F-type ATPase (see Fig. 11–39; Table 11–3) similar in structure and mechanism to the ATP synthases of chloroplasts and eubacteria. This large enzyme complex of the inner mitochondrial membrane catalyzes the formation of ATP

Efraim Racker,
1913–1991

from ADP and $P_i$, accompanied by the flow of protons from the P to the N side of the membrane (Eqn 19–10). ATP synthase, also called Complex V, has two distinct components: $F_1$, a peripheral membrane protein, and $F_o$ (o denoting oligomycin-sensitive), which is integral to the membrane. $F_1$, the first factor recognized as essential for oxidative phosphorylation, was identified and purified by Efraim Racker and his colleagues in the early 1960s.

In the laboratory, small membrane vesicles formed from inner mitochondrial membranes carry out ATP synthesis coupled to electron transfer. When $F_1$ is gently extracted, the "stripped" vesicles still contain intact respiratory chains and the $F_o$ portion of ATP synthase. The vesicles can catalyze electron transfer from NADH to $O_2$ but cannot produce a proton gradient: $F_o$ has a proton pore through which protons leak as fast as they are pumped by electron transfer, and without a proton gradient the $F_1$-depleted vesicles cannot make ATP. Isolated $F_1$ catalyzes ATP hydrolysis (the reversal of synthesis) and was therefore originally called **$F_1$ATPase.** When purified $F_1$ is added back to the depleted vesicles, it reassociates with $F_o$, plugging its proton pore and restoring the membrane's capacity to couple electron transfer and ATP synthesis.

## ATP Is Stabilized Relative to ADP on the Surface of $F_1$

Isotope exchange experiments with purified $F_1$ reveal a remarkable fact about the enzyme's catalytic mechanism: on the enzyme surface, the reaction ADP + $P_i$ $\rightleftharpoons$ ATP + $H_2O$ is readily reversible—the free-energy change for ATP synthesis is close to zero! When ATP is hydrolyzed by $F_1$ in the presence of $^{18}O$-labeled water, the $P_i$ released contains an $^{18}O$ atom. Careful measurement of the $^{18}O$ content of $P_i$ formed in vitro by $F_1$-catalyzed hydrolysis of ATP reveals that the $P_i$ has not one, but three or four $^{18}O$ atoms (Fig. 19–21). This indicates that the terminal pyrophosphate bond in ATP is cleaved and re-formed repeatedly before $P_i$ leaves the enzyme surface. With $P_i$ free to tumble in its binding site, each hydrolysis inserts $^{18}O$ randomly at one of the

**(a)**

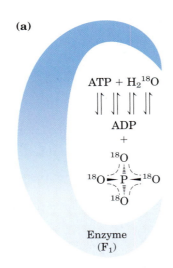

**(b)**

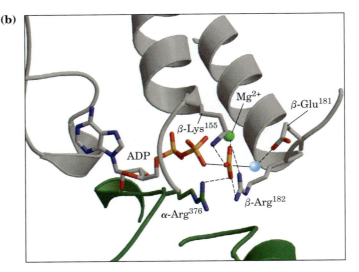

**FIGURE 19–21 Catalytic mechanism of $F_1$. (a)** $^{18}O$-exchange experiment. $F_1$ solubilized from mitochondrial membranes is incubated with ATP in the presence of $^{18}O$-labeled water. At intervals, a sample of the solution is withdrawn and analyzed for the incorporation of $^{18}O$ into the $P_i$ produced from ATP hydrolysis. In minutes, the $P_i$ contains three or four $^{18}O$ atoms, indicating that both ATP hydrolysis and ATP synthesis have occurred several times during the incubation. **(b)** The likely transition state complex for ATP hydrolysis and synthesis in ATP synthase (derived from PDB ID 1BMF). The $\alpha$ subunit is shown in green, $\beta$ in gray. The positively charged residues $\beta$-Arg$^{182}$ and $\alpha$-Arg$^{376}$ coordinate two oxygens of the pentavalent phosphate intermediate; $\beta$-Lys$^{155}$ interacts with a third oxygen, and the $Mg^{2+}$ ion (green sphere) further stabilizes the intermediate. The blue sphere represents the leaving group ($H_2O$). These interactions result in the ready equilibration of ATP and ADP + $P_i$ in the active site.

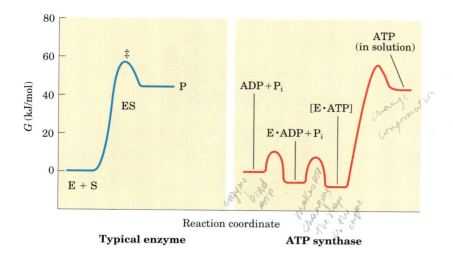

**FIGURE 19–22 Reaction coordinate diagrams for ATP synthase and for a more typical enzyme.** In a typical enzyme-catalyzed reaction (left), reaching the transition state (‡) between substrate and product is the major energy barrier to overcome. In the reaction catalyzed by ATP synthase (right), release of ATP from the enzyme, not formation of ATP, is the major energy barrier. The free-energy change for the formation of ATP from ADP and $P_i$ in aqueous solution is large and positive, but on the enzyme surface, the very tight binding of ATP provides sufficient binding energy to bring the free energy of the enzyme-bound ATP close to that of ADP + $P_i$, so the reaction is readily reversible. The equilibrium constant is near 1. The free energy required for the release of ATP is provided by the proton-motive force.

four positions in the molecule. This exchange reaction occurs in unenergized $F_oF_1$ complexes (with no proton gradient) and with isolated $F_1$—the exchange does not require the input of energy.

Kinetic studies of the initial rates of ATP synthesis and hydrolysis confirm the conclusion that $\Delta G'^\circ$ for ATP synthesis on the enzyme is near zero. From the measured rates of hydrolysis ($k_1 = 10$ s$^{-1}$) and synthesis ($k_{-1} = 24$ s$^{-1}$), the calculated equilibrium constant for the reaction

$$\text{Enz-ATP} \rightleftharpoons \text{Enz–(ADP + P}_i)$$

is

$$K'_{eq} = \frac{k_{-1}}{k_1} = \frac{24 \text{ s}^{-1}}{10 \text{ s}^{-1}} = 2.4$$

From this $K'_{eq}$, the calculated apparent $\Delta G'^\circ$ is close to zero. This is much different from the $K'_{eq}$ of about $10^5$ ($\Delta G'^\circ = -30.5$ kJ/mol) for the hydrolysis of ATP free in solution (not on the enzyme surface).

What accounts for the huge difference? ATP synthase stabilizes ATP relative to ADP + $P_i$ by binding ATP more tightly, releasing enough energy to counterbalance the cost of making ATP. Careful measurements of the binding constants show that $F_oF_1$ binds ATP with very high affinity ($K_d \leq 10^{-12}$ M) and ADP with much lower affinity ($K_d \approx 10^{-5}$ M). The difference in $K_d$ corresponds to a difference of about 40 kJ/mol in binding energy, and this binding energy drives the equilibrium toward formation of the product ATP.

## The Proton Gradient Drives the Release of ATP from the Enzyme Surface

Although ATP synthase equilibrates ATP with ADP + $P_i$, in the absence of a proton gradient the newly synthesized ATP does not leave the surface of the enzyme.

It is the proton gradient that causes the enzyme to release the ATP formed on its surface. The reaction coordinate diagram of the process (Fig. 19–22) illustrates the difference between the mechanism of ATP synthase and that of many other enzymes that catalyze endergonic reactions.

For the continued synthesis of ATP, the enzyme must cycle between a form that binds ATP very tightly and a form that releases ATP. Chemical and crystallographic studies of the ATP synthase have revealed the structural basis for this alternation in function.

## Each β Subunit of ATP Synthase Can Assume Three Different Conformations

Mitochondrial $F_1$ has nine subunits of five different types, with the composition $\alpha_3\beta_3\gamma\delta\varepsilon$. Each of the three β subunits has one catalytic site for ATP synthesis. The crystallographic determination of the $F_1$ structure by John E. Walker and colleagues revealed structural details very helpful in explaining the catalytic mechanism of the enzyme. The knoblike portion of $F_1$ is a flattened sphere, 8 nm high and 10 nm across, consisting of alternating α and β subunits arranged like the sections of an orange (Fig. 19–23a–c). The polypeptides that make up the stalk in the $F_1$ crystal structure are asymmetrically arranged, with one domain of the single γ subunit making up a central shaft that passes through $F_1$, and another domain of γ associated primarily with one of the three β subunits, designated β-empty (Fig. 19–23c). Although the amino acid sequences of the three β subunits are identical, *their conformations differ*, in part because of the association of the γ subunit with just one of the three. The structures of the δ and ε subunits are not revealed in these crystallographic studies.

The conformational differences among β subunits extend to differences in their ATP/ADP-binding sites.

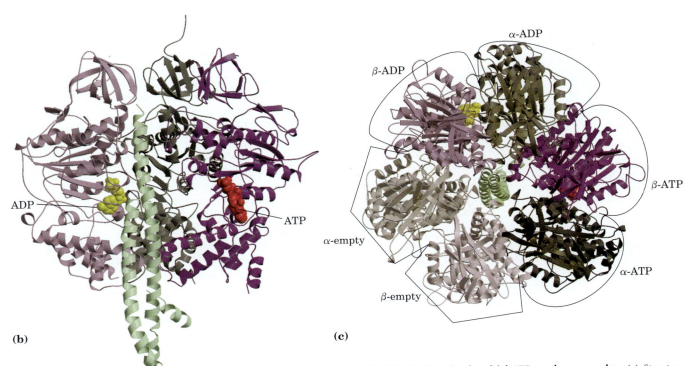

**(b)**

**(a)**

John E. Walker

When researchers crystallized the protein in the presence of ADP and App(NH)p, a close structural analog of ATP that cannot be hydrolyzed by the ATPase activity of $F_1$, the binding site of one of the three $\beta$ subunits was filled with App(NH)p, the second was filled with

App(NH)p ($\beta,\gamma$-imidoadenosine 5'-triphosphate)

**(c)**

**FIGURE 19–23 Mitochondrial ATP synthase complex. (a)** Structure of the $F_1$ complex, deduced from crystallographic and biochemical studies. In $F_1$, three $\alpha$ and three $\beta$ subunits are arranged like the segments of an orange, with alternating $\alpha$ (shades of gray) and $\beta$ (shades of purple) subunits around a central shaft, the $\gamma$ subunit (green). **(b)** Crystal structure of bovine $F_1$ (PDB ID 1BMF), viewed from the side. Two $\alpha$ subunits and one $\beta$ subunit have been omitted to reveal the central shaft ($\gamma$ subunit) and the binding sites for ATP (red) and ADP (yellow) on the $\beta$ subunits. The $\delta$ and $\varepsilon$ subunits are not shown here. **(c)** $F_1$ viewed from above (that is, from the N side of the membrane), showing the three $\beta$ and three $\alpha$ subunits and the central shaft ($\gamma$ subunit, green). Each $\beta$ subunit, near its interface with the neighboring $\alpha$ subunit, has a nucleotide-binding site critical to the catalytic activity. The single $\gamma$ subunit associates primarily with one of the three $\alpha\beta$ pairs, forcing each of the three $\beta$ subunits into slightly different conformations, with different nucleotide-binding sites. In the crystalline enzyme, one subunit ($\beta$-ADP) has ADP (yellow) in its binding site, the next ($\beta$-ATP) has ATP (red), and the third ($\beta$-empty) has no bound nucleotide. **(d)** Side view of the $F_oF_1$ structure. This is a composite, in which the crystallographic coordinates of bovine mitochondrial $F_1$ (shades of purple and gray) have been combined with those of yeast mitochondrial $F_o$ (shades of yellow and orange) (PDB ID 1QO1). Subunits a, b, $\delta$, and $\varepsilon$ were not part of the crystal structure shown here. **(e)** The $F_oF_1$ structure, viewed end-on in the direction P side to N side. The major structures visible in this cross section are the two transmembrane helices of each of ten c subunits arranged in concentric circles. **(f)** Diagram of the $F_oF_1$ complex, deduced from biochemical and crystallographic studies. The two b subunits of $F_o$ associate firmly with the $\alpha$ and $\beta$ subunits of $F_1$, holding them fixed relative to the membrane. In $F_o$, the membrane-embedded cylinder of c subunits is attached to the shaft made up of $F_1$ subunits $\gamma$ and $\varepsilon$. As protons flow through the membrane from the P side to the N side through $F_o$, the cylinder and shaft rotate, and the $\beta$ subunits of $F_1$ change conformation as the $\gamma$ subunit associates with each in turn.

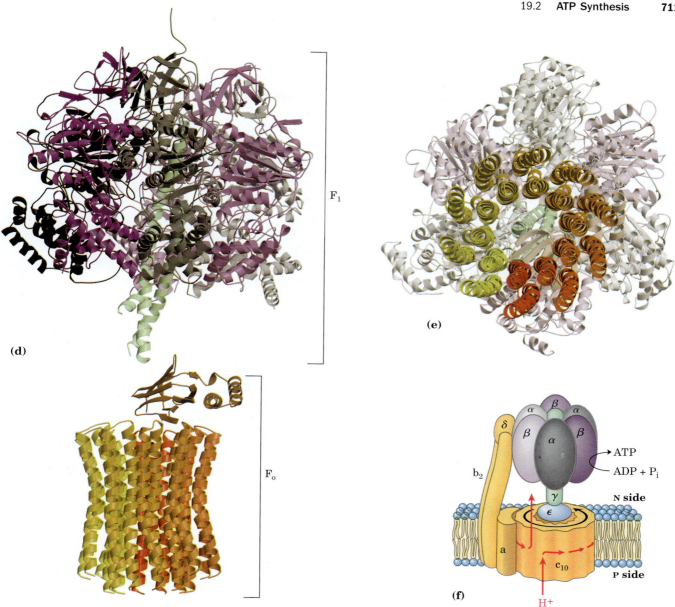

**(d)**

F$_1$

F$_0$

**(e)**

**(f)**

ATP

ADP + P$_i$

N side

P side

H$^+$

ADP, and the third was empty. The corresponding $\beta$ subunit conformations are designated $\beta$-ATP, $\beta$-ADP, and $\beta$-empty (Fig. 19–23c). This difference in nucleotide binding among the three subunits is critical to the mechanism of the complex.

The F$_0$ complex making up the proton pore is composed of three subunits, a, b, and c, in the proportion ab$_2$c$_{10-12}$. Subunit c is a small ($M_r$ 8,000), very hydrophobic polypeptide, consisting almost entirely of two transmembrane helices, with a small loop extending from the matrix side of the membrane. The crystal structure of the yeast F$_0$F$_1$, solved in 1999, shows the arrangement of the c subunits. The yeast complex has ten c subunits, each with two transmembrane helices roughly perpendicular to the plane of the membrane and arranged in two concentric circles (Fig. 19–23d, e). The inner circle is made up of the amino-terminal helices of each c subunit; the outer circle, about 55 Å in diameter, is made up of the carboxyl-terminal helices. The $\varepsilon$ and $\gamma$ subunits of F$_1$ form a leg-and-foot that projects from the bottom (membrane) side of F$_1$ and stands firmly on the ring of c subunits. The schematic drawing in Figure 19–23f combines the structural information from studies of bovine F$_1$ and yeast F$_0$F$_1$.

## Rotational Catalysis Is Key to the Binding-Change Mechanism for ATP Synthesis

On the basis of detailed kinetic and binding studies of the reactions catalyzed by F$_0$F$_1$, Paul Boyer proposed a **rotational catalysis** mechanism in which the three active sites of F$_1$ take turns catalyzing ATP synthesis

Paul Boyer

(Fig. 19–24). A given $\beta$ subunit starts in the $\beta$-ADP conformation, which binds ADP and $P_i$ from the surrounding medium. The subunit now changes conformation, assuming the $\beta$-ATP form that tightly binds and stabilizes ATP, bringing about the ready equilibration of ADP + $P_i$ with ATP on the enzyme surface. Finally, the subunit changes to the $\beta$-empty conformation, which has very low affinity for ATP, and the newly synthesized ATP leaves the enzyme surface. Another round of catalysis begins when

this subunit again assumes the $\beta$-ADP form and binds ADP and $P_i$.

The conformational changes central to this mechanism are driven by the passage of protons through the $F_o$ portion of ATP synthase. The streaming of protons through the $F_o$ "pore" causes the cylinder of c subunits and the attached $\gamma$ subunit to rotate about the long axis of $\gamma$, which is perpendicular to the plane of the membrane. The $\gamma$ subunit passes through the center of the $\alpha_3\beta_3$ spheroid, which is held stationary relative to the membrane surface by the $b_2$ and $\delta$ subunits (Fig. 19–23f). With each rotation of 120°, $\gamma$ comes into contact with a different $\beta$ subunit, and the contact forces that $\beta$ subunit into the $\beta$-empty conformation.

The three $\beta$ subunits interact in such a way that when one assumes the $\beta$-empty conformation, its neighbor to one side *must* assume the $\beta$-ADP form, and the other neighbor the $\beta$-ATP form. Thus one complete rotation of the $\gamma$ subunit causes each $\beta$ subunit to cycle through all three of its possible conformations, and for each rotation, three ATP are synthesized and released from the enzyme surface.

One strong prediction of this binding-change model is that the $\gamma$ subunit should rotate in one direction when $F_oF_1$ is synthesizing ATP and in the opposite direction when the enzyme is hydrolyzing ATP. This prediction was confirmed in elegant experiments in the laboratories of Masasuke Yoshida and Kazuhiko Kinosita, Jr. The rotation of $\gamma$ in a single $F_1$ molecule was observed microscopically by attaching a long, thin, fluorescent actin polymer to $\gamma$ and watching it move relative to $\alpha_3\beta_3$ immobilized on a microscope slide, as ATP was hydrolyzed. When the entire $F_oF_1$ complex (not just $F_1$) was used in a similar experiment, the entire ring of c subunits rotated with $\gamma$ (Fig. 19–25). The "shaft" rotated in the predicted direction through 360°. The rotation was not smooth, but occurred in three discrete steps of 120°. As calculated from the known rate of ATP hydrolysis by one $F_1$ molecule and from the frictional drag on the long actin polymer, the efficiency of this mechanism in converting chemical energy into motion is close to 100%. It is, in Boyer's words, "a splendid molecular machine!"

## Chemiosmotic Coupling Allows Nonintegral Stoichiometries of $O_2$ Consumption and ATP Synthesis

Before the general acceptance of the chemiosmotic model for oxidative phosphorylation, the assumption was that the overall reaction equation would take the following form:

$$x\text{ADP} + x\text{P}_i + \tfrac{1}{2}\text{O}_2 + \text{H}^+ + \text{NADH} \longrightarrow$$
$$x\text{ATP} + \text{H}_2\text{O} + \text{NAD}^+ \quad (19\text{–}11)$$

with the value of $x$—sometimes called the **P/O ratio** or the **P/2e$^-$ ratio**—always an integer. When intact mito-

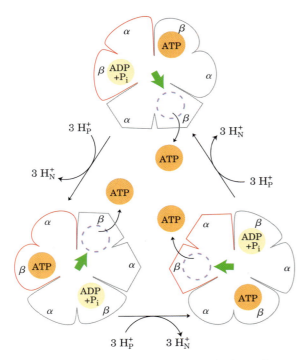

**FIGURE 19–24 Binding-change model for ATP synthase.** The $F_1$ complex has three nonequivalent adenine nucleotide–binding sites, one for each pair of $\alpha$ and $\beta$ subunits. At any given moment, one of these sites is in the $\beta$-ATP conformation (which binds ATP tightly), a second is in the $\beta$-ADP (loose-binding) conformation, and a third is in the $\beta$-empty (very-loose-binding) conformation. The proton-motive force causes rotation of the central shaft—the $\gamma$ subunit, shown as a green arrowhead—which comes into contact with each $\alpha\beta$ subunit pair in succession. This produces a cooperative conformational change in which the $\beta$-ATP site is converted to the $\beta$-empty conformation, and ATP dissociates; the $\beta$-ADP site is converted to the $\beta$-ATP conformation, which promotes condensation of bound ADP + $P_i$ to form ATP; and the $\beta$-empty site becomes a $\beta$-ADP site, which loosely binds ADP + $P_i$ entering from the solvent. This model, based on experimental findings, requires that at least two of the three catalytic sites alternate in activity; ATP cannot be released from one site unless and until ADP and $P_i$ are bound at the other.

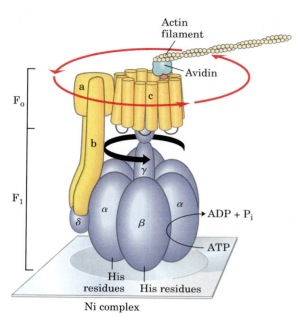

His residues   His residues

Ni complex

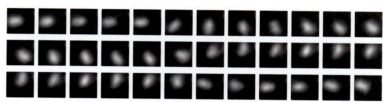

**FIGURE 19–25 Rotation of $F_o$ and $\gamma$ experimentally demonstrated.** $F_1$ genetically engineered to contain a run of His residues adheres tightly to a microscope slide coated with a Ni complex; biotin is covalently attached to a c subunit of $F_o$. The protein avidin, which binds biotin very tightly, is covalently attached to long filaments of actin labeled with a fluorescent probe. Biotin-avidin binding now attaches the actin filaments to the c subunit. When ATP is provided as substrate for the ATPase activity of $F_1$, the labeled filament is seen to rotate continuously in one direction, proving that the $F_o$ cylinder of c subunits rotates. In another experiment, a fluorescent actin filament was attached directly to the $\gamma$ subunit. The series of fluorescence micrographs shows the position of the actin filament at intervals of 133 ms. Note that as the filament rotates, it makes a discrete jump about every eleventh frame. Presumably the cylinder and shaft move as one unit.

chondria are suspended in solution with an oxidizable substrate such as succinate or NADH and are provided with $O_2$, ATP synthesis is readily measurable, as is the decrease in $O_2$. Measurement of P/O, however, is complicated by the fact that intact mitochondria consume ATP in many reactions taking place in the matrix, and they consume $O_2$ for purposes other than oxidative phosphorylation. Most experiments have yielded P/O (ATP to $\frac{1}{2}O_2$) ratios of between 2 and 3 when NADH was the electron donor, and between 1 and 2 when succinate was the donor. Given the assumption that P/O should have an integral value, most experimenters agreed that the P/O ratios must be 3 for NADH and 2 for succinate, and for years those values have appeared in research papers and textbooks.

With introduction of the chemiosmotic paradigm for coupling ATP synthesis to electron transfer, there was no theoretical requirement for P/O to be integral. The relevant questions about stoichiometry became, how many protons are pumped outward by electron transfer from one NADH to $O_2$, and how many protons must flow inward through the $F_oF_1$ complex to drive the synthesis of one ATP? The measurement of proton fluxes is technically complicated; the investigator must take into account the buffering capacity of mitochondria, nonproductive leakage of protons across the inner membrane, and use of the proton gradient for functions other than ATP synthesis, such as driving the transport of substrates across the inner mitochondrial membrane (described below). The consensus values for number of protons pumped out per pair of electrons are 10 for NADH

and 6 for succinate. The most widely accepted experimental value for number of protons required to drive the synthesis of an ATP molecule is 4, of which 1 is used in transporting $P_i$, ATP, and ADP across the mitochondrial membrane (see below). If 10 protons are pumped out per NADH and 4 must flow in to produce 1 ATP, the proton-based P/O ratio is 2.5 for NADH as the electron donor and 1.5 (6/4) for succinate. We use the P/O values of 2.5 and 1.5 throughout this book, but the values 3.0 and 2.0 are still common in the biochemical literature. The final word on proton stoichiometry will probably not be written until we know the full details of the $F_oF_1$ reaction mechanism.

## The Proton-Motive Force Energizes Active Transport

Although the primary role of the proton gradient in mitochondria is to furnish energy for the synthesis of ATP, the proton-motive force also drives several transport processes essential to oxidative phosphorylation. The inner mitochondrial membrane is generally impermeable to charged species, but two specific systems transport ADP and $P_i$ into the matrix and ATP out to the cytosol (Fig. 19–26).

The **adenine nucleotide translocase,** integral to the inner membrane, binds $ADP^{3-}$ in the intermembrane space and transports it into the matrix in exchange for an $ATP^{4-}$ molecule simultaneously transported outward (see Fig. 13–1 for the ionic forms of ATP and ADP). Because this antiporter moves four negative charges out for every three moved in, its activity is favored by the

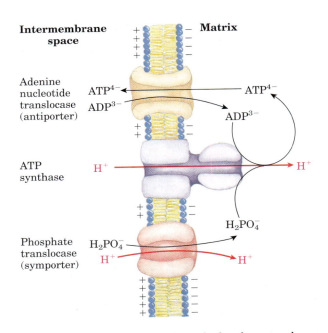

**Intermembrane space**

**Matrix**

Adenine nucleotide translocase (antiporter)

$ATP^{4-}$    $ATP^{4-}$

$ADP^{3-}$    $ADP^{3-}$

ATP synthase

$H^+$       $H^+$

$H_2PO_4^-$

Phosphate translocase (symporter)

$H_2PO_4^-$

$H^+$      $H^+$

**FIGURE 19–26 Adenine nucleotide and phosphate translocases.** Transport systems of the inner mitochondrial membrane carry ADP and $P_i$ into the matrix and newly synthesized ATP into the cytosol. The adenine nucleotide translocase is an antiporter; the same protein moves ADP into the matrix and ATP out. The effect of replacing $ATP^{4-}$ with $ADP^{3-}$ is the net efflux of one negative charge, which is favored by the charge difference across the inner membrane (outside positive). At pH 7, $P_i$ is present as both $HPO_4^{2-}$ and $H_2PO_4^-$; the phosphate translocase is specific for $H_2PO_4^-$. There is no net flow of charge during symport of $H_2PO_4^-$ and $H^+$, but the relatively low proton concentration in the matrix favors the inward movement of $H^+$. Thus the proton-motive force is responsible both for providing the energy for ATP synthesis and for transporting substrates (ADP and $P_i$) in and product (ATP) out of the mitochondrial matrix. All three of these transport systems can be isolated as a single membrane-bound complex (ATP synthasome).

transmembrane electrochemical gradient, which gives the matrix a net negative charge; the proton-motive force drives ATP-ADP exchange. Adenine nucleotide translocase is specifically inhibited by atractyloside, a toxic glycoside formed by a species of thistle. If the transport of ADP into and ATP out of mitochondria is inhibited, cytosolic ATP cannot be regenerated from ADP, explaining the toxicity of atractyloside.

A second membrane transport system essential to oxidative phosphorylation is the **phosphate translocase,** which promotes symport of one $H_2PO_4^-$ and one $H^+$ into the matrix. This transport process, too, is favored by the transmembrane proton gradient (Fig. 19–26). Notice that the process requires movement of one proton from the P to the N side of the inner membrane, consuming some of the energy of electron transfer. A complex of the ATP synthase and both translocases, the **ATP synthasome,** can be isolated from

mitochondria by gentle dissection with detergents, suggesting that the functions of these three proteins are very tightly integrated.

## Shuttle Systems Indirectly Convey Cytosolic NADH into Mitochondria for Oxidation

The NADH dehydrogenase of the inner mitochondrial membrane of animal cells can accept electrons only from NADH in the matrix. Given that the inner membrane is not permeable to NADH, how can the NADH generated by glycolysis in the cytosol be reoxidized to $NAD^+$ by $O_2$ via the respiratory chain? Special shuttle systems carry reducing equivalents from cytosolic NADH into mitochondria by an indirect route. The most active NADH shuttle, which functions in liver, kidney, and heart mitochondria, is the **malate-aspartate shuttle** (Fig. 19–27). The reducing equivalents of cytosolic NADH are first transferred to cytosolic oxaloacetate to yield malate, catalyzed by cytosolic malate dehydrogenase. The malate thus formed passes through the inner membrane via the malate–α-ketoglutarate transporter. Within the matrix the reducing equivalents are passed to $NAD^+$ by the action of matrix malate dehydrogenase, forming NADH; this NADH can pass electrons directly to the respiratory chain. About 2.5 molecules of ATP are generated as this pair of electrons passes to $O_2$. Cytosolic oxaloacetate must be regenerated by transamination reactions and the activity of membrane transporters to start another cycle of the shuttle.

Skeletal muscle and brain use a different NADH shuttle, the **glycerol 3-phosphate shuttle** (Fig. 19–28). It differs from the malate-aspartate shuttle in that it delivers the reducing equivalents from NADH to ubiquinone and thus into Complex III, not Complex I (Fig. 19–8), providing only enough energy to synthesize 1.5 ATP molecules per pair of electrons.

The mitochondria of plants have an *externally* oriented NADH dehydrogenase that can transfer electrons directly from cytosolic NADH into the respiratory chain at the level of ubiquinone. Because this pathway bypasses the NADH dehydrogenase of Complex I and the associated proton movement, the yield of ATP from cytosolic NADH is less than that from NADH generated in the matrix (Box 19–1).

### *SUMMARY 19.2* ATP Synthesis

- The flow of electrons through Complexes I, III, and IV results in pumping of protons across the inner mitochondrial membrane, making the matrix alkaline relative to the intermembrane space. This proton gradient provides the energy (in the form of the proton-motive force) for ATP synthesis from ADP and $P_i$ by ATP synthase ($F_oF_1$ complex) in the inner membrane.

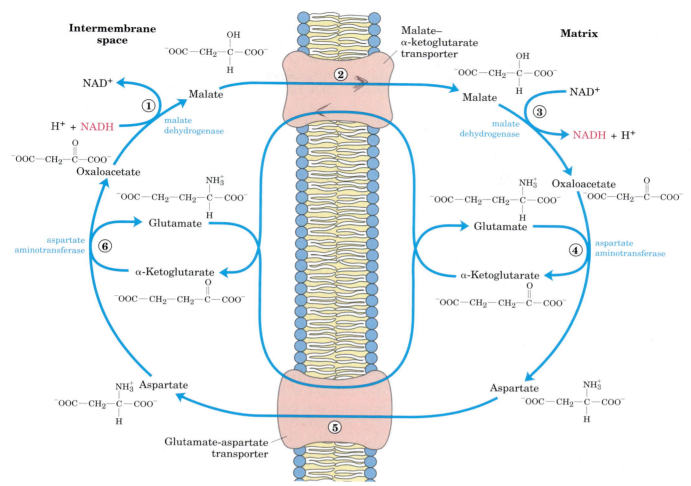

**FIGURE 19–27 Malate-aspartate shuttle.** This shuttle for transporting reducing equivalents from cytosolic NADH into the mitochondrial matrix is used in liver, kidney, and heart. ① NADH in the cytosol (intermembrane space) passes two reducing equivalents to oxaloacetate, producing malate. ② Malate crosses the inner membrane via the malate–α-ketoglutarate transporter. ③ In the matrix, malate passes two reducing equivalents to NAD⁺, and the resulting NADH is oxidized by the respiratory chain. The oxaloacetate formed from malate cannot pass directly into the cytosol. ④ It is first transaminated to aspartate, which ⑤ can leave via the glutamate-aspartate transporter. ⑥ Oxaloacetate is regenerated in the cytosol, completing the cycle.

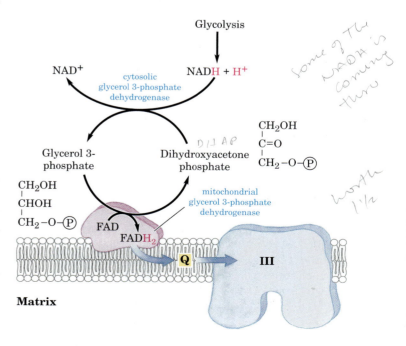

**FIGURE 19–28 Glycerol 3-phosphate shuttle.** This alternative means of moving reducing equivalents from the cytosol to the mitochondrial matrix operates in skeletal muscle and the brain. In the cytosol, dihydroxyacetone phosphate accepts two reducing equivalents from NADH in a reaction catalyzed by cytosolic glycerol 3-phosphate dehydrogenase. An isozyme of glycerol 3-phosphate dehydrogenase bound to the outer face of the inner membrane then transfers two reducing equivalents from glycerol 3-phosphate in the intermembrane space to ubiquinone. Note that this shuttle does not involve membrane transport systems.

■ ATP synthase carries out "rotational catalysis," in which the flow of protons through $F_o$ causes each of three nucleotide-binding sites in $F_1$ to cycle from (ADP + $P_i$)–bound to ATP-bound to empty conformations.

■ ATP formation on the enzyme requires little energy; the role of the proton-motive force is to push ATP from its binding site on the synthase.

■ The ratio of ATP synthesized per $\frac{1}{2}O_2$ reduced to $H_2O$ (the P/O ratio) is about 2.5 when electrons enter the respiratory chain at Complex I, and 1.5 when electrons enter at CoQ.

■ Energy conserved in a proton gradient can drive solute transport uphill across a membrane.

■ The inner mitochondrial membrane is impermeable to NADH and $NAD^+$, but NADH equivalents are moved from the cytosol to the matrix by either of two shuttles. NADH equivalents moved in by the malate-aspartate shuttle enter the respiratory chain at Complex I and yield a P/O ratio of 2.5; those moved in by the glycerol 3-phosphate shuttle enter at CoQ and give a P/O ratio of 1.5.

## 19.3 Regulation of Oxidative Phosphorylation

Oxidative phosphorylation produces most of the ATP made in aerobic cells. Complete oxidation of a molecule of glucose to $CO_2$ yields 30 or 32 ATP (Table 19–5). By comparison, glycolysis under anaerobic conditions (lactate fermentation) yields only 2 ATP per glucose. Clearly, the evolution of oxidative phosphorylation provided a tremendous increase in the energy efficiency of catabolism. Complete oxidation to $CO_2$ of the coenzyme A derivative of palmitate (16:0), which also occurs in the mitochondrial matrix, yields 108 ATP per palmitoyl-CoA (see Table 17–1). A similar calculation can be made for the ATP yield from oxidation of each of the amino acids (Chapter 18). Aerobic oxidative pathways that result in electron transfer to $O_2$ accompanied by oxidative phosphorylation therefore account for the vast majority of the ATP produced in catabolism, so the regulation of ATP production by oxidative phosphorylation to match the cell's fluctuating needs for ATP is absolutely essential.

### Oxidative Phosphorylation Is Regulated by Cellular Energy Needs

The rate of respiration ($O_2$ consumption) in mitochondria is tightly regulated; it is generally limited by the availability of ADP as a substrate for phosphorylation. Dependence of the rate of $O_2$ consumption on the availability of the $P_i$ acceptor ADP (Fig. 19–18b), the **acceptor control** of respiration, can be remarkable. In some animal tissues, the **acceptor control ratio,** the ratio of the maximal rate of ADP-induced $O_2$ consumption to the basal rate in the absence of ADP, is at least ten.

The intracellular concentration of ADP is one measure of the energy status of cells. Another, related measure is the **mass-action ratio** of the ATP-ADP system, [ATP]/([ADP][$P_i$]). Normally this ratio is very high, so the ATP-ADP system is almost fully phosphorylated. When the rate of some energy-requiring process (protein synthesis, for example) increases, the rate of breakdown of ATP to ADP and $P_i$ increases, lowering the mass-action ratio. With more ADP available for oxidative phosphorylation, the rate of respiration increases, causing regeneration of ATP. This continues until the mass-action ratio returns to its normal high level, at which point respiration slows again. The rate of oxidation of cellular fuels is regulated with such sensitivity and precision that the [ATP]/([ADP][$P_i$]) ratio fluctuates only slightly in most tissues, even during extreme variations in energy demand. In short, ATP is formed only as fast as it is used in energy-requiring cellular activities.

| **TABLE 19–5** | **ATP Yield from Complete Oxidation of Glucose** | |
|---|---|---|
| *Process* | *Direct product* | *Final ATP* |
| Glycolysis | 2 NADH (cytosolic) | 3 or 5[*] |
| | 2 ATP | 2 |
| Pyruvate oxidation (two per glucose) | 2 NADH (mitochondrial matrix) | 5 |
| Acetyl-CoA oxidation in citric acid cycle | 6 NADH (mitochondrial matrix) | 15 |
| (two per glucose) | 2 FADH$_2$ | 3 |
| | 2 ATP or 2 GTP | 2 |
| Total yield per glucose | | 30 or 32 |

[*]The number depends on which shuttle system transfers reducing equivalents into the mitochondrion.

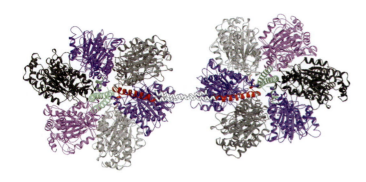

**FIGURE 19–29 Structure of bovine F$_1$-ATPase in a complex with its regulatory protein IF$_1$.** (Derived from PDB ID 1OHH) Two F$_1$ molecules are viewed here as in Figure 19–23c. The inhibitor IF$_1$ (red) binds to the $\alpha\beta$ interface of the subunits in the diphosphate (ADP) conformation ($\alpha$ADP and $\beta$ADP), freezing the two F$_1$ complexes and thereby blocking ATP hydrolysis (and synthesis). (Parts of IF$_1$ that failed to resolve in crystals of F$_1$ are shown in white outline as they occur in crystals of isolated IF$_1$.) This complex is stable only at the low cytosolic pH characteristic of cells that are producing ATP by glycolysis; when aerobic metabolism resumes, the cytosolic pH rises, the inhibitor is destabilized, and ATP synthase becomes active.

## An Inhibitory Protein Prevents ATP Hydrolysis during Ischemia

We have already encountered ATP synthase as an ATP-driven proton pump (see Fig. 11–39; Table 11–3), catalyzing the reverse of ATP synthesis. When a cell is ischemic (deprived of oxygen), as in a heart attack or stroke, electron transfer to oxygen ceases, and so does the pumping of protons. The proton-motive force soon collapses. Under these conditions, the ATP synthase could operate in reverse, hydrolyzing ATP to pump protons outward and causing a disastrous drop in ATP levels. This is prevented by a small (84 amino acids) protein inhibitor, IF$_1$, which simultaneously binds to two ATP synthase molecules, inhibiting their ATPase activity (Fig. 19–29). IF$_1$ is inhibitory only in its dimeric form, which is favored at pH lower than 6.5. In a cell starved for oxygen, the main source of ATP becomes glycolysis, and the pyruvic or lactic acid thus formed lowers the pH in the cytosol and the mitochondrial matrix. This favors IF$_1$ dimerization, leading to inhibition of the ATPase activity of ATP synthase, thereby preventing wasteful hydrolysis of ATP. When aerobic metabolism resumes, production of pyruvic acid slows, the pH of the cytosol rises, the IF$_1$ dimer is destabilized, and the inhibition of ATP synthase is lifted.

## Uncoupled Mitochondria in Brown Fat Produce Heat

There is a remarkable and instructive exception to the general rule that respiration slows when the ATP supply is adequate. Most newborn mammals, including humans, have a type of adipose tissue called **brown fat** in which fuel oxidation serves not to produce ATP but to generate heat to keep the newborn warm. This specialized adipose tissue is brown because of the presence of large numbers of mitochondria and thus large amounts of cytochromes, whose heme groups are strong absorbers of visible light.

The mitochondria of brown fat are like those of other mammalian cells in all respects, except that they have a unique protein in their inner membrane. **Thermogenin,** also called the **uncoupling protein** (Table 19–4), provides a path for protons to return to the matrix without passing through the F$_o$F$_1$ complex (Fig. 19–30).

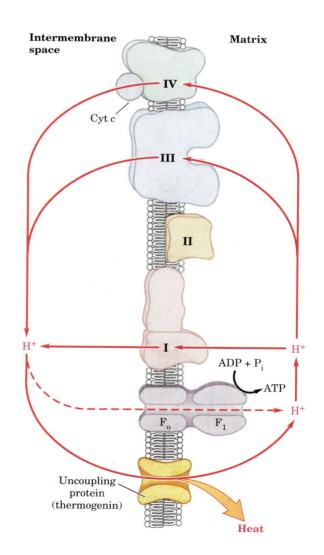

**FIGURE 19–30 Heat generation by uncoupled mitochondria.** The uncoupling protein (thermogenin) of brown fat mitochondria, by providing an alternative route for protons to reenter the mitochondrial matrix, causes the energy conserved by proton pumping to be dissipated as heat.

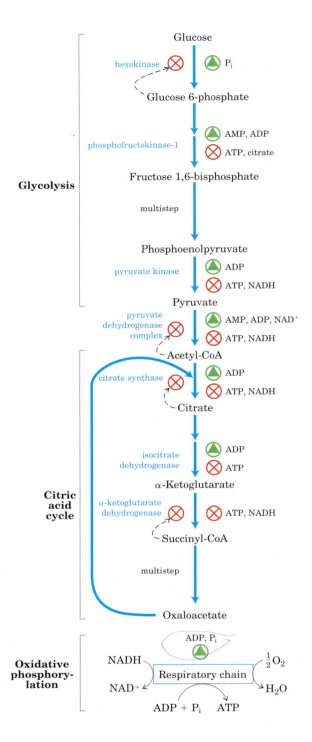

**FIGURE 19–31 Regulation of the ATP-producing pathways.** This diagram shows the interlocking regulation of glycolysis, pyruvate oxidation, the citric acid cycle, and oxidative phosphorylation by the relative concentrations of ATP, ADP, and AMP, and by NADH. High [ATP] (or low [ADP] and [AMP]) produces low rates of glycolysis, pyruvate oxidation, acetate oxidation via the citric acid cycle, and oxidative phosphorylation. All four pathways are accelerated when the use of ATP and the formation of ADP, AMP, and $P_i$ increase. The interlocking of glycolysis and the citric acid cycle by citrate, which inhibits glycolysis, supplements the action of the adenine nucleotide system. In addition, increased levels of NADH and acetyl-CoA also inhibit the oxidation of pyruvate to acetyl-CoA, and a high [NADH]/[NAD$^+$] ratio inhibits the dehydrogenase reactions of the citric acid cycle (see Fig. 16–18).

## ATP-Producing Pathways Are Coordinately Regulated

The major catabolic pathways have interlocking and concerted regulatory mechanisms that allow them to function together in an economical and self-regulating manner to produce ATP and biosynthetic precursors. The relative concentrations of ATP and ADP control not only the rates of electron transfer and oxidative phosphorylation but also the rates of the citric acid cycle, pyruvate oxidation, and glycolysis (Fig. 19–31). Whenever ATP consumption increases, the rate of electron transfer and oxidative phosphorylation increases. Simultaneously, the rate of pyruvate oxidation via the citric acid cycle increases, increasing the flow of electrons into the respiratory chain. These events can in turn evoke an increase in the rate of glycolysis, increasing the rate of pyruvate formation. When conversion of ADP to ATP lowers the ADP concentration, acceptor control slows electron transfer and thus oxidative phosphorylation. Glycolysis and the citric acid cycle are also slowed, because ATP is an allosteric inhibitor of the glycolytic enzyme phosphofructokinase-1 (see Fig. 15–18) and of pyruvate dehydrogenase (see Fig. 16–18).

Phosphofructokinase-1 is also inhibited by citrate, the first intermediate of the citric acid cycle. When the cycle is "idling," citrate accumulates within mitochondria, then spills into the cytosol. When the concentrations of both ATP and citrate rise, they produce a concerted allosteric inhibition of phosphofructokinase-1 that is greater than the sum of their individual effects, slowing glycolysis.

As a result of this short-circuiting of protons, the energy of oxidation is not conserved by ATP formation but is dissipated as heat, which contributes to maintaining the body temperature of the newborn. Hibernating animals also depend on uncoupled mitochondria of brown fat to generate heat during their long dormancy (see Box 17–1).

## SUMMARY 19.3 Regulation of Oxidative Phosphorylation

■ Oxidative phosphorylation is regulated by cellular energy demands. The intracellular [ADP] and the mass-action ratio [ATP]/([ADP][$P_i$]) are measures of a cell's energy status.

- In ischemic (oxygen-deprived) cells, a protein inhibitor blocks ATP hydrolysis by the ATP synthase operating in reverse, preventing a drastic drop in [ATP].

- In brown fat, which is specialized for the production of metabolic heat, electron transfer is uncoupled from ATP synthesis and the energy of fatty acid oxidation is dissipated as heat.

- ATP and ADP concentrations set the rate of electron transfer through the respiratory chain via a series of interlocking controls on respiration, glycolysis, and the citric acid cycle.

# 19.4 Mitochondrial Genes: Their Origin and the Effects of Mutations

Mitochondria contain their own genome, a circular, double-stranded DNA molecule. Each of the hundreds or thousands of mitochondria in a typical cell has about five copies of this genome. The human mitochondrial chromosome (Fig. 19–32) contains 37 genes (16,569 bp), including 13 that encode subunits of proteins of the respiratory chain (Table 19–6); the remaining genes code for rRNA and tRNA molecules essential to the protein-synthesizing machinery of mitochondria. About 900 different mitochondrial proteins are encoded by nuclear genes, synthesized on cytoplasmic ribosomes, then imported and assembled within the mitochondria (Chapter 27).

## Mutations in Mitochondrial Genes Cause Human Disease

A growing number of human diseases can be attributed to mutations in mitochondrial genes. Many of these diseases, those known as the **mitochondrial encephalomyopathies,** affect primarily the brain and skeletal muscle (both heavily dependent on an abundant supply of ATP). These diseases are invariably inherited from the mother, because a developing embryo derives all its mitochondria from the mother's egg. The rare disease **Leber's hereditary optic neuropathy (LHON)** affects the central nervous system, including the optic nerves, causing bilateral loss of vision in early adulthood. A single base change in the mitochondrial gene *ND4* (Fig. 19–32a) changes an Arg residue to a His residue in a polypeptide of Complex I, and the result is mitochondria partially defective in electron transfer from NADH to ubiquinone. Although these mitochondria can produce some ATP by electron transfer from succinate, they apparently cannot supply sufficient ATP to support the very active metabolism of neurons. One result is damage to the optic nerve, leading to blindness. A single base change in the mitochondrial gene for cytochrome *b,* a component of Complex III, also produces LHON, demonstrating that the pathology results from a general reduction of mitochondrial function, not specifically from a defect in electron transfer through Complex I.

**Myoclonic epilepsy and ragged-red fiber disease (MERRF)** is caused by a mutation in the mitochondrial gene that encodes a transfer RNA specific for lycine (lysyl-tRNA). This disease, characterized by uncontrollable muscular jerking, apparently results from defective production of several of the proteins whose synthesis involves mitochondrial tRNAs. Skeletal muscle fibers of individuals with MERRF have abnormally shaped mitochondria that sometimes contain paracrystalline structures (Fig. 19–32b). Mutations in the mitochondrial lysyl-tRNA gene are also one of the causes of adult-onset (type II) diabetes mellitus. Other mutations in mitochondrial genes are believed to be responsible for the progressive muscular weakness that characterizes mitochondrial myopathy and for enlargement and deterioration of the heart muscle in hypertrophic cardiomyopathy. According to one hypothesis on the progressive changes that accompany aging, the accumulation of mutations in mitochondrial DNA during a lifetime of exposure to DNA-damaging agents such as $\cdot O_2^-$ (see below) results in mitochondria that cannot supply sufficient ATP for normal cellular function. Mitochondrial disease can also result from mutations in any of the 900 nuclear genes that encode mitochondrial proteins. ■

**TABLE 19–6    Respiratory Proteins Encoded by Mitochondrial Genes in Humans**

| Complex | Number of subunits | Number of subunits encoded by mitochondrial DNA |
|---|---|---|
| I NADH dehydrogenase | 43 | 7 |
| II Succinate dehydrogenase | 4 | 0 |
| III Ubiquinone:cytochrome *c* oxidoreductase | 11 | 1 |
| IV Cytochrome oxidase | 13 | 3 |
| V ATP synthase | 8 | 2 |

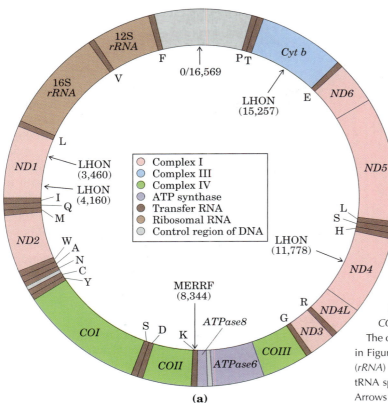

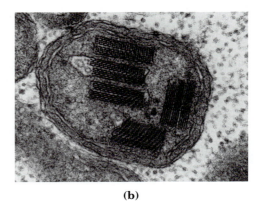

**(b)**

**FIGURE 19-32 Mitochondrial genes and mutations.**
**(a)** Map of human mitochondrial DNA, showing the genes that encode proteins of Complex I, the NADH dehydrogenase (*ND1* to *ND6*); the cytochrome *b* of Complex III (*Cyt b*); the subunits of cytochrome oxidase (Complex IV) (*COI* to *COIII*); and two subunits of ATP synthase (*ATPase6* and *ATPase8*). The colors of the genes correspond to those of the complexes shown in Figure 19–7. Also included here are the genes for ribosomal RNAs (*rRNA*) and for a number of mitochondrion-specific transfer RNAs; tRNA specificity is indicated by the one-letter codes for amino acids. Arrows indicate the positions of mutations that cause Leber's hereditary optic neuropathy (LHON) and myoclonic epilepsy and ragged-red fiber disease (MERRF). Numbers in parentheses indicate the position of the altered nucleotides (nucleotide 1 is at the top of the circle and numbering proceeds counterclockwise). **(b)** Electron micrograph of an abnormal mitochondrion from the muscle of an individual with MERRF, showing the paracrystalline protein inclusions sometimes present in the mutant mitochondria.

**(a)**

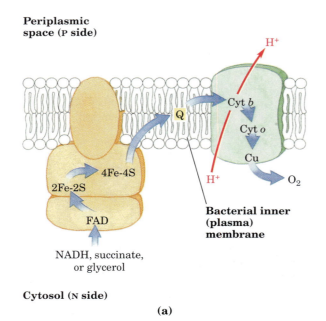

**FIGURE 19-33 Bacterial respiratory chain. (a)** Shown here are the respiratory carriers of the inner membrane of *E. coli*. Eubacteria contain a minimal form of Complex I, containing all the prosthetic groups normally associated with the mitochondrial complex but only 14 polypeptides. This plasma membrane complex transfers electrons from NADH to ubiquinone or to **(b)** menaquinone, the bacterial equivalent of ubiquinone, while pumping protons outward and creating an electrochemical potential that drives ATP synthesis.

## Mitochondria Evolved from Endosymbiotic Bacteria

The existence of mitochondrial DNA, ribosomes, and tRNAs supports the hypothesis of the endosymbiotic origin of mitochondria (see Fig. 1–36), which holds that the first organisms capable of aerobic metabolism, including respiration-linked ATP production, were prokaryotes. Primitive eukaryotes that lived anaerobically (by fermentation) acquired the ability to carry out oxidative phosphorylation when they established a symbiotic relationship with bacteria living in their cytosol. After much evolution and the movement of many bacterial genes into the nucleus of the "host" eukaryote, the endosymbiotic bacteria eventually became mitochondria.

This hypothesis presumes that early free-living prokaryotes had the enzymatic machinery for oxidative phosphorylation and predicts that their modern prokaryotic descendants must have respiratory chains closely similar to those of modern eukaryotes. They do. Aerobic bacteria carry out NAD-linked electron transfer from substrates to $O_2$, coupled to the phosphorylation of cytosolic ADP. The dehydrogenases are located in the bacterial cytosol and the respiratory chain in the plasma membrane. The electron carriers are similar to some mitochondrial electron carriers (Fig. 19–33). They translocate protons outward across the plasma membrane as electrons are transferred to $O_2$. Bacteria such as *Escherichia coli* have $F_oF_1$ complexes in their plasma membranes; the $F_1$ portion protrudes into the cytosol and catalyzes ATP synthesis from ADP and $P_i$ as protons flow back into the cell through the proton channel of $F_o$.

The respiration-linked extrusion of protons across the bacterial plasma membrane also provides the driving force for other processes. Certain bacterial transport systems bring about uptake of extracellular nutrients (lactose, for example) against a concentration gradient, in symport with protons (see Fig. 11–42). And the rotary motion of bacterial flagella is provided by "proton turbines," molecular rotary motors driven not by ATP but directly by the transmembrane electrochemical potential generated by respiration-linked proton pumping (Fig. 19–34). It appears likely that the chemiosmotic mechanism evolved early, before the emergence of eukaryotes.

### SUMMARY 19.4 Mitochondrial Genes: Their Origin and the Effects of Mutations

- A small proportion of human mitochondrial proteins (13 proteins) are encoded in the mitochondrial genome and synthesized within mitochondria. About 900 mitochondrial proteins are encoded in nuclear genes and imported into mitochondria after their synthesis.

- Mutations in the genes that encode components of the respiratory chain, whether in the mitochondrial genes or in the nuclear genes that encode mitochondrial proteins, cause a variety of human diseases, which often affect muscle and brain most severely.

- Mitochondria most likely arose from aerobic prokaryotes that entered into an endosymbiotic relationship with ancestral eukaryotes.

## 19.5 The Role of Mitochondria in Apoptosis and Oxidative Stress

Besides their central role in ATP synthesis, mitochondria also participate in processes associated with cellular damage and death. **Apoptosis** is a controlled process by which cells die for the good of the organism, while the organism conserves the molecular components (amino acids, nucleotides, and so forth) of the dead cells. Apoptosis may be triggered by an external signal, acting at a receptor in the plasma membrane, or by internal events such as a viral infection. When a cell receives a signal for apoptosis, one consequence is an increase in the permeability of the outer mitochondrial membrane, allowing escape of the cytochrome $c$ normally confined in the intermembrane space (see Fig. 12–50). The released cytochrome $c$ activates one of the proteolytic enzymes (caspase 9) responsible for protein degradation during apoptosis. This is a dramatic case of

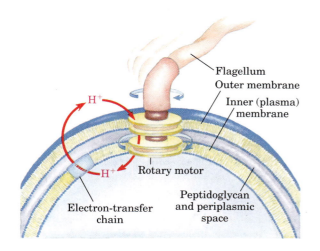

**FIGURE 19–34 Rotation of bacterial flagella by proton-motive force.** The shaft and rings at the base of the flagellum make up a rotary motor that has been called a "proton turbine." Protons ejected by electron transfer flow back into the cell through the turbine, causing rotation of the shaft of the flagellum. This motion differs fundamentally from the motion of muscle and of eukaryotic flagella and cilia, for which ATP hydrolysis is the energy source.

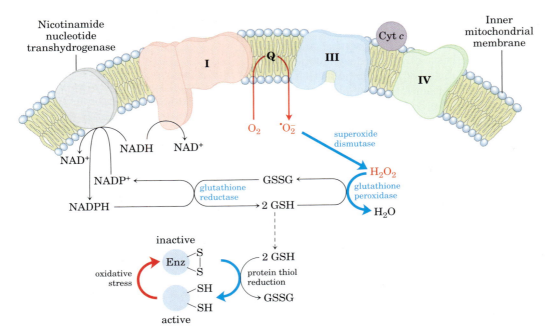

**FIGURE 19–35 Mitochondrial production and disposal of super-oxide.** Superoxide radical, $\cdot O_2^-$, is formed in side reactions at Complexes I and III, as the partially reduced ubiquinone radical ($\cdot Q^-$) donates an electron to $O_2$. The reactions shown in blue defend the cell against the damaging effects of superoxide. Reduced glutathione (GSH; see Fig. 22–27) donates electrons for the reduction of hydrogen peroxide ($H_2O_2$) and of oxidized Cys residues (—S—S—) in proteins, and GSH is regenerated from the oxidized form (GSSG) by reduction with NADPH.

one protein (cytochrome *c*) playing two very different roles in the cell.

Mitochondria are also involved in the cell's response to oxidative stress. As we have seen, several steps in the path of oxygen reduction in mitochondria have the potential to produce highly reactive free radicals that can damage cells. The passage of electrons from $QH_2$ to cytochrome $b_L$ through Complex III, and passage of electrons from Complex I to $QH_2$, involve the radical $\cdot Q^-$ as an intermediate. The $\cdot Q^-$ can, with a low probability, pass an electron to $O_2$ in the reaction

$$O_2 + e^- \longrightarrow \cdot O_2^-$$

The superoxide free radical thus generated, $\cdot O_2^-$, is very reactive and can damage enzymes, membrane lipids, and nucleic acids. Antimycin A, an inhibitor of Complex III, may act by occupying the $Q_N$ site (Fig. 19–11), thus blocking the Q cycle and prolonging the binding of $\cdot Q^-$ to the $Q_P$ site; this would increase the likelihood of superoxide radical formation and cellular damage. From 0.1% to as much as 4% of the $O_2$ used by actively respiring mitochondria forms $\cdot O_2^-$—more than enough to have lethal effects on a cell unless the free radical is quickly disposed of.

To prevent oxidative damage by $\cdot O_2^-$, cells have several forms of the enzyme **superoxide dismutase,** which catalyzes the reaction

$$2 \cdot O_2^- + 2H^+ \longrightarrow H_2O_2 + O_2$$

The hydrogen peroxide ($H_2O_2$) generated by this reaction is rendered harmless by the action of **glutathione peroxidase** (Fig. 19–35). This enzyme is remarkable for the presence of a selenocysteine residue (see Fig. 3–8a), in which an atom of selenium replaces the sulfur atom normally present in the thiol of the side chain. The selenol group (—SeH) is more acidic than the thiol (—SH); its pKa is about 5, so at neutral pH, the selenocysteine side chain is essentially fully ionized (—$CH_2Se^-$). Glutathione reductase recycles oxidized glutathione to its reduced form, using electrons from the NADPH formed by nicotinamide nucleotide transhydrogenase or by the pentose phosphate pathway (see Fig. 14–20). Reduced glutathione also serves in keeping protein sulfhydryl groups in their reduced state, preventing some of the deleterious effects of oxidative stress (Fig. 19–35).

## SUMMARY 19.5 The Role of Mitochondria in Apoptosis and Oxidative Stress

■ Mitochondrial cytochrome *c*, released into the cytosol, participates in activation of one of the proteases (caspase 9) involved in apoptosis.

■ Reactive oxygen species produced in mitochondria are inactivated by a set of protective enzymes, including superoxide dismutase and glutathione peroxidase.

# PHOTOSYNTHESIS: HARVESTING LIGHT ENERGY

We now turn to another reaction sequence in which the flow of electrons is coupled to the synthesis of ATP: light-driven phosphorylation. The capture of solar energy by photosynthetic organisms and its conversion to the chemical energy of reduced organic compounds is the ultimate source of nearly all biological energy. Photosynthetic and heterotrophic organisms live in a balanced steady state in the biosphere (Fig. 19–36). Photosynthetic organisms trap solar energy and form ATP and NADPH, which they use as energy sources to make carbohydrates and other organic compounds from $CO_2$ and $H_2O$; simultaneously, they release $O_2$ into the atmosphere. Aerobic heterotrophs (humans, for example, as well as plants during dark periods) use the $O_2$ so formed to degrade the energy-rich organic products of photosynthesis to $CO_2$ and $H_2O$, generating ATP. The $CO_2$ returns to the atmosphere, to be used again by photosynthetic organisms. Solar energy thus provides the driving force for the continuous cycling of $CO_2$ and $O_2$ through the biosphere and provides the reduced substrates—fuels, such as glucose—on which nonphotosynthetic organisms depend.

Photosynthesis occurs in a variety of bacteria and in unicellular eukaryotes (algae) as well as in vascular plants. Although the process in these organisms differs in detail, the underlying mechanisms are remarkably similar, and much of our understanding of photosynthesis in vascular plants is derived from studies of simpler organisms. The overall equation for photosynthesis in vascular plants describes an oxidation-reduction reaction in which $H_2O$ donates electrons (as hydrogen) for the reduction of $CO_2$ to carbohydrate ($CH_2O$):

$$CO_2 + H_2O \xrightarrow{\text{light}} O_2 + (CH_2O)$$

# 19.6 General Features of Photophosphorylation

Unlike NADH (the major electron donor in oxidative phosphorylation), $H_2O$ is a poor donor of electrons; its standard reduction potential is 0.816 V, compared with −0.320 V for NADH. Photophosphorylation differs from oxidative phosphorylation in requiring the input of energy in the form of light to *create* a good electron donor and a good electron acceptor. In photophosphorylation, electrons flow through a series of membrane-bound carriers including cytochromes, quinones, and iron-sulfur proteins, while protons are pumped across a membrane to create an electrochemical potential. Electron transfer and proton pumping are catalyzed by membrane complexes homologous in structure and function to Complex III of mitochondria. The electrochemical potential they produce is the driving force for ATP synthesis from ADP and $P_i$, catalyzed by a membrane-bound ATP synthase complex closely similar to that of oxidative phosphorylation.

Photosynthesis in plants encompasses two processes: the **light-dependent reactions,** or **light reactions,** which occur only when plants are illuminated, and the **carbon-assimilation reactions** (or **carbon-fixation reactions**), sometimes misleadingly called the dark reactions, which are driven by products of the light reactions (Fig. 19–37). In the light reactions, chlorophyll and other pigments of photosynthetic cells absorb light energy and conserve it as ATP and NADPH; simultaneously, $O_2$ is evolved. In the carbon-assimilation reactions, ATP and NADPH are used to reduce $CO_2$ to form triose phosphates, starch, and sucrose, and other products derived from them. In this chapter we are concerned only with the light-dependent reactions that lead to the synthesis of ATP and NADPH. The reduction of $CO_2$ is described in Chapter 20.

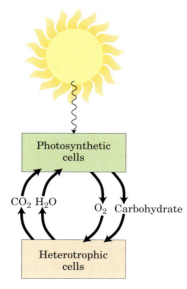

**FIGURE 19–36  Solar energy as the ultimate source of all biological energy.** Photosynthetic organisms use the energy of sunlight to manufacture glucose and other organic products, which heterotrophic cells use as energy and carbon sources.

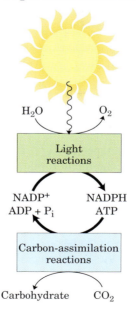

**FIGURE 19–37  The light reactions of photosynthesis generate energy-rich NADPH and ATP at the expense of solar energy.** These products are used in the carbon-assimilation reactions, which occur in light or darkness, to reduce $CO_2$ to form trioses and more complex compounds (such as glucose) derived from trioses.

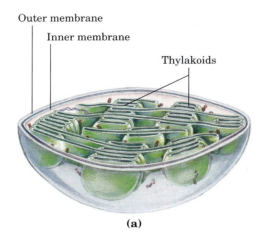

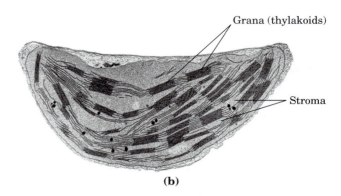

**FIGURE 19–38  Chloroplast. (a)** Schematic diagram. **(b)** Electron micrograph at high magnification showing grana, stacks of thylakoid membranes.

## Photosynthesis in Plants Takes Place in Chloroplasts

In photosynthetic eukaryotic cells, both the light-dependent and the carbon-assimilation reactions take place in the chloroplasts (Fig. 19–38), membrane-bounded intracellular organelles that are variable in shape and generally a few micrometers in diameter. Like mitochondria, they are surrounded by two membranes, an outer membrane that is permeable to small molecules and ions, and an inner membrane that encloses the internal compartment. This compartment contains many flattened, membrane-surrounded vesicles or sacs, the **thylakoids,** usually arranged in stacks called **grana** (Fig. 19–38b). Embedded in the thylakoid membranes (commonly called **lamellae**) are the photosynthetic pigments and the enzyme complexes that carry out the light reactions and ATP synthesis. The **stroma** (the aqueous phase enclosed by the inner membrane) contains most of the enzymes required for the carbon-assimilation reactions.

## Light Drives Electron Flow in Chloroplasts

In 1937 Robert Hill found that when leaf extracts containing chloroplasts were illuminated, they (1) evolved $O_2$ and (2) reduced a nonbiological electron acceptor added to the medium, according to the **Hill reaction:**

$$2H_2O + 2A \xrightarrow{\text{light}} 2AH_2 + O_2$$

where A is the artificial electron acceptor, or **Hill reagent.** One Hill reagent, the dye 2,6-dichlorophenolindophenol, is blue when oxidized (A) and colorless when reduced ($AH_2$), making the reaction easy to follow. When a leaf extract supplemented with the dye was illuminated, the blue dye became colorless and $O_2$ was evolved. In the dark, neither $O_2$ evolution nor dye reduction took place. This was the first evidence that absorbed light energy causes electrons to flow from $H_2O$ to an electron acceptor. Moreover, Hill found that $CO_2$

was neither required nor reduced to a stable form under these conditions; $O_2$ evolution could be dissociated from $CO_2$ reduction. Several years later Severo Ochoa showed that $NADP^+$ is the biological electron acceptor in chloroplasts, according to the equation

$$2H_2O + 2NADP^+ \xrightarrow{\text{light}} 2NADPH + 2H^+ + O_2$$

To understand this photochemical process, we must first consider the more general topic of the effects of light absorption on molecular structure.

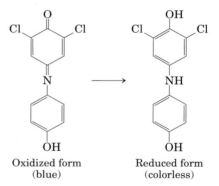

Oxidized form (blue) → Reduced form (colorless)

Dichlorophenolindophenol

## SUMMARY 19.6 General Features of Photophosphorylation

- The light reactions of photosynthesis are those directly dependent on the absorption of light; the resulting photochemistry takes electrons from $H_2O$ and drives them through a series of membrane-bound carriers, producing NADPH and ATP.

- The carbon-assimilation reactions of photosynthesis reduce $CO_2$ with electrons from NADPH and energy from ATP.

## 19.7 Light Absorption

Visible light is electromagnetic radiation of wavelengths 400 to 700 nm, a small part of the electromagnetic spectrum (Fig. 19–39), ranging from violet to red. The energy of a single **photon** (a quantum of light) is greater at the violet end of the spectrum than at the red end; shorter wavelength (and higher frequency) corresponds to higher energy. The energy, $E$, in a "mole" of photons (1 einstein, or $6 \times 10^{23}$ photons) of visible light is 170 to 300 kJ, as given by the Planck equation:

$$E = h\nu$$

where $h$ is Planck's constant ($6.626 \times 10^{-34}$ J·s) and $\nu$ is the wavelength. These amounts of energy are almost an order of magnitude greater than the 30 to 50 kJ required to synthesize a mole of ATP from ADP and $P_i$.

When a photon is absorbed, an electron in the absorbing molecule (chromophore) is lifted to a higher energy level. This is an all-or-nothing event; to be absorbed, the photon must contain a quantity of energy (a **quantum**) that exactly matches the energy of the electronic transition. A molecule that has absorbed a photon is in an **excited state,** which is generally unstable. An electron lifted into a higher-energy orbital usually returns rapidly to its normal lower-energy orbital; the excited molecule decays to the stable **ground state,** giving up the absorbed quantum as light or heat or using it to do chemical work. Light emission accompanying decay of excited molecules (called **fluorescence**) is always at a longer wavelength (lower energy) than that of the absorbed light (see Box 12–2). An alternative mode of decay important in photosynthesis involves direct transfer of excitation energy from an excited molecule to a neighboring molecule. Just as the photon is a quantum of light energy, so the **exciton** is a quantum of energy passed from an excited molecule to another molecule in a process called **exciton transfer.**

### Chlorophylls Absorb Light Energy for Photosynthesis

The most important light-absorbing pigments in the thylakoid membranes are the **chlorophylls,** green pigments with polycyclic, planar structures resembling the protoporphyrin of hemoglobin (see Fig. 5–1), except that $Mg^{2+}$, not $Fe^{2+}$, occupies the central position (Fig. 19–40). The four inward-oriented nitrogen atoms of chlorophyll are coordinated with the $Mg^{2+}$. All chlorophylls have a long **phytol** side chain, esterified to a carboxyl-group substituent in ring IV, and chlorophylls also have a fifth five-membered ring not present in heme.

The heterocyclic five-ring system that surrounds the $Mg^{2+}$ has an extended polyene structure, with alternating single and double bonds. Such polyenes characteristically show strong absorption in the visible region of the spectrum (Fig. 19–41); the chlorophylls have unusually high molar extinction coefficients (see Box 3–1) and are therefore particularly well-suited for absorbing visible light during photosynthesis.

Chloroplasts always contain both chlorophyll $a$ and chlorophyll $b$ (Fig. 19–40a). Although both are green, their absorption spectra are sufficiently different (Fig. 19–41) that they complement each other's range of light absorption in the visible region. Most plants contain about twice as much chlorophyll $a$ as chlorophyll $b$. The pigments in algae and photosynthetic bacteria include chlorophylls that differ only slightly from the plant pigments.

Chlorophyll is always associated with specific binding proteins, forming **light-harvesting complexes (LHCs)** in which chlorophyll molecules are fixed in relation to each other, to other protein complexes, and to the membrane. The detailed structure of one light-harvesting complex is known from x-ray crystallography (Fig. 19–42). It contains seven molecules of chlorophyll $a$, five of chlorophyll $b$, and two of the accessory pigment lutein (see below).

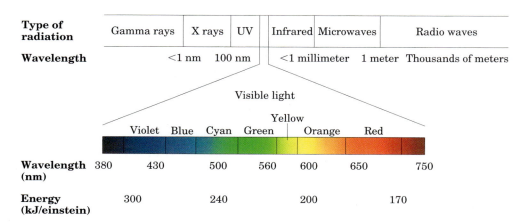

**FIGURE 19–39 Electromagnetic radiation.** The spectrum of electromagnetic radiation, and the energy of photons in the visible range of the spectrum. One einstein is $6 \times 10^{23}$ photons.

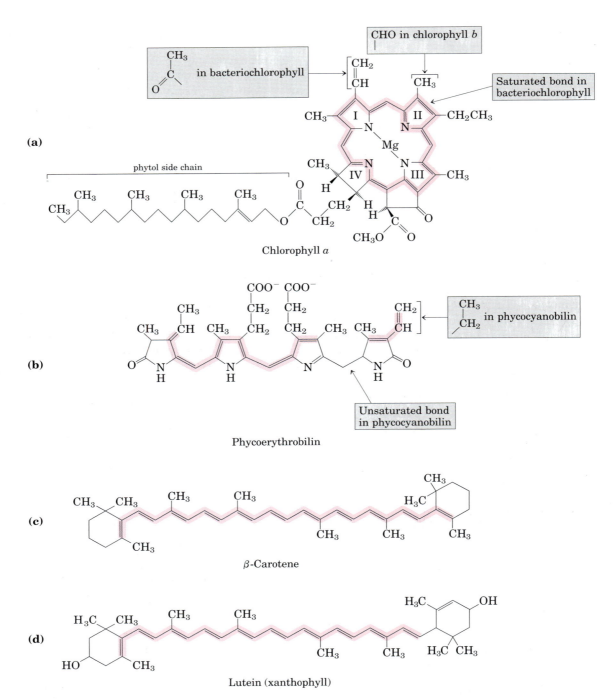

**FIGURE 19–40 Primary and secondary photopigments. (a)** Chlorophylls *a* and *b* and bacteriochlorophyll are the primary gatherers of light energy. **(b)** Phycoerythrobilin and phycocyanobilin (phycobilins) are the antenna pigments in cyanobacteria and red algae. **(c)** β-Carotene (a carotenoid) and **(d)** lutein (a xanthophyll) are accessory pigments in plants. The areas shaded pink are the conjugated systems (alternating single and double bonds) that largely account for the absorption of visible light.

Cyanobacteria and red algae employ **phycobilins** such as phycoerythrobilin and phycocyanobilin (Fig. 19–40b) as their light-harvesting pigments. These open-chain tetrapyrroles have the extended polyene system found in chlorophylls, but not their cyclic structure or central $Mg^{2+}$. Phycobilins are covalently linked to specific binding proteins, forming **phycobiliproteins,** which associate in highly ordered complexes called phycobilisomes (Fig. 19–43) that constitute the primary light-harvesting structures in these microorganisms.

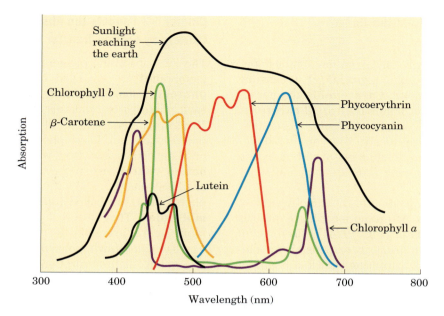

**FIGURE 19–41 Absorption of visible light by photopigments.** Plants are green because their pigments absorb light from the red and blue regions of the spectrum, leaving primarily green light to be reflected or transmitted. Compare the absorption spectra of the pigments with the spectrum of sunlight reaching the earth's surface; the combination of chlorophylls (*a* and *b*) and accessory pigments enables plants to harvest most of the energy available in sunlight.

The relative amounts of chlorophylls and accessory pigments are characteristic of a particular plant species. Variation in the proportions of these pigments is responsible for the range of colors of photosynthetic organisms, from the deep blue-green of spruce needles, to the greener green of maple leaves, to the red, brown, or purple color of some species of multicellular algae and the leaves of some foliage plants favored by gardeners.

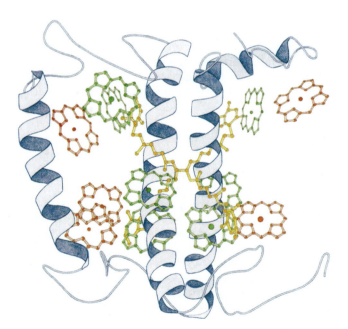

**FIGURE 19–42 A light-harvesting complex, LHCII.** The functional unit is an LHC trimer, with 36 chlorophyll and 6 lutein molecules. Shown here is a monomer, viewed in the plane of the membrane, with its three transmembrane α-helical segments, seven chlorophyll *a* molecules (green), five chlorophyll *b* molecules (red), and two molecules of the accessory pigment lutein (yellow), which form an internal cross-brace.

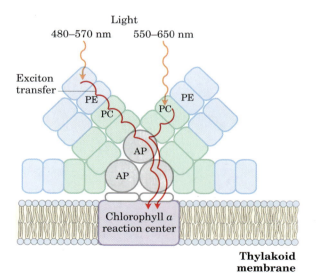

**FIGURE 19–43 A phycobilisome.** In these highly structured assemblies found in cyanobacteria and red algae, phycobilin pigments bound to specific proteins form complexes called phycoerythrin (PE), phycocyanin (PC), and allophycocyanin (AP). The energy of photons absorbed by PE or PC is conveyed through AP (a phycocyanobilin-binding protein) to chlorophyll *a* of the reaction center by exciton transfer, a process discussed in the text.

## Accessory Pigments Extend the Range of Light Absorption

In addition to chlorophylls, thylakoid membranes contain secondary light-absorbing pigments, or **accessory pigments,** called carotenoids. **Carotenoids** may be yellow, red, or purple. The most important are **β-carotene,** which is a red-orange isoprenoid, and the yellow carotenoid **lutein** (Fig. 19–40c, d). The carotenoid pigments absorb light at wavelengths not absorbed by the chlorophylls (Fig. 19–41) and thus are supplementary light receptors.

Experimental determination of the effectiveness of light of different colors in promoting photosynthesis yields an **action spectrum** (Fig. 19–44), often useful in identifying the pigment primarily responsible for a biological effect of light. By capturing light in a region of the spectrum not used by other organisms, a photosynthetic organism can claim a unique ecological niche. For example, the phycobilins in red algae and cyanobacteria absorb light in the range 520 to 630 nm (Fig. 19–41), allowing them to occupy niches where light of lower or higher wavelength has been filtered out by the pigments of other organisms living in the water above them, or by the water itself.

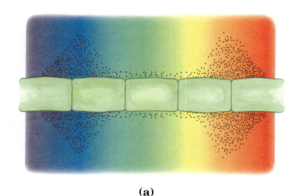

**(a)**

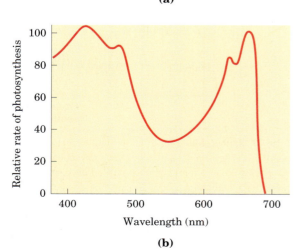

**(b)**

## Chlorophyll Funnels the Absorbed Energy to Reaction Centers by Exciton Transfer

The light-absorbing pigments of thylakoid or bacterial membranes are arranged in functional arrays called **photosystems.** In spinach chloroplasts, for example, each photosystem contains about 200 chlorophyll and 50 carotenoid molecules. All the pigment molecules in a photosystem can absorb photons, but only a few chlorophyll molecules associated with the **photochemical reaction center** are specialized to transduce light into chemical energy. The other pigment molecules in a photosystem are called **light-harvesting** or **antenna molecules.** They absorb light energy and transmit it rapidly and efficiently to the reaction center (Fig. 19–45).

The chlorophyll molecules in light-harvesting complexes have light-absorption properties that are subtly different from those of free chlorophyll. When isolated chlorophyll molecules in vitro are excited by light, the absorbed energy is quickly released as fluorescence and heat, but when chlorophyll in intact leaves is excited by visible light (Fig. 19–46, step ①), very little fluorescence is observed. Instead, the excited antenna chlorophyll transfers energy directly to a neighboring chlorophyll molecule, which becomes excited as the first molecule returns to its ground state (step ②). This transfer of energy, exciton transfer, extends to a third, fourth, or subsequent neighbor, until one of a special pair of chlorophyll *a* molecules at the photochemical reaction center is excited (step ③). In this excited chlorophyll molecule, an electron is promoted to a higher-energy orbital. This electron then passes to a nearby electron acceptor that is part of the electron-transfer chain, leaving the reaction-center chlorophyll with a

**FIGURE 19–44 Two ways to determine the action spectrum for photosynthesis. (a)** Results of a classic experiment performed by T. W. Englemann in 1882 to determine the wavelength of light that is most effective in supporting photosynthesis. Englemann placed cells of a filamentous photosynthetic alga on a microscope slide and illuminated them with light from a prism, so that one part of the filament received mainly blue light, another part yellow, another red. To determine which algal cells carried out photosynthesis most actively, Englemann also placed on the microscope slide bacteria known to migrate toward regions of high $O_2$ concentration. After a period of illumination, the distribution of bacteria showed highest $O_2$ levels (produced by photosynthesis) in the regions illuminated with violet and red light.

**(b)** Results of a similar experiment that used modern techniques (an oxygen electrode) for the measurement of $O_2$ production. An action spectrum (as shown here) describes the relative rate of photosynthesis for illumination with a constant number of photons of different wavelengths. An action spectrum is useful because, by comparison with absorption spectra (such as those in Fig. 19–41), it suggests which pigments can channel energy into photosynthesis.

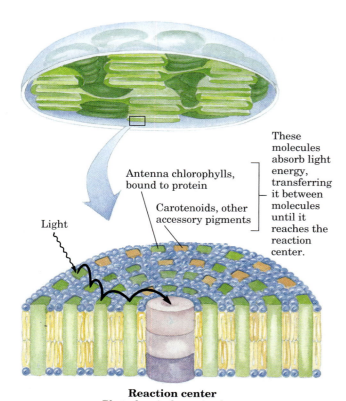

**Reaction center**
Photochemical reaction here
converts the energy of a photon
into a separation of charge,
initiating electron flow.

**FIGURE 19–45 Organization of photosystems in the thylakoid membrane.** Photosystems are tightly packed in the thylakoid membrane, with several hundred antenna chlorophylls and accessory pigments surrounding a photoreaction center. Absorption of a photon by any of the antenna chlorophylls leads to excitation of the reaction center by exciton transfer (black arrow). Also embedded in the thylakoid membrane are the cytochrome $b_6f$ complex and ATP synthase (see Fig. 19–52).

missing electron (an "electron hole," denoted by + in Fig. 19–46) (step ④). The electron acceptor acquires a negative charge in this transaction. The electron lost by the reaction-center chlorophyll is replaced by an electron from a neighboring electron-donor molecule (step ⑤), which thereby becomes positively charged. In this way, *excitation by light causes electric charge separation and initiates an oxidation-reduction chain.*

**FIGURE 19–46 Exciton and electron transfer.** This generalized scheme shows conversion of the energy of an absorbed photon into separation of charges at the reaction center. The steps are further described in the text. Note that step ① may repeat between successive antenna molecules until the exciton reaches a reaction-center chlorophyll. The asterisk (*) represents the excited state of an antenna molecule.

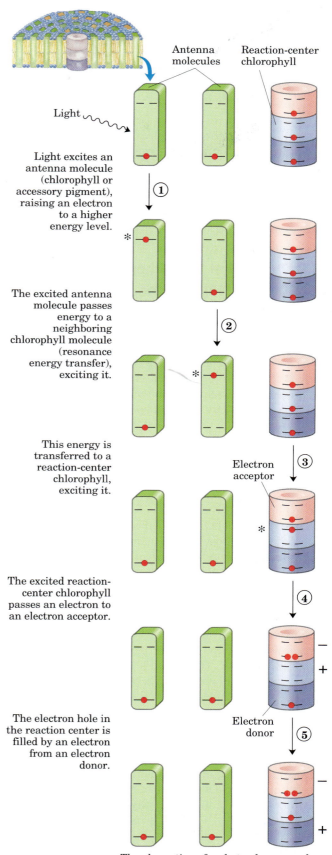

Light excites an antenna molecule (chlorophyll or accessory pigment), raising an electron to a higher energy level.

The excited antenna molecule passes energy to a neighboring chlorophyll molecule (resonance energy transfer), exciting it.

This energy is transferred to a reaction-center chlorophyll, exciting it.

The excited reaction-center chlorophyll passes an electron to an electron acceptor.

The electron hole in the reaction center is filled by an electron from an electron donor.

The absorption of a photon has caused separation of charge in the reaction center.

## *SUMMARY 19.7* Light Absorption

- Photophosphorylation in the chloroplasts of green plants and in cyanobacteria involves electron flow through a series of membrane-bound carriers.

- In the light reactions of plants, absorption of a photon excites chlorophyll molecules and other (accessory) pigments, which funnel the energy into reaction centers in the thylakoid membranes. In the reaction centers, photo-excitation results in a charge separation that produces a strong electron donor (reducing agent) and a strong electron acceptor.

## 19.8 The Central Photochemical Event: Light-Driven Electron Flow

Light-driven electron transfer in plant chloroplasts during photosynthesis is accomplished by multienzyme systems in the thylakoid membrane. Our current picture of photosynthetic mechanisms is a composite, drawn from studies of plant chloroplasts and a variety of bacteria and algae. Determination of the molecular structures of bacterial photosynthetic complexes (by x-ray crystallography) has given us a much improved understanding of the molecular events in photosynthesis in general.

### Bacteria Have One of Two Types of Single Photochemical Reaction Center

One major insight from studies of photosynthetic bacteria came in 1952 when Louis Duysens found that illumination of the photosynthetic membranes of the purple bacterium *Rhodospirillum rubrum* with a pulse of light of a specific wavelength (870 nm) caused a temporary decrease in the absorption of light at that wavelength; a pigment was "bleached" by 870 nm light. Later studies by Bessel Kok and Horst Witt showed similar bleaching of plant chloroplast pigments by light of 680 and 700 nm. Furthermore, addition of the (nonbiological) electron acceptor $[Fe(CN)_6]^{3-}$ (ferricyanide) caused bleaching at these wavelengths *without illumination*. These findings indicated that bleaching of the pigments was due to the loss of an electron from a photochemical reaction center. The pigments were named for the wavelength of maximum bleaching: P870, P680, and P700.

Photosynthetic bacteria have relatively simple phototransduction machinery, with one of two general types of reaction center. One type (found in purple bacteria) passes electrons through **pheophytin** (chlorophyll lacking the central $Mg^{2+}$ ion) to a quinone. The other (in green sulfur bacteria) passes electrons through a quinone to an iron-sulfur center. Cyanobacteria and plants have two photosystems (PSI, PSII), one of each type, acting in tandem. Biochemical and biophysical studies have revealed many of the molecular details of reaction centers of bacteria, which therefore serve as prototypes for the more complex phototransduction systems of plants.

***The Pheophytin-Quinone Reaction Center (Type II Reaction Center)*** The photosynthetic machinery in purple bacteria consists of three basic modules (Fig. 19–47a): a single reaction center (P870), a cytochrome $bc_1$ electron-transfer complex similar to Complex III of the mitochondrial electron-transfer chain, and an ATP synthase, also similar to that of mitochondria. Illumination drives electrons through pheophytin and a quinone to the cytochrome $bc_1$ complex; after passing through the complex, electrons flow through cytochrome $c_2$ back to the reaction center, restoring its preillumination state. This light-driven cyclic flow of electrons provides the energy for proton pumping by the cytochrome $bc_1$ complex. Powered by the resulting proton gradient, ATP synthase produces ATP, exactly as in mitochondria.

The three-dimensional structures of the reaction centers of purple bacteria (*Rhodopseudomonas viridis* and *Rhodobacter sphaeroides*), deduced from x-ray crystallography, shed light on how phototransduction takes place in a pheophytin-quinone reaction center. The *R. viridis* reaction center (Fig. 19–48a) is a large protein complex containing four polypeptide subunits and 13 cofactors: two pairs of bacterial chlorophylls, a pair of pheophytins, two quinones, a nonheme iron, and four hemes in the associated *c*-type cytochrome.

The extremely rapid sequence of electron transfers shown in Figure 19–48b has been deduced from physical studies of the bacterial pheophytin-quinone centers, using brief flashes of light to trigger phototransduction and a variety of spectroscopic techniques to follow the flow of electrons through several carriers. A pair of bacteriochlorophylls—the "special pair," designated $(Chl)_2$—is the site of the initial photochemistry in the bacterial reaction center. Energy from a photon absorbed by one of the many antenna chlorophyll molecules surrounding the reaction center reaches $(Chl)_2$ by exciton transfer. When these two chlorophyll molecules—so close that their bonding orbitals overlap—absorb an exciton, the redox potential of $(Chl)_2$ is shifted, by an amount equivalent to the energy of the photon, converting the special pair to a very strong electron donor. The $(Chl)_2$ donates an electron that passes through a neighboring chlorophyll monomer to pheophytin (Pheo). This produces two radicals, one positively charged (the special pair of chlorophylls) and one negatively charged (the pheophytin):

$$(Chl)_2 + 1 \text{ exciton} \longrightarrow (Chl)_2^* \qquad \text{(excitation)}$$

$$(Chl)_2^* + Pheo \longrightarrow \cdot(Chl)_2^+ + \cdot Pheo^- \quad \text{(charge separation)}$$

The pheophytin radical now passes its electron to a tightly bound molecule of quinone ($Q_A$), converting it to a semiquinone radical, which immediately donates its

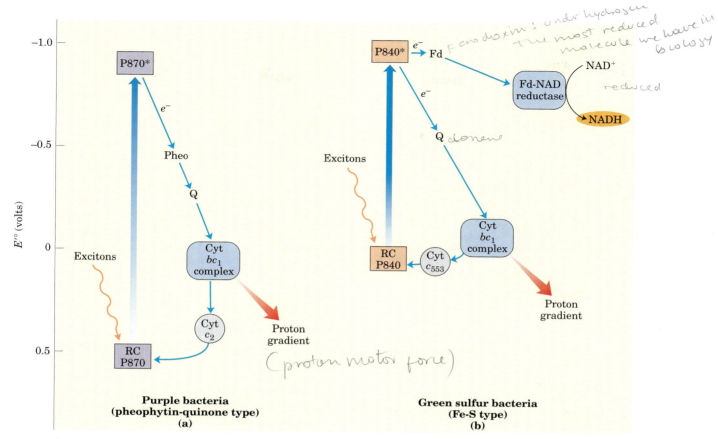

*Handwritten annotations on figure (b):* "Ferredoxin: under hydrogen The most reduced molecule we have in biology", "reduced", "Q donene", and below figure (a) "(proton motor force)"

**FIGURE 19–47 Functional modules of photosynthetic machinery in purple bacteria and green sulfur bacteria. (a)** In purple bacteria, light energy drives electrons from the reaction center P870 through pheophytin (Pheo), a quinone (Q), and the cytochrome $bc_1$ complex, then through cytochrome $c_2$ back to the reaction center. Electron flow through the cytochrome $bc_1$ complex causes proton pumping, creating an electrochemical potential that powers ATP synthesis. **(b)** Green sulfur bacteria have two routes for electrons driven by excitation of P840: a cyclic route passes through a quinone to the cytochrome $bc_1$ complex and back to the reaction center via cytochrome $c$, and a noncyclic route from the reaction center through the iron-sulfur protein ferredoxin (Fd), then to $NAD^+$ in a reaction catalyzed by ferredoxin:NAD reductase.

extra electron to a second, loosely bound quinone ($Q_B$). Two such electron transfers convert $Q_B$ to its fully reduced form, $Q_BH_2$, which is free to diffuse in the membrane bilayer, away from the reaction center:

$$2 \cdot Pheo^- + 2H^+ + Q_B \longrightarrow 2\ Pheo + Q_BH_2$$

$$\text{(quinone reduction)}$$

The hydroquinone ($Q_BH_2$), carrying in its chemical bonds some of the energy of the photons that originally excited P870, enters the pool of reduced quinone ($QH_2$) dissolved in the membrane and moves through the lipid phase of the bilayer to the cytochrome $bc_1$ complex.

Like the homologous Complex III in mitochondria, the cytochrome $bc_1$ complex of purple bacteria carries electrons from a quinol donor ($QH_2$) to an electron acceptor, using the energy of electron transfer to pump protons across the membrane, producing a protonmotive force. The path of electron flow through this complex is believed to be very similar to that through mitochondrial Complex III, involving a Q cycle (Fig. 19–12) in which protons are consumed on one side of the membrane and released on the other. The ultimate electron acceptor in purple bacteria is the electron-depleted form of P870, $\cdot(Chl)_2^+$ (Fig. 19–47a). Electrons move from the cytochrome $bc_1$ complex to P870 via a soluble $c$-type cytochrome, cytochrome $c_2$. The electron-transfer process completes the cycle, returning the reaction center to its unbleached state, ready to absorb another exciton from antenna chlorophyll.

A remarkable feature of this system is that all the chemistry occurs in the *solid state*, with reacting species held close together in the right orientation for reaction. The result is a very fast and efficient series of reactions.

**The Fe-S Reaction Center (Type I Reaction Center)**  Photosynthesis in green sulfur bacteria involves the same three modules as in purple bacteria, but the process differs in several respects and involves additional enzymatic reactions (Fig. 19–47b). Excitation causes an electron to move from the reaction center to the cytochrome $bc_1$ complex via a quinone carrier. Electron transfer through this complex powers proton transport and creates the proton-motive force used for ATP synthesis, just as in purple bacteria and in mitochondria.

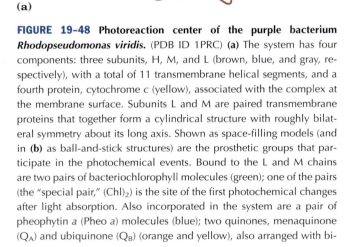

**(a)**

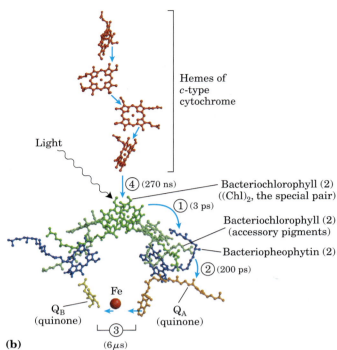

Hemes of
*c*-type
cytochrome

Light

④ (270 ns)

① (3 ps)

— Bacteriochlorophyll (2)
(($Chl)_2$, the special pair)

Bacteriochlorophyll (2)
(accessory pigments)

Bacteriopheophytin (2)

② (200 ps)

Fe

$Q_B$
(quinone)

$Q_A$
(quinone)

③

**(b)**

(6 μs)

**FIGURE 19–48 Photoreaction center of the purple bacterium** ***Rhodopseudomonas viridis.*** (PDB ID 1PRC) **(a)** The system has four components: three subunits, H, M, and L (brown, blue, and gray, respectively), with a total of 11 transmembrane helical segments, and a fourth protein, cytochrome *c* (yellow), associated with the complex at the membrane surface. Subunits L and M are paired transmembrane proteins that together form a cylindrical structure with roughly bilateral symmetry about its long axis. Shown as space-filling models (and in **(b)** as ball-and-stick structures) are the prosthetic groups that participate in the photochemical events. Bound to the L and M chains are two pairs of bacteriochlorophyll molecules (green); one of the pairs (the "special pair," $(Chl)_2$) is the site of the first photochemical changes after light absorption. Also incorporated in the system are a pair of pheophytin *a* (Pheo *a*) molecules (blue); two quinones, menaquinone ($Q_A$) and ubiquinone ($Q_B$) (orange and yellow), also arranged with bi-

lateral symmetry; and a single nonheme Fe (red) located approximately on the axis of symmetry between the quinones. Shown at the top of the figure are four heme groups (red) associated with the *c*-type cytochrome of the reaction center. The reaction center of another purple bacterium, *Rhodobacter sphaeroides,* is very similar, except that cytochrome *c* is not part of the crystalline complex.

**(b)** Sequence of events following excitation of the special pair of bacteriochlorophylls (all components colored as in **(a)**), with the time scale of the electron transfers in parentheses. ① The excited special pair passes an electron to pheophytin, ② from which the electron moves rapidly to the tightly bound menaquinone, $Q_A$. ③ This quinone passes electrons much more slowly to the diffusible ubiquinone, $Q_B$, through the nonheme Fe. Meanwhile, ④ the "electron hole" in the special pair is filled by an electron from a heme of cytochrome *c*.

However, in contrast to the cyclic flow of electrons in purple bacteria, some electrons flow from the reaction center to an iron-sulfur protein, **ferredoxin,** which then passes electrons via ferredoxin:NAD reductase to $NAD^+$, producing NADH. The electrons taken from the reaction center to reduce $NAD^+$ are replaced by the oxidation of $H_2S$ to elemental S, then to $SO_4^{2-}$, in the reaction that defines the green sulfur bacteria. This oxidation of $H_2S$ by bacteria is chemically analogous to the oxidation of $H_2O$ by oxygenic plants.

### Kinetic and Thermodynamic Factors Prevent the Dissipation of Energy by Internal Conversion

The complex construction of reaction centers is the product of evolutionary selection for efficiency in the photosynthetic process. The excited state $(Chl)_2^*$ could in principle decay to its ground state by internal conversion, a very rapid process (10 picoseconds; 1 ps = $10^{-12}$ s) in which the energy of the absorbed photon is

converted to heat (molecular motion). Reaction centers are constructed to prevent the inefficiency that would result from internal conversion. The proteins of the reaction center hold the bacteriochlorophylls, bacteriopheophytins, and quinones in a fixed orientation relative to each other, allowing the photochemical reactions to take place in a virtually solid state. This accounts for the high efficiency and rapidity of the reactions; nothing is left to chance collision or random diffusion. Exciton transfer from antenna chlorophyll to the special pair of the reaction center is accomplished in less than 100 ps with >90% efficiency. Within 3 ps of the excitation of P870, pheophytin has received an electron and become a negatively charged radical; less than 200 ps later, the electron has reached the quinone $Q_B$ (Fig. 19–48b). The electron-transfer reactions not only are fast but are thermodynamically "downhill"; the excited special pair $(Chl)_2^*$ is a very good electron donor ($E'^\circ$ −1 V), and each successive electron transfer is to an acceptor of substantially less negative $E'^\circ$. The standard free-energy change for the process is therefore negative and large; recall from Chapter 13 that $\Delta G'^{\circ} = -n\mathcal{F}\Delta E'^{\circ}$; here, $\Delta E'^{\circ}$ is the difference between the standard reduction potentials of the two half-reactions

$$(1)\ (Chl)_2^* \longrightarrow \cdot(Chl)_2^+ + e^- \qquad E'^{\circ} \approx -1.0\ V$$
$$(2)\ Q + 2H^+ + 2e^- \longrightarrow QH_2 \qquad E'^{\circ} = -0.045\ V$$

Thus

$$\Delta E'^{\circ} = -0.045\ V - (-1.0\ V) \approx 0.95\ V$$

and

$$\Delta G'^{\circ} = -2(96.5\ kJ/V \cdot mol)(0.95\ V) = -180\ kJ/mol$$

The combination of fast kinetics and favorable thermodynamics makes the process virtually irreversible and highly efficient. The overall energy yield (the percentage of the photon's energy conserved in $QH_2$) is >30%, with the remainder of the energy dissipated as heat.

### In Plants, Two Reaction Centers Act in Tandem

The photosynthetic apparatus of modern cyanobacteria, algae, and vascular plants is more complex than the one-center bacterial systems, and it appears to have evolved through the combination of two simpler bacterial photocenters. The thylakoid membranes of chloroplasts have two different kinds of photosystems, each with its own type of photochemical reaction center and set of antenna molecules. The two systems have distinct and complementary functions (Fig. 19–49). **Photosystem II (PSII)** is a pheophytin-quinone type of system (like the single photosystem of purple bacteria) containing roughly equal amounts of chlorophylls $a$ and $b$. Excitation of its reaction center P680 drives electrons through the cytochrome $b_6f$ complex with concomitant movement of protons across the thylakoid membrane. **Photosystem I (PSI)** is structurally and functionally related to the type I reaction center of green sulfur bacteria. It has a reaction center designated P700 and a high ratio of chlorophyll $a$ to chlorophyll $b$. Excited P700 passes electrons to the Fe-S protein ferredoxin, then to $NADP^+$, producing NADPH. The thylakoid membranes of a single spinach chloroplast have many hundreds of each kind of photosystem.

These two reaction centers in plants act in tandem to catalyze the light-driven movement of electrons from $H_2O$ to $NADP^+$ (Fig. 19–49). Electrons are carried between the two photosystems by the soluble protein **plastocyanin,** a one-electron carrier functionally similar to cytochrome $c$ of mitochondria. To replace the electrons that move from PSII through PSI to $NADP^+$, cyanobacteria and plants oxidize $H_2O$ (as green sulfur

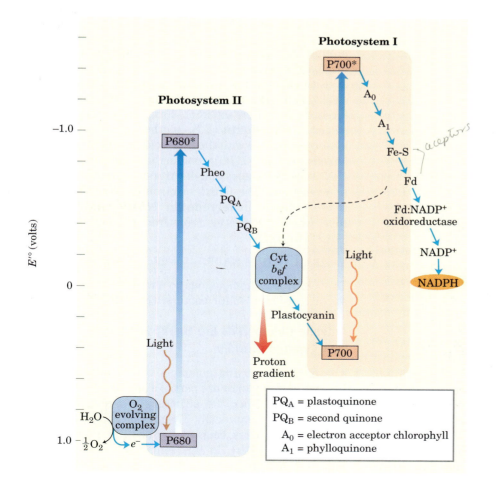

**FIGURE 19–49 Integration of photosystems I and II in chloroplasts.** This "Z scheme" shows the pathway of electron transfer from $H_2O$ (lower left) to $NADP^+$ (far right) in noncyclic photosynthesis. The position on the vertical scale of each electron carrier reflects its standard reduction potential. To raise the energy of electrons derived from $H_2O$ to the energy level required to reduce $NADP^+$ to NADPH, each electron must be "lifted" twice (heavy arrows) by photons absorbed in PSII and PSI. One photon is required per electron in each photosystem. After excitation, the high-energy electrons flow "downhill" through the carrier chains shown. Protons move across the thylakoid membrane during the water-splitting reaction and during electron transfer through the cytochrome $b_6f$ complex, producing the proton gradient that is central to ATP formation. The dashed arrow is the path of cyclic electron transfer (discussed later in the text), which involves only PSI; electrons return via the cyclic pathway to PSI, instead of reducing $NADP^+$ to NADPH.

$PQ_A$ = plastoquinone
$PQ_B$ = second quinone
$A_0$ = electron acceptor chlorophyll
$A_1$ = phylloquinone

bacteria oxidize $H_2S$), producing $O_2$ (Fig. 19–49, bottom left). This process is called **oxygenic photosynthesis** to distinguish it from the anoxygenic photosynthesis of purple and green sulfur bacteria. All $O_2$-evolving photosynthetic cells—those of plants, algae, and cyanobacteria—contain both PSI and PSII; organisms with only one photosystem do not evolve $O_2$. The diagram in Figure 19–49, often called the **Z scheme** because of its overall form, outlines the pathway of electron flow between the two photosystems and the energy relationships in the light reactions. The Z scheme thus describes the complete route by which electrons flow from $H_2O$ to $NADP^+$, according to the equation

$$2H_2O + 2NADP^+ + 8 \text{ photons} \longrightarrow O_2 + 2NADPH + 2H^+$$

For every two photons absorbed (one by each photosystem), one electron is transferred from $H_2O$ to $NADP^+$. To form one molecule of $O_2$, which requires transfer of four electrons from two $H_2O$ to two $NADP^+$, a total of eight photons must be absorbed, four by each photosystem.

The mechanistic details of the photochemical reactions in PSII and PSI are essentially similar to those of the two bacterial photosystems, with several important additions. In PSII, two very similar proteins, D1 and D2, form an almost symmetrical dimer, to which all the electron-carrying cofactors are bound (Fig. 19–50). Excitation of P680 in PSII produces P680*, an excellent electron donor that, within picoseconds, transfers an electron to pheophytin, giving it a negative charge ($\cdot Pheo^-$). With the loss of its electron, P680* is transformed into a radical cation, $P680^+$. $\cdot Pheo^-$ very rapidly passes its extra electron to a protein-bound **plastoquinone, $PQ_A$** (or $Q_A$), which in turn passes its electron to another, more loosely bound plastoquinone, $PQ_B$ (or $Q_B$). When $PQ_B$ has acquired two electrons in two such transfers from $PQ_A$ and two protons from the solvent water, it is in its fully reduced quinol form, $PQ_BH_2$. The overall reaction initiated by light in PSII is

$$4P680 + 4H^+ + 2PQ_B + 4 \text{ photons} \longrightarrow$$
$$4P680^+ + 2PQ_BH_2 \qquad (19\text{–}12)$$

Eventually, the electrons in $PQ_BH_2$ pass through the cytochrome $b_6f$ complex (Fig. 19–49). The electron initially removed from P680 is replaced with an electron obtained from the oxidation of water, as described below. The binding site for plastoquinone is the point of action of many commercial herbicides that kill plants by blocking electron transfer through the cytochrome $b_6f$ complex and preventing photosynthetic ATP production.

The photochemical events that follow excitation of PSI at the reaction center P700 are formally similar to those in PSII. The excited reaction center P700* loses an electron to an acceptor, $A_0$ (believed to be a special form of chlorophyll, functionally homologous to the pheophytin of PSII), creating $A_0^-$ and $P700^+$ (Fig. 19–49, right side); again, excitation results in charge separation

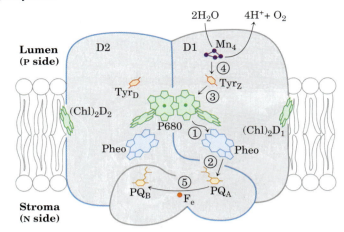

**FIGURE 19–50 Photosystem II of the cyanobacterium *Synechococcus elongates.*** The monomeric form of the complex shown here has two major transmembrane proteins, D1 and D2, each with its set of cofactors. Although the two subunits are nearly symmetric, electron flow occurs through only one of the two branches of cofactors, that on the right (on D1). The arrows show the path of electron flow from the Mn ion cluster ($Mn_4$, purple) of the water-splitting enzyme to the quinone $PQ_B$ (orange). The photochemical events occur in the sequence indicated by the step numbers. Notice the close similarity between the positions of cofactors here and the positions in the bacterial photoreaction center shown in Figure 19–48. The role of the Tyr residues is discussed later in the text.

at the photochemical reaction center. $P700^+$ is a strong oxidizing agent, which quickly acquires an electron from plastocyanin, a soluble Cu-containing electron-transfer protein. $A_0^-$ is an exceptionally strong reducing agent that passes its electron through a chain of carriers that leads to $NADP^+$. First, **phylloquinone** ($A_1$) accepts an electron and passes it to an iron-sulfur protein (through three Fe-S centers in PSI). From here, the electron moves to **ferredoxin** (Fd), another iron-sulfur protein loosely associated with the thylakoid membrane. Spinach ferredoxin ($M_r$ 10,700) contains a 2Fe-2S center (Fig. 19–5) that undergoes one-electron oxidation and reduction reactions. The fourth electron carrier in the chain is the flavoprotein **ferredoxin:$NADP^+$ oxidoreductase,** which transfers electrons from reduced ferredoxin ($Fd_{red}$) to $NADP^+$:

$$2Fd_{red} + 2H^+ + NADP^+ \longrightarrow 2Fd_{ox} + NADPH + H^+$$

This enzyme is homologous to the ferredoxin:NAD reductase of green sulfur bacteria (Fig. 19–47b).

## Antenna Chlorophylls Are Tightly Integrated with Electron Carriers

The electron-carrying cofactors of PSI and the light-harvesting complexes are part of a supramolecular complex (Fig. 19–51a), the structure of which has been solved crystallographically. The protein consists of three identical complexes, each composed of 11 different proteins (Fig. 19–51b). In this remarkable structure the many antenna chlorophyll and carotenoid molecules are

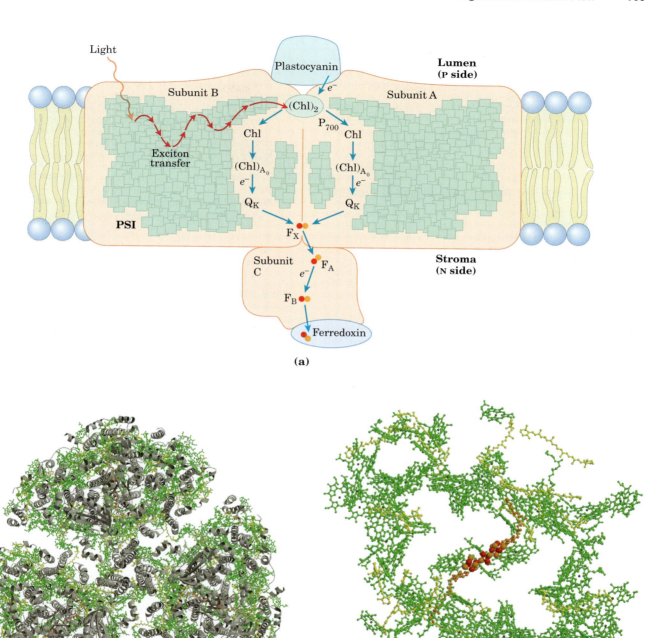

**FIGURE 19–51 The supramolecular complex of PSI and its associated antenna chlorophylls.** (a) Schematic drawing of the essential proteins and cofactors in a single unit of PSI. A large number of antenna chlorophylls surround the reaction center and convey to it (red arrows) the energy of photons they have absorbed. The result is excitation of the pair of chlorophyll molecules that constitute P700. Excitation of P700 greatly decreases its reduction potential, and it passes an electron through two nearby chlorophylls to phylloquinone ($Q_K$; also called $A_1$). Reduced phylloquinone is reoxidized as it passes two electrons, one at a time, to an Fe-S protein ($F_X$) near the N side of the membrane. From $F_X$, electrons move through two more Fe-S centers ($F_A$ and $F_B$), then to the Fe-S protein ferredoxin in the stroma. Ferredoxin then donates electrons to $NADP^+$ (not shown), reducing it to NADPH, one of the forms in which the energy of photons is trapped in chloroplasts. (b) The trimeric structure (derived from PDB ID 1JBO), viewed from the thylakoid lumen perpendicular to the membrane, showing all protein subunits (gray) and cofactors (described in (c)). (c) A monomer of PSI with all the proteins omitted, revealing the antenna and reaction center chlorophylls (green with dark green $Mg^{2+}$ ions in the center), carotenoids (yellow), and Fe-S centers of the reaction center (space-filling red and orange structures). The proteins in the complex hold the components rigidly in orientations that maximize efficient exciton transfers between excited antenna molecules and the reaction center.

precisely arrayed around the reaction center (Fig. 19–51c). The reaction center's electron-carrying cofactors are therefore tightly integrated with antenna chlorophylls. This arrangement allows very rapid and efficient exciton transfer from antenna chlorophylls to the reaction center. In contrast to the single path of electrons in PSII, the electron flow initiated by absorption of a photon is believed to occur through both branches of carriers in PSI.

### Spatial Separation of Photosystems I and II Prevents Exciton Larceny

The energy required to excite PSI (P700) is less than that needed to excite PSII (P680) (shorter wavelength corresponds to higher energy). If PSI and PSII were physically contiguous, excitons originating in the antenna system of PSII would migrate to the reaction center of PSI, leaving PSII chronically underexcited and interfering with the operation of the two-center system. This "exciton larceny" is prevented by separation of PSI and PSII in the thylakoid membrane (Fig. 19–52). PSII is located almost exclusively in the tightly appressed membrane stacks of thylakoid grana (granal lamellae); its associated light-harvesting complex (LHCII) mediates the tight association of adjacent membranes in the grana. PSI and the ATP synthase complex are located almost exclusively in the thylakoid membranes that are not appressed (the stromal lamellae), where both have access to the contents of the stroma, including ADP and $NADP^+$. The cytochrome $b_6f$ complex is present throughout the thylakoid membrane.

The association of LHCII with PSII is regulated by light intensity and wavelength. In bright sunlight (with a large component of blue light), PSII absorbs more light

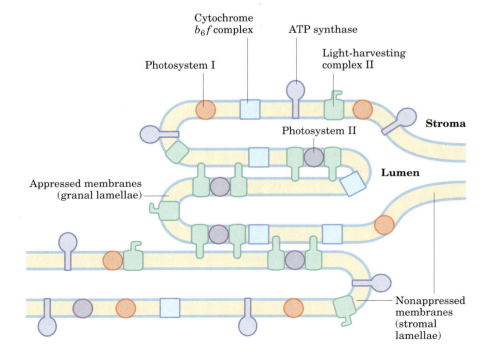

**FIGURE 19–53 Equalization of electron flow in PSI and PSII by modulation of granal stacking.** A hydrophobic domain of LHCII in one thylakoid lamella inserts into the neighboring lamella and closely appresses the two membranes. Accumulation of plastoquinol (not shown) stimulates a protein kinase that phosphorylates a Thr residue in the hydrophobic domain of LHCII, which reduces its affinity for the neighboring thylakoid membrane and converts appressed regions (granal lamellae) to nonappressed regions (stromal lamellae). A specific protein phosphatase reverses this regulatory phosphorylation when the [PQ]/[PQH$_2$] ratio increases.

than PSI and produces reduced plastoquinone (plastoquinol, $PQH_2$) faster than PSI can oxidize it. The resulting accumulation of $PQH_2$ activates a protein kinase that phosphorylates a Thr residue on LHCII (Fig. 19–53). Phosphorylation weakens the interaction of LHCII with PSII, and some LHCII dissociates and moves to the stromal lamellae; here it captures photons for PSI, speeding the oxidation of $PQH_2$ and reversing the imbalance between electron flow in PSI and PSII. In less intense light (in the shade, with more red light), PSI oxidizes $PQH_2$ faster than PSII can make it, and the resulting increase in PQ concentration triggers dephosphorylation of LHCII, reversing the effect of phosphorylation.

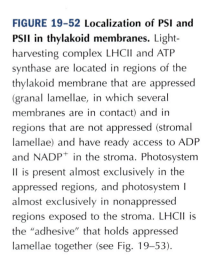

**FIGURE 19–52 Localization of PSI and PSII in thylakoid membranes.** Light-harvesting complex LHCII and ATP synthase are located in regions of the thylakoid membrane that are appressed (granal lamellae, in which several membranes are in contact) and in regions that are not appressed (stromal lamellae) and have ready access to ADP and NADP$^+$ in the stroma. Photosystem II is present almost exclusively in the appressed regions, and photosystem I almost exclusively in nonappressed regions exposed to the stroma. LHCII is the "adhesive" that holds appressed lamellae together (see Fig. 19–53).

## The Cytochrome $b_6f$ Complex Links Photosystems II and I

Electrons temporarily stored in plastoquinol as a result of the excitation of P680 in PSII are carried to P700 of PSI via the cytochrome $b_6f$ complex and the soluble protein plastocyanin (Fig. 19–49, center). Like Complex III

of mitochondria, the cytochrome $b_6f$ complex (Fig. 19–54) contains a $b$-type cytochrome with two heme groups (designated $b_H$ and $b_L$), a Rieske iron-sulfur protein ($M_r$ 20,000), and cytochrome $f$ (for the Latin *frons*, "leaf"). Electrons flow through the cytochrome $b_6f$ complex from $PQ_BH_2$ to cytochrome $f$, then to plastocyanin, and finally to P700, thereby reducing it.

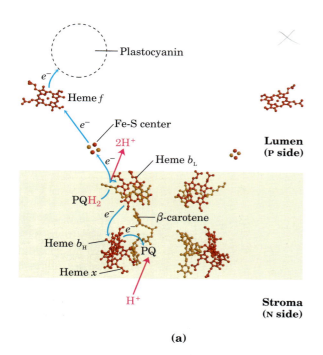

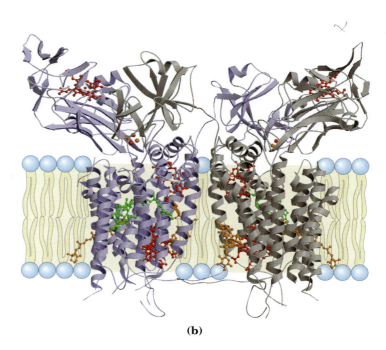

**(a)** **(b)**

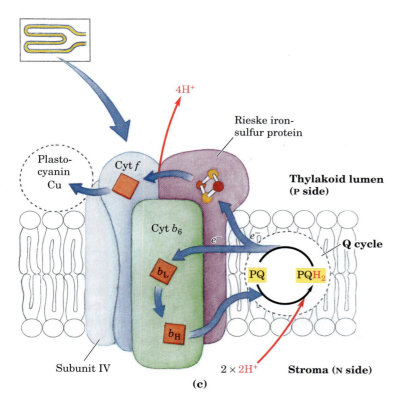

**(c)**

**FIGURE 19–54 Electron and proton flow through the cytochrome $b_6f$ complex. (a)** The crystal structure of the complex (PDB ID 1UM3) reveals the positions of the cofactors involved in electron transfers. In addition to the hemes of cytochrome $b$ (heme $b_H$ and $b_L$; also called heme $b_N$ and $b_P$, respectively, because of their proximity to the N and P sides of the bilayer) and that of cytochrome $f$ (heme $f$), there is a fourth (heme $x$) near heme $b_H$, and there is a $\beta$-carotene of unknown function. Two sites bind plastoquinone: the $PQH_2$ site near the P side of the bilayer, and the PQ site near the N side. The Fe-S center of the Rieske protein lies just outside the bilayer on the P side, and the heme $f$ site is on a protein domain that extends well into the thylakoid lumen. **(b)** The complex is a homodimer arranged to create a cavern connecting the $PQH_2$ and PQ sites (compare with the structure of mitochondrial Complex III in Fig. 19–12). This cavern allows plastoquinone movement between the sites of its oxidation and reduction.

**(c)** Plastoquinol ($PQH_2$) formed in PSII is oxidized by the cytochrome $b_6f$ complex in a series of steps like those of the Q cycle in the cytochrome $bc_1$ complex (Complex III) of mitochondria (see Fig. 19–11). One electron from $PQH_2$ passes to the Fe-S center of the Rieske protein (purple), the other to heme $b_L$ of cytochrome $b_6$ (green). The net effect is passage of electrons from $PQH_2$ to the soluble protein plastocyanin, which carries them to PSI.

Like Complex III of mitochondria, cytochrome $b_6f$ conveys electrons from a reduced quinone—a mobile, lipid-soluble carrier of two electrons (Q in mitochondria, $PQ_B$ in chloroplasts)—to a water-soluble protein that carries one electron (cytochrome $c$ in mitochondria, plastocyanin in chloroplasts). As in mitochondria, the function of this complex involves a Q cycle (Fig. 19–12) in which electrons pass, one at a time, from $PQ_BH_2$ to cytochrome $b_6$. This cycle results in the pumping of protons across the membrane; in chloroplasts, the direction of proton movement is from the stromal compartment to the thylakoid lumen, up to four protons moving for each pair of electrons. The result is production of a proton gradient across the thylakoid membrane as electrons pass from PSII to PSI. Because the volume of the flattened thylakoid lumen is small, the influx of a small number of protons has a relatively large effect on lumenal pH. The measured difference in pH between the stroma (pH 8) and the thylakoid lumen (pH 5) represents a 1,000-fold difference in proton concentration—a powerful driving force for ATP synthesis.

## Cyanobacteria Use the Cytochrome $b_6f$ Complex and Cytochrome $c_6$ in Both Oxidative Phosphorylation and Photophosphorylation

Cyanobacteria can synthesize ATP by oxidative phosphorylation or by photophosphorylation, although they have neither mitochondria nor chloroplasts. The enzymatic machinery for both processes is in a highly convoluted plasma membrane (see Fig. 1–6). Two protein components function in both processes (Fig. 19–55). The proton-pumping cytochrome $b_6f$ complex carries electrons from plastoquinone to cytochrome $c_6$ in photosynthesis, and also carries electrons from ubiquinone to cytochrome $c_6$ in oxidative phosphorylation—the role played by cytochrome $bc_1$ in mitochondria. Cytochrome $c_6$, homologous to mitochondrial cytochrome $c$, carries electrons from Complex III to Complex IV in cyanobacteria; it can also carry electrons from the cytochrome $b_6f$ complex to PSI—a role performed in plants by plastocyanin. We therefore see the functional homology between the cyanobacterial cytochrome $b_6f$ complex and the mitochondrial cytochrome $bc_1$ complex, and between cyanobacterial cytochrome $c_6$ and plant plastocyanin.

## Water Is Split by the Oxygen-Evolving Complex

The ultimate source of the electrons passed to NADPH in plant (oxygenic) photosynthesis is water. Having given up an electron to pheophytin, $P680^+$ (of PSII) must acquire an electron to return to its ground state in preparation for capture of another photon. In principle, the required electron might come from any number of organic or inorganic compounds. Photosynthetic bacteria use a variety of electron donors for this purpose—acetate, succinate, malate, or sulfide—depending on

what is available in a particular ecological niche. About 3 billion years ago, evolution of primitive photosynthetic bacteria (the progenitors of the modern cyanobacteria) produced a photosystem capable of taking electrons from a donor that is always available—water. Two water molecules are split, yielding four electrons, four protons, and molecular oxygen:

$$2H_2O \longrightarrow 4H^+ + 4e^- + O_2$$

A single photon of visible light does not have enough energy to break the bonds in water; four photons are required in this photolytic cleavage reaction.

The four electrons abstracted from water do not pass directly to $P680^+$, which can accept only one electron at a time. Instead, a remarkable molecular device, the **oxygen-evolving complex** (also called the **water-splitting complex**), passes four electrons *one at a*

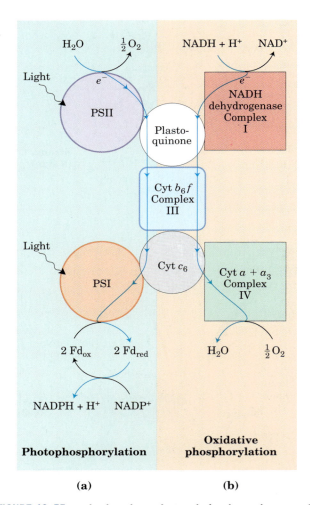

**(a)**                          **(b)**

**FIGURE 19–55 Dual roles of cytochrome $b_6f$ and cytochrome $c_6$ in cyanobacteria.** Cyanobacteria use cytochrome $b_6f$, cytochrome $c_6$, and plastoquinone for both oxidative phosphorylation and photophosphorylation. **(a)** In photophosphorylation, electrons flow (top to bottom) from water to $NADP^+$. **(b)** In oxidative phosphorylation, electrons flow from NADH to $O_2$. Both processes are accompanied by proton movement across the membrane, accomplished by a Q cycle.

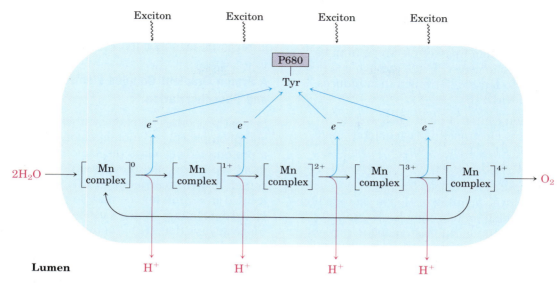

**FIGURE 19–56 Water-splitting activity of the oxygen-evolving complex.** Shown here is the process that produces a four-electron oxidizing agent—believed to be a multinuclear center with several Mn ions—in the water-splitting complex of PSII. The sequential absorption of four photons (excitons), each absorption causing the loss of one elec-tron from the Mn center, produces an oxidizing agent that can remove four electrons from two molecules of water, producing $O_2$. The electrons lost from the Mn center pass one at a time to an oxidized Tyr residue in a PSII protein, then to $P680^+$.

*time* to $P680^+$ (Fig. 19–56). The immediate electron donor to $P680^+$ is a Tyr residue (often designated Z or $Tyr_z$) in protein subunit D1 of the PSII reaction center. The Tyr residue loses both a proton and an electron, generating the electrically neutral Tyr free radical, $^\bullet$Tyr:

$$4P680^+ + 4\,\text{Tyr} \longrightarrow 4P680 + 4\,^\bullet\text{Tyr} \quad (19\text{–}13)$$

The Tyr radical regains its missing electron and proton by oxidizing a cluster of four manganese ions in the water-splitting complex. With each single-electron transfer, the Mn cluster becomes more oxidized; four single-electron transfers, each corresponding to the absorption of one photon, produce a charge of +4 on the Mn complex (Fig. 19–56):

$$4\,^\bullet\text{Tyr} + [\text{Mn complex}]^0 \longrightarrow$$
$$4\,\text{Tyr} + [\text{Mn complex}]^{4+} \quad (19\text{–}14)$$

In this state, the Mn complex can take four electrons from a pair of water molecules, releasing 4 $H^+$ and $O_2$:

$$[\text{Mn complex}]^{4+} + 2H_2O \longrightarrow$$
$$[\text{Mn complex}]^0 + 4H^+ + O_2 \quad (19\text{–}15)$$

Because the four protons produced in this reaction are released into the thylakoid lumen, the oxygen-evolving complex acts as a proton pump, driven by electron transfer. The sum of Equations 19–12 through 19–15 is

$$2H_2O + 2PQ_B + 4\text{ photons} \longrightarrow O_2 + 2PQ_BH_2 \quad (19\text{–}16)$$

The water-splitting activity associated with the PSII reaction center has proved exceptionally difficult to purify. A peripheral membrane protein ($M_r$ 33,000) on the lumenal side of the thylakoid membrane is believed to stabilize the Mn complex. In the crystal structure (PDB ID 1FE1; see Fig. 19–50), four Mn ions are clustered with precise geometry near a Tyr residue on the D1 sub-unit, presumably the one involved in water oxidation. Manganese can exist in stable oxidation states from +2 to +7, so a cluster of four Mn ions can certainly donate or accept four electrons. The mechanism shown in Figure 19–56 is consistent with the experimental facts, but until the exact chemical structures of all the intermediates of the Mn cluster are known, the detailed mechanism remains elusive.

## SUMMARY 19.8 The Central Photochemical Event: Light-Driven Electron Flow

- Bacteria have a single reaction center; in purple bacteria, it is of the pheophytin-quinone type, and in green sulfur bacteria, the Fe-S type.

- Structural studies of the reaction center of a purple bacterium have provided information about light-driven electron flow from an excited special pair of chlorophyll molecules, through pheophytin, to quinones. Electrons then pass from quinones through the cytochrome $bc_1$ complex, and back to the photoreaction center.

- An alternative path, in green sulfur bacteria, sends electrons from reduced quinones to $NAD^+$.

- Cyanobacteria and plants have two different photoreaction centers, arranged in tandem.

- Plant photosystem I passes electrons from its excited reaction center, P700, through a series of carriers to ferredoxin, which then reduces $NADP^+$ to NADPH.

- The reaction center of plant photosystem II, P680, passes electrons to plastoquinone, and the electrons lost from P680 are replaced by electrons from $H_2O$ (electron donors other than $H_2O$ are used in other organisms).

- The light-driven splitting of $H_2O$ is catalyzed by a Mn-containing protein complex; $O_2$ is produced. The reduced plastoquinone carries electrons to the cytochrome $b_6f$ complex; from here they pass to plastocyanin, and then to P700 to replace those lost during its photoexcitation.

- Electron flow through the cytochrome $b_6f$ complex drives protons across the plasma membrane, creating a proton-motive force that provides the energy for ATP synthesis by an ATP synthase.

# 19.9 ATP Synthesis by Photophosphorylation

The combined activities of the two plant photosystems move electrons from water to $NADP^+$, conserving some of the energy of absorbed light as NADPH (Fig. 19–49). Simultaneously, protons are pumped across the thylakoid membrane and energy is conserved as an electrochemical potential. We turn now to the process by which this proton gradient drives the synthesis of ATP, the other energy-conserving product of the light-dependent reactions.

In 1954 Daniel Arnon and his colleagues discovered that ATP is generated from ADP and $P_i$ during photosynthetic electron transfer in illuminated spinach chloroplasts. Support for these findings came from the work of Albert Frenkel, who detected light-dependent ATP production in pigment-containing membranous structures called **chromatophores,** derived from photosynthetic

Daniel Arnon, 1910–1994

bacteria. Investigators concluded that some of the light energy captured by the photosynthetic systems of these organisms is transformed into the phosphate bond en-

ergy of ATP. This process is called **photophosphorylation,** to distinguish it from oxidative phosphorylation in respiring mitochondria.

## A Proton Gradient Couples Electron Flow and Phosphorylation

Several properties of photosynthetic electron transfer and photophosphorylation in chloroplasts indicate that a proton gradient plays the same role as in mitochondrial oxidative phosphorylation. (1) The reaction centers, electron carriers, and ATP-forming enzymes are located in a proton-impermeable membrane—the thylakoid membrane—which must be intact to support photophosphorylation. (2) Photophosphorylation can be uncoupled from electron flow by reagents that promote the passage of protons through the thylakoid membrane. (3) Photophosphorylation can be blocked by venturicidin and similar agents that inhibit the formation of ATP from ADP and $P_i$ by the mitochondrial ATP synthase (Table 19–4). (4) ATP synthesis is catalyzed by $F_oF_1$ complexes, located on the outer surface of the thylakoid membranes, that are very similar in structure and function to the $F_oF_1$ complexes of mitochondria.

Electron-transferring molecules in the chain of carriers connecting PSII and PSI are oriented asymmetrically in the thylakoid membrane, so photoinduced electron flow results in the net movement of protons across the membrane, from the stromal side to the thylakoid lumen (Fig. 19–57). In 1966 André Jagendorf showed that a pH gradient across the thylakoid membrane (alkaline outside) could furnish the driving force to generate ATP. His early observations provided some of the most important experimental evidence in support of Mitchell's chemiosmotic hypothesis.

André Jagendorf

Jagendorf incubated chloroplasts, in the dark, in a pH 4 buffer; the buffer slowly penetrated into the inner compartment of the thylakoids, lowering their internal pH. He added ADP and $P_i$ to the dark suspension of chloroplasts and then suddenly raised the pH of the outer medium to 8, momentarily creating a large pH gradient across the membrane. As protons moved out of the thylakoids into the medium, ATP was generated from ADP and $P_i$. Because the formation of ATP occurred in the dark (with no input of energy from light), this experiment showed that a pH gradient across the membrane is a high-energy state that, as in mitochondrial oxidative phosphorylation, can mediate the transduction of energy from electron transfer into the chemical energy of ATP.

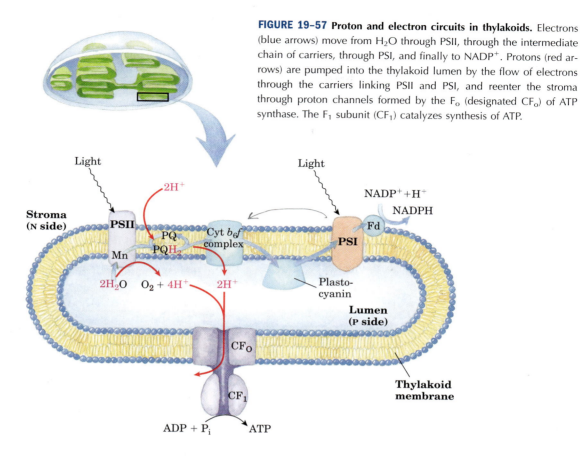

**FIGURE 19–57 Proton and electron circuits in thylakoids.** Electrons (blue arrows) move from $H_2O$ through PSII, through the intermediate chain of carriers, through PSI, and finally to $NADP^+$. Protons (red arrows) are pumped into the thylakoid lumen by the flow of electrons through the carriers linking PSII and PSI, and reenter the stroma through proton channels formed by the $F_o$ (designated $CF_o$) of ATP synthase. The $F_1$ subunit ($CF_1$) catalyzes synthesis of ATP.

## The Approximate Stoichiometry of Photophosphorylation Has Been Established

As electrons move from water to $NADP^+$ in plant chloroplasts, about 12 $H^+$ move from the stroma to the thylakoid lumen per four electrons passed (that is, per $O_2$ formed). Four of these protons are moved by the oxygen-evolving complex, and up to eight by the cytochrome $b_6f$ complex. The measurable result is a 1,000-fold difference in proton concentration across the thylakoid membrane ($\Delta pH = 3$). Recall that the energy stored in a proton gradient (the electrochemical potential) has two components: a proton concentration difference ($\Delta pH$) and an electrical potential ($\Delta\psi$) due to charge separation. In chloroplasts, $\Delta pH$ is the dominant component; counterion movement apparently dissipates most of the electrical potential. In illuminated chloroplasts, the energy stored in the proton gradient per mole of protons is

$$\Delta G = 2.3RT\,\Delta pH + Z\mathcal{F}\,\Delta\psi = -17 \text{ kJ/mol}$$

so the movement of 12 mol of protons across the thylakoid membrane represents conservation of about 200 kJ of energy—enough energy to drive the synthesis of several moles of ATP ($\Delta G'^\circ = 30.5$ kJ/mol). Experimental measurements yield values of about 3 ATP per $O_2$ produced.

At least eight photons must be absorbed to drive four electrons from $H_2O$ to NADPH (one photon per electron at each reaction center). The energy in eight photons of visible light is more than enough for the synthesis of three molecules of ATP.

ATP synthesis is not the only energy-conserving reaction of photosynthesis in plants; the NADPH formed in the final electron transfer is (like its close analog NADH) also energetically rich. The overall equation for noncyclic photophosphorylation (a term explained below) is

$$2H_2O + 8 \text{ photons} + 2NADP^+ + {\sim}3ADP + {\sim}3P_i \longrightarrow$$
$$O_2 + {\sim}3ATP + 2NADPH \quad (19\text{–}17)$$

## Cyclic Electron Flow Produces ATP but Not NADPH or $O_2$

Using an alternative path of light-induced electron flow, plants can vary the ratio of NADPH to ATP formed in the light; this path is called **cyclic electron flow** to differentiate it from the normally unidirectional or **noncyclic electron flow** from $H_2O$ to $NADP^+$, as discussed thus far. Cyclic electron flow (Fig. 19–49) involves only PSI. Electrons passing from P700 to ferredoxin do not continue to $NADP^+$, but move back through the cytochrome $b_6f$ complex to plastocyanin. The path of electrons matches that in green sulfur bacteria (Fig. 19–47b). Plastocyanin donates electrons to P700, which transfers them to ferredoxin when the plant is illuminated. Thus, in the light, PSI can cause electrons to cycle continuously out of and back into the reaction center of PSI, each electron propelled around the cycle by the energy yielded by the absorption of one photon. Cyclic electron flow is not accompanied by net formation of NADPH or evolution of $O_2$. However, it *is* accompanied by proton pumping by the cytochrome $b_6f$ complex and by phosphorylation of ADP to ATP, referred to as **cyclic photophosphorylation.** The overall equation for cyclic electron flow and photophosphorylation is simply

$$ADP + P_i \xrightarrow{\text{light}} ATP + H_2O$$

By regulating the partitioning of electrons between $NADP^+$ reduction and cyclic photophosphorylation, a plant adjusts the ratio of ATP to NADPH produced in the light-dependent reactions to match its needs for these products in the carbon-assimilation reactions and other biosynthetic processes. As we shall see in Chapter 20, the carbon-assimilation reactions require ATP and NADPH in the ratio 3:2.

## The ATP Synthase of Chloroplasts Is Like That of Mitochondria

The enzyme responsible for ATP synthesis in chloroplasts is a large complex with two functional components, $CF_o$ and $CF_1$ (C denoting its location in chloroplasts). $CF_o$ is a transmembrane proton pore composed of several integral membrane proteins and is homologous to mitochondrial $F_o$. $CF_1$ is a peripheral

membrane protein complex very similar in subunit composition, structure, and function to mitochondrial $F_1$.

Electron microscopy of sectioned chloroplasts shows ATP synthase complexes as knoblike projections on the *outside* (stromal or N) surface of thylakoid membranes; these complexes correspond to the ATP synthase complexes seen to project on the *inside* (matrix or N) surface of the inner mitochondrial membrane. Thus the relationship between the orientation of the ATP synthase and the direction of proton pumping is the same in chloroplasts and mitochondria. In both cases, the $F_1$ portion of ATP synthase is located on the more alkaline (N) side of the membrane through which protons flow down their concentration gradient; the direction of proton flow relative to $F_1$ is the same in both cases: P to N (Fig. 19–58).

The mechanism of chloroplast ATP synthase is also believed to be essentially identical to that of its mitochondrial analog; ADP and $P_i$ readily condense to form ATP on the enzyme surface, and the release of this enzyme-bound ATP requires a proton-motive force. Rotational catalysis sequentially engages each of the three $\beta$ subunits of the ATP synthase in ATP synthesis, ATP release, and $ADP + P_i$ binding (Figs 19–24, 19–25).

## Chloroplasts Evolved from Endosymbiotic Bacteria

Like mitochondria, chloroplasts contain their own DNA and protein-synthesizing machinery. Some of the polypeptides of chloroplast proteins are encoded by chloroplast genes and synthesized in the chloroplast; others are encoded by nuclear genes, synthesized outside the chloroplast, and imported (Chapter 27). When plant cells grow and divide, chloroplasts give rise to new

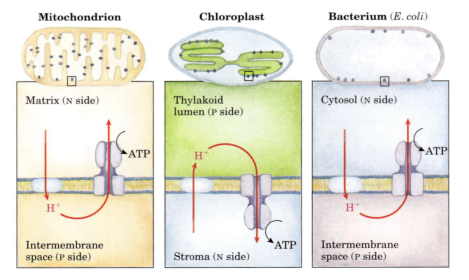

**FIGURE 19–58** Comparison of the topology of proton movement and ATP synthase orientation in the membranes of mitochondria, chloroplasts, and the bacterium *E. coli.* In each case, orientation of the proton gradient relative to ATP synthase activity is the same.

chloroplasts by division, during which their DNA is replicated and divided between daughter chloroplasts. The machinery and mechanism for light capture, electron flow, and ATP synthesis in photosynthetic bacteria are similar in many respects to those in the chloroplasts of plants. These observations led to the now widely accepted hypothesis that the evolutionary progenitors of modern plant cells were primitive eukaryotes that engulfed photosynthetic bacteria and established stable endosymbiotic relationships with them (see Fig. 1–36).

## Diverse Photosynthetic Organisms Use Hydrogen Donors Other Than Water

At least half of the photosynthetic activity on Earth occurs in microorganisms—algae, other photosynthetic eukaryotes, and photosynthetic bacteria. Cyanobacteria have PSII and PSI in tandem, and the PSII has an associated water-splitting activity resembling that of plants. However, the other groups of photosynthetic bacteria have single reaction centers and do not split $H_2O$ or produce $O_2$. Many are obligate anaerobes and cannot tolerate $O_2$; they must use some compound other than $H_2O$ as electron donor. Some photosynthetic bacteria use inorganic compounds as electron (and hydrogen) donors. For example, green sulfur bacteria use hydrogen sulfide:

$$2H_2S + CO_2 \xrightarrow{\text{light}} (CH_2O) + H_2O + 2S$$

These bacteria, instead of producing molecular $O_2$, form elemental sulfur as the oxidation product of $H_2S$. (They further oxidize the S to $SO_4^{2-}$.) Other photosynthetic bacteria use organic compounds such as lactate as electron donors:

$$2 \text{ Lactate} + CO_2 \xrightarrow{\text{light}} (CH_2O) + H_2O + 2 \text{ pyruvate}$$

The fundamental similarity of photosynthesis in plants and bacteria, despite the differences in the electron donors they employ, becomes more obvious when the equation of photosynthesis is written in the more general form

$$2H_2D + CO_2 \xrightarrow{\text{light}} (CH_2O) + H_2O + 2D$$

in which $H_2D$ is an electron (and hydrogen) donor and D is its oxidized form. $H_2D$ may be water, hydrogen sulfide, lactate, or some other organic compound, depending on the species. Most likely, the bacteria that first developed photosynthetic ability used $H_2S$ as their electron source, and only after the later development of oxygenic photosynthesis (about 2.3 billion years ago) did oxygen become a significant proportion of the earth's atmosphere. With that development, the evolution of electron-transfer systems that used $O_2$ as their ultimate electron acceptor became possible, leading to the highly efficient energy extraction of oxidative phosphorylation.

## In Halophilic Bacteria, a Single Protein Absorbs Light and Pumps Protons to Drive ATP Synthesis

The halophilic ("salt-loving") bacterium *Halobacterium salinarum*, an archaebacterium derived from very ancient evolutionary progenitors, traps the energy of sunlight in a process very different from the photosynthetic mechanisms we have described so far. This bacterium lives only in brine ponds and salt lakes (Great Salt Lake and the Dead Sea, for example), where the high salt concentration—which can exceed 4 M—results from water loss by evaporation; indeed, halobacteria cannot live in NaCl concentrations lower than 3 M. These organisms are aerobes and normally use $O_2$ to oxidize organic fuel molecules. However, the solubility of $O_2$ is so low in brine ponds that sometimes oxidative metabolism must be supplemented by sunlight as an alternative source of energy.

The plasma membrane of *H. salinarum* contains patches of the light-absorbing pigment **bacteriorhodopsin,** which contains retinal (the aldehyde derivative of vitamin A; see Fig. 10–21) as a prosthetic group. When the cells are illuminated, all-*trans*-retinal bound to the bacteriorhodopsin absorbs a photon and undergoes photoisomerization to 13-*cis*-retinal. The restoration of all-*trans*-retinal is accompanied by the outward movement of protons through the plasma membrane. Bacteriorhodopsin, with only 247 amino acid residues, is the simplest light-driven proton pump known. The difference in the three-dimensional structure of bacteriorhodopsin in the dark and after illumination (Fig. 19–59a) suggests a pathway by which a concerted series of proton "hops" could effectively move a proton across the membrane. The chromophore retinal is bound through a Schiff-base linkage to the $\varepsilon$-amino group of a Lys residue. In the dark, the N of this Schiff base is protonated; in the light, photoisomerization of retinal lowers the $pK_a$ of this group and it releases its proton to a nearby Asp residue, triggering a series of proton hops that ultimately result in the release of a proton at the outer surface of the membrane (Fig. 19–59b).

The electrochemical potential across the membrane drives protons back into the cell through a membrane ATP synthase complex very similar to that of mitochondria and chloroplasts. Thus, when $O_2$ is limited, halobacteria can use light to supplement the ATP synthesized by oxidative phosphorylation. Halobacteria do not evolve $O_2$, nor do they carry out photoreduction of $NADP^+$; their phototransducing machinery is therefore much simpler than that of cyanobacteria or plants. Nevertheless, the proton-pumping mechanism used by this simple protein may prove to be prototypical for the many other, more complex, ion pumps. **Bacteriorhodopsin** ■

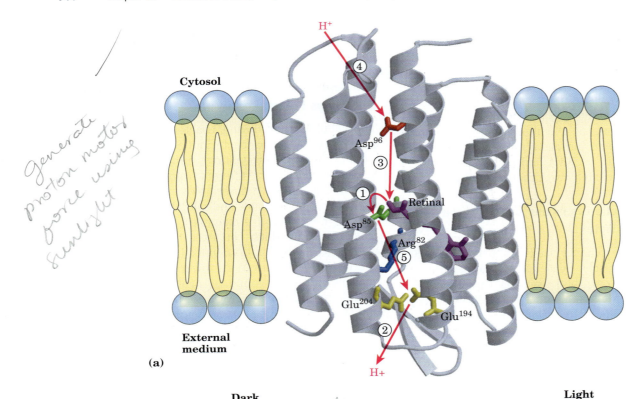

*Generate proton motor force using sunlight* (handwritten)

**(a)**

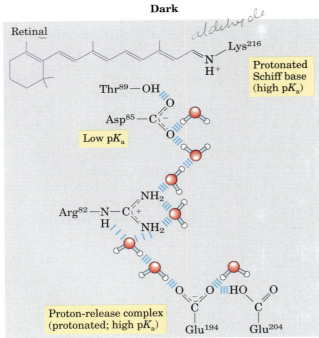

**(b)**

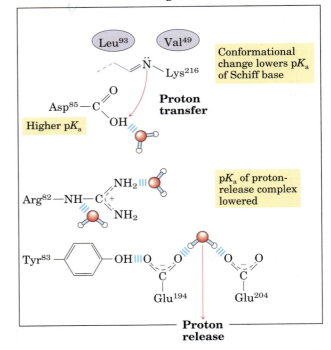

**FIGURE 19–59 Light-driven proton pumping by bacteriorhodopsin.**
**(a)** Bacteriorhodopsin ($M_r$ 26,000) has seven membrane-spanning $\alpha$ helices (PDB ID 1C8R). The chromophore all-*trans*-retinal (purple) is covalently attached via a Schiff base to the $\varepsilon$-amino group of a Lys residue deep in the membrane interior. Running through the protein are a series of Asp and Glu residues and a series of closely associated water molecules that together provide the transmembrane path for protons (red arrows). Steps ① through ⑤ indicate proton movements, described below.

**(b)** In the dark (left panel), the Schiff base is protonated. Illumination (right panel) photoisomerizes the retinal, forcing subtle confor-

mational changes in the protein that alter the distance between the Schiff base and its neighboring amino acid residues. Interaction with these neighbors lowers the $pK_a$ of the protonated Schiff base, and the base gives up its proton to a nearby carboxyl group on Asp85 (step ① in (a)). This initiates a series of concerted proton hops between water molecules (see Fig. 2–14) in the interior of the protein, which ends with ② the release of a proton that was shared by Glu194 and Glu204 near the extracellular surface. ③ The Schiff base reacquires a proton from Asp96, which ④ takes up a proton from the cytosol. ⑤ Finally, Asp85 gives up its proton, leading to reprotonation of the Glu204-Glu194 pair. The system is now ready for another round of proton pumping.

## SUMMARY 19.9 ATP Synthesis by Photophosphorylation

- In plants, both the water-splitting reaction and electron flow through the cytochrome $b_6f$ complex are accompanied by proton pumping across the thylakoid membrane. The proton-motive force thus created drives ATP synthesis by a $CF_oCF_1$ complex similar to the mitochondrial $F_oF_1$ complex.

- Flow of electrons through the photosystems produces NADPH and ATP, in the ratio of about 2:3. A second type of electron flow (cyclic flow) produces ATP only and allows variability in the proportions of NADPH and ATP formed.

- The localization of PSI and PSII between the granal and stromal lamellae can change and is indirectly controlled by light intensity, optimizing the distribution of excitons between PSI and PSII for efficient energy capture.

- Chloroplasts, like mitochondria, evolved from bacteria living endosymbiotically within early eukaryotic cells. The ATP synthases of eubacteria, cyanobacteria, mitochondria, and chloroplasts share a common evolutionary precursor and a common enzymatic mechanism.

- Many photosynthetic microorganisms obtain electrons for photosynthesis not from water but from donors such as $H_2S$.

## Key Terms

*Terms in bold are defined in the glossary.*

**chemiosmotic theory**   690
nicotinamide nucleotide–linked dehydrogenases   692
**flavoprotein**   692
**reducing equivalent**   693
ubiquinone (coenzyme Q, Q)   693
**cytochromes**   693
**iron-sulfur protein**   693
Complex I   696
**vectorial metabolism**   697
Complex II   698
Complex III   699
cytochrome $bc_1$ complex   699
Q cycle   700
Complex IV   700
cytochrome oxidase   700
**proton-motive force**   703

**ATP synthase**   704
$F_1ATPase$   708
**rotational catalysis**   711
**P/O ratio**   712
$P/2e^-$ ratio   712
**acceptor control**   716
**mass-action ratio**   716
**light-dependent reactions**   723
light reactions   723
**carbon-assimilation reactions**   723
**carbon-fixation reaction**   723
**thylakoid**   724
**stroma**   724
exciton transfer   725
**chlorophylls**   725

light-harvesting complexes (LHCs)   725
**accessory pigments**   728
**photosystem**   728
**photochemical reaction center**   728
light-harvesting (antenna) molecules   728
photosystem II (PSII)   733
photosystem I (PSI)   733
oxygenic photosynthesis   734
oxygen-evolving complex (water-splitting complex)   738
**photophosphorylation** 740
**cyclic electron flow**   741
**noncyclic electron flow**   741
**cyclic photophosphorylation** 742

## Further Reading

### History and General Background

**Arnon, D.I.** (1984) The discovery of photosynthetic phosphorylation. *Trends Biochem. Sci.* **9**, 258–262.

**Beinert, H.** (1995) These are the moments when we live! From Thunberg tubes and manometry to phone, fax and FedEx. In *Selected Topics in the History of Biochemistry: Personal Recollections,* Comprehensive Biochemistry, Vol. 38, Elsevier Science Publishing Co., Inc., New York.
  An engaging personal account of the exciting period when the biochemistry of respiratory electron transfer was worked out.

**Blankenship, R.E.** (2002) *Molecular Mechanisms of Photosynthesis,* Blackwell Science Inc., London.
  An intermediate-level discussion of all aspects of photosynthesis.

**Gray, M.W., Berger, G., & Lang, B.F.** (1999) Mitochondrial evolution. *Science* **283**, 1476–1481.

  Compact review of the endosymbiotic origin hypothesis and the evidence for and against it.

**Harold, F.M.** (1986) *The Vital Force: A Study in Bioenergetics,* W. H. Freeman and Company, New York.
  A very readable synthesis of the principles of bioenergetics and their application to energy transductions.

**Heldt, H.-W.** (1997) *Plant Biochemistry and Molecular Biology,* Oxford University Press, Oxford.
  A textbook of plant biochemistry with excellent discussions of photophosphorylation.

**Keilin, D.** (1966) *The History of Cell Respiration and Cytochrome,* Cambridge University Press, London.
  An authoritative and absorbing account of the discovery of cytochromes and their roles in respiration, written by the man who discovered cytochromes.

**Mitchell, P.** (1979) Keilin's respiratory chain concept and its chemiosmotic consequences. *Science* **206,** 1148–1159.

Mitchell's Nobel lecture, outlining the evolution of the chemiosmotic hypothesis.

**Nicholls, D.G. & Ferguson, S.J.** (2002) *Bioenergetics 3,* Academic Press, Amsterdam.

Up-to-date, comprehensive, well-illustrated treatment of all aspects of mitochondrial and chloroplast energy transductions. A comprehensive, advanced treatise.

**Scheffler, I.E.** (1999) *Mitochondria,* Wiley-Liss, New York.

An excellent survey of mitochondrial structure and function.

**Slater, E.C.** (1987) The mechanism of the conservation of energy of biological oxidations. *Eur. J. Biochem.* **166,** 489–504.

A clear and critical account of the evolution of the chemiosmotic model.

# OXIDATIVE PHOSPHORYLATION

## Respiratory Electron Flow

**Babcock, G.T. & Wickström, M.** (1992) Oxygen activation and the conservation of energy in cell respiration. *Nature* **356,** 301–309.

Advanced discussion of the reduction of water and pumping of protons by cytochrome oxidase.

**Brandt, U.** (1997) Proton-translocation by membrane-bound NADH:ubiquinone-oxidoreductase (complex I) through redox-gated ligand conduction. *Biochim. Biophys. Acta* **1318,** 79–91.

Advanced discussion of models for electron movement through Complex I.

**Brandt, U. & Trumpower, B.** (1994) The protonmotive Q cycle in mitochondria and bacteria. *Crit. Rev. Biochem. Mol. Biol.* **29,** 165–197.

**Crofts, A.R. & Berry, E.A.** (1998) Structure and function of the cytochrome $bc_1$ complex of mitochondria and photosynthetic bacteria. *Curr. Opin. Struct. Biol.* **8,** 501–509.

**Michel, H., Behr, J., Harrenga, A., & Kannt, A.** (1998) Cytochrome *c* oxidase: structure and spectroscopy. *Annu. Rev. Biophys. Biomol. Struct.* **27,** 329–356.

Advanced review of Complex IV structure and function.

**Rottenberg, H.** (1998) The generation of proton electrochemical potential gradient by cytochrome *c* oxidase. *Biochim. Biophys. Acta* **1364,** 1–16.

**Tielens, A.G.M., Rotte, C., van Hellemond, J.J., & Martin, W.** (2002) Mitochondria as we don't know them. *Trends Biochem. Sci.* **27,** 564–572.

Intermediate-level discussion of the many organisms in which mitochondria do not depend on oxygen as the final electron donor.

**Tsukihara, T., Aoyama, H., Yamashita, E., Tomizaki, T., Yamaguchi, H., Shinzawa-Itoh, K., Nakashima, R., Yaono, R., & Yoshikawa, S.** (1996) The whole structure of the 13-subunit oxidized cytochrome *c* oxidase at 2.8 Å. *Science* **272,** 1136–1144.

The solution by x-ray crystallography of the structure of this huge membrane protein.

**Xia, D., Yu, C.-A., Kim, H., Xia, J.-Z., Kachurin, A.M., Zhang, L., Yu, L., & Deisenhofer, J.** (1997) Crystal structure of the cytochrome $bc_1$ complex from bovine heart mitochondria. *Science* **277,** 60–66.

Report revealing the crystallographic structure of Complex III.

**Yankovskaya, V., Horsefield, R., Törnroth, S., Luna-Chavez, C., Myoshi, H., Léger, C., Byrne, B., Cecchini, G., & Iwata, S.** (2003) Architecture of succinate dehydrogenase and reactive oxygen species generation. *Science* **299,** 700–704.

## Coupling ATP Synthesis to Respiratory Electron Flow

**Abrahams, J.P., Leslie, A.G.W., Lutter, R., & Walker, J.E.** (1994) The structure of $F_1$-ATPase from bovine heart mitochondria determined at 2.8 Å resolution. *Nature* **370,** 621–628.

**Bianchet, M.A., Hullihen, J., Pedersen, P.L., & Amzel, L.M.** (1998) The 2.80 Å structure of rat liver $F_1$-ATPase: configuration of a critical intermediate in ATP synthesis-hydrolysis. *Proc. Natl. Acad. Sci. USA* **95,** 11,065–11,070.

Research paper that provided important structural detail in support of the catalytic mechanism.

**Boyer, P.D.** (1997) The ATP synthase—a splendid molecular machine. *Annu. Rev. Biochem.* **66,** 717–749.

An account of the historical development and current state of the binding-change model, written by its principal architect.

**Cabezón, E., Montgomery, M.G., Leslie, A.G.W., & Walker, J.E.** (2003) The structure of bovine $F_1$-ATPase in complex with its regulatory protein $IF_1$ *Nat. Struct. Biol.* **10,** 744–750.

**Hinkle, P.C., Kumar, M.A., Resetar, A., & Harris, D.L.** (1991) Mechanistic stoichiometry of mitochondrial oxidative phosphorylation. *Biochemistry* **30,** 3576–3582.

A careful analysis of experimental results and theoretical considerations on the question of nonintegral P/O ratios.

**Khan, S.** (1997) Rotary chemiosmotic machines. *Biochim. Biophys. Acta* **1322,** 86–105.

Detailed review of the structures that underlie proton-driven rotary motion of ATP synthase and bacterial flagella.

**Sambongi, Y., Iko, Y., Tanabe, M., Omote, H., Iwamoto-Kihara, A., Ueda, I., Yanagida, T., Wada, Y., & Futai, M.** (1999) Mechanical rotation of the c subunit oligomer in ATP synthase ($F_oF_1$): direct observation. *Science* **286,** 1722–1724.

The experimental evidence for rotation of the entire cylinder of c subunits in $F_oF_1$.

**Stock, D., Leslie, A.G.W., & Walker, J.E.** (1999) Molecular architecture of the rotary motor in ATP synthase. *Science* **286,** 1700–1705.

The first crystallographic view of the $F_o$ subunit, in the yeast $F_oF_1$. See also R. H. Fillingame's editorial comment in the same issue of *Science.*

**Weber, J. & Senior, A.E.** (1997) Catalytic mechanism of $F_1$-ATPase. *Biochim. Biophys. Acta* **1319,** 19–58.

An advanced review of kinetic, structural, and biochemical evidence for the ATP synthase mechanism.

**Yasuda, R., Noji, H., Kinosita, K., Jr., & Yoshida, M.** (1998) $F_1$-ATPase is a highly efficient molecular motor that rotates with discrete 120° steps. *Cell* **93,** 1117–1124.

Graphical demonstration of the rotation of ATP synthase.

## Regulation of Mitochondrial Oxidative Phosphorylation

**Brand, M.D. & Murphy, M.P.** (1987) Control of electron flux through the respiratory chain in mitochondria and cells. *Biol. Rev. Camb. Philos. Soc.* **62,** 141–193.

An advanced description of respiratory control.

**Harris, D.A. & Das, A.M.** (1991) Control of mitochondrial ATP synthesis in the heart. *Biochem. J.* **280,** 561–573.

> Advanced discussion of the regulation of ATP synthase by $Ca^{2+}$ and other factors.

**Klingenberg, M. & Huang, S.-G.** (1999) Structure and function of the uncoupling protein from brown adipose tissue. *Biochim. Biophys. Acta* **1415,** 271–296.

### Apoptosis and Mitochondrial Diseases

**Kroemer, G.** (2003) Mitochondrial control of apoptosis: an introduction. *Biochem. Biophys. Res. Commun.* **304,** 433–435.

> A short introduction to a collection of excellent papers in this journal issue.

**McCord, J.M.** (2002) Superoxide dismutase in aging and disease: an overview. *Meth. Enzymol.* **349,** 331–341.

**Newmeyer, D.D. & Ferguson-Miller, S.** (2003) Mitochondria: releasing power for life and unleashing the machineries of death. *Cell* **112,** 481–490.

> Intermediate-level review of the possible roles of mitochondria in apoptosis.

**Schapira, A.H.V.** (2002) Primary and secondary defects of the mitochondrial respiratory chain. *J. Inher. Metab. Dis.* **25,** 207–214.

**Wallace, D.C.** (1999) Mitochondrial disease in man and mouse. *Science* **283,** 1482–1487.

## PHOTOSYNTHESIS

### Harvesting Light Energy

**Cogdell, R.J., Isaacs, N.W., Howard, T.D., McLuskey, K., Fraser, N.J., & Prince, S.M.** (1999) How photosynthetic bacteria harvest solar energy. *J. Bacteriol.* **181,** 3869–3879.

> A short, intermediate-level review of the structure and function of the light-harvesting complex of the purple bacteria and exciton flow to the reaction center.

**Green, B.R., Pichersky, E., & Kloppstech, K.** (1991) Chlorophyll *a/b*-binding proteins: an extended family. *Trends Biochem. Sci.* **16,** 181–186.

> An intermediate-level description of the proteins that orient chlorophyll molecules in chloroplasts.

**Kargul, J., Nield, J., & Barber, J.** (2003) Three-dimensional reconstruction of a light-harvesting Complex I–Photosystem I (LHCI-PSI) supercomplex from the green alga *Chlamydomonas reinhardtii. J. Biol. Chem.* **278,** 16,135–16,141.

**Zuber, H.** (1986) Structure of light-harvesting antenna complexes of photosynthetic bacteria, cyanobacteria and red algae. *Trends Biochem. Sci.* **11,** 414–419.

### Light-Driven Electron Flow

**Barber, J.** (2002) Photosystem II: a multisubunit membrane protein that oxidizes water. *Curr. Opin. Struct. Biol.* **12,** 523–530.

> A short, intermediate-level summary of the structure of PSII.

**Barber, J. & Anderson, J.M. (eds)** (2002) Photosystem II: molecular structure and function. *Philos. Trans. R. Soc. (Biol. Sci.)* **357,** 1321–1512.

> A collection of 16 papers on photosystem II.

**Chitnis, P.R.** (2001) Photosystem I: function and physiology. *Annu. Rev. Plant Physiol. Plant Mol. Biol.* **52,** 593–626.

> An advanced and lengthy review.

**Deisenhofer, J. & Michel, H.** (1991) Structures of bacterial photosynthetic reaction centers. *Annu. Rev. Cell Biol.* **7,** 1–23.

> Description of the structure of the reaction center of purple bacteria and implications for the function of bacterial and plant reaction centers.

**Heathcote, P., Fyfe, P.K., & Jones, M.R.** (2002) Reaction centres: the structure and evolution of biological solar power. *Trends. Biochem Sci.* **27,** 79–87.

> Intermediate-level review of photosystems I and II.

**Huber, R.** (1990) A structural basis of light energy and electron transfer in biology. *Eur. J. Biochem.* **187,** 283–305.

> Huber's Nobel lecture, describing the physics and chemistry of phototransductions; an exceptionally clear and well-illustrated discussion, based on crystallographic studies of reaction centers.

**Jagendorf, A.T.** (1967) Acid-base transitions and phosphorylation by chloroplasts. *Fed. Proc.* **26,** 1361–1369.

> Report of the classic experiment establishing the ability of a proton gradient to drive ATP synthesis in the dark.

**Jordan, P., Fromme, P., Witt, H.T., Klukas, O., Saenger, W., & Krauss, N.** (2001) Three-dimensional structure of cyanobacterial photosystem I at 2.5 Å. *Nature* **411,** 909–917.

**Kamiya, N. & Shen, J.-R.** (2003) Crystal structure of oxygen-evolving photosystem II from *Thermosynechococcus vulcanus* at 3.7 Å resolution. *Proc. Natl. Acad. Sci. USA* **100,** 98–103.

**Kargul, J., Nield, J., & Barber, J.** (2003) Three-dimensional reconstruction of a light-harvesting complex I–photosystem I (LHCI-PSI) supercomplex from the green alga *Chlamydomonas reinhardtii:* insights into light harvesting for PSI. *J. Biol. Chem.* **278,** 16,135–16,141.

**Leister, D.** (2003) Chloroplast research in the genomic age. *Trends Genet.* **19,** 47–56.

**Zouni, A., Witt, H.-T., Kern, J., Fromme, P., Krauss, N., Saenger, W., & Orth, P.** (2001) Crystal structure of photosystem II from *Synechococcus elongates* at 3.8 Å resolution. *Nature* **409,** 739–743.

### Water-Splitting Complex

**Rögner, M., Boekema, E.J., & Barber, J.** (1996) How does photosystem 2 split water? The structural basis of efficient energy conversion. *Trends Biochem. Sci.* **21,** 44–49.

**Szalai, V.A. & Brudvig, G.W.** (1998) How plants produce dioxygen. *Am. Sci.* **86,** 542–551.

> A well-illustrated introduction to the oxygen-evolving complex of plants.

### Bacteriorhodopsin

**Luecke, H.** (2000) Atomic resolution structures of bacteriorhodopsin photocycle intermediates: the role of discrete water molecules in the function of this light-driven ion pump. *Biochim. Biophys. Acta* **1460,** 133–156.

> Advanced review of a proton pump that employs an internal chain of water molecules.

**Luecke, H., Schobert, B., Richter, H.-T., Cartailler, J.-P., & Lanyi, J.K.** (1999) Structural changes in bacteriorhodopsin during ion transport at 2 angstrom resolution. *Science* **286,** 255–264.

> This paper, accompanied by an editorial comment in the same *Science* issue, describes the model for $H^+$ translocation by proton hopping.

# Problems

**1. Oxidation-Reduction Reactions** The NADH dehydrogenase complex of the mitochondrial respiratory chain promotes the following series of oxidation-reduction reactions, in which $Fe^{3+}$ and $Fe^{2+}$ represent the iron in iron-sulfur centers, Q is ubiquinone, $QH_2$ is ubiquinol, and E is the enzyme:

(1)   $NADH + H^+ + E\text{-}FMN \longrightarrow NAD^+ + E\text{-}FMNH_2$

(2)   $E\text{-}FMNH_2 + 2Fe^{3+} \longrightarrow E\text{-}FMN + 2Fe^{2+} + 2H^+$

(3)   $2Fe^{2+} + 2H^+ + Q \longrightarrow 2Fe^{3+} + QH_2$

*Sum:* $NADH + H^+ + Q \longrightarrow NAD^+ + QH_2$

For each of the three reactions catalyzed by the NADH dehydrogenase complex, identify (a) the electron donor, (b) the electron acceptor, (c) the conjugate redox pair, (d) the reducing agent, and (e) the oxidizing agent.

**2. All Parts of Ubiquinone Have a Function** In electron transfer, only the quinone portion of ubiquinone undergoes oxidation-reduction; the isoprenoid side chain remains unchanged. What is the function of this chain?

**3. Use of FAD Rather Than $NAD^+$ in Succinate Oxidation** All the dehydrogenases of glycolysis and the citric acid cycle use $NAD^+$ ($E'^\circ$ for $NAD^+$/NADH is $-0.32$ V) as electron acceptor except succinate dehydrogenase, which uses covalently bound FAD ($E'^\circ$ for FAD/$FADH_2$ in this enzyme is 0.050 V). Suggest why FAD is a more appropriate electron acceptor than $NAD^+$ in the dehydrogenation of succinate, based on the $E'^\circ$ values of fumarate/succinate ($E'^\circ = 0.031$), $NAD^+$/NADH, and the succinate dehydrogenase FAD/$FADH_2$.

**4. Degree of Reduction of Electron Carriers in the Respiratory Chain** The degree of reduction of each carrier in the respiratory chain is determined by conditions in the mitochondrion. For example, when NADH and $O_2$ are abundant, the steady-state degree of reduction of the carriers decreases as electrons pass from the substrate to $O_2$. When electron transfer is blocked, the carriers before the block become more reduced and those beyond the block become more oxidized (see Fig. 19–6). For each of the conditions below, predict the state of oxidation of ubiquinone and cytochromes $b$, $c_1$, $c$, and $a + a_3$.

(a) Abundant NADH and $O_2$, but cyanide added

(b) Abundant NADH, but $O_2$ exhausted

(c) Abundant $O_2$, but NADH exhausted

(d) Abundant NADH and $O_2$

**5. Effect of Rotenone and Antimycin A on Electron Transfer** Rotenone, a toxic natural product from plants, strongly inhibits NADH dehydrogenase of insect and fish mitochondria. Antimycin A, a toxic antibiotic, strongly inhibits the oxidation of ubiquinol.

(a) Explain why rotenone ingestion is lethal to some insect and fish species.

(b) Explain why antimycin A is a poison.

(c) Given that rotenone and antimycin A are equally effective in blocking their respective sites in the electron-transfer chain, which would be a more potent poison? Explain.

**6. Uncouplers of Oxidative Phosphorylation** In normal mitochondria the rate of electron transfer is tightly coupled to the demand for ATP. When the rate of use of ATP is relatively low, the rate of electron transfer is low; when demand for ATP increases, electron-transfer rate increases. Under these conditions of tight coupling, the number of ATP molecules produced per atom of oxygen consumed when NADH is the electron donor—the P/O ratio—is about 2.5.

(a) Predict the effect of a relatively low and a relatively high concentration of uncoupling agent on the rate of electron transfer and the P/O ratio.

(b) Ingestion of uncouplers causes profuse sweating and an increase in body temperature. Explain this phenomenon in molecular terms. What happens to the P/O ratio in the presence of uncouplers?

(c) The uncoupler 2,4-dinitrophenol was once prescribed as a weight-reducing drug. How could this agent, in principle, serve as a weight-reducing aid? Uncoupling agents are no longer prescribed, because some deaths occurred following their use. Why might the ingestion of uncouplers lead to death?

**7. Effects of Valinomycin on Oxidative Phosphorylation** When the antibiotic valinomycin is added to actively respiring mitochondria, several things happen: the yield of ATP decreases, the rate of $O_2$ consumption increases, heat is released, and the pH gradient across the inner mitochondrial membrane increases. Does valinomycin act as an uncoupler or an inhibitor of oxidative phosphorylation? Explain the experimental observations in terms of the antibiotic's ability to transfer $K^+$ ions across the inner mitochondrial membrane.

**8. Mode of Action of Dicyclohexylcarbodiimide (DCCD)** When DCCD is added to a suspension of tightly coupled, actively respiring mitochondria, the rate of electron transfer (measured by $O_2$ consumption) and the rate of ATP production dramatically decrease. If a solution of 2,4-dinitrophenol is now added to the preparation, $O_2$ consumption returns to normal but ATP production remains inhibited.

(a) What process in electron transfer or oxidative phosphorylation is affected by DCCD?

(b) Why does DCCD affect the $O_2$ consumption of mitochondria? Explain the effect of 2,4-dinitrophenol on the inhibited mitochondrial preparation.

(c) Which of the following inhibitors does DCCD most resemble in its action: antimycin A, rotenone, or oligomycin?

**9. Compartmentalization of Citric Acid Cycle Components** Isocitrate dehydrogenase is found only in the mitochondrion, but malate dehydrogenase is found in both the cytosol and mitochondrion. What is the role of cytosolic malate dehydrogenase?

**10. The Malate–$\alpha$-Ketoglutarate Transport System** The transport system that conveys malate and $\alpha$-ketoglutarate across the inner mitochondrial membrane (see Fig. 19–27) is inhibited by $n$-butylmalonate. Suppose $n$-butylmalonate is added to an aerobic suspension of kidney cells using glucose exclusively as fuel. Predict the effect of this inhibitor on (a) glycolysis, (b) oxygen consumption, (c) lactate formation, and (d) ATP synthesis.

**11. Cellular ADP Concentration Controls ATP Formation** Although both ADP and $P_i$ are required for the synthesis of ATP, the rate of synthesis depends mainly on the concentration of ADP, not $P_i$. Why?

**12. The Pasteur Effect** When $O_2$ is added to an anaerobic suspension of cells consuming glucose at a high rate, the rate of glucose consumption declines greatly as the $O_2$ is used up, and accumulation of lactate ceases. This effect, first observed by Louis Pasteur in the 1860s, is characteristic of most cells capable of both aerobic and anaerobic glucose catabolism.

(a) Why does the accumulation of lactate cease after $O_2$ is added?

(b) Why does the presence of $O_2$ decrease the rate of glucose consumption?

(c) How does the onset of $O_2$ consumption slow down the rate of glucose consumption? Explain in terms of specific enzymes.

**13. Respiration-Deficient Yeast Mutants and Ethanol Production** Respiration-deficient yeast mutants ($p^-$; "petites") can be produced from wild-type parents by treatment with mutagenic agents. The mutants lack cytochrome oxidase, a deficit that markedly affects their metabolic behavior. One striking effect is that fermentation is not suppressed by $O_2$—that is, the mutants lack the Pasteur effect (see Problem 12). Some companies are very interested in using these mutants to ferment wood chips to ethanol for energy use. Explain the advantages of using these mutants rather than wild-type yeast for large-scale ethanol production. Why does the absence of cytochrome oxidase eliminate the Pasteur effect?

**14. How Many Protons in a Mitochondrion?** Electron transfer translocates protons from the mitochondrial matrix to the external medium, establishing a pH gradient across the inner membrane (outside more acidic than inside). The tendency of protons to diffuse back into the matrix is the driving force for ATP synthesis by ATP synthase. During oxidative phosphorylation by a suspension of mitochondria in a medium of pH 7.4, the pH of the matrix has been measured as 7.7.

(a) Calculate [$H^+$] in the external medium and in the matrix under these conditions.

(b) What is the outside-to-inside ratio of [$H^+$]? Comment on the energy inherent in this concentration difference. (Hint: See Eqn 11–3, p. 398.)

(c) Calculate the number of protons in a respiring liver mitochondrion, assuming its inner matrix compartment is a sphere of diameter 1.5 $\mu m$.

(d) From these data, is the pH gradient alone sufficient to generate ATP?

(e) If not, suggest how the necessary energy for synthesis of ATP arises.

**15. Rate of ATP Turnover in Rat Heart Muscle** Rat heart muscle operating aerobically fills more than 90% of its ATP needs by oxidative phosphorylation. Each gram of tissue consumes $O_2$ at the rate of 10.0 $\mu mol/min$, with glucose as the fuel source.

(a) Calculate the rate at which the heart muscle consumes glucose and produces ATP.

(b) For a steady-state concentration of ATP of 5.0 $\mu mol/g$ of heart muscle tissue, calculate the time required (in seconds) to completely turn over the cellular pool of ATP. What does this result indicate about the need for tight regulation of ATP production? (Note: Concentrations are expressed as

micromoles per gram of muscle tissue because the tissue is mostly water.)

**16. Rate of ATP Breakdown in Flight Muscle** ATP production in the flight muscle of the fly *Lucilia sericata* results almost exclusively from oxidative phosphorylation. During flight, 187 ml of $O_2/hr \cdot g$ of body weight is needed to maintain an ATP concentration of 7.0 $\mu mol/g$ of flight muscle. Assuming that flight muscle makes up 20% of the weight of the fly, calculate the rate at which the flight-muscle ATP pool turns over. How long would the reservoir of ATP last in the absence of oxidative phosphorylation? Assume that reducing equivalents are transferred by the glycerol 3-phosphate shuttle and that $O_2$ is at 25 °C and 101.3 kPa (1 atm).

**17. Transmembrane Movement of Reducing Equivalents** Under aerobic conditions, extramitochondrial NADH must be oxidized by the mitochondrial electron-transfer chain. Consider a preparation of rat hepatocytes containing mitochondria and all the cytosolic enzymes. If [4-$^3$H]NADH is introduced, radioactivity soon appears in the mitochondrial matrix. However, if [7-$^{14}$C]NADH is introduced, no radioactivity appears in the matrix. What do these observations reveal about the oxidation of extramitochondrial NADH by the electron-transfer chain?

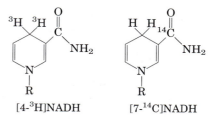

[4-$^3$H]NADH          [7-$^{14}$C]NADH

**18. NAD Pools and Dehydrogenase Activities** Although both pyruvate dehydrogenase and glyceraldehyde 3-phosphate dehydrogenase use $NAD^+$ as their electron acceptor, the two enzymes do not compete for the same cellular NAD pool. Why?

**19. Photochemical Efficiency of Light at Different Wavelengths** The rate of photosynthesis, measured by $O_2$ production, is higher when a green plant is illuminated with light of wavelength 680 nm than with light of 700 nm. However, illumination by a combination of light of 680 nm and 700 nm gives a higher rate of photosynthesis than light of either wavelength alone. Explain.

**20. Balance Sheet for Photosynthesis** In 1804 Theodore de Saussure observed that the total weights of oxygen and dry organic matter produced by plants is greater than the weight of carbon dioxide consumed during photosynthesis. Where does the extra weight come from?

**21. Role of $H_2S$ in Some Photosynthetic Bacteria** Illuminated purple sulfur bacteria carry out photosynthesis in the presence of $H_2O$ and $^{14}CO_2$, but only if $H_2S$ is added and $O_2$ is absent. During the course of photosynthesis, measured by formation of [$^{14}$C]carbohydrate, $H_2S$ is converted to elemental sulfur, but no $O_2$ is evolved. What is the role of the conversion of $H_2S$ to sulfur? Why is no $O_2$ evolved?

**22. Boosting the Reducing Power of Photosystem I by Light Absorption** When photosystem I absorbs red light at 700 nm, the standard reduction potential of P700 changes from 0.40 V to about $-1.2$ V. What fraction of the absorbed light is trapped in the form of reducing power?

**23. Limited ATP Synthesis in the Dark** In a laboratory experiment, spinach chloroplasts are illuminated in the absence of ADP and $P_i$, then the light is turned off and ADP and $P_i$ are added. ATP is synthesized for a short time in the dark. Explain this finding.

**24. Mode of Action of the Herbicide DCMU** When chloroplasts are treated with 3-(3,4-dichlorophenyl)-1,1-dimethylurea (DCMU, or diuron), a potent herbicide, $O_2$ evolution and photophosphorylation cease. Oxygen evolution, but not photophosphorylation, can be restored by addition of an external electron acceptor, or Hill reagent. How does DCMU act as a weed killer? Suggest a location for the inhibitory action of this herbicide in the scheme shown in Figure 19–49. Explain.

**25. Bioenergetics of Photophosphorylation** The steady-state concentrations of ATP, ADP, and $P_i$ in isolated spinach chloroplasts under full illumination at pH 7.0 are 120.0, 6.0, and 700.0 $\mu m$, respectively.

    (a) What is the free-energy requirement for the synthesis of 1 mol of ATP under these conditions?

    (b) The energy for ATP synthesis is furnished by light-induced electron transfer in the chloroplasts. What is the minimum voltage drop necessary (during transfer of a pair of electrons) to synthesize ATP under these conditions? (You may need to refer to Eqn 13–6, p. 510.)

**26. Light Energy for a Redox Reaction** Suppose you have isolated a new photosynthetic microorganism that oxidizes $H_2S$ and passes the electrons to $NAD^+$. What wavelength of light would provide enough energy for $H_2S$ to reduce $NAD^+$ under standard conditions? Assume 100% efficiency in the photochemical event, and use $E'^\circ$ of $-243$ mV for $H_2S$ and $-320$ mV for $NAD^+$. See Figure 19–39 for energy equivalents of wavelengths of light.

**27. Equilibrium Constant for Water-Splitting Reactions** The coenzyme $NADP^+$ is the terminal electron acceptor in chloroplasts, according to the reaction

$$2H_2O + 2NADP^+ \longrightarrow 2NADPH + 2H^+ + O_2$$

Use the information in Table 19–2 to calculate the equilibrium constant for this reaction at 25 °C. (The relationship between $K'_{eq}$ and $\Delta G'^\circ$ is discussed on p. 492.) How can the chloroplast overcome this unfavorable equilibrium?

**28. Energetics of Phototransduction** During photosynthesis, eight photons must be absorbed (four by each photosystem) for every $O_2$ molecule produced:

$$2H_2O + 2NADP^+ + 8 \text{ photons} \longrightarrow 2NADPH + 2H^+ + O_2$$

Assuming that these photons have a wavelength of 700 nm (red) and that the absorption and use of light energy are 100% efficient, calculate the free-energy change for the process.

**29. Electron Transfer to a Hill Reagent** Isolated spinach chloroplasts evolve $O_2$ when illuminated in the presence of potassium ferricyanide (a Hill reagent), according to the equation

$$2H_2O + 4Fe^{3+} \longrightarrow O_2 + 4H^+ + 4Fe^{2+}$$

where $Fe^{3+}$ represents ferricyanide and $Fe^{2+}$, ferrocyanide. Is NADPH produced in this process? Explain.

**30. How Often Does a Chlorophyll Molecule Absorb a Photon?** The amount of chlorophyll $a$ ($M_r$ 892) in a spinach leaf is about 20 $\mu g/cm^2$ of leaf. In noonday sunlight (average energy 5.4 $J/cm^2 \cdot min$), the leaf absorbs about 50% of the radiation. How often does a single chlorophyll molecule absorb a photon? Given that the average lifetime of an excited chlorophyll molecule in vivo is 1 ns, what fraction of the chlorophyll molecules is excited at any one time?

**31. Effect of Monochromatic Light on Electron Flow** The extent to which an electron carrier is oxidized or reduced during photosynthetic electron transfer can sometimes be observed directly with a spectrophotometer. When chloroplasts are illuminated with 700 nm light, cytochrome $f$, plastocyanin, and plastoquinone are oxidized. When chloroplasts are illuminated with 680 nm light, however, these electron carriers are reduced. Explain.

**32. Function of Cyclic Photophosphorylation** When the $[NADPH]/[NADP^+]$ ratio in chloroplasts is high, photophosphorylation is predominantly cyclic (see Fig. 19–49). Is $O_2$ evolved during cyclic photophosphorylation? Is NADPH produced? Explain. What is the main function of cyclic photophosphorylation?

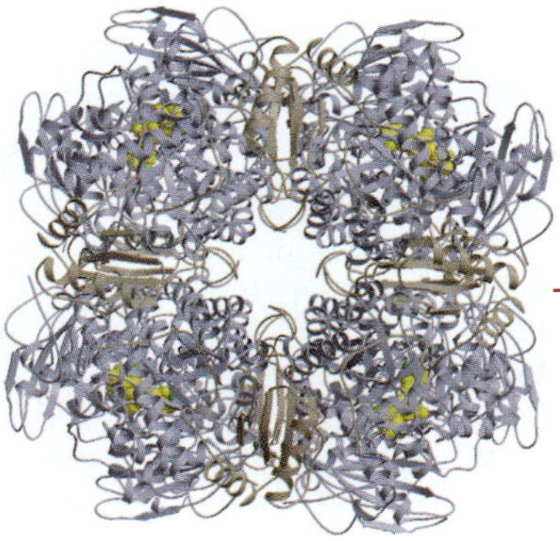

# CARBOHYDRATE BIOSYNTHESIS IN PLANTS AND BACTERIA

... the discovery of the long-lived isotope of carbon, carbon-14, by Samuel Ruben and Martin Kamen in 1940 provided the ideal tool for the tracing of the route along which carbon dioxide travels on its way to carbohydrate.

—*Melvin Calvin, Nobel Address, 1961*

We have now reached a turning point in our study of cellular metabolism. Thus far in Part II we have described how the major metabolic fuels—carbohydrates, fatty acids, and amino acids—are degraded through converging *catabolic* pathways to enter the citric acid cycle and yield their electrons to the respiratory chain, and how this exergonic flow of electrons to oxygen is coupled to the endergonic synthesis of ATP. We now turn to *anabolic* pathways, which use chemical energy in the form of ATP and NADH or NADPH to synthesize cellular components from simple precursor molecules. Anabolic pathways are generally reductive rather than oxidative. Catabolism and anabolism proceed simultaneously in a dynamic steady state, so the energy-yielding degradation of cellular components is counterbalanced by biosynthetic processes, which create and maintain the intricate orderliness of living cells.

Plants must be especially versatile in their handling of carbohydrates, for several reasons. First, plants are autotrophs, able to convert inorganic carbon (as $CO_2$) into organic compounds. Second, biosynthesis occurs primarily in plastids, membrane-bounded organelles unique to plants, and the movement of intermediates between cellular compartments is an important aspect of metabolism. Third, plants are not motile: they cannot move to find better supplies of water, sunlight, or nutrients. They must have sufficient metabolic flexibility to allow them to adapt to changing conditions in the place where they are rooted. Finally, plants have thick cell walls made of carbohydrate polymers, which must be assembled outside the plasma membrane and which constitute a significant proportion of the cell's carbohydrate.

The chapter begins with a description of the process by which $CO_2$ is assimilated into trioses and hexoses, then considers photorespiration, an important side reaction during $CO_2$ fixation, and the ways in which certain plants avoid this side reaction. We then look at how the biosynthesis of sucrose (for sugar transport) and starch (for energy storage) is accomplished by mechanisms analogous to those employed by animal cells to make glycogen. The next topic is the synthesis of the cellulose of plant cell walls and the peptidoglycan of bacterial cell walls, illustrating the problems of energy-dependent biosynthesis outside the plasma membrane. Finally, we discuss how the various pathways that share pools of common intermediates are segregated within organelles yet integrated with one another.

## 20.1 Photosynthetic Carbohydrate Synthesis

The synthesis of carbohydrates in animal cells always employs precursors having at least three carbons, all of which are less oxidized than the carbon in $CO_2$. Plants

751

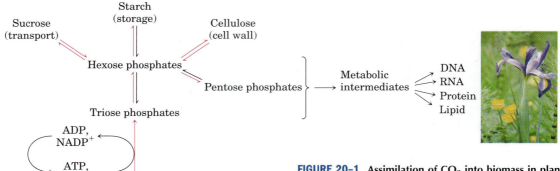

**FIGURE 20–1 Assimilation of CO$_2$ into biomass in plants.** The light-driven synthesis of ATP and NADPH, described in Chapter 19, provides energy and reducing power for the fixation of CO$_2$ into trioses, from which all the carbon-containing compounds of the plant cell are synthesized. The processes shown with red arrows are the focus of this chapter.

and photosynthetic microorganisms, by contrast, can synthesize carbohydrates from CO$_2$ and water, reducing CO$_2$ at the expense of the energy and reducing power furnished by the ATP and NADPH that are generated by the light-dependent reactions of photosynthesis (Fig. 20–1). Plants (and other autotrophs) can use CO$_2$ as the sole source of the carbon atoms required for the biosynthesis of cellulose and starch, lipids and proteins, and the many other organic components of plant cells. By contrast, heterotrophs cannot bring about the net reduction of CO$_2$ to achieve a net synthesis of glucose.

Green plants contain in their chloroplasts unique enzymatic machinery that catalyzes the conversion of CO$_2$ to simple (reduced) organic compounds, a process called **CO$_2$ assimilation.** This process has also been called **CO$_2$ fixation** or **carbon fixation,** but we reserve these terms for the specific reaction in which CO$_2$ is incorporated (fixed) into a three-carbon organic compound, the triose phosphate 3-phosphoglycerate. This simple product of photosynthesis is the precursor of more complex biomolecules, including sugars, polysaccharides, and the metabolites derived from them, all of which are synthesized by metabolic pathways similar to those of animal tissues. Carbon dioxide is assimilated via a cyclic pathway, its key intermediates constantly regenerated. The pathway was elucidated in the early 1950s by Melvin Calvin, Andrew Benson, and James A. Bassham, and is often called the **Calvin cycle** or, more descriptively, the **photosynthetic carbon reduction cycle.**

Melvin Calvin,
1911–1997

Carbohydrate metabolism is more complex in plant cells than in animal cells or in nonphotosynthetic microorganisms. In addition to the universal pathways of glycolysis and gluconeogenesis, plants have the unique reaction sequences for reduction of CO$_2$ to triose phosphates and the associated reductive pentose phosphate pathway—all of which must be coordinately regulated to ensure proper allocation of carbon to energy production and synthesis of starch and sucrose. Key enzymes are regulated, as we shall see, by (1) reduction of disulfide bonds by electrons flowing from photosystem I and (2) changes in pH and Mg$^{2+}$ concentration that result from illumination. When we look at other aspects of plant carbohydrate metabolism, we also find enzymes that are modulated by (3) conventional allosteric regulation by one or more metabolic intermediates and (4) covalent modification (phosphorylation).

## Plastids Are Organelles Unique to Plant Cells and Algae

Most of the biosynthetic activities in plants (including CO$_2$ assimilation) occur in **plastids,** a family of self-reproducing organelles bounded by a double membrane and containing a small genome that encodes some of their proteins. Most proteins destined for plastids are encoded in nuclear genes, which are transcribed and translated like other nuclear genes; then the proteins are imported into plastids. Plastids reproduce by binary fission, replicating their genome (a single circular DNA molecule) and using their own enzymes and ribosomes to synthesize the proteins encoded by that genome. **Chloroplasts** (see Fig. 19–38) are the sites of CO$_2$ assimilation. The enzymes for this process are contained in the stroma, the soluble phase bounded by the inner chloroplast membrane. **Amyloplasts** are colorless plastids (that is, they lack chlorophyll and other pig-

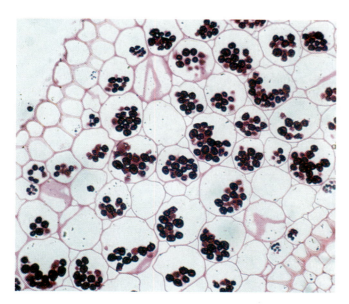

**FIGURE 20-2** **Amyloplasts filled with starch (dark granules) are stained with iodine in this section of *Ranunculus* root cells.** Starch granules in various tissues range from 1 to 100 $\mu m$ in diameter.

ments found in chloroplasts). They have no internal membranes analogous to the photosynthetic membranes (thylakoids) of chloroplasts, and in plant tissues rich in starch these plastids are packed with starch granules (Fig. 20–2). Chloroplasts can be converted to **proplastids** by the loss of their internal membranes and chlorophyll, and proplastids are interconvertible with amyloplasts (Fig. 20–3). In turn, both amyloplasts and proplastids can develop into chloroplasts. The relative proportions of the plastid types depend on the type of plant tissue and on the intensity of illumination. Cells of green leaves are rich in chloroplasts, whereas amyloplasts dominate in nonphotosynthetic tissues that store starch in large quantities, such as potato tubers.

The inner membranes of all types of plastids are impermeable to polar and charged molecules. Traffic across these membranes is mediated by sets of specific transporters.

## Carbon Dioxide Assimilation Occurs in Three Stages

The first stage in the assimilation of $CO_2$ into biomolecules (Fig. 20–4) is the **carbon-fixation reaction:** condensation of $CO_2$ with a five-carbon acceptor, **ribulose 1,5-bisphosphate,** to form two molecules of **3-phosphoglycerate.** In the second stage, the 3-phosphoglycerate is reduced to triose phosphates. Overall, three molecules of $CO_2$ are fixed to three molecules of ribulose 1,5-bisphosphate to form six molecules of glyceraldehyde 3-phosphate (18 carbons) in equilibrium with dihydroxyacetone phosphate. In the third stage, five of the six molecules of triose phosphate (15 carbons) are used to regenerate three molecules of ribu-

lose 1,5-bisphosphate (15 carbons), the starting material. The sixth molecule of triose phosphate, the net product of photosynthesis, can be used to make hexoses for fuel and building materials, sucrose for transport to nonphotosynthetic tissues, or starch for storage. Thus the overall process is cyclical, with the continuous conversion of $CO_2$ to triose and hexose phosphates. Fructose 6-phosphate is a key intermediate in stage 3 of $CO_2$ assimilation; it stands at a branch point, leading either to regeneration of ribulose 1,5-bisphosphate or to synthesis of starch. The pathway from hexose phosphate to pentose bisphosphate involves many of the same reactions used in animal cells for the conversion of pentose phosphates to hexose phosphates during the nonoxidative phase of the **pentose phosphate pathway** (see Fig. 14–22). In the photosynthetic assimilation of $CO_2$, essentially the same set of reactions operates in the other direction, converting hexose phosphates to pentose phosphates. This **reductive pentose phosphate cycle** uses the same enzymes as the oxidative pathway, and several more enzymes that make the reductive cycle irreversible. All 13 enzymes of the pathway are in the chloroplast stroma.

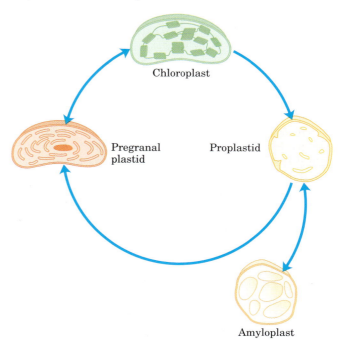

**FIGURE 20-3** **Plastids: their origins and interconversions.** All types of plastids are bounded by a double membrane, and some (notably the mature chloroplast) have extensive internal membranes. The internal membranes can be lost (when a mature chloroplast becomes a proplastid) and resynthesized (as a proplastid gives rise to a pregranal plastid and then a mature chloroplast). Proplastids in nonphotosynthetic tissues (such as root) give rise to amyloplasts, which contain large quantities of starch. All plant cells have plastids, and these organelles are the site of other important processes, including the synthesis of essential amino acids, thiamine, pyridoxal phosphate, flavins, and vitamins A, C, E, and K.

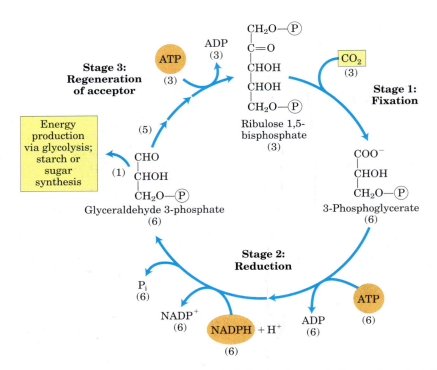

**FIGURE 20–4** **The three stages of $CO_2$ assimilation in photosynthetic organisms.** Stoichiometries of three key intermediates (numbers in parentheses) reveal the fate of carbon atoms entering and leaving the cycle. As shown here, three $CO_2$ are fixed for the net synthesis of one molecule of glyceraldehyde 3-phosphate. This cycle is the photosynthetic carbon reduction cycle, or the Calvin cycle.

***Stage 1: Fixation of $CO_2$ into 3-Phosphoglycerate*** An important clue to the nature of the $CO_2$-assimilation mechanisms in photosynthetic organisms came in the late 1940s. Calvin and his associates illuminated a suspension of green algae in the presence of radioactive carbon dioxide ($^{14}CO_2$) for just a few seconds, then quickly killed the cells, extracted their contents, and with the help of chromatographic methods searched for the metabolites in which the labeled carbon first appeared. The first compound that became labeled was **3-phosphoglycerate,** with the $^{14}C$ predominantly located in the carboxyl carbon atom. These experiments strongly suggested that 3-phosphoglycerate is an early intermediate in photosynthesis. The many plants in which this three-carbon compound is the first intermediate are called **$C_3$ plants,** in contrast with the $C_4$ plants described below.

The enzyme that catalyzes incorporation of $CO_2$ into an organic form is **ribulose 1,5-bisphosphate carboxylase/oxygenase,** a name mercifully shortened to **rubisco.** As a carboxylase, rubisco catalyzes the covalent attachment of $CO_2$ to the five-carbon sugar ribulose 1,5-bisphosphate and cleavage of the unstable six-carbon intermediate to form two molecules of 3-phosphoglycerate, one of which bears the carbon introduced as $CO_2$ in its carboxyl group (Fig. 20–4). The enzyme's oxygenase activity is discussed in Section 20.2.

Plant rubisco, the crucial enzyme in the production of biomass from $CO_2$, has a complex structure (Fig. 20–5a), with eight identical large subunits ($M_r$ 53,000; encoded in the chloroplast genome, or plastome), each containing a catalytic site, and eight identical small subunits ($M_r$ 14,000; encoded in the nuclear genome) of uncertain function. The rubisco of photosynthetic bacteria is simpler in structure, having two subunits that in many respects resemble the large subunits of the plant enzyme (Fig. 20–5b). This similarity is consistent with the endosymbiont hypothesis for the origin of chloroplasts (p. 35). The plant enzyme has an exceptionally low turnover number; only three molecules of $CO_2$ are fixed per second per molecule of rubisco at 25 °C. To achieve high rates of $CO_2$ fixation, plants therefore need large amounts of this enzyme. In fact, rubisco makes up almost 50% of soluble protein in chloroplasts and is probably one of the most abundant enzymes in the biosphere.

Central to the proposed mechanism for plant rubisco is a carbamoylated Lys side chain with a bound $Mg^{2+}$ ion. The $Mg^{2+}$ ion brings together and orients the reactants at the active site (Fig. 20–6) and polarizes the $CO_2$, opening it to nucleophilic attack by the five-carbon enediolate reaction intermediate formed on the enzyme (Fig. 20–7). The resulting six-carbon intermediate breaks down to yield two molecules of 3-phosphoglycerate.

Top view

Side view

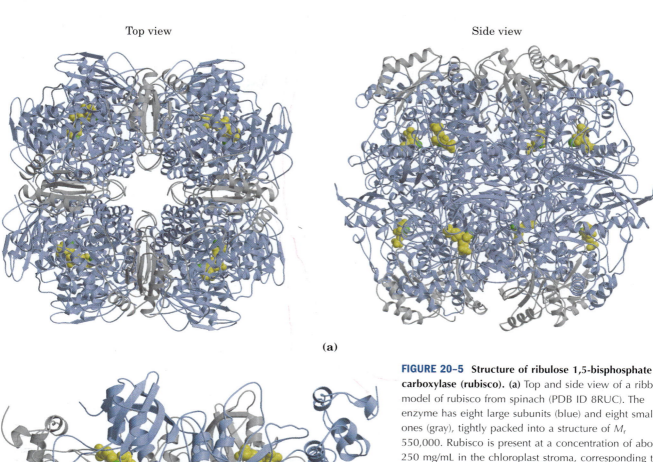

**(a)**

**(b)**

**FIGURE 20–5 Structure of ribulose 1,5-bisphosphate carboxylase (rubisco).** **(a)** Top and side view of a ribbon model of rubisco from spinach (PDB ID 8RUC). The enzyme has eight large subunits (blue) and eight small ones (gray), tightly packed into a structure of $M_r$ 550,000. Rubisco is present at a concentration of about 250 mg/mL in the chloroplast stroma, corresponding to an extraordinarily high concentration of active sites (~4 mM). Amino acid residues of the active site are shown in yellow, $Mg^{2+}$ in green. **(b)** Ribbon model of rubisco from the bacterium *Rhodospirillum rubrum* (PDB ID 9RUB). The subunits are in gray and blue. A Lys residue at the active site that is carboxylated to a carbamate in the active enzyme is shown in red. The substrate, ribulose 1,5-bisphosphate, is yellow; $Mg^{2+}$ is green.

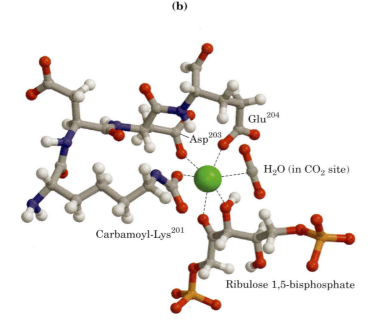

Glu$^{204}$

Asp$^{203}$

$H_2O$ (in $CO_2$ site)

Carbamoyl-Lys$^{201}$

Ribulose 1,5-bisphosphate

**FIGURE 20–6 Central role of $Mg^{2+}$ in the catalytic mechanism of rubisco.** (Derived from PDB ID 1RXO) $Mg^{2+}$ is coordinated in a roughly octahedral complex with six oxygen atoms: one oxygen in the carbamate on Lys$^{201}$; two in the carboxyl groups of Glu$^{204}$ and Asp$^{203}$; two at C-2 and C-3 of the substrate, ribulose 1,5-bisphosphate; and one in the other substrate, $CO_2$. A water molecule occupies the $CO_2$-binding site in this crystal structure. (Residue numbers refer to the spinach enzyme.)

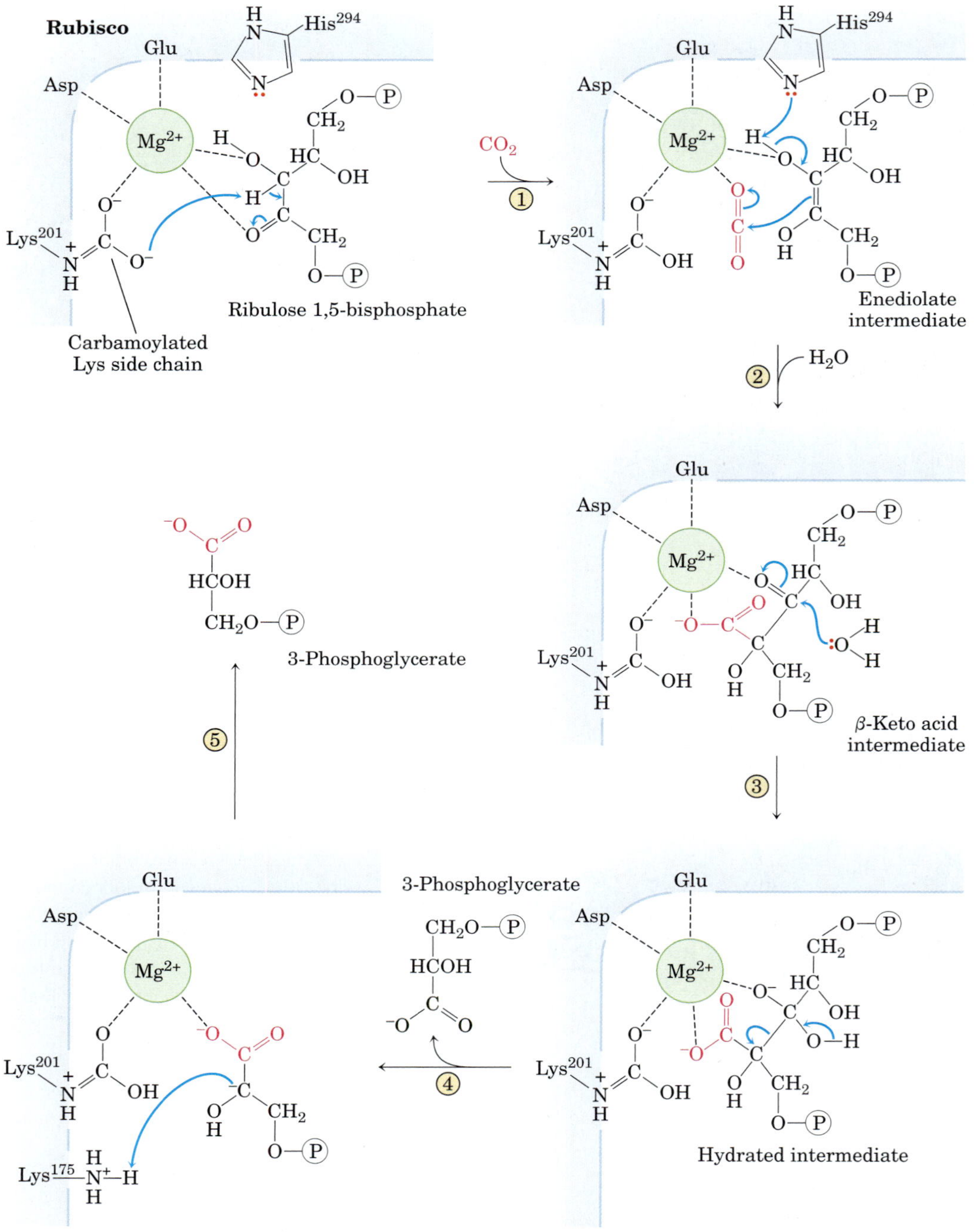

**MECHANISM FIGURE 20–7 First stage of $CO_2$ assimilation: rubisco's carboxylase activity.** The $CO_2$-fixation reaction is catalyzed by ribulose 1,5-bisphosphate carboxylase/oxygenase (rubisco). ① Ribulose 1,5-bisphosphate forms an enediolate at the active site. ② $CO_2$, polarized by the proximity of the $Mg^{2+}$ ion, undergoes nucleophilic attack by the enediolate, producing a branched six-carbon sugar. ③ Hydroxylation at C-3 of this sugar, followed by aldol cleavage ④, forms one molecule of 3-phosphoglycerate, which leaves the enzyme active site.

⑤ The carbanion of the remaining three-carbon fragment is protonated by the nearby side chain of $Lys^{175}$, generating a second molecule of 3-phosphoglycerate. The overall reaction therefore accomplishes the combination of one $CO_2$ and one ribulose 1,5-bisphosphate to form two molecules of 3-phosphoglycerate, one of which contains the carbon atom from $CO_2$ (red). **Rubisco Mechanism; Rubisco Tutorial**

**FIGURE 20–8 Role of rubisco activase in the carbamoylation of Lys$^{201}$ of rubisco.** When the substrate ribulose 1,5-bisphosphate is bound to the active site, Lys$^{201}$ is not accessible. Rubisco activase couples ATP hydrolysis to expulsion of the bound sugar bisphosphate, exposing Lys$^{201}$; this Lys residue can now be carbamoylated with $CO_2$ in a reaction that is apparently not enzyme-mediated. $Mg^{2+}$ is attracted to and binds to the negatively charged carbamoyl-Lys, and the enzyme is thus activated.

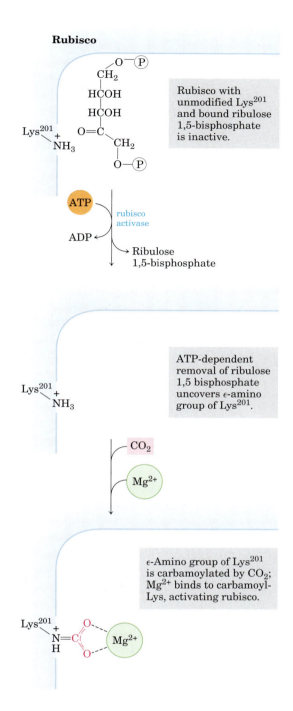

As the catalyst for the first step of photosynthetic $CO_2$ assimilation, rubisco is a prime target for regulation. The enzyme is inactive until carbamoylated on the $\varepsilon$ amino group of Lys$^{201}$ (Fig. 20–8). Ribulose 1,5-bisphosphate inhibits carbamoylation by binding tightly to the active site and locking the enzyme in the "closed" conformation, in which Lys$^{201}$ is inaccessible. **Rubisco activase** overcomes the inhibition by promoting ATP-dependent release of the ribulose 1,5-bisphosphate, exposing the Lys amino group to nonenzymatic carbamoylation by $CO_2$; this is followed by $Mg^{2+}$ binding, which activates the rubisco. Rubisco activase in some species is activated by light through a redox mechanism (see Fig. 20–19).

Another regulatory mechanism involves the "nocturnal inhibitor" 2-carboxyarabinitol 1-phosphate, a naturally occurring transition-state analog (see Box 6–3) with a structure similar to that of the $\beta$-keto acid intermediate of the rubisco reaction (Fig. 20–7; see also Fig. 20–20). This compound, synthesized in the dark in some plants, is a potent inhibitor of carbamoylated rubisco. It is either broken down when light returns or is expelled by rubisco activase, activating the rubisco.

$$CH_2-O-PO_3^{2-}$$
$$HO-C-C{\overset{O}{\underset{O^-}{}}}$$
$$H-C-OH$$
$$H-C-OH$$
$$CH_2OH$$

2-Carboxyarabinitol 1-phosphate

### Stage 2: Conversion of 3-Phosphoglycerate to Glyceraldehyde 3-Phosphate

The 3-phosphoglycerate formed in stage 1 is converted to glyceraldehyde 3-phosphate in two steps that are essentially the reversal of the corresponding steps in glycolysis, with one exception: the nucleotide cofactor for the reduction of 1,3-bisphosphoglycerate is NADPH rather than NADH (Fig. 20–9). The chloroplast stroma contains all the glycolytic enzymes except phosphoglycerate mutase. The stromal and cytosolic enzymes are isozymes; both sets of enzymes catalyze the same reactions, but they are the products of different genes.

In the first step of stage 2, the stromal **3-phosphoglycerate kinase** catalyzes the transfer of a phosphoryl group from ATP to 3-phosphoglycerate, yielding 1,3-bisphosphoglycerate. Next, NADPH donates electrons in a reduction catalyzed by the chloroplast-specific isozyme of **glyceraldehyde 3-phosphate dehydrogenase,** producing glyceraldehyde 3-phosphate and $P_i$. Triose phosphate isomerase then interconverts glyceraldehyde 3-phosphate and dihydroxyacetone phosphate.

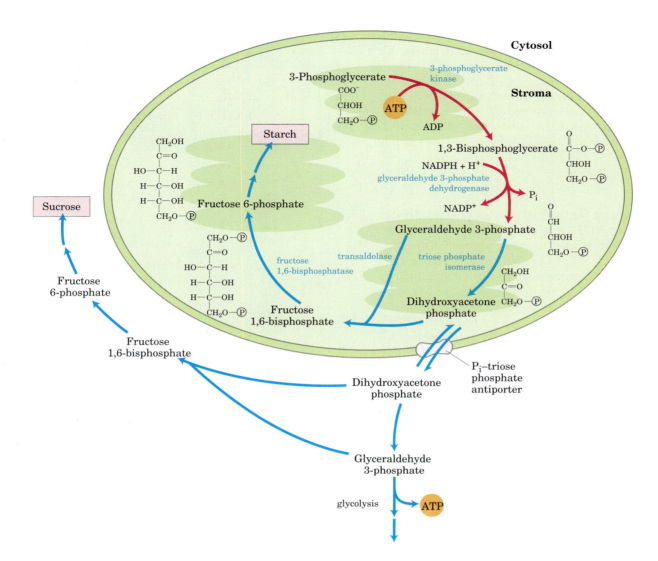

**FIGURE 20–9 Second stage of $CO_2$ assimilation.** 3-Phosphoglycerate is converted to glyceraldehyde 3-phosphate (red arrows). Also shown are the alternative fates of the fixed carbon of glyceraldehyde 3-phosphate (blue arrows). Most of the glyceraldehyde 3-phosphate is recycled to ribulose 1,5-bisphosphate as shown in Figure 20–10. A small fraction of the "extra" glyceraldehyde 3-phosphate may be used immediately as a source of energy, but most is converted to sucrose for transport or is stored in the chloroplast as starch. In the latter case, glyceraldehyde 3-phosphate condenses with dihydroxyacetone phosphate in the stroma to form fructose 1,6-bisphosphate, a precursor of starch. In other situations the glyceraldehyde 3-phosphate is converted to dihydroxyacetone phosphate, which leaves the chloroplast via a specific transporter (see Fig. 20–15) and, in the cytosol, can be degraded glycolytically to provide energy or used to form fructose 6-phosphate and hence sucrose.

Most of the triose phosphate thus produced is used to regenerate ribulose 1,5-bisphosphate; the rest is either converted to starch in the chloroplast and stored for later use or immediately exported to the cytosol and converted to sucrose for transport to growing regions of the plant. In developing leaves, a significant portion of the triose phosphate may be degraded by glycolysis to provide energy.

**Stage 3: Regeneration of Ribulose 1,5-Bisphosphate from Triose Phosphates**    The first reaction in the assimilation of $CO_2$ into triose phosphates consumes ribulose 1,5-bisphosphate and, for continuous flow of $CO_2$ into carbohydrate, ribulose 1,5-bisphosphate must be constantly regenerated. This is accomplished in a series of reactions (Fig. 20–10) that, together with stages 1 and 2, constitute the cyclic pathway shown in Figure 20–4. The product of

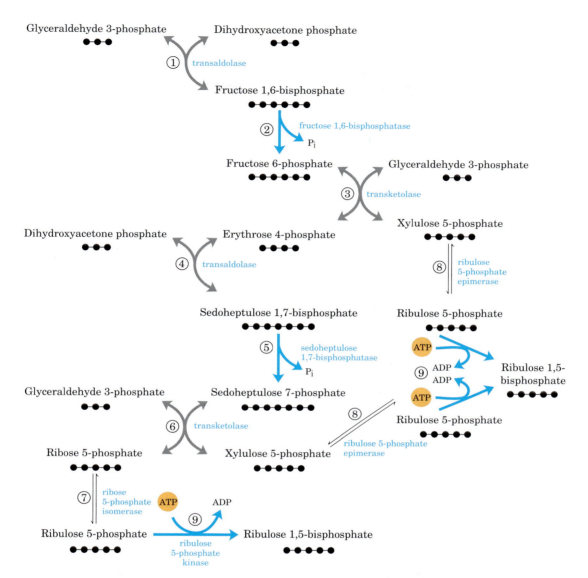

**FIGURE 20–10 Third stage of CO₂ assimilation.** This schematic diagram shows the interconversions of triose phosphates and pentose phosphates. Black dots represent the number of carbons in each compound. The starting materials are glyceraldehyde 3-phosphate and dihydroxyacetone phosphate. Reactions catalyzed by transaldolase (①and ④) and transketolase (③and ⑥) produce pentose phosphates that are converted to ribulose 1,5-bisphosphate—ribose 5-phosphate by ribose 5-phosphate isomerase (⑦) and xylulose 5-phosphate by ribulose 5-phosphate epimerase (⑧). In step ⑨, ribulose 5-phosphate is phosphorylated, regenerating ribulose 1,5-bisphosphate. The steps with blue arrows are exergonic and make the whole process irreversible: steps ②fructose 1,6-bisphosphatase, ⑤ sedoheptulose bisphosphatase, and ⑨ ribulose 5-phosphate kinase.

the first assimilation reaction (3-phosphoglycerate) thus undergoes transformations that regenerate ribulose 1,5-bisphosphate. The intermediates in this pathway include three-, four-, five-, six-, and seven-carbon sugars. In the following discussion, all step numbers refer to Figure 20–10.

Steps ①and ④ are catalyzed by the same enzyme, **transaldolase.** It first catalyzes the reversible conden-

sation of glyceraldehyde 3-phosphate with dihydroxyacetone phosphate, yielding fructose 1,6-bisphosphate (step ①); this is cleaved to fructose 6-phosphate and Pᵢ by fructose 1,6-bisphosphatase (FBPase-1) in step ②. The reaction is strongly exergonic and essentially irreversible. Step ③ is catalyzed by **transketolase,** which contains thiamine pyrophosphate (TPP) as its prosthetic group (see Fig. 14–13a) and requires $Mg^{2+}$.

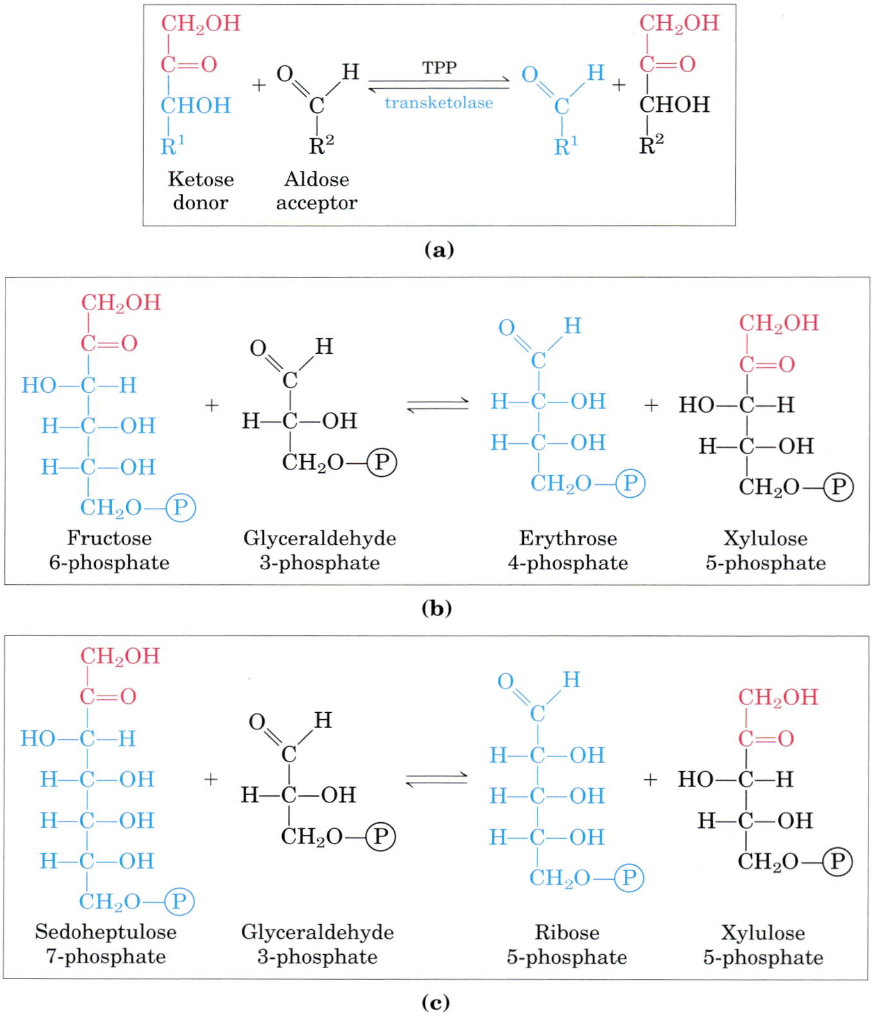

**(a)**

**(b)**

**(c)**

**FIGURE 20–11 Transketolase-catalyzed reactions of the Calvin cycle. (a)** General reaction catalyzed by transketolase: the transfer of a two-carbon group, carried temporarily on enzyme-bound TPP, from a ketose donor to an aldose acceptor. **(b)** Conversion of a hexose and a triose to a four-carbon and a five-carbon sugar (step ③ of Fig. 20–10). **(c)** Conversion of seven-carbon and three-carbon sugars to two pentoses (step ⑥ of Fig. 20–10).

Transketolase catalyzes the reversible transfer of a 2-carbon ketol group ($CH_2OH$—CO—) from a ketose phosphate donor, fructose 6-phosphate, to an aldose phosphate acceptor, glyceraldehyde 3-phosphate (Fig. 20–11a, b), forming the pentose xylulose 5-phosphate and the tetrose erythrose 4-phosphate. In step ④, transaldolase acts again, combining erythrose 4-phosphate with dihydroxyacetone phosphate to form the seven-carbon **sedoheptulose 1,7-bisphosphate.** An enzyme unique to plastids, sedoheptulose 1,7-bisphosphatase, converts the bisphosphate to sedoheptulose 7-phosphate (step ⑤); this is the second irreversible reaction in the pathway. Transketolase now acts again, converting sedoheptulose 7-phosphate and glyceralde-hyde 3-phosphate to two pentose phosphates in step ⑥ (Fig. 20–11c). Figure 20–12 shows how a two-carbon fragment is temporarily carried on the transketolase cofactor TPP and condensed with the three carbons of glyceraldehyde 3-phosphate in step ⑥.

The pentose phosphates formed in the transketolase reactions—ribose 5-phosphate and xylulose 5-phosphate—are converted to **ribulose 5-phosphate** (steps ⑦ and ⑧), which in the final step (⑨) of the cycle is phosphorylated to ribulose 1,5-bisphosphate by ribulose 5-phosphate kinase (Fig. 20–13). This is the third very exergonic reaction of the pathway, as the phosphate anhydride bond in ATP is swapped for a phosphate ester in ribulose 1,5-bisphosphate.

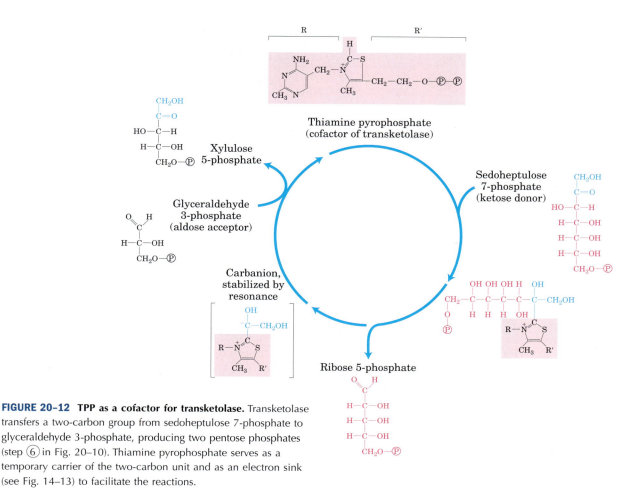

**FIGURE 20–12 TPP as a cofactor for transketolase.** Transketolase transfers a two-carbon group from sedoheptulose 7-phosphate to glyceraldehyde 3-phosphate, producing two pentose phosphates (step ⑥ in Fig. 20–10). Thiamine pyrophosphate serves as a temporary carrier of the two-carbon unit and as an electron sink (see Fig. 14–13) to facilitate the reactions.

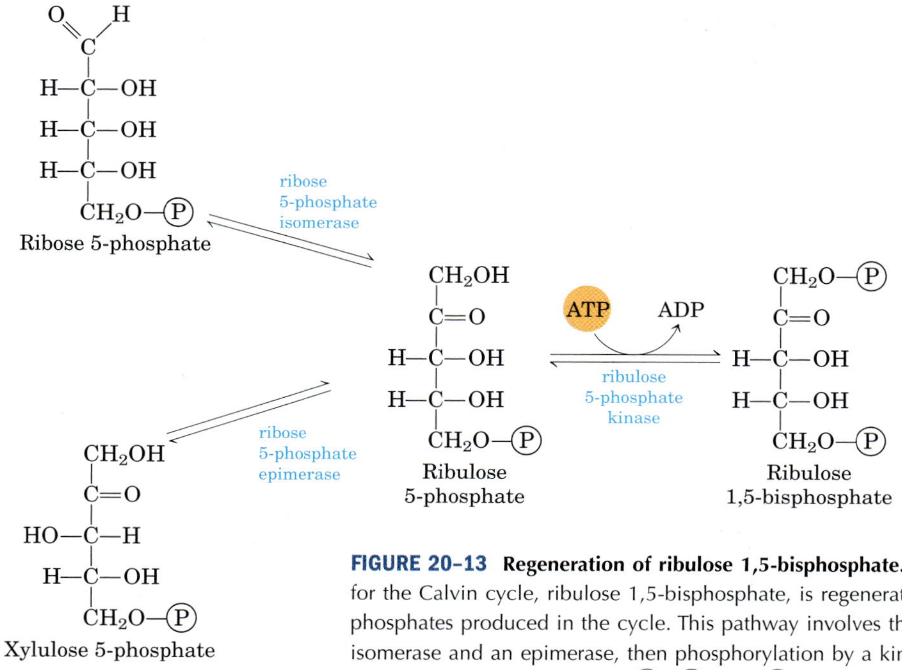

**FIGURE 20–13 Regeneration of ribulose 1,5-bisphosphate.** The starting material for the Calvin cycle, ribulose 1,5-bisphosphate, is regenerated from two pentose phosphates produced in the cycle. This pathway involves the action of an isomerase and an epimerase, then phosphorylation by a kinase, with ATP as phosphate group donor (steps ⑦, ⑧, and ⑨ of Fig. 20–10).

## Synthesis of Each Triose Phosphate from CO₂ Requires Six NADPH and Nine ATP

The net result of three turns of the Calvin cycle is the conversion of three molecules of $CO_2$ and one molecule of phosphate to a molecule of triose phosphate. The stoichiometry of the overall path from $CO_2$ to triose phosphate, with regeneration of ribulose 1,5-bisphosphate, is shown in Figure 20–14. Three molecules of ribulose 1,5-bisphosphate (a total of 15 carbons) condense with three $CO_2$ (3 carbons) to form six molecules of 3-phosphoglycerate (18 carbons). These six molecules of 3-phosphoglycerate are reduced to six molecules of glyceraldehyde 3-phosphate (which is in equilibrium with dihydroxyacetone phosphate), with the expenditure of six ATP (in the synthesis of 1,3-bisphosphoglycerate) and six NADPH (in the reduction of 1,3-bisphosphoglycerate to glyceraldehyde 3-phosphate). The isozyme of glyceraldehyde 3-phosphate dehydrogenase present in chloroplasts can use NADP as its electron carrier and normally functions in the direction of 1,3-bisphosphoglycerate reduction. The cytosolic isozyme uses NAD, as does the glycolytic enzyme of animals and other eukaryotes, and in the dark this isozyme acts in glycolysis to oxidize glyceraldehyde 3-phosphate. Both glyceraldehyde 3-phosphate dehydrogenase isozymes, like all enzymes, catalyze the reaction in both directions.

One molecule of glyceraldehyde 3-phosphate is the net product of the carbon assimilation pathway. The other five triose phosphate molecules (15 carbons) are rearranged in steps ① to ⑨ of Figure 20–10 to form three molecules of ribulose 1,5-bisphosphate (15 carbons). The last step in this conversion requires one ATP per ribulose 1,5-bisphosphate, or a total of three ATP. Thus, in summary, for every molecule of triose phosphate produced by photosynthetic $CO_2$ assimilation, six NADPH and nine ATP are required.

NADPH and ATP are produced in the light-dependent reactions of photosynthesis in about the same ratio (2:3) as they are consumed in the Calvin cycle. Nine ATP molecules are converted to ADP and phosphate in the generation of a molecule of triose phosphate; eight of the phosphates are released as $P_i$ and combined with eight ADP to regenerate ATP. The ninth phosphate is incorporated into the triose phosphate itself. To convert the ninth ADP to ATP, a molecule of $P_i$ must be imported from the cytosol, as we shall see.

In the dark, the production of ATP and NADPH by photophosphorylation, and the incorporation of $CO_2$ into triose phosphate (by the so-called dark reactions),

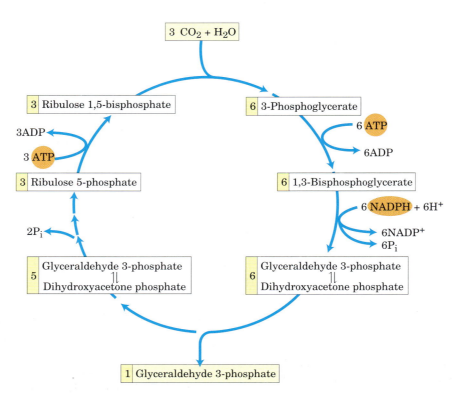

**FIGURE 20–14** **Stoichiometry of CO₂ assimilation in the Calvin cycle.** For every three $CO_2$ molecules fixed, one molecule of triose phosphate (glyceraldehyde 3-phosphate) is produced and nine ATP and six NADPH are consumed.

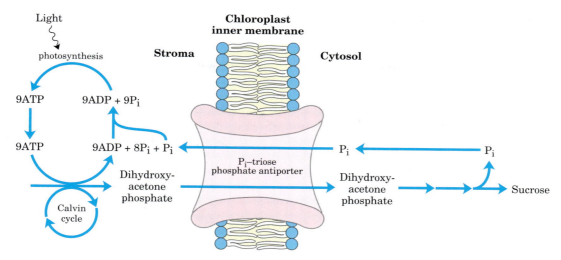

**FIGURE 20–15** **The $P_i$–triose phosphate antiport system of the inner chloroplast membrane.** This transporter facilitates the exchange of cytosolic $P_i$ for stromal dihydroxyacetone phosphate. The products of photosynthetic carbon assimilation are thus moved into the cytosol where they serve as a starting point for sucrose biosynthesis, and $P_i$ required for photophosphorylation is moved into the stroma. This same antiporter can transport 3-phosphoglycerate and acts in the shuttle for exporting ATP and reducing equivalents (see Fig. 20–16).

cease. The "dark reactions" of photosynthesis were so named to distinguish them from the *primary* light-driven reactions of electron transfer to $NADP^+$ and synthesis of ATP, described in Chapter 19. They do not, in fact, occur at significant rates in the dark and are thus more appropriately called the **carbon-assimilation reactions.** Later in this section we describe the regulatory mechanisms that turn on carbon assimilation in the light and turn it off in the dark.

The chloroplast stroma contains all the enzymes necessary to convert the triose phosphates produced by $CO_2$ assimilation (glyceraldehyde 3-phosphate and dihydroxyacetone phosphate) to starch, which is temporarily stored in the chloroplast as insoluble granules. Aldolase condenses the trioses to fructose 1,6-bisphosphate; fructose 1,6-bisphosphatase produces fructose 6-phosphate; phosphohexose isomerase yields glucose 6-phosphate; and phosphoglucomutase produces glucose 1-phosphate, the starting material for starch synthesis (see Section 20.3).

All the reactions of the Calvin cycle except those catalyzed by rubisco, sedoheptulose 1,7-bisphosphatase, and ribulose 5-phosphate kinase also take place in animal tissues. Lacking these three enzymes, animals cannot carry out net conversion of $CO_2$ to glucose.

## A Transport System Exports Triose Phosphates from the Chloroplast and Imports Phosphate

The inner chloroplast membrane is impermeable to most phosphorylated compounds, including fructose 6-phosphate, glucose 6-phosphate, and fructose 1,6-bisphosphate. It does, however, have a specific antiporter that catalyzes the one-for-one exchange of $P_i$ with a triose phosphate, either dihydroxyacetone phosphate or 3-phosphoglycerate (Fig. 20–15; see also Fig. 20–9). This antiporter simultaneously moves $P_i$ into the chloroplast, where it is used in photophosphorylation, and moves triose phosphate into the cytosol, where it can be used to synthesize sucrose, the form in which the fixed carbon is transported to distant plant tissues.

Sucrose synthesis in the cytosol and starch synthesis in the chloroplast are the major pathways by which the excess triose phosphate from photosynthesis is "harvested." Sucrose synthesis (described below) releases four $P_i$ molecules from the four triose phosphates required to make sucrose. For every molecule of triose phosphate removed from the chloroplast, one $P_i$ is transported into the chloroplast, providing the ninth $P_i$ mentioned above, to be used in regenerating ATP. If this exchange were blocked, triose phosphate synthesis would quickly deplete the available $P_i$ in the chloroplast, slowing ATP synthesis and suppressing assimilation of $CO_2$ into starch.

The $P_i$–triose phosphate antiport system serves one additional function. ATP and reducing power are needed in the cytosol for a variety of synthetic and energy-requiring reactions. These requirements are met to an as yet undetermined degree by mitochondria, but a second potential source of energy is the ATP and NADPH generated in the chloroplast stroma during the light reactions. However, neither ATP nor NADPH can cross the chloroplast membrane. The $P_i$–triose phosphate antiport system has the indirect effect of moving ATP

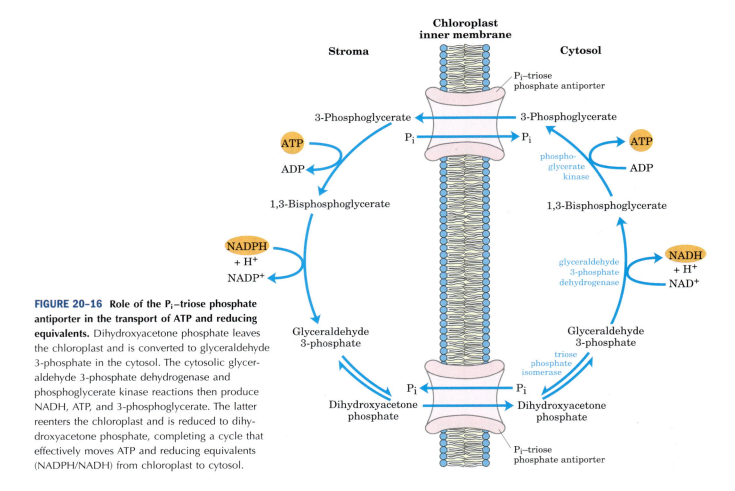

**FIGURE 20–16 Role of the $P_i$–triose phosphate antiporter in the transport of ATP and reducing equivalents.** Dihydroxyacetone phosphate leaves the chloroplast and is converted to glyceraldehyde 3-phosphate in the cytosol. The cytosolic glyceraldehyde 3-phosphate dehydrogenase and phosphoglycerate kinase reactions then produce NADH, ATP, and 3-phosphoglycerate. The latter reenters the chloroplast and is reduced to dihydroxyacetone phosphate, completing a cycle that effectively moves ATP and reducing equivalents (NADPH/NADH) from chloroplast to cytosol.

equivalents and reducing equivalents from the chloroplast to the cytosol (Fig. 20–16). Dihydroxyacetone phosphate formed in the stroma is transported to the cytosol, where it is converted by glycolytic enzymes to 3-phosphoglycerate, generating ATP and NADH. 3-Phosphoglycerate reenters the chloroplast, completing the cycle.

## Four Enzymes of the Calvin Cycle Are Indirectly Activated by Light

The reductive assimilation of $CO_2$ requires a lot of ATP and NADPH, and their stromal concentrations increase when chloroplasts are illuminated (Fig. 20–17). The light-induced transport of protons across the thylakoid membrane (Chapter 19) also increases the stromal pH from about 7 to about 8, and it is accompanied by a flow of $Mg^{2+}$ from the thylakoid compartment into the stroma, raising the $[Mg^{2+}]$ from 1 to 3 mM to 3 to 6 mM. Several stromal enzymes have evolved to take advantage of these light-induced conditions, which signal the availability of ATP and NADPH: the enzymes are more active in an alkaline environment and at high $[Mg^{2+}]$. For example, activation of rubisco by formation of the

carbamoyllysine is faster at alkaline pH, and high stromal $[Mg^{2+}]$ favors formation of the enzyme's active $Mg^{2+}$ complex. Fructose 1,6-bisphosphatase requires $Mg^{2+}$ and is very dependent on pH (Fig. 20–18); its activity increases more than 100-fold when pH and $[Mg^{2+}]$ rise during chloroplast illumination.

Four Calvin cycle enzymes are subject to a special type of regulation by light. Ribulose 5-phosphate kinase, fructose 1,6-bisphosphatase, sedoheptulose 1,7-bisphosphatase, and glyceraldehyde 3-phosphate dehydrogenase are activated by light-driven reduction of disulfide bonds between two Cys residues critical to their catalytic activities. When these Cys residues are disulfide-bonded (oxidized), the enzymes are inactive; this is the normal situation in the dark. With illumination, electrons flow from photosystem I to ferredoxin (see Fig. 19–49), which passes electrons to a small, soluble, disulfide-containing protein called **thioredoxin** (Fig. 20–19), in a reaction catalyzed by **ferredoxin-thioredoxin reductase**. Reduced thioredoxin donates electrons for the reduction of the disulfide bonds of the light-activated enzymes, and these reductive cleavage reactions are accompanied by conformational changes that increase enzyme activities. At nightfall, the Cys residues in the

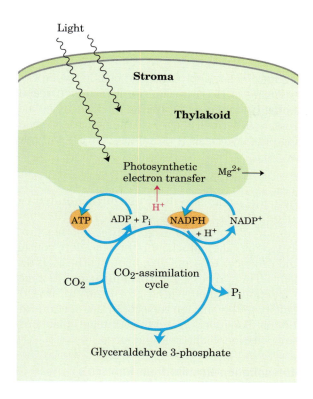

**FIGURE 20–17  Source of ATP and NADPH.** ATP and NADPH produced by the light reactions are essential substrates for the reduction of $CO_2$. The photosynthetic reactions that produce ATP and NADPH are accompanied by movement of protons (red) from the stroma into the thylakoid, creating alkaline conditions in the stroma. Magnesium ions pass from the thylakoid into the stroma, increasing the stromal $[Mg^{2+}]$.

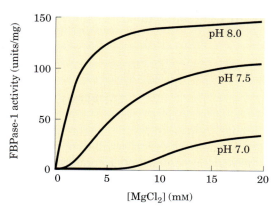

**FIGURE 20–18  Activation of chloroplast fructose 1,6-bisphosphatase.** Reduced fructose 1,6-bisphosphatase (FBPase-1) is activated by light and by the combination of high pH and high $[Mg^{2+}]$ in the stroma, both of which are produced by illumination.

four enzymes are reoxidized to their disulfide forms, the enzymes are inactivated, and ATP is not expended in $CO_2$ assimilation. Instead, starch synthesized and stored during the daytime is degraded to fuel glycolysis at night.

Glucose 6-phosphate dehydrogenase, the first enzyme in the *oxidative* pentose phosphate pathway, is also regulated by this light-driven reduction mechanism, but in the opposite sense. During the day, when photosynthesis produces plenty of NADPH, this enzyme is not needed for NADPH production. Reduction of a critical disulfide bond by electrons from ferredoxin *inactivates* the enzyme.

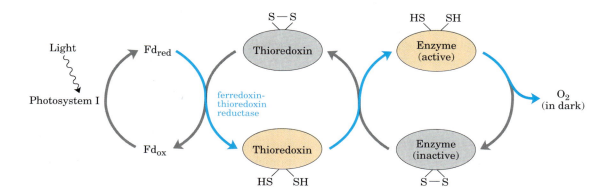

**FIGURE 20–19  Light activation of several enzymes of the Calvin cycle.** The light activation is mediated by thioredoxin, a small, disulfide-containing protein. In the light, thioredoxin is reduced by electrons moving from photosystem I through ferredoxin (Fd) (blue arrows), then thioredoxin reduces critical disulfide bonds in each of the enzymes sedoheptulose 1,7-bisphosphatase, fructose 1,6-bisphosphatase, ribulose 5-phosphate kinase, and glyceraldehyde 3-phosphate dehydrogenase, activating these enzymes. In the dark, the —SH groups undergo reoxidation to disulfides, inactivating the enzymes.

*SUMMARY 20.1* Photosynthetic Carbohydrate Synthesis

- Photosynthesis in vascular plants takes place in chloroplasts. In the $CO_2$-assimilating reactions (the Calvin cycle), ATP and NADPH are used to reduce $CO_2$ to triose phosphates. These reactions occur in three stages: the fixation reaction itself, catalyzed by rubisco; reduction of the resulting 3-phosphoglycerate to glyceraldehyde 3-phosphate; and regeneration of ribulose 1,5-bisphosphate from triose phosphates.

- Rubisco condenses $CO_2$ with ribulose 1,5-bisphosphate, forming an unstable hexose bisphosphate that splits into two molecules of 3-phosphoglycerate. Rubisco is activated by covalent modification (carbamoylation of $Lys^{201}$) catalyzed by rubisco activase and is inhibited by a natural transition-state analog, whose concentration rises in the dark and falls during daylight.

- Stromal isozymes of the glycolytic enzymes catalyze reduction of 3-phosphoglycerate to glyceraldehyde 3-phosphate; each molecule reduced requires one ATP and one NADPH.

- Stromal enzymes, including transketolase and transaldolase, rearrange the carbon skeletons of triose phosphates, generating intermediates of three, four, five, six, and seven carbons and eventually yielding pentose phosphates. The pentose phosphates are converted to ribulose 5-phosphate, then phosphorylated to ribulose 1,5-bisphosphate to complete the Calvin cycle.

- The cost of fixing three $CO_2$ into one triose phosphate is nine ATP and six NADPH, which are provided by the light-dependent reactions of photosynthesis.

- An antiporter in the inner chloroplast membrane exchanges $P_i$ in the cytosol for 3-phosphoglycerate or dihydroxyacetone phosphate produced by $CO_2$ assimilation in the stroma. Oxidation of dihydroxyacetone phosphate in the cytosol generates ATP and NADH, thus moving ATP and reducing equivalents from the chloroplast to the cytosol.

- Four enzymes of the Calvin cycle are activated indirectly by light and are inactive in the dark, so that hexose synthesis does not compete with glycolysis—which is required to provide energy in the dark.

# 20.2 Photorespiration and the $C_4$ and CAM Pathways

As we have seen, photosynthetic cells produce $O_2$ (by the splitting of $H_2O$) during the light-driven reactions (Chapter 19) and use $CO_2$ during the light-independent processes (described above), so the net gaseous change during photosynthesis is the uptake of $CO_2$ and release of $O_2$:

$$CO_2 + H_2O \longrightarrow O_2 + (CH_2O)$$

In the dark, plants also carry out **mitochondrial respiration,** the oxidation of substrates to $CO_2$ and the conversion of $O_2$ to $H_2O$. And there is another process in plants that, like mitochondrial respiration, consumes $O_2$ and produces $CO_2$ and, like photosynthesis, is driven by light. This process, **photorespiration,** is a costly side reaction of photosynthesis, a result of the lack of specificity of the enzyme rubisco. In this section we describe this side reaction and the strategies plants use to minimize its metabolic consequences.

### Photorespiration Results from Rubisco's Oxygenase Activity

Rubisco is not absolutely specific for $CO_2$ as a substrate. Molecular oxygen ($O_2$) competes with $CO_2$ at the active site, and about once in every three or four turnovers, rubisco catalyzes the condensation of $O_2$ with ribulose 1,5-bisphosphate to form 3-phosphoglycerate and **2-phosphoglycolate** (Fig. 20–20), a metabolically useless product. This is the oxygenase activity referred to in the full name of the enzyme: ribulose 1,5-bisphosphate carboxylase/oxygenase. The reaction with $O_2$ results in no fixation of carbon and appears to be a net liability to the cell; salvaging the carbons from 2-phosphoglycolate (by the pathway outlined below) consumes significant amounts of cellular energy and releases some previously fixed $CO_2$.

Given that the reaction with oxygen is deleterious to the organism, why did the evolution of rubisco produce an active site unable to discriminate well between $CO_2$ and $O_2$? Perhaps much of this evolution occurred before the time, about 2.5 billion years ago, when production of $O_2$ by photosynthetic organisms started to raise the oxygen content of the atmosphere. Before that time, there was no selective pressure for rubisco to discriminate between $CO_2$ and $O_2$. The $K_m$ for $CO_2$ is about 9 $\mu M$, and that for $O_2$ is about 350 $\mu M$. The modern atmosphere contains about 20% $O_2$ and only 0.04% $CO_2$, so an aqueous solution in equilibrium with air at room temperature contains about 250 $\mu M$ $O_2$ and 11 $\mu M$ $CO_2$—concentrations that allow significant $O_2$ "fixation" by rubisco and thus a significant waste of energy. The temperature dependence of the solubilities of $O_2$ and $CO_2$ is

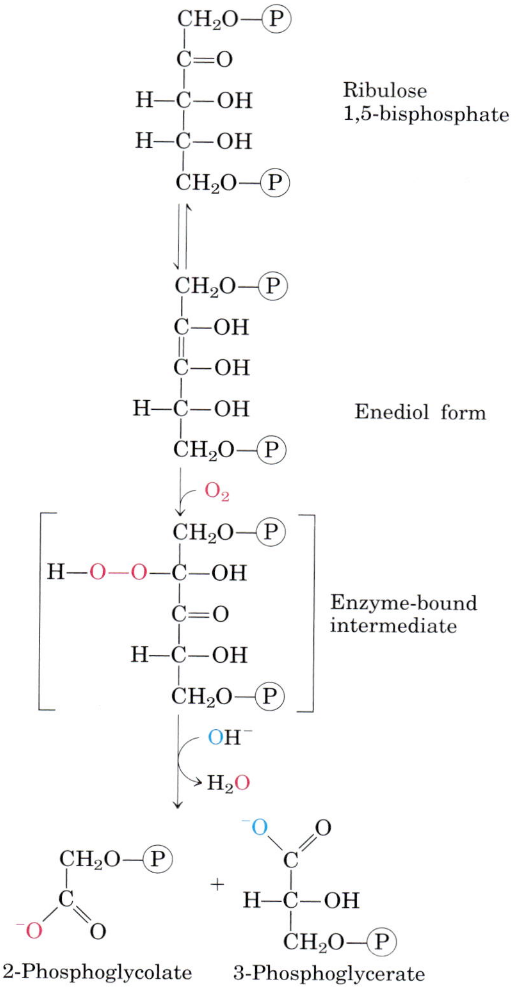

**FIGURE 20–20 Oxygenase activity of rubisco.** Rubisco can incorporate $O_2$ rather than $CO_2$ into ribulose 1,5-bisphosphate. The unstable intermediate thus formed splits into 2-phosphoglycolate (recycled as described in Fig. 20–21) and 3-phosphoglycerate, which can reenter the Calvin cycle.

such that at higher temperatures, the ratio of $O_2$ to $CO_2$ in solution increases. In addition, the affinity of rubisco for $CO_2$ decreases with increasing temperature, exacerbating its tendency to catalyze the wasteful oxygenase reaction. And as $CO_2$ is consumed in the assimilation reactions, the ratio of $O_2$ to $CO_2$ in the air spaces of a leaf increases, further favoring the oxygenase reaction.

## The Salvage of Phosphoglycolate Is Costly

The **glycolate pathway** converts two molecules of 2-phosphoglycolate to a molecule of serine (three carbons) and a molecule of $CO_2$ (Fig. 20–21). In the chloroplast, a phosphatase converts 2-phosphoglycolate to glycolate, which is exported to the peroxisome. There, glycolate is

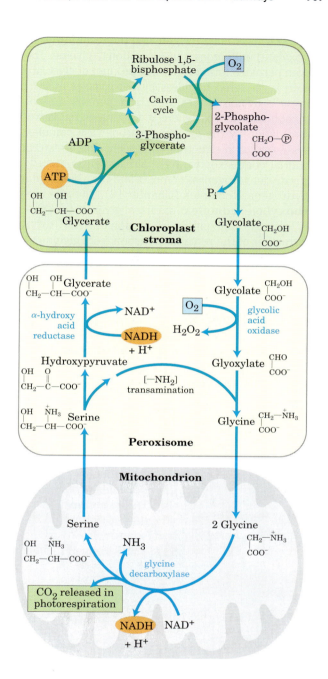

**FIGURE 20–21 Glycolate pathway.** This pathway, which salvages 2-phosphoglycolate (shaded pink) by its conversion to serine and eventually 3-phosphoglycerate, involves three cellular compartments. Glycolate formed by dephosphorylation of 2-phosphoglycolate in chloroplasts is oxidized to glyoxylate in peroxisomes and then transaminated to glycine. In mitochondria, two glycine molecules condense to form serine and the $CO_2$ released during photorespiration (shaded green). This reaction is catalyzed by glycine decarboxylase, an enzyme present at very high levels in the mitochondria of $C_3$ plants (see text). The serine is converted to hydroxypyruvate and then to glycerate in peroxisomes; glycerate reenters the chloroplasts to be phosphorylated, rejoining the Calvin cycle. Oxygen (shaded blue) is consumed at two steps during photorespiration.

oxidized by molecular oxygen, and the resulting aldehyde (glyoxylate) undergoes transamination to glycine. The hydrogen peroxide formed as a side product of glycolate oxidation is rendered harmless by peroxidases in the peroxisome. Glycine passes from the peroxisome to the mitochondrial matrix, where it undergoes oxidative decarboxylation by the glycine decarboxylase complex, an enzyme similar in structure and mechanism to two mitochondrial complexes we have already encountered: the pyruvate dehydrogenase complex and the α-ketoglutarate dehydrogenase complex (Chapter 16). The **glycine decarboxylase complex** oxidizes glycine to $CO_2$ and $NH_3$, with the concomitant reduction of $NAD^+$ to NADH and transfer of the remaining carbon from glycine to the cofactor tetrahydrofolate (Fig. 20–22). The one-carbon

unit carried on tetrahydrofolate is then transferred to a second glycine by serine hydroxymethyltransferase, producing serine. The net reaction catalyzed by the glycine decarboxylase complex and serine hydroxymethyltransferase is

$$2 \text{ Glycine} + NAD^+ + H_2O \longrightarrow$$
$$\text{serine} + CO_2 + NH_3 + NADH + H^+$$

The serine is converted to hydroxypyruvate, to glycerate, and finally to 3-phosphoglycerate, which is used to regenerate ribulose 1,5-bisphosphate, completing the long, expensive cycle (Fig. 20–21).

In bright sunlight, the flux through the glycolate salvage pathway can be very high, producing about five times more $CO_2$ than is typically produced by all the ox-

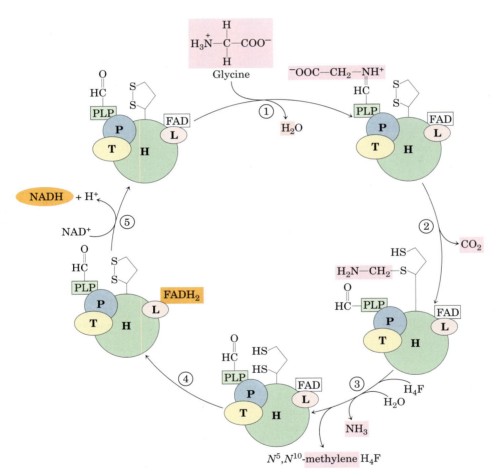

**FIGURE 20–22 The glycine decarboxylase system.** Glycine decarboxylase in plant mitochondria is a complex of four types of subunits, with the stoichiometry $P_4H_{27}T_9L_2$. Protein H has a covalently attached lipoic acid residue that can undergo reversible oxidation. Step ① is formation of a Schiff base between pyridoxal phosphate (PLP) and glycine, catalyzed by protein P (named for its bound PLP). In step ②, protein P catalyzes oxidative decarboxylation of glycine, releasing $CO_2$; the remaining methylamine group is attached to one of the —SH groups of reduced lipoic acid. ③ Protein T (which uses tetrahydrofolate ($H_4F$) as cofactor) now releases $NH_3$ from the methylamine

moiety and transfers the remaining one-carbon fragment to tetrahydrofolate, producing $N^5,N^{10}$-methylene tetrahydrofolate. ④ Protein L oxidizes the two —SH groups of lipoic acid to a disulfide, passing electrons through FAD to $NAD^+$ ⑤, thus completing the cycle. The $N^5,N^{10}$-methylene tetrahydrofolate formed in this process is used by serine hydroxymethyltransferase to convert a molecule of glycine to serine, regenerating the tetrahydrofolate that is essential for the reaction catalyzed by protein T. The L subunit of glycine decarboxylase is identical to the dihydrolipoyl dehydrogenase ($E_3$) of pyruvate dehydrogenase and α-ketoglutarate dehydrogenase (see Fig. 16–6).

idations of the citric acid cycle. To generate this large flux, mitochondria contain prodigious amounts of the glycine decarboxylase complex: the four proteins of the complex make up *half* of all the protein in the mitochondrial matrix in the leaves of pea and spinach plants! In nonphotosynthetic parts of a plant, such as potato tubers, mitochondria have very low concentrations of the glycine decarboxylase complex.

The combined activity of the rubisco oxygenase and the glycolate salvage pathway consumes $O_2$ and produces $CO_2$—hence the name **photorespiration.** This pathway is perhaps better called the **oxidative photosynthetic carbon cycle** or $C_2$ **cycle**, names that do not invite comparison with respiration in mitochondria. Unlike mitochondrial respiration, "photorespiration" does not conserve energy and may actually inhibit net biomass formation as much as 50%. This inefficiency has led to evolutionary adaptations in the carbon-assimilation processes, particularly in plants that have evolved in warm climates.

## In $C_4$ Plants, $CO_2$ Fixation and Rubisco Activity Are Spatially Separated

In many plants that grow in the tropics (and in temperate-zone crop plants native to the tropics, such as maize, sugarcane, and sorghum) a mechanism has evolved to circumvent the problem of wasteful photorespiration. The step in which $CO_2$ is fixed into a three-carbon product, 3-phosphoglycerate, is preceded by several steps, one of which is temporary fixation of $CO_2$ into a four-carbon compound. Plants that use this process are referred to as $C_4$ **plants,** and the assimilation process as $C_4$ **metabolism** or the $C_4$ **pathway.** Plants that use the carbon-assimilation method we have described thus far, in which the *first step* is reaction of $CO_2$ with ribulose 1,5-bisphosphate to form 3-phosphoglycerate, are called $C_3$ **plants.**

The $C_4$ plants, which typically grow at high light intensity and high temperatures, have several important characteristics: high photosynthetic rates, high growth rates, low photorespiration rates, low rates of water loss, and a specialized leaf structure. Photosynthesis in the leaves of $C_4$ plants involves two cell types: mesophyll and bundle-sheath cells (Fig. 20–23a). There are three variants of $C_4$ metabolism, worked out in the 1960s by Marshall Hatch and Rodger Slack (Fig. 20–23b).

In plants of tropical origin, the first intermediate into which $^{14}CO_2$ is fixed is oxaloacetate, a four-carbon compound. This reaction, which occurs in the cytosol of leaf mesophyll cells, is catalyzed by **phosphoenolpyruvate carboxylase,** for which the substrate is $HCO_3^-$, not $CO_2$. The oxaloacetate thus formed is either reduced to malate at the expense of NADPH (as shown in Fig. 20–23b) or converted to aspartate by transamination:

Oxaloacetate + $\alpha$-amino acid $\longrightarrow$ L-aspartate + $\alpha$-keto acid

The malate or aspartate formed in the mesophyll cells then passes into neighboring bundle-sheath cells through plasmodesmata, protein-lined channels that connect two plant cells and provide a path for movement of metabolites and even small proteins between cells. In the bundle-sheath cells, malate is oxidized and decarboxylated to yield pyruvate and $CO_2$ by the action of **malic enzyme,** reducing NADP$^+$. In plants that use aspartate as the $CO_2$ carrier, aspartate arriving in bundle-sheath cells is transaminated to form oxaloacetate and reduced to malate, then the $CO_2$ is released by malic enzyme or PEP carboxykinase. As labeling experiments show, the free $CO_2$ released in the bundle-sheath cells is the same $CO_2$ molecule originally fixed into oxaloacetate in the mesophyll cells. This $CO_2$ is now fixed again, this time by rubisco, in exactly the same reaction that occurs in $C_3$ plants: incorporation of $CO_2$ into C-1 of 3-phosphoglycerate.

The pyruvate formed by decarboxylation of malate in bundle-sheath cells is transferred back to the mesophyll cells, where it is converted to PEP by an unusual enzymatic reaction catalyzed by **pyruvate phosphate dikinase** (Fig. 20–23b). This enzyme is called a dikinase because two different molecules are simultaneously phosphorylated by one molecule of ATP: pyruvate to PEP, and phosphate to pyrophosphate. The pyrophosphate is subsequently hydrolyzed to phosphate, so two high-energy phosphate groups of ATP are used in regenerating PEP. The PEP is now ready to receive another molecule of $CO_2$ in the mesophyll cell.

The PEP carboxylase of mesophyll cells has a high affinity for $HCO_3^-$ (which is favored relative to $CO_2$ in aqueous solution and can fix $CO_2$ more efficiently than can rubisco). Unlike rubisco, it does not use $O_2$ as an alternative substrate, so there is no competition between $CO_2$ and $O_2$. The PEP carboxylase reaction, then, serves to fix and concentrate $CO_2$ in the form of malate. Release of $CO_2$ from malate in the bundle-sheath cells yields a sufficiently high local concentration of $CO_2$ for rubisco to function near its maximal rate, and for suppression of the enzyme's oxygenase activity.

Once $CO_2$ is fixed into 3-phosphoglycerate in the bundle-sheath cells, the other reactions of the Calvin cycle take place exactly as described earlier. Thus in $C_4$ plants, mesophyll cells carry out $CO_2$ assimilation by the $C_4$ pathway and bundle-sheath cells synthesize starch and sucrose by the $C_3$ pathway.

Three enzymes of the $C_4$ pathway are regulated by light, becoming more active in daylight. Malate dehydrogenase is activated by the thioredoxin-dependent reduction mechanism shown in Figure 20–19; PEP carboxylase is activated by phosphorylation of a Ser residue; and pyruvate phosphate dikinase is activated by dephosphorylation. In the latter two cases, the details of how light effects phosphorylation or dephosphorylation are not known.

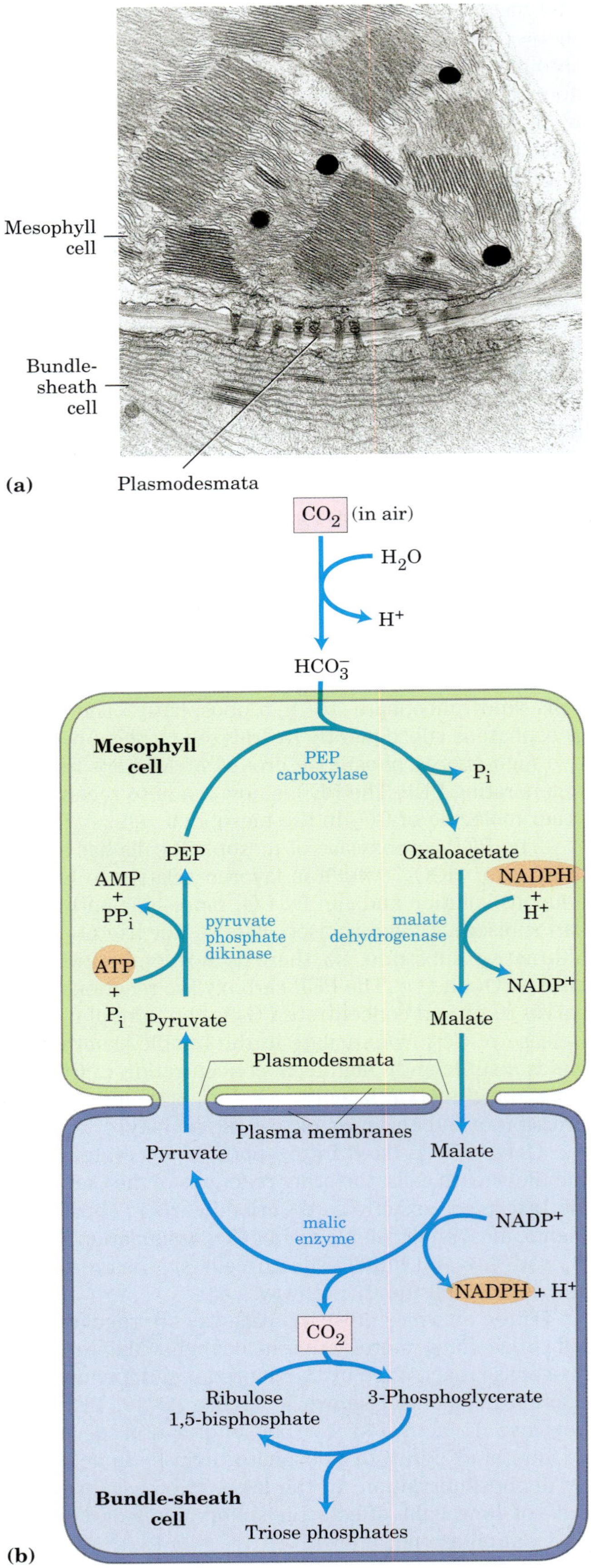

(a)

(b)

The pathway of $CO_2$ assimilation has a greater energy cost in $C_4$ plants than in $C_3$ plants. For each molecule of $CO_2$ assimilated in the $C_4$ pathway, a molecule of PEP must be regenerated at the expense of two high-energy phosphate groups of ATP. Thus $C_4$ plants need five ATP molecules to assimilate one molecule of $CO_2$, whereas $C_3$ plants need only three (nine per triose phosphate). As the temperature increases (and the affinity of rubisco for $CO_2$ decreases, as noted above), a point is reached (at about 28 to 30 °C) at which the gain in efficiency from the elimination of photorespiration more than compensates for this energetic cost. $C_4$ plants (crabgrass, for example) outgrow most $C_3$ plants during the summer, as any experienced gardener can attest.

## In CAM Plants, $CO_2$ Capture and Rubisco Action Are Temporally Separated

Succulent plants such as cactus and pineapple, which are native to very hot, very dry environments, have another variation on photosynthetic $CO_2$ fixation, which reduces loss of water vapor through the pores (stomata) by which $CO_2$ and $O_2$ must enter leaf tissue. Instead of separating the initial trapping of $CO_2$ and its fixation by rubisco across space (as do the $C_4$ plants), they separate these two events over time. At night, when the air is cooler and moister, the stomata open to allow entry of $CO_2$, which is then fixed into oxaloacetate by PEP carboxylase. The oxaloacetate is reduced to malate and stored in the vacuoles, to protect cytosolic and plastid enzymes from the low pH produced by malic acid dissociation. During the day the stomata close, preventing the water loss that would result from high daytime temperatures, and the $CO_2$ trapped overnight in malate is released as $CO_2$ by the NADP-linked malic enzyme. This $CO_2$ is now assimilated by the action of rubisco and the Calvin cycle enzymes. Because this method of $CO_2$ fixation was first discovered in stonecrops, perennial flowering plants of the family *Crassulaceae*, it is called *crassulacean acid metabolism*, and the plants are called **CAM plants.**

**FIGURE 20–23 Carbon assimilation in $C_4$ plants.** The $C_4$ pathway, involving mesophyll cells and bundle-sheath cells, predominates in plants of tropical origin. **(a)** Electron micrograph showing chloroplasts of adjacent mesophyll and bundle-sheath cells. The bundle-sheath cell contains starch granules. Plasmodesmata connecting the two cells are visible. **(b)** The $C_4$ pathway of $CO_2$ assimilation, which occurs through a four-carbon intermediate.

**SUMMARY 20.2** Photorespiration and the C$_4$ and CAM Pathways

- When rubisco uses O$_2$ rather than CO$_2$ as substrate, the 2-phosphoglycolate so formed is disposed of in an oxygen-dependent pathway. The result is increased consumption of O$_2$—photorespiration or, more accurately, the oxidative photosynthetic carbon cycle or C$_2$ cycle. The 2-phosphoglycolate is converted to glyoxylate, to glycine, and then to serine in a pathway that involves enzymes in the chloroplast stroma, the peroxisome, and the mitochondrion.

- In C$_4$ plants, the carbon-assimilation pathway minimizes photorespiration: CO$_2$ is first fixed in mesophyll cells into a four-carbon compound, which passes into bundle-sheath cells and releases CO$_2$ in high concentrations. The released CO$_2$ is fixed by rubisco, and the remaining reactions of the Calvin cycle occur as in C$_3$ plants.

- In CAM plants, CO$_2$ is fixed into malate in the dark and stored in vacuoles until daylight, when the stomata are closed (minimizing water loss) and malate serves as a source of CO$_2$ for rubisco.

## 20.3 Biosynthesis of Starch and Sucrose

During active photosynthesis in bright light, a plant leaf produces more carbohydrate (as triose phosphates) than it needs for generating energy or synthesizing precursors. The excess is converted to sucrose and transported to other parts of the plant, to be used as fuel or stored. In most plants, starch is the main storage form, but in a few plants, such as sugar beet and sugarcane, sucrose is the primary storage form. The synthesis of sucrose and starch occurs in different cellular compartments (cytosol and plastids, respectively), and these processes are coordinated by a variety of regulatory mechanisms that respond to changes in light level and photosynthetic rate.

### ADP-Glucose Is the Substrate for Starch Synthesis in Plant Plastids and for Glycogen Synthesis in Bacteria

Starch, like glycogen, is a high molecular weight polymer of D-glucose in ($\alpha1\rightarrow4$) linkage. It is synthesized in chloroplasts for temporary storage as one of the stable end products of photosynthesis, and for long-term storage it is synthesized in the amyloplasts of the nonphotosynthetic parts of plants—seeds, roots, and tubers (underground stems).

The mechanism of glucose activation in starch synthesis is similar to that in glycogen synthesis. An activated **nucleotide sugar,** in this case **ADP-glucose,** is formed by condensation of glucose 1-phosphate with ATP in a reaction made essentially irreversible by the presence in plastids of inorganic pyrophosphatase (p. 502). **Starch synthase** then transfers glucose residues from ADP-glucose to preexisting starch molecules. Although it has generally been assumed that glucose is added to the *nonreducing* end of starch, as in glycogen synthesis (see Fig. 15–8), evidence now suggests that starch synthase has two equivalent active sites that alternate in inserting a glucosyl residue onto the *reducing* end of the growing chain. This end remains covalently attached to the enzyme, first at one active site, then at the other (Fig. 20–24). Attachment to one active site effectively activates the reducing end of the growing chain for nucleophilic displacement of the enzyme by the attacking C-4 hydroxyl of a glucosyl moiety bound to the other active site, forming the ($\alpha1\rightarrow4$) linkage characteristic of starch.

The amylose of starch is unbranched, but amylopectin has numerous ($\alpha1\rightarrow6$)-linked branches (see Fig. 7–15). Chloroplasts contain a branching enzyme, similar to glycogen-branching enzyme (see Fig. 15–9), that introduces the ($\alpha1\rightarrow6$) branches of amylopectin. Taking into account the hydrolysis by inorganic pyrophosphatase of the PP$_i$ produced during ADP-glucose synthesis, the overall reaction for starch formation from glucose 1-phosphate is

$$\text{Starch}_n + \text{glucose 1-phosphate} + \text{ATP} \longrightarrow$$
$$\text{starch}_{n+1} + \text{ADP} + 2\text{P}_i$$
$$\Delta G'^\circ = -50 \text{ kJ/mol}$$

Starch synthesis is regulated at the level of ADP-glucose formation, as discussed below.

Many types of bacteria store carbohydrate in the form of glycogen (essentially highly branched starch), which they synthesize in a reaction analogous to that catalyzed by glycogen synthase in animals. Bacteria, like plant plastids, use ADP-glucose as the activated form of glucose, whereas animal cells use UDP-glucose. Again, the similarity between plastid and bacterial metabolism is consistent with the endosymbiont hypothesis for the origin of organelles (see Fig. 1–36).

### UDP-Glucose Is the Substrate for Sucrose Synthesis in the Cytosol of Leaf Cells

Most of the triose phosphate generated by CO$_2$ fixation in plants is converted to sucrose (Fig. 20–25) or starch. In the course of evolution, sucrose may have been selected as the transport form of carbon because of its unusual linkage between the anomeric C-1 of glucose and the anomeric C-2 of fructose. This bond is not hydrolyzed by amylases or other common carbohydrate-cleaving

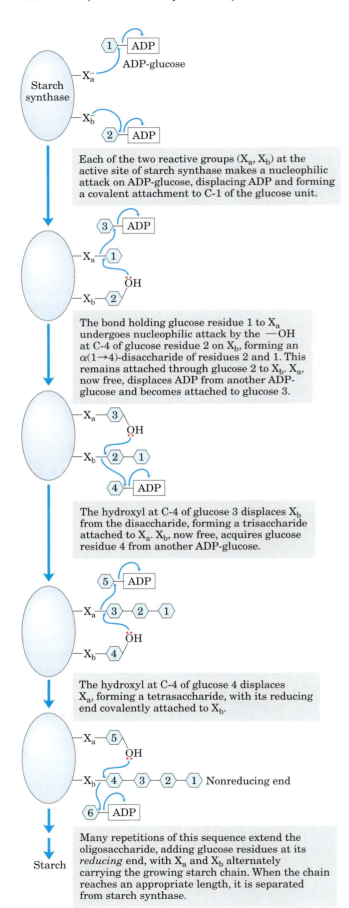

**FIGURE 20–24 Starch synthesis.** Starch synthesis proceeds by a two-site insertion mechanism, with ADP-glucose as the initial glucosyl donor. The two identical active sites on starch synthase alternate in displacing the growing chain from each other, and new glucosyl units are inserted at the reducing end of the growing chain.

Each of the two reactive groups ($X_a$, $X_b$) at the active site of starch synthase makes a nucleophilic attack on ADP-glucose, displacing ADP and forming a covalent attachment to C-1 of the glucose unit.

The bond holding glucose residue 1 to $X_a$ undergoes nucleophilic attack by the —OH at C-4 of glucose residue 2 on $X_b$, forming an $\alpha(1{\rightarrow}4)$-disaccharide of residues 2 and 1. This remains attached through glucose 2 to $X_b$. $X_a$, now free, displaces ADP from another ADP-glucose and becomes attached to glucose 3.

The hydroxyl at C-4 of glucose 3 displaces $X_b$ from the disaccharide, forming a trisaccharide attached to $X_a$. $X_b$, now free, acquires glucose residue 4 from another ADP-glucose.

The hydroxyl at C-4 of glucose 4 displaces $X_a$, forming a tetrasaccharide, with its reducing end covalently attached to $X_b$.

Many repetitions of this sequence extend the oligosaccharide, adding glucose residues at its *reducing* end, with $X_a$ and $X_b$ alternately carrying the growing starch chain. When the chain reaches an appropriate length, it is separated from starch synthase.

enzymes, and the unavailability of the anomeric carbons prevents sucrose from reacting nonenzymatically (as does glucose) with amino acids and proteins.

Sucrose is synthesized in the cytosol, beginning with dihydroxyacetone phosphate and glyceraldehyde 3-phosphate exported from the chloroplast. After condensation of two triose phosphates to form fructose 1,6-bisphosphate (catalyzed by aldolase), hydrolysis by fructose 1,6-bisphosphatase yields fructose 6-phosphate. **Sucrose 6-phosphate synthase** then catalyzes the reaction of fructose 6-phosphate with **UDP-glucose** to form **sucrose 6-phosphate** (Fig. 20–25). Finally, **sucrose 6-phosphate phosphatase** removes the phosphate group, making sucrose available for export to other tissues. The reaction catalyzed by sucrose 6-phosphate synthase is a low-energy process ($\Delta G'^\circ = -5.7$ kJ/mol), but the hydrolysis of sucrose 6-phosphate to sucrose is sufficiently exergonic ($\Delta G'^\circ = -16.5$ kJ/mol) to make the overall synthesis of sucrose essentially irreversible. Sucrose synthesis is regulated and closely coordinated with starch synthesis, as we shall see.

One remarkable difference between the cells of plants and animals is the absence in the plant cell cytosol of the enzyme inorganic pyrophosphatase, which catalyzes the reaction

$$PP_i + H_2O \longrightarrow 2P_i \qquad \Delta G'^\circ = -19.2 \text{ kJ/mol}$$

For many biosynthetic reactions that liberate $PP_i$, pyrophosphatase activity makes the process more favorable energetically, tending to make these reactions irreversible. In plants, this enzyme is present in plastids but absent from the cytosol. As a result, the cytosol of leaf cells contains a substantial concentration of $PP_i$—enough (~0.3 mM) to make reactions such as that catalyzed by UDP-glucose pyrophosphorylase (Fig. 15–7) readily reversible. Recall from Chapter 14 (p. 527) that the cytosolic isozyme of phosphofructokinase in plants uses $PP_i$, not ATP, as the phosphoryl donor.

## Conversion of Triose Phosphates to Sucrose and Starch Is Tightly Regulated

Triose phosphates produced by the Calvin cycle in bright sunlight, as we have noted, may be stored temporarily in the chloroplast as starch, or converted to sucrose and exported to nonphotosynthetic parts of the plant, or both. The balance between the two processes is tightly regulated, and both must be coordinated with the rate of carbon fixation. Five-sixths of the triose

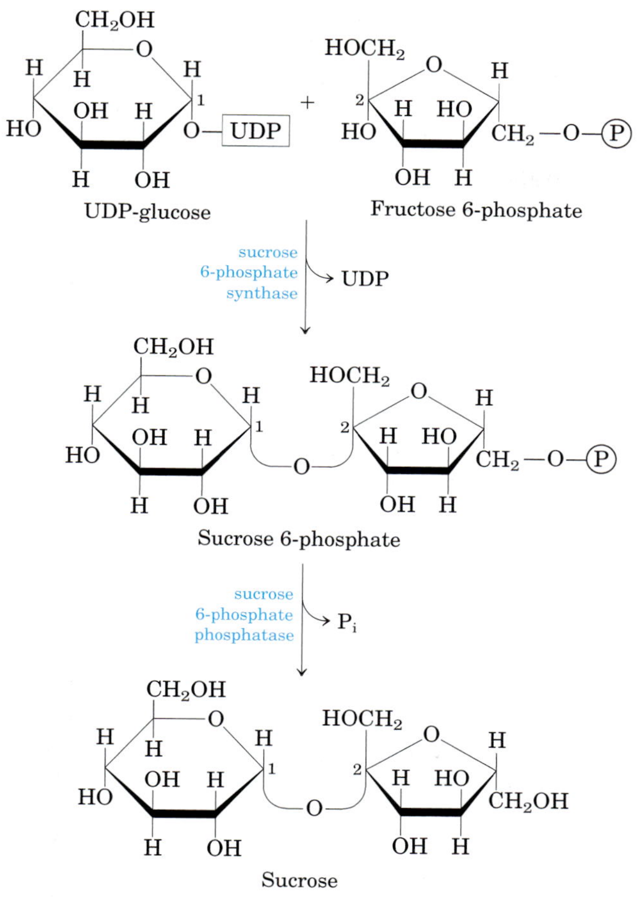

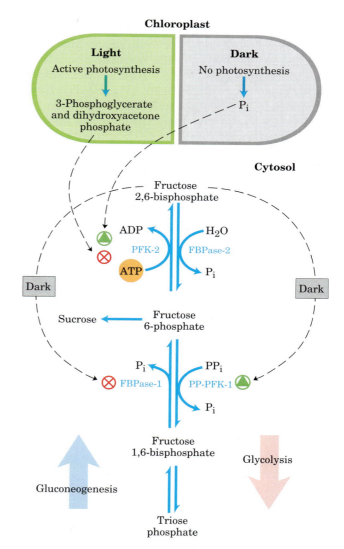

**FIGURE 20–25 Sucrose synthesis.** Sucrose is synthesized from UDP-glucose and fructose 6-phosphate, which are synthesized from triose phosphates in the plant cell cytosol by pathways shown in Figures 15–7 and 20–9. The sucrose 6-phosphate synthase of most plant species is allosterically regulated by glucose 6-phosphate and $P_i$.

phosphate formed in the Calvin cycle must be recycled to ribulose 1,5-bisphosphate (Fig. 20–14); if more than one-sixth of the triose phosphate is drawn out of the cycle to make sucrose and starch, the cycle will slow or stop. However, *insufficient* conversion of triose phosphate to starch or sucrose would tie up phosphate, leaving a chloroplast deficient in $P_i$, which is also essential for operation of the Calvin cycle.

The flow of triose phosphates into sucrose is regulated by the activity of fructose 1,6-bisphosphatase (FBPase-1) and the enzyme that effectively reverses its action, $PP_i$-dependent phosphofructokinase (PP-PFK-1; p. 527). These enzymes are therefore critical points for determining the fate of triose phosphates produced by photosynthesis. Both enzymes are regulated by **fructose 2,6-bisphosphate (F2,6BP)**, which inhibits FBPase-1 and stimulates PP-PFK-1. In vascular plants, the concentration of F2,6BP varies inversely with the rate of photosynthesis (Fig. 20–26). Phosphofructokinase-2,

responsible for F2,6BP synthesis, is inhibited by dihydroxyacetone phosphate or 3-phosphoglycerate and stimulated by fructose 6-phosphate and $P_i$. During active photosynthesis, dihydroxyacetone phosphate is produced and $P_i$ is consumed, resulting in inhibition of PFK-2 and lowered concentrations of F2,6BP. This

**FIGURE 20–26 Fructose 2,6-bisphosphate as regulator of sucrose synthesis.** The concentration of the allosteric regulator fructose 2,6-bisphosphate in plant cells is regulated by the products of photosynthetic carbon assimilation and by $P_i$. Dihydroxyacetone phosphate and 3-phosphoglycerate produced by $CO_2$ assimilation inhibit phosphofructokinase-2 (PFK-2), the enzyme that synthesizes the regulator; $P_i$ stimulates PFK-2. The concentration of the regulator is therefore inversely proportional to the rate of photosynthesis. In the dark, the concentration of fructose 2,6-bisphosphate increases and stimulates the glycolytic enzyme $PP_i$-dependent phosphofructokinase-1 (PP-PFK-1), while inhibiting the gluconeogenic enzyme fructose 1,6-bisphosphatase (FBPase-1). When photosynthesis is active (in the light), the concentration of the regulator drops and the synthesis of fructose 6-phosphate and sucrose is favored.

favors greater flux of triose phosphate into fructose 6-phosphate formation and sucrose synthesis. With this regulatory system, sucrose synthesis occurs when the level of triose phosphate produced by the Calvin cycle exceeds that needed to maintain the operation of the cycle.

Sucrose synthesis is also regulated at the level of sucrose 6-phosphate synthase, which is allosterically activated by glucose 6-phosphate and inhibited by $P_i$. This enzyme is further regulated by phosphorylation and dephosphorylation; a protein kinase phosphorylates the enzyme on a specific Ser residue, making it less active, and a phosphatase reverses this inactivation by removing the phosphate (Fig. 20–27). Inhibition of the kinase by glucose 6-phosphate, and of the phosphatase by $P_i$, strengthens the effects of these two compounds on sucrose synthesis. When hexose phosphates are abundant, sucrose 6-phosphate synthase is activated by glucose 6-phosphate; when $P_i$ is elevated (as when photosynthesis is slow), sucrose synthesis is slowed. During active photosynthesis, triose phosphates are converted to fructose 6-phosphate, which is rapidly equilibrated with glucose 6-phosphate by

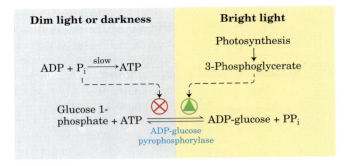

**FIGURE 20–28 Regulation of ADP-glucose phosphorylase by 3-phosphoglycerate and $P_i$.** This enzyme, which produces the precursor for starch synthesis, is rate-limiting in starch production. The enzyme is stimulated allosterically by 3-phosphoglycerate (3-PGA) and inhibited by $P_i$; in effect, the ratio [3-PGA]/[$P_i$], which rises with increasing rates of photosynthesis, controls starch synthesis at this step.

phosphohexose isomerase. Because the equilibrium lies far toward glucose 6-phosphate, as soon as fructose 6-phosphate accumulates, the level of glucose 6-phosphate rises and sucrose synthesis is stimulated.

The key regulatory enzyme in starch synthesis is **ADP-glucose pyrophosphorylase** (Fig. 20–28); it is activated by 3-phosphoglycerate (which accumulates during active photosynthesis) and inhibited by $P_i$ (which accumulates when light-driven condensation of ADP and $P_i$ slows). When sucrose synthesis slows, 3-phosphoglycerate formed by $CO_2$ fixation accumulates, activating this enzyme and stimulating the synthesis of starch.

## SUMMARY 20.3 Biosynthesis of Starch and Sucrose

- Starch synthase in chloroplasts and amyloplasts catalyzes the addition of single glucose residues, donated by ADP-glucose, to the reducing end of a starch molecule by a two-step insertion mechanism. Branches in amylopectin are introduced by a second enzyme.

- Sucrose is synthesized in the cytosol in two steps from UDP-glucose and fructose 1-phosphate.

- The partitioning of triose phosphates between sucrose synthesis and starch synthesis is regulated by fructose 2,6-bisphosphate (F2,6BP), an allosteric effector of the enzymes that determine the level of fructose 6-phosphate. F2,6BP concentration varies inversely with the rate of photosynthesis, and F2,6BP inhibits the synthesis of fructose 6-phosphate, the precursor to sucrose.

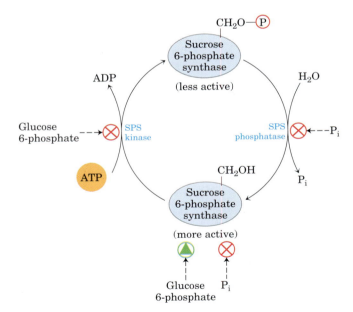

**FIGURE 20–27 Regulation of sucrose phosphate synthase by phosphorylation.** A protein kinase (SPS kinase) specific for sucrose phosphate synthase (SPS) phosphorylates a Ser residue in SPS, inactivating it; a specific phosphatase (SPS phosphatase) reverses this inhibition. The kinase is inhibited allosterically by glucose 6-phosphate, which also activates SPS allosterically. The phosphatase is inhibited by Pi, which also inhibits SPS directly. Thus when the concentration of glucose 6-phosphate is high as a result of active photosynthesis, SPS is activated and produces sucrose phosphate. A high $P_i$ concentration, which occurs when photosynthetic conversion of ADP to ATP is slow, inhibits sucrose phosphate synthesis.

# 20.4 Synthesis of Cell Wall Polysaccharides: Plant Cellulose and Bacterial Peptidoglycan

Cellulose is a major constituent of plant cell walls, providing strength and rigidity and preventing the swelling of the cell and rupture of the plasma membrane that might result when osmotic conditions favor water entry into the cell. Each year, worldwide, plants synthesize more than $10^{11}$ metric tons of cellulose, making this simple polymer one of the most abundant compounds in the biosphere. The structure of cellulose is simple: linear polymers of thousands of $\beta(1{\rightarrow}4)$-linked D-glucose units, assembled into bundles of about 36 chains, which aggregate side by side to form a microfibril (Fig. 20–29).

The biosynthesis of cellulose is less well understood than that of glycogen or starch. As a major component of the plant cell wall, cellulose must be synthesized from intracellular precursors but deposited and assembled outside the plasma membrane. The enzymatic machinery for initiation, elongation, and export of cellulose chains is more complicated than that needed to synthesize starch or glycogen (which are not exported). Bacteria face a similar set of problems when they synthesize the complex polysaccharides that make up their cell walls, and they may employ some of the same mechanisms to solve these problems.

## Cellulose Is Synthesized by Supramolecular Structures in the Plasma Membrane

The complex enzymatic machinery that assembles cellulose chains spans the plasma membrane, with one part positioned to bind the substrate, UDP-glucose, in the cytosol and another part extending to the outside, responsible for elongating and crystallizing cellulose molecules in the extracellular space. Freeze-fracture electron microscopy shows these **terminal complexes**, also called **rosettes,** to be composed of six large particles arranged in a regular hexagon (Fig. 20–30). Several proteins, including the catalytic subunit of **cellulose synthase,** make up the terminal complex. Cellulose synthase has not been isolated in its active form, but its amino acid sequence has been determined from the nucleotide sequence of the gene that encodes it. From the primary structure we can use hydropathy plots (see Fig. 11–11) to deduce that the enzyme has eight transmembrane segments, connected by short loops on the outside, and several longer loops exposed to the cytosol. Much of the recent progress in understanding cellulose synthesis stems from genetic and molecular genetic studies of the plant *Arabidopsis thaliana*, which is especially amenable to genetic dissection and whose genome has been sequenced.

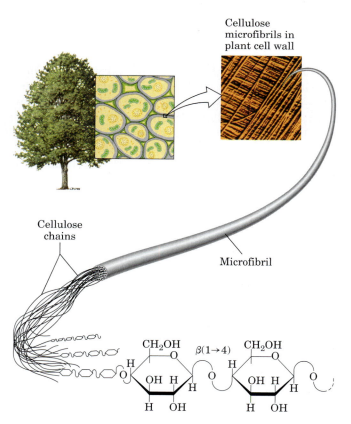

Cellulose microfibrils in plant cell wall

Cellulose chains

Microfibril

$CH_2OH$  $\beta(1{\rightarrow}4)$  $CH_2OH$

**FIGURE 20–29 Cellulose structure.** The plant cell wall is made up in part of cellulose molecules arranged side by side to form paracrystalline arrays—cellulose microfibrils. Many microfibrils combine to form a cellulose fiber, seen in the scanning electron microscope as a structure 5 to 12 nm in diameter, laid down on the cell surface in several layers distinguishable by the different orientations of their fibers.

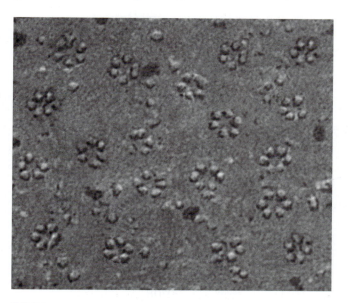

**FIGURE 20–30 Rosettes.** The outside surface of the plant plasma membrane in a freeze-fractured sample, viewed here with electron microscopy, contains many hexagonal arrays of particles about 10 nm in diameter, believed to be composed of cellulose synthase molecules and associated enzymes.

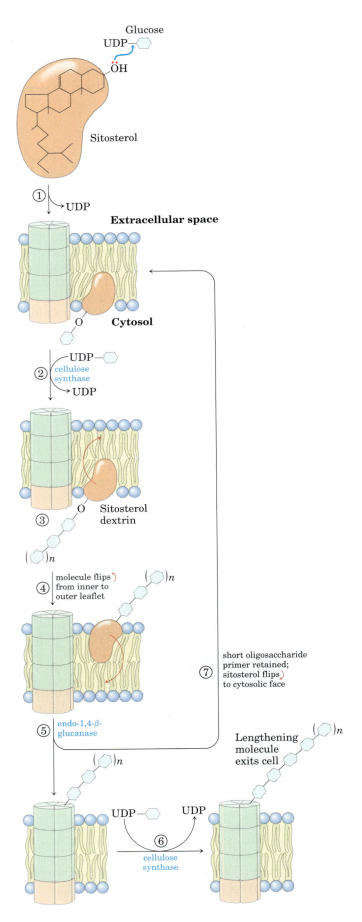

New cellulose chains appear to be initiated by the formation of a lipid-linked intermediate unlike anything involved in starch or glycogen synthesis. Glucose is transferred from UDP-glucose to a membrane lipid, probably the plant sterol sitosterol (Fig. 20–31), on the inner face of the plasma membrane. Here, intracellular cellulose synthase adds several more glucose residues to the first one, in ($\beta 1 \rightarrow 4$) linkage, forming a short oligosaccharide chain attached to the sitosterol (sitosterol dextrin). Next, the whole sitosterol dextrin flips across to the outer face of the plasma membrane, where most of the polysaccharide chain is removed by endo-1,4-$\beta$-glucanase. The shortened sitosterol dextrin primer now associates, perhaps covalently, with another form of cellulose synthase. Presumably this entire process occurs in the rosettes. Whether each of the 36 cellulose chains is initiated on its own lipid primer, or the primer recycles to start a number of chains, is not yet clear. In either case, the second form of cellulose synthase extends the polymer to 500 to 15,000 glucose units, extruding it onto the outer surface of the cell. The action of the enzyme is processive: one enzyme molecule adds many glucose units before releasing the growing cellulose chain. The direction of chain growth (whether addition occurs at the reducing end or at the nonreducing end) has not been established.

The finished cellulose is in the form of crystalline microfibrils (Fig. 20–29), each consisting of 36 separate cellulose chains lying side by side, all with the same (parallel) orientation of nonreducing and reducing ends. It seems likely that each particle in the rosette synthesizes six separate cellulose chains simultaneously and in parallel with the chains made by the other five particles, so that 36 polymers arrive together on the outer surface of the cell, already aligned and ready to crystallize as a microfibril of the cell wall. When the 36 polymers reach some critical length, their synthesis is terminated by an unknown mechanism; crystallization into a microfibril follows.

In addition to its catalytic subunit, cellulose synthase may have subunits that mediate extrusion of the polysaccharide chain (the pore subunit) and crystallization of the polysaccharide chains outside the cell

**FIGURE 20–31 Lipid primer for cellulose synthesis.** This proposed pathway begins with ① the transfer of a glucosyl residue from UDP-glucose to a lipid "primer" (probably sitosterol) in the inner leaflet of the plasma membrane. After this initiation, ② the chain of carbohydrate is elongated by transfer of glucosyl residues from UDP-glucose, until ③ a critical length of oligosaccharide is reached. ④ The sitosterol with its attached oligosaccharide now flips from the inner leaflet to the outer leaflet. ⑤ An endo-1,4-$\beta$-glucanase separates the growing chain from a short oligonucleotide still attached to the lipid. As it is pushed out of the cell, ⑥ the lipid-free polymer of glucosyl residues (the glucan acceptor) is further extended by the addition of glucosyl residues from UDP-glucose, catalyzed by cellulose synthase. ⑦ The lipid-linked oligosaccharide returns to serve as the primer for another chain of cellulose.

(the crystallization subunit). The potent herbicide CGA 325′615, which specifically inhibits cellulose synthesis, causes rosettes to fall apart; the small amount of cellulose still synthesized remains tightly, perhaps covalently, bound to the catalytic subunit of cellulose synthase. The inhibitor may act by dissociating the catalytic subunit from the pore and crystallization subunits, preventing the later stages of cellulose synthesis.

The UDP-glucose used for cellulose synthesis is generated from sucrose produced during photosynthesis, by the reaction catalyzed by sucrose synthase (named for the reverse reaction):

$$\text{Sucrose} + \text{UDP} \longrightarrow \text{UDP-glucose} + \text{fructose}$$

In one proposed model, cellulose synthase spans the plasma membrane and uses cytosolic UDP-glucose as the precursor for extracellular cellulose synthesis. In another, a membrane-bound form of sucrose synthase forms a complex with cellulose synthase, feeding UDP-glucose from sucrose directly into cell wall synthesis (Fig. 20–32).

In the activated precursor of cellulose (UDP-glucose), the glucose is $\alpha$-linked to the nucleotide, but in the product (cellulose), glucose residues are $\beta(1\rightarrow4)$-linked, so there is an inversion of configuration at the anomeric carbon (C-1) as the glycosidic bond forms. Glycosyltransferases that invert configuration are generally assumed to use a single-displacement mechanism, with nucleophilic attack by the acceptor species at the anomeric carbon of the donor sugar (UDP-glucose).

Certain bacteria (*Acetobacter*, *Agrobacteria*, *Rhizobia*, and *Sarcina*) and many simple eukaryotes also carry out cellulose synthesis, apparently by a mechanism similar to that in plants. If the bacteria use a membrane lipid to initiate new chains, it cannot be a sterol—bacteria do not contain sterols.

## Lipid-Linked Oligosaccharides Are Precursors for Bacterial Cell Wall Synthesis

Like plants, many bacteria have thick, rigid extracellular walls that protect them from osmotic lysis. The **peptidoglycan** that gives bacterial envelopes their strength and rigidity is an alternating linear copolymer of *N*-acetylglucosamine (GlcNAc) and *N*-acetylmuramic acid (Mur2Ac), linked by $(\beta1\rightarrow4)$ glycosidic bonds and cross-linked by short peptides attached to the Mur2Ac (Fig. 20–33). During assembly of the polysaccharide backbone of this complex macromolecule, both GlcNAc and Mur2Ac are activated by attachment of a uridine

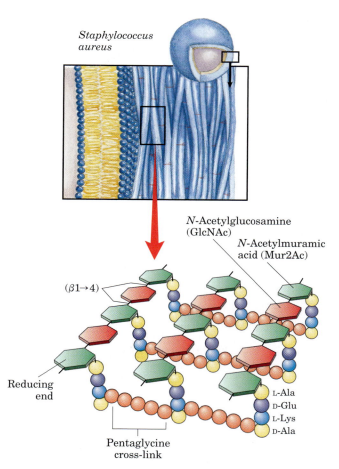

**FIGURE 20–33 Peptidoglycan structure.** This is the peptidoglycan of the cell wall of *Staphylococcus aureus*, a gram-positive bacterium. Peptides (strings of colored spheres) covalently link *N*-acetylmuramic acid residues in neighboring polysaccharide chains. Note the mixture of L and D amino acids in the peptides. Gram-positive bacteria such as *S. aureus* have a pentaglycine chain in the cross-link. Gram-negative bacteria, such as *E. coli*, lack the pentaglycine; instead, the terminal D-Ala residue of one tetrapeptide is attached directly to a neighboring tetrapeptide through either L-Lys or a lysine-like amino acid, diaminopimelic acid.

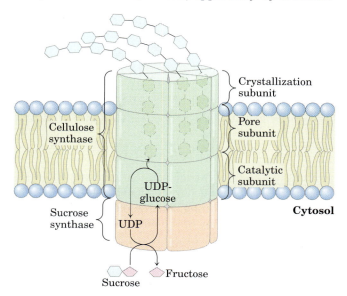

**FIGURE 20–32 A plausible model for the structure of cellulose synthase.** The enzyme complex includes a catalytic subunit with eight transmembrane segments and several other subunits that are presumed to act in threading cellulose chains through the catalytic site and out of the cell, and in the crystallization of 36 cellulose strands into the paracrystalline microfibrils shown in Figure 20–29.

**FIGURE 20–34 Synthesis of bacterial peptidoglycan.**
In the early steps of this pathway (① through ④),
*N*-acetylglucosamine (GlcNAc) and *N*-acetylmuramic
acid (Mur2Ac) are activated by attachment of a uridine
nucleotide (UDP) to their anomeric carbons and, in
the case of Mur2Ac, of a long-chain isoprenyl alcohol
(dolichol) through a phosphodiester bond. These activating groups participate in the formation of glycosidic
linkages; they serve as excellent leaving groups. After
⑤, ⑥ assembly of a disaccharide with a peptide side
chain (10 amino acid residues), ⑦ this precursor is
transferred to the nonreducing end of an existing peptidoglycan chain, which serves as a primer for the
polymerization reaction. Finally, ⑧ in a transpeptidation reaction between the peptide side chains on two
different peptidoglycan molecules, a Gly residue at the
end of one chain displaces a terminal D-Ala in the
other chain, forming a cross-link. This transpeptidation
reaction is inhibited by the penicillins, which kill
bacteria by weakening their cell walls.

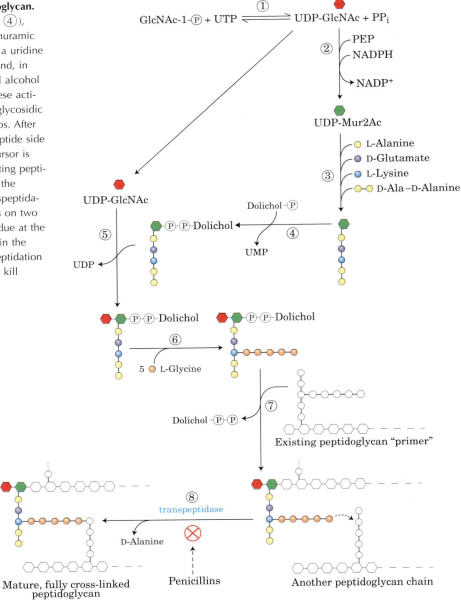

nucleotide at their anomeric carbons. First, GlcNAc 1-
phosphate condenses with UTP to form UDP-GlcNAc
(Fig. 20–34, step ①), which reacts with phosphoenolpyruvate to form UDP-Mur2Ac (step ②); five amino
acids are then added (step ③). The Mur2Ac-pentapep-
tide moiety is transferred from the uridine nucleotide
to the membrane lipid dolichol, a long-chain isoprenoid
alcohol (see Fig. 10–22f) (step ④), and a GlcNAc
residue is donated by UDP-GlcNAc (step ⑤). In many
bacteria, five glycines are added in peptide linkage to
the amino group of the Lys residue of the pentapeptide
(step ⑥). Finally, this disaccharide decapeptide is added
to the nonreducing end of an existing peptidoglycan

molecule (step ⑦). A transpeptidation reaction cross-
links adjacent polysaccharide chains (step ⑧), contributing to a huge, strong, macromolecular wall around
the bacterial cell. Many of the most effective antibiotics
in use today act by inhibiting reactions in the synthesis
of the peptidoglycan (Box 20–1).

Many other oligosaccharides and polysaccharides
are synthesized by similar routes in which sugars are activated for subsequent reactions by attachment to nucleotides. In the glycosylation of proteins, for example
(see Fig. 27–34), the precursors of the carbohydrate
moieties include sugar nucleotides and lipid-linked
oligosaccharides.

BOX 20–1 BIOCHEMISTRY IN MEDICINE

## The Magic Bullet versus the Bulletproof Vest: Penicillin and β-Lactamase

Because peptidoglycans are unique to bacterial cell walls, with no known homologous structures in mammals, the enzymes responsible for their synthesis are ideal targets for antibiotic action. Antibiotics that hit specific bacterial targets are sometimes called "magic bullets." Penicillin and its many synthetic analogs have been used to treat bacterial infections since these drugs came into wide application in World War II.

Penicillins and related antibiotics contain the β-lactam ring (Fig. 1), variously modified. All penicillins have a thiazolidine ring attached to the β-lactam, but they differ in the substituent at position 6, which accounts for the different pharmacological properties of the penicillins. For example, penicillin V is acid stable and can be administered orally, but methicillin is acid labile and must be given intravenously or intramuscularly. However, methicillin resists breakdown by bacterial enzymes (β-lactamases) whereas many other penicillins do not. The β-lactams have many of the properties that make a good drug. First, they target a metabolic pathway present in bacteria but not in people. Second, they have half-lives in the body long enough to be clinically useful. Third, they reach thera-

peutic concentrations in most, if not all, tissues and organs. Finally, they are effective against a broad range of bacterial species.

Penicillins block formation of the peptide cross-links in peptidoglycans, acting as mechanism-based (suicide) inhibitors. The normal catalytic mechanism of the target enzyme activates the inhibitor, which then covalently modifies a critical residue in the active site. Transpeptidases employ a reaction mechanism (involving Ser residues) similar to that of chymotrypsin (see Fig. 6–21); the reaction activates β-lactams such as penicillin, which in turn inactivate the transpeptidases. After penicillin enters the transpeptidase active site, the proton on the hydroxyl group of an active-site Ser residue is abstracted to the nitrogen of the β-lactam ring, and the activated oxygen of the Ser hydroxyl attacks the carbonyl carbon at position 7 of the β-lactam, opening the ring and forming a stable penicilloyl-enzyme derivative that inactivates the enzyme (Fig. 2a).

Widespread use of antibiotics has driven the selection and evolution of antibiotic resistance in many pathogenic bacteria. The most important mechanism of resistance is inactivation of the antibiotic by enzymatic hydrolysis of the lactam ring, catalyzed by bacterial

(continued on next page)

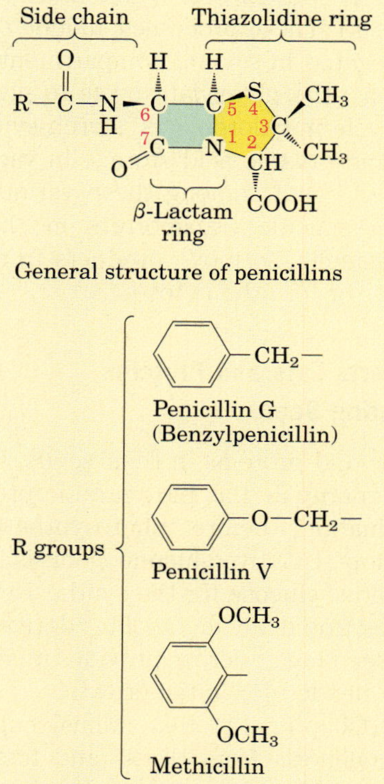

FIGURE 1

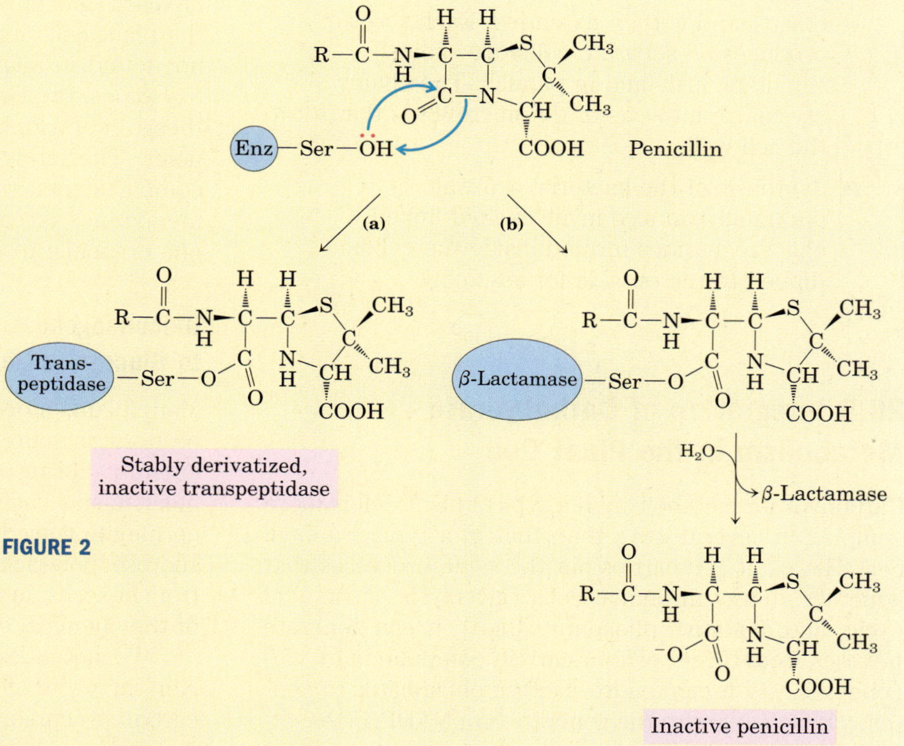

FIGURE 2

BOX 20–1   **BIOCHEMISTRY IN MEDICINE** *(continued from previous page)*

β-lactamases, which provide bacteria with a bullet-proof vest (Fig. 2b). A β-lactamase forms a temporary covalent adduct with the carboxyl group of the opened β-lactam ring, which is immediately hydrolyzed, regenerating active enzyme. One approach to circumventing antibiotic resistance of this type is to synthesize penicillin analogs, such as methicillin, that are poor substrates for β-lactamases. Another approach is to administer along with antibiotics a β-lactamase inhibitor such as clavulanate or sulbactam.

Antibiotic resistance is a significant threat to public health. Some bacterial infections are now essen-tially untreatable with antibiotics. By the early 1990s, 20% to 40% of *Staphylococcus aureus* (the causative agent of "staph" infections) was resistant to methicillin, and 32% of *Neisseria gonorrhoeae* (the causative agent of gonorrhea) was resistant to penicillin. By 1986, 32% of *Shigella* (a pathogen responsible for se-vere forms of dysentery, some with a lethality of up to 15%) was resistant to ampicillin. Significantly, many of these pathogens are also resistant to many other antibiotics. In the future, we will need to develop new drugs that circumvent the bacterial resistance mech-anisms or that act on different bacterial targets.

## SUMMARY 20.4   Synthesis of Cell Wall Polysaccharides: Plant Cellulose and Bacterial Peptidoglycan

- Cellulose synthesis takes place in terminal complexes (rosettes) in the plasma membrane. Each cellulose chain begins as a sitosterol dextrin formed inside the cell. It then flips to the outside, where the oligosaccharide portion is transferred to cellulose synthase in the rosette and is then extended. Each rosette produces 36 separate cellulose chains simultaneously and in parallel. The chains crystallize into one of the microfibrils that form the cell wall.

- Synthesis of the bacterial cell wall peptidoglycan also involves lipid-linked oligosaccharides formed inside the cell and flipped to the outside for assembly.

## 20.5 Integration of Carbohydrate Metabolism in the Plant Cell

Carbohydrate metabolism in a typical plant cell is more complex in several ways than that in a typical animal cell. The plant cell carries out the same processes that generate energy in animal cells (glycolysis, citric acid cycle, and oxidative phosphorylation); it can generate hexoses from three- or four-carbon compounds by glu-coneogenesis; it can oxidize hexose phosphates to pen-tose phosphates with the generation of NADPH (the ox-idative pentose phosphate pathway); and it can produce a polymer of ($\alpha1\rightarrow4$)-linked glucose (starch) and de-grade it to generate hexoses. But besides these carbo-hydrate transformations that it shares with animal cells, the photosynthetic plant cell can fix $CO_2$ into organic compounds (the rubisco reaction); use the products of fixation to generate trioses, hexoses, and pentoses (the Calvin cycle); and convert acetyl-CoA generated from fatty acid breakdown to four-carbon compounds (the glyoxylate cycle) and the four-carbon compounds to hexoses (gluconeogenesis). These processes, unique to the plant cell, are segregated in several compartments not found in animal cells: the glyoxylate cycle in gly-oxysomes, the Calvin cycle in chloroplasts, starch syn-thesis in amyloplasts, and organic acid storage in vac-uoles. The integration of events among these various compartments requires specific transporters in the membranes of each organelle, to move products from one organelle to another or into the cytosol.

### Gluconeogenesis Converts Fats and Proteins to Glucose in Germinating Seeds

Many plants store lipids and proteins in their seeds, to be used as sources of energy and as biosynthetic pre-cursors during germination, before photosynthetic mechanisms have developed. Active gluconeogenesis in germinating seeds provides glucose for the synthesis of sucrose, polysaccharides, and many metabolites derived from hexoses. In plant seedlings, sucrose provides much of the chemical energy needed for initial growth.

We noted earlier (Chapter 14) that animal cells can carry out gluconeogenesis from three- and four-carbon precursors, but not from the two acetyl carbons of acetyl-CoA. Because the pyruvate dehydrogenase reaction is effectively irreversible (pp. 602–603), animal cells have no way to convert acetyl-CoA to pyruvate or oxaloacetate. Unlike animals, plants and some microorganisms *can* convert acetyl-CoA derived from fatty acid oxidation to glucose. Some of the enzymes essential to this conversion are sequestered in glyoxysomes, where glyoxysome-specific isozymes of β-oxidation break down fatty acids to acetyl-CoA (see Fig. 16–22). The physical separation of the glyoxylate cycle and β-oxidation enzymes from the mitochondrial citric acid cycle enzymes prevents further oxidation of acetyl-CoA to $CO_2$. Instead, the acetyl-CoA is converted to succinate in the glyoxylate cycle (see Fig. 16–20). The succinate passes into the mitochondrial matrix, where it is converted by citric acid cycle enzymes to oxaloacetate, which moves into the cytosol. Cytosolic oxaloacetate is converted by gluconeogenesis to fructose 6-phosphate, the precursor of sucrose. Thus the integration of reaction sequences in three subcellular compartments is required for the production of fructose 6-phosphate or sucrose from stored lipids. Because only three of the four carbons in each molecule of oxaloacetate are converted to hexose in the cytosol, about 75% of the carbon in the fatty acids stored as seed lipids is converted to carbohydrate by the combined pathways of Figure 20–35. The other 25% is lost as $CO_2$ in the conversion of oxaloacetate to phosphoenolpyruvate. Hydrolysis of storage triacylglycerols also produces glycerol 3-phosphate, which can enter the gluconeogenic pathway after its oxidation to dihydroxyacetone phosphate (Fig. 20–36).

Glucogenic amino acids (see Table 14–4) derived from the breakdown of stored seed proteins also yield precursors for gluconeogenesis, following transamination and oxidation to succinyl-CoA, pyruvate, oxaloacetate, fumarate, and α-ketoglutarate (Chapter 18)—all good starting materials for gluconeogenesis.

## Pools of Common Intermediates Link Pathways in Different Organelles

Although we have described metabolic transformations in plant cells in terms of individual pathways, these pathways interconnect so completely that we should instead consider pools of metabolic intermediates shared among these pathways and connected by readily reversible reactions (Fig. 20–37). One such **metabolite pool** includes the hexose phosphates glucose 1-phosphate, glucose 6-phosphate, and fructose 6-phosphate; a second includes the 5-phosphates of the pentoses ribose, ribulose, and xylulose; a third includes the triose phosphates dihydroxyacetone phosphate and glyceraldehyde 3-phosphate. Metabolite fluxes through these

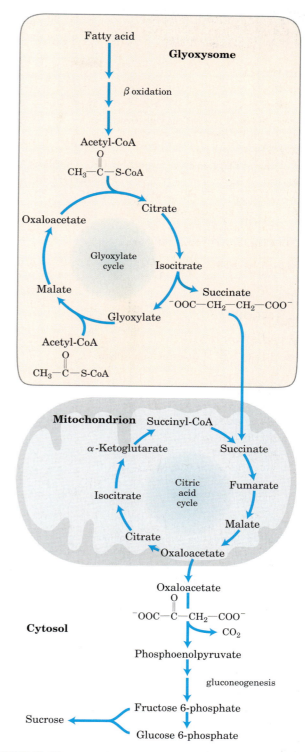

**FIGURE 20–35 Conversion of stored fatty acids to sucrose in germinating seeds.** This pathway begins in glyoxysomes. Succinate is produced and exported to mitochondria, where it is converted to oxaloacetate by enzymes of the citric acid cycle. Oxaloacetate enters the cytosol and serves as the starting material for gluconeogenesis and for the synthesis of sucrose, the transport form of carbon in plants.

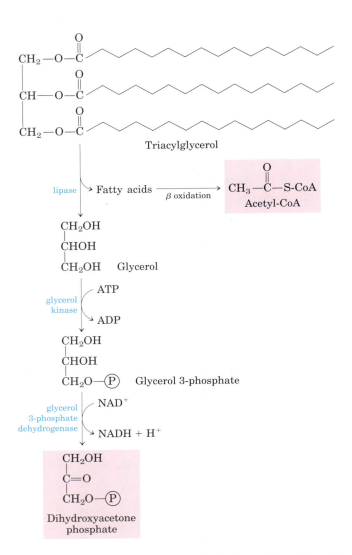

**FIGURE 20-36 Conversion of the glycerol moiety of triacylglycerols to sucrose in germinating seeds.** The glycerol of triacylglycerols is oxidized to dihydroxyacetone phosphate, which enters the gluconeogenic pathway at the triose phosphate isomerase reaction.

pools change in magnitude and direction in response to changes in the circumstances of the plant, and they vary with tissue type. Transporters in the membranes of each organelle move specific compounds in and out, and the regulation of these transporters presumably influences the degree to which the pools mix.

During daylight hours, triose phosphates produced in leaf tissue by the Calvin cycle move out of the chloroplast and into the cytosolic hexose phosphate pool, where they are converted to sucrose for transport to nonphotosynthetic tissues. In these tissues, sucrose is converted to starch for storage or is used as an energy source via glycolysis. In growing plants, hexose phosphates are also withdrawn from the pool for the synthesis of cell walls. At night, starch is metabolized by glycolysis to provide energy, essentially as in nonphotosynthetic organisms, and NADPH and ribose 5-phosphate are obtained through the oxidative pentose phosphate pathway.

## SUMMARY 20.5 Integration of Carbohydrate Metabolism in the Plant Cell

- Plants can synthesize sugars from acetyl-CoA, the product of fatty acid breakdown, by the combined actions of the glyoxylate cycle and gluconeogenesis.

- The individual pathways of carbohydrate metabolism in plants overlap extensively; they share pools of common intermediates, including hexose phosphates, pentose phosphates, and triose phosphates. Transporters in the membranes of chloroplasts, mitochondria, amyloplasts, and peroxisomes mediate the movement of sugar phosphates between organelles. The direction of metabolite flow through the pools changes from day to night.

**FIGURE 20-37 Pools of pentose phosphates, triose phosphates, and hexose phosphates.** The compounds in each pool are readily interconvertible by reactions that have small standard free-energy changes. When one component of the pool is temporarily depleted, a new equilibrium is quickly established to replenish it. Movement of the sugar phosphates between intracellular compartments is limited; specific transporters must be present in an organelle membrane.

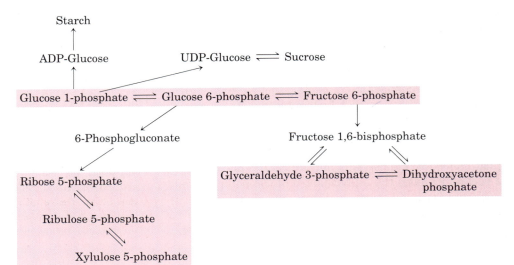

## Key Terms

*Terms in bold are defined in the glossary.*

Calvin cycle   752
**plastids**   752
chloroplast   752
amyloplast   752
**carbon-fixation reaction**   753
ribulose 1,5-bisphosphate   753
3-phosphoglycerate   753
**pentose phosphate pathway**   753
reductive pentose phosphate
   cycle   753
$C_3$ plants   754
**ribulose 1,5-bisphosphate
   carboxylase/oxygenase
   (rubisco)**   754
rubisco activase   757

transaldolase   759
transketolase   759
sedoheptulose 1,7-bisphosphate   760
ribulose 5-phosphate   760
**carbon-assimilation
   reactions**   763
thioredoxin   764
ferredoxin-thioredoxin
   reductase   764
**photorespiration**   766
2-phosphoglycolate   766
**glycolate pathway**   767
oxidative photosynthetic carbon cycle
   ($C_2$ cycle)   769
**$C_4$ plants**   769

phosphoenolpyruvate
   carboxylase   769
malic enzyme   769
pyruvate phosphate dikinase   769
**CAM plants**   770
nucleotide sugars   771
ADP-glucose   771
starch synthase   771
sucrose 6-phosphate synthase   772
fructose 2,6-bisphosphate   773
ADP-glucose pyrophosphorylase   774
cellulose synthase   775
peptidoglycan   777
metabolite pools   781

## Further Reading

### General References

**Blankenship, R.E.** (2002) *Molecular Mechanisms of Photosynthesis,* Blackwell Science, Oxford.
   Very readable, well-illustrated, intermediate-level treatment of all aspects of photosynthesis, including the carbon metabolism covered in this chapter and the light-driven reactions described in Chapter 19.

**Buchanan, B.B., Gruissem, W., & Jones, R.L. (eds)** (2000) *Biochemistry and Molecular Biology of Plants,* American Society of Plant Physiology, Rockville, MD.
   This wonderful book—up to date and authoritative—covers all aspects of plant biochemistry and molecular biology. The following chapters cover carbohydrate synthesis in greater depth: **Malkin, R. & Niyogi, K.,** Chapter 12, Photosynthesis (pp. 568–629); **Dennis, D.T. & Blakeley, S.D.,** Chapter 13, Carbohydrate Metabolism (pp. 630–675); **Siedow, J.N. & Day, D.A.,** Chapter 14, Respiration and Photorespiration (pp. 676–729).

**Heldt, H.-W.** (1997) *Plant Biochemistry and Molecular Biology,* Oxford University Press, Oxford.
   Another excellent textbook of plant biochemistry. Especially useful are Chapter 6, Photosynthetic $CO_2$ Assimilation by the Calvin Cycle; Chapter 7, Photorespiration; and Chapter 9, Polysaccharides.

### Photosynthetic Carbohydrate Synthesis

**Andersson, I., Knight, S., Schneider, G., Lindqvist, Y., Lundqvist, T., Brändén, C.-I., & Lorimer, G.H.** (1989) Crystal structure of the active site of ribulose-bisphosphate carboxylase. *Nature* **337,** 229–234.

**Benson, A.A.** (2002) Following the path of carbon in photosynthesis: a personal story—history of photosynthesis. *Photosynth. Res.* **73,** 31–49.

**Cleland, W.W., Andrews, T.J., Gutteridge, S., Hartman, F.C., & Lorimer, G.H.** (1998) Mechanism of rubisco—the carbamate as general base. *Chem. Rev.* **98,** 549–561.
   Review with a special focus on the carbamate at the active site.

**Dietz, K.J., Link, G., Pistorius, E.K., & Scheibe, R.** (2002) Redox regulation in oxygenic photosynthesis. *Prog. Botany* **63,** 207–245.

**Flügge, U.-I.** (1999) Phosphate translocaters in plastids. *Annu. Rev. Plant Physiol. Plant Mol. Biol.* **50,** 27–45.
   Review of the transporters that carry $P_i$ and various sugar phosphates across plastid membranes.

**Fridlyand, L.E. & Scheibe R.** (1999) Homeostatic regulation upon changes of enzyme activities in the Calvin cycle as an example for general mechanisms of flux control: what can we expect from transgenic plants? *Photosynth. Res.* **61,** 227–239.

**Hartman, F.C. & Harpel, M.R.** (1994) Structure, function, regulation and assembly of D-ribulose-1,5-bisphosphate carboxylase/oxygenase. *Annu. Rev. Biochem.* **63,** 197–234.

**Horecker, B.L.** (2002) The pentose phosphate pathway. *J. Biol. Chem.* **277,** 47,965–47,971.

**Portis, A.R., Jr.** (2003) Rubisco activase: rubisco's catalytic chaperon. *Photosynth. Res.* **75,** 11–27.
   Structure, regulation, mechanism, and importance of rubisco activase.

**Raven, J. A. & Girard-Bascou, J.** (2001) Algal model systems and the elucidation of photosynthetic metabolism. *J. Phycol.* **37,** 943–950.
   Short, intermediate-level review.

**Schneider, G., Lindqvist, Y., Brändén, C.-I., & Lorimer, G.** (1986) Three-dimensional structure of ribulose-1,5-bisphosphate carboxylase/oxygenase from *Rhodospirillum rubrum* at 2.9 Å resolution. *EMBO J.* **5,** 3409–3415.

**Smith, A.M., Denyer, K., & Martin, C.** (1997) The synthesis of the starch granule. *Annu. Rev. Plant Physiol. Plant Mol. Biol.* **48,** 67–87.
   Review of the role of ADP-glucose pyrophosphorylase in the synthesis of amylose and amylopectin in starch granules.

**Spreitzer, R.J. & Salvucci, M.E.** (2002). Rubisco: structure, regulatory interactions, and possibilities for a better enzyme. *Annu. Rev. Plant Biol.* **53,** 449–475.

Advanced review on rubisco and rubisco activase.

**Stitt, M.** (1995) Regulation of metabolism in transgenic plants. *Annu. Rev. Plant Physiol. Plant Mol. Biol.* **46,** 341–368.

Review of the genetic approaches to defining points of regulation in vivo.

**Tabita, F.R.** (1999) Microbial ribulose 1,5-bisphosphate carboxylase/oxygenase: a different perspective. *Photosynth. Res.* **60,** 1–28.

Discussion of biochemical and genetic studies of the microbial rubisco, and comparison with the enzyme from plants.

**Wolosiuk, R., Ballicora, M., & Hagelin, K.** (1993) The reductive pentose phosphate cycle for photosynthetic $CO_2$ assimilation: enzyme modulation. *FASEB J.* **7,** 622–637.

### Photorespiration and the $C_4$ and CAM Pathways

**Douce, R., Bourguignon, J., Neuburger, M., & Rébeillé, F.** (2001) The glycine decarboxylase system: a fascinating complex. *Trends Plant Sci.* **6,** 167–176.

Intermediate-level description of the structure and the reaction mechanism of the enzyme.

**Douce, R. & Neuberger, M.** (1999) Biochemical dissection of photorespiration. *Curr. Opin. Plant Biol.* **2,** 214–222.

**Hatch, M.C.** (1987) $C_4$ photosynthesis: a unique blend of modified biochemistry, anatomy and ultrastructure. *Biochim. Biophys. Acta* **895,** 81–106.

Intermediate-level review by one of the discoverers of the $C_4$ pathway.

**Tolbert, N.E.** (1997) The $C_2$ oxidative photosynthetic carbon cycle. *Annu. Rev. Plant Physiol. Plant Mol. Biol.* **48,** 1–25.

A fascinating personal account of the development of our understanding of photorespiration, by one who was central in this development.

### Biosynthesis of Starch and Sucrose

**Ball, S.G., van de Wal, M.H.B.J., & Visser, R.G.F.** (1998) Progress in understanding the biosynthesis of amylose. *Trends Plant Sci.* **3,** 462–467.

**Delmer, D.P.** (1999) Cellulose biosynthesis: exciting times for a difficult field of study. *Annu. Rev. Plant Physiol. Plant Mol. Biol.* **50,** 245–276.

**Doblin, M.S., Kurek, I., Javob-Wilk, D., & Delmer, D.P.** (2002) Cellulose biosynthesis in plants: from genes to rosettes. *Plant Cell Physiol.* **43,** 1407–1420.

**Errington, J., Daniel, R.A., & Scheffers, D.J.** (2003) Cytokinesis in bacteria. *Microbiol. Mol. Biol. Rev.* **67,** 52–65.

**Fernie, A.R., Willmitzer, L., & Trethewey, R.N.** (2002) Sucrose to starch: a transition in molecular plant physiology. *Trends Plant Sci.* **7,** 35–41.

Intermediate-level review of the genes and proteins involved in starch synthesis in plant tubers.

**Huber, S.C. & Huber, J.L.** (1996) Role and regulation of sucrose-phosphate synthase in higher plants. *Annu. Rev. Plant Physiol. Plant Mol. Biol.* **47,** 431–444.

Short review of factors that regulate this critical enzyme.

**Leloir, L.F.** (1971) Two decades of research on the biosynthesis of saccharides. *Science* **172,** 1299–1303.

Leloir's Nobel address, including a discussion of the role of sugar nucleotides in metabolism.

**Mukerjea, R., Yu, L., & Robyt, J.F.** (2002) Starch biosynthesis: mechanism for the elongation of starch chains. *Carbohydrate Res.* **337,** 1015–1022.

Recent evidence that the starch chain grows at its *reducing* end.

### Synthesis of Cellulose and Peptidoglycan

**Delmer, D.P.** (1999) Cellulose biosynthesis: exciting times for a difficult field of study. *Annu. Rev. Plant Physiol. Plant Mol. Biol.* **50,** 245–276.

**Dhugga, K.S.** (2001) Building the wall: genes and enzyme complexes for polysaccharide synthases. *Curr. Opin. Plant Biol.* **4,** 488–493.

**Peng, L., Kawagoe, Y., Hogan, P., & Delmer, D.** (2002) Sitosterol-$\beta$-glucoside as primer for cellulose synthesis in plants. *Science* **295,** 147–150.

Recent evidence for the lipid-oligosaccharide intermediate in cellulose synthesis.

**Reid, J.S.G.** (2000) Cementing the wall: cell wall polysaccharide synthesising enzymes. *Curr. Opin. Plant Biol.* **3,** 512–516.

**Saxena, I.M. & Brown, R.M., Jr.** (2000) Cellulose synthases and related enzymes. *Curr. Opin. Plant Biol.* **3,** 523–531.

**Williamson, R.E., Burn, J.E., & Hocart, C.H.** (2002) Towards the mechanism of cellulose synthesis. *Trends Plant Sci.* **7,** 461–467.

## Problems

**1. Segregation of Metabolism in Organelles** What are the advantages to the plant cell of having different organelles to carry out different reaction sequences that share intermediates?

**2. Phases of Photosynthesis** When a suspension of green algae is illuminated in the absence of $CO_2$ and then incubated with $^{14}CO_2$ in the dark, $^{14}CO_2$ is converted to [$^{14}C$]glucose for a brief time. What is the significance of this observation with regard to the $CO_2$-assimilation process, and how is it related to the light reactions of photosynthesis? Why does the conversion of $^{14}CO_2$ to [$^{14}C$]glucose stop after a brief time?

**3. Identification of Key Intermediates in $CO_2$ Assimilation** Calvin and his colleagues used the unicellular green alga *Chlorella* to study the carbon-assimilation reactions of photosynthesis. They incubated $^{14}CO_2$ with illuminated suspensions of algae and followed the time course of appearance of $^{14}C$ in two compounds, X and Y, under two sets of conditions. Suggest the identities of X and Y, based on your understanding of the Calvin cycle.

(a) Illuminated *Chlorella* were grown with unlabeled $CO_2$, then the light was turned off and $^{14}CO_2$ was added (vertical dashed line in the graph below). Under these conditions, X was the first compound to become labeled with $^{14}C$; Y was unlabeled.

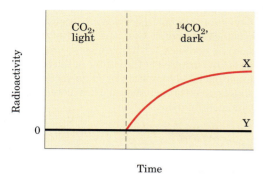

(b) Illuminated *Chlorella* cells were grown with $^{14}CO_2$. Illumination was continued until all the $^{14}CO_2$ had disappeared (vertical dashed line in the graph below). Under these conditions, X became labeled quickly but lost its radioactivity with time, whereas Y became more radioactive with time.

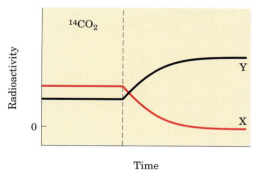

**4. Regulation of the Calvin Cycle**  Iodoacetate reacts irreversibly with the free —SH groups of Cys residues in proteins.

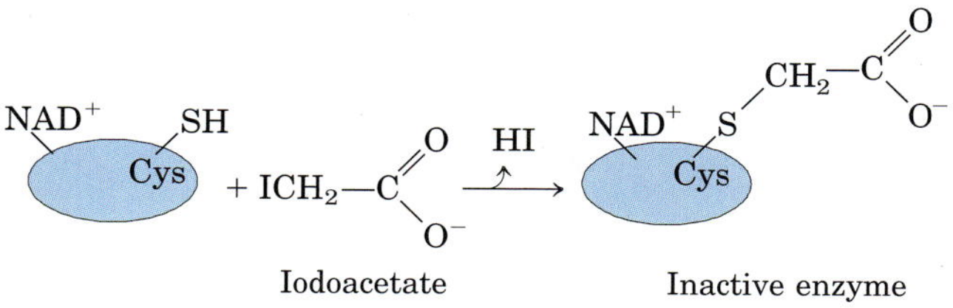

$$NAD^+ \quad SH \qquad \qquad NAD^+ \quad S-CH_2-C\begin{smallmatrix}O\\O^-\end{smallmatrix}$$
$$\bigcirc\!\!Cys + ICH_2-C\begin{smallmatrix}O\\O^-\end{smallmatrix} \xrightarrow{HI} \bigcirc\!\!Cys$$

Iodoacetate                    Inactive enzyme

Predict which Calvin cycle enzyme(s) would be inhibited by iodoacetate, and explain why.

**5. Thioredoxin in Regulation of Calvin Cycle Enzymes**  Motohashi and colleagues[*] used thioredoxin as a hook to fish out from plant extracts the proteins that are activated by thioredoxin. To do this, they prepared a mutant thioredoxin in which one of the reactive Cys residues was replaced with a Ser. Explain why this modification was necessary for their experiments.

**6. Comparison of the Reductive and Oxidative Pentose Phosphate Pathways**  The *reductive* pentose phosphate pathway generates a number of intermediates identical to those of the *oxidative* pentose phosphate pathway (Chapter 14). What role does each pathway play in cells where it is active?

**7. Photorespiration and Mitochondrial Respiration**  Compare the oxidative photosynthetic carbon cycle ($C_2$ cycle), also called *photorespiration*, with the *mitochondrial respiration* that drives ATP synthesis. Why are both processes referred to as respiration? Where in the cell do they occur, and under what circumstances? What is the path of electron flow in each?

**8. Rubisco and the Composition of the Atmosphere**  N. E. Tolbert[†] has argued that the dual specificity of rubisco for $CO_2$ and $O_2$ is not simply a leftover from evolution in a low-oxygen environment. He suggests that the relative activities of the carboxylase and oxygenase activities of rubisco actually have set, and now maintain, the ratio of $CO_2$ to $O_2$ in the earth's atmosphere. Discuss the pros and cons of this hypothesis, in molecular terms and in global terms. How does the existence of $C_4$ organisms bear on the hypothesis?

**9. Role of Sedoheptulose 1,7-Bisphosphatase**  What effect on the cell and the organism might result from a defect in sedoheptulose 1,7-bisphosphatase in (a) a human hepatocyte and (b) the leaf cell of a green plant?

**10. Pathway of $CO_2$ Assimilation in Maize**  If a maize (corn) plant is illuminated in the presence of $^{14}CO_2$, after about 1 second, more than 90% of all the radioactivity incorporated in the leaves is found at C-4 of malate, aspartate, and oxaloacetate. Only after 60 seconds does $^{14}C$ appear at C-1 of 3-phosphoglycerate. Explain.

**11. Identifying CAM Plants**  Given some $^{14}CO_2$ and all the tools typically present in a biochemistry research lab, how would you design a simple experiment to determine whether a plant was a typical $C_4$ plant or a CAM plant?

**12. Chemistry of Malic Enzyme: Variation on a Theme**  Malic enzyme, found in the bundle-sheath cells of $C_4$ plants, carries out a reaction that has a counterpart in the citric acid cycle. What is the analogous reaction? Explain your choice.

**13. The Cost of Storing Glucose as Starch**  Write the sequence of steps and the net reaction required to calculate the cost, in ATP molecules, of converting a molecule of cytosolic glucose 6-phosphate to starch and back to glucose 6-phosphate. What fraction of the maximum number of ATP molecules available from complete catabolism of glucose 6-phosphate to $CO_2$ and $H_2O$ does this cost represent?

[*]Motohashi, K., Kondoh, A., Stumpp, M.T., & Hisabori, T. (2001) Comprehensive survey of proteins targeted by chloroplast thioredoxin. *Proc. Natl. Acad. Sci. USA* **98**, 11,224–11,229.

[†]Tolbert, N.E. (1994) The role of photosynthesis and photorespiration in regulating atmospheric $CO_2$ and $O_2$. In *Regulation of Atmospheric $CO_2$ and $O_2$ by Photosynthetic Carbon Metabolism* (Tolbert, N.E. & Preiss, J., eds), pp. 8–33, Oxford University Press, New York.

**14. Inorganic Pyrophosphatase** The enzyme inorganic pyrophosphatase contributes to making many biosynthetic reactions that generate inorganic pyrophosphate essentially irreversible in cells. By keeping the concentration of $PP_i$ very low, the enzyme "pulls" these reactions in the direction of $PP_i$ formation. The synthesis of ADP-glucose in chloroplasts is one reaction that is pulled in the forward direction by this mechanism. However, the synthesis of UDP-glucose in the plant cytosol, which produces $PP_i$, is readily reversible in vivo. How do you reconcile these two facts?

**15. Regulation of Starch and Sucrose Synthesis** Sucrose synthesis occurs in the cytosol and starch synthesis in the chloroplast stroma, yet the two processes are intricately balanced. What factors shift the reactions in favor of (a) starch synthesis and (b) sucrose synthesis?

**16. Regulation of Sucrose Synthesis** In the regulation of sucrose synthesis from the triose phosphates produced during photosynthesis, 3-phosphoglycerate and $P_i$ play critical roles (see Fig. 20–26). Explain why the concentrations of these two regulators reflect the rate of photosynthesis.

**17. Sucrose and Dental Caries** The most prevalent infection in humans worldwide is dental caries, which stems from the colonization and destruction of tooth enamel by a variety of acidifying microorganisms. These organisms synthesize and live within a water-insoluble network of dextrans, called dental plaque, composed of ($\alpha 1 \rightarrow 6$)-linked polymers of glucose with many ($\alpha 1 \rightarrow 3$) branch points. Polymerization of dextran requires dietary sucrose, and the reaction is catalyzed by a bacterial enzyme, dextran-sucrose glucosyltransferase.

(a) Write the overall reaction for dextran polymerization.

(b) In addition to providing a substrate for the formation of dental plaque, how does dietary sucrose also provide oral bacteria with an abundant source of metabolic energy?

**18. Differences between $C_3$ and $C_4$ Plants** The plant genus *Atriplex* includes some $C_3$ and some $C_4$ species. From the data in the plots below (species 1, black curve; species 2, red curve), identify which is a $C_3$ plant and which is a $C_4$ plant. Justify your answer in molecular terms that account for the data in all three plots.

**19. $C_4$ Pathway in a Single Cell** In typical $C_4$ plants, the initial capture of $CO_2$ occurs in one cell type, and the Calvin cycle reactions occur in another (see Fig. 20–23). Voznesenskaya and colleagues[††] have described a plant, *Bienertia cycloptera*—which grows in salty depressions of semidesert in Central Asia—that shows the biochemical properties of a $C_4$ plant but unlike typical $C_4$ plants does not segregate the reactions of $CO_2$ fixation into two cell types. PEP carboxylase and rubisco are present in the same cell. However, the cells have two types of chloroplasts, which are localized differently, as shown in the micrograph. One type, relatively poor in grana (thylakoids), is confined to the periphery; the more typical chloroplasts are clustered in the center of the cell, separated from the peripheral chloroplasts by large vacuoles. Thin cytosolic bridges pass through the vacuoles, connecting the peripheral and central cytosol.

10 $\mu m$

In this plant, where would you expect to find (a) PEP carboxylase, (b) rubisco, and (c) starch granules? Explain your answers with a model for $CO_2$ fixation in these $C_4$ cells.

[††]Voznesenskaya, E.V., Fraceschi, V.R., Kiirats, O., Artyusheva, E.G., Freitag, H., & Edwards, G.E. (2002) Proof of $C_4$ photosynthesis without Kranz anatomy in *Bienertia cycloptera* (Chenopodiaceae). *Plant J.* **31**, 649–662.

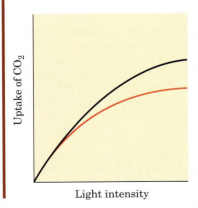

Uptake of $CO_2$ — Light intensity

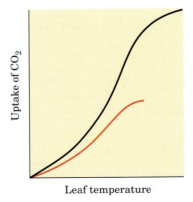

Uptake of $CO_2$ — Leaf temperature

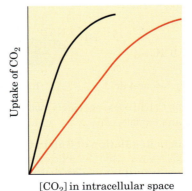

Uptake of $CO_2$ — [$CO_2$] in intracellular space

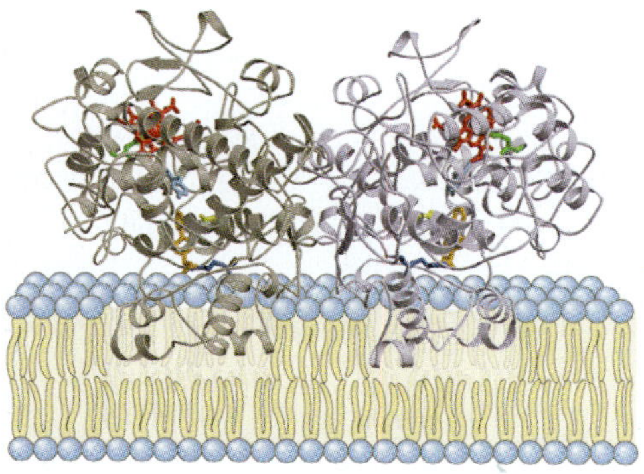

# LIPID BIOSYNTHESIS

How the division of "spoils" came about I do not recall—it may have been by drawing lots. At any rate, David Shemin "drew" amino acid metabolism, which led to his classical work on heme biosynthesis. David Rittenburg was to continue his interest in protein synthesis and turnover, and lipids were to be my territory.

—*Konrad Bloch, on how his career turned to problems of lipid metabolism after the death of his mentor, Rudolf Schoenheimer; article in* Annual Review of Biochemistry, *1987*

Lipids play a variety of cellular roles, some only recently recognized. They are the principal form of stored energy in most organisms and major constituents of cellular membranes. Specialized lipids serve as pigments (retinal, carotene), cofactors (vitamin K), detergents (bile salts), transporters (dolichols), hormones (vitamin D derivatives, sex hormones), extracellular and intracellular messengers (eicosanoids, phosphatidylinositol derivatives), and anchors for membrane proteins (covalently attached fatty acids, prenyl groups, and phosphatidylinositol). The ability to synthesize a variety of lipids is essential to all organisms. This chapter describes the biosynthetic pathways for some of the most common cellular lipids, illustrating the strategies employed in assembling these water-insoluble products from water-soluble precursors such as acetate. Like other biosynthetic pathways, these reaction sequences are endergonic and reductive. They use ATP as a source of metabolic energy and a reduced electron carrier (usually NADPH) as reductant.

We first describe the biosynthesis of fatty acids, the primary components of both triacylglycerols and phospholipids, then examine the assembly of fatty acids into triacylglycerols and the simpler membrane phospholipids. Finally, we consider the synthesis of cholesterol, a component of some membranes and the precursor of steroids such as the bile acids, sex hormones, and adrenocortical hormones.

## 21.1 Biosynthesis of Fatty Acids and Eicosanoids

After the discovery that fatty acid oxidation takes place by the oxidative removal of successive two-carbon (acetyl-CoA) units (see Fig. 17–8), biochemists thought the biosynthesis of fatty acids might proceed by a simple reversal of the same enzymatic steps. However, as they were to find out, fatty acid biosynthesis and breakdown occur by different pathways, are catalyzed by different sets of enzymes, and take place in different parts of the cell. Moreover, biosynthesis requires the participation of a three-carbon intermediate, **malonyl-CoA,** that is not involved in fatty acid breakdown.

We focus first on the pathway of fatty acid synthesis, then turn our attention to regulation of the pathway and to the biosynthesis of longer-chain fatty acids, unsaturated fatty acids, and their eicosanoid derivatives.

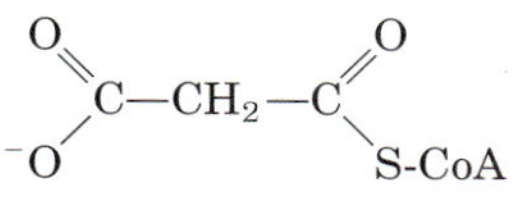

$$\overset{O}{\overset{\|}{\underset{-O}{C}}}-CH_2-\overset{O}{\overset{\|}{\underset{S\text{-CoA}}{C}}}$$

Malonyl-CoA

### Malonyl-CoA Is Formed from Acetyl-CoA and Bicarbonate

The formation of malonyl-CoA from acetyl-CoA is an irreversible process, catalyzed by **acetyl-CoA carboxylase.** The bacterial enzyme has three separate polypeptide subunits (Fig. 21–1); in animal cells, all three

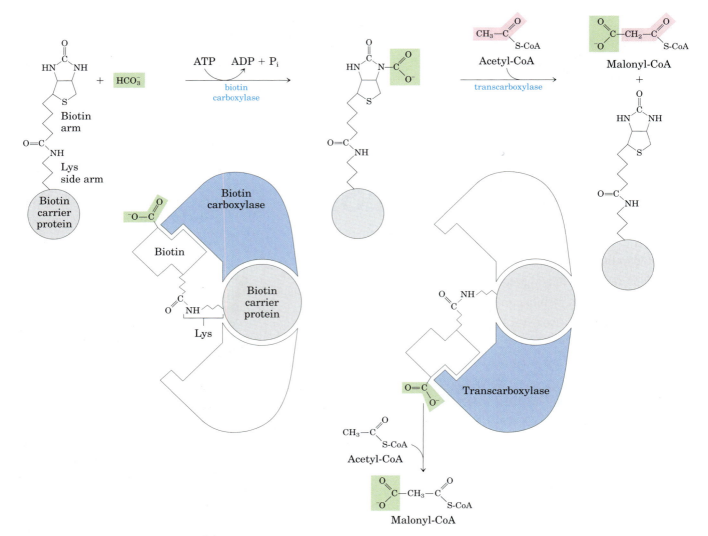

**FIGURE 21–1 The acetyl-CoA carboxylase reaction.** Acetyl-CoA carboxylase has three functional regions: biotin carrier protein (gray); biotin carboxylase, which activates $CO_2$ by attaching it to a nitrogen in the biotin ring in an ATP-dependent reaction (see Fig. 16–16); and transcarboxylase, which transfers activated $CO_2$ (shaded green) from biotin to acetyl-CoA, producing malonyl-CoA. The long, flexible biotin arm carries the activated $CO_2$ from the biotin carboxylase region to the transcarboxylase active site, as shown in the diagrams below the reaction arrows. The active enzyme in each step is shaded blue.

activities are part of a single multifunctional polypeptide. Plant cells contain both types of acetyl-CoA carboxylase. In all cases, the enzyme contains a biotin prosthetic group covalently bound in amide linkage to the $\varepsilon$-amino group of a Lys residue in one of the three polypeptides or domains of the enzyme molecule. The two-step reaction catalyzed by this enzyme is very similar to other biotin-dependent carboxylation reactions, such as those catalyzed by pyruvate carboxylase (see Fig. 16–16) and propionyl-CoA carboxylase (see Fig. 17–11). The carboxyl group, derived from bicarbonate ($HCO_3^-$), is first transferred to biotin in an ATP-dependent reaction. The biotinyl group serves as a tem-

porary carrier of $CO_2$, transferring it to acetyl-CoA in the second step to yield malonyl-CoA.

## Fatty Acid Synthesis Proceeds in a Repeating Reaction Sequence

The long carbon chains of fatty acids are assembled in a repeating four-step sequence (Fig. 21–2). A saturated acyl group produced by this set of reactions becomes the substrate for subsequent condensation with an activated malonyl group. With each passage through the cycle, the fatty acyl chain is extended by two carbons. When the chain length reaches 16 carbons, the product

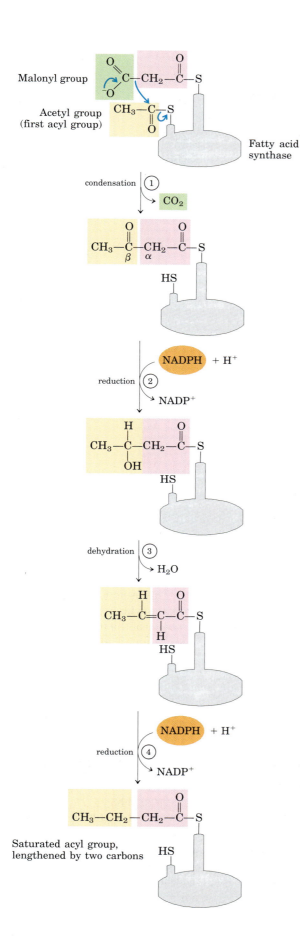

(palmitate, 16:0; see Table 10–1) leaves the cycle. Carbons C-16 and C-15 of the palmitate are derived from the methyl and carboxyl carbon atoms, respectively, of an acetyl-CoA used directly to prime the system at the outset (Fig. 21–3); the rest of the carbon atoms in the chain are derived from acetyl-CoA via malonyl-CoA.

Both the electron-carrying cofactor and the activating groups in the reductive anabolic sequence differ from those in the oxidative catabolic process. Recall that in $\beta$ oxidation, $NAD^+$ and FAD serve as electron acceptors and the activating group is the thiol (—SH) group of coenzyme A (see Fig. 17–8). By contrast, the reducing agent in the synthetic sequence is NADPH and the activating groups are two different enzyme-bound —SH groups, as described below.

All the reactions in the synthetic process are catalyzed by a multienzyme complex, **fatty acid synthase.** Although the details of enzyme structure differ in prokaryotes such as *Escherichia coli* and in eukaryotes, the four-step process of fatty acid synthesis is the same in all organisms. We first describe the process as it occurs in *E. coli*, then consider differences in enzyme structure in other organisms.

## The Fatty Acid Synthase Complex Has Seven Different Active Sites

The core of the *E. coli* fatty acid synthase system consists of seven separate polypeptides (Table 21–1), and at least three others act at some stage of the process. The proteins act together to catalyze the formation of fatty acids from acetyl-CoA and malonyl-CoA. Throughout the process, the intermediates remain covalently attached as thioesters to one of two thiol groups of the synthase complex. One point of attachment is the —SH group of a Cys residue in one of the seven synthase proteins ($\beta$-ketoacyl-ACP synthase); the other is the —SH group of acyl carrier protein.

**FIGURE 21–2 Addition of two carbons to a growing fatty acyl chain: a four-step sequence.** Each malonyl group and acetyl (or longer acyl) group is activated by a thioester that links it to fatty acid synthase, a multienzyme complex described later in the text. ① Condensation of an activated acyl group (an acetyl group from acetyl-CoA is the first acyl group) and two carbons derived from malonyl-CoA, with elimination of $CO_2$ from the malonyl group, extends the acyl chain by two carbons. The mechanism of the first step of this reaction is given to illustrate the role of decarboxylation in facilitating condensation. The $\beta$-keto product of this condensation is then reduced in three more steps nearly identical to the reactions of $\beta$ oxidation, but in the reverse sequence: ② the $\beta$-keto group is reduced to an alcohol, ③ elimination of $H_2O$ creates a double bond, and ④ the double bond is reduced to form the corresponding saturated fatty acyl group.

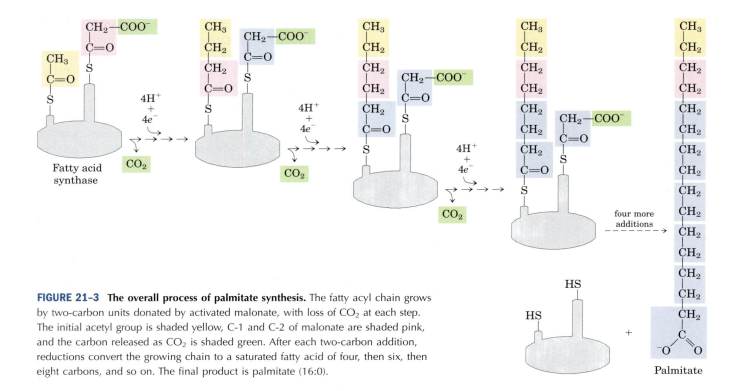

**FIGURE 21–3  The overall process of palmitate synthesis.** The fatty acyl chain grows by two-carbon units donated by activated malonate, with loss of $CO_2$ at each step. The initial acetyl group is shaded yellow, C-1 and C-2 of malonate are shaded pink, and the carbon released as $CO_2$ is shaded green. After each two-carbon addition, reductions convert the growing chain to a saturated fatty acid of four, then six, then eight carbons, and so on. The final product is palmitate (16:0).

**Acyl carrier protein (ACP)** of *E. coli* is a small protein ($M_r$ 8,860) containing the prosthetic group **4′-phosphopantetheine** (Fig. 21–4; compare this with the panthothenic acid and $\beta$-mercaptoethylamine moiety of coenzyme A in Fig. 8–41). Hydrolysis of thioesters is highly exergonic, and the energy released helps to make two different steps (① and ⑤ in Fig. 21–5) in fatty acid synthesis (condensation) thermodynamically favorable. The 4′-phosphopantetheine prosthetic group of ACP is believed to serve as a flexible arm, tethering the growing fatty acyl chain to the surface of the fatty acid synthase complex while carrying the reaction intermediates from one enzyme active site to the next.

## Fatty Acid Synthase Receives the Acetyl and Malonyl Groups

Before the condensation reactions that build up the fatty acid chain can begin, the two thiol groups on the enzyme complex must be charged with the correct acyl groups (Fig. 21–5, top). First, the acetyl group of acetyl-CoA is transferred to the Cys —SH group of the $\beta$-ketoacyl-ACP synthase. This reaction is catalyzed by **acetyl-CoA–ACP transacetylase** (AT in Fig. 21–5). The second reaction, transfer of the malonyl group from malonyl-CoA to the —SH group of ACP, is catalyzed by **malonyl-CoA–ACP transferase** (MT), also part of the complex. In the charged synthase complex, the acetyl

| TABLE 21–1   Proteins of the Fatty Acid Synthase Complex of *E. coli* | |
|---|---|
| Component | Function |
| Acyl carrier protein (ACP) | Carries acyl groups in thioester linkage |
| Acetyl-CoA–ACP transacetylase (AT) | Transfers acyl group from CoA to Cys residue of KS |
| $\beta$-Ketoacyl-ACP synthase (KS) | Condenses acyl and malonyl groups (KS has at least three isozymes) |
| Malonyl-CoA–ACP transferase (MT) | Transfers malonyl group from CoA to ACP |
| $\beta$-Ketoacyl-ACP reductase (KR) | Reduces $\beta$-keto group to $\beta$-hydroxyl group |
| $\beta$-Hydroxyacyl-ACP dehydratase (HD) | Removes $H_2O$ from $\beta$-hydroxyacyl-ACP, creating double bond |
| Enoyl-ACP reductase (ER) | Reduces double bond, forming saturated acyl-ACP |

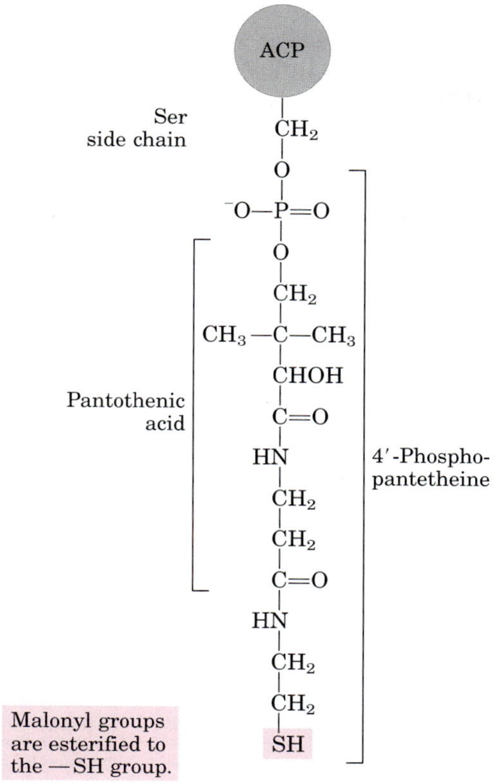

FIGURE 21–4 **Acyl carrier protein (ACP).** The prosthetic group is 4'-phosphopantetheine, which is covalently attached to the hydroxyl group of a Ser residue in ACP. Phosphopantetheine contains the B vitamin pantothenic acid, also found in the coenzyme A molecule. Its —SH group is the site of entry of malonyl groups during fatty acid synthesis.

and malonyl groups are very close to each other and are activated for the chain-lengthening process. The first four steps of this process are now considered in some detail; all step numbers refer to Figure 21–5.

**Step ① Condensation** The first reaction in the formation of a fatty acid chain is condensation of the activated acetyl and malonyl groups to form **acetoacetyl-ACP,** an acetoacetyl group bound to ACP through the phosphopantetheine —SH group; simultaneously, a molecule of $CO_2$ is produced. In this reaction, catalyzed by **β-ketoacyl-ACP synthase** (KS), the acetyl group is transferred from the Cys —SH group of the enzyme to the malonyl group on the —SH of ACP, becoming the methyl-terminal two-carbon unit of the new acetoacetyl group.

The carbon atom of the $CO_2$ formed in this reaction is the same carbon originally introduced into malonyl-CoA from $HCO_3^-$ by the acetyl-CoA carboxylase reaction (Fig. 21–1). Thus $CO_2$ is only transiently in covalent linkage during fatty acid biosynthesis; it is removed as each two-carbon unit is added.

Why do cells go to the trouble of adding $CO_2$ to make a malonyl group from an acetyl group, only to lose the $CO_2$ during the formation of acetoacetate? Recall that in the β oxidation of fatty acids (see Fig. 17–8), cleavage of the bond between two acyl groups (cleavage of an acetyl unit from the acyl chain) is highly exergonic, so the simple condensation of two acyl groups (two acetyl-CoA molecules, for example) is highly endergonic. The use of activated malonyl groups rather than acetyl groups is what makes the condensation reactions thermodynamically favorable. The methylene carbon (C-2) of the malonyl group, sandwiched between carbonyl and carboxyl carbons, is chemically situated to act as a good nucleophile. In the condensation step (step ①), decarboxylation of the malonyl group facilitates the nucleophilic attack of the methylene carbon on the thioester linking the acetyl group to β-ketoacyl-ACP synthase, displacing the enzyme's —SH group. Coupling the condensation to the decarboxylation of the malonyl group renders the overall process highly exergonic. A similar carboxylation-decarboxylation sequence facilitates the formation of phosphoenolpyruvate from pyruvate in gluconeogenesis (see Fig. 14–17).

By using activated malonyl groups in the synthesis of fatty acids and activated acetate in their degradation, the cell makes both processes energetically favorable, although one is effectively the reversal of the other. The extra energy required to make fatty acid synthesis favorable is provided by the ATP used to synthesize malonyl-CoA from acetyl-CoA and $HCO_3^-$ (Fig. 21–1).

**Step ② Reduction of the Carbonyl Group** The acetoacetyl-ACP formed in the condensation step now undergoes reduction of the carbonyl group at C-3 to form D-β-hydroxybutyryl-ACP. This reaction is catalyzed by **β-ketoacyl-ACP reductase** (KR) and the electron donor is NADPH. Notice that the D-β-hydroxybutyryl group does not have the same stereoisomeric form as the L-β-hydroxyacyl intermediate in fatty acid oxidation (see Fig. 17–8).

**Step ③ Dehydration** The elements of water are now removed from C-2 and C-3 of D-β-hydroxybutyryl-ACP to yield a double bond in the product, **trans-$\Delta^2$- butenoyl-ACP.** The enzyme that catalyzes this dehydration is **β-hydroxyacyl-ACP dehydratase** (HD).

**Step ④ Reduction of the Double Bond** Finally, the double bond of trans-$\Delta^2$-butenoyl-ACP is reduced (saturated) to form **butyryl-ACP** by the action of **enoyl-ACP reductase** (ER); again, NADPH is the electron donor.

## The Fatty Acid Synthase Reactions Are Repeated to Form Palmitate

Production of the four-carbon, saturated fatty acyl–ACP completes one pass through the fatty acid synthase

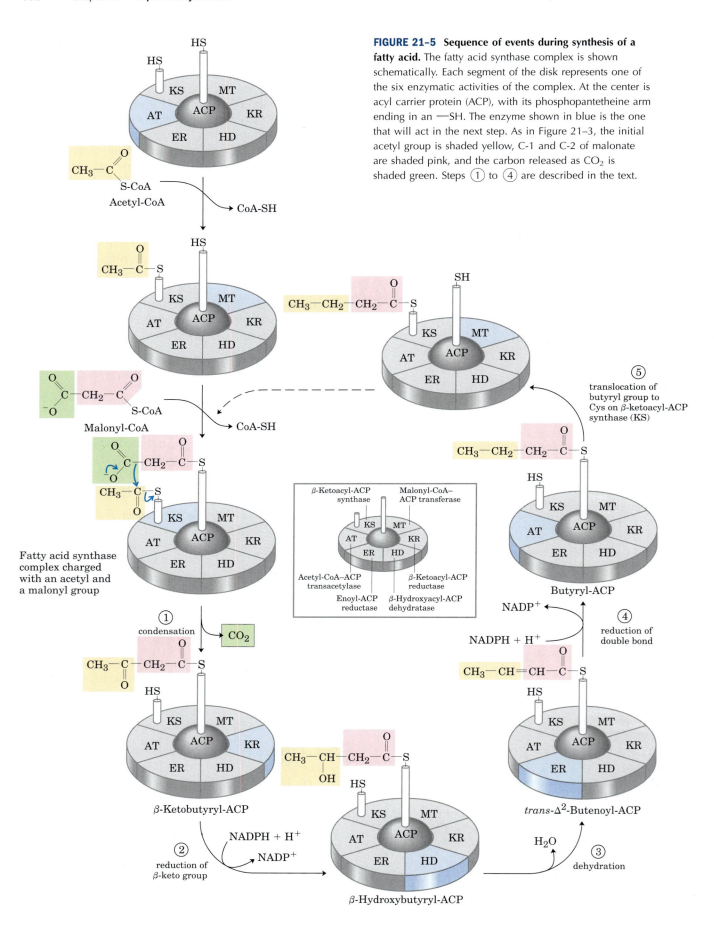

**FIGURE 21–5 Sequence of events during synthesis of a fatty acid.** The fatty acid synthase complex is shown schematically. Each segment of the disk represents one of the six enzymatic activities of the complex. At the center is acyl carrier protein (ACP), with its phosphopantetheine arm ending in an —SH. The enzyme shown in blue is the one that will act in the next step. As in Figure 21–3, the initial acetyl group is shaded yellow, C-1 and C-2 of malonate are shaded pink, and the carbon released as $CO_2$ is shaded green. Steps ① to ④ are described in the text.

complex. The butyryl group is now transferred from the phosphopantetheine —SH group of ACP to the Cys —SH group of β-ketoacyl-ACP synthase, which initially bore the acetyl group (Fig. 21–5). To start the next cycle of four reactions that lengthens the chain by two more carbons, another malonyl group is linked to the now unoccupied phosphopantetheine —SH group of ACP (Fig. 21–6). Condensation occurs as the butyryl group, acting like the acetyl group in the first cycle, is linked to two carbons of the malonyl-ACP group with concurrent loss of $CO_2$. The product of this condensation is a six-carbon acyl group, covalently bound to the phosphopantetheine —SH group. Its β-keto group is reduced in the next three steps of the synthase cycle to yield the saturated acyl group, exactly as in the first round of reactions—in this case forming the six-carbon product.

Seven cycles of condensation and reduction produce the 16-carbon saturated palmitoyl group, still bound to ACP. For reasons not well understood, chain elongation by the synthase complex generally stops at this point and free palmitate is released from the ACP by a hydrolytic activity in the complex. Small amounts of longer fatty acids such as stearate (18:0) are also formed. In certain plants (coconut and palm, for example) chain termination occurs earlier; up to 90% of the fatty acids in the oils of these plants are between 8 and 14 carbons long.

We can consider the overall reaction for the synthesis of palmitate from acetyl-CoA in two parts. First, the formation of seven malonyl-CoA molecules:

$$7 \text{ Acetyl-CoA} + 7CO_2 + 7ATP \rightarrow$$
$$7 \text{ malonyl-CoA} + 7ADP + 7P_i \quad (21\text{–}1)$$

then seven cycles of condensation and reduction:

$$\text{Acetyl-CoA} + 7 \text{ malonyl-CoA} + 14NADPH + 14H^+ \rightarrow$$
$$\text{palmitate} + 7CO_2 + 8 \text{ CoA} + 14NADP^+ + 6H_2O \quad (21\text{–}2)$$

The overall process (the sum of Eqns 21–1 and 21–2) is

$$8 \text{ Acetyl-CoA} + 7ATP + 14NADPH + 14H^+ \rightarrow$$
$$\text{palmitate} + 8 \text{ CoA} + 7ADP + 7P_i + 14NADP^+ + 6H_2O$$
$$(21\text{–}3)$$

**FIGURE 21–6 Beginning of the second round of the fatty acid synthesis cycle.** The butyryl group is on the Cys —SH group. The incoming malonyl group is first attached to the phosphopantetheine —SH group. Then, in the condensation step, the entire butyryl group on the Cys —SH is exchanged for the carboxyl group of the malonyl residue, which is lost as $CO_2$ (green). This step is analogous to step ① in Figure 21–5. The product, a six-carbon β-ketoacyl group, now contains four carbons derived from malonyl-CoA and two derived from the acetyl-CoA that started the reaction. The β-ketoacyl group then undergoes steps ② through ④, as in Figure 21–5.

The biosynthesis of fatty acids such as palmitate thus requires acetyl-CoA and the input of chemical energy in two forms: the group transfer potential of ATP and the reducing power of NADPH. The ATP is required to attach $CO_2$ to acetyl-CoA to make malonyl-CoA; the NADPH is required to reduce the double bonds. We return to the sources of acetyl-CoA and NADPH soon, but first let's consider the structure of the remarkable enzyme complex that catalyzes the synthesis of fatty acids.

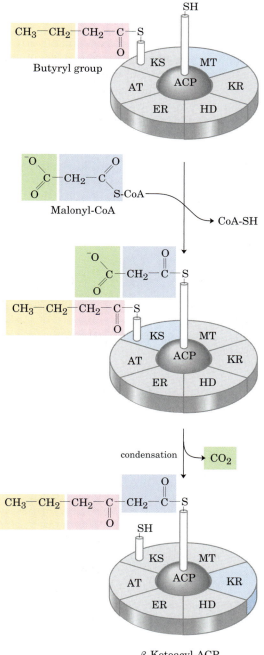

## The Fatty Acid Synthase of Some Organisms Consists of Multifunctional Proteins

In *E. coli* and some plants, the seven active sites for fatty acid synthesis (six enzymes and ACP) reside in seven separate polypeptides (Fig. 21–7, top). In these complexes, each enzyme is positioned with its active site near that of the preceding and succeeding enzymes of the sequence. The flexible pantetheine arm of ACP can reach all the active sites, and it carries the growing fatty acyl chain from one site to the next; intermediates are not released from the enzyme complex until it has formed the finished product. As we have seen in earlier chapters, this channeling of intermediates from one active site to the next increases the efficiency of the overall process.

The fatty acid synthases of yeast and of vertebrates are also multienzyme complexes, and their integration is even more complete than in *E. coli* and plants. In yeast, the seven distinct active sites reside in two large, multifunctional polypeptides, with three activities on the α subunit and four on the β subunit. In vertebrates, a single large polypeptide ($M_r$ 240,000) contains all seven enzymatic activities as well as a hydrolytic activity that cleaves the finished fatty acid from the ACP-like part of the enzyme complex. The vertebrate enzyme functions as a dimer ($M_r$ 480,000) in which the two identical subunits lie head-to-tail. The subunits appear to function independently. When all the active sites in one subunit are inactivated by mutation, palmitate synthesis is only modestly reduced.

## Fatty Acid Synthesis Occurs in the Cytosol of Many Organisms but in the Chloroplasts of Plants

In most higher eukaryotes, the fatty acid synthase complex is found exclusively in the cytosol (Fig. 21–8), as are the biosynthetic enzymes for nucleotides, amino acids, and glucose. This location segregates synthetic processes from degradative reactions, many of which take place in the mitochondrial matrix. There is a corresponding segregation of the electron-carrying cofactors used in anabolism (generally a reductive process) and those used in catabolism (generally oxidative).

Usually, NADPH is the electron carrier for anabolic reactions, and NAD⁺ serves in catabolic reactions. In hepatocytes, the [NADPH]/[NADP⁺] ratio is very high (about 75) in the cytosol, furnishing a strongly reducing environment for the reductive synthesis of fatty acids and other biomolecules. The cytosolic [NADH]/[NAD⁺] ratio is much smaller (only about $8 \times 10^{-4}$), so the NAD⁺-dependent oxidative catabolism of glucose can take place in the same compartment, and at the same time, as fatty acid synthesis. The [NADH]/[NAD⁺] ratio in the mitochondrion is much higher than in the cytosol, because of the flow of electrons to NAD⁺ from the oxidation of fatty acids, amino acids, pyruvate, and acetyl-CoA. This high mitochondrial [NADH]/[NAD⁺] ratio favors the reduction of oxygen via the respiratory chain.

In hepatocytes and adipocytes, cytosolic NADPH is largely generated by the pentose phosphate pathway (see Fig. 14–21) and by **malic enzyme** (Fig. 21–9a). The NADP-linked malic enzyme that operates in the carbon-assimilation pathway of C₄ plants (see Fig. 20–23) is unrelated in function. The pyruvate produced in the reaction shown in Figure 21–9a reenters the mitochondrion. In hepatocytes and in the mammary gland of lactating animals, the NADPH required for fatty acid biosynthesis is supplied primarily by the pentose phosphate pathway (Fig. 21–9b).

In the photosynthetic cells of plants, fatty acid synthesis occurs not in the cytosol but in the chloroplast stroma (Fig. 21–8). This makes sense, given that NADPH is produced in chloroplasts by the light reactions of photosynthesis:

$$H_2O + NADP^+ \xrightarrow{\text{light}} \tfrac{1}{2}O_2 + NADPH + H^+$$

Again, the resulting high [NADPH]/[NADP⁺] ratio provides the reducing environment that favors reductive anabolic processes such as fatty acid synthesis.

## Acetate Is Shuttled out of Mitochondria as Citrate

In nonphotosynthetic eukaryotes, nearly all the acetyl-CoA used in fatty acid synthesis is formed in mito-

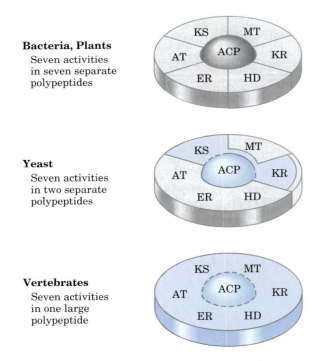

**Bacteria, Plants**
Seven activities in seven separate polypeptides

**Yeast**
Seven activities in two separate polypeptides

**Vertebrates**
Seven activities in one large polypeptide

**FIGURE 21–7  Structure of fatty acid synthases.** The fatty acid synthase of bacteria and plants is a complex of at least seven different polypeptides. In yeast, all seven activities reside in only two polypeptides; the vertebrate enzyme is a single large polypeptide.

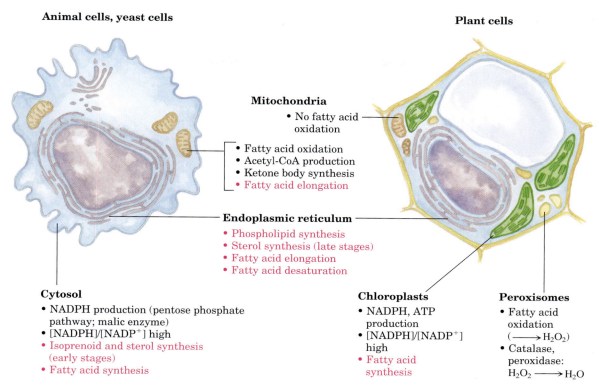

**FIGURE 21–8 Subcellular localization of lipid metabolism.** Yeast and vertebrate cells differ from higher plant cells in the compartmentation of lipid metabolism. Fatty acid synthesis takes place in the compart- ment in which NADPH is available for reductive synthesis (i.e., where the [NADPH]/[NADP$^+$] ratio is high). Processes in red type are cov- ered in this chapter.

chondria from pyruvate oxidation and from the catabo- lism of the carbon skeletons of amino acids. Acetyl-CoA arising from the oxidation of fatty acids is not a signifi- cant source of acetyl-CoA for fatty acid biosynthesis in animals, because the two pathways are reciprocally reg- ulated, as described below.

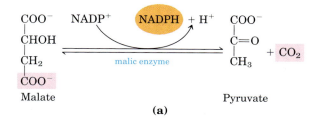

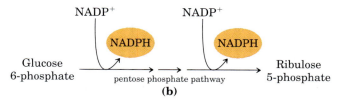

**FIGURE 21–9 Production of NADPH.** Two routes to NADPH, cat- alyzed by **(a)** malic enzyme and **(b)** the pentose phosphate pathway.

The mitochondrial inner membrane is impermeable to acetyl-CoA, so an indirect shuttle transfers acetyl group equivalents across the inner membrane (Fig. 21–10). Intramitochondrial acetyl-CoA first reacts with oxaloacetate to form citrate, in the citric acid cycle re- action catalyzed by **citrate synthase** (see Fig. 16–7). Citrate then passes through the inner membrane on the **citrate transporter.** In the cytosol, citrate cleavage by **citrate lyase** regenerates acetyl-CoA in an ATP- dependent reaction. Oxaloacetate cannot return to the mitochondrial matrix directly, as there is no oxaloacetate transporter. Instead, cytosolic malate dehydrogenase re- duces the oxaloacetate to malate, which returns to the mitochondrial matrix on the malate–$\alpha$-ketoglutarate transporter in exchange for citrate. In the matrix, malate is reoxidized to oxaloacetate to complete the shuttle. Al- ternatively, the malate produced in the cytosol is used to generate cytosolic NADPH through the activity of malic enzyme (Fig. 21–9a).

## Fatty Acid Biosynthesis Is Tightly Regulated

When a cell or organism has more than enough meta- bolic fuel to meet its energy needs, the excess is gen- erally converted to fatty acids and stored as lipids such as triacylglycerols. The reaction catalyzed by acetyl-CoA

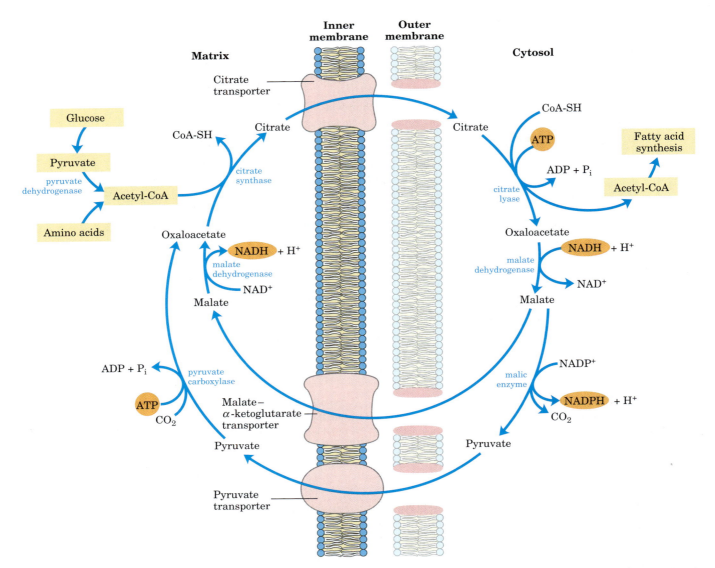

**FIGURE 21–10 Shuttle for transfer of acetyl groups from mitochondria to the cytosol.** The mitochondrial outer membrane is freely permeable to all these compounds. Pyruvate derived from amino acid catabolism in the mitochondrial matrix, or from glucose by glycolysis in the cytosol, is converted to acetyl-CoA in the matrix. Acetyl groups pass out of the mitochondrion as citrate; in the cytosol they are delivered as acetyl-CoA for fatty acid synthesis. Oxaloacetate is reduced to malate, which returns to the mitochondrial matrix and is converted to oxaloacetate. An alternative fate for cytosolic malate is oxidation by malic enzyme to generate cytosolic NADPH; the pyruvate produced returns to the mitochondrial matrix.

carboxylase is the rate-limiting step in the biosynthesis of fatty acids, and this enzyme is an important site of regulation. In vertebrates, palmitoyl-CoA, the principal product of fatty acid synthesis, is a feedback inhibitor of the enzyme; citrate is an allosteric activator (Fig. 21–11a), increasing $V_{max}$. Citrate plays a central role in diverting cellular metabolism from the consumption (oxidation) of metabolic fuel to the storage of fuel as fatty acids. When the concentrations of mitochondrial acetyl-CoA and ATP increase, citrate is transported out of mitochondria; it then becomes both the precursor of cytosolic acetyl-CoA and an allosteric signal for the activation

of acetyl-CoA carboxylase. At the same time, citrate inhibits the activity of phosphofructokinase-1 (see Fig. 15–18), reducing the flow of carbon through glycolysis.

Acetyl-CoA carboxylase is also regulated by covalent modification. Phosphorylation, triggered by the hormones glucagon and epinephrine, inactivates the enzyme and reduces its sensitivity to activation by citrate, thereby slowing fatty acid synthesis. In its active (dephosphorylated) form, acetyl-CoA carboxylase polymerizes into long filaments (Fig. 21–11b); phosphorylation is accompanied by dissociation into monomeric subunits and loss of activity.

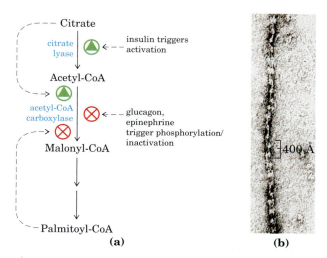

**(a)**    **(b)**

**FIGURE 21–11 Regulation of fatty acid synthesis. (a)** In the cells of vertebrates, both allosteric regulation and hormone-dependent covalent modification influence the flow of precursors into malonyl-CoA. In plants, acetyl-CoA carboxylase is activated by the changes in $[Mg^{2+}]$ and pH that accompany illumination (not shown here). **(b)** Filaments of acetyl-CoA carboxylase (the active, dephosphorylated form) as seen with the electron microscope.

The acetyl-CoA carboxylase of plants and bacteria is not regulated by citrate or by a phosphorylation-dephosphorylation cycle. The plant enzyme is activated by an increase in stromal pH and $[Mg^{2+}]$, which occurs on illumination of the plant (see Fig. 20–18). Bacteria do not use triacylglycerols as energy stores. In *E. coli*, the primary role of fatty acid synthesis is to provide precursors for membrane lipids; the regulation of this process is complex, involving guanine nucleotides (such as ppGpp) that coordinate cell growth with membrane formation (see Figs 8–42, 28–24).

In addition to the moment-by-moment regulation of enzymatic activity, these pathways are regulated at the level of gene expression. For example, when animals ingest an excess of certain polyunsaturated fatty acids, the expression of genes encoding a wide range of lipogenic enzymes in the liver is suppressed. The detailed mechanism by which these genes are regulated is not yet clear.

If fatty acid synthesis and $\beta$ oxidation were to proceed simultaneously, the two processes would constitute a futile cycle, wasting energy. We noted earlier (see Fig. 17–12) that $\beta$ oxidation is blocked by malonyl-CoA, which inhibits carnitine acyltransferase I. Thus during fatty acid synthesis, the production of the first intermediate, malonyl-CoA, shuts down $\beta$ oxidation at the level of a transport system in the mitochondrial inner membrane. This control mechanism illustrates another advantage of segregating synthetic and degradative pathways in different cellular compartments.

## Long-Chain Saturated Fatty Acids Are Synthesized from Palmitate

Palmitate, the principal product of the fatty acid synthase system in animal cells, is the precursor of other long-chain fatty acids (Fig. 21–12). It may be lengthened to form stearate (18:0) or even longer saturated fatty acids by further additions of acetyl groups, through the action of **fatty acid elongation systems** present in the smooth endoplasmic reticulum and in mitochondria. The more active elongation system of the ER extends the 16-carbon chain of palmitoyl-CoA by two carbons, forming stearoyl-CoA. Although different enzyme systems are involved, and coenzyme A rather than ACP is the acyl carrier in the reaction, the mechanism of elongation in the ER is otherwise identical to that in palmitate synthesis: donation of two carbons by malonyl-CoA, followed by reduction, dehydration, and reduction to the saturated 18-carbon product, stearoyl-CoA.

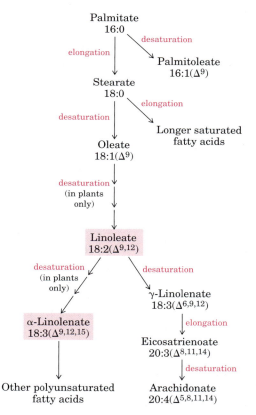

**FIGURE 21–12 Routes of synthesis of other fatty acids.** Palmitate is the precursor of stearate and longer-chain saturated fatty acids, as well as the monounsaturated acids palmitoleate and oleate. Mammals cannot convert oleate to linoleate or $\alpha$-linolenate (shaded pink), which are therefore required in the diet as essential fatty acids. Conversion of linoleate to other polyunsaturated fatty acids and eicosanoids is outlined. Unsaturated fatty acids are symbolized by indicating the number of carbons and the number and position of the double bonds, as in Table 10–1.

## Desaturation of Fatty Acids Requires a Mixed-Function Oxidase

Palmitate and stearate serve as precursors of the two most common monounsaturated fatty acids of animal tissues: palmitoleate, $16:1(\Delta^9)$, and oleate, $18:1(\Delta^9)$; both of these fatty acids have a single cis double bond between C-9 and C-10 (see Table 10–1). The double bond is introduced into the fatty acid chain by an oxidative reaction catalyzed by **fatty acyl–CoA desaturase** (Fig. 21–13), a **mixed-function oxidase** (Box 21–1). Two different substrates, the fatty acid and NADH or NADPH, simultaneously undergo two-electron oxidations. The path of electron flow includes a cytochrome (cytochrome $b_5$) and a flavoprotein (cytochrome $b_5$ reductase), both of which, like fatty acyl–CoA desaturase, are in the smooth ER. Bacteria have two cytochrome $b_5$ reductases, one NADH-dependent and the other NADPH-dependent; which of these is the main electron donor in vivo is unclear. In plants, oleate is produced by a stearoyl-ACP desaturase in the chloroplast stroma that uses reduced ferredoxin as the electron donor.

Mammalian hepatocytes can readily introduce double bonds at the $\Delta^9$ position of fatty acids but cannot introduce additional double bonds between C-10 and the

---

### BOX 21–1    THE WORLD OF BIOCHEMISTRY

#### Mixed-Function Oxidases, Oxygenases, and Cytochrome P-450

In this chapter we encounter several enzymes that carry out oxidation-reduction reactions in which molecular oxygen is a participant. The reaction that introduces a double bond into a fatty acyl chain (see Fig. 21–13) is one such reaction.

The nomenclature for enzymes that catalyze reactions of this general type is often confusing to students, as is the mechanism of the reactions. **Oxidase** is the general name for enzymes that catalyze oxidations in which molecular oxygen is the electron acceptor but oxygen atoms do not appear in the oxidized product (however, there is an exception to this "rule," as we shall see!). The enzyme that creates a double bond in fatty acyl–CoA during the oxidation of fatty acids in peroxisomes (see Fig. 17–13) is an oxidase of this type; a second example is the cytochrome oxidase of the mitochondrial electron-transfer chain (see Fig. 19–14). In the first case, the transfer of two electrons to $H_2O$ produces hydrogen peroxide, $H_2O_2$; in the second, two electrons reduce $\frac{1}{2}O_2$ to $H_2O$. Many, but not all, oxidases are flavoproteins.

**Oxygenases** catalyze oxidative reactions in which oxygen atoms *are* directly incorporated into the substrate molecule, forming a new hydroxyl or carboxyl group, for example. **Dioxygenases** catalyze reactions in which both oxygen atoms of $O_2$ are incorporated into the organic substrate molecule. An example of a dioxygenase is tryptophan 2, 3-dioxygenase, which catalyzes the opening of the five-membered ring of tryptophan in the catabolism of this amino acid. When this reaction takes place in the presence of $^{18}O_2$, the isotopic oxygen atoms are found in the two carbonyl groups of the product (shown in red).

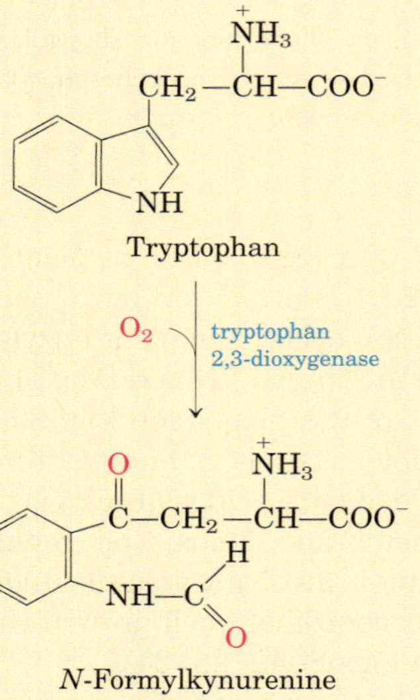

Tryptophan

tryptophan 2,3-dioxygenase

*N*-Formylkynurenine

**Monooxygenases,** more abundant and more complex in their action, catalyze reactions in which only one of the two oxygen atoms of $O_2$ is incorporated into the organic substrate, the other being reduced to $H_2O$. Monooxygenases require two substrates to serve as reductants of the two oxygen atoms of $O_2$. The main substrate accepts one of the two oxygen atoms, and a cosubstrate furnishes hydrogen atoms to reduce the other oxygen atom to $H_2O$. The general reaction equation for monooxygenases is

$$AH + BH_2 + O{-}O \longrightarrow A{-}OH + B + H_2O$$

where AH is the main substrate and $BH_2$ the cosubstrate. Because most monooxygenases catalyze reactions in which the main substrate becomes hydroxylated, they are also called **hydroxylases.** They are

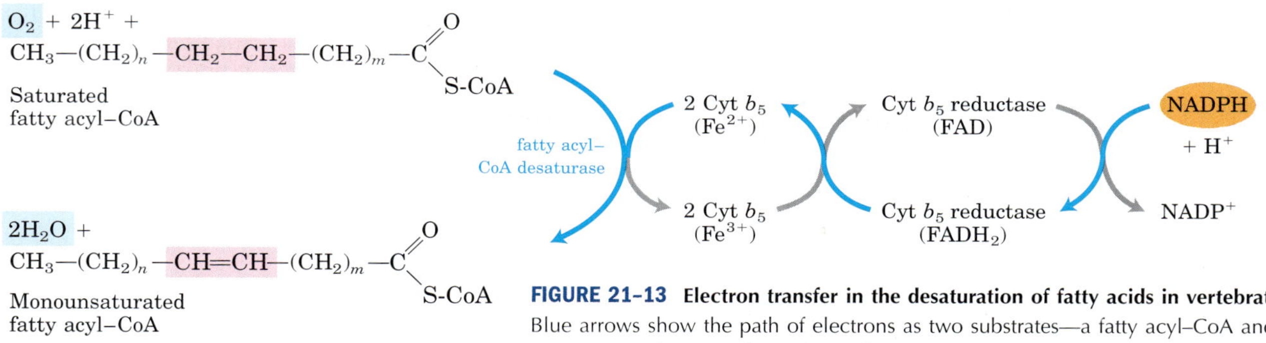

$O_2 + 2H^+ +$
$CH_3$—$(CH_2)_n$—$CH_2$—$CH_2$—$(CH_2)_m$—$\overset{\displaystyle O}{\underset{\displaystyle S\text{-CoA}}{C}}$

**Saturated fatty acyl–CoA**

2 Cyt $b_5$ (Fe$^{2+}$) — Cyt $b_5$ reductase (FAD) — NADPH

$+ H^+$

2 Cyt $b_5$ (Fe$^{3+}$) — Cyt $b_5$ reductase (FADH$_2$) — NADP$^+$

fatty acyl–CoA desaturase

$2H_2O +$
$CH_3$—$(CH_2)_n$—$CH$=$CH$—$(CH_2)_m$—$\overset{\displaystyle O}{\underset{\displaystyle S\text{-CoA}}{C}}$

**Monounsaturated fatty acyl–CoA**

**FIGURE 21–13 Electron transfer in the desaturation of fatty acids in vertebrates.** Blue arrows show the path of electrons as two substrates—a fatty acyl–CoA and NADPH—undergo oxidation by molecular oxygen. These reactions take place on the lumenal face of the smooth ER. A similar pathway, but with different electron carriers, occurs in plants.

also sometimes called **mixed-function oxidases** or **mixed-function oxygenases,** to indicate that they oxidize two different substrates simultaneously. (Note here the use of "oxidase"—a deviation from the general meaning of this term noted above.)

There are different classes of monooxygenases, depending on the nature of the cosubstrate. Some use reduced flavin nucleotides (FMNH$_2$ or FADH$_2$), others use NADH or NADPH, and still others use $\alpha$-ketoglutarate as the cosubstrate. The enzyme that hydroxylates the phenyl ring of phenylalanine to form tyrosine is a monooxygenase for which tetrahydrobiopterin serves as cosubstrate (see Fig. 18–23). This is the enzyme that is defective in the human genetic disease phenylketonuria.

The most numerous and most complex monooxygenation reactions are those employing a type of heme protein called **cytochrome P-450.** This cytochrome is usually present in the smooth ER rather than the mitochondria. Like mitochondrial cytochrome oxidase, cytochrome P-450 can react with $O_2$ and bind carbon monoxide, but it can be differentiated from cytochrome oxidase because the carbon monoxide complex of its reduced form absorbs light strongly at 450 nm—thus the name P-450.

Cytochrome P-450 catalyzes hydroxylation reactions in which an organic substrate, RH, is hydroxylated to R—OH, incorporating one oxygen atom of $O_2$; the other oxygen atom is reduced to $H_2O$ by reducing equivalents that are furnished by NADH or NADPH but are usually passed to cytochrome P-450 by an iron-sulfur protein. Figure 1 shows a simplified outline of the action of cytochrome P-450, which has intermediate steps not yet fully understood.

Cytochrome P-450 is actually a family of similar proteins; several hundred members of this protein

family are known, each with a different substrate specificity. In the adrenal cortex, for example, a specific cytochrome P-450 participates in the hydroxylation of steroids to yield the adrenocortical hormones (see Fig. 21–47). Cytochrome P-450 is also important in the hydroxylation of many different drugs, such as barbiturates and other xenobiotics (substances foreign to the organism), particularly if they are hydrophobic and relatively insoluble. The environmental carcinogen benzo[a]pyrene (found in cigarette smoke) undergoes cytochrome P-450–dependent hydroxylation during detoxification. Hydroxylation of xenobiotics makes them more soluble in water and allows their excretion in the urine. Unfortunately, hydroxylation of some compounds converts them to toxic substances, subverting the detoxification system.

Reactions described in this chapter that are catalyzed by mixed-function oxidases are those involved in fatty acyl–CoA desaturation (Fig. 21–13), leukotriene synthesis (Fig. 21–16), plasmalogen synthesis (Fig. 21–30), conversion of squalene to cholesterol (Fig. 21–37), and steroid hormone synthesis (Fig. 21–47).

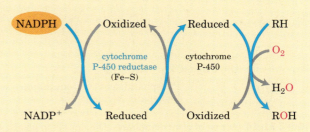

NADPH — Oxidized — Reduced — RH

$O_2$

cytochrome P-450 reductase (Fe–S)   cytochrome P-450

$H_2O$

NADP$^+$ — Reduced — Oxidized — ROH

**FIGURE 1**

methyl-terminal end. Thus mammals cannot synthesize linoleate, $18:2(\Delta^{9,12})$, or $\alpha$-linolenate, $18:3(\Delta^{9,12,15})$. Plants, however, can synthesize both; the desaturases that introduce double bonds at the $\Delta^{12}$ and $\Delta^{15}$ positions are located in the ER and the chloroplast. The ER enzymes act not on free fatty acids but on a phospholipid, phosphatidylcholine, that contains at least one oleate linked to the glycerol (Fig. 21–14). Both plants and bacteria must synthesize polyunsaturated fatty acids to ensure membrane fluidity at reduced temperatures.

Because they are necessary precursors for the synthesis of other products, linoleate and linolenate are **essential fatty acids** for mammals; they must be obtained from dietary plant material. Once ingested, linoleate may be converted to certain other polyunsaturated acids, particularly $\gamma$-linolenate, eicosatrienoate, and arachidonate (eicosatetraenoate), all of which can

be made only from linoleate (Fig. 21–12). Arachidonate, $20:4(\Delta^{5,8,11,14})$, is an essential precursor of regulatory lipids, the eicosanoids. The 20-carbon fatty acids are synthesized from linoleate (and linolenate) by fatty acid elongation reactions analogous to those described on page 797.

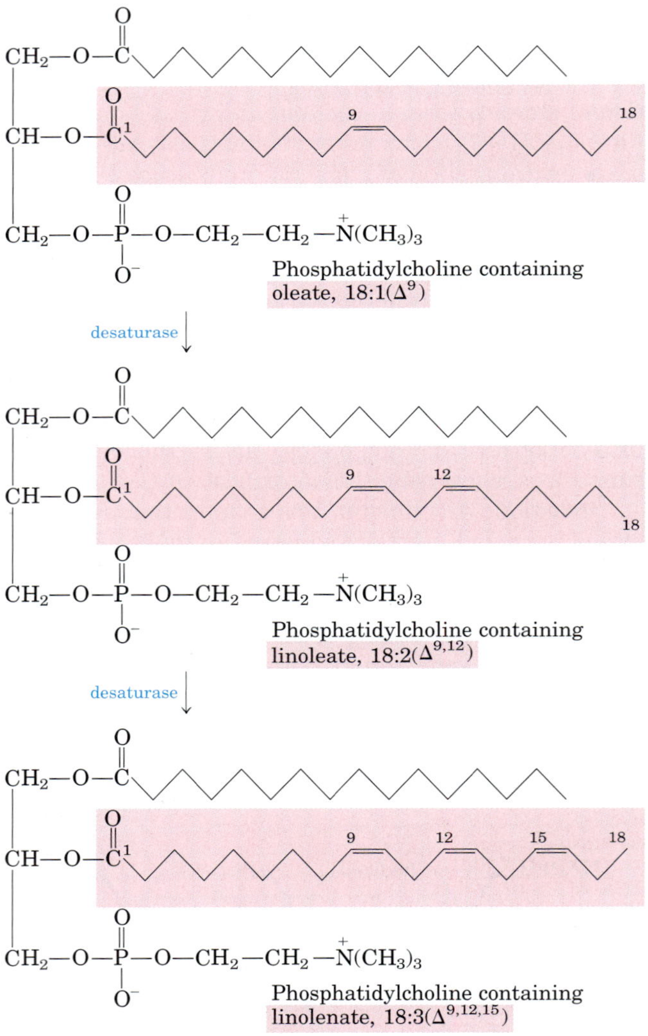

## Eicosanoids Are Formed from 20-Carbon Polyunsaturated Fatty Acids

Eicosanoids are a family of very potent biological signaling molecules that act as short-range messengers, affecting tissues near the cells that produce them. In response to hormonal or other stimuli, phospholipase $A_2$, present in most types of mammalian cells, attacks membrane phospholipids, releasing arachidonate from the middle carbon of glycerol. Enzymes of the smooth ER then convert arachidonate to **prostaglandins,** beginning with the formation of prostaglandin $H_2$ ($PGH_2$), the immediate precursor of many other prostaglandins and of thromboxanes (Fig. 21–15a). The two reactions that lead to $PGH_2$ are catalyzed by a bifunctional enzyme, **cyclooxygenase (COX),** also called **prostaglandin $H_2$ synthase.** In the first of two steps, the cyclooxygenase activity introduces molecular oxygen to convert arachidonate to $PGG_2$. The second step, catalyzed by the peroxidase activity of COX, converts $PGG_2$ to $PGH_2$.

Aspirin (acetylsalicylate; Fig. 21–15b) irreversibly inactivates the cyclooxygenase activity of COX by acetylating a Ser residue and blocking the enzyme's active site, thus inhibiting the synthesis of prostaglandins and thromboxanes. Ibuprofen, a widely used nonsteroidal antiinflammatory drug (NSAID; Fig. 21–15c), inhibits the same enzyme. The recent discovery that there are two isozymes of COX has led to the development of more precisely targeted NSAIDs with fewer undesirable side effects (Box 21–2).

**Thromboxane synthase,** present in blood platelets (thrombocytes), converts $PGH_2$ to thromboxane $A_2$, from which other **thromboxanes** are derived (Fig. 21–15a). Thromboxanes induce constriction of blood vessels and platelet aggregation, early steps in blood clotting. Low doses of aspirin, taken regularly, reduce the probability of heart attacks and strokes by reducing thromboxane production. ∎

Thromboxanes, like prostaglandins, contain a ring of five or six atoms; the pathway from arachidonate to these two classes of compounds is sometimes called the "cyclic" pathway, to distinguish it from the "linear" pathway that leads from arachidonate to the **leukotrienes,** which are linear compounds (Fig. 21–16). Leukotriene synthesis begins with the action of several lipoxygenases that catalyze the incorporation of molecular oxygen into arachidonate. These enzymes, found in leukocytes and in heart, brain, lung, and spleen, are mixed-function oxidases that use cytochrome P-450 (Box 21–1). The various leukotrienes differ in the position of the peroxide

**FIGURE 21–14 Action of plant desaturases.** Desaturases in plants oxidize phosphatidylcholine-bound oleate to polyunsaturated fatty acids. Some of the products are released from the phosphatidylcholine by hydrolysis.

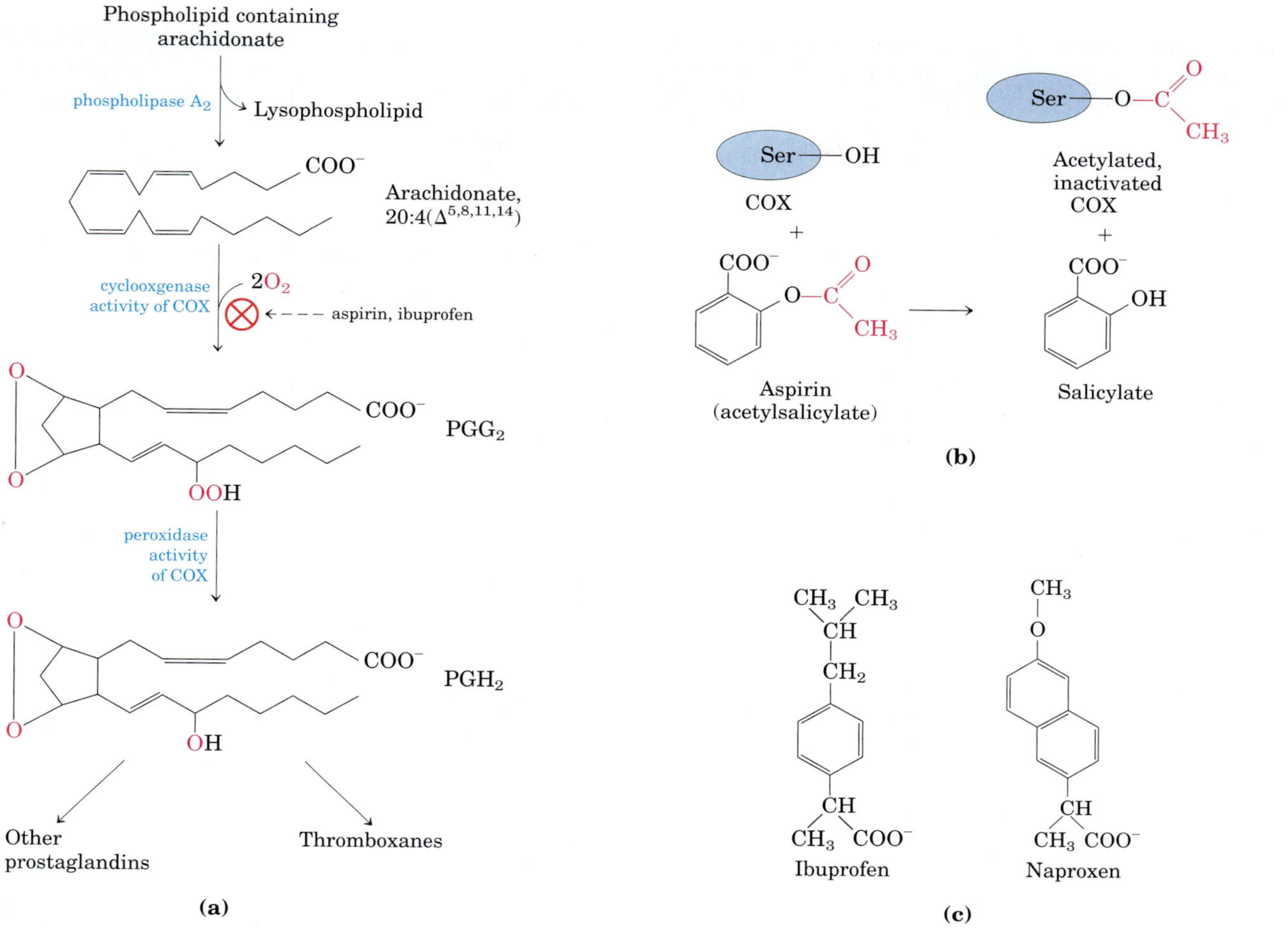

**FIGURE 21–15  The "cyclic" pathway from arachidonate to prostaglandins and thromboxanes. (a)** After arachidonate is released from phospholipids by the action of phospholipase $A_2$, the cyclooxygenase and peroxidase activities of COX (also called prostaglandin $H_2$ synthase) catalyze the production of $PGH_2$, the precursor of other prostaglandins and thromboxanes. **(b)** Aspirin inhibits the first reaction by acetylating an essential Ser residue on the enzyme. **(c)** Ibuprofen and naproxen inhibit the same step, probably by mimicking the structure of the substrate or an intermediate in the reaction.

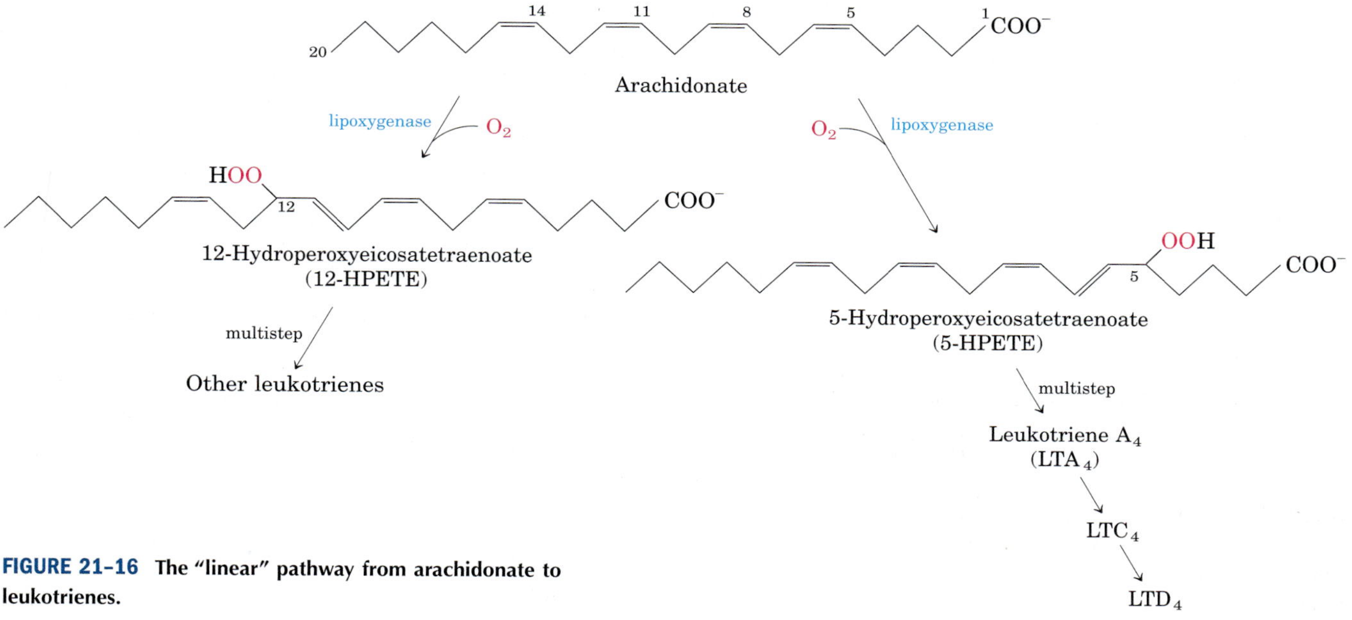

**FIGURE 21–16  The "linear" pathway from arachidonate to leukotrienes.**

BOX 21–2    BIOCHEMISTRY IN MEDICINE

## Relief Is in (the Active) Site: Cyclooxygenase Isozymes and the Search for a Better Aspirin

Each year, several *thousand tons* of aspirin (acetylsalicylate) are consumed around the world for the relief of headaches, sore muscles, inflamed joints, and fever. Because aspirin inhibits platelet aggregation and blood clotting, it is also used in low doses to treat patients at risk of heart attacks. The medicinal properties of the compounds known as salicylates, including aspirin, were first described by western science in 1763, when Edmund Stone of England noted that bark of the willow tree *Salix alba* was effective against fevers, aches, and pains. By the 1830s, German chemists had purified the active components from willow and from another plant rich in salicylates, the meadowsweet, *Spiraea ulmaria*. However, salicylate itself was bitter-tasting and its use had some unpleasant side effects, including severe stomach irritation in some cases. To address these problems, Felix Hoffmann and Arthur Eichengrun synthesized acetylsalicylate at the Bayer company in Germany in 1897. The new compound, with fewer side effects than salicylate, was marketed in 1899 under the trade name Aspirin (from *a* for acetyl and *spir* for *Spirsaüre*, the German word for the acid prepared from *Spiraea*). Within a few years, aspirin was in widespread use.

Aspirin (now a generic name) is one of a number of nonsteroidal antiinflammatory drugs (NSAIDs); others include ibuprofen and naproxen (see Fig. 21–15), all now sold over the counter. Unfortunately, aspirin reduces but does not eliminate the side effects of salicylates. In some patients, aspirin itself can produce stomach bleeding, kidney failure, and, in extreme cases, death. New NSAIDs with the beneficial effects of aspirin but without its side effects would be medically valuable.

Aspirin and other NSAIDs inhibit the cyclooxygenase activity of prostaglandin $H_2$ synthase (also called COX, for cyclooxygenase), which adds molecular oxygen to arachidonate to initiate prostaglandin synthesis (see Fig. 21–15a). Prostaglandins regulate many physiological processes, including platelet aggrega-

tion, uterine contractions, pain, inflammation, and the secretion of mucins that protect the gastric mucosa from acid and proteolytic enzymes in the stomach. The stomach irritation that is a common side effect of aspirin use results from the drug's interference with the secretion of gastric mucin.

Mammals have two isozymes of prostaglandin $H_2$ synthase, COX-1 and COX-2. These have different functions but closely similar amino acid sequences (60% to 65% sequence identity) and similar reaction mechanisms at both of their catalytic centers. COX-1 is responsible for the synthesis of the prostaglandins that regulate the secretion of gastric mucin, and COX-2 for the prostaglandins that mediate inflammation, pain, and fever. Aspirin inhibits both isozymes about equally, so a dose sufficient to reduce inflammation also risks stomach irritation. Much research is aimed at developing new NSAIDs that inhibit COX-2 specifically, and several such drugs have become available.

The development of COX-2–specific inhibitors has been helped immensely by knowledge of the detailed three-dimensional structures of COX-1 and COX-2 (Fig. 1). Both proteins are homodimers. Each monomer ($M_r$ 70,000) has an amphipathic domain that penetrates but does not span the ER; this anchors the enzyme on the lumenal side of the ER (a very unusual topology—generally the hydrophobic regions of integral membrane proteins span the entire bilayer). Both catalytic sites are on the globular domain protruding into the ER lumen.

COX-1 and COX-2 have virtually identical tertiary and quaternary structures, but they differ subtly in a long, thin hydrophobic channel extending from the membrane interior to the lumenal surface. The channel includes both catalytic sites and is presumed to be the binding site for the hydrophobic substrate, arachidonate. Both COX-1 and COX-2 have been crystallized in the presence of several different bound NSAID compounds, defining the NSAID-binding site (Fig. 1). The bound drugs block the hydrophobic channel and prevent arachidonate entry. The subtle differences between the channels of COX-1 and COX-2 have guided

---

group introduced by the lipoxygenases. This linear pathway from arachidonate, unlike the cyclic pathway, is not inhibited by aspirin or other NSAIDs.

Plants also derive important signaling molecules from fatty acids. As in animals, a key step in the initiation of signaling involves activation of a specific phospholipase. In plants, the fatty acid substrate that is re-

leased is $\alpha$-linolenate. A lipoxygenase then catalyzes the first step in a pathway that converts linolenate to jasmonate, a substance known to have signaling roles in insect defense, resistance to fungal pathogens, and pollen maturation. Jasmonate (see Fig. 12–28) also affects seed germination, root growth, and fruit and seed development.

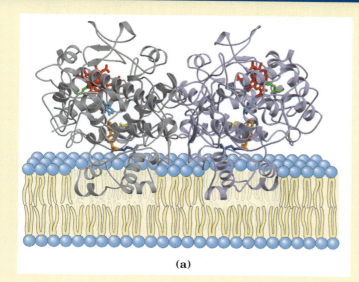

(a)

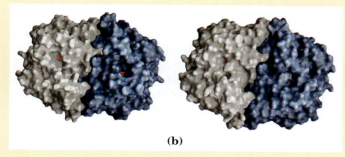

(b)

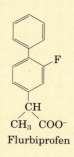

**FIGURE 1 Structures of COX-1 and COX-2. (a)** COX-1 with an NSAID inhibitor (flurbiprofen, orange) bound (PDB ID 3PGH). The enzyme consists of two identical monomers (gray and blue) each with three domains: a membrane anchor consisting of four amphipathic helices; a second domain that somewhat resembles a domain of the epidermal growth factor; and the catalytic domain, which contains the cyclooxygenase and peroxidase activities, as well as the hydrophobic channel in which the substrate (arachidonate) binds. The heme that is part of the peroxidase active sites is shown in red; Tyr$^{385}$, a key residue in the cyclooxygenase site, is turquoise. Other catalytically important residues include Arg$^{120}$ (dark blue), His$^{388}$ (green), and Ser$^{530}$ (yellow). Flurbiprofen blocks access to the enzyme active site.

F

CH

CH$_3$  COO$^-$

**Flurbiprofen**

**(b)** A look at COX-1 and COX-2 side by side. The COX-1 enzyme from sheep is shown at left, with bound ibuprofen (PDB ID 1EQG). The COX-2 enzyme from mouse is shown at right, with a similar inhibitor, Sc-558, bound (PDB ID 6COX). The inhibitors (red) are partially buried within the structures in these representations, which emphasize surface contours. COX-1 and COX-2 are very similar in structure, but enzymologists exploit the small differences in the structures of the active sites and the channels leading to them to design inhibitors specific for one enzyme or the other.

the design of NSAIDs that selectively fit COX-2 and therefore inhibit COX-2 more effectively than COX-1. Two of these drugs have been approved for use worldwide: celecoxib (Celebrex) for osteoarthritis and rheumatoid arthritis, and rofecoxib (Vioxx) for osteoarthritis and acute musculoskeletal pain (Fig. 2). In clinical trials, these drugs have proven effective while significantly reducing the stomach irritation and other side effects of aspirin and other NSAIDs. The use of precise structural information about an enzyme's active site is a powerful tool in the development of better, more specific drugs.

H$_3$C

N—N

CF$_3$

O

H$_2$N

S

O

**Celecoxib
(Celebrex)**

O

O

H$_3$C

S

O

**Rofecoxib
(Vioxx)**

**FIGURE 2 Two COX-2–specific drugs that bind to COX-2 about 1,000 times better than to COX-1.**

## SUMMARY 21.1 Biosynthesis of Fatty Acids and Eicosanoids

■ Long-chain saturated fatty acids are synthesized from acetyl-CoA by a cytosolic complex of six enzyme activities plus acyl carrier protein (ACP). The fatty acid synthase complex, which in some organisms consists of multifunctional polypeptides, contains two types of —SH groups (one furnished by the phosphopantetheine of ACP, the other by a Cys residue of β-ketoacyl-ACP synthase) that function as carriers of the fatty acyl intermediates.

- Malonyl-ACP, formed from acetyl-CoA (shuttled out of mitochondria) and $CO_2$, condenses with an acetyl bound to the Cys—SH to yield acetoacetyl-ACP, with release of $CO_2$. This is followed by reduction to the D-$\beta$-hydroxy derivative, dehydration to the $trans$-$\Delta^2$-unsaturated acyl-ACP, and reduction to butyryl-ACP. NADPH is the electron donor for both reductions. Fatty acid synthesis is regulated at the level of malonyl-CoA formation.

- Six more molecules of malonyl-ACP react successively at the carboxyl end of the growing fatty acid chain to form palmitoyl-ACP—the end product of the fatty acid synthase reaction. Free palmitate is released by hydrolysis.

- Palmitate may be elongated to the 18-carbon stearate. Palmitate and stearate can be desaturated to yield palmitoleate and oleate, respectively, by the action of mixed-function oxidases.

- Mammals cannot make linoleate and must obtain it from plant sources; they convert exogenous linoleate to arachidonate, the parent compound of eicosanoids (prostaglandins, thromboxanes, and leukotrienes), a family of very potent signaling molecules.

## 21.2 Biosynthesis of Triacylglycerols

Most of the fatty acids synthesized or ingested by an organism have one of two fates: incorporation into triacylglycerols for the storage of metabolic energy or incorporation into the phospholipid components of membranes. The partitioning between these alternative fates depends on the organism's current needs. During rapid growth, synthesis of new membranes requires the production of membrane phospholipids; when an organism has a plentiful food supply but is not actively growing, it shunts most of its fatty acids into storage fats. Both pathways begin at the same point: the formation of fatty acyl esters of glycerol. In this section we examine the route to triacylglycerols and its regulation, and the production of glycerol 3-phosphate in the process of glyceroneogenesis.

### Triacylglycerols and Glycerophospholipids Are Synthesized from the Same Precursors

Animals can synthesize and store large quantities of triacylglycerols, to be used later as fuel (see Box 17–1). Humans can store only a few hundred grams of glycogen in liver and muscle, barely enough to supply the body's energy needs for 12 hours. In contrast, the total amount of stored triacylglycerol in a 70-kg man of average build is about 15 kg, enough to support basal energy needs for as long as 12 weeks (see Table 23–5). Triacylglycerols have the highest energy content of all stored nutrients—more than 38 kJ/g. Whenever carbohydrate is ingested in excess of the organism's capacity to store glycogen, the excess is converted to triacylglycerols and stored in adipose tissue. Plants also manufacture triacylglycerols as an energy-rich fuel, mainly stored in fruits, nuts, and seeds.

In animal tissues, triacylglycerols and glycerophospholipids such as phosphatidylethanolamine share two precursors (fatty acyl–CoA and L-glycerol 3-phosphate) and several biosynthetic steps. The vast majority of the glycerol 3-phosphate is derived from the glycolytic intermediate dihydroxyacetone phosphate (DHAP) by the action of the cytosolic NAD-linked **glycerol 3-phosphate dehydrogenase;** in liver and kidney, a small amount of glycerol 3-phosphate is also formed from glycerol by the action of **glycerol kinase** (Fig. 21–17). The other precursors of triacylglycerols are fatty acyl–CoAs, formed from fatty acids by **acyl-CoA synthetases,** the same enzymes responsible for the activation of fatty acids for $\beta$ oxidation (see Fig. 17–5).

The first stage in the biosynthesis of triacylglycerols is the acylation of the two free hydroxyl groups of L-glycerol 3-phosphate by two molecules of fatty acyl–CoA to yield **diacylglycerol 3-phosphate,** more commonly called **phosphatidic acid** or phosphatidate (Fig. 21–17). Phosphatidic acid is present in only trace amounts in cells but is a central intermediate in lipid biosynthesis; it can be converted either to a triacylglycerol or to a glycerophospholipid. In the pathway to triacylglycerols, phosphatidic acid is hydrolyzed by **phosphatidic acid phosphatase** to form a 1,2-diacylglycerol (Fig. 21–18). Diacylglycerols are then converted to triacylglycerols by transesterification with a third fatty acyl–CoA.

### Triacylglycerol Biosynthesis in Animals Is Regulated by Hormones

In humans, the amount of body fat stays relatively constant over long periods, although there may be minor short-term changes as caloric intake fluctuates. Carbohydrate, fat, or protein consumed in excess of energy needs is stored in the form of triacylglycerols that can be drawn upon for energy, enabling the body to withstand periods of fasting.

Biosynthesis and degradation of triacylglycerols are regulated such that the favored path depends on the metabolic resources and requirements of the moment. The rate of triacylglycerol biosynthesis is profoundly altered by the action of several hormones. Insulin, for example, promotes the conversion of carbohydrate to triacylglycerols (Fig. 21–19). People with severe diabetes mellitus, due to failure of insulin secretion or action, not only are unable to use glucose properly but also fail to synthesize fatty acids from carbohydrates or amino acids. If the diabetes is untreated, these individuals have increased rates of fat oxidation and ketone body formation (Chapter 17) and therefore lose weight. ■

An additional factor in the balance between biosynthesis and degradation of triacylglycerols is that approximately 75% of all fatty acids released by lipolysis are reesterified to form triacylglycerols rather than used for fuel. This ratio persists even under starvation conditions, when energy metabolism is shunted from the use of carbohydrate to the oxidation of fatty acids. Some of this fatty acid recycling takes place in adipose tissue, with the reesterification occurring before release into the bloodstream; some takes place via a systemic cycle in which free fatty acids are transported to liver, recycled to triacylglycerol, exported again to the blood (transport of lipids in the blood is discussed in Section 21.4), and taken up again by adipose tissue after release from triacylglycerol by extracellular lipoprotein lipase

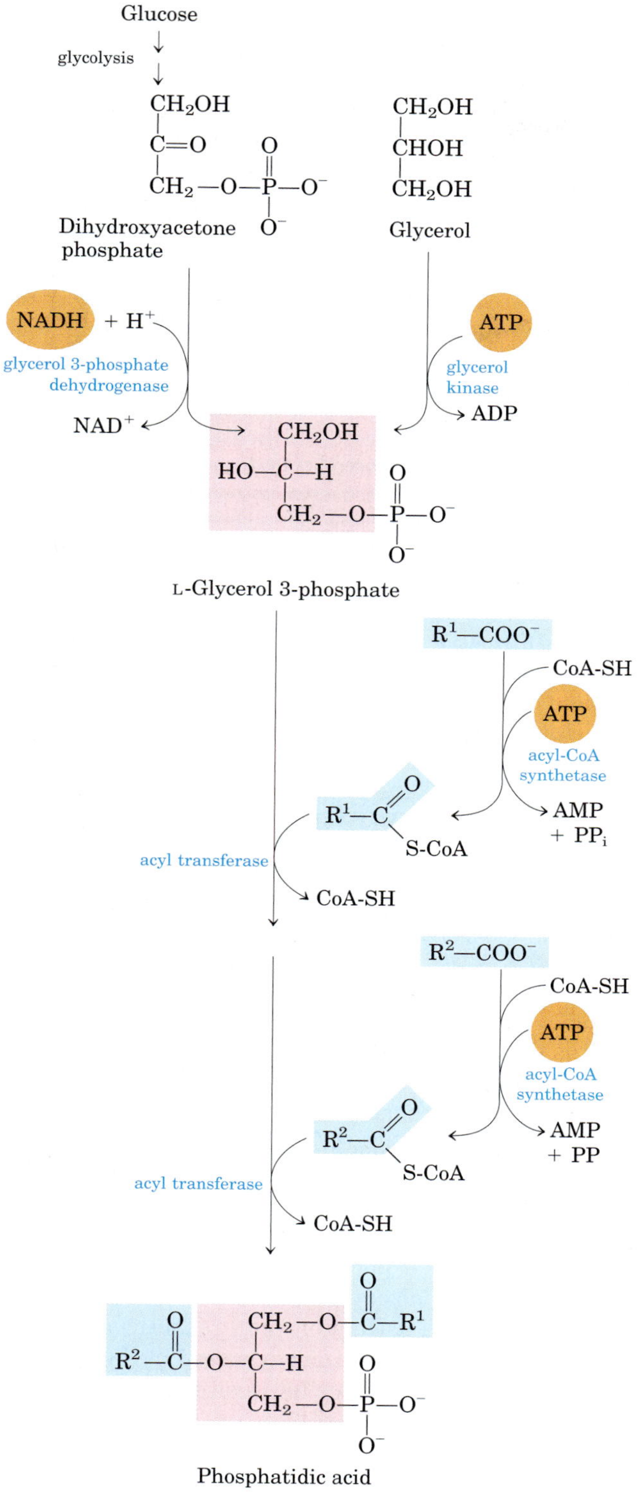

**FIGURE 21–17 Biosynthesis of phosphatidic acid.** A fatty acyl group is activated by formation of the fatty acyl–CoA, then transferred to ester linkage with L-glycerol 3-phosphate, formed in either of the two ways shown. Phosphatidic acid is shown here with the correct stereochemistry at C-2 of the glycerol molecule. To conserve space in subsequent figures (and in Fig. 21–14), both fatty acyl groups of glycerophospholipids, and all three acyl groups of triacylglycerols, are shown projecting to the right.

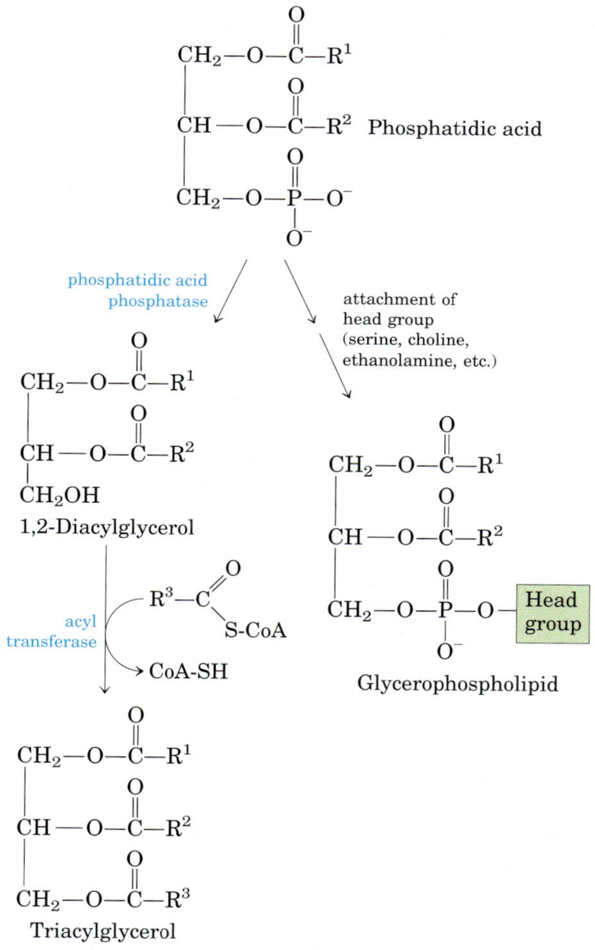

**FIGURE 21–18 Phosphatidic acid in lipid biosynthesis.** Phosphatidic acid is the precursor of both triacylglycerols and glycerophospholipids. The mechanisms for head-group attachment in phospholipid synthesis are described later in this section.

(Fig. 21–20; see also Fig. 17–1). Flux through this **triacylglycerol cycle** between adipose tissue and liver may be quite low when other fuels are available and the release of fatty acids from adipose tissue is limited, but as noted above, the proportion of released fatty acids that are reesterified remains roughly constant at 75% under all metabolic conditions. The level of free fatty acids in the blood thus reflects both the rate of release of fatty acids and the balance between the synthesis and breakdown of triacylglycerols in adipose tissue and liver.

When the mobilization of fatty acids is required to meet energy needs, release from adipose tissue is stimulated by the hormones glucagon and epinephrine (see Figs 17–3, 17–12). Simultaneously, these hormonal signals decrease the rate of glycolysis and increase the rate of gluconeogenesis in the liver (providing glucose for the brain, as further elaborated in Chapter 23). The released fatty acid is taken up by a number of tissues, including muscle, where it is oxidized to provide energy. Much of the fatty acid taken up by liver is not oxidized but is recycled to triacylglycerol and returned to adipose tissue.

The function of the apparently futile triacylglycerol cycle (futile cycles are discussed in Chapter 15) is not well understood. However, as we learn more about how the triacylglycerol cycle is sustained via metabolism in two separate organs and is coordinately regulated, some possibilities emerge. For example, the excess capacity in the triacylglycerol cycle (the fatty acid that is even-

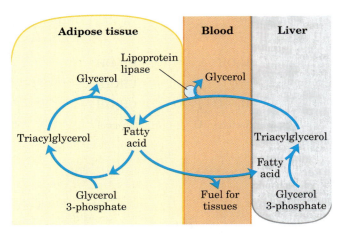

**FIGURE 21–20 The triacylglycerol cycle.** In mammals, triacylglycerol molecules are broken down and resynthesized in a triacylglycerol cycle during starvation. Some of the fatty acids released by lipolysis of triacylglycerol in adipose tissue pass into the bloodstream, and the remainder are used for resynthesis of triacylglycerol. Some of the fatty acids released into the blood are used for energy (in muscle, for example), and some are taken up by the liver and used in triacylglycerol synthesis. The triacylglycerol formed in the liver is transported in the blood back to adipose tissue, where the fatty acid is released by extracellular lipoprotein lipase, taken up by adipocytes, and reesterified into triacylglycerol.

tually reconverted to triacylglycerol rather than oxidized as fuel) could represent an energy reserve in the bloodstream during fasting, one that would be more rapidly mobilized in a "fight or flight" emergency than would stored triacylglycerol.

The constant recycling of triacylglycerols in adipose tissue even during starvation raises a second question: what is the source of the glycerol 3-phosphate required for this process? As noted above, glycolysis is suppressed in these conditions by the action of glucagon and epinephrine, so little DHAP is available, and glycerol released during lipolysis cannot be converted directly to glycerol 3-phosphate in adipose tissue, because these cells lack glycerol kinase (Fig. 21–17). So, how is sufficient glycerol 3-phosphate produced? The answer lies in a pathway discovered more than three decades ago and given little attention until recently, a pathway intimately linked to the triacylglycerol cycle and, in a larger sense, to the balance between fatty acid and carbohydrate metabolism.

## Adipose Tissue Generates Glycerol 3-phosphate by Glyceroneogenesis

**Glyceroneogenesis** is a shortened version of gluconeogenesis, from pyruvate to DHAP (see Fig. 14–16), followed by conversion of the DHAP to glycerol 3-phosphate by cytosolic NAD-linked glycerol 3-phosphate dehydrogenase (Fig. 21–21). Glycerol 3-phosphate is subsequently used in triacylglycerol synthesis. Glycero-

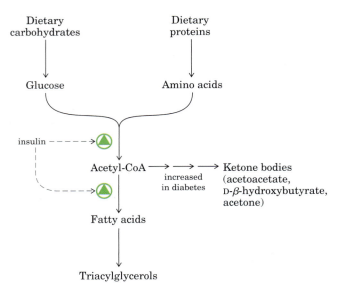

**FIGURE 21–19 Regulation of triacylglycerol synthesis by insulin.** Insulin stimulates conversion of dietary carbohydrates and proteins to fat. Individuals with diabetes mellitus lack insulin; in uncontrolled disease, this results in diminished fatty acid synthesis, and the acetyl-CoA arising from catabolism of carbohydrates and proteins is shunted instead to ketone body production. People in severe ketosis smell of acetone, so the condition is sometimes mistaken for drunkenness (p. 909).

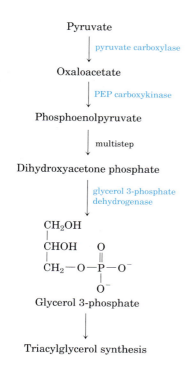

**FIGURE 21–21 Glyceroneogenesis.** The pathway is essentially an abbreviated version of gluconeogenesis, from pyruvate to dihydroxyacetone phosphate (DHAP), followed by conversion of DHAP to glycerol 3-phosphate, which is used for the synthesis of triacylglycerol.

neogenesis was discovered in the 1960s by Lea Reshef, Richard Hanson, and John Ballard, and simultaneously by Eleazar Shafrir and his coworkers, who were intrigued by the presence of two gluconeogenic enzymes, pyruvate carboxylase and phosphoenolpyruvate (PEP) carboxykinase, in adipose tissue, where glucose is not synthesized. After a long period of inattention, interest in this pathway has been renewed by the demonstration of a link between glyceroneogenesis and late-onset (type 2) diabetes, as we shall see.

Glyceroneogenesis has multiple roles. In adipose tissue, glyceroneogenesis coupled with reesterification of free fatty acids controls the rate of fatty acid release to the blood. In brown adipose tissue, the same pathway may control the rate at which free fatty acids are delivered to mitochondria for use in thermogenesis (see Fig. 19–30). And in fasting humans, glyceroneogenesis in the liver alone supports the synthesis of enough glycerol 3-phosphate to account for up to 65% of fatty acids reesterified to triacylglycerol.

Flux through the triacylglycerol cycle between liver and adipose tissue is controlled to a large degree by the activity of PEP carboxykinase, which limits the rate of both gluconeogenesis and glyceroneogenesis. Glucocorticoid hormones such as cortisol (a biological steroid derived from cholesterol; see Fig. 21–46) and dexamethasone (a synthetic glucocorticoid) regulate the

levels of PEP carboxykinase reciprocally in the liver and adipose tissue. Acting through the glucocorticoid receptor, these steroid hormones increase the expression of the gene encoding PEP carboxykinase in the liver, thus increasing gluconeogenesis and glyceroneogenesis (Fig. 21–22).

Cortisol

Dexamethasone

Stimulation of glyceroneogenesis leads to an increase in the synthesis of triacylglycerol molecules in the liver and their release into the blood. At the same time, glucocorticoids suppress the expression of the gene encoding PEP carboxykinase in adipose tissue. This results in a decrease in glyceroneogenesis in adipose tissue; recycling of fatty acids declines as a result, and more free fatty acids are released into the blood. Thus glyceroneogenesis is regulated reciprocally in the liver and adipose tissue, affecting lipid metabolism in opposite ways: a lower rate of glyceroneogenesis in adipose tissue leads to more fatty acid release (rather than recycling), whereas a higher rate in the liver leads to more synthesis and export of triacylglycerols. The net result is an increase in flux through the triacylglycerol cycle. When the glucocorticoids are no longer present, flux through the cycle declines as the expression of PEP carboxykinase increases in adipose tissue and decreases in the liver.

The recent attention given to glyceroneogenesis has arisen in part from the connection between this pathway and diabetes. High levels of free fatty acids in the blood interfere with glucose utilization in muscle and promote the insulin resistance that leads to type 2 diabetes. A new class of drugs called **thiazolidinediones** have been shown to reduce the levels of fatty acids circulating in the blood and increase sensitivity to

Rosiglitazone (Avandia)

Pioglitazone (Actos)
**Thiazolidinediones**

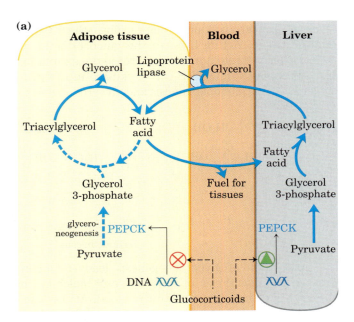

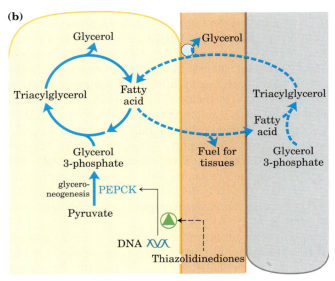

**FIGURE 21–22 Regulation of glyceroneogenesis. (a)** Glucocorticoid hormones stimulate glyceroneogenesis and gluconeogenesis in the liver, while suppressing glyceroneogenesis in the adipose tissue (by reciprocal regulation of the gene expressing PEP carboxykinase (PEPCK) in the two tissues); this increases the flux through the triacylglycerol cycle. The glycerol freed by the breakdown of triacylglycerol in adipose tissue is released to the blood and transported to the liver, where it is primarily converted to glucose, although some is converted to glycerol 3-phosphate by glycerol kinase.

**(b)** A class of drugs called thiazolidinediones are now used to treat type 2 diabetes. In this disease, high levels of free fatty acids in the blood interfere with glucose utilization in muscle and promote insulin resistance. Thiazolidinediones activate a nuclear receptor called peroxisome proliferator-activated receptor γ (PPARγ), which induces the activity of PEP carboxykinase. Therapeutically, thiazolidinediones increase the rate of glyceroneogenesis, thus increasing the resynthesis of triacylglycerol in adipose tissue and reducing the amount of free fatty acid in the blood.

insulin. Thiazolidinediones bind to and activate a nuclear hormone receptor called peroxisome proliferator-activated receptor γ (PPARγ), leading to the induction in adipose tissue of PEP carboxykinase (Fig. 21–22); a higher activity of PEP carboxykinase then leads to increased synthesis of the precursors of glyceroneogenesis. The therapeutic effect of thiazolidinediones is thus due, at least in part, to the increase in glyceroneogenesis, which in turn increases the resynthesis of triacylglycerol in adipose tissue and reduces the release of free fatty acid from adipose tissue into the blood. ■

### SUMMARY 21.2 Biosynthesis of Triacylglycerols

- Triacylglycerols are formed by reaction of two molecules of fatty acyl–CoA with glycerol 3-phosphate to form phosphatidic acid; this product is dephosphorylated to a diacylglycerol, then acylated by a third molecule of fatty acyl–CoA to yield a triacylglycerol.

- The synthesis and degradation of triacylglycerol are hormonally regulated.

- Mobilization and recycling of triacylglycerol molecules results in a triacylglycerol cycle. Triacylglycerols are resynthesized from free fatty acids and glycerol 3-phosphate even during starvation. The dihydroxyacetone phosphate precursor of glycerol 3-phosphate is derived from pyruvate via glyceroneogenesis.

## 21.3 Biosynthesis of Membrane Phospholipids

In Chapter 10 we introduced two major classes of membrane phospholipids: glycerophospholipids and sphingolipids. Many different phospholipid species can be constructed by combining various fatty acids and polar head groups with the glycerol or sphingosine backbone (see Figs 10–8, 10–12). All the biosynthetic pathways follow a few basic patterns. In general, the assembly of phospholipids from simple precursors requires (1) synthesis of the backbone molecule (glycerol or sphingosine); (2) attachment of fatty acid(s) to the backbone through an ester or amide linkage; (3) addition of a hydrophilic head group to the backbone through a phosphodiester linkage; and, in some cases, (4) alteration or exchange of the head group to yield the final phospholipid product.

In eukaryotic cells, phospholipid synthesis occurs primarily on the surfaces of the smooth endoplasmic reticulum and the mitochondrial inner membrane. Some newly formed phospholipids remain at the site of synthesis, but most are destined for other cellular locations.

The process by which water-insoluble phospholipids move from the site of synthesis to the point of their eventual function is not fully understood, but we conclude this section by discussing some mechanisms that have emerged in recent years.

## Cells Have Two Strategies for Attaching Phospholipid Head Groups

The first steps of glycerophospholipid synthesis are shared with the pathway to triacylglycerols (Fig. 21–17): two fatty acyl groups are esterified to C-1 and C-2 of L-glycerol 3-phosphate to form phosphatidic acid. Commonly but not invariably, the fatty acid at C-1 is saturated and that at C-2 is unsaturated. A second route to phosphatidic acid is the phosphorylation of a diacylglycerol by a specific kinase.

The polar head group of glycerophospholipids is attached through a phosphodiester bond, in which each of two alcohol hydroxyls (one on the polar head group and one on C-3 of glycerol) forms an ester with phosphoric acid (Fig. 21–23). In the biosynthetic process, one of the hydroxyls is first activated by attachment of a nucleotide, cytidine diphosphate (CDP). Cytidine monophosphate (CMP) is then displaced in a nucleophilic attack by the other hydroxyl (Fig. 21–24). The CDP is attached either to the diacylglycerol, forming the activated phosphatidic acid **CDP-diacylglycerol**

(strategy 1), or to the hydroxyl of the head group (strategy 2). Eukaryotic cells employ both strategies, whereas prokaryotes use only the first. The central importance of cytidine nucleotides in lipid biosynthesis was discovered by Eugene P. Kennedy in the early 1960s.

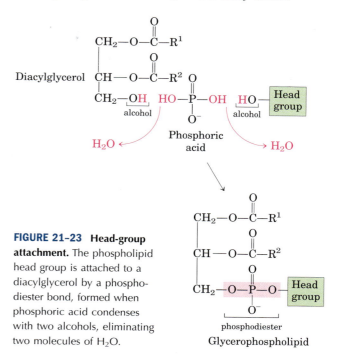

**FIGURE 21–23 Head-group attachment.** The phospholipid head group is attached to a diacylglycerol by a phosphodiester bond, formed when phosphoric acid condenses with two alcohols, eliminating two molecules of $H_2O$.

Eugene P. Kennedy

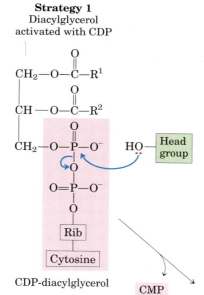

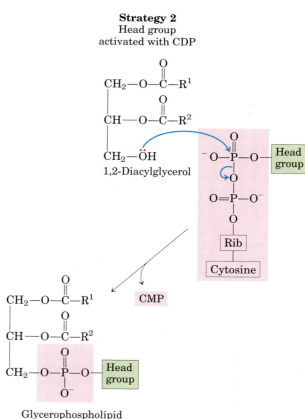

**FIGURE 21–24 Two general strategies for forming the phosphodiester bond of phospholipids.** In both cases, CDP supplies the phosphate group of the phosphodiester bond.

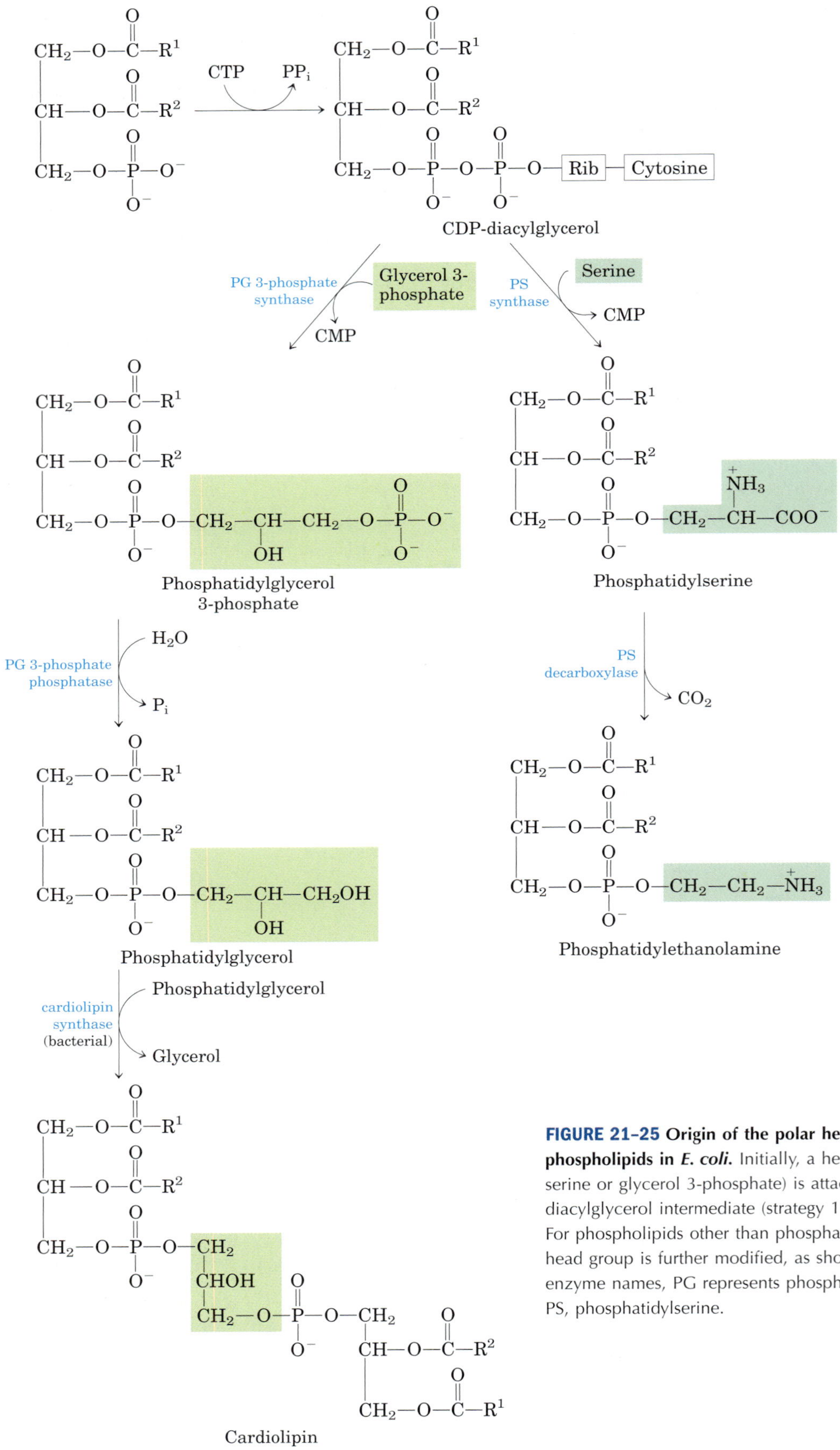

The figure shows chemical structures and reactions:

$$CH_2-O-C(=O)-R^1$$
$$CH-O-C(=O)-R^2$$
$$CH_2-O-P(=O)(O^-)-O^-$$

CTP → PP$_i$

CDP-diacylglycerol

$$CH_2-O-C(=O)-R^1$$
$$CH-O-C(=O)-R^2$$
$$CH_2-O-P(=O)(O^-)-O-P(=O)(O^-)-O-[Rib]-[Cytosine]$$

PG 3-phosphate synthase — Glycerol 3-phosphate → CMP

PS synthase — Serine → CMP

**Phosphatidylglycerol 3-phosphate**

$$CH_2-O-C(=O)-R^1$$
$$CH-O-C(=O)-R^2$$
$$CH_2-O-P(=O)(O^-)-O-CH_2-CH(OH)-CH_2-O-P(=O)(O^-)-O^-$$

**Phosphatidylserine**

$$CH_2-O-C(=O)-R^1$$
$$CH-O-C(=O)-R^2$$
$$CH_2-O-P(=O)(O^-)-O-CH_2-CH(\overset{+}{N}H_3)-COO^-$$

PG 3-phosphate phosphatase — H$_2$O → P$_i$

PS decarboxylase → CO$_2$

**Phosphatidylglycerol**

$$CH_2-O-C(=O)-R^1$$
$$CH-O-C(=O)-R^2$$
$$CH_2-O-P(=O)(O^-)-O-CH_2-CH(OH)-CH_2OH$$

**Phosphatidylethanolamine**

$$CH_2-O-C(=O)-R^1$$
$$CH-O-C(=O)-R^2$$
$$CH_2-O-P(=O)(O^-)-O-CH_2-CH_2-\overset{+}{N}H_3$$

cardiolipin synthase (bacterial) — Phosphatidylglycerol → Glycerol

**Cardiolipin**

$$CH_2-O-C(=O)-R^1$$
$$CH-O-C(=O)-R^2$$
$$CH_2-O-P(=O)(O^-)-O-CH_2$$
$$CHOH$$
$$CH_2-O-P(=O)(O^-)-O-CH_2$$
$$CH-O-C(=O)-R^2$$
$$CH_2-O-C(=O)-R^1$$

**FIGURE 21–25 Origin of the polar head groups of phospholipids in *E. coli*.** Initially, a head group (either serine or glycerol 3-phosphate) is attached via a CDP-diacylglycerol intermediate (strategy 1 in Fig. 21–24). For phospholipids other than phosphatidylserine, the head group is further modified, as shown here. In the enzyme names, PG represents phosphatidylglycerol; PS, phosphatidylserine.

## Phospholipid Synthesis in *E. coli* Employs CDP-Diacylglycerol

The first strategy for head-group attachment is illustrated by the synthesis of phosphatidylserine, phosphatidylethanolamine, and phosphatidylglycerol in *E. coli*. The diacylglycerol is activated by condensation of phosphatidic acid with cytidine triphosphate (CTP) to form CDP-diacylglycerol, with the elimination of pyrophosphate (Fig. 21–25). Displacement of CMP through nucleophilic attack by the hydroxyl group of serine or by the C-1 hydroxyl of glycerol 3-phosphate yields **phosphatidylserine** or phosphatidylglycerol 3-phosphate, respectively. The latter is processed further by cleavage of the phosphate monoester (with release of P$_i$) to yield **phosphatidylglycerol.**

Phosphatidylserine and phosphatidylglycerol can serve as precursors of other membrane lipids in bacteria (Fig. 21–25). Decarboxylation of the serine moiety in phosphatidylserine, catalyzed by phosphatidylserine decarboxylase, yields **phosphatidylethanolamine.** In *E. coli*, condensation of two molecules of phosphatidylglycerol, with elimination of one glycerol, yields **cardiolipin,** in which two diacylglycerols are joined through a common head group.

## Eukaryotes Synthesize Anionic Phospholipids from CDP-Diacylglycerol

In eukaryotes, phosphatidylglycerol, cardiolipin, and the phosphatidylinositols (all anionic phospholipids; see Fig. 10–8) are synthesized by the same strategy used for phospholipid synthesis in bacteria. Phosphatidylglycerol is made exactly as in bacteria. Cardiolipin synthesis in eukaryotes differs slightly: phosphatidylglycerol condenses with CDP-diacylglycerol (Fig. 21–26), not another molecule of phosphatidylglycerol as in *E. coli* (Fig. 21–25).

Phosphatidylinositol is synthesized by condensation of CDP-diacylglycerol with inositol (Fig. 21–26). Specific **phosphatidylinositol kinases** then convert phosphatidylinositol to its phosphorylated derivatives (see Fig. 10–17). Phosphatidylinositol and its phosphorylated products in the plasma membrane play a central role in signal transduction in eukaryotes (see Figs 12–8, 12–19).

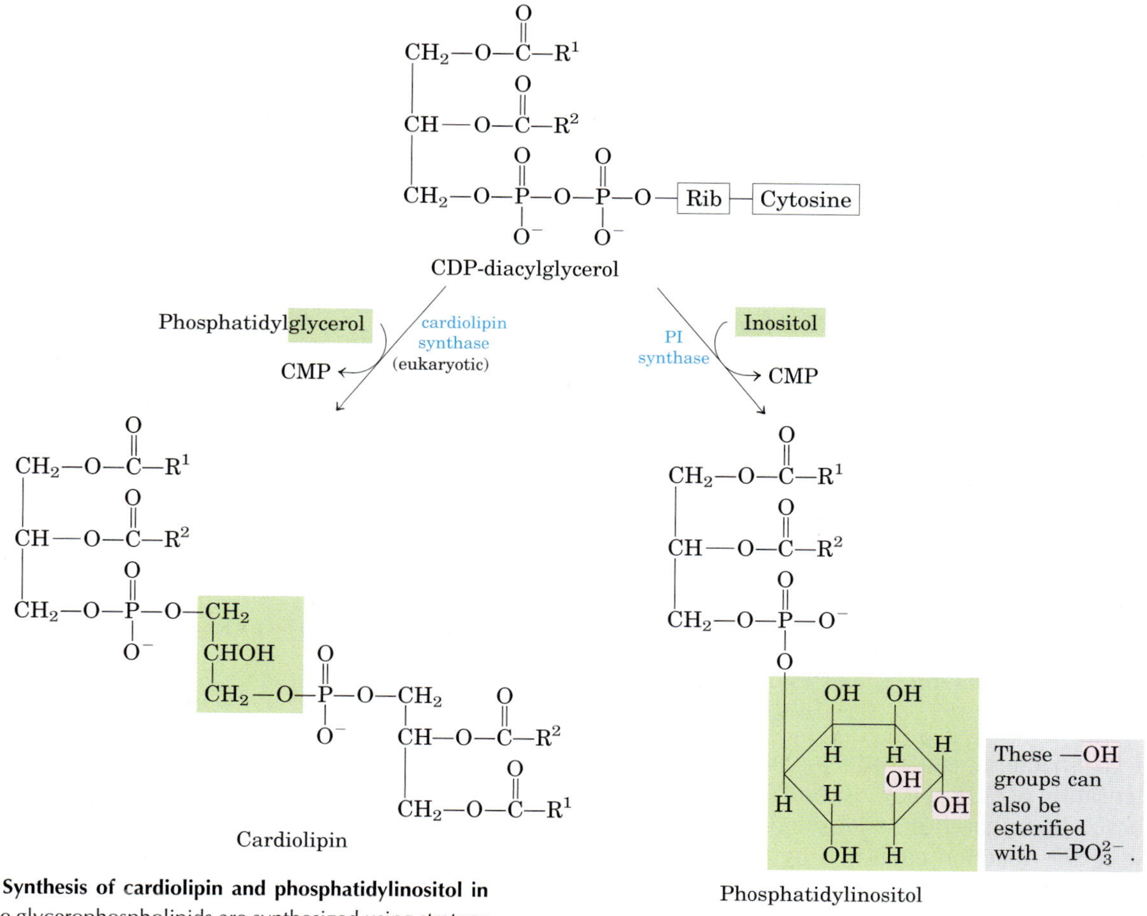

**FIGURE 21–26 Synthesis of cardiolipin and phosphatidylinositol in eukaryotes.** These glycerophospholipids are synthesized using strategy 1 in Figure 21-24. Phosphatidylglycerol is synthesized as in bacteria (see Fig. 21-25). PI represents phosphatidylinositol.

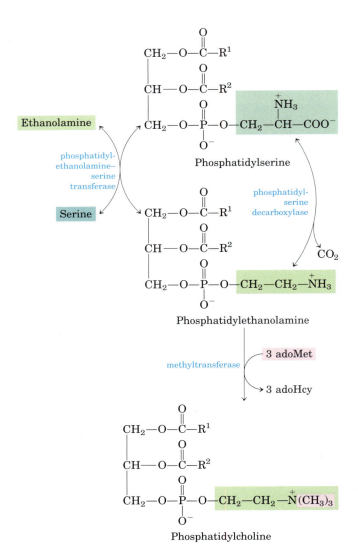

FIGURE 21–27 The "salvage" pathway from phosphatidylserine to phosphatidylethanolamine and phosphatidylcholine in yeast. Phosphatidylserine and phosphatidylethanolamine are interconverted by a reversible head-group exchange reaction. In mammals, phosphatidylserine is derived from phosphatidylethanolamine by a reversal of this reaction; adoMet is S-adenosylmethionine; adoHcy, S-adenosylhomocysteine.

## Eukaryotic Pathways to Phosphatidylserine, Phosphatidylethanolamine, and Phosphatidylcholine Are Interrelated

Yeast, like bacteria, can produce phosphatidylserine by condensation of CDP-diacylglycerol and serine, and can synthesize phosphatidylethanolamine from phosphatidylserine in the reaction catalyzed by phosphatidylserine decarboxylase (Fig. 21–27). In mammalian cells, an alternative route to phosphatidylserine is a head-group

exchange reaction, in which free serine displaces ethanolamine. Phosphatidylethanolamine may also be converted to **phosphatidylcholine** (lecithin) by the addition of three methyl groups to its amino group; *S*-adenosylmethionine is the methyl group donor (see Fig. 18–18) for all three methylation reactions.

In mammals, phosphatidylserine is not synthesized from CDP-diacylglycerol; instead, it is derived from phosphatidylethanolamine via the head-group exchange reaction (Fig. 21–27). Synthesis of phosphatidylethanolamine and phosphatidylcholine in mammals occurs by strategy 2 of Figure 21–24: phosphorylation and activa-

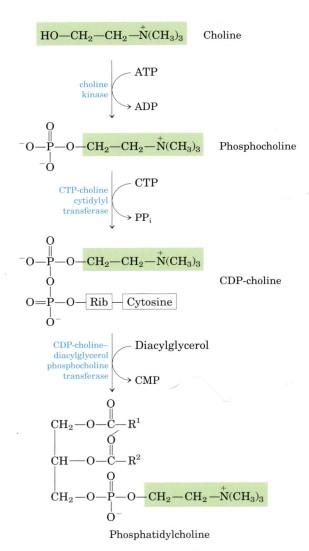

FIGURE 21–28 Pathway for phosphatidylcholine synthesis from choline in mammals. The same strategy shown here (strategy 2 in Fig. 21–24) is also used for salvaging ethanolamine in phosphatidylethanolamine synthesis.

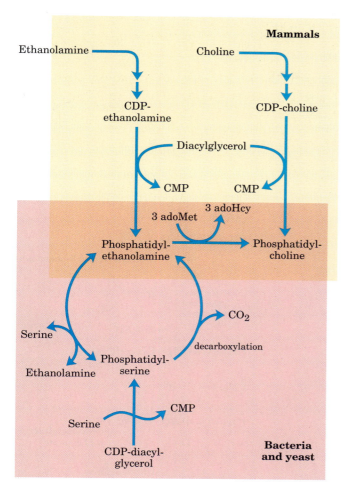

**FIGURE 21–29 Summary of the pathways to phosphatidylcholine and phosphatidylethanolamine.** Conversion of phosphatidylethanolamine to phosphatidylcholine in mammals takes place only in the liver.

tion of the head group, followed by condensation with diacylglycerol. For example, choline is reused ("salvaged") by being phosphorylated then converted to CDP-choline by condensation with CTP. A diacylglycerol displaces CMP from CDP-choline, producing phosphatidylcholine (Fig. 21–28). An analogous salvage pathway converts ethanolamine obtained in the diet to phosphatidylethanolamine. In the liver, phosphatidylcholine is also produced by methylation of phosphatidylethanolamine (with *S*-adenosylmethionine, as described above), but in all other tissues phosphatidylcholine is produced only by condensation of diacylglycerol and CDP-choline. The pathways to phosphatidylcholine and phosphatidylethanolamine in various organisms are summarized in Figure 21–29.

Although the role of lipid composition in membrane function is not entirely understood, changes in composition can produce dramatic effects. Researchers have isolated fruit flies with mutations in the gene that encodes ethanolamine kinase (analogous to choline kinase; Fig. 21–28). Lack of this enzyme eliminates one pathway for phosphatidylethanolamine synthesis, thereby reducing the amount of this lipid in cellular membranes. Flies with this mutation—those with the genotype *easily shocked*—exhibit transient paralysis following electrical stimulation or mechanical shock that would not affect wild-type flies.

## Plasmalogen Synthesis Requires Formation of an Ether-Linked Fatty Alcohol

The biosynthetic pathway to ether lipids, including **plasmalogens** and the **platelet-activating factor** (see Fig. 10–9), involves the displacement of an esterified fatty acyl group by a long-chain alcohol to form the ether linkage (Fig. 21–30). Head-group attachment follows, by mechanisms essentially like those used in formation of the common ester-linked phospholipids. Finally, the characteristic double bond of plasmalogens (shaded blue in Fig. 21–30) is introduced by the action of a mixed-function oxidase similar to that responsible for desaturation of fatty acids (Fig. 21–13). The peroxisome is the primary site of plasmalogen synthesis.

## Sphingolipid and Glycerophospholipid Synthesis Share Precursors and Some Mechanisms

The biosynthesis of sphingolipids takes place in four stages: (1) synthesis of the 18-carbon amine **sphinganine** from palmitoyl-CoA and serine; (2) attachment of a fatty acid in amide linkage to yield **N-acylsphinganine;** (3) desaturation of the sphinganine moiety to form **N-acylsphingosine** (ceramide); and (4) attachment of a head group to produce a sphingolipid such as a **cerebroside** or **sphingomyelin** (Fig. 21–31). The pathway shares several features with the pathways leading to glycerophospholipids: NADPH provides reducing power, and fatty acids enter as their activated CoA derivatives. In cerebroside formation, sugars enter as their activated nucleotide derivatives. Head-group attachment in sphingolipid synthesis has several novel aspects. Phosphatidylcholine, rather than CDP-choline, serves as the donor of phosphocholine in the synthesis of sphingomyelin.

In glycolipids, the cerebrosides and **gangliosides** (see Fig. 10–12), the head-group sugar is attached directly to the C-1 hydroxyl of sphingosine in glycosidic linkage rather than through a phosphodiester bond. The sugar donor is a UDP-sugar (UDP-glucose or UDP-galactose).

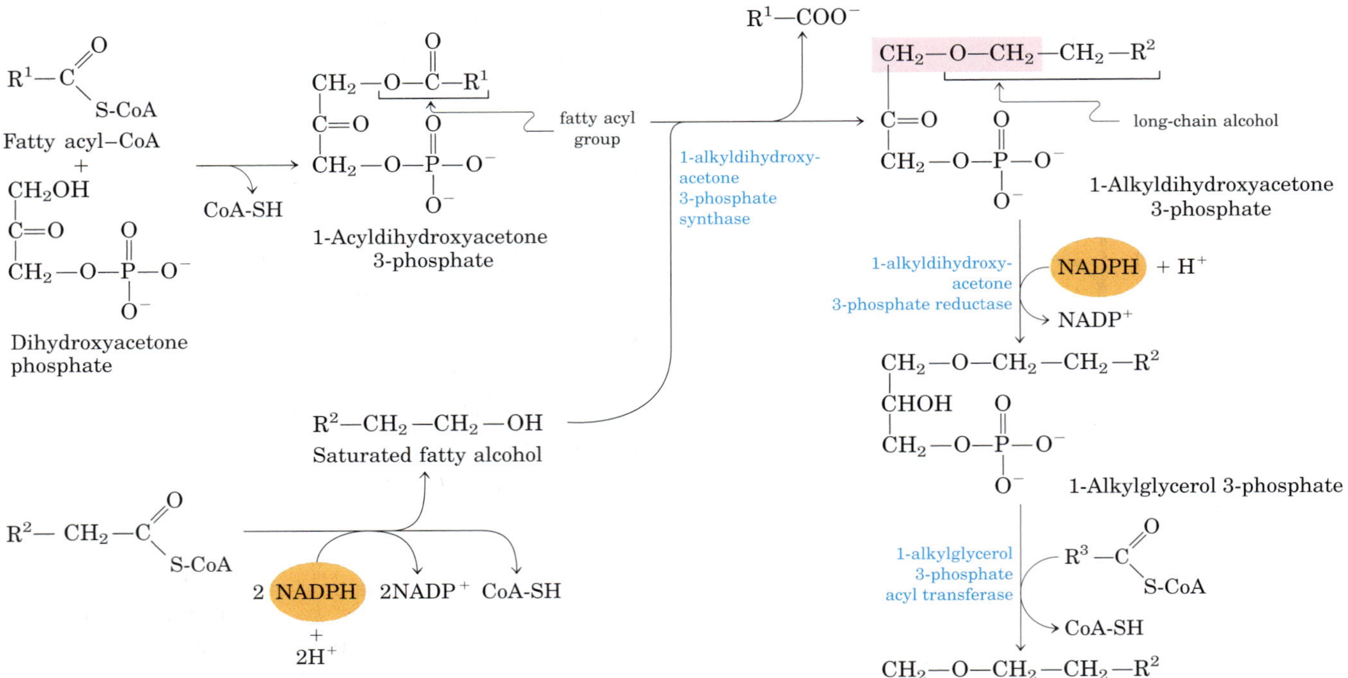

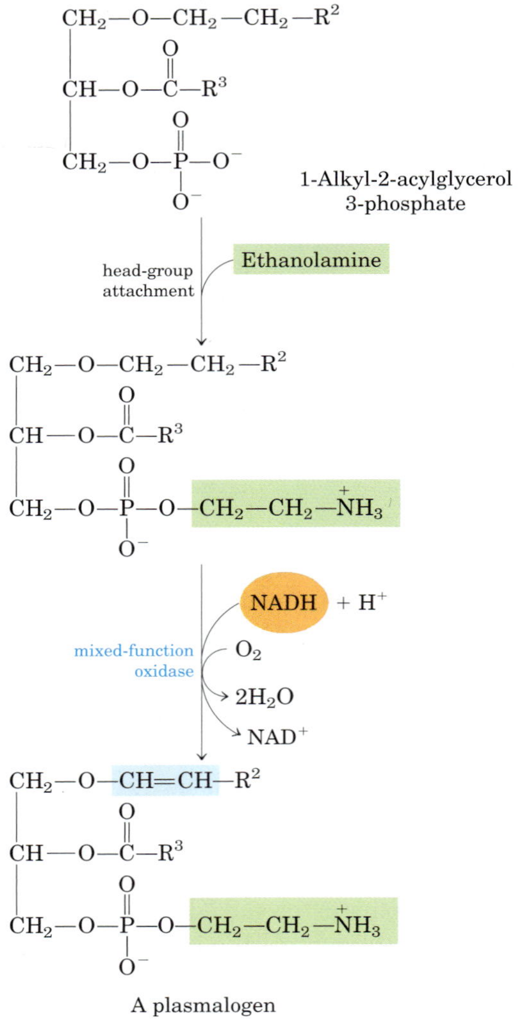

**FIGURE 21–30 Synthesis of ether lipids and plasmalogens.** The newly formed ether linkage is shaded pink. The intermediate 1-alkyl-2-acyl-glycerol 3-phosphate is the ether analog of phosphatidic acid. Mechanisms for attaching head groups to ether lipids are essentially the same as for their ester-linked analogs. The characteristic double bond of plasmalogens (shaded blue) is introduced in a final step by a mixed-function oxidase system similar to that shown in Figure 21–13.

## Polar Lipids Are Targeted to Specific Cellular Membranes

After synthesis on the smooth ER, the polar lipids, including the glycerophospholipids, sphingolipids, and glycolipids, are inserted into specific cellular membranes in specific proportions, by mechanisms not yet understood. Membrane lipids are insoluble in water, so they cannot simply diffuse from their point of synthesis (the ER) to their point of insertion. Instead, they are delivered in membrane vesicles that bud from the Golgi complex then move to and fuse with the target membrane (see Fig. 11–23). Cytosolic proteins also bind phospholipids and sterols and transport them between cellular membranes. These mechanisms contribute to the establishment of the characteristic lipid compositions of organelle membranes (see Fig. 11–2).

**FIGURE 21-31 Biosynthesis of sphingolipids.** Condensation of palmitoyl-CoA and serine followed by reduction with NADPH yields sphinganine, which is then acylated to *N*-acylsphinganine (a ceramide). In animals, a double bond (shaded pink) is created by a mixed-function oxidase, before the final addition of a head group: phosphatidylcholine, to form sphingomyelin; glucose, to form a cerebroside.

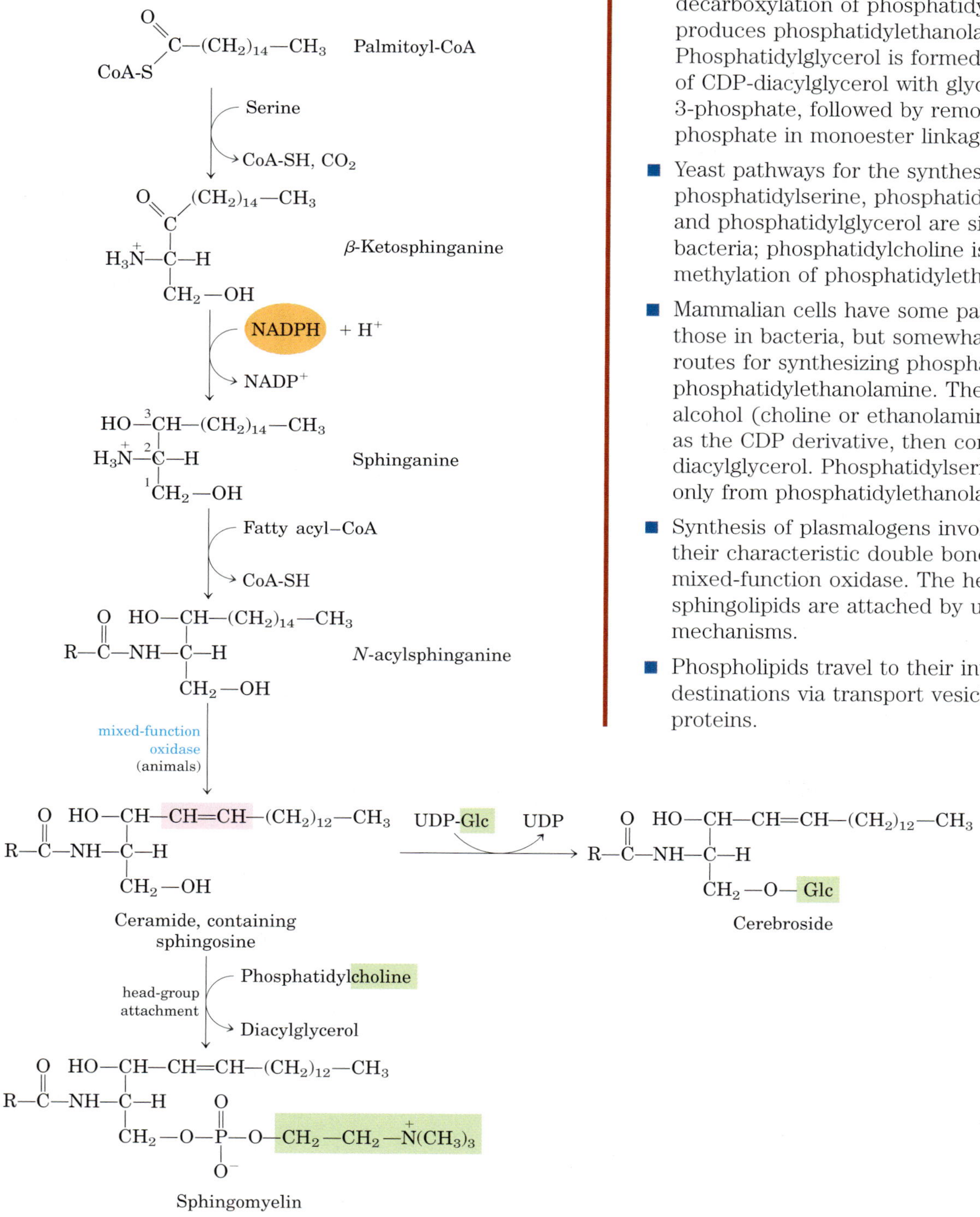

Palmitoyl-CoA

Serine

CoA-SH, CO₂

β-Ketosphinganine

NADPH + H⁺

NADP⁺

Sphinganine

Fatty acyl–CoA

CoA-SH

*N*-acylsphinganine

mixed-function oxidase (animals)

Ceramide, containing sphingosine

UDP-Glc   UDP

Cerebroside

head-group attachment

Phosphatidylcholine

Diacylglycerol

Sphingomyelin

## SUMMARY 21.3 Biosynthesis of Membrane Phospholipids

- Diacylglycerols are the principal precursors of glycerophospholipids.

- In bacteria, phosphatidylserine is formed by the condensation of serine with CDP-diacylglycerol; decarboxylation of phosphatidylserine produces phosphatidylethanolamine. Phosphatidylglycerol is formed by condensation of CDP-diacylglycerol with glycerol 3-phosphate, followed by removal of the phosphate in monoester linkage.

- Yeast pathways for the synthesis of phosphatidylserine, phosphatidylethanolamine, and phosphatidylglycerol are similar to those in bacteria; phosphatidylcholine is formed by methylation of phosphatidylethanolamine.

- Mammalian cells have some pathways similar to those in bacteria, but somewhat different routes for synthesizing phosphatidylcholine and phosphatidylethanolamine. The head-group alcohol (choline or ethanolamine) is activated as the CDP derivative, then condensed with diacylglycerol. Phosphatidylserine is derived only from phosphatidylethanolamine.

- Synthesis of plasmalogens involves formation of their characteristic double bond by a mixed-function oxidase. The head groups of sphingolipids are attached by unique mechanisms.

- Phospholipids travel to their intracellular destinations via transport vesicles or specific proteins.

## 21.4 Biosynthesis of Cholesterol, Steroids, and Isoprenoids

Cholesterol is doubtless the most publicized lipid, notorious because of the strong correlation between high levels of cholesterol in the blood and the incidence of human cardiovascular diseases. Less well advertised is cholesterol's crucial role as a component of cellular membranes and as a precursor of steroid hormones and bile acids. Cholesterol is an essential molecule in many animals, including humans, but is not required in the mammalian diet—all cells can synthesize it from simple precursors.

The structure of this 27-carbon compound suggests a complex biosynthetic pathway, but all of its carbon atoms are provided by a single precursor—acetate (Fig. 21–32). The **isoprene** units that are the essential intermediates in the pathway from acetate to cholesterol are also precursors to many other natural lipids, and the mechanisms by which isoprene units are polymerized are similar in all these pathways.

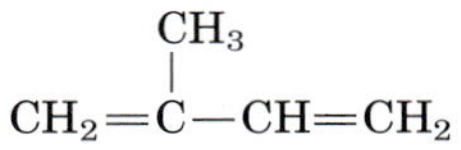

$$CH_2{=}\overset{\overset{\displaystyle CH_3}{|}}{C}{-}CH{=}CH_2$$
Isoprene

We begin with an account of the main steps in the biosynthesis of cholesterol from acetate, then discuss the transport of cholesterol in the blood, its uptake by cells, the normal regulation of cholesterol synthesis, and its regulation in those with defects in cholesterol uptake or transport. We next consider other cellular components derived from cholesterol, such as bile acids and steroid hormones. Finally, an outline of the biosynthetic pathways to some of the many compounds derived from isoprene units, which share early steps with the pathway to cholesterol, illustrates the extraordinary versatility of isoprenoid condensations in biosynthesis.

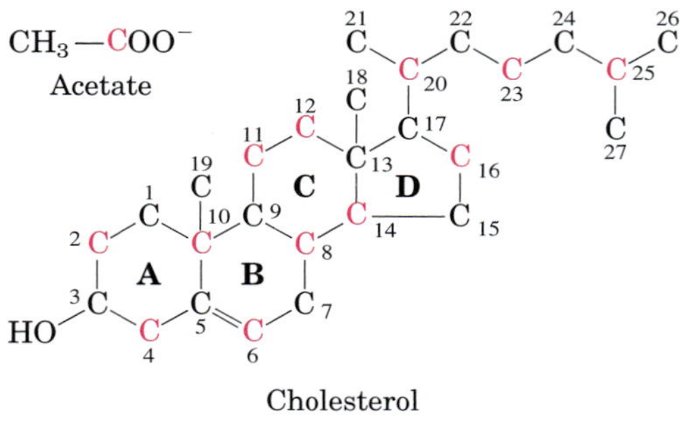

### Cholesterol Is Made from Acetyl-CoA in Four Stages

Cholesterol, like long-chain fatty acids, is made from acetyl-CoA, but the assembly plan is quite different. In early experiments, animals were fed acetate labeled with $^{14}C$ in either the methyl carbon or the carboxyl carbon. The pattern of labeling in the cholesterol isolated from the two groups of animals (Fig. 21–32) provided the blueprint for working out the enzymatic steps in cholesterol biosynthesis.

Synthesis takes place in four stages, as shown in Figure 21–33: ① condensation of three acetate units to form a six-carbon intermediate, mevalonate; ② conversion of mevalonate to activated isoprene units; ③ polymerization of six 5-carbon isoprene units to form the 30-carbon linear squalene; and ④ cyclization of squalene to form the four rings of the steroid nucleus, with a further series of changes (oxidations, removal or migration of methyl groups) to produce cholesterol.

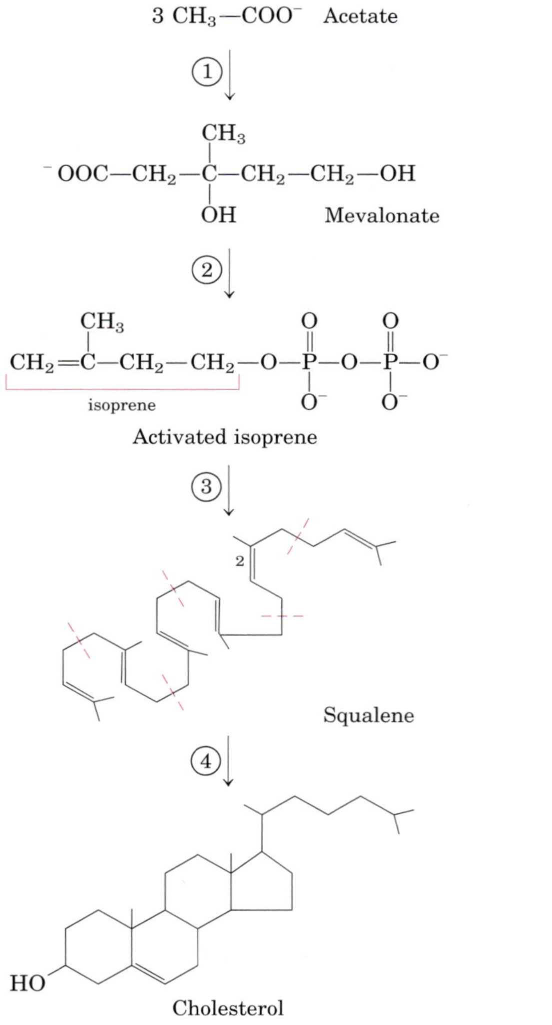

**FIGURE 21–33 Summary of cholesterol biosynthesis.** The four stages are discussed in the text. Isoprene units in squalene are set off by red dashed lines.

**FIGURE 21–32 Origin of the carbon atoms of cholesterol.** This can be deduced from tracer experiments with acetate labeled in the methyl carbon (black) or the carboxyl carbon (red). The individual rings in the fused-ring system are designated A through D.

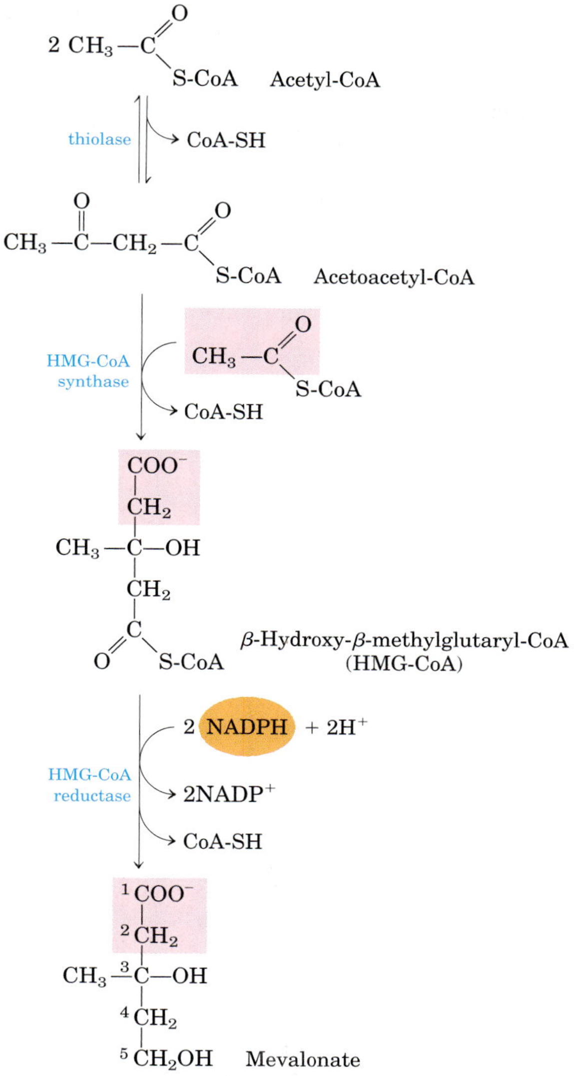

**FIGURE 21-34 Formation of mevalonate from acetyl-CoA.** The origin of C-1 and C-2 of mevalonate from acetyl-CoA is shown in pink.

**Stage ① Synthesis of Mevalonate from Acetate** The first stage in cholesterol biosynthesis leads to the intermediate **mevalonate** (Fig. 21–34). Two molecules of acetyl-CoA condense to form acetoacetyl-CoA, which condenses with a third molecule of acetyl-CoA to yield the six-carbon compound **β-hydroxy-β-methylglutaryl-CoA (HMG-CoA).** These first two reactions are catalyzed by **thiolase** and **HMG-CoA synthase,** respectively. The cytosolic HMG-CoA synthase in this pathway is distinct from the mitochondrial isozyme that catalyzes HMG-CoA synthesis in ketone body formation (see Fig. 17–18).

The third reaction is the committed and rate-limiting step: reduction of HMG-CoA to mevalonate, for which each of two molecules of NADPH donates two electrons. **HMG-CoA reductase,** an integral membrane protein of the smooth ER, is the major point of regulation on the pathway to cholesterol, as we shall see.

**Stage ② Conversion of Mevalonate to Two Activated Isoprenes**
In the next stage of cholesterol synthesis, three phosphate groups are transferred from three ATP molecules to mevalonate (Fig. 21–35). The phosphate attached to the C-3 hydroxyl group of mevalonate in the intermediate 3-phospho-5-pyrophosphomevalonate is a good

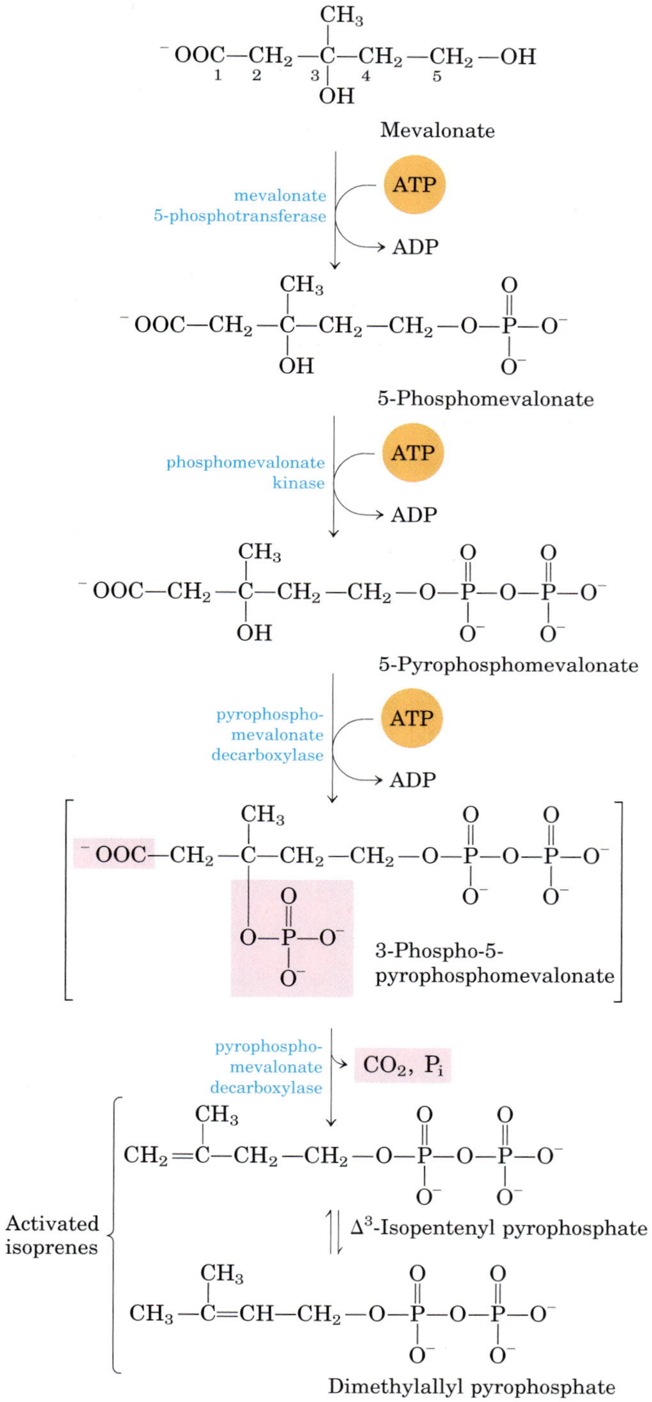

**FIGURE 21-35 Conversion of mevalonate to activated isoprene units.** Six of these activated units combine to form squalene (see Fig. 21–36). The leaving groups of 3-phospho-5-pyrophosphomevalonate are shaded pink. The bracketed intermediate is hypothetical.

leaving group; in the next step, both this phosphate and the nearby carboxyl group leave, producing a double bond in the five-carbon product, $\Delta^3$-**isopentenyl pyrophosphate.** This is the first of the two activated isoprenes central to cholesterol formation. Isomerization of $\Delta^3$-isopentenyl pyrophosphate yields the second activated isoprene, **dimethylallyl pyrophosphate.** Synthesis of isopentenyl pyrophosphate in the cytoplasm of plant cells follows the pathway described here. However, plant chloroplasts and many bacteria use a mevalonate-independent pathway. This alternative pathway does not occur in animals, so it is an attractive target for the development of new antibiotics.

**Stage ③ Condensation of Six Activated Isoprene Units to Form Squalene** Isopentenyl pyrophosphate and dimethylallyl pyrophosphate now undergo a head-to-tail condensation, in which one pyrophosphate group is displaced and a 10-carbon chain, **geranyl pyrophosphate,** is formed (Fig. 21–36). (The "head" is the end to which pyrophosphate is joined.) Geranyl pyrophosphate undergoes another head-to-tail condensation with isopentenyl pyro-

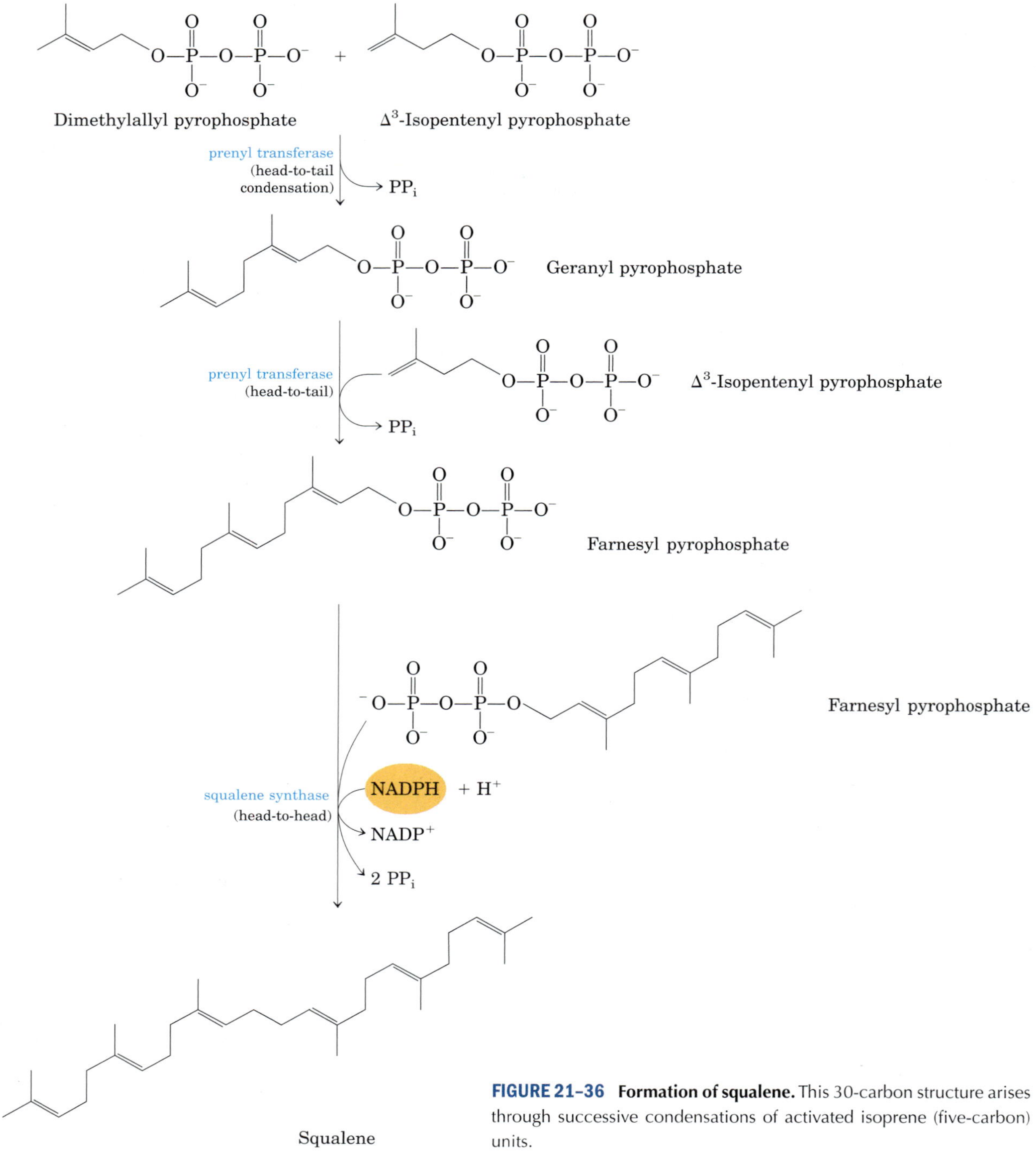

**FIGURE 21–36 Formation of squalene.** This 30-carbon structure arises through successive condensations of activated isoprene (five-carbon) units.

phosphate, yielding the 15-carbon intermediate **farne-syl pyrophosphate.** Finally, two molecules of farnesyl pyrophosphate join head to head, with the elimination of both pyrophosphate groups, to form **squalene.**

The common names of these intermediates derive from the sources from which they were first isolated. Geraniol, a component of rose oil, has the aroma of geraniums, and farnesol is an aromatic compound found in the flowers of the Farnese acacia tree. Many natural scents of plant origin are synthesized from isoprene units. Squalene, first isolated from the liver of sharks (genus *Squalus*), has 30 carbons, 24 in the main chain and 6 in the form of methyl group branches.

**Stage ④** *Conversion of Squalene to the Four-Ring Steroid Nucleus* When the squalene molecule is represented as in Figure 21–37, the relationship of its linear structure to the cyclic structure of the sterols becomes apparent. All

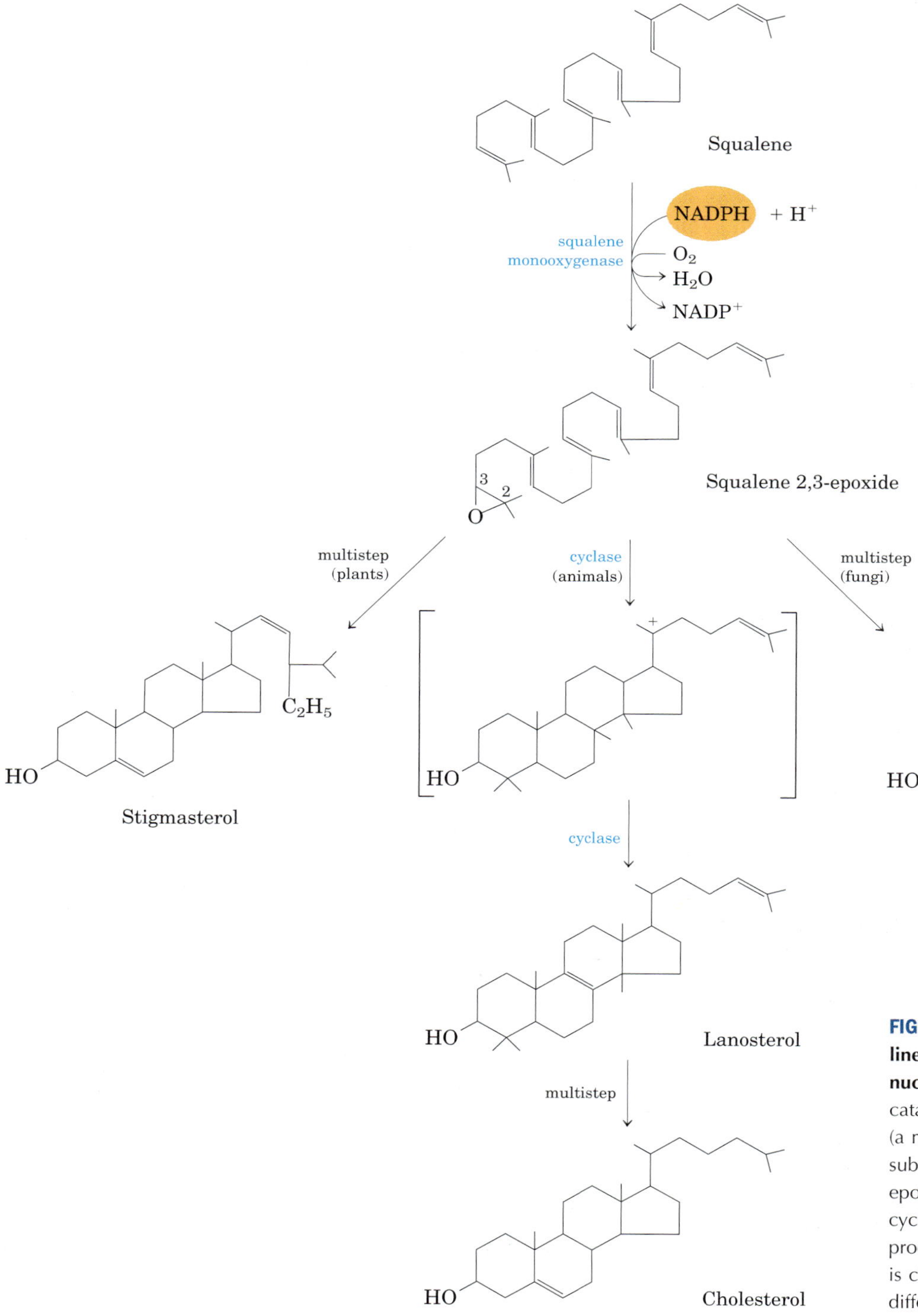

**FIGURE 21–37  Ring closure converts linear squalene to the condensed steroid nucleus.** The first step in this sequence is catalyzed by a mixed-function oxidase (a monooxygenase), for which the co-substrate is NADPH. The product is an epoxide, which in the next step is cyclized to the steroid nucleus. The final product of these reactions in animal cells is cholesterol; in other organisms, slightly different sterols are produced, as shown.

sterols have the four fused rings that form the steroid nucleus, and all are alcohols, with a hydroxyl group at C-3—thus the name "sterol." The action of **squalene monooxygenase** adds one oxygen atom from $O_2$ to the end of the squalene chain, forming an epoxide. This enzyme is another mixed-function oxidase (Box 21–1); NADPH reduces the other oxygen atom of $O_2$ to $H_2O$. The double bonds of the product, **squalene 2,3-epoxide,** are positioned so that a remarkable concerted reaction can convert the linear squalene epoxide to a cyclic structure. In animal cells, this cyclization results in the formation of **lanosterol,** which contains the four rings characteristic of the steroid nucleus. Lanosterol is finally converted to cholesterol in a series of about 20 reactions that include the migration of some methyl groups and the removal of others. Elucidation of this extraordinary biosynthetic pathway, one of the most complex known, was accomplished by Konrad Bloch, Feodor Lynen, John Cornforth, and George Popják in the late 1950s.

Cholesterol is the sterol characteristic of animal cells; plants, fungi, and protists make other, closely related sterols instead. They use the same synthetic pathway as far as squalene 2,3-epoxide, at which point the pathways diverge slightly, yielding other sterols, such as stigmasterol in many plants and ergosterol in fungi (Fig. 21–37).

Konrad Bloch, 1912–2000

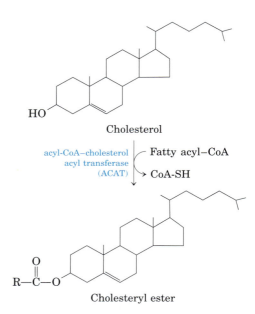

Feodor Lynen, 1911–1979

John Cornforth

George Popják

**FIGURE 21–38 Synthesis of cholesteryl esters.** Esterification converts cholesterol to an even more hydrophobic form for storage and transport.

## Cholesterol Has Several Fates

Much of the cholesterol synthesis in vertebrates takes place in the liver. A small fraction of the cholesterol made there is incorporated into the membranes of hepatocytes, but most of it is exported in one of three forms: biliary cholesterol, bile acids, or cholesteryl esters. **Bile acids** and their salts are relatively hydrophilic cholesterol derivatives that are synthesized in the liver and aid in lipid digestion (see Fig. 17–1). **Cholesteryl esters** are formed in the liver through the action of **acyl-CoA–cholesterol acyl transferase (ACAT).** This enzyme catalyzes the transfer of a fatty acid from coenzyme A to the hydroxyl group of cholesterol (Fig. 21–38), converting the cholesterol to a more hydrophobic form. Cholesteryl esters are transported in secreted lipoprotein particles to other tissues that use cholesterol, or they are stored in the liver.

All growing animal tissues need cholesterol for membrane synthesis, and some organs (adrenal gland and gonads, for example) use cholesterol as a precursor for steroid hormone production (discussed below). Cholesterol is also a precursor of vitamin D (see Fig. 10–20a).

## Cholesterol and Other Lipids Are Carried on Plasma Lipoproteins

Cholesterol and cholesteryl esters, like triacylglycerols and phospholipids, are essentially insoluble in water, yet must be moved from the tissue of origin to the tissues in which they will be stored or consumed. They are carried in the blood plasma as **plasma lipoproteins,**

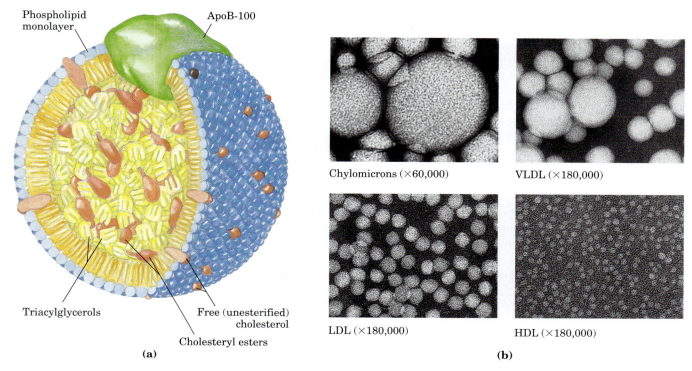

**FIGURE 21–39 Lipoproteins. (a)** Structure of a low-density lipoprotein (LDL). Apolipoprotein B-100 (apoB-100) is one of the largest single polypeptide chains known, with 4,636 amino acid residues ($M_r$ 513,000). **(b)** Four classes of lipoproteins, visualized in the electron microscope after negative staining. Clockwise from top left: chylomicrons, 50 to 200 nm in diameter; VLDL, 28 to 70 nm; HDL, 8 to 11 nm; and LDL, 20 to 25 nm. For properties of lipoproteins, see Table 21–2.

macromolecular complexes of specific carrier proteins, **apolipoproteins,** with various combinations of phospholipids, cholesterol, cholesteryl esters, and triacylglycerols.

Apolipoproteins ("apo" designates the protein in its lipid-free form) combine with lipids to form several classes of lipoprotein particles, spherical complexes with hydrophobic lipids in the core and hydrophilic amino acid side chains at the surface (Fig. 21–39a). Different combinations of lipids and proteins produce particles of different densities, ranging from chylomicrons to high-density lipoproteins. These particles can be separated by ultracentrifugation (Table 21–2) and visualized by electron microscopy (Fig. 21–39b).

Each class of lipoprotein has a specific function, determined by its point of synthesis, lipid composition, and apolipoprotein content. At least nine different apolipoproteins are found in the lipoproteins of human plasma (Table 21–3), distinguishable by their size, their reactions with specific antibodies, and their characteristic distribution in the lipoprotein classes. These protein components act as signals, targeting lipoproteins to specific tissues or activating enzymes that act on the lipoproteins.

**Chylomicrons,** discussed in Chapter 17 in connection with the movement of dietary triacylglycerols from the intestine to other tissues, are the largest of the lipoproteins and the least dense, containing a high

**TABLE 21–2  Major Classes of Human Plasma Lipoproteins: Some Properties**

| Lipoprotein | Density (g/mL) | Composition (wt %) | | | | |
| --- | --- | --- | --- | --- | --- | --- |
| | | Protein | Phospholipids | Free cholesterol | Cholesteryl esters | Triacylglycerols |
| Chylomicrons | <1.006 | 2 | 9 | 1 | 3 | 85 |
| VLDL | 0.95–1.006 | 10 | 18 | 7 | 12 | 50 |
| LDL | 1.006–1.063 | 23 | 20 | 8 | 37 | 10 |
| HDL | 1.063–1.210 | 55 | 24 | 2 | 15 | 4 |

**Source:** Modified from Kritchevsky, D. (1986) Atherosclerosis and nutrition. *Nutr. Int.* **2**, 290–297.

proportion of triacylglycerols (see Fig. 17–2). Chylomicrons are synthesized in the ER of epithelial cells that line the small intestine, then move through the lymphatic system and enter the bloodstream via the left subclavian vein. The apolipoproteins of chylomicrons include apoB-48 (unique to this class of lipoproteins), apoE, and apoC-II (Table 21–3). ApoC-II activates lipoprotein lipase in the capillaries of adipose, heart, skeletal muscle, and lactating mammary tissues, allowing the release of free fatty acids to these tissues. Chylomicrons thus carry dietary fatty acids to tissues where they will be consumed or stored as fuel (Fig. 21–40). The remnants of chylomicrons (depleted of most of their triacylglycerols but still containing cholesterol, apoE, and apoB-48) move through the bloodstream to the liver. Receptors in the liver bind to the apoE in the chylomicron remnants and mediate their uptake by endocytosis. In the liver, the remnants release their cholesterol and are degraded in lysosomes.

When the diet contains more fatty acids than are needed immediately as fuel, they are converted to triacylglycerols in the liver and packaged with specific apolipoproteins into **very-low-density lipoprotein (VLDL).** Excess carbohydrate in the diet can also be converted to triacylglycerols in the liver and exported as VLDLs (Fig. 21–40a). In addition to triacylglycerols, VLDLs contain some cholesterol and cholesteryl esters, as well as apoB-100, apoC-I, apoC-II, apoC-III, and apoE (Table 21–3). These lipoproteins are transported in the blood from the liver to muscle and adipose tissue, where activation of lipoprotein lipase by apoC-II causes the release of free fatty acids from the VLDL triacylglycerols. Adipocytes take up these fatty acids, reconvert them to triacylglycerols, and store the products in intracellular lipid droplets; myocytes, in contrast, primarily oxidize the fatty acids to supply energy. Most VLDL remnants are removed from the circulation by hepatocytes. The uptake, like that for chylomicrons, is

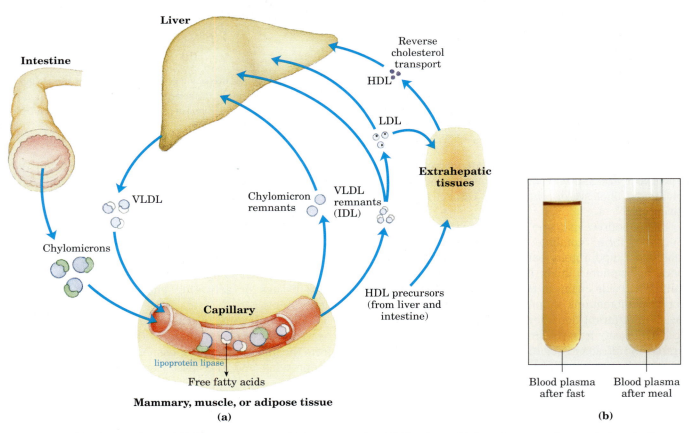

**FIGURE 21–40 Lipoproteins and lipid transport. (a)** Lipids are transported in the bloodstream as lipoproteins, which exist as several variants that have different functions, different protein and lipid compositions (see Tables 21–2, 21–3), and thus different densities. Dietary lipids are packaged into chylomicrons; much of their triacylglycerol content is released by lipoprotein lipase to adipose and muscle tissues during transport through capillaries. Chylomicron remnants (containing largely protein and cholesterol) are taken up by the liver. Endogenous lipids and cholesterol from the liver are delivered to adipose and muscle tissue by VLDL. Extraction of lipid from VLDL (along with loss of some apolipoproteins) gradually converts some of it to LDL, which delivers cholesterol to extrahepatic tissues or returns to the liver. The liver takes up LDL, VLDL remnants, and chylomicron remnants by receptor-mediated endocytosis. Excess cholesterol in extrahepatic tissues is transported back to the liver as HDL. In the liver, some cholesterol is converted to bile salts.

**(b)** Blood plasma samples collected after a fast (left) and after a high-fat meal (right). Chylomicrons produced after a fatty meal give the plasma a milky appearance.

**TABLE 21-3 Apolipoproteins of the Human Plasma Lipoproteins**

| Apolipoprotein | Molecular weight | Lipoprotein association | Function (if known) |
|---|---|---|---|
| ApoA-I | 28,331 | HDL | Activates LCAT; interacts with ABC transporter |
| ApoA-II | 17,380 | HDL | |
| ApoA-IV | 44,000 | Chylomicrons, HDL | |
| ApoB-48 | 240,000 | Chylomicrons | |
| ApoB-100 | 513,000 | VLDL, LDL | Binds to LDL receptor |
| ApoC-I | 7,000 | VLDL, HDL | |
| ApoC-II | 8,837 | Chylomicrons, VLDL, HDL | Activates lipoprotein lipase |
| ApoC-III | 8,751 | Chylomicrons, VLDL, HDL | Inhibits lipoprotein lipase |
| ApoD | 32,500 | HDL | |
| ApoE | 34,145 | Chylomicrons, VLDL, HDL | Triggers clearance of VLDL and chylomicron remnants |

**Source:** Modified from Vance, D.E. & Vance, J.E. (eds) (1985) *Biochemistry of Lipids and Membranes*. The Benjamin/Cummings Publishing Company, Menlo Park, CA.

receptor-mediated and depends on the presence of apoE in the VLDL remnants (Box 21–3 describes a link between apoE and Alzheimer's disease).

The loss of triacylglycerol converts some VLDL to VLDL remnants (also called intermediate density lipoprotein, IDL); further removal of triacylglycerol from VLDL produces **low-density lipoprotein (LDL)** (Table 21–2). Very rich in cholesterol and cholesteryl esters and containing apoB-100 as their major apolipoprotein, LDLs carry cholesterol to extrahepatic tissues that have specific plasma membrane receptors that recognize apoB-100. These receptors mediate the uptake of cholesterol and cholesteryl esters in a process described below.

The fourth major lipoprotein type, **high-density lipoprotein (HDL),** originates in the liver and small intestine as small, protein-rich particles that contain relatively little cholesterol and no cholesteryl esters (Fig. 21–40). HDLs contain apoA-I, apoC-I, apoC-II, and other apolipoproteins (Table 21–3), as well as the enzyme **lecithin-cholesterol acyl transferase (LCAT),** which catalyzes the formation of cholesteryl esters from lecithin (phosphatidylcholine) and cholesterol (Fig. 21–41). LCAT on the surface of nascent (newly forming) HDL particles converts the cholesterol and phosphatidylcholine of chylomicron and VLDL remnants to cholesteryl esters, which begin to form a core, transforming the disk-shaped nascent HDL to a mature, spherical HDL particle. This cholesterol-rich lipoprotein then returns to the liver, where the cholesterol is unloaded; some of this cholesterol is converted to bile salts.

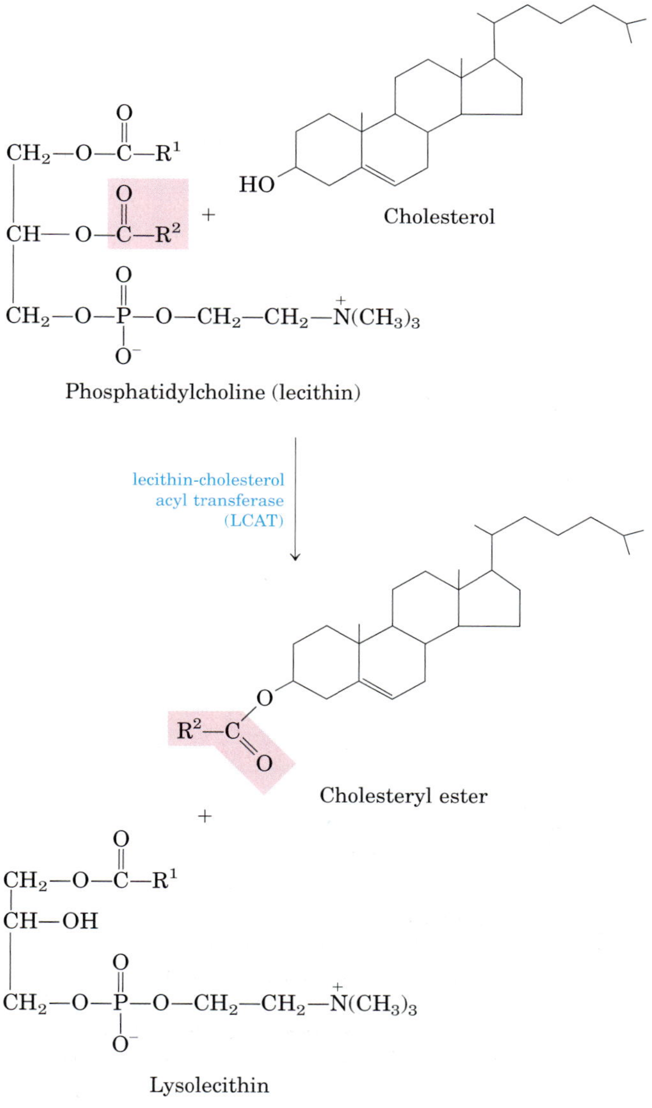

**FIGURE 21-41 Reaction catalyzed by lecithin-cholesterol acyl transferase (LCAT).** This enzyme is present on the surface of HDL and is stimulated by the HDL component apoA-I. Cholesteryl esters accumulate within nascent HDLs, converting them to mature HDLs.

## BOX 21–3 BIOCHEMISTRY IN MEDICINE

### ApoE Alleles Predict Incidence of Alzheimer's Disease

In the human population there are three common variants, or alleles, of the gene encoding apolipoprotein E. The most common, accounting for about 78% of human apoE alleles, is *APOE3;* alleles *APOE4* and *APOE2* account for 15% and 7%, respectively. The *APOE4* allele is particularly common in humans with Alzheimer's disease, and the link is highly predictive. Individuals who inherit *APOE4* have an increased risk of late-onset Alzheimer's disease. Those who are homozygous for *APOE4* have a 16-fold increased risk of developing the disease; for those who do, the mean age of onset is just under 70 years. For people who inherit two copies of *APOE3,* by contrast, the mean age of onset of Alzheimer's disease exceeds 90 years.

The molecular basis for the association between apoE4 and Alzheimer's disease is not yet known. Speculation has focused on a possible role for apoE in stabilizing the cytoskeletal structure of neurons. The apoE2 and apoE3 proteins bind to a number of proteins associated with neuronal microtubules, whereas apoE4 does not. This may accelerate the death of neurons. Whatever the mechanism proves to be, these observations promise to expand our understanding of the biological functions of apolipoproteins.

HDL may be taken up in the liver by receptor-mediated endocytosis, but at least some of the cholesterol in HDL is delivered to other tissues by a novel mechanism. HDL can bind to plasma membrane receptor proteins called SR-BI in hepatic and steroidogenic tissues such as the adrenal gland. These receptors mediate not endocytosis but a partial and selective transfer of cholesterol and other lipids in HDL into the cell. Depleted HDL then dissociates to recirculate in the bloodstream and extract more lipids from chylomicron and VLDL remnants. Depleted HDL can also pick up cholesterol stored in extrahepatic tissues and carry it to the liver, in **reverse cholesterol transport** pathways (Fig. 21–40). In one reverse transport path, interaction of nascent HDL with SR-BI receptors in cholesterol-rich cells triggers passive movement of cholesterol from the cell surface into HDL, which then carries it back to the liver. In a second pathway, apoA-I in depleted HDL interacts with an active transporter, the ABC1 protein, in a cholesterol-rich cell. The apoA-I (and presumably the HDL) is taken up by endocytosis, then resecreted with a load of cholesterol, which it transports to the liver.

The ABC1 protein is a member of a large family of multidrug transporters, sometimes called ABC transporters because they all have *A*TP-*b*inding *c*assettes; they also have two transmembrane domains with six transmembrane helices (Chapter 11). These proteins actively transport a variety of ions, amino acids, vitamins, steroid hormones, and bile salts across plasma membranes. The CFTR protein that is defective in cystic fibrosis (see Box 11–3) is another member of this ABC family of multidrug transporters.

### Cholesteryl Esters Enter Cells by Receptor-Mediated Endocytosis

Each LDL particle in the bloodstream contains apoB-100, which is recognized by specific surface receptor proteins, **LDL receptors,** on cells that need to take up cholesterol. The binding of LDL to an LDL receptor initiates endocytosis, which conveys the LDL and its receptor into the cell within an endosome (Fig. 21–42). The endosome eventually fuses with a lysosome, which contains enzymes that hydrolyze the cholesteryl esters, releasing cholesterol and fatty acid into the cytosol. The apoB-100 of LDL is also degraded to amino acids that are released to the cytosol, but the LDL receptor escapes degradation and is returned to the cell surface, to function again in LDL uptake. ApoB-100 is also present in VLDL, but its receptor-binding domain is not available for binding to the LDL receptor; conversion of VLDL to LDL exposes the receptor-binding domain of apoB-100. This pathway for the transport of cholesterol in blood and its **receptor-mediated endocytosis** by target tissues was elucidated by Michael Brown and Joseph Goldstein.

Michael Brown and Joseph Goldstein

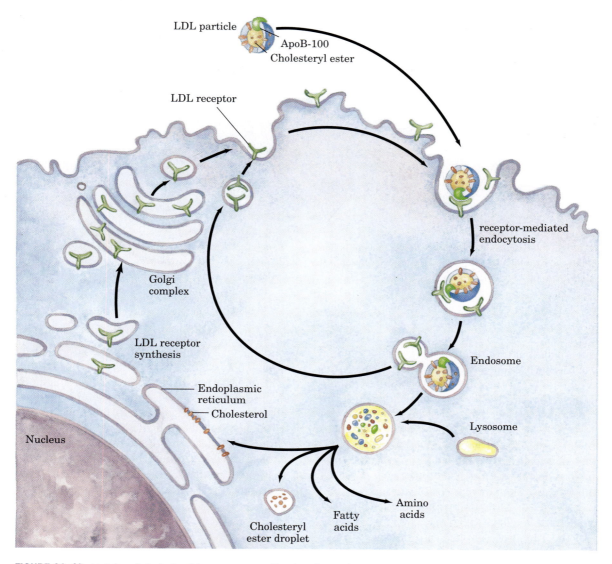

**FIGURE 21–42** Uptake of cholesterol by receptor-mediated endocytosis.

Cholesterol that enters cells by this path may be incorporated into membranes or reesterified by ACAT (Fig. 21–38) for storage within cytosolic lipid droplets. Accumulation of excess intracellular cholesterol is prevented by reducing the rate of cholesterol synthesis when sufficient cholesterol is available from LDL in the blood.

The LDL receptor also binds to apoE and plays a significant role in the hepatic uptake of chylomicrons and VLDL remnants. However, if LDL receptors are unavailable (as, for example, in a mouse strain that lacks the gene for the LDL receptor), VLDL remnants and chylomicrons are still taken up by the liver even though LDL is not. This indicates the presence of a back-up system for receptor-mediated endocytosis of VLDL remnants and chylomicrons. One back-up receptor is *l*ipoprotein *r*eceptor-related *p*rotein (LRP), which binds to apoE as well as to a number of other ligands.

## Cholesterol Biosynthesis Is Regulated at Several Levels

Cholesterol synthesis is a complex and energy-expensive process, so it is clearly advantageous to an organism to regulate the biosynthesis of cholesterol to complement dietary intake. In mammals, cholesterol production is regulated by intracellular cholesterol concentration and by the hormones glucagon and insulin. The rate-limiting step in the pathway to cholesterol (and a major site of regulation) is the conversion of HMG-CoA to mevalonate (Fig. 21–34), the reaction catalyzed by HMG-CoA reductase.

Regulation in response to cholesterol levels is mediated by an elegant system of transcriptional regulation of the gene encoding HMG-CoA reductase. This gene, along with more than 20 other genes encoding enzymes that mediate the uptake and synthesis of cholesterol and

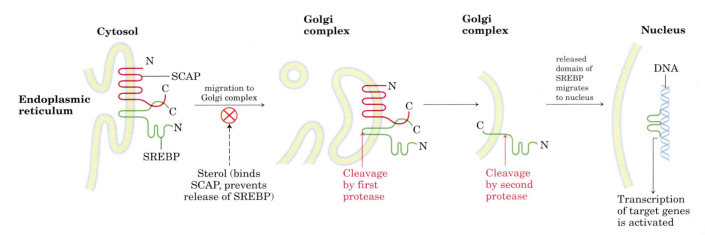

**Cytosol**       **Golgi complex**       **Golgi complex**       **Nucleus**

**FIGURE 21–43 SREBP activation.** Sterol regulatory element-binding proteins (SREBPs, shown in green) are embedded in the ER when first synthesized, in a complex with the protein SREBP cleavage-activating protein (SCAP, red). (N and C represent the amino and carboxyl termini of the proteins.) When bound to SCAP, SREBPs are inactive. When sterol levels decline, the complex migrates to the Golgi complex, and SREBP is cleaved by two different proteases in succession. The liberated amino-terminal domain of SREBP migrates to the nucleus, where it activates transcription of sterol-regulated genes.

unsaturated fatty acids, is controlled by a small family of proteins called sterol regulatory element-binding proteins (SREBPs). When newly synthesized, these proteins are embedded in the ER. Only the soluble amino-terminal domain of an SREBP functions as a transcriptional activator, using mechanisms discussed in Chapter 28. However, this domain has no access to the nucleus and cannot participate in gene activation while it remains part of the SREBP molecule. To activate transcription of the HMG-CoA reductase gene and other genes, the transcriptionally active domain is separated from the rest of the SREBP by proteolytic cleavage. When cholesterol levels are high, SREBPs are inactive, secured to the ER in a complex with another protein called SREBP cleavage-activating protein (SCAP) (Fig. 21–43). It is SCAP that binds cholesterol and a number of other sterols, thus acting as a sterol sensor. When sterol levels are high, the SCAP-SREBP complex probably interacts with another protein that retains the entire complex in the ER. When the level of sterols in the cell declines, a conformational change in SCAP causes release of the SCAP-SREBP complex from the ER-retention activity, and the complex migrates within vesicles to the Golgi complex. In the Golgi complex, SREBP is cleaved twice by two different proteases, the second cleavage releasing the amino-terminal domain into the cytosol. This domain travels to the nucleus and activates transcription of its target genes. The amino-terminal domain of SREBP has a short half-life and is rapidly degraded by proteasomes (see Fig. 27–42). When sterol levels increase sufficiently, the proteolytic release of SREBP amino-terminal domains is again blocked, and proteasome degradation of the existing active domains results in a rapid shut-down of the gene targets.

Several other mechanisms also regulate cholesterol synthesis (Fig. 21–44). Hormonal control is mediated by covalent modification of HMG-CoA reductase itself. The enzyme exists in phosphorylated (inactive) and dephosphorylated (active) forms. Glucagon stimulates phosphorylation (inactivation), and insulin promotes dephosphorylation, activating the enzyme and favoring cholesterol synthesis. High intracellular concentrations of cholesterol activate ACAT, which increases esterification of cholesterol for storage. Finally, a high cellular cholesterol level diminishes transcription of the gene that encodes the LDL receptor, reducing production of the receptor and thus the uptake of cholesterol from the blood.

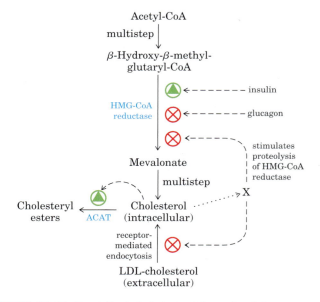

**FIGURE 21–44 Regulation of cholesterol formation balances synthesis with dietary uptake.** Glucagon promotes phosphorylation (inactivation) of HMG-CoA reductase; insulin promotes dephosphorylation (activation). X represents unidentified metabolites of cholesterol that stimulate proteolysis of HMG-CoA reductase.

Unregulated cholesterol production can lead to serious human disease. When the sum of cholesterol synthesized and cholesterol obtained in the diet exceeds the amount required for the synthesis of membranes, bile salts, and steroids, pathological accumulations of cholesterol in blood vessels (atherosclerotic plaques) can develop, resulting in obstruction of blood vessels (**atherosclerosis**). Heart failure due to occluded coronary arteries is a leading cause of death in industrialized societies. Atherosclerosis is linked to high levels of cholesterol in the blood, and particularly to high levels of LDL-bound cholesterol; there is a *negative* correlation between HDL levels and arterial disease.

In familial hypercholesterolemia, a human genetic disorder, blood levels of cholesterol are extremely high and severe atherosclerosis develops in childhood. These individuals have a defective LDL receptor and lack receptor-mediated uptake of cholesterol carried by LDL. Consequently, cholesterol is not cleared from the blood; it accumulates and contributes to the formation of atherosclerotic plaques. Endogenous cholesterol synthesis continues despite the excessive cholesterol in the blood, because extracellular cholesterol cannot enter the cell to regulate intracellular synthesis (Fig. 21–44). Two products derived from fungi, **lovastatin** and **compactin,** are used to treat patients with familial hypercholesterolemia. Both these compounds, and several synthetic analogs, resemble mevalonate (Fig. 21–45) and are competitive inhibitors of HMG-CoA reductase, thus inhibiting cholesterol synthesis. Lovastatin treatment lowers serum cholesterol by as much as 30% in individuals having one defective copy of the gene for the LDL receptor. When combined with an edible resin that binds bile acids and prevents their reabsorption from the intestine, the drug is even more effective.

In familial HDL deficiency, HDL levels are very low; they are almost undetectable in Tangier disease. Both genetic disorders are the result of mutations in the ABC1 protein. Cholesterol-depleted HDL cannot take up cholesterol from cells that lack ABC1 protein, and cholesterol-poor HDL is rapidly removed from the blood and destroyed. Both familial HDL deficiency and Tangier disease are very rare (worldwide, fewer than 100 families with Tangier disease are known), but the existence of these diseases establishes a role for ABC1 protein in the regulation of plasma HDL levels. Because low plasma HDL levels correlate with a high incidence of coronary artery disease, the ABC1 protein may prove a useful target for drugs to control HDL levels. ■

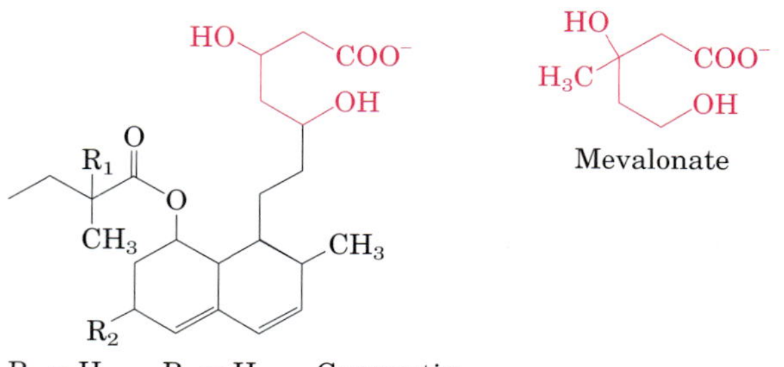

## Steroid Hormones Are Formed by Side-Chain Cleavage and Oxidation of Cholesterol

Humans derive all their steroid hormones from cholesterol (Fig. 21–46). Two classes of steroid hormones are synthesized in the cortex of the adrenal gland: **mineralocorticoids,** which control the reabsorption of inorganic ions ($Na^+$, $Cl^-$, and $HCO_3^-$) by the kidney, and **glucocorticoids,** which help regulate gluconeogenesis and reduce the inflammatory response. Sex hormones are produced in male and female gonads and the placenta. They include **progesterone,** which regulates the female reproductive cycle, and **androgens** (such as testosterone) and **estrogens** (such as estradiol), which influence the development of

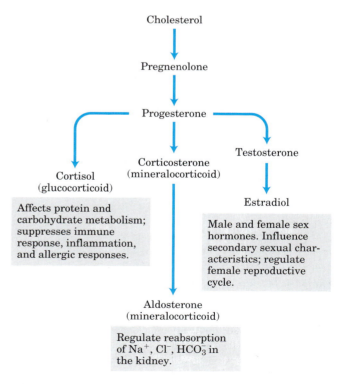

**FIGURE 21–45 Inhibitors of HMG-CoA reductase.** A comparison of the structures of mevalonate and four pharmaceutical compounds that inhibit HMG-CoA reductase.

Mevalonate

| $R_1$ = H | $R_2$ = H | Compactin |
|---|---|---|
| $R_1$ = $CH_3$ | $R_2$ = $CH_3$ | Simvastatin (Zocor) |
| $R_1$ = H | $R_2$ = OH | Pravastatin (Pravachol) |
| $R_1$ = H | $R_2$ = $CH_3$ | Lovastatin (Mevacor) |

Progesterone

Cholesterol
↓
Pregnenolone
↓
Progesterone

Cortisol (glucocorticoid)

Affects protein and carbohydrate metabolism; suppresses immune response, inflammation, and allergic responses.

Corticosterone (mineralocorticoid)

Testosterone
↓
Estradiol

Male and female sex hormones. Influence secondary sexual characteristics; regulate female reproductive cycle.

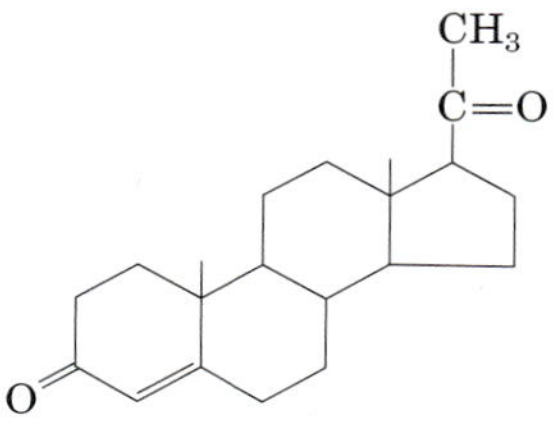

Aldosterone (mineralocorticoid)

Regulate reabsorption of $Na^+$, $Cl^-$, $HCO_3^-$ in the kidney.

**FIGURE 21–46 Some steroid hormones derived from cholesterol.** The structures of some of these compounds are shown in Figure 10–19.

secondary sexual characteristics in males and females, respectively. Steroid hormones are effective at very low concentrations and are therefore synthesized in relatively small quantities. In comparison with the bile salts, their production consumes relatively little cholesterol.

Synthesis of steroid hormones requires removal of some or all of the carbons in the "side chain" on C-17

of the D ring of cholesterol. Side-chain removal takes place in the mitochondria of steroidogenic tissues. Removal involves the hydroxylation of two adjacent carbons in the side chain (C-20 and C-22) followed by cleavage of the bond between them (Fig. 21–47). Formation of the various hormones also involves the introduction of oxygen atoms. All the hydroxylation and oxygenation reactions in steroid biosynthesis are catalyzed by mixed-function oxidases (Box 21–1) that use NADPH, $O_2$, and mitochondrial cytochrome P-450.

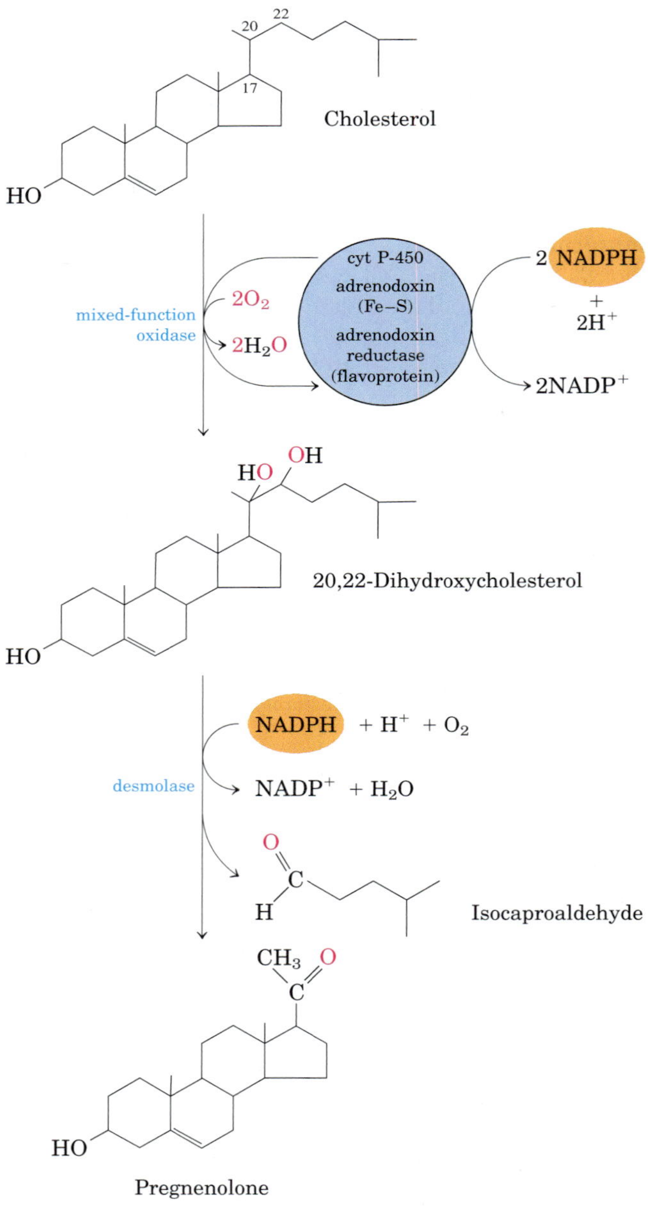

## Intermediates in Cholesterol Biosynthesis Have Many Alternative Fates

In addition to its role as an intermediate in cholesterol biosynthesis, isopentenyl pyrophosphate is the activated precursor of a huge array of biomolecules with diverse biological roles (Fig. 21–48). They include vitamins A, E, and K; plant pigments such as carotene and the phytol chain of chlorophyll; natural rubber; many essential oils (such as the fragrant principles of lemon oil, eucalyptus, and musk); insect juvenile hormone, which controls metamorphosis; dolichols, which serve as lipid-soluble carriers in complex polysaccharide synthesis; and ubiquinone and plastoquinone, electron carriers in mitochondria and chloroplasts. Collectively, these molecules are called isoprenoids. More than

**FIGURE 21–47 Side-chain cleavage in the synthesis of steroid hormones.** Cytochrome P-450 acts as electron carrier in this mixed-function oxidase system that oxidizes adjacent carbons. The process also requires the electron-transferring proteins adrenodoxin and adrenodoxin reductase. This system for cleaving side chains is found in mitochondria of the adrenal cortex, where active steroid production occurs. Pregnenolone is the precursor of all other steroid hormones (see Fig. 21–46).

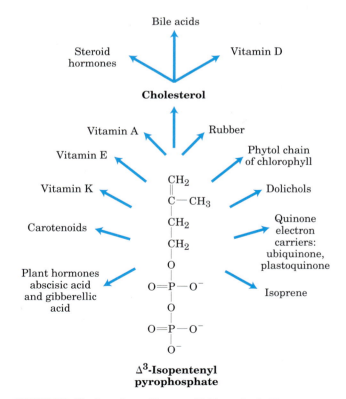

**FIGURE 21–48 Overview of isoprenoid biosynthesis.** The structures of most of the end products shown here are given in Chapter 10.

20,000 different isoprenoid molecules have been discovered in nature, and hundreds of new ones are reported each year.

Prenylation (covalent attachment of an isoprenoid; see Fig. 27–30) is a common mechanism by which proteins are anchored to the inner surface of cellular membranes in mammals (see Fig. 11–14). In some of these proteins the attached lipid is the 15-carbon farnesyl group; others have the 20-carbon geranylgeranyl group. Different enzymes attach the two types of lipids. It is possible that prenylation reactions target proteins to different membranes, depending on which lipid is attached. Protein prenylation is another important role for the isoprene derivatives of the pathway to cholesterol.

## SUMMARY 21.4 Biosynthesis of Cholesterol, Steroids, and Isoprenoids

- Cholesterol is formed from acetyl-CoA in a complex series of reactions, through the intermediates β-hydroxy-β-methylglutaryl-CoA, mevalonate, and two activated isoprenes, dimethylallyl pyrophosphate and isopentenyl pyrophosphate. Condensation of isoprene units produces the noncyclic squalene, which is cyclized to yield the steroid ring system and side chain.
- Cholesterol synthesis is under hormonal control and is also inhibited by elevated concentrations of intracellular cholesterol, which acts through covalent modification and transcriptional regulation mechanisms.
- Cholesterol and cholesteryl esters are carried in the blood as plasma lipoproteins. VLDL carries cholesterol, cholesteryl esters, and triacylglycerols from the liver to other tissues, where the triacylglycerols are degraded by lipoprotein lipase, converting VLDL to LDL. The LDL, rich in cholesterol and its esters, is taken up by receptor-mediated endocytosis, in which the apolipoprotein B-100 of LDL is recognized by receptors in the plasma membrane. HDL removes cholesterol from the blood, carrying it to the liver. Dietary conditions or genetic defects in cholesterol metabolism may lead to atherosclerosis and heart disease.
- The steroid hormones (glucocorticoids, mineralocorticoids, and sex hormones) are produced from cholesterol by alteration of the side chain and introduction of oxygen atoms into the steroid ring system. In addition to cholesterol, a wide variety of isoprenoid compounds are derived from mevalonate through condensations of isopentenyl pyrophosphate and dimethylallyl pyrophosphate.
- Prenylation of certain proteins targets them for association with cellular membranes and is essential for their biological activity.

## Key Terms

***Terms in bold are defined in the glossary.***

acetyl-CoA carboxylase 787
fatty acid synthase 789
acyl carrier protein (ACP) 790
fatty acyl-CoA desaturase 798
**mixed-function oxidases** 799
mixed-function oxygenases 799
cytochrome P-450 799
essential fatty acids 800
**prostaglandins** 800
cyclooxygenase (COX) 800
prostaglandin $H_2$ synthase 800
**thromboxane synthase** 800
thromboxanes 800
leukotrienes 800
glycerol 3-phosphate dehydrogenase 804
triacylglycerol cycle 806
**glyceroneogenesis** 806
thiazolidinediones 807
phosphatidylserine 811
phosphatidylglycerol 811
phosphatidylethanolamine 811
**cardiolipin** 811
phosphatidylcholine 812
**plasmalogen** 813
platelet-activating factor 813
cerebroside 813
sphingomyelin 813
**gangliosides** 813
**isoprene** 816
mevalonate 817
β-hydroxy-β-methylglutaryl-CoA (HMG-CoA) 817
thiolase 817
HMG-CoA synthase 817
HMG-CoA reductase 817
**bile acids** 820
cholesteryl esters 820
**apolipoproteins** 821
**chylomicron** 821
very-low-density lipoprotein (VLDL) 822
low-density lipoprotein (LDL) 823
high-density lipoprotein (HDL) 823
reverse cholesterol transport 824
LDL receptors 824
receptor-mediated endocytosis 824
atherosclerosis 827
lovastatin 827
mineralocorticoids 827
glucocorticoids 827
progesterone 827
androgens 827
estrogens 827

# Further Reading

*The general references in Chapters 10 and 17 are also useful.*

## General

**Bell, S.J., Bradley, D., Forse, R.A., & Bistrian, B.R.** (1997) The new dietary fats in health and disease. *J. Am. Dietetic Assoc.* **97,** 280–286.

**Gotto, A.M., Jr. (ed.)** (1987) *Plasma Lipoproteins,* New Comprehensive Biochemistry, Vol. 14 (Neuberger, A. & van Deenen, L.L.M., series eds), Elsevier Biomedical Press, Amsterdam.

Twelve reviews cover the structure, synthesis, and metabolism of lipoproteins, regulation of cholesterol synthesis, and the enzymes LCAT and lipoprotein lipase.

**Hajjar, D.P. & Nicholson, A.C.** (1995) Atherosclerosis. *Am. Sci.* **83,** 460–467.

A good description of the molecular basis of this disease and prospects for therapy.

**Hawthorne, J.N. & Ansell, G.B. (eds)** (1982) *Phospholipids,* New Comprehensive Biochemistry, Vol. 4 (Neuberger, A. & van Deenen, L.L.M., series eds), Elsevier Biomedical Press, Amsterdam.

This volume has excellent reviews of biosynthetic pathways to glycerophospholipids and sphingolipids, phospholipid transfer proteins, and bilayer assembly.

**Ohlrogge, J. & Browse, J.** (1995) Lipid biosynthesis. *Plant Cell* **7,** 957–970.

A good summary of pathways for lipid biosynthesis in plants.

**Vance, D.E. & Vance, J.E. (eds)** (1996) *Biochemistry of Lipids, Lipoproteins, and Membranes,* New Comprehensive Biochemistry, Vol. 31, Elsevier Science Publishing Co., Inc., New York.

Excellent reviews of lipid structure, biosynthesis, and function.

## Biosynthesis of Fatty Acids and Eicosanoids

**Capdevila, J.H., Falck, J.R., & Estabrook, R.W.** (1992) Cytochrome P450 and the arachidonate cascade. *FASEB J.* **6,** 731–736.

This issue contains 20 articles on the structure and function of various types of cytochrome P-450.

**Creelman, R.A. & Mullet, J.E.** (1997) Biosynthesis and action of jasmonates in plants. *Annu. Rev. Plant Physiol. Plant Mol. Biol.* **48,** 355–381.

**DeWitt, D.L.** (1999) Cox-2-selective inhibitors: the new super aspirins. *Mol. Pharmacol.* **55,** 625–631.

A short, clear review of the topic discussed in Box 21–2.

**Drazen, J.M., Israel, E., & O'Byrne, P.M.** (1999) Drug therapy: treatment of asthma with drugs modifying the leukotriene pathway. *New Engl. J. Med.* **340,** 197–206.

**Lands, W.E.M. (1991)** Biosynthesis of prostaglandins. *Annu. Rev. Nutr.* **11,** 41–60.

Discussion of the nutritional requirement for unsaturated fatty acids and recent biochemical work on pathways from arachidonate to prostaglandins; advanced level.

**Munday, M.R.** (2002) Regulation of mammalian acetyl-CoA carboxylase. *Biochem. Soc. Trans.* **30,** 1059–1064.

**Reshef, L., Olswang, Y., Cassuto, H., Blum, B., Croniger, C.M., Kalhan, S.C., Tilghman, S.M., & Hanson, R.M.** (2003) Glyceroneogenesis and the triglyceride/fatty acid cycle. *J. Biol. Chem.* **278,** 30,413–30,416.

**Smith, S.** (1994) The animal fatty acid synthase: one gene, one polypeptide, seven enzymes. *FASEB J.* **8,** 1248–1259.

**Smith, W.L., Garavito, R.M., & DeWitt, D.L.** (1996) Prostaglandin endoperoxide H synthases (cyclooxygenases)-1 and -2. *J. Biol. Chem.* **271,** 33,157–33,160.

A concise review of the properties and roles of COX-1 and COX-2.

## Biosynthesis of Membrane Phospholipids

**Bishop, W.R. & Bell, R.M.** (1988) Assembly of phospholipids into cellular membranes: biosynthesis, transmembrane movement and intracellular translocation. *Annu. Rev. Cell Biol.* **4,** 579–610.

Advanced review of the enzymology and cell biology of phospholipid synthesis and targeting.

**Dowhan, W.** (1997) Molecular basis for membrane phospholipid diversity: why are there so many lipids? *Annu. Rev. Biochem.* **66,** 199–232.

**Kennedy, E.P.** (1962) The metabolism and function of complex lipids. *Harvey Lect.* **57,** 143–171.

A classic description of the role of cytidine nucleotides in phospholipid synthesis.

**Pavlidis, P., Ramaswami, M., & Tanouye, M.A.** (1994) The *Drosophila easily shocked* gene: a mutation in a phospholipid synthetic pathway causes seizure, neuronal failure, and paralysis. *Cell* **79,** 23–33.

Description of the fascinating effects of changing the composition of membrane lipids in fruit flies.

**Raetz, C.R.H. & Dowhan, W.** (1990) Biosynthesis and function of phospholipids in *Escherichia coli. J. Biol. Chem.* **265,** 1235–1238.

A brief review of bacterial biosynthesis of phospholipids and lipopolysaccharides.

## Biosynthesis of Cholesterol, Steroids, and Isoprenoids

**Bittman, R. (ed.)** (1997) *Subcellular Biochemistry,* Vol. 28: *Cholesterol: Its Functions and Metabolism in Biology and Medicine,* Plenum Press, New York.

**Bloch, K.** (1965) The biological synthesis of cholesterol. *Science* **150,** 19–28.

The author's Nobel address; a classic description of cholesterol synthesis in animals.

**Brown, M.S., & Goldstein, J.L.** (1999) A proteolytic pathway that controls the cholesterol content of membranes, cells, and blood. *Proc. Natl. Acad. Sci. USA* **96,** 11,041–11,048.

**Chang, T.Y., Chang, C.C.Y., & Cheng, D.** (1997) Acyl–coenzyme A: cholesterol acyltransferase. *Annu. Rev. Biochem.* **66,** 613–638.

**Edwards, P.A. & Ericsson, J.** (1999) Sterols and isoprenoids: signaling molecules derived from the cholesterol biosynthetic pathway. *Annu. Rev. Biochem.* **68,** 157–185.

**Gimpl, G., Burger, K., & Fahrenholz, F.** (2002) A closer look at the cholesterol sensor. *Trends Biochem. Sci.* **27**, 596–599.

**Goldstein, J.L. & Brown, M.S.** (1990) Regulation of the mevalonate pathway. *Nature* **343**, 425–430.

Description of the allosteric and covalent regulation of the enzymes of the mevalonate pathway; includes a short discussion of the prenylation of Ras and other proteins.

**Knopp, R.H.** (1999) Drug therapy: drug treatment of lipid disorders. *New Engl. J. Med.* **341**, 498–511.

Review of the use of HMG-CoA inhibitors and bile acid–binding resins to reduce serum cholesterol.

**Krieger, M.** (1999) Charting the fate of the "good cholesterol": identification and characterization of the high-density lipoprotein receptor SR-BI. *Annu. Rev. Biochem.* **68**, 523–558.

**McGarvey, D.J. & Croteau, R.** (1995) Terpenoid metabolism. *Plant Cell* **7**, 1015–1026.

A description of the amazing diversity of isoprenoids in plants.

**Olson, R.E.** (1998) Discovery of the lipoproteins, their role in fat transport and their significance as risk factors. *J. Nutr.* **128** (2 Suppl.), 439S–443S.

Brief, clear, historical background on studies of lipoprotein function.

**Russell, D.W.** (2003) The enzymes, regulation, and genetics of bile acid synthesis. *Annu. Rev. Biochem.* **72**, 137–174.

**Young, S.G. & Fielding, C.J.** (1999) The ABCs of cholesterol efflux. *Nat. Genet.* **22**, 316–318.

A brief review of three papers in this journal issue that establish mutations in ABC1 as the cause of Tangier disease and familial HDL deficiency.

# Problems

**1. Pathway of Carbon in Fatty Acid Synthesis**  Using your knowledge of fatty acid biosynthesis, provide an explanation for the following experimental observations:

(a) Addition of uniformly labeled [$^{14}$C]acetyl-CoA to a soluble liver fraction yields palmitate uniformly labeled with $^{14}$C.

(b) However, addition of a *trace* of uniformly labeled [$^{14}$C]acetyl-CoA in the presence of an excess of unlabeled malonyl-CoA to a soluble liver fraction yields palmitate labeled with $^{14}$C only in C-15 and C-16.

**2. Synthesis of Fatty Acids from Glucose**  After a person has ingested large amounts of sucrose, the glucose and fructose that exceed caloric requirements are transformed to fatty acids for triacylglycerol synthesis. This fatty acid synthesis consumes acetyl-CoA, ATP, and NADPH. How are these substances produced from glucose?

**3. Net Equation of Fatty Acid Synthesis**  Write the net equation for the biosynthesis of palmitate in rat liver, starting from mitochondrial acetyl-CoA and cytosolic NADPH, ATP, and $CO_2$.

**4. Pathway of Hydrogen in Fatty Acid Synthesis**  Consider a preparation that contains all the enzymes and cofactors necessary for fatty acid biosynthesis from added acetyl-CoA and malonyl-CoA.

(a) If [2-$^2$H]acetyl-CoA (labeled with deuterium, the heavy isotope of hydrogen)

$$\underset{\overset{|}{^2H}}{\overset{\overset{^2H}{|}}{^2H-C}}-\underset{\text{S-CoA}}{\overset{O}{C}}$$

and an excess of unlabeled malonyl-CoA are added as substrates, how many deuterium atoms are incorporated into every molecule of palmitate? What are their locations? Explain.

(b) If unlabeled acetyl-CoA and [2-$^2$H]malonyl-CoA

$$^-OOC-\underset{\overset{|}{^2H}}{\overset{\overset{^2H}{|}}{C}}-\underset{\text{S-CoA}}{\overset{O}{C}}$$

are added as substrates, how many deuterium atoms are incorporated into every molecule of palmitate? What are their locations? Explain.

**5. Energetics of $\beta$-Ketoacyl-ACP Synthase**  In the condensation reaction catalyzed by $\beta$-ketoacyl-ACP synthase (see Fig. 21–5), a four-carbon unit is synthesized by the combination of a two-carbon unit and a three-carbon unit, with the release of $CO_2$. What is the thermodynamic advantage of this process over one that simply combines two two-carbon units?

**6. Modulation of Acetyl-CoA Carboxylase**  Acetyl-CoA carboxylase is the principal regulation point in the biosynthesis of fatty acids. Some of the properties of the enzyme are described below.

(a) Addition of citrate or isocitrate raises the $V_{max}$ of the enzyme as much as 10-fold.

(b) The enzyme exists in two interconvertible forms that differ markedly in their activities:

Protomer (inactive) $\rightleftharpoons$ filamentous polymer (active)

Citrate and isocitrate bind preferentially to the filamentous form, and palmitoyl-CoA binds preferentially to the protomer.

Explain how these properties are consistent with the regulatory role of acetyl-CoA carboxylase in the biosynthesis of fatty acids.

**7. Shuttling of Acetyl Groups across the Mitochondrial Inner Membrane**  The acetyl group of acetyl-CoA, produced by the oxidative decarboxylation of pyruvate in the mitochondrion, is transferred to the cytosol by the acetyl group shuttle outlined in Figure 21-10.

(a) Write the overall equation for the transfer of one acetyl group from the mitochondrion to the cytosol.

(b) What is the cost of this process in ATPs per acetyl group?

(c) In Chapter 17 we encountered an acyl group shuttle in the transfer of fatty acyl–CoA from the cytosol to the mitochondrion in preparation for $\beta$ oxidation (see Fig. 17–6). One result of that shuttle was separation of the mitochondrial and cytosolic pools of CoA. Does the acetyl group shuttle also accomplish this? Explain.

**8. Oxygen Requirement for Desaturases** The biosynthesis of palmitoleate (see Fig. 21–12), a common unsaturated fatty acid with a cis double bond in the $\Delta^9$ position, uses palmitate as a precursor. Can this be carried out under strictly anaerobic conditions? Explain.

**9. Energy Cost of Triacylglycerol Synthesis** Use a net equation for the biosynthesis of tripalmitoylglycerol (tripalmitin) from glycerol and palmitate to show how many ATPs are required per molecule of tripalmitin formed.

**10. Turnover of Triacylglycerols in Adipose Tissue** When [$^{14}$C]glucose is added to the balanced diet of adult rats, there is no increase in the total amount of stored triacylglycerols, but the triacylglycerols become labeled with $^{14}$C. Explain.

**11. Energy Cost of Phosphatidylcholine Synthesis** Write the sequence of steps and the net reaction for the biosynthesis of phosphatidylcholine by the salvage pathway from oleate, palmitate, dihydroxyacetone phosphate, and choline. Starting from these precursors, what is the cost (in number of ATPs) of the synthesis of phosphatidylcholine by the salvage pathway?

**12. Salvage Pathway for Synthesis of Phosphatidylcholine** A young rat maintained on a diet deficient in methionine fails to thrive unless choline is included in the diet. Explain.

**13. Synthesis of Isopentenyl Pyrophosphate** If 2-[$^{14}$C]acetyl-CoA is added to a rat liver homogenate that is synthesizing cholesterol, where will the $^{14}$C label appear in $\Delta^3$-isopentenyl pyrophosphate, the activated form of an isoprene unit?

**14. Activated Donors in Lipid Synthesis** In the biosynthesis of complex lipids, components are assembled by transfer of the appropriate group from an activated donor. For example, the activated donor of acetyl groups is acetyl-CoA. For each of the following groups, give the form of the activated donor: (a) phosphate; (b) D-glucosyl; (c) phosphoethanolamine; (d) D-galactosyl; (e) fatty acyl; (f) methyl; (g) the two-carbon group in fatty acid biosynthesis; (h) $\Delta^3$-isopentenyl.

**15. Importance of Fats in the Diet** When young rats are placed on a totally fat-free diet, they grow poorly, develop a scaly dermatitis, lose hair, and soon die—symptoms that can be prevented if linoleate or plant material is included in the diet. What makes linoleate an essential fatty acid? Why can plant material be substituted?

**16. Regulation of Cholesterol Biosynthesis** Cholesterol in humans can be obtained from the diet or synthesized de novo. An adult human on a low-cholesterol diet typically synthesizes 600 mg of cholesterol per day in the liver. If the amount of cholesterol in the diet is large, de novo synthesis of cholesterol is drastically reduced. How is this regulation brought about?

# BIOSYNTHESIS OF AMINO ACIDS, NUCLEOTIDES, AND RELATED MOLECULES

Time passes rapidly when you are having fun. The thrill of seeing people get well who might otherwise have died of disease . . . cannot be described in words. The Nobel Prize was only the icing on the cake.

—*Gertrude Elion, quoted in an article in* Science, *2002*

Nitrogen ranks behind only carbon, hydrogen, and oxygen in its contribution to the mass of living systems. Most of this nitrogen is bound up in amino acids and nucleotides. In this chapter we address all aspects of the metabolism of these nitrogen-containing compounds except amino acid catabolism, which is covered in Chapter 18.

Discussing the biosynthetic pathways for amino acids and nucleotides together is a sound approach, not only because both classes of molecules contain nitrogen (which arises from common biological sources) but because the two sets of pathways are extensively intertwined, with several key intermediates in common. Certain amino acids or parts of amino acids are incorporated into the structure of purines and pyrimidines, and in one case part of a purine ring is incorporated into an amino acid (histidine). The two sets of pathways also share

much common chemistry, in particular a preponderance of reactions involving the transfer of nitrogen or one-carbon groups.

The pathways described here can be intimidating to the beginning biochemistry student. Their complexity arises not so much from the chemistry itself, which in many cases is well understood, but from the sheer number of steps and variety of intermediates. These pathways are best approached by maintaining a focus on metabolic principles we have already discussed, on key intermediates and precursors, and on common classes of reactions. Even a cursory look at the chemistry can be rewarding, for some of the most unusual chemical transformations in biological systems occur in these pathways; for instance, we find prominent examples of the rare biological use of the metals molybdenum, selenium, and vanadium. The effort also offers a practical dividend, especially for students of human or veterinary medicine. Many genetic diseases of humans and animals have been traced to an absence of one or more enzymes of amino acid and nucleotide metabolism, and many pharmaceuticals in common use to combat infectious diseases are inhibitors of enzymes in these pathways—as are a number of the most important agents in cancer chemotherapy.

Regulation is crucial in the biosynthesis of the nitrogen-containing compounds. Because each amino acid and each nucleotide is required in relatively small amounts, the metabolic flow through most of these pathways is not nearly as great as the biosynthetic flow leading to carbohydrate or fat in animal tissues. Because the different amino acids and nucleotides must be made in the correct ratios and at the right time for protein and nucleic acid synthesis, their biosynthetic pathways must be accurately regulated and coordinated with each other. And because amino acids and nucleotides are charged molecules, their levels must be regulated to maintain electrochemical balance in the cell. As discussed in earlier chapters, pathways can be controlled by changes in either the activity or the amounts of specific enzymes. The pathways we encounter in this chapter provide some of the best-understood examples of the regulation of enzyme activity. Control of the *amounts* of different enzymes in a cell (that is, of their synthesis and degradation) is a topic covered in Chapter 28.

## 22.1 Overview of Nitrogen Metabolism

The biosynthetic pathways leading to amino acids and nucleotides share a requirement for nitrogen. Because soluble, biologically useful nitrogen compounds are generally scarce in natural environments, most organisms maintain strict economy in their use of ammonia, amino acids, and nucleotides. Indeed, as we shall see, free amino acids, purines, and pyrimidines formed during metabolic turnover of proteins and nucleic acids are often salvaged and reused. We first examine the pathways by which nitrogen from the environment is introduced into biological systems.

### The Nitrogen Cycle Maintains a Pool of Biologically Available Nitrogen

The most important source of nitrogen is air, which is four-fifths molecular nitrogen ($N_2$). However, relatively few species can convert atmospheric nitrogen into forms useful to living organisms. In the biosphere, the metabolic processes of different species function interdependently to salvage and reuse biologically available nitrogen in a vast **nitrogen cycle** (Fig. 22–1). The first step in the cycle is **fixation** (reduction) of atmospheric nitrogen by nitrogen-fixing bacteria to yield ammonia ($NH_3$ or $NH_4^+$). Although ammonia can be used by most living organisms, soil bacteria that derive their energy by oxidizing ammonia to nitrite ($NO_2^-$) and ultimately nitrate ($NO_3^-$) are so abundant and active that nearly all ammonia reaching the soil is oxidized to nitrate. This process is known as **nitrification.** Plants and many bacteria can take up and readily reduce nitrate and nitrite through the action of nitrate and nitrite reductases. The ammonia so formed is incorporated into amino acids by plants. Animals then use plants as a source of amino acids, both nonessential and essential, to build their proteins. When organisms die, microbial degradation of their proteins returns ammonia to the soil, where nitrifying bacteria again convert it to nitrite and nitrate. A balance is maintained between fixed nitrogen and atmospheric nitrogen by bacteria that convert nitrate to

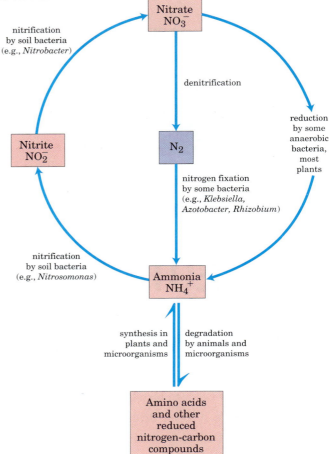

**FIGURE 22–1** The nitrogen cycle. The total amount of nitrogen fixed annually in the biosphere exceeds $10^{11}$ kg.

$N_2$ under anaerobic conditions, a process called **denitrification** (Fig. 22–1). These soil bacteria use $NO_3^-$ rather than $O_2$ as the ultimate electron acceptor in a series of reactions that (like oxidative phosphorylation) generates a transmembrane proton gradient, which is used to synthesize ATP.

Now let's examine the process of nitrogen fixation, the first step in the nitrogen cycle.

### Nitrogen Is Fixed by Enzymes of the Nitrogenase Complex

Only certain prokaryotes can fix atmospheric nitrogen. These include the cyanobacteria of soils and fresh and salt waters, other kinds of free-living soil bacteria such as *Azotobacter* species, and the nitrogen-fixing bacteria that live as **symbionts** in the root nodules of leguminous plants. The first important product of nitrogen fixation is ammonia, which can be used by all organisms either directly or after its conversion to other soluble compounds such as nitrites, nitrates, or amino acids.

The reduction of nitrogen to ammonia is an exergonic reaction:

$$N_2 + 3H_2 \longrightarrow 2NH_3 \qquad \Delta G'^\circ = -33.5 \text{ kJ/mol}$$

The N≡N triple bond, however, is very stable, with a bond energy of 930 kJ/mol. Nitrogen fixation therefore has an extremely high activation energy, and atmospheric nitrogen is almost chemically inert under normal conditions. Ammonia is produced industrially by the Haber process (named for its inventor, Fritz Haber), which requires temperatures of 400 to 500 °C and nitrogen and hydrogen at pressures of tens of thousands of kilopascals (several hundred atmospheres) to provide the necessary activation energy. Biological nitrogen fixation, however, must occur at biological temperatures and at 0.8 atm of nitrogen, and the high activation barrier is overcome by other means. This is accomplished, at least in part, by the binding and hydrolysis of ATP. The overall reaction can be written

$$N_2 + 10H^+ + 8e^- + 16ATP \longrightarrow 2NH_4^+ + 16ADP + 16P_i + H_2$$

Biological nitrogen fixation is carried out by a highly conserved complex of proteins called the **nitrogenase complex** (Fig. 22–2), the crucial components of which are **dinitrogenase reductase** and **dinitrogenase** (Fig. 22–3). Dinitrogenase reductase ($M_r$ 60,000) is a dimer of two identical subunits. It contains a single 4Fe-4S redox center (see Fig. 19–5), bound between the subunits, and can be oxidized and reduced by one electron. It also has two binding sites for ATP/ADP (one site on each subunit). Dinitrogenase ($M_r$ 240,000), a tetramer with two copies of two different subunits, contains both iron and molybdenum; its redox centers have a total of 2 Mo, 32 Fe, and 30 S per tetramer. About half of the iron and sulfur is present as two bridged pairs of 4Fe-4S centers called P clusters; the remainder is present as part of a novel iron-molybdenum cofactor. A form of nitrogenase that contains vanadium rather than molybdenum has been discovered, and some bacterial species can produce both types of nitrogenase systems. The vanadium-containing enzyme may be the primary nitrogen-fixing system under some environmental conditions, but it is not yet as well characterized as the molybdenum-dependent enzyme.

Nitrogen fixation is carried out by a highly reduced form of dinitrogenase and requires eight electrons: six for the reduction of $N_2$ and two to produce one molecule of $H_2$ as an obligate part of the reaction mechanism. Dinitrogenase is reduced by the transfer of electrons from dinitrogenase reductase (Fig. 22–2). The dinitrogenase tetramer has two binding sites for the reductase. The required eight electrons are transferred from reductase to dinitrogenase one at a time: a reduced reductase molecule binds to the dinitrogenase and transfers a single electron, then the oxidized reductase dissociates from dinitrogenase, in a repeating cycle. Each turn of the cycle requires the hydrolysis of two ATP molecules by the dimeric reductase. The immediate source of electrons to reduce dinitrogenase reductase varies, with reduced **ferredoxin** (p. 733; see also Fig. 19–5), reduced flavodoxin, and perhaps other sources playing a role. In at least one species, the ultimate source of electrons to reduce ferredoxin is pyruvate (Fig. 22–2).

The role of ATP in this process is somewhat unusual. As you will recall, ATP can contribute not only chemical energy, through the hydrolysis of one or more of its phosphoanhydride bonds, but also binding energy (pp. 196, 301), through noncovalent interactions that lower the activation energy. In the reaction carried out by dinitrogenase reductase, both ATP binding

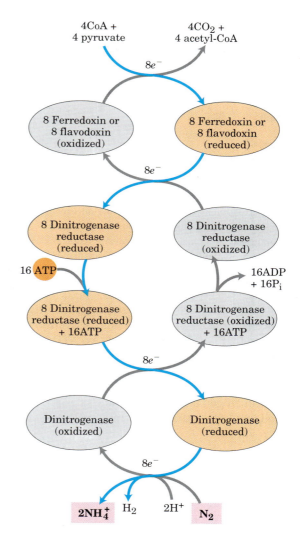

**FIGURE 22–2 Nitrogen fixation by the nitrogenase complex.** Electrons are transferred from pyruvate to dinitrogenase via ferredoxin (or flavodoxin) and dinitrogenase reductase. Dinitrogenase reductase reduces dinitrogenase one electron at a time, with at least six electrons required to fix one molecule of $N_2$. An additional two electrons are used to reduce 2 $H^+$ to $H_2$ in a process that obligatorily accompanies nitrogen fixation in anaerobes, making a total of eight electrons required per $N_2$ molecule. The subunit structures and metal cofactors of the dinitrogenase reductase and dinitrogenase proteins are described in the text and in Figure 22–3.

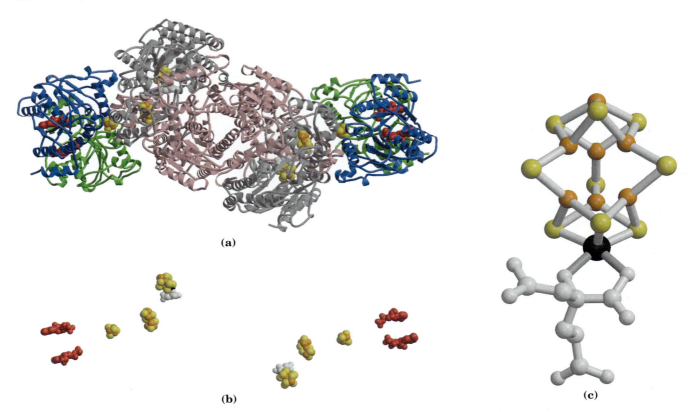

(a)

(b)

(c)

**FIGURE 22–3 Enzymes and cofactors of the nitrogenase complex.** (PDB ID 1N2C) **(a)** In this ribbon diagram, the dinitrogenase subunits are shown in gray and pink, the dinitrogenase reductase subunits in blue and green. The bound ADP is red. Note the 4Fe-4S complex (Fe atoms orange, S atoms yellow) and the iron-molybdenum cofactor (Mo black, homocitrate light gray). The P clusters (bridged pairs of 4Fe-4S complexes) are also shown. **(b)** The dinitrogenase complex cofactors without the protein (colors as in **(a)**). **(c)** The iron-molybdenum cofactor contains 1 Mo (black), 7 Fe (orange), 9 S (yellow), and one molecule of homocitrate (gray).

and ATP hydrolysis bring about protein conformational changes that help overcome the high activation energy of nitrogen fixation. The binding of two ATP molecules to the reductase shifts the reduction potential ($E'^\circ$) of this protein from $-300$ to $-420$ mV, an enhancement of its reducing power that is required to transfer electrons to dinitrogenase. The ATP molecules are then hydrolyzed just before the actual transfer of one electron to dinitrogenase.

Another important characteristic of the nitrogenase complex is an extreme lability in the presence of oxygen. The reductase is inactivated in air, with a half-life of 30 seconds; dinitrogenase has a half-life of 10 minutes in air. Free-living bacteria that fix nitrogen cope with this problem in a variety of ways. Some live only anaerobically or repress nitrogenase synthesis when oxygen is present. Some aerobic species, such as *Azotobacter vinelandii*, partially uncouple electron transfer from ATP synthesis so that oxygen is burned off as rapidly as it enters the cell (see Box 19–1). When fixing nitrogen, cultures of these bacteria actually increase in temperature as a result of their efforts to rid themselves of oxygen.

The symbiotic relationship between leguminous plants and the nitrogen-fixing bacteria in their root nodules (Fig. 22–4) takes care of both the energy requirements and the oxygen lability of the nitrogenase complex. The energy required for nitrogen fixation was probably the evolutionary driving force for this plant-bacteria association. The bacteria in root nodules have access to a large reservoir of energy in the form of abundant carbohydrate and citric acid cycle intermediates made available by the plant. This may allow the bacteria to fix hundreds of times more nitrogen than their free-living cousins can fix under conditions generally encountered in soils. To solve the oxygen-toxicity problem, the bacteria in root nodules are bathed in a solution of the oxygen-binding heme protein **leghemoglobin,** produced by the plant (although the heme may be contributed by the bacteria). Leghemoglobin binds all available oxygen so that it cannot interfere with nitrogen fixation, and efficiently delivers the oxygen to the bacterial electron-transfer system. The benefit to the plant, of course, is a ready supply of reduced nitrogen. The efficiency of the symbiosis between plants and bacteria is evident in the enrichment of soil nitrogen brought about by leguminous plants. This enrichment is the basis of crop rotation methods, in which plantings of nonleguminous plants (such as maize) that extract fixed nitrogen from

(a)

(b)    2 μm

**FIGURE 22–4    Nitrogen-fixing nodules. (a)** Root nodules of bird's-foot trefoil, a legume. The flower of this common plant is shown in the inset. **(b)** Artificially colorized electron micrograph of a thin section through a pea root nodule. Symbiotic nitrogen-fixing bacteria, or bacteroids (red), live inside the nodule cells, surrounded by the peribacteroid membrane (blue). Bacteroids produce the nitrogenase complex that converts atmospheric nitrogen ($N_2$) to ammonium ($NH_4^+$); with-out the bacteroids, the plant is unable to utilize $N_2$. The infected root cells provide some factors essential for nitrogen fixation, including leghemoglobin; this heme protein has a very high binding affinity for oxygen, which strongly inhibits nitrogenase. (The cell nucleus is shown in yellow/green. Not visible in this micrograph are other organelles of the infected root cell that are normally found in plant cells.)

the soil are alternated every few years with plantings of legumes such as alfalfa, peas, or clover.

Nitrogen fixation is the subject of intense study, because of its immense practical importance. Industrial production of ammonia for use in fertilizers requires a large and expensive input of energy, and this has spurred a drive to develop recombinant or transgenic organisms that can fix nitrogen. Recombinant DNA techniques (Chapter 9) are being used to transfer the DNA that encodes the enzymes of nitrogen fixation into non-nitrogen-fixing bacteria and plants. Success in these efforts will depend on overcoming the problem of oxygen toxicity in any cell that produces nitrogenase.

## Ammonia Is Incorporated into Biomolecules through Glutamate and Glutamine

Reduced nitrogen in the form of $NH_4^+$ is assimilated into amino acids and then into other nitrogen-containing biomolecules. Two amino acids, **glutamate** and **glutamine**, provide the critical entry point. Recall that these same two amino acids play central roles in the catabolism of ammonia and amino groups in amino acid oxidation (Chapter 18). Glutamate is the source of amino groups for most other amino acids, through transamination reactions (the reverse of the reaction shown in Fig. 18–4). The amide nitrogen of glutamine is a source of amino groups in a wide range of biosynthetic processes. In most types of cells, and in extracellular fluids in higher organisms, one or both of these amino acids are present at higher concentrations—sometimes an order of magnitude or more higher—than other amino acids. An *Escherichia coli* cell requires so much glutamate that this amino acid is one of the primary solutes in the cytosol. Its concentration is regulated not only in response to the cell's nitrogen requirements but also to maintain an osmotic balance between the cytosol and the external medium.

The biosynthetic pathways to glutamate and glutamine are simple, and all or some of the steps occur in most organisms. The most important pathway for the assimilation of $NH_4^+$ into glutamate requires two reactions. First, **glutamine synthetase** catalyzes the reaction of glutamate and $NH_4^+$ to yield glutamine. This reaction takes place in two steps, with enzyme-bound γ-glutamyl phosphate as an intermediate (see Fig. 18–8):

(1)  Glutamate + ATP ⟶ γ-glutamyl phosphate + ADP

(2)  γ-Glutamyl phosphate + $NH_4^+$ ⟶ glutamine + $P_i$ + $H^+$

_____

*Sum:* Glutamate + $NH_4^+$ + ATP ⟶

$$\text{glutamine} + \text{ADP} + \text{Pi} + H^+ \qquad (22\text{–}1)$$

Glutamine synthetase is found in all organisms. In addition to its importance for $NH_4^+$ assimilation in bacteria, it has a central role in amino acid metabolism in mammals, converting toxic free $NH_4^+$ to glutamine for transport in the blood (Chapter 18).

In bacteria and plants, glutamate is produced from glutamine in a reaction catalyzed by **glutamate synthase.** α-Ketoglutarate, an intermediate of the citric acid cycle, undergoes reductive amination with glutamine as nitrogen donor:

α-Ketoglutarate + glutamine + NADPH + $H^+$ ⟶

$$2 \text{ glutamate} + \text{NADP}^+ \qquad (22\text{–}2)$$

The net reaction of glutamine synthetase and glutamate synthase (Eqns 22–1 and 22–2) is

α-Ketoglutarate + $NH_4^+$ + NADPH + ATP ⟶

$$\text{L-glutamate} + \text{NADP}^+ + \text{ADP} + P_i$$

Glutamate synthase is not present in animals, which, instead, maintain high levels of glutamate by processes such as the transamination of α-ketoglutarate during amino acid catabolism.

Glutamate can also be formed in yet another, albeit minor, pathway: the reaction of α-ketoglutarate and $NH_4^+$ to form glutamate in one step. This is catalyzed by L-glutamate dehydrogenase, an enzyme present in all organisms. Reducing power is furnished by NADPH:

α-Ketoglutarate + $NH_4^+$ + NADPH ⟶

$$\text{L-glutamate} + \text{NADP}^+ + H_2O$$

We encountered this reaction in the catabolism of amino acids (see Fig. 18–7). In eukaryotic cells, L-glutamate dehydrogenase is located in the mitochondrial matrix. The reaction equilibrium favors reactants, and the $K_m$ for $NH_4^+$ (~1 mM) is so high that the reaction probably makes only a modest contribution to $NH_4^+$ assimilation into amino acids and other metabolites. (Recall that the glutamate dehydrogenase reaction, in reverse (see Fig. 18–10), is one source of $NH_4^+$ destined for the urea cycle.) Concentrations of $NH_4^+$ high enough for the glutamate dehydrogenase reaction to make a significant contribution to glutamate levels generally occur only when $NH_3$ is added to the soil or when organisms are grown in a laboratory in the presence of high $NH_3$ concentrations. In general, soil bacteria and plants rely on the two-enzyme pathway outlined above (Eqns 22–1, 22–2).

## Glutamine Synthetase Is a Primary Regulatory Point in Nitrogen Metabolism

The activity of glutamine synthetase is regulated in virtually all organisms—not surprising, given its central metabolic role as an entry point for reduced nitrogen. In enteric bacteria such as *E. coli*, the regulation is unusually complex. The enzyme has 12 identical subunits of $M_r$ 50,000 (Fig. 22–5) and is regulated both allosterically and by covalent modification. Alanine, glycine, and at least six end products of glutamine metabolism are allosteric inhibitors of the enzyme (Fig. 22–6). Each inhibitor alone produces only partial inhibition, but the effects of multiple inhibitors are more than additive, and all eight together virtually shut down the enzyme. This control mechanism provides a constant adjustment of glutamine levels to match immediate metabolic requirements.

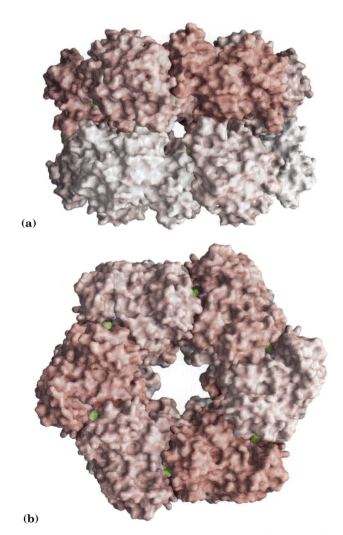

**(a)**

**(b)**

**FIGURE 22–5 Subunit structure of glutamine synthetase as determined by x-ray diffraction.** (PDB ID 2GLS) **(a)** Side view. The 12 subunits are identical; they are differently colored to illustrate packing and placement. **(b)** Top view, showing active sites (green).

Superimposed on the allosteric regulation is inhibition by adenylylation of (addition of AMP to) $Tyr^{397}$, located near the enzyme's active site (Fig. 22–7). This covalent modification increases sensitivity to the allosteric inhibitors, and activity decreases as more subunits are adenylylated. Both adenylylation and deadenylylation are promoted by **adenylyltransferase** (AT in Fig. 22–7), part of a complex enzymatic cascade that responds to levels of glutamine, $\alpha$-ketoglutarate, ATP, and $P_i$. The activity of adenylyltransferase is modulated by binding to a regulatory protein called $P_{II}$, and the activity of $P_{II}$, in turn, is regulated by covalent modification (uridylylation), again at a Tyr residue. The adenylyltransferase complex with uridylylated $P_{II}$ ($P_{II}$-UMP) stimulates deadenylylation, whereas the same complex

**FIGURE 22–6 Allosteric regulation of glutamine synthetase.** The enzyme undergoes cumulative regulation by six end products of glutamine metabolism. Alanine and glycine probably serve as indicators of the general status of amino acid metabolism in the cell.

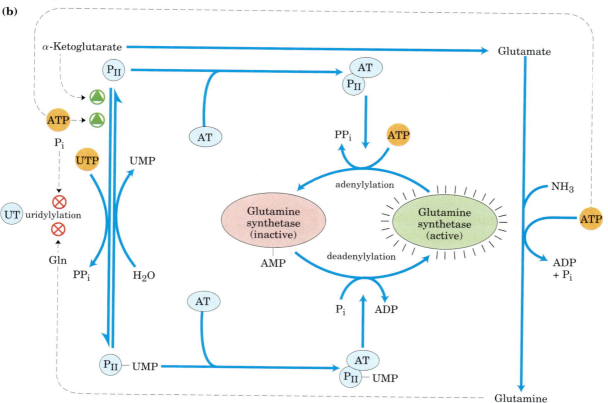

**(a)**

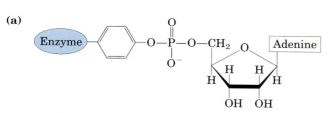

◀ **FIGURE 22–7  Second level of regulation of glutamine synthetase: covalent modifications.** **(a)** An adenylylated Tyr residue. **(b)** Cascade leading to adenylylation (inactivation) of glutamine synthetase. AT represents adenylyltransferase; UT, uridylyltransferase. Details of this cascade are discussed in the text.

with deuridylylated $P_{II}$ stimulates adenylylation of glutamine synthetase. Both uridylylation and deuridylylation of $P_{II}$ are brought about by a single enzyme, **uridylyltransferase.** Uridylylation is inhibited by binding of glutamine and $P_i$ to uridylyltransferase and is stimulated by binding of $\alpha$-ketoglutarate and ATP to $P_{II}$.

The regulation does not stop there. The uridylylated $P_{II}$ also mediates the activation of transcription of the gene encoding glutamine synthetase, thus increasing the cellular concentration of the enzyme; the deuridylylated $P_{II}$ brings about a decrease in transcription of the same gene. This mechanism involves an interaction of $P_{II}$ with additional proteins involved in gene regulation, of a type described in Chapter 28. The net result of this elaborate system of controls is a decrease in glutamine synthetase activity when glutamine levels are high, and an increase in activity when glutamine levels are low and $\alpha$-ketoglutarate and ATP (substrates for the synthetase reaction) are available. The multiple layers of regulation permit a sensitive response in which glutamine synthesis is tailored to cellular needs.

## Several Classes of Reactions Play Special Roles in the Biosynthesis of Amino Acids and Nucleotides

The pathways described in this chapter include a variety of interesting chemical rearrangements. Several of these recur and deserve special note before we progress to the pathways themselves. These are (1) transamination reactions and other rearrangements promoted by enzymes containing pyridoxal phosphate; (2) transfer of one-carbon groups, with either tetrahydrofolate (usually at the —CHO and —$CH_2OH$ oxidation levels) or $S$-adenosylmethionine (at the —$CH_3$ oxidation level) as cofactor; and (3) transfer of amino groups derived from the amide nitrogen of glutamine. Pyridoxal phosphate (PLP), tetrahydrofolate ($H_4$ folate), and $S$-adenosylmethionine (adoMet) were described in some detail in Chapter 18 (see Figs 18–6, 18–17, and 18–18). Here we focus on amino group transfer involving the amide nitrogen of glutamine.

More than a dozen known biosynthetic reactions use glutamine as the major physiological source of amino groups, and most of these occur in the pathways outlined in this chapter. As a class, the enzymes catalyzing these reactions are called **glutamine amidotransferases.** All have two structural domains: one binding glutamine, the other binding the second substrate, which serves as amino group acceptor (Fig. 22–8). A conserved Cys residue in the glutamine-binding domain is believed to act as a nucleophile, cleaving the amide bond of glutamine and forming a covalent glutamyl-enzyme intermediate. The $NH_3$ produced in this reaction is not released, but instead is transferred through an "ammonia channel" to a second active site, where it reacts with the second substrate to form the aminated product. The

covalent intermediate is hydrolyzed to the free enzyme and glutamate. If the second substrate must be activated, the usual method is the use of ATP to generate an acyl phosphate intermediate (R—OX in Fig. 22–8, with X as a phosphoryl group). The enzyme glutaminase acts in a similar fashion but uses $H_2O$ as the second substrate, yielding $NH_4^+$ and glutamate (see Fig. 18–8).

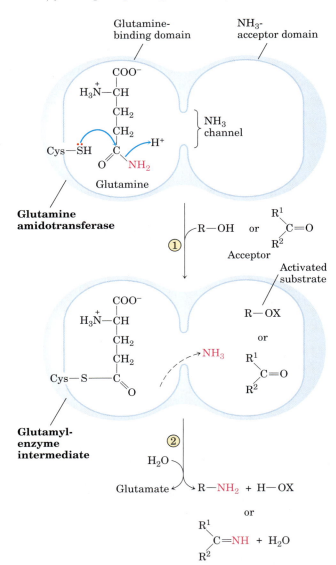

**MECHANISM FIGURE 22–8 Proposed mechanism for glutamine amidotransferases.** Each enzyme has two domains. The glutamine-binding domain contains structural elements conserved among many of these enzymes, including a Cys residue required for activity. The $NH_3$-acceptor (second-substrate) domain varies. ① The $\gamma$-amido nitrogen of glutamine (red) is released as $NH_3$ in a reaction that probably involves a covalent glutamyl-enzyme intermediate. The $NH_3$ travels through a channel to the second active site, where ② it reacts with any of several acceptors. Two types of amino acceptors are shown. X represents an activating group, typically a phosphoryl group derived from ATP, that facilitates displacement of a hydroxyl group from R—OH by $NH_3$.

### SUMMARY 22.1 Overview of Nitrogen Metabolism

- The molecular nitrogen that makes up 80% of the earth's atmosphere is unavailable to most living organisms until it is reduced. This fixation of atmospheric $N_2$ takes place in certain free-living bacteria and in symbiotic bacteria in the root nodules of leguminous plants.

- The nitrogen cycle entails formation of ammonia by bacterial fixation of $N_2$, nitrification of ammonia to nitrate by soil organisms, conversion of nitrate to ammonia by higher plants, synthesis of amino acids from ammonia by all organisms, and conversion of nitrate to $N_2$ by denitrifying soil bacteria.

- Fixation of $N_2$ as $NH_3$ is carried out by the nitrogenase complex, in a reaction that requires ATP. The nitrogenase complex is highly labile in the presence of $O_2$.

- In living systems, reduced nitrogen is incorporated first into amino acids and then into a variety of other biomolecules, including nucleotides. The key entry point is the amino acid glutamate. Glutamate and glutamine are the nitrogen donors in a wide variety of biosynthetic reactions. Glutamine synthetase, which catalyzes the formation of glutamine from glutamate, is a main regulatory enzyme of nitrogen metabolism.

- The amino acid and nucleotide biosynthetic pathways make repeated use of the biological cofactors pyridoxal phosphate, tetrahydrofolate, and $S$-adenosylmethionine. Pyridoxal phosphate is required for transamination reactions involving glutamate and for other amino acid transformations. One-carbon transfers require $S$-adenosylmethionine and tetrahydrofolate. Glutamine amidotransferases catalyze reactions that incorporate nitrogen derived from glutamine.

## 22.2 Biosynthesis of Amino Acids

All amino acids are derived from intermediates in glycolysis, the citric acid cycle, or the pentose phosphate pathway (Fig. 22–9). Nitrogen enters these pathways by way of glutamate and glutamine. Some pathways are simple, others are not. Ten of the amino acids are just one or several steps removed from the common metabolite from which they are derived. The biosynthetic pathways for others, such as the aromatic amino acids, are more complex.

Organisms vary greatly in their ability to synthesize the 20 common amino acids. Whereas most bacteria and plants can synthesize all 20, mammals can synthesize only about half of them—generally those with simple pathways. These are the **nonessential amino acids,** not needed in the diet (see Table 18–1). The remainder, the **essential amino acids,** must be obtained from food. Unless otherwise indicated, the pathways for the 20 common amino acids presented below are those operative in bacteria.

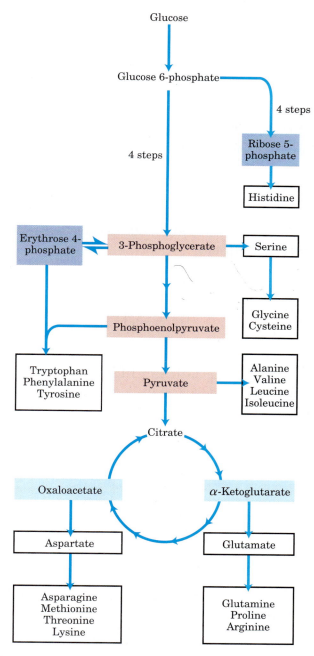

**FIGURE 22–9  Overview of amino acid biosynthesis.** The carbon skeleton precursors derive from three sources: glycolysis (pink), the citric acid cycle (blue), and the pentose phosphate pathway (purple).

A useful way to organize these biosynthetic pathways is to group them into six families corresponding to their metabolic precursors (Table 22–1), and we use this approach to structure the detailed descriptions that follow. In addition to these six precursors, there is a notable intermediate in several pathways of amino acid and nucleotide synthesis—**5-phosphoribosyl-1-pyrophosphate (PRPP):**

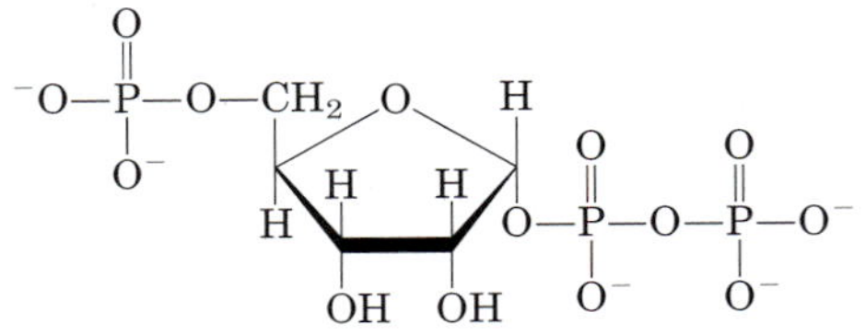

PRPP is synthesized from ribose 5-phosphate derived from the pentose phosphate pathway (see Fig. 14–21), in a reaction catalyzed by **ribose phosphate pyrophosphokinase:**

$$\text{Ribose 5-phosphate} + \text{ATP} \longrightarrow \text{5-phosphoribosyl-1-pyrophosphate} + \text{AMP}$$

This enzyme is allosterically regulated by many of the biomolecules for which PRPP is a precursor.

**TABLE 22–1 Amino Acid Biosynthetic Families, Grouped by Metabolic Precursor**

| | |
|---|---|
| **α-Ketoglutarate** | **Pyruvate** |
| Glutamate | Alanine |
| Glutamine | Valine* |
| Proline | Leucine* |
| Arginine | Isoleucine* |
| **3-Phosphoglycerate** | **Phosphoenolpyruvate and erythrose 4-phosphate** |
| Serine | Tryptophan* |
| Glycine | Phenylalanine* |
| Cysteine | Tyrosine† |
| **Oxaloacetate** | **Ribose 5-phosphate** |
| Aspartate | Histidine* |
| Asparagine | |
| Methionine* | |
| Threonine* | |
| Lysine* | |

*Essential amino acids.
†Derived from phenylalanine in mammals.

## α-Ketoglutarate Gives Rise to Glutamate, Glutamine, Proline, and Arginine

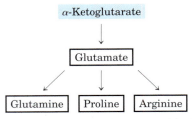

We have already described the biosynthesis of **glutamate** and **glutamine. Proline** is a cyclized derivative of glutamate (Fig. 22–10). In the first step of proline synthesis, ATP reacts with the γ-carboxyl group of glutamate to form an acyl phosphate, which is reduced by NADPH or NADH to glutamate γ-semialdehyde. This intermediate undergoes rapid spontaneous cyclization and is then reduced further to yield proline.

**Arginine** is synthesized from glutamate via ornithine and the urea cycle in animals (Chapter 18). In principle, ornithine could also be synthesized from glutamate γ-semialdehyde by transamination, but the spontaneous cyclization of the semialdehyde in the proline pathway precludes a sufficient supply of this intermediate for ornithine synthesis. Bacteria have a de novo biosynthetic pathway for ornithine (and thus arginine) that parallels some steps of the proline pathway but includes two additional steps that avoid the problem of the spontaneous cyclization of glutamate γ-semialdehyde (Fig. 22–10). In the first step, the α-amino group of glutamate is blocked by an acetylation requiring acetyl-CoA; then, after the transamination step, the acetyl group is removed to yield ornithine.

The pathways to proline and arginine are somewhat different in mammals. Proline can be synthesized by the pathway shown in Figure 22–10, but it is also formed from arginine obtained from dietary or tissue protein. Arginase, a urea cycle enzyme, converts arginine to ornithine and urea (see Figs 18–10, 18–26). The ornithine is converted to glutamate γ-semialdehyde by the enzyme **ornithine δ-aminotransferase** (Fig. 22–11). The semialdehyde cyclizes to $\Delta^1$-pyrroline-5-carboxylate, which is then converted to proline (Fig. 22–10). The pathway for arginine synthesis shown in Figure 22–10 is absent in mammals. When arginine from dietary intake or protein turnover is insufficient for protein synthesis, the ornithine δ-aminotransferase reaction operates in the direction of ornithine formation. Ornithine is then converted to citrulline and arginine in the urea cycle.

## Serine, Glycine, and Cysteine Are Derived from 3-Phosphoglycerate

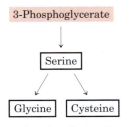

The major pathway for the formation of **serine** is the same in all organisms (Fig. 22–12). In the first step, the hydroxyl group of 3-phosphoglycerate is oxidized by a

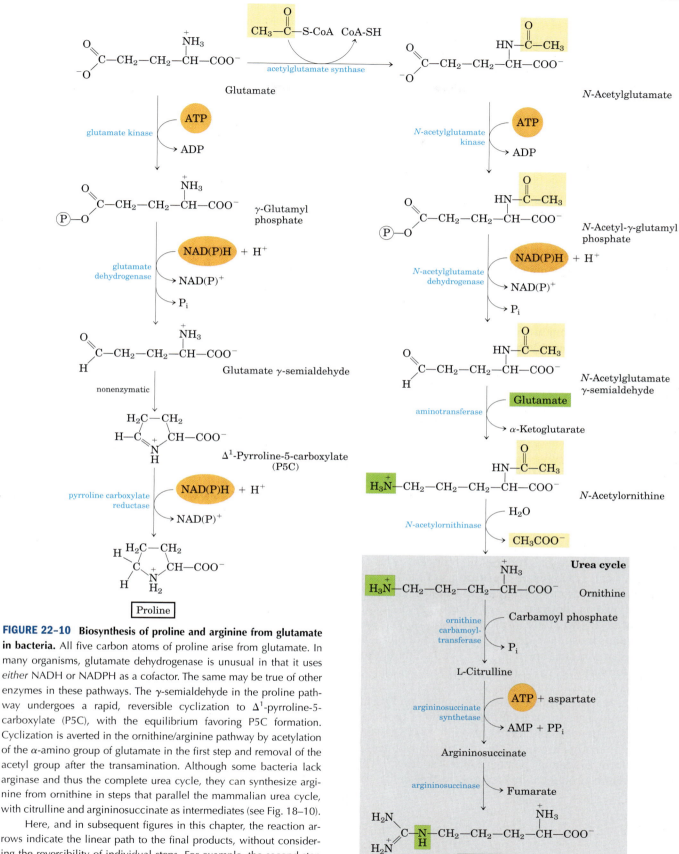

**FIGURE 22–10 Biosynthesis of proline and arginine from glutamate in bacteria.** All five carbon atoms of proline arise from glutamate. In many organisms, glutamate dehydrogenase is unusual in that it uses *either* NADH or NADPH as a cofactor. The same may be true of other enzymes in these pathways. The γ-semialdehyde in the proline pathway undergoes a rapid, reversible cyclization to Δ$^1$-pyrroline-5-carboxylate (P5C), with the equilibrium favoring P5C formation. Cyclization is averted in the ornithine/arginine pathway by acetylation of the α-amino group of glutamate in the first step and removal of the acetyl group after the transamination. Although some bacteria lack arginase and thus the complete urea cycle, they can synthesize arginine from ornithine in steps that parallel the mammalian urea cycle, with citrulline and argininosuccinate as intermediates (see Fig. 18–10).

Here, and in subsequent figures in this chapter, the reaction arrows indicate the linear path to the final products, without considering the reversibility of individual steps. For example, the second step of the pathway leading to arginine, catalyzed by *N*-acetylglutamate dehydrogenase, is chemically similar to the glyceraldehyde 3-phosphate dehydrogenase reaction in glycolysis (see Fig. 14–7) and is readily reversible.

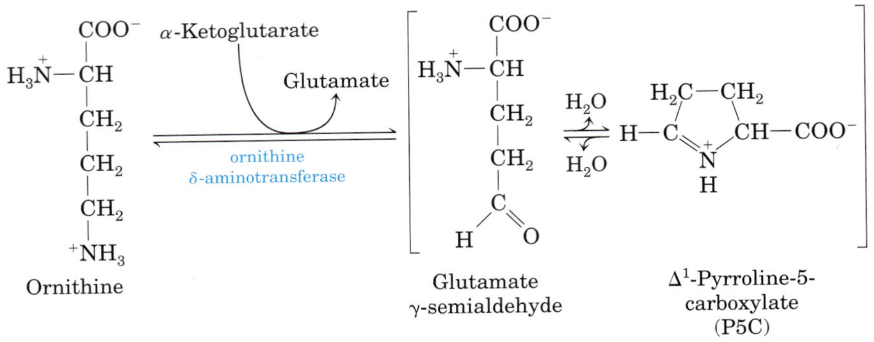

**FIGURE 22–11 Ornithine δ-aminotransferase reaction: a step in the mammalian pathway to proline.** This enzyme is found in the mitochondrial matrix of most tissues. Although the equilibrium favors P5C formation, the reverse reaction is the only mammalian pathway for synthesis of ornithine (and thus arginine) when arginine levels are insufficient for protein synthesis.

dehydrogenase (using $NAD^+$) to yield 3-phosphohydroxypyruvate. Transamination from glutamate yields 3-phosphoserine, which is hydrolyzed to free serine by phosphoserine phosphatase.

Serine (three carbons) is the precursor of **glycine** (two carbons) through removal of a carbon atom by **serine hydroxymethyltransferase** (Fig. 22–12). Tetrahydrofolate accepts the $\beta$ carbon (C-3) of serine, which forms a methylene bridge between N-5 and N-10 to yield $N^5,N^{10}$-methylenetetrahydrofolate (see Fig. 18–17). The overall reaction, which is reversible, also requires pyridoxal phosphate. In the liver of vertebrates, glycine can be made by another route: the reverse of the reaction shown in Figure 18–20c, catalyzed by **glycine synthase** (also called **glycine cleavage enzyme**):

$$CO_2 + NH_4^+ + N^5,N^{10}\text{-methylenetetrahydrofolate} +$$
$$NADH + H^+ \longrightarrow$$
$$\text{glycine} + \text{tetrahydrofolate} + NAD^+$$

Plants and bacteria produce the reduced sulfur required for the synthesis of **cysteine** (and methionine, described later) from environmental sulfates; the pathway is shown on the right side of Figure 22–13. Sulfate is activated in two steps to produce 3-phosphoadenosine 5′-phosphosulfate (PAPS), which undergoes an eight-electron reduction to sulfide. The sulfide is then used in formation of cysteine from serine in a two-step pathway. Mammals synthesize cysteine from two amino acids: methionine furnishes the sulfur atom and serine furnishes the carbon skeleton. Methionine is first converted to $S$-adenosylmethionine (see Fig. 18–18), which can lose its methyl group to any of a number of acceptors to form $S$-adenosylhomocysteine (adoHcy). This demethylated product is hydrolyzed to free homocys-

teine, which undergoes a reaction with serine, catalyzed by **cystathionine $\beta$-synthase,** to yield cystathionine (Fig. 22–14). Finally, **cystathionine $\gamma$-lyase,** a PLP-requiring enzyme, catalyzes removal of ammonia and cleavage of cystathionine to yield free cysteine.

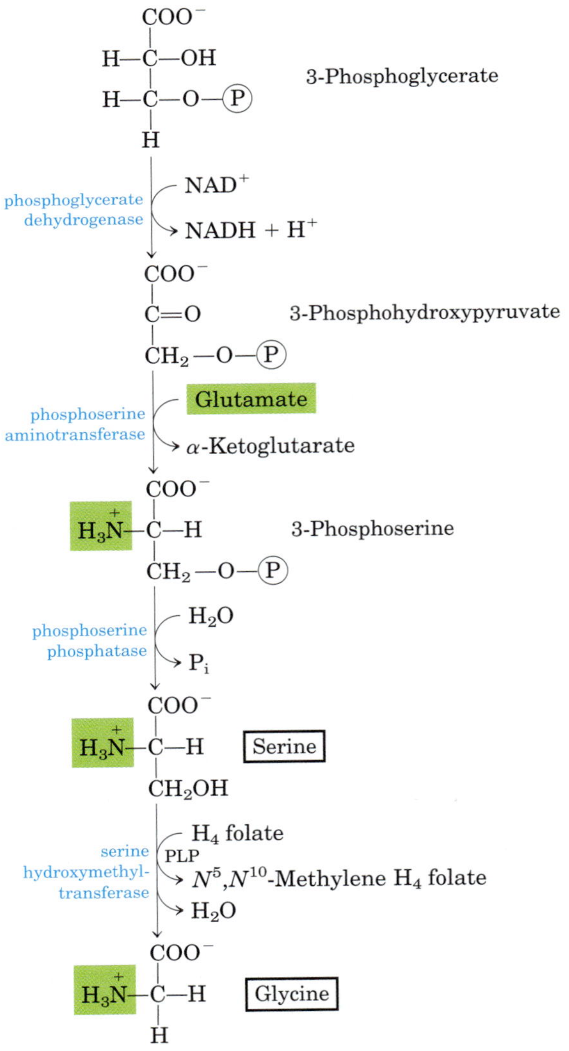

**FIGURE 22–12 Biosynthesis of serine from 3-phosphoglycerate and of glycine from serine in all organisms.** Glycine is also made from $CO_2$ and $NH_4^+$ by the action of glycine synthase, with $N^5,N^{10}$-methylenetetrahydrofolate as methyl group donor (see text).

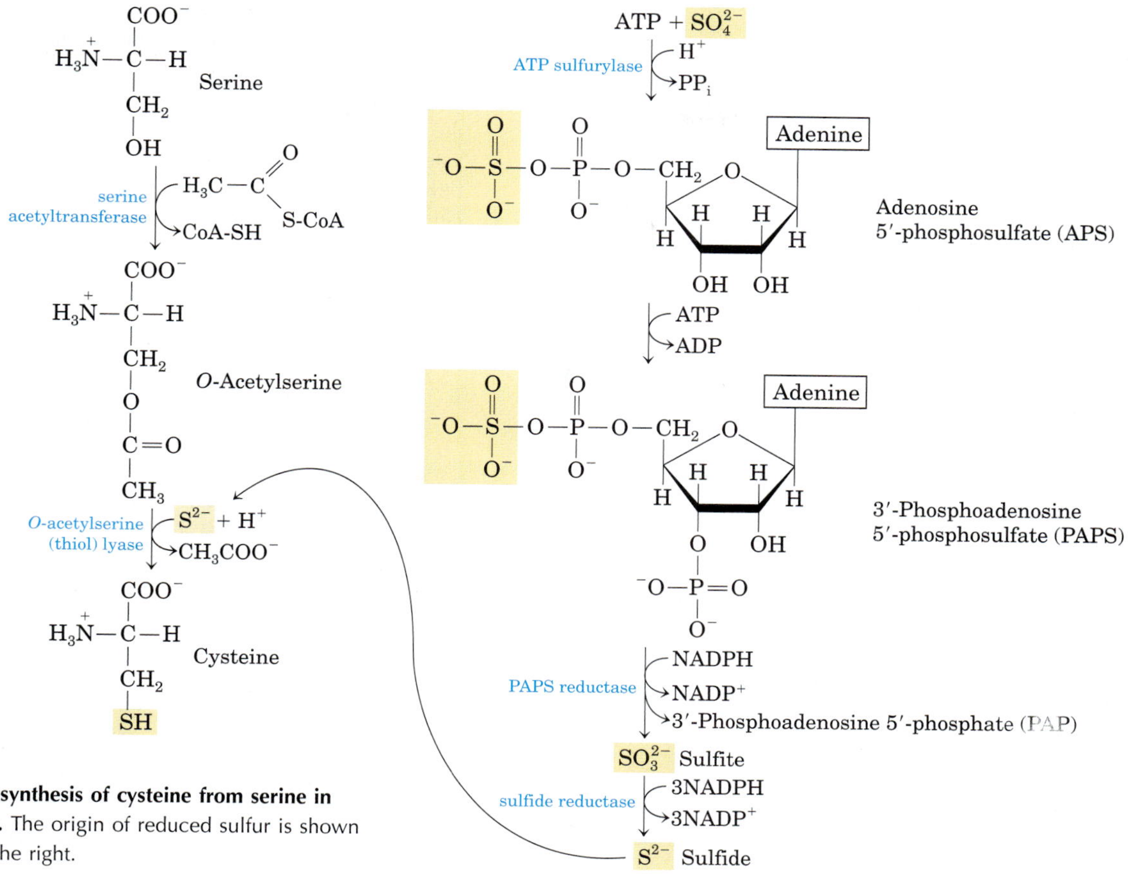

**FIGURE 22–13 Biosynthesis of cysteine from serine in bacteria and plants.** The origin of reduced sulfur is shown in the pathway on the right.

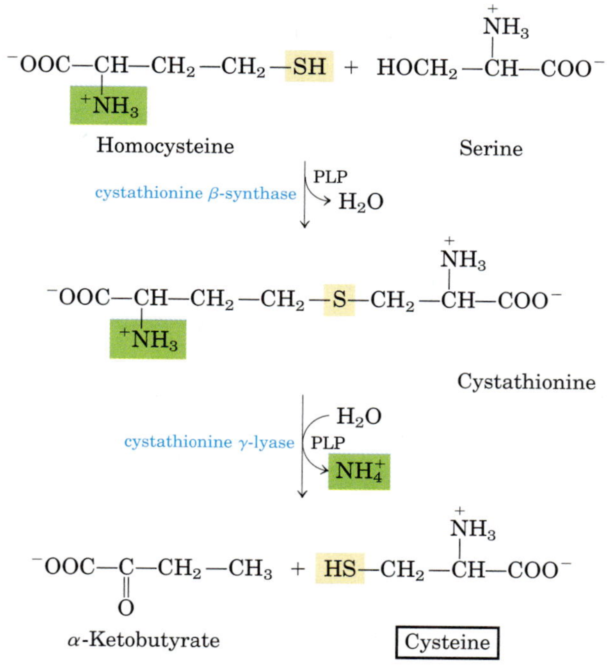

**FIGURE 22–14 Biosynthesis of cysteine from homocysteine and serine in mammals.** The homocysteine is formed from methionine, as described in the text.

## Three Nonessential and Six Essential Amino Acids Are Synthesized from Oxaloacetate and Pyruvate

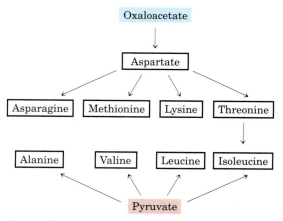

**Alanine** and **aspartate** are synthesized from pyruvate and oxaloacetate, respectively, by transamination from glutamate. **Asparagine** is synthesized by amidation of aspartate, with glutamine donating the $NH_4^+$. These are nonessential amino acids, and their simple biosynthetic pathways occur in all organisms.

Methionine, threonine, lysine, isoleucine, valine, and leucine are essential amino acids. Their biosynthetic pathways are complex and interconnected (Fig. 22–15).

**FIGURE 22-15 Biosynthesis of six essential amino acids from oxalo-acetate and pyruvate in bacteria: methionine, threonine, lysine, isoleucine, valine, and leucine.** Here, and in other multistep pathways, the enzymes are listed in the key. Note that L,L-α,ε-diaminopimelate, the product of step ⑭, is symmetric. The carbons derived from pyruvate (and the amino group derived from glutamate) are not traced beyond this point, because subsequent reactions may place them at either end of the lysine molecule.

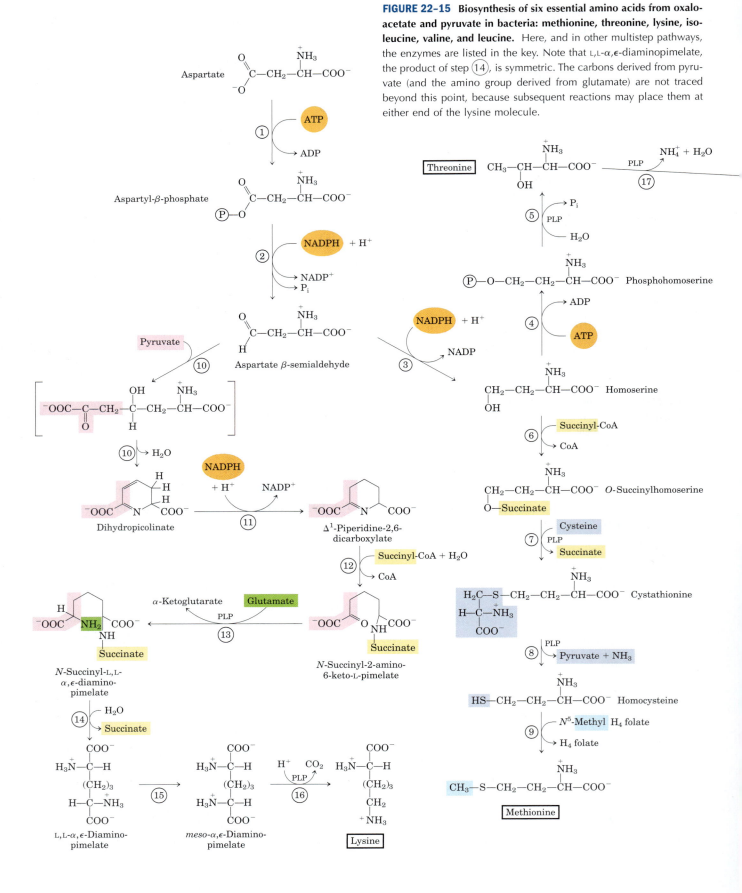

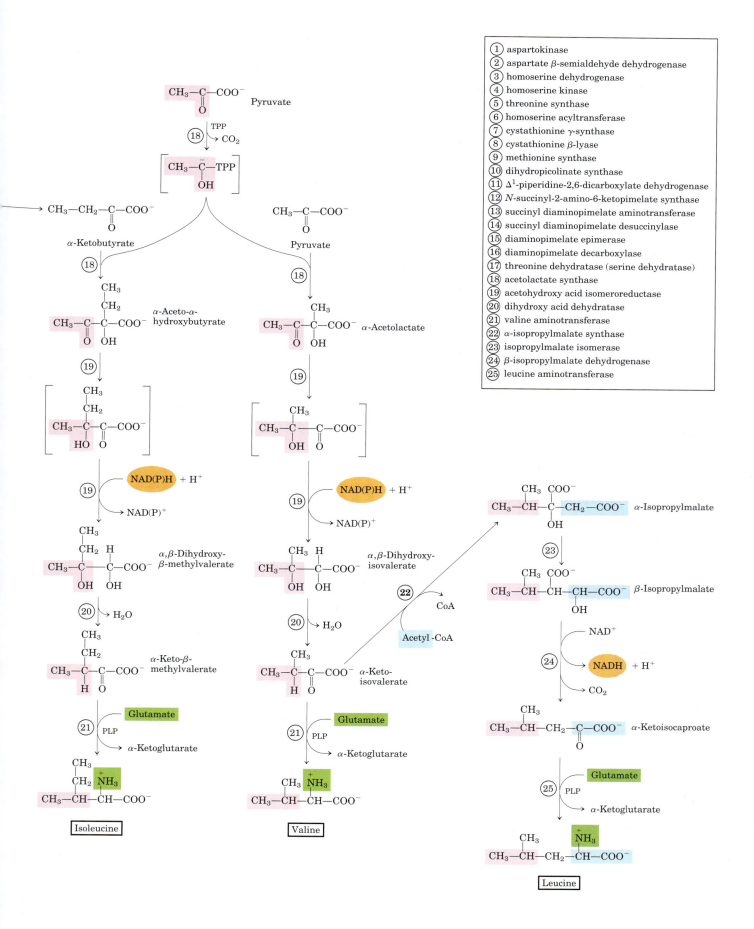

In some cases, the pathways in bacteria, fungi, and plants differ significantly. The bacterial pathways are outlined in Figure 22–15.

Aspartate gives rise to **methionine, threonine, and lysine.** Branch points occur at aspartate β-semialdehyde, an intermediate in all three pathways, and at homoserine, a precursor of threonine and methionine. Threonine, in turn, is one of the precursors of isoleucine. The **valine** and **isoleucine** pathways share four enzymes (Fig. 22–15, steps 18 to 21). Pyruvate gives rise

to valine and isoleucine in pathways that begin with condensation of two carbons of pyruvate (in the form of hydroxyethyl thiamine pyrophosphate; see Fig. 14–13) with another molecule of pyruvate (valine path) or with α-ketobutyrate (isoleucine path). The α-ketobutyrate is derived from threonine in a reaction that requires pyridoxal phosphate (Fig. 22–15, step 17). An intermediate in the valine pathway, α-ketoisovalerate, is the starting point for a four-step branch pathway leading to **leucine** (steps 22 to 25).

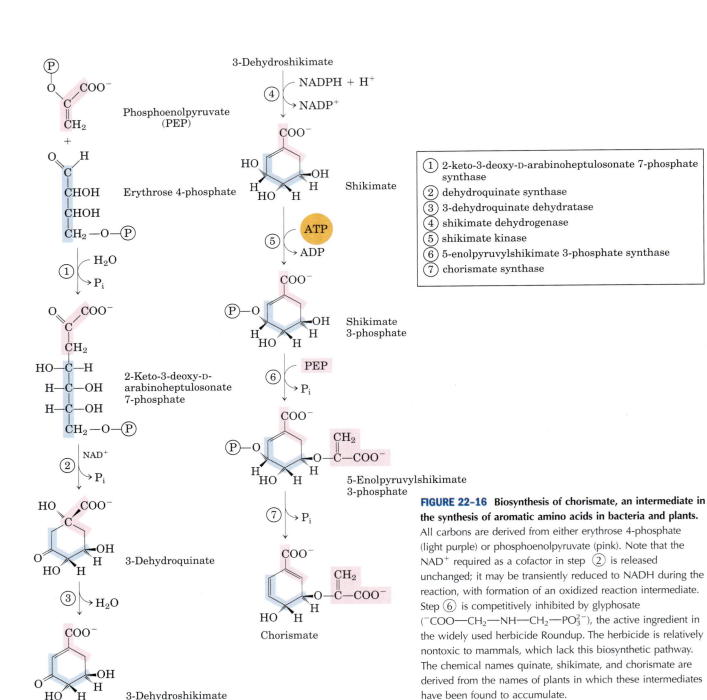

1. 2-keto-3-deoxy-D-arabinoheptulosonate 7-phosphate synthase
2. dehydroquinate synthase
3. 3-dehydroquinate dehydratase
4. shikimate dehydrogenase
5. shikimate kinase
6. 5-enolpyruvylshikimate 3-phosphate synthase
7. chorismate synthase

**FIGURE 22–16 Biosynthesis of chorismate, an intermediate in the synthesis of aromatic amino acids in bacteria and plants.** All carbons are derived from either erythrose 4-phosphate (light purple) or phosphoenolpyruvate (pink). Note that the NAD$^+$ required as a cofactor in step 2 is released unchanged; it may be transiently reduced to NADH during the reaction, with formation of an oxidized reaction intermediate. Step 6 is competitively inhibited by glyphosate ($^-COO$—$CH_2$—$NH$—$CH_2$—$PO_3^{2-}$), the active ingredient in the widely used herbicide Roundup. The herbicide is relatively nontoxic to mammals, which lack this biosynthetic pathway. The chemical names quinate, shikimate, and chorismate are derived from the names of plants in which these intermediates have been found to accumulate.

## Chorismate Is a Key Intermediate in the Synthesis of Tryptophan, Phenylalanine, and Tyrosine

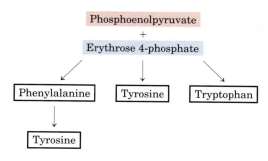

Aromatic rings are not readily available in the environment, even though the benzene ring is very stable. The branched pathway to tryptophan, phenylalanine, and tyrosine, occurring in bacteria, fungi, and plants, is the main biological route of aromatic ring formation. It proceeds through ring closure of an aliphatic precursor followed by stepwise addition of double bonds. The first four steps produce shikimate, a seven-carbon molecule derived from erythrose 4-phosphate and phosphoenolpyruvate (Fig. 22–16). Shikimate is converted to chorismate in three steps that include the addition of three more carbons from another molecule of phosphoenolpyruvate. Chorismate is the first branch point of the pathway, with one branch leading to tryptophan, the other to phenylalanine and tyrosine.

In the **tryptophan** branch (Fig. 22–17), chorismate is converted to anthranilate in a reaction in which glutamine donates the nitrogen that will become part of the indole ring. Anthranilate then condenses with PRPP. The indole ring of tryptophan is derived from the ring carbons and amino group of anthranilate plus two carbons derived from PRPP. The final reaction in the sequence is catalyzed by **tryptophan synthase.** This enzyme has an $\alpha_2\beta_2$ subunit structure and can be dissociated into two $\alpha$ subunits and a $\beta_2$ subunit that catalyze different parts of the overall reaction:

$$\text{Indole-3-glycerol phosphate} \xrightarrow[\alpha \text{ subunit}]{}$$
$$\text{indole} + \text{glyceraldehyde 3-phosphate}$$

$$\text{Indole} + \text{serine} \xrightarrow[\beta_2 \text{ subunit}]{} \text{tryptophan} + H_2O$$

---

① anthranilate synthase
② anthranilate phosphoribosyltransferase
③ $N$-(5'-phosphoribosyl)-anthranilate isomerase
④ indole-3-glycerol phosphate synthase
⑤ tryptophan synthase

**FIGURE 22–17 Biosynthesis of tryptophan from chorismate in bacteria and plants.** In *E. coli,* enzymes catalyzing steps ① and ② are subunits of a single complex.

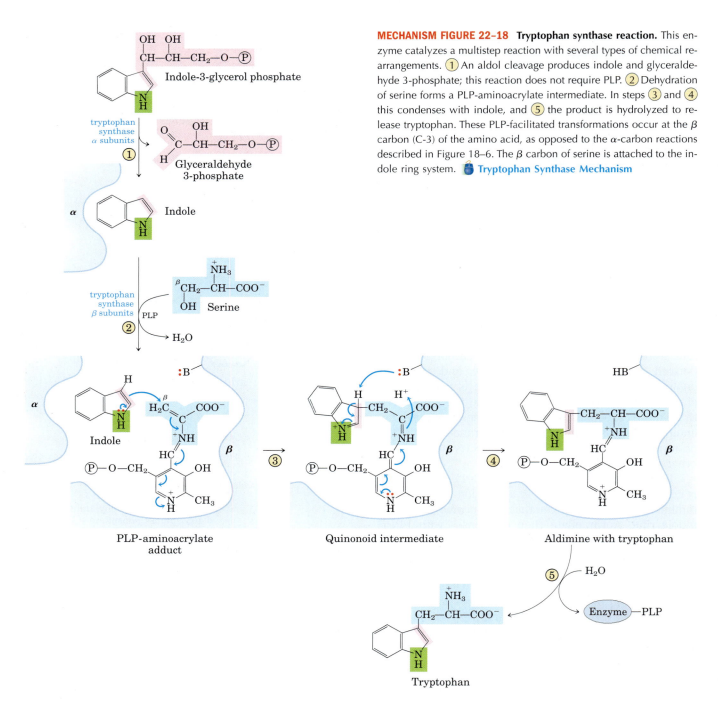

**MECHANISM FIGURE 22–18 Tryptophan synthase reaction.** This enzyme catalyzes a multistep reaction with several types of chemical rearrangements. ① An aldol cleavage produces indole and glyceraldehyde 3-phosphate; this reaction does not require PLP. ② Dehydration of serine forms a PLP-aminoacrylate intermediate. In steps ③ and ④ this condenses with indole, and ⑤ the product is hydrolyzed to release tryptophan. These PLP-facilitated transformations occur at the β carbon (C-3) of the amino acid, as opposed to the α-carbon reactions described in Figure 18–6. The β carbon of serine is attached to the indole ring system. 🔲 **Tryptophan Synthase Mechanism**

The second part of the reaction requires pyridoxal phosphate (Fig. 22–18). Indole formed in the first part is not released by the enzyme, but instead moves through a channel from the α-subunit active site to the β-subunit active site, where it condenses with a Schiff base intermediate derived from serine and PLP. Intermediate channeling of this type may be a feature of the entire pathway from chorismate to tryptophan. Enzyme active sites catalyzing different steps (sometimes not sequential steps) of the pathway to tryptophan are found on single polypeptides in some species of fungi and bacteria, but are separate proteins in others. In addition, the activity of some of these enzymes requires a noncovalent association with other enzymes of the pathway. These observations suggest that all the pathway enzymes are components of a large, multienzyme complex in both prokaryotes and eukaryotes. Such complexes are generally not preserved intact when the enzymes are isolated using traditional biochemical methods, but evidence for the existence of multienzyme complexes is accumulating for this and a number of other metabolic pathways (p. 605).

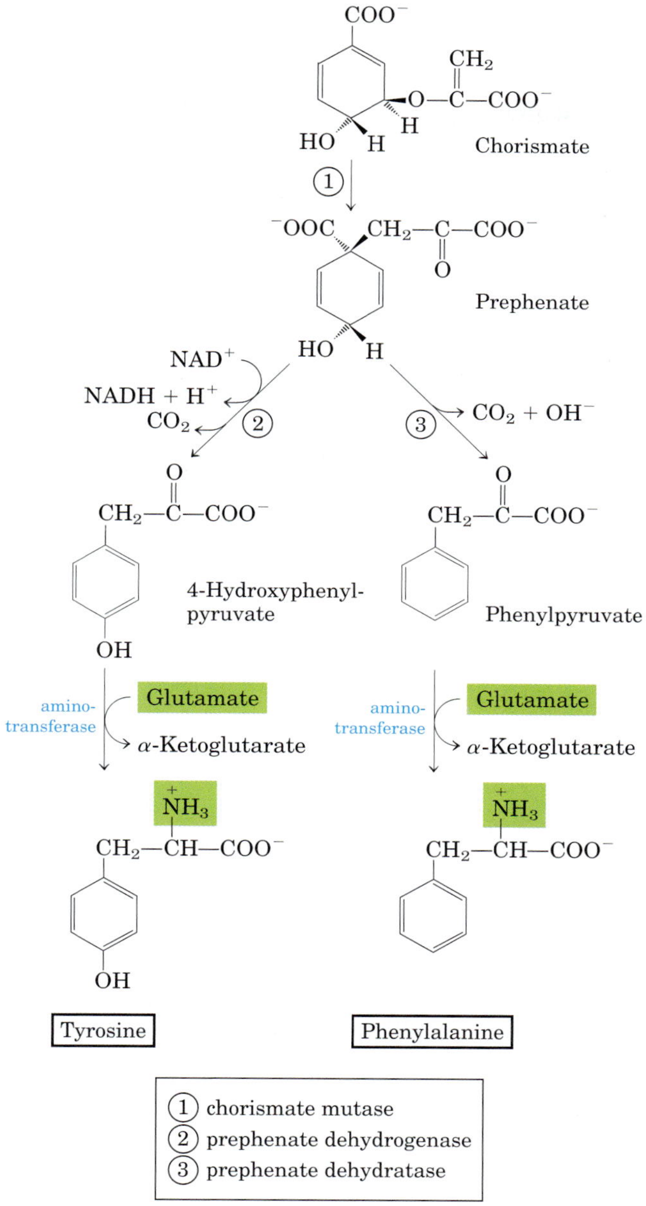

## Biosynthesis of phenylalanine and tyrosine

COO⁻

Chorismate

① chorismate mutase

Prephenate

NAD⁺
NADH + H⁺
$CO_2$
②

③
$CO_2 + OH^-$

4-Hydroxyphenyl-
pyruvate

Phenylpyruvate

OH

amino-
transferase
Glutamate
α-Ketoglutarate

amino-
transferase
Glutamate
α-Ketoglutarate

$NH_3$

$NH_3$

OH

Tyrosine

Phenylalanine

① chorismate mutase
② prephenate dehydrogenase
③ prephenate dehydratase

**FIGURE 22–19 Biosynthesis of phenylalanine and tyrosine from chorismate in bacteria and plants.** Conversion of chorismate to prephenate is a rare biological example of a Claisen rearrangement.

In plants and bacteria, **phenylalanine** and **tyrosine** are synthesized from chorismate in pathways much less complex than the tryptophan pathway. The common intermediate is prephenate (Fig. 22–19). The final step in both cases is transamination with glutamate.

Animals can produce tyrosine directly from phenylalanine through hydroxylation at C-4 of the phenyl group by **phenylalanine hydroxylase;** this enzyme also participates in the degradation of phenylalanine

(see Figs 18–23, 18–24). Tyrosine is considered a conditionally essential amino acid, or as nonessential insofar as it can be synthesized from the essential amino acid phenylalanine.

### Histidine Biosynthesis Uses Precursors of Purine Biosynthesis

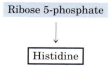

The pathway to **histidine** in all plants and bacteria differs in several respects from other amino acid biosynthetic pathways. Histidine is derived from three precursors (Fig. 22–20): PRPP contributes five carbons, the purine ring of ATP contributes a nitrogen and a carbon, and glutamine supplies the second ring nitrogen. The key steps are condensation of ATP and PRPP, in which N-1 of the purine ring is linked to the activated C-1 of the ribose of PRPP (step ① in Fig. 22–20); purine ring opening that ultimately leaves N-1 and C-2 of adenine linked to the ribose (step ③); and formation of the imidazole ring, a reaction in which glutamine donates a nitrogen (step ⑤). The use of ATP as a metabolite rather than a high-energy cofactor is unusual— but not wasteful, because it dovetails with the purine biosynthetic pathway. The remnant of ATP that is released after the transfer of N-1 and C-2 is 5-aminoimidazole-4-carboxamide ribonucleotide (AICAR), an intermediate of purine biosynthesis (see Fig. 22–33) that is rapidly recycled to ATP.

### Amino Acid Biosynthesis Is under Allosteric Regulation

The most responsive regulation of amino acid synthesis takes place through feedback inhibition of the first reaction in a sequence by the end product of the pathway. This first reaction is usually irreversible and catalyzed by an allosteric enzyme. As an example, Figure 22–21 shows the allosteric regulation of isoleucine synthesis from threonine (detailed in Fig. 22–15). The end product, isoleucine, is an allosteric inhibitor of the first reaction in the sequence. In bacteria, such allosteric modulation of amino acid synthesis occurs as a minute-to-minute response.

Allosteric regulation can be considerably more complex. An example is the remarkable set of allosteric controls exerted on glutamine synthetase of *E. coli* (Fig. 22–6). Six products derived from glutamine serve as negative feedback modulators of the enzyme, and the overall effects of these and other modulators are more than additive. Such regulation is called **concerted inhibition.**

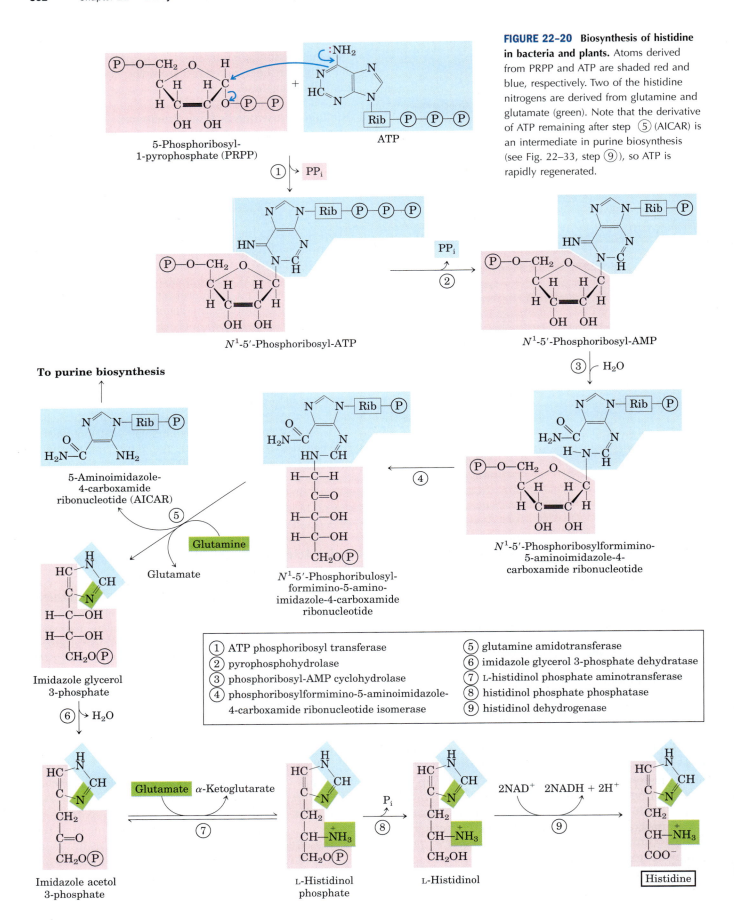

**FIGURE 22–20 Biosynthesis of histidine in bacteria and plants.** Atoms derived from PRPP and ATP are shaded red and blue, respectively. Two of the histidine nitrogens are derived from glutamine and glutamate (green). Note that the derivative of ATP remaining after step ⑤ (AICAR) is an intermediate in purine biosynthesis (see Fig. 22–33, step ⑨), so ATP is rapidly regenerated.

5-Phosphoribosyl-1-pyrophosphate (PRPP)

ATP

① → $PP_i$

$N^1$-5'-Phosphoribosyl-ATP

② $PP_i$

$N^1$-5'-Phosphoribosyl-AMP

③ ⧵ $H_2O$

$N^1$-5'-Phosphoribosylformimino-5-aminoimidazole-4-carboxamide ribonucleotide

④

$N^1$-5'-Phosphoribulosyl-formimino-5-amino-imidazole-4-carboxamide ribonucleotide

**To purine biosynthesis**

5-Aminoimidazole-4-carboxamide ribonucleotide (AICAR)

⑤

Glutamine

Glutamate

Imidazole glycerol 3-phosphate

⑥ ⧵ $H_2O$

Imidazole acetol 3-phosphate

Glutamate    α-Ketoglutarate

⑦

L-Histidinol phosphate

$P_i$

⑧

L-Histidinol

2NAD⁺   2NADH + 2H⁺

⑨

Histidine

① ATP phosphoribosyl transferase
② pyrophosphohydrolase
③ phosphoribosyl-AMP cyclohydrolase
④ phosphoribosylformimino-5-aminoimidazole-4-carboxamide ribonucleotide isomerase
⑤ glutamine amidotransferase
⑥ imidazole glycerol 3-phosphate dehydratase
⑦ L-histidinol phosphate aminotransferase
⑧ histidinol phosphate phosphatase
⑨ histidinol dehydrogenase

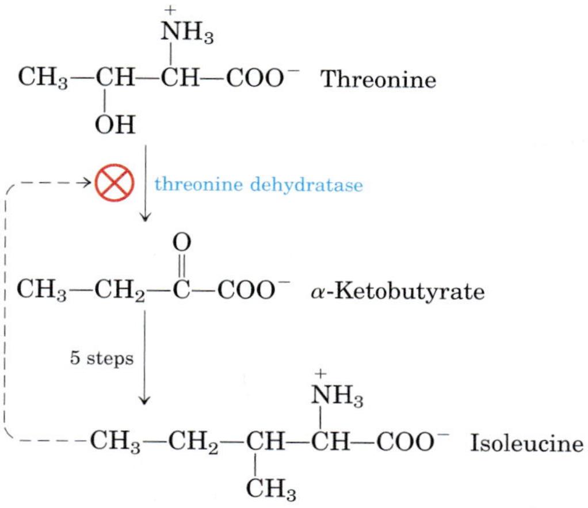

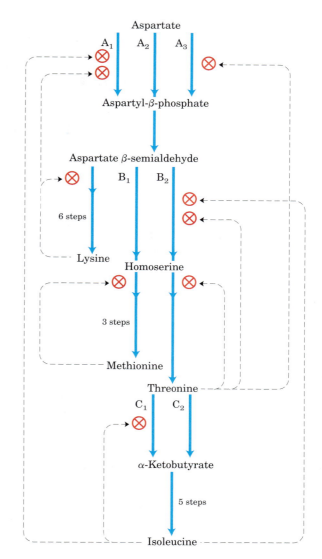

**FIGURE 22–21 Allosteric regulation of isoleucine biosynthesis.** The first reaction in the pathway from threonine to isoleucine is inhibited by the end product, isoleucine. This was one of the first examples of allosteric feedback inhibition to be discovered. The steps from α-ketobutyrate to isoleucine correspond to steps ⑱ through ㉑ in Figure 22–15 (five steps because ⑲ is a two-step reaction).

Because the 20 common amino acids must be made in the correct proportions for protein synthesis, cells have developed ways not only of controlling the rate of synthesis of individual amino acids but also of coordinating their formation. Such coordination is especially well developed in fast-growing bacterial cells. Figure 22–22 shows how *E. coli* cells coordinate the synthesis of lysine, methionine, threonine, and isoleucine, all made from aspartate. Several important types of inhibition patterns are evident. The step from aspartate to aspartyl-β-phosphate is catalyzed by three isozymes, each independently controlled by different modulators. This **enzyme multiplicity** prevents one biosynthetic end product from shutting down key steps in a pathway when other products of the same pathway are required. The steps from aspartate β-semialdehyde to homoserine and from threonine to α-ketobutyrate (detailed in Fig. 22–15) are also catalyzed by dual, independently controlled isozymes. One isozyme for the conversion of aspartate to aspartyl-β-phosphate is allosterically inhibited by two different modulators, lysine and isoleucine, whose action is more than additive—another example of concerted inhibition. The sequence from aspartate to isoleucine undergoes multiple, overlapping negative feedback inhibition; for example, isoleucine inhibits the conversion of threonine to α-ketobutyrate (as described above), and threonine inhibits its own formation at three points: from homoserine, from aspartate β-semialdehyde, and from aspartate (steps ④, ③, and ① in Fig. 22–15). This overall regulatory mechanism is called **sequential feedback inhibition.**

Similar patterns are evident in the pathways leading to the aromatic amino acids. The first step of the

**FIGURE 22–22 Interlocking regulatory mechanisms in the biosynthesis of several amino acids derived from aspartate in *E. coli*.** Three enzymes (A, B, C) have either two or three isozyme forms, indicated by numerical subscripts. In each case, one isozyme ($A_2$, $B_1$, and $C_2$) has no allosteric regulation; these isozymes are regulated by changes in the amount synthesized (Chapter 28). Synthesis of isozymes $A_2$ and $B_1$ is repressed when methionine levels are high, and synthesis of isozyme $C_2$ is repressed when isoleucine levels are high. Enzyme A is aspartokinase; B, homoserine dehydrogenase; C, threonine dehydratase.

early pathway to the common intermediate chorismate is catalyzed by the enzyme 2-keto-3-deoxy-D-arabino-heptulosonate 7-phosphate (DAHP) synthase (step ① in Fig. 22–16). Most microorganisms and plants have three DAHP synthase isozymes. One is allosterically inhibited (feedback inhibition) by phenylalanine, another by tyrosine, and the third by tryptophan. This scheme helps the overall pathway to respond to cellular requirements for one or more of the aromatic amino acids. Additional regulation takes place after the pathway branches at chorismate. For example, the enzymes catalyzing the first two steps of the tryptophan branch are subject to allosteric inhibition by tryptophan.

### SUMMARY 22.2 Biosynthesis of Amino Acids

- Plants and bacteria synthesize all 20 common amino acids. Mammals can synthesize about half; the others are required in the diet (essential amino acids).

- Among the nonessential amino acids, glutamate is formed by reductive amination of $\alpha$-ketoglutarate and serves as the precursor of glutamine, proline, and arginine. Alanine and aspartate (and thus asparagine) are formed from pyruvate and oxaloacetate, respectively, by transamination. The carbon chain of serine is derived from 3-phosphoglycerate. Serine is a precursor of glycine; the $\beta$-carbon atom of serine is transferred to tetrahydrofolate. In microorganisms, cysteine is produced from serine and from sulfide produced by the reduction of environmental sulfate. Mammals produce cysteine from methionine and serine by a series of reactions requiring S-adenosylmethionine and cystathionine.

- Among the essential amino acids, the aromatic amino acids (phenylalanine, tyrosine, and tryptophan) form by a pathway in which chorismate occupies a key branch point. Phosphoribosyl pyrophosphate is a precursor of tryptophan and histidine. The pathway to histidine is interconnected with the purine synthetic pathway. Tyrosine can also be formed by hydroxylation of phenylalanine (and thus is considered conditionally essential). The pathways for the other essential amino acids are complex.

- The amino acid biosynthetic pathways are subject to allosteric end-product inhibition; the regulatory enzyme is usually the first in the sequence. Regulation of the various synthetic pathways is coordinated.

## 22.3 Molecules Derived from Amino Acids

In addition to their role as the building blocks of proteins, amino acids are precursors of many specialized biomolecules, including hormones, coenzymes, nucleotides, alkaloids, cell wall polymers, porphyrins, antibiotics, pigments, and neurotransmitters. We describe here the pathways to a number of these amino acid derivatives.

### Glycine Is a Precursor of Porphyrins

The biosynthesis of **porphyrins,** for which glycine is a major precursor, is our first example, because of the central importance of the porphyrin nucleus in heme proteins such as hemoglobin and the cytochromes. The porphyrins are constructed from four molecules of the monopyrrole derivative **porphobilinogen,** which itself is derived from two molecules of $\delta$-aminolevulinate. There are two major pathways to $\delta$-aminolevulinate. In higher eukaryotes (Fig. 22–23a), glycine reacts with succinyl-CoA in the first step to yield $\alpha$-amino-$\beta$-ketoadipate, which is then decarboxylated to $\delta$-aminolevulinate. In plants, algae, and most bacteria, $\delta$-aminolevulinate is formed from glutamate (Fig. 22–23b). The glutamate is first esterified to glutamyl-tRNA$^{Glu}$ (see Chapter 27 on the topic of transfer RNAs); reduction by NADPH converts the glutamate to glutamate 1-semialdehyde, which is cleaved from the tRNA. An aminotransferase converts the glutamate 1-semialdehyde to $\delta$-aminolevulinate.

In all organisms, two molecules of $\delta$-aminolevulinate condense to form porphobilinogen and, through a series of complex enzymatic reactions, four molecules of porphobilinogen come together to form **protoporphyrin** (Fig. 22–24). The iron atom is incorporated after the protoporphyrin has been assembled, in a step catalyzed by ferrochelatase. Porphyrin biosynthesis is regulated in higher eukaryotes by the concentration of the heme product, which serves as a feedback inhibitor of early steps in the synthetic pathway. Genetic defects in the biosynthesis of porphyrins can lead to the accumulation of pathway intermediates, causing a variety of human diseases known collectively as **porphyrias** (Box 22–1).

### Heme Is the Source of Bile Pigments

The iron-porphyrin (heme) group of hemoglobin, released from dying erythrocytes in the spleen, is degraded to yield free $Fe^{3+}$ and, ultimately, **bilirubin.** This pathway is arresting for its capacity to inject color into human biochemistry.

The first step in the two-step pathway, catalyzed by heme oxygenase (HO), converts heme to biliverdin, a linear (open) tetrapyrrole derivative (Fig. 22–25). The other products of the reaction are free $Fe^{2+}$ and CO. The $Fe^{2+}$ is quickly bound by ferritin. Carbon monoxide is a poison that binds to hemoglobin (see Box 5–1), and the production of CO by heme oxygenase ensures that, even in the absence of environmental exposure, about 1% of an individual's heme is complexed with CO.

Biliverdin is converted to bilirubin in the second step, catalyzed by biliverdin reductase. You can monitor this reaction colorimetrically in a familiar in situ experiment. When you are bruised, the black and/or purple color results from hemoglobin released from damaged erythrocytes. Over time, the color changes to the green of biliverdin, and then to the yellow of bilirubin. Biliru-

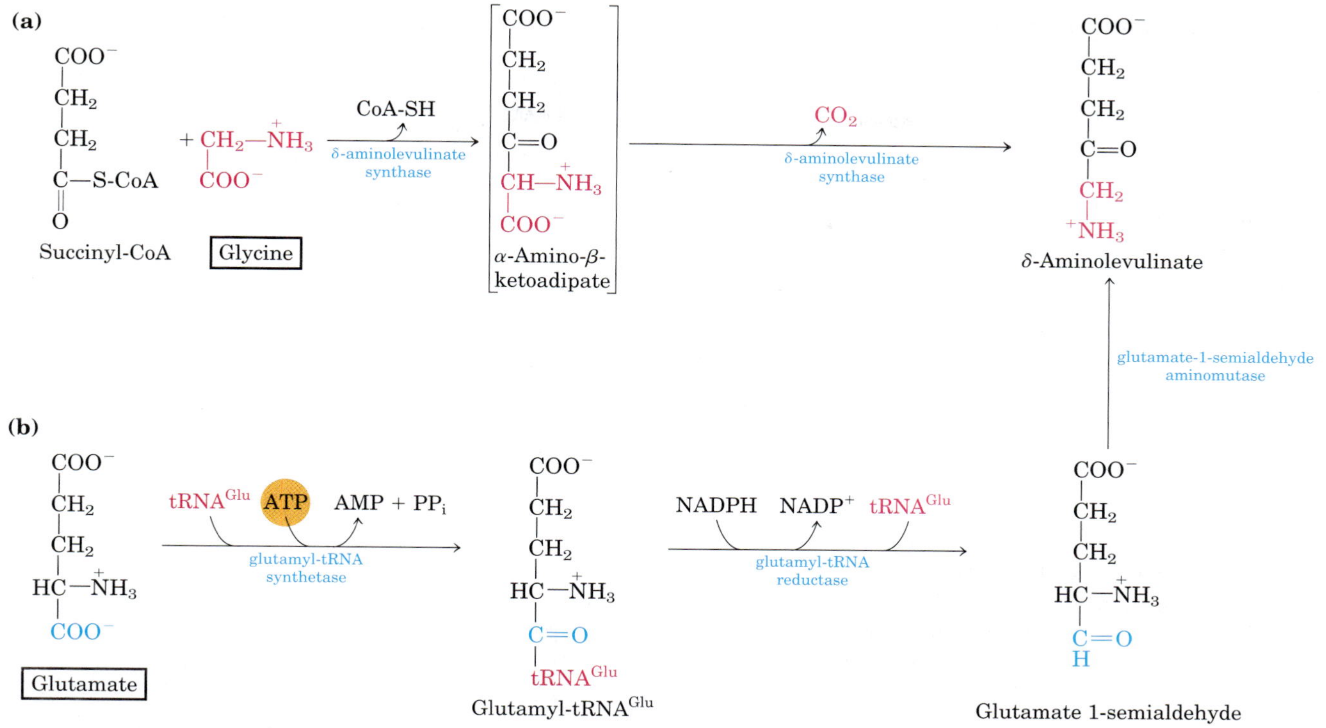

**(a)**

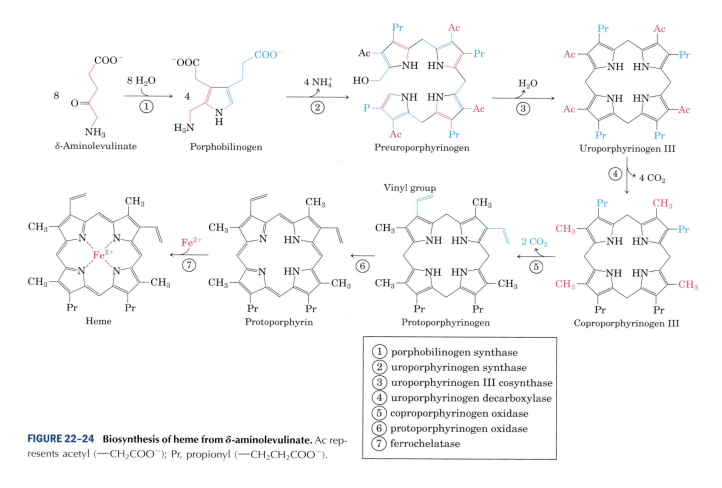

**FIGURE 22–23** Biosynthesis of δ-aminolevulinate. **(a)** In most animals, including mammals, δ-aminolevulinate is synthesized from glycine and succinyl-CoA. The atoms furnished by glycine are shown in red. **(b)** In bacteria and plants, the precursor of δ-aminolevulinate is glutamate.

**FIGURE 22–24** Biosynthesis of heme from δ-aminolevulinate. Ac represents acetyl (—CH₂COO⁻); Pr, propionyl (—CH₂CH₂COO⁻).

① porphobilinogen synthase
② uroporphyrinogen synthase
③ uroporphyrinogen III cosynthase
④ uroporphyrinogen decarboxylase
⑤ coproporphyrinogen oxidase
⑥ protoporphyrinogen oxidase
⑦ ferrochelatase

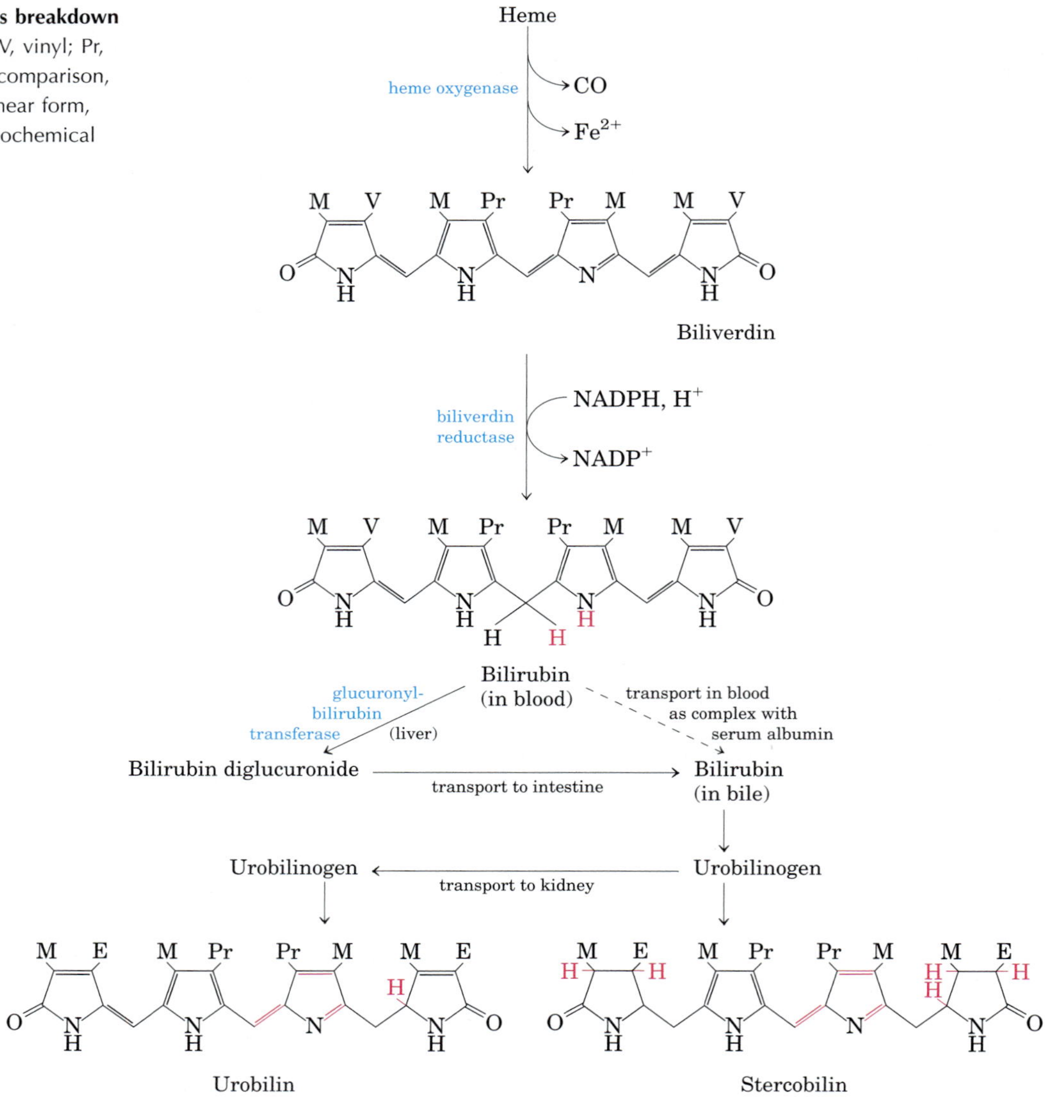

**FIGURE 22–25  Bilirubin and its breakdown products.** M represents methyl; V, vinyl; Pr, propionyl; E, ethyl. For ease of comparison, these structures are shown in linear form, rather than in their correct stereochemical conformations.

bin is largely insoluble, and it travels in the bloodstream as a complex with serum albumin. In the liver, bilirubin is transformed to the bile pigment bilirubin diglucuronide. This product is sufficiently water-soluble to be secreted with other components of bile into the small intestine, where microbial enzymes convert it to several products, predominantly urobilinogen. Some urobilinogen is reabsorbed into the blood and transported to the kidney, where it is converted to urobilin, the compound that gives urine its yellow color (Fig. 22–25, left branch). Urobilinogen remaining in the intestine is converted (in another microbe-dependent reaction) to stercobilin (Fig. 22–25, right branch), which imparts the red-brown color to feces.

Impaired liver function or blocked bile secretion causes bilirubin to leak from the liver into the blood, resulting in a yellowing of the skin and eyeballs, a con-

dition called jaundice. In cases of jaundice, determination of the concentration of bilirubin in the blood may be useful in the diagnosis of underlying liver disease. Newborn infants sometimes develop jaundice because they have not yet produced enough glucuronyl bilirubin transferase to process their bilirubin. A traditional treatment to reduce excess bilirubin, exposure to a fluorescent lamp, causes a photochemical conversion of bilirubin to compounds that are more soluble and easily excreted.

These pathways of heme breakdown play significant roles in protecting cells from oxidative damage and in regulating certain cellular functions. The CO produced by heme oxygenase is toxic at high concentrations, but at the very low concentrations generated during heme degradation it appears to have some regulatory and/or signaling functions. It acts as a vasodilator, much the

## BOX 22–1    BIOCHEMISTRY IN MEDICINE

### Biochemistry of Kings and Vampires

Porphyrias (listed at right) are a group of genetic diseases in which, because of defects in enzymes of the biosynthetic pathway from glycine to porphyrins, specific porphyrin precursors accumulate in erythrocytes, body fluids, and the liver. The most common form is acute intermittent porphyria. Most affected individuals are heterozygotes and are usually asymptomatic, because the single copy of the normal gene provides a sufficient level of enzyme function. However, certain nutritional or environmental factors (as yet poorly understood) can cause a buildup of δ-aminolevulinate and porphobilinogen, leading to attacks of acute abdominal pain and neurological dysfunction. King George III, British monarch during the American Revolution, suffered several episodes of apparent madness that tarnished the record of this otherwise accomplished man. The symptoms of his condition suggest that George III suffered from acute intermittent porphyria.

One of the rarer porphyrias results in an accumulation of uroporphyrinogen I, an abnormal isomer of a protoporphyrin precursor. This compound stains the urine red, causes the teeth to fluoresce strongly in ultraviolet light, and makes the skin abnormally sensitive to sunlight. Many individuals with this porphyria are anemic, because insufficient heme is synthesized. This genetic condition may have given rise to the vampire myths of folk legend.

The symptoms of most porphyrias are now readily controlled with dietary changes or the administration of heme or heme derivatives.

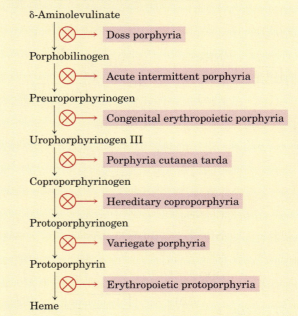

same as (but less potent than) nitric oxide (discussed below). Low levels of CO also have some regulatory effects on neurotransmission. Bilirubin is the most abundant antioxidant in mammalian tissues and is responsible for most of the antioxidant activity in serum. Its protective effects appear to be especially important in the developing brain of newborn infants. The cell toxicity associated with jaundice may be due to bilirubin levels in excess of the serum albumin needed to solubilize it.

Given these varied roles of heme degradation products, the degradative pathway is subject to regulation, mainly at the first step. Humans have at least three isozymes of heme oxygenase. HO-1 is highly regulated; the expression of its gene is induced by a wide range of stress conditions (shear stress, angiogenesis (uncontrolled development of blood vessels), hypoxia, hyperoxia, heat shock, exposure to ultraviolet light, hydrogen peroxide, and many other metabolic insults). HO-2 is found mainly in brain and testes, where it is continuously expressed. The third isozyme, HO-3, is not yet well characterized. ∎

### Amino Acids Are Precursors of Creatine and Glutathione

**Phosphocreatine,** derived from **creatine,** is an important energy buffer in skeletal muscle (see Fig. 13–5). Creatine is synthesized from glycine and arginine (Fig. 22–26); methionine, in the form of S-adenosylmethionine, acts as methyl group donor.

**Glutathione (GSH),** present in plants, animals, and some bacteria, often at high levels, can be thought of as a redox buffer. It is derived from glycine, glutamate, and cysteine (Fig. 22–27). The γ-carboxyl group of glutamate is activated by ATP to form an acyl phosphate intermediate, which is then attacked by the α-amino group of cysteine. A second condensation reaction follows, with the α-carboxyl group of cysteine activated to an acyl phosphate to permit reaction with glycine. The oxidized form of glutathione (GSSG), produced in the course of its redox activities, contains two glutathione molecules linked by a disulfide bond.

Glutathione probably helps maintain the sulfhydryl groups of proteins in the reduced state and the iron of heme in the ferrous ($Fe^{2+}$) state, and it serves as a reducing agent for glutaredoxin in deoxyribonucleotide synthesis (see Fig. 22–39). Its redox function is also used to remove toxic peroxides formed in the normal course of growth and metabolism under aerobic conditions:

$$2GSH + R{-}O{-}O{-}H \longrightarrow GSSG + H_2O + R{-}OH$$

This reaction is catalyzed by **glutathione peroxidase,** a remarkable enzyme in that it contains a covalently

bound selenium (Se) atom in the form of selenocysteine (see Fig. 3–8a), which is essential for its activity.

## D-Amino Acids Are Found Primarily in Bacteria

Although D-amino acids do not generally occur in proteins, they do serve some special functions in the structure of bacterial cell walls and peptide antibiotics. Bacterial peptidoglycans (see Fig. 20–23) contain both D-alanine and D-glutamate. D-Amino acids arise directly from the L isomers by the action of amino acid racemases, which have pyridoxal phosphate as cofactor (see Fig. 18–6). Amino acid racemization is uniquely important to bacterial metabolism, and enzymes such as

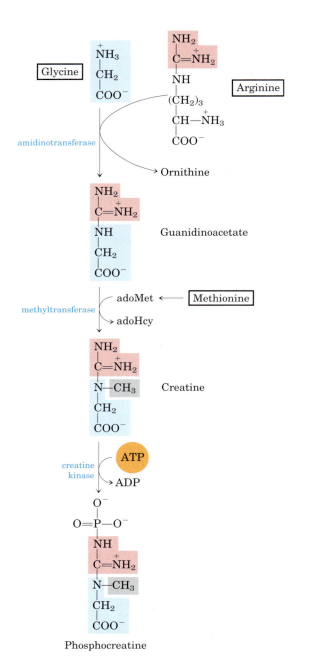

**FIGURE 22–26 Biosynthesis of creatine and phosphocreatine.** Creatine is made from three amino acids: glycine, arginine, and methionine. This pathway shows the versatility of amino acids as precursors of other nitrogenous biomolecules.

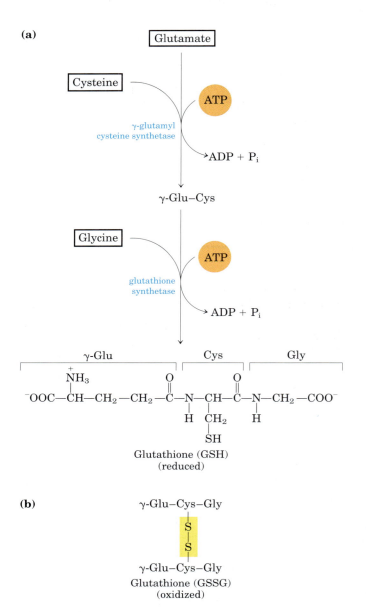

**FIGURE 22–27 Glutathione metabolism. (a)** Biosynthesis of glutathione. **(b)** Reduced form of glutathione.

alanine racemase are prime targets for pharmaceutical agents. One such agent, **L-fluoroalanine,** is being tested as an antibacterial drug. Another, **cycloserine,** is used to treat tuberculosis. Because these inhibitors also affect some PLP-requiring human enzymes, however, they have potentially undesirable side effects. ■

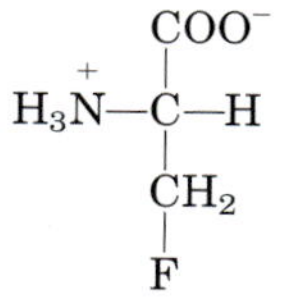

L-Fluoroalanine 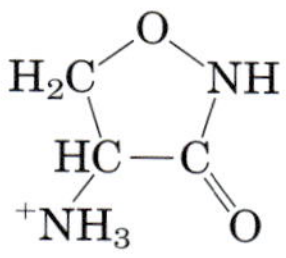         Cycloserine

## Aromatic Amino Acids Are Precursors of Many Plant Substances

Phenylalanine, tyrosine, and tryptophan are converted to a variety of important compounds in plants. The rigid polymer **lignin,** derived from phenylalanine and tyrosine, is second only to cellulose in abundance in plant tissues. The structure of the lignin polymer is complex and not well understood. Tryptophan is also the precursor of the plant growth hormone indole-3-acetate, or **auxin** (Fig. 22–28a), which has been implicated in the regulation of a wide range of biological processes in plants.

Phenylalanine and tyrosine also give rise to many commercially significant natural products, including the tannins that inhibit oxidation in wines; alkaloids such as morphine, which have potent physiological effects; and the flavoring of cinnamon oil (Fig. 22–28b), nutmeg, cloves, vanilla, cayenne pepper, and other products.

## Biological Amines Are Products of Amino Acid Decarboxylation

Many important neurotransmitters are primary or secondary amines, derived from amino acids in simple pathways. In addition, some polyamines that form complexes with DNA are derived from the amino acid ornithine, a component of the urea cycle. A common denominator of many of these pathways is amino acid decarboxylation, another PLP-requiring reaction (see Fig. 18–6).

The synthesis of some neurotransmitters is illustrated in Figure 22–29. Tyrosine gives rise to a family of catecholamines that includes **dopamine, norepinephrine,** and **epinephrine.** Levels of catecholamines are correlated with, among other things, changes in blood pressure. The neurological disorder Parkinson's disease is associated with an underproduction of dopamine, and it has traditionally been treated by administering L-dopa. Overproduction of dopamine in the brain may be linked to psychological disorders such as schizophrenia.

Glutamate decarboxylation gives rise to **γ-aminobutyrate (GABA),** an inhibitory neurotransmitter. Its underproduction is associated with epileptic seizures.

GABA analogs are used in the treatment of epilepsy and hypertension. Levels of GABA can also be increased by administering inhibitors of the GABA-degrading enzyme GABA aminotransferase. Another important neurotransmitter, **serotonin,** is derived from tryptophan in a two-step pathway.

Histidine undergoes decarboxylation to **histamine,** a powerful vasodilator in animal tissues. Histamine is released in large amounts as part of the allergic response, and it also stimulates acid secretion in the stomach. A growing array of pharmaceutical agents are being designed to interfere with either the synthesis or the action of histamine. A prominent example is the histamine receptor antagonist **cimetidine** (Tagamet), a structural analog of histamine:

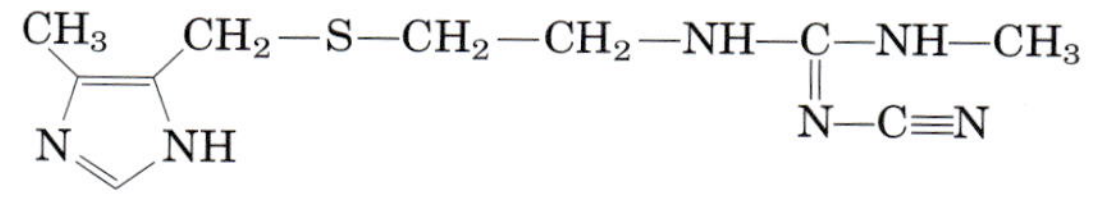

It promotes the healing of duodenal ulcers by inhibiting secretion of gastric acid.

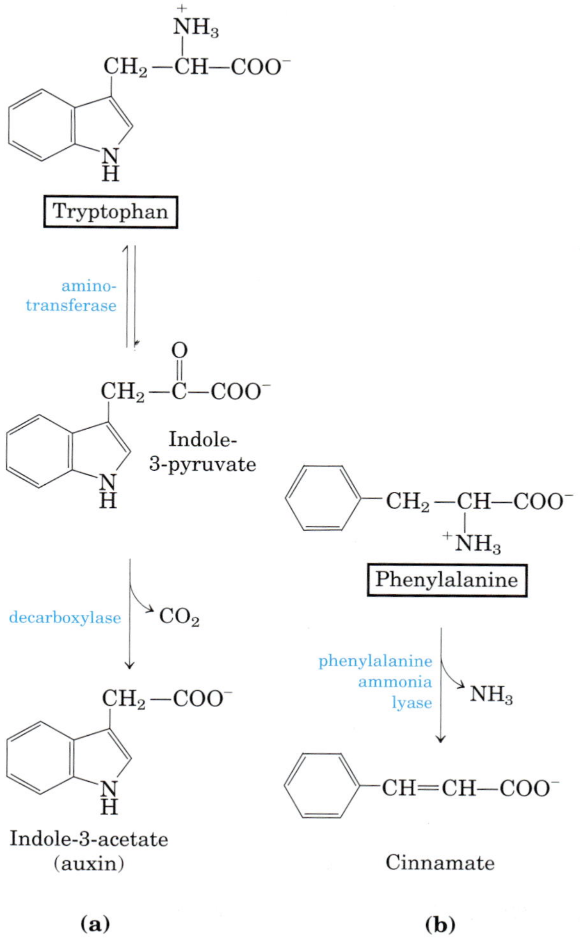

**FIGURE 22–28 Biosynthesis of two plant substances from amino acids.** **(a)** Indole-3-acetate (auxin) and **(b)** cinnamate (cinnamon flavor).

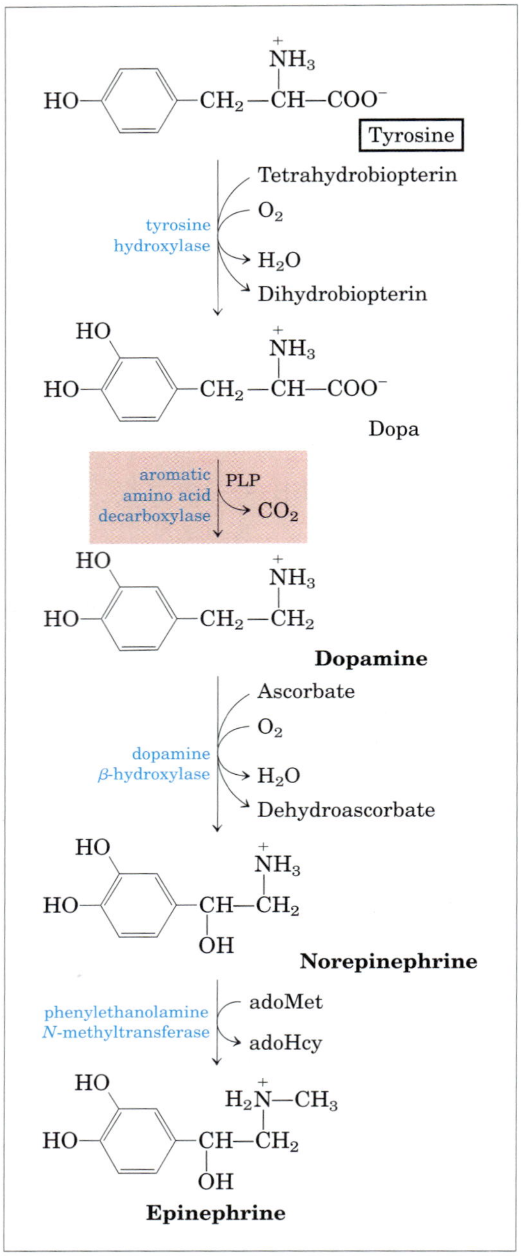

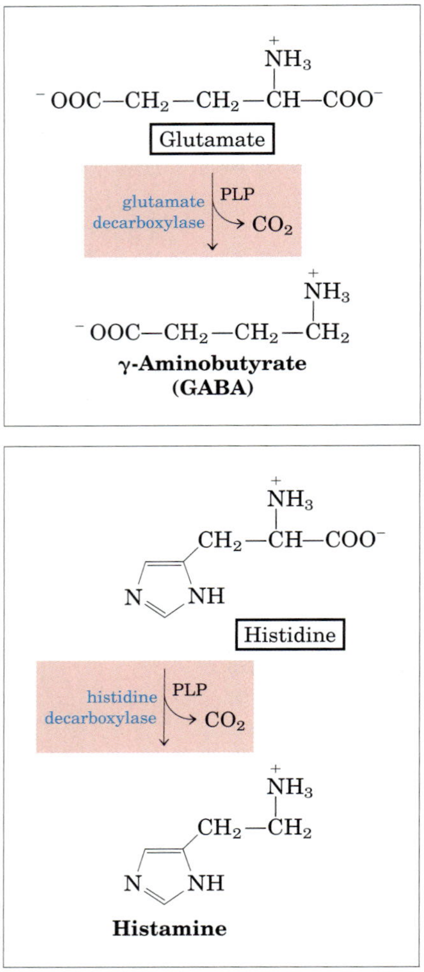

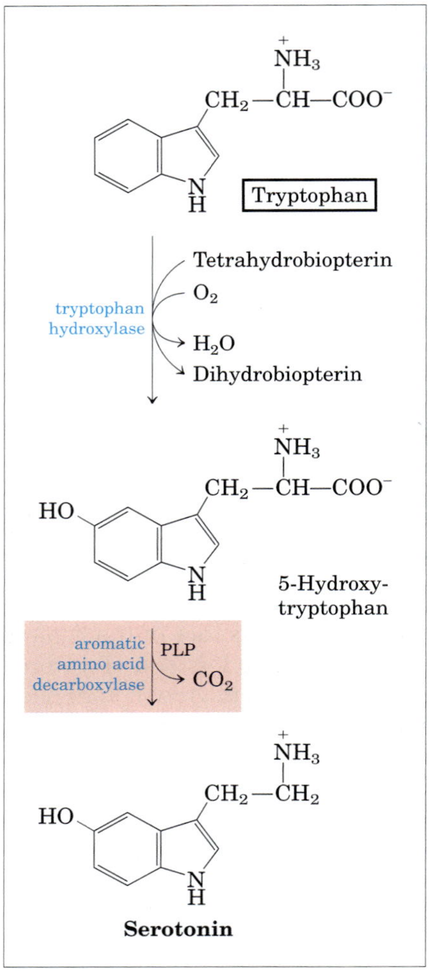

**FIGURE 22–29 Biosynthesis of some neurotransmitters from amino acids.** The key step is the same in each case: a PLP-dependent decarboxylation (shaded in pink).

Polyamines such as **spermine** and **spermidine,** involved in DNA packaging, are derived from methionine and ornithine by the pathway shown in Figure 22–30. The first step is decarboxylation of ornithine, a precursor of arginine (Fig. 22–10). **Ornithine decarboxylase,** a PLP-requiring enzyme, is the target of several powerful inhibitors used as pharmaceutical agents (Box 22–2). ■

### Arginine Is the Precursor for Biological Synthesis of Nitric Oxide

A surprise finding in the mid-1980s was the role of nitric oxide (NO)—previously known mainly as a component of smog—as an important biological messenger. This simple gaseous substance diffuses readily through membranes, although its high reactivity limits its range of diffusion to about a 1 mm radius from the site of synthesis. In humans NO plays a role in a range of physiological processes, including neurotransmission, blood clotting, and the control of blood pressure. Its mode of action is described in Chapter 12 (p. 434).

Nitric oxide is synthesized from arginine in an NADPH-dependent reaction catalyzed by nitric oxide synthase (Fig. 22–31), a dimeric enzyme structurally related to NADPH cytochrome P-450 reductase (see Box 21–1). The reaction is a five-electron oxidation. Each subunit of the enzyme contains one bound molecule of each of four different cofactors: FMN, FAD, tetrahydrobiopterin, and $Fe^{3+}$ heme. NO is an unstable molecule and cannot be stored. Its synthesis is stimulated by interaction of nitric oxide synthase with $Ca^{2+}$-calmodulin (see Fig. 12–21).

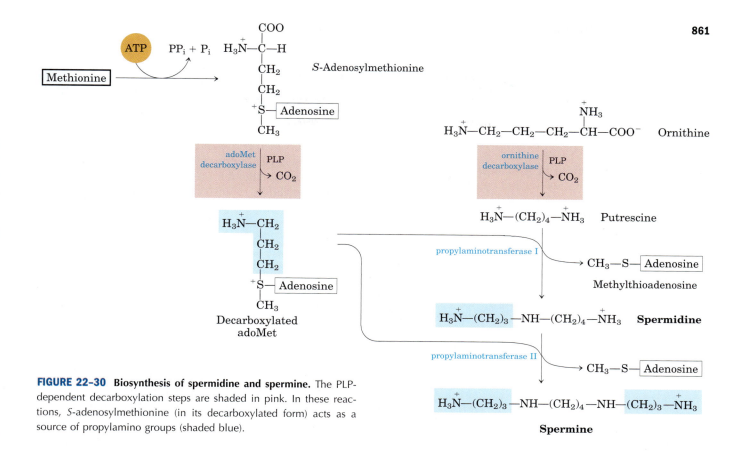

**FIGURE 22–30 Biosynthesis of spermidine and spermine.** The PLP-dependent decarboxylation steps are shaded in pink. In these reactions, S-adenosylmethionine (in its decarboxylated form) acts as a source of propylamino groups (shaded blue).

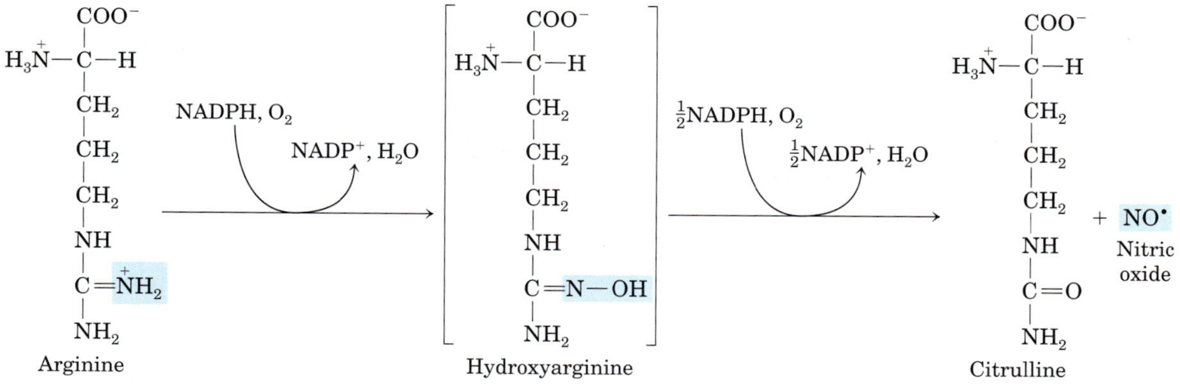

**FIGURE 22–31 Biosynthesis of nitric oxide.** Both steps are catalyzed by nitric oxide synthase. The nitrogen of the NO is derived from the guanidino group of arginine.

## SUMMARY 22.3 Molecules Derived from Amino Acids

■ Many important biomolecules are derived from amino acids. Glycine is a precursor of porphyrins. Degradation of iron-porphyrin (heme) generates bilirubin, which is converted to bile pigments, with several physiological functions.

■ Glycine and arginine give rise to creatine and phosphocreatine, an energy buffer. Glutathione, formed from three amino acids, is an important cellular reducing agent.

■ Bacteria synthesize D-amino acids from L-amino acids in racemization reactions requiring pyridoxal phosphate. D-amino acids are commonly found in certain bacterial walls and certain antibiotics.

■ The aromatic amino acids give rise to many plant substances. The PLP-dependent decarboxylation of some amino acids yields important biological amines, including neurotransmitters.

■ Arginine is the precursor of nitric oxide, a biological messenger.

## BOX 22–2 BIOCHEMISTRY IN MEDICINE

### Curing African Sleeping Sickness with a Biochemical Trojan Horse

African sleeping sickness, or African trypanosomiasis, is caused by protists (single-celled eukaryotes) called trypanosomes (Fig. 1). This disease (and related trypanosome-caused diseases) is medically and economically significant in many developing nations. Until recently, the disease was virtually incurable. Vaccines are ineffective, because the parasite has a novel mechanism to evade the host immune system.

**FIGURE 1** *Trypanosoma brucei rhodesiense,* one of several trypanosomes known to cause African sleeping sickness.

The cell coat of trypanosomes is covered with a single protein, which is the antigen to which the immune system responds. Every so often, however, by a process of genetic recombination (see Table 28–1), a few cells in the population of infecting trypanosomes switch to a new protein coat, not recognized by the immune system. This process of "changing coats" can occur hundreds of times. The result is a chronic cyclic infection: the human host develops a fever, which subsides as the immune system beats back the first infection; trypanosomes with changed coats then become the seed for a second infection, and the fever recurs. This cycle can repeat for weeks, and the weakened person eventually dies.

Some modern approaches to treating African sleeping sickness have been based on an understanding of enzymology and metabolism. In at least one such approach, this involves pharmaceutical agents designed as mechanism-based enzyme inactivators (suicide inactivators; p. 211). A vulnerable point in trypanosome metabolism is the pathway of polyamine biosynthesis. The polyamines spermine and spermidine, used in DNA packaging, are required in large amounts in rapidly dividing cells. The first step in their synthesis is catalyzed by ornithine decarboxylase, a PLP-requiring enzyme (see Fig.

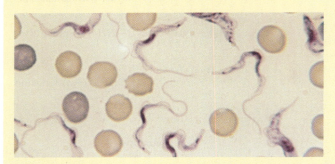

**FIGURE 2** Mechanism of ornithine decarboxylase reaction.

## 22.4 Biosynthesis and Degradation of Nucleotides

As discussed in Chapter 8, nucleotides play a variety of important roles in all cells. They are the precursors of DNA and RNA. They are essential carriers of chemical energy—a role primarily of ATP and to some extent GTP. They are components of the cofactors NAD, FAD, *S*-adenosylmethionine, and coenzyme A, as well as of activated biosynthetic intermediates such as UDP-glucose and CDP-diacylglycerol. Some, such as cAMP and cGMP, are also cellular second messengers.

Two types of pathways lead to nucleotides: the **de novo pathways** and the **salvage pathways.** De novo synthesis of nucleotides begins with their metabolic precursors: amino acids, ribose 5-phosphate, $CO_2$, and $NH_3$. Salvage pathways recycle the free bases and nucleosides released from nucleic acid breakdown. Both types of

22–30). In mammalian cells, ornithine decarboxylase undergoes rapid turnover—that is, a constant round of enzyme degradation and synthesis. In some trypanosomes, however, the enzyme—for reasons not well understood—is stable, not readily replaced by newly synthesized enzyme. An inhibitor of ornithine decarboxylase that binds permanently to the enzyme would thus have little effect on human cells, which could rapidly replace inactivated enzyme, but would adversely affect the parasite.

The first few steps of the normal reaction catalyzed by ornithine decarboxylase are shown in Figure 2. Once $CO_2$ is released, the electron movement is reversed and putrescine is produced (see Fig. 22–30). Based on this mechanism, several suicide inactivators have been designed, one of which is difluoromethylornithine (DFMO). DFMO is relatively inert in solution. When it binds to ornithine decarboxylase, however, the enzyme is quickly inactivated (Fig. 3). The inhibitor acts by providing an alternative electron sink in the form of two strategically placed fluorine atoms, which are excellent leaving groups. Instead of electrons moving into the ring structure of PLP, the reaction results in displacement of a fluorine atom. The S of a Cys residue at the enzyme's active site then forms a covalent complex with the highly reactive PLP-inhibitor adduct in an essentially irreversible reaction. In this way, the inhibitor makes use of the enzyme's own reaction mechanisms to kill it.

DFMO has proved highly effective against African sleeping sickness in clinical trials and is now used to treat African sleeping sickness caused by *T. brucei gambiense*. Approaches such as this show great promise for treating a wide range of diseases. The design of drugs based on enzyme mechanism and structure can complement the more traditional trial-and-error methods of developing pharmaceuticals.

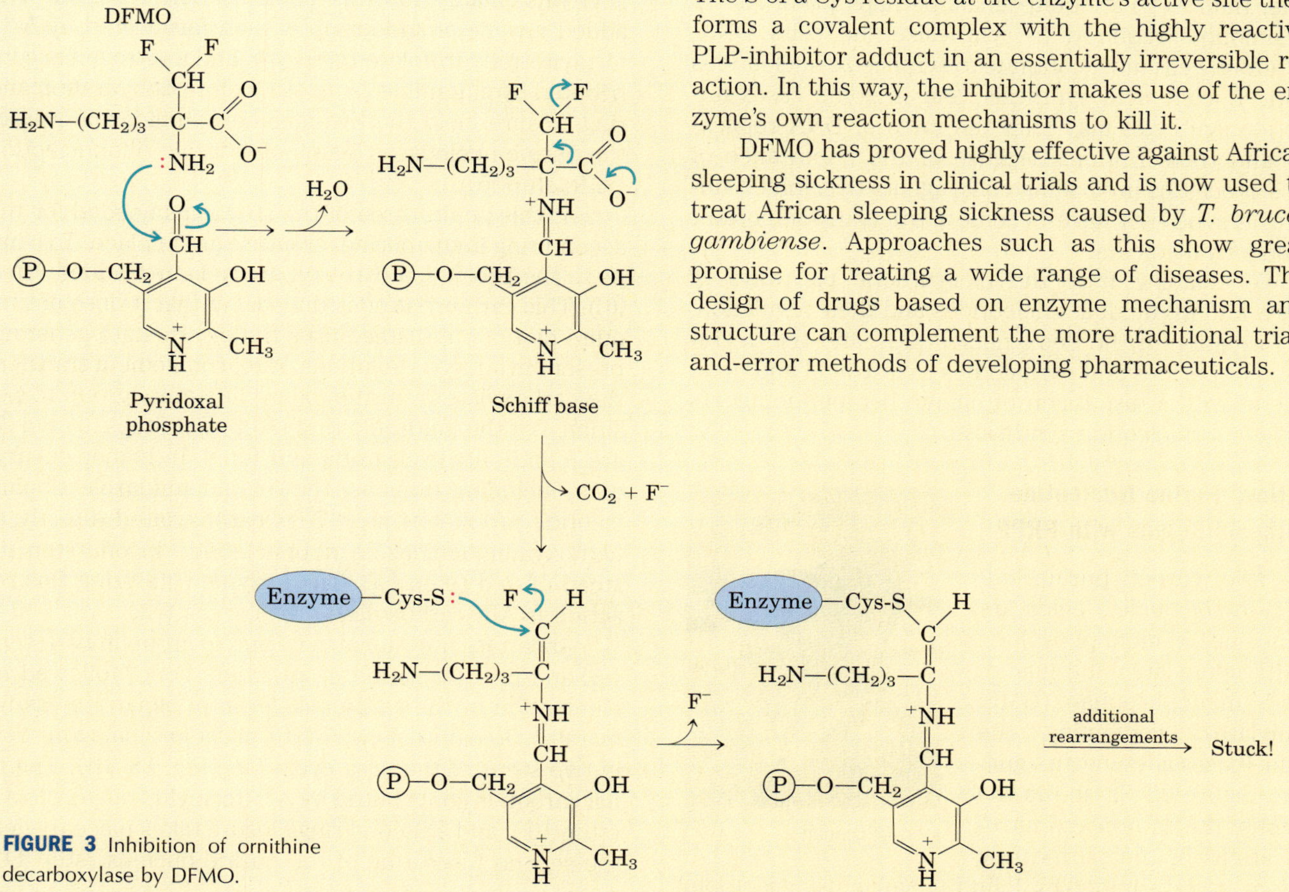

FIGURE 3 Inhibition of ornithine decarboxylase by DFMO.

pathways are important in cellular metabolism and both are presented in this section.

The de novo pathways for purine and pyrimidine biosynthesis appear to be nearly identical in all living organisms. Notably, the free bases guanine, adenine, thymine, cytidine, and uracil are *not* intermediates in these pathways; that is, the bases are not synthesized and then attached to ribose, as might be expected. The purine ring structure is built up one or a few atoms at a time, attached to ribose throughout the process. The pyrimidine ring is synthesized as **orotate,** attached to ribose phosphate, and then converted to the common pyrimidine nucleotides required in nucleic acid synthesis. Although the free bases are not intermediates in the de novo pathways, they are intermediates in some of the salvage pathways.

Several important precursors are shared by the de novo pathways for synthesis of pyrimidines and purines.

Phosphoribosyl pyrophosphate (PRPP) is important in both, and in these pathways the structure of ribose is retained in the product nucleotide, in contrast to its fate in the tryptophan and histidine biosynthetic pathways discussed earlier. An amino acid is an important precursor in each type of pathway: glycine for purines and aspartate for pyrimidines. Glutamine again is the most important source of amino groups—in five different steps in the de novo pathways. Aspartate is also used as the source of an amino group in the purine pathways, in two steps.

Two other features deserve mention. First, there is evidence, especially in the de novo purine pathway, that the enzymes are present as large, multienzyme complexes in the cell, a recurring theme in our discussion of metabolism. Second, the cellular pools of nucleotides (other than ATP) are quite small, perhaps 1% or less of the amounts required to synthesize the cell's DNA. Therefore, cells must continue to synthesize nucleotides during nucleic acid synthesis, and in some cases nucleotide synthesis may limit the rates of DNA replication and transcription. Because of the importance of these processes in dividing cells, agents that inhibit nucleotide synthesis have become particularly important to modern medicine.

We examine here the biosynthetic pathways of purine and pyrimidine nucleotides and their regulation, the formation of the deoxynucleotides, and the degradation of purines and pyrimidines to uric acid and urea. We end with a discussion of chemotherapeutic agents that affect nucleotide synthesis.

## De Novo Purine Nucleotide Synthesis Begins with PRPP

John Buchanan

The two parent purine nucleotides of nucleic acids are adenosine 5′-monophosphate (AMP; adenylate) and guanosine 5′-monophosphate (GMP; guanylate), containing the purine bases adenine and guanine. Figure 22–32 shows the origin of the carbon and nitrogen atoms of the purine ring system, as determined by John Buchanan using isotopic tracer experiments in birds. The detailed pathway of purine biosynthesis was worked out primarily by Buchanan and G. Robert Greenberg in the 1950s.

In the first committed step of the pathway, an amino group donated by glutamine is attached at C-1 of PRPP (Fig. 22–33). The resulting **5-phosphoribosylamine** is highly unstable, with a half-life of 30 seconds at pH 7.5. The purine ring is subsequently built up on this structure. The pathway described here is identical in all organisms, with the exception of one step that differs in higher eukaryotes as noted below.

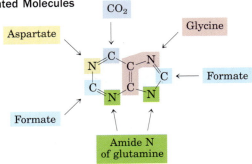

**FIGURE 22–32 Origin of the ring atoms of purines.** This information was obtained from isotopic experiments with $^{14}$C- or $^{15}$N-labeled precursors. Formate is supplied in the form of $N^{10}$-formyltetrahydrofolate.

The second step is the addition of three atoms from glycine (Fig. 22–33, step ②). An ATP is consumed to activate the glycine carboxyl group (in the form of an acyl phosphate) for this condensation reaction. The added glycine amino group is then formylated by $N^{10}$-formyltetrahydrofolate (step ③), and a nitrogen is contributed by glutamine (step ④), before dehydration and ring closure yield the five-membered imidazole ring of the purine nucleus, as 5-aminoimidazole ribonucleotide (AIR; step ⑤).

At this point, three of the six atoms needed for the second ring in the purine structure are in place. To complete the process, a carboxyl group is first added (step ⑥). This carboxylation is unusual in that it does not require biotin, but instead uses the bicarbonate generally present in aqueous solutions. A rearrangement transfers the carboxylate from the exocyclic amino group to position 4 of the imidazole ring (step ⑦). Steps ⑥ and ⑦ are found only in bacteria and fungi. In higher eukaryotes, including humans, the 5-aminoimidazole ribonucleotide product of step ⑤ is carboxylated directly to carboxyaminoimidazole ribonucleotide in one step instead of two (step ⑥a). The enzyme catalyzing this reaction is AIR carboxylase.

Aspartate now donates its amino group in two steps (⑧ and ⑨): formation of an amide bond, followed by elimination of the carbon skeleton of aspartate (as fumarate). Recall that aspartate plays an analogous role in two steps of the urea cycle (see Fig. 18–10). The final carbon is contributed by $N^{10}$-formyltetrahydrofolate (step ⑩), and a second ring closure takes place to yield the second fused ring of the purine nucleus (step ⑪).

**FIGURE 22–33** (facing page) **De novo synthesis of purine nucleotides: construction of the purine ring of inosinate (IMP).** Each addition to the purine ring is shaded to match Figure 22–32. After step ②, R symbolizes the 5-phospho-D-ribosyl group on which the purine ring is built. Formation of 5-phosphoribosylamine (step ①) is the first committed step in purine synthesis. Note that the product of step ⑨, AICAR, is the remnant of ATP released during histidine biosynthesis (see Fig. 22–20, step ⑤). Abbreviations are given for most intermediates to simplify the naming of the pathway enzymes. Step ⑥a is the alternative path from AIR to CAIR occurring in higher eukaryotes.

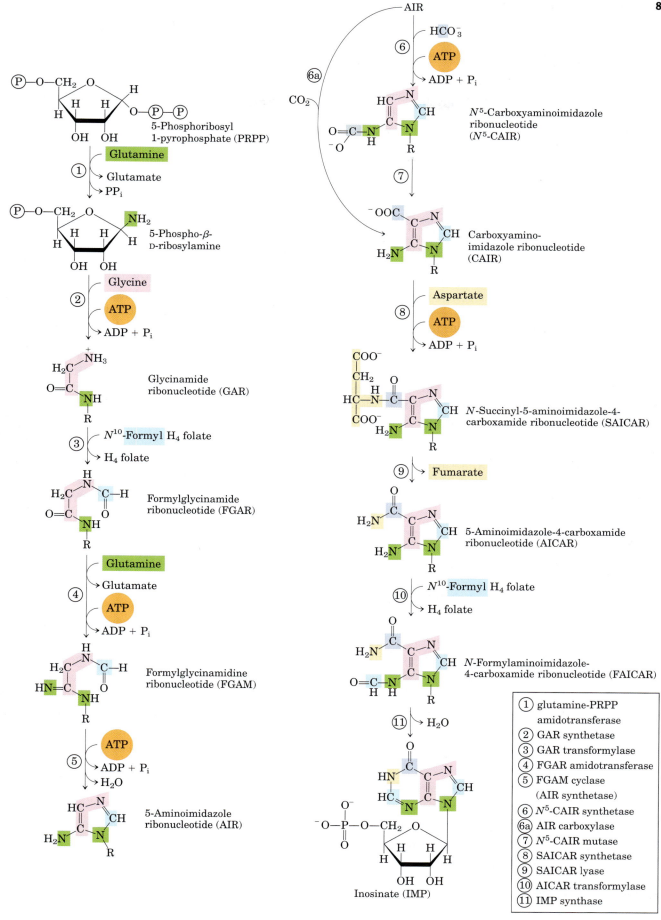

5-Phosphoribosyl 1-pyrophosphate (PRPP)

① Glutamine → Glutamate, PP$_i$

5-Phospho-$\beta$-D-ribosylamine

② Glycine, ATP → ADP + P$_i$

Glycinamide ribonucleotide (GAR)

③ $N^{10}$-Formyl H$_4$ folate → H$_4$ folate

Formylglycinamide ribonucleotide (FGAR)

④ Glutamine → Glutamate, ATP → ADP + P$_i$

Formylglycinamidine ribonucleotide (FGAM)

⑤ ATP → ADP + P$_i$, H$_2$O

5-Aminoimidazole ribonucleotide (AIR)

AIR

⑥ HCO$_3^-$, ATP → ADP + P$_i$

$N^5$-Carboxyaminoimidazole ribonucleotide ($N^5$-CAIR)

⑥a CO$_2$

⑦

Carboxyaminoimidazole ribonucleotide (CAIR)

⑧ Aspartate, ATP → ADP + P$_i$

$N$-Succinyl-5-aminoimidazole-4-carboxamide ribonucleotide (SAICAR)

⑨ Fumarate

5-Aminoimidazole-4-carboxamide ribonucleotide (AICAR)

⑩ $N^{10}$-Formyl H$_4$ folate → H$_4$ folate

$N$-Formylaminoimidazole-4-carboxamide ribonucleotide (FAICAR)

⑪ H$_2$O

Inosinate (IMP)

① glutamine-PRPP amidotransferase
② GAR synthetase
③ GAR transformylase
④ FGAR amidotransferase
⑤ FGAM cyclase (AIR synthetase)
⑥ $N^5$-CAIR synthetase
⑥a AIR carboxylase
⑦ $N^5$-CAIR mutase
⑧ SAICAR synthetase
⑨ SAICAR lyase
⑩ AICAR transformylase
⑪ IMP synthase

The first intermediate with a complete purine ring is **inosinate (IMP).**

As in the tryptophan and histidine biosynthetic pathways, the enzymes of IMP synthesis appear to be organized as large, multienzyme complexes in the cell. Once again, evidence comes from the existence of single polypeptides with several functions, some catalyzing nonsequential steps in the pathway. In eukaryotic cells ranging from yeast to fruit flies to chickens, steps ①, ③, and ⑤ in Figure 22–33 are catalyzed by a multifunctional protein. An additional multifunctional protein catalyzes steps ⑩ and ⑪. In humans, a multifunctional enzyme combines the activities of AIR carboxylase and SAICAR synthetase (steps ⑥a and ⑧). In bacteria, these activities are found on separate proteins, but a large noncovalent complex may exist in these cells. The channeling of reaction intermediates from one enzyme to the next permitted by these complexes is probably especially important for unstable intermediates such as 5-phosphoribosylamine.

Conversion of inosinate to adenylate requires the insertion of an amino group derived from aspartate (Fig. 22–34); this takes place in two reactions similar to those used to introduce N-1 of the purine ring (Fig. 22–33, steps ⑧ and ⑨). A crucial difference is that GTP rather than ATP is the source of the high-energy phosphate in synthesizing adenylosuccinate. Guanylate is formed by the $NAD^+$-requiring oxidation of inosinate at C-2, followed by addition of an amino group derived from glutamine. ATP is cleaved to AMP and $PP_i$ in the final step (Fig. 22–34).

## Purine Nucleotide Biosynthesis Is Regulated by Feedback Inhibition

Three major feedback mechanisms cooperate in regulating the overall rate of de novo purine nucleotide synthesis and the relative rates of formation of the two end products, adenylate and guanylate (Fig. 22–35). The first mechanism is exerted on the first reaction that is unique to purine synthesis—transfer of an amino group to PRPP to form 5-phosphoribosylamine. This reaction is catalyzed by the allosteric enzyme glutamine-PRPP amidotransferase, which is inhibited by the end products IMP, AMP, and GMP. AMP and GMP act synergistically in this concerted inhibition. Thus, whenever either AMP or GMP accumulates to excess, the first step in its biosynthesis from PRPP is partially inhibited.

In the second control mechanism, exerted at a later stage, an excess of GMP in the cell inhibits formation of xanthylate from inosinate by IMP dehydrogenase, without affecting the formation of AMP (Fig. 22–35). Conversely, an accumulation of adenylate inhibits formation of adenylosuccinate by adenylosuccinate synthetase, without affecting the biosynthesis of GMP. In the third mechanism, GTP is required in the conversion of IMP to AMP (Fig. 22–34, step ①), whereas ATP is required for conversion of IMP to GMP (step ④), a reciprocal arrangement that tends to balance the synthesis of the two ribonucleotides.

The final control mechanism is the inhibition of PRPP synthesis by the allosteric regulation of ribose phosphate pyrophosphokinase. This enzyme is inhibited

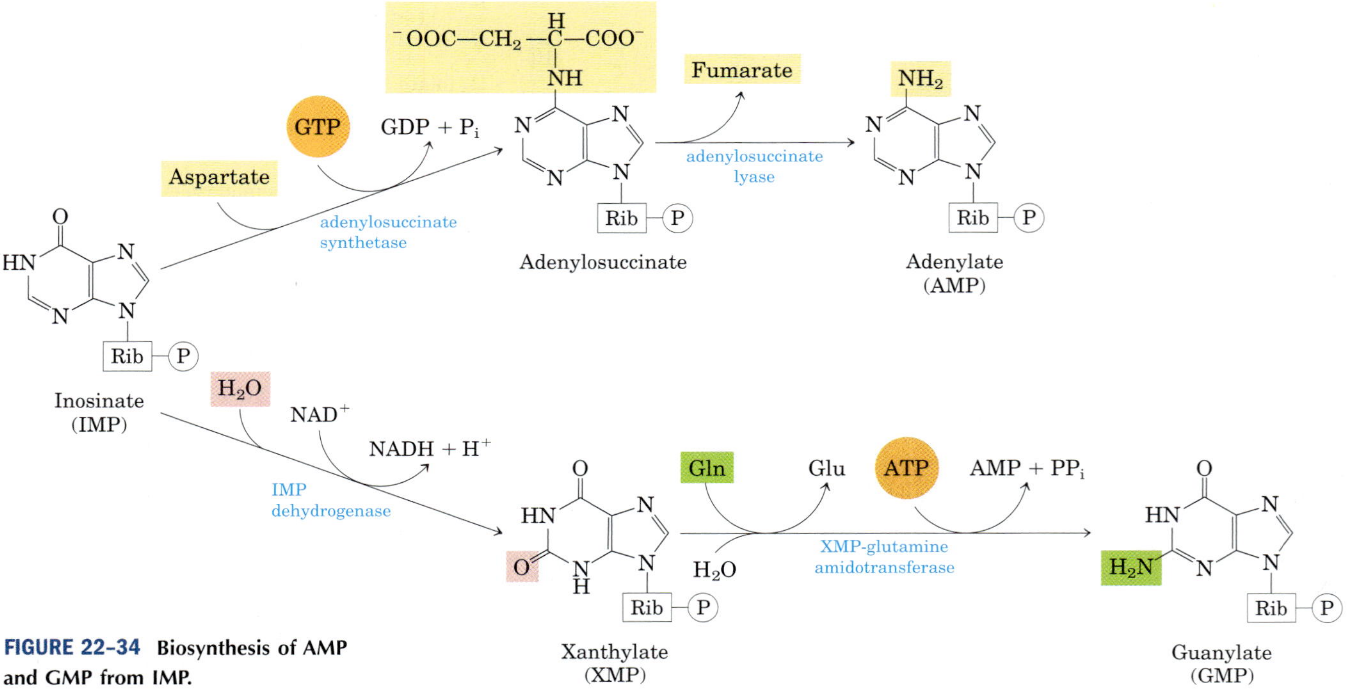

**FIGURE 22–34  Biosynthesis of AMP and GMP from IMP.**

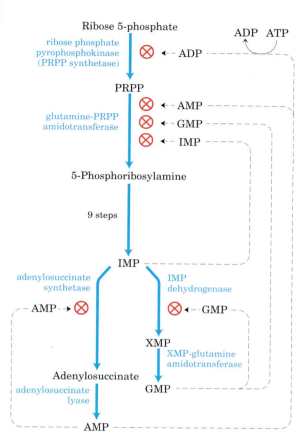

**FIGURE 22–35 Regulatory mechanisms in the biosynthesis of adenine and guanine nucleotides in _E. coli._** Regulation of these pathways differs in other organisms.

by ADP and GDP, in addition to metabolites from other pathways of which PRPP is a starting point.

## Pyrimidine Nucleotides Are Made from Aspartate, PRPP, and Carbamoyl Phosphate

The common pyrimidine ribonucleotides are cytidine 5′-monophosphate (CMP; cytidylate) and uridine 5′-monophosphate (UMP; uridylate), which contain the pyrimidines cytosine and uracil. De novo pyrimidine nucleotide biosynthesis (Fig. 22–36) proceeds in a somewhat different manner from purine nucleotide synthesis; the six-membered pyrimidine ring is made first and then attached to ribose 5-phosphate. Required in this process is carbamoyl phosphate, also an intermediate in the urea cycle (see Fig. 18–10). However, as we noted

**FIGURE 22–36 De novo synthesis of pyrimidine nucleotides: biosynthesis of UTP and CTP via orotidylate.** The pyrimidine is constructed from carbamoyl phosphate and aspartate. The ribose 5-phosphate is then added to the completed pyrimidine ring by orotate phosphoribosyltransferase. The first step in this pathway (not shown here; see Fig. 18–11a) is the synthesis of carbamoyl phosphate from $CO_2$ and $NH_4^+$, catalyzed in eukaryotes by carbamoyl phosphate synthetase II.

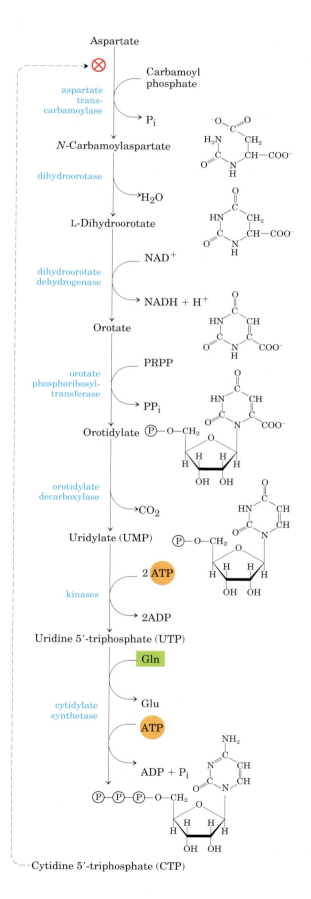

in Chapter 18, in animals the carbamoyl phosphate required in urea synthesis is made in mitochondria by carbamoyl phosphate synthetase I, whereas the carbamoyl phosphate required in pyrimidine biosynthesis is made in the cytosol by a different form of the enzyme, **carbamoyl phosphate synthetase II.** In bacteria, a single enzyme supplies carbamoyl phosphate for the synthesis of arginine and pyrimidines. The bacterial enzyme has three separate active sites, spaced along a channel nearly 100 Å long (Fig. 22–37). Bacterial carbamoyl phosphate synthetase provides a vivid illustration of the channeling of unstable reaction intermediates between active sites.

Carbamoyl phosphate reacts with aspartate to yield N-carbamoylaspartate in the first committed step of pyrimidine biosynthesis (Fig. 22–36). This reaction is catalyzed by **aspartate transcarbamoylase.** In bacteria, this step is highly regulated, and bacterial aspartate transcarbamoylase is one of the most thoroughly studied allosteric enzymes (see below). By removal of water from N-carbamoylaspartate, a reaction catalyzed by **dihydroorotase,** the pyrimidine ring is closed to form L-dihydroorotate. This compound is oxidized to the pyrimidine derivative orotate, a reaction in which $NAD^+$ is the ultimate electron acceptor. In eukaryotes, the first three enzymes in this pathway—carbamoyl phosphate synthetase II, aspartate transcarbamoylase, and dihydroorotase—are part of a single trifunctional protein. The protein, known by the acronym CAD, contains three identical polypeptide chains (each of $M_r$ 230,000), each with active sites for all three reactions. This suggests that large, multienzyme complexes may be the rule in this pathway.

Once orotate is formed, the ribose 5-phosphate side chain, provided once again by PRPP, is attached to yield orotidylate (Fig. 22–36). Orotidylate is then decarboxylated to uridylate, which is phosphorylated to UTP. CTP is formed from UTP by the action of **cytidylate synthetase,** by way of an acyl phosphate intermediate (consuming one ATP). The nitrogen donor is normally glutamine, although the cytidylate synthetases in many species can use $NH_4^+$ directly.

## Pyrimidine Nucleotide Biosynthesis Is Regulated by Feedback Inhibition

Regulation of the rate of pyrimidine nucleotide synthesis in bacteria occurs in large part through aspartate transcarbamoylase (ATCase), which catalyzes the first reaction in the sequence and is inhibited by CTP, the end product of the sequence (Fig. 22–36). The bacterial ATCase molecule consists of six catalytic subunits and six regulatory subunits (see Fig. 6–27). The catalytic subunits bind the substrate molecules, and the allosteric subunits bind the allosteric inhibitor, CTP. The entire ATCase molecule, as well as its subunits, exists in two conformations, active and inactive. When CTP is

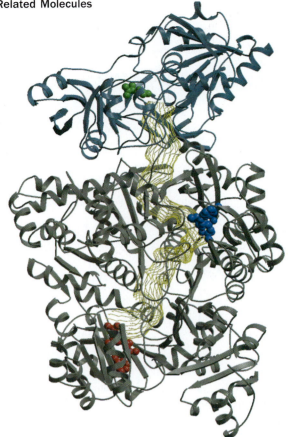

**FIGURE 22–37 Channeling of intermediates in bacterial carbamoyl phosphate synthetase.** (Derived from PDB ID 1M6V.) The reaction catalyzed by this enzyme is illustrated in Figure 18–11a. The large and small subunits are shown in gray and blue, respectively; the channel between active sites (almost 100 Å long) is shown as a yellow mesh. A glutamine molecule (green) binds to the small subunit, donating its amido nitrogen as $NH_4^+$ in a glutamine amidotransferase–type reaction. The $NH_4^+$ enters the channel, which takes it to a second active site, where it combines with bicarbonate in a reaction requiring ATP (bound ADP in blue). The carbamate then reenters the channel to reach the third active site, where it is phosphorylated to carbamoyl phosphate (bound ADP in red).

not bound to the regulatory subunits, the enzyme is maximally active. As CTP accumulates and binds to the regulatory subunits, they undergo a change in conformation. This change is transmitted to the catalytic subunits, which then also shift to an inactive conformation. ATP prevents the changes induced by CTP. Figure 22–38 shows the effects of the allosteric regulators on the activity of ATCase.

## Nucleoside Monophosphates Are Converted to Nucleoside Triphosphates

Nucleotides to be used in biosynthesis are generally converted to nucleoside triphosphates. The conversion pathways are common to all cells. Phosphorylation of AMP to ADP is promoted by **adenylate kinase,** in the reaction

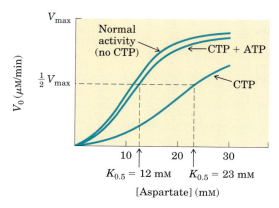

**FIGURE 22–38 Allosteric regulation of aspartate transcarbamoylase by CTP and ATP.** Addition of 0.8 mM CTP, the allosteric inhibitor of ATCase, increases the $K_{0.5}$ for aspartate (lower curve) and the rate of conversion of aspartate to N-carbamoylaspartate. ATP at 0.6 mM fully reverses this effect (middle curve).

$$ATP + AMP \rightleftharpoons 2ADP$$

The ADP so formed is phosphorylated to ATP by the glycolytic enzymes or through oxidative phosphorylation.

ATP also brings about the formation of other nucleoside diphosphates by the action of a class of enzymes called **nucleoside monophosphate kinases.** These enzymes, which are generally specific for a particular base but nonspecific for the sugar (ribose or deoxyribose), catalyze the reaction

$$ATP + NMP \rightleftharpoons ADP + NDP$$

The efficient cellular systems for rephosphorylating ADP to ATP tend to pull this reaction in the direction of products.

Nucleoside diphosphates are converted to triphosphates by the action of a ubiquitous enzyme, **nucleoside diphosphate kinase,** which catalyzes the reaction

$$NTP_D + NDP_A \rightleftharpoons NDP_D + NTP_A$$

This enzyme is notable in that it is not specific for the base (purines or pyrimidines) or the sugar (ribose or deoxyribose). This nonspecificity applies to both phosphate acceptor (A) and donor (D), although the donor ($NTP_D$) is almost invariably ATP, because it is present in higher concentration than other nucleoside triphosphates under aerobic conditions.

## Ribonucleotides Are the Precursors of Deoxyribonucleotides

Deoxyribonucleotides, the building blocks of DNA, are derived from the corresponding ribonucleotides by direct reduction at the 2′-carbon atom of the D-ribose to form the 2′-deoxy derivative. For example, adenosine diphosphate (ADP) is reduced to 2′-deoxyadenosine

diphosphate (dADP), and GDP is reduced to dGDP. This reaction is somewhat unusual in that the reduction occurs at a nonactivated carbon; no closely analogous chemical reactions are known. The reaction is catalyzed by **ribonucleotide reductase,** best characterized in *E. coli,* in which its substrates are ribonucleoside diphosphates.

The reduction of the D-ribose portion of a ribonucleoside diphosphate to 2′-deoxy-D-ribose requires a pair of hydrogen atoms, which are ultimately donated by NADPH via an intermediate hydrogen-carrying protein, **thioredoxin.** This ubiquitous protein serves a similar redox function in photosynthesis (see Fig. 20–19) and other processes. Thioredoxin has pairs of —SH groups that carry hydrogen atoms from NADPH to the ribonucleoside diphosphate. Its oxidized (disulfide) form is reduced by NADPH in a reaction catalyzed by **thioredoxin reductase** (Fig. 22–39), and reduced thioredoxin is then used by ribonucleotide reductase to reduce the nucleoside diphosphates (NDPs) to deoxyribonucleoside diphosphates (dNDPs). A second source of reducing equivalents for ribonucleotide reductase is glutathione (GSH). Glutathione serves as the reductant for a protein closely related to thioredoxin,

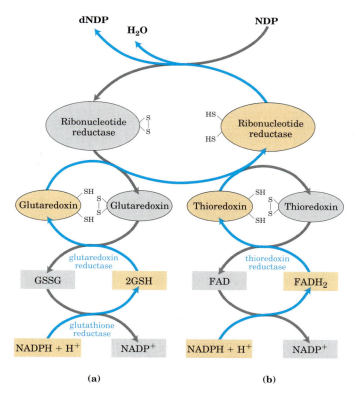

**FIGURE 22–39 Reduction of ribonucleotides to deoxyribonucleotides by ribonucleotide reductase.** Electrons are transmitted (blue arrows) to the enzyme from NADPH by **(a)** glutaredoxin or **(b)** thioredoxin. The sulfide groups in glutaredoxin reductase are contributed by two molecules of bound glutathione (GSH; GSSG indicates oxidized glutathione). Note that thioredoxin reductase is a flavoenzyme, with FAD as prosthetic group.

**glutaredoxin,** which then transfers the reducing power to ribonucleotide reductase (Fig. 22–39).

Ribonucleotide reductase is notable in that its reaction mechanism provides the best-characterized example of the involvement of free radicals in biochemical transformations, once thought to be rare in biological systems. The enzyme in *E. coli* and most eukaryotes is a dimer, with subunits designated R1 and R2 (Fig. 22–40). The R1 subunit contains two kinds of regulatory sites, as described below. The two active sites of the enzyme are formed at the interface between the R1 and R2 subunits. At each active site, R1 contributes two sulfhydryl groups required for activity and R2 contributes a stable tyrosyl radical. The R2 subunit also has a binuclear iron ($Fe^{3+}$) cofactor that helps generate and stabilize the tyrosyl radicals (Fig. 22–40). The tyrosyl radical is too far from the active site to interact directly with the site, but it generates another radical at the active site that functions in catalysis.

A likely mechanism for the ribonucleotide reductase reaction is illustrated in Figure 22–41. The 3′-ribonucleotide radical formed in step ① helps stabilize the cation formed at the 2′ carbon after the loss of $H_2O$ (steps ② and ③). Two one-electron transfers accompanied by oxidation of the dithiol reduce the radical cation (step ④). Step ⑤ is the reverse of step ①, regenerating the active site radical (ultimately, the tyrosyl radical) and forming the deoxy product. The oxidized dithiol is reduced to complete the cycle (step ⑥). In *E. coli*, likely sources of the required reducing equivalents for this reaction are thioredoxin and glutaredoxin, as noted above.

Four classes of ribonucleotide reductase have been reported. Their mechanisms (where known) generally conform to the scheme in Figure 22–41, but they differ in the identity of the group supplying the active-site radical and in the cofactors used to generate it. The *E. coli* enzyme (class I) requires oxygen to regenerate the tyrosyl radical if it is quenched, so this enzyme functions only in an aerobic environment. Class II enzymes, found in other microorganisms, have 5′-deoxyadenosylcobalamin (see Box 17–2) rather than a binuclear iron center. Class III enzymes have evolved to function in an anaerobic environment. *E. coli* contains a separate class III ribonucleotide reductase when grown anaerobically; this enzyme contains an iron-sulfur cluster (structurally distinct from the binuclear iron center of the class I enzyme) and requires NADPH and *S*-adenosylmethionine for activity. It uses nucleoside triphosphates rather than nucleoside diphosphates as substrates. A class IV ribonucleotide reductase, containing a binuclear manganese center, has been reported in some microorganisms. The evolution of different classes of ribonucleotide reductase for production of DNA precursors in different environments reflects the importance of this reaction in nucleotide metabolism.

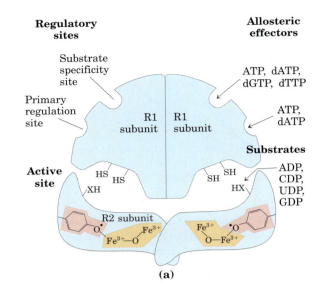

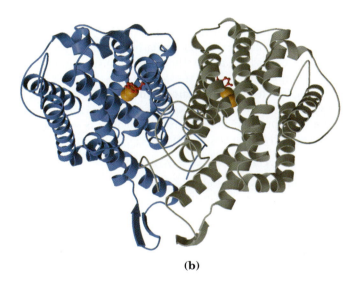

**FIGURE 22–40 Ribonucleotide reductase. (a)** Subunit structure. The functions of the two regulatory sites are explained in Figure 22–42. Each active site contains two thiols and a group (—XH) that can be converted to an active-site radical; this group is probably the —SH of $Cys^{439}$, which functions as a thiyl radical. **(b)** The R2 subunits of *E. coli* ribonucleotide reductase (PDB ID 1PFR). The Tyr residue that acts as the tyrosyl radical is shown in red; the binuclear iron center is orange. **(c)** The tyrosyl radical functions to generate the active-site radical (—X⋅), which is used in the mechanism shown in Figure 22–41.

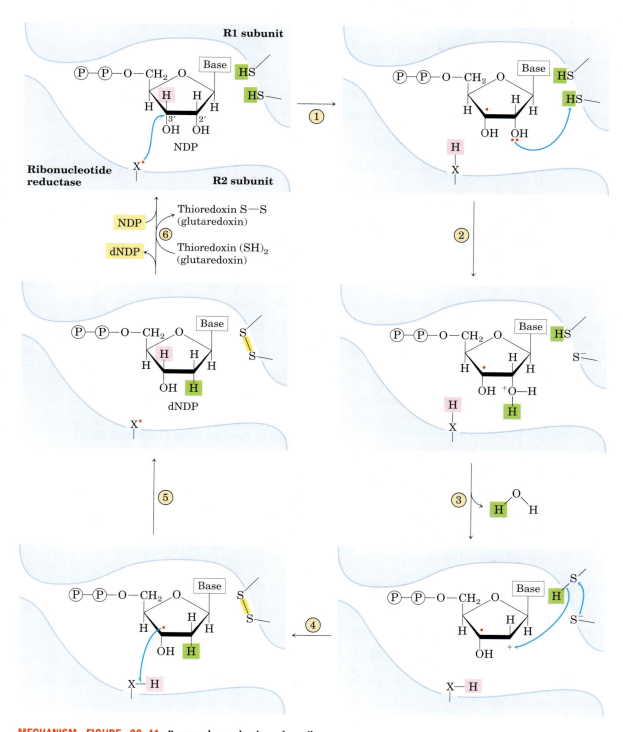

**MECHANISM FIGURE 22–41 Proposed mechanism for ribonucleotide reductase.** In the enzyme of *E. coli* and most eukaryotes, the active thiol groups are on the R1 subunit; the active-site radical (—X˙) is on the R2 subunit and in *E. coli* is probably a thiyl radical of Cys[439] (see Fig. 22–40). Steps ① through ⑥ are described in the text.

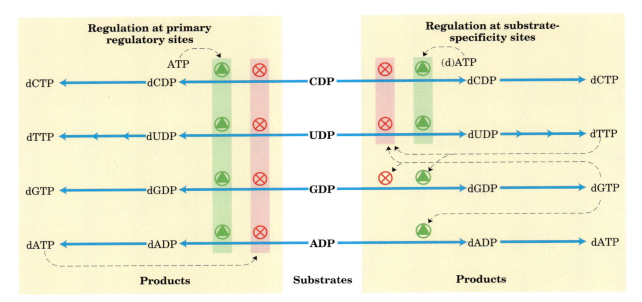

**FIGURE 22–42 Regulation of ribonucleotide reductase by deoxynucleoside triphosphates.** The overall activity of the enzyme is affected by binding at the primary regulatory site (left). The substrate specificity of the enzyme is affected by the nature of the effector molecule bound at the second type of regulatory site (right). The diagram indicates inhibition or stimulation of enzyme activity with the four different substrates. The pathway from dUDP to dTTP is described later (see Figs 22–43, 22–44).

Regulation of *E. coli* ribonucleotide reductase is unusual in that not only its *activity* but its *substrate specificity* is regulated by the binding of effector molecules. Each R1 subunit has two types of regulatory site (Fig. 22–40). One type affects overall enzyme activity and binds either ATP, which activates the enzyme, or dATP, which inactivates it. The second type alters substrate specificity in response to the effector molecule—ATP, dATP, dTTP, or dGTP—that is bound there (Fig. 22–42). When ATP or dATP is bound, reduction of UDP and CDP is favored. When dTTP or dGTP is bound, reduction of GDP or ADP, respectively, is stimulated. The scheme is designed to provide a balanced pool of precursors for DNA synthesis. ATP is also a general activator for biosynthesis and ribonucleotide reduction. The presence of dATP in small amounts increases the reduction of pyrimidine nucleotides. An oversupply of the pyrimidine dNTPs is signaled by high levels of dTTP, which shifts the specificity to favor reduction of GDP. High levels of dGTP, in turn, shift the specificity to ADP reduction, and high levels of dATP shut the enzyme down. These effectors are thought to induce several distinct enzyme conformations with altered specificities.

### Thymidylate Is Derived from dCDP and dUMP

DNA contains thymine rather than uracil, and the de novo pathway to thymine involves only deoxyribonucleotides. The immediate precursor of thymidylate (dTMP) is dUMP. In bacteria, the pathway to dUMP begins with formation of dUTP, either by deamination of dCTP or by phosphorylation of dUDP (Fig. 22–43). The dUTP is converted to dUMP by a dUTPase. The latter reaction must be efficient to keep dUTP pools low and prevent incorporation of uridylate into DNA.

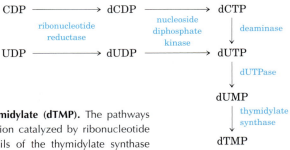

**FIGURE 22–43 Biosynthesis of thymidylate (dTMP).** The pathways are shown beginning with the reaction catalyzed by ribonucleotide reductase. Figure 22–44 gives details of the thymidylate synthase reaction.

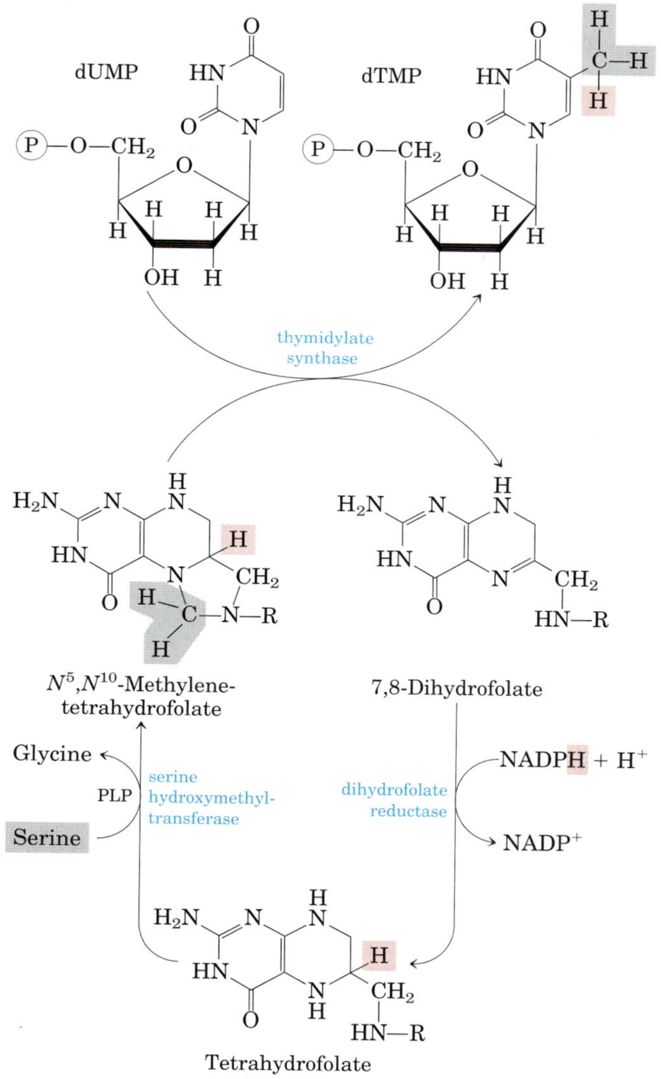

**FIGURE 22–44 Conversion of dUMP to dTMP by thymidylate synthase and dihydrofolate reductase.** Serine hydroxymethyltransferase is required for regeneration of the $N^5,N^{10}$-methylene form of tetrahydrofolate. In the synthesis of dTMP, all three hydrogens of the added methyl group are derived from $N^5,N^{10}$-methylenetetrahydrofolate (pink and gray).

Conversion of dUMP to dTMP is catalyzed by **thymidylate synthase.** A one-carbon unit at the hydroxymethyl ($-CH_2OH$) oxidation level (see Fig. 18–17) is transferred from $N^5,N^{10}$-methylenetetrahydrofolate to dUMP, then reduced to a methyl group (Fig. 22–44). The reduction occurs at the expense of oxidation of tetrahydrofolate to dihydrofolate, which is unusual in tetrahydrofolate-requiring reactions. (The mechanism of this reaction is shown in Fig. 22–50.) The dihydrofolate is reduced to tetrahydrofolate by **dihydrofolate reductase**—a regeneration that is essential for the many processes that require tetrahydrofolate. In plants and at least one protist, thymidylate synthase and dihydrofolate reductase reside on a single bifunctional protein.

## Degradation of Purines and Pyrimidines Produces Uric Acid and Urea, Respectively

Purine nucleotides are degraded by a pathway in which they lose their phosphate through the action of **5′-nucleotidase** (Fig. 22–45). Adenylate yields adenosine, which is deaminated to inosine by **adenosine deaminase,** and inosine is hydrolyzed to hypoxanthine (its purine base) and D-ribose. Hypoxanthine is oxidized successively to xanthine and then uric acid by **xanthine oxidase,** a flavoenzyme with an atom of molybdenum and four iron-sulfur centers in its prosthetic group. Molecular oxygen is the electron acceptor in this complex reaction.

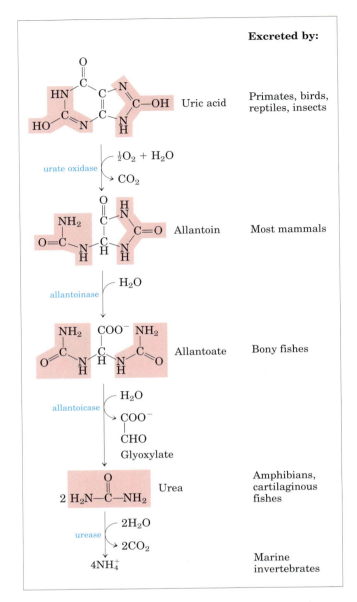

**FIGURE 22–45 Catabolism of purine nucleotides.** Note that primates excrete much more nitrogen as urea via the urea cycle (Chapter 18) than as uric acid from purine degradation. Similarly, fish excrete much more nitrogen as $NH_4^+$ than as urea produced by the pathway shown here.

GMP catabolism also yields uric acid as end product. GMP is first hydrolyzed to guanosine, which is then cleaved to free guanine. Guanine undergoes hydrolytic removal of its amino group to yield xanthine, which is converted to uric acid by xanthine oxidase (Fig. 22–45).

Uric acid is the excreted end product of purine catabolism in primates, birds, and some other animals. A healthy adult human excretes uric acid at a rate of about 0.6 g/24 h; the excreted product arises in part from ingested purines and in part from turnover of the purine nucleotides of nucleic acids. In most mammals and many other vertebrates, uric acid is further degraded to **allantoin** by the action of **urate oxidase.** In other organisms the pathway is further extended, as shown in Figure 22–45.

The pathways for degradation of pyrimidines generally lead to $NH_4^+$ production and thus to urea synthesis. Thymine, for example, is degraded to methylmalonylsemialdehyde (Fig. 22–46), an intermediate of valine catabolism. It is further degraded through propionyl-CoA and methylmalonyl-CoA to succinyl-CoA (see Fig. 18–27).

Genetic aberrations in human purine metabolism have been found, some with serious consequences. For example, **adenosine deaminase (ADA) deficiency** leads to severe immunodeficiency disease in which T lymphocytes and B lymphocytes do not develop properly. Lack of ADA leads to a 100-fold increase in the cellular concentration of dATP, a strong inhibitor of ribonucleotide reductase (Fig. 22–42). High levels

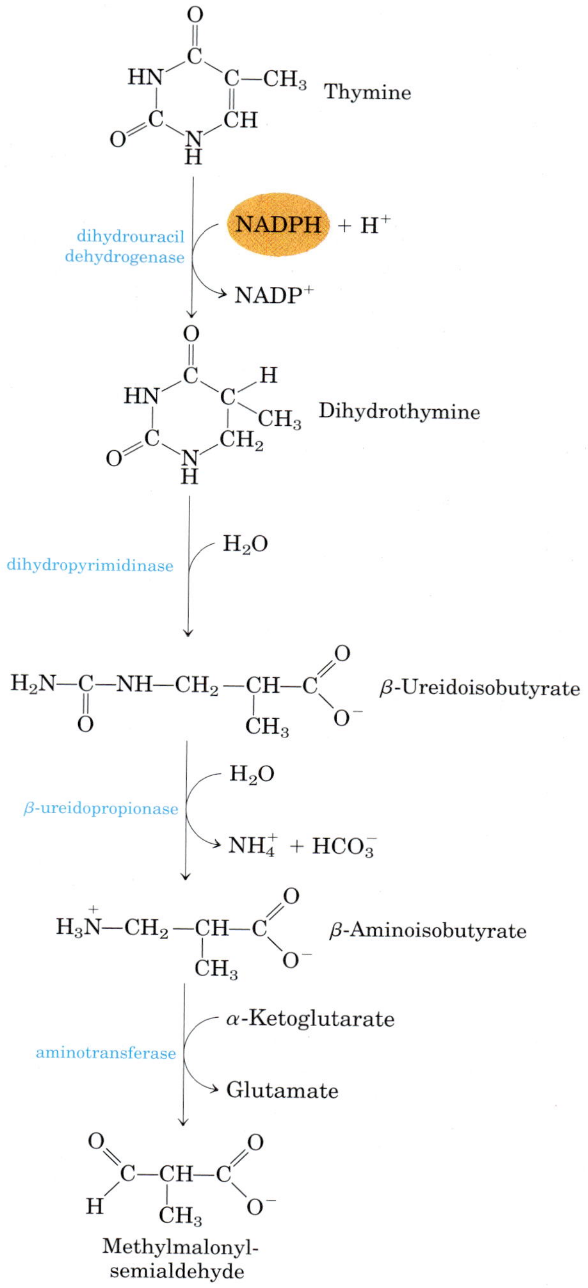

**FIGURE 22–46 Catabolism of a pyrimidine.** Shown here is the pathway for thymine. The methylmalonylsemialdehyde is further degraded to succinyl-CoA.

of dATP produce a general deficiency of other dNTPs in T lymphocytes. The basis for B-lymphocyte toxicity is less clear. Individuals with ADA deficiency lack an effective immune system and do not survive unless isolated in a sterile "bubble" environment. ADA deficiency is one of the first targets of human gene therapy trials (see Box 9–2). ■

## Purine and Pyrimidine Bases Are Recycled by Salvage Pathways

Free purine and pyrimidine bases are constantly released in cells during the metabolic degradation of nucleotides. Free purines are in large part salvaged and reused to make nucleotides, in a pathway much simpler than the de novo synthesis of purine nucleotides described earlier. One of the primary salvage pathways consists of a single reaction catalyzed by **adenosine phosphoribosyltransferase,** in which free adenine reacts with PRPP to yield the corresponding adenine nucleotide:

$$\text{Adenine} + \text{PRPP} \longrightarrow \text{AMP} + \text{PP}_i$$

Free guanine and hypoxanthine (the deamination product of adenine; Fig. 22–45) are salvaged in the same way by **hypoxanthine-guanine phosphoribosyltransferase.** A similar salvage pathway exists for pyrimidine bases in microorganisms, and possibly in mammals.

A genetic lack of hypoxanthine-guanine phosphoribosyltransferase activity, seen almost exclusively in male children, results in a bizarre set of symptoms called **Lesch-Nyhan syndrome.** Children with this genetic disorder, which becomes manifest by the age of 2 years, are sometimes poorly coordinated and mentally retarded. In addition, they are extremely hostile and show compulsive self-destructive tendencies: they mutilate themselves by biting off their fingers, toes, and lips.

The devastating effects of Lesch-Nyhan syndrome illustrate the importance of the salvage pathways. Hypoxanthine and guanine arise constantly from the breakdown of nucleic acids. In the absence of hypoxanthine-guanine phosphoribosyltransferase, PRPP levels rise and purines are overproduced by the de novo pathway, resulting in high levels of uric acid production and gout-like damage to tissue (see below). The brain is especially dependent on the salvage pathways, and this may account for the central nervous system damage in children with Lesch-Nyhan syndrome. This syndrome is another target of early trials in gene therapy (see Box 9–2). ■

## Excess Uric Acid Causes Gout

Long thought, erroneously, to be due to "high living," gout is a disease of the joints caused by an elevated concentration of uric acid in the blood and tissues. The joints become inflamed, painful, and arthritic, owing to the abnormal deposition of sodium urate crystals. The kidneys are also affected, as excess uric acid is deposited in the kidney tubules. Gout occurs predominantly in males. Its precise cause is not known, but it often involves an underexcretion of urate. A genetic deficiency of one or another enzyme of purine metabolism may also be a factor in some cases.

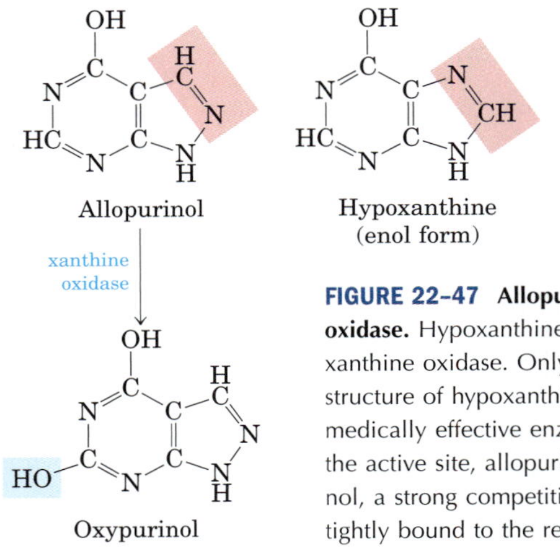

Allopurinol

Hypoxanthine
(enol form)

xanthine
oxidase

Oxypurinol

**FIGURE 22–47 Allopurinol, an inhibitor of xanthine oxidase.** Hypoxanthine is the normal substrate of xanthine oxidase. Only a slight alteration in the structure of hypoxanthine (shaded pink) yields the medically effective enzyme inhibitor allopurinol. At the active site, allopurinol is converted to oxypurinol, a strong competitive inhibitor that remains tightly bound to the reduced form of the enzyme.

Gertrude Elion (1918–1999) and
George Hitchings (1905–1998)

Gout is effectively treated by a combination of nutritional and drug therapies. Foods especially rich in nucleotides and nucleic acids, such as liver or glandular products, are withheld from the diet. Major alleviation of the symptoms is provided by the drug **allopurinol** (Fig. 22–47), which inhibits xanthine oxidase, the enzyme that catalyzes the conversion of purines to uric acid. Allopurinol is a substrate of xanthine oxidase, which converts allopurinol to oxypurinol (alloxanthine). Oxypurinol inactivates the reduced form of the enzyme by remaining tightly bound in its active site. When xanthine oxidase is inhibited, the excreted products of purine metabolism are xanthine and hypoxanthine, which are more water-soluble than uric acid and less likely to form crystalline deposits. Allopurinol was developed by Gertrude Elion and George Hitchings, who also developed acyclovir, used in treating people with AIDS, and other purine analogs used in cancer chemotherapy. ■

## Many Chemotherapeutic Agents Target Enzymes in the Nucleotide Biosynthetic Pathways

The growth of cancer cells is not controlled in the same way as cell growth in most normal tissues. Cancer cells have greater requirements for nucleotides as precursors of DNA and RNA, and consequently are generally more sensitive than normal cells to inhibitors of nucleotide biosynthesis. A growing array of important chemotherapeutic agents—for cancer and other diseases—act by inhibiting one or more enzymes in these pathways. We describe here several well-studied examples that illustrate productive approaches to treatment and help us understand how these enzymes work.

The first set of agents includes compounds that inhibit glutamine amidotransferases. Recall that glutamine is a nitrogen donor in at least half a dozen separate reactions in nucleotide biosynthesis. The binding sites for glutamine and the mechanism by which $NH_4^+$ is extracted are quite similar in many of these enzymes. Most are strongly inhibited by glutamine analogs such as **aza-**

**serine** and **acivicin** (Fig. 22–48). Azaserine, characterized by John Buchanan in the 1950s, was one of the first examples of a mechanism-based enzyme inactivator (suicide inactivator; p. 211 and Box 22–2). Acivicin shows promise as a cancer chemotherapeutic agent.

Other useful targets for pharmaceutical agents are thymidylate synthase and dihydrofolate reductase, enzymes that provide the only cellular pathway for thymine synthesis (Fig. 22–49). One inhibitor that acts on thymidylate synthase, **fluorouracil,** is an important chemotherapeutic agent. Fluorouracil itself is not the enzyme inhibitor. In the cell, salvage pathways convert it to the deoxynucleoside monophosphate FdUMP, which binds to and inactivates the enzyme. Inhibition by FdUMP (Fig. 22–50) is a classic example of mechanism-based enzyme inactivation. Another prominent chemotherapeutic agent, **methotrexate,** is an inhibitor of dihydrofolate reductase. This folate analog acts as a competitive inhibitor; the enzyme binds methotrexate with about 100 times higher affinity than dihydrofolate. **Aminopterin** is a related compound that acts similarly.

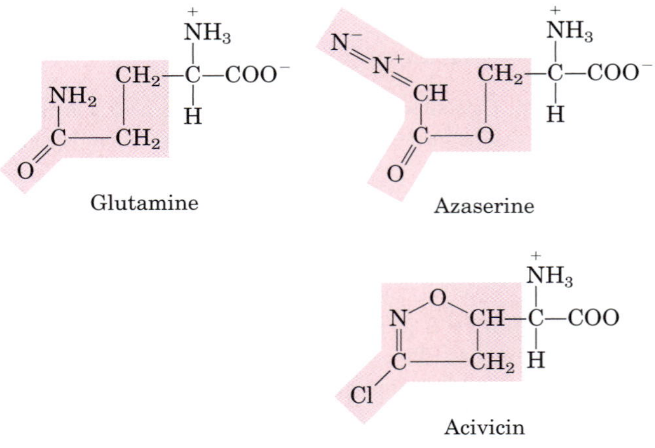

Glutamine

Azaserine

Acivicin

**FIGURE 22–48 Azaserine and acivicin, inhibitors of glutamine amidotransferases.** These analogs of glutamine interfere in a number of amino acid and nucleotide biosynthetic pathways.

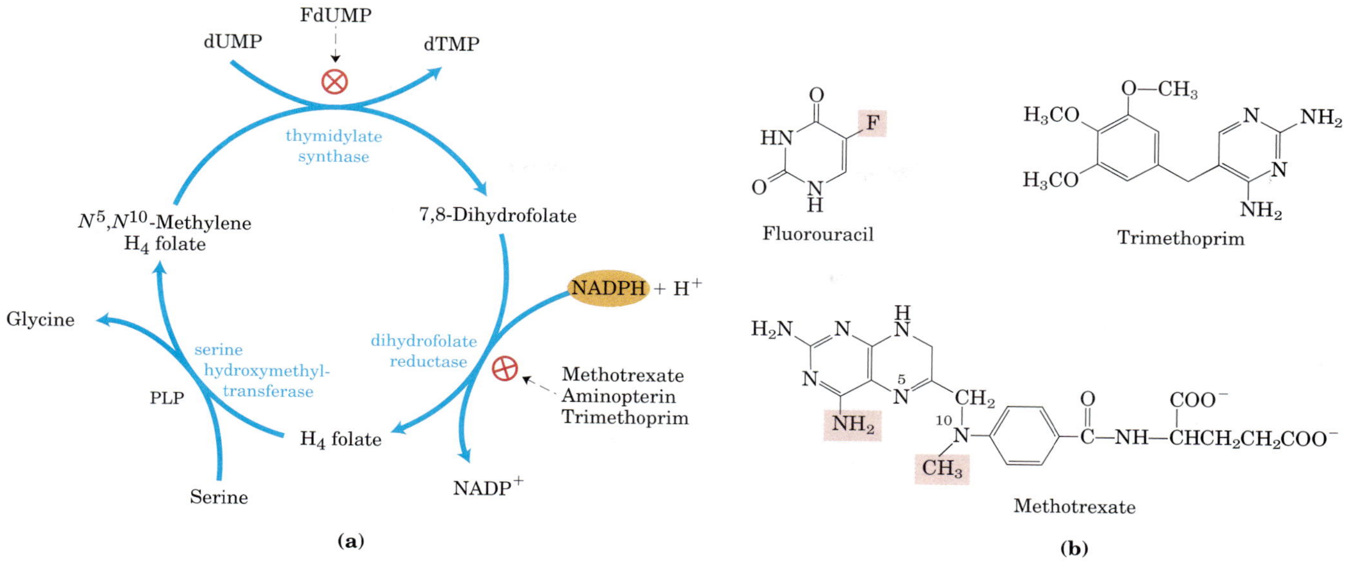

**FIGURE 22–49 Thymidylate synthesis and folate metabolism as targets of chemotherapy. (a)** During thymidylate synthesis, $N^5,N^{10}$-methylenetetrahydrofolate is converted to 7,8-dihydrofolate; the $N^5,N^{10}$-methylenetetrahydrofolate is regenerated in two steps (see Fig. 22–44). This cycle is a major target of several chemotherapeutic agents. **(b)** Fluorouracil and methotrexate are important chemotherapeutic agents. In cells, fluorouracil is converted to FdUMP, which inhibits thymidylate synthase. Methotrexate, a structural analog of tetrahydrofolate, inhibits dihydrofolate reductase; the shaded amino and methyl groups replace a carbonyl oxygen and a proton, respectively, in folate (see Fig. 22–44). Another important folate analog, aminopterin, is identical to methotrexate except that it lacks the shaded methyl group. Trimethoprim, a tight-binding inhibitor of bacterial dihydrofolate reductase, was developed as an antibiotic.

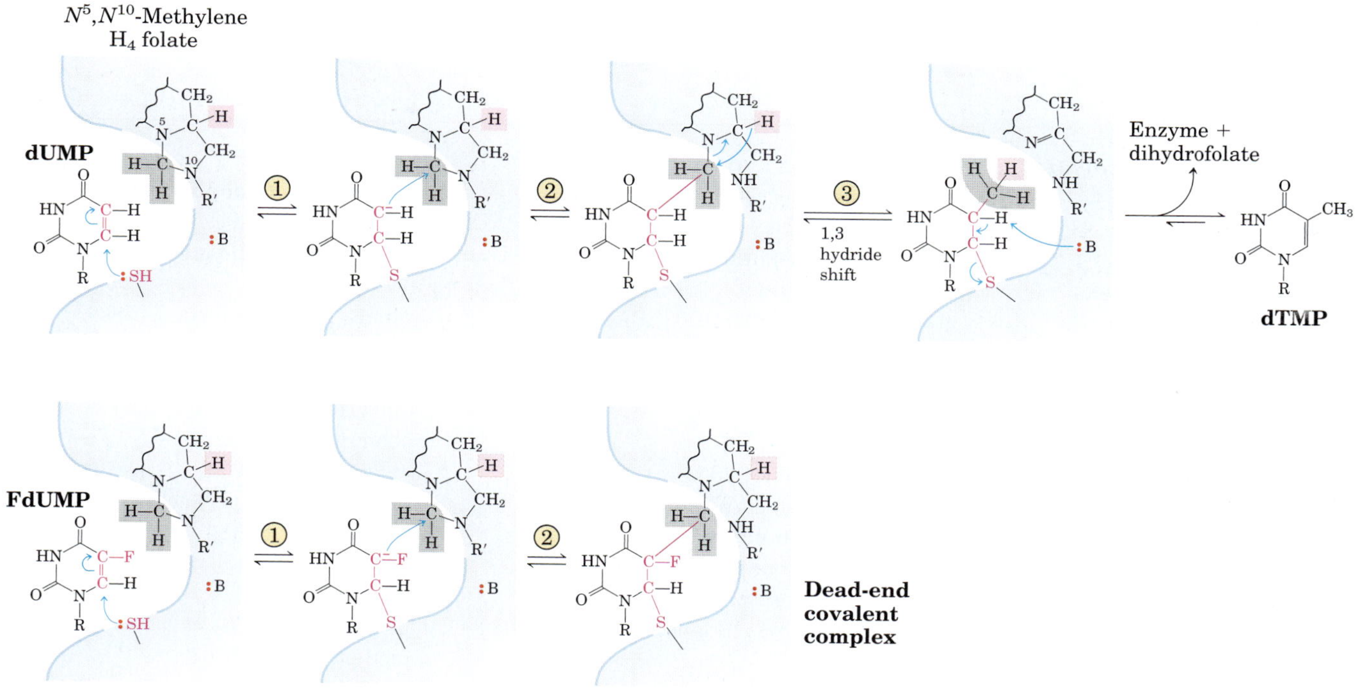

**MECHANISM FIGURE 22–50 Conversion of dUMP to dTMP and its inhibition by FdUMP.** The top row is the normal reaction mechanism of thymidylate synthase. The nucleophilic sulfhydryl group contributed by the enzyme in step ① and the ring atoms of dUMP taking part in the reaction are shown in red; :B denotes an amino acid side chain that acts as a base to abstract a proton after step ③. The hydrogens derived from the methylene group of $N^5,N^{10}$-methylenetetrahydrofolate are shaded in gray. A novel feature of this reaction mechanism is a 1,3 hydride shift (step ③), which moves a hydride ion (shaded pink) from C-6 of $H_4$ folate to the methyl group of thymidine, resulting in the oxidation of tetrahydrofolate to dihydrofolate. It is this hydride shift that apparently does not occur when FdUMP is the substrate (bottom). Steps ① and ② proceed normally, but result in a stable complex—with FdUMP linked covalently to the enzyme and to tetrahydrofolate—that inactivates the enzyme. 🔵 **Thymidylate Synthase Mechanism**

The medical potential of inhibitors of nucleotide biosynthesis is not limited to cancer treatment. All fast-growing cells (including bacteria and protists) are potential targets. **Trimethoprim,** an antibiotic developed by Hitchings and Elion, binds to bacterial dihydrofolate reductase nearly 100,000 times better than to the mammalian enzyme. It is used to treat certain urinary and middle ear bacterial infections. Parasitic protists, such as the trypanosomes that cause African sleeping sickness (African trypanosomiasis), lack pathways for de novo nucleotide biosynthesis and are particularly sensitive to agents that interfere with their scavenging of nucleotides from the surrounding environment using salvage pathways. Allopurinol (Fig. 22–47) and a number of related purine analogs have shown promise for the treatment of African trypanosomiasis and related afflictions. See Box 22–2 for another approach to combating African trypanosomiasis, made possible by advances in our understanding of metabolism and enzyme mechanism. ■

## SUMMARY 22.4 Biosynthesis and Degradation of Nucleotides

- The purine ring system is built up step-by-step beginning with 5-phosphoribosylamine. The amino acids glutamine, glycine, and aspartate furnish all the nitrogen atoms of purines. Two ring-closure steps form the purine nucleus.

- Pyrimidines are synthesized from carbamoyl phosphate and aspartate, and ribose 5-phosphate is then attached to yield the pyrimidine ribonucleotides.

- Nucleoside monophosphates are converted to their triphosphates by enzymatic phosphorylation reactions. Ribonucleotides are converted to deoxyribonucleotides by ribonucleotide reductase, an enzyme with novel mechanistic and regulatory characteristics. The thymine nucleotides are derived from dCDP and dUMP.

- Uric acid and urea are the end products of purine and pyrimidine degradation.

- Free purines can be salvaged and rebuilt into nucleotides. Genetic deficiencies in certain salvage enzymes cause serious disorders such as Lesch-Nyhan syndrome and ADA deficiency.

- Accumulation of uric acid crystals in the joints, possibly caused by another genetic deficiency, results in gout.

- Enzymes of the nucleotide biosynthetic pathways are targets for an array of chemotherapeutic agents used to treat cancer and other diseases.

## Key Terms

*Terms in bold are defined in the glossary.*

**nitrogen cycle**   834
**nitrogen fixation**   834
**symbionts**   834
**nitrogenase complex**   835
leghemoglobin   836
glutamine synthetase   838
glutamate synthase   838
glutamine amidotransferases   840
5-phosphoribosyl-1-
    pyrophosphate (PRPP)   842
tryptophan synthase   849
**porphyrin**   854
**porphyria**   854
bilirubin   854
phosphocreatine   857
creatine   857
glutathione (GSH)   857

**auxin**   859
dopamine   859
norepinephrine   859
epinephrine   859
γ-aminobutyrate (GABA)   859
serotonin   859
histamine   859
cimetidine   859
spermine   860
spermidine   860
ornithine decarboxylase   860
**de novo pathway**   862
**salvage pathway**   862
inosinate (IMP)   866
carbamoyl phosphate synthetase II
    868
aspartate transcarbamoylase   868

**nucleoside monophosphate kinase**
    869
**nucleoside diphosphate kinase**
    869
ribonucleotide reductase   869
thioredoxin   869
thymidylate synthase   873
dihydrofolate reductase   873
adenosine deaminase
    deficiency   874
Lesch-Nyhan syndrome   875
allopurinol   876
azaserine   876
acivicin   876
fluorouracil   876
methotrexate   876
aminopterin   876

## Further Reading

### Nitrogen Fixation

**Burris, R.H.** (1995) Breaking the N–N bond. *Annu. Rev. Plant Physiol. Plant Mol. Biol.* **46,** 1–19.

**Igarishi, R.Y. & Seefeldt, L.C.** (2003) Nitrogen fixation: the mechanism of the Mo-dependent nitrogenase. *Crit. Rev. Biochem. Mol. Biol.,* **38,** 351–384.

**Patriarca, E.J., Tate, R., & Iaccarino, M.** (2002) Key role of bacterial NH$_4^+$ metabolism in rhizobium-plant symbiosis. *Microbiol. Mol. Biol. Rev.* **66,** 203–222.

A good overview of ammonia assimilation in bacterial systems and its regulation.

**Sinha, S.C. & Smith, J.L.** (2001) The PRT protein family. *Curr. Opin. Struct. Biol.* **11,** 733–739.

Description of a protein family that includes many amidotransferases, with channels for the movement of NH$_3$ from one active site to another.

**Ye, R.W. & Thomas, S.M.** (2001) Microbial nitrogen cycles: physiology, genomics and applications. *Curr. Opin. Microbiol.* **4,** 307–312.

### Amino Acid Biosynthesis

**Abeles, R.H., Frey, P.A., & Jencks, W.P.** (1992) *Biochemistry,* Jones and Bartlett Publishers, Boston.

This book includes excellent accounts of reaction mechanisms, including one-carbon metabolism and pyridoxal phosphate enzymes.

**Bender, D.A.** (1985) *Amino Acid Metabolism,* 2nd edn, Wiley-Interscience, New York.

**Neidhardt, F.C. (ed.)** (1996) Escherichia coli *and* Salmonella: *Cellular and Molecular Biology,* 2nd edn, ASM Press, Washington, DC.

Volume 1 of this two-volume set has 13 chapters devoted to detailed descriptions of amino acid and nucleotide biosynthesis in bacteria. The web-based version, *EcoSal,* is updated regularly. A valuable resource.

**Pan P., Woehl, E., & Dunn, M.F.** (1997) Protein architecture, dynamics and allostery in tryptophan synthase channeling. *Trends Biochem. Sci.* **22,** 22–27.

### Compounds Derived from Amino Acids

**Bredt, D.S. & Snyder, S.H.** (1994) Nitric oxide: a physiologic messenger molecule. *Annu. Rev. Biochem.* **63,** 175–195.

**Meister, A. & Anderson, M.E.** (1983) Glutathione. *Annu. Rev. Biochem.* **52,** 711–760.

**Morse, D. & Choi, A.M.K.** (2002) Heme oxygenase-1—the "emerging molecule" has arrived. *Am. J. Resp. Cell Mol. Biol.* **27,** 8–16.

**Rondon, M.R., Trzebiatowski, J.R., & Escalante-Semerena, J.C.** (1997) Biochemistry and molecular genetics of cobalamin biosynthesis. *Prog. Nucleic Acid Res. Mol. Biol.* **56,** 347–384.

**Stadtman, T.C.** (1996) Selenocysteine. *Annu. Rev. Biochem.* **65,** 83–100.

### Nucleotide Biosynthesis

**Blakley, R.L. & Benkovic, S.J.** (1985) *Folates and Pterins,* Vol. 2: *Chemistry and Biochemistry of Pterins,* Wiley-Interscience, New York.

**Carreras, C.W. & Santi, D.V.** (1995) The catalytic mechanism and structure of thymidylate synthase. *Annu. Rev. Biochem.* **64,** 721–762.

**Eliasson, R., Pontis, E., Sun, X., & Reichard, P.** (1994) Allosteric control of the substrate specificity of the anaerobic ribonucleotide reductase from *Escherichia coli. J. Biol. Chem.* **269,** 26,052–26,057.

**Holmgren, A.** (1989) Thioredoxin and glutaredoxin systems. *J. Biol. Chem.* **264,** 13,963–13,966.

**Jordan, A. & Reichard P.** (1998) Ribonucleotide reductases. *Annu. Rev. Biochem.* **67,** 71–98.

**Kappock, T.J., Ealick, S.E., & Stubbe, J.** (2000) Modular evolution of the purine biosynthetic pathway. *Curr. Opin. Chem. Biol.* **4,** 567–572.

**Kornberg, A. & Baker, T.A.** (1991) *DNA Replication,* 2nd edn, W. H. Freeman and Company, New York.

This text includes a good summary of nucleotide biosynthesis.

**Lee, L., Kelly, R.E., Pastra-Landis, S.C., & Evans, D.R.** (1985) Oligomeric structure of the multifunctional protein CAD that initiates pyrimidine biosynthesis in mammalian cells. *Proc. Natl. Acad. Sci. USA* **82,** 6802–6806.

**Licht, S., Gerfen, G.J., & Stubbe, J.** (1996) Thiyl radicals in ribonucleotide reductases. *Science* **271,** 477–481.

**Schachman, H.K.** (2000) Still looking for the ivory tower. *Annu. Rev. Biochem.* **69,** 1–29.

A lively description of research on aspartate transcarbamoylase, accompanied by delightful tales of science and politics.

**Stubbe, J. & Riggs-Gelasco, P.** (1998) Harnessing free radicals: formation and function of the tyrosyl radical in ribonucleotide reductase. *Trends Biochem. Sci.* **23,** 438–443.

**Villafranca, J.E., Howell, E.E., Voet, D.H., Strobel, M.S., Ogden, R.C., Abelson, J.N., & Kraut, J.** (1983) Directed mutagenesis of dihydrofolate reductase. *Science* **222,** 782–788.

A report of structural studies on this important enzyme.

### Genetic Diseases

**Scriver, C.R., Beaudet, A.L., Valle, D., Sly, W.S., Childs, B., Kinzler, L.W., & Vogelstein, B. (eds)** (2001) *The Metabolic and Molecular Bases of Inherited Disease,* 8th edn, McGraw-Hill Professional, New York.

This four-volume set has good chapters on disorders of amino acid, porphyrin, and heme metabolism. See also the chapters on inborn errors of purine and pyrimidine metabolism.

## Problems

**1. ATP Consumption by Root Nodules in Legumes** Bacteria residing in the root nodules of the pea plant consume more than 20% of the ATP produced by the plant. Suggest why these bacteria consume so much ATP.

**2. Glutamate Dehydrogenase and Protein Synthesis** The bacterium *Methylophilus methylotrophus* can synthesize protein from methanol and ammonia. Recombinant DNA techniques have improved the yield of protein by introducing into *M. methylotrophus* the glutamate dehydrogenase gene from *E. coli*. Why does this genetic manipulation increase the protein yield?

**3. Transformation of Aspartate to Asparagine** There are two routes for transforming aspartate to asparagine at the expense of ATP. Many bacteria have an asparagine synthetase that uses ammonium ion as the nitrogen donor. Mammals have an asparagine synthetase that uses glutamine as the nitrogen donor. Given that the latter requires an extra ATP (for the synthesis of glutamine), why do mammals use this route?

**4. Equation for the Synthesis of Aspartate from Glucose** Write the net equation for the synthesis of aspartate (a nonessential amino acid) from glucose, carbon dioxide, and ammonia.

**5. Phenylalanine Hydroxylase Deficiency and Diet** Tyrosine is normally a nonessential amino acid, but individuals with a genetic defect in phenylalanine hydroxylase require tyrosine in their diet for normal growth. Explain.

**6. Cofactors for One-Carbon Transfer Reactions** Most one-carbon transfers are promoted by one of three cofactors: biotin, tetrahydrofolate, or $S$-adenosylmethionine (Chapter 18). $S$-Adenosylmethionine is generally used as a methyl group donor; the transfer potential of the methyl group in $N^5$-methyltetrahydrofolate is insufficient for most biosynthetic reactions. However, one example of the use of $N^5$-methyltetrahydrofolate in methyl group transfer is in methionine formation by the methionine synthase reaction (step ⑨ of Fig. 22–15); methionine is the immediate precursor of $S$-adenosylmethionine (see Fig. 18–18). Explain how the methyl group of $S$-adenosylmethionine can be derived from $N^5$-methyltetrahydrofolate, even though the transfer potential of the methyl group in $N^5$-methyltetrahydrofolate is one-thousandth of that in $S$-adenosylmethionine.

**7. Concerted Regulation in Amino Acid Biosynthesis** The glutamine synthetase of *E. coli* is independently modulated by various products of glutamine metabolism (see Fig. 22–6). In this concerted inhibition, the extent of enzyme inhibition is greater than the sum of the separate inhibitions caused by each product. For *E. coli* grown in a medium rich in histidine, what would be the advantage of concerted inhibition?

**8. Relationship between Folic Acid Deficiency and Anemia** Folic acid deficiency, believed to be the most common vitamin deficiency, causes a type of anemia in which hemoglobin synthesis is impaired and erythrocytes do not mature properly. What is the metabolic relationship between hemoglobin synthesis and folic acid deficiency?

**9. Nucleotide Biosynthesis in Amino Acid Auxotrophic Bacteria** Normal *E. coli* cells can synthesize all 20 common amino acids, but some mutants, called amino acid auxotrophs, are unable to synthesize a specific amino acid and require its addition to the culture medium for optimal growth. Besides their role in protein synthesis, some amino acids are also precursors for other nitrogenous cell products. Consider the three amino acid auxotrophs that are unable to synthesize glycine, glutamine, and aspartate, respectively. For each mutant, what nitrogenous products other than proteins would the cell fail to synthesize?

**10. Inhibitors of Nucleotide Biosynthesis** Suggest mechanisms for the inhibition of (a) alanine racemase by L-fluoroalanine and (b) glutamine amidotransferases by azaserine.

**11. Mode of Action of Sulfa Drugs** Some bacteria require $p$-aminobenzoate in the culture medium for normal growth, and their growth is severely inhibited by the addition of sulfanilamide, one of the earliest sulfa drugs. Moreover, in the presence of this drug, 5-aminoimidazole-4-carboxamide ribonucleotide (AICAR; see Fig. 22–33) accumulates in the culture medium. These effects are reversed by addition of excess $p$-aminobenzoate.

$p$-Aminobenzoate                    Sulfanilamide

(a) What is the role of $p$-aminobenzoate in these bacteria? (Hint: See Fig. 18–16).

(b) Why does AICAR accumulate in the presence of sulfanilamide?

(c) Why are the inhibition and accumulation reversed by addition of excess $p$-aminobenzoate?

**12. Pathway of Carbon in Pyrimidine Biosynthesis** Predict the locations of $^{14}C$ in orotate isolated from cells grown on a small amount of uniformly labeled [$^{14}C$]succinate. Justify your prediction.

**13. Nucleotides as Poor Sources of Energy** Under starvation conditions, organisms can use proteins and amino acids as sources of energy. Deamination of amino acids produces carbon skeletons that can enter the glycolytic pathway and the citric acid cycle to produce energy in the form of ATP. Nucleotides, on the other hand, are not similarly degraded for use as energy-yielding fuels. What observations about cellular physiology support this statement? What aspect of the structure of nucleotides makes them a relatively poor source of energy?

**14. Treatment of Gout** Allopurinol (see Fig. 22–47), an inhibitor of xanthine oxidase, is used to treat chronic gout. Explain the biochemical basis for this treatment. Patients treated with allopurinol sometimes develop xanthine stones in the kidneys, although the incidence of kidney damage is much lower than in untreated gout. Explain this observation in the light of the following solubilities in urine: uric acid, 0.15 g/L; xanthine, 0.05 g/L; and hypoxanthine, 1.4 g/L.

**15. Inhibition of Nucleotide Synthesis by Azaserine** The diazo compound $O$-(2-diazoacetyl)-L-serine, known also as azaserine (see Fig. 22–48), is a powerful inhibitor of glutamine amidotransferases. If growing cells are treated with azaserine, what intermediates of nucleotide biosynthesis would accumulate? Explain.

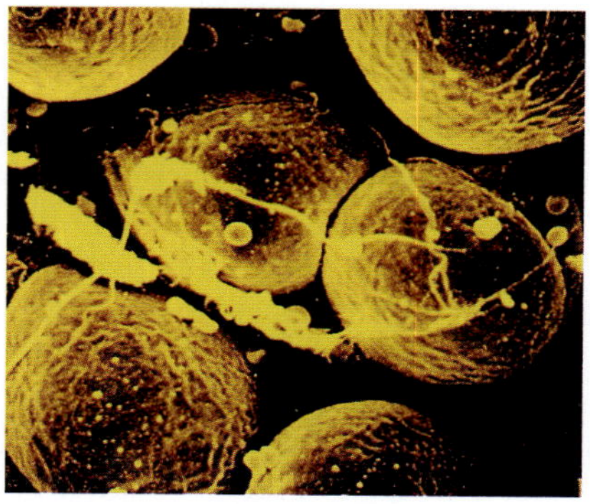

## chapter 23

# HORMONAL REGULATION AND INTEGRATION OF MAMMALIAN METABOLISM

We recognize that each tissue and, more generally, each cell of the organism secretes . . . special products or ferments into the blood which thereby influence all the other cells thus integrated with each other by a mechanism other than the nervous system.

—*Charles Édouard Brown-Séquard and J. d'Arsonval, article in Comptes Rendus de la Société de Biologie, 1891*

In Chapters 13 through 22 we have discussed metabolism at the level of the individual cell, emphasizing central pathways common to almost all cells, prokaryotic and eukaryotic. We have seen how metabolic processes within cells are regulated at the level of individual enzyme reactions, by substrate availability, by allosteric mechanisms, and by phosphorylation or other covalent modifications of enzymes.

To appreciate fully the significance of individual metabolic pathways and their regulation, we must view these pathways in the context of the whole organism. An essential characteristic of multicellular organisms is cell differentiation and division of labor. The specialized functions of the tissues and organs of complex organisms such as humans impose characteristic fuel requirements and patterns of metabolism. Hormonal signals integrate and coordinate the metabolic activities of different tissues and optimize the allocation of fuels and precursors to each organ.

In this chapter we focus on mammals, looking at the specialized metabolism of several major organs and tissues and the integration of metabolism in the whole organism. We begin by examining the broad range of hormones and hormonal mechanisms, then turn to the tissue-specific functions regulated by these mechanisms. We discuss the distribution of nutrients to various organs—emphasizing the central role played by the liver—and the metabolic cooperation among these organs. To illustrate the integrative role of hormones, we describe the interplay of insulin, glucagon, and epinephrine in coordinating fuel metabolism in muscle, liver, and adipose tissue. The metabolic disturbances in diabetes further illustrate the importance of hormonal regulation of metabolism. Finally we discuss the long-term hormonal regulation of body mass.

## 23.1 Hormones: Diverse Structures for Diverse Functions

Virtually every process in a complex organism is regulated by one or more hormones: maintenance of blood pressure, blood volume, and electrolyte balance; embryogenesis; sexual differentiation, development, and reproduction; hunger, eating behavior, digestion, and fuel allocation—to name but a few. We examine here the methods for detecting and measuring hormones and their interaction with receptors, and consider a representative selection of hormone types.

The coordination of metabolism in mammals is achieved by the **neuroendocrine system.** Individual cells in one tissue sense a change in the organism's circumstances and respond by secreting a chemical messenger that passes to another cell in the same or different tissue, where it binds to a receptor molecule and triggers a change in this second cell. In neuronal signaling (Fig. 23–1a), the chemical messenger (neurotransmitter; acetylcholine, for example) may travel only a fraction of a micrometer, across the synaptic cleft to the next neuron in a network. In hormonal signaling, the messengers—hormones—are carried in the bloodstream to neighboring cells or to distant organs and tissues; they may travel a meter or more before encountering

their target cell (Fig. 23–1b). Except for this anatomic difference, these two chemical signaling mechanisms are remarkably similar. Epinephrine and norepinephrine, for example, serve as neurotransmitters in certain synapses of the brain and smooth muscle and as hormones that regulate fuel metabolism in liver and muscle. The following discussion of cellular signaling emphasizes hormone action, drawing on discussions of fuel metabolism in earlier chapters, but most of the fundamental mechanisms described here also occur in neurotransmitter action.

## The Discovery and Purification of Hormones Requires a Bioassay

How is a hormone discovered and isolated? First, researchers find that a physiological process in one tissue depends on a signal that originates in another tissue. Insulin, for example, was first recognized as a substance that is produced in the pancreas and affects the volume and composition of urine (Box 23–1). Once a physiological effect of the putative hormone is discovered, a quantitative bioassay for the hormone can be developed. In the case of insulin, the assay consisted of injecting extracts of pancreas (a crude source of insulin) into experimental animals deficient in insulin, then quantifying the resulting changes in glucose concentration in blood and urine. To isolate a hormone, the biochemist fractionates extracts containing the putative hormone, with the same techniques used to purify other biomolecules (solvent fractionation, chromatography, and electrophoresis), and then assays each fraction for hormone activity. Once the chemical has been purified, its composition and structure can be determined.

This protocol for hormone characterization is deceptively simple. Hormones are extremely potent and are produced in very small amounts. Obtaining sufficient hormone to allow its chemical characterization often involves biochemical isolations on a heroic scale. When Roger Guillemin and Andrew Schally independently purified and characterized thyrotropin-releasing hormone (TRH) from the hypothalamus, Schally's group processed about 20 tons of hypothalamus from nearly two million sheep, and Guillemin's group extracted the

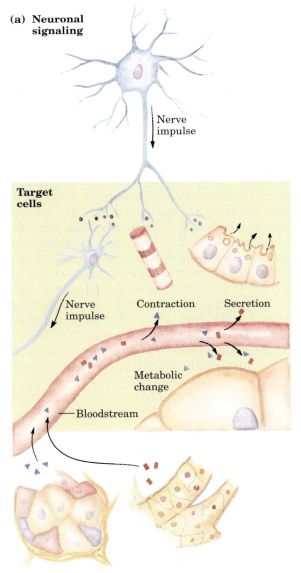

**(a) Neuronal signaling**

Nerve impulse

Target cells

Nerve impulse

Contraction

Secretion

Metabolic change

Bloodstream

**(b) Endocrine signaling**

**FIGURE 23–1 Signaling by the neuroendocrine system. (a)** In neuronal signaling, electrical signals (nerve impulses) originate in the cell body of a neuron and travel very rapidly over long distances to the axon tip, where neurotransmitters are released and diffuse to the target cell. The target cell (another neuron, a myocyte, or a secretory cell) is only a fraction of a micrometer or a few micrometers away from the site of neurotransmitter release. **(b)** In the endocrine system, hormones are secreted into the bloodstream, which carries them throughout the body to target tissues that may be a meter or more away from the secreting cell. Both neurotransmitters and hormones interact with specific receptors on or in their target cells, triggering responses.

## BOX 23–1    BIOCHEMISTRY IN MEDICINE

### How Is a Hormone Discovered? The Arduous Path to Purified Insulin

Millions of people with type I (insulin-dependent) diabetes mellitus inject themselves daily with pure insulin to compensate for the lack of production of this critical hormone by their own pancreatic $\beta$ cells. Insulin injection is not a cure for diabetes, but it allows people who otherwise would have died young to lead long and productive lives. The discovery of insulin, which began with an accidental observation, illustrates the combination of serendipity and careful experimentation that led to the discovery of many of the hormones.

In 1889, Oskar Minkowski, a young assistant at the Medical College of Strasbourg, and Josef von Mering, at the Hoppe-Seyler Institute in Strasbourg, had a friendly disagreement about whether the pancreas, known to contain lipases, was important in fat digestion in dogs. To resolve the issue, they began an experiment on fat digestion. They surgically removed the pancreas from a dog, but before their experiment got any farther, Minkowski noticed that the dog was now producing far more urine than normal (a common symptom of untreated diabetes). Also, the dog's urine had glucose levels far above normal (another symptom of diabetes). These findings suggested that lack of some pancreatic product caused diabetes.

Minkowski tried unsuccessfully to prepare an extract of dog pancreas that would reverse the effect of removing the pancreas—that is, would lower the urinary or blood glucose levels. We now know that insulin is a protein, and that the pancreas is very rich in proteases (trypsin and chymotrypsin), normally released directly into the small intestine to aid in digestion. These proteases doubtless degraded the insulin in the pancreatic extracts in Minkowski's experiments.

Despite considerable effort, no significant progress was made in the isolation or characterization of the "antidiabetic factor" until the summer of 1921,

when Frederick G. Banting, a young scientist working in the laboratory of J. J. R. MacLeod at the University of Toronto, and a student assistant, Charles Best, took up the problem. By that time, several lines of evidence pointed to a group of specialized cells in the pancreas (the islets of Langerhans; see Fig. 23–24) as the source of the antidiabetic factor, which came to be called insulin (from Latin *insula,* "island").

Taking precautions to prevent proteolysis, Banting and Best (later aided by biochemist J. B. Collip) succeeded in December 1921 in preparing a purified pancreatic extract that cured the symptoms of experimental diabetes in dogs. On January 25, 1922 (just one month later!), their insulin preparation was injected into Leonard Thompson, a 14-year-old boy severely ill with diabetes mellitus. Within days, the levels of ketone bodies and glucose in Thompson's urine dropped dramatically; the extract saved his life. In 1923, Banting and MacLeod won the Nobel Prize for their isolation of insulin. Banting immediately announced that he would share his prize with Best; MacLeod shared his with Collip.

By 1923, pharmaceutical companies were supplying thousands of patients throughout the world with insulin extracted from porcine pancreas. With the development of genetic engineering techniques in the 1980s (Chapter 9), it became possible to produce unlimited quantities of human insulin by inserting the cloned human gene for insulin in a microorganism, which was then cultured on an industrial scale. Some patients with diabetes are now fitted with implanted insulin pumps, which release adjustable amounts of insulin on demand to meet changing needs at meal times and during exercise. There is a reasonable prospect that, in the future, transplantation of pancreatic tissue will provide diabetic patients with a source of insulin that responds as well as normal pancreas, releasing insulin into the bloodstream only when blood glucose rises.

Frederick G. Banting, 1891–1941

J. J. R. MacLeod, 1876–1935

Charles Best, 1899–1978

J. B. Collip, 1892–1965

Pyroglutamate  Histidine  Prolylamide

**pyroGlu–His–Pro–NH₂**

**FIGURE 23–2 The structure of thyrotropin-releasing hormone (TRH).** Purified (by heroic efforts) from extracts of hypothalamus, TRH proved to be a derivative of the tripeptide Glu–His–Pro. The side-chain carboxyl group of the amino-terminal Glu forms an amide (red bond) with the residue's $\alpha$-amino group, creating pyroglutamate, and the carboxyl group of the carboxyl-terminal Pro is converted to an amide (red —NH₂). Such modifications are common among the small peptide hormones. In a typical protein of $M_r$ ~50, the charges on the amino- and carboxyl-terminal groups contribute relatively little to the overall charge on the protein, but in a tripeptide these two charges dominate the properties of the peptide. Formation of the amide derivatives removes these charges.

hypothalamus from about a million pigs! TRH proved to be a simple derivative of the tripeptide Glu–His–Pro (Fig. 23–2). Once the structure of the hormone was known, it could be chemically synthesized in large quantities for use in physiological and biochemical studies.

For their work on hypothalamic hormones, Schally and Guillemin shared the Nobel Prize in Physiology or Medicine in 1977, along with Rosalyn Yalow, who (with Solomon A. Berson) developed the extraordinarily sensitive **radioimmunoassay (RIA)** for peptide hormones and used it to study hormone action. RIA revolutionized hormone research by making possible the rapid, quantitative, and specific measurement of hormones in minute amounts.

Hormone-specific antibodies are the key to the radioimmunoassay. Purified hormone, injected into rabbits, elicits antibodies that bind to that hormone with very high affinity and specificity. When a constant amount of isolated antibody is incubated with a fixed amount of the radioactively labeled hormone, a certain fraction of the radioactive hormone binds to the antibody (Fig. 23–3). If, in addition to the radiolabeled hormone, unlabeled hormone is also present, the unlabeled hormone competes with and displaces some of the labeled hormone from its binding site on the antibody. This binding competition can be quantified by reference to a standard curve obtained with known amounts of unlabeled hormone. The degree to which labeled hormone is displaced from antibody is a measure of the amount of unlabeled hormone in a sample of blood or tissue extract. By using very highly radioactive hormone, researchers can make the assay sensitive to picograms of hormone. A newer variation of this technique, enzyme-linked immunosorbent assay (ELISA), is illustrated in Figure 5–28b.

## Hormones Act through Specific High-Affinity Cellular Receptors

As we saw in Chapter 12, all hormones act through highly specific receptors in hormone-sensitive target cells, to which the hormones bind with high affinity (see Fig. 12–2). Each cell type has its own combination of hormone receptors, which define the range of its hormone responsiveness. Moreover, two cell types with the same type of receptor may have different intracellular targets of hormone action and thus may respond differently to the same hormone. The specificity of hormone action results from structural complementarity between the hormone and its receptor; this interaction is extremely selective, so structurally similar hormones can have different effects. The high affinity of the interaction allows cells to respond to very low concentrations of hormone. In the design of drugs intended to intervene in hormonal regulation, we need to know the relative specificity and affinity of the drug and the natural hormone. Recall that hormone-receptor interactions can be quantified by **Scatchard analysis** (see Box 12–1), which, under favorable conditions, yields a quantitative measure of affinity (the dissociation constant for the complex) and the number of hormone-binding sites in a preparation of receptor.

Roger Guillemin

Andrew V. Schally

Rosalyn S. Yalow

**(a)**

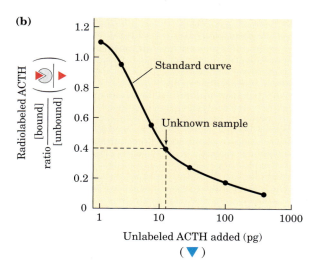

**(b)**

FIGURE 23–3 **Radioimmunoassay (RIA). (a)** A low concentration of radiolabeled hormone (red) is incubated with ① a fixed amount of antibody specific for that hormone or ② a fixed amount of antibody and various concentrations of unlabeled hormone (blue). In the latter case, unlabeled hormone competes with labeled hormone for binding to the antibody; the amount of labeled hormone bound varies inversely with the concentration of unlabeled hormone present. **(b)** A radioimmunoassay for adrenocorticotropic hormone (ACTH). A standard curve of the ratio [bound] to [unbound radiolabeled ACTH] vs. [unlabeled ACTH added] is constructed and used to determine the amount of (unlabeled) ACTH in an unknown sample. If an aliquot containing an unknown quantity of unlabeled hormone gives, say, a value of 0.4 for the ratio [bound]/[unbound] (see arrow), the aliquot must contain about 20 pg of ACTH.

The locus of the encounter between hormone and receptor may be extracellular, cytosolic, or nuclear, depending on the hormone type. The intracellular consequences of hormone-receptor interaction are of at least six general types: (1) a change in membrane potential results from the opening or closing of a hormone-gated

ion channel; (2) a receptor enzyme is activated by the extracellular hormone; (3) a second messenger (such as cAMP or inositol trisphosphate) generated inside the cell acts as an allosteric regulator of one or more enzymes; (4) a receptor with no intrinsic enzyme activity recruits a soluble protein kinase in the cytosol, which passes on the signal; (5) an adhesion receptor on the cell surface interacts with molecules in the extracellular matrix and conveys information to the cytoskeleton; or (6) a steroid or steroidlike molecule causes a change in the level of expression (transcription of DNA into mRNA) of one or more genes, mediated by a nuclear hormone receptor protein (see Fig. 12–2).

Water-soluble peptide and amine hormones (insulin and epinephrine, for example) act extracellularly by binding to cell surface receptors that span the plasma membrane (Fig. 23–4). When the hormone binds to its extracellular domain, the receptor undergoes a conformational change analogous to that produced in an allosteric enzyme by binding of an effector molecule. The conformational change triggers the downstream effects of the hormone.

A single hormone molecule, in forming a hormone-receptor complex, activates a catalyst that produces many molecules of second messenger, so the receptor

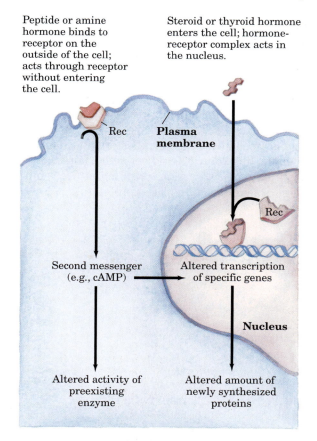

FIGURE 23–4 **Two general mechanisms of hormone action.** The peptide and amine hormones are faster acting than steroid and thyroid hormones.

serves not only as a signal transducer but also as a signal amplifier. The signal may be further amplified by a signaling cascade, a series of steps in which a catalyst activates a catalyst, resulting in very large amplifications of the original signal. A cascade of this type occurs in the regulation of glycogen synthesis and breakdown by epinephrine (see Fig. 12–16). Epinephrine activates (through its receptor) adenylyl cyclase, which produces many molecules of cAMP for each molecule of receptor-bound hormone. Cyclic AMP in turn activates cAMP-dependent protein kinase, which activates phosphory-lase kinase, which activates glycogen phosphorylase. The result is signal amplification: one epinephrine molecule causes the production of many thousands of molecules of glucose 1-phosphate from glycogen.

Water-insoluble hormones (steroid, retinoid, and thyroid hormones) readily pass through the plasma membrane of their target cells to reach their receptor proteins in the nucleus (Fig. 23–4). With this class of hormones, the hormone-receptor complex itself carries the message; it interacts with DNA to alter the expression of specific genes, changing the enzyme complement of the cell and thereby changing cellular metabolism (see Fig. 12–40).

Hormones that act through plasma membrane receptors generally trigger very rapid physiological or biochemical responses. Just seconds after the adrenal medulla secretes epinephrine into the bloodstream, skeletal muscle responds by accelerating the breakdown of glycogen. By contrast, the thyroid hormones and the sex (steroid) hormones promote maximal responses in their target tissues only after hours or even days. These differences in response time correspond to different modes of action. In general, the fast-acting hormones lead to a change in the activity of one or more preexisting enzymes in the cell, by allosteric mechanisms or covalent modification. The slower-acting hormones generally alter gene expression, resulting in the synthesis of more or less of the regulated protein(s).

## Hormones Are Chemically Diverse

Mammals have several classes of hormones, distinguishable by their chemical structures and their modes of action (Table 23–1). Peptide, amine, and eicosanoid hormones act from outside the target cell via surface receptors. Steroid, vitamin D, retinoid, and thyroid hormones enter the cell and act through nuclear receptors. Nitric oxide also enters the cell, but activates a cytosolic enzyme, guanylyl cyclase (see Fig. 12–10).

Hormones can also be classified by the way they get from the point of their release to their target tissue. **Endocrine** (from the Greek *endon,* "within," and *krinein,* "to release") hormones are released into the blood and carried to target cells throughout the body (insulin is an example). **Paracrine** hormones are released into the extracellular space and diffuse to neighboring target cells (the eicosanoid hormones are of this type). **Autocrine** hormones are released by and affect the same cell, binding to receptors on the cell surface.

Mammals are hardly unique in possessing hormonal signaling systems. Insects and nematode worms have highly developed systems for hormonal regulation, with fundamental mechanisms similar to those in mammals. Plants, too, use hormonal signals to coordinate the activities of their various tissues (Chapter 12). The study of hormone action is not as advanced in plants as in animals, but we do know that some mechanisms are shared. To illustrate the structural diversity and range of action of mammalian hormones, we consider representative examples of each major class listed in Table 23–1.

*Peptide Hormones*   Peptide hormones may have from 3 to 200 or more amino acid residues. They include the pancreatic hormones insulin, glucagon, and somatostatin, the parathyroid hormone, calcitonin, and all the hormones of the hypothalamus and pituitary (described below). These hormones are synthesized on ribosomes in the form of longer precursor proteins (prohormones),

**TABLE 23–1   Classes of Hormones**

| Type | Example | Synthetic path | Mode of action |
|------|---------|----------------|----------------|
| Peptide | Insulin, glucagon | Proteolytic processing of prohormone | Plasma membrane receptors; second messengers |
| Catecholamine | Epinephrine | From tyrosine | |
| Eicosanoid | PGE$_1$ | From arachidonate (20:4 fatty acid) | |
| Steroid | Testosterone | From cholesterol | Nuclear receptors; transcriptional regulation |
| Vitamin D | 1,25-Dihydroxycholecalciferol | From cholesterol | |
| Retinoid | Retinoic acid | From vitamin A | |
| Thyroid | Triiodothyronine (T$_3$) | From Tyr in thyroglobulin | |
| Nitric oxide | Nitric oxide | From arginine + O$_2$ | Cytosolic receptor (guanylate cyclase) and second messenger (cGMP) |

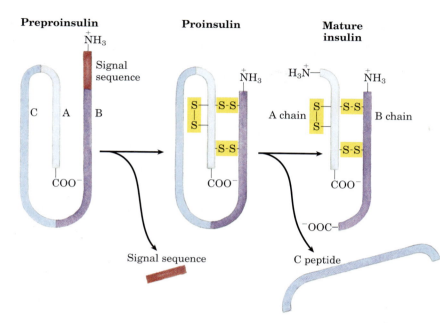

**Preproinsulin**     **Proinsulin**     **Mature insulin**

**FIGURE 23–5 Insulin.** Mature insulin is formed from its larger precursor preproinsulin by proteolytic processing. Removal of a 23 amino acid segment (the signal sequence) at the amino terminus of preproinsulin and formation of three disulfide bonds produces proinsulin. Further proteolytic cuts remove the C peptide from proinsulin to produce mature insulin, composed of A and B chains. The amino acid sequence of bovine insulin is shown in Figure 3–24.

then packaged into secretory vesicles and proteolytically cleaved to form the active peptides. **Insulin** is a small protein ($M_r$ 5,800) with two polypeptide chains, A and B, joined by two disulfide bonds. It is synthesized in the pancreas as an inactive single-chain precursor, preproinsulin (Fig. 23–5), with an amino-terminal "signal sequence" that directs its passage into secretory vesicles. (Signal sequences are discussed in Chapter 27; see Fig. 27–33.) Proteolytic removal of the signal sequence and formation of three disulfide bonds produces proinsulin, which is stored in secretory granules in pancreatic $\beta$ cells. When elevated blood glucose triggers in-

sulin secretion, proinsulin is converted to active insulin by specific proteases, which cleave two peptide bonds to form the mature insulin molecule.

In some cases, prohormone proteins yield a single peptide hormone, but often several active hormones are carved out of the same prohormone. Pro-opiomelanocortin (POMC) is a spectacular example of multiple hormones encoded by a single gene. The POMC gene encodes a large polypeptide that is progressively carved up into at least nine biologically active peptides (Fig. 23–6). The terminal residues of peptide hormones are often modified, as in TRH (Fig. 23–2).

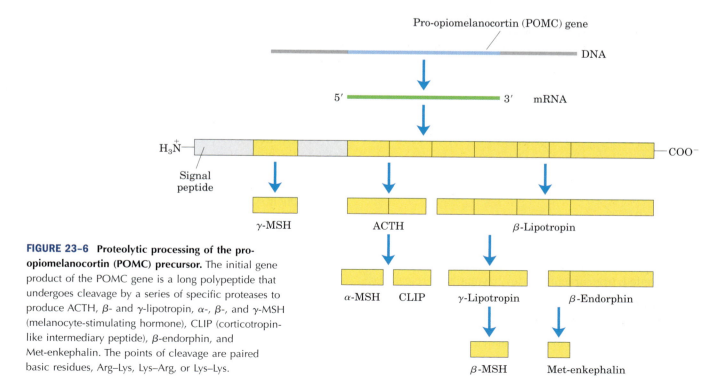

**FIGURE 23–6 Proteolytic processing of the pro-opiomelanocortin (POMC) precursor.** The initial gene product of the POMC gene is a long polypeptide that undergoes cleavage by a series of specific proteases to produce ACTH, $\beta$- and $\gamma$-lipotropin, $\alpha$-, $\beta$-, and $\gamma$-MSH (melanocyte-stimulating hormone), CLIP (corticotropin-like intermediary peptide), $\beta$-endorphin, and Met-enkephalin. The points of cleavage are paired basic residues, Arg–Lys, Lys–Arg, or Lys–Lys.

The concentration of peptide hormones within secretory granules is so high that the vesicle contents are virtually crystalline; when the contents are released by exocytosis, a large amount of hormone is released suddenly. The capillaries that serve peptide-producing endocrine glands are fenestrated (and thus permeable to peptides), so the hormone molecules readily enter the bloodstream for transport to target cells elsewhere. As noted earlier, all peptide hormones act by binding to receptors in the plasma membrane. They cause the generation of a second messenger in the cytosol, which changes the activity of an intracellular enzyme, thereby altering the cell's metabolism.

**Catecholamine Hormones** The water-soluble compounds **epinephrine (adrenaline)** and **norepinephrine (noradrenaline)** are **catecholamines,** named for the structurally related compound catechol. They are synthesized from tyrosine.

$$\text{Tyrosine} \longrightarrow \text{L-DOPA} \longrightarrow \text{Dopamine} \longrightarrow$$
$$\text{Norepinephrine} \longrightarrow \text{Epinephrine}$$

Catecholamines produced in the brain and in other neural tissues function as neurotransmitters, but epinephrine and norepinephrine are also hormones, synthesized and secreted by the adrenal glands. Like the peptide hormones, catecholamines are highly concentrated within secretory vesicles and released by exocytosis, and they act through surface receptors to generate intracellular second messengers. They mediate a wide variety of physiological responses to acute stress (see Table 23–6).

**Eicosanoids** The eicosanoid hormones (prostaglandins, thromboxanes, and leukotrienes) are derived from the 20-carbon polyunsaturated fatty acid arachidonate.

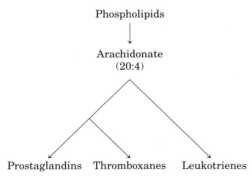

Unlike the hormones described above, they are not synthesized in advance and stored; they are produced, when needed, from arachidonate enzymatically released from membrane phospholipids by phospholipase $A_2$ (see Fig. 10–18). The enzymes of the pathway leading to prostaglandins and thromboxanes (see Fig. 21–15) are very widely distributed in mammalian tissues; most cells can produce these signals, and cells of many tissues can respond to them through specific plasma membrane receptors. The eicosanoid hormones are paracrine hormones, secreted into the interstitial fluid (not primarily into the blood) and acting on nearby cells.

Prostaglandins promote the contraction of smooth muscle, including that of the intestine and uterus (and can therefore be used medically to induce labor). They also mediate pain and inflammation in all tissues. Many antiinflammatory drugs act by inhibiting steps in the prostaglandin synthetic pathway (see Box 21–2). Thromboxanes regulate platelet function and therefore blood clotting. Leukotrienes $LTC_4$ and $LTD_4$ act through plasma membrane receptors to stimulate contraction of smooth muscle in the intestine, pulmonary airways, and trachea. They are mediators of the severe immune response called anaphylaxis. ∎

**Steroid Hormones** The steroid hormones (adrenocortical hormones and sex hormones) are synthesized from cholesterol in several endocrine tissues.

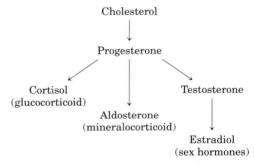

They travel to their target cells through the bloodstream, bound to carrier proteins. More than 50 corticosteroid hormones are produced in the adrenal cortex by reactions that remove the side chain from the D ring of cholesterol and introduce oxygen to form keto and hydroxyl groups. Many of these reactions involve cytochrome P-450 enzymes (see Box 21–1). The steroid hormones are of two general types. Glucocorticoids (such as cortisol) primarily affect the metabolism of carbohydrates; mineralocorticoids (such as aldosterone) regulate the concentrations of electrolytes in the blood. Androgens (testosterone) and estrogens (such as estradiol; see Fig. 10–19) are synthesized in the testes and ovaries. Their synthesis also involves cytochrome P-450 enzymes that cleave the side chain of cholesterol and introduce oxygen atoms. These hormones affect sexual development, sexual behavior, and a variety of other reproductive and nonreproductive functions.

All steroid hormones act through nuclear receptors to change the level of expression of specific genes (p. 465). Recent evidence indicates that they also have more rapid effects, mediated by receptors localized in the plasma membrane.

***Vitamin D Hormone*** Calcitriol (1,25-dihydroxycholecalciferol) is produced from vitamin D by enzyme-catalyzed hydroxylation in the liver and kidneys (see Fig. 10–20a).

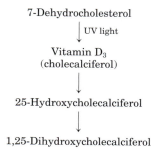

7-Dehydrocholesterol

↓ UV light

Vitamin $D_3$
(cholecalciferol)

↓

25-Hydroxycholecalciferol

↓

1,25-Dihydroxycholecalciferol

Vitamin D is obtained in the diet or by photolysis of 7-dehydrocholesterol in skin exposed to sunlight. Calcitriol works in concert with parathyroid hormone in $Ca^{2+}$ homeostasis, regulating $[Ca^{2+}]$ in the blood and the balance between $Ca^{2+}$ deposition and $Ca^{2+}$ mobilization from bone. Acting through nuclear receptors, calcitriol activates the synthesis of an intestinal $Ca^{2+}$-binding protein essential for uptake of dietary $Ca^{2+}$. Inadequate dietary vitamin D or defects in the biosynthesis of calcitriol result in serious diseases such as rickets, in which bones are weak and malformed (see Fig. 10–20b).

***Retinoid Hormones*** Retinoids are potent hormones that regulate the growth, survival, and differentiation of cells via nuclear retinoid receptors. The prohormone retinol is synthesized from vitamin A, primarily in liver (see Fig. 10–21), and many tissues convert retinol to the hormone retinoic acid (RA).

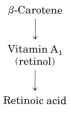

$\beta$-Carotene

↓

Vitamin $A_1$
(retinol)

↓

Retinoic acid

All tissues are retinoid targets, as all cell types have at least one form of nuclear retinoid receptor. In adults, the most significant targets include cornea, skin, epithelia of the lungs and trachea, and the immune system. RA regulates the synthesis of proteins essential for growth or differentiation. Excessive vitamin A can cause birth defects, and pregnant women are advised not to use the retinoid creams that have been developed for treatment of severe acne.

***Thyroid Hormones*** The thyroid hormones $T_4$ (thyroxine) and $T_3$ (triiodothyronine) are synthesized from the precursor protein thyroglobulin ($M_r$ 660,000). Up to 20 Tyr residues in thyroglobulin are enzymatically iodinated in the thyroid gland, then two iodotyrosine residues condense to form the precursor to thyroxine. When needed, thyroxine is released by proteolysis. Condensation of monoiodotyrosine with diiodotyrosine produces $T_3$, which is also an active hormone released by proteolysis.

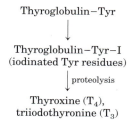

Thyroglobulin–Tyr

↓

Thyroglobulin–Tyr–I
(iodinated Tyr residues)

↓ proteolysis

Thyroxine ($T_4$),
triiodothyronine ($T_3$)

The thyroid hormones act through nuclear receptors to stimulate energy-yielding metabolism, especially in liver and muscle, by increasing the expression of genes encoding key catabolic enzymes.

***Nitric Oxide (NO)*** Nitric oxide is a relatively stable free radical synthesized from molecular oxygen and the guanidino nitrogen of arginine (see Fig. 22–31) in a reaction catalyzed by **NO synthase.**

$$\text{Arginine} + 1\tfrac{1}{2}\text{ NADPH} + 2O_2 \longrightarrow$$
$$\text{NO} + \text{citrulline} + 2H_2O + 1\tfrac{1}{2}\text{ NADP}^+$$

This enzyme is found in many tissues and cell types: neurons, macrophages, hepatocytes, myocytes of smooth muscle, endothelial cells of the blood vessels, and epithelial cells of the kidney. NO acts near its point of release, entering the target cell and activating the cytosolic enzyme guanylyl cyclase, which catalyzes the formation of the second messenger cGMP (see Fig. 12–10).

## Hormone Release Is Regulated by a Hierarchy of Neuronal and Hormonal Signals

The changing levels of specific hormones regulate specific cellular processes, but what regulates the level of each hormone? The brief answer is that the central nervous system receives input from many internal and external sensors—signals about danger, hunger, dietary intake, blood composition and pressure, for example—and orchestrates the production of appropriate hormonal signals by the endocrine tissues. For a more complete answer, we must look at the hormone-producing systems of the human body and some of their functional interrelationships.

Figure 23–7 shows the anatomic location of the major endocrine glands in humans, and Figure 23–8 represents the "chain of command" in the hormonal signaling hierarchy. The **hypothalamus,** a small region of the brain (Fig. 23–9), is the coordination center of the

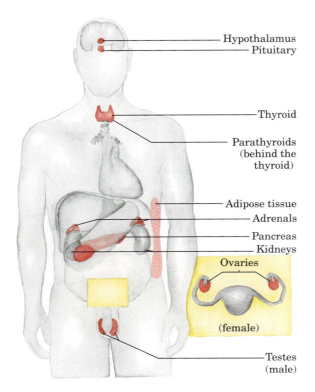

**FIGURE 23–7 The major endocrine glands.** The glands are shaded dark pink.

endocrine system; it receives and integrates messages from the central nervous system. In response to these messages, the hypothalamus produces regulatory hormones (releasing factors) that pass directly to the nearby pituitary gland, through special blood vessels and neurons that connect the two glands (Fig. 23–9b). The pituitary gland has two functionally distinct parts. The **posterior pituitary** contains the axonal endings of many neurons that originate in the hypothalamus. These neurons produce the short peptide hormones oxytocin and vasopressin (Fig. 23–10), which then move down the axon to the nerve endings in the pituitary, where they are stored in secretory granules to await the signal for their release.

The **anterior pituitary** responds to hypothalamic hormones carried in the blood, producing **tropic hormones,** or **tropins** (from the Greek *tropos,* "turn"). These relatively long polypeptides activate the next rank of endocrine glands (Fig. 23–8), which includes the adrenal cortex, thyroid gland, ovaries, and testes. These glands in turn secrete their specific hormones, which are carried in the bloodstream to the receptors of cells in the target tissues. For example, corticotropin-releasing hormone from the hypothalamus stimulates the anterior pituitary to release ACTH, which travels to the zona fasciculata of the adrenal cortex and triggers the release of

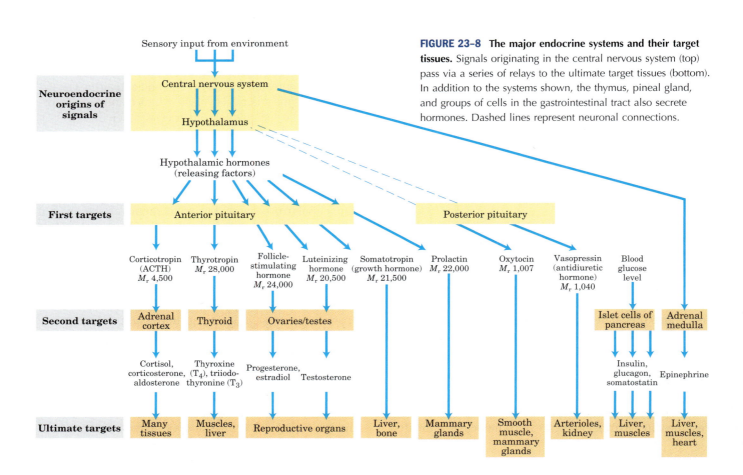

**FIGURE 23–8 The major endocrine systems and their target tissues.** Signals originating in the central nervous system (top) pass via a series of relays to the ultimate target tissues (bottom). In addition to the systems shown, the thymus, pineal gland, and groups of cells in the gastrointestinal tract also secrete hormones. Dashed lines represent neuronal connections.

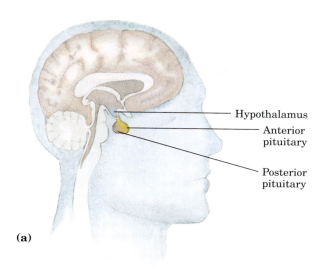

**(a)**

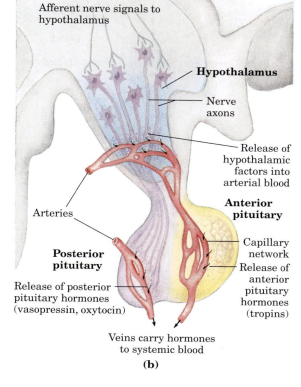

**FIGURE 23–9  Neuroendocrine origins of hormone signals. (a)** Location of the hypothalamus and pituitary gland. **(b)** Details of the hypothalamus-pituitary system. Signals from connecting neurons stimulate the hypothalamus to secrete releasing factors into a blood vessel that carries the hormones directly to a capillary network in the anterior pituitary. In response to each hypothalamic releasing factor, the anterior pituitary releases the appropriate hormone into the general circulation. Posterior pituitary hormones are synthesized in neurons arising in the hypothalamus, transported along axons to nerve endings in the posterior pituitary, and stored there until released into the blood in response to a neuronal signal.

cortisol. Cortisol, the ultimate hormone in this cascade, acts through its receptor in many types of target cells to alter their metabolism. In hepatocytes, one effect of cortisol is to increase the rate of gluconeogenesis.

Hormonal cascades such as those responsible for the release of cortisol and epinephrine result in large amplifications of the initial signal and allow exquisite fine-tuning of the output of the ultimate hormone (Fig. 23–11). At each level in the cascade, a small signal elicits a larger response. The initial electrical signal to the hypothalamus results in the release of a few *nanograms* of corticotropin-releasing hormone, which elicits the release of a few *micrograms* of corticotropin. Corticotropin acts on the adrenal cortex to cause the release of *milligrams* of cortisol, for an overall amplification of at least a millionfold.

At each level of a hormonal cascade, feedback inhibition of earlier steps in the cascade is possible; an unnecessarily elevated level of the ultimate hormone or of one of the intermediate hormones inhibits the release of earlier hormones in the cascade. These feedback mechanisms accomplish the same end as those that limit the output of a biosynthetic pathway (compare Fig. 23–11 with Fig. 6–28): a product is synthesized (or released) only until the necessary concentration is reached.

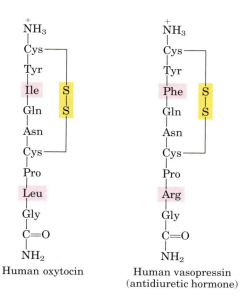

**FIGURE 23–10  Two hormones of the posterior pituitary gland.** The carboxyl-terminal residues are glycinamide, $-NH-CH_2-CONH_2$ (as noted in Fig. 23–2, amidation of the carboxyl terminus is common in short peptide hormones). These two hormones, identical in all but two residues (shaded), have very different biological effects. Oxytocin acts on the smooth muscles of the uterus and mammary gland, causing uterine contractions during labor and promoting milk release during lactation. Vasopressin (also called antidiuretic hormone) increases water reabsorption in the kidney and promotes the constriction of blood vessels, thereby increasing blood pressure.

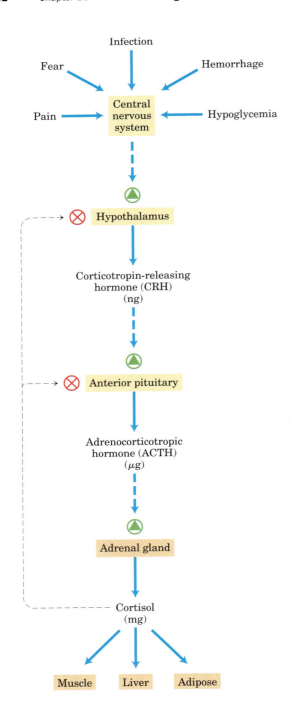

Fear

Infection

Hemorrhage

Pain → Central nervous system ← Hypoglycemia

⊗ Hypothalamus

Corticotropin-releasing hormone (CRH) (ng)

⊗ Anterior pituitary

Adrenocorticotropic hormone (ACTH) (μg)

Adrenal gland

Cortisol (mg)

Muscle    Liver    Adipose

**FIGURE 23–11 Cascade of hormone release following central nervous system input to the hypothalamus.** In each endocrine tissue along the pathway, a stimulus from the level above is received, amplified, and transduced into the release of the next hormone in the cascade. The cascade is sensitive to regulation at several levels through feedback inhibition by the ultimate hormone. The product therefore regulates its own production, as in feedback inhibition of biosynthetic pathways within a single cell.

- Hormonal cascades, in which catalysts activate catalysts, amplify the initial stimulus by several orders of magnitude, often in a very short time (seconds).

- Nerve impulses stimulate the hypothalamus to send specific hormones to the pituitary gland, thus stimulating (or inhibiting) the release of tropic hormones. The anterior pituitary hormones in turn stimulate other endocrine glands (thyroid, adrenals, pancreas) to secrete their characteristic hormones, which in turn stimulate specific target tissues.

- Peptide, amine, and eicosanoid hormones act outside the target cell on specific receptors in the plasma membrane, altering the level of an intracellular second messenger.

- Steroid, vitamin D, retinoid, and thyroid hormones enter target cells and alter gene expression by interacting with specific nuclear receptors.

## 23.2 Tissue-Specific Metabolism: The Division of Labor

Each tissue of the human body has a specialized function, reflected in its anatomy and metabolic activity (Fig. 23–12). Skeletal muscle allows directed motion; adipose tissue stores and releases energy in the form of fats, which serve as fuel throughout the body; the brain pumps ions across plasma membranes to produce electrical signals. The liver plays a central processing and distributing role in metabolism and furnishes all other organs and tissues with an appropriate mix of nutrients via the bloodstream. The functional centrality of the liver is indicated by the common reference to all other tissues and organs as "extrahepatic" or "peripheral." We therefore begin our discussion of the division of metabolic labor by considering the transformations of carbohydrates, amino acids, and fats in the mammalian liver. This is followed by brief descriptions of the primary metabolic functions of adipose tissue, muscle, brain, and the medium that interconnects all others: the blood.

## SUMMARY 23.1 Hormones: Diverse Structures for Diverse Functions

- Hormones are chemical messengers secreted by certain tissues into the blood or interstitial fluid, serving to regulate the activity of other cells or tissues.

- Radioimmunoassay (RIA) and ELISA are two very sensitive techniques for detecting and quantifying hormones.

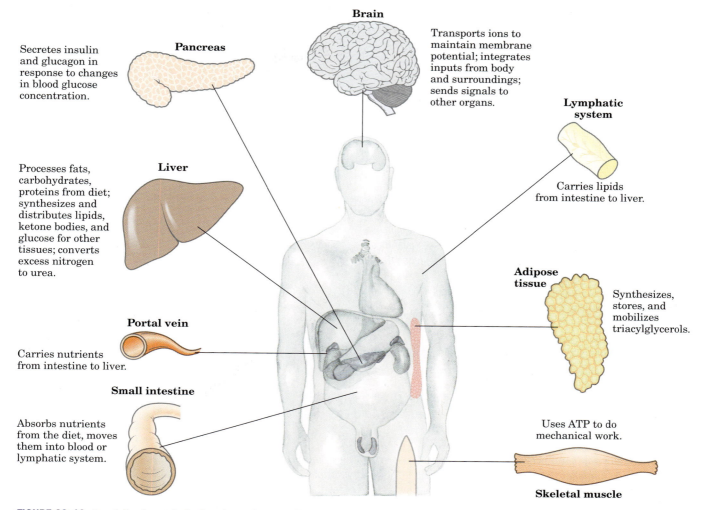

Secretes insulin and glucagon in response to changes in blood glucose concentration.

**Pancreas**

**Brain**

Transports ions to maintain membrane potential; integrates inputs from body and surroundings; sends signals to other organs.

**Lymphatic system**

Carries lipids from intestine to liver.

Processes fats, carbohydrates, proteins from diet; synthesizes and distributes lipids, ketone bodies, and glucose for other tissues; converts excess nitrogen to urea.

**Liver**

**Adipose tissue**

Synthesizes, stores, and mobilizes triacylglycerols.

**Portal vein**

Carries nutrients from intestine to liver.

**Small intestine**

Absorbs nutrients from the diet, moves them into blood or lymphatic system.

Uses ATP to do mechanical work.

**Skeletal muscle**

**FIGURE 23–12** Specialized metabolic functions of mammalian tissues.

## The Liver Processes and Distributes Nutrients

During digestion in mammals, the three main classes of nutrients (carbohydrates, proteins, and fats) undergo enzymatic hydrolysis into their simple constituents. This breakdown is necessary because the epithelial cells lining the intestinal lumen absorb only relatively small molecules. Many of the fatty acids and monoacylglycerols released by digestion of fats in the intestine are reassembled within these epithelial cells into triacylglycerols (TAGs).

After being absorbed, most sugars and amino acids and some TAGs travel in the bloodstream to the liver; the remaining TAGs enter adipose tissue via the lymphatic system. The portal vein is a direct route from the digestive organs to the liver, and liver therefore has first access to ingested nutrients. The liver has two main cell types. Kupffer cells are phagocytes, important in immune function. **Hepatocytes,** of primary interest here,

transform dietary nutrients into the fuels and precursors required by other tissues and export them via the blood. The kinds and amounts of nutrients supplied to the liver vary with several factors, including the diet and the time between meals. The demand of extrahepatic tissues for fuels and precursors varies among organs and with the level of activity and overall nutritional state of the individual.

To meet these changing circumstances, the liver has remarkable metabolic flexibility. For example, when the diet is rich in protein, hepatocytes supply themselves with high levels of enzymes for amino acid catabolism and gluconeogenesis. Within hours after a shift to a high-carbohydrate diet, the levels of these enzymes begin to drop and the hepatocytes increase their synthesis of enzymes essential to carbohydrate metabolism and fat synthesis. Liver enzymes turn over (are synthesized and degraded) at five to ten times the rate of enzyme turnover in other tissues, such as muscle. Extrahepatic

**TABLE 23–2 Pathways of Carbohydrate, Amino Acid, and Fat Metabolism Illustrated in Earlier Chapters**

| Pathway | Figure reference |
|---|---|
| *Citric acid cycle:* acetyl-CoA $\longrightarrow$ $2CO_2$ | 16–7 |
| *Oxidative phosphorylation:* ATP synthesis | 19–17 |
| **Carbohydrate catabolism** | |
| *Glycogenolysis:* glycogen $\longrightarrow$ glucose 1-phosphate $\longrightarrow$ blood glucose | 15–3; 15–4 |
| *Hexose entry into glycolysis:* fructose, mannose, galactose $\longrightarrow$ glucose 6-phosphate | 14–9 |
| *Glycolysis:* glucose $\longrightarrow$ pyruvate | 14–2 |
| *Pyruvate dehydrogenase reaction:* pyruvate $\longrightarrow$ acetyl-CoA | 16–2 |
| *Lactic acid fermentation:* glucose $\longrightarrow$ lactate + 2ATP | 14–3 |
| *Pentose phosphate pathway:* glucose 6-phosphate $\longrightarrow$ pentose phosphates + NADPH | 14–21 |
| **Carbohydrate anabolism** | |
| *Gluconeogenesis:* citric acid cycle intermediates $\longrightarrow$ glucose | 14–16 |
| *Glucose-alanine cycle:* glucose $\longrightarrow$ pyruvate $\longrightarrow$ alanine $\longrightarrow$ glucose | 18–9 |
| *Glycogen synthesis:* glucose 6-phosphate $\longrightarrow$ glucose 1-phosphate $\longrightarrow$ glycogen | 15–8 |
| **Amino acid and nucleotide metabolism** | |
| *Amino acid degradation:* amino acids $\longrightarrow$ acetyl-CoA, citric acid cycle intermediates | 18–15 |
| *Amino acid synthesis* | 22–9 |
| *Urea cycle:* $NH_3$ $\longrightarrow$ urea | 18–10 |
| *Glucose-alanine cycle:* alanine $\longrightarrow$ glucose | 18–9 |
| *Nucleotide synthesis:* amino acids $\longrightarrow$ purines, pyrimidines | 22–33; 22–36 |
| *Hormone and neurotransmitter synthesis* | 22–29 |
| **Fat catabolism** | |
| *β Oxidation of fatty acids:* fatty acid $\longrightarrow$ acetyl-CoA | 17–8 |
| *Oxidation of ketone bodies:* β-hydroxybutyrate $\longrightarrow$ acetyl-CoA $\longrightarrow$ $CO_2$ citric acid cycle | 17–19 |
| **Fat anabolism** | |
| *Fatty acid synthesis:* acetyl-CoA $\longrightarrow$ fatty acids | 21–5 |
| *Triacylglycerol synthesis:* acetyl-CoA $\longrightarrow$ fatty acids $\longrightarrow$ triacylglycerol | 21–18; 21–19 |
| *Ketone body formation:* acetyl-CoA $\longrightarrow$ acetoacetate, β-hydroxybutyrate | 17–18 |
| *Cholesterol and cholesteryl ester synthesis:* acetyl-CoA $\longrightarrow$ cholesterol $\longrightarrow$ cholesteryl esters | 21–33 to 21–37 |
| *Phospholipid synthesis:* fatty acids $\longrightarrow$ phospholipids | 21–17; 21–23 to 21–28 |

tissues also can adjust their metabolism to prevailing conditions, but none is as adaptable as the liver, and none is so central to the organism's overall metabolism. What follows is a survey of the possible fates of sugars, amino acids, and lipids that enter the liver from the bloodstream. To help you recall the metabolic transformations discussed here, Table 23–2 shows the major pathways and processes to which we refer and indicates by figure number where each pathway is presented in detail. Here, we present summaries of the pathways, referring to the step numbers in Figures 23–13 to 23–15.

***Sugars*** The glucose transporter in hepatocytes (GLUT2) is so effective that the concentration of glucose within a hepatocyte is essentially the same as that in the blood. Glucose entering hepatocytes is phosphorylated by hexokinase IV (glucokinase) to yield glucose 6-phosphate. Glucokinase has a much higher $K_m$ for glucose (10 mM) than do the hexokinase isozymes in other cells (p. 578) and, unlike these other isozymes, it is not inhibited by

its product, glucose 6-phosphate. The presence of glucokinase allows hepatocytes to continue phosphorylating glucose when the glucose concentration rises well above levels that would overwhelm other hexokinases. The high $K_m$ of glucokinase also ensures that the phosphorylation of glucose in hepatocytes is minimal when the glucose concentration is low, preventing the liver from consuming glucose as fuel via glycolysis. This spares glucose for other tissues. Fructose, galactose, and mannose, all absorbed from the small intestine, are also converted to glucose 6-phosphate by enzymatic pathways examined in Chapter 14. Glucose 6-phosphate is at the crossroads of carbohydrate metabolism in the liver. It may take any of several major metabolic routes (Fig. 23–13), depending on the current metabolic needs of the organism. By the action of various allosterically regulated enzymes, and through hormonal regulation of enzyme synthesis and activity, the liver directs the flow of glucose into one or more of these pathways.

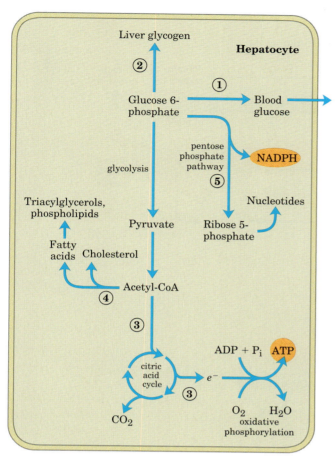

**FIGURE 23–13 Metabolic pathways for glucose 6-phosphate in the liver.** Here and in Figures 23–14 and 23–15, anabolic pathways are shown leading upward, catabolic pathways leading downward, and distribution to other organs horizontally. The numbered processes in each figure are described in the text.

① Glucose 6-phosphate is dephosphorylated by glucose 6-phosphatase to yield free glucose (p. 547), which is exported to replenish blood glucose. Export is the predominant pathway when glucose 6-phosphate is in limited supply, because the blood glucose concentration must be kept sufficiently high (4 mM) to provide adequate energy for the brain and other tissues. ② Glucose 6-phosphate not immediately needed to form blood glucose is converted to liver glycogen, or it has one of several other fates. Following glucose 6-phosphate breakdown by glycolysis and decarboxylation of the pyruvate (by the pyruvate dehydrogenase reaction), ③ the acetyl-CoA so formed can be oxidized for energy production by the citric acid cycle, with ensuing electron transfer and oxidative phosphorylation yielding ATP. (Normally, however, fatty acids are the preferred fuel for energy production in hepatocytes.) ④ Acetyl-CoA can also serve as the precursor of fatty acids, which are incorporated into TAGs and phospholipids, and cholesterol. Much of the lipid synthesized in the liver is

transported to other tissues by blood lipoproteins. ⑤ Finally, glucose 6-phosphate can enter the pentose phosphate pathway, yielding both reducing power (NADPH), needed for the biosynthesis of fatty acids and cholesterol, and D-ribose 5-phosphate, a precursor for nucleotide biosynthesis. NADPH is also an essential cofactor in the detoxification and elimination of many drugs and other xenobiotics metabolized in the liver.

**Amino Acids**    Amino acids that enter the liver follow several important metabolic routes (Fig. 23–14). ① They are precursors for protein synthesis, a process discussed in Chapter 27. The liver constantly renews its own proteins, which have a relatively high turnover rate (average half-life of only a few days), and is also the site of biosynthesis of most plasma proteins. ② Alternatively, amino acids pass in the bloodstream to other organs, to be used in the synthesis of tissue proteins. ③ Other amino acids are precursors in the biosynthesis of nucleotides, hormones, and other nitrogenous compounds in the liver and other tissues. ④a Amino acids not needed as biosynthetic precursors are transaminated or deaminated and degraded to yield pyruvate and citric acid cycle intermediates, with various fates; ④b the ammonia released is converted to the excretory product urea. ⑤ Pyruvate can be converted to glucose and glycogen via gluconeogenesis or ⑥ it can be converted to acetyl-CoA, which has several possible fates. ⑦ It can be oxidized via the citric acid cycle and ⑧ oxidative phosphorylation to produce ATP, or ⑨ converted to lipids for storage. ⑩ Citric acid cycle intermediates can be siphoned off into glucose synthesis by gluconeogenesis.

The liver also metabolizes amino acids that arrive intermittently from other tissues. The blood is adequately supplied with glucose just after the digestion and absorption of dietary carbohydrate or, between meals, by the conversion of liver glycogen to blood glucose. During the interval between meals, especially if prolonged, some muscle protein is degraded to amino acids. These amino acids donate their amino groups (by transamination) to pyruvate, the product of glycolysis, to yield alanine, which ⑪ is transported to the liver and deaminated. Hepatocytes convert the resulting pyruvate to blood glucose (via gluconeogenesis ⑤), and the ammonia to urea for excretion. One benefit of this glucose-alanine cycle (see Fig. 18–9) is the smoothing out of fluctuations in blood glucose between meals. The amino acid deficit incurred in the muscles is made up after the next meal by incoming dietary amino acids.

**Lipids**    The fatty acid components of the lipids entering hepatocytes also have several different fates (Fig. 23–15). ① Some are converted to liver lipids. ② Under most circumstances, fatty acids are the primary oxidative fuel in the liver. Free fatty acids may be activated

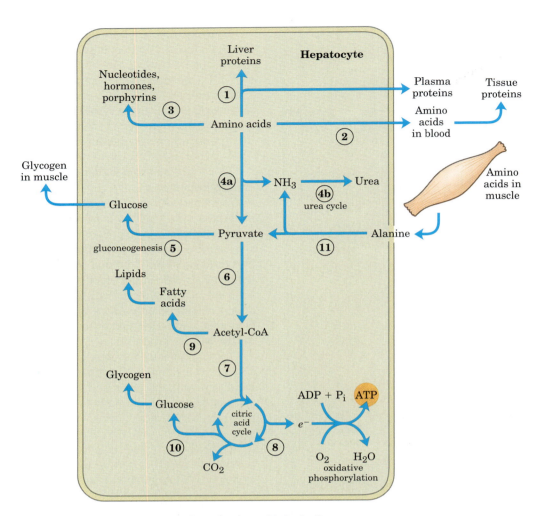

**FIGURE 23–14** Metabolism of amino acids in the liver.

and oxidized to yield acetyl-CoA and NADH. ③ and ④ The acetyl-CoA is further oxidized via the citric acid cycle, and the oxidations in the cycle drive the synthesis of ATP by oxidative phosphorylation. ⑤ Excess acetyl-CoA released by oxidation of fatty acids and not required by the liver is converted to the ketone bodies, acetoacetate and $\beta$-hydroxybutyrate; these circulate in the blood to other tissues, to be used as fuel for the citric acid cycle. Ketone bodies may be regarded as a transport form of acetyl groups. They can supply a significant fraction of the energy in some extrahepatic tissues—up to one-third in the heart; and as much as 60% to 70% in the brain during prolonged fasting. ⑥ Some of the acetyl-CoA derived from fatty acids (and from glucose) is used for the biosynthesis of cholesterol, which is required for membrane synthesis. Cholesterol is also the precursor of all steroid hormones and of the bile salts, which are essential for the digestion and absorption of lipids.

The final two metabolic fates of lipids involve specialized mechanisms for the transport of insoluble lipids in the blood. ⑦ Fatty acids are converted to the phospholipids and TAGs of plasma lipoproteins, which carry lipids to adipose tissue for storage as TAGs. ⑧ Some free fatty acids become bound to serum albumin and are carried to the heart and skeletal muscles, which absorb and oxidize free fatty acids as a major fuel. Serum albumin is the most abundant plasma protein; one molecule of serum albumin can carry up to 10 molecules of free fatty acid to the tissues where the fatty acids are released and consumed.

The liver thus serves as the body's distribution center, exporting nutrients in the correct proportions to other organs, smoothing out fluctuations in metabolism caused by intermittent food intake, and processing excess amino groups into urea and other products to be disposed of by the kidneys. Certain nutrients are stored in the liver, including Fe ions and vitamin A. The liver also detoxifies foreign organic compounds, such as drugs, food additives, preservatives, and other possibly harmful agents with no food value. Detoxification often involves the cytochrome P-450–dependent hydroxylation of relatively insoluble organic compounds, making them sufficiently soluble for further breakdown and excretion (see Box 21–1).

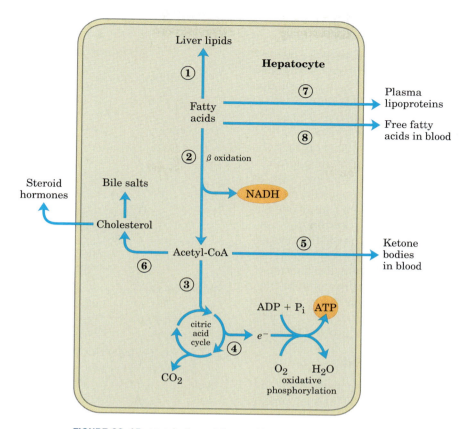

**FIGURE 23–15** Metabolism of fatty acids in the liver.

## Adipose Tissue Stores and Supplies Fatty Acids

Adipose tissue, which consists of **adipocytes** (fat cells) (Fig. 23–16), is amorphous and widely distributed in the body: under the skin, around the deep blood vessels, and in the abdominal cavity. It typically makes up about 15% of the mass of a young adult human, with approximately 65% of this mass in the form of triacylglycerols. Adipocytes are metabolically very active, responding quickly to hormonal stimuli in a metabolic interplay with the liver, skeletal muscles, and the heart.

Like other cell types, adipocytes have an active glycolytic metabolism, use the citric acid cycle to oxidize pyruvate and fatty acids, and carry out oxidative phosphorylation. During periods of high carbohydrate intake, adipose tissue can convert glucose (via pyruvate and acetyl-CoA) to fatty acids, convert the fatty acids to TAGs, and store them as large fat globules—although, in humans, much of the fatty acid synthesis occurs in hepatocytes. Adipocytes store TAGs arriving from the liver (carried in the blood as VLDLs; see Fig. 21–40a) and from the intestinal tract (carried in chylomicrons), particularly after meals rich in fat.

When fuel demand rises, lipases in adipocytes hydrolyze stored TAGs to release free fatty acids, which can travel in the bloodstream to skeletal muscles and the heart. The release of fatty acids from adipocytes is greatly accelerated by epinephrine, which stimulates the cAMP-dependent phosphorylation of perilipin; this gives triacylglycerol lipase access to TAGs in the lipid droplet. The hormone-sensitive lipase is also stimulated by phosphorylation, but this is not the main cause of increased lipolysis (see Fig. 17–3). Insulin counterbalances this effect of epinephrine, decreasing the activity of triacylglycerol lipase.

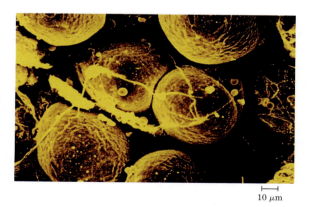

10 μm

**FIGURE 23–16** Scanning electron micrograph of human adipocytes. In fat tissues, capillaries and collagen fibers form a supporting network around spherical adipocytes. Almost the entire volume of these metabolically active cells is taken up by fat droplets.

The breakdown and synthesis of TAGs in adipose tissues constitute a substrate cycle; up to 70% of the fatty acids released by triacylglycerol lipase are reesterified in adipocytes, re-forming TAGs. Recall from Chapter 15 that such substrate cycles allow fine regulation of the rate and direction of flow of intermediates through a bidirectional pathway. In adipose tissue, glycerol liberated by triacylglycerol lipase cannot be reused in the synthesis of TAGs, because adipocytes lack the enzyme glycerol kinase. Instead, the glycerol phosphate required for TAG synthesis is made from pyruvate by glyceroneogenesis, involving the cytosolic enzyme PEP carboxykinase (see Fig. 21–22). This enzyme is one target of the drugs (thiazolidinediones) used in the treatment of type II diabetes, raising the possibility that defective regulation of cytosolic PEP carboxykinase in fat tissue may be a causative factor in type II diabetes.

Human infants, and many hibernating animals, have adipose tissue called brown fat, which is specialized to generate heat rather than ATP during oxidation of fatty acids. Adult humans have very little brown fat tissue.

## Muscles Use ATP for Mechanical Work

Metabolism in the cells of skeletal muscle—the **myocytes**—is specialized to generate ATP as the immediate source of energy for contraction. Moreover, skeletal muscle is adapted to do its mechanical work in an intermittent fashion, on demand. Sometimes skeletal muscles must work at their maximum capacity for a short time, as in a 100 m sprint; at other times more prolonged work is required, as in running a marathon or extended physical labor.

There are two general classes of muscle tissue, which differ in physiological role and fuel utilization. **Slow-twitch muscle,** also called red muscle, provides relatively low tension but is highly resistant to fatigue. It produces ATP by the relatively slow but steady process of oxidative phosphorylation. Red muscle is very rich in mitochondria and is served by very dense networks of blood vessels, which bring the oxygen essential to ATP production. It is the cytochromes in mitochondria and the hemoglobin in blood that give the tissue its characteristic red color. **Fast-twitch muscle,** or white muscle, has fewer mitochondria than red muscle and is less well supplied with blood vessels, but it can develop greater tension, and do so faster. White muscle is quicker to fatigue, because when active, it uses ATP faster than it can replace it. There is a genetic component to the proportion of red and white muscle in any individual; with training, the endurance of fast-twitch muscle can be improved.

Skeletal muscle can use free fatty acids, ketone bodies, or glucose as fuel, depending on the degree of muscular activity (Fig. 23–17). In resting muscle, the primary fuels are free fatty acids from adipose tissue and

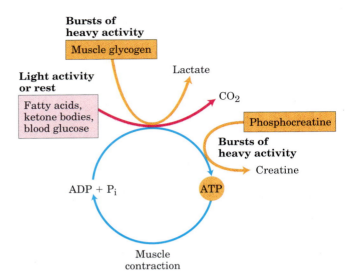

**FIGURE 23–17 Energy sources for muscle contraction.** Different fuels are used for ATP synthesis during bursts of heavy activity and during light activity or rest. Phosphocreatine can rapidly supply ATP.

ketone bodies from the liver. These are oxidized and degraded to yield acetyl-CoA, which enters the citric acid cycle for oxidation to $CO_2$. The ensuing transfer of electrons to $O_2$ provides the energy for ATP synthesis by oxidative phosphorylation. Moderately active muscle uses blood glucose in addition to fatty acids and ketone bodies. The glucose is phosphorylated, then degraded by glycolysis to pyruvate, which is converted to acetyl-CoA and oxidized via the citric acid cycle and oxidative phosphorylation.

In maximally active fast-twitch muscles, the demand for ATP is so great that the blood flow cannot provide $O_2$ and fuels fast enough to supply sufficient ATP by aerobic respiration alone. Under these conditions, stored muscle glycogen is broken down to lactate by fermentation (p. 523). Each glucose unit degraded yields three ATP, because phosphorolysis of glycogen produces glucose 6-phosphate, sparing the ATP normally consumed in the hexokinase reaction. Lactic acid fermentation thus responds to an increased need for ATP more quickly than does oxidative phosphorylation, supplementing basal ATP production that results from aerobic oxidation of other fuels via the citric acid cycle and respiratory chain. The use of blood glucose and muscle glycogen as fuels for muscular activity is greatly enhanced by the secretion of epinephrine, which stimulates both the release of glucose from liver glycogen and the breakdown of glycogen in muscle tissue.

The relatively small amount of glycogen in skeletal muscle (about 1% of its total weight) limits the amount of glycolytic energy available during all-out exertion. Moreover, the accumulation of lactate and consequent decrease in pH in maximally active muscles reduces their efficiency. Skeletal muscle, however, contains another

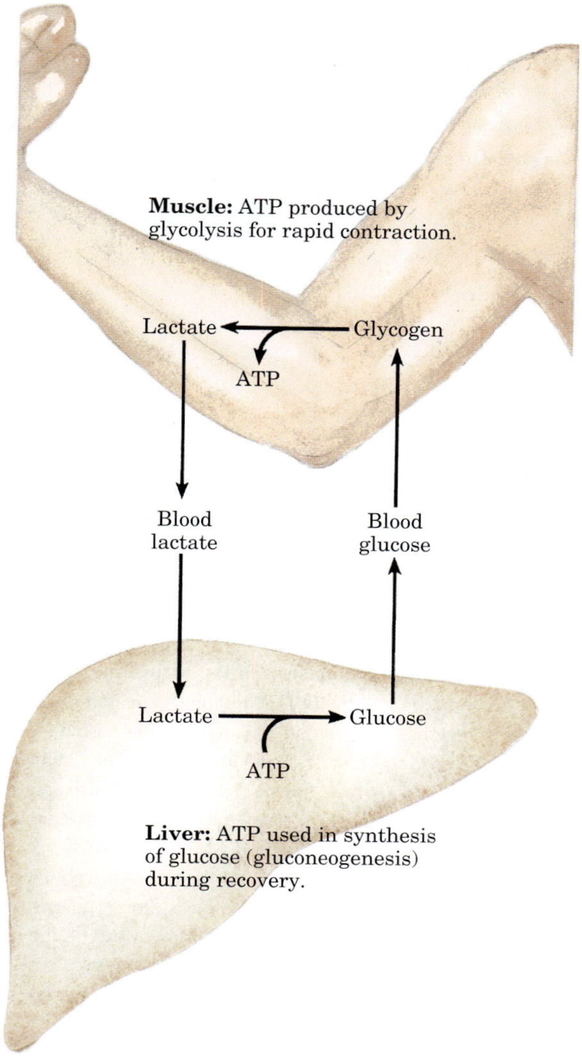

**Muscle:** ATP produced by glycolysis for rapid contraction.

Lactate ← Glycogen

ATP

Blood lactate

Blood glucose

Lactate → Glucose

ATP

**Liver:** ATP used in synthesis of glucose (gluconeogenesis) during recovery.

**FIGURE 23–18 Metabolic cooperation between skeletal muscle and the liver.** Extremely active muscles use glycogen as energy source, generating lactate via glycolysis. During recovery, some of this lactate is transported to the liver and converted to glucose via gluconeogenesis. This glucose is released to the blood and returned to the muscles to replenish their glycogen stores. The overall pathway (glucose → lactate → glucose) constitutes the Cori cycle.

source of ATP, in the form of phosphocreatine (10 to 30 mM), which can rapidly regenerate ATP from ADP by the creatine kinase reaction:

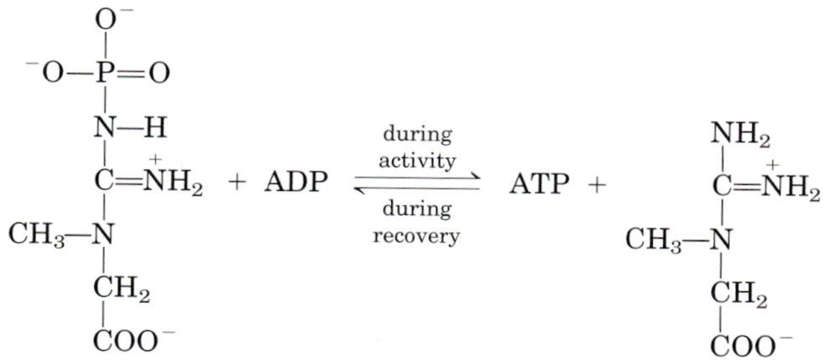

Phosphocreatine + ADP $\underset{\text{during recovery}}{\overset{\text{during activity}}{\rightleftharpoons}}$ ATP + Creatine

During periods of active contraction and glycolysis, this reaction proceeds predominantly in the direction of ATP synthesis; during recovery from exertion, the same enzyme resynthesizes phosphocreatine from creatine at the expense of ATP.

After a period of intense muscular activity, the individual continues breathing heavily for some time, using much of the extra $O_2$ for oxidative phosphorylation in the liver. The ATP produced is used for gluconeogenesis from lactate that has been carried in the blood from the muscles. The glucose thus formed returns to the muscles to replenish their glycogen, completing the Cori cycle (Fig. 23–18; see also Box 15–1).

Heart muscle differs from skeletal muscle in that it is continuously active in a regular rhythm of contraction and relaxation, and it has a completely aerobic metabolism at all times. Mitochondria are much more abundant in heart muscle than in skeletal muscle, making up almost half the volume of the cells (Fig. 23–19). The heart uses as its fuel mainly free fatty acids, but also some glucose and ketone bodies taken up from the blood; these fuels are oxidized via the citric acid cycle and oxidative phosphorylation to generate ATP. Like skeletal muscle, heart muscle does not store lipids or glycogen in large amounts. It does have small amounts of reserve energy in the form of phosphocreatine, enough for a few seconds of contraction. Because the heart is normally aerobic and obtains its energy from oxidative phosphorylation, the failure of $O_2$ to reach a portion of the heart muscle when the blood vessels are blocked by lipid deposits (atherosclerosis) or blood clots (coronary thrombosis) can cause that region of the heart muscle to die. This is what happens in myocardial infarction, more commonly known as a heart attack. ■

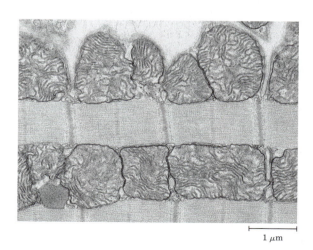

1 μm

**FIGURE 23–19 Electron micrograph of heart muscle.** In the profuse mitochondria of heart tissue, pyruvate, fatty acids, and ketone bodies are oxidized to drive ATP synthesis. This steady aerobic metabolism allows the human heart to pump blood at a rate of nearly 6 L/min, or about 350 L/hr—or $200 \times 10^6$ L over 70 years.

## The Brain Uses Energy for Transmission of Electrical Impulses

The metabolism of the brain is remarkable in several respects. The neurons of the adult mammalian brain normally use only glucose as fuel (Fig. 23–20). (Astrocytes, the other major cell type in the brain, can oxidize fatty acids.) The brain has a very active respiratory metabolism (Fig. 23–21); it uses $O_2$ at a fairly constant rate, accounting for almost 20% of the total $O_2$ consumed by the body at rest. Because the brain contains very little glycogen, it is constantly dependent on incoming glucose from the blood. Should blood glucose fall significantly below a critical level for even a short time, severe and sometimes irreversible changes in brain function may result.

Although the neurons of the brain cannot directly use free fatty acids or lipids from the blood as fuels, they can, when necessary, use $\beta$-hydroxybutyrate (a ketone body), which is formed from fatty acids in the liver. The capacity of the brain to oxidize $\beta$-hydroxybutyrate via acetyl-CoA becomes important during prolonged fasting or starvation, after liver glycogen has been depleted, because it allows the brain to use body fat as an energy source. This spares muscle proteins—until they become the brain's ultimate source of glucose (via gluconeogenesis in the liver) during severe starvation.

Neurons oxidize glucose by glycolysis and the citric acid cycle, and the flow of electrons from these oxidations through the respiratory chain provides almost all the ATP used by these cells. Energy is required to create and maintain an electrical potential across the neuronal plasma membrane. The membrane contains an electrogenic ATP-driven antiporter, the $Na^+K^+$ ATPase, which simultaneously pumps 2 $K^+$ ions into and 3 $Na^+$ ions out of the neuron (see Fig. 11–37). The resulting

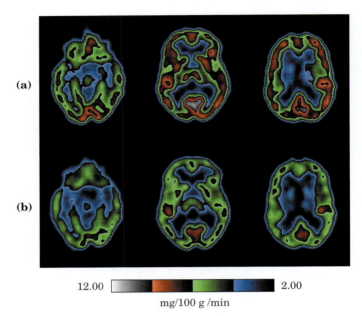

12.00 ▭▬▬▬▬▬▬ 2.00
mg/100 g /min

**FIGURE 23–21 Glucose metabolism in the brain.** The technique of positron emission tomography (PET) scanning shows metabolic activity in specific regions of the brain. PET scans allow visualization of isotopically labeled glucose in precisely localized regions of the brain of a living person, in real time. A positron-emitting glucose analog (2-[$^{18}$F]-fluoro-2-deoxy-D-glucose) is injected into the bloodstream; a few seconds later, a PET scan shows how much of the glucose has been taken up by each region of the brain—a measure of metabolic activity. Shown here are PET scans of front-to-back cross sections of the brain at three levels, from the top (at the left) downward (to the right). The scans compare glucose metabolism (in mg/100 g/min) when the experimental subject **(a)** is rested and **(b)** has been deprived of sleep for 48 hours.

transmembrane potential changes transiently as an electrical signal (action potential) sweeps from one end of a neuron to the other (see Fig. 12–5). Action potentials are the chief mechanism of information transfer in the nervous system, so a depletion of ATP in neurons has disastrous effects on all activities coordinated by neuronal signaling.

## Blood Carries Oxygen, Metabolites, and Hormones

Blood mediates the metabolic interactions among all tissues. It transports nutrients from the small intestine to the liver, and from the liver and adipose tissue to other organs; it also transports waste products from the tissues to the kidneys for excretion. Oxygen moves in the bloodstream from the lungs to the tissues, and $CO_2$ generated by tissue respiration returns via the bloodstream to the lungs for exhalation. Blood also carries hormonal signals from one tissue to another. In its role as signal carrier, the circulatory system resembles the nervous system; both regulate and integrate the activities of different organs.

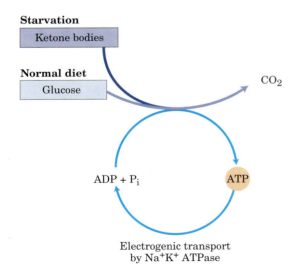

**FIGURE 23–20 Energy sources in the brain vary with nutritional state.** The ketone body used by the brain is $\beta$-hydroxybutyrate.

The average adult human has 5 to 6 L of blood. Almost half of this volume is occupied by three types of blood cells (Fig. 23–22): **erythrocytes** (red cells), filled with hemoglobin and specialized for carrying $O_2$ and $CO_2$; much smaller numbers of **leukocytes** (white cells) of several types (including **lymphocytes**, also found in lymphatic tissue), which are central to the immune system that defends against infections; and **platelets,** which help to mediate blood clotting. The liquid portion is the **blood plasma,** which is 90% water and 10% solutes. Dissolved or suspended in the plasma is a large variety of proteins, lipoproteins, nutrients, metabolites, waste products, inorganic ions, and hormones. More than 70% of the plasma solids are **plasma proteins** (Fig. 23–22), primarily immunoglobulins (circulating antibodies), serum albumin, apolipoproteins involved in the transport of lipids, transferrin (for iron transport), and blood-clotting proteins such as fibrinogen and prothrombin.

The ions and low molecular weight solutes in blood plasma are not fixed components but are in constant flux between blood and various tissues. Dietary uptake of the inorganic ions that are the predominant electrolytes of blood and cytosol ($Na^+$, $K^+$, and $Ca^{2+}$) is, in general, counterbalanced by their excretion in the urine. For many blood components, something near a dynamic steady state is achieved; the concentration of the component changes little, although a continuous flux occurs between the digestive tract, blood, and urine. The plasma levels of $Na^+$, $K^+$, and $Ca^{2+}$ remain close to 140, 5, and 2.5 mM, respectively, with little change in response to dietary intake. Any significant departure from these values can result in serious illness or death. The kidneys play an especially important role in maintaining ion balance by selectively filtering waste products and excess ions out of the blood while preventing the loss of essential nutrients and ions.

The concentration of glucose in the plasma is also subject to tight regulation. We have noted the constant requirement of the brain for glucose and the role of the liver in maintaining blood glucose in the normal range of 60 to 90 mg/100 mL. When blood glucose in a human drops to 40 mg/100 mL (the hypoglycemic condition), the person experiences discomfort and mental confusion (Fig. 23–23); further reductions lead to coma, convulsions, and in extreme hypoglycemia, death.

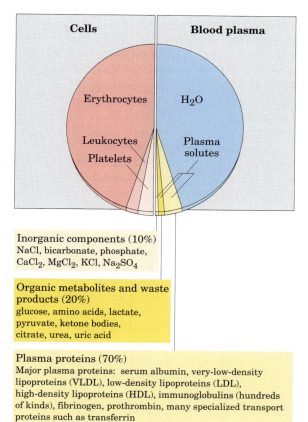

**FIGURE 23–22  The composition of blood.** Whole blood can be separated into blood plasma and cells by centrifugation. About 10% of blood plasma is solutes, of which about 10% consists of inorganic salts, 20% small organic molecules, and 70% plasma proteins. The major dissolved components are listed. Blood contains many other substances, often in trace amounts. These include other metabolites, enzymes, hormones, vitamins, trace elements, and bile pigments. Measurements of the concentrations of components in blood plasma are important in the diagnosis and treatment of many diseases.

**Cells**    **Blood plasma**

Erythrocytes    $H_2O$

Leukocytes
Platelets    Plasma solutes

Inorganic components (10%)
NaCl, bicarbonate, phosphate, $CaCl_2$, $MgCl_2$, KCl, $Na_2SO_4$

Organic metabolites and waste products (20%)
glucose, amino acids, lactate, pyruvate, ketone bodies, citrate, urea, uric acid

Plasma proteins (70%)
Major plasma proteins: serum albumin, very-low-density lipoproteins (VLDL), low-density lipoproteins (LDL), high-density lipoproteins (HDL), immunoglobulins (hundreds of kinds), fibrinogen, prothrombin, many specialized transport proteins such as transferrin

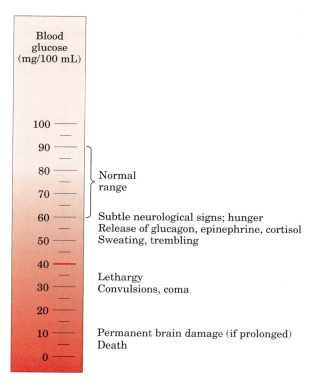

Blood
glucose
(mg/100 mL)

100
90
80          Normal
            range
70

60          Subtle neurological signs; hunger
            Release of glucagon, epinephrine, cortisol
50          Sweating, trembling

40

30          Lethargy
            Convulsions, coma

20

10          Permanent brain damage (if prolonged)
            Death
0

**FIGURE 23–23  Physiological effects of low blood glucose in humans.** Blood glucose levels of 40 mg/100 mL and below constitute severe hypoglycemia.

Maintaining the normal concentration of glucose in the blood is therefore a very high priority of the organism, and a variety of regulatory mechanisms have evolved to achieve that end. Among the most important regulators of blood glucose are the hormones insulin, glucagon, and epinephrine, as discussed in Section 23.3. ■

## SUMMARY 23.2 Tissue-Specific Metabolism: The Division of Labor

- In mammals there is a division of metabolic labor among specialized tissues and organs. The liver is the central distributing and processing organ for nutrients. Sugars and amino acids produced in digestion cross the intestinal epithelium and enter the blood, which carries them to the liver. Some triacylglycerols derived from ingested lipids also make their way to the liver, where the constituent fatty acids are used in a variety of processes.

- Glucose 6-phosphate is the key intermediate in carbohydrate metabolism. It may be polymerized into glycogen, dephosphorylated to blood glucose, or converted to fatty acids via acetyl-CoA. It may undergo oxidation by glycolysis, the citric acid cycle, and respiratory chain to yield ATP, or enter the pentose phosphate pathway to yield pentoses and NADPH.

- Amino acids are used to synthesize liver and plasma proteins, or their carbon skeletons are converted to glucose and glycogen by gluconeogenesis; the ammonia formed by deamination is converted to urea.

- The liver converts fatty acids to triacylglycerols, phospholipids, or cholesterol and its esters, for transport as plasma lipoproteins to adipose tissue for storage. Fatty acids can also be oxidized to yield ATP or to form ketone bodies, which are circulated to other tissues.

- Skeletal muscle is specialized to produce and use ATP for mechanical work. During strenuous muscular activity, glycogen is the ultimate fuel, supplying ATP through lactic acid fermentation. During recovery, the lactate is reconverted (through gluconeogenesis) to glycogen and glucose in the liver. Phosphocreatine is an immediate source of ATP during active contraction.

- Heart muscle obtains nearly all its ATP from oxidative phosphorylation.

- The neurons of the brain use only glucose and β-hydroxybutyrate as fuels, the latter being important during fasting or starvation. The brain uses most of its ATP for the active transport of $Na^+$ and $K^+$ and maintenance of the electrical potential across the neuronal membrane.

- The blood carries nutrients, waste products, and hormonal signals among the organs.

# 23.3 Hormonal Regulation of Fuel Metabolism

The minute-by-minute adjustments that keep the blood glucose level near 4.5 mM involve the combined actions of insulin, glucagon, epinephrine, and cortisol on metabolic processes in many body tissues, but especially in liver, muscle, and adipose tissue. Insulin signals these tissues that blood glucose is higher than necessary; as a result, cells take up excess glucose from the blood and convert it to the storage compounds glycogen and triacylglycerol. Glucagon signals that blood glucose is too low, and tissues respond by producing glucose through glycogen breakdown and (in liver) gluconeogenesis and by oxidizing fats to reduce the use of glucose. Epinephrine is released into the blood to prepare the muscles, lungs, and heart for a burst of activity. Cortisol mediates the body's response to longer-term stresses. We discuss these hormonal regulations in the context of three normal metabolic states—well-fed, fasted, and starving—and look at the metabolic consequences of diabetes mellitus, which results from derangements in the signaling pathways that control glucose metabolism.

## The Pancreas Secretes Insulin or Glucagon in Response to Changes in Blood Glucose

When glucose enters the bloodstream from the intestine after a carbohydrate-rich meal, the resulting increase in blood glucose causes increased secretion of **insulin** (and decreased secretion of glucagon). Insulin release by the pancreas is largely regulated by the level of glucose in the blood supplied to the pancreas. The peptide hormones insulin, glucagon, and somatostatin are produced by clusters of specialized pancreatic cells, the islets of Langerhans (Fig. 23–24). Each cell type of the islets produces a single hormone: α cells produce glucagon; β cells, insulin; and δ cells, somatostatin.

When blood glucose rises, GLUT2 transporters carry glucose into the β cells, where it is immediately converted to glucose 6-phosphate by hexokinase IV (glucokinase) and enters glycolysis (Fig. 23–25). The increased rate of glucose catabolism raises [ATP], causing the closing of ATP-gated $K^+$ channels in the plasma membrane. Reduced efflux of $K^+$ depolarizes the membrane, thereby opening voltage-sensitive $Ca^{2+}$ channels in the plasma membrane. The resulting influx of $Ca^{2+}$

triggers the release of insulin by exocytosis. Stimuli from the parasympathetic and sympathetic nervous systems also stimulate and inhibit insulin release, respectively. A simple feedback loop limits hormone release: insulin lowers blood glucose by stimulating glucose uptake by the tissues; the reduced blood glucose is detected by the β cell as a diminished flux through the hexokinase reaction; this slows or stops the release of insulin. This feedback regulation holds blood glucose concentration nearly constant despite large fluctuations in dietary intake.

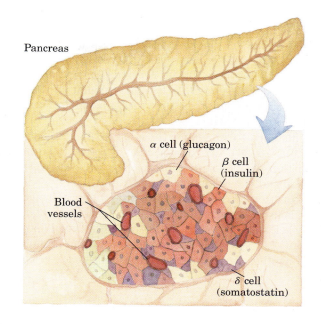

**FIGURE 23–24** **The endocrine system of the pancreas.** In addition to the exocrine cells (see Fig. 18–3b), which secrete digestive enzymes in the form of zymogens, the pancreas contains endocrine tissue, the islets of Langerhans. The islets contain α, β, and δ cells (also known as A, B, and D cells, respectively), each cell type secreting a specific polypeptide hormone.

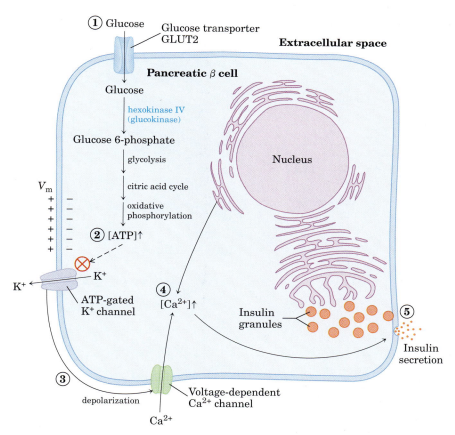

**FIGURE 23–25** **Glucose regulation of insulin secretion by pancreatic β cells.** When the blood glucose level is high, active metabolism of glucose in the β cell raises intracellular [ATP], which leads to closing of $K^+$ channels in the plasma membrane, depolarizing the membrane. In response to the change in membrane potential, voltage-gated $Ca^{2+}$ channels in the plasma membrane open, allowing $Ca^{2+}$ to flow into the cell; this raises the cytosolic $[Ca^{2+}]$ enough to trigger insulin release by exocytosis.

## Insulin Counters High Blood Glucose

Insulin stimulates glucose uptake by muscle and adipose tissue (Table 23–3), where the glucose is converted to glucose 6-phosphate. In the liver, insulin also activates glycogen synthase and inactivates glycogen phosphorylase, so that much of the glucose 6-phosphate is channeled into glycogen.

Insulin also stimulates the storage of excess fuel as fat (Fig. 23–26). In the liver, insulin activates both the oxidation of glucose 6-phosphate to pyruvate via glycolysis and the oxidation of pyruvate to acetyl-CoA. If not oxidized further for energy production, this acetyl-CoA is used for fatty acid synthesis in the liver, and the fatty acids are exported as the TAGs of plasma lipoproteins (VLDLs) to the adipose tissue. Insulin stimulates TAG synthesis in adipocytes, from fatty acids released

from the VLDL triacylglycerols. These fatty acids are ultimately derived from the excess glucose taken up from the blood by the liver. In summary, the effect of insulin is to favor the conversion of excess blood glucose to two storage forms: glycogen (in the liver and muscle) and triacylglycerols (in adipose tissue) (Table 23–3).

## Glucagon Counters Low Blood Glucose

Several hours after the intake of dietary carbohydrate, blood glucose levels fall slightly because of the ongoing oxidation of glucose by the brain and other tissues. Lowered blood glucose triggers secretion of glucagon and decreases insulin release (Fig. 23–27).

Glucagon causes an increase in blood glucose concentration in several ways (Table 23–4). Like epinephrine, it stimulates the net breakdown of liver glycogen

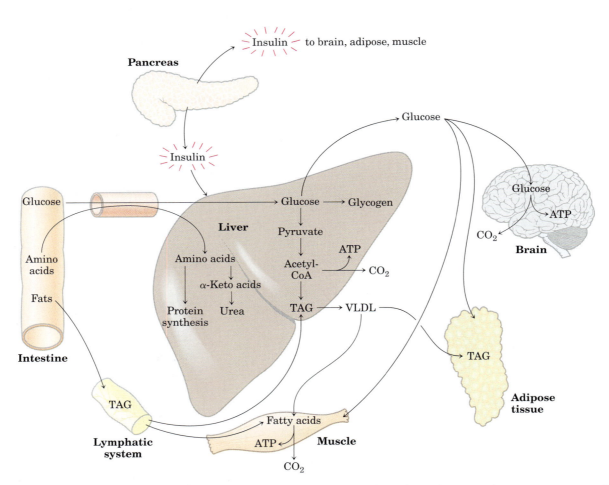

**FIGURE 23-26 The well-fed state: the lipogenic liver.** Immediately after a calorie-rich meal, glucose, fatty acids, and amino acids enter the liver. Insulin released in response to the high blood glucose concentration stimulates glucose uptake by the tissues. Some glucose is exported to the brain for its energetic needs, and some to fat and muscle tissue. In the liver, excess glucose is oxidized to acetyl-CoA, which is used to synthesize fatty acids for export as triacylglycerols in

VLDLs to fat and muscle tissue. The NADPH necessary for lipid synthesis is obtained by oxidation of glucose in the pentose phosphate pathway. Excess amino acids are converted to pyruvate and acetyl-CoA, which are also used for lipid synthesis. Dietary fats move via the lymphatic system, as chylomicrons, from the intestine to muscle and fat tissues.

**TABLE 23–3  Effects of Insulin on Blood Glucose: Uptake of Glucose by Cells and Storage as Triacylglycerols and Glycogen**

| Metabolic effect | Target enzyme |
|---|---|
| ↑ Glucose uptake (muscle, adipose) | ↑ Glucose transporter (GLUT4) |
| ↑ Glucose uptake (liver) | ↑ Glucokinase (increased expression) |
| ↑ Glycogen synthesis (liver, muscle) | ↑ Glycogen synthase |
| ↓ Glycogen breakdown (liver, muscle) | ↓ Glycogen phosphorylase |
| ↑ Glycolysis, acetyl-CoA production (liver, muscle) | ↑ PFK-1 (by ↑ PFK-2) |
|  | ↑ Pyruvate dehydrogenase complex |
| ↑ Fatty acid synthesis (liver) | ↑ Acetyl-CoA carboxylase |
| ↑ Triacylglycerol synthesis (adipose tissue) | ↑ Lipoprotein lipase |

by activating glycogen phosphorylase and inactivating glycogen synthase; both effects are the result of phosphorylation of the regulated enzymes, triggered by cAMP. Glucagon inhibits glucose breakdown by glycolysis in the liver and stimulates glucose synthesis by gluconeogenesis. Both effects result from lowering the concentration of fructose 2,6-bisphosphate, an allosteric inhibitor of the gluconeogenic enzyme fructose 1,6-bisphosphatase (FBPase-1) and an activator of phosphofructokinase-1. Recall that [fructose 2,6-bisphosphate] is ultimately controlled by a cAMP-dependent protein phosphorylation reaction (see Fig. 15–23). Glucagon also inhibits the glycolytic enzyme pyruvate kinase (by promoting its cAMP-dependent phosphorylation), thus blocking the conversion of phosphoenolpyruvate to pyruvate and preventing oxidation of pyruvate via the citric acid cycle. The resulting accumulation of phosphoenolpyruvate favors gluconeogenesis. This effect is augmented by glucagon's stimulation of the synthesis of the gluconeogenic enzyme PEP carboxykinase. By stimulating glycogen breakdown, preventing glycolysis, and promoting gluconeogenesis in hepatocytes, glucagon

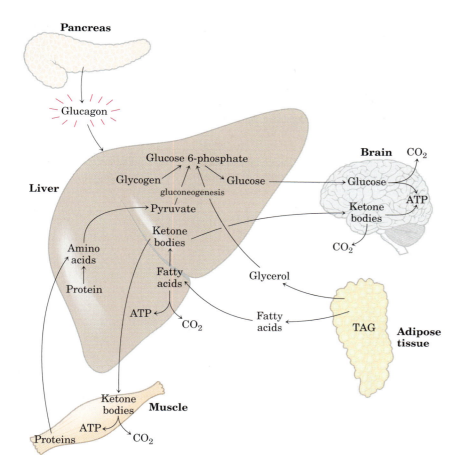

**FIGURE 23–27  The fasting state: the glucogenic liver.** After some hours without a meal, the liver becomes the principal source of glucose for the brain. Liver glycogen is broken down, and the glucose 1-phosphate produced is converted to glucose 6-phosphate, then to free glucose, which is released into the bloodstream. Amino acids from the degradation of proteins and glycerol from the breakdown of TAGs in adipose tissue are used for gluconeogenesis. The liver uses fatty acids as its principal fuel, and excess acetyl-CoA is converted to ketone bodies for export to other tissues for fuel; the brain is especially dependent on this fuel when glucose is in short supply.

**TABLE 23-4    Effects of Glucagon on Blood Glucose: Production and Release of Glucose by the Liver**

| Metabolic effect | Effect on glucose metabolism | Target enzyme |
|---|---|---|
| ↑ Glycogen breakdown (liver) | Glycogen ⟶ glucose | ↑ Glycogen phosphorylase |
| ↓ Glycogen synthesis (liver) | Less glucose stored as glycogen | ↓ Glycogen synthase |
| ↓ Glycolysis (liver) | Less glucose used as fuel in liver | ↓ PFK-1 |
| ↑ Gluconeogenesis (liver) | Amino acids ⎫<br>Glycerol  ⎬ ⟶ glucose<br>Oxaloacetate ⎭ | ↑ FBPase-2<br>↓ Pyruvate kinase<br>↑ PEP carboxykinase |
| ↑ Fatty acid mobilization (adipose tissue) | Less glucose used as fuel by liver, muscle | ↑ Triacylglycerol lipase<br>    Perilipin phosphorylation |
| ↑ Ketogenesis | Provides alternative to glucose as<br>    energy source for brain | ↑ Acetyl-CoA carboxylase |

enables the liver to export glucose, restoring blood glucose to its normal level.

Although its primary target is the liver, glucagon (like epinephrine) also affects adipose tissue, activating TAG breakdown by causing cAMP-dependent phosphorylation of perilipin and triacylglycerol lipase. The activated lipase liberates free fatty acids, which are exported to the liver and other tissues as fuel, sparing glucose for the brain. The net effect of glucagon is therefore to stimulate glucose synthesis and release by the liver and to mobilize fatty acids from adipose tissue, to be used instead of glucose as fuel for tissues other than the brain (Table 23–4). All these effects of glucagon are mediated by cAMP-dependent protein phosphorylation.

## During Fasting and Starvation, Metabolism Shifts to Provide Fuel for the Brain

The fuel reserves of a healthy adult human are of three types: glycogen stored in the liver and, in relatively small quantities, in muscles; large quantities of triacylglycerols in adipose tissues; and tissue proteins, which can be degraded when necessary to provide fuel (Table 23–5).

In the first few hours after a meal, the blood glucose level is diminished slightly, and tissues receive glucose released from liver glycogen. There is little or no synthesis of lipids. By 24 hours after a meal, blood glucose has fallen further, insulin secretion has slowed, and glucagon secretion has increased. These hormonal signals

**TABLE 23-5    Available Metabolic Fuels in a Normal-Weight 70 kg Man and in an Obese 140 kg Man at the Beginning of a Fast**

| Type of fuel | Weight (kg) | Caloric equivalent (thousands of kcal (kJ)) | Estimated survival (months)[*] |
|---|---|---|---|
| **Normal-weight, 70 kg man** | | | |
| Triacylglycerols (adipose tissue) | 15 | 141 (589) | |
| Proteins (mainly muscle) | 6 | 24 (100) | |
| Glycogen (muscle, liver) | 0.225 | 0.90 (3.8) | |
| Circulating fuels (glucose, fatty acids, triacylglycerols, etc.) | 0.023 | 0.10 (0.42) | |
| Total | | 166 (694) | 3 |
| **Obese, 140 kg man** | | | |
| Triacylglycerols (adipose tissue) | 80 | 752 (3,140) | |
| Proteins (mainly muscle) | 8 | 32 (134) | |
| Glycogen (muscle, liver) | 0.23 | 0.92 (3.8) | |
| Circulating fuels | 0.025 | 0.11 (0.46) | |
| Total | | 785 (3,280) | 14 |

[*]Survival time is calculated on the assumption of a basal energy expenditure of 1,800 kcal/day.

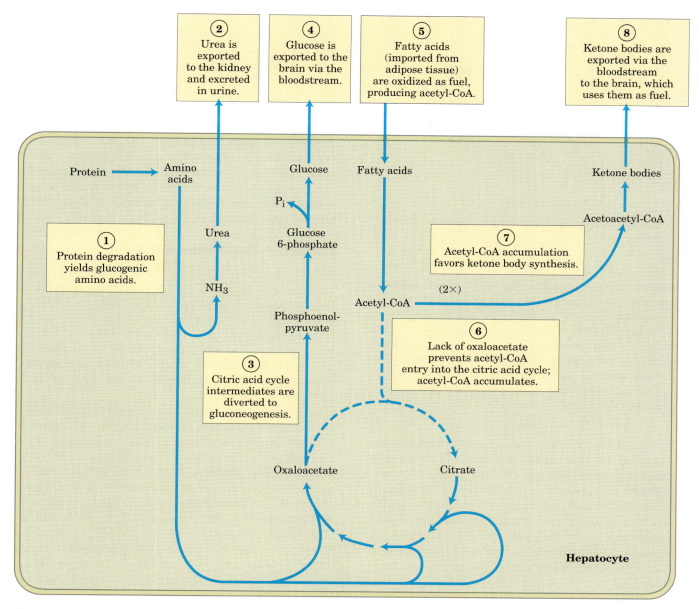

**FIGURE 23–28 Fuel metabolism in the liver during prolonged fasting or in uncontrolled diabetes mellitus.** After depletion of stored carbohydrates, ① to ④ proteins become an important source of glucose, produced from glucogenic amino acids by gluconeogenesis. ⑤ mobilize triacylglycerols, which now become the primary fuel for muscle and liver. Figure 23–28 shows the responses to prolonged fasting. ① To provide glucose for the brain, the liver degrades certain proteins—those most expendable in an organism not ingesting food. Their nonessential amino acids are transaminated or deaminated (Chapter 18), and ② the extra amino groups are converted to urea, which is exported via the bloodstream to the kidney and excreted.

Also in the liver, ③ the carbon skeletons of glucogenic amino acids are converted to pyruvate or inter-

to ⑧ Fatty acids imported from adipose tissue are converted to ketone bodies for export to the brain. Broken arrows represent reactions with reduced flux under these conditions. The steps are further described in the text.

mediates of the citric acid cycle. ④ These intermediates, as well as the glycerol ⑤ derived from triacylglycerols in adipose tissue, provide the starting materials for gluconeogenesis in the liver, yielding glucose for the brain. Eventually the use of citric acid cycle intermediates for gluconeogenesis depletes oxaloacetate, inhibiting entry of acetyl-CoA into the citric acid cycle. ⑥ Acetyl-CoA produced by fatty acid oxidation now accumulates, favoring ⑦ the formation of acetoacetyl-CoA and ketone bodies in the liver. After a few days of fasting, the levels of ketone bodies in the blood rise

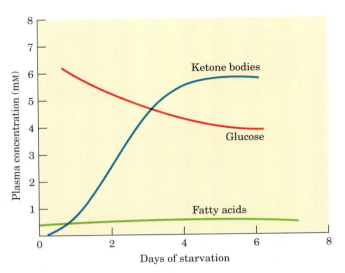

**FIGURE 23–29 Concentrations of fatty acids, glucose, and ketone bodies in the plasma during the first week of starvation.** Despite the hormonal mechanisms for maintaining the level of glucose in the blood, it begins to diminish after two days of fasting. The level of ketone bodies, almost immeasurable before the fast, rises dramatically after 2 to 4 days of fasting. These water-soluble ketones, acetoacetate and β-hydroxybutyrate, supplement glucose as an energy source during a long fast. Fatty acids cannot serve as a fuel for the brain; they do not cross the blood-brain barrier.

Triacylglycerols stored in the adipose tissue of a normal-weight adult could provide enough fuel to maintain a basal rate of metabolism for about three months; a very obese adult has enough stored fuel to endure a fast of more than a year (Table 23–5). When fat reserves are gone, the degradation of essential proteins begins; this leads to loss of heart and liver function, and eventually death. Stored fat can provide adequate energy (calories) during a fast or rigid diet, but vitamins and minerals must be provided, and sufficient dietary glucogenic amino acids are needed to replace those being used for gluconeogenesis. Rations for those on a weight-reduction diet are therefore commonly fortified with vitamins, minerals, and amino acids or proteins.

### Epinephrine Signals Impending Activity

When an animal is confronted with a stressful situation that requires increased activity—fighting or fleeing, in the extreme case—neuronal signals from the brain trigger the release of epinephrine and norepinephrine from the adrenal medulla. Both hormones dilate the respiratory passages to facilitate the uptake of $O_2$, increase the rate and strength of the heartbeat, and raise the blood pressure, thereby promoting the flow of $O_2$ and fuels to the tissues (Table 23–6).

Epinephrine acts primarily on muscle, adipose, and liver tissues. It activates glycogen phosphorylase and inactivates glycogen synthase by cAMP-dependent phosphorylation of the enzymes, thus stimulating the conversion of liver glycogen to blood glucose, the fuel for anaerobic muscular work. Epinephrine also promotes the anaerobic breakdown of muscle glycogen by lactic acid fermentation, stimulating glycolytic ATP formation. The stimulation of glycolysis is accomplished by raising the concentration of fructose 2,6-bisphosphate, a potent allosteric activator of the key glycolytic enzyme phosphofructokinase-1 (see Figs 15–22, 15–23). Epinephrine

(Fig. 23–29) as these fuels are exported from the liver to the heart, skeletal muscle, and brain, which use them instead of glucose (⑧).

Acetyl-CoA is a critical regulator of the fate of pyruvate; it allosterically inhibits pyruvate dehydrogenase and stimulates pyruvate carboxylase (see Fig. 15–20). In these ways acetyl-CoA prevents it own further production from pyruvate while stimulating the conversion of pyruvate to oxaloacetate, the first step in gluconeogenesis.

**TABLE 23–6    Physiological and Metabolic Effects of Epinephrine: Preparation for Action**

| Immediate effect | Overall effect |
|---|---|
| **Physiological** | |
| ↑ Heart rate | |
| ↑ Blood pressure | Increase delivery of $O_2$ to tissues (muscle) |
| ↑ Dilation of respiratory passages | |
| **Metabolic** | |
| ↑ Glycogen breakdown (muscle, liver) | |
| ↓ Glycogen synthesis (muscle, liver) | Increase production of glucose for fuel |
| ↑ Gluconeogenesis (liver) | |
| ↑ Glycolysis (muscle) | Increases ATP production in muscle |
| ↑ Fatty acid mobilization (adipose tissue) | Increases availability of fatty acids as fuel |
| ↑ Glucagon secretion | Reinforce metabolic effects of epinephrine |
| ↓ Insulin secretion | |

also stimulates fat mobilization in adipose tissue, activating (by cAMP-dependent phosphorylation) both perilipin and triacylglycerol lipase (see Fig. 17–3). Finally, epinephrine stimulates glucagon secretion and inhibits insulin secretion, reinforcing its effect of mobilizing fuels and inhibiting fuel storage.

## Cortisol Signals Stress, Including Low Blood Glucose

A variety of stressors (anxiety, fear, pain, hemorrhage, infections, low blood glucose, starvation) stimulate release of the corticosteroid hormone **cortisol** from the adrenal cortex. Cortisol acts on muscle, liver, and adipose tissue to supply the organism with fuel to withstand the stress. Cortisol is a relatively slow-acting hormone that alters metabolism by changing the kinds and amounts of certain enzymes synthesized in its target cell, rather than by regulating the activity of existing enzyme molecules.

In adipose tissue, cortisol leads to an increase in the release of fatty acids from stored TAGs. The fatty acids are exported to serve as fuel for other tissues, and the glycerol is used for gluconeogenesis in the liver. Cortisol stimulates the breakdown of muscle proteins and the export of amino acids to the liver, where they serve as precursors for gluconeogenesis. In the liver, cortisol promotes gluconeogenesis by stimulating synthesis of the key enzyme PEP carboxykinase (see Fig. 14–17b); glucagon has the same effect, whereas insulin has the opposite effect. Glucose produced in this way is stored in the liver as glycogen or exported immediately to tissues that need glucose for fuel. The net effect of these metabolic changes is to restore blood glucose to its normal level and to increase glycogen stores, ready to support the fight-or-flight response commonly associated with stress. The effects of cortisol therefore counterbalance those of insulin.

## Diabetes Mellitus Arises from Defects in Insulin Production or Action

**Diabetes mellitus,** caused by a deficiency in the secretion or action of insulin, is a relatively common disease: nearly 6% of the United States population shows some degree of abnormality in glucose metabolism that is indicative of diabetes or a tendency toward the condition. There are two major clinical classes of diabetes mellitus: **type I diabetes,** or insulin-dependent diabetes mellitus (IDDM), and **type II diabetes,** or non-insulin-dependent diabetes mellitus (NIDDM), also called insulin-resistant diabetes.

In type I diabetes, the disease begins early in life and quickly becomes severe. This disease responds to insulin injection, because the metabolic defect stems from a paucity of pancreatic $\beta$ cells and a consequent inability to produce sufficient insulin. IDDM requires insulin therapy and careful, lifelong control of the balance between dietary intake and insulin dose. Characteristic symptoms of type I (and type II) diabetes are excessive thirst and frequent urination (polyuria), leading to the intake of large volumes of water (polydipsia) ("diabetes mellitus" means "excessive excretion of sweet urine"). These symptoms are due to the excretion of large amounts of glucose in the urine, a condition known as **glucosuria.**

Type II diabetes is slow to develop (typically in older, obese individuals), and the symptoms are milder and often go unrecognized at first. This is really a group of diseases in which the regulatory activity of insulin is defective: insulin is produced, but some feature of the insulin-response system is defective. These individuals are insulin-resistant. The connection between type II diabetes and obesity (discussed below) is an active area of research.

Individuals with either type of diabetes are unable to take up glucose efficiently from the blood; recall that insulin triggers the movement of GLUT4 glucose transporters to the plasma membrane of muscle and adipose tissue (see Fig. 12–8). Another characteristic metabolic change in diabetes is excessive but incomplete oxidation of fatty acids in the liver. The acetyl-CoA produced by $\beta$ oxidation cannot be completely oxidized by the citric acid cycle, because the high [NADH]/[NAD$^+$] ratio produced by $\beta$ oxidation inhibits the cycle (recall that three steps convert NAD$^+$ to NADH). Accumulation of acetyl-CoA leads to overproduction of the ketone bodies acetoacetate and $\beta$-hydroxybutyrate, which cannot be used by extrahepatic tissues as fast as they are made in the liver. In addition to $\beta$-hydroxybutyrate and acetoacetate, the blood of diabetics also contains acetone, which results from the spontaneous decarboxylation of acetoacetate:

$$CH_3-\overset{\overset{\displaystyle O}{\|}}{C}-CH_2-COO^- + H_2O \longrightarrow CH_3-\overset{\overset{\displaystyle O}{\|}}{C}-CH_3 + HCO_3^-$$

Acetoacetate                Acetone

Acetone is volatile and is exhaled, and in uncontrolled diabetes, the breath has a characteristic odor sometimes mistaken for ethanol. A diabetic individual who is experiencing mental confusion due to high blood glucose is occasionally misdiagnosed as intoxicated, an error that can be fatal. The overproduction of ketone bodies, called **ketosis,** results in greatly increased concentrations of ketone bodies in the blood (ketonemia) and urine (ketonuria).

The ketone bodies are carboxylic acids, which ionize, releasing protons. In uncontrolled diabetes this acid production can overwhelm the capacity of the blood's bicarbonate buffering system and produce a lowering of blood pH called **acidosis** or, in combination with ketosis, **ketoacidosis,** a potentially life-threatening condition.

Biochemical measurements on blood and urine samples are essential in the diagnosis and treatment of diabetes. A sensitive diagnostic criterion is provided by the **glucose-tolerance test.** The patient fasts overnight, then drinks a test dose of 100 g of glucose dissolved in a glass of water. The blood glucose concentration is measured before the test dose and at 30 min intervals for several hours thereafter. A healthy individual assimilates the glucose readily, the blood glucose rising to no more than about 9 or 10 mM; little or no glucose appears in the urine. Diabetic individuals assimilate the test dose of glucose poorly; their blood glucose level far exceeds the kidney threshold (about 10 mM), causing glucose to appear in the urine. ■

### SUMMARY 23.3 Hormonal Regulation of Fuel Metabolism

- The concentration of glucose in blood is hormonally regulated. Fluctuations in blood glucose level (normally 60 to 90 mg/100 mL, or about 4.5 mM) due to dietary intake or vigorous exercise are counterbalanced by a variety of hormonally triggered changes in the metabolism of several organs.

- High blood glucose elicits the release of insulin, which speeds the uptake of glucose by tissues and favors the storage of fuels as glycogen and triacylglycerols, while inhibiting fatty acid mobilization in adipose tissue.

- Low blood glucose triggers release of glucagon, which stimulates glucose release from liver glycogen and shifts fuel metabolism in liver and muscle to fatty acid oxidation, sparing glucose for use by the brain. In prolonged fasting, triacylglycerols become the principal fuel; the liver converts the fatty acids to ketone bodies for export to other tissues, including the brain.

- Epinephrine prepares the body for increased activity by mobilizing blood glucose from glycogen and other precursors.

- Cortisol, released in response to a variety of stressors (including low blood glucose), stimulates gluconeogenesis from amino acids and glycerol in the liver, thus raising blood glucose and counterbalancing the effects of insulin.

- In diabetes, insulin is either not produced or not recognized by the tissues, and the uptake of blood glucose is compromised. When blood glucose levels are high, glucose is excreted. Tissues then depend on fatty acids for fuel (producing ketone bodies) and degrade cellular proteins to provide glucogenic amino acids for glucose synthesis. Uncontrolled diabetes is characterized by high glucose levels in the blood and urine and the production and excretion of ketone bodies.

## 23.4 Obesity and the Regulation of Body Mass

In the United States population, 30% of adults are obese and another 35% are overweight. (Obesity is defined in terms of body mass index (BMI): BMI = weight in kg/(height in m)$^2$. A BMI below 25 is considered normal; 25 to 30 is overweight, and greater than 30, obese.) Obesity is life-threatening. It significantly increases the chances of developing type II diabetes as well as heart attack, stroke, and cancers of the colon, breast, prostate, and endometrium. Consequently, there is great interest in understanding how body mass and the storage of fats in adipose tissue are regulated. ■

To a first approximation, obesity is the result of taking in more calories in the diet than are expended by the body's energy-consuming activities. The body can deal with an excess of dietary calories in three ways: (1) convert excess fuel to fat and store it in adipose tissue, (2) burn excess fuel by extra exercise, and (3) "waste" fuel by diverting it to heat production (**thermogenesis**) in uncoupled mitochondria. In mammals, a complex set of hormonal and neuronal signals act to keep fuel intake and energy expenditure in balance, so as to hold the amount of adipose tissue at a suitable level. Dealing effectively with obesity requires understanding these various checks and balances under normal conditions, and how these homeostatic mechanisms fail in obesity.

### The Lipostat Theory Predicts the Feedback Regulation of Adipose Tissue

The **lipostat theory** postulates a mechanism that inhibits eating behavior and increases energy consumption whenever body weight exceeds a certain value (the set point); the inhibition is relieved when body weight drops below the set point (Fig. 23–30). This theory predicts that a feedback signal originating in adipose tissue influences the brain centers that control eating behavior and activity (metabolic and motor). The first such factor, leptin, was discovered in 1994, and several others are now known.

**Leptin** (Greek *leptos,* "thin") is a small protein (167 amino acids) that is produced in adipocytes and moves through the blood to the brain, where it acts on receptors in the hypothalamus to curtail appetite. Leptin was first identified as the product of a gene designated *OB* (obese) in laboratory mice. Mice with two defective copies of this gene (*ob/ob* genotype; lowercase letters signify a mutant form of the gene) show the behavior and physiology of animals in a constant state of starvation: their serum cortisol levels are elevated; they are unable to stay warm, they grow abnormally, do not reproduce, and exhibit unrestrained appetite. As a consequence of the last effect, they become severely obese,

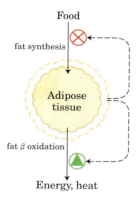

**FIGURE 23–30 Set-point model for maintaining constant mass.** When the mass of adipose tissue increases, released leptin inhibits feeding and fat synthesis and stimulates oxidation of fatty acids. When the mass of adipose tissue decreases, a lowered leptin production favors a greater food intake and less fatty acid oxidation.

weighing as much as three times more than normal mice (Fig. 23–31). They also have metabolic disturbances very similar to those of diabetic animals, and they are insulin-resistant. When leptin is injected into *ob/ob* mice, they lose weight and increase their locomotor activity and thermogenesis.

A second mouse gene, designated *DB* (diabetic), has also been found to have a role in appetite regulation. Mice with two defective copies (*db/db*) are obese and diabetic. The *DB* gene encodes the **leptin receptor.** When the leptin receptor is defective, the signaling function of leptin is lost.

The leptin receptor is expressed primarily in regions of the brain known to regulate feeding behavior— neurons of the arcuate nucleus of the hypothalamus

**FIGURE 23–31 Obesity caused by defective leptin production.** Both these mice, which are the same age, have defects in the *OB* gene. The mouse on the right was provided with purified leptin by daily injection, and weighs 35 g. The mouse on the left got no leptin, consequently ate more food and was less active, and weighs 67 g.

(Fig. 23–32a). Leptin carries the message that fat reserves are sufficient, and it promotes a reduction in fuel intake and increased expenditure of energy. Leptin-receptor interaction in the hypothalamus alters the release of neuronal signals to the region of the brain that affects appetite. Leptin also stimulates the sympathetic nervous system, increasing blood pressure, heart rate, and thermogenesis by uncoupling electron transfer from ATP synthesis in the mitochondria of adipocytes (Fig. 23–32b). Recall that **thermogenin,** also called uncoupling protein (UCP), forms a channel in the inner mitochondrial membrane that allows protons to reenter the mitochondrial matrix without passing through the ATP synthase complex (see Fig. 19–30). This permits continual oxidation of fuel (fatty acids in an adipocyte) without ATP synthesis, dissipating energy as heat and consuming dietary calories or stored fats in potentially very large amounts.

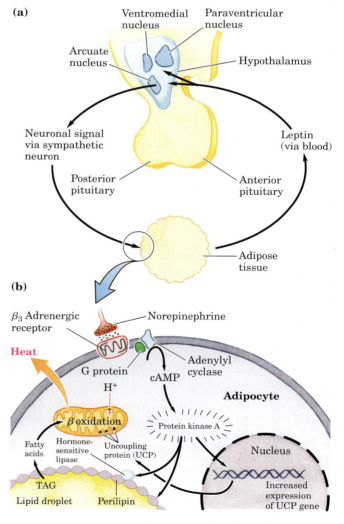

**FIGURE 23–32 Hypothalamic regulation of food intake and energy expenditure. (a)** Anatomy of the hypothalamus. **(b)** Interactions between the hypothalamus and an adipocyte, described later in text.

## Leptin Stimulates Production of Anorexigenic Peptide Hormones

Two types of neurons in the arcuate nucleus control fuel intake and metabolism (Fig. 23–33). The **orexigenic** (appetite-stimulating) neurons stimulate eating by producing and releasing **neuropeptide Y (NPY),** which causes the next neuron in the circuit to send the signal to the brain, Eat! The blood level of NPY rises during starvation, and is elevated in both *ob/ob* and *db/db* mice. The high NPY concentration presumably underlies the obesity of these mice, who eat voraciously.

The **anorexigenic** (appetite-suppressing) neurons in the arcuate nucleus produce α-melanocyte-stimulating hormone **(α-MSH),** formed from its polypeptide precursor pro-opiomelanocortin (POMC; Fig. 23–6). Release of α-MSH causes the next neuron in the circuit to send the signal to the brain, Stop eating!

The amount of leptin released by adipose tissue depends on both the number and the size of adipocytes. When weight loss decreases the mass of lipid tissue, leptin levels in the blood decrease, the production of NPY is diminished, and the processes in adipose tissue shown in Figure 23–32 are reversed. Uncoupling is diminished,

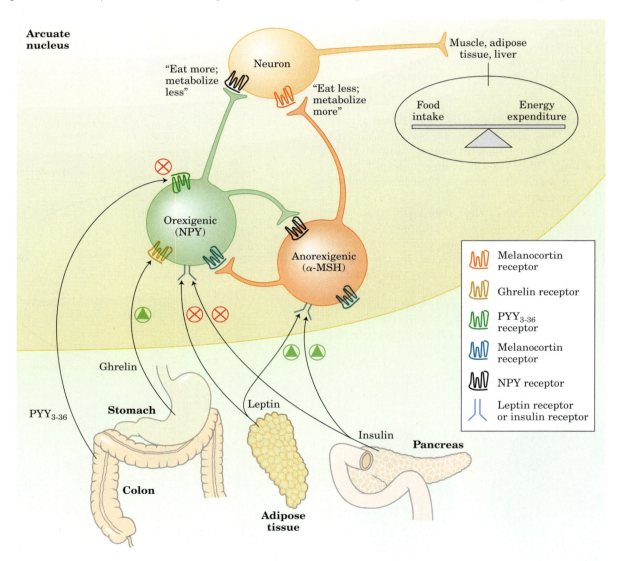

**FIGURE 23–33 Hormones that control eating.** In the arcuate nucleus, two sets of neurosecretory cells receive hormonal input and relay neuronal signals to the cells of muscle, adipose tissue, and liver. Leptin and insulin are released from adipose tissue and pancreas, respectively, in proportion to the mass of body fat. The two hormones act on anorexigenic neurosecretory cells (red) to trigger release of α-MSH; this produces neuronal signals to eat less and metabolize more fuel. Leptin and insulin also act on orexigenic neurosecretory cells (green) to inhibit the release of NPY, reducing the "eat" signal sent to the tissues. As described later in the text, the gastric hormone ghrelin *stimulates* appetite by activating the NPY-expressing cells; PYY₃₋₃₆, released from the colon, *inhibits* these neurons and thereby decreases appetite. Each of the two types of neurosecretory cells inhibits hormone production by the other, so any stimulus that activates orexigenic cells inactivates anorexigenic cells, and vice versa. This strengthens the effect of stimulatory inputs.

slowing thermogenesis and saving fuel, and fat mobilization slows in response to reduced signaling by cAMP. Consumption of more food combined with more efficient utilization of fuel results in replenishment of the fat reserve in adipose, bringing the system back into balance.

## Leptin Triggers a Signaling Cascade That Regulates Gene Expression

The leptin signal is transduced by a mechanism also used by receptors for interferon and growth factors, the JAK-STAT system (Fig. 23–34; see Fig. 12–9). The leptin receptor, which has a single transmembrane segment, dimerizes when leptin binds to the extracellular domain of two monomers. Both monomers are phosphorylated on a Tyr residue of the intracellular domain by a *Janus kinase (JAK).* The Ⓟ–Tyr residues become docking sites for three proteins that are *s*ignal *t*ransducers and *a*ctivators of *t*ranscription (**STATs** 3, 5, and 6, sometimes called fat-STATS). The docked STATs are then phosphorylated on Tyr residues by the

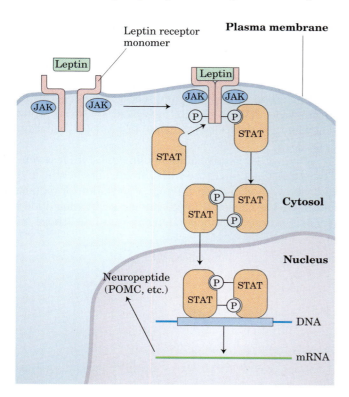

**FIGURE 23–34  The JAK-STAT mechanism of leptin signal transduction in the hypothalamus.** Leptin binding induces dimerization of the leptin receptor, followed by phosphorylation of Tyr residues of the receptor, catalyzed by Janus kinase (JAK). STATs bound to the phosphorylated leptin receptor through their SH2 domains are now phosphorylated on Tyr residues by a separate activity of JAK. The STATs dimerize, binding each other's Ⓟ–Tyr residues, and enter the nucleus. Here, they bind specific regulatory regions in the DNA and alter the expression of certain genes. The products of these genes ultimately influence the organism's feeding behavior and energy expenditure.

same JAK. After phosphorylation, the STATs dimerize then move to the nucleus, where they bind to specific DNA sequences and stimulate the expression of target genes, including the gene for POMC, from which $\alpha$-MSH is produced.

The increased catabolism and thermogenesis triggered by leptin are due in part to increased synthesis of the mitochondrial uncoupling protein UCP in adipocytes. Leptin stimulates the synthesis of this uncoupling protein by altering synaptic transmissions from neurons in the arcuate nucleus to adipose and other tissues. In these tissues, leptin causes increased release of norepinephrine, which acts through $\beta_3$-adrenergic receptors to stimulate transcription of the gene for UCP. The resulting uncoupling of electron transfer from oxidative phosphorylation consumes fat and is **thermogenic** (Fig. 23–32).

Might human obesity be the result of insufficient leptin production, and therefore treatable by the injection of leptin? Blood levels of leptin are in fact usually much *higher* in obese animals (including humans) than in animals of normal body mass (except, of course, in *ob/ob* animals, which cannot make leptin). Some downstream element in the leptin response system must be defective in obese individuals, and the elevation in leptin is the result of an (unsuccessful) attempt to overcome the leptin resistance. In those very rare humans with extreme obesity who have a defective leptin gene (*OB*), leptin injection does result in dramatic weight loss. In the vast majority of obese individuals, however, the *OB* gene is intact. In clinical trials, the injection of leptin did not have the weight-reducing effect observed in obese *ob/ob* mice. Clearly, most cases of human obesity involve one or more factors in addition to leptin.

## The Leptin System May Have Evolved to Regulate the Starvation Response

Although much of the initial interest in leptin resulted from its possible role in preventing obesity, the leptin system probably evolved to adjust an animal's activity and metabolism during periods of fasting and starvation, not to restrict weight. The *reduction* in leptin level triggered by nutritional deficiency reverses the thermogenic processes illustrated in Figure 23–32, allowing fuel conservation. Leptin activates **AMP-dependent protein kinase (AMPK)**, which regulates many aspects of fuel metabolism. Leptin also triggers decreased production of thyroid hormone (slowing basal metabolism), decreased production of sex hormones (preventing reproduction), and increased production of glucocorticoids (mobilizing the body's fuel-generating resources). By minimizing energy expenditures and maximizing the use of endogenous reserves of energy, these leptin-mediated responses may allow an animal to survive periods of severe nutritional deprivation.

## Insulin Acts in the Arcuate Nucleus to Regulate Eating and Energy Conservation

Insulin secretion reflects both the size of fat reserves (adiposity) and the current energy balance (blood glucose level). Insulin acts on insulin receptors in the hypothalamus to inhibit eating (Fig. 23–33). Insulin receptors in the orexigenic neurons of the arcuate nucleus *inhibit* the release of NPY, and insulin receptors in the anorexigenic neurons *stimulate* α-MSH production, thereby decreasing fuel intake and increasing thermogenesis. By mechanisms discussed in Section 23.3, insulin also signals muscle, liver, and adipose tissues to increase catabolic reactions, including fat oxidation, which results in weight loss.

Leptin makes the cells of liver and muscle more sensitive to insulin. One hypothesis to explain this effect suggests cross-talk between the protein tyrosine kinases activated by leptin and those activated by insulin (Fig. 23–35); common second messengers in the two signaling pathways allow leptin to trigger some of the same downstream events that are triggered by insulin, through insulin receptor substrate-2 (IRS-2) and phosphoinositide 3-kinase (PI-3K) (Chapter 12).

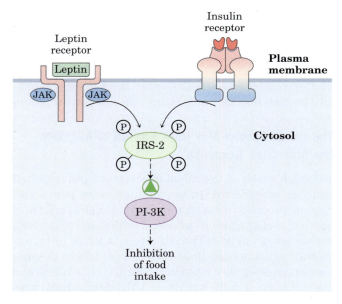

**FIGURE 23–35 A possible mechanism for cross-talk between receptors for insulin and leptin.** The insulin receptor has intrinsic Tyr kinase activity (see Fig. 12–6), and the leptin receptor, when occupied by its ligand, is phosphorylated by a soluble Tyr kinase (JAK). One possible explanation for the observed interaction between leptin and insulin is that both may phosphorylate the same substrate—in the case shown here, insulin receptor substrate-2 (IRS-2). When phosphorylated, IRS-2 activates PI-3K, which has downstream consequences that include inhibition of food intake. IRS-2 serves here as an integrator of the input from two receptors.

## Adiponectin Acts through AMPK

**Adiponectin** is a peptide hormone (224 amino acids) produced almost exclusively in adipose tissue. It circulates in the blood and powerfully affects the metabolism of fatty acids and carbohydrates in liver and muscle. Adiponectin increases the uptake of fatty acids from the blood by myocytes and the rate at which fatty acids undergo β oxidation in the muscle. It also blocks fatty acid synthesis and gluconeogenesis in hepatocytes, and it stimulates glucose uptake and catabolism in muscle and liver (Fig. 23–36). These effects of adiponectin occur indirectly, through activation of the key regulatory enzyme AMPK by increased cytosolic [AMP]. Increased [AMP] also results from ATP consumption during intense muscular activity, but it can be brought about by adiponectin through other, unknown mechanisms. When activated, AMPK phosphorylates a number of target proteins critical to the metabolism of fatty acids and carbohydrates, with profound effects on the metabolism of the whole animal.

One enzyme regulated by AMPK is acetyl-CoA carboxylase, which produces malonyl-CoA, the first intermediate committed to fatty acid synthesis. Malonyl-CoA is a powerful inhibitor of the enzyme carnitine acyltransferase I, which starts the process of β oxidation by transporting fatty acids into the mitochondrion (see Fig. 17–6). By phosphorylating and inactivating acetyl-CoA carboxylase, AMPK inhibits fatty acid synthesis while relieving the inhibition (by malonyl-CoA) of β oxidation (Fig. 23–37).

Mice with defective adiponectin genes are less sensitive to insulin than those with normal adiponectin, and they show poor glucose tolerance; ingestion of dietary carbohydrate causes a long-lasting rise in their blood glucose. These metabolic defects resemble those of

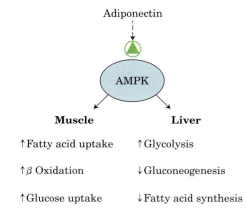

**FIGURE 23–36 Effects of adiponectin on muscle and adipose tissue.** By interacting with its receptors on the surface of myocytes and hepatocytes, adiponectin activates their AMPK. The activated kinase phosphorylates key metabolic enzymes (see Fig. 23–37, for example), shifting metabolism toward oxidation of fatty acids and away from lipid and glucose synthesis.

**FIGURE 23–37 Regulation of fatty acid synthesis and β oxidation by AMPK action on acetyl-CoA carboxylase.** When activated by elevated 5′-AMP, AMPK phosphorylates a Thr residue on acetyl-CoA carboxylase (ACC), inactivating it. This prevents the synthesis of malonyl-CoA, the first intermediate in fatty acid synthesis, and reduction in [malonyl-CoA] relieves the inhibition of carnitine acyltransferase I, allowing fatty acids to enter the mitochondrial matrix to undergo β oxidation.

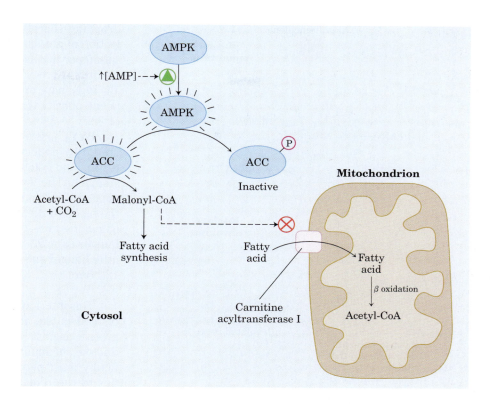

humans with type II diabetes, who also are **insulin-insensitive** and clear glucose from the blood only slowly. Indeed, individuals with obesity or type II diabetes have lower blood adiponectin levels than nondiabetic controls. Moreover, the drugs used in treatment of type II diabetes—the thiazolidinediones, such as rosiglitazone (Avandia) and pioglitazone (Actos) (p. 807)—increase the expression of adiponectin mRNA in adipose tissue and increase blood adiponectin levels in experimental animals; they also activate AMPK. It appears that adiponectin, acting through AMPK, modulates the sensitivity of cells and tissues to insulin. Perhaps this hormone will prove to be one of the links between type II diabetes and its most important predisposing factor, obesity.

Three factors improve the health of individuals with type II diabetes: regular exercise, use of thiazolidinediones, and dietary restriction. We have seen that exercise activates AMPK, as does adiponectin, and that thiazolidinediones increase the concentration of adiponectin in plasma, increasing insulin sensitivity. Dietary restriction may act by regulating the expression of genes that encode proteins involved in fatty acid oxidation and in energy expenditure via thermogenesis.

### Diet Regulates the Expression of Genes Central to Maintaining Body Mass

Proteins in a family of ligand-activated transcription factors, the **peroxisome proliferator-activated receptors (PPARs)**, respond to changes in dietary lipid by altering the expression of genes involved in fat and carbohydrate metabolism. These transcription factors were first recognized for their roles in peroxisome synthesis—thus their name. Their normal ligands are fatty acids or fatty acid derivatives, but they can also bind synthetic agonists and can be activated in the laboratory by genetic manipulation. PPARα, PPARδ, and PPARγ are members of the nuclear receptor superfamily. They act in the nucleus by forming heterodimers with another nuclear receptor, RXR (retinoid X receptor), binding to regulatory regions of DNA near the genes under their control and changing the rate of transcription of those genes (Fig. 23–38).

**PPARγ,** expressed primarily in liver and adipose tissue, is involved in turning on genes necessary to the differentiation of fibroblasts into adipocytes and genes that encode proteins required for lipid synthesis and storage in adipocytes. PPARγ is activated by drugs of the thiazolidinedione class, which are used to treat type II diabetes. **PPARα** in hepatocytes turns on the genes necessary for β oxidation of fatty acids and formation of ketone bodies during fasting.

**PPARδ** is a key regulator of fat oxidation, which acts by sensing changes in dietary lipid. It acts in liver and muscle, stimulating the transcription of at least nine genes encoding proteins for β oxidation and for energy dissipation through uncoupling of mitochondria. Normal mice overfed on high-fat diets accumulate massive amounts of both brown and white fat, and fat droplets accumulate in the liver. But when the same overfeeding experiment is done with mice that have a genetically

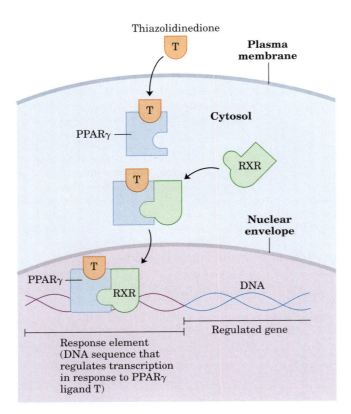

**FIGURE 23–38 Mode of action of PPARs.** PPARs, when bound to their cognate ligand, form heterodimers with the nuclear receptor RXR. The dimer binds specific regions of DNA, response elements, stimulating transcription of genes in those regions.

altered, always active PPARδ, this fat accumulation is prevented. In mice with a nonfunctioning leptin receptor (*db/db*), activated PPARδ prevents the development of obesity that would otherwise occur (see Fig. 23–31). By stimulating fatty acid breakdown in uncoupled mitochondria, PPARδ causes fat depletion, weight loss, and thermogenesis. Seen in this light, thermogenesis is both a means of keeping warm and a defense against obesity. Clearly, PPARδ is a potential target for drugs to treat obesity.

## Short-Term Eating Behavior Is Set by Ghrelin and PYY₃₋₃₆

**Ghrelin** is a peptide hormone (28 amino acids) produced in cells lining the stomach. It was originally recognized as the stimulus for the release of growth hormone (*ghre* is the Proto-Indo-European root of "grow"), then subsequently shown to be a powerful appetite stimulant that works on a shorter time scale (between meals) than leptin and insulin. Ghrelin receptors are located in the pituitary gland (presumably mediating growth hormone release) and in the hypothalamus (affecting appetite), as well as in heart muscle and adipose tissue. The concentration of ghrelin in the blood varies strikingly between meals, peaking just before a meal and

dropping sharply just after the meal (Fig. 23–39). Injection of ghrelin into humans produces immediate sensations of intense hunger. Individuals with Prader-Willi syndrome, whose blood levels of ghrelin are exceptionally high, have an uncontrollable appetite, leading to extreme obesity that often results in death before the age of 30.

**PYY₃₋₃₆** is a peptide hormone (34 amino acids) secreted by endocrine cells in the lining of the small intestine and colon in response to food entering from the stomach. The level of PYY₃₋₃₆ in the blood rises after a meal and remains high for some hours. It is carried in the blood to the arcuate nucleus, where it acts on orexigenic neurons, inhibiting NPY release and reducing hunger (Fig. 23–33). Humans injected with PYY₃₋₃₆ feel little hunger and eat less than normal amounts for about 12 hours.

This interlocking system of neuroendocrine controls of food intake and metabolism presumably evolved to protect against starvation and to eliminate counterproductive accumulation of fat (extreme obesity). The difficulty most people face in trying to lose weight testifies to the remarkable effectiveness of these controls.

**(a)**

**(b)**

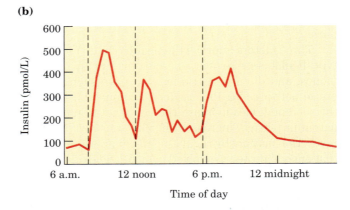

**FIGURE 23–39 Variations in ghrelin and insulin relative to meal times. (a)** Plasma levels of ghrelin rise sharply just *before* the normal time for meals (7 a.m. breakfast, 12 noon lunch, 5:30 p.m. dinner) and drop precipitously just after meals, paralleling the subjective feelings of hunger. **(b)** Insulin levels rise immediately *after* each meal, in response to the increase in blood glucose concentration.

## *SUMMARY 23.4* Obesity and the Regulation of Body Mass

- Obesity is increasingly common in the developed countries and predisposes people toward several life-threatening conditions.

- Adipose tissue produces leptin, a hormone that regulates feeding behavior and energy expenditure so as to maintain adequate reserves of fat. Leptin production and release increase with the number and size of adipocytes.

- Leptin acts on receptors in the arcuate nucleus of the hypothalamus, causing the release of anorexigenic peptides, including $\alpha$-MSH, that act in the brain to inhibit eating. Leptin also stimulates sympathetic nervous system action on adipocytes, leading to uncoupling of mitochondrial oxidative phosphorylation, with consequent thermogenesis.

- The signal-transduction mechanism for leptin involves phosphorylation of the JAK-STAT system. On phosphorylation by JAK, STATs can bind to regulatory regions in nuclear DNA and alter the expression of genes for the proteins that set the level of metabolic activity and determine feeding behavior. Insulin acts on receptors in the arcuate nucleus, with results similar to those caused by leptin.

- The hormone adiponectin stimulates fatty acid uptake and oxidation and inhibits fatty acid synthesis. Its actions are mediated by AMPK.

- Ghrelin, a hormone produced in the stomach, acts on orexigenic neurons in the arcuate nucleus to produce hunger before a meal. PYY$_{3-36}$, a peptide hormone of the intestine, acts at the same site to lessen hunger after a meal.

## Key Terms

*Terms in bold are defined in the glossary.*

neuroendocrine system   882
**radioimmunoassay (RIA)**   884
Scatchard analysis   884
endocrine glands   886
paracrine   886
autocrine   886
insulin   887
epinephrine   888
norepinephrine   888
**catecholamines**   888
**eicosanoid hormones**   888
**steroid hormones**   888
vitamin D hormone   889
retinoid hormones   889

thyroid hormones   889
nitric oxide (NO)   889
NO synthase   889
hypothalamus   889
posterior pituitary   890
anterior pituitary   890
**tropic hormone**   890
**tropin**   890
**hepatocyte**   893
**adipocyte**   897
**myocyte**   898
**erythrocyte**   901
**leukocyte**   901
**lymphocyte**   901
**platelets**   901
blood plasma   901
**plasma proteins**   901

cortisol   909
**diabetes mellitus**   909
**type I diabetes**   909
**type II diabetes**   909
glucosuria   909
**ketosis**   909
**acidosis**   909
ketoacidosis   909
glucose-tolerance test   910
**thermogenesis**   910
leptin   910
**thermogenin (uncoupling protein)**   911
orexigenic   912
neuropeptide Y (NPY)   912

anorexigenic   912
$\alpha$-melanocyte-stimulating hormone ($\alpha$-MSH)   912
Janus kinase (JAK)   913
STAT (signal transducer and activator of transcription)   913
**thermogenic**   913
AMP-dependent protein kinase (AMPK)   913
adiponectin   914
PPAR (peroxisome proliferator-activated receptor)   915
ghrelin   916
PYY$_{3-36}$   916

## Further Reading

### General Background and History

**Crapo, L.** (1985) *Hormones: The Messengers of Life,* W. H. Freeman and Company, New York.

> Short, entertaining account of the history and results of hormone research.

**Litwack, G. & Norman, A.W.** (1997) *Hormones,* 2nd edn, Academic Press, Orlando, FL.

> An excellent, authoritative, well-illustrated, clear introduction to hormone structure, synthesis, and action.

**Yalow, R.S.** (1978) Radioimmunoassay: a probe for the fine structure of biologic systems. *Science* **200,** 1236–1245.

> History of the development of radioimmunoassays; the author's Nobel lecture.

### Hormone Structure and Function

**Litwack, G. & Schmidt, R.J.** (2001) Biochemistry of hormones I: polypeptide hormones. In *Textbook of Biochemistry with Clinical Correlations,* 5th edn (Devlin, T.M., ed.), pp. 905–958. John Wiley & Sons, Inc., New York.

Litwack, G. & Schmidt, R.J. (2001) Biochemistry of hormones II: steroid hormones. In *Textbook of Biochemistry with Clinical Correlations,* 5th edn (Devlin, T.M., ed.), pp. 959–989. John Wiley & Sons, Inc., New York.

**Tissue-Specific Metabolism**

Arias, I.M., Boyer, J.L., Chisari, F.V., Fausto, N., Schachter, D., & Shafritz, D.A. (2001) *The Liver: Biology and Pathobiology,* 4th edn, Lippincott, Williams & Wilkins, Philadelphia.

Advanced text; includes chapters on the metabolism of carbohydrates, fats, and proteins in the liver.

Nosadini, R., Avogaro, A., Doria, A., Fioretto, P., Trevisan, R., & Morocutti, A. (1989) Ketone body metabolism: a physiological and clinical overview. *Diabetes Metab. Rev.* **5,** 299–319.

Randle, P.J. (1995) Metabolic fuel selection: general integration at the whole-body level. *Proc. Nutr. Soc.* **54,** 317–327.

**Hormonal Regulation of Fuel Metabolism**

Attie, A.D. & Raines, R.T. (1995) Analysis of receptor-ligand interactions. *J. Chem. Ed.* **72,** 119–123.

Carling, D. (2004) The AMP-activated protein kinase cascade— a unifying system for energy control. *Trends Biochem. Sci.* **29,** [in press].

Elia, M. (1995) General integration and regulation of metabolism at the organ level. *Proc. Nutr. Soc.* **54,** 213–234.

Holloszy, J.O. & Kohrt, W.M. (1996) Regulation of carbohydrate and fat metabolism during and after exercise. *Annu. Rev. Nutr.* **16,** 121–138.

Pilkis, S.J. & Claus, T.H. (1991) Hepatic gluconeogenesis/glycolysis: regulation and structure/function relationships of substrate cycle enzymes. *Annu. Rev. Nutr.* **11,** 465–515.

Advanced review.

Snyder, S.H. (1985) The molecular basis of communication between cells. *Sci. Am.* **253** (October), 132–141.

Introductory discussion of the human endocrine system.

Zammit, V.A. (1996) Role of insulin in hepatic fatty acid partitioning: emerging concepts. *Biochem. J.* **314,** 1–14.

**Control of Body Mass**

Auwerx, J. & Staels, B. (1998) Leptin. *Lancet* **351,** 737–742.

Brief overview of the leptin system and JAK-STAT signal transductions.

Elmquist, J.K., Maratos-Flier, E., Saper, C.B., & Flier, J.S. (1998) Unraveling the central nervous system pathways underlying responses to leptin. *Nat. Neurosci.* **1,** 445–450.

Short, excellent review.

Flier, J.S. & Maratos-Flier, E. (1998) Obesity and the hypothalamus: novel peptides for new pathways. *Cell* **92,** 437–440.

Freake, H.C. (1998) Uncoupling proteins: beyond brown adipose tissue. *Nutr. Rev.* **56,** 185–189.

Review of the structure, function, and role of uncoupling proteins.

Friedman, J.M. (2002) The function of leptin in nutrition, weight, and physiology. *Nutr. Rev.* **60,** S1–S14.

Intermediate-level review of all aspects of leptin biology.

Inui, A. (1999) Feeding and body-weight regulation by hypothalamic neuropeptides—mediation of the actions of leptin. *Trends Neurosci.* **22,** 62–67.

Short, intermediate-level review of the leptin system.

Jequier, E. & Tappy, L. (1999) Regulation of body weight in humans. *Physiol. Rev.* **79,** 451–480.

Detailed review of the role of leptin in body-weight regulation, the control of food intake, and the roles of white and brown adipose tissues in energy expenditure.

Korner, J. & Aronne, L.J. (2003) The emerging science of body weight regulation and its impact on obesity treatment. *J. Clin. Invest.* **111,** 565–570.

Marx, J. (2003) Cellular warriors at the battle of the bulge. *Science* **299,** 846–849.

Short review of biochemistry of weight control, and introduction to several papers in the same issue of *Science* on obesity in humans.

Unger, R.H. (2003) The physiology of cellular liporegulation. *Annu. Rev. Physiol.* **65,** 333–347.

Advanced review.

Wang, Y.-X., Lee, C.-H., Tiep, S., Yu, R.T., Ham, J., Kang, H., & Evans, R.M. (2003) Peroxisome-proliferator-activated receptor activates fat metabolism to prevent obesity. *Cell* **113,** 159–170.

Describes the role of PPARδ.

Wood, S.C., Seeley, R.J., Porte, D., Jr., & Schwartz, M.W. (1998) Signals that regulate food intake and energy homeostasis. *Science* **280,** 1378–1383.

Review of the roles of leptin, insulin, and other neuropeptides in the regulation of feeding and catabolism.

## Problems

**1. ATP and Phosphocreatine as Sources of Energy for Muscle** During muscle contraction, the concentration of phosphocreatine in skeletal muscle drops while the concentration of ATP remains fairly constant. However, in a classic experiment, Robert Davies found that if he first treated muscle with 1-fluoro-2,4-dinitrobenzene (p. 97), the concentration of ATP declined rapidly while the concentration of phosphocreatine remained unchanged during a series of contractions. Suggest an explanation.

**2. Metabolism of Glutamate in the Brain** Brain tissue takes up glutamate from the blood, transforms it into glutamine, then releases it into the blood. What is accomplished by this metabolic conversion? How does it take place? The amount of glutamine produced in the brain can actually exceed the amount of glutamate entering from the blood. How does this extra glutamine arise? (Hint: You may want to review amino acid catabolism in Chapter 18; recall that $NH_4^+$ is very toxic to the brain.)

**3. Absence of Glycerol Kinase in Adipose Tissue** Glycerol 3-phosphate is required for the biosynthesis of triacylglycerols. Adipocytes, specialized for the synthesis and degradation of triacylglycerols, cannot use glycerol directly, because they lack glycerol kinase, which catalyzes the reaction

$$\text{Glycerol} + \text{ATP} \longrightarrow \text{glycerol 3-phosphate} + \text{ADP}$$

How does adipose tissue obtain the glycerol 3-phosphate necessary for triacylglycerol synthesis?

**4. Oxygen Consumption during Exercise** A sedentary adult consumes about 0.05 L of $O_2$ in 10 seconds. A sprinter, running a 100 m race, consumes about 1 L of $O_2$ in 10 seconds. After finishing the race, the sprinter continues to breathe at an elevated (but declining) rate for some minutes, consuming an extra 4 L of $O_2$ above the amount consumed by the sedentary individual.

(a) Why does the need for $O_2$ increase dramatically during the sprint?

(b) Why does the demand for $O_2$ remain high after the sprint is completed?

**5. Thiamine Deficiency and Brain Function** Individuals with thiamine deficiency show some characteristic neurological signs and symptoms, including loss of reflexes, anxiety, and mental confusion. Why might thiamine deficiency be manifested by changes in brain function?

**6. Potency of Hormones** Under normal conditions, the human adrenal medulla secretes epinephrine ($C_9H_{13}NO_3$) at a rate sufficient to maintain a concentration of $10^{-10}$ M in circulating blood. To appreciate what that concentration means, calculate the diameter of a round swimming pool, with a water depth of 2.0 m, that would be needed to dissolve 1.0 g (about 1 teaspoon) of epinephrine to a concentration equal to that in blood.

**7. Regulation of Hormone Levels in the Blood** The half-life of most hormones in the blood is relatively short. For example, when radioactively labeled insulin is injected into an animal, half of the labeled hormone disappears from the blood within 30 min.

(a) What is the importance of the relatively rapid inactivation of circulating hormones?

(b) In view of this rapid inactivation, how is the level of circulating hormone kept constant under normal conditions?

(c) In what ways can the organism make rapid changes in the level of a circulating hormone?

**8. Water-Soluble versus Lipid-Soluble Hormones** On the basis of their physical properties, hormones fall into one of two categories: those that are very soluble in water but relatively insoluble in lipids (e.g., epinephrine) and those that are relatively insoluble in water but highly soluble in lipids (e.g., steroid hormones). In their role as regulators of cellular activity, most water-soluble hormones do not enter their target cells. The lipid-soluble hormones, by contrast, do enter their target cells and ultimately act in the nucleus. What

is the correlation between solubility, the location of receptors, and the mode of action of these two classes of hormones?

**9. Metabolic Differences between Muscle and Liver in a "Fight or Flight" Situation** During a "fight or flight" situation, the release of epinephrine promotes glycogen breakdown in the liver, heart, and skeletal muscle. The end product of glycogen breakdown in the liver is glucose; the end product in skeletal muscle is pyruvate.

(a) What is the reason for the different products of glycogen breakdown in the two tissues?

(b) What is the advantage to an organism that must fight or flee of these specific glycogen breakdown routes?

**10. Excessive Amounts of Insulin Secretion: Hyperinsulinism** Certain malignant tumors of the pancreas cause excessive production of insulin by the $\beta$ cells. Affected individuals exhibit shaking and trembling, weakness and fatigue, sweating, and hunger.

(a) What is the effect of hyperinsulinism on the metabolism of carbohydrates, amino acids, and lipids by the liver?

(b) What are the causes of the observed symptoms? Suggest why this condition, if prolonged, leads to brain damage.

**11. Thermogenesis Caused by Thyroid Hormones** Thyroid hormones are intimately involved in regulating the basal metabolic rate. Liver tissue of animals given excess thyroxine shows an increased rate of $O_2$ consumption and increased heat output (thermogenesis), but the ATP concentration in the tissue is normal. Different explanations have been offered for the thermogenic effect of thyroxine. One is that excess thryroxine causes uncoupling of oxidative phosphorylation in mitochondria. How could such an effect account for the observations? Another explanation suggests that the thermogenesis is due to an increased rate of ATP utilization by the thyroxine-stimulated tissue. Is this a reasonable explanation? Why?

**12. Function of Prohormones** What are the possible advantages in the synthesis of hormones as prohormones?

**13. Sources of Glucose during Starvation** The typical human adult uses about 160 g of glucose per day, 120 g of which is used by the brain. The available reserve of glucose (~20 g of circulating glucose and ~190 g of glycogen) is adequate for about one day. After the reserve has been depleted during starvation, how would the body obtain more glucose?

**14. Parabiotic *ob/ob* Mice** By careful surgery, researchers can connect the circulatory systems of two mice so that the same blood circulates through both animals. In these **parabiotic** mice, products released into the blood by one animal reach the other animal via the shared circulation. Both animals are free to eat independently. If an *ob/ob* mouse (both copies of the *OB* gene are defective) and a normal *OB/OB* mouse (two good copies of the *OB* gene) were made parabiotic, what would happen to the weight of each mouse?

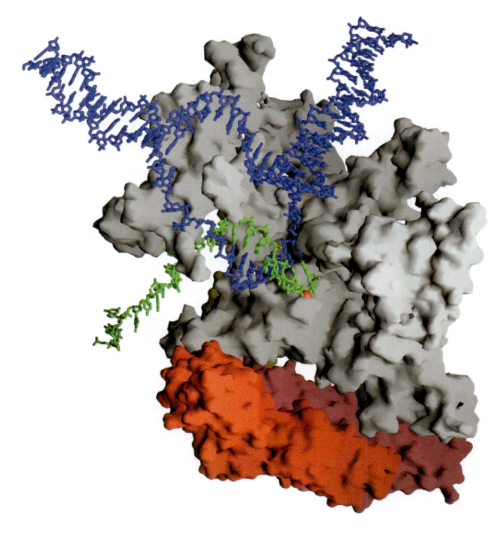

The bacterial RNA polymerase

# INFORMATION PATHWAYS

The third and final part of this book explores the biochemical mechanisms underlying the apparently contradictory requirements for both genetic continuity and the evolution of living organisms. What is the molecular nature of genetic material? How is genetic information transmitted from one generation to the next with high fidelity? How do the rare changes in genetic material that are the raw material of evolution arise? How is genetic information ultimately expressed in the amino acid sequences of the astonishing variety of protein molecules in a living cell?

The fundamental unit of information in living systems is the **gene.** A gene can be defined biochemically as a segment of DNA (or, in a few cases, RNA) that encodes the information required to produce a functional biological product. The final product is usually a protein, so much of the material in Part III concerns genes that encode proteins. A functional gene product might also be one of several classes of RNA molecules. The storage, maintenance, and metabolism of these informational units form the focal points of our discussion in Part III.

Modern biochemical research on gene structure and function has brought to biology a revolution comparable to that stimulated by the publication of Darwin's theory on the origin of species nearly 150 years ago. An understanding of how information is stored and used in cells has brought penetrating new insights to some of the most fundamental questions about cellular structure and function. A comprehensive conceptual framework for biochemistry is now unfolding.

Today's understanding of information pathways has arisen from the convergence of genetics, physics, and chemistry in modern biochemistry. This was epitomized by the discovery of the double-helical structure of DNA, postulated by James Watson and Francis Crick in 1953 (see Fig. 8–15). Genetic theory contributed the concept of coding by genes. Physics permitted the determination of molecular structure by x-ray diffraction analysis. Chemistry revealed the composition of DNA. The profound impact of the Watson-Crick hypothesis arose from its ability to account for a wide range of observations derived from studies in these diverse disciplines.

This revolution in our understanding of the structure of DNA inevitably stimulated questions about its function. The double-helical structure itself clearly suggested how DNA might be copied so that the information it contains can be transmitted from one generation to the next. Clarification of how the information in DNA is converted into functional proteins came with the discovery of both messenger RNA and transfer RNA and with the deciphering of the genetic code.

These and other major advances gave rise to the central dogma of molecular biology, comprising the three major processes in the cellular utilization of genetic information. The first is **replication,** the copying of parental DNA to form daughter DNA molecules with identical nucleotide sequences. The second is **transcription,** the process by which parts of the genetic message encoded in DNA are copied precisely into RNA. The third is **translation,** whereby the genetic message encoded in messenger RNA is translated on the ribosomes into a polypeptide with a particular sequence of amino acids.

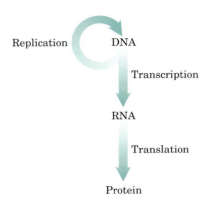

Replication    DNA

Transcription

RNA

Translation

Protein

The central dogma of molecular biology, showing the general pathways of information flow via replication, transcription, and translation. The term "dogma" is a misnomer. Introduced by Francis Crick at a time when little evidence supported these ideas, the dogma has become a well-established principle.

Part III explores these and related processes. In Chapter 24 we examine the structure, topology, and packaging of chromosomes and genes. The processes underlying the central dogma are elaborated in Chapters 25 through 27. Finally, we turn to regulation, examining how the expression of genetic information is controlled (Chapter 28).

A major theme running through these chapters is the added complexity inherent in the biosynthesis of macromolecules that contain information. Assembling nucleic acids and proteins with particular sequences of nucleotides and amino acids represents nothing less than preserving the faithful expression of the template upon which life itself is based. We might expect the formation of phosphodiester bonds in DNA or peptide bonds in proteins to be a trivial feat for cells, given the arsenal of enzymatic and chemical tools described in Part II. However, the framework of patterns and rules established in our examination of metabolic pathways thus far must be enlarged considerably to take into account molecular information. Bonds must be formed between *particular* subunits in informational biopolymers, avoiding either the occurrence or the persistence of sequence errors. This has an enormous impact on the thermodynamics, chemistry, and enzymology of the biosynthetic processes. Formation of a peptide bond requires an energy input of only about 21 kJ/mol of bonds and can be catalyzed by relatively simple enzymes. But to synthesize a bond between two specific amino acids at a particular point in a polypeptide, the cell invests about 125 kJ/mol while making use of more than 200 enzymes, RNA molecules, and specialized proteins. The chemistry involved in peptide bond formation does not change because of this requirement, but additional processes are layered over the basic reaction to ensure that the peptide bond is formed between particular amino acids. Information is expensive.

The dynamic interaction between nucleic acids and proteins is another central theme of Part III. With the important exception of a few catalytic RNA molecules (discussed in Chapters 26 and 27), the processes that make up the pathways of cellular information flow are catalyzed and regulated by proteins. An understanding of these enzymes and other proteins can have practical as well as intellectual rewards, because they form the basis of recombinant DNA technology (introduced in Chapter 9).

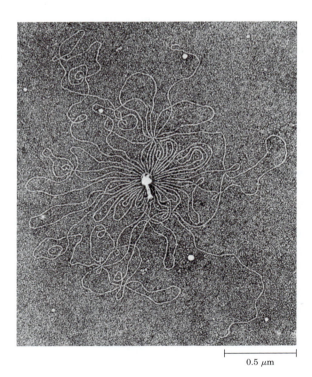

chapter **24**

# GENES AND CHROMOSOMES

DNA topoisomerases are the magicians of the DNA world. By allowing DNA strands or double helices to pass through each other, they can solve all of the topological problems of DNA in replication, transcription and other cellular transactions.

> —*James Wang, article in* Nature Reviews in Molecular Cell Biology, *2002*

Supercoiling, in fact, does more for DNA than act as an executive enhancer; it keeps the unruly, spreading DNA inside the cramped confines that the cell has provided for it.

> —*Nicholas Cozzarelli,* Harvey Lectures, *1993*

Almost every cell of a multicellular organism contains the same complement of genetic material—its **genome.** Just look at any human individual for a hint of the wealth of information contained in each human cell. **Chromosomes,** the nucleic acid molecules that are the repository of an organism's genetic information, are the largest molecules in a cell and may contain thousands of genes as well as considerable tracts of intergenic DNA. The 16 chromosomes in the relatively small genome of the yeast *Saccharomyces cerevisiae* have molecular masses ranging from $1.5 \times 10^8$ to $1 \times 10^9$ daltons, corresponding to DNA molecules with 230,000 to 1,532,000 contiguous base pairs (bp). Human chromosomes range up to 279 million bp.

The very size of DNA molecules presents an interesting biological puzzle, given that they are generally much longer than the cells or viral packages that con-

tain them (Fig. 24–1). In this chapter we shift our focus from the secondary structure of DNA, considered in Chapter 8, to the extraordinary degree of organization required for the tertiary packaging of DNA into chromosomes. We first examine the elements within viral and cellular chromosomes, then assess their size and organization. We next consider DNA topology, providing a

|— 0.5 μm —|

**FIGURE 24–1 Bacteriophage T2 protein coat surrounded by its single, linear molecule of DNA.** The DNA was released by lysing the bacteriophage particle in distilled water and allowing the DNA to spread on the water surface. An undamaged T2 bacteriophage particle consists of a head structure that tapers to a tail by which the bacteriophage attaches itself to the outer surface of a bacterial cell. All the DNA shown in this electron micrograph is normally packaged inside the phage head.

description of the coiling of DNA molecules. Finally, we discuss the protein-DNA interactions that organize chromosomes into compact structures.

## 24.1 Chromosomal Elements

Cellular DNA contains genes and intergenic regions, both of which may serve functions vital to the cell. The more complex genomes, such as those of eukaryotic cells, demand increased levels of chromosomal organization, and this is reflected in the chromosome's structural features. We begin by considering the different types of DNA sequences and structural elements within chromosomes.

### Genes Are Segments of DNA That Code for Polypeptide Chains and RNAs

Our understanding of genes has evolved tremendously over the last century. Classically, a gene was defined as a portion of a chromosome that determines or affects a single character or **phenotype** (visible property), such as eye color. George Beadle and Edward Tatum proposed a molecular definition of a gene in 1940. After exposing spores of the fungus *Neurospora crassa* to x rays and other agents known to damage DNA and cause alterations in DNA sequence **(mutations),** they detected mutant fungal strains that lacked one or another specific enzyme, sometimes resulting in the failure of an entire metabolic pathway. Beadle and Tatum concluded that a gene is a segment of genetic material that determines or codes for one enzyme: the **one gene–one enzyme** hypothesis. Later this concept was broadened to **one gene–one polypeptide,** because many genes code for proteins that are not enzymes or for one polypeptide of a multisubunit protein.

The modern biochemical definition of a gene is even more precise. A gene is all the DNA that encodes the primary sequence of some final gene product, which can be either a polypeptide or an RNA with a structural or catalytic function. DNA also contains other segments or sequences that have a purely regulatory function. **Regulatory sequences** provide signals that may denote the beginning or the end of genes, or influence the transcription of genes, or function as initiation points for replication or recombination (Chapter 28). Some genes can be expressed in different ways to generate multiple gene products from one segment of DNA. The special transcriptional and translational mechanisms that allow this are described in Chapters 26 through 28.

We can make direct estimations of the minimum overall size of genes that encode proteins. As described in detail in Chapter 27, each amino acid of a polypeptide chain is coded for by a sequence of three consecutive nucleotides in a single strand of DNA (Fig. 24–2), with these "codons" arranged in a sequence that corresponds to the sequence of amino acids in the polypeptide that the gene encodes. A polypeptide chain of 350 amino acid residues (an average-size chain) corre-

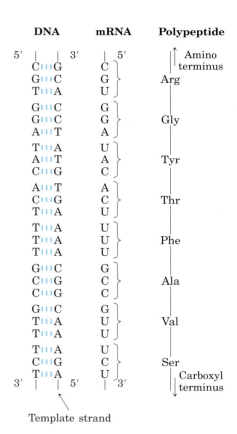

**FIGURE 24–2 Colinearity of the coding nucleotide sequences of DNA and mRNA and the amino acid sequence of a polypeptide chain.** The triplets of nucleotide units in DNA determine the amino acids in a protein through the intermediary mRNA. One of the DNA strands serves as a template for synthesis of mRNA, which has nucleotide triplets (codons) complementary to those of the DNA. In some bacterial and many eukaryotic genes, coding sequences are interrupted at intervals by regions of noncoding sequences (called introns).

George W. Beadle, 1903–1989

Edward L. Tatum, 1909–1975

sponds to 1,050 bp. Many genes in eukaryotes and a few in prokaryotes are interrupted by noncoding DNA segments and are therefore considerably longer than this simple calculation would suggest.

How many genes are in a single chromosome? The *Escherichia coli* chromosome, one of the prokaryotic genomes that has been completely sequenced, is a circular DNA molecule (in the sense of an endless loop rather than a perfect circle) with 4,639,221 bp. These base pairs encode about 4,300 genes for proteins and another 115 genes for stable RNA molecules. Among eukaryotes, the approximately 3.2 billion base pairs of the human genome include 30,000 to 35,000 genes on 24 different chromosomes.

## DNA Molecules Are Much Longer Than the Cellular Packages That Contain Them

Chromosomal DNAs are often many orders of magnitude longer than the cells or viruses in which they are found (Fig. 24–1; Table 24–1). This is true of every class of organism or parasite.

**Viruses**  Viruses are not free-living organisms; rather, they are infectious parasites that use the resources of a host cell to carry out many of the processes they require to propagate. Many viral particles consist of no more than a genome (usually a single RNA or DNA molecule) surrounded by a protein coat.

Almost all plant viruses and some bacterial and animal viruses have RNA genomes. These genomes tend to be particularly small. For example, the genomes of mammalian retroviruses such as HIV are about 9,000 nucleotides long, and that of the bacteriophage $Q\beta$ has 4,220 nucleotides. Both types of viruses have single-stranded RNA genomes.

The genomes of DNA viruses vary greatly in size (Table 24–1). Many viral DNAs are circular for at least part of their life cycle. During viral replication within a host cell, specific types of viral DNA called **replicative forms** may appear; for example, many linear DNAs become circular and all single-stranded DNAs become double-stranded. A typical medium-sized DNA virus is bacteriophage $\lambda$ (lambda), which infects *E. coli*. In its replicative form inside cells, $\lambda$ DNA is a circular double helix. This double-stranded DNA contains 48,502 bp and has a contour length of 17.5 $\mu$m. Bacteriophage $\phi$X174 is a much smaller DNA virus; the DNA in the viral particle is a single-stranded circle, and the double-stranded replicative form contains 5,386 bp. Although viral genomes are small, the contour lengths of their DNAs are much greater than the long dimensions of the viral particles that contain them. The DNA of bacteriophage T4, for example, is about 290 times longer than the viral particle itself (Table 24–1).

**Bacteria**  A single *E. coli* cell contains almost 100 times as much DNA as a bacteriophage $\lambda$ particle. The chromosome of an *E. coli* cell is a single double-stranded circular DNA molecule. Its 4,639,221 bp have a contour length of about 1.7 mm, some 850 times the length of the *E. coli* cell (Fig. 24–3). In addition to the very large, circular DNA chromosome in their nucleoid, many bacteria contain one or more small circular DNA molecules that are free in the cytosol. These extrachromosomal elements are called **plasmids** (Fig. 24–4; see also p. 311). Most plasmids are only a few thousand base pairs long, but some contain more than 10,000 bp. They carry genetic information and undergo replication to yield daughter plasmids, which pass into the daughter cells at cell division. Plasmids have been found in yeast and other fungi as well as in bacteria.

In many cases plasmids confer no obvious advantage on their host, and their sole function appears to be self-propagation. However, some plasmids carry genes that are useful to the host bacterium. For example, some plasmid genes make a host bacterium resistant to antibacterial agents. Plasmids carrying the gene for the enzyme $\beta$-lactamase confer resistance to $\beta$-lactam antibiotics such as penicillin and amoxicillin (see Box 20–1). These and similar plasmids may pass from an antibiotic-resistant cell to an antibiotic-sensitive cell of the same or another bacterial species, making the recipient cell antibiotic resistant. The extensive use of antibiotics

**TABLE 24–1  The Sizes of DNA and Viral Particles for Some Bacterial Viruses (Bacteriophages)**

| Virus | Size of viral DNA (bp) | Length of viral DNA (nm) | Long dimension of viral particle (nm) |
|---|---|---|---|
| $\phi$X174 | 5,386 | 1,939 | 25 |
| T7 | 39,936 | 14,377 | 78 |
| $\lambda$ (lambda) | 48,502 | 17,460 | 190 |
| T4 | 168,889 | 60,800 | 210 |

**Note:** Data on size of DNA are for the replicative form (double-stranded). The contour length is calculated assuming that each base pair occupies a length of 3.4 Å (see Fig. 8–15).

**FIGURE 24–3** The length of the *E. coli* chromosome (1.7 mm) depicted in linear form relative to the length of a typical *E. coli* cell (2 μm).

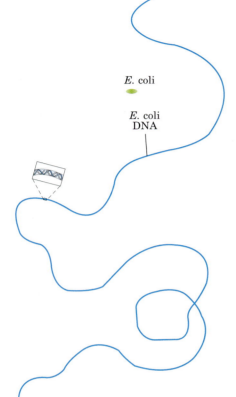

*E.* coli

*E.* coli
DNA

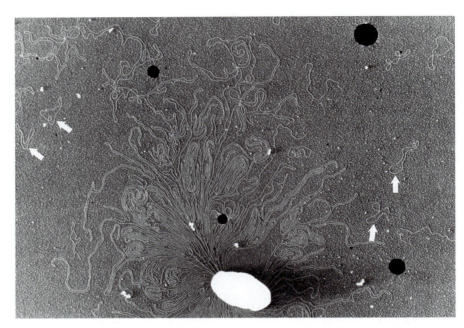

**FIGURE 24–4 DNA from a lysed *E. coli* cell.** In this electron micrograph several small, circular plasmid DNAs are indicated by white arrows. The black spots and white specks are artifacts of the preparation.

in some human populations has served as a strong selective force, encouraging the spread of antibiotic resistance–coding plasmids (as well as transposable elements, described below, that harbor similar genes) in disease-causing bacteria and creating bacterial strains that are resistant to several antibiotics. Physicians are becoming increasingly reluctant to prescribe antibiotics unless a clear clinical need is confirmed. For similar reasons, the widespread use of antibiotics in animal feeds is being curbed.

***Eukaryotes*** A yeast cell, one of the simplest eukaryotes, has 2.6 times more DNA in its genome than an *E. coli* cell (Table 24–2). Cells of *Drosophila*, the fruit fly used in classical genetic studies, contain more than 35 times as much DNA as *E. coli* cells, and human cells have almost 700 times as much. The cells of many plants and amphibians contain even more. The genetic material of eukaryotic cells is apportioned into chromosomes, the diploid ($2n$) number depending on the species (Table 24–2). A human somatic cell, for example, has 46 chro-

mosomes (Fig. 24–5). Each chromosome of a eukaryotic cell, such as that shown in Figure 24–5a, contains a single, very large, duplex DNA molecule. The DNA molecules in the 24 different types of human chromosomes (22 matching pairs plus the X and Y sex chromosomes) vary in length over a 25-fold range. Each type of chromosome in eukaryotes carries a characteristic set of genes. Interestingly, the number of genes does not vary nearly as much as does genome size (see Chapter 9 for a discussion of the types of sequences, besides genes, that contribute to genome size).

The DNA of one human genome (22 chromosomes plus X and Y or two X chromosomes), placed end to end, would extend for about a meter. Most human cells are diploid and each cell contains a total of 2 m of DNA. An adult human body contains approximately $10^{14}$ cells and thus a total DNA length of $2 \times 10^{11}$ km. Compare this with the circumference of the earth ($4 \times 10^4$ km) or the distance between the earth and the sun ($1.5 \times 10^8$ km)—a dramatic illustration of the extraordinary degree of DNA compaction in our cells.

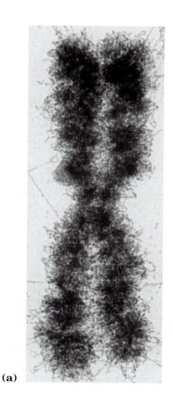

(b)

(a)

**FIGURE 24–5 Eukaryotic chromosomes. (a)** A pair of linked and condensed sister chromatids from a human chromosome. Eukaryotic chromosomes are in this state after replication and at metaphase during mitosis. **(b)** A complete set of chromosomes from a leukocyte from one of the authors. There are 46 chromosomes in every normal human somatic cell.

Eukaryotic cells also have organelles, mitochondria (Fig. 24–6) and chloroplasts, that contain DNA. Mitochondrial DNA (mtDNA) molecules are much smaller than the nuclear chromosomes. In animal cells, mtDNA contains fewer than 20,000 bp (16,569 bp in human mtDNA) and is a circular duplex. Each mitochondrion typically has two to ten copies of this mtDNA molecule, and the number can rise to hundreds in certain cells when an embryo is undergoing cell differentiation. In a few organisms (trypanosomes, for example) each mitochondrion contains thousands of copies of mtDNA, organized into a complex and interlinked matrix known as a kinetoplast. Plant cell mtDNA ranges in size from 200,000 to 2,500,000 bp. Chloroplast DNA (cpDNA) also exists as circular duplexes and ranges in size from 120,000 to 160,000 bp. The evolutionary origin of mitochondrial and chloroplast DNAs has been the subject of much speculation. A widely accepted view is that they are vestiges of the chromosomes of ancient bacteria that gained access to the cytoplasm of host cells and became the precursors of these organelles (see Fig. 1–36).

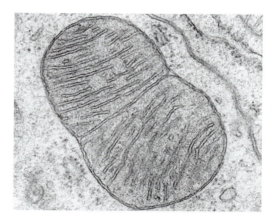

**FIGURE 24–6 A dividing mitochondrion.** Some mitochondrial proteins and RNAs are encoded by one of the copies of the mitochondrial DNA (none of which are visible here). The DNA (mtDNA) is replicated each time the mitochondrion divides, before cell division.

**TABLE 24-2** DNA, Gene, and Chromosome Content in Some Genomes

| | Total DNA (bp) | Number of chromosomes[*] | Approximate number of genes |
|---|---|---|---|
| Bacterium (*Escherichia coli*) | 4,639,221 | 1 | 4,405 |
| Yeast (*Saccharomyces cerevisiae*) | 12,068,000 | 16[†] | 6,200 |
| Nematode (*Caenorhabditis elegans*) | 97,000,000 | 12[‡] | 19,000 |
| Plant (*Arabidopsis thaliana*) | 125,000,000 | 10 | 25,500 |
| Fruit fly (*Drosophila melanogaster*) | 180,000,000 | 18 | 13,600 |
| Plant (*Oryza sativa*; rice) | 480,000,000 | 24 | 57,000 |
| Mouse (*Mus musculus*) | 2,500,000,000 | 40 | 30,000–35,000 |
| Human (*Homo sapiens*) | 3,200,000,000 | 46 | 30,000–35,000 |

**Note:** This information is constantly being refined. For the most current information, consult the websites for the individual genome projects.

[*]The diploid chromosome number is given for all eukaryotes except yeast.

[†]Haploid chromosome number. Wild yeast strains generally have eight (octoploid) or more sets of these chromosomes.

[‡]Number for females, with two X chromosomes. Males have an X but no Y, thus 11 chromosomes in all.

Mitochondrial DNA codes for the mitochondrial tRNAs and rRNAs and for a few mitochondrial proteins. More than 95% of mitochondrial proteins are encoded by nuclear DNA. Mitochondria and chloroplasts divide when the cell divides. Their DNA is replicated before and during division, and the daughter DNA molecules pass into the daughter organelles.

## Eukaryotic Genes and Chromosomes Are Very Complex

Many bacterial species have only one chromosome per cell and, in nearly all cases, each chromosome contains only one copy of each gene. A very few genes, such as those for rRNAs, are repeated several times. Genes and regulatory sequences account for almost all the DNA in prokaryotes. Moreover, almost every gene is precisely colinear with the amino acid sequence (or RNA sequence) for which it codes (Fig. 24–2).

The organization of genes in eukaryotic DNA is structurally and functionally much more complex. The study of eukaryotic chromosome structure, and more recently the sequencing of entire eukaryotic genomes, has yielded many surprises. Many, if not most, eukaryotic genes have a distinctive and puzzling structural feature: their nucleotide sequences contain one or more intervening segments of DNA that do not code for the amino acid sequence of the polypeptide product. These nontranslated inserts interrupt the otherwise colinear relationship between the nucleotide sequence of the gene and the amino acid sequence of the polypeptide it encodes. Such nontranslated DNA segments in genes are called **intervening sequences** or **introns,** and the coding segments are called **exons.** Few prokaryotic genes contain introns.

In higher eukaryotes, the typical gene has much more intron sequence than sequences devoted to exons. For example, in the gene coding for the single polypeptide chain of the avian egg protein ovalbumin (Fig. 24–7), the introns are much longer than the exons; altogether, seven introns make up 85% of the gene's DNA. In the gene for the $\beta$ subunit of hemoglobin, a single intron contains more than half of the gene's DNA. The gene for the muscle protein titin is the intron champion, with 178 introns. Genes for histones appear to have no introns. In most cases the function of introns is not clear. In total, only about 1.5% of human DNA is "coding" or exon DNA, carrying information for protein or RNA products. However, when the much larger introns are included in the count, as much as 30% of the human genome consists of genes.

The relative paucity of genes in the human genome leaves a lot of DNA unaccounted for. Figure 24–8 provides a summary of sequence types. Much of the nongene DNA is in the form of repeated sequences of several kinds. Perhaps most surprising, about half the human genome is made up of moderately repeated sequences that are derived from transposable elements—segments of DNA, ranging from a few hundred to several thousand base pairs long, that can move from one location to another in the genome. Transposable elements (transposons) are a kind of molecular parasite, efficiently making a home within the host genome. Many have genes encoding proteins that catalyze the transposition process, described in more detail in Chapters 25 and 26. Some transposons in the human genome are active, moving at a low frequency, but most are inactive relics, evolutionarily altered by mutations. Although these elements generally do not encode proteins or RNAs that are used in human cells, they have played a

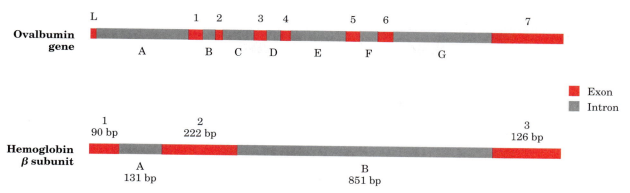

FIGURE 24–7 **Introns in two eukaryotic genes.** The gene for ovalbumin has seven introns (A to G), splitting the coding sequences into eight exons (L, and 1 to 7). The gene for the β subunit of hemoglobin has two introns and three exons, including one intron that alone contains more than half the base pairs of the gene.

major role in human evolution: movement of transposons can lead to the redistribution of other genomic sequences.

Another 3% or so of the human genome consists of **highly repetitive** sequences, also referred to as **simple-sequence DNA** or **simple sequence repeats (SSR).** These short sequences, generally less than 10 bp long, are sometimes repeated millions of times per cell. The simple-sequence DNA has also been called **satellite DNA,** so named because its unusual base composition often causes it to migrate as "satellite" bands (separated from the rest of the DNA) when fragmented cellular DNA samples are centrifuged in a cesium chloride density gradient. Studies suggest that simple-sequence DNA does not encode proteins or RNAs. Unlike the transposable elements, the highly repetitive DNA can have identifiable functional importance in human cellular metabolism, because much of it is associated with two defining features of eukaryotic chromosomes: centromeres and telomeres.

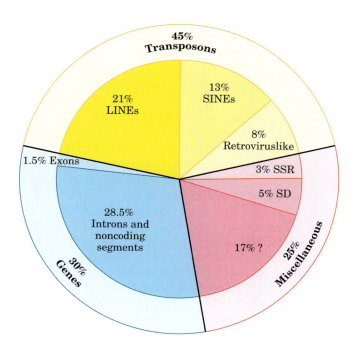

FIGURE 24–8 **Types of sequences in the human genome.** This pie chart divides the genome into transposons (transposable elements), genes, and miscellaneous sequences. There are four main classes of transposons. Long interspersed elements (LINEs), 6 to 8 kbp long (1 kbp = 1,000 bp), typically include a few genes encoding proteins that catalyze transposition. The genome has about 850,000 LINEs. Short interspersed elements (SINEs) are about 100 to 300 bp long. Of the 1.5 million in the human genome more than 1 million are Alu elements, so called because they generally include one copy of the recognition sequence for *Alu*I, a restriction endonuclease (see Fig. 9–3). The genome also contains 450,000 copies of retroviruslike transposons, 1.5 to 11 kbp long. Although these are "trapped" in the genome and cannot move from one cell to another, they are evolutionarily related to the retroviruses (Chapter 26), which include HIV. A final class of transposons (making up <1% and not shown here) consists of a variety of transposon remnants that differ greatly in length.

About 30% of the genome consists of sequences included in genes for proteins, but only a small fraction of this DNA is in exons (coding sequences). Miscellaneous sequences include simple-sequence repeats (SSR) and large segmental duplications (SD), the latter being segments that appear more than once in different locations. Among the unlisted sequence elements (denoted by a question mark) are genes encoding RNAs (which can be harder to identify than genes for proteins) and remnants of transposons that have been evolutionarily altered so that they are now hard to identify.

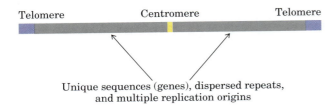

FIGURE 24–9 **Important structural elements of a yeast chromosome.**

The **centromere** (Fig. 24–9) is a sequence of DNA that functions during cell division as an attachment point for proteins that link the chromosome to the mitotic spindle. This attachment is essential for the equal and orderly distribution of chromosome sets to daughter cells. The centromeres of *Saccharomyces cerevisiae* have been isolated and studied. The sequences essential to centromere function are about 130 bp long and are very rich in A=T pairs. The centromeric sequences of higher eukaryotes are much longer and, unlike those of yeast, generally contain simple-sequence DNA, which consists of thousands of tandem copies of one or a few short sequences of 5 to 10 bp, in the same orientation. The precise role of simple-sequence DNA in centromere function is not yet understood.

**Telomeres** (Greek *telos,* "end") are sequences at the ends of eukaryotic chromosomes that help stabilize the chromosome. The best-characterized telomeres are those of the simpler eukaryotes. Yeast telomeres end with about 100 bp of imprecisely repeated sequences of the form

$$(5')(T_xG_y)_n$$
$$(3')(A_xC_y)_n$$

where $x$ and $y$ are generally between 1 and 4. The number of telomere repeats, $n$, is in the range of 20 to 100 for most single-celled eukaryotes and generally more than 1,500 in mammals. The ends of a linear DNA molecule cannot be routinely replicated by the cellular replication machinery (which may be one reason why bacterial DNA molecules are circular). Repeated telomeric sequences are added to eukaryotic chromosome ends primarily by the enzyme telomerase (see Fig. 26–35).

Artificial chromosomes (Chapter 9) have been constructed as a means of better understanding the functional significance of many structural features of eukaryotic chromosomes. A reasonably stable artificial linear chromosome requires only three components: a centromere, telomeres at each end, and sequences that allow the initiation of DNA replication. Yeast artificial chromosomes (YACs; see Fig. 9–8) have been developed as a research tool in biotechnology. Similarly, human artificial chromosomes (HACs) are being developed for the treatment of genetic diseases by somatic gene therapy.

## SUMMARY 24.1 Chromosomal Elements

- Genes are segments of a chromosome that contain the information for a functional polypeptide or RNA molecule. In addition to genes, chromosomes contain a variety of regulatory sequences involved in replication, transcription, and other processes.

- Genomic DNA and RNA molecules are generally orders of magnitude longer than the viral particles or cells that contain them.

- Many genes in eukaryotic cells, and a few in bacteria, are interrupted by noncoding sequences called introns. The coding segments separated by introns are called exons.

- Less than one-third of human genomic DNA consists of genes. Much of the remainder consists of repeated sequences of various types. Nucleic acid parasites known as transposons account for about half of the human genome.

- Eukaryotic chromosomes have two important special-function repetitive DNA sequences: centromeres, which are attachment points for the mitotic spindle, and telomeres, located at the ends of chromosomes.

## 24.2 DNA Supercoiling

Cellular DNA, as we have seen, is extremely compacted, implying a high degree of structural organization. The folding mechanism must not only pack the DNA but also permit access to the information in the DNA. Before considering how this is accomplished in processes such as replication and transcription, we need to examine an important property of DNA structure known as **supercoiling.**

Supercoiling means the coiling of a coil. A telephone cord, for example, is typically a coiled wire. The path taken by the wire between the base of the phone and the receiver often includes one or more supercoils (Fig. 24–10). DNA is coiled in the form of a double helix, with both strands of the DNA coiling around an axis. The further coiling of that axis upon itself (Fig. 24–11) produces DNA supercoiling. As detailed below, DNA supercoiling is generally a manifestation of structural strain. When there is no net bending of the DNA axis upon itself, the DNA is said to be in a **relaxed** state.

We might have predicted that DNA compaction involved some form of supercoiling. Perhaps less predictable is that replication and transcription of DNA also affect and are affected by supercoiling. Both processes

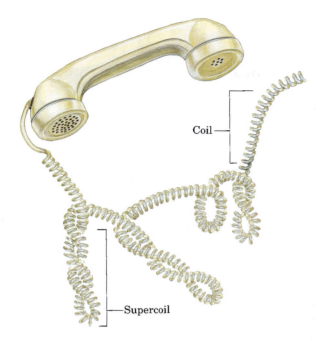

FIGURE 24–10 **Supercoils.** A typical phone cord is coiled like a DNA helix, and the coiled cord can itself coil in a supercoil. The illustration is especially appropriate because an examination of phone cords helped lead Jerome Vinograd and his colleagues to the insight that many properties of small circular DNAs can be explained by supercoiling. They first detected DNA supercoiling, in small circular viral DNAs, in 1965.

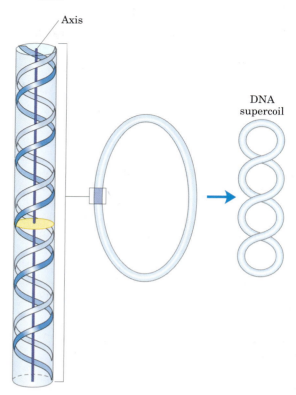

FIGURE 24–11 **Supercoiling of DNA.** When the axis of the DNA double helix is coiled on itself, it forms a new helix (superhelix). The DNA superhelix is usually called a supercoil.

require a separation of DNA strands—a process complicated by the helical interwinding of the strands (as demonstrated in Fig. 24–12).

That DNA would bend on itself and become supercoiled in tightly packaged cellular DNA would seem logical, then, and perhaps even trivial, were it not for one additional fact: many circular DNA molecules remain highly supercoiled even after they are extracted and purified, freed from protein and other cellular components. This indicates that supercoiling is an intrinsic property of DNA tertiary structure. It occurs in all cellular DNAs and is highly regulated by each cell.

A number of measurable properties of supercoiling have been established, and the study of supercoiling has provided many insights into DNA structure and function. This work has drawn heavily on concepts derived from a branch of mathematics called **topology,** the study of the properties of an object that do not change under continuous deformations. For DNA, continuous deformations include conformational changes due to thermal motion or an interaction with proteins or other molecules; discontinuous deformations involve DNA strand breakage. For circular DNA molecules, a topological property is one that is unaffected by deformations

FIGURE 24–12 **Supercoiling induced by separating the strands of a helical structure.** Twist two linear strands of rubber band into a right-handed double helix as shown. Fix one end by having a friend hold onto it, then pull apart the two strands at the other end. The resulting strain will produce supercoiling.

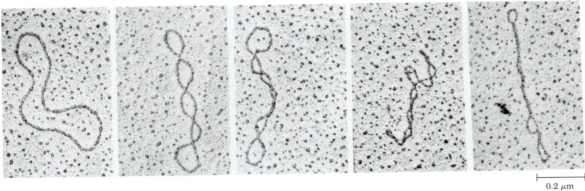

**FIGURE 24–13** **Relaxed and supercoiled plasmid DNAs.** The molecule in the leftmost electron micrograph is relaxed; the degree of supercoiling increases from left to right.

0.2 μm

of the DNA strands as long as no breaks are introduced. Topological properties are changed only by breakage and rejoining of the backbone of one or both DNA strands.

We now examine the fundamental properties and physical basis of supercoiling.

## Most Cellular DNA Is Underwound

To understand supercoiling we must first focus on the properties of small circular DNAs such as plasmids and small viral DNAs. When these DNAs have no breaks in either strand, they are referred to as **closed-circular DNAs.** If the DNA of a closed-circular molecule conforms closely to the B-form structure (the Watson-Crick structure; see Fig. 8–15), with one turn of the double helix per 10.5 bp, the DNA is relaxed rather than supercoiled (Fig. 24–13). Supercoiling results when DNA is subject to some form of structural strain. Purified closed-circular DNA is rarely relaxed, regardless of its biological origin. Furthermore, DNAs derived from a given cellular source have a characteristic degree of supercoiling. DNA structure is therefore strained in a manner that is regulated by the cell to induce the supercoiling.

In almost every instance, the strain is a result of **underwinding** of the DNA double helix in the closed circle. In other words, the DNA has *fewer* helical turns than would be expected for the B-form structure. The effects of underwinding are summarized in Figure 24–14. An 84 bp segment of a circular DNA in the relaxed state would contain eight double-helical turns, or one for every 10.5 bp. If one of these turns were removed, there would be (84 bp)/7 = 12.0 bp per turn, rather than the 10.5 found in B-DNA (Fig. 24–14b). This is a deviation from the most stable DNA form, and the molecule is thermodynamically strained as a result. Generally, much of this strain would be accommodated by coiling the axis of the DNA on itself to form a supercoil (Fig. 24–14c; some of the strain in this 84 bp segment would simply become dispersed in the untwisted struc-

ture of the larger DNA molecule). In principle, the strain could also be accommodated by separating the two DNA strands over a distance of about 10 bp (Fig. 24–14d). In isolated closed-circular DNA, strain introduced by underwinding is generally accommodated by supercoiling rather than strand separation, because coiling the axis of the DNA usually requires less energy than breaking the hydrogen bonds that stabilize paired bases. Note, however, that the underwinding of DNA in vivo makes

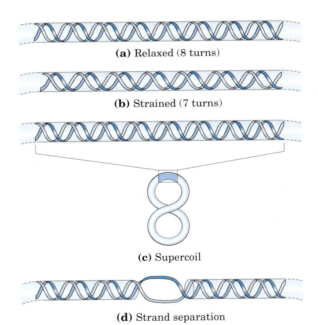

**(a)** Relaxed (8 turns)

**(b)** Strained (7 turns)

**(c)** Supercoil

**(d)** Strand separation

**FIGURE 24–14** **Effects of DNA underwinding.** **(a)** A segment of DNA within a closed-circular molecule, 84 bp long, in its relaxed form with eight helical turns. **(b)** Removal of one turn induces structural strain. **(c)** The strain is generally accommodated by formation of a supercoil. **(d)** DNA underwinding also makes the separation of strands somewhat easier. In principle, each turn of underwinding should facilitate strand separation over about 10 bp, as shown. However, the hydrogen-bonded base pairs would generally preclude strand separation over such a short distance, and the effect becomes important only for longer DNAs and higher levels of DNA underwinding.

it easier to separate DNA strands, giving access to the information they contain.

Every cell actively underwinds its DNA with the aid of enzymatic processes (described below), and the resulting strained state represents a form of stored energy. Cells maintain DNA in an underwound state to facilitate its compaction by coiling. The underwinding of DNA is also important to enzymes of DNA metabolism that must bring about strand separation as part of their function.

The underwound state can be maintained only if the DNA is a closed circle or if it is bound and stabilized by proteins so that the strands are not free to rotate about each other. If there is a break in one strand of an isolated, protein-free circular DNA, free rotation at that point will cause the underwound DNA to revert spontaneously to the relaxed state. In a closed-circular DNA molecule, however, the number of helical turns cannot be changed without at least transiently breaking one of the DNA strands. The number of helical turns in a DNA molecule therefore provides a precise description of supercoiling.

## DNA Underwinding Is Defined by Topological Linking Number

The field of topology provides a number of ideas that are useful to this discussion, particularly the concept of **linking number.** Linking number is a topological property of double-stranded DNA, because it does not vary when the DNA is bent or deformed, as long as both DNA strands remain intact. Linking number ($Lk$) is illustrated in Figure 24–15.

Let's begin by visualizing the separation of the two strands of a double-stranded circular DNA. If the two strands are linked as shown in Figure 24–15a, they are effectively joined by what can be described as a topological bond. Even if all hydrogen bonds and base-stacking interactions were abolished such that the strands were not in physical contact, this topological bond would still link the two strands. Visualize one of the circular strands as the boundary of a surface (such as a soap film spanning the space framed by a circular wire before you blow a soap bubble). The linking number can be defined as the number of times the second strand pierces this surface. For the molecule in Figure 24–15a, $Lk = 1$; for that in Figure 24–15b, $Lk = 6$. The linking number for a closed-circular DNA is always an integer. By convention, if the links between two DNA strands are arranged so that the strands are interwound in a right-handed helix, the linking number is defined as positive ($+$); for strands interwound in a left-handed helix, the linking number is negative ($-$). Negative linking numbers are, for all practical purposes, not encountered in DNA.

We can now extend these ideas to a closed-circular DNA with 2,100 bp (Fig. 24–16a). When the molecule

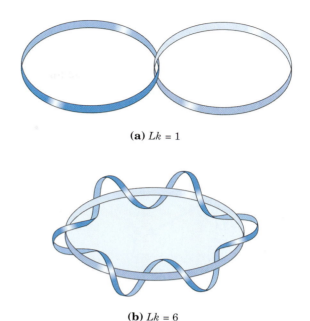

**(a)** $Lk = 1$

**(b)** $Lk = 6$

**FIGURE 24–15 Linking number, Lk.** Here, as usual, each blue ribbon represents one strand of a double-stranded DNA molecule. For the molecule in **(a)**, $Lk = 1$. For the molecule in **(b)**, $Lk = 6$. One of the strands in **(b)** is kept untwisted for illustrative purposes, to define the border of an imaginary surface (shaded blue). The number of times the twisting strand penetrates this surface provides a rigorous definition of linking number.

is relaxed, the linking number is simply the number of base pairs divided by the number of base pairs per turn, which is close to 10.5; so in this case, $Lk = 200$. For a circular DNA molecule to have a topological property such as linking number, neither strand may contain a break. If there is a break in either strand, the strands can, in principle, be unraveled and separated completely. In this case, no topological bond exists and $Lk$ is undefined (Fig. 24–16b).

We can now describe DNA underwinding in terms of changes in the linking number. The linking number in relaxed DNA, $Lk_0$, is used as a reference. For the molecule shown in Figure 24–16a, $Lk_0 = 200$; if two turns are removed from this molecule, $Lk = 198$. The change can be described by the equation

$$\Delta Lk = Lk - Lk_0 = 198 - 200 = -2$$

It is often convenient to express the change in linking number in terms of a quantity that is independent of the length of the DNA molecule. This quantity, called the **specific linking difference** ($\sigma$), or **superhelical density,** is a measure of the number of turns removed relative to the number present in relaxed DNA:

$$\sigma = \frac{\Delta Lk}{Lk_0}$$

In the example in Figure 24–16c, $\sigma = -0.01$, which means that 1% (2 of 200) of the helical turns present

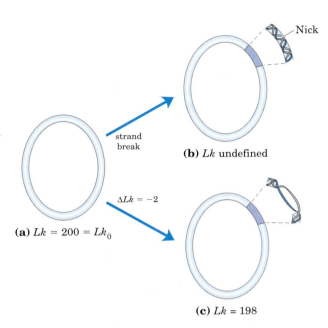

**(a)** $Lk = 200 = Lk_0$

**(b)** $Lk$ undefined

**(c)** $Lk = 198$

**FIGURE 24–16 Linking number applied to closed-circular DNA molecules.** A 2,100 bp circular DNA is shown in three forms: **(a)** relaxed, $Lk = 200$; **(b)** relaxed with a nick (break) in one strand, $Lk$ undefined; and **(c)** underwound by two turns, $Lk = 198$. The underwound molecule generally exists as a supercoiled molecule, but underwinding also facilitates the separation of DNA strands.

in the DNA (in its B form) have been removed. The degree of underwinding in cellular DNAs generally falls in the range of 5% to 7%; that is, $\sigma = -0.05$ to $-0.07$. The negative sign indicates that the change in linking number is due to underwinding of the DNA. The supercoiling induced by underwinding is therefore defined as negative supercoiling. Conversely, under some conditions DNA can be overwound, resulting in positive supercoiling. Note that the twisting path taken by the axis of the DNA helix when the DNA is underwound (negative supercoiling) is the mirror image of that taken when the DNA is overwound (positive supercoiling) (Fig. 24–17). Supercoiling is not a random process; the path of the supercoiling is largely prescribed by the torsional strain imparted to the DNA by decreasing or increasing the linking number relative to B-DNA.

Linking number can be changed by ±1 by breaking one DNA strand, rotating one of the ends 360° about the unbroken strand, and rejoining the broken ends. This change has no effect on the number of base pairs or the number of atoms in the circular DNA molecule. Two forms of a circular DNA that differ only in a topological property such as linking number are referred to as **topoisomers.**

Linking number can be broken down into two structural components called writhe ($Wr$) and twist ($Tw$) (Fig. 24–18). These are more difficult to describe than linking number, but writhe may be thought of as a measure of the coiling of the helix axis and twist as deter-

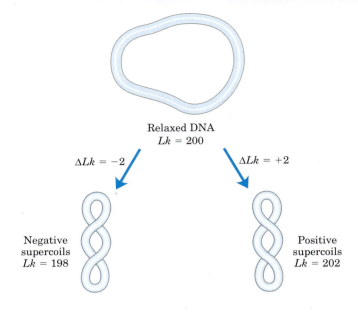

Relaxed DNA
$Lk = 200$

$\Delta Lk = -2$     $\Delta Lk = +2$

Negative supercoils
$Lk = 198$

Positive supercoils
$Lk = 202$

**FIGURE 24–17 Negative and positive supercoils.** For the relaxed DNA molecule of Figure 24–16a, underwinding or overwinding by two helical turns ($Lk = 198$ or $202$) will produce negative or positive supercoiling, respectively. Note that the DNA axis twists in opposite directions in the two cases.

mining the local twisting or spatial relationship of neighboring base pairs. When the linking number changes, some of the resulting strain is usually compensated for by writhe (supercoiling) and some by changes in twist, giving rise to the equation

$$Lk = Tw + Wr$$

$Tw$ and $Wr$ need not be integers. Twist and writhe are geometric rather than topological properties, because they may be changed by deformation of a closed-circular DNA molecule.

In addition to causing supercoiling and making strand separation somewhat easier, the underwinding of

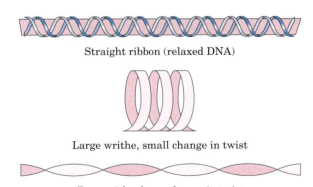

Straight ribbon (relaxed DNA)

Large writhe, small change in twist

Zero writhe, large change in twist

**FIGURE 24–18 Ribbon model for illustrating twist and writhe.** The pink ribbon represents the axis of a relaxed DNA molecule. Strain introduced by twisting the ribbon (underwinding the DNA) can be manifested as writhe or twist. Changes in linking number are usually accompanied by changes in both writhe and twist.

DNA facilitates a number of structural changes in the molecule. These are of less physiological importance but help illustrate the effects of underwinding. Recall that a cruciform (see Fig. 8–21) generally contains a few unpaired bases; DNA underwinding helps to maintain the required strand separation (Fig. 24–19). Underwinding of a right-handed DNA helix also facilitates the formation of short stretches of left-handed Z-DNA in regions where the base sequence is consistent with the Z form (Chapter 8).

## Topoisomerases Catalyze Changes in the Linking Number of DNA

DNA supercoiling is a precisely regulated process that influences many aspects of DNA metabolism. Every cell has enzymes with the sole function of underwinding and/or relaxing DNA. The enzymes that increase or decrease the extent of DNA underwinding are **topoisomerases;** the property of DNA that they change is the linking number. These enzymes play an especially im-

portant role in processes such as replication and DNA packaging. There are two classes of topoisomerases. **Type I topoisomerases** act by transiently breaking one of the two DNA strands, passing the unbroken strand through the break, and rejoining the broken ends; they change $Lk$ in increments of 1. **Type II topoisomerases** break both DNA strands and change $Lk$ in increments of 2.

The effects of these enzymes can be demonstrated using agarose gel electrophoresis (Fig. 24–20). A population of identical plasmid DNAs with the same linking number migrates as a discrete band during electrophoresis. Topoisomers with $Lk$ values differing by as little as 1 can be separated by this method, so changes in linking number induced by topoisomerases are readily detected.

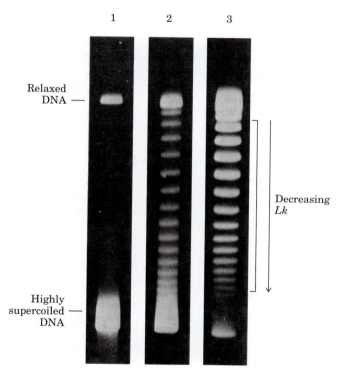

**FIGURE 24–20 Visualization of topoisomers.** In this experiment, all DNA molecules have the same number of base pairs but exhibit some range in the degree of supercoiling. Because supercoiled DNA molecules are more compact than relaxed molecules, they migrate more rapidly during gel electrophoresis. The gels shown here separate topoisomers (moving from top to bottom) over a limited range of superhelical density. In lane 1, highly supercoiled DNA migrates in a single band, even though different topoisomers are probably present. Lanes 2 and 3 illustrate the effect of treating the supercoiled DNA with a type I topoisomerase; the DNA in lane 3 was treated for a longer time than that in lane 2. As the superhelical density of the DNA is reduced to the point where it corresponds to the range in which the gel can resolve individual topoisomers, distinct bands appear. Individual bands in the region indicated by the bracket next to lane 3 each contain DNA circles with the same linking number; the linking number changes by 1 from one band to the next.

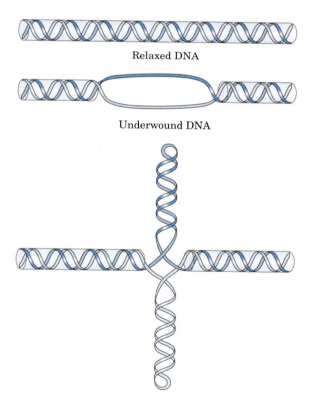

**FIGURE 24–19 Promotion of cruciform structures by DNA underwinding.** In principle, cruciforms can form at palindromic sequences (see Fig. 8–21), but they seldom occur in relaxed DNA because the linear DNA accommodates more paired bases than does the cruciform structure. Underwinding of the DNA facilitates the partial strand separation needed to promote cruciform formation at appropriate sequences.

**(a)**

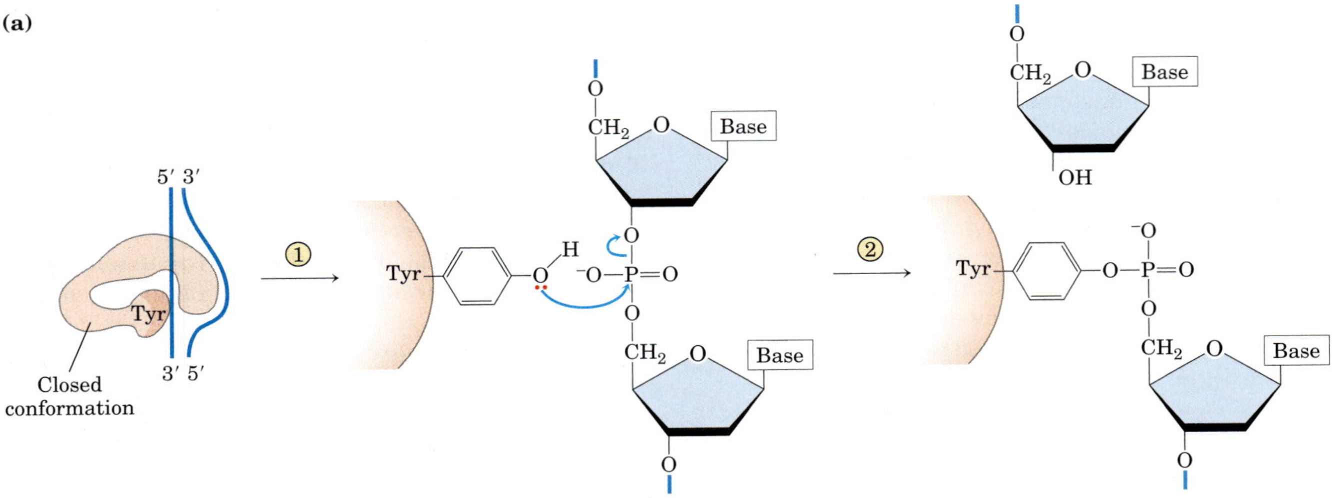

After DNA binds (step ①), an active-site Tyr attacks a phosphodiester bond on one DNA strand in step ②, cleaving it, creating a covalent 5'-Ⓟ–Tyr protein-DNA linkage, and liberating the 3'-hydroxyl group of the adjacent nucleotide.

**(b)**

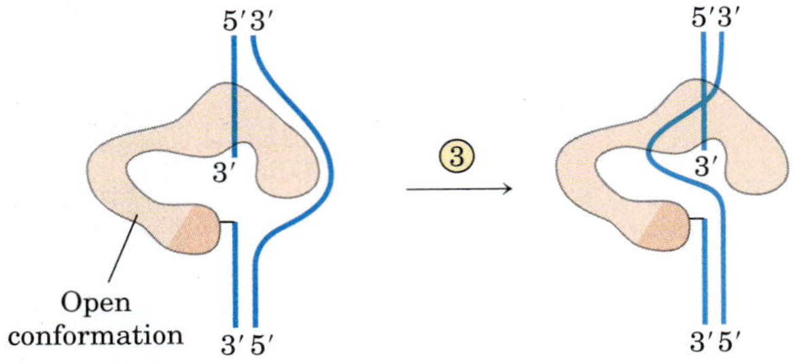

In step ③ the enzyme switches to its open conformation, and the unbroken DNA strand passes through the break in the first strand.

**(c)**

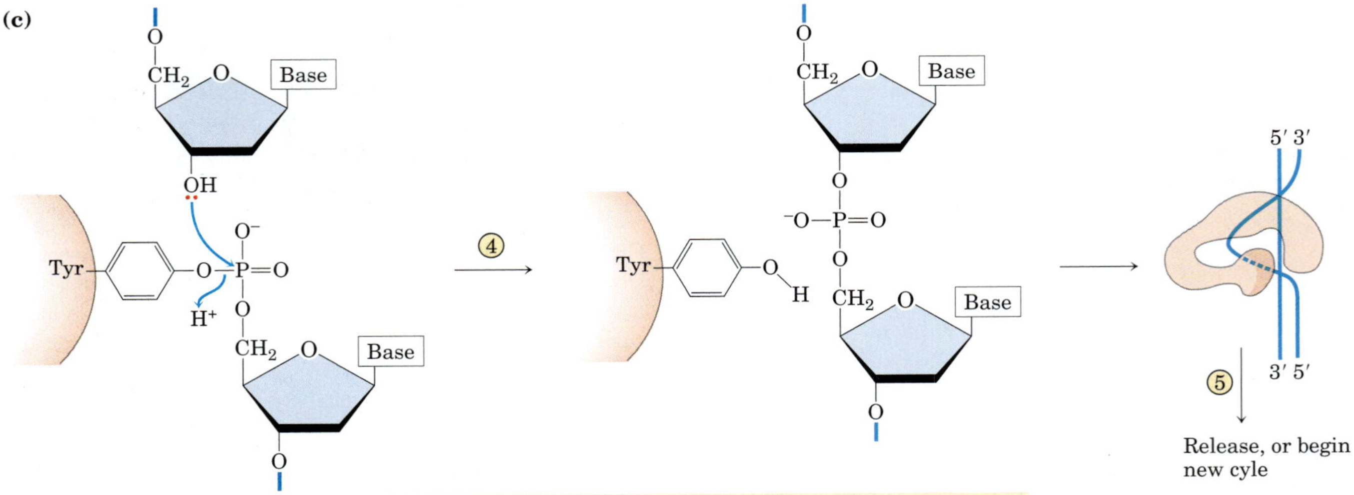

With the enzyme in the closed conformation, the liberated 3'-hydroxyl group attacks the 5'-Ⓟ–Tyr protein-DNA linkage in step ④ to religate the cleaved DNA strand.

**MECHANISM FIGURE 24–21 Bacterial type I topoisomerases alter linking number.** A proposed reaction sequence for the bacterial topoisomerase I is illustrated. The enzyme has closed and open conformations. **(a)** A DNA molecule binds to the closed conformation and one DNA strand is cleaved. **(b)** The enzyme changes to its open conformation, and the other DNA strand moves through the break in the first strand. **(c)** In the closed conformation, the DNA strand is religated.

*E. coli* has at least four different individual topoisomerases (I through IV). Those of type I (topoisomerases I and III) generally relax DNA by removing negative supercoils (increasing $Lk$). The way in which bacterial type I topoisomerases change linking number is illustrated in Figure 24–21. A bacterial type II enzyme, called either topoisomerase II or DNA gyrase, can introduce negative supercoils (decrease $Lk$). It uses the energy of ATP to accomplish this. To alter DNA linking number, type II topoisomerases cleave both strands of a DNA molecule and pass another duplex through the break. The degree of supercoiling of bacterial DNA is maintained by regulation of the net activity of topoisomerases I and II.

Eukaryotic cells also have type I and type II topoisomerases. The type I enzymes are topoisomerases I and III; the type II enzymes are topoisomerases II$\alpha$ and II$\beta$. The eukaryotic type II topoisomerases cannot underwind DNA (introduce negative supercoils), but they can relax both positive and negative supercoils. We consider one probable origin of negative supercoils in eukaryotic cells in our discussion of chromatin in Section 24.3. The process catalyzed by eukaryotic type II topoisomerases is illustrated in Figure 24–22.

## DNA Compaction Requires a Special Form of Supercoiling

Supercoiled DNA molecules are uniform in a number of respects. The supercoils are right-handed in a negatively supercoiled DNA molecule (Fig. 24–17), and they tend to be extended and narrow rather than compacted, often with multiple branches (Fig. 24–23). At the superhelical densities normally encountered in cells, the length of the supercoil axis, including branches, is about 40% of the length of the DNA. This type of supercoiling is referred to as **plectonemic** (from the Greek *plektos*, "twisted," and *nema*, "thread"). This term can be applied to any structure with strands intertwined in some simple and regular way, and it is a good description of the general structure of supercoiled DNA in solution.

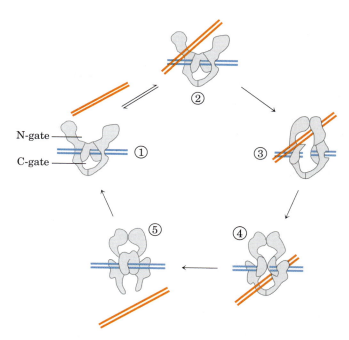

**FIGURE 24–22 Proposed mechanism for the alteration of linking number by eukaryotic type IIA topoisomerases.** ① The multisubunit enzyme binds one DNA molecule (blue). Gated cavities above and below the bound DNA are called the N-gate and the C-gate. ② A second segment of the same DNA molecule (red) is bound at the N-gate and ③ trapped. Both strands of the first DNA are now cleaved (the chemistry is similar to that in Fig. 24–20b), and ④ the second DNA segment is passed through the break. ⑤ The broken DNA is re-ligated, and the second DNA segment is released through the C-gate. Two ATPs are bound and hydrolyzed during this cycle; it is likely that one is hydrolyzed in the step leading to the complex in step ④. Additional details of the ATP hydrolysis component of the reaction remain to be worked out.

Plectonemic supercoiling, the form observed in isolated DNAs in the laboratory, does not produce sufficient compaction to package DNA in the cell. A second form of supercoiling, **solenoidal** (Fig. 24–24), can be adopted by an underwound DNA. Instead of the

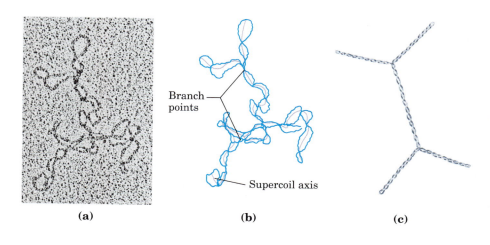

**(a)**        **(b)**        **(c)**

Branch points

Supercoil axis

**FIGURE 24–23 Plectonemic supercoiling. (a)** Electron micrograph of plectonemically supercoiled plasmid DNA and **(b)** an interpretation of the observed structure. The purple lines show the axis of the supercoil; note the branching of the supercoil. **(c)** An idealized representation of this structure.

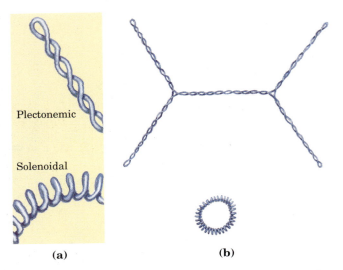

**FIGURE 24–24 Plectonemic and solenoidal supercoiling. (a)** Plectonemic supercoiling takes the form of extended right-handed coils. Solenoidal negative supercoiling takes the form of tight left-handed turns about an imaginary tubelike structure. The two forms are readily interconverted, although the solenoidal form is generally not observed unless certain proteins are bound to the DNA. **(b)** Plectonemic (top) and solenoidal supercoiling of the same DNA molecule, drawn to scale. Solenoidal supercoiling provides a much greater degree of compaction.

extended right-handed supercoils characteristic of the plectonemic form, solenoidal supercoiling involves tight left-handed turns, similar to the shape taken up by a garden hose neatly wrapped on a reel. Although their structures are dramatically different, plectonemic and solenoidal supercoiling are two forms of negative supercoiling that can be taken up by the *same* segment of underwound DNA. The two forms are readily interconvertible. Although the plectonemic form is more stable in solution, the solenoidal form can be stabilized by protein binding and is the form found in chromatin. It provides a much greater degree of compaction (Fig. 24–24b). Solenoidal supercoiling is the mechanism by which underwinding contributes to DNA compaction.

## SUMMARY 24.2 DNA Supercoiling

- Most cellular DNAs are supercoiled. Underwinding decreases the total number of helical turns in the DNA relative to the relaxed, B form. To maintain an underwound state, DNA must be either a closed circle or bound to protein. Underwinding is quantified by a topological parameter called linking number, $Lk$.

- Underwinding is measured in terms of specific linking difference, $\sigma$ (also called superhelical

density), which is $(Lk - Lk_0)/Lk_0$. For cellular DNAs, $\sigma$ is typically $-0.05$ to $-0.07$, which means that approximately 5% to 7% of the helical turns in the DNA have been removed. DNA underwinding facilitates strand separation by enzymes of DNA metabolism.

- DNAs that differ only in linking number are called topoisomers. Enzymes that underwind and/or relax DNA, the topoisomerases, catalyze changes in linking number. The two classes of topoisomerases, type I and type II, change $Lk$ in increments of 1 or 2, respectively, per catalytic event.

## 24.3 The Structure of Chromosomes

The term "chromosome" is used to refer to a nucleic acid molecule that is the repository of genetic information in a virus, a bacterium, a eukaryotic cell, or an organelle. It also refers to the densely colored bodies seen in the nuclei of dye-stained eukaryotic cells, as visualized using a light microscope.

### Chromatin Consists of DNA and Proteins

The eukaryotic cell cycle (see Fig. 12–41) produces remarkable changes in the structure of chromosomes (Fig. 24–25). In nondividing eukaryotic cells (in G0) and those in interphase (G1, S, and G2), the chromosomal material, **chromatin,** is amorphous and appears to be randomly dispersed in certain parts of the nucleus. In the S phase of interphase the DNA in this amorphous state replicates, each chromosome producing two sister chromosomes (called sister chromatids) that remain associated with each other after replication is complete. The chromosomes become much more condensed during prophase of mitosis, taking the form of a species-specific number of well-defined pairs of sister chromatids (Fig. 24–5).

Chromatin consists of fibers containing protein and DNA in approximately equal masses, along with a small amount of RNA. The DNA in the chromatin is very tightly associated with proteins called **histones,** which package and order the DNA into structural units called **nucleosomes** (Fig. 24–26). Also found in chromatin are many nonhistone proteins, some of which help maintain chromosome structure, others that regulate the expression of specific genes (Chapter 28). Beginning with nucleosomes, eukaryotic chromosomal DNA is packaged into a succession of higher-order structures that ultimately yield the compact chromosome seen with the light microscope. We now turn to a description of this structure in eukaryotes and compare it with the packaging of DNA in bacterial cells.

**FIGURE 24–25 Changes in chromosome structure during the eukaryotic cell cycle.** Cellular DNA is uncondensed throughout interphase. The interphase period can be subdivided (see Fig. 12–41) into the G1 (gap) phase; the S (synthesis) phase, when the DNA is replicated; and the G2 phase, in which the replicated chromosomes cohere to one another. The DNA undergoes condensation in the prophase of mitosis. Cohesins (green) and condensins (red) are proteins involved in cohesion and condensation (discussed later in the chapter). The architecture of the cohesin-condensin-DNA complex is not yet established, and the interactions shown here are figurative, simply suggesting their role in condensation of the chromosome. During metaphase, the condensed chromosomes line up along a plane halfway between the spindle poles. One chromosome of each pair is linked to each spindle pole via microtubules that extend between the spindle and the centromere. The sister chromatids separate at anaphase, each drawn toward the spindle pole to which it is connected. After cell division is complete, the chromosomes decondense and the cycle begins anew.

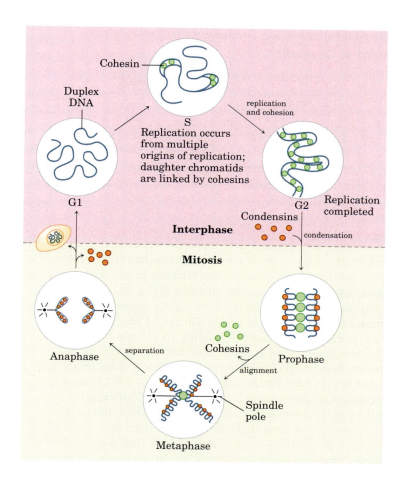

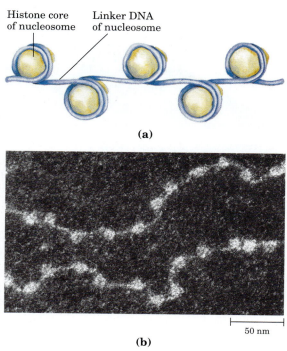

**FIGURE 24–26 Nucleosomes.** Regularly spaced nucleosomes consist of histone complexes bound to DNA. **(a)** Schematic illustration and **(b)** electron micrograph.

## Histones Are Small, Basic Proteins

Found in the chromatin of all eukaryotic cells, histones have molecular weights between 11,000 and 21,000 and are very rich in the basic amino acids arginine and lysine (together these make up about one-fourth of the amino acid residues). All eukaryotic cells have five major classes of histones, differing in molecular weight and amino acid composition (Table 24–3). The H3 histones are nearly identical in amino acid sequence in all eukaryotes, as are the H4 histones, suggesting strict conservation of their functions. For example, only 2 of 102 amino acid residues differ between the H4 histone molecules of peas and cows, and only 8 differ between the H4 histones of humans and yeast. Histones H1, H2A, and H2B show less sequence similarity among eukaryotic species.

Each type of histone has variant forms, because certain amino acid side chains are enzymatically modified by methylation, ADP-ribosylation, phosphorylation, glycosylation, or acetylation. Such modifications affect the net electric charge, shape, and other properties of histones, as well as the structural and functional properties of the chromatin, and they play a role in the regulation of transcription (Chapter 28).

## Nucleosomes Are the Fundamental Organizational Units of Chromatin

The eukaryotic chromosome depicted in Figure 24–5 represents the compaction of a DNA molecule about $10^5$ μm long into a cell nucleus that is typically 5 to 10 μm in diameter. This compaction involves several levels of highly organized folding. Subjection of chromosomes to treatments that partially unfold them reveals a structure in which the DNA is bound tightly to beads of protein, often regularly spaced (Fig. 24–26). The beads in this "beads-on-a-string" arrangement are complexes of histones and DNA. The bead plus the connecting DNA that leads to the next bead form the nucleosome, the fundamental unit of organization upon which the higher-order packing of chromatin is built. The bead of each nucleosome contains eight histone molecules: two copies each of H2A, H2B, H3, and H4. The spacing of the nucleosome beads provides a repeating unit typically of about 200 bp, of which 146 bp are bound tightly around the eight-part histone core and the remainder serve as linker DNA between nucleosome beads. Histone H1 binds to the linker DNA. Brief treatment of chromatin with enzymes that digest DNA causes preferential degradation of the linker DNA, releasing histone particles containing 146 bp of bound DNA that have been protected from digestion. Researchers have crystallized nucleosome cores obtained in this way, and x-ray diffraction analysis reveals a particle made up of the eight histone molecules with the DNA wrapped around it in the form of a left-handed solenoidal supercoil (Fig. 24–27).

A close inspection of this structure reveals why eukaryotic DNA is underwound even though eukaryotic cells lack enzymes that underwind DNA. Recall that the solenoidal wrapping of DNA in nucleosomes is but one form of supercoiling that can be taken up by underwound (negatively supercoiled) DNA. The tight wrapping of DNA around the histone core requires the removal of about one helical turn in the DNA. When the protein core of a nucleosome binds in vitro to a relaxed, closed-circular DNA, the binding introduces a negative supercoil. Because this binding process does not break the DNA or change the linking number, the formation of a negative solenoidal supercoil must be accompanied by a compensatory positive supercoil in the unbound region of the DNA (Fig. 24–28). As mentioned earlier, eukaryotic topoisomerases can relax positive supercoils. Relaxing the unbound positive supercoil leaves the negative supercoil fixed (through its binding to the nucleosome histone core) and results in an overall decrease in linking number. Indeed, topoisomerases have proved necessary for assembling chromatin from purified histones and closed-circular DNA in vitro.

Another factor that affects the binding of DNA to histones in nucleosome cores is the sequence of the

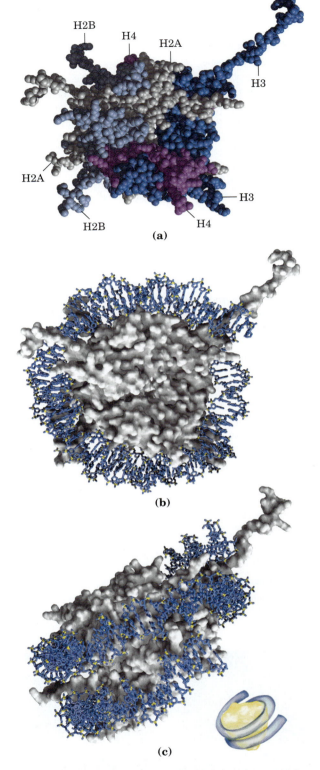

**FIGURE 24–27 DNA wrapped around a nucleosome core. (a)** Space-filling representation of the nucleosome protein core, with different colors for the different histones (PDB ID 1AOI). **(b)** Top and **(c)** side views of the crystal structure of a nucleosome with 146 bp of bound DNA. The protein is depicted as a gray surface contour, with the bound DNA in blue. The DNA binds in a left-handed solenoidal supercoil that circumnavigates the histone complex 1.8 times. A schematic drawing is included in **(c)** for comparison with other figures depicting nucleosomes.

**TABLE 24–3 Types and Properties of Histones**

| Histone | Molecular weight | Number of amino acid residues | Content of basic amino acids (% of total) | |
|---------|------------------|-------------------------------|----------|----------|
| | | | Lys | Arg |
| H1* | 21,130 | 223 | 29.5 | 11.3 |
| H2A* | 13,960 | 129 | 10.9 | 19.3 |
| H2B* | 13,774 | 125 | 16.0 | 16.4 |
| H3 | 15,273 | 135 | 19.6 | 13.3 |
| H4 | 11,236 | 102 | 10.8 | 13.7 |

*The sizes of these histones vary somewhat from species to species. The numbers given here are for bovine histones.

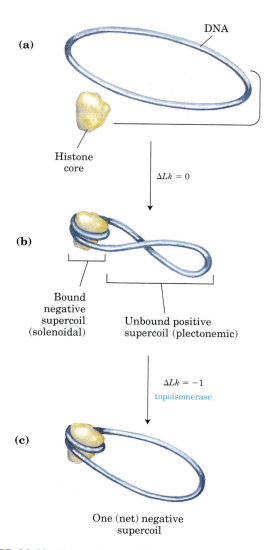

**FIGURE 24–28 Chromatin assembly. (a)** Relaxed, closed-circular DNA. **(b)** Binding of a histone core to form a nucleosome induces one negative supercoil. In the absence of any strand breaks, a positive supercoil must form elsewhere in the DNA ($\Delta Lk = 0$). **(c)** Relaxation of this positive supercoil by a topoisomerase leaves one net negative supercoil ($\Delta Lk = -1$).

bound DNA. Histone cores do not bind randomly to DNA; rather, they tend to position themselves at certain locations. This positioning is not fully understood but in some cases appears to depend on a local abundance of A≡T base pairs in the DNA helix where it is in contact with the histones (Fig. 24–29). The tight wrapping of the DNA around the nucleosome's histone core requires compression of the minor groove of the helix at these points, and a cluster of two or three A≡T base pairs makes this compression more likely.

Other proteins are required for the positioning of some nucleosome cores on DNA. In several organisms, certain proteins bind to a specific DNA sequence and then facilitate the formation of a nucleosome core nearby. Precise positioning of nucleosome cores can play a role in the expression of some eukaryotic genes (Chapter 28).

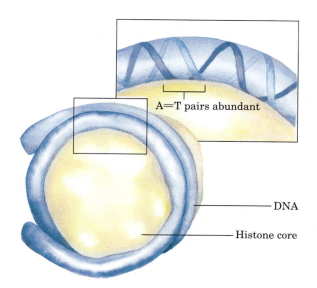

**FIGURE 24–29** Positioning of a nucleosome to make optimal use of A≡T base pairs where the histone core is in contact with the minor groove of the DNA helix.

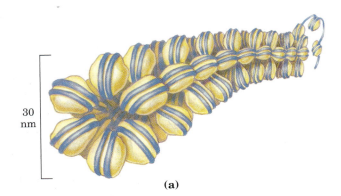

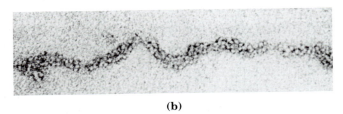

**(b)**

**FIGURE 24–30** **The 30 nm fiber, a higher-order organization of nucleosomes.** **(a)** Schematic illustration of the probable structure of the fiber, showing nucleosome packing. **(b)** Electron micrograph.

## Nucleosomes Are Packed into Successively Higher Order Structures

Wrapping of DNA around a nucleosome core compacts the DNA length about sevenfold. The overall compaction in a chromosome, however, is greater than 10,000-fold—ample evidence for even higher orders of structural organization. In chromosomes isolated by very gentle methods, nucleosome cores appear to be organized into a structure called the **30 nm fiber** (Fig. 24–30). This packing requires one molecule of histone H1 per nucleosome core. Organization into 30 nm fibers does not extend over the entire chromosome but is punctuated by regions bound by sequence-specific (nonhistone) DNA-binding proteins. The 30 nm structure also appears to depend on the transcriptional activity of the particular region of DNA. Regions in which genes are being transcribed are apparently in a less-ordered state that contains little, if any, histone H1.

The 30 nm fiber, a second level of chromatin organization, provides an approximately 100-fold compaction of the DNA. The higher levels of folding are not yet understood, but it appears that certain regions of DNA associate with a nuclear scaffold (Fig. 24–31). The scaffold-associated regions are separated by loops of DNA with perhaps 20 to 100 kbp. The DNA in a loop may contain a set of related genes. For example, in *Drosophila* complete sets of histone-coding genes seem to cluster together in loops that are bounded by scaffold attachment sites (Fig. 24–32). The scaffold itself appears to contain several proteins, notably large

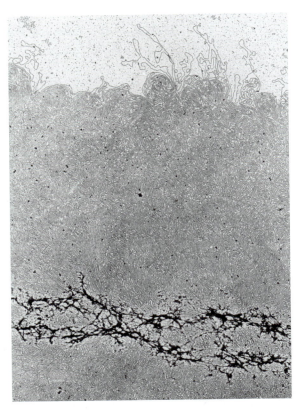

**FIGURE 24–31** A partially unraveled human chromosome, revealing numerous loops of DNA attached to a scaffoldlike structure.

amounts of histone H1 (located in the interior of the fiber) and topoisomerase II. The presence of topoisomerase II further emphasizes the relationship between DNA underwinding and chromatin structure. Topoisomerase II is so important to the maintenance of chromatin structure that inhibitors of this enzyme can kill

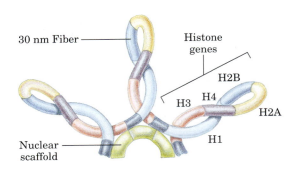

**FIGURE 24–32** **Loops of chromosomal DNA attached to a nuclear scaffold.** The DNA in the loops is packaged as 30 nm fibers, so the loops are the next level of organization. Loops often contain groups of genes with related functions. Complete sets of histone-coding genes, as shown in this schematic illustration, appear to be clustered in loops of this kind. Unlike most genes, histone genes occur in multiple copies in many eukaryotic genomes.

rapidly dividing cells. Several drugs used in cancer chemotherapy are topoisomerase II inhibitors that allow the enzyme to promote strand breakage but not the resealing of the breaks.

Evidence exists for additional layers of organization in eukaryotic chromosomes, each dramatically enhancing the degree of compaction. One model for achieving this compaction is illustrated in Figure 24–33. Higher-order chromatin structure probably varies from chromosome to chromosome, from one region to the next in a single chromosome, and from moment to moment in the life of a cell. No single model can adequately describe these structures. Nevertheless, the principle is clear: DNA compaction in eukaryotic chromosomes is likely to involve coils upon coils upon coils . . . Three-Dimensional Packaging of Nuclear Chromosomes

### Condensed Chromosome Structures Are Maintained by SMC Proteins

A third major class of chromatin proteins, in addition to the histones and topoisomerases, is the **SMC proteins** (structural *m*aintenance of *c*hromosomes). The primary structure of SMC proteins consists of five distinct domains (Fig. 24–34a). The amino- and carboxyl-terminal globular domains, N and C, each of which has part of an ATP hydrolytic site, are connected by two regions of α-helical coiled-coil motifs (see Fig. 4–11) that are joined by a hinge domain. The proteins are generally dimeric, forming a V-shaped complex that is thought to be tied together through their hinge domains (Fig. 24–34b). One N and one C domain come together to form a complete ATP hydrolytic site at each end of the V.

Proteins in the SMC family are found in all types of organisms, from bacteria to humans. Eukaryotes have two major types, cohesins and condensins (Fig. 24–25). The **cohesins** play a substantial role in linking together sister chromatids immediately after replication and keeping them together as the chromosomes condense to metaphase. This linkage is essential if chromosomes are to segregate properly at cell division. The detailed mechanism by which cohesins link sister chromosomes, and the role of ATP hydrolysis, are not yet understood. The **condensins** are essential to the condensation of chromosomes as cells enter mitosis. In the laboratory, condensins bind to DNA in a manner that creates positive supercoils; that is, condensin binding causes the DNA to become overwound, in contrast to the underwinding induced by the binding of nucleosomes. It is not yet clear how this helps to compact the chromatin, although one possibility is presented in Figure 24–35.

### Bacterial DNA Is Also Highly Organized

We now turn briefly to the structure of bacterial chromosomes. Bacterial DNA is compacted in a structure called the **nucleoid,** which can occupy a significant

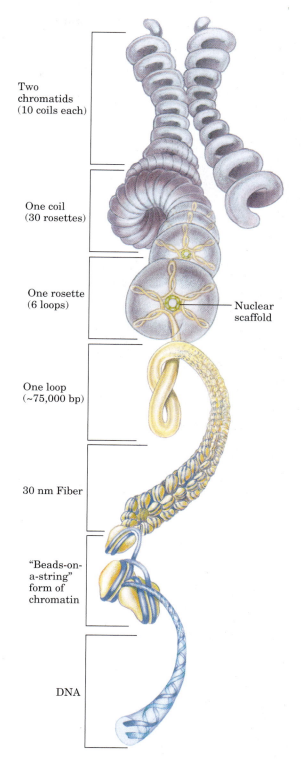

Two chromatids (10 coils each)

One coil (30 rosettes)

One rosette (6 loops) — Nuclear scaffold

One loop (~75,000 bp)

30 nm Fiber

"Beads-on-a-string" form of chromatin

DNA

**FIGURE 24–33** Compaction of DNA in a eukaryotic chromosome. Model for levels of organization that could provide DNA compaction in the chromosomes of eukaryotes. The levels take the form of coils upon coils. In cells, the higher-order structures (above the 30 nm fibers) are unlikely to be as uniform as depicted here.

**(a)**

N            Hinge             C

Coiled coil         Coiled coil

**(b)**

**(c)**

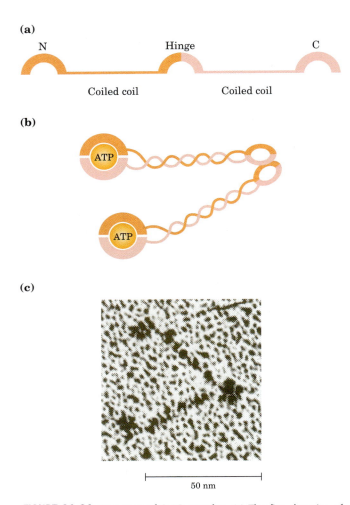

50 nm

**FIGURE 24–34 Structure of SMC proteins. (a)** The five domains of the SMC primary structure. N and C denoted the amino-terminal and carboxyl-terminal domains, respectively. **(b)** Each polypeptide is folded so that the two coiled-coil domains wrap around each other and the N and C domains come together to form a complete ATP-binding site. Two of these domains are linked at the hinge region to form the dimeric V-shaped molecule. **(c)** Electron micrograph of SMC proteins from *Bacillus subtilis*.

fraction of the cell volume (Fig. 24–36). The DNA appears to be attached at one or more points to the inner surface of the plasma membrane. Much less is known about the structure of the nucleoid than of eukaryotic chromatin. In *E. coli,* a scaffoldlike structure appears to organize the circular chromosome into a series of looped domains, as described above for chromatin. Bacterial DNA does not seem to have any structure comparable to the local organization provided by nucleosomes in eukaryotes. Histonelike proteins are abundant in *E. coli*—the best-characterized example is a two-subunit protein called HU ($M_r$ 19,000)—but these proteins bind and dissociate within minutes, and no regular, stable DNA-histone structure has been found. The bacterial chromosome is a relatively dy-

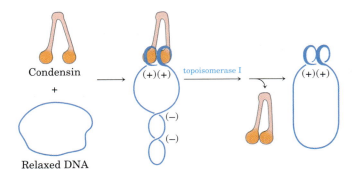

Condensin     (+)(+)    topoisomerase I    (+)(+)

+         (−)

              (−)

Relaxed DNA

**FIGURE 24–35 Model for the effect of condensins on DNA supercoiling.** Binding of condensins to a closed-circular DNA in the presence of topoisomerase I leads to the production of positive supercoils (+). Wrapping of the DNA about the condensin introduces positive supercoils because it wraps in the opposite sense to a solenoidal supercoil (see Fig. 24–24). The compensating negative supercoils (−) that appear elsewhere in the DNA are then relaxed by topoisomerase I. In the chromosome, it is the wrapping of the DNA about condensin that may contribute to DNA condensation.

namic molecule, possibly reflecting a requirement for more ready access to its genetic information. The bacterial cell division cycle can be as short as 15 min, whereas a typical eukaryotic cell may not divide for hours or even months. In addition, a much greater fraction of prokaryotic DNA is used to encode RNA and/or protein products. Higher rates of cellular metabolism in bacteria mean that a much higher proportion of the DNA is being transcribed or replicated at a given time than in most eukaryotic cells.

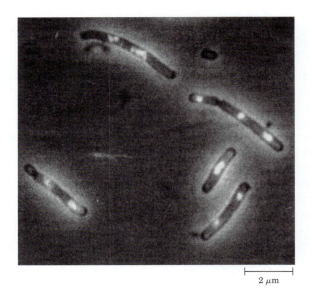

2 μm

**FIGURE 24–36 *E. coli* cells showing nucleoids.** The DNA is stained with a dye that fluoresces when exposed to UV light. The light area defines the nucleoid. Note that some cells have replicated their DNA but have not yet undergone cell division and hence have multiple nucleoids.

With this overview of the complexity of DNA structure, we are now ready to turn, in the next chapter, to a discussion of DNA metabolism.

### SUMMARY 24.3 The Structure of Chromosomes

- The fundamental unit of organization in the chromatin of eukaryotic cells is the nucleosome, which consists of histones and a 200 bp segment of DNA. A core protein particle containing eight histones (two copies each of histones H2A, H2B, H3, and H4) is encircled by a segment of DNA (about 146 bp) in the form of a left-handed solenoidal supercoil.

- Nucleosomes are organized into 30 nm fibers, and the fibers are extensively folded to provide the 10,000-fold compaction required to fit a typical eukaryotic chromosome into a cell nucleus. The higher-order folding involves attachment to a nuclear scaffold that contains histone H1, topoisomerase II, and SMC proteins.

- Bacterial chromosomes are also extensively compacted into the nucleoid, but the chromosome appears to be much more dynamic and irregular in structure than eukaryotic chromatin, reflecting the shorter cell cycle and very active metabolism of a bacterial cell.

## Key Terms

*Terms in bold are defined in the glossary.*

| | | | |
|---|---|---|---|
| **gene**  921 | **exon**  928 | underwinding  932 | solenoidal  937 |
| **genome**  923 | simple-sequence | **linking number**  933 | **chromatin**  938 |
| **chromosome**  923 | DNA  929 | specific linking difference | **histones**  938 |
| **phenotype**  924 | **satellite DNA**  929 | ($\sigma$)  933 | **nucleosome**  938 |
| **mutation**  924 | **centromere**  930 | **superhelical** | 30 nm fiber  942 |
| **regulatory** | **telomere**  930 | **density**  933 | SMC proteins  943 |
| **sequence**  924 | **supercoil**  930 | **topoisomers**  934 | cohesins  943 |
| **plasmid**  925 | **relaxed DNA**  930 | **topoisomerases**  935 | condensins  943 |
| **intron**  928 | **topology**  931 | **plectonemic**  937 | **nucleoid**  943 |

## Further Reading

### General

**Blattner, F.R., Plunkett, G., III, Bloch, C.A., Perna, N.T., Burland, V., Riley, M., Collado-Vides, J., Glasner, J.D., Rode, C.K., Mayhew, G.F., et al.** (1997) The complete genome sequence of *Escherichia coli* K-12. *Science* **277,** 1453–1474.

New secrets of this common laboratory organism are revealed.

**Cozzarelli, N.R. & Wang, J.C. (eds)** (1990) *DNA Topology and Its Biological Effects,* Cold Spring Harbor Laboratory Press, Cold Spring Harbor, NY.

**Kornberg, A. & Baker, T.A.** (1991) *DNA Replication,* 2nd edn, W. H. Freeman & Company, New York.

A good place to start for further information on the structure and function of DNA.

**Lodish, H., Berk, A., Matsudaira, P., Kaiser, C.A., Krieger, M., Scott, M.P., Zipursky, S.L., & Darnell, J.** (2003) *Molecular Cell Biology,* 5th edn, W. H. Freeman & Company, New York.

Another excellent general reference.

### Genes and Chromosomes

**Bromham, L.** (2002) The human zoo: endogenous retroviruses in the human genome. *Trends Ecol. Evolut.* **17,** 91–97.

A thorough description of one of the transposon classes that makes up a large part of the human genome.

**Goffeau, A., Barrell, B.G., Bussey, H., Davis, R.W., Dujon, B., Feldmann, H., Galibert, F., Hoheisel, J.D., Jacq, C., Johnston, M., et al.** (1996) Life with 6000 genes. *Science* **274,** 546, 563–567.

Report of the first complete sequence of a eukaryotic genome, the yeast *Saccharomyces cerevisiae.*

**Greider, C.W. & Blackburn, E.H.** (1996) Telomeres, telomerase and cancer. *Sci. Am.* **274** (February), 92–97.

**Huxley, C.** (1997) Mammalian artificial chromosomes and chromosome transgenics. *Trends Genet.* **13,** 345–347.

**Lander, E.S., Linton, L.M., Birren, B., Nusbaum, C., Zody, M.C., Baldwin, J., Devon, K., Dewar, K., Doyle, M., FitzHugh, W., et al.** (2001) Initial sequencing and analysis of the human genome. *Nature* **409,** 860–921.

One of the first reports on the draft sequence of the human genome, with lots of analysis and many associated articles.

**Long, M., de Souza, S.J., & Gilbert, W.** (1995) Evolution of the intron-exon structure of eukaryotic genes. *Curr. Opin. Genet. Dev.* **5,** 774–778.

**McEachern, M.J., Krauskopf, A., & Blackburn, E.H.** (2000) Telomeres and their control. *Annu. Rev. Genet.* **34,** 331–358.

**Schmid, C.W.** (1996) Alu: structure, origin, evolution, significance and function of one-tenth of human DNA. *Prog. Nucleic Acid Res. Mol. Biol.* **53**, 283–319.

**Tyler-Smith, C. & Floridia, G.** (2000) Many paths to the top of the mountain: diverse evolutionary solutions to centromere structure. *Cell* **102**, 5–8.

Details of the diversity of centromere structures from different organisms, as currently understood.

**Zakian, V.A.** (1996) Structure, function, and replication of *Saccharomyces cerevisiae* telomeres. *Annu. Rev. Genet.* **30**, 141–172.

### Supercoiling and Topoisomerases

**Berger, J.M.** (1998) Type II DNA topoisomerases. *Curr. Opin. Struct. Biol.* **8**, 26–32.

**Boles, T.C., White, J.H., & Cozzarelli, N.R.** (1990) Structure of plectonemically supercoiled DNA. *J. Mol. Biol.* **213**, 931–951.

A study that defines several fundamental features of supercoiled DNA.

**Champoux, J.J.** (2001) DNA topoisomerases: structure, function, and mechanism. *Annu. Rev. Biochem.* **70**, 369–413.

An excellent summary of the topoisomerase classes.

**Cozzarelli, N.R., Boles, T.C., & White, J.H.** (1990) Primer on the topology and geometry of DNA supercoiling. In *DNA Topology and Its Biological Effects* (Cozzarelli, N.R. & Wang, J.C., eds), pp. 139–184, Cold Spring Harbor Laboratory Press, Cold Spring Harbor, NY.

A more advanced and thorough discussion.

**Lebowitz, J.** (1990) Through the looking glass: the discovery of supercoiled DNA. *Trends Biochem. Sci.* **15**, 202–207.

A short and interesting historical note.

**Wang, J.C.** (2002) Cellular roles of DNA topoisomerases: a molecular perspective. *Nat. Rev. Mol. Cell Biol.* **3**, 430–440.

### Chromatin and Nucleosomes

**Filipski, J., Leblanc, J., Youdale, T., Sikorska, M., & Walker, P.R.** (1990) Periodicity of DNA folding in higher order chromatin structures. *EMBO J.* **9**, 1319–1327.

**Hirano, T.** (2002) The ABCs of SMC proteins: two-armed ATPases for chromosome condensation, cohesion and repair. *Genes Dev.* **16**, 399–414.

Description of the rapid advances in understanding of this interesting class of proteins.

**Kornberg, R.D.** (1974) Chromatin structure: a repeating unit of histones and DNA. *Science* **184**, 868–871.

A classic paper that introduced the subunit model for chromatin.

**Nasmyth, K.** (2002) Segregating sister genomes: the molecular biology of chromosome separation. *Science* **297**, 559–565.

**Wyman, C. & Kanaar, R.** (2002) Chromosome organization: reaching out to embrace new models. *Curr. Biol.* **12**, R446–R448.

A good, short summary of chromosome structure and the roles of SMC proteins within it.

**Zlatanova, J. & van Holde, K.** (1996) The linker histones and chromatin structure: new twists. *Prog. Nucleic Acid Res. Mol. Biol.* **52**, 217–259.

# Problems

**1. Packaging of DNA in a Virus** Bacteriophage T2 has a DNA of molecular weight $120 \times 10^6$ contained in a head about 210 nm long. Calculate the length of the DNA (assume the molecular weight of a nucleotide pair is 650) and compare it with the length of the T2 head.

**2. The DNA of Phage M13** The base composition of phage M13 DNA is A, 23%; T, 36%; G, 21%; C, 20%. What does this tell you about the DNA of phage M13?

**3. The *Mycoplasma* Genome** The complete genome of the simplest bacterium known, *Mycoplasma genitalium*, is a circular DNA molecule with 580,070 bp. Calculate the molecular weight and contour length (when relaxed) of this molecule. What is $Lk_0$ for the *Mycoplasma* chromosome? If $\sigma = -0.06$, what is $Lk$?

**4. Size of Eukaryotic Genes** An enzyme isolated from rat liver has 192 amino acid residues and is coded for by a gene with 1,440 bp. Explain the relationship between the number of amino acid residues in the enzyme and the number of nucleotide pairs in its gene.

**5. Linking Number** A closed-circular DNA molecule in its relaxed form has an $Lk$ of 500. Approximately how many base pairs are in this DNA? How is the linking number altered (increases, decreases, doesn't change, becomes undefined) when (a) a protein complex is bound to form a nucleosome, (b) one DNA strand is broken, (c) DNA gyrase and ATP are added to the DNA solution, or (d) the double helix is denatured by heat?

**6. Superhelical Density** Bacteriophage $\lambda$ infects *E. coli* by integrating its DNA into the bacterial chromosome. The success of this recombination depends on the topology of the *E. coli* DNA. When the superhelical density ($\sigma$) of the *E. coli* DNA is greater than $-0.045$, the probability of integration is $<20\%$; when $\sigma$ is less than $-0.06$, the probability is $>70\%$. Plasmid DNA isolated from an *E. coli* culture is found to have a length of 13,800 bp and an $Lk$ of 1,222. Calculate $\sigma$ for this DNA and predict the likelihood that bacteriophage $\lambda$ will be able to infect this culture.

**7. Altering Linking Number** (a) What is the $Lk$ of a 5,000 bp circular duplex DNA molecule with a nick in one strand? (b) What is the $Lk$ of the molecule in (a) when the nick is sealed (relaxed)? (c) How would the $Lk$ of the molecule in (b) be affected by the action of a single molecule of *E. coli* topoisomerase I? (d) What is the $Lk$ of the molecule in (b) after eight enzymatic turnovers by a single molecule of DNA gyrase in the presence of ATP? (e) What is the $Lk$ of the molecule in (d) after four enzymatic turnovers by a single molecule of bacterial type I topoisomerase? (f) What is the $Lk$ of the molecule in (d) after binding of one nucleosome?

**8. Chromatin**   Early evidence that helped researchers define nucleosome structure is illustrated by the agarose gel below, in which the thick bands represent DNA. It was generated by briefly treating chromatin with an enzyme that degrades DNA, then removing all protein and subjecting the purified DNA to electrophoresis. Numbers at the side of the gel denote the position to which a linear DNA of the indicated size would migrate. What does this gel tell you about chromatin structure? Why are the DNA bands thick and spread out rather than sharply defined?

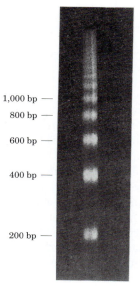

**9. DNA Structure**   Explain how the underwinding of a B-DNA helix might facilitate or stabilize the formation of Z-DNA.

**10. Maintaining DNA Structure**   (a) Describe two structural features required for a DNA molecule to maintain a negatively supercoiled state. (b) List three structural changes that become more favorable when a DNA molecule is negatively supercoiled. (c) What enzyme, with the aid of ATP, can generate negative superhelicity in DNA? (d) Describe the physical mechanism by which this enzyme acts.

**11. Yeast Artificial Chromosomes (YACs)**   YACs are used to clone large pieces of DNA in yeast cells. What three types of DNA sequences are required to ensure proper replication and propagation of a YAC in a yeast cell?

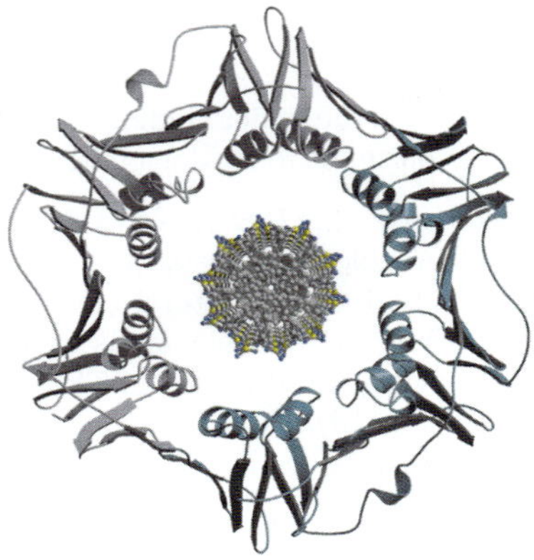

# DNA METABOLISM

I have never known a dull enzyme.

—*Arthur Kornberg, in the essay "For Love of Enzymes," 1975*

As the repository of genetic information, DNA occupies a unique and central place among biological macromolecules. The nucleotide sequences of DNA encode the primary structures of all cellular RNAs and proteins and, through enzymes, indirectly affect the synthesis of all other cellular constituents. This passage of information from DNA to RNA and protein guides the size, shape, and functioning of every living thing.

DNA is a marvelous device for the stable storage of genetic information. The phrase "stable storage," however, conveys a static and misleading picture. It fails to capture the complexity of processes by which genetic information is preserved in an uncorrupted state and then transmitted from one generation of cells to the next. DNA metabolism comprises both the process by which copies of DNA molecules are faithfully made (replication) and the processes that affect the inherent structure of the information (repair and recombination). Together, these activities are the focus of this chapter.

The metabolism of DNA is shaped by the requirement for an exquisite degree of accuracy. The chemistry of joining one nucleotide to the next in DNA replication is elegant and simple, almost deceptively so. As we shall see, complexity arises in the form of enzymatic devices to ensure that the genetic information is transmitted intact. Uncorrected errors that arise during DNA synthesis can have dire consequences, not only because they can permanently affect or eliminate the function of a gene but also because the change is inheritable.

The enzymes that synthesize DNA may copy DNA molecules that contain millions of bases. They do so with extraordinary fidelity and speed, even though the DNA substrate is highly compacted and bound with other proteins. Formation of phosphodiester bonds to link nucleotides in the backbone of a growing DNA strand is therefore only one part of an elaborate process that requires myriad proteins and enzymes.

Maintaining the integrity of genetic information lies at the heart of DNA repair. As detailed in Chapter 8, DNA is susceptible to many types of damaging reactions. Such reactions are infrequent but significant nevertheless, because of the very low biological tolerance for changes in DNA sequence. DNA is the only macromolecule for which repair systems exist; the number, diversity, and complexity of DNA repair mechanisms reflect the wide range of insults that can harm DNA.

Cells can rearrange their genetic information by processes collectively called recombination—seemingly undermining the principle that the stability and integrity of genetic information are paramount. However, most DNA rearrangements in fact play constructive roles in maintaining genomic integrity, contributing in special ways to DNA replication, DNA repair, and chromosome segregation.

Special emphasis is given in this chapter to the *enzymes* of DNA metabolism. They merit careful study, both for their increasing importance in medicine and for their everyday use as reagents in a wide range of modern biochemical technologies. Many of the seminal discoveries in DNA metabolism have been made with *Escherichia coli,* so its well-understood enzymes are generally used to illustrate the ground rules. A quick look at some relevant genes on the *E. coli* genetic map (Fig. 25–1) provides just a hint of the complexity of the enzymatic systems involved in DNA metabolism.

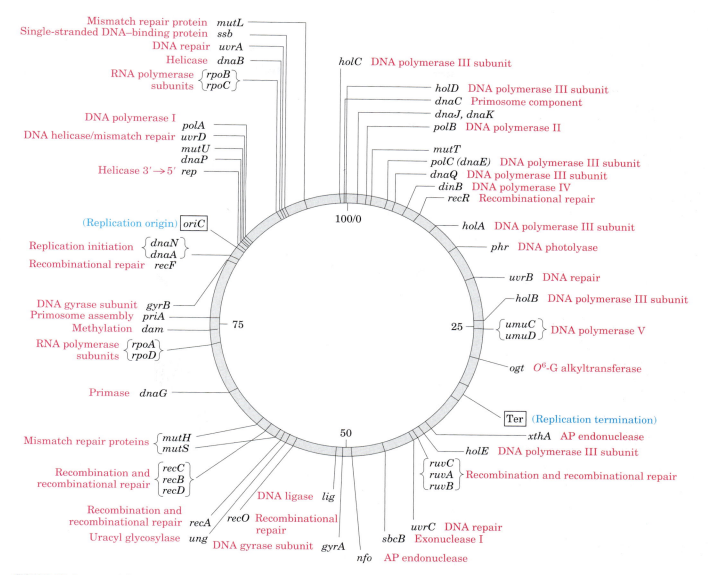

**FIGURE 25–1 Map of the *E. coli* chromosome.** The map shows the relative positions of genes encoding many of the proteins important in DNA metabolism. The number of genes known to be involved provides a hint of the complexity of these processes. The numbers 0 to 100 inside the circular chromosome denote a genetic measurement called minutes. Each minute corresponds to ~40,000 bp along the DNA molecule of *E. coli*. The three-letter names of genes and other elements generally reflect some aspect of their function. These include *mut*, mutagenesis; *dna*, DNA replication; *pol*, DNA polymerase; *rpo*, RNA polymerase; *uvr*, UV resistance; *rec*, recombination; *dam*, DNA adenine methylation; *lig*, DNA ligase; Ter, termination of replication; and ori, origin of replication.

***A Word about Terminology*** Before beginning to look closely at replication, we must make a short digression into the use of abbreviations in naming genes and proteins. By convention, bacterial genes generally are named using three lowercase, italicized letters that often reflect their apparent function. For example, the *dna*, *uvr*, and *rec* genes affect *DNA* replication, *resistance* to the damaging effects of *UV* radiation, and *recombination*, respectively. Where several genes affect the same process, the letters *A*, *B*, *C*, and so forth, are added—as in *dnaA*, *dnaB*, *dnaQ*, for example—usually reflecting their order of discovery rather than their order in a reaction sequence.

During genetic investigations, the protein product of each gene is usually isolated and characterized. Many bacterial genes have been identified and named before the roles of their protein products are understood in detail. Sometimes the gene product is found to be a previously isolated protein, and some renaming occurs. Often the product turns out to be an as yet unknown protein, with an activity not easily described by a simple enzyme name. In a practice that can be confusing, these bacterial proteins often retain the name of their genes. When referring to the protein, roman type is used and the first letter is capitalized: for example, the *dnaA* and *recA* gene products are called the DnaA and RecA proteins, respectively. You will encounter many such examples in this chapter.

Similar conventions exist for the naming of eukaryotic genes, although the exact form of the abbreviations may vary with the species and no single convention applies to all eukaryotic systems.

## 25.1 DNA Replication

Long before the structure of DNA was known, scientists wondered at the ability of organisms to create faithful copies of themselves and, later, at the ability of cells to produce many identical copies of large and complex macromolecules. Speculation about these problems centered around the concept of a **template,** a structure that would allow molecules to be lined up in a specific order and joined, to create a macromolecule with a unique sequence and function. The 1940s brought the revelation that DNA was the genetic molecule, but not until James Watson and Francis Crick deduced its structure did the way in which DNA could act as a template for the replication and transmission of genetic information become clear: *one strand is the complement of the other.* The strict base-pairing rules mean that each strand provides the template for a sister strand with a predictable and complementary sequence (see Figs 8–16, 8–17). 🖱 **Nucleotides: Building Blocks of Nucleic Acids**

The fundamental properties of the DNA replication process and the mechanisms used by the enzymes that catalyze it have proved to be essentially identical in all species. This mechanistic unity is a major theme as we proceed from general properties of the replication process, to *E. coli* replication enzymes, and, finally, to replication in eukaryotes.

### DNA Replication Follows a Set of Fundamental Rules

Early research on bacterial DNA replication and its enzymes helped to establish several basic properties that have proven applicable to DNA synthesis in every organism.

***DNA Replication Is Semiconservative*** Each DNA strand serves as a template for the synthesis of a new strand, producing two new DNA molecules, each with one new strand and one old strand. This is **semiconservative replication.**

Watson and Crick proposed the hypothesis of semiconservative replication soon after publication of their 1953 paper on the structure of DNA, and the hypothesis was proved by ingeniously designed experiments carried out by Matthew Meselson and Franklin Stahl in

1957. Meselson and Stahl grew *E. coli* cells for many generations in a medium in which the sole nitrogen source ($NH_4Cl$) contained $^{15}N$, the "heavy" isotope of nitrogen, instead of the normal, more abundant "light" isotope, $^{14}N$. The DNA isolated from these cells had a density about 1% greater than that of normal [$^{14}N$]DNA (Fig. 25–2a). Although this is only a small difference, a mixture of heavy [$^{15}N$]DNA and light [$^{14}N$]DNA can be separated by centrifugation to equilibrium in a cesium chloride density gradient.

The *E. coli* cells grown in the $^{15}N$ medium were transferred to a fresh medium containing only the $^{14}N$ isotope, where they were allowed to grow until the cell population had just doubled. The DNA isolated from these first-generation cells formed a *single* band in the CsCl gradient at a position indicating that the double-helical DNA molecules of the daughter cells were hybrids containing one new $^{14}N$ strand and one parent $^{15}N$ strand (Fig. 25–2b).

This result argued against conservative replication, an alternative hypothesis in which one progeny DNA

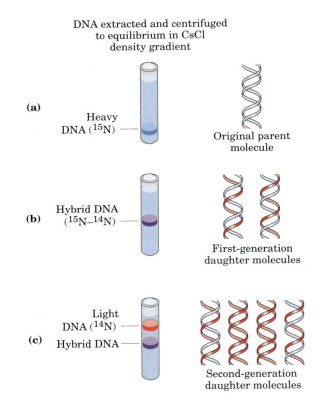

DNA extracted and centrifuged to equilibrium in CsCl density gradient

**(a)** Heavy DNA ($^{15}N$) — Original parent molecule

**(b)** Hybrid DNA ($^{15}N$–$^{14}N$) — First-generation daughter molecules

**(c)** Light DNA ($^{14}N$) / Hybrid DNA — Second-generation daughter molecules

**FIGURE 25–2 The Meselson-Stahl experiment. (a)** Cells were grown for many generations in a medium containing only heavy nitrogen, $^{15}N$, so that all the nitrogen in their DNA was $^{15}N$, as shown by a single band (blue) when centrifuged in a CsCl density gradient. **(b)** Once the cells had been transferred to a medium containing only light nitrogen, $^{14}N$, cellular DNA isolated after one generation equilibrated at a higher position in the density gradient (purple band). **(c)** Continuation of replication for a second generation yielded two hybrid DNAs and two light DNAs (red), confirming semiconservative replication.

molecule would consist of two newly synthesized DNA strands and the other would contain the two parent strands; this would not yield hybrid DNA molecules in the Meselson-Stahl experiment. The semiconservative replication hypothesis was further supported in the next step of the experiment (Fig. 25–2c). Cells were again allowed to double in number in the $^{14}$N medium. The isolated DNA product of this second cycle of replication exhibited *two* bands in the density gradient, one with a density equal to that of light DNA and the other with the density of the hybrid DNA observed after the first cell doubling.

### Replication Begins at an Origin and Usually Proceeds Bidirectionally

Following the confirmation of a semiconservative mechanism of replication, a host of questions arose. Are the parent DNA strands completely unwound before each is replicated? Does replication begin at random places or at a unique point? After initiation at any point in the DNA, does replication proceed in one direction or both?

An early indication that replication is a highly coordinated process in which the parent strands are simultaneously unwound and replicated was provided by John Cairns, using autoradiography. He made *E. coli* DNA radioactive by growing cells in a medium containing thymidine labeled with tritium ($^{3}$H). When the DNA was carefully isolated, spread, and overlaid with a photographic emulsion for several weeks, the radioactive thymidine residues generated "tracks" of silver grains in the emulsion, producing an image of the DNA molecule. These tracks revealed that the intact chromosome of *E. coli* is a single huge circle, 1.7 mm long. Radioactive DNA isolated from cells during replication showed an extra loop (Fig. 25–3a). Cairns concluded that the loop resulted from the formation of two radioactive daughter strands, each complementary to a parent strand. One or both ends of the loop are dynamic points, termed **replication forks,** where parent DNA is being unwound and the separated strands quickly replicated. Cairns's results demonstrated that both DNA strands are replicated simultaneously, and a variation on his experiment (Fig. 25–3b) indicated that replication of bacterial chromosomes is bidirectional: both ends of the loop have active replication forks.

The determination of whether the replication loops originate at a unique point in the DNA required landmarks along the DNA molecule. These were provided

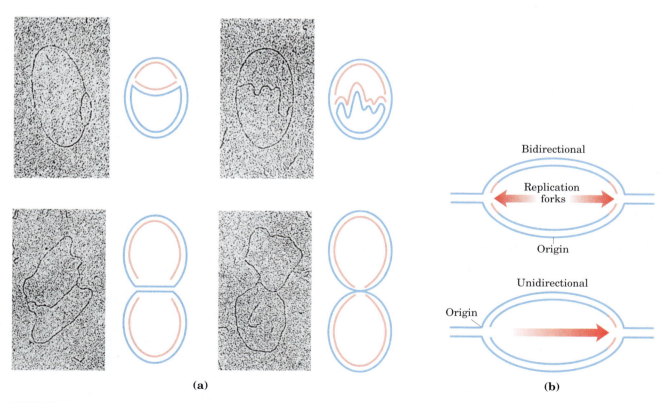

**(a)**

**(b)**

**FIGURE 25–3 Visualization of bidirectional DNA replication.** Replication of a circular chromosome produces a structure resembling the Greek letter theta ($\theta$). **(a)** Labeling with tritium ($^{3}$H) shows that both strands are replicated at the same time (new strands shown in red). The electron micrographs illustrate the replication of a circular *E. coli* plasmid as visualized by autoradiography. **(b)** Addition of $^{3}$H for a short period just before the reaction is stopped allows a distinction to be made between unidirectional and bidirectional replication, by determining whether label (red) is found at one or both replication forks in autoradiograms. This technique has revealed bidirectional replication in *E. coli, Bacillus subtilis,* and other bacteria.

by a technique called **denaturation mapping,** developed by Ross Inman and colleagues. Using the 48,502 bp chromosome of bacteriophage λ, Inman showed that DNA could be selectively denatured at sequences unusually rich in A=T base pairs, generating a reproducible pattern of single-strand bubbles (see Fig. 8–31). Isolated DNA containing replication loops can be partially denatured in the same way. This allows the position and progress of the replication forks to be measured and mapped, using the denatured regions as points of reference. The technique revealed that in this system the replication loops always initiate at a unique point, which was termed an **origin.** It also confirmed the earlier observation that replication is usually bidirectional. For circular DNA molecules, the two replication forks meet at a point on the side of the circle opposite to the origin. Specific origins of replication have since been identified and characterized in bacteria and lower eukaryotes.

### DNA Synthesis Proceeds in a 5′→3′ Direction and Is Semidiscontinuous

A new strand of DNA is always synthesized in the 5′→3′ direction, with the free 3′ OH as the point at which the DNA is elongated (the 5′ and 3′ ends of a DNA strand are defined in Fig. 8–7). Because the two DNA strands are antiparallel, the strand serving as the template is read from its 3′ end toward its 5′ end.

If synthesis always proceeds in the 5′→3′ direction, how can both strands be synthesized simultaneously? If both strands were synthesized *continuously* while the replication fork moved, one strand would have to undergo 3′→5′ synthesis. This problem was resolved by Reiji Okazaki and colleagues in the 1960s. Okazaki found that one of the new DNA strands is synthesized in short pieces, now called **Okazaki fragments.** This work ultimately led to the conclusion that one strand is synthesized continuously and the other discontinuously (Fig. 25–4). The continuous strand, or **leading strand,** is the one in which 5′→3′ synthesis proceeds in the *same* direction as replication fork movement. The discontinuous strand, or **lagging strand,** is the one in which 5′→3′ synthesis proceeds in the direction *opposite* to the direction of fork movement. Okazaki fragments range in length from a few hundred to a few thousand nucleotides, depending on the cell type. As we shall see later, leading and lagging strand syntheses are tightly coordinated.

### DNA Is Degraded by Nucleases

To explain the enzymology of DNA replication, we first introduce the enzymes that degrade DNA rather than synthesize it. These enzymes are known as **nucleases,** or **DNases** if they are specific for DNA rather than RNA. Every cell contains several different nucleases, belonging to two broad classes: exonucleases and endonucleases. **Exonucleases** degrade nucleic acids from one

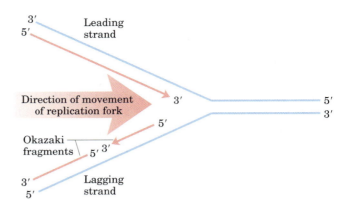

**FIGURE 25–4 Defining DNA strands at the replication fork.** A new DNA strand (red) is always synthesized in the 5′→3′ direction. The template is read in the opposite direction, 3′→5′. The leading strand is continuously synthesized in the direction taken by the replication fork. The other strand, the lagging strand, is synthesized discontinuously in short pieces (Okazaki fragments) in a direction opposite to that in which the replication fork moves. The Okazaki fragments are spliced together by DNA ligase. In bacteria, Okazaki fragments are ~1,000 to 2,000 nucleotides long. In eukaryotic cells, they are 150 to 200 nucleotides long.

end of the molecule. Many operate in only the 5′→3′ or the 3′→5′ direction, removing nucleotides only from the 5′ or the 3′ end, respectively, of one strand of a double-stranded nucleic acid or of a single-stranded DNA. **Endonucleases** can begin to degrade at specific internal sites in a nucleic acid strand or molecule, reducing it to smaller and smaller fragments. A few exonucleases and endonucleases degrade only single-stranded DNA. There are a few important classes of endonucleases that cleave only at specific nucleotide sequences (such as the restriction endonucleases that are so important in biotechnology; see Chapter 9, Fig. 9–3). You will encounter many types of nucleases in this and subsequent chapters.

### DNA Is Synthesized by DNA Polymerases

The search for an enzyme that could synthesize DNA began in 1955. Work by Arthur Kornberg and colleagues led to the purification and characterization of DNA polymerase from *E. coli* cells, a single-polypeptide enzyme now called **DNA polymerase I** ($M_r$ 103,000; encoded by the *polA* gene). Much later, investigators found that *E. coli* contains at least four other distinct DNA polymerases, described below.

Arthur Kornberg

Detailed studies of DNA polymerase I revealed features of the DNA synthetic process that are now known to be common to all DNA polymerases. The fundamental reaction is a phosphoryl group transfer. The nucleophile is the 3'-hydroxyl group of the nucleotide at the 3' end of the growing strand. Nucleophilic attack occurs at the α phosphorus of the incoming deoxynucleoside 5'-triphosphate (Fig. 25–5). Inorganic pyrophosphate is released in the reaction. The general reaction is

$$(dNMP)_n + dNTP \longrightarrow (dNMP)_{n+1} + PP_i \qquad (25\text{–}1)$$
$$\text{DNA} \qquad\qquad \text{Lengthened} \qquad\qquad\qquad$$
$$\text{DNA} \qquad\qquad\qquad\qquad$$

where dNMP and dNTP are deoxynucleoside 5'-mono-

phosphate and 5'-triphosphate, respectively. The reaction appears to proceed with only a minimal change in free energy, given that one phosphodiester bond is formed at the expense of a somewhat less stable phosphate anhydride. However, noncovalent base-stacking and base-pairing interactions provide additional stabilization to the lengthened DNA product relative to the free nucleotide. Also, the formation of products is facilitated in the cell by the 19 kJ/mol generated in the subsequent hydrolysis of the pyrophosphate product by the enzyme pyrophosphatase.

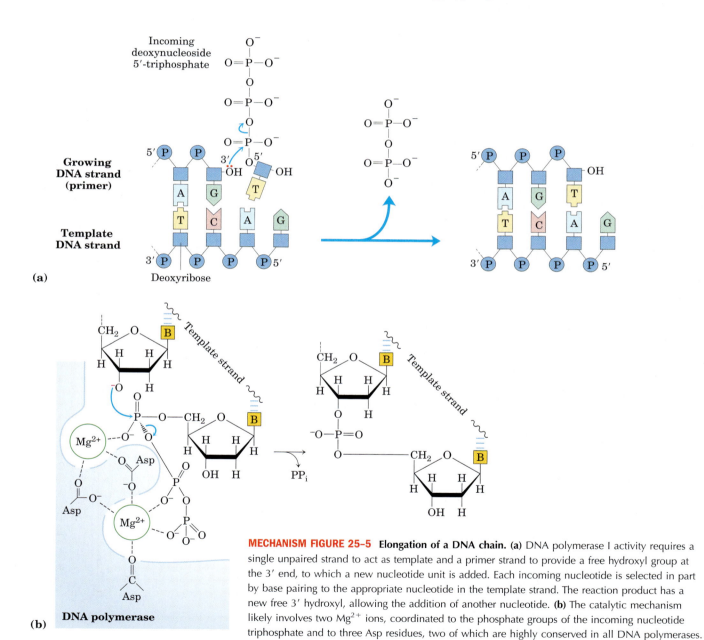

**MECHANISM FIGURE 25–5 Elongation of a DNA chain. (a)** DNA polymerase I activity requires a single unpaired strand to act as template and a primer strand to provide a free hydroxyl group at the 3' end, to which a new nucleotide unit is added. Each incoming nucleotide is selected in part by base pairing to the appropriate nucleotide in the template strand. The reaction product has a new free 3' hydroxyl, allowing the addition of another nucleotide. **(b)** The catalytic mechanism likely involves two $Mg^{2+}$ ions, coordinated to the phosphate groups of the incoming nucleotide triphosphate and to three Asp residues, two of which are highly conserved in all DNA polymerases. The top $Mg^{2+}$ ion in the figure facilitates attack of the 3'-hydroxyl group of the primer on the α phosphate of the nucleotide triphosphate; the lower $Mg^{2+}$ ion facilitates displacement of the pyrophosphate. Both ions stabilize the structure of the pentacovalent transition state. RNA polymerases use a similar mechanism (See Fig. 26–1b). ● **Nucleic Acid Synthesis**

Early work on DNA polymerase I led to the definition of two central requirements for DNA polymerization. First, all DNA polymerases require a **template.** The polymerization reaction is guided by a template DNA strand according to the base-pairing rules predicted by Watson and Crick: where a guanine is present in the template, a cytosine deoxynucleotide is added to the new strand, and so on. This was a particularly important discovery, not only because it provided a chemical basis for accurate semiconservative DNA replication but also because it represented the first example of the use of a template to guide a biosynthetic reaction.

Second, the polymerases require a **primer.** A primer is a strand segment (complementary to the template) with a free 3'-hydroxyl group to which a nucleotide can be added; the free 3' end of the primer is called the **primer terminus.** In other words, part of the new strand must already be in place: all DNA polymerases can only add nucleotides to a preexisting strand. Most primers are oligonucleotides of RNA rather than DNA, and specialized enzymes synthesize primers when and where they are required.

After adding a nucleotide to a growing DNA strand, a DNA polymerase either dissociates or moves along the template and adds another nucleotide. Dissociation and reassociation of the polymerase can limit the overall polymerization rate—the process is generally faster when a polymerase adds more nucleotides without dissociating from the template. The average number of nucleotides added before a polymerase dissociates defines its **processivity.** DNA polymerases vary greatly in processivity; some add just a few nucleotides before dissociating, others add many thousands. 🖱 **Nucleotide Polymerization by DNA Polymerase**

## Replication Is Very Accurate

Replication proceeds with an extraordinary degree of fidelity. In *E. coli*, a mistake is made only once for every $10^9$ to $10^{10}$ nucleotides added. For the *E. coli* chromosome of ~$4.6 \times 10^6$ bp, this means that an error occurs only once per 1,000 to 10,000 replications. During polymerization, discrimination between correct and incorrect nucleotides relies not just on the hydrogen bonds that specify the correct pairing between complementary bases but also on the common geometry of the standard A=T and G≡C base pairs (Fig. 25–6). The active site of DNA polymerase I accommodates only base pairs with this geometry. An incorrect nucleotide may be able to hydrogen-bond with a base in the template, but it generally will not fit into the active site. Incorrect bases can be rejected before the phosphodiester bond is formed.

The accuracy of the polymerization reaction itself, however, is insufficient to account for the high degree of fidelity in replication. Careful measurements in vitro have shown that DNA polymerases insert one incorrect nucleotide for every $10^4$ to $10^5$ correct ones. These

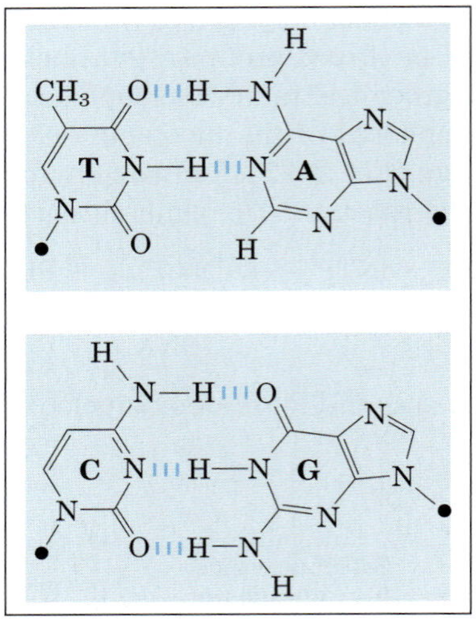

(a)

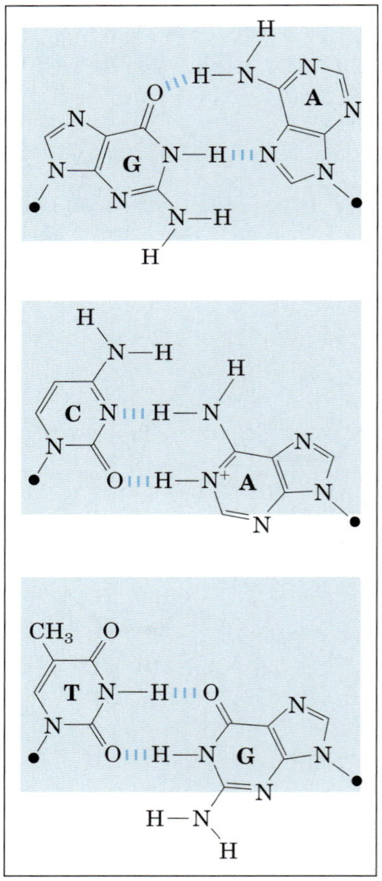

(b)

**FIGURE 25–6 Contribution of base-pair geometry to the fidelity of DNA replication. (a)** The standard A=T and G≡C base pairs have very similar geometries, and an active site sized to fit one (blue shading) will generally accommodate the other. **(b)** The geometry of incorrectly paired bases can exclude them from the active site, as occurs on DNA polymerase I.

mistakes sometimes occur because a base is briefly in an unusual tautomeric form (see Fig. 8–9), allowing it to hydrogen-bond with an incorrect partner. In vivo, the error rate is reduced by additional enzymatic mechanisms.

One mechanism intrinsic to virtually all DNA polymerases is a separate $3' \rightarrow 5'$ exonuclease activity that double-checks each nucleotide after it is added. This nuclease activity permits the enzyme to remove a newly added nucleotide and is highly specific for mismatched base pairs (Fig. 25–7). If the polymerase has added the wrong nucleotide, translocation of the enzyme to the position where the next nucleotide is to be added is inhibited. This kinetic pause provides the opportunity for a correction. The $3' \rightarrow 5'$ exonuclease activity removes the mispaired nucleotide, and the polymerase begins again. This activity, known as **proofreading,** is not simply the reverse of the polymerization reaction (Eqn 25–1), because pyrophosphate is not involved. The polymerizing and proofreading activities of a DNA polymerase can be measured separately. Proofreading improves the inherent accuracy of the polymerization reaction $10^2$- to $10^3$-fold. In the monomeric DNA polymerase I, the polymerizing and proofreading activities have separate active sites within the same polypeptide.

When base selection and proofreading are combined, DNA polymerase leaves behind one net error for every $10^6$ to $10^8$ bases added. Yet the measured accuracy of replication in *E. coli* is higher still. The additional accuracy is provided by a separate enzyme system that repairs the mismatched base pairs remaining after replication. We describe this mismatch repair, along with other DNA repair processes, in Section 25.2.

## *E. coli* Has at Least Five DNA Polymerases

More than 90% of the DNA polymerase activity observed in *E. coli* extracts can be accounted for by DNA polymerase I. Soon after the isolation of this enzyme in 1955, however, evidence began to accumulate that it is not suited for replication of the large *E. coli* chromosome. First, the rate at which it adds nucleotides (600 nucleotides/min) is too slow (by a factor of 100 or more) to account for the rates at which the replication fork moves in the bacterial cell. Second, DNA polymerase I has a relatively low processivity. Third, genetic studies have demonstrated that many genes, and therefore many proteins, are involved in replication: DNA polymerase I clearly does not act alone. Fourth, and most important, in 1969 John Cairns isolated a bacterial strain with an altered gene for DNA polymerase I that produced an inactive enzyme. Although this strain was abnormally sensitive to agents that damaged DNA, it was nevertheless viable!

A search for other DNA polymerases led to the discovery of *E. coli* **DNA polymerase II** and **DNA polymerase III** in the early 1970s. DNA polymerase II

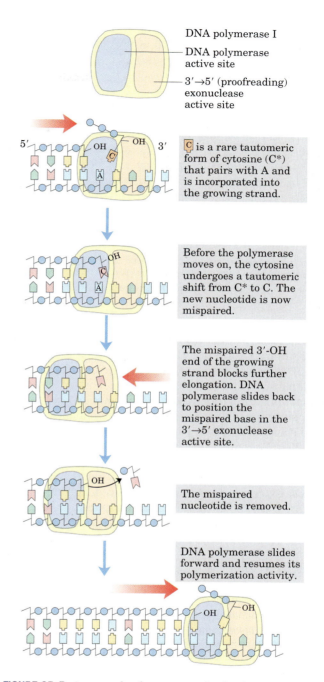

**FIGURE 25–7 An example of error correction by the $3' \rightarrow 5'$ exonuclease activity of DNA polymerase I.** Structural analysis has located the exonuclease activity ahead of the polymerase activity as the enzyme is oriented in its movement along the DNA. A mismatched base (here, a C–A mismatch) impedes translocation of DNA polymerase I to the next site. Sliding backward, the enzyme corrects the mistake with its $3' \rightarrow 5'$ exonuclease activity, then resumes its polymerase activity in the $5' \rightarrow 3'$ direction.

is an enzyme involved in one type of DNA repair (Section 25.3). DNA polymerase III is the principal replication enzyme in *E. coli*. The properties of these three DNA polymerases are compared in Table 25–1. DNA

**TABLE 25-1 Comparison of DNA Polymerases of *E. coli***

| | DNA polymerase | | |
| --- | :---: | :---: | :---: |
| | I | II | III |
| Structural gene* | *polA* | *polB* | *polC (dnaE)* |
| Subunits (number of different types) | 1 | 7 | ≥10 |
| $M_r$ | 103,000 | 88,000† | 791,500 |
| $3' \rightarrow 5'$ Exonuclease (proofreading) | Yes | Yes | Yes |
| $5' \rightarrow 3'$ Exonuclease | Yes | No | No |
| Polymerization rate (nucleotides/s) | 16–20 | 40 | 250–1,000 |
| Processivity (nucleotides added before polymerase dissociates) | 3–200 | 1,500 | ≥500,000 |

*For enzymes with more than one subunit, the gene listed here encodes the subunit with polymerization activity. Note that *dnaE* is an earlier designation for the gene now referred to as *polC*.

†Polymerization subunit only. DNA polymerase II shares several subunits with DNA polymerase III, including the β, γ, δ, δ', χ, and ψ subunits (see Table 25–2).

polymerases IV and V, identified in 1999, are involved in an unusual form of DNA repair (Section 25.2).

DNA polymerase I, then, is not the primary enzyme of replication; instead it performs a host of clean-up functions during replication, recombination, and repair. The polymerase's special functions are enhanced by its $5' \rightarrow 3'$ exonuclease activity. This activity, distinct from the $3' \rightarrow 5'$ proofreading exonuclease (Fig. 25–7), is located in a structural domain that can be separated from the enzyme by mild protease treatment. When the $5' \rightarrow 3'$ exonuclease domain is removed, the remaining fragment ($M_r$ 68,000), the **large fragment** or **Klenow fragment** (Fig. 25–8), retains the polymerization and

proofreading activities. The $5' \rightarrow 3'$ exonuclease activity of intact DNA polymerase I can replace a segment of DNA (or RNA) paired to the template strand, in a process known as nick translation (Fig. 25–9). Most other DNA polymerases lack a $5' \rightarrow 3'$ exonuclease activity.

DNA polymerase III is much more complex than DNA polymerase I, having ten types of subunits (Table 25–2). Its polymerization and proofreading activities reside in its α and ε (epsilon) subunits, respectively. The θ subunit associates with α and ε to form a core polymerase, which can polymerize DNA but with limited processivity. Two core polymerases can be linked by

**TABLE 25-2 Subunits of DNA Polymerase III of *E. coli***

| Subunit | Number of subunits per holoenzyme | $M_r$ of subunit | Gene | Function of subunit | |
| --- | :---: | :---: | --- | --- | --- |
| α | 2 | 129,900 | *polC (dnaE)* | Polymerization activity | ⎫ |
| ε | 2 | 27,500 | *dnaQ (mutD)* | $3' \rightarrow 5'$ Proofreading exonuclease | ⎬ Core polymerase |
| θ | 2 | 8,600 | *holE* | | ⎭ |
| τ | 2 | 71,100 | *dnaX* | Stable template binding; core enzyme dimerization | ⎫ |
| γ | 1 | 47,500 | *dnaX** | Clamp loader | ⎪ Clamp-loading (γ) complex that |
| δ | 1 | 38,700 | *holA* | Clamp opener | ⎬ loads β subunits on lagging |
| δ' | 1 | 36,900 | *holB* | Clamp loader | ⎭ strand at each Okazaki fragment |
| χ | 1 | 16,600 | *holC* | Interaction with SSB | |
| ψ | 1 | 15,200 | *holD* | Interaction with γ and χ | |
| β | 4 | 40,600 | *dnaN* | DNA clamp required for optimal processivity | |

*The γ subunit is encoded by a portion of the gene for the τ subunit, such that the amino-terminal 66% of the τ subunit has the same amino acid sequence as the γ subunit. The γ subunit is generated by a translational frameshifting mechanism (see Box 27–1) that leads to premature translational termination.

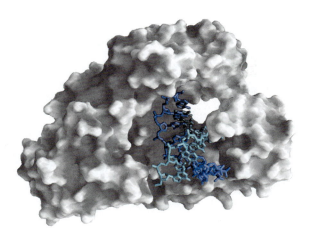

**FIGURE 25–8 Large (Klenow) fragment of DNA polymerase I.** This polymerase is widely distributed in bacteria. The Klenow fragment, produced by proteolytic treatment of the polymerase, retains the polymerization and proofreading activities of the enzyme. The Klenow fragment shown here is from the thermophilic bacterium *Bacillus stearothermophilus* (PDB ID 3BDP). The active site for addition of nucleotides is deep in the crevice at the far end of the bound DNA. The dark blue strand is the template.

another set of subunits, a clamp-loading complex, or $\gamma$ complex, consisting of five subunits of four different types, $\tau_2\gamma\delta\delta'$. The core polymerases are linked through the $\tau$ (tau) subunits. Two additional subunits, $\chi$ (chi) and $\psi$ (psi), are bound to the clamp-loading complex. The entire assembly of 13 protein subunits (nine different types) is called DNA polymerase III* (Fig. 25–10a).

DNA polymerase III* can polymerize DNA, but with a much lower processivity than one would expect for the organized replication of an entire chromosome. The necessary increase in processivity is provided by the addition of the $\beta$ subunits, four of which complete the DNA polymerase III holoenzyme. The $\beta$ subunits associate in pairs to form donut-shaped structures that encircle the DNA and act like clamps (Fig. 25–10b). Each dimer associates with a core subassembly of polymerase III* (one dimeric clamp per core subassembly) and slides along the DNA as replication proceeds. The $\beta$ sliding clamp prevents the dissociation of DNA polymerase III from DNA, dramatically increasing processivity—to greater than 500,000 (Table 25–1).

## DNA Replication Requires Many Enzymes and Protein Factors

Replication in *E. coli* requires not just a single DNA polymerase but 20 or more different enzymes and proteins, each performing a specific task. The entire complex has been termed the **DNA replicase system** or **replisome.** The enzymatic complexity of replication reflects the constraints imposed by the structure of DNA

and by the requirements for accuracy. The main classes of replication enzymes are considered here in terms of the problems they overcome.

Access to the DNA strands that are to act as templates requires separation of the two parent strands. This is generally accomplished by **helicases,** enzymes that move along the DNA and separate the strands, using chemical energy from ATP. Strand separation creates topological stress in the helical DNA structure (see Fig. 24–12), which is relieved by the action of **topoisomerases.** The separated strands are stabilized by **DNA-binding proteins.** As noted earlier, before DNA polymerases can begin synthesizing DNA, primers must be present on the template—generally short segments

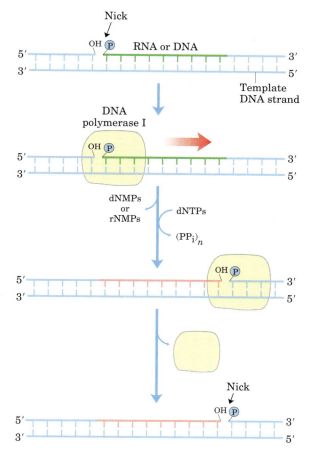

**FIGURE 25–9 Nick translation.** In this process, an RNA or DNA strand paired to a DNA template is simultaneously degraded by the 5'→3' exonuclease activity of DNA polymerase I and replaced by the polymerase activity of the same enzyme. These activities have a role in both DNA repair and the removal of RNA primers during replication (both described later). The strand of nucleic acid to be removed (either DNA or RNA) is shown in green, the replacement strand in red. DNA synthesis begins at a nick (a broken phosphodiester bond, leaving a free 3' hydroxyl and a free 5' phosphate). Polymerase I extends the nontemplate DNA strand and moves the nick along the DNA—a process called nick translation. A nick remains where DNA polymerase I dissociates, and is later sealed by another enzyme.

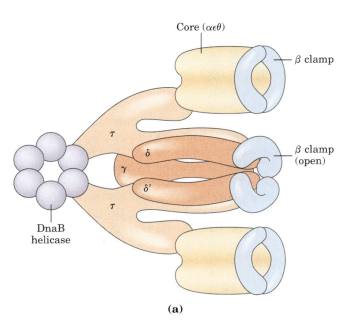

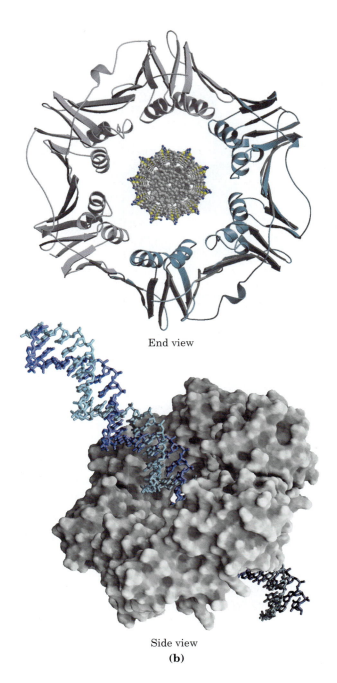

End view

Side view
**(b)**

**FIGURE 25–10 DNA polymerase III. (a)** Architecture of bacterial DNA polymerase III. Two core domains, composed of subunits $\alpha$, $\varepsilon$, and $\theta$, are linked by a five-subunit $\gamma$ complex (also known as the clamp-loading complex) with the composition $\tau_2\gamma\delta\delta'$. The $\gamma$ and $\tau$ subunits are encoded by the same gene. The $\gamma$ subunit is a shortened version of $\tau$; the $\tau$ subunit thus contains a domain identical to $\gamma$, along with an additional segment that interacts with the core polymerase. The other two subunits of DNA polymerase III*, $\chi$ and $\psi$ (not shown), also bind to the $\gamma$ complex. Two $\beta$ clamps interact with the two-core subassembly, each clamp a dimer of the $\beta$ subunit. The complex interacts with the DnaB helicase through the $\tau$ subunit. **(b)** Two $\beta$ subunits of *E. coli* polymerase III form a circular clamp that surrounds the DNA. The clamp slides along the DNA molecule, increasing the processivity of the polymerase III holoenzyme to greater than 500,000 by preventing its dissociation from the DNA. The end-on view shows the two $\beta$ subunits as gray and light-blue ribbon structures surrounding a space-filling model of DNA. In the side view, surface contour models of the $\beta$ subunits (gray) surround a stick representation of a DNA double helix (light and dark blue) (derived from PDB ID 2POL).

of RNA synthesized by enzymes known as **primases.** Ultimately, the RNA primers are removed and replaced by DNA; in *E. coli,* this is one of the many functions of DNA polymerase I. After an RNA primer is removed and the gap is filled in with DNA, a nick remains in the DNA backbone in the form of a broken phosphodiester bond. These nicks are sealed by **DNA ligases.** All these processes require coordination and regulation, an interplay best characterized in the *E. coli* system.

### Replication of the *E. coli* Chromosome Proceeds in Stages

The synthesis of a DNA molecule can be divided into three stages: initiation, elongation, and termination, distinguished both by the reactions taking place and by the enzymes required. As you will find here and in the next two chapters, synthesis of the major information-containing biological polymers—DNAs, RNAs, and proteins—can be understood in terms of these same three stages, with the stages of each pathway having unique characteristics. The events described below reflect information derived primarily from in vitro experiments using purified *E. coli* proteins, although the principles are highly conserved in all replication systems.

***Initiation*** The *E. coli* replication origin, *oriC,* consists of 245 bp; it bears DNA sequence elements that are highly conserved among bacterial replication origins. The general arrangement of the conserved sequences is

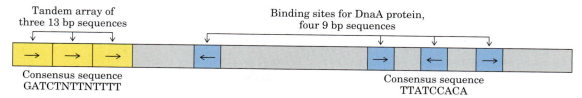

Tandem array of three 13 bp sequences

Binding sites for DnaA protein, four 9 bp sequences

Consensus sequence
GATCTNTTNTTTT

Consensus sequence
TTATCCACA

**FIGURE 25–11** Arrangement of sequences in the *E. coli* replication origin, *oriC*. Although the repeated sequences (shaded in color) are not identical, certain nucleotides are particularly common in each po-

sition, forming a consensus sequence. In positions where there is no consensus, N represents any of the four nucleotides. The arrows indicate the orientations of the nucleotide sequences.

illustrated in Figure 25–11. The key sequences of interest here are two series of short repeats: three repeats of a 13 bp sequence and four repeats of a 9 bp sequence.

At least nine different enzymes or proteins (summarized in Table 25–3) participate in the initiation phase of replication. They open the DNA helix at the origin and establish a prepriming complex for subsequent reactions. The crucial component in the initiation process is the DnaA protein. A single complex of four to five DnaA protein molecules binds to the four 9 bp repeats in the origin (Fig. 25–12, step ①), then recognizes and successively denatures the DNA in the region of the three 13 bp repeats, which are rich in A═T pairs (step ②). This process requires ATP and the bacterial histonelike protein HU. The DnaC protein then loads the DnaB protein onto the unwound region. Two ring-shaped hexamers of DnaB, one loaded onto each DNA strand, act as helicases, unwinding the DNA bidirectionally and creating two potential replication forks. If the *E. coli* single-stranded DNA–binding protein (SSB) and DNA gyrase (DNA topoisomerase II) are now added in vitro, thousands of base pairs are rapidly unwound by the DnaB helicase, proceeding out from the origin. Many molecules of SSB bind cooperatively to single-stranded DNA, stabilizing the separated strands and preventing renaturation while gyrase relieves the topological stress produced by the DnaB helicase. When additional replication proteins are included in the in vitro system, the DNA unwinding mediated by DnaB is coupled to replication, as described below.

Initiation is the only phase of DNA replication that is known to be regulated, and it is regulated such that replication occurs only once in each cell cycle. The mechanism of regulation is not yet well understood, but genetic and biochemical studies have provided a few insights.

The timing of replication initiation is affected by DNA methylation and interactions with the bacterial plasma membrane. The *oriC* DNA is methylated by the Dam methylase (Table 25–3), which methylates the $N^6$ position of adenine within the palindromic sequence (5′)GATC. (Dam is not a biochemical expletive; it stands for *D*NA *a*denine *m*ethylation.) The *oriC* region of *E. coli* is highly enriched in GATC sequences—it has 11 of them in its 245 bp, whereas the average frequency of GATC in the *E. coli* chromosome as a whole is 1 in 256 bp.

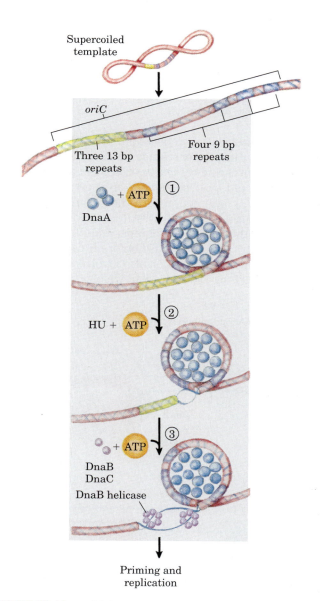

Supercoiled template

*oriC*

Three 13 bp repeats

Four 9 bp repeats

DnaA + ATP ①

HU + ATP ②

DnaB DnaC + ATP ③

DnaB helicase

Priming and replication

**FIGURE 25–12** Model for initiation of replication at the *E. coli* origin, *oriC*. ① About 20 DnaA protein molecules, each with a bound ATP, bind at the four 9 bp repeats. The DNA is wrapped around this complex. ② The three A═T-rich 13 bp repeats are denatured sequentially. ③ Hexamers of the DnaB protein bind to each strand, with the aid of DnaC protein. The DnaB helicase activity further unwinds the DNA in preparation for priming and DNA synthesis.

**TABLE 25-3   Proteins Required to Initiate Replication at the *E. coli* Origin**

| Protein | $M_r$ | Number of subunits | Function |
|---|---|---|---|
| DnaA protein | 52,000 | 1 | Recognizes ori sequence; opens duplex at specific sites in origin |
| DnaB protein (helicase) | 300,000 | 6* | Unwinds DNA |
| DnaC protein | 29,000 | 1 | Required for DnaB binding at origin |
| HU | 19,000 | 2 | Histonelike protein; DNA-binding protein; stimulates initiation |
| Primase (DnaG protein) | 60,000 | 1 | Synthesizes RNA primers |
| Single-stranded DNA–binding protein (SSB) | 75,600 | 4* | Binds single-stranded DNA |
| RNA polymerase | 454,000 | 5 | Facilitates DnaA activity |
| DNA gyrase (DNA topoisomerase II) | 400,000 | 4 | Relieves torsional strain generated by DNA unwinding |
| Dam methylase | 32,000 | 1 | Methylates (5′)GATC sequences at *oriC* |

*Subunits in these cases are identical.

Immediately after replication, the DNA is hemi-methylated: the parent strands have methylated *oriC* sequences but the newly synthesized strands do not. The hemimethylated *oriC* sequences are now sequestered for a period by interaction with the plasma membrane (the mechanism is unknown). After a time, *oriC* is released from the plasma membrane, and it must be fully methylated by Dam methylase before it can again bind DnaA. Regulation of initiation also involves the slow hydrolysis of ATP by DnaA protein, which cycles the protein between active (with bound ATP) and inactive (with bound ADP) forms on a timescale of 20 to 40 minutes.

***Elongation***   The elongation phase of replication includes two distinct but related operations: leading strand synthesis and lagging strand synthesis. Several enzymes at the replication fork are important to the synthesis of both strands. Parent DNA is first unwound by DNA helicases, and the resulting topological stress is relieved by topoisomerases. Each separated strand is then stabilized by

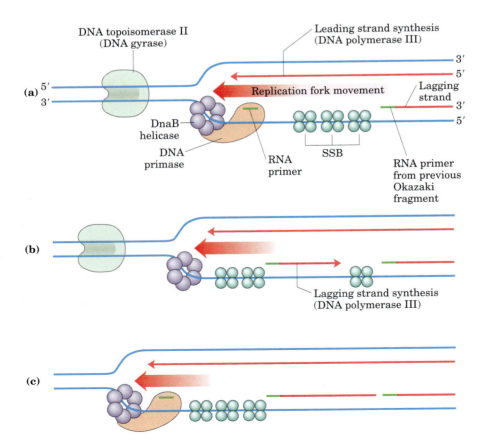

**FIGURE 25–13   Synthesis of Okazaki fragments. (a)** At intervals, primase synthesizes an RNA primer for a new Okazaki fragment. Note that if we consider the two template strands as lying side by side, lagging strand synthesis formally proceeds in the opposite direction from fork movement. **(b)** Each primer is extended by DNA polymerase III. **(c)** DNA synthesis continues until the fragment extends as far as the primer of the previously added Okazaki fragment. A new primer is synthesized near the replication fork to begin the process again.

SSB. From this point, synthesis of leading and lagging strands is sharply different.

Leading strand synthesis, the more straightforward of the two, begins with the synthesis by primase (DnaG protein) of a short (10 to 60 nucleotide) RNA primer at the replication origin. Deoxyribonucleotides are added to this primer by DNA polymerase III. Leading strand synthesis then proceeds continuously, keeping pace with the unwinding of DNA at the replication fork.

Lagging strand synthesis, as we have noted, is accomplished in short Okazaki fragments. First, an RNA primer is synthesized by primase and, as in leading strand synthesis, DNA polymerase III binds to the RNA primer and adds deoxyribonucleotides (Fig. 25–13). On this level, the synthesis of each Okazaki fragment seems straightforward, but the reality is quite complex. The complexity lies in the *coordination* of leading and lagging strand synthesis: both strands are produced by a *single* asymmetric DNA polymerase III dimer, which is accomplished by looping the DNA of the lagging strand as shown in Figure 25–14, bringing together the two points of polymerization.

**FIGURE 25–14 DNA synthesis on the leading and lagging strands.** Events at the replication fork are coordinated by a single DNA polymerase III dimer, in an integrated complex with DnaB helicase. This figure shows the replication process already underway (parts (a) through (e) are discussed in the text). The lagging strand is looped so that DNA synthesis proceeds steadily on both the leading and lagging strand templates at the same time. Red arrows indicate the 3′ end of the two new strands and the direction of DNA synthesis. Black arrows show the direction of movement of the parent DNA through the complex. An Okazaki fragment is being synthesized on the lagging strand.

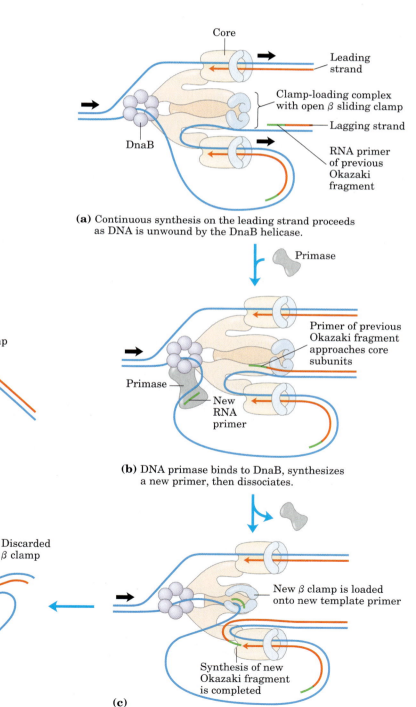

(a) Continuous synthesis on the leading strand proceeds as DNA is unwound by the DnaB helicase.

(b) DNA primase binds to DnaB, synthesizes a new primer, then dissociates.

(c) New β clamp is loaded onto new template primer. Synthesis of new Okazaki fragment is completed.

(d)

(e) The next β clamp is readied

The synthesis of Okazaki fragments on the lagging strand entails some elegant enzymatic choreography. The DnaB helicase and DnaG primase constitute a functional unit within the replication complex, the **primosome**. DNA polymerase III uses one set of its core subunits (the core polymerase) to synthesize the leading strand continuously, while the other set of core subunits cycles from one Okazaki fragment to the next on the looped lagging strand. The DnaB helicase unwinds the DNA at the replication fork (Fig. 25–14a) as it travels along the lagging strand template in the 5'→3' direction. DNA primase occasionally associates with DnaB helicase and synthesizes a short RNA primer (Fig. 25–14b). A new β sliding clamp is then positioned at the primer by the clamp-loading complex of DNA polymerase III (Fig. 25–14c). When synthesis of an Okazaki fragment has been completed, replication halts, and the core subunits of DNA polymerase III dissociate from their β sliding clamp (and from the completed Okazaki fragment) and associate with the new clamp (Fig. 25–14d, e). This initiates synthesis of a new Okazaki fragment. As noted earlier, the entire complex responsible for coordinated DNA synthesis at a replication fork is a **replisome**. The proteins acting at the replication fork are summarized in Table 25–4.

The replisome promotes rapid DNA synthesis, adding ~1,000 nucleotides/s to each strand (leading and lagging). Once an Okazaki fragment has been completed, its RNA primer is removed and replaced with DNA by DNA polymerase I, and the remaining nick is sealed by DNA ligase (Fig. 25–15).

DNA ligase catalyzes the formation of a phosphodiester bond between a 3' hydroxyl at the end of one DNA strand and a 5' phosphate at the end of another strand. The phosphate must be activated by adenylylation. DNA ligases isolated from viruses and eukaryotes use ATP for this purpose. DNA ligases from bacteria are unusual in that they generally use $NAD^+$—a cofactor that normally functions in hydride transfer reactions

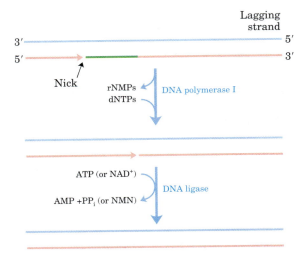

**FIGURE 25–15 Final steps in the synthesis of lagging strand segments.** RNA primers in the lagging strand are removed by the 5'→3' exonuclease activity of DNA polymerase I and replaced with DNA by the same enzyme. The remaining nick is sealed by DNA ligase. The role of ATP or $NAD^+$ is shown in Figure 25–16.

(see Fig. 13–15)—as the source of the AMP activating group (Fig. 25–16). DNA ligase is another enzyme of DNA metabolism that has become an important reagent in recombinant DNA experiments (see Fig. 9–1).

**Termination** Eventually, the two replication forks of the circular *E. coli* chromosome meet at a terminus region containing multiple copies of a 20 bp sequence called Ter (for *ter*minus) (Fig. 25–17a). The Ter sequences are arranged on the chromosome to create a sort of trap that a replication fork can enter but cannot leave. The Ter sequences function as binding sites for a protein called Tus (*t*erminus *u*tilization *s*ubstance). The Tus-Ter complex can arrest a replication fork from only one direction. Only one Tus-Ter complex functions per replication cycle—the complex first encountered by either

**TABLE 25–4  Proteins at the *E. coli* Replication Fork**

| Protein | $M_r$ | Number of subunits | Function |
|---|---|---|---|
| SSB | 75,600 | 4 | Binding to single-stranded DNA |
| DnaB protein (helicase) | 300,000 | 6 | DNA unwinding; primosome constituent |
| Primase (DnaG protein) | 60,000 | 1 | RNA primer synthesis; primosome constituent |
| DNA polymerase III | 791,500 | 17 | New strand elongation |
| DNA polymerase I | 103,000 | 1 | Filling of gaps; excision of primers |
| DNA ligase | 74,000 | 1 | Ligation |
| DNA gyrase (DNA topoisomerase II) | 400,000 | 4 | Supercoiling |

Modified from Kornberg, A. (1982) *Supplement to DNA Replication*, Table S11-2, W. H. Freeman and Company, New York.

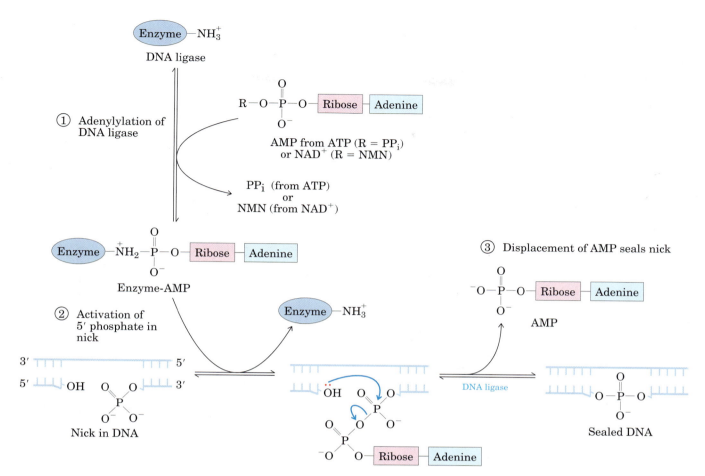

**FIGURE 25–16 Mechanism of the DNA ligase reaction.** In each of the three steps, one phosphodiester bond is formed at the expense of another. Steps ① and ② lead to activation of the 5′ phosphate in the nick. An AMP group is transferred first to a Lys residue on the enzyme and then to the 5′ phosphate in the nick. In step ③, the 3′-hydroxyl group attacks this phosphate and displaces AMP, producing a phosphodiester bond to seal the nick. In the *E. coli* DNA ligase reaction, AMP is derived from NAD⁺. The DNA ligases isolated from a number of viral and eukaryotic sources use ATP rather than NAD⁺, and they release pyrophosphate rather than nicotinamide mononucleotide (NMN) in step ①.

replication fork. Given that opposing replication forks generally halt when they collide, Ter sequences do not seem essential, but they may prevent overreplication by one replication fork in the event that the other is delayed or halted by an encounter with DNA damage or some other obstacle.

So, when either replication fork encounters a functional Tus-Ter complex, it halts; the other fork halts when it meets the first (arrested) fork. The final few hundred base pairs of DNA between these large protein complexes are then replicated (by an as yet unknown mechanism), completing two topologically interlinked (catenated) circular chromosomes (Fig. 25–17b). DNA circles linked in this way are known as **catenanes.** Separation of the catenated circles in *E. coli* requires topoisomerase IV (a type II topoisomerase). The separated chromosomes then segregate into daughter cells at cell division. The terminal phase of replication of other circular chromosomes, including many of the DNA viruses that infect eukaryotic cells, is similar.

## Bacterial Replication Is Organized in Membrane-Bound Replication Factories

The replication of a circular bacterial chromosome is highly organized. Once bidirectional replication is initiated at the origin, the two replisomes do not travel away from each other along the DNA. Instead, the replisomes are linked together and tethered to one point on the bacterial inner membrane, and the DNA substrate is fed through this "replication factory" (Fig. 25–18a). The tethering point is at the center of the elongated bacterial cell. After initiation, each of the two newly synthesized replication origins is partitioned into one half of

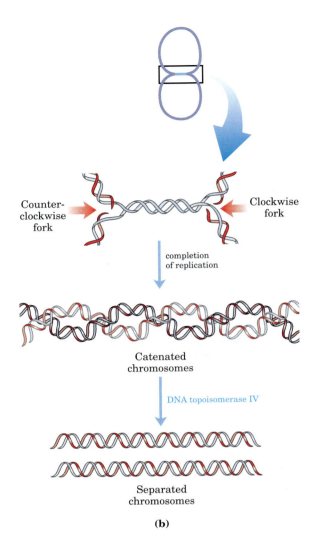

**FIGURE 25-17 Termination of chromosome replication in *E. coli*. (a)** The Ter sequences are positioned on the chromosome in two clusters with opposite orientations. **(b)** Replication of the DNA separating the opposing replication forks leaves the completed chromosomes joined as catenanes, or topologically interlinked circles. The circles are not covalently linked, but because they are interwound and each is covalently closed, they cannot be separated—except by the action of topoisomerases. In *E. coli*, a type II topoisomerase known as DNA topoisomerase IV plays the primary role in the separation of catenated chromosomes, transiently breaking both DNA strands of one chromosome and allowing the other chromosome to pass through the break.

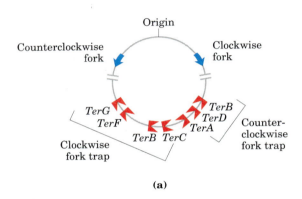

**(a)**

**(b)**

the cell, and continuing replication extrudes each new chromosome into that half (Fig. 25–18b). The elaborate spatial organization of the newly replicated chromosomes is orchestrated and maintained by many proteins, including bacterial homologs of the SMC proteins and topoisomerases (Chapter 24). Once replication is terminated, the cell divides, and the chromosomes sequestered in the two halves of the original cell are accurately partitioned into the daughter cells. When replication commences in the daughter cells, the origin of replication is sequestered in new replication factories formed at a point on the membrane at the center of the cell, and the entire process is repeated.

## Replication in Eukaryotic Cells Is More Complex

The DNA molecules in eukaryotic cells are considerably larger than those in bacteria and are organized into complex nucleoprotein structures (chromatin; p. 938). The essential features of DNA replication are the same in eukaryotes and prokaryotes, and many of the protein complexes are functionally and structurally conserved. However, some interesting variations on the general principles discussed above promise new insights into the regulation of replication and its link with the cell cycle.

Origins of replication, called *autonomously replicating sequences* (ARS) or **replicators,** have been identified and best studied in yeast. Yeast replicators span ~150 bp and contain several essential conserved sequences. About 400 replicators are distributed among the 16 chromosomes in a haploid yeast genome. Initiation of replication in all eukaryotes requires a multisubunit protein, the *origin recognition complex* (ORC), which binds to several sequences within the replicator. ORC interacts with and is regulated by a number of other proteins involved in control of the eukaryotic cell cycle. Two other proteins, CDC6 (discovered in a screen for genes affecting the *cell division cycle*) and CDT1 (*Cdc10-dependent transcript 1*), bind to ORC and mediate the loading of a heterohexamer of *minichromosome maintenance* proteins (MCM2 to MCM7). The MCM complex is a ring-shaped replicative helicase, analogous to the bacterial DnaB helicase. The CDC6 and CDT1 proteins have a role comparable to that of the bacterial DnaC protein, loading the MCM helicase onto the DNA near the replication origin.

The rate of replication fork movement in eukaryotes (~50 nucleotides/s) is only one-twentieth that observed in *E. coli*. At this rate, replication of an average human chromosome proceeding from a single origin

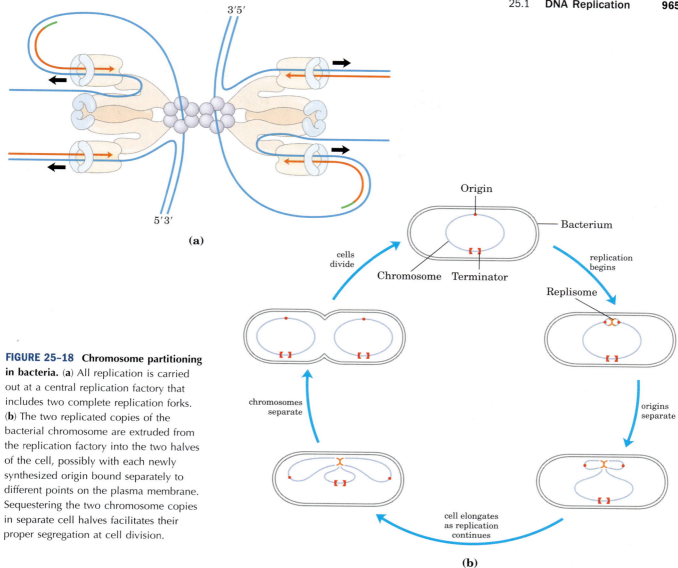

**(a)**

**(b)**

**FIGURE 25–18 Chromosome partitioning in bacteria.** (a) All replication is carried out at a central replication factory that includes two complete replication forks. (b) The two replicated copies of the bacterial chromosome are extruded from the replication factory into the two halves of the cell, possibly with each newly synthesized origin bound separately to different points on the plasma membrane. Sequestering the two chromosome copies in separate cell halves facilitates their proper segregation at cell division.

would take more than 500 hours. Replication of human chromosomes in fact proceeds bidirectionally from many origins, spaced 30,000 to 300,000 bp apart. Eukaryotic chromosomes are almost always much larger than bacterial chromosomes, so multiple origins are probably a universal feature in eukaryotic cells.

Like bacteria, eukaryotes have several types of DNA polymerases. Some have been linked to particular functions, such as the replication of mitochondrial DNA. The replication of nuclear chromosomes involves DNA polymerase $\alpha$, in association with DNA polymerase $\delta$. **DNA polymerase $\alpha$** is typically a multisubunit enzyme with similar structure and properties in all eukaryotic cells. One subunit has a primase activity, and the largest subunit ($M_r$ ~180,000) contains the polymerization activity. However, this polymerase has no proofreading $3' \rightarrow 5'$ exonuclease activity, making it unsuitable for high-fidelity DNA replication. DNA polymerase $\alpha$ is believed to function only in the synthesis of short primers (containing either RNA or DNA) for Okazaki fragments on the lagging strand. These primers

are then extended by the multisubunit **DNA polymerase $\delta$.** This enzyme is associated with and stimulated by a protein called *proliferating cell nuclear antigen* (PCNA; $M_r$ 29,000), found in large amounts in the nuclei of proliferating cells. The three-dimensional structure of PCNA is remarkably similar to that of the $\beta$ subunit of *E. coli* DNA polymerase III (Fig. 25–10b), although primary sequence homology is not evident. PCNA has a function analogous to that of the $\beta$ subunit, forming a circular clamp that greatly enhances the processivity of the polymerase. DNA polymerase $\delta$ has a $3' \rightarrow 5'$ proofreading exonuclease activity and appears to carry out both leading and lagging strand synthesis in a complex comparable to the dimeric bacterial DNA polymerase III.

Yet another polymerase, **DNA polymerase $\varepsilon$,** replaces DNA polymerase $\delta$ in some situations, such as in DNA repair. DNA polymerase $\varepsilon$ may also function at the replication fork, perhaps playing a role analogous to that of the bacterial DNA polymerase I, removing the primers of Okazaki fragments on the lagging strand.

Many DNA viruses encode their own DNA polymerases, and some of these have become targets for pharmaceuticals. For example, the DNA polymerase of the herpes simplex virus is inhibited by acyclovir, a compound developed by Gertrude Elion (p. 876). Acyclovir consists of guanine attached to an incomplete ribose ring. It is phosphorylated by a virally encoded thymidine kinase; acyclovir binds to this viral enzyme with an affinity 200-fold greater than its binding to the cellular thymidine kinase. This ensures that phosphorylation occurs mainly in virus-infected cells. Cellular kinases convert the resulting acyclo-GMP to acyclo-GTP, which is both an inhibitor and a substrate of DNA polymerases, and which competitively inhibits the herpes DNA polymerase more strongly than cellular DNA polymerases. Because it lacks a 3′ hydroxyl, acyclo-GTP also acts as a chain terminator when incorporated into DNA. Thus viral replication is inhibited at several steps.

Acyclovir

Two other protein complexes also function in eukaryotic DNA replication. RPA (*replication protein A*) is a eukaryotic single-stranded DNA–binding protein, equivalent in function to the *E. coli* SSB protein. RFC (*replication factor C*) is a clamp loader for PCNA and facilitates the assembly of active replication complexes. The subunits of the RFC complex have significant sequence similarity to the subunits of the bacterial clamp-loading ($\gamma$) complex.

The termination of replication on linear eukaryotic chromosomes involves the synthesis of special structures called **telomeres** at the ends of each chromosome, as discussed in the next chapter.

### SUMMARY 25.1 DNA Replication

- Replication of DNA occurs with very high fidelity and at a designated time in the cell cycle. Replication is semiconservative, each strand acting as template for a new daughter strand. It is carried out in three identifiable phases: initiation, elongation, and termination. The reaction starts at the origin and usually proceeds bidirectionally.

- DNA is synthesized in the 5′→3′ direction by DNA polymerases. At the replication fork, the leading strand is synthesized continuously in the same direction as replication fork movement; the lagging strand is synthesized discontinuously as Okazaki fragments, which are subsequently ligated.

- The fidelity of DNA replication is maintained by (1) base selection by the polymerase, (2) a 3′→5′ proofreading exonuclease activity that is part of most DNA polymerases, and (3) specific repair systems for mismatches left behind after replication.

- Most cells have several DNA polymerases. In *E. coli*, DNA polymerase III is the primary replication enzyme. DNA polymerase I is responsible for special functions during replication, recombination, and repair.

- Replication of the *E. coli* chromosome involves many enzymes and protein factors organized in replication factories, in which template DNA is spooled through two replisomes tethered to the bacterial plasma membrane.

- Replication is similar in eukaryotic cells, but eukaryotic chromosomes have many replication origins.

## 25.2 DNA Repair

A cell generally has only one or two sets of genomic DNA. Damaged proteins and RNA molecules can be quickly replaced by using information encoded in the DNA, but DNA molecules themselves are irreplaceable. Maintaining the integrity of the information in DNA is a cellular imperative, supported by an elaborate set of DNA repair systems. DNA can become damaged by a variety of processes, some spontaneous, others catalyzed by environmental agents (Chapter 8). Replication itself can very occasionally damage the information content in DNA when errors introduce mismatched base pairs (such as G paired with T).

The chemistry of DNA damage is diverse and complex. The cellular response to this damage includes a wide range of enzymatic systems that catalyze some of the most interesting chemical transformations in DNA metabolism. We first examine the effects of alterations in DNA sequence and then consider specific repair systems.

### Mutations Are Linked to Cancer

The best way to illustrate the importance of DNA repair is to consider the effects of *unrepaired* DNA damage (a lesion). The most serious outcome is a change in the base sequence of the DNA, which, if replicated and transmitted to future cell generations, becomes permanent. A permanent change in the nucleotide sequence of DNA is called a **mutation.** Mutations can involve the replacement of one base pair with another (substitution mutation) or the addition or deletion of one or more base pairs (insertion or deletion mutations). If the mutation affects nonessential DNA or if it has a negligible

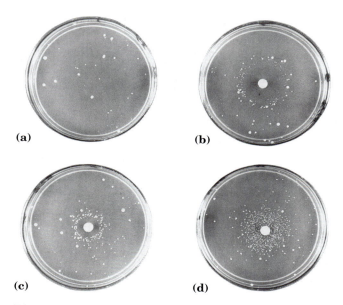

**(a)**   **(b)**

**(c)**   **(d)**

**FIGURE 25–19 Ames test for carcinogens, based on their mutagenicity.** A strain of *Salmonella typhimurium* having a mutation that inactivates an enzyme of the histidine biosynthetic pathway is plated on a histidine-free medium. Few cells grow. **(a)** The few small colonies of *S. typhimurium* that do grow on a histidine-free medium carry spontaneous back-mutations that permit the histidine biosynthetic pathway to operate. Three identical nutrient plates **(b), (c),** and **(d)** have been inoculated with an equal number of cells. Each plate then receives a disk of filter paper containing progressively lower concentrations of a mutagen. The mutagen greatly increases the rate of back-mutation and hence the number of colonies. The clear areas around the filter paper indicate where the concentration of mutagen is so high that it is lethal to the cells. As the mutagen diffuses away from the filter paper, it is diluted to sublethal concentrations that promote back-mutation. Mutagens can be compared on the basis of their effect on mutation rate. Because many compounds undergo a variety of chemical transformations after entering a cell, compounds are sometimes tested for mutagenicity after first incubating them with a liver extract. Some substances have been found to be mutagenic only after this treatment.

The genome of a typical mammalian cell accumulates many thousands of lesions during a 24-hour period. However, as a result of DNA repair, fewer than 1 in 1,000 becomes a mutation. DNA is a relatively stable molecule, but in the absence of repair systems, the cumulative effect of many infrequent but damaging reactions would make life impossible.

## All Cells Have Multiple DNA Repair Systems

The number and diversity of repair systems reflect both the importance of DNA repair to cell survival and the diverse sources of DNA damage (Table 25–5). Some common types of lesions, such as pyrimidine dimers (see Fig. 8–34), can be repaired by several distinct systems. Many DNA repair processes also appear to be extraordinarily inefficient energetically—an exception to

effect on the function of a gene, it is known as a **silent mutation.** Rarely, a mutation confers some biological advantage. Most nonsilent mutations, however, are deleterious.

In mammals there is a strong correlation between the accumulation of mutations and cancer. A simple test developed by Bruce Ames measures the potential of a given chemical compound to promote certain easily detected mutations in a specialized bacterial strain (Fig. 25–19). Few of the chemicals that we encounter in daily life score as mutagens in this test. However, of the compounds known to be carcinogenic from extensive animal trials, more than 90% are also found to be mutagenic in the Ames test. Because of this strong correlation between mutagenesis and carcinogenesis, the Ames test for bacterial mutagens is widely used as a rapid and inexpensive screen for potential human carcinogens.

**TABLE 25–5 Types of DNA Repair Systems in *E. coli***

| Enzymes/proteins | Type of damage |
|---|---|
| **Mismatch repair** | |
| Dam methylase | Mismatches |
| MutH, MutL, MutS proteins | |
| DNA helicase II | |
| SSB | |
| DNA polymerase III | |
| Exonuclease I | |
| Exonuclease VII | |
| RecJ nuclease | |
| Exonuclease X | |
| DNA ligase | |
| **Base-excision repair** | |
| DNA glycosylases | Abnormal bases (uracil, hypoxanthine, xanthine); alkylated bases; in some other organisms, pyrimidine dimers |
| AP endonucleases | |
| DNA polymerase I | |
| DNA ligase | |
| **Nucleotide-excision repair** | |
| ABC excinuclease | DNA lesions that cause large structural changes (e.g., pyrimidine dimers) |
| DNA polymerase I | |
| DNA ligase | |
| **Direct repair** | |
| DNA photolyases | Pyrimidine dimers |
| $O^6$-Methylguanine-DNA methyltransferase | $O^6$-Methylguanine |
| AlkB protein | 1-Methylguanine, 3-methylcytosine |

the pattern observed in the metabolic pathways, where every ATP is generally accounted for and used optimally. When the integrity of the genetic information is at stake, the amount of chemical energy invested in a repair process seems almost irrelevant.

DNA repair is possible largely because the DNA molecule consists of two complementary strands. DNA damage in one strand can be removed and accurately replaced by using the undamaged complementary strand as a template. We consider here the principal types of repair systems, beginning with those that repair the rare nucleotide mismatches that are left behind by replication.

***Mismatch Repair*** Correction of the rare mismatches left after replication in *E. coli* improves the overall fidelity of replication by an additional factor of $10^2$ to $10^3$. The mismatches are nearly always corrected to reflect the information in the old (template) strand, so the repair system must somehow discriminate between the template and the newly synthesized strand. The cell accomplishes this by tagging the template DNA with methyl groups to distinguish it from newly synthesized strands. The mismatch repair system of *E. coli* includes at least 12 protein components (Table 25–5) that function either in strand discrimination or in the repair process itself.

The strand discrimination mechanism has not been worked out for most bacteria or eukaryotes, but is well understood for *E. coli* and some closely related bacteria. In these prokaryotes, strand discrimination is based on the action of Dam methylase (Table 25–3), which, as you will recall, methylates DNA at the $N^6$ position of all adenines within (5′)GATC sequences. Immediately after passage of the replication fork, there is a short period (a few seconds or minutes) during which the template strand is methylated but the newly synthesized strand is not (Fig. 25–20). The transient unmethylated state of GATC sequences in the newly synthesized strand permits the new strand to be distinguished from the template strand. Replication mismatches in the vicinity of a hemimethylated GATC sequence are then repaired according to the information in the methylated parent (template) strand. Tests in vitro show that if both strands are methylated at a GATC sequence, few mismatches are repaired; if neither strand is methylated, repair occurs but does not favor either strand. The cell's

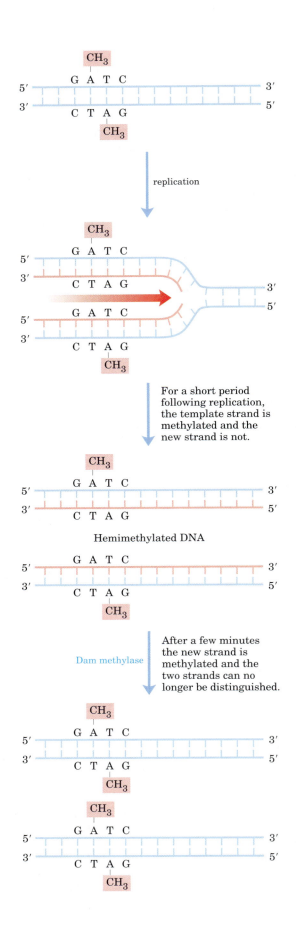

**FIGURE 25–20 Methylation and mismatch repair.** Methylation of DNA strands can serve to distinguish parent (template) strands from newly synthesized strands in *E. coli* DNA, a function that is critical to mismatch repair (see Fig. 25–21). The methylation occurs at the $N^6$ of adenines in (5′)GATC sequences. This sequence is a palindrome (see Fig. 8–20), present in opposite orientations on the two strands.

methyl-directed mismatch repair system efficiently repairs mismatches up to 1,000 bp from a hemimethylated GATC sequence. For many bacterial species, the mechanism of strand discrimination during mismatch repair has not been determined.

How is the mismatch correction process directed by relatively distant GATC sequences? A mechanism is illustrated in Figure 25–21. MutL protein forms a complex with MutS protein, and the complex binds to all mismatched base pairs (except C–C). MutH protein binds to MutL and to GATC sequences encountered by the MutL-MutS complex. DNA on both sides of the mismatch is threaded through the MutL-MutS complex, creating a DNA loop; simultaneous movement of both legs of the loop through the complex is equivalent to the complex moving in both directions at once along the DNA. MutH has a site-specific endonuclease activity that is inactive until the complex encounters a hemimethylated GATC sequence. At this site, MutH catalyzes cleavage of the unmethylated strand on the 5' side of the G in GATC, which marks the strand for repair. Further steps in the pathway depend on where the mismatch is located relative to this cleavage site (Fig. 25–22).

When the mismatch is on the 5' side of the cleavage site, the unmethylated strand is unwound and degraded in the 3'→5' direction from the cleavage site through the mismatch, and this segment is replaced with new DNA. This process requires the combined action of DNA helicase II, SSB, exonuclease I or exonuclease X (both of which degrade strands of DNA in the 3'→5' direction), DNA polymerase III, and DNA ligase. The pathway for repair of mismatches on the 3' side of the cleavage site is similar, except that the exonuclease is either exonuclease VII (which degrades single-stranded DNA in the 5'→3' or 3'→5' direction) or RecJ nuclease (which degrades single-stranded DNA in the 5'→3' direction).

Mismatch repair is a particularly expensive process for *E. coli* in terms of energy expended. The mismatch may be 1,000 bp or more from the GATC sequence. The degradation and replacement of a strand segment of this length require an enormous investment in activated deoxynucleotide precursors to repair a *single* mismatched base. This again underscores the importance to the cell of genomic integrity.

All eukaryotic cells have several proteins structurally and functionally analogous to the bacterial MutS and MutL (but not MutH) proteins. Alterations in human genes encoding proteins of this type produce some of the most common inherited cancer-susceptibility syndromes (Box 25–1), further demonstrating the value to the organism of DNA repair systems. The main MutS homologs in most eukaryotes, from yeast to humans, are MSH2 (*MutS homolog 2*), MSH3, and MSH6. Heterodimers of MSH2 and MSH6 generally bind to single

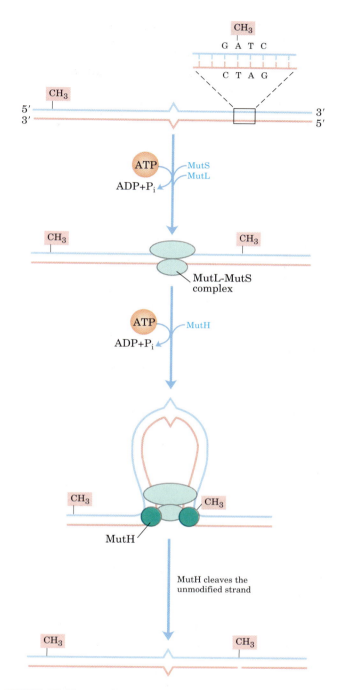

**FIGURE 25–21 A model for the early steps of methyl-directed mismatch repair.** The proteins involved in this process in *E. coli* have been purified (see Table 25–5). Recognition of the sequence (5')GATC and of the mismatch are specialized functions of the MutH and MutS proteins, respectively. The MutL protein forms a complex with MutS at the mismatch. DNA is threaded through this complex such that the complex moves simultaneously in both directions along the DNA until it encounters a MutH protein bound at a hemimethylated GATC sequence. MutH cleaves the unmethylated strand on the 5' side of the G in this sequence. A complex consisting of DNA helicase II and one of several exonucleases then degrades the unmethylated DNA strand from that point toward the mismatch (see Fig. 25–22).

BOX 25–1 BIOCHEMISTRY IN MEDICINE

## DNA Repair and Cancer

Human cancer develops when certain genes that regulate normal cell division (oncogenes and tumor suppressor genes; Chapter 12) fail to function, are activated at the wrong time, or are altered. As a consequence, cells may grow out of control and form a tumor. The genes controlling cell division can be damaged by spontaneous mutation or overridden by the invasion of a tumor virus (Chapter 26). Not surprisingly, alterations in DNA-repair genes that result in an increase in the rate of mutation can greatly increase an individual's susceptibility to cancer. Defects in the genes encoding the proteins involved in nucleotide-excision repair, mismatch repair, recombinational repair, and error-prone translesion synthesis have all been linked to human cancers. Clearly, DNA repair can be a matter of life and death.

Nucleotide-excision repair requires a larger number of proteins in humans than in bacteria, although the overall pathways are very similar. Genetic defects that inactivate nucleotide-excision repair have been associated with several genetic diseases, the best-studied of which is xeroderma pigmentosum, or XP. Because nucleotide-excision repair is the sole repair pathway for pyrimidine dimers in humans, people with XP are extremely light sensitive and readily develop sunlight-induced skin cancers. Most people with XP also have neurological abnormalities, presumably because of their inability to repair certain lesions caused by the high rate of oxidative metabolism in neurons. Defects in the genes encoding any of at least seven different protein components of the nucleotide-excision repair system can result in XP, giving rise to seven different genetic groups denoted XPA to XPG. Several of these proteins (notably XPB, XPD, and XPG) also play roles in transcription-coupled base-excision repair of oxidative lesions, described in Chapter 26.

Most microorganisms have redundant pathways for the repair of cyclobutane pyrimidine dimers—making use of DNA photolyase and sometimes base-excision repair as alternatives to nucleotide-excision repair—but humans and other placental mammals do not. This lack of a back-up to nucleotide-excision repair for the removal of pyrimidine dimers has led to speculation that early mammalian evolution involved small, furry, nocturnal animals with little need to repair UV damage. However, mammals do have a path-

way for the translesion bypass of cyclobutane pyrimidine dimers, which involves DNA polymerase η. This enzyme preferentially inserts two A residues opposite a T–T pyrimidine dimer, minimizing mutations. People with a genetic condition in which DNA polymerase η function is missing exhibit an XP-like illness known as XP-variant or XP-V. Clinical manifestations of XP-V are similar to those of the classic XP diseases, although mutation levels are higher when cells are exposed to UV light. Apparently, the nucleotide-excision repair system works in concert with DNA polymerase η in normal human cells, repairing and/or bypassing pyrimidine dimers as needed to keep cell growth and DNA replication going. Exposure to UV light introduces a heavy load of pyrimidine dimers, requiring that some be bypassed by translesion synthesis to keep replication on track. When either system is missing, it is partly compensated for by the other. A loss of polymerase η activity leads to stalled replication forks and bypass of UV lesions by different, and more mutagenic, translesion synthesis (TLS) polymerases. As when other DNA repair systems are absent, the resulting increase in mutations often leads to cancer.

One of the most common inherited cancer-susceptibility syndromes is hereditary nonpolyposis colon cancer, or HNPCC. This syndrome has been traced to defects in mismatch repair. Human and other eukaryotic cells have several proteins analogous to the bacterial MutL and MutS proteins (see Fig. 25–21). Defects in at least five different mismatch repair genes can give rise to HNPCC. The most prevalent are defects in the *hMLH1* (human MutL homolog 1) and *hMSH2* (human MutS homolog 2) genes. In individuals with HNPCC, cancer generally develops at an early age, with colon cancers being most common.

Most human breast cancer occurs in women with no known predisposition. However, about 10% of cases are associated with inherited defects in two genes, *BRCA1* and *BRCA2*. BRCA1 and BRCA2 are large proteins (human BRCA1 and BRCA2 are 1834 and 3418 amino acid residues long, respectively). They both interact with a wide range of other proteins involved in transcription, chromosome maintenance, DNA repair, and control of the cell cycle. However, the precise molecular function of BRACA1 and BRCA2 in these various cellular processes is not yet clear. Women with defects in either the *BRCA1* or *BRCA2* gene have a greater than 80% chance of developing breast cancer.

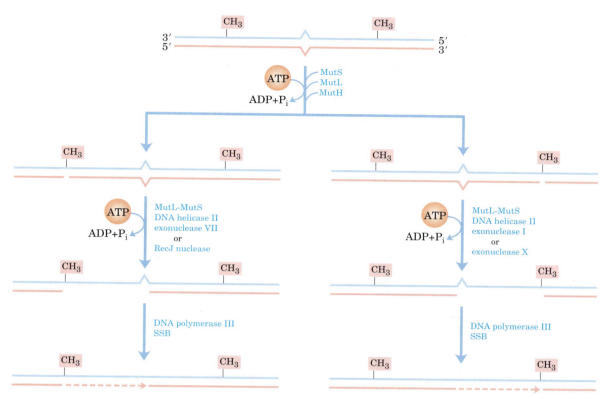

**FIGURE 25–22 Completing methyl-directed mismatch repair.** The combined action of DNA helicase II, SSB, and one of four different exonucleases removes a segment of the new strand between the MutH cleavage site and a point just beyond the mismatch. The exonuclease

that is used depends on the location of the cleavage site relative to the mismatch. The resulting gap is filled in by DNA polymerase III, and the nick is sealed by DNA ligase (not shown).

base-pair mismatches, and bind less well to slightly longer mispaired loops. In many organisms the longer mismatches (2 to 6 bp) may be bound instead by a heterodimer of MSH2 and MSH3, or are bound by both types of heterodimers in tandem. Homologs of MutL, predominantly a heterodimer of MLH1 and PMS1 (*post-meiotic segregation*), bind to and stabilize the MSH complexes. Many details of the subsequent events in eukaryotic mismatch repair remain to be worked out. In particular, we do not know the mechanism by which newly synthesized DNA strands are identified, although research has revealed that this strand identification does not involve GATC sequences.

**Base-Excision Repair** Every cell has a class of enzymes called **DNA glycosylases** that recognize particularly common DNA lesions (such as the products of cytosine and adenine deamination; see Fig. 8–33a) and remove the affected base by cleaving the *N*-glycosyl bond. This cleavage creates an apurinic or apyrimidinic site in the DNA, commonly referred to as an **AP site** or **abasic**

**site.** Each DNA glycosylase is generally specific for one type of lesion.

Uracil DNA glycosylases, for example, found in most cells, specifically remove from DNA the uracil that results from spontaneous deamination of cytosine. Mutant cells that lack this enzyme have a high rate of G≡C to A=T mutations. This glycosylase does not remove uracil residues from RNA or thymine residues from DNA. The capacity to distinguish thymine from uracil, the product of cytosine deamination—necessary for the selective repair of the latter—may be one reason why DNA evolved to contain thymine instead of uracil (p. 293).

Bacteria generally have just one type of uracil DNA glycosylase, whereas humans have at least four types, with different specificities—an indicator of the importance of uracil removal from DNA. The most abundant human uracil glycosylase, UNG, is associated with the human replisome, where it eliminates the occasional U residue inserted in place of a T during replication. The deamination of C residues is 100-fold faster in single-stranded DNA than in double-stranded DNA, and

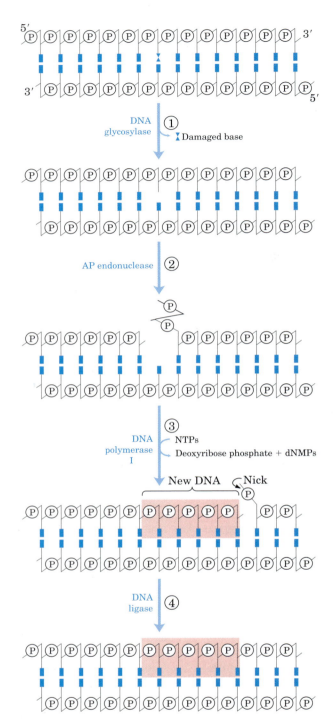

**FIGURE 25–23 DNA repair by the base-excision repair pathway.** ① A DNA glycosylase recognizes a damaged base and cleaves between the base and deoxyribose in the backbone. ② An AP endonuclease cleaves the phosphodiester backbone near the AP site. ③ DNA polymerase I initiates repair synthesis from the free 3′ hydroxyl at the nick, removing (with its 5′→3′ exonuclease activity) a portion of the damaged strand and replacing it with undamaged DNA. ④ The nick remaining after DNA polymerase I has dissociated is sealed by DNA ligase.

humans have the enzyme hSMUG1, which removes any U residues that occur in single-stranded DNA during replication or transcription. Two other human DNA glycosylases, TDG and MBD4, remove either U or T residues paired with G, generated by deamination of cytosine or 5-methylcytosine, respectively.

Other DNA glycosylases recognize and remove a variety of damaged bases, including formamidopyrimidine and 8-hydroxyguanine (both arising from purine oxidation), hypoxanthine (arising from adenine deamination), and alkylated bases such as 3-methyladenine and 7-methylguanine. Glycosylases that recognize other lesions, including pyrimidine dimers, have also been identified in some classes of organisms. Remember that AP sites also arise from the slow, spontaneous hydrolysis of the N-glycosyl bonds in DNA (see Fig. 8–33b).

Once an AP site has formed, another group of enzymes must repair it. The repair is *not* made by simply inserting a new base and re-forming the N-glycosyl bond. Instead, the deoxyribose 5′-phosphate left behind is removed and replaced with a new nucleotide. This process begins with **AP endonucleases,** enzymes that cut the DNA strand containing the AP site. The position of the incision relative to the AP site (5′ or 3′ to the site) varies with the type of AP endonuclease. A segment of DNA including the AP site is then removed, DNA polymerase I replaces the DNA, and DNA ligase seals the remaining nick (Fig. 25–23). In eukaryotes, nucleotide replacement is carried out by specialized polymerases, as described below.

***Nucleotide-Excision Repair*** DNA lesions that cause large distortions in the helical structure of DNA generally are repaired by the nucleotide-excision system, a repair pathway critical to the survival of all free-living organisms. In nucleotide-excision repair (Fig. 25–24), a multisubunit enzyme hydrolyzes two phosphodiester bonds, one on either side of the distortion caused by the lesion. In *E. coli* and other prokaryotes, the enzyme system hydrolyzes the fifth phosphodiester bond on the 3′ side and the eighth phosphodiester bond on the 5′ side to generate a fragment of 12 to 13 nucleotides (depending on whether the lesion involves one or two bases). In humans and other eukaryotes, the enzyme system hydrolyzes the sixth phosphodiester bond on the 3′ side and the twenty-second phosphodiester bond on the 5′ side, producing a fragment of 27 to 29 nucleotides. Following the dual incision, the excised oligonucleotides are released from the duplex and the resulting gap is filled—by DNA polymerase I in *E. coli* and DNA polymerase $\varepsilon$ in humans. DNA ligase seals the nick.

In *E. coli*, the key enzymatic complex is the ABC excinuclease, which has three subunits, UvrA ($M_r$ 104,000), UvrB ($M_r$ 78,000), and UvrC ($M_r$ 68,000). The

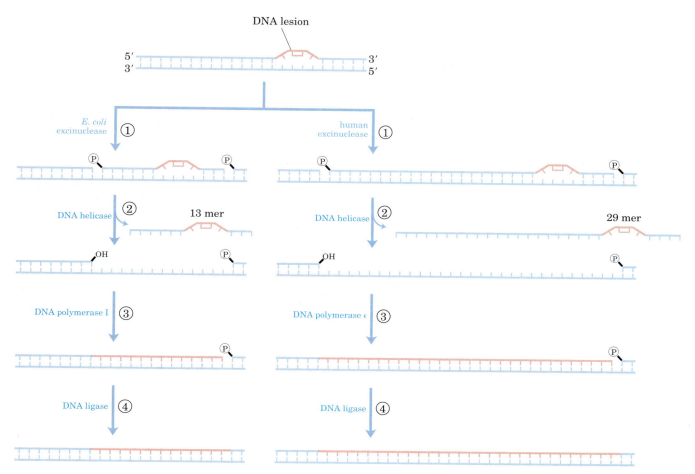

**FIGURE 25–24 Nucleotide-excision repair in *E. coli* and humans.** The general pathway of nucleotide-excision repair is similar in all organisms. ① An excinuclease binds to DNA at the site of a bulky lesion and cleaves the damaged DNA strand on either side of the lesion. ② The DNA segment—of 13 nucleotides (13 mer) or 29 nucleotides (29 mer)—is removed with the aid of a helicase. ③ The gap is filled in by DNA polymerase, and ④ the remaining nick is sealed with DNA ligase.

term "excinuclease" is used to describe the unique capacity of this enzyme complex to catalyze two specific endonucleolytic cleavages, distinguishing this activity from that of standard endonucleases. A complex of the UvrA and UvrB proteins ($A_2B$) scans the DNA and binds to the site of a lesion. The UvrA dimer then dissociates, leaving a tight UvrB-DNA complex. UvrC protein then binds to UvrB, and UvrB makes an incision at the fifth phosphodiester bond on the 3′ side of the lesion. This is followed by a UvrC-mediated incision at the eighth phosphodiester bond on the 5′ side. The resulting 12 to 13 nucleotide fragment is removed by UvrD helicase. The short gap thus created is filled in by DNA polymerase I and DNA ligase. This pathway is a primary repair route for many types of lesions, including cyclo-butane pyrimidine dimers, 6-4 photoproducts (see Fig. 8–34), and several other types of base adducts including benzo[*a*]pyrene-guanine, which is formed in DNA by exposure to cigarette smoke. The nucleolytic activity of the ABC excinuclease is novel in the sense that two cuts are made in the DNA (Fig. 25–24).

The mechanism of eukaryotic excinucleases is quite similar to that of the bacterial enzyme, although 16 polypeptides with no similarity to the *E. coli* excinuclease subunits are required for the dual incision. As described in Chapter 26, some of the nucleotide-excision repair and base-excision repair in eukaryotes is closely tied to transcription. Genetic deficiencies in nucleotide-excision repair in humans give rise to a variety of serious diseases (Box 25–1).

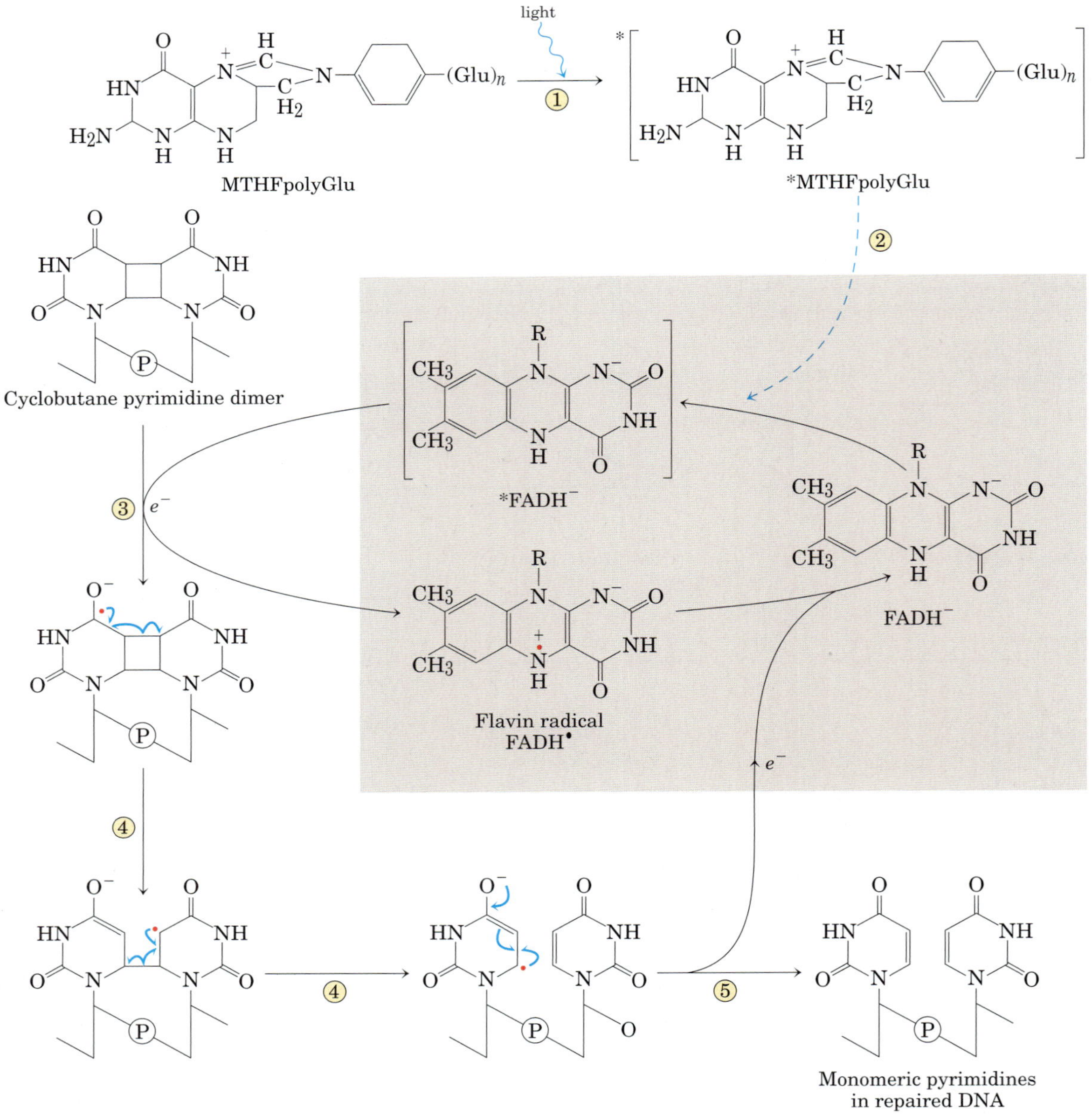

**MECHANISM FIGURE 25–25 Repair of pyrimidine dimers with photolyase.** Energy derived from absorbed light is used to reverse the photoreaction that caused the lesion. The two chromophores in *E. coli* photolyase ($M_r$ 54,000), $N^5,N^{10}$-methenyltetrahydrofolylpolyglutamate (MTHFpolyGlu) and FADH$^-$, perform complementary functions. On binding of photolyase to a pyrimidine dimer, repair proceeds as follows. ① A blue-light photon (300 to 500 nm wavelength) is absorbed by the MTHFpolyGlu, which functions as a photoantenna. ② The excitation energy passes to FADH$^-$ in the active site of the enzyme. ③ The excited flavin (*FADH$^-$) donates an electron to the pyrimidine dimer (shown here in a simplified representation) to generate an unstable dimer radical. ④ Electronic rearrangement restores the monomeric pyrimidines, and ⑤ the electron is transferred back to the flavin radical to regenerate FADH$^-$.

**Direct Repair** Several types of damage are repaired without removing a base or nucleotide. The best-characterized example is direct photoreactivation of cyclobutane pyrimidine dimers, a reaction promoted by **DNA photolyases.** Pyrimidine dimers result from an ultraviolet light–induced reaction, and photolyases use energy derived from absorbed light to reverse the damage (Fig. 25–25). Photolyases generally contain two cofactors that serve as light-absorbing agents, or chromophores. One of the chromophores is always FADH$^-$. In *E. coli* and yeast, the other chromophore is a folate. The reaction mechanism entails the generation of free radicals. DNA photolyases are not present in the cells of placental mammals (which include humans).

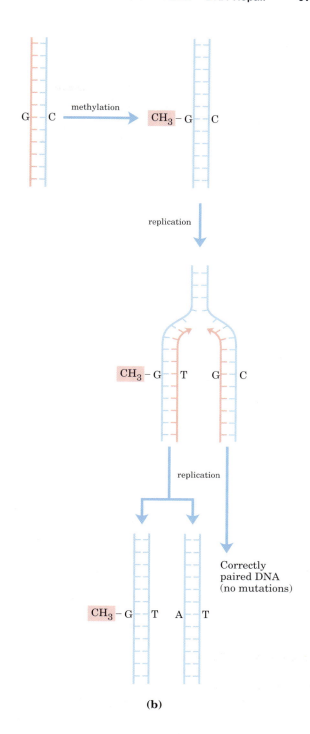

**FIGURE 25–26 Example of how DNA damage results in mutations.**
**(a)** The methylation product $O^6$-methylguanine pairs with thymine rather than cytosine. **(b)** If not repaired, this leads to a G≡C to A=T mutation after replication.

Additional examples can be seen in the repair of nucleotides with alkylation damage. The modified nucleotide $O^6$-methylguanine forms in the presence of alkylating agents and is a common and highly mutagenic lesion (p. 295). It tends to pair with thymine rather than cytosine during replication, and therefore causes G≡C to A=T mutations (Fig. 25–26). Direct repair of $O^6$-methylguanine is carried out by $O^6$-methylguanine-DNA methyltransferase, a protein that catalyzes transfer of the methyl group of $O^6$-methylguanine to one of its own Cys residues. This methyltransferase is not strictly an enzyme, because a single methyl transfer event permanently methylates the protein, making it inactive in this pathway. The consumption of an entire protein molecule to correct a single damaged base is another vivid illustration of the priority given to maintaining the integrity of cellular DNA.

A very different but equally direct mechanism is used to repair 1-methyladenine and 3-methylcytosine. The amino groups of A and C residues are sometimes methylated when the DNA is single-stranded, and the methylation directly affects proper base pairing. In *E. coli,* oxidative demethylation of these alkylated nucleotides is mediated by the AlkB protein, a member of the α-ketoglutarate-$Fe^{2+}$–dependent dioxygenase superfamily (Fig. 25–27). (See Box 4–3 for a description of another member of this enzyme family.)

Cys—SH
active

Cys—S—CH$_3$
inactive

methyltransferase

$O^6$-Methylguanine nucleotide

Guanine nucleotide

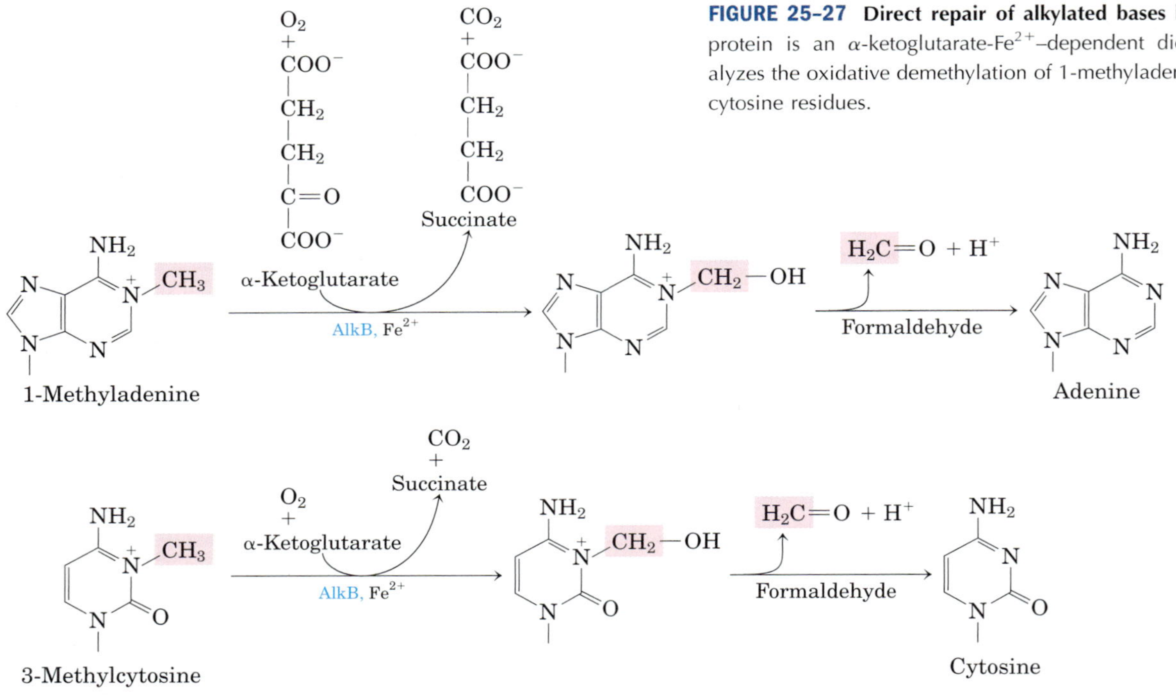

**FIGURE 25–27 Direct repair of alkylated bases by AlkB.** The AlkB protein is an $\alpha$-ketoglutarate-$Fe^{2+}$–dependent dioxygenase. It catalyzes the oxidative demethylation of 1-methyladenine and 3-methylcytosine residues.

## The Interaction of Replication Forks with DNA Damage Can Lead to Error-Prone Translesion DNA Synthesis

The repair pathways considered to this point generally work only for lesions in double-stranded DNA, the undamaged strand providing the correct genetic information to restore the damaged strand to its original state. However, in certain types of lesions, such as double-strand breaks, double-strand cross-links, or lesions in a single-stranded DNA, the complementary strand is itself damaged or is absent. Double-strand breaks and lesions in single-stranded DNA most often arise when a replication fork encounters an unrepaired DNA lesion (Fig. 25–28). Such lesions and DNA cross-links can also result from ionizing radiation and oxidative reactions.

At a stalled bacterial replication fork, there are two avenues for repair. In the absence of a second strand, the information required for accurate repair must come from a separate, homologous chromosome. The repair system thus involves homologous genetic recombination. This **recombinational DNA repair** is considered in detail in Section 25.3. Under some conditions, a second repair pathway, **error-prone translesion DNA synthesis** (often abbreviated TLS), becomes available. When this pathway is active, DNA repair becomes significantly less accurate and a high mutation rate can result. In bacteria, error-prone translesion DNA synthesis is part of a cellular stress response to extensive DNA damage known, appropriately enough, as the **SOS response.** Some SOS proteins, such as the UvrA and UvrB proteins already described (Table 25–6), are normally

present in the cell but are induced to higher levels as part of the SOS response. Additional SOS proteins participate in the pathway for error-prone repair; these include the UmuC and UmuD proteins ("Umu" from *unmut*able; lack of the *umu* gene function eliminates error-prone repair). The UmuD protein is cleaved in an SOS-regulated process to a shorter form called UmuD′, which forms a complex with UmuC to create a specialized DNA polymerase (DNA polymerase V) that can replicate past many of the DNA lesions that would normally block replication. Proper base pairing is often impossible at the site of such a lesion, so this translesion replication is error-prone.

Given the emphasis on the importance of genomic integrity throughout this chapter, the existence of a system that increases the rate of mutation may seem incongruous. However, we can think of this system as a desperation strategy. The *umuC* and *umuD* genes are fully induced only late in the SOS response, and they are not activated for translesion synthesis initiated by UmuD cleavage unless the levels of DNA damage are particularly high and all replication forks are blocked. The mutations resulting from DNA polymerase V–mediated replication kill some cells and create deleterious mutations in others, but this is the biological price an organism pays to overcome an otherwise insurmountable barrier to replication, as it permits at least a few mutant cells to survive.

In addition to DNA polymerase V, translesion replication requires the RecA protein, SSB, and some subunits derived from DNA polymerase III. Yet another DNA polymerase, DNA polymerase IV, is also induced during

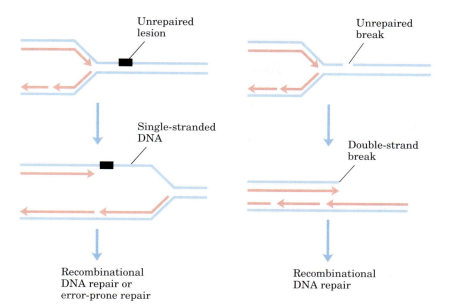

Unrepaired lesion

Single-stranded DNA

Recombinational DNA repair or error-prone repair

Unrepaired break

Double-strand break

Recombinational DNA repair

**FIGURE 25-28  DNA damage and its effect on DNA replication.** If the replication fork encounters an unrepaired lesion or strand break, replication generally halts and the fork may collapse. A lesion is left behind in an unreplicated, single-stranded segment of the DNA; a strand break becomes a double-strand break. In each case, the damage to one strand cannot be repaired by mechanisms described earlier in this chapter, because the complementary strand required to direct accurate repair is damaged or absent. There are two possible avenues for repair: recombinational DNA repair (described in Fig. 25–37) or, when lesions are unusually numerous, error-prone repair. The latter mechanism involves a novel DNA polymerase (DNA polymerase V, encoded by the *umuC* and *umuD* genes) that can replicate, albeit inaccurately, over many types of lesions. The repair mechanism is referred to as error-prone because mutations often result.

the SOS response. Replication by DNA polymerase IV, a product of the *dinB* gene, is also highly error-prone. The bacterial DNA polymerases IV and V are part of a family of TLS polymerases found in all organisms. These enzymes lack a proofreading exonuclease activity, and the fidelity of replicative base selection can be reduced by a factor of $10^2$, lowering overall replication fidelity to one error in ~1,000 nucleotides.

Mammals have many low-fidelity DNA polymerases of the TLS polymerase family. However, the presence of these enzymes does not necessarily translate into an unacceptable mutational burden, because most of the

**TABLE 25–6  Genes Induced as Part of the SOS Response *in E. coli***

| Gene name | Protein encoded and/or role in DNA repair |
|---|---|
| **Genes of known function** | |
| *polB (dinA)* | Encodes polymerization subunit of DNA polymerase II, required for replication restart in recombinational DNA repair |
| *uvrA* / *uvrB* | Encode ABC excinuclease subunits UvrA and UvrB |
| *umuC* / *umuD* | Encode DNA polymerase V |
| *sulA* | Encodes protein that inhibits cell division, possibly to allow time for DNA repair |
| *recA* | Encodes RecA protein, required for error-prone repair and recombinational repair |
| *dinB* | Encodes DNA polymerase IV |
| **Genes involved in DNA metabolism, but role in DNA repair unknown** | |
| *ssb* | Encodes single-stranded DNA–binding protein (SSB) |
| *uvrD* | Encodes DNA helicase II (DNA-unwinding protein) |
| *himA* | Encodes subunit of integration host factor (IHF), involved in site-specific recombination, replication, transposition, regulation of gene expression |
| *recN* | Required for recombinational repair |
| **Genes of unknown function** | |
| *dinD* | |
| *dinF* | |

Note: Some of these genes and their functions are further discussed in Chapter 28.

enzymes also have specialized functions in DNA repair. DNA polymerase $\eta$ (eta), for example, is a TLS polymerase found in all eukaryotes. It promotes translesion synthesis primarily across cyclobutane T–T dimers. Few mutations result in this case, because the enzyme preferentially inserts two A residues across from the linked T residues. Several other low-fidelity polymerases, including DNA polymerases $\beta$, $\iota$ (iota), and $\lambda$, have specialized roles in eukaryotic base-excision repair. Each of these enzymes has a 5′-deoxyribose phosphate lyase activity in addition to its polymerase activity. After base removal by a glycosylase and backbone cleavage by an AP endonuclease, these enzymes remove the abasic site (a 5′-deoxyribose phosphate) and fill in the very short gap with their polymerase activity. The frequency of mutations due to DNA polymerase $\eta$ activity is minimized by the very short lengths (often one nucleotide) of DNA synthesized.

What emerges from research into cellular DNA repair systems is a picture of a DNA metabolism that maintains genomic integrity with multiple and often redundant systems. In the human genome, more than 130 genes encode proteins dedicated to the repair of DNA. In many cases, the loss of function of one of these proteins results in genomic instability and an increased occurrence of oncogenesis (Box 25–1). These repair systems are often integrated with the DNA replication systems and are complemented by the recombination systems that we turn to next.

### SUMMARY 25.2 DNA Repair

■ Cells have many systems for DNA repair. Mismatch repair in *E. coli* is directed by transient nonmethylation of (5′)GATC sequences on the newly synthesized strand.

■ Base-excision repair systems recognize and repair damage caused by environmental agents (such as radiation and alkylating agents) and spontaneous reactions of nucleotides. Some repair systems recognize and excise only damaged or incorrect bases, leaving an AP (abasic) site in the DNA. This is repaired by excision and replacement of the DNA segment containing the AP site.

■ Nucleotide-excision repair systems recognize and remove a variety of bulky lesions and pyrimidine dimers. They excise a segment of the DNA strand including the lesion, leaving a gap that is filled in by DNA polymerase and ligase activities.

■ Some DNA damage is repaired by direct reversal of the reaction causing the damage:

pyrimidine dimers are directly converted to monomeric pyrimidines by a photolyase, and the methyl group of $O^6$-methylguanine is removed by a methyltransferase.

■ In bacteria, error-prone translesion DNA synthesis, involving TLS DNA polymerases, occurs in response to very heavy DNA damage. In eukaryotes, similar polymerases have specialized roles in DNA repair that minimize the introduction of mutations.

## 25.3 DNA Recombination

The rearrangement of genetic information within and among DNA molecules encompasses a variety of processes, collectively placed under the heading of genetic recombination. The practical applications of DNA rearrangements in altering the genomes of increasing numbers of organisms are now being explored (Chapter 9).

Genetic recombination events fall into at least three general classes. **Homologous genetic recombination** (also called general recombination) involves genetic exchanges between any two DNA molecules (or segments of the same molecule) that share an extended region of nearly identical sequence. The actual sequence of bases is irrelevant, as long as it is similar in the two DNAs. In **site-specific recombination,** the exchanges occur only at a *particular* DNA sequence. **DNA transposition** is distinct from both other classes in that it usually involves a short segment of DNA with the remarkable capacity to move from one location in a chromosome to another. These "jumping genes" were first observed in maize in the 1940s by Barbara McClintock. There is in addition a wide range of unusual genetic rearrangements for which no mechanism or purpose has yet been proposed. Here we focus on the three general classes.

Barbara McClintock, 1902–1992

The functions of genetic recombination systems are as varied as their mechanisms. They include roles in specialized DNA repair systems, specialized activities in DNA replication, regulation of expression of certain genes, facilitation of proper chromosome segregation during eukaryotic cell division, maintenance of genetic diversity, and implementation of programmed genetic rearrangements during embryonic development. In most cases, genetic recombination is closely integrated with other processes in DNA metabolism, and this becomes a theme of our discussion.

## Homologous Genetic Recombination Has Several Functions

In bacteria, homologous genetic recombination is primarily a DNA repair process and in this context (as noted in Section 25.2) is referred to as **recombinational DNA repair.** It is usually directed at the reconstruction of replication forks stalled at the site of DNA damage. Homologous genetic recombination can also occur during conjugation (mating), when chromosomal DNA is transferred from a donor to a recipient bacterial cell. Recombination during conjugation, although rare in wild bacterial populations, contributes to genetic diversity.

In eukaryotes, homologous genetic recombination can have several roles in replication and cell division, including the repair of stalled replication forks. Recombination occurs with the highest frequency during **meiosis,** the process by which diploid germ-line cells with two sets of chromosomes divide to produce haploid gametes—sperm cells or ova in higher eukaryotes—each gamete having only one member of each chromosome pair (Fig. 25–29). Meiosis begins with replication of the DNA in the germ-line cell so that each DNA molecule is present in four copies. The cell then goes through two rounds of cell division without an intervening round of DNA replication. This reduces the DNA content to the haploid level in each gamete.

After the DNA is replicated during prophase of the first meiotic division, the resulting sister chromatids remain associated at their centromeres. At this stage, each set of four homologous chromosomes exists as two pairs of chromatids. Genetic information is now exchanged between the closely associated homologous chromatids

by homologous genetic recombination, a process involving the breakage and rejoining of DNA (Fig. 25–30). This exchange, also referred to as crossing over, can be observed with the light microscope. Crossing over links the two pairs of sister chromatids together at points called chiasmata (singular, chiasma).

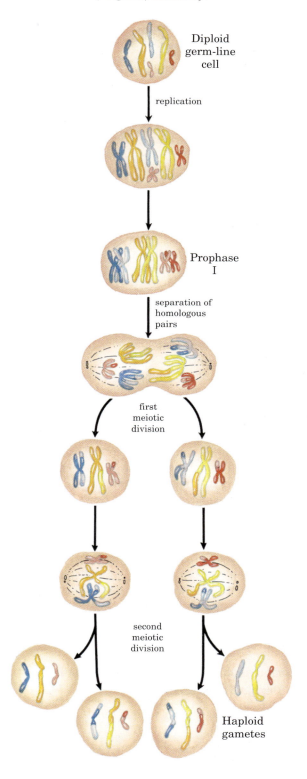

**FIGURE 25–29 Meiosis in eukaryotic germ-line cells.** The chromosomes of a hypothetical diploid germ-line cell (six chromosomes; three homologous pairs) replicate and are held together at their centromeres. Each replicated double-stranded DNA molecule is called a chromatid (sister chromatid). In prophase I, just before the first meiotic division, the three homologous sets of chromatids align to form tetrads, held together by covalent links at homologous junctions (chiasmata). Crossovers occur within the chiasmata (see Fig. 25–30). These transient associations between homologs ensure that the two tethered chromosomes segregate properly in the next step, when they migrate toward opposite poles of the dividing cell in the first meiotic division. The products of this division are two daughter cells, each with three pairs of chromatids. The pairs now line up across the equator of the cell in preparation for separation of the chromatids (now called chromosomes). The second meiotic division produces four haploid daughter cells that can serve as gametes. Each has three chromosomes, half the number of the diploid germ-line cell. The chromosomes have resorted and recombined.

Diploid germ-line cell

replication

Prophase I

separation of homologous pairs

first meiotic division

second meiotic division

Haploid gametes

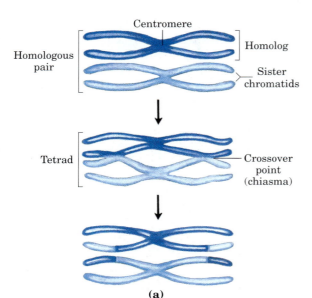

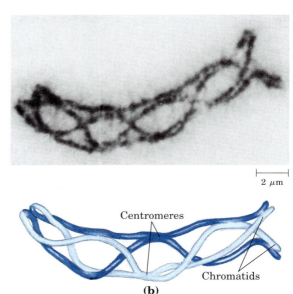

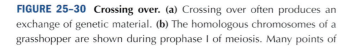

**(a)**

**(b)**

**FIGURE 25–30 Crossing over. (a)** Crossing over often produces an exchange of genetic material. **(b)** The homologous chromosomes of a grasshopper are shown during prophase I of meiosis. Many points of joining (chiasmata) are evident between the two homologous pairs of chromatids. These chiasmata are the physical manifestation of prior homologous recombination (crossing over) events.

Crossing over effectively links together all four homologous chromatids, a linkage that is essential to the proper segregation of chromosomes in the subsequent meiotic cell divisions. Crossing over is not an entirely random process, and "hot spots" have been identified on many eukaryotic chromosomes. However, the assumption that crossing over can occur with equal probability at almost any point along the length of two homologous chromosomes remains a reasonable approximation in many cases, and it is this assumption that permits the genetic mapping of genes. The frequency of homologous recombination in any region separating two points on a chromosome is roughly proportional to the distance between the points, and this allows determination of the relative positions of and distances between different genes.

Homologous recombination thus serves at least three identifiable functions: (1) it contributes to the repair of several types of DNA damage; (2) it provides, in eukaryotic cells, a transient physical link between chromatids that promotes the orderly segregation of chromosomes at the first meiotic cell division; and (3) it enhances genetic diversity in a population.

### Recombination during Meiosis Is Initiated with Double-Strand Breaks

A likely pathway for homologous recombination during meiosis is outlined in Figure 25–31a. The model has four key features. First, homologous chromosomes are aligned. Second, a double-strand break in a DNA mole-

cule is enlarged by an exonuclease, leaving a single-strand extension with a free 3'-hydroxyl group at the broken end (step ①). Third, the exposed 3' ends invade the intact duplex DNA, and this is followed by **branch migration** (Fig. 25–32) and/or replication to create a pair of crossover structures, called Holliday junctions (Fig. 25–31a, steps ② to ④). Fourth, cleavage of the two crossovers creates two complete recombinant products (step ⑤).

In this **double-strand break repair model** for recombination, the 3' ends are used to initiate the genetic exchange. Once paired with the complementary strand on the intact homolog, a region of hybrid DNA is created that contains complementary strands from two different parent DNAs (the product of step ② in Fig. 25–31a). Each of the 3' ends can then act as a primer for DNA replication. The structures thus formed, **Holliday intermediates** (Fig. 25–31b), are a feature of homologous genetic recombination pathways in all organisms.

Homologous recombination can vary in many details from one species to another, but most of the steps outlined above are generally present in some form. There are two ways to cleave, or "resolve," the Holliday intermediate so that the two recombinant products carry genes in the same linear order as in the substrates—the original, unrecombined chromosomes (step ⑤ of Fig. 25–31a). If cleaved one way, the DNA flanking the region containing the hybrid DNA is not recombined; if cleaved the other way, the flanking DNA is recombined. Both outcomes are observed in vivo in eukaryotes and prokaryotes.

The homologous recombination illustrated in Figure 25–31 is a very elaborate process with subtle molecular consequences for the generation of genetic diversity. To understand how this process contributes to diversity, we should keep in mind that the two homologous chromo-

somes that undergo recombination are not necessarily *identical.* The linear array of genes may be the same, but the base sequences in some of the genes may differ slightly (in different alleles). In a human, for example, one chromosome may contain the allele for hemoglobin A

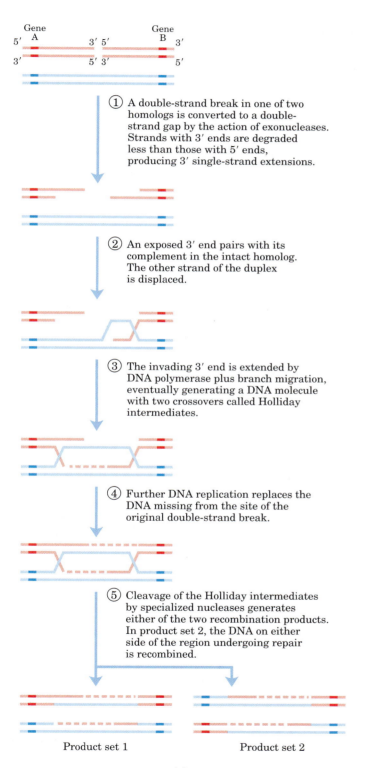

① A double-strand break in one of two homologs is converted to a double-strand gap by the action of exonucleases. Strands with 3' ends are degraded less than those with 5' ends, producing 3' single-strand extensions.

② An exposed 3' end pairs with its complement in the intact homolog. The other strand of the duplex is displaced.

③ The invading 3' end is extended by DNA polymerase plus branch migration, eventually generating a DNA molecule with two crossovers called Holliday intermediates.

④ Further DNA replication replaces the DNA missing from the site of the original double-strand break.

⑤ Cleavage of the Holliday intermediates by specialized nucleases generates either of the two recombination products. In product set 2, the DNA on either side of the region undergoing repair is recombined.

Product set 1          Product set 2

**(a)**

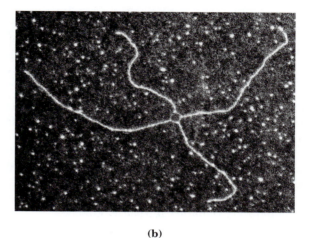

**(b)**

**FIGURE 25–31 Recombination during meiosis.**
**(a)** Model of double-strand break repair for homologous genetic recombination. The two homologous chromosomes involved in this recombination event have similar sequences. Each of the two genes shown has different alleles on the two chromosomes. The DNA strands and alleles are colored differently so that their fate is evident. The steps are described in the text.
**(b)** A Holliday intermediate formed between two bacterial plasmids in vivo, as seen with the electron microscope. The intermediates are named for Robin Holliday, who first proposed their existence in 1964.

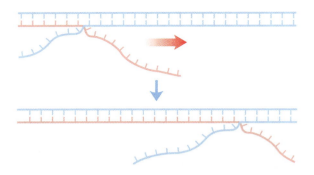

**FIGURE 25–32 Branch migration.** When a template strand pairs with two different complementary strands, a branch is formed at the point where the three complementary strands meet. The branch "migrates" when base pairing to one of the two complementary strands is broken and replaced with base pairing to the other complementary strand. In the absence of an enzyme to direct it, this process can move the branch spontaneously in either direction. Spontaneous branch migration is blocked wherever one of the otherwise complementary strands has a sequence nonidentical to the other strand.

(normal hemoglobin) while the other contains the allele for hemoglobin S (the sickle-cell mutation). The difference may consist of no more than one base pair among millions. Homologous recombination does not change the linear array of genes, but it can determine which alleles become linked together on a single chromosome.

## Recombination Requires a Host of Enzymes and Other Proteins

Enzymes that promote various steps of homologous recombination have been isolated from both prokaryotes and eukaryotes. In *E. coli*, the *recB*, *recC*, and *recD* genes encode the RecBCD enzyme, which has both helicase and nuclease activities. The RecA protein pro-

motes all the central steps in the homologous recombination process: the pairing of two DNAs, formation of Holliday intermediates, and branch migration (as described below). The RuvA and RuvB proteins (*r*epair of *UV* damage) form a complex that binds to Holliday intermediates, displaces RecA protein, and promotes branch migration at higher rates than does RecA. Nucleases that specifically cleave Holliday intermediates, often called resolvases, have been isolated from bacteria and yeast. The RuvC protein is one of at least two such nucleases in *E. coli*.

The RecBCD enzyme binds to linear DNA at a free (broken) end and moves inward along the double helix, unwinding and degrading the DNA in a reaction coupled to ATP hydrolysis (Fig. 25–33). The activity of the enzyme is altered when it interacts with a sequence referred to as chi, (5′)GCTGGTGG. From that point, degradation of the strand with a 3′ terminus is greatly reduced, but degradation of the 5′-terminal strand is increased. This process creates a single-stranded DNA with a 3′ end, which is used during subsequent steps in recombination (Fig. 25–31). The 1,009 chi sequences scattered throughout the *E. coli* genome enhance the frequency of recombination about five- to tenfold within 1,000 bp of the chi site. The enhancement declines as the distance from the site increases. Sequences that enhance recombination frequency have also been identified in several other organisms.

RecA is unusual among the proteins of DNA metabolism in that its active form is an ordered, helical filament of up to several thousand RecA monomers that assemble cooperatively on DNA (Fig. 25–34). This fila-

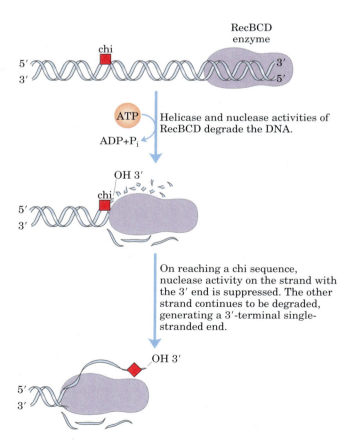

Helicase and nuclease activities of RecBCD degrade the DNA.

On reaching a chi sequence, nuclease activity on the strand with the 3′ end is suppressed. The other strand continues to be degraded, generating a 3′-terminal single-stranded end.

**FIGURE 25–33 Helicase and nuclease activities of the RecBCD enzyme.** Entering at a double-stranded end, RecBCD unwinds and degrades the DNA until it encounters a chi sequence. The interaction with chi alters the activity of RecBCD so that it generates a single-stranded DNA with a 3′ end, suitable for subsequent steps in recombination. Movement of the enzyme requires ATP hydrolysis. This enzyme is believed to help initiate homologous genetic recombination in *E. coli*. It is also involved in the repair of double-strand breaks at collapsed replication forks.

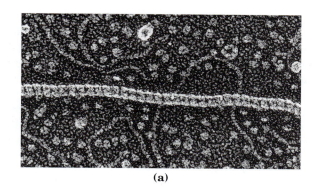

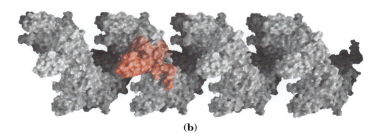

**FIGURE 25–34 RecA. (a)** Nucleoprotein filament of RecA protein on single-stranded DNA, as seen with the electron microscope. The striations indicate the right-handed helical structure of the filament. **(b)** Surface contour model of a 24-subunit RecA filament. The filament has six subunits per turn. One subunit is colored red to provide perspective (derived from PDB ID 2REB).

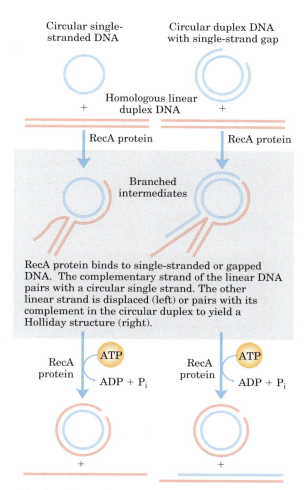

Continued branch migration yields a circular duplex with a nick and a displaced linear strand (left) or a partially single-stranded linear duplex (right).

**FIGURE 25–35 DNA strand-exchange reactions promoted by RecA protein in vitro.** Strand exchange involves the separation of one strand of a duplex DNA from its complement and transfer of the strand to an alternative complementary strand to form a new duplex (heteroduplex) DNA. The transfer forms a branched intermediate. Formation of the final product depends on branch migration, which is facilitated by RecA. The reaction can involve three strands (left) or a reciprocal exchange between two homologous duplexes—four strands in all (right). When four strands are involved, the branched intermediate that results is a Holliday intermediate. RecA protein promotes the branch-migration phases of these reactions, using energy derived from ATP hydrolysis.

ment normally forms on single-stranded DNA, such as that produced by the RecBCD enzyme. The filament will also form on a duplex DNA with a single-strand gap; in this case, the first RecA monomers bind to the single-stranded DNA in the gap, after which the assembled filament rapidly envelops the neighboring duplex. The RecF, RecO, and RecR proteins regulate the assembly and disassembly of RecA filaments.

A useful model to illustrate the recombination activities of the RecA filament is the in vitro DNA strand exchange reaction (Fig. 25–35). A single strand of DNA is first bound by RecA to establish the nucleoprotein filament. The RecA filament then takes up a homologous duplex DNA and aligns it with the bound single strand. Strands are then exchanged between the two DNAs to create hybrid DNA. The exchange occurs at a rate of 6 bp/s and progresses in the 5′→3′ direction relative to the single-stranded DNA within the RecA filament. This reaction can involve either three or four strands (Fig. 25–35); in the latter case, a Holliday intermediate forms during the process.

As the duplex DNA is incorporated within the RecA filament and aligned with the bound single-stranded DNA over regions of hundreds of base pairs, one strand of the duplex switches pairing partners (Fig. 25–36,

step ②). Because DNA is a helical structure, continued strand exchange requires an ordered rotation of the two aligned DNAs. This brings about a spooling action (steps ③ and ④) that shifts the branch point along the helix. ATP is hydrolyzed by RecA protein during this reaction.

Once a Holliday intermediate has formed, a host of enzymes—topoisomerases, the RuvAB branch migration protein, a resolvase, other nucleases, DNA polymerase

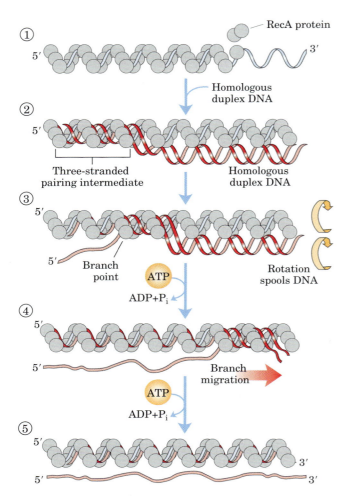

**FIGURE 25–36 Model for DNA strand exchange mediated by RecA protein.** A three-strand reaction is shown. The balls representing RecA protein are undersized relative to the thickness of DNA to clarify the fate of the DNA strands. ① RecA protein forms a filament on the single-stranded DNA. ② A homologous duplex incorporates into this complex. ③ As spooling shifts the three-stranded region from left to right, one of the strands in the duplex is transferred to the single strand originally bound in the filament. The other strand of the duplex is displaced, and a new duplex forms within the filament. As rotation continues (④ and ⑤), the displaced strand separates entirely. In this model, hydrolysis of ATP by RecA protein rotates the two DNA molecules relative to each other and thus directs the strand exchange from left to right as shown.

I or III, and DNA ligase—are required to complete recombination. The RuvC protein ($M_r$ 20,000) of *E. coli* cleaves Holliday intermediates to generate full-length, unbranched chromosome products.

## All Aspects of DNA Metabolism Come Together to Repair Stalled Replication Forks

Like all cells, bacteria sustain high levels of DNA damage even under normal growth conditions. Most DNA lesions are repaired rapidly by base-excision repair, nucleotide-excision repair, and the other pathways described earlier. Nevertheless, almost every bacterial replication fork encounters an unrepaired DNA lesion or break at some point in its journey from the replication origin to the terminus (Fig. 25–28). DNA polymerase III cannot proceed past many types of DNA lesions, and these encounters tend to leave the lesion in a single-strand gap. An encounter with a DNA strand break creates a double-strand break. Both situations require recombinational DNA repair (Fig. 25–37). Under normal growth conditions, stalled replication forks are reactivated by an elaborate repair pathway encompassing recombinational DNA repair, the restart of replication, and the repair of any lesions left behind. All aspects of DNA metabolism come together in this process.

After a replication fork has been halted, it can be restored by at least two major paths, both of which require the RecA protein. The repair pathway for lesion-containing DNA gaps also requires the RecF, RecO, and RecR proteins. Repair of double-strand breaks requires the RecBCD enzyme (Fig. 25–37). Additional recombination steps are followed by a process called **origin-independent restart of replication,** in which the replication fork reassembles with the aid of a complex of seven proteins (PriA, B, and C, and DnaB, C, G, and T). This complex, originally discovered as a component required for the replication of ϕX174 DNA in vitro, is now termed the **replication restart primosome.** Restart of the replication fork also requires DNA polymerase II, in a role not yet defined; this polymerase II activity gives way to DNA polymerase III for the extensive replication generally required to complete the chromosome.

The repair of stalled replication forks entails a coordinated transition from replication to recombination and back to replication. The recombination steps function to fill the DNA gap or rejoin the broken DNA branch to recreate the branched DNA structure at the replication fork. Lesions left behind in what is now duplex DNA are repaired by pathways such as base-excision or nucleotide-excision repair. Thus a wide range of enzymes encompassing every aspect of DNA metabolism ultimately take part in the repair of a stalled replication fork. This type of repair process is clearly a primary function of the homologous recombination system of every cell, and defects in recombinational DNA repair play an important role in human disease (Box 25–1).

## Site-Specific Recombination Results in Precise DNA Rearrangements

Homologous genetic recombination can involve any two homologous sequences. The second general type of recombination, site-specific recombination, is a very different type of process: recombination is limited to spe-

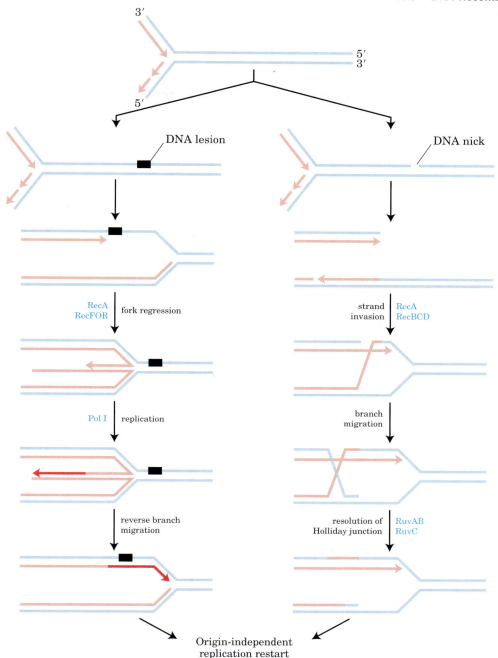

**FIGURE 25–37 Models for recombinational DNA repair of stalled replication forks.** The replication fork collapses on encountering a DNA lesion (left) or strand break (right). Recombination enzymes promote the DNA strand transfers needed to repair the branched DNA structure at the replication fork. A lesion in a single-strand gap is repaired in a reaction requiring the RecF, RecO, and RecR proteins. Double-strand breaks are repaired in a pathway requiring the RecBCD enzyme. Both pathways require RecA. Recombination intermediates are processed by additional enzymes (e.g., RuvA, RuvB, and RuvC, which process Holliday intermediates). Lesions in double-stranded DNA are repaired by nucleotide-excision repair or other pathways. The replication fork re-forms with the aid of enzymes catalyzing origin-independent replication restart, and chromosomal replication is completed. The overall process requires an elaborate coordination of all aspects of bacterial DNA metabolism.

cific sequences. Recombination reactions of this type occur in virtually every cell, filling specialized roles that vary greatly from one species to another. Examples include regulation of the expression of certain genes and promotion of programmed DNA rearrangements in em-

bryonic development or in the replication cycles of some viral and plasmid DNAs. Each site-specific recombination system consists of an enzyme called a recombinase and a short (20 to 200 bp), unique DNA sequence where the recombinase acts (the recombination site). One or

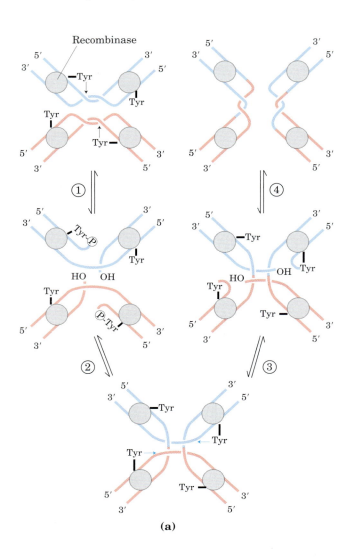

**(a)**

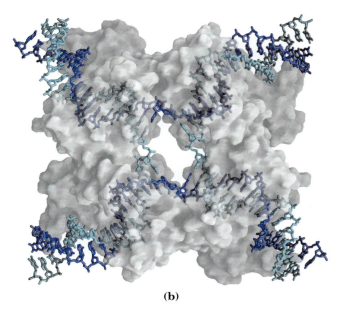

**(b)**

**FIGURE 25–38** **A site-specific recombination reaction. (a)** The reaction shown here is for a common class of site-specific recombinases called integrase-class recombinases (named after bacteriophage λ integrase, the first recombinase characterized). The reaction is carried out within a tetramer of identical subunits. Recombinase subunits bind to a specific sequence, often called simply the recombination site. ① One strand in each DNA is cleaved at particular points within the sequence. The nucleophile is the OH group of an active-site Tyr residue, and the product is a covalent phosphotyrosine link between protein and DNA. ② The cleaved strands join to new partners, producing a Holliday intermediate. Steps ③ and ④ complete the reaction by a process similar to the first two steps. The original sequence of the recombination site is regenerated after recombining the DNA flanking the site. These steps occur within a complex of multiple recombinase subunits that sometimes includes other proteins not shown here. **(b)** A surface contour model of a four-subunit integrase-class recombinase called the Cre recombinase, bound to a Holliday intermediate (shown with light blue and dark blue helix strands). The protein has been rendered transparent so that the bound DNA is visible (derived from PDB ID 3CRX).

more auxiliary proteins may regulate the timing or outcome of the reaction.

In vitro studies of many site-specific recombination systems have elucidated some general principles, including the fundamental reaction pathway (Fig. 25–38a). A separate recombinase recognizes and binds to each of two recombination sites on two different DNA molecules or within the same DNA. One DNA strand in each site is cleaved at a specific point within the site, and the recombinase becomes covalently linked to the DNA at the cleavage site through a phosphotyrosine (or phosphoserine) bond (step ①). The transient protein-DNA linkage preserves the phosphodiester bond that is lost in cleaving the DNA, so high-energy cofactors such as ATP are unnecessary in subsequent steps. The cleaved DNA strands are rejoined to new partners to form a Holliday intermediate, with new phosphodiester bonds created at the expense of the protein-DNA linkage (step ②). To complete the reaction, the process must be repeated at a second point within each of the two recombination sites (steps ③ and ④). In some systems, both strands of each recombination site are cut concurrently and rejoined to new partners without the Holliday intermediate. The exchange is always reciprocal and precise, regenerating the recombination sites when the reaction is complete. We can view a recombinase as a site-specific endonuclease and ligase in one package.

The sequences of the recombination sites recognized by site-specific recombinases are partially asymmetric (nonpalindromic), and the two recombining sites align in the same orientation during the recombinase reaction. The outcome depends on the location and orientation of the recombination sites (Fig. 25–39). If the two sites are on the same DNA molecule, the reaction either inverts or deletes the intervening DNA, determined by whether the recombination sites have the opposite or the same

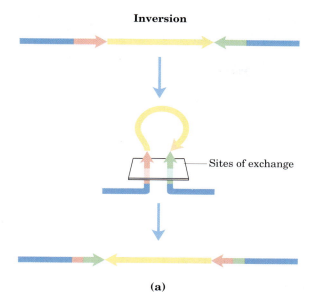

**Inversion**

**Deletion and insertion**

Sites of exchange

insertion ‖ deletion

+

**(a)**

**(b)**

**FIGURE 25–39 Effects of site-specific recombination.** The outcome of site-specific recombination depends on the location and orientation of the recombination sites (red and green) in a double-stranded DNA molecule. Orientation here (shown by arrowheads) refers to the order of nucleotides in the recombination site, not the 5′→3′ direction.

**(a)** Recombination sites with opposite orientation in the same DNA molecule. The result is an inversion. **(b)** Recombination sites with the same orientation, either on one DNA molecule, producing a deletion, or on two DNA molecules, producing an insertion.

orientation, respectively. If the sites are on different DNAs, the recombination is intermolecular; if one or both DNAs are circular, the result is an insertion. Some recombinase systems are highly specific for one of these reaction types and act only on sites with particular orientations.

The first site-specific recombination system studied in vitro was that encoded by bacteriophage λ. When λ phage DNA enters an *E. coli* cell, a complex series of regulatory events commits the DNA to one of two fates.

The λ DNA either replicates and produces more bacteriophages (destroying the host cell) or integrates into the host chromosome, replicating passively along with the chromosome for many cell generations. Integration is accomplished by a phage-encoded recombinase (λ integrase) that acts at recombination sites on the phage and bacterial DNAs—at attachment sites attP and attB, respectively (Fig. 25–40). The role of site-specific recombination in regulating gene expression is considered in Chapter 28.

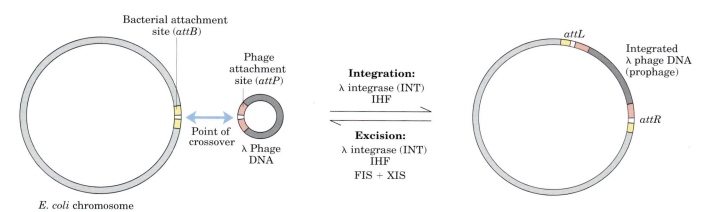

**FIGURE 25–40 Integration and excision of bacteriophage λ DNA at the chromosomal target site.** The attachment site on the λ phage DNA (attP) shares only 15 bp of complete homology with the bacterial site (attB) in the region of the crossover. The reaction generates two new attachment sites (attR and attL) flanking the integrated phage DNA.

The recombinase is the λ integrase (or INT protein). Integration and excision use different attachment sites and different auxiliary proteins. Excision uses the proteins XIS, encoded by the bacteriophage, and FIS, encoded by the bacterium. Both reactions require the protein IHF (*integration host factor*), encoded by the bacterium.

## Complete Chromosome Replication Can Require Site-Specific Recombination

Recombinational DNA repair of a circular bacterial chromosome, while essential, sometimes generates deleterious byproducts. The resolution of a Holliday junction at a replication fork by a nuclease such as RuvC, followed by completion of replication, can give rise to one of two products: the usual two monomeric chromosomes or a contiguous dimeric chromosome (Fig. 25–41). In the latter case, the covalently linked chromosomes cannot be segregated to daughter cells at cell division and the dividing cells become "stuck." A specialized site-specific recombination system in *E. coli*, the XerCD system, converts the dimeric chromosomes to monomeric chromosomes so that cell division can proceed. The reaction is a site-specific deletion reaction (Fig. 25–39b). This is another example of the close coordination between DNA recombination processes and other aspects of DNA metabolism.

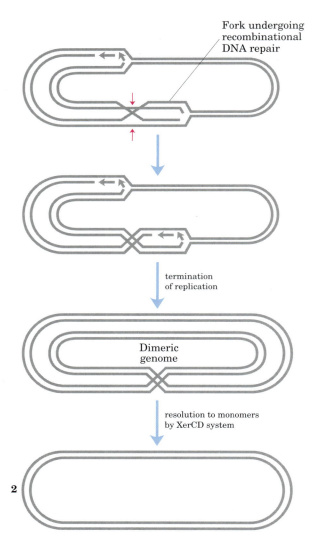

**FIGURE 25–41  DNA deletion to undo a deleterious effect of recombinational DNA repair.** The resolution of a Holliday intermediate during recombinational DNA repair (if cut at the points indicated by red arrows) can generate a contiguous dimeric chromosome. A specialized site-specific recombinase in *E. coli*, XerCD, converts the dimer to monomers, allowing chromosome segregation and cell division to proceed.

## Transposable Genetic Elements Move from One Location to Another

We now consider the third general type of recombination system: recombination that allows the movement of transposable elements, or **transposons.** These segments of DNA, found in virtually all cells, move, or "jump," from one place on a chromosome (the donor site) to another on the same or a different chromosome (the target site). DNA sequence homology is not usually required for this movement, called **transposition;** the new location is determined more or less randomly. Insertion of a transposon in an essential gene could kill the cell, so transposition is tightly regulated and usually very infrequent. Transposons are perhaps the simplest of molecular parasites, adapted to replicate passively within the chromosomes of host cells. In some cases they carry genes that are useful to the host cell, and thus exist in a kind of symbiosis with the host.

Bacteria have two classes of transposons. **Insertion sequences** (simple transposons) contain only the sequences required for transposition and the genes for proteins (transposases) that promote the process. **Complex transposons** contain one or more genes in addition to those needed for transposition. These extra genes might, for example, confer resistance to antibiotics and thus enhance the survival chances of the host cell. The spread of antibiotic-resistance elements among disease-causing bacterial populations that is rendering some antibiotics ineffectual (pp. 925–926) is mediated in part by transposition.

Bacterial transposons vary in structure, but most have short repeated sequences at each end that serve as binding sites for the transposase. When transposition occurs, a short sequence at the target site (5 to 10 bp) is duplicated to form an additional short repeated sequence that flanks each end of the inserted transposon (Fig. 25–42). These duplicated segments result from the cutting mechanism used to insert a transposon into the DNA at a new location.

There are two general pathways for transposition in bacteria. In direct or simple transposition (Fig. 25–43, left), cuts on each side of the transposon excise it, and the transposon moves to a new location. This leaves a double-strand break in the donor DNA that must be

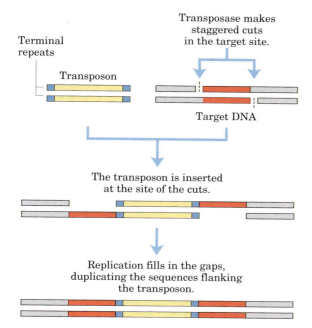

**FIGURE 25–42 Duplication of the DNA sequence at a target site when a transposon is inserted.** The duplicated sequences are shown in red. These sequences are generally only a few base pairs long, so their size (compared with that of a typical transposon) is greatly exaggerated in this drawing.

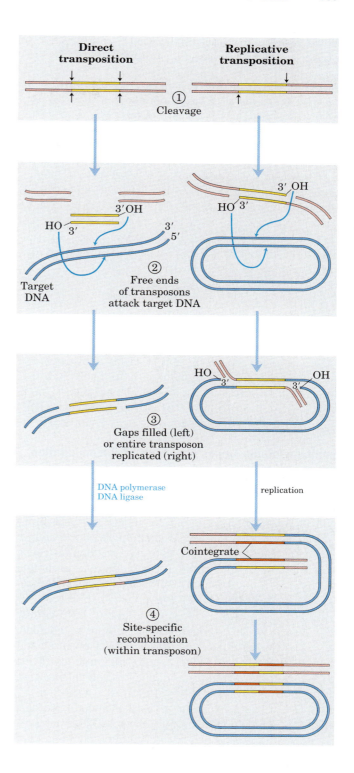

repaired. At the target site, a staggered cut is made (as in Fig. 25–42), the transposon is inserted into the break, and DNA replication fills in the gaps to duplicate the target site sequence. In replicative transposition (Fig. 25–43, right), the entire transposon is replicated, leaving a copy behind at the donor location. A **cointegrate** is an intermediate in this process, consisting of the donor region covalently linked to DNA at the target site. Two complete copies of the transposon are present in the cointegrate, both having the same relative orientation in the DNA. In some well-characterized transposons, the cointegrate intermediate is converted to products by site-specific recombination, in which specialized recombinases promote the required deletion reaction.

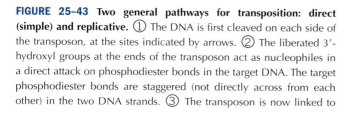

**FIGURE 25–43 Two general pathways for transposition: direct (simple) and replicative.** ① The DNA is first cleaved on each side of the transposon, at the sites indicated by arrows. ② The liberated 3'-hydroxyl groups at the ends of the transposon act as nucleophiles in a direct attack on phosphodiester bonds in the target DNA. The target phosphodiester bonds are staggered (not directly across from each other) in the two DNA strands. ③ The transposon is now linked to the target DNA. In direct transposition, replication fills in gaps at each end. In replicative transposition, the entire transposon is replicated to create a cointegrate intermediate. ④ The cointegrate is often resolved later, with the aid of a separate site-specific recombination system. The cleaved host DNA left behind after direct transposition is either repaired by DNA end-joining or degraded (not shown). The latter outcome can be lethal to an organism.

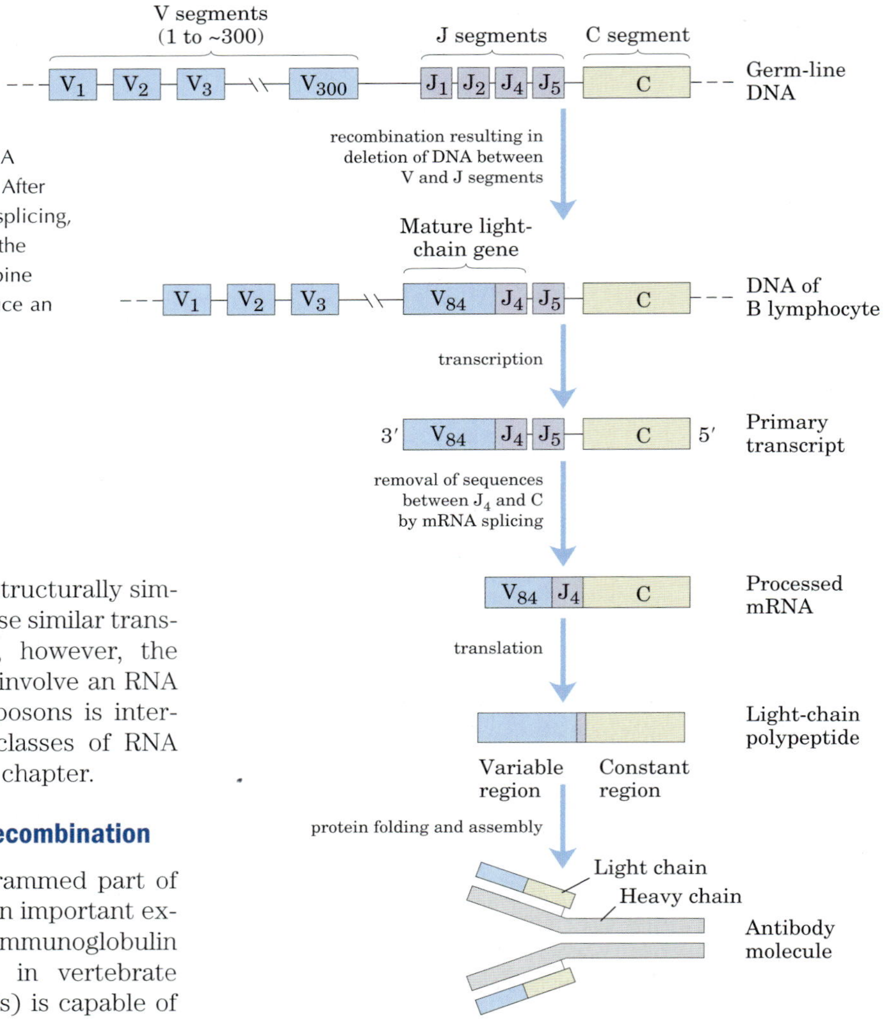

**FIGURE 25–44 Recombination of the V and J gene segments of the human IgG kappa light chain.** This process is designed to generate antibody diversity. At the top is shown the arrangement of IgG-coding sequences in a bone marrow stem cell. Recombination deletes the DNA between a particular V segment and a J segment. After transcription, the transcript is processed by RNA splicing, as described in Chapter 26; translation produces the light-chain polypeptide. The light chain can combine with any of 5,000 possible heavy chains to produce an antibody molecule.

Eukaryotes also have transposons, structurally similar to bacterial transposons, and some use similar transposition mechanisms. In other cases, however, the mechanism of transposition appears to involve an RNA intermediate. Evolution of these transposons is intertwined with the evolution of certain classes of RNA viruses. Both are described in the next chapter.

## Immunoglobulin Genes Assemble by Recombination

Some DNA rearrangements are a programmed part of development in eukaryotic organisms. An important example is the generation of complete immunoglobulin genes from separate gene segments in vertebrate genomes. A human (like other mammals) is capable of producing *millions* of different immunoglobulins (antibodies) with distinct binding specificities, even though the human genome contains only ~35,000 genes. Recombination allows an organism to produce an extraordinary diversity of antibodies from a limited DNA-coding capacity. Studies of the recombination mechanism reveal a close relationship to DNA transposition and suggest that this system for generating antibody diversity may have evolved from an ancient cellular invasion of transposons.

We can use the human genes that encode proteins of the immunoglobulin G (IgG) class to illustrate how antibody diversity is generated. Immunoglobulins consist of two heavy and two light polypeptide chains (see Fig. 5–23). Each chain has two regions, a variable region, with a sequence that differs greatly from one immunoglobulin to another, and a region that is virtually constant within a class of immunoglobulins. There are also two distinct families of light chains, kappa and lambda, which differ somewhat in the sequences of their constant regions. For all three types of polypeptide chain (heavy chain, and kappa and lambda light chains), diversity in the variable regions is generated by a simi-

lar mechanism. The genes for these polypeptides are divided into segments, and the genome contains clusters with multiple versions of each segment. The joining of one version of each of the segments creates a complete gene.

Figure 25–44 depicts the organization of the DNA encoding the kappa light chains of human IgG and shows how a mature kappa light chain is generated. In undifferentiated cells, the coding information for this polypeptide chain is separated into three segments. The V (variable) segment encodes the first 95 amino acid residues of the variable region, the J (joining) segment encodes the remaining 12 residues of the variable region, and the C segment encodes the constant region. The genome contains ~300 different V segments, 4 different J segments, and 1 C segment.

As a stem cell in the bone marrow differentiates to form a mature B lymphocyte, one V segment and one J segment are brought together by a specialized recombination system (Fig. 25–44). During this programmed DNA deletion, the intervening DNA is discarded. There are about $300 \times 4 = 1,200$ possible V–J combinations.

The recombination process is not as precise as the site-specific recombination described earlier, so additional variation occurs in the sequence at the V–J junction. This increases the overall variation by a factor of at least 2.5, thus the cells can generate about $2.5 \times 1{,}200 = 3{,}000$ different V–J combinations. The final joining of the V–J combination to the C region is accomplished by an RNA-splicing reaction after transcription, a process described in Chapter 26.

The recombination mechanism for joining the V and J segments is illustrated in Figure 25–45. Just beyond each V segment and just before each J segment lie recombination signal sequences (RSS). These are bound by proteins called RAG1 and RAG2 (*recombination activating gene*). The RAG proteins catalyze the formation of a double-strand break between the signal sequences and the V (or J) segments to be joined. The V and J segments are then joined with the aid of a second complex of proteins.

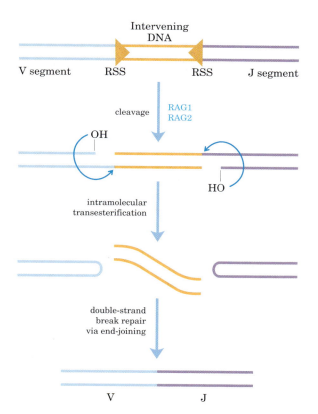

**FIGURE 25–45 Mechanism of immunoglobulin gene rearrangement.** The RAG1 and RAG2 proteins bind to the recombination signal sequences (RSS) and cleave one DNA strand between the RSS and the V (or J) segments to be joined. The liberated 3′ hydroxyl then acts as a nucleophile, attacking a phosphodiester bond in the other strand to create a double-strand break. The resulting hairpin bends on the V and J segments are cleaved, and the ends are covalently linked by a complex of proteins specialized for end-joining repair of double-strand breaks. The steps in the generation of the double-strand break catalyzed by RAG1 and RAG2 are chemically related to steps in transposition reactions.

The genes for the heavy chains and the lambda light chains form by similar processes. Heavy chains have more gene segments than light chains, with more than 5,000 possible combinations. Because any heavy chain can combine with any light chain to generate an immunoglobulin, each human has at least $3{,}000 \times 5{,}000 = 1.5 \times 10^7$ possible IgGs. And additional diversity is generated by high mutation rates (of unknown mechanism) in the V sequences during B-lymphocyte differentiation. Each mature B lymphocyte produces only one type of antibody, but the range of antibodies produced by different cells is clearly enormous.

Did the immune system evolve in part from ancient transposons? The mechanism for generation of the double-strand breaks by RAG1 and RAG2 does mirror several reaction steps in transposition (Fig. 25–45). In addition, the deleted DNA, with its terminal RSS, has a sequence structure found in most transposons. In the test tube, RAG1 and RAG2 can associate with this deleted DNA and insert it, transposonlike, into other DNA molecules (probably a rare reaction in B lymphocytes). Although we cannot know for certain, the properties of the immunoglobulin gene rearrangement system suggest an intriguing origin in which the distinction between host and parasite has become blurred by evolution.

### SUMMARY 25.3 DNA Recombination

- DNA sequences are rearranged in recombination reactions, usually in processes tightly coordinated with DNA replication or repair.

- Homologous genetic recombination can take place between any two DNA molecules that share sequence homology. In meiosis (in eukaryotes), this type of recombination helps to ensure accurate chromosomal segregation and create genetic diversity. In both bacteria and eukaryotes it serves in the repair of stalled replication forks. A Holliday intermediate forms during homologous recombination.

- Site-specific recombination occurs only at specific target sequences, and this process can also involve a Holliday intermediate. Recombinases cleave the DNA at specific points and ligate the strands to new partners. This type of recombination is found in virtually all cells, and its many functions include DNA integration and regulation of gene expression.

- In virtually all cells, transposons use recombination to move within or between chromosomes. In vertebrates, a programmed recombination reaction related to transposition joins immunoglobulin gene segments to form immunoglobulin genes during B-lymphocyte differentiation.

# Key Terms

*Terms in bold are defined in the glossary.*

**template** 950
semiconservative
    replication 950
**replication fork** 951
**origin** 952
Okazaki fragments 952
**leading strand** 952
**lagging strand** 952
**nucleases** 952
**exonuclease** 952
**endonuclease** 952
DNA polymerase I 952
**primer** 954
**primer terminus** 954

**processivity** 954
**proofreading** 955
DNA polymerase III
    955
**replisome** 957
**helicases** 957
**topoisomerases** 957
**primases** 958
**DNA ligase** 958
**primosome** 962
**catenane** 963
DNA polymerase $\alpha$ 965
DNA polymerase $\delta$ 965
DNA polymerase $\varepsilon$ 965

**mutation** 966
base-excision repair
    971
DNA glycosylases 971
AP site 971
AP endonucleases 972
DNA photolyases 974
**recombinational DNA
    repair** 976
error-prone translesion
    DNA synthesis 976
**SOS response** 976
**homologous genetic re-
    combination** 978

**site-specific recombina-
    tion** 978
DNA transposition 978
**meiosis** 979
**branch migration** 980
double-strand break repair
    model 980
**Holliday intermediate**
    980
**transposons** 988
**transposition** 988
**insertion sequence**
    988
**cointegrate** 989

# Further Reading

## General

**Friedberg, E.C., Walker, G.C., & Siede, W.** (1995) *DNA Repair and Mutagenesis,* American Society for Microbiology, Washington, DC.

A thorough treatment of DNA metabolism and a good place to start exploring this field.

**Kornberg, A. & Baker, T.A.** (1991) *DNA Replication,* 2nd edn, W. H. Freeman and Company, New York.

Excellent primary source for all aspects of DNA metabolism.

## DNA Replication

**Benkovic, S.J., Valentine, A.M., & Salinas, F.** (2001) Replisome-mediated DNA replication. *Annu. Rev. Biochem.* **70,** 181–208.

This review describes the similar strategies and enzymes of DNA replication in different classes of organisms.

**Boye, E., Lobner-Olesen, A., & Skarstad, K.** (2000) Limiting DNA replication to once and only once. *EMBO Rep.* **1,** 479–483.

Good summary of the mechanisms by which replication initiation is regulated.

**Davey, M.J. & O'Donnell, M.** (2000) Mechanisms of DNA replication. *Curr. Opin. Chem. Biol.* **4,** 581–586.

**Ellison, V. & Stillman, B.** (2001) Opening of the clamp: an intimate view of an ATP-driven biological machine. *Cell* **106,** 655–660.

**Frick, D.N. & Richardson, C.C.** (2001) DNA primases. *Annu. Rev. Biochem.* **70,** 39–80.

**Hübscher, U., Maga, G., & Spadari, S.** (2002) Eukaryotic DNA polymerases. *Annu. Rev. Biochem.* **71,** 133–163.

Good summary of the properties and roles of the more than one dozen known eukaryotic DNA polymerases.

**Jeruzalmi, D., O'Donnell, M., & Kuriyan, J.** (2002) Clamp loaders and sliding clamps. *Curr. Opin. Struct. Biol.* **12,** 217–224.

Summary of some of the elegant work elucidating how clamp loaders function.

**Kamada, K., Horiuchi, T., Ohsumi, K., Shimamoto, N., & Morikawa, K.** (1996) Structure of a replication-terminator protein complexed with DNA. *Nature* **383,** 598–603.

The report revealing the structure of the Tus-Ter complex.

**Katayama, T.** (2001) Feedback controls restrain the initiation of *Escherichia coli* chromosomal replication. *Mol. Microbiol.* **41,** 9–17.

**Kool, E.T.** (2002) Active site tightness and substrate fit in DNA replication. *Annu. Rev. Biochem.* **71,** 191–219.

Excellent summary of the molecular basis of replication fidelity by a DNA polymerase—base-pair geometry as well as hydrogen bonding.

**Lemon, K.P. & Grossman, A.D.** (2001) The extrusion-capture model for chromosome partitioning in bacteria. *Genes Dev.* **15,** 2031–2041.

Report describing the replication factory model for bacterial DNA replication.

**Nishitani, H. & Lygerou, Z.** (2002) Control of DNA replication licensing in a cell cycle. *Genes Cells* **7,** 523–534.

A good summary of recent advances in the understanding of how eukaryotic DNA replication is initiated.

**Toyn, J.H., Toone, M.W., Morgan, B.A., & Johnston, L.H.** (1995) The activation of DNA replication in yeast. *Trends Biochem. Sci.* **20,** 70–73.

## DNA Repair

**Begley, T.J. & Samson, L.D.** (2003) AlkB mystery solved: oxidative demethylation of N1-methyladenine and N3-methylcytosine adducts by a direct reversal mechanism. *Trends Biochem. Sci.* **28,** 2–5.

**Friedberg, E.C., Fischhaber, P.L., & Kisker, C.** (2001) Error-prone DNA polymerases: novel structures and the benefits of infidelity. *Cell* **107,** 9–12.

**Goodman, M.F.** (2002) Error-prone repair DNA polymerases in prokaryotes and eukaryotes. *Annu. Rev. Biochem.* **71,** 17–50.

Review of a class of DNA polymerases that continues to grow.

**Kolodner, R.D.** (1995) Mismatch repair: mechanisms and relationship to cancer susceptibility. *Trends Biochem. Sci.* **20,** 397–401.

**Lindahl, T. & Wood, R.D.** (1999) Quality control by DNA repair. *Science* **286,** 1897–1905.

**Marnett, L.J. & Plastaras, J.P.** (2001) Endogenous DNA damage and mutation. *Trends Genet.* **17,** 214–221.

**McCullough, A.K., Dodson, M.L., & Lloyd, R.S.** (1999) Initiation of base excision repair: glycosylase mechanisms and structures. *Annu. Rev. Biochem.* **68,** 255–286.

**Modrich, P. & Lahue, R.** (1996) Mismatch repair in replication fidelity, genetic recombination, and cancer biology. *Annu. Rev. Biochem.* **65,** 101–133.

**Sancar, A.** (1996) DNA excision repair. *Annu. Rev. Biochem.* **65,** 43–81.

**Sutton, M.D., Smith, B.T., Godoy, V.G., & Walker, G.C.** (2000) The SOS response: recent insights into umuDC-dependent mutagenesis and DNA damage tolerance. *Annu. Rev. Genet.* **34,** 479–497.

**Wood, R.D., Mitchell, M., Sgouros, J., & Lindahl T.** (2001) Human DNA repair genes. *Science* **291,** 1284–1289.

Description of what an early look at the human genome reveals about DNA repair.

### DNA Recombination

**Cox, M.M.** (2001) Historical overview: searching for replication help in all of the rec places. *Proc. Natl. Acad. Sci. USA* **98,** 8173–8180.

A review of how recombination was shown to be a replication fork repair process.

**Craig, N.L.** (1995) Unity in transposition reactions. *Science* **270,** 253–254.

**Eggleston, A.K. & West, S.C.** (1996) Exchanging partners: recombination in *E. coli. Trends Genet.* **12,** 20–26.

**Gellert, M.** (2002) V(D)J recombination: RAG proteins, repair factors, and regulation. *Annu. Rev. Biochem.* **71,** 101–132.

**Hallet, B. & Sherratt, D.J.** (1997) Transposition and site-specific recombination: adapting DNA cut-and-paste mechanisms to a variety of genetic rearrangements. *FEMS Microbiol. Rev.* **21,** 157–178.

**Kogoma, T.** (1996) Recombination by replication. *Cell* **85,** 625–627.

**Lieber, M.** (1996) Immunoglobulin diversity: rearranging by cutting and repairing. *Curr. Biol.* **6,** 134–136.

**Lusetti, S.L. & Cox, M.M.** (2002) The bacterial RecA protein and the recombinational DNA repair of stalled replication forks. *Annu. Rev. Biochem.* **71,** 71–100.

**Marians, K.J.** (2000) PriA-directed replication fork restart in *Escherichia coli. Trends Biochem. Sci.* **25,** 185–189.

**Pâques, F. & Haber, J.E.** (1999) Multiple pathways of recombination induced by double-strand breaks in *Saccharomyces cerevisiae. Microbiol. Mol. Biol. Rev.* **63,** 349–404.

**Van Duyne, G.D.** (2001) A structural view of Cre-loxP site-specific recombination. *Annu. Rev. Biophys. Biomol. Struct.* **30,** 87–104.

## Problems

**1. Conclusions from the Meselson-Stahl Experiment** The Meselson-Stahl experiment (see Fig. 25–2) proved that DNA undergoes semiconservative replication in *E. coli.* In the "dispersive" model of DNA replication, the parent DNA strands are cleaved into pieces of random size, then joined with pieces of newly replicated DNA to yield daughter duplexes. In the Meselson-Stahl experiment, each strand would contain random segments of heavy and light DNA. Explain how the results of Meselson and Stahl's experiment ruled out such a model.

**2. Heavy Isotope Analysis of DNA Replication** A culture of *E. coli* growing in a medium containing $^{15}NH_4Cl$ is switched to a medium containing $^{14}NH_4Cl$ for three generations (an eightfold increase in population). What is the molar ratio of hybrid DNA ($^{15}N-^{14}N$) to light DNA ($^{14}N-^{14}N$) at this point?

**3. Replication of the *E. coli* Chromosome** The *E. coli* chromosome contains 4,639,221 bp.

(a) How many turns of the double helix must be unwound during replication of the *E. coli* chromosome?

(b) From the data in this chapter, how long would it take to replicate the *E. coli* chromosome at 37 °C if two replica-tion forks proceeded from the origin? Assume replication occurs at a rate of 1,000 bp/s. Under some conditions *E. coli* cells can divide every 20 min. How might this be possible?

(c) In the replication of the *E. coli* chromosome, about how many Okazaki fragments would be formed? What factors guarantee that the numerous Okazaki fragments are assembled in the correct order in the new DNA?

**4. Base Composition of DNAs Made from Single-Stranded Templates** Predict the base composition of the total DNA synthesized by DNA polymerase on templates provided by an equimolar mixture of the two complementary strands of bacteriophage øX174 DNA (a circular DNA molecule). The base composition of one strand is A, 24.7%; G, 24.1%; C, 18.5%; and T, 32.7%. What assumption is necessary to answer this problem?

**5. DNA Replication** Kornberg and his colleagues incubated soluble extracts of *E. coli* with a mixture of dATP, dTTP, dGTP, and dCTP, all labeled with $^{32}P$ in the α-phosphate group. After a time, the incubation mixture was treated with trichloroacetic acid, which precipitates the DNA but not the nucleotide precursors. The precipitate was collected, and the extent of precursor incorporation into DNA was determined from the amount of radioactivity present in the precipitate.

(a) If any one of the four nucleotide precursors were omitted from the incubation mixture, would radioactivity be found in the precipitate? Explain.

(b) Would $^{32}$P be incorporated into the DNA if only dTTP were labeled? Explain.

(c) Would radioactivity be found in the precipitate if $^{32}$P labeled the $\beta$ or $\gamma$ phosphate rather than the $\alpha$ phosphate of the deoxyribonucleotides? Explain.

**6. Leading and Lagging Strands** Prepare a table that lists the names and compares the functions of the precursors, enzymes, and other proteins needed to make the leading versus lagging strands during DNA replication in *E. coli*.

**7. Function of DNA Ligase** Some *E. coli* mutants contain defective DNA ligase. When these mutants are exposed to $^3$H-labeled thymine and the DNA produced is sedimented on an alkaline sucrose density gradient, two radioactive bands appear. One corresponds to a high molecular weight fraction, the other to a low molecular weight fraction. Explain.

**8. Fidelity of Replication of DNA** What factors promote the fidelity of replication during the synthesis of the leading strand of DNA? Would you expect the lagging strand to be made with the same fidelity? Give reasons for your answers.

**9. Importance of DNA Topoisomerases in DNA Replication** DNA unwinding, such as that occurring in replication, affects the superhelical density of DNA. In the absence of topoisomerases, the DNA would become overwound ahead of a replication fork as the DNA is unwound behind it. A bacterial replication fork will stall when the superhelical density ($\sigma$) of the DNA ahead of the fork reaches $+0.14$ (see Chapter 24).

Bidirectional replication is initiated at the origin of a 6,000 bp plasmid in vitro, in the absence of topoisomerases. The plasmid initially has a $\sigma$ of $-0.06$. How many base pairs will be unwound and replicated by each replication fork before the forks stall? Assume that each fork travels at the same rate and that each includes all components necessary for elongation except topoisomerase.

**10. The Ames Test** In a nutrient medium that lacks histidine, a thin layer of agar containing $\sim10^9$ *Salmonella typhimurium* histidine auxotrophs (mutant cells that require histidine to survive) produces $\sim13$ colonies over a two-day incubation period at 37 °C (see Fig. 25–19). How do these colonies arise in the absence of histidine? The experiment is repeated in the presence of 0.4 $\mu$g of 2-aminoanthracene. The number of colonies produced over two days exceeds 10,000. What does this indicate about 2-aminoanthracene? What can you surmise about its carcinogenicity?

**11. DNA Repair Mechanisms** Vertebrate and plant cells often methylate cytosine in DNA to form 5-methylcytosine (see Fig. 8–5a). In these same cells, a specialized repair system recognizes G–T mismatches and repairs them to G≡C base pairs. How might this repair system be advantageous to the cell? (Explain in terms of the presence of 5-methylcytosine in the DNA.)

**12. DNA Repair in People with Xeroderma Pigmentosum** The condition known as xeroderma pigmentosum (XP) arises from mutations in at least seven different human genes. The deficiencies are generally in genes encoding enzymes involved in some part of the pathway for human nucleotide-excision repair. The various types of XP are labeled A through G (XPA, XPB, etc.), with a few additional variants lumped under the label XPV.

Cultures of cells from healthy individuals and from patients with XPG are irradiated with ultraviolet light. The DNA is isolated and denatured, and the resulting single-stranded DNA is characterized by analytical ultracentrifugation.

(a) Samples from the normal fibroblasts show a significant reduction in the average molecular weight of the single-stranded DNA after irradiation, but samples from the XPG fibroblasts show no such reduction. Why might this be?

(b) If you assume that a nucleotide-excision repair system is operative, which step might be defective in the fibroblasts from the patients with XPG? Explain.

**13. Holliday Intermediates** How does the formation of Holliday intermediates in homologous genetic recombination differ from their formation in site-specific recombination?

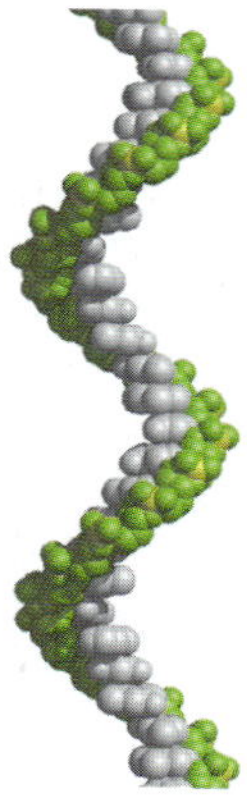

# chapter 26

# RNA METABOLISM

The RNA of the cell is partly in the nucleus, partly in particles in the cytoplasm and partly as the "soluble" RNA of the cell sap; many workers have shown that all these three fractions turn over differently. It is very important to realize in any discussion of the role of RNA in the cell that it is very inhomogeneous metabolically, and probably of more than one type.

—*Francis H. C. Crick, article in* Symposia of the Society for Experimental Biology, *1958*

Expression of the information in a gene generally involves production of an RNA molecule transcribed from a DNA template. Strands of RNA and DNA may seem quite similar at first glance, differing only in that RNA has a hydroxyl group at the 2′ position of the aldopentose and uracil instead of thymine. However, unlike DNA, most RNAs carry out their functions as single strands, strands that fold back on themselves and have the potential for much greater structural diversity than DNA (Chapter 8). RNA is thus suited to a variety of cellular functions.

RNA is the only macromolecule known to have a role both in the storage and transmission of information and in catalysis, which has led to much speculation about its possible role as an essential chemical intermediate in the development of life on this planet. The discovery of catalytic RNAs, or ribozymes, has changed the very definition of an enzyme, extending it beyond the domain of proteins. Proteins nevertheless remain essential to RNA and its functions. In the modern cell, all nucleic acids, including RNAs, are complexed with proteins. Some of these complexes are quite elaborate, and RNA can assume both structural and catalytic roles within complicated biochemical machines.

All RNA molecules except the RNA genomes of certain viruses are derived from information permanently stored in DNA. During **transcription,** an enzyme system converts the genetic information in a segment of double-stranded DNA into an RNA strand with a base sequence complementary to one of the DNA strands. Three major kinds of RNA are produced. **Messenger RNAs (mRNAs)** encode the amino acid sequence of one or more polypeptides specified by a gene or set of genes. **Transfer RNAs (tRNAs)** read the information encoded in the mRNA and transfer the appropriate amino acid to a growing polypeptide chain during protein synthesis. **Ribosomal RNAs (rRNAs)** are constituents of ribosomes, the intricate cellular machines that synthesize proteins. Many additional specialized RNAs have regulatory or catalytic functions or are precursors to the three main classes of RNA.

During replication the entire chromosome is usually copied, but transcription is more selective. Only particular genes or groups of genes are transcribed at any one time, and some portions of the DNA genome are never transcribed. The cell restricts the expression of genetic information to the formation of gene products needed at any particular moment. Specific regulatory sequences mark the beginning and end of the DNA segments to be transcribed and designate which strand in duplex DNA is to be used as the template. The regulation of transcription is described in detail in Chapter 28.

In this chapter we examine the synthesis of RNA on a DNA template and the postsynthetic processing and turnover of RNA molecules. In doing so we encounter many of the specialized functions of RNA, including catalytic functions. Interestingly, the substrates for RNA enzymes are often other RNA molecules. We also describe systems in which RNA is the template and DNA the product, rather than vice versa. The information pathways thus come full circle, revealing that template-dependent nucleic acid synthesis has standard rules regardless of the nature of template or product (RNA or DNA). This examination of the biological interconversion of DNA and RNA as information carriers leads to a discussion of the evolutionary origin of biological information.

## 26.1 DNA-Dependent Synthesis of RNA

Our discussion of RNA synthesis begins with a comparison between transcription and DNA replication (Chapter 25). Transcription resembles replication in its fundamental chemical mechanism, its polarity (direction of synthesis), and its use of a template. And like replication, transcription has initiation, elongation, and termination phases—though in the literature on transcription, initiation is further divided into discrete phases of DNA binding and initiation of RNA synthesis. Transcription differs from replication in that it does not require a primer and, generally, involves only limited segments of a DNA molecule. Additionally, within transcribed segments only one DNA strand serves as a template.

### RNA Is Synthesized by RNA Polymerases

The discovery of DNA polymerase and its dependence on a DNA template spurred a search for an enzyme that synthesizes RNA complementary to a DNA strand. By 1960, four research groups had independently detected an enzyme in cellular extracts that could form an RNA polymer from ribonucleoside 5'-triphosphates. Subsequent work on the purified *Escherichia coli* RNA polymerase helped to define the fundamental properties of transcription (Fig. 26–1). **DNA-dependent RNA polymerase** requires, in addition to a DNA template, all four ribonucleoside 5'-triphosphates (ATP, GTP, UTP, and CTP) as precursors of the nucleotide units of RNA, as well as $Mg^{2+}$. The protein also binds one $Zn^{2+}$. The chemistry and mechanism of RNA synthesis closely resemble those used by DNA polymerases (see Fig. 25–5). RNA polymerase elongates an RNA strand by adding ribonucleotide units to the 3'-hydroxyl end, building RNA in the $5' \rightarrow 3'$ direction. The 3'-hydroxyl group acts as a nucleophile, attacking the $\alpha$ phosphate of the incoming ribonucleoside triphosphate (Fig. 26–1b) and releasing pyrophosphate. The overall reaction is

$$\underset{\text{RNA}}{(NMP)_n} + NTP \longrightarrow \underset{\text{Lengthened RNA}}{(NMP)_{n+1}} + PP_i$$

RNA polymerase requires DNA for activity and is most active when bound to a double-stranded DNA. As noted above, only one of the two DNA strands serves as a template. The template DNA strand is copied in the $3' \rightarrow 5'$ direction (antiparallel to the new RNA strand), just as in DNA replication. Each nucleotide in the newly formed RNA is selected by Watson-Crick base-pairing interac-

tions; U residues are inserted in the RNA to pair with A residues in the DNA template, G residues are inserted to pair with C residues, and so on. Base-pair geometry (see Fig. 25–6) may also play a role in base selection.

Unlike DNA polymerase, RNA polymerase does not require a primer to initiate synthesis. Initiation occurs when RNA polymerase binds at specific DNA sequences called promoters (described below). The 5'-triphosphate group of the first residue in a nascent (newly formed) RNA molecule is not cleaved to release $PP_i$, but instead remains intact throughout the transcription process. During the elongation phase of transcription, the growing end of the new RNA strand base-pairs temporarily with the DNA template to form a short hybrid

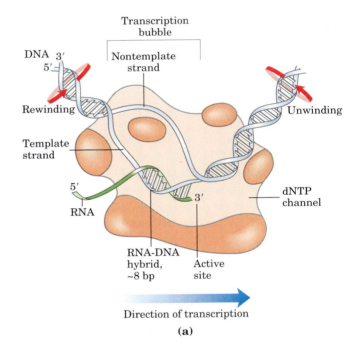

**MECHANISM FIGURE 26–1 Transcription by RNA polymerase in *E. coli.*** For synthesis of an RNA strand complementary to one of two DNA strands in a double helix, the DNA is transiently unwound. **(a)** About 17 bp are unwound at any given time. RNA polymerase and the bound transcription bubble move from left to right along the DNA as shown, facilitating RNA synthesis. The DNA is unwound ahead and rewound behind as RNA is transcribed. Red arrows show the direction in which the DNA must rotate to permit this process. As the DNA is rewound, the RNA-DNA hybrid is displaced and the RNA strand extruded. The RNA polymerase is in close contact with the DNA ahead of the transcription bubble, as well as with the separated DNA strands and the RNA within and immediately behind the bubble. A channel in the protein funnels new nucleoside triphosphates (NTPs) to the polymerase active site. The polymerase footprint encompasses about 35 bp of DNA during elongation.

**(b)** Catalytic mechanism of RNA synthesis by RNA polymerase. Note that this is essentially the same mechanism used by DNA poly-

RNA-DNA double helix, estimated to be 8 bp long (Fig. 26–1a). The RNA in this hybrid duplex "peels off" shortly after its formation, and the DNA duplex re-forms.

To enable RNA polymerase to synthesize an RNA strand complementary to one of the DNA strands, the DNA duplex must unwind over a short distance, forming a transcription "bubble." During transcription, the *E. coli* RNA polymerase generally keeps about 17 bp unwound. The 8 bp RNA-DNA hybrid occurs in this unwound region. Elongation of a transcript by *E. coli* RNA polymerase proceeds at a rate of 50 to 90 nucleotides/s. Because DNA is a helix, movement of a transcription bubble requires considerable strand rotation of the nucleic acid molecules. DNA strand rotation is restricted

in most DNAs by DNA-binding proteins and other structural barriers. As a result, a moving RNA polymerase generates waves of positive supercoils ahead of the transcription bubble and negative supercoils behind (Fig. 26–1c). This has been observed both in vitro and in vivo (in bacteria). In the cell, the topological problems caused by transcription are relieved through the action of topoisomerases (Chapter 24).

The two complementary DNA strands have different roles in transcription. The strand that serves as template for RNA synthesis is called the **template strand.** The DNA strand complementary to the template, the **nontemplate strand,** or **coding strand,** is identical in base sequence to the RNA transcribed from the gene,

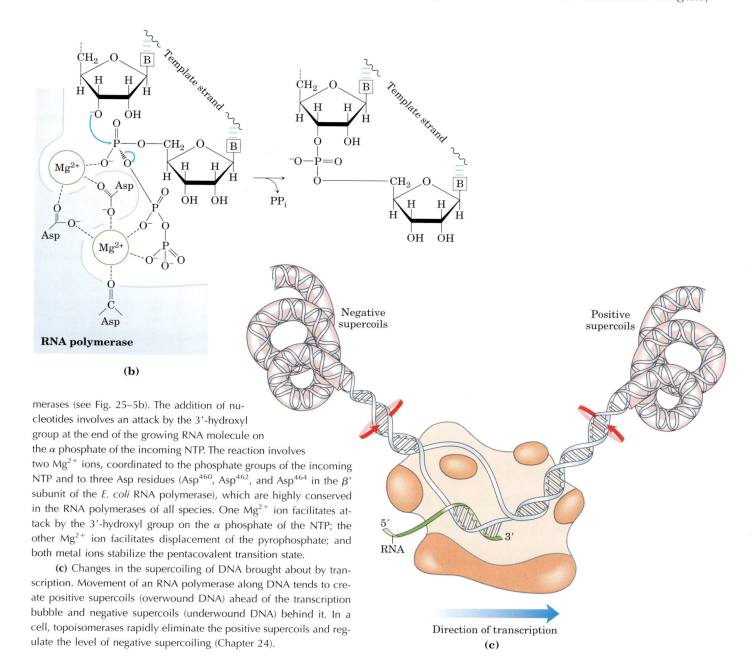

**(b)**

merases (see Fig. 25–5b). The addition of nucleotides involves an attack by the 3'-hydroxyl group at the end of the growing RNA molecule on the α phosphate of the incoming NTP. The reaction involves two $Mg^{2+}$ ions, coordinated to the phosphate groups of the incoming NTP and to three Asp residues ($Asp^{460}$, $Asp^{462}$, and $Asp^{464}$ in the β' subunit of the *E. coli* RNA polymerase), which are highly conserved in the RNA polymerases of all species. One $Mg^{2+}$ ion facilitates attack by the 3'-hydroxyl group on the α phosphate of the NTP; the other $Mg^{2+}$ ion facilitates displacement of the pyrophosphate; and both metal ions stabilize the pentacovalent transition state.

**(c)** Changes in the supercoiling of DNA brought about by transcription. Movement of an RNA polymerase along DNA tends to create positive supercoils (overwound DNA) ahead of the transcription bubble and negative supercoils (underwound DNA) behind it. In a cell, topoisomerases rapidly eliminate the positive supercoils and regulate the level of negative supercoiling (Chapter 24).

(5') C G C T A T A G C G T T T (3')      DNA nontemplate (coding) strand

(3') G C G A T A T C G C A A A (5')      DNA template strand

(5') C G C U A U A G C G U U U (3')      RNA transcript

**FIGURE 26–2** **Template and nontemplate (coding) DNA strands.** The two complementary strands of DNA are defined by their function in transcription. The RNA transcript is synthesized on the template strand and is identical in sequence (with U in place of T) to the nontemplate strand, or coding strand.

with U in the RNA in place of T in the DNA (Fig. 26–2). The coding strand for a particular gene may be located in either strand of a given chromosome (as shown in Fig. 26–3 for a virus). The regulatory sequences that control transcription (described later in this chapter) are by convention designated by the sequences in the coding strand.

The DNA-dependent RNA polymerase of *E. coli* is a large, complex enzyme with five core subunits ($\alpha_2\beta\beta'\omega$; $M_r$ 390,000) and a sixth subunit, one of a group designated $\sigma$, with variants designated by size (molecular weight). The $\sigma$ subunit binds transiently to the core and directs the enzyme to specific binding sites on the DNA (described below). These six subunits constitute the RNA polymerase holoenzyme (Fig. 26–4). The RNA polymerase holoenzyme of *E. coli* thus exists in several forms, depending on the type of $\sigma$ subunit. The most common subunit is $\sigma^{70}$ ($M_r$ 70,000), and the upcoming discussion focuses on the corresponding RNA polymerase holoenzyme.

RNA polymerases lack a separate proofreading $3' \rightarrow 5'$ exonuclease active site (such as that of many DNA polymerases), and the error rate for transcription is higher than that for chromosomal DNA replication—approximately one error for every $10^4$ to $10^5$ ribonucleotides incorporated into RNA. Because many copies of an RNA are generally produced from a single gene and all RNAs are eventually degraded and replaced, a mistake in an RNA molecule is of less consequence to the cell than a mistake in the permanent information stored in DNA. Many RNA polymerases, including bacterial RNA polymerase and the eukaryotic RNA polymerase II (discussed below), do pause when a mispaired base is added during transcription, and they can remove

mismatched nucleotides from the 3' end of a transcript by direct reversal of the polymerase reaction. But we do not yet know whether this activity is a true proofreading function and to what extent it may contribute to the fidelity of transcription.

## RNA Synthesis Begins at Promoters

Initiation of RNA synthesis at random points in a DNA molecule would be an extraordinarily wasteful process. Instead, an RNA polymerase binds to specific sequences in the DNA called **promoters,** which direct the transcription of adjacent segments of DNA (genes). The sequences where RNA polymerases bind can be quite variable, and much research has focused on identifying the particular sequences that are critical to promoter function.

In *E. coli*, RNA polymerase binding occurs within a region stretching from about 70 bp before the transcription start site to about 30 bp beyond it. By convention, the DNA base pairs that correspond to the beginning of an RNA molecule are given positive numbers, and those preceding the RNA start site are given negative numbers. The promoter region thus extends between positions $-70$ and $+30$. Analyses and comparisons of the most common class of bacterial promoters (those recognized by an RNA polymerase holoenzyme containing $\sigma^{70}$) have revealed similarities in two short sequences centered about positions $-10$ and $-35$ (Fig. 26–5). These sequences are important interaction sites for the $\sigma^{70}$ subunit. Although the sequences are not identical for all bacterial promoters in this class, certain nucleotides that are particularly common at each position form a **consensus sequence** (recall the *E. coli*

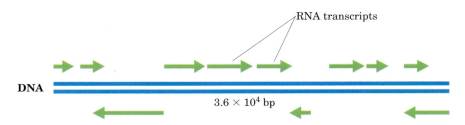

RNA transcripts

DNA

$3.6 \times 10^4$ bp

**FIGURE 26–3** **Organization of coding information in the adenovirus genome.** The genetic information of the adenovirus genome (a conveniently simple example) is encoded by a double-stranded DNA molecule of 36,000 bp, both strands of which encode proteins. The information for most proteins is encoded by the top strand—by convention, the strand transcribed from left to right—but some is encoded by the bottom strand, which is transcribed in the opposite

direction. Synthesis of mRNAs in adenovirus is actually much more complex than shown here. Many of the mRNAs shown for the upper strand are initially synthesized as a single, long transcript (25,000 nucleotides), which is then extensively processed to produce the separate mRNAs. Adenovirus causes upper respiratory tract infections in some vertebrates.

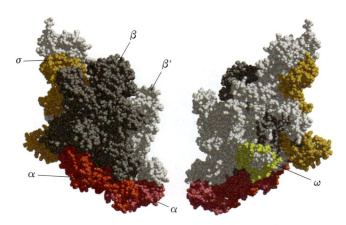

**FIGURE 26–4** **Structure of the RNA polymerase holoenzyme of the bacterium *Thermus aquaticus*.** (Derived from PDB ID 1IW7.) The overall structure of this enzyme is very similar to that of the *E. coli* RNA polymerase; no DNA or RNA is shown here. The $\beta$ subunit is in gray, the $\beta'$ subunit is white; the two $\alpha$ subunits are different shades of red; the $\omega$ subunit is yellow; the $\sigma$ subunit is orange. The image on the left is oriented as in Figure 26–6. When the structure is rotated 180° about the *y* axis (right) the small $\omega$ subunit is visible.

tain highly expressed genes. The UP element is bound by the $\alpha$ subunit of RNA polymerase. The efficiency with which an RNA polymerase binds to a promoter and initiates transcription is determined in large measure by these sequences, the spacing between them, and their distance from the transcription start site.

Many independent lines of evidence attest to the functional importance of the sequences in the −35 and −10 regions. Mutations that affect the function of a given promoter often involve a base pair in these regions. Variations in the consensus sequence also affect the efficiency of RNA polymerase binding and transcription initiation. A change in only one base pair can decrease the rate of binding by several orders of magnitude. The promoter sequence thus establishes a basal level of expression that can vary greatly from one *E. coli* gene to the next. A method that provides information about the interaction between RNA polymerase and promoters is illustrated in Box 26–1.

The pathway of transcription initiation is becoming much better defined (Fig. 26–6a). It consists of two major parts, binding and initiation, each with multiple steps. First, the polymerase binds to the promoter, forming, in succession, a closed complex (in which the bound DNA is intact) and an open complex (in which the bound DNA is intact and partially unwound near the −10 sequence). Second, transcription is initiated within the complex, leading to a conformational change that converts the complex to the elongation form, followed by movement of the transcription complex away from

*oriC* consensus sequence; see Fig. 25–11). The consensus sequence at the −10 region is (5′)TATAAT(3′); the consensus sequence at the −35 region is (5′)TTGACA(3′). A third AT-rich recognition element, called the UP (upstream promoter) element, occurs between positions −40 and −60 in the promoters of cer-

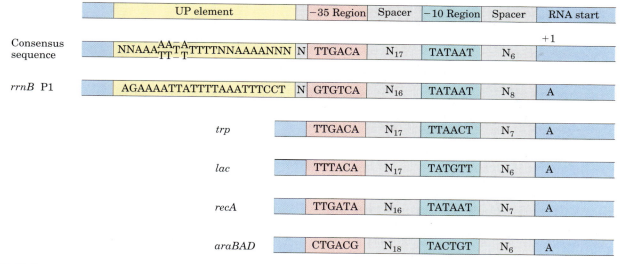

| | UP element | | −35 Region | Spacer | −10 Region | Spacer | RNA start |
|---|---|---|---|---|---|---|---|
| **Consensus sequence** | NNAAA$^{AA-A}_{TT-T}$TTTTNNAAAANNN | N | TTGACA | $N_{17}$ | TATAAT | $N_6$ | +1 |
| *rrnB* P1 | AGAAAATTATTTTAAATTTCCT | N | GTGTCA | $N_{16}$ | TATAAT | $N_8$ | A |
| *trp* | | | TTGACA | $N_{17}$ | TTAACT | $N_7$ | A |
| *lac* | | | TTTACA | $N_{17}$ | TATGTT | $N_6$ | A |
| *recA* | | | TTGATA | $N_{16}$ | TATAAT | $N_7$ | A |
| *araBAD* | | | CTGACG | $N_{18}$ | TACTGT | $N_6$ | A |

**FIGURE 26–5** **Typical *E. coli* promoters recognized by an RNA polymerase holoenzyme containing $\sigma^{70}$.** Sequences of the nontemplate strand are shown, read in the 5′→3′ direction, as is the convention for representations of this kind. The sequences vary from one promoter to the next, but comparisons of many promoters reveal similarities, particularly in the −10 and −35 regions. The sequence element UP, not present in all *E. coli* promoters, is shown in the P1 promoter for the highly expressed rRNA gene *rrnB*. UP elements, generally occur-

ring in the region between −40 and −60, strongly stimulate transcription at the promoters that contain them. The UP element in the *rrnB* P1 promoter encompasses the region between −38 and −59. The consensus sequence for *E. coli* promoters recognized by $\sigma^{70}$ is shown second from the top. Spacer regions contain slightly variable numbers of nucleotides (N). Only the first nucleotide coding the RNA transcript (at position +1) is shown.

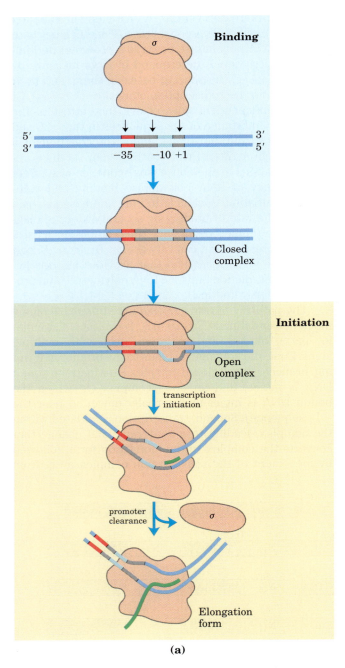

**(a)**

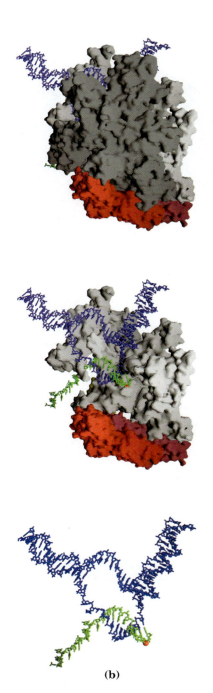

**(b)**

**FIGURE 26–6 Transcription initiation and elongation by *E. coli* RNA polymerase. (a)** Initiation of transcription requires several steps generally divided into two phases, binding and initiation. In the binding phase, the initial interaction of the RNA polymerase with the promoter leads to formation of a closed complex, in which the promoter DNA is stably bound but not unwound. A 12 to 15 bp region of DNA—from within the −10 region to position +2 or +3—is then unwound to form an open complex. Additional intermediates (not shown) have been detected in the pathways leading to the closed and open complexes, along with several changes in protein conformation. The initiation phase encompasses transcription initiation and promoter clearance. Once the first 8 or 9 nucleotides of a new RNA are synthesized, the σ subunit is released and the polymerase leaves the promoter and becomes committed to elongation of the RNA.

**(b)** Structure of the RNA core polymerase from *E. coli*. RNA and DNA are included here to illustrate a polymerase in the elongation phase. Subunit coloring matches Figure 26–4: the β and β' subunits are light gray and white; the α subunits, shades of red. The ω subunit is on the opposite side of the complex and is not visible in this view. The σ subunit is not present in this complex, having dissociated after the initiation steps. The top panel shows the entire complex. The active site for transcription is in a cleft between the β and β' subunits. In the middle panel, the β subunit has been removed, exposing the active site and the DNA-RNA hybrid region. The active site is marked in part by a $Mg^{2+}$ ion (red). In the bottom panel, all the protein has been removed to reveal the circuitous path taken by the DNA and RNA through the complex.

the promoter (promoter clearance). Any of these steps can be affected by the specific makeup of the promoter sequences. The $\sigma$ subunit dissociates as the polymerase enters the elongation phase of transcription (Fig. 26–6a).

E. coli has other classes of promoters, bound by RNA polymerase holoenzymes with different $\sigma$ subunits. An example is the promoters of the heat-shock genes. The products of this set of genes are made at higher levels when the cell has received an insult, such as a sudden increase in temperature. RNA polymerase binds to the promoters of these genes only when $\sigma^{70}$ is replaced with the $\sigma^{32}$ ($M_r$ 32,000) subunit, which is specific for the heat-shock promoters (see Fig. 28–3). By using different $\sigma$ subunits the cell can coordinate the expression of sets of genes, permitting major changes in cell physiology.

## Transcription Is Regulated at Several Levels

Requirements for any gene product vary with cellular conditions or developmental stage, and transcription of each gene is carefully regulated to form gene products only in the proportions needed. Regulation can occur at any step in transcription, including elongation and termination. However, much of the regulation is directed at the polymerase binding and transcription initiation steps outlined in Figure 26–6. Differences in promoter sequences are just one of several levels of control.

The binding of proteins to sequences both near to and distant from the promoter can also affect levels of gene expression. Protein binding can *activate* transcription by facilitating either RNA polymerase binding or steps further along in the initiation process, or it can *repress* transcription by blocking the activity of the polymerase. In E. coli, one protein that activates transcription is the **cAMP receptor protein (CRP)**, which increases the transcription of genes coding for enzymes that metabolize sugars other than glucose when cells are grown in the absence of glucose. **Repressors** are proteins that block the synthesis of RNA at specific genes. In the case of the Lac repressor (Chapter 28), transcription of the genes for the enzymes of lactose metabolism is blocked when lactose is unavailable.

Transcription is the first step in the complicated and energy-intensive pathway of protein synthesis, so much of the regulation of protein levels in both bacterial and eukaryotic cells is directed at transcription, particularly its early stages. In Chapter 28 we describe many mechanisms by which this regulation is accomplished.

## Specific Sequences Signal Termination of RNA Synthesis

RNA synthesis is processive (that is, the RNA polymerase has high processivity; p. 954)—necessarily so, because if an RNA polymerase released an RNA transcript prematurely, it could not resume synthesis of the same RNA but instead would have to start over. However, an encounter with certain DNA sequences results in a pause in RNA synthesis, and at some of these sequences transcription is terminated. The process of termination is not yet well understood in eukaryotes, so our focus is again on bacteria. E. coli has at least two classes of termination signals: one class relies on a protein factor called $\rho$ (rho) and the other is $\rho$-independent.

Most $\rho$-independent terminators have two distinguishing features. The first is a region that produces an RNA transcript with self-complementary sequences, permitting the formation of a hairpin structure (see Fig. 8–21a) centered 15 to 20 nucleotides before the projected end of the RNA strand. The second feature is a highly conserved string of three A residues in the template strand that are transcribed into U residues near the 3′ end of the hairpin. When a polymerase arrives at a termination site with this structure, it pauses (Fig. 26–7). Formation of the hairpin structure in the RNA disrupts several A=U base pairs in the RNA-DNA hybrid segment and may disrupt important interactions

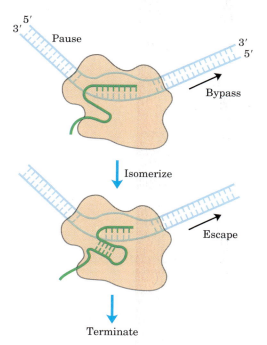

**FIGURE 26–7** Model for $\rho$-independent termination of transcription in E. coli. RNA polymerase pauses at a variety of DNA sequences, some of which are terminators. One of two outcomes is then possible: the polymerase bypasses the site and continues on its way, or the complex undergoes a conformational change (isomerization). In the latter case, intramolecular pairing of complementary sequences in the newly formed RNA transcript may form a hairpin that disrupts the RNA-DNA hybrid and/or the interactions between the RNA and the polymerase, resulting in isomerization. An A=U hybrid region at the 3′ end of the new transcript is relatively unstable, and the RNA dissociates completely, leading to termination and dissociation of the RNA molecule. This is the usual outcome at terminators. At other pause sites, the complex may escape after the isomerization step to continue RNA synthesis.

**BOX 26–1    WORKING IN BIOCHEMISTRY**

## RNA Polymerase Leaves Its Footprint on a Promoter

**Footprinting,** a technique derived from principles used in DNA sequencing, identifies the DNA sequences bound by a particular protein. Researchers isolate a DNA fragment thought to contain sequences recognized by a DNA-binding protein and radiolabel one end of one strand (Fig. 1). They then use chemical or enzymatic reagents to introduce random breaks in the DNA fragment (averaging about one per molecule). Separation of the labeled cleavage products (broken fragments of various lengths) by high-resolution electrophoresis produces a ladder of radioactive bands. In a separate tube, the cleavage procedure is repeated on copies of the same DNA fragment in the presence of the DNA-binding protein. The researchers then subject the two sets of cleavage products to electrophoresis and compare them side by side. A gap ("footprint") in the series of radioactive bands derived from the DNA-protein sample, attributable to protection of the DNA by the bound protein, identifies the sequences that the protein binds.

The precise location of the protein-binding site can be determined by directly sequencing (see Fig. 8–37) copies of the same DNA fragment and including the sequencing lanes (not shown here) on the same gel with the footprint. Figure 2 shows footprinting results for the binding of RNA polymerase to a DNA fragment containing a promoter. The polymerase covers 60 to 80 bp; protection by the bound enzyme includes the −10 and −35 regions.

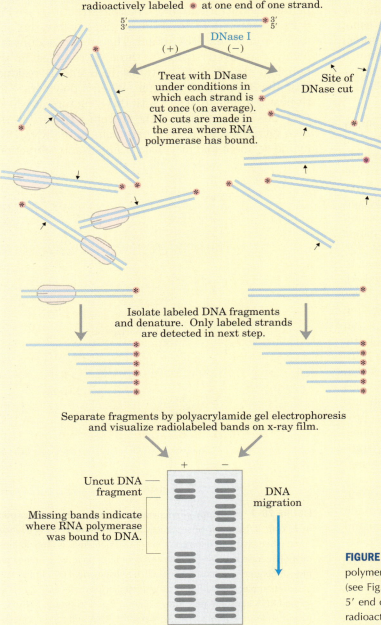

Solution of identical DNA fragments radioactively labeled ✳ at one end of one strand.

Treat with DNase under conditions in which each strand is cut once (on average). No cuts are made in the area where RNA polymerase has bound.

Site of DNase cut

Isolate labeled DNA fragments and denature. Only labeled strands are detected in next step.

Separate fragments by polyacrylamide gel electrophoresis and visualize radiolabeled bands on x-ray film.

Uncut DNA fragment

DNA migration

Missing bands indicate where RNA polymerase was bound to DNA.

**FIGURE 1** Footprint analysis of the RNA polymerase–binding site on a DNA fragment. Separate experiments are carried out in the presence (+) and absence (−) of the polymerase.

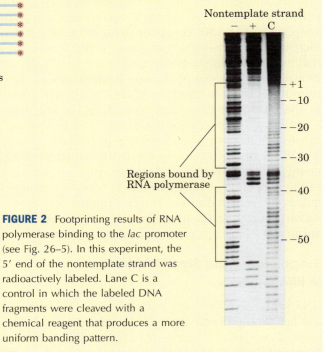

Nontemplate strand

Regions bound by RNA polymerase

**FIGURE 2** Footprinting results of RNA polymerase binding to the *lac* promoter (see Fig. 26–5). In this experiment, the 5′ end of the nontemplate strand was radioactively labeled. Lane C is a control in which the labeled DNA fragments were cleaved with a chemical reagent that produces a more uniform banding pattern.

between RNA and the RNA polymerase, facilitating dissociation of the transcript.

The ρ-dependent terminators lack the sequence of repeated A residues in the template strand but usually include a CA-rich sequence called a *rut* (*r*ho *ut*ilization) element. The ρ protein associates with the RNA at specific binding sites and migrates in the 5′→3′ direction until it reaches the transcription complex that is paused at a termination site. Here it contributes to release of the RNA transcript. The ρ protein has an ATP-dependent RNA-DNA helicase activity that promotes translocation of the protein along the RNA, and ATP is hydrolyzed by ρ protein during the termination process. The detailed mechanism by which the protein promotes the release of the RNA transcript is not known.

## Eukaryotic Cells Have Three Kinds of Nuclear RNA Polymerases

The transcriptional machinery in the nucleus of a eukaryotic cell is much more complex than that in bacteria. Eukaryotes have three RNA polymerases, designated I, II, and III, which are distinct complexes but have certain subunits in common. Each polymerase has a specific function and is recruited to a specific promoter sequence.

RNA polymerase I (Pol I) is responsible for the synthesis of only one type of RNA, a transcript called preribosomal RNA (or pre-rRNA), which contains the precursor for the 18S, 5.8S, and 28S rRNAs (see Fig. 26–22). Pol I promoters vary greatly in sequence from one species to another. The principal function of RNA polymerase II (Pol II) is synthesis of mRNAs and some specialized RNAs. This enzyme can recognize thousands of promoters that vary greatly in sequence. Many Pol II promoters have a few sequence features in common, including a TATA box (eukaryotic consensus sequence TATAAA) near base pair −30 and an Inr sequence (initiator) near the RNA start site at +1 (Fig. 26–8).

RNA polymerase III (Pol III) makes tRNAs, the 5S rRNA, and some other small specialized RNAs. The promoters recognized by Pol III are well characterized. Interestingly, some of the sequences required for the regulated initiation of transcription by Pol III are located within the gene itself, whereas others are in more conventional locations upstream of the RNA start site (Chapter 28).

## RNA Polymerase II Requires Many Other Protein Factors for Its Activity

RNA polymerase II is central to eukaryotic gene expression and has been studied extensively. Although this polymerase is strikingly more complex than its bacterial counterpart, the complexity masks a remarkable conservation of structure, function, and mechanism. Pol II is a huge enzyme with 12 subunits. The largest subunit (RBP1) exhibits a high degree of homology to the β′ subunit of bacterial RNA polymerase. Another subunit (RBP2) is structurally similar to the bacterial β subunit, and two others (RBP3 and RBP11) show some structural homology to the two bacterial α subunits. Pol II must function with genomes that are more complex and with DNA molecules more elaborately packaged than in bacteria. The need for protein-protein contacts with the numerous other protein factors required to navigate this labyrinth accounts in large measure for the added complexity of the eukaryotic polymerase.

The largest subunit of Pol II also has an unusual feature, a long carboxyl-terminal tail consisting of many repeats of a consensus heptad amino acid sequence –YSPTSPS–. There are 27 repeats in the yeast enzyme (18 exactly matching the consensus) and 52 (21 exact) in the mouse and human enzymes. This carboxyl-terminal domain (CTD) is separated from the main body of the enzyme by an unstructured linker sequence. The CTD has many important roles in Pol II function, as outlined below.

RNA polymerase II requires an array of other proteins, called **transcription factors,** in order to form the active transcription complex. The **general transcription factors** required at every Pol II promoter

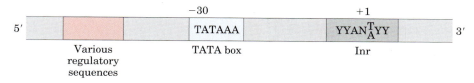

FIGURE 26–8 **Common sequences in promoters recognized by eukaryotic RNA polymerase II.** The TATA box is the major assembly point for the proteins of the preinitiation complexes of Pol II. The DNA is unwound at the initiator sequence (Inr), and the transcription start site is usually within or very near this sequence. In the Inr consensus sequence shown here, N represents any nucleotide; Y, a pyrimidine nucleotide. Many additional sequences serve as binding sites for a wide variety of proteins that affect the activity of Pol II. These sequences are important in regulating Pol II promoters and vary greatly in type and number, and in general the eukaryotic promoter is much more complex than suggested here. Many of the sequences are located within a few hundred base pairs of the TATA box on the 5′ side; others may be thousands of base pairs away. The sequence elements summarized here are more variable among the Pol II promoters of eukaryotes than among the E. coli promoters (see Fig. 26–5). Many Pol II promoters lack a TATA box or a consensus Inr element or both. Additional sequences around the TATA box and downstream (to the right as drawn) of Inr may be recognized by one or more transcription factors.

(factors usually designated TFII with an additional identifier) are highly conserved in all eukaryotes (Table 26–1). The process of transcription by Pol II can be described in terms of several phases—assembly, initiation, elongation, termination—each associated with characteristic proteins (Fig. 26–9). The step-by-step pathway described below leads to active transcription in vitro. In the cell, many of the proteins may be present in larger, preassembled complexes, simplifying the pathways for assembly on promoters. As you read about this process, consult Figure 26–9 and Table 26–1 to help keep track of the many participants.

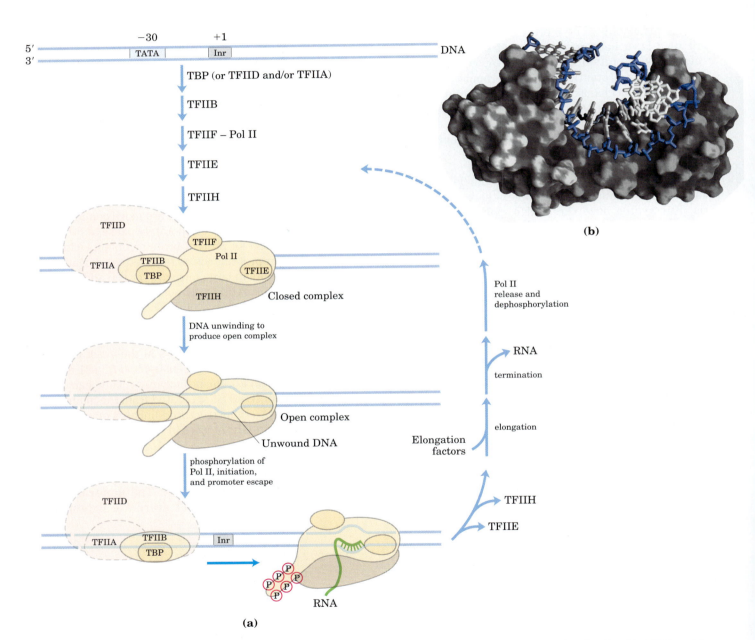

**(a)**

**(b)**

**FIGURE 26–9  Transcription at RNA polymerase II promoters. (a)** The sequential assembly of TBP (often with TFIIA), TFIIB, TFIIF plus Pol II, TFIIE, and TFIIH results in a closed complex. TBP often binds as part of a larger complex, TFIID. Some of the TFIID subunits play a role in transcription regulation (see Fig. 28–30). Within the complex, the DNA is unwound at the Inr region by the helicase activity of TFIIH and perhaps of TFIIE, creating an open complex. The carboxyl-terminal domain of the largest Pol II subunit is phosphorylated by TFIIH, and the polymerase then escapes the promoter and begins transcription. Elongation is accompanied by the release of many transcription factors and is also enhanced by elongation factors (see Table 26–1). After termination, Pol II is released, dephosphorylated, and recycled. **(b)** The structure of human TBP (gray) bound to DNA (blue and white) (PDB ID 1TGH).

**TABLE 26–1 Proteins Required for Initiation of Transcription at the RNA Polymerase II (Pol II) Promoters of Eukaryotes**

| Transcription protein | Number of subunits | Subunit(s) $M_r$ | Function(s) |
|---|---|---|---|
| **Initiation** | | | |
| Pol II | 12 | 10,000–220,000 | Catalyzes RNA synthesis |
| TBP (TATA-binding protein) | 1 | 38,000 | Specifically recognizes the TATA box |
| TFIIA | 3 | 12,000, 19,000, 35,000 | Stabilizes binding of TFIIB and TBP to the promoter |
| TFIIB | 1 | 35,000 | Binds to TBP; recruits Pol II–TFIIF complex |
| TFIIE | 2 | 34,000, 57,000 | Recruits TFIIH; has ATPase and helicase activities |
| TFIIF | 2 | 30,000, 74,000 | Binds tightly to Pol II; binds to TFIIB and prevents binding of Pol II to nonspecific DNA sequences |
| TFIIH | 12 | 35,000–89,000 | Unwinds DNA at promoter (helicase activity); phosphorylates Pol II (within the CTD); recruits nucleotide-excision repair proteins |
| **Elongation\*** | | | |
| ELL[†] | 1 | 80,000 | |
| p-TEFb | 2 | 43,000, 124,000 | Phosphorylates Pol II (within the CTD) |
| SII (TFIIS) | 1 | 38,000 | |
| Elongin (SIII) | 3 | 15,000, 18,000, 110,000 | |

*The function of all elongation factors is to suppress the pausing or arrest of transcription by the Pol II–TFIIF complex.

[†]Name derived from *eleven-nineteen lysine-rich leukemia*. The gene for ELL is the site of chromosomal recombination events frequently associated with acute myeloid leukemia.

**Assembly of RNA Polymerase and Transcription Factors at a Promoter** The formation of a closed complex begins when the TATA-binding protein (TBP) binds to the TATA box (Fig. 26–9b). TBP is bound in turn by the transcription factor TFIIB, which also binds to DNA on either side of TBP. TFIIA binding, although not always essential, can stabilize the TFIIB-TBP complex on the DNA and can be important at nonconsensus promoters where TBP binding is relatively weak. The TFIIB-TBP complex is next bound by another complex consisting of TFIIF and Pol II. TFIIF helps target Pol II to its promoters, both by interacting with TFIIB and by reducing the binding of the polymerase to nonspecific sites on the DNA. Finally, TFIIE and TFIIH bind to create the closed complex. TFIIH has DNA helicase activity that promotes the unwinding of DNA near the RNA start site (a process requiring the hydrolysis of ATP), thereby creating an open complex. Counting all the subunits of the various essential factors (excluding TFIIA), this minimal active assembly has more than 30 polypeptides.

**RNA Strand Initiation and Promoter Clearance** TFIIH has an additional function during the initiation phase. A kinase activity in one of its subunits phosphorylates Pol II at many places in the CTD (Fig. 26–9). Several other protein kinases, including CDK9 (cyclin-dependent kinase 9), which is part of the complex pTEFb (*positive transcription elongation factor b*), also phosphorylate the CTD. This causes a conformational change in the overall complex, initiating transcription. Phosphorylation of the CTD is also important during the subsequent elongation phase, and it affects the interactions between the transcription complex and other enzymes involved in processing the transcript (as described below).

During synthesis of the initial 60 to 70 nucleotides of RNA, first TFIIE and then TFIIH is released, and Pol II enters the elongation phase of transcription.

**Elongation, Termination, and Release** TFIIF remains associated with Pol II throughout elongation. During this stage, the activity of the polymerase is greatly enhanced by proteins called elongation factors (Table 26–1). The elongation factors suppress pausing during transcription and also coordinate interactions between protein complexes involved in the posttranscriptional processing of mRNAs. Once the RNA transcript is completed, transcription is terminated. Pol II is dephosphorylated and recycled, ready to initiate another transcript (Fig. 26–9).

**Regulation of RNA Polymerase II Activity** Regulation of transcription at Pol II promoters is quite elaborate. It involves the interaction of a wide variety of other proteins with the preinitiation complex. Some of these regulatory proteins interact with transcription factors, others with Pol II itself. Many interact through TFIID, a complex of about 12 proteins, including TBP and certain

*TBP-associated factors*, or TAFs. The regulation of transcription is described in more detail in Chapter 28.

**Diverse Functions of TFIIH**  In eukaryotes, the repair of damaged DNA (see Table 25–5) is more efficient within genes that are actively being transcribed than for other damaged DNA, and the template strand is repaired somewhat more efficiently than the nontemplate strand. These remarkable observations are explained by the alternative roles of the TFIIH subunits. Not only does TFIIH participate in the formation of the closed complex during assembly of a transcription complex (as described above), but some of its subunits are also essential components of the separate nucleotide-excision repair complex (see Fig. 25–24).

When Pol II transcription halts at the site of a DNA lesion, TFIIH can interact with the lesion and recruit the entire nucleotide-excision repair complex. Genetic loss of certain TFIIH subunits can produce human diseases. Some examples are xeroderma pigmentosum (see Box 25–1) and Cockayne's syndrome, which is characterized by arrested growth, photosensitivity, and neurological disorders. ■

## DNA-Dependent RNA Polymerase Undergoes Selective Inhibition

The elongation of RNA strands by RNA polymerase in both bacteria and eukaryotes is inhibited by the antibiotic **actinomycin D** (Fig. 26–10). The planar portion of this molecule inserts (intercalates) into the double-helical DNA between successive G≡C base pairs, deforming the DNA. This prevents movement of the polymerase along the template. Because actinomycin D

inhibits RNA elongation in intact cells as well as in cell extracts, it is used to identify cell processes that depend on RNA synthesis. **Acridine** inhibits RNA synthesis in a similar fashion (Fig. 26–10).

**Rifampicin** inhibits bacterial RNA synthesis by binding to the $\beta$ subunit of bacterial RNA polymerases, preventing the promoter clearance step of transcription (Fig. 26–6). It is sometimes used as an antibiotic.

The mushroom *Amanita phalloides* has evolved a very effective defense mechanism against predators. It produces **$\alpha$-amanitin,** which disrupts mRNA formation in animal cells by blocking Pol II and, at higher concentrations, Pol III. Neither Pol I nor bacterial RNA polymerase is sensitive to $\alpha$-amanitin—nor is the RNA polymerase II of *A. phalloides* itself!

## SUMMARY 26.1  DNA-Dependent Synthesis of RNA

- Transcription is catalyzed by DNA-dependent RNA polymerases, which use ribonucleoside 5'-triphosphates to synthesize RNA complementary to the template strand of duplex DNA. Transcription occurs in several phases: binding of RNA polymerase to a DNA site called a promoter, initiation of transcript synthesis, elongation, and termination.

- Bacterial RNA polymerase requires a special subunit to recognize the promoter. As the first committed step in transcription, binding of RNA polymerase to the promoter and initiation of transcription are closely regulated. Transcription stops at sequences called terminators.

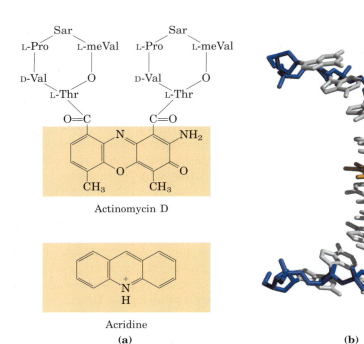

**FIGURE 26–10  Actinomycin D and acridine, inhibitors of DNA transcription.** (a) The shaded portion of actinomycin D is planar and intercalates between two successive G≡C base pairs in duplex DNA. The two cyclic peptide structures of actinomycin D bind to the minor groove of the double helix. Sarcosine (Sar) is *N*-methylglycine; meVal is methylvaline. Acridine also acts by intercalation in DNA. (b) A complex of actinomycin D with DNA (PDB ID 1DSC). The DNA backbone is shown in blue, the bases are white, the intercalated part of actinomycin (shaded in (a)) is orange, and the remainder of the actinomycin is red. The DNA is bent as a result of the actinomycin binding.

- Eukaryotic cells have three types of RNA polymerases. Binding of RNA polymerase II to its promoters requires an array of proteins called transcription factors. Elongation factors participate in the elongation phase of transcription. The largest subunit of Pol II has a long carboxyl-terminal domain, which is phosphorylated during the initiation and elongation phases.

## 26.2 RNA Processing

Many of the RNA molecules in bacteria and virtually all RNA molecules in eukaryotes are processed to some degree after synthesis. Some of the most interesting molecular events in RNA metabolism occur during this postsynthetic processing. Intriguingly, several of the enzymes that catalyze these reactions consist of RNA rather than protein. The discovery of these catalytic RNAs, or **ribozymes,** has brought a revolution in thinking about RNA function and about the origin of life.

A newly synthesized RNA molecule is called a **primary transcript.** Perhaps the most extensive processing of primary transcripts occurs in eukaryotic mRNAs and in tRNAs of both bacteria and eukaryotes.

The primary transcript for a eukaryotic mRNA typically contains sequences encompassing one gene, although the sequences encoding the polypeptide may not be contiguous. Noncoding tracts that break up the coding region of the transcript are called introns, and the coding segments are called exons (see the discussion of introns and exons in DNA in Chapter 24). In a process called **splicing,** the introns are removed from the primary transcript and the exons are joined to form a con-

tinuous sequence that specifies a functional polypeptide. Eukaryotic mRNAs are also modified at each end. A modified residue called a 5′ cap (p. 1008) is added at the 5′ end. The 3′ end is cleaved, and 80 to 250 A residues are added to create a poly(A) "tail." The sometimes elaborate protein complexes that carry out each of these three mRNA-processing reactions do not operate independently. They appear to be organized in association with each other and with the phosphorylated CTD of Pol II; each complex affects the function of the others. Other proteins involved in mRNA transport to the cytoplasm are also associated with the mRNA in the nucleus, and the processing of the transcript is coupled to its transport. In effect, a eukaryotic mRNA, as it is synthesized, is ensconced in an elaborate complex involving dozens of proteins. The composition of the complex changes as the primary transcript is processed, transported to the cytoplasm, and delivered to the ribosome for translation. These processes are outlined in Figure 26–11 and described in more detail below.

The primary transcripts of prokaryotic and eukaryotic tRNAs are processed by the removal of sequences from each end (cleavage) and in a few cases by the removal of introns (splicing). Many bases and sugars in tRNAs are also modified; mature tRNAs are replete with unusual bases not found in other nucleic acids (see Fig. 26–24).

The ultimate fate of any RNA is its complete and regulated degradation. The rate of turnover of RNAs plays a critical role in determining their steady-state levels and the rate at which cells can shut down expression of a gene whose product is no longer needed. During the development of multicellular organisms, for example, certain proteins must be expressed at one stage only, and the mRNA encoding such a protein must be made and destroyed at the appropriate times.

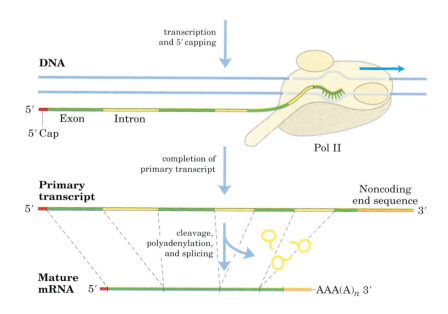

**FIGURE 26–11** Formation of the primary transcript and its processing during maturation of mRNA in a eukaryotic cell. The 5′ cap (red) is added before synthesis of the primary transcript is complete. A noncoding sequence following the last exon is shown in orange. Splicing can occur either before or after the cleavage and polyadenylation steps. All the processes shown here take place within the nucleus.

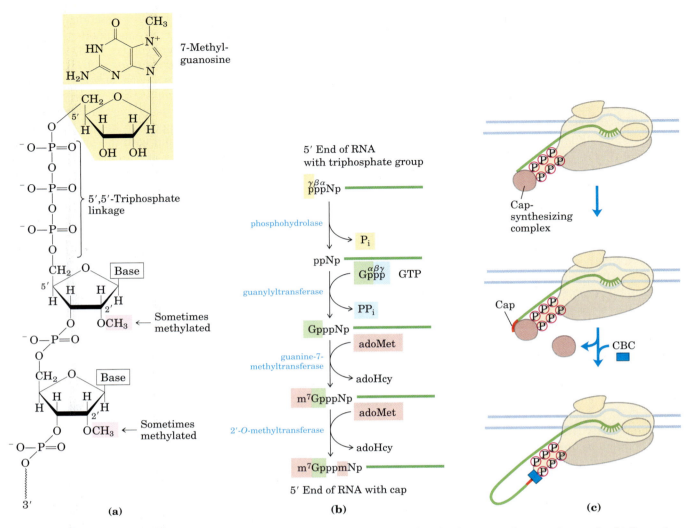

**FIGURE 26–12** **The 5′ cap of mRNA.** (a) 7-Methylguanosine is joined to the 5′ end of almost all eukaryotic mRNAs in an unusual 5′,5′-triphosphate linkage. Methyl groups (pink) are often found at the 2′ position of the first and second nucleotides. RNAs in yeast cells lack the 2′-methyl groups. The 2′-methyl group on the second nucleotide is generally found only in RNAs from vertebrate cells. (b) Generation of the 5′ cap involves four to five separate steps (adoHcy is *S*-adenosylhomocysteine). (c) Synthesis of the cap is carried out by enzymes tethered to the CTD of Pol II. The cap remains tethered to the CTD through an association with the cap-binding complex (CBC).

## Eukaryotic mRNAs Are Capped at the 5′ End

Most eukaryotic mRNAs have a **5′ cap,** a residue of 7-methylguanosine linked to the 5′-terminal residue of the mRNA through an unusual 5′,5′-triphosphate linkage (Fig. 26–12). The 5′ cap helps protect mRNA from ribonucleases. The cap also binds to a specific cap-binding complex of proteins and participates in binding of the mRNA to the ribosome to initiate translation (Chapter 27).

The 5′ cap is formed by condensation of a molecule of GTP with the triphosphate at the 5′ end of the transcript. The guanine is subsequently methylated at N-7, and additional methyl groups are often added at the 2′ hydroxyls of the first and second nucleotides adjacent to the cap (Fig. 26–12). The methyl groups are derived from *S*-adenosylmethionine. All these reactions occur very early in transcription, after the first 20 to 30 nucleotides of the transcript have been added. All three of the capping enzymes, and through them the 5′ end of the transcript itself, are associated with the RNA polymerase II CTD until the cap is synthesized. The capped 5′ end is then released from the capping enzymes and bound by the cap-binding complex (Fig. 26–12c).

## Both Introns and Exons Are Transcribed from DNA into RNA

In bacteria, a polypeptide chain is generally encoded by a DNA sequence that is colinear with the amino acid sequence, continuing along the DNA template without interruption until the information needed to specify the polypeptide is complete. However, the notion that *all* genes are continuous was disproved in 1977 when

Phillip Sharp and Richard Roberts independently discovered that many genes for polypeptides in eukaryotes are interrupted by noncoding sequences (introns).

The vast majority of genes in vertebrates contain introns; among the few exceptions are those that encode histones. The occurrence of introns in other eukaryotes varies. Many genes in the yeast *Saccharomyces cerevisiae* lack introns, although in some other yeast species introns are more common. Introns are also found in a few eubacterial and archaebacterial genes. Introns in DNA are transcribed along with the rest of the gene by RNA polymerases. The introns in the primary RNA transcript are then spliced, and the exons are joined to form a mature, functional RNA. In eukaryotic mRNAs, most exons are less than 1,000 nucleotides long, with many in the 100 to 200 nucleotide size range, encoding stretches of 30 to 60 amino acids within a longer polypeptide. Introns vary in size from 50 to 20,000 nucleotides. Genes of higher eukaryotes, including humans, typically have much more DNA devoted to introns than to exons. Many genes have introns; some genes have dozens of them.

## RNA Catalyzes the Splicing of Introns

There are four classes of introns. The first two, the group I and group II introns, differ in the details of their splicing mechanisms but share one surprising characteristic: they are *self-splicing*—no protein enzymes are involved. Group I introns are found in some nuclear, mitochondrial, and chloroplast genes coding for rRNAs, mRNAs, and tRNAs. Group II introns are generally found in the primary transcripts of mitochondrial or chloroplast mRNAs in fungi, algae, and plants. Group I and group II introns are also found among the rarer examples of introns in bacteria. Neither class requires a high-energy cofactor (such as ATP) for splicing. The splicing mechanisms in both groups involve two transesterification reaction steps (Fig. 26–13). A ribose 2'- or 3'-hydroxyl group makes a nucleophilic attack on a phosphorus and, in each step, a new phosphodiester bond is formed at the expense of the old, maintaining the balance of energy. These reactions are very similar to the DNA breaking and rejoining reactions promoted by topoisomerases (see Fig. 24–21) and site-specific recombinases (see Fig. 25–38).

The group I splicing reaction requires a guanine nucleoside or nucleotide cofactor, but the cofactor is not used as a source of energy; instead, the 3'-hydroxyl group of guanosine is used as a nucleophile in the first step of the splicing pathway. The guanosine 3'-hydroxyl group forms a normal 3',5'-phosphodiester bond with the 5' end of the intron (Fig. 26–14). The 3' hydroxyl of the exon that is displaced in this step then acts as a nucleophile in a similar reaction at the 3' end of the intron. The result is precise excision of the intron and ligation of the exons.

In group II introns the reaction pattern is similar except for the nucleophile in the first step, which in this case is the 2'-hydroxyl group of an A residue *within* the intron (Fig. 26–15). A branched lariat structure is formed as an intermediate.

Self-splicing of introns was first revealed in 1982 in studies of the splicing mechanism of the group I rRNA intron from the ciliated protozoan *Tetrahymena thermophila*, conducted by Thomas Cech and colleagues. These workers transcribed isolated *Tetrahymena* DNA (including the intron) in vitro using purified bacterial RNA polymerase. The resulting RNA spliced itself accurately without any protein enzymes from *Tetrahymena*. The discovery that RNAs could have catalytic functions was a milestone in our understanding of biological systems.

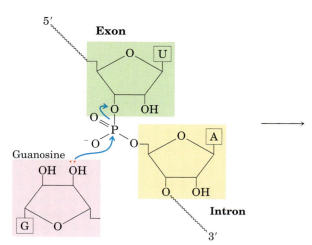

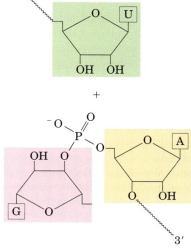

**FIGURE 26–13 Transesterification reaction.** This is the first step in the splicing of group I introns. Here, the 3' OH of a guanosine molecule acts as nucleophile.

Thomas Cech

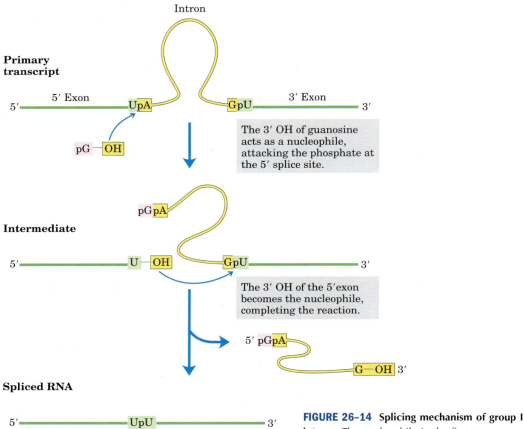

**FIGURE 26–14 Splicing mechanism of group I introns.** The nucleophile in the first step may be guanosine, GMP, GDP, or GTP. The spliced intron is eventually degraded.

Most introns are *not* self-splicing, and these types are not designated with a group number. The third and largest class of introns includes those found in nuclear mRNA primary transcripts. These are called **spliceosomal introns,** because their removal occurs within and is catalyzed by a large protein complex called a **spliceosome.** Within the spliceosome, the introns undergo splicing by the same lariat-forming mechanism as the group II introns. The spliceosome is made up of specialized RNA-protein complexes, *small nuclear ribonucleoproteins* (snRNPs, often pronounced "snurps"). Each snRNP contains one of a class of eukaryotic RNAs, 100 to 200 nucleotides long, known as **small nuclear RNAs (snRNAs).** Five snRNAs (U1, U2, U4, U5, and U6) involved in splicing reactions are generally found in abundance in eukaryotic nuclei. The RNAs and proteins in snRNPs are highly conserved in eukaryotes from yeasts to humans.  **mRNA Splicing**

Spliceosomal introns generally have the dinucleotide sequence GU and AG at the 5′ and 3′ ends, respectively, and these sequences mark the sites where splicing occurs. The U1 snRNA contains a sequence complementary to sequences near the 5′ splice site of nuclear mRNA introns (Fig. 26–16a), and the U1 snRNP

binds to this region in the primary transcript. Addition of the U2, U4, U5, and U6 snRNPs leads to formation of the spliceosome (Fig. 26–16b). The snRNPs together contribute five RNAs and about 50 proteins to the spliceosome, a supramolecular assembly nearly as complex as the ribosome (described in Chapter 27). ATP is required for assembly of the spliceosome, but the RNA cleavage-ligation reactions do not seem to require ATP. Some mRNA introns are spliced by a less common type of spliceosome, in which the U1 and U2 snRNPs are replaced by the U11 and U12 snRNPs. Whereas U1- and U2-containing spliceosomes remove introns with (5′)GU and AG(3′) terminal sequences, as shown in Figure 26–16, the U11- and U12-containing spliceosomes remove a rare class of introns that have (5′)AU and AC(3′) terminal sequences to mark the intronic splice sites. The spliceosomes used in nuclear RNA splicing may have evolved from more ancient group II introns, with the snRNPs replacing the catalytic domains of their self-splicing ancestors.

Some components of the splicing apparatus appear to be tethered to the CTD of RNA polymerase II, suggesting an interesting model for the splicing reaction. As the first splice junction is synthesized, it is bound by

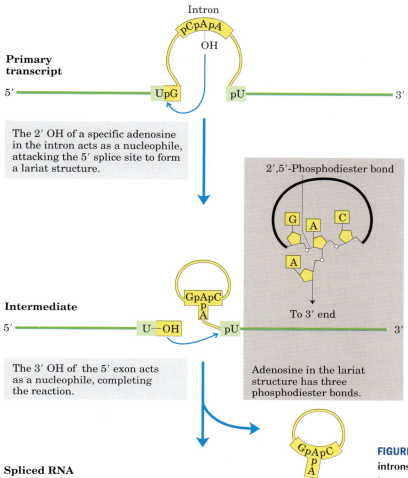

**FIGURE 26–15 Splicing mechanism of group II introns.** The chemistry is similar to that of group I intron splicing, except for the identity of the nucleophile in the first step and formation of a lariatlike intermediate, in which one branch is a 2′,5′-phosphodiester bond.

a tethered spliceosome. The second splice junction is then captured by this complex as it passes, facilitating the juxtaposition of the intron ends and the subsequent splicing process (Fig. 26–16c). After splicing, the intron remains in the nucleus and is eventually degraded.

The fourth class of introns, found in certain tRNAs, is distinguished from the group I and II introns in that the splicing reaction requires ATP and an endonuclease. The splicing endonuclease cleaves the phosphodiester bonds at both ends of the intron, and the two exons are joined by a mechanism similar to the DNA ligase reaction (see Fig. 25–16).

Although spliceosomal introns appear to be limited to eukaryotes, the other intron classes are not. Genes with group I and II introns have now been found in both bacteria and bacterial viruses. Bacteriophage T4, for example, has several protein-encoding genes with group I introns. Introns appear to be more common in archaebacteria than in eubacteria.

## Eukaryotic mRNAs Have a Distinctive 3′ End Structure

At their 3′ end, most eukaryotic mRNAs have a string of 80 to 250 A residues, making up the **poly(A) tail.** This tail serves as a binding site for one or more specific proteins. The poly(A) tail and its associated proteins probably help protect mRNA from enzymatic destruction. Many prokaryotic mRNAs also acquire poly(A) tails, but these tails stimulate decay of mRNA rather than protecting it from degradation.

The poly(A) tail is added in a multistep process. The transcript is extended beyond the site where the poly(A) tail is to be added, then is cleaved at the poly(A) addition site by an endonuclease component of a large enzyme complex, again associated with the CTD of RNA polymerase II (Fig. 26–17). The mRNA site where cleavage occurs is marked by two sequence elements: the highly conserved sequence (5′)AAUAAA(3′), 10 to 30

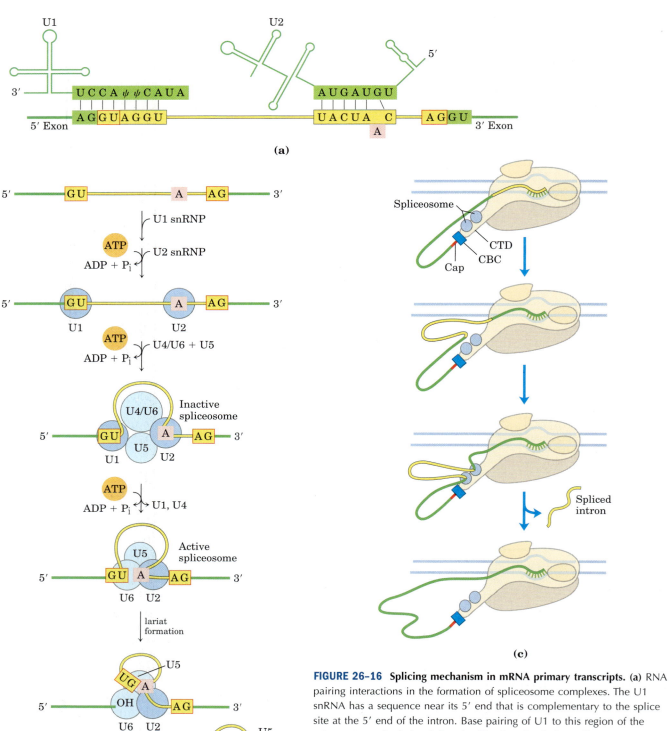

**FIGURE 26–16 Splicing mechanism in mRNA primary transcripts. (a)** RNA pairing interactions in the formation of spliceosome complexes. The U1 snRNA has a sequence near its 5′ end that is complementary to the splice site at the 5′ end of the intron. Base pairing of U1 to this region of the primary transcript helps define the 5′ splice site during spliceosome assembly (Ψ is pseudouridine; see Fig. 26–24). U2 is paired to the intron at a position encompassing the A residue (shaded pink) that becomes the nucleophile during the splicing reaction. Base pairing of U2 snRNA causes a bulge that displaces and helps to activate the adenylate, whose 2′ OH will form the lariat structure through a 2′,5′-phosphodiester bond.

**(b)** Assembly of spliceosomes. The U1 and U2 snRNPs bind, then the remaining snRNPs (the U4/U6 complex and U5) bind to form an inactive spliceosome. Internal rearrangements convert this species to an active spliceosome in which U1 and U4 have been expelled and U6 is paired with both the 5′ splice site and U2. This is followed by the catalytic steps, which parallel those of the splicing of group II introns (see Fig. 26–15).

**(c)** Coordination of splicing with transcription provides an attractive mechanism for bringing the two splice sites together. See the text for details.

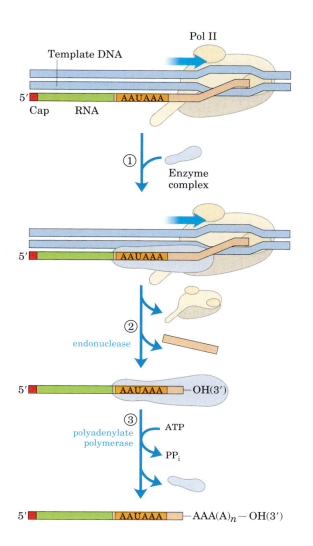

**FIGURE 26–17 Addition of the poly(A) tail to the primary RNA transcript of eukaryotes.** Pol II synthesizes RNA beyond the segment of the transcript containing the cleavage signal sequences, including the highly conserved upstream sequence (5')AAUAAA. ① The cleavage signal sequence is bound by an enzyme complex that includes an endonuclease, a polyadenylate polymerase, and several other multisubunit proteins involved in sequence recognition, stimulation of cleavage, and regulation of the length of the poly(A) tail. ② The RNA is cleaved by the endonuclease at a point 10 to 30 nucleotides 3' to (downstream of) the sequence AAUAAA. ③ The polyadenylate polymerase synthesizes a poly(A) tail 80 to 250 nucleotides long, beginning at the cleavage site.

nucleotides on the 5' side (upstream) of the cleavage site, and a less well-defined sequence rich in G and U residues, 20 to 40 nucleotides downstream of the cleavage site. Cleavage generates the free 3'-hydroxyl group that defines the end of the mRNA, to which A residues are immediately added by **polyadenylate polymerase,** which catalyzes the reaction

$$RNA + nATP \longrightarrow RNA\text{–}(AMP)_n + nPP_i$$

where $n = 80$ to 250. This enzyme does not require a template but does require the cleaved mRNA as a primer.

The overall processing of a typical eukaryotic mRNA is summarized in Figure 26–18. In some cases the polypeptide-coding region of the mRNA is also modified by RNA "editing" (see Box 27–1 for details). This editing includes processes that add or delete bases in the coding regions

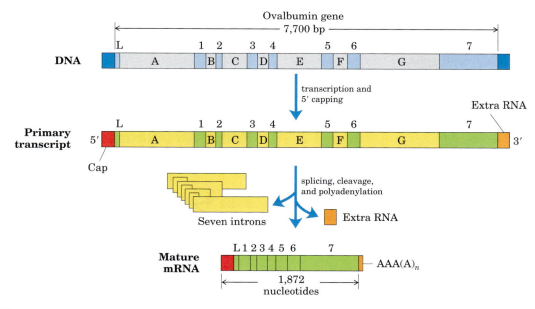

**FIGURE 26–18 Overview of the processing of a eukaryotic mRNA.** The ovalbumin gene, shown here, has introns A to G and exons 1 to 7 and L (L encodes a signal peptide sequence that targets the protein for export from the cell; see Fig. 27–34). About three-quarters of the RNA is removed during processing. Pol II extends the primary transcript well beyond the cleavage and polyadenylation site ("extra RNA") before terminating transcription. Termination signals for Pol II have not yet been defined.

of primary transcripts or that change the sequence (by, for example, enzymatic deamination of a C residue to create a U residue). A particularly dramatic example occurs in trypanosomes, which are parasitic protozoa: large regions of an mRNA are synthesized without any uridylate, and the U residues are inserted later by RNA editing.

## A Gene Can Give Rise to Multiple Products by Differential RNA Processing

The transcription of introns seems to consume cellular resources and energy without returning any benefit to the organism, but introns may confer an advantage not yet fully appreciated by scientists. Introns may be vestiges of a molecular parasite not unlike transposons (Chapter 25). Although the benefits of introns are not yet clear in most cases, cells have evolved to take advantage of the splicing pathways to alter the expression of certain genes.

Most eukaryotic mRNA transcripts produce only one mature mRNA and one corresponding polypeptide, but some can be processed in more than one way to produce *different* mRNAs and thus different polypeptides. The primary transcript contains molecular signals for all the alternative processing pathways, and the pathway favored in a given cell is determined by processing factors, RNA-binding proteins that promote one particular path.

Complex transcripts can have either more than one site for cleavage and polyadenylation or alternative splicing patterns, or both. If there are two or more sites for cleavage and polyadenylation, use of the one closest to the 5′ end will remove more of the primary transcript sequence (Fig. 26–19a). This mechanism, called poly(A) site choice, generates diversity in the variable domains of immunoglobulin heavy chains. Alternative splicing patterns (Fig. 26–19b) produce, from a common primary transcript, three different forms of the myosin heavy chain at different stages of fruit fly development. *Both* mechanisms come into play when a single RNA transcript is processed differently to produce two different hormones: the calcium-regulating hormone calcitonin in rat thyroid and calcitonin-gene-related peptide (CGRP) in rat brain (Fig. 26–20).

## Ribosomal RNAs and tRNAs Also Undergo Processing

Posttranscriptional processing is not limited to mRNA. Ribosomal RNAs of both prokaryotic and eukaryotic cells are made from longer precursors called **preribosomal RNAs,** or pre-rRNAs, synthesized by Pol I. In bacteria, 16S, 23S, and 5S rRNAs (and some tRNAs, although most tRNAs are encoded elsewhere) arise from a single 30S RNA precursor of about 6,500 nucleotides. RNA at both ends of the 30S precursor and segments between the rRNAs are removed during processing (Fig. 26–21).

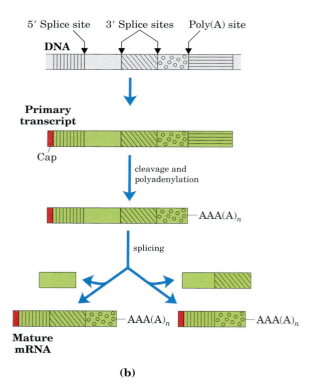

**(a)**

**(b)**

**FIGURE 26–19 Two mechanisms for the alternative processing of complex transcripts in eukaryotes. (a)** Alternative cleavage and polyadenylation patterns. Two poly(A) sites, $A_1$ and $A_2$, are shown.

**(b)** Alternative splicing patterns. Two different 3′ splice sites are shown. In both mechanisms, different mature mRNAs are produced from the same primary transcript.

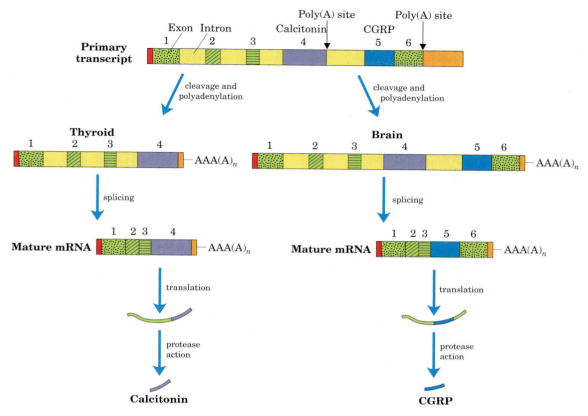

**FIGURE 26–20 Alternative processing of the calcitonin gene transcript in rats.** The primary transcript has two poly(A) sites; one predominates in the brain, the other in the thyroid. In the brain, splicing eliminates the calcitonin exon (exon 4); in the thyroid, this exon is re-

tained. The resulting peptides are processed further to yield the final hormone products: calcitonin-gene-related peptide (CGRP) in the brain and calcitonin in the thyroid.

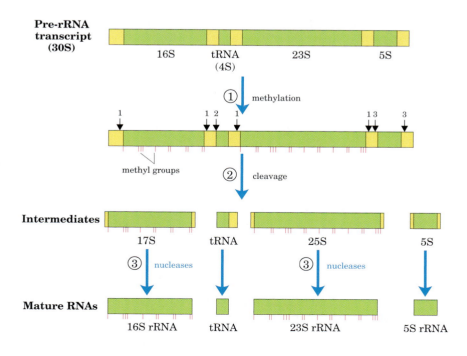

**FIGURE 26–21 Processing of pre-rRNA transcripts in bacteria.** ① Before cleavage, the 30S RNA precursor is methylated at specific bases. ② Cleavage liberates precursors of rRNAs and tRNA(s). Cleavage at the points labeled 1, 2, and 3 is carried out by the enzymes RNase III, RNase P, and RNase E, respectively. As discussed later in the text, RNase P is a ribozyme. ③ The final 16S, 23S, and 5S rRNA products result from the action of a variety of specific nucleases. The seven copies of the gene for pre-rRNA in the *E. coli* chromosome differ in the number, location, and identity of tRNAs included in the primary transcript. Some copies of the gene have additional tRNA gene segments between the 16S and 23S rRNA segments and at the far 3′ end of the primary transcript.

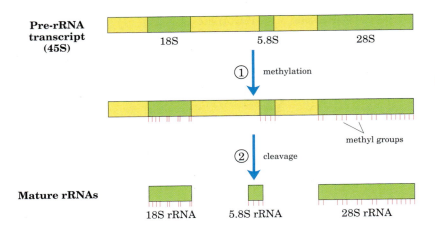

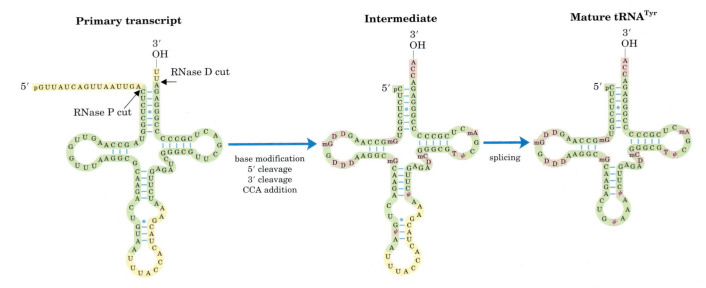

**FIGURE 26–22 Processing of pre-rRNA transcripts in vertebrates.** In step ①, the 45S precursor is methylated at more than 100 of its 14,000 nucleotides, mostly on the 2'-OH groups of ribose units retained in the final products. ② A series of enzymatic cleavages produces the 18S, 5.8S, and 28S rRNAs. The cleavage reactions require RNAs found in the nucleolus, called small nucleolar RNAs (snoRNAs), within protein complexes reminiscent of spliceosomes. The 5S rRNA is produced separately.

The genome of *E. coli* encodes seven pre-rRNA molecules. All these genes have essentially identical rRNA-coding regions, but they differ in the segments between these regions. The segment between the 16S and 23S rRNA genes generally encodes one or two tRNAs, with different tRNAs arising from different pre-rRNA transcripts. Coding sequences for tRNAs are also found on the 3' side of the 5S rRNA in some precursor transcripts.

In eukaryotes, a 45S pre-rRNA transcript is processed in the nucleolus to form the 18S, 28S, and 5.8S rRNAs characteristic of eukaryotic ribosomes (Fig. 26–22). The 5S rRNA of most eukaryotes is made as a completely separate transcript by a different polymerase (Pol III instead of Pol I).

Most cells have 40 to 50 distinct tRNAs, and eukaryotic cells have multiple copies of many of the tRNA genes. Transfer RNAs are derived from longer RNA precursors by enzymatic removal of nucleotides from the 5' and 3' ends (Fig. 26–23). In eukaryotes, introns are present in a few tRNA transcripts and must be excised. Where two or more different tRNAs are contained in a single primary transcript, they are separated by enzymatic cleavage. The endonuclease RNase P, found in all organisms, removes RNA at the 5' end of tRNAs. This enzyme contains both protein and RNA. The RNA component is essential for activity, and in bacterial cells it can carry out its processing function with precision even without the protein component. RNase P is therefore another example of a catalytic RNA, as described in more detail below. The 3' end of tRNAs is processed by one or more nucleases, including the exonuclease RNase D.

**FIGURE 26–23 Processing of tRNAs in bacteria and eukaryotes.** The yeast tRNA^Tyr (the tRNA specific for tyrosine binding; see Chapter 27) is used to illustrate the important steps. The nucleotide sequences shown in yellow are removed from the primary transcript. The ends are processed first, the 5' end before the 3' end. CCA is then added to the 3' end, a necessary step in processing eukaryotic tRNAs and those bacterial tRNAs that lack this sequence in the primary transcript. While the ends are being processed, specific bases in the rest of the transcript are modified (see Fig. 26–24). For the eukaryotic tRNA shown here, the final step is splicing of the 14-nucleotide intron. Introns are found in some eukaryotic tRNAs but not in bacterial tRNAs.

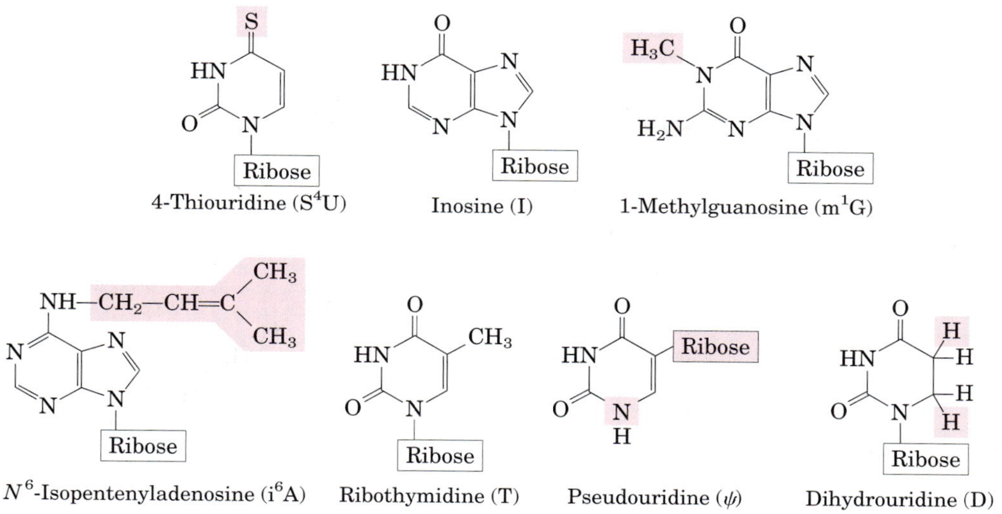

**FIGURE 26–24 Some modified bases of tRNAs, produced in posttranscriptional reactions.** The standard symbols (used in Fig. 26–23) are shown in parentheses. Note the unusual ribose attachment point in pseudouridine.

Transfer RNA precursors may undergo further post-transcriptional processing. The 3′-terminal trinucleotide CCA(3′) to which an amino acid will be attached during protein synthesis (Chapter 27) is absent from some bacterial and all eukaryotic tRNA precursors and is added during processing (Fig. 26–23). This addition is carried out by tRNA nucleotidyltransferase, an unusual enzyme that binds the three ribonucleoside triphosphate precursors in separate active sites and catalyzes formation of the phosphodiester bonds to produce the CCA(3′) sequence. The creation of this defined sequence of nucleotides is therefore not dependent on a DNA or RNA template—the template is the binding site of the enzyme.

The final type of tRNA processing is the modification of some of the bases by methylation, deamination, or reduction (Fig. 26–24). In the case of pseudouridine ($\Psi$), the base (uracil) is removed and reattached to the sugar through C-5. Some of these modified bases occur at characteristic positions in all tRNAs (Fig. 26–23).

## RNA Enzymes Are the Catalysts of Some Events in RNA Metabolism

The study of posttranscriptional processing of RNA molecules led to one of the most exciting discoveries in modern biochemistry—the existence of RNA enzymes. The best-characterized ribozymes are the self-splicing group I introns, RNase P, and the hammerhead ribozyme (discussed below). Most of the activities of these ribozymes are based on two fundamental reactions: transesterification (Fig. 26–13) and phosphodiester bond hydrolysis (cleavage). The substrate for ribozymes is often an RNA molecule, and it may even be part of the ribozyme itself. When its substrate is RNA, an RNA catalyst can make use of base-pairing interactions to align the substrate for the reaction.

Ribozymes vary greatly in size. A self-splicing group I intron may have more than 400 nucleotides. The hammerhead ribozyme consists of two RNA strands with only 41 nucleotides in all (Fig. 26–25). As with protein enzymes, the three-dimensional structure of ribozymes is important for function. Ribozymes are inactivated by heating above their melting temperature or by addition of denaturing agents or complementary oligonucleotides, which disrupt normal base-pairing patterns. Ribozymes can also be inactivated if essential nucleotides are changed. The secondary structure of a self-splicing group I intron from the 26S rRNA precursor of *Tetrahymena* is shown in detail in Figure 26–26.

***Enzymatic Properties of Group I Introns*** Self-splicing group I introns share several properties with enzymes besides accelerating the reaction rate, including their kinetic behaviors and their specificity. Binding of the guanosine cofactor (Fig. 26–13) to the *Tetrahymena* group I rRNA intron (Fig. 26–26) is saturable ($K_m \approx 30\ \mu$M) and can be competitively inhibited by 3′-deoxyguanosine. The intron is very precise in its excision reaction, largely due to a segment called the **internal guide sequence** that can base-pair with exon sequences near the 5′ splice site (Fig. 26–26). This pairing promotes the alignment of specific bonds to be cleaved and rejoined.

Because the intron itself is chemically altered during the splicing reaction—its ends are cleaved—it may appear to lack one key enzymatic property: the ability to catalyze multiple reactions. Closer inspection has shown that after excision, the 414 nucleotide intron from *Tetrahymena* rRNA can, in vitro, act as a true enzyme (but in vivo it is quickly degraded). A series of

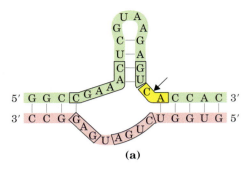

**(a)**

**FIGURE 26–25 Hammerhead ribozyme.** Certain viruslike elements called virusoids have small RNA genomes and usually require another virus to assist in their replication and/or packaging. Some virusoid RNAs include small segments that promote site-specific RNA cleavage reactions associated with replication. These segments are called hammerhead ribozymes, because their secondary structures are shaped like the head of a hammer. Hammerhead ribozymes have been defined and studied separately from the much larger viral RNAs. **(a)** The minimal sequences required for catalysis by the ribozyme. The boxed nucleotides are highly conserved and are required for catalytic function. The arrow indicates the site of self-cleavage. **(b)** Three-dimensional structure (PDB ID 1MME). The strands are colored as in **(a).** The hammerhead ribozyme is a metalloenzyme; $Mg^{2+}$ ions are required for activity. The phosphodiester bond at the site of self-cleavage is indicated by an arrow. 🖱 **Hammerhead Ribozyme**

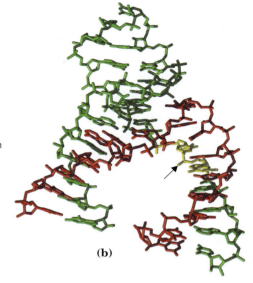

**(b)**

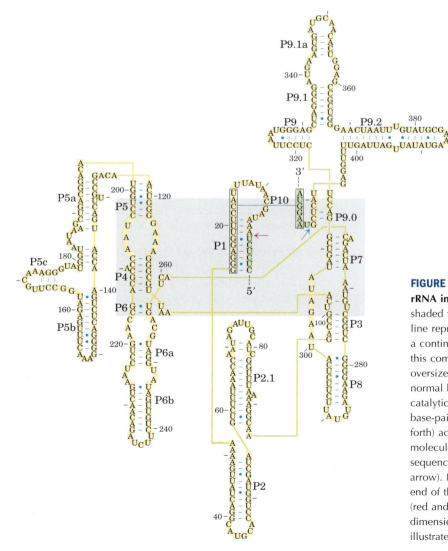

**FIGURE 26–26 Secondary structure of the self-splicing rRNA intron from *Tetrahymena.*** Intron sequences are shaded yellow, exon sequences green. Each thick yellow line represents a bond between neighboring nucleotides in a continuous sequence (a device necessitated by showing this complex molecule in two dimensions; similarly an oversize blue line between a C and G residue indicates normal base pairing); all nucleotides are shown. The catalytic core of the self-splicing activity is shaded. Some base-paired regions are labeled (P1, P3, P2.1, P5a, and so forth) according to an established convention for this RNA molecule. The P1 region, which contains the internal guide sequence (boxed), is the location of the 5' splice site (red arrow). Part of the internal guide sequence pairs with the end of the 3' exon, bringing the 5' and 3' splice sites (red and blue arrows) into close proximity. The three-dimensional structure of a large segment of this intron is illustrated in Figure 8–28c.

intramolecular cyclization and cleavage reactions in the excised intron leads to the loss of 19 nucleotides from its 5′ end. The remaining 395 nucleotide, linear RNA—referred to as L-19 IVS—promotes nucleotidyl transfer reactions in which some oligonucleotides are lengthened at the expense of others (Fig. 26–27). The best substrates are oligonucleotides, such as a synthetic $(C)_5$ oligomer, that can base-pair with the same guanylate-rich internal guide sequence that held the 5′ exon in place for self-splicing.

The enzymatic activity of the L-19 IVS ribozyme results from a cycle of transesterification reactions mechanistically similar to self-splicing. Each ribozyme molecule can process about 100 substrate molecules per hour and is not altered in the reaction; therefore the intron acts as a catalyst. It follows Michaelis-Menten kinetics, is specific for RNA oligonucleotide substrates, and can be competitively inhibited. The $k_{cat}/K_m$ (specificity constant) is $10^3 \ M^{-1} \ s^{-1}$, lower than that of many enzymes, but the ribozyme accelerates hydrolysis by a factor of $10^{10}$ relative to the uncatalyzed reaction. It makes use of substrate orientation, covalent catalysis, and metal-ion catalysis—strategies used by protein enzymes.

**Characteristics of Other Ribozymes** *E. coli* RNase P has both an RNA component (the M1 RNA, with 377 nu-

cleotides) and a protein component ($M_r$ 17,500). In 1983 Sidney Altman and Norman Pace and their coworkers discovered that under some conditions, the M1 RNA alone is capable of catalysis, cleaving tRNA precursors at the correct position. The protein component apparently serves to stabilize the RNA or facilitate its function in vivo. The RNase P ribozyme recognizes the three-dimensional shape of its pre-tRNA substrate, along with the CCA sequence, and thus can cleave the 5′ leaders from diverse tRNAs (Fig. 26–23).

The known catalytic repertoire of ribozymes continues to expand. Some virusoids, small RNAs associated with plant RNA viruses, include a structure that promotes a self-cleavage reaction; the hammerhead ribozyme illustrated in Figure 26–25 is in this class, catalyzing the hydrolysis of an internal phosphodiester bond. The splicing reaction that occurs in a spliceosome seems to rely on a catalytic center formed by the U2, U5, and U6 snRNAs (Fig. 26–16). And perhaps most important, an RNA component of ribosomes catalyzes the synthesis of proteins (Chapter 27).

Exploring catalytic RNAs has provided new insights into catalytic function in general and has important implications for our understanding of the origin and evolution of life on this planet, a topic discussed in Section 26.3.

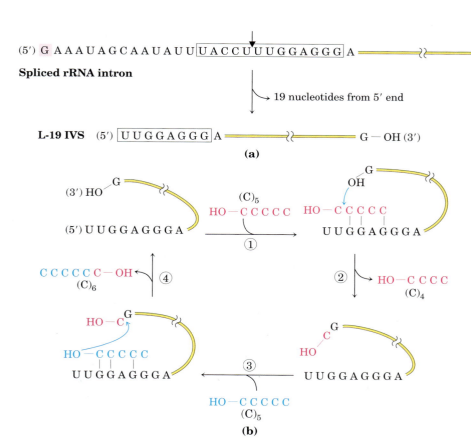

(5′) G A A A U A G C A A U A U U ⬚U A C C U U U G G A G G G⬚ A ══════ ⟩⟩ ══════ G—OH (3′)

**Spliced rRNA intron**

↳ 19 nucleotides from 5′ end

**L-19 IVS** (5′) ⬚U U G G A G G G⬚ A ══════ ⟩⟩ ══════ G — OH (3′)

**(a)**

**(b)**

**FIGURE 26–27 In vitro catalytic activity of L-19 IVS. (a)** L-19 IVS is generated by the autocatalytic removal of 19 nucleotides from the 5′ end of the spliced *Tetrahymena* intron. The cleavage site is indicated by the arrow in the internal guide sequence (boxed). The G residue (shaded pink) added in the first step of the splicing reaction (see Fig. 26–14) is part of the removed sequence. A portion of the internal guide sequence remains at the 5′ end of L-19 IVS. **(b)** L-19 IVS lengthens some RNA oligonucleotides at the expense of others in a cycle of transesterification reactions (steps ① through ④). The 3′ OH of the G residue at the 3′ end of L-19 IVS plays a key role in this cycle (note that this is *not* the G residue added in the splicing reaction). $(C)_5$ is one of the ribozyme's better substrates because it can base-pair with the guide sequence remaining in the intron. Although this catalytic activity is probably irrelevant to the cell, it has important implications for current hypotheses on evolution, discussed at the end of this chapter.

## Cellular mRNAs Are Degraded at Different Rates

The expression of genes is regulated at many levels. A crucial factor governing a gene's expression is the cellular concentration of its associated mRNA. The concentration of any molecule depends on two factors: its rate of synthesis and its rate of degradation. When synthesis and degradation of an mRNA are balanced, the concentration of the mRNA remains in a steady state. A change in either rate will lead to net accumulation or depletion of the mRNA. Degradative pathways ensure that mRNAs do not build up in the cell and direct the synthesis of unnecessary proteins.

The rates of degradation vary greatly for mRNAs from different eukaryotic genes. For a gene product that is needed only briefly, the half-life of its mRNA may be only minutes or even seconds. Gene products needed constantly by the cell may have mRNAs that are stable over many cell generations. The average half-life of a vertebrate cell mRNA is about 3 hours, with the pool of each type of mRNA turning over about ten times per cell generation. The half-life of bacterial mRNAs is much shorter—only about 1.5 min—perhaps because of regulatory requirements.

Messenger RNA is degraded by ribonucleases present in all cells. In *E. coli*, the process begins with one or a few cuts by an endoribonuclease, followed by $3' \rightarrow 5'$ degradation by exoribonucleases. In lower eukaryotes, the major pathway involves first shortening the poly(A) tail, then decapping the 5' end and degrading the mRNA in the $5' \rightarrow 3'$ direction. A $3' \rightarrow 5'$ degradative pathway also exists and may be the major path in higher eukaryotes. All eukaryotes have a complex of up to ten conserved $3' \rightarrow 5'$ exoribonucleases, called the exosome, which is involved in the processing of the 3' end of rRNAs and tRNAs as well as the degradation of mRNAs.

A hairpin structure in bacterial mRNAs with a $\rho$-independent terminator (Fig. 26–7) confers stability against degradation. Similar hairpin structures can make some parts of a primary transcript more stable, leading to nonuniform degradation of transcripts. In eukaryotic cells, both the 3' poly(A) tail and the 5' cap are important to the stability of many mRNAs. 🖱 **Life Cycle of an mRNA**

## Polynucleotide Phosphorylase Makes Random RNA-like Polymers

In 1955, Marianne Grunberg-Manago and Severo Ochoa discovered the bacterial enzyme **polynucleotide phosphorylase,** which in vitro catalyzes the reaction

$$(\text{NMP})_n + \text{NDP} \rightleftharpoons \underset{\substack{\text{Lengthened} \\ \text{polynucleotide}}}{(\text{NMP})_{n+1}} + \text{P}_i$$

Polynucleotide phosphorylase was the first nucleic acid–synthesizing enzyme discovered (Arthur Kornberg's discovery of DNA polymerase followed soon thereafter).

The reaction catalyzed by polynucleotide phosphorylase differs fundamentally from the polymerase activities discussed so far in that it is not template-dependent. The enzyme uses the 5'-diphosphates of ribonucleosides as substrates and cannot act on the homologous 5'-triphosphates or on deoxyribonucleoside 5'-diphosphates. The RNA polymer formed by polynucleotide phosphorylase contains the usual 3',5'-phosphodiester linkages, which can be hydrolyzed by ribonuclease. The reaction is readily reversible and can be pushed in the direction of breakdown of the polyribonucleotide by increasing the phosphate concentration. The probable function of this enzyme in the cell is the degradation of mRNAs to nucleoside diphosphates.

Because the polynucleotide phosphorylase reaction does not use a template, the polymer it forms does not have a specific base sequence. The reaction proceeds equally well with any or all of the four nucleoside diphosphates, and the base composition of the resulting polymer reflects nothing more than the relative concentrations of the 5'-diphosphate substrates in the medium.

Polynucleotide phosphorylase can be used in the laboratory to prepare RNA polymers with many different base sequences and frequencies. Synthetic RNA polymers of this sort were critical for deducing the genetic code for the amino acids (Chapter 27).

## SUMMARY 26.2 RNA Processing

- Eukaryotic mRNAs are modified by addition of a 7-methylguanosine residue at the 5' end and by cleavage and polyadenylation at the 3' end to form a long poly(A) tail.

- Many primary mRNA transcripts contain introns (noncoding regions), which are removed by splicing. Excision of the group I introns found in some rRNAs requires a guanosine cofactor. Some group I and group II introns are capable of self-splicing; no protein enzymes are required. Nuclear mRNA precursors have a third class (the largest class) of introns, which are spliced

Marianne Grunberg-Manago

Severo Ochoa, 1905–1993

with the aid of RNA-protein complexes called snRNPs, assembled into spliceosomes. A fourth class of introns, found in some tRNAs, is the only class known to be spliced by protein enzymes.

- Ribosomal RNAs and transfer RNAs are derived from longer precursor RNAs, trimmed by nucleases. Some bases are modified enzymatically during the maturation process.

- The self-splicing introns and the RNA component of RNase P (which cleaves the 5′ end of tRNA precursors) are two examples of ribozymes. These biological catalysts have the properties of true enzymes. They generally promote hydrolytic cleavage and transesterification, using RNA as substrate. Combinations of these reactions can be promoted by the excised group I intron of *Tetrahymena* rRNA, resulting in a type of RNA polymerization reaction.

- Polynucleotide phosphorylase reversibly forms RNA-like polymers from ribonucleoside 5′-diphosphates, adding or removing ribonucleotides at the 3′-hydroxyl end of the polymer. The enzyme degrades RNA in vivo.

## 26.3 RNA-Dependent Synthesis of RNA and DNA

In our discussion of DNA and RNA synthesis up to this point, the role of the template strand has been reserved for DNA. However, some enzymes use an RNA template for nucleic acid synthesis. With the very important exception of viruses with an RNA genome, these enzymes play only a modest role in information pathways. RNA viruses are the source of most RNA-dependent polymerases characterized so far.

The existence of RNA replication requires an elaboration of the central dogma (Fig. 26–28; contrast this with the diagram on p. 922). The enzymes involved in RNA replication have profound implications for investigations into the nature of self-replicating molecules that may have existed in prebiotic times.

### Reverse Transcriptase Produces DNA from Viral RNA

Certain RNA viruses that infect animal cells carry within the viral particle an RNA-dependent DNA polymerase called **reverse transcriptase.** On infection, the single-stranded RNA viral genome (~10,000 nucleotides) and the enzyme enter the host cell. The reverse transcriptase first catalyzes the synthesis of a DNA strand complementary to the viral RNA (Fig. 26–29), then degrades the RNA strand of the viral RNA-DNA hybrid and replaces it with DNA. The resulting duplex DNA often becomes incorporated into the genome of the eukaryotic host cell. These integrated (and dormant) viral genes can be activated and transcribed, and the gene products—viral proteins and the viral RNA genome itself—packaged as new viruses. The RNA viruses that contain reverse transcriptases are known as **retroviruses** (*retro* is the Latin prefix for "backward").

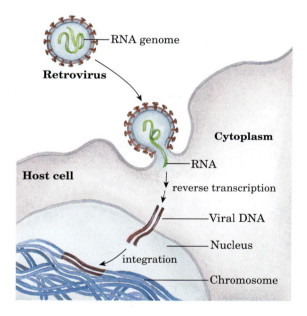

**FIGURE 26–29 Retroviral infection of a mammalian cell and integration of the retrovirus into the host chromosome.** Viral particles entering the host cell carry viral reverse transcriptase and a cellular tRNA (picked up from a former host cell) already base-paired to the viral RNA. The tRNA facilitates immediate conversion of viral RNA to double-stranded DNA by the action of reverse transcriptase, as described in the text. Once converted to double-stranded DNA, the DNA enters the nucleus and is integrated into the host genome. The integration is catalyzed by a virally encoded integrase. Integration of viral DNA into host DNA is mechanistically similar to the insertion of transposons in bacterial chromosomes (see Fig. 25–43). For example, a few base pairs of host DNA become duplicated at the site of integration, forming short repeats of 4 to 6 bp at each end of the inserted retroviral DNA (not shown).

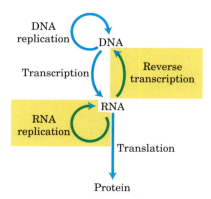

**FIGURE 26–28** Extension of the central dogma to include RNA-dependent synthesis of RNA and DNA.

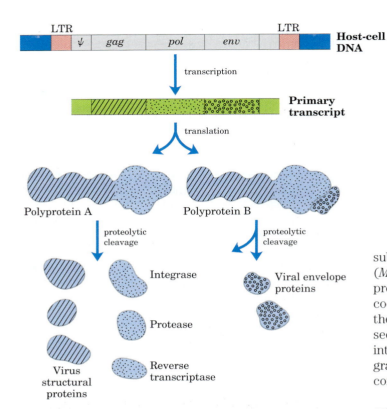

Howard Temin,
1934–1994

David Baltimore

**FIGURE 26–30 Structure and gene products of an integrated retroviral genome.** The long terminal repeats (LTRs) have sequences needed for the regulation and initiation of transcription. The sequence denoted Ψ is required for packaging of retroviral RNAs into mature viral particles. Transcription of the retroviral DNA produces a primary transcript encompassing the *gag*, *pol*, and *env* genes. Translation (Chapter 27) produces a polyprotein, a single long polypeptide derived from the *gag* and *pol* genes, which is cleaved into six distinct proteins. Splicing of the primary transcript yields an mRNA derived largely from the *env* gene, which is also translated into a polyprotein, then cleaved to generate viral envelope proteins.

The existence of reverse transcriptases in RNA viruses was predicted by Howard Temin in 1962, and the enzymes were ultimately detected by Temin and, independently, by David Baltimore in 1970. Their discovery aroused much attention as dogma-shaking proof that genetic information can flow "backward" from RNA to DNA.

Retroviruses typically have three genes: *gag* (derived from the historical designation *g*roup *a*ssociated *a*ntigen), *pol*, and *env* (Fig. 26–30). The transcript that contains *gag* and *pol* is translated into a long "polyprotein," a single large polypeptide that is cleaved into six proteins with distinct functions. The proteins derived from the *gag* gene make up the interior core of the viral particle. The *pol* gene encodes the protease that cleaves the long polypeptide, an integrase that inserts the viral DNA into the host chromosomes, and reverse transcriptase. Many reverse transcriptases have two

subunits, $\alpha$ and $\beta$. The *pol* gene specifies the $\beta$ subunit ($M_r$ 90,000), and the $\alpha$ subunit ($M_r$ 65,000) is simply a proteolytic fragment of the $\beta$ subunit. The *env* gene encodes the proteins of the viral envelope. At each end of the linear RNA genome are long terminal repeat (LTR) sequences of a few hundred nucleotides. Transcribed into the duplex DNA, these sequences facilitate integration of the viral chromosome into the host DNA and contain promoters for viral gene expression.

Reverse transcriptases catalyze three different reactions: (1) RNA-dependent DNA synthesis, (2) RNA degradation, and (3) DNA-dependent DNA synthesis. Like many DNA and RNA polymerases, reverse transcriptases contain $Zn^{2+}$. Each transcriptase is most active with the RNA of its own virus, but each can be used experimentally to make DNA complementary to a variety of RNAs. The DNA and RNA synthesis and RNA degradation activities use separate active sites on the protein. For DNA synthesis to begin, the reverse transcriptase requires a primer, a cellular tRNA obtained during an earlier infection and carried within the viral particle. This tRNA is base-paired at its 3' end with a complementary sequence in the viral RNA. The new DNA strand is synthesized in the 5'→3' direction, as in all RNA and DNA polymerase reactions. Reverse transcriptases, like RNA polymerases, do not have 3'→5' proofreading exonucleases. They generally have error rates of about 1 per 20,000 nucleotides added. An error rate this high is extremely unusual in DNA replication and appears to be a feature of most enzymes that replicate the genomes of RNA viruses. A consequence is a higher mutation rate and faster rate of viral evolution, which is a factor in the frequent appearance of new strains of disease-causing retroviruses.

Reverse transcriptases have become important reagents in the study of DNA-RNA relationships and in DNA cloning techniques. They make possible the synthesis of DNA complementary to an mRNA template, and synthetic DNA prepared in this manner, called **complementary DNA (cDNA),** can be used to clone cellular genes (see Fig. 9–14).

**FIGURE 26–31 Rous sarcoma virus genome.** The *src* gene encodes a tyrosine-specific protein kinase, one of a class of enzymes known to function in systems that affect cell division, cell-cell interactions, and intercellular communication (Chapter 12). The same gene is found in chicken DNA (the usual host for this virus) and in the genomes of many other eukaryotes, including humans. When associated with the Rous sarcoma virus, this oncogene is often expressed at abnormally high levels, contributing to unregulated cell division and cancer.

## Some Retroviruses Cause Cancer and AIDS

Retroviruses have featured prominently in recent advances in the molecular understanding of cancer. Most retroviruses do not kill their host cells but remain integrated in the cellular DNA, replicating when the cell divides. Some retroviruses, classified as RNA tumor viruses, contain an oncogene that can cause the cell to grow abnormally (see Fig. 12–47). The first retrovirus of this type to be studied was the Rous sarcoma virus (also called avian sarcoma virus; Fig. 26–31), named for F. Peyton Rous, who studied chicken tumors now known to be caused by this virus. Since the initial discovery of oncogenes by Harold Varmus and Michael Bishop, many dozens of such genes have been found in retroviruses.

The human immunodeficiency virus (HIV), which causes acquired immune deficiency syndrome (AIDS), is a retrovirus. Identified in 1983, HIV has an RNA genome with standard retroviral genes along with several other unusual genes (Fig. 26–32). Unlike many other retroviruses, HIV kills many of the cells it infects (principally T lymphocytes) rather than causing tumor formation. This gradually leads to suppression of the immune system in the host organism. The reverse transcriptase of HIV is even more error prone than other known reverse transcriptases—ten times more so—resulting in high mutation rates in this virus. One or more errors are generally made every time the viral genome is replicated, so any two viral RNA molecules are likely to differ.

Many modern vaccines for viral infections consist of one or more coat proteins of the virus, produced by methods described in Chapter 9. These proteins are not infectious on their own but stimulate the immune system to recognize and resist subsequent viral invasions (Chapter 5). Because of the high error rate of the HIV reverse transcriptase, the *env* gene in this virus (along with the rest of the genome) undergoes very rapid mutation, complicating the development of an effective vaccine. However, repeated cycles of cell invasion and replication are needed to propagate an HIV infection, so inhibition of viral enzymes offers promise as an effective therapy. The HIV protease is targeted by a class of drugs called protease inhibitors (see Box 6–3). Reverse transcriptase is the target of some additional drugs widely used to treat HIV-infected individuals (Box 26–2).

## Many Transposons, Retroviruses, and Introns May Have a Common Evolutionary Origin

Some well-characterized eukaryotic DNA transposons from sources as diverse as yeast and fruit flies have a structure very similar to that of retroviruses; these are sometimes called retrotransposons (Fig. 26–33). Retrotransposons encode an enzyme homologous to the retroviral reverse transcriptase, and their coding regions are flanked by LTR sequences. They transpose from one position to another in the cellular genome by means of an RNA intermediate, using reverse transcriptase to make a DNA copy of the RNA, followed by integration of the DNA at a new site. Most transposons in eukaryotes use this mechanism for transposition, distinguishing them from bacterial transposons, which move as DNA directly from one chromosomal location to another (see Fig. 25–43).

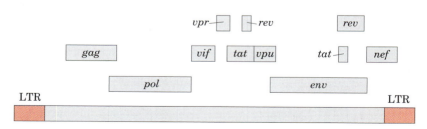

**FIGURE 26–32 The genome of HIV, the virus that causes AIDS.** In addition to the typical retroviral genes, HIV contains several small genes with a variety of functions (not identified here, and not all known). Some of these genes overlap (see Box 27–1). Alternative splicing mechanisms produce many different proteins from this small ($9.7 \times 10^3$ nucleotides) genome.

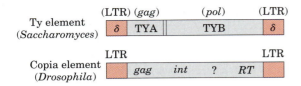

**FIGURE 26–33 Eukaryotic transposons.** The Ty element of the yeast *Saccharomyces* and the copia element of the fruit fly *Drosophila* serve as examples of eukaryotic transposons, which often have a structure similar to retroviruses but lack the *env* gene. The δ sequences of the Ty element are functionally equivalent to retroviral LTRs. In the copia element, *int* and *RT* are homologous to the integrase and reverse transcriptase segments, respectively, of the *pol* gene.

Retrotransposons lack an *env* gene and so cannot form viral particles. They can be thought of as defective viruses, trapped in cells. Comparisons between retroviruses and eukaryotic transposons suggest that reverse transcriptase is an ancient enzyme that predates the evolution of multicellular organisms.

Interestingly, many group I and group II introns are also mobile genetic elements. In addition to their self-splicing activities, they encode DNA endonucleases that promote their movement. During genetic exchanges between cells of the same species, or when DNA is introduced into a cell by parasites or by other means, these endonucleases promote insertion of the intron into an identical site in another DNA copy of a homologous gene that does not contain the intron, in a process termed **homing** (Fig. 26–34). Whereas group I intron homing is DNA-based, group II intron homing occurs through an RNA intermediate. The endonucleases of the group II introns have associated reverse transcriptase activity. The proteins can form complexes with the intron RNAs themselves, after the introns are spliced from the primary transcripts. Because the homing process involves insertion of the RNA intron into DNA and reverse transcription of the intron, the movement of these introns has been called retrohoming. Over time, every copy of a particular gene in a population may acquire the intron.

---

 **BOX 26–2    BIOCHEMISTRY IN MEDICINE**

### Fighting AIDS with Inhibitors of HIV Reverse Transcriptase

Research into the chemistry of template-dependent nucleic acid biosynthesis, combined with modern techniques of molecular biology, has elucidated the life cycle and structure of the human immunodeficiency virus, the retrovirus that causes AIDS. A few years after the isolation of HIV, this research resulted in the development of drugs capable of prolonging the lives of people infected by HIV.

The first drug to be approved for clinical use was AZT, a structural analog of deoxythymidine. AZT was first synthesized in 1964 by Jerome P. Horwitz. It failed as an anticancer drug (the purpose for which it was made), but in 1985 it was found to be a useful treatment for AIDS. AZT is taken up by T lymphocytes, immune system cells that are particularly vulnerable to HIV infection, and converted to AZT triphosphate. (AZT triphosphate taken directly would be ineffective, because it cannot cross the plasma membrane.) HIV's reverse transcriptase has a higher affinity for AZT triphosphate than for dTTP, and binding of AZT triphosphate to this enzyme competitively inhibits dTTP binding. When AZT is added to the 3' end of the growing DNA strand, lack of a 3' hydroxyl means that the DNA strand is terminated prematurely and viral DNA synthesis grinds to a halt.

AZT triphosphate is not as toxic to the T lymphocytes themselves, because *cellular* DNA polymerases have a lower affinity for this compound than for dTTP. At concentrations of 1 to 5 μM, AZT affects HIV reverse transcription but not most cellular DNA replication. Unfortunately, AZT appears to be toxic to the bone marrow cells that are the progenitors of erythrocytes, and many individuals taking AZT develop anemia. AZT can increase the survival time of people with advanced AIDS by about a year, and it delays the onset of AIDS in those who are still in the early stages of HIV infection. Some other AIDS drugs, such as dideoxyinosine (DDI), have a similar mechanism of action. Newer drugs target and inactivate the HIV protease. Because of the high error rate of HIV reverse transcriptase and the resulting rapid evolution of HIV, the most effective treatments of HIV infections use a combination of drugs directed at both the protease and the reverse transcriptase.

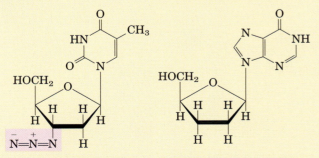

3'-Azido-2',3'-dideoxy-thymidine (AZT)

2',3'-Dideoxyinosine (DDI)

**FIGURE 26–34 Introns that move: homing and retrohoming.** Certain introns include a gene (shown in red) for enzymes that promote homing (type I introns) or retrohoming (type II introns). **(a)** The gene within the spliced intron is bound by a ribosome and translated. Type I homing introns specify a site-specific endonuclease, called a homing endonuclease. Type II retrohoming introns specify a protein with both endonuclease and reverse transcriptase activities.

**(b)** Homing. Allele *a* of a gene *X* containing a type I homing intron is present in a cell containing allele *b* of the same gene, which lacks the intron. The homing endonuclease produced by *a* cleaves *b* at the position corresponding to the intron in *a*, and double-strand break repair (recombination with allele *a*; see Fig. 25–31a) then creates a new copy of the intron in *b*. **(c)** Retrohoming. Allele *a* of gene *Y* contains a retrohoming type II intron; allele *b* lacks the intron. The spliced intron inserts itself into the coding strand of *b* in a reaction that is the reverse of the splicing that excised the intron from the primary transcript (see Fig. 26–15), except that here the insertion is into DNA rather than RNA. The noncoding DNA strand of *b* is then cleaved by the intron-encoded endonuclease/reverse transcriptase. This same enzyme uses the inserted RNA as a template to synthesize a complementary DNA strand. The RNA is then degraded by cellular ribonucleases and replaced with DNA.

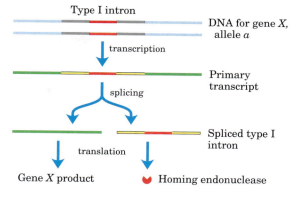

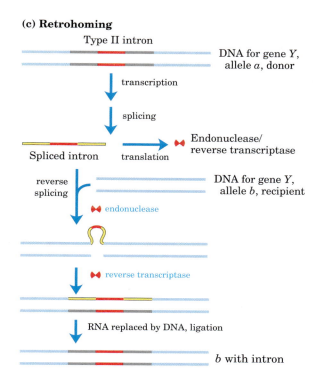

Much more rarely, the intron may insert itself into a new location in an unrelated gene. If this event does not kill the host cell, it can lead to the evolution and distribution of an intron in a new location. The structures and mechanisms used by mobile introns support the idea that at least some introns originated as molecular parasites whose evolutionary past can be traced to retroviruses and transposons.

## Telomerase Is a Specialized Reverse Transcriptase

Telomeres, the structures at the ends of linear eukaryotic chromosomes (see Fig. 24–9), generally consist of many tandem copies of a short oligonucleotide sequence. This sequence usually has the form $T_xG_y$ in one strand and $C_yA_x$ in the complementary strand, where $x$ and $y$ are typically in the range of 1 to 4 (p. 930). Telomeres vary in length from a few dozen base pairs in some ciliated protozoans to tens of thousands of base pairs in mammals. The TG strand is longer than its complement, leaving a region of single-stranded DNA of up to a few hundred nucleotides at the 3′ end.

The ends of a linear chromosome are not readily replicated by cellular DNA polymerases. DNA replication requires a template and primer, and beyond the end of a linear DNA molecule no template is available for the pairing of an RNA primer. Without a special mechanism for replicating the ends, chromosomes would be shortened somewhat in each cell generation. The enzyme **telomerase** solves this problem by adding telomeres to chromosome ends.

Although the existence of this enzyme may not be surprising, the mechanism by which it acts is remarkable and unprecedented. Telomerase, like some other enzymes described in this chapter, contains both RNA and protein components. The RNA component is about 150 nucleotides long and contains about 1.5 copies of the appropriate $C_yA_x$ telomere repeat. This region of the RNA acts as a template for synthesis of the $T_xG_y$ strand of the telomere. Telomerase thereby acts as a cellular reverse transcriptase that provides the active site for RNA-dependent DNA synthesis. Unlike retroviral reverse transcriptases, telomerase copies only a small segment of RNA that it carries within itself. Telomere synthesis requires the 3' end of a chromosome as primer and proceeds in the usual 5'→3' direction. Having syn-

thesized one copy of the repeat, the enzyme repositions to resume extension of the telomere (Fig. 26–35a).

After extension of the $T_xG_y$ strand by telomerase, the complementary $C_yA_x$ strand is synthesized by cellular DNA polymerases, starting with an RNA primer (see Fig. 25–13). The single-stranded region is protected by specific binding proteins in many lower eukaryotes, especially those species with telomeres of less than a few hundred base pairs. In higher eukaryotes (including mammals) with telomeres many thousands of base pairs long, the single-stranded end is sequestered in a specialized structure called a **T loop.** The single-stranded end is folded back and paired with its complement in the double-stranded portion of the telomere. The formation of a T loop involves invasion of the 3' end

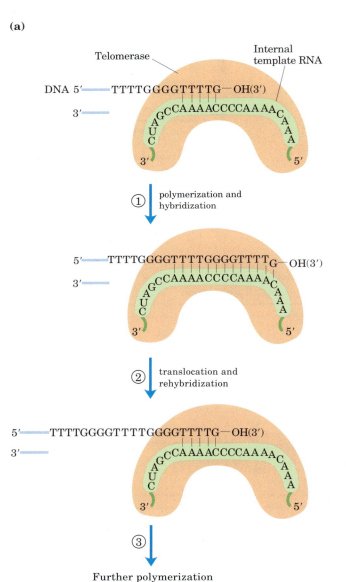

**(a)**

① polymerization and hybridization

② translocation and rehybridization

③

Further polymerization

**FIGURE 26–35 The TG strand and T loop of telomeres.** The internal template RNA of telomerase binds to and base-pairs with the DNA's TG primer (T$x$G$y$). ① Telomerase adds more T and G residues to the TG primer, then ② repositions the internal template RNA to allow ③ the addition of more T and G residues. The complementary strand is synthesized by cellular DNA polymerases (not shown). **(b)** Proposed structure of T loops in telomeres. The single-stranded tail synthesized by telomerase is folded back and paired with its complement in the duplex portion of the telomere. The telomere is bound by several telomere-binding proteins, including TRF1 and TRF2 (telomere repeat binding factors). **(c)** Electron micrograph of a T loop at the end of a chromosome isolated from a mouse hepatocyte. The bar at the bottom of the micrograph represents a length of 5,000 bp.

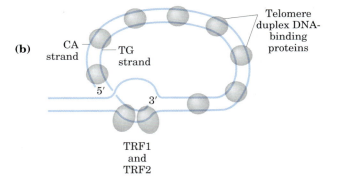

**(b)**

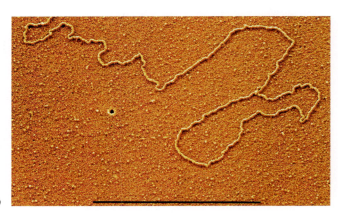

**(c)**

of the telomere's single strand into the duplex DNA, perhaps by a mechanism similar to the initiation of homologous genetic recombination (see Fig. 25–31). In mammals, the looped DNA is bound by two proteins, TRF1 and TRF2, with the latter protein involved in formation of the T loop. T loops protect the 3′ ends of chromosomes, making them inaccessible to nucleases and the enzymes that repair double-strand breaks (Fig. 26–35b).

In protozoans (such as *Tetrahymena*), loss of telomerase activity results in a gradual shortening of telomeres with each cell division, ultimately leading to the death of the cell line. A similar link between telomere length and cell senescence (cessation of cell division) has been observed in humans. In germ-line cells, which contain telomerase activity, telomere lengths are maintained; in somatic cells, which lack telomerase, they are not. There is a linear, inverse relationship between the length of telomeres in cultured fibroblasts and the age of the individual from whom the fibroblasts were taken: telomeres in human somatic cells gradually shorten as an individual ages. If the telomerase reverse transcriptase is introduced into human somatic cells in vitro, telomerase activity is restored and the cellular life span increases markedly.

Is the gradual shortening of telomeres a key to the aging process? Is our natural life span determined by the length of the telomeres we are born with? Further research in this area should yield some fascinating insights.

### Some Viral RNAs Are Replicated by RNA-Dependent RNA Polymerase

Some *E. coli* bacteriophages, including f2, MS2, R17, and Qβ, as well as some eukaryotic viruses (including influenza and Sindbis viruses, the latter associated with a form of encephalitis) have RNA genomes. The single-stranded RNA chromosomes of these viruses, which also function as mRNAs for the synthesis of viral proteins, are replicated in the host cell by an **RNA-dependent RNA polymerase (RNA replicase).** All RNA viruses—with the exception of retroviruses—must encode a protein with RNA-dependent RNA polymerase activity because the host cells do not possess this enzyme.

The RNA replicase of most RNA bacteriophages has a molecular weight of ~210,000 and consists of four subunits. One subunit ($M_r$ 65,000) is the product of the replicase gene encoded by the viral RNA and has the active site for replication. The other three subunits are host proteins normally involved in host-cell protein synthesis: the *E. coli* elongation factors Tu ($M_r$ 30,000) and Ts ($M_r$ 45,000) (which ferry amino acyl–tRNAs to the ribosomes) and the protein S1 (an integral part of the 30S ribosomal subunit).

These three host proteins may help the RNA replicase locate and bind to the 3′ ends of the viral RNAs.

RNA replicase isolated from Qβ-infected *E. coli* cells catalyzes the formation of an RNA complementary to the viral RNA, in a reaction equivalent to that catalyzed by DNA-dependent RNA polymerases. New RNA strand synthesis proceeds in the 5′→3′ direction by a chemical mechanism identical to that used in all other nucleic acid synthetic reactions that require a template. RNA replicase requires RNA as its template and will not function with DNA. It lacks a separate proofreading endonuclease activity and has an error rate similar to that of RNA polymerase. Unlike the DNA and RNA polymerases, RNA replicases are specific for the RNA of their own virus; the RNAs of the host cell are generally not replicated. This explains how RNA viruses are preferentially replicated in the host cell, which contains many other types of RNA.

### RNA Synthesis Offers Important Clues to Biochemical Evolution

The extraordinary complexity and order that distinguish living from inanimate systems are key manifestations of fundamental life processes. Maintaining the living state requires that *selected* chemical transformations occur very rapidly—especially those that use environmental energy sources and synthesize elaborate or specialized cellular macromolecules. Life depends on powerful and selective catalysts—enzymes—and on informational systems capable of both securely storing the blueprint for these enzymes and accurately reproducing the blueprint for generation after generation. Chromosomes encode the blueprint not for the cell but for the enzymes that construct and maintain the cell. The parallel demands for information and catalysis present a classic conundrum: what came first, the information needed to specify structure or the enzymes needed to maintain and transmit the information?

The unveiling of the structural and functional complexity of RNA led Carl Woese, Francis Crick, and Leslie Orgel to propose in the 1960s that this macromolecule might serve as both information carrier and catalyst. The discovery of catalytic RNAs took this proposal from

Carl Woese

Francis Crick

Leslie Orgel

conjecture to hypothesis and has led to widespread speculation that an "RNA world" might have been important in the transition from prebiotic chemistry to life (see Fig. 1–34). The parent of all life on this planet, in the sense that it could reproduce itself across the generations from the origin of life to the present, might have been a self-replicating RNA or a polymer with equivalent chemical characteristics.

How might a self-replicating polymer come to be? How might it maintain itself in an environment where the precursors for polymer synthesis are scarce? How could evolution progress from such a polymer to the modern DNA-protein world? These difficult questions can be addressed by careful experimentation, providing clues about how life on Earth began and evolved.

The probable origin of purine and pyrimidine bases is suggested by experiments designed to test hypotheses about prebiotic chemistry (pp. 32–33). Beginning with simple molecules thought to be present in the early atmosphere ($CH_4$, $NH_3$, $H_2O$, $H_2$), electrical discharges such as lightning generate, first, more reactive molecules such as HCN and aldehydes, then an array of amino acids and organic acids (see Fig. 1–33). When molecules such as HCN become abundant, purine and pyrimidine bases are synthesized in detectable amounts. Remarkably, a concentrated solution of ammonium cyanide, refluxed for a few days, generates adenine in yields of up to 0.5% (Fig. 26–36). Adenine may well have been the first and most abundant nucleotide constituent to appear on Earth. Intriguingly, most enzyme cofactors contain adenosine as part of their structure, although it plays no direct role in the cofactor function (see Fig. 8–41). This may suggest an evolutionary relationship, based on the simple synthesis of adenine from cyanide.

The RNA world hypothesis requires a nucleotide polymer to reproduce itself. Can a ribozyme bring about its own synthesis in a template-directed manner? The self-splicing rRNA intron of *Tetrahymena* (Fig. 26–26) catalyzes the reversible attack of a guanosine residue on the 5′ splice junction (Fig. 26–37). If the 5′ splice site and the internal guide sequence are removed from the intron, the rest of the intron can bind RNA strands paired with short oligonucleotides. Part of the remaining intact intron effectively acts as a template for the alignment and ligation of the short oligonucleotides. The reaction is in essence a reversal of the attack of guanosine on the 5′ splice junction, but the result is the synthesis of long RNA polymers from short ones, with the sequence of the product defined by an RNA template.

A self-replicating polymer would quickly use up available supplies of precursors provided by the relatively slow processes of prebiotic chemistry. Thus, from an early stage in evolution, metabolic pathways would be required to generate precursors efficiently, with the synthesis of precursors presumably catalyzed by ribozymes. The extant ribozymes found in nature have a limited repertoire of catalytic functions, and of the ribozymes that may once have existed, no trace is left. To explore the RNA world hypothesis more deeply, we need to know whether RNA has the potential to catalyze the many different reactions needed in a primitive system of metabolic pathways.

The search for RNAs with new catalytic functions has been aided by the development of a method that rapidly searches pools of random polymers of RNA and extracts those with particular activities: SELEX is nothing less than accelerated evolution in a test tube (Box 26–3). It has been used to generate RNA molecules that bind to amino acids, organic dyes, nucleotides, cyanocobalamin, and other molecules. Researchers have isolated ribozymes that catalyze ester and amide bond formation, $S_N2$ reactions, metallation of (addition of metal ions to) porphyrins, and carbon–carbon bond formation. The evolution of enzymatic cofactors with nucleotide "handles" that facilitate their binding to ribozymes might have further expanded the repertoire of chemical processes available to primitive metabolic systems.

As we shall see in the next chapter, some natural RNA molecules catalyze the formation of peptide bonds, offering an idea of how the RNA world might have been transformed by the greater catalytic potential of proteins. The synthesis of proteins would have been a major event in the evolution of the RNA world, but would also have hastened its demise. The information-carrying role of RNA may have passed to DNA because DNA is chemically more stable. RNA replicase and reverse transcriptase may be modern versions of enzymes that once played important roles in making the transition to the modern DNA-based system.

Molecular parasites may also have originated in an RNA world. With the appearance of the first inefficient self-replicators, transposition could have been a potentially important alternative to replication as a strategy for successful reproduction and survival. Early parasitic RNAs would simply hop into a self-replicating molecule via catalyzed transesterification, then passively undergo replication. Natural selection would have driven transposition to become site-specific, targeting sequences that did not interfere with the catalytic activities of the

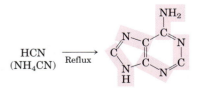

**FIGURE 26–36 Possible prebiotic synthesis of adenine from ammonium cyanide.** Adenine is derived from five molecules of cyanide, denoted by shading.

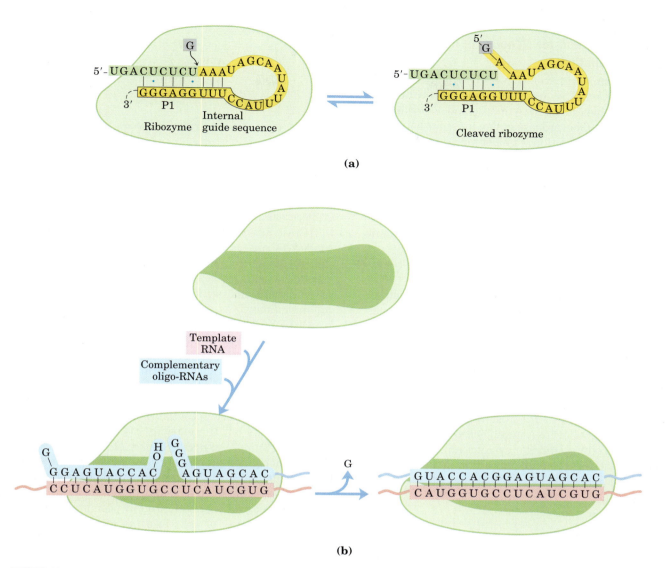

**(a)**

**(b)**

**FIGURE 26–37  RNA-dependent synthesis of an RNA polymer from oligonucleotide precursors. (a)** The first step in the removal of the self-splicing group I intron of the rRNA precursor of *Tetrahymena* is reversible attack of a guanosine residue on the 5′ splice site. Only P1, the region of the ribozyme that includes the internal guide sequence (boxed) and the 5′ splice site, is shown in detail; the rest of the ribozyme is represented as a green blob. The complete secondary structure of the ribozyme is shown in Figure 26–26. **(b)** If P1 is removed (shown as the darker green "hole"), the ribozyme retains both its three-dimensional shape and its catalytic capacity. A new RNA molecule added in vitro can bind to the ribozyme in the same manner as does the internal guide sequence of P1 in **(a).** This provides a template for further RNA polymerization reactions when oligonucleotides complementary to the added RNA base-pair with it. The ribozyme can link these oligonucleotides in a process equivalent to the *reversal* of the reaction in **(a).** Although only one such reaction is shown in **(b),** repeated binding and catalysis can result in the RNA-dependent synthesis of long RNA polymers.

host RNA. Replicators and RNA transposons could have existed in a primitive symbiotic relationship, each contributing to the evolution of the other. Modern introns, retroviruses, and transposons may all be vestiges of a "piggy-back" strategy pursued by early parasitic RNAs. These elements continue to make major contributions to the evolution of their hosts.

Although the RNA world remains a hypothesis, with many gaps yet to be explained, experimental evidence supports a growing list of its key elements. Further experimentation should increase our understanding. Important clues to the puzzle will be found in the workings of fundamental chemistry, in living cells, and perhaps on other planets.

**BOX 26-3**    **WORKING IN BIOCHEMISTRY**

## The SELEX Method for Generating RNA Polymers with New Functions

SELEX (*systematic evolution of ligands by exponential enrichment*) is used to generate **aptamers,** oligonucleotides selected to tightly bind a specific molecular target. The process is generally automated to allow rapid identification of one or more aptamers with the desired binding specificity.

Figure 1 illustrates how SELEX is used to select an RNA species that binds tightly to ATP. In step ①, a random mixture of RNA polymers is subjected to "unnatural selection" by passing it through a resin to which ATP is attached. The practical limit for the complexity of an RNA mixture in SELEX is about $10^{15}$ different sequences, which allows for the complete randomization of 25 nucleotides ($4^{25} = 10^{15}$). When longer RNAs are used, the RNA pool used to initiate the search does not include all possible sequences. ② RNA polymers that pass through the column are discarded; ③ those that bind to ATP are washed from the column with salt solution and collected. ④ The collected RNA polymers are amplified by reverse transcriptase to make many DNA complements to the selected RNAs; then an RNA polymerase makes many RNA complements of the resulting DNA molecules. ⑤ This new pool of RNA is subjected to the same selection procedure, and the cycle is repeated a dozen or more times. At the end, only a few aptamers, in this

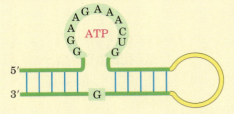

**FIGURE 2** RNA aptamer that binds ATP. The shaded nucleotides are those required for the binding activity.

case RNA sequences with considerable affinity for ATP, remain.

Critical sequence features of an RNA aptamer that binds ATP are shown in Figure 2; molecules with this general structure bind ATP (and other adenosine nucleotides) with $K_d < 50$ $\mu$M. Figure 3 presents the three-dimensional structure of a 36 nucleotide RNA aptamer (shown as a complex with AMP) generated by SELEX. This RNA has the backbone structure shown in Figure 2.

In addition to its use in exploring the potential functionality of RNA, SELEX has an important practical side in identifying short RNAs with pharmaceutical uses. Finding an aptamer that binds specifically to every potential therapeutic target may be impossible, but the capacity of SELEX to rapidly select and amplify a specific oligonucleotide sequence from a highly complex pool of sequences makes this a promising approach for the generation of new therapies. For example, one could select an RNA that binds tightly to a receptor protein prominent in the plasma membrane of cells in a particular cancerous tumor. Blocking the activity of the receptor, or targeting a toxin to the tumor cells by attaching it to the aptamer, would kill the cells. SELEX also has been used to select DNA aptamers that detect anthrax spores. Many other promising applications are under development. ■

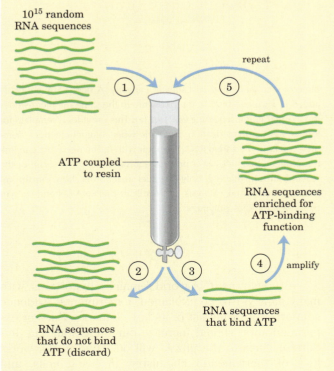

**FIGURE 1** The SELEX procedure.

10$^{15}$ random RNA sequences

repeat

①

⑤

ATP coupled to resin

RNA sequences enriched for ATP-binding function

②

③

④ amplify

RNA sequences that bind ATP

RNA sequences that do not bind ATP (discard)

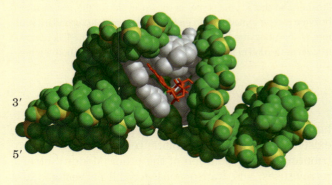

**FIGURE 3** (Derived from PDB ID 1RAW.) RNA aptamer bound to AMP. The bases of the conserved nucleotides (forming the binding pocket) are white; the bound AMP is red.

## SUMMARY 26.3 RNA-Dependent Synthesis of RNA and DNA

- RNA-dependent DNA polymerases, also called reverse transcriptases, were first discovered in retroviruses, which must convert their RNA genomes into double-stranded DNA as part of their life cycle. These enzymes transcribe the viral RNA into DNA, a process that can be used experimentally to form complementary DNA.

- Many eukaryotic transposons are related to retroviruses, and their mechanism of transposition includes an RNA intermediate.

- Telomerase, the enzyme that synthesizes the telomere ends of linear chromosomes, is a specialized reverse transcriptase that contains an internal RNA template.

- RNA-dependent RNA polymerases, such as the replicases of RNA bacteriophages, are template-specific for the viral RNA.

- The existence of catalytic RNAs and pathways for the interconversion of RNA and DNA has led to speculation that an important stage in evolution was the appearance of an RNA (or an equivalent polymer) that could catalyze its own replication. The biochemical potential of RNAs can be explored by SELEX, a method for rapidly selecting RNA sequences with particular binding or catalytic properties.

## Key Terms

*Terms in bold are defined in the glossary.*

| | | |
|---|---|---|
| **transcription**  995 | **repressor**   1001 | **reverse transcriptase**   1021 |
| **messenger RNA (mRNA)**   995 | **footprinting**   1002 | **retrovirus**   1021 |
| **transfer RNA (tRNA)**   995 | transcription factors   1003 | **complementary DNA (cDNA)**   1022 |
| **ribosomal RNA (rRNA)**   995 | **ribozymes**   1007 | homing   1024 |
| DNA-dependent RNA polymerase   996 | **primary transcript**   1007 | telomerase   1025 |
| **promoter**   998 | **RNA splicing**   1007 | RNA-dependent RNA polymerase (RNA replicase)   1027 |
| **consensus sequence**   998 | 5′ cap   1008 | **aptamer**   1030 |
| **cAMP receptor protein (CRP)**   1001 | **spliceosome**   1010 | |
| | **poly(A) tail**   1011 | |

## Further Reading

### General

**Jacob, F. & Monod, J.** (1961) Genetic regulatory mechanisms in the synthesis of proteins. *J. Mol. Biol.* **3,** 318–356.

A classic article that introduced many important ideas.

**Lodish, H., Berk, A., Matsudaira, P., Kaiser, C.A., Krieger, M., Scott, M.P., Zipursky, S.L., & Darnell, J.** (2003) *Molecular Cell Biology,* 5th edn, W. H. Freeman & Company, New York.

### DNA-Directed RNA Synthesis

**Conaway, J.W. & Conaway, R.C.** (1999) Transcription elongation and human disease. *Annu. Rev. Biochem.* **68,** 301–320.

**Conaway, J.W., Shilatifard, A., Dvir, A., & Conaway, R.C.** (2000) Control of elongation by RNA polymerase II. *Trends Biochem. Sci.* **25,** 375–380.

A particularly good summary of what is known about elongation factors.

**DeHaseth, P.L., Zupancic, M.L., & Record, M.T., Jr.** (1998) RNA polymerase-promoter interactions: the comings and goings of RNA polymerase. *J. Bacteriol.* **180,** 3019–3025.

**Friedberg, E.C.** (1996) Relationships between DNA repair and transcription. *Annu. Rev. Biochem.* **65,** 15–42.

**Kornberg, R.D.** (1996) RNA polymerase II transcription control. *Trends Biochem. Sci.* **21,** 325–327.

Introduction to an issue of *Trends in Biochemical Sciences* that is devoted to RNA polymerase II.

**Mooney, R.A., Artsimovitch, I., & Landick, R.** (1998) Informational processing by RNA polymerase: recognition of regulatory signals during RNA chain elongation. *J. Bacteriol.* **180,** 3265–3275.

**Murakami, K.S. & Darst, S.A.** (2003) Bacterial RNA polymerases: the whole story. *Curr. Opin. Struct. Biol.* **13,** 31–39.

This article and the two listed below explore the wealth of new structural information and what it tells us about RNA polymerase function.

**Woychik, N.A. & Hampsey, M.** (2002) The RNA polymerase II machinery: structure illuminates function. *Cell* **108,** 453–463.

**Young, B.A., Gruber, T.M., & Gross, C.A.** (2002) Views of transcription initiation. *Cell* **109,** 417–420.

### RNA Processing

**Beelman, C.A. & Parker, R.** (1995) Degradation of mRNA in eukaryotes. *Cell* **81,** 179–183.

Brow, D.A. (2002) Allosteric cascade of spliceosome activation. *Annu. Rev. Genet.* **36,** 333–360.

Chevalier, B.S. & Stoddard, B.L. (2001) Homing endonucleases: structural and functional insight into the catalysts of intron/intein mobility. *Nucleic Acid Res.* **29,** 3757–3774.

Curcio, M.J. & Belfort, M. (1996) Retrohoming: cDNA-mediated mobility of group II introns requires a catalytic RNA. *Cell* **84,** 9–12.

Frank, D.N. & Pace, N.R. (1998) Ribonuclease P: unity and diversity in a tRNA-processing ribozyme. *Annu. Rev. Biochem.* **67,** 153–180.

Jensen, T.H., Dower, K., Libri, D., & Rosbash, M. (2003) Early formation of mRNP: license for export or quality control? *Mol. Cell* **11,** 1129–1138.

   A good summary of current ideas about the coupled processing and transport of eukaryotic mRNAs.

Kushner, S.R. (2002) mRNA decay in *Escherichia coli* comes of age. *J. Bacteriol.* **184,** 4658–4665.

Narlikar, G.J. & Herschlag, D. (1997) Mechanistic aspects of enzymatic catalysis: lessons from comparison of RNA and protein enzymes. *Annu. Rev. Biochem.* **66,** 19–59.

Proudfoot, N.J., Furger, A., & Dye, M.J. (2002) Integrating mRNA processing with transcription. *Cell* **108,** 501–512.

   A description of current evidence for how processing is linked to the CTD of RNA polymerase II.

Sarkar, N. (1997) Polyadenylation of mRNA in prokaryotes. *Annu. Rev. Biochem.* **66,** 173–197.

Staley, J.P. & Guthrie, C. (1988) Mechanical devices of the spliceosome—motors, clocks, springs, and things. *Cell* **92,** 315–326.

**RNA-Directed RNA or DNA Synthesis**

Bishop, J.M. (1991) Molecular themes in oncogenesis. *Cell* **64,** 235–248.

   A good overview of oncogenes; it introduces a series of more detailed reviews included in the same issue of *Cell.*

Blackburn, E.H. (1992) Telomerases. *Annu. Rev. Biochem.* **61,** 113–129.

Boeke, J.D. & Devine, S.E. (1998) Yeast retrotransposons: finding a nice, quiet neighborhood. *Cell* **93,** 1087–1089.

Collins, K. (1999) Ciliate telomerase biochemistry. *Annu. Rev. Biochem.* **68,** 187–218.

Frankel, A.D. & Young, J.A.T. (1998) HIV-1: fifteen proteins and an RNA. *Annu. Rev. Biochem.* **67,** 1–25.

Greider, C.W. (1996) Telomere length regulation. *Annu. Rev. Biochem.* **65,** 337–365.

Griffith, J.D., Comeau, L., Rosenfield, S., Stansel, R.M., Bianchi, A., Moss, H., & de Lange, T. (1999) Mammalian telomeres end in a large duplex loop. *Cell* **97,** 503–514.

Lingner, J. & Cech, T.R. (1998) Telomerase and chromosome end maintenance. *Curr. Opin. Genet. Dev.* **8,** 226–232.

Temin, H.M. (1976) The DNA provirus hypothesis: the establishment and implications of RNA-directed DNA synthesis. *Science* **192,** 1075–1080.

   Discussion of the original proposal for reverse transcription in retroviruses.

Zakian, V.A. (1995) Telomeres: beginning to understand the end. *Science* **270,** 1601–1607.

**Ribozymes and Evolution**

Bittker, J.A., Phillips, K.J., & Liu, D.R. (2002) Recent advances in the in vitro evolution of nucleic acids. *Curr. Opin. Chem. Biol.* **6,** 367–374.

DeRose, V.J. (2002) Two decades of RNA catalysis. *Chem. Biol.* **9,** 961–969.

Johnston, W.K., Unrau, P.J., Lawrence, M.S., Glasner, M.E., & Bartel, D.P. (2001) RNA-catalyzed RNA polymerization: accurate and general RNA-templated primer extension. *Science* **292,** 1319–1325.

   Review of progress toward the laboratory evolution of a self-replicating RNA.

Joyce, G.F. (2002) The antiquity of RNA-based evolution. *Nature* **418,** 214–221.

Wilson, D.S. & Szostak, J.W. (1999) In vitro selection of functional nucleic acids. *Annu. Rev. Biochem.* **68,** 611–648.

Yarus, M. (2002) Primordial genetics: phenotype of the ribocyte. *Annu. Rev. Genet.* **36,** 125–151.

   Detailed speculations about what an RNA-based life form might have been like, and a good summary of the research behind the speculations.

## Problems

**1. RNA Polymerase** (a) How long would it take for the *E. coli* RNA polymerase to synthesize the primary transcript for the *E. coli* genes encoding the enzymes for lactose metabolism (the 5,300 bp *lac* operon, considered in Chapter 28)? (b) How far along the DNA would the transcription "bubble" formed by RNA polymerase move in 10 seconds?

**2. Error Correction by RNA Polymerases** DNA polymerases are capable of editing and error correction, whereas the capacity for error correction in RNA polymerases appears to be quite limited. Given that a single base error in either replication or transcription can lead to an error in protein synthesis, suggest a possible biological explanation for this striking difference.

**3. RNA Posttranscriptional Processing** Predict the likely effects of a mutation in the sequence (5′)AAUAAA in a eukaryotic mRNA transcript.

**4. Coding versus Template Strands** The RNA genome of phage Qβ is the nontemplate or coding strand, and when introduced into the cell it functions as an mRNA. Suppose the RNA replicase of phage Qβ synthesized primarily template-strand RNA and uniquely incorporated this, rather than nontemplate strands, into the viral particles. What would be the fate of the template strands when they entered a new cell? What enzyme would such a template-strand virus need to include in the viral particles for successful invasion of a host cell?

**5. The Chemistry of Nucleic Acid Biosynthesis** Describe three properties common to the reactions catalyzed by DNA polymerase, RNA polymerase, reverse transcriptase, and RNA replicase. How is the enzyme polynucleotide phosphorylase similar to and different from these three enzymes?

**6. RNA Splicing** What is the minimum number of transesterification reactions needed to splice an intron from an mRNA transcript? Explain.

**7. RNA Genomes** The RNA viruses have relatively small genomes. For example, the single-stranded RNAs of retroviruses have about 10,000 nucleotides and the Qβ RNA is only 4,220 nucleotides long. Given the properties of reverse transcriptase and RNA replicase described in this chapter, can you suggest a reason for the small size of these viral genomes?

**8. Screening RNAs by SELEX** The practical limit for the number of different RNA sequences that can be screened in a SELEX experiment is $10^{15}$. (a) Suppose you are working with oligonucleotides 32 nucleotides in length. How many sequences exist in a randomized pool containing every sequence possible? (b) What percentage of these can be screened in a SELEX experiment? (c) Suppose you wish to select an RNA molecule that catalyzes the hydrolysis of a particular ester. From what you know about catalysis (Chapter 6), propose a SELEX strategy that might allow you to select the appropriate catalyst.

**9. Slow Death** The death cap mushroom, *Amanita phalloides,* contains several dangerous substances, including the lethal α-amanitin. This toxin blocks RNA elongation in consumers of the mushroom by binding to eukaryotic RNA polymerase II with very high affinity; it is deadly in concentrations as low as $10^{-8}$ M. The initial reaction to ingestion of the mushroom is gastrointestinal distress (caused by some of the other toxins). These symptoms disappear, but about 48 hours later, the mushroom-eater dies, usually from liver dysfunction. Speculate on why it takes this long for α-amanitin to kill.

**10. Detection of Rifampicin-Resistant Strains of Tuberculosis** Rifampicin is an important antibiotic used to treat tuberculosis, as well as other mycobacterial diseases.

Some strains of *Mycobacterium tuberculosis,* the causative agent of tuberculosis, are resistant to rifampicin. These strains become resistant through mutations that alter the *rpoB* gene, which encodes the β subunit of the RNA polymerase. Rifampicin cannot bind to the mutant RNA polymerase and so is unable to block the initiation of transcription. DNA sequences from a large number of rifampicin-resistant *M. tuberculosis* strains have been found to have mutations in a specific 69 bp region of *rpoB*. One well-characterized strain with rifampicin resistance has a single base pair alteration in *rpoB* that results in a single amino acid substitution in the β subunit: a His residue is replaced by an Asp residue.

(a) Based on your knowledge of protein chemistry (Chapters 3 and 4), suggest a technique that would allow detection of the rifampicin-resistant strain containing this particular mutant protein.

(b) Based on your knowledge of nucleic acid chemistry (Chapter 8), suggest a technique to identify the mutant form of *rpoB.*

## Biochemistry on the Internet

**11. The Ribonuclease Gene** Human pancreatic ribonuclease has 128 amino acid residues.

(a) What is the minimum number of nucleotide pairs required to code for this protein?

(b) The mRNA expressed in human pancreatic cells was copied with reverse transcriptase to create a "library" of human DNA. The sequence of the mRNA coding for human pancreatic ribonuclease was determined by sequencing the complementary DNA (cDNA) from this library that included an open reading frame for the protein. Use the Entrez database system (www.ncbi.nlm.nih.gov/Entrez) to find the published sequence of this mRNA (search the nucleotide database for accession number D26129). What is the length of this mRNA?

(c) How can you account for the discrepancy between the size you calculated in (a) and the actual length of the mRNA?

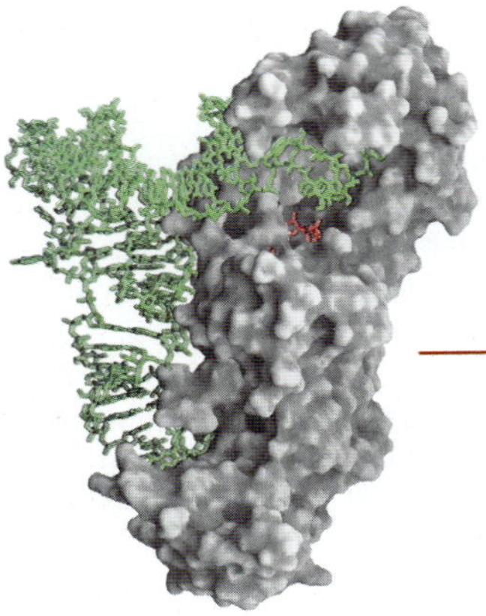

# PROTEIN METABOLISM

Obviously, Harry [Noller]'s finding doesn't speak to how life started, and it doesn't explain what came before RNA. But as part of the continually growing body of circumstantial evidence that there was a life form before us on this planet, from which we emerged—boy, it's very strong!

—*Gerald Joyce, quoted in commentary in* Science, *1992*

**P**roteins are the end products of most information pathways. A typical cell requires thousands of different proteins at any given moment. These must be synthesized in response to the cell's current needs, transported (targeted) to their appropriate cellular locations, and degraded when no longer needed.

An understanding of protein synthesis, the most complex biosynthetic process, has been one of the greatest challenges in biochemistry. Eukaryotic protein synthesis involves more than 70 different ribosomal proteins; 20 or more enzymes to activate the amino acid precursors; a dozen or more auxiliary enzymes and other protein factors for the initiation, elongation, and termination of polypeptides; perhaps 100 additional enzymes for the final processing of different proteins; and 40 or more kinds of transfer and ribosomal RNAs. Overall, almost 300 different macromolecules cooperate to synthesize polypeptides. Many of these macromolecules are organized into the complex three-dimensional structure of the ribosome.

To appreciate the central importance of protein synthesis, consider the cellular resources devoted to this process. Protein synthesis can account for up to 90% of

the chemical energy used by a cell for all biosynthetic reactions. Every prokaryotic and eukaryotic cell contains from several to thousands of copies of many different proteins and RNAs. The 15,000 ribosomes, 100,000 molecules of protein synthesis–related protein factors and enzymes, and 200,000 tRNA molecules in a typical bacterial cell can account for more than 35% of the cell's dry weight.

Despite the great complexity of protein synthesis, proteins are made at exceedingly high rates. A polypeptide of 100 residues is synthesized in an *Escherichia coli* cell (at 37 °C) in about 5 seconds. Synthesis of the thousands of different proteins in a cell is tightly regulated, so that just enough copies are made to match the current metabolic circumstances. To maintain the appropriate mix and concentration of proteins, the targeting and degradative processes must keep pace with synthesis. Research is gradually uncovering the finely coordinated cellular choreography that guides each protein to its proper cellular location and selectively degrades it when it is no longer required.

The study of protein synthesis offers another important reward: a look at a world of RNA catalysts that may have existed before the dawn of life "as we know it." Researchers have elucidated the structure of bacterial ribosomes, revealing the workings of cellular protein synthesis in beautiful molecular detail. And what did they find? Proteins are synthesized by a gigantic RNA enzyme!

## 27.1 The Genetic Code

Three major advances set the stage for our present knowledge of protein biosynthesis. First, in the early 1950s, Paul Zamecnik and his colleagues designed a set of experiments to investigate where in the cell proteins are synthesized. They injected radioactive amino acids into rats and, at different time intervals after the injec-

Paul Zamecnik

tion, removed the liver, homogenized it, fractionated the homogenate by centrifugation, and examined the subcellular fractions for the presence of radioactive protein. When hours or days were allowed to elapse after injection of the labeled amino acids, *all* the subcellular fractions contained labeled proteins. However, when only minutes had elapsed, labeled protein appeared only in a fraction containing small ribonucleoprotein particles. These particles, visible in animal tissues by electron microscopy, were therefore identified as the site of protein synthesis from amino acids, and later were named ribosomes (Fig. 27–1).

The second key advance was made by Mahlon Hoagland and Zamecnik, when they found that amino acids were "activated" when incubated with ATP and the cytosolic fraction of liver cells. The amino acids became attached to a heat-stable soluble RNA of the type that had been discovered and characterized by Robert Holley and later called transfer RNA (tRNA), to form **aminoacyl-tRNAs.** The enzymes that catalyze this process are the **aminoacyl-tRNA synthetases.**

The third advance resulted from Francis Crick's reasoning on how the genetic information encoded in the 4-letter language of nucleic acids could be translated into

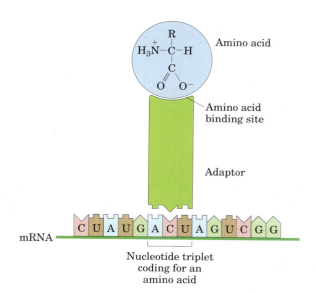

FIGURE 27–2 Crick's adaptor hypothesis. Today we know that the amino acid is covalently bound at the 3' end of a tRNA molecule and that a specific nucleotide triplet elsewhere in the tRNA interacts with a particular triplet codon in mRNA through hydrogen bonding of complementary bases.

the 20-letter language of proteins. A small nucleic acid (perhaps RNA) could serve the role of an adaptor, one part of the adaptor molecule binding a specific amino acid and another part recognizing the nucleotide sequence encoding that amino acid in an mRNA (Fig. 27–2). This idea was soon verified. The tRNA adaptor "translates" the nucleotide sequence of an mRNA into the amino acid sequence of a polypeptide. The overall process of mRNA-guided protein synthesis is often referred to simply as **translation.**

These three developments soon led to recognition of the major stages of protein synthesis and ultimately to the elucidation of the genetic code that specifies each amino acid.

## The Genetic Code Was Cracked Using Artificial mRNA Templates

By the 1960s it had long been apparent that at least three nucleotide residues of DNA are necessary to encode each amino acid. The four code letters of DNA (A, T, G, and C) in groups of two can yield only $4^2 = 16$ different combinations, insufficient to encode 20 amino acids. Groups of three, however, yield $4^3 = 64$ different combinations.

Several key properties of the genetic code were established in early genetic studies (Figs 27–3, 27–4). A **codon** is a triplet of nucleotides that codes for a specific amino acid. Translation occurs in such a way that these nucleotide triplets are read in a successive, nonoverlapping fashion. A specific first codon in the

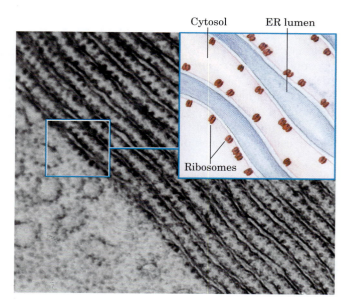

FIGURE 27–1 Ribosomes and endoplasmic reticulum. Electron micrograph and schematic drawing of a portion of a pancreatic cell, showing ribosomes attached to the outer (cytosolic) face of the endoplasmic reticulum (ER). The ribosomes are the numerous small dots bordering the parallel layers of membranes.

Nonoverlapping code
A U A C G A G U C _ _ _ _
1    2    3

Overlapping code
A U A C G A G U C
1
2
3

**FIGURE 27–3 Overlapping versus nonoverlapping genetic codes.** In a nonoverlapping code, codons (numbered consecutively) do not share nucleotides. In an overlapping code, some nucleotides in the mRNA are shared by different codons. In a triplet code with maximum overlap, many nucleotides, such as the third nucleotide from the left (A), are shared by three codons. Note that in an overlapping code, the triplet sequence of the first codon limits the possible sequences for the second codon. A nonoverlapping code provides much more flexibility in the triplet sequence of neighboring codons and therefore in the possible amino acid sequences designated by the code. The genetic code used in all living systems is now known to be nonoverlapping.

sequence establishes the **reading frame,** in which a new codon begins every three nucleotide residues. There is no punctuation between codons for successive amino acid residues. The amino acid sequence of a protein is defined by a linear sequence of contiguous triplets. In principle, any given single-stranded DNA or mRNA sequence has three possible reading frames. Each reading frame gives a different sequence of codons (Fig. 27–5), but only one is likely to encode a given protein. A key question remained: what were the three-letter code words for each amino acid?

In 1961 Marshall Nirenberg and Heinrich Matthaei reported the first breakthrough. They incubated synthetic polyuridylate, poly(U), with an *E. coli* extract, GTP, ATP, and a mixture of the 20 amino acids in 20 different tubes, each tube containing a different radioactively labeled amino acid. Because poly(U) mRNA is made up of many successive UUU triplets, it should promote the synthesis of a polypeptide containing only the amino acid encoded

Marshall Nirenberg

by the triplet UUU. A radioactive polypeptide was indeed formed in only one of the 20 tubes, the one containing radioactive phenylalanine. Nirenberg and Matthaei therefore concluded that the triplet codon UUU encodes phenylalanine. The same approach revealed that polycytidylate, poly(C), encodes a polypeptide containing only proline (polyproline), and polyadenylate, poly(A), encodes polylysine. Polyguanylate did not generate any polypeptide in this experiment because it spontaneously forms tetraplexes (see Fig. 8–22) that cannot be bound by ribosomes.

The synthetic polynucleotides used in such experiments were prepared with polynucleotide phosphorylase (p. 1020), which catalyzes the formation of RNA polymers starting from ADP, UDP, CDP, and GDP. This enzyme requires no template and makes polymers with a base composition that directly reflects the relative concentrations of the nucleoside 5′-diphosphate precursors in the medium. If polynucleotide phosphorylase is presented with UDP only, it makes only poly(U). If it is presented with a mixture of five parts ADP and one part CDP, it makes a polymer in which about five-sixths of the residues are adenylate and one-sixth are cytidylate. This random polymer is likely to have many triplets of the sequence AAA, smaller numbers of AAC, ACA, and CAA triplets, relatively few ACC, CCA, and CAC triplets, and very few CCC triplets (Table 27–1). Using a variety of artificial mRNAs made by polynucleotide phosphorylase from different starting mixtures of ADP, GDP, UDP, and CDP, investigators soon identified the base compositions of the triplets coding for almost all the amino acids. Although these experiments revealed the base composition of the coding triplets, they could not reveal the sequence of the bases.

**FIGURE 27–4 The triplet, nonoverlapping code.** Evidence for the general nature of the genetic code came from many types of experiments, including genetic experiments on the effects of deletion and insertion mutations. Inserting or deleting one base pair (shown here in the mRNA transcript) alters the sequence of triplets in a nonoverlapping code; all amino acids coded by the mRNA following the change are affected. Combining insertion and deletion mutations affects some amino acids but can eventually restore the correct amino acid sequence. Adding or subtracting three nucleotides (not shown) leaves the remaining triplets intact, providing evidence that a codon has three, rather than four or five, nucleotides. The triplet codons shaded in gray are those transcribed from the original gene; codons shaded in blue are new codons resulting from the insertion or deletion mutations.

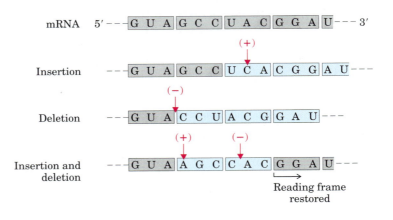

Reading frame 1  5′ ---U U C U C G G A C C U G G A G A U U C A C A G U --- 3′

Reading frame 2  ---U U C U C G G A C C U G G A G A U U C A C A G U ---

Reading frame 3  ---U U C U C G G A C C U G G A G A U U C A C A G U ---

**FIGURE 27–5 Reading frames in the genetic code.** In a triplet, nonoverlapping code, all mRNAs have three potential reading frames, shaded here in different colors. The triplets, and hence the amino acids specified, are different in each reading frame.

---

**TABLE 27–1  Incorporation of Amino Acids into Polypeptides in Response to Random Polymers of RNA**

| Amino acid | Observed frequency of incorporation (Lys = 100) | Tentative assignment for nucleotide composition* of corresponding codon | Expected frequency of incorporation based on assignment (Lys = 100) |
|---|---|---|---|
| Asparagine | 24 | $A_2C$ | 20 |
| Glutamine | 24 | $A_2C$ | 20 |
| Histidine | 6 | $AC_2$ | 4 |
| Lysine | 100 | AAA | 100 |
| Proline | 7 | $AC_2$, CCC | 4.8 |
| Threonine | 26 | $A_2C$, $AC_2$ | 24 |

Note: Presented here is a summary of data from one of the early experiments designed to elucidate the genetic code. A synthetic RNA containing only A and C residues in a 5:1 ratio directed polypeptide synthesis, and both the identity and the quantity of incorporated amino acids were determined. Based on the relative abundance of A and C residues in the synthetic RNA, and assigning the codon AAA (the most likely codon) a frequency of 100, there should be three different codons of composition $A_2C$, each at a relative frequency of 20; three of composition $AC_2$, each at a relative frequency of 4.0; and CCC at a relative frequency of 0.8. The CCC assignment was based on information derived from prior studies with poly(C). Where two tentative codon assignments are made, both are proposed to code for the same amino acid.

*These designations of nucleotide composition contain no information on nucleotide sequence (except, of course, AAA and CCC).

---

In 1964 Nirenberg and Philip Leder achieved another experimental breakthrough. Isolated *E. coli* ribosomes would bind a specific aminoacyl-tRNA in the presence of the corresponding synthetic polynucleotide messenger. (By convention, the identity of a tRNA is indicated by a superscript, such as tRNA[Ala], and the aminoacylated tRNA by a hyphenated name: alanyl-tRNA[Ala] or Ala-tRNA[Ala].) For example, ribosomes incubated with poly(U) and phenylalanyl-tRNA[Phe] (Phe-tRNA[Phe]) bind both RNAs, but if the ribosomes are incubated with poly(U) and some other aminoacyl-tRNA, the aminoacyl-tRNA is not bound, because it does not recognize the UUU triplets in poly(U) (Table 27–2). Even trinucleotides could promote specific binding of appropriate tRNAs, so these experiments could be carried out with chemically synthesized small oligonucleotides. With this technique researchers determined which aminoacyl-tRNA bound to about 50 of the 64 possible triplet codons. For some codons, either no aminoacyl-tRNA or more than one would bind. Another method was needed to complete and confirm the entire genetic code.

**TABLE 27–2  Trinucleotides That Induce Specific Binding of Aminoacyl-tRNAs to Ribosomes+**

| Trinucleotide | Relative increase in [14]C-labeled aminoacyl-tRNA bound to ribosome* | | |
|---|---|---|---|
| | Phe-tRNA[Phe] | Lys-tRNA[Lys] | Pro-tRNA[Pro] |
| UUU | 4.6 | 0 | 0 |
| AAA | 0 | 7.7 | 0 |
| CCC | 0 | 0 | 3.1 |

Source: Modified from Nirenberg, M. & Leder, P. (1964) RNA code words and protein synthesis. *Science* **145,** 1399.

*Each number represents the factor by which the amount of bound [14]C increased when the indicated trinucleotide was present, relative to a control with no trinucleotide.

At about this time, a complementary approach was provided by H. Gobind Khorana, who developed chemical methods to synthesize poly-ribonucleotides with defined, repeating sequences of two to four bases. The polypeptides produced by these mRNAs had one or a few amino acids in repeating patterns. These patterns, when combined with information from the random

H. Gobind Khorana

polymers used by Nirenberg and colleagues, permitted unambiguous codon assignments. The copolymer $(AC)_n$, for example, has alternating ACA and CAC codons: ACACACACACACACA. The polypeptide synthesized on this messenger contained equal amounts of threonine and histidine. Given that a histidine codon has one A and two Cs (Table 27–1), CAC must code for histidine and ACA for threonine.

Consolidation of the results from many experiments permitted the assignment of 61 of the 64 possible codons. The other three were identified as termination codons, in part because they disrupted amino acid coding patterns when they occurred in a synthetic RNA polymer (Fig. 27–6). Meanings for all the triplet codons (tabulated in Fig. 27–7) were established by 1966 and have been verified in many different ways. The cracking of the genetic code is regarded as one of the most important scientific discoveries of the twentieth century.

Codons are the key to the translation of genetic information, directing the synthesis of specific proteins. The reading frame is set when translation of an mRNA molecule begins, and it is maintained as the synthetic machinery reads sequentially from one triplet to the next. If the initial reading frame is off by one or two bases, or if translation somehow skips a nucleotide in the mRNA, all the subsequent codons will be out of register; the result is usually a "missense" protein with a garbled amino acid sequence. There are a few unusual but interesting exceptions to this rule (Box 27–1).

Several codons serve special functions (Fig. 27–7). The **initiation codon** AUG is the most common signal for the beginning of a polypeptide in all cells (some rare

alternatives are discussed in Box 27–2), in addition to coding for Met residues in internal positions of polypeptides. The **termination codons** (UAA, UAG, and UGA), also called stop codons or nonsense codons, normally signal the end of polypeptide synthesis and do not code for any known amino acids.

As described in Section 27.2, initiation of protein synthesis in the cell is an elaborate process that relies on initiation codons and other signals in the mRNA. In retrospect, the experiments of Nirenberg and Khorana to identify codon function should not have worked in the absence of initiation codons. Serendipitously, experimental conditions caused the normal initiation require-

First letter of codon (5′ end)

Second letter of codon →

|   | U | | C | | A | | G | |
|---|---|---|---|---|---|---|---|---|
| **U** | UUU | Phe | UCU | Ser | UAU | Tyr | UGU | Cys |
| | UUC | Phe | UCC | Ser | UAC | Tyr | UGC | Cys |
| | UUA | Leu | UCA | Ser | UAA | Stop | UGA | Stop |
| | UUG | Leu | UCG | Ser | UAG | Stop | UGG | Trp |
| **C** | CUU | Leu | CCU | Pro | CAU | His | CGU | Arg |
| | CUC | Leu | CCC | Pro | CAC | His | CGC | Arg |
| | CUA | Leu | CCA | Pro | CAA | Gln | CGA | Arg |
| | CUG | Leu | CCG | Pro | CAG | Gln | CGG | Arg |
| **A** | AUU | Ile | ACU | Thr | AAU | Asn | AGU | Ser |
| | AUC | Ile | ACC | Thr | AAC | Asn | AGC | Ser |
| | AUA | Ile | ACA | Thr | AAA | Lys | AGA | Arg |
| | AUG | Met | ACG | Thr | AAG | Lys | AGG | Arg |
| **G** | GUU | Val | GCU | Ala | GAU | Asp | GGU | Gly |
| | GUC | Val | GCC | Ala | GAC | Asp | GGC | Gly |
| | GUA | Val | GCA | Ala | GAA | Glu | GGA | Gly |
| | GUG | Val | GCG | Ala | GAG | Glu | GGG | Gly |

**FIGURE 27–7** "Dictionary" of amino acid code words in mRNAs. The codons are written in the 5′→3′ direction. The third base of each codon (in bold type) plays a lesser role in specifying an amino acid than the first two. The three termination codons are shaded in pink, the initiation codon AUG in green. All the amino acids except methionine and tryptophan have more than one codon. In most cases, codons that specify the same amino acid differ only at the third base.

Reading frame 1   5′ --- G U A A G U A A G U A A G U A A G U A A --- 3′

Reading frame 2   --- G U A A G U A A G U A A G U A A G U A A ---

Reading frame 3   --- G U A A G U A A G U A A G U A A G U A A ---

**FIGURE 27–6** Effect of a termination codon in a repeating tetranucleotide. Termination codons (pink) are encountered every fourth codon in three different reading frames (shown in different colors). Dipeptides or tripeptides are synthesized, depending on where the ribosome initially binds.

ments for protein synthesis to be relaxed. Diligence combined with chance to produce a breakthrough—a common occurrence in the history of biochemistry.

In a random sequence of nucleotides, 1 in every 20 codons in each reading frame is, on average, a termination codon. In general, a reading frame without a termination codon among 50 or more codons is referred to as an **open reading frame (ORF).** Long open reading frames usually correspond to genes that encode proteins. In the analysis of sequence databases, sophisticated programs are used to search for open reading frames in order to find genes among the often huge background of nongenic DNA. An uninterrupted gene coding for a typical protein with a molecular weight of 60,000 would require an open reading frame with 500 or more codons.

A striking feature of the genetic code is that an amino acid may be specified by more than one codon, so the code is described as **degenerate.** This does *not* suggest that the code is flawed: although an amino acid may have two or more codons, each codon specifies only one amino acid. The degeneracy of the code is not uniform. Whereas methionine and tryptophan have single codons, for example, three amino acids (Leu, Ser, Arg) have six codons, five amino acids have four, isoleucine has three, and nine amino acids have two (Table 27–3).

The genetic code is nearly universal. With the intriguing exception of a few minor variations in mitochondria, some bacteria, and some single-celled eukaryotes (Box 27–2), amino acid codons are identical in all species examined so far. Human beings, *E. coli,* tobacco plants, amphibians, and viruses share the same genetic code. Thus it would appear that all life forms have a common evolutionary ancestor, whose genetic code has been preserved throughout biological evolution. Even the variations (Box 27–2) reinforce this theme.

## Wobble Allows Some tRNAs to Recognize More than One Codon

When several different codons specify one amino acid, the difference between them usually lies at the third base position (at the 3′ end). For example, alanine is coded by the triplets GCU, GCC, GCA, and GCG. The codons for most amino acids can be symbolized by $XY^A_G$ or $XY^U_C$. The first two letters of each codon are the primary determinants of specificity, a feature that has some interesting consequences.

Transfer RNAs base-pair with mRNA codons at a three-base sequence on the tRNA called the **anticodon.** The first base of the codon in mRNA (read in the 5′→3′ direction) pairs with the third base of the anticodon (Fig. 27–8a). If the anticodon triplet of a tRNA recognized only one codon triplet through Watson-Crick base pairing at all three positions, cells would have a different tRNA for each amino acid codon. This is not the case, however, because the anticodons in some tRNAs include the nucleotide inosinate (designated I), which contains the uncommon base hypoxanthine (see Fig. 8–5b). Inosinate can form hydrogen bonds with three different nucleotides (U, C, and A; Fig. 27–8b), although

| TABLE 27–3 | Degeneracy of the Genetic Code | | |
|---|---|---|---|
| Amino acid | Number of codons | Amino acid | Number of codons |
| Met | 1 | Tyr | 2 |
| Trp | 1 | Ile | 3 |
| Asn | 2 | Ala | 4 |
| Asp | 2 | Gly | 4 |
| Cys | 2 | Pro | 4 |
| Gln | 2 | Thr | 4 |
| Glu | 2 | Val | 4 |
| His | 2 | Arg | 6 |
| Lys | 2 | Leu | 6 |
| Phe | 2 | Ser | 6 |

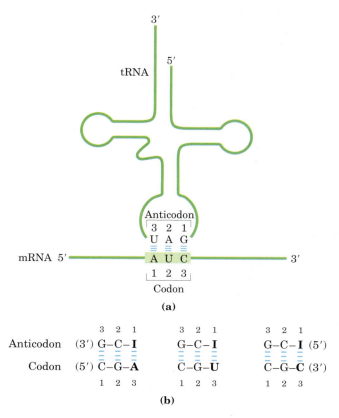

**FIGURE 27–8  Pairing relationship of codon and anticodon. (a)** Alignment of the two RNAs is antiparallel. The tRNA is shown in the traditional cloverleaf configuration. **(b)** Three different codon pairing relationships are possible when the tRNA anticodon contains inosinate.

## BOX 27–1    WORKING IN BIOCHEMISTRY

### Changing Horses in Midstream: Translational Frameshifting and mRNA Editing

Once the reading frame has been set during protein synthesis, codons are translated without overlap or punctuation until the ribosomal complex encounters a termination codon. The other two possible reading frames usually contain no useful genetic information, but a few genes are structured so that ribosomes "hiccup" at a certain point in the translation of their mRNAs, changing the reading frame from that point on. This appears to be a mechanism either to allow two or more related but distinct proteins to be produced from a single transcript or to regulate the synthesis of a protein.

One of the best-documented examples occurs in translation of the mRNA for the overlapping *gag* and *pol* genes of the Rous sarcoma virus (see Fig. 26–31). The reading frame for *pol* is offset to the left by one base pair (−1 reading frame) relative to the reading frame for *gag* (Fig. 1).

The product of the *pol* gene (reverse transcriptase) is translated as a larger polyprotein, on the same mRNA that is used for the *gag* protein alone (see Fig. 26–30). The polyprotein, or *gag-pol* protein, is then trimmed to the mature reverse transcriptase by proteolytic digestion. Production of the polyprotein requires a translational frameshift in the overlap region to allow the ribosome to bypass the UAG termination codon at the end of the *gag* gene (shaded pink in Fig. 1).

Frameshifts occur during about 5% of translations of this mRNA, and the *gag-pol* polyprotein (and ulti-

mately reverse transcriptase) is synthesized at about one-twentieth the frequency of the *gag* protein, a level that suffices for efficient reproduction of the virus. In some retroviruses, another translational frameshift allows translation of an even larger polyprotein that includes the product of the *env* gene fused to the *gag* and *pol* gene products (see Fig. 26–30). A similar mechanism produces both the $\tau$ and $\gamma$ subunits of *E. coli* DNA polymerase III from a single *dnaX* gene transcript (see Table 25–2).

This mechanism also occurs in the gene for *E. coli* release factor 2 (RF-2), discussed in Section 27.2, which is required for termination of protein synthesis at the termination codons UAA and UGA. The twenty-sixth codon in the transcript of the gene for RF-2 is UGA, which would normally halt protein synthesis. The remainder of the gene is in the +1 reading frame (offset one base pair to the right) relative to this UGA codon. Translation pauses at this codon, but termination does not occur unless RF-2 is bound to the codon (the lower the level of RF-2, the less likely the binding). The absence of bound RF-2 prevents the termination of protein synthesis at UGA and allows time for a frameshift to occur. The UGA plus the C that follows it (UGAC) is therefore read as GAC, which translates to Asp. Translation then proceeds in the new reading frame to complete synthesis of RF-2. In this way, RF-2 regulates its own synthesis in a feedback loop.

Some mRNAs are edited before translation. The initial transcripts of the genes that encode cytochrome oxidase subunit II in some protist mitochondria do not correspond precisely to the sequence needed at the

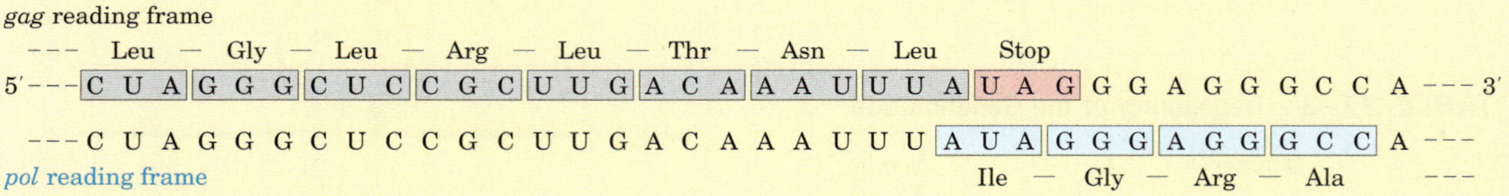

*gag* reading frame

   ---   Leu  —  Gly  —  Leu  —  Arg  —  Leu  —  Thr  —  Asn  —  Leu    Stop

5′ --- C U A G G G C U C C G C U U G A C A A A U U U A U A G G G A G G G C C A --- 3′

   --- C U A G G G C U C C G C U U G A C A A A U U U A U A G G G A G G G C C A ---

*pol* reading frame                                Ile  —  Gly  —  Arg  —  Ala     ---

**FIGURE 1** The *gag-pol* overlap region in Rous sarcoma virus RNA.

these pairings are much weaker than the hydrogen bonds of Watson-Crick base pairs (G≡C and A=U). In yeast, one tRNA$^{Arg}$ has the anticodon (5′)ICG, which recognizes three arginine codons: (5′)CGA, (5′)CGU, and (5′)CGC. The first two bases are identical (CG) and form strong Watson-Crick base pairs with the corresponding bases of the anticodon, but the third base

(A, U, or C) forms rather weak hydrogen bonds with the I residue at the first position of the anticodon.

Examination of these and other codon-anticodon pairings led Crick to conclude that the third base of most codons pairs rather loosely with the corresponding base of its anticodon; to use his picturesque word, the third base of such codons (and the first base of their corre-

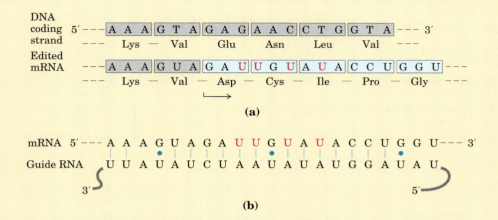

**(a)**

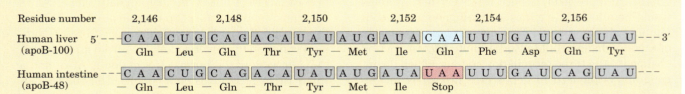

**(b)**

**FIGURE 2** RNA editing of the transcript of the cytochrome oxidase subunit II gene from *Trypanosoma brucei* mitochondria. **(a)** Insertion of four U residues (pink) produces a revised reading frame. **(b)** A special class of guide RNAs, complementary to the edited product, may act as templates for the editing process.

carboxyl terminus of the protein product. A posttranscriptional editing process inserts four U residues that shift the translational reading frame of the transcript. Figure 2a shows the added U residues in the small part of the transcript that is affected by editing. Neither the function nor the mechanism of this editing process is understood. Investigators have detected a special class of RNA molecules encoded by these mitochondria, with sequences complementary to the edited mRNAs. These so-called guide RNAs (Fig. 2b) appear to act as templates for the editing process. Note that the base pairing involves a number of G≡U base pairs (blue dots), which are common in RNA molecules.

A distinct form of RNA editing occurs in the gene for the apolipoprotein B component of low-density lipoprotein in vertebrates. One form of apolipoprotein B, apoB-100 ($M_r$ 513,000), is synthesized in the liver; a second form, apoB-48 ($M_r$ 250,000), is synthesized in the intestine. Both are encoded by an mRNA produced from the gene for apoB-100. A cytosine deaminase enzyme found only in the intestine binds to the mRNA at the codon for amino acid residue 2,153 (CAA = Gln) and converts the C to a U, to introduce the termination codon UAA. The apoB-48 produced in the intestine from this modified mRNA is simply an abbreviated form (corresponding to the amino-terminal half) of apoB-100 (Fig. 3). This reaction permits tissue-specific synthesis of two different proteins from one gene.

| Residue number | 2,146 | | 2,148 | | 2,150 | | 2,152 | | 2,154 | | 2,156 | |
|---|---|---|---|---|---|---|---|---|---|---|---|---|
| Human liver (apoB-100) 5′--- | CAA | CUG | CAG | ACA | UAU | AUG | AUA | CAA | UUU | GAU | CAG | UAU ---3′ |
| | — Gln — | Leu — | Gln — | Thr — | Tyr — | Met — | Ile — | Gln — | Phe — | Asp — | Gln — | Tyr — |
| Human intestine --- (apoB-48) | CAA | CUG | CAG | ACA | UAU | AUG | AUA | UAA | UUU | GAU | CAG | UAU --- |
| | — Gln — | Leu — | Gln — | Thr — | Tyr — | Met — | Ile | Stop | | | | |

**FIGURE 3** RNA editing of the transcript of the gene for the apolipoprotein B-100 component of LDL. Deamination, which occurs only in the intestine, converts a specific cytosine to uracil, changing a Gln codon to a stop codon and producing a truncated protein.

sponding anticodons) "wobbles." Crick proposed a set of four relationships called the **wobble hypothesis:**

1. The first two bases of an mRNA codon always form strong Watson-Crick base pairs with the corresponding bases of the tRNA anticodon and confer most of the coding specificity.

2. The first base of the anticodon (reading in the 5′→3′ direction; this pairs with the third base of the codon) determines the number of codons recognized by the tRNA. When the first base of the anticodon is C or A, base pairing is specific and only one codon is recognized by that tRNA. When the first base is U or G, binding is less

BOX 27–2    WORKING IN BIOCHEMISTRY

## Exceptions That Prove the Rule: Natural Variations in the Genetic Code

In biochemistry, as in other disciplines, exceptions to general rules can be problematic for instructors and frustrating for students. At the same time, though, they teach us that life is complex and inspire us to search for more surprises. Understanding the exceptions can even reinforce the original rule in surprising ways.

One would expect little room for variation in the genetic code. Even a single amino acid substitution can have profoundly deleterious effects on the structure of a protein. Nevertheless, variations in the code do occur in some organisms, and they are both interesting and instructive. The types of variation and their rarity provide powerful evidence for a common evolutionary origin of all living things.

To alter the code, changes must occur in one or more tRNAs, with the obvious target for alteration being the anticodon. Such a change would lead to the systematic insertion of an amino acid at a codon that, according to the normal code (see Fig. 27–7), does not specify that amino acid. The genetic code, in effect, is defined by two elements: (1) the anticodons on tRNAs (which determine where an amino acid is placed in a growing polypeptide) and (2) the specificity of the enzymes—the aminoacyl-tRNA synthetases—that charge the tRNAs, which determines the identity of the amino acid attached to a given tRNA.

Most sudden changes in the code would have catastrophic effects on cellular proteins, so code alterations are more likely where relatively few proteins would be affected—such as in small genomes encoding only a few proteins. The biological consequences of a code change could also be limited by restricting changes to the three termination codons, which do not generally occur *within* genes (see Box 27–4 for exceptions to *this* rule). This pattern is in fact observed.

Of the very few variations in the genetic code that we know of, most occur in mitochondrial DNA (mtDNA), which encodes only 10 to 20 proteins. Mitochondria have their own tRNAs, so their code variations do not affect the much larger cellular genome.

The most common changes in mitochondria (and the only code changes that have been observed in cellular genomes) involve termination codons. These changes affect termination in the products of only a subset of genes, and sometimes the effects are minor because the genes have multiple (redundant) termination codons.

In mitochondria, these changes can be viewed as a kind of genomic streamlining. Vertebrate mtDNAs have genes that encode 13 proteins, 2 rRNAs, and 22 tRNAs (see Fig. 19–32). An unusual set of wobble rules allows the 22 tRNAs to decode all 64 possible codon triplets; not all of the 32 tRNAs required for the normal code are needed. Four codon families (in which the amino acid is determined entirely by the first two nucleotides) are decoded by a single tRNA with a U residue in the first (or wobble) position in the anticodon. Either the U pairs somehow with any of the four possible bases in the third position of the codon or a "two out of three" mechanism is used—that is, no base pairing is needed at the third position. Other tRNAs recognize codons with either A or G in the third position, and yet others recognize U or C, so that virtually all the tRNAs recognize either two or four codons.

In the normal code, only two amino acids are specified by single codons: methionine and tryptophan (see Table 27–3). If all mitochondrial tRNAs recognize two codons, we would expect additional Met and Trp codons in mitochondria. And we find that the single most common code variation is the normal termination codon UGA specifying tryptophan. The tRNA^Trp recognizes and inserts a Trp residue at either UGA or the normal Trp codon, UGG. The second most common variation is conversion of AUA from an Ile codon to a Met codon; the normal Met codon is AUG, and a single tRNA recognizes both codons. The known coding variations in mitochondria are summarized in Table 1.

Turning to the much rarer changes in the codes for cellular (as distinct from mitochondrial) genomes, we find that the only known variation in a prokaryote is again the use of UGA to encode Trp residues, oc-

specific and two different codons may be read. When inosine (I) is the first (wobble) nucleotide of an anticodon, three different codons can be recognized—the maximum number for any tRNA. These relationships are summarized in Table 27–4.

**3.** When an amino acid is specified by several different codons, the codons that differ in either of the first two bases require different tRNAs.

**4.** A minimum of 32 tRNAs are required to translate all 61 codons (31 to encode the amino acids and 1 for initiation).

curring in the simplest free-living cell, *Mycoplasma capricolum*. Among eukaryotes, the only known extramitochondrial coding changes occur in a few species of ciliated protists, in which both termination codons UAA and UAG can specify glutamine.

Changes in the code need not be absolute; a codon might not always encode the same amino acid. In *E. coli* we find two examples of amino acids being inserted at positions not specified in the normal code. The first is the occasional use of GUG (Val) as an initiation codon. This occurs only for those genes in which the GUG is properly located relative to particular mRNA sequences that affect the initiation of translation (as discussed in Section 27.2).

The second *E. coli* example also involves contextual signals that alter coding patterns. A few proteins in all cells (such as formate dehydrogenase in bacteria and glutathione peroxidase in mammals) require the element selenium for their activity, generally in the form of the modified amino acid selenocysteine. Although modified amino acids are generally produced in posttranslational reactions (described in Section 27.3), in *E. coli* selenocysteine is introduced into formate dehydrogenase during translation, in response to an in-frame UGA codon. A special type of serine tRNA, present at lower levels than other Ser-tRNAs, recognizes UGA and no other codons. This tRNA is charged with serine, and the serine is enzymatically converted to selenocysteine before its use at the ribosome. The charged tRNA does not recognize just any UGA codon; some contextual signal in the mRNA, still to be identified, ensures that this tRNA recognizes only the few UGA codons, within certain genes, that specify selenocysteine. In effect, *E. coli* has 21 common amino acids, and UGA doubles as a codon for both termination and (sometimes) selenocysteine.

These variations tell us that the code is not quite as universal as once believed, but that its flexibility is severely constrained. The variations are obviously derivatives of the normal code, and no example of a completely different code has been found. The limited scope of code variants strengthens the principle that all life on this planet evolved on the basis of a single (slightly flexible) genetic code.

## TABLE 1 Known Variant Codon Assignments in Mitochondria

| | Codons* | | | | |
|---|---|---|---|---|---|
| | UGA | AUA | AGA AGG | CUN | CGG |
| Normal code assignment | Stop | Ile | Arg | Leu | Arg |
| Animals | | | | | |
|   Vertebrates | Trp | Met | Stop | + | + |
|   *Drosophila* | Trp | Met | Ser | + | + |
| Yeasts | | | | | |
|   *Saccharomyces cerevisiae* | Trp | Met | + | Thr | + |
|   *Torulopsis glabrata* | Trp | Met | + | Thr | ? |
|   *Schizosaccharomyces pombe* | Trp | + | + | + | + |
| Filamentous fungi | Trp | + | + | + | + |
| Trypanosomes | Trp | + | + | + | + |
| Higher plants | + | + | + | + | Trp |
| *Chlamydomonas reinhardtii* | ? | + | + | + | ? |

*N indicates any nucleotide; +, codon has the same meaning as in the normal code; ?, codon not observed in this mitochondrial genome.

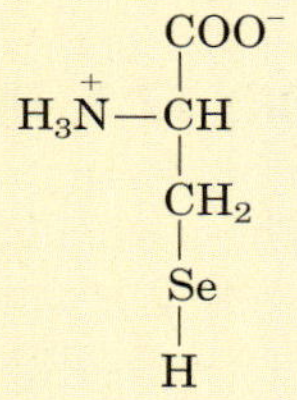

$$H_3\overset{+}{N}-\underset{|}{\overset{|}{C}}H$$

COO⁻
|
H₃N⁺—CH
|
CH₂
|
Se
|
H

Selenocysteine

The wobble (or third) base of the codon contributes to specificity, but, because it pairs only loosely with its corresponding base in the anticodon, it permits rapid dissociation of the tRNA from its codon during protein synthesis. If all three bases of a codon engaged in strong Watson-Crick pairing with the three bases of the anticodon, tRNAs would dissociate too slowly and this would severely limit the rate of protein synthesis. Codon-anticodon interactions balance the requirements for accuracy and speed.

The genetic code tells us how protein sequence information is stored in nucleic acids and provides some

**TABLE 27–4    How the Wobble Base of the Anticodon Determines the Number of Codons a tRNA Can Recognize**

1. One codon recognized:

| Anticodon | (3′) X–Y–**C** (5′) | (3′) X–Y–**A** (5′) |
|---|---|---|
| Codon | (5′) Y–X–**G** (3′) | (5′) Y–X–**U** (3′) |

2. Two codons recognized:

| Anticodon | (3′) X–Y–**U** (5′) | (3′) X–Y–**G** (5′) |
|---|---|---|
| Codon | (5′) Y–X–**A/G** (3′) | (5′) Y–X–**C/U** (3′) |

3. Three codons recognized:

| Anticodon | (3′) X–Y–**I** (5′) |
|---|---|
| Codon | (5′) Y–X–**A/U/C** (3′) |

Note: X and Y denote bases complementary to and capable of strong Watson-Crick base pairing with X′ and Y′, respectively. Wobble bases—in the 3′ position of codons and 5′ position of anticodons—are shaded in pink.

clues about how that information is translated into protein. We now turn to the molecular mechanisms of the translation process.

## SUMMARY 27.1 The Genetic Code

- The particular amino acid sequence of a protein is constructed through the translation of information encoded in mRNA. This process is carried out by ribosomes.

- Amino acids are specified by mRNA codons consisting of nucleotide triplets. Translation requires adaptor molecules, the tRNAs, that recognize codons and insert amino acids into their appropriate sequential positions in the polypeptide.

- The base sequences of the codons were deduced from experiments using synthetic mRNAs of known composition and sequence.

- The codon AUG signals initiation of translation. The triplets UAA, UAG, and UGA are signals for termination.

- The genetic code is degenerate: it has multiple code words for almost every amino acid.

- The standard genetic code words are universal in all species, with some minor deviations in mitochondria and a few single-celled organisms.

- The third position in each codon is much less specific than the first and second and is said to wobble.

## 27.2 Protein Synthesis

As we have seen for DNA and RNA (Chapters 25 and 26), the synthesis of polymeric biomolecules can be considered in terms of initiation, elongation, and termination stages. These fundamental processes are typically bracketed by two additional stages: activation of precursors before synthesis and postsynthetic processing of the completed polymer. Protein synthesis follows the same pattern. The activation of amino acids before their incorporation into polypeptides and the posttranslational processing of the completed polypeptide play particularly important roles in ensuring both the fidelity of synthesis and the proper function of the protein product. The cellular components involved in the five stages of protein synthesis in *E. coli* and other bacteria are listed in Table 27–5; the requirements in eukaryotic cells are quite similar, although the components are in some cases more numerous. An initial overview of the stages of protein synthesis provides a useful outline for the discussion that follows.

### Protein Biosynthesis Takes Place in Five Stages

**Stage 1: Activation of Amino Acids**    For the synthesis of a polypeptide with a defined sequence, two fundamental chemical requirements must be met: (1) the carboxyl group of each amino acid must be activated to facilitate formation of a peptide bond, and (2) a link must be established between each new amino acid and the information in the mRNA that encodes it. Both these requirements are met by attaching the amino acid to a tRNA in the first stage of protein synthesis. Attaching the right amino acid to the right tRNA is critical. This reaction takes place in the cytosol, not on the ribosome. Each of the 20 amino acids is covalently attached to a specific tRNA at the expense of ATP energy, using $Mg^{2+}$-dependent activating enzymes known as aminoacyl-tRNA synthetases. When attached to their amino acid (aminoacylated) the tRNAs are said to be "charged."

**Stage 2: Initiation**    The mRNA bearing the code for the polypeptide to be made binds to the smaller of two ribosomal subunits and to the initiating aminoacyl-tRNA. The large ribosomal subunit then binds to form an initiation complex. The initiating aminoacyl-tRNA base-pairs with the mRNA codon AUG that signals the beginning of the polypeptide. This process, which requires GTP, is promoted by cytosolic proteins called initiation factors.

**Stage 3: Elongation**    The nascent polypeptide is lengthened by covalent attachment of successive amino acid units, each carried to the ribosome and correctly positioned by its tRNA, which base-pairs to its corresponding codon in the mRNA. Elongation requires cytosolic proteins known as elongation factors. The binding of each incoming aminoacyl-tRNA and the movement of

**TABLE 27-5    Components Required for the Five Major Stages of Protein Synthesis in *E. coli***

| Stage | Essential components |
|---|---|
| 1. Activation of amino acids | 20 amino acids<br>20 aminoacyl-tRNA synthetases<br>32 or more tRNAs<br>ATP<br>$Mg^{2+}$ |
| 2. Initiation | mRNA<br>*N*-Formylmethionyl-tRNA$^{fmet}$<br>Initiation codon in mRNA (AUG)<br>30S ribosomal subunit<br>50S ribosomal subunit<br>Initiation factors (IF-1, IF-2, IF-3)<br>GTP<br>$Mg^{2+}$ |
| 3. Elongation | Functional 70S ribosome (initiation complex)<br>Aminoacyl-tRNAs specified by codons<br>Elongation factors (EF-Tu, EF-Ts, EF-G)<br>GTP<br>$Mg^{2+}$ |
| 4. Termination and release | Termination codon in mRNA<br>Release factors (RF-1, RF-2, RF-3) |
| 5. Folding and posttranslational processing | Specific enzymes, cofactors, and other components for removal of initiating residues and signal sequences, additional proteolytic processing, modification of terminal residues, and attachment of phosphate, methyl, carboxyl, carbohydrate, or prosthetic groups |

Masayasu Nomura

the ribosome along the mRNA are facilitated by the hydrolysis of GTP as each residue is added to the growing polypeptide.

**Stage 4: Termination and Release**    Completion of the polypeptide chain is signaled by a termination codon in the mRNA. The new polypeptide is released from the ribosome, aided by proteins called release factors.

**Stage 5: Folding and Posttranslational Processing**    In order to achieve its biologically active form, the new polypeptide must fold into its proper three-dimensional conformation. Before or after folding, the new polypeptide may undergo enzymatic processing, including removal of one or more amino acids (usually from the amino terminus); addition of acetyl, phosphoryl, methyl, carboxyl, or other groups to certain amino acid residues; proteolytic cleavage; and/or attachment of oligosaccharides or prosthetic groups.

Before looking at these five stages in detail, we must examine two key components in protein biosynthesis: the ribosome and tRNAs.

## The Ribosome Is a Complex Supramolecular Machine

Each *E. coli* cell contains 15,000 or more ribosomes, making up almost a quarter of the dry weight of the cell. Bacterial ribosomes contain about 65% rRNA and 35% protein; they have a diameter of about 18 nm and are composed of two unequal subunits with sedimentation coefficients of 30S and 50S and a combined sedimentation coefficient of 70S. Both subunits contain dozens of ribosomal proteins and at least one large rRNA (Table 27–6).

Following Zamecnik's discovery that ribosomes are the complexes responsible for protein synthesis, and following elucidation of the genetic code, the study of ribosomes accelerated. In the late 1960s Masayasu Nomura and colleagues demonstrated that both ribosomal subunits can be broken down into their RNA and protein components, then reconstituted in vitro. Under appropriate experimental conditions, the RNA and protein spontaneously reassemble to form 30S or 50S subunits nearly identical in structure and activity to native subunits. This breakthrough fueled decades of research into the function and structure of ribosomal RNAs and proteins. At the same time, increasingly sophisticated structural methods revealed more and more details about ribosome structure.

The dawn of a new millennium brought with it the elucidation of the first high-resolution structures of bacterial ribosomal subunits. The bacterial ribosome is complex, with a combined molecular weight of ~2.7 million, and it is providing a wealth of surprises (Fig. 27–9). First, the traditional focus on the protein components of ribosomes was shifted. The ribosomal subunits are huge RNA molecules. In the 50S subunit, the 5S and 23S rRNAs form the structural core. The proteins are secondary elements in the complex, decorating the surface. Second and most important, there is no protein within 18 Å of the active site for peptide bond formation. The high-resolution structure thus confirms what many had suspected for more than a decade: the ribosome is a ribozyme. In addition to the insight they provide into the mechanism of protein synthesis (as elaborated below), the detailed

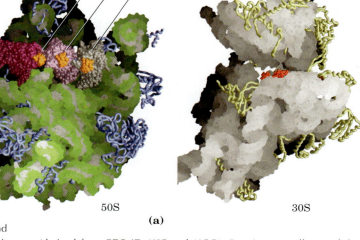

**FIGURE 27–9 Ribosomes.** Our understanding of ribosome structure took a giant step forward with the publication in 2000 of the high-resolution structure of the 50S ribosomal subunit of the bacterium *Haloarcula marismortui* by Thomas Steitz, Peter Moore, and their colleagues. This was followed by additional high-resolution structures of the ribosomal subunits from several different bacterial species, and models of the corresponding complete ribosomes. A sampling of that progress is presented here.

**(a)** The 50S and 30S bacterial subunits, split apart to visualize the surfaces that interact in the active ribosome. The structure on the left is the 50S subunit (derived from PDB ID 1JJ2 and 1GIY), with tRNAs (purple, mauve, and gray); bound to sites E, P, and A, described later in the text; the tRNA anticodons are in orange. Proteins appear as blue wormlike structures; the rRNA as a blended space-filling representation designed to highlight surface features, with the bases in white and the backbone in green. The structure on the right is the 30S subunit (derived from PDB ID 1J5E and 1JGO). Proteins are yellow and the rRNA white. The part of the mRNA that interacts with the tRNA anticodons is shown in red. The rest of the mRNA winds through grooves or channels on the 30S subunit surface.

**(b)** A model of a complete active bacterial ribosome (derived from PDB ID 1J5E, 1JJ2, 1JGO, and 1GIY). All components are colored as in **(a).** This is a view down into the groove separating the subunits. A second view (inset) is from the same angle, but with the tRNAs removed to give a better sense of the cleft where protein synthesis occurs.

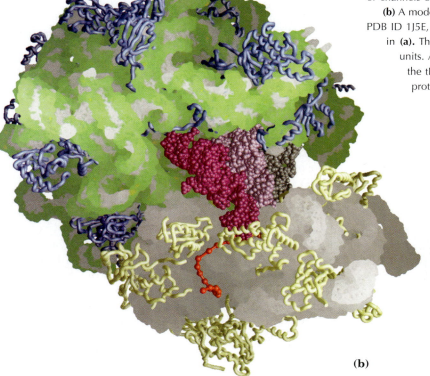

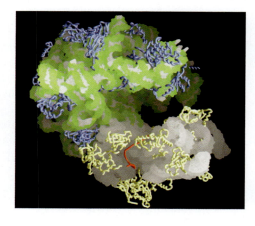

**(b)**

**TABLE 27–6** RNA and Protein Components of the *E. coli* Ribosome

| Subunit | Number of different proteins | Total number of proteins | Protein designations | Number and type of rRNAs |
|---------|------------------------------|--------------------------|----------------------|--------------------------|
| 30S | 21 | 21 | S1–S21 | 1 (16S rRNA) |
| 50S | 33 | 36 | L1–L36* | 2 (5S and 23S rRNAs) |

*The L1 to L36 protein designations do not correspond to 36 different proteins. The protein originally designated L7 is in fact a modified form of L12, and L8 is a complex of three other proteins. Also, L26 proved to be the same protein as S20 (and not part of the 50S subunit). This gives 33 different proteins in the large subunit. There are four copies of the L7/L12 protein, with the three extra copies bringing the total protein count to 36.

structures of the ribosome and its subunits have stimulated a new look at the evolution of life (Box 27–3).

The two irregularly shaped ribosomal subunits fit together to form a cleft through which the mRNA passes as the ribosome moves along it during translation (Fig. 27–9b). The 55 proteins in bacterial ribosomes vary enormously in size and structure. Molecular weights range from about 6,000 to 75,000. Most of the proteins have globular domains arranged on the ribosome surface. Some also have snakelike protein extensions that protrude into the rRNA core of the ribosome, stabilizing its structure. The functions of some of these proteins have not yet been elucidated in detail, although a structural role seems evident for many of them.

The sequences of the rRNAs of many organisms are now known. Each of the three single-stranded rRNAs of

**(c)** Structure of the 50S bacterial ribosome subunit (PDB ID 1Q7Y). The subunit is again viewed from the side that attaches to the 30S subunit, but is tilted down slightly compared to its orientation in **(a).** The active site for peptide bond formation (the peptidyl transferase activity), deep within a surface groove and far away from any protein, is marked by a bound inhibitor, puromycin (red).

**(d)** Summary of the composition and mass of ribosomes in prokaryotes and eukaryotes. Ribosomal subunits are identified by their S (Svedberg unit) values, sedimentation coefficients that refer to their rate of sedimentation in a centrifuge. The S values are not necessarily additive when subunits are combined, because rates of sedimentation are affected by shape as well as mass.

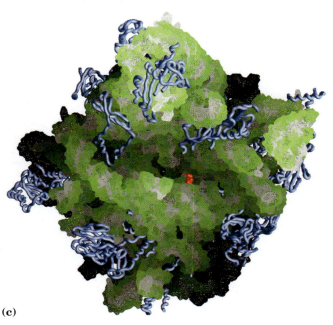

**(c)**

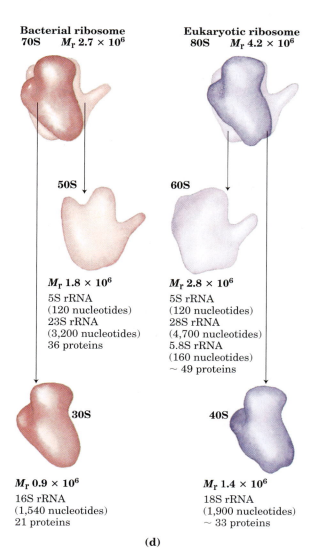

**Bacterial ribosome**
70S $M_r$ 2.7 × 10⁶

**Eukaryotic ribosome**
80S $M_r$ 4.2 × 10⁶

**50S**

**60S**

$M_r$ 1.8 × 10⁶
5S rRNA
(120 nucleotides)
23S rRNA
(3,200 nucleotides)
36 proteins

$M_r$ 2.8 × 10⁶
5S rRNA
(120 nucleotides)
28S rRNA
(4,700 nucleotides)
5.8S rRNA
(160 nucleotides)
~ 49 proteins

**30S**

**40S**

$M_r$ 0.9 × 10⁶
16S rRNA
(1,540 nucleotides)
21 proteins

$M_r$ 1.4 × 10⁶
18S rRNA
(1,900 nucleotides)
~ 33 proteins

**(d)**

## From an RNA World to a Protein World

Extant ribozymes generally promote one of two types of reactions: hydrolytic cleavage of phosphodiester bonds or phosphoryl transfers (Chapter 26). In both cases, the substrates of the reactions are also RNA molecules. The ribosomal RNAs provide an important expansion of the catalytic range of known ribozymes. Coupled to the laboratory exploration of potential RNA catalytic function (see Box 26–3), the idea of an RNA world as a precursor to current life forms becomes increasingly attractive.

A viable RNA world would require an RNA capable of self-replication, a primitive metabolism to generate the needed ribonucleotide precursors, and a cell boundary to aid in concentrating the precursors and sequestering them from the environment. The requirements for catalysis of reactions involving a growing range of metabolites and macromolecules could have led to larger and more complex RNA catalysts. The many negatively charged phosphoryl groups in the RNA backbone limit the stability of very large RNA molecules. In an RNA world, divalent cations or other positively charged groups could be incorporated into the structures to augment stability.

Certain peptides could stabilize large RNA molecules. For example, many ribosomal proteins in modern eukaryotic cells have long extensions, lacking secondary structure, that snake into the rRNAs and help stabilize them (Fig. 1). Ribozyme-catalyzed synthesis of peptides could thus initially have evolved as part of a general solution to the structural maintenance of large RNA molecules. The synthesis of peptides may have helped stabilize large ribozymes, but this advance also marked the beginning of the end for the RNA world. Once peptide synthesis was possible, the greater catalytic potential of proteins would have set in motion an irreversible transition to a protein-dominated metabolic system.

Most enzymatic processes, then, were eventually surrendered to the proteins—but not all. In every organism, the critical task of synthesizing the proteins remains, even now, a ribozyme-catalyzed process.

There appears to be only one good arrangement (or just a very few) of nucleotide residues in a ribozyme active site that can catalyze peptide synthesis. The rRNA residues that seem to be involved in the peptidyl transferase activity of ribosomes are highly conserved in the large-subunit rRNAs of all species. Using in vitro evolution (SELEX; see Box 26–3), investigators have isolated artificial ribozymes that promote peptide synthesis. Intriguingly, most of them include the ribonucleotide octet (5′)AUAACAGG(3′), a highly conserved sequence found at the peptidyl transferase active site in the ribosomes of all cells. There may be just one optimal solution to the overall chemical problem of ribozyme-catalyzed synthesis of proteins of defined sequence. Evolution found this solution once, and no life form has notably improved on it.

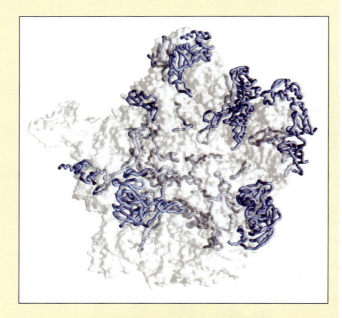

**FIGURE 1** The 50S subunit of a bacterial ribosome (PDB ID 1NKW). The protein backbones are shown as blue wormlike structures; the rRNA components are transparent. The unstructured extensions of many of the ribosomal proteins snake into the rRNA structures, helping to stabilize them.

---

*E. coli* has a specific three-dimensional conformation featuring extensive intrachain base pairing. The predicted secondary structure of the rRNAs (Fig. 27–10) has largely been confirmed in the high-resolution models, but fails to convey the extensive network of tertiary interactions evident in the complete structure.

The ribosomes of eukaryotic cells (other than mitochondrial and chloroplast ribosomes) are larger and more complex than bacterial ribosomes (Fig. 27–9d), with a diameter of about 23 nm and a sedimentation coefficient of about 80S. They also have two subunits, which vary in size among species but on average are 60S

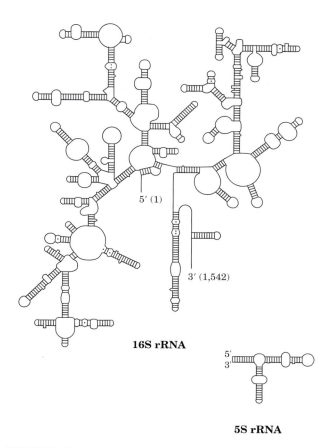

**16S rRNA**

**5S rRNA**

**FIGURE 27–10 Bacterial rRNAs.** Diagrams of the secondary structure of *E. coli* 16S and 5S rRNAs. The first (5′ end) and final (3′ end) ribonucleotide residues of the 16S rRNA are numbered.

and 40S. Altogether, eukaryotic ribosomes contain more than 80 different proteins. The ribosomes of mitochondria and chloroplasts are somewhat smaller and simpler than bacterial ribosomes. Nevertheless, ribosomal structure and function are strikingly similar in all organisms and organelles.

## Transfer RNAs Have Characteristic Structural Features

To understand how tRNAs can serve as adaptors in translating the language of nucleic acids into the language of proteins, we must first examine their structure in more detail. Transfer RNAs are relatively small and consist of a single strand of RNA folded into a precise three-dimensional structure (see Fig. 8–28a). The tRNAs in bacteria and in the cytosol of eukaryotes have between 73 and 93 nucleotide residues, corresponding to molecular weights of 24,000 to 31,000. Mitochondria and chloroplasts contain distinctive, somewhat smaller tRNAs. Cells have at least one kind of tRNA for each amino acid; at least 32 tRNAs are required to recognize all the amino acid codons (some recognize more than one codon), but some cells use more than 32.

Yeast alanine tRNA (tRNA<sup>Ala</sup>), the first nucleic acid to be completely sequenced (Fig. 27–11), contains 76 nucleotide residues, 10 of which have modified bases. Comparisons of tRNAs from various species have revealed many common denominators of structure (Fig. 27–12). Eight or more of the nucleotide residues have modified bases and sugars, many of which are methylated derivatives of the principal bases. Most tRNAs have a guanylate (pG) residue at the 5′ end, and all have the trinucleotide sequence CCA(3′) at the 3′ end. When

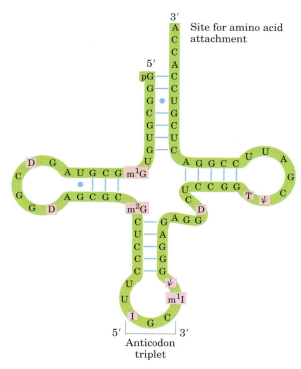

**FIGURE 27–11 Nucleotide sequence of yeast tRNA<sup>Ala</sup>.** This structure was deduced in 1965 by Robert W. Holley and his colleagues; it is shown in the cloverleaf conformation in which intrastrand base pairing is maximal. The following symbols are used for the modified nucleotides (shaded pink): ψ, pseudouridine; I, inosine; T, ribothymidine; D, 5,6-dihydrouridine; m¹I, 1-methylinosine; m¹G, 1-methylguanosine; m²G, *N*²-dimethylguanosine (see Fig. 26–24). Blue lines between parallel sections indicate Watson-Crick base pairs. The anticodon can recognize three codons for alanine (GCA, GCU, and GCC). Other features of tRNA structure are shown in Figures 27–12 and 27–13. Note the presence of two G=U base pairs, signified by a blue dot to indicate non-Watson-Crick pairing. In RNAs, guanosine is often base-paired with uridine, although the G=U pair is not as stable as the Watson-Crick G≡C pair (Chapter 8).

Robert W. Holley, 1922–1993

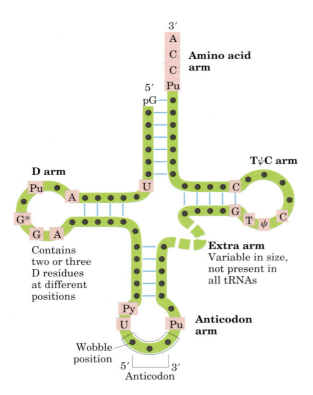

**FIGURE 27–12  General cloverleaf secondary structure of tRNAs.** The large dots on the backbone represent nucleotide residues; the blue lines represent base pairs. Characteristic and/or invariant residues common to all tRNAs are shaded in pink. Transfer RNAs vary in length from 73 to 93 nucleotides. Extra nucleotides occur in the extra arm or in the D arm. At the end of the anti-codon arm is the anticodon loop, which always contains seven unpaired nucleotides. The D arm contains two or three D (5,6-dihydrouridine) residues, depending on the tRNA. In some tRNAs, the D arm has only three hydrogen-bonded base pairs. In addition to the symbols explained in Figure 27–11: Pu, purine nucleotide; Py, pyrimidine nucleotide; G*, guanylate *or* 2'-*O*-methylguanylate.

drawn in two dimensions, the hydrogen-bonding pattern of all tRNAs forms a cloverleaf structure with four arms; the longer tRNAs have a short fifth arm, or extra arm (Fig. 27–12). In three dimensions, a tRNA has the form of a twisted L (Fig. 27–13).

Two of the arms of a tRNA are critical for its adaptor function. The **amino acid arm** can carry a specific amino acid esterified by its carboxyl group to the 2'- or 3'-hydroxyl group of the A residue at the 3' end of the tRNA. The **anticodon arm** contains the anticodon. The other major arms are the **D arm,** which contains the un-usual nucleotide dihydrouridine (D), and the **TψC arm,** which contains ribothymidine (T), not usually present in RNAs, and pseudouridine (ψ), which has an unusual carbon–carbon bond between the base and ribose (see Fig. 26–24). The D and TψC arms contribute important

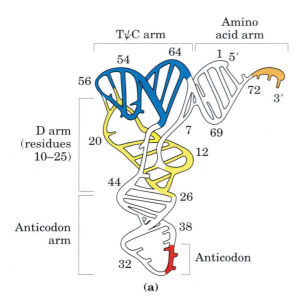

**(a)**

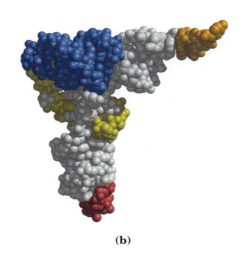

**(b)**

**FIGURE 27–13  Three-dimensional structure of yeast tRNA^Phe deduced from x-ray diffraction analysis.** The shape resembles a twisted L. **(a)** Schematic diagram with the various arms identified in Figure 27–12 shaded in different colors. **(b)** A space-filling model, with the same color coding (PDB ID 4TRA).The CCA sequence at the 3' end (orange) is the attachment point for the amino acid.

interactions for the overall folding of tRNA molecules, and the TψC arm interacts with the large-subunit rRNA.

Having looked at the structures of ribosomes and tRNAs, we now consider in detail the five stages of protein synthesis.

## Stage 1: Aminoacyl-tRNA Synthetases Attach the Correct Amino Acids to Their tRNAs

During the first stage of protein synthesis, taking place in the cytosol, aminoacyl-tRNA synthetases esterify the 20 amino acids to their corresponding tRNAs. Each enzyme is specific for one amino acid and one or more corresponding tRNAs. Most organisms have one aminoacyl-tRNA synthetase for each amino acid. For amino acids with two or more corresponding tRNAs, the same enzyme usually aminoacylates all of them.

The structures of all the aminoacyl-tRNA synthetases of *E. coli* have been determined. Researchers have divided them into two classes (Table 27–7) based on substantial differences in primary and tertiary structure and in reaction mechanism (Fig. 27–14); these two classes are the same in all organisms. There is no evidence for a common ancestor, and the biological, chemical, or evolutionary reasons for two enzyme classes for essentially identical processes remain obscure.

The reaction catalyzed by an aminoacyl-tRNA synthetase is

$$\text{Amino acid} + \text{tRNA} + \text{ATP} \underset{}{\overset{\text{Mg}^{2+}}{\rightleftharpoons}}$$

$$\text{aminoacyl-tRNA} + \text{AMP} + \text{PP}_i$$

This reaction occurs in two steps in the enzyme's active site. In step ① (Fig. 27–14) an enzyme-bound intermediate, aminoacyl adenylate (aminoacyl-AMP), forms when the carboxyl group of the amino acid reacts with the α-phosphoryl group of ATP to form an anhydride linkage, with displacement of pyrophosphate. In the sec-

ond step the aminoacyl group is transferred from enzyme-bound aminoacyl-AMP to its corresponding specific tRNA. The course of this second step depends on the class to which the enzyme belongs, as shown by pathways ②a and ②b in Figure 27–14. The resulting ester linkage between the amino acid and the tRNA (Fig. 27–15) has a highly negative standard free energy of hydrolysis ($\Delta G'^\circ = -29$ kJ/mol). The pyrophosphate formed in the activation reaction undergoes hydrolysis to phosphate by inorganic pyrophosphatase. Thus *two* high-energy phosphate bonds are ultimately expended for each amino acid molecule activated, rendering the overall reaction for amino acid activation essentially irreversible:

$$\text{Amino acid} + \text{tRNA} + \text{ATP} \xrightarrow{\text{Mg}^{2+}}$$

$$\text{aminoacyl-tRNA} + \text{AMP} + 2\text{P}_i$$
$$\Delta G'^\circ \approx -29 \text{ kJ/mol}$$

***Proofreading by Aminoacyl-tRNA Synthetases*** The aminoacylation of tRNA accomplishes two ends: (1) activation of an amino acid for peptide bond formation and (2) attachment of the amino acid to an adaptor tRNA that ensures appropriate placement of the amino acid in a growing polypeptide. The identity of the amino acid attached to a tRNA is not checked on the ribosome, so attachment of the correct amino acid to the tRNA is essential to the fidelity of protein synthesis.

As you will recall from Chapter 6, enzyme specificity is limited by the binding energy available from enzyme-substrate interactions. Discrimination between two similar amino acid substrates has been studied in detail in the case of Ile-tRNA synthetase, which distinguishes between valine and isoleucine, amino acids that differ by only a single methylene group (—CH2—). Ile-tRNA synthetase favors activation of isoleucine (to form Ile-AMP) over valine by a factor of 200—as we would expect, given the amount by which a methylene group (in Ile) could enhance substrate binding. Yet valine is erroneously incorporated into proteins in positions normally occupied by an Ile residue at a frequency of only about 1 in 3,000. How is this greater than tenfold increase in accuracy brought about? Ile-tRNA synthetase, like some other aminoacyl-tRNA synthetases, has a proofreading function.

Recall a general principle from the discussion of proofreading by DNA polymerases (p. 955): if available binding interactions do not provide sufficient discrimination between two substrates, the necessary specificity can be achieved by substrate-specific binding in *two successive* steps. The effect of forcing the system through two successive filters is multiplicative. In the case of Ile-tRNA synthetase, the first filter is the initial binding of the amino acid to the enzyme and its activation to aminoacyl-AMP. The second is the binding of any

**TABLE 27–7 The Two Classes of Aminoacyl-tRNA Synthetases**

| Class I | | Class II | |
|---------|-----|----------|-----|
| Arg | Leu | Ala | Lys |
| Cys | Met | Asn | Phe |
| Gln | Trp | Asp | Pro |
| Glu | Tyr | Gly | Ser |
| Ile | Val | His | Thr |

Note: Here, Arg represents arginyl-tRNA synthetase, and so forth. The classification applies to all organisms for which tRNA synthetases have been analyzed and is based on protein structural distinctions and on the mechanistic distinction outlined in Figure 27–14.

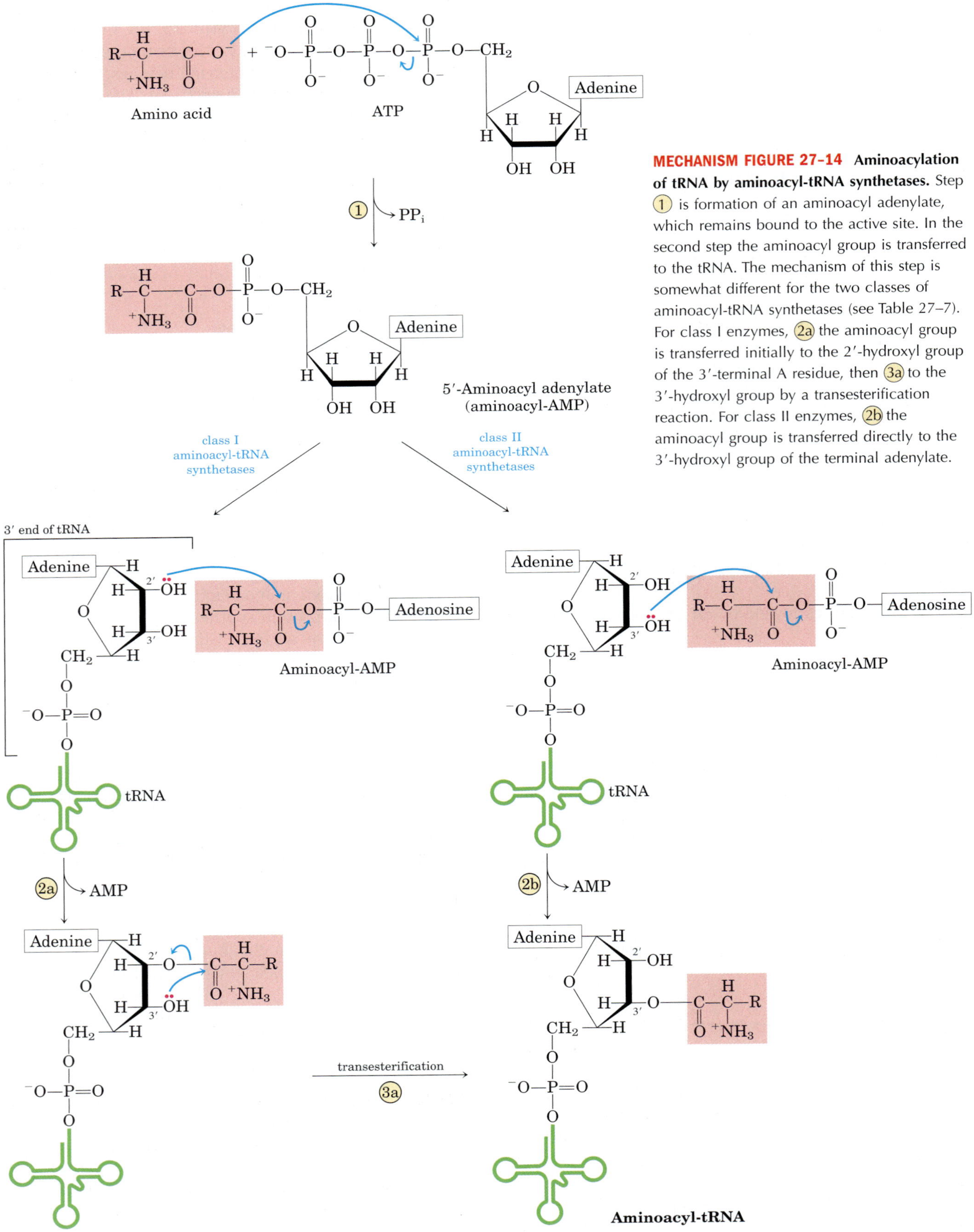

**MECHANISM FIGURE 27–14** **Aminoacylation of tRNA by aminoacyl-tRNA synthetases.** Step ① is formation of an aminoacyl adenylate, which remains bound to the active site. In the second step the aminoacyl group is transferred to the tRNA. The mechanism of this step is somewhat different for the two classes of aminoacyl-tRNA synthetases (see Table 27–7). For class I enzymes, ②a the aminoacyl group is transferred initially to the 2'-hydroxyl group of the 3'-terminal A residue, then ③a to the 3'-hydroxyl group by a transesterification reaction. For class II enzymes, ②b the aminoacyl group is transferred directly to the 3'-hydroxyl group of the terminal adenylate.

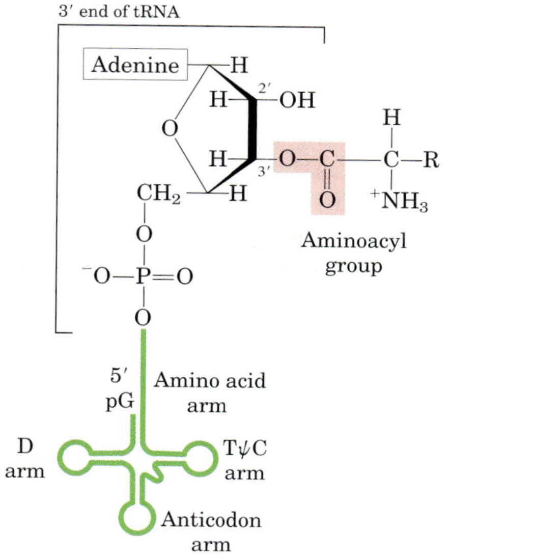

FIGURE 27–15 **General structure of aminoacyl-tRNAs.** The aminoacyl group is esterified to the 3′ position of the terminal A residue. The ester linkage that both activates the amino acid and joins it to the tRNA is shaded pink.

low as that of DNA replication. Because flaws in a protein are eliminated when the protein is degraded and are not passed on to future generations, they have less biological significance. The degree of fidelity in protein synthesis is sufficient to ensure that most proteins contain no mistakes and that the large amount of energy required to synthesize a protein is rarely wasted. One defective protein molecule is usually unimportant when many correct copies of the same protein are present.

***Interaction between an Aminoacyl-tRNA Synthetase and a tRNA: A "Second Genetic Code"*** An individual aminoacyl-tRNA synthetase must be specific not only for a single amino acid but for certain tRNAs as well. Discriminating among dozens of tRNAs is just as important for the overall fidelity of protein biosynthesis as is distinguishing among amino acids. The interaction between aminoacyl-tRNA synthetases and tRNAs has been referred to as the "second genetic code," reflecting its critical role in maintaining the accuracy of protein synthesis. The "coding" rules appear to be more complex than those in the "first" code.

Figure 27–16 summarizes what we know about the nucleotides involved in recognition by some aminoacyl-tRNA synthetases. Some nucleotides are conserved in all tRNAs and therefore cannot be used for discrimination.

*incorrect* aminoacyl-AMP products to a separate active site on the enzyme; a substrate that binds in this second active site is hydrolyzed. The R group of valine is slightly smaller than that of isoleucine, so Val-AMP fits the hydrolytic (proofreading) site of the Ile-tRNA synthetase but Ile-AMP does not. Thus Val-AMP is hydrolyzed to valine and AMP in the proofreading active site, and tRNA bound to the synthetase does not become aminoacylated to the wrong amino acid.

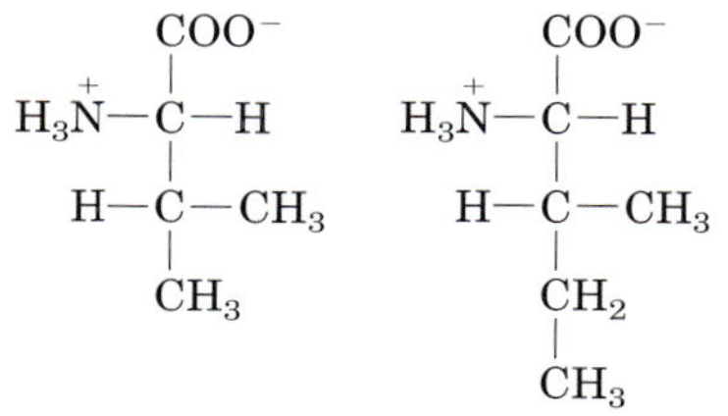

In addition to proofreading after formation of the aminoacyl-AMP intermediate, most aminoacyl-tRNA synthetases can also hydrolyze the ester linkage between amino acids and tRNAs in the aminoacyl-tRNAs. This hydrolysis is greatly accelerated for incorrectly charged tRNAs, providing yet a third filter to enhance the fidelity of the overall process. The few aminoacyl-tRNA synthetases that activate amino acids with no close structural relatives (Cys-tRNA synthetase, for example) demonstrate little or no proofreading activity; in these cases, the active site for aminoacylation can sufficiently discriminate between the proper substrate and any incorrect amino acid.

The overall error rate of protein synthesis (~1 mistake per $10^4$ amino acids incorporated) is not nearly as

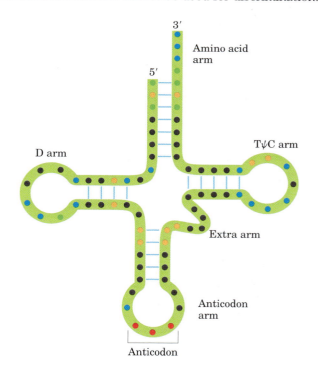

FIGURE 27–16 **Nucleotide positions in tRNAs that are recognized by aminoacyl-tRNA synthetases.** Some positions (blue dots) are the same in all tRNAs and therefore cannot be used to discriminate one from another. Other positions are known recognition points for one (orange) or more (green) aminoacyl-tRNA synthetases. Structural features other than sequence are important for recognition by some of the synthetases.

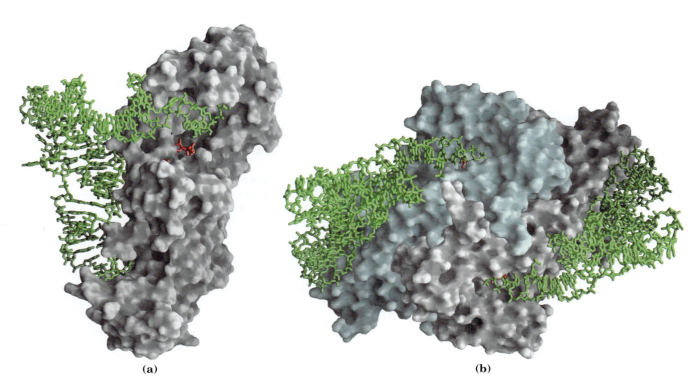

**(a)**                                    **(b)**

**FIGURE 27–17 Aminoacyl-tRNA synthetases.** Both synthetases are complexed with their cognate tRNAs (green stick structures). Bound ATP (red) pinpoints the active site near the end of the aminoacyl arm.

**(a)** Gln-tRNA synthetase from *E. coli,* a typical monomeric type I synthetase (PDB ID 1QRT). **(b)** Asp-tRNA synthetase from yeast, a typical dimeric type II synthetase (PDB ID 1ASZ).

By observing changes in nucleotides that alter substrate specificity, researchers have identified nucleotide positions that are involved in discrimination by the aminoacyl-tRNA synthetases. These nucleotide positions seem to be concentrated in the amino acid arm and the anticodon arm, including the nucleotides of the anticodon itself, but are also located in other parts of the tRNA molecule. Determination of the crystal structures of aminoacyl-tRNA synthetases complexed with their cognate tRNAs and ATP has added a great deal to our understanding of these interactions (Fig. 27–17).

Ten or more specific nucleotides may be involved in recognition of a tRNA by its specific aminoacyl-tRNA synthetase. But in a few cases the recognition mechanism is quite simple. Across a range of organisms from bacteria to humans, the primary determinant of tRNA recognition by the Ala-tRNA synthetases is a single G=U base pair in the amino acid arm of tRNA$^{Ala}$ (Fig. 27–18a). A short RNA with as few as 7 bp arranged in a simple hairpin minihelix is efficiently aminoacylated by the Ala-tRNA synthetase, as long as the RNA contains the critical G=U (Fig. 27–18b). This relatively simple alanine system may be an evolutionary relic of a period when RNA oligonucleotides, ancestors to tRNA, were aminoacylated in a primitive system for protein synthesis.

## Stage 2: A Specific Amino Acid Initiates Protein Synthesis

Protein synthesis begins at the amino-terminal end and proceeds by the stepwise addition of amino acids to the carboxyl-terminal end of the growing polypeptide, as determined by Howard Dintzis in 1961 (Fig. 27–19). The AUG initiation codon thus specifies an *amino-terminal* methionine residue. Although methionine has only one codon, (5′)AUG, all organisms have two tRNAs for methionine. One is used exclusively when (5′)AUG is the initiation codon for protein synthesis. The other is used to code for a Met residue in an internal position in a polypeptide.

The distinction between an initiating (5′)AUG and an internal one is straightforward. In bacteria, the two types of tRNA specific for methionine are designated tRNA$^{Met}$ and tRNA$^{fMet}$. The amino acid incorporated in response to the (5′)AUG initiation codon is *N*-formylmethionine (fMet). It arrives at the ribosome as *N*-formylmethionyl-tRNA$^{fMet}$ (fMet-tRNA$^{fMet}$), which is formed in two successive reactions. First, methionine is attached to tRNA$^{fMet}$ by the Met-tRNA synthetase (which in *E. coli* aminoacylates both tRNA$^{fMet}$ and tRNA$^{Met}$):

Methionine + tRNA$^{fMet}$ + ATP $\longrightarrow$

Met-tRNA$^{fMet}$ + AMP + PP$_i$

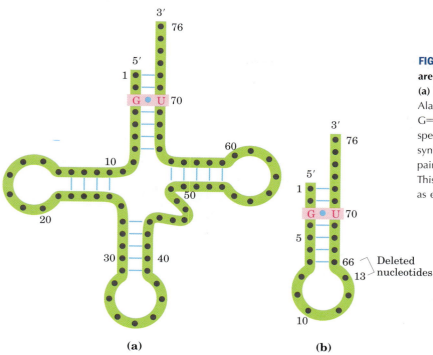

**(a)**

**(b)**

**FIGURE 27–18 Structural elements of tRNA^Ala that are required for recognition by Ala-tRNA synthetase.** (a) The tRNA^Ala structural elements recognized by the Ala-tRNA synthetase are unusually simple. A single G≡U base pair (pink) is the only element needed for specific binding and aminoacylation. (b) A short synthetic RNA minihelix, with the critical G≡U base pair but lacking most of the remaining tRNA structure. This is specifically aminoacylated with alanine almost as efficiently as the complete tRNA^Ala.

Next, a transformylase transfers a formyl group from $N^{10}$-formyltetrahydrofolate to the amino group of the Met residue:

$$N^{10}\text{-Formyltetrahydrofolate} + \text{Met-tRNA}^{\text{fMet}} \longrightarrow$$
$$\text{tetrahydrofolate} + \text{fMet-tRNA}^{\text{fMet}}$$

The transformylase is more selective than the Met-tRNA synthetase; it is specific for Met residues attached to tRNA^fMet, presumably recognizing some unique structural feature of that tRNA. By contrast, Met-tRNA^Met inserts methionine in interior positions in polypeptides.

Addition of the $N$-formyl group to the amino group of methionine by the transformylase prevents fMet from entering interior positions in a polypeptide while also allowing fMet-tRNA^fMet to be bound at a specific ribosomal initiation site that accepts neither Met-tRNA^Met nor any other aminoacyl-tRNA.

$$
\begin{array}{c}
\phantom{H-C-}\text{H}\phantom{-}\text{COO}^- \\
\phantom{H-C-}| \phantom{---}| \\
\text{H}-\text{C}-\text{N}-\text{C}-\text{H} \\
\phantom{H-}\|\phantom{-----}| \\
\phantom{H-}\text{O}\phantom{-----}\text{CH}_2 \\
\phantom{H-C-N-C}| \\
\phantom{H-C-N-C}\text{CH}_2 \\
\phantom{H-C-N-C}| \\
\phantom{H-C-N-C}\text{S} \\
\phantom{H-C-N-C}| \\
\phantom{H-C-N-C}\text{CH}_3
\end{array}
$$

*N*-Formylmethionine

In eukaryotic cells, all polypeptides synthesized by cytosolic ribosomes begin with a Met residue (rather than fMet), but, again, the cell uses a specialized initiating

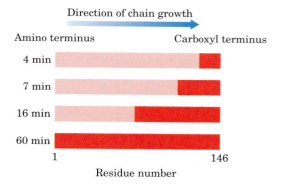

Direction of chain growth

Amino terminus — Carboxyl terminus

4 min

7 min

16 min

60 min

1 — 146

Residue number

**FIGURE 27–19 Proof that polypeptides grow by addition of amino acid residues to the carboxyl end: the Dintzis experiment.** Reticulocytes (immature erythrocytes) actively synthesizing hemoglobin were incubated with radioactive leucine (selected because it occurs frequently in both the α- and β-globin chains). Samples of completed α chains were isolated from the reticulocytes at various times afterward, and the distribution of radioactivity was determined. The dark red zones show the portions of completed α-globin chains containing radioactive Leu residues. At 4 min, only a few residues at the carboxyl end of α-globin were labeled, because the only *complete* globin chains with incorporated label after 4 min were those that had nearly completed synthesis at the time the label was added. With longer incubation times, successively longer segments of the polypeptide contained labeled residues, always in a block at the carboxyl end of the chain. The unlabeled end of the polypeptide (the amino terminus) was thus defined as the initiating end, which means that polypeptides grow by successive addition of amino acids to the carboxyl end.

tRNA that is distinct from the tRNA$^{Met}$ used at (5′)AUG codons at interior positions in the mRNA. Polypeptides synthesized by mitochondrial and chloroplast ribosomes, however, begin with *N*-formylmethionine. This strongly supports the view that mitochondria and chloroplasts originated from bacterial ancestors that were symbiotically incorporated into precursor eukaryotic cells at an early stage of evolution (see Fig. 1–36).

How can the single (5′)AUG codon distinguish between the starting *N*-formylmethionine (or methionine, in eukaryotes) and interior Met residues? The details of the initiation process provide the answer.

**The Three Steps of Initiation** The **initiation** of polypeptide synthesis in bacteria requires (1) the 30S ribosomal subunit, (2) the mRNA coding for the polypeptide to be made, (3) the initiating fMet-tRNA$^{fMet}$, (4) a set of three proteins called initiation factors (IF-1, IF-2, and IF-3), (5) GTP, (6) the 50S ribosomal subunit, and (7) Mg$^{2+}$. Formation of the initiation complex takes place in three steps (Fig. 27–20).

In step ① the 30S ribosomal subunit binds two initiation factors, IF-1 and IF-3. Factor IF-3 prevents the 30S and 50S subunits from combining prematurely. The mRNA then binds to the 30S subunit. The initiating (5′)AUG is guided to its correct position by the **Shine-Dalgarno sequence** (named for Australian researchers John Shine and Lynn Dalgarno, who identified it) in the mRNA. This consensus sequence is an initiation signal of four to nine purine residues, 8 to 13 bp to the 5′ side of the initiation codon (Fig. 27–21a). The sequence base-pairs with a complementary pyrimidine-rich sequence near the 3′ end of the 16S rRNA of the 30S ribosomal subunit (Fig. 27–21b). This mRNA-rRNA interaction positions the initiating (5′)AUG sequence of the mRNA in the precise position on the 30S subunit where it is required for initiation of translation. The particular (5′)AUG where fMet-tRNA$^{fMet}$ is to be bound is distinguished from other methionine codons by its proximity to the Shine-Dalgarno sequence in the mRNA.

Bacterial ribosomes have three sites that bind aminoacyl-tRNAs, the **aminoacyl (A) site**, the **peptidyl (P) site,** and the **exit (E) site.** Both the 30S and the 50S subunits contribute to the characteristics of the A and P sites, whereas the E site is largely confined to the 50S subunit. The initiating (5′)AUG is positioned at the P site, the only site to which fMet-tRNA$^{fMet}$ can bind (Fig. 27–20). The fMet-tRNA$^{fMet}$ is the only aminoacyl-tRNA that binds first to the P site; during the subsequent elongation stage, all other incoming aminoacyl-tRNAs (including the Met-tRNA$^{Met}$ that binds to interior AUG codons) bind first to the A site and only subsequently to the P and E sites. The E site is the site from which the "uncharged" tRNAs leave during elongation. Factor IF-1 binds at the A site and prevents tRNA binding at this site during initiation.

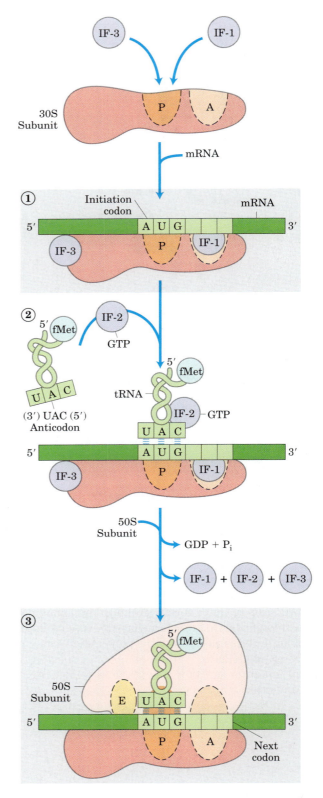

**FIGURE 27–20 Formation of the initiation complex in bacteria.** The complex forms in three steps (described in the text) at the expense of the hydrolysis of GTP to GDP and P$_i$. IF-1, IF-2, and IF-3 are initiation factors. P designates the peptidyl site, A the aminoacyl site, and E the exit site. Here the anticodon of the tRNA is oriented 3′ to 5′, left to right, as in Figure 27–8 but opposite to the orientation in Figures 27–16 and 27–18.

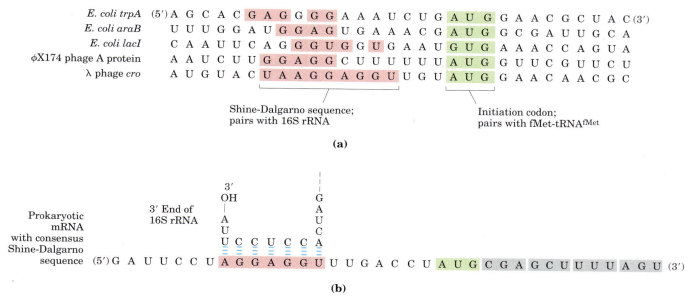

| | | |
|---|---|---|
| *E. coli trpA* | (5') A G C A C | **G A G G G G** A A A U C U G **A U G** G A A C G C U A C (3') |
| *E. coli araB* | U U U G G A U | **G G A G** U G A A A C G **A U G** G C G A U U G C A |
| *E. coli lacI* | C A A U U C A G | **G G U G G** U G A A U **G U G** A A A C C A G U A |
| φX174 phage A protein | A A U C U U | **G G A G G** C U U U U U U **A U G** G U U C G U U C U |
| λ phage *cro* | A U G U A C | **U A A G G A G G U** U G U **A U G** G A A C A A C G C |

Shine-Dalgarno sequence;
pairs with 16S rRNA

Initiation codon;
pairs with fMet-tRNA^fMet

**(a)**

Prokaryotic
mRNA
with consensus
Shine-Dalgarno
sequence

3' End of
16S rRNA

3'
OH
|
A
U
U C C U C C A

G
A
U
C

(5') G A U U C C U **A G G A G G U** U U G A C C U **A U G C G A G C U U U U A G U** (3')

**(b)**

**FIGURE 27–21 Messenger RNA sequences that serve as signals for initiation of protein synthesis in bacteria. (a)** Alignment of the initiating AUG (shaded in green) at its correct location on the 30S ribosomal subunit depends in part on upstream Shine-Dalgarno sequences (pink). Portions of the mRNA transcripts of five prokaryotic genes are shown. Note the unusual example of the *E. coli* LacI protein, which initiates with a GUG (Val) codon (see Box 27–2). **(b)** The Shine-Dalgarno sequence of the mRNA pairs with a sequence near the 3' end of the 16S rRNA.

In step ② of the initiation process (Fig. 27–20), the complex consisting of the 30S ribosomal subunit, IF-3, and mRNA is joined by both GTP-bound IF-2 and the initiating fMet-tRNA^fMet. The anticodon of this tRNA now pairs correctly with the mRNA's initiation codon.

In step ③ this large complex combines with the 50S ribosomal subunit; simultaneously, the GTP bound to IF-2 is hydrolyzed to GDP and $P_i$, which are released from the complex. All three initiation factors depart from the ribosome at this point.

Completion of the steps in Figure 27–20 produces a functional 70S ribosome called the **initiation complex,** containing the mRNA and the initiating fMet-tRNA^fMet. The correct binding of the fMet-tRNA^fMet to the P site in the complete 70S initiation complex is assured by at least three points of recognition and attachment: the codon-anticodon interaction involving the initiation AUG fixed in the P site; interaction between the Shine-Dalgarno sequence in the mRNA and the 16S rRNA; and binding interactions between the ribosomal P site and the fMet-tRNA^fMet. The initiation complex is now ready for elongation.

### Initiation in Eukaryotic Cells

Translation is generally similar in eukaryotic and bacterial cells; most of the significant differences are in the mechanism of initiation. Eukaryotic mRNAs are bound to the ribosome as a complex with a number of specific binding proteins. Several of these tie together the 5' and 3' ends of the message. At the 3' end, the mRNA is bound by the poly(A) binding protein (PAB). Eukaryotic cells have at least nine initiation factors. A complex called eIF4F, which includes the proteins eIF4E, eIF4G, and eIF4A, binds to the 5' cap (see Fig. 26–12) through eIF4E. The protein eIF4G binds to both eIF4E and PAB, effectively tying them together (Fig. 27–22). The protein eIF4A has an RNA helicase activity. It is the eIF4F complex that associates

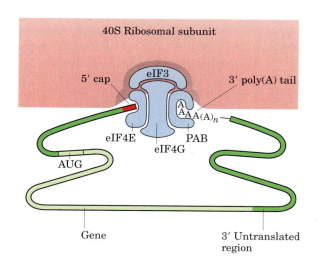

**FIGURE 27–22 Protein complexes in the formation of a eukaryotic initiation complex.** The 3' and 5' ends of eukaryotic mRNAs are linked by a complex of proteins that includes several initiation factors and the poly(A) binding protein (PAB). The factors eIF4E and eIF4G are part of a larger complex called eIF4F. This complex binds to the 40S ribosomal subunit.

**TABLE 27–8** Protein Factors Required for Initiation of Translation in Bacterial and Eukaryotic Cells

| Factor | Function |
|---|---|
| **Bacterial** | |
| IF-1 | Prevents premature binding of tRNAs to A site |
| IF-2 | Facilitates binding of fMet-tRNA$^{fMet}$ to 30S ribosomal subunit |
| IF-3 | Binds to 30S subunit; prevents premature association of 50S subunit; enhances specificity of P site for fMet-tRNA$^{fMet}$ |
| **Eukaryotic*** | |
| eIF2 | Facilitates binding of initiating Met-tRNA$^{Met}$ to 40S ribosomal subunit |
| eIF2B, eIF3 | First factors to bind 40S subunit; facilitate subsequent steps |
| eIF4A | RNA helicase activity removes secondary structure in the mRNA to permit binding to 40S subunit; part of the eIF4F complex |
| eIF4B | Binds to mRNA; facilitates scanning of mRNA to locate the first AUG |
| eIF4E | Binds to the 5′ cap of mRNA; part of the eIF4F complex |
| eIF4G | Binds to eIF4E and to poly(A) binding protein (PAB); part of the eIF4F complex |
| eIF5 | Promotes dissociation of several other initiation factors from 40S subunit as a prelude to association of 60S subunit to form 80S initiation complex |
| eIF6 | Facilitates dissociation of inactive 80S ribosome into 40S and 60S subunits |

*The prefix "e" identifies these as eukaryotic factors.

with another factor, eIF3, and with the 40S ribosomal subunit. The efficiency of translation is affected by many properties of the mRNA and proteins in this complex, including the length of the 3′ poly(A) tract (in most cases, longer is better). The end-to-end arrangement of the eukaryotic mRNA facilitates translational regulation of gene expression, considered in Chapter 28.

The initiating (5′)AUG is detected within the mRNA not by its proximity to a Shine-Dalgarno-like sequence but by a scanning process: a scan of the mRNA from the 5′ end until the first AUG is encountered, signaling the beginning of the reading frame. The eIF4F complex is probably involved in this process, perhaps using the RNA helicase activity of eIF4A to eliminate secondary structure in the 5′ untranslated portion of the mRNA. Scanning is also facilitated by another protein, eIF4B.

The roles of the various bacterial and eukaryotic initiation factors in the overall process are summarized in Table 27–8. The mechanism by which these proteins act is an important area of investigation.

## Stage 3: Peptide Bonds Are Formed in the Elongation Stage

The third stage of protein synthesis is **elongation.** Again, our initial focus is on bacterial cells. Elongation requires (1) the initiation complex described above, (2) aminoacyl-tRNAs, (3) a set of three soluble cytosolic proteins called **elongation factors** (EF-Tu, EF-Ts, and EF-G in bacteria), and (4) GTP. Cells use three steps to add each amino acid residue, and the steps are repeated as many times as there are residues to be added.

***Elongation Step 1: Binding of an Incoming Aminoacyl-tRNA*** In the first step of the elongation cycle (Fig. 27–23), the appropriate incoming aminoacyl-tRNA binds to a complex of GTP-bound EF-Tu. The resulting aminoacyl-tRNA–EF-Tu–GTP complex binds to the A site of the 70S initiation complex. The GTP is hydrolyzed and an EF-Tu–GDP complex is released from the 70S ribosome. The EF-Tu–GTP complex is regenerated in a process involving EF-Ts and GTP.

***Elongation Step 2: Peptide Bond Formation*** A peptide bond is now formed between the two amino acids bound by their tRNAs to the A and P sites on the ribosome. This occurs by the transfer of the initiating *N*-formylmethionyl group from its tRNA to the amino group of the second amino acid, now in the A site (Fig. 27–24). The α-amino group of the amino acid in the A site acts as a nucleophile, displacing the tRNA in the P site to form the peptide bond. This reaction produces a dipeptidyl-tRNA in the A site, and the now "uncharged" (deacylated) tRNA$^{fMet}$ remains bound to the P site. The tRNAs then shift to a hybrid binding state, with elements of each spanning two different sites on the ribosome, as shown in Figure 27–24.

The enzymatic activity that catalyzes peptide bond formation has historically been referred to as **peptidyl transferase** and was widely assumed to be intrinsic to one or more of the proteins in the large ribosomal subunit. We now know that this reaction is catalyzed by the 23S rRNA (Fig. 27–9), adding to the known catalytic repertoire of ribozymes. This discovery has interesting implications for the evolution of life (Box 27–3).

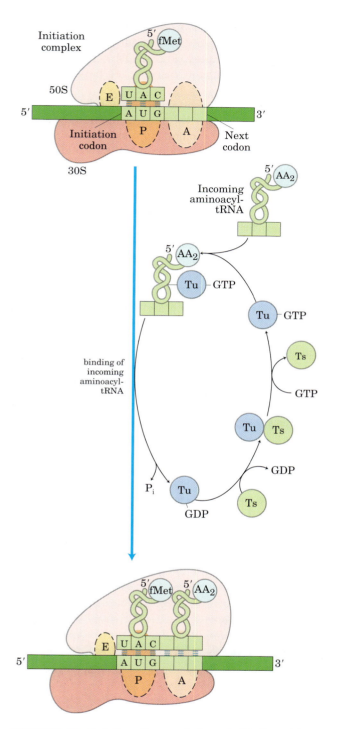

**FIGURE 27–23 First elongation step in bacteria: binding of the second aminoacyl-tRNA.** The second aminoacyl-tRNA enters the A site of the ribosome bound to EF-Tu (shown here as Tu), which also contains GTP. Binding of the second aminoacyl-tRNA to the A site is accompanied by hydrolysis of the GTP to GDP and Pi and release of the EF-Tu–GDP complex from the ribosome. The bound GDP is released when the EF-Tu–GDP complex binds to EF-Ts, and EF-Ts is subsequently released when another molecule of GTP binds to EF-Tu. This recycles EF-Tu and makes it available to repeat the cycle.

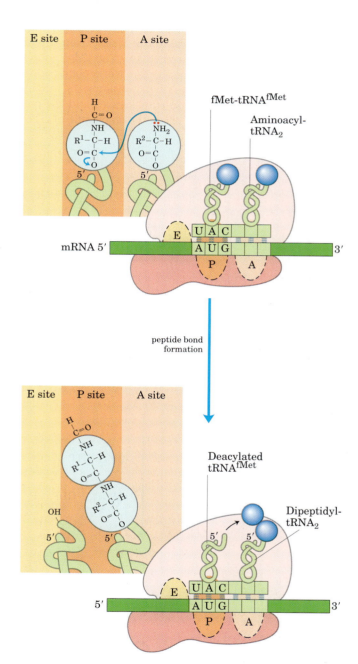

**FIGURE 27–24 Second elongation step in bacteria: formation of the first peptide bond.** The peptidyl transferase catalyzing this reaction is the 23S rRNA ribozyme. The *N*-formylmethionyl group is transferred to the amino group of the second aminoacyl-tRNA in the A site, forming a dipeptidyl-tRNA. At this stage, both tRNAs bound to the ribosome shift position in the 50S subunit to take up a hybrid binding state. The uncharged tRNA shifts so that its 3′ and 5′ ends are in the E site. Similarly, the 3′ and 5′ ends of the peptidyl tRNA shift to the P site. The anticodons remain in the A and P sites.

***Elongation Step 3: Translocation*** In the final step of the elongation cycle, **translocation,** the ribosome moves one codon toward the 3' end of the mRNA (Fig. 27–25a). This movement shifts the anticodon of the dipeptidyl-tRNA, which is still attached to the second codon of the mRNA, from the A site to the P site, and shifts the deacylated tRNA from the P site to the E site, from where the tRNA is released into the cytosol. The third codon of the mRNA now lies in the A site and the second codon in the P site. Movement of the ribosome along the mRNA requires EF-G (also known as translocase) and the energy provided by hydrolysis of another molecule of GTP.

A change in the three-dimensional conformation of the entire ribosome results in its movement along the mRNA. Because the structure of EF-G mimics the structure of the EF-Tu–tRNA complex (Fig. 27–25b), EF-G can bind the A site and presumably displace the peptidyl-tRNA.

The ribosome, with its attached dipeptidyl-tRNA and mRNA, is now ready for the next elongation cycle and attachment of a third amino acid residue. This process occurs in the same way as addition of the second residue (as shown in Figs 27–23, 27–24, and 27–25). For each amino acid residue correctly added to the growing polypeptide, two GTPs are hydrolyzed to GDP and $P_i$ as the ribosome moves from codon to codon along the mRNA toward the 3' end.

The polypeptide remains attached to the tRNA of the most recent amino acid to be inserted. This association maintains the functional connection between the information in the mRNA and its decoded polypeptide output. At the same time, the ester linkage between this tRNA and the carboxyl terminus of the growing polypeptide activates the terminal carboxyl group for nucleophilic attack by the incoming amino acid to form a new peptide bond (Fig. 27–24). As the existing ester linkage between the polypeptide and tRNA is broken during

**FIGURE 27–25 Third elongation step in bacteria: translocation.**
**(a)** The ribosome moves one codon toward the 3' end of the mRNA, using energy provided by hydrolysis of GTP bound to EF-G (translocase). The dipeptidyl-tRNA is now entirely in the P site, leaving the A site open for the incoming (third) aminoacyl-tRNA. The uncharged tRNA dissociates from the E site, and the elongation cycle begins again. **(b)** The structure of EF-G mimics the structure of EF-Tu complexed with tRNA. Shown here are (left) EF-Tu complexed with tRNA (green) (PDB ID 1B23) and (right) EF-G complexed with GDP (red) (PDB ID 1DAR). The carboxyl-terminal part of EF-G (dark gray) mimics the structure of the anticodon loop of tRNA in both shape and charge distribution.

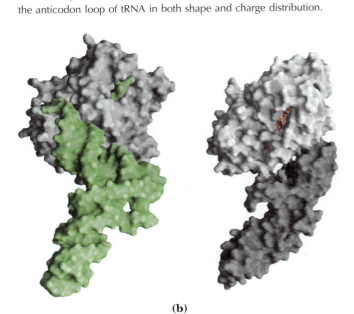

**(b)**

**(a)**

peptide bond formation, the linkage between the polypeptide and the information in the mRNA persists, because each newly added amino acid is still attached to its tRNA.

The elongation cycle in eukaryotes is quite similar to that in prokaryotes. Three eukaryotic elongation factors (eEF1$\alpha$, eEF1$\beta\gamma$, and eEF2) have functions analogous to those of the bacterial elongation factors (EF-Tu, EF-Ts, and EF-G, respectively). Eukaryotic ribosomes do not have an E site; uncharged tRNAs are expelled directly from the P site.

**Proofreading on the Ribosome** The GTPase activity of EF-Tu during the first step of elongation in bacterial cells (Fig. 27–23) makes an important contribution to the rate and fidelity of the overall biosynthetic process. Both the EF-Tu–GTP and EF-Tu–GDP complexes exist for a few milliseconds before they dissociate. These two intervals provide opportunities for the codon-anticodon interactions to be proofread. Incorrect aminoacyl-tRNAs normally dissociate from the A site during one of these periods. If the GTP analog guanosine 5′-O-(3-thiotriphosphate) (GTP$\gamma$S) is used in place of GTP, hydrolysis is slowed, improving the fidelity (by increasing the proofreading intervals) but reducing the rate of protein synthesis.

Guanosine 5′-*O*-(3-thiotriphosphate)
(GTP$\gamma$S)

The process of protein synthesis (including the characteristics of codon-anticodon pairing already described) has clearly been optimized through evolution to balance the requirements of both speed and fidelity. Improved fidelity might diminish speed, whereas increases in speed would probably compromise fidelity. Note that the proofreading mechanism on the ribosome establishes only that the proper codon-anticodon pairing has taken place. The identity of the amino acid attached to a tRNA is not checked on the ribosome. If a tRNA is successfully aminoacylated with the wrong amino acid (as can be done experimentally), this incorrect amino acid is efficiently incorporated into a protein in response to whatever codon is normally recognized by the tRNA.

## Stage 4: Termination of Polypeptide Synthesis Requires a Special Signal

Elongation continues until the ribosome adds the last amino acid coded by the mRNA. **Termination,** the fourth stage of polypeptide synthesis, is signaled by the presence of one of three termination codons in the mRNA (UAA, UAG, UGA), immediately following the final coded amino acid. Mutations in a tRNA anticodon that allow an amino acid to be inserted at a termination codon are generally deleterious to the cell (Box 27–4).

In bacteria, once a termination codon occupies the ribosomal A site, three **termination factors,** or **release factors**—the proteins RF-1, RF-2, and RF-3—contribute to (1) hydrolysis of the terminal peptidyl-tRNA bond; (2) release of the free polypeptide and the last tRNA, now uncharged, from the P site; and (3) dissociation of the 70S ribosome into its 30S and 50S subunits, ready to start a new cycle of polypeptide synthesis (Fig. 27–26). RF-1 recognizes the termination codons UAG and UAA, and RF-2 recognizes UGA and UAA. Either RF-1 or RF-2 (depending on which codon is present) binds at a termination codon and induces peptidyl transferase to transfer the growing polypeptide to a water molecule rather than to another amino acid. The release factors have domains thought to mimic the structure of tRNA, as shown for the elongation factor EF-G in Figure 27–25b. The specific function of RF-3 has not been firmly established, although it is thought to release the ribosomal subunit. In eukaryotes, a single release factor, eRF, recognizes all three termination codons.

**Energy Cost of Fidelity in Protein Synthesis** Synthesis of a protein true to the information specified in its mRNA requires energy. Formation of each aminoacyl-tRNA uses two high-energy phosphate groups. An additional ATP is consumed each time an incorrectly activated amino acid is hydrolyzed by the deacylation activity of an aminoacyl-tRNA synthetase, as part of its proofreading activity. A GTP is cleaved to GDP and P$_i$ during the first elongation step, and another during the translocation step. Thus, on average, the energy derived from the hydrolysis of more than four NTPs to NDPs is required for the formation of each peptide bond of a polypeptide.

This represents an exceedingly large thermodynamic "push" in the direction of synthesis: at least $4 \times 30.5$ kJ/mol = 122 kJ/mol of phosphodiester bond energy to generate a peptide bond, which has a standard free energy of hydrolysis of only about $-21$ kJ/mol. The net free-energy change during peptide bond synthesis is thus $-101$ kJ/mol. Proteins are information-containing polymers. The biochemical goal is not simply the formation of a peptide bond but the formation of a peptide bond between two *specified* amino acids. Each of the high-energy phosphate compounds expended in this process plays a critical role in maintaining proper alignment between each new codon in the mRNA and its associated amino acid at the growing end of the polypeptide. This energy permits very high fidelity in the biological translation of the genetic message of mRNA into the amino acid sequence of proteins.

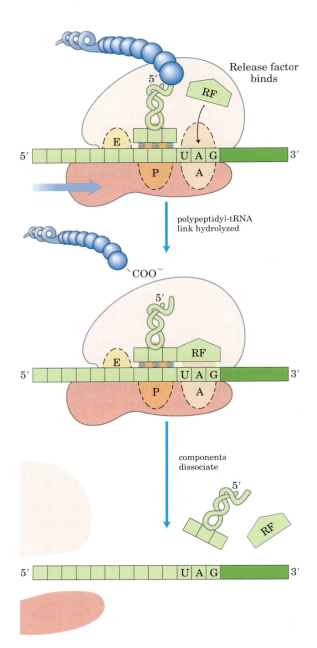

**FIGURE 27–26 Termination of protein synthesis in bacteria.** Termination occurs in response to a termination codon in the A site. First, a release factor, RF (RF-1 or RF-2, depending on which termination codon is present), binds to the A site. This leads to hydrolysis of the ester linkage between the nascent polypeptide and the tRNA in the P site and release of the completed polypeptide. Finally, the mRNA, deacylated tRNA, and release factor leave the ribosome, and the ribosome dissociates into its 30S and 50S subunits.

***Rapid Translation of a Single Message by Polysomes*** Large clusters of 10 to 100 ribosomes that are very active in protein synthesis can be isolated from both eukaryotic and bacterial cells. Electron micrographs show a fiber between adjacent ribosomes in the cluster, which is called a **polysome** (Fig. 27–27). The connecting strand is a single molecule of mRNA that is being translated simultaneously by many closely spaced ribosomes, allowing the highly efficient use of the mRNA.

In bacteria, transcription and translation are tightly coupled. Messenger RNAs are synthesized and translated in the same 5′→3′ direction. Ribosomes begin translating the 5′ end of the mRNA before transcription is complete (Fig. 27–28). The situation is quite different in eukaryotic cells, where newly transcribed mRNAs must leave the nucleus before they can be translated.

Bacterial mRNAs generally exist for just a few minutes (p. 1020) before they are degraded by nucleases. In order to maintain high rates of protein synthesis, the mRNA for a given protein or set of proteins must be made continuously and translated with maximum efficiency. The short lifetime of mRNAs in bacteria allows a rapid cessation of synthesis when the protein is no longer needed.

## Stage 5: Newly Synthesized Polypeptide Chains Undergo Folding and Processing

In the final stage of protein synthesis, the nascent polypeptide chain is folded and processed into its biologically active form. During or after its synthesis, the polypeptide progressively assumes its native conformation, with the formation of appropriate hydrogen bonds and van der Waals, ionic, and hydrophobic interactions. In this way the linear, or one-dimensional, genetic message in the mRNA is converted into the three-dimensional structure of the protein. Some newly made proteins, both prokaryotic and eukaryotic, do not attain their final biologically active conformation until they have been altered by one or more processing reactions called **posttranslational modifications.**

***Amino-Terminal and Carboxyl-Terminal Modifications*** The first residue inserted in all polypeptides is *N*-formylmethionine (in bacteria) or methionine (in eukaryotes). However, the formyl group, the amino-terminal Met residue, and often additional amino-terminal (and, in some cases, carboxyl-terminal) residues may be removed enzymatically in formation of the final functional protein. In as many as 50% of eukaryotic proteins, the amino group of the amino-terminal residue is *N*-acetylated after translation. Carboxyl-terminal residues are also sometimes modified.

***Loss of Signal Sequences*** As we shall see in Section 27.3, the 15 to 30 residues at the amino-terminal end of some proteins play a role in directing the protein to its ultimate destination in the cell. Such **signal sequences** are ultimately removed by specific peptidases.

***Modification of Individual Amino Acids*** The hydroxyl groups of certain Ser, Thr, and Tyr residues of some proteins are enzymatically phosphorylated by ATP (Fig.

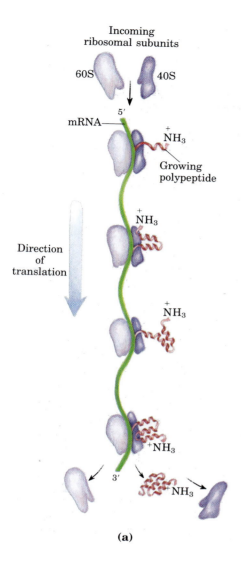

**(a)**

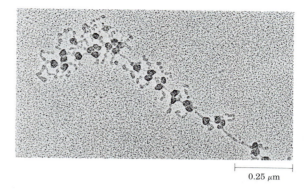

0.25 μm

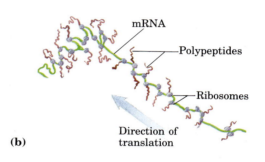

**(b)**

**FIGURE 27–27** **Polysome.** **(a)** Four ribosomes translating a eukaryotic mRNA molecule simultaneously, moving from the 5′ end to the 3′ end and synthesizing a polypeptide from the amino terminus to the carboxyl terminus. **(b)** Electron micrograph and explanatory diagram of a polysome from the silk gland of a silkworm larva. The mRNA is being translated by many ribosomes simultaneously. The nascent polypeptides become longer as the ribosomes move toward the 3′ end of the mRNA. The final product of this process is silk fibroin.

27–29a); the phosphate groups add negative charges to these polypeptides. The functional significance of this modification varies from one protein to the next. For example, the milk protein casein has many phospho-serine groups that bind $Ca^{2+}$. Calcium, phosphate, and amino acids are all valuable to suckling young, so casein efficiently provides three essential nutrients. And as we have seen in numerous instances, phosphorylation-dephosphorylation cycles regulate the activity of many enzymes and regulatory proteins.

Extra carboxyl groups may be added to Glu residues of some proteins. For example, the blood-clotting protein prothrombin contains a number of γ-carboxygluta-mate residues (Fig. 27–29b) in its amino-terminal region, introduced by an enzyme that requires vitamin K. These carboxyl groups bind $Ca^{2+}$, which is required to initiate the clotting mechanism.

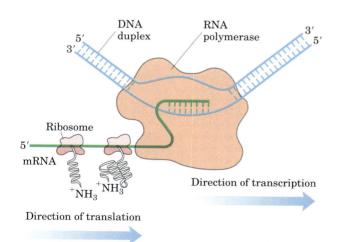

**FIGURE 27–28** **Coupling of transcription and translation in bacteria.** The mRNA is translated by ribosomes while it is still being transcribed from DNA by RNA polymerase. This is possible because the mRNA in bacteria does not have to be transported from a nucleus to the cytoplasm before encountering ribosomes. In this schematic diagram the ribosomes are depicted as smaller than the RNA polymerase. In reality the ribosomes ($M_r$ $2.7 \times 10^6$) are an order of magnitude larger than the RNA polymerase ($M_r$ $3.9 \times 10^5$).

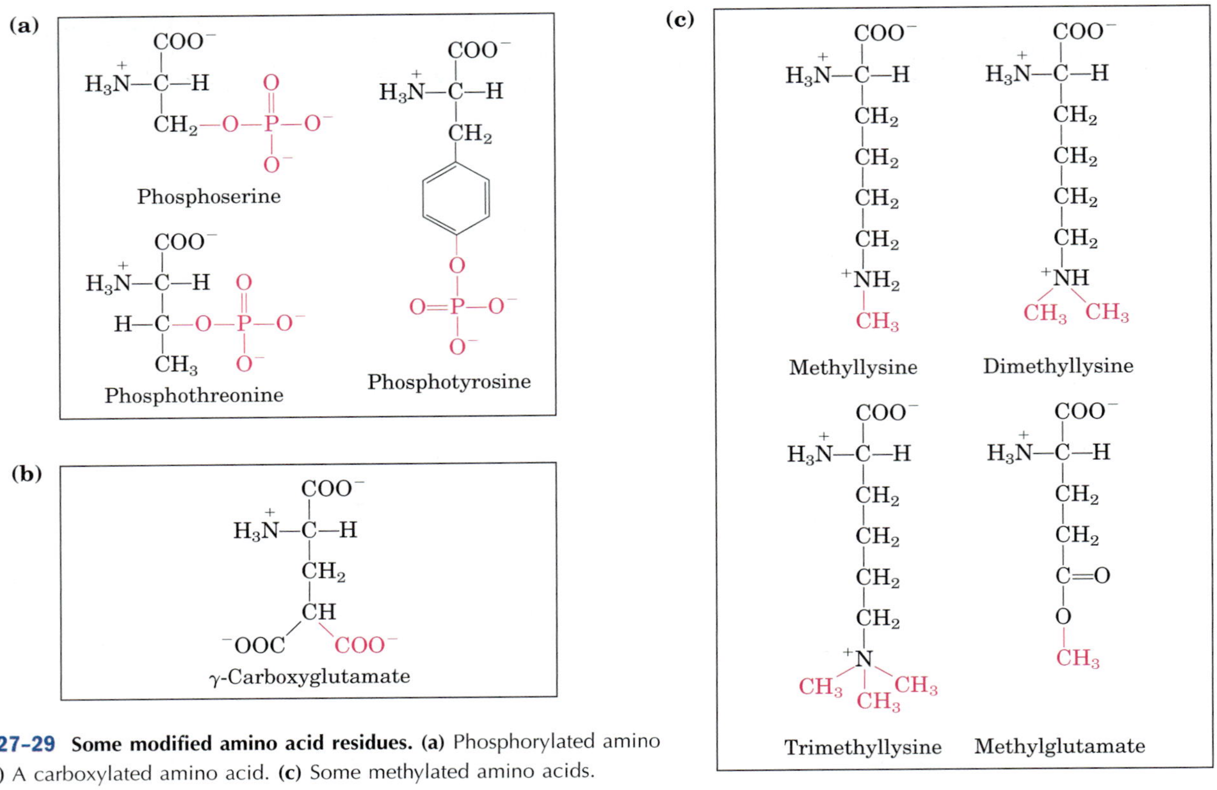

**(a)**

Phosphoserine

Phosphothreonine

Phosphotyrosine

**(b)**

γ-Carboxyglutamate

**(c)**

Methyllysine        Dimethyllysine

Trimethyllysine    Methylglutamate

**FIGURE 27–29 Some modified amino acid residues. (a)** Phosphorylated amino acids. **(b)** A carboxylated amino acid. **(c)** Some methylated amino acids.

Monomethyl- and dimethyllysine residues (Fig. 27–29c) occur in some muscle proteins and in cytochrome *c*. The calmodulin of most species contains one trimethyllysine residue at a specific position. In other proteins, the carboxyl groups of some Glu residues undergo methylation, removing their negative charge.

**Attachment of Carbohydrate Side Chains** The carbohydrate side chains of glycoproteins are attached covalently during or after synthesis of the polypeptide. In some glycoproteins, the carbohydrate side chain is attached enzymatically to Asn residues (*N*-linked oligosaccharides), in others to Ser or Thr residues (*O*-linked oligosaccharides) (see Fig. 7–31). Many proteins that function extracellularly, as well as the lubricating proteoglycans that coat mucous membranes, contain oligosaccharide side chains (see Fig. 7–29).

**Addition of Isoprenyl Groups** A number of eukaryotic proteins are modified by the addition of groups derived from isoprene (isoprenyl groups). A thioether bond is formed between the isoprenyl group and a Cys residue of the protein (see Fig. 11–14). The isoprenyl groups are derived from pyrophosphorylated intermediates of the cholesterol biosynthetic pathway (see Fig. 21–33), such as farnesyl pyrophosphate (Fig. 27–30). Proteins modified in this way include the Ras proteins, products of the *ras* oncogenes and proto-oncogenes, and G proteins (both discussed in Chapter 12), and lamins, pro-

teins found in the nuclear matrix. The isoprenyl group helps to anchor the protein in a membrane. The transforming (carcinogenic) activity of the *ras* oncogene is lost when isoprenylation of the Ras protein is blocked, a finding that has stimulated interest in identifying inhibitors of this posttranslational modification pathway for use in cancer chemotherapy.

**Addition of Prosthetic Groups** Many prokaryotic and eukaryotic proteins require for their activity covalently bound prosthetic groups. Two examples are the biotin

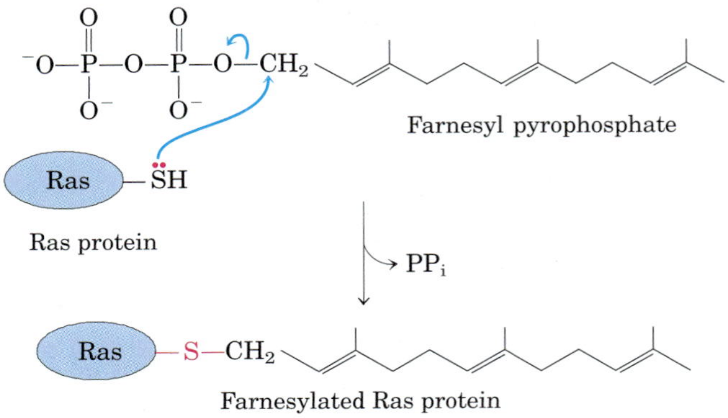

Farnesyl pyrophosphate

Ras protein

Farnesylated Ras protein

**FIGURE 27–30 Farnesylation of a Cys residue.** The thioether linkage is shown in red. The Ras protein is the product of the *ras* oncogene.

BOX 27–4 WORKING IN BIOCHEMISTRY

## Induced Variation in the Genetic Code: Nonsense Suppression

When a mutation introduces a termination codon in the interior of a gene, translation is prematurely halted and the incomplete polypeptide is usually inactive. These are called nonsense mutations. The gene can be restored to normal function if a second mutation either (1) converts the misplaced termination codon to a codon specifying an amino acid or (2) suppresses the effects of the termination codon. Such restorative mutations are called **nonsense suppressors;** they generally involve mutations in tRNA genes to produce altered (suppressor) tRNAs that can recognize the termination codon and insert an amino acid at that position. Most known suppressor tRNAs have single base substitutions in their anticodons.

Suppressor tRNAs constitute an experimentally induced variation in the genetic code to allow the reading of what are usually termination codons, much like the naturally occurring code variations described in Box 27–2. Nonsense suppression does not completely disrupt normal information transfer in a cell, because the cell usually has several copies of each tRNA gene; some of these duplicate genes are weakly expressed and account for only a minor part of the cellular pool of a particular tRNA. Suppressor mutations usually involve a "minor" tRNA, leaving the major tRNA to read its codon normally.

For example, *E. coli* has three identical genes for tRNA$^{Tyr}$, each producing a tRNA with the anticodon

(5′)GUA. One of these genes is expressed at relatively high levels and thus its product represents the major tRNA$^{Tyr}$ species; the other two genes are transcribed in only small amounts. A change in the anticodon of the tRNA product of one of these duplicate tRNA$^{Tyr}$ genes, from (5′)GUA to (5′)CUA, produces a minor tRNA$^{Tyr}$ species that will insert tyrosine at UAG stop codons. This insertion of tyrosine at UAG is carried out inefficiently, but it can produce enough full-length protein from a gene with a nonsense mutation to allow the cell to survive. The major tRNA$^{Tyr}$ continues to translate the genetic code normally for the majority of proteins.

The mutation that leads to creation of a suppressor tRNA does not always occur in the anticodon. The suppression of UGA nonsense codons generally involves the tRNA$^{Trp}$ that normally recognizes UGG. The alteration that allows it to read UGA (and insert Trp residues at these positions) is a G to A change at position 24 (in an arm of the tRNA somewhat removed from the anticodon); this tRNA can now recognize *both* UGG and UGA. A similar change is found in tRNAs involved in the most common naturally occurring variation in the genetic code (UGA = Trp; see Box 27–2).

Suppression should lead to many abnormally long proteins, but this does not always occur. We understand only a few details of the molecular events in translation termination and nonsense suppression.

molecule of acetyl-CoA carboxylase and the heme group of hemoglobin or cytochrome *c*.

***Proteolytic Processing*** Many proteins are initially synthesized as large, inactive precursor polypeptides that are proteolytically trimmed to form their smaller, active forms. Examples include proinsulin, some viral proteins, and proteases such as chymotrypsinogen and trypsinogen (see Fig. 6–33).

***Formation of Disulfide Cross-Links*** After folding into their native conformations, some proteins form intrachain or interchain disulfide bridges between Cys residues. In eukaryotes, disulfide bonds are common in proteins to be exported from cells. The cross-links formed in this way help to protect the native conformation of the protein molecule from denaturation in the extracellular environment, which can differ greatly from intracellular conditions and is generally oxidizing.

## Protein Synthesis Is Inhibited by Many Antibiotics and Toxins

Protein synthesis is a central function in cellular physiology and is the primary target of many naturally occurring antibiotics and toxins. Except as noted, these antibiotics inhibit protein synthesis in bacteria. The differences between bacterial and eukaryotic protein synthesis, though in some cases subtle, are sufficient that most of the compounds discussed below are relatively harmless to eukaryotic cells. Natural selection has favored the evolution of compounds that exploit minor differences in order to affect bacterial systems selectively, such that these biochemical weapons are synthesized by some microorganisms and are extremely toxic to others. Because nearly every step in protein synthesis can be specifically inhibited by one antibiotic or another, antibiotics have become valuable tools in the study of protein biosynthesis.

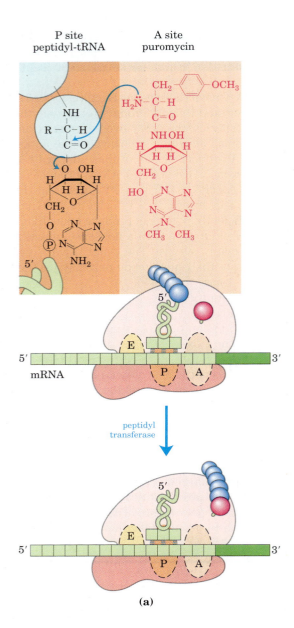

(a)

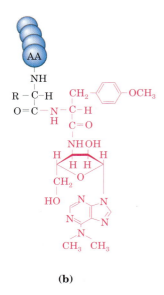

(b)

**FIGURE 27–31** **Disruption of peptide bond formation by puromycin.**
**(a)** The antibiotic puromycin resembles the aminoacyl end of a charged tRNA, and it can bind to the ribosomal A site and participate in peptide bond formation. The product of this reaction, instead of being translocated to the P site, dissociates from the ribosome, causing premature chain termination. **(b)** Peptidyl puromycin.

**Puromycin,** made by the mold *Streptomyces alboniger,* is one of the best-understood inhibitory antibiotics. Its structure is very similar to the 3′ end of an aminoacyl-tRNA, enabling it to bind to the ribosomal A site and participate in peptide bond formation, producing peptidyl-puromycin (Fig. 27–31). However, because puromycin resembles only the 3′ end of the tRNA, it does not engage in translocation and dissociates from the ribosome shortly after it is linked to the carboxyl terminus of the peptide. This prematurely terminates polypeptide synthesis.

**Tetracyclines** inhibit protein synthesis in bacteria by blocking the A site on the ribosome, preventing the binding of aminoacyl-tRNAs. **Chloramphenicol** inhibits protein synthesis by bacterial (and mitochondrial

Tetracycline

Chloramphenicol

and chloroplast) ribosomes by blocking peptidyl transfer; it does not affect cytosolic protein synthesis in eukaryotes. Conversely, **cycloheximide** blocks the peptidyl transferase of 80S eukaryotic ribosomes but not that of 70S bacterial (and mitochondrial and chloroplast) ribosomes. **Streptomycin,** a basic trisaccharide, causes misreading of the genetic code (in bacteria) at relatively low concentrations and inhibits initiation at higher concentrations.

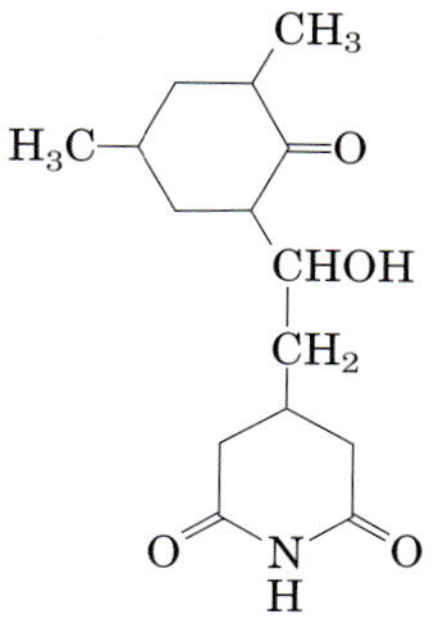

Cycloheximide

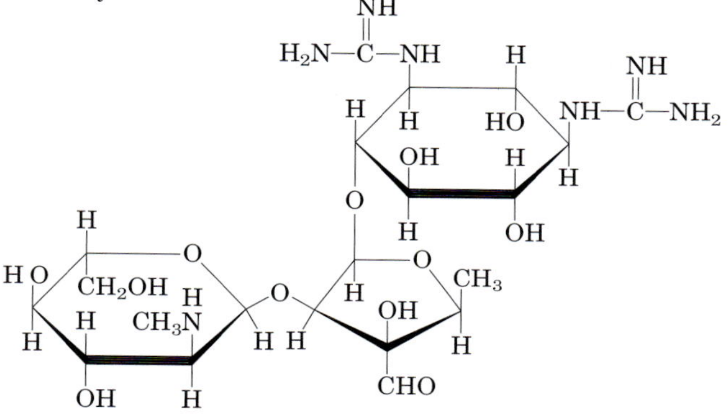

Streptomycin

Several other inhibitors of protein synthesis are notable because of their toxicity to humans and other mammals. **Diphtheria toxin** ($M_r$ 58,330) catalyzes the ADP-ribosylation of a diphthamide (a modified histidine) residue of eukaryotic elongation factor eEF2, thereby inactivating it. **Ricin** ($M_r$ 29,895), an extremely toxic protein of the castor bean, inactivates the 60S subunit of eukaryotic ribosomes by depurinating a specific adenosine in 23S rRNA.

## SUMMARY 27.2 Protein Synthesis

■ Protein synthesis occurs on the ribosomes, which consist of protein and rRNA. Bacteria have 70S ribosomes, with a large (50S) and a small (30S) subunit. Eukaryotic ribosomes are significantly larger (80S) and contain more proteins.

■ Transfer RNAs have 73 to 93 nucleotide residues, some of which have modified bases.

Each tRNA has an amino acid arm with the terminal sequence CCA(3′) to which an amino acid is esterified, an anticodon arm, a TψC arm, and a D arm; some tRNAs have a fifth arm. The anticodon is responsible for the specificity of interaction between the aminoacyl-tRNA and the complementary mRNA codon.

■ The growth of polypeptides on ribosomes begins with the amino-terminal amino acid and proceeds by successive additions of new residues to the carboxyl-terminal end.

■ Protein synthesis occurs in five stages.

*1.* Amino acids are activated by specific aminoacyl-tRNA synthetases in the cytosol. These enzymes catalyze the formation of aminoacyl-tRNAs, with simultaneous cleavage of ATP to AMP and $PP_i$. The fidelity of protein synthesis depends on the accuracy of this reaction, and some of these enzymes carry out proofreading steps at separate active sites. In bacteria, the initiating aminoacyl-tRNA in all proteins is $N$-formylmethionyl-tRNA$^{fMet}$.

*2.* Initiation of protein synthesis involves formation of a complex between the 30S ribosomal subunit, mRNA, GTP, fMet-tRNA$^{fMet}$, three initiation factors, and the 50S subunit; GTP is hydrolyzed to GDP and $P_i$.

*3.* In the elongation steps, GTP and elongation factors are required for binding the incoming aminoacyl-tRNA to the A site on the ribosome. In the first peptidyl transfer reaction, the fMet residue is transferred to the amino group of the incoming aminoacyl-tRNA. Movement of the ribosome along the mRNA then translocates the dipeptidyl-tRNA from the A site to the P site, a process requiring hydrolysis of GTP. Deacylated tRNAs dissociate from the ribosomal E site.

*4.* After many such elongation cycles, synthesis of the polypeptide is terminated with the aid of release factors. At least four high-energy phosphate equivalents (from ATP and GTP) are required to generate each peptide bond, an energy investment required to guarantee fidelity of translation.

*5.* Polypeptides fold into their active, three-dimensional forms. Many proteins are further processed by posttranslational modification reactions.

■ Many well-studied antibiotics and toxins inhibit some aspect of protein synthesis.

# 27.3 Protein Targeting and Degradation

The eukaryotic cell is made up of many structures, compartments, and organelles, each with specific functions that require distinct sets of proteins and enzymes. These proteins (with the exception of those produced in mitochondria and plastids) are synthesized on ribosomes in the cytosol, so how are they directed to their final cellular destinations?

We are now beginning to understand this complex and fascinating process. Proteins destined for secretion, integration in the plasma membrane, or inclusion in lysosomes generally share the first few steps of a pathway that begins in the endoplasmic reticulum. Proteins destined for mitochondria, chloroplasts, or the nucleus use three separate mechanisms. And proteins destined for the cytosol simply remain where they are synthesized.

The most important element in many of these targeting pathways is a short sequence of amino acids called a **signal sequence,** whose function was first postulated by Günter Blobel and colleagues in 1970. The signal sequence directs a protein to its appropriate location in the cell and, for many proteins, is removed during transport or after the protein has reached its final destination. In proteins slated for transport into mitochondria, chloroplasts, or the ER, the signal sequence is at the amino terminus of a newly synthesized polypeptide. In many cases, the targeting capacity of particular signal sequences has been confirmed by fusing the signal sequence from one protein to a second protein and showing that the signal directs the second protein to the location where the first protein is normally found. The selective degradation of proteins no longer needed by the cell also relies largely on a set of molecular signals embedded in each protein's structure.

In this concluding section we examine protein targeting and degradation, emphasizing the underlying signals and molecular regulation that are so crucial to cellular metabolism. Except where noted, the focus is now on eukaryotic cells.

## Posttranslational Modification of Many Eukaryotic Proteins Begins in the Endoplasmic Reticulum

Perhaps the best-characterized targeting system begins in the ER. Most lysosomal, membrane, or secreted proteins have an amino-terminal signal sequence (Fig. 27–32) that marks them for translocation into the lumen of the ER; hundreds of such signal sequences have been determined. The carboxyl terminus of the signal sequence is defined by a cleavage site, where protease action removes the sequence after the protein is imported into the ER. Signal sequences vary in length from 13 to 36 amino acid residues, but all have the following features: (1) about 10 to 15 hydrophobic amino acid residues; (2) one or more positively charged residues, usually near the amino terminus, preceding the hydrophobic sequence; and (3) a short sequence at the carboxyl terminus (near the cleavage site) that is relatively polar, typically having amino acid residues with short side chains (especially Ala) at the positions closest to the cleavage site.

As originally demonstrated by George Palade, proteins with these signal sequences are synthesized on ribosomes attached to the ER. The signal sequence itself helps to direct the ribosome to the ER, as illustrated by

Günter Blobel

George Palade

| | cleavage site |
|---|---|
| Human influenza virus A | Met Lys Ala Lys Leu Leu Val Leu Leu Tyr Ala Phe Val Ala Gly ↓ Asp Gln -- |
| Human preproinsulin | Met Ala Leu Trp Met Arg Leu Leu Pro Leu Leu Ala Leu Leu Ala Leu Trp Gly Pro Asp Pro Ala Ala Ala ↓ Phe Val -- |
| Bovine growth hormone | Met Met Ala Ala Gly Pro Arg Thr Ser Leu Leu Leu Ala Phe Ala Leu Leu Cys Leu Pro Trp Thr Gln Val Val Gly ↓ Ala Phe -- |
| Bee promellitin | Met Lys Phe Leu Val Asn Val Ala Leu Val Phe Met Val Val Tyr Ile Ser Tyr Ile Tyr Ala ↓ Ala Pro -- |
| *Drosophila* glue protein | Met Lys Leu Leu Val Val Ala Val Ile Ala Cys Met Leu Ile Gly Phe Ala Asp Pro Ala Ser Gly ↓ Cys Lys -- |

**FIGURE 27–32** **Translocation into the ER directed by amino-terminal signal sequences of some eukaryotic proteins.** The hydrophobic core (yellow) is preceded by one or more basic residues (blue). Note the polar and short-side-chain residues immediately preceding (to the left of, as shown here) the cleavage sites (indicated by red arrows).

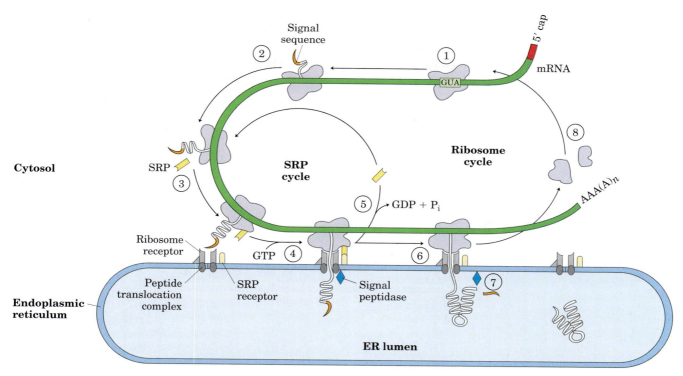

**FIGURE 27–33 Directing eukaryotic proteins with the appropriate signals to the endoplasmic reticulum.** This process involves the SRP cycle and translocation and cleavage of the nascent polypeptide. The steps are described in the text. SRP is a rod-shaped complex containing a 300 nucleotide RNA (7SL-RNA) and six different proteins (combined $M_r$ 325,000). One protein subunit of SRP binds directly to the signal sequence, inhibiting elongation by sterically blocking the entry of aminoacyl-tRNAs and inhibiting peptidyl transferase. Another protein subunit binds and hydrolyzes GTP. The SRP receptor is a heterodimer of $\alpha$ ($M_r$ 69,000) and $\beta$ ($M_r$ 30,000) subunits, both of which bind and hydrolyze multiple GTP molecules during this process.

steps ① through ⑧ in Figure 27–33. ① The targeting pathway begins with initiation of protein synthesis on free ribosomes. ② The signal sequence appears early in the synthetic process, because it is at the amino terminus, which as we have seen is synthesized first. ③ As it emerges from the ribosome, the signal sequence—and the ribosome itself—are bound by the large **signal recognition particle (SRP);** SRP then binds GTP and halts elongation of the polypeptide when it is about 70 amino acids long and the signal sequence has completely emerged from the ribosome. ④ The GTP-bound SRP now directs the ribosome (still bound to the mRNA) and the incomplete polypeptide to GTP-bound SRP receptors in the cytosolic face of the ER; the nascent polypeptide is delivered to a **peptide translocation complex** in the ER, which may interact directly with the ribosome. ⑤ SRP dissociates from the ribosome, accompanied by hydrolysis of GTP in both SRP and the SRP receptor. ⑥ Elongation of the polypeptide now resumes, with the ATP-driven translocation complex feeding the growing polypeptide into the ER lumen until the complete protein has been synthesized. ⑦ The signal sequence is removed by a signal peptidase within the ER lumen; ⑧ the ribosome dissociates and is recycled.

## Glycosylation Plays a Key Role in Protein Targeting

In the ER lumen, newly synthesized proteins are further modified in several ways. Following the removal of signal sequences, polypeptides are folded, disulfide bonds formed, and many proteins glycosylated to form glycoproteins. In many glycoproteins the linkage to their oligosaccharides is through Asn residues. These N-linked oligosaccharides are diverse (Chapter 7), but the pathways by which they form have a common first step. A 14 residue core oligosaccharide is built up in a stepwise fashion, then transferred from a dolichol phosphate donor molecule to certain Asn residues in the protein (Fig. 27–34). The transferase is on the lumenal face of the ER and thus cannot catalyze glycosylation of cytosolic proteins. After transfer, the core oligosaccharide is trimmed and elaborated in different ways on different

Dolichol phosphate
($n = 9$–$22$)

- ⬡ $N$-Acetylglucosamine (GlcNAc)
- ⬡ Mannose (Man)
- ⬡ Glucose (Glc)

**FIGURE 27–34 Synthesis of the core oligosaccharide of glycoproteins.** The core oligosaccharide is built up by the successive addition of monosaccharide units. ①, ② The first steps occur on the cytosolic face of the ER. ③ Translocation moves the incomplete oligosaccharide across the membrane (mechanism not shown), and ④ completion of the core oligosaccharide occurs within the lumen of the ER. The precursors that contribute additional mannose and glucose residues to the growing oligosaccharide in the lumen are dolichol phosphate derivatives. In the first step in the construction of the $N$-linked oligosaccharide moiety of a glycoprotein, ⑤, ⑥ the core oligosaccharide is transferred from dolichol phosphate to an Asn residue of the protein within the ER lumen. The core oligosaccharide is then further modified in the ER and the Golgi complex in pathways that differ for different proteins. The five sugar residues shown surrounded by a beige screen (after step ⑦) are retained in the final structure of all $N$-linked oligosaccharides. ⑧ The released dolichol pyrophosphate is again translocated so that the pyrophosphate is on the cytosolic face of the ER, then ⑨ a phosphate is hydrolytically removed to regenerate dolichol phosphate.

proteins, but all $N$-linked oligosaccharides retain a pentasaccharide core derived from the original 14 residue oligosaccharide. Several antibiotics act by interfering with one or more steps in this process and have aided in elucidating the steps of protein glycosylation. The best-characterized is **tunicamycin,** which mimics the structure of UDP-$N$-acetylglucosamine and blocks the first step of the process (Fig. 27–34, step ①). A few proteins are $O$-glycosylated in the ER, but most $O$-glycosylation occurs in the Golgi complex or in the cytosol (for proteins that do not enter the ER).

Suitably modified proteins can now be moved to a variety of intracellular destinations. Proteins travel from the ER to the Golgi complex in transport vesicles (Fig. 27–35). In the Golgi complex, oligosaccharides are $O$-linked to some proteins, and $N$-linked oligosaccharides are further modified. By mechanisms not yet fully understood, the Golgi complex also sorts proteins and sends them to their final destinations. The processes that segregate proteins targeted for secretion from those targeted for the plasma membrane or lysosomes must distinguish among these proteins on the basis of structural features other than signal sequences, which were removed in the ER lumen.

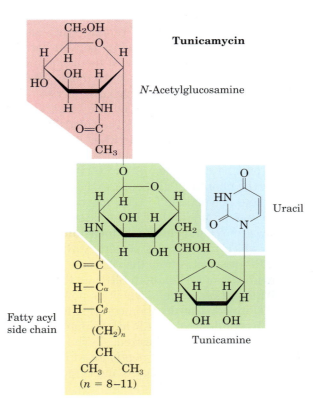

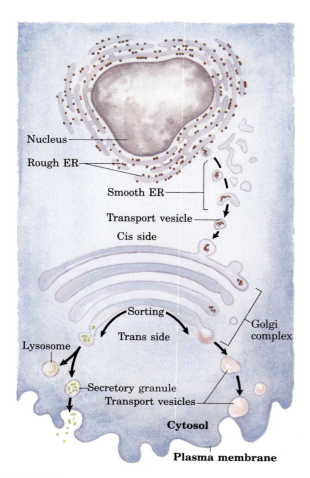

**FIGURE 27–35 Pathway taken by proteins destined for lysosomes, the plasma membrane, or secretion.** Proteins are moved from the ER to the cis side of the Golgi complex in transport vesicles. Sorting occurs primarily in the trans side of the Golgi complex.

Labels in figure: Nucleus, Rough ER, Smooth ER, Transport vesicle, Cis side, Sorting, Trans side, Lysosome, Golgi complex, Secretory granule, Transport vesicles, Cytosol, **Plasma membrane**

This sorting process is best understood in the case of hydrolases destined for transport to lysosomes. On arrival of a hydrolase (a glycoprotein) in the Golgi complex, an as yet undetermined feature (sometimes called a signal patch) of the three-dimensional structure of the hydrolase is recognized by a phosphotransferase, which phosphorylates certain mannose residues in the oligosaccharide (Fig. 27–36). The presence of one or more mannose 6-phosphate residues in its N-linked oligosaccharide is the structural signal that targets the protein to lysosomes. A receptor protein in the membrane of the Golgi complex recognizes the mannose 6-phosphate signal and binds the hydrolase so marked. Vesicles containing these receptor-hydrolase complexes bud from the trans side of the Golgi complex and make their way to sorting vesicles. Here, the receptor-hydrolase complex dissociates in a process facilitated by the lower pH in the vesicle and by phosphatase-catalyzed removal of phosphate groups from the mannose 6-phosphate residues. The receptor is then recycled to the Golgi complex, and vesicles containing the hydrolases bud from

the sorting vesicles and move to the lysosomes. In cells treated with tunicamycin (Fig. 27–34, step ①), hydrolases that should be targeted for lysosomes are instead secreted, confirming that the N-linked oligosaccharide plays a key role in targeting these enzymes to lysosomes.

The pathways that target proteins to mitochondria and chloroplasts also rely on amino-terminal signal sequences. Although mitochondria and chloroplasts contain DNA, most of their proteins are encoded by nuclear DNA and must be targeted to the appropriate organelle. Unlike other targeting pathways, however, the mitochondrial and chloroplast pathways begin only *after* a precursor protein has been completely synthesized and released from the ribosome. Precursor proteins destined for mitochondria or chloroplasts are bound by cytosolic chaperone proteins and delivered to receptors on the exterior surface of the target organelle. Specialized translocation mechanisms then transport the protein to its final destination in the organelle, after which the signal sequence is removed.

## Signal Sequences for Nuclear Transport Are Not Cleaved

Molecular communication between the nucleus and the cytosol requires the movement of macromolecules through nuclear pores. RNA molecules synthesized in the nucleus are exported to the cytosol. Ribosomal proteins synthesized on cytosolic ribosomes are imported into the nucleus and assembled into 60S and 40S ribosomal subunits in the nucleolus; completed subunits are then exported back to the cytosol. A variety of nuclear proteins (RNA and DNA polymerases, histones, topoisomerases, proteins that regulate gene expression, and so forth) are synthesized in the cytosol and imported into the nucleus. This traffic is modulated by a complex system of molecular signals and transport proteins that is gradually being elucidated.

In most multicellular eukaryotes, the nuclear envelope breaks down at each cell division, and once division is completed and the nuclear envelope reestablished, the dispersed nuclear proteins must be reimported. To allow this repeated nuclear importation, the signal sequence that targets a protein to the nucleus—the nuclear localization sequence, NLS—is not removed after the protein arrives at its destination. An NLS, unlike other signal sequences, may be located almost anywhere along the primary sequence of the protein. NLSs can vary considerably, but many consist of four to eight amino acid residues and include several consecutive basic (Arg or Lys) residues.

Nuclear importation is mediated by a number of proteins that cycle between the cytosol and the nucleus (Fig. 27–37), including importin $\alpha$ and $\beta$ and a small GTPase known as Ran. A heterodimer of importin $\alpha$ and $\beta$ functions as a soluble receptor for proteins targeted to the nucleus, with the $\alpha$ subunit binding NLS-bearing

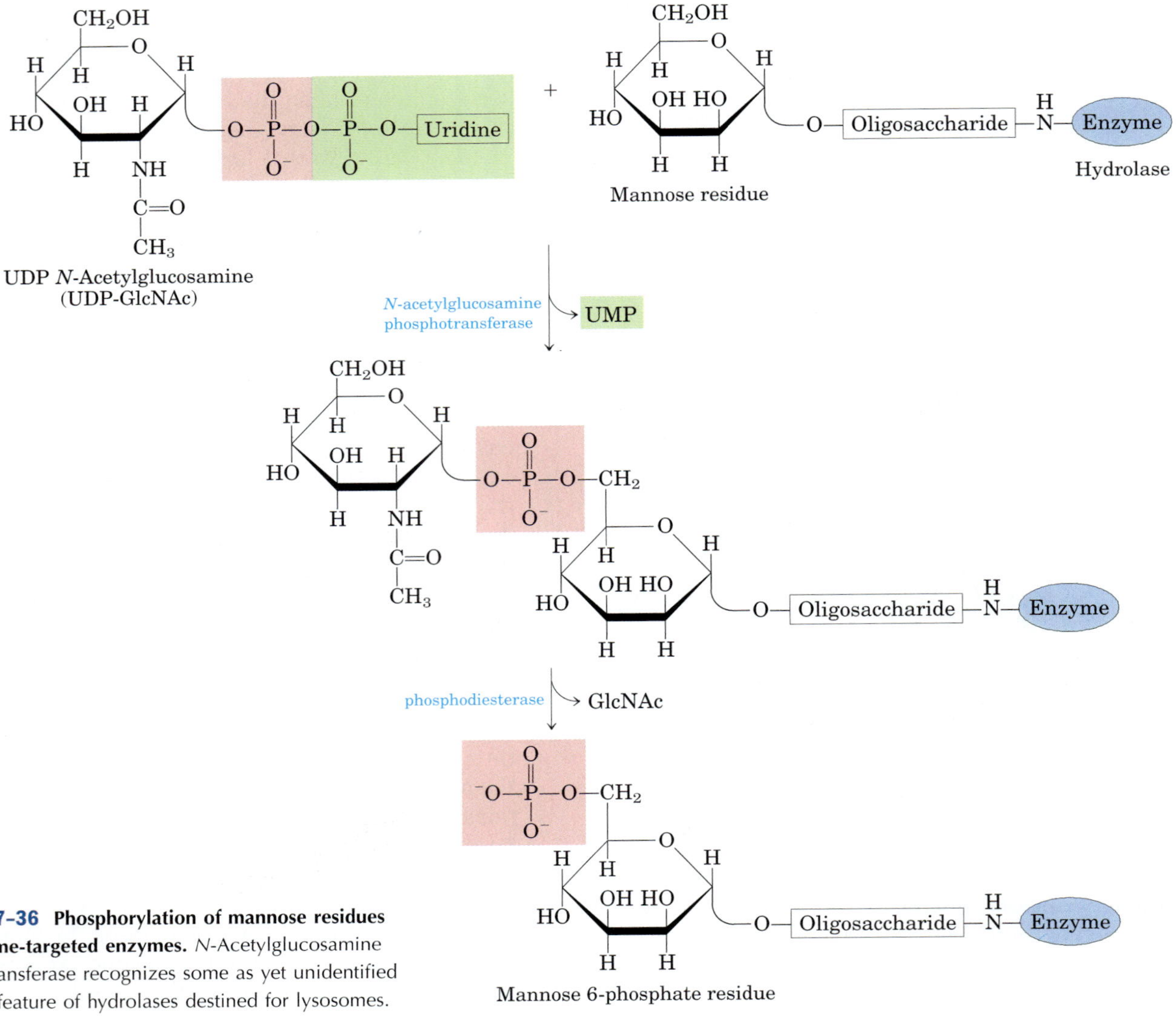

**FIGURE 27–36 Phosphorylation of mannose residues on lysosome-targeted enzymes.** N-Acetylglucosamine phosphotransferase recognizes some as yet unidentified structural feature of hydrolases destined for lysosomes.

proteins in the cytosol. The complex of the NLS-bearing protein and the importin docks at a nuclear pore and is translocated through the pore by an energy-dependent mechanism that requires the Ran GTPase. The two importin subunits separate during the translocation, and the NLS-bearing protein dissociates from importin $\alpha$ inside the nucleus. Importin $\alpha$ and $\beta$ are then exported from the nucleus to repeat the process. How importin $\alpha$ remains dissociated from the many NLS-bearing proteins inside the nucleus is not yet clear.

## Bacteria Also Use Signal Sequences for Protein Targeting

Bacteria can target proteins to their inner or outer membranes, to the periplasmic space between these membranes, or to the extracellular medium. They use signal sequences at the amino terminus of the proteins (Fig.

27–38), much like those on eukaryotic proteins targeted to the ER, mitochondria, and chloroplasts.

Most proteins exported from *E. coli* make use of the pathway shown in Figure 27–39. Following translation, a protein to be exported may fold only slowly, the amino-terminal signal sequence impeding the folding. The soluble chaperone protein SecB binds to the protein's signal sequence or other features of its incompletely folded structure. The bound protein is then delivered to SecA, a protein associated with the inner surface of the plasma membrane. SecA acts as both a receptor and a translocating ATPase. Released from SecB and bound to SecA, the protein is delivered to a translocation complex in the membrane, made up of SecY, E, and G, and is translocated stepwise through the membrane at the SecYEG complex in lengths of about 20 amino acid residues. Each step is facilitated by the hydrolysis of ATP, catalyzed by SecA.

**(a)**

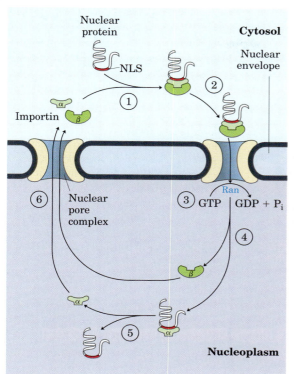

**(b)**

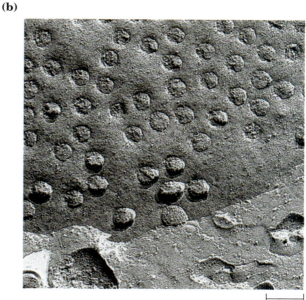

0.2 μm

**FIGURE 27–37 Targeting of nuclear proteins. (a)** ① A protein with an appropriate nuclear localization signal (NLS) is bound by a complex of importin α and β. ② The resulting complex binds to a nuclear pore, and ③ translocation is mediated by the Ran GTPase. ④ Inside the nucleus, importin β dissociates from importin α, and ⑤ importin α then releases the nuclear protein. ⑥ Importin α and β are transported out of the nucleus and recycled. **(b)** Scanning electron micrograph of the surface of the nuclear envelope, showing numerous nuclear pores.

**Inner membrane proteins**

Phage fd, major coat protein  Met Lys Lys Ser Leu Val Leu Lys Ala Ser Val Ala Val Ala Thr Leu Val Pro Met Leu Ser Phe Ala↓Ala Glu --

Phage fd, minor coat protein  Met Lys Lys Leu Leu Phe Ala Ile Pro Leu Val Val Pro Phe Tyr Ser His Ser↓Ala Glu --

**Periplasmic proteins**

Alkaline phosphatase  Met Lys Gln Ser Thr Ile Ala Leu Ala Leu Leu Pro Leu Leu Phe Thr Pro Val Thr Lys Ala↓Arg Thr --

Leucine-specific binding protein  Met Lys Ala Asn Ala Lys Thr Ile Ile Ala Gly Met Ile Ala Leu Ala Ile Ser His Thr Ala Met Ala↓Asp Asp --

β-Lactamase of pBR322  Met Ser Ile Gln His Phe Arg Val Ala Leu Ile Pro Phe Phe Ala Ala Phe Cys Leu Pro Val Phe Ala↓His Pro --

**Outer membrane proteins**

Lipoprotein  Met Lys Ala Thr Lys Leu Val Leu Gly Ala Val Ile Leu Gly Ser Thr Leu Leu Ala Gly↓Cys Ser --

LamB  Leu Arg Lys Leu Pro Leu Ala Val Ala Val Ala Ala Gly Val Met Ser Ala Gln Ala Met Ala↓Val Asp --

OmpA  Met Met Ile Thr Met Lys Lys Thr Ala Ile Ala Ile Ala Val Ala Leu Ala Gly Phe Ala Thr Val Ala Gln Ala↓Ala Pro --

cleavage site

**FIGURE 27–38 Signal sequences that target proteins to different locations in bacteria.** Basic amino acids (blue) near the amino terminus and hydrophobic core amino acids (yellow) are highlighted. The cleavage sites marking the ends of the signal sequences are indicated by red arrows. Note that the inner bacterial cell membrane (see Fig. 1–6) is where phage fd coat proteins and DNA are assembled into phage particles. OmpA is outer membrane protein A; LamB is a cell surface receptor protein for bacteriophage lambda.

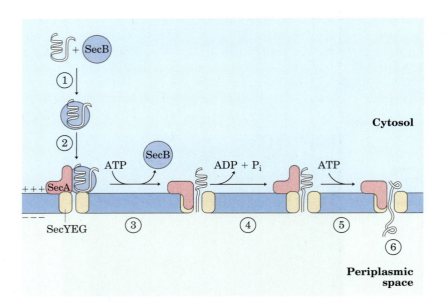

**FIGURE 27–39  Model for protein export in bacteria.** ① A newly translated polypeptide binds to the cytosolic chaperone protein SecB, which ② delivers it to SecA, a protein associated with the translocation complex (SecYEG) in the bacterial cell membrane. ③ SecB is released, and SecA inserts itself into the membrane, forcing about 20 amino acid residues of the protein to be exported through the translocation complex. ④ Hydrolysis of an ATP by SecA provides the energy for a conformational change that causes SecA to withdraw from the membrane, releasing the polypeptide. ⑤ SecA binds another ATP, and the next stretch of 20 amino acid residues is pushed across the membrane through the translocation complex. Steps ④ and ⑤ are repeated until ⑥ the entire protein has passed through and is released to the periplasm. The electrochemical potential across the membrane (denoted by + and −) also provides some of the driving force required for protein translocation.

An exported protein is thus pushed through the membrane by a SecA protein located on the cytoplasmic surface, rather than being pulled through the membrane by a protein on the periplasmic surface. This difference may simply reflect the need for the translocating ATPase to be where the ATP is. The transmembrane electrochemical potential can also provide energy for translocation of the protein, by an as yet unknown mechanism.

Although most exported bacterial proteins use this pathway, some follow an alternative pathway that uses signal recognition and receptor proteins homologous to components of the eukaryotic SRP and SRP receptor (Fig. 27–33).

## Cells Import Proteins by Receptor-Mediated Endocytosis

Some proteins are imported into cells from the surrounding medium; examples in eukaryotes include low-density lipoprotein (LDL), the iron-carrying protein transferrin, peptide hormones, and circulating proteins destined for degradation. The proteins bind to receptors in invaginations of the membrane called **coated pits,** which concentrate endocytic receptors in preference to other cell-surface proteins. The pits are coated on their cytosolic side with a lattice of the protein **clathrin,** which forms closed polyhedral structures (Fig. 27–40). The clathrin lattice grows as more recep-

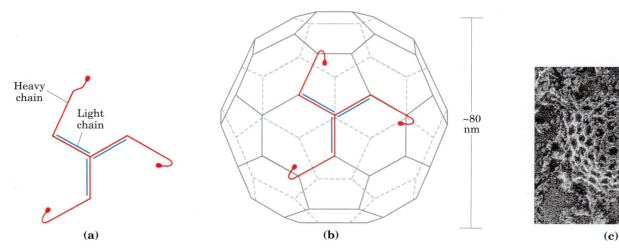

**FIGURE 27–40  Clathrin. (a)** Three light (L) chains ($M_r$ 35,000) and three heavy (H) chains ($M_r$ 180,000) of the $(HL)_3$ clathrin unit, organized as a three-legged structure called a triskelion. **(b)** Triskelions tend to assemble into polyhedral lattices. **(c)** Electron micrograph of a coated pit on the cytosolic face of the plasma membrane of a fibroblast.

tors are occupied by target proteins, until a complete membrane-bounded endocytic vesicle buds off the plasma membrane and enters the cytoplasm. The clathrin is quickly removed by uncoating enzymes, and the vesicle fuses with an endosome. ATPase activity in the endosomal membranes reduces the pH therein, facilitating dissociation of receptors from their target proteins.

The imported proteins and receptors then go their separate ways, their fates varying with the cell and protein type. Transferrin and its receptor are eventually recycled. Some hormones, growth factors, and immune complexes, after eliciting the appropriate cellular response, are degraded along with their receptors. LDL is degraded after the associated cholesterol has been delivered to its destination, but the LDL receptor is recycled (see Fig. 21–42).

Receptor-mediated endocytosis is exploited by some toxins and viruses to gain entry to cells. Influenza virus (see Fig. 11–24), diphtheria toxin, and cholera toxin all enter cells in this way.

## Protein Degradation Is Mediated by Specialized Systems in All Cells

Protein degradation prevents the buildup of abnormal or unwanted proteins and permits the recycling of amino acids. The half-lives of eukaryotic proteins vary from 30 seconds to many days. Most proteins turn over rapidly relative to the lifetime of a cell, although a few (such as hemoglobin) can last for the life of the cell (about 110 days for an erythrocyte). Rapidly degraded proteins include those that are defective because of incorrectly inserted amino acids or because of damage accumulated during normal functioning. And enzymes that act at key regulatory points in metabolic pathways often turn over rapidly.

Defective proteins and those with characteristically short half-lives are generally degraded in both bacterial and eukaryotic cells by selective ATP-dependent cytosolic systems. A second system in vertebrates, operating in lysosomes, recycles the amino acids of membrane proteins, extracellular proteins, and proteins with characteristically long half-lives.

In *E. coli,* many proteins are degraded by an ATP-dependent protease called Lon (the name refers to the "long form" of proteins, observed only when this protease is absent). The protease is activated in the presence of defective proteins or those slated for rapid turnover; two ATP molecules are hydrolyzed for every peptide bond cleaved. The precise role of this ATP hydrolysis is not yet clear. Once a protein has been reduced to small inactive peptides, other ATP-independent proteases complete the degradation process.

The ATP-dependent pathway in eukaryotic cells is quite different, involving the protein **ubiquitin,** which,

as its name suggests, occurs throughout the eukaryotic kingdoms. One of the most highly conserved proteins known, ubiquitin (76 amino acid residues) is essentially identical in organisms as different as yeasts and humans. Ubiquitin is covalently linked to proteins slated for destruction via an ATP-dependent pathway involving three separate enzymes (E1, E2, and E3 in Fig. 27–41).

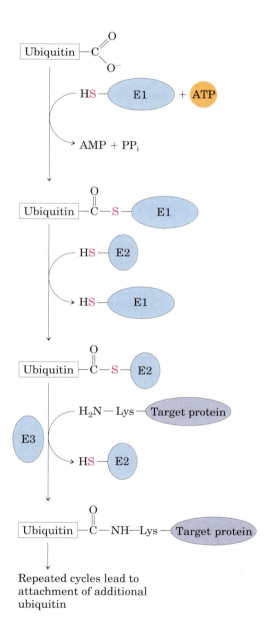

**FIGURE 27–41 Three-step cascade pathway by which ubiquitin is attached to a protein.** Two different enzyme-ubiquitin intermediates are involved. The free carboxyl group of ubiquitin's carboxyl-terminal Gly residue is ultimately linked through an amide (isopeptide) bond to an $\epsilon$-amino group of a Lys residue of the target protein. Additional cycles produce polyubiquitin, a covalent polymer of ubiquitin subunits that targets the attached protein for destruction in eukaryotes.

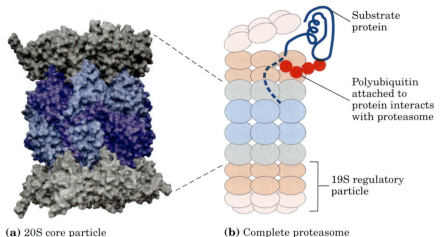

**(a)** 20S core particle    **(b)** Complete proteasome

**FIGURE 27–42 Three-dimensional structure of the eukaryotic proteasome.** The 26S proteasome is highly conserved in all eukaryotes. The two subassemblies are the 20S core particle and the 19S regulatory particle. **(a)** (PDB ID 1IRU) The core particle consists of four rings arranged to form a barrel-like structure. Each of the inner rings has seven different β subunits (light blue), three of which have protease activities (dark blue). The outer rings each have seven different α subunits (gray). **(b)** A regulatory particle forms a cap on each end of the core particle. The core particle is colored as in **(a)**. The base and lid segments of each regulatory particle are presented in different shades of red. The regulatory particle unfolds ubiquitinated proteins (blue) and translocates them into the core particle, as shown.

Ubiquitinated proteins are degraded by a large complex known as the **26S proteasome** ($M_r$ $2.5 \times 10^6$) (Fig. 27–42). The proteasome consists of two copies each of at least 32 different subunits, most of which are highly conserved from yeasts to humans. The proteasome contains two main types of subcomplexes, a barrel-like core particle and regulatory particles on either end of the barrel. The 20S core particle consists of four rings; the outer rings are formed from seven α subunits, and the inner rings from seven β subunits. Three of the

seven subunits in each β ring have protease activities, each with different substrate specificities. The stacked rings of the core particle form the barrel-like structure within which target proteins are degraded. The 19S regulatory particle on each end of the core particle contains 18 subunits, including some that recognize and bind to ubiquitinated proteins. Six of the subunits are ATPases that probably function in unfolding the ubiquitinated proteins and translocating the unfolded polypeptide into the core particle for degradation.

Although we do not yet understand all the signals that trigger ubiquitination, one simple signal has been found. For many proteins, the identity of the first residue that remains after removal of the amino-terminal Met residue, and any other posttranslational proteolytic processing of the amino-terminal end, has a profound influence on half-life (Table 27–9). These amino-terminal signals have been conserved over billions of years of evolution, and are the same in bacterial protein degradation systems and in the human ubiquitination pathway. More complex signals, such as the destruction box discussed in Chapter 12 (see Fig. 12–44), are also being identified.

Ubiquitin-dependent proteolysis is as important for the regulation of cellular processes as for the elimination of defective proteins. Many proteins required at only one stage of the eukaryotic cell cycle are rapidly degraded by the ubiquitin-dependent pathway after completing their function. The same pathway also processes and presents class I MHC antigens (see Fig. 5–22). Ubiquitin-dependent destruction of cyclin is critical to cell-cycle regulation (see Fig. 12–44). The E2 and E3 components of the ubiquitination cascade pathway

**TABLE 27–9 Relationship between Protein Half-Life and Amino-Terminal Amino Acid Residue**

| Amino-terminal residue | Half-life* |
|---|---|
| **Stabilizing** | |
| Met, Gly, Ala, Ser, Thr, Val | >20 h |
| **Destabilizing** | |
| Ile, Gln | ~30 min |
| Tyr, Glu | ~10 min |
| Pro | ~7 min |
| Leu, Phe, Asp, Lys | ~3 min |
| Arg | ~2 min |

Source: Modified from Bachmair, A., Finley, D., & Varshavsky, A. (1986) In vivo half-life of a protein is a function of its amino-terminal residue. *Science* **234**, 179–186.

*Half-lives were measured in yeast for the β-galactosidase protein modified so that in each experiment it had a different amino-terminal residue. (See Chapter 9 for a discussion of techniques used to engineer proteins with altered amino acid sequences.) Half-lives may vary for different proteins and in different organisms, but this general pattern appears to hold for all organisms.

(Fig. 27–41) are in fact two large families of proteins. Different E2 and E3 enzymes exhibit different specificities for target proteins and thus regulate different cellular processes. Some E2 and E3 enzymes are highly localized in certain cellular compartments, reflecting a specialized function.

Not surprisingly, defects in the ubiquitination pathway have been implicated in a wide range of disease states. An inability to degrade certain proteins that activate cell division (the products of oncogenes) can lead to tumor formation, whereas a too-rapid degradation of proteins that act as tumor suppressors can have the same effect. The ineffective or overly rapid degradation of cellular proteins also appears to play a role in a range of other conditions: renal diseases, asthma, neurodegenerative disorders such as Alzheimer's and Parkinson's diseases (associated with the formation of characteristic proteinaceous structures in neurons), cystic fibrosis (caused in some cases by a too-rapid degradation of a chloride ion channel, with resultant loss of function; see Box 11–3), Liddle's syndrome (in which a sodium channel in the kidney is not degraded, leading to excessive Na$^+$ absorption and early-onset hypertension)—and many other disorders. Drugs designed to inhibit proteasome function are being developed as potential treatments for some of these conditions. In a changing metabolic environment, protein degradation is as important to a cell's survival as is protein synthesis, and much remains to be learned about these interesting pathways. ■

## SUMMARY 27.3 Protein Targeting and Degradation

- After synthesis, many proteins are directed to particular locations in the cell. One targeting mechanism involves a peptide signal sequence, generally found at the amino terminus of a newly synthesized protein.

- In eukaryotic cells, one class of signal sequences is recognized by the signal recognition particle (SRP), which binds the signal sequence as soon as it appears on the ribosome and transfers the entire ribosome and incomplete polypeptide to the ER. Polypeptides with these signal sequences are moved into the ER lumen as they are synthesized; once in the lumen they may be modified and moved to the Golgi complex, then sorted and sent to lysosomes, the plasma membrane, or transport vesicles.

- Proteins targeted to mitochondria and chloroplasts in eukaryotic cells, and those destined for export in bacteria, also make use of an amino-terminal signal sequence.

- Proteins targeted to the nucleus have an internal signal sequence that is not cleaved once the protein is successfully targeted.

- Some eukaryotic cells import proteins by receptor-mediated endocytosis.

- All cells eventually degrade proteins, using specialized proteolytic systems. Defective proteins and those slated for rapid turnover are generally degraded by an ATP-dependent system. In eukaryotic cells, the proteins are first tagged by linkage to ubiquitin, a highly conserved protein. Ubiquitin-dependent proteolysis is carried out by proteasomes, also highly conserved, and is critical to the regulation of many cellular processes.

## Key Terms

*Terms in bold are defined in the glossary.*

**aminoacyl-tRNA** 1035
**aminoacyl-tRNA synthetases** 1035
**translation** 1035
**codon** 1035
**reading frame** 1036
**initiation codon** 1038
**termination codons** 1038
**open reading frame (ORF)** 1039
**anticodon** 1039
**wobble** 1041

initiation 1056
**Shine-Dalgarno sequence** 1056
aminoacyl (A) site 1056
peptidyl (P) site 1056
exit (E) site 1056
**initiation complex** 1057
elongation 1058
**elongation factors** 1058
**peptidyl transferase** 1058

translocation 1060
termination 1061
**release factors** 1061
**polysome** 1062
**posttranslational modification** 1062
**nonsense suppressor** 1065
**puromycin** 1066
tetracyclines 1066
chloramphenicol 1066
cycloheximide 1067
streptomycin 1067

diphtheria toxin 1067
ricin 1067
**signal sequence** 1068
signal recognition particle (SRP) 1069
tunicamycin 1070
coated pits 1074
clathrin 1074
**ubiquitin** 1075
**proteasome** 1076

## Further Reading

### Genetic Code

**Bass, B.L.** (2002) RNA editing by adenosine deaminases that act on RNA. *Annu. Rev. Biochem.* **71**, 817–846.

**Blanc, V. & Davidson, N.O.** (2003) C-to-U RNA editing: mechanisms leading to genetic diversity *J. Biol. Chem.* **278**, 1395–1398.

**Crick, F.H.C.** (1966) The genetic code: III. *Sci. Am.* **215** (October), 55–62.

An insightful overview of the genetic code at a time when the code words had just been worked out.

**Fox, T.D.** (1987) Natural variation in the genetic code. *Annu. Rev. Genet.* **21**, 67–91.

**Hatfield, D. & Oroszlan, S.** (1990) The *where, what* and *how* of ribosomal frameshifting in retroviral protein synthesis. *Trends Biochem. Sci.* **15**, 186–190.

**Klobutcher, L.A. & Farabaugh, P.J.** (2002) Shifty ciliates: frequent programmed translational frameshifting in Euplotids. *Cell* **111**, 763–766.

**Knight, R.D., Freeland, S.J., & Landweber, L.F.** (2001) Rewiring the keyboard: evolvability of the genetic code. *Nat. Rev. Genet.* **2**, 49–58.

**Maas, S., Rich, A., & Nishikura, K.** (2003) A-to-I RNA editing: recent news and residual mysteries. *J. Biol. Chem.* **278**, 1391–1394.

**Nirenberg, M.W.** (1963) The genetic code: II. *Sci. Am.* **208** (March), 80–94.

A description of the original experiments.

**Stadtman, T.C.** (1996) Selenocysteine. *Annu. Rev. Biochem.* **65**, 83–100.

### Protein Synthesis

**Ban, N., Nissen, P., Hansen, J., Moore, P.B., & Steitz, T.A.** (2000) The complete atomic structure of the large ribosomal subunit at 2.4 angstrom resolution. *Science* **289**, 905–920.

The first high-resolution structure of a major ribosomal subunit.

**Björk, G.R., Ericson, J.U., Gustafsson, C.E.D., Hagervall, T.G., Jönsson, Y.H., & Wikström, P.M.** (1987) Transfer RNA modification. *Annu. Rev. Biochem.* **56**, 263–288.

**Chapeville, F., Lipmann, F., von Ehrenstein, G., Weisblum, B., Ray, W.J., Jr., & Benzer, S.** (1962) On the role of soluble ribonucleic acid in coding for amino acids. *Proc. Natl. Acad. Sci. USA* **48**, 1086–1092.

Classic experiments providing proof for Crick's adaptor hypothesis and showing that amino acids are not checked after they are linked to tRNAs.

**Dintzis, H.M.** (1961) Assembly of the peptide chains of hemoglobin. *Proc. Natl. Acad. Sci. USA* **47**, 247–261.

A classic experiment establishing that proteins are assembled beginning at the amino terminus.

**Giege, R., Sissler, M., & Florentz, C.** (1998) Universal rules and idiosyncratic features in tRNA identity. *Nucleic Acid Res.* **26**, 5017–5035.

**Gingras, A.-C., Raught, B., & Sonenberg, N.** (1999) eIF4 initiation factors: effectors of mRNA recruitment to ribosomes and regulators of translation. *Annu. Rev. Biochem.* **68**, 913–964.

**Gray, N.K. & Wickens, M.** (1998) Control of translation initiation in animals. *Annu. Rev. Cell Dev. Biol.* **14**, 399–458.

**Green, R. & Noller, J.F.** (1997) Ribosomes and translation. *Annu. Rev. Biochem.* **66**, 679–716.

**Ibba, M. & Soll, D.** (2000) Aminoacyl-tRNA synthesis. *Annu. Rev. Biochem.* **69**, 617–650.

**Maden, B.E.H.** (1990) The numerous modified nucleotides in eukaryotic ribosomal RNA. *Prog. Nucleic Acid Res. Mol. Biol.* **39**, 241–303.

**Moore, P.B. & Steitz, T.A.** (2003) The structural basis of large ribosomal subunit function. *Annu. Rev. Biochem.* **72**, 813–850.

**Ramakrishnan, V.** (2002) Ribosome structure and the mechanism of translation. *Cell* **108**, 557–572.

A good overview, incorporating structural advances.

**Rodnina, M.V. & Wintermeyer, W.** (2001) Fidelity of aminoacyl-tRNA selection on the ribosome: kinetic and structural mechanisms. *Annu. Rev. Biochem.* **70**, 415–435.

**Sprinzl, M.** (1994) Elongation factor Tu: a regulatory GTPase with an integrated effector. *Trends Biochem. Sci.* **19**, 245–250.

**Woese, C.R., Olsen, G.J., Ibba, M., & Soll, D.** (2000) Aminoacyl-tRNA synthetases, the genetic code, and the evolutionary process. *Microbiol. Mol. Biol. Rev.* **64**, 202–236.

### Protein Targeting and Secretion

**Görlich, D. & Mattaj, I.W.** (1996) Nucleocytoplasmic transport. *Science* **271**, 1513–1518.

**Hartmann-Petersen, R., Seeger, M., & Gordon C.** (2003) Transferring substrates to the 26S proteasome. *Trends Biochem. Sci.* **28**, 26–31.

**Higgins, M.K. & McMahon, H.T.** (2002) Snap-shots of clathrin-mediated endocytosis. *Trends Biochem. Sci.* **27**, 257–263.

**Neupert, W.** (1997) Protein import into mitochondria. *Annu. Rev. Biochem.* **66**, 863–917.

**Pryer, N.K., Wuestehube, L.J., & Schekman, R.** (1992) Vesicle-mediated protein sorting. *Annu. Rev. Biochem.* **61**, 471–516.

**Rapoport, T.A., Jungnickel, B., & Kutay, U.** (1996) Protein transport across the eukaryotic endoplasmic reticulum and bacterial inner membranes. *Annu. Rev. Biochem.* **65**, 271–303.

**Schatz, G. & Dobberstein, B.** (1996) Common principles of protein translocation across membranes. *Science* **271**, 1519–1525.

**Schekman, R. & Orci, L.** (1996) Coat proteins and vesicle budding. *Science* **271**, 1526–1532.

**Schmid, S.L.** (1997) Clathrin-coated vesicle formation and protein sorting: an integrated process. *Annu. Rev. Biochem.* **66**, 511–548.

**Varshavsky, A.** (1997) The ubiquitin system. *Trends Biochem. Sci.* **22**, 383–387.

**Voges, D., Zwickl, P., & Baumeister, W.** (1999) The 26S proteasome: a molecular machine designed for controlled proteolysis. *Annu. Rev. Biochem.* **68**, 1015–1057.

**Ward, W.H.J.** (1987) Diphtheria toxin: a novel cytocidal enzyme. *Trends Biochem. Sci.* **12**, 28–31.

# Problems

**1. Messenger RNA Translation** Predict the amino acid sequences of peptides formed by ribosomes in response to the following mRNA sequences, assuming that the reading frame begins with the first three bases in each sequence.

(a) GGUCAGUCGCUCCUGAUU

(b) UUGGAUGCGCCAUAAUUUGCU

(c) CAUGAUGCCUGUUGCUAC

(d) AUGGACGAA

**2. How Many Different mRNA Sequences Can Specify One Amino Acid Sequence?** Write all the possible mRNA sequences that can code for the simple tripeptide segment Leu–Met–Tyr. Your answer will give you some idea about the number of possible mRNAs that can code for one polypeptide.

**3. Can the Base Sequence of an mRNA Be Predicted from the Amino Acid Sequence of Its Polypeptide Product?** A given sequence of bases in an mRNA will code for one and only one sequence of amino acids in a polypeptide, if the reading frame is specified. From a given sequence of amino acid residues in a protein such as cytochrome *c*, can we predict the base sequence of the unique mRNA that coded it? Give reasons for your answer.

**4. Coding of a Polypeptide by Duplex DNA** The template strand of a segment of double-helical DNA contains the sequence

(5′)CTTAACACCCCTGACTTCGCGCCGTCG(3′)

(a) What is the base sequence of the mRNA that can be transcribed from this strand?

(b) What amino acid sequence could be coded by the mRNA in (a), starting from the 5′ end?

(c) If the complementary (nontemplate) strand of this DNA were transcribed and translated, would the resulting amino acid sequence be the same as in (b)? Explain the biological significance of your answer.

**5. Methionine Has Only One Codon** Methionine is one of two amino acids with only one codon. How does the single codon for methionine specify both the initiating residue and interior Met residues of polypeptides synthesized by *E. coli*?

**6. Synthetic mRNAs** The genetic code was elucidated with polyribonucleotides synthesized either enzymatically or chemically in the laboratory. Given what we now know about the genetic code, how would you make a polyribonucleotide that could serve as an mRNA coding predominantly for many Phe residues and a small number of Leu and Ser residues? What other amino acid(s) would be coded for by this polyribonucleotide, but in smaller amounts?

**7. Energy Cost of Protein Biosynthesis** Determine the minimum energy cost, in terms of ATP equivalents expended, required for the biosynthesis of the β-globin chain of hemoglobin (146 residues), starting from a pool including all necessary amino acids, ATP, and GTP. Compare your answer with the direct energy cost of the biosynthesis of a linear glycogen chain of 146 glucose residues in (α1→4) linkage, starting from a pool including glucose, UTP, and ATP (Chapter 15). From your data, what is the *extra* energy cost

of making a protein, in which all the residues are ordered in a specific sequence, compared with the cost of making a polysaccharide containing the same number of residues but lacking the informational content of the protein?

In addition to the direct energy cost for the synthesis of a protein, there are indirect energy costs—those required for the cell to make the necessary enzymes for protein synthesis. Compare the magnitude of the indirect costs to a eukaryotic cell of the biosynthesis of linear (α1→4) glycogen chains and the biosynthesis of polypeptides, in terms of the enzymatic machinery involved.

**8. Predicting Anticodons from Codons** Most amino acids have more than one codon and attach to more than one tRNA, each with a different anticodon. Write all possible anticodons for the four codons of glycine: (5′)GGU, GGC, GGA, and GGG.

(a) From your answer, which of the positions in the anticodons are primary determinants of their codon specificity in the case of glycine?

(b) Which of these anticodon-codon pairings has/have a wobbly base pair?

(c) In which of the anticodon-codon pairings do all three positions exhibit strong Watson-Crick hydrogen bonding?

**9. Effect of Single-Base Changes on Amino Acid Sequence** Much important confirmatory evidence on the genetic code has come from assessing changes in the amino acid sequence of mutant proteins after a single base has been changed in the gene that encodes the protein. Which of the following amino acid replacements would be consistent with the genetic code if the replacements were caused by a single base change? Which cannot be the result of a single-base mutation? Why?

(a) Phe→Leu        (e) Ile→Leu

(b) Lys→Ala        (f) His→Glu

(c) Ala→Thr        (g) Pro→Ser

(d) Phe→Lys

**10. Basis of the Sickle-Cell Mutation** Sickle-cell hemoglobin has a Val residue at position 6 of the β-globin chain, instead of the Glu residue found in normal hemoglobin A. Can you predict what change took place in the DNA codon for glutamate to account for replacement of the Glu residue by Val?

**11. Importance of the "Second Genetic Code"** Some aminoacyl-tRNA synthetases do not recognize and bind the anticodon of their cognate tRNAs but instead use other structural features of the tRNAs to impart binding specificity. The tRNAs for alanine apparently fall into this category.

(a) What features of tRNA^Ala are recognized by Ala-tRNA synthetase?

(b) Describe the consequences of a C→G mutation in the third position of the anticodon of tRNA^Ala.

(c) What other kinds of mutations might have similar effects?

(d) Mutations of these types are never found in natural populations of organisms. Why? (Hint: Consider what might happen both to individual proteins and to the organism as a whole.)

**12. Maintaining the Fidelity of Protein Synthesis**    The chemical mechanisms used to avoid errors in protein synthesis are different from those used during DNA replication. DNA polymerases use a $3' \rightarrow 5'$ exonuclease proofreading activity to remove mispaired nucleotides incorrectly inserted into a growing DNA strand. There is no analogous proofreading function on ribosomes and, in fact, the identity of an amino acid attached to an incoming tRNA and added to the growing polypeptide is never checked. A proofreading step that hydrolyzed the previously formed peptide bond after an incorrect amino acid had been inserted into a growing polypeptide (analogous to the proofreading step of DNA polymerases) would be impractical. Why? (Hint: Consider how the link between the growing polypeptide and the mRNA is maintained during elongation; see Figs 27–24 and 27–25.)

**13. Predicting the Cellular Location of a Protein**    The gene for a eukaryotic polypeptide 300 amino acid residues long is altered so that a signal sequence recognized by SRP occurs at the polypeptide's amino terminus and a nuclear localization signal (NLS) occurs internally, beginning at residue 150. Where is the protein likely to be found in the cell?

**14. Requirements for Protein Translocation across a Membrane**    The secreted bacterial protein OmpA has a precursor, ProOmpA, which has the amino-terminal signal sequence required for secretion. If purified ProOmpA is denatured with 8 M urea and the urea is then removed (such as by running the protein solution rapidly through a gel filtration column) the protein can be translocated across isolated bacterial inner membranes in vitro. However, translocation becomes impossible if ProOmpA is first allowed to incubate for a few hours in the absence of urea. Furthermore, the capacity for translocation is maintained for an extended period if ProOmpA is first incubated in the presence of another bacterial protein called trigger factor. Describe the probable function of this factor.

**15. Protein-Coding Capacity of a Viral DNA**    The 5,386 bp genome of bacteriophage $\phi$X174 includes genes for 10 proteins, designated A to K, with sizes given in the table below. How much DNA would be required to encode these 10 proteins? How can you reconcile the size of the $\phi$X174 genome with its protein-coding capacity?

| Protein | Number of amino acid residues | Protein | Number of amino acid residues |
|---------|-------------------------------|---------|-------------------------------|
| A | 455 | F | 427 |
| B | 120 | G | 175 |
| C | 86 | H | 328 |
| D | 152 | J | 38 |
| E | 91 | K | 56 |

# REGULATION OF GENE EXPRESSION

The fundamental problem of chemical physiology and of embryology is to understand why tissue cells do not all express, all the time, all the potentialities inherent in their genome.

—*François Jacob and Jacques Monod,*
*article in* Journal of Molecular Biology, *1961*

Of the 4,000 or so genes in the typical bacterial genome, or the perhaps 35,000 genes in the human genome, only a fraction are expressed in a cell at any given time. Some gene products are present in very large amounts: the elongation factors required for protein synthesis, for example, are among the most abundant proteins in bacteria, and ribulose 1,5-bisphosphate carboxylase/oxygenase (rubisco) of plants and photosynthetic bacteria is, as far as we know, the most abundant enzyme in the biosphere. Other gene products occur in much smaller amounts; for instance, a cell may contain only a few molecules of the enzymes that repair rare DNA lesions. Requirements for some gene products change over time. The need for enzymes in certain metabolic pathways may wax and wane as food sources change or are depleted. During development of a multicellular organism, some proteins that influence cellular differentiation are present for just a brief time in only a few cells. Specialization of cellular function can dramatically affect the need for various gene products; an example is the uniquely high concentration of a single protein—hemoglobin—in erythrocytes. Given the high cost of protein synthesis, regulation of gene expression is essential to making optimal use of available energy.

The cellular concentration of a protein is determined by a delicate balance of at least seven processes, each having several potential points of regulation:

1. Synthesis of the primary RNA transcript (transcription)
2. Posttranscriptional modification of mRNA
3. Messenger RNA degradation
4. Protein synthesis (translation)
5. Posttranslational modification of proteins
6. Protein targeting and transport
7. Protein degradation

These processes are summarized in Figure 28–1. We have examined several of these mechanisms in previous chapters. Posttranscriptional modification of mRNA, by processes such as alternative splicing patterns (see Fig. 26–19b) or RNA editing (see Box 27–1), can affect which proteins are produced from an mRNA transcript and in what amounts. A variety of nucleotide sequences in an mRNA can affect the rate of its degradation (p. 1020). Many factors affect the rate at which an mRNA is translated into a protein, as well as the posttranslational modification, targeting, and eventual degradation of that protein (Chapter 27).

This chapter focuses primarily on the regulation of transcription initiation, although aspects of posttranscriptional and translational regulation are also described. Of the regulatory processes illustrated in Figure 28–1, those operating at the level of transcription initiation are the best documented and probably the most

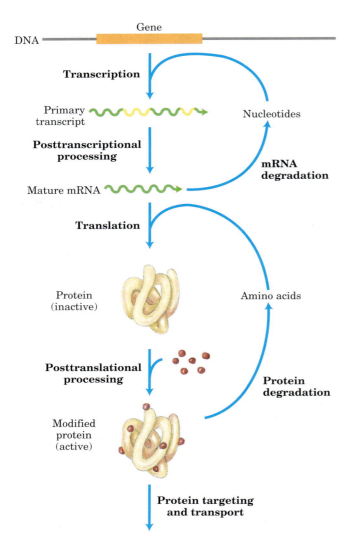

DNA — Gene

**Transcription**

Primary transcript

Nucleotides

**Posttranscriptional processing**

**mRNA degradation**

Mature mRNA

**Translation**

Protein (inactive)

Amino acids

**Posttranslational processing**

**Protein degradation**

Modified protein (active)

**Protein targeting and transport**

**FIGURE 28–1 Seven processes that affect the steady-state concentration of a protein.** Each process has several potential points of regulation.

common. As in all biochemical processes, an efficient place for regulation is at the beginning of the pathway. Because synthesis of informational macromolecules is so extraordinarily expensive in terms of energy, elaborate mechanisms have evolved to regulate the process. Researchers continue to discover complex and sometimes surprising regulatory mechanisms. Increasingly, posttranscriptional and translational regulation are proving to be among the more important of these processes, especially in eukaryotes. In fact, the regulatory processes themselves can involve a considerable investment of chemical energy.

Control of transcription initiation permits the synchronized regulation of multiple genes encoding products with interdependent activities. For example, when their DNA is heavily damaged, bacterial cells require a coordinated increase in the levels of the many DNA repair enzymes. And perhaps the most sophisticated form

of coordination occurs in the complex regulatory circuits that guide the development of multicellular eukaryotes, which can involve many types of regulatory mechanisms.

We begin by examining the interactions between proteins and DNA that are the key to transcriptional regulation. We next discuss the specific proteins that influence the expression of specific genes, first in prokaryotic and then in eukaryotic cells. Information about posttranscriptional and translational regulation is included in the discussion, where relevant, to provide a more complete overview of the rich complexity of regulatory mechanisms.

## 28.1 Principles of Gene Regulation

Genes for products that are required at all times, such as those for the enzymes of central metabolic pathways, are expressed at a more or less constant level in virtually every cell of a species or organism. Such genes are often referred to as **housekeeping genes.** Unvarying expression of a gene is called **constitutive gene expression.**

For other gene products, cellular levels rise and fall in response to molecular signals; this is **regulated gene expression.** Gene products that increase in concentration under particular molecular circumstances are referred to as **inducible;** the process of increasing their expression is **induction.** The expression of many of the genes encoding DNA repair enzymes, for example, is induced by high levels of DNA damage. Conversely, gene products that decrease in concentration in response to a molecular signal are referred to as **repressible,** and the process is called **repression.** For example, in bacteria, ample supplies of tryptophan lead to repression of the genes for the enzymes that catalyze tryptophan biosynthesis.

Transcription is mediated and regulated by protein-DNA interactions, especially those involving the protein components of RNA polymerase (Chapter 26). We first consider how the activity of RNA polymerase is regulated, and proceed to a general description of the proteins participating in this process. We then examine the molecular basis for the recognition of specific DNA sequences by DNA-binding proteins.

### RNA Polymerase Binds to DNA at Promoters

RNA polymerases bind to DNA and initiate transcription at promoters (see Fig. 26–5), sites generally found near points at which RNA synthesis begins on the DNA template. The regulation of transcription initiation often entails changes in how RNA polymerase interacts with a promoter.

The nucleotide sequences of promoters vary considerably, affecting the binding affinity of RNA polymerases and thus the frequency of transcription initiation. Some

**FIGURE 28-2 Consensus sequence for many E. coli promoters.** Most base substitutions in the −10 and −35 regions have a negative effect on promoter function. Some promoters also include the UP (upstream promoter) element (see Fig. 26–5). By convention, DNA sequences are shown as they exist in the nontemplate strand, with the 5′ terminus on the left. Nucleotides are numbered from the transcription start site, with positive numbers to the right (in the direction of transcription) and negative numbers to the left. N indicates any nucleotide.

*Escherichia coli* genes are transcribed once per second, others less than once per cell generation. Much of this variation is due to differences in promoter sequence. In the absence of regulatory proteins, differences in promoter sequences may affect the frequency of transcription initiation by a factor of 1,000 or more. Most *E. coli* promoters have a sequence close to a consensus (Fig. 28–2). Mutations that result in a shift away from the consensus sequence usually decrease promoter function; conversely, mutations toward consensus usually enhance promoter function.

Although housekeeping genes are expressed constitutively, the cellular concentrations of the proteins they encode vary widely. For these genes, the RNA polymerase–promoter interaction strongly influences the rate of transcription initiation; differences in promoter sequence allow the cell to synthesize the appropriate level of each housekeeping gene product.

The basal rate of transcription initiation at the promoters of nonhousekeeping genes is also determined by the promoter sequence, but expression of these genes is further modulated by regulatory proteins. Many of these proteins work by enhancing or interfering with the interaction between RNA polymerase and the promoter.

The sequences of eukaryotic promoters are more variable than their prokaryotic counterparts (see Fig. 26–8). The three eukaryotic RNA polymerases usually require an array of general transcription factors in order to bind to a promoter. Yet, as with prokaryotic gene expression, the basal level of transcription is determined by the effect of promoter sequences on the function of RNA polymerase and its associated transcription factors.

## Transcription Initiation Is Regulated by Proteins That Bind to or near Promoters

At least three types of proteins regulate transcription initiation by RNA polymerase: **specificity factors** alter the specificity of RNA polymerase for a given promoter or set of promoters; **repressors** impede access of RNA polymerase to the promoter; and **activators** enhance the RNA polymerase–promoter interaction.

We introduced prokaryotic specificity factors in Chapter 26 (see Fig. 26–5), although we did not refer to them by that name. The $\sigma$ subunit of the *E. coli* RNA polymerase holoenzyme is a specificity factor that mediates promoter recognition and binding. Most *E. coli* promoters are recognized by a single $\sigma$ subunit ($M_r$ 70,000), $\sigma^{70}$. Under some conditions, some of the $\sigma^{70}$ subunits are replaced by another specificity factor. One notable case arises when the bacteria are subjected to heat stress, leading to the replacement of $\sigma^{70}$ by $\sigma^{32}$ ($M_r$ 32,000). When bound to $\sigma^{32}$, RNA polymerase is directed to a specialized set of promoters with a different consensus sequence (Fig. 28–3). These promoters control the expression of a set of genes that encode the heat-shock response proteins. Thus, through changes in the binding affinity of the polymerase that direct it to different promoters, a set of genes involved in related processes is coordinately regulated. In eukaryotic cells, some of the general transcription factors, in particular the TATA-binding protein (TBP; see Fig. 26–8), may be considered specificity factors.

Repressors bind to specific sites on the DNA. In prokaryotic cells, such binding sites, called **operators,** are generally near a promoter. RNA polymerase binding,

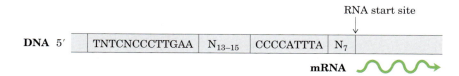

**FIGURE 28-3 Consensus sequence for promoters that regulate expression of the E. coli heat-shock genes.** This system responds to temperature increases as well as some other environmental stresses, resulting in the induction of a set of proteins. Binding of RNA polymerase to heat-shock promoters is mediated by a specialized $\sigma$ subunit of the polymerase, $\sigma^{32}$, which replaces $\sigma^{70}$ in the RNA polymerase initiation complex.

or its movement along the DNA after binding, is blocked when the repressor is present. Regulation by means of a repressor protein that blocks transcription is referred to as **negative regulation.** Repressor binding to DNA is regulated by a molecular signal (or **effector**), usually a small molecule or a protein, that binds to the repressor and causes a conformational change. The interaction between repressor and signal molecule either increases or decreases transcription. In some cases, the conformational change results in dissociation of a DNA-bound repressor from the operator (Fig. 28–4a). Transcription initiation can then proceed unhindered. In other cases, interaction between an inactive repressor and the signal molecule causes the repressor to bind to the operator (Fig. 28–4b). In eukaryotic cells, the binding site for a repressor may be some distance from the promoter; binding has the same effect as in bacterial cells: inhibit-

ing the assembly or activity of a transcription complex at the promoter.

Activators provide a molecular counterpoint to repressors; they bind to DNA and *enhance* the activity of RNA polymerase at a promoter; this is **positive regulation.** Activator binding sites are often adjacent to promoters that are bound weakly or not at all by RNA polymerase alone, such that little transcription occurs in the absence of the activator. Some eukaryotic activators bind to DNA sites, called **enhancers,** that are quite distant from the promoter, affecting the rate of transcription at a promoter that may be located thousands of base pairs away. Some activators are normally bound to DNA, enhancing transcription until dissociation of the activator is triggered by the binding of a signal molecule (Fig. 28–4c). In other cases the activator binds to DNA only after interaction with a signal molecule

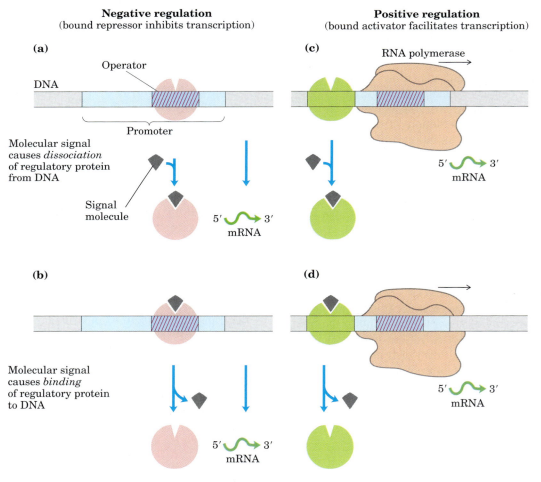

**FIGURE 28–4 Common patterns of regulation of transcription initiation.** Two types of negative regulation are illustrated. **(a)** Repressor (pink) binds to the operator in the absence of the molecular signal; the external signal causes dissociation of the repressor to permit transcription. **(b)** Repressor binds in the presence of the signal; the repressor dissociates and transcription ensues when the signal is removed. Positive regulation is mediated by gene activators. Again, two types are shown. **(c)** Activator (green) binds in the absence of the mo-

lecular signal and transcription proceeds; when the signal is added, the activator dissociates and transcription is inhibited. **(d)** Activator binds in the presence of the signal; it dissociates only when the signal is removed. Note that "positive" and "negative" regulation refer to the type of regulatory protein involved: the bound protein either facilitates or inhibits transcription. In either case, addition of the molecular signal may increase or decrease transcription, depending on its effect on the regulatory protein.

(Fig. 28–4d). Signal molecules can therefore increase or decrease transcription, depending on how they affect the activator. Positive regulation is particularly common in eukaryotes, as we shall see.

## Many Prokaryotic Genes Are Clustered and Regulated in Operons

Bacteria have a simple general mechanism for coordinating the regulation of genes encoding products that participate in a set of related processes: these genes are clustered on the chromosome and are transcribed together. Many prokaryotic mRNAs are polycistronic—multiple genes on a single transcript—and the single promoter that initiates transcription of the cluster is the site of regulation for expression of all the genes in the cluster. The gene cluster and promoter, plus additional sequences that function together in regulation, are called an **operon** (Fig. 28–5). Operons that include two to six genes transcribed as a unit are common; some operons contain 20 or more genes.

   Many of the principles of prokaryotic gene expression were first defined by studies of lactose metabolism in *E. coli*, which can use lactose as its sole carbon source. In 1960, François Jacob and Jacques Monod published a short paper in the *Proceedings* of the French Academy of Sciences that described how two adjacent genes involved in lactose metabolism were coordinately regulated by a genetic element located at one end of the gene cluster. The genes were those for β-galactosidase, which cleaves lactose to galactose and glucose, and galactoside permease, which transports lactose into the cell (Fig. 28–6). The terms "operon" and "operator" were first introduced in this paper. With the operon model, gene regulation could, for the first time, be considered in molecular terms.

## The *lac* Operon Is Subject to Negative Regulation

The lactose (*lac*) operon (Fig. 28–7a) includes the genes for β-galactosidase (*Z*), galactoside permease (*Y*), and thiogalactoside transacetylase (*A*). The last of these enzymes appears to modify toxic galactosides to facilitate their removal from the cell. Each of the three genes is preceded by a ribosome binding site (not shown in Fig. 28–7) that independently directs the translation

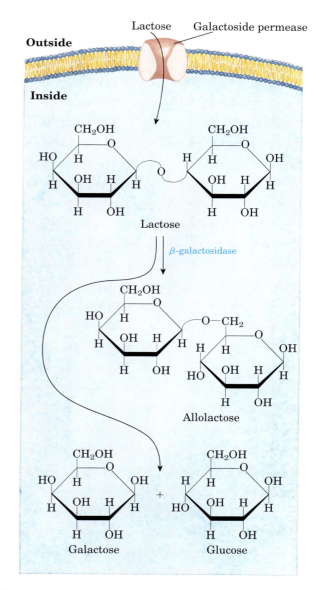

**FIGURE 28–6 Lactose metabolism in *E. coli*.** Uptake and metabolism of lactose require the activities of galactoside permease and β-galactosidase. Conversion of lactose to allolactose by transglycosylation is a minor reaction also catalyzed by β-galactosidase.

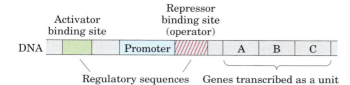

**FIGURE 28–5 Representative prokaryotic operon.** Genes A, B, and C are transcribed on one polycistronic mRNA. Typical regulatory sequences include binding sites for proteins that either activate or repress transcription from the promoter.

François Jacob

Jacques Monod, 1910–1976

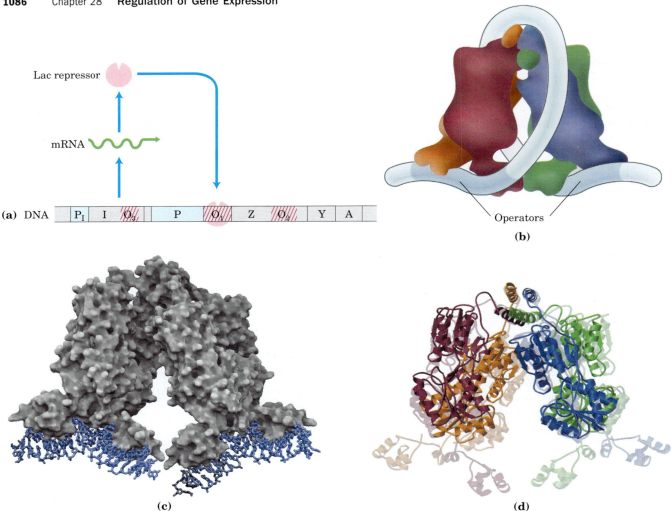

**FIGURE 28–7 The *lac* operon. (a)** The *lac* operon in the repressed state. The *I* gene encodes the Lac repressor. The *lac Z, Y,* and *A* genes encode β-galactosidase, galactoside permease, and thiogalactoside transacetylase, respectively. P is the promoter for the *lac* genes, and $P_I$ is the promoter for the *I* gene. $O_1$ is the main operator for the *lac* operon; $O_2$ and $O_3$ are secondary operator sites of lesser affinity for the Lac repressor. **(b)** The Lac repressor binds to the main operator and $O_2$ or $O_3$, apparently forming a loop in the DNA that might wrap around the repressor as shown. **(c)** Lac repressor bound to DNA (derived from PDB ID 1LBG). This shows the protein (gray) bound to short, discontinuous segments of DNA (blue). **(d)** Conformational change in the Lac repressor caused by binding of the artificial inducer isopropylthiogalactoside, IPTG (derived from PDB ID 1LBH and 1LBG). The structure of the tetrameric repressor is shown without IPTG bound (transparent image) and with IPTG bound (overlaid solid image; IPTG not shown). The DNA bound when IPTG is absent (transparent structure) is not shown. When IPTG is bound and DNA is not bound, the repressor's DNA-binding domains are too disordered to be defined in the crystal structure.

of that gene (Chapter 27). Regulation of the *lac* operon by the *lac* repressor protein (Lac) follows the pattern outlined in Figure 28–4a.

The study of *lac* operon mutants has revealed some details of the workings of the operon's regulatory system. In the absence of lactose, the *lac* operon genes are repressed. Mutations in the operator or in another gene, the *I* gene, result in constitutive synthesis of the gene products. When the *I* gene is defective, repression can be restored by introducing a functional *I* gene into the cell on another DNA molecule, demonstrating that the *I* gene encodes a diffusible molecule that causes gene repression. This molecule proved to be a protein, now called the Lac repressor, a tetramer of identical monomers. The operator to which it binds most tightly ($O_1$) abuts the transcription start site (Fig. 28–7a). The *I* gene is transcribed from its own promoter ($P_I$) independent of the *lac* operon genes. The *lac* operon has two secondary binding sites for the Lac repressor. One ($O_2$) is centered near position +410, within the gene encoding β-galactosidase (*Z*); the other ($O_3$) is near position −90, within the *I* gene. To repress the operon, the Lac repressor appears to bind to both the main operator and one of the two secondary sites, with the intervening DNA looped out (Fig. 28–7b, c). Either binding arrangement blocks transcription initiation.

Despite this elaborate binding complex, repression is not absolute. Binding of the Lac repressor reduces the rate of transcription initiation by a factor of $10^3$. If the $O_2$ and $O_3$ sites are eliminated by deletion or mutation, the binding of repressor to $O_1$ alone reduces transcription by a factor of about $10^2$. Even in the repressed state, each cell has a few molecules of $\beta$-galactosidase and galactoside permease, presumably synthesized on the rare occasions when the repressor transiently dissociates from the operators. This basal level of transcription is essential to operon regulation.

When cells are provided with lactose, the *lac* operon is induced. An inducer (signal) molecule binds to a specific site on the Lac repressor, causing a conformational change (Fig. 28–7d) that results in dissociation of the repressor from the operator. The inducer in the *lac* operon system is not lactose itself but allolactose, an isomer of lactose (Fig. 28–6). After entry into the *E. coli* cell (via the few existing molecules of permease), lactose is converted to allolactose by one of the few existing $\beta$-galactosidase molecules. Release of the operator by Lac repressor, triggered as the repressor binds to allolactose, allows expression of the *lac* operon genes and leads to a $10^3$-fold increase in the concentration of $\beta$-galactosidase.

Several $\beta$-galactosides structurally related to allolactose are inducers of the *lac* operon but are not substrates for $\beta$-galactosidase; others are substrates but not inducers. One particularly effective and nonmetabolizable inducer of the *lac* operon that is often used experimentally is isopropylthiogalactoside (IPTG):

Isopropylthiogalactoside
(IPTG)

An inducer that cannot be metabolized allows researchers to explore the physiological function of lactose as a carbon source for growth, separate from its function in the regulation of gene expression.

In addition to the multitude of operons now known in bacteria, a few polycistronic operons have been found in the cells of lower eukaryotes. In the cells of higher eukaryotes, however, almost all protein-encoding genes are transcribed separately.

The mechanisms by which operons are regulated can vary significantly from the simple model presented in Figure 28–7. Even the *lac* operon is more complex than indicated here, with an activator also contributing to the overall scheme, as we shall see in Section 28.2. Before any further discussion of the layers of regulation of gene expression, however, we examine the critical molecular interactions between DNA-binding proteins (such as repressors and activators) and the DNA sequences to which they bind.

## Regulatory Proteins Have Discrete DNA-Binding Domains

Regulatory proteins generally bind to specific DNA sequences. Their affinity for these target sequences is roughly $10^4$ to $10^6$ times higher than their affinity for any other DNA sequences. Most regulatory proteins have discrete DNA-binding domains containing substructures that interact closely and specifically with the DNA. These binding domains usually include one or more of a relatively small group of recognizable and characteristic structural motifs.

To bind specifically to DNA sequences, regulatory proteins must recognize surface features on the DNA. Most of the chemical groups that differ among the four bases and thus permit discrimination between base pairs are hydrogen-bond donor and acceptor groups exposed in the major groove of DNA (Fig. 28–8), and most of the protein-DNA contacts that impart specificity are hydrogen bonds. A notable exception is the nonpolar surface

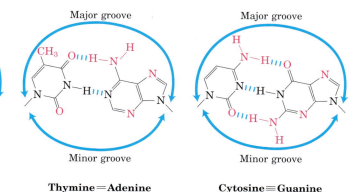

**FIGURE 28–8 Groups in DNA available for protein binding.** Shown here are functional groups on all four base pairs that are displayed in the major and minor grooves of DNA. Groups that can be used for base-pair recognition by proteins are shown in red.

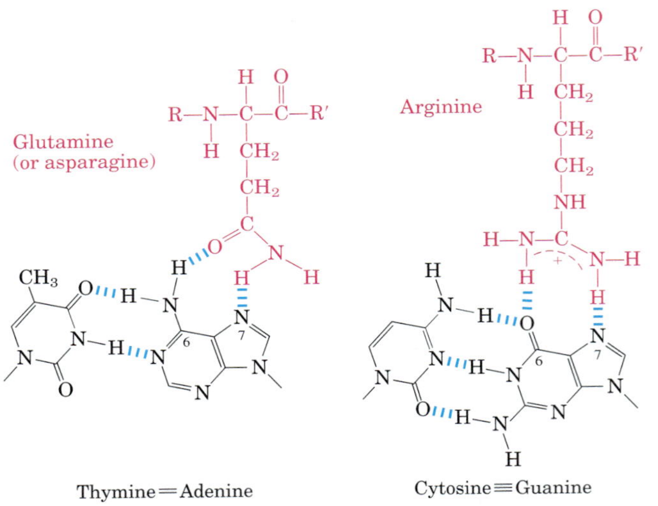

Glutamine
(or asparagine)

Arginine

Thymine═Adenine

Cytosine≡Guanine

**FIGURE 28–9** Two examples of specific amino acid–base pair interactions that have been observed in DNA-protein binding.

near C-5 of pyrimidines, where thymine is readily distinguished from cytosine by its protruding methyl group. Protein-DNA contacts are also possible in the minor groove of the DNA, but the hydrogen-bonding patterns here generally do not allow ready discrimination between base pairs.

Within regulatory proteins, the amino acid side chains most often hydrogen-bonding to bases in the DNA are those of Asn, Gln, Glu, Lys, and Arg residues. Is there a simple recognition code in which a particular amino acid always pairs with a particular base? The two hydrogen bonds that can form between Gln or Asn and the $N^6$ and N-7 positions of adenine cannot form with any other base. And an Arg residue can form two hydrogen bonds with N-7 and $O^6$ of guanine (Fig. 28–9). Examination of the structures of many DNA-binding proteins, however, has shown that a protein can recognize each base pair in more than one way, leading to the conclusion that there is no simple amino acid–base code. For some proteins, the Gln-adenine interaction can specify A═T base pairs, but in others a van der Waals pocket for the methyl group of thymine can recognize A═T base pairs. Researchers cannot yet examine the structure of a DNA-binding protein and infer the DNA sequence to which it binds.

To interact with bases in the major groove of DNA, a protein requires a relatively small structure that can stably protrude from the protein surface. The DNA-binding domains of regulatory proteins tend to be small (60 to 90 amino acid residues), and the structural motifs within these domains that are actually in contact with the DNA are smaller still. Many small proteins are unstable because of their limited capacity to form layers of structure to bury hydrophobic groups (p. 118). The DNA-binding motifs provide either a very compact stable structure or a way of allowing a segment of protein to protrude from the protein surface.

The DNA-binding sites for regulatory proteins are often inverted repeats of a short DNA sequence (a palindrome) at which multiple (usually two) subunits of a regulatory protein bind cooperatively. The Lac repressor is unusual in that it functions as a tetramer, with two dimers tethered together at the end distant from the DNA-binding sites (Fig. 28–7b). An *E. coli* cell normally contains about 20 tetramers of the Lac repressor. Each of the tethered dimers separately binds to a palindromic operator sequence, in contact with 17 bp of a 22 bp region in the *lac* operon (Fig. 28–10). And each of the tethered dimers can independently bind to an operator sequence, with one generally binding to $O_1$ and the other to $O_2$ or $O_3$ (as in Fig. 28–7b). The symmetry of the $O_1$ operator sequence corresponds to the twofold axis of symmetry of two paired Lac repressor subunits. The tetrameric Lac repressor binds to its operator sequences in vivo with an estimated dissociation constant of about $10^{-10}$ M. The repressor discriminates between the operators and other sequences by a factor of about $10^6$, so binding to these few base pairs among the 4.6 million or so of the *E. coli* chromosome is highly specific.

Several DNA-binding motifs have been described, but here we focus on two that play prominent roles in the binding of DNA by regulatory proteins: the **helix-turn-helix** and the **zinc finger.** We also consider a type of DNA-binding domain—the homeodomain—found in some eukaryotic proteins.

*Helix-Turn-Helix*    This DNA-binding motif is crucial to the interaction of many prokaryotic regulatory proteins with DNA, and similar motifs occur in some eukaryotic regulatory proteins. The helix-turn-helix motif comprises about 20 amino acids in two short α-helical segments,

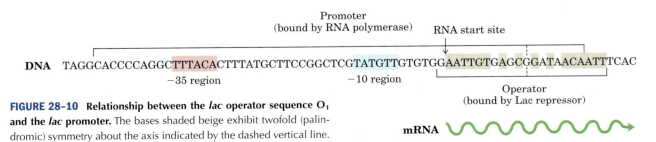

Promoter
(bound by RNA polymerase)

RNA start site

**DNA**    TAGGCACCCCAGGCTTTACACTTTATGCTTCCGGCTCGTATGTTGTGTGGAATTGTGAGCGGATAACAATTTCAC

−35 region

−10 region

Operator
(bound by Lac repressor)

**FIGURE 28–10 Relationship between the *lac* operator sequence $O_1$ and the *lac* promoter.** The bases shaded beige exhibit twofold (palindromic) symmetry about the axis indicated by the dashed vertical line.

**mRNA**

each seven to nine amino acid residues long, separated by a β turn (Fig. 28–11). This structure generally is not stable by itself; it is simply the reactive portion of a somewhat larger DNA-binding domain. One of the two α-helical segments is called the recognition helix, because it usually contains many of the amino acids that interact with the DNA in a sequence-specific way. This α helix is stacked on other segments of the protein structure so that it protrudes from the protein surface. When bound to DNA, the recognition helix is positioned in or nearly in the major groove. The Lac repressor has this DNA-binding motif (Fig. 28–11).

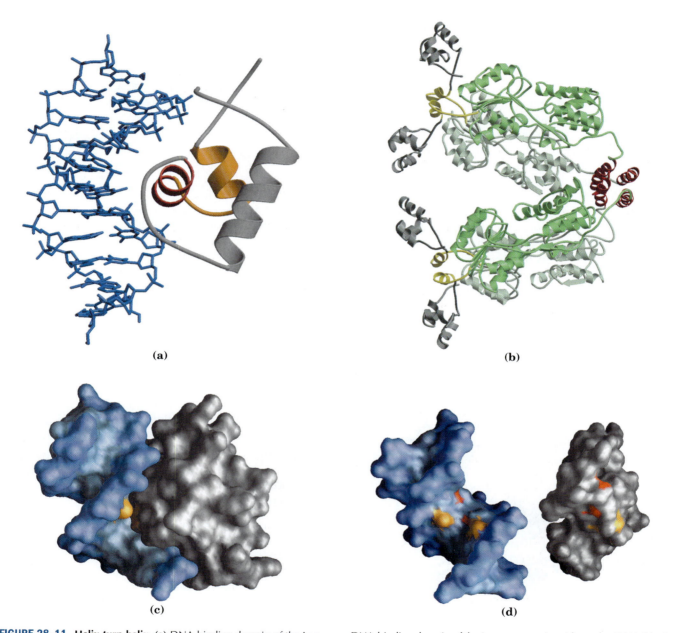

(a)

(b)

(c)

(d)

**FIGURE 28–11** **Helix-turn-helix.** (a) DNA-binding domain of the Lac repressor (PDB ID 1LCC). The helix-turn-helix motif is shown in red and orange; the DNA recognition helix is red. (b) Entire Lac repressor (derived from PDB ID 1LBG). The DNA-binding domains are gray, and the α helices involved in tetramerization are red. The remainder of the protein (shades of green) has the binding sites for allolactose. The allolactose-binding domains are linked to the DNA-binding domains through linker helices (yellow). (c) Surface rendering of the DNA-binding domain of the Lac repressor (gray) bound to DNA (blue). (d) The same DNA-binding domain as in (c), but separated from the DNA, with the binding interaction surfaces shown. Some groups on the protein and DNA that interact through hydrogen-bonding are shown in red; some groups that interact through hydrophobic interactions are in orange. This model shows only a few of the groups involved in sequence recognition. The complementary nature of the two surfaces is evident.

***Zinc Finger*** In a zinc finger, about 30 amino acid residues form an elongated loop held together at the base by a single $Zn^{2+}$ ion, which is coordinated to four of the residues (four Cys, or two Cys and two His). The zinc does not itself interact with DNA; rather, the coordination of zinc with the amino acid residues stabilizes this small structural motif. Several hydrophobic side chains in the core of the structure also lend stability. Figure 28–12 shows the interaction between DNA and three zinc fingers of a single polypeptide from the mouse regulatory protein Zif268.

Many eukaryotic DNA-binding proteins contain zinc fingers. The interaction of a single zinc finger with DNA is typically weak, and many DNA-binding proteins, like Zif268, have multiple zinc fingers that substantially enhance binding by interacting simultaneously with the DNA. One DNA-binding protein of the frog *Xenopus* has 37 zinc fingers. There are few known examples of the zinc finger motif in prokaryotic proteins.

The precise manner in which proteins with zinc fingers bind to DNA differs from one protein to the next. Some zinc fingers contain the amino acid residues that are important in sequence discrimination, whereas others appear to bind DNA nonspecifically (the amino acids required for specificity are located elsewhere in the protein). Zinc fingers can also function as RNA-binding motifs—for example, in certain proteins that bind eukaryotic mRNAs and act as translational repressors. We discuss this role later (Section 28.3).

***Homeodomain*** Another type of DNA-binding domain has been identified in a number of proteins that function as transcriptional regulators, especially during eukaryotic

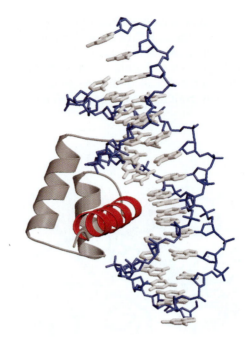

**FIGURE 28–13 Homeodomain.** Shown here is a homeodomain bound to DNA; one of the α helices (red), stacked on two others, can be seen protruding into the major groove (PDB ID 1B8I). This is only a small part of the much larger protein Ultrabithorax (Ubx), active in the regulation of development in fruit flies.

development. This domain of 60 amino acids—called the **homeodomain,** because it was discovered in homeotic genes (genes that regulate the development of body patterns)—is highly conserved and has now been identified in proteins from a wide variety of organisms, including humans (Fig. 28–13). The DNA-binding segment of the domain is related to the helix-turn-helix motif. The DNA sequence that encodes this domain is known as the **homeobox.**

## Regulatory Proteins Also Have Protein-Protein Interaction Domains

Regulatory proteins contain domains not only for DNA binding but also for protein-protein interactions—with RNA polymerase, other regulatory proteins, or other subunits of the same regulatory protein. Examples include many eukaryotic transcription factors that function as gene activators, which often bind as dimers to the DNA, using DNA-binding domains that contain zinc fingers. Some structural domains are devoted to the interactions required for dimer formation, which is generally a prerequisite for DNA binding. Like DNA-binding motifs, the structural motifs that mediate protein-protein interactions tend to fall within one of a few common categories. Two important examples are the **leucine zipper** and the **basic helix-loop-helix.** Structural motifs such as

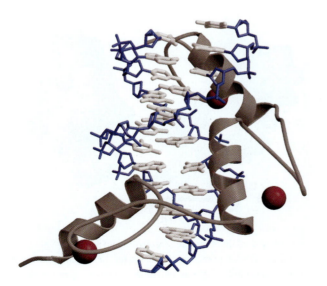

**FIGURE 28–12 Zinc fingers.** Three zinc fingers (gray) of the regulatory protein Zif268, complexed with DNA (blue and white) (PDB ID 1A1L). Each $Zn^{2+}$ (maroon) coordinates with two His and two Cys residues (not shown).

these are the basis for classifying some regulatory proteins into structural families.

**Leucine Zipper**   This motif is an amphipathic α helix with a series of hydrophobic amino acid residues concentrated on one side (Fig. 28–14), with the hydrophobic surface forming the area of contact between the two polypeptides of a dimer. A striking feature of these α helices is the occurrence of Leu residues at every seventh position, forming a straight line along the hydrophobic surface. Although researchers initially thought the Leu residues interdigitated (hence the name "zipper"), we now know that they line up side by side as the interacting α helices coil around each other (forming a coiled coil; Fig. 28–14b). Regulatory proteins with leucine zippers often have a separate DNA-binding domain with a high concentration of basic (Lys or Arg) residues that can interact with the negatively charged phosphates of the DNA backbone. Leucine zippers have been found in many eukaryotic and a few prokaryotic proteins.

**Basic Helix-Loop-Helix**   Another common structural motif occurs in some eukaryotic regulatory proteins implicated in the control of gene expression during the development of multicellular organisms. These proteins share a conserved region of about 50 amino acid residues important in both DNA binding and protein dimerization. This region can form two short amphipathic α helices linked by a loop of variable length, the helix-loop-helix (distinct from the helix-turn-helix motif associated with DNA binding). The helix-loop-helix motifs of two polypeptides interact to form dimers (Fig. 28–15). In these proteins, DNA binding is mediated by an adjacent short amino acid sequence rich in basic residues, similar to the separate DNA-binding region in proteins containing leucine zippers.

**Subunit Mixing in Eukaryotic Regulatory Proteins**   Several families of eukaryotic transcription factors have been defined based on close structural similarities. Within each family, dimers can sometimes form between two identical proteins (a homodimer) or between two different members of the family (a heterodimer). A hypothetical family of four different leucine-zipper proteins could thus form up to ten different dimeric species. In many cases, the different combinations appear to have distinct regulatory and functional properties.

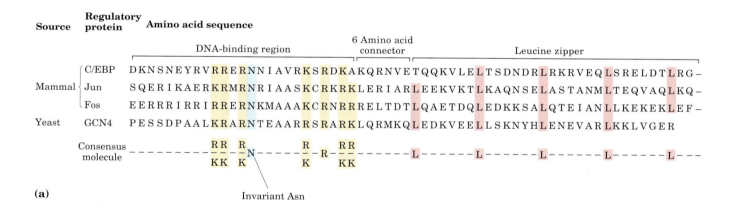

(a)

Invariant Asn

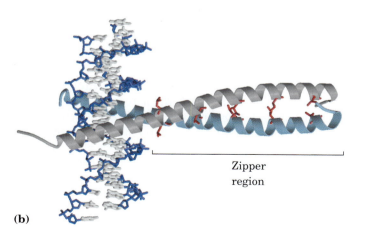

(b)

Zipper region

**FIGURE 28–14 Leucine zippers. (a)** Comparison of amino acid sequences of several leucine zipper proteins. Note the Leu (L) residues at every seventh position in the zipper region, and the number of Lys (K) and Arg (R) residues in the DNA-binding region. **(b)** Leucine zipper from the yeast activator protein GCN4 (PDB ID 1YSA). Only the "zippered" α helices (gray and light blue), derived from different subunits of the dimeric protein, are shown. The two helices wrap around each other in a gently coiled coil. The interacting Leu residues are shown in red.

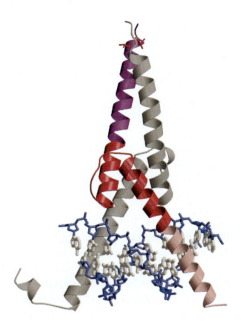

**FIGURE 28–15 Helix-loop-helix.** The human transcription factor Max, bound to its DNA target site (PDB ID 1HLO). The protein is dimeric; one subunit is colored. The DNA-binding segment (pink) merges with the first helix of the helix-loop-helix (red). The second helix merges with the carboxyl-terminal end of the subunit (purple). Interaction of the carboxyl-terminal helices of the two subunits describes a coiled coil very similar to that of a leucine zipper (see Fig. 28–14b), but with only one pair of interacting Leu residues (red side chains near the top) in this particular example. The overall structure is sometimes called a helix-loop-helix/leucine zipper motif.

In addition to structural domains devoted to DNA binding and dimerization (or oligomerization), many regulatory proteins must interact with RNA polymerase, with unrelated regulatory proteins, or with both. At least three different types of additional domains for protein-protein interaction have been characterized (primarily in eukaryotes): glutamine-rich, proline-rich, and acidic domains, the names reflecting the amino acid residues that are especially abundant.

Protein-DNA binding interactions are the basis of the intricate regulatory circuits fundamental to gene function. We now turn to a closer examination of these gene regulatory schemes, first in prokaryotic, then in eukaryotic systems.

### *SUMMARY 28.1* Principles of Gene Regulation

- The expression of genes is regulated by processes that affect the rates at which gene products are synthesized and degraded. Much of this regulation occurs at the level of transcription initiation, mediated by regulatory proteins that either repress transcription (negative regulation) or activate transcription (positive regulation) at specific promoters.

- In bacteria, genes that encode products with interdependent functions are often clustered in an operon, a single transcriptional unit. Transcription of the genes is generally blocked by binding of a specific repressor protein at a DNA site called an operator. Dissociation of the repressor from the operator is mediated by a specific small molecule, an inducer. These principles were first elucidated in studies of the lactose (*lac*) operon. The Lac repressor dissociates from the *lac* operator when the repressor binds to its inducer, allolactose.

- Regulatory proteins are DNA-binding proteins that recognize specific DNA sequences; most have distinct DNA-binding domains. Within these domains, common structural motifs that bind DNA are the helix-turn-helix, zinc finger, and homeodomain.

- Regulatory proteins also contain domains for protein-protein interactions, including the leucine zipper and helix-loop-helix, which are involved in dimerization, and other motifs involved in activation of transcription.

## 28.2 Regulation of Gene Expression in Prokaryotes

As in many other areas of biochemical investigation, the study of the regulation of gene expression advanced earlier and faster in bacteria than in other experimental organisms. The examples of bacterial gene regulation presented here are chosen from among scores of well-studied systems, partly for their historical significance, but primarily because they provide a good overview of the range of regulatory mechanisms employed in prokaryotes. Many of the principles of prokaryotic gene regulation are also relevant to understanding gene expression in eukaryotic cells.

We begin by examining the lactose and tryptophan operons; each system has regulatory proteins, but the overall mechanisms of regulation are very different. This is followed by a short discussion of the SOS response in *E. coli*, illustrating how genes scattered throughout the genome can be coordinately regulated. We then describe two prokaryotic systems of quite different types, illustrating the diversity of gene regulatory mechanisms: regulation of ribosomal protein synthesis at the level of translation, with many of the regulatory proteins binding to RNA (rather than DNA), and regulation of a process called phase variation in *Salmonella*, which results from genetic recombination. First, we return to the *lac* operon to examine its features in greater detail.

## The *lac* Operon Undergoes Positive Regulation

The operator-repressor-inducer interactions described earlier for the *lac* operon (Fig. 28–7) provide an intuitively satisfying model for an on/off switch in the regulation of gene expression. In truth, operon regulation is rarely so simple. A bacterium's environment is too complex for its genes to be controlled by one signal. Other factors besides lactose affect the expression of the *lac* genes, such as the availability of glucose. Glucose, metabolized directly by glycolysis, is *E. coli*'s preferred energy source. Other sugars can serve as the main or sole nutrient, but extra steps are required to prepare them for entry into glycolysis, necessitating the synthesis of additional enzymes. Clearly, expressing the genes for proteins that metabolize sugars such as lactose or arabinose is wasteful when glucose is abundant.

What happens to the expression of the *lac* operon when both glucose and lactose are present? A regulatory mechanism known as **catabolite repression** restricts expression of the genes required for catabolism of lactose, arabinose, and other sugars in the presence of glucose, even when these secondary sugars are also present. The effect of glucose is mediated by cAMP, as a coactivator, and an activator protein known as **cAMP receptor protein**, or **CRP** (the protein is sometimes called CAP, for *c*atabolite gene *a*ctivator *p*rotein). CRP is a homodimer (subunit $M_r$ 22,000) with binding sites for DNA and cAMP. Binding is mediated by a helix-turn-helix motif within the protein's DNA-binding domain (Fig. 28–16). When glucose is absent, CRP-cAMP binds to a site near the *lac* promoter (Fig. 28–17a) and stimulates RNA transcription 50-fold. CRP-cAMP is therefore a positive regulatory element responsive to glucose levels, whereas the Lac repressor is a negative regulatory element responsive to lactose. The two act in con-

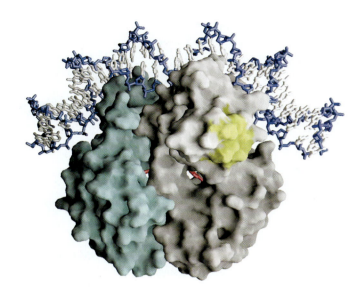

**FIGURE 28–16 CRP homodimer.** (PDB ID 1RUN) Bound molecules of cAMP are shown in red. Note the bending of the DNA around the protein. The region that interacts with RNA polymerase is shaded yellow.

cert. CRP-cAMP has little effect on the *lac* operon when the Lac repressor is blocking transcription, and dissociation of the repressor from the *lac* operator has little effect on transcription of the *lac* operon unless CRP-cAMP is present to facilitate transcription; when CRP is not bound, the wild-type *lac* promoter is a relatively weak promoter (Fig. 28–17b). The open complex of RNA polymerase and the promoter (see Fig. 26–6) does not form readily unless CRP-cAMP is present. CRP interacts directly with RNA polymerase (at the region shown in Fig. 28–16) through the polymerase's $\alpha$ subunit.

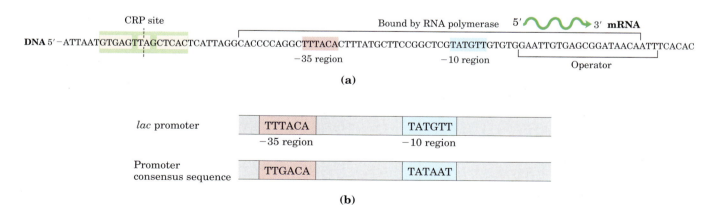

**(a)**

**(b)**

**FIGURE 28–17 Activation of transcription of the *lac* operon by CRP.** **(a)** The binding site for CRP-cAMP is near the promoter. As in the case of the *lac* operator, the CRP site has twofold symmetry (bases shaded beige) about the axis indicated by the dashed line. **(b)** Sequence of the *lac* promoter compared with the promoter consensus sequence. The differences mean that RNA polymerase binds relatively weakly to the *lac* promoter until the polymerase is activated by CRP-cAMP.

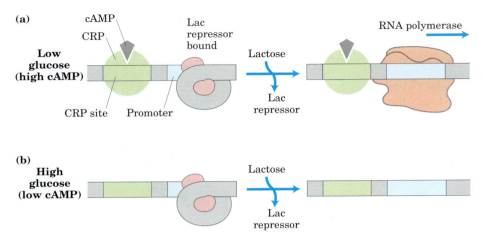

**FIGURE 28–18 Combined effects of glucose and lactose on expression of the *lac* operon. (a)** High levels of transcription take place only when glucose concentrations are low (so cAMP levels are high and CRP-cAMP is bound) and lactose concentrations are high (so the Lac repressor is not bound). **(b)** Without bound activator (CRP-cAMP), the *lac* promoter is poorly transcribed even when lactose concentrations are high and the Lac repressor is not bound.

The effect of glucose on CRP is mediated by the cAMP interaction (Fig. 28–18). CRP binds to DNA most avidly when cAMP concentrations are high. In the presence of glucose, the synthesis of cAMP is inhibited and efflux of cAMP from the cell is stimulated. As [cAMP] declines, CRP binding to DNA declines, thereby decreasing the expression of the *lac* operon. Strong induction of the *lac* operon therefore requires both lactose (to inactivate the *lac* repressor) and a lowered concentration of glucose (to trigger an increase in [cAMP] and increased binding of cAMP to CRP).

CRP and cAMP are involved in the coordinated regulation of many operons, primarily those that encode enzymes for the metabolism of secondary sugars such as lactose and arabinose. A network of operons with a common regulator is called a **regulon.** This arrangement, which allows for coordinated shifts in cellular functions that can require the action of hundreds of genes, is a major theme in the regulated expression of dispersed networks of genes in eukaryotes. Other bacterial regulons include the heat-shock gene system that responds to changes in temperature (p. 1083) and the genes induced in *E. coli* as part of the SOS response to DNA damage, described later.

## Many Genes for Amino Acid Biosynthetic Enzymes Are Regulated by Transcription Attenuation

The 20 common amino acids are required in large amounts for protein synthesis, and *E. coli* can synthesize all of them. The genes for the enzymes needed to synthesize a given amino acid are generally clustered in an operon and are expressed whenever existing supplies of that amino acid are inadequate for cellular requirements. When the amino acid is abundant, the biosynthetic enzymes are not needed and the operon is repressed.

The *E. coli* tryptophan (*trp*) operon (Fig. 28–19) includes five genes for the enzymes required to convert chorismate to tryptophan. Note that two of the enzymes catalyze more than one step in the pathway. The mRNA from the *trp* operon has a half-life of only about 3 min, allowing the cell to respond rapidly to changing needs for this amino acid. The Trp repressor is a homodimer, each subunit containing 107 amino acid residues (Fig. 28–20). When tryptophan is abundant it binds to the Trp repressor, causing a conformational change that permits the repressor to bind to the *trp* operator and inhibit expression of the *trp* operon. The *trp* operator site overlaps the promoter, so binding of the repressor blocks binding of RNA polymerase.

Once again, this simple on/off circuit mediated by a repressor is not the entire regulatory story. Different cellular concentrations of tryptophan can vary the rate of synthesis of the biosynthetic enzymes over a 700-fold range. Once repression is lifted and transcription begins, the rate of transcription is fine-tuned by a second regulatory process, called **transcription attenuation,** in which transcription is initiated normally but is abruptly halted *before* the operon genes are transcribed. The frequency with which transcription is attenuated is regulated by the availability of tryptophan and relies on the very close coupling of transcription and translation in bacteria.

The *trp* operon attenuation mechanism uses signals encoded in four sequences within a 162 nucleotide **leader** region at the 5′ end of the mRNA, preceding the initiation codon of the first gene (Fig. 28–21a). Within the leader lies a region known as the **attenuator,** made up of sequences 3 and 4. These sequences base-pair to

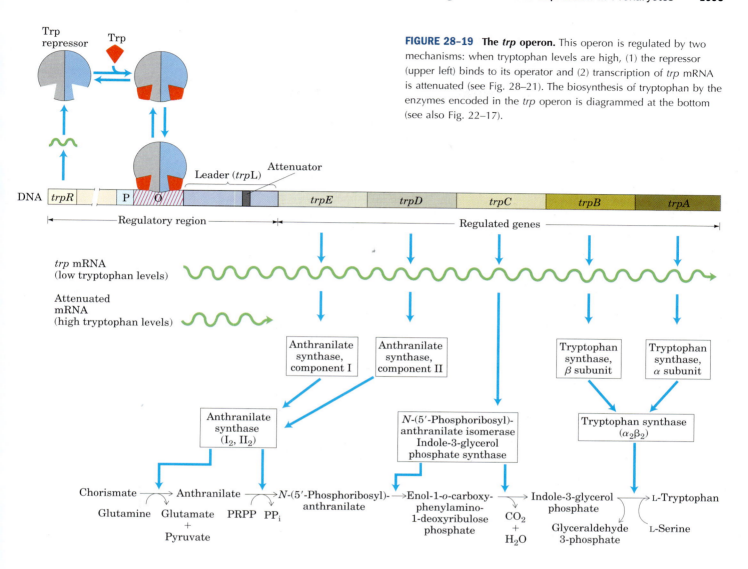

FIGURE 28–19 **The *trp* operon.** This operon is regulated by two mechanisms: when tryptophan levels are high, (1) the repressor (upper left) binds to its operator and (2) transcription of *trp* mRNA is attenuated (see Fig. 28–21). The biosynthesis of tryptophan by the enzymes encoded in the *trp* operon is diagrammed at the bottom (see also Fig. 22–17).

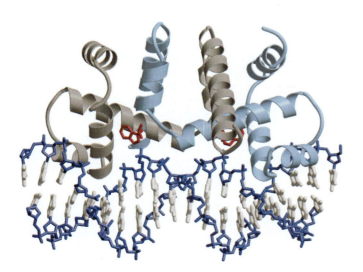

FIGURE 28–20 **Trp repressor.** The repressor is a dimer, with both subunits (gray and light blue) binding the DNA at helix-turn-helix motifs (PDB ID 1TRO). Bound molecules of tryptophan are in red.

form a G≡C-rich stem-and-loop structure closely followed by a series of U residues. The attenuator structure acts as a transcription terminator (Fig. 28–21b). Sequence 2 is an alternative complement for sequence 3 (Fig. 28–21c). If sequences 2 and 3 base-pair, the attenuator structure cannot form and transcription continues into the *trp* biosynthetic genes; the loop formed by the pairing of sequences 2 and 3 does not obstruct transcription.

Regulatory sequence 1 is crucial for a tryptophan-sensitive mechanism that determines whether sequence 3 pairs with sequence 2 (allowing transcription to continue) or with sequence 4 (attenuating transcription). Formation of the attenuator stem-and-loop structure depends on events that occur during *translation* of regulatory sequence 1, which encodes a leader peptide (so called because it is encoded by the leader region of the mRNA) of 14 amino acids, two of which are Trp residues. The leader peptide has no other known cellular function; its synthesis is simply an operon regulatory device.

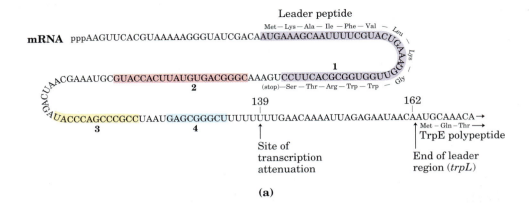

**(a)**

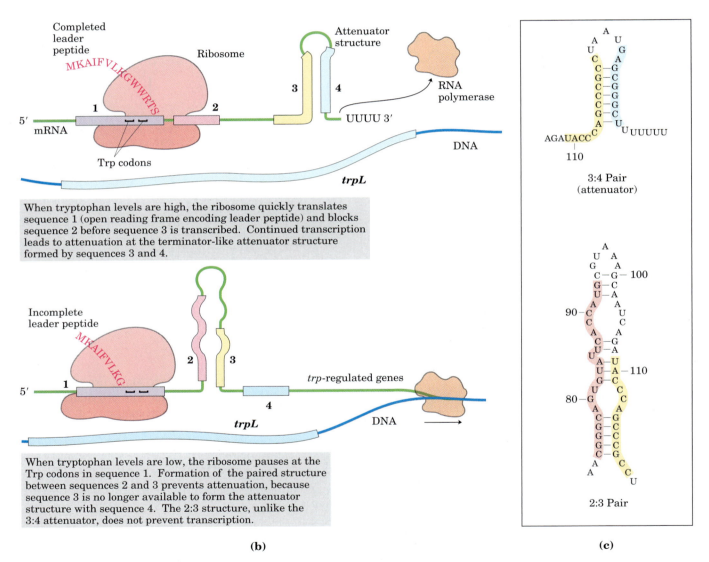

**(b)**

**(c)**

**FIGURE 28–21 Transcriptional attenuation in the *trp* operon.** Transcription is initiated at the beginning of the 162 nucleotide mRNA leader encoded by a DNA region called *trpL* (see Fig. 28-19). A regulatory mechanism determines whether transcription is attenuated at the end of the leader or continues into the structural genes. **(a)** The *trp* mRNA leader (*trpL*). The attenuation mechanism in the *trp* operon involves sequences 1 to 4 (highlighted). **(b)** Sequence 1 encodes a small peptide, the leader peptide, containing two Trp residues (W); it is translated immediately after transcription begins. Sequences 2 and 3 are complementary, as are sequences 3 and 4. The attenuator structure forms by the pairing of sequences 3 and 4 (top). Its structure and function are similar to those of a transcription terminator (see Fig. 26–7). Pairing of sequences 2 and 3 (bottom) prevents the attenuator structure from forming. Note that the leader peptide has no other cellular function. Translation of its open reading frame has a purely regulatory role that determines which complementary sequences (2 and 3 or 3 and 4) are paired. **(c)** Base-pairing schemes for the complementary regions of the *trp* mRNA leader.

This peptide is translated immediately after it is transcribed, by a ribosome that follows closely behind RNA polymerase as transcription proceeds.

When tryptophan concentrations are high, concentrations of charged tryptophan tRNA (Trp-tRNA$^{Trp}$) are also high. This allows translation to proceed rapidly past the two Trp codons of sequence 1 and into sequence 2, before sequence 3 is synthesized by RNA polymerase. In this situation, sequence 2 is covered by the ribosome and unavailable for pairing to sequence 3 when sequence 3 is synthesized; the attenuator structure (sequences 3 and 4) forms and transcription halts (Fig. 28–21b, top). When tryptophan concentrations are low, however, the ribosome stalls at the two Trp codons in sequence 1, because charged tRNA$^{Trp}$ is less available. Sequence 2 remains free while sequence 3 is synthesized, allowing these two sequences to base-pair and permitting transcription to proceed (Fig. 28–21b, bottom). In this way, the proportion of transcripts that are attenuated declines as tryptophan concentration declines.

Many other amino acid biosynthetic operons use a similar attenuation strategy to fine-tune biosynthetic enzymes to meet the prevailing cellular requirements. The 15 amino acid leader peptide produced by the *phe* operon contains seven Phe residues. The *leu* operon leader peptide has four contiguous Leu residues. The leader peptide for the *his* operon contains seven contiguous His residues. In fact, in the *his* operon and a number of others, attenuation is sufficiently sensitive to be the *only* regulatory mechanism.

### Induction of the SOS Response Requires Destruction of Repressor Proteins

Extensive DNA damage in the bacterial chromosome triggers the induction of many distantly located genes. This response, called the SOS response (p. 976), provides another good example of coordinated gene regulation. Many of the induced genes are involved in DNA repair (see Table 25–6). The key regulatory proteins are the RecA protein and the LexA repressor.

The LexA repressor ($M_r$ 22,700) inhibits transcription of all the SOS genes (Fig. 28–22), and induction of the SOS response requires removal of LexA. This is not a simple dissociation from DNA in response to binding of a small molecule, as in the regulation of the *lac* operon described above. Instead, the LexA repressor is

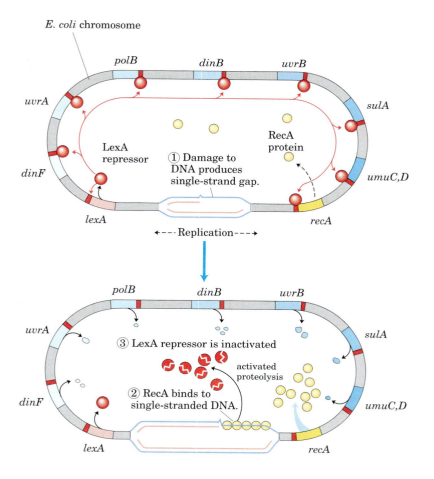

**FIGURE 28–22 SOS response in *E. coli*.** See Table 25–6 for the functions of many of these proteins. The LexA protein is the repressor in this system, which has an operator site (red) near each gene. Because the *recA* gene is not entirely repressed by the LexA repressor, the normal cell contains about 1,000 RecA monomers. ① When DNA is extensively damaged (e.g., by UV light), DNA replication is halted and the number of single-strand gaps in the DNA increases. ② RecA protein binds to this damaged, single-stranded DNA, activating the protein's coprotease activity. ③ While bound to DNA, the RecA protein facilitates cleavage and inactivation of the LexA repressor. When the repressor is inactivated, the SOS genes, including *recA*, are induced; RecA levels increase 50- to 100-fold.

inactivated when it catalyzes its own cleavage at a specific Ala–Gly peptide bond, producing two roughly equal protein fragments. At physiological pH, this autocleavage reaction requires the RecA protein. RecA is not a protease in the classical sense, but its interaction with LexA facilitates the repressor's self-cleavage reaction. This function of RecA is sometimes called a coprotease activity.

The RecA protein provides the functional link between the biological signal (DNA damage) and induction of the SOS genes. Heavy DNA damage leads to numerous single-strand gaps in the DNA, and only RecA that is bound to single-stranded DNA can facilitate cleavage of the LexA repressor (Fig. 28–22, bottom). Binding of RecA at the gaps eventually activates its coprotease activity, leading to cleavage of the LexA repressor and SOS induction.

During induction of the SOS response in a severely damaged cell, RecA also cleaves and thus inactivates the repressors that otherwise allow propagation of certain viruses in a dormant lysogenic state within the bacterial host. This provides a remarkable illustration of evolutionary adaptation. These repressors, like LexA, also undergo self-cleavage at a specific Ala–Gly peptide bond, so induction of the SOS response permits replication of the virus and lysis of the cell, releasing new viral particles. Thus the bacteriophage can make a hasty exit from a compromised bacterial host cell.

## Synthesis of Ribosomal Proteins Is Coordinated with rRNA Synthesis

In bacteria, an increased cellular demand for protein synthesis is met by increasing the number of ribosomes rather than altering the activity of individual ribosomes. In general, the number of ribosomes increases as the cellular growth rate increases. At high growth rates, ribosomes make up approximately 45% of the cell's dry weight. The proportion of cellular resources devoted to making ribosomes is so large, and the function of ribosomes so important, that cells must coordinate the synthesis of the ribosomal components: the ribosomal proteins (r-proteins) and RNAs (rRNAs). This regulation is distinct from the mechanisms described so far, because it occurs largely at the level of *translation*.

The 52 genes that encode the r-proteins occur in at least 20 operons, each with 1 to 11 genes. Some of these operons also contain the genes for the subunits of DNA primase (see Fig. 25–13), RNA polymerase (see Fig. 26–4), and protein synthesis elongation factors (see Fig. 27–23)—revealing the close coupling of replication, transcription, and protein synthesis during cell growth.

The r-protein operons are regulated primarily through a translational feedback mechanism. One r-protein encoded by each operon also functions as a **translational repressor,** which binds to the mRNA

transcribed from that operon and blocks translation of all the genes the messenger encodes (Fig. 28–23). In general, the r-protein that plays the role of repressor also binds directly to an rRNA. Each translational repressor r-protein binds with higher affinity to the appropriate rRNA than to its mRNA, so the mRNA is bound and translation repressed only when the level of the r-protein exceeds that of the rRNA. This ensures that translation of the mRNAs encoding r-proteins is repressed only when synthesis of these r-proteins exceeds that needed to make functional ribosomes. In this way, the rate of r-protein synthesis is kept in balance with rRNA availability.

The mRNA binding site for the translational repressor is near the translational start site of one of the genes in the operon, usually the first gene (Fig. 28–23). In other operons this would affect only that one gene, because in bacterial polycistronic mRNAs most genes have independent translation signals. In the r-protein operons, however, the translation of one gene depends on the translation of all the others. The mechanism of this translational coupling is not yet understood in detail. However, in some cases the translation of multiple genes appears to be blocked by folding of the mRNA into an elaborate three-dimensional structure that is stabilized both by internal base-pairing (as in Fig. 8–26) and by binding of the translational repressor protein. When the translational repressor is absent, ribosome binding and translation of one or more of the genes disrupts the folded structure of the mRNA and allows all the genes to be translated.

Because the synthesis of r-proteins is coordinated with the available rRNA, the regulation of ribosome production reflects the regulation of rRNA synthesis. In *E. coli*, rRNA synthesis from the seven rRNA operons responds to cellular growth rate and to changes in the availability of crucial nutrients, particularly amino acids. The regulation coordinated with amino acid concentrations is known as the **stringent response** (Fig. 28–24). When amino acid concentrations are low, rRNA synthesis is halted. Amino acid starvation leads to the binding of uncharged tRNAs to the ribosomal A site; this triggers a sequence of events that begins with the binding of an enzyme called **stringent factor** (RelA protein) to the ribosome. When bound to the ribosome, stringent factor catalyzes formation of the unusual nucleotide guanosine tetraphosphate (ppGpp; see Fig. 8–42); it adds pyrophosphate to the 3′ position of GTP, in the reaction

$$\text{GTP} + \text{ATP} \longrightarrow \text{pppGpp} + \text{AMP}$$

then a phosphohydrolase cleaves off one phosphate to form ppGpp. The abrupt rise in ppGpp level in response to amino acid starvation results in a great reduction in rRNA synthesis, mediated at least in part by the binding of ppGpp to RNA polymerase.

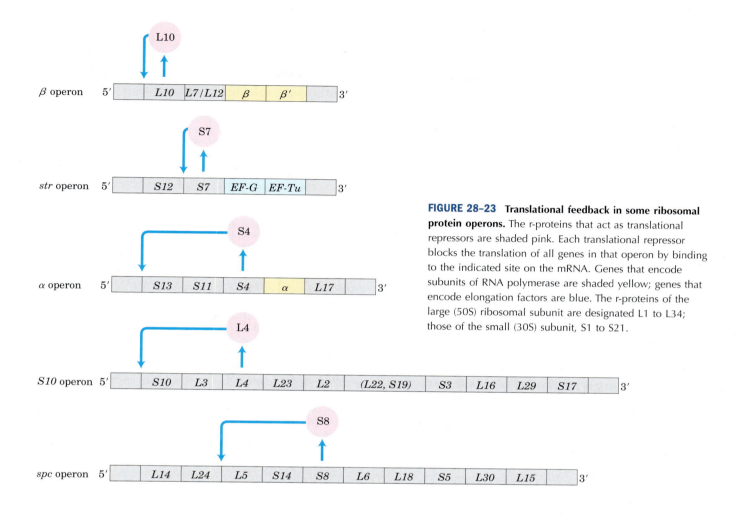

**FIGURE 28–23 Translational feedback in some ribosomal protein operons.** The r-proteins that act as translational repressors are shaded pink. Each translational repressor blocks the translation of all genes in that operon by binding to the indicated site on the mRNA. Genes that encode subunits of RNA polymerase are shaded yellow; genes that encode elongation factors are blue. The r-proteins of the large (50S) ribosomal subunit are designated L1 to L34; those of the small (30S) subunit, S1 to S21.

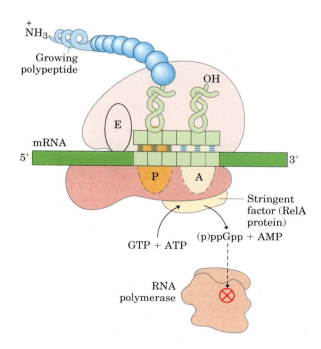

**FIGURE 28–24 Stringent response in *E. coli*.** This response to amino acid starvation is triggered by binding of an uncharged tRNA in the ribosomal A site. A protein called stringent factor binds to the ribosome and catalyzes the synthesis of pppGpp, which is converted by a phosphohydrolase to ppGpp. The signal ppGpp reduces transcription of some genes and increases that of others, in part by binding to the β subunit of RNA polymerase and altering the enzyme's promoter specificity. Synthesis of rRNA is reduced when ppGpp levels increase.

The nucleotide ppGpp, along with cAMP, belongs to a class of modified nucleotides that act as cellular second messengers (p. 302). In *E. coli*, these two nucleotides serve as starvation signals; they cause large changes in cellular metabolism by increasing or decreasing the transcription of hundreds of genes. In eukaryotic cells, similar nucleotide second messengers also have multiple regulatory functions. The coordination of cellular metabolism with cell growth is highly complex, and further regulatory mechanisms undoubtedly remain to be discovered.

## Some Genes Are Regulated by Genetic Recombination

*Salmonella typhimurium*, which inhabits the mammalian intestine, moves by rotating the flagella on its cell surface (Fig. 28–25). The many copies of the protein flagellin ($M_r$ 53,000) that make up the flagella are prominent targets of mammalian immune systems. But *Salmonella* cells have a mechanism that evades the immune response: they switch between two distinct flagellin proteins (FljB and FliC) roughly once every 1,000 generations, using a process called **phase variation.**

The switch is accomplished by periodic inversion of a segment of DNA containing the promoter for a flagellin gene. The inversion is a site-specific recombination reaction (see Fig. 25–39) mediated by the Hin recombinase at specific 14 bp sequences (*hix* sequences)

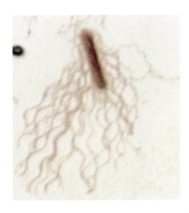

**FIGURE 28–25** *Salmonella typhimurium*, with flagella evident.

at either end of the DNA segment. When the DNA segment is in one orientation, the gene for FljB flagellin and the gene encoding a repressor (FljA) are expressed (Fig. 28–26a); the repressor shuts down expression of the gene for FliC flagellin. When the DNA segment is inverted (Fig. 28–26b), the *fljA* and *fljB* genes are no longer transcribed, and the *fliC* gene is induced as the repressor becomes depleted. The Hin recombinase, encoded by the *hin* gene in the DNA segment that undergoes inversion, is expressed when the DNA segment is in either orientation, so the cell can always switch from one state to the other.

This type of regulatory mechanism has the advantage of being absolute: gene expression is impossible

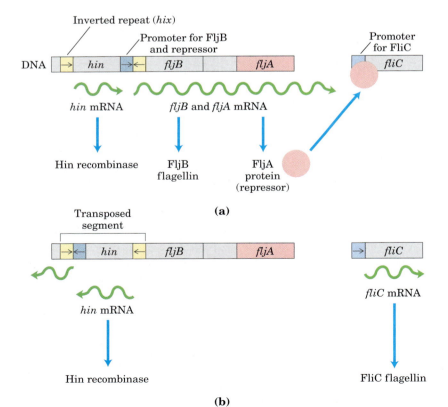

**FIGURE 28–26** **Regulation of flagellin genes in *Salmonella*: phase variation.** The products of genes *fliC* and *fljB* are different flagellins. The *hin* gene encodes the recombinase that catalyzes inversion of the DNA segment containing the *fljB* promoter and the *hin* gene. The recombination sites (inverted repeats) are called *hix* (yellow). **(a)** In one orientation, *fljB* is expressed along with a repressor protein (product of the *fljA* gene) that represses transcription of the *fliC* gene. **(b)** In the opposite orientation only the *fliC* gene is expressed; the *fljA* and *fljB* genes cannot be transcribed. The interconversion between these two states, known as phase variation, also requires two other nonspecific DNA-binding proteins (not shown), HU (histonelike protein from *U*13, a strain of *E. coli*) and FIS (factor for inversion stimulation).

## TABLE 28–1 Examples of Gene Regulation by Recombination

| System | Recombinase/ recombination site | Type of recombination | Function |
|---|---|---|---|
| Phase variation (*Salmonella*) | Hin/*hix* | Site-specific | Alternative expression of two flagellin genes allows evasion of host immune response. |
| Host range (bacteriophage $\mu$) | Gin/*gix* | Site-specific | Alternative expression of two sets of tail fiber genes affects host range. |
| Mating-type switch (yeast) | HO endonuclease, RAD52 protein, other proteins/*MAT* | Nonreciprocal gene conversion[*] | Alternative expression of two mating types of yeast, a and $\alpha$, creates cells of different mating types that can mate and undergo meiosis. |
| Antigenic variation (trypanosomes)[†] | Varies | Nonreciprocal gene conversion[*] | Successive expression of different genes encoding the variable surface glycoproteins (VSGs) allows evasion of host immune response. |

[*]In nonreciprocal gene conversion (a class of recombination events not discussed in Chapter 25), genetic information is moved from one part of the genome (where it is silent) to another (where it is expressed). The reaction is similar to replicative transposition (see Fig. 25–43).

[†]Trypanosomes cause African sleeping sickness and other diseases (see Box 22–2). The outer surface of a trypanosome is made up of multiple copies of a single VSG, the major surface antigen. A cell can change surface antigens to more than 100 different forms, precluding an effective defense by the host immune system.

when the gene is physically separated from its promoter (note the position of the *fljB* promoter in Fig. 28–26b). An absolute on/off switch may be important in this system (even though it affects only one of the two flagellin genes), because a flagellum with just one copy of the wrong flagellin might be vulnerable to host antibodies against that protein. The *Salmonella* system is by no means unique. Similar regulatory systems occur in a number of other bacteria and in some bacteriophages, and recombination systems with similar functions have been found in eukaryotes (Table 28–1). Gene regulation by DNA rearrangements that move genes and/or promoters is particularly common in pathogens that benefit by changing their host range or by changing their surface proteins, thereby staying ahead of host immune systems.

## SUMMARY 28.2 Regulation of Gene Expression in Prokaryotes

- In addition to repression by the Lac repressor, the *E. coli lac* operon undergoes positive regulation by the cAMP receptor protein (CRP). When [glucose] is low, [cAMP] is high and CRP-cAMP binds to a specific site on the DNA, stimulating transcription of the *lac* operon and production of lactose-metabolizing enzymes. The presence of glucose depresses [cAMP], decreasing expression of *lac* and other genes involved in metabolism of secondary sugars. A group of coordinately regulated operons is referred to as a regulon.

- Operons that produce the enzymes of amino acid synthesis have a regulatory circuit called attenuation, which uses a transcription termination site (the attenuator) in the mRNA. Formation of the attenuator is modulated by a mechanism that couples transcription and translation while responding to small changes in amino acid concentration.

- In the SOS system, multiple unlinked genes repressed by a single repressor are induced simultaneously when DNA damage triggers RecA protein–facilitated autocatalytic proteolysis of the repressor.

- In the synthesis of ribosomal proteins, one protein in each r-protein operon acts as a translational repressor. The mRNA is bound by the repressor, and translation is blocked only when the r-protein is present in excess of available rRNA. Some genes are regulated by genetic recombination processes that move promoters relative to the genes being regulated. Regulation can also take place at the level of translation. These diverse mechanisms permit very sensitive cellular responses to environmental change.

## 28.3 Regulation of Gene Expression in Eukaryotes

Initiation of transcription is a crucial regulation point for both prokaryotic and eukaryotic gene expression. Although some of the same regulatory mechanisms are used in both systems, there is a fundamental difference in the regulation of transcription in eukaryotes and bacteria.

We can define a transcriptional ground state as the inherent activity of promoters and transcriptional machinery in vivo in the absence of regulatory sequences. In bacteria, RNA polymerase generally has access to every promoter and can bind and initiate transcription at some level of efficiency in the absence of activators or repressors; the transcriptional ground state is therefore nonrestrictive. In eukaryotes, however, strong promoters are generally inactive in vivo in the absence of regulatory proteins; that is, the transcriptional ground state is restrictive. This fundamental difference gives rise to at least four important features that distinguish the regulation of gene expression in eukaryotes from that in bacteria.

First, access to eukaryotic promoters is restricted by the structure of chromatin, and activation of transcription is associated with many changes in chromatin structure in the transcribed region. Second, although eukaryotic cells have both positive and negative regulatory mechanisms, positive mechanisms predominate in all systems characterized so far. Thus, given that the transcriptional ground state is restrictive, virtually every eukaryotic gene requires activation to be transcribed. Third, eukaryotic cells have larger, more complex multimeric regulatory proteins than do bacteria. Finally, transcription in the eukaryotic nucleus is separated from translation in the cytoplasm in both space and time.

The complexity of regulatory circuits in eukaryotic cells is extraordinary, as the following discussion shows. We conclude the section with an illustrated description of one of the most elaborate circuits: the regulatory cascade that controls development in fruit flies.

### Transcriptionally Active Chromatin Is Structurally Distinct from Inactive Chromatin

The effects of chromosome structure on gene regulation in eukaryotes have no clear parallel in prokaryotes. In the eukaryotic cell cycle, interphase chromosomes appear, at first viewing, to be dispersed and amorphous (see Figs 12–41, 24–25). Nevertheless, several forms of chromatin can be found along these chromosomes. About 10% of the chromatin in a typical eukaryotic cell is in a more condensed form than the rest of the chromatin. This form, **heterochromatin,** is transcriptionally inactive. Heterochromatin is generally associated with particular chromosome structures—the centromeres, for example. The remaining, less condensed chromatin is called **euchromatin.**

Transcription of a eukaryotic gene is strongly repressed when its DNA is condensed within heterochromatin. Some, but not all, of the euchromatin is transcriptionally active. Transcriptionally active chromosomal regions can be detected based on their increased sensitivity to nuclease-mediated degradation. Nucleases such as DNase I tend to cleave the DNA of carefully isolated chromatin into fragments of multiples of about 200 bp, reflecting the regular repeating structure of the nucleosome (see Fig. 24–26). In actively transcribed regions, the fragments produced by nuclease activity are smaller and more heterogeneous in size. These regions contain **hypersensitive sites,** sequences especially sensitive to DNase I, which consist of about 100 to 200 bp within the 1,000 bp flanking the 5′ ends of transcribed genes. In some genes, hypersensitive sites are found farther from the 5′ end, near the 3′ end, or even within the gene itself.

Many hypersensitive sites correspond to binding sites for known regulatory proteins, and the relative absence of nucleosomes in these regions may allow the binding of these proteins. Nucleosomes are entirely absent in some regions that are very active in transcription, such as the rRNA genes. Transcriptionally active chromatin tends to be deficient in histone H1, which binds to the linker DNA between nucleosome particles.

Histones within transcriptionally active chromatin and heterochromatin also differ in their patterns of covalent modification. The core histones of nucleosome particles (H2A, H2B, H3, H4; see Fig. 24–27) are modified by irreversible methylation of Lys residues, phosphorylation of Ser or Thr residues, acetylation (see below), or attachment of ubiquitin (see Fig. 27–41). Each of the core histones has two distinct structural domains. A central domain is involved in histone-histone interaction and the wrapping of DNA around the nucleosome. A second, lysine-rich amino-terminal domain is generally positioned near the exterior of the assembled nucleosome particle; the covalent modifications occur at specific residues concentrated in this amino-terminal domain. The patterns of modification have led some researchers to propose the existence of a histone code, in which modification patterns are recognized by enzymes that alter the structure of chromatin. Modifications associated with transcriptional activation would be recognized by enzymes that make the chromatin more accessible to the transcription machinery.

5-Methylation of cytosine residues of CpG sequences is common in eukaryotic DNA (p. 296), but DNA in transcriptionally active chromatin tends to be undermethylated. Furthermore, CpG sites in particular genes are more often undermethylated in cells from tissues where the genes are expressed than in those where the genes are not expressed. The overall pattern suggests that active chromatin is prepared for transcription by the removal of potential structural barriers.

## Chromatin Is Remodeled by Acetylation and Nucleosomal Displacements

The detailed mechanisms for transcription-associated structural changes in chromatin, called **chromatin remodeling,** are now coming to light, including identification of a variety of enzymes directly implicated in the process. These include enzymes that covalently modify the core histones of the nucleosome and others that use the chemical energy of ATP to remodel nucleosomes on the DNA (Table 28–2).

The acetylation and deacetylation of histones figure prominently in the processes that activate chromatin for transcription. As noted above, the amino-terminal domains of the core histones are generally rich in Lys residues. Particular Lys residues are acetylated by histone acetyltransferases (HATs). Cytosolic (type B) HATs acetylate newly synthesized histones before the histones are imported into the nucleus. The subsequent assembly of the histones into chromatin is facilitated by additional proteins: CAF1 for H3 and H4, and NAP1 for H2A and H2B. (See Table 28–2 for an explanation of some of these abbreviated names.)

Where chromatin is being activated for transcription, the nucleosomal histones are further acetylated by nuclear (type A) HATs. The acetylation of multiple Lys residues in the amino-terminal domains of histones H3 and H4 can reduce the affinity of the entire nucleosome for DNA. Acetylation may also prevent or promote interactions with other proteins involved in transcription or its regulation. When transcription of a gene is no longer required, the acetylation of nucleosomes in that vicinity is reduced by the activity of histone deacetylases, as part of a general gene-silencing process that restores the chromatin to a transcriptionally inactive state. In addition to the removal of certain acetyl groups, new covalent modification of histones marks chromatin as transcriptionally inactive. As an example, the Lys residue at position 9 in histone H3 is often methylated in heterochromatin.

Chromatin remodeling also requires protein complexes that actively move or displace nucleosomes, hydrolyzing ATP in the process (Table 28–2). The enzyme complex SWI/SNF found in all eukaryotic cells, contains 11 polypeptides (total $M_r$ $2 \times 10^6$) that together create hypersensitive sites in the chromatin and stimulate the binding of transcription factors. SWI/SNF is not required for the transcription of every gene. NURF is another ATP-dependent enzyme complex that remodels chromatin in ways that complement and overlap the activity of SWI/SNF. These enzyme complexes play an important role in preparing a region of chromatin for active transcription.

## Many Eukaryotic Promoters Are Positively Regulated

As already noted, eukaryotic RNA polymerases have little or no intrinsic affinity for their promoters; initiation of transcription is almost always dependent on the action of multiple activator proteins. One important reason for the apparent predominance of positive regulation seems obvious: the storage of DNA within chromatin effectively renders most promoters inaccessible, so genes are normally silent in the absence of other regulation. The structure of chromatin affects access to some promoters more than others, but repressors that

**TABLE 28–2 Some Enzyme Complexes Catalyzing Chromatin Structural Changes Associated with Transcription**

| Enzyme complex[*] | Oligomeric structure (number of polypeptides) | Source | Activities |
|---|---|---|---|
| GCN5-ADA2-ADA3 | 3 | Yeast | GCN5 has type A HAT activity |
| SAGA/PCAF | >20 | Eukaryotes | Includes GCN5-ADA2-ADA3 |
| SWI/SNF | 11; total $M_r$ $2 \times 10^6$ | Eukaryotes | ATP-dependent nucleosome remodeling |
| NURF | 4; total $M_r$ 500,000 | Drosophila | ATP-dependent nucleosome remodeling |
| CAFI | >2 | Humans; Drosophila | Responsible for binding histones H3 and H4 to DNA |
| NAP1 | 1; $M_r$ 125,000 | Widely distributed in eukaryotes | Responsible for binding histones H2A and H2B to DNA |

[*]The abbreviations for eukaryotic genes and proteins are often more confusing or obscure than those used for bacteria. The complex of GCN5 (general control nonderepressible) and ADA (alteration/deficiency activation) proteins was discovered during investigation of the regulation of nitrogen metabolism genes in yeast. These proteins can be part of the larger SAGA complex (SPF, ADA2,3, GCN5, acetyltransferase) in yeasts. The equivalent of SAGA in humans is PCAF (p300/CBP-associated factor). SWI (switching) was discovered as a protein required for expression of certain genes involved in mating-type switching in yeast, and SNF (sucrose nonfermenting) as a factor for expression of the yeast gene for sucrase. Subsequent studies revealed multiple SWI and SNF proteins that acted in a complex. The SWI/SNF complex has a role in the expression of a wide range of genes and has been found in many eukaryotes, including humans. NURF is nuclear remodeling factor; CAF1, chromatin assembly factor; and NAP1, nucleosome assembly protein.

bind to DNA so as to preclude access of RNA polymerase (negative regulation) would often be simply redundant. Other factors are at play in the use of positive regulation, and speculation generally centers around two: the large size of eukaryotic genomes and the greater efficiency of positive regulation.

First, nonspecific DNA binding of regulatory proteins becomes a more important problem in the much larger genomes of higher eukaryotes. And the chance that a single specific binding sequence will occur randomly at an inappropriate site also increases with genome size. Specificity for transcriptional activation can be improved if each of several positive-regulatory proteins must bind specific DNA sequences and then form a complex in order to become active. The average number of regulatory sites for a gene in a multicellular organism is probably at least five. The requirement for binding of several positive-regulatory proteins to specific DNA sequences vastly reduces the probability of the random occurrence of a functional juxtaposition of all the necessary binding sites. In principle, a similar strategy could be used by multiple negative-regulatory elements, but this brings us to the second reason for the use of positive regulation: it is simply more efficient. If the 30,000 to 35,000 genes in the human genome were negatively regulated, each cell would have to synthesize, at all times, this same number of different repressors (or many times this number if multiple regulatory elements were used at each promoter) in concentrations sufficient to permit specific binding to each "unwanted" gene. In positive regulation, most of the genes are normally inactive (that is, RNA polymerases do not bind to the promoters) and the cell synthesizes only the activator proteins needed to promote transcription of the subset of genes required in the cell at that time. These arguments notwithstanding, there are examples of negative regulation in eukaryotes, from yeast to humans, as we shall see.

### DNA-Binding Transactivators and Coactivators Facilitate Assembly of the General Transcription Factors

To continue our exploration of the regulation of gene expression in eukaryotes, we return to the interactions between promoters and RNA polymerase II (Pol II), the enzyme responsible for the synthesis of eukaryotic mRNAs. Although most (but not all) Pol II promoters include the TATA box and Inr (initiator) sequences, with their standard spacing (see Fig. 26–8), they vary greatly in both the number and the location of additional sequences required for the regulation of transcription. These additional regulatory sequences are usually called **enhancers** in higher eukaryotes and **upstream activator sequences (UASs)** in yeast. A typical enhancer may be found hundreds or even thousands of base pairs

upstream from the transcription start site, or may even be downstream, within the gene itself. When bound by the appropriate regulatory proteins, an enhancer increases transcription at nearby promoters regardless of its orientation in the DNA. The UASs of yeast function in a similar way, although generally they must be positioned upstream and within a few hundred base pairs of the transcription start site. An average Pol II promoter may be affected by a half-dozen regulatory sequences of this type, and even more complex promoters are quite common.

Successful binding of active RNA polymerase II holoenzyme at one of its promoters usually requires the action of other proteins (Fig. 28–27), of three types: (1) **basal transcription factors** (see Fig. 26–9, Table 26–1), required at every Pol II promoter; (2) **DNA-binding transactivators,** which bind to enhancers or UASs and facilitate transcription; and (3) **coactivators.** The latter group act indirectly—not by binding to the DNA—and are required for essential communication between the DNA-binding transactivators and the complex composed of Pol II and the general transcription factors. Furthermore, a variety of repressor proteins can interfere with communication between the RNA polymerase and the DNA-binding transactivators, resulting in repression of transcription (Fig. 28–27b). Here we focus on the protein complexes shown in Figure 28–27 and on how they interact to activate transcription.

**TATA-Binding Protein**   The first component to bind in the assembly of a preinitiation complex at the TATA box of a typical Pol II promoter is the **TATA-binding protein (TBP).** The complete complex includes the basal (or general) transcription factors TFIIB, TFIIE, TFIIF, TFIIH; Pol II; and perhaps TFIIA (not all of the factors are shown in Fig. 28–27). This minimal preinitiation complex, however, is often insufficient for the initiation of transcription and generally does not form at all if the promoter is obscured within chromatin. Positive regulation leading to transcription is imposed by the transactivators and coactivators.

**DNA-Binding Transactivators**   The requirements for transactivators vary greatly from one promoter to another. A few transactivators are known to facilitate transcription at hundreds of promoters, whereas others are specific for a few promoters. Many transactivators are sensitive to the binding of signal molecules, providing the capacity to activate or deactivate transcription in response to a changing cellular environment. Some enhancers bound by DNA-binding transactivators are quite distant from the promoter's TATA box. How do the transactivators function at a distance? The answer in most cases seems to be that, as indicated earlier, the intervening DNA is looped so that the various protein complexes can interact directly. The looping is promoted by certain non-

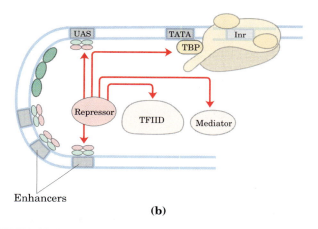

HMG proteins

Transcription

**(a)**

Enhancers

DNA-binding transactivators

**(b)**

Enhancers

**FIGURE 28–27 Eukaryotic promoters and regulatory proteins.** RNA polymerase II and its associated general transcription factors form a preinitiation complex at the TATA box and Inr site of the cognate promoters, a process facilitated by DNA-binding transactivators, acting through TFIID and/or mediator. **(a)** A composite promoter with typical sequence elements and protein complexes found in both yeast and higher eukaryotes. The carboxyl-terminal domain (CTD) of Pol II (see Fig. 26–9) is an important point of interaction with mediator and other protein complexes. Not shown are the protein complexes required for histone acetylation and chromatin remodeling. For the DNA-binding transactivators, DNA-binding domains are shown in green, activation domains in pink. The interactions symbolized by blue arrows are discussed in the text. **(b)** A wide variety of eukaryotic transcriptional repressors function by a range of mechanisms. Some bind directly to DNA, displacing a protein complex required for activation; others interact with various parts of the transcription or activation complexes to prevent activation. Possible points of interaction are indicated with red arrows.

histone proteins that are abundant in chromatin and bind nonspecifically to DNA. These high mobility group (HMG) proteins (Fig. 28–27; "high mobility" refers to their electrophoretic mobility in polyacrylamide gels) play an important structural role in chromatin remodeling and transcriptional activation.

**Coactivator Protein Complexes** Most transcription requires the presence of additional protein complexes. Some major regulatory protein complexes that interact with Pol II have been defined both genetically and biochemically. These coactivator complexes act as intermediaries between the DNA-binding transactivators and the Pol II complex.

The best-characterized coactivator is the transcription factor TFIID (Fig. 28–27). In eukaryotes, TFIID is a large complex that includes TBP and ten or more TBP-associated factors (TAFs). Some TAFs resemble histones and may play a role in displacing nucleosomes during the activation of transcription. Many DNA-binding transactivators aid in transcription initiation by interacting with one or more TAFs. The requirement for TAFs to initiate transcription can vary greatly from one gene to another. Some promoters require TFIID, some do not, and some require only subsets of the TFIID TAF subunits.

Another important coactivator consists of 20 or more polypeptides in a protein complex called **mediator** (Fig. 28–27); the 20 core polypeptides are highly conserved from fungi to humans. Mediator binds tightly to the carboxyl-terminal domain (CTD) of the largest subunit of Pol II. The mediator complex is required for both basal and regulated transcription at promoters used by Pol II, and it also stimulates the phosphorylation of the CTD by TFIIH. Both mediator and TFIID are required at some promoters. As with TFIID, some DNA-binding transactivators interact with one or more components of the mediator complex. Coactivator complexes function at or near the promoter's TATA box.

**Choreography of Transcriptional Activation** We can now begin to piece together the sequence of transcriptional activation events at a typical Pol II promoter. First, crucial remodeling of the chromatin takes place in stages. Some DNA-binding transactivators have significant affinity for their binding sites even when the sites are within condensed chromatin. Binding of one transactivator may facilitate the binding of others, gradually displacing some nucleosomes.

The bound transactivators can then interact directly with HATs or enzyme complexes such as SWI/SNF (or both), accelerating the remodeling of the surrounding chromatin. In this way a bound transactivator can draw in other components necessary for further chromatin remodeling to permit transcription of specific genes. The bound transactivators, generally acting through complexes such as TFIID or mediator (or both), stabilize the binding of Pol II and its associated transcription factors and greatly facilitate formation of the preinitiation transcription complex. Complexity in these regulatory circuits is the rule rather than the exception, with multiple DNA-bound transactivators promoting transcription.

The script can change from one promoter to another, but most promoters seem to require a precisely ordered assembly of components to initiate transcription. The assembly process is not always fast. At some genes it may take minutes; at certain genes in higher eukaryotes the process can take days.

***Reversible Transcriptional Activation***    Although rarer, some eukaryotic regulatory proteins that bind to Pol II promoters can act as repressors, inhibiting the formation of active preinitiation complexes (Fig. 28–27b). Some transactivators can adopt different conformations, enabling them to serve as transcriptional activators or repressors. For example, some steroid hormone receptors (described later) function in the nucleus as DNA-binding transactivators, stimulating the transcription of certain genes when a particular steroid hormone signal is present. When the hormone is absent, the receptor proteins revert to a repressor conformation, *preventing* the formation of preinitiation complexes. In some cases this repression involves interaction with histone deacetylases and other proteins that help restore the surrounding chromatin to its transcriptionally inactive state.

## The Genes of Galactose Metabolism in Yeast Are Subject to Both Positive and Negative Regulation

Some of the general principles described above can be illustrated by one well-studied eukaryotic regulatory circuit (Fig. 28–28). The enzymes required for the importation and metabolism of galactose in yeast are encoded by genes scattered over several chromosomes (Table 28–3). Each of the *GAL* genes is transcribed separately, and yeast cells have no operons like those in bacteria. However, all the *GAL* genes have similar promoters and are regulated coordinately by a common set of proteins. The promoters for the *GAL* genes consist of the TATA box and Inr sequences, as well as an upstream activator sequence (UAS$_G$) recognized by a DNA-binding transcriptional activator known as Gal4 protein (Gal4p). Regulation of gene expression by galactose entails an interplay between Gal4p and two other proteins, Gal80p and Gal3p (Fig. 28–28). Gal80p forms a complex with Gal4p, preventing Gal4p from functioning as an activator of the *GAL* promoters. When galactose is present, it binds Gal3p, which then interacts with Gal80p, allowing Gal4p to function as an activator at the various *GAL* promoters.

Other protein complexes also have a role in activating transcription of the *GAL* genes. These may include the SAGA complex for histone acetylation, the SWI/SNF complex for nucleosome remodeling, and the mediator complex. Figure 28–29 provides an idea of the complexity of protein interactions in the overall process of transcriptional activation in eukaryotic cells.

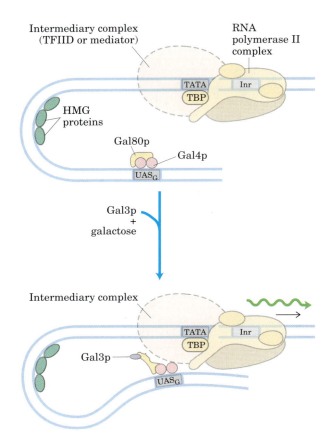

**OFIGURE 28–28 Regulation of transcription at genes of galactose metabolism in yeast.** Galactose is imported into the cell and converted to galactose 6-phosphate by a pathway involving six enzymes whose genes are scattered over three chromosomes (see Table 28–3). Transcription of these genes is regulated by the combined actions of the proteins Gal4p, Gal80p, and Gal3p, with Gal4p playing the central role of DNA-binding transactivator. The Gal4p-Gal80p complex is inactive in gene activation. Binding of galactose to Gal3p and its interaction with Gal80p produce a conformational change in Gal80p that allows Gal4p to function in transcription activation.

Glucose is the preferred carbon source for yeast, as it is for bacteria. When glucose is present, most of the *GAL* genes are repressed—whether galactose is present or not. The *GAL* regulatory system described above is effectively overridden by a complex catabolite repression system that includes several proteins (not depicted in Fig. 28–29).

## DNA-Binding Transactivators Have a Modular Structure

DNA-binding transactivators typically have a distinct structural domain for specific DNA binding and one or more additional domains for transcriptional activation or for interaction with other regulatory proteins. Interaction of two regulatory proteins is often mediated by domains containing leucine zippers (Fig. 28–14) or helix-loop-helix motifs (Fig. 28–15). We consider here three

**TABLE 28-3  Genes of Galactose Metabolism in Yeast**

| | Protein function | Chromosomal location | Protein size (number of residues) | Relative protein expression in different carbon sources | | |
| --- | --- | --- | --- | --- | --- | --- |
| | | | | Glucose | Glycerol | Galactose |
| **Regulated genes** | | | | | | |
| GAL1 | Galactokinase | II | 528 | − | − | +++ |
| GAL2 | Galactose permease | XII | 574 | − | − | +++ |
| PGM2 | Phosphoglucomutase | XIII | 569 | + | + | ++ |
| GAL7 | Galactose 1-phosphate uridylyltransferase | II | 365 | − | − | +++ |
| GAL10 | UDP-glucose 4-epimerase | II | 699 | − | − | +++ |
| MEL1 | α-Galactosidase | II | 453 | − | + | ++ |
| **Regulatory genes** | | | | | | |
| GAL3 | Inducer | IV | 520 | − | + | ++ |
| GAL4 | Transcriptional activator | XVI | 881 | +/− | + | + |
| GAL80 | Transcriptional inhibitor | XIII | 435 | + | + | ++ |

Source: Adapted from Reece, R. & Platt, A. (1997) Signaling activation and repression of RNA polymerase II transcription in yeast. *Bioessays* 19, 1001–1010.

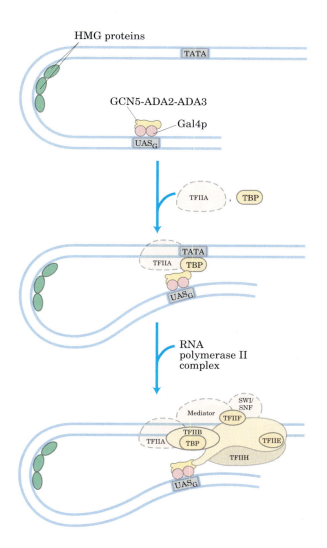

**FIGURE 28–29 Protein complexes involved in transcription activation of a group of related eukaryotic genes.** The *GAL* system illustrates the complexity of this process, but not all these protein complexes are yet known to affect *GAL* gene transcription. Note that many of the complexes (such as SWI/SNF, GCN5-ADA2-ADA3, and mediator) affect the transcription of many genes. The complexes assemble stepwise. First the DNA-binding transactivators bind, then the additional protein complexes needed to remodel the chromatin and allow transcription to begin.

distinct types of structural domains used in activation by DNA-binding transactivators (Fig. 28–30a): Gal4p, Sp1, and CTF1.

Gal4p contains a zinc fingerlike structure in its DNA-binding domain, near the amino terminus; this domain has six Cys residues that coordinate two $Zn^{2+}$. The protein functions as a homodimer (with dimerization mediated by interactions between two coiled coils) and binds to $UAS_G$, a palindromic DNA sequence about 17 bp long. Gal4p has a separate activation domain with many acidic amino acid residues. Experiments that substitute a variety of different peptide sequences for the **acidic activation domain** of Gal4p suggest that the acidic nature of this domain is critical to its function, although its precise amino acid sequence can vary considerably.

Sp1 ($M_r$ 80,000) is a DNA-binding transactivator for a large number of genes in higher eukaryotes. Its DNA binding site, the GC box (consensus sequence

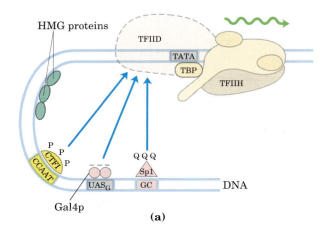

**(a)**

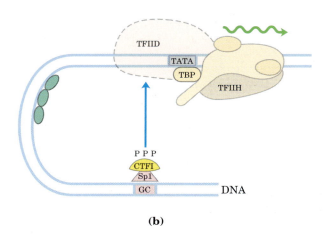

**(b)**

**FIGURE 28–30  DNA-binding transactivators. (a)** Typical DNA-binding transactivators such as CTF1, Gal4p, and Sp1 have a DNA-binding domain and an activation domain. The nature of the activation domain is indicated by symbols: − − −, acidic; Q Q Q, glutamine-rich; P P P, proline-rich. Some or all of these proteins may activate transcription by interacting with intermediary complexes such as TFIID or mediator. Note that the binding sites illustrated here are not generally found together near a single gene. **(b)** A chimeric protein containing the DNA-binding domain of Sp1 and the activation domain of CTF1 activates transcription if a GC box is present.

GGGCGG), is usually quite near the TATA box. The DNA-binding domain of the Sp1 protein is near its carboxyl terminus and contains three zinc fingers. Two other domains in Sp1 function in activation, and are notable in that 25% of their amino acid residues are Gln. A wide variety of other activator proteins also have these **glutamine-rich domains.**

CCAAT-binding transcription factor 1 (CTF1) belongs to a family of DNA-binding transactivators that bind a sequence called the CCAAT site (its consensus sequence is $TGGN_6GCCAA$, where N is any nucleotide). The DNA-binding domain of CTF1 contains many basic amino acid residues, and the binding region is probably arranged as an $\alpha$ helix. This protein has neither a helix-turn-helix nor a zinc finger motif; its DNA-binding mechanism is not yet clear. CTF1 has a **proline-rich activation domain,** with Pro accounting for more than 20% of the amino acid residues.

The discrete activation and DNA-binding domains of regulatory proteins often act completely independently, as has been demonstrated in "domain-swapping" experiments. Genetic engineering techniques (Chapter 9) can join the proline-rich activation domain of CTF1 to the DNA-binding domain of Sp1 to create a protein that, like normal Sp1, binds to GC boxes on the DNA and activates transcription at a nearby promoter (as in Fig. 28–30b). The DNA-binding domain of Gal4p has similarly been replaced experimentally with the DNA-binding domain of the prokaryotic LexA repressor (of the SOS response; Fig. 28–22). This chimeric protein neither binds at $UAS_G$ nor activates the yeast *GAL* genes (as would normal Gal4p) unless the $UAS_G$ sequence in the DNA is replaced by the LexA recognition site.

## Eukaryotic Gene Expression Can Be Regulated by Intercellular and Intracellular Signals

The effects of steroid hormones (and of thyroid and retinoid hormones, which have the same mode of action) provide additional well-studied examples of the modulation of eukaryotic regulatory proteins by direct interaction with molecular signals (see Fig. 12–40). Unlike other types of hormones, steroid hormones do not have to bind to plasma membrane receptors. Instead, they can interact with intracellular receptors that are themselves transcriptional transactivators. Steroid hormones too hydrophobic to dissolve readily in the blood (estrogen, progesterone, and cortisol, for example) travel on specific carrier proteins from their point of release to their target tissues. In the target tissue, the hormone passes through the plasma membrane by simple diffusion and binds to its specific receptor protein in the nucleus. The hormone-receptor complex acts by binding to highly specific DNA sequences called **hormone response elements (HREs),** thereby altering gene expression. Hormone binding triggers changes in the conformation of the receptor proteins so that they become capable of interacting with additional transcription factors. The bound hormone-receptor complex can either enhance or suppress the expression of adjacent genes.

The DNA sequences (HREs) to which hormone-receptor complexes bind are similar in length and arrangement, but differ in sequence, for the various steroid hormones. Each receptor has a consensus HRE sequence (Table 28–4) to which the hormone-receptor complex binds well, with each consensus consisting of two six-nucleotide sequences, either contiguous or separated by three nucleotides, in tandem or in a palindromic arrangement. The hormone receptors have a highly conserved DNA-binding domain with two zinc fingers

**TABLE 28–4 Hormone Response Elements (HREs) Bound by Steroid-Type Hormone Receptors**

| Receptor | Consensus sequence bound[*] |
|---|---|
| Androgen | GG(A/T)ACAN$_2$TGTTCT |
| Glucocorticoid | GGTACAN$_3$TGTTCT |
| Retinoic acid (some) | AGGTCAN$_5$AGGTCA |
| Vitamin D | AGGTCAN$_3$AGGTCA |
| Thyroid hormone | AGGTCAN$_3$AGGTCA |
| RX[†] | AGGTCANAGGTCANAGGTCANAGGTCA |

[*]N represents any nucleotide.

[†]Forms a dimer with the retinoic acid receptor or vitamin D receptor.

(Fig. 28–31). The hormone-receptor complex binds to the DNA as a dimer, with the zinc finger domains of each monomer recognizing one of the six-nucleotide sequences. The ability of a given hormone to act through the hormone-receptor complex to alter the expression of a specific gene depends on the exact sequence of the HRE, its position relative to the gene, and the number of HREs associated with the gene.

Unlike the DNA-binding domain, the ligand-binding region of the receptor protein—always at the carboxyl terminus—is quite specific to the particular receptor. In the ligand-binding region, the glucocorticoid receptor is only 30% similar to the estrogen receptor and 17% similar to the thyroid hormone receptor. The size of the ligand-binding region varies dramatically; in the vitamin D receptor it has only 25 amino acid residues, whereas in the mineralocorticoid receptor it has 603 residues. Mutations that change one amino acid in these regions can result in loss of responsiveness to a specific hormone.

Some humans unable to respond to cortisol, testosterone, vitamin D, or thyroxine have mutations of this type.

### Regulation Can Result from Phosphorylation of Nuclear Transcription Factors

We noted in Chapter 12 that the effects of insulin on gene expression are mediated by a series of steps leading ultimately to the activation of a protein kinase in the nucleus that phosphorylates specific DNA-binding proteins and thereby alters their ability to act as transcription factors (see Fig. 12–6). This general mechanism mediates the effects of many nonsteroid hormones. For example, the $\beta$-adrenergic pathway that leads to elevated levels of cytosolic cAMP, which acts as a second messenger in eukaryotes as well as in prokaryotes (see Figs 12–12, 28–18), also affects the transcription of a set of genes, each of which is located near a specific DNA sequence called a cAMP response element (CRE). The catalytic subunit of protein kinase A, released when cAMP levels rise (see Fig. 12–15), enters the nucleus and phosphorylates a nuclear protein, the CRE-binding protein (CREB). When phosphorylated, CREB binds to CREs near certain genes and acts as a transcription factor, turning on the expression of these genes.

### Many Eukaryotic mRNAs Are Subject to Translational Repression

Regulation at the level of translation assumes a much more prominent role in eukaryotes than in bacteria and is observed in a range of cellular situations. In contrast to the tight coupling of transcription and translation in bacteria, the transcripts generated in a eukaryotic nucleus

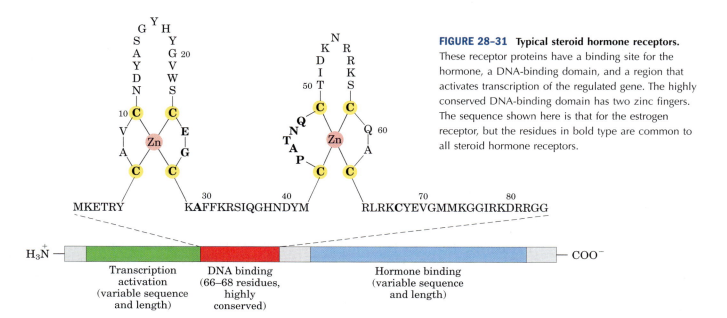

**FIGURE 28–31 Typical steroid hormone receptors.** These receptor proteins have a binding site for the hormone, a DNA-binding domain, and a region that activates transcription of the regulated gene. The highly conserved DNA-binding domain has two zinc fingers. The sequence shown here is that for the estrogen receptor, but the residues in bold type are common to all steroid hormone receptors.

must be processed and transported to the cytoplasm before translation. This can impose a significant delay on the appearance of a protein. When a rapid increase in protein production is needed, a translationally repressed mRNA already in the cytoplasm can be activated for translation without delay. Translational regulation may play an especially important role in regulating certain very long eukaryotic genes (a few are measured in the millions of base pairs), for which transcription and mRNA processing can require many hours. Some genes are regulated at both the transcriptional and translational stages, with the latter playing a role in the fine-tuning of cellular protein levels. In some anucleate cells, such as reticulocytes (immature erythrocytes), transcriptional control is entirely unavailable and translational control of stored mRNAs becomes essential. As described below, translational controls can also have spatial significance during development, when the regulated translation of prepositioned mRNAs creates a local gradient of the protein product.

Eukaryotes have at least three main mechanisms of translational regulation.

1. Initiation factors are subject to phosphorylation by a number of protein kinases. The phosphorylated forms are often less active and cause a general depression of translation in the cell.

2. Some proteins bind directly to mRNA and act as translational repressors, many of them binding at specific sites in the 3′ untranslated region (3′UTR). So positioned, these proteins interact with other translation initiation factors bound to the mRNA or with the 40S ribosomal subunit to prevent translation initiation (Fig. 28–32; compare this with Fig. 27–22).

3. Binding proteins, present in eukaryotes from yeast to mammals, disrupt the interaction between eIF4E and eIF4G (see Fig. 27–22). The mammalian versions are known as 4E-BPs (eIF4E binding proteins). When cell growth is slow, these proteins limit translation by binding to the site on eIF4E that normally interacts with eIF4G. When cell growth resumes or increases in response to growth factors or other stimuli, the binding proteins are inactivated by protein kinase–dependent phosphorylation.

The variety of translational regulation mechanisms provides flexibility, allowing focused repression of a few mRNAs or global regulation of all cellular translation.

Translational regulation has been particularly well studied in reticulocytes. One such mechanism in these cells involves eIF2, the initiation factor that binds to the initiator tRNA and conveys it to the ribosome; when Met-tRNA has bound to the P site, the factor eIF2B

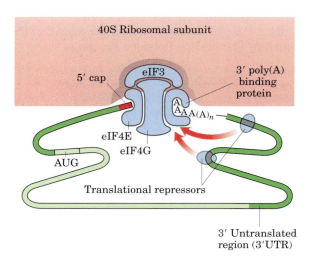

**FIGURE 28–32 Translational regulation of eukaryotic mRNA.** One of the most important mechanisms for translational regulation in eukaryotes involves the binding of translational repressors (RNA-binding proteins) to specific sites in the 3′ untranslated region (3′UTR) of the mRNA. These proteins interact with eukaryotic initiation factors or with the ribosome (see Fig. 27–22) to prevent or slow translation.

binds to eIF2, recycling it with the aid of GTP binding and hydrolysis. The maturation of reticulocytes includes destruction of the cell nucleus, leaving behind a plasma membrane packed with hemoglobin. Messenger RNAs deposited in the cytoplasm before the loss of the nucleus allow for the replacement of hemoglobin. When reticulocytes become deficient in iron or heme, the translation of globin mRNAs is repressed. A protein kinase called HCR (*h*emin-*c*ontrolled *r*epressor) is activated, catalyzing the phosphorylation of eIF2. In its phosphorylated form, eIF2 forms a stable complex with eIF2B that sequesters the eIF2, making it unavailable for participation in translation. In this way, the reticulocyte coordinates the synthesis of globin with the availability of heme.

Many additional examples of translational regulation have been found in studies of the development of multicellular organisms, as discussed in more detail below.

### Posttranscriptional Gene Silencing Is Mediated by RNA Interference

In higher eukaryotes, including nematodes, fruit flies, plants, and mammals, a class of small RNAs has been discovered that mediates the silencing of particular genes. The RNAs function by interacting with mRNAs, often in the 3′UTR, resulting in either mRNA degradation or translation inhibition. In either case, the mRNA, and thus the gene that produces it, is silenced. This form of gene regulation controls developmental timing in at least some organisms. It is also used as a mechanism to protect against invading RNA viruses (particularly important in plants, which lack an immune system) and to control the activity of transposons. In addition, small RNA molecules may play a critical (but still undefined) role in the formation of heterochromatin.

The small RNAs are sometimes called micro-RNAs (miRNAs). Many are present only transiently during development, and these are sometimes referred to as small temporal RNAs (stRNAs). Hundreds of different miRNAs have been identified in higher eukaryotes. They are transcribed as precursor RNAs about 70 nucleotides long, with internally complementary sequences that form hairpinlike structures (Fig. 28–33). The precursors are cleaved by endonucleases to form short duplexes about 20 to 25 nucleotides long. The best-characterized nuclease goes by the delightfully suggestive name Dicer; endonucleases in the Dicer family are widely distributed in higher eukaryotes. One strand of the processed miRNA is transferred to the target mRNA (or to a viral or transposon RNA), leading to inhibition of translation or degradation of the RNA (Fig. 28–33a).

This gene regulation mechanism has an interesting and very useful practical side. If an investigator introduces into an organism a duplex RNA molecule corresponding in sequence to virtually any mRNA, the Dicer endonuclease cleaves the duplex into short segments,

called small interfering RNAs (siRNAs). These bind to the mRNA and silence it (Fig. 28–33b). The process is known as **RNA interference (RNAi).** In plants, virtually any gene can be effectively shut down in this way. In nematodes, simply introducing the duplex RNA into the worm's diet produces very effective suppression of the target gene. The technique has rapidly become an important tool in the ongoing efforts to study gene function, because it can disrupt gene function without creating a mutant organism. The procedure can be applied to humans as well. Laboratory-produced siRNAs have already been used to block HIV and poliovirus infections in cultured human cells for a week or so at a time. Although this work is in its infancy, the rapid progress makes RNA interference a field to watch for future medical advances.

### Development Is Controlled by Cascades of Regulatory Proteins

For sheer complexity and intricacy of coordination, the patterns of gene regulation that bring about development of a zygote into a multicellular animal or plant have no peer. Development requires transitions in morphology and protein composition that depend on tightly coordinated changes in expression of the genome. More genes are expressed during early development than in any other part of the life cycle. For example, in the sea urchin, an oocyte has about 18,500 *different* mRNAs, compared with about 6,000 different mRNAs in the cells of a typical differentiated tissue. The mRNAs in the oocyte give rise to a cascade of events that regulate the expression of many genes across both space and time.

Several animals have emerged as important model systems for the study of development, because they are easy to maintain in a laboratory and have relatively short generation times. These include nematodes, fruit flies, zebra fish, mice, and the plant *Arabidopsis*. This discussion focuses on the development of fruit flies. Our understanding of the molecular events during development of *Drosophila melanogaster* is particularly well advanced and can be used to illustrate patterns and principles of general significance.

The life cycle of the fruit fly includes complete metamorphosis during its progression from an embryo to an adult (Fig. 28–34). Among the most important characteristics of the embryo are its **polarity** (the anterior and posterior parts of the animal are readily distinguished, as are its dorsal and ventral parts) and its **metamerism** (the embryo body is made up of serially repeating segments, each with characteristic features). During development, these segments become organized into a head, thorax, and abdomen. Each segment of the adult thorax has a different set of appendages. Development of this complex pattern is under genetic control, and a variety of pattern-regulating genes have been

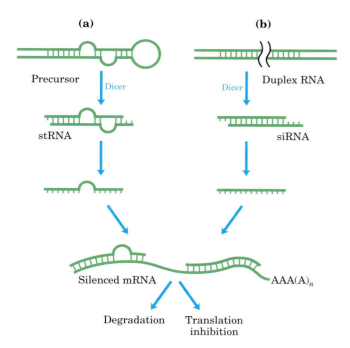

**FIGURE 28–33 Gene silencing by RNA interference. (a)** Small temporal RNAs (stRNAs) are generated by Dicer-mediated cleavage of longer precursors that fold to create duplex regions. The stRNAs then bind to mRNAs, leading to degradation of mRNA or inhibition of translation. **(b)** Double-stranded RNAs can be constructed and introduced into a cell. Dicer processes the duplex RNAs into small interfering RNAs (siRNAs), which interact with the target mRNA. Again, the mRNA is either degraded or its translation inhibited.

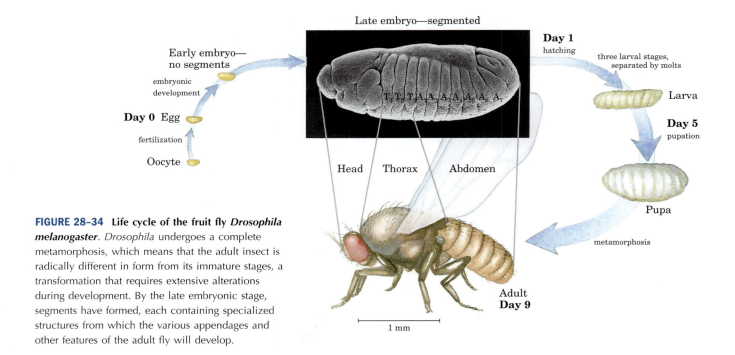

**FIGURE 28–34** **Life cycle of the fruit fly *Drosophila melanogaster*.** *Drosophila* undergoes a complete metamorphosis, which means that the adult insect is radically different in form from its immature stages, a transformation that requires extensive alterations during development. By the late embryonic stage, segments have formed, each containing specialized structures from which the various appendages and other features of the adult fly will develop.

discovered that dramatically affect the organization of the body.

The *Drosophila* egg, along with 15 nurse cells, is surrounded by a layer of follicle cells (Fig. 28–35). As the egg cell forms (before fertilization), mRNAs and proteins originating in the nurse and follicle cells are deposited in the egg cell, where some play a critical role in development. Once a fertilized egg is laid, its nucleus divides and the nuclear descendants continue to divide in synchrony every 6 to 10 min. Plasma membranes are not formed around the nuclei, which are distributed within the egg cytoplasm (or syncytium). Between the eighth and eleventh rounds of nuclear division, the nuclei migrate to the outer layer of the egg, forming a monolayer of nuclei surrounding the common yolk-rich cytoplasm; this is the syncytial blastoderm. After a few additional divisions, membrane invaginations surround the nuclei to create a layer of cells that form the cellular blastoderm. At this stage, the mitotic cycles in the various cells lose their synchrony. The developmental fate of the cells is determined by the mRNAs and proteins originally deposited in the egg by the nurse and follicle cells.

Proteins that, through changes in local concentration or activity, cause the surrounding tissue to take up a particular shape or structure are sometimes referred to as **morphogens;** they are the products of pattern-regulating genes. As defined by Christiane Nüsslein-Volhard, Edward B. Lewis, and Eric F. Wieschaus, three major classes of pattern-regulating genes—maternal, segmentation, and homeotic genes—function in successive stages of development to specify the basic features of the *Drosophila* embryo's body. **Maternal genes** are expressed in the unfertilized egg, and the resulting **maternal mRNAs** remain dormant until fertilization. These provide most of the proteins needed in very early development, until the cellular blastoderm is formed. Some of the proteins encoded by maternal mRNAs direct the spatial organization of the developing embryo at early stages, establishing its polarity. **Segmentation genes,** transcribed after fertilization, direct the formation of the proper number of body segments. At least three subclasses of segmentation genes act at successive stages: **gap genes** divide the developing embryo into several broad regions, and **pair-rule genes** together with **segment polarity genes** define 14 stripes that become the 14 segments of a normal embryo. **Homeotic genes** are expressed still later; they specify which organs and appendages will develop in particular body segments.

The many regulatory genes in these three classes direct the development of an adult fly, with a head, thorax, and abdomen, with the proper number of segments, and with the correct appendages on each segment. Although embryogenesis takes about a day to complete, all these genes are activated during the first four hours. Some mRNAs and proteins are present for only a few minutes at specific points during this period. Some of the genes code for transcription factors that affect the expression of other genes in a kind of developmental cascade. Regulation at the level of translation also occurs, and many of the regulatory genes encode translational repressors, most of which bind to the 3′UTR of the mRNA (Fig. 28–32). Because many mRNAs are

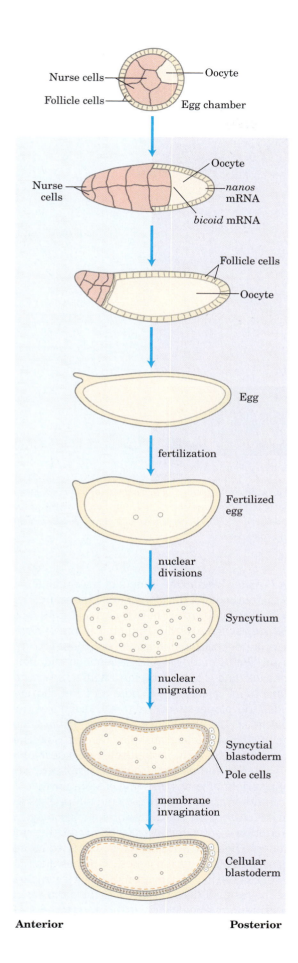

**Anterior**                                    **Posterior**

deposited in the egg long before their translation is required, translational repression provides an especially important avenue for regulation in developmental pathways.

***Maternal Genes*** Some maternal genes are expressed within the nurse and follicle cells, and some in the egg itself. Within the unfertilized *Drosophila* egg, the maternal gene products establish two axes—anterior-posterior and dorsal-ventral—and thus define which regions of the radially symmetric egg will develop into the head and abdomen and the top and bottom of the adult fly. A key event in very early development is establishment of mRNA and protein gradients along the body axes. Some maternal mRNAs have protein products that diffuse through the cytoplasm to create an asymmetric distribution in the egg. Different cells in the cellular blastoderm therefore inherit different amounts of these proteins, setting the cells on different developmental paths. The products of the maternal mRNAs include transcriptional activators or repressors as well as translational repressors, all regulating the expression of other pattern-regulating genes. The resulting specific patterns and sequences of gene expression therefore differ between cell lineages, ultimately orchestrating the development of each adult structure.

The anterior-posterior axis in *Drosophila* is defined at least in part by the products of the *bicoid* and *nanos* genes. The *bicoid* gene product is a major anterior morphogen, and the *nanos* gene product is a major posterior morphogen. The mRNA from the *bicoid* gene is synthesized by nurse cells and deposited in the unfertilized egg near its anterior pole. Nüsslein-Volhard found that this mRNA is translated soon after fertilization, and the Bicoid protein diffuses through

Christiane Nüsslein-Volhard

**FIGURE 28–35 Early development in *Drosophila*.** During development of the egg, maternal mRNAs (including the *bicoid* and *nanos* gene transcripts, discussed in the text) and proteins are deposited in the developing oocyte (unfertilized egg cell) by nurse cells and follicle cells. After fertilization, the two nuclei of the fertilized egg divide in synchrony within the common cytoplasm (syncytium), then migrate to the periphery. Membrane invaginations surround the nuclei to create a monolayer of cells at the periphery; this is the cellular blastoderm stage. During the early nuclear divisions, several nuclei at the far posterior become pole cells, which later become the germ-line cells.

**(a)**

Normal egg

**(b)**

*bcd⁻/bcd⁻* egg

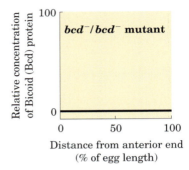

Normal larva

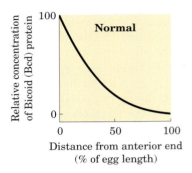

Double-posterior larva

**FIGURE 28–36 Distribution of a maternal gene product in a *Drosophila* egg. (a)** Micrograph of an immunologically stained egg, showing distribution of the *bicoid* (*bcd*) gene product. The graph measures stain intensity. This distribution is essential for normal develop-ment of the anterior structures of the animal. **(b)** If the *bcd* gene is not expressed by the mother (*bcd⁻/bcd⁻* mutant) and thus no *bicoid* mRNA is deposited in the egg, the resulting embryo has two posteriors (and soon dies).

the cell to create, by the seventh nuclear division, a concentration gradient radiating out from the anterior pole (Fig. 28–36a). The Bicoid protein is a transcription factor that activates the expression of a number of segmentation genes; the protein contains a homeodomain (p. 1090). Bicoid is also a translational repressor that inactivates certain mRNAs. The amounts of Bicoid protein in various parts of the embryo affect the subsequent expression of a number of other genes in a threshold-dependent manner. Genes are transcriptionally activated or translationally repressed only where the Bicoid protein concentration exceeds the threshold. Changes in the shape of the Bicoid concentration gradient have dramatic effects on the body pattern. Lack of Bicoid protein results in development of an embryo with two abdomens but neither head nor thorax (Fig. 28–36b); however, embryos without Bicoid will develop normally if an adequate amount of *bicoid* mRNA is injected into the egg at the appropriate end. The *nanos* gene has an analogous role, but its mRNA is deposited at the posterior end of the egg and the anterior-posterior protein gradient peaks at the posterior pole. The Nanos protein is a translational repressor.

A broader look at the effects of maternal genes reveals the outline of a developmental circuit. In addition to the *bicoid* and *nanos* mRNAs, which are deposited in the egg asymmetrically, a number of other maternal mRNAs are deposited uniformly throughout the egg cytoplasm. Three of these mRNAs encode the Pumilio, Hunchback, and Caudal proteins, all affected by *nanos* and *bicoid* (Fig. 28–37). Caudal and Pumilio are involved in development of the posterior end of the fly. Caudal is a transcriptional activator with a homeodomain; Pumilio is a translational repressor. Hunchback protein plays an important role in the development of the anterior end and is also a transcriptional regulator of a variety of genes, in some cases a positive regulator, in other cases negative. Bicoid suppresses translation of *caudal* in the anterior and also acts as a transcriptional activator of *hunchback* in the cellular blastoderm. Because *hunchback* is expressed both from maternal mRNAs and from genes in the developing egg, it is considered both a maternal and a segmentation gene. The result of the activities of Bicoid is an increased concentration of Hunchback at the anterior end of the

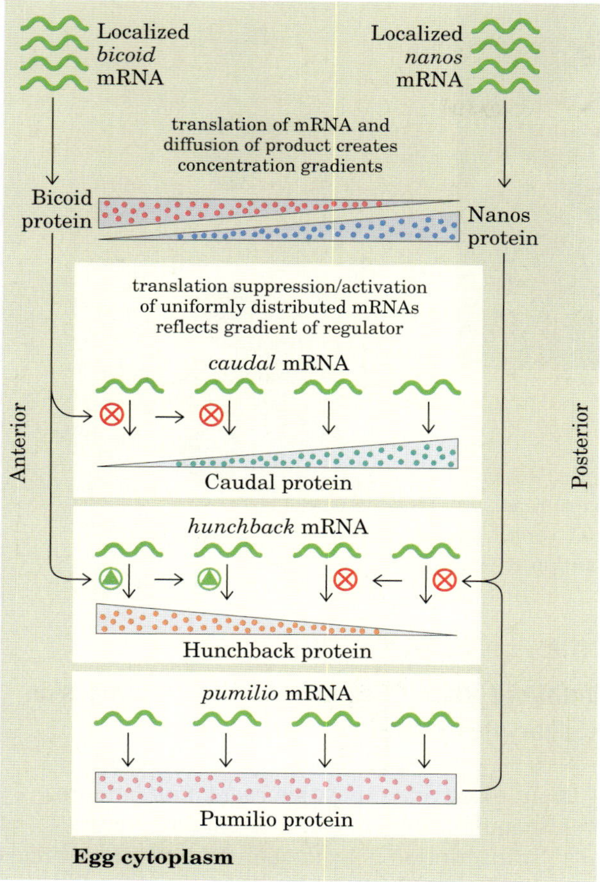

## Figure 28-37 region

Localized *bicoid* mRNA

Localized *nanos* mRNA

translation of mRNA and diffusion of product creates concentration gradients

Bicoid protein

Nanos protein

translation suppression/activation of uniformly distributed mRNAs reflects gradient of regulator

*caudal* mRNA

Caudal protein

*hunchback* mRNA

Hunchback protein

*pumilio* mRNA

Pumilio protein

Anterior

Posterior

**Egg cytoplasm**

**FIGURE 28–37 Regulatory circuits of the anterior-posterior axis in a *Drosophila* egg.** The *bicoid* and *nanos* mRNAs are localized near the anterior and posterior poles, respectively. The *caudal, hunchback,* and *pumilio* mRNAs are distributed throughout the egg cytoplasm. The gradients of Bicoid (Bcd) and Nanos proteins lead to accumulation of Hunchback protein in the anterior and Caudal protein in the posterior of the egg. Because Pumilio protein requires Nanos protein for its activity as a translational repressor of *hunchback*, it functions only at the posterior end.

egg. The Nanos and Pumilio proteins act as translational repressors of *hunchback*, suppressing synthesis of its protein near the posterior end of the egg. Pumilio does not function in the absence of the Nanos protein, and the gradient of Nanos expression confines the activity of both proteins to the posterior region. Translational repression of the *hunchback* gene leads to degradation of *hunchback* mRNA near the posterior end. However, lack of Bicoid protein in the posterior leads to expression of *caudal*. In this way, the Hunchback and Caudal proteins become asymmetrically distributed in the egg.

**Segmentation Genes** Gap genes, pair-rule genes, and segment polarity genes, three subclasses of segmentation genes in *Drosophila,* are activated at successive

stages of embryonic development. Expression of the gap genes is generally regulated by the products of one or more maternal genes. At least some of the gap genes encode transcription factors that affect the expression of other segmentation or (later) homeotic genes.

One well-characterized segmentation gene is *fushi tarazu* (*ftz*), of the pair-rule subclass. When *ftz* is deleted, the embryo develops 7 segments instead of the normal 14, each segment twice the normal width. The Fushi-tarazu protein (Ftz) is a transcriptional activator with a homeodomain. The mRNAs and proteins derived from the normal *ftz* gene accumulate in a striking pattern of seven stripes that encircle the posterior two-thirds of the embryo (Fig. 28–38). The stripes demarcate the positions of segments that develop later; these segments are eliminated if *ftz* function is lost. The Ftz protein and a few similar regulatory proteins directly or indirectly regulate the expression of vast numbers of genes in the continuing developmental cascade.

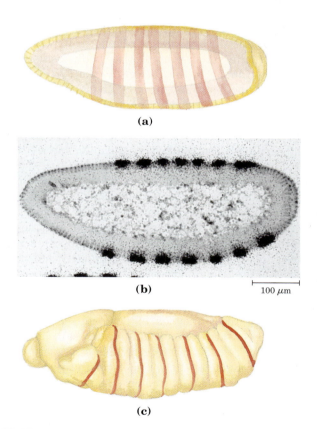

(a)

(b)

100 μm

(c)

**FIGURE 28–38 Distribution of the *fushi tarazu* (*ftz*) gene product in early *Drosophila* embryos. (a)** In the normal embryo, the gene product can be detected in seven bands around the circumference of the embryo (shown schematically). These bands **(b)** appear as dark spots (generated by a radioactive label) in a cross-sectional autoradiograph and **(c)** demarcate the anterior margins of the segments in the late embryo (marked in red).

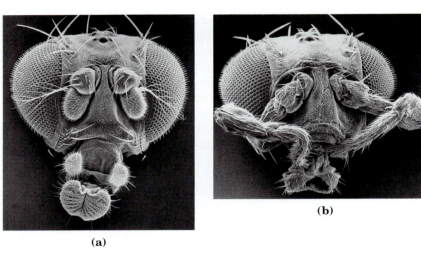

(a)

(b)

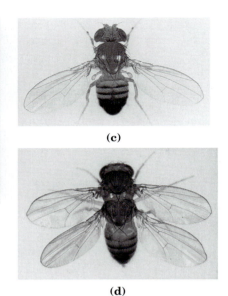

(c)

(d)

**FIGURE 28–39  Effects of mutations in homeotic genes in *Drosophila*. (a)** Normal head. **(b)** Homeotic mutant (*antennapedia*) in which antennae are replaced by legs. **(c)** Normal body structure. **(d)** Homeotic mutant (*bithorax*) in which a segment has developed incorrectly to produce an extra set of wings.

**Homeotic Genes**  Loss of homeotic genes by mutation or deletion causes the appearance of a normal appendage or body structure at an inappropriate body position. An important example is the *ultrabithorax* (*ubx*) gene. When Ubx function is lost, the first abdominal segment develops incorrectly, having the structure of the third thoracic segment. Other known homeotic mutations cause the formation of an extra set of wings, or two legs at the position in the head where the antennae are normally found (Fig. 28–39).

The homeotic genes often span long regions of DNA. The *ubx* gene, for example, is 77,000 bp long. More than 73,000 bp of this gene are in introns, one of which is more than 50,000 bp long. Transcription of the *ubx* gene takes nearly an hour. The delay this imposes on *ubx* gene expression is believed to be a timing mechanism involved in the temporal regulation of subsequent steps in development. The Ubx protein is yet another transcriptional activator with a homeodomain (Fig. 28–13).

Many of the principles of development outlined above apply to eukaryotes from nematodes to humans. Some of the regulatory proteins themselves are conserved. For example, the products of the homeobox-containing genes *HOX 1.1* in mouse and *antennapedia* in fruit fly differ in only one amino acid residue. Of course, although the molecular regulatory mechanisms may be similar, many of the ultimate developmental events are not conserved (humans do not have wings or antennae). The discovery of structural determinants with identifiable molecular functions is the first step in understanding the molecular events underlying development. As more genes and their protein products are discovered, the biochemical side of this vast puzzle will be elucidated in increasingly rich detail.

## SUMMARY 28.3  Regulation of Gene Expression in Eukaryotes

■ In eukaryotes, positive regulation is more common than negative regulation, and transcription is accompanied by large changes in chromatin structure. Promoters for Pol II typically have a TATA box and Inr sequence, as well as multiple binding sites for DNA-binding transactivators. The latter sites, sometimes located hundreds or thousands of base pairs away from the TATA box, are called upstream activator sequences in yeast and enhancers in higher eukaryotes.

■ Large complexes of proteins are generally required to regulate transcriptional activity. The effects of DNA-binding transactivators on Pol II are mediated by coactivator protein complexes such as TFIID or mediator. The modular structures of the transactivators have distinct activation and DNA-binding domains. Other protein complexes, including histone acetyltransferases such as GCN5-ADA2-ADA3 and ATP-dependent complexes such as SWI/SNF and NURF, reversibly remodel chromatin structure.

■ Hormones affect the regulation of gene expression in one of two ways. Steroid hormones interact directly with intracellular receptors that are DNA-binding regulatory proteins; binding of the hormone has either positive or negative effects on the transcription of genes targeted by the hormone. Nonsteroid hormones bind to cell-surface receptors, triggering a signaling pathway that can lead to phosphorylation of a regulatory protein, affecting its activity.

■ Development of a multicellular organism presents the most complex regulatory challenge. The fate of cells in the early embryo is determined by establishment of anterior-posterior and dorsal-ventral gradients of proteins that act as transcriptional transactivators or translational repressors, regulating the genes required for the development of structures appropriate to a particular part of the organism. Sets of regulatory genes operate in temporal and spatial succession, transforming given areas of an egg cell into predictable structures in the adult organism.

## Key Terms

*Terms in bold are defined in the glossary.*

| | | | |
|---|---|---|---|
| housekeeping genes 1082 | **leucine zipper** 1090 | chromatin remodeling 1103 | RNA interference (RNAi) 1111 |
| **induction** 1082 | basic helix-loop-helix 1090 | **enhancers** 1104 | **polarity** 1111 |
| **repression** 1082 | catabolite repression 1093 | upstream activator sequences (UASs) 1104 | **metamerism** 1111 |
| specificity factor 1083 | **cAMP receptor protein (CRP)** 1093 | basal transcription factors 1104 | morphogens 1112 |
| **repressor** 1083 | **regulon** 1094 | DNA-binding transactivators 1104 | maternal genes 1112 |
| **activator** 1083 | transcription attenuation 1094 | coactivators 1104 | maternal mRNAs 1112 |
| **operator** 1083 | **translational repressor** 1098 | TATA-binding protein (TBP) 1104 | segmentation genes 1112 |
| negative regulation 1084 | stringent response 1098 | mediator 1105 | gap genes 1112 |
| positive regulation 1084 | phase variation 1100 | **hormone response elements (HREs)** 1108 | pair-rule genes 1112 |
| **operon** 1085 | hypersensitive sites 1102 | | segment polarity genes 1112 |
| helix-turn-helix 1088 | | | |
| **zinc finger** 1088 | | | **homeotic genes** 1112 |
| **homeodomain** 1090 | | | |
| **homeobox** 1090 | | | |

## Further Reading

### General

**Hershey, J.W.B., Mathews, M.B., & Sonenberg, N.** (1996) *Translational Control,* Cold Spring Harbor Laboratory Press, Cold Spring Harbor, NY.

Many detailed reviews cover all aspects of this topic.

**Müller-Hill, B.** (1996) *The lac Operon: A Short History of a Genetic Paradigm,* Walter de Gruyter, New York.

An excellent detailed account of the investigation of this important system.

**Neidhardt, F.C. (ed.)** (1996) Escherichia coli *and* Salmonella typhimurium, 2nd edn, Vol. 1: *Cellular and Molecular Biology* (Curtiss, R., Ingraham, J.L., Lin, E.C.C., Magasanik, B., Low, K.B., Reznikoff, W.S., Riley, M., Schaechter, M., & Umbarger, H.E., vol. eds), American Society for Microbiology, Washington, DC.

An excellent source for reviews of many bacterial operons. The Web-based version, *EcoSal,* is updated regularly.

**Pabo, C.O. & Sauer, R.T.** (1992) Transcription factors: structural factors and principles of DNA recognition. *Annu. Rev. Biochem.* **61,** 1053–1095.

**Schleif, R.** (1993) *Genetics and Molecular Biology,* 2nd edn, The Johns Hopkins University Press, Baltimore.

Provides an excellent account of the experimental basis of important concepts of prokaryotic gene regulation.

### Regulation of Gene Expression in Prokaryotes

**Condon, C., Squires, C., & Squires, C.L.** (1995) Control of rRNA transcription in *Escherichia coli. Microbiol. Rev.* **59,** 623–645.

**Gourse, R.L., Gaal, T., Bartlett, M.S., Appleman, J.A., & Ross, W.** (1996) rRNA transcription and growth rate–dependent regulation of ribosome synthesis in *Escherichia coli. Annu. Rev. Microbiol.* **50,** 645–677.

**Jacob, F. & Monod, J.** (1961) Genetic regulatory mechanisms in the synthesis of proteins. *J. Mol. Biol.* **3,** 318–356.

The operon model and the concept of messenger RNA, first proposed in the *Proceedings* of the French Academy of Sciences in 1960, are presented in this historic paper.

**Johnson, R.C.** (1991) Mechanism of site-specific DNA inversion in bacteria. *Curr. Opin. Genet. Dev.* **1,** 404–411.

**Kolb, A., Busby, S., Buc, H., Garges, S., & Adhya, S.** (1993) Transcriptional regulation by cAMP and its receptor protein. *Annu. Rev. Biochem.* **62,** 749–795.

**Romby, P. & Springer, M.** (2003) Bacterial translational control at atomic resolution. *Trends Genet.* **19,** 155–161.

**Yanofsky, C., Konan, K.V., & Sarsero, J.P.** (1996) Some novel transcription attenuation mechanisms used by bacteria. *Biochimie* **78,** 1017–1024.

## Regulation of Gene Expression in Eukaryotes

**Agami, R.** (2002) RNAi and related mechanisms and their potential use for therapy. *Curr. Opin. Chem. Biol.* **6,** 829–834.

**Bashirullah, A., Cooperstock, R.L., & Lipshitz, H.D.** (1998) RNA localization in development. *Annu. Rev. Biochem.* **67,** 335–394.

**Becker, P.B. & Horz W.** (2002) ATP-dependent nucleosome remodeling. *Annu. Rev. Biochem.* **71,** 247–273.

**Boube, M., Joulia, L., Cribbs, D.L., & Bourbon, H.M.** (2002) Evidence for a mediator of RNA polymerase II transcriptional regulation conserved from yeast to man. *Cell* **110,** 143–151.

**Cerutti, H.** (2003) RNA interference: traveling in the cell and gaining functions? *Trends Genet.* **19,** 9–46.

**Conaway, R.C., Brower, C.S., & Conaway, J.W.** (2002) Gene expression—emerging roles of ubiquitin in transcription regulation. *Science* **296,** 1254–1258.

**Cosma, M.P.** (2002) Ordered recruitment: gene-specific mechanism of transcription activation. *Mol. Cell* **10,** 227–236.

**Dean, K.A., Aggarwal, A.K., & Wharton, R.P.** (2002) Translational repressors in *Drosophila. Trends Genet.* **18,** 572–577.

**DeRobertis, E.M., Oliver, G., & Wright, C.V.E.** (1990) Homeobox genes and the vertebrate body plan. *Sci. Am.* **263** (July), 46–52.

**Edmondson, D.G. & Roth, S.Y.** (1996) Chromatin and transcription. *FASEB J.* **10,** 1173–1182.

**Gingras, A.-C., Raught, B., & Sonenberg, N.** (1999) eIF4 initiation factors: effectors of mRNA recruitment to ribosomes and regulators of translation. *Annu. Rev. Biochem.* **68,** 913–963.

**Gray, N.K. & Wickens, M.** (1998) Control of translation initiation in animals. *Annu. Rev. Cell Dev. Biol.* **14,** 399–458.

**Hannon, G.J.** (2002) RNA interference. *Nature* **418,** 244–251.

**Luger, K.** (2003) Structure and dynamic behavior of nucleosomes. *Curr. Opin. Genet. Dev.* **13,** 127–135.

**Mannervik, M., Nibu, Y., Zhang, H., & Levine, M.** (1999) Transcriptional coregulators in development. *Science* **284,** 606–609.

**Martens, J.A. & Winston, F.** (2003) Recent advances in understanding chromatin remodeling by Swi/Snf complexes. *Curr. Opin. Genet. Dev.* **13,** 136–142.

**McKnight, S.L.** (1991) Molecular zippers in gene regulation. *Sci. Am.* **264** (April), 54–64.
   A good description of leucine zippers.

**Melton, D.A.** (1991) Pattern formation during animal development. *Science* **252,** 234–241.

**Muller, W.A.** (1997) *Developmental Biology,* Springer, New York.
   A good elementary text.

**Myers, L.C. & Kornberg, R.D.** (2000) Mediator of transcriptional regulation. *Annu. Rev. Biochem.* **69,** 729–749.

**Reese, J.C.** (2003) Basal transcription factors. *Curr. Opin. Genet. Dev.* **13,** 114–118.

**Rivera-Pomar, R. & Jackle, H.** (1996) From gradients to stripes in *Drosophila* embryogenesis: filling in the gaps. *Trends Genet.* **12,** 478–483.

**Struhl, K.** (1999) Fundamentally different logic of gene regulation in eukaryotes and prokaryotes. *Cell* **98,** 1–4.

**Waterhouse, P.M. & Helliwell, C.A.** (2003) Exploring plant genomes by RNA-induced gene silencing. *Nat. Rev. Genet.* **4,** 29–38.

## Problems

**1. Effect of mRNA and Protein Stability on Regulation** *E. coli* cells are growing in a medium with glucose as the sole carbon source. Tryptophan is suddenly added. The cells continue to grow, and divide every 30 min. Describe (qualitatively) how the amount of tryptophan synthase activity in the cells changes with time under the following conditions:

   (a) The *trp* mRNA is stable (degraded slowly over many hours).

   (b) The *trp* mRNA is degraded rapidly, but tryptophan synthase is stable.

   (c) The *trp* mRNA and tryptophan synthase are both degraded rapidly.

**2. Negative Regulation** Describe the probable effects on gene expression in the *lac* operon of a mutation in (a) the *lac* operator that deletes most of $O_1$; (b) the *lac*I gene that inactivates the repressor; and (c) the promoter that alters the region around position $-10$.

**3. Specific DNA Binding by Regulatory Proteins** A typical prokaryotic repressor protein discriminates between its specific DNA binding site (operator) and nonspecific DNA by a factor of $10^4$ to $10^6$. About 10 molecules of repressor per cell are sufficient to ensure a high level of repression. Assume that a very similar repressor existed in a human cell, with a similar specificity for its binding site. How many copies of the repressor would be required to elicit a level of repression similar to that in the prokaryotic cell? (Hint: The *E. coli* genome contains about 4.6 million bp; the human haploid genome has about 3.2 billion bp.)

**4. Repressor Concentration in *E. coli*** The dissociation constant for a particular repressor-operator complex is very low, about $10^{-13}$ M. An *E. coli* cell (volume $2 \times 10^{-12}$ mL) contains 10 copies of the repressor. Calculate the cellular concentration of the repressor protein. How does this value compare with the dissociation constant of the repressor-operator complex? What is the significance of this result?

**5. Catabolite Repression** *E. coli* cells are growing in a medium containing lactose but no glucose. Indicate whether each of the following changes or conditions would increase, decrease, or not change the expression of the *lac* operon. It may be helpful to draw a model depicting what is happening in each situation.

(a) Addition of a high concentration of glucose

(b) A mutation that prevents dissociation of the Lac repressor from the operator

(c) A mutation that completely inactivates β-galactosidase

(d) A mutation that completely inactivates galactoside permease

(e) A mutation that prevents binding of CRP to its binding site near the *lac* promoter

**6. Transcription Attenuation** How would transcription of the *E. coli trp* operon be affected by the following manipulations of the leader region of the *trp* mRNA?

(a) Increasing the distance (number of bases) between the leader peptide gene and sequence 2

(b) Increasing the distance between sequences 2 and 3

(c) Removing sequence 4

(d) Changing the two Trp codons in the leader peptide gene to His codons

(e) Eliminating the ribosome-binding site for the gene that encodes the leader peptide

(f) Changing several nucleotides in sequence 3 so that it can base-pair with sequence 4 but not with sequence 2

**7. Repressors and Repression** How would the SOS response in *E. coli* be affected by a mutation in the *lexA* gene that prevented autocatalytic cleavage of the LexA protein?

**8. Regulation by Recombination** In the phase variation system of *Salmonella,* what would happen to the cell if the Hin recombinase became more active and promoted recombination (DNA inversion) several times in each cell generation?

**9. Initiation of Transcription in Eukaryotes** A new RNA polymerase activity is discovered in crude extracts of cells derived from an exotic fungus. The RNA polymerase initiates transcription only from a single, highly specialized promoter. As the polymerase is purified its activity declines, and the purified enzyme is completely inactive unless crude extract is added to the reaction mixture. Suggest an explanation for these observations.

**10. Functional Domains in Regulatory Proteins** A biochemist replaces the DNA-binding domain of the yeast Gal4 protein with the DNA-binding domain from the Lac repressor, and finds that the engineered protein no longer regulates transcription of the *GAL* genes in yeast. Draw a diagram of the different functional domains you would expect to find in the Gal4 protein and in the engineered protein. Why does the engineered protein no longer regulate transcription of the *GAL* genes? What might be done to the DNA-binding site recognized by this chimeric protein to make it functional in activating transcription of *GAL* genes?

**11. Inheritance Mechanisms in Development** A *Drosophila* egg that is *bcd⁻/bcd⁻* may develop normally but as an adult will not be able to produce viable offspring. Explain.

### Biochemistry on the Internet

**12. TATA Binding Protein and the TATA Box** To examine the interactions between transcription factors and DNA, go to the Protein Data Bank (www.rcsb.org/pdb) and download the PDB file 1TGH. This file models the interactions between a human TATA-binding protein and a segment of double-stranded DNA. Use the Noncovalent Bond Finder at the Chime Resources website (www.umass.edu/microbio/chime) to examine the roles of hydrogen bonds and hydrophobic interactions involved in the binding of this transcription factor to the TATA box.

Within the Noncovalent Bond Finder program, load the PDB file and display the protein in Spacefill mode and the DNA in Wireframe mode.

(a) Which of the base pairs in the DNA form hydrogen bonds with the protein? Which of these contribute to the specific recognition of the TATA box by this protein? (Hydrogen-bond length between hydrogen donor and hydrogen acceptor ranges from 2.5 to 3.3 Å.)

(b) Which amino acid residues in the protein interact with these base pairs? On what basis did you make this determination? Do these observations agree with the information presented in the text?

(c) What is the sequence of the DNA in this model and which portions of the sequence are recognized by the TATA-binding protein?

(d) Can you identify any hydrophobic interactions in this complex? (Hydrophobic interactions usually occur with interatomic distances of 3.3 to 4.0 Å.)

# Glossary

## a

**ABC transporter:** A member of a large family of membrane transporters that have sequences that make up *ATP-Binding Cassettes*. These transporters move a large variety of substrates, including inorganic ions, lipids, and nonpolar drugs, outward across the plasma membrane, using ATP as the energy source.

**absolute configuration:** The configuration of four different substituent groups around an asymmetric carbon atom, in relation to D- and L-glyceraldehyde.

**absorption:** Transport of the products of digestion from the intestinal tract into the blood.

**acceptor control:** The regulation of the rate of respiration by the availability of ADP as phosphate group acceptor.

**accessory pigments:** Visible light-absorbing pigments (carotenoids, xanthophyll, and phycobilins) in plants and photosynthetic bacteria that complement chlorophylls in trapping energy from sunlight.

**acidosis:** A metabolic condition in which the capacity of the body to buffer $H^+$ is diminished; usually accompanied by decreased blood pH.

**actin:** A protein making up the thin filaments of muscle; also an important component of the cytoskeleton of many eukaryotic cells.

**activation energy ($\Delta G^{\ddagger}$):** The amount of energy (in joules) required to convert all the molecules in 1 mole of a reacting substance from the ground state to the transition state.

**activator:** (1) A DNA-binding protein that positively regulates the expression of one or more genes; that is, transcription rates increase when an activator is bound to the DNA. (2) A positive modulator of an allosteric enzyme.

**active site:** The region of an enzyme surface that binds the substrate molecule and catalytically transforms it; also known as the catalytic site.

**active transport:** Energy-requiring transport of a solute across a membrane in the direction of increasing concentration.

**activity:** The true thermodynamic activity or potential of a substance, as distinct from its molar concentration.

**activity coefficient:** The factor by which the numerical value of the concentration of a solute must be multiplied to give its true thermodynamic activity.

**acyl phosphate:** Any molecule with the general chemical form $R-C-OPO_3^{2-}$.
$$\underset{O}{\overset{\parallel}{}}$$

**adenosine 3′,5′-cyclic monophosphate:** *See* cyclic AMP.

**adenosine diphosphate:** *See* ADP.

**adenosine triphosphate:** *See* ATP.

**adipocyte:** An animal cell specialized for the storage of fats (triacylglycerols).

**adipose tissue:** Connective tissue specialized for the storage of large amounts of triacylglycerols.

**ADP (adenosine diphosphate):** A ribonucleoside 5′-diphosphate serving as phosphate group acceptor in the cell energy cycle.

**aerobe:** An organism that lives in air and uses oxygen as the terminal electron acceptor in respiration.

**aerobic:** Requiring or occurring in the presence of oxygen.

**alcohol fermentation:** *See* ethanol fermentation.

**aldose:** A simple sugar in which the carbonyl carbon atom is an aldehyde; that is, the carbonyl carbon is at one end of the carbon chain.

**alkalosis:** A metabolic condition in which the capacity of the body to buffer $OH^-$ is diminished; usually accompanied by an increase in blood pH.

**allosteric enzyme:** A regulatory enzyme, with catalytic activity modulated by the noncovalent binding of a specific metabolite at a site other than the active site.

**allosteric protein:** A protein (generally with multiple subunits) with multiple ligand-binding sites, such that ligand binding at one site affects ligand binding at another.

**allosteric site:** The specific site on the surface of an allosteric enzyme molecule to which the modulator or effector molecule is bound.

**α helix:** A helical conformation of a polypeptide chain, usually right-handed, with maximal intrachain hydrogen bonding; one of the most common secondary structures in proteins.

**α oxidation:** An alternative path for the oxidation of β-methyl fatty acids in peroxisomes.

**Ames test:** A simple bacterial test for carcinogens, based on the assumption that carcinogens are mutagens.

**amino acid activation:** ATP-dependent enzymatic esterification of the carboxyl group of an amino acid to the 3′-hydroxyl group of its corresponding tRNA.

**amino acids:** α-Amino-substituted carboxylic acids, the building blocks of proteins.

**amino-terminal residue:** The only amino acid residue in a polypeptide chain with a free α-amino group; defines the amino terminus of the polypeptide.

**aminoacyl-tRNA:** An aminoacyl ester of a tRNA.

**aminoacyl-tRNA synthetases:** Enzymes that catalyze synthesis of an aminoacyl-tRNA at the expense of ATP energy.

**aminotransferases:** Enzymes that catalyze the transfer of amino groups from α-amino to α-keto acids; also called transaminases.

**ammonotelic:** Excreting excess nitrogen in the form of ammonia.

**amphibolic pathway:** A metabolic pathway used in both catabolism and anabolism.

**amphipathic:** Containing both polar and nonpolar domains.

**ampholyte:** A substance that can act as either a base or an acid.

**amphoteric:** Capable of donating and accepting protons, thus able to serve as an acid or a base.

**anabolism:** The phase of intermediary metabolism concerned with the energy-requiring biosynthesis of cell components from smaller precursors.

**anaerobe:** An organism that lives without oxygen. Obligate anaerobes die when exposed to oxygen.

**anaerobic:** Occurring in the absence of air or oxygen.

**analyte:** A molecule to be analyzed by mass spectrometry.

**anaplerotic reaction:** An enzyme-catalyzed reaction that can replenish the supply of intermediates in the citric acid cycle.

**angstrom (Å):** A unit of length ($10^{-8}$ cm) used to indicate molecular dimensions.

**anhydride:** The product of the condensation of two carboxyl or phosphate groups in which the elements of water are eliminated to form a compound with the general structure $R-X-O-X-R$, where X is either carbon
$$\underset{O}{\overset{\parallel}{}} \quad \underset{O}{\overset{\parallel}{}}$$
or phosphorus.

**anion-exchange resin:** A polymeric resin with fixed cationic groups; used in the chromatographic separation of anions.

**anomers:** Two stereoisomers of a given sugar that differ only in the configuration about the carbonyl (anomeric) carbon atom.

**antibiotic:** One of many different organic compounds that are formed and secreted by various species of microorganisms and plants, are toxic to other species, and presumably have a defensive function.

**antibody:** A defense protein synthesized by the immune system of vertebrates. *See also* immunoglobulin.

**anticodon:** A specific sequence of three nucleotides in a tRNA, complementary to a codon for an amino acid in an mRNA.

**antigen:** A molecule capable of eliciting the synthesis of a specific antibody in vertebrates.

**antiparallel:** Describes two linear polymers that are opposite in polarity or orientation.

**antiport:** Cotransport of two solutes across a membrane in opposite directions.

**apoenzyme:** The protein portion of an enzyme, exclusive of any organic or inorganic cofactors or prosthetic groups that might be required for catalytic activity.

**apolipoprotein:** The protein component of a lipoprotein.

**apoprotein:** The protein portion of a protein, exclusive of any organic or inorganic cofactors or prosthetic groups that might be required for activity.

**apoptosis:** (app′-a-toe′-sis) Programmed cell death, in which a cell brings about its own death and lysis, signaled from outside or programmed in its genes, by systematically degrading its own macromolecules.

**aptamer:** Oligonucleotide that binds specifically to one molecular target, usually selected by an iterative cycle of affinity-based enrichment (SELEX).

**archaebacteria:** One of five kingdoms of living organisms; includes many species that thrive in extreme environments of high ionic strength, high temperature, or low pH.

**arrestins:** A family of proteins that bind to the phosphorylated carboxyl-terminal region of serpentine receptors, preventing their interactions with G proteins and thereby terminating the signal through those receptors.

**asymmetric carbon atom:** A carbon atom that is covalently bonded to four different groups and thus may exist in two different tetrahedral configurations.

**ATP (adenosine triphosphate):** A ribonucleoside 5′-triphosphate functioning as a phosphate group donor in the cell energy cycle; carries chemical energy between metabolic pathways by serving as a shared intermediate coupling endergonic and exergonic reactions.

**ATP synthase:** An enzyme complex that forms ATP from ADP and phosphate during oxidative phosphorylation in the inner mitochondrial membrane or the bacterial plasma membrane, and during photophosphorylation in chloroplasts.

**ATPase:** An enzyme that hydrolyzes ATP to yield ADP and phosphate; usually coupled to some process requiring energy.

**attenuator:** An RNA sequence involved in regulating the expression of certain genes; functions as a transcription terminator.

**autophosphorylation:** Strictly, the phosphorylation of an amino acid residue in a protein, catalyzed by the same molecule. Often extended to include phosphorylation of one subunit of a homodimer by the other subunit.

**autotroph:** An organism that can synthesize its own complex molecules from very simple carbon and nitrogen sources, such as carbon dioxide and ammonia.

**auxin:** A plant growth hormone.

**auxotrophic mutant (auxotroph):** A mutant organism defective in the synthesis of a given biomolecule, which must therefore be supplied for the organism's growth.

**Avogadro's number ($N$):** The number of molecules in a gram molecular weight (a mole) of any compound ($6.02 \times 10^{23}$).

 **b**

**back-mutation:** A mutation that causes a mutant gene to regain its wild-type base sequence.

**bacteriophage (phage):** A virus capable of replicating in a bacterial cell.

**basal metabolic rate:** The rate of oxygen consumption by an animal's body at complete rest, long after a meal.

**base pair:** Two nucleotides in nucleic acid chains that are paired by hydrogen bonding of their bases; for example, A with T or U, and G with C.

**$\beta$ conformation:** An extended, zigzag arrangement of a polypeptide chain; a common secondary structure in proteins.

**$\beta$ oxidation:** Oxidative degradation of fatty acids into acetyl-CoA by successive oxidations at the $\beta$-carbon atom.

**$\beta$ turn:** A type of secondary structure in polypeptides consisting of four amino acid residues arranged in a tight turn so that the polypeptide turns back on itself.

**bilayer:** A double layer of oriented amphipathic lipid molecules, forming the basic structure of biological membranes. The hydrocarbon tails face inward to form a continuous nonpolar phase.

**bile salts:** Amphipathic steroid derivatives with detergent properties, participating in digestion and absorption of lipids.

**binding energy:** The energy derived from noncovalent interactions between enzyme and substrate or receptor and ligand.

**binding site:** The crevice or pocket on a protein in which a ligand binds.

**bioassay:** A method for measuring the amount of a biologically active substance (such as a hormone) in a sample by quantifying the biological response to aliquots of that sample.

**biocytin:** The conjugate amino acid residue arising from covalent attachment of biotin, through an amide linkage, to a Lys residue.

**biomolecule:** An organic compound normally present as an essential component of living organisms.

**biopterin:** An enzymatic cofactor derived from pterin and involved in certain oxidation-reduction reactions.

**biosphere:** All the living matter on or in the earth, the seas, and the atmosphere.

**biotin:** A vitamin; an enzymatic cofactor involved in carboxylation reactions.

**B lymphocyte (B cell):** One of a class of blood cells (lymphocytes), responsible for the production of circulating antibodies.

**bond energy:** The energy required to break a bond.

**branch migration:** Movement of the branch point in branched DNA formed from two DNA molecules with identical sequences. *See also* Holliday intermediate.

**buffer:** A system capable of resisting changes in pH, consisting of a conjugate acid-base pair in which the ratio of proton acceptor to proton donor is near unity.

 **c**

**calorie:** The amount of heat required to raise the temperature of 1.0 g of water from 14.5 to 15.5 °C. One calorie (cal) equals 4.18 joules (J).

**Calvin cycle:** The cyclic pathway used by plants to fix carbon dioxide and produce triose phosphates.

**cAMP:** *See* cyclic AMP.

**cAMP receptor protein (CRP):** A specific regulatory protein that controls initiation of transcription of the genes producing the enzymes required for a bacterial cell to use some other nutrient when glucose is lacking. Also called catabolite gene activator protein (CAP).

**CAP:** *See* cAMP receptor protein.

**capsid:** The protein coat of a virion or virus particle.

**carbanion:** A negatively charged carbon atom.

**carbocation:** A positively charged carbon atom; also called a carbonium ion.

**carbon-assimilation reactions:** Reaction sequences in which atmospheric $CO_2$ is converted into organic compounds.

**carbon-fixation reaction:** The reaction catalyzed by rubisco during photosynthesis, or by other carboxylases, in which atmospheric $CO_2$ is initially incorporated into an organic compound.

**carboxyl-terminal residue:** The only amino acid residue in a polypeptide chain with a free $\alpha$-carboxyl group; defines the carboxyl terminus of the polypeptide.

**carnitine shuttle:** Mechanism for moving fatty acids from the cytosol to the mitochondrial matrix as fatty esters of carnitine.

**carotenoids:** Lipid-soluble photosynthetic pigments made up of isoprene units.

**cascade:** A series of events in which a stimulus activates an enzyme molecule, which acts catalytically to activate many molecules of another enzyme, each of which in turn activates many molecules of a third, and so on. The result is a large amplification of the original signal.

**catabolism:** The phase of intermediary metabolism concerned with the energy-yielding degradation of nutrient molecules.

**catabolite gene activator protein (CAP):** *See* cAMP receptor protein.

**catalytic site:** *See* active site.

**catecholamines:** Hormones, such as epinephrine, that are amino derivatives of catechol.

**catenane:** Circular polymeric molecules with a noncovalent topological link resembling the links of a chain.

**cation-exchange resin:** An insoluble polymer with fixed negative charges; used in the chromatographic separation of cationic substances.

**cDNA:** *See* complementary DNA.

**cDNA library:** DNA library consisting entirely of cloned cDNAs from a particular organism or cell type.

**central dogma:** The organizing principle of molecular biology: genetic information flows from DNA to RNA to protein.

**centromere:** A specialized site within a chromosome, serving as the attachment point for the mitotic or meiotic spindle.

**cerebroside:** Sphingolipid containing one sugar residue as a head group.

**channeling:** The direct transfer of a reaction product (common intermediate) from the active site of one enzyme to the active site of a different enzyme catalyzing the next step in a sequential pathway.

**chemiosmotic coupling:** Coupling of ATP synthesis to electron transfer via a transmembrane difference in charge and pH.

**chemiosmotic theory:** The theory that energy derived from electron transfer reactions is temporarily stored as a transmembrane difference in charge and pH, which subsequently drives the formation of ATP in oxidative phosphorylation and photophosphorylation.

**chemotaxis:** A cell's sensing of and movement toward, or away from, a specific chemical agent.

**chemotroph:** An organism that obtains energy by metabolizing organic compounds derived from other organisms.

**chiral center:** An atom with substituents arranged so that the molecule is not superimposable on its mirror image.

**chiral compound:** A compound that contains an asymmetric center (chiral atom or chiral center) and thus can occur in two nonsuperimposable mirror-image forms (enantiomers).

**chlorophylls:** A family of green pigments functioning as receptors of light energy in photosynthesis; magnesium-porphyrin complexes.

**chloroplasts:** Chlorophyll-containing photosynthetic organelles in some eukaryotic cells.

**chromatin:** A filamentous complex of DNA, histones, and other proteins, constituting the eukaryotic chromosome.

**chromatography:** A process in which complex mixtures of molecules are separated by many repeated partitionings between a flowing (mobile) phase and a stationary phase.

**chromosome:** A single large DNA molecule and its associated proteins, containing many genes; stores and transmits genetic information.

**chylomicron:** A plasma lipoprotein consisting of a large droplet of triacylglycerols stabilized by a coat of protein and phospholipid; carries lipids from the intestine to the tissues.

**chymotrypsin:** A well-studied protease that cleaves on the carbonyl side of amino acid residues with aromatic side chains.

**chymotrypsinogen:** The zymogen precursor of chymotrypsin.

**cis and trans isomers:** *See* geometric isomers.

**cistron:** A unit of DNA or RNA corresponding to one gene.

**citric acid cycle:** A cyclic system of enzymatic reactions for the oxidation of acetyl residues to carbon dioxide, in which formation of citrate is the first step; also known as the Krebs cycle or tricarboxylic acid cycle.

**clones:** The descendants of a single cell.

**cloning:** The production of large numbers of identical DNA molecules, cells, or organisms, from a single ancestral DNA molecule, cell, or organism.

**closed system:** A system that exchanges neither matter nor energy with the surroundings. *See also* system.

**cobalamin:** *See* coenzyme $B_{12}$.

**codon:** A sequence of three adjacent nucleotides in a nucleic acid that codes for a specific amino acid.

**coenzyme:** An organic cofactor required for the action of certain enzymes; often contains a vitamin as a component.

**coenzyme A:** A pantothenic acid-containing coenzyme serving as an acyl group carrier in certain enzymatic reactions.

**coenzyme $B_{12}$:** An enzymatic cofactor derived from the vitamin cobalamin, involved in certain types of carbon skeletal rearrangements.

**cofactor:** An inorganic ion or a coenzyme required for enzyme activity.

**cognate:** Describes two biomolecules that normally interact; for example, an enzyme

and its normal substrate, or a receptor and its normal ligand.

**cohesive ends:** *See* sticky ends.

**cointegrate:** An intermediate in the migration of certain DNA transposons in which the donor DNA and target DNA are covalently attached.

**colligative properties:** Properties of solutions that depend on the number of solute particles per unit volume; for example, freezing-point depression.

**common intermediate:** A chemical compound common to two chemical reactions, as a product of one and a reactant in the other.

**competitive inhibition:** A type of enzyme inhibition reversed by increasing the substrate concentration; a competitive inhibitor generally competes with the normal substrate or ligand for a protein's binding site.

**complementary:** Having a molecular surface with chemical groups arranged to interact specifically with chemical groups on another molecule.

**complementary DNA (cDNA):** A DNA used in DNA cloning, usually made by reverse transcriptase; complementary to a given mRNA.

**configuration:** The spatial arrangement of an organic molecule that is conferred by the presence of either (1) double bonds, about which there is no freedom of rotation, or (2) chiral centers, around which substituent groups are arranged in a specific sequence. Configurational isomers cannot be interconverted without breaking one or more covalent bonds.

**conformation:** The spatial arrangement of substituent groups that are free to assume different positions in space, without breaking any bonds, because of the freedom of bond rotation.

**conformation, $\beta$:** *See* $\beta$ conformation.

**conjugate acid-base pair:** A proton donor and its corresponding deprotonated species; for example, acetic acid (donor) and acetate (acceptor).

**conjugate redox pair:** An electron donor and its corresponding electron acceptor form; for example, $Cu^+$ (donor) and $Cu^{2+}$ (acceptor), or NADH (donor) and $NAD^+$ (acceptor).

**conjugated protein:** A protein containing one or more prosthetic groups.

**consensus sequence:** A DNA or amino acid sequence consisting of the residues that occur most commonly at each position within a set of similar sequences.

**conservative substitution:** Replacement of an amino acid residue in a polypeptide by another residue with similar properties; for example, substitution of Glu by Asp.

**constitutive enzymes:** Enzymes required at all times by a cell and present at some constant level; for example, many enzymes of the central metabolic pathways. Sometimes called housekeeping enzymes.

**contig:** A series of overlapping clones or a continuous sequence defining an uninterrupted section of a chromosome.

**contour length:** The length of a helical polymeric molecule as measured along the molecule's helical axis.

**corticosteroids:** Steroid hormones formed by the adrenal cortex.

**cotransport:** The simultaneous transport, by a single transporter, of two solutes across a membrane. *See* antiport, symport.

**coupled reactions:** Two chemical reactions that have a common intermediate and thus a means of energy transfer from one to the other.

**covalent bond:** A chemical bond that involves sharing of electron pairs.

**cristae:** Infoldings of the inner mitochondrial membrane.

**CRP:** *See* cAMP receptor protein.

**cruciform:** Secondary structure in double-stranded RNA or DNA, in which the double helix is denatured at palindromic repeat sequences in each strand, and each separated strand is paired internally to form opposing hairpin structures. *See also* hairpin.

**cyclic AMP (cAMP):** A second messenger within cells; its formation by adenylyl cyclase is stimulated by certain hormones or other molecular signals.

**cyclic electron flow:** In chloroplasts, the light-induced flow of electrons originating from and returning to photosystem I.

**cyclic photophosphorylation:** ATP synthesis driven by cyclic electron flow through photosystem I.

**cyclin:** One of a family of proteins that activate cyclin-dependent protein kinases and thereby regulate the cell cycle.

**cytochromes:** Heme proteins serving as electron carriers in respiration, photosynthesis, and other oxidation-reduction reactions.

**cytokine:** One of a family of small secreted proteins (such as interleukins or interferons) that activate cell division or differentiation by binding to plasma membrane receptors in sensitive's cells.

**cytokinesis:** The final separation of daughter cells following mitosis.

**cytoplasm:** The portion of a cell's contents outside the nucleus but within the plasma membrane; includes organelles such as mitochondria.

**cytoskeleton:** The filamentous network providing structure and organization to the cytoplasm; includes actin filaments, microtubules, and intermediate filaments.

**cytosol:** The continuous aqueous phase of the cytoplasm, with its dissolved solutes; excludes the organelles such as mitochondria.

## d

**dalton:** The weight of a single hydrogen atom ($1.66 \times 10^{-24}$ g).

**dark reactions:** *See* carbon-assimilation reactions.

**de novo pathway:** Pathway for synthesis of a biomolecule, such as a nucleotide, from simple precursors; as distinct from a salvage pathway.

**deamination:** The enzymatic removal of amino groups from biomolecules such as amino acids or nucleotides.

**degenerate code:** A code in which a single element in one language is specified by more than one element in a second language.

**dehydrogenases:** Enzymes catalyzing the removal of pairs of hydrogen atoms from their substrates.

**deletion mutation:** A mutation resulting from the deletion of one or more nucleotides from a gene or chromosome.

**denaturation:** Partial or complete unfolding of the specific native conformation of a polypeptide chain, protein, or nucleic acid.

**denatured protein:** A protein that has lost its native conformation by exposure to a destabilizing agent such as heat or detergent.

**deoxyribonucleic acid:** *See* DNA.

**deoxyribonucleotides:** Nucleotides containing 2-deoxy-D-ribose as the pentose component.

**desaturases:** Enzymes that catalyze the introduction of double bonds into the hydrocarbon portion of fatty acids.

**desensitization:** Universal process by which sensory mechanisms cease to respond after prolonged exposure to the specific stimulus they detect.

**desolvation:** In aqueous solution, the release of bound water surrounding a solute.

**dextrorotatory isomer:** A stereoisomer that rotates the plane of plane-polarized light clockwise.

**diabetes mellitus:** A metabolic disease resulting from insulin deficiency; characterized by a failure in glucose transport from the blood into cells at normal glucose concentrations.

**dialysis:** Removal of small molecules from a solution of a macromolecule by allowing them to diffuse through a semipermeable membrane into a buffered solution.

**differential centrifugation:** Separation of cell organelles or other particles of different size by their different rates of sedimentation in a centrifugal field.

**differentiation:** Specialization of cell structure and function during embryonic growth and development.

**diffusion:** The net movement of molecules in the direction of lower concentration.

**digestion:** Enzymatic hydrolysis of major nutrients in the gastrointestinal system to yield their simpler components.

**diploid:** Having two sets of genetic information; describes a cell with two chromosomes of each type.

**dipole:** A molecule having both positive and negative charges.

**diprotic acid:** An acid having two dissociable protons.

**disaccharide:** A carbohydrate consisting of two covalently joined monosaccharide units.

**dissociation constant:** (1) An equilibrium constant ($K_d$) for the dissociation of a complex of two or more biomolecules into its components; for example, dissociation of a substrate from an enzyme. (2) The dissociation constant ($K_a$) of an acid, describing its dissociation into its conjugate base and a proton.

**disulfide bridge:** A covalent cross link between two polypeptide chains formed by a cystine residue (two Cys residues).

**DNA (deoxyribonucleic acid):** A polynucleotide having a specific sequence of deoxyribonucleotide units covalently joined through $3',5'$-phosphodiester bonds; serves as the carrier of genetic information.

**DNA chimera:** DNA containing genetic information derived from two different species.

**DNA chip:** An informal term for a DNA microarray, referring to the small size of typical microarrays.

**DNA cloning:** *See* cloning.

**DNA library:** A collection of cloned DNA fragments.

**DNA ligase:** An enzyme that creates a phosphodiester bond between the $3'$ end of one DNA segment and the $5'$ end of another.

**DNA looping:** The interaction of proteins bound at distant sites on a DNA molecule so that the intervening DNA forms a loop.

**DNA microarray:** A collection of DNA sequences immobilized on a solid surface, with individual sequences laid out in patterned arrays that can be probed by hybridization.

**DNA polymerase:** An enzyme that catalyzes template-dependent synthesis of DNA from its deoxyribonucleoside $5'$-triphosphate precursors.

**DNA replicase system:** The entire complex of enzymes and specialized proteins required in biological DNA replication.

**DNA supercoiling:** The coiling of DNA upon itself, generally as a result of bending, underwinding, or overwinding of the DNA helix.

**DNA transposition:** *See* transposition.

**domain:** A distinct structural unit of a polypeptide; domains may have separate functions and may fold as independent, compact units.

**double helix:** The natural coiled conformation of two complementary, antiparallel DNA chains.

**double-reciprocal plot:** A plot of $1/V_0$ versus $1/[S]$, which allows a more accurate determination of $V_{max}$ and $K_m$ than a plot of $V_0$ versus $[S]$; also called the Lineweaver-Burk plot.

### e

$E'^{\circ}$: *See* standard reduction potential.

***E. coli (Escherichia coli):*** A common bacterium found in the small intestine of vertebrates; the most well-studied organism.

**electrochemical gradient:** The sum of the gradients of concentration and of electric charge of an ion across a membrane; the driving force for oxidative phosphorylation and photophosphorylation.

**electrochemical potential:** The energy required to maintain a separation of charge and of concentration across a membrane.

**electrogenic:** Contributing to an electrical potential across a membrane.

**electron acceptor:** A substance that receives electrons in an oxidation-reduction reaction.

**electron carrier:** A protein, such as a flavoprotein or a cytochrome, that can reversibly gain and lose electrons; functions in the transfer of electrons from organic nutrients to oxygen or some other terminal acceptor.

**electron donor:** A substance that donates electrons in an oxidation-reduction reaction.

**electron transfer:** Movement of electrons from substrates to oxygen via the carriers of the respiratory (electron-transfer) chain.

**electrophile:** An electron-deficient group with a strong tendency to accept electrons from an electron-rich group (nucleophile).

**electrophoresis:** Movement of charged solutes in response to an electrical field; often used to separate mixtures of ions, proteins, or nucleic acids.

**electroporation:** Introduction of macromolecules into cells after rendering the cells transiently permeable by the application of a high-voltage pulse.

**elongation factors:** (1) Proteins that function in the elongation phase of eukaryotic transcription. (2) Specific proteins required in the elongation of polypeptide chains by ribosomes.

**eluate:** The effluent from a chromatographic column.

**enantiomers:** Stereoisomers that are nonsuperimposable mirror images of each other.

**end-product inhibition:** *See* feedback inhibition.

**endergonic reaction:** A chemical reaction that consumes energy (that is, for which $\Delta G$ is positive).

**endocrine glands:** Groups of cells specialized to synthesize hormones and secrete them into the blood to regulate other types of cells.

**endocytosis:** The uptake of extracellular material by its inclusion within a vesicle (endosome) formed by an invagination of the plasma membrane.

**endonuclease:** An enzyme that hydrolyzes the interior phosphodiester bonds of a nucleic acid; that is, it acts at points other than the terminal bonds.

**endoplasmic reticulum:** An extensive system of double membranes in the cytoplasm of eukaryotic cells; it encloses secretory channels and is often studded with ribosomes (rough endoplasmic reticulum).

**endothermic reaction:** A chemical reaction that takes up heat (that is, for which $\Delta H$ is positive).

**energy charge:** The fractional degree to which the ATP/ADP/AMP system is filled with high-energy phosphate groups.

**energy coupling:** The transfer of energy from one process to another.

**enhancers:** DNA sequences that facilitate the expression of a given gene; may be located a few hundred, or even thousand, base pairs away from the gene.

**enthalpy ($H$):** The heat content of a system.

**enthalpy change ($\Delta H$):** For a reaction, is approximately equal to the difference between the energy used to break bonds and the energy gained by the formation of new ones.

**entropy ($S$):** The extent of randomness or disorder in a system.

**enzyme:** A biomolecule, either protein or RNA, that catalyzes a specific chemical reaction. It does not affect the equilibrium of the catalyzed reaction; it enhances the rate of a reaction by providing a reaction path with a lower activation energy.

**enzyme cascade:** A series of reactions, often involved in regulatory events, in which one enzyme activates another (often by phosphorylation), which activates a third, and so on. The effect of a catalyst activating a catalyst is a large amplification of the signal that initiated the cascade.

**epimerases:** Enzymes that catalyze the reversible interconversion of two epimers.

**epimers:** Two stereoisomers differing in configuration at one asymmetric center, in a compound having two or more asymmetric centers.

**epithelial cell:** Any cell that forms part of the outer covering of an organism or organ.

**epitope:** An antigenic determinant; the particular chemical group or groups within a macromolecule (antigen) to which a given antibody binds.

**epitope tag:** A protein sequence or domain bound by some well-characterized antibody.

**equilibrium:** The state of a system in which no further net change is occurring; the free energy is at a minimum.

**equilibrium constant ($K_{eq}$):** A constant, characteristic for each chemical reaction; relates the specific concentrations of all reactants and products at equilibrium at a given temperature and pressure.

**erythrocyte:** A cell containing large amounts of hemoglobin and specialized for oxygen transport; a red blood cell.

***Escherichia coli:*** *See E. coli.*

**essential amino acids:** Amino acids that cannot be synthesized by humans (and other vertebrates) and must be obtained from the diet.

**essential fatty acids:** The group of polyunsaturated fatty acids produced by plants, but not by humans; required in the human diet.

**ethanol fermentation:** The anaerobic conversion of glucose to ethanol via glycolysis. Also called alcohol fermentation. *See also* fermentation.

**eubacteria:** One of the five kingdoms of living organisms. Eubacteria have a plasma membrane but no internal organelles or nucleus.

**eukaryote:** A unicellular or multicellular organism with cells having a membrane-bounded nucleus, multiple chromosomes, and internal organelles.

**excited state:** An energy-rich state of an atom or molecule; produced by the absorption of light energy.

**exergonic reaction:** A chemical reaction that proceeds with the release of free energy (that is, for which $\Delta G$ is negative).

**exocytosis:** The fusion of an intracellular vesicle with the plasma membrane, releasing the vesicle contents to the extracellular space.

**exon:** The segment of a eukaryotic gene that encodes a portion of the final product of the gene; a portion that remains after posttranscriptional processing and is transcribed into a protein or incorporated into the structure of an RNA. *See* intron.

**exonuclease:** An enzyme that hydrolyzes only those phosphodiester bonds that are in the terminal positions of a nucleic acid.

**exothermic reaction:** A chemical reaction that releases heat (that is, for which $\Delta H$ is negative).

**expressed sequence tag (EST):** A specific type of sequence-tagged site representing a gene that is expressed.

**expression vector:** *See* vector.

**extrahepatic:** Term used to refer to all tissues outside the liver; implies the centrality of the liver in metabolism.

## f

**facilitated diffusion:** Diffusion of a polar substance across a biological membrane through a protein transporter; also called passive diffusion or passive transport.

**facultative cells:** Cells that can live in the presence or absence of oxygen.

**FAD (flavin adenine dinucleotide):** The coenzyme of some oxidation-reduction enzymes; contains riboflavin.

**fatty acid:** A long-chain aliphatic carboxylic acid found in natural fats and oils; also a component of membrane phospholipids and glycolipids.

**feedback inhibition:** Inhibition of an allosteric enzyme at the beginning of a metabolic sequence by the end product of the sequence; also known as end-product inhibition.

**fermentation:** Energy-yielding anaerobic breakdown of a nutrient molecule, such as glucose, without net oxidation; yields lactate, ethanol, or some other simple product.

**fibroblast:** A cell of the connective tissue that secretes connective tissue proteins such as collagen.

**fibrous proteins:** Insoluble proteins that serve in a protective or structural role; contain polypeptide chains that generally share a common secondary structure.

**fingerprinting:** *See* peptide mapping.

**first law of thermodynamics:** The law stating that in all processes, the total energy of the universe remains constant.

**Fischer projection formulas:** *See* projection formulas.

**5′ end:** The end of a nucleic acid that lacks a nucleotide bound at the 5′ position of the terminal residue.

**flagellum:** A cell appendage used in propulsion. Bacterial flagella have a much simpler structure than eukaryotic flagella, which are similar to cilia.

**flavin-linked dehydrogenases:** Dehydrogenases requiring one of the riboflavin coenzymes, FMN or FAD.

**flavin nucleotides:** Nucleotide coenzymes (FMN and FAD) containing riboflavin.

**flavoprotein:** An enzyme containing a flavin nucleotide as a tightly bound prosthetic group.

**flippases:** Membrane proteins that catalyze the movement of lipids from one leaf of a membrane bilayer to the other.

**fluid mosaic model:** A model describing biological membranes as a fluid lipid bilayer with embedded proteins; the bilayer exhibits both structural and functional asymmetry.

**fluorescence:** Emission of light by excited molecules as they revert to the ground state.

**FMN (flavin mononucleotide):** Riboflavin phosphate, a coenzyme of certain oxidation-reduction enzymes.

**fold:** A distinct folding pattern for elements of secondary structure, observed in one or more proteins; also called a motif or super-secondary structure.

**footprinting:** A technique for identifying the nucleic acid sequence bound by a DNA- or RNA-binding protein.

**fractionation:** The process of separating the proteins or other components of a complex molecular mixture into fractions based on differences in their physical properties, such as size, net charge, and solubility.

**frame shift:** A mutation caused by insertion or deletion of one or more paired nucleotides, changing the reading frame of codons during protein synthesis; the polypeptide product has a garbled amino acid sequence beginning at the mutated codon.

**FRAP (Fluorescence return after photobleaching):** A technique used to quantify the diffusion of membrane components (lipids or proteins) in the plane of the bilayer.

**free energy ($G$):** The component of the total energy of a system that can do work at constant temperature and pressure.

**free energy of activation ($\Delta G^{\ddagger}$):** *See* activation energy.

**free-energy change ($\Delta G$):** The amount of free energy released (negative $\Delta G$) or absorbed (positive $\Delta G$) in a reaction at constant temperature and pressure.

**free radical:** *See* radical.

**functional group:** The specific atom or group of atoms that confers a particular chemical property on a biomolecule.

**furanose:** A simple sugar containing the five-membered furan ring.

**fusion protein:** (1) A family of proteins that facilitate membrane fusion. (2) The protein product of a gene created by the fusion of two distinct genes or portions of genes.

**futile cycle:** A set of enzyme-catalyzed cyclic reactions that results in release of thermal energy by the hydrolysis of ATP.

## g

**G protein–coupled receptors (GPCR):** *See* serpentine receptors.

**G proteins:** A family of heterotrimeric GTP-binding proteins that act in intracellular signaling pathways. Commonly, ligand binding to a serpentine receptor induces the exchange of GTP for bound GDP, enabling the G protein to activate a downstream enzyme in a signaling pathway. G proteins have intrinsic GTPase activity, and therefore self-inactivate.

**$\Delta G^{\prime \circ}$:** *See* standard free-energy change.

**gametes:** Reproductive cells with a haploid gene content; sperm or egg cells.

**gangliosides:** Sphingolipids containing complex oligosaccharides as head groups; especially common in nervous tissue.

**gel filtration:** *See* size-exclusion chromatography.

**gene:** A chromosomal segment that codes for a single functional polypeptide chain or RNA molecule.

**gene expression:** Transcription, and in the case of proteins, translation, to yield the product of a gene; a gene is expressed when its biological product is present and active.

**gene splicing:** The enzymatic attachment of one gene, or part of a gene, to another.

**general acid-base catalysis:** Catalysis involving proton transfer(s) to or from a molecule other than water.

**genetic code:** The set of triplet code words in DNA (or mRNA) coding for the amino acids of proteins.

**genetic information:** The hereditary information contained in a sequence of nucleotide bases in chromosomal DNA or RNA.

**genetic map:** A diagram showing the relative sequence and position of specific genes along a chromosome.

**genome:** All the genetic information encoded in a cell or virus.

**genomic library:** A DNA library containing DNA segments representing all (or most) of the sequences in an organism's genome.

**genomics:** A science devoted broadly to the understanding of cellular and organism genomes.

**genotype:** The genetic constitution of an organism, as distinct from its physical characteristics, or phenotype.

**geometric isomers:** Isomers related by rotation about a double bond; also called cis and trans isomers.

**germ-line cell:** A type of animal cell that is formed early in embryogenesis and may multiply by mitosis or may produce, by meiosis, cells that develop into gametes (egg or sperm cells).

**globular proteins:** Soluble proteins with a globular (somewhat rounded) shape.

**glucogenic amino acids:** Amino acids with carbon chains that can be metabolically converted into glucose or glycogen via gluconeogenesis.

**gluconeogenesis:** The biosynthesis of a carbohydrate from simpler, noncarbohydrate precursors such as oxaloacetate or pyruvate.

**GLUT:** Designation for a family of membrane proteins that transport glucose.

**glycan:** Another term for polysaccharide; a polymer of monosaccharide units joined by glycosidic bonds.

**glycerophospholipid:** An amphipathic lipid with a glycerol backbone; fatty acids are ester-linked to C-1 and C-2 of glycerol, and a polar alcohol is attached through a phosphodiester linkage to C-3.

**glycoconjugate:** A compound containing a carbohydrate component bound covalently to a protein or lipid, forming a glycoprotein or glycolipid.

**glycogenin:** The protein that both primes the synthesis of new glycogen chains and catalyzes the polymerization of the first few sugar residues of each chain before glycogen synthase continues the extension.

**glycogenesis:** The process of converting glucose to glycogen.

**glycolipid:** A lipid containing a carbohydrate group.

**glycolysis:** The catabolic pathway by which a molecule of glucose is broken down into two molecules of pyruvate.

**glycoprotein:** A protein containing a carbohydrate group.

**glycosaminoglycan:** A heteropolysaccharide of two alternating units: one is either *N*-acetylglucosamine or *N*-acetylgalactosamine; the other is a uronic acid (usually glucuronic acid). Formerly called mucopolysaccharide.

**glycosidic bonds:** Bonds between a sugar and another molecule (typically an alcohol, purine, pyrimidine, or sugar) through an intervening oxygen.

**glyoxylate cycle:** A variant of the citric acid cycle, for the net conversion of acetate into succinate and, eventually, new carbohydrate; present in bacteria and some plant cells.

**glyoxysome:** A specialized peroxisome containing the enzymes of the glyoxylate cycle; found in cells of germinating seeds.

**Golgi complex:** A complex membranous organelle of eukaryotic cells; functions in the posttranslational modification of proteins and their secretion from the cell or incorporation into the plasma membrane or organellar membranes.

**gram molecular weight:** The weight in grams of a compound that is numerically equal to its molecular weight; the weight of 1 mole.

**grana:** Stacks of thylakoids, flattened membranous sacs or disks, in chloroplasts.

**green fluorescent protein (GFP):** A small protein from a marine organism that produces a bright fluorescence in the green region of the visible spectrum. Fusion proteins with GFP are commonly used to determine the subcellular location of the fused protein by fluorescence microscopy.

**ground state:** The normal, stable form of an atom or molecule; as distinct from the excited state.

**group transfer potential:** A measure of the ability of a compound to donate an activated group (such as a phosphate or acyl group); generally expressed as the standard free energy of hydrolysis.

**growth factors:** Proteins or other molecules that act from outside a cell to stimulate cell growth and division.

#  h

**half-life:** The time required for the disappearance or decay of one-half of a given component in a system.

**hairpin:** Secondary structure in single-stranded RNA or DNA, in which complementary parts of a palindromic repeat fold back and are paired to form an antiparallel duplex helix that is closed at one end.

**haploid:** Having a single set of genetic information; describes a cell with one chromosome of each type.

**hapten:** A small molecule which, when linked to a larger molecule, elicits an immune response.

**Haworth perspective formulas:** A method for representing cyclic chemical structures so as to define the configuration of each substituent group; the method commonly used for representing sugars.

**helicase:** An enzyme that catalyzes the separation of strands in a DNA molecule before replication.

**helix, α:** *See* α helix.

**heme:** The iron-porphyrin prosthetic group of heme proteins.

**heme protein:** A protein containing a heme as a prosthetic group.

**hemoglobin:** A heme protein in erythrocytes; functions in oxygen transport.

**Henderson-Hasselbalch equation:** An equation relating the pH, the $pK_a$, and the ratio of the concentrations of the proton-acceptor ($A^-$) and proton-donor (HA) species in a solution.

**hepatocyte:** The major cell type of liver tissue.

**heteroduplex DNA:** Duplex DNA containing complementary strands derived from two different DNA molecules with similar sequences, often as a product of genetic recombination.

**heteropolysaccharide:** A polysaccharide containing more than one type of sugar.

**heterotroph:** An organism that requires complex nutrient molecules, such as glucose, as a source of energy and carbon.

**heterotropic:** Describes an allosteric modulator that is distinct from the normal ligand.

**heterotropic enzyme:** An allosteric enzyme requiring a modulator other than its substrate.

**hexose:** A simple sugar with a backbone containing six carbon atoms.

**hexose monophosphate pathway:** *See* phophogluconate pathway.

**high-energy compound:** A compound that on hydrolysis undergoes a large decrease in free energy under standard conditions.

**high-performance liquid chromatography (HPLC):** Chromatographic procedure, often conducted at relatively high pressures, using automated equipment that permits refined and highly reproducible profiles.

**Hill coefficient:** A measure of cooperative interaction between protein subunits.

**Hill reaction:** The evolution of oxygen and the photoreduction of an artificial electron acceptor by a chloroplast preparation in the absence of carbon dioxide.

**histones:** The family of five basic proteins that associate tightly with DNA in the chromosomes of all eukaryotic cells.

**Holliday intermediate:** An intermediate in genetic recombination in which two double-stranded DNA molecules are joined by virtue of a reciprocal crossover involving one strand of each molecule.

**holoenzyme:** A catalytically active enzyme including all necessary subunits, prosthetic groups, and cofactors.

**homeobox:** A conserved DNA sequence of 180 base pairs encoding a protein domain found in many proteins that play a regulatory role in development.

**homeodomain:** The protein domain encoded by the homeobox.

**homeostasis:** The maintenance of a dynamic steady state by regulatory mechanisms that compensate for changes in external circumstances.

**homeotic genes:** Genes that regulate the development of the pattern of segments in the *Drosophila* body plan; similar genes are found in most vertebrates.

**homologous genetic recombination:** Recombination between two DNA molecules of similar sequence, occurring in all cells; occurs during meiosis and mitosis in eukaryotes.

**homologous proteins:** Proteins having sequences and functions similar in different species; for example, the hemoglobins.

**homopolysaccharide:** A polysaccharide made up of only one type of monosaccharide unit.

**homotropic:** Describes an allosteric modulator that is identical to the normal ligand.

**homotropic enzyme:** An allosteric enzyme that uses its substrate as a modulator.

**hormone:** A chemical substance synthesized in small amounts by an endocrine tissue and carried in the blood to another tissue, where it acts as a messenger to regulate the function of the target tissue or organ.

**hormone receptor:** A protein in, or on the surface of, target cells that binds a specific hormone and initiates the cellular response.

**hormone response element (HRE):** A short (12 to 20 bp) DNA sequence to which receptors for steroid, retinoid, thyroid, and vitamin D hormones bind, altering the expression of the contiguous genes. For each hormone, there is a consensus sequence preferred by the cognate receptor.

**hyaluronic acid:** A high molecular weight, acidic polysaccharide typically composed of the alternating disaccharide GlcUA($\beta$→3) GlcNAc. Hyaluronic acid is a major component of the extracellular matrix, and forms larger complexes (proteoglycans) with proteins and other acidic polysaccharides.

**hydrogen bond:** A weak electrostatic attraction between one electronegative atom (such as oxygen or nitrogen) and a hydrogen atom covalently linked to a second electronegative atom.

**hydrolases:** Enzymes (proteases, lipases, phosphatases, nucleases, for example) that catalyze hydrolysis reactions.

**hydrolysis:** Cleavage of a bond, such as an anhydride or peptide bond, by the addition of the elements of water, yielding two or more products.

**hydronium ion:** The hydrated hydrogen ion ($H_3O^+$).

**hydropathy index:** A scale that expresses the relative hydrophobic and hydrophilic tendencies of a chemical group.

**hydrophilic:** Polar or charged; describes molecules or groups that associate with (dissolve easily in) water.

**hydrophobic:** Nonpolar; describes molecules or groups that are insoluble in water.

**hydrophobic interactions:** The association of nonpolar groups, or compounds, with each other in aqueous systems, driven by the tendency of the surrounding water molecules to seek their most stable (disordered) state.

**hyperchromic effect:** The large increase in light absorption at 260 nm occurring as a double-helical DNA is melted (unwound).

**hypoxia:** The metabolic condition in which the supply of oxygen is severely limited.

## i

**immune response:** The capacity of a vertebrate to generate antibodies to an antigen, a macromolecule foreign to the organism.

**immunoglobulin:** An antibody protein generated against, and capable of binding specifically to, an antigen.

**in vitro:** "In glass"; that is, in the test tube.

**in vivo:** "In life"; that is, in the living cell or organism.

**induced fit:** A change in the conformation of an enzyme in response to substrate binding that renders the enzyme catalytically active; also used to denote changes in the conformation of any macromolecule in response to ligand binding such that the binding site of the macromolecule better conforms to the shape of the ligand.

**inducer:** A signal molecule that, when bound to a regulatory protein, produces an increase in the expression of a given gene.

**induction:** An increase in the expression of a gene in response to a change in the activity of a regulatory protein.

**informational macromolecules:** Biomolecules containing information in the form of specific sequences of different monomers; for example, many proteins, lipids, polysaccharides, and nucleic acids.

**inhibitory G protein (G$_i$):** A trimeric GTP-binding protein that, when activated by an associated plasma membrane receptor, acts to *inhibit* a neighboring membrane enzyme such as adenylyl cyclase. G$_i$ has effects opposite those of G$_s$.

**initiation codon:** AUG (sometimes GUG in prokaryotes); codes for the first amino acid in a polypeptide sequence: *N*-formylmethionine in prokaryotes, and methionine in eukaryotes.

**initiation complex:** A complex of a ribosome with an mRNA and the initiating Met-tRNA$^{Met}$ or fMet-tRNA$^{fMet}$, ready for the elongation steps.

**inorganic pyrophosphatase:** An enzyme that hydrolyzes a molecule of inorganic pyrophosphate to yield two molecules of (ortho) phosphate; also known as pyrophosphatase.

**insertion mutation:** A mutation caused by insertion of one or more extra bases, or a mutagen, between successive bases in DNA.

**insertion sequence:** Specific base sequences at either end of a transposable segment of DNA.

**integral proteins:** Proteins firmly bound to a membrane by hydrophobic interactions; as distinct from peripheral proteins.

**integrin:** One of a large family of heterodimeric transmembrane proteins that mediate adhesion of cells to other cells or to the extracellular matrix.

**intercalating mutagen:** A mutagen that inserts itself between successive bases in a nucleic acid, causing a frame-shift mutation.

**intercalation:** Insertion between stacked aromatic or planar rings; for example, the insertion of a planar molecule between two successive bases in a nucleic acid.

**interferons:** A class of glycoproteins with antiviral activities.

**intermediary metabolism:** In cells, the enzyme-catalyzed reactions that extract chemical energy from nutrient molecules and utilize it to synthesize and assemble cell components.

**intron (intervening sequence):** A sequence of nucleotides in a gene that is transcribed but excised before the gene is translated.

**ion channel:** An integral protein that provides for the regulated transport of a specific ion, or ions, across a membrane.

**ion-exchange resin:** A polymeric resin that contains fixed charged groups; used in chromatographic columns to separate ionic compounds.

**ion product of water ($K_W$):** The product of the concentrations of $H^+$ and $OH^-$ in pure water: $K_w = [H^+][OH^-] = 1 \times 10^{-14}$ at 25 °C.

**ionizing radiation:** A type of radiation, such as x rays, that causes loss of electrons from some organic molecules, thus making them more reactive.

**ionophore:** A compound that binds one or more metal ions and is capable of diffusing across a membrane, carrying the bound ion.

**iron-sulfur center:** A prosthetic group of certain redox proteins involved in electron transfers; $Fe^{2+}$ or $Fe^{3+}$ is bound to inorganic sulfur and to Cys groups in the protein.

**iron-sulfur protein:** One of a large family of electron-transfer proteins in which the electron carrier is one or more iron ions associated with two or more sulfur atoms from cysteine residues or from inorganic sulfide ion.

**isoelectric focusing:** An electrophoretic method for separating macromolecules on the basis of their isoelectric pH.

**isoelectric pH (isoelectric point):** The pH at which a solute has no net electric charge and thus does not move in an electric field.

**isoenzymes:** *See* isozymes.

**isomerases:** Enzymes that catalyze the transformation of compounds into their positional isomers.

**isomers:** Any two molecules with the same molecular formula but a different arrangement of molecular groups.

**isoprene:** The hydrocarbon 2-methyl-1, 3-butadiene, a recurring structural unit of the terpenoid biomolecules.

**isoprenoid:** Any of a large number of natural products synthesized by enzymatic polymerization of two or more isoprene units; also called terpenoids.

**isothermal:** Occurring at constant temperature.

**isotopes:** Stable or radioactive forms of an element that differ in atomic weight but are otherwise chemically identical to the naturally abundant form of the element; used as tracers.

**isozymes:** Multiple forms of an enzyme that catalyze the same reaction but differ from each other in their amino acid sequence, substrate affinity, $V_{max}$, and/or regulatory properties; also called isoenzymes.

## k

**keratins:** Insoluble protective or structural proteins consisting of parallel polypeptide chains in $\alpha$-helical or $\beta$ conformations.

**ketogenic amino acids:** Amino acids with carbon skeletons that can serve as precursors of the ketone bodies.

**ketone bodies:** Acetoacetate, D-$\beta$-hydroxybutyrate, and acetone; water-soluble fuels normally exported by the liver but overproduced during fasting or in untreated diabetes mellitus.

**ketose:** A simple monosaccharide in which the carbonyl group is a ketone.

**ketosis:** A condition in which the concentration of ketone bodies in the blood, tissues, and urine is abnormally high.

**kinases:** Enzymes that catalyze the phosphorylation of certain molecules by ATP.

**kinetics:** The study of reaction rates.

**Krebs cycle:** *See* citric acid cycle.

## l

**lagging strand:** The DNA strand that, during replication, must be synthesized in the direction opposite to that in which the replication fork moves.

**law of mass action:** The law stating that the rate of any given chemical reaction is proportional to the product of the activities (or concentrations) of the reactants.

**leader:** A short sequence near the amino terminus of a protein or the 5′ end of an RNA that has a specialized targeting or regulatory function.

**leading strand:** The DNA strand that, during replication, is synthesized in the same direction in which the replication fork moves.

**leaky mutant:** A mutant gene that gives rise to a product with a detectable level of biological activity.

**leaving group:** The departing or displaced molecular group in a unimolecular elimination or a bimolecular substitution reaction.

**lectin:** A protein that binds a carbohydrate, commonly an oligosaccharide, with very high affinity and specificity, mediating cell-cell interactions.

**lethal mutation:** A mutation that inactivates a biological function essential to the life of the cell or organism.

**leucine zipper:** A protein structural motif involved in protein-protein interactions in many eukaryotic regulatory proteins; consists of two interacting $\alpha$ helices in which Leu residues in every seventh position are a prominent feature of the interacting surfaces.

**leukocyte:** White blood cell; involved in the immune response in mammals.

**leukotrienes:** A family of molecules derived from arachidonate; muscle contractants that constrict air passages in the lungs and are involved in asthma.

**levorotatory isomer:** A stereoisomer that rotates the plane of plane-polarized light counterclockwise.

**ligand:** A small molecule that binds specifically to a larger one; for example, a hormone is the ligand for its specific protein receptor.

**ligases:** Enzymes that catalyze condensation reactions in which two atoms are joined

using the energy of ATP or another energy-rich compound.

**light-dependent reactions:** The reactions of photosynthesis that require light and cannot occur in the dark; also known as the light reactions.

**Lineweaver-Burk equation:** An algebraic transform of the Michaelis-Menten equation, allowing determination of $V_{max}$ and $K_m$ by extrapolation of [S] to infinity.

**linking number:** The number of times one closed circular DNA strand is wound about another; the number of topological links holding the circles together.

**lipases:** Enzymes that catalyze the hydrolysis of triacylglycerols.

**lipid:** A small water-insoluble biomolecule generally containing fatty acids, sterols, or isoprenoid compounds.

**lipoate (lipoic acid):** A vitamin for some microorganisms; an intermediate carrier of hydrogen atoms and acyl groups in $\alpha$-keto acid dehydrogenases.

**lipoprotein:** A lipid-protein aggregate that serves to carry water-insoluble lipids in the blood. The protein component alone is an apolipoprotein.

**liposome:** A small, spherical vesicle composed of a phospholipid bilayer, which forms spontaneously when phospholipids are suspended in an aqueous buffer.

**low-energy phosphate compound:** A phosphorylated compound with a relatively small standard free energy of hydrolysis.

**lyases:** Enzymes that catalyze the removal of a group from a molecule to form a double bond, or the addition of a group to a double bond.

**lymphocytes:** A subclass of leukocytes involved in the immune response. B lymphocytes synthesize and secrete antibodies; T lymphocytes either play a regulatory role in immunity or kill foreign and virus-infected cells.

**lysis:** Destruction of a cell's plasma membrane or of a bacterial cell wall, releasing the cellular contents and killing the cell.

**lysosome:** A membrane-bounded organelle in the cytoplasm of eukaryotic cells; it contains many hydrolytic enzymes and serves as a degrading and recycling center for unneeded components.

## m

**macromolecule:** A molecule having a molecular weight in the range of a few thousand to many millions.

**mass-action ratio ($Q$):** For the reaction $aA + bB \rightleftharpoons cC + dD$, the ratio: $\dfrac{[C]^c[D]^d}{[A]^a[B]^b}$ .

**matrix:** The aqueous contents of a cell or organelle (the mitochondrion, for example) with dissolved solutes.

**meiosis:** A type of cell division in which diploid cells give rise to haploid cells destined to become gametes.

**membrane potential ($V_m$):** The difference in electrical potential across a biological membrane, commonly measured by the insertion of a microelectrode. Typical membrane potentials vary from $-25$ mV (by convention, the negative sign indicates that the inside is negative relative to the outside) to greater than $-100$ mV across some plant vacuole membranes.

**membrane transport:** Movement of a polar solute across a membrane via a specific membrane protein (a transporter).

**messenger RNA (mRNA):** A class of RNA molecules, each of which is complementary to one strand of DNA; carries the genetic message from the chromosome to the ribosomes.

**metabolic control:** The mechanisms by which the flux through a metabolic pathway is changed to reflect a cell's altered circumstances.

**metabolic regulation:** The mechanisms by which a cell resists changes in the concentration of individual metabolites that would otherwise occur when metabolic control mechanisms alter the flux through a pathway.

**metabolism:** The entire set of enzyme-catalyzed transformations of organic molecules in living cells; the sum of anabolism and catabolism.

**metabolite:** A chemical intermediate in the enzyme-catalyzed reactions of metabolism.

**metabolon:** A supramolecular assembly of sequential metabolic enzymes.

**metalloprotein:** A protein having a metal ion as its prosthetic group.

**metamerism:** Division of the body into segments, as in insects, for example.

**micelle:** An aggregate of amphipathic molecules in water, with the nonpolar portions in the interior and the polar portions at the exterior surface, exposed to water.

**Michaelis constant ($K_m$):** The substrate concentration at which an enzyme-catalyzed reaction proceeds at one-half its maximum velocity.

**Michaelis-Menten equation:** The equation describing the hyperbolic dependence of the initial reaction velocity, $V_0$, on substrate concentration, [S], in many enzyme-catalyzed reactions: $V_0 = \dfrac{V_{max}[S]}{K_m + [S]}$ .

**Michaelis-Menten kinetics:** A kinetic pattern in which the initial rate of an enzyme-catalyzed reaction exhibits a hyperbolic dependence on substrate concentration.

**microbodies:** Cytoplasmic, membrane-bounded vesicles containing peroxide-forming and peroxide-destroying enzymes; include lysosomes, peroxisomes, and glyoxysomes.

**microfilaments:** Thin filaments composed of actin, found in the cytoplasm of eukaryotic cells; serve in structure and movement.

**microsomes:** Membranous vesicles formed by fragmentation of the endoplasmic reticulum of eukaryotic cells; recovered by differential centrifugation.

**microtubules:** Thin tubules assembled from two types of globular tubulin subunits; present in cilia, flagella, centrosomes, and other contractile or motile structures.

**mismatch:** A base pair in a nucleic acid that cannot form normal Watson-Crick pairs.

**mismatch repair:** An enzymatic system for repairing base mismatches in DNA.

**mitochondrion:** Membrane-bounded organelle in the cytoplasm of eukaryotes; contains the enzyme systems required for the citric acid cycle, fatty acid oxidation, electron transfer, and oxidative phosphorylation.

**mitosis:** The multistep process in eukaryotic cells that results in the replication of chromosomes and cell division.

**mixed-function oxidases:** Enzymes (a monooxygenase, for example) that catalyze reactions in which two reductants, one of which is generally NADPH, the other the substrate, are oxidized. One oxygen atom is incorporated into the product, the other is reduced to $H_2O$. These enzymes often employ cytochrome P-450 to carry electrons from NADPH to $O_2$.

**mixed inhibition:** The reversible inhibition pattern resulting when an inhibitor molecule can bind to either the free enzyme or to the enzyme-substrate complex (not necessarily with the same affinity).

**modulator:** A metabolite that, when bound to the allosteric site of an enzyme, alters its kinetic characteristics.

**molar solution:** One mole of solute dissolved in water to give a total volume of 1,000 mL.

**mole:** One gram molecular weight of a compound. *See* Avogadro's number.

**monocistronic mRNA:** An mRNA that can be translated into only one protein.

**monoclonal antibodies:** Antibodies produced by a cloned hybridoma cell, which therefore are identical and directed against the same epitope of the antigen.

**monolayer:** A single layer of oriented lipid molecules.

**monoprotic acid:** An acid having only one dissociable proton.

**monosaccharide:** A carbohydrate consisting of a single sugar unit.

**motif:** A distinct folding pattern for elements of secondary structure, observed in one or more proteins; also called a fold or supersecondary structure.

**mRNA:** *See* messenger RNA.

**mucopolysaccharide:** An older name for a glycosaminoglycan.

**multienzyme system:** A group of related enzymes participating in a given metabolic pathway.

**mutarotation:** The change in specific rotation of a pyranose or furanose sugar or glycoside accompanying the equilibrium of its $\alpha$- and $\beta$-anomeric forms.

**mutases:** Enzymes that catalyze the transposition of functional groups.

**mutation:** An inheritable change in the nucleotide sequence of a chromosome.

**myocyte:** A muscle cell.

**myofibril:** A unit of thick and thin filaments of muscle fibers.

**myosin:** A contractile protein; the major component of the thick filaments of muscle and other actin-myosin systems.

# n

**NAD, NADP (nicotinamide adenine dinucleotide, nicotinamide adenine dinucleotide phosphate):** Nicotinamide-containing coenzymes functioning as carriers of hydrogen atoms and electrons in some oxidation-reduction reactions.

**native conformation:** The biologically active conformation of a macromolecule.

**negative cooperativity:** A phenomenon of some multisubunit enzymes or proteins in which binding of a ligand or substrate to one subunit impairs binding to another subunit.

**negative feedback:** Regulation of a biochemical pathway achieved when a reaction product inhibits an earlier step in the pathway.

**neuron:** A cell of nervous tissue specialized for transmission of a nerve impulse.

**neurotransmitter:** A low molecular weight compound (usually containing nitrogen) secreted from the terminal of a neuron and bound by a specific receptor in the next neuron; serves to transmit a nerve impulse.

**nicotinamide adenine dinucleotide, nicotinamide adenine dinucleotide phosphate:** *See* NAD, NADP.

**ninhydrin reaction:** A color reaction given by amino acids and peptides on heating with ninhydrin; widely used for their detection and estimation.

**nitrogen cycle:** The cycling of various forms of biologically available nitrogen through the plant, animal, and microbial worlds, and through the atmosphere and geosphere.

**nitrogen fixation:** Conversion of atmospheric nitrogen ($N_2$) into a reduced, biologically available form by nitrogen-fixing organisms.

**nitrogenase complex:** A system of enzymes capable of reducing atmospheric nitrogen to ammonia in the presence of ATP.

**noncyclic electron flow:** The light-induced flow of electrons from water to $NADP^+$ in oxygen-evolving photosynthesis; it involves both photosystems I and II.

**nonessential amino acids:** Amino acids that can be made by humans and other vertebrates from simpler precursors, and are thus not required in the diet.

**nonheme iron proteins:** Proteins, usually acting in oxidation-reduction reactions, containing iron but no porphyrin groups.

**nonpolar:** Hydrophobic; describes molecules or groups that are poorly soluble in water.

**nonsense codon:** A codon that does not specify an amino acid, but signals the termination of a polypeptide chain.

**nonsense mutation:** A mutation that results in the premature termination of a polypeptide chain.

**nonsense suppressor:** A mutation, usually in the gene for a tRNA, that causes an amino acid to be inserted into a polypeptide in response to a termination codon.

**nuclear magnetic resonance (NMR) spectroscopy:** A technique that utilizes certain quantum mechanical properties of atomic nuclei to study the structure and dynamics of the molecules of which they are a part.

**nucleases:** Enzymes that hydrolyze the internucleotide (phosphodiester) linkages of nucleic acids.

**nucleic acids:** Biologically occurring polynucleotides in which the nucleotide residues are linked in a specific sequence by phosphodiester bonds; DNA and RNA.

**nucleoid:** In bacteria, the nuclear zone that contains the chromosome but has no surrounding membrane.

**nucleolus:** A densely staining structure in the nucleus of eukaryotic cells; involved in rRNA synthesis and ribosome formation.

**nucleophile:** An electron-rich group with a strong tendency to donate electrons to an electron-deficient nucleus (electrophile); the entering reactant in a bimolecular substitution reaction.

**nucleoplasm:** The portion of a cell's contents enclosed by the nuclear membrane; also called the nuclear matrix.

**nucleoside:** A compound consisting of a purine or pyrimidine base covalently linked to a pentose.

**nucleoside diphosphate kinase:** An enzyme that catalyzes the transfer of the terminal phosphate of a nucleoside 5'-triphosphate to a nucleoside 5'-diphosphate.

**nucleoside diphosphate sugar:** A coenzymelike carrier of a sugar molecule, functioning in the enzymatic synthesis of polysaccharides and sugar derivatives.

**nucleoside monophosphate kinase:** An enzyme that catalyzes the transfer of the terminal phosphate of ATP to a nucleoside 5'-monophosphate.

**nucleosome:** Structural unit for packaging chromatin; consists of a DNA strand wound around a histone core.

**nucleotide:** A nucleoside phosphorylated at one of its pentose hydroxyl groups.

**nucleus:** In eukaryotes, a membrane-bounded organelle that contains chromosomes.

# o

**oligomer:** A short polymer, usually of amino acids, sugars, or nucleotides; the definition of "short" is somewhat arbitrary, but usually less than 50 subunits.

**oligomeric protein:** A multisubunit protein having two or more identical polypeptide chains.

**oligonucleotide:** A short polymer of nucleotides (usually less than 50).

**oligopeptide:** A few amino acids joined by peptide bonds.

**oligosaccharide:** Several monosaccharide groups joined by glycosidic bonds.

**ω oxidation:** An alternative mode of fatty acid oxidation in which the initial oxidation is at the carbon most distant from the carboxyl carbon.

**oncogene:** A cancer-causing gene; any of several mutant genes that cause cells to exhibit rapid, uncontrolled proliferation. *See also* proto-oncogene.

**open reading frame:** A group of contiguous nonoverlapping nucleotide codons in a DNA or RNA molecule that does not include a termination codon.

**open system:** A system that exchanges matter and energy with its surroundings. *See also* system.

**operator:** A region of DNA that interacts with a repressor protein to control the expression of a gene or group of genes.

**operon:** A unit of genetic expression consisting of one or more related genes and the operator and promoter sequences that regulate their transcription.

**optical activity:** The capacity of a substance to rotate the plane of plane-polarized light.

**optimum pH:** The characteristic pH at which an enzyme has maximal catalytic activity.

**organelles:** Membrane-bounded structures found in eukaryotic cells; contain enzymes and other components required for specialized cell functions.

**origin:** The nucleotide sequence or site in DNA where DNA replication is initiated.

**orthologs:** Genes in different organisms that possess a clear sequence and functional relationship to each other.

**osmosis:** Bulk flow of water through a semipermeable membrane into another aqueous compartment containing solute at a higher concentration.

**osmotic pressure:** Pressure generated by the osmotic flow of water through a semipermeable membrane into an aqueous compartment containing solute at a higher concentration.

**oxidases:** Enzymes that catalyze oxidation reactions in which molecular oxygen serves as the electron acceptor, but neither of the oxygen atoms is incorporated into the product. *Compare* oxygenases.

**oxidation:** The loss of electrons from a compound. *See also* α oxidation, β oxidation, and ω oxidation.

**oxidation-reduction reaction:** A reaction in which electrons are transferred from a donor to an acceptor molecule; also called a redox reaction.

**oxidative phosphorylation:** The enzymatic phosphorylation of ADP to ATP coupled to electron transfer from a substrate to molecular oxygen.

**oxidizing agent (oxidant):** The acceptor of electrons in an oxidation-reduction reaction.

**oxygen debt:** The extra oxygen (above the normal resting level) consumed in the recovery period after strenuous physical exertion.

**oxygenases:** Enzymes that catalyze reactions in which oxygen atoms are directly incorporated into the product, forming a hydroxyl or carboxyl group. In reactions catalyzed by a monooxygenase, only one of the two O atoms is incorporated; the other is reduced to $H_2O$; in reactions catalyzed by a dioxygenase, both O atoms are incorporated into the product. *Compare* oxidases.

# p

**palindrome:** A segment of duplex DNA in which the base sequences of the two strands exhibit twofold rotational symmetry about an axis.

**paradigm:** In biochemistry, an experimental model or example.

**partition coefficient:** A constant that expresses the ratio in which a given solute will be partitioned or distributed between two given immiscible liquids at equilibrium.

**pathogenic:** Disease-causing.

**pentose:** A simple sugar with a backbone containing five carbon atoms.

**pentose phosphate pathway:** A pathway that serves to interconvert hexoses and pentoses and is a source of reducing equivalents and pentoses for biosynthetic processes; present in most organisms. Also called the phosphogluconate pathway and the hexose monophosphate pathway.

**peptidases:** Enzymes that hydrolyze peptide bonds.

**peptide:** Two or more amino acids covalently joined by peptide bonds.

**peptide bond:** A substituted amide linkage between the α-amino group of one amino acid and the α-carboxyl group of another, with the elimination of the elements of water.

**peptide mapping:** The characteristic two-dimensional pattern (on paper or gel) formed by the separation of a mixture of peptides resulting from partial hydrolysis of a protein; also known as peptide fingerprinting.

**peptidoglycan:** A major component of bacterial cell walls; generally consists of parallel heteropolysaccharides cross-linked by short peptides.

**peptidyl transferase:** Enzymatic activity in which the peptide bonds of proteins are synthesized on ribosomes. The peptidyl transferase activity is part of the rRNA of the large ribosomal subunit, and is thus a ribozyme.

**peripheral proteins:** Proteins that are loosely or reversibly bound to a membrane by hydrogen bonds or electrostatic forces; generally water soluble once released from the membrane.

**permeases:** *See* transporters.

**peroxisome:** Membrane-bounded organelle in the cytoplasm of eukaryotic cells; contains peroxide-forming and peroxide-destroying enzymes.

**pH:** The negative logarithm of the hydrogen ion concentration of an aqueous solution.

**phage:** *See* bacteriophage.

**phenotype:** The observable characteristics of an organism.

**phosphatases:** Enzymes that hydrolyze a phosphate ester or anhydride, releasing inorganic phosphate, $P_i$.

**phosphodiester linkage:** A chemical grouping that contains two alcohols esterified to one molecule of phosphoric acid, which thus serves as a bridge between them.

**phosphogluconate pathway:** An oxidative pathway beginning with glucose 6-phosphate and leading, via 6-phosphogluconate, to pentose phosphates and yielding NADPH. Also called the pentose phosphate pathway and the hexose monophosphate pathway.

**phospholipid:** A lipid containing one or more phosphate groups.

**phosphorolysis:** Cleavage of a compound with phosphate as the attacking group; analogous to hydrolysis.

**phosphorylases:** Enzymes that catalyze phosphorolysis.

**phosphorylation:** Formation of a phosphate derivative of a biomolecule, usually by enzymatic transfer of a phosphoryl group from ATP.

**phosphorylation potential ($\Delta G_p$):** The actual free-energy change of ATP hydrolysis under the nonstandard conditions prevailing within a cell.

**photochemical reaction center:** The part of a photosynthetic complex where the energy of an absorbed photon causes charge separation, initiating electron transfer.

**photon:** The ultimate unit (a quantum) of light energy.

**photophosphorylation:** The enzymatic formation of ATP from ADP coupled to the light-dependent transfer of electrons in photosynthetic cells.

**photoreduction:** The light-induced reduction of an electron acceptor in photosynthetic cells.

**photorespiration:** Oxygen consumption occurring in illuminated temperate-zone plants, largely due to oxidation of phosphoglycolate.

**photosynthesis:** The use of light energy to produce carbohydrates from carbon dioxide and a reducing agent such as water.

**photosynthetic phosphorylation:** *See* photophosphorylation.

**photosystem:** In photosynthetic cells, a functional set of light-absorbing pigments and its reaction center.

**phototroph:** An organism that can use the energy of light to synthesize its own fuels from simple molecules such as carbon dioxide, oxygen, and water; as distinct from a chemotroph.

**p$K_a$:** The negative logarithm of an equilibrium constant.

**plasma membrane:** The exterior membrane surrounding the cytoplasm of a cell.

**plasma proteins:** The proteins present in blood plasma.

**plasmalogen:** A phospholipid with an alkenyl ether substituent on the C-1 of glycerol.

**plasmid:** An extrachromosomal, independently replicating, small circular DNA molecule; commonly employed in genetic engineering.

**plastid:** In plants, a self-replicating organelle; may differentiate into a chloroplast.

**platelets:** Small, enucleated cells that initiate blood clotting; they arise from cells called megakaryocytes in the bone marrow. Also known as thrombocytes.

**pleated sheet:** The side-by-side, hydrogen-bonded arrangement of polypeptide chains in the extended $\beta$ conformation.

**plectonemic:** Describes a structure in a molecular polymer in which there is a net twisting of strands about each other in some simple and regular way.

**polar:** Hydrophilic, or "water-loving"; describes molecules or groups that are soluble in water.

**polarity:** (1) In chemistry, the nonuniform distribution of electrons in a molecule; polar molecules are usually soluble in water. (2) In molecular biology, the distinction between the 5′ and 3′ ends of nucleic acids.

**poly(A) tail:** A length of adenosine residues added to the 3′ ends of many mRNAs in eukaryotes (and sometimes in bacteria).

**polycistronic mRNA:** A contiguous mRNA with more than two genes that can be translated into proteins.

**polyclonal antibodies:** A heterogeneous pool of antibodies produced in an animal by a number of different B lymphocytes in response to an antigen. Different antibodies in the pool recognize different parts of the antigen.

**polylinker:** A short, often synthetic, fragment of DNA containing recognition sequences for several restriction endonucleases.

**polymerase chain reaction (PCR):** A repetitive procedure that results in a geometric amplification of a specific DNA sequence.

**polymorphic:** Describing a protein for which amino acid sequence variants exist in a population of organisms, but the variations do not destroy the protein's function.

**polynucleotide:** A covalently linked sequence of nucleotides in which the 3′ hydroxyl of the pentose of one nucleotide residue is joined by a phosphodiester bond to the 5′ hydroxyl of the pentose of the next residue.

**polypeptide:** A long chain of amino acids linked by peptide bonds; the molecular weight is generally less than 10,000.

**polyribosome:** *See* polysome.

**polysaccharide:** A linear or branched polymer of monosaccharide units linked by glycosidic bonds.

**polysome:** A complex of an mRNA molecule and two or more ribosomes; also called polyribosome.

**P/O ratio:** The number of moles of ATP formed in oxidative phosphorylation per $\frac{1}{2}O_2$ reduced (thus, per pair of electrons passed to $O_2$). Experimental values used in this text are 2.5 for passage of electrons from NADH to $O_2$, and 1.5 for passage of electrons from FADH to $O_2$. Some textbooks use the integral values of 3.0 and 2.0.

**porphyria:** Genetic condition resulting from the lack of one or more enzymes required to synthesize porphyrins.

**porphyrin:** Complex nitrogenous compound containing four substituted pyrroles covalently joined into a ring; often complexed with a central metal atom.

**positive cooperativity:** A phenomenon of some multisubunit enzymes or proteins in which binding of a ligand or substrate to one subunit facilitates binding to another subunit.

**posttranscriptional processing:** The enzymatic processing of the primary RNA transcript, producing functional mRNA, tRNA, and/or rRNA molecules.

**posttranslational modification:** Enzymatic processing of a polypeptide chain after translation from its mRNA.

**primary structure:** A description of the covalent backbone of a polymer (macromolecule), including the sequence of monomeric subunits and any interchain and intrachain covalent bonds.

**primary transcript:** The immediate RNA product of transcription before any posttranscriptional processing reactions.

**primase:** An enzyme that catalyzes the formation of RNA oligonucleotides used as primers by DNA polymerases.

**primer:** A short oligomer (of sugars or nucleotides, for example) to which an enzyme adds additional monomeric subunits.

**primer terminus:** The end of the primer to which monomeric subunits are added.

**priming:** In protein phosphorylation, refers to the phosphorylation of an amino acid residue that becomes the binding site and point of reference for phosphorylation of other residues in the same protein.

**primosome:** An enzyme complex that synthesizes the primers required for lagging strand DNA synthesis.

**probe:** A labeled fragment of nucleic acid containing a nucleotide sequence complementary to a gene or genomic sequence that one wishes to detect in a hybridization experiment.

**processivity:** For any enzyme that catalyzes the synthesis of a biological polymer, the property of adding multiple subunits to the polymer without dissociating from the substrate.

**prochiral molecule:** A symmetric molecule that can react asymmetrically with an enzyme having an asymmetric active site, generating a chiral product.

**projection formulas:** A method for representing molecules to show the configuration of groups around chiral centers; also known as Fischer projection formulas.

**prokaryote:** A bacterium; a unicellular organism with a single chromosome, no nuclear envelope, and no membrane-bounded organelles.

**promoter:** A DNA sequence at which RNA polymerase may bind, leading to initiation of transcription.

**proofreading:** The correction of errors in the synthesis of an information-containing biopolymer by removing incorrect monomeric subunits after they have been covalently added to the growing polymer.

**prostaglandins:** A class of lipid-soluble, hormonelike regulatory molecules derived from arachidonate and other polyunsaturated fatty acids.

**prosthetic group:** A metal ion or an organic compound (other than an amino acid) that is covalently bound to a protein and is essential to its activity.

**proteasome:** Supramolecular assembly of enzymatic complexes that function in the degradation of damaged or unneeded cellular proteins.

**protein:** A macromolecule composed of one or more polypeptide chains, each with a characteristic sequence of amino acids linked by peptide bonds.

**protein kinases:** Enzymes that transfer the terminal phosphoryl group of ATP or another nucleoside triphosphate to a Ser, Thr, Tyr, Asp, or His side chain in a target protein, thereby regulating the activity or other properties of that protein.

**protein targeting:** The process by which newly synthesized proteins are sorted and transported to their proper locations in the cell.

**proteoglycan:** A hybrid macromolecule consisting of a heteropolysaccharide joined to a polypeptide; the polysaccharide is the major component.

**proteome:** The full complement of proteins expressed in a given cell, or the complete complement of proteins that can be expressed by a given genome.

**proteomics:** Broadly, the study of the protein complement of a cell or organism.

**protomer:** A general term describing any repeated unit of one or more stably associated protein subunits within a larger protein structure. If a protomer has multiple subunits, the subunits may be identical or different.

**proto-oncogene:** A cellular gene, usually encoding a regulatory protein, that can be converted into an oncogene by mutation.

**proton acceptor:** An anionic compound capable of accepting a proton from a proton donor; that is, a base.

**proton donor:** The donor of a proton in an acid-base reaction; that is, an acid.

**proton-motive force:** The electrochemical potential inherent in a transmembrane gradient of $H^+$ concentration; used in oxidative phosphorylation and photophosphorylation to drive ATP synthesis.

**protoplasm:** A general term referring to the entire contents of a living cell.

**purine:** A nitrogenous heterocyclic base found in nucleotides and nucleic acids; containing fused pyrimidine and imidazole rings.

**puromycin:** An antibiotic that inhibits polypeptide synthesis by being incorporated into a growing polypeptide chain, causing its premature termination.

**pyranose:** A simple sugar containing the six-membered pyran ring.

**pyridine nucleotide:** A nucleotide coenzyme containing the pyridine derivative nicotinamide; NAD or NADP.

**pyridoxal phosphate:** A coenzyme containing the vitamin pyridoxine (vitamin $B_6$); functions in reactions involving amino group transfer.

**pyrimidine:** A nitrogenous heterocyclic base found in nucleotides and nucleic acids.

**pyrimidine dimer:** A covalently joined dimer of two adjacent pyrimidine residues in DNA, induced by absorption of UV light; most commonly derived from two adjacent thymines (a thymine dimer).

**pyrophosphatase:** *See* inorganic pyrophosphatase.

## q

**quantum:** The ultimate unit of energy.

**quaternary structure:** The three-dimensional structure of a multisubunit protein; particularly the manner in which the subunits fit together.

## r

**R group:** (1) Formally, an abbreviation denoting any alkyl group. (2) Occasionally, used in a more general sense to denote virtually any organic substituent (the R groups of amino acids, for example).

**racemic mixture (racemate):** An equimolar mixture of the D and L stereoisomers of an optically active compound.

**radical:** An atom or group of atoms possessing an unpaired electron; also called a free radical.

**radioactive isotope:** An isotopic form of an element with an unstable nucleus that stabilizes itself by emitting ionizing radiation.

**radioimmunoassay (RIA):** A sensitive and quantitative method for detecting trace amounts of a biomolecule, based on its capacity to displace a radioactive form of the molecule from combination with its specific antibody.

**rate constant:** The proportionality constant that relates the velocity of a chemical reaction to the concentration(s) of the reactant(s).

**rate-limiting step:** (1) Generally, the step in an enzymatic reaction with the greatest activation energy or the transition state of highest free energy. (2) The slowest step in a metabolic pathway.

**reaction intermediate:** Any chemical species in a reaction pathway that has a finite chemical lifetime.

**reading frame:** A contiguous and nonoverlapping set of three-nucleotide codons in DNA or RNA.

**recombinant DNA:** DNA formed by the joining of genes into new combinations.

**recombination:** Any enzymatic process by which the linear arrangement of nucleic acid sequences in a chromosome is altered by cleavage and rejoining.

**recombinational DNA repair:** recombinational processes that are directed at the repair of DNA strand breaks or cross-links, especially at inactivated replication forks.

**redox pair:** An electron donor and its corresponding oxidized form; for example, NADH and $NAD^+$.

**redox reaction:** *See* oxidation-reduction reaction.

**reducing agent (reductant):** The electron donor in an oxidation-reduction reaction.

**reducing end:** The end of a polysaccharide having a terminal sugar with a free anomeric carbon; the terminal residue can act as a reducing sugar.

**reducing equivalent:** A general or neutral term for an electron or an electron equivalent in the form of a hydrogen atom or a hydride ion.

**reducing sugar:** A sugar in which the carbonyl (anomeric) carbon is not involved in a glycosidic bond and can therefore undergo oxidation.

**reduction:** The gain of electrons by a compound or ion.

**regulatory enzyme:** An enzyme having a regulatory function through its capacity to undergo a change in catalytic activity by allosteric mechanisms or by covalent modification.

**regulatory gene:** A gene that gives rise to a product involved in the regulation of the expression of another gene; for example, a gene coding for a repressor protein.

**regulatory sequence:** A DNA sequence involved in regulating the expression of a gene; for example, a promoter or operator.

**regulon:** A group of genes or operons that are coordinately regulated even though some, or all, may be spatially distant within the chromosome or genome.

**relaxed DNA:** Any DNA that exists in its most stable and unstrained structure, typically B form under most cellular conditions.

**release factors:** Protein factors of the cytosol required in releasing a completed polypeptide chain from a ribosome; also known as termination factors.

**releasing factors:** Hypothalamic hormones that stimulate release of other hormones by the pituitary gland.

**renaturation:** Refolding of an unfolded (denatured) globular protein so as to restore native structure and protein function.

**replication:** Synthesis of daughter nucleic acid molecules identical to the parental nucleic acids.

**replication fork:** The Y-shaped structure generally found at the point where DNA is being synthesized.

**replicative form:** Any of the full-length structural forms of a viral chromosome that serve as distinct replication intermediates.

**replisome:** The multiprotein complex that promotes DNA synthesis at the replication fork.

**repressible enzyme:** In bacteria, an enzyme whose synthesis is inhibited when its reaction product is readily available to the cell.

**repression:** A decrease in the expression of a gene in response to a change in the activity of a regulatory protein.

**repressor:** The protein that binds to the regulatory sequence or operator for a gene, blocking its transcription.

**residue:** A single unit within a polymer; for example, an amino acid within a polypeptide chain. The term reflects the fact that sugars, nucleotides, and amino acids lose a few atoms (generally the elements of water) when incorporated in their respective polymers.

**respiration:** Any metabolic process that leads to the uptake of oxygen and the release of $CO_2$.

**respiration-linked phosphorylation:** ATP formation from ADP and $P_i$, driven by electron flow through a series of membrane-bound carriers, with a proton gradient as the direct source of energy driving rotational catalysis by ATP synthase.

**respiratory chain:** The electron-transfer chain; a sequence of electron-carrying proteins that transfers electrons from substrates to molecular oxygen in aerobic cells.

**restriction endonucleases:** Site-specific endodeoxyribonucleases causing cleavage of both strands of DNA at points within or near the specific site recognized by the enzyme; important tools in genetic engineering.

**restriction fragment:** A segment of double-stranded DNA produced by the action of a restriction endonuclease on a larger DNA.

**restriction fragment length polymorphisms (RFLPs):** Variations, among individuals in a population, in the length of certain restriction fragments within which certain genomic sequences occur. These variations result from rare sequence changes that create or destroy restriction sites in the genome.

**retrovirus:** An RNA virus containing a reverse transcriptase.

**reverse transcriptase:** An RNA-directed DNA polymerase in retroviruses; capable of making DNA complementary to an RNA.

**reversible inhibition:** Inhibition by a molecule that binds reversibly to the enzyme, such that the enzyme activity returns when the inhibitor is no longer present.

**ribonuclease:** A nuclease that catalyzes the hydrolysis of certain internucleotide linkages of RNA.

**ribonucleic acid:** *See* RNA.

**ribonucleotide:** A nucleotide containing D-ribose as its pentose component.

**ribosomal RNA (rRNA):** A class of RNA molecules serving as components of ribosomes.

**ribosome:** A supramolecular complex of rRNAs and proteins, approximately 18 to 22 nm in diameter; the site of protein synthesis.

**ribozymes:** Ribonucleic acid molecules with catalytic activities; RNA enzymes.

**Rieske iron-sulfur protein:** A type of iron-sulfur protein in which two of the ligands to the central iron ion are His side chains. These proteins act in many electron-transfer sequences, including oxidative phosphorylation and photophosphorylation.

**RNA (ribonucleic acid):** A polyribonucleotide of a specific sequence linked by successive 3′,5′-phosphodiester bonds.

**RNA polymerase:** An enzyme that catalyzes the formation of RNA from ribonucleoside 5′-triphosphates, using a strand of DNA or RNA as a template.

**RNA splicing:** Removal of introns and joining of exons in a primary transcript.

**rRNA:** *See* ribosomal RNA.

### S

**S-adenosylmethionine (adoMet):** An enzymatic cofactor involved in methyl group transfers.

**salvage pathway:** Synthesis of a biomolecule, such as a nucleotide, from intermediates in the degradative pathway for the biomolecule; a recycling pathway, as distinct from a de novo pathway.

**saponification:** Alkaline hydrolysis of triacylglycerols to yield fatty acids as soaps.

**sarcomere:** A functional and structural unit of the muscle contractile system.

**satellite DNA:** Highly repeated, nontranslated segments of DNA in eukaryotic chromosomes; most often associated with the centromeric region. Its function is not clear.

**saturated fatty acid:** A fatty acid containing a fully saturated alkyl chain.

**second law of thermodynamics:** The law stating that in any chemical or physical process, the entropy of the universe tends to increase.

**second messenger:** An effector molecule synthesized within a cell in response to an external signal (first messenger) such as a hormone.

**secondary metabolism:** Pathways that lead to specialized products not found in every living cell.

**secondary structure:** The residue-by-residue conformation of the backbone of a polymer.

**sedimentation coefficient:** A physical constant specifying the rate of sedimentation of a particle in a centrifugal field under specified conditions.

**selectins:** A large family of membrane proteins, lectins that bind oligosaccharides on other cells tightly and specifically, and serve to carry signals across the plasma membrane.

**SELEX:** A method for rapid experimental identification of nucleic acid sequences (usually RNA) that have particular catalytic or ligand-binding properties.

**sequence-tagged site (STS):** Any known sequence that has been mapped within a chromosome and/or clones derived from it.

**serpentine receptors:** A large family of membrane receptor proteins with seven transmembrane helical segments. These receptors often associate with G proteins to transduce an extracellular signal into a change in cellular metabolism. Also called G protein–coupled receptors (GPCR) and 7 transmembrane segment (7tm) receptors.

**7 transmembrane segment (7tm) receptors:** *See* serpentine receptors.

**Shine-Dalgarno sequence:** A sequence in an mRNA required for binding prokaryotic ribosomes.

**SH2 domain:** A protein domain that binds tightly to a phosphotyrosine residue in certain proteins such as the receptor tyrosine kinases, initiating the formation of a multiprotein complex that acts in a signaling pathway.

**shuttle vector:** A recombinant DNA vector that can be replicated in two or more different host species. *See also* vector.

**sickle-cell anemia:** A human disease characterized by defective hemoglobin molecules; caused by a homozygous allele coding for the $\beta$ chain of hemoglobin.

**sickle-cell trait:** A human condition recognized by the sickling of erythrocytes when exposed to low oxygen tension; occurs in individuals heterozygous for the allele responsible for sickle-cell anemia.

**signal sequence:** An amino acid sequence, often at the amino terminus, that signals the cellular fate or destination of a newly synthesized protein.

**signal transduction:** The process by which an extracellular signal (chemical, mechanical, or electrical) is amplified and converted to a cellular response.

**silent mutation:** A mutation in a gene that causes no detectable change in the biological characteristics of the gene product.

**simple diffusion:** The movement of solute molecules across a membrane to a region of lower concentration, unassisted by a protein transporter.

**simple protein:** A protein yielding only amino acids on hydrolysis.

**single nucleotide polymorphisms (SNPs):** Single genomic base pair changes that distinguish one individual from another within a given species.

**site-directed mutagenesis:** A set of methods used to create specific alterations in the sequence of a gene.

**site-specific recombination:** A type of genetic recombination that occurs only at specific sequences.

**size-exclusion chromatography:** A procedure for the separation of a mixture of molecules on the basis of size, based on the capacity of porous polymers to exclude solutes above a certain size. Also called gel filtration.

**small nuclear RNA (snRNA):** Any of several small RNA molecules in the nucleus; most have a role in the splicing reactions that remove introns from mRNA, tRNA, and rRNA molecules.

**somatic cells:** All body cells except the germ-line cells.

**SOS response:** In bacteria, a coordinated induction of a variety of genes as a response to high levels of DNA damage.

**Southern blot:** A DNA hybridization procedure in which one or more specific DNA fragments are detected in a larger population by means of hybridization to a complementary, labeled nucleic acid probe.

**specific acid-base catalysis:** Acid or base catalysis involving the constituents of water (hydroxide or hydronium ions).

**specific activity:** The number of micromoles ($\mu$mol) of a substrate transformed by an enzyme preparation per minute per milligram of protein at 25 °C; a measure of enzyme purity.

**specific heat:** The amount of energy (in joules or calories) needed to raise the temperature of 1 g of a pure substance by 1 °C.

**specific rotation:** The rotation, in degrees, of the plane of plane-polarized light (D-line of sodium) by an optically active compound at 25 °C, with a specified concentration and light path.

**specificity:** The ability of an enzyme or receptor to discriminate among competing substrates or ligands.

**sphingolipid:** An amphipathic lipid with a sphingosine backbone to which are attached a long-chain fatty acid and a polar alcohol.

**spliceosome:** A complex of RNAs and proteins involved in the splicing of mRNAs in eukaryotic cells.

**splicing:** *See* gene splicing, RNA splicing.

**standard free-energy change ($\Delta G°$):** The free-energy change for a reaction occurring under a set of standard conditions: temperature, 298 K; pressure, 1 atm or 101.3 kPa; and all solutes at 1 M concentration. $\Delta G'°$ denotes the standard free-energy change at pH 7.0 in 55.5 M water.

**standard reduction potential ($E'°$):** The electromotive force exhibited at an electrode by 1 M concentrations of a reducing agent and its oxidized form at 25 °C and pH 7.0; a measure of the relative tendency of the reducing agent to lose electrons.

**steady state:** A nonequilibrium state of a system through which matter is flowing and in which all components remain at a constant concentration.

**stem cells:** The common, self-regenerating cells in bone marrow that give rise to

differentiated blood cells such as erythrocytes and lymphocytes.

**stereoisomers:** Compounds that have the same composition and the same order of atomic connections but different molecular arrangements.

**sterols:** A class of lipids containing the steroid nucleus.

**sticky ends:** Two DNA ends in the same DNA molecule, or in different molecules, with short overhanging single-stranded segments that are complementary to one another, facilitating ligation of the ends; also known as cohesive ends.

**stimulatory G protein ($G_s$):** A trimeric regulatory GTP-binding protein that, when activated by an associated plasma membrane receptor, stimulates a neighboring membrane enzyme such as adenylyl cyclase. $G_s$ has effects opposite those of $G_i$.

**stop codons:** *See* termination codons.

**stroma:** The space and aqueous solution enclosed within the inner membrane of a chloroplast, not including the contents within the thylakoid membranes.

**structural gene:** A gene coding for a protein or RNA molecule; as distinct from a regulatory gene.

**substitution mutation:** A mutation caused by the replacement of one base by another.

**substrate:** The specific compound acted upon by an enzyme.

**substrate channeling:** Movement of the chemical intermediates in a series of enzyme-catalyzed reactions from the active site of one enzyme to that of the next enzyme in the pathway without leaving the surface of a protein complex that includes both enzymes.

**substrate-level phosphorylation:** Phosphorylation of ADP or some other nucleoside 5'-diphosphate coupled to the dehydrogenation of an organic substrate; independent of the electron-transfer chain.

**suicide inactivator:** A relatively inert molecule that is transformed by an enzyme, at its active site, into a reactive substance that irreversibly inactivates the enzyme.

**supercoil:** The twisting of a helical (coiled) molecule on itself; a coiled coil.

**supercoiled DNA:** DNA that twists upon itself because it is under- or overwound (and thereby strained) relative to B-form DNA.

**superhelical density:** In a helical molecule such as DNA, the number of supercoils (superhelical turns) relative to the number of coils (turns) in the relaxed molecule.

**supersecondary structure:** A distinct folding pattern for elements of secondary structure, observed in one or more proteins; also called a motif or fold.

**suppressor mutation:** A mutation that totally or partially restores a function lost by a primary mutation; located at a site different from the site of the primary mutation.

**Svedberg (S):** A unit of measure of the rate at which a particle sediments in a centrifugal field.

**symbionts:** Two or more organisms that are mutually interdependent; usually living in physical association.

**symport:** Cotransport of solutes across a membrane in the same direction.

**synteny:** Conserved gene order along the chromosomes of different species.

**synthases:** Enzymes that catalyze condensation reactions in which no nucleoside triphosphate is required as an energy source.

**synthetases:** Enzymes that catalyze condensation reactions using ATP or another nucleoside triphosphate as an energy source.

**system:** An isolated collection of matter; all other matter in the universe apart from the system is called the surroundings.

 **t**

**telomere:** Specialized nucleic acid structure found at the ends of linear eukaryotic chromosomes.

**template:** A macromolecular mold or pattern for the synthesis of an informational macromolecule.

**template strand:** A strand of nucleic acid used by a polymerase as a template to synthesize a complementary strand.

**terminal transferase:** An enzyme that catalyzes the addition of nucleotide residues of a single kind to the 3' end of DNA chains.

**termination codons:** UAA, UAG, and UGA; in protein synthesis, these codons signal the termination of a polypeptide chain. Also known as stop codons.

**termination factors:** *See* release factors.

**termination sequence:** A DNA sequence that appears at the end of a transcriptional unit and signals the end of transcription.

**terpenes:** Organic hydrocarbons or hydrocarbon derivatives constructed from recurring isoprene units. They produce some of the scents and tastes of plant products.

**tertiary structure:** The three-dimensional conformation of a polymer in its native folded state.

**tetrahydrobiopterin:** The reduced coenzyme form of biopterin.

**tetrahydrofolate:** The reduced, active coenzyme form of the vitamin folate.

**thermogenesis:** The biological generation of heat by muscle activity (shivering), uncoupled oxidative phosphorylation, or the operation of futile cycles.

**thermogenin:** A protein of the inner mitochondrial membrane of brown fat mitochondria that allows transmembrane movement of protons, short-circuiting the normal use of protons to drive ATP synthesis, and dissipating the energy of substrate oxidation as heat. Also called uncoupling protein (UCP).

**thiamine pyrophosphate (TPP):** The active coenzyme form of vitamin $B_1$; involved in aldehyde transfer reactions.

**thioester:** An ester of a carboxylic acid with a thiol or mercaptan.

**3′ end:** The end of a nucleic acid that lacks a nucleotide bound at the 3′ position of the terminal residue.

**thrombocytes:** *See* platelets.

**thromboxanes:** A class of molecules derived from arachidonate and involved in platelet aggregation during blood clotting.

**thylakoid:** Closed cisterna, or disk, formed by the pigment-bearing internal membranes of chloroplasts.

**thymine dimer:** *See* pyrimidine dimer.

**tissue culture:** Method by which cells derived from multicellular organisms are grown in liquid media.

**titration curve:** A plot of the pH versus the equivalents of base added during titration of an acid.

**T lymphocyte (T cell):** One of a class of blood cells (lymphocytes) of thymic origin, involved in cell-mediated immune reactions.

**tocopherols:** Forms of vitamin E.

**topoisomerases:** Enzymes that introduce positive or negative supercoils in closed, circular duplex DNA.

**topoisomers:** Different forms of a covalently closed, circular DNA molecule that differ only in their linking number.

**topology:** The study of the properties of an object that do not change under continuous deformations such as twisting or bending.

**toxins:** Proteins produced by some organisms and toxic to certain other species.

**trace element:** A chemical element required by an organism in only trace amounts.

**transaminases:** *See* aminotransferases.

**transamination:** Enzymatic transfer of an amino group from an $\alpha$-amino acid to an $\alpha$-keto acid.

**transcription:** The enzymatic process whereby the genetic information contained in one strand of DNA is used to specify a complementary sequence of bases in an mRNA chain.

**transcriptional control:** The regulation of a protein's synthesis by regulation of the formation of its mRNA.

**transduction:** (1) Generally, the conversion of energy or information from one form to another. (2) The transfer of genetic information from one cell to another by means of a viral vector.

**transfer RNA (tRNA):** A class of RNA molecules ($M_r$ 25,000 to 30,000), each of which combines covalently with a specific amino acid as the first step in protein synthesis.

**transformation:** Introduction of an exogenous DNA into a cell, causing the cell to acquire a new phenotype.

**transgenic:** Describes an organism that has genes from another organism incorporated within its genome as a result of recombinant DNA procedures.

**transition state:** An activated form of a molecule in which the molecule has undergone a partial chemical reaction; the highest point on the reaction coordinate.

**translation:** The process in which the genetic information present in an mRNA molecule specifies the sequence of amino acids during protein synthesis.

**translational control:** The regulation of a protein's synthesis by regulation of the rate of its translation on the ribosome.

**translational repressor:** A repressor that binds to an mRNA, blocking translation.

**translocase:** (1) An enzyme that catalyzes membrane transport. (2) An enzyme that causes a movement such as the movement of a ribosome along an mRNA.

**transpiration:** Passage of water from the roots of a plant to the atmosphere via the vascular system and the stomata of the leaves.

**transporters:** Proteins that span a membrane and transport specific nutrients, metabolites, ions, or proteins across the membrane; sometimes called permeases.

**transposition:** The movement of a gene or set of genes from one site in the genome to another.

**transposon (transposable element):** A segment of DNA that can move from one position in the genome to another.

**triacylglycerol:** An ester of glycerol with three molecules of fatty acid; also called a triglyceride or neutral fat.

**tricarboxylic acid cycle:** *See* citric acid cycle.

**triose:** A simple sugar with a backbone containing three carbon atoms.

**tRNA:** *See* transfer RNA.

**tropic hormone (tropin):** A peptide hormone that stimulates a specific target gland to secrete its hormone; for example, thyrotropin produced by the pituitary stimulates secretion of thyroxine by the thyroid.

**tumor suppressor gene:** One of a class of genes that encode proteins that normally suppress the division of cells. When defective, the normal gene becomes a tumor suppressor gene, and when both copies are defective, the cell is allowed to continue dividing without limitation; it becomes a tumor cell.

**turnover number:** The number of times an enzyme molecule transforms a substrate molecule per unit time, under conditions giving maximal activity at substrate concentrations that are saturating.

**two-component signaling systems:** Signal-transducing systems found in prokaryotes and plants, composed of a receptor His kinase that phosphorylates an internal His residue when occupied by its ligand. It then catalyzes phosphoryl transfer to a second component, the response regulator, which, when phosphorylated, alters the output of the signaling system.

**u**

**ubiquitin:** A small, highly conserved protein that targets an intracellular protein for degradation by proteasomes. Several ubiquitin molecules are covalently attached in tandem to a Lys residue in the target protein by a specific ubiquitinating enzyme.

**ultraviolet (UV) radiation:** Electromagnetic radiation in the region of 200 to 400 nm.

**uncompetitive inhibition:** The reversible inhibition pattern resulting when an inhibitor molecule can bind to the enzyme-substrate complex but not to the free enzyme.

**uncoupling agent:** A substance that uncouples phosphorylation of ADP from electron transfer; for example, 2,4-dinitrophenol.

**uniport:** A transport system that carries only one solute, as distinct from cotransport.

**unsaturated fatty acid:** A fatty acid containing one or more double bonds.

**urea cycle:** A metabolic pathway in vertebrates, for the synthesis of urea from amino groups and carbon dioxide; occurs in the liver.

**ureotelic:** Excreting excess nitrogen in the form of urea.

**uricotelic:** Excreting excess nitrogen in the form of urate (uric acid).

**v**

$V_{max}$**:** The maximum velocity of an enzymatic reaction when the binding site is saturated with substrate.

**vector:** A DNA molecule known to replicate autonomously in a host cell, to which a segment of DNA may be spliced to allow its replication; for example, a plasmid or an artificial chromosome.

**vectorial:** Refers to an enzymatic reaction or transport process in which the protein has a specific orientation in a biological membrane such that the substrate is moved from one side of the membrane to the other as it is converted into product.

**vectorial metabolism:** Metabolic transformations in which the location (not the chemical composition) of a substrate changes relative to a cellular membrane dividing two compartments. Transporters catalyze vectorial reactions, as do the proton pumps of oxidative and photophosphorylation.

**viral vector:** A viral DNA altered so that it can act as a vector for recombinant DNA.

**virion:** A virus particle.

**virus:** A self-replicating, infectious, nucleic acid–protein complex that requires an intact host cell for its replication; its genome is either DNA or RNA.

**vitamin:** An organic substance required in small quantities in the diet of some species; generally functions as a component of a coenzyme.

## W

**wild type:** The normal (unmutated) phenotype.

**wobble:** The relatively loose base pairing between the base at the 3′ end of a codon and the complementary base at the 5′ end of the anticodon.

## X

**x-ray crystallography:** The analysis of x-ray diffraction patterns of a crystalline compound, used to determine the molecule's three-dimensional structure.

## Z

**zinc finger:** A specialized protein motif involved in DNA recognition by some DNA-binding proteins; characterized by a single atom of zinc coordinated to four Lys residues or to two His and two Lys residues.

**zwitterion:** A dipolar ion, with spatially separated positive and negative charges.

**zymogen:** An inactive precursor of an enzyme; for example, pepsinogen, the precursor of pepsin.

# Credits

## Molecular Modeling

MOLECULAR GRAPHICS Unless indicated, all molecular graphics were produced by Jean-Yves Sgro, Ph.D., at the Department of Biochemistry, University of Wisconsin–Madison, using a Silicon Graphics Unix workstation and the following software.

GRASP software from Anthony Nicholls. See Nicholls, A., Sharp, K.A., & Honig, B. (1991) Protein folding and association: insights from interfacial and thermodynamic properties of hydrocarbons. *Proteins Struct. Funct. Genet.* 11, 281–296. (http://trantor.bioc.columbia.edu/grasp/)

MidasPlus software from the Computer Graphics Laboratory, University of California, San Francisco. See Ferrin, T.E., et al. (1988) The MIDAS display system. *J. Mol. Graphics* 6, 13. (www.cgl.ucsf.edu/Outreach/midasplus)

MOLMOL See Koradi, R., Billeter, M., & Wüthrich, K. (1996) MOLMOL: a program for display and analysis of macromolecular structures. *J. Mol. Graphics* 14, 51–55. (www.mol.biol.ethz.ch/wuthrich/software/molmol)

Molscript (versions 1.4 and 2.1) from Per Kraulis. See Kraulis, P.J. (1991) MOLSCRIPT: a program to produce both detailed and schematic plots of protein structures. *J. Appl. Crystallogr.* 24, 946–950. (www.avatar.se/molscript)

RasMol 2.6b and 2.7.2 for Unix (©Roger Sayle, 1992–1994; Bernstein et al., 2003) was used to create template scripts for Molscript/Raster3D renderings subsequently altered to add specific colors or renderings. See www.umass.edu/microbio/rasmol/ and www.bernstein-plus-sons.com/software/rasmol/.

Raster3D (v. 2.2 & 2.7). See Merritt, E.A., & Murphy, M.E.P. (1994) Raster3D version 2.0, a program for photorealistic molecular graphics. *Acta Cryst.* D50, 869–873.

WebLabViewerLite3.2 software from Molecular Simulation, Inc., www.msi.com. In some cases images were superimposed with PhotoShop (Adobe, www.adobe.com) on a Macintosh computer or with Image Magick (www.wizards.dupont.com/cristy/ImageMagick.html) on a Silicon Graphics Unix workstation.

ATOMIC COORDINATES Unless indicated, all atomic coordinates were obtained from the Protein Data Bank at the Research Collaboratory for Structural Bioinformatics at http://www.rcsb.org/pdb/. See Berman, H.M., Westbrook, J., Feng, Z., Gilliland, G., Bhat, T.N., Weissig, H., Shindyalov, I.N., Bourne, P.E. (2000) The Protein Data Bank *Nucl. Acids Res.* 28, 235–242; Bernstein, F.C., Koetzle, T.F., Williams, G.J.B., Meyer, E.F., Jr., Brice, M.D., Rodgers, J.R., Kennard, O., Shimanouchi, T., & Tasumi, M. (1977) Protein data bank. A computer-based archival file for macromolecular structures. *J. Mol. Biol.* 112, 535; and Abola, E.E., Bernstein, F.C., Bryant, S.H., Koetzle, T.F., & Weng, J. (1987) Protein data bank. In *Crystallographic Databases—Information Content, Software Systems, Scientific Applications* (Allen, F.H., Bergerhoff, G., & Sievers, R., eds), p. 107, Data Commission of the International Union of Crystallography, Bonn.

Some structures (noted below) were generated using Sybyl 6.2, Tripos Inc. http://www.tripos.com/

**CHAPTER 1 p. 1** François Jacob (1970) *The Logic of Life: A History of Heredity.* Pantheon Books, Inc., New York, 1973; originally published (1970) as *La logique du vivant: une histoire de l'hérédité.* Editions Gallimard, Paris; Charles Darwin (1871) *The Descent of Man, and Selection in Relation to Sex.* John Murray, London; **Figure 1–1a** Ron Boardman/Frank Lane Agency/Corbis; **Figure 1–1b** W. Perry Conway/Corbis; **Figure 1–1c** Karen Tweedy-Holmes/Corbis; **Figure 1–2** The Bridgeman Art Library; **Figure 1–4** Adapted from Woese, C.R. (1987) Bacterial evolution. *Microbiol. Rev.* 51, 221, Fig. 4; **Figure 1–6 (top left)** T.J. Beveridge/Biological Photo Service; **(top right, bottom left)** Biological Photo Service; **(bottom right)** Norma J. Lange/Biological Photo Service; **Figure 1–8** Adapted from Alberts, B., Bray, D., Lewis, J., Raff, M., Roberts, K., & Watson, J.D. (1989) *Molecular Biology of the Cell,* 2nd edn, pp. 165–166, Garland Publishing, Inc., New York; **Figure 1–9a (top)** Robert Goldman, Northwestern University Medical School, Department of Cell Biology and Anatomy; **(bottom)** M. Schliwa/Visuals Unlimited; **Figure 1–9b (top)** Dr. Gopal Murti/Science Photo Library/Photo Researchers; **(bottom)** K.G. Murti/Visuals Unlimited; **Figure 1–9c (top)** Dr. Peter Dawson/Science Photo Library/Photo Researchers; **(bottom)** Ueli Aebi, Maurice E. Müller-Institut für Hochauflösende Elektronenmikroskopie am Biozentrum der Universität Basel; **Figure 1–11** Adapted from Becker, W.M., & Deamer, D.W. (1991) *The World of the Cell,* 2nd edn, Fig. 2–15,

Benjamin/Cummings Publishing Company, Menlo Park, CA; **Figure 1–17b,c** Coordinates generated by Sybyl; **Figure 1–20** Adapted from Carroll, F. (1998) *Perspectives on Structure and Mechanism in Organic Chemistry,* p. 63, Brooks/Cole Publishing Co., Pacific Grove, CA; **Box 1–2** (Pasteur) The Granger Collection; **Figure 1–22** PDB ID 1AKX; Brodsky, A.S., & Williamson, J.R. (1997) Solution structure of the HIV-2 Tar-argininamide complex. *J. Mol. Biol.* 267, 624; **p. 23** (Gibbs) Historical Pictures Service/Stock Montage; **Figure 1–29a** Erich Lessing/Art Resource, NY; **Figure 1–29b** Dr. Gopal Murti-CNRI/Phototake NYC; **p. 31** Theodosius Dobzhansky (1973). Nothing in biology makes sense except in the light of evolution, *The American Biology Teacher* (March) 35, 125–129. **p. 36** (Linnaeus, Darwin) Corbis/Bettmann.

**PART I pp. 44, 45** PDB ID 7GCH; see citation for Figure 6–1; **p. 45** A. Tiselius (1946) Presentation speech for the award of the Nobel Prize in Chemistry to James B. Sumner, John H. Northrop, and Wendell M. Stanley. In *Nobel Lectures Including Presentation Speeches and Laureates' Biographies: Chemistry, 1942–1962.* Elsevier Publishing Co., Amsterdam, 1964.

**CHAPTER 2 p. 47** Linus Pauling (1939) *The Nature of the Chemical Bond and the Structure of Molecules and Crystals: An Introduction to Modern Structural Chemistry.* Cornell University Press, Ithaca, New York. **Figure 2–9** PDB ID 1A3N; Tame, J., & Vallone, B. (1998) Deoxy human hemoglobin (primary citation not available); **Figure 2–10** Adapted from Nicolls, P. (2000) Introduction: the biology of the water molecule. *Cell. Mol. Life Sci.* 57, 987, Fig. 6a (redrawn from information in the PDB and a Kinemage file published by Martinez, S.E., Huang, D., Ponomarev, M., Cramer, W.A., & Smith, J.L. (1996) The heme redox center of chloroplast cytochrome *f* is linked to a buried five-water chain. *Protein Sci.* 5, 1081); **Box 2–1 Figure 1a,b** Runk/Schoenberger/Grant Heilman Photography; **Box 2–1 Figure 2a,b** Adrian P. Davies/Bruce Coleman; **p. 70** Jon Bertsch/Visuals Unlimited.

**CHAPTER 3 p. 75** J.J. Berzelius (1838) Letter to G.J. Mulder. In H.B. Vickery (1950) The origin of the word protein. *Yale Journal of Biology and Medicine* 22, 387–393. **Figure 3–1a** Runk/Schoenburger/Grant Heilman Photography; **Figure 3–1b** Bill Longcore/Photo Researchers; **Figure 3–1c** Animals Animals; **Figure 3–16** Adapted from Dickerson, R.E., & Geis I. (1969) *The Structure and Action of Proteins,* Benjamin/Cummings Publishing Company, Menlo Park, CA. © 1969 by Irving Geis; **Figure 3–19b** Richard R. Burgess & Jerome J. Jendrias, University of Wisconsin–Madison, Biotechnology Center; **Figure 3–22b** Patrick H. O'Farrell, University of California Medical Center, San Francisco, Department of Biochemistry and Biophysics; **p. 97** UPI/Corbis-Bettmann; **Box 3–2 Figures 1, 2** Adapted from Mann, M., & Wilm, M. (1995) Electrospray mass spectrometry for protein characterization. *Trends Biochem Sci.* 20, 219; **p. 105** Corbis/Bettmann; **Figure 3–30** Adapted from Gupta, R.S. (1998) Protein phylogenies and signature sequences: A reappraisal of evolutionary relationships among archaebacteria, eubacteria, and eukaryotes. *Microbiol. Mol. Biol. Rev.* 62, 1435, Fig. 2; **Figure 3–31** Adapted from Henikoff, S., & Henikoff, J.G. (1993) Performance evaluation of amino acid substitution matrices. *Proteins* 17, 49; **Figure 3–32** see citation for Figure 3–30, Fig. 7; **Figure 3–33** see citation for Figure 3–30, Fig. 11.

**CHAPTER 4 p. 116** Personal communication from Yong Duan and Peter A. Kollman, University of California, San Francisco, Department of Pharmaceutical Science; J.C. Kendrew et al. (1958) A three-dimensional model of the myoglobin molecule obtained by x-ray analysis. *Nature* 181, 662–666; **Figure 4–1** PDB ID 6GCH; Brady, K., Wei, A., Ringe, D., & Abeles, R.H. (1990) Structure of chymotrypsin-trifluoromethyl ketone inhibitor complexes: comparison of slowly and rapidly equilibrating inhibitors. *Biochemistry* 29, 7600; glycine coordinates from Sybyl; **Figure 4–3,** Adapted from Creighton, T.E. (1984) *Proteins,* p. 166. © 1984 by W. H. Freeman and Company. Reprinted by permission; **p. 120** (Pauling) Corbis/Bettmann; (Corey) AP/Wide World Photos; **Figures 4–4c,d** and **4–5** PDB ID 4TNC; Satyshur, K.A., Rao, S.T., Pyzalska, D., Drendel, W., Greaser, M., & Sundaralingam, M. (1988) Refined structure of chicken skeletal muscle troponin C in the two-calcium state at 2-angstroms resolution. *J. Biol. Chem.* 263, 1628; **Figure 4–9a** See citation for Figure 4–3; **Figure 4–9b** Personal communication from Hazel Holden, Department of Biochemistry and Enzyme Institute, University of Wisconsin–Madison; **Figure 4–10** From Zubay, G. (1988) *Biochemistry,* p. 34. © 1988 by Macmillan Publishing Company. Reprinted by permission; **Figure 4–12** PDB ID 1CGD (modified); Bella, J., Brodsky, B., & Berman, H.M. (1995)

Hydration structure of a collagen peptide. *Structure* 3, 893; **Figure 4–13** Science Source/Photo Researchers; **Figure 4–14b** Dr. Dennis Kunkel/ Phototake NYC; **p. 130** Ethel Wedgwood (1906) *The Memoirs of the Lord of Joinville: A New English Version.* E.P. Dutton and Company, New York. **Box 4–3 Figure 1** National Archives of Canada; **Box 4–3 Figure 2** Courtesy of the Royal College of Physicians of Edinburgh; **Figure 4–16** PDB ID 1MBO; Phillips, S.E.V. (1980) Structure and refinement of oxymyoglobin at 1.6 angstroms resolution. *J. Mol. Biol.* 142, 531; **Figure 4–18 (left)** PDB ID 1CCR; Ochi, H., Hata, Y., Tanaka, N., Kakudo, M., Sakurai, T., Aihara, S., & Morita, Y. (1983) Structure of rice ferricytochrome *c* at 2.0 angstroms resolution. *J. Mol. Biol.* 166, 407; **Figure 4–18 (center)** PDB ID 3LYM; Kundrot, C.E., & Richards, F.M. (1987) Crystal structure of hen egg-white lysozyme at a hydrostatic pressure of 1000 atmospheres. *J. Mol. Biol.*; **Figure 4–18 (right)** PDB ID 3RN3; Howlin, B., Moss, D.S., & Harris, G.W. (1989) Segmented anisotropic refinement of bovine ribonuclease A by the application of the rigid-body TLS model. *Acta Crystallogr., Sect. A* 45, 851; **Box 4–4 Figure 1a,b,c** George N. Phillips, Jr., University of Wisconsin–Madison, Department of Biochemistry; **Box 4–4 Figure 1d** PDB ID 2MBW; Brucker, E.A., Olson, J.S., Phillips, G.N., Jr., Dou, Y., & Ikeda-Saito, M. (1996) High resolution crystal structures of the deoxy-, oxy-, and aquomet- forms of cobalt myoglobin. *J. Biol. Chem.* 271, 25,419–425,422; **Box 4–4 Figures 2,3a** Volkman, B.F., Alam, S.L., Satterlee, J.D., & Markley, J.L. (1998) Solution structure and backbone dynamics of component IV-glycera dibranchiata monomeric hemoglobin-CO. *Biochemistry* 37, 10,906; **Box 4–4 Figure 3b,c** Created by Brian Volkman, National Magnetic Resonance Facility at Madison, using MOLMOL; PDB ID 1VRF **(b)** and 1VRE **(c)**; Volkman, B.F., Alam, S.L., Satterlee, J.D., & Markley, J.L. (1998) Solution structure and backbone dynamics of component IV-glycera dibranchiata monomeric hemoglobin-CO. *Biochemistry* 37, 10,906; **Figure 4–19** PDB ID 4TNC; see citation at Figure 4–4c,d; **Figure 4–20d (left)** PDB ID 7AHL; Song, L., Hobaugh, M.R., Shustak, C., Cheley, S., Bayley, H., & Gouaux, J.E. (1996) Structure of staphylococcal α hemolysin, a heptameric transmembrane pore. *Science* 274, 1859; **Figure 4–20d (right)** PDB ID 1DNP; Park, H.W., Kim, S.T., Sancar, A., & Deisenhofer, J. (1995) Crystal structure of DNA photolyase from *Escherichia coli*. *Science* 268, 1866; **Figure 4–21** PDB ID 1PKN; Larsen, T.M., Laughlin, L.T., Holden, H.M., Rayment, I., & Reed, G.H. (1994) Structure of rabbit muscle pyruvate kinase complexed with $Mn^{2+}$, $K^+$, and pyruvate. *Biochemistry* 33, 6301; **Figure 4–22 (all α)** PDB ID 1A06; Sugio, S., Kashima, A., Mochizuki, S., Noda, M., & Kobayashi, K. (1999) Crystal structure of human serum albumin at 2.5 angstrom resolution. *Protein Eng.* 12, 439; PDB ID 1BCF; Frolow, F., Kalb (Gilboa), A.J., & Yariv, J. (1994) Structure of a unique, two-fold symmetric, haem-binding site. *Nat. Struct. Biol.* 1, 453; PDB ID 1GAI; Aleshin, A.E., Stoffer, B., Firsov, L.M., Svensson, B., & Honzatko, R. B. (1996) Crystallo-graphic complexes of glucoamylase with maltooligosaccharide analogs: relationship of stereochemical distortions at the nonreducing end to the catalytic mechanism. *Biochemistry* 35, 8319; PDB ID 1ENH; Clarke, N.D., Kissinger, C.R., Desjarlais, J., Gilliland, G.L., & Pabo, C.O. (1994) Structural studies of the engrailed homeodomain. *Protein Sci.* 3, 1779; **(all β)** PDB ID 1HOE; Pflugrath, J.W., Wiegand, G., Huber, R., & Vertesy, L. (1986) Crystal structure determination, refinement and the molecular model of the α-amylase inhibitor Hoe-467A. *J. Mol. Biol.* 189, 383; PDB ID 1LXA; Raetz, C.R.H., & Roderick, S.L. (1995) A left-handed parallel β-helix in the structure of UDP *N*-acetylglucosamine acyltransferase. *Science* 270, 997; PDB ID 1PEX; Gomis-Ruth, F.X., Gohlke, U., Betz, M., Knauper, V., Murphy, G., Lopez-Otin, C., & Bode, W. (1996) The helping hand of collagenase-3 (Mmp-13): 2.7 Å crystal structure of its C-terminal haemopexin-like domain. *J. Mol. Biol.* 264, 556; PDB ID 1JPC; Wright, C.S., & Hester, G. (1996) 2Å Structure of a cross-linked complex between snowdrop lectin and a branched penta-mannoside: evidence for two unique binding modes. *Structure* 4, 1339–1352; PDB ID 1CD8; Leahy, D.J., Axel, R., & Hendrickson, W.A. (1992) Crystal structure of a soluble form of the human T cell co-receptor Cd8 at 2.6 angstroms resolution. *Cell* 68, 1145; **(α/β)** PDB ID 1DEH; Davis, G.J., Stone, C.J., Bosron, W.F., & Hurley, T.D. (1996) X-ray structure of human $\beta_3\beta_3$ alcohol dehydrogenase: the contribution of ionic interactions to coenzyme binding. *J. Biol. Chem.* 271, 17,057; PDB ID 1DUB; Engel, C.K., Mathieu, M., Zeelen, J.P., Hiltunen, J.K., & Wierenga, R. K. (1996) Crystal structure of enoyl-coenzyme A (CoA) hydratase at 2.5 angstroms resolution: a spiral fold defines the CoA-binding pocket. *EMBO J.* 15, 5135; PDB ID 1PFK; Shirakihara, Y., & Evans, P.R. (1988) Crystal structure of the complex of phosphofructokinase from *Escherichia coli* with its reaction products. *J. Mol. Biol.* 204, 973; **(α + β)** PDB ID 2PIL; Forest, K.T., Dunham, S.A., Koomey, M., & Tainer, J.A. (1999) Crystallographic structure of phosphorylated pilin from *Neisseria*: phosphoserine sites modify type IV pilus surface chemistry and morphology. *Mol. Microbiol.* 31, 743; PDB ID 1SYN; Stout, T.J., & Stroud, R.M. (1996) The complex of the anti-cancer therapeutic, BW1843U89, with thymidylate synthase at 2.0 Å resolution: implications for a new mode of inhibition. *Structure* 4, 67–77; PDB ID 1EMA; Ormo, M., Cubitt, A.B., Kallio, K., Gross, L.A., Tsien, R.Y., & Remington, S.J. (1996) Crystal structure of the *Aequorea victoria* green fluorescent protein. *Science* 273, 1392; PDB ID 1U9A; Tong, H., Hateboer, G., Perrakis, A., Bernards, R., & Sixma, T.K. (1997) Crystal structure of murine/human Ubc9 provides insight into the variability of the ubiquitin-conjugating system. *J. Biol. Chem.* 272, 21,381; **p. 144** Corbis/Hulton Deutsch Collection; **Figure 4–23** PDB ID 2HHB; Fermi, G., Perutz, M.F., Shaanan, B., & Fourme, R. (1984) The crystal structure of human deoxyhaemoglobin at 1.74 angstroms resolution. *J. Mol. Biol.* 175, 159; **Figure 4–25a** PDB ID 2PLV; Filman, D.J., Syed, R., Chow, M., Macadam, A.J., Minor, P.D., & Hogle, J.M. (1989) Structural factors that control conformational transitions and serotype specificity in type 3 poliovirus. *EMBO J.* 8, 1567; **Figure 4–25b** Science Source/Photo Researchers; PDB ID 1VTM; Pattanayek, R., & Stubbs, G. (1992) Structure of the U2 strain of tobacco mosaic virus refined at 3.5 angstroms resolution using x-ray fiber diffraction. *J. Mol. Biol.* 228, 516; **Figure 4–26a** Data from Sendak, R.A., Rothwarf, D.M., Wedemeyer, W.J., Houry, W.A., & Scheraga, H.A. (1996) Kinetic and thermodynamic studies of the folding/unfolding of a tryptophan-containing mutant of ribonuclease A. *Biochemistry* 35, 12,978; Nishii, I. Kataoka, M., & Goto, Y. (1995) Thermodynamic stability of the molten globule states of apomyoglobin. *J. Mol. Biol.* 250, 223; **Figure 4–26b** Data from Houry, W.A., Rothwarf, D.M., & Scheraga, H.A. (1996) Circular dichroism evidence for the presence of burst-phase intermediates on the conformational folding pathway of ribonuclease A. *Biochemistry* 35, 10,125; **Figure 4–28** Personal communication from Yong Duan and Peter A. Kollman, University of California, San Francisco, Department of Pharamaceutical Science; **Figure 4–29** Adapted from Wolynes, P.G., Onuchic, J.N., & Thirumalai, D. (1995) Navigating the folding routes. *Science* 267, 1619; **Box 4–5 Figure 1** Stephen J. DeArmond; **Box 4–5 Figure 2** PDB ID 1QM0; Zahn, R., Liu, A., Luhrs, T., Riek, R., Von Schroetter, C., Garcia, F.L., Billeter, M., Calzolai, L., Wider, G., & Wuthrich, K. (2000) NMR solution structure of the human prion protein *Proc. Natl. Acad. Sci. USA* 97, 145; **Figure 4–31** PDB ID 1AON; Xu, Z., Horwich, A.L., & Sigler, P.B. (1997) The crystal structure of the asymmetric GroEL-GroES-(ADP)7 chaperonin complex. *Nature* 388, 741.

**CHAPTER 5 p. 157** See citation for Figure 5–19a; Friedrich Ludwig Hünefeld (1840) *Der Chemismus in der thierischen Organisation* (The Chemistry of Animal Organization). Brockhaus, Leipzig, p. 160; Emil Fischer (1906) Untersuchen über Aminosäuren, Polypeptide und Proteine (Studies of Amino Acids, Polypeptides and Proteins). *Berichte der deutschen chemischen Gesellschaft zu Berlin* 39, 530; **Figure 5–1c** Heme extracted from PDB ID 1CCR; see citation for Figure 4–18 (left); **Figures 5–3, 5–5c, 5–6 (left)** PDB ID 1MBO; Phillips, S.E.V. (1980) Structure and refinement of oxymyoglobin at 1.6 angstroms resolution. *J. Mol. Biol.* 142, 531; **Figures 5–6 (right), 5–8, 5–9a, 5–10 (T state), 5–11 (T state)** PDB ID 1HGA; Liddington, R., Derewenda, Z., Dodson, E., Hubbard, R., & Dodson, G. (1992) High resolution crystal structures and comparisons of T state deoxyhaemoglobin and two liganded T-state haemoglobins: T(α-oxy) haemoglobin and T(met)haemoglobin. *J. Mol. Biol.* 228, 551; **Figures 5–10 (R state), 5–11 (R state)** PDB ID 1BBB; Silva, M.M., Rogers, P.H., & Arnone, A. (1992) A third quaternary structure of human hemoglobin at 1.7-angstroms resolution. *J. Biol. Chem.* 267, 17,248; in Fig. 5–11 R-state was modified to represent $O_2$ instead of CO; **Box 5–1 Figure 1** Adapted from Coburn, R.F., Forster, R.E., & Kane, P.B. (1965) Considerations of the physiological variables that determine the blood carboxyhemoglobin concentration in man. *J. Clin. Invest.* 44, 1899; **Box 5–1 Figure 2** Adapted from Roughton, F.J.W., & Darling, R.C. (1944) The effect of carbon monoxide on the oxyhemoglobin dissociation curve. *Am. J. Physiol.* 141, 17; **Figure 5–18** PDB ID 1HGA; see citation for Figure 5–6 (right). Coordinates for 2,3-bisphosphoglycerate obtained from Paul Reisberg, Wellesley College, Department of Chemistry; **Figure 5–19a** Andrew Syred/Science Photo Library/Custom Medical Stock Photo; **Figure 5–19b** Custom Medical Stock Photo; **Figure 5–22** PDB ID 1DDH; Li, H., Natarajan, K., Malchiodi, E.L., Margulies, D.H., & Mariuzza, R.A. (1998) Three-dimensional structure of H-2Dd complexed with an immunodominant peptide from human immunodeficiency virus envelope glycoprotein 120. *J. Mol. Biol.* 283, 179; **Figure 5–23b** PDB ID 1IGT; Harris, L.J., Larson, S.B., Hasel, K.W., & McPherson, A. (1997) Refined structure of an intact IgG2A monoclonal antibody. *Biochemistry* 36, 1581; **Figure 5–27a** PDB ID 1GGC; Stanfield, R.L., Takimoto-Kamimura, M., Rini, J.M., Profy, A.T., & Wilson, I.A. (1993) Major antigen-induced domain rearrangements in an antibody. *Structure* 1, 83; **Figure 5–27b,c** PDB ID 1GGI; Rini, J.M., Stanfield, R.L., Stura, E.A., Salinas, P.A., Profy, A.T., & Wilson, I.A. (1993) Crystal structure of an human immunodeficiency virus type 1 neutralizing antibody, 50.1, in complex with its V3 loop peptide antigen. *Proc. Natl. Acad. Sci. USA* 90, 6325; **p. 181** (Kohler and Milstein) Corbis/UPI/Bettmann; **Figure 5–28b** State of Wisconsin Laboratory of Hygiene, Madison, WI; **Figure 5–28c** Son, M., Gundersen, R.E., & Nelson, D.L. (1993) A second member of the novel Ca²⁺-dependent protein kinase family from *Paramecium tetraurelia*: Purification and characterization. *J. Biol. Chem.* 268, 5940; **Figure 5–29a** David Shotton/University of Oxford, Department of Zoology; **Figure 5–29c** Personal communication from Ivan Rayment, Enzyme Institute and Department of Biochemistry, University of Wisconsin–Madison (see also PDB ID 2MYS; Rayment, I., Rypniewski, W.R., Schmidt-Base, K., Smith, R., Tomchick, D.R., Benning, M.M., Winkelmann, D.A., Wesenberg, G., & Holden, H.M. (1993) Three-dimensional structure of myosin subfragment-1: a molecular motor. *Science* 261, 50); **Figure 5–30a** Eisaku Katayama, Institute of Medical Science, The University of Tokyo, Department of Fine Morphology; **Figure 5–30b** Roger Craig, University of Massachusetts Medical School; **Figure 5–30c** See citation for Figure 5–29c; **Figure 5–31b,c** James E. Dennis/Phototake NYC.

**CHAPTER 6    p. 190** PDB ID 1ONE; see citation for Figure 6–23b; J.B.S. Haldane (1930) *Enzymes.* Longmans, Green & Co., London, p. 182; William P. Jencks (1975) Binding energy, specificity, and enzymic catalysis: the Circe effect. *Advances in Enzymology* 43, 219–410; **p. 191** (Buchner) Science Photo Library/Photo Researchers; (Sumner and Haldane) AP/Wide World Photos; **Figure 6–1** PDB ID 7GCH; Brady, K., Wei, A., Ringe, D., & Abeles, R.H. (1990) Structure of chymotrypsin-trifluoromethyl ketone inhibitor complexes. Comparison of slowly and rapidly equilibrating inhibitors. *Biochemistry* 29, 7600; **Figure 6–4** PDB ID 1RA2; Sawaya, M.R., & Kraut, J. (1997) Loop and subdomain movements in the mechanism of *Escherichia coli* dihydrofolate reductase: crystallographic evidence. *Biochemistry* 36, 586; **p. 203** (Michaelis) Rockefeller University Archive Center; (Menten) Courtesy Dorothy C. Craig; **Figure 6–18b,c,d** PDB ID 7GCH; see citation for Figure 6–1; **Figure 6–22a** PDB ID 2YHX; Anderson, C.M., Stenkamp, R.E., & Steitz, T.A. (1978) Sequencing a protein by x-ray crystallography: II. Refinement of yeast hexokinase B coordinates and sequence at 2.1 angstroms. *J. Mol. Biol.* 123, 15; **Figure 6–22b** PDB ID 1HKG; Steitz, T.A., Shoham, M., & Bennett, W.S., Jr. (1981) Structural dynamics of yeast hexokinase during catalysis. *Philos. Trans. R. Soc. London Ser. B* 293, 43; glucose (GLC) coordinates transformed from PDB ID 1GLK; St. Charles, R., Harrison, R.W., Bell, G.I., Pilkis, S.J., & Weber, I.T. (1994) Molecular model of human beta-cell glucokinase built by analogy to the crystal structure of yeast hexokinase B. *Diabetes* 43, 784; **Figure 6–23b** PDB ID 1ONE; Larsen, T.M., Wedekind, J.E., Rayment, I., & Reed, G.H. (1996) A carboxylate oxygen of the substrate bridges the magnesium ions at the active site of enolase: structure of the yeast enzyme complexed with the equilibrium mixture of 2-phosphoglycerate and phosphoenolpyruvate at 1.8 Å resolution. *Biochemistry* 35, 4349; **Box 6–3 Figure 1** Adapted from Fersht, A. (1985) *Enzyme Structure and Mechanism*, 2nd edn, p. 318. © 1977, 1985 by W. H. Freeman and Company. Reprinted by permission; **Figure 6–24a** PDB ID 1LZE; Maenaka, K., Matsushima, M., Song, H., Sunada, F., Watanabe, K., & Kumagai, I. (1995) Dissection of protein-carbohydrate interactions in mutant hen egg-white lysozyme complexes and their hydrolytic activity. *J. Mol. Biol.* 247, 281; **Figure 6–25a,b** PDB ID 1H6M; Vocadlo, D.J., Davies, G.J., Laine, R., & Withers, S.G. (2001) Catalysis by hen egg-white lysozyme proceeds via a covalent intermediate *Nature* 412, 835; **Figure 6–27** PDB ID 2AT2; Stevens, R.C., Reinisch, K.M., & Lipscomb, W.N. (1991) Molecular structure of *Bacillus subtilis* aspartate transcarbamoylase at 3.0 angstroms resolution. *Proc. Natl. Acad. Sci. USA* 88, 6087. Thank you to Raymond C. Stevens (University of California, Berkeley) for providing symmetry-related coordinates.

**CHAPTER 7    p. 238** Rida Johnson Young (lyrics) and Victor Herbert (music) (1910) Ah! Sweet Mystery of Life; E.B. White (1977) Coon Tree. In *Essays of E. B. White.* HarperCollins, Perennial Classics edn, New York, 1999, pp. 47–48; **Figure 7–14a** Biophoto Associates/Photo Researchers; **Figure 7–14b** Barry King/Biological Photo Service; **Figure 7–16b** Adapted from Lehninger, A. (1982) *Principles of Biochemistry*, p. 290, Worth Publishers, New York. © Irving Geis; **Figure 7–17** Courtesy of Jack Scheper, Floridata.com; **Figure 7–18b** Leroy Somon/Visuals Unlimited; **Figure 7–20** Courtesy of H.-J. Gabius and Herbert Kaltner, University of Munich, from a figure provided by C.-W. von der Lieth, Heidelberg; **Figure 7–21b** See citation for Figure 7–16b; **Figure 7–25** PDB ID 1E0O; Pellegrini, L., Burke, D.F., von Delft, F., Mulloy, B., & Blundell, T.L. (2000) Crystal structure of fibroblast growth factor receptor ectodomain bound to ligand and heparin. *Nature* 407, 1029; **Figures 7–27b, 7–28** Adapted from Turnbull, J., Powell, A., & Guimond, S. (2001) Heparan sulfate: decoding a dynamic multifunctional cell regulator. *Trends Cell Biol.* 11, 75; **Figure 7–32b** Lipopolysaccharide (LPS) from PDB ID 2FCP; Ferguson, A.D., Hofmann, E., Coulton, J.W., Diederichs,

K., & Welte, W. (1998) Siderophore-mediated iron transport: crystal structure of FhuA with bound lipopolysaccharide. *Science* 282, 2215; **Figure 7–34** R.M. Genta and D.Y. Graham, Veterans Affairs Medical Center, Houston, TX; **Figure 7–35a,b** PDB ID 1QFO; May, A.P., Robinson, R.C., Vinson, M., Crocker, P.R., & Jones, E.Y. (1998) Crystal structure of the *N*-terminal domain of sialoadhesin in complex with 3′ sialyllactose at 1.85 Å resolution. *Mol. Cell* 1, 719; **Figure 7–35c,d** PDB ID 1M6P; Roberts, D.L., Weix, D.J., Dahms, N.M., & Kim, J.J. (1998) Molecular basis of lysosomal enzyme recognition: three-dimensional structure of the cation-dependent mannose 6-phosphate receptor. *Cell* 93, 639; **Figure 7–36** Adapted from a figure provided by Dr. C.-W. von der Lieth, Heidelberg, in Gabius, H.-J. (2000) Biological information transfer beyond the genetic code: the sugar code. *Naturwissenschaften* 87, 108, Fig. 6; **Figure 7–37** Adapted from Sharon, N., & Lis, H. (1993) Carbohydrates in cell recognition. *Sci. Am.* 268 (January), 82.

**CHAPTER 8    p. 273** James D. Watson (1968) *The Double Helix: A Personal Account of the Discovery of the Structure of DNA.* Atheneum, New York; **p. 279** (Watson and Crick) Corbis/UPI/Bettmann; **p. 282** (Franklin) Science Photo Library/Photo Researchers; (Wilkins) Corbis/UPI/Bettmann; **Figure 8–14** Science Source/Photo Researchers; **Figures 8–15b,c, 8–19** Coordinates generated by Sybyl; **Figure 8–22b** PDB ID 1BCE; Asensio, J.L., Brown, T., & Lane, A.N. (1998) Comparison of the solution structures of intramolecular DNA triple helices containing adjacent and non-adjacent CG.C+ triplets. *Nucl. Acids Res.* 26, 3677; **Figure 8–22d** PDB ID 1QDG; Marathias, V.M., Wang, K.Y., Kumar, S., Swaminathan, S., & Bolton, P.H. (1996) Determination of the number and location of the manganese binding sites of DNA quadruplexes in solution by EPR and NMR in the presence and absence of thrombin. *J. Mol. Biol.* 260, 378; **Figure 8–23a** Htun, H., & Dahlberg, J.E. (1988) Single strands, triple strands, and kinks in H-DNA. *Science* 241, 1793. © 1988 by the American Association for the Advancement of Science; **Figure 8–23b** Htun, H., & Dahlberg, J.E. (1989) Topology and formation of triple-stranded H-DNA. *Science* 243, 1571. © 1989 by the American Association for the Advancement of Science; **Figure 8–25** Coordinates generated by Sybyl; **Figure 8–27** James, B., Olsen, G.J., Liu, J., & Pace, N.R. (1988) The secondary structure of ribonuclease P RNA, the catalytic element of a ribonucleoprotein enzyme. *Cell* 52, 19–26; **Figure 8–28a** PDB ID 1TRA; Westhof, E., & Sundaralingam, M. (1986) Restrained refinement of the monoclinic form of yeast phenylalanine transfer RNA: temperature factors and dynamics, coordinated waters, and base-pair propeller twist angles. *Biochemistry* 25, 4868; **Figure 8–28b** PDB ID 1MME; Scott, W.G., Finch, J.T., & Klug, A. (1995) The crystal structure of an all-RNA hammerhead ribozyme: a proposed mechanism for RNA catalytic cleavage. *Cell* 81, 991; **Figure 8–28c** PDB ID 1GRZ; Golden, B.L., Gooding, A.R., Podell, E.R., & Cech, T.R. (1998) A preorganized active site in the crystal structure of the *Tetrahymena* ribozyme. *Science* 282, 259; **Figure 8–30b** After Marmur, J., & Doty, P. (1962) Determination of the base composition of deoxyribonucleic acid from its thermal denaturation temperature. *J. Mol. Biol.* 5, 109; **Figure 8–31** Ross B. Inman, University of Wisconsin–Madison, Department of Molecular Biology; **Figure 8–36c** Lloyd Smith, University of Wisconsin–Madison, Department of Chemistry; **Figure 8–37** Data provided by Lloyd Smith, Department of Chemistry, University of Wisconsin–Madison.

**CHAPTER 9    p. 306** Emile Zuckerkandl and Linus Pauling (1965) Molecules as documents of evolutionary history. *Journal of Theoretical Biology* 8, 357–366; **p. 306** (Berg) Courtesy Stanford Visual Art Services; (Boyer) Courtesy Genentech, Inc.; (Cohen) Courtesy Stanford Visual Art Services; **Figure 9–2** PDB ID 1RVC; Kostrewa, D., & Winkler, F.K. (1995) Mg²⁺ binding to the active site of EcoRV endonuclease: a crystallographic study of complexes with substrate and product DNA at 2 angstroms resolution. *Biochemistry* 34, 683; **Figure 9–5** Elizabeth A. Wood, University of Wisconsin–Madison, Department of Biochemistry; **Figure 9–15a** Courtesy of Kevin Strange and Michael Christensen, Department of Pharmacology, Vanderbilt University Medical Center; **Box 9–1 Figure 1** Phil Borden/PhotoEdit; **p. 322** (Collins) Alex Wong/ Newsmakers; (Venter) Mike Theiler/ Reuters; **Figure 9–20** Adapted from Wolfsberg, T.G., McEntyre, J., & Schuler, G.D. (2001) Guide to the draft human genome. *Nature* 409, 824, Fig. 1; **Figure 9–23** Patrick O. Brown, Stanford University School of Medicine, Department of Biochemistry; **Figure 9–25** Adapted from Pandey, A., & Mann, M. (2000) Proteomics to study genes and genomes. *Nature* 405, 837, Fig. 7; **Figure 9–29** Keith Wood, University of California, San Diego, Department of Biology; **Figure 9–30** From Fischhoff, D.A., & Bowdish, K.S. (1987) Insect tolerant transgenic tomato plants. *BioTechnology* 5, 807. Photo courtesy David Fischhoff, Monsanto Company; **Figure 9–31a,b** Chris Boerboom, University of Wisconsin–Madison, Department of Agronomy; **Figure 9–33** R.L. Brinster and R.E. Hammer, University of Pennsylvania, School of Veterinary Medicine;.

**CHAPTER 10    p. 343** Coordinates from Sybyl; Michel-Eugéne Chevreul (1814) Chemical inquiries into the nature of several fatty substances, and particularly on their combinations with the alkalis. Of a new substance, called margarine, obtained from the soap made from the fat of pork and potash. *Philosophical Magazine* 44, 193; **Figures 10–1a,b, 10–2** Coordinates from Sybyl; **Figure 10–3a** Biological Photo Service; **Figure 10–3b** Courtesy Howard Goodman, Department of Genetics, Harvard Medical School; **Figure 10–5b** William J. Weber/Visuals Unlimited; **p. 353** From Drabkin, D.L. (1958) *Thudichum: Chemist of the Brain*, University of Pennsylvania Press, credited to Thudichum, J.L.W. (1898) *Briefe über öffentliche Gesundheitspflege ihre bisherigen Leistungen und heutigen Aufgaben*, F. Pietzcker, Tübingen; **Figure 10–13** Coordinates from Sybyl; **Figure 10–16** PDB coordinates from Dave Woodcock, Okanagan University College, Kelowna, British Columbia, Department of Chemistry; **Box 10–2 Figure 2** Herbert A. Fischler, Isaac Albert Research Institute of the Kingsbrook Jewish Medical Center; **p. 359** (Vane, Bergström, and Samuelsson) Ira Wyman/Sygma; **Figure 10–20b** Courtesy Dr. Russell Chesney, Chairman, Department of Pediatrics, University of Tennessee, Memphis, in collaboration with H.F. DeLuca, Department of Biochemistry, University of Wisconsin–Madison; **p. 363** (Dam and Doisy) AP/Wide World Photos; **Figure 10–24** Christie, W.W. (1996) Beginners' guide to mass spectrometry of fatty acids. 2. General purpose derivatives. *Lipid Technology* 8, 64.

**CHAPTER 11    p. 369** PDB ID 7AHL; see citation at Figure 11–13; Robert Frost (1914) Mending Wall. In *North of Boston*. David Nutt, London; **Figure 11–1** Courtesy J.D. Robertson; **Figure 11–5** Data from Zachowski, A. (1993) Phospholipids in animal eukaryotic membranes: transverse asymmetry and movement. *Biochem. J.* 294, 1; **Figure 11–7** Adapted from Marchesi, V.T., Furthmayr, H., & Tomita, M. (1976) The red cell membrane. *Annu. Rev. Biochem.* 45, 667; **Figure 11–8** PDB ID 2AT9; Mitsuoka, K., Hirai, T., Murata, K., Miyazawa, A., Kidera, A., Kimura, Y., & Fujiyoshi, Y. (1999) The structure of bacteriorhodopsin at 3.0 Å resolution based on electron crystallography: implication of the charge distribution. *J. Mol. Biol.* 286, 861; **Figure 11–9** PDB ID 1PRC; Deisenhofer, J., Epp, O., Sinning, I., & Michel, H. (1995) Crystallographic refinement at 2.3 angstroms resolution and refined model of the photosynthetic reaction center from *Rhodopseudomonas viridis. J. Mol. Biol.* 246, 4299; **Figure 11–12** PDB ID 1BL8; Doyle, D.A., Morais Cabral, J., Pfuetzner, R.A., Kuo, A., Gulbis, J.M., Cohen, S.L., Chait, B.T., & MacKinnon, R. (1998) The structure of the potassium channel: molecular basis of K$^+$ conduction and selectivity. *Science* 280, 69; PDB ID 1AF6; Wang, Y.F., Dutzler, R., Rizkallah, P.J., Rosenbusch, J.P., & Schirmer, T. (1997) Channel specificity: structural basis for sugar discrimination and differential flux rates in maltoporin. *J. Mol. Biol.* 272, 56; PDB ID 1QD5; Snijder, H.J., Ubarretxena-Belandia, I., Blaauw, M., Kalk, K.H., Verheij, H.M., Egmond, M.R., Dekker, N., & Dijkstra, B.W. (1999) Structural evidence for dimerization-regulated activation of an integral membrane phospholipase. *Nature* 401, 717; PDB ID 1QJ9; Vogt, J., & Schulz, G.E. (1999) The structure of the outer membrane protein OmpX from *Escherichia coli* reveals mechanisms of virulence. *Structure* 7, 1301; PDB ID 1PHO; Cowan, S.W., Schirmer, T., Rummel, G., Steiert, M., Ghosh, R., Pauptit, R.A., Jansonius, J.N., & Rosenbusch, J.P. (1992) Crystal structures explain functional properties of two *E. coli* porins. *Nature* 358, 727; **Figure 11–13** PDB ID 1FEP; Buchanan, S.K., Smith, B. S., Venkatramani, L., Xia, D., Esser, L., Palnitkar, M., Chakraborty, R., van der Helm, D., & Deisenhofer, J. (1999) Crystal structure of the outer membrane active transporter FepA from *Escherichia coli. Nat. Struct. Biol.* 6, 56; PDB ID 1QD5; Snijder, H.J., Ubarretxena-Belandia, I., Blaauw, M., Kalk, K.H., Verheij, H.M., Egmond, M.R., Dekker, N., & Dijkstra, B.W. (1999) Structural evidence for dimerization-regulated activation of an integral membrane phospholipase *Nature* 401, 717; PDB ID 1MAL; Schirmer, T., Keller, T.A., Wang, Y.F., & Rosenbusch, J.P. (1995) Structural basis for sugar translocation through maltoporin channels at 3.1 Å resolution. *Science* 267, 512; PDB ID 1EK9; Koronakis, V., Sharff, A.J., Koronakis, E., Luisi, B., & Hughes, C. (2000) Crystal structure of the bacterial membrane protein TolC central to multidrug efflux and protein export. *Nature* 405, 914; PDB ID 7AHL; Song, L., Hobaugh, M.R., Shustak, C., Cheley, S., Bayley, H., & Gouaux, J. E. (1996) Structure of staphylococcal α-hemolysin, a heptameric transmembrane pore. *Science* 274, 1859; **Figure 11–15** Heller, H, Schaefer, M., & Schulten, K. (1993) Molecular dynamics simulation of a bilayer of 200 lipids, in the gel and in the liquid-crystal phases. *J. Phys. Chem.* 97, 8343; **Figure 11–18** Courtesy of Takahiro Fujiwara, Ken Ritchie, Hideji Murakoshi, Ken Jacobson, & Akihiro Kusumi; **Box 11–1 Figure 2a** Courtesy of Daniel J. Müeller, Ueli Aebi, and Andreas Engel/M.E. Müller Institute (MSB), Biozentrum, University of Basel; **Box 11–1 Figure 2b** Courtesy of S. Scheuring, D. Fotiadis, C. Möller, S.A. Müller, A. Engel, & D.J. Müeller; **Box 11–1 Figure 2c** Courtesy of Daniel J. Müeller, Max-Planck Institute of Molecular Cell Biology and Genetic; **Figure 11–20b** Courtesy of J.M.

Edwardson, Department of Pharmacology, University of Cambridge, UK; **Figure 11–24** Adapted from Cross, K.J., Burleigh, L.M., & Steinhauer, D.A. (2001) Mechanisms of cell entry by influenza virus. *Expert Rev. Mol. Med.* 3, 1; **Figure 11–25** Adapted from Chen, Y.A., & Scheller, R.H. (2001) SNARE-mediated membrane fusion. *Nature* 2, 98; **Figure 11–30** Adapted from Mueckler, M. (1994) Facilitative glucose transporters, *Eur. J. Biochem.* 219, 713; **Box 11–2 Figure 1** Adapted from Lienhard, F.E., Slot, J.W., James, D.E., & Mueckler, M.M. (1992) How cells absorb glucose. *Sci. Am.* 266 (January), 86; **p. 398** (Skou) Courtesy Information Office, University of Aarhus, Denmark; **Figure 11–38** PDB ID 1EUL Toyoshima, C., Nakasako, M., Nomura, H., & Ogawa, H. (2000) Crystal structure of the calcium pump of sarcoplasmic reticulum at 2.6 angstrom resolution. *Nature* 405, 647; **Figure 11–41a** PDB ID 1JSQ; Chang, G., & Roth, C.B. (2001) Structure of Msba from *E. coli:* A homolog of the multidrug resistance ATP binding cassette (ABC) transporters *Science* 293, 1793; **Figure 11–41b** PDB ID 1L7V; Locher, K.P., Lee, A.T., & Rees, D.C. (2002) The *E. coli* BtuCD structure: A framework for ABC transporter architecture and mechanism. *Science* 296, 1091; **Box 11–3 Figure 2** Tom Moninger, University of Iowa, Ames; **Figure 11–43a** PDB ID 1PVL; Abramson, J., Smirnova, I., Kasho, V., Verner, G., Kaback, H.R., & Iwata, S. (2003) Structure and mechanism of the lactose permease of *Escherichia coli. Science* 301, 610; **Figure 11–43b** Coordinates provided prior to publication courtesy of Ron Kaback and So Iwata; **p. 406** (Agre) Courtesy of the Royal Swedish Academy of Sciences; **Figure 11–45** Coordinates prepared for The Virtual Museum of Minerals and Molecules, www.soils.wisc.edu/virtual_museum, by Phillip Barak, University of Wisconsin–Madison, Department of Soil Science, using data from Neupert-Laves, K., & Dobler, M. (1975) The crystal structure of a K$^+$ complex of valinomycin. *Helv. Chim. Acta* 58, 432; **Figure 11–46** PDB ID 1J4N; Sui, H., Han, B.-G., Lee, J.K., Walian, P., & Jap, B.K. (2001) Structural basis of water-specific transport through the AQP1 water channel *Nature* 414, 872; **p. 409** (Sakmann, Neher)The Nobel Foundation; (MacKinnon) Courtesy of the Royal Swedish Academy of Sciences; **Figure 11–48a,b** PDB ID 1BL8; Doyle, D.A., Cabral, J.M., Pfuetzner, R.A., Kuo, A., Gulbis, J.M., Cohen, S.L., Chait, B.T., & Mackinnon, R. (1998) The structure of the potassium channel: molecular basis of K$^+$ conduction and selectivity. *Science* 280, 69; **Figure 11–49** PDB ID 1J95; Zhou, M., Morais-Cabral, J.H., Mann, S., & MacKinnon, R. (2001) Potassium channel receptor site for the inactivation gate and quaternary amine inhibitors. *Nature* 411, 657; **Figure 11–50** Adapted from Taylor, R. (1994) Evolutions: the voltage-gated sodium channel. *J. NIH Res.* 6, 112; **Figure 11–51** Adapted from Changeux, J.P. (1993) Chemical signaling in the brain. *Sci. Am.* 269 (November), 58.

**CHAPTER 12    p. 421** PDB ID 1JST; see citation at Figure 12–42c; Earl W. Sutherland (1971) Nobel Address, www.nobel.se/medicine/laureates/1971/sutherland-lecture.html. **Figure 12–7a** PDB ID 1IRK; Hubbard, S.R., Wei, L., Ellis, L., Hendrickson, W.A. (1994) Crystal structure of the tyrosine kinase domain of the human insulin receptor. *Nature* 372, 746; **Figure 12–7b** PDB ID 1IR3; Hubbard, S. R. (1997) Crystal structure of the activated insulin receptor tyrosine kinase in complex with peptide substrate and ATP analog. *EMBO J.* 16, 5572; **p. 436** (Gilman) Office of News and Publications, The University of Texas Southwestern Medical Center at Dallas; (Rodbell) Courtesy Andrew Rodbell; **Figure 12–13** PDB ID 1AZS; Tesmer, J.J., Sunahara, R.K., Gilman, A.G., & Sprang, S.R. (1997) Crystal structure of the catalytic domains of adenylyl cyclase in a complex with G$_{s\alpha}$·GTPγS. *Science* 278, 1907; **Figure 12–15b** PDB ID 1JBP; Madhusudan, Trafny, E.A., Xuong, N.H., Adams, J.A., Ten Eyck, L.F., Taylor, S.S., & Sowadski, J.M. (1994) cAMP-Dependent protein kinase: crystallographic insights into substrate recognition and phosphotransfer. *Protein Sci.* 3, 176; **Figure 12–20a** Courtesy Michael D. Cahalan, Department of Physiology and Biophysics, University of California, Irvine; **Figure 12–20b** Rooney, T.A., Sass, E.J., & Thomas, A.P. (1989) Characterization of cytosolic calcium oscillations induced by phenylephrine and vasopressin in single fura-2-loaded hepatocytes. *J. Biol. Chem.* 264, 17,131; **Figure 12–21a** PDB ID 1CLL; Chattopadhyaya, R., Meador, W.E., Means, A.R., & Quiocho, F. (1992) A calmodulin structure refined at 1.7 angstroms resolution. *J. Mol. Biol.* 228, 1177; **Figure 12–21b,c** PDB ID 1CDL; Meador, W.E., Means, A.R., & Quiocho, F.A. (1992) Target enzyme recognition by calmodulin: 2.4 angstroms structure of a calmodulin-peptide complex. *Science* 257, 1251; **Box 12–2 Figure 1** Chris Parks/ImageQuest Marine; **Box 12–2 Figure 2** PDB ID 1GFL; Yang, F., Moss, L.G., & Phillips, G.N., Jr. (1996) The molecular structure of green fluorescent protein. *Nat. Biotechnol.* 14, 1246; **Figure 12–22** PDB ID 1SHC; Zhou, M.M., Ravichandran, K.S., Olejniczak, E.F., Petros, A.M., Meadows, R.P., Sattler, M., Harlan, J.E., Wade, W.S., Burakoff, S.J., & Fesik, S.W. (1995) Structure and ligand recognition of the phosphotyrosine binding domain of Shc. *Nature* 378, 584; **Figure 12–24** Adapted from Pawson, T., Gish, G.D., & Nash, P. (2001) SH2 domains, interaction modules and cellular wiring. *Trends*

*Cell Biol.* 11, 504, Fig. 5; **Figure 12–29** Adapted from Ouaked, F., Rozhon, W., Lecourieux, D., & Hirt, H. (2003) A MAPK pathway mediates ethylene signaling in plants. *EMBO J.* 22, 1282; **Figure 12–30** Adapted from Tichtinsky, G., Vanoosthuyse, V., Cock, J.M., & Gaude, T. (2003) Making inroads into plant receptor kinase signalling pathways. *Trends Plant Sci.* 8, 231, Fig. 1; **Figure 12–33** PDB ID 1BAC; Chou, K.-C., Carlacci, L., Maggiora, G.M., Parodi, L.A., & Schulz, M.W. (1992) An energy-based approach to packing the 7-helix bundle of bacteriorho-dopsin. *Protein Sci.* 1, 810; **Figure 12–35** Adapted from Nathans, J. (1989) The genes for color vision. *Sci. Am.* 260 (February), 42; **Box 12–3 Figure 1** Courtesy Professor J.D. Mollon, Department of Experimental Psychology, Cambridge University; **Figure 12–40** O'Malley, B.W., & Tsai, M.-J. (1992) Molecular pathways of steroid receptor action. *Biol. Reprod.* 46, 163; **Figure 12–42a** PDB ID 1HCK; Schulze-Gahmen, U., De Bondt, H.L., & Kim, S.-H. (1996) High-resolution crystal structures of human cyclin-dependent kinase 2 with and without ATP: bound waters and natural ligand as guides for inhibitor design. *J. Med. Chem.* 39, 4540; **Figure 12–42b** PDB ID 1FIN; Jeffrey, P.D., Russo, A.A., Polyak, K., Gibbs, E., Hurwitz, J., Massague, J., & Pavletich, N.P. (1995) Mechanism of Cdk activation revealed by the structure of a cyclin a-Cdk2 complex. *Nature* 376, 313; **Figure 12–42c** PDB ID 1JST; Russo, A.A., Jeffrey, P.D., & Pavletich, N.P. (1996) Structural basis of cyclin-dependent kinase activation by phos-phorylation. *Nat. Struct. Biol.* 3, 696. **Figure 12–43** Data from Pines, J. (1999) Four-dimensional control of the cell cycle. *Nat. Cell Biol.* 1, E73; **Figure 12–49** Adapted from Kinzler, K.W., & Vogelstein, B. (1996) Lessons from hereditary colorectal cancer. *Cell* 87, 159.

**PART II    p. 480** PDB ID 1JBO; see citation for Figure 19–51b; **p. 481,** PDB ID 5CTS; see citation at Figure 16–8b.

**CHAPTER 13    p. 489** ATP extracted from PDB ID 1MO8; Hilge, M., Siegal, G., Vuister, G.W., Guentert, P., Gloor, S.M., & Abrahams, J.P. (2003) ATP-induced conformational changes of the nucleotide-binding domain of Na,K-ATPase *Nat. Struct. Biol.* 10, 468; Rudolf Clausius (1865) Über Verschiedene für die Anwendung bequeme Formen der Hauptgleichungen der mechanischen Wärmetheorie (On several convenient forms of the second fundamental equations of the mechanical theory of heat). In *The Mechanical Theory of Heat with Its Applications to the Steam-Engine and to the Physical Properties of Bodies* (J. Tyndall, trans.). John van Voorst, London, p. 365; François Jacob (1970) *The Logic of Life: A History of Heredity*. Pantheon Books, Inc., New York, 1973; originally published (1970) as *La logique du vivant: une histoire de l'hérédité.* Editions Gallimard, Paris; (Lavoisier) The Granger Collection; **p. 490** © Sidney Harris; **Box 13–2 Figure 1** E.R. Degginger/Photo Researchers; **Figure 13–16** PDB ID 1LDH; White, J.L., Hackert, M.L., Buehner, M., Adams, M.J., Ford, G.C., Lentz, P.J., Jr., Smiley, I.E., Steindel, S.J., & Rossmann, M.G. (1976) A comparison of the structures of apo dogfish M4 lactate dehydrogenase and its ternary complexes. *J. Mol. Biol.* 102, 759; **p. 515** (Strong) Courtesy of the Department of Biochemistry, University of Wisconsin–Madison, (Woolley) Rockefeller Archive Center, (Elvehjem) Courtesy of the Department of Biochemistry, University of Wisconsin–Madison.

**CHAPTER 14    p. 521** Arthur Harden (1923) *Alcoholic Fermentation,* 3d edn. Longmans, Green and Co., London; **p. 522** (von Euler) Austrian Archives/ Corbis, (Embden) Courtesy of Institut für Biochemie I: Molekulare Bioenergetik, Universitaetsklinikum Frankfurt, ZBC, (Meyerhoff) Hulton-Seutsch Collection/ Corbis; **p. 525** (Harden) Hulton Getty/Liaison; (Young) Courtesy Medical History Museum, The University of Melbourne; **p. 534** (Warburg) Hulton Archive by Getty Images; **Box 14–1** Animals Animals; **Figure 14–14** New Brunswick Scientific Co., Inc..

**CHAPTER 15    p. 560** PDB ID 1LL2; see citation at Figure 15–10; CF. Cori and G.T. Cori (1929) Glycogen formation in the liver from *d*- and *l*-lactic acid. *Journal of Biological Chemistry* 81, 389–403; Alfred A. Knopf, New York; **Figure 15–1** Courtesy of Minoru Kanehisa, Institute for Chemical Research, Kyoto University, Japan; **Figure 15–2** BCC Microimaging, reproduced with permission; **p. 565** (Leloir) The Nobel Foundation; **Box 15–1** AP; **Figure 15–10** PDB ID 1LL2; Gibbons, B.J., Roach, P.J., & Hurley, T.D. (2002) Crystal structure of the autocatalytic initiator of glycogen biosynthesis, glycogenin. *J. Mol. Biol.* 319, 463; **Figure 15–18a** PDB ID 1PFK; Shirakihara, Y., & Evans, P.R. (1988) Crystal structure of the complex of phosphofructokinase from *Escherichia coli* with its reaction products. *J. Mol. Biol.* 204, 973; **p. 584** (Sutherland) The Nobel Foundation; **p. 591** H.L. Mencken (1949) *A Mencken Chrestomathy;* (Buchner) The Nobel Foundation; **Figure 15–33** Data from Torres, N.V., Mateo, F., Melendez-Hevia, E., & Kacser, H. (1986) Kinetics of metabolic pathways. A system in vitro to study the control of flux. *Biochem. J.* 234, 169; **Box 15–3 Figure 2** Fell, D. (1997) *Understanding the Control of Metabolism*, Portland, London, p. 103.

**CHAPTER 16    p. 601** See citation for Figure 16–5; H.A. Krebs and W.A. Johnson (1937) The role of citric acid in intermediate metabolism in animal tissues. *Enzymologia* 4, 148; (Krebs) Courtesy Public Relations Office, University of Sheffield; **Figure 16–5a,b** Courtesy of Dr. Z. Hong Zhou, Department of Pathology and Laboratory Medicine, University of Texas–Houston Medical School; **Figure 16–8a** PDB ID 5CSC; Liao, D.-I., Karpusas, M., & Remington, S.J. (1991) Crystal structure of an open conformation of citrate synthase from chicken heart at 2.8-Å resolution. *Biochemistry* 30, 6031; **Figure 16–8b** PDB ID 5CTS; Karpusas, M., Branchaud, B., & Remington, S.J. (1990) Proposed mechanism for the condensation reaction of citrate synthase: 1.9-angstroms structure of the ternary complex with oxaloac-etate and carboxymethyl coenzyme A. *Biochemistry* 29, 2213; **Figure 16–9** Adapted from Remington, J.S. (1992) Mechanisms of citrate synthase and related enzymes (triose phosphate isomerase and mandelate racemase). *Curr. Opin. Struct. Biol.* 2, 730; **Figure 16–12b** PDB ID 1SCU; Wolodko, W.T., Fraser, M.E., James, M.N.G., & Bridger, W.A. (1994) The crystal structure of succinyl-CoA synthetase from *Escherichia coli* at 2.5 angstroms resolution. *J. Biol. Chem.* 289, 10,883; **Figure 16–21** Richard N. Trelease, Arizona State University, Department of Botany.

**CHAPTER 17    p. 631** John Clarke (1639) *Paroemiologia Anglo-Latina (Proverbs English and Latin).* Imprinted by F. Kyngston for R. Mylbourne, London; **Box 17–1** Stouffer Productions/Animals Animals; **Box 17–2** (Hodgkin)The Nobel Foundation.

**CHAPTER 18    p. 656** PDB ID 1AJS; see citation at Figure 18–5c, d, e; Hans Krebs (1970) The history of the tricarboxylic acid cycle. *Perspectives in Biology and Medicine* 14, 154–170, 161–162; **Figure 18–5c,d,e** PDB ID 1AJS; Rhee, S., Silva, M.M., Hyde, C.C., Rogers, P.H., Metzler, C.M., Metzler, D.E., & Arnone, A. (1997) Refinement and comparisons of the crystal struc-tures of pig cytosolic aspartate aminotransferase and its complex with 2-methylaspartate. *J. Biol. Chem.* 272, 17,293.

**CHAPTER 19    p. 690** PDB ID 1BMF; see citation for Figure 19–23c; Claude Bernard (1865) *An Introduction to the Study of Experimental Medicine* (H.C. Greene, trans.) Harry Schumann, Paris; reprinted by Dover Publications, New York, 1957; Peter Mitchell (1979) Keilin's respiratory chain concept and its chemiosmotic consequences. *Science* 206, 1148–1159; **p. 691** (Lehninger) Alan Mason Chesney Medical Archives of Johns Hopkins Medical Institutions; **p. 693** (Beinert) Courtesy of Helmut Beinert, University of Wisconsin–Madison; **Figure 19–5d** PDB ID 1FRD; Jacobson, B.L., Chae, Y.K., Markley, J.L., Rayment, I., & Holden, H.M. (1993) Molecular structure of the oxidized, recombinant, heterocyst (2Fe-2S) ferredoxin from Anabaena 7120 determined to 1.7 ångstroms resolution. *Biochemistry* 32, 6788; **Figure 19–10** PDB ID 1NEK; Yankovskaya, V., Horsefield, R., Tornroth, S., Luna-Chavez, C., Miyoshi, H., Leger, C., Byrne, B., Cecchini, G., & Iwata, S. (2003) Architecture of succinate dehydrogenase and reactive oxygen species generation. *Science* 299, 700; **Figure 19–11a** PDB ID 1BJY; Iwata, S., Lee, J.W., Okada, K., Lee, J.K., Iwata, M., Rasmussen, B., Link, T.A., Ramaswamy, S., & Jap, B.K. (1998) Complete structure of the 11-subunit bovine mitochondrial cytochrome $bc_1$ complex. *Science* 281, 64; **Figure 19–13** PDB ID 1OCC; Tsukihara, T., Aoyama, H., Yamashita, E., Tomizaki, T., Yamaguchi, H., Shinzawa-Itoh, K., Nakashima, R., Yaono, R., & Yoshikawa, S. (1996) The whole structure of the 13-subunit oxidized cytochrome *c* oxidase at 2.8 Å. *Science* 272, 1136; **Figure 19–14** Adapted from Williams, R.J.P. (1995) *Nature* 378, 235, a correction to Williams, R.J.P. (1995) Purpose of proton pathways. *Nature* 376, 643; **p. 704** (Mitchell) AP/Wide World Photos; **Box 19–1 Figure 1** D. Cavagnaro/Visuals Unlimited; **p. 708** (Racker) Courtesy E. Racker; **Figure 19–21b** PDB ID 1BMF; Abrahams, J.P., Leslie, A.G., Lutter, R., & Walker, J.E. (1994) Structure at 2.8 Å resolution of $F_1$-ATPase from bovine heart mitochondria. *Nature* 370, 621; **Figure 19–23b,c** PDB ID 1BMF; see citation for Figure 19–21b; **p. 710** (Walker) The Nobel Foundation; **Figure 19–23d,e** PDB ID 1QO1; Stock, D., Leslie, A. G.W., & Walker, J.E. (1999) Molecular architecture of the rotary motor in ATP synthase. *Science* 286, 1700; protein backbones were calculated from the $C_\alpha$ PDB entry using MIDAS-PLUS and Sybyl; **Figure 19–23f** Adapted from Fillingame, R.H. (1999) Protein structure: molecular rotary motors. *Science* 286, 1687; **p. 712** (Boyer) The Nobel Foundation; **Figure 19–25 (left)** Adapted from Sambongi, Y., Iko, Y., Tanabe, M., Omote, H., Iwamoto-Kihara, A., Ueda, I., Yanagida, T., Wada, Y., & Futai, M. (1999) Mechanical rotation of the c subunit oligomer in ATP synthase ($F_oF_1$): direct observation. *Science* 286, 1722–1724; **Figure 19–25 (right)** Courtesy Ryohei Yasuda and Kazuhiko Kinosita, from Yasuda, R., Noji, H., Kinosita, K., Jr., & Yoshida, M. (1998) $F_1$-ATPase is a highly efficient molecular motor that rotates with discrete 120° steps. *Cell* 93, 1117; **Figure 19–29** PDB ID 1OHH; Cabezon, E., Montgomery, M.G., Leslie, A.G.W., & Walker, J.E. (2003) The structure of bovine $F_1$-ATPase in complex with its regulatory protein IF$_1$ *Nat. Struct. Biol.* 10, 744; **Figure 19–32a** Morris, M.A. (1990) Mitomutations in neuro-ophthalmological diseases: a review. *J. Clin. Neuroophthalmol.* 10, 159; **Figure 19–32b** From Wallace, D., Zheng, X., Lott, M.T., Shoffner, J.M., Hodge, J.A., Kelley, R.I., Epstein, C.M., & Hopkins, L.C. (1988) Familial mitochondrial encephalomyopathy (MERRF): genetic, pathophysiological, and biochemical characterization of a mitochondrial DNA disease. *Cell* 55, 601; **Figure 19–33** Adapted from Harris, D.A. (1995) *Bioenergetics at a Glance.* Blackwell Science, London, p. 36; **Figure 19–38b** Biological Photo Service; **Figure 19–42** Kühlbrandt, W., Wang, D.N., & Fujiyoshi, Y. (1994) Atomic model of plant light-harvesting complex by electron crystallography. *Nature* 367, 614; **Figures 19–43, 19–47a, b, 19–52, 19–53, 19–55** Adapted from Heldt, H.-W. (1997) *Plant Biochemistry and Molecular Biology.* Oxford University Press, Oxford, pp. 57, 62, 63, 100, 101, 133; **Figure 19–48** PDB ID 1PRC; Deisenhofer, J., Epp, O., Sinning, I., & Michel, H. (1995) Crystallographic refinement at 2.3 angstroms resolution and refined model of the photosynthetic reaction center from *Rhodopseudomonas viridis. J. Mol. Biol.* 246, 429; **Figure 19–50** Adapted from A.W. Rutherford, A.W., & Faller, P. (2001) The heart of photosynthesis in glorious 3D. *Trends Biochem. Sci.* 26, 341, Fig. 1; **Figure 19–51a** Adapted from Kuhlbrandt, W. (2001) Structural biology: Chlorophylls galore. *Nature* 411, 896, Fig. 1; **Figure 19–51b,c** PDB ID 1JBO; Nield, J., Rizkallah, P.J., Barber, J., & Chayen, N.E. (2003) The 1.45Å three-dimensional structure of c-phycocyanin from the thermophilic cyanobacterium *Synechococcus elongatus. J. Struct. Biol.* 141, 149; **Figure 19–54a,b** PDB ID 1UM3; Kurisu, G., Zhang, H., Smith, J. L., & Cramer, W. A. (2003) Structure of the cytochrome *b6f* complex of oxygenic photosynthesis: Tuning the cavity *Science* 302, 1009; **p. 740** (Arnon) University of California, Berkeley; (Jagendorf) Cornell University; **Figure 19–59a** PDB ID 1C8R; Luecke, H., Schobert, B., Richter, H.-T., Cartailler, J.-P., & Lanyi, J.K. (1999) Structural changes in bacteriorhodopsin during ion transport at 2 Å resolution. *Science* 286, 255; **Figure 19–59b** Adapted from Gennis, R.B., & Ebrey, T.G. (1999) Proton pump caught in the act. *Science* 286, 252.

**CHAPTER 20    p. 751** PDB ID 8RUC; see citation for Figure 20–2; Melvin Calvin (1961) Nobel Address, www.nobel.se/chemistry/laureates/1961/calvin-lecture.html; **Figure 20–1** PhotoDisc; **p. 752** (Calvin) The Nobel Foundation; **Figure 20–2** Ken Wagner/ Visuals Unlimited; **Figure 20–5a** PDB ID 8RUC; Andersson, I. (1996) Large structures at high resolution: the 1.6 angstroms crystal structure of spinach ribulose-1,5-bisphosphate carboxylase/oxygenase complexed with 2-carboxyarabinitol bisphosphate. *J. Mol. Biol.* 259, 160; **Figure 20–5b** PDB ID 9RUB; Lundqvist, T., & Schneider, G. (1991) Crystal structure of activated ribulose-1,5-bisphosphate carboxylase complexed with its substrate, ribulose-1,5-bisphosphate. *J. Biol. Chem.* 266, 12,604; **Figure 20–6** PDB ID 1RCX Taylor, T.C., & Andersson, I. (1997) The structure of the complex between rubisco and its natural substrate ribulose 1,5-bisphosphate. *J. Mol. Biol.* 265, 432; **Figure 20–18** Halliwell, B. (1984) *Chloroplast Metabolism: The Structure and Function of Chloroplasts in Green Leaf Cells,* Clarendon Press, Oxford, p. 97; **Figure 20–23a** Ray Evert, University of Wisconsin–Madison, Department of Botany; **Figure 20–29** (cellulose) Ken Wagner/ Visuals Unlimited; art adapted from Becker, W. M., & Deamer, D.W. (1991) *The World of the Cell,* 2nd edn, The Benjamin/Cummings Publishing Company, Inc., Redwood City, CA, Fig. 3–20, p. 60; **Figure 20–30** Courtesy of Mark J. Grimson, Texas Tech University, & Candace H. Haigler, North Carolina State University; **Figure 20–32** Adapted from Buchanan, B. B., Gruissem, W., & Jones, R.L. (2000) *Biochemistry & Molecular Biology of Plants,* American Society of Plant Physiologists, Rockville, MD, Fig. 13.24, p. 649; **p. 786** **(problem 11)** Courtesy of Elena V. Voznesenskaya, Vincent R. Franceschi, Olavi Kiirats, Elena G. Artyusheva, Helmut Freitag, & Gerald E. Edwards.

**CHAPTER 21    p. 787** PDB ID 3PGH; see citation for Box 21–2 Figure 1a; Konrad Bloch (1987) Summing up. *Annual Review of Biochemistry* 56, 10; **Figure 21–11b** Daniel Lane, Johns Hopkins University, School of Medicine; **Box 21–2 Figure 1a** PDB ID 3PGH; Kurumbail, R.G., Stevens, A.M., Gierse, J.K., McDonald, J.J., Stegeman, R.A., Pak, J.Y., Gildehaus, D., Miyashiro, J.M., Penning, T.D., Seibert, K., Isakson, P.C., & Stallings, W.C. (1996) Structural basis for selective inhibition of cyclooxygenase-2 by anti-inflammatory agents. *Nature* 384, 644; **Box 21–2 Figure 1b** PDB ID 1EQG; Selinsky, B.S., Gupta, K., Sharkey, C.T., & Loll, P.J. (2001) Structural analysis of NSAID binding by prostaglandin H2 synthase: Time-dependent and time-independent inhibitors elicit identical enzyme conformations. *Biochemistry* 40, 5172; **p. 809** (Kennedy) Harvard Medical School; **p. 820** (Bloch, Lynen, Cornforth) AP/Wide World Photos; (Popjak) Corbis/UPI/Bettmann; **Figure 21–39b** Courtesy Robert L. Hamilton and the Arteriosclerosis Specialized Center of Research, University of California, San Francisco; **Figure 21–40b** Alan Attie, Department of Biochemistry, University of Wisconsin–Madison; **p. 824** Corbis/UPI/Bettmann.

**CHAPTER 22    p. 833** Ric Ergenbright/Corbis; Gertrude Elion, quoted in S.B. McGrayne (2002) Damn the torpedoes. Full speed ahead. *Science* 296, **851**–852; **Figure 22–3** PDB ID 1N2C; Schindelin, H., Kisker, C., Schlessman, J.L., Howard, J.B., & Rees, D.C. (1997) Structure of ADP X Alf4⁻-stabilized nitrogenase complex and its implications for signal transduction. *Nature* 387, 370; **Figure 22–4a** C.P. Vance/Visuals Unlimited; **(inset)** Ric Ergenbright/ Corbis; **Figure 22–4b** Jeremy Burgess/Photo Researchers; **Figure 22–5** PDB ID 2GLS; Yamashita, M.M., Almassy, R.J., Janson, C.A., Cascio, D., & Eisenberg, D. (1989) Refined atomic model of glutamine synthetase at 3.5 angstroms resolution. *J. Biol. Chem.* 264, 17,681; **Box 22–2 Figure 1** John Mansfield, University of Wisconsin–Madison, Department of Veterinary Science; **p. 864** (Buchanan) Massachusetts Institute of Technology Museum Collection; **Figure 22–37** PDB ID 1M6V; Thoden, J.B., Huang, X., Raushel, F.M., Holden, H.M. (2002) Carbamoyl-phosphate synthetase: creation of an escape route for ammonia. *J. Biol. Chem.* 277, 39,722; Jim Thoden and Hazel Holden, University of Wisconsin–Madison, Department of Biochemistry and Enzyme Institute, provided preliminary data for the channel path; **Figure 22–40a** Thelander, L., & Reichard, P. (1979) Reduction of ribonucleotides. *Annu. Rev. Biochem.* 48, 133; **Figure 22–40b** PDB ID 1PFR; Logan, D.T., Su, X.D., Aberg, A., Regnstrom, K., Hajdu, J., Eklund, H., & Nordlund, P. (1996) Crystal structure of reduced protein r2 of ribonucleotide reductase: the structural basis for oxygen activation at a dinuclear iron site. *Structure* 4, 1053. **p. 876** Courtesy Kathy Bendo Hitchings.

**CHAPTER 23    p. 881** Fred Hossler/Visuals Unlimited; Charles Édouard Brown-Séquard and J. d'Arsonval (1891) Additions à une note sur l'injection des extraits liquides de divers organes, comme méthode thérapeutique. (Additions to a note on the injection of aqueous extracts of various organs, as a therapeutic method). *Comptes Rendus de la Société de Biologie* (9)3, 267; **Box 23–1** (Banting, MacLeod, Best, Collip) Corbis/UPI/Bettmann; **p. 884** (Guillemin) Corbis/UPI/Bettmann; (Schally) AP/Wide World Photos; (Yalow) Courtesy Rosalyn S. Yalow, Veterans Affairs Medical Center, NYC; **Figure 23–3** Adapted from Orth, D.N. (1975) General considerations for radioimmunoassay of peptide hormones. *Meth. Enzymol.* 37, 22; **Figure 23–16** Fred Hossler/Visuals Unlimited; **Figure 23–19** D. W. Fawcett/Photo Researchers; **Figure 23–21** Courtesy M. L. Thomas, H. C. Sing, G. Belenky, Division of Neuropsychiatry, Walter Reed Army Institute of Research, U.S. Army Medical Research Materiel Command; **Figure 23–31** John Sholtis, The Rockefeller University, New York; **Figure 23–32** Adapted from Ezzell, C. (1995) Fat times for obesity research: tons of new information, but how does it all fit together? *J. NIH Res.* 7, 39; **Figure 23–33** Adapted from Schwartz, M.W., & Morton, G.J. (2002) Obesity: Keeping hunger at bay. *Nature* 418, 595, Fig. 1; **Figure 23–34** Adapted from Auwerx, J., & Staels, B. (1998) Leptin. *Lancet* 351, 737; **Figure 23–39** Adapted from Cummings, D.E., Purnell, J.Q., Frayo, R.S., Schmidova, K., Wisse, B.E., & Weigle, D.S. (2001) A preprandial rise in plasma ghrelin levels suggests a role in meal initiation in humans. *Diabetes* 50, 1714, Fig. 1.

**PART III    p. 920** Derived from coordinates supplied by Seth Darst, The Rockefeller University; **p. 921** PDB ID 1Q7Y; see citation for Figure 27–9c.

**CHAPTER 24    p. 923** G.F. Bahr/Biological Photo Service; J.C. Wang (2002) Cellular roles of DNA topoisomerases: a molecular perspective. *Nature Review of Molecular and Cell Biology* 3, 430–440; N.R. Cozzarelli (1993) The structure and function of DNA supercoiling and catenanes. *Harvey Lectures* 87, 35–55; **Figure 24–1** From Kleinschmidt, A.K., Land, D., Jackerts, D., & Zahn, R.K. (1962) Darstellung und Längenmessungen des gesamten Desoxyribonucleinsäure-Inhalter von T₂-Bakteriophagen. *Biochim. Biophys. Acta* 61, pp. 857–864; **p. 924** (Beadle) Archive Photos; (Tatum) Corbis/UPI/Bettmann; **Figure 24–4** Huntington Potter and David Dressler, Harvard Medical School, Department of Neurobiology; **Figure 24–5a,b** G.F. Bahr/Biological Photo Service; **Figure 24–6** D.W. Fawcett/Photo Researchers; **Figures 24–11** Adapted from Cozzarelli, N.R., Boles, T.C., & White, J.H. (1990) Primer on the topology and geometry of DNA supercoiling. In *DNA Topology and Its Biological Effects* (Cozzarelli, N.R., & Wang, J.C., eds.), pp. 139–184. Cold Spring Harbor Laboratory Press, Cold Spring Harbor, NY; **Figure 24–12** Adapted from Saenger, W. (1984) *Principles of Nucleic Acid Structure,* Springer-Verlag, New York, p. 452; **Figure 24–13** Laurien Polder, from Kornberg, A. (1980) *DNA Replication,* W. H. Freeman & Company, New York, p. 29; **Figures 24–14, 24–15** see citation for Figure 24–11; **Figure 24–20** Keller, W. (1975) Characterization of purified DNA-relaxing enzyme from human tissue culture cells. *Proc. Natl. Acad. Sci. USA* 72, 2553; **Figures 24–21, 24–22** Adapted from Champoux, J.J. (2001) DNA topoisomerases: structure, function, and mechanism. *Annu. Rev. Biochem.* 70, 369, Figs 3, 11; **Figure 24–23a** James H. White, T. Christian Boles, & N.R. Cozzarelli, University of California, Berkeley, Department of Molecular and Cell Biology; **Figure 24–26b** Ada L. Olins & Donald E. Olins, Oak Ridge

National Laboratory; **Figure 24–27** PDB ID 1AOI; Luger, K., Maeder, A.W., Richmond, R.K., Sargent, D.F., & Richmond, T.J. (1997) Crystal structure of the nucleosome core particle at 2.8 Å resolution. *Nature* 389, 251; **Figure 24–30b** Barbara Hamkalo, University of California, Irvine, Department of Molecular Biology and Biochemistry; **Figure 24–31** D. W. Fawcett/Visuals Unlimited; **Figure 24–34a,b** Adapted from Hirano, T. (2002) The ABCs of SMC proteins: two-armed ATPases for chromosome condensation, cohesion, and repair. *Genes Dev.* 16, 399; **Figure 24–34c** Thomas E. Melby, Charles N. Ciampaglio, Gina Briscoe, & Harold P. Erickson; **Figure 24–35** Adapted from Bazett-Jones, D.P., Kimura, K., & Hirano, T. (2002) Efficient supercoiling of DNA by a single condensin complex as revealed by electron spectroscopic imaging. *Mol. Cell* 9, 1183, Fig. 5; **p. 947 (problem 8)** Roger Kornberg, MRC Laboratory of Molecular Biology.

**CHAPTER 25   p. 948** See citation for Figure 25–10b; Arthur Kornberg (1975) For the love of enzymes. In *In Honour of Severo Ochoa* (A. Kornberg et al., eds), 1st edn. Pergamon Press, New York, 1976, pp. 243–251; **Figure 25–3a** Bernard Hirt, Institut Suisse de Recherches Experimentales sur le Cancer; **Figure 25–3c** Cairns, J. (1963) The chromosomes of *E. coli*. *Cold Spring Harbor Symp. Quant. Biol.* 28, 44; **p. 952** (Kornberg) AP/Wide World Photos; **Figure 25–8** PDB ID 3BDP; Kiefer, J.R., Mao, C., Braman, J.C., & Beese, L.S. (1998) Visualizing DNA replication in a catalytically active *Bacillus* DNA polymerase crystal. *Nature* 391, 304; **Figure 25–10b** PDB ID 2POL; Kong, X.-P., Onrust, R., O'Donnell, M., & Kuriyan, J. (1992) Three-dimensional structure of the β subunit of *Escherichia coli* DNA polymerase III holoenzyme: a sliding DNA clamp. *Cell* 69, 425; **Figure 25–12** Bramhill, D., & Kornberg, A. (1988) A model for initiation at origins of DNA replication. *Cell* 54, 915; **Figure 25–14** Adapted from an animation kindly provided by Mike O'Donnell, The Rockefeller University; **Figure 25–17** Wake, R.G., & King, G.F. (1997) A tale of two terminators: crystal structures sharpen the debate on DNA replication fork arrest mechanisms *Structure* 5, 1; **Figure 25–18b** Adapted from Lemon, K.P., & Grossman, A.D. (2001) The extrusion-capture model for chromosome partitioning in bacteria. *Genes Dev.* 15, 2031; **Figure 25–19** Bruce N. Ames, University of California, Berkeley, Department of Biochemistry & Molecular Biology; **Figure 25–21** Adapted from a figure provided by Paul Modrich; **Figure 25–22** Adapted from Grilley, M., Griffith, J., & Modrich, P. (1993) Bidirectional excision in methyl-directed mismatch repair. *J. Biol. Chem.* 268, 11,830; **Figure 25–23** Watson, J.D., Hopkins, N.H., Roberts, J.W., Steitz, J.A., & Weiner, A.M. (1987) *Molecular Biology of the Gene,* 4th edn, The Benjamin/Cummings Publishing Company, Menlo Park, CA, p. 350; **Figure 25–24** Adapted from a figure provided by Aziz Sancar; **p. 978** (McClintock) AP/Wide World Photos; **Figure 25–30b** Bernard John, The Australian National University; **Figure 25–31b** Huntington Potter and David Dressler, Harvard Medical School, Department of Neurobiology; **Figure 25–34a** John Heuser, Washington University Medical School, Department of Biochemistry; **Figure 25–34b** PDB ID 2REB; Story, R.M., Weber, I.T., & Steitz, T.A. (1992) The structure of the *E. coli* RecA protein monomer and polymer. *Nature* 355, 318; **Figure 25–36** Roca, A.I., & Cox, M.M. (1990) The RecA protein: structure and function. *Crit. Rev. Biochem. Mol. Biol.* 25, 415; **Figure 25–38a** Adapted from Guo, F., Gopaul, D.N., & Van Duyne, G.D. (1997) Structure of Cre recombinase complexed with DNA in a site-specific recombination synapse. *Nature* 389, 40; **Figure 25–38b** PDB ID 3CRX; Gopaul, D.N., Guo, F., & Van Duyne, G.D. (1998) Structure of the Holliday junction intermediate in Cre-Loxp site-specific recombination. *EMBO J.* 17, 4175.

**CHAPTER 26   p. 995** Coordinates generated by Sybyl; Francis H.C. Crick (1958) The biological replication of macromolecules. *Symposium of the Society of Experimental Biology* 12, 146; **Figure 26–4** PDB ID 1IW7; Vassylyev, D.G., Sekine, S., Laptenko, O., Lee, J., Vassylyeva, M.N., Borukhov, S., Yokoyama, S. (2002) Crystal structure of a bacterial RNA polymerase holoenzyme at 2.6 Å resolution. *Nature* 417, 712; **Figure 26–6b** Derived from coordinates supplied by Seth Darst, The Rockefeller University; **Figure 26–7** Landick, R. (1997) RNA polymerase slides home: pause and termination site recognition. *Cell* 88, 741; **Box 26–1 Figure 2** Carol Gross, University of California, San Francisco, Department of Stomatology; **Figure 26–9b** PDB ID 1TGH; Juo, Z.S., Chiu, T.K., Leiberman, P.M., Baikalov, I., Berk, A.J., & Dickerson, R.E. (1996) How proteins recognize the TATA-box. *J. Mol. Biol.* 261, 239; **Figure 26–10b** PDB ID 1DSC; Lian, C., Robinson, H., & Wang, A.H.-J. (1996) Structure of actinomycin D bound with (GAAGCTTC)2 and (GATGCTTC)2 and its binding to the (CAG)N:(CTG)N triplet sequence by NMR analysis. *J. Am. Chem. Soc.* 118, 8791; **Figure 26–12a** Pierre Chambon, Laboratorie de Génétique des Eucaryotes, Faculté de Médecine (CNRS); **Figure 26–12b,c** Chambon, P. (1981) Split genes. *Sci. Am.* 244 (May), 60; **p. 1009** (Cech) Corbis/UPI/Bettmann; **Figure 26–15** Cech, T.R. (1986) RNA as an enzyme. *Sci. Am.* 255 (November), 64; **Figure 26–16a**

Kramer, A. (1996) The structure and function of proteins involved in mammalian pre-mRNA splicing. *Annu. Rev. Biochem.* 65, 367; **Figure 26–25b** PDB ID 1MME; see citation for Figure 8–28b; **Figure 26–26** Cech, T.R., Damberger, S.H., & Gutell, R.R. (1994) Representation of the secondary and tertiary structure of group I introns. *Struct. Biol.* 1, 273; **Figure 26–27** Cech, T.R. (1986) RNA as an enzyme. *Sci. Am.* 255 (November), 64; **p. 1020** (Grunberg-Manago) Courtesy Marianne Grunberg-Manago; (Ochoa) AP/Wide World Photos; **p. 1022** (Temin) Corbis/UPI/Bettmann; (Baltimore) AP/Wide World Photos; **Figure 26–32** Haseltine, W.A., & Wong-Staal, F. (1988) The molecular biology of the AIDS virus. *Sci. Am.* 259 (October), 52; **Figure 26–33** Kingsman, A.J., & Kingsman, S.M. (1988) Ty: a retroelement moving forward. *Cell* 53, 333; **Figure 26–35a** Boeke, J.D. (1990) Reverse transcriptase, the end of the chromosome, and the end of life. *Cell* 61, 193; **Figure 26–35c** Jack Griffith, University of North Carolina at Chapel Hill, Comprehensive Cancer Center; **p. 1027** (Woese, Crick, Orgel) AP/Wide World Photos; **Box 26–3, Figure 3** PDB ID 1RAW; Dieckmann, T., Suzuki, E., Nakamura, G.K., Feigon, J. (1996) Solution structure of an ATP-binding RNA aptamer reveals a novel fold. *RNA* 2, 628.

**CHAPTER 27   p. 1034** PDB ID 1QRT; see citation for Figure 27–17a; Gerald Joyce, quoted in Waldrop, M.M. (1992) Finding RNA makes proteins gives "RNA World" a big boost. *Science* 256, 1396–1397; **p. 1035** (Zamecnik) News Office, Massachusetts General Hospital; **Figure 27–1** D.W. Fawcett/ Visuals Unlimited; **p. 1037** (Nirenberg) AP/Wide World Photos; **p. 1038** (Khorana) Courtesy Archives, University of Wisconsin–Madison; **p. 1045** Courtesy Masayasu Nomura; **Figure 27–9a (left)** PDB ID 1JJ2; Klein, D.J., Schmeing, T.M., Moore, P.B., & Steitz, T.A. (2001) The kink-turn: A new RNA secondary structure motif *EMBO J.* 20, 4214; and PDB ID 1GIY; Yusupov, M.M., Yusupova, G.Z., Baucom, A., Lieberman, K., Earnest, T.N., Cate, J.H.D., & Noller, H.F. (2001) Crystal structure of the ribosome at 5.5 Å resolution *Science* 292, 883; **Figure 27–9a (right)** PDB ID 1J5E; Wimberly, B.T., Brodersen, D.E., Clemons W.M., Jr., Morgan-Warren, R., Carter, A.P., Vonrhein, C., Hartsch, T., & Ramakrishnan, V. (2000) Structure of the 30S ribosomal subunit *Nature* 407, 327; and PDB ID 1JGO; Yusupova, G.Z., Yusupov, M.M., Cate, J.H., & Noller, H.F. (2001) The path of messenger RNA through the ribosome. *Cell* 106, 233; **Figure 27–9b** PDB IDs 1J5E, 1JJ2, 1JGO, 1GIY; see citations for Figure 27–8a; **Figure 27–9c** PDB ID 1Q7Y; Hansen, J.L., Schmeing, T.M., Moore, P.B., & Steitz, T.A. (2002) Structural insights into peptide bond formation. *Proc. Natl. Acad. Sci. USA* 99, 11,670; **Box 27–3 Figure 1** PDB ID 1NKW; Harms, J.M., Schluenzen, F., Zarivach, R., Bashan, A., Gat, S., Agmon, I., Bartels, H., Franceschi, F., Yonath, A. (2001) High resolution structure of the large ribosomal subunit from a mesophilic eubacterium. *Cell* 107, 679; **p. 1049** (Holley) Corbis/UPI/ Bettmann; **Figure 27–13b** PDB ID 4TRA; Westhof, E., Dumas, P., Moras, D. (1988) Restrained refinement of two crystalline forms of yeast aspartic acid and phenylalanine transfer RNA crystals. *Acta Cryst. A* 44, 112; **Figure 27–17a** PDB ID 1QRT; Arnez, J.G., & Steitz, T.A. (1996) Crystal structures of three misacylating mutants of *Escherichia coli* glutaminyl-tRNA synthetase complexed with tRNA(Gln) and ATP. *Biochemistry* 35, 14,725; **Figure 27–17b** PDB ID 1ASZ; Cavarelli, J., Eriani, G., Rees, B., Ruff, M., Boeglin, M., Mitschler, A., Martin, F., Gangloff, J., Thierry, J.C., & Moras, D. (1994) The active site of yeast aspartyl-tRNA synthetase: structural and functional aspects of the aminoacylation reaction. *EMBO J.* 13, 327; **Figure 27–25b (left)** PDB ID 1B23; Nissen, P., Thirup, S., Kjeldgaard, M., & Nyborg, J. (1999) The crystal structure of Cys-tRNACys-EF-Tu-GDPNP reveals general and specific features in the ternary complex and in tRNA. *Struct. Fold Des.* 7, 143; **Figure 27–25b (right)** PDB ID 1DAR; al-Karadaghi, S., Aevarsson, A., Garber, M., Zheltonosova, J., Liljas, A. (1996) The structure of elongation factor G in complex with GDP: conformational flexibility and nucleotide exchange. *Structure* 4, 555; **Figure 27–27b** Steven L. McKnight and Oscar L. Miller, University of Virginia, Department of Biology; **p. 1068** (Blobel) Courtesy Günter Blobel, Rockefeller University; (Palade) AP/Wide World Photos; **Figure 27–37b** Don W. Fawcett/Photo Researchers; **Figure 27–40c** John Heuser, Washington University Medical School, Department of Biochemistry; **Figure 27–42a** PDB ID 1IRU; Unno, M., Mizushima, T., Morimoto, Y., Tomisugi, Y., Tanaka, K., Yasuoka, N., & Tsukihara, T. (2002) The structure of the mammalian 20S proteasome at 2.75 Å resolution. *Structure* 10, 609.

**CHAPTER 28   p. 1081** See citation for Figure 28–39d; François Jacob and Jacques Monod (1961) Genetic regulatory mechanisms in the synthesis of proteins. *Journal of Molecular Biology* 3, 318–356; **p. 1085** (Jacob) Corbis/Bettmann; (Monod) Corbis/Bettmann; **Figure 28–7b** Personal communication from M. Thomas Record and Oleg Tsodikov, Department of Biochemistry, University of Wisconsin–Madison; **Figure 28–7c,d** PDB ID 1LBG and PDB ID 1LBH; Lewis, M., Chang, G., Horton, N.C., Kercher, M.A., Pace, H.C., Schumacher, M.A., Brennan, R.G., & Lu, P. (1996) Crystal structure of the lactose operon repressor and its complexes with DNA and inducer. *Science* 271, 1247. Side chains were added with software programs MIDAS-PLUS and Sybyl to the $C_\alpha$-only PDB entry; **Figure 28–11a,c,d** PDB ID 1LCC; Chuprina, V.P., Rullmann, J.A.C., Lamerichs, R.M.J.N., Van Boom, J.H., Boelens, R., & Kaptein, R. (1993) Structure of the complex of Lac repressor headpiece and an 11 base-pair half-operator determined by nuclear magnetic resonance spectroscopy and restrained molecular dynamics. *J. Mol. Biol.* 234, 446; **Figure 28–11b** PDB ID 1LBG; see citation for Figure 28–7c; **Figure 28–12** PDB ID 1A1L; Elrod-Erickson, M., Benson, T.E., & Pabo, C.O. (1998) High-resolution structures of variant Zif268-DNA complexes: implications for understanding zinc finger-DNA recognition. *Structure* 6, 451; **Figure 28–13** PDB ID 1B8I; Passner, J.M., Ryoo, H.-D., Shen, L., Mann, R.S., & Aggarwal, A.K. (1999) Structure of a DNA-bound ultrabithorax-extradenticle homeo-domain complex. *Nature* 397, 714; **Figure 28–14a** McKnight, S.L. (1991) Molecular zippers in gene regulation. *Sci. Am.* 264 (April), 54–64; **Figure 28–14b** PDB ID 1YSA; Ellenberger, T.E., Brandl, C.J., Struhl, K., & Harrison, S.C. (1992) The GCN4 basic region leucine zipper binds DNA as a dimer of uninterrupted $\alpha$ helices: crystal structure of the protein-DNA complex. *Cell* 71, 1223; **Figure 28–15** PDB ID 1HLO; Brownlie, P., Ceska, T.A., Lamers, M., Romier, C., Theo, H., & Suck, D. (1997) The crystal structure of an intact human max-DNA complex: new insights into mechanisms of transcriptional control. *Structure* 5, 509; **Figure 28–16** PDB ID 1RUN; Parkinson, G., Gunasekera, A., Vojtechovsky, J., Zhang, X., Kunkel, T.A., Berman, H., &

Ebright, R.H. (1996) Aromatic hydrogen bond in sequence-specific protein-DNA recognition. *Nat. Struct. Biol.* 3, 837; **Figure 28–20** PDB ID 1TRO; Otwinowski, Z., Schevitz, R.W., Zhang, R.-G., Lawson, C.L., Joachimiak, A.J., Marmorstein, R., Luisi, B.F., & Sigler, P.B. (1988) Crystal structure of Trp repressor operator complex at atomic resolution. *Nature* 335, 321; **Figure 28–21a** Watson, J.D., Hopkins, N.H., Roberts, J.W., Steitz, J.A., & Weiner, A.M. (1987) *Molecular Biology of the Gene,* 4th edn, The Benjamin/Cummings Publishing Company, Menlo Park, CA, p. 487; **Figure 28–23** Nomura, M., Gourse, R., & Baughman, G. (1984) Regulation of the synthesis of ribosomes and ribosomal components. *Annu. Rev. Biochem.* 53, 75; **Figure 28–25** John D. Cunningham/Visuals Unlimited; **Figure 28–31** Schwabe, J.W.R., & Rhodes, D. (1991) Beyond zinc fingers: steroid hormone receptors have a novel structural motif for DNA recognition. *Trends Biochem. Sci.* 16, 291; **Figure 28–34** F.R. Turner, University of Indiana, Bloomington, Department of Biology; **p. 1113** (Nüsslein-Volhard) AP/Wide World Photos/Mark Lennihan; **Figure 28–36** Wolfgang Driever and Christiane Nüsslein-Volhard, Max-Planck-Institut; **Figure 28–37** Adapted from a figure provided by Marv Wickens, University of Wisconsin–Madison, Department of Biochemistry; **Figure 28–38b** Courtesy Phillip Ingham, Imperial Cancer Research Fund, Oxford University; **Figure 28–39a,b** F.R. Turner, University of Indiana, Bloomington, Department of Biology; **Figure 28–39c,d** E.B. Lewis, California Institute of Technology, Division of Biology.

# Appendix A    Common Abbreviations in the Biochemical Research Literature

| | |
|---|---|
| Å | angstrom |
| A | adenine, adenosine, or adenylate |
| $A$ | absorbance |
| Ab | antibody |
| ACh | acetylcholine |
| ACP | acyl carrier protein |
| ACTH | adrenocorticotropic hormone |
| Acyl-CoA | acyl derivatives of coenzyme A (also, acyl-S-CoA) |
| ADH | alcohol dehydrogenase |
| adoHcy | $S$-adenosylhomocysteine |
| adoMet | $S$-adenosylmethionine (also, SAM) |
| Ag | antigen |
| AIDS | acquired immunodeficiency syndrome |
| AKAP | PKA-anchoring protein |
| Ala | alanine |
| $[\alpha]_D^{25°C}$ | specific rotation |
| AMP, ADP, ATP | adenosine 5′-mono-, di-, triphosphate |
| AQP | aquaporin |
| Arg | arginine |
| ARS | autonomously replicating sequence |
| Asn | asparagine |
| Asp | aspartate |
| ATCase | aspartate transcarbamoylase |
| atm | atmosphere |
| ATPase | adenosine triphosphatase |
| $B_{12}$ | coenzyme $B_{12}$, cobalamin |
| BAC | bacterial artificial chromosome |
| BMR | basal metabolic rate |
| bp | base pair |
| 1,3-BPG | 1,3-bisphosphoglycerate |
| C | cytosine, cytidine, or cytidylate |
| CaM | calmodulin |
| cAMP, cGMP | 3′,5′-cyclic AMP, 3′,5′-cyclic GMP |
| CDK | cyclin-dependent protein kinase |
| cDNA | complementary DNA |
| CFTR | cystic fibrosis transmembrane conductance regulator |
| Chl | chlorophyll |
| CL | cardiolipin |
| CMP, CDP, CTP | cytidine 5′-mono-, di-, triphosphate |
| CoA | coenzyme A (also, CoASH) |
| CoQ | coenzyme Q (ubiquinone; also, UQ) |
| CRP | cAMP receptor protein |
| Cys | cysteine |
| D | dihydrouridine |
| $D$ | diffusion coefficient |
| $d$ | density |
| dADP, dGDP, etc. | deoxyadenosine 5′-diphosphate, deoxyguanosine 5′-diphosphate, etc. |
| dAMP, dGMP, etc. | deoxyadenosine 5′-monophosphate, deoxyguanosine 5′-monophosphate, etc. |
| dATP, dGTP, etc. | deoxyadenosine 5′-triphosphate, deoxyguanosine 5′-triphosphate, etc. |
| DEAE | diethylaminoethyl |
| DFP (DIFP) | diisopropylfluorophosphate |
| DHAP | dihydroxyacetone phosphate |
| DHF | dihydrofolate (also, $H_2$ folate) |
| DMS | dimethyl sulfate |
| DNA | deoxyribonucleic acid |
| DNase | deoxyribonuclease |
| DNP | 2,4-dinitrophenol |
| Dol | dolichol |
| DOPA | dihydroxyphenylalanine |
| $E$ | electrical potential |
| E.C. | Enzyme Commission (followed by numbers indicating the formal classification of an enzyme) |
| EDTA | ethylenediaminetetraacetate |
| EF | elongation factor |
| EGF | epidermal growth factor |
| ELISA | enzyme-linked immunosorbent assay |
| EM | electron microscopy |
| $\epsilon$ | molar absorption coefficient |
| ER | endoplasmic reticulum |
| EST | expressed sequence tag |
| $\eta$ | viscosity |
| $\mathscr{F}$ | faraday |
| $f$ | frictional coefficient |
| FA | fatty acid |
| FAD, $FADH_2$ | flavin adenine dinucleotide, and its reduced form |
| FBPase-1 | fructose-1,6-bisphosphatase |
| FBPase-2 | fructose-2,6-bisphosphatase |
| Fd | ferredoxin |
| FDNB (DNFB) | 1-fluoro-2,4-dinitrobenzene |
| FFA | free fatty acid |
| FH | familial hypercholesterolemia |
| fMet | $N$-formylmethionine |
| FMN, $FMNH_2$ | flavin mononucleotide, and its reduced form |
| $F_oF_1$ | mitochondrial ATP synthase |
| FP | flavoprotein |
| F1P | fructose 1-phosphate |
| F6P | fructose 6-phosphate |
| Fru | D-fructose |
| $\Delta G$ | free-energy change |
| $\Delta G'°$ | standard free-energy change |
| $\Delta G^{\ddagger}$ | activation energy |
| $\Delta G_B$ | binding energy |
| $\Delta G_p$ | free-energy change of ATP hydrolysis under cellular conditions |
| G | guanine, guanosine, or guanylate |
| $g$ | acceleration due to gravity |
| GABA | δ-aminobutyric acid |
| Gal | D-galactose |

| | | | |
|---|---|---|---|
| GalN | D-galactosamine | $k$ | rate constant |
| GalNAc | N-acetyl-D-galactosamine | $k_{cat}$ | turnover number |
| GAP | glyceraldehyde 3-phosphate; G protein–activating factor | kb | kilobase |
| | | kbp | kilobase pair |
| GDH | glutamate dehydrogenase | $\alpha$-KG | $\alpha$-ketoglutarate |
| GH | growth hormone | lac | lactose |
| GLC | gas-liquid chromatography | $\lambda$ | wavelength |
| Glc | D-glucose | LDH | lactate dehydrogenase |
| GlcA | D-gluconic acid | LDL | low-density lipoprotein |
| GlcN | D-glucosamine | Leu | leucine |
| GlcNAc | N-acetyl-D-glucosamine (also, NAG) | LH | luteinizing hormone |
| GlcUA | D-glucuronic acid | $Lk$ | linking number |
| Gln | glutamine | ln | logarithm to the base $e$ |
| Glu | glutamate | log | logarithm to the base 10 |
| Gly | glycine | LT | leukotriene |
| GMP, GDP, GTP | guanosine 5′-mono-, di-, triphosphate | LTR | long terminal repeat |
| G1P | glucose 1-phosphate | Lys | lysine |
| GPCR | G protein–coupled receptor | $M_r$ | relative molecular mass |
| GPI | glycosylated derivatives of phosphatidylinositol | Man | D-mannose |
| | | MAPK | mitosis-associated protein kinase |
| GRK | G protein–coupled receptor kinase | Mb, MbO$_2$ | myoglobin, oxymyoglobin |
| G6P | glucose 6-phosphate | Met | methionine |
| GSH, GSSG | glutathione, and its oxidized form | miRNA | micro-RNA |
| $\Delta H$ | enthalpy change | mRNA | messenger RNA |
| Hb, HbO$_2$, HbCO | hemoglobin, oxyhemoglobin, carbon monoxide hemoglobin | MSH | melanocyte-stimulating hormone |
| | | mDNA | mitochondrial DNA |
| HDL | high-density lipoprotein | $\mu$ | electrophoretic mobility |
| H$_2$ folate | dihydrofolate (also, DHF) | Mur | muramic acid |
| H$_4$ folate | tetrahydrofolate (also, THF) | Mur2Ac | N-acetylmuramic acid (also, NAM) |
| His | histidine | NAD$^+$, NADH | nicotinamide adenine dinucleotide, and its reduced form |
| HIV | human immunodeficiency virus | | |
| HMG-CoA | $\beta$-hydroxy-$\beta$-methylglutaryl-CoA | NADP$^+$, NADPH | nicotinamide adenine dinucleotide phosphate, and its reduced form |
| hnRNA | heterogeneous nuclear RNA | | |
| HPLC | high-performance liquid chromatography | NAG | N-acetylglucosamine (also, GlcNAc) |
| HRE | hormone response element | NAM | N-acetylmuramic acid (also, MurNAc) |
| HTH | helix-turn-helix | Neu5Ac | N-acetylneuraminic acid (sialic acid) |
| Hyp | hydroxyproline | NMN$^+$, NMNH | nicotinamide mononucleotide, and its reduced form |
| I | inosine | | |
| IF | initiation factor | NMP, NDP, NTP | nucleoside mono-, di-, and triphosphate |
| Ig | immunoglobulin | NMR | nuclear magnetic resonance |
| IgG | immunoglobulin G | NO | nitric oxide |
| Ile | isoleucine | OAA | oxaloacetate |
| IMP, IDP, ITP | inosine 5′-mono-, di-, triphosphate | ORF | open reading frame |
| IP$_3$ | inositol 1,4,5-trisphosphate | $P$ | pressure |
| IR | infrared | P$_i$ | inorganic orthophosphate |
| IS | insertion sequence | pO$_2$ | partial pressure of oxygen |
| J | joule | PAB or PABA | p-aminobenzoate |
| JAK | Janus kinase | PAGE | polyacrylamide gel electrophoresis |
| $K_a$ | acid dissociation constant | PC | plastocyanin; phosphatidylcholine |
| $K_d$ | dissociation constant | PCR | polymerase chain reaction |
| $K_D$ | diffusion constant | PDE | cyclic nucleotide phosphodiesterase |
| $K_{eq}$ | equilibrium constant | PE | phosphatidylethanolamine |
| $K'_{eq}$ | equilibrium constant under standard conditions | PEP | phosphoenolpyruvate |
| | | PFK | phosphofructokinase |
| $K_I$ | inhibition constant | PG | prostaglandin |
| $K_m$ | Michaelis constant | 2PGA | 2-phosphoglycerate |

| | |
|---|---|
| 3PGA | 3-phosphoglycerate |
| pH | log $1/[H^+]$ |
| Phe | phenylalanine |
| pI | isoelectric point |
| PI | phosphatidylinositol |
| PIP$_2$ | phosphatidylinositol 4,5-bisphosphate |
| PK | protein kinase; pyruvate kinase |
| p$K$ | log $1/K$ |
| PKA | cAMP-dependent protein kinase |
| PKC | $Ca^{2+}$/calmodulin-dependent protein kinase |
| PKG | cGMP-dependent protein kinase |
| PLC | phospholipase C |
| PLP | pyridoxal-5-phosphate |
| Pol | polymerase (DNA or RNA) |
| PP$_i$ | inorganic pyrophosphate |
| PQ | plastoquinone |
| pRb | retinoblastoma protein |
| Pro | proline |
| PRPP | 5-phosphoribosyl-1-pyrophosphate |
| PS | phosphatidylserine |
| $\Delta\varphi$ | transmembrane electrical potential |
| PUFA | polyunsaturated fatty acid |
| Rb | retinoblastoma gene, a tumor suppressor gene |
| RC | photoreaction center |
| RER | rough endoplasmic reticulum |
| RF | release factor; replicative form |
| RFLP | restriction-fragment length polymorphism |
| RGS | regulator of G protein signaling, a protein |
| RIA | radioimmunoassay |
| Rib | D-ribose |
| RNA | ribonucleic acid |
| RNase | ribonuclease |
| rRNA | ribosomal RNA |
| RNAi | RNA interference |
| RSV | Rous sarcoma virus |
| RTK | receptor tyrosine kinase |
| $\Delta S$ | entropy change |
| SAM | $S$-adenosylmethionine (also, adoMet) |
| SDS | sodium dodecyl sulfate |
| SER | smooth endoplasmic reticulum |
| Ser | serine |

| | |
|---|---|
| SMC | structural maintenance of chromosomes |
| SNARE | synaptosome-associated protein receptor |
| SNP | single nucleotide polymorphism |
| snRNA | small nuclear RNA |
| snRNP | small nuclear ribonucleoprotein |
| SRP | signal recognition particle |
| STAT | signal transducer and activator of transcription |
| STP | standard temperature and pressure |
| stRNA | small temporal RNA |
| STS | sequence-tagged site |
| T | thymine, thymidine, or thymidylate |
| $T$ | absolute temperature |
| THF | tetrahydrofolate (also, H$_4$ folate) |
| Thr | threonine |
| TLC | thin layer chromatography |
| TMP, TDP, TTP | thymidine 5′-mono-, di-, triphosphate |
| TMV | tobacco mosaic virus |
| TPI | triose phosphate isomerase |
| TPP | thiamine pyrophosphate |
| tRNA | transfer RNA |
| Trp | tryptophan |
| TX | thromboxane |
| Tyr | tyrosine |
| U | uracil, uridine, or uridylate |
| UCP | uncoupling protein of the mitochondrial inner membrane |
| UDP-Gal | uridine diphosphate galactose (also, UDP-galactose) |
| UDP-Glc | uridine diphosphate glucose (also, UDP-glucose) |
| UMP, UDP, UTP | uridine 5′-mono-, di-, triphosphate |
| UQ | coenzyme Q (ubiquinone; also, CoQ) |
| UV | ultraviolet |
| $V_m$ | transmembrane electrical potential |
| $V_{max}$ | maximum velocity |
| $V_0$ | initial velocity |
| Val | valine |
| VLDL | very low-density lipoprotein |
| YAC | yeast artificial chromosome |
| $Z$ | net charge |

# Appendix B  Abbreviated Solutions to Problems

## Chapter 1

1. **(a)** Diameter of magnified cell = 500 mm **(b)** $2.7 \times 10^{12}$ actin molecules **(c)** 36,000 mitochondria **(d)** $3.9 \times 10^{10}$ glucose molecules **(e)** 50 glucose molecules per hexokinase molecule

2. **(a)** $1 \times 10^{-12}$ g = 1 pg **(b)** 10% **(c)** 5%

3. **(a)** 1.6 mm; 800 times longer than the cell; DNA must be tightly coiled. **(b)** 4,000 proteins

4. **(a)** Metabolic rate is limited by diffusion, which is limited by surface area. **(b)** 12 $\mu m^{-1}$ for the bacterium; 0.04 $\mu m^{-1}$ for the amoeba; surface-to-volume ratio 300 times higher in the bacterium.

5. **(a)** $2 \times 10^6$ s (about 23 days) **(b)** Fast, directed transport allows unidirectional movement of molecules, larger cell size, localization of cellular functions to discrete regions, and regulation of molecular movement.

6. The vitamin molecules from the two sources are identical; the body cannot distinguish the source; only associated impurities might vary with the source.

7.

**(a)** H₃N⁺—C—C—OH  Amino  Hydroxyl

**(b)** H—C—OH, H—C—OH, H—C—OH, Hydroxyls

**(c)** HO—P—O⁻ Phosphoryl; C=C—COO⁻ Carboxyl

**(d)** COO⁻ Carboxyl; Amino H₃N⁺—C—H; H—C—OH Hydroxyl; CH₃ Methyl

**(e)** ⁻O—C—O Carboxyl; CH₂; CH₂; NH Amido; C=O; H—C—OH Hydroxyl; Methyl CH₃—C—CH₃ Methyl; CH₂OH Hydroxyl

**(f)** H—C=O Aldehyde; H—C—NH₃⁺ Amino; Hydroxyl HO—C—H; H—C—OH; H—C—OH Hydroxyls; CH₂OH

8.

The two enantiomers have different interactions with a chiral biological "receptor" (a protein).

9. **(a)** Only the amino acids have amino groups; separation could be based on the charge or binding affinity of these groups. Fatty acids are less soluble in water than amino acids, and the two types of molecules also differ in size and shape—either of these property differences could be the basis for separation. **(b)** Glucose is a smaller molecule than a nucleotide; separation could be based on size. The nitrogenous base and/or the phosphate group also endows nucleotides with characteristics (solubility, charge) that could be used for separation from glucose.

10. **(a)** It is improbable that silicon could serve as the central organizing element for life, especially in an $O_2$-containing atmosphere such as that of Earth. Long chains of silicon atoms are not readily synthesized; the polymeric macromolecules necessary for more complex functions would not readily form. Oxygen disrupts bonds between silicon atoms, and silicon–oxygen bonds are extremely stable and difficult to break, preventing the breaking and making of bonds that is essential to life processes.

11. Only one enantiomer of the drug was physiologically active. Dexedrine consisted of the single enantiomer; Benzedrine consisted of a racemic mixture.

12. **(a)** 3 Phosphoric acid groups; α-D-ribose; guanine **(b)** Choline; phosphoric acid; glycerol; oleic acid; palmitic acid **(c)** Tyrosine; 2 glycines; phenylalanine; methionine

13. **(a)** $CH_2O$; $C_3H_6O_3$
**(b)**

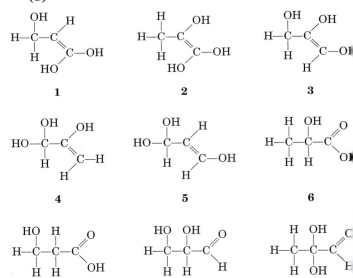

**(c)** X contains a chiral center; eliminates all but **6** and **8**. **(d)** X contains an acidic functional group; eliminates **8**; structure **6** is consistent with all data. **(e)** Structure **6**; we cannot distinguish between the two possible enantiomers.

# Chapter 2

**1.** 3.3 mL

**2.** pH 1.1

**3.** $1.7 \times 10^{-9}$ mol of acetylcholine

**4.** **(a)** 460 mM $Na^+$, 10 mM $K^+$, 590 mM $Cl^-$, 10 mM $Ca^{2+}$, 50 mM $Mg^{2+}$; the seawater contains 1,120 mosmol/L. **(b)** Cells contain 235 mosmol/L. **(c)** Ions (notably $Na^+$ and $Cl^-$) diffuse down their concentration gradients into cells; water moves down its concentration gradient out of cells. The cells become severely dehydrated. **(d)** 2.6 g **(e)** 5.3 g

**5.** **(a)** pH 8.6 to 10.6 **(b)** 4/5 **(c)** 10 mL **(d)** pH = p$K$a − 2

**6.** **(a)** 0.1 M HCl **(b)** 0.1 M NaOH **(c)** 0.1 M NaOH

**7.** **(d)** Bicarbonate, a weak base, titrates —OH to —O⁻, making the compound more polar and more water-soluble.

**8.** Stomach; the neutral form of aspirin present at the lower pH is less polar and passes through the membrane more easily.

**9.** $NaH_2PO_4 \cdot H_2O$, 5.8 g/L; $Na_2HPO_4$, 8.2 g/L

**10.** **(a)** 4.76 **(b)** 9.19 **(c)** 4.0 **(d)** 4.82

**11.** **(a)** $1.51 \times 10^{-4}$ M **(b)** $2.95 \times 10^{-7}$ M **(c)** $7.76 \times 10^{-12}$ M

**12.** **(a)** 5.06 **(b)** 4.28 **(c)** 5.46 **(d)** 4.76 **(e)** 3.76

**13.** **(a)** 4.6 **(b)** 0.1 pH unit **(c)** 4.3 pH units

**14.** 8.8

**15.** 7.3

**16.** **(a)** Blood pH is controlled by the carbon dioxide–bicarbonate buffer system, $CO_2 + H_2O \rightleftharpoons H^+ + HCO_3^-$. During *hypoventilation*, $[CO_2]$ increases in the lungs and arterial blood, driving the equilibrium to the right, raising $[H^+]$ and lowering pH. **(b)** During *hyperventilation*, $[CO_2]$ decrease in the lungs and arterial blood, reducing $[H^+]$ and increasing pH above the normal 7.4 value. **(c)** Lactate is a moderately strong acid, completely dissociating under physiological conditions and thus lowering the pH of blood and muscle tissue. Hyperventilation removes $H^+$, raising the pH of blood and tissues in anticipation of the acid buildup.

# Chapter 3

**1.** L; determine the absolute configuration at the $\alpha$ carbon and compare it with D- and L-glyceraldehyde.

**2.** **(a)** I **(b)** II **(c)** IV **(d)** II **(e)** IV **(f)** II and IV **(g)** III **(h)** III **(i)** V **(j)** III **(k)** V **(l)** II **(m)** III **(n)** V **(o)** I, III, and V

**3.** **(a)** pI > p$K$a of the $\alpha$-carboxyl group and pI < p$K$a of the $\alpha$-amino group, so both groups are charged (ionized). **(b)** 1 in $2.19 \times 10^7$. The pI of alanine is 6.01. From Table 3-1 and the Henderson-Hasselbalch equation, 1/4,680 carboxyl groups and 1/4,680 amino groups are uncharged. The fraction of alanine molecules with both groups uncharged is the product of these fractions.

**4.** **(a)–(c)**

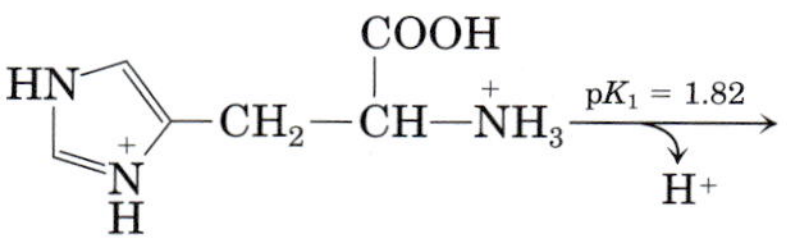

**1**

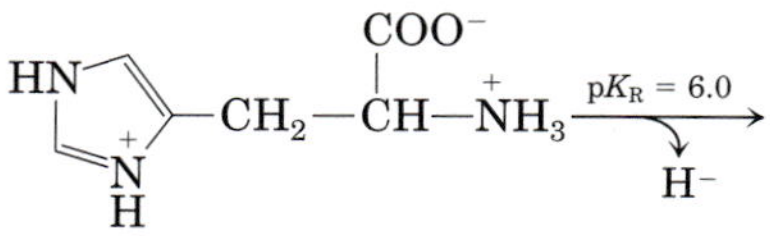

**2**

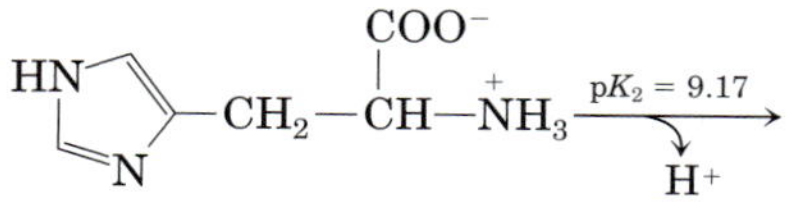

**3**

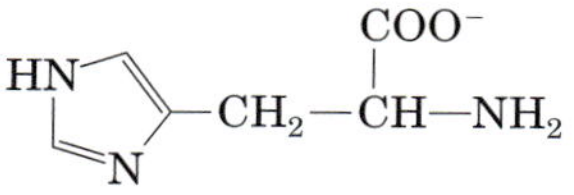

**4**

| pH | Structure | Net charge | Migrates toward |
|----|-----------|------------|-----------------|
| 1  | **1**     | +2         | Cathode         |
| 4  | **2**     | +1         | Cathode         |
| 8  | **3**     | 0          | Does not migrate |
| 12 | **4**     | −1         | Anode           |

**5.** **(a)** Asp **(b)** Met **(c)** Glu **(d)** Gly **(e)** Ser

**6.** **(a)** 2 **(b)** 4 **(c)**

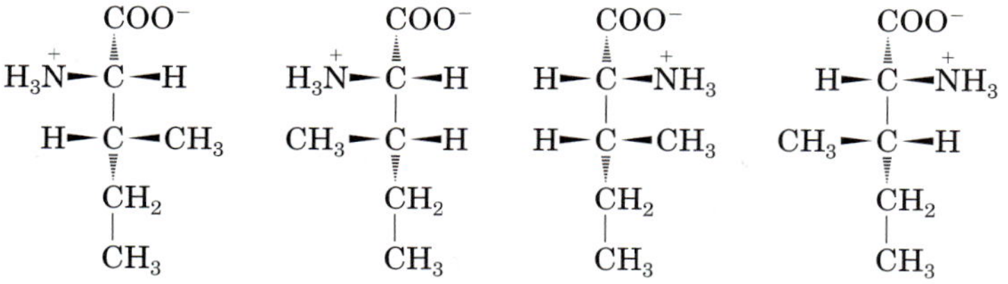

**7.** **(a)** Structure at pH 7:

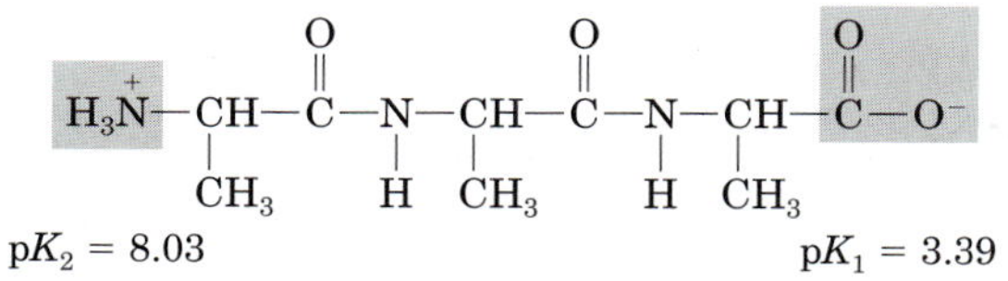

**(b)** Electrostatic interaction between the carboxylate anion and the protonated amino group of the alanine zwitterion favorably affects the ionization of the carboxyl group. This favorable electrostatic interaction decreases as the length of the poly(Ala) increases, resulting in an increase in p$K$1.
**(c)** Ionization of the protonated amino group destroys the favorable electrostatic interaction noted in (b). With increasing distance between the charged groups, removal of the proton from the amino group in poly(Ala) becomes easier and thus p$K_2$ is lower.

**8.** 75,000

**9.** **(a)** 32,000 **(b)** 2

**10.** **(a)** at pH 3, +2; at pH 8, 0; at pH 11, −1 **(b)** pI = 7.8

**11.** pI ≈ 1; carboxylate groups; Asp and Glu

**12.** Lys, His, Arg; negatively charged phosphate groups in DNA interact with positively charged side groups in histones.

**13.** **(a)** (Glu)$_{20}$ **(b)** (Lys–Ala)$_3$ **(c)** (Asn–Ser–His)$_5$ **(d)** (Asn–Ser–His)$_5$

**14.** **(a)** Specific activity after step 1 is 200 units/mg; step 2, 600 units/mg; step 3, 250 units/mg; step 4, 4,000 units/mg; step 5, 15,000 units/mg; step 6, 15,000 units/mg **(b)** Step 4 **(c)** Step 3 **(d)** Yes. Specific activity did not increase in step 6; SDS polyacrylamide gel electrophoresis

**15.** Tyr–Gly–Gly–Phe–Leu

16.

The arrows correspond to the orientation of the peptide bonds, —CO → NH—.

17. 88%, 97%

18. **(a)** The protein to be isolated (citrate synthase, CS) is a relatively small fraction of the total cellular protein. Cold temperatures reduce protein degradation; sucrose provides an isotonic environment that preserves the integrity of organelles during homogenization. **(b)** This step separates organelles on the basis of relative size. **(c)** The first addition of ammonium sulfate removes some unwanted proteins from the homogenate. Additional ammonium sulfate precipitates CS. **(d)** To resuspend (solubilize) CS, ammonium sulfate must be removed under conditions of pH and ionic strength that support the native conformation. **(e)** CS molecules are larger than the pore size of the chromatographic gel. Protein is detectable at 280 nm because of absorption at this wavelength by Try and Trp residues. **(f)** CS has a positive charge and thus binds to the negatively charged cation-exchange column. After the neutral and negatively charged proteins pass through, CS is displaced from the column using the washing solution of higher pH, which alters the charge on CS. **(g)** Different proteins can have the same pI. The SDS gel confirmed that only a single protein was purified. SDS is difficult to remove completely from a protein, and its presence distorts the acid-base properties of the protein, including pI.

## Chapter 4

1. **(a)** Shorter bonds have a higher bond order (are multiple rather than single) and are stronger. The peptide C—N bond is stronger than a single bond and is midway between a single and a double bond in character. **(b)** Rotation about the peptide bond is difficult at physiological temperatures because of its partial double-bond character.

2. **(a)** The principal structural units in the wool fiber polypeptide ($\alpha$-keratin) are successive turns of the $\alpha$ helix, at 5.4 Å intervals; coiled coils produce the 5.2 Å spacing. Steaming and stretching the fiber yields an extended polypeptide chain with the $\beta$ conformation, with a distance between adjacent R groups of about 7.0 Å. As the polypeptide reassumes an $\alpha$-helical structure, the fiber shortens. **(b)** Processed wool shrinks when polypeptide chains are converted from an extended $\beta$ conformation to the native $\alpha$-helical conformation in the presence of moist heat. The structure of silk—$\beta$-pleated sheets, with their small, closely packed amino acid side chains—is more stable than that of wool.

3. ~42 peptide bonds per second

4. At pH > 6, the carboxyl groups of poly(Glu) are deprotonated; repulsion among negatively charged carboxylate groups leads to unfolding. Similarly, at pH < 9, the amino groups of poly(Lys) are protonated; repulsion among these positively charged groups also leads to unfolding.

5. **(a)** Disulfide bonds are covalent bonds, which are much stronger than the noncovalent interactions that stabilize most proteins. They cross-link protein chains, increasing their stiffness, mechanical strength, and hardness. **(b)** Cystine residues (disulfide bonds) prevent the complete unfolding of the protein.

6. **(a)** Bends are most likely at residues 7 and 19; Pro residues in the cis configuration accommodate turns well. **(b)** The Cys residues at positions 13 and 24 can form disulfide bonds. **(c)** External surface: polar and charged residues (Asp, Gln, Lys); interior: nonpolar and aliphatic residues (Ala, Ile); Thr, though polar, has a hydropathy index near zero and thus can be found either on the external surface or in the interior of the protein.

7. 30 amino acid residues; 0.87

8. The bacterial enzyme is a collagenase; it destroys the connective-tissue barrier of the host, allowing the bacterium to invade the tissues. Bacteria do not contain collagen.

9. **(a)** The number of moles of DNP-valine formed per mole of protein equals the number of amino termini and thus the number of polypeptide chains. **(b)** 4 **(c)** Different chains would likely run as discrete bands on an SDS polyacrylamide gel.

10. **(a)** NF-$\kappa$-B transcription factor **(b)** No. You will obtain similar results, but with additional related proteins listed. **(c)** The protein has two subunits. There are multiple variants of the subunits, with the best-characterized being 50, 52, or 65 kD. These pair with each other to form a variety of homodimers and heterodimers. The structures of a number of different variants can be found in the PDB database. **(d)** The NF-$\kappa$-B transcription factor is a dimeric protein that binds specific DNA sequences, enhancing transcription of nearby genes. One such gene is the immunoglobulin $\kappa$ light chain, from which the transcription factor gets its name.

## Chapter 5

1. Protein B has a higher affinity for ligand X; it will be half-saturated at a much lower concentration of X than will protein A. Protein A has $K_a = 10^6$ $M^{-1}$; protein B has $K_a = 10^9$ $M^{-1}$.

2. **(a)**, **(b)**, and **(c)** all have $^nH < 1.0$. Apparent negative cooperativity in ligand binding can be caused by the presence of two or more types of ligand-binding sites with different affinities for the ligand on the same or different proteins in the same solution. Apparent negative cooperativity is also commonly observed in heterogeneous protein preparations. There are few well-documented cases of true negative cooperativity.

3. Effect on myoglobin's affinity for $O_2$: **(a)** none **(b)** none **(c)** none. Effect on hemoglobin's affinity for $O_2$ : **(a)** decreases **(b)** increases **(c)** decreases.

4. The cooperative behavior of hemoglobin arises from subunit interactions.

5. **(a)** The observation that hemoglobin A (HbA; maternal) is about 60% saturated when the $pO_2$ is 4 kPa, whereas hemoglobin F (HbF; fetal) is more than 90% saturated under the same physiological conditions, indicates that HbF has a higher $O_2$ affinity than does HbA. **(b)** The higher $O_2$ affinity of HbF ensures that oxygen will flow from maternal blood to fetal blood in the placenta. Fetal blood approaches full saturation where the $O_2$ affinity of HbA is low. **(c)** The observation that the $O_2$ -saturation curve of HbA undergoes a larger shift on BPG binding than does that of HbF suggests that HbA binds BPG more tightly than does HbF. Differential binding of BPG to the two hemoglobins may determine the difference in their $O_2$ affinities.

6. **(a)** Hb Memphis **(b)** HbS, Hb Milwaukee, Hb Providence, possibly Hb Cowtown **(c)** Hb Providence

7. **(a)** $1.2 \times 10^{-8}$ M **(b)** $5.0 \times 10^{-8}$ M **(c)** $7.5 \times 10^{-8}$ M **(d)** $2.0 \times 10^{-7}$ M. Note that a rearrangement of Eqn 5-8 gives $[L] = \theta K_d/(1 - \theta)$.

8. The epitope is likely to be a structure that is buried when G-actin polymerizes to F-actin.

9. Many pathogens, including HIV, have evolved mechanisms by which they can repeatedly alter the surface proteins to which

immune system components initially bind. Thus the host organism regularly faces new antigens and requires time to mount an immune response to each one. As the immune system responds to one variant, new variants are created.

**10.** Binding of ATP to myosin triggers dissociation of myosin from the actin thin filament. In the absence of ATP, actin and myosin bind tightly to each other.

**11.**

   (a)        (b)        (c)        (d)

**12. (a)** Chain L is the light chain and chain H is the heavy chain of the Fab fragment of this antibody molecule. Chain Y is lysozyme. **(b)** $\beta$ Structures are predominant in the variable and constant regions of the fragment. **(c)** Fab heavy-chain fragment, 218 amino acid residues; light-chain fragment, 214; lysozyme, 129. Less than 15% of the lysozyme molecule is in contact with the Fab fragment. **(d)** In the H chain, residues that seem to be in contact with lysozyme include $Gly^{31}$, $Tyr^{32}$, $Arg^{99}$, $Asp^{100}$, and $Tyr^{101}$. In the L chain the residues that appear to be in contact with lysozyme include $Tyr^{32}$, $Tyr^{49}$, $Tyr^{50}$, and $Trp^{92}$. In lysozyme, residues $Asn^{19}$, $Gly^{22}$, $Tyr^{23}$, $Ser^{24}$, $Lys^{116}$, $Gly^{117}$, $Thr^{118}$, $Asp^{119}$, $Gln^{121}$, and $Arg^{125}$ appear to be situated at the antigen-antibody interface. Not all these residues are adjacent in the primary structure. Folding of the polypeptide chain into higher levels of structure brings the non-consecutive residues together to form the antigen-binding site.

**13. (a)** The protein with a $K_d$ of 5 $\mu$M will have the highest affinity for ligand L. When the $K_d$ is 10 $\mu$M, doubling [L] from 0.2 to 0.4 $\mu$M (values well below the $K_d$) will nearly double $\theta$ (the actual increase factor is 1.96). This is a property of the hyperbolic curve; at low ligand concentrations, the $\theta$ is an almost linear function of [L]. On the other hand, doubling [L] from 40 to 80 $\mu$M (well above the $K_d$, where the binding curve is approaching its asymptotic limit) will increase $\theta$ by a factor of only 1.1. The increase factors are identical for the curve generated from Eqn 5-11. **(b)** $\theta = 0.998$. **(c)** A variety of answers will be obtained depending on the values entered for the different parameters.

## Chapter 6

**1.** The activity of the enzyme that converts sugar to starch is destroyed by heat denaturation.

**2.** $2.4 \times 10^{-6}$ M

**3.** $9.5 \times 10^8$ years

**4.** The enzyme-substrate complex is more stable than the enzyme alone.

**5. (a)** 190 Å **(b)** Three-dimensional folding of the enzyme brings the amino acid residues into proximity.

**6.** The reaction rate can be measured by following the decrease in absorption by NADH (at 340 nm) as the reaction proceeds. Determine the $K_m$ value; using substrate concentrations well above the $K_m$, measure initial rate (rate of NADH disappearance with time, measured spectrophotometrically) at several known enzyme concentrations, and make a plot of initial rate versus concentration of enzyme. The plot should be linear, with a slope that provides a measure of LDH concentration.

**7. (a)** $1.7 \times 10^{-3}$ M **(b)** 0.33; 0.67; 0.91

**8.** $V_{max} \approx 140$ $\mu$M/min; $K_m \approx 1 \times 10^{-5}$ M

**9. (a)** $V_{max} = 51.5$ mM/min, $K_m = 0.59$ mM **(b)** Competitive inhibition

**10.** $K_m = 2.2$ mM; $V_{max} = 0.50$ $\mu$mol/min

**11.** Curve A

**12.** $k_{cat} = 2.0 \times 10^7$ min$^{-1}$

**13.** The basic assumptions of the Michaelis-Menten equation still hold. The reaction is at steady state, and the rate is determined by $V_0 = k_2$ [ES]. The equations needed to solve for [ES] are

$$[E_t] = [E] + [ES] + [EI] \quad \text{and} \quad [EI] = \frac{[E][I]}{K_1}$$

[E] can be obtained by rearranging Eqn 6-19. The rest follows the pattern of the Michaelis-Menten equation derivation in the text.

**14.** Minimum $M_r = 29,000$

**15.** Activity of the prostate enzyme equals total phosphatase activity in a blood sample minus phosphatase activity in the presence of enough tartrate to completely inhibit the prostate enzyme.

**16.** The inhibition is mixed. Because $K_m$ appears not to change appreciably, this could be the special case of mixed inhibition called noncompetitive.

**17.** The [S] at which $V_0 = V_{max}/2\alpha'$ is obtained when all terms except $V_{max}$ on the right side of Eqn 6-30—that is, [S]/$\alpha K_m$ + $\alpha'$[S]—equal 1/2 $\alpha'$. Begin with [S]/$\alpha K_m$ + $\alpha'$[S] = 1/2 $\alpha'$ and solve for [S].

**18.** The optimum activity occurs when $Glu^{35}$ is protonated and $Asp^{52}$ is unprotonated.

**19. (a)** Increase factor = 1.96; $V_0 = 50$ $\mu$M/s; increase factor = 1.048. **(b)** When $\alpha = 2.0$, the curve is shifted to the right as the $K_m$ is increased by a factor of 2. When $\alpha' = 3.0$, the asymptote of the curve (the $V_{max}$) declines by a factor of 3. When $\alpha = 2.0$ and $\alpha' = 3.0$, the curve briefly rises above the curve where both $\alpha$ and $\alpha' = 1.0$, due to a decline in $K_m$. However, the asymptote is lower, because $V_{max}$ declines by a factor of 3. **(c)** When $\alpha = 2.0$, the $x$ intercept moves to the right. When $\alpha = 2.0$ and $\alpha' = 3.0$, the $x$ intercept moves to the left.

## Chapter 7

**1.** Yes. The empirical formula is $CH_2O$, typical of a carbohydrate.

**2.** With reduction of the carbonyl oxygen to a hydroxyl group, the chemistry at C-1 and C-3 is the same; the glycerol molecule is not chiral.

**3.** Osazone formation destroys the configuration around C-2 of aldoses, so aldoses differing only at the C-2 configuration give the same derivative, with the same melting point.

**4. (a)**

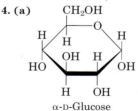 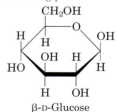

   $\alpha$-D-Glucose        $\beta$-D-Glucose

**(b)** A fresh solution of $\alpha$-D-glucose, or of $\beta$-D-glucose, undergoes mutarotation to an equilibrium mixture of the $\alpha$ and $\beta$ forms. **(c)** 36% $\alpha$ form; 64% $\beta$ form

**5.** Fructose cyclizes to either the pyranose or the furanose structure. Increasing the temperature shifts the equilibrium in the direction of the furanose, the less sweet form.

**6.** The rate of mutarotation is sufficiently high that, as the enzyme consumes $\beta$-D-glucose, more $\alpha$-D-glucose is converted to the β form and, eventually, all the glucose is oxidized. Glucose oxidase is specific for glucose and does not detect other reducing sugars (such as galactose) that react with Fehling's reagent.

7. **(a)** Measure the change in optical rotation with time. **(b)** The optical rotation of the mixture is negative (inverted) relative to that of the sucrose solution. **(c)** $-2.0°$

8. Prepare a slurry of sucrose and water for the core; add a small amount of invertase; immediately coat with chocolate.

9. Sucrose has no free anomeric carbon to undergo mutarotation.

10. Native cellulose consists of glucose units linked by $(\beta1\rightarrow4)$ glycosidic bonds, which force the polymer chain into an extended conformation. Parallel series of these extended chains form intermolecular hydrogen bonds, aggregating into long, tough, insoluble fibers. Glycogen consists of glucose units linked by $(\alpha1\rightarrow4)$ glycosidic bonds, which cause bends in the chain and prevent formation of long fibers. In addition, glycogen is highly branched and, because many of its hydroxyl groups are exposed to water, is highly hydrated and disperses in water.

  Cellulose is a structural material in plants, consistent with its side-by-side aggregation into insoluble fibers. Glycogen is a storage fuel in animals. Highly hydrated glycogen granules with their many nonreducing ends are rapidly hydrolyzed by glycogen phosphorylase to release glucose 1-phosphate.

11. 6,000 residues/s

12. 11 s

13. The negative charges on chondroitin sulfate repel each other and force the molecule into an extended conformation. The polar molecule attracts many water molecules, increasing the molecular volume. In the dehydrated solid, each negative charge is counterbalanced by a positive ion, and the molecule condenses.

14. Positively charged amino acid residues would bind the highly negatively charged groups on heparin. In fact, Lys residues of antithrombin III interact with heparin.

15. Oligosaccharides; their subunits can be combined in more ways than the amino acid subunits of oligopeptides. Each hydroxyl group can participate in glycosidic bonds, and the configuration of each glycosidic bond can be either $\alpha$ or $\beta$. The polymer can be linear or branched.

16. **(a)** Branch-point residues yield 2,3-di-O-methylglucose; the unbranched residues yield 2,3,6-tri-O-methylglucose. **(b)** 3.75%

17. Chains of $(1\rightarrow6)$-linked D-glucose residues with occasional $(1\rightarrow3)$-linked branches, with about one branch every 20 residues

## Chapter 8

1. N-3 and N-7

2. (5′)GCGCAATATTTTGAGAAATATTGCGC(3′); it contains a palindrome. The individual strands can form hairpin structures; the two strands can form a cruciform.

3. $9.4 \times 10^{-4}$ g

4. **(a)** 40° **(b)** 0°

5. The RNA helix is in the A conformation; the DNA helix is generally in the B conformation.

6. In eukaryotic DNA, about 5% of C residues are methylated. 5-Methylcytosine can spontaneously deaminate to form thymine; the resulting G–T pair is one of the most common mismatches in eukaryotic cells.

7. Base stacking in nucleic acids tends to reduce the absorption of UV light. Denaturation involves loss of base stacking, and UV absorption increases.

8. 0.35 mg/mL

9. One DNA contains 32% A, 32% T, 18% G, and 18% C; the other 17% A, 17% T, 33% G, and 33% C. This assumes that both are double-stranded. The DNA with 33% G and 33% C most likely came from the thermophilic bacterium; its higher G≡C content makes it much more stable to heat.

10.

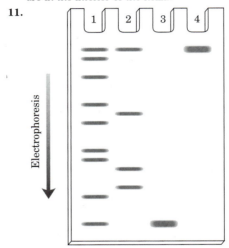

Deoxyribose        Guanine        Phosphate

Solubilities: phosphate > deoxyribose > guanine. The highly polar phosphate groups and sugar moieties are on the outside of the double helix, exposed to water; the hydrophobic bases are in the interior of the helix.

11.

12. (5′)P—GCGCCAUUG(3′)—OH
 (5′)P—GCGCCAUU(3′)—OH
 (5′)P—GCGCCAU(3′)—OH
 (5′)P—GCGCCA(3′)—OH
 (5′)P—GCGCC(3′)—OH
 (5′)P—GCGC(3′)—OH
 (5′)P—GCG(3′)—OH
 (5′)P—GC(3′)—OH

 and the nucleoside 5′-phosphates

13. **(a)** Water is a participant in most biological reactions, including those that cause mutations. The low water content in endospores reduces the activity of mutation-causing enzymes and slows the rate of nonenzymatic depurination reactions, which are hydrolysis reactions. **(b)** UV light induces formation of cyclobutane pyrimidine dimers. Because *B. subtilis* is a soil organism, spores can be lofted to the top of the soil or into the air, where they may be subject to prolonged UV exposure.

14. **(a)** The DNA backbone can be identified using "Backbone" display in Rasmol or Chime. In Chime, individual bases are named at the bottom of the frame by clicking on bases in the ball-and-stick mode. In Rasmol, each base that you click on is identified in the separate text window. The base at the 5′ end is adenine. This is a right-handed helix. **(b)** This is a left-handed helix. **(c)** See the Study Guide or website (www.whfreeman.com/lehninger) for additional tips on viewing structures in stereo.

# Chapter 9

**1. (a)** (5′)---G(3′)   and   (5′)AATTC---(3′)
   (5′)---CTTAA          (5′)G---
**(b)** (5′)---GAATT(3′) and (5′)AATTC---(3′)
   (5′)---CTTAA          (5′)TTAAG---
**(c)** (5′)---GAATTAATTC---(3′)
   (5′)---CTTAATTAAG---
**(d)** (5′)---G(3′)   and   (5′)C---(3′)
   (5′)---C          (5′)G---
**(e)** (5′)---GAATTC---(3′)
   (5′)---CTTAAG---
**(f)** (5′)---CAG(3′)   and   (5′)CTG---(3′)
   (5′)---GTC          (5′)GAC---
**(g)** (5′)---CAGAATTC---(3′)
   (5′)---GTCTTAAG---
**(h)** Method 1: cut the DNA with *Eco*RI as in (a). At this point, one could treat the DNA as in (b) or (d), then ligate a synthetic DNA fragment with the *Bam*HI recognition sequence between the two resulting blunt ends. Method 2 (more efficient): synthesize a DNA fragment with the structure

(5′)AATTGGATCC
(5′)CCTAGGTTAA

This would ligate efficiently to the sticky ends generated by *Eco*RI cleavage, would introduce a *Bam*HI site, but would not regenerate the *Eco*RI site.
**(i)** The four fragments (with N = any nucleotide), in order of discussion in the problem, are

(5′)AATTCNNNNCTGCA
GNNNNG
(5′)AATTCNNNNGTGCA
GNNNNC
(5′)AATTGNNNNCTGCA
CNNNNG
(5′)AATTGNNNNGTGCA
CNNNNC

**2.** λ phage DNA can be packaged into infectious phage particles only if it is between 40,000 and 53,000 bp in length. Since bacteriophage vectors generally include about 30,000 bp (in two pieces), they will not be packaged into phage particles unless they contain a sufficient length of inserted DNA (10,000 to 23,000 bp).

**3. (a)** Plasmids in which the original pBR322 was regenerated without insertion of a foreign DNA fragment; these would retain resistance to ampicillin. Also, two or more molecules of pBR322 might be ligated together with or without insertion of foreign DNA. **(b)** The clones in lanes 1 and 2 each have one DNA fragment inserted in different orientations. The clone in lane 3 has two DNA fragments, ligated such that the *Eco*RI proximal ends are joined.

**4.** Focus on the amino acids with the fewest codons: Met and Trp. The best possibility is the span of DNA from the codon for the first Trp residue to the first two nucleotides of the codon for Ile. The sequence of the probe would be:

(5′)UGGUA(U/C)UG(U/C)AUGGA(U/C)UGGAU

The synthesis would be designed to incorporate either U or C where indicated, producing a mixture of eight 20-nucleotide probes that differ only at one or more of these positions.

**5.** Your test would require DNA primers, a heat-stable DNA polymerase, deoxynucleoside triphosphates, and a PCR machine (thermal cycler). The primers would be designed to simplify a DNA segment encompassing the CAG repeat. The DNA strand shown is the coding strand, oriented 5′→3′ left to right. The

primer targeted to DNA to the left of the repeat would be identical to any 25-nucleotide sequence shown in the region to the left of the CAG repeat. The primer on the right side must be *complementary* and *antiparallel* to a 25-nucleotide sequence to the right of the CAG repeat. Using the primers, DNA including the CAG repeat would be amplified by PCR, and its size would be determined by comparison to size markers after electrophoresis. The length of the DNA would reflect the length of the CAG repeat, providing a simple test for the disease.

**6.** Design PCR primers complementary to DNA in the deleted segment, but which would direct DNA synthesis away from each other. No PCR product will be generated unless the ends of the deleted segment are joined to create a circle.

**7.** None of the children can be excluded. All have one band that could be derived from the father.

**8.**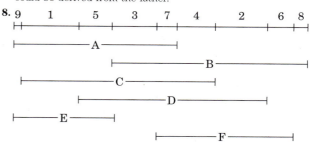

**9.** Simply for convenience; the 200,000 bp Ti plasmid, even when the T DNA is removed, is too large to isolate in quantity and manipulate in vitro. It is also too large to reintroduce into a cell by standard transformation techniques. The *vir* genes will facilitate the transfer of any DNA between the T DNA repeats, even if they are on a separate plasmid. The second plasmid in the two-plasmid system, because it requires only the T DNA repeats and a few sequences necessary for plasmid selection and propagation, is relatively small, easily isolated, and easily manipulated (foreign DNA easily added and/or altered).

**10.**

(gel electrophoresis figure: lanes M, A (1 2), B (1 2), C (1 2), with size markers 1–10)

**11.** Cover spot 4, add solution containing activated T, irradiate, wash.

1. A–T          2. G–T          3. A–T          4. G–C

Cover spots 2 and 4, add solution containing activated G, irradiate, wash.

1. A–T–G          2. G–T          3. A–T–G          4. G–C

Cover spot 3, add solution containing activated C, irradiate, wash.

1. A–T–G–C     2. G–T–C          3. A–T–G          4. G–C–C

Cover spots 1, 3, and 4, add solution containing activated C, irradiate, wash.

1. A–T–G–C     2. G–T–C–C     3. A–T–G          4. G–C–C

Cover spots 1 and 2, add solution containing activated G, irradiate, wash.

1. A–T–G–C     2. G–T–C–C     3. A–T–G–C     4. G–C–C–C

**12.** The vectors must be introduced into a cell infected with a helper virus that can provide the necessary replication and packaging functions but cannot itself be packaged. Once recombinant DNA is integrated into the chromosome of the target cell with these vectors, the lack of recombination and packaging functions makes the integration very stable by preventing the deletion or replication of the integrated DNA.

## Chapter 10

**1.** The term "lipid" does not specify a particular chemical structure. Compounds are categorized as lipids based on their greater solubility in organic solvents than in water.

**2. (a)** The number of cis double bonds. Each cis double bond causes a bend in the hydrocarbon chain, lowering the melting temperature.
**(b)** Six different triacylglycerols can be constructed, in order of increasing melting points:

$$OOO < OOP = OPO < PPO = POP < PPP$$

where O = oleic and P = palmitic acid. The greater the content of saturated fatty acid, the higher is the melting point.
**(c)** Branched-chain fatty acids increase the fluidity of membranes because they decrease the extent of membrane lipid packing.

**3.** Lecithin, an amphipathic compound, is an emulsifying agent, facilitating the solubilization of butter.

**4.** *Hydrophobic units:* **(a)** 2 fatty acids **(b)**, **(c)**, and **(d)** 1 fatty acid and the hydrocarbon chain of sphingosine **(e)** steroid nucleus and acyl side chain. *Hydrophilic units:* **(a)** phosphoethanolamine **(b)** phosphocholine **(c)** D-galactose **(d)** several sugar molecules **(e)** alcohol group (OH)

**5.** The triacylglycerols of animal fats (grease) are hydrolyzed by NaOH (saponified) to form soaps, which are much more soluble in water than are triacylglycerols.

**6. (a)** The free —OH group on C-2 and the phosphorylcholine head group on C-3 are hydrophilic; the fatty acid on C-1 of lysolecithin is hydrophobic. **(b)** Certain steroids such as prednisone inhibit the action of phospholipase $A_2$, inhibiting the release of arachidonic acid from C-2. Arachidonic acid is converted to a variety of eicosanoids, some of which cause inflammation and pain. **(c)** Phospholipase $A_2$ releases arachidonic acid, a precursor of other eicosanoids with vital protective functions in the body; it also breaks down dietary glycerophospholipids.

**7.** Diacylglycerol is hydrophobic and remains in the membrane lipid. Inositol 1,4,5-trisphosphate is highly polar, very soluble in water, and more readily diffusible in the cytosol. Both are second messengers.

**8.** Water-soluble vitamins are more rapidly excreted in the urine and are not stored effectively. Fat-soluble vitamins have very low solubility in water and are stored in body lipids.

**9. (a)** Glycerol and the sodium salts of palmitic and stearic acids. **(b)** D-Glycerol 3-phosphorylcholine and the sodium salts of palmitic and oleic acids.

**10.** Solubilities in water: monoacylglycerol > diacylglycerol > triacylglycerol.

**11.** First eluted to last eluted: cholesteryl palmitate and triacylglycerol; cholesterol and *n*-tetradecanol; phosphatidylcholine and phosphatidylethanolamine; sphingomyelin; phosphatidylserine and palmitate.

**12. (a)** Subject acid hydrolysates of each compound to chromatography (GLC or silica gel TLC) and compare the result with known standards. *Sphingomyelin hydrolysate:* sphingosine, fatty acids, phosphocholine, choline, and phosphate; *cerebroside hydrolysate:* sphingosine, fatty acids, sugars, but no phosphate. **(b)** Strong alkaline hydrolysis of sphingomyelin yields sphingosine; phosphatidylcholine yields glycerol. Detect hydrolysate components on thin-layer chromatograms by comparing with standards or by their differential reaction with FDNB (only sphingosine reacts to form a colored product). Treatment with phospholipase $A_1$ or $A_2$ releases free fatty acids from phosphatidylcholine, but not from sphingomyelin.

**13.** Phosphatidylethanolamine and phosphatidylserine.

## Chapter 11

**1.** From the known amount of lipid used, its molecular weight, and the area occupied by a monolayer (determined as shown), the area per molecule can be calculated.

**2.** The data support a bilayer of lipid in the dog erythrocytes: a single cell, with surface area 98 $\mu m^2$, has a lipid monolayer area of 200 $\mu m^2$. In the case of sheep and human erythrocytes, the data suggest a monolayer, not a bilayer. In fact, significant experimental errors occurred in these early experiments; recent, more accurate measurements support a bilayer in all cases.

**3.** 63 SDS molecules per micelle

**4. (a)** Lipids that form bilayers are amphipathic molecules: they contain a hydrophilic and a hydrophobic region. To minimize the hydrophobic area exposed to the water surface, these lipids form two-dimensional sheets with the hydrophilic regions exposed to water and the hydrophobic regions buried in the interior of the sheet. Furthermore, to avoid exposing the hydrophobic edges of the sheet to water, lipid bilayers close upon themselves. **(b)** These sheets form the closed membrane surfaces that envelop cells and compartments within cells (organelles).

**5.** 2 nm. Two palmitates placed end to end span about 4 nm, approximately the width of a typical bilayer.

**6.** Decrease. Movement of individual lipids in bilayers occurs much faster at 37 °C, when the lipids are in the "fluid" phase, than at 10 °C, when they are in the "solid" phase.

**7.** 35 kJ/mol, neglecting the effects of transmembrane electrical potential; 0.6 mol.

**8.** 13 kJ/mol.

**9.** Most of the $O_2$ consumed by a tissue is for oxidative phosphorylation, the source of most of the ATP. Therefore, about two-thirds of the ATP synthesized by the kidney is used for pumping $K^+$ and $Na^+$.

**10.** No. The symporter may carry more than one equivalent of $Na^+$ for each mole of glucose transported.

**11.** Salt extraction indicates a peripheral location, and inaccessibility to protease in intact cells indicates an internal location. X resembles a peripheral membrane protein.

**12.** The interactions among membrane lipids are noncovalent and reversible and form spontaneously.

13. The temperature of body tissues at the extremities is lower than that of tissues closer to the center of the body. If lipid is to remain fluid at this lower temperature, it must contain a higher proportion of unsaturated fatty acids; unsaturated fatty acids lower the melting point of lipid mixtures.

14. The energetic cost of moving the highly polar, sometimes charged, head group through the hydrophobic interior of the bilayer is prohibitive.

15. At pH 7, tryptophan bears a positive and a negative charge, but indole is uncharged. The movement of the less polar indole through the hydrophobic core of the bilayer is energetically more favorable.

16. $3 \times 10^{-2}$ s

17. Treat a suspension of cells with unlabeled NEM in the presence of excess lactose, remove the lactose, then add radiolabeled NEM. Use SDS-PAGE to determine the $M_r$ of the radiolabeled band (the transporter).

18. Construct a hydropathy plot; hydrophobic regions of 20 or more residues suggest transmembrane segments. Determine whether the protein in intact erythrocytes reacts with a membrane-impermeant reagent specific for primary amines; if so, the transporter is of type I.

19. The leucine transporter is specific for the L isomer, but the binding site can accommodate either L-leucine or L-valine. Reduction of $V_{max}$ in the absence of $Na^+$ indicates that leucine (or valine) is transported by symport with $Na^+$.

20. $V_{max}$ reduced; $K_t$ unaffected.

21. ~1%; estimated by calculating the surface area of the cell and of 10,000 transporter molecules (using the dimensions of hemoglobin (p. 163) as a model globular protein).

22. **(a)** The rise-per-residue for an $\alpha$ helix (Chapter 4) is about 1.5 Å = 0.15 nm. To span a 4 nm bilayer, an $\alpha$ helix must contain about 27 residues; thus for seven spans, about 190 residues are required. A protein of $M_r$ 64,000 has about 580 residues. **(b)** A hydropathy plot is used to locate transmembrane regions. **(c)** Because about half of this portion of the epinephrine receptor consists of charged residues, it probably represents an intracellular loop that connects two adjacent membrane-spanning regions of the protein. **(d)** Because this helix is composed mostly of hydrophobic residues, this portion of the receptor is probably one of the membrane-spanning regions of the protein.

## Chapter 12

1. Albuterol raises [cAMP], leading to relaxation and dilation of the bronchi and bronchioles. Because $\beta$-adrenergic receptors control many other processes, this drug would have undesirable side effects. To minimize them, find an agonist specific for the subtype of $\beta$-adrenergic receptors found in the bronchial smooth muscle.

2. Amplification results as one molecule of a catalyst activates many molecules of another catalyst, in an amplification cascade involving, in order, insulin receptor, IRS-1, Raf, MEK, ERK; ERK activates a transcription factor, which stimulates mRNA production.

3. Hormone degradation; hydrolysis of GTP bound to a G protein; degradation, metabolism, or sequestration of second messenger; receptor desensitization; removal of receptor from the cell surface.

4. Two cells expressing the same surface receptor may have different complements of target proteins for protein phosphorylation.

5. **(a)** If $V_m$ were set by permeability to (primarily) $K^+$, the Nernst equation would predict a $V_m$ of $-90$ mV, not the observed $-95$ mV, so some other conductance must contribute

to $V_m$. **(b)** Chloride ion is probably the determinant of $V_m$; the predicted $E_{Cl}^-$ is $-94$ mV.

6. **(a)** $V_m$ of the oocyte membrane changes from $-60$ mV to $-10$ mV—that is, the membrane is depolarized. **(b)** The effect of KCl depends on influx of $Ca^{2+}$ from the extracellular medium.

7. Hyperpolarization results in the closing of voltage-dependent $Ca^{2+}$ channels in the presynaptic region of the rod cell. The resulting decrease in $[Ca^{2+}]_{in}$ diminishes release of an inhibitory neurotransmitter that suppresses activity in the next neuron of the visual circuit. When this inhibition is removed in response to a light stimulus, the circuit becomes active and visual centers in the brain are excited.

8. X is cAMP; its production is stimulated by epinephrine. **(a)** Centrifugation sediments adenylyl cyclase (which catalyzes cAMP formation) in the particulate fraction. **(b)** Added cAMP stimulates glycogen phosphorylase. **(c)** cAMP is heat stable; it can be prepared by treating ATP with barium hydroxide.

9. **(a)** It increases [cAMP]. **(b)** cAMP regulates $Na^+$ permeability. **(c)** Replace lost body fluids and electrolytes.

10. Unlike cAMP, dibutyryl-cAMP passes readily through the plasma membrane.

11. $G_s$ remains in its activated form when the nonhydrolyzable analog is bound. The analog therefore prolongs the effect of epinephrine on the injected cell.

12. *Shared properties of Ras and $G_s$:* both bind either GDP or GTP; both activated by GTP; both, when active, activate a downstream enzyme; both have intrinsic GTPase activity that shuts them off after a short period of activation. *Differences between Ras and $G_s$:* Ras is a small, monomeric protein; $G_s$ is heterotrimeric. *Functional difference between $G_s$ and $G_i$:* $G_s$ activates adenylyl cyclase, $G_i$ inhibits it.

13. Vasopressin acts by elevating cytosolic $[Ca^{2+}]$ to $10^{-6}$ M, activating protein kinase C. EGTA injection blocks vasopressin action, but should not affect the response to glucagon, which uses cAMP, *not* $Ca^{2+}$, as second messenger.

14. Individuals with Oguchi's disease might have a defect in rhodopsin kinase or in arrestin.

15. **(a)** The mutation makes R unable to bind and inhibit C, so C is constantly active. **(b)** The mutation prevents cAMP binding to R, leaving C inhibited by bound R.

16. *Kinase (factor in parentheses):* PKA (cAMP); PKG (cGMP); PKC ($Ca^{2+}$, DAG); $Ca^{2+}$/CaM kinase ($Ca^{2+}$, CaM); cyclin-dependent kinase (cyclin); protein Tyr kinase (ligand for the receptor, such as insulin); MAPK (Raf); Raf (Ras); glycogen phosphorylase kinase (PKA).

17. A normal tumor suppressor gene encodes a protein that restrains cell division. A mutant form of the protein fails to suppress cell division, but if either of the two alleles encodes normal protein, normal function will continue. A normal oncogene encodes a regulator protein that triggers cell division, but only when an appropriate signal (growth factor) is present. The mutant version of the oncogene product constantly sends the signal to divide, whether or not growth factors are present.

18. In a child who develops multiple tumors in both eyes, every retinal cell had a defective copy of the *Rb* gene at birth. Early in the child's life, several cells independently undergo a second mutation that damages the one good *Rb* allele, producing a tumor. A child who develops a single tumor had, at birth, two good copies of the *Rb* gene in every cell; mutation in both *Rb* alleles in one cell (extremely rare) causes a single tumor.

19. A mutation in Ras that inactivates its GTPase activity creates a protein that, once activated by the binding of GTP, continues to give, through Raf, the insulin-response signal.

## Chapter 13

1. Consider the developing chick as the system; the nutrients, egg shell, and outside world are the surroundings. Transformation of the single cell into a chick drastically reduces the entropy of the system. Initially, the parts of the egg outside the embryo (the surroundings) contain complex fuel molecules (a low-entropy condition). During incubation, some of these complex molecules are converted to large numbers of $CO_2$ and $H_2O$ molecules (high entropy). This increase in the entropy of the surroundings is larger than the decrease in entropy of the chick (the system).

2. **(a)** $-4.8$ kJ/mol **(b)** 7.56 kJ/mol **(c)** $-13.7$ kJ/mol

3. **(a)** 262 M **(b)** 608 M **(c)** 0.30

4. $K'_{eq} = 21$; $\Delta G'^{\circ} = -7.6$ kJ/mol

5. $-31$ kJ/mol

6. **(a)** $-1.68$ kJ/mol **(b)** $-4.4$ kJ/mol **(c)** At a given temperature, the value of $\Delta G'^{\circ}$ for any reaction is fixed and is defined for standard conditions (here, both fructose 6-phosphate and glucose 6-phosphate at 1 M). In contrast, $\Delta G$ is a variable that can be calculated for any set of reactant and product concentrations.

7. Less. The overall equation for ATP hydrolysis can be approximated as

$$ATP^{4-} + H_2O \longrightarrow ADP^{3-} + HPO_4^{2-} + H^+$$

(This is only an approximation, because the ionized species shown here are the major, but not the only, forms present.) Under standard conditions (i.e., [ATP] = [ADP] = [$P_i$] = 1 M), the concentration of water is 55 M and does not change during the reaction. Because $H^+$ ions are produced in the reaction, at a higher [$H^+$] (pH 5.0) the equilibrium would be shifted to the left and less free energy would be released.

8. 10

9. **(a)** $3.85 \times 10^{-3}$ $M^{-1}$; [glucose 6-phosphate] = $8.9 \times 10^{-8}$ M; no. **(b)** 14 M; because the maximum solubility of glucose is less than 1 M, this is not a reasonable step. **(c)** 837 ($\Delta G'^{\circ} = -16.7$ kJ/mol); [glucose] = $1.2 \times 10^{-7}$ M; yes. **(d)** No. This would require such high [$P_i$] that the phosphate salts of divalent cations would precipitate. **(e)** By directly transferring the phosphoryl group from ATP to glucose, the phosphoryl group transfer potential ("tendency" or "pressure") of ATP is utilized without generating high concentrations of intermediates. The essential part of this transfer is, of course, the enzymatic catalysis.

10. **(a)** $-12.5$ kJ/mol **(b)** $-14.6$ kJ/mol

11. **(a)** $3 \times 10^{-4}$ **(b)** 68.7 **(c)** $7.4 \times 10^4$

12. $-13$ kJ/mol

13. 46.1 kJ/mol

14. **(a)** 46 kJ/mol **(b)** 46 kg; 68% **(c)** ATP is synthesized as it is needed, then broken down to ADP and $P_i$; its concentration is maintained in a steady state.

15. The ATP system is in a dynamic steady state; [ATP] remains constant because the rate of ATP consumption equals its rate of synthesis. ATP consumption involves release of the terminal ($\gamma$) phosphoryl group; synthesis of ATP from ADP involves replacement of this phosphoryl group. Hence the terminal phosphate undergoes rapid turnover. In contrast, the central ($\beta$) phosphate undergoes only relatively slow turnover.

16. **(a)** 1.7 kJ/mol **(b)** Inorganic pyrophosphatase catalyzes the hydrolysis of pyrophosphate and drives the net reaction toward the synthesis of acetyl-CoA.

17. 34 kJ/mol

18. **(a)** $NAD^+/NADH$ **(b)** Pyruvate/lactate **(c)** Lactate formation **(d)** $-26.1$ kJ/mol **(e)** $3.63 \times 10^4$

19. **(a)** 1.14 V **(b)** $-220$ kJ/mol **(c)** $\sim4$

20. **(a)** $-0.35$ V **(b)** $-0.320$ V **(c)** $-0.29$ V

21. In order of increasing tendency: **(a)**, **(d)**, **(b)**, **(c)**

22. **(c)** and **(d)**

## Chapter 14

1. Net equation: Glucose + 2ATP → 2 glyceraldehyde 3-phosphate + 2ADP + $2H^+$; $\Delta G'^{\circ} = 2.1$ kJ/mol

2. Net equation: 2 Glyceraldehyde 3-phosphate + 4ADP + $2P_i$ → 2 lactate + $2NAD^+$; $\Delta G'^{\circ} = -114$ kJ/mol

3. **(a)** $^{14}CH_3CH_2OH$ **(b)** [3-$^{14}$C]glucose or [4-$^{14}$C]glucose

4. Soybeans and wheat contain starch, a polymer of glucose. The microorganisms break down starch to glucose, glucose to pyruvate via glycolysis, and—because the process is carried out in the absence of $O_2$ (i.e., it is a fermentation)—pyruvate to lactic acid and ethanol. If $O_2$ were present, pyruvate would be oxidized to acetyl-CoA, then to $CO_2$ and $H_2O$. Some of the acetyl-CoA, however, would also be hydrolyzed to acetic acid (vinegar) in the presence of oxygen.

5. C-1. This experiment demonstrates the reversibility of the aldolase reaction. The C-1 of glyceraldehyde 3-phosphate is equivalent to C-4 of fructose 1,6-bisphosphate (see Fig. 14-6). The starting glyceraldehyde 3-phosphate must have been labeled at C-1. The C-3 of dihydroxyacetone phosphate becomes labeled through the triose phosphate isomerase reaction, thus giving rise to $^{14}$[C-3]fructose 1,6-bisphosphate.

6. No. There would be no anaerobic production of ATP; aerobic ATP production would be diminished only slightly.

7. No. Lactate dehydrogenase is required to recycle $NAD^+$ from the NADH formed during the oxidation of glyceraldehyde 3-phosphate.

8. The transformation of glucose to lactate occurs when myocytes are low in oxygen, and it provides a means of generating ATP under $O_2$-deficient conditions. Because lactate can be oxidized to pyruvate, glucose is not wasted; pyruvate is oxidized by aerobic reactions when $O_2$ becomes plentiful. This metabolic flexibility gives the organism a greater capacity to adapt to its environment.

9. It rapidly removes the 1,3-bisphosphoglycerate in a favorable subsequent step, catalyzed by phosphoglycerate kinase.

10. **(a)** 3-Phosphoglycerate is the product. **(b)** In the presence of arsenate there is no net ATP synthesis under anaerobic conditions.

11. **(a)** Alcohol fermentation requires 2 mol of $P_i$ per mole of glucose. **(b)** Ethanol is the reduced product formed during reoxidation of NADH to $NAD^+$, and $CO_2$ is the byproduct of the conversion of pyruvate to ethanol. Yes; pyruvate must be converted to ethanol, to produce a continuous supply of $NAD^+$ for the oxidation of glyceraldehyde 3-phosphate. Fructose 1,6-bisphosphate accumulates; it is formed as an intermediate in glycolysis. **(c)** Arsenate replaces $P_i$ in the glyceraldehyde 3-phosphate dehydrogenase reaction to yield an acyl arsenate, which spontaneously hydrolyzes. This prevents formation of ATP, but 3-phosphoglycerate continues through the pathway.

12. Dietary niacin is used to synthesize $NAD^+$. Oxidations carried out by $NAD^+$ are part of cyclic processes, with $NAD^+$ as electron carrier (reducing agent); one molecule of $NAD^+$ can oxidize many thousands of molecules of glucose, and thus the dietary requirement for the precursor vitamin (niacin) is relatively small.

13. Net equation: Glycerol + $2NAD^+$ + ADP + $P_i$ → pyruvate + 2NADH + ATP + $2H^+$

14. *Galactokinase deficiency:* galactose (less toxic); *UDP-glucose: galactose 1-phosphate uridylyl deficiency:* galactose 1-phosphate (more toxic).

15. The proteins have been degraded to amino acids and used for gluconeogenesis.

16. **(a)** In the pyruvate carboxylase reaction, $^{14}CO_2$ is added to pyruvate, but PEP carboxykinase removes the *same* $CO_2$ in the next step. Thus, $^{14}C$ is not (initially) incorporated into glucose.
**(b)**

1-$^{14}$C-Pyruvate → Oxaloacetate → Phosphoenolpyruvate

1,3-Bisphospho-glycerate ← 3-Phospho-glycerate ← 2-Phospho-glycerate

Glyceraldehyde 3-phosphate ⇌ Dihydroxyacetone phosphate

Fructose 1,6-bisphosphate → → 3,4-$^{14}$C-Glucose

17. Pyruvate carboxylase is a mitochondrial enzyme. The [$^{14}$C]oxaloacetate formed mixes with the oxaloacetate pool of the citric acid cycle and is equilibrated with the cycle intermediates to form a mixture of [1-$^{14}$C] and [4-$^{14}$C]oxaloacetate. Oxaloacetate labeled at C-1 leads to the formation of [3,4-$^{14}$C]glucose (see Problem 16).

18. 4 ATP equivalents per glucose molecule

19. **(a), (b), (d)** are glucogenic; **(c) (e)** are not.

20. Consumption of alcohol forces competition for $NAD^+$ between ethanol metabolism and gluconeogenesis. The problem is compounded by strenuous exercise and lack of food, because at these times the level of blood glucose is already low.

21. **(a)** The rapid increase in glycolysis; the rise in pyruvate and NADH results in a rise in lactate. **(b)** Lactate is transformed to glucose via pyruvate; this is a slower process, because formation of pyruvate is limited by $NAD^+$ availability, the LDH equilibrium is in favor of lactate, and conversion of pyruvate to glucose is energy-requiring. **(c)** The equilibrium for the LDH reaction is in favor of lactate formation.

22. Lactate is transformed to glucose in the liver by gluconeogenesis (see Figs 14–15, 14–16). A defect in FBPase-1 would prevent entry of lactate into the gluconeogenic pathway in hepatocytes, causing lactate to accumulate in the blood.

23. Succinate is transformed to oxaloacetate, which passes into the cytosol and is converted to PEP by PEP carboxykinase. Two moles of PEP are then required to produce a mole of glucose by the route outlined in Figure 14–16.

24. If the catabolic and anabolic pathways of glucose metabolism are operating simultaneously, futile cycling of ATP occurs, with extra $O_2$ consumption.

## Chapter 15

1. **(a)** 0.029 **(b)** 308 **(c)** No. $Q$ is much lower than $K'_{eq}$, indicating that the PFK-1 reaction is far from equilibrium in cells; this reaction is slower than the subsequent reactions in glycolysis. Flux through the glycolytic pathway is largely determined by the activity of PFK-1.

2. In the absence of $O_2$, the ATP needs are met by anaerobic glucose metabolism (fermentation to lactate). Because aerobic oxidation of glucose produces far more ATP than does fermentation, less glucose is needed to produce the same amount of ATP.

3. **(a)** There are two binding sites for ATP: a catalytic site and a regulatory site. Binding of ATP to a regulatory site inhibits PFK-1, by reducing $V_{max}$ or increasing $K_m$ for ATP at the catalytic site. **(b)** Glycolytic flux is reduced when ATP is plentiful. **(c)** The graph indicates that increased [ADP] suppresses the inhibition of ATP. Because the adenine nucleotide pool is fairly constant, consumption of ATP leads to an increase in [ADP]. The data show that the activity of PFK-1 may be regulated by the [ATP]/[ADP] ratio.

4. **(a)** $1.4 \times 10^{-9}$ M **(b)** The physiological concentration (0.023 mM) is 16,000 times the equilibrium concentration; this reaction does not reach equilibrium in the cell. Many reactions in the cell are not at equilibrium.

5. The phosphate group on glucose 6-phosphate is completely ionized at pH 7, giving the molecule an overall negative charge. Because membranes are generally impermeable to electrically charged molecules, glucose 6-phosphate cannot pass from the bloodstream into cells and hence cannot enter the glycolytic pathway and generate ATP. (This is why glucose, once phosphorylated, cannot escape from the cell.)

6. **(a)** *In muscle:* glycogen breakdown supplies energy (ATP) via glycolysis. Glycogen phosphorylase catalyzes the conversion of stored glycogen to glucose 1-phosphate, which is converted to glucose 6-phosphate, an intermediate in glycolysis. During strenuous activity, skeletal muscle requires large quantities of glucose 6-phosphate. *In the liver:* glycogen breakdown maintains a steady level of blood glucose between meals (glucose 6-phosphate is converted to free glucose). **(b)** In actively working muscle, ATP flux requirements are very high and glucose 1-phosphate must be produced rapidly, requiring a high $V_{max}$.

7. **(a)** [$P_i$]/[glucose 1-phosphate] = 3.3/1 **(b)** and **(c)** The value of this ratio in the cell (>100:1) indicates that [glucose 1-phosphate] is far below the equilibrium value. The rate at which glucose 1-phosphate is removed (through entry into glycolysis) is greater than its rate of production (by the glycogen phosphorylase reaction), so metabolite flow is from glycogen to glucose 1-phosphate. The glycogen phosphorylase reaction is probably the regulatory step in glycogen breakdown.

8. **(a)** Increases **(b)** Decreases **(c)** Increases

9. *Resting:* [ATP] high; [AMP] low; [acetyl-CoA] and [citrate] intermediate. *Running:* [ATP] intermediate; [AMP] high; [acetyl-CoA] and [citrate] low. Glucose flux through glycolysis increases during the anaerobic sprint because (1) the ATP inhibition of glycogen phosphorylase and PFK-1 is partially relieved, (2) AMP stimulates both enzymes, and (3) lower citrate and acetyl-CoA levels relieve their inhibitory effects on PFK-1 and pyruvate kinase, respectively.

10. The migrating bird relies on the highly efficient aerobic oxidation of fats, rather than the anaerobic metabolism of glucose used by a sprinting rabbit. The bird reserves its muscle glycogen for short bursts of energy during emergencies.

11. *Case A:* (f), (3); *Case B:* (c), (3); *Case C:* (h), (4); *Case D:* (d), (6)

## Chapter 16

1. **(a)**

   ① *Citrate synthase:*
   Acetyl-CoA + oxaloacetate + $H_2O$ → citrate + CoA
   ② *Aconitase:*
   Citrate → isocitrate
   ③ *Isocitrate dehydrogenase:*
   Isocitrate + $NAD^+$ → α-ketoglutarate + $CO_2$ + NADH
   ④ *α-Ketoglutarate dehydrogenase:*
   α-Ketoglutarate + $NAD^+$ + CoA → succinyl-CoA + $CO_2$ + NADH
   ⑤ *Succinyl-CoA synthetase:*
   Succinyl-CoA + $P_i$ + GDP → succinate + GTP + CoA
   ⑥ *Succinate dehydrogenase:*
   Succinate + FAD → fumarate + $FADH_2$
   ⑦ *Fumarase:*
   Fumarate + $H_2O$ → malate
   ⑧ *Malate dehydrogenase:*
   Malate + $NAD^+$ → oxaloacetate + NADH + $H^+$

   **(b), (c)** ① CoA, condensation; ② none, isomerization; ③ $NAD^+$, oxidative decarboxylation; ④ $NAD^+$, CoA, and thiamine pyrophosphate, oxidative decarboxylation; ⑤ CoA, substrate-level phosphorylation; ⑥ FAD, oxidation; ⑦ none, hydration; ⑧ $NAD^+$, oxidation

   **(d)** Acetyl-CoA + $3NAD^+$ + FAD + GDP + $P_i$ + $2H_2O$ →
   $2CO_2$ + 3NADH + $FADH_2$ + GTP + $2H^+$ + CoA

2. **(a)** Oxidation; methanol → formaldehyde + [H—H]
   **(b)** Oxidation; formaldehyde → formate + [H—H]
   **(c)** Reduction; $CO_2$ + [H—H] → formate + $H^+$
   **(d)** Reduction; glycerate + $H^+$ + [H—H] → glyceraldehyde + $H_2O$
   **(e)** Oxidation; glycerol → dihydroxyacetone + [H—H]
   **(f)** Oxidation; $2H_2O$ + toluene → benzoate + $H^+$ + 3[H—H]
   **(g)** Oxidation; succinate → fumarate + [H—H]
   **(h)** Oxidation; pyruvate + $H_2O$ → acetate + $CO_2$ + [H—H]

3. From the structural formulas, we see that the carbon-bound H/C ratio of hexanoic acid (11/6) is higher than that of glucose (7/6). Hexanoic acid is more reduced and yields more energy on complete combustion to $CO_2$ and $H_2O$.

4. **(a)** Oxidized; ethanol + $NAD^+$ → acetaldehyde + NADH + $H^+$
   **(b)** Reduced; 1,3-bisphosphoglycerate + NADH + $H^+$ →
   glyceraldehyde 3-phosphate + $NAD^+$ + $HPO_4^{2-}$
   **(c)** Unchanged; pyruvate + $H^+$ → acetaldehyde + $CO_2$
   **(d)** Oxidized; pyruvate + $NAD^+$ →
   acetate + $CO_2$ + NADH + $H^+$
   **(e)** Reduced; oxaloacetate + NADH + $H^+$ → malate + $NAD^+$
   **(f)** Unchanged; acetoacetate + $H^+$ → acetone + $CO_2$

5. Oxygen consumption is a measure of the activity of the first two stages of cellular respiration: glycolysis and the citric acid cycle. The addition of oxaloacetate or malate stimulates the cit-

ric acid cycle and thus stimulates respiration. The added oxaloacetate or malate serves a catalytic role, because it is regenerated in the latter part of the citric acid cycle.

6. **(a)** $5.6 \times 10^{-6}$ **(b)** $1.1 \times 10^{-8}$ M **(c)** 28 molecules

7. The terminal phosphoryl group in GTP can be transferred to ADP in a reaction catalyzed by nucleoside diphosphate kinase, with an equilibrium constant of 1.0: GTP + ADP → GDP + ATP

8. **(a)** $^-$OOC—$CH_2$—$CH_2$—COO$^-$ (succinate) **(b)** Malonate is a competitive inhibitor of succinate dehydrogenase. **(c)** A block in the citric acid cycle stops NADH formation, which stops electron transfer, which stops respiration. **(d)** A large excess of succinate (substrate) overcomes the competitive inhibition.

9. **(a)** Add uniformly labeled [$^{14}$C]glucose and check for the release of $^{14}CO_2$. **(b)** Equally distributed in C-2 and C-3 of oxaloacetate; an infinite number

10. **(a)** C-1 **(b)** C-3 **(c)** C-3 **(d)** C-2 (methyl group) **(e)** C-4 **(f)** C-4 **(g)** Equally distributed in C-2 and C-3

11. Thiamine is required for the synthesis of thiamine pyrophosphate (TPP), a prosthetic group in the pyruvate dehydrogenase and α-ketoglutarate dehydrogenase complexes. A thiamine deficiency reduces the activity of these enzyme complexes and causes the observed accumulation of precursors.

12. No. For every two carbons that enter as acetate, two leave the cycle as $CO_2$; thus there is no net synthesis of oxaloacetate. Net synthesis of oxaloacetate occurs by the carboxylation of pyruvate, an anaplerotic reaction.

13. **(a)** Inhibition of aconitase **(b)** Fluorocitrate; competes with citrate; by a large excess of citrate **(c)** Citrate and fluorocitrate are inhibitors of PFK-1. **(d)** All catabolic processes necessary for ATP production are shut down.

14. *Glycolysis:*
   Glucose + $2P_i$ + 2ADP + $2NAD^+$ →
   2 pyruvate + 2ATP + 2NADH + $2H^+$ + $2H_2O$
   *Pyruvate carboxylase reaction:*
   2 Pyruvate + $2CO_2$ + 2ATP + $2H_2O$ →
   2 oxaloacetate + 2ADP + $2P_i$ + $4H^+$
   *Malate dehydrogenase reaction:*
   2 Oxaloacetate + 2NADH + $2H^+$ → 2 L-malate + $2NAD^+$
   This recycles nicotinamide coenzymes under anaerobic conditions. The overall reaction is
   Glucose + $2CO_2$ → 2 L-malate + $4H^+$
   This produces four $H^+$ per glucose, increasing the acidity and thus the tartness of the wine.

15. Net reaction: 2 Pyruvate + ATP + $2NAD^+$ + $H_2O$ →
   α-ketoglutarate + $CO_2$ + ADP + $P_i$ + 2NADH + $3H^+$

16. **(a)** Decreases **(b)** Increases **(c)** Decreases

17. **(a)** Citrate is produced through the action of citrate synthase on oxaloacetate and acetyl-CoA. Citrate synthase can be used for net synthesis of citrate when (1) there is a continuous influx of new oxaloacetate and acetyl-CoA and (2) isocitrate synthesis is restricted, as in a medium low in $Fe^{3+}$. Aconitase requires $Fe^{3+}$, so an $Fe^{3+}$-restricted medium restricts the synthesis of aconitase.
   **(b)** Sucrose + $H_2O$ → glucose + fructose
   Glucose + $2P_i$ + 2ADP + $2NAD^+$ →
   2 pyruvate + 2ATP + 2NADH + $2H^+$ + $2H_2O$
   Fructose + $2P_i$ + 2ADP + $2NAD^+$ →
   2 pyruvate + 2ATP + 2NADH + $2H^+$ + $2H_2O$
   2 Pyruvate + $2NAD^+$ + 2CoA →
   2 acetyl-CoA + 2NADH + $2H^+$ + $2CO_2$
   2 Pyruvate + $2CO_2$ + 2ATP + $2H_2O$ →
   2 oxaloacetate + 2ADP + $2P_i$ + $4H^+$
   2 Acetyl-CoA + 2 oxaloacetate + $2H_2O$ → 2 citrate + 2CoA
   The overall reaction is

Sucrose + $H_2O$ + $2P_i$ + 2ADP + 6NAD$^+$ →
2 citrate + 2ATP + 6NADH + 10H$^+$

**(c)** The overall reaction consumes NAD$^+$. Because the cellular pool of this oxidized coenzyme is limited, it must be recycled by the electron-transfer chain with consumption of $O_2$. Consequently, the overall conversion of sucrose to citric acid is an aerobic process and requires molecular oxygen.

**18.** Succinyl-CoA is an intermediate of the citric acid cycle; its accumulation signals reduced flux through the cycle, calling for reduced entry of acetyl-CoA into the cycle. Citrate synthase, by regulating the primary oxidative pathway of the cell, regulates the supply of NADH and thus the flow of electrons from NADH to $O_2$.

**19.** Fatty acid catabolism increases [acetyl-CoA], which stimulates pyruvate carboxylase. The resulting increase in [oxaloacetate] stimulates acetyl-CoA consumption by the citric acid cycle, and [citrate] rises, inhibiting glycolysis at the level of PFK-1. In addition, increased [acetyl-CoA] inhibits the pyruvate dehydrogenase complex, slowing the utilization of pyruvate from glycolysis.

**20.** Oxygen is needed to recycle NAD$^+$ from the NADH produced by the oxidative reactions of the citric acid cycle. Reoxidation of NADH occurs during mitochondrial oxidative phosphorylation.

**21.** Toward citrate; $\Delta G$ for the citrate synthase reaction under these conditions is about $-8.$ kJ/mol.

**22.** Steps ④ and ⑤ are essential in the reoxidation of the enzyme's reduced lipoamide cofactor.

# Chapter 17

**1.** The fatty acid portion; the carbons in fatty acids are more reduced than those in glycerol.

**2.** **(a)** $4.0 \times 10^5$ kJ ($9.6 \times 10^4$ kcal) **(b)** 48 days **(c)** 0.48 lb/day

**3.** The first step in fatty acid oxidation is analogous to the conversion of succinate to fumarate; the second step, to the conversion of fumarate to malate; the third step, to the conversion of malate to oxaloacetate.

**4.** **(a)** R—COO$^-$ +ATP → acyl-AMP + PP$_i$
Acyl-AMP + CoA → acyl-CoA + AMP
**(b)** Irreversible hydrolysis of PP$_i$ to 2P$_i$ by cellular inorganic pyrophosphatase

**5.** Yes. Some of the tritium is removed from palmitate during the dehydrogenation reactions of $\beta$ oxidation. The removed tritium appears as tritiated water.

**6.** Fatty acyl groups condensed with CoA in the cytosol are first transferred to carnitine, releasing CoA, then transported into the mitochondrion, where they are again condensed with CoA. The cytosolic and mitochondrial pools of CoA are thus kept separate, and no radioactive CoA from the cytosolic pool enters the mitochondrion.

**7.** **(a)** In the pigeon, $\beta$ oxidation predominates; in the pheasant, anaerobic glycolysis of glycogen predominates. **(b)** Pigeon muscle would consume more $O_2$. **(c)** Fat contains more energy per gram than glycogen does. In addition, the anaerobic breakdown of glycogen is limited by the tissue's tolerance to lactate buildup. Thus the pigeon, operating on the oxidative catabolism of fats, is the long-distance flyer. **(d)** These enzymes are the regulatory enzymes of their respective pathways and thus limit ATP production rates.

**8.** **(a)** The carnitine-mediated entry of fatty acids into mitochondria is the rate-limiting step in fatty acid oxidation. Carnitine deficiency slows fatty acid oxidation; added carnitine increases the rate. **(b)** All increase the metabolic need for fatty acid oxidation. **(c)** Carnitine deficiency might result from a deficiency

of lysine, its precursor, or from a defect in one of the enzymes in the biosynthesis of carnitine.

**9.** Oxidation of fats releases metabolic water; 1.4 L of water per kg of tripalmitoylglycerol (ignores the small contribution of glycerol to the mass)

**10.** The bacteria can be used to completely oxidize hydrocarbons to $CO_2$ and $H_2O$. However, contact between hydrocarbons and bacterial enzymes may be difficult to achieve. Bacterial nutrients such as nitrogen and phosphorus may be limiting and inhibit growth.

**11.** **(a)** $M_r$ 136; phenylacetic acid **(b)** Even

**12.** Because the mitochondrial pool of CoA is small, CoA must be recycled from acetyl-CoA via the formation of ketone bodies. This allows the operation of the $\beta$-oxidation pathway, necessary for energy production.

**13.** **(a)** Glucose yields pyruvate via glycolysis, and pyruvate is the main source of oxaloacetate. Without glucose in the diet, [oxaloacetate] drops and the citric acid cycle slows. **(b)** Odd-numbered; propionate conversion to succinyl-CoA provides intermediates for the citric acid cycle and four-carbon precursors for gluconeogenesis.

**14.** $\beta$ Oxidation of $\omega$-fluorooleate forms fluoroacetyl-CoA, which enters the citric acid cycle and produces fluorocitrate, a powerful inhibitor of aconitase. Inhibition of aconitase shuts down the citric acid cycle. Without reducing equivalents from the citric acid cycle, oxidative phosphorylation (ATP synthesis) is fatally slowed.

**15.** Enz-FAD, having a more positive standard reduction potential, is a better electron acceptor than NAD$^+$, and the reaction is driven in the direction of fatty acyl–CoA oxidation. This more favorable equilibrium is obtained at the cost of 1 ATP; only 1.5 ATP are produced per FADH$_2$ oxidized in the respiratory chain (vs. 2.5 per NADH).

**16.** 9 turns; arachidic acid, a 20-carbon saturated fatty acid, yields 10 molecules of acetyl-CoA, the last two formed in the ninth turn.

**17.** See Fig. 17-11. [3-$^{14}$C]Succinyl-CoA is formed, which gives rise to oxaloacetate labeled at C-2 and C-3.

**18.** ATP hydrolysis in the energy-requiring reactions of a cell takes up water in the reaction ATP + $H_2O$ → ADP + P$_i$; thus, in the steady state, there is no *net* production of $H_2O$.

**19.** Methylmalonyl-CoA mutase requires the cobalt-containing cofactor formed from vitamin B$_{12}$.

**20.** Mass lost per day is about 0.66 kg, or about 140 kg in 7 months. Ketosis could be avoided by degradation of nonessential body proteins to supply amino acid skeletons for gluconeogenesis.

# Chapter 18

**1.**

**(a)**  $^-$OOC—CH$_2$—$\overset{\overset{\displaystyle O}{\|}}{C}$—COO$^-$          Oxaloacetate

**(b)**  $^-$OOC—CH$_2$—CH$_2$—$\overset{\overset{\displaystyle O}{\|}}{C}$—COO$^-$          $\alpha$-Ketoglutarate

**(c)**  CH$_3$—$\overset{\overset{\displaystyle O}{\|}}{C}$—COO$^-$          Pyruvate

**(d)**  ⬡—CH$_2$—$\overset{\overset{\displaystyle O}{\|}}{C}$—COO$^-$          Phenylpyruvate

**2.** This is a coupled-reaction assay. The product of the slow transamination (pyruvate) is rapidly consumed in the subsequent "indicator reaction" catalyzed by lactate dehydrogenase, which consumes NADH. Thus the rate of disappearance of NADH is a measure of the rate of the aminotransferase reaction. The indicator reaction is monitored by observing the decrease in absorption of NADH at 340 nm with a spectrophotometer.

**3.** No. The nitrogen in alanine can be transferred to oxaloacetate via transamination, to form aspartate.

**4.** **(a)** Phenylalanine hydroxylase; a low-phenylalanine diet **(b)** The normal route of phenylalanine metabolism via hydroxylation to tyrosine is blocked, and phenyalanine accumulates. **(c)** Phenylalanine is transformed to phenylpyruvate by transamination, and then to phenyllactate by reduction. The transamination reaction has an equilibrium constant of 1.0, and phenylpyruvate is formed in significant amounts when phenylalanine accumulates. **(d)** Because of the deficiency in production of tyrosine, which is a precursor of melanin, the pigment normally present in hair.

**5.** Catabolism of the carbon skeletons of valine, methionine, and isoleucine is impaired because a functional methylmalonyl-CoA mutase (a coenzyme $B_{12}$ enzyme) is absent. The physiological effects of loss of this enzyme are described in Table 18-2 and Box 18-2.

**6.** 15 mol of ATP per mole of lactate; 13 mol of ATP per mole of alanine, when nitrogen removal is included

**7.** **(a)** $^{15}NH_2{-}CO{-}^{15}NH_2$

**(b)** $^-OO^{14}C{-}CH_2{-}CH_2{-}^{14}COO^-$

**(c)** R—NH—$\overset{\overset{\displaystyle ^{15}NH}{\|}}{C}$—$^{15}NH_2$

**(d)** R—NH—$\overset{\overset{\displaystyle O}{\|}}{C}$—$^{15}NH_2$

**(e)** No label

**(f)** $^-OO^{14}C{-}\overset{\overset{\displaystyle ^{15}NH_2}{|}}{\underset{\underset{\displaystyle H}{|}}{C}}{-}CH_2{-}^{14}COO^-$

**8.** **(a)** Isoleucine $\overset{①}{\longrightarrow}$ II $\overset{②}{\longrightarrow}$ IV $\overset{③}{\longrightarrow}$ I $\overset{④}{\longrightarrow}$ V $\overset{⑤}{\longrightarrow}$ III $\overset{⑥}{\longrightarrow}$ acetyl-CoA + propionyl-CoA **(b)** Step ① transamination, no analogous reaction, PLP; ② oxidative decarboxylation, analogous to the pyruvate dehydrogenase reaction, NAD$^+$, TPP, lipoate, FAD; ③ oxidation, analogous to the succinate dehydrogenase reaction, FAD; ④ hydration, analogous to the fumarase reaction, no cofactor; ⑤ oxidation, analogous to the malate dehydrogenase reaction, NAD$^+$; ⑥ thiolysis (reverse aldol condensation), analogous to the thiolase reaction, CoA.

**9.** A likely mechanism is:

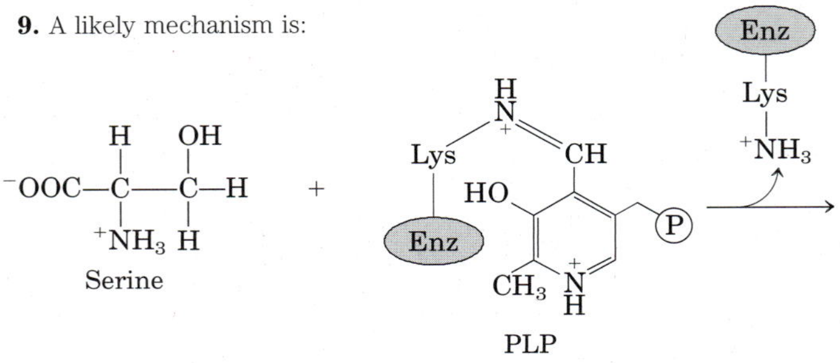

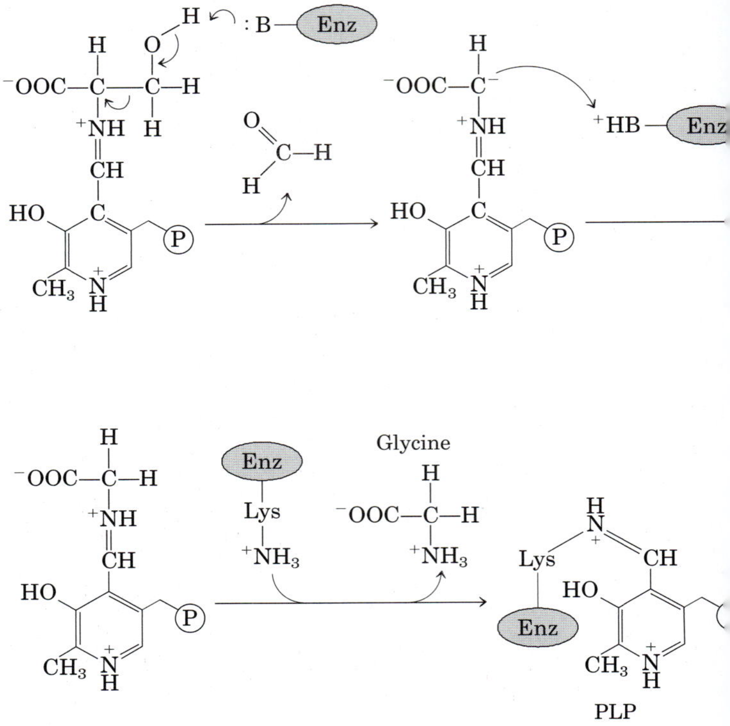

The formaldehyde (HCHO) produced in the second step reacts rapidly with tetrahydrofolate at the enzyme active site to produce $N^5, N^{10}$-methylenetetrahydrofolate (see Fig. 18-17).

**10.** **(a)** Transamination; no analogies; PLP
**(b)** Oxidative decarboxylation; analogous to oxidative decarboxylation of pyruvate to acetyl-CoA prior to entry into the citric acid cycle, and of $\alpha$-ketoglutarate to succinyl-CoA in the citric acid cycle; NAD$^+$, FAD, lipoate, and TPP
**(c)** Dehydrogenation (oxidation); analogous to dehydrogenation of succinate to fumarate in the citric acid cycle, and of fatty acyl–CoA to enoyl-CoA in $\beta$ oxidation; FAD
**(d)** Carboxylation; no analogous in citric acid cycle or $\beta$ oxidation; ATP and biotin
**(e)** Hydration; analogous to hydration of fumarate to malate in the citric acid cycle, and of enoyl-CoA to 3-hydroxyacyl-CoA in $\beta$ oxidation; no cofactors
**(f)** Reverse aldol reaction; analogous to reverse of citrate synthase reaction in the citric acid cycle

**11.** **(a)** Fasting resulted in low blood glucose; subsequent administration of the experimental diet led to rapid catabolism of glucogenic amino acids. **(b)** Oxidative deamination caused the rise in NH$_3$ levels; the absence of arginine (an intermediate in the urea cycle) prevented conversion of NH$_3$ to urea; arginine is not synthesized in sufficient quantities in the cat to meet the needs imposed by the stress of the experiment. This suggests that arginine is an essential amino acid in the cat's diet. **(c)** Ornithine is converted to arginine by the urea cycle.

**12.** $H_2O$ + glutamate + NAD$^+$ $\rightarrow$
$\qquad\qquad\alpha$-ketoglutarate + NH$_4^+$ + NADH + H$^+$
$NH_4^+$ + 2ATP + $H_2O$ + $CO_2$ $\rightarrow$
$\qquad\qquad$ carbamoyl phosphate + 2ADP + P$_i$ + 3H$^+$
Carbamoyl phosphate + ornithine $\rightarrow$ citrulline + P$_i$ + H$^+$
Citrulline + aspartate + ATP $\rightarrow$
$\qquad\qquad$ argininosuccinate + AMP + PP$_i$ + H$^+$
Argininosuccinate $\rightarrow$ arginine + fumarate
Fumarate + $H_2O$ $\rightarrow$ malate

Malate + NAD$^+$ → oxaloacetate + NADH + H$^+$

Oxaloacetate + glutamate → aspartate + α-ketoglutarate

Arginine + H$_2$O → urea + ornithine

2 Glutamate + CO$_2$ + 4H$_2$O + 2NAD$^+$ + 3ATP →

2 α-ketoglutarate + 2NADH + 7H$^+$ + urea + 2ADP +

$$\text{AMP} + \text{PP}_i + 2\text{P}_i \qquad (1)$$

Additional reactions that need to be considered:

$$\text{AMP} + \text{ATP} \rightarrow 2\text{ADP} \qquad (2)$$

O$_2$ + 8H$^+$ + 2NADH + 6ADP + 6P$_i$ →

$$2\text{NAD}^+ + 6\text{ATP} + 8\text{H}_2\text{O} \qquad (3)$$

$$\text{H}_2\text{O} + \text{PP}_i \rightarrow 2\text{P}_i + \text{H}^+ \qquad (4)$$

Summing equations (1) through (4),

2 Glutamate + CO$_2$ + O$_2$ + 2ADP + 2P$_i$ →

$$2 \text{ α-ketoglutarate} + \text{urea} + 3\text{H}_2\text{O} + 2\text{ATP}$$

**13.** The second amino group introduced into urea is transferred from aspartate, which is generated during the transamination of glutamate to oxaloacetate, a reaction catalyzed by aspartate aminotransferase. Approximately one-half of all the amino groups excreted as urea must pass through the aspartate aminotransferase reaction, making this the most highly active aminotransferase.

**14. (a)** A person on a diet consisting only of protein must use amino acids as the principal source of metabolic fuel. Because the catabolism of amino acids requires the removal of nitrogen as urea, the process consumes abnormally large quantities of water to dilute and excrete the urea in the urine. Furthermore, electrolytes in the "liquid protein" must be diluted with water and excreted. If the daily water loss through the kidney is not balanced by a sufficient water intake, a net loss of body water results. **(b)** When considering the nutritional benefits of protein, one must keep in mind the total amount of amino acids needed for protein synthesis and the distribution of amino acids in the dietary protein. Gelatin contains a nutritionally unbalanced distribution of amino acids. As large amounts of gelatin are ingested and the excess amino acids are catabolized, the capacity of the urea cycle may be exceeded, leading to ammonia toxicity. This is further complicated by the dehydration that may result from excretion of large quantities of urea. A combination of these two factors could produce coma and death.

**15.** Alanine and glutamine play special roles in the transport of amino groups from muscle and from other nonhepatic tissues, respectively, to the liver.

# Chapter 19

**1.** *Reaction (1):* **(a), (d)** NADH; **(b), (e)** E-FMN; **(c)** NAD$^+$/NADH and E-FMN/FMNH$_2$

*Reaction (2):* **(a), (d)** E-FMNH$_2$; **(b), (e)** Fe$^{3+}$; **(c)** E-FMN/FMNH$_2$ and Fe$^{3+}$/Fe$^{2+}$

*Reaction (3):* **(a), (d)** Fe$^{2+}$; **(b), (e)** Q; **(c)** Fe$^{3+}$/Fe$^{2+}$ and Q/QH$_2$

**2.** The side chain makes ubiquinone soluble in lipids and allows it to diffuse in the semifluid membrane.

**3.** From the difference in standard reduction potential ($\Delta E'^\circ$) for each pair of half-reactions, one can calculate $\Delta G'^\circ$. The oxidation of succinate by FAD is favored by the negative standard free-energy change ($\Delta G'^\circ = -3.7$ kJ/mol). Oxidation by NAD$^+$ would require a large, positive, standard free-energy change ($\Delta G'^\circ = 68$ kJ/mol).

**4. (a)** All carriers reduced; CN$^-$ blocks the reduction of O$_2$ catalyzed by cytochrome oxidase. **(b)** All carriers reduced; in the absence of O$_2$, the reduced carriers are not reoxidized. **(c)** All carriers oxidized. **(d)** Early carriers more reduced; later carriers more oxidized.

**5. (a)** Inhibition of NADH dehydrogenase by rotenone decreases the rate of electron flow through the respiratory chain, which in turn decreases the rate of ATP production. If this reduced rate is unable to meet the organism's ATP requirements, the organism dies. **(b)** Antimycin A strongly inhibits the oxidation of Q in the respiratory chain, reducing the rate of electron transfer and leading to the consequences described in (a). **(c)** Because antimycin A blocks *all* electron flow to oxygen, it is a more potent poison than rotenone, which blocks electron flow from NADH but not from FADH$_2$.

**6. (a)** The rate of electron transfer necessary to meet the ATP demand increases, and thus the P/O ratio decreases. **(b)** High concentrations of uncoupler produce P/O ratios near zero. The P/O ratio decreases, and more fuel must be oxidized to generate the same amount of ATP. The extra heat released by this oxidation raises the body temperature. **(c)** Increased activity of the respiratory chain in the presence of an uncoupler requires the degradation of additional fuel. By oxidizing more fuel (including fat reserves) to produce the same amount of ATP, the body loses weight. When the P/O ratio approaches zero, the lack of ATP results in death.

**7.** Valinomycin acts as an uncoupler. It combines with K$^+$ to form a complex that passes through the inner mitochondrial membrane, dissipating the membrane potential. ATP synthesis decreases, which causes the rate of electron transfer to increase. This results in an increase in the H$^+$ gradient, O$_2$ consumption, and amount of heat released.

**8. (a)** The formation of ATP is inhibited. **(b)** The formation of ATP is tightly coupled to electron transfer; 2,4-dinitrophenol is an uncoupler of oxidative phosphorylation. **(c)** Oligomycin

**9.** Cytosolic malate dehydrogenase plays a key role in the transport of reducing equivalents across the inner mitochondrial membrane via the malate-aspartate shuttle.

**10. (a)** Glycolysis becomes anaerobic. **(b)** Oxygen consumption ceases. **(c)** Lactate formation increases. **(d)** ATP synthesis decreases to 2 ATP/glucose.

**11.** The steady-state concentration of P$_i$ in the cell is much higher than that of ADP. The P$_i$ released by ATP hydrolysis changes total [P$_i$] very little.

**12. (a)** NADH is reoxidized via electron transfer instead of lactic acid fermentation. **(b)** Oxidative phosphorylation is more efficient. **(c)** The high mass-action ratio of the ATP system inhibits phosphofructokinase-1.

**13.** Fermentation to ethanol could be accomplished in the presence of O$_2$, which is an advantage because strict anaerobic conditions are difficult to maintain. The Pasteur effect is not observed, since the citric acid cycle and electron-transfer chain are inactive.

**14. (a)** External medium: $4.0 \times 10^{-8}$ M; matrix: $2.0 \times 10^{-8}$ M **(b)** 2:1; 1.7 kJ/mol **(c)** 21 **(d)** No **(e)** From the transmembrane potential

**15. (a)** 0.91 $\mu$mol/s · g **(b)** 5.5 s; to provide a constant level of ATP, regulation of ATP production must be tight and rapid.

**16.** 53 $\mu$mol/s · g. With a steady state [ATP] of 7.0 $\mu$mol/g, this is equivalent to 10 turnovers of the ATP pool per second; the reservoir would last about 0.13 s.

**17.** The inner mitochondrial membrane is impermeable to NADH, but the reducing equivalents of NADH are transferred (shuttled) through the membrane indirectly: they are transferred to oxaloacetate in the cytosol, the resulting malate is transported into the matrix, and mitochondrial NAD$^+$ is reduced to NADH.

**18.** Pyruvate dehydrogenase is located in mitochondria; glyceraldehyde 3-phosphate dehydrogenase in the cytosol. The NAD pools are separated by the inner mitochondrial membrane.

**19.** For the maximum photosynthetic rate, PSI (which absorbs light of 700 nm) and PSII (which absorbs light of 680 nm) must be operating simultaneously.

**20.** The extra weight comes from the water consumed in the overall reaction.

**21.** Purple sulfur bacteria use $H_2S$ as the hydrogen donor in photosynthesis. No $O_2$ is evolved, because the single photosystem lacks the manganese-containing water-splitting complex.

**22.** 0.44

**23.** During illumination, a proton gradient is established. When ADP and $P_i$ are added, ATP synthesis is driven by the gradient, which becomes exhausted in the absence of light.

**24.** DCMU blocks electron transfer between PSII and the first site of ATP production.

**25.** **(a)** 55 kJ/mol **(b)** 0.29 V

**26.** From the difference in reduction potentials, one can calculate that $\Delta G'^{\circ} = 15$ kJ for the redox reaction. Figure 19–39 shows that the energy of photons in any region of the visible spectrum is more than sufficient to drive this endergonic reaction.

**27.** $1.3 \times 10^{-77}$; the reaction is highly unfavorable! In chloroplasts, the input of light energy overcomes this barrier.

**28.** $-920$ kJ/mol

**29.** No. The electrons from $H_2O$ flow to the artificial electron acceptor $Fe^{3+}$, not to $NADP^+$.

**30.** About once every 0.1 s; 1 in $10^8$ is excited.

**31.** Light of 700 nm excites PSI but not PSII; electrons flow from P700 to $NADP^+$, but no electrons flow from P680 to replace them. When light of 680 nm excites PSII, electrons tend to flow to PSI, but the electron carriers between the two photosystems quickly become completely reduced.

**32.** No. The excited electron from P700 returns to refill the electron "hole" created by illumination. PSII is not needed to supply electrons, and no $O_2$ is evolved from $H_2O$. NADPH is not formed, because the excited electron returns to P700.

## Chapter 20

**1.** Within subcellular organelles, concentrations of specific enzymes and metabolites are elevated; separate pools of cofactors and intermediates are maintained; regulatory mechanisms affect only one set of enzymes and pools.

**2.** This observation suggests that ATP and NADPH are generated in the light and are essential for $CO_2$ fixation; conversion stops as the supply of ATP and NADPH becomes exhausted. Furthermore, some enzymes are switched off in the dark.

**3.** X is 3-phosphoglycerate; Y is ribulose 1,5-bisphosphate.

**4.** Ribulose 5-phosphate kinase, fructose 1,6-bisphosphatase, sedoheptulose 1,7-bisphosphatase, and glyceraldehyde 3-phosphate dehydrogenase; all are activated by reduction of a critical disulfide bond to a pair of sulfhydryls; iodoacetate reacts irreversibly with free sulfhydryls.

**5.** To carry out the disulfide exchange reaction that activates the Calvin cycle enzymes, thioredoxin needs both of its sulfhydryl groups.

**6.** Reductive pentose phosphate pathway regenerates ribulose 1,5-bisphosphate from triose phosphates produced during photosynthesis. Oxidative pentose phosphate pathway provides NADPH for reductive biosynthesis and pentose phosphates for nucleotide synthesis.

**7.** Both types of "respiration" consume $O_2$ and produce $CO_2$. Mitochondrial respiration occurs continuously; electrons derived from various fuels are passed through a chain of carriers in the inner mitochondrial membrane to $O_2$. Photorespiration occurs in chloroplasts, peroxisomes, and mitochondria. Photorespiration occurs during the daytime, when photosynthetic carbon fixation is occurring; mitochondrial respiration occurs primarily at night, or during cloudy days. Path of electron flow in photorespiration is shown in 20-21; that for mitochondrial respiration, in Fig.19–15.

**8.** This hypothesis assumes directed evolution, or evolution with a purpose—ideas not generally accepted by evolutionary biologists.

**9.** **(a)** Without production of NADPH by the pentose phosphate pathway, cells would be unable to synthesize lipids and other reduced products. **(b)** Without generation of ribulose 1,5-bisphosphate, the Calvin cycle is effectively blocked.

**10.** In maize, $CO_2$ is fixed by the $C_4$ pathway elucidated by Hatch and Slack, in which PEP is carboxylated rapidly to oxaloacetate (some of which undergoes transamination to aspartate) and reduced to malate. Only after subsequent decarboxylation does the $CO_2$ enter the Calvin cycle.

**11.** Measure the rate of fixation of $^{14}C$ carbon dioxide in the light (daytime) and the dark. Greater fixation in the dark identifies the CAM plant. One could also determine the titratable acidity; acids stored in the vacuole during the night can be quantified in this way.

**12.** Isocitrate dehydrogenase reaction

**13.** Storage consumes 1 mol of ATP per mole of glucose 6-phosphate; this represents 3.3% of the total ATP available from glucose 6-phosphate metabolism (i.e., the efficiency of storage is 96.7%).

**14.** $[PP_i]$ is high in the cytosol because the cytosol lacks inorganic pyrophosphatase.

**15.** **(a)** Low $[P_i]$ in the cytosol and high [triose phosphate] in the chloroplast **(b)** High [triose phosphate] in the cytosol

**16.** 3-Phosphoglycerate is the primary product of photosynthesis; $[P_i]$ rises when light-driven synthesis of ATP from ADP and $P_i$ slows.

**17.** **(a)** Sucrose + $(glucose)_n \rightarrow (glucose)_{n+1}$ + fructose **(b)** Fructose generated in the synthesis of dextran is readily imported and metabolized by the bacteria.

**18.** Species 1 is C4; species 2, C3.

**19.** **(a)** In peripheral chloroplasts; for both **(b)** and **(c)**, In central sphere

## Chapter 21

**1.** **(a)** The 16 carbons of palmitate are derived from 8 acetyl groups of 8 acetyl-CoA molecules. The $^{14}C$-labeled acetyl-CoA gives rise to malonyl-CoA labeled at C-1 and C-2. **(b)** The metabolic pool of malonyl-CoA, the source of all palmitate carbons except C-16 and C-15, does not become labeled with small amounts of $^{14}C$-labeled acetyl-CoA. Hence, only [15,16-$^{14}C$] palmitate is formed.

**2.** Both glucose and fructose are degraded to pyruvate in glycolysis. Pyruvate is converted to acetyl-CoA by the pyruvate dehydrogenase complex. Some of this acetyl-CoA enters the citric acid cycle, which produces reducing equivalents (NADH and NADPH). Mitochondrial electron transfer to $O_2$ yields ATP.

**3.** 8 Acetyl-CoA + 15ATP + 14NADPH + 9$H_2O \rightarrow$
    palmitate + 8CoA + 15ADP + 15$P_i$ + 14$NADP^+$ + 2$H^+$

**4.** **(a)** 3 deuteriums per palmitate; all located on C-16; all other two-carbon units are derived from unlabeled malonyl-CoA. **(b)** 7 deuteriums per palmitate; located on all *even*-numbered carbons except C-16.

**5.** By using the three-carbon unit malonyl-CoA, the activated form of acetyl-CoA (recall that malonyl-CoA synthesis requires ATP), metabolism is driven in the direction of fatty acid synthesis by the exergonic release of $CO_2$.

6. The rate-limiting step in fatty acid synthesis is carboxylation of acetyl-CoA, catalyzed by acetyl-CoA carboxylase. High [citrate] and [isocitrate] indicate that conditions are favorable for fatty acid synthesis: an active citric acid cycle is providing a plentiful supply of ATP, reduced pyridine nucleotides, and acetyl-CoA. Citrate stimulates (increases the $V_{max}$ of) acetyl-CoA carboxylase **(a)**. Because citrate binds more tightly to the filamentous form of the enzyme (the active form), high [citrate] drives the protomer $\rightleftharpoons$ filament equilibrium in the direction of the active form **(b)**. In contrast, palmitoyl-CoA (the end product of fatty acid synthesis) drives the equilibrium in the direction of the inactive (protomer) form. Hence, when the end product of fatty acid synthesis accumulates, the biosynthetic path slows.

7. **(a)** Acetyl-CoA$_{(mit)}$ + ATP + CoA$_{(cyt)}$ → acetyl-CoA$_{(cyt)}$ + ADP + P$_i$ + CoA$_{(mit)}$ **(b)** 1 ATP per acetyl group

   **(c)** Yes

8. The double bond in palmitoleate is introduced by an oxidation catalyzed by fatty acyl–CoA desaturase, a mixed-function oxidase that requires O$_2$ as a cosubstrate.

9. 3 Palmitate + glycerol + 7ATP + 4H$_2$O →

   tripalmitin + 7ADP + 7P$_i$ + 7H$^+$

10. In adult rats, stored triacylglycerols are maintained at a steady-state level through a balance of the rates of degradation and biosynthesis. Hence, the triacylglycerols of adipose (fat) tissue are constantly turned over, which explains the incorporation of $^{14}$C label from dietary glucose.

11. Net reaction:

    Dihydroxyacetone phosphate + NADH + palmitate + oleate + 3ATP + CTP + choline + 4H$_2$O →

    phosphatidylcholine + NAD$^+$ + 2AMP + ADP + H$^+$ + CMP
    + 5Pi

    7 ATP per molecule of phosphatidylcholine

12. Methionine deficiency reduces the level of adoMet, which is required for the de novo synthesis of phosphatidylcholine. The salvage pathway does not employ adoMet, but uses available choline. Thus phosphatidylcholine can be synthesized even when the diet is deficient in methionine, as long as choline is available.

13. $^{14}$C label appears in three places in the activated isoprene:

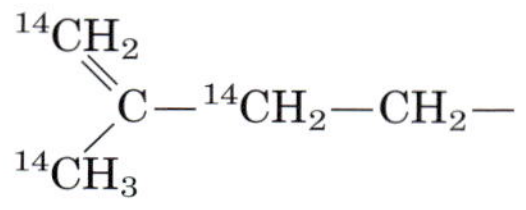

14. **(a)** ATP **(b)** UDP-glucose **(c)** CDP-ethanolamine **(d)** UDP-galactose **(e)** Fatty acyl–CoA **(f)** S-Adenosylmethionine **(g)** Malonyl-CoA **(h)** Isopentenyl pyrophosphate

15. Linoleate is required in the synthesis of prostaglandins. Animals are unable to transform oleate to linoleate, so linoleate is an essential fatty acid. Plants can transform oleate to linoleate, and they provide animals with the required linoleate (see Fig. 21–12).

16. The rate-determining step in the biosynthesis of cholesterol is the synthesis of mevalonate, catalyzed by hydroxymethylglutaryl-CoA reductase. This enzyme is allosterically regulated by mevalonate and cholesterol derivatives. High intracellular [cholesterol] also reduces transcription of the gene encoding HMG-CoA reductase.

## Chapter 22

1. In their symbiotic relationship with the plant, bacteria supply ammonium ion by reducing atmospheric nitrogen, which requires large quantities of ATP.

2. The transfer of nitrogen from NH$_3$ to carbon skeletons can be catalyzed by (1) glutamine synthetase and (2) glutamate dehydrogenase. The latter enzyme produces glutamate, the amino group donor in all transamination reactions, necessary to the formation of amino acids for protein synthesis.

3. In the mammalian route, toxic ammonium ions are transformed to glutamine, reducing toxic effects on the brain.

4. Glucose + 2CO$_2$ + 2NH$_3$ → 2 aspartate + 2H$^+$ + 2H$_2$O

5. If phenylalanine hydroxylase is defective, the biosynthetic route to tyrosine is blocked and tyrosine must be obtained from the diet.

6. In adoMet synthesis, triphosphate is released from ATP. Hydrolysis of the triphosphate renders the reaction thermodynamically more favorable.

7. If the inhibition of glutamine synthase were not concerted, saturating concentrations of histidine would shut down the enzyme and cut off production of glutamine, which the bacterium needs to synthesize other products.

8. Folic acid is a precursor of tetrahydrofolate (Fig. 18–16), required in the biosynthesis of glycine (Fig. 22–12), a precursor of porphyrins. A folic acid deficiency therefore impairs hemoglobin synthesis.

9. *For glycine auxotrophs:* adenine and guanine; *for glutamine auxotrophs:* adenine, guanine, and cytosine; *for aspartate auxotrophs:* adenine, guanine, cytosine, and uridine.

10. **(a)** See Figure 18–6, step ②, for the reaction mechanism of amino acid racemization. The F atom of fluoroalanine is an excellent leaving group. Fluoroalanine causes irreversible (covalent) inhibition of alanine racemase. One plausible mechanism is (Nuc denotes any nucleophilic amino acid side chain in the enzyme active site):

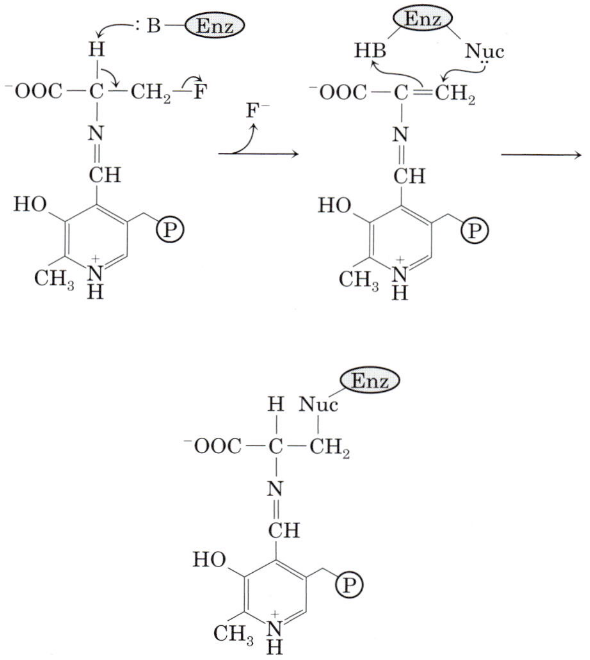

**(b)** Azaserine (see Fig. 22–48) is an analog of glutamine. The diazoacetyl group is highly reactive and forms covalent bonds with nucleophiles at the active site of a glutamine amidotransferase.

11. **(a)** As shown in Figure 18–16, p-aminobenzoate is a component of $N^5,N^{10}$-methylenetetrahydrofolate, the cofactor involved in the transfer of one-carbon units. **(b)** In the presence of sulfanilamide, a structural analog of p-aminobenzoate, bacteria are

unable to synthesize tetrahydrofolate, a cofactor necessary for converting AICAR to FAICAR; thus AICAR accumulates.

**(c)** The competitive inhibition by sulfanilamide of the enzyme involved in tetrahydrofolate biosynthesis is overcome by the addition of excess substrate (*p*-aminobenzoate).

**12.** The $^{14}$C-labeled orotate arises from the following pathway (the first three steps are part of the citric acid cycle):

$$^-OO^{14}C—^{14}CH_2—^{14}CH_2—^{14}COO^-$$
Succinate

↓

$$^-OO^{14}C \quad \backslash \quad H$$
$$^{14}C=^{14}C$$
$$H \quad / \quad ^{14}COO^-$$
Fumarate

↓

$$^-OO^{14}C—^{14}\underset{H}{\overset{OH}{C}}—^{14}\underset{H}{\overset{H}{C}}—^{14}COO^-$$
Malate

↓

$$^-OO^{14}C—^{14}\overset{O}{\overset{\|}{C}}—^{14}\underset{H}{\overset{H}{C}}—^{14}COO^-$$
Oxaloacetate

↓ transamination

$$^-OO^{14}C—^{14}\underset{H}{\overset{\overset{+}{N}H_3}{C}}—^{14}CH_2—^{14}COO^-$$
Aspartate

↓

↓

↓

Orotate

**13.** Organisms do not store nucleotides to be used as fuel, and they do not completely degrade them, but rather hydrolyze them to release the bases, which can be recovered in salvage pathways. The low C:N ratio of nucleotides makes them poor sources of energy.

**14.** Treatment with allopurinol has two consequences. (1) It inhibits conversion of hypoxanthine to uric acid, causing accumulation of hypoxanthine, which is more soluble and more readily excreted; this alleviates the clinical problems associated with AMP degradation. (2) It inhibits conversion of guanine to uric acid, causing accumulation of xanthine, which is less soluble than uric acid; this is the source of xanthine stones. Because the amount of GMP degradation is low relative to AMP degradation, the kidney damage caused by xanthine stones is less than that caused by untreated gout.

**15.** 5-Phosphoribosyl-1-pyrophosphate; this is the first $NH_3$ acceptor in the purine biosynthetic pathway.

## Chapter 23

**1.** Steady-state levels of ATP are maintained by phosphoryl group transfer to ADP from phosphocreatine. 1-Fluoro-2,4-dinitrobenzene inhibits creatine kinase.

**2.** Ammonia is very toxic to nervous tissue, especially the brain. Excess $NH_3$ is removed by transformation of glutamate to glutamine, which travels to the liver and is subsequently transformed to urea. The additional glutamine arises from the transformation of glucose to α-ketoglutarate, transamination of α-ketoglutarate to glutamate, and conversion of glutamate to glutamine.

**3.** From glucose, by the following route: Glucose → dihydroxyacetone phosphate (in glycolysis); dihydroxyacetone phosphate + NADH + H$^+$ → glycerol 3-phosphate + NAD$^+$ (glycerol 3-phosphate dehydrogenase reaction)

**4. (a)** Increased muscular activity increases the demand for ATP, which is met by increased $O_2$ consumption. **(b)** After the sprint, lactate produced by anaerobic glycolysis is converted to glucose and glycogen, which requires ATP and therefore $O_2$ .

**5.** Glucose is the primary fuel of the brain. TPP-dependent oxidative decarboxylation of pyruvate to acetyl-CoA is essential to complete glucose metabolism.

**6.** 190 m

**7. (a)** Inactivation provides a rapid means to change hormone concentrations. **(b)** Insulin level is maintained by equal rates of synthesis and degradation. **(c)** Changes in the rate of release from storage, rate of transport, and rate of conversion from prohormone to active hormone.

**8.** Water-soluble hormones bind to receptors on the outer surface of the cell, triggering the formation of a second messenger (e.g., cAMP) inside the cell. Lipid-soluble hormones can pass through the plasma membrane to act on target molecules or receptors directly.

**9. (a)** Heart and skeletal muscle lack glucose 6-phosphatase. Any glucose 6-phosphate produced enters the glycolytic pathway, and under $O_2$-deficient conditions is converted to lactate via pyruvate. **(b)** In a "fight or flight" situation, the concentration of glycolytic precursors must be high in preparation for muscular activity. Phosphorylated intermediates cannot escape from the cell, because the membrane is not permeable to charged species, and glucose 6-phosphate is not exported on the glucose transporter. The liver, by contrast, must release the glucose necessary to maintain blood glucose level; glucose is formed from glucose 6-phosphate and enters the bloodstream.

10. **(a)** Excessive uptake and use of blood glucose by the liver, leading to hypoglycemia; shutdown of amino acid and fatty acid catabolism **(b)** Little circulating fuel is available for ATP requirements. Brain damage results because glucose is the main source of fuel for the brain.

11. Thyroxine acts as an uncoupler of oxidative phosphorylation. Uncouplers lower the P/O ratio, and the tissue must increase respiration to meet the normal ATP demands. Thermogenesis could also be due to the increased rate of ATP utilization by the thyroid-stimulated tissue, as increased ATP demands are met by increased oxidative phosphorylation and thus respiration.

12. Because prohormones are inactive, they can be stored in quantity in secretory granules. Rapid activation is achieved by enzymatic cleavage in response to an appropriate signal.

13. In animals, glucose can be synthesized from many precursors (see Fig. 14-15). In humans, the principal precursors are glycerol from triacylglycerols and glucogenic amino acids from protein.

14. The *ob/ob* mouse, which is initially obese, will lose weight. The *OB/OB* mouse will retain its normal body weight.

# Chapter 24

1. $6.1 \times 10^4$ nm; 290 times longer than the T2 phage head

2. The number of A residues does not equal the number of T residues, nor does the number of G equal the number of C, so the DNA is not a base-paired double helix; the M13 DNA is single-stranded.

3. $M_r = 3.8 \times 10^8$; length $= 200$ $\mu$m; $Lk_0 = 55,200$; $Lk = 51,900$

4. The exons contain $3 \times 192 = 576$ bp. The remaining 864 bp are in introns, possibly in a leader or signal sequence, and/or in other noncoding DNA.

5. 5,000 bp. **(a)** Doesn't change; $Lk$ cannot change without breaking and re-forming the covalent backbone of the DNA. **(b)** Becomes undefined; a circular DNA with a break in one strand has, by definition, no $Lk$. **(c)** Decreases; in the presence of ATP, gyrase underwinds DNA. **(d)** Doesn't change; this assumes that neither of the DNA strands is broken in the heating process.

6. $\sigma = -0.067$; >70% probability

7. **(a)** Undefined; the strands of a nicked DNA could be separated and thus have no $Lk$. **(b)** 476 **(c)** 476; the DNA is already relaxed, so the topoisomerase does not cause a net change. **(d)** 460; gyrase plus ATP reduces the $Lk$ in increments of 2. **(e)** 464; eukaryotic type I topoisomerases increase the $Lk$ of underwound or negatively supercoiled DNA in increments of 1. **(f)** 460; nucleosome binding does not break any DNA strands and thus cannot change $Lk$.

8. A fundamental structural unit in chromatin repeats about every 200 bp; the DNA is accessible to the nuclease only at 200 bp intervals. The brief treatment was insufficient to cleave the DNA at every accessible point, so a ladder of DNA bands is created in which the DNA fragments are multiples of 200 bp. The thickness of the DNA bands suggests that the distance between cleavage sites varies somewhat. For instance, not all the fragments in the lowest band are exactly 200 bp long.

9. A right-handed helix has a positive $Lk$; a left-handed helix (such as Z-DNA) has a negative $Lk$. Decreasing the $Lk$ of a closed circular B-DNA by underwinding it facilitates formation of regions of Z-DNA within certain sequences. (See Chapter 8, p. 285, for a description of sequences that permit the formation of Z-DNA.)

10. **(a)** Both strands must be covalently closed, and the molecule must be either circular or constrained at both ends. **(b)** Formation of cruciforms, left-handed Z DNA, plectonemic or solenoidal supercoils, and unwinding of the DNA are favored. **(c)** *E. coli* DNA topoisomerase II or DNA gyrase. **(d)** It binds the DNA at a point where it crosses on itself, cleaves both strands of one of the crossing segments, passes the other segment through the break, then reseals the break. The result is a change in $Lk$ of $-2$.

11. Centromere, telomeres, and an autonomous replicating sequence or replication origin

# Chapter 25

1. In random, dispersive replication, in the second generation, all the DNAs would have the same density and would appear as a single band, not the two bands observed in the Meselson-Stahl experiment.

2. In this extension of the Meselson-Stahl experiment, after three generations the molar ratio of $^{15}$N–$^{14}$N DNA to $^{14}$N–$^{14}$N DNA is $2/6 = 0.33$.

3. **(a)** $4.42 \times 10^5$ turns; **(b)** 40 min. In cells dividing every 20 min, a replicative cycle is initiated every 20 min, each cycle beginning before the prior one is complete. **(c)** 2,000 to 5,000 Okazaki fragments. The fragments are 1,000 to 2,000 nucleotides long and are firmly bound to the template strand by base pairing. Each fragment is quickly joined to the lagging strand, thus preserving the correct order of the fragments.

4. A 28.7%; G 21.3%; C 21.3%; T 28.7%. The DNA strand made from the template strand: A 32.7%; G 18.5%; C 24.1%; T 24.7%; the DNA strand made from the complementary template strand: A 24.7%; G 24.1%; C 18.5%; T 32.7%. It is assumed that the two template strands are replicated completely.

5. **(a)** No. Incorporation of $^{32}$P into DNA results from the synthesis of new DNA, which requires the presence of *all four* nucleotide precursors. **(b)** Yes. Although all four nucleotide precursors must be present for DNA synthesis, only one of them has to be radioactive in order for radioactivity to appear in the new DNA. **(c)** No. Radioactivity is incorporated only if the $^{32}$P label is in the $\alpha$ phosphate; DNA polymerase cleaves off pyrophosphate—i.e., the $\beta$ and $\gamma$ phosphate groups.

6. *Leading strand:* Precursors: dATP, dGTP, dCTP, dTTP (also needs a template DNA strand and DNA primer); enzymes and other proteins: DNA gyrase, helicase, single-stranded DNA–binding protein, DNA polymerase III, topoisomerases, and pyrophosphatase. *Lagging strand:* Precursors: ATP, GTP, CTP, UTP, dATP, dGTP, dCTP, dTTP (also needs an RNA primer); enzymes and other proteins: DNA gyrase, helicase, single-stranded DNA–binding protein, primase, DNA polymerase III, DNA polymerase I, DNA ligase, topoisomerases, and pyrophosphatase. NAD$^+$ is also required as a cofactor for DNA ligase.

7. Mutants with defective DNA ligase produce a DNA duplex in which one of the strands remains in pieces (as Okazaki fragments). When this duplex is denatured, sedimentation results in one fraction containing the intact single strand (the high molecular weight band) and one fraction containing the unspliced fragments (the low molecular weight band).

8. Watson-Crick base pairing between template and leading strand; proofreading and removal of wrongly inserted nucleotides by the 3'-exonuclease activity of DNA polymerase III.

Yes—perhaps. Because the factors ensuring fidelity of replication are operative in both the leading and the lagging strands, the lagging strand would probably be made with the same fidelity. However, the greater number of distinct chemical operations involved in making the lagging strand might provide a greater opportunity for errors to arise.

**9.** ~1,200 bp (600 in each direction)

**10.** A small fraction (13 of $10^9$ cells) of the histidine-requiring mutants spontaneously undergo back-mutation and regain their capacity to synthesize histidine. 2-Aminoanthracene increases the rate of back-mutations about 1800-fold and is therefore mutagenic. Since most carcinogens are mutagenic, 2-aminoanthracene is probably carcinogenic.

**11.** Spontaneous deamination of 5-methylcytosine (see p. 293) produces thymine, and thus a G–T mismatched pair. These are among the most common mismatches in the DNA of eukaryotes. The specialized repair system restores the G≡C pair.

**12. (a)** Ultraviolet irradiation produces pyrimidine dimers; in normal fibroblasts these are excised by cleavage of the damaged strand by a special excinuclease. Thus the denatured single-stranded DNA contains the many fragments caused by the cleavage, and the average molecular weight is lowered. These fragments of single-stranded DNA are absent from the XPG samples, as indicated by the unchanged average molecular weight. **(b)** The absence of fragments in the single-stranded DNA from the XPG cells after irradiation suggests the special excinuclease is defective or missing.

**13.** During homologous genetic recombination, a Holliday intermediate may be formed almost anywhere within the two paired, homologous chromosomes; the branch point of the intermediate can move extensively by branch migration. In site-specific recombination, the Holliday intermediate is formed between two specific sites, and branch migration is generally restricted by heterologous sequences on either side of the recombination sites.

## Chapter 26

**1. (a)** 60 to 100s; **(b)** 500 to 900 nucleotides

**2.** A single base error in DNA replication, if not corrected, would cause one of the two daughter cells, and all its progeny, to have a mutated chromosome. A single base error in RNA transcription would not affect the chromosome; it would lead to formation of some defective copies of one protein, but because mRNAs turn over rapidly, most copies of the protein would not be defective. The progeny of this cell would be normal.

**3.** Normal posttranscriptional processing at the 3′ end (cleavage and polyadenylation) would be inhibited or blocked.

**4.** Because the template-strand RNA does not encode the enzymes needed to initiate viral infection, it would probably be inert or simply degraded by cellular ribonucleases. Replication of the template-strand RNA and propagation of the virus could occur only if intact RNA replicase (RNA-dependent RNA polymerase) were introduced into the cell along with the template strand.

**5.** (1) Use of a template strand of nucleic acid; (2) synthesis in the 3′→5′ direction; (3) use of nucleoside triphosphate substrates, with formation of a phosphodiester bond and displacement of $PP_i$

**6.** Generally two: one to cleave the phosphodiester bond at one intron-exon junction; the other to link the resulting free exon end to the exon at the other end of the intron. If the nucleophile in the first step were water, this step would be a hydrolytic event and only one transesterification step would be required to complete the splicing process.

**7.** These enzymes lack a 3′→5′ proofreading exonuclease and have a high error rate; the likelihood of a replication error that would inactivate the virus is much less in a small genome than in a large one.

**8. (a)** $4^{32} = 1.8 \times 10^{19}$ **(b)** 0.006% **(c)** For the "unnatural selection" step, use a chromatographic resin with a bound molecule that is a transition-state analog of the ester hydrolysis reaction (e.g., an appropriate phosphonate compound; see Box 6-3).

**9.** Though RNA synthesis is quickly halted by α-amanitin toxin, it takes several days for the critical mRNAs and the proteins in the liver to degrade, causing liver dysfunction and death.

**10. (a)** After lysis of the cells and partial purification of the contents, the protein extract could be subjected to isoelectric focusing. The β subunit could be detected by an antibody-based assay. The difference in amino acid residues between the normal β subunit and the mutated form (i.e., the different charges on the amino acids) would alter the electrophoretic mobility of the mutant protein in an isoelectric focusing gel, relative to the protein from a nonresistant strain. **(b)** Direct DNA sequencing (by the Sanger method).

**11. (a)** 384 nucleotide pairs **(b)** 1,620 nucleotide pairs **(c)** Most of the nucleotides are untranslated regions at the 3′ and 5′ ends of the mRNA. Also, most mRNAs code for a signal sequence (Chapter 27) in their protein products, which is eventually cleaved off to produce the mature and functional protein.

## Chapter 27

**1. (a)** Gly–Gln–Ser–Leu–Leu–Ile **(b)** Leu–Asp–Ala–Pro **(c)** His–Asp–Ala–Cys–Cys–Tyr **(d)** Met–Asp–Glu in eukaryotes; fMet–Asp–Glu in prokaryotes

**2.** UUAAUGUAU, UUGAUGUAU, CUUAUGUAU, CUCAUGUAU, CUAAUGUAU, CUGAUGUAU, UUAAUGUAC, UUGAUGUAC, CUUAUGUAC, CUCAUGUAC, CUAAUGUAC, CUGAUGUAC

**3.** No. Because nearly all the amino acids have more than one codon (e.g., Leu has six), any given polypeptide can be coded for by a number of different base sequences. However, some amino acids are encoded by only one codon and those with multiple codons often share the same nucleotide at two of the three positions, so *certain parts* of the mRNA sequence encoding a protein of known amino acid sequence can be predicted with high certainty.

**4. (a)** (5′)CGACGGCGCGAAGUCAGGGGUGUUAAG(3′) **(b)** Arg–Arg–Arg–Glu–Val–Arg–Gly–Val–Lys **(c)** No. The complementary antiparallel strands in double-helical DNA do not have the same base sequence in the 5′→3′ direction. RNA is transcribed from only one specific strand of duplex DNA. The RNA polymerase must therefore recognize and bind to the correct strand.

**5.** There are two tRNAs for methionine: tRNA^fMet, which is the initiating tRNA, and tRNA^Met, which can insert a Met residue in interior positions in a polypeptide. Only fMet-tRNA^fMet is recognized by the initiation factor IF-2 and is aligned with the initiating AUG positioned at the ribosomal P site in the initiation complex. AUG codons in the interior of the mRNA can bind and incorporate only Met-tRNA^Met.

**6.** Allow polynucleotide phosphorylase to act on a mixture of UDP and CDP in which UDP has, say, five times the concentration of CDP. The result would be a synthetic RNA polymer with many UUU triplets (coding for Phe), a smaller number of UUC (also Phe), UCU (Ser), and CUU (Leu), a much smaller number of UCC (also Ser), CUC (also Leu), and CCU (Pro), and the smallest number of CCC (Pro).

**7.** A minimum of 583 ATP equivalents (based on 4 per amino acid residue added, except that there are only 145 translocation steps). Correction of each error requires 2 ATP equivalents. For glycogen synthesis, 292 ATP equivalents. The extra energy cost for $\beta$-globin synthesis reflects the cost of the information content of the protein. At least 20 activating enzymes, 70 ribosomal proteins, 4 rRNAs, 32 or more tRNAs, an mRNA, and 10 or more auxiliary enzymes must be made by the eukaryotic cell in order to synthesize a protein from amino acids. The synthesis of an $(\alpha1\rightarrow4)$ chain of glycogen from glucose requires only 4 or 5 enzymes (Chapter 15).

**8.**

| Glycine codons | Anticodons |
| --- | --- |
| (5')GGU | (5')ACC, GCC, ICC |
| (5')GGC | (5')GCC, ICC |
| (5')GGA | (5')UCC, ICC |
| (5')GGG | (5')CCC, UCC |

**(a)** The 3' and middle position **(b)** Pairings with anticodons (5')GCC, ICC, and UCC **(c)** Pairings with anticodons (5')ACC and CCC

**9. (a), (c), (e),** and **(g)** only; (b), (d), and (f) cannot be the result of single-base mutations: (b) and (f) would require substitutions of two bases, and (d) would require substitutions of all three bases.

**10.** The two DNA codons for Glu are GAA and GAG, and the four DNA codons for Val are GTT, GTC, GTA, and GTG. A single base change in GAA to form GTA or in GAG to form GTG could account for the Glu $\rightarrow$ Val replacement in sickle-cell hemoglobin. Much less likely are two-base changes, from GAA to GTG, GTT, or GTC; and from GAG to GTA, GTT, or GTC.

**11. (a)** The Ala-tRNA synthetase recognizes the $G^3$–$U^{70}$ base pair in the amino acid arm of tRNA$^{Ala}$. **(b)** The mutant tRNA$^{Ala}$ would insert Ala residues at codons encoding Pro. **(c)** A mutation that might have similar effects is an alteration in tRNA$^{Pro}$ that allowed it to be recognized and aminoacylated by Ala-tRNA synthetase. **(d)** Most of the proteins in the cell would be inactivated, so these would be lethal mutations and hence never observed. This represents a powerful selective pressure for maintaining the genetic code.

**12.** The amino acid most recently added to a growing polypeptide chain is the only one covalently attached to a tRNA and thus is the only link between the polypeptide and the mRNA encoding it. A proofreading activity would sever this link, halting synthesis of the polypeptide and releasing it from the mRNA.

**13.** The protein would be directed into the ER, and from there the targeting would depend on additional signals. SRP binds the amino-terminal signal early in protein synthesis and directs the nascent polypeptide and ribosome to receptors in the ER. Because the protein is translocated into the lumen of the ER as it is synthesized, the NLS is never accessible to the proteins involved in nuclear targeting.

**14.** Trigger factor is a molecular chaperone that stabilizes an unfolded and translocation-competent conformation of ProOmpA.

**15.** DNA with a minimum of 5,784 bp; some of the coding sequences must be nested or overlapping.

## Chapter 28

**1. (a)** Tryptophan synthase levels remain high in spite of the presence of tryptophan. **(b)** Levels again remain high. **(c)** Levels rapidly decrease, preventing wasteful synthesis of tryptophan.

**2. (a)** Constitutive, low-level expression of the operon; most mutations in the operator would make the repressor less likely to bind. **(b)** Either constitutive expression, as in (a), or constant repression, if the mutation destroyed the capability to bind to lactose and related compounds and hence the response to inducers. **(c)** Either increased or decreased expression of the operon (under conditions in which it is induced), depending on whether the mutation made the promoter more or less similar, respectively, to the consensus *E. coli* promoter.

**3.** 7,000 copies

**4.** $8 \times 10^{-9}$ M, about $10^5$ times greater than the dissociation constant. With 10 copies of active repressor in the cell, the operator site is always bound by the repressor molecule.

**5. (a)–(e)** Each condition decreases expression of *lac* operon genes.

**6. (a)** Less attenuation of transcription. The ribosome completing the translation of sequence 1 would no longer overlap and block sequence 2; sequence 2 would always be available to pair with sequence 3, preventing formation of the attenuator structure. **(b)** More attenuation of transcription. Sequence 2 would pair less efficiently with sequence 3; the attenuator structure would be formed more often, even when sequence 2 was not blocked by a ribosome. **(c)** No attenuation of transcription. The only regulation would be that afforded by the Trp repressor. **(d)** Attenuation loses its sensitivity to Trp tRNA. It might become sensitive to His tRNA. **(e)** Attenuation would rarely, if ever, occur. Sequences 2 and 3 always block formation of the attenuator. **(f)** Constant attenuation of transcription. Attenuator always forms, regardless of the availability of tryptophan.

**7.** Induction of the SOS response could not occur, making the cells more sensitive to high levels of DNA damage.

**8.** Each *Salmonella* cell would have flagella made up of both types of flagellar protein, and the cell would be vulnerable to antibodies generated in response to either protein.

**9.** A dissociable factor necessary for activity (e.g., a specificity factor similar to the $\sigma$ subunit of the *E. coli* enzyme) may have been lost during purification of the polymerase.

**10.**

**Gal4 protein**

| Gal4 DNA-binding domain | Gal4 activator domain |
| --- | --- |

**Engineered protein**

| Lac repressor DNA-binding domain | Gal4 activator domain |
| --- | --- |

The engineered protein cannot bind to the Gal4 binding site in the *GAL* gene (UAS$_G$), because it lacks the Gal4 DNA-binding domain. Modify the Gal4p DNA binding site to give it the nucleotide sequence to which the Lac repressor normally binds (using methods described in Chapter 9).

**11.** The *bcd* mRNA needed for development is contributed to the egg by the mother. The egg develops normally even if its genotype is *bcd⁻/bcd⁻*, as long as the mother has one normal *bcd* gene and the *bcd⁻* allele is recessive. However, the adult *bcd⁻/bcd⁻* female will be sterile because she has no normal *bcd* mRNA to contribute to her own eggs.

**12. (a)** Hydrogen bonds form between the protein and DNA backbone at A106, A110, A118, T119, and A122, and between the protein and DNA bases at A106, T107, A118, and T119. The latter four nucleotides contribute directly to DNA

# Index

Key: f = figures; t = tables; s = structural formulas; **boldface** = boldfaced terms in text

## a

A bands, 184, 184f
Abasic site, in base-excision repair, **971,** 972f
ABC1 protein, 824, 827
ABC exinuclease, 972–973
ABC transporter, **402,** 402f, 824
A (aminoacyl) binding site, ribosomal, 1056, 1506f
Abiotic synthesis, 32, 33f
Absolute configuration, **77**
Absolute temperature, units of, 491t
Absorbance *(A),* **82**
Absorption
  of fat, in small intestine, 632, 633f
  of light, 725–730, 725f. *See also* Light,
    absorption of
Absorption spectra
  of cytochrome *c,* 693, 694f
  of nucleotides, 278, 278f
  of opsins, 460, 460f
ACAT (acyl-CoA-cholesterol acyl transferase),
  **820,** 825
Acceptor control, **716**
Acceptor control ratio, **716**
Accessory pigments, **728**
  in light absorption, 728, 728f
Acetaldehyde, 509s
Acetals, **241,** 242f
Acetate, 499s
  activated, in citric acid cycle, 602–606, 602f,
    603f, 604f, 605f
  in cholesterol synthesis, 816, 816f
  in fatty acid synthesis, 794–795, 796f
  oxidation of, 615–616, 617f
  as source of phosphoenolpyruvate, 623–624
  transport of, 794–795, 796f
Acetic acid, 499s, 509s
  p$K_a$ of, 83, 83f
  titration curve for, 64, 64f
Acetic acid-acetate buffer system, 66, 66f
Acetoacetate, **650,** 909f
Acetoacetate decarboxylase, **651**
Acetoacetyl-ACP, 790t, **791,** 792f, 793f
Acetoacetyl-CoA, 301
Acetone, 509s, **650,** 909f
  in diabetic ketoacidosis, 909f
Acetylcholine receptor, **411**–415, 413f
  defective, 415t
  open/closed conformation of, 427, 427f
  in signaling, 426–427, 427f, 428f
  synaptic aggregation of, 383
Acetyl-CoA, 15s, 603s
  amino acid degradation to, 677–678, 678f, 679f
  AMPK and, 914
  $\beta$ oxidation yielding, 638f, 639
  in cholesterol synthesis, 816–820, 816f
  decarboxylation of pyruvate to, 605, 605f
  in fatty acid synthesis, 787–789, 788f, 789f, 793,
    794, 904
  in glucose metabolism, 904, 904f, 907–908, 907f
  hepatic metabolism of, 896
  hydrolysis of, 499, 499s, 499t
  oxidation of, in citric acid cycle, 639, 640t
  oxidation of pyruvate to, 602–603, 602f
  in plant gluconeogenesis, 781
  production of, 794–795
    in citric acid cycle, 602–606, 602f, 603f, 604f,
      605f

by pyruvate dehydrogenase complex, 621–622,
  621f
Acetyl-CoA–ACP transacetylase, **790**–791, 790t,
  792f, 793f
Acetyl-CoA carboxylase
  in bacteria, 797
  in fatty acid synthesis, **787**–789, 788f, 789f,
    795–797, 797f
  in plants, 797
Acetyl-coenzyme A. *See* Acetyl-CoA
*N*-Acetyl-$\beta$-D-glucosamine, 244s
Acetylene, 509s
*N*-Acetylgalactosamine, in glycosaminoglycans,
  253, 253f
*N*-Acetylglucosamine (GlcNAc), 243, 244f
  in bacterial cell walls, 252, 252f
  in glycosaminoglycans, 253, 253f
  in peptidoglycan synthesis, 777–778, 778f
*N*-Acetylglutamate, **669**
*N*-Acetylglutamate synthase, **669**
Acetyl groups
  in fatty acid synthesis, 795, 796f
  transport of, 795, 796f
*N*-Acetylmuramic acid (Mur2Ac), 243, 244f
  in bacterial cell walls, 252, 252f
  in peptidoglycan synthesis, 777–778, 778f
*N*-Acetylneuraminic acid (Neu5Ac), 243, 244s,
  262–263, 262t, 263s, 353s
  in gangliosides, 352–353, 353s
Acetylsalicylic acid, 800, 801s, 802–803
Achiral molecules, **17,** 18f
Acid(s). *See also specific acid, e.g.,* Acetic acid
  amino acids as, 84–85
  as buffers, 66–69, 67f, 68f
  definition of, 63
  dissociation constant ($K_a$) of, **63**–65, 63f–65f
  fatty. *See* Fatty acid(s)
  relative strength of. *See* p$K_a$
  strong, 63
  titration curve for, 64–65, 64f, 65f
    Henderson-Hasselbalch equation for, **66**–67,
      84
Acid anhydride, standard free-energy changes of,
  493t
Acid-base catalysis
  general, **200,** 201–202, 201f, 202f
  specific, **200**
Acid-base pairs, conjugate, **63**–64, 63f
  as buffer systems, 66–68, 66f
Acid-base titration, 64–65, 64f, 65f, 82–84, 83f
Acidemia
  argininosuccinic, 677t
  methylmalonic, 677t, 684
Acidic activation domain, **1107,** 1108f
Acidic sugars, 244s
Acidosis, **652**
  diabetic, **909**
Acivicin, **876,** 876f
Aconitase, **608**
*cis*-Aconitate, **608,** 609s
  isocitrate formation via, 608–609, 610f
Aconitate hydratase, **608**
ACP (acyl carrier protein), **790,** 790t, 791f, 792f,
  794
Acridine, **1006,** 1006f
Actin, **183**–184, 183f
  in ATP hydrolysis, 184, 185, 186f
  in muscle contraction, 185–186, 185f, 186f
  structure of, 183–184
  in thin filaments, 183–184, 183f, 185, 185f
Actin filaments, 9, 9f

Actin-myosin complex
  ATP in, 504–505
  phosphorylation of, 470
Actin-myosin interactions, 185–186, 185f, 186f
Actinomycin D, **1006,** 1006f
Action spectrum, **728**
Activation barrier, 26, 27f
Activation energy ($\Delta G^{\ddagger}$), **26,** 27f, 196
  of enzymatic reactions, **194**–195, 196
  in membrane transport, 391, 391f
  rate constant and, 196
Activators, **1083,** 1084–1085, 1084f
Active site, 158, **193,** 193f
  transition-state complementarity and, 220–221,
    221f
Active transport, ATP and, 504, 713–714
Actos, **807**–808, 808t, 915
Acyclovir, 966
Acyl-carnitine/carnitine transporter, **636**
Acyl carrier protein (ACP), **790,** 790t, 791f, 792f,
  794
Acyl-CoA, fatty, **635**
  conversion of fatty acids into, 635–636, 635f
Acyl-CoA acetyltransferase, **638**
Acyl-CoA–cholesterol acyl transferase (ACAT),
  **820,** 825
Acyl-CoA dehydrogenase, **637, 699**
  medium-chain, **643**
Acyl-CoA synthetases, **635**
  in triacylglycerol synthesis, **804,** 805f
Acyl-enzyme intermediate, in chymotrypsin
  mechanism, 213–215, 215f, 216f
Acyl phosphate, **530**
*N*-Acylsphinganine, **813,** 815f
*N*-Acylsphingosine, **813,** 815f
Adaptor hypothesis, 1035, 1035f
Adenine, 10s, **274,** 275f, 275t, 497s, 513s. *See
  also* Purine bases
  deamination of, 293, 294f
  evolutionary significance of, 1028, 1028f
Adenine nucleotides
  biosynthesis of, regulatory mechanisms in, 866,
    867f
  cellular, 497t
Adenine nucleotide translocase, **713**
Adenosine, 275s
  anti form of, 284, 284f
  as enzyme cofactor, 301–302, 301f
  evolutionary significance of, 301–302
  methylation of, 296
  syn form of, 284, 284f
Adenosine 2,3′-cyclic monophosphate, 276s
Adenosine 2′ monophosphate, 276s
Adenosine 3,′5′-cyclic monophosphate. *See* cAMP
  (adenosine 3,′5′-cyclic monophosphate)
Adenosine 3′-monophosphate, 276s
Adenosine 5′-monophosphate, 276s. *See also* AMP
  (adenosine monophosphate)
Adenosine 5′-triphosphate. *See* ATP (adenosine
  triphosphate)
Adenosine deaminase, **873**
Adenosine deaminase deficiency, **874**
  gene therapy for, 336–337
Adenosine diphosphate (ADP). *See* ADP
  (adenosine diphosphate)
Adenosine monophosphate (AMP). *See* AMP
  (adenosine monophosphate)
Adenosine phosphoribosyltransferase, **875**
Adenosine triphosphate (ATP). *See* ATP
  (adenosine triphosphate)
*S*-Adenosylhomocysteine, **674**